2014 International Power Electronics Conference

(IPEC-Hiroshima 2014 ECCE-ASIA)

AA002293

Hiroshima, Japan
18-21 May 2014

Pages 3190-3965

IEEE Catalog Number:	CFP14CPB-POD
ISBN:	978-1-4799-2706-7

**Copyright © 2014 by the Institute of Electrical and Electronic Engineers, Inc
All Rights Reserved**

Copyright and Reprint Permissions: Abstracting is permitted with credit to the source. Libraries are permitted to photocopy beyond the limit of U.S. copyright law for private use of patrons those articles in this volume that carry a code at the bottom of the first page, provided the per-copy fee indicated in the code is paid through Copyright Clearance Center, 222 Rosewood Drive, Danvers, MA 01923.

For other copying, reprint or republication permission, write to IEEE Copyrights Manager, IEEE Service Center, 445 Hoes Lane, Piscataway, NJ 08854. All rights reserved.

***This publication is a representation of what appears in the IEEE Digital Libraries. Some format issues inherent in the e-media version may also appear in this print version.**

IEEE Catalog Number: CFP14CPB-POD
ISBN 13:   978-1-4799-2706-7

Additional Copies of This Publication Are Available From:

Curran Associates, Inc
57 Morehouse Lane
Red Hook, NY 12571 USA
Phone: (845) 758-0400
Fax: (845) 758-2633
E-mail: curran@proceedings.com
Web: www.proceedings.com

TABLE OF CONTENTS

A NOVEL CONTROL SCHEME FOR THREE-LEVEL FULL-BRIDGE CONVERTER ACHIEVING LOW THD OUTPUT VOLTAGE 66
Liu, Jilong ; Xiao, Fei ; Chen, Wei ; Yang, Guorun

PARALLEL CONNECTED THREE PHASE INVERTERS BASED ON MODULAR DESIGN AND DISTRIBUTED CONTROL 72
Xiao, Fei ; Chen, Wei ; Liu, Jilong ; Wang, Hengli

EFFICIENCY INVESTIGATIONS OF A 3KW T-TYPE INVERTER FOR SWITCHING FREQUENCIES UP TO 100 KHZ 78
Anthon, Alexander ; Zhang, Zhe ; Andersen, Michael A.E. ; Franke, Toke

MINIATURIZATION OF THE BOOST-UP TYPE ACTIVE BUFFER CIRCUIT IN A SINGLE-PHASE INVERTER 84
Watanabe, Hiroki ; Koiwa, Kazuhiro ; Itoh, Jun-ichi ; Ohnuma, Yoshiya ; Miyawaki, Satoshi

TESTING FACILITY USING LARGE CAPACITY INVERTER 92
Ishimaru, Yusuke ; Adachi, Mitsuo ; Tsukakoshi, Masahiko ; Nakamura, Ritaka ; Masuda, Hiroyuki ; Ogashi, Yoshihiro ; Tsuboi, Yuichi

PERFORMANCE EVALUATION UNDER THE ACTUAL OPERATING CONDITION OF A LARGE CAPACITY VSI INVERTER FOR STEEL MILL APPLICATIONS 97
Mamun, Mostafa ; Yoshizawa, Daisuke ; Mukunoki, Makoto

A SOFT-SWITCHING SINGLE-PHASE UNIFIED POWER QUALITY CONDITIONER 105
Jiang, Maoh-Chin ; Chang, Kai-Chi ; Lu, Kao-Yi ; Shih, Bing-Jyun ; Liu, Tai-Chun

NOVEL THREE-PHASE PWM AC-AC CONVERTERS SOLVING COMMUTATION PROBLEM 110
Khan, Ashraf Ali ; Shin, Hyunhak ; Cha, Honnyong ; Kim, Heung-Geun

EXPERIMENTAL INVESTIGATION OF NORMALLY-ON TYPE BIDIRECTIONAL SWITCH FOR INDIRECT MATRIX CONVERTERS 117
Sung, Kyungmin ; Iijima, Ryuji ; Nishizawa, Shinichi ; Norigoe, Isami ; Ohashi, Hiromichi

VISUALIZATION OF PWM WAVEFORMS OF OUTPUT VOLTAGE AND INPUT CURRENT FOR A DIRECT MATRIX CONVERTER 123
Asai, Inami ; Takeshita, Takaharu

SPACE VECTOR MODULATION BASED ON VIRTUAL INDIRECT CONTROL FOR HIGH FREQUENCY AC-LINKED MATRIX CONVERTER 130
Inoue, Keita ; Shioda, Masashi ; Katade, Motohumi ; Goto, Akira ; Morishita, Shin ; Itoh, Junichi ; Koiwa, Kazuhiro

A FUNDAMENTAL VERIFICATION OF A SINGLE-PHASE TO THREE-PHASE MATRIX CONVERTER WITH A PDM CONTROL BASED ON SPACE VECTOR MODULATION 138
Nakata, Yuki ; Itoh, Jun-ichi

STEADY STATE CHARACTERISTICS OF THE BOOST-TYPE MATRIX CONVERTER FOR STAND-ALONE POWER SOURCE 146
Nagano, Y. ; Yamamura, N. ; Ishida, M. ; Hirokado, K.

DESIGN PROCEDURE FOR OUTPUT CURRENT CONTROL AND DAMPING CONTROL OF MATRIX CONVERTER 152
Takahashi, Hiroki ; Itoh, Jun-ichi

A NOVEL LCL FILTER PARAMETER DESIGN METHOD BASING ON RESONANT FREQUENCY OPTIMIZATION OF THREE-LEVEL NPC GRID CONNECTED INVERTER 160
Li, Ning ; Wang, Yue ; Niu, Ruigen ; Guo, Wei ; Lei, Wanjun ; Wang, Zhao'An

DESIGN AND ANALYSIS OF ISOLATED BI-DIRECTIONAL DC/DC CONVERTER USING QUASI-RESONANT ZVS 166
Noh, Yong-Su ; Won, Chung-Yuen ; Oh, Min-Seok ; Jeon, Jin-Yong ; Jung, Yong-Chae

AN ACTIVE-CLAMPING ZVS FLYBACK CONVERTER WITH INTEGRATED TRANSFORMER 172
Lin, Jing-Yuan ; Lo, Yu-Kang ; Chiu, Huang-Jen ; Wang, Chao-Fu ; Lin, Chien-Yu

PFM AND PWM HYBRID CONTROLLED LLC CONVERTER 177
Yamamoto, Junichi ; Zaitsu, Toshiyuki ; Abe, Seiya ; Ninomiya, Tamotsu

DISCUSSIONS ON VARIOUS VOLTAGE EQUALIZERS FOR EDLCS USING CW CIRCUIT 183
Khant, Hlaing Kyi Pyar ; Matsui, Keiju ; Hasegawa, Masaru ; Yasubayashi, Mikio ; Umeno, Masayoshi ; Ooishi, Eiji

ISOLATION SYSTEM WITH WIRELESS POWER TRANSFER FOR MULTIPLE GATE DRIVER SUPPLIES OF A MEDIUM VOLTAGE INVERTER 191
Kusaka, Keisuke ; Orikawa, Koji ; Itoh, Jun-ichi ; Morita, Kazunori ; Hirao, Kuniaki

STUDY AND IMPLEMENTATION OF A 15-W POWER AMPLIFIER FOR PIEZOELECTRIC ACTUATOR 199
Lo, Yu-Kang ; Chiu, Huang-Jen ; Liu, Yu-Chen ; Lin, Chung-Yi ; Cheng, Shih-Jen ; Yang, CS

ISOLATED VOLTAGE-BOOSTING CONVERTER 204
Hwu, K.I. ; Jiang, W.Z. ; Shieh, Jenn-Jong

HIGH VOLTAGE CONVERSION RATIO CASCADE BOOST CONVERTER WITH DC SNUBBER 208
Lee, Yuang-Shung ; Yu, Ling-Chia ; Chou, Tzu-Han

DESIGN-ORIENTED ANALYSIS OF RESONANCE DAMPING AND HARMONIC COMPENSATION FOR LCL-FILTERED VOLTAGE SOURCE CONVERTERS 216
Wang, Xiongfei ; Blaabjerg, Frede ; Loh, Poh Chiang

STATE-SPACE AVERAGE MODELING OF BIDIRECTIONAL DC-DC CONVERTER FOR BATTERY CHARGER USING LCLC FILTER ... 224

Moon, Sang-Ho ; Jou, Sung-Tak ; Lee, Kyo-Beum

A NEW SVPWM STRATEGY FOR INPUT SWITCHED MULTILEVEL CONVERTER ... 230

Xiong, Li ; Prasanna, U.R. ; Bilal, Akin ; Rajashekara, Kaushik

ESD RELIABILITY INFLUENCE OF A 60 V POWER LDMOS BY THE FOD-BASED (& DOTTED-OD) DRAIN ... 236

Chen, Shen-Li ; Lee, Min-Hua

ENHANCED TRANSVERSE-FLUX MOTOR WITH TORUS COILS ... 240

Tanaka, Junya ; Sakai, Kazuto

THE INFLUENCE OF MAGNETIC PROPERTIES OF PERMANENT MAGNET ON THE PERFORMANCE OF IPMSM FOR AUTOMOTIVE APPLICATION ... 246

Yoshioka, S. ; Morimoto, S. ; Sanada, M. ; Inoue, Y.

CHARACTERISTICS OF INTERIOR PERMANENT MAGNET SYNCHRONOUS MOTOR WITH IMPERFECT MAGNETS ... 252

Shinagawa, Syuhei ; Ishikawa, Takeo ; Kurita, Nobuyuki

STUDY OF STATOR STRUCTURE TO IMPROVE RELUCTANCE TORQUE FOR IPMSM WITH CONCENTRATED WINDING ... 258

Morikawa, R. ; Sanada, M. ; Morimoto, S. ; Inoue, Y.

DEVELOPMENT AND VERIFICATION OF ENERGY-ACCURATE SIMULATION MODELS FOR PERMANENT MAGNET SYNCHRONOUS MOTORS IN AUTOMATION SYSTEMS ... 264

Blank, Frederic ; Roth-Stielow, Jorg

COMPARISON OF THE RESISTANCE- AND INDUCTANCE-BASED SALIENCY OF A PMSM DUE TO A SHORT-CIRCUITED ROTOR WINDING ... 270

Graus, Johannes ; Rambetius, Alexander ; Hahn, Ingo

DESIGN AND OPTIMIZATION OF HIGH-SPEED SWITCHED RELUCTANCE MOTOR USING SOFT MAGNETIC COMPOSITE MATERIAL ... 278

Gaing, Zwe-Lee ; Kuo, Kuan-Yi ; Hu, Jia-Sheng ; Hsieh, Min-Fu ; Tsai, Ming-Hsiao

INFLUENCE OF PULSE WIDTH MODULATION (PWM) ON THE IRON LOSSES OF ELECTRICAL STEEL ... 283

Boehm, Andreas ; Hahn, Ingo

INVESTIGATION ON IRON LOSS CHARACTERISTICS IN STAR-CONNECTION AND DELTA-CONNECTION UNDER THREE PHASE PWM INVERTER EXCITATION ... 289

Odawara, Shunya ; Fujisaki, Keisuke ; Fukuhara, Shuhei

OPTIMIZATION ON ARRANGEMENT OF PERMANENT MAGNETS FOR MAGNETIC LEVITATION SYSTEM FOR THIN STEEL PLATE (FUNDAMENTAL CONSIDERATION ON LEVITATION PROBABILITY) ... 294

Ishii, Hirotaka ; Hasegawa, Shinya ; Narita, Takayoshi ; Oshinoya, Yasuo

EFFECT OF A MAGNETIC FIELD FROM THE HORIZONTAL DIRECTION ON A MAGNETICALLY LEVITATED STEEL PLATE (FUNDAMENTAL CONSIDERATIONS ON THE SHAPE ANALYSIS OF ULTRATHIN STEEL PLATE) ... 299

Kurihara, Takeshi ; Hasegawa, Shinya ; Narita, Takayoshi ; Oshinoya, Yasuo

NOVEL MAGNETIC STRUCTURE OF INTEGRATED DIFFERENTIAL-MODE AND COMMON-MODE INDUCTORS TO SUPPRESS DC SATURATION ... 304

Umetani, Kazuhiro ; Tera, Takahiro ; Shirakawa, Kazuhiro

A NOVEL CONTROL METHOD IN FLUX-WEAKENING REGION FOR EFFICIENT OPERATION OF INTERIOR PERMANENT MAGNET SYNCHRONOUS MOTOR ... 312

Ueda, K. ; Morimoto, S. ; Inoue, Y. ; Sanada, M.

IMPLEMENTATION OF THE MTPA AND MTPV CONTROL WITH ONLINE PARAMETER IDENTIFICATION FOR A HIGH SPEED IPMSM USED AS TRACTION DRIVE ... 318

Nguyen, Quoc Khanh ; Petrich, Matthias ; Roth-Stielow, Jorg

CORRECTION OF REFERENCE FLUX FOR MTPA CONTROL IN DIRECT TORQUE CONTROLLED INTERIOR PERMANENT MAGNET SYNCHRONOUS MOTOR DRIVES ... 324

Shinohara, Atsushi ; Inoue, Yukinori ; Morimoto, Shigeo ; Sanada, Masayuki

VOLTAGE REGULATION AND MAXIMUM OUTPUT POWER TRACKING OF A 4.5KW PERMANENT-MAGNET SYNCHRONOUS GENERATOR ... 330

Chang, Yuan-Chih ; Chang, Hsiu-Feng ; Dai, Wei-Fu ; Wu, Chun-Wei

A NOVEL FLUX-WEAKENING CONTROL METHOD BASED ON SINGLE CURRENT REGULATOR FOR PERMANENT MAGNET SYNCHRONOUS MOTOR ... 335

Fang, Xiaocun ; Hu, Taiyuan ; Lin, Fei ; Yang, Zhongping

PREDICTIVE CURRENT CONTROL METHOD IN INDUCTION MOTOR SPEED SENSORLESS DRIVE ... 341

Wei, Sun ; Yong, Yu ; Dianguo, Xu ; Jin, Xu ; Li, Ding

REAL-TIME IMPLEMENTATION OF AN ONLINE MODEL PREDICTIVE CONTROL FOR IPMSM USING PARALLEL COMPUTING ON FPGA ... 346

Leuer, Michael ; Bocker, Joachim

AN INTEGRAL SLIDING-MODE CONTROLLER FOR ENERGY EFFICIENCY IMPROVEMENT IN AC POWER SOURCE SUPPLIED AC MACHINE DRIVES ... 351

Shieh, Hsin-Jang ; Chen, Ying-Zuo

PERFORMANCE IMPROVEMENT OF ULTRA-HIGH-SPEED PMSM DRIVE SYSTEM BASED ON DTC BY USING SIC INVERTER ... 356

Togashi, Ryo ; Inoue, Yukinori ; Morimoto, Shigeo ; Sanada, Masayuki

MATHEMATICAL MODEL FOR HIGH-EFFICIENCY CONTROL OF PERMANENT-MAGNET SYNCHRONOUS MOTOR IN STATOR FLUX LINKAGE SYNCHRONOUS FRAME 363
Inoue, Tatsuki ; Inoue, Yukinori ; Morimoto, Shigeo ; Sanada, Masayuki

WIDE-SPEED-RANGE OPERATION OF DTC-BASED PMSM DRIVE SYSTEM USING MTPF CONTROL 370
Inoue, Yukinori ; Ichiya, Takahiro ; Morimoto, Shigeo ; Sanada, Masayuki

AN INDUSTRIAL LOW-VOLTAGE INVERTER FOR PRM CONTROL 376
Nakamura, M. ; Oka, T. ; Oishi, K.

OPTIMAL PULSE PATTERN DETERMINATION BASED ON PULSE HARMONIC MODULATION 383
Furukawa, Kimihisa ; Ajima, Toshiyuki ; Miyazaki, Hideki

METHOD FOR AUTO-TUNING OF CURRENT AND SPEED CONTROLLER IN IPMSM DRIVE SYSTEM BASED ON PARAMETER IDENTIFICATION 390
Tadokoro, D. ; Morimoto, S. ; Inoue, Y. ; Sanada, M.

COMPARATIVE STUDY OF PWM STRATEGIES FOR THREE-PHASE OPEN-END WINDING INDUCTION MOTOR DRIVES 395
Zhu, B. ; Prasanna, U.R. ; Rajashekara, K. ; Kubo, H.

10MW,3.3MWH ENERGY STORAGE SYSTEM CONSISTING OF 4000 FLYWHEELS CONTROLLED BY ICT NETWORK FOR SHORT CYCLE POWER FLUCTUATION COMPENSATION 403
Kato, Koji ; Ishigma, Satoru ; Nakajima, Yoichiro ; Arai, Haruki ; Ueda, Tetsuya ; Iwata, Tetsuki ; Ito, Yoichi ; Sugao, Kazumi

VERSATILE POWER TRANSFER STRATEGIES OF PV-BATTERY HYBRID SYSTEM FOR RESIDENTIAL USE WITH ENERGY MANAGEMENT SYSTEM 409
Choi, Seong-Chon ; Sin, Min-ho ; Kim, Dong-Rak ; Won, Chung-Yuen ; Jung, Yong-Chae

HIGH-EFFICIENCY AND COST-MINIMIZATION METHOD OF ENERGY STORAGE SYSTEM WITH MULTI STORAGE DEVICES FOR GRID CONNECTION 415
Haga, Hitoshi ; Shimao, Toshihiro ; Kondo, Seiji ; Kato, Koji ; Itoh, Youichi ; Arimatsu, Kenji ; Matsuda, Katsuhiro

BIDIRECTIONAL DC-DC CONVERTER WITH MULTIPLE SWITCHED-CAPACITOR CELLS 421
Lee, Yuang-Shung ; Huang, Hsin-Wei ; Chou, Tzu-Han

SWITCHED-CAPACITOR CHARGE EQUALIZATION CIRCUIT FOR SERIES-CONNECTED BATTERIES 429
Hsieh, Yao-Ching ; Cai, Zheng-Xiu ; Wu, Wen-Zhe

PERFORMANCE ANALYSIS OF UNITL-H6 INVERTER WITH SIC MOSFETS 433
Barater, Davide ; Buticchi, Giampaolo ; Concari, Carlo ; Franceschini, Giovanni ; Gurpinar, Emre ; De, Dipankar ; Castellazzi, Alberto

MAXIMUM POWER POINT TRACKING OF GRID-TIED PHOTOVOLTAIC POWER SYSTEMS 440
Lee, Ya-Ting ; Chiu, Chian-Song ; Chiu, Tse-Wei

A NEW VOLTAGE TYPE MAGNETICALLY COUPLED T-SOURCE INVERTER 446
Tran, Q.V. ; Low, K.S.

A HIGH EFFICIENCY HYBRID 7-LEVEL INVERTER WITH SINGLE DC SOURCE 452
Yanhong, Zhang ; Kazuya, Ogura ; Oi, Kazunobu

OPTIMAL IDLING CONTROL STRATEGY FOR THREE-PORT FULL-BRIDGE CONVERTER 458
Jiang, Yongjie ; Liu, Fuxin ; Ruan, Xinbo ; Wang, Lipeng

FILTER DESIGN FOR THREE-LEVEL GRID-CONNECTED INVERTER WITH LOW SWITCHING FREQUENCY 465
Ren, Kangle ; Zhang, Xing ; Wang, Fusheng ; Tu, Yunwu ; Wang, Lingxiang ; Deng, Lirong

A NOVEL EFFICIENT T TYPE THREE LEVEL NEUTRAL-POINT-CLAMPED INVERTER FOR RENEWABLE ENERGY SYSTEM 470
Wu, Wenlong ; Wang, Fei ; Wang, Yong

A NOVEL NEUTRAL POINT VOLTAGE AUTOMATIC BALANCING CARRIER-BASED MODULATION STRATEGY OF THREE-LEVEL NPC CONVERTER 475
Li, Ning ; Wang, Yue ; Niu, Ruigen ; Guo, Wei ; Lei, Wanjun ; Wang, Zhao'An

A HIGH VOLTAGE GAIN SWITCHED-COUPLED-INDUCTOR QUASI-Z-SOURCE INVERTER 480
Ahmed, Furqan ; Cha, Honnyong ; Kim, Su-Han ; Kim, Heung-Geun

A NOVEL CONTROL STRATEGY TO SUPPRESS DC CURRENT INJECTION TO THE GRID FOR THREE-PHASE PV INVERTER 485
Zhang, Tao ; He, Guofeng ; Chen, Min ; Xu, Dehong

CLC FILTER DESIGN OF A FLYBACK-INVERTER FOR PHOTOVOLTAIC SYSTEMS 493
Shin, Yesl ; Lee, June-Hee ; Lee, June-Seok ; Lee, Kyo-Beum

THREE-PHASE INVERTER TOPOLOGIES FOR GRID-CONNECTED PHOTOVOLTAIC SYSTEMS 498
Ozkan, Ziya ; Hava, Ahmet M.

A THREE-PORT TOPOLOGY COMPARISON FOR A LOW POWER STAND-ALONE PHOTOVOLTAIC SYSTEM 506
Mira, Maria C. ; Knott, Arnold ; Andersen, Michael A.E.

EFFECT OF CONVENTIONAL GRID-VOLTAGE FEEDFORWARD ON THE OUTPUT IMPEDANCE OF A THREE-PHASE PHOTOVOLTAIC INVERTER 514
Messo, T. ; Jokipii, J. ; Suntio, T.

POWER AMPLIFIER SUITABLE FOR PHOTOVOLTAIC CELL BOOSTER 522
Kohama, Teruhiko ; Sogawa, Yuki ; Tsuji, Satoshi

REALIZATION STUDY OF INTERLEAVED PV MICROINVERTER BY QUADRATURE-PHASE-SHIFT SPWM CONTROL 526
Hsieh, Hung-I ; Hsieh, Guan-Cyun ; Hou, Jiaxin

CURRENT SENSORLESS MPPT METHOD FOR A PV FLYBACK MICROINVERTERS USING A DUAL-MODE ... 532
Lee, June-Hee ; Lee, June-Seok ; Lee, Kyo-Beum

A NOVEL METHOD OF SUPPRESSING INRUSH CURRENTS OF SQUIRREL-CAGE INDUCTION MACHINE USING MATRIX CONVERTER IN WIND POWER GENERATION SYSTEMS ... 538
Yamada, Hiroaki ; Hanamoto, Tsuyoshi

NONLINEAR PITCH CONTROL DESIGN FOR LOAD REDUCTION ON WIND TURBINES ... 543
Xiao, Shuai ; Yang, Geng ; Geng, Hua

DEVICE LOADING OF MODULAR MULTILEVEL CONVERTER MMC IN WIND POWER APPLICATION ... 548
Popova, L. ; Pyrhonen, J. ; Ma, K. ; Blaabjerg, F.

A NOVEL OPTIMAL DESIGN OF DFIG CROWBAR RESISTOR DURING GRID FAULTS ... 555
Hu, Sheng ; Zou, XuDong ; Kang, Yong

DC-VOLTAGE REGULATION OF A FIVE LEVELS NEUTRAL POINT CLAMPED CASCADED CONVERTER FOR WIND ENERGY CONVERSION SYSTEM ... 560
Merahi, Farid ; Mekhilef, Saad ; Berkouk, El Madjid

A REACTIVE POWER SHARING METHOD BASED ON VIRTUAL CAPACITOR IN ISLANDING MICROGRID ... 567
Xu, Haizhen ; Zhang, Xing ; Liu, Fang ; Shi, Rongliang ; Yu, Changzhou ; Zhao, Wei ; Yu, Yong ; Cao, Wei

STORAGE CAPACITY PERFORMANCE FOR HYBRID PV/DIESEL SYSTEM IN SABAH MALAYSIA ... 573
Hidayat, Nabil M ; Kari, Mat Nasir ; Mohd Arif, Mohd Johari

NEW TECHNIQUES FOR MEASURING ISLANDED MICROGRID IMPEDANCE CHARACTERISTICS BASED ON CURRENT INJECTION ... 577
Hou, Lixiang ; Liu, Baoquan ; Shi, Hongtao ; Yi, Hao ; Zhuo, Fang

A GENERAL FRAMEWORK TO DESIGN OPERATION MODES OF DC MICROGRIDS WITHOUT COMMUNICATION LINKS ... 582
Pan, Miao ; Shen, Na ; Yang, Geng ; Morita, Kazunori ; Ogura, Kazuya ; Wu, Weiyang

IMPLEMENTATION DESIGN OF THE CONVERTER-BASED GALVANIC ISOLATION FOR LOW VOLTAGE DC DISTRIBUTION ... 587
Mattsson, A. ; Vaisanen, V. ; Nuutinen, P. ; Kaipia, T. ; Lana, A. ; Peltoniemi, P. ; Silventoinen, P. ; Partanen, J.

PEAK DETECTION METHOD USING TWO-DELTA OPERATION FOR SINGLE VOLTAGE SAG ... 595
Lee, Woo-Cheol ; Lee, Taeck-Kie

LINE LOSS MINIMIZATION IN RADIAL DISTRIBUTION SYSTEM USING MULTIPLE STATCOMS AND STATIC CAPACITORS ... 601
Miyazaki, Kensuke ; Takeshita, Takaharu

A NOVEL CONTROL METHOD FOR INDIVIDUAL DC VOLTAGE BALANCING IN H-BRIDGE CASCADED STATCOM ... 609
Xu, Rong ; Yu, Yong ; Yang, Rongfeng ; Qu, Lizhi ; Sun, Wei ; Xu, Dianguo

RESEARCH ON THE CONTROL STRATEGY OF STATCOM BASED ON MODULAR MULTILEVEL CONVERTER ... 614
Zhang, Wei ; Gao, Qiang ; Su, Bonan ; Jin, Miaoxin ; Xu, Dianguo ; Liu, Jianyu

FAULT DIAGNOSIS IN LARGE FORMAT LIFEPO4 ESS APPLICATION THROUGH DWT-BASED MRA ... 619
Kim, Jonghoon

COMPARISON OF DIFFERENT IGBT BASED DESIGNS OF POWER ELECTRONIC TRANSFORMER ... 624
Wang, Xinyu ; Ouyang, Shaodi ; Liu, Jinjun ; Meng, Fei ; Javed, Riffat

SEMI-ADAPTIVE HARMONIC CONTROL FOR POWER BALANCING DEVICE FOR AC TRACTION ... 629
Akagi, Masataka ; Tsuruta, Hironori ; Oso, Hiroshi

RESEARCH OF EFFICIENT MAIN POWER EQUIPMENT USING SIC POWER DEVICE ... 634
Shinbo, Mitsuo ; Sonoda, Hideki ; Ishida, Takahito ; Abiko, Hiroshi ; Shibanuma, Kenichi ; Chiba, Yoshinori

A HIGH PERFORMANCE CONTROL STRATEGY FOR THREE-LEVEL NPC EMU CONVERTERS ... 640
Song Kejian ; Wu Mingli ; Wang Hui ; Agelidis, Vassilios Georgios

A DESIGN OF INRUSH CURRENT IDENTIFICATION SYSTEM FOR HIGH-SPEED TRAIN'S TRACTION TRANSFORMER ... 647
Yu, Weikai ; Liu, Xiankai ; Zhang, Yuzhuo ; Cao, Yuan ; Ma, Weigang ; Hei, Xinhong ; Huang, Zhenhui ; Jiang, Dawang

CURRENT SOURCE INVERTER BASED CASCADED SOLID STATE TRANSFORMER FOR AC TO DC POWER CONVERSION ... 651
Roy, Sudhin ; De, Ankan ; Bhattacharya, Subhashish

EVALUATION OF HIGH VOLTAGE 15 KV SIC IGBT AND 10 KV SIC MOSFET FOR ZVS AND ZCS HIGH POWER DC -DC CONVERTERS ... 656
Moballegh, Shiva ; Madhusoodhanan, Sachin ; Bhattacharya, Subhashish

THE DIRECT YAW-MOMENT CONTROL TO FOLLOW THE NEUTRAL STEERING PATH REGARDLESS OF VELOCITY ... 664
Jang, Young-Jin ; Nam, Kwang-Hee

NEXT-GENERATION IGBT MODULE STRUCTURE FOR HYBRID VEHICLE WITH HIGH COOLING PERFORMANCE AND HIGH TEMPERATURE OPERATION ... 671
Morozumi, Akira ; Gohara, Hiromichi ; Momose, Fumihiko ; Saito, Takashi ; Nishimura, Yoshitaka ; Mochizuki, Eiji ; Takahashi, Yoshikazu

INTEGRATION OF PLUG-IN ELECTRIC VEHICLES IN POWER SYSTEMS USING CHARGING MODE SWITCHING ... 677
Wen-Tai Li ; Wen, Chao-Kai ; Chen, Jung-Chieh ; Teng, Jen-Hao ; Ting, Pangan

A NOVEL COMPENSATION METHOD FOR A MOTOR PHASE CURRENT SENSOR OFFSET ERROR VARIED DURING A VSI-MOTOR DRIVE 682
Tamura, Hiroshi ; Noto, Yasuo ; Ajima, Toshiyuki ; Itoh, Jun-ichi

INVESTIGATION OF CALCULATION METHOD OF LOSSES IN PWM INVERTER WITH VOLTAGE BOOSTER USING BOTH DC LINK VOLTAGE CONTROL AND FLUX WEAKENING CONTROL 689
Imakiire, Akihiro ; Hikita, Masayuki ; Yamamoto, Kichiro ; Yonemori, Ryo

DYNAMIC AND STEADY-STATE BEHAVIOR OF A PARALLELING THREE-PHASE AC-TO-DC CONVERTER WITH REDUCED DC BUS CAPACITOR 694
Kamnarn, Uthen ; Kanthaphayao, Yutthana ; Chunkag, Viboon

REACTIVE POWER LOSS OPTIMIZATION METHOD FOR BI-DIRECTIONAL ISOLATED DC-DC CONVERTERS 702
Wen, Huiqing

POWER SUPPLY FOR A WIRELESS SENSOR NETWORK: AIRLINER FLIGHT TEST CASE STUDY 707
Durand Estebe, P. ; Boitier, V. ; Bafleur, M. ; Dilhac, J-M. ; Berhouet, S.

A CONFIGURABLE THREE-PHASED INVERTER FOR TEACHING POWER ELECTRONICS 712
Kern, Ansgar

A BACHELOR-STUDENT PROJECT: BUCK-BOOST OPERATION OF AN INTEGRATED H-BRIDGE FOR VARIABLE-SPEED ENERGY STORAGE SYSTEMS USING MEASUREMENT COILS IN THE STATOR OF A DC-MACHINE 718
De Belie, Frederik ; Darba, Araz ; Melkebeek, Jan

DEVELOPMENT OF A WEB-BASED REMOTE EXPERIMENT SYSTEM FOR ELECTRICAL MACHINERY LEARNERS 724
Ishibashi, Makoto ; Fukumoto, Hisao ; Furukawa, Tatsuya ; Itoh, Hideaki ; Ohchi, Masashi

DEVELOPMENT OF POWER MEASUREMENT SYSTEM IN SIMULATED MICRO GRID SYSTEM FOR EDUCATION 730
Hira, Yuki ; Furukawa, Tatsuya ; Yakabe, Seichiro ; Fukumoto, Hisao ; Itoh, Hideaki ; Ohchi, Masashi

POWER ELECTRONIC TECHNOLOGIES FOR FLEXIBLE DC DISTRIBUTION GRIDS 736
De Doncker, Rik W.

2.5KV, 200KW BI-DIRECTIONAL ISOLATED DC/DC CONVERTER FOR MEDIUM-VOLTAGE APPLICATIONS 744
Matsuoka, Yuji ; Wada, Keiji ; Nakahara, Mizuki ; Takao, Kazuto ; Kyungmin Sung ; Ohashi, Hiromichi ; Nishizawa, Shinichi

POWER-LOSS BREAKDOWN OF A 750-V, 100-KW, 20-KHZ BIDIRECTIONAL ISOLATED DC-DC CONVERTER USING SIC-MOSFET/SBD DUAL MODULES 750
Akagi, Hirofumi ; Yamagishi, Tatsuya ; Tan, Nadia M.L. ; Kinouchi, Shin-ichi ; Miyazaki, Yuji ; Koyama, Masato

DESIGN CONSIDERATIONS OF A 15KV SIC IGBT ENABLED HIGH-FREQUENCY ISOLATED DC-DC CONVERTER 758
Tripathi, Awneesh ; Mainali, Krishna ; Patel, Dhaval ; Kadavelugu, Arun ; Hazra, Samir ; Bhattacharya, Subhashish ; Hatua, Kamalesh

COMMON-MODE CURRENTS IN MULTI-CELL SOLID-STATE TRANSFORMERS 766
Huber, Jonas E. ; Kolar, Johann W.

SINGLE-STAGE RECONFIGURABLE DC/DC CONVERTER FOR WIDE INPUT VOLTAGE RANGE OPERATION IN HEVS 774
Zeljkovic, Sandra ; Reiter, Tomas ; Gerling, Dieter

A TWO STAGE DC/DC CONVERTER WITH WIDE INPUT RANGE FOR EV 782
Peng Wen ; Changsheng Hu ; Haitao Yang ; Longlong Zhang ; Cheng Deng ; Yashun Li ; Dehong Xu

INTERMEDIATE AND LIGHT LOAD EFFICIENCY IMPROVEMENT OF A HIGH-POWER DENSITY BIDIRECTIONAL DC-DC CONVERTER IN HYBRID ELECTRIC VEHICLES WITH MR FLUID GAP INDUCTOR 790
Ahmed, Furqan ; Su-Han Kim ; Cha, Honnyong ; Kim, Dong-Hun ; Heung-Geun Kim

REGENERATIVE CONTROL OF BI-DIRECTIONAL DC-DC CONVERTER CONTROLLING VARIABLE DC-LINK FOR FCEV 796
Il-Kuen Won ; An-Yeol Ko ; Do-Yun Kim ; Chung-Yuen Won ; Young-Ryul Kim

LARGE DRIVING RANGE INCREASE OF SERIES CHOPPER BASED POWER TRAIN USING MOTOR TEST BENCH 801
Hosoyamada, Yu ; Takeda, Masashi ; Motoi, Naoki ; Kawamura, Atsuo

THE POWER ELECTRONICS PROGRAM AT BEIJING JIAOTONG UNIVERSITY 807
Fei Lin ; Zhongping Yang ; Zheng, T.Q.

EFFORTS FOR POWER ELECTRONICS EDUCATION IN A START-UP COMPANY 811
Hattori, Fumiya ; Imaoka, Jun ; Ishitobi, Manabu ; Nagai, Shinichiroh ; Yamamoto, Masayoshi

EDUCATION FOR THE ENGINEERS OF TRACTION POWER SUPPLY DIVISION IN EAST JAPAN RAILWAY COMPANY 817
Takino, Toshiaki ; Iwakami, Tetsuro

SUCCESSFUL ONLINE EDUCATION - GECKOCIRCUITS AS OPEN-SOURCE SIMULATION PLATFORM 821
Musing, Andreas ; Kolar, Johann W.

AN ELECTRIC VEHICLE PROJECT FOR ECO-RUN RACE 829
Yamagata, Shinichi ; Oda, Yoshinori ; Tanai, Masanobu ; Sung, Kyungmin

MULTI-LOOP CONTROLLER DESIGN FOR DIODE-ASSISTED BUCK-BOOST VOLTAGE SOURCE INVERTER 835
Yan Zhang ; Jinjun Liu ; Xiaolong Ma ; Junjie Feng

VOLUME 2

REAL-TIME SIMULATION OF WIND TURBINE CONVERTER-GRID SYSTEMS 843
Shah, Shahil ; Vieto, Ignacio ; Nian Heng ; Sun, Jian

TECHNOLOGIES FOR MITIGATING FLUCTUATION CAUSED BY RENEWABLE ENERGY SOURCES 850
Katoh, Shuji ; Ohara, Shinya ; Itoh, Tomomichi

RELIABILITY-ORIENTED ENERGY STORAGE SIZING IN WIND POWER SYSTEMS 857
Zian Qin ; Liserre, Marco ; Blaabjerg, Frede ; Poh Chiang Loh

A MULTI-LEVEL VIRTUAL CONDUCTOR AS A BACKBONE OF A DC POWER ROUTING SYSTEM 863
Ramadan, Husam A. ; Imamura, Yasutaka ; Kawachi, Konosuke ; Yang, Sihun ; Shoyama, Masahito

SEMI-NUMERICAL METHOD FOR LOSS-CALCULATION IN FOIL-WINDINGS EXPOSED TO AN AIR-GAP FIELD 868
Leuenberger, D. ; Biela, J.

LOSS REDUCTION OF LAMINATED CORE INDUCTOR USED IN ON-BOARD CHARGER FOR EVS 876
Tera, Takahiro ; Taki, Hiroshi ; Shimizu, Toshihisa

FEASIBLE EVALUATIONS OF COUPLED MULTILAYER CHIP INDUCTOR FOR POL CONVERTER 883
Imaoka, Jun ; Kimura, Shota ; Itoh, Yuki ; Yamamoto, Masayoshi ; Suzuki, Michiaki ; Kawano, Kenji

OPTIMAL INDUCTOR DESIGN FOR 3-PHASE VOLTAGE-SOURCE PWM CONVERTERS CONSIDERING DIFFERENT MAGNETIC MATERIALS AND A WIDE SWITCHING FREQUENCY RANGE 891
Burkart, Ralph M. ; Uemura, Hirofumi ; Kolar, Johann W.

COMPARATIVE ANALYSIS OF INDUCTOR CONCEPTS FOR HIGH PEAK LOAD LOW DUTY CYCLE OPERATION 899
Leibl, Michael ; Kolar, Johann W.

INITIAL POSITION ESTIMATION FOR IPMSMS USING COMB FILTERS AND EFFECTS ON VARIOUS INJECTED SIGNAL FREQUENCIES 907
Suzuki, Toshiki ; Tomita, Mutuwo ; Hasegawa, Masaru ; Doki, Shinji

ADAPTIVE SIGNAL INJECTION METHOD COMBINED WITH EEMF BASED POSITION SENSORLESS CONTROL OF IPMSM DRIVES 914
Ohnuma, Takumi ; Makaino, Yuki ; Saitoh, Ryoh

STUDY OF LOW SPEED SENSORLESS DRIVES FOR SPMSM BY CONTROLLING ELLIPTICAL INDUCTANCE 919
Maekawa, Sari ; Hinata, Toshifumi ; Suzuki, Nobuyuki ; Kubota, Hisao

SUPPRESSION OF INJECTION VOLTAGE DISTURBANCE FOR HIGH FREQUENCY SQUARE-WAVE INJECTION SENSORLESS DRIVE WITH REGULATION OF INDUCED HIGH FREQUENCY CURRENT RIPPLE 925
Dongouk Kim ; Yong-Cheol Kwon ; Seung-Ki Sul ; Jang-Hwan Kim ; Rae-Sung Yu

APPLICATION TREND OF SALIENCY-BASED SENSORLESS DRIVES 933
Yamazaki, Akira ; Ide, Kozo

SWITCHING-LEVEL SIMULATION MODEL OF MMC-BASED BACK-TO-BACK CONVERTER FOR HVDC APPLICATION 937
Byung Moon Han ; Jong kyou Jeong

POWER-CELL SWITCHING-CYCLE CAPACITOR VOLTAGE CONTROL FOR THE MODULAR MULTILEVEL CONVERTERS 944
Wang, Jun ; Burgos, Rolando ; Boroyevich, Dushan ; Bo Wen

A COMPARISON OF MODULAR MULTILEVEL ENERGY CONVERSION PROCESSES: DC/AC VERSUS DC/DC 951
Kish, Gregory J. ; Lehn, Peter W.

A NOVEL TOPOLOGY OF WIND POWER PLANT SUITABLE FOR DC POWER TRANSMISSION SYSTEMS 959
Nishikata, Shoji ; Tatsuta, Fujio ; Suzuki, Katsumi

AN IMPEDANCE-BASED APPROACH TO HVDC SYSTEM STABILITY ANALYSIS AND CONTROL DEVELOPMENT 967
Liu, Hanchao ; Shah, Shahil ; Sun, Jian

TOPOLOGY EVALUATION OF SLOTLESS BEARINGLESS MOTORS WITH TOROIDAL WINDINGS 975
Steinert, Daniel ; Nussbaumer, Thomas ; Kolar, Johann W.

WINDING ARRANGEMENT IN SINGLE-DRIVE BEARINGLESS MOTOR WITH RADIAL GAP 982
Sugimoto, Hiroya ; Tanaka, Seiyu ; Chiba, Akira ; Rahman, M.A.

DEVELOPMENT OF A ONE-AXIS ACTIVELY REGULATED BEARINGLESS MOTOR WITH A REPULSIVE TYPE PASSIVE MAGNETIC BEARING 988
Asama, Junichi ; Watanabe, Daisuke ; Oiwa, Takaaki ; Chiba, Akira

CONTROL CHARACTERISTICS OF 8/10 AND 12/14 BEARINGLESS SWITCHED RELUCTANCE MOTOR 994
Zhenyao Xu ; Dong-Hee Lee ; Jin-Woo Ahn

BASIC CHARACTERISTIC OF A TWO-UNIT OUTER ROTOR TYPE BEARINGLESS MOTOR WITH CONSEQUENT POLE PERMANENT MAGNET STRUCTURE 1000
Takemoto, Masatsugu

VOLTAGE RIPPLE ELIMINATION IN INDUCTOR-LESS AC-TO-AC CONVERTERS FOR MULTI-POLE PERMANENT MAGNET SYNCHRONOUS GENERATORS 1006

Tanaka, Koutaro ; Fujita, Hideaki

A NEW SVM METHOD TO REDUCE COMMON-MODE VOLTAGE IN DIRECT MATRIX CONVERTER 1013

Huu-Nhan Nguyen ; Hong-Hee Lee

EXPERIMENTAL VERIFICATION OF HIGH FREQUENCY LINK DC-AC CONVERTER USING PULSE DENSITY MODULATION AT SECONDARY MATRIX CONVERTER 1021

Itoh, Jun-ichi ; Oshima, Ryo ; Takahashi, Hiroki

LOSS ANALYSIS AND DESIGN METHOD FOR HIGH EFFICIENCY MATRIX CONVERTER 1028

Koiwa, Kazuhiro ; Goh Teck Chiang ; Itoh, Jun-ichi

CAPACITOR CLAMPED MULTI-LEVEL MATRIX CONVERTER 1036

Raju, Siddharth ; Mohan, Ned

EUROPEAN TRENDS AND TECHNOLOGIES IN TRACTION 1043

Drofenik, Uwe ; Canales, Francisco

CO-PHASE POWER SUPPLY SYSTEM FOR HSR 1050

Qunzhan Li ; Wei Liu ; Zeliang Shu ; Shaofeng Xie ; Fulin Zhou

THE APPLICATION OF ELECTRONIC FREQUENCY CONVERTER TO THE SHINKANSEN RAILYARD POWER SUPPLY 1054

Shimizu, Toshimasa ; Kunomura, Ken ; Kai, Masahiko ; Onishi, Mitsuru ; Masuzawa, Hiroshi ; Miyajima, Hiroki ; Otsuki, Midori ; Tsuruma, Yoshinori

APPLICATION EXAMPLES OF ENERGY SAVING MEASURES IN JAPANESE DC FEEDING SYSTEM 1062

Suzuki, Takashi ; Hayashiya, Hitoshi ; Yamanoi, Takashi ; Kawahara, Keiji

LITHIUM ION BATTERY APPLICATION IN TRACTION POWER SUPPLY SYSTEM 1068

Teshima, Masato ; Takahashi, Hirotaka

INTEGRATED ISOLATION AND VOLTAGE BALANCING LINK OF 3-PHASE 3-LEVEL PWM RECTIFIER AND INVERTER SYSTEMS 1073

Boillat, David O. ; Kolar, Johann W.

VOLTAGE STEP-UP CONVERTER BASED ON MULTISTAGE STACKED BOOST ARCHITECTURE (MSBA) 1081

Rufer, Alfred ; Barrade, Philippe ; Steinke, Gina

COMPARISON OF CASCADED MULTILEVEL CONVERTER TOPOLOGIES FOR AC/AC CONVERSION 1087

Ilves, Kalle ; Bessegato, Luca ; Norrga, Staffan

EVALUATION OF ISOLATED THREE-PHASE AC-DC CONVERTER USING MODULAR MULTILEVEL CONVERTER TOPOLOGY 1095

Nakanishi, Toshiki ; Itoh, Jun-ichi

SELF-DECOUPLED DUAL PICK-UP COILS WITH LARGE LATERAL TOLERANCE FOR ROADWAY POWERED ELECTRIC VEHICLES 1103

Choi, Su Y. ; Lee, Sung W. ; Lee, Eun S. ; Jeong, Seog Y. ; Gu, Beom W. ; Rim, Chun T.

CONTACTLESS POWER TRANSFER SYSTEM SUITABLE FOR LOW VOLTAGE AND LARGE CURRENT CHARGING FOR EDLCS 1109

Kudo, Takahiro ; Toi, Takahiro ; Kaneko, Yasuyoshi ; Abe, Shigeru

EXCITATION SYSTEM BY CONTACTLESS POWER TRANSFER SYSTEM WITH THE PRIMARY SERIES CAPACITOR METHOD 1115

Nozawa, Ryosuke ; Kobayashi, Ryota ; Tanifuji, Hikaru ; Kaneko, Yasuyoshi ; Abe, Shigeru

DESIGN OF FERRITE CORES OF INDUCTIVE POWER COLLECTION COILS FOR MOVING VEHICLES 1122

Shimode, Daisuke ; Murai, Toshiaki ; Sawada, Tadashi

TORQUE/CURRENT RATIO IMPROVEMENT AND VIBRATION REDUCTION OF SWITCHED RELUCTANCE MOTORS USING MULTI-STAGE STRUCTURE 1128

Matsui, Ryota ; Nakao, Noriya ; Akatsu, Kan

IMPROVEMENT OF EFFICIENCY BY STEPPED-SKEWING ROTOR FOR SWITCHED RELUCTANCE MOTORS 1135

Sugiura, Makoto ; Ishihara, Yuji ; Ishikawa, Hiroki ; Naitoh, Haruo

A SINGLE PHASE SRM DRIVEN BY COMMERCIAL AC POWER SUPPLY 1141

Aiso, Kohei ; Nakao, Noriya ; Akatsu, Kan

FAST ANALYTICAL MODEL OF SWITCHED RELUCTANCE MACHINE 1148

Smaka, Senad ; Masic, Semsudin ; Cosovic, Mirsad

DETAILED ANALYSIS AND A GENERAL DESIGN PROCEDURE OF DAMPED LCL FILTERS IN THREE PHASE VOLTAGE SOURCE CONVERTERS 1155

Baoquan Liu ; Shaohui Zhong ; Yixin Zhu ; Hao Yi ; Fang Zhuo

70 KHZ, 15 KW SILICON-CARBIDE MOSFET INVERTER FOR INDUSTRIAL INDUCTION HEATING SYSTEMS 1160

Komeda, Shohei ; Tsuboi, Yoshiki ; Fujita, Hideaki

A STUDY ON EFFICIENCY IMPROVEMENT OF HIGH-FREQUENCY CURRENT OUTPUT INVERTER BASED ON IMMITTANCE CONVERSION ELEMENT 1166

Suzuki, Shun ; Shimizu, Toshihisa

HIGH-SPEED SWITCHING METHOD OF MOSFET USING VOLTAGE BOOST AUXILIARY CIRCUIT FED BY GATE DRIVE POWER SUPPLY 1173

Noguchi, Toshihiko ; Murata, Munehiro

OPERATING STRATEGY FOR BI-DIRECTIONAL LLC RESONANT CONVERTER WITH SEAMLESS OPERATION 1179
Abe, Seiya ; Yamamoto, Junichi ; Zaitsu, Toshiyuki ; Ninomiya, Tamotsu

NEGATIVE SEQUENCE CURRENT INJECTION CONTROL ALGORITHM COMPENSATING FOR UNBALANCED PCC VOLTAGE IN MEDIUM VOLTAGE PMSG WIND TURBINES 1185
Jayoon Kang ; Daesu Han ; Suh, Yongsug ; Byoungchang Jung ; Jeongjoong Kim ; Jonghyung Park ; Youngjoon Choi

OPTIMIZATION OF AN OFF-GRID HYBRID SYSTEM FOR SUPPLYING OFFSHORE PLATFORMS IN ARCTIC CLIMATES 1193
Kalogera, Maria ; Bauer, Pavol

ACTIVE DAMPING CONTROL OF LLCL FILTERS FOR THREE-LEVEL T-TYPE GRID CONVERTERS 1201
Alemi, Payam ; Lee, Dong-Choon

DEVELOPING A NEW TOPOLOGY FOR THE DC-DC CONVERTER USED IN FUEL CELL-ELECTRIC DOUBLE LAYER CAPACITOR HYBRID POWER SOURCE SYSTEM FOR MOBILE DEVICES 1207
Tosaka, Shuhei ; Yamanaka, Tatsuya ; Katayama, Noboru ; Hayase, Masanori ; Dowaki, Kiyoshi ; Kogoshi, Sumio

MULTIPLE OUTPUT CHARGER BASED ON PHASE SHIFT FULL BRIDGE CONVERTER WITH NOVEL TIME DIVISION MULTIPLE CONTROL TECHNIQUE 1214
Van-Long Tran ; Woojin Choi

DC-BREAKER FOR A MULTI-MEGAWATT BATTERY ENERGY STORAGE SYSTEM 1220
Demetriades, Georgios D. ; Hermansson, Willy ; Svensson, Jan R ; Papastergiou, Konstantinos ; Larsson, Tomas

ENERGY MANAGEMENT METHOD USING THE IIR FILTER FOR PEMFC-SUPERCAPACITOR HYBRID POWER SOURCE 1227
Yamanaka, Tatsuya ; Katayama, Noboru ; Tosaka, Shuhei ; Kogoshi, Sumio

ADVANCED TORQUE AND CURRENT CONTROL TECHNIQUES FOR PMSMS WITH A REAL-TIME SIMULATOR INSTALLED BEHAVIOR MOTOR MODEL 1234
Tanabe, Ryo ; Akatsu, Kan

COMPENSATION OF THE CURRENT MEASUREMENT ERROR WITH PERIODIC DISTURBANCE OBSERVER FOR MOTOR DRIVE 1242
Yamaguchi, Takashi ; Tadano, Yugo ; Hoshi, Nobukazu

RAPID AND STABLE SPEED CONTROL OF SPMSM BASED ON CURRENT DIFFERENTIAL SIGNAL 1247
Kitajima, Jun ; Ohishi, Kiyoshi

PARALLEL CONNECTED MULTIPLE DRIVE SYSTEM USING SMALL AUXILIARY INVERTER FOR NUMBERS OF PMSM 1253
Nagano, Tsuyoshi ; Itoh, Jun-chi

A TRANSFORMER INRUSH REDUCTION TECHNIQUE FOR LOW-VOLTAGE RIDE-THROUGH OPERATION OF RENEWABLE CONVERTERS 1261
Hsin-Chih Chen ; Ping-Heng Wu ; Cheng, Po-Tai

A CELL CAPACITOR ENERGY BALANCING CONTROL OF MODULAR MULTILEVEL CONVERTER CONSIDERING THE UNBALANCED AC GRID CONDITIONS 1268
Jung, Jae-Jung ; Shenghui Cui ; Kim, Sungmin ; Sul, Seung-Ki

FAULT CURRENT LIMITATION USING THYRISTOR BASED DEVICES 1276
Komatsu, Wilson ; Giaretta, Antonio Ricardo ; de Miranda, Rubens Domingos ; Jardini, Jose Antonio ; Casolari, Ronaldo Pedro ; Vasquez-Arnez, Ricardo Leon ; Hojo, Toshiaki ; Carvalho, Eden Luiz ; Maezono, Paulo Koiti

DC-DC BOOST CONVERTER BASED MSHE-PWM CASCADED MULTILEVEL INVERTER CONTROL FOR STATCOM SYSTEMS 1283
Law, Kah Haw ; Dahidah, Mohamed S.A.

NOVEL PRINCIPLE FOR FLUX SENSING IN THE APPLICATION OF A DC + AC CURRENT SENSOR 1291
Schrittwieser, L. ; Mauerer, M. ; Bortis, D. ; Ortiz, G. ; Kolar, J.W.

UTILIZING VOLTAGE MEASUREMENT OF FET SWITCH FOR MPPT OF DC ENERGY SOURCE 1299
Kimura, Noriyuki ; Niijima, Koji ; Morizane, Toshimitsu ; Omori, Hideki

HIGH FREQUENCY TRANSFORMER BASED ON A COUPLED INDUCTOR TOPOLOGY WITH DIELECTRIC ISOLATION 1303
Amanci, Adrian Z. ; Dawson, Francis P. ; Ruda, Harry E.

CONCEPT AND EXPERIMENTAL EVALUATION OF A NOVEL DC- 100MHZ WIRELESS OSCILLOSCOPE 1309
Lobsiger, Yanick ; Ortiz, Gabriel ; Bortis, Dominik ; Kolar, Johann W.

INTRODUCTION AND EFFECTIVENESS OF STATCOM TO THE INDEPENDENT POWER SYSTEM OF JR EAST 1317
Omi, Masataro ; Kotegawa, Ryo ; Ando, Masato ; Masui, Takeshi ; Horita, Yasuhisa

THE ANALYSIS OF TIME-VARYING RESONANCES IN THE POWER SUPPLY LINE OF HIGH SPEED TRAINS 1322
Chu, Xi ; Lin, Fei ; Yang, Zhongping

FUZZY FEED-FORWARD CHARGE/DISCHARGE CONTROL OF STATIONARY ENERGY STORAGE SYSTEMS FOR DC ELECTRIC RAILWAYS 1328
Kikuchi, Takuya ; Taga, Hironori ; Takagi, Ryo

TRAIN GROUP CONTROL FOR ENERGY-SAVING DC-ELECTRIC RAILWAY OPERATION 1334
Watanabe, Shoichiro ; Koseki, Takafumi

TRANSFORMER-LESS UNIFIED POWER FLOW CONTROLLER USING THE CASCADE MULTILEVEL INVERTER 1342
Fang Zheng ; Shao Zhang ; Shuitao Yang ; Gunasekaran, Deepak ; Karki, Ujjwal

A NEW POWER FLOW CONTROLLER USING SIX MULTILEVEL CASCADED CONVERTERS FOR DISTRIBUTION SYSTEMS..1350
Tsuruta, Ryoji ; Hosaka, Tatsuya ; Fujita, Hideaki

A PROPOSAL OF MODULAR MULTILEVEL CONVERTER APPLYING THREE WINDING TRANSFORMER..1357
Tamada, Shunsuke ; Nakazawa, Yosuke ; Irokawa, Shoichi

BACK-TO-BACK SYSTEM FOR FIVE-LEVEL CONVERTER WITH COMMON FLYING CAPACITORS.........................1365
Hasegawa, Isamu ; Urushibata, Shota ; Kondo, Takeshi ; Hirao, Kuniaki ; Kodama, Takashi ; Hui Zhang

HARMONIC MODELING OF A VEHICLE TRACTION CIRCUIT TOWARDS THE DC BUS..1373
Haghbin, Saeid ; Karvonen, Andreas ; Thiringer, Torbjorn

AC/DC CONVERTER BASED ON INSTANTANEOUS POWER BALANCE CONTROL FOR REDUCING DC-LINK CAPACITANCE..1379
Tokumasu, Akira ; Taki, Hiroshi ; Shirakawa, Kazuhiro ; Wada, Keiji

MODULAR CONVERTER ARCHITECTURE FOR MEDIUM VOLTAGE ULTRA FAST EV CHARGING STATIONS: DUAL HALF-BRIDGE-BASED ISOLATION STAGE...1386
Vasiladiotis, Michail ; Bahrani, Behrooz ; Burger, Niklaus ; Rufer, Alfred

NEW INTERLEAVED CURRENT-FED RESONANT CONVERTER WITH SIGNIFICANTLY REDUCED HIGH CURRENT OUTPUT FILTER FOR EV AND HEV APPLICATION...1394
Moon, Dongok ; Park, Junsung ; Choi, Sewan

15 PHASE INDUCTION MOTOR DRIVE WITH 1:3:5 SPEED RATIOS USING POLE PHASE MODULATION...................1400
Umesh B S ; Sivakumar K

MATHEMATICAL MODEL OF NOVEL WOUND-FIELD SYNCHRONOUS MOTOR SELF-EXCITED BY SPACE HARMONICS..1405
Aoyama, Masahiro ; Noguchi, Toshihiko

DUAL PURPOSE NO VOLTAGE WINDING DESIGN FOR THE BEARINGLESS AC HOMOPOLAR AND CONSEQUENT POLE MOTORS..1412
Severson, Eric ; Nilssen, Robert ; Undeland, Tore ; Mohan, Ned

HARVESTING ENERGY FROM SHIP ROLLING USING AN ECCENTRIC DISK REVOLVING IN A HULA-HOOP MOTION..1420
Yu-Jen Wang

LOAD-INDEPENDENT CURRENT OUTPUT OF INDUCTIVE POWER TRANSFER CONVERTERS WITH OPTIMIZED EFFICIENCY..1425
Zhang, Wei ; Wong, Siu-Chung ; Tse, Chi K. ; Chen, Qianhong

VOLTAGE CONTROL OF INDUCTIVE CONTACTLESS POWER TRANSFER SYSTEM WITH COAXIAL CORELESS TRANSFORMER FOR DC POWER DISTRIBUTION...1430
Miiura, Yushi ; Ojika, Satoshi ; Ise, Tomofumi

CONTACTLESS HIGH POWER TRANSFORMER TECHNOLOGIES FOR RAILWAY VEHICLES...............................1438
Kondo, Keiichiro ; Yamamoto, Kohei ; Kitazawa, Satochi

TWO-SWITCH VOLTAGE EQUALIZER BASED ON HALF-BRIDGE CONVERTER WITH MULTI-STACKED CURRENT DOUBLERS FOR SERIES-CONNECTED BATTERIES...1444
Uno, Masatoshi ; Kukita, Akio

OPTIMAL ENERGY STORAGE SYSTEM PLANNING FOR MICROGRIDS WITH CONTRACT CAPACITY CONSTRAINT..1452
Shu-Hung Liao ; Jen-Hao Teng ; Yung-Ching Huang ; Dong-Jing Lee

OPTIMAL ZERO SEQUENCE INJECTION IN MULTILEVEL CASCADED H-BRIDGE CONVERTER UNDER UNBALANCED PHOTOVOLTAIC POWER GENERATION...1458
Yu, Yifan ; Konstantinou, Georgios ; Hredzak, Branislav ; Agelidis, Vassilios G.

SIMPLE METHOD FOR MEASURING OUTPUT IMPEDANCE OF A THREE-PHASE INVERTER IN DQ-DOMAIN..1466
Jokipii, Juha ; Messo, Tuomas ; Suntio, Teuvo

ANALYSIS AND DESIGN OF POWER MANAGEMENT SCHEME FOR AN ON-BOARD SOLAR ENERGY STORAGE SYSTEM..1471
Jiang, W. ; Yu, F.Y. ; Lin, Z.Y. ; Wu, G.F. ; Chen, H. ; Hashimoto, S

LVRT CONTROL STRATEGY OF CSC-DPMSG-WGS UNDER UNBALANCED GRID FAULTS.................................1476
Meiqin Mao ; Yong Ding ; Shiting Weng ; Liuchen Chang

A NEW CURRENT CONTROL DROOP STRATEGY FOR VSI-BASED ISLANDED MICROGRIDS..............................1482
Shoeiby, B. ; Davoodnezhad, R. ; Holmes, D.G. ; McGrath, B.P.

POWER EXCHANGE USING PFC FOR MICRO GRID..1490
Sakai, Tomoyasu ; Takeda, Takashi ; Yukita, Kazuto ; Goto, Yasuyuki ; Ichiyanagi, Katsuhiro ; Morita, Hiroshi

DETERMINATION OF ROTOR TEMPERATURE FOR AN INTERIOR PERMANENT MAGNET SYNCHRONOUS MACHINE USING A PRECISE FLUX OBSERVER...1501
Specht, Andreas ; Wallscheid, Oliver ; Bocker, Joachim

MONITORING CRITICAL TEMPERATURES IN PERMANENT MAGNET SYNCHRONOUS MOTORS USING LOW-ORDER THERMAL MODELS..1508
Huber, Tobias ; Peters, Wilhelm ; Bocker, Joachim

ROBUST CURRENT CONTROL INSENSITIVE TO GAIN DEVIATION AND OFFSET OF INVERTER DC-LINK CURRENT SENSOR FOR SPMSM..1516
Matsuura, Kei ; Ando, Itaru ; Ohishi, Kiyoshi ; Matsuhashi, Masataka

AUTO-TUNING METHOD OF INDUCTANCES FOR PERMANENT MAGNET SYNCHRONOUS MOTORS.....................1522
Nomura, Naofumi ; Higuchi, Shinichi

AN IMPEDANCE-BASED STABILITY ANALYSIS METHOD FOR PARALLELED VOLTAGE SOURCE CONVERTERS 1529
Wang, Xiongfei ; Blaabjerg, Frede ; Loh, Poh Chiang

DYNAMIC CHARACTERISTICS AND STABILITY COMPARISONS BETWEEN VIRTUAL SYNCHRONOUS GENERATOR AND DROOP CONTROL IN INVERTER-BASED DISTRIBUTED GENERATORS 1536
Jia Liu ; Miura, Yushi ; Ise, Toshifumi

EMBEDDED LIMITATIONS AND PROTECTIONS FOR DROOP-BASED CONTROL SCHEMES WITH CASCADED LOOPS IN THE SYNCHRONOUS REFERENCE FRAME 1544
D'Arco, Salvatore ; Guidi, Giuseppe ; Suul, Jon Are

VIRTUAL SYNCHRONOUS GENERATOR CONTROL WITH DOUBLE DECOUPLED SYNCHRONOUS REFERENCE FRAME FOR SINGLE-PHASE INVERTER 1552
Hirase, Yuko ; Noro, Osamu ; Yoshimura, Eiji ; Nakagawa, Hidehiko ; Sakimoto, Kenichi ; Shindo, Yuji

CONTACTLESS DC CONNECTOR BASED ON GAN LLC CONVERTER FOR NEXT GENERATION DATA CENTERS 1560
Hayashi, Yusuke ; Toyoda, Hajime ; Ise, Toshifumi ; Matsumoto, Akira

ANALYSIS OF MIS-INTERRUPTION OF SEMICONDUCTOR BREAKER IN DC POWER FEEDING SYSTEM 1567
Murai, Kensuke ; Kanai, Yasuyuki ; Asakimori, Koki ; Babasaki, Tadatoshi

A RELIABLE ELECTRONIC CHOKE WITH NO NEED OF GAIN ADJUSTMENT FOR WIRE COMMUNICATION SYSTEM 1575
Katsuki, Akihiko ; Nakamura, Tatsuya ; Mizuki, Tatsuya ; Shibahara, Kohei ; Abe, Tomohiko ; Ikeda, Tomohiko ; Maeyama, Shigetaka

DESIGN OF NEW CONTROL STRATEGIES FOR A FOUR-LEG THREE-PHASE INVERTER TO ELIMINATE THE NEUTRAL CURRENT UNDER UNBALANCED LOADS 1580
Zhao-Qin Guo ; Panda, Sanjib Kumar ; Prasanna, I.V.

RESEARCH TRENDS OF MODULAR MULTILEVEL CASCADE INVERTER (MMCI-DSCC)-BASED MEDIUM-VOLTAGE MOTOR DRIVES IN A LOW-SPEED RANGE 1586
Okazaki, Yuhei ; Matsui, Hitoshi ; Hagiwara, Makoto ; Akagi, Hirofumi

AN INPUT SWITCHED MULTILEVEL INVERTER FOR OPEN-END WINDING INDUCTION MOTOR DRIVE 1594
Zhu, B. ; Jia, Y. ; Prasanna, U.R. ; Rajashekara, K. ; Kubo, H.

VARIABLE CARRIER FREQUENCY MIXED PWM TECHNIQUE BASED ON CURRENT RIPPLE PREDICTION FOR REDUCED SWITCHING LOSS 1601
Kubo, Hajime ; Yamamoto, Yasuhiro

SLIDING MODE PWM FOR EFFECTIVE CURRENT CONTROL IN SWITCHED RELUCTANCE MACHINE DRIVES 1606
Manolas, Iakovos ; Papafotiou, Georgios ; Manias, Stefanos N.

EXPERIMENTAL VERIFICATION OF AN EMC FILTER USED FOR PWM INVERTER WITH WIDE BAND-GAP DEVICES 1613
Itoh, Jun-ichi ; Araki, Takahiro ; Orikawa, Koji

PACKAGING FOR SIC POWER DEVICE 1621
Funaki, Tsuyoshi

SOLID STATE TRANSFORMER AND MV GRID TIE APPLICATIONS ENABLED BY 15 KV SIC IGBTS AND 10 KV SIC MOSFETS BASED MULTILEVEL CONVERTERS 1626
Madhusoodhanan, Sachin ; Tripathi, Awneesh ; Patel, Dhaval ; Mainali, Krishna ; Kadavelugu, Arun ; Hazra, Samir ; Bhattacharya, Subhashish ; Hatua, Kamalesh

VOLUME 3

GENERALIZED MODULAR MULTILEVEL CONVERTER AND MODULATION 1634
Hui Liu ; Loh, Poh Chiang ; Blaabjerg, Frede

AVERAGE POWER CONTROL OF DC BUS VOLTAGES OF CASCADED H-BRIDGE MULTILEVEL CONVERTERS 1639
Lee, Chia-Tse ; Chen, Hsin-Chih ; Ching-Wei Wang ; Ching-Hsiang Yang ; Cheng, Po-Tai

ANALYSIS AND COMPARISON OF HIGH POWER SEMICONDUCTOR DEVICE LOSSES IN 5MW PMSG MV WIND TURBINES 1646
Kihyun Lee ; Kyungsub Jung ; Seunghoo Song ; Suh, Yongsug ; Changwoo Kim ; Hyoyol Yoo ; Sunsoon Park

APPLICATION OF MODULAR MATRIX CONVERTER TO WIND TURBINE GENERATOR 1654
Inomata, Kentaro ; Hara, Hidenori ; Morimoto, Shinya ; Fujii, Junji ; Takeda, Kotaro ; Yamamoto, Eiji

FREE MOTION MECHANICAL POWER FACTOR; COMPARISON BETWEEN ROBOTS IN DIFFERENT STRUCTURE AND COORDINATE 1660
Mizoguchi, Takahiro ; Nozaki, Takahiro ; Ohnishi, Kouhei

ANALYSIS OF SETTLING BEHAVIOR AND DESIGN OF CASCADED PRECISE POSITIONING CONTROL IN PRESENCE OF NONLINEAR FRICTION 1665
Ruderman, Michael ; Iwasaki, Makoto

FIELD AND BENCH TEST EVALUATION OF RANGE EXTENSION CONTROL SYSTEM FOR ELECTRIC VEHICLES BASED ON FRONT AND REAR DRIVING-BRAKING FORCE DISTRIBUTIONS 1671
Fujimoto, Hiroshi ; Harada, Shingo ; Goto, Yuichi ; Kawano, Daisuke ; Sato, Koji ; Matsuo, Yusuke

VIBRATION SUPPRESSION OF INTEGRATED RESONANT AND TIME DELAY SYSTEM BY REFLECTED WAVE REJECTION 1679
Saito, Eiichi ; Oboe, Roberto ; Katsura, Seiichiro

THRUST CHARACTERISTICS IMPROVEMENT OF A CIRCULAR SHAFT MOTOR FOR DIRECT-DRIVE APPLICATIONS 1685
Omura, Mototsugu ; Shimono, Tomoyuki ; Fujimoto, Yasutaka

DESIGN OF A BEARINGLESS FLUX-SWITCHING SLICE MOTOR 1691
Gruber, Wolfgang ; Radman, Karlo ; Schob, Reto.T.

PROPOSAL OF A PERMANENT MAGNET HYBRID TYPE AXIAL MAGNETICALLY LEVITATED MOTOR 1697
Kurita, Nobuyuki ; Ishikawa, Takeo ; Takada, Hiromu ; Suzuki, Genri

COMPARISON OF HIGH SPEED BEARINGLESS DRIVE TOPOLOGIES WITH COMBINED WINDINGS 1701
Mitterhofer, Hubert ; Mrak, Branimir ; Gruber, Wolfgang

HIGH-SPEED MAGNETICALLY LEVITATED REACTION WHEEL DEMONSTRATOR 1707
Zwyssig, Christof ; Baumgartner, Thomas ; Kolar, Johann W.

STABILIZED SUSPENSION CONTROL CONSIDERING ARMATURE REACTION IN A D-Q AXIS CURRENT CONTROL BEARINGLESS MOTOR 1715
Ooshima, Masahide ; Kumakura, Yoshito

ANALYSIS AND DESIGN OF A HIGH-FREQUENCY ISOLATED DUAL-TANK LCL RESONANT AC-DC CONVERTER 1721
Du, Yimian ; Bhat, Ashoka K.S.

VERIFICATION OF LLC RESONANT CONVERTER APPLIED A CURRENT-BALANCING HIGH-FREQUENCY TRANSFORMER WITH MULTI-OUTPUT WINDINGS 1728
Araki, Jun ; Shinozaki, Ikki ; Funato, Hirohito ; Ogasawara, Satoshi ; Murakami, Daichi ; Hirota, Yukitsugu ; Mihara, Teruyoshi ; Mouri, Masayuki ; Okazaki, Fumihiro

LIGHT-LOAD EFFICIENCY IMPROVEMENT STRATEGY FOR LLC RESONANT CONVERTER UTILIZING A STEP-GAP TRANSFORMER 1734
Huang, Wen-Nan ; Lee, Shiu-Hui ; Chen, Ching-Guo

A NOVEL ACCURATE PRIMARY SIDE CONTROL (PSC) METHOD FOR HALF-BRIDGE (HB) LLC CONVERTER 1738
Jae-Bum Lee ; Kim, Chong-Eun ; Jae-Hyun Kim ; Cheol-O Yeon ; Young-Do Kim ; Moon, Gun-Woo

A SIMPLE CONTROL SCHEME FOR IMPROVING LIGHT-LOAD EFFICIENCY IN A FULL-BRIDGE LLC RESONANT CONVERTER 1743
Kim, Jae-Hyun ; Kim, Chong-Eun ; Lee, Jae-Bum ; Young-Do Kim ; Han-Shin Youn ; Moon, Gun-Woo

POWER CONDITIONER FOR STABILIZING POWER DISTURBANCE CAUSED OF WIND TURBINE GENERATOR SYSTEM 1748
Saga, Yasunao ; Fujii, Kansuke ; Yoda, Kazuyuki

A FRONT-TO-FRONT (FTF) SYSTEM CONSISTING OF MULTIPLE MODULAR MULTILEVEL CASCADE CONVERTERS FOR OFFSHORE WIND FARMS 1761
Sasongko, Firman ; Hagiwara, Makoto ; Akagi, Hirofumi

MODELLING, DESIGN AND CONTROL OF GRID CONNECTED CONVERTER FOR HIGH ALTITUDE WIND POWER APPLICATION 1775
Adhikari, Jeevan ; Rathore, Akshay K. ; Panda, S K

PRACTICAL STUDY OF A HIGH STEP-DOWN CONVERTER 1781
Jinno, Masahito ; Su, Hong-Wei ; Tsai, Jiung-Lin ; Matsuo, Hirofumi

GENERALIZED MODELING AND OPTIMIZATION OF A BIDIRECTIONAL DUAL ACTIVE BRIDGE DC-DC CONVERTER INCLUDING FREQUENCY VARIATION 1788
Jauch, Felix ; Biela, Jurgen

BALANCED DISCHARGING OF POWER BANK WITH BUCK-BOOST BATTERY POWER MODULES 1796
Moo, Chin-Sien ; Wu, Tsung-Hsi ; Hou, Chih-Hao ; Hsieh, Yao-Ching

Y-SOURCE IMPEDANCE-NETWORK-BASED ISOLATED BOOST DC/DC CONVERTER 1801
Siwakoti, Yam P. ; Town, Graham E. ; Loh, Poh Chiang ; Blaabjerg, Frede

MULTI-PHASE DC-DC CONVERTER WITH RIPPLE-LESS OPERATION FOR THERMO-ELECTRIC GENERATOR 1806
Kimura, Noriyuki ; Niijima, Koji ; Morizane, Toshimitsu ; Omori, Hideki

POSITION SENSORLESS START-UP METHOD OF SURFACE PERMANENT MAGNET SYNCHRONOUS MOTOR USING NONLINEAR ROTOR POSITION OBSERVER 1811
Hanamoto, Tsuyoshi ; Yamada, Hiroaki ; Okuyama, Yoshihiro

SENSORLESS CONTROL OF PMSM FOR THE WHOLE SPEED RANGE USING TWO-DEGREE-OF-FREEDOM CURRENT CONTROL AND HF TEST CURRENT INJECTION FOR LOW SPEED RANGE 1816
Seilmeier, Markus ; Piepenbreier, Bernhard

ELLIPSE-TRAJECTORY-ORIENTED VECTOR CONTROL FOR ENERGY EFFICIENT/WIDE-SPEED-RANGE DRIVES OF SENSORLESS PMSM 1824
Shinnaka, Shinji ; Amano, Yuki

DEVELOPMENT OF POSITION SENSORLESS CONTROL FOR PERMANENT-MAGNET SYNCHRONOUS GENERATOR DRIVE 1832
Chang, Yuan-Chih ; Lin, Chia-Yu ; Dai, Wei-Fu ; Wu, Chun-Wei

CONTROL OF A 750KW PERMANENT MAGNET SYNCHRONOUS MOTOR 1837
Liping Zheng ; Dong Le

REGIONAL SMART GRID OF ISLAND IN CHINA WITH MULTIFOLD RENEWABLE ENERGY 1842
Xu Cai ; Zheng Li

STABILIZING SMALL ISLAND POWER SYSTEM WITH RENEWABLES BY USE OF POWER CONDITIONING SYSTEMS - JAPANESE ISLAND SYSTEM CASE - 1849
Baba, Jumpei

POWER ELECTRONICS SOLUTIONS APPLIED TO A VARIETY OF DEMONSTRATIVE MICROGRID PROJECTS 1855
Ueda, Yoshinobu

MOVING TOWARDS THE SMART GRID: THE NORWEGIAN CASE 1861
Fosso, Olav B. ; Molinas, Marta ; Sand, Kjell ; Coldevin, Grete H.

POWER ELECTRONICS TECHNOLOGY IN SMART GRID PROJECTS -APPLICATIONS AND EXPERIENCES- 1868
Kobayashi, Takenori

EV AND HEV MOTOR DEVELOPMENT IN TOSHIBA 1874
Arata, Masanori ; Kurihara, Yoshihiro ; Misu, Daisuke ; Matsubara, Masakatsu

MOTOR STATOR WITH THICK RECTANGULAR WIRE LAP WINDING FOR HEVS 1880
Ishigami, Takashi ; Tanaka, Yuichiro ; Homma, Hiroshi

COMPARISON STUDY OF VARIOUS MOTORS FOR EVS AND THE POTENTIALITY OF A FERRITE MAGNET MOTOR 1886
Matsuhashi, Daiki ; Matsuo, Keisuke ; Okitsu, Takashi ; Ashikaga, Tadashi ; Mizuno, Takayuki

OPTIMAL FIELD EXCITATION CONTROL OF A CLAW POLE MOTOR FOR HYBRID ELECTRIC VEHICLE 1892
Azuma, M. ; Hazeyama, M. ; Morita, M. ; Kuroda, Y. ; Daikoku, A. ; Inoue, M.

A WIDE SPEED RANGE HIGH EFFICIENCY EV DRIVE SYSTEM USING WINDING CHANGEOVER TECHNIQUE AND SIC DEVICES 1898
Takatsuka, Yushi ; Hara, Hidenori ; Yamada, Kenji ; Maemura, Akihiko ; Kume, Tsuneo

PERFORMANCE COMPARISON OF A GAN GIT AND A SI IGBT FOR HIGH-SPEED DRIVE APPLICATIONS 1904
Tuysuz, Arda ; Bosshard, Roman ; Kolar, Johann W.

WIDE-BAND GAP DEVICES IN PV SYSTEMS - OPPORTUNITIES AND CHALLENGES 1912
Sintamarean, C. ; Eni, E. ; Blaabjerg, F. ; Teodorescu, R. ; Wang, H.

POWER ELECTRONICS EQUIPMENTS APPLYING NOVEL SIC POWER SEMICONDUCTOR MODULES 1920
Mino, Kazuaki ; Yamada, Ryuji ; Kimura, Hiroshi ; Matsumoto, Yasushi

EMI PREDICTION METHOD FOR SIC INVERTER BY THE MODELING OF STRUCTURE AND THE ACCURATE MODEL OF POWER DEVICE 1929
Maekawa, Sari ; Tsuda, Junichi ; Kuzumaki, Atsuhiko ; Matsumoto, Shuhei ; Mochikawa, Hiroshi ; Kubota, Hisao

SYSTEM INTEGRATION OF GAN TECHNOLOGY 1935
Ferreira, J.A. ; Popovic, J. ; van Wyk, J.D. ; Pansier, F.

POWER LOSSES OF MULTILEVEL CONVERTERS IN TERMS OF THE NUMBER OF THE OUTPUT VOLTAGE LEVELS 1943
Kashihara, Yugo ; Itoh, Jum-ichi

A LARGE CAPACITY 3-LEVEL IEGT INVERTER 1950
Yoshizawa, Daisuke ; Mukunoki, Makoto ; Omote, Kenichiro ; Hayashi, Makoto ; Isida, Takashi

VIBRATION SUPPRESSING CONTROL METHOD OF ANGULAR TRANSMISSION ERROR OF CYCLOID GEAR FOR INDUSTRIAL ROBOTS 1956
Yoshioka, Takashi ; Hirano, Yosei ; Ohishi, Kiyoshi ; Miyazaki, Toshimasa ; Yokokura, Yuki

AN ADVANCED POSITION CONTROL OF OVERHEAD CRANE BY SWAY SUPPRESSION METHOD EMULATING NATURAL DAMPING 1962
Kurabayashi, Toshiyuki ; Yang Chuan ; Murakami, Toshiyuki

A ROBOTIC CANE FOR WALKING ASSISTANCE 1968
Shimizu, Kyohei ; Smadi, Issam ; Fujimoto, Yasutaka

HAND POSITION ESTIMATION IN BINOCULAR VISUAL SPACE USING LINEAR APPROXIMATION OF KINEMATICS 1974
Komada, Satoshi ; Turpin, Santiago ; Hashimoto, Kento ; Yashiro, Daisuke ; Hirai, Junji

CONTACT STATE RECOGNITION BASED ON HAPTIC SIGNAL PROCESSING FOR ROBOTIC TOOL USE 1978
Matsuzaki, Ryohei ; Okuma, Jun ; Sakaino, Sho ; Tsuji, Toshiaki

RECENT TECHNICAL TRENDS IN MAGNETIC MATERIALS 1984
Wajima, Kiyoshi ; Toda, Hiroaki ; Kosaka, Takashi ; Marukawa, Yasuhiro ; Ishihara, Chio

MULTI-DOMAIN CO-SIMULATION WITH NUMERICALLY IDENTIFIED PMSM INTERWORKING AT HILS FOR ELECTRIC PROPULSION 1990
Park, Gyeong-Jae ; Jung, Hochang ; Kim, Yong-Jae ; Jung, Sang-Yong

RECENT TECHNICAL TRENDS IN PMSM 1997
Morimoto, Shigeo ; Asano, Yoshinari ; Kosaka, Takashi ; Enomoto, Yuji

RECENT TECHNICAL TRENDS IN SRM AND FSM 2004
Kano, Yoshiaki

RECENT TECHNICAL TRENDS IN VARIABLE FLUX MOTORS 2011
Toba, Akio ; Daikoku, Akihiro ; Nishiyama, Noriyoshi ; Yoshikawa, Yuichi ; Kawazoe, Yosuke

A GENERAL DISCRETE TIME MODEL TO EVALUATE ACTIVE DAMPING OF GRID CONVERTERS WITH LCL FILTERS 2019
Parker, S.G. ; McGrath, B.P. ; Holmes, D.G.

ANALYSIS AND REDUCTION OF POWER LOSSES IN PV CONVERTERS FOR GRID CONNECTION TO LOW-VOLTAGE THREE-PHASE THREE-WIRE SYSTEMS 2027
Amma, Ryosuke ; Fujita, Hideaki

DESIGN OF GRID CONNECTED PWM CONVERTERS CONSIDERING TOPOLOGY AND PWM METHODS FOR LOW-VOLTAGE RENEWABLE ENERGY APPLICATIONS..2034
Kantar, Emre ; Hava, Ahmet M.

PERFORMANCE OF DEAD TIME COMPENSATION METHODS IN THREE-PHASE GRID-CONNECTION CONVERTERS 2042
Mannen, Tomoyuki ; Fujita, Hideaki

D-S DIGITAL CONTROL FOR THREE-PHASE BI-DIRECTIONAL INVERTERS 2050
Wu, T.-F. ; Chang, C.-H. ; Lin, L.-C.

EXPECTATIONS OF NEXT-GENERATION POWER DEVICES FOR HOME AND CONSUMER APPLIANCES..2058
Kanouda, Akihiko ; Shoji, Hiroyuki ; Shimada, Takae ; Okubo, Toshikazu

APPLICATION TREND AND FORESIGHT OF SIC POWER DEVICES TO AIR CONDITIONERS....................2064
Kamikura, Mamoru ; Murata, Yuichiro ; Kutsuki, Tomohiro ; Saito, Katsuhiko

RECENT TECHNICAL TRENDS AND FUTURE PROSPECTS OF IGBTS AND POWER MOSFETS 2068
Ogura, Tsuneo

RECENT DEVELOPMENT AND FUTURE PROSPECTS OF POWER SIC DEVICES 2074
Nakamura, T. ; Nakano, Y. ; Aketa, M. ; Hanada, T.

RECENT ADVANCES AND FUTURE PROSPECTS ON GAN-BASED POWER DEVICES 2075
Ueda, Tetsuzo

SCALING AND BALANCING OF MULTI-CELL CONVERTERS...2079
Kasper, Matthias ; Bortis, Dominik ; Kolar, Johann W.

HYBRID MODULATED UNIVERSAL SOFT-SWITCHING CURRENT-FED DC/DC CONVERTER FOR WIDE VOLTAGE REGULATION FOR PV/FUEL CELLS/BATTERY APPLICATIONS 2087
Moorthy, Radha Sree Krishna ; Rathore, Akshay Kumar

HIGH EFFICIENCY POWER CONVERTERS FOR BATTERY ENERGY STORAGE SYSTEMS 2095
Kawakami, Noriko ; Iijima, Yukihia ; Li, Haiqing ; Ota, Satoru

IMPLEMENTATION OF BRIDGELESS CUK POWER FACTOR CORRECTOR WITH POSITIVE OUTPUT VOLTAGE 2100
Yang, Hong-Tzer ; Chiang, Hsin-Wei

A NOVEL SYNCHRONOUS RECTIFIER METHOD FOR A LLC RESONANT CONVERTER WITH VOLTAGE-DOUBLER RECTIFIER 2108
Murata, Koji ; Kurokawa, Fujio

LATEST DEVELOPMENTS IN INCREASING THE POWER DENSITY OF TRACTION DRIVES 2113
Bakran, Mark-M. ; Marz, Andreas ; Laska, Bernd ; Krafft, Eberhard ; Korner, Olaf ; Nagel, Andreas

CATENARY AND STORAGE BATTERY HYBRID SYSTEM FOR ELECTRIC RAILCAR SERIES EV-E301 2120
Kono, Y. ; Shiraki, N. ; Yokoyama, H. ; Furuta, R.

TECHNOLOGY FOR ENERGY-SAVING RAILWAY OPERATION THROUGH POWER-LIMITING BRAKES—A CASE STUDY AT AN URBAN RAILWAY 2126
Koseki, Takafumi ; Watanabe, Shoichiro ; Hamazaki, Yasuhiro ; Kondo, Keiichiro ; Hasegawa, Tomonori ; Mizuma, Takeshi

AN OVERVIEW ON BRAKING ENERGY REGENERATION TECHNOLOGIES IN CHINESE URBAN RAILWAY TRANSPORTATION 2133
Yang, Zhongping ; Xia, Huan ; Wang, Bin ; Lin, Fei

TRACTION INVERTER THAT APPLIES COMPACT 3.3 KV / 1200 A SIC HYBRID MODULE................2140
Ishikawa, Katsumi ; Yukutake, Seigo ; Kono, Yasuhiko ; Ogawa, Kazutoshi ; Kameshiro, Norifumi

POWER ELECTRONIC-BASED PROTECTION FOR DIRECT-CURRENT POWER DISTRIBUTION IN MICRO-GRIDS...2145
Tseng, K.J. ; Luo, Guomin

A CONCEPT OF HIGH POWER DC/DC CONVERTER WITH DOUBLE LOW POWER OUTPUTS 2152
Hojo, Masahide ; Nishioka, Tomoya ; Yamanaka, Kenji

PERFORMANCE EVALUATION FOR GRID IMPEDANCE BASED ISLANDING DETECTION METHOD 2156
Liu, Ning ; Aljankawey, A.S. ; Diduch, C.P. ; Chang, L. ; Mao, Meiqin ; Yazdkhasti, Pegah ; Su, Jianhui

IDENTIFYING NATURAL DEGRADATION/AGING IN POWER MOSFETS IN A LIVE GRID-TIED PV INVERTER USING SPREAD SPECTRUM TIME DOMAIN REFLECTOMETRY 2161
Li, Qian ; Khan, Faisal H.

CONTROL METHOD FOR INDUCTIVE POWER TRANSFER WITH HIGH PARTIAL-LOAD EFFICIENCY AND RESONANCE TRACKING 2167
Bosshard, R. ; Kolar, J.W. ; Wunsch, B.

STANDARD MODELS FOR SMART GRID SIMULATIONS 2175
Noda, Taku ; Nagashima, Tomohiro ; Sekisue, Takayuki ; Kabasawa, Yuichiro ; Kato, Shinji ; Sekiba, Yoichi ; Tokuda, Hirokazu ; Kounoto, Masaaki

MODEL DEVELOPMENT FOR MOTOR DRIVE SYSTEM SIMULATIONS.....................................2183
Ishikawa, Hiroki ; Abe, Takashi ; Kato, Toshiji ; Kubota, Yutaka ; Shimomura, Junichi ; Kohno, Yusuke ; Ikeda, Masahiro ; Umeda, Nobuhiro ; Kimura, Noriyuki ; Shigematsu, Koichi ; Inoue, Yukinori

PRACTICAL SIMULATION EXAMPLES OF AUTOMOTIVE AND POWER SUPPLY SYSTEMS.................................2189
Abe, Takashi ; Fukushima, Kentaro ; Sekisue, Takayuki ; Shigematsu, Koichi ; Ichihara, Junichi ; Kato, Toshiji ; Ishikawa, Hiroki ; Kouno, Yusuke ; Konoto, Masaaki ; Saito, Ryoji ; Nishida, Yasuyuki

ADMITTANCE MATRICES OF VOLTAGE SOURCE CONVERTERS FOR DISTRIBUTED GENERATORS 2195
Lian, K.L. ; Huang, T.D.

FPGA-BASED SIMULATION OF POWER ELECTRONICS USING ITERATIVE METHODS.................................2202
Zhang, Huiguo ; Sun, Jian

GALLIUM ARSENIDE IC TECHNOLOGY FOR POWER SUPPLIES ON CHIP.................................2208
Pala, Vipindas ; Peng, Han ; Hella, Mona ; Chow, T.Paul

SILICON ON NANOCRYSTALLINE AND MICROCRYSTALLINE DIAMOND STACKING STRUCTURE FOR POWER SUPPLY ON CHIP.................................2212
Yamada, Takatoshi ; Hasegawa, Masataka

A NOVEL LOAD REGULATION TECHNIQUE FOR POWER-SOC WITH PARALLEL CONNECTED POLS.................................2216
Abe, Seiya ; Matsumoto, Satoshi ; Hidaka, Akira ; Rikitake, Jungo ; Ninomiya, Tamotsu

MATRIX-POL ARCHITECTURE FOR INTEGRATED POWER SUPPLY 2222
Ishizuka, Yoichi ; Shibahara, Ryota ; Ninomiya, Tamotsu ; Tanaka, Kiminori ; Abe, Seiya

ON-CHIP BUCK CONVERTER WITH SPIRAL FERRITE INDUCTOR AND REDUCING IR DROP IN 3D STACKED INTEGRATION.................................2228
Fuketa, Hiroshi ; Shinozuka, Yasuhiro ; Ishida, Koichi ; Takamiya, Makoto ; Sakurai, Takayasu

DCM ANALYSIS OF A SINGLE SIC SWITCH BASED ZVZCS TAPPED BOOST CONVERTER 2232
Choi, Bo H. ; Lee, Eun S. ; Kim, Ji H. ; Rim, Chun T.

EFFECT OF INPUT AND OUTPUT TERMINAL SOURCES ON DYNAMIC BEHAVIOR OF SWITCHED-MODE CONVERTERS.................................2240
Suntio, T. ; Viinamaki, J. ; Jokipii, J. ; Messo, T. ; Sitbon, M. ; Kuperman, A.

A FULLY SOFT-SWITCHED MULTIPHASE DC-DC CONVERTER WITH REDUCED SWITCH COUNT FOR HIGH POWER APPLICATION 2247
Kim, Minjae ; Yang, Daeki ; Choi, Sewan

A STATIC CHARACTERISTIC ANALYSIS OF PROPOSED BI-DIRECTIONAL DUAL ACTIVE BRIDGE DC-DC CONVERTER.................................2252
Nagata, Shun ; Takasaki, Mika ; Furukawa, Yutaka ; Hirose, Toshiro ; Ishizuka, Yoichi

HYBRID BATTERY CHARGING SYSTEM COMBINING OBC WITH LDC FOR ELECTRIC VEHICLES 2260
Kim, Sconghyc ; Kang, Feel soon

TRANSIENT BEHAVIOR OF THE DUAL ACTIVE BRIDGE CONVERTER IN HIGH EFFICIENT ENERGY CONVERSION SYSTEM 2266
Aoyama, Kohei ; Motoi, Naoki ; Tsuruta, Yukinori ; Kawamura, Atsuo

STATE-OF-CHARGE ESTIMATION FOR LITHIUM-ION BATTERY PACK USING RECONSTRUCTED OPEN-CIRCUIT-VOLTAGE CURVE.................................2272
Chun, Chang Yoon ; Seo, Gab-Su ; Yoon, Sung Hyun ; Cho, Bo-Hyung

SYSTEM DESIGN OF ELECTRIC ASSISTED BICYCLE USING EDLCS AND WIRELESS CHARGER.................................2277
Itoh, Jun-ichi ; Noguchi, Kenji ; Orikawa, Koji

STUDY ON LOW-LOSS GATE DRIVE CIRCUIT FOR HIGH EFFICIENCY SERVER POWER SUPPLY USING NORMALLY-OFF SIC-JFET.................................2285
Katoh, Kaoru ; Ishikawa, Katsumi ; Hatanaka, Ayumu ; Ogawa, Kazutoshi ; Akiyama, Satoru ; Ogawa, Takashi ; Yokoyama, Natsuki ; Maru, Naoki ; Takahashi, Osamu ; Nishisu, Koji

A SHORT CIRCUIT PROTECTION METHOD BASED ON A GATE CHARGE CHARACTERISTIC.................................2290
Horiguchi, Takeshi ; Kinouchi, Shin-ichi ; Nakayama, Yasushi ; Oi, Takeshi ; Urushibata, Hiroaki ; Okamoto, Shoji ; Tominaga, Shinji ; Akagi, Hirofumi

HIGHLY RELIABLE 1200-V P-TYPE MOSFET FOR LEVEL-SHIFT CIRCUIT USED IN DRIVER IC.................................2297
Sakurai, Naoki ; Hakutou, Takuma ; Yura, Masashi

A NEW LEVEL UP SHIFTER FOR HVICS WITH HIGH NOISE TOLERANCE.................................2302
Akahane, Masashi ; Jonishi, Akihiro ; Yamaji, Masaharu ; Kanno, Hiroshi ; Tanaka, Takahide ; Nishio, Haruhiko ; Sumida, Hitoshi

OUTPUT RIPPLE MINIMIZATION OF SINGLE-STAGE POWER FACTOR CORRECTED BI-DIRECTIONAL BUCK AC/DC CONVERTER.................................2310
Veerasamy, Balaji ; Kitagawa, Wataru ; Takeshita, Takaharu

THREE-PHASE ISOLATED FULL-BRIDGE BOOST PFC WITH FLYBACK PASSIVE AUXILIARY CONVERTER.................................2318
Meng, Tao ; Yu, Shuai ; Ben, Hongqi ; Wei, Guo ; Sun, Shaohua

CONTROL AND EXPERIMENT OF A MODULAR PUSH-PULL PWM CONVERTER FOR A BATTERY ENERGY STORAGE SYSTEM.................................2323
Hagiwara, Makoto ; Akagi, Hirofumi

ACTIVE FRONT-END TOPOLOGY FOR 5 LEVEL MEDIUM VOLTAGE DRIVE SYSTEM WITH ISOLATED DC BUS.................................2330
Oka, Toshiaki ; Kusunoki, Hironobu ; Tsukakoshi, Masahiko ; Kleinecke, John ; Daskalos, Mike

A DUAL ACTIVE BRIDGE DC-DC CONVERTER WITH OPTIMAL DC-LINK VOLTAGE SCALING AND FLYBACK MODE FOR ENHANCED LOW-POWER OPERATION IN HYBRID PV/STORAGE SYSTEMS.................................2336
Poshtkouhi, Shahab ; Trescases, Olivier

NOVEL MODULAR MULTIPLE-INPUT BIDIRECTIONAL DC-DC POWER CONVERTER (MIPC) 2343
Hintz, Andrew ; Prasanna, Udupi.R. ; Rajashekara, Kaushik

SINGLE-SWITCH PWM CONVERTER INTEGRATING VOLTAGE EQUALIZER FOR PHOTOVOLTAIC MODULES UNDER PARTIAL SHADING..2351
Uno, Masatoshi ; Kukita, Akio

NEW DC RAIL SIDE SOFT-SWITCHING PWM DC-DC CONVERTER WITH VOLTAGE DOUBLER RECTIFIER FOR PV GENERATION INTERFACE..2359
Sayed, Khairy ; Kwon, Soon-Kurl ; Nishida, Katsumi ; Nakaoka, Mutsuo

MODELING METHOD OF STRAY MAGNETIC COUPLINGS IN AN EMC FILTER FOR A SIC SOLAR INVERTER..2366
Masuzawa, Takashi ; Hoene, Eckart ; Hoffmann, Stefan ; Lang, Klaus-Dieter

DC BUS VOLTAGE EMI MITIGATION IN THREE-PHASE ACTIVE RECTIFIERS USING A VIRTUAL NEUTRAL FILTER..2372
Parker, S.G. ; Segaran, D.S. ; Holmes, D.G. ; McGrath, B.P.

EFFECTS OF TRANSFORMER STRUCTURES ON THE NOISE BALANCING AND CANCELLATION MECHANISMS OF SWITCHING POWER CONVERTERS..2380
Hsieh, Hung-I ; Shih, Sheng-Fang

A NOVEL TECHNIQUE FOR REDUCING LEAKAGE CURRENT BY APPLICATION OF ZERO-SEQUENCE VOLTAGE..2385
Ayano, Hideki ; Murakami, Kouhei ; Matsui, Yoshihiro

AC-CHOPPERS USING INSTANTANEOUS VOLTAGE CONTROL TECHNIQUE TO SOLVE VOLTAGE SAG PROBLEMS..2392
Khomfoi, Surin

VOLTAGE REGULATION IN DISTRIBUTION SYSTEM USING THE COMBINED DVR..2400
Nakamura, Sota ; Aoki, Mutsumi ; Ukai, Hiroyuki

NONLINEAR CONTROL OF THREE-PHASE FOUR-WIRE DYNAMIC VOLTAGE RESTORERS FOR DISTRIBUTION SYSTEM..2406
Jeong, Seon-Yeong ; Nguyen, Thanh Hai ; Lee, Dong-Choon ; Kim, Jang-Mok

VOLUME 4

DISTURBANCE CALCULATION BASED ON SPACE VECTOR DOT PRODUCT: APPLICATIONS TO COMPENSATORS..2413
de Carvalho, Kelly Caroline Mingorancia ; Ama, Naji Rajai Nasri ; Komatsu, Wilson ; Martinz, Fernando Ortiz ; Figueredo, Ricardo Souza ; Matakas, Lourenco

PROPOSAL OF 6TH RADIAL FORCE CONTROL BASED ON FLUX LINKAGE..2421
Kanematsu, Masato ; Miyajima, Takayuki ; Fujimoto, Hiroshi ; Hori, Yoichi ; Enomoto, Toshio ; Kondou, Masahiko ; Komiya, Hiroshi ; Yoshimoto, Kantaro ; Miyakawa, Takayuki

AIR GAP CONTROL OF MULTI-PHASE TRANSVERSE FLUX PERMANENT MAGNET LINEAR SYNCHRONOUS MOTOR BY USING INDEPENDENT VECTOR CONTROL..2427
Hwang, Seon-Hwan ; Bang, Deok-Je ; Kim, Ji-Won

MODIFIED DIRECT INSTANTANEOUS TORQUE CONTROL OF SWITCHED RELUCTANCE MOTOR WITH HIGH TORQUE PER AMPERE AND REDUCED SOURCE CURRENT RIPPLE..2433
Suryadevara, Rohit ; Fernandes, B.G.

CONTROL OF WOUND FIELD SYNCHRONOUS MOTOR INTEGRATED WITH ZSI..2438
Tajima, G. ; Kosaka, T. ; Matsui, N. ; Tonogi, K. ; Minoshima, N. ; Yoshida, T.

A NOVEL IPMSM MODEL FOR ROBUST POSITION SENSORLESS CONTROL TO MAGNETIC SATURATION..2445
Matsumoto, Atsushi ; Hasegawa, Masaru ; Doki, Shinji

MOTOR DRIVE SYSTEM USING NONLINEAR MATHEMATICAL MODEL FOR PERMANENT MAGNET SYNCHRONOUS MOTORS..2451
Iwaji, Yoshitaka ; Nakatsugawa, Junnosuke ; Sakai, Toshifumi ; Aoyagi, Shigehisa ; Nagura, Hirokazu

SENSORLESS-ORIENTED DESIGN OF IPMSM..2457
Kano, Yoshiaki

NOISE REDUCTION METHOD BY INJECTED FREQUENCY CONTROL FOR POSITION SENSORLESS CONTROL OF PERMANENT MAGNET SYNCHRONOUS MOTOR..2465
Taniguchi, Shun ; Yasui, Kazuya ; Yuki, Kazuaki

FORCE SENSORLESS BILATERAL CONTROL USING A DYNAMICAL ASYMMETRIC COMPENSATOR..2470
Hama, Ryota ; Imai, Jun ; Takahashi, Akiko ; Funabiki, Shigeyuki

DESIGN OF M-IPD CONTROLLER OF MULTI-INERTIA SYSTEM USING DIFFERENTIAL EVOLUTION..2476
Ikeda, Hidehiro ; Tsuyoshi, Hanamoto

A GUIDE TO DESIGN DISTURBANCE OBSERVER BASED MOTION CONTROL SYSTEMS..2483
Sariyildiz, Emre ; Ohnishi, Kouhei

IDENTIFICATION OF TWO-MASS MECHANICAL SYSTEMS USING TORQUE EXCITATION: DESIGN AND EXPERIMENTAL EVALUATION..2489
Saarakkala, Seppo E. ; Hinkkanen, Marko

INDUCTOR LOSS CALCULATION OF COUPLED INDUCTORS FOR HIGH POWER DENSITY BOOST CONVERTER..2497
Itoh, Yuki ; Kimura, Shota ; Imaoka, Jun ; Yamamoto, Masayoshi

1.2KW DUAL-ACTIVE BRIDGE CONVERTER USING SIC POWER MOSFETS AND PLANAR MAGNETICS..2503
De, D. ; Castellazzi, A. ; Lamantia, A.

ANALYSIS OF HYSTERESIS AND EDDY-CURRENT LOSSES FOR A MEDIUM-FREQUENCY TRANSFORMER IN AN ISOLATED DC-DC CONVERTER..2511
Nakahara, Mizuki ; Wada, Keiji

EXPERIMENTAL VERIFICATION OF CAPACITIVE POWER TRANSFER USING ONE PULSE SWITCHING ACTIVE CAPACITOR FOR PRACTICAL USE..2517
Kitabayashi, Tatsuaki ; Funato, Hirohito ; Kobayashi, Hiroya ; Yamaichi, Katsuya

A SINGLE-STAGE HIGH-PF DRIVER FOR SUPPLYING A T8-TYPE LED LAMP................................2523
Cheng, Chun-An ; Chang, Chien-Hsuan ; Cheng, Hung-Liang ; Chung, Tsung-Yuan

ELIMINATION OF ELECTROLYTIC CAPACITOR IN AC-DC SYSTEM OF LED DRIVER................2529
Mustapa, Rijalul Fahmi ; Hidayat, Nabil M ; Tukiman, Rahayu

A NOVEL BRIDGELESS BOOST HALF-BRIDGE ZVS-PWM SINGLE-STAGE UTILITY FREQUENCY AC-HIGH FREQUENCY AC RESONANT CONVERTER FOR DOMESTIC INDUCTION HEATERS................2533
Mishima, Tomoakzu ; Nakagawa, Yuki ; Nakaoka, Mutsuo

APPLICATION OF VIRTUAL VALIDATION SYSTEM FOR INVERTER HEAT PUMP SYSTEM................2541
Kanamori, Masaki ; Noda, Koji ; Endo, Takahisa ; Suzuki, Nobuyuki

TEST SETUP FOR ACCELERATED TEST OF HIGH POWER IGBT MODULES WITH ONLINE MONITORING OF VCE AND VF VOLTAGE DURING CONVERTER OPERATION................................2547
de Vega, Angel Ruiz ; Ghimire, Pramod ; Pedersen, Kristian Bonderup ; Trintis, Ionut ; Beczckowski, Szymon ; Munk-Nielsen, Stig ; Rannestad, Bjorn ; Thogersen, Paul

DESIGN OF HIGH-SPEED IGBT-BASED SWITCHING MODULES FOR PULSED POWER APPLICATIONS................2554
Kluge, Andreas ; Goehler, Lutz ; Gueldner, Henry ; Trompa, Thomas ; Mory, David ; Segsa, Karl-Heinz

COMPARATIVE SUITABILITY EVALUATION OF REVERSE-BLOCKING IGBTS FOR CURRENT-SOURCE BASED CONVERTER..2562
De, Ankan ; Roy, Sudhin ; Bhattacharya, Subhashish

NEW REVERSE-CONDUCTING IGBT (1200V) WITH REVOLUTIONARY COMPACT PACKAGE................2569
Takahashi, K. ; Yoshida, S. ; Noguchi, S. ; Kuribayashi, H. ; Nashida, N. ; Kobayashi, Y. ; Kobayashi, H. ; Mochizuki, K. ; Ikeda, Y. ; Ikawa, O.

AN IMPROVED MODULATED CARRIER CONTROL OF SINGLE-PHASE CCM BOOST PFC CONVERTER................2575
Kim, Hyejin ; Cho, Bo-Hyung ; Choi, Hangseok

MODIFIED INTERLEAVED CURRENT SENSORLESS CONTROL FOR THREE-LEVEL BOOST PFC CONVERTER WITH ASYMMETRIC LOADS..2580
Chen, Hung-Chi ; Liao, Jhen-Yu

A NOVEL CRITICAL-CONDUCTION-MODE BRIDGELESS INTERLEAVED BOOST PFC RECTIFIER................2587
Cao, Guoen ; Kim, Hee-Jun

ANALYSIS AND DESIGN OF A PUSH-PULL SINGLE-STAGE FLYBACK POWER FACTOR CORRECTOR................2593
Lo, Yu-Kang ; Chiu, Huang-Jen ; Liu, Yu-Chen ; Lin, Chung-Yi ; Cheng, Shih-Jen ; Yang, CS

LINEAR OVER-MODULATION STRATEGY FOR CURRENT CONTROL IN PHOTOVOLTAIC INVERTER................2598
Park, Yongsoon ; Sul, Seung-Ki ; Hong, Ki-Nam

DESIGN OF DECENTRALIZED VOLTAGE CONTROL FOR PV INVERTERS TO MITIGATE VOLTAGE RISE IN DISTRIBUTION POWER SYSTEM WITHOUT COMMUNICATION................2606
Lee, Tzung-Lin ; Yang, Shih-Sian ; Hu, Shang-Hung

STABILITY ANALYSIS AND ACTIVE DAMPING FOR LLCL-FILTER BASED GRID-CONNECTED INVERTERS..2610
Huang, Min ; Blaabjerg, Frede ; Loh, Poh Chiang ; Wu, Weimin

INTEGRATED COMMON AND DIFFERENTIAL MODE FILTER APPLIED TO A SINGLE-PHASE TRANSFORMERLESS PV MICROINVERTER WITH LOW LEAKAGE CURRENT................2618
Figueredo, Ricardo Souza ; de Carvalho, Kelly Caroline Mingorancia ; Matakas, Lourenco

DESIGN AND INTEGRATION OF INTERPHASE INDUCTORS FOR INTERLEAVED THREE PHASE VOLTAGE-SOURCE-INVERTERS IN DC-FED MOTOR DRIVE SYSTEMS................2626
Zhang, Xuning ; Boroyevich, Dushan ; Burgos, Rolando

A NOVEL TRANSFORMER MODEL USING MAGNETIC CIRCUIT..2632
Nakamurame, Fuminori ; Ise, Toshifumi

HARDWARE-IN-THE-LOOP SIMULATION OF A MACHINE MODEL WITH REAL-TIME ANIMATION................2638
Xiaojie Zhuang ; Hibino, Shinya ; Harakawa, Masaya ; Terabe, Ryosuke ; Ozaki, Takayuki ; Nagano, Tetsuaki

DEVELOPMENT OF REAL TIME DIGITAL SIMULATOR FOR SELF-COMMUTATED SVC TO SUPPRESS VOLTAGE FLICKER..2644
Terao, Yutaka ; Shishida, Yasuhiro ; Tsuruma, Yoshinori ; Ishizuka, Tomotsugu ; Aoyama, Fumio ; Yoshino, Teruo ; Kato, Yutaka ; Belanger, Jean

OPERATIONAL ASPECTS AND POWER ARCHITECTURE DESIGN FOR A MICROGRID TO INCREASE THE USE OF RENEWABLE ENERGY IN WIRELESS COMMUNICATION NETWORKS................2649
Kwasinski, Alexis ; Kwasinski, Andres

P+ MULTIPLE RESONANT CONTROL FOR OUTPUT VOLTAGE REGULATION OF MICROGRID WITH UNBALANCED AND NONLINEAR LOADS..2656
Kyungbae Lim ; Jaeho Choi ; Juyoung Jang ; Junghum Lee ; Jaesig Kim

130MVA-STATCOM FOR TRANSIENT STABILITY IMPROVEMENT..2663
Imanishi, Takao ; Nagatomo, Yoshinobu ; Iwasaki, Shinya ; Masaki, Kenji ; Fujii, Toshiyuki ; Ieda, Jun

IMPROVED DROOP CONTROLLER FOR MICROGRID INVERTER CONSIDERING THE LINE IMPEDANCE MISMATCHING..2668
Du Yan ; Liuchen Chang ; Meiqin Mao ; Jianhui Su ; Ning Liu

SUPPRESSION CONTROL METHOD FOR IRON LOSS OF MATRIX MOTOR UNDER FLUX WEAKENING UTILIZING INDIVIDUAL WINDING CURRENT CONTROL 2673
Hijikata, Hiroki ; Akatsu, Kan ; Miyama, Yoshihiro ; Arita, Hideaki ; Daikoku, Akihiro

PERFORMANCE ANALYSIS OF A NEW CONCENTRATEDWINDING INTERIOR PERMANENT MAGNET SYNCHRONOUS MACHINE UNDER FIELD ORIENTED CONTROL 2679
Nguyen, D. ; Dutta, R. ; Fletcher, J. ; Rahman, F. ; Lovatt, Howard

ONLINE PARTICLE SWARM OPTIMIZATION FOR SENSORLESS IPMSM DRIVES CONSIDERING PARAMETER VARIATION 2686
Song, Z.Q. ; Xiao, D. ; Rahman, M.F.

A DTC-PWM CONTROL SCHEME OF PMSM BASED ON 12-SECTORS DIVISION AND SPEED INFORMATION 2693
Yunchang Kwak ; Jin-Woo Ahn ; Dong-Hee Lee

CONTROL OF POWER FLOW BETWEEN THE WIND GENERATOR AND NETWORK 2700
Stumpf, Peter ; Nagy, Istvan ; Vajk, Istvan

ADVANCES IN NANOGRID TECHNOLOGY AND ITS INTEGRATION INTO RURAL ELECTRIFICATION IN INDIA 2707
Mishra, Santanu ; Ray, Olive

STUDY AND IMPLEMENTATION OF SEVEN-LEVEL INVERTER USING COUPLED INDUCTOR AND SWITCHED-CAPACITOR 2714
Yi-Chun Lin ; Jiann-Fuh Chen ; Wen-Chien Hsu ; Sheng-Kai Kao

CASCADED MULTILEVEL CONVERTER BASED BIDIRECTIONAL INDUCTIVE POWER TRANSFER (BIPT) SYSTEM 2722
Bac Xuan Nguyen ; Vilathgamuwa, D.M. ; Foo, Gilbert ; Ong, Andrew ; Sampath, Prasad K. ; Madawala, Udaya K.

UNDERSAMPLING CONTROL OF A BIDIRECTIONAL CASCADED BUCK+BOOST DC-DC CONVERTER 2729
Rosekeit, Martin ; Joebges, Philipp ; Lelie, Markus ; Sauer, Dirk Uwe ; De Doncker, Rik W.

SUB-MICROSECOND RESPONSE DIGITAL CONTROLLER FOR POL 2737
Nonaka, Hirotaka ; Ishizuka, Yoichi ; Mii, Kenji ; Takenami, Fumiaki ; Kanemoto, Daisuke

GAIN CONTROLLED HIGH EFFICIENCY POWER FACTOR CORRECTION CIRCUIT 2745
Yonezawa, Yu ; Nakao, Hiroshi ; Sasaki, Tomotake ; Matsui, Yoshinobu ; Nakashima, Yoshiyasu ; Kaneko, Junji ; Shimamori, Hiroshi ; Yoshino, Yukio ; Hisato, Hosoyama ; Atsushi, Manabe ; Motizuki, Shun ; Yamashita, Shigeharu

DESIGN OF QUASI-RESONANT FLYBACK CONVERTER CONTROL IC WITH DCM AND CCM OPERATION 2750
Kai-Hui Chen ; Tsorng-Juu Liang

LOAD TRANSIENT RESPONSE IMPROVEMENT BASED ON PID CONTROL 2754
Yau, Y.T. ; Hwu, K.I.

AN ACTIVE-CLAMPING FORWARD CONVERTER WITH NON-LINEAR STEP-DOWN CONVERSION 2758
Jing-Yuan Lin ; Yu-Kang Lo ; Huang-Jen Chiu ; Chao-Fu Wang ; Chien-Yu Lin

SWITCHING LOSS MINIMIZATION OF 3-PHASE INTERLEAVED BIDIRECTIONAL DC-DC CONVERTER 2763
Eui-Cheol Nho ; Jae-Hun Jung ; Hak-Soo Kim ; In-Dong Kim ; Heung-Geun Kim ; Tae-Won Chun

MODIFIED THREE-PHASE THREE-LEVEL DC-DC CONVERTER -ADOPTING ASYMMETRICAL DUTY CYCLE CONTROL 2768
Yue Chen ; Xuling Chen ; Liu, Fuxin ; Ruan, Xinbo

DEADBEAT CONTROL OF POWER LEVELING UNIT WITH BIDIRECTIONAL BUCK/BOOST DC/DC CONVERTER 2775
Hamasaki, Shin-ichi ; Mukai, Ryosuke ; Yano, Yoshihiro ; Tsuji, Mineo

DESIGN OF OPTIMIZED ON-OFF CONTROL TO IMPROVE EFFICIENCY OF PARALLELED CONVERTER SYSTEM 2781
Kohama, Teruhiko ; Sogawa, Yuki ; Tsuji, Satoshi

EFFICIENCY IMPROVEMENTS IN A SINGLE ACTIVE BRIDGE MODULAR DC-DC CONVERTER WITH SNUBBER CAPACITANCE OPTIMISATION 2787
Ting, Yeh ; de Haan, Sjoerd ; Ferreira, Jan A.

A WIRELESS POWER TRANSFER SYSTEM OPTIMIZED FOR HIGH EFFICIENCY AND HIGH POWER APPLICATIONS 2794
Bani Shamseh, Mohammad ; Kawamura, Atsuo ; Yuzurihara, Itsuo ; Takayanagi, Atsushi

NON-ITERATIVE LCL FILTER DESIGN FOR THREE-PHASE TWO-LEVEL VOLTAGE-SOURCE PWM CONVERTERS 2802
Byung-Geuk Cho ; Seung-Ki Sul

DSP-BASED INTERLEAVED BUCK POWER FACTOR CORRECTOR 2810
Yu-Chen Liu ; Tsan Chen ; Po-Jung Tseng ; Yu-Kang Lo ; Huang-Jen Chiu

THE AVERAGE MODEL OF A THREE-PHASE THREE-STAGE POWER ELECTRONIC TRANSFORMER 2815
Shaodi Ouyang ; Liu, Jinjun ; Wang, Xinyu ; Wang, Xiaojian ; Fei Meng ; Riffat, Javid

A MULTI-CARRIER PWM FOR AC-DC-AC CONVERTER WITHOUT DC LINK ELECTROLYTIC CAPACITOR 2821
Chung-Chuan Hou ; Hsin-Ping Su

A DECOUPLING OFFSET-BASED PWM CONTROL FOR A MULTILEVEL INVERTER UNDER DC VOLTAGE UNBALANCE 2826
Nho Van Nguyen ; Tam Khanh Tu Nguyen ; Lee, Hong-Hee

?-? PARETO OPTIMIZATION OF 3-PHASE 3-LEVEL T-TYPE AC-DC-AC CONVERTER COMPRISING SI AND SIC HYBRID POWER STAGE..2834
Uemura, Hirofumi ; Krismer, Florian ; Okuma, Yasuhiro ; Kolar, Johann W.

PRACTICAL INVESTIGATION OF THE GATE BIAS EFFECT ON THE REVERSE RECOVERY BEHAVIOR OF THE BODY DIODE IN POWER MOSFETS 2842
Lindberg-Poulsen, Kristian ; Petersen, Lars Press ; Ouyang, Ziwei ; Andersen, Michael A.E.

AN ONLINE VCE MEASUREMENT AND TEMPERATURE ESTIMATION METHOD FOR HIGH POWER IGBT MODULE IN NORMAL PWM OPERATION 2850
Ghimire, Pramod ; de Vega, Angel Ruiz ; Beczkowski, Szymon ; Munk-Nielsen, Stig ; Rannested, Bjorn ; Thogersen, Paul Bach

EVALUATION ON IRON LOSS CHARACTERISTICS IN SERIES CONNECTION AND PARALLEL CONNECTION OF LOADS WITH INVERTER EXCITATION 2856
Odawara, Shunya ; Fujisaki, Keisuke

LOSS AND THERMAL MODEL FOR POWER SEMICONDUCTORS INCLUDING DEVICE RATING INFORMATION 2862
Ma, K. ; Bahman, A.S. ; Beczkowski, S.M. ; Blaabjerg, F.

IMPROVING RELIABILITY OF IGBT SURFACE ELECTRODE FOR 200 C OPERATION 2870
Nishimura, Tomohiro ; Ikeda, Yoshinari ; Hokazono, Hiroaki ; Mochizuki, Eiji ; Takahashi, Yoshikazu

INFLUENCE OF CARRIER FREQUENCY ON IRON LOSS TAKING ACCOUNT OF DEAD TIME EFFECT 2874
Kogi, Ryosuke ; Odawara, Shunya ; Fujisaki, Keisuke

DECREASE OF SIC-BJT DRIVER LOSSES BY ONE-STEP COMMUTATION 2881
Barth, Henry ; Hofmann, Wilfried

POWER PROFILE BASED SELECTION AND OPERATION OPTIMIZATION OF PARALLEL-CONNECTED POWER CONVERTER COMBINATIONS 2887
Vogt, T. ; Peters, A. ; Frohleke, N. ; Bocker, J. ; Kempen, S.

A NOVEL POWER LOSS CALCULATION METHOD FOR IGBTS IN POWER CONVERTERS VIA CHAOTIC SPWM CONTROL 2893
Boyu Wang ; Li, Hong ; Xiaojie You ; Trillion Zheng

LOSS ANALYSIS AND SOFT-SWITCHING CHARACTERISTICS OF FLYBACK-FORWARD HIGH GAIN DC/DC CONVERTER WITH GAN FET 2899
Zhang Yajing ; Zheng, Trillion Q. ; Li Yan

INSULATED METAL SUBSTRATE FOR POWER MODULES USING ANODIC OXIDE FILM OF ALUMINUM 2904
Tokuyama, Takeshi ; Kusukawa, Jumpei ; Nakatsu, Kinya

A FAST-TRANSIENT-RESPONSE BUCK CONVERTER WITH SPLIT-TYPE III COMPENSATION AND CHARGE-PUMP CIRCUIT TECHNIQUE 2910
Chen, Jiann-Jong ; Wei-Ting Hsu ; Jih-Hua Yu ; Hwang, Yuh-Shyan ; Cheng-Chieh Yu

ADVANTAGES OF LOW PARASITIC INDUCTANCE PACKAGES OF POWER MOSFET FOR SERVER POWER APPLICATIONS 2914
Wonsuk Choi ; Dongkook Son ; Dongwook Kim

MODULAR INTEGRATION OF A MATRIX CONVERTER 2920
Solomon, Adane Kassa ; Skuriat, Robert ; Castellazzi, Alberto ; Wheeler, Pat

A MODULAR NANOSECOND PULSE GENERATION SYSTEM FOR PLASMA-ASSISTED IGNITION 2926
Peng Gao ; Fletcher, John ; O'Byrne, Sean

DEVELOPMENT OF A SINGLE SWITCH CELL FOR MODULAR NANOSECOND PULSE GENERATION SYSTEMS 2932
Peng Gao ; Fletcher, John ; O'Byrne, Sean

ADVANTAGE OF SUPER JUNCTION MOSFET FOR POWER SUPPLY APPLICATION 2939
Tabira, K. ; Watanabe, S. ; Shimatou, T. ; Watashima, T. ; Takenoiri, S.

STUDY ON AN ACCURATE CALCULATION OF THE CONDUCTED EMI NOISE OF THE POWER CONVERTERS 2944
Omata, Shinpei ; Shimizu, Toshihisa

AN EXACT DISCRETE-TIME MODEL CONSIDERING DEAD-TIME NONLINEARITY FOR AN H-BRIDGE GRID-CONNECTED INVERTER 2950
Xie, Ruiliang ; Hao, Xiang ; Yang, Xu ; Chen, Wenjie ; Huang, Lang ; Chao Wang

THEORETICAL ANALYSIS OF THE DUALITY PRINCIPLE APPLIED TO INTERLEAVED TOPOLOGIES 2954
Caris, M.L.A. ; Huisman, H. ; Duarte, J.L.

A NEW IMPEDANCE MEASUREMENT METHOD BASED ON HIGH FREQUENCY COMPENSATION 2960
Yue, Xiaolong ; Zhuo, Fang ; Hao Yi

NUMERICAL AND EXPERIMENTAL INVESTIGATION OF PARASITIC EDGE CAPACITANCE FOR PHOTOVOLTAIC PANEL 2967
Wenjie Chen ; Xiaomei Song ; Hao Huang ; Xu Yang

VEHICLE INTERIOR NOISE CONTROL OF ULTRA-COMPACT ELECTRIC VEHICLE (FUNDAMENTAL CONSIDERATION USING RECTANGULAR ENCLOSURE) 2972
Kato, Taro ; Kato, Hideaki ; Oshinoya, Yasuo ; Suzuki, Ryosuke ; Hasegawa, Shinya

CONSIDERATION FOR THE PROPAGATION PATH OF CONDUCTIVE NOISE IN AIR CONDITIONERS 2977
Tokiwa, Tsuyoshi ; Kanamori, Masaki ; Endo, Takahisa ; Iida, Mikiya ; Ogasawara, Satoshi ; Yizhanyi Tang

IRON LOSS EVALUATION OF IRON POWDER CORE SUITABLE FOR INDUCTOR USED IN POWER CONVERTERS 2983
Mori, Tomohiro ; Igarashi, Kazunori ; Kanagawa, Kinji ; Yamashita, Nobuyuki ; Shimizu, Toshihisa ; Bizen, Yosio

OPTIMIZED TUNING METHOD OF STATIONARY FRAME PROPORTIONAL RESONANT CURRENT CONTROLLERS 2988

Martinz, Fernando Ortiz ; de Carvalho, Kelly Caroline Mingorancia ; Ama, Naji Rajai Nasri ; Komatsu, Wilson ; Matakas, Lourenco

INSTANTANEOUS POWER THEORY APPLIED TO POWER CONDITIONING UNDER DISTORTED MAINS VOLTAGES: A MATLAB/SIMULINK APPROACH....................2996

Nicolae, Petre-Marian ; Popa, Lucian-Dinut ; Nicolae, Marian-Stefan ; Nicolae, Ileana-Diana

THE RESEARCH ON RELIABILITY AND REAL-TIME OF THE SCHEME OF PROCESS LAYER GOOSE NETWORK IN SMART SUBSTATION BASED ON ARTIFICIAL COBWEB TOPOLOGY STRUCTURE 3002

Liu, Xiaosheng ; Zhu, Honglin ; Xu, Dianguo ; Li, Yanxiang

EFFICIENCY IMPROVEMENT OF A SELF-START TYPE PERMANENT MAGNET SYNCHRONOUS MOTOR....................3007

Saikusa, H. ; Arikawa, S. ; Higuchi, T. ; Yokoi, Y. ; Abe, T.

CONSIDERATION OF OPTIMAL NUMBER OF POLES AND FREQUENCY FOR HIGH-EFFICIENCY PERMANENT MAGNET MOTOR 3012

Misu, Daisuke ; Matsushita, Makoto ; Takeuchi, Katsutoku ; Oishi, Koji ; Kawamura, Mitsuhiro

BASIC STUDY ON THE SUITABLE STRUCTURE OF A PERMANENT MAGNET SYNCHRONOUS MOTOR WITH A POWDER MAGNETIC CORE 3018

Hashimoto, Shizuka ; Sanada, Masayuki ; Morimoto, Shigeo ; Inoue, Yukinori

CHARACTERISTICS OF A HALF-WAVE RECTIFIED BRUSHLESS SYNCHRONOUS GENERATOR....................3024

Hirakawa, Yuki ; Higuchi, Tsuyoshi ; Yokoi, Yuichi ; Abe, Takashi

MODELING OF WOUND ROTOR SYNCHRONOUS MACHINES CONSIDERING HARMONICS, GEOMETRIC SALIENCIES AND SATURATION INDUCED SALIENCIES 3029

Rambetius, Alexander ; Luthardt, Sven ; Piepenbreier, Bernhard

DESIGN AND COMPARISON OF HIGH FREQUENCY TRANSFORMERS USING FOIL AND ROUND WINDINGS 3037

Iyer, Kartik V ; Robbins, William P ; Mohan, Ned

A METHOD TO CALCULATE THE PERFORMANCE OF LINEAR INDUCTION MOTORS USING SIMPLE TWO-PHASE MODEL 3044

Hirahara, Hideaki ; Yamamoto, Shu ; Ara, Takahiro ; Shimizu, Toshihisa

AN ESP DOWNHOLE PARAMETERS MONITORING SYSTEM BASED ON CURRENT LOOP TRANSMISSION METHOD 3050

Jin Miaoxin ; Zhang Wei ; Gao Qiang ; Xu Dianguo

BENDING MAGNETIC LEVITATION CONTROL FOR THIN STEEL PLATE (EXPERIMENTAL CONSIDERATION USING SLIDING MODE CONTROL)....................3055

Yonezawa, Hikaru ; Narita, Takayoshi ; Oshinoya, Yasuo ; Marumori, Hiroki ; Hasegawa, Shinya

TRANSFORMER WINDING LOSSES WITH ROUND CONDUCTORS FOR DUTY-CYCLE REGULATED SQUARE WAVES....................3061

Iyer, Kartik V ; Robbins, William P ; Basu, Kaushik ; Mohan, Ned

SIMULATION OF RESIN MOLDED TYPE SENSOR IN POLE SWITCH FOR POWER DELIBERY SYSTEMS 3067

Furukawa, Tatsuya ; Muta, Shoichiro ; Fukumoto, Hisao ; Itoh, Hideaki ; Ohchi, Masashi

ROBUST STARTUP CONTROL OF SENSORLESS PMSM DRIVES WITH SELF-COMMISSIONING 3072

Lin, Chiao-Chien ; Tzou, Ying-Yu

POSITION SENSORLESS CONTROL OF PMSM WITH A LOW-FREQUENCY SIGNAL INJECTION 3079

Nimura, Tomohiro ; Doki, Shinji ; Fujitsuna, Masami

A COMPARISON OF DIFFERENT SENSORLESS POSITION ACQUISITION METHODS AT LOW SPEEDS FOR A PERMANENT MAGNET SYNCHRONOUS MACHINE IN VEHICLE APPLICATIONS....................3085

Lehmann, Oliver ; Zehelein, Matthias ; Schuster, Johannes ; Roth-Stielow, Jorg

STABILITY COMPARISON OF IPMSM SENSORLESS VECTOR CONTROL SYSTEMS USING EXTENDED EMF 3093

Tsuji, Mineo ; Mizusaki, Hiroshi ; Hamasaki, Sin-ichi

INDUCTION MACHINE BASED FLYWHEEL SPEED ESTIMATION AT STAND-BY MODE....................3099

Liu, Rongqiang ; Xu, David

SYMMETRICAL SIGNALING SYSTEM FOR SENSOR-LESS SRM DRIVE 3106

Yamamoto, Kenji ; Takahashi, Hisashi ; Ushiro, Nobumasa ; Shirasawa, Koki

DIGITAL INTEGRATORS FOR CONDITION MONITORING: A DC AND MULTITONE SIGNAL ANALYSIS 3111

Peretti, L.

AUDIBLE NOISE REDUCTION METHOD IN IPMSM POSITION SENSORLESS CONTROL BASED ON HIGH-FREQUENCY CURRENT INJECTION 3119

Tauchi, Yuki ; Kubota, Hisao

A NOVEL DESIGN FOR INDUCTION MOTOR FLUX ESTIMATION USING IMPULSIVE OBSERVER....................3124

Peng Wang ; Yan Li ; Jianwen Zhang ; Xu Cai ; Zhengzhi Han

LOAD TORQUE AND INERTIA SIMULATION BASED ON DOUBLE-STATOR PERMANENT-MAGNET SYNCHRONOUS MOTOR 3129

Zhe Wang ; Mingyan Wang ; Ben Guo ; Chai Feng

INDEPENDENT SPEED AND POSITION CONTROL OF TWO PERMANENT MAGNET SYNCHRONOUS MOTORS FED BY A FOUR-LEG INVERTER 3134

Kubo, Yuji ; Moroi, Takayuki ; Kouki, Matsuse ; Kubota, Hisao ; Rajashekara, Kaushik

MINIMIZATION OF STATOR CURRENTS FOR MONO INVERTER DUAL PARALLEL PMSM DRIVE SYSTEM..3140
Yongjae Lee ; Ha, Jung-Ik

PERFORMANCE COMPARISON OF INVERTER AND DRIVE CONFIGURATIONS WITH OPEN-END AND STAR-CONNECTED WINDINGS...3145
Neubert, Markus ; Koschik, Stefan ; De Doncker, Rik W.

INPUT CURRENT HARMONICS REDUCTION CONTROL FOR ELECTROLYTIC CAPACITOR LESS INVERTER BASED IPMSM DRIVE SYSTEM...3153
Abe, Kodai ; Ohishi, Kiyoshi ; Haga, Hitoshi

NONCONTACT GUIDE SYSTEM FOR TRAVELING ELASTIC STEEL PLATES (THEORETICAL STUDY ON THE SHAPE OF TRAVELING STEEL PLATE)...3159
Sakaba, Kouichi ; Hasegawa, Shinya ; Narita, Takayoshi ; Oshinoya, Yasuo

ACTIVE SEAT SUSPENSION FOR ULTRA-COMPACT VEHICLE (FUNDAMENTAL CONSIDERATION ON ELECTROMYOGRAM WHEN FALL FROM THE BUMP)...3162
Mashino, Masahiro ; Sunaga, Keita ; Hasegawa, Shinya ; Ishida, Masaki ; Kato, Hideaki ; Oshinoya, Yasuo

ADAPTIVE CURRENT TRACKING OF THREE-PHASE ACTIVE POWER FILTER USING BACKSTEPPING CONTROL..3168
Yunmei Fang ; Juntao Fei ; Shixi Hou ; Weili Dai

FAST IDENTIFICATION OF RESONANCE CHARACTERISTIC FOR 2-MASS SYSTEM WITH ELASTIC LOAD...3174
Ming Yang ; Liang Hao ; Dianguo Xu

AUTONOMOUS NAVIGATION SYSTEM BASED ON COLLISION DANGER-DEGREE FOR UNMANNED GROUND VEHICLE...3179
Yasuno, Takashi ; Tanaka, Daiki ; Kuwahara, Akinobu

A HIGH-PERFORMANCE BIDIRECTIONAL DC-DC CONVERTER FOR DC MICRO-GRID SYSTEM APPLICATION..3185
Shu-Wei Kuo ; Yu-Kang Lo ; Huang-Jen Chiu ; Shih-Jen Cheng ; Chung-Yi Lin ; Yang, CS

VOLUME 5

IMPROVEMENT IN EFFICIENCY OF LED LIGHTING SYSTEM..3190
Hwu, K.I. ; Jiang, W.Z. ; Jenn-Jong Shieh

COMPARISON AND EVALUATION OF VIBRATION-BASED PIEZOELECTRIC POWER GENERATORS.............3194
Basari, Amat A. ; Awaji, Sosuke ; Hashimoto, Seiji ; Kasai, Makoto ; Suto, Kenji ; Kumagai, Shunji ; Kasai, Makoto ; Suto, Kenji ; Wei Jiang ; Shuren Wang

BATTERY SELECTION FOR HYBRID ENERGY SYSTEMS AND THERMAL MANAGEMENT IN ARCTIC CLIMATES..3200
Kalogera, Maria ; Bauer, Pavol

100KW PV PCS WITH NATURAL CONVECTION COOLING FOR OUTDOOR INSTALLATION.........................3207
Jin, Yasuhiro ; Matsuoka, Kazumasa ; Takahashi, Takehiro ; Takahashi, Nobuhiro

A NEW PLL BASED ON FAST POSITIVE AND NEGATIVE SEQUENCE DECOMPOSITION ALGORITHM WITH MATRIX OPERATION UNDER DISTORTED GRID CONDITIONS...3213
Shaohua Sun ; Hongqi Ben ; Tao Meng ; Jinyong Zhang

PERFORMANCE IMPROVEMENT OF PHOTOVOLTAIC POWER GENERATION SYSTEMS USING ON-OFF CONTROL METHODS...3218
Kenji, Matsumoto ; Nomura, Shinichi

LOW VOLTAGE PV POWER INTEGRATION INTO MEDIUM VOLTAGE GRID USING HIGH VOLTAGE SIC DEVICES...3225
Chattopadhyay, Ritwik ; Bhattacharya, Subhashish ; Foureaux, Nicole C. ; Silva, Sidelmo M. ; Braz Cardoso, F. ; de Paula, Helder ; Pires, Igor A. ; Cortizio, Porfirio C. ; Moraes, Lenin ; de S.Brito, Jose A.

A NOVEL GLOBAL MAXIMUM POWER POINT TRACKING METHOD FOR PHOTOVOLTAIC GENERATION SYSTEM OPERATING UNDER PARTIALLY SHADED CONDITION...3233
Jing-Hsiao Chen ; Yu-Shan Cheng ; Shun-Chung Wang ; Huang, Jia-Wei ; Liu, Yi-Hua

AN APPLICATION OF Z-SOURCE CONVERTER TO BATTERIES CHARGE WITH A PHOTOVOLTAIC SYSTEM..3239
Razik, H. ; Zitouni, Y. ; Maret, C.

PCS WITH SCANNING-TYPE MPPT CONTROL FOR INDUSTRIAL GRID-CONNECTED PV POWER GENERATION SYSTEM...3244
Itako, Kazutaka

FEASIBLE METHOD OF CALCULATING LEAKAGE REACTANCE OF 9-WINDING TRANSFORMER FOR HIGH-VOLTAGE INVERTER SYSTEM...3249
Fukumoto, Hisao ; Furukawa, Tatsuya ; Itoh, Hideaki ; Ohchi, Masashi

HIGH POWER HVDC-DC CONVERTERS FOR THE INTERCONNECTION OF HVDC LINES WITH DIFFERENT LINE TOPOLOGIES...3255
Schon, Andre ; Bakran, Mark-M.

CHARACTERIZATION OF A CURRENT SHUNT AND AN INDUCTIVE VOLTAGE DIVIDER FOR PMU CALIBRATION..3263
Kon, Saytaro ; Yamada, Tatsuji

DISTRIBUTED SERIES/HYBRID-SHUNT COMPENSATION FOR HARMONIC MITIGATION IN COMMERCIAL FACILITIES..3270
Diniz, Rogerio Azevedo ; Pires, Igor A. ; Franca, Gleisson J. ; Cardoso, Braz J.

ROBUST CONTROL DESIGN FOR THE VOLTAGE TRACKING LOOP OF A DVR................................3278
Ferrari, Bruno Augusto ; Ama, Naji Rajai Nasri ; de Carvalho, Kelly Caroline Mingorancia ; Martinz, Fernando Ortiz ; Matakas, Lourenco

MULTI-PORT SOLID STATE TRANSFORMER FOR INTER-GRID POWER FLOW CONTROL 3286
Roy, Sudhin ; De, Ankan ; Bhattacharya, Subhashish

REACTIVE POWER CONTROL STRATEGY BASED ON DC CAPACITOR VOLTAGE CONTROL FOR ACTIVE LOAD BALANCER IN THREE-PHASE FOUR-WIRE DISTRIBUTION SYSTEMS 3292
Tint Soe Win ; Hisada, Yoshihiro ; Tanaka, Toshihiko ; Hiraki, Eiji ; Okamoto, Masayuki ; Lee, Seong Ryong

VOLTAGE SAG RIDE-THROUGH PERFORMANCE OF VIRTUAL SYNCHRONOUS GENERATOR.................3298
Alipoor, Jaber ; Miura, Yushi ; Ise, Toshifumi

CONTROL OF DISTRIBUTED GENERATION SYSTEMS UNDER UNBALANCED VOLTAGE CONDITIONS 3306
Kabiri, R. ; Holmes, D.G. ; McGrath, B.P.

STABILITY ANALYSIS OF GRID-CONNECTED INVERTERS WITH LCL-FILTER BASED ON HARMONIC BALANCE AND FLOQUET THEORY..3314
Jing Bian ; Hong Li ; Zheng, Trillion Q.

COMPARATIVE EVALUATION OF PASSIVE DAMPING TOPOLOGIES FOR PARALLEL GRID-CONNECTED CONVERTERS WITH LCL FILTERS 3320
Beres, Remus ; Wang, Xiongfei ; Blaabjerg, Frede ; Bak, Claus Leth ; Liserre, Marco

STUDY AND IMPLEMENTATION OF A SEPIC LED DRIVER WITH ADJUSTABLE OUTPUT VOLTAGE 3328
Po-Jung Tseng ; Yu-Chen Liu ; Yu-Kang Lo ; Chiu, Huang-Jen ; Yun-Chu Chiu

AN INTERLEAVED SINGLE-STAGE LLC RESONANT CONVERTER USED FOR MULTI-CHANNEL LED DRIVING 3333
Chang, Chien-Hsuan ; Cheng, Chun-An ; Jinno, Masahito ; Cheng, Hung-Liang

A NOVEL TYPE OF WIRELESS V2H SYSTEM WITH BIDIRECTIONAL RESONANT SINGLE-ENDED INVERTER 3341
Fukuoka, Hiroki ; Iga, Yuichi ; Omori, Hideki ; Morizane, Tosimitsu ; Kimura, Noriyuki ; Nakaoka, Mutuo

DESIGN AND IMPLEMENTATION OF AN INTERLEAVED BCM BOOST PFC CONTROL IC 3346
Kuan-Hsien Chou ; Tsorng-Juu Liang ; Kai-Hui Chen ; Ji-Shiang Lee

LOW CAPACITIVE INDUCTORS FOR FAST SWITCHING DEVICES IN ACTIVE POWER FACTOR CORRECTION APPLICATIONS 3352
Hernandez, Juan C. ; Petersen, Lars P. ; Andersen, Michael A.E.

TEMPERATURE-ROBUST LC3 LED DRIVER WITH LOW THD, HIGH EFFICIENCY, AND LONG LIFE.........................3358
Lee, Eun S. ; Choi, Bo H. ; Cheon, Jun P. ; Kim, Bong C. ; Rim, Chun T.

OPTIMIZING REPULSIVE LORENTZ FORCES FOR A LEVITATING INDUCTION COOKER.................................3365
Zingerli, Claudius M. ; Nussbaumer, Thomas ; Kolar, Johann W.

DESIGN OF A MODULAR RESONANT CONVERTER FOR 25KV-8A DC POWER SUPPLY OF RF CAVITIES..3371
Siemaszko, Daniel ; Pittet, Serge ; Aguglia, Davide ; de Mallac, Louis

A NOVEL TRANSFORMER-LESS INTERLEAVED FOUR-PHASE HIGH STEP-DOWN DC CONVERTER WITH LOW SWITCH VOLTAGE STRESS 3379
Ching-Tasi Pan ; Chen-Feng Chuang ; Chia-Chi Chu ; Hao-Chien Cheng

EFFICIENCY IMPROVEMENT OF POWER SUPPLY WITH TRANSIENT CURRENT CIRCUIT USING DIGITAL CONTROL..3386
Takashita, Haruomi ; Shoyama, Masahito ; Yonezawa, Yu ; Nakashima, Yoshiyasu

ULTRA HIGH STEP-DOWN CONVERTER 3392
Yau, Y.T. ; Hwu, K.I.

DIGITAL CONTROL OF PWM INVERTER USING ULTRA HIGH SPEED NETWORK FOR FEEDBACK SIGNALS WITH COMMUNICATION DISTURBANCE OBSERVER BASED ON ROCKET I/O PROTOCOL 3397
Saito, Ryo ; Tsuchida, Kazuo ; Yokoyama, Tomoki

100 KHZ DC CHOPPER DIGITALLY GATE CONTROLLED WITH PARTIAL TURN- OFF SWITCHING USING SIC-MOSFET AND FPGA..3403
Tsuruta, Yukinori ; Kawamura, Atsuo

VARIABLE CARRIER DEADBEAT CONTROL WITH DIGITAL HYSTERESIS METHOD USING SOC-FPGA FOR UTILITY INTERACTIVE INVERTER 3410
Ohashi, Shunsuke ; Yoshida, Morito ; Yokoyama, Tomoki

A SPACE VECTOR MODULATION STRATEGY FOR THREE-LEVEL OPERATION BASED ON DUAL TWO-LEVEL VOLTAGE SOURCE INVERTERS 3417
Kumsuwan, Yuttana ; Srirattanawichaikul, Watcharin

INVESTIGATION ON THE PARALLEL OPERATION OF ALL-GAN POWER MODULE AND THERMAL PERFORMANCE EVALUATION 3425
Cheng, Stone ; Po-Chien Chou

FULL SILICON CARBIDE BOOST CHOPPER MODULE FOR HIGH FREQUENCY AND HIGH TEMPERATURE OPERATION..3432
Pettersson, Sami ; Kicin, Slavo ; Holm, Toni ; Bianda, Enea ; Canales, Francisco

DEVELOPMENT OF ULTRAHIGH VOLTAGE SIC POWER DEVICES 3440

Fukuda, Kenji ; Okamoto, Dai ; Harada, Shinsuke ; Tanaka, Yasunori ; Yonezawa, Yoshiyuki ; Deguchi, Tadayoshi ; Katakami, Shuji ; Ishimori, Hitoshi ; Takasu, Shinji ; Arai, Manabu ; Takenaka, Kensuke ; Fujisawa, Hiroyuki ; Takei, Manabu ; Matsumoto, Kazushi ; Ohse, Naoyuki ; Ryo, Mina ; Ota, Chiharu ; Takao, Kazuto ; Mizukami, Makoto ; Kato, Tomohisa ; Izumi, Toru ; Hayashi, Toshihiko ; Nakayama, Koji ; Asano, Katsunori ; Okumura, Hajime ; Kimoto, Tsunenobu

HIGH SWITCHING PERFORMANCE OF 1.7KV, 50A SIC POWER MOSFET OVER SI IGBT FOR ADVANCED POWER CONVERSION APPLICATIONS 3447

Hazra, Samir ; De, Ankan ; Bhattacharya, Subhashish ; Lin Cheng ; Palmour, John ; Schupbach, Marcelo ; Hull, Brett ; Allen, Scott

CONTROL METHOD FOR FIVE LEVEL CONVERTER WITH COMMON FLYING CAPACITORS TO AVOID VOLTAGE LEVEL SKIP 3455

Wei Yan ; Hui Zhang ; Ogura, Kazuya ; Urushibata, Shota

LOW-COMPLEXITY ANALYTICAL APPROXIMATIONS OF SWITCHING FREQUENCY HARMONICS OF 3-PHASE N-LEVEL VOLTAGE-SOURCE PWM CONVERTERS 3460

Burkart, Ralph M. ; Kolar, Johann W.

DYNAMIC VOLTAGE BALANCING ALGORITHM FOR MODULAR MULTILEVEL CONVERTER WITH THREE-LEVEL FLYING CAPACITOR SUBMODULES 3468

Dekka, Apparao ; Wu, Bin ; Zargari, Navid R.

MODULAR MEDIUM VOLTAGE DRIVE FOR DEMANDING APPLICATIONS 3476

Dujic, Drazen ; Wahlstroem, Jonas ; Marrero Sosa, Juan Alberto ; Fritz, Dominik

ASYMMETRICAL FAULT RIDE-THROUGH OF THREE-PHASE PV SYSTEMS USING FOUR-WIRE DC-AC CONVERTERS 3482

Iyer, Shivkumar ; Bin Wu ; Yunwei Li ; Singh, B.N.

OPERATION MODE ANALYSIS FOR SOLVING THE PARTIAL SHADOW IN A NOVEL PV POWER GENERATION SYSTEM 3489

Qi Zhang ; Xiangdong Sun ; Yanru Zhong ; Lie Guo ; Matsui, Mikihiko

ANALYSIS OF PARTIAL POWER PROCESSING DISTRIBUTED MPPT FOR A PV POWERED ELECTRIC AIRCRAFT 3496

Marzouk, Ahmad Diab ; Fournier-Bidoz, Sebastien ; Yablecki, Jessica ; McLean, Kenneth ; Trescases, Olivier

IMPACTS OF RECTIFIER CIRCUIT LOADS ON ISLANDING DETECTION OF PHOTOVOLTAIC SYSTEMS 3503

Yoshida, Yoshiaki ; Suzuki, Hirokazu

INDUCTION MOTOR MADE OF SMC 3509

Morimoto, Masayuki ; Inamori, Mamiko

ESTIMATION AND COMPARISON OF THE WINDAGE LOSS OF A 60 KW SWITCHED RELUCTANCE MOTOR FOR HYBRID ELECTRIC VEHICLES 3513

Kiyota, Kyohei ; Kakishima, Takeo ; Chiba, Akira

DEVELOPMENT OF HIGH-POWER PMASYNRM USING FERRITE MAGNETS FOR REDUCING RARE-EARTH MATERIAL USE 3519

Sanada, Masayuki ; Morimoto, Shigeo ; Inoue, Yukinori

CONSIDERATION OF 10KW IN-WHEEL TYPE AXIAL-GAP MOTOR USING FERRITE PERMANENT MAGNETS 3525

Sone, Kodai ; Takemoto, Masatsugu ; Ogasawara, Satoshi ; Takezaki, Kenichi ; Hino, Wataru

POWER CONTROL METHOD FOR MULTI-PARALLEL DC DISTRIBUTION SYSTEM THROUGH THE EQUIVALENT CIRCUIT MODEL 3532

Seok-Jin Hong ; Soo-Cheol Shin ; Hee-Jun Lee ; Chung-Yuen Won ; Taeck-Kie Lee

A COMMUNICATION-LESS DISTRIBUTED VOLTAGE CONTROL STRATEGY FOR A MULTI-BUS AC ISLANDED MICROGRID 3538

Wang, Yanbo ; Yongdong Tan ; Chen, Zhe ; Wang, Xiongfei ; Tian, Yanjun

AN ENHANCED LOAD POWER SHARING STRATEGY FOR LOW-VOLTAGE MICROGRIDS BASED ON INVERSE-DROOP CONTROL METHOD 3546

Yixin Zhu ; Fang Zhuo ; Baoquan Liu ; Hao Yi

ADDING VIRTUAL RESISTANCE IN SOURCE SIDE CONVERTERS FOR STABILIZATION OF CASCADED CONNECTED TWO STAGE CONVERTER SYSTEMS WITH CONSTANT POWER LOADS IN DC MICROGRIDS 3553

Mingfei Wu ; Lu, Dylan D.C.

EXPANSION OF OPERATING RANGE AND IMPROVEMENT OF TORQUE RESPONSE OF PMSM DRIVE BY USING MODEL PREDICTIVE CONTROL 3557

N/A

NONLINEAR MODEL PREDICTIVE TORQUE CONTROL OF A LOAD COMMUTATED INVERTER AND SYNCHRONOUS MACHINE 3563

Almer, Stefan ; Besselmann, Thomas ; Ferreau, Joachim

MODEL PREDICTIVE CURRENT CONTROL FOR PMSM CONSIDERING NUMBER OF SWITCHING OPERATIONS 3568

Zanma, Tadanao ; Yasumura, Yuji ; Liu, KangZhi

PREDICTIVE INDIRECT MATRIX CONVERTER FED TORQUE RIPPLE MINIMIZATION WITH WEIGHTING FACTOR OPTIMIZATION 3574

Uddin, Muslem ; Mekhilef, Saad ; Rivera, Marco ; Rodriguez, Jose

HIGH-POWER DENSITY HYBRID CONVERTER TOPOLOGIES FOR LOW-POWER DC-DC SMPS 3582

Radic, Aleksandar ; Ahssanuzzaman, S.M. ; Mahdavikhah, Behzad ; Prodic, Aleksandar

COUPLED INDUCTOR BASED CURRENT-FED SWITCHED INVERTER FOR LOW VOLTAGE RENEWABLE INTERFACE 3587
Nag, Soumya Shubhra ; Mishra, Santanu Kumar

A SEMI-ISOLATED MULTI-INPUT CONVERTER FOR HYBRID PV/WIND POWER CHARGER SYSTEM 3592
Cheng-Wei Chen ; Kun-Hung Chen ; Chen, Yaow-Ming

HFL PV MICRO-INVERTER WITH FRONT-END CURRENT-FED CONVERTER AND HALF-WAVE CYCLOCONVERTER 3598
Nayanasiri, D.R. ; Vilathgamuwa, D.M. ; Maskell, D.L.

COMPREHENSIVE STUDY ABOUT STABILITY ISSUES OF MULTI-MODULE DISTRIBUTED SYSTEM 3604
Liu, Fangcheng ; Liu, Jinjun ; Zhang, Haodong ; Xue, Danhong ; Dou, Qinyun

CHARACTERISTICS STUDY OF NEURAL NETWORK AIDED DIGITAL CONTROL FOR DC-DC CONVERTER 3611
Maruta, Hidenori ; Motomura, Masashi ; Kurokawa, Fujio

ZERO CURRENT SWITCHING CURRENT-FED PARALLEL RESONANT PUSH-PULL (CFPRPP) CONVERTER 3616
Moorthy, Radha Sree Krishna ; Rathore, Akshay Kumar

CHARACTERISTICS OF TRANSMISSION CARRIER IN A NEW WIRE COMMUNICATION SYSTEM BY THE USE OF HIGH-RIPPLE DC-DC CONVERTER 3624
Katsuki, Akihiko ; Mizuki, Tatsuya ; Shibahara, Kohei ; Morita, Kosuke ; Masutomo, Kazufumi ; Maeyama, Shigetaka

5MHZ PWM-CONTROLLED CURRENT-MODE RESONANT DC-DC CONVERTER USING GAN-FETS 3630
Hariya, Akinori ; Yanagi, Hiroshige ; Ishizuka, Yoichi ; Matsuura, Ken ; Tomioka, Satoshi ; Ninomiya, Tamotsu

DESIGN AND PERFORMANCE EVALUATION OF DIGITAL CONTROL FOR LLC SERIES RESONANT DC-TO-DC CONVERTERS 3638
Pidaparthy, Syam Kumar ; Choi, Byungcho ; Jang, Jinhaeng

EXPERIMENTAL VERIFICATION OF NOISELESS SAMPLING FOR BUCK CHOPPER CIRCUIT WITH CURRENT CONTROL 3646
Takeuchi, Shun ; Wada, Keiji

CONTROL CHARACTERISTICS IMPROVEMENT OF FULL-BRIDGE DC-DC CONVERTER WITH SNUBBER CAPACITOR 3652
Domoto, Kazuhide ; Ishizuka, Yoichi ; Abe, Seiya ; Ninomiya, Tamotsu

DCM CONTROL METHOD OF BOOST CONVERTER BASED ON CONVENTIONAL CCM CONTROL 3659
Le Hoai Nam ; Orikawa, Koji ; Itoh, Jun-ichi

TECHNICAL ASSESSMENT OF LOAD COMMUTATION SWITCH IN HYBRID HVDC BREAKER 3667
Hassanpoor, Arman ; Hafner, Jurgen ; Jacobson, Bjorn

CONTROL OF HEXAGONAL MODULAR MULTILEVEL CONVERTER FOR 3-PHASE BTB SYSTEM 3674
Hamasaki, Shin-ichi ; Okamura, Kazuki ; Tsubakidani, Takashi ; Tsuji, Mineo

A SYNTHESIZED CAPACITORS VOLTAGE CONTROL FOR MODULAR MULTILEVEL CONVERTER IN HVDC APPLICATION 3680
Rongfeng Yang ; Shunke Sui ; Binbin Li ; Wei Wang ; Dianguo Xu

OPERATING PHASE AND FREQUENCY SELECTION OF LOW FREQUENCY AC TRANSMISSION SYSTEM USING CYCLOCONVERTERS 3687
Achara, Pichetjamroen ; Ise, Toshifumi

FAST ACTING DC CIRCUIT BREAKER FOR HVDC TRANSMISSION LINE BASED ON DC/DC CHOPPER 3695
Liangyi Tang ; Bin Wu ; Yaramasu, Venkata ; Weirong Chen ; Athab, Hussain S.

1700V SI-IGBT AND SIC-SBD HYBRID MODULE FOR AC690V INVERTER SYSTEM 3702
Haining Wang ; Ikawa, O. ; Miyashita, S. ; Nishimura, T. ; Igarashi, S.

SWITCHING SIMULATION OF SIC HIGH-POWER MODULE WITH LOW PARASITIC INDUCTANCE 3707
Yamamoto, Takashi ; Hasegawa, Kohei ; Ishida, Masaaki ; Takao, Kazuto

SWITCHING PERFORMANCE OF PARALLEL-CONNECTED POWER MODULES WITH SIC MOSFETS 3712
Colmenares, Juan ; Peftitsis, Dimosthenis ; Nee, Hans-Peter ; Rabkowski, Jacek

BUILT-IN RELIABILITY DESIGN OF A HIGH-FREQUENCY SIC MOSFET POWER MODULE 3718
Jianfeng Li ; Gurpinar, Emre ; Lopez-Arevalo, Saul ; Castellazzi, Alberto ; Mills, Liam

EXPERIMENTAL SWITCHING FREQUENCY LIMITS OF 15 KV SIC N-IGBT MODULE 3726
Kadavelugu, Arun ; Bhattacharya, Subhashish ; Ryu, Sei-Hyung ; Van Brunt, Edward ; Grider, Dave ; Leslie, Scott

SELECTION OF SUITABLE CARRIER-BASED PWM METHOD FOR MODULAR MULTILEVEL CONVERTER 3734
Ciftci, Baris ; Erturk, Feyzullah ; Hava, Ahmet M.

CONTROL AND EXPERIMENT OF A 380-V, 15-KW MOTOR DRIVE USING MODULAR MULTILEVEL CASCADE CONVERTER BASED ON TRIPLE-STAR BRIDGE CELLS (MMCC-TSBC) 3742
Kawamura, Wataru ; Hagiwara, Makoto ; Akagi, Hirofumi

A POWER ELECTRONIC TRANSFORMER WITH SINUSOIDAL VOLTAGES AND CURRENTS USING MODULAR MULTILEVEL CONVERTER 3750
Sahoo, Ashish Kumar ; Mohan, Ned

VARYING AND UNEQUAL CARRIER FREQUENCY PWM TECHNIQUES FOR MODULAR MULTILEVEL CONVERTERS 3758
Konstantinou, Georgios ; Darus, Rosheila ; Pou, Josep ; Ceballos, Salvador ; Agelidis, Vassilios G.

COMPARISON OF PHASE-SHIFTED AND LEVEL-SHIFTED PWM IN THE MODULAR MULTILEVEL CONVERTER 3764
Darus, Rosheila ; Konstantinou, Georgios ; Pou, Josep ; Ceballos, Salvador ; Agelidis, Vassilios G.

A SINGLE-PHASE POWER CONDITIONER WITH A BUCK-BOOST-TYPE POWER DECOUPLING CIRCUIT 3771
Yamaguchi, Shota ; Shimizu, Toshihisa

A NOVEL ASYMMETRICAL FLC-BASED MPPT TECHNIQUE FOR PHOTOVOLTAIC GENERATION SYSTEM 3778
Yi-Hsun Chiu ; Yu-Shan Cheng ; Yi-Hua Liu ; Shun-Chung Wang ; Zong-Zhen Yang

A NOVEL CURRENT LINK DISTRIBUTED MPPT PV SYSTEM - OVERALL SYSTEM PROTOTYPING AND EVALUATION 3784
Mikihiko ; Toru ; Akira ; Xiang-Dong Sun ; Byung-Gyu Yu

POWER FLOW CONTROL AND MPPT PARAMETER SELECTION FOR RESIDENTIAL GRID-CONNECTED PV SYSTEMS WITH BATTERY STORAGE 3789
Chokchai, Chuenwattanapraniti

A MAXIMUM POWER POINT TRACKING METHOD WITH RIPPLE CURRENT ORIENTATION 3796
Moo, Chin-Sien ; Wu, Gwo-Bin

OUTPUT CHARACTERISTICS OF A SURFACE PERMANENT MAGNET-TYPE VERNIER MOTOR - COMPARISON OF TEST RESULTS AND CALCULATION 3801
Kataoka, Yasuhiro ; Takayama, Masakazu ; Anazawa, Yoshihisa ; Matsushima, Yoshitarou

TOPOLOGY OPTIMIZATION FOR SKEW OF SPMSM BY USING MULTI-STEP PARALLEL GA 3809
Kitagawa, Wataru ; Takeshita, Takaharu

LOSS MINIMIZATION DESIGN USING MAGNETIC EQUIVALENT CIRCUIT FOR A PERMANENT MAGNET SYNCHRONOUS MOTOR 3815
Sato, Daisuke ; Itoh, Jun-ichi

THE PROPOSAL OF A NEW MOTOR WHICH HAS A HIGH WINDING FACTOR AND A HIGH SLOT FILL FACTOR 3823
Makita, Shinji ; Ito, Yasuhide ; Aoyama, Tomohiro ; Doki, Shinji

VARIABLE LEAKAGE FLUX INTERIOR PERMANENT MAGNET SYNCHRONOUS MACHINE FOR IMPROVING EFFICIENCY ON DUTY CYCLE 3828
Minowa, Masanao ; Hijikata, Hiroki ; Akatsu, Kan ; Kato, Takashi

HISTORY AND TRENDS OF CONVERTER TECHNOLOGY FOR DC AND AC TRANSMISSION IN JAPAN 3834
Yoshino, Teruo

ACCURATE OUTPUT POWER CONTROL OF CONVERTERS FOR MICROGRIDS BASED ON LOCAL MEASUREMENT AND UNIFIED CONTROL 3842
Meiqin Mao ; Zheng Dong ; Yong Ding ; Liuchen Chang

IMPEDANCE-BASED ANALYSIS OF ACTIVE FREQUENCY DRIFT ISLANDING DETECTION METHOD FOR GRID-TIED INVERTER SYSTEM 3850
Wen, Bo ; Boroyevich, Dushan ; Burgos, Rolando ; Shen, Zhiyu ; Mattavelli, Paolo

DEVELOPMENT OF 200-MVAR CLASS THYRISTOR SWITCHED CAPACITOR SUPPORTING FAULT RIDE-THROUGH 3857
Ohtake, Asuka ; Fei Zhang ; Fujimoto, Takafumi ; Nakayama, Naoyuki

DETAILED ANALYSIS AND DESIGN OF A THREE-PHASE PHASE-MODULAR ISOLATED MATRIX-TYPE PFC RECTIFIER 3864
Cortes, Patricio ; Fassler, Lukas ; Bortis, Dominik ; Kolar, Johann W. ; Silva, Marcelo

AN ENERGY SAVING DRIVE METHOD OF AN INDUCTION MOTOR WITH THE SUPPRESSION OF SUDDEN ACCELERATION AND DECELERATION 3872
Asano, Yuji ; Inoue, Kaoru ; Kotera, Keito ; Kato, Toshiji

FIELD ORIENTED CONTROL OF SENSORLESS LINEAR INDUCTION MOTOR USING MATRIX CONVERTER 3877
Sayed, Mahmoud A. ; Mohamed, Essam Ebaid ; Mohamed, Tarek Hassan ; Takeshita, Takaharu

A STATOR-EQUATION-BASED REDUCED-ORDER OBSERVER FOR POSITION-SENSORLESS VECTOR CONTROL SYSTEM OF DOUBLY-FED INDUCTION MACHINES 3885
Smiththisomboon, Somrat ; Suwankawin, Surapong

INPUT CURRENT RIPPLE ANALYSIS OF INVERTER FED DUAL THREE-PHASE AC MOTORS 3893
Dahono, Pekik Argo ; Satria, Andri

OFFLINE EXTRACTION OF INDUCTION MACHINE PARAMETERS FOR CONTROL STRATEGY SYNTHESIS 3898
Koschik, Stefan ; Bauer, Florian ; De Doncker, R.W.

HIGH CURRENT PLANAR TRANSFORMER FOR VERY HIGH EFFICIENCY ISOLATED BOOST DC-DC CONVERTERS 3905
Pittini, Riccardo ; Zhe Zhang ; Andersen, Michael A.E.

HIGH VOLTAGE-GAIN INTERLEAVED BOOST DC-DC CONVERTER DISCARDED ELECTROLYTIC CAPACITOR 3913
Nha, Quang Trong ; Huang-Jen Chiu ; Yu-Kang Lo ; Pham Phu Hieu

PARALLEL BI-DIRECTIONAL DC-DC CONVERTER FOR ENERGY STORAGE SYSTEM 3920
Ouchi, Takayuki ; Kanoda, Akihiko ; Takahashi, Naoya

CHARGING SCENARIO OF SERIAL BATTERY POWER MODULES WITH BUCK-BOOST CONVERTERS 3928
Jhen-Yu Jian ; Chu-Shen Chang ; Moo, Chin-Sien ; Hau-Chen Yen

COMPARATIVE THERMAL PERFORMANCE EVALUATION OF SIC MOSFETS AND SI MOSFET FOR 1.2 KW 300 KHZ DC-DC BOOST CONVERTER AS A SOLAR PV PRE-REGULATOR 3933
Taekyun Kim ; Minsoo Jang ; Agelidis, Vassilios G.

TOLERANCE ANALYSIS OF A CONSTANT-ON TIME CURRENT-MODE VOLTAGE REGULATOR WITH ADAPTIVE VOLTAGE POSITION FEATURE 3938
Chih Wei Chen ; Dan Chen ; Shin Shiung Wang

FPGA-BASED DIGITAL-CONTROLLED POWER CONVERTER DESIGNED WITH UNIVERSAL INPUT MEETING 80 PLUS PLATINUM EFFICIENCY CODE AND STANDBY POWER CODE FOR SEVER POWER APPLICATIONS 3942
Lai, Yen-Shin ; Ho, Kung-Min

STATIC AND DYNAMIC ANALYSES OF DIGITAL PEAK CURRENT MODE DC-DC CONVERTER 3950
Kajiwara, Kazuhiro ; Kurokawa, Fujio ; Shibata, Yuichiro

EXTENDED DISCRETE CONTROL OF CLASS E AMPLIFIER IN ORDER TO ACHIEVE NOMINAL OPERATION 3955
Suetsugu, Tadashi ; Xiuqin Wei ; Kuga, Shotaro

ADAPTIVE POWER EFFICIENCY CONTROL BY COMPUTER POWER CONSUMPTION PREDICTION USING PERFORMANCE COUNTERS 3959
Kawaguchi, Shinichi ; Yachi, Toshiaki

Author Index

Improvement in Efficiency of LED Lighting System

K. I. Hwu, W. Z. Jiang
Department of Electrical Engineering
National Taipei University of Technology
Taipei, Taiwan
eaglehwu@ntut.edu.tw, newjerusalem333@gmail.com

Jenn-Jong Shieh
Department of Electrical and Electronic Engineering
Ta Hwa University of Science and Technology
Hsinchu, Taiwan
eesjj@tust.edu.tw

Abstract—**In this paper, a LED dimming circuit, together with the KY converter, is presented, which is controlled based on the field-programmable gate array (FPGA). By a given dimming command and the proposed maximum gate voltage detector, the voltage across the MOSFET in the linear current regulator can be reduced so as to upgrade the efficiency of the overall system. Aside from this, each LED string takes level dimming, and is powered by the KY converter, which has an output inductor and hence upgrades the life of the output capacitor. Furthermore, via some experimental results, the efficiency based on the proposed control method is higher than that based on the traditional control one.**

Keywords—Efficiency; LED lighting; triode region.

I. Introduction

Recently, the light emitting diode (LED) has got more and more attracted in the world. In the future, such a light source will replace the other light sources. As compared to the other light sources, the LED has some advantages [1], such as no Hg corresponding to environment protection, small size with resistance to vibration and pressure, high-speed response, etc. In [2], a linear current regulator for a single LED string is presented, and this can be extended and applied to multiple strings. This circuit is basically constructed by one operational amplifier, one MOSFET and one current-limiting resistor R_s. Via the virtual ground of the operational amplifier, the voltage across R_s is equal to a given reference voltage, and hence the current flowing through the LED strings can be adjusted by such a voltage. However, during the dimming period, the less the current flowing through the LED string is, the more the voltage on MOSFET, and hence the corresponding efficiency of the overall system is deteriorated [3]. Therefore, the literature [4] has presented a method to improve the efficiency of the overall system, especially at light load. However, there is still some room to improve the efficiency based on this method. Consequently, in this paper, the feedback voltage is subtracted from a variable voltage reference, which is controlled by the dimming command and the maximum gate voltage detector, so as to make the voltage across the MOSFET in the linear current regulator as minimum as possible for any load. For the power stage to feed LED strings to be considered, the KY converter [5] is employed herein, which is suitable for low-power applications and always operates in the continuous conduction mode (CCM). Furthermore, the output inductor current is non-pulsating, and hence not only reduces the output voltage ripple but also prolongs the life of the output capacitor. Therefore, the level dimming circuit, together with the KY converter, is adopted herein.

II. Overall System Configuration

Fig. 1 shows the proposed overall system configuration, which is built up by the KY converter used to provide a desired voltage for five LED strings, one maximum gate voltage detector, two analog-to-digital converters (ADCs), one voltage divider constructed by two resistors R_1 and R_2, one Butterworth filter to get the DC value of the control force created from the FPGA, two gate drivers, and five linear current regulators with each constructed by one operational amplifier, one MOSFET switch and one current limiting resistor. Sequentially, how to reduce the voltage across the MOSFET is described as follows.

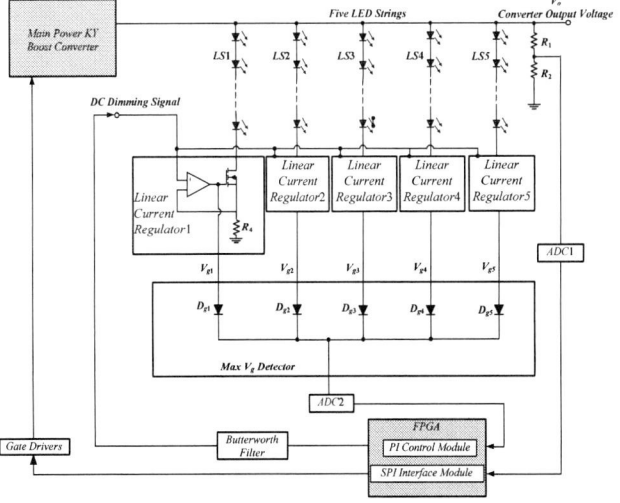

Fig. 1. Proposed LED lighting system.

III. REDUCTION OF THE VOLTAGE ON MOSFET

As the dimming command is changed, the voltages across the LED strings will be varied and hence the voltages across the MOSFETs in the individual linear current regulators will be altered, also. As shown in Fig. 2(a), if the MOSFET operates in the saturation region, the current in the LED string, I_{LS}, is not changed with the voltage between the drain and the source, V_{ds}, but is varied with the voltage between the gate and the source, V_{gs}, whereas if the MOSFET operates in the triode region, then in order to satisfy the virtual ground of the operational amplifier in the linear current regulator, variations in V_{gs} will follow variations in V_{ds} so as to make I_{LS} constant. However, variations in I_{LS} are affected significantly by V_{ds} in the triode region such that I_{LS} is not so easy to control. Consequently, based on the mention above, in order to make V_{ds} as minimum as possible and I_{LS} easy to control, V_{ds} is controlled near the knee of the saturation region. In the following, how to do this will be illustrated, and after this, some experimental results will be given to verify the proposed method.

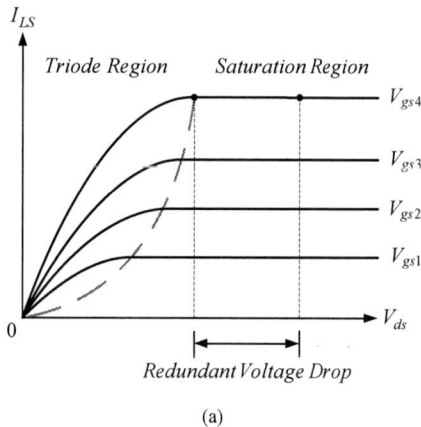

(a)

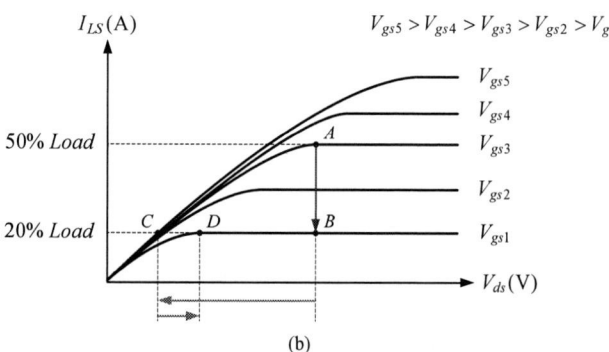

(b)

Fig. 2. (a) Curve of MOSFET characteristics; (b) curve of V_{ds} versus V_{gs} for the down dimming from 50% to 20% load.

In Fig. 2(b), for the down dimming from 50% to 20% load, the output voltage of the KY converter is reduced step by step such that V_{ds} goes down simultaneously. At the same time, the gate voltage goes down from V_{gs3} at point A to V_{gs1} at point B,

then abruptly to V_{gs5} at point C to keep I_{LS} constant in the triode region, and eventually back to V_{gs1} at point D to make sure that the MOSFET operates near the knee of the saturation region. As V_{ds} goes from point B to point C, the corresponding signal inside the FPGA will be changed from the low level to the high level immediately. Therefore, the FPGA will make the output voltage of the KY converter stop reducing and then increase a bit, so as to let the MOSFET operate at point D.

IV. EXPERIMENTAL RESULTS

Before this section is taken up, there are some specifications to be given as follows: (i) the input voltage of the KY converter is 14V; (ii) the output voltage of the KY converter is set to 20V; (iii) the switching frequency for the KY converter is 100kHz; (iv) the LED module consists of five LED strings connected in parallel with five LEDs per LED string; and (v) the dimming current range for each LED string is from 0.438A to 1.75A, i.e., from 25% to 100% of the rated output current.

Figs. 3 to 5 show the waveforms for v_{gs}, V_o, i_L and I_o of the KY converter under the proposed dimming topology, at 50%, 75% and 100% load, respectively. From Figs. 3 to 5, it can be seen that the proposed LED lighting system can stably operate under various dimming power levels. Furthermore, based on the proposed dimming topology, Table I shows the voltages across five MOSFETs, v_{ds1}, v_{ds2}, v_{ds3}, v_{ds4} and v_{ds5}, the output voltage of the KY converter, V_o, and the efficiency, η, under 25%, 50%, 75% and 100% of the rated load, whereas based on the traditional dimming topology, Table II shows the same measured items as Table I. From Tables I and II, it can be seen that the values of efficiency based on the proposed dimming topology are much higher than those based on the traditional dimming topology. This is because the former makes the output voltage of the KY converter vary with the load but the latter does not.

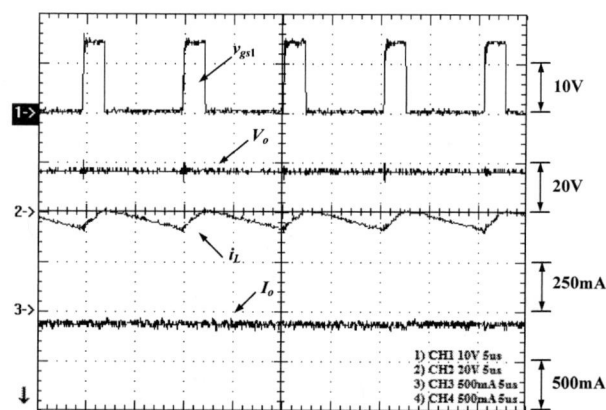

Fig. 3. Waveforms for v_{gs}, V_o, i_L and I_o of the KY converter at 50% load under the proposed dimming topology.

V. Conclusion

In this paper, the proposed LED dimming strategy, together with the KY converter used to power LEDs, is verified by experimental results, with the efficiency of the proposed LED lighting system upgraded significantly, particularly under low dimming power level.

Acknowledgment

The authors would like to thank the National Science Council for supporting this work under Grant NSC 101-2221-E-027-107-MY2; NSC 99-2632-E-233-001-MY3.

References

[1] S. Peralta and H. Ruda, "Application for advanced solid-state lamps," *IEEE Industry Application Magazine*, vol. 4, no. 4, pp. 31-42, 1998.

[2] H. J. Chiu, H. M. Huang, H. T. Yang and S. J. Cheng, "An improved single-stage Flyback PFC converter for high-luminance lighting LED lamps," *Int. J. Circuit Theory Appl*, vol. 36, no. 2, 2008, pp. 205-210.

[3] Y. Hu and M. M. Jovanovic, "LED driver with self-adaptive drive voltage," *IEEE Trans. Power Electron.*, vol. 23, no. 6, pp. 3116-3125, 2008.

[4] K. I. Hwu and W. C. Tu, "LED dimming with efficiency considered," *IET Electronics Letters*, vol. 47, no. 7, pp. 457-459, 2011.

[5] K. I. Hwu and Y. T. Yau, "KY converter and its derivatives," *IEEE Trans. Power Electron.*, vol. 24, no. 1, pp. 128-136, 2009.

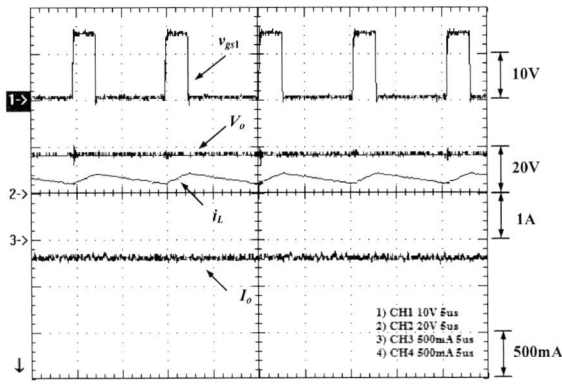

Fig. 4. Waveforms for v_{gs}, V_o, i_L and I_o of the KY converter at 75% load under the proposed dimming topology.

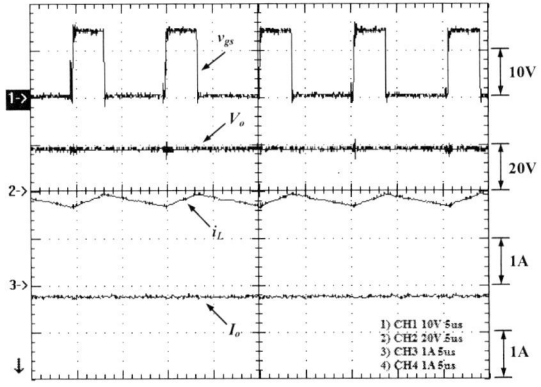

Fig. 5. Waveforms for v_{gs}, V_o, i_L and I_o of the KY converter at 100% load under the proposed dimming topology.

Table I Values of v_{ds1}, v_{ds2}, v_{ds3}, v_{ds4} and v_{ds5}, V_o and η under 25%, 50%, 75% and 100% of the rated load, based on the proposed topology

Load	v_{ds1} (V)	v_{ds2} (V)	v_{ds3} (V)	v_{ds4} (V)	v_{ds5} (V)	V_o (V)	η (%)
25%	0.343	0.229	0.197	0.161	0.119	15.40	98.63
50%	0.420	0.319	0.306	0.208	0.264	16.40	98.15
75%	0.603	0.458	0.477	0.326	0.473	16.55	97.17
100%	0.762	0.593	0.620	0.410	0.832	17.41	96.3

The 2014 International Power Electronics Conference

Table II Values of v_{ds1}, v_{ds2}, v_{ds3}, v_{ds4} and v_{ds5}, V_o and η under 25%, 50%, 75% and 100% of the rated load, based on the traditional topology

Load	v_{ds1} (V)	v_{ds2} (V)	v_{ds3} (V)	v_{ds4} (V)	v_{ds5} (V)	V_o (V)	η (%)
25%	4.924	4.835	4.408	4.760	4.061	19.99	76.60
50%	4.519	4.407	4.417	4.322	4.188	20.02	78.16
75%	3.947	3.860	3.910	3.775	4.267	20.06	80.30
100%	3.572	3.510	3.596	3.419	5.083	19.99	80.85

Comparison and Evaluation of Vibration-Based Piezoelectric Power Generators

Amat A. Basari, Sosuke Awaji
Yunshun Zhang, Song Wang
and Seiji Hashimoto
Div. of Electronics and Informatics
Gunma University
Kiryu, Gunma, Japan
hashimotos@gunma-u.ac.jp

Shunji Kumagai
Makoto Kasai and Kenji Suto
Research and Dev. Dept.
Mitsuba Corporation
Hirosawa-cho
Kiryu-shi, Japan

Wei Jiang
and Shuren Wang
School of Energy and Power Eng.
Yangzhou University
196 Huayangxilu Rd, Yangzhou
China
jiangwei@yzu.edu.cn

Abstract—Vibration energy regeneration with lead zirconate titanate piezoelectric (PZT) device has been reported by many researchers for many years. Various factors influence the energy regeneration efficiency of the PZT devices in converting the mechanical vibration energy to the electrical energy. This paper presents the analytical and experimental evaluation of energy regeneration efficiency of PZT devices through impedance matching method, energy regeneration efficiency of different shape of PZT devices, and effects of forced vibration input with concentrated load applied on the device to the PZT device's energy regeneration ability. The results show that the impedance matching method has increased the energy regeneration efficiency to between 4% and 88% while triangular shape of PZT device produce a stable efficiency in the energy regeneration. Besides that, it becomes clear that energy regeneration of triangular PZT device as well as its efficiency is optimum with auxiliary mass of 2.6g.

Keywords—Energy regeneration efficiency, Forced vibration, Impedance matching, PZT device.

I. INTRODUCTION

In recent years, solutions for the environmental and energy problems correspond to the increment of power demand all over the world have become a great interest of research. One of the solution methods that highly recommended by literature is by doubling our efforts in research and development on renewable energy technology. Wind, solar, thermal, vibration and many other types of energies are always available and they do not adversely affect the environment. To date, for example, solar, thermal and wind energy have successfully been converted to usable electrical energy and used in various industrial and home appliance products [1]-[3], while vibration energy has yet to become an alternative source of energy for self-powered system. Conversion of vibration to electrical energy is possible using piezoelectric, electromagnetic and electrostatic based device [4]. It is stated in [5] that, generally, electrostatic device is able to produce electrical energy up to 2% of efficiency while electromagnetic device according to [6], it can generate electrical power up to 1.4mW with 25% efficiency.

This paper presents the evaluation of vibration energy regeneration efficiency of PZT devices which is produced through impedance matching approach and by different shape of PZT devices. In addition to these, energy regeneration efficiency of PZT device with forced vibration input and concentrated load applied on it is also evaluated analytically and experimentally.

II. IMPEDANCE MATCHING OF PZT DEVICE

A. Power and Energy Generation with PZT Devices

Impedance matching is one of the methods to increase power and energy output of PZT device power generator. Theoretically, by matching the load impedance with input impedance, maximum power can be delivered from source to the load. Prior to the evaluation on the impedance matching of the four selected bimorph PZT devices, discussion on power and energy generation by all PZT devices will be presented.

To evaluate the power and energy generation by all PZT devices, experimental setup as shown in Fig. 1 was built. The experiments were conducted based on the experimental conditions as listed in Table I. Relationship between the resonant frequency, matching impedance and energy is described in the following section.

B. Impedance Matching for Maximum Power and Energy Generation

In the previous section, discussion on power generation by the PZT devices with fixed load resistor of 10kΩ has been presented. This section will discuss the impedance matching of PZT devices so that maximum power and energy can be generated. For PZT device A, load resistor was increased gradually from 100Ω to 3MΩ. The results of power dissipation by each resistor are shown in Fig. 2. From the figure, maximum power was

TABLE I.   EXPERIMENTAL CONDITIONS

Parameter	Details
1) Input signal	step signal
2) Amplitude	0.5mm
3) Measurement signals	PZT voltage and displacement of PZT device's free end
4) Sampling time	0.2ms
5) Displ. resolution	3.5μm
6) ADC resolution	16 bit

978-1-4799-2706-7/14 $31.00 © 2014 IEEE

The 2014 International Power Electronics Conference

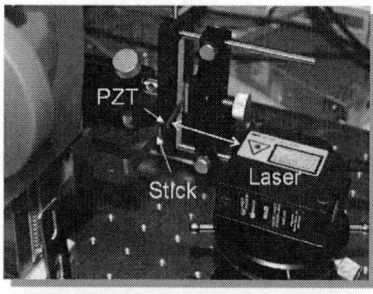

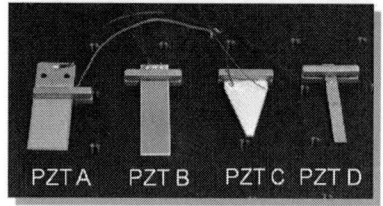

Fig. 1.   Experimental setup and PZT devices.

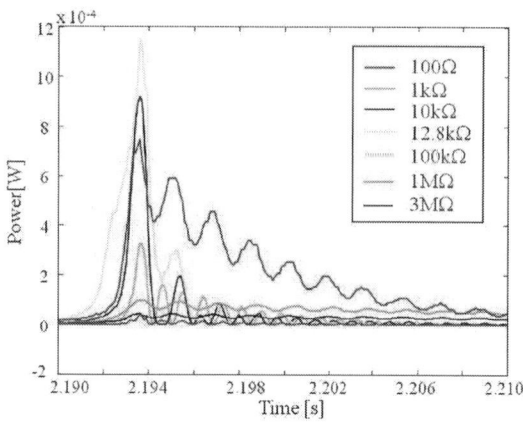

Fig. 2.   Power dissipation by resistors - PZT device A.

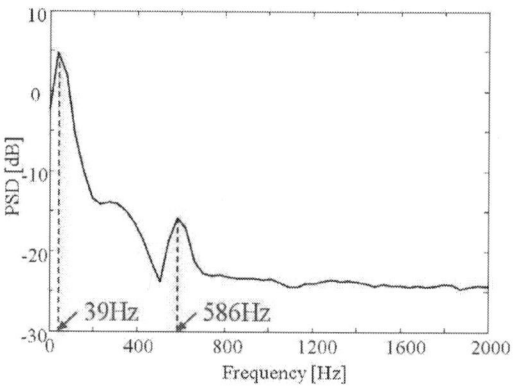

Fig. 3.   Voltage power spectral density of 100kΩ resistor.

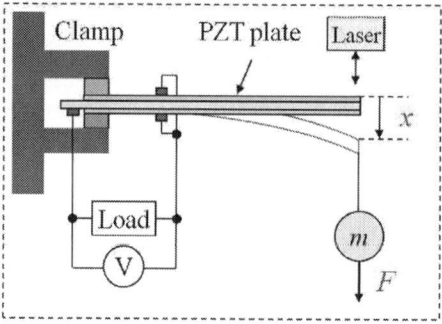

Fig. 4.   Experimental setup for energy measurement.

dissipated by the 12.8kΩ resistor and maximum energy was generated when resistor is 100kΩ. Notice that the maximum power and energy are dependent on different value of resistors. From this observation we know that PZT device A has two resonant frequencies. Next, power spectral density (PSD) of 100kΩ resistor was plotted. It is clearly can be seen in Fig. 3, the resonant frequency of PZT device A appears at two different frequencies; at 39Hz and 586Hz.

For PZT device B, C and D, same experiments were conducted and their dissipated voltage's power spectral densities were analyzed. Maximum power and energy for each PZT device were found to be dependent on the load resistors.

C. Energy Regeneration Efficiency

To evaluate the energy regeneration efficiency of each PZT device, experimental set up as shown in Fig. 4 was constructed and ratio of output to input energy was calculated. Note that one end of the cantilever beam was clamped and the other end was free and attached with mass $m=0.1$kg by string. At the rest position, the displacement of the free end is x_0. The displacement of the free end after the string is cut was measured using laser measuring device while the voltage drop was measured using voltmeter. Without considering the impedance matching results, the load resistor is set to 10kΩ for all PZT devices. Input energy, output energy and energy efficiency were calculated based on the equation (1), (2) and (3). Results are summarized in Table II.

$$W_i = \int F \mathrm{d}x \approx \frac{1}{2} mgx_0 \qquad (1)$$

$$W_0 = \frac{1}{R} \int v^2 \mathrm{d}t \qquad (2)$$

$$\eta = \frac{W_0}{W_i} \times 100\% \qquad (3)$$

Now, based on the impedance matching results, load resistor was replaced with the impedance matching resistor that produced maximum energy for each PZT device. The displacement and voltage drop were measured and recorded. As shown in Table II, it is clear that energy regeneration efficiency has increased between 4% to 88% for all PZT devices with the impedance matching resistor.

TABLE II.   ENERGY REGENERATION EFFICIENCY

PZT	Efficiency-with 10kΩ resistor (%)	Efficiency-with impedance matching resistor (%)	Increment in efficiency
A	3.0	5.4	80%
B	4.8	5.0	4%
C	3.1	4.9	58%
D	1.7	3.2	88%

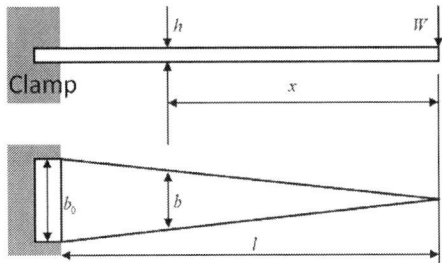

Fig. 5.   Cantilever beam of PZT device C.

III.   ENERGY REGENERATION EFFICIENCY OF DIFFERENT SHAPE PZT DEVICES

This section will discuss the effect of the shape of PZT device to the energy regeneration efficiency. In the previous section, four different size PZT devices have been tested. Basically, PZT device A, B and D have the same shape. Thus, in this section, to evaluate the effect of the PZT shape on the energy regeneration efficiency, PZT device A and C which having different shapes were used.

A. Stress, Deflection and Elastic Energy Analysis

Dimension of PZT device C in the cantilever beam structure is shown in Fig. 5. In this figure, W represents the concentrated load that applied on the free end of the cantilever beam. Theoretically, in bending mode, maximum stress is calculated using equation (4).

$$\sigma_{max} = \frac{M}{Z} = -\frac{6Wx}{bh^2} \tag{4}$$

where, M is bending moment, Z is section modulus, x is distance from the load, b is the width and h is the thickness of the PZT plate. From equation (4), thickness h is a fixed value, and if ratio of x/b is also fixed, the maximum stress at all points of the PZT plate will become constant. Based on this fact, different from PZT device A which having the maximum stress at the clamped point, PZT device C is the one with the constant value of maximum stress at all points of the plate for which the width b will increase as distance from the free end x increases.

Further analysis on deflection and elastic energy experienced by both PZT devices was performed. Based on the above analysis, the curvature of PZT device C will be a constant value when concentrated load is applied. Moreover, if compared to PZT device A, the deflection

angle of the PZT C is double and this will increase its deflection and the input energy W_i, to 1.5 times of that the PZT device A has. The equation that denoted the input energy is shown in equation 1. By looking at the equation, it can be expected that energy regeneration efficiency will become better than what PZT device A has. Another merit point that PZT device C has here is if we look to its volume. By having the same durability, the cost also can be reduced as its volume is half of the PZT device A.

Meanwhile, in terms of elastic energy per unit volume for PZT device A and C, they are denoted by equation (5) and (6) respectively.

$$U = \frac{1}{2}Wy_0\frac{1}{V} = \frac{\sigma^2}{18E} \tag{5}$$

$$U = \frac{1}{2}Wy_0\frac{1}{V} = \frac{\sigma^2}{6E} \tag{6}$$

where y_0 is the maximum deflection and V is the volume. As mentioned in above, with 1.5 times of maximum deflection and half of the volume, it is clearly can be seen from the equation that PZT device C will be able to absorb 3 times greater externally applied elastic energy than of that the PZT device A would absorb.

B. Experimental Analysis

Energy regeneration by PZT device is very much related to the stress that can be generated by the PZT device where we know that deflection is the source for the stress. An increment in the deflection simply can be done by increasing the concentrated load. Therefore, to observe the effects of the shape of PZT plate to the energy regeneration efficiency, an experimental setup as shown in Fig. 4 was constructed. The experiments were conducted with load mass which was used as the varying parameter. It was increased gradually from 50g to 350g. The output was connected to the impedance matching resistor of each case. Input and output energy data were measured and recorded.

From Fig. 6, it can be seen that both input energy of PZT device A and C increase with quadratic function of the load mass. Besides that, the output energy of PZT

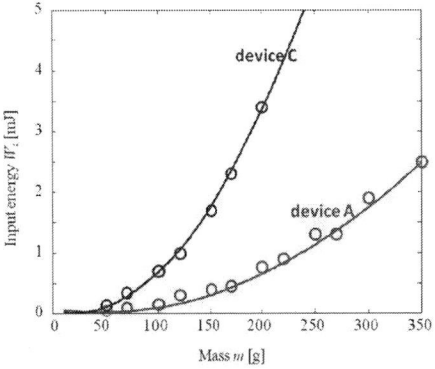

Fig. 6.   Input energy of PZT cantilever beam versus load mass.

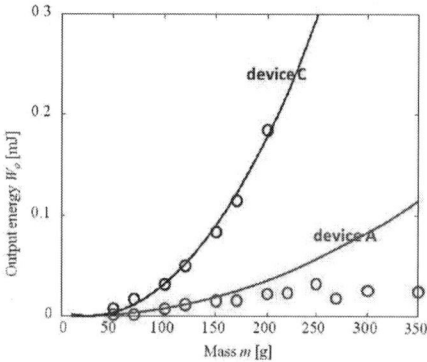

Fig. 7. Output energy of PZT cantilever beam versus load mass.

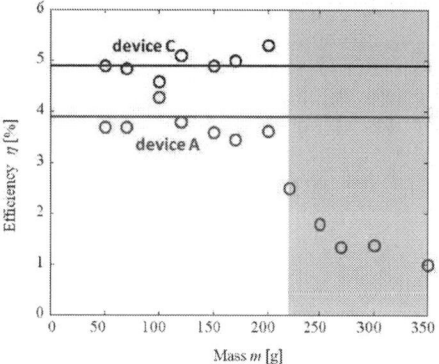

Fig. 8. Energy regeneration efficiency of PZT cantilever beam versus load mass.

device C, as shown in Fig. 7 also increases with the same function. Contradict to the rest, the output energy generated by the PZT device A is saturated at low level even the load mass is increased. In terms of energy regeneration efficiency, as can be seen in Fig. 8, the PZT device C marked a stable efficiency at about 5% as the load mass increases when compared to that of the PZT device A.

Next, further analysis on the energy regeneration of PZT device C was conducted. Experimental setup as shown in Fig. 9 was constructed. This time, mass was directly fixed on free end of PZT device as shown in Fig. 9. Auxiliary mass was used as the varying parameter and step signal was used as the input (0.5mm). The results are shown in Table III. Based on the results, it is obvious that with auxiliary mass applied on the free end of the device, power and energy generated by the device is higher than without the auxiliary mass. However, in case of maximum power, as the auxiliary mass bigger than 2.6g, it decrease gradually. From our analysis, with a bigger auxiliary mass attached to the device, resonant frequency of the device will decrease and thus contribute to the higher matching impedance value. Therefore, power generation also decreased for which it is denoted by $P(t) = v^2(t)/Z$. In terms of energy, we can see that, as damping time of the response become longer,

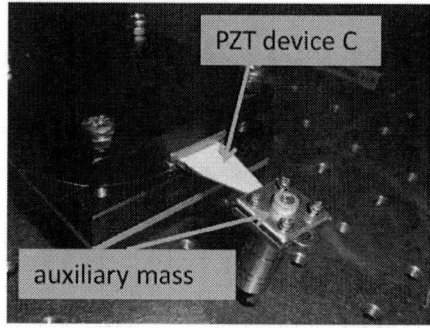

Fig. 9. Experimental setup - Input is a step signal of 0.5mm.

TABLE III.   EXPERIMENTAL RESULTS OF PZT DEVICE C

Auxiliary mass [g]	Max. power [mW]	Energy [μJ]	Damping time [s]
0	0.34	0.058	0.2
2.6	**10.6**	96.0	0.4
9.5	5.6	100.0	0.9
20.0	3.6	**109.2**	1.6

the output energy also increased. In this case, energy generated by the device with mass of 20g has a longest damping time of 1.6s and it has generated 109.2μJ of energy.

IV.   EFFECT OF FORCED VIBRATION INPUT ON ENERGY REGENERATION EFFICIENCY

A. Analytical Analysis on Energy Regeneration Efficiency

To evaluate the energy regeneration efficiency of PZT device C subject to the forced vibration input, analytical as well as experimental analyses were conducted. This section dedicated to the analytical analysis and the experimental analysis and results is highlighted in the next section. The analytical analysis was based on [7]. Lets consider an experimental setup and its equivalent circuit as shown in Figs. 10 and 11. The circuit parameters are summarized in Table IV. Input vibration y is denoted by equation (7) and oscillation of cantilever beam at free end, which behaves like a spring is governed by equation (8).

$$y[m] = Y\cos(\omega_n t) \qquad (7)$$

$$x[m] = X\cos(\omega_n t + \frac{\pi}{2}) \qquad (8)$$

ω_n in the equations is natural frequency of the cantilever beam, Y is maximum amplitude of the input vibration and X is maximum amplitude of the free end of the cantilever beam. When cantilever beam is excited by the vibrator, PZT plate will be deflected accordingly. This will generate a potential different and eventually current I flows to the load resistor. Energy balance equation within the characteristic period of time can be written as equation (9).

$$kXY\pi = c\omega_n(X^2 + Y^2)\pi + \frac{\pi}{\omega_n}(R_p + R_1)I^2 \qquad (9)$$

978-1-4799-2706-7/14 $31.00 © 2014 IEEE   3197

TABLE IV. EQUIVALENT CIRCUIT PARAMETERS.

k	Cantilever beam equivalent stiffness constant
c	Cantilever beam equivalent damping constant
θ	Electrical mechanical coupling coefficient
R_p	PZT internal resistor
R_1	Load resistor
C_p	PZT capacitance
I	Load current

TABLE V. EXPERIMENTAL RESULTS WITH FORCED VIBRATION INPUT

Auxiliary mass [g]	Power [mW]	Energy [μJ]	Efficiency [%]
0	–	0.0047	1.2
1.0	**6.49**	35.0	30.6
2.6	5.79	**37.2**	**37.2**
5.0	3.99	36.6	30.9
8.0	2.73	33.1	34.0
11.0	1.61	25.4	23.9
15.0	1.50	23.8	17.6
20.0	0.93	20.9	17.6

where in this equation, $(kXY\pi)$ represents the input vibration energy, $(c\omega_n(X^2 + Y^2)\pi)$ is the loss of energy due to the attenuation of the cantilever beam and $(\pi/\omega_n(R_p + R_1)I^2)$ represents the energy dissipation by the combined resistor.

The first term of the output equation, the energy that dissipated by the attenuation of the cantilever beam is considered as the mechanical energy loss for which it can not be retrieved. Same with the energy that dissipated by the internal resistance R_p which cannot be used too. Hence, the actual energy efficiency calculation will be based on equation (10)

$$\eta_{eff} = \frac{W_0}{W_i} = \frac{\frac{\pi}{\omega_n}R_1 I^2}{kXY\pi} \times 100[\%] \qquad (10)$$

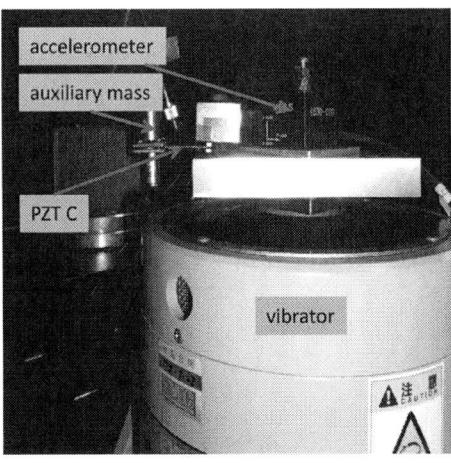

Fig. 10. Experimental setup for forced vibration mode of PZT device C.

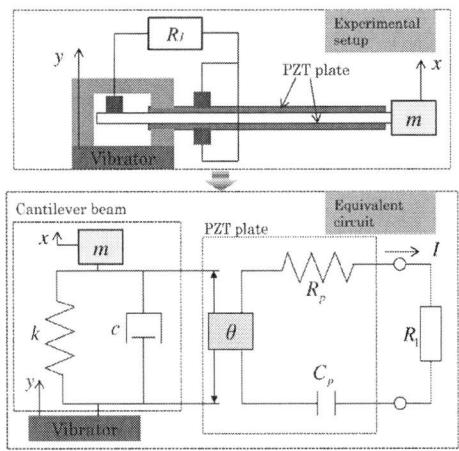

Fig. 11. Experimental setup and its equivalent circuit.

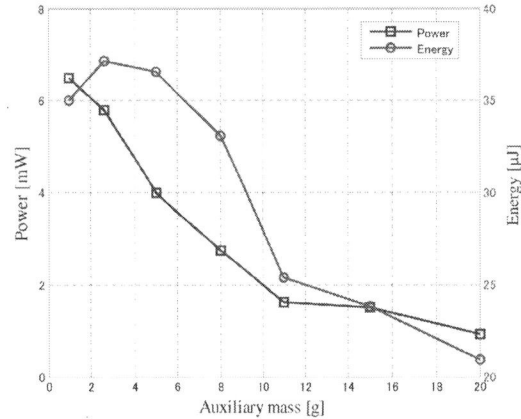

Fig. 12. Output power and energy versus auxiliary mass.

B. Experimental Analysis of Energy Regeneration Efficiency of PZT Device C with Auxiliary Mass

In the above section, analytical analysis on the electrical energy regeneration efficiency by cantilever beam of PZT device has been presented. In this section, experimental study and results will be presented and discussed. Experimental setup and its equivalent circuit are shown in Figs. 10 and 11. Forced vibration input used in this experiment was a sin wave which denoted by equation (7) and input energy was set to be constant. Results are listed and plotted in Table V and Fig. 12 Obviously from the table that energy regeneration with auxiliary mass applied on the device is much better than without the auxiliary mass. In terms of output power, the results is in line with the finding in the previous section, where it become less with increment in the auxiliary mass. Whereas, energy generation has decreased with the increment of auxiliary mass which is contradict to the earlier finding. The main reason for this is calculation of energy was based on the one period of time. Therefore, shorter sampling time for the energy calculation results in small energy generation can be obtained. Maximum energy regeneration of 37.2% can be seen when auxiliary mass is 2.6g. In terms of energy regeneration efficiency, the highest percentage is 37.2% for which the auxiliary mass is also 2.6g.

V. CONCLUSION

This paper has successfully evaluated how impedance matching, shape of devices, concentrated mass and forced vibration contribute to the variation of the energy regeneration efficiency of PZT devices. Input and load impedance matching has shown an increment in the energy regeneration efficiency to between 4% and 88%. On the other hand, a stable 5% energy regeneration efficiency was produced by a triangular plate of PZT device while the rectangular plate's efficiency has worsened as load mass increased.

Further analysis on the output of triangular PZT device shows that increment of auxiliary mass will reduce the maximum power that can be harvested even though the output energy is increased. Results from the forced vibration mode analysis has supported the above finding where increment in the auxiliary mass will worsen the harvesting power. Finally based on the results from both experimental study, it can be concluded that energy regeneration efficiency is optimum with auxiliary mass of 2.6g.

REFERENCES

[1] J. J. Rizza, "Solar-Driven $LiBr/H_2O$ Air Conditioning System With a R-123 Heat Pump Assist," Journal of Solar Energy Engineering, 136(1), pp. 1-5, 2013.

[2] D. G. Beshore, F. A. Jaeger, E. M. Gartner, "Thermal Energy Storage/Waste Heat Recovery Application in the Cement Industry," Proc. of First Industrial Energy Technology Conference, pp. 747-756, 1979.

[3] N. Hatziargyriou, A. Zervos, "Wind Power Development in Europe," Proc. of the IEEE, pp. 1765-1782, 2001.

[4] R. D. Hulst, T. Sterken, R. Puers, G. Deconinck, J. Driesen, "Power Processing Circuits for Piezoelectric Vibration-Based Energy Harvesters," IEEE Trans. on Industrial Electronics, vol. 57, no. 12, pp. 4170-4177, 2010.

[5] J. O. M. Miranda, "Electrostatic Vibration-to-Electric Energy Conversion," PHD thesis, Massachusetts Institute of Technology, 2004.

[6] G. Robert, O. Radu, "Harvesting Vibration Energy by Electromagnetic Induction," Electrical Engineering Series, pp. 7-12, 2011.

[7] K. Adachi, T. Sakamoto, "Study on Energy Transfer Efficiency Analysis of Cantilever Type of Piezocomposit Vibration Energy Harvester," 2012 The Japan Society of Mechanical Engineers, pp. 271-281, 2012.

[8] N. Kong, D. S. Ha, A. Erturk, D. J. Inman, "Resistive Impedance Matching Circuit for Piezoelectric Energy Harvesting," Journal of Intelligent Material Systems and Structures, pp. 1-10, 2010.

[9] M. I. Friswell, S. Adhikari, "Sensor Shape Design for Piezoelectric cantilever Beams to Harvest Vibration Energy," Journal of Applied Physics, 108, 1/7, 2010.

[10] S. Hashimoto, N. Nagai, Y. Fujikura, J. Takahashi, S. Kumagai, M. Kasai, K. Suto, H. Okada, "Multi-Mode Vibration-Based Power Generation for Automobiles," Proc of the 2012 IEEE-IAS Annual Meeting, 1/5, 2012.

[11] M. Young, "The PWM Strategy on DC-DC Converter," IEEE Journal of Industry Applications, vol. 28, no. 15, pp. 123-129, 1989.

[12] G. Eason, B. Noble, and I. N. Sneddon, "On Certain Integrals of Lipschitz-Hankel Type Involving Products of Bessel Function," IEEE Trans. on Power Electronics, vol. 247, no. 8, pp. 529-551, 1995.

[13] J. C. Maxwell, "A Treatise on Electricity and Magnetism," IEEE Trans. on Industry Applications, vol. 589, no. 2, pp. 68-73, 2010.

Battery Selection for Hybrid Energy Systems and Thermal Management in Arctic Climates

Maria Kalogera*, Pavol Bauer, *Senior Member IEEE* **
Delft University of Technology
*e-mail: kalogera.maria@gmail.com
**e-mail: p.bauer@tudelft.nl

Abstract— Hybrid energy systems are considered as a promising solution for meeting the power needs of offshore platforms due to the advances in renewable energy technologies and the need for reduction of dependency on fossil fuels. However, the unpredictable behavior resulting from the fluctuation and intermittence of the renewable energy sources and their dependence on local meteorological conditions, executes considerable impacts on normal operation of the hybrid system. As such, the energy storage is crucial for incessant energy supply and presents a preferable solution. The proposal explored in this paper is to give a full scope review of the principal energy storage technologies being developed so far, and the features and benefits of energy storage systems. The harsh Arctic conditions are also accounted, with a view to proposing a thermal management system.

Keywords— *Arctic, battery selection, microgrid, sustainable energy, thermal management.*

I. INTRODUCTION

The rising energy demands worldwide coupled with the depleting energy resources are intensifying the demand for -and consumption of new energy supplies [1], [2]. These imperative issues spur Arctic growth onwards, and attention has turned there as the most promising area for petroleum development. New opportunities and challenges are emerging across the Arctic and it is considered that the vast Arctic region is probably the last remaining unexplored source of hydrocarbons in the planet. Energy companies do not shy away from these issues and are on their way for exploring ways to expand and diversify their energy supply. If responsibly developed, Arctic energy resources can help offset supply constraints and maintain energy security for consumers throughout the world.

Hybrid energy systems constitute a promising solution for power supplying offshore-platforms in Arctic climates. As such, this paper focuses on the selection of the appropriate energy storage system. Due to the harsh conditions that Arctic presents, a thermal management system takes over part of the paper.

This paper is developed in the following steps. Initially, the first section expands on the introductory observations and discusses the need for battery storage in hybrid systems. Later on, the classification of energy storage techniques is presented which is followed by the selection of the appropriate battery type for Arctic climates. Ultimately, the selected battery model is extensively reviewed and finally a mathematical model for the thermal management of the batteries in harsh Arctic conditions is shown. In the end, conclusions are drawn upon the specification of this proposed thermal management model.

II. ENERGY STORAGE

A. Background

Renewable energy sources essentially have random behavior and cannot have accurate prediction. Continuous sunny /windy days give abundant PV/ Wind power because of which the battery banks are underutilized. On the other hand, cloudy/ still days can discharge the batteries beyond the lower discharge limit [3]. Thus, it comes to the point that energy storage is an essential part of the hybrid energy system. Storage of electrical and thermal energy is compelling so as to cope with the fluctuations of the natural resources. If energy is sufficiently available, the energy is stored to be used later on. Likewise, the energy storage system is the enabling power element to compensate for the lack of power when the production of renewable energy resources is not adequate.

TABLE I
CLASSIFICATION OF ELECTRICAL ENERGY STORAGE SYSTEMS ACCORDING TO ENERGY FORM

Electrical Energy Storage	Electrostatic energy storage including capacitors and supercapacitors
	Magnetic/current energy storage including SMES
Mechanical energy storage	Kinetic energy storage (flywheels); Potential energy storage (PHES and CAES).
Chemical Energy Storage	Electrochemical energy storage (conventional batteries)
	Chemical energy storage (fuel cells, Molten-Carbonate Fuel Cells – MCFCs and Metal-Air batteries)
	Thermochemical energy storage (solar hydrogen, solar metal, solar ammonia dissociation–recombination and solar methane dissociation–recombination).
Thermal Energy Storage	Low temperature energy storage (Aquiferous cold energy storage, cryogenic energy storage);
	High temperature energy storage (sensible heat systems such as steam or hot water accumulators)

The 2014 International Power Electronics Conference

TABLE II
COMPARISON OF THE TECHNICAL SPECIFICATIONS OF VARIOUS ENERGY STORAGE TECHNIQUES [7], [8], [9], [10], [11]

Technology	Power Rating	Discharge Time	Self Discharge per day	Main advantages	Main Disadvantages	Power Quality Applications	Energy Management Applications	Common Appication
PHES	100-500MW	Several Days	Very Small	High capacity, high energy efficiency, Low cost	Requires special topology	↓	↑↑	Long period energy management and peak shaving
CAES	50-300 MW	Several days	Small	High capacity, Low cost	Requires special topology	↓	↑↑	Long period energy management and peak shaving
Lead-Acid	Several kW-20 MW	Seconds – Days	0.1-0.3%	Low cost, technical maturity	Low lifetime under deep discharges	↑↑	→	Peak shaving
NiCd	Several W – 20 MW	Seconds - Days	0.2-0.6%	High power and energy density, high charge & discharge rate	Memory effect, high Cost	↑↑	↑	Small and medium period energy management
NaS	50 kW- 10 MW	Seconds - Days	~ 20%	High power and energy density, High pulse power, High efficiency	High cost, High operating temperature	↑↑	↑↑	Small and medium period energy management
Li-ion	Several W – 100 kW	Minutes - Hours	0.1-0.3%	High power and energy density, High efficiency, Long lifetime under high DOD	High cost, special charge circuit	↑↑	→	Portable devices
VRB	30 kW-3MW	Seconds- hours	Small	Independent power and energy design, Long lifetime	Low energy and power density	↑	↑↑	Small and medium period energy management
ZnBr	50 kW-2MW	Seconds- hours	Small	Independent power and energy design, Long lifetime	Low energy and power density	↑	↑↑	Small and medium period energy management
PSB	1-15 MW	Seconds - Hours	Small	Independent power and energy design, Long lifetime	Low energy and power density	↑	↑↑	Small and medium period energy management
Fuel Cell	0-50 MW	Seconds- several Days	Almost zero	High energy and power density	High cost, low efficiency	↑↑	↑↑	Small and medium period energy management
Metal Air	0-10 kW	Seconds- several Days	Very small	Low cost, Environmentally friendly	Short lifetime, low efficiency, difficult recharging	↓	↑↑	Bridging power
Solar Fuel	0-10 MW	Several days	Almost zero	Environmentally friendly, transport capability	R&D stage	↑↑	↑↑	Energy management
SMES	100 kW-10 MW	ms-8 sec	10-15%	High efficiency, high power output	High cost, low energy density	↑↑	↓	Power Quality
Flywheel	0-10 MW	ms-15 min	-	High power output	Low energy density	↑↑	→	Power quality
Capacitor/ Supercapacit or	0-300 kW	ms-60 min	20-40%	High efficiency	Low energy density	↑↑	↑	Power quality
AL-TES	0-5 MW	1-8 h	0.5%	-	-	→	↑↑	Energy management
CES	100kW- 300 MW	1-8h	0.5-1%	-	-	→	↑↑	Energy management
HT-TES	0-60 MW	Several days	0.05-1%	-	-	→	↑↑	Energy management

978-1-4799-2706-7/14 $31.00 © 2014 IEEE

B. Classification of energy storage techniques

There are many energy storage techniques. According to their form they can be classified as follows [4], [5], [6] and are listed in Table I. They are compared in Table II. The various energy storage techniques have diverse applications in power systems such as [5]:

- Rapid reserve
- Area control and frequency responsive reserve
- Commodity storage
- Transmission system stability
- Transmission voltage regulation
- Transmission and Distribution facility deferral
- Renewable energy management
- Customer energy management
- Power quality and reliability

Each technology has its own particular strengths and operational characteristics. The selection of the appropriate storage method is based upon the following criteria:

- Reliability which is the capability of the system to satisfy the load needs at all times.
- Efficiency defined by the ability to use the components in a way of minimizing the losses.
- Life Cycle Cost which includes the capital cost and the maintenance costs over the lifespan of the system.
- Technical Maturity regarding the commercial availability and proven reliability of the technologies used.
- Environmental Impact including the impact that each component and the overall system have on the surrounding area.

Almost always there is no "ideal" specific energy storage technique. Conversely, each time the type selected is a tradeoff among the criteria of the decision maker.

What is more, a storage system needs to be intimately adapted to the application type which may include low to mid power isolated areas, etc and to the production type (permanent, portable, renewable etc). In this thesis, the battery bank as energy storage device is only considered. Chemical Storage offers the best trade for efficiency, reliability, maturity and cost. With efficiencies ranging between 75% and 85% and costs lying on a low scale compared to other methods, the battery bank as energy storage is the one that will be dealt in the framework of this thesis [12], [13].

However, due to their restricted lifetimes, batteries represent one of the weakest parts of energy systems. Consequently, the study of their performance is of high importance when exploring the performances of such systems.

In short the strengths and weaknesses of the batteries are shown in Table III. Some of the characteristics of the commonly used batteries are summarized in Table IV.

TABLE III
STRENGTHS AND WEAKNESSES OF BATTERY ENERGY STORAGE

Strengths	Weaknesses
Reliable providing quickly back-up power	Batteries should be controlled and regulated as there is risk of overcharging etc.
Early charge-discharge High Power Density	Possible release of toxic materials Reduced capacity in low temperatures
No greenhouse gas emissions	Low recharging ability in low temperatures
Prolong lifetime in cold climates as a result of lower self-discharge rates.	

C. Selection of battery type

There is an assortment of key issues that have to be taken into consideration when selecting the appropriate type of battery system. The selection of such type is a compromise as no battery fulfills all the characteristics for a fully satisfactory solution. Among the different

TABLE IV
CHARACTERISTICS OF COMMONLY USED RECHARGEABLE BATTERIES [14], [15], [16], [17], [18], [19]

	Vanadium redox flow battery	Nickel – Cadmium (NiCd)	Nickel-metal-hybrid (NiMH)	Valve Regulated Lead-acid (VRLA)	Lithium -ion (Li-ion)
Energy density [Wh/kg]	25	45-80	60-120	30-50	100-200
Power density [Wh/kg]		175	250-1000	180	1800
Internal Resistance (in mΩ)		100-200 6V pack	200-300 6V pack	<100 12V pack	150-250 7.2V pack
Cycle Life	10000	1500	300-500	200-300	300-500
Overcharge Tolerance (discharge only)	High	Moderate	Low	High	Very low
Fast Charge Time		1h typical	2-4h	8-16h	2-4h
Operating Temperature (°C)	0…60	-40…60	-20…60	-20…60	-20…60
Maintenance Requirement		30 to 60 days	60 to 90 days	3 to 6 months	Not req.
Typical Cost ($)	500 $/ kWh	50 (7.2V)	60 (7.2V)	25 (6V)	100 (7.2V)
Toxicity	Very low	High	Relatively low	High	Low
Commercial date	1980	1950	1990	1970	1991
Remarks	Long lifetime, Easy capacity extension, Can be charged or mechanically refueled, Unlimited shell life	memory effect, robust, long term storage possible	limited discharge current, high maintenance, complex charge algorithm	charged storage, lifetime is temperature sensitive no maintenance	protection needed, not fully mature technology

types of batteries, Nickel Cadmium (NiCd) batteries, score the best for Arctic climates. Their use is desired and considered to be the best option for storage of energy in the case studied.

They are preferred based on the following characteristics:

• They have been used before in Arctic climates with reasonable results [11].

• NiCd batteries are rechargeable and achieve full power relatively quickly, giving them prolonged practical life than lithium non-rechargeable or alkaline batteries.

• They are simple to use and incorporate an inexpensive technology over the course of their useful life [20].

• NiCd have a renowned reputation for reliability and robustness. They operate well in a wide temperature range which is the benchmark technology for harsh and challenging applications. However, cold and high heat reduce charge acceptance and the battery must be brought into moderate conditions before charging.

• They can be designed for a different speed of charge.

D. Basic Operation of NiCd Batteries

The three basic elements of a NiCd cell are the positive nickel electrode (cathode), the negative cadmium electrode, immersed in a concentrated KOH electrolyte solution. Small quantities of lithium hydroxide (LiOH) are also included in the electrolyte for enhancing life and temperature performance. The electrolyte is used only for ion transfer and is not chemically charged or degraded during the charge/discharge cycle. Both electrodes are separated by a thin porous layer of insulating material, namely the separator, to inhibit electrical contact between them.

The charge/discharge reaction of a nickel cadmium battery is as follows:

$2 \text{NiOOH} + 2 \text{H}_2\text{0} + \text{Cd} \rightarrow$	$\leftarrow 2 \text{Ni(OH)}_2 + \text{Cd (OH)}_2$
Charged	Discharged

As illustrated from the above reaction, during discharge, the nickel oxyhydroxide combines with water and produces nickel hydroxide and a hydroxide ion while cadmium hydroxide is produced at the negative electrode.

To charge the battery the process can be reversed. When the charging current is applied to the NiCd battery, the negative plates lose oxygen and begin forming metallic cadmium, while nickel-hydroxide becomes more oxidized. This continuous process stops when the charging current is applied or until all the oxygen is removed from the negative plates and only cadmium remains [21]. Nevertheless, during charging, oxygen can be produced at the positive electrode and hydrogen can be produced at the negative electrode. As a consequence some venting and water addition is required, but much less than required for a LA battery [22].

E. Charactertistics of NiCd batteries

NiCd batteries differ from typical lead-acid or alkaline batteries in the following crucial aspects; the typical batteries (alkaline or lead-acid) have cell voltage of around 1.5V which drops off as it is depleted. Unlike these batteries, NiCd keep a steady voltage of 1.2V, enabling them to deliver full power till the end of discharge. What is more, NiCd batteries incorporate low internal resistance, thus facilitating the quick charge/discharge uptake. These two facts allow a high amount of amperage at higher voltage than alkaline batteries [23].

Besides, NiCd charge rate ranges between 1.2-1.45 V/cell, thus demanding slight variation in voltage and almost steady amperage, conversely with lead-acid batteries. Normally, the charge rate for a NiCd battery is 10 percent of the capacity, unless speed chargers are used. In addition, NiCd batteries can reach the 115 percent of their capacity, undergoing only a small decline of their life time. What is more, NiCd batteries incorporate higher energy density, smaller size and lighter characteristics than that of lead-acid batteries [24]. This is the reason why they are mostly preferable in cases of size and weight limitation (i.e. aircraft) [25].

NiCd cells have lower capacity compared to alkaline cells and are more costly [26]. However, taking into account that the alkaline battery's chemical reaction is not reversible, a reusable NiCd battery has a considerably longer lifetime.

However, a common drawback of the Nickel Cadmium batteries is the memory effect. They tend to suffer from the chemical memory that is developed and causes fast discharge upon reaching a certain point [27]. The memory effect occurs due to the abnormal growth of large crystals on cadmium electrode, resulting in subsequent reduction of surface area. In turn, this results in increasing the battery's effective internal resistance. This phenomenon tends to appear more quickly when the batteries are overcharged after leaving them connected to the charger for extensive periods of time without being able to discharge them properly. In order to maintain the rated capacity of these batteries, a time-consuming recharging regime is required [28].

Worst of all, cadmium is a poisonous heavy metal that can contaminate the environment [29]. It is highly toxic, and listed among the 25 hazardous substances that pose the most severe potential hazard to the human health [30], [31].

F. Performance of NiCd batteries in Arctic

Generally, extended exposure to low temperatures, leads to limited performance mostly because of the change in the cell chemistry. The latter is defined as the electrochemical characteristics of the active chemicals used in the cell, due to which the nominal voltage of a galvanic cell is fixed. The terminals' voltage at each time is linked with the load current and the internal impedance of the cell. These in turn, depend on the temperature, the

978-1-4799-2706-7/14 $31.00 © 2014 IEEE

state of charge and the age of the cell. Cell performance can change radically with temperature.

The optimum performance of the batteries takes place at room temperature [16]. Although, NiCd batteries can go as low as the frigid point of -40°C, the discharge rate is limited to 0.2C [32]. It is essential to select an appropriate temperature for optimizing charging performances. Temperature range within 10-45°C yield the highest efficiency, which drops at above 45°C. On the other hand, repetitive charging at less than 0°C may cause cell internal pressure build-up, causing electrolyte leakage as in high temperature conditions. As a result, batteries can be charged at temperature ranges between 0-45°C under standard charging conditions, but preferably at 10-45°C under fast charging conditions [33].

NiCd batteries are not more susceptible to the freezing point at low levels of charge. However, this does not automatically mean that they can also be charged under those conditions. Most batteries need to be brought up to temperatures above the freezing point for charging. The NiCd can be recharged at below freezing provided the charge rate is reduced to 0.1C.

In Fig. 1, the decrease in charge efficiency of NiCd is demonstrated. As it is clearly illustrated in Fig. 1, the battery can only accept 70% of its full capacity at 45°C while at 60°C the charge acceptance is further reduced to 45%.

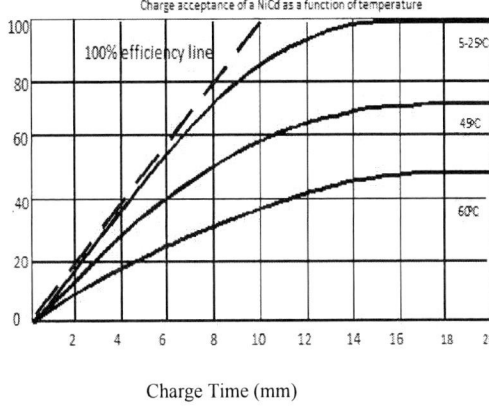

Fig. 1. NiCd charge acceptance as a function of temperature

III. Thermal Management of the Batteries

A. Thermal Management Concept

In the framework of this paper, a thermal management model including insulation for the batteries enclosure and an internal heater so as to ensure that the temperature never drops below a predefined model will be applied.

B. Thermal Management Model

1) Insulation Material

There are various materials used for thermal insulation. Among them, the most popular are the polystyrene, polyurethane, glass wool etc. The expanded polystyrene and the polyurethane foam are indicative for insulation at low temperature ranges. For temperature exceeding the 80°C range, these kinds of insulation materials are not considered as they are susceptible to degradation by heat action. However, this is not an issue for the Beaufort Sea and an aluminum box insulated with extruded polystyrene is considered for the insulation of the batteries.

2) Mathematical Model

Assuming that the heat generation of a battery cell is uniform, the heat generation rate per unit volume for discharging a NiCd cell is expressed as follows [34], [35] and [36]

$$Q = I\,(E_{oc} - E)$$

While the Heat Generation Rate per Unit Volume for charging a NiCd cell is:

$$Q = I\,(E - E_{oc})\eta + I\,E\,(1 - \eta)$$

Where:

I	Current passing through the cell	A
E	Actual cell voltage	V
E_{oc}	Cell open circuit voltage	V
T	Cell temperature	K
η	Cell charge efficiency	$Dimensionless$

The open circuit voltage of a cell is considered to be constant over the optimum operating temperature for Ni-Cd Cells ($0 - 20°C$). For each Ni-Cd cell some typical parameter values from [35] are considered. These include:

Open circuit voltage	$E_{oc} = 1.461$	V
Depth of discharge	25%	$Dimensionless$
Efficiency	$\eta = 0.8$	$Dimensionless$

Based on the assumption that the temperature of the enclosure exterior is the same as the ambient air temperature T_amb, and the predominant heat transfer phenomenon between the interior and the exterior is conduction, the energy balance in the system is formulated as follows [37]:

$$Q_{cond} + q = c_{p,bat}m_{bat}(T - T_o) \Rightarrow$$
$$\frac{dQ_{cond}}{dt} = c_{p,bat}m_{bat}\frac{dT}{dt} - q \qquad (1)$$

From Fourier's law for heat conduction

$$\frac{dQ_{cond}}{dt} = -\frac{kA}{L}(T - T_{amb}) \qquad (2)$$

Combining the equations (3.1) and (3.2) the following differential equation is derived.

$$T = -\frac{Lc_{p,bat}m_{bat}}{kA}\frac{dT}{dt} + \frac{qL}{kA} \qquad (3)$$

Where:

Q_{cond}	Heat conducted in the enclosure	
Q	Heat generated by thermal losses in the Battery	
m_{pcm}/m_{bat}	Mass of PCM/batteries respectively	
q	Average heat generation due to battery inefficiencies	
A	Area of the wall	
Q	Total heat generation	

For the thermal management model the differential equation (3.3) along with the following assumptions were implemented.

Thermal Conductivity of Polysterene	0.029	$\dfrac{W}{m^2 \cdot K}$
Thickness of the polystyrene insulator	0.1	m
Polysterene density	1050	kg/m^3

The heat generation from the batteries inefficiencies is very low and is shown in Fig. 2. It ranges up to $3.3W$.

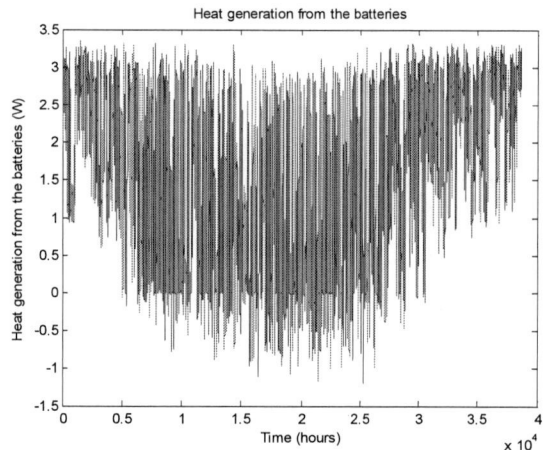

Fig.2. Heat generation of the batteries in one year

Implementing the polystyrene insulation for the thermal management and an initial battery temperature of 0°C would result in the temperature profile of the battery enclosure shown in Fig. 3 (marked with black). As illustrated, at some point the battery temperature approaches −12°C which is not desired for the batteries performance.

Thus, for further improvement of the model, a thermostat is also implemented to ensure that the temperature never drops below 0°C. In order to achieve so, a heating of the enclosure is also required. After testing the model for various values and testing its behavior an extra heating element (i.e. resistor) of 100W was considered sufficient. The results of the model are shown in Fig. 3. As shown the battery temperature fluctuates around 0°C resulting in a successful operation of the thermal management system.

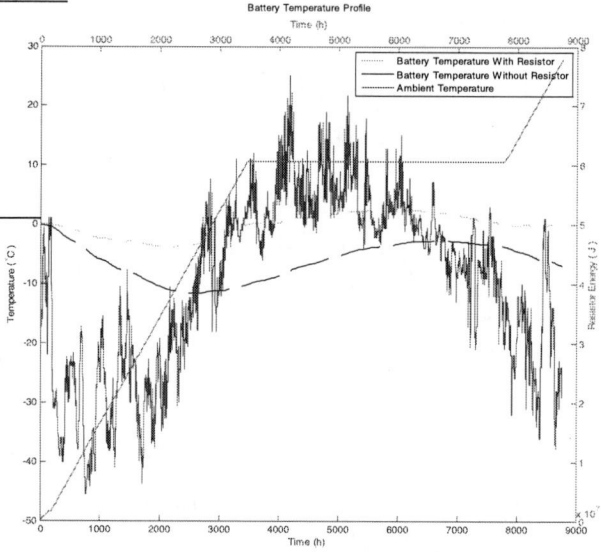

Fig. 3. Thermal Management Profile

IV. CONCLUSION

Power supplying of offshore platforms in Arctic climates can be made feasible with hybrid energy systems. These systems incorporate renewable energy sources which are fluctuating and do not have predictable behavior. As a result, energy storage is crucial for continuous energy supply and presents a preferable solution. In this paper, the chemical storage was considered as a possibility as it presents the best trade-off for efficiency, reliability, maturity and cost. From the various chemical storage types, NiCd batteries were selected based on various criteria. However, even if NiCd batteries operate well in a wide temperature range, their ability to efficiently operate in very low temperatures is limited. This was the primary reason for the battery thermal management system which was developed in the paper. The proposed model includes insulation for the batteries enclosure and an internal heater so as to make sure that the temperature never drops below a predefined model. The results of this model were shown in a graph which verified the validity of the model.

REFERENCES

[1] P.Bauer A.Kaas, "Design approaches for a sustainable off-grid power system," in IECON 2011 37th Annual conference of the IEEE industrial electronics society, Melbourne, 2011.

[2] P.Bauer E.Stamati, "Green Energy for On-road Charging of Electric Vehicles," in Invited paper 15th Mechatronika 2012, Prague, 2012.

[3] Vaishalee Dash Prabodh Bajpai, "Hybrid Renewable energy systems for power generation in stand-alone applications: A review ," Elsevier- Renewable and Sustainable Energy Reviews, pp. 2926-2939, 2012.

[4] J. F. Baalbergen, "System design and power management of a generator-set with energy storage for a 4Q drive," Delft , 2007.

[5] Adrian Ilincaa, and Jean Perronb Hussein Ibrahimab, "Comparison and Analysis of Different Energy Storage

Techniques Based on their Performance Index," in IEEE Electrical Power Conference Canada , Canada, 2007, pp. 393-398.

[6] International Electrotechnical Commission, "Electrical Energy Storage," IEC, White Paper 2011.

[7] A. Ilinca, J. Perron H. Ibrahim, "Energy storage systems— Characteristics and comparisons," Renewable and Sustainable Energy Reviews, vol. ISSN 1364-0321, pp. 1221-1250, June 2008.

[8] D. Zafirakis, K. Kavadias J.K. Kaldellis, "Techno-economic comparison of energy storage systems for island autonomous electrical networks," Renewable and Sustainable Energy Reviews, vol. ISSN 1364-0321, pp. 378-392, February 2009.

[9] Jian X. Jin, "HTS energy storage techniques for use in distributed generation systems," Physica C: Superconductivity, vol. ISSN 0921-4534, pp. 1449-1450, September 2007.

[10] Vladimir Strezov, Tim J. Evans Annette Evans, "Assessment of utility energy storage options for increased renewable energy penetration," Renewable and Sustainable Energy Reviews, vol. ISSN 1364-0321, pp. 4141-4147, August 2012.

[11] A., Ó'Gallachóir, B., McKeogh, E. & Lynch, Gonzalez, "Study of Electricity Storage Technologies and Their potential to address wind energy intermittency in Ireland. ," Sustainable Energy Ireland, http://www.sei.ie/Grants/Renewable_Energy_RD_D/Projects_fun ded_to_date/Wind/Study_of_Elec_ 2004.

[12] K.C. Divya J. Ostergaard, "Battery energy storage technology for power systems," Electric Power Systems Res., vol. 79, pp. 511-520, 2009.

[13] K. Magnago, A. da Rosa Abaide and B. Wottric L. N. Canha, "Multicriteria decision making for management of storage energy technologies on renewable hybrid systems-the analytic hierarchy process and the fuzzy logic," in 6th International Conference on the European, 2009, p. 166.

[14] A. De Broe, R. De Coninck K. Van Rattinghe, "Design Specification batteries and auxiliary equipment," Princess Elisabeth Antarctic Research Station, 2007.

[15] G. Hayman, C.T. Avelar, J.G.McGowan, U.Abdulwahid, K.Wu A. Rogers, "A hybrid system simulation model – theory manual," University of Massachusetts, 2006.

[16] "http://www.batteryuniversity.com/," Battery University,.

[17] "Energy Storage Systems," http://www.vrbpower.com/,.

[18] "Vanadium Redox Battery," http://www.vrb.unsw.edu.au/,.

[19] "British Antarctic Survey ," http://www.antarctica.ac.uk/,.

[20] W.S Kruijt, P.H.L Notten H. J Bergveld, "Electronic Network modelling of rechargeable NiCd cells and its application to the design of battery management systems," Journal of Power Sources, vol. 77, no. 2, pp. 143-158, February 1999.

[21] Mike Tooley Llloyd Dingle, "NiCd Cells ," in Aircraft Engineering Principles. Chennai, India: Charon Tec Pvt. Ltd , 2004, p. 329.

[22] David Connolly, "A Review of Energy Storage Technologies for the integration of fluctuating renewable energy ," University of Limerick, 2010.

[23] Terry L. Duran, "Secondary Cells ," in Armed Services Vocational Apritude Battery. N.Y, USA: Barron's Educational Series , 2006, p. 214.

[24] Andreas Poullikkas, Venizelos Efthimiou Ioannis Hadjipaschalis, "Overview of current and future energy storage technologies for electric power applications," in Renewable and Sustainable Energy Reviews.: ISSN 1364-0321 10.1016/j.rser.2008.09.028, August–September 2009, pp. 1513-1522.

[25] F. Putois, "Market for nickel-cadmium batteries," Journal of Power Sources, vol. 57, no. ISSN 0378-7753 10.1016/0378-7753(95)02243-0 http://www.sciencedirect.com/science/article/pii/03787753950224 30, pp. 67-70, September–December 1995.

[26] SF Fullam EC Anderson, "System and method for managing utilization of a battery," Apple Computer, Inc., US Patent 5 5963255, 1999.

[27] Emmett Dulaney, Toby Skandier Quentin Docter, , Jeff Kellum, Ed. Indianapolis, Indiana: Wiley Publishing, 2009, pp. 2-91.

[28] A.S. Arico, V. Antonucci A.K Shuklaa, "An appraisal of electric automobile power sources ," in Renewable and Sustainable Energy Reviews.: ISSN 1364-0321 10.1016/S1364-0321(00)00011-3, 2001, pp. 137-155.

[29] Jacob Østergaard K.C. Divya, "Battery Energy Storage Technology for power systems-An overview," in Electric Power Systems Research. Denmark: ISSN 0378-7796 10.1016/j.epsr.2008.09.017 http://www.sciencedirect.com/science/article/pii/S0378779608002 642, 2009, pp. 511-520.

[30] A. Oudalov, "Value Analysis of Battery Energy Storage Applications in Power Systems," in Power Systems Conference and Exposition, 2006. PSCE '06. 2006 IEEE PES, Atlanta, GA, 2006, pp. 2206 - 2211.

[31] JI-HOON KIM, LIEN DUONG THI, TAHIR IMRAN QURESHI YOUNG-JU KIM, "Recycling of NiCd Batteries by Hydrometallurgical Process on Small Scale," Journal of the Chemical Society of Pakistan, vol. 33, March 2011.

[32] Discharging at High and Low Temperatures, http://www.buchmann.ca/Chap5-page5.asp,.

[33] Online, Nickel Cadmium Technical Book.: http://www.gpina.com/pdf/NiCd.pdf.

[34] P. Stangerup P. Montalenti, "Thermal Simulation of NiCd Batteries for Spacecraft," Journal Of Power Sources, pp. 147-162, February 1977-1978.

[35] A. El-Dib A. Megahed, "Thermal Design and Analysis for Battery Module for a Remote Sensing Satellite," Journal of Spacecraft and Rockets, vol. 44, August 2007.

[36] A. Atef M. Zahran, "Electric and Thermal Properties of NiCd Batteries for Satellite," in 6th WSEAS International Conference on Power Systems, Lisbon, Portugal, 2006, pp. 122-130.

[37] Michael Ross, "Estimating wintertime battery temperature in stand-alone photovoltaic systems with insulated battery enclosures," in Renewable Energy Technologies in Cold Climates, Quebec, Canada , May 1998, pp. 344-350

[38] Long Lam; Bauer, P., "Practical Capacity Fading Model for Li-Ion Battery Cells in Electric Vehicles," Power Electronics, IEEE Transactions on , vol.28, no.12, pp.5910,5918, Dec. 2013

[39] Kumar, P.; Bauer, P., "Parameter extraction of battery models using multiobjective optimization genetic algorithms," Power Electronics and Motion Control Conference (EPE/PEMC), 2010 14th International , vol., no., pp.T9-106,T9-110, 6-8 Sept. 2010.

[40] Bauer, P.; Stembridge, N.; Doppler, J.; Kumar, P., "Battery modeling and fast charging of EV," Power Electronics and Motion Control Conference (EPE/PEMC), 2010 14th International , vol., no., pp.S11-39,S11-45, 6-8 Sept. 2010.

The 2014 International Power Electronics Conference

100kW PV PCS with Natural Convection Cooling for Outdoor Installation

Yasuhiro Jin, Kazumasa Matsuoka, Takehiro Takahashi, Nobuhiro Takahashi

Power Electronics Systems Division

TOSHIBA MITSUBISHI-ELECTRIC INDUSTRIAL SYSTEMS CORPORATION (TMEIC)

Tokyo, Japan

Abstract-The paper introduces development of the power inverter for the Power Conditioning System (PCS) for the middle scale Photo Voltaic (PV) power plants. The paper introduces a 100kW-rated PCS for outdoor installation. The feature of PCSs is in cooling. The 100kW PCS is cooled only by natural convection and it is the world first PCS eliminating the cooling fans, in this capacity range for outdoor installation. In the development, the heat transfers both from inside and outside of the enclosure have been minimized. Furthermore, the PCS has been developed by solving contradictory requirements where good cooling needs sufficient air flow but good protection from weather needs narrow openings. The features of the 100kW PCS offer short installation term at PV plants and also economical maintenance in the life cycle.

Keywords— Natural convection cooling, Outdoor installation, Photo voltaic power generation (PV), Power Conditioning System (PCS)

I. INTRODUCTION

The photovoltaic (PV) power generation is now increasing capacity in Japan and also in the world. [1][2] In order to promote renewable energy, the government of Japan started the feed-in tariff scheme in 2011. The scheme includes, photo-voltaic, wind, geo-thermal and small-scale hydraulic generations. By the statistics of Japan Photovoltaic Energy Association, PV generation capacity jumps up almost triple compared before and after introduction of the scheme as shown in Fig. 1. [3] The statistics also indicates a considerably large increase in utility scale PV generation for non-residential use in addition to capacity increase for residential application. The increase in 2009 was made because of another promotion scheme for residential PV generation.

For non-residential use, large utility scale plants rated at several tens of MW are frequently reported and well known. However, the PV generation facility is now found in spare spaces like rooftops of office buildings, warehouses and so on. For rooftops of business buildings, typical capacity of PV generation lies around 100kW, which is in middle between residential application and large utility scale application. Then, 100kW PCSs are developed to cover the capacity range. The developed PCS, of course, equipped with efficient control systems including the maximum power point tracking. [4]

In general, high system efficiency and low running

cost are keys for successful PV plant operation considering long life cycle. From these viewpoints, a 100kW PCS has been developed to be cooled only by natural convection and eliminating cooling fans. Furthermore, for easy installation at site, the PCS is sometimes required to stand outside without a house for protection from weather. Although the capacity is large, the circuits in PCS are made of electronic components, the circuits should be protected from high temperature and rain water of outside. However, the natural convection cooling and the protection from weather are contradictory requirements. The convection cooling needs good air flow in the PCS. On the contrary, for protection from weather, the openings should be narrow to avoid rain water penetration to the electronic component.

In this paper, at first, a theoretical consideration is made in order to confirm if the natural convection cooling is feasible or not for 100kW PCS. Then, computer aided 3-dimension design (3D-CAD) for PCS is performed combined with thermal simulations with finite element method to find detailed practical solution. For proving the design, factory tests are performed for cooling aspects in addition to electric performance tests. Weather proof tests are also performed outdoor for protection from rain water and the solar radiation.

The developed PCSs offer shorter installation term at site and less maintenance works compared with the conventional PCSs, which are for indoor installation and with the cooling fans and its associated components.

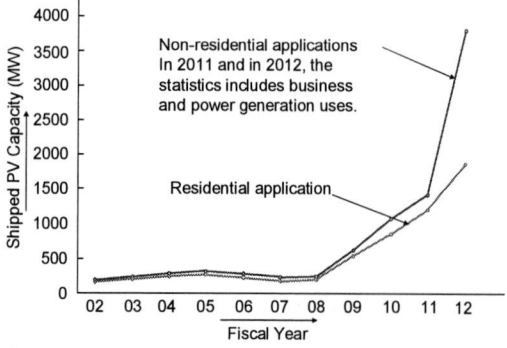

Fig. 1. PV capacity shipped for domestic market in Japan

978-1-4799-2706-7/14 $31.00 © 2014 IEEE   3207

II. OUTLINE OF DEVELOPED 100kW PCS

Before the detailed description, the outline of developed 100kW PCS is introduced. The specifications and the ratings are shown in Table I. The external view is shown in Fig. 2. As seen from Fig.2, the developed 100kW PCS for outdoor installation has been designed considering the following factors. The details will be described later.

- The enclosure is designed considering better thermal insulation between inside and outside.
- The canopy at the roof is selected fulfilling both the protection from rain falls and air flow for cooling.
- The inlet at the base the enclosures designed considering both protection from dusts or particles from outside and the intake air flow.

TABLE I
SPECIFICATIONS AND RATINGS OF PCS

Item	Specifications and ratings	
Installation	Outdoor	Indoor
Input voltage	DC 350V	
Output voltage	AC200/210/415/440V 3-Phase; 3-Wire	
Frequency	50/60Hz	
Power factor	More than 0.95	
Current THD	Less than 5%	
Weight	1000kg	900kg
Footprint	1000×1000mm	800×1000mm
Height	2445mm	1900mm
Temperature Range	-5°C to 40°C	
Efficiency	96.2% max at 50% load	
Anti-islanding	Passive methods: Voltage phase jump detection Active methods: Slip mode frequency shift	
Cooling method	Natural Convection	

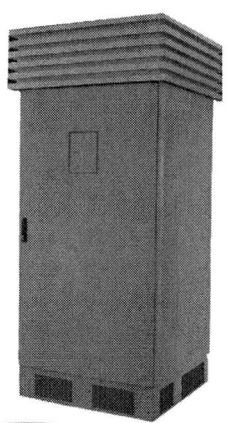 

(a) for outdoor installation   (b) for indoor installation

Fig. 2. The external view of developed 100kW PCS

100kW PCS circuit configuration is shown in Fig. 3. It mainly consists of circuit breakers (MCCBs) for input and output, an inverter (INV), an isolation transformer (TR), a magnetic contactor (MC) and an inductor to connect with low-voltage distribution system. All of 100kW PCS main circuit is placed in an enclosure, which is cooled only by natural convection.

The same circuit configuration is applied for both of outdoor and indoor installations although the enclosure is different. The PCS operates within the specification of temperature range shown in Table I.

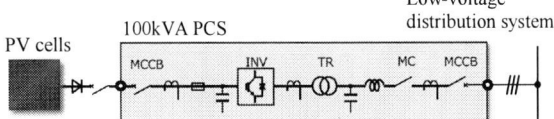

Fig. 3. 100kW PCS circuit configuration for both of outdoor and indoor installation

III. FEATURES OF DEVELOPED 100kW PCS

Comparisons of PCS features based on cooling method are shown in Table II. MTBF is expected to be improved compared with our conventional cooling PCS because the cooling fan and its peripheral control circuit are eliminated. No frequent maintenance will be necessary because no rotating mechanical parts are inside. As seen from Table II, many advantages are expected with PCS without cooling fans.

On the other hand, it is obvious to confront many challenges to cool the PCS. Therefore, this paper introduces the challenges and the developed solutions.

TABLE II
COMPARISONS OF PCS FEATURES BASED ON COOLING METHOD

Item	Comparisons	
Product	Developed PCS	Conventional PCS
Cooling	Natural Convection	Forced air cooling
Design	Advanced air flow design	Conventional cooling fans
Reliability	Better MTBF due to fan elimination	Typical MTBF due to cooling fans
Maintenance	Less frequent due to no fan replacement	Every 5 years in general due to fan replacement
Efficiency	Higher efficiency due to no cooling power consumption	Usual efficiency due to conventional cooling system
Environment	Improved condition because of slow movement of air	Getting dusty due to forced air cooling
Enclosure	Added a canopy and designed with thermal protection for outdoor use	Used normal frame for indoor use

From the cooling viewpoint, one of the major challenges is to reduce the loss of inverter portion. The first step is adopting 6th generation IGBT, the latest technology for the inverter circuit. Switching loss and conducting loss of the latest IGBT are decreased compared with the former generations of IGBT. In addition to use the latest IGBT, a high-efficiency transformer is introduced in the second step. Core material and total transformer design are studied carefully with various tests for several combinations. As a result, the maximum efficiency reaches 96.2% and high efficiency is kept for wide range of output as shown in Fig.4.

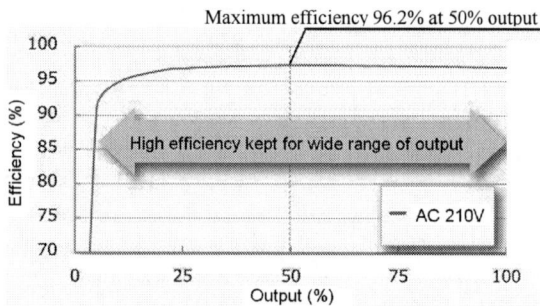

Fig. 4. 100kW PCS efficiency characteristics

Other challenge is to improve air flow in the enclosure. The development of natural convection cooling for 100kW PCS is discussed in detail in the following chapter.

IV. FEASIBILITY OF NATURAL CONVECTION COOLING FOR 100KW PCS

The first step of the development is feasibility check by calculating theoretically the air flow rate by natural convection if it is enough for cooling. In this chapter, the PCS enclosure is assumed to be a simple cylinder as shown in Fig. 5. The bottom of the cylinder corresponds to the air inlet of the enclosure while the top corresponds to the outlet.

The pressure difference (Pa or N/m^2) between the inside air and the outside air can be expressed by equation (1).

$$\Delta P = g(\rho_{out} - \rho_{in})h \qquad (1)$$

where, ρ_{out} and ρ_{in} are density of outside air and density of inside air respectively, h is height between outlet and inlet air (m) and g is acceleration of gravity (9.81 m/s^2).

The pressure difference will balance with the pressure loss by the air flow friction. The pressure loss is proportional to square of the air flow velocity v as shown by equation (2).[5]

$$\Delta P = \lambda \frac{l}{d_h} \rho_{in} \frac{v^2}{2} \qquad (2)$$

where, λ (dimensionless) is the friction factor, l (m) is the length of the air path, d_h (m) is hydraulic equivalent diameter of the air path of the enclosure.

From equations (1) and (2), the air velocity (m/s) is obtained by equation (3).

$$v = \sqrt{\frac{2g(\rho_{out} - \rho_{in})h \cdot d_h}{\lambda \cdot l \cdot \rho_{in}}} \qquad (3)$$

Then, air volume rate q_n (m^3/s) made by natural convection can be calculated by equation (4) considering

the air flows through the circle of diameter d_h.

$$q_n = \pi \left(\frac{d_h}{2}\right)^2 \sqrt{\frac{2g(\rho_{out} - \rho_{in})h \cdot d_h}{\lambda \cdot l \cdot \rho_{in}}} \qquad (4)$$

Considering that the air density ρ is proportional to the absolute temperature, equation (4) can be written as equation (5).

$$q_n = \pi \left(\frac{d_h}{2}\right)^2 \sqrt{\frac{2g \cdot h \cdot d_h}{\lambda \cdot l} \frac{\Delta T}{T_{in}}} \qquad (5)$$

where, ΔT is the temperature difference between inside and outside and T_{in} is the absolute temperature of inside calculated as $T_{in} = 273 + T_c$ from Celsius temperature T_c.

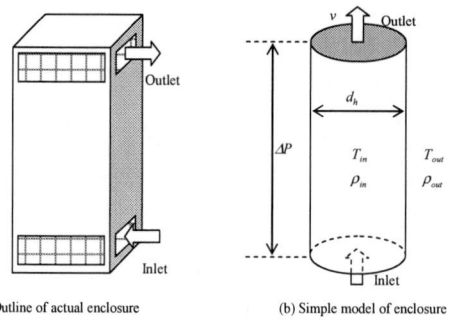

(a) Outline of actual enclosure   (b) Simple model of enclosure

Fig. 5. Simple hydraulic model of enclosure for feasibility check

The second step is to calculate heat transfer from the power loss P_l in the enclosure to the air. When the air flow volume q is given, the thermal capacity of the air flow per second is $C \cdot \rho_{in} \cdot q$ where C is the specific heat of the air. Assuming the temperature rise ΔT_e, the heat transfer balance can be expressed by equation (6).

$$P_l = C \cdot \rho_{in} \cdot q \cdot \Delta T_e \qquad (6)$$

The feasibility is checked if the air flow rate q_n by equation (5) is enough larger than that for heat transfer in equation (6). The preliminary calculation is performed with values based on experiences. The calculation yields $q_n \approx 27 m^3/\text{min}$ with temperature difference $\Delta T = 15 \deg$. The power loss P_l is estimated from the design efficiency η of the PCS as $P_l = 100kW \cdot (1 - \eta) \approx 3kW$. The very good efficiency is realized by the IGBT development. [6] From equation (6), the flow rate q is required to be larger than $10.6 m^3/\text{min}$. Then, q_n is confirmed to be sufficient for cooling. After the feasibility check, the challenge to eliminate cooling fan has started.

V. ACTUAL NATURAL CONVECTION COOLING DESIGN WITH CONCURRENT FEM SIMULATION

The feasibility check is performed based on the simple model where the PCS enclosure is treated as a cylinder. However, the structure of the enclosure is complicated since it accommodates many components for electrical circuits associated with mechanical components.

Then, the enclosure design has been done with help of 3D-CAD of which data can be used in the FEM (Finite Element Method) simulation for air flow and thermal analysis. Try and error cycles have been made in a desk top computer for rapid design cycle to find optimum cooling path considering the actual component and structure arrangement in the PCS enclosure. An example is shown in Fig. 6. The air flow path is shown and its velocity distribution is indicated by colors. The air flow passes through all components necessary for cooling.

Through the design and simulation process, the technical challenge to eliminate fans has been confirmed to be realized.

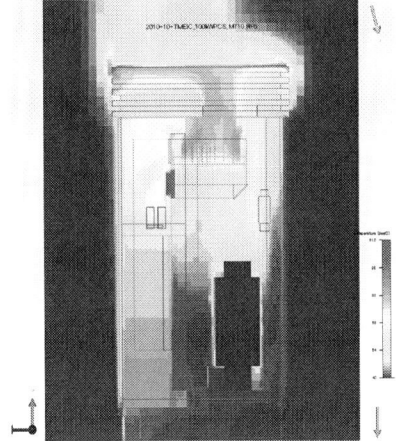

Fig.6. PCS design and concurrent thermal flow analysis example

VI. ENCLOSURE TESTS AT FACTORY

The first 100kW PCS for outdoor type has been manufactured and tested at the factory. In addition to the electrical characteristics and performances, the thermal performance and the protection performance have also evaluated. The full load test has been done as the most severe thermal condition.

At the first test, the temperature near the top of enclosure was found to be higher than expected. It was from difficulty for outdoor installation. The first canopy design was based on the enclosure for the indoor enclosure which had successfully passed the tests. For outdoor installation, the canopy should protect the PCS from rain water. Thus, the canopy for indoor type was modified to be surrounded by rain covers. This was contradictory to cooling performance since the covers could be obstacles for air flow. As considered, the investigation showed that the first canopy shape was not sufficient for cooling. Then, the canopy shape was modified to be taller with multi eaves as shown in Fig. 2 to satisfy both better cooling performance and weather protection at the same time. After the modification, the temperature rise tests are repeated and the PCS showed good cooling performance.

After the temperature rise test, the tests for weather protection are performed outside the factory as shown in Fig. 7 which shows the water shower test.

In the international standard IEC 60529, the enclosure protection test is defined in detail. For outdoor installation, the protection level should be greater than "IP X4". "IP" stands for "ingress protection" and the protection level is indicated by two indices. The first index "X" stands for protection from solid particles. The

second index "4" stands for liquid ingress. The test for proving "X4" level is defined as follows;

- Water splashing from all directions
- Duration : 5 min
- Water flow rate: 10 l/min
- Pressure : 80kPa to 100kPa

Fig.7. Weather proof test of the outdoor type PCS

VII. PROTECTION FROM SOLAR RADIATION HEAT

The other important requirement to the PCS for PV plant is the protection from heat generated by solar radiation in addition to the internal power loss from the electrical components of the PCS. As the PV cells generate electric poser from the sun the PV PCS is also exposed to solar radiation when it is installed outside. It is the same phenomena which many car drivers experience. The cars under the sunshine absorb the solar radiation and the surface temperature increases to several tens of Centigrade. Under such condition, if the door or walls of the PCS are made of a single steel plate, the heat transfer P_C through the plate can be expressed as equation (7).

$$P_C = h_C A (T_s - T_{in})_e \qquad (7)$$

where, h_C is heat transfer factor of the steel, A is the area and T_s is the surface temperature.

When no countermeasure is taken to prevent the heat penetration to the enclosure, the sum $P_C + P_l$, the solar radiation heat and the internal power loss, may exceed the capability of the natural convection cooling. Then, the protection from the solar radiation is also necessary for the outdoor installation. For the purpose, the natural convection principle shown by Fig. 5 (b) is repeated to be applied to the structure of the door and the side plates. Actually, the door and the plates are made of two steel layers and have air inlets at the bottom and outlets at the top. The structure offers thermal insulation for heat conduction by making an air layer between two steel layers.

The air layer has much smaller heat transfer factor compared with the steel. The structure also offers air flow by natural convection and keeps the air layer temperature low taking fresh air from the inlet. [7]

The protection performance from the solar radiation is performed as shown in Fig. 8. outside of the factory on a

978-1-4799-2706-7/14 $31.00 © 2014 IEEE

sunny day. Temperatures inside and at surface of the PCS are measured for several hours. Fig. 9 shows the recorded temperature chart. As shown in Fig. 9, the inside temperature is kept almost equal to the outside atmosphere temperature. On the contrary, the temperature at the door surface rises as the sun radiation increases along with the time.

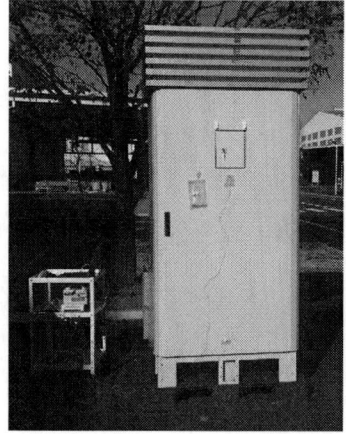

Fig.8. Solar radiation test of outdoor type PCS

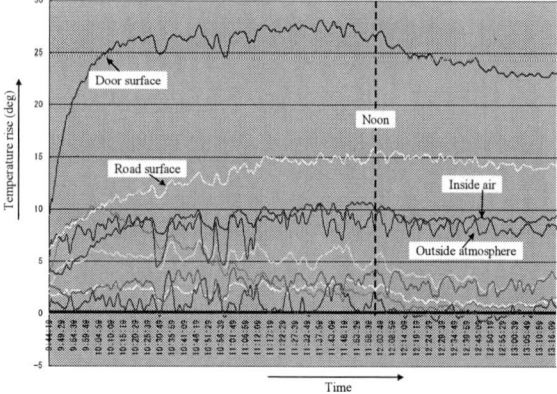

Fig.9. Measured temperatures at solar radiation test

The outdoor enclosure is built with durable materials, which is chosen from those for the house construction, considering the expected life of the PCS. The enclosure is designed considering also the maximum regional wind speed which is recorded in the past and typical snow accumulation in Japan. Then, the outdoor PCS can be installed in most parts of Japan. As the exhaust outlet is located on the side of canopy, the cooling capability of developed PCS will not be affected even if the snow is on top of the enclosure as shown in Fig.10.

A typical operation waveform of developed PCS is shown in Fig. 11, which is recorded at the development test. An example of PCS operation on a sunny day in April is shown in Fig. 12, which shows the sum of outputs from two types of PCS. One is the developed 100kW PCS and the other is a smaller PCS rated at 10 kW. As shown in Fig.12, the developed PCS successfully operate to convert the power from PV cells to the distribution system. Fig. 12 is from the field test record performed at the factory.

Fig. 10. PCS working for more than a year in the field

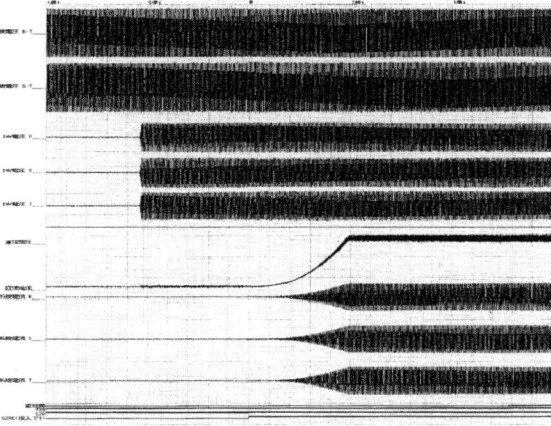

Fig.11. Full load operation test started with soft output ramp-up.

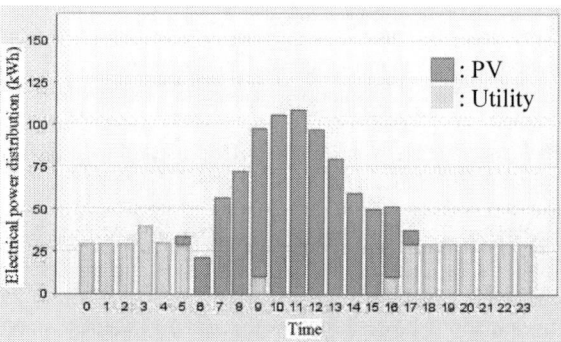

Fig.12. PCS operation on a sunny day in April

VIII. SHORT INSTALLATION WORK AND SMALL MAINTENANCE WORK

The developed PCS for outdoor installation does not need any housing at the PV plant site. The PCS is designed considering practical aspects of outdoor environment and the restrictions to the site are also small. Then, the installation work is easy and done in short time.

The PCS do not have any cooling fans and air filters, which require frequent replacement during expected life of the PCS. Then, the developed 100kW PCSs do not require such frequent maintenance as shown in Fig.13. Furthermore, the slow movement of air flow is not expected to introduce contaminants into the PCS from outside environment. Then, high reliability is expected for the developed 100kW PCSs.

These benefits are very effective especially when the PCSs are located sparsely or remote from the maintenance centers.

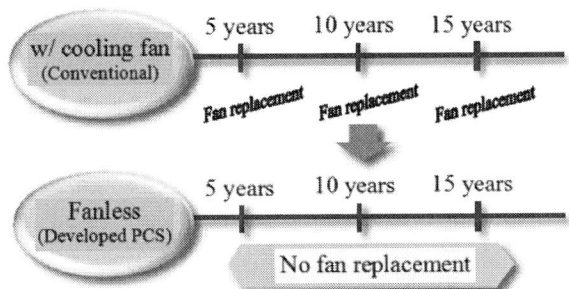

Fig.13. Comparison of typical maintenance period

IX. CONCLUSION

The 100kW rated PCSs have been successfully developed considering various kinds of requirement necessary for outdoor installation for the middle scale PV plants. The key points were the application of the latest technology of cooling design realizing efficient natural convection cooling. The development was successfully done by overcoming high technical challenges. The efficient cooling system for the outdoor installations has been successfully developed with help of the concurrent design process by the 3D CAD and the air flow and thermal analysis. The designs were verified by factory tests including the water shower test and the solar radiation test.

The natural convection cooling PCSs for outdoor installation offers easy installation at site. It also reduces frequent maintenance works and expected to contribute high operation efficiency. The advantages are well understood widely and the PCSs have been already delivered and working in the fields and show successful operation experiences. The developed PCSs are hoped to contribute to make PV generation much more popular and feasible now and in future.

REFERENCES

[1] T. Yoshino et al., "Power Electronics Conserves the Environment for Better Future", Proceedings of PESC2008, pp. XCIX-CIV, June 2008.

[2] T. Yoshino et al., "MW-Rated Power Electronics for Sustainable and Low Carbon Industrial Revolution", Proceedings of ISIE2010, pp. 3811 – 3816, 4-7 July 2010.

[3] http://www.jpea.gr.jp/en/statistic/index.html, Web page of Japan Photovoltaic Power Generation

[4] K.S. Lee et al., "Development of a 250kW PV PCS and adaptive MPPT method," Proceedings of IPEC2010, pp. 2598 - 2602, 21-24 June 2010.

[5] T. Takahashi et al., "100kW High-Power PV PCS with No Cooling Fans - High Efficiency and Low Running Cost –", Proceedings of ECCE Asia Down Under, June 2013.

[6] R.B. Bird, W.E. Stewart, E.N. Lightfoot: Transport Phenomena, Second Edition, p. 179, Wiley (2002)

[7] Peter Brackett, "Beat the Sun - Keep Your Equipment Cool", Proc. of AREMA 2003 Annual Conference, Chicago, IL - October 5-8, 2003

A New PLL Based on Fast Positive and Negative Sequence Decomposition Algorithm with Matrix Operation under Distorted Grid Conditions

Shaohua Sun, Hongqi Ben, Tao Meng, Jinyong Zhang
School of Electrical Engineering and Automation
Harbin Institute of Technology
Harbin, China

Abstract—This paper investigates and presents a new phase locked loop (PLL) based on fast positive and negative sequence decomposition algorithm with matrix operation under distorted grid conditions. In renewable energy integration filed, some information of unbalanced and distorted grid voltages, such as frequency and phase angle of fundamental wave, should be exactly and quickly detected for reliable operation. Due to the unfavorable transient response and very poor phase angle tracking of PLL based on the positive and negative sequence components decomposition with a quarter cycle delayed α variable and β variable in a stationary α-β reference frame under distorted grid conditions, a fast and exact PLL method by matrix operation of α variable and β variable of grid voltage is proposed. This process is achieved by transforming the positive, negative and low-order harmonics into multiple synchronous reference frames, and then by a matrix operation, the frequency and phase information are extracted. Thus the positive and negative sequences are separated after 5 sampling periods. The paper presents a detailed description and derivation of the proposed PLL method. Simulations and experimental results based on digital signal processor chip TMS320F2812 verify and validate the excellent performance of the new PLL method.

Keywords—phase locked loop, positive and negative sequence decomposition algorithm, matrix operation, unbalanced and distorted grid conditions

I. INTRODUCTION

In renewable energy integration field, some information of the fundamental frequency and phase of the three-phase voltages must be detected to feed back to the grid-connected converter for reliable operation [1], and the positive and negative sequence components are also needed for converter control. An ideal phase locked loop (PLL) should provide the fast and accurate synchronization information with a high degree of immunity and insensitivity to disturbances, harmonics of the grid voltage.

In the three-phase systems, the PLL based on the synchronous reference frame (SRF-PLL) has become a conventional grid synchronization technique [2]-[7]. It is widely used under ideal grid conditions. However, in case the grid voltage is distorted with harmonics, the SRF-PLL can still work if its bandwidth is reduced to

reject and cancel out the effect of the harmonics at the cost of the PLL response speed reduction.

In [8], a new multiple-complex coefficient-filter (MCCF) based synchronization technique without the symmetrical component, the fundamental positive and negative information and other harmonic components can be accurately estimated under distorted grid conditions. But it needs some complicated coordinate transformations of large trigonometric functions which increase the computing time. In [9], a decoupled double synchronous reference frame (DDSRF) based PLL is proposed to improve the accuracy and the response, using two dq frames to insulate the positive and negative sequence of grid voltages with unbalanced utility voltages. To suppress the harmonic components, another DDSRF is needed for every undesired harmonic component increasing the processing time.

Other solutions are based on the delayed signal cancellation (DSC) method. In [10], one-quarter of the fundamental period is delayed, in such a way that some harmonic components are rejected, but the time response is a quarter cycle of the grid fundamental period, and some harmonic components are still existed in the positive sequence component. To overcome its drawbacks, a second-order generalized integrator (SOGI) filter is used to suppress harmonics [11]. A dual second-order generalized integrator is presented in [12], by using two set of SOGI, negative sequence component is cancelled. Although the advanced synchronization systems are capable of making an accurate estimation of the fundamental components of the voltages by means of SOGI, for its correct operation, SOGI needs the information about the grid frequency that can be obtained by using a FLL in a feedback closed loop to guarantee good performance.

This paper proposes a new PLL based on fast positive and negative sequence decomposition algorithm with matrix operation, based on that, the PLL can be realized in 5 sampling periods by simple matrix operation. As following, the main building block of the proposed PLL method is presented and its performance is evaluated by simulations and experiments. Compared with the conventional method mentioned in [10], the new PLL improves the speed response and harmonic rejection.

978-1-4799-2706-7/14 $31.00 © 2014 IEEE

II. CONVENTIONAL PLL UNDER DISTORTED GRID CONDITIONS

In some ideal conditions, there are only fundamental positive sequence component in three-phase grid voltages. In practice, however, the voltages may contain negative sequence components and harmonics. When considering a three-phase three-wire system, where zero sequence is omitted, and to simplify the derivation, only the 5th, 7th, 11th and 13th harmonic components are considered, the three-phase voltages can be generally represented as in (1),

$$
\begin{bmatrix} E_a \\ E_b \\ E_c \end{bmatrix} = E^+ \begin{bmatrix} \cos(\omega t+\psi_+) \\ \cos(\omega t-\frac{2\pi}{3}+\psi_+) \\ \cos(\omega t+\frac{2\pi}{3}+\psi_+) \end{bmatrix} + E^- \begin{bmatrix} \cos(\omega t+\psi_-) \\ \cos(\omega t+\frac{2\pi}{3}+\psi_-) \\ \cos(\omega t-\frac{2\pi}{3}+\psi_-) \end{bmatrix} + E^{5-} \begin{bmatrix} \cos(5\omega t+\psi_5) \\ \cos(5\omega t+\frac{2\pi}{3}+\psi_5) \\ \cos(5\omega t-\frac{2\pi}{3}+\psi_5) \end{bmatrix}
$$

$$
+E^{7+} \begin{bmatrix} \cos(7\omega t+\psi_7) \\ \cos(7\omega t-\frac{2\pi}{3}+\psi_7) \\ \cos(7\omega t+\frac{2\pi}{3}+\psi_7) \end{bmatrix} + E^{11-} \begin{bmatrix} \cos(11\omega t+\psi_{11}) \\ \cos(11\omega t+\frac{2\pi}{3}+\psi_{11}) \\ \cos(11\omega t-\frac{2\pi}{3}+\psi_{11}) \end{bmatrix} + E^{13+} \begin{bmatrix} \cos(13\omega t+\psi_{13}) \\ \cos(13\omega t-\frac{2\pi}{3}+\psi_{13}) \\ \cos(13\omega t+\frac{2\pi}{3}+\psi_{13}) \end{bmatrix}
$$

(1)

Where a-b-c represents the three phases of the ac grid, E and Ψ are the amplitude and initial phase angle of voltages respectively. ω is the angle frequency of grid voltage. So the grid voltage can be represented as the sum of positive-, negative-, 5th, 7th and 11th and 13th harmonic components (represented by the symbols+, -, 5-, 7+, 11-, and13+, respectively).

By using Clarke transformation, equation (1) can be expressed as

$$
E_{\alpha\beta} = \begin{bmatrix} E_\alpha \\ E_\beta \end{bmatrix} = T_{3s/2s} \begin{bmatrix} E_a \\ E_b \\ E_c \end{bmatrix} = \begin{bmatrix} E_\alpha^+ \\ E_\beta^+ \end{bmatrix} + \begin{bmatrix} E_\alpha^- \\ E_\beta^- \end{bmatrix}
$$

$$
+\begin{bmatrix} E_\alpha^{5-} \\ E_\beta^{5-} \end{bmatrix} + \begin{bmatrix} E_\alpha^{7+} \\ E_\beta^{7+} \end{bmatrix} + \begin{bmatrix} E_\alpha^{11-} \\ E_\beta^{11-} \end{bmatrix} + \begin{bmatrix} E_\alpha^{13+} \\ E_\beta^{13+} \end{bmatrix}
$$

(2)

Where

$$
T_{3s/2s} = \sqrt{\frac{2}{3}} \begin{bmatrix} 1 & -\frac{1}{2} & -\frac{1}{2} \\ 0 & -\frac{\sqrt{3}}{2} & \frac{\sqrt{3}}{2} \end{bmatrix}
$$

In multiple synchronous reference frames, the voltages can also be given in equation (3) below.

$$
E_{\alpha\beta} = E_{dq}^+(t)e^{j(\omega t+\psi_+)} + E_{dq}^-(t)e^{-j(\omega t+\psi_-)}
$$

$$
+E_{dq}^{5-}(t)e^{-j(5\omega t+\psi_5)} + E_{dq}^{7+}(t)e^{j(7\omega t+\psi_7)}
$$

$$
+E_{dq}^{11-}(t)e^{-j(11\omega t+\psi_{11})} + E_{dq}^{13+}(t)e^{j(13\omega t+\psi_{13})}
$$

(3)

T is the grid period, time delayed $E_{\alpha\beta}$ by $T/4$, and equation (4) is obtained as follow.

$$
E_{\alpha\beta}(t-\frac{T}{4}) = E_{dq}^+(t-\frac{T}{4})e^{j(\omega t+\psi_+)}e^{-j\frac{\pi}{2}} + E_{dq}^-(t-\frac{T}{4})e^{-j(\omega t+\psi_-)}e^{j\frac{\pi}{2}}
$$

$$
+E_{dq}^{5-}(t-\frac{T}{4})e^{-j(5\omega t+\psi_5)}e^{j\frac{\pi}{2}} + E_{dq}^{7+}(t-\frac{T}{4})e^{j(7\omega t+\psi_7)}e^{-j\frac{\pi}{2}}
$$

(4)

$$
+E_{dq}^{11-}(t-\frac{T}{4})e^{-j(11\omega t+\psi_{11})}e^{-j\frac{\pi}{2}} +E_{dq}^{13+}(t-\frac{T}{4})e^{j(13\omega t+\psi_{13})}e^{-j\frac{\pi}{2}}
$$

In this equation, E_{dq}^+, E_{dq}^-, E_{dq}^{5-}, E_{dq}^{7+}, E_{dq}^{11-} and E_{dq}^{13+} are the magnitudes of positive , negative , 5th , 7th,

11th and 13th harmonic components in their rotating synchronous reference frames respectively. They vary slowly with time. So equation (4) can be approximated by equation (5).

$$
E_{\alpha\beta}(t-T/4) = E_{dq}^+(t)e^{j(\omega t+\psi_+)}e^{-j\frac{\pi}{2}} + E_{dq}^-(t)e^{-j(\omega t+\psi_-)}e^{j\frac{\pi}{2}}
$$

$$
+E_{dq}^{5-}(t)e^{-j(5\omega t+\psi_5)}e^{j\frac{\pi}{2}} + E_{dq}^{7+}(t)e^{j(7\omega t+\psi_7)}e^{j\frac{\pi}{2}}
$$

$$
+E_{dq}^{11-}(t)e^{-j(11\omega t+\psi_{11})}e^{-j\frac{\pi}{2}} + E_{dq}^{13+}(t)e^{j(13\omega t+\psi_{13})}e^{-j\frac{\pi}{2}}
$$

(5)

From equations (3) and (5) we can derive equations (6) and (7) below.

$$
E_\alpha^+ = \frac{1}{2}(E_\alpha(t) - E_\beta(t-\frac{T}{4})) - E_\alpha^{11-} - E_\alpha^{13+}
$$

(6)

$$
E_\beta^+ = \frac{1}{2}(E_\beta(t) + E_\alpha(t-\frac{T}{4})) - E_\beta^{11-} - E_\beta^{13+}
$$

(7)

Considered that the distorted grid voltages only contain the positive, the negative sequence component, 5th and 7th harmonic components, so E_α^{11-}, E_β^{11-}, E_α^{13+} and E_β^{13+} in equations (6) and (7) are zero, based on that, the positive sequence of grid voltages can be extracted exactly by equations (6) and (7). Then the positive sequence components are translated from stationary reference frame to the dq rotating reference frame by using Park's transformation. The frequency and angle can be realized by a feedback loop which regulates the q component to zero. The block diagram of PLL based on extracting the positive sequence in a quarter cycles delayed is shown in Fig.1.

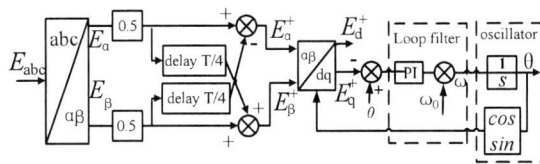

Fig.1. the block diagram of PLL based on the positive and negative component decomposition in a quarter cycles delayed under distorted grid conditions

However, when the 11th and 13th harmonics are contained in the grid voltage, only an approximation not the true amplitude and phase angle of the positive sequence component are detected, and the dynamic response of the system is significantly reduced because of the inherent time delays of a quarter cycle of the grid period. Therefore, in applications requiring high accuracy and a good dynamic response under distorted voltages, the conventional PLL technique does not present the most appropriate solution.

III. PROPOSED PLL UNDER DISTORTED GRID CONDITIONS

T_s is the sampling period, time delayed $E_{\alpha\beta}$ by T_s, equation (3) can be changed to equation (8) as follow.

$$
E_{\alpha\beta}(t-T_s) = E_{dq}^+(t-T_s)e^{j(\omega t+\psi_+)}e^{-j\omega T_s} + E_{dq}^-(t-T_s)e^{-j(\omega t+\psi_-)}e^{j\omega T_s}
$$

$$
+E_{dq}^{5-}(t-T_s)e^{-j(5\omega t+\psi_5)}e^{j5\omega T_s} + E_{dq}^{7+}(t-T_s)e^{j(7\omega t+\psi_7)}e^{-j7\omega T_s}
$$

$$
+E_{dq}^{11-}(t-T_s)e^{-j(11\omega t+\psi_{11})}e^{j11\omega T_s} + E_{dq}^{13+}(t-T_s)e^{j(13\omega t+\psi_{13})}e^{-j13\omega T_s}
$$

(8)

T_s is small, so equation (8) can be approximated by equation (9).

$$E_{\alpha\beta}(t-T_s) = E_{dq}^+(t)e^{j(\omega t+\psi_+)}e^{-j\omega T_S} + E_{dq}^-(t)e^{-j(\omega t+\psi_-)}e^{j\omega T_S}$$
$$+ E_{dq}^{5-}(t)e^{-j(5\omega t+\psi_5)}e^{j5\omega T_S} + E_{dq}^{7+}(t)e^{j(7\omega t+\psi_7)}e^{-j7\omega T_S}$$
$$+ E_{dq}^{11-}(t)e^{-j(11\omega t+\psi_{11})}e^{j11\omega T_S} + E_{dq}^{13+}(t)e^{j(13\omega t+\psi_{13})}e^{-j13\omega T_S}$$
(9)

Similarly, time delayed $E_{\alpha\beta}$ by $2T_s$, $3T_s$, $4T_s$ and $5T_s$, equations (10), (11), (12) and (13) are obtained as follow.

$$E_{\alpha\beta}(t-2T_s) = E_{dq}^+(t)e^{j(\omega t+\psi_+)}e^{-j2\omega T_S} + E_{dq}^-(t)e^{-j(\omega t+\psi_-)}e^{j2\omega T_S}$$
$$+ E_{dq}^{5-}(t)e^{-j(5\omega t+\psi_5)}e^{j10\omega T_S} + E_{dq}^{7+}(t)e^{j(7\omega t+\psi_7)}e^{-j14\omega T_S}$$
$$+ E_{dq}^{11-}(t)e^{-j(11\omega t+\psi_{11})}e^{j22\omega T_S} + E_{dq}^{13+}(t)e^{j(13\omega t+\psi_{13})}e^{-j26\omega T_S}$$
(10)

$$E_{\alpha\beta}(t-3T_s) = E_{dq}^+(t)e^{j(\omega t+\psi_+)}e^{-j3\omega T_S} + E_{dq}^-(t)e^{-j(\omega t+\psi_-)}e^{j3\omega T_S}$$
$$+ E_{dq}^{5-}(t)e^{-j(5\omega t+\psi_5)}e^{j15\omega T_S} + E_{dq}^{7+}(t)e^{j(7\omega t+\psi_7)}e^{-j21\omega T_S}$$
$$+ E_{dq}^{11-}(t)e^{-j(11\omega t+\psi_{11})}e^{j33\omega T_S} + E_{dq}^{13+}(t)e^{j(13\omega t+\psi_{13})}e^{-j39\omega T_S}$$
(11)

$$E_{\alpha\beta}(t-4T_s) = E_{dq}^+(t)e^{j(\omega t+\psi_+)}e^{-j4\omega T_S} + E_{dq}^-(t)e^{-j(\omega t+\psi_-)}e^{j4\omega T_S}$$
$$+ E_{dq}^{5-}(t)e^{-j(5\omega t+\psi_5)}e^{j20\omega T_S} + E_{dq}^{7+}(t)e^{j(7\omega t+\psi_7)}e^{-j28\omega T_S}$$
$$+ E_{dq}^{11-}(t)e^{-j(11\omega t+\psi_{11})}e^{j44\omega T_S} + E_{dq}^{13+}(t)e^{j(13\omega t+\psi_{13})}e^{-j52\omega T_S}$$
(12)

$$E_{\alpha\beta}(t-5T_s) = E_{dq}^+(t)e^{j(\omega t+\psi_+)}e^{-j5\omega T_S} + E_{dq}^-(t)e^{-j(\omega t+\psi_-)}e^{j5\omega T_S}$$
$$+ E_{dq}^{5-}(t)e^{-j(5\omega t+\psi_5)}e^{j25\omega T_S} + E_{dq}^{7+}(t)e^{j(7\omega t+\psi_7)}e^{-j35\omega T_S}$$
$$+ E_{dq}^{11-}(t)e^{-j(11\omega t+\psi_{11})}e^{j55\omega T_S} + E_{dq}^{13+}(t)e^{j(13\omega t+\psi_{13})}e^{-j65\omega T_S}$$
(13)

Let

$$E_{dq}^+(t)e^{j(\omega t+\psi_+)} = A, \quad E_{dq}^-(t)e^{-j(\omega t+\psi_-)} = B,$$
$$E_{dq}^{5-}(t)e^{-j(5\omega t+\psi_5)} = C, \quad E_{dq}^{7+}(t_s)e^{j(7\omega t+\psi_7)} = D,$$
$$E_{dq}^{11-}(t_s)e^{-j(11\omega t+\psi_{11})} = F, \quad E_{dq}^{13+}(t_s)e^{j(13\omega t+\psi_{13})} = H$$
$$e^{j\omega T_S} = k_1, \quad e^{-j\omega T_S} = k_2$$

From equations (3), (9), (10), (11), (12) and (13), we can derive equations (14) below.

$$\begin{cases} E_{\alpha\beta}(t) = A + B + C + D + F + H \\ E_{\alpha\beta}(t-T_s) = Ak_2 + Bk_1 + Ck_1^5 + Dk_2^7 + Fk_1^{11} + Hk_2^{13} \\ E_{\alpha\beta}(t-2T_s) = Ak_2^2 + Bk_1^2 + Ck_1^{10} + Dk_2^{14} + Fk_1^{22} + Hk_2^{26} \\ E_{\alpha\beta}(t-3T_s) = Ak_2^3 + Bk_1^3 + Ck_1^{15} + Dk_2^{21} + Fk_1^{33} + Hk_2^{39} \\ E_{\alpha\beta}(t-4T_s) = Ak_2^4 + Bk_1^4 + Ck_1^{20} + Dk_2^{28} + Fk_1^{44} + Hk_2^{42} \\ E_{\alpha\beta}(t-5T_s) = Ak_2^5 + Bk_1^5 + Ck_1^{25} + Dk_2^{35} + Fk_1^{55} + Hk_2^{65} \end{cases}$$
(14)

Equations (14) can also be described by matrix, so equations (15) are obtained below.

$$\begin{bmatrix} E_{\alpha\beta}(t) \\ E_{\alpha\beta}(t-T_s) \\ E_{\alpha\beta}(t-2T_s) \\ E_{\alpha\beta}(t-3T_s) \\ E_{\alpha\beta}(t-4T_s) \\ E_{\alpha\beta}(t-5T_s) \end{bmatrix} = \begin{bmatrix} 1 & 1 & 1 & 1 & 1 & 1 \\ k_2 & k_1 & k_1^5 & k_2^7 & k_1^{11} & k_2^{13} \\ k_2^2 & k_1^2 & k_1^{10} & k_2^{14} & k_1^{22} & k_2^{26} \\ k_2^3 & k_1^3 & k_1^{15} & k_2^{21} & k_1^{33} & k_2^{39} \\ k_2^4 & k_1^4 & k_1^{20} & k_2^{28} & k_1^{44} & k_2^{52} \\ k_2^5 & k_1^5 & k_1^{25} & k_2^{35} & k_1^{55} & k_2^{65} \end{bmatrix} \begin{bmatrix} A \\ B \\ C \\ D \\ F \\ H \end{bmatrix}$$
(15)

By matrix inversion, the positive and negative sequence components and harmonic components can be deduced as follow.

$$\begin{bmatrix} A \\ B \\ C \\ D \\ F \\ H \end{bmatrix} = \begin{bmatrix} 1 & 1 & 1 & 1 & 1 & 1 \\ k_2 & k_1 & k_1^5 & k_2^7 & k_1^{11} & k_2^{13} \\ k_2^2 & k_1^2 & k_1^{10} & k_2^{14} & k_1^{22} & k_2^{26} \\ k_2^3 & k_1^3 & k_1^{15} & k_2^{21} & k_1^{33} & k_2^{39} \\ k_2^4 & k_1^4 & k_1^{20} & k_2^{28} & k_1^{44} & k_2^{52} \\ k_2^5 & k_1^5 & k_1^{25} & k_2^{35} & k_1^{55} & k_2^{65} \end{bmatrix}^{-1} \begin{bmatrix} E_{\alpha\beta}(t) \\ E_{\alpha\beta}(t-T_s) \\ E_{\alpha\beta}(t-2T_s) \\ E_{\alpha\beta}(t-3T_s) \\ E_{\alpha\beta}(t-4T_s) \\ E_{\alpha\beta}(t-5T_s) \end{bmatrix}$$
(16)

Where $\gamma = \omega T_s$, the following equations are obtained.

$$\begin{bmatrix} E_\alpha^+ \\ E_\beta^+ \\ E_\alpha^- \\ E_\beta^- \\ E_\alpha^{5-} \\ E_\beta^{5-} \\ E_\alpha^{7+} \\ E_\beta^{7+} \\ E_\alpha^{11-} \\ E_\beta^{11-} \\ E_\alpha^{13+} \\ E_\beta^{13+} \end{bmatrix} = K^{-1}(\omega) \begin{bmatrix} E_\alpha(t) \\ E_\beta(t) \\ E_\alpha(t-T_s) \\ E_\beta(t-T_s) \\ E_\alpha(t-2T_s) \\ E_\beta(t-2T_s) \\ E_\alpha(t-3T_s) \\ E_\beta(t-3T_s) \\ E_\alpha(t-4T_s) \\ E_\beta(t-4T_s) \\ E_\alpha(t-5T_s) \\ E_\beta(t-5T_s) \end{bmatrix}$$
(17)

Let

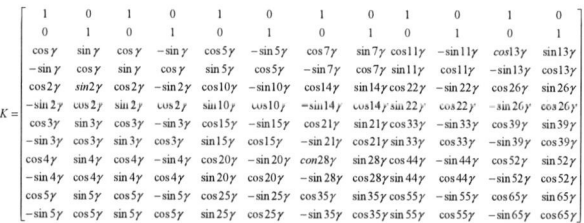

As analyzed above, Fig.2 shows the block diagram of the fast PLL algorithm based on matrix operation under distorted grid conditions.

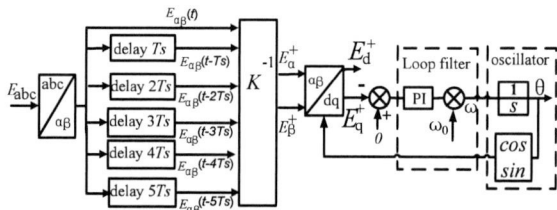

Fig.2. the block diagram of the fast PLL algorithm based on matrix operation under distorted grid conditions

Since γ is known, K is constant coefficient matrix, and the inversion of K can be calculated beforehand. So implementation of K^{-1} can not consume DSP resources. From equations (17), it is obviously that the positive sequence components in a stationary α-β reference frame can be extracted in 5 sampling periods by simple operation with $E_{\alpha\beta}$ and the variables of delayed $E_{\alpha\beta}$. After that, the frequency and angle can be realized by using loop filter and oscillator in a feedback closed loop which regulates q axis voltage component to zero. The response speed of the proposed PLL method is increased.

IV. SIMULATION RESULTS

To verify the effectiveness of the proposed algorithm, some significant cases of the conventional PLL and the proposed PLL have been simulated. The simulations have been performed with Matlab-Simulink software.

The main objective of the performed tests is to verify the high detecting speed and the accuracy of new PLL when the three-phase supply voltages contain asymmetries and harmonics. In simulations, the initial parameters are as follows: three-phase grid voltages are balanced and its RMS value of phase voltage and fundamental frequency are 220V and 50Hz, respectively. The sampling frequency is 8 kHz. At 0.1s, the voltage of phase C to neutral drops to 85% of the initial voltage, and fifth harmonic, seventh harmonic, eleventh harmonic and thirtieth harmonic components are superposed to the initial voltages with the values of 3.41%, 2.55%, 2.63% and 1.65% of the initial voltages, respectively. Fig.3 gives the grid voltage waveforms when the grid voltages change from balanced to distorted conditions; Fig.4 and Fig.5 show the simulation results of the conventional PLL and the proposed PLL method under the given grid conditions.

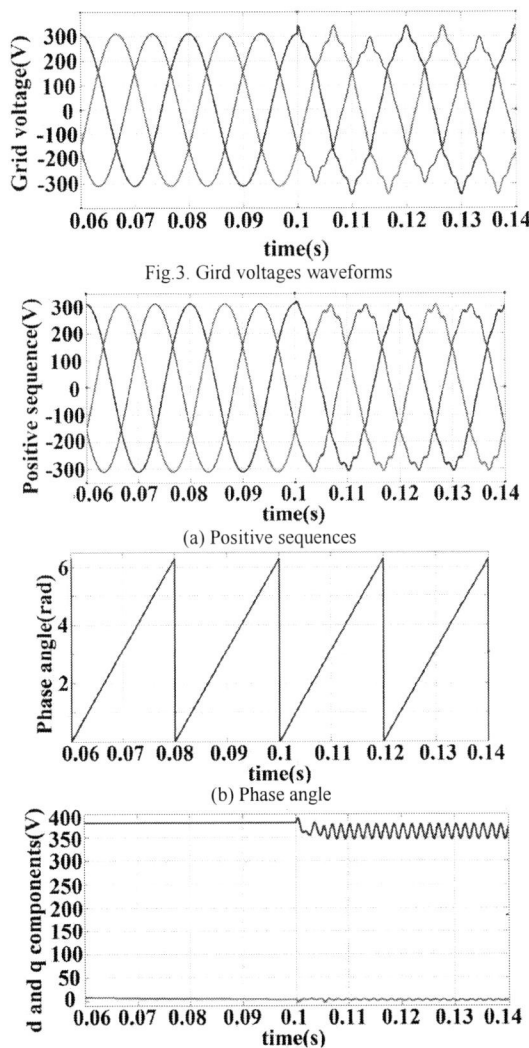

Fig.3. Gird voltages waveforms

(a) Positive sequences

(b) Phase angle

(c) The d and q components

Fig.4. Simulation results of the conventional PLL

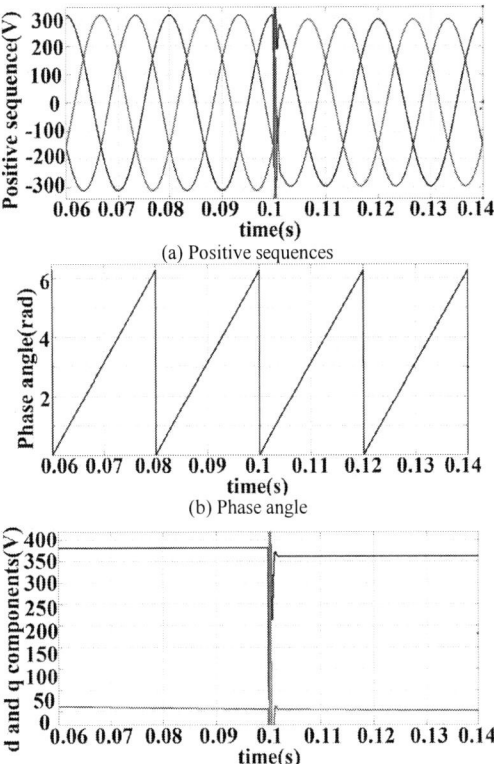

(a) Positive sequences

(b) Phase angle

(c) The d and q components

Fig.5. simulation results of the proposed PLL

As you can see from simulation waveforms, when the grid voltages change from balanced to distorted conditions, the response time in Fig.5 (c) is reduced to about 2msec compared with that of Fig.4. (c), which is about 5msec. And the ripples of the phase angle shown in Fig.5 (b) and the d voltage component shown in Fig.5 (c) estimated by the proposed PLL are smaller than that of the conventional PLL method shown in Fig.4 (b) and (c). The positive sequences extracted by the proposed PLL shown in Fig.5 (a) are more accurate than that of the conventional PLL shown in Fig.4 (a).

V. EXPERIMENTAL VERIFICATION

Experiments use a Chroma-61511 programmable AC source to generate the same simulated distorted grid voltages. The three-phase grid signals were acquired using 12 bit AD converters and processed in a floating point Texas TMS320F2812 DSP. In order to plot the positive sequences and the d axis voltage component, the DSP variables to be examined were sent to TLV5614, which is a quadruple 12-bit voltage output DAC with 4-wire serial interface to SPI of DSP. The experimental results are shown in Fig.6 and Fig.7.

The 2014 International Power Electronics Conference

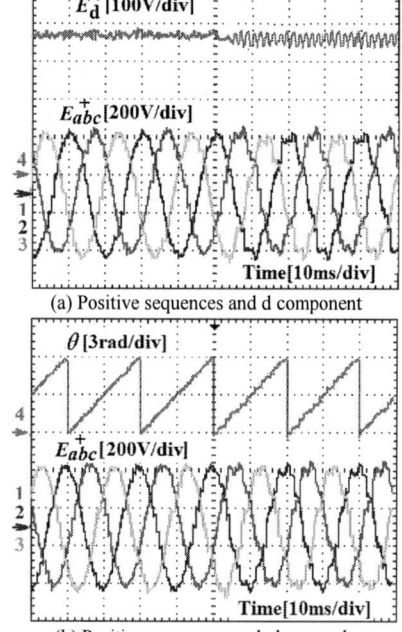

(a) Positive sequences and d component

(b) Positive sequences and phase angle

Fig.6. Experimental results obtained by the conventional PLL

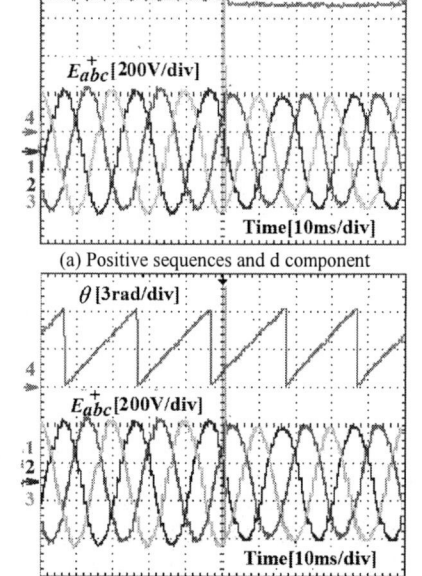

(a) Positive sequences and d component

(b) Positive sequences and phase angle

Fig.7.Experimental results obtained by the proposed PLL

When the grid voltages change from balanced to distorted conditions, from Fig.6 (a) the positive sequences which extracted by the conventional PLL method are not accurate, which contains 12th harmonic component caused by 11th and 13th harmonics, and the transient response time is about 5ms. But from Fig.7 (a), the accurate positive sequences without harmonic component can be achieved, and the transient response time is less than 2msec. From Fig.6 (b), there is clearly oscillation appear in the phase angle which detected by the conventional PLL. As a consequence of Fig.7 (b), the Proposed PLL estimates the phase angle of grid distorted voltage with better performance than that of the

conventional PLL method. Similarly, compared Fig.6 (a) with Fig.7 (a), it can be seen the d component which achieved by the proposed PLL has smaller ripple than that of the conventional PLL which is about 40V.

VI. CONCLUSIONS

From the aforementioned analysis and results, it can be seen that the proposed method shows fast and precise characterization under unbalanced and distorted conditions, without making the structure of PLL sophisticated, so it is very promising in the grid-connected inverter applications.

REFERENCES

[1] Blaabjerg F, Teodorescuk R, Liserre M, et al. "Overview of control and grid synchronization for distributed power generation systems," *IEEE Transactions on Industrial Electronics*, vol. 53, no. 5, pp. 1398-1409, 2006.

[2] Chung SeKyo, "A phase tracking system for three phase utility interface inverters," *IEEE Transactions on Power electronics*, vol.15, no. 3, pp. 431-438, 2000.

[3] Vikram Kaura, Vladimir Blasko, "Operation of a phase locked loop system under distorted utility conditions", *IEEE Transactions on Industry Applications*, vol. 33, no. 1, pp. 58-63, 1997.

[4] Chung S K, "Phase-locked loop for grid-connected three-phase power conversion systems", *Electric Power Applications*, vol.147, no. 3, pp. 213-219, 2000.

[5] Chong H Ng, Li Ran, Jim Bumby, "Unbalanced-grid-fault ride-through control for a wind turbine inverter", *IEEE Transactions on Industry Applications*, vol.44, no.3, pp. 845-856, 2008.

[6] M. A. Perez, J. R. Espinoza, L. A. Moran, M. A. Torres, and E. A. Araya,"A robust phase-locked loop algorithm to synchronize static-power converters with polluted AC systems," *IEEE Trans. Power Electron.*, vol. 55,no. 5, pp. 2185–2192, 2008.

[7] P. Li, L. Xue, P. Hazucha, T. Karnik, and R. Bashirullah, "A delay-locked loop synchronization scheme for high-frequency multiphase hysteretic DC-DC converters," *IEEE J. Solid-State Circuits*, vol. 44, no. 11,pp. 3131–3145, 2009.

[8] Guo Xiaoqiang, Wu Weiyang, "Multiplecomplex coefficient-filter-based phase-locked loop and synchronization technique for three-phase grid-interfaced converters in distributed utility networks". *IEEE Trans.on Industrial Electronics*, vol.58, no 4, pp. 1194-1204. 2011.

[9] Rodriguez P, Pou J, Bergas J, et al, "Decoupled double synchronous reference frame PLL for power converters control", *IEEE Trans. on Power Electronics*, vol. 22, no. 2, pp. 584-592.

[10] Alper AKdag, Susumu Tadakuma, Hideaki Minakata, "Load Balancing Control by Symmetrical Coordinates Frame for PWM Inverter Based Reactive Power Compensators", *T IEE Japan*, Vol. 121-D, No. I, pp. 43-51, 2001.

[11] Ciobotaru M, Teodorescu R, Blaabjerg F, "A new single-phase PLL structure based on second order generalized integrator", IEEE *37th Power Electronics Specialists Conference*, pp. 1-6, 2006.

[12] Rodriguez P, Teodorescu R Candela I et al, "New positive-sequence voltage detector for grid synchronization of power converters under faulty grid conditions", *37th IEEE Power Electronics Specialists Conference*, pp. 1-7, 2006.

978-1-4799-2706-7/14 $31.00 © 2014 IEEE

Performance Improvement of Photovoltaic Power Generation Systems Using On-Off Control Methods

Matsumoto Kenji, Shinichi Nomura
Meiji University
1-1-1 Higashimita, Tama-ku, Kawasaki, Kanagawa, Japan
nomuras@meiji.ac.jp

Abstract- **The purpose of this work is to investigate the output power improvement of photovoltaic power generation systems using ON-OFF control methods. In the photovoltaic power generation system, the disturbance of the solar illuminance between the solar cells causes the imbalance of the internal resistance in each solar cell, and lowers the performance of the MPPT operation of the total photovoltaic power generation system. To overcome this problem, the authors investigate the feasibility of ON-OFF control methods in order to improve the MPPT operation. In the ON-OFF control methods, the solar cell module, which has a higher internal resistance due to a lower solar illuminance, can be disconnected in order to avoid the decrease in the total output power of the photovoltaic power generation system. Moreover, the disconnected solar cell module by the ON-OFF control method can be connected in parallel in order to extract the higher output power from the total photovoltaic power generation system. The authors demonstrated the validity of the ON-OFF control method using a small experimental model of a domestic photovoltaic power generation system.**

I. INTRODUCTION

Photovoltaic power generation is one of the feasible options in the recent electric power system. In order to extract the output power from the solar cells, the maximum power point tracking (MPPT) operation is very important. However, the disturbance of the solar illuminance between the solar cells causes the imbalance of the internal resistance in each solar cell, and lowers the performance of the MPPT operation of the total photovoltaic power generation system. To overcome this problem, the feasibility of the individual MPPT operation is discussed [1]. The individual MPPT operation requires the DC-DC converter connected to each solar cell module to compensate the imbalance of the internal resistance in each solar cell. This configuration may make the photovoltaic power generation system complex.

In this work, instead of the individual MPPT operation, the authors discuss the feasibility of ON-OFF control methods in order to improve the MPPT operation even when the disturbance of the solar illuminance between the solar cells occurs. Since the typical load factor of the

photovoltaic power generation system is around 10%, the solar cell module, which has a higher internal resistance due to a lower solar illuminance, can be disconnected in order to avoid the decrease in the total output power of the photovoltaic power generation system.

Although a simple configuration for the ON-OFF control method is to use traditional bypass diodes, the output power of the photovoltaic power generation system can not be controlled depending on the solar illuminance. Recently, the feasibility of active bypass circuits is also proposed to enhance the higher power production [2].

As a first step of this work, the authors evaluate the relationship between the disturbance of the solar illuminance and the decrease in the MPPT operation performance based on the experimental results. Using a small photovoltaic power generation system, the authors demonstrate the validity of the ON-OFF control method applied to a domestic photovoltaic power generation system. Moreover, in order to extract the higher output power from the solar cells, the feasibility of a series-parallel control method of the solar cells is also discussed.

II. ON-OFF CONTROL CONCEPT FOR A DOMESTIC PHOTOVOLTAIC POWER GENERATION SYSTEM

Fig. 1 shows schematic diagrams of the ON-OFF control concept applied to a domestic photovoltaic power generation system. Most domestic photovoltaic power generation system is composed of east, south and west side solar cell modules. In order to obtain a higher output voltage for the grid connection, the solar cell modules should be connected in series. In this work, a switch box for the ON-OFF control is installed between the solar cell modules as shown in Fig. 1.

In the morning hours, the solar illuminance of the west side module is lower than that of the east and south side modules. In this case, the internal resistance of the west side solar cells will be increased. This imbalance may cause to reduce the total output power if all the solar cell modules are connected. Therefore, the west side module can be disconnected in order to avoid the degradation of the MPPT operation performance of the total

978-1-4799-2706-7/14 $31.00 © 2014 IEEE

The 2014 International Power Electronics Conference

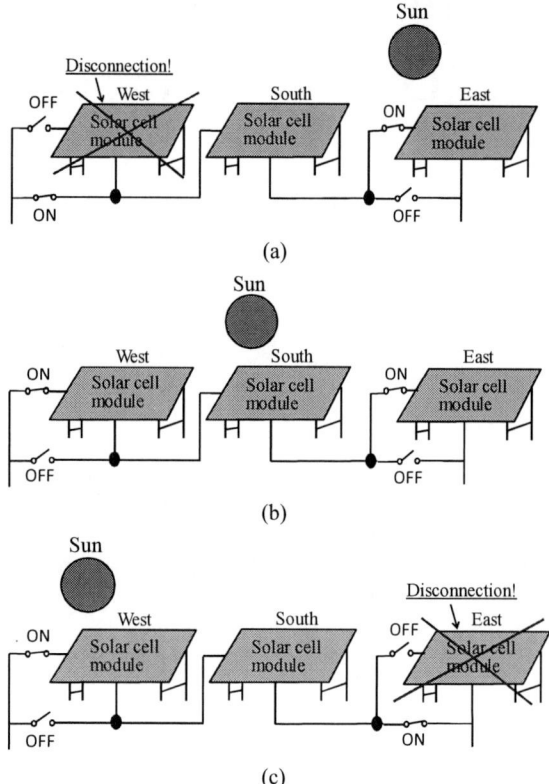

(a)

(b)

(c)

Fig. 1. Schematic diagram of the ON-OFF control concept for a domestic photovoltaic power generation system. In the morning hours, the west side of the solar cells is disconnected due to the lower illuminance from the sun (a). If the enough illuminance is obtained, all sides of the solar cells are connected in series (b). In the evening hours, the east side of the solar cells is disconnected (c).

photovoltaic power generation system as shown in Fig. 1 (a). Similarly, the east side module can be disconnected in the evening hours as shown in Fig. 1 (c).

III. EVALUATION OF MPPT OPERATION PERFORMANCES DUE TO DISTURBANCES OF THE SOLAR ILLUMINANCE

In the ON-OFF control method, the connecting or disconnecting control of the solar cell module is depending on the disturbance of the solar illuminance. In order to investigate this control method, the authors evaluated the relationship between the disturbance of the solar illuminance and the decrease in the MPPT operation performance.

Table I summaries the main parameter of the sample solar battery for the evaluation. The sample solar battery is made of amorphous-silicon solar cells with a dimension of 15 cm×16.5 cm. For the assumption of a domestic photovoltaic power generation system, the sample solar battery is set on the board with a tilted angle of 30 degree as shown in Fig. 2.

TABLE I
MAIN PARAMETERS OF THE SAMPLE SOLAR BATTERY.

Open-circuit voltage	8.6 V (50 kLx)
Short-circuit current	126.9 mA (50 kLx)
Maximum power	702 mW (50 kLx, 6.6 V, 106.3 mA)

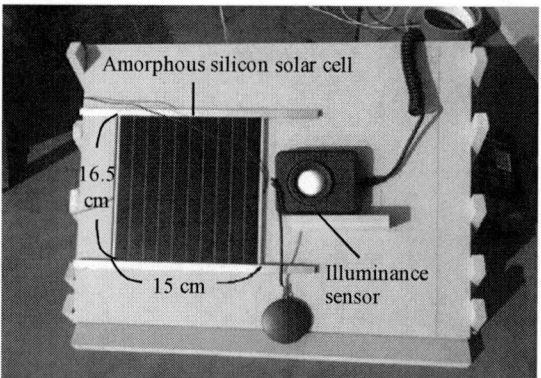

(a)

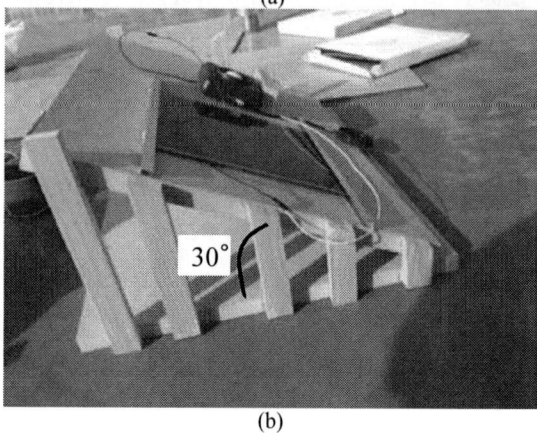

(b)

Fig. 2. Experimental layout of the sample solar battery for the demonstration of the ON-OFF control concept. The illuminance on the solar battery is evaluated by the illuminance sensor (a). Assuming that the photovoltaic power generation system on the roof, the solar battery is set on the board with a tilted angle of 30 degree (b).

Fig. 3 summaries the power curves of the sample solar battery. The maximum power point is depending on the illuminance. Using two solar batteries, the authors investigated the influence of the maximum power point due to the disturbance of the illuminance.

Fig. 4 compares the power curves between the series connection of the solar batteries and the parallel connection of those. In the experiments, one solar battery is applied 100 kLx of the illuminance, and another solar battery is applied 50 kLx. From the results, the output voltage in the series connection case is higher than that in the parallel connection case. However, since the solar

978-1-4799-2706-7/14 $31.00 © 2014 IEEE

The 2014 International Power Electronics Conference

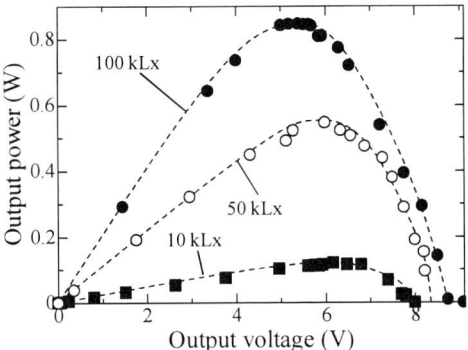

Fig. 3. The power curve dependences on the illuminance of the sample solar battery.

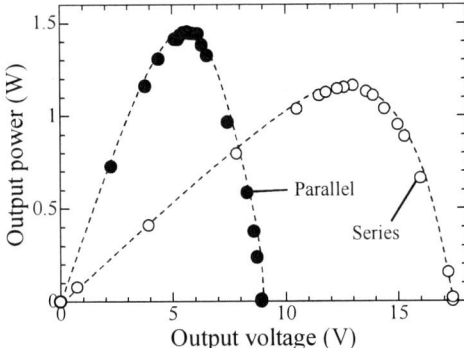

Fig. 4. Comparison of the power curves between the series connection of two sample solar batteries and the parallel connection of those with a difference of 50 kLx. The higher illuminance is 100 kLx. Without the disturbance of the illuminance, the total output power will be about 1.8W at 100 kLx.

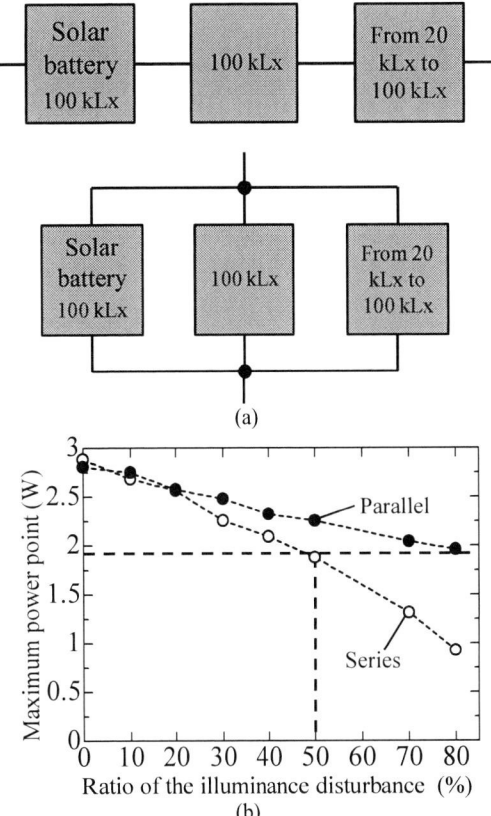

(a)

(b)

Fig. 5. Experimental conditions (a) and maximum power point as a function of the illuminance disturbance (b). The constant illuminance of 100 kLx applies to two sample solar batteries, and the illuminance of the other is varied from 20 kLx to 100 kLx. The experiments compare the series connection of three sample solar batteries with the parallel connection of those. The output power from the two sample batteries is 1.8 W with the uniform illuminance of 100 kLx.

battery with a lower illuminance has a higher internal resistance, the output current in the series connection case is decreased. Due to this, the maximum power point is lowered especially in the series connection case.

In order to estimate the influence of the illuminance disturbance in the case of a domestic photovoltaic power generation system, the authors investigate the maximum power point properties depending on the series or parallel connections by using three sample solar batteries as shown in Fig. 5(a).

In this work, the ratio of the illuminance disturbance is defined as

$$(\frac{E_{vH} - E_{vL}}{E_{vH}}) \times 100 \ (\%), \qquad (1)$$

where E_{vH} and E_{vL} are the highest illuminance and the lowest illuminance applied to the solar batteries. In the experiments, the same illuminance of 100 kLx applies to two of the sample batteries, which corresponds to the highest illuminance E_{vH}. The illuminance applied to the other E_{vL} is varied from 20 kLx to 100 kLx.

Fig. 5(b) shows the maximum power point dependence on the ratio of the illuminance disturbance. When the ratio of the illuminance disturbance is 0% (100 kLx of the uniform illuminance), the output power from three sample solar batteries is about 2.8 W.

From the results, the maximum power point decreases with the ratio of the illuminance disturbance. Especially, in the case of the series connection, the decrease in the maximum power point is drastically. If the ratio of the illuminance disturbance is over 50%, the maximum power from three series connected solar batteries is lower than that from two series connected solar batteries with 100 kLx of the uniform illuminance. This result means that the solar battery with the lowest illuminance can be disconnected in order to avoid the degradation of the MPPT operation performance of the total photovoltaic power generation system.

978-1-4799-2706-7/14 $31.00 © 2014 IEEE

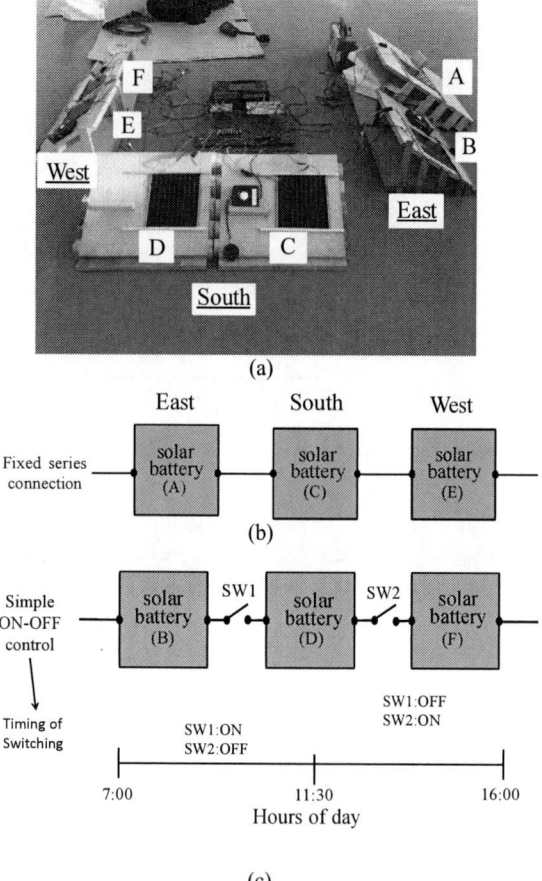

(a)

East　　　　South　　　　West

Fixed series connection

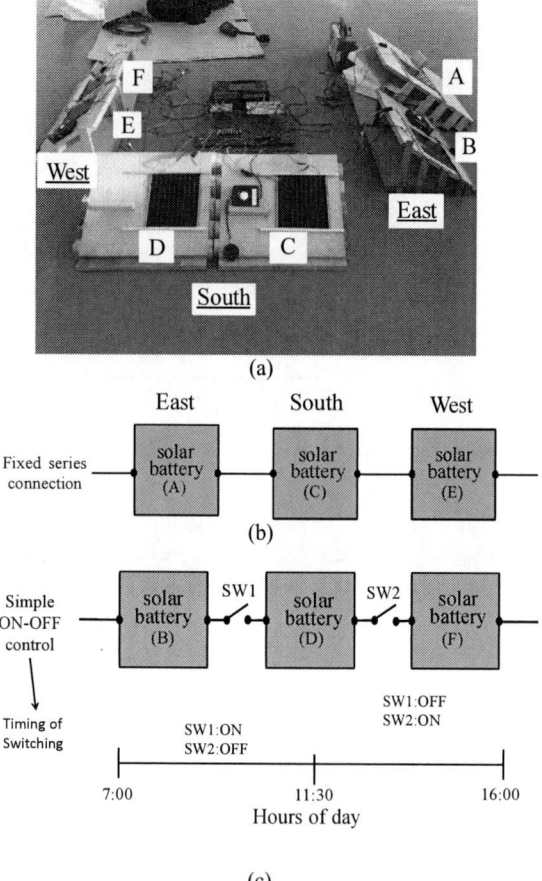

(b)

Simple ON-OFF control

Timing of Switching

(c)

Fig. 6. Experimental layout for the demonstration of the ON-OFF control concept applied to a domestic photovoltaic power generation system (a). Two sets of the independent circuits are compared. The sample solar batteries A, C and E are connected in series (b). The sample solar batteries B, D and F are also connected in series through the switch boxes for the ON-OFF control (c). From the results in Fig. 7, the sample F (West side) was disconnected from 7:00 to 11:30, and the sample B (East side) was disconnected from 11:30 to 16:00.

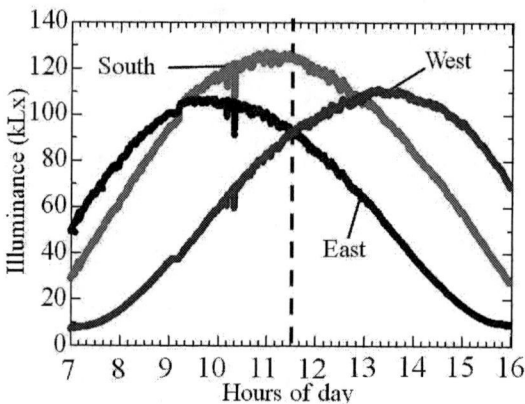

Fig. 7. Example of the transition of the solar illuminance depending on the directions (September 18, 2013).

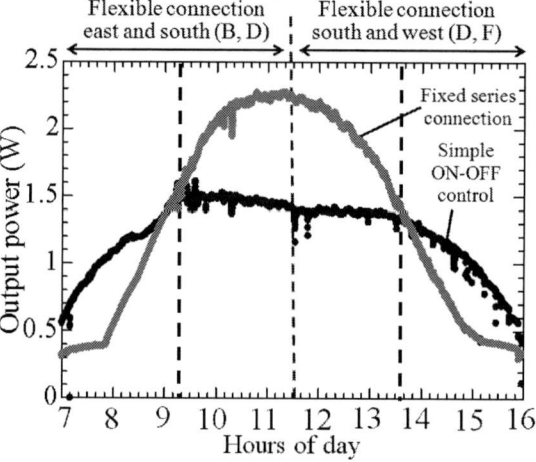

Fig. 8. Comparison of the maximum power point tracking operations between the fixed series connection of the solar batteries and the flexible series connection based on the ON-OFF control method (September 18.2013).

IV. DEMONSTRATION OF THE ON-OFF CONTROL CONCEPT FOR A DOMESTIC PHOTOVOLTAIC SYSTEM

In order to demonstrate the validity of the ON-OFF control method, the authors developed a small model of a domestic photovoltaic power generation system by using the solar batteries as shown in Fig. 6(a).

Two types of the solar battery circuits are compared in this demonstration. One circuit is composed of three solar batteries based on the fixed series connection as shown in Fig. 6(b). Another circuit is also composed of three solar batteries based on the flexible series connection through the switch boxes for the ON-OFF control operation as shown in Fig. 6(c).

Fig. 7 shows an example of the transition of the solar illuminance depending on the directions. Before 11:30, the illuminance from the east is higher than that from the west, meaning that the electric power will be mainly provided from the east side and the south side solar batteries.

On the other hand, after 11:30, the illuminance from the west is higher than that from the east. This means that the south side and the west side solar batteries will mainly provide the electric power. From these results, in the ON-OFF control method, the solar battery facing the west is disconnected before 11:30, and the solar battery facing the east is disconnected after 11:30 as shown in Fig. 6(c).

Fig. 8 compares the MPPT operation performance between the fixed series connection and the flexible series connection based on the ON-OFF control method. The output power of the fixed series connection is larger than that of the ON-OFF control method from 09:15 to 13:40. However, before 09:15, the output power of the ON-OFF control method is higher than that of the fixed connection. From Fig. 7, the ratio of illuminance

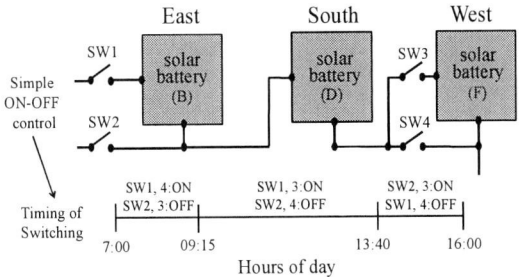

Fig. 9. An optimal MPPT operation by the ON-OFF control method.

disturbance between the east and the west is over 50%. Compare to the result in Fig. 5, the solar battery facing the west can be disconnected in order to extract the output power from the total solar battery system.

On the other hand, after 13:40, the ratio of the illuminance disturbance between the east and the west is also over 50%. Due to this, the solar battery facing the east can be disconnected in order to obtain the higher output power from the total solar battery system. These results show the validity of the ON-OFF control method.

An optimal MPPT operation by the ON-OFF control method is summarized in Fig. 9. Based on the results in Fig. 8, the ON-OFF control method can be expected to obtain 10% larger electrical energy from the photovoltaic power generation system than that of the conventional MPPT operation case with the fixed series connection of the solar cell modules.

V. SERIES-PARALLEL CONTROL OF THE SOLAR CELL MODULES BASED ON THE ON-OFF CONTROL CONCEPT

A. Concept of the Series-Parallel Control Method

In the previous section, the authors demonstrated that the solar cell module with a lowest illuminance can be disconnected in order to avoid the degradation of the MPPT operation performance of the total photovoltaic power generation system. However, from the results in Fig. 5, the parallel connected solar cell modules can generate the higher output power even when the ratio of the illuminance disturbance is higher. This property means that the disconnected solar cell module by the ON-OFF control method can be connected in parallel in order to extract the higher output power from the total photovoltaic power generation system.

Fig. 10 shows the concept of the series-parallel control method based on the ON-OFF control concept for a domestic photovoltaic power generation system. The switch box for the series-parallel control can be also installed between the solar cell modules

In the morning hours, the solar illuminance of the west side module is lower than that of the east and south side modules. In the simple ON-OFF control method, the west side module should be disconnected in order to avoid the degradation of the MPPT operation performance. Instead

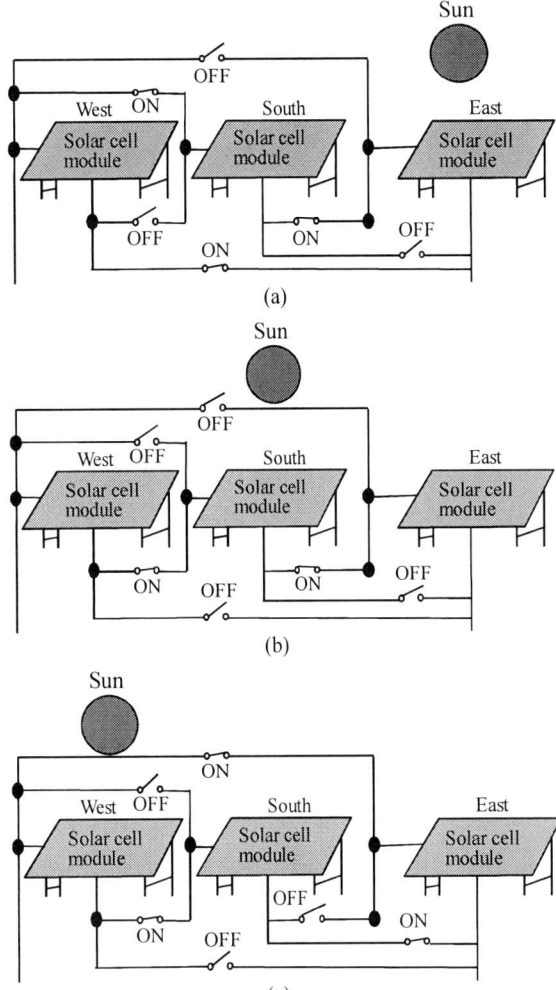

Fig. 10. Schematic diagram of the series-parallel control method based on the ON-OFF control concept for a domestic photovoltaic power generation system. In the morning hours, the west side of the solar cells is connected in parallel to the east and south solar cell modules (a). If the enough illuminance is obtained, all sides of the solar cells are connected in series (b). In the evening hours, the east side of the solar cells is connected in parallel to the south and west solar cell modules (c).

of this method, in the series-parallel control method, the west side module is connected in parallel to the east and west side modules in order to extract the higher output power from the total photovoltaic power generation system as shown in Fig. 10 (a). Similarly, the east side module is connected in parallel to the south and west side modules in the evening hours as shown in Fig. 10 (c). If the enough illuminance is obtained, all sides of the solar cells are connected in series as shown in Fig. 10 (b).

B. Demonstration of the Series-Parallel Control Method

In order to demonstrate the validity of the series-parallel control method, the authors developed a small

The 2014 International Power Electronics Conference

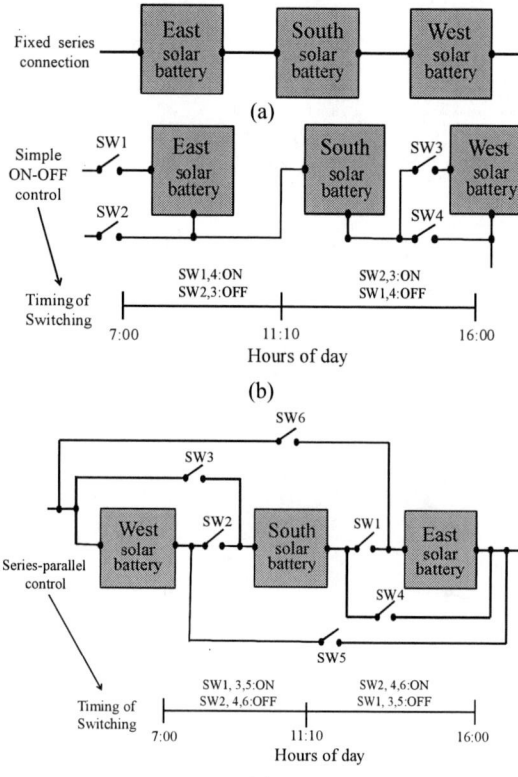

(a)

(b)

(c)

Fig. 11. Three types of the solar battery circuits are compared. First type consists of three sample solar batteries connected in series (a). The second type is also composed of three sample solar batteries connected in series through the switch boxes for the ON-OFF control (b). From the results in Fig. 12, the west side sample solar battery was disconnected from 7:00 to 11:10, and the east side sample solar battery was disconnected from 11:10 to 16:00. The other type is based on the series-parallel control method using three sample solar batteries (c). From the results in Fig. 12, the west side sample was connected in parallel to the east and south side sample from 7:00 to 11:10, and the east side sample was connected in parallel to the south and west sample from 11:10 to 16:00.

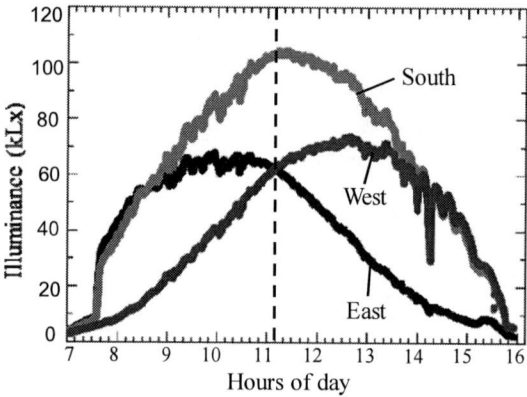

Fig. 12. Example of the transition of the solar illuminance depending on the directions (November 26, 2013).

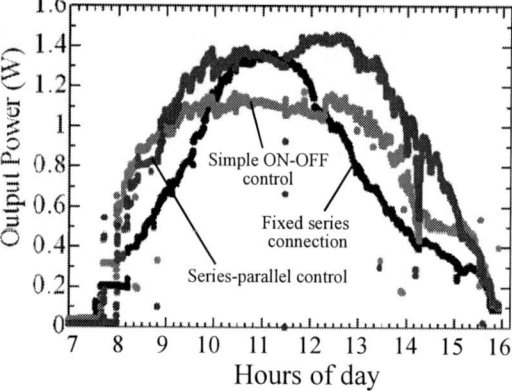

Fig. 13. Comparison of the maximum power point tracking operations between the fixed series connection of the solar batteries and the flexible series connection based on the ON-OFF control method and combine series with parallel method (November 26.2013).

model of a domestic photovoltaic power generation system which is described in the section IV.

Three types of the solar battery circuits are compared in this demonstration as shown in Fig. 11. The first type of the experimental circuit is the fixed series connection of the sample solar batteries as shown in Fig. 11 (a). The second type is the flexible series connection of the sample solar batteries based on the simple ON-OFF control method as shown in Fig. 11 (b). The other type is also composed of three sample solar batteries, and based on the series-parallel control method as shown in Fig. 11 (c).

Fig. 12 shows an example of the transition of the solar illuminance depending on the directions. Before 11:10, the illuminance from the east is higher than that from the west, meaning that the electric power will be mainly provided from the east side and the south side solar batteries. On the other hand, after 11:10, the illuminance

from the west is higher than that from the east. This means that the south side and the west side solar batteries will mainly provide the electric power. From these results, in the series-parallel control method, the solar battery facing the west is connected to parallel at the solar battery facing the east and south before 11:10, and the solar battery facing the east is connected to parallel at the solar battery facing the south and east after 11:10 as shown in Fig. 11 (c).

Fig. 13 compares the MPPT operation performance among the fixed series connection, the simple ON-OFF control method, and the series-parallel control based on the ON-OFF control method. From these results, the series-parallel control method generates the highest output power among three, which fact demonstrate the validity of the series-parallel control method instead of the disconnection of the solar battery with the lowest illuminance based on the simple ON-OFF control method.

Fig. 14 summaries an optimal MPPT operation by the series-parallel control method. From the results in Fig. 13,

978-1-4799-2706-7/14 $31.00 © 2014 IEEE

The 2014 International Power Electronics Conference

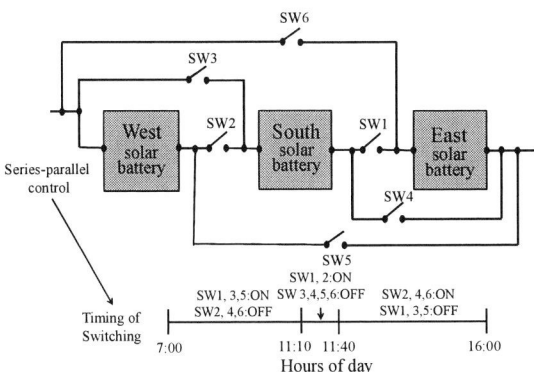

Fig. 14. An optimal MPPT operation by the series-parallel control based on the ON-OFF control method.

the series-parallel control method can be expected to obtain 35% larger electrical energy from the photovoltaic power generation system than that of the conventional MPPT operation case with the fixed series connection of the solar cell modules. Addition to this, the series-parallel control method can be also expected to extract 18% larger electrical energy than that of the simple ON-OFF control case.

VI. CONCLUSIONS

In order to improve the MPPT operation performance of the photovoltaic power generation system, the feasibility of the ON-OFF control method and series-parallel control method depending on the solar illuminance has been discussed. Using a small experimental model, the authors evaluated the

relationship between the disturbance of the solar illuminance and the decrease in the MPPT operation performance. From the experimental results, when the disturbance of the solar illuminance between the solar batteries is increased, the solar battery with the lowest illuminance can be disconnected in order to avoid the degradation of the MPPT operation performance of the total photovoltaic power generation system. Moreover, if the disconnected solar battery is connected in parallel, the total photovoltaic power generation system can be expected to extract the higher electrical energy.

However, especially in the large scale photovoltaic power generation system, the switching circuit for the series-parallel control method may be more complex. Therefore, further optimization of the switch box for the ON-OFF control method is required. As a future work, the authors should investigate an optimal configuration of the ON-OFF control method including power conditioning system for the grid connection, and demonstrate the validity of the ON-OFF control method using solar cell modules.

REFERENCES

[1] Eduardo Roman, Ricardo Alonso, Pedro Ibanez, Sabino Elorduizapatarietxe, and Damian Goitia, "Intelligent PV Module for Grid- Connected PV system," *IEEE Trans. on Industrial Electronics*, vol.53, pp. 1066- 1073, 2006.

[2] Vincenzo d'Alessandro, Pierluigi Guerriero, and Santolo Daliento "A Simple Bipolar Transistor-Based Bypass Approach for Photovoltaic Modules." *IEEE JOURNAL OF PHOTOVOLTAICS, VOL. 4, No. 1*, pp. 405-413 JANUARY 2014.

978-1-4799-2706-7/14 $31.00 © 2014 IEEE

Low Voltage PV Power Integration into Medium Voltage Grid Using High Voltage SiC Devices

Ritwik Chattopadhyay, Subhashish
Bhattacharya
FREEDM Systems Centre
North Carolina State University, Dept. of
ECE Raleigh, North Carolina, USA.

Nicole C. Foureaux, Sidelmo M. Silva, Braz Cardoso F., Helder
de Paula, Igor A. Pires, Porfírio C. Cortizio, Lênin Moraes
TESLA Power Engineering
Federal University of Minas Gerais, Dept. of EE
Belo Horizonte, Minas Gerais, BRA

José A. de S. Brito
COELBA
Salvador, BA

Abstract - **High voltage high power semiconductor devices are being used for grid integration of renewable energy sources. 1200V, 100A SiC Mosfets, 10 kV SiC Mosfets and 10kV SiC JBS Diodes have proven to be beneficial for high voltage application. High Voltage SiC devices enable high switching frequency operation thus reducing size of passive elements. Scope of this paper focuses on an alternative approach for 0.9 MW PV power plant, which is currently being constructed in Brazil. Use of high power SiC devices for PV power plant for integration into 13.8 kV grid provides higher efficiency, reduction in size and volume.**

I. INTRODUCTION

PV power plants have been considered to be one of the key factors for increased amount of renewable energy production in order to reduce fossil fuel consumption. PV power plants demand efficient power conversion process, to integrate the produced energy into utility grid. Traditional methods of PV power conversion employ solar modules, responsible for generating electrical energy, along with power electronics for converting dc power into usable ac power and transformers for stepping up the ac voltage to connect to grid. This paper reports the work based on a 0.9 MW solar power plant, which is under development in Brazil. Since this PV plant has just been commissioned, it is assumed to represent the current state of the art at its power rating. A thorough description of the PV plant is presented and analyzed to point out its limitations as well as opportunities for new technologies. Based on such studies, a new alternative configuration for the 0.9 MW solar power plant with new age power electronics to eliminate the large power line transformers and to enable a more efficient power conversion process.

II. PV POWER PLANT

A. Description

This study is based on an 840 kW solar power plant installed in Brazil. This occupies an area of 15135 m² and it is located close to a soccer stadium in the northeast region of Brazil. A picture of the PV power plant considered in this paper is shown in figure 1.

In this power plant, the 3360 PV panels are grouped in four arrays. In each array, 20 modules are connected in series and 14 of such series strings are connected in parallel. The PV panels are then organized in 12 arrays. The total installed capacity is 890 kWp(peak power),

(assuming STC, Standard Test Conditions). A representation of the PV arrays is presented in figure 2.

The PV panels employed in this power plant were supplied by Yingli: model Panda 265C – 30B (Si mono-crystalline). The PV panels characteristics and technical specifications are shown in figure 3 and table 1, respectively.

The power available from the PV arrays is converted by a single 840kW central inverter composed by 4 individual inverters. Figure 4 shows a block diagram representation of the central inverter. It consists of 4 two-level inverters with independent MPPT trackers (with master-slave configuration). Three PV panels arrays are connected to each inverter. All of the inverters have three individual fused isolating switches (one for each PV array), overvoltage protection elements in dc side and overvoltage protection elements in ac side. Table 2 summarizes the technical specifications of the PV panels arrays and inverters.

Figure 1 – PV power plant location.

All inverters outputs are connected to a common ac bus. The PV plant is then connected to the grid through a step up transformer (1 MW, 380V/13.8kV, 60 Hz). The grid impedances at the point of common coupling are presented in table 3.

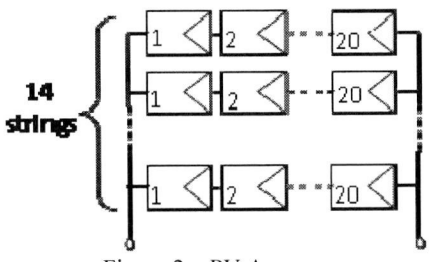

Figure 2 – PV Arrays

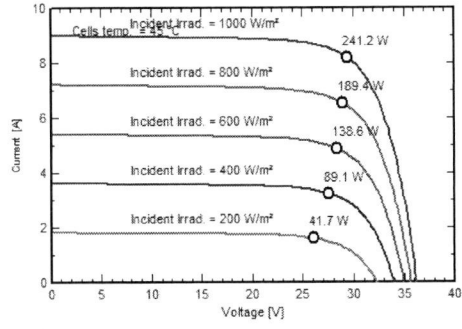

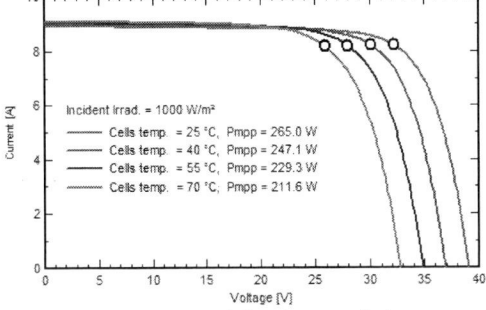

Figure 3 – Module Characteristics

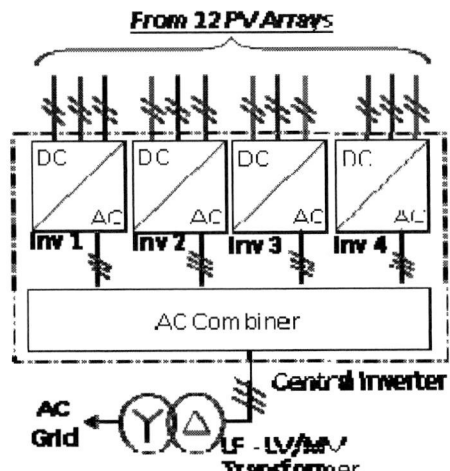

Figure 4 – PV Panels and Grid connection DC –AC Output

TABLE 1 – PV Panel Technical Specifications

Pmpp	Vmpp	Impp	Voc	Isc
265W	39V	8.93A	31V	8.55A

TABLE 2- PV Panel and Inverter Technical specifications

Single array		Inverter			
DC – Output		DC- Input		AC - Output	
Pmpp	74200W	Pmpp	222600W	Vout	230V
Vmpp	620V	Vmpp	620V	Triphase	
Impp	119.7A	Impp	359.1A		
Voc	780V	Voc	780V		
Isc	125.02A	Isc	375.06A		

Table 3 – Grid Impedances at PCC

Z1	0.4277+j1.9077
Z2	0.4277+j1.9077
Z0	0.7144+j4.3664

B. Performance simulation

A simplified model of the installation and its main elements was built and simulated using *PV Syst* software. Preliminary results are presented in table 4. The most important losses are presented. In special, the PV array losses due to temperature are very significant. It is important to keep in mind that Brazil is a tropical country and the PV power plant is located at lat 8° 2' 24" S. High temperatures are expected throughout the year (27 °C average, with highs at 38 °C), far from standard test temperature conditions.

It is important also to emphasize that the performance ratio shows that a good percentage of the available solar energy is converted into electricity, considering the high temperatures at the site. For comparison, current performance ratios for PV plants in Germany range from 80% to 90%. Inverter losses are also listed and the figures refer to the energy at the inverter input. Grid connection transformer losses were not computed. Typical efficiencies for similar power transformers are higher than 99%.

Table 4 – Preliminary simulation results

System Production	
Performance Ratio (calculated)	**78.3%**
Produced Energy (calculated)	**1671 MWh/year**

System Losses	
Array nominal energy @STC	**2069 MWh/year**
PV arrays losses	
Irradiance levels losses	-2.2%
Temperature losses	-10.3%
Soiling losses	-2.1%
Module quality losses	-1.6%
Module array mismatch losses	-2.1%
Ohmic wiring	-1.1%
Inverter losses	**-1,3%**

C. Limitations and opportunities

The conventional central inverter solution has many disadvantages that should be evaluated in order to increase the conversion efficiency and to indicate an improved one.

Central inverter has low voltage and low frequency output, which implies in a LV/MV low frequency transformer. Despite the high efficiency of power transformers, they are bulky and heavy and have also maintenance and reliability issues. Transformer isolation in solar power plants is necessary due to energy security standards (required in few countries) and to avoid DC current injection into the grid [1-2]. The transformer's role in PV power plants is also aimed at the mitigation of leakage currents, mainly due to the panel's parasite capacitances to the ground [3-4]. Transformer-less applications do not limit such currents, which contribute to conduced and irradiated EMI, harmonics and power losses [3-4]. Regardless of the converter's topology, it is not possible to connect PV plants to the MV grid without transformer isolation.

The presented central inverter has only 4 MPP trackers and this can be seen as another disadvantage due to its poor overall maximum power point tracking, mainly during partial shading conditions. In order to obtain the maximum power output from the panels, the ideal tracking system would be one MPP tracker per panel or one per string of panels. This solution is not suitable to PV power plants, due to its high number of converters, sensors and controllers. Also, a LV/MV transformer is needed for the MV grid connection [5-6]. An ideal number of trackers for PV power plants should be figured out to achieve the optimal solution.

Considering the presented disadvantages, the main opportunities in central inverters technology are: minimized weight and volume and a large number of MPP trackers [7], while achieving the same requirements of galvanic isolation. In the next sections a new technology addressing such issues is presented.

III. PROPOSED CONFIGURATION FOR PV PLANT

A. Proposed Configuration

With the invention of new age power electronics, like advent of High Voltage SiC devices, grid integration of renewable energy sources at high power and high voltage levels is much less complex and highly feasible. The main use of the step up transformer is voltage boost, which can be done in a more efficient way using high frequency transformer based power electronic converters. With the advent of 10kV SiC-MOSFET[8][9], it is possible to use a cascaded inverter structure to feed power from renewable energy sources into medium voltage grids such as 13.8 kV. The high voltage 10 kV/120A SiC Module, shown in figure 5[8], from [8-10], is capable of withstanding 5-6kV voltages at converter level operation. A co-pack module of 10 kV/10A SiC Mosfet and SiC-JBS diode is shown in figure 6(a)[10] and a single 10 kV/10A JBS diode picture is shown in figure 6(b)[10]. The switching characteristics and losses

for the co-pack module is determined using double pulse test(DPT) procedure using the schematic circuit of figure 7(a)[10] and hardware setup of figure 7(b)[10]. The switching characteristics, switching losses and conduction losses are shown in figure 8[10], 9[10] and 10[9], for a switching voltage of 6 kV using 100Ω turn-on gate resistance and 10Ω turn-off gate resistance[10].

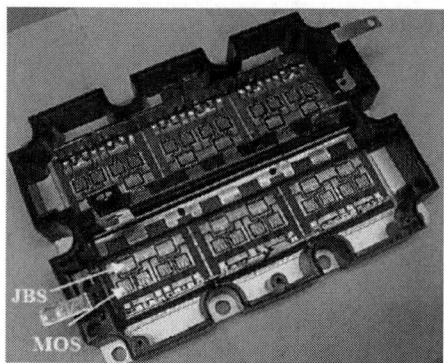

Fig. 5. Internal Structure of 10kV/120A Mosfet and 10 kV JBS diode module [8]

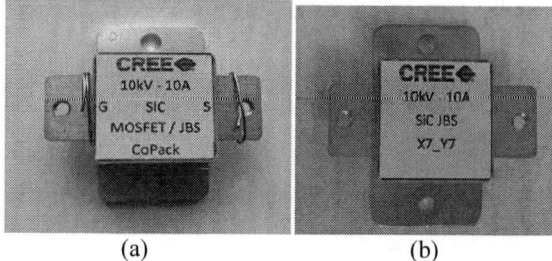

(a)   (b)

Fig. 6.(a) 10kV/10A SiC Mosfet/JBS diode Co-pack Module [10], (b) 10kV/10A SiC JBS diode [10]

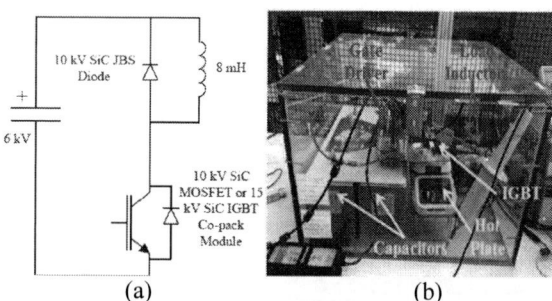

(a)   (b)

Fig. 7.(a) Double pulse test schematic circuit[10], (b) Double pulse test hardware setup [10]

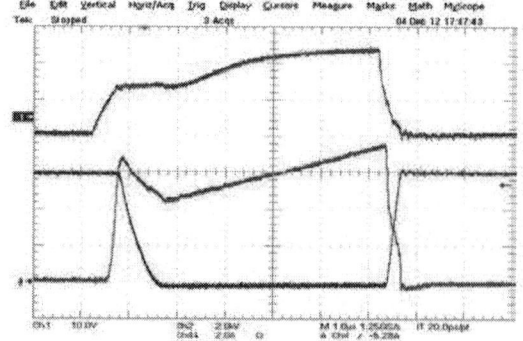

Fig. 8. Switching characteristics of 10 kV SiC Mosfet[10]

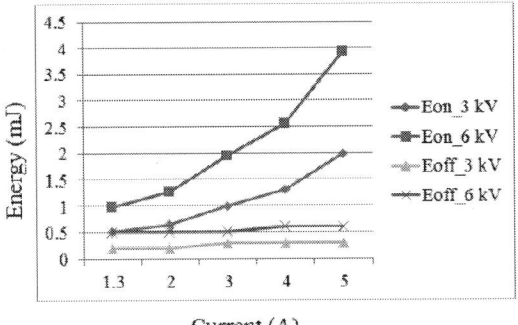

Fig. 9. Switching losses for turn-on and turn-off[10]

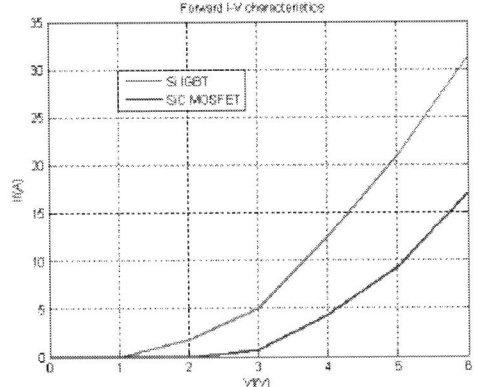

Fig. 10. Forward characteistics for 10 kV Sic Mosfet[9]

The conduction loss data for 10 kV JBS diode is shown is shown in figure 11[8].

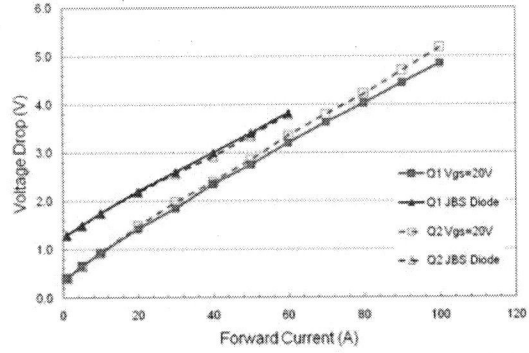

Fig. 11. Forward characteistics for 10 kV JBS diode[8]

For a VSI to feed power into 13.8 kV medium voltage grid, it is required to have a topology with series connected devices or cascaded structure with modular approach. For the PV plant application a cascaded three level PWM inverter topology is considered for the application. Each phase of the grid is fed by two cascaded single phase three level PWM inverter, using 10 kV SiC Mosfet as switching device. The dc links of the PWM three level inverters, each withstanding a voltage of 7kV, are fed by a dc-dc converter which has low voltage PV at its input and it generates three high voltage dc links. The three high voltage dc links are connected to the dc link of three level inverters for the three phases. Another such dc-dc converter is connected in same manner to the lower set of three level inverters. The topology schematic is shown in figure 12.

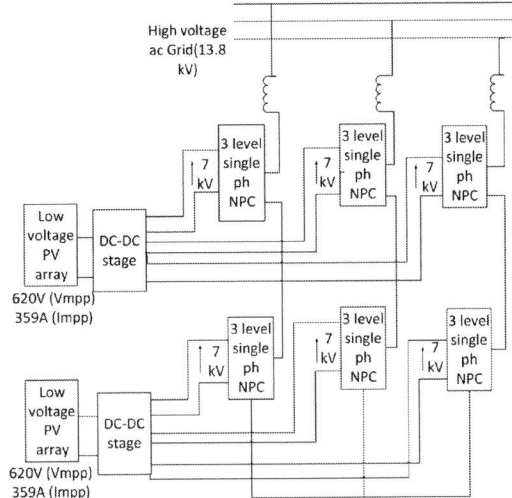

Fig. 12. System Schematic for Proposed Configuration

B. DC-DC Converter Topology

The proposed dc-dc topology shown in figure 13, has a PV input power same as that referenced in table 2. However, the total power can be increased by using parallel dc-dc converters connected to the same three dc links as done by the other dc link. The dc-dc stage is a high frequency transformer isolated three phase dual active bridge converter. The dc-dc stage has three high frequency power transformers, each having four outer limbs and one central limb, each limb carrying a winding. One winding from each of the transformer with one winding each from other two transformers are connected in star connection and one winding from each of the transformer with one winding each from other two transformers are connected in delta connection. Since each transformer has four outer limbs, hence two star connections and two delta connections can be formed. Four three phase two level inverters, using 1200V/100A SiC Mosfet[14] as the switching device, are used, which are connected to the two star connection and two delta connection.

The inverters are connected to a common dc link which is fed from PV source. The inverters operate at high frequency with constant duty cycle of 50% and the delta connected inverters have a phase shift of -30 degree to produce a winding voltage which is in phase with that produced in the star connected winding. The voltages produced by the star and delta windings add up in the middle limb and produce a stepped wave of high frequency, which has high fundamental component and relatively lower other harmonic components. This configuration enables more power to be added up in the central limb. Each of the three middle limb windings are connected to a three level single phase NPC converter through an inductor which acts as power transferring element. The structure of the dc-dc converter is shown in figure 8. A practical high frequency transformer of 35 kVA from [11], is shown in figure 14 to give a volumetric idea about the smaller size transformers.

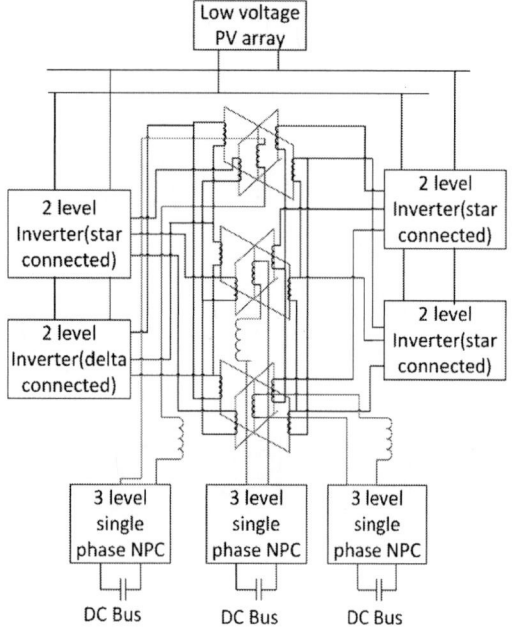

Fig. 13. DC-DC Stage Configuration using three phase dual active bridge having high power density.

Fig. 14. 35 kVA, 10 kHz power transformer[11]

C. Transformer Structure

The structure of a four leg high frequency transformer is shown in figure 15, where each outer winding is connected in either star or delta, the voltages applied on the outer windings are in phase with each other, which is realized by switching the respective converters w.r.t same reference wave. The fluxes produced by the four outer winding voltages are in phase with each other and they add up at the middle limb. The flux linking the central limb winding is given as $\_c = 1 + 2 + 3 + 4$. The voltage thus induced at the middle winding is given as follows

$$V = n\frac{d\_c}{dt} = n\frac{d}{dt}(1 + 2 + 3 + 4)$$

$V = n*(V_1 + V_2 + V_3 + V_4)$, where V_1, V_2, V_3, V_4 are the applied voltages on the four limbs.

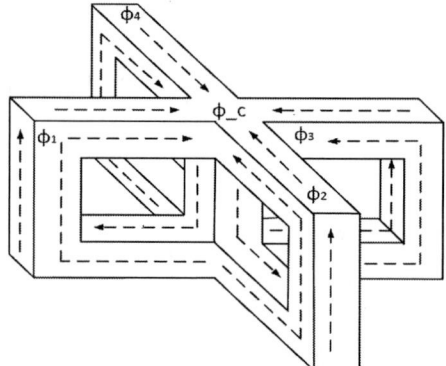

Fig. 15. Transformer Structure

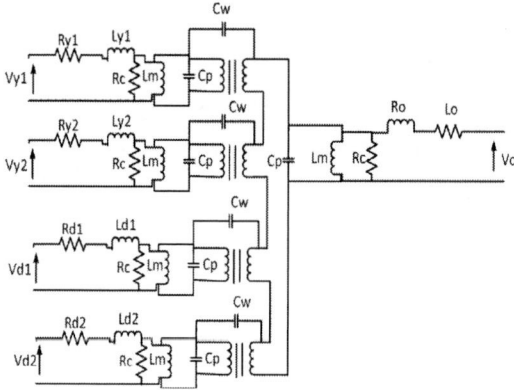

Fig. 16. Electrical Equivalent Circuit of the Transformer Structure

It can observed that if the applied voltages at the outer windings are equal, then the induced voltage at middle winding is nearly equal to turns ratio times sum of the outer winding voltages. For high power and high voltage operation of transformers, realization of the equivalent circuit of the transformer is important. Since the transformer has the windings placed in different limbs, main flux linkage takes place between the outer windings and the inner winding. The flux linkage among the different outer windings is neglected as each outer winding has a source voltage applied to it, hence the effect of induced voltage due to other sources is negligible, hence are not considered. In [13], an equivalent circuit of three winding single phase transformer is shown, where the windings are placed on the same limb. However, in this construction, due to placement of windings on separate limbs, the inter winding capacitive effect among outer limb windings are neglected. Since the induced voltage at the middle limb is nearly equal to turns ratio times sum of the outer winding voltages, the equivalent circuit can be realized as shown in figure 16, where the leakage capacitive effect is present between outer and middle limb windings and across each winding.

D. DC-DC Converter Operation

The NPC converters operate at high frequency to

produce a stepped five level waveform. Accurate switching can produce small amounts of third, fifth and seventh harmonic components in the output voltage of NPC converter. The small amounts of harmonics produced by NPC converter reduces the effect of harmonic components in the middle limb winding voltage, thus the required inductance for power transfer is reduced and the inductor current is nearly sinusoidal at high frequency. Also, the structure of the transformer, similar to that in [11-13], adds on significant amount of leakage inductance which in effect reduces the required inductance value. For example for the rating shown in figure 6, the turns ratio for the outer limb to inner limb of the transformers is 1:3 and the DAB inductance value is around 2.5 mH. The dc buses connected to 3 level NPC converters are assumed to have a constant dc voltage maintained by the grid side inverter. High voltage dc bus capacitance is chosen as 2000μF and PV side dc bus capacitance is chosen as 5000 μF.

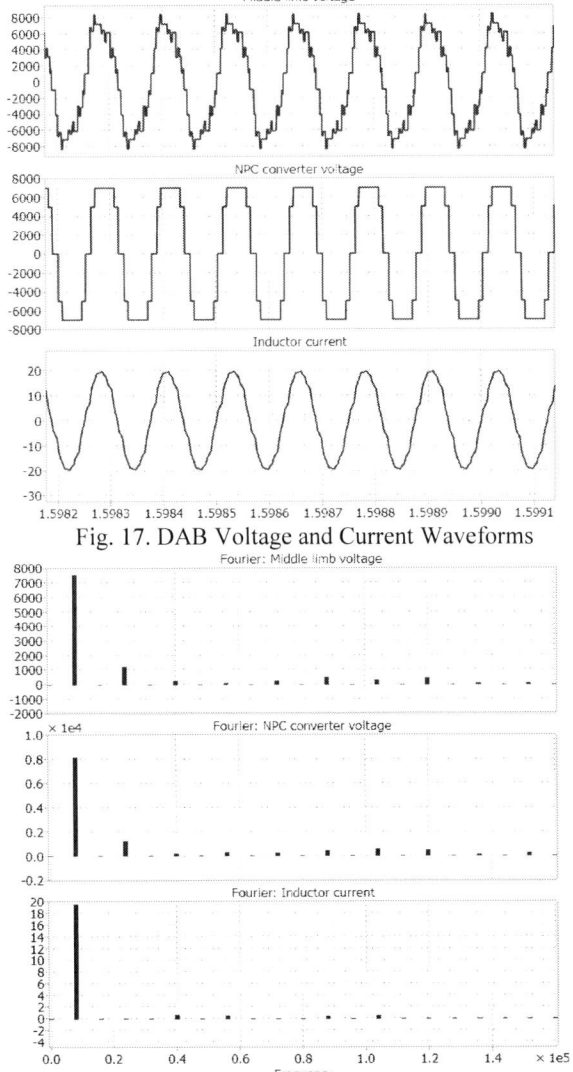

Fig. 17. DAB Voltage and Current Waveforms

Fig. 18. Frequency Spectrum for DAB Voltage and Current Waveforms

Simulation waveforms for the middle limb voltage, the NPC converter voltage and the DAB inductor current along with frequency spectrum are shown in figure 17 and 18. It can be observed that the inductor current is nearly sinusoidal at high frequency(8 kHz). The middle limb voltage distortion is due to the parasitic leakage of the transformer.

E. Control of Cascaded Converters and DC-DC converter with Simulation Results

In the proposed configuration of Fig. 12, the power available from PV modules has to be fed to the grid. The isolated dc-dc converter here acts as MPPT converter and follows the voltage reference from MPPT controller for input dc bus. The three level NPC converters regulate the individual dc bus voltages by operating at a fixed modulation index and adjusting the power angle δ. The control loop is shown in Fig. 19.

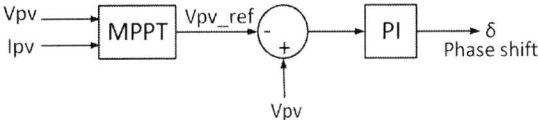

Fig. 19. Control Loop of DC-DC Converter

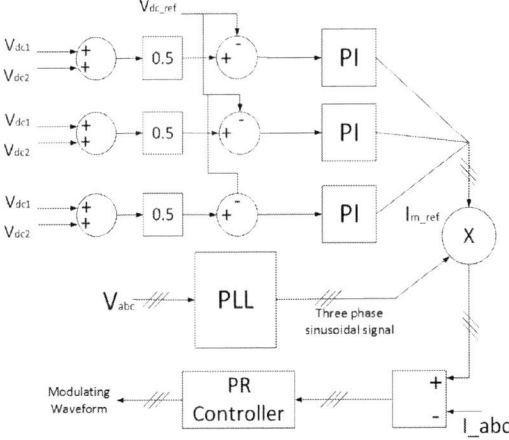

Fig. 20. Control of Cascaded Inverter

For the cascaded inverter control, it is considered that the power output of both the dc-dc converters are equal. The control block diagram for the cascaded inverter is shown in figure 20. PR controller tuned at grid frequency is used as current controller here, since PR controller is capable to provide high gain at resonance frequency. The grid inductance is chosen as 10 mH, each of the 7kV dc buses have a capacitance of 2000 μF, the DAB inductance is around 2-2.5 mH and the PV side capacitance is 5000 μF. According to table 2, the PV voltage at MPP is 620V. In simulation PV array is considered as a current source of 359A, according to table 2, the input DC link is regulated at 620V to emulate the PV MPP operating condition. PV voltage regulation is shown in figure 21, the two 7 kV dc bus voltage regulation and the grid voltage(phase) and line current waveforms using L filter, are shown in figure 22 and 23 respectively.

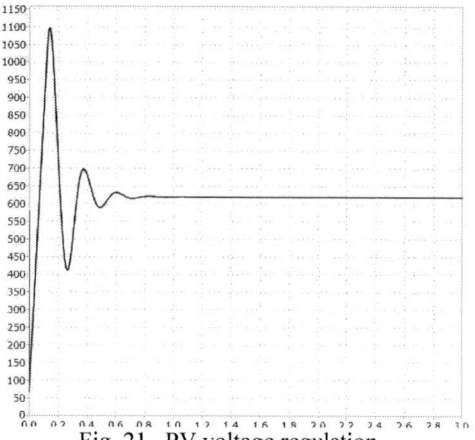

Fig. 21. PV voltage regulation

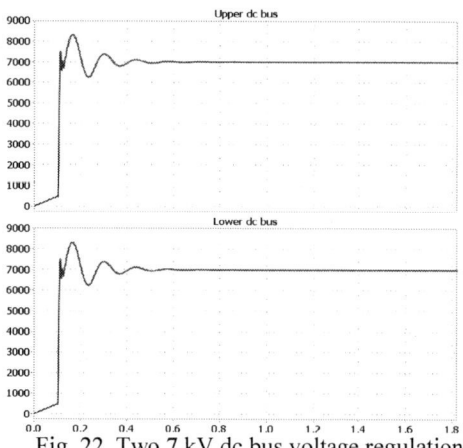

Fig. 22. Two 7 kV dc bus voltage regulation

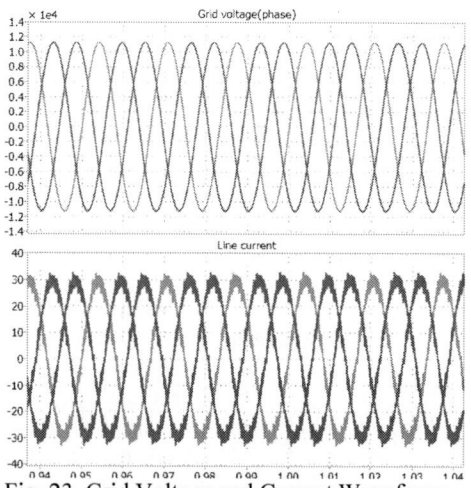

Fig. 23. Grid Voltage and Current Waveforms

F. Total Device Loss Estimation

Use of high voltage devices enables power converters to get rid of big and heavy power line transformers, but performance of these devices vary from different operating conditions. In the proposed topology, the low voltage side converter is made up of 1200V/100A SiC Mosfet[14] and the high voltage side converter uses 10kV SiC Mosfet and 10 kV JBS diode co-pack module[8-10] devices. From the configuration of figure 12, the star and delta connected windings have same currents, but the line current for delta connected inverter is higher than star connected inverter. The high voltage side NPC converters use 10 kV SiC Mosfet and 10 kV JBS diode from [8-10]. Depending on the operating power levels as from table 2, the star connected two level three phase inverter uses two 1200V/100A Mosfet devices in parallel as a single switching element and the delta connected inverter uses three 1200V/100A Mosfet devices in parallel as a single switching element. The NPC converter uses three 10kV/10A SiC Mosfet co-pack module[10] in parallel as a single switching element along with three 10 kV JBS diodes[8] as the neutral connected diodes of NPC. The switching and conduction loss data of 10kV SiC Mosfet and 10kV JBS diode are obtained from figures 9,10,11[8-10], the loss data for 1200V/100A SiC Mosfet data is obtained from [14]. A MATLAB/PLECS model of the proposed system of figure 12 is built to analyze the switching loss data and conduction loss data for the proposed system. The whole system has eight two level inverters made of 1200V/100A SiC Mosfet, six three level NPC converters for DAB and six three level NPC PWM converters for grid integration. Simulation is run in MATLAB/PLECS environment with the loss data of the devices, and the devices losses are measured in watts in a time scale averaging manner. Device loss data for eight low voltage side two level converters, is given in table 3, device loss data for six three level NPC converter of three phase DAB is shown in table 4 and device loss data for six three level PWM converters is shown in table 5. Total device loss for the converters is shown in table 6.

TABLE 3- Two level inverter device losses

Sw. Freq (kHz)	Star Connected Inverters(Watts)		Delta Connected Inverters(Watts)		Total Loss (Watts)
	Sw. Loss	Cond Loss	Sw. Loss	Cond. Loss	
5	62	445	106	845	1458
8	84	445	143	845	1517
10	100	445	163	845	1553

TABLE 4- DAB NPC converter device losses

Sw. Freq. (kHz)	Three level NPC converter for DAB Losses(Watts)			Total Loss (Watts)
	10kV Mosfet cond. Loss	10kV JBS diode cond. loss	10kV Mosfet sw. loss	
5	560	60	215	835
8	560	60	325	945

978-1-4799-2706-7/14 $31.00 © 2014 IEEE

| 10 | 560 | 60 | 355 | 975 |

TABLE 5- Grid side converter device losses

Sw. Freq. (kHz)	Three level NPC converter for Grid side converter losses(Watts)			Total Loss
	10kV Mosfet cond. Loss	10kV JBS diode cond. loss	10kV Mosfet sw. loss	
5	732	67	245	1044
8	732	67	364	1163
10	732	67	406	1205

TABLE 6- Total device losses

Sw. Freq.	Total device losses(Watts)	Percentage of total input power
5 kHz	3337 W	0.75%
8 kHz	3625 W	0.815%
10 kHz	3733 W	0.84%

IV. CONCLUSION

It can be observed that using the SiC devices, the total device losses for the system shown in figure 6 are lesser than one percent. Proper designing of high frequency inductors and transformers, busbars, use of low leakage and low ESR capacitors can limit the overall converter efficiency to an acceptable limit. Compared to the system shown in section II, the topology presented in section III is much lower sized due to absence of line frequency transformers, less heavy and much lower device losses.

ACKNOWLEDGMENT

This project is funded by COELBA through the ANEEL R&D Program "Chamada ANEEL: 013/2011", Research Grant number PD-0047-0060/2011

This work made use of FREEDM ERC shared facilities supported by National Science Foundation.

REFERENCES

[1] IEEE Std. 929-2000, IEEE Recommended Practice for Utility Interface of Distributed (PV) Systems.
[2] Automatic disconnecting facility for photovoltaic installations with a rated output ≤ 4,6 kVA and a single-phase parallel feed by means of an inverter into the public low-voltage mains, Draft standard DIN VDEOI26, 2004.
[3] Hurng-Liahng Jou Wen-Jung Chiang ; Jinn-Chang Wu "Voltage-Mode Grid-Connected Solar Inverter with High Frequency Isolated Transformer", *Industrial Electronics, 2009. ISIE 2009. IEEE International Symposium on, 5-8 July 2009 Pg 1087 – 1092.*
[4] T. Tran-Quoc, C. LE Thi Minh, H. Colin, C. Duvauchelle, S. Bacha, S. Aissanou, B. Gaiddon, C. Kieny, G. Moine, Y. Tanguy "TRANSFORMERLESS INVERTERS AND RCD: WHAT'S THE PROBLEM?", *25th European Photovoltaic*

Solar Energy Conference and Exhibition /5th World Conference on Photovoltaic Energy Conversion, 6-10 September 2010, Valencia, Spain
[5] Elasser, A. Agamy, M. ; Sabate, J. ; Steigerwald, R. ; Fisher, R. ; Harfman-Todorovic, M. "A Comparative Study of Central and Distributed MPPT Architectures for Megawatt Utility and Large Scale Commercial Photovoltaic Plants", *in Proc. IECON 2010 - 36th Annual Conference on IEEE Industrial Electronics Society, 2010 IEEE, pp. 2753 – 2758.*
[6] Villarejo, J.A. Molina-Garcia, A. ; De Jodar, E. "Comparison of Central vs Distributed Inverters: Application to Photovoltaic Systems" *Industrial Electronics (ISIE), 2011 IEEE International Symposium on 27-30 June 2011 1741 – 1746.*
[7] Krishnamoorthy, H.S. Essakiappan, S. ; Enjeti, P.N. ; Balog, R.S. ; Ahmed, S. "A New Multilevel Converter for Megawatt Scale Solar Photovoltaic Utility Integration", *in Proc. Applied Power Electronics Conference and Exposition (APEC), 2012 Twenty-Seventh Annual IEEE, pp.1431 – 1438.*
[8] Das, M.K; Capell, C.; Grider, D.E.; Raju, R.; Schutten, M.; Nasadoski, J.; Leslie, S.; Ostop, J.; Hefner, A. "10 kV, 120 A SiC half H-bridge power MOSFET modules suitable for high frequency, medium voltage applications", *in Proc. Energy Conversion Congress and Exposition (ECCE), 2011 IEEE, pp. 2689-2692.*
[9] G. Wang, X. Huang, J. Wang, T. Zhao, S. Bhattacharya, A.Q. Huang, "Comparisons of 6.5kV 25A Si IGBT and 10-kV SiC MOSFET in Solid-State Transformer application", ", *in Proc. Energy Conversion Congress and Exposition (ECCE), 2010 IEEE, pp. 100-104.*
[10] Shiva Moballegh, Sachin Madhusoodhanan, Subhashish Bhattacharya, "Evaluation of High Voltage 15 kV SiC IGBT and 10 kV SiC MOSFET for ZVS and ZCS High Power DC - DC Converters", *accepted and to be presented at International Power Electronic Conference(IPEC) 2014.*
[11] A.K. Tripathi, K. Mainali, D. Patel, S. Bhattacharya, K. Hatua, "Closed Loop D-Q Control of High-Voltage High-Power Three-Phase Dual Active Bridge Converter in Presence of Real Transformer Parasitic Parameters", *presented at Energy Conversion Congress and Exposition (ECCE), 2013 IEEE.*
[12] K. Mainali, A.K. Tripathi, D. Patel, S. Bhattacharya, T. Challita, "Design, Measurement and Equivalent Circuit Synthesis of High Power HF Transformer for Three-Phase Composite Dual Active Bridge Topology", *to be presented at . Applied Power Electronics Conference and Exposition (APEC), 2014, IEEE.*
[13] K. Hatua, S. Dutta, A. Tripathi, Seunghun Baek, G. Karimi, S. Bhattacharya, "Transformer less Intelligent Power Substation design with 15kV SiC IGBT for grid interconnection", *in Proc. Energy Conversion Congress and Exposition (ECCE), 2011 IEEE, pp. 4225-4232.*
[14] Datasheet of CAS100H12AM1, 1200V, 100A Silicon Carbide Half bridge module, *from Cree,Inc.*

A novel global maximum power point tracking method for photovoltaic generation system operating under partially shaded condition

Jing-Hsiao Chen
Department of Electrical
Engineering, NTUST
Taipei, Taiwan, R.O.C.
D10107201@mail.ntust.edu.tw

Yu-Shan Cheng
Department of Electrical
Engineering, NTUST
Taipei, Taiwan, R.O.C.
D10107206@mail.ntust.edu.tw

Shun-Chung Wang
Department of Electrical
Engineering, LHU
Taoyuan, Taiwan, R.O.C.
scwang.hinet@msa.hinet.net

Jia-Wei Huang
Photovoltaic Technology
Division System Application
Department, Green Energy
and Environment Research
Laboratories, ITRI
Hsinchu, Taiwan, R.O.C.
J.W.Huang@itri.org.tw

Yi-Hua Liu
Department of Electrical
Engineering, NTUST
Taipei, Taiwan, R.O.C.
yhliu@mail.ntust.edu.tw

Abstract—Solar cells have nonlinear P-V curves that vary with insolation and temperature. Therefore, maximum power point tracking (MPPT) technique is used to operate and keep a solar cell at its maximum power output. In practical application, solar cells are usually influenced by partially shaded conditions (PSC). There is a bypass diode that prevents damage to the solar cell, which causes the P-V characteristic curve to have multiple peak values. As the traditional MPPT technology only tracks local maximum value, techniques which can find the global maximum value should be developed to maintain MPPT. This study proposed a new type of global maximum power point tracking (GMPPT) algorithm and experimentally proved that the global maximum power point (GMPP) could be tracked rapidly and successfully in different shaded conditions.

Keywords— *solar energy; global maximum power point tracking; partially shaded conditions*

I. Introduction

Environmental considerations have risen due to the greenhouse effect, and the exhaustion of energy sources has attracted wide attention to renewable energy. Among various renewable energy sources, solar power generation is a promising energy source [1]. The P-V characteristic curve of a solar cell is nonlinear, and maximum power can be obtained only if the system is operated at the peak point of the P-V curve. This method is known as the maximum power point tracking (MPPT). Numerous MPPT methods have been proposed, including the straight-line approximation method, open-circuit voltage, short-circuit voltage, perturb & observe (P&O), and incremental conductance (INC) [2, 3]. These methods have a good tracking performance when the insolation is uniform. However, series-connected solar cell arrays are sometimes shaded by clouds, buildings, and trees, resulting in non-uniform insolation. This causes the P-V curve to have multiple peak values at different heights. If the maximum power point (MPP) is tracked using the aforesaid methods, the found MPP may be a local maximum power point (LMPP) instead of the global maximum power point (GMPP). Therefore, an accurate and rapid GMPPT method is necessary.

Many GMPPT methods for handling partially shaded conditions (PSC) have been proposed in the literatures. [4] used the P&O method to search for all LMPPs in order to find the GMPP, which led to a long tracking time. [5] used dividing rectangles (DIRECT) technology to reduce the search area gradually and find the GMPP; however, this method cannot guarantee to find the GMPP. [6] used constant power operation to determine the section where the GMPP is located, and then used the P&O method to move the operating point to the GMPP. However, this method cannot find the GMPP correctly if the power command variation is too large. [7] used system characteristic line to quickly move the operating point to somewhere near the GMPP when the PSC occurred, and then used the P&O method to move the operating point to the GMPP. This method also cannot guarantee to find the GMPP. [8] proposed a particle swarm optimization (PSO) based GMPPT method; however, as the PSO is characterized by a random number, the characteristics in each execution cannot be guaranteed.

This paper proposes a new type of GMPPT method, in which the section containing the GMPP is determined by sectional scanning. The variable-step P&O method is then used to find the GMPP. According to the experimental results, this method could find the GMPP effectively in different PSCs.

II. The effect of PSC on PGS

A. The basic characteristics of a solar cell

A solar cell can be modeled as the equivalent circuit as shown in Fig. 1. The equivalent circuit includes a current source I_{SC}, a diode D, a series equivalent resistance R_S and a parallel equivalent resistance R_{SH}.

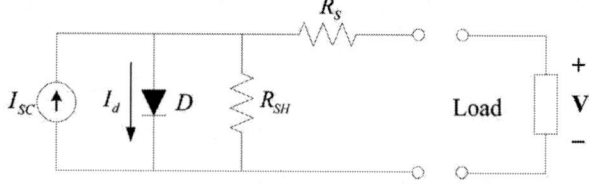

Fig. 1 Equivalent circuit of a solar cell

The relationship of the output voltage and output current of a solar cell can then be represented as

$$I = I_{SC} - I_O \cdot \left[\exp\left(\frac{q \cdot (V + R_S \cdot I)}{n \cdot k \cdot T_K} \right) - 1 \right] - \frac{V + R_S \cdot I}{R_{SH}} \quad (1)$$

where n is the quality factor, k is the Boltzmann's constant, q is the electron charge, T_K is the temperature in Kelvin and I_{SC}, I and I_O are the photogenerated current, panel current and saturation currents, respectively. R_S has higher effect on the I-V curve when solar cell operates in the voltage source region, and R_{SH} has stronger effect on the I-V curve when solar cell operates in the current source region.

B. The effect of PSC

As shown in Fig.2, bypass diodes are be added to protect the solar cell from hot-spot effect caused by partially shading condition. From Eq. (1), a shaded solar cell is not able to produce as much current as a non-shaded cell. Therefore, the bypass diode connected in parallel to the shaded cell will conduct to allow current from non-shaded cell to pass by the shaded cell, this will prevent the shaded cell to go into reverse bias. However, the bypass diodes will result in multiple steps in the I-V curves and multiple peaks in the P-V curves, as shown in Fig. 3.

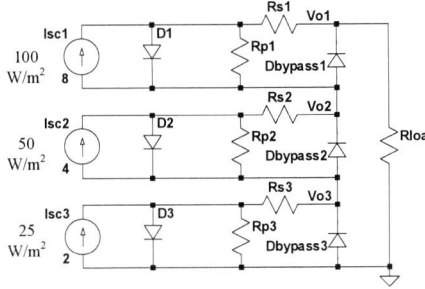

Fig. 2 Solar cells connect in series

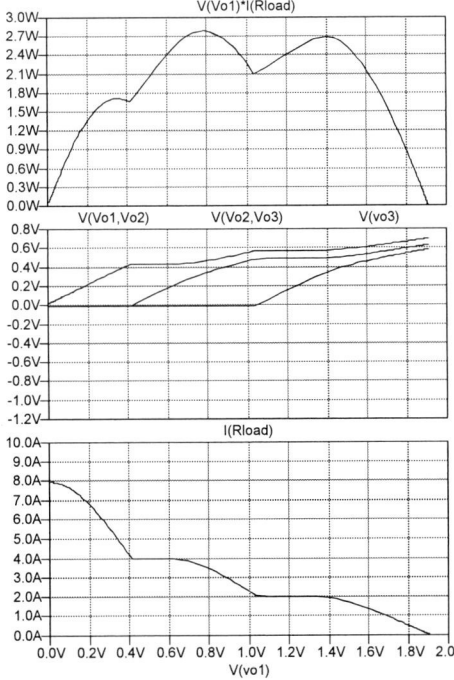

Fig. 3 The I-V curve and the P-V curve of a PV array under PSC

III. New type of GMPPT method applicable to PSC

The open-circuit voltage (OCV) of one single solar cell can be known beforehand, and when multiple solar cells are connected in series, the total OCV can be measured. Therefore, as long as the measured total OCV is divided into enough sections, there will not be more than one peak value in each section interval. If the power is sampled at a specific position in each interval, this power can represent the interval. Afterwards, the representative powers of various intervals can be compared, and the P&O method can be used to find the GMPP in the interval with maximum power. In order to further increase the tracking speed, this study used the variable-step P&O method. When the distance to the GMPP is longer, a larger step size is used for perturbation; when the distance to the GMPP is shorter, a smaller step size is used instead. Therefore, this method pays attention to both tracking speed and accuracy. Fig. 4 is a schematic diagram of the proposed segment search.

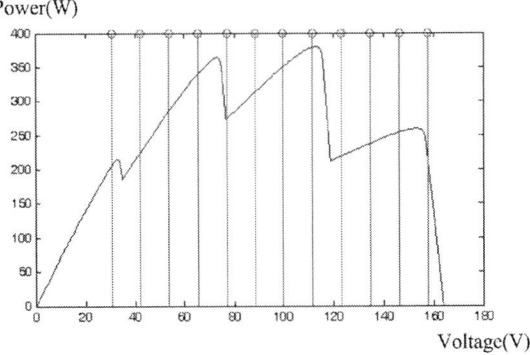

Fig. 4 Schematic diagram of the proposed method

The design procedure of the proposed method is briefly described below:

Step 1: The total open-circuit voltage V_{OC} of the series-connected solar cells is obtained and the upper and lower range V_a, V_b is determined, in which $V_a = 0.1 \cdot V_{OC}$, and $V_b = 0.9 \cdot V_{OC}$. The number of sections N is set as greater than the number of panels in the series connection.

Step 2: The power of each section point is measured as the representative power P_i of the interval, and the interval with maximum representative power is determined as the global maximum power P_m.

Step 3: The variable-step P&O method is used for the interval with the GMPP. First, a larger step size is adopted, and then once the GMPP is approached, a smaller step size is adopted for perturbation.

Step 4: When the search is completed, the variation of the power value with time is checked, so as to ensure there is no change in the insolation.

In the flow chart, V_{GP} represents the voltage corresponding to the GMPP. When the variation of power ΔP exceeds the preset ΔP_{crit}, it can be regarded as a change in the insolation or solar cell shading, meaning the maximum power interval has probably changed, so the representative power of various intervals must be checked again. This flow chart is shown in Fig. 5.

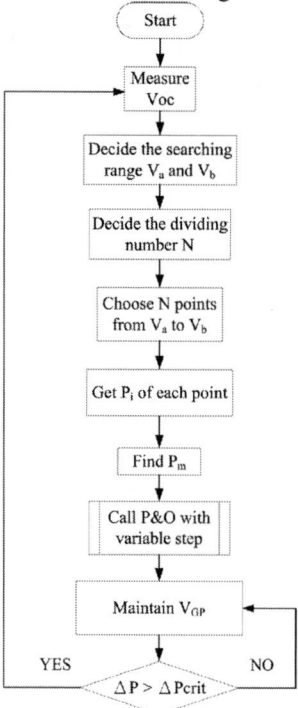

Fig. 5 Flowchart of the proposed method

IV. Variable-step P&O method

The variable-step P&O method adopted in this study was similar to the variable-step INC proposed in [9], in which the ratio of the power variation ΔP of the solar cell to the input voltage variation ΔV and the proportional factor M_V designed for the system was used to determine the voltage command variation, which are used to determine the next voltage command ($V(n+1)$). The calculation of the next voltage command is expressed as Eq. (2) and (3), and the flow chart of the variable-step P&O method used in this paper is shown in Fig. 6.

$$\Delta V_{(n+1)} = M_v \times \frac{\left|P_{(n)} - P_{(n-1)}\right|}{\left|V_{(n)} - V_{(n-1)}\right|} = M_v \times \frac{|\Delta P|}{|\Delta V|} \quad (2)$$

$$V_{(n+1)} = V_{(n)} \pm \Delta V_{(n+1)} \quad (3)$$

where

$\Delta V(n+1)$ is the voltage command variation. $V(n+1)$ $V(n)$ and $V(n-1)$ is the voltage command of the next, present and previous time interval, respectively. Similarly, $P(n)$ and $P(n-1)$ is the calculated power of the present and previous time interval, respectively.

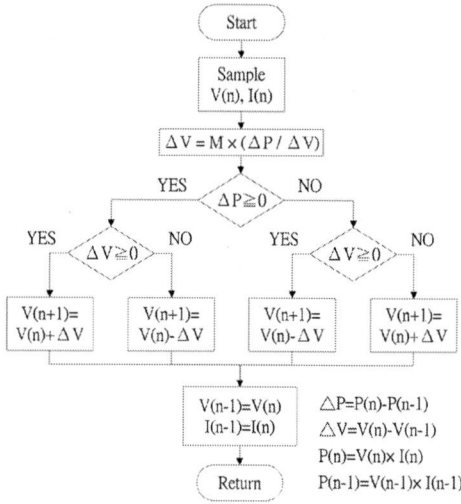

Fig. 6 Flowchart of the utilized variable-step P&O method

V. Experimental Results

Since shading patterns are complex and changeful in real world, whether the proposed method can work correctly with uniform insolation, as well as whether the GMPP can be tracked and converged rapidly and correctly under different shading patterns, must be considered in order to validate the GMPPT method. Five tests are performed to prove that the proposed method could find and converge at the GMPP. Finally, considering the dynamic nature of insolation and shading patterns in a true environment, the shading pattern was changed continuously during testing, so as to prove that the proposed method could not only detect the change in insolation but also successfully track the position after the change and converge correctly.

In this paper, an experimental platform as shown in Fig. 7 is built-up to validate the effectiveness of the proposed GMPPT method, the photo of the realized experimental platform is shown in Fig. 8. In this paper, solar array simulator 62050H-600S from Chroma is adopted to provide the P-V curve, and the proposed GMPPT algorithm is implemented in a low cost digital signal controller (DSC) dsPIC33FJ16GS502 from Microchip corp. A DC electric load 63202 from Chroma operating in constant voltage mode is utilized in this paper to simulate a battery load. The rating of the implemented power circuit is 2.5 kW.

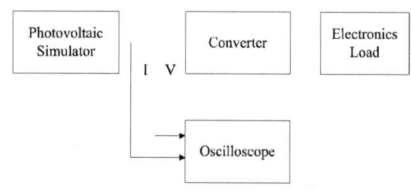

Fig. 7 Block diagram of the experimental platform

Fig. 8 Photo of the experimental platform

The experimental results for all the tests are presneted in Fig. 9 through Fig. 20. For all the measured waveforms, the upper waveform shows the PV voltage, the middle waveform shows the PV current and the lower waveform shows the calculated PV power. Details of the experimental results will be provided as follows:

Test 1 (unifom irradiation): The I-V curve and P-V curve for Test 1 is shown in Fig. 9, and the measured waveform for Test 1 is displayed in Fig. 10. From Fig. 9, the target GMPP is 2244.4 W (V=308.08 V, I=7.285 A). From Fig. 10, it takes 2.8 sec for the proposed method to acquire the segment where the GMPP locates, and the total tracking time required is 8 sec, the obtained tracking accuracy is 99.96 %.

Test 2-4 (non-unifom irradiation): Test 2-4 provides the experimental results of applying the proposed method to P-V curves with three peaks. In Test 2 through 4, the GMPP locates at the left side, middle and the right side of the P-V curve, respectively. The I-V curve and P-V curve for Test 2 is shown in Fig. 11, and the measured waveform for Test 2 is displayed in Fig. 12. From Fig. 11, the target GMPP is 1205.6 W (V=181.24 V, I=6.652 A). From Fig. 12, it takes 2.8 sec for the proposed method to acquire the segment where the GMPP locates, and the total tracking time required is 9 sec, the obtained tracking accuracy is 99.81 %. The I-V curve and P-V curve for Test 3 is shown in Fig. 13, and the measured waveform for Test 3 is displayed in Fig. 14. From Fig. 13, the target GMPP is 1243.9 W (V=232.89 V, I=5.341 A). From Fig. 14, it takes 2.8 sec for the proposed method to acquire the segment where the GMPP locates, and the total tracking time required is 8 sec, the obtained tracking accuracy is 99.81 %. The I-V curve and P-V curve for Test 4 is shown in Fig. 15, and the measured waveform for Test 4 is displayed in Fig. 16. From Fig. 15, the target GMPP is 1250.4 W (V=291.82 V, I=4.285 A). From Fig. 16, it takes 2.8 sec for the proposed method to acquire the segment where the GMPP locates, and the total tracking time required is 9 sec, the obtained tracking accuracy is 99.37 %.

From the test results of Test 1 through Test 4, the proposed method can successfully acquire the GMPP under both uniform and non-uniform irradiation condition. In order to validate that the proposed method can be applied to P-V curve with more complex characteristics, Test 5 is also conducted. Test 5 provides the experimental results of applying the proposed method to P-V curves with five peaks. The I-V curve and P-V curve for Test 5 is shown in Fig. 17, and the measured waveform for Test 5 is displayed

in Fig. 18. From Fig. 17, the target GMPP is 865.5 W (V=245.32 V, I=3.528 A). From Fig. 18, it takes 2.8 sec for the proposed method to acquire the segment where the GMPP locates, and the total tracking time required is 9 sec, the obtained tracking accuracy is 99.93 %. Finally, Test 6 is conducted to validate the dynamic tracking performance of the proposed system. Fig. 19 shows the changing sequence of the tested shading patterns. Fig. 20 shows the measured waveforms of Test 6. From Fig. 20, the proposed algorithm can successfully detect the shading pattern changes and reinitialize the MPPT process accordingly. All the experimental results is summarized in Table 1.

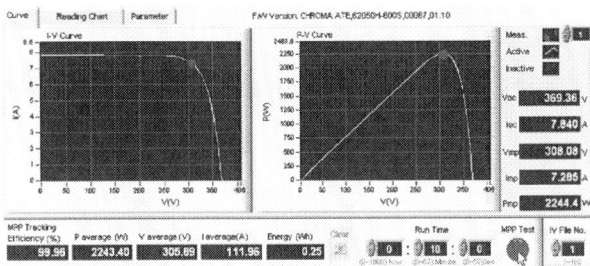

Fig. 9 I-V curve and P-V curve of Test 1

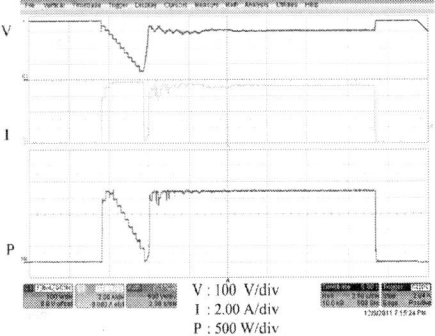

Fig. 10 Measured waveforms of Test 1

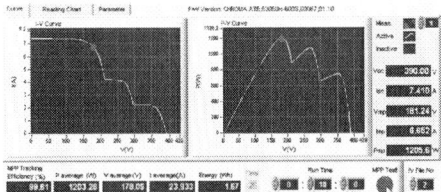

Fig. 11 I-V curve and P-V curve of Test 2

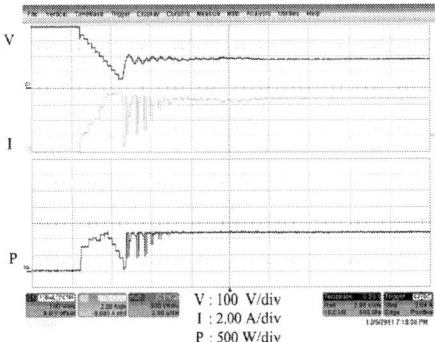

Fig. 12 Measured waveforms of Test 2

The 2014 International Power Electronics Conference

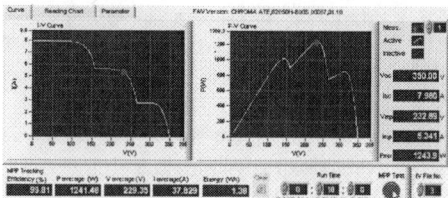

Fig. 13 I-V curve and P-V curve of Test 3

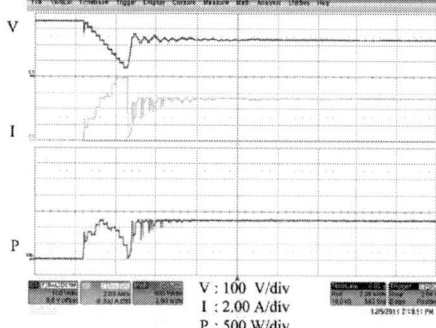

V : 100 V/div
I : 2.00 A/div
P : 500 W/div

Fig. 14 Measured waveforms of Test 3

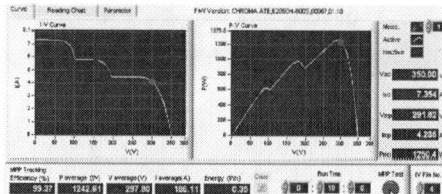

Fig. 15 I-V curve and P-V curve of Test 4

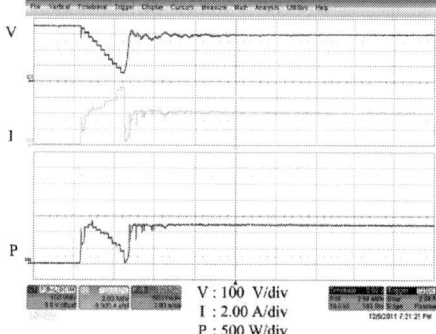

V : 100 V/div
I : 2.00 A/div
P : 500 W/div

Fig. 16 Measured waveforms of Test 4

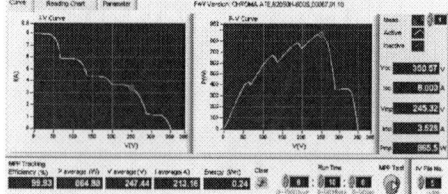

Fig. 17 I-V curve and P-V curve of Test 5

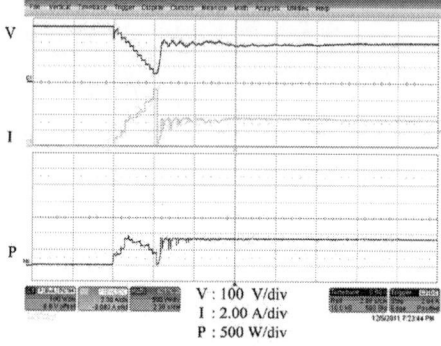

V : 100 V/div
I : 2.00 A/div
P : 500 W/div

Fig. 18 Measured waveforms of Test 5

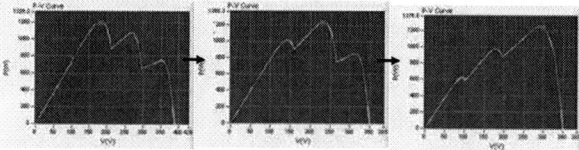

Fig. 19 I-V curve and P-V curve of Test 6

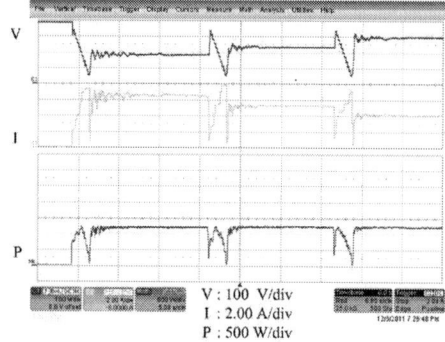

V : 100 V/div
I : 2.00 A/div
P : 500 W/div

Fig. 20 Measured waveforms of Test 6

Table 1 Summary of the experimental results

Test Condition	The Time of reached the global maximum point	The Total Processes before the convergence time
Test 1	2.8 Sec.	8 Sec.
Test 2	2.8 Sec.	9 Sec.
Test 3	2.8 Sec.	8 Sec.
Test 4	2.8 Sec.	9 Sec.
Test 5	2.8 Sec.	10 Sec.

VI. Conclusion and Discussion

This study proposed a new type of GMPPT algorithm. The proposed method contained two stages. In Stage I, the segmentation at fixed intervals was used to search for the interval of the GMPP. In Stage II, the variable-step P&O method was used to find the exact position of the GMPP. The architecture of the proposed method is simple and can be implemented using a low-cost microcontroller. It can also easily be integrated into traditional PGS firmware. In order to validate the accuracy of the proposed method, this study used a solar simulator to generate multiple different P-V characteristic curves, and operated the electronic load

978-1-4799-2706-7/14 $31.00 © 2014 IEEE

at a constant voltage to simulate the battery load for the experiment. According to the experimental results, the proposed GMPPT method could track the GMPP of P-V curve of multiple peak values successfully and could identify changes in the characteristic curve. The tracking speed of the proposed method was less than 10 sec, and the tracking accuracy was higher than 99%.

Acknowledgment

The research underlying this paper was supported by The Proposal for PV Universalization Environment Construction and Promotion Project (102-D0305). The authors are grateful for the Bureau of Energy.

References

[1] Bidram A., Davoudi A., Balog R. S. Control and Circuit Techniques to Mitigate Partial Shading Effects in Photovoltaic Arrays. IEEE Journal of Photovoltaics 2012;2:532-46.

[2] Ishaque K., and Salam Z. A review of maximum power point tracking techniques of PV system for uniform insolation and partial shading condition. Renewable and Sustainable Energy Reviews 2013;19:475-88.

[3] Salam Z., Ahmed Jubaer., Merugu B. S. The application of soft computing methods for MPPT of PV system: A technological and status review. Applied Energy 2013;107:135-148.

[4] Patel H., Agarwal V. Maximum Power Point Tracking Scheme for PV Systems Operating Under Partially Shaded Conditions. IEEE Transactions on Industrial Electronics 2008;55:1689-98.

[5] Nguyen T. L., Low K. S. A Global Maximum Power Point Tracking Scheme Employing DIRECT Search Algorithm for Photovoltaic Systems. IEEE Transactions on Industrial Electronics 2010;57:3456-67.

[6] Koutroulis E., Blaabjerg F. A New Technique for Tracking the Global Maximum Power Point of PV Arrays Operating Under Partial-Shading Conditions. IEEE Journal of Photovoltaics 2012;2:184-90.

[7] Ji Y. H., Jung D. Y., Kim J. G., Kim J. H., Lee T. W., Won C. Y. A Real Maximum Power Point Tracking Method for Mismatching Compensation in PV Array Under Partially Shaded Conditions. IEEE Transactions on Power Electronics 2011;26:1001-9.

[8] Liu Y. H., Huang S. C., Huang J. W., Liang W. C. A Particle Swarm Optimization-Based Maximum Power Point Tracking Algorithm for PV Systems Operating Under Partially Shaded Conditions. IEEE Transactions on Energy Conversion 2012;27:1027-35.

[9] Liu, F., Duan S., Liu F., Liu B., and Kang Y. A Variable Step Size INC MPPT Method for PV Systems. IEEE Transactions on Industrial Electronics 2008; 55: 2622-8.

An application of Z-Source converter to batteries charge with a photovoltaic system

H. Razik, Y. Zitouni
Université Claude Bernard Lyon 1
Laboratoire AMPERE - UMR 5005
F - 69622 Villeurbanne, France
E-mail: [Hubert.Razik, Younes.Zitouni]@univ-lyon1.fr

C. Maret
APAVE SUDEUROPE
33 avenue Georges Lévy, BP 116
69634 Venissieux, France
E-mail: Cyril.maret@apave.com

Abstract—**This paper deals with a photovoltaic system based on a Z-source DC-DC converter. This type of structure differs from a classical converter by the presence of a shoot-through duty cycle in comparison with the standard duty cycle. This recent topology, involving an X-shape LC impedance network, is attractive. So, in this application, the photovoltaic energy is stored in lead acid batteries thanks to a Z-source DC-DC converter. A prototype is designed and several experimental results are included to show the interest of the proposed system.**

Keywords—Energy storage, Photovoltaic, Z-Source converter.

I. INTRODUCTION

Renewable energy is one of the current major concerns of our society. The guidelines are numerous at national levels as well as European and international levels. The side effects of the use of primary sources of energy lead to explore other sources such as photovoltaic. A look at the distribution of the production capacity of solar power in 2011 induces the following conclusion: Germany (35.6%), Italy (18.3%), USA (5.7% in the fifth position) and France (4.1% in the seventh position) [1]. Energy production in Europe in 2011 is 188TWh for wind, hydro and 360TWh for 37TWh for solar (22TWh in 2010).

The demand of energy from solar or wind power is increasing from year to year [2]. Unfortunately, the production is not constant because it is highly dependent on daily weather conditions. Numerous studies have been conducted and are still on the hybridization / combination of energy sources into a single system. The consequences induced deal with the management of this energy which is intermittent, the energy storing and the availability of this one. The decrease in the price of electronic components led to consider interconnect topologies solar panels with individual control of the MPPT (Maximum Power Point Tracking). Therefore, these primary energy sources could be managed through networks of communication between each source of production [3].

In this paper, a topology of the Z-source DC-DC converter is applied for the storage of energy. This kind of structure, the DC-DC converter in a single stage has a booster effect and its topology is relatively novel [4]. This is due to the introduction of a passive X-shape LC impedance network. Consequently, this type of structure

can buck and boost its output voltage without any active switch. In fact, the converter is based on a Z-source network placed between the supply and the load. In 2007, an application for fuel cell vehicles had been studied [5]. This converter topology has a boost capability [6], moreover this structure is immunized from short circuits produced by the conduction of both transistors. However this structure has an increased cost in comparison to classical DC-DC converters. As a consequence, the efficiency is reduced too [7].

This paper is organized as follows. Section II deals with the Z-source converter topology. Section III includes the circuit analysis. Section IV shows the main characteristics of a solar panel. Section V includes the solar characteristic. Section VI shows the control strategy. Section VII includes experimental results and finally, section VIII presents the conclusion.

II. THE Z-SOURCE CONVERTER

Fig. 1 shows the configuration of the Z-source converter which allows to charge the batteries from a solar panel. This structure is composed of elements connected in X-shape:

- 2 separate inductances L1 and L2;

- 2 capacitors C1 and C2.

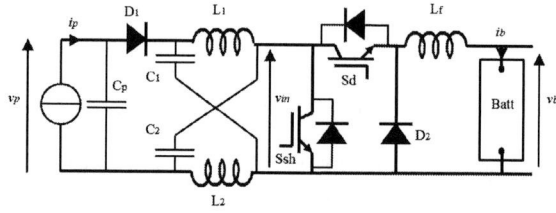

Fig. 1. Configuration of the Z-source DC-DC converter.

In continuous conduction mode there are three structures [4]:

- The null mode;

- The shoot-through mode;

- The non shoot-through mode.

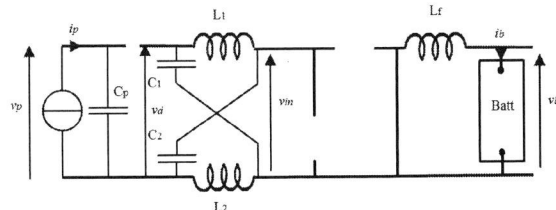

Fig. 2. Structure of the Z-source DC-DC converter: null mode.

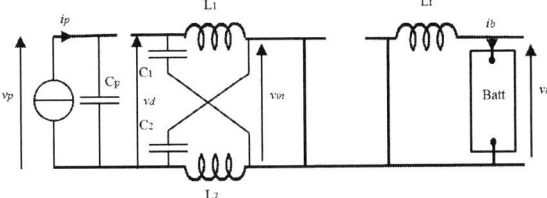

Fig. 3. Structure of the Z-source DC-DC converter: shoot-through mode.

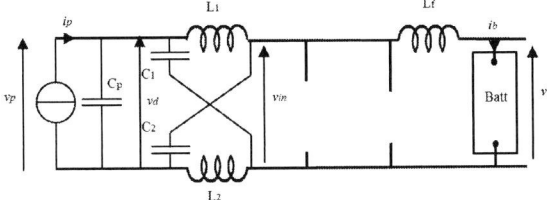

Fig. 4. Structure of the Z-source DC-DC converter: the non-shoot through mode.

The first mode is the null mode (Fig. 2). The Z-network is isolated. The second mode is the shoot-through mode (Fig. 3) which is the mode ST. In this configuration, the two inductances are charged and produce a voltage boost. During this duration the input diode D1 is reverse-biased. The third mode is the non shoot-through mode (Fig. 4) where the energy is transferred to batteries.

III. CIRCUIT ANALYSIS

Considering that inductances are identical ($L1 = L2 = L$) and capacitors also ($C1 = C2 = C$), a relation between the output and the input voltage can be expressed in the CCM.

As the circuit is symmetrical, voltage relations are obtained as follows:

$$V_{C1} = V_{C2} = V_C \\ V_{L1} = V_{L2} = V_L \qquad (1)$$

During T_{sh} the time of shoot-through mode, the voltage V_{in} is equal to zero. Consequently, $V_d = 2V_C = 2V_L$.

During $T - d$ the time of non shoot-through mode, the voltages are: $V_{in} = V_p - 2V_L$, $V_C = V_{in} + V_L$ and $V_{in} = 2V_C - V_p$.

After some calculus, the average voltage applied to batteries is given by:

$$\overline{v_b} = \frac{d_d}{1 - 2d_{st}} V_p \qquad (2)$$

where $d_d = \frac{T_d}{T}$, $d_{st} = \frac{T_{sh}}{T}$ are duty cycles.

One constraint is $T_o + T_d + T_{sh} = T$ where T_o is the time of application of the null mode and T the switching period. It is obvious that the shoot-through mode is equivalent to a boost phenomenon and the non shoot-through mode is equivalent to a classical buck converter.

IV. PHOTOVOLTAIC GENERATOR MODELING

The solar cell, used in this application, could be represented by a simple equivalent circuit as shown in Fig. 5. Several equivalent circuits exist from simplistic models to heavy models. In fact, knowing that a solar panel is composed of many solar in parallel and series, one can define a system as a generator of DC-current with several other components like resistances and diodes [8].

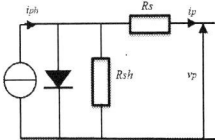

Fig. 5. Equivalent circuit of the solar cell.

Datasheet gives the output circuit voltage V_{oc}, the short circuit current I_{sc}, the maximum-power current and voltage at specific conditions, i.e. under an irradiance $G = 1000W/m^2$ and 25^oC. Consequently, the current-voltage relationship of a panel I_p is written as follows:

$$i_p = N_p I_{ph} - N_p I_{rs} \left\{ \exp^{\frac{q}{AkT_c}\left(\frac{V_p}{N_s} + \frac{i_p r_s}{N_p}\right)} - 1 \right\} \\ - \frac{N_p}{r_p}\left(\frac{V_p}{N_s} + \frac{i_p r_s}{N_p}\right) \qquad (3)$$

where: N_p is the number of parallel combination, N_s the number of series combination, r_s the cell intrinsic series resistance, r_p the cell intrinsic shunt resistance, I_{ph} is the solar cell photocurrent, I_{rs} is the solar cell reverse saturation current. A is the P-N junction ideal factor, k the Boltzmann's constant (1.380658e-23 J/K), q the electron charge (1.60217733e-19 Cb), T_c the solar cell temperature (K).

Figure 5 shows a model of a solar panel where the shunt resistance R_{sh} is in parallel with the source of current I_{ph}. The serial resistance R_s has to be considered due to the wiring resistance from the panel to the converter (15 meters in our application).

V. SOLAR CHARACTERISTIC

The maximum power generated by a solar cell is characterized by a point on the current-voltage where the product VI is maximum. This point is known as the

978-1-4799-2706-7/14 $31.00 © 2014 IEEE

The 2014 International Power Electronics Conference

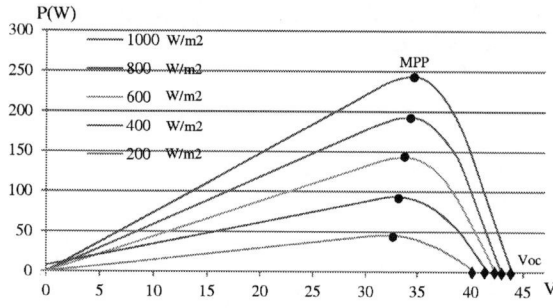

Fig. 6. Characteristic curves of a solar panel.

Maximum Power Point (MPP) as showed in Fig. 6. where the previous points are represented.

The influence of the irradiance level on the voltage-power (V-P) is obvious. Thus, the MPP changes all day long and it has to be tracked.

The maximum power point tracking based on the incremental conductance technic is the most commonly technic used [9], [10]. This one is based on null slop of the variation of the power dP/dV at MPP.

$$\frac{P}{V} = \frac{VI}{V} = I + V\frac{dI}{dV} \quad (4)$$

Thus, $dP/dV < 0$ at right of MMP, $dP/dV = 0$ at MMP and $dP/dV > 0$ at left of MMP. The most common algorithm is described as follows:

Algorithm 1 The incremental conductance algorithm

Measure: V(t) and I(t)
Calculate:
$\triangle I = I(t) - I(t - \triangle t)$ and $\triangle V = V(t) - V(t - \triangle t)$
If $\triangle I$=0 then
    If $\triangle V \neq 0$
        If $\triangle V > 0$; Increase Iref.
        else Decrease Iref.
else
    If $\triangle V/\triangle I \neq -V/I$
        If $\triangle V/\triangle I > -V/I$; Increase Iref.
        else Decrease Iref.
end

VI. CONTROL STRATEGY

The control strategy is shown in Fig. 7. To achieve the current regulation, a hysteresis comparator is used. This method is well suited for slowly varying setup. The different levels at which switchings occur are the high and low threshold current in this kind of application. So the output gives the main switching frequency. The shoot-through mode is achieved by using an impulse generator (monostable). Its output will drive the Ssh transistor. Of course, the time constraint has to be considered. The standard switching frequency in steady state must be restricted due to the shoot-through time.

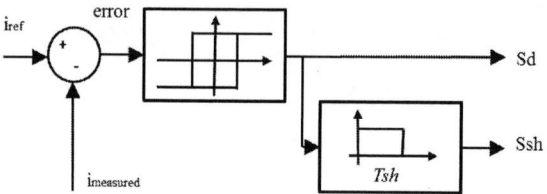

Fig. 7. Control scheme.

VII. EXPERIMENTAL RESULTS

Experiments have been performed to confirm the analysis. This prototype is designed for a switch frequency in the range from 5 kHz to 10 kHz. The Z-source parameters are given in Table I. Constraints are the voltage panel close to V_{oc} (33 V) and the maximum current which is given by I_{sc} (8.5 A) as given in Table II.

Fig. 8. Setup of the Z-converter.

TABLE I. Z-SOURCE NETWORK PARAMETERS

Symbol	Value
$L_1 = L_2 = L$	2.9mH
$C_1 = C_2 = C$	360μF
L_F	2mH
C_F	730μF
$D_1 = D_2$	STPS16150
$S_D = S_{SH}$	IRL3215
Photovoltaic panel	TE 210-54M
Battery	GEL 12-110 Victron Energy

TABLE II. SOLAR PANEL CHARACTERISTICS

Symbol	Value
V_{oc}	32.9V
I_{SC}	8.5A
V_{MP}	26.25V
I_{MP}	8A
P_{MP}	210W
G	1000W/m^2
T	25°C

To demonstrate the converter performance, 5 figures show voltages and currents and the last one the conversion of the solar energy under 2 irradiance conditions. The switching frequency is 5 kHz and the shoot-through mode is applied during 20 μs in this experiment. Experiments are made with Td in the range: 0 μs to 170 μs knowing that $Td + Tsh$ must be lesser than the switching period T. In this experiment, Td is equal to 100 μs.

Fig. 9 shows the instantaneous commands of the switch Ssh and Sd which are shown in all 4 figures.

978-1-4799-2706-7/14 $31.00 © 2014 IEEE

The 2014 International Power Electronics Conference

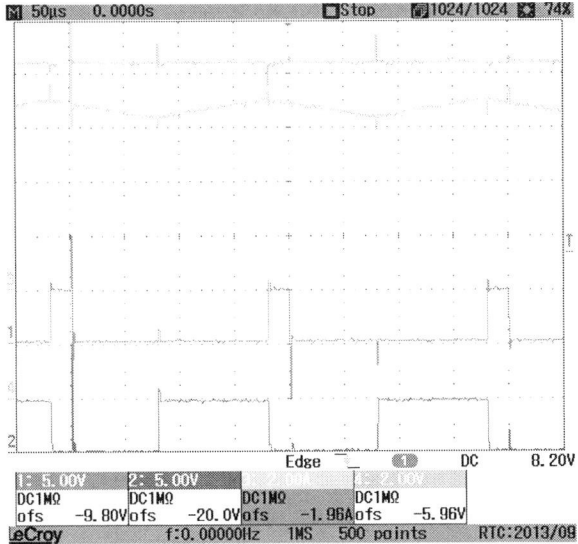

Fig. 9. Experiment: Chanel 1 - T_{sh} signal, Chanel 2 - T_d signal, Chanel 3 - I_b Current in batteries, Chanel 4 - V_b Voltage of batteries.

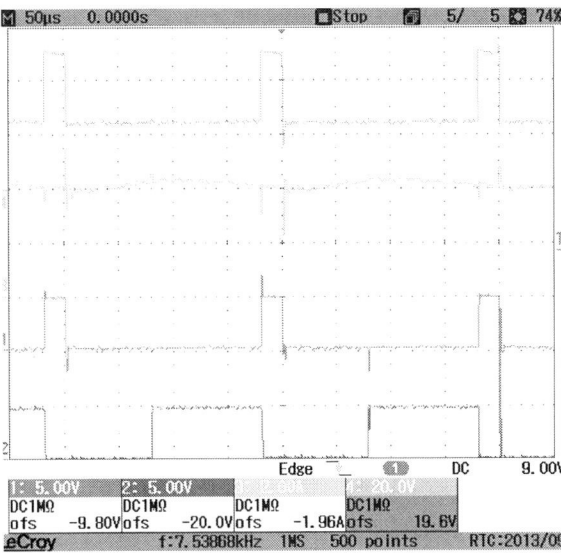

Fig. 11. Experiment: Chanel 1 - T_{sh} signal, Chanel 2 - T_d signal, Chanel 3 - I_b, Chanel 4 - V_d.

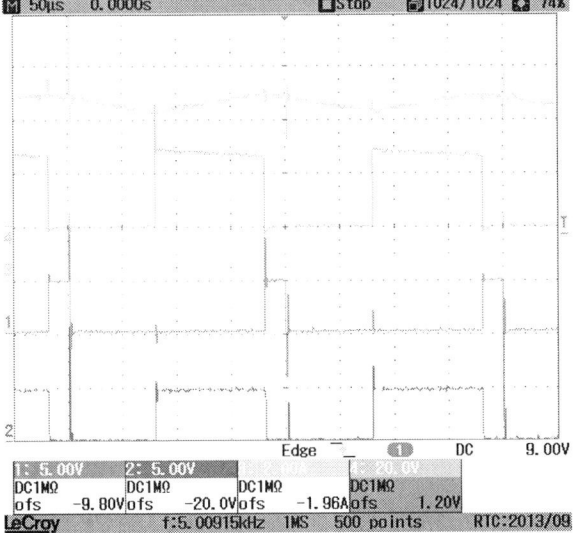

Fig. 10. Experiment: Chanel 1 - T_{sh} signal, Chanel 2 - T_d signal, Chanel 3 - I_b, Chanel 4 - V_{DZ}.

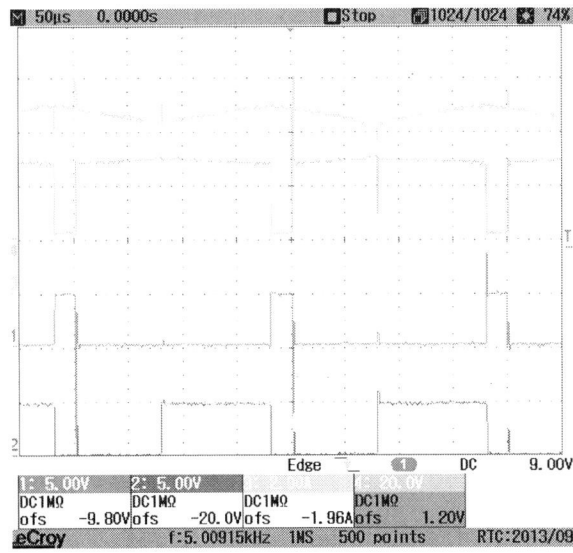

Fig. 12. Experiment: Chanel 1 - T_{sh} signal, Chanel 2 - T_d signal, Chanel 3 - I_D , Chanel 4 - V_{in}.

Chanel 4 shows the instantaneous voltage of the batteries as 2 batteries are connected in parallel. Chanel 3 shows the current with an average value close to 6.5 A. During the shoot-through time as shown in Fig. 2, the current I_b decreases. During the standard mode as shown in Fig. 3 the current I_b increases. During the rest, the structure is in the null mode as shown in Fig. 1.

In Fig. 10 can be seen the instantaneous voltage across the diode D_2 and its stress. This voltage is higher than V_b.

Fig. 11 shows the current flowing in batteries and the voltage across the X-shape LC impedance network. It can be seen that during the shoot-through mode, the value is very high in comparison with the voltage across the panel

V_D. Its value is almost constant during both other modes.

In Fig. 12 the voltage across the switch T_d is illustrated. We can see the increasing value of the voltage V_{in} that corresponds to the transfer of energy into the LC impedance network. This is the null mode. During both other modes, V_{in} is decreasing as expected.

In Fig 13 the voltage across the solar panel is illustrated. The value of the capacitor C_p was selected in order to have 1 V of peak to peak ripple in the worst case. In this test, its value is close to 25 V and it is almost constant.

In Fig. 14 are illustrated the voltage across the solar panel, the current flowing inside it and the power generated. Two tests were made. The first one was made at 9 am under a "low" irradiance, the second one was

978-1-4799-2706-7/14 $31.00 © 2014 IEEE

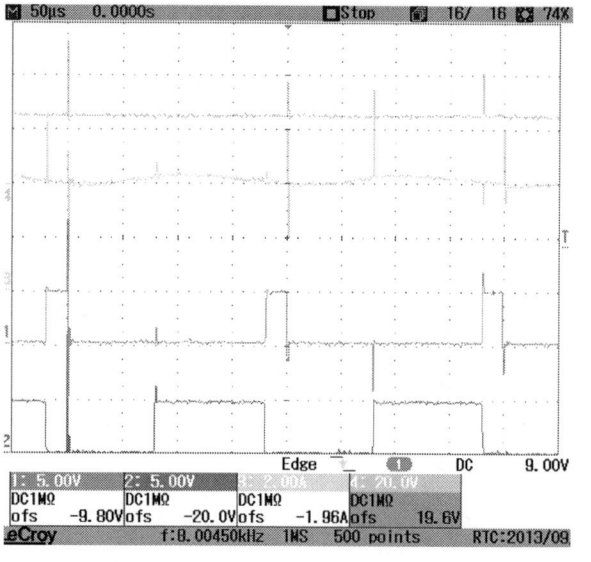

Fig. 13. Experiment: Chanel 1 - T_{sh} signal, Chanel 2 - T_d signal, Chanel 3 - I_D, Chanel 4 - V_D.

made at 2 pm under a quasi optimal irradiance. During them both, the standard mode varies from 0 μs to 170 μs. At "low" irradiance, it appears from the $Vp\,l$ curve that a maximum exists. This is the main goal of MPPT (Maximum Power Point Tracking) algorithm. We can see that it is not possible for the "optimal" irradiance. So it is necessary to modify the shoot-through time because they would no longer apply the null mode.

VIII. CONCLUSION

This paper has presented the design of a Z-source converter as a possible solution for photovoltaic source. In this application, the converter is supplied by a solar panel. Even if the number of components is higher in comparison with standard structures, the reliability is better, due to the X-shape. This paper has shown the interest of this type of structure for the transfer of solar energy from a solar source to batteries. The hardware setup presented in this paper is under development in the laboratory.

REFERENCES

[1] http://www.iea.org/etp/tracking/renewables/

[2] R.A. Messenger, and J. Ventre, *Photovoltaic Systems Engineering*, CRC Press, 2010.

[3] A. Khaligh, and O.C. Onar, *Solar, Wind, and Ocean Energy Conversion Systems*, CRC Press, 2009.

[4] Fang Zheng Peng, "Z-Source Inverter", *IEEE Trans. on Industry Applications*, vol. 39, no. 2, pp. 504-510, 2003.

[5] M. Shen, A. Joseph, J. Wang, F. Peng, and D. Adams, "Comparison of traditional inverters and z-source inverter for fuel cell vehicles", *IEEE Trans. On Power Electronics*, vol. 22, no. 4, pp. 1453-1463, 2007.

[6] S.J. Amodeo, H.G. Chiacchiarini, A. Oliva, C.A. Busada, and M.B. D'Amico, "Enhanced-performance control of a DC-DC Z-Source converter", *in proc. IEMDC '09*, pp. 363-368, 2009.

[7] Q.V. Tran, T.W. Chun, J.R. Ahn, and H.H. Lee, "Algorithms for Controlling Both the DC Boost and AC Output Voltage of Z-Source Inverter", *IEEE Trans. On Industrial Electronics*, vol. 54, no. 5, pp. 2745-2750, 2007.

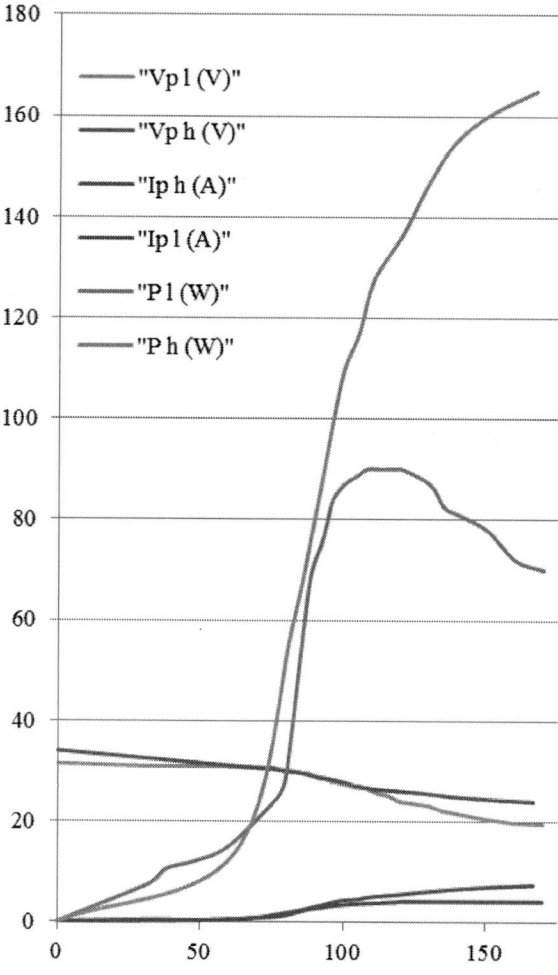

Fig. 14. Experiment: Voltage Vp at high irradiance (Vp h), low irradiance (Vp l), current Ip at high irradiance (Ip h), low irradiance (Ip l), power at high irradiance (P h), low irradiance (P l), x axis T_d in μs.

[8] Roger A. Messenger, J. Ventre, *Photovoltaic Systems Engineering*, third edition, CRC Press, 2010.

[9] Hussein KH, Zhao G. "Maximum photovoltaic power tracking: an algorithm for rapidly changing atmospheric conditions", *in proc. IEE Proceedings of Generation, Transmission, Distribution '95*, pp. 59-64, 1995.

[10] T. Esram, P.L. Chapman, "Comparison of Photovoltaic Array Maximum Power Point Tracking Techniques", *IEEE Trans. on Energy Conversion*, vol. 22, no. 2, pp. 439-449, June 2007.

The 2014 International Power Electronics Conference

PCS with Scanning-type MPPT Control for Industrial Grid-Connected PV Power Generation System

Kazutaka Itako

Department of Electronic & Electrical Engineering
Kanagawa Institute of Technology
1030 Shimo-Ogino, Atsugi, Kanagawa 243-0292, Japan
e-mail:itako@ele.kanagawa-it.ac.jp

Abstract— **In a standalone photovoltaic (PV) power generation system, the author previously proposed a new type of MPPT (Maximum Power Point Tracking) control method in which the I–V characteristics are scanned with detection interval control at specified intervals and the maximum power point is monitored. The author obtained satisfactory results using this new MPPT control method. This paper investigates applying the new MPPT control method to power conditioning system (PCS) in an industrial grid-connected type PV power generation system. The experimental results clearly demonstrate that the developed PCS offers outstanding effectiveness in partially shaded environments.**

Keywords— PCS, MPPT, PV generation system, DC-DC converter

I. INTRODUCTION

In recent years, renewable energy has been attracting interest for reducing greenhouse gas emissions. Photovoltaic power generation has attracted particular attention due to sunlight being a nearly unlimited energy source and the ease of installing photovoltaic cells. The maximum power point of photovoltaic cells fluctuates with weather conditions since they affect the solar radiation intensity and the panel temperature. It is thus necessary to use maximum power point tracking (MPPT) control to track the maximum power point and to efficiently extract power from photovoltaic cells.

The author has proposed an I–V characteristics scanning-type MPPT control method (hereafter, simply referred to as the scanning method) as a new type of MPPT control for a stand-alone photovoltaic power generation system. This method scans the I–V characteristics at regular intervals and detects the maximum power point. In a stand-alone photovoltaic power generation system, the scanning method has been clearly shown to significantly increase the power generated when the panel is partially shaded, during mixed use, or similar conditions, relative to that generated when the conventional MPPT control method of hill climbing (the P & O method) is used. However, a new system for purchasing photovoltaic power was introduced in Japan in July 2012. Because this system permits the purchase of all photovoltaic power for power generation, industrial photovoltaic power generating facilities are expected to rapidly increase.

Accordingly, this paper describes the development of a power conditioning system (PCS) that applies this scanning method to industrial photovoltaic power generation systems and compares the power generated when the panel is partially shaded to that for the P & O method.

II. OVERALL CONFIGURATION OF THE SYSTEM

A. System configuration

Fig. 1 shows the proposed industrial photovoltaic power generation system. With this control method, MPPT control (the scanning method) is implemented using a DC–DC converter to control the input voltage V_D and the power factor P.F. of a three-phase inverter and to export power to the grid. When applying this control method to a photovoltaic power generation system, it is necessary to confirm the effect of the scanning operation (which is a characteristic of the scanning method) on the inverter unit.

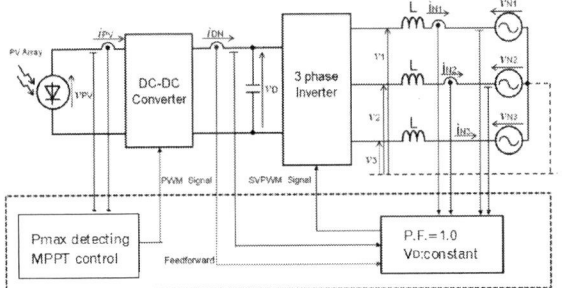

Fig. 1. Industrial grid-connected photovoltaic power generation system.

B. DC–DC converter unit (scanning method)

In the DC–DC converter unit, MPPT control of the photovoltaic cell is performed using the scanning method for which the I–V characteristics of the photovoltaic cell are instantaneously scanned at regular intervals to detect the maximum power point. The MPPT control method causes the photovoltaic cell to operate at the optimal operating voltage V_{OP}.

Fig. 2 shows a schematic diagram of the operation of the

978-1-4799-2706-7/14 $31.00 © 2014 IEEE

The 2014 International Power Electronics Conference

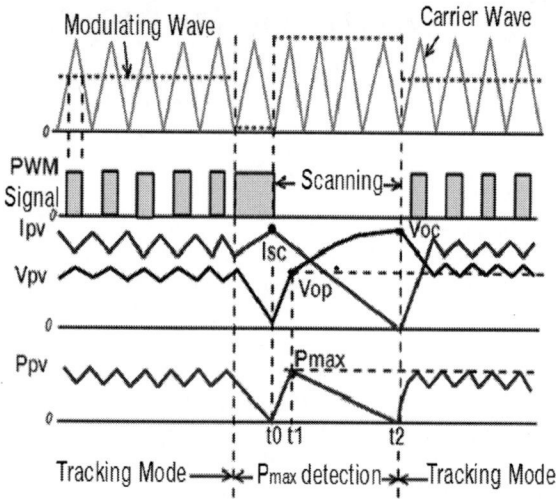

Fig. 2. Schematic diagram of operation of scanning method.

scanning method. During the P_{max} detection interval, the photovoltaic cell is in the open state; it subsequently changes to the short-circuit state. Using the change in the photovoltaic cell voltage and the current during the interval t0 to t2, the point of maximum power P_{max} (t = t1) is detected and the optimal operating voltage V_{OP} at that time is determined. Control is implemented during the tracking operation interval so that the photovoltaic cell voltage V_{PV} will become the optimal operating voltage V_{OP} that is determined during the detection interval. The detection interval and the tracking operation interval are repeated at regular intervals to realize MPPT control.

.

C. Three-phase inverter

In this circuit, a typical space vector PWM method is used to switch the inverter unit. The inverter unit controls the input voltage V_D and the power factor PF. Fig. 3 shows a block diagram of the inverter control. With this control, the input current i_{DN} is fed forward and incorporated in the control to improve the response to fluctuations in the solar radiation intensity. In Fig. 3, the three-phase/two-phase conversion of phase voltages V_{N1}, V_{N2}, and V_{N3} and line currents i_{N1}, i_{N2}, and i_{N3} is performed according to the following equation.

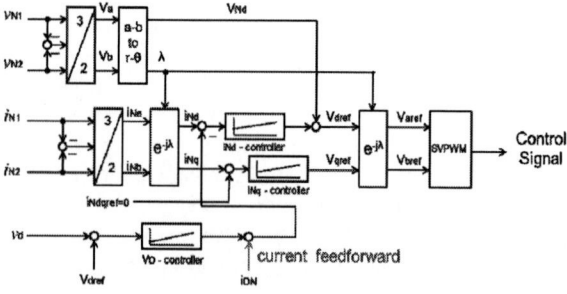

Fig. 3 Block diagram of inverter control

$$\begin{pmatrix} v_a \\ v_b \end{pmatrix} = \frac{2}{3}\begin{pmatrix} 1 & -\dfrac{1}{2} & -\dfrac{1}{2} \\ 0 & \dfrac{\sqrt{3}}{2} & -\dfrac{\sqrt{3}}{2} \end{pmatrix}\begin{pmatrix} v_{N1} \\ v_{N2} \\ v_{N3} \end{pmatrix} \tag{1}$$

$$\begin{pmatrix} i_{Na} \\ i_{Nb} \end{pmatrix} = \frac{2}{3}\begin{pmatrix} 1 & -\dfrac{1}{2} & -\dfrac{1}{2} \\ 0 & \dfrac{\sqrt{3}}{2} & -\dfrac{\sqrt{3}}{2} \end{pmatrix}\begin{pmatrix} i_{N1} \\ i_{N2} \\ i_{N3} \end{pmatrix} \tag{2}$$

Additionally, Va and Vb obtained using Eq. (1) are converted to polar coordinates using Eq. (3).

$$v_{Nd} = \sqrt{v_a^2 + v_b^2}$$

$$\lambda = \tan^{-1}\frac{v_b}{v_a} \tag{3}$$

From the results of Eqs. (2) and (3), the effective current component i_{Nd} and the reactive current component i_{Nq} can be calculated using the following equation.

$$\begin{pmatrix} i_{Nd} \\ i_{Nq} \end{pmatrix} = \begin{pmatrix} \cos\lambda & \sin\lambda \\ \sin\lambda & \cos\lambda \end{pmatrix}\begin{pmatrix} i_{Na} \\ i_{Nb} \end{pmatrix} \tag{4}$$

Fig. 4 shows a space vector diagram at this time. From Fig. 4, if the power factor setting value of the inverter is denoted by P.F.$_{ref}$, the value of the reactive current component i_{Nq} can be expressed by the following equation.

$$i_{Nq} = i_{Nd}\tan\left(\cos^{-1}\left(P.F._{ref}\right)\right) \tag{5}$$

With this control method, the provision of an intermediate voltage setting value V_{Dref} and the inverter power factor setting value P.F.$_{ref}$ enables the input voltage V_D and the reactive power to be freely controlled.

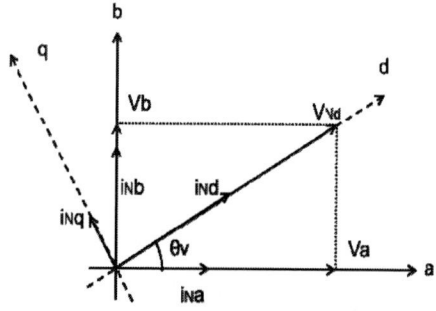

Fig. 4 Space vector diagram of inverter

III. EXPERIMENTAL RESULTS

A. Experimental conditions

Fig. 5 shows the circuit used in this experiment. Four series-connected GT133S (monocrystalline silicon: open voltage V_{OC} = 20.5 V; short-circuit current I_{SC} = 3.35 A; nominal

978-1-4799-2706-7/14 $31.00 © 2014 IEEE

maximum output operating voltage V_{OP} = 16.4 V; nominal maximum output operating current I_{OP} = 3.05 A; nominal maximum output power: 50 W for solar radiation intensity G = 1000 W/m² at 25℃) photovoltaic cells (200 W) were used. The following settings were used: grid line-to-line rms voltage V_N = 200 V, frequency f = 50 Hz, input voltage setting V_{Dref}= 400 V, power factor setting P.F.$_{ref}$ = 1.0, detection period of scanning method Td = 1.0 s, coils L_0 = 7.5 mH and L = 280 mH, and switching frequency fc = 10 kHz.

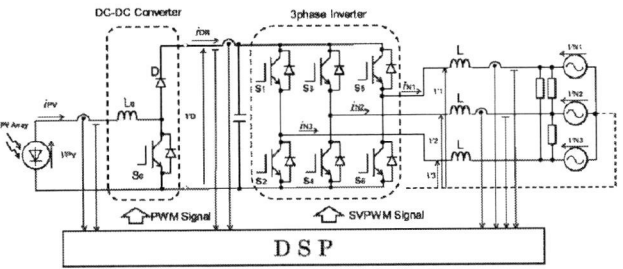

Fig. 5 Experimental circuit diagram

B. Confirmation of operation with simulated power supply

(1) Steady-state characteristics

Using a photovoltaic cell simulated power supply device, four photovoltaic cell panels were connected in series, as shown in Fig. 6. Operation when two of the panels were partially shaded was then examined. The partial shading was set such that the solar radiation intensity on a photovoltaic panel would be 1/4 of its usual value; Fig. 7 shows the P–V characteristics at this time.

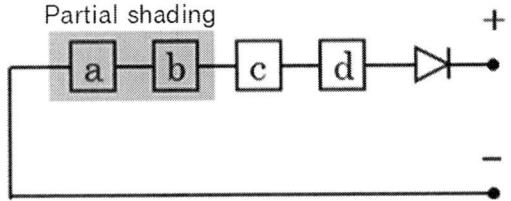

Fig. 6 Experimental conditions of photovoltaic panel

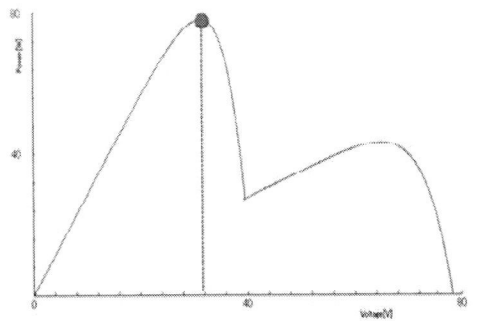

Fig. 7 P–V characteristics during partial shading

Fig.8(a) shows the operating waveform when using the scanning method. It shows that there are no fluctuations in the input voltage V_D or disturbances in the line current i_{N1} during the scanning operation. Moreover, the line current i_{N1} and the line voltage V_{N1} can be confirmed to be stable and in the same phase as when set. Figs. 8(b) and (c) show the three phases of grid phase voltage and the phase current; they show that they are highly stable.

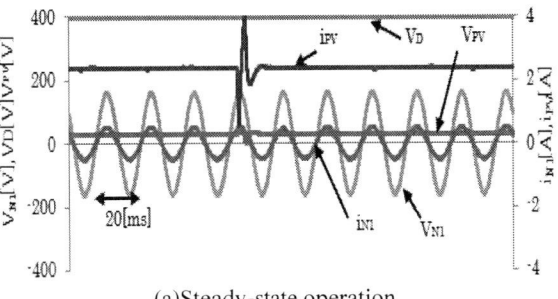

(a)Steady-state operation

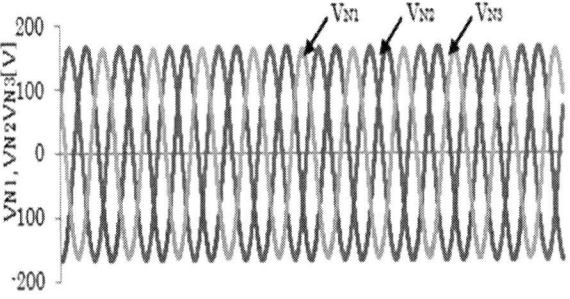

(b)Line voltages V_{N1}, V_{N2}, and V_{N3}

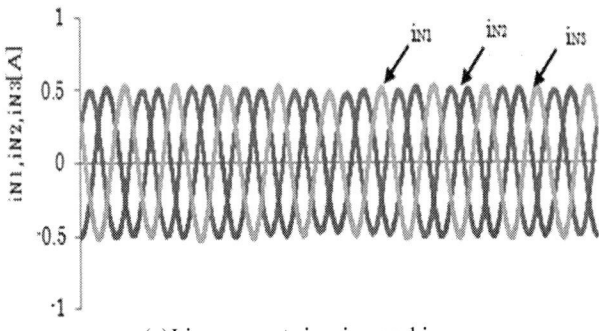

(c)Line currents i_{N1}, i_{N2}, and i_{N3}

Fig. 8 Operating waveform

Fig. 9 shows the response waveforms at startup for the scanning and P & O methods. For the scanning method, Fig. 9(a) shows that the optimal operating voltage is detected and that steady-state operation is reached instantaneously. However, Fig. 9(b) shows that it takes some time to reach the optimal operating voltage for the P & O method. In addition, the peak point on the high-power low-voltage side is tracked in the scanning method, whereas the peak point on the low-power high-voltage side is tracked in the P & O method. The

978-1-4799-2706-7/14 $31.00 © 2014 IEEE

The 2014 International Power Electronics Conference

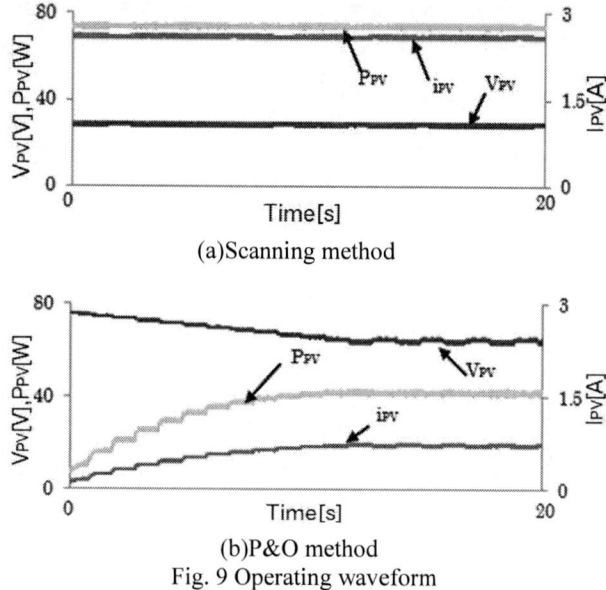

(a)Scanning method

(b)P&O method

Fig. 9 Operating waveform

powers acquired at these times were 76.8 and 44.2 W for the scanning and P & O methods, respectively.

(2) Transient response

Next, Fig. 10 shows the step response waveforms when the solar radiation intensity at two of the four serially connected panels is varied every 20 s from 1000 W/m² → 250 W/m² → 1000 W/m². Fig. 10(a) shows the photovoltaic cell current and Fig. 10(b) shows the instantaneous inverter output current waveform when the solar radiation intensity is varied. Even if the solar radiation intensity changes suddenly when partial shading is applied, the inverter output current maintains a sinusoidal profile and it can be confirmed to have regained a steady state after about one period.

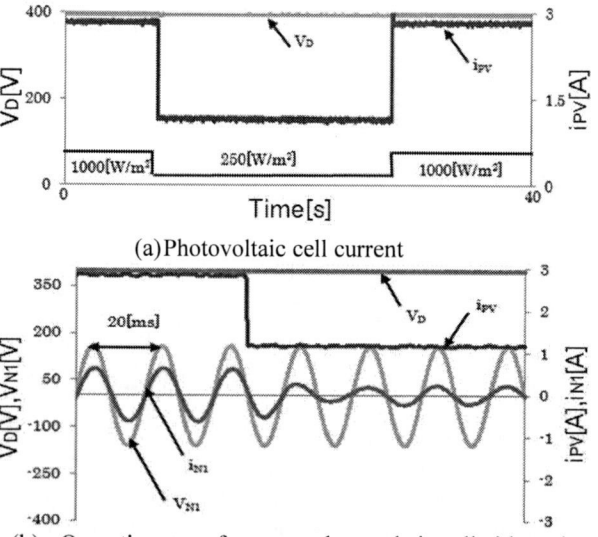

(a)Photovoltaic cell current

(b) Operating waveforms on photovoltaic cell side and on grid side

Fig. 10 Step response

C. Experimental results for photovoltaic cells

The following experiment was conducted using an actual photovoltaic cell. Fig. 11 shows the operating waveform when using the scanning method. Even during scanning operation, the input voltage V_D is stable and there were no disturbances in the phase current i_{N1}. In addition, the power factor P.F. was 0.99 at this time.

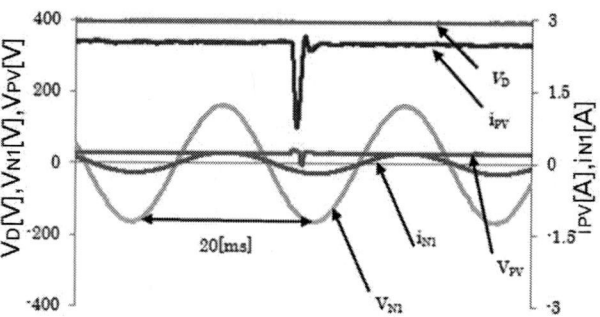

Fig. 11 Steady-state operation

Next, Fig. 12 shows the operating waveform when partial shading (3/4) is applied to two of the four serially connected photovoltaic panels. With the newly developed PCS, Fig. 12(a) shows that the maximum power point on the low-voltage side is tracked when partial shading is applied. However, with the conventional PCS, Fig. 12(b) shows that the low peak of power is tracked on the high-voltage side, even when partial shading is applied. The powers generated during partial shading were 86.5 and 59.4 W with the newly developed PCS and the conventional PCS, respectively.

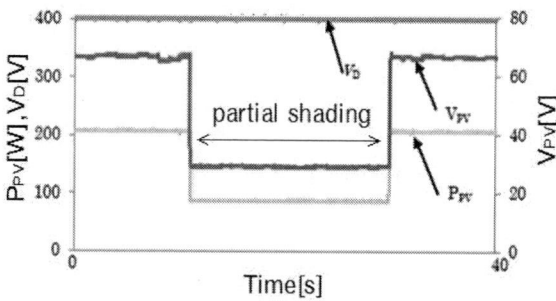

Fig. 12 Operation when partial shading is applied

D. Power factor control

Because of the problem of fluctuating grid voltage, reactive power control must be controlled in an industrial grid-connected system. Accordingly, using Eq.(5), the response was examined when the power factor setting P.F.ref was varied stepwise from 1.0 to 0.8. Fig. 13 shows the operating waveform. The operation reaches steady state about two periods after providing a command value; at that time, the power factor PF is 0.80 and reactive power control of the inverter operates normally.

978-1-4799-2706-7/14 $31.00 © 2014 IEEE

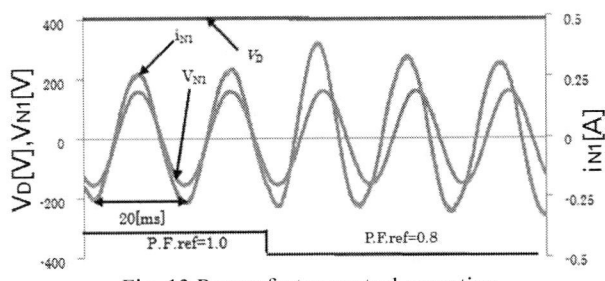

Fig. 13 Power factor control operation

IV. CONCLUSIONS

In this study, the author developed a prototype of the proposed industrial PCS that employs the scanning method. We were able to operate this prototype without any problems during scanning and when the solar radiation intensity fluctuated rapidly. Moreover, this newly developed PCS enables more power to be generated during partial shading relative to that generated using a conventional PCS employing the P & O method.

The above results demonstrate the effectiveness of the newly developed industrial PCS employing the scanning method. In the future, we intend to conduct experiments on a real scale and to conduct studies for practical applications.

REFERENCES

[1] K.Itako and T.Mori: "A New Current Sensorless MPPT Control Method for PV Generation Systems"Proceedings of 11th European Conference on Power Electronics and Applications 2005.(CD-ROM)

[2] K.Itako and T.Mori: "A Current Sensorless MPPT Control Method for a Stand-Alone-Type PV Generation System" Electrical Engineering in Japan, WILEY(USA), Vol.157,No.2,pp. 65〜71, 2006.

[3] K.Itako and T.Mori : "A Single Sensor Type MPPT Control Method for PV Generation Systems"Proceedings of 12th EuropeanConference on Power Electronics and Applications 2007. (CD-ROM)

[4] K.Itako,S.Daidouji and T.Mori: "Characteristics in Partial Shadow of MPPT Control With I-V Characteristics Scanning" Proceedings of Renewable Energy 2010. (CD-ROM)

[5] K.Itako: "A Detecting Interval Control in MPPT Control with I-V Characteristics Scanning for a PV Generation System"Proceedings of The International Conference on Electrical Engineering 2012.(CD-ROM)

[6] N.Femia, G.Petrone, G.Spagnuolo and M.Vitelli,"Optimization of Perturb and Observe Maximum Power Point Tracking Method"IEEE Trans. Power Electronics, vol.20, No.4, pp.963-973, July 2005

[7] K.Itako K.itako and T.Agari :High Perfomance Three-Phase SVPWM Converter System with Feedforward Control:proceedings of The International Conference on Electrical Engineering 2011.(CD-ROM)

The 2014 International Power Electronics Conference

Feasible Method of Calculating Leakage Reactance of 9-Winding Transformer for High-Voltage Inverter System

Hisao Fukumoto, Tatsuya Furukawa, Hideaki Itoh
Graduate School of Science and Engineering
Saga University
Saga, JAPAN

Masashi Ohchi
Faculty of Engineering
Chiba Institute of Technology
Chiba, JAPAN

Abstract—**Calculating leakage reactance of transformers is a very important factor in designing transformers. In the traditional two-winding transformer, the calculation methods of leakage reactance have been presented using the inductance deduced by the geometry mean length between two windings. However, it is difficult to calculate the leakage reactance in multi-winding transformers, because its structure is more complex. Finite element method (FEM) is useful to obtain the characterization of the electromagnetic behavior of the magnetic components. FEM is an accurate method but its computational cost is very high. In this paper, the leakage reactance of 9-winding transformer was calculated by using an ordinary three dimensional FEM (3D FEM) and hybrid FEM which is a quasi-three dimensional analysis method. The comparison results show that the hybrid FEM is able to calculate the leakage reactance with a low computational cost.**

Keywords—*Multi-Winding, transformer, leakage inductance, finite element method.*

I. INTRODUCTION

Recently, the power conditioner systems have been widely used in power systems. A multi-winding type transformer has been used for the harmonic suppression of high-voltage inverters. Estimating precise leakage reactance of transformer is a very important factor in designing such transformer and power conditioners. In the traditional two-winding transformer design, the calculating method of leakage reactance has been presented using the inductance deduced by the geometry mean length between two windings[1], [2]. However, it is difficult to calculate the leakage reactance in multi-winding transformers, because its structure is more complex.

An equivalent circuit model matched terminal-leakage inductance measurements for three-winding transformers has been presented [3]. The generalization model that extended a model of [3] to multi-winding transformers has been presented as unifying terminal models with duality derived models [4].

The short-circuit condition of multi-winding transformers was discussed in [5]. Under the short-circuit condition, it is necessary to consider the distribution of leakage flux field. These investigations were concerning concentric winding transformers especially. Few papers

have reported the analysis of leakage inductance of no-concentric winding transformers.

The authors have been studying leakage reactance of power transformers using FEM [6]. In the FEM analysis of magnetic leakage flux, two-dimensional analyses using Cartesian coordinate system can not consider the three dimensional distribution of magnetic leakage flux. In the axis symmetry three dimensional analysis, it is able to calculate the flux density around the core, however, the flux through the yoke of the transformer is not considered. Therefore, the results are not precise. In this paper, the leakage reactance of a 9-winding transformer was calculated by using an ordinary three dimensional FEM (3D FEM) and the hybrid FEM[7] which is a quasi-three dimensional analysis method devised by Emeritus Professor K.Nakata et al., Okayama University. The leakage reactance of the transformer is calculated by above the two methods and compared with the experimental values.

II. METHOD

A. 9-winding transformer

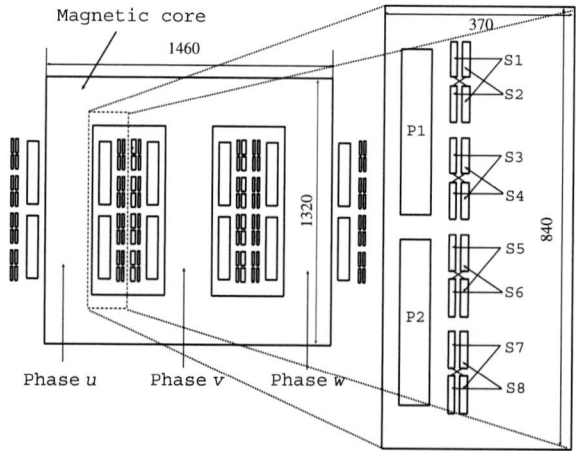

Fig. 1. Structure of windings of 9-winding transformer.

In this study, the target transformer was a three-limb and three-phase transformer. This multi-winding transformer is used for the harmonic suppression of high-voltage inverters. Fig. 1 shows the structure of the 9-

978-1-4799-2706-7/14 $31.00 © 2014 IEEE

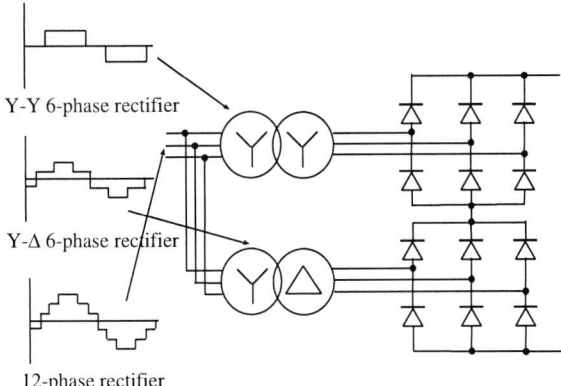

Fig. 2. Example of a 12-phase rectifier using this transformer.

winding transformer. The high-voltage winding sections (P1 and P2) are in parallel in this transformer. The second winding consists of eight-windings (from S1 to S8). The primary winding is connected in Y configuration. The secondary winding of S1, S3, S5 and S7 are connected in Delta configuration. And other secondary windings of S2, S4, S6 and S8 are connected in Y configuration. Using this connected transformer, it is able to reduce the effect of harmonics produced by the converter to the power system. Fig. 2 shows the 12 pulse converter which is connected in series of two three-phase bridge converters. This converter is able to reduce the harmonic noise for shifting phase of secondary voltage by 30 degrees.

B. FEM Analysis

In order to calculate leakage reactance in the 9-winding transformer, the authors assumed the followings.

- Ignore the eddy current. That is, the core is a laminated structure.

- Ignore the displacement current. That is, quasi-stationary analysis was conducted.

- Ignore the magnetic saturation of the core. The permeability of the core is constant.

The magnetostatic equation using the magnetic vector potential is

$$\text{rot}(\nu \text{rot} \boldsymbol{A}) = \boldsymbol{J} \quad (1)$$

where $\boldsymbol{A}$ is the vector potential, $\boldsymbol{J}$ is the current density and ν is the reluctivity as the inverse number of the permeability (μ).

C. Hybrid FEM

The hybrid FEM has been presented to perform analyses using the combined two-dimensional analysis and axis symmetry three dimensional analysis. The hybrid FEM adopts mixed two-dimensional Cartesian (xy) and axis symmetric ($r\zeta$) coordinate systems shown in Fig. 3, where Φ_{xy} and $\Phi_{r\zeta}$ are the fluxes in each coordinate system, respectively and they are equal to each other

at the interfaces of both the core ends of the coil, the axis symmetric region only for such as a coil is taken into account as a three-dimensional area so that the flux linkage across a coil could be successfully simulated in the region by good precision [7]. This method does not require high computational cost, e.g. CPU power, memory size and high priced FEM software.

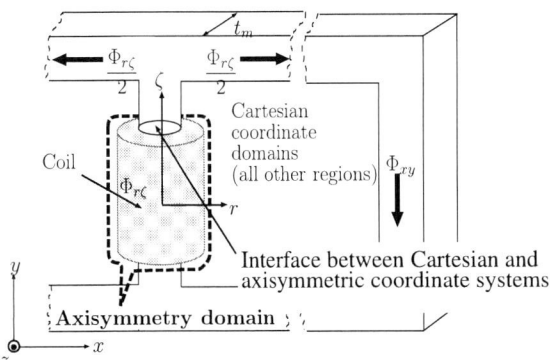

Fig. 3. Model of a transformer in the hybrid FEM analysis.

D. 3D FEM

The geometry of the 9-winding transformer to calculate leakage reactance between the primary windings (P1, P2) and secondary winding of S1 is shown in Fig. 4. This model is made and analyzed by a CAE tool. The grid of tetrahedral elements has been used for the 3D FEM.

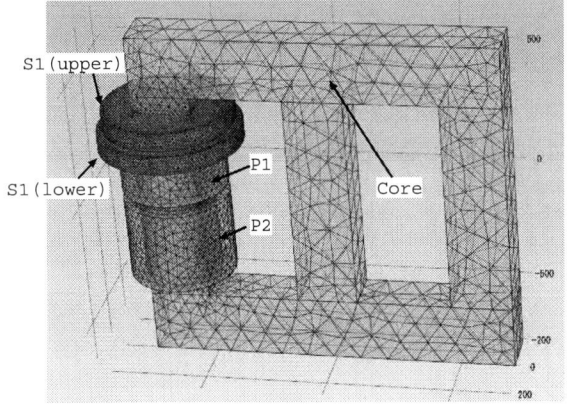

Fig. 4. 3D-model of 9-winding transformer (P-S1 analysis).

E. Leakage reactance

The leakage reactance is measured by a short-circuit test in which one winding is energized with a second winding that is short-circuited while keeping other windings open. The example of the short-circuit is shown in Fig. 5. In this case, the ampere-turn is balanced in the primary winding and second winding. It adopts the same amplitude and inversed direction of ampere-turns between the primary winding and secondary winding for the FEM analysis. That is

$$n_1 i_1 = -n_2 i_2 \quad (2)$$

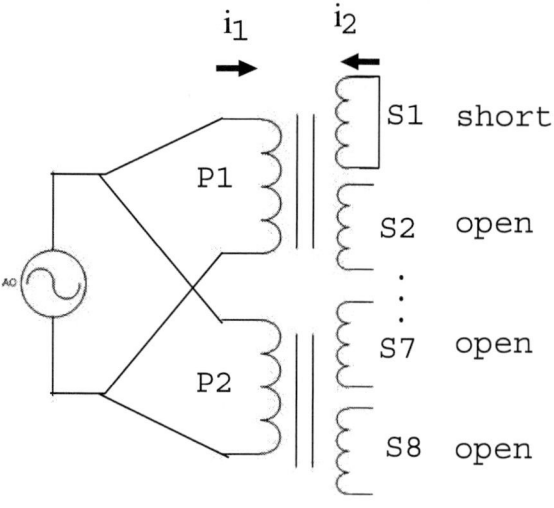

Fig. 5. Circuit of short-circuit test for measuring leakage inductance.

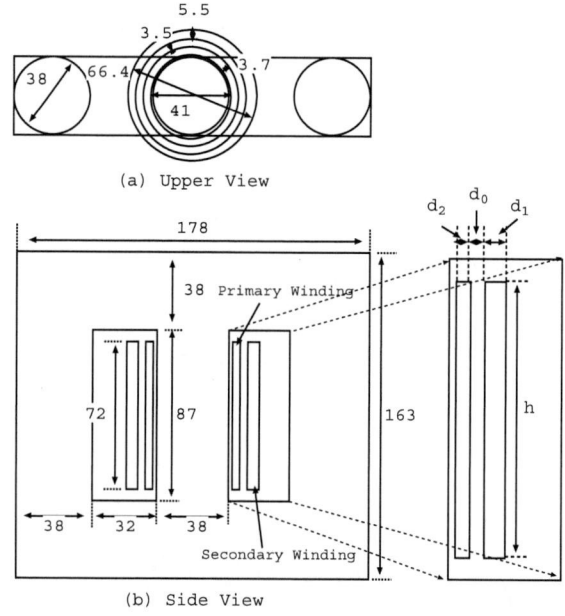

(a) Upper View

(b) Side View

Fig. 6. Schematics of the two-winding transformer.

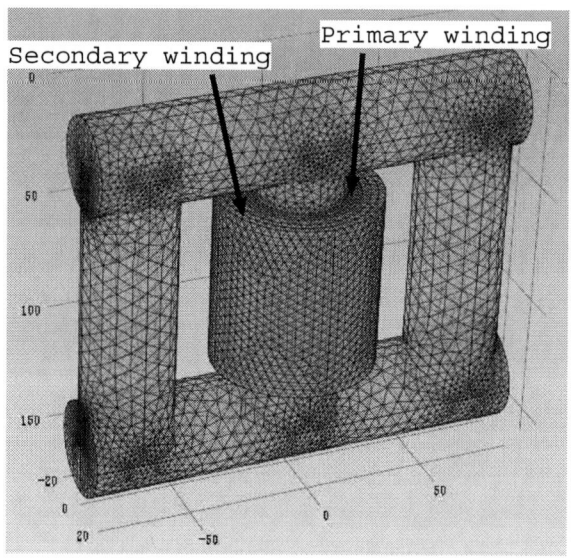

Fig. 7. Mesh pattern of the two-winding transformer.

where n_1 and n_2 are turn numbers of each winding, i_l and i_2 are the primary current and the secondary current. In this case, the all magnetic flux of the transformer is magnetic leakage flux. The total energy stored in the magnetic field of the model is calculated as

$$W = \int_V \frac{\boldsymbol{B} \cdot \boldsymbol{H}}{2} dV = \int_V \frac{|\boldsymbol{B}|^2}{2\mu} dV \qquad (3)$$

where $\boldsymbol{B}$ is flux density, $\boldsymbol{H}$ is the magnetic field strength and μ is permeability [8].

Using this W, the leakage inductance to the secondary side yields is obtained as

$$L_l = \left(\frac{n_2}{n_1}\right)^2 L_{l1} + L_{l2} = \frac{2W}{i_2^2} \qquad (4)$$

where n_1 and n_2 are turn numbers of each winding, L_{l1} and L_{l2} are the primary leakage inductance and the secondary inductance. The leakage reactance is calculated as $X_l = 2\pi f L_l$, where f is the frequency of power supply.

III. SIMULATION RESULTS

The simulation of the described FEM analysis was performed on a PC with an Intel Core-i7-2600 CPU 1.6GHz and 16GB RAM. The OS was Ubuntu 13.04 and the linux kernel version was 3.8.0-31.

A. Normal two-winding transformer

The normal two-winding transformer with specifications given in Fig. 6 [2] is simulated. In this simulation, the leakage reactance using total electromagnetic energy in the 3D-FEM analysis is evaluated by known leakage reactance. In this transformer, the leakage inductance L is obtained as [2]

$$L = \mu_0 K \frac{T^2 U_m}{h}\left(d_0 + \frac{d_1 + d_2}{3}\right) \qquad (5)$$

where μ_0 is absolute permeability, K is correction factor for consideration of the flux fringing in the top and bottom of the windings, T is number of winding turns, U_m is mean length per turn for the whole arrangement of the coils, h is the height of the windings, d_0 is the width of the space between primary and secondary windings, and d_1 and d_2 are the width of each primary winding and secondary winding. The turns number of the primary winding and secondary winding are $N_1 = 981$, and $N_2 = 178$, respectively.

Fig. 7 shows the mesh pattern of the model in 3D-FEM. The relative permeability of the core is 4,000. The obtained result is shown in Fig. 8, which presents the distribution of magnetic flux density in the transformer. Using the equation (5) and 3D FEM analysis, leakage %reactance has been calculated and the results are compared to each other in Table I. The leakage %reactances of 3D-FEM and what is calculated by equation (5) agree very well, in spite of the simulation is not considered the eddy current and the magnetic saturation of the core.

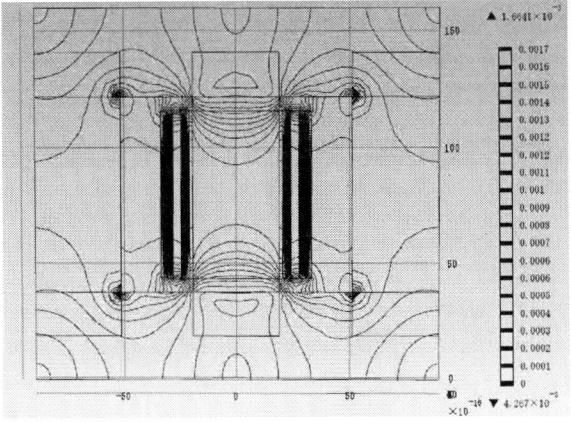

Fig. 8.   Distribution of flux density in two-winding transformer.

TABLE I.   RESULTS COMPARISON.

	Equation (5)	3D-FEM	% Difference
%Leakage reactance	6.83 [%]	6.63 [%]	-2.93 [%]

B. 9-winding transformer

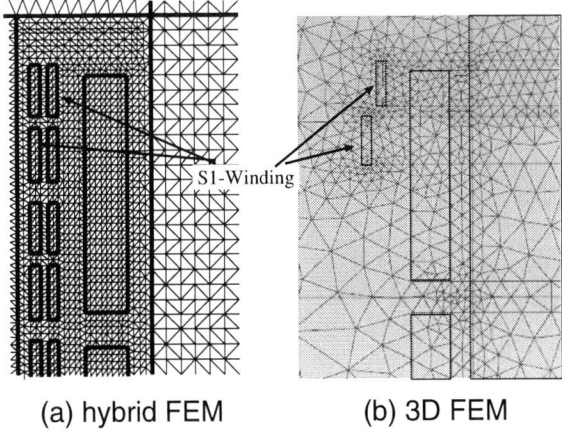

(a) hybrid FEM   (b) 3D FEM

Fig. 9.   Mesh patterns in the hybrid FEM and 3D FEM.

In this section, the 9-winding transformer is simulated.

In Fig. 9, the meshed patterns show the hybrid FEM and 3D FEM around the second winding of S1. The mesh pattern in the hybrid FEM was more fine. The important area of the analysis is meshed in detail, because the two-dimensional mesh is adopted in the hybrid FEM. On the other hand, the mesh size is automated according to the structure in the 3D FEM. The mesh size in 3D FEM is mostly automated but can be chosen manually to some

extent, whereas the mesh size in hybrid FEM can be chosen more arbitrarily.

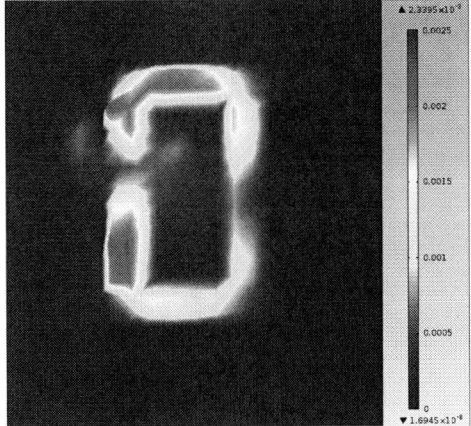

Fig. 10.   Distribution of flux density in hybrid FEM.

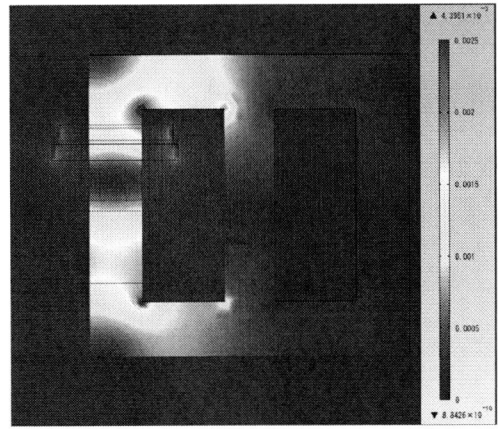

Fig. 11.   Distribution of magnetic flux density in 3D FEM.

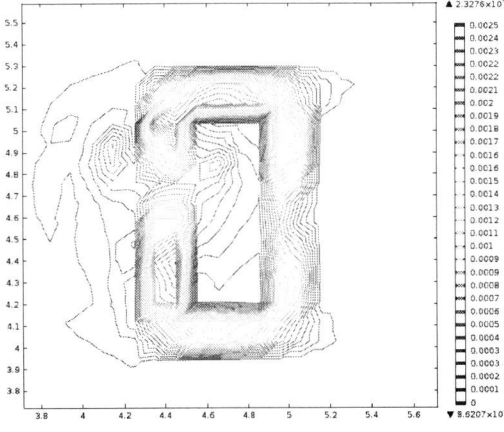

Fig. 12.   Contour graph of magnetic flux density in hybrid FEM.

The obtained results are shown in Figs. 10 and 11, which present the distribution of magnetic flux density in the transformer. In Figs. 12 and 13, the contour pattern of magnetic flux density are shown. The relative

The 2014 International Power Electronics Conference

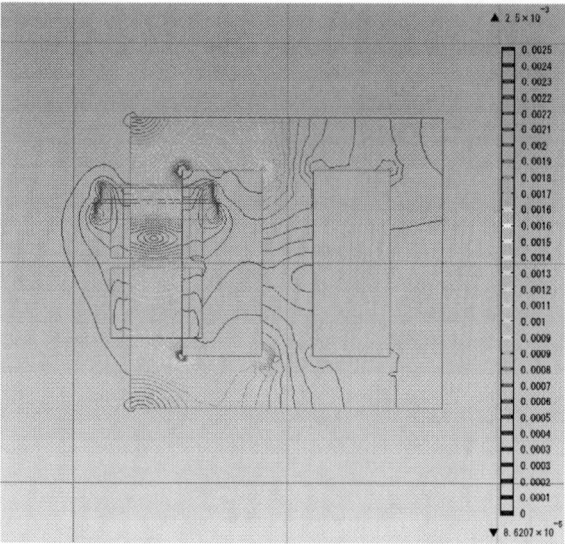

Fig. 13.   Contour graph of magnetic flux density in 3D FEM.

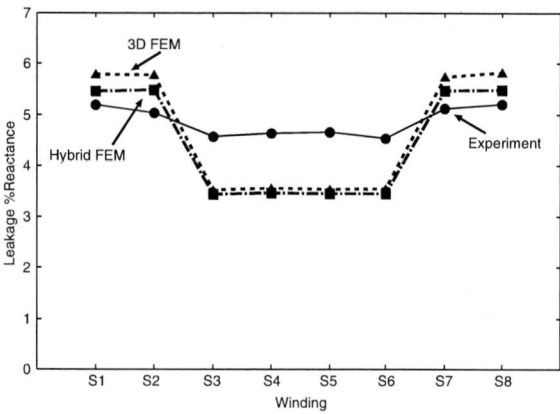

Fig. 14.   Difference of leakage %reactance in each Method.

permeability of the core is 4,000 in hybrid FEM and 3D-FEM analysis. These results are similar to the distribution and amplitude of the magnetic flux density.

Using the hybrid FEM analysis and 3D FEM analysis, leakage %reactance has been calculated and results are compared to experimental results in Table II and III. The %difference in the tables is calculated based on the experimental result. The obtained leakage %reactance values agree very well in the hybrid FEM and 3D FEM analysis. However, there is a difference between the results of each FEM analyses and the experimental result. The difference was in the range from 75% to 114% of the experimental result. It is considered that these differences are caused by ignoring the magnetic saturation of the core.

Fig. 14 shows the relation between second winding position and leakage %reactance. In this figure, the leakage reactance is larger in outer windings than in inner windings. Table IV shows the comparison of the calculation cost between the hybrid FEM and 3D FEM. It is clear that the hybrid FEM is of very low computational

TABLE II.   LEAKAGE %REACTANCE OF HYBRID FEM

	Hybrid FEM	Experiment	%Difference
S1	5.46	5.19	5.20
S2	5.48	5.04	8.73
S3	3.43	4.57	-25.0
S4	3.46	4.63	-25.3
S5	3.45	4.66	-26.0
S6	3.44	4.54	-24.2
S7	5.47	5.12	6.84
S8	5.48	5.20	5.38

TABLE III.   LEAKAGE %REACTANCE OF 3D-FEM

	3D FEM	Experiment	%Difference
S1	5.79	5.19	11.6
S2	5.77	5.04	14.5
S3	3.52	4.57	-23.0
S4	3.55	4.63	-23.3
S5	3.53	4.66	-24.2
S6	3.54	4.54	-22.0
S7	5.74	5.12	12.1
S8	5.82	5.20	11.9

TABLE IV.   COMPARISON OF CALCULATION COSTS.

	3D FEM	Hybrid FEM
number of elements	393,187	54,688
memory	4.4 [GB]	194 [MB]
CPU time	2456 [s]	7 [s]

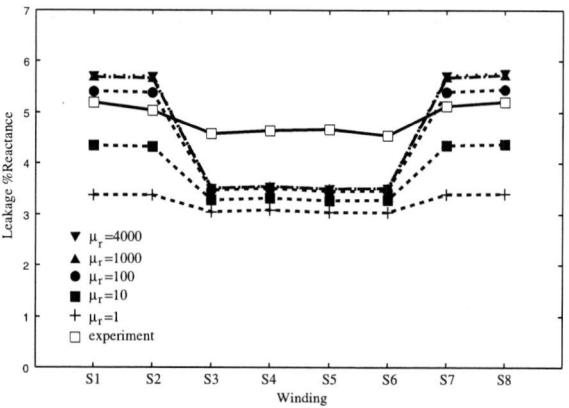

Fig. 15.   Relationship of %reactance and relative permeability.

cost, though the results showed the same accuracy.

Fig. 15 shows the relationship of the leakage %reactance and relative permeability of the core in the 3D-FEM analysis. The leakage %reactances of S1, S2, S7 and S8 winding change significantly as the relative permeability grows up. In the S3, S4, S5 and S6 winding, the leakage %reactances do not change very much. In these simulations, the relative permeability of the core is determined to be $\mu_r = 4,000$, because the results are almost the same as those of the $\mu_r = 1,000$.

IV.   CONCLUSION

In this study, the leakage reactance of a 9-winding transformer is evaluated by a hybrid FEM and 3D FEM.

978-1-4799-2706-7/14 $31.00 © 2014 IEEE

The analysis by the hybrid FEM and 3D FEM are obtained similar values of leakage reactance. The hybrid FEM was better with regard to the computational cost. In future work, the authors will investigate the difference between experimental leakage reactance and the calculated one. It is necessary to consider effects of eddy current and magnetic saturation of the core.

ACKNOWLEDGEMENT

Thanks to Mr. M. Hashiguchi, Keisoku Engineering System Co. Ltd. for useful discussion of the FEM analysis.

REFERENCES

[1] Y. Ikeda, I. Tanaka and S. Zaizen, "Calculation of leakage reactance in transformer windings using geometric mean distance", Proceedings of the United Assembly of Four Electric Society, No.527, 1968 (in Japanese),

[2] T. Takeuchi, "Design of electrical machinery", Ohmsha, 1993 (in Japanese).

[3] F. Leon and J.Martinez, "Dual three-winding transformer equivalent circuit matching leakage measurements", *IEEE Trans. on Power Delivery*, vol. 24, no. 1, pp. 160–168, 2008.

[4] C.A. Marino, F. Leon and M.L. Fernandez, "Equivalent circuit for the leakage inductance of multiwinding transformers: Unification of terminal and duality models", *IEEE Trans. on Power Delivery*, vol. 27, no. 1, pp. 353–361, 2012.

[5] G.B. Kumbhar and S.V.Kulkarni, "Analysis of short-circuit performance of split-winding transformer using coupled field-circuit approach", *IEEE Trans. on Power Delivery*, vol. 22, no. 2, pp. 936–943, 2007.

[6] M. Hata, T. Furukawa, H. Itoh, H. Fukumoto, H. Wakuya and M. Ohchi, "Methods for estimation of leakage reactance in multi-wound type transformer for harmonic suppression of high-voltage inverter", *IEEJ Technical Meeting on Magnetics Application* MAG–10–193, pp.7–12, 2010 (in Japanese).

[7] T. Nakata and Y. Kawase, "Finite element analysis of magnetic circuits composed of many axisymmetric and rectangular regions", *IEEJ Technical Meeting on Static Apparatus and Rotating Machinery*, RM–85–34, SA–85–43, pp.33–42, 1985 (in Japanese).

[8] I. Hernandez, F. Leon and P.Gomez, "Design formulas for the leakage inductance of troidal distribution transformers", *IEEE Trans. on Power Delivery*, vol. 26, no. 4, pp. 2197–2204, 2011.

The 2014 International Power Electronics Conference

High Power HVDC-DC converters for the interconnection of HVDC lines with different line topologies

André Schön
University of Bayreuth
Department of Mechatronics
95447 Bayreuth, Germany
Email: andre.schoen@uni-bayreuth.de

Mark-M. Bakran
University of Bayreuth
Department of Mechatronics
95447 Bayreuth, Germany

Abstract—The structural change in energy generation and distribution will demand that large amounts of power are transported over long onshore distances. High Voltage Direct Current (HVDC) power transmission is most suitable for that task since, compared to High Voltage Alternating Current (HVAC) power transmission, there are no stability issues and the transmission losses are much lower. However the junction points of a segmented HVDC transmission corridor still have to be realized as AC nodes, since missing components, like DC breakers and or efficient HVDC-DC converters prevent the direct interconnection of transmission segments to a radial or even meshed DC grid. With the new HVDC auto transformer topology the direct interconnection of HVDC lines with different voltage levels becomes feasible. In this paper the ability of this new HVDC-DC converter topology to connect different DC line topologies and to preserve the transmission redundancy in case of a DC pole malfunction is investigated.

Keywords—HVDC-DC Converter, HVDC Power Transmission, Modular Multilevel Converter, Multi-Terminal HVDC Grid

I. INTRODUCTION

Long distance HVAC power transmission brings an AC grid to its limits. The AC line impedance causes stability issues, the transmission losses, compared to HVDC, are significantly higher and the reactive power causes even more losses. Therefore, HVDC is the preferable choice for long distance power transmission. Within the AC grid, the segmentation in several voltage levels, e.g. the highest voltage level for long distance power transmission and lower voltage levels for power distribution is well established. Within a DC backbone, the same segmentation, e.g. a lower collection and distribution level, matched to the actual power, and a higher transmission level for bulk power transportation between the generation and the load center, can be reasonable, too. To avoid multiple AC-DC power conversion high power HVDC-DC converters are needed to allow the power transfer between the distinct voltage levels. A new topology, the HVDC auto transformer has been presented in [1]. Also its ability to suppress DC fault currents, respectively to protect the fault free DC level from a fault on the other DC level has been investigated in [2]. In this paper, the ability of the new HVDC auto transformer topology to connect

not simply different voltage levels, but also different HVDC transmission topologies is investigated. Regarding the scenario in Fig. 1, the transmission reliability of the bulk transmission line, connecting the generation and the load center, is of special interest. Therefore, the focus lies on the preservation of the transmission redundancy of a bipolar DC transmission configuration, as it would be used for HVDC bulk power transportation.

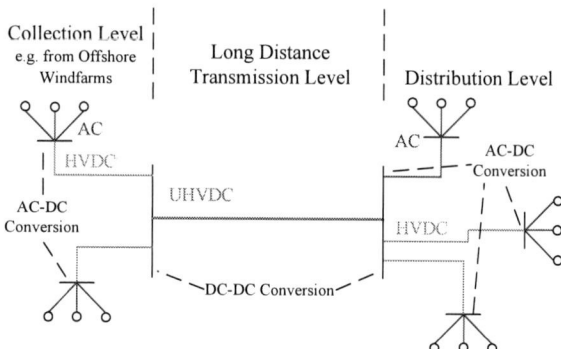

Fig. 1. A HVDC power transmission and distribution scheme with two DC voltage levels

II. BIDIRECTIONAL HVDC-DC CONVERTERS

Most of the previously presented HVDC-DC converters are based on a two stage topology (e.g. [3] [4] [5] [6]). The DC power of the first DC voltage level is inverted, the voltage level transformed by an AC transformer and again rectified into the second DC voltage level. The only advantage of this dual active bridge topology is the galvanic separation of both DC sides. However, both AC-DC converters as well as the AC transformer have to be dimensioned for the full DC power. Every AC-DC converter has to be dimensioned for its DC voltage level. In terms of efficiency, there is no difference if the DC-DC conversion is done by one single topology, e.g. a dual active bridge with a specially designed AC link or, as proposed in [7], in a "grid of independent lines" by two separate AC-DC converters linked through the AC grid.

This "grid of independent lines" not only gives an outlook into a first DC grid, it also represents the state of

978-1-4799-2706-7/14 $31.00 © 2014 IEEE

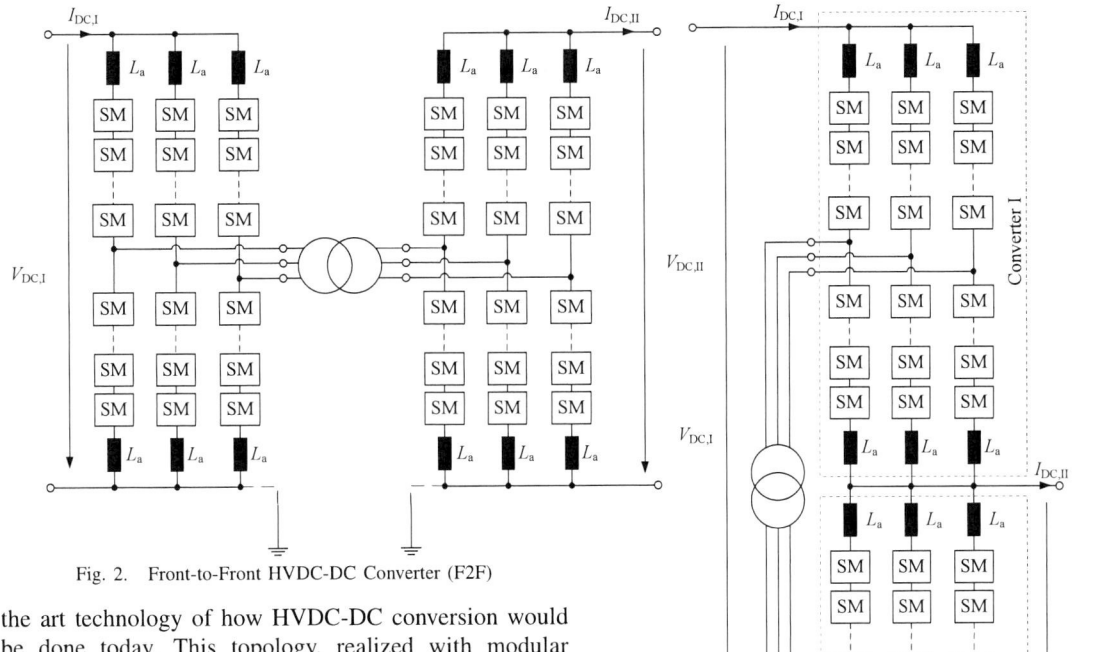

Fig. 2.   Front-to-Front HVDC-DC Converter (F2F)

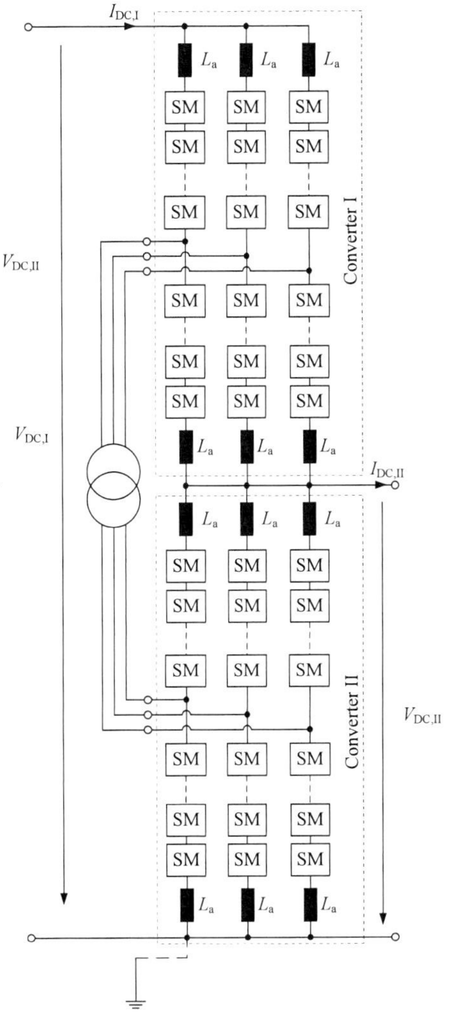

Fig. 3.   HVDC-DC auto transformer (HVDC-AT)

the art technology of how HVDC-DC conversion would be done today. This topology, realized with modular multilevel converters (MMC), is shown in Fig. 2. It consists of two converter stations connected by an AC node. Based on the terminology for the Back-to-Back topology, where two AC grids are connected through a DC link, the name "Front-to-Front" HVDC-DC converter (F2F) for the connection of two DC lines via an AC link was introduced.

In contrast to the many two stage topologies, only few single stage HVDC-DC converters without galvanic separation have been presented (e.g. [1] [8] [9]). The focus of this paper is on the HVDC auto transformer (HVDC-AT) presented in [1]. This single stage HVDC-DC converter consists of two series connected voltage source converters per DC pole. The sum of both converter voltages forms the high level DC voltage and the lower converter forms the low level DC voltage. Both converters are still linked through an AC connection. The HVDC auto transformer is also realized with MMC subconverters and pictured in Fig. 3.

The realization with MMC type subconverters allows a good comparability between the F2F topology and the new HVDC-AT and also a good scalability for different voltage transformation ratios. The power rating of a MMC is mostly determined by its DC voltage and the current rating of the used semiconductors in the MMC cells. For a better comparability of the regarded topologies it is assumed, that there are identical MMC cells used in all pictured subconverters. The number of cells in each subconverter is only determined by its DC voltage. The power rating of each MMC is therefore given by its DC voltage and the maximum DC current of the used cells:

$$P_{\mathrm{MMC}} = V_{\mathrm{DC}} \cdot I_{\mathrm{DC,max}} \overset{\mathrm{Def.}}{=} P_{\mathrm{DC,max}} \qquad (1)$$

III. COMPARISON OF THE F2F AND THE NEW HVDC AUTO TRANSFORMER TOPOLOGY

With the assumption, that identical cells are used in both converter stations of the F2F topology, its transferable DC power is only given by the voltage rating of the

low level subconverter. Hence, it declines with a rising transformation ratio proportional to $1/n$.

$$P_{\mathrm{DC,F2F}} = V_{\mathrm{DC,II}} \cdot I_{\mathrm{DC,max}} \qquad (2)$$

$$= \frac{V_{\mathrm{DC,I}}}{n} I_{\mathrm{DC,max}} = \frac{P_{\mathrm{DC,max}}}{n} \qquad (3)$$

$$\text{with } n = V_{\mathrm{DC,I}}/V_{\mathrm{DC,II}} \quad n \in \mathbb{R}, n > 1 \qquad (4)$$

The number of submodules for that topology is given by the sum of cells of both substations:

$$m_{\mathrm{F2F}} = m_{\mathrm{F2F,CI}} + m_{\mathrm{F2F,CII}} \qquad (5)$$

$$= k_{\mathrm{phaselegs}} \cdot \left(\frac{2V_{\mathrm{DC,I}}}{V_{\mathrm{m}}} + \frac{2V_{\mathrm{DC,II}}}{V_{\mathrm{m}}} \right) \qquad (6)$$

$$= k_{\mathrm{phaselegs}} \cdot \frac{2V_{\mathrm{DC,I}}}{V_{\mathrm{m}}} \cdot \frac{n+1}{n} \qquad (7)$$

The new HVDC auto transformer consist of two series connected MMCs and both subconverters together have to be rated for the high level DC voltage. The number of submodules for this topology is therefore given by

$$m_{\mathrm{AT}} = k_{\mathrm{phaselegs}} \cdot \frac{2V_{\mathrm{DC,I}}}{V_{\mathrm{m}}} \qquad (8)$$

and is constant for all transformation ratios.

The DC power, due to the DC voltage at the terminals of the upper subconverter (Converter I in Fig. 3) and the high level DC current is inverted and transferred to the lower subconverter, where it is rectified and fed into the low level DC side[1]. Only a fraction of the total DC power has to be transferred through the AC link. The power rating of the AC link, and therewith the power rating of the transformer is given with Eq.(10).

$$P_{\mathrm{Tr}} = P_{\mathrm{DC,CI}} = I_{\mathrm{DC,I}} \cdot (V_{\mathrm{DC,I}} - V_{\mathrm{DC,II}}) \qquad (9)$$

$$= P_{\mathrm{DC}} \cdot \frac{n-1}{n} \quad \text{with } n = \frac{V_{\mathrm{DC,I}}}{V_{\mathrm{DC,II}}} \qquad (10)$$

Hence, both subconverter only see a fraction of the total DC power. The current through the upper subconverter is equal to the high level DC current. The current through the lower subconverter equals the difference between the high and the low level DC current.

$$I_{\mathrm{DC,CI}} = I_{\mathrm{DC,I}} \qquad (11)$$

$$I_{\mathrm{DC,CII}} = I_{\mathrm{DC,I}} - I_{\mathrm{DC,II}} \qquad (12)$$

$$= I_{\mathrm{DC,I}} \cdot (1 - n) \qquad (13)$$

As long as the low level DC current is less then twice the high level DC current (meaning $n \leq 2$), the upper subconverter sees the higher current and the transferable DC power is limited by that converter. For a transformation ratio of $n > 2$ the lower subconverter sees the higher current and therewith limits the transferable DC power. With the maximum allowed DC current through a subconverter $I_{\mathrm{DC,max}}$ the currents and the DC power of the HVDC-AT topology is defined as follows:

$$|I_{\mathrm{DC,CI}}| = \begin{cases} I_{\mathrm{DC,max}} & \text{for } n \leq 2 \\ \frac{I_{\mathrm{DC,max}}}{n-1} & \text{else} \end{cases} \qquad (14)$$

$$|I_{\mathrm{DC,CII}}| = \begin{cases} I_{\mathrm{DC,max}} \cdot (n-1) & \text{for } n \leq 2 \\ I_{\mathrm{DC,max}} & \text{else} \end{cases} \qquad (15)$$

$$P_{\mathrm{DC,I}} = \begin{cases} P_{\mathrm{DC,max}} & \text{for } n \leq 2 \\ \frac{P_{\mathrm{DC,max}}}{n-1} & \text{else} \end{cases} \qquad (16)$$

Eq.(16) shows, that with the new HVDC auto transformer topology, up to a transformation ratio of $n = 2$, the full DC power of a comparable MMC AC-DC converter can actually be transferred between both DC sides. Only at higher transformation ratios the transferable DC power declines. However, the transferable DC power and therewith the utilization of the installed converter power of the new topology is always higher than that of the conventional two stage topology.

Fig. 4 shows the transferable DC power of both topologies, scaled to the maximum DC power of the high level subconverter of Fig. 2. At a transformation ratio of $n = 2$ the new HVDC auto transformer topology is able to transfer twice the DC power of the conventional two stage topology. The advantage of the new HVDC auto transformer topology becomes even more obvious in Fig. 5. Here, the transferable DC power is scaled to the installed converter power (respectively to the necessary

---

[1]Given the pictured load direction in Fig. 3. The load direction can also be reversed.

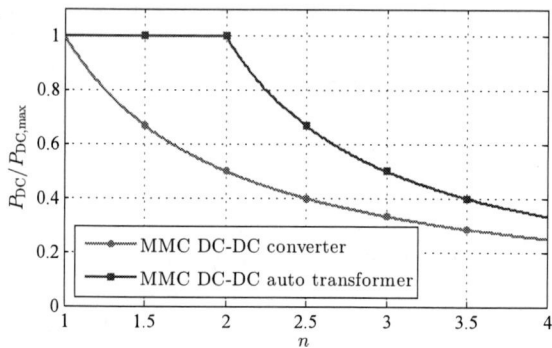

Fig. 4. Transferable DC power of both considered topologies, scaled to the maximum power of the high level subconverter of Fig. 2

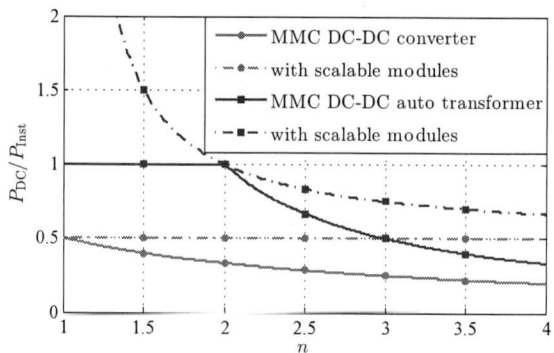

Fig. 5. Converter utilization: Transferable DC power scaled to the installed converter power

number of submodules). By using identical submodules in all subconverters, at a transformation ratio of $n = 2$, the HVDC auto transformer topology is able to transfer twice the DC power of the conventional two stage topology by using only two thirds of its submodules. If perfectly scalable submodules are available, in theory the new HVDC auto transformer topology can, at a transformation ratio of $n < 2$, transfer even more DC power, than one would have to install in conversion power.

IV. INTERCONNECTION OF DIFFERENT DC LINE TOPOLOGIES

There are three different HVDC transmission configurations, that differ in the power rating, the dimensioning of the AC transformer, the number of DC conductors and their redundancy options. The most basic variant is the asymmetrical, monopolar configuration. A single converter supplies, related to the ground potential, a single DC pole. The ground return is not necessarily a metallic conductor, but can also be a water or a true earth return path, if that is allowed. The AC side transformer has to be isolated not only for the AC, but also for the DC voltage. The second DC transmission configuration is called a symmetrical monopole. The relation to the ground potential is supplied by the AC side transformer with a high resistive connection of the star-point to ground. It no longer has to be isolated for the DC voltage. With the virtual, not loadable, ground relation, the converter supplies two DC poles with $\pm V_{\mathrm{DC}}/2$. Hence, given the same DC pole to ground voltage twice as much DC

The 2014 International Power Electronics Conference

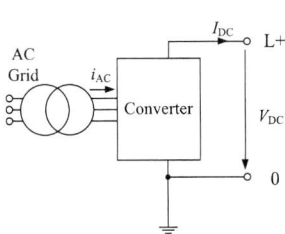 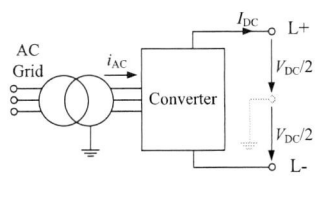 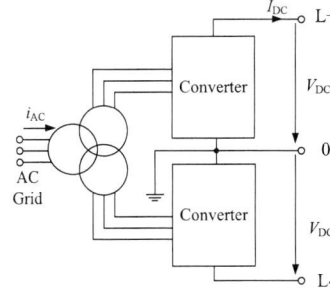

(a) asymmetrical, monopolar HVDC transmission with loadable, metallic ground return

(b) symmetrical, monopolar HVDC transmission without loadable ground return

(c) symmetrical, bipolar HVDC transmission with loadable, metallic ground return

Fig. 6. HVDC transmission configurations [10] [11]

power can be transferred. The third DC transmission configuration is the symmetrical bipole. It consists of two asymmetrical monopoles, each forming one DC pole with $\pm V_{DC}$. During normal operation, the ground return sees no load and can be omitted. All three configurations are shown in Fig. 6.

In case of a converter or a DC line fault both monopolar topologies have to be turned off, since there is no alternative current path. However the loadable ground return of the bipolar configuration can be used as an additional current path. Hence, this DC transmission configuration offers a redundancy where, during such a fault scenario, half of the original DC power can be transmitted. Regarding the scenario pictured in Fig. 1 it is most likely that this redundancy is wanted for the ultra-high voltage DC line, as this feature simply increases the transmission security of the bulk power transmission. However, existing and planned HVDC lines, especially for offshore connection are mostly monopolar configurations. Therefore, one key feature of a HVDC-DC converter for the pictured scenario is its ability to connect monopolar and bipolar configurations by keeping the transmission redundancy of the bipolar DC line.

Because of the full AC link the Front-to-Front topology inherently allows the interconnection of any DC transmission configuration. E.g. the incoming monopolar transmission line ends at a monopolar converter topology, the power is fed into the AC link and then again rectified by a bipolar converter topology. This topology also inherently allows the connection to an AC grid. But, as already shown in [1] this topology is very inefficient and expensive. It will be shown that, with a few modifications, the new HVDC auto transformer topology also has the ability to interconnect any DC transmission configuration.

Fig. 7(a) to (c) show the conventional interconnection of monopolar and bipolar HVDC lines with different voltage levels. Due to the full AC link the interconnection to an AC grid is also possible at this point. This "grid of independent lines" [7] represents the state of the art technology. Every AC link stands for a node in the DC grid.

With the new HVDC-AT topology, DC lines can be interconnected directly, resulting in the topologies of Fig. 7(e) and Fig. 7(d). The basic HVDC auto transformer topology is a asymmetrical, monopolar topology. Hence, for a bipolar transmission and a symmetrical, monopolar

configuration one HVDC-AT per DC pole is needed.

If both DC sides are symmetrical, monopolar configurations the inner subconverters of both DC poles can be summarized to one single subconverter as shown in Fig. 7(d). This configuration also leads to an upgrade option for an already existing HVDC link. With the exiting inner converter station only the outer subconverters and a new transformer would have to be added to enable an energy transfer to a new HVDC line with a higher voltage level.

In a bipolar transmission configuration, during symmetrical, fault free operation both poles can operate independently from each other and the third conductor is current free. However, due to different, polarity dependent leakage currents a small unbalance occurs even during normal operation. To enable an energy transfer between both poles and therefor equalize this unbalance, the use of three winding transformers is suggested.

The resulting additional AC link between both HVDC-ATs also allows the direct connection to an AC grid for additional symmetrical power transfer from or to both DC lines. Even if one DC side has to be disconnected, e.g. because of DC line failure or maintenance, the full power of the high level DC side or a fraction of the power of the low level DC side (limited by the power rating of the lower subconverter) can be transferred to the AC grid. However, the pictured connection to an AC grid in both topologies (Fig. 7(e) and (d)) is optional and not necessary for the DC-DC conversion.

In case of a converter or DC pole failure, the bipolar DC transmission configuration allows a redundant operation using the third conductor. The following two subsections explain how the new HVDC-AT can be enabled to feature such an asymmetrical operation of one DC side by simultaneously maintaining the symmetric operation of the other DC side. Due to the symmetry of the topology, the following considerations apply for both load directions and also for a failure of the other DC pole.

A. Operation of the HVDC-AT during a DC Pole Failure on the Low Level DC Side

If the positive terminal of the low level DC side is not available, this side degenerates to an asymmetrical monopole consisting of the negative and the ground terminal. This case is pictured in Fig. 8(a). The in comparison to the normal, bipolar operation changed variables are

978-1-4799-2706-7/14 $31.00 © 2014 IEEE   3258

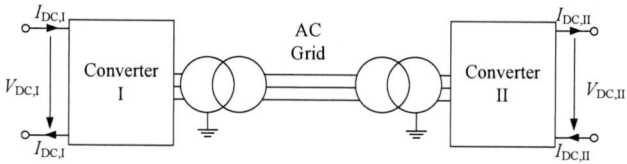

(a) F2F converter for the interconnection of two monopolar DC lines

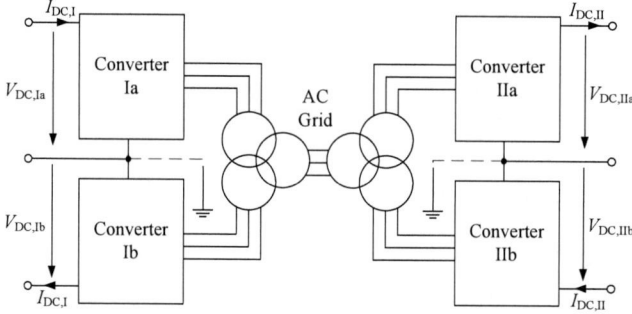

(b) F2F converter for the interconnection of two bipolar DC lines

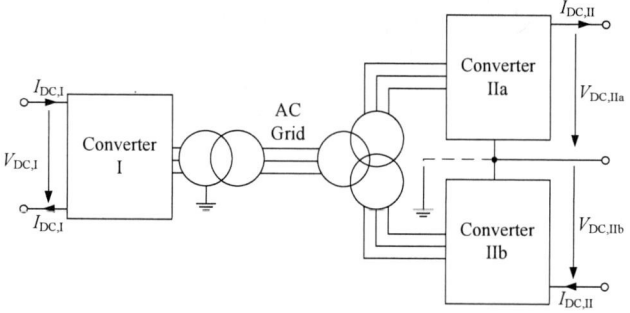

(c) F2F converter for the interconnection of one monopolar with one bipolar DC line

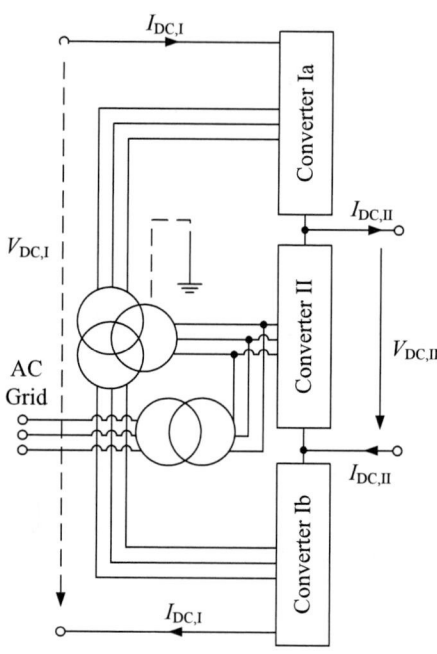

(d) HVDC AT for the interconnection of two monopolar DC lines

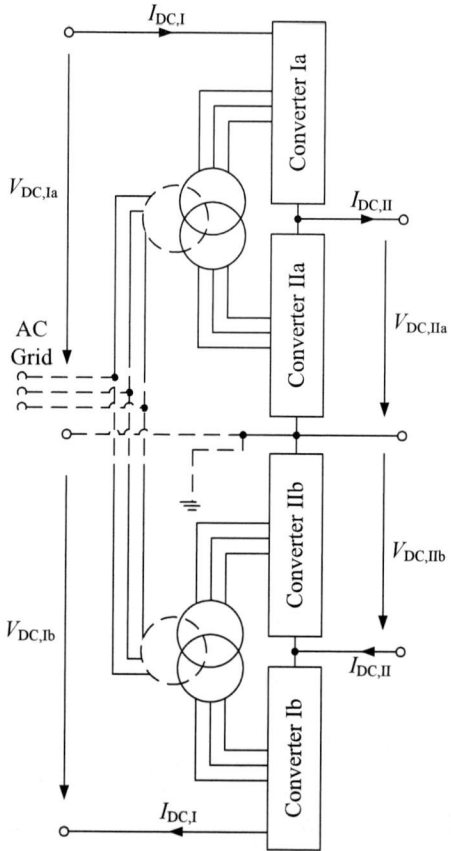

(e) HVDC AT for the interconnection of monopolar und bipolar DC lines

Fig. 7. Implementation of the considered HVDC-DC converter topologies for different DC transmission configurations

marked with a star. Considering the pictured load flow direction, the subconverters Ia, Ib and IIa work as an inverter, feeding the AC link. Subconverter IIb continues to work as a rectifier. The DC voltage at subconverter IIa is still present. Therefore, the faulty terminal has to be disconnected mechanically. Since the effective voltage of the low level DC side is halved, the transferred DC power has to be halved as well. For the further considerations it is assumed that:

$$P_{DC,I}^* = P_{DC,I}/2 \qquad (17)$$

$$I_{DC,I}^* = I_{DC,I}/2 \qquad (18)$$

$$I_{DC,II}^* = I_{DC,II} \qquad (19)$$

During symmetrical operation, per DC pole $P_{DC,I}/2$ is transferred. Hence, during the asymmetrical operation of the low level DC side $P_{DC,I}/4$ have to be transferred by each DC pole.

Compared to the normal operation, in this asymmetrical operation point the load of AC transformers is higher. The upper three winding transformer sees, independent from the actual transformation ration one fourth of the previous DC power $P_{DC,I}$. In addition to that power, the lower three winding transformer has to transfer the inverted DC power of subconverter Ib. The total power rating of the lower transformer during a line fault on the low level DC side is given by Eq.(21).

978-1-4799-2706-7/14 $31.00 © 2014 IEEE   3259

The 2014 International Power Electronics Conference

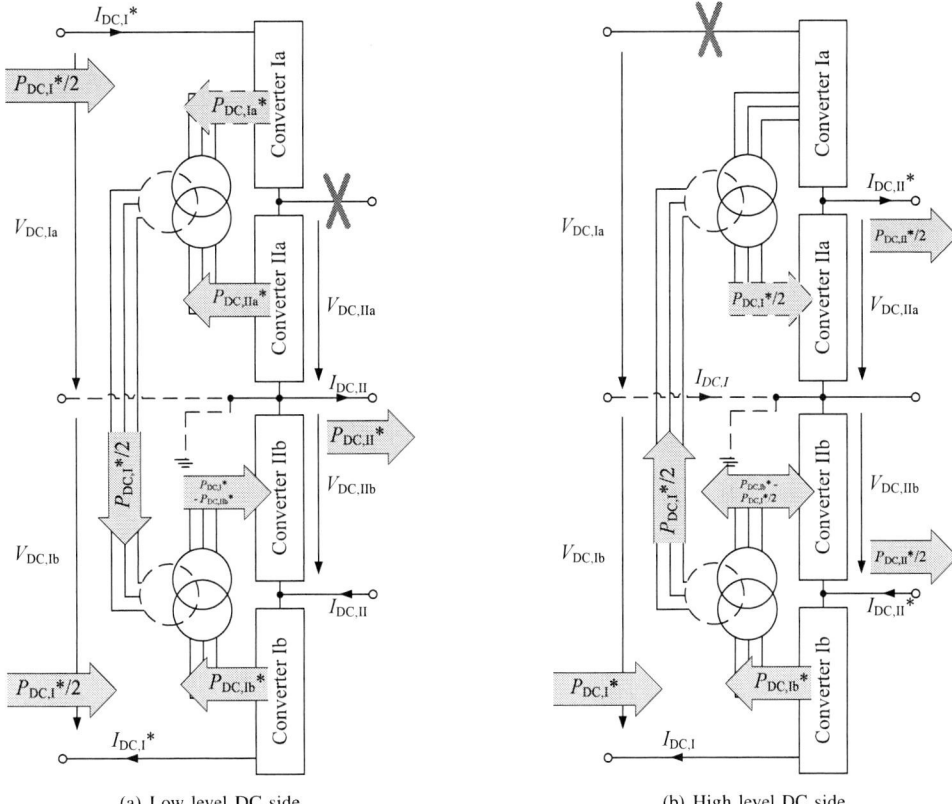

(a) Low level DC side   (b) High level DC side

Fig. 8. Load flow during a DC pole failure

$$P^*_{\text{DC,CIb}} = \frac{P_{\text{DC,I}}}{4} \frac{n-1}{n} \qquad (20)$$

$$P^*_{\text{Tr,b}} = \frac{P_{\text{DC,I}}}{4} \frac{2n-1}{n} \qquad (21)$$

Considering the load of the four subconverters in this asymmetrical operation point, subconverter Ia and Ib see less load, since the high level side DC current is only half of its nominal value. Hence, both subconverts are already rated for the new operation point. The DC current through subconverter IIa during the asymmetrical operation of the low level DC side is identical to the high level DC current and therewith, for a transformation ratio of $n < 1.5$ higher than during normal, symmetrical operation. However, given the assumption that identical cells are being used for all subconverters, subconverter IIa is still in a save operation point since $I^*_{\text{DC,I}}$ is always less than or equal to the maximum allowed DC current $I_{\text{DC,max}}$.

Subconverter IIb rectifies the total power fed into the lower transformer.

$$P_{\text{DC,CIIb}} = P_{\text{DC,I}}/4 + P_{\text{DC,CIb}} \qquad (22)$$

$$= \frac{P_{\text{DC,I}}}{4} \frac{2n-1}{n} \qquad (23)$$

The necessary power rating for all subconverters and both AC transformers is given in Fig. 11. The theoretically necessary power rating of subconverter IIb (solid red line) is for any transformation ratio higher than the nominal power of that subconverter (solid blue line). Assuming, that the cells power rating is fixed and designed

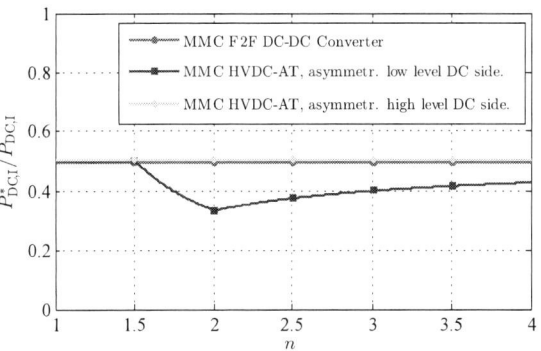

Fig. 9. Transferable DC power, during a single side pole failure, scaled to the nominal DC power during symmetrical operation

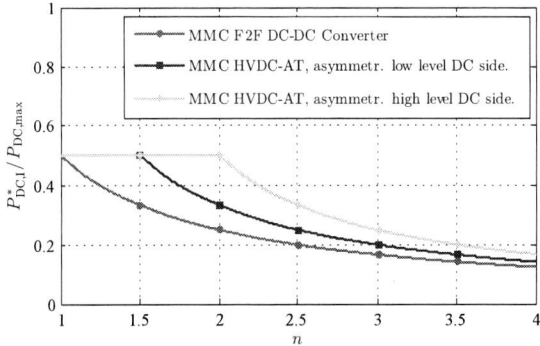

Fig. 10. Transferable DC power, during a single side pole failure, scaled to the maximum power of the high level subconverter of Fig. 2

The 2014 International Power Electronics Conference

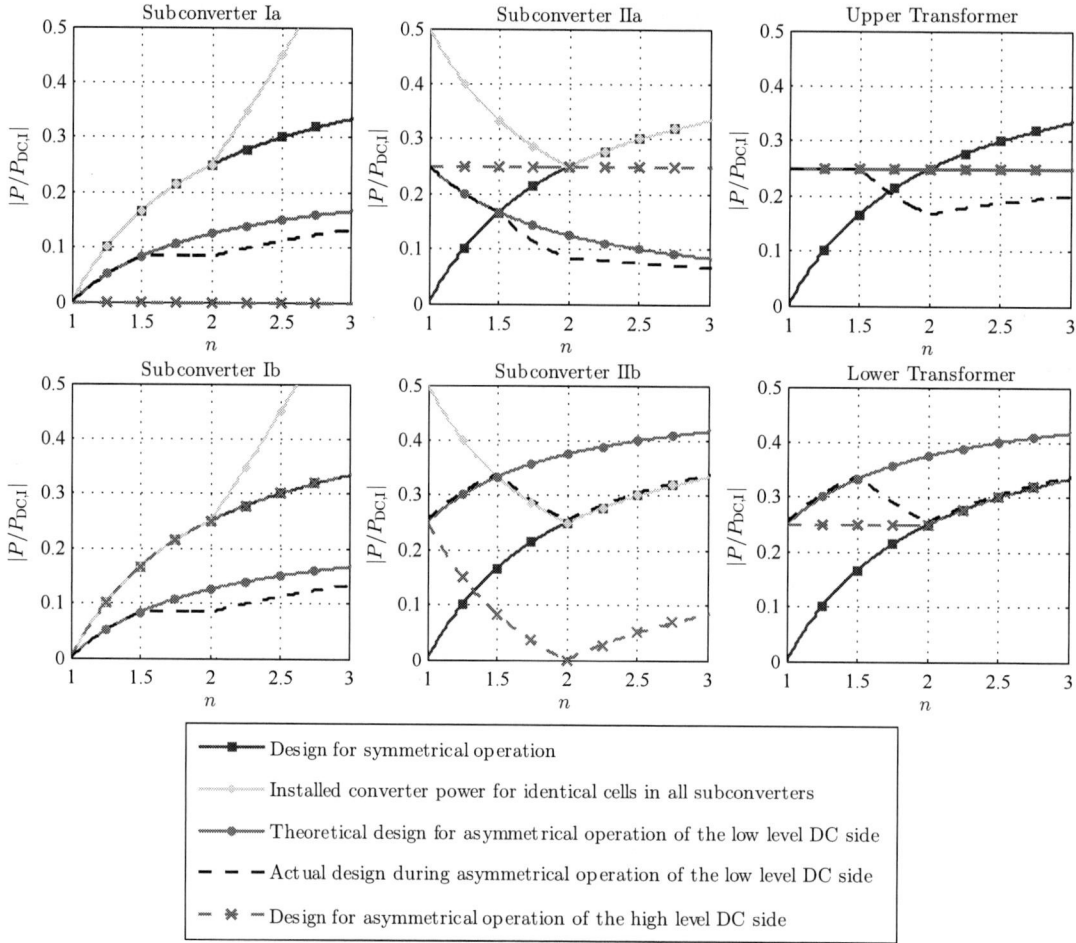

Fig. 11. Power rating of the subconverters of the HVDC-AT topology for single side DC pole failures

for the normal DC-DC converter operation, the installed converter power (solid green line) of subconverter IIb limits the transferable DC power for any transformation ratio greater than 1.5. Scaled to the nominal power of the symmetrical operation the transferable power during a low level DC pole failure is given in Eq.(24) and in Fig. 9.

$$\frac{P_{DC,I}^*}{P_{DC,I}} = \begin{cases} \frac{1}{2} & \text{for } n \leq 1.5 \\ \frac{1}{2n-1} & \text{for } 1.5 < n \leq 2 \\ \frac{n-1}{2n-1} & \text{else} \end{cases} \quad (24)$$

For a transformation ratio of $n > 1.5$, the transferable power during a low level DC side pole failure has to be reduced to less than half of the nominal DC power $P_{DC,I}$. However, in relation to the maximum DC power of an AC-DC converter with the same DC voltage rating, the transferable DC power is still higher than that of a comparable F2F DC-DC converter (Fig. 10).

The reduced transferable DC power of subconverter IIb also leads to a lower utilization of all other subconverters (dashed black line in Fig. 11) and both transformers. If the reduced transferable DC power is considered, the power rating of both AC transformers has to be increased only for $n < 2$.

B. Operation of the HVDC-AT during a DC Pole Failure on the High Level DC Side

If the positive terminal of the high level DC side is not available this side also degenerates to an asymmetrical monopole consisting of the negative and the ground terminal. This case is pictured in Fig. 8(b). Considering the pictured load flow direction, subconverter IIa continues to work as a rectifier and subconverter Ib continues to work as an inverter. Depending on the transformation ratio n subconverter IIb either works as a rectifier($n > 2$) or as an inverter($n < 2$). Subconverter Ia is directly connected to the faulty terminal and sees no load. For the further considerations it is assumed that:

$$P_{DC,I}^* = P_{DC,I}/2 \quad (25)$$
$$I_{DC,I}^* = I_{DC,I} \quad (26)$$
$$I_{DC,II}^* = I_{DC,II}/2 \quad (27)$$

The AC power of the upper transformer, and therewith the AC power of subconverter IIa, is independent of the actual transformation ration equal to $P_{DC,I}^*/2$ and therewith, for a transformation ratio of $n < 2$, higher than the rated power during the symmetrical operation of both DC sides. Assuming identical cells are being used in all subconverters, subconverter IIa is still in a save operation area.

978-1-4799-2706-7/14 $31.00 © 2014 IEEE

For $n < 2$ the lower transformer also has to be rated for $P_{DC,I}^*/2$. At a higher transformation ratio, the AC power of the lower transformer is given by the DC power of subconverter Ib:

$$P_{DC,CIb}^* = I_{DC,I} \cdot (V_{DC,I} - V_{DC,I}) \qquad (28)$$

$$= P_{DC,I}^* \frac{n-1}{n} \qquad (29)$$

The load of subconverter Ib during the asymmetrical operation of is therewith identical to that during symmetrical operation of both DC sides.

The DC current through subconverter IIb is given by the difference of the high and the low level DC current:

$$I_{DC,CIIb}^* = I_{DC,I} - I_{DC,II}^* \qquad (30)$$

$$= \begin{cases} I_{DC,max}\left(1 - \frac{n}{2}\right) & \text{for } n \leq 2 \\ \frac{I_{DC,max}}{2}\frac{2-n}{n-1} & \text{else} \end{cases} \qquad (31)$$

$$P_{DC,CIIb}^* = \begin{cases} \frac{P_{DC,max}}{n}\left(1 - \frac{n}{2}\right) & \text{for } n \leq 2 \\ \frac{P_{DC,max}}{2n}\frac{2-n}{n-1} & \text{else} \end{cases} \qquad (32)$$

Scaled to the converter power during symmetrical operation this leads to

$$\frac{P_{DC,CIIb}^*}{P_{DC,CIIb}} = \frac{2-n}{2(n-1)} \qquad (33)$$

Hence, in comparison to the symmetrical operation, the power of subconverter IIb during asymmetrical operation is only for a transformation ratio of $n < 4/3$ higher. However, given that identical cells are being used in all subconverters, the power of subconverter IIb is always lower than the installed converter power.

Fig. 11 shows the necessary power rating of all subconverters and both transformers during a pole failure of the high level DC side. For the asymmetrical operation of the high level DC side, only the power rating the two AC transformers and only for a transformation ratio of less than two, has to be increased. The nominal power rating of all subconverters is sufficient to continue transferring one half of the previous symmetrical DC power.

V. CONCLUSION

In this paper, the ability of the new HVDC auto transformer topology to interconnect different HVDC transmission topologies has been discussed. It has been shown, that it is possible to connect all known DC transmission topologies with the HVDC-AT. The new topology supports all essential features of a HVDC-DC converter without limitations, except for the galvanic separation of both DC sides. The comparison to the state of the art technology, of connecting two HVDC transmission lines via two distinct converter stations and a full AC link shows, that with the new topology both the transformation losses and the investment cost for such a HVDC-DC converter can be massively reduced.

Additionally this new topology also allows the connection to an AC grid to transfer power to, or draw power from both DC lines at the interconnection point.

The integrated partial AC link also supports an asymmetric operation of a bipolar DC transmission configuration during a single side DC pole failure. During a DC pole failure of the high level DC side one half of the original designed DC power can be transferred from the bipolar low level DC side to the asymmetrical monopolar high level DC side and vice versa. Given the scenario of Fig. 1, this is the most likely application, where the redundancy of bipolar DC lines is necessary.

During a DC pole failure of the low level DC side, depending on the actual transformation ratio, the transferable DC power has to be reduced below one half of original designed DC power to a minimum of one third at a transformation ratio of $n = 2$. If this is not acceptable, the current rating of the subconverters IIa and b must be increased.

In both cases of DC pole failures and especially for a transformation ratio of $n < 2$, the AC transformers must be rated for a higher power, since the power through the AC link increases in the asymmetrical operation points.

However, even with the higher rating of subconverter IIa and b and both AC transformers, the installed converter power is still lower than that of a state of the art F2F HVDC-DC converter of the same power rating.

REFERENCES

[1] A.Schön and M.-M.Bakran, "A new HVDC-DC converter for the efficient connection of HVDC networks," in *Proceedings of the PCIM Europe*. VDE Verlag GmbH, 2013, pp. 525–532.

[2] A.Schön and M.-M.Bakran, "A new HVDC-DC converter with inherent fault clearing capability," in *Proceedings of the 15th European Conference on Power Electronics and Applications (EPE)*, 2013, pp. 1–10.

[3] C.Barker, C.Davidson, D. R.Trainer, and R. S.Whitehouse, "Requirements of DC-DC Converters to facilitate large DC Grids: Alstom Grid UK Ltd," *Cigré Session 2012*, 2012.

[4] D.Jovcic, D.vanHertem, K.Linden, J.-P.Taisne, and W.Grieshaber, "Feasibility of DC transmission networks," *IEEE 2nd PES Innovative Smart Grid Technologies Europe*, pp. 1–8, 2011.

[5] D.Jovcic and J.Zhang, "High power IGBT-based DC/DC converter with DC fault tolerance," in *Proceedings of the 15th International Power Electronics and Motion Control Conference (PEMC)*, 2012, pp. DS3b.6–1–DS3b.6–6.

[6] M.Jimenez Carrizosa, A.Benchaib, P.Alou, and G.Damm, "DC transformer for DC/DC connection in HVDC network," in *Proceedings of the 15th European Conference on Power Electronics and Applications (EPE)*, 2013, pp. 1–10.

[7] N.Ahmed, A.Haider, D.vanHertem, Lidong Zhang, and H.-P.Nee, "Prospects and challenges of future HVDC SuperGrids with modular multilevel converters," in *Proceedings of the 14th European Conference on Power Electronics and Applications (EPE)*, 2011, pp. 1–10.

[8] J. A.Ferreira, "The Multilevel Modular DC Converter," *IEEE Transactions on Power Electronics*, vol. 28, no. 10, pp. 4460–4465, 2013.

[9] G. J.Kish and P. W.Lehn, "A modular bidirectional DC power flow controller with fault blocking capability for DC networks," in *14th Workshop on: Control and Modeling for Power Electronics (COMPEL)*, 2013, pp. 1–7.

[10] Cigré Working Group B4-52, Ed., *HVDC Grid Feasibility Study*, 2012.

[11] Friends of the Supergrid, WG 2. Technological, "Roadmap to the Supergrid Technologies: Update Report," 2013.

Characterization of a current shunt and an inductive voltage divider for PMU calibration

Saytaro Kon and Tatsuji Yamada

National Metrology Institute of Japan (NMIJ)
National Institute of Advanced Industrial Science and Technology (AIST)
Tsukuba, JAPAN
seitaro-kon@aist.go.jp

Abstract— This paper describes a phasor-measurement-unit (PMU) calibration method and presents the measurement results of the characteristics of an inductive voltage divider (IVD) and a current shunt. The frequency characteristics of an IVD and a current shunt were evaluated up to 10 kHz in terms of PMUs. The in-phase components of the IVD were less than 0.39 ppm, whereas its quadrature components reached 0.53 ppm at 10 kHz with small measurement uncertainties. The measurement uncertainty of the current shunt reached approximately 14 ppm at 10 kHz. These results imply that the frequency characteristics of the IVD and current shunt were found to be sufficient for the evaluation of PMUs.

Keywords— *Phasor measurement unit (PMU), shunt resistor, inductive voltage divider (IVD), calibration.*

I. INTRODUCTION

One instrument that provides visualization of the electrical power grid is a phasor measurement unit (PMU) [1]. PMUs are important tools for making electric power grids stable and reliable. As PMUs measure the synchronized voltage and current at a given point by accurate clock signals and time stamps based on the Global Positioning System (GPS), they are useful in monitoring and controlling the system state over wide areas of an electrical power grid. On the basis of their utility, the use of PMUs has increased sharply, particularly in China and the U.S., and is set to grow further around the world in the coming years.

Although the PMU specifications are defined in IEEE standard C37.118 [2], development of new calibration methods and algorithms [3-6] are needed to produce higher-accuracy PMUs. In this context, a prospective system for testing PMUs using the IEEE 1588 network precision time protocol for time synchronization is also under consideration [7].

In this paper, a system for calibrating a PMU by means of a GPS clock and the frequency and harmonic characteristics of key system component voltage and current transducers are described. Evaluation of the harmonic distortion is desired up to 50th harmonic at least, and if possible, the 100th harmonic. Therefore, the measurement frequency range was set up to 10 kHz.

II. SYSTEM CONFIGURATION

The synchrophasor $X(n)$ described in IEEE C37.118 [2] is given by

$$X(n) = X_r(n) + jX_i(n) = \frac{X_m(n)}{\sqrt{2}} e^{j\theta(n)} \tag{1}$$

where $X_m(n)$ is the amplitude of the signal $X(n)$, and $\theta(n)$ is its phase angle at the system frequency ω synchronized to the coordinated universal time (UTC). The IEEE standard defines the measurement requirements for the PMU in terms of the total vector error (TVE). The TVE is an expression of the difference between the theoretical synchrophasor and the reading of the PMU at the same instant of time. The TVE is defined as

$$\text{TVE}(n) \equiv \sqrt{\frac{(\hat{X}_r(n) - X(n)_r)^2 + (\hat{X}_i(n) - X(n)_i)^2}{(X_r(n))^2 + (X_i(n))^2}} \tag{2}$$

where $\hat{X}_r(n)$ and $\hat{X}_i(n)$ are the measurement values given by the PMU (device under test), and $X_r(n)$ and $X_i(n)$ are the reference values given by the input signal. The standard also defines the measurement requirements for the frequency errors (FEs) and the rate of change of frequency (ROCOF) error (RFE). The ROCOF is defined as

$$\text{ROCOF}(t) = \frac{df(t)}{dt} \tag{3}$$

where $f(t)$ is the frequency. The FE and RFE are the absolute values of the difference between the theoretical values and the actual values.

Fig. 1 shows the calibration system for the PMUs. A PMU measures the parameters of the synchrophasor, and a signal that has been synchronized with a GPS reference clock is applied to three voltage amplifiers and three transconductance amplifiers (TCAs) for calibration of a PMU. The signals from the GPS are a one-pulse-per-second (1 PPS) signal, an Inter-Range Instrumentation Group Time Code Format B (IRIG-B) [8], and 10-MHz signals. To calculate the theoretical phasor parameters, the voltage applied to the PMU is divided by a

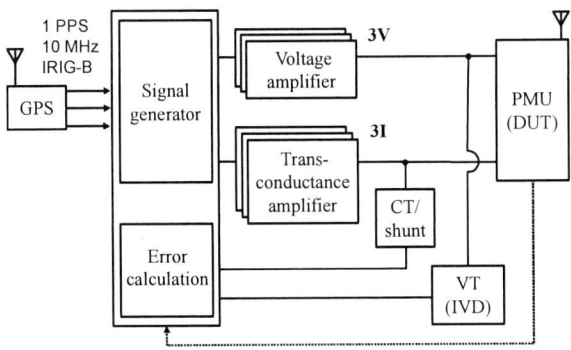

Fig. 1.   Schematic of the proposed calibration system for a PMU.

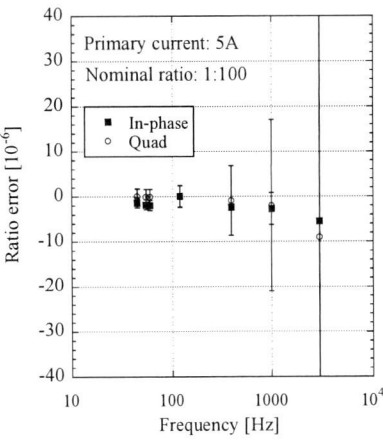

Fig. 2.   Frequency characteristics of the in-phase and quadrature components of the current transformer up to 3 kHz.

pre-calibrated voltage transducer, and the current is measured using a current transducer such as a current transformer (CT) or an ac shunt [9]. Finally, the TVE, FE, and RFE are calculated by comparing the theoretical values with the PMU measurement values. As their measurement stabilities are important to PMU calibration, the voltage and current transducers need to be evaluated accurately. If harmonic characterization is required, the signal generator can be calibrated by a non-sinusoidal power measurement standard [10].

The measurement uncertainty of the TVE(n) is given by

$$u^2(\mathrm{TVE}) = \sum_l^N (\frac{\partial(\mathrm{TVE})}{\partial x_l})u^2(x_l) \qquad (4)$$

where $x_l = \hat{X}_r, \hat{X}_i, X_r, X_i$. Therefore, the generation of a stable and accurate reference signal and the accuracy of the synchrophasor measurements by current and voltage transducers are quite important to reduce the measurement uncertainty of the system.

III.   CHARACTERIZATION OF THE TRANSDUCERS

A. Current transducers

A two-stage current transformer (CT) [11] with a compensation circuit that had been calibrated using the ampere-turn balance method by means of a self-calibrated current comparator [12] was used to measure the applied current. The frequency characteristics of the CT were evaluated at frequencies up to 3 kHz for a primary current of 5 A, as shown in Fig. 2. The in-phase component remains constant up to 1 kHz; however, it reaches a few parts per million at 3 kHz, and the quadrature component varies with frequency. The error bars in Fig. 2 represent the expanded measurement uncertainties (coverage factor: k = 2) [13] of each measurement. As can be seen in Fig. 2, the expanded uncertainties of the CT measurements sharply increase with frequency. The values of the uncertainties of the in-phase and quadrature components are respectively 3.5 ppm and 19 μrad at 1 kHz and 55 ppm and 77 μrad at 3 kHz. The ratio errors of

the CT increase with frequency because of the leakage current that is derived from the stray capacitance among the windings and a decrease in the magnetic permeability of the core and magnetic shielding. In addition, the difference in the load may have harmful effects on these measurement uncertainties and ratio errors. The PMU evaluations need to be improved by tightening the method for calibrating the ac current ratio standards at high frequencies by using current transducers that can be calibrated with small measurement uncertainty. Measurement uncertainties of less than 50 ppm at 10 kHz are preferred. Therefore, it is also better to consider another current transducer such as an ac shunt that allows for the most precise measurement of current.

These shunts play an important role in several research areas in which accurate current measurements are necessary. For example, evaluation of any new current sensor requires an accurate and stable current standard. We used a cage-type standard shunt resistor that consists of many resistor elements connected in parallel [14]. This parallel configuration reduces the harmful effects of inductance and heat due to the input current on the resistance. Specifically, the current shunt (Fluke, model A40B) with nominal values of 0.8 Ω, 0.16 Ω, 0.08 Ω, and 0.008 Ω, and the standard shunt that consists of one hundred 10-Ω metal-foil resistors in parallel with a nominal value of 0.1 Ω have been used.

The influence of current on the properties of an ac shunt is important issue because there may be a variation in the output of ac shunt due to current heating. To measure this effect, a voltage source, a TCA, shunt resistors, thermography, and an ac voltmeter are used. The thermal distribution of the ac shunt with a nominal value of 0.08 Ω was measured by thermography at an input current of 10 A and an ambient temperature of 23 °C ± 1 °C. The surrounding temperature of the resistor elements arranged in parallel is higher than the temperature of the other parts. The temperature increased to approximately 41 °C 30 min after the start of current input. In addition, the maximum temperature of the ac shunt with a nominal value of 0.8 Ω at 1 A is 30 °C and that of the ac shunt with a nominal value of 0.008 Ω at 100 A is 72 °C. The temperatures and output voltages of the shunts continue to change for approximately 1000 s after the initial flow of the input current, as shown in Fig. 3. The thermal time constants are approximately 200 s for all shunts, and these values do not depend on the frequency of the input current. To separate the warm-up effects, a TCA was warmed up with a dummy load over a period of 1 h before each measurement.

The several cage-type ac shunts [14] with different rated currents are calibrated by a combination of a dc resistance calibration and ac–dc difference measurements using a thermal voltage converter (TVC) [15]. As shown in Fig. 4, the frequency characteristics of the ac shunts are flat up to 10 kHz. The expanded measurement uncertainties also remain about the same. The expanded uncertainties of the shunts at 10 kHz are 14 ppm at 1 A and 25 ppm at 10 A. The combination method for the evaluation of shunt resistors is suitable for the frequency characterization of the ac resistance of the shunts. However, the method cannot provide the phase angle of the shunts because of thermoelectric conversion by using a TVC.

Therefore, another method that can provide both the ac resistance and the phase angle of the shunts is needed. Thus, an ac shunt calibration system using a current-bridge method is developed [9] as shown in Fig. 5. The current-bridge method uses a CT, current comparator (CC), and standard ac resistor calibrated by the ac current ratio standard and ac resistance standard in advance. A standard ac resistor with a nominal value of 1 kΩ is used in the system, and the nominal values of the shunt resistors are small, such as 0.1 Ω, for example. In this case, the resistance ratio between the standard ac resistor and the shunt is 10000. Therefore, a current ratio of 10000 is needed by a combination of a CT and a CC to compare the standard ac resistor with the shunt. The in-phase and quadrature components of the CT, CC, and standard ac resistor are obtained by their calibration results. Therefore, the ac resistance and phase angle of the shunts can be obtained by the current-bridge method.

Fig. 6 shows the comparison of the calibration results obtained with the current-bridge method and with the combination method of dc resistance and ac–dc measurements. These results are consistent with each other within the measurement uncertainties. The measurement uncertainty of the current-bridge method is smaller than that of the combination method.

Fig. 7 shows the frequency characteristics of the phase angle of a shunt. The phase angle varies proportionally to the frequency. The results for the phase angle are reasonable assuming the basic equivalent circuit model of an ac resistor, which consists of a combination of a resistor and an inductor in series and a capacitor in parallel. The values of the phase angle of the shunt are small, and they appear to have a very small effect on the PMU evaluation in this study; therefore, the values of shunts have enough accuracy to evaluate the PMUs.

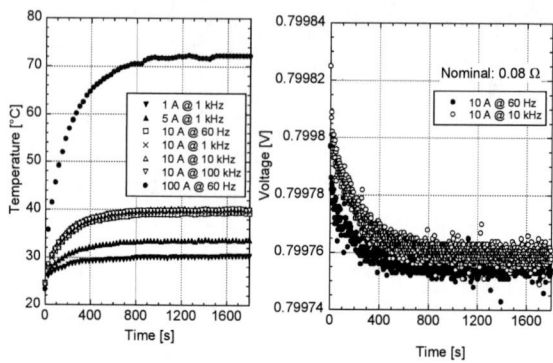

Fig. 3.　Variations in surface temperature of the shunt resistor elements and in the output voltage of the shunts with time.

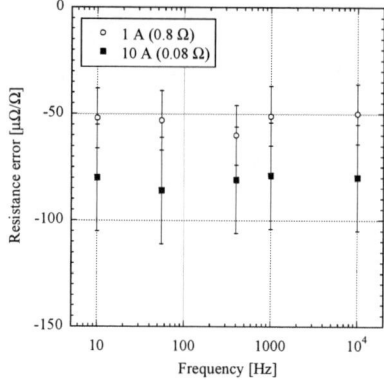

Fig. 4.　Frequency characteristics of the ac shunts obtained using the dc resistance and ac–dc difference method up to 10 kHz.

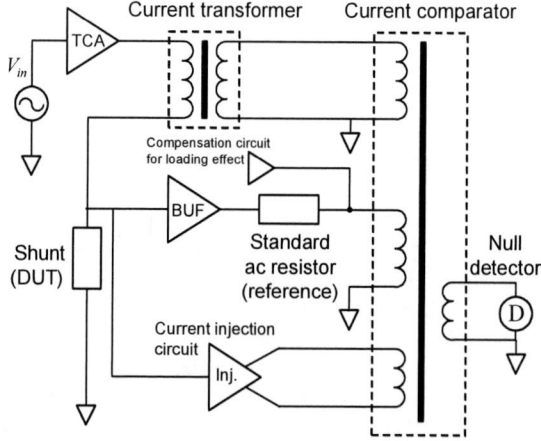

Fig. 5.　Schematic of the ac shunt calibration system using the current-bridge method.

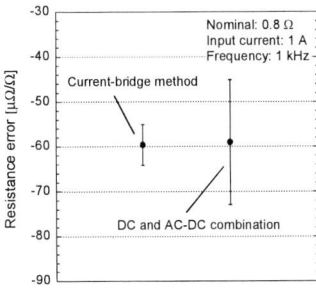

Fig. 6.   Comparison of the calibration results obtained using the current-bridge method with those obtained using the dc resistance and ac–dc difference methods.

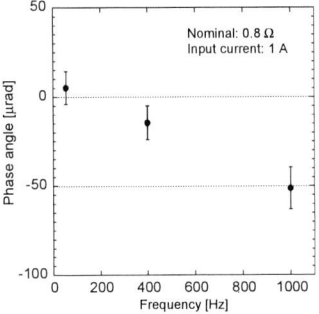

Fig. 7.   Frequency characteristic of the phase angle of a shunt obtained using the current-bridge method.

B. Voltage transducer

An inductive voltage divider (IVD) is used for a voltage transducer. IVDs are instruments that divide the input ac voltage into a voltage depending on the setting ratios using the electromagnetic induction phenomenon. An IVD consists of a high magnetic permeability core, several dozen to hundreds of windings, and some output terminals (taps). Further, an IVD can provide accurate voltage dividing by only changing the winding ratios. Therefore, IVDs play an important role in several research areas where accurate impedance or power measurements are required. The ratios of the IVDs are not susceptible to the influence of ambient conditions such as temperature and humidity. The ratio error of IVDs can be less than 1×10^{-7} by using a two-stage structure [16] and magnetic shielding. The two-stage IVD used to evaluate the PMU is fabricated with a primary-stage toroidal core with dimensions of $100 \times 60 \times 30$ mm (outer diameter × internal diameter × height), and a secondary-stage toroidal core with dimensions of $60 \times 45 \times 10$ mm. These cores are made of nano-crystals, and Copper wire with a diameter of 0.8-mm is used for windings. The turn number for the excitation windings is 80, and that of ratio windings is four by using 22 twisted wires. The magnetic and electrostatic shielding is added between the excitation windings and the ratio windings to reduce the

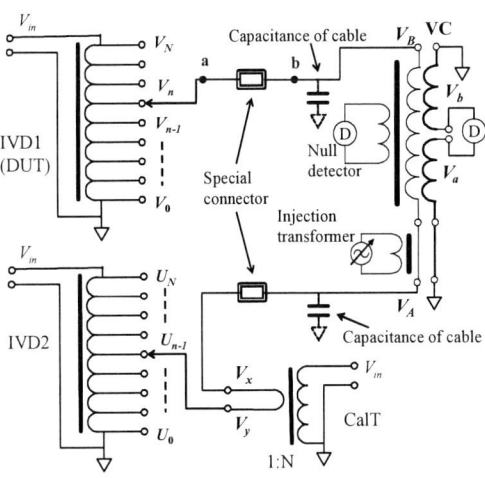

Fig. 8.   Schematic of the IVD calibration system using the build-up method.

leakage flux. The two-stage IVD has 1:20 step ratios.

Fig. 8 shows the IVD calibration system using the build-up method [17] based on Thompson's method [18]. IVD1 is the device under test (DUT), IVD2 is used for voltage adjustment, and CalT is a transformer that has a ratio between the secondary and primary windings of 1:N. The build-up method can be expected to provide precise calibration results because the potentials of each tap of IVD1 are kept stable. The voltage difference between every two adjacent taps, $V_n - V_{n-1}$, in IVD1 is measured to determine the voltage ratio of each tap of IVD1 using the voltage difference, $V_x - V_y$, on the secondary side of CalT as a reference.

Therefore, small voltage differences such as the voltage differences between V_n or V_{n-1} and V_x or V_y have to be measured accurately to calibrate the IVD. A voltage comparator (VC) is used to achieve accurate voltage measurements and is based on the principle of a coaxial difference transformer. A coaxial cable is wound on a toroidal core, the outer conductor normally used for a shield is cut in a half, and the output terminals are connected to the outer conductor. Therefore, the inner conductor and outer conductor constitute a transformer. The output of the differential transformer can be written as

$$V_a - V_b = -(V_A - V_B) \qquad (5)$$

where V_A and V_B are input voltages for the differential transformer. However, an error in the transformer is generated in practice. Therefore, detector windings are added to the core of a differential transformer to reduce the errors, and an injection transformer is added to the input terminal side of the core of a differential transformer. The detector windings are connected to the detector, and the injection transformer can provide same voltages to the inner and outer conductors at the same time. The voltage difference between V_A and V_B induces

The 2014 International Power Electronics Conference

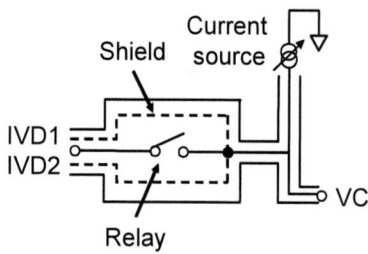

Fig. 9.   Schematic of the special connector

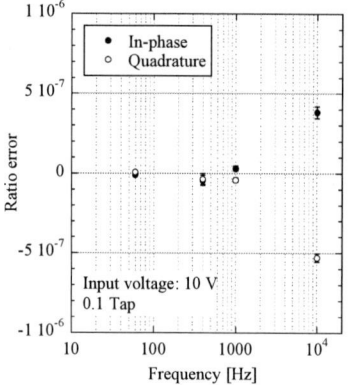

Fig. 10.   Frequency characteristics of the IVD obtained using the build-up method up to 10 kHz.

a voltage to the detector windings, and the detector measures a value. The injection transformer is adjusted such that the detector measures zero. Further, these procedures can provide a three-digit accuracy improvement for the voltage difference measurements in the system.

The current flow from IVD1 to the VC is also quite small because the voltage differences between V_n or V_{n-1} and V_x or V_y are small. However, the cable used for connecting IVD1/IVD2 and the VC has some capacitance; therefore, a capacitive current flows from IVD1/IVD2 to ground. The influence of the capacitive current depends on the output impedance of the IVDs and the measurement frequency of the system. Special connectors are used to reduce the effects of the capacitive current and consist of a relay switch and shield, as shown in Fig. 9. Current injection to the special connector is also possible by using an adjustable current source. Further, the output current from the IVD can become zero by using the special connector. Specifically, the relay is first turned on, and the injection transformer is adjusted such that the null detector of the VC is zero. Next, the relay is turned off, and the null detector of the VC does not show zero at this time. The current source connected to the special connector is then adjusted such that the null detector of the VC is zero. Next, the relay is turned on, and the injection transformer is adjusted such that the null detector is zero again. When these adjustment procedures are repeated, the null detector finally shows a constant value that is independent of the status of the relay

because the potentials at nodes a and b are the same at this time. This means that the current derived from the IVD is zero, even if the relay turns on. The potential of the shield of the special connector is also same as nodes a and b at this time. Therefore, the special connector blocks the capacitive current from the inner conductor to the outer conductor perfectly. If there is no shield in the special connector, a capacitive current flows from the inner conductor to the outer conductor via the stray capacitance between the inner and outer conductors, even if the potentials at nodes a and b are the same. Therefore, the IVD can be calibrated without current flows.

IVD1 was calibrated with a build-up method based on Thompson's method, as mentioned earlier. To allow for characterization up to the 100th harmonic, the frequency characteristics of the IVD are evaluated up to 10 kHz. The frequency characteristics of IVD1 (0.1 tap) for an input voltage of 10 V from 60 Hz to 10 kHz in Fig. 10 reveal that the deviation from the nominal ratio (in-phase component) remains constant up to 1 kHz and subsequently reaches approximately 0.39 ppm at 10 kHz. However, the quadrature component varies over the entire frequency range, reaching approximately 0.04 ppm and then 0.53 ppm at 1 kHz and 10 kHz, respectively. The expanded measurement uncertainties of IVD1 are also very small, and these values possess satisfactory accuracy for the evaluation of PMUs.

However, these values were obtained using the build-up method under approximately unloaded conditions. When calibrated IVDs are used in some applications such as the evaluation of a PMU, the IVD ratio error may deviate from the calibrated values because of the relationship between the IVD output impedance and the loads. Therefore, the loading effects on the IVDs should be evaluated, even though the output impedance of a two-stage IVD is practically quite small.

The IVD load characteristics were evaluated by comparing the measured values under resistive, capacitive, and inductive loads and without load. The IVD load characteristics are evaluated by measuring the output voltage differences between IVD1 and IVD2. To detect the very small variation with loads, IVD2 was used to obtain an output voltage similar to that of IVD1. The input impedance of the detector affected the evaluation of the IVD load characteristics. Therefore, we evaluated the effect of the detector input impedance. The ratio errors due to the input impedance at 10 kHz were less than 1×10^{-6} for the in-phase and quadrature components.

Fig. 11 shows the frequency characteristics of IVD1 with resistive, capacitive, and inductive loads. The in-phase components with a resistive load have a flat frequency response. In addition, the in-phase components with capacitive loads were stable up to 1 kHz and increased sharply at 10 kHz. The quadrature components with resistive and capacitive loads were proportional to the frequency, and the in-phase and quadrature components with inductive loads were quite large. The measurement uncertainties of the in-phase and quadrature components depended on the load value and frequency and

978-1-4799-2706-7/14 $31.00 © 2014 IEEE   3267

were estimated to be 0.1×10^{-6} with a 10-kΩ load, 0.3×10^{-6} with a 1-kΩ load, and 2.1×10^{-5} with a 100-Ω load, respectively. The measurement uncertainties of the quadrature values are 0.1×10^{-6} with a 10-kΩ load, 0.2×10^{-6} with a 1-kΩ load, and 1.5×10^{-5} with a 100-Ω load.

To check the load characteristic measurements, the positions of IVD1 and IVD2 were swapped, and IVD2 was also evaluated using same method. The load characteristics of IVD2 were similar to IVD1. Therefore, the effects of the loads on the IVDs were significant for both the IVD in-phase and quadrature components. It is preferable that the input impedance of the instruments connected to the IVD should be greater than 10 kΩ at least.

IV. CONCLUSION

The characteristics of some key components for PMU evaluation, such as the CT, ac shunt, and IVD were evaluated. The results of this evaluation showed that the IVD is accurate enough if the load value is greater than 10 kΩ, and the ac shunt is also accurate enough if the temperature and output voltage of a shunt are stable after the initial flow of the input current for PMU evaluation. Our future work will evaluate the PMU by using the proposed method and estimate its measurement uncertainties.

REFERENCES

[1] A.G. Phadke and J.S. Thorp, "Synchronized Phasor Measurements and Their Applications," Power Electronics and Power Systems, Springer, 2010.
[2] "IEEE Standard for Synchrophasor Measurements for Power System," IEEE Std C37.118.1-2011, pp. 1–61, 2011.
[3] G. Rietveld, J.P. Braun, P.S. Wright, and N. Zisky, "Realization of a Smart Grid Metrology Infrastructure in Europe," CPEM2012 Digest, pp. 408–409, 2012.
[4] U. Pogliano, J.P. Braun, B. Volj and R. Lapuh, "Software Platform for PMU Algorithms Testing," CPEM2012 Digest, pp. 412–413, 2012.
[5] J.P. Braun and C. Mester, "Reference Grade Calibrator for the Testing of the Dynamic Behavior of Phasor Measurement Units," CPEM2012 Digest, pp. 410–411, 2012.
[6] Y. Tang and G.N. Stenbakken, "Calibration of Phasor Measurement Unit at NIST," CPEM2012 Digest, pp. 414–415, 2012.
[7] J. Amelot and G. Stenbakken, "Testing Phasor Measurement Units using IEEE 1588 Precision Time Protocol," CPEM2012 Digest, pp. 230–231, 2012.
[8] IRIG Std., 200-04, Sep. 2004.
[9] S. Kon and T. Yamada, "Uncertainty Evaluations of an AC Shunt Calibration System with a Load Effect Reduction Circuit," IEEE Trans. Instrum. Meas., vol. 60, no. 7, pp. 2286–2291, 2011.
[10] T. Yamada, S. Kon, N. Sakamoto and H. Kato, "Uncertainty Estimation and Performance Evaluation of a Non-Sinusoidal Power Measurement Standard," IEEJ Trans. Fundam. Mater., vol. 132, no. 3, pp. 251–256, 2012.
[11] T.M. Souders, "Wide-Band Two-Stage Current Transformers of High Accuracy," IEEE Trans. Instrum. Meas., vol. 21, no. 4, pp. 340–345, 1972.
[12] W.J.M. Moore and P.N. Miljianic, "The Current Comparator," IEE Electrical Measurement Series 4, Institution of Engineering and Technology, 1988.
[13] International Organization for Standardization, "Guide to the Expression of Uncertainty in Measurement," 1993.
[14] K. Lind, T. Sorsdal, and H. Slinde, "Design, Modeling, and Verification of High-Performance AC-DC Current Shunts from Inexpensive Components," IEEE Trans. Instrum. Meas., vol. 57, no. 1, pp. 176–181, 2008.
[15] J.R. Kinard, T.E. Lipe, and C.B. Childers, "AC-DC Difference Relationships for Current Shunt and Thermal Converter Combinations," IEEE Trans. Instr. Meas., vol. 40, no. 2, pp. 352–355, 1991.
[16] T.A. Deacon and J.J. Hill, "Two-stage inductive voltage divider," Proc. IEEE, vol. 115, no. 6, pp. 888–892, 1968.

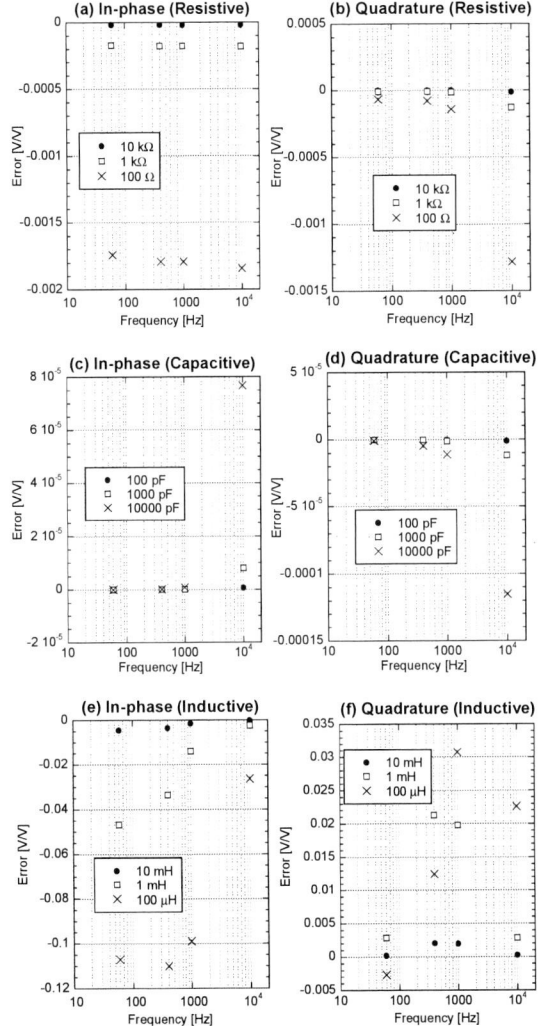

Fig. 11. Loading effect of the IVD up to 10 kHz. Frequency characteristics of the ratio errors of in-phase and quadrature components with resistive, capacitive, and inductive loads.

[17] Y. Nakamura, "Calibration and Uncertainty Estimation of a Two-staged Inductive Voltage Divider," AIST Bull. Metrol., vol. 4, no. 1, pp. 45–52, 2005.

[18] A.M. Thompson, "Precise calibration of ratio transformers," IEEE Trans. Instrum. Meas., vol. 32, no. 1, pp. 47–50, 1983.

Distributed Series/Hybrid-Shunt Compensation for Harmonic Mitigation in Commercial Facilities

Rogério Azevedo Diniz
Diretoria de Apoio aos Negócios e Operações
Banco do Brasil
Brasília, Brazil
rogeriodiniz@bb.com.br

Igor A. Pires*, Gleisson J. França, Braz J. Cardoso**
Electronic Engineering Departament*, Electrical Engineering Departament**
Universidade Federal de Minas Gerais
Belo Horizonte, Brazil
iap@ufmg.br, gleissonjf@ig.com.br, cardosob@ufmg.br

Abstract— **This work presents the development and application of an electronically controlled harmonic mitigation device based on the insertion of active series harmonic impedances in the electric network. These active harmonic impedances are inserted at specific frequency (or frequencies). Important advantages of the proposed system are the reduced converter rating and the utilization of coaxial transformers with single turn secondary. In this paper, the active harmonic impedance device is applied to implement distributed harmonic mitigation strategy, leading to a flexible and cost-effective solution for industrial and commercial power systems applications. The desired overall results are investigated through modeling and simulation of an actual commercial facility, based on the PTW - Power Tools for Windows/SKM. Harmonic distortions and harmonic sources data are measured in the considered facility, characterizing the situation before and after the implementation of series-shunt compensation for harmonic mitigation.**

Keywords - **Active Harmonic Impedance, *H*armonics, Power Factor Correction, Distributed Compensation.**

I. INTRODUCTION

The advance of power electronic applications has been stimulated by continuing advances on semiconductor device technologies and demands for energy savings and productivity increase in commercial and industrial facilities. Machinery and processes using static converters, arc furnaces, switching power supplies, reactive compensators (SVC's) are employed in growing numbers, increasing the presence of nonlinear loads in industrial parks, and on a smaller scale, but no less relevant, in commercial building facilities. Such loads imply new challenges in the electrical power system, demanding attention to the harmonic distortion of voltage and current, which can cause problems such as:

✓ Overloading power equipment and systems;
✓ Transformers: temperature rise due to the increase of losses in iron (voltage harmonics) and copper (current harmonics) into harmonic frequencies; audible noise emission;
✓ Cables: temperature rise due to increased losses; skin effect losses; voltage stress and corona from dielectric failure;

✓ Capacitors: possibility of resonance due to high levels of voltage and current, increased heating, dielectric stress, reduction of life;
✓ Reduced power factor.

The alternatives normally used for harmonic mitigation are grouped as passive, active and hybrid solutions. Passive solutions use only inductors, capacitors and resistors on its construction. The major difficulties of the passive solutions are the reduced flexibility and performance strongly dependent on system impedance and load configuration. Active solutions are based on series and/or parallel active filters. Finally, hybrid solutions are a composition of active and passive solutions. The main difficulties of active and hybrid solutions application are the implementation cost, mainly in high current and/or voltage systems, and reliability [1].

Recognizing the fact that it is only necessary to reduce the harmonic content of the electric system to levels compatible with the existing equipment, this work proposes a series active harmonic impedance insertion system where impedances are coupled only at specific desired harmonic frequencies. An important feature of the proposed system is the use of a coaxial transformer for harmonic impedance coupling. The harmonic impedances can be used as series or hybrid shunt compensation with the objective to control the harmonic current flow, directing it to paths that do not cause damage to the system components.

In this paper, Series Active Harmonic Impedance – SAHI – devices are considered for distributed harmonic mitigation in a commercial building facility. The selected commercial building presents a high density of non-linear loads with relevant harmonic currents flow. Such high harmonic content causes significant electrical losses, especially in the secondary of the transformer substation, and cabling. The desired overall results are investigated through modeling and simulation based on the PTW - Power Tools for Windows/SKM. Harmonic distortions and harmonic sources data are all obtained from field measurement in the considered facility, characterizing the situation before and after the implementation of series-shunt compensation for harmonic mitigation.

II. SERIES ACTIVE HARMONIC IMPEDANCE COMPENSATION

A. General Description

The series active harmonic impedance proposed here is essentially a controlled voltage source based on single phase static DC/AC converter. The harmonic voltage induced in the main power cables is determined from a desired proportion between the harmonic current and voltage. This relationship is defined independently from the fundamental frequency, for one or more harmonic frequencies in the electrical system.

An important characteristic of the active harmonic impedance proposed in this work is the flexibility introduced by the utilization of coaxial transformers, as can be shown in Figure 1. This solution allows the use of existing power cables as the secondary winding. In [2], a similar coupling system was discussed for application in distributed fundamental frequency power flow control, proposed in order to obtain flexibility to improve renewable energy market.

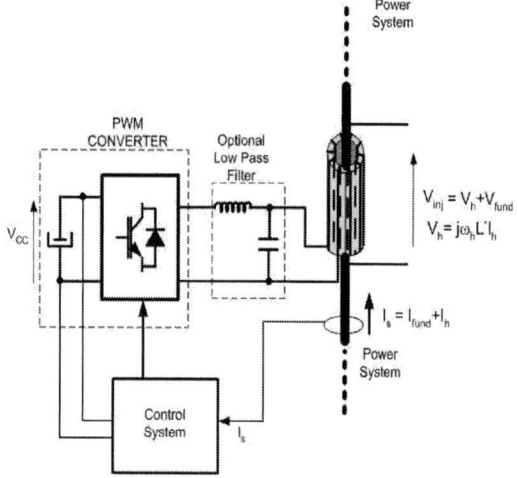

Fig. 1. General diagram of the proposed active harmonic impedance.

Active harmonic impedance devices are independent single phase low power units. They can be cascaded to achieve the desired total harmonic impedance at one point or distributed to control harmonic flow. The flexibility inherent to the system allows its use for detuning (or tuning) a capacitor bank composed of several stages, as an alternative to the use of one reactor per stage for detuning or the implementation of shunt filter banks. Since electrical loads change along the day, typical automatic capacitor banks used for reactive compensation and filtering uses up to 12 capacitor stages in order to obtain an efficient compensation at any load demand. Each capacitor stage uses a dedicated reactor in detuned and tuned systems which lead to costly and heavy reactive compensation systems. The single phase approach gives the advantage to treat unbalanced harmonic conditions since each phase is considered and compensated independently from the others.

Figure 2 shows a block diagram of the active harmonic impedance cell. The transformer secondary current (electrical system power cable) I_s is measured and a PLL algorithm is used to obtain the fundamental frequency current component amplitude and phase. The estimated fundamental current - I_f is subtracted from the measured current and the result – I_h (harmonic component) is directed to another PLL input, tuned to estimate the amplitude and phase of the desired harmonic frequency.

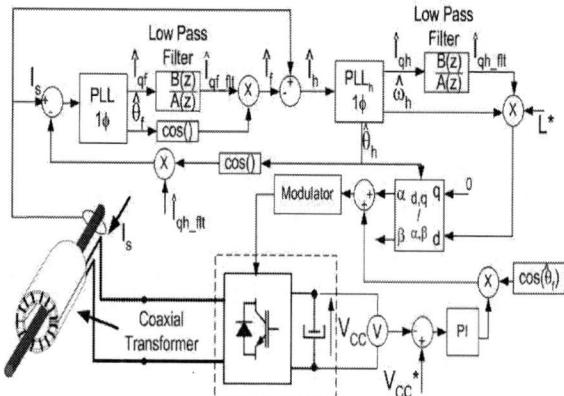

Fig. 2. Block diagram of the active harmonic impedance cell.

In order to apply the proposed system for the compensation of additional harmonic components it is necessary to introduce additional PLL algorithms and inverse Park transforms for each additional harmonic. In practice, the compensation of the 5th, 7th and 11th harmonic is sufficient to obtain satisfactory results on most industrial systems.

In this work the harmonic impedance synthesized is a pure inductive reactance, represented by the inductance L^*, according to the equation (1).

$$V_h = L^* \frac{dI_h}{dt} \tag{1}$$

Equation (1) is implemented considering that the current Is, which flows on the coaxial transformer, corresponds to the α component used in Clark transformation, on the single phase PLL.

B. Active harmonic Impedance Implementation

As the PLL algorithms, (1) is implemented considering the emulation of a balanced three-phase system. The calculation of the voltage to be synthesized from equation (1) is carried out in the synchronous reference frame based on the following development, observing the reference axis of Figure 3. Park's transformation applied to the components α and β is given by:

$$I_q = I_\alpha \cos \theta_h + I_\beta \operatorname{sen} \theta_h \tag{2}$$

$$I_d = I_\alpha \operatorname{sen} \theta_h - I_\beta \cos \theta_h \tag{3}$$

where:

$$\theta_h = \omega_h t + \phi_h \tag{4}$$

978-1-4799-2706-7/14 $31.00 © 2014 IEEE   3271

Differentiation of the equations (2) and (3) with respect to time results in:

$$\frac{dI_q}{dt} = \left(\frac{dI_\alpha}{dt} \cos\theta_h + \frac{dI_\beta}{dt} sen\theta_h \right) - \omega_h \left(I_\alpha sen\theta_h - I_\beta \cos\theta_h \right) \quad (5)$$

$$\frac{dI_d}{dt} = \left(\frac{dI_\alpha}{dt} sen\theta_h - \frac{dI_\beta}{dt} \cos\theta_h \right) + \omega_h \left(I_\alpha \cos\theta_h + I_\beta sen\theta_h \right) \quad (6)$$

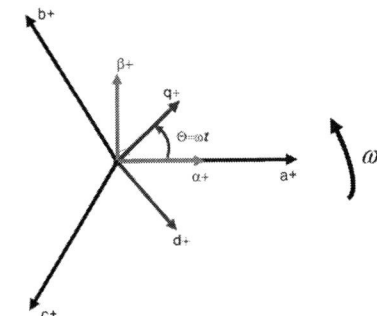

Fig. 3. Axis references for axis transformations.

Considering a steady state condition:

$$\frac{dI_q}{dt} = \frac{dI_d}{dt} = 0 \quad (7)$$

$$\left(\frac{dI_\alpha}{dt} \cos\theta_h + \frac{dI_\beta}{dt} sen\theta_h \right) = \omega_h I_d \quad (8)$$

$$\left(\frac{dI_\alpha}{dt} sen\theta_h - \frac{dI_\beta}{dt} \cos\theta_h \right) = -\omega_h I_q \quad (9)$$

Based on the development represented by (2) to (9), and recognizing that in (5) and (6), the terms of the first parentheses after the equal sign represent the Park transformation of the derivatives of the components α and β of the desired currents [7], the components of the voltage to be synthesized are calculated on the synchronous reference frame as follows:

$$V_{qh} = \omega_h L^* I_{dh} \quad (10)$$

$$V_{dh} = -\omega_h L^* I_{qh} \quad (11)$$

Applying the inverse Park transform on the results from (10) and (11) and taking the α-axis component, (12) is obtained representing the open loop voltage to be synthesized by the PWM modulator, seen at the secondary side of the coupling transformer, responsible for injecting the desired active harmonic impedance.

$$V_h = V_\alpha = -\omega L^* I_{qh} sen\theta_h \quad (12)$$

In Section III, simulation results from application of equation (12) to the implementation of the proposed active harmonic impedance are presented. Experimental results are shown in section V.

C. PLL Algorithms

The PLL algorithms are based on the Park transformations for treatment of desired variables on a synchronous reference frame. The PLL tracking system is based on a PI that objectives to cancel the d axis component resulting from the Park transformation [3].

The proposed system requires at least two separate PLL algorithms: one for obtaining the amplitude, frequency and phase of the fundamental current component and one PLL in order to obtain the parameters for each harmonic component of interest. It can be found on the literature several alternatives for implementation of the PLL algorithms, but the method of coordinate transformations for the use of synchronous reference has been widely applied. Detailed analysis of PLL algorithms is given in [3].

D. Coupling Coaxial Transformer Considerations

The design of the coaxial transformer, shown in Figure 4, should take into consideration the maximum voltage to be injected on the desired frequency and also to ensure the use of the magnetic core at a flux density level lower than the saturation level of the magnetic material used on its construction, as stated by eq. (2). Also, maximum flux density choosing is critical, as it is directly related to the iron losses on the transformer core. Another important factor is the primary coil number of turns, which should be defined taking into account the voltage and current ratings of commercially available power semiconductors to achieve a good cost-benefit relationship.

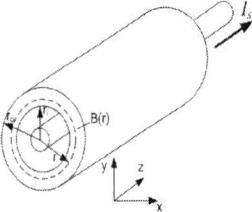

Fig. 4. Coaxial transformer used for active impedance coupling.

$$B(t) = B_1(t) + \sum_h B_h(t) < B_{sat\,max} \quad (13)$$

The effective harmonic voltage induced in the single turn secondary of the transformer is given by equation (13). Considering the harmonic induced voltage as a design input, (13) can be used for calculation of the minimum sectional area of the magnetic core necessary to induce the desired harmonic voltage. The magnetic core should be chosen satisfying simultaneously the equations (13) and (14).

$$V_{h2\_rms} = \frac{B_h \omega h A_e}{\sqrt{2}} \quad (14)$$

III. SIMULATION RESULTS

A. PLL Algorithms

Figures 5 and 6 shows the simulation results obtained from the application of the PLL algorithms to a signal composed by a fundamental component plus fifth and seventh harmonic components, with amplitudes of 1 pu, 1 pu and 0.1 pu respectively.

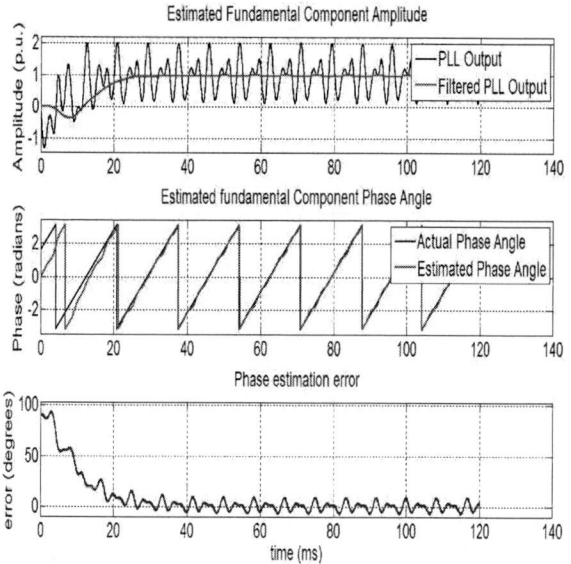

Fig. 5. Amplitude (I_q) and phase angle (θ_f) response obtained from PLL - fundamental component estimation applied to a signal composed by: 1 p.u. (fundamental component) + 1 p.u. (5th harmonic) + 0.1 p.u. (7th harmonic).

The PLL simulation results shows the effectiveness of the implemented algorithms, as the fundamental and the fifth harmonic components were correctly tracked even in the presence of an additional harmonic component.

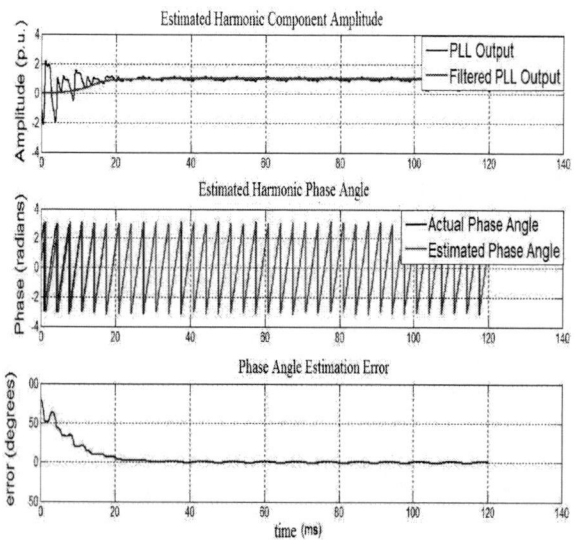

Fig. 6. Amplitude (I_{q5h}) and phase angle (θ_{5h}) response of the PLL used to estimate the fifth harmonic component.

IV. CASE STUDY: COMMERCIAL BUILDING

In typical commercial facilities the harmonic sources are distributed in the system. The allocation of filters would be as close as possible to all relevant harmonic sources. By doing so, the installed system would be able to achieve the overall reduction in energy consumption in kW and standardized levels of harmonic distortion. However, this method often becomes costly due to the number of filters to be installed and the difficulties of adapting the existing infrastructure to the solution. Proper selection of the installation points of the active harmonic impedance device and sizing of such units is the aim of this case study.

A. Building characterization

➢ Type: commercial building;
➢ Floors: 11;
➢ Constructed area: 9.000 m²;
➢ Population: 1.200 people;
➢ Installed power: 1.500 kW;
➢ Demanded power: 655 kW;
➢ Profile loads: air conditioning 54%, non-linear loads 36%, others 10%;
➢ Typical single phase non linear loads -127(V);
➢ Availability of all power plants and loads list.

B. Methodology

The methodology adopted for this case study, in progress, is based on the steps below:

i. The power quality survey of the entire plant via quality analyzer (Fluke 434), registering the higher incidence points of harmonic distortion. So, already at this stage, one can observe the technical feasibility of the application of active harmonic impedance devices, object of this paper.

ii. Modeling of the entire installation, on the software PTW. All components are properly configured and specified with real data from manufacturers.

iii. Modeling the active harmonic impedance device in PTW. Such model must present performance equivalent to harmonic compensation technology, proving its effectiveness in simulation studies.

iv. Investigation of the best installation points and sizing of the active harmonic impedance devices as required in this study case. Those two points identified: the input PFC and the BUS-TC Rooms and Elev, concentrates high level of harmonic distortion. The last one represents 69% of facilities non-linear critical loads. This condition has been consolidated and are shown on Table I and Figure 11 - *Quality energy diagram*. The implementation of SAHI TC at this point presents some advantages: contributed to reduce involved current levels, consequently the SAHI dimensions and costs; it represents the best facility place to install the equipment, ensuring harmonic mitigation to upstream loads, including the essential energy bus - BUS EE.

v. Practical implementation of the active harmonic impedance devices and installation at the points identified in the previous steps; perform new measurements of the entire installation.

V. MODELING AND SIMULATION OF THE FACILITY

First of all, the entire installation has been modeled, on the software PTW, shown in Appendix and described on item A – *Building characterization*. All components are properly configured and specified with real data from manufacturers. The simulation performance of the electrical system ratified the aforementioned results on the primary stage, identifying two high current harmonic distortion points by field measurements, Fluke analyzer and PTW simulations: BUS-TC Rooms and Elev data are registered at Table I and Figure 8 below, and BUS-En+EE at Table II and Figure 10 ahead:

TABLE I

Table of THDi, obtained at BW-QGBT/EE, demanded by BUS-TC Rooms and Elev, with quality analyzer equipment, before optimization.

Amp	A	B	C	N
THD%f	26,4	24,9	25,6	419
H3%f	21,9	22,4	22,7	418
H5%f	11,6	9,6	10,8	30,2
H7%f	7,7	3,8	3,8	26,6

TABLE II

Table of THDi, obtained at BW-SE/QGBT, demanded by BUS-En+EE and a lard contribution of Capacitor Bank loads mainly at 5th harmonic and 7th harmonic, obtained with quality analyzer equipment, before optimization.

Amp	A	B	C	N
THD%f	13	11,2	9,9	307,6
H3%f	7,9	8,7	7,4	305,8
H5%f	6,8	4,7	4,1	27
H7%f	6,3	4,4	4,1	22,6

TABLE III

Table of THDi, obtained at BW-GMG/QGBT, demanded by BUS-EE loads, obtained with quality analyzer equipment, before optimization.

Amp	A	B	C	N
THD%f	20.6	20.8	19.1	1026
H3%f	17.8	19.1	17.8	1033
H5%f	7.6	6.9	5.7	80.6
H7%f	5.7	3.4	2.6	55.2

Tables II and III, revels the propagation and impact of harmonics high level generated by nonlinear loads located downstream facility and named: TC Rooms and Elev.

Above, Figure 7, shows the parameterized voltage waveform on PTW and equivalent injected voltage synthesized by the PWM modulator, seen at the secondary side of the coupling transformer, and responsible for injecting the desired active harmonic impedance. As shown in TABLE IV, the relevance of harmonic order 3rd and 5th is evident, so deserving attenuation by the compensator.

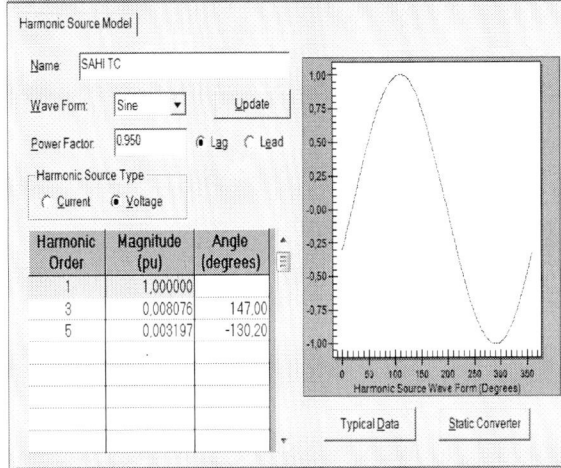

Fig. 7. PTW´s SAHI TC parameterized wave form

The values sizing of magnitude and angle defined in PTW and illustrated in Figure 7, were based on the values calculated by the software in the situation without optimization, ie, the amplitude of the harmonics and their respective angles of square phase shifted with respect the current.

TABLE IV

This table consolidates the facilities scenario object of these studies in the light of PTW before implementation of SAHI, showing numbers of power losses caused by respective harmonics distortion. Once more high THDi indices are measured.

HARMONIC CURRENT SPECTRUM REPORT SAHI OUT				
BW-QGBT/EE	kWLoss	kVARLoss	Harmonic Order/THDi%	IEEE-519 Limit
SAHI out	0,453	4,327	1/20,95	12
			3/18,851	10
			5/8,079	10
BW-GMG/QGI	kWLoss	kVARLoss	Harmonic Order/THDi%	IEEE-519 Limit
SAHI out	0,418	3,334	1/15,53	12
			3/13,492	10
			5/6,738	10
BW-SE/QGBT	kWLoss	kVARLoss	Harmonic Order/THDi%	IEEE-519 Limit
SAHI out	0,338	4,166	1/17,58	15
			3/9,578	12

V. PRACTICAL RESULTS

On the following three graphs presented, called "Waveform Distortion" and part of Figures 8, 09 and 10, we observe the behavior of the injected voltage signal wave induced current, where the angular displacement and previously parameterized amplitude in the desired harmonic, shows the effective compensation of their precursors unwanted signal. As a consequence, the positive result is materialized in a significant decline in the rates of both THDi as 3rd in 5th order, in all three evaluated situations.

It is noteworthy that these results were achieved exclusively and sufficiently with the implementation of SAHI-TC, ie, mitigation of harmonic distortion in the center of applied loads identified as being of higher harmonic contributions to the installation, further downstream. Its effectiveness can be observed at BW-SE/QGBT, with the exclusively actuation of SAHI-TC.

At this moment there is no contribution of active compensator implemented on BUS-CB yet. It is possible to observe the important influence of Capacitor Bank increasing harmonics 7th order. However specific studies should be further add to this paper.

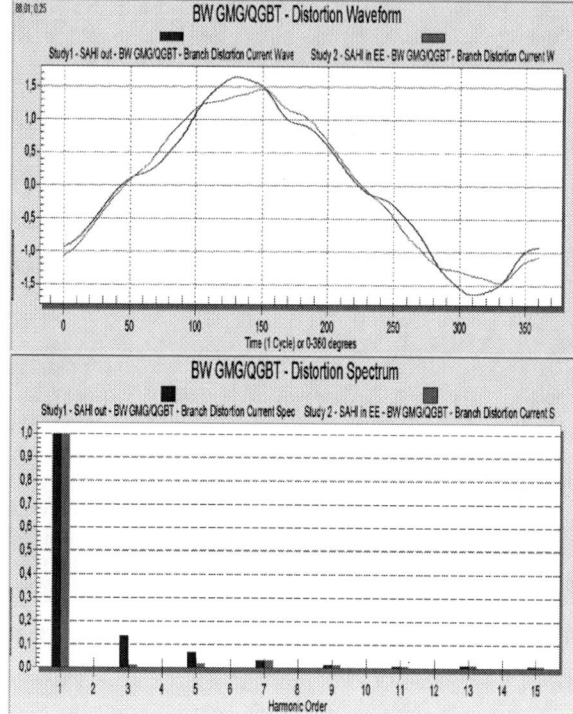

Fig. 9. Current Harmonic distortion spectrum, obtained from PTW simulation, at BW-GMG/QGBT, demanded by BUS-EE loads, before and after SAHI TC implementation.

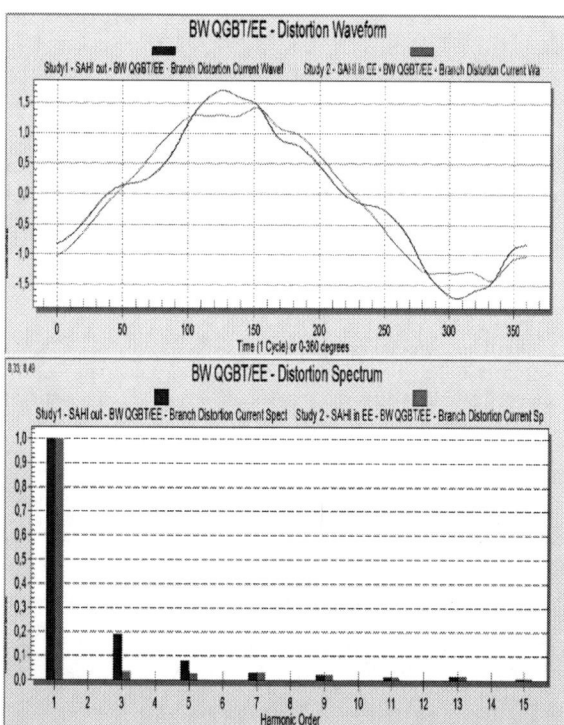

Fig. 8. Current Harmonic distortion spectrum, obtained from PTW simulation, at BW-QGBT/EE, demanded by BUS-TC Rooms and Elev loads, before and after SAHI TC implementation.

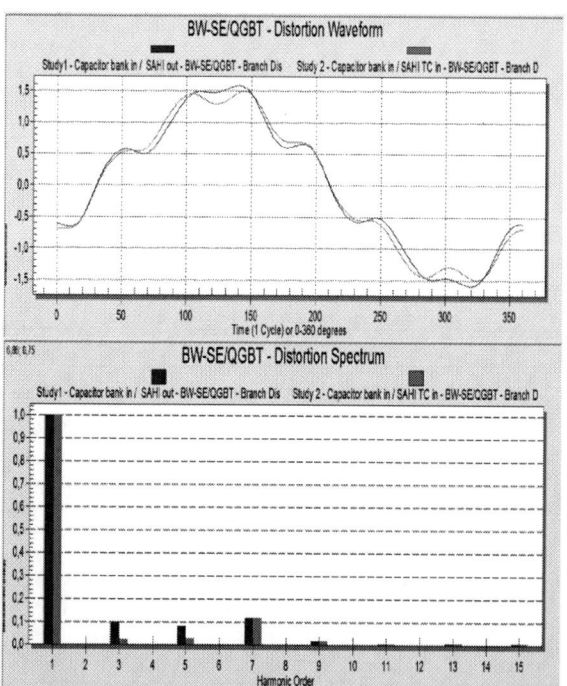

Fig. 10. Current Harmonic distortion spectrum, obtained from PTW simulation, at BW-SE/QGBT, demanded by BUS-En+EE, before and after SAHI TC implementation.

TABLE V

This table represents the significant mitigation of THDi harmonics levels on the facility, reaching values below the limits established by the IEEE-519 standard. Moreover, the power losses reduction saves 0,143% kW and 2,4% kVAR .

HARMONIC CURRENT SPECTRUM REPORT SAHI IN

BW-QGBT/EE	kWLoss	kVARLoss	Harmonic Order/THDi%	IEEE-519 Limit
SAHI in	0,248	1,371	1/6,03	12
			3/3,319	10
			5/2,648	10
BW-GMG/QGB	kWLoss	kVARLoss	Harmonic Order/THDi%	IEEE-519 Limit
SAHI in	0,288	1,46	1/4,21	12
			3/0,980	10
			5/1,733	10
BW-SE/QGBT	kWLoss	kVARLoss	Harmonic Order/THDi%	IEEE-519 Limit
SAHI in	0,278	2,681	1/12,66	15
			3/2,348	12
			5/2,66	12

Economical feasibility: The low values of "Voltage Magnitude" registered on Figure 7, above, confirm one of the most powerful SAHI's advantage, its competitive cost and volume.

VI.   CONCLUSION

This paper described a flexible harmonic compensation system that can be used to implement series or hybrid harmonic mitigation solutions. The use of a coaxial coupling transformer gives flexibility to the system allowing a fast field solution evaluation that is especially useful for detecting the better configuration in existing power systems.

The active harmonic impedance device performance and practical application is investigated based on a case study. A commercial building was selected and the harmonic flow in its power network was characterized based on field measurements. The building was modeled in PTW for further studies involving the application of the active harmonic impedance devices, aiming at the selection of the installation points and sizing of the harmonic mitigation devices required.

Aligned with the fact that it is only necessary to reduce the harmonic content of the electric systems to the compatibility levels of the equipment that shares it, and not completely eliminates the harmonic content, this work provide recommendations for reducing harmonic distortion, improving system capacity, availability and reliability while evaluating economic feasibility.

This is an ongoing project. The economic feasibility can be increased with the conclusion of details studies based on the electrical losses caused by harmonic distortion and can also add the function of limiting the harmonic currents through the capacitors without the need to raise their voltage levels and to control the harmonic flow, with the objective of distribute harmonics currents through patches that do not harm the system components.

REFERENCES

[1] D. J. Carnovale, T. J. Dionise, and T. M. Blooming, "Price and Performance Considerations for Harmonic solutions". *Power Systems World, Power Quality 2003 Conference*, Long Beach, Califórnia.

[2] H. Johal, D. Divan.," Design Considerations for Series-Connected Distributed FACTS Converters", *IEEE Trans. on Industry Applications*, vol, 43, n° 6, pgs. 1609-1618, November/December 2007.

[3] L. N. Arruda, B. J. Cardoso Filho, S. M. Silva, S. A. C. Diniz, "Wide Bandwidth Single and Three-Phase PLL Structures for Grid-Tied PV Systems", Photovoltaic Specialists Conference, 2000. Conference Record of the Twent€ y-Eighth IEEE, pgs. 1660-1663.

[4] G.J. França, B.J. Cardoso Filho, "Series-shunt distributed compensation for harmonic mitigation and dynamic power factor correction". Doctorate Thesis, PPGEE - Federal University of Minas Gerais, Belo Horizonte MG, 2012.

[5] G.J. França, B.J. Cardoso Filho. Series-Shunt Compensation for mitigation and dynamic power factor correction. Eletrônica de Potência (Printed), v. 17, p. 641-650, 2012.

[6] G.J. França, B.J. Cardoso Filho. Method and equipment for selective harmonic mitigation and its use. Pat. Req. BR1020130257079, 2013, INPI, Brazil.

[7] Bhattacharya S., P. Chen, D. Divan, "Hybrid Solutions for Improving Passive Filter Performance in High Power Applications", *IEEE Trans. on Industry Applications*, vol. 33, n° 3, May/June 1997.

[8] PTW, "Power Tools for Windows", SKM Power Tool Version 6.0.2.1

The 2014 International Power Electronics Conference

APPENDIX

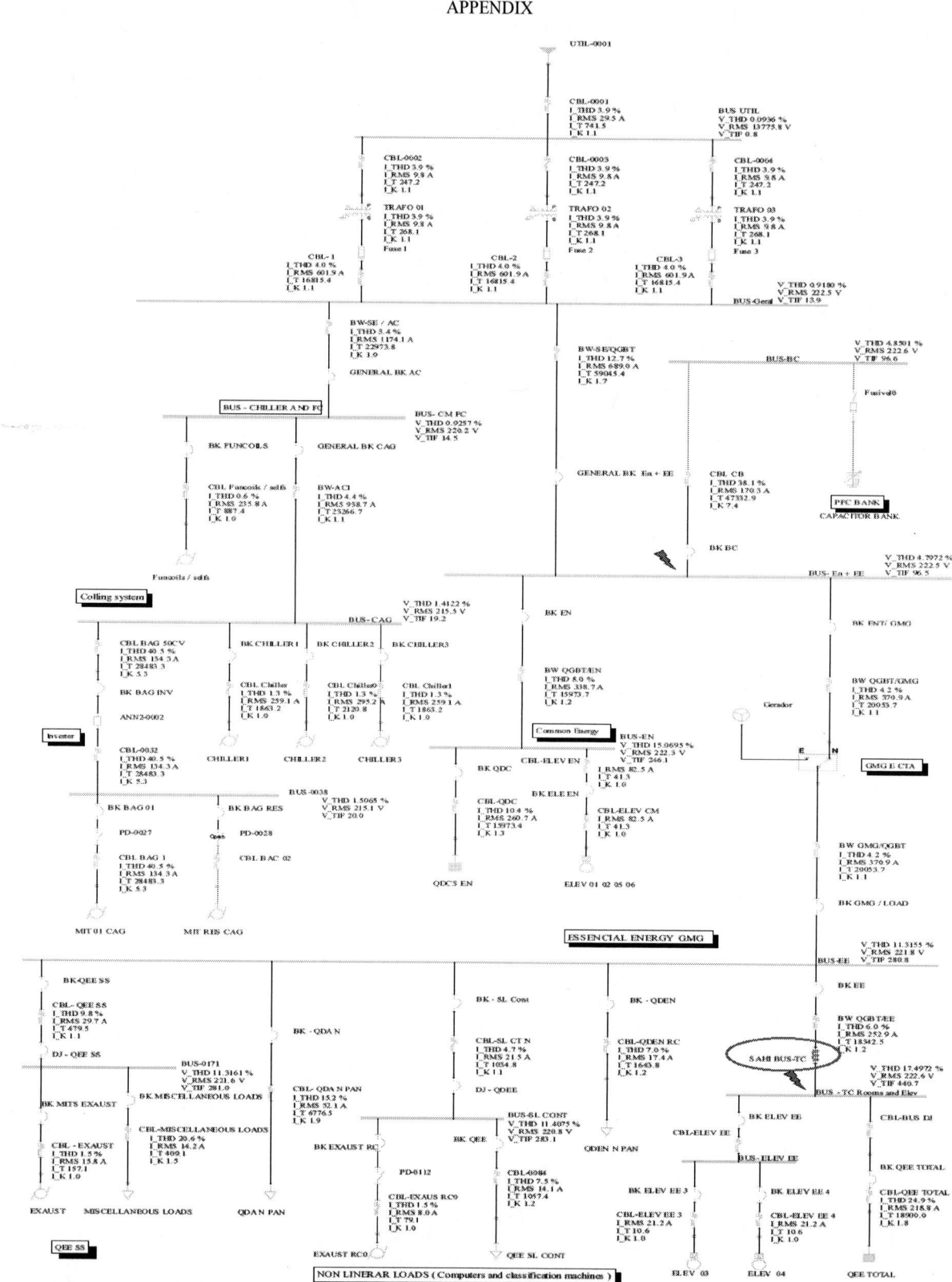

Fig. 11. Optimizided quality energy diagram on both essencial points

Robust Control Design for the Voltage Tracking Loop of a DVR

Bruno Augusto Ferrari, Naji Rajai Nasri Ama, Kelly Caroline Mingorancia de Carvalho, Fernando Ortiz Martinz, Lourenço Matakas Junior

Electrical Energy and Automation Department
Polytechnic School of the University of São Paulo
São Paulo, Brazil
bruno.ferrari@usp.br - matakas@pea.usp.br

Abstract-**The Dynamic Voltage Restorer (DVR) is a solution for protection of sensitive loads from voltage sags and swells. Its basic function is to inject voltages in the grid line in order to mitigate voltage sags and/or swells. Typically the controller structure for a DVR is composed by an inner current loop and an outer voltage loop. Usually a proportional or proportional integral controller is used for the current loop and a resonant controller is used for the voltage loop. This paper presents the design of a robust controller for the voltage tracking loop of a DVR that guaranties the robust stability against load parameters variation, and assures the tracking of a sinusoidal voltage waveform and the rejection of the non linear load current influence with a pre specified error. The voltage controller design is based on H$_\infty$ parameter specification approach. All the performance and robustness requirements are specified and analyzed based on the frequency response plot of closed loop transfer function. The proposed controller performance is validated by using PSIM software simulation and by experiments carried out on a DVR.**

I. INTRODUCTION

Power quality has been obtained increasing attention by utilities, as well as by both industrial and commercial electrical consumers. The Dynamic Voltage Restorer (DVR) can compensate mainly sags, swells and unbalances present at mains voltages, maintaining the load voltage at a desired magnitude and phase. In order to achieve this goal, a DVR must inject the required voltages in series with the line.

The DVR tracking performance is affected by the load current especially by distorted ones. Thus, the DVR controller must track the reference signal and adequately reject the disturbance caused by load variation and load current harmonics on the injected voltages. Additionally stability robustness must be assured for load variation.

Robust control [1][2][3] is a convenient tool to cope with the above listed issues. [1][2][3] use the H$_\infty$ mix-sensitivity approach. [1] considers robustness against variations on plant parameters, which does not includes the load model. [2] considers an UPS application, and includes a RL model of the load. [1],[2] and [3] do not quantify the tracking error nor the rejection of the load current disturbance, nor considers the influence of discretization and associated delays caused by the sampled PWM and the processor calculation time. They also consider the weighting function for tracking error

performance as a second order one, similar to a resonant controller, to provide high gain at the vicinity of the line frequency. This approach does not consider the influence of higher frequency harmonics of the load current on the converter output voltage.

This paper quantifies the performance for tracking error at fundamental frequency and disturbance rejection for the high frequency components of the load current, in the design of the robust controller. Considering that the load current is affected by the grid and the DVR voltage, the load model is not included in the nominal plant, but as a multiplicative uncertainty of the real plant.

This paper is organized as follows. Section II discuss the structure of the controller and the plant model taking into account the delays due to the sampled PWM and processor calculation time. Section III presents the formulation of the robust control problem including: i) the performance specification in terms of barriers in the frequency response plot; ii) weighting functions specification for closed loop system shaping and iii) discretization of the continuous time controller transfer function. Section IV and V presents the simulation and experimental results respectively.

II. SYSTEM MODELING

The DVR and its controller simplified block diagram are presented in Fig. 1. The plant is basically composed of a voltage source inverter and a second order LC filter, where the capacitor is connected to the secondary of the transformer. The controller consists of an inner current loop based on filter inductor current feedback and an outer voltage loop that tracks the voltage on filter capacitor. The DVR is modeled in continuous time domain including a delay block $e^{-T_d s}$ [4][5]. The time delay T_d is composed of two parts: i) the delay of $T_s / 2$ introduced by the sampled PWM where T_s is the sampling period; ii) the delay of T_s, related to the processor calculation time.

The internal current loop has the basic functions of attenuate the plant resonance frequency [1][2], and improve the capability to reject the disturbance caused by the load current.

The robust voltage controller $K_V(s)$ guarantees: i) the stability for all possible load conditions; ii) a pre defined tracking error performance and iii) the disturbance rejection requirement. Next section details the voltage controller specification and design.

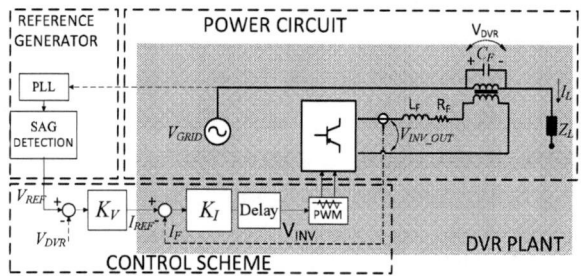

Fig. 1. DVR topology and controller block diagram.

Fig. 2 presents the linearized model of the DVR plant of Fig. 1. The proportional current controller K_I is considered part of the plant for the synthesis of the voltage controller $K_V(s)$ [1].

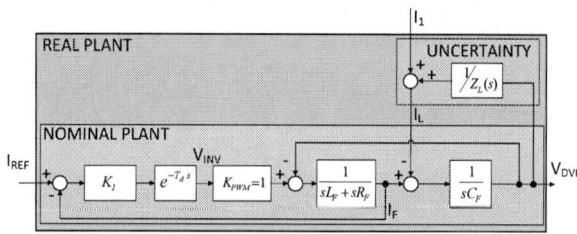

Fig. 2. DVR plant including the internal current loop.

The parameters in Fig. 2 are:

R_F: Filter resistance[1]. I_{REF}: Current reference.

L_F: Filter inductance[2]. V_{INV}: Inverter voltage.

C_F: Filter capacitor. I_F: Filter current.

$Z_L(s)$: Load. V_{DVR}: Output voltage.

K_I: Current controller. I_L: Load Current.

The time delay $e^{-T_d s}$ is approximated by a first order Padé approximation.

$$e^{-sT_d} \cong \frac{1 - sT_d/2}{1 + sT_d/2} \tag{1}$$

The H$_\infty$ formulation problem requires the block diagram of Fig. 2 to be rearranged to explicitly present the blocks $G_n(s)$ and $D(s)$ as shown in Fig. 3 and (2), (3) and (4).

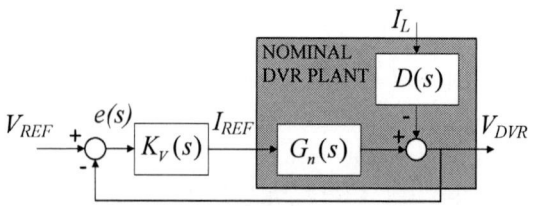

Fig. 3. Complete nominal control system.

$$V_{DVR} = G_n(s)I_{REF} - D(s)I_L \tag{2}$$

[1,2] Considers the sum of the filter and the transformer parameters

$$G_n = \frac{-K_I s + K_I \, 2/T_d}{s^3 C_F L_F + s^2 C_F (L_F \, 2/T_d + R_F.K_I) + s(C_F R_F \, 2/T_d + C_F K_I \, 2/T_d + 1) + 2/T_d} \tag{3}$$

$$D(s) = \frac{s^2 L_F + s(R_F + L_F \, 2/T_d - K_I) + R_F \, 2/T_d + K_I \, 2/T_d}{s^3 C_F L_F + s^2 C_F (L_F \, 2/T_d + R_F.K_I) + s(C_F R_F \, 2/T_d + C_F K_I \, 2/T_d + 1) + 2/T_d} \tag{4}$$

This paper models the load as a multiplicative uncertainty of the plant. In this sense, the dynamics of the load (dashed-line in the Fig. 2) is not considered in the nominal plant $G_n(s)$, but it is considered in the real plant $G_R(s)$.

For the real DVR plant, the total disturbance current I_L is a sum of two parts, one directly depending on line voltage (I_1), and the other depending on the DVR voltage, according to (5).

$$I_L = I_1 + I_2 = I_1 + \frac{V_{DVR}}{Z_L(s)} \tag{5}$$

So, once we consider the real plant, we have the dynamic of the load $Z_L(s)$ as a part of it. Substituting (5) in (2), results in the expression (6), that contains the real plant transfer function $G_R(s)$:

$$V_{DVR} = \underbrace{\frac{G_n(s)}{1 + D(s)A(s)}}_{G_R(s)} I_{REF} - \frac{D(s)}{1 + D(s)A(s)} I_1, \tag{6}$$

where:

$$Z_L(s) = R_L + sL_L, \tag{7}$$

$$A(s) = \frac{1}{Z_L(s)}, \tag{8}$$

and R_L and L_L are the load parameters; $G_R(s)$ is the real plant; $G_n(s)$ is the nominal plant.

Since the current controller is part of the nominal plant, the gain K_I must be defined before starting the robust controller design. According to [6] the maximum permissible gain for proportional current controller is $K_I \cong 19$ for 2-level, double update PWM converter and for the parameters defined in Table I. The effect of variation of K_I on $D(s)$, for the parameters in Table I is presented in Fig. 4.

The behavior of $D(s)$ affects the disturbance rejection specifications (explained in next section). The commitment of current controller specification in this paper is to reduce the natural resonance gain without increasing the low frequencies gain of $D(s)$. Taking into account the maximum current controller gain and Fig. 4, the preliminary choice for K_I is 10.

TABLE I - DVR SYSTEM PARAMETERS.

Parameter	Value	Dimension
C_F	20	μF
L_F	2.144	mH
R_F	0.235	Ω
DC Bus Voltage	175	V
Line voltage	110	V_{RMS}
Transformer ratio	1:1	-
Transformer leakage inductance	0.252	mH
Transformer series resistance	0.034	Ω
Switching frequency.	12	kHz
T_d^*	1/24000	s

* In this paper only calculation delay was considered.

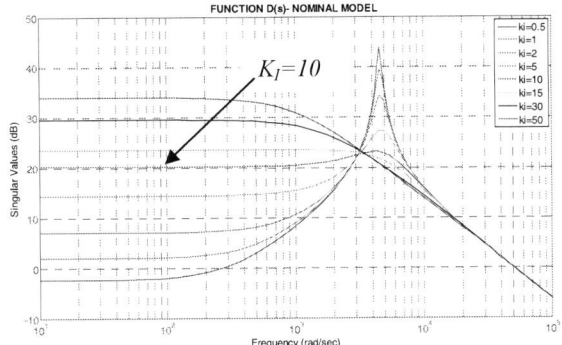

Fig. 4. D(s) frequency response for variation of K_I.

III. ROBUST CONTROL PROBLEM DEFINITION

This paper does not explore the generation of the voltage reference signal. It concentrates on the design of $K_V(s)$ that matches the performance and robustness specifications.

The H$_\infty$ problem formulation is defined in terms of sensitivity $S(s)$ and complementary sensitivity $T(s)$ transfer functions, presented in (9) and (10) respectively.

$$S(s) = \frac{1}{1 + G_n(s)K_V(s)} \quad (9)$$

$$T(s) = \frac{G_n(s)K_V(s)}{1 + G_n(s)K_V(s)} \quad (10)$$

The nominal plant $G_n(s)$ and the real plant $G_R(s)$ transfer functions were defined in section II. The plant multiplicative uncertainty $\Delta(s)$ is defined in (11).

$$\Delta(s) = \frac{G_R(s) - G_n(s)}{G_n(s)} \quad (11)$$

The problem formulation consists of specifying barriers in the frequency response curves that should be respected by the sensitivity and complementary sensitivity transfer functions of the closed loop system. Three different kinds of barriers are defined: i) barriers for robust stability against load parameters variation; ii) barriers for tracking error performance and iii) barriers for disturbance rejection performance. The H$_\infty$ method is used to obtain the robust controller $K_V(s)$ that respects the pre-defined barriers. It has the standard configuration

shown in Fig. 5, where the function $G_n(s)$ represents the nominal plant and $W_1(s)$, $W_3(s)$ and $W_2(s)$ are the weighting functions for tracking error performance, robust stability and control signal effort respectively.

These weighting functions should be defined such that the augmented plant $P(s)$ (with inputs u_1 and u_2 and outputs y_{1a}, y_{1b}, y_{1c} and y_2) carries out all the information about performance and robustness that the controller $K_V(s)$ will follow. The H$_\infty$ controller synthesis is conducted by using the mixed-sensitivity approach. The H$_\infty$ problem consists of finding a controller $K_V(s)$ that stabilizes the system and respects condition (12) (in terms of H$_\infty$ norm).

$$\left\| \begin{matrix} W_1(s)S(s) \\ W_2(s)K_V(s)S(s) \\ W_3(s)T(s) \end{matrix} \right\|_\infty < 1 \quad (12)$$

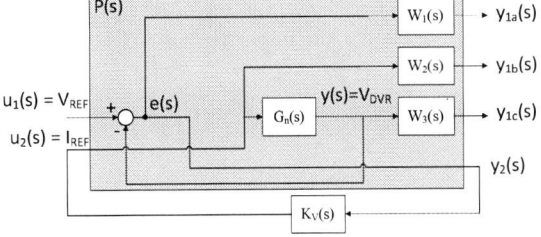

Fig. 5. Block diagram of H$_\infty$ control system.

Control effort is not relevant in this paper. It concentrates to analyze and to specify the tracking error and disturbance rejection performances. A small value of 0.1 is assigned to weighting function $W_2(s)$.

A. Weighting function definition for tracking error and disturbance rejection performance

Assuming that the maximum tracking error $e(s)$ allowed in the system is δ_r. Based on the block diagram of Fig. 3, the performance definition in terms of sensitivity function $S(s)$ is deduced as follow:

$$e(s) = \frac{1}{1 + G_n(s)K_V(s)} V_{REF} \quad (13)$$

$$\left| \frac{e(s)}{V_{REF}} \right| = \left| \frac{1}{1 + G_n(s)K_V(s)} \right| \leq \delta_r \quad (14)$$

$$|S(s)| \leq \delta_r, \quad \text{or} \quad \frac{1}{|S(s)|} \geq \frac{1}{\delta_r}. \quad (15)$$

Defining a function $W_1(s)$ such that $|W_1(s)| \geq \frac{1}{\delta_r}$, equation (16) is a sufficient condition for (15).

$$|W_1(s)S(s)| \leq 1 \quad (16)$$

978-1-4799-2706-7/14 $31.00 © 2014 IEEE

This is equivalent to $\left\| W_1(s)S(s) \right\|_\infty \leq 1$ in terms of H$_\infty$ norm.

By inspection of the block diagram of the complete control system shown in Fig. 3, the expression for the error $e(s)$ caused by the disturbance I_L is:

$$e(s) = -\frac{D(s)}{1+G_n(s)K_V(s)}I_L \qquad (17)$$

If δ_d is the maximum error caused by the disturbance, then:

$$\left| \frac{e(s)}{I_L} \right| = \left| \frac{D(s)}{1+G_n(s)K_V(s)} \right| \leq \delta_d \qquad (18)$$

that is the same as:

$$\left| \frac{1+G_n(s)K_V(s)}{D(s)} \right| \geq \frac{1}{\delta_d}. \qquad (19)$$

Thus, equation (20) presents the disturbance rejection performance in terms of the sensitivity function.

$$\frac{1}{|S(s)|} \geq \frac{1}{\delta_d}|D(s)| \qquad (20)$$

Since the performance specification for disturbance rejection (20) is in terms of $S(s)$, and defining the same function $W_1(s)$ such that $|W_1(s)| \geq \frac{1}{\delta_d}|D(s)|$, equation (16) is a sufficient condition for (20).

Taking into account that the same weighting function $W_1(s)$ is applied for disturbance rejection and tracking error, the definition of $W_1(s)$ is:

$$|W_1(s)| \begin{cases} \geq \dfrac{1}{\delta_r} \\[2mm] \geq \dfrac{1}{\delta_d}|D(s)| \end{cases} \qquad (21)$$

It is inferred from equations (20) and (21) that $D(s)$ influences the definition of the barriers and consequently the choice of weighting function $W_1(s)$.

In this paper the barrier δ_r is specified in per unit value. Additional care is required to specify δ_d, where a disturbance in the load current causes a voltage error in the DVR output. In other words δ_d has a dimension of Ohms while δ_r is dimensionless.

For each frequency $h\omega$ an harmonic load current I_{L_h} causes a voltage error E_h. The value of δ_{d_h} for frequency $h\omega$ can be calculated by (18), resulting in (22).

$$\delta_{d\_h} = \frac{E_h}{I_{L\_h}} \qquad (22)$$

Considering that the DVR nominal injected voltage is U_{nom} and the nominal load current is I_{L_nom}, it is possible to define the per unit values $E_h^{p.u.}$ and $I_{L\_h}^{p.u.}$ as (23).

$$\begin{cases} E_h = E_h^{p.u.} \cdot U_{nom} \\[2mm] I_{L\_h} = I_{L\_h}^{p.u.} I_{L\_nom} \end{cases} \qquad (23)$$

Rewriting (22) in terms of (23)

$$\delta_{d\_h} = \frac{E_h^{p.u.} \cdot U_{nom}}{I_{L\_h}^{p.u.} I_{L\_nom}} \qquad (24)$$

Defining the nominal DVR impedance as $Z_{nom} = U_{nom}/I_{L\_nom}$, δ_{d_h} can be easily obtained by (25) based on per unit values of load current and voltage error spectra.

$$\delta_{d\_h} = \frac{E_h^{p.u.}}{I_{L\_h}^{p.u.}}Z_{nom} \qquad (25)$$

The disturbance rejection barrier is defined not only for the nominal frequency but also for the most typical expressive harmonics (3rd order, 5th order, 7th order and 9th order), taking into account the existence of distorted load current.

Considering as a design example a small scale DVR, rated for 0.55 kVA, with nominal output voltage of $U_{nom} = 110V_{RMS}$ and nominal current of 5 A$_{RMS}$ ($Z_{nom} = 22\Omega$) and a non linear load with current spectrum defined by the first row of Table II, it is possible to define barriers δ_r and δ_d as follows:

1) This design example proposes to track only sinusoidal reference voltage at fundamental mains frequency. Consequently, $\delta_r = 0.05$ is defined only for fundamental frequency (red barrier with triangle mask in Fig. 6).

2) Defining $E_{rms}^{p.u.} = 0.05$ as the RMS value of the error caused by the load current and assuming that the 1 p.u. fundamental current causes 0.01 p.u. fundamental voltage error ($E_{h=1}^{p.u.} = 0.01$) and the remaining current harmonics causes voltage errors with the same amplitude $E_{h=3}^{p.u.} = E_{h=5}^{p.u.} = E_{h=7}^{p.u.} = E_{h=9}^{p.u.}$, it is possible to evaluate the $E_{h=3,5,7,9}^{p.u.}$ by (26), resulting in $E_{h=3,5,7,9}^{p.u.} = 0.0245$ (2nd row of Table II).

$$0.05 = \sqrt{(E_{h=1}^{p.u.})^2 + (E_{h=3}^{p.u.})^2 + (E_{h=5}^{p.u.})^2 + (E_{h=7}^{p.u.})^2 + (E_{h=9}^{p.u.})^2} \qquad (26)$$

TABLE II : DISTURBANCE REJECTION PERFORMANCE SPECIFICATION

	60 Hz	180 Hz	300 Hz	420 Hz	540 Hz
Load current harmonics in p.u.*	1	0.478	0.316	0.117	0.042
Maximum error caused by current load in p.u.	0.01	0.0245	0.0245	0.0245	0.0245
Calculated δ_{d_h}	0.22	1.127	1.701	4.583	12.634

*Non linear load current is based on diode rectifier with a capacitive filter.

Considering the per unit spectrum of a single phase rectifier (1st row of Table II), the permissible disturbance voltage error (2nd row of Table II), δ_{d_h} is calculated for each frequency based on equation (25) (3rd row of Table II). Using (21), the barriers $\left.\dfrac{1}{\delta_d}\left|D(s)\right|\right|_{(db)}$ are calculated and plotted in black without triangle mask in Fig. 6.

Considering these specifications for performance at low frequencies (tracking error and disturbance rejection), a convenient choice for $W_1(s)$ is shown in (27) and plotted in Fig. 6.

$$W_1(s) = \frac{4.494x10^5}{s^2 + 226.2s + 1.421x10^5} \qquad (27)$$

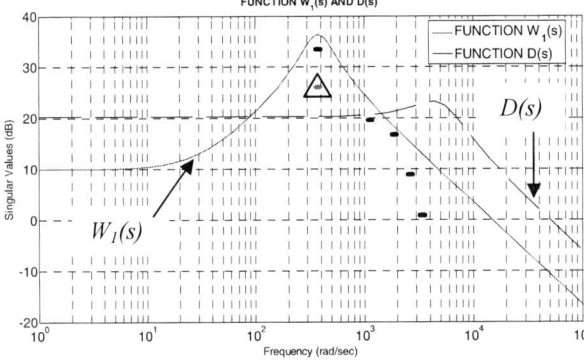

Fig. 6. Function $W_1(s)$. over the specified barriers.

B. Weighting function definition for robust stability against load parameters variation

As discussed before, the designed robust controller $K_V(s)$ should be suitable for operating with all possible load conditions. The load was modeled as a multiplicative uncertainty of the plant (11), where the load parameters R_L and L_L are chosen to obtain a impedance Z_L whose amplitude varies in the range $[22, \infty]\Omega$ and angle varies in the range $[0, 53.3°]$.

Considering the family of curves of $\Delta(s)$ obtained for each possible value of Z_L , $\Delta_{max} = \max\left|\Delta(s)\right|$ is the superior envelope of these $\Delta(s)$ curves. To achieve the desired robust stability, the condition (28) in terms of the complementary sensitivity function must be satisfied.

$$\left|T(s)\right| < \frac{1}{\Delta_{max}} \qquad (28)$$

Choosing a function $W_3(s)$ such that $\left|W_3(s)\right| > \Delta_{max}(s)$, (29) is a sufficient condition for (28).

$$\left|W_3(s)T(s)\right| < 1 \qquad (29)$$

This is equivalent to $\left\|W_3(s)T(s)\right\|_\infty < 1$ in terms of H$_\infty$ norm.

Fig. 7 shows the uncertainty Δ_{max} and the chosen weighting function $W_3(s)$ defined in (30).

$$W_3(s) = \frac{0.2917}{0.0001429\,s + 1} \qquad (30)$$

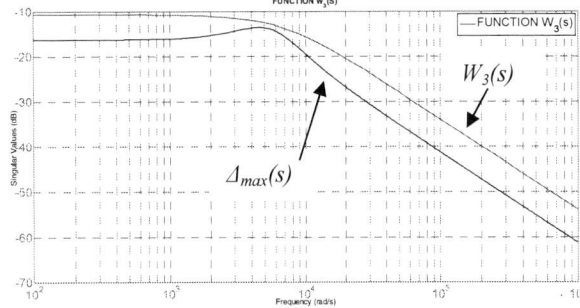

Fig. 7. $W_3(s)$ over the specified barrier $\Delta_{max}(s)$.

There is a commitment between the choices of the weighting functions. They cannot be chosen arbitrarily. Equations (16) and (29) results in (31).

$$\frac{1}{\left|W_1(s)\right|} \geq \left|S(s)\right| \quad \text{and} \quad \frac{1}{\left|W_3(s)\right|} > \left|T(s)\right| \qquad (31)$$

Taking into account (32) and (33) the inequality (34) must be satisfied.

$$\left|S(s) + T(s)\right| = 1 \qquad (32)$$

$$\left|S(s) + T(s)\right| \leq \left|S(s)\right| + \left|T(s)\right| \qquad (33)$$

$$\frac{1}{\left|W_1(s)\right|} + \frac{1}{\left|W_3(s)\right|} > 1 \qquad (34)$$

C. Voltage H$_\infty$ controller synthesis

Having defined the necessary weighting functions, the voltage H$_\infty$ controller $K_V(s)$ can be synthesized by using the Control Toolbox in MATLAB . The order of $K_V(s)$ depends on the order of the augmented plant (sum of the order of the nominal plant and the weighting functions). In this case synthesized $K_V(s)$ is a 6th order transfer function given by (35).

$$K_V(s) = \frac{a_0s^6 + a_1s^5 + a_2s^4 + a_3s^3 + a_4s^2 + a_5s^1 + a_6}{b_0s^6 + b_1s^5 + b_2s^4 + b_3s^3 + b_4s^2 + b_5s^1 + b_6} \qquad (35)$$

The parameters for $K_V(s)$ are:

$a_0 =$	0	$b_0 =$	1.0
$a_1 =$	6.137×10^6	$b_1 =$	7.236×10^5
$a_2 =$	3.127×10^{11}	$b_2 =$	7.268×10^{10}
$a_3 =$	3.285×10^{15}	$b_3 =$	4.589×10^{15}
$a_4 =$	1.595×10^{19}	$b_4 =$	2.905×10^{19}
$a_5 =$	4.336×10^{22}	$b_5 =$	6.985×10^{21}
$a_6 =$	8.874×10^{23}	$b_6 =$	3.981×10^{24}

Fig. 8 presents the sensitivity and complementary sensitivity function of the closed loop system together with $|1/W_1(s)|$ and $|1/W_3(s)|$. The plot shows that the requirements were achieved by the H$_\infty$ controller $K_V(s)$. The controller shaped the sensitivity and complementary sensitivity function properly to satisfy performance and robustness specification (31).

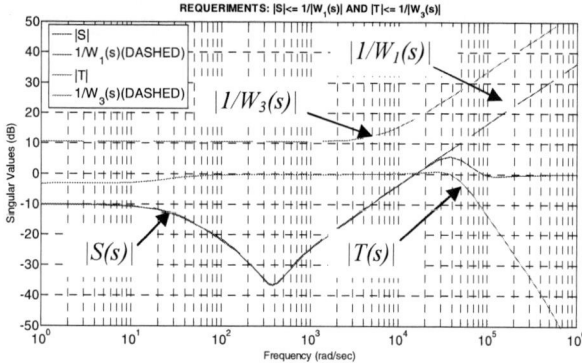

Fig. 8. Frequency response for functions $|1/W_1(s)|$ and $|1/W_3(s)|$ and Sensitivity $|S(s)|$ and Complementary Sensitivity $|T(s)|$ functions.

The voltage controller was reduced to a 5th order transfer function using the Schur method model reduction (MATLAB function *schbal*). The 5th order controller is then converted to a discrete time using Bilinear (Tustin) approximation with a sampling time $T_S = 1/24000$ seconds resulting in (36)

$$K_{V\_d}(z) = \frac{\alpha_0 + \alpha_1 z^{-1} + \alpha_2 z^{-2} + \alpha_3 z^{-3} + \alpha_4 z^{-4} + \alpha_5 z^{-5}}{\beta_0 + \beta_1 z^{-1} + \beta_2 z^{-2} + \beta_3 z^{-3} + \beta_4 z^{-4} + \beta_5 z^{-5}} \quad (36)$$

where the parameters are:

$\alpha_0 =$	3.1335164712625	$\beta_0 =$	1.0
$\alpha_1 =$	-5.8486467047783	$\beta_1 =$	-0.4990627568967
$\alpha_2 =$	-0.0390346055478	$\beta_2 =$	-1.1256865291907
$\alpha_3 =$	5.5846593475772	$\beta_3 =$	0.0452203699248
$\alpha_4 =$	-3.0943811073108	$\beta_4 =$	0.3191686221708
$\alpha_5 =$	0.2640881156050	$\beta_5 =$	0.2612459337884

IV. SIMULATION RESULTS

The performance of the proposed control scheme for a DVR was first verified by computer simulation considering the discrete time controller. The simulation was conducted using PSIM software. The parameters used in the simulation are shown in the Table I.

Several load conditions were tested in order to verify the tracking error performance and disturbance load current rejection such as: i) no load operation; ii) linear load operation, mainly to verify the performance of the controller with the disturbance in the fundamental frequency and iii) with non linear load operation, mainly to verify the harmonics load current disturbance rejection . The DVR reference voltages are calculated according to [7].

The simulation with no load operation is presented in Fig. 9, and summarized in the first column of Table III. The voltage tracking error is 1.53 % and is lower than the maximum specified tracking error $\delta_r = 5\%$.

Fig. 10 shows simulation results with a resistive load ($R_L = 33\Omega$). It was simulated with a sag of 58% and line voltage of 127 V$_{RMS}$. Fig. 11 shows simulation results with a series RL inductive load ($R_L = 13.4\Omega$ and $L_L = 46mH$). In this case the reference voltage is of 110 V$_{RMS}$. Both figures show the steady state waveforms immediately after the voltage sag. The calculated RMS amplitudes of V_{REF}, V_{DVR}, $E_{h=1}$ (voltage error at fundamental frequency) and I_L, and percent voltage error are given in Table III.

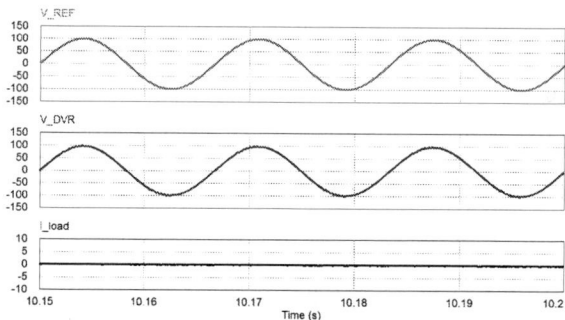

Fig. 9. Simulation results of no load operation. The voltage reference is in red (top), DVR output voltage is in blue (middle) and the load current is in black (bottom).

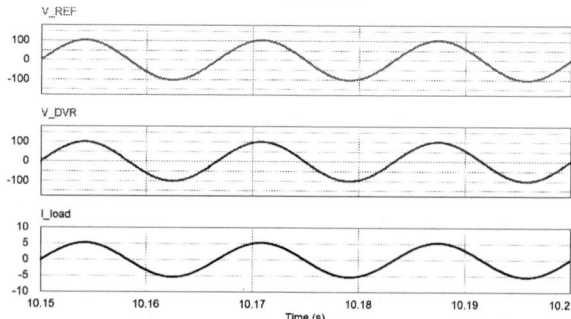

Fig. 10. Simulation results with resistive load. The voltage reference is in red (top), DVR output voltage is in blue (middle) and the load current is in black (bottom).

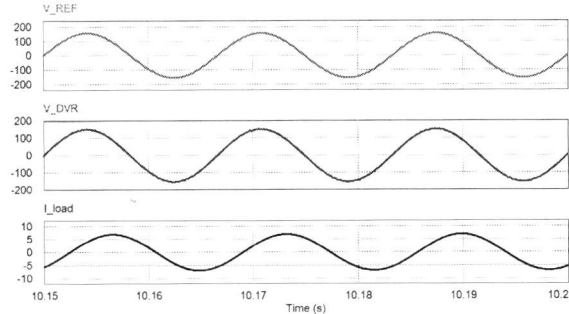

Fig. 11. Simulation results with inductive load. The voltage reference is in red (top), DVR output voltage is in blue (middle) and the load current is in black (bottom).

It is not possible to separate the tracking and disturbance errors. The propose design specifies 5% of tracking error and 1% disturbance error (at 1 p.u. of load current).The worst total error would be 6%. The total error obtained in all cases is significantly lower than 6%.

TABLE III - SIMULATION RESULTS WITH LINEAR LOAD.

	No-Load	R_L=33 Ω	R_L=33 Ω and L_L=46 mH
V_{REF} [V_{RMS}]	70.39	73.17	109.43
V_{DVR} [V_{RMS}]	69.23	71.41	107.19
$E_{h=1}$ [V_{RMS}]	1.08	1.68	2.25
$E_{h=1}$ [%]	1.53	2.29	2.05
I_L [A_{RMS}]	0	3.77	4.88

The simulation with non linear load (diode rectifier with capacitive filter) is shown in Fig. 12. In this case the reference voltage is of 110 V_{RMS}. The calculated RMS amplitudes of voltage error E_h, load current I_L and the specified and simulated values of disturbance gain δ_d are given in Table IV.

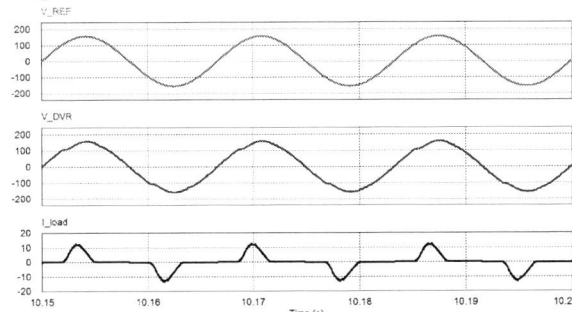

Fig. 12.Simulation results with non linear load. The voltage reference is in red (top), DVR output voltage is in blue (middle) and the load current is in black (bottom).

All the simulated values of δ_d are lower than the values specified in Table II (calculated δ_{d_h}).

TABLE IV - SIMULATION RESULTS WITH NON LINEAR LOAD.
DISTURBANCE REJECTION PERFORMANCE.

h	3	5	7	9
f	180 Hz	300 Hz	420 Hz	540 Hz
E_h [V_{RMS}]	1.89	2.34	1.41	0.55
I_L [A_{RMS}]	2.79	1.74	0.74	0.19
Specified δ_d [Ω]	1.127	1.701	4.583	12.634
Simulated δ_d [Ω]	0.67	1.34	1.90	2.89

V. EXPERIMENTAL RESULTS

The experimental setup is based on Fig. 1 and uses the same parameters of the simulation tests given in Table I. The voltage sag generator of Fig. 13 generates the V_{GRID} in Fig. 1. The control loop is implemented in a Texas TMS 320F28335 DSP. Voltages and current of the control loop are measured by means of Hall effect sensors (LEM). All the plotted signals are acquired using a DSO6014 Agilent oscilloscope with an Agilent N2775 current and differential voltage probes.

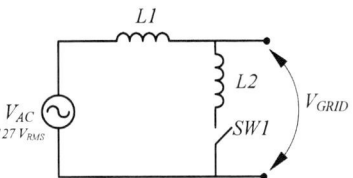

Fig. 13. Voltage sag generator.

Three experiments were carried out to validate the previous simulations. Fig. 14 shows the experimental waveforms of V_{REF}, V_{DVR} and voltage error for a voltage reference of 100 V_{peak} without load current. These signals were obtained from the DSP's D/A outputs.

Fig. 15 shows the experimental waveforms for a voltage sag of 58% of V_{GRID} (127 V_{RMS}). The calculated RMS amplitudes of V_{REF}, V_{DVR}, $E_{h=1}$ and I_L, and percent voltage error are given in Table V.

Table V shows good agreement with the simulation results of Table III. The experimental tracking error at no current load is lower than the specified value of 5%. In the case of the resistive load, the total error is also lower than the specified maximum error of 6% (worst case).

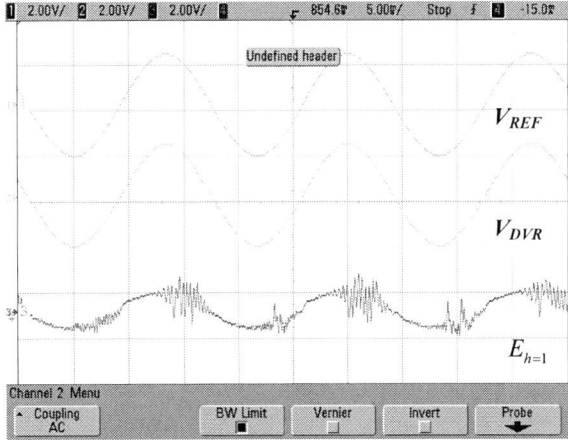

Fig. 14.Experimental results with no load. The voltage reference is in yellow (top), DVR output voltage is in green (middle) and the voltage error is in purple (bottom). Vertical scales: V_{REF} and V_{DVR}: 88V/div; E_{h-1}: 4V/div.

It is worth noting that V_{DVR} has harmonics (Fig. 15 and Fig. 16). It happens because the reference generator proposed in [7] also includes the active filter function. In the real experiment V_{REF} detects the V_{GRID} harmonics, shown in Fig. 15 and Fig. 16 (before the sag). The V_{GRID} is distorted however the load voltage is no so distorted.

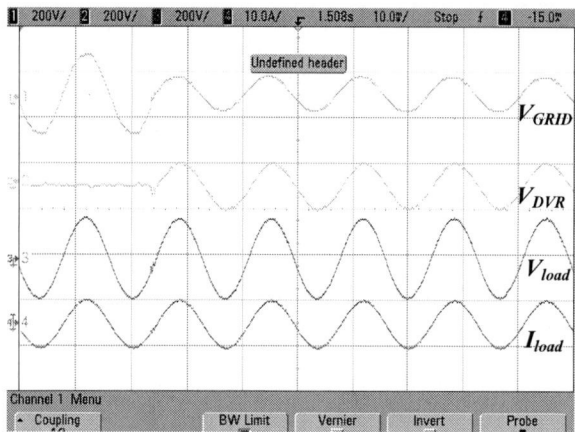

Fig. 15.Experimental results with resistive load. The grid voltage is in yellow (CH1), DVR output voltage is in green (CH2), load voltage is in purple (CH3) and the load current is in pink (CH4).

TABLE V - EXPERIMENTAL RESULTS WITH LINEAR LOAD.

	No-Load	$R_L = 33\ \Omega$
$V_{REF}\ [V_{RMS}]$	69.32	71.67
$V_{DVR}\ [V_{RMS}]$	69.05	71.02
$E_{h=1}\ [V_{RMS}]$	1.09	1.81
$E_{h=1}\ [\%]$	1.56	2.52
$I_L\ [A_{RMS}]$	0	3.54

The experimental waveforms with non linear load are shown in Fig. 16. It was tested with a voltage sag of 58% of V_{GRID} (127 V_{RMS}). The reference generator of [7] was modified to exclude the active voltage filter function, so as to obtain sinusoidal reference voltages to not corrupt the evaluation of disturbance rejection performance. The calculated RMS amplitudes of voltage error E_h, load current I_L and the specified and measured values of disturbance gain δ_d are given in Table VI.

Table VI shows good agreement with the simulation results of Table IV. Again, it is possible to see that all the measured values of δ_d are lower than the specified ones.

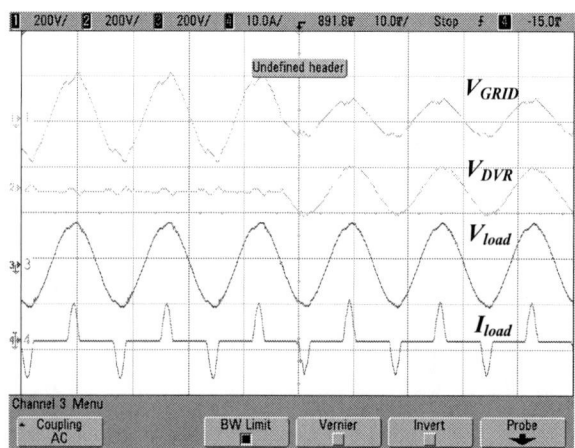

Fig. 16.Experimental results with non linear load. The grid voltage is in yellow (CH1), DVR output voltage is in green (CH2), load voltage is in purple (CH3) and the load current is in pink (CH4).

TABLE VI - EXPERIMENTAL RESULTS WITH NON LINEAR LOAD. DISTURBANCE REJECTION PERFORMANCE.

	h	3	5	7	9
	f	180 Hz	300 Hz	420 Hz	540 Hz
$E_h\ [V_{RMS}]$		1.18	1.38	0.86	0.12
$I_L\ [A_{RMS}]$		1.47	0.94	0.41	0.05
Specified $\delta_d\ [\Omega]$		1.127	1.701	4.583	12.634
Measured $\delta_d\ [\Omega]$		0.80	1.461	2.097	2.082

VI. CONCLUSION

A voltage control strategy for DVR applications using H_∞ synthesis was presented. An inner proportional current loop was used and included in the plant. Disturbance rejection for non linear load was included in the controller design procedure. A large range of load variation was modeled as an uncertainty of the plant. With proper weighting functions selection, the voltage controller synthesized by H_∞ method achieved all the performance and robustness specifications. DSP computation time delay was considered in the continuous time model.

The validation of the effectiveness of the proposed controller was conducted by simulation and experimental tests. The tracking and disturbance rejection performance of the controller was verified for several load conditions including non linear load.

ACKNOWLEDGMENT

The authors are grateful to Texas Instruments Academic Program, for donating the DSP Evaluation Kit and respective software, and to CNPq and FDTE for the financial support.

REFERENCES

[1] Y. W.Li, F. Blaabjerg, P. C. Loh "A Robust Control Scheme for Medium-Voltage-Level DVR Implementation," *IEEJ Transactions Electronics*, vol.54, no. 4, pp. 2249-2261, 2007.

[2] T.-S. Lee, S.-J. Chiang, J.-M. Chang "H∞ Loop-Shaping Designs for the Single-Phase UPS Inverter"*IEEE Trans. on Power Electronics*, vol. 16, no. 4, pp. 473-481, 2001.

[3] Yun Wei Li; Vilathgamuwa, D.M.; Poh Chiang Loh, "Robust Control Scheme for a Microgrid With PFC Capacitor Connected," Industry Applications, IEEE Transactions on , vol.43, no.5, pp.1172,1182, Sept.-oct. 2007.

[4] S. Buso, and P. Mattavelli, "Digital Control in Power Electronics," 1st ed., Morgan & Claypool, 2006.

[5] D.G. Holmes, T.A. Lipo, B.P. McGrath, and W.Y. Kong, "Optimized Design of Stationary Frame Three Phase AC Current Regulators," IEEE Transactions on Power Electronics, vol. 24, no. 11, pp. 2417-2426, 2009.

[6] F.O. Martinz, R.D. Miranda, W. Komatsu, and L. Matakas Jr., "Gain limits for current loop controllers of single and three-phase PWM converters," Proceedings of 2010 IPEC, Sapporo, vol.1, pp. 201-208, 2010.

[7] Terrazas, T.M.; Marafao, F.P.; Monteiro, T.C.; Giaretta, A.R.; Matakas, L.; Komatsu, W., "Reference generator for voltage controlled power conditioners," Power Electronics Conference (COBEP), 2011 Brazilian , vol., no., pp.513,519, 11-15 Sept. 2011

Multi-Port Solid State Transformer for Inter-Grid Power Flow Control

Sudhin Roy[1], Ankan De[2]
Department of Electrical and Computer Engineering,
North Carolina State University, USA
[1]sudhin.roy@gmail.com; [2]ade@ncsu.edu

Subhashish Bhattacharya
Department of Electrical and Computer Engineering,
North Carolina State University, USA
sbhattacharya@ncsu.edu

Abstract— **This paper discusses about a multi-port solid state transformer (SST) which can be used as an inter-grid power flow controller. The objective of the converter is to integrate renewable energy sources with the existing grid along with controlled power flow between the grids. As an example, this paper describes a three terminal isolated SST with an integrated battery bank connected to it. The energy extracted from the renewable energy sources can be stored in the battery bank or can be directly fed to the utility grid. A suitable converter structure is chosen to activate bidirectional power flow between the energy sources. The extracted energy can also be used to supply the local demand. A laboratory prototype of the converter has been built and the functionality of the converter is validated with detailed experimental results.**

Keywords— *Solid state transformer, renewable energy integration and current source converter.*

I. INTRODUCTION

The term "smart grid" [1] considered as next generation power grid, has become a topic of interest where the most power electronic research is focused. In the existing power grid, the power flow is unidirectional, i.e., from the generating station to the user end. This structure of power grid is not going to serve the purpose in the near future, as the conventional energy sources will not be available. As a solution to this problem, people have started to include non-conventional energy sources in the existing grid in a distributed manner [2]. Therefore, the power flow in the grid needs to be bi-directional. At the point of penetration of renewable energy sources, the transformer is a must which will interlink the high voltage transmission line and low/medium voltage sources. The demand of transformers goes high with the number of distributed sources. It is very obvious that the line frequency transformers are heavy and bulky. Thus, the need of small and light weight transformers (solid state transformers) came into picture. This paper discusses a current source inverter based solid state transformer with integrated renewable energy sources.

There have been many topologies based on voltage source converters as a replacement of the line frequency transformers. The basic solid state transformer (SST) concept is discussed in [3]-[4]. [5] shows a detailed comparative study of different topologies for the solid state transformer. Voltage source converter based topologies are not suitable for medium and high voltage systems because of the lack of control on short circuit current during fault conditions. Current source converters are therefore preferred over the voltage source converters.

This paper discusses a current source converter based solid state transformer. [6] (Chapter 10) discusses about a fly-back cyclo-converter with isolation. A new topology for bi-directional SST is presented in [7]. The SST topology used in this paper works on same principle as the fly-back cyclo-converter. The isolation between the input and the output sides are taken care with an extra winding of the fly-back inductor. This can also be used to step up or down the voltage. In this particular SST topology, another winding is added to the fly-back inductor to interface a battery-bank with the grid. It is charged from the grid and during peak power demand, the stored energy is fed back to the grid. Moreover, non-conventional energy sources (like solar PV cell) can also be interfaced to the grid through this port. The stored energy in the battery-bank can also be used to meet the local requirement.

II. CONVERTER STRUCTURE

Fig. 1(a) shows the schematic of the solid state transformer with the integrated battery storage system. This storage system is controlled in such a way that it does not affect the main power flow between the AC sources. It is interfaced with the AC sources with a bidirectional converter. The main converter (shown in Fig. 1(b)) which is used for the main grid to grid power flow consists of two cascaded full bridge inverters and one high frequency fly-back inductor.

The inductor with multiple windings provides electrical isolation between the sources. It also allows controlled power flow between the sources with unequal voltages. As shown in Fig.1 (b), the converter consists of a clamp circuit which restricts the inductor voltage to rise beyond a safe limit. The two cascaded converters are realized using an IGBT in series with a diode (Fig.1 (a)). The diode is used to block the reverse voltage across the switch. The switch allows unidirectional current and it can block bi-directional voltage.

978-1-4799-2706-7/14 $31.00 © 2014 IEEE

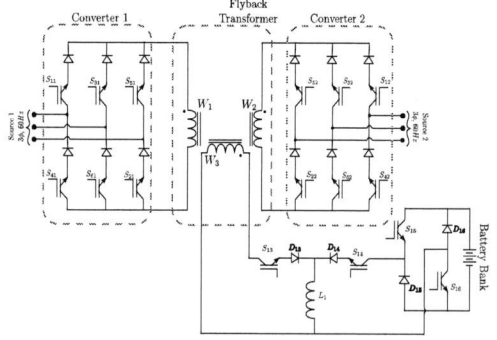

(a) With integrated battery energy storage

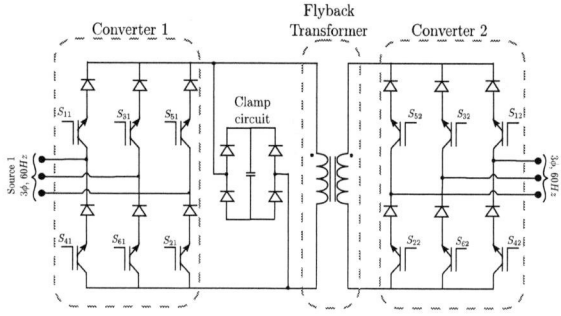

(b) Without the battery energy storage

Figure 1: Structure of the solid state transformer with and without the integrated battery energy storage system

III. PRINCIPLE OF OPERATION

It is mentioned earlier that the converter consists of four sections as (i) front end converter, (ii) fly-back inductor, (iii) rear end converter and (iv) battery charge/discharge circuit. The front end converter is used to store energy in the fly-back inductor depending on the two AC source side voltages. For a particular instant, switches corresponding to any two phases of the front end converter are turned ON to store energy in the fly-back inductor while the other converter remains in sleeping mode. For example, the switches S_{11} and S_{12} (Fig. 2) can be turned ON to store energy in the inductor drawing current from A and C phases. At this instant the other switches of the converter are in OFF condition. The path of the current flow is shown in Fig. 2(a). These switches are turned OFF at an instant when the inductor current reaches a particular value decided by the source voltages. The stored energy in the fly-back inductor is then delivered to the other source by turning ON the proper switches of the rear end converter. Fig. 2(b) shows one of the possible paths to de-energize the fly-back inductor. During this time the front end converter goes to sleeping mode like the rear end converter in charging mode. In practice, there should be an overlapping time between the two converters to avoid high voltage due to high L(di/dt). The moment the charging converter switches are turned OFF, the discharging converter switches should be ready

to provide path for the inductor current to flow. The inductor charging current flows through the primary winding while the discharging current through the secondary winding. Therefore, it is obvious that the coupling between the windings needs to be good in order to avoid voltage spikes due to the leakage flux between the windings. However, some sort of leakage inductance is unavoidable and it produces voltage spikes during the current change over instance from primary to secondary winding. The role of the clamp circuit is to absorb the voltage spikes which may appear due to many reasons other than the leakage flux effect. The clamp circuit consists of diode bridge rectifier and a DC capacitor as shown in Fig. 1(b). The energy stored by the capacitor due to the voltage spikes appearing on the inductor side has to be removed in order the clamp circuit to work effectively. A suitable resistance can be connected across the capacitor to dissipate the stored energy, however, it is not always recommended as it is associated with extra losses. The stored energy in the DC capacitor can be used to supply the local low power electronics via suitable bi-directional DC-DC converters.

The charging and discharging of the inductor is performed within a short period compared to the fundamental period of the source voltages. The pair of switches to be selected for this job are decided based on the specific amount of power to draw from a particular phase of the source. For example, the switches corresponding to the maximum and minimum phase voltages are turned ON to energize and then de-energize the inductor respectively. In this way power can be transferred from one source to the other without much control complexity. The only flaw in this case is that the shape of the current on both the converter grid sides becomes non-sinusoidal. The energizing and the de-energizing of the inductor can be controlled so as to draw sinusoidal and unity power factor currents from the sources. However, this brings complexity in the control technique.

A battery-bank is connected to the system via a bidirectional converter as shown in Fig.1. The converter is kept isolated from the system using a third winding in the fly-back transformer (winding W_3 as in Fig1). The number of turns for this winding is selected based on the grid voltage and the voltage of the battery bank. There are two operating modes for the converter, viz. (i) charging mode and (ii) discharging mode. The winding W_3 of the fly-back transformer always acts as a secondary winding of a two winding transformer. During charging mode power is first transferred from the winding W_1 to the inductor L_1 based on simple transformer action. In this duration the switch S_{13} is turned ON and all other switches of the battery charger remain OFF. The stored energy in the inductor L_1 is then transferred to the battery bank by turning switch S_{14} ON. The diodes D_{14}, D_{15} and D_{16} conduct during this period. The charging current to the battery bank can be controlled by varying the charging time for the inductor L_1. The charging of L_1 happens for a portion of the charging time of the main fly-back inductor. The

978-1-4799-2706-7/14 $31.00 © 2014 IEEE   3287

discharging of L_1 to the battery bank happens during the discharge time of the main fly-back inductor. In other words, during the charging time of the main fly-back inductor a portion of the drawn power from source 1 is stored in the fly-back inductor air gap and the other portion is transferred to the inductor L_1. The stored energy in the main fly-back transformer goes to source 2 and the stored energy in the inductor L_1 is delivered to the battery bank.

The discharge mode is selected for the battery bank when it is required to transfer power to the AC grids. In this mode the inductor L_1 is first charged from the battery bank and then the stored energy is discharged to the grid. This charging happens during the charging of the main fly-back inductor and the discharge of L_1 happens during the discharge of main fly-back inductor. Switches S_{14}, S_{15} and S_{16} are turned ON during the charging of L_1 from the battery bank. During the discharging mode S_{14} is turned OFF and S_{13} is turned ON. This connects the inductor L_1 in parallel with the main fly-back inductor and both the inductors discharges to the AC source.

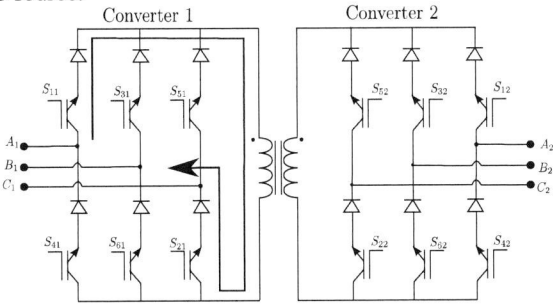

(a) During charging of the fly-back inductor

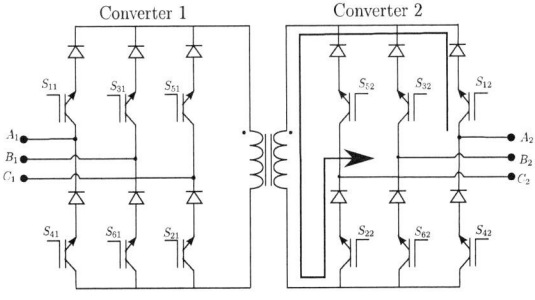

(b) During discharging of the fly-back inductor

Figure 2: Schematic of the power converter while the fly-back inductor is energized and de-energized respectively with the line voltage VAC

IV. CONTROL ALGORITHM

Fig. 3 shows a switching time interval with the inductor currents and voltage referred to primary winding. The power flow is happening from A_1-B_1-C_1 grid to A_2-B_2-C_2 grid. The switching time interval is divided into three parts, i.e., charging time (0 to T_1), discharging time (T_1 to

T_2) and the dead time. During the charging time the fly-back inductor is charged via converter 1 and it is discharged to the other grid via converter 2. The dead time is a small time duration when all the converter switches are OFF and it ensures discontinuous current of the fly-back inductor. It is well-known fact that instead of continuous current through the fly-back inductor, the discontinuous current gives requires smaller size of magnetic circuit.

The grid voltages are sensed and converter 1 switches corresponding to the maximum phase voltage are selected during each of the charging intervals. Thus the inductor current ramps up from 0 to peak value during the charging time interval, i.e., 0 to T_1. After T_1, converter 2 is turned ON (switches corresponding to minimum phase voltages) to discharge the stored energy in the fly-back inductor to the other grid. The time intervals are so selected that the inductor current comes back to 0 at time T_2. This is clearly depicted in the Fig. 3 below.

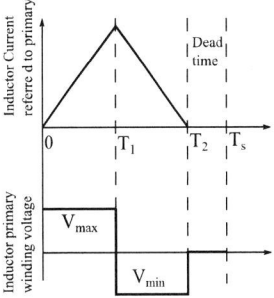

Figure 3: Fly-back inductor primary winding voltage and current during a switching time interval.

The control algorithm for the main power flow remains untouched even in the presence of the battery charger:-

$$P = V_{max} * I_{avg}$$

The time duration (t_{on}) is so chosen to appropriately allow a particular throughput power (P),

$$I_{avg} = \frac{\int_0^{t_{on}} i(\text{t}).dt}{(1/F)}$$

Where,

$$i(\text{t}) = \frac{V_{max}}{L} t$$

'L' is the magnetizing inductance of the fly-back transformer. The overall expression for ton to allow power 'P' to transfer from one grid to other is:

$$t_{on} = \sqrt{\frac{P}{V_{max}^2} * \frac{2L}{F}}$$

The following paragraphs explain the control topology during the charging and discharging mode of the battery.

The battery in charging mode:-

Fig. 4(a) shows the fly-back inductor and the inductor L_1 voltage and current while the battery is in charging mode. The control scheme for the main converters remains untouched. As explained above, L_1 is charged while the fly-back inductor is storing energy from the A_1-B_1-C_1 grid. The stored energy in L_1 is then discharged to the battery bank. The time interval T_3 is selected depending on the charging status of the battery.

The battery in dis-charging mode:-

Fig. 4(b) shows the fly-back inductor and the inductor L_1 voltage and current while the battery stored energy is discharged through grid A_2-B_2-C_2. During the charging interval of the main fly-back inductor, L_1 is also charged from the battery bank and during the dis-charging mode it is discharged to the grid via converter 2.

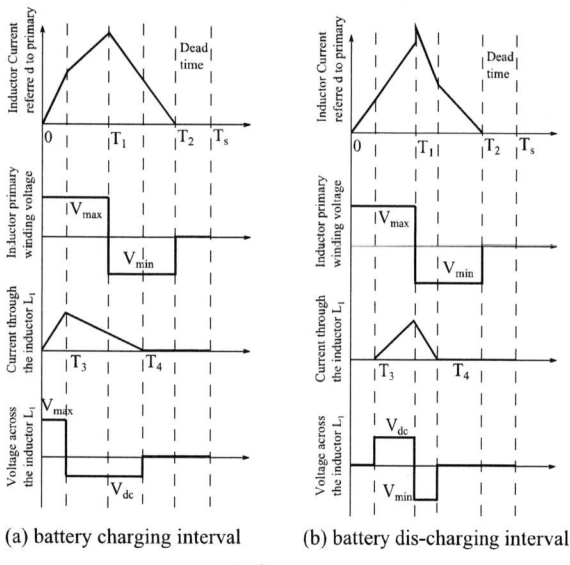

(a) battery charging interval   (b) battery dis-charging interval

Figure 4: Fly-back inductor and the battery charging inductor voltage and current during a switching time interval.

V. EXPERIMENTAL RESULTS

The above mentioned converter operation is experimentally verified as shown below. The control logic is implemented using a DSP platform (TMS320F2812). Gate pulses are generated for the IGBTs corresponding to the maximum and minimum phase voltages as explained in the above section.

Fig.5 (a) and (b) show the primary winding voltage and current of the fly-back transformer with and without the battery charger operation respectively. It is clear that with the battery charger ON, the primary winding (W_1 winding) current has two different slopes. The initial higher slope is because of the current drawn to charge the main inductor as well as to charge the inductor L_1 (looking at Fig. 1(a)) and the final slope is due to only the main fly-back inductor charging current. During this interval current is drawn only to store energy in the main fly-back inductor. This shape of the primary winding

current (two different slopes) is missing while the battery charger is OFF (Fig. 5(b)). In the discharging period of the switching time Ts, the secondary winding of the main fly-back inductor carries the current and the stored energy is fed to the other grid. Also, the stored energy in the inductor L_1 (looking at Fig. 1(a)), is used to charge the battery bank. The duration for which the inductor is charged depends on the state of the battery and the grid side line voltage which is selected for charging the battery-bank. If the battery is fully charged the inductor L_1 is disconnected electrically from the circuit with the help of the switches S_{13} and S_{14} as shown in Fig. 1 (a).

The stored energy in the battery-bank can be delivered to the grid in case of emergency. In this mode of operation, the inductor L_1 is first charged from the battery bank and then that energy is delivered back to the grid in the discharging period of the main fly-back inductor. In this particular case, the winding W_1 is used to charge the battery bank and to discharge it the winding W_2 is used. Fig. 6(a) and (b) show the voltage and current of the winding W_2 with and without the battery in discharge mode respectively. Similar to charging mode, here also the current has two slopes during battery discharging.

Fig. 7(a) and (b) show the voltage and current of W_3 winding while the battery charger is used in charging and discharging mode respectively.

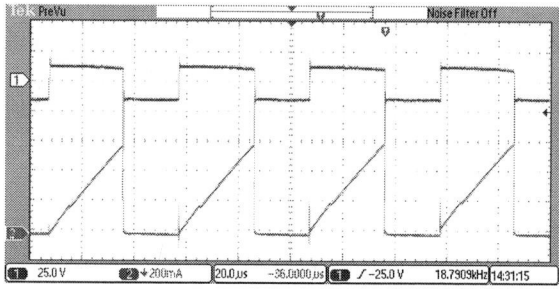

(a) With the battery charger ON

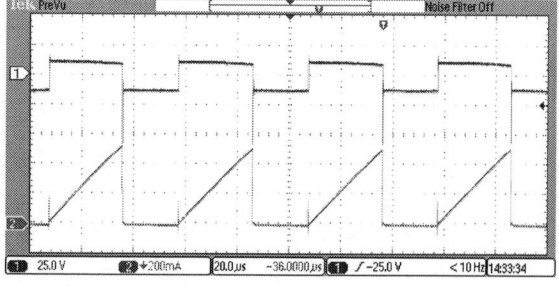

(b) With the battery charger OFF

Figure 5: Primary winding voltage and current of the fly-back transformer with and without the battery charger (ch1: voltage; ch2: current); (voltage scale: 250V/div; current scale: 4A/div; time scale: 20us/div)

20us/div, (b) 40us/div)

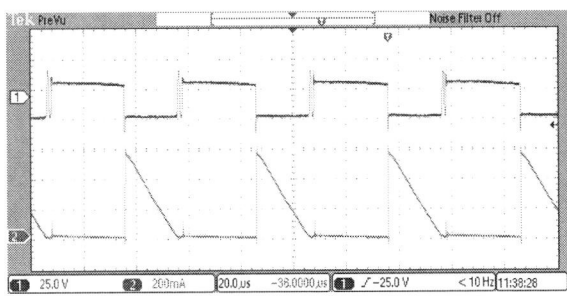

(a) With the battery in charging mode

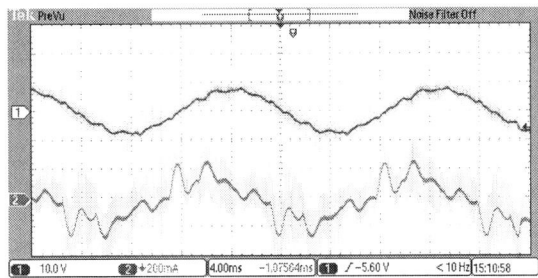

(a) Input side grid voltage and current

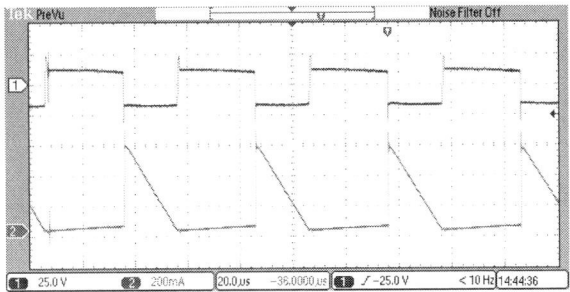

(b) Without the battery in discharging mode

Figure 6: Secondary winding voltage and current of the fly-back transformer with and without the battery in discharging mode (ch1: voltage; ch2: current); (voltage scale: 250V/div; current scale: 4A/div; time scale: 20us/div)

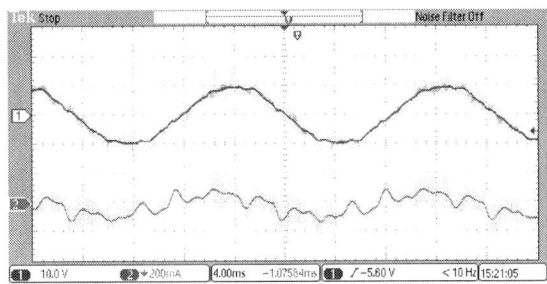

(b) Output side grid voltage and current

Figure 8: Input and output side grid voltage and current (R-phase) (ch1: voltage; ch2: current); (voltage scale: 100V/div; Current scale: 4A/div; time scale: 4ms/div)

Fig.8 (a) and (b) show the R-phase grid voltage and current of the input and output side respectively.

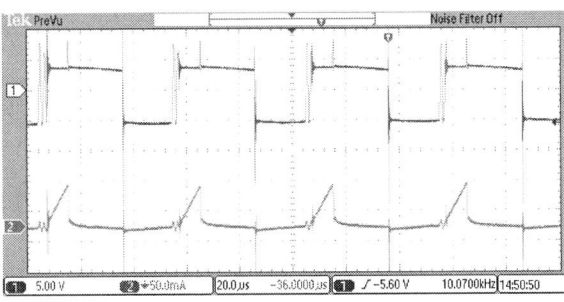

(a) With the battery in charging mode

VI. CONCLUSIONS

The solid state transformer topology presented in this paper has low loss compared to its voltage source converter based counterpart. The battery interface circuit adds an interesting feature of adding renewable energy sources to the existing power network. The extracted energy can be stored in the battery bank and the excess amount can be fed directly to the grid. The battery bank can also be charged from a renewable energy source and during peak demand that energy can be fed to the grid via the converter shown in this paper. Thus the new era of green energy can be accelerated. In this case an experimental prototype of 2kW has been built to verify the operation of the system. From the experimental results presented in this paper it is clear that the solid state transformer operates as expected.

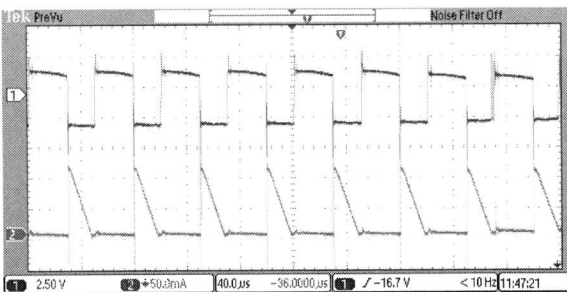

(b) Without the battery in discharging mode

Figure 7: Third winding (W3) voltage and current of the fly-back transformer with the battery in charging and discharging mode respectively (ch1: voltage; ch2: current); (voltage scale: (a) 50V/div (b) 25V/div; current scale: 1A/div; time scale: (a)

VII. APPENDIX

The laboratory prototype is built for the following specifications.

Power Rating: 2kW
Voltage Rating: 300V
Inductance of the main fly-back inductor: 300mH

Number of turns of the three winding fly-back inductor: 40 : 40 : 10 (N1 : N2 : N3)
Inductance of the battery charger inductor (L1): 700mH
Part No of the switches: IGBT (IXA12IF1200HB); Diode (ISL9R30120G2)

ACKNOWLEDGMENT

This work made use of FREEDM ERC shared facilities supported by National Science Foundation under Award Number EEC-0812121.

REFERENCES

[1] R. Hassan and R. G, "Survey on smart grid," *IEEE SoutheastCon 2010 (SoutheastCon), Proceedings of the, March 2010, pp. 210–213.*

[2] A. Q. Huang and B. J, "Freedm system: Role of power electronics and power semiconductors in developing an energy internet," *Power Semiconductor Devices & IC's, 2009. ISPSD 2009. 21st International Symposium on, June 2009, pp. 9–12.*

[3] J. van der Merwe and H. du T. Mouton, "The solid-state transformer concept: A new era in power distribution," *AFRICON, 2009. AFRICON '09., 2009, pp. 1–6.*

[4] L. Heinemann and G. Mauthe, "The universal power electronics based distribution transformer, an unified approach," Specialists Conference, 2001. PESC, vol. 2, 2001, pp. 504–509.

[5] S. Falcones, X. Mao, and R. Ayyanar, "Topology comparison for solid state transformer implementation," Power and Energy Society General Meeting, 2010 IEEE, 2010, pp. 1–8.

[6] K. D. T. Ngo, "Topology and analysis in pwm inversion, rectification and cycloconversion," Ph.D. dissertation, California Institute of Technology, May 1984.

[7] US Patent Publication # US20130201733 A1, Aug 8, 2013. "Isolated dynamic current converters", by Deepakraj M. Divan, Anish Prasai, Hao Chen.

Reactive Power Control Strategy Based on DC Capacitor Voltage Control for Active Load Balancer in Three-Phase Four-Wire Distribution Systems

Tint Soe Win[1], Yoshihiro Hisada[1], Toshihiko Tanaka[1*], Eiji Hiraki[2], Masayuki Okamoto[3] and Seong Ryong Lee[4]

[1]Yamaguchi University, Japan
[2]Okayama University, Japan
[3]Ube National College of Technology, Japan
[4]Kunsan National University, Korea
[1*]Email: totanaka@yamaguchi-u.ac.jp

Abstract—This paper proposes a reactive power control strategy for the active load balancer (ALB) in three-phase four-wire distribution systems. The proposed reactive power control strategy is based on constant DC capacitor voltage control, which is always used in active power line conditioners. Therefore, the proposed reactive power control strategy does not require active, reactive calculation blocks of load currents for the reference source current calculation. Balanced source currents with a predefined power factor can be achieved without unbalanced active and reactive components detection of the load currents. The basic principle of the reactive power control strategy for the ALB is discussed in detail, and then confirmed by digital computer simulation using PSIM software. A prototype experimental model is constructed and tested to validate the feasibility of the proposed control algorithm. Simulation and experimental results demonstrate that balanced source currents with a predefined power factor are achieved in three-phase four-wire distribution systems.

Keywords—*active load balancer, DC capacitor voltage control, reactive power control, three-phase four-wire distribution system.*

I. Introduction

A wide variety of distribution systems are used worldwide for residential and commercial power supply. Among them, three-phase four-wire wye-connected distribution systems are widely used in South Korea, Myanmar, and other countries. Both three-phase and single-phase power can be fed simultaneously using these distribution systems, as shown in Fig. 1. However, single-phase loads on a three-phase feeder result in unbalanced load conditions in the distribution systems. The unbalanced load currents cause unbalanced voltages, and affect other loads connected at the same point of common coupling (PCC). These unbalanced load conditions cause excessive neutral current, which results in transformer overheating, power losses, and lower system efficiency. Therefore, load balancing is necessary in three-phase four-wire distribution systems to improve the power quality and efficiency of the distribution system. To solve these power quality problems, active power line conditioners with several control algorithms have been proposed for three-phase four-wire distribution systems [1]–[6]. Some control algorithms are based on instantaneous active-reactive power theory and its extensions for the reference

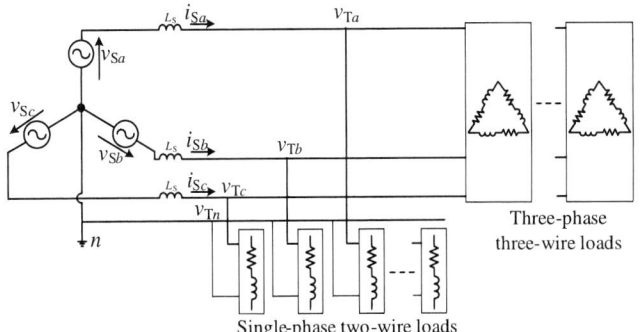

Fig. 1. Three-phase four-wire distribution systems.

current calculation [1]–[3]. Decomposition in the time domain of the load currents, the sample and hold circuit method, and instantaneous symmetrical component theory are also used in the control algorithm [4]–[6]. All these reference current calculation methods require a significant number of computation steps. To reduce the number of computation steps, we have proposed a simple control algorithm for the active load balancer (ALB), which is based on constant DC capacitor voltage control [7]. Balanced source currents with unity power factor can be achieved without any active and reactive calculations of the load currents. Therefore, the number of calculation steps can be considerably reduced using the simple control algorithm. However, a large power rating is required for the ALB to achieve a unity power factor in the source-side currents. A reduced ALB power rating is necessary for practical applications.

In this paper, we propose a reactive power control strategy based on constant DC capacitor voltage control to reduce the power rating of the ALB. Balanced source currents with a power factor of 0.9, which is acceptable in Japanese power distribution systems [8], are achieved without any calculation blocks of unbalanced active and reactive components in the load currents. The basic principle of reactive power control strategy based on constant DC capacitor voltage control for the ALB is discussed in detail, and then confirmed with digital computer simulation using PSIM software. A prototype

The 2014 International Power Electronics Conference

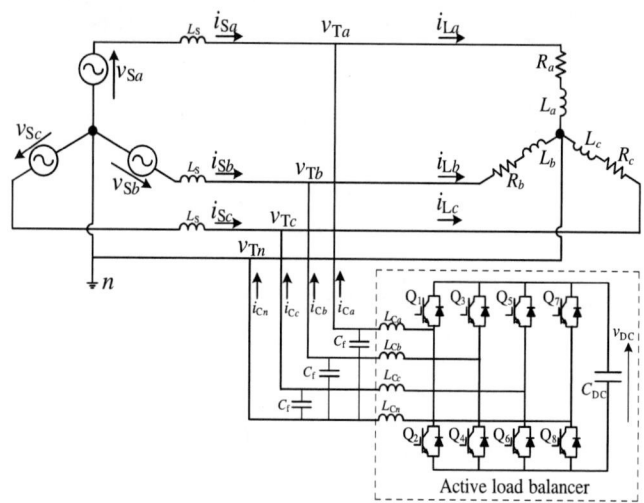

Fig. 2. Power circuit diagram of the proposed active load balancer.

experimental model is constructed and tested to validate the feasibility of the proposed control algorithm. Simulation and experimental results demonstrate that the proposed reactive power control strategy is highly applicable for the ALB in three-phase four-wire distribution systems.

II. REACTIVE POWER CONTROL STRATEGY BASED ON DC CAPACITOR VOLTAGE CONTROL

A. Power Circuit Configuration

Fig. 2 shows a power circuit diagram of the proposed ALB for a three-phase four-wire distribution system. The three unbalanced single-phase loads are used for the unbalanced load conditions. These three single-phase unbalanced loads are connected through a three-phase four-wire distribution feeder. The ALB, which is constructed with four-leg power-switching devices with a common DC capacitor, is connected in parallel to three single-phase loads. Thus, the three legs of the ALB are connected to each phase of the distribution transformer, and the fourth leg is connected to the neutral-line. The unbalanced active and reactive currents drawn by the three single-phase loads are compensated by the ALB, which provides balanced source currents with a predefined power factor in the distribution transformer.

B. Proposed Control Algorithm

Fig. 3 shows a block diagram of the proposed reactive power control strategy for the ALB in three-phase four-wire distribution systems. The principle of reactive power control strategy is discussed. The three-phase terminal voltages v_{Ta}, v_{Tb}, and v_{Tc} in Fig. 2 are expressed as

$$
\begin{aligned}
v_{Ta} &= \sqrt{2}V_T\cos(\omega t), \\
v_{Tb} &= \sqrt{2}V_T\cos(\omega t - \frac{2\pi}{3}), \\
v_{Tc} &= \sqrt{2}V_T\cos(\omega t - \frac{4\pi}{3}).
\end{aligned}
\tag{1}
$$

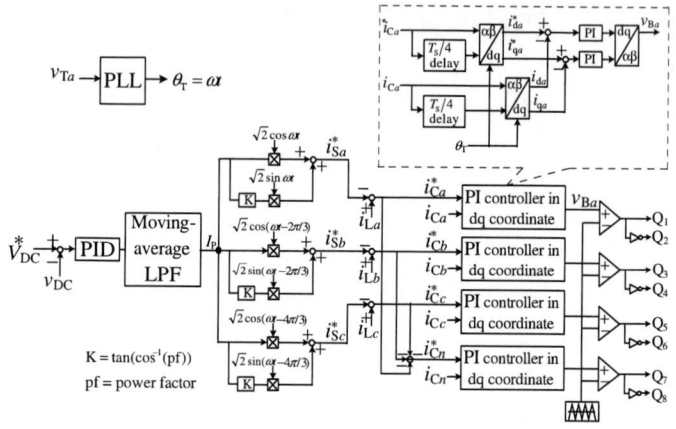

Fig. 3. Proposed reactive power control strategy for the active load balancer.

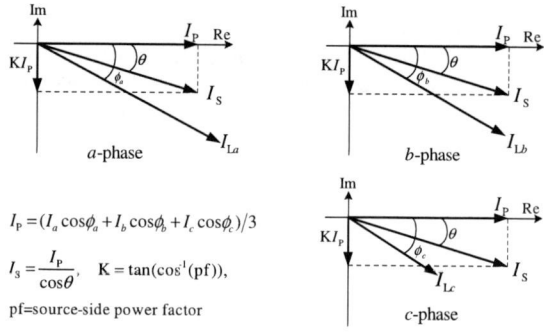

Fig. 4. Phasor diagrams of the proposed control strategy.

The load currents i_{La}, i_{Lb}, and i_{Lc} drawn by each single-phase load in Fig. 2 are also expressed as

$$
\begin{aligned}
i_{La} &= \sqrt{2}I_a\cos(\omega t - \phi_a), \\
i_{Lb} &= \sqrt{2}I_b\cos(\omega t - \frac{2\pi}{3} - \phi_b), \\
i_{Lc} &= \sqrt{2}I_c\cos(\omega t - \frac{4\pi}{3} - \phi_c).
\end{aligned}
\tag{2}
$$

Let us assume that the three-phase source currents i_{Sa}, i_{Sb} and i_{Sc} are balanced with a power factor of $\cos\theta$ after compensating the unbalanced active components with reactive power control. The three-phase source currents, therefore, can be expressed as

$$
\begin{aligned}
i_{Sa} &= \sqrt{2}I_S\cos(\omega t - \theta), \\
i_{Sb} &= \sqrt{2}I_S\cos(\omega t - \frac{2\pi}{3} - \theta), \\
i_{Sc} &= \sqrt{2}I_S\cos(\omega t - \frac{4\pi}{3} - \theta),
\end{aligned}
\tag{3}
$$

where $I_S = (I_a\cos\phi_a + I_b\cos\phi_b + I_c\cos\phi_c)/(3\cos\theta)$ as shown in Fig. 4. I_S is the theoretical rms value of the balanced source current with a power factor of $\cos\theta$ for each phase.

From (2) and (3), the compensation currents of the ALB

978-1-4799-2706-7/14 $31.00 © 2014 IEEE

are calculated as

$$
\begin{aligned}
i_{Ca} &= i_{La} - i_{Sa} \\
&= (I_a\cos\phi_a - I_S\cos\theta)\,\sqrt{2}\cos(\omega t) + \\
&\quad (I_a\sin\phi_a - I_S\sin\theta)\,\sqrt{2}\sin(\omega t), \\
i_{Cb} &= i_{Lb} - i_{Sb} \\
&= (I_b\cos\phi_b - I_S\cos\theta)\,\sqrt{2}\cos(\omega t - \tfrac{2\pi}{3}) + \\
&\quad (I_b\sin\phi_b - I_S\sin\theta)\,\sqrt{2}\sin(\omega t - \tfrac{2\pi}{3}), \\
i_{Cc} &= i_{Lc} - i_{Sc} \\
&= (I_c\cos\phi_c - I_S\cos\theta)\,\sqrt{2}\cos(\omega t - \tfrac{4\pi}{3}) + \\
&\quad (I_c\sin\phi_c - I_S\sin\theta)\,\sqrt{2}\sin(\omega t - \tfrac{4\pi}{3}). \quad (4)
\end{aligned}
$$

The instantaneous power P_C flowing to the ALB can be calculated as

$$
\begin{aligned}
P_C &= v_{Ta}\cdot i_{Ca} + v_{Tb}\cdot i_{Cb} + v_{Tc}\cdot i_{Cc} \\
&= \Big(2I_a\cos\phi_a - I_b\cos\phi_b - I_c\cos\phi_c + \sqrt{3}I_b\sin\phi_b - \\
&\quad \sqrt{3}I_c\sin\phi_c\Big)\frac{1}{2}V_T\cos(2\omega t) + \\
&\quad \Big(2I_a\sin\phi_a - I_b\sin\phi_b - I_c\sin\phi_c - \sqrt{3}I_b\cos\phi_b + \\
&\quad \sqrt{3}I_c\cos\phi_c\Big)\frac{1}{2}V_T\sin(2\omega t). \quad (5)
\end{aligned}
$$

The mean value of the instantaneous power P_C in (5) is zero, while the source currents in (3) are balanced with the same phase angle θ. Thus, maintaining a constant DC capacitor voltage in the ALB means that a balanced condition with a predefined power factor $\cos\theta$ is achieved on the source side. Therefore, the constant DC capacitor voltage control can ideally be used for the reactive power control strategy of the ALB in three-phase four-wire distribution systems. In practical applications, the instantaneous DC capacitor voltage is not constant owing to the 2ω components caused by the unbalanced load conditions. Thus, the constant mean value of the DC capacitor voltage is controlled in the proposed method.

The DC capacitor voltage v_{DC} is detected in Fig. 2. Then the difference between the detected DC capacitor voltage v_{DC} and the reference DC capacitor voltage V_{DC}^* is amplified by the PID controller as shown in Fig. 3. The output value of the PID controller is input into a moving-average low-pass filter (LPF). The moving-average LPF is designed to remove the 2ω frequency components. The transfer function of the moving-average LPF is expressed as

$$
H(z) = \frac{1}{N}\sum_{n=0}^{N-1} z^{-n}, \quad (6)
$$

where N is the number of samples. After filtering with the moving-average LPF, the effective value I_P of the source-side active current is obtained by performing constant DC capacitor voltage control. To calculate the reference compensation currents for the ALB, the a-phase terminal voltage v_{Ta} is detected, and then the electrical angle ($\theta_T = \omega t$) is generated using a single-phase phased-lock loop (PLL) [9]. Then, $\sqrt{2}\cos(\omega t)$, $\sqrt{2}\cos(\omega t - \tfrac{2\pi}{3})$, $\sqrt{2}\cos(\omega t - \tfrac{4\pi}{3})$, $\sqrt{2}\sin(\omega t)$, $\sqrt{2}\sin(\omega t - \tfrac{2\pi}{3})$, and $\sqrt{2}\sin(\omega t - \tfrac{4\pi}{3})$ are calculated using θ_T. Using these calculated values and the effective value I_P, the reference currents for each phase are calculated as

$$
\begin{aligned}
i_{Sa}^* &= \sqrt{2}I_P\cos(\omega t) + K\sqrt{2}I_P\sin(\omega t), \\
i_{Sb}^* &= \sqrt{2}I_P\cos(\omega t - \tfrac{2\pi}{3}) + K\sqrt{2}I_P\sin(\omega t - \tfrac{2\pi}{3}), \\
i_{Sc}^* &= \sqrt{2}I_P\cos(\omega t - \tfrac{4\pi}{3}) + K\sqrt{2}I_P\sin(\omega t - \tfrac{4\pi}{3}), \quad (7)
\end{aligned}
$$

where $K = \tan(\cos^{-1}(pf))$ as shown in Fig. 4. Finally, the compensation reference signals for the ALB are expressed as

$$
\begin{aligned}
i_{Ca}^* &= i_{La} - i_{Sa}^*, \\
i_{Cb}^* &= i_{Lb} - i_{Sb}^*, \\
i_{Cc}^* &= i_{Lc} - i_{Sc}^*, \\
i_{Cn}^* &= -\big(i_{Ca}^* + i_{Cb}^* + i_{Cc}^*\big). \quad (8)
\end{aligned}
$$

The PI controllers in dq coordinates are used to control the compensation currents i_{Ca}, i_{Cb}, i_{Cc} and i_{Cn} of the ALB [10]. The operating principle is the same for the phases and neutral, as shown in Fig. 3. In the a-phase compensation current control, for example, the a-phase reference compensation current i_{Ca}^* is delayed by $T_S/4$, where T_S is the cycle of the a-phase terminal voltage. i_{Ca}^* corresponds to the α-component, and the delayed current through the $T_S/4$ delay block corresponds to the β-component. Using θ_T, the electrical angle of the a-phase terminal voltage generated by the PLL, the α- and β-components are transformed into i_{da}^* and i_{qa}^*, respectively. The compensation output current i_{Ca} is also transformed into i_{da} and i_{qa} in the same way. The differences between the reference currents i_{da}^*, i_{qa}^* and the detected currents i_{da}, i_{qa} are amplified by the PI controller in dq coordinates. The amplified values are retransformed into the a-phase components. Then, using the pulse width modulation (PWM) technique, the gate signals for the power switching devices of the ALB are generated.

III. Simulation results

To confirm the validity and practicability of the proposed reactive power control strategy for the ALB, digital computer simulation is implemented using PSIM software. The rating of the distribution transformer is three-phase, 380 V, 21.5 kVA, and 60 Hz. A line-to-line voltage of 380 V is used in the simulation for Korea's three-phase four-wire distribution systems. The base power rating is 7.1 kVA for each single-phase system. Three single-phase load parameters and unbalanced load conditions are shown in Table I. The load conditions for the a-phase include two different loads, while the b-phase and c-phase loads are kept constant. The unbalanced load percentage is calculated as the ratio of the negative-sequence to positive-sequence values in accordance with the International Electrotechnical Commission (IEC) standard. Table II shows the circuit constants for the ALB in Fig. 2. The capacity of the DC capacitor voltage is 2200 μF. The reference value of the DC capacitor voltage V_{DC}^* is set to 780 V.

978-1-4799-2706-7/14 $31.00 © 2014 IEEE

TABLE I. LOAD CONDITIONS OF THREE-PHASE FOUR-WIRE DISTRIBUTION SYSTEM.

Item	Symbol	Value
a-phase load (large-load condition) (0.9pu, power factor 0.8)	R_a	6.1 Ω
	L_a	12 mH
a-phase load (small-load condition) (0.2pu, power factor 0.8)	R_a	25 Ω
	L_a	50 mH
b-phase load (0.5pu, power factor 0.8)	R_b	10 Ω
	L_b	20 mH
c-phase load (0.25pu, power factor 0.8)	R_c	20 Ω
	L_c	40 mH
unbalanced load percentage (with a-phase large-load)		31%
unbalanced load percentage (with a-phase small-load)		30%

TABLE II. CIRCUIT CONSTANTS OF ACTIVE LOAD BALANCER IN THE SIMULATION.

Item	Symbol	Value
Reference DC capacitor voltage	V_{DC}^*	780 V
Capacity of capacitor	C_{DC}	2200 μF
Compensation inductance	$L_{Ca}, L_{Cb}, L_{Cc}, L_{Cn}$	2.5 mH
Switching frequency	f_{sw}	12 kHz

Fig. 5 shows simulation waveforms for the ALB in Fig. 2 with the proposed reactive power control strategy. v_{Ta}, v_{Tb}, and v_{Tc} are the a-phase, b-phase, and c-phase terminal voltages, respectively. The a-phase load current i_{La} is varied from 0.9 pu to 0.2 pu, while the b-phase and c-phase load currents, i_{Lb} and i_{Lc}, are kept constant. The unbalanced load percentage is 31% before the load variation and 30% after the load variation. The power factor is set to 0.9 in the control algorithm in accordance with the Japanese guidelines [8]. Before and after the load current variation, the three source currents i_{Sa}, i_{Sb}, and i_{Sc} are balanced as shown in Fig. 5. The power factor $\cos\theta$ is 0.9 under the heavy-load condition and 0.91 under the light-load condition. A slight variation in the power factor occured because of the filter capacitor effect of the ALB. The DC capacitor voltage v_{DC} is well controlled with respect to its reference value V_{DC}^* in both the transient and steady states. The ripple amount of the DC capacitor voltage is 2.8% in the transient state and less than ±1% in the steady state.

Fig. 6 shows the simulation results for the ALB in Fig. 2 with the proposed reactive power control strategy. The a-phase load current i_{La} is changed from 0.2 pu to 0.9 pu, while the b-phase load i_{Lb} and c-phase load i_{Lc} are kept constant. Before and after the load current variation, the source currents i_{Sa}, i_{Sb}, and i_{Sc} are balanced. The DC capacitor voltage v_{DC} closely follows its reference value V_{DC}^* in this light-load to heavy-load variation. The amount of ripple of the DC capacitor voltage is less than 3.2% in the transient state.

IV. EXPERIMENTAL RESULTS

A reduced-scale experimental model of the ALB is constructed and tested to demonstrate the validity and practicability of the proposed reactive power control algorithm. Fig. 7 shows a block diagram of the constructed experimental model. A Δ-Y connected distribution transformer is used in the experimental setup. Electric power utilities in Japan distribute 200 V for commercial supply. Therefore, the voltage rating

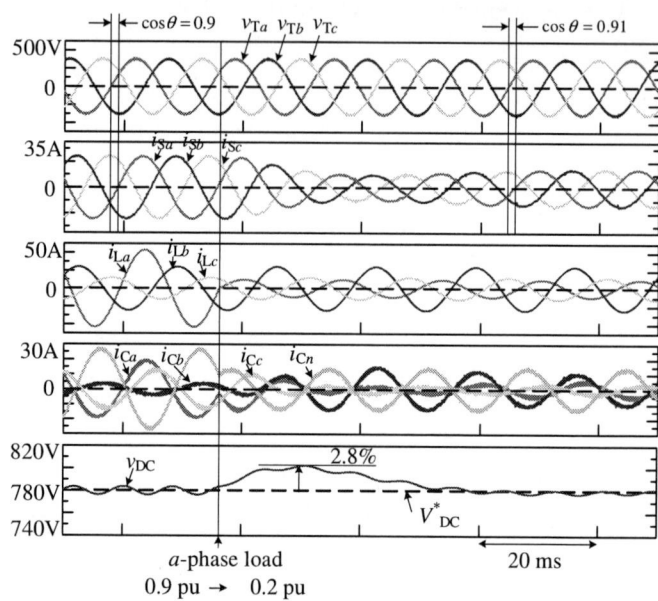

Fig. 5. Simulation waveforms using the proposed reactive power control strategy with heavy-load to light-load variation (power factor is set to 0.9).

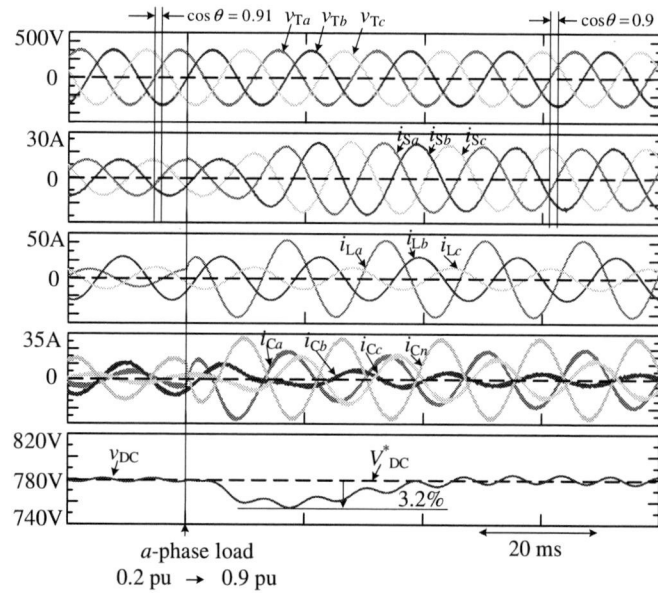

Fig. 6. Simulation waveforms using the proposed reactive power control strategy with light-load to heavy-load variation (power factor is set to 0.9).

(line-to-line) is 200 V in the experiment instead of 380 V. The rating of the transformer is three phase, 200 V, 6 kVA, and 60 Hz on both the primary and secondary sides. The line-to-neutral voltage is 115 V for each single phase system in the secondary side. The circuit constants of the ALB are shown in Table III. The reference DC capacitor voltage V_{DC}^* is 385 V in the experiment. The load conditions of the distribution system are the same as those in Table I.

A digital signal processor (DSP: TMS320C6713, 225 MHz) is used in the experimental setup. The line-to-neutral voltage v_{Ta}; the three load currents i_{La}, i_{Lb}, and i_{Lc};

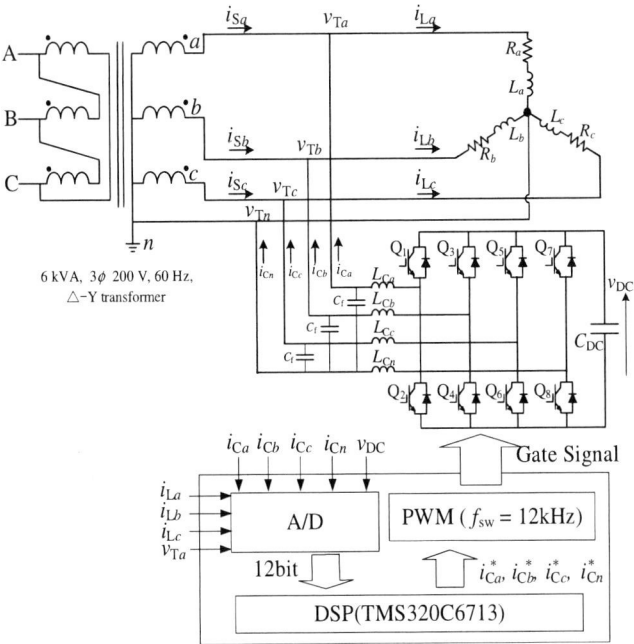

Fig. 7. Constructed experimental model for Fig. 2.

TABLE III. CIRCUIT CONSTANTS OF ACTIVE LOAD BALANCER IN THE EXPERIMENT.

Item	Symbol	Value
Reference DC capacitor voltage	V_{DC}^*	385 V
Capacity of capacitor	C_{DC}	2200 μF
Compensation inductance	$L_{Ca}, L_{Cb}, L_{Cc}, L_{Cn}$	1.5 mH
Switching frequency	f_{sw}	12 kHz

the four compensation output currents i_{Ca}, i_{Cb}, i_{Cc}, and i_{Cn}; and the DC capacitor voltage v_{DC} are detected. These detected signals are input into the DSP through 12-bit A/D converters as shown in Fig. 7. In the DSP, the reference compensation currents i_{Ca}^*, i_{Cb}^*, i_{Cc}^*, and i_{Cn}^* are calculated using the proposed reactive power control strategy as in (8). The sine-triangle intercept technique is used to control the output currents i_{Ca}, i_{Cb}, i_{Cc}, and i_{Cn}. These compensation output currents are detected by the DSP for current feedback control, where the PI controller in dq coordinates is also constructed. A Yokogawa SL1000 high-speed data acquisition unit with a sampling rate of 5 μs is used for waveform detection.

Fig. 8 shows the experimental results for the ALB in Fig. 7 with the proposed reactive power control strategy based on constant DC capacitor voltage control. The a-phase load current i_{La} is changed from 0.9 pu to 0.2 pu, while the b-phase load i_{Lb} and c-phase load i_{Lc} are kept constant. The unbalanced load percentage is 31% before the load variation and 30% after the load variation. The power factor is set to 0.9 in the control algorithm in accordance with the Japanese guidelines [8]. Before and after the load current variation, the source currents i_{Sa}, i_{Sb}, and i_{Sc} are balanced. The power factor $\cos\theta$ is 0.9 under the heavy-load condition and 0.91 under the light-load condition. A slight variation in the power factor occured because of the filter capacitor effect of the ALB. The DC capacitor voltage v_{DC} closely follows its reference value

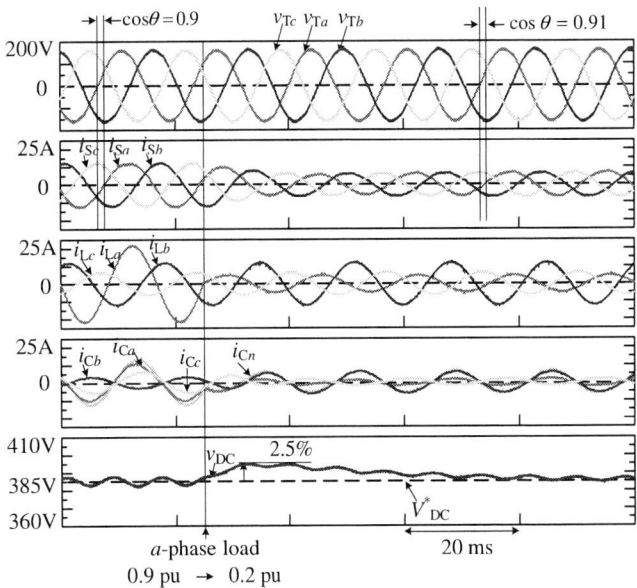

Fig. 8. Experimental results of the proposed reactive power control strategy with heavy-load to light-load variation (power factor is set to 0.9).

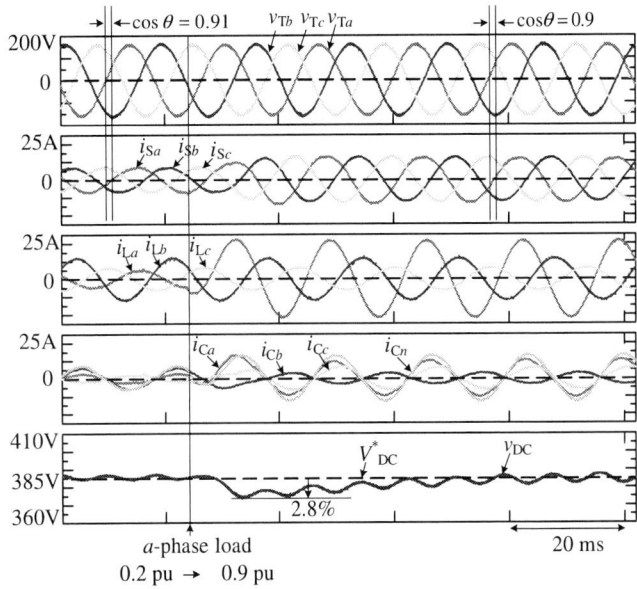

Fig. 9. Experimental results of the proposed reactive power control strategy with light-load to heavy-load variation (power factor is set to 0.9).

V_{DC}^* in the transient and steady states. The amount of ripple of the DC capacitor voltage is less than 2.5% in both the transient and steady states.

Fig. 9 shows the experimental results for the ALB in Fig. 7 with the proposed reactive power control strategy. The a-phase load current i_{La} is changed from 0.2 pu to 0.9 pu, while the b-phase load i_{Lb} and c-phase load i_{Lc} are kept constant. Before and after the load current variation, the source currents i_{Sa}, i_{Sb}, and i_{Sc} are balanced. The DC capacitor voltage v_{DC} closely follows its reference value V_{DC}^* in the transient state in this light-load to heavy-load variation. The amount of ripple of the

DC capacitor voltage is less than 2.8% in the transient state.

V. REQUIRED POWER RATING OF ALB

The required power rating S_C of the ALB with four-leg switching devices is discussed. Fig. 10 shows the equivalent circuit of the ALB in three-phase four-wire distribution systems. Each leg of the ALB is represented by a current source. From the equivalent circuit, the neutral-leg current should be included in the required power rating calculation of the ALB. The power rating of ALB S_C is given by

$$S_C = V_{Ln}(I_{Ca} + I_{Cb} + I_{Cc} + I_{Cn}). \qquad (9)$$

where V_{Ln} is the rms line-to-neutral voltage, and I_{Ca}, I_{Cb}, I_{Cc}, and I_{Cn} are the rms compensation currents of the ALB. Fig. 11 shows a comparison of the experimental results using two control strategies under the same load condition. Fig. 11(a) shows the experimental results with a unity power factor control strategy in [7]. The compensation current flowing in each phase is $I_{Ca} = 10.642$ A, $I_{Cb} = 6.113$ A, $I_{Cc} = 5.035$ A and $I_{Cn} = 8.712$ A. Thus, the required power rating of the ALB is 3.5 kVA. Fig. 11(b) shows the experimental results with the proposed reactive power control strategy. The compensation current flowing in each phase is $I_{Ca} = 7.367$ A, $I_{Cb} = 2.356$ A, $I_{Cc} = 4.098$ A and $I_{Cn} = 8.970$ A. The required power rating of the ALB is 2.6 kVA. Therefore, the proposed reactive power control strategy reduces the required power rating of the ALB by 26% as compared to that with the already proposed control strategy.

VI. CONCLUSION

In this paper, we have proposed a simple reactive power control strategy based on DC capacitor voltage control for an ALB in three-phase four-wire distribution systems. Balanced source side currents with a power factor of 0.9 can be achieved without any calculation blocks of active, reactive components of load currents. A digital computer simulation and experimental results confirm that the proposed reactive power control strategy is well suited to reactive power control to reduce the power rating of the ALB. From the experimental analysis, the reactive power control strategy, which achieves a power factor of 0.9, reduces the ALB's power rating by 26% compared to the control strategy, which is with a unity power factor. Therefore, we are confident that we have offered the simplest and most efficient control algorithm for an ALB in three-phase four-wire distribution systems.

REFERENCES

[1] H. Akagi, Y. Kanazawa, and A. Nabe, "Instantaneous reactive power compensators comprising switching devices without energy storage components," *IEEE Trans. Ind. Appl.*, Vol. IA-20, No. 3, May/June 1984.

[2] A. Nava-Segura and G. Mino-Aguilar, "Four-branches-inverter-based-active-filter for unbalanced 3-phase 4-wires electrical distribution systems," *Proc. IEEE Ind. Appl. Conf.*, Vol. 4, pp. 2503-2508, 2000.

[3] A. Adya, A.P. Mittal, and J.R.P. Gupta, "Modeling and control of DSTATCOM for three-phase, four-wire distribution systems," *IEEE-IAS Ann. Meeting*, Vol. 4, pp. 2428-2434, 2005.

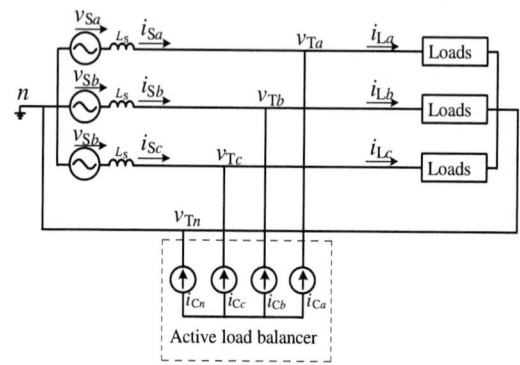

Fig. 10. Equivalent circuit of the ALB in three-phase four-wire distribution systems.

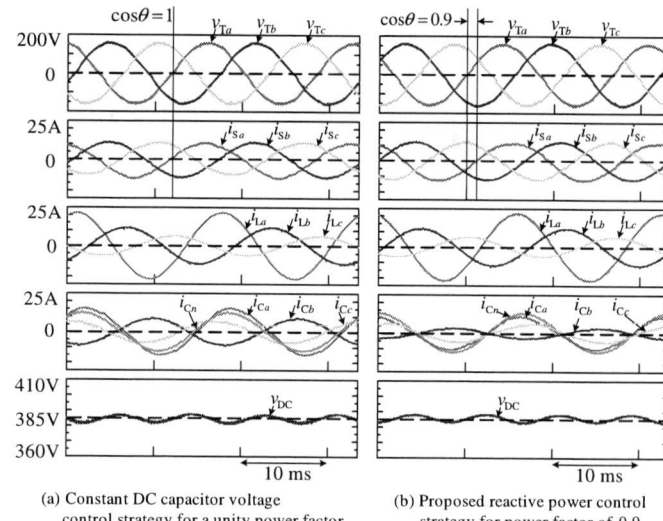

(a) Constant DC capacitor voltage control strategy for a unity power factor

(b) Proposed reactive power control strategy for power factor of 0.9

Fig. 11. Comparison of two control strategies for the ALB.

[4] G. Garcera, M. Pascual, and E. Figueres, "A new current controller applied to four-branch inverter shunt active filters with UPF control method," *IEEE 32nd Ann. Power Electron. Specialists Conf.*, Vol. 3, pp. 1402-1407, 2001.

[5] L.W. Dixon, J.J. Garcia, and L. Moran, "Control system for three-phase active power filter which simultaneously compensates power factor and unbalanced loads," *IEEE Trans. Ind. Electron.*, Vol. 42, pp. 636-641, 1995.

[6] N. Geddada, S.B. Karanki, M.K. Mishra, and B.K. Kumar, "Modified four leg DSTATCOM topology for compensation of unbalanced and nonlinear loads in three phase four wire system," *Proc. of EPE'11*, pp. 1-10, 2011.

[7] Tint Soe Win, E. Hiraki, M. Okamoto, S.R. Lee and T. Tanaka, "Constant DC capacitor voltage control based strategy for active load balancer in three-phase four-wire distribution system," *Proc. of ICEMS 2013 International Conf.*, pp. 1560-1565, 2013.

[8] Japan Electric Association, Indoor Wiring Guidelines, JESC E0005, p. 32 (2005) (in Japanese).

[9] L.N. Arruda, S.M. Silva, and B.J.C. Filho, "PLL structure for utility connected systems," *Conf. Rec. 36th IEEE-IAS Ann. Meeting*, pp. 2655-2660, Chicago, USA, 2001.

[10] F. S. Zhang, "Control of single phase power converter in D-Q rotating coordinates," U.S Patent 6 621 252 B2, Sep. 16, 2003.

Voltage Sag Ride-through Performance of Virtual Synchronous Generator

Jaber Alipoor, Yushi Miura, Toshifumi Ise,
Division of Electrical, Electronic and Information Engineering, Osaka University
Suita, Japan
alipoor@pe.eei.eng.osaka-u.ac.jp

Abstract— **Virtual Synchronous Generator (VSG) is an inverter control structure that supports power system stability by imitating a synchronous machine. Because of the limitation in inverter power and current, their operation under disturbances should be evaluated and enhanced. In this paper, the VSG unit response to different types of faults at grid side is assessed. Besides, a theoretical analysis that traces the trajectory of state variable of the system during voltage sag is represented to justify the effect of the characteristics of symmetrical and unsymmetrical voltage sags. Knowing the critical characteristics (duration and initial point-on-wave) of each type of voltage sag created by fault, proper measures can be embedded to eliminate the hazardous consequences of voltage sags. Furthermore, three additional controllers for voltage sag ride-through enhancement are implemented and tested by experiments.**

Keywords— *distributed generation, inverter, voltage sag.*

I. INTRODUCTION

Conventional enormous synchronous generators comprise rotating inertia due to their rotating parts. These generators are capable of injecting the kinetic energy preserved in their rotating parts to power grid during disturbances or sudden changes. Therefore the system is robust against instability. On the other hand, penetration of Distributed Generating (DG) units in power systems is increasing rapidly. A power system with a big portion of inverter based DGs is prone to instability due to lack of adequate balancing energy injection within the proper time interval. The solution can be found in the control scheme of inverter-based DGs. By controlling the switching pattern of an inverter, it can emulate the behavior of a real synchronous machine. In Virtual Synchronous Generator (VSG) concept, the power electronics interface of DG unit is controlled in a way to exhibit a reaction similar to that of a synchronous machine to a change or disturbance [1].

VSG concept and application were investigated in [2, 3]. The same concept under the title of Synchronverters is described in [4]. The VSG systems addressed in [5-7] are designed to connect only an energy storage unit to the main grid. Reference [8] implements a linear and ideal model of synchronous machine to produce current reference signals for hysteresis controller of inverter. In this Virtual Synchronous Machine (VISMA), authors

added an algorithm for small disturbance compensation to improve voltage quality of grid. Reference [9] introduces a mechanism for voltage, frequency and active and reactive power flow control of VSG. Our research group introduced an other VSG design [10] and added reactive power control to it to have a constant voltage at VSG terminals [11].

Faults on transmission lines cause voltage drops in several points in power system that affects electrical equipment. Voltage sags (drops) are classified in several types based on the fault type. Three-phase fault on power line produces symmetrical voltage sag, while other fault types cause various unsymmetrical sags.

Several works addressed the effect of inverter based DG units on voltage sags in power system focusing on compensation effect of DGs [12,13]. Reference [14] compared the performance of two control strategies in voltage sag ride-through improvement of a single phase converter-connected DG.

Since the VSG is inherently a power electronics instrument, it is sensitive to grid side faults and disturbances. Consequently, before practical usage, their operation must be evaluated under disturbances conditions. In this paper, VSG is tested under various types of voltage sags and the influence of the characteristics of voltage sags on the VSG transient current is investigated. To verify the results mathematically, current trajectory analysis in phase plane is introduced. The trajectory of system state variable (VSG current) is monitored in phase plane and its equations are extracted for during and after the sag. The effects of the characteristic of voltage sag can be observed clearly by this method. High magnitude transient current is the major hazardous consequence of voltage sags on inverter based DGs. To limit the overcurrent, three additional controllers of voltage amplitude control, output power control and virtual inertia control are embedded in the model and tested. Experiments were performed on a 10 kVA VSG controlled inverter and the results showed that the additional controllers enhanced the voltage sag ride-though capability of the VSG system.

The structure of VSG system is reviewed in Sections II. In Section III, the voltage sag types and characteristics are explained. In section IV, the consequences of voltages sags on the VSG unit is monitored by

978-1-4799-2706-7/14 $31.00 © 2014 IEEE

simulation. In section V, the effect of the characteristics of voltages sags on its consequences is clarified by a theoretical analysis. In section VI, tree additional controllers are introduced to improve the voltage sag ride-through performance of the VSG system. Experimental results are represented in Section VII. Finally, conclusion is given in Section VIII.

II. VIRTUAL SYNCHRONOUS GENERATOR STRUCTURE

Fig. 1 shows the control block diagram of the VSG system. In this scheme, a distributed resource (DR) is connected to the main power system via an inverter controlled based on VSG concept. The model of synchronous generator which is used in this paper is a cylindrical-rotor type synchronous generator connected to an infinite bus. The well-known swing equation of synchronous generator is used as the heart of VSG model:

$$P_{in} - P_{out} = J\omega_m \frac{d\omega_m}{dt} + D\Delta\omega \qquad (1)$$

where P_{in} is input power (as same as the prime mover power in a synchronous generator), P_{out} is output power, J is the moment of inertia of the rotor, ω_m is the virtual angular velocity of the virtual rotor and D is the damping factor. $\Delta\omega$ is given by $\Delta\omega = \omega_m - \omega_{grid}$, ω_{grid} being the grid frequency or the reference frequency when the grid is not available. Using voltage and current signals measured at the VSG terminals, its output power and frequency are calculated.

A governor model shown in Fig. 2 is implemented to tune the input power command based on the frequency deviation. Having the essential parameters, (1) can be solved by numerical integration. By solving this equation in each control cycle, the momentary ω_m is calculated and by passing through an integrator, the virtual mechanical phase angle, θ_m is produced. This phase angle and a voltage magnitude reference are used as the VSG output voltage angle and magnitude commands for generating PWM pulses.

III. VOLTAGE SAGS TYPES AND CHARACTERISTICS

Voltage sag (dip) is a momentary drop in RMS voltage of at least one of the three phase or line voltages. Voltage sags originate from short-term increase in current in power system due to fault or starting large loads [15]. Severity of the effects of voltage sag on equipment depends on the sag characteristics. Sag magnitude and duration are major characteristics. In this paper, we assume that the implemented sags are rectangular. At the instant when the sag begins, voltage of phase "a", v_a is expressed by:

$$v_a(t)\big|_{t=t_{sag}} = V_m \sin(\theta_0) \qquad (2)$$

where V_m is the phase voltage amplitude. Voltage phase angle at the sag beginning instant, θ_0 is called point-on-wave. θ_0 is considered as a characteristic of voltage sag

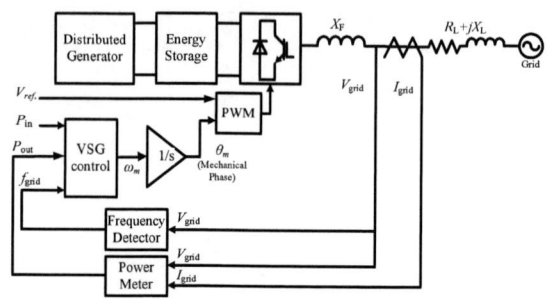

Fig. 1. Block diagram of VSG unit.

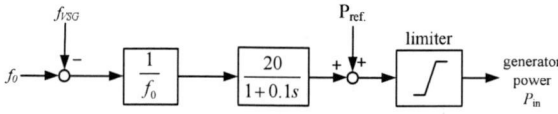

Fig. 2. Governor diagram.

that affects the response of appliance to voltage sag. For both symmetrical and unsymmetrical voltage sags, sag duration is the time interval between sag beginning and ending.

Voltage sags can be either symmetrical or unsymmetrical. If the individual phase voltages are equal and the phase angle relationship is 120°, the sag is symmetrical. Otherwise, the sag is unsymmetrical. Reference [16] classified the main voltage sags in 7 types.

Three-phase short circuits can produce symmetrical sags classified as type A. In this type, voltage magnitudes of three phases decrease equally. For symmetrical sags, the magnitude of sag is the remained RMS voltage in per unit or percent of rated voltage that expressed by a parameter denoted as "h".

Type B voltage sag is one kind of unbalanced voltage sag that originates from single phase to ground fault. Single-phase-to-ground (after passing through a "Δ-Y" transformer) and phase-to-phase faults on power lines produce voltage sags of type C. In this type, relative phase angles between voltages vectors also change. If the voltage sag of type C passes through a "Δ-Y" transformer, it will be converted to type D voltage sag.

For unsymmetrical sags, "h" is a coefficient that appears in voltage equations and causes difference in their magnitude and/or phase angles as shown in Fig. 3. Lightest and severest characteristics of each voltage sag type are summarized in Table I.

TABLE I
LIGHTEST AND SEVEREST CHARACTERISTICS OF VOLTAGE SAGS

Sag type	Lightest characteristics		severest characteristics	
	duration	θ_0	duration	θ_0
A	k^*T^{**}	-	$(k-1/2)T$	-
B	kT	$\pi/2, 3\pi/2$	$(k-1/2)T$	$0, \pi$
C	kT	$0, \pi$	$(k-1/2)T$	$\pi/2, 3\pi/2$
D	kT	$\pi/2, 3\pi/2$	$(k-1/2)T$	$0, \pi$

$*. k \in \mathbb{N}$

$**. T$: One cycle time period

The 2014 International Power Electronics Conference

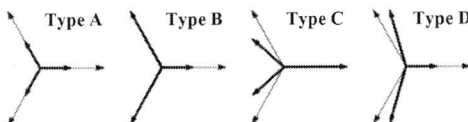

Fig. 3. Phasor diagram of voltage sag types for $h = 0.5$.

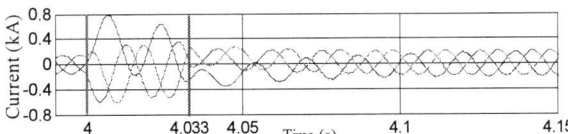

Fig. 4. Currents of VSG subjected to a voltage sag of type A with the duration of 2 cycles and h=0.1.

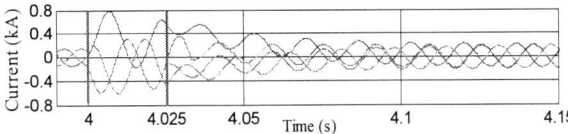

Fig. 5. Currents of VSG subjected to a voltage sag of type A with the duration of 1.5 cycles and h=0.1.

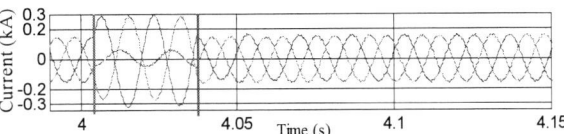

Fig. 6. Currents of VSG subjected to a voltage sag of type B with the duration of 2 cycles, initial point-on-wave of $\pi/2$ and h=0.1

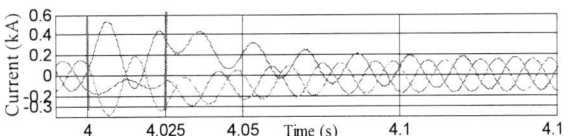

Fig. 7. Currents of VSG subjected to a voltage sag of type B with the duration of 1.5 cycles, initial point-on-wave of 0 and h=0.1

IV. VSG SUBJECTED TO VOLTAGE SAGS

The VSG model discussed in section II was simulated in PSCAD/EMTDC software. The parameters of the system are: $P_{base} = 35$ kW, $f_{base} = 60$ Hz, $R_L = 12.5\%$, $X_L = 33.0\%$, $X_F = 42.4\%$, $J = 2.46$ kgm^2 and $D = 17$ pu. Voltage sags were applied to the VSG and its transient currents were investigated. In this section, results are summarized to voltage sags of types A and B. For all cases of simulation the voltage sag intensity parameter, h is set to 0.1.

Fig. 4 shows the current waveforms of VSG subjected to voltage sag of type A with the duration of 2 cycles. Since this is the lightest characteristic of voltage sag, overcurrent happens only during voltage sag. Whereas, as shown in Fig. 5, for voltage sag of type A with the duration of 1.5 cycles there are current oscillations which appear after voltage recovery.

Initial point-of-wave must be taken into account for voltage sag of type B. Fig. 6 shows the VSG response to a voltage sag of type B with the duration of 2 cycles and initial point-on-wave of $\pi/2$ that is the kindest condition. Slight overcurrent happens in two phases during voltage sag. On the other hand, for the severest characteristics, i.e., the duration of 1.5 cycles and initial point-on-wave of zero there are significant current oscillations during and after the voltage sag observed in Fig. 7.

V. STATE VARIABLE ANALYSIS IN PHASE PLANE

The difference between the VSG response to various sags and their characteristics can be illustrated by VSG state variable (current) analysis in phase plane.

A. Symmetrical Voltage Sag (Type A)

Inverter output current during voltage sag is calculated by integrating the voltage in polar coordinate in stationary frame. Gird voltage during voltage sag of type A is expressed as:

$$V_g = hV_m e^{j(\theta_0 - \pi/2)} e^{j\omega_{grid} t} \tag{3}$$

where hV_m is the remained voltage amplitude and θ_0 is the initial point-on-wave. As mentioned before, h is the voltage sag intensity parameter varies from 0 (zero remained voltage magnitude) to 1 (normal voltage). The fault current is calculated by integrating the voltage difference as:

$$I_{sag} = \frac{1}{L}\int_0^{t_{sag}} (1-h)V_m e^{j(\omega_{grid}t + \theta_0 - \pi/2)} dt \tag{4}$$

Where I_{sag}, L and t_{sag} are the fault current in the stationary reference frame, the interconnecting inductance and the sag ending time, respectively. Solving (4) and transferring to the synchronous reference frame yields:

$$I_{sag}^{synch} = (1-h)\frac{V_m}{\omega_{grid}L} e^{j(\theta_0 - \pi)}(1 - e^{-j\omega_{grid}t}) \tag{5}$$

Based on (5), during voltage sag, current vector is circulating in the synchronous dq-frame phase plane with fixed radius as shown in Fig. 8 (the damping terms is neglected). The radius of the circle, $(1-h)V_m/L\omega_{grid}$, determines the oscillation magnitude during voltage sag. It is independent of the θ_0 obviously.

If the duration of voltage sag is a multiple of full cycles, the transient current will have zero magnitude at voltage recovery point. In the other words, state variable position at the sag ending moment will be close to its initial point. Therefore, minimum oscillations happen after voltage sag. Whereas if the duration is half a cycle more than any number of full cycles, transient current has its maximum value based on (5). It means that the state variable position has maximum distance to its normal point. The transient current must settle in the origin of coordinate (current before fault is neglected) after voltage recovery. The second transient can be expressed as:

$$I_{after\ sag}^{synch} = 2(1-h)\frac{V_m}{\omega_{grid}L} e^{j(\theta_0 - \pi)} e^{-j\omega t} \tag{6}$$

Equation (6) is a circular trajectory with the center coinciding with the origin. The radius of trajectory after

The 2014 International Power Electronics Conference

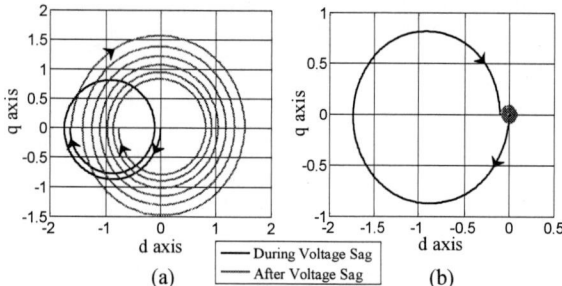

Fig. 8. VSG current trajectory in phase plane during (solid line) and after (dotted line) voltage sag.

voltage sag has its maximum possible value proving that the assumed duration is the severest one. Fig. 7 includes the current paths in phase plane. Part (a) of this figure is related to sag with the duration of 1.5 cycles. At the sag ending point the state variable goes through circles with a large radius (dotted line) that causes severe current transient. While in part (b) of the figure, since there is a small distance between the position of the variable after and before the sag, light transient happens.

B. Unsymmetrical Voltage Sag

The same analysis with more intense mathematical work is performed for voltage sag of type B as the representative of unsymmetrical voltage sags. The three-phase phasor expressions of voltage sag type B is as follows:

$$
\begin{cases}
V_a = hV_{rms} \\
V_b = -\dfrac{1}{2}V_{rms} - j\dfrac{\sqrt{3}}{2}V_{rms} \\
V_c = -\dfrac{1}{2}V_{rms} - j\dfrac{\sqrt{3}}{2}V_{rms}
\end{cases}
\tag{7}
$$

Transforming into polar form in the stationary coordinate the yields:

$$
V_{sag} = \frac{V_m}{3}\left((h+2)e^{j(\omega_{grid}t+\theta_0-\pi/2)} \right.
$$
$$
\left. + (h-1)e^{-j(\omega_{grid}t+\theta_0-\pi/2)}\right)
\tag{8}
$$

After integrating the voltage difference between the inverter and grid, the fault current in the synchronous frame is given as follows:

$$
I_{sag}^{synch} = \frac{V_m(1-h)}{3L\omega_{grid}}\left(-e^{j\theta_0} \right.
$$
$$
\left. + 2\cos\theta_0 e^{-j\omega_{grid}t} - e^{-j\theta_0}e^{-2j\omega_{grid}t}\right)
\tag{9}
$$

It is observed that the coefficient $\cos\theta_0$ appears in the current equation and affects its magnitude during oscillation. The form of (9) on the phase plane is not exactly circular and depends on θ_0. The characteristics of the voltage sags that result in the maximum absolute value of (9) are the critical characteristics for this type of voltage sag. $\theta_0 = 0$, π brings forth the maximum

oscillations during voltage sag type B and with this value of θ_0, when the duration of the sag is a multiple of a half-cycle, the state variable position has maximum distance to its normal point. When voltage is recovered at this moment, the state variable returns to its normal point with severest oscillation similar to one of the symmetrical voltage sag, i.e., the circular trace in the phase plane. The terms $e^{-2j\omega_{grid}t}$ in (9) indicates that the current oscillates with twice of the system frequency during unsymmetrical voltage sags. The mildest characteristics of voltage sag type B can be deduced from (9) as well. $\theta_0 = \pi/2$ results in lightest oscillation during voltage sag. For this initial point-on-wave, the duration of a multiple of a half-cycle and the duration of an odd multiple of a quarter-cycle have the mildest and severest consequences after sag, respectively (only the oscillatory component with double system frequency exists).

VI. VOLTAGE SAG RIDE-THROUGH ENHANCEMENT

Three approaches are added to the VSG system to limit the overcurrent during and after voltage sag. Fig. 8 shows the VSG system with the additional controls.

A. Voltage Amplitude Control

Subsystem A in Fig. 9 calculates the RMS value of grid voltage and uses it as reference voltage for inverter output. By this control, when grid voltage drops during voltage sag, inverter output voltage also will be reduced to prevent overcurrent.

B. Output Power Control

Since inverter output power is proportional to product of inverter and grid voltages, once the output voltage follows the grid voltage, inverter output power must be limited proportional to the square of grid voltage. Subsystem B calculates the output power reference of VSG proportional to the square of grid voltage.

C. Virtual Inertia Control

This control was introduced by authors in [17] to impose additional damping and stabilize the VSG after a disturbance. After a disturbance, the angular velocity generated by VSG oscillates in sinusoidal form and causes power oscillations in VSG terminal as same as the

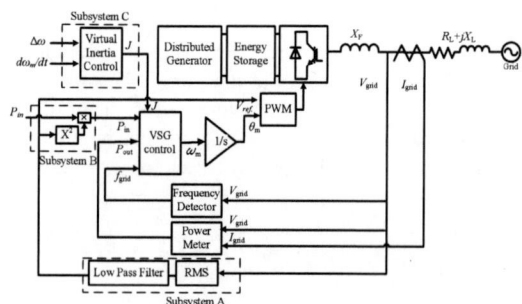

Fig. 9. VSG unit with additional controls for voltage sag ride-through enhancement.

power-angle oscillations of a synchronous machine. During this oscillation, system transient energy is damped until energy balance is achieved after a change or disturbance.

Consider the power-angle curve of Fig. 10. When a fault happens, power-angle curve follows the dashed line curve and load angle moves to the point δ_2. After the fault clearance, operating point moves along the original power-angle curve and load angle oscillates around the equilibrium point δ_1. Machine condition during each phase of an oscillation cycle is summarized in Table II. It should be noted that the sign of the $d\omega_m/dt$ does not determine acceleration or deceleration by itself; whereas, its sign respect to the sign of the relative angular velocity defines the acceleration or deceleration. For example, in segment ③ of Fig. 4, during transition from points "c" to "b", both of $d\omega_m/dt$ and $\Delta\omega$ are positive and act in the same direction, therefore it is an acceleration period. On the other hand, when they have opposite signs like segment ④, it is a deceleration period.

The derivative of angular velocity, $d\omega_m/dt$ indicates the rate of acceleration or deceleration. Considering (1), it is observed that this rate has a reverse relation to the moment of inertia, J. Based on this fact, one can select a large value of J during acceleration phases ("a" to "b" and "c" to "b") to reduce the acceleration and a small value of J during deceleration phases ("b" to "c" and "b" to "a") to boost the deceleration. The big moment of inertia J_{big} and the small one J_{small} can be chosen within a wide range depending on the rated power so that the difference between J_{big} and J_{small} determines the damped power in each half-cycle of oscillation. The subsystem C in Fig. 8 adopts the value of the virtual inertia based on the stated algorithm summarized in Table II.

VII. EXPERIMENTAL RESULTS

A laboratory scale test system is used to investigate the voltage sag ride-through performance of the VSG. The overall system configuration is depicted in Fig. 11 and main parameters of the system are presented in TABLE III. The transmission unit (TU) in Fig. 11 simulates the π model of a 40 km transmission line shown in Fig. 12.

Due to the strict overcurrent limitation of the inverter unit, when a deep voltage sag happens, the VSG unit fails quickly. Therefore, only mild voltage sags with the magnitude bigger than 90% can be tested on this system. For light voltage sags, since the state variable position in phase plane does not change considerably, the characteristics do not affect the severity of transient oscillation. Because of this practical limitation, the voltage sag ride-through performance of the VSG system subjected to the mild voltage sags with the duration of 10 cycles was evaluated.

A. Symmetrical Voltage Sag

The voltage sag shown in Fig. 13 appears due to three-phase short circuit happened at the fault point and measured at the Point of Common Coupling (PCC) indicated in Fig. 11. The VSG was subjected to this voltage sag and voltage sag ride-through performance of VSG unit with and without the additional controllers was investigated.

To assess the performance of the additional controls, first the VSG unit without the controllers is subjected to the voltage sag presented in Fig. 13. As stated formerly, two transient states happen: the transient during voltage sag and the one after voltage recovery. To see the

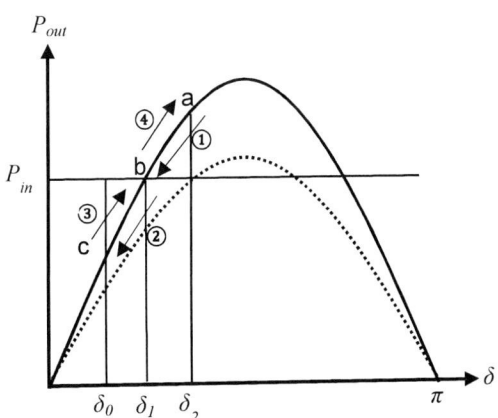

Fig. 10. Power-angle curve of a typical synchronous machine.

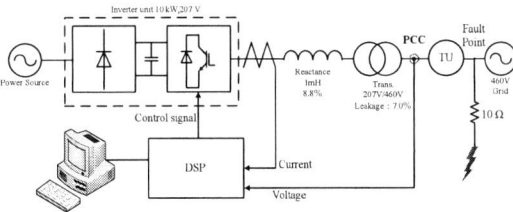

Fig. 11. Experimental system.

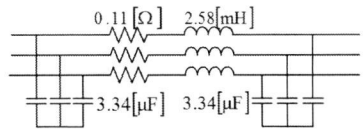

Fig. 12. Transmission unit.

TABLE II
MACHINE MODES DURING OSCILLATION.

Segment	$\Delta\omega$	$d\omega_m/dt$	Mode	J
a→b	$\Delta\omega < 0$	$d\omega_m/dt < 0$	Accelerating	J_{big}
b→c	$\Delta\omega < 0$	$d\omega_m/dt > 0$	Decelerating	J_{small}
c→b	$\Delta\omega > 0$	$d\omega_m/dt > 0$	Accelerating	J_{big}
b→a	$\Delta\omega > 0$	$d\omega_m/dt < 0$	Decelerating	J_{small}

TABLE III
SPECIFICATIONS OF THE VSG UNIT.

Base power P_{base}	10 kW
Base voltage V	207 V
Base Frequency	60 Hz
Interconnecting reactance X_L	7.0%
Damping factor	17 pu
Moment of inertia J	0.563 kgm^2
J_{big}	0.563 kgm^2
J_{small}	0.1 kgm^2

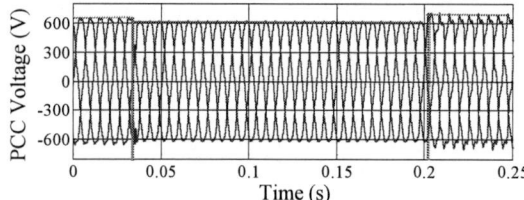

Fig. 13. Type A (symmetrical) voltage sag at PCC due to symmetrical 3-phase fault.

importance of the transient after voltage sag recovery, the VSG was subjected to the voltage sag while injecting 1 kW power to the grid and its current in synchronous d-q coordinate was monitored. Fig. 14 shows the currents and dc-link voltage in this condition. The oscillations after voltage sag caused power and current oscillation which went to negative value. This revers current increased the dc-link voltage and resulted in the failure.

During the voltage sag, current is raised by the voltage difference and results in an overcurrent failure. Since the voltage sag magnitude in this research is high (i.e. mild voltage sag), the initial current of the VSG affects the fault ride-through performance. When the VSG unit was loaded at 2.6 kW, the voltage sag type A was applied. Fig. 15 shows that the VSG current raised sharply and VSG stopped. It can be concluded that for lower loading levels, the oscillations after fault recovery causes the failure and for higher loading levels, the overcurrent during voltage sag results in the failure.

In the next step, only the voltage amplitude control and output power control are activated. When the VSG output power was the inconsiderable value of 1 kW, the type A voltage sag happened. As it is observed in Fig. 16, the oscillation after voltage sag fell to the negative level and activated the dc-link voltage protection.

The VSG current and dc-link voltage affected by voltage sag when the VSG power is 2.6 kW is shown in Fig. 17. In this condition, the voltage amplitude and power control banned the overcurrent to some extent and the VSG was able to pass-through the fault. When the voltage sag starts, it is expected that the voltage amplitude controller calculate the grid RMS-voltage and apply it as the voltage reference of the inverter. Although the overcurrent magnitude was reduced slightly; however, severe current transient happened. It is because of the delay in sensing, filtering, and applying the voltage reference. In spite of this technical shortcoming, the VSG unit was able to ride-through this voltage sag. It should be noted that the RMS value of I_d and I_q is concerned as the threatening overcurrent.

In the next experiment, all of the additional controllers were activated. First the VSG with the low output power of 1kW was subjected to the voltage sag type A. As it is observed in Fig. 18, the inertia control suppressed the after sag oscillations by imposing a damping effect. It will be shown in the unsymmetrical voltage sag experiment that the virtual inertia control is so effective that even for very lower output power (200 W), the oscillation after voltage recovery does not cause a failure.

In high loading condition, the overcurrent at the sag

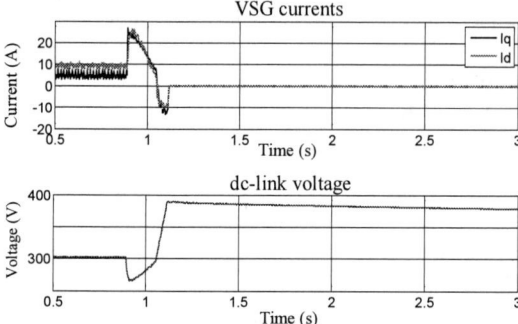

Fig. 14. Currents and dc-link voltage of the VSG with 1 kW output power and without additional controller subjected to voltage sag type A.

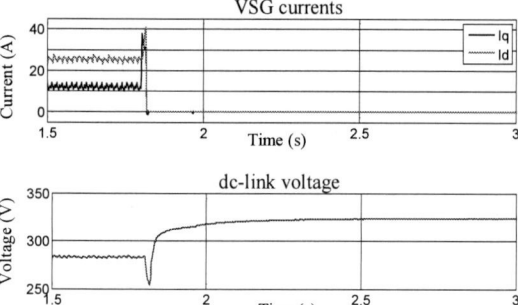

Fig. 15. Currents and dc-link voltage of the VSG with 2.6 kW output power and without additional controller subjected to voltage sag type A.

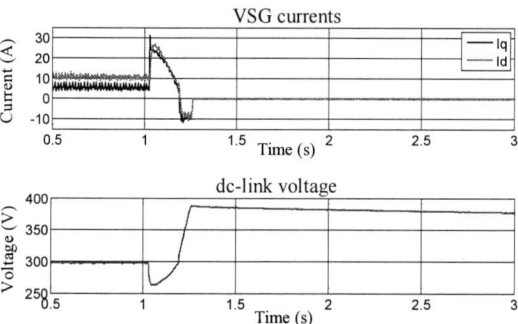

Fig. 16. Currents and dc-link voltage of the VSG with 1 kW output power and with voltage amplitude and output power controller subjected to voltage sag type A.

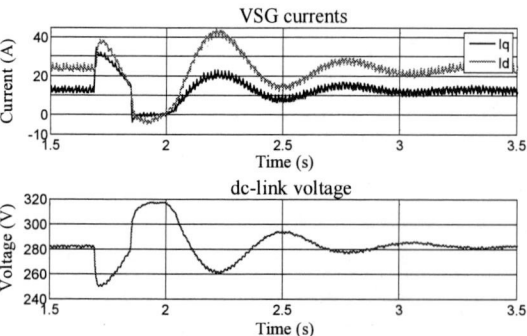

Fig. 17. Currents and dc-link voltage of the VSG with 2.6 kW output power and with voltage amplitude and output power controller subjected to voltage sag type A.

starting moment is perilous. Fig. 18 shows the currents and dc-link voltage of the VSG with 3 kW output power equipped with the additional controllers subjected to the voltage sag type A. Because of the delay in voltage control, output voltage did not follow the grid voltage promptly; therefore, overcurrent appears right away when sag starts. However, it fell down quickly and the VSG passed through the voltage sag even with high output power of 3 kW.

B. Unsymmetrical Voltage Sag

The VSG unit was also evaluated under unsymmetrical voltage sag condition. A phase-to-phase fault with the line-to-line resistance of 20 Ω happened at the fault point of Fig. 11 which produced type C voltage sag at PCC as shown in Fig. 20.

Figs. 21 and 22 show the experiment results of VSG subjected to voltage sag type C at 1kW and 2.6 kW output power, respectively. In these cases, the behavior of the VSG without any additional controller was similar to that of the symmetrical voltage sag. The only difference is the form of overcurrent during voltage sag that has an oscillatory component with twice of the system frequency. The VSG unit with output power less than 1 kW failed due to the transients after voltage sag and it failed due to overcurrent during the sag when the output power is higher than 2.6 kW.

Fig. 23 depicts the current and dc-link voltage of the VSG with voltage sag ride-through controllers, at low output power of 200 W, subjected to the unsymmetrical

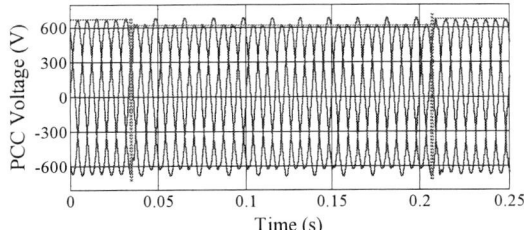

Fig. 20. Type C voltage sag at PCC due to phase-to-phase fault.

voltage sag type C. It is observed that when the additional controllers were activated, the oscillation after voltage recovery was eliminated and hence, the failure because of the dc-link overvoltage was prevented.

The same experiment was done when the VSG works at 2.6 kW output power and the results are represented in Fig. 24. It is observed that the additional controllers improved the voltage sag ride-through capability of the VSG system effectively. However, there might be a problem in calculating the RMS voltage of the grid during unsymmetrical voltage sag. Since the qd-component voltages oscillate with double system frequency during unsymmetrical voltage sag, the RMS value calculated by $V_{rms} = \sqrt{V_d^2 + V_q^2}$ will be oscillatory as well. As a result, the VSG voltage reference applied by the voltage amplitude controller might be oscillatory. This problem did not emerge since the low pass filter of the voltage amplitude controller removes this oscillation. In accordance with the theoretical analysis of section V, the transient current during unsymmetrical voltage sag with double system frequency component is observed in Figs. 21 to 24.

VIII. CONCLUSION

VSG has been invented to support grid stability by inserting virtual inertia into power system. VSG may be vulnerable in fault condition because of its power electronics basis. Evaluation of voltage sag consequences is the first step of VSG protection in fault condition. Various types of voltage sags were applied to VSG and its response monitored by checking the current transients. If symmetrical voltage sags lasts a half cycle more than a multiple of full cycles, severest current oscillation appears after voltage recovery. For unsymmetrical voltage sags, initial point-on-wave is also important. Initial point-on-wave of 0 and π are the severest ones for both voltage sag types of B and D. whereas for type C, π/2 and 3π/2 are the critical ones. These results were verified by analyzing the current trajectory in phase plane. Transient current vector circulates in phase plane. By extracting the radius of circulations, the overcurrent severity can be evaluated. Finally, three subsystems of voltage amplitude control, output power control, and virtual inertia control were added to the VSG system to suppress the overcurrent during and after voltage sags. Experimental results showed that the additional controls improved voltage sag ride-through capability of VSG effectively for symmetrical and unsymmetrical sag types.

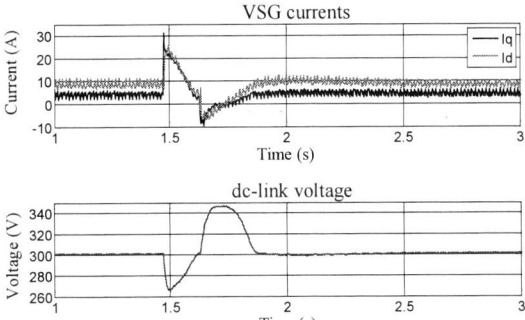

Fig. 18. Currents and dc-link voltage of the VSG with 1 kW output power and with voltage amplitude, output power, and virtual inertia controller subjected to voltage sag type A.

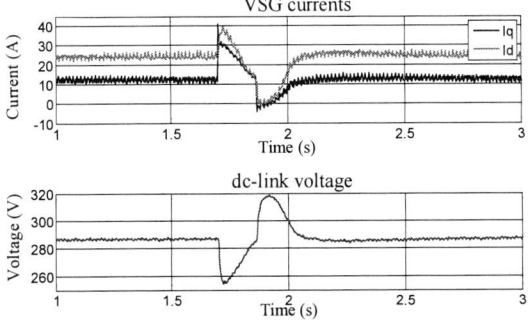

Fig. 19. Currents and dc-link voltage of the VSG with 3 kW output power and with voltage amplitude, output power, and virtual inertia controller subjected to voltage sag type A.

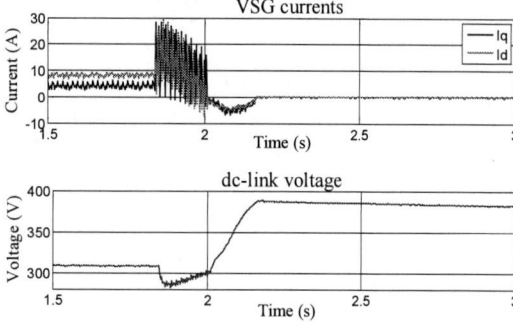

Fig. 21. Currents and dc-link voltage of the VSG with 1 kW output power and without additional controller subjected to voltage sag type C.

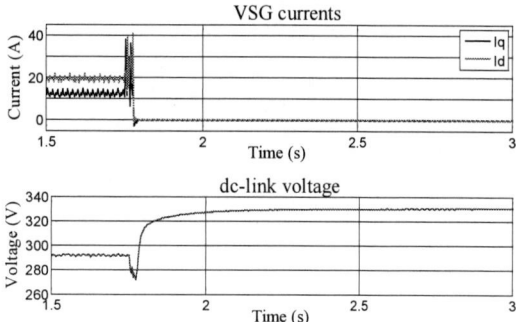

Fig. 22. Currents and dc-link voltage of the VSG with 2.6 kW output power and without additional controller subjected to voltage sag type C.

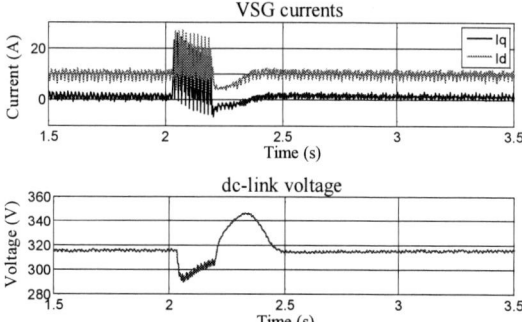

Fig. 23. Currents and dc-link voltage of the VSG with 200 W output power and with voltage amplitude, output power, and virtual inertia controller subjected to voltage sag type C.

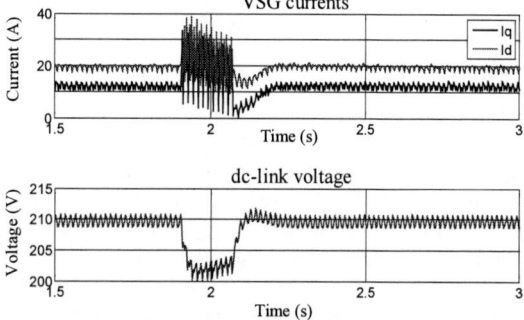

Fig. 24. Currents and dc-link voltage of the VSG with 2.6 kW output power and with voltage amplitude, output power, and virtual inertia controller subjected to voltage sag type C.

REFERENCES

[1] J. Alipoor, Y. Miura, T. Ise, "Evaluation of Virtual Synchronous Generator (VSG) Operation under Different Voltage Sag Conditions," in IEEJ Conference on Power Technology and power systems, Tokyo, JAPAN, pp. 41-46, PE-12-60, PSE-12-76, 2012.

[2] J. Driesen and K. Visscher, "Virtual Synchronous Generators," in IEEE Power and Energy Society General Meeting - Conversion and Delivery of Electrical Energy in the 21st Century, pp.1-3, 2008.

[3] T. Loix, S. De Breucker, P. Vanassche, J. Van den Keybus, J. Driesen, and K. Visscher , "Layout and Performance of the Power Electronic Converter Platform for the VSYNC Project," in IEEE Powertech Conference, pp.1-8, 2009.

[4] Q. C. Zhong, G. Weiss, "Synchronverters: Inverters That Mimic Synchronous Generators," IEEE Transactions on Industrial Electronics, Vol. 58, No. 4, pp.1259-1267, 2011.

[5] M.P.N. van Wesenbeeck, S.W.H. de Haan, P. Varela and K. Visscher, "Grid Tied Converter with Virtual Kinetic Storage," in IEEE Powertech Conference, Bucharest, pp.1-7, 2009.

[6] M. Torres, L.A.C. Lopes, "Virtual synchronous generator control in autonomous wind-diesel power systems," in IEEE Electrical Power & Energy Conference (EPEC), Montreal, pp.1-6, 2009.

[7] V. Karapanos, S. de Haan, and K. Zwetsloot, "Real Time Simulation of a Power System with VSG Hardware in the Loop," in 37th Annual Conference on IEEE Industrial Electronics Society (IECON), Melbourne, pp.3748-3754, 2011.

[8] R. Hesse, D. Turschner, and H.-P. Beck, "Micro grid stabilization using the virtual synchronous machine," in International Conference on Renewable Energies and Power Quality (ICREPQ'09), Spain, 2009.

[9] Y. Xiang-zhen, S. Jian-hui, D. Ming, L. Jin-wei, D. Yan, "Control Strategy for Virtual Synchronous Generator in Microgrid," in 4th Int. Conference on Electric Utility Deregulation and Restructuring and Power Technologies (DRPT), pp. 1633-1637, 2011.

[10] K. Sakimoto, Y. Miura and T. Ise, "Stabilization of a Power System with a Distributed Generators by a Virtual Synchronous Generator function," in 8th IEEE international conference on Power electronics- ECCE Asia, Jeju, Korea, pp. 1498-1505, 2011.

[11] T. Shintai, Y. Miura, T. Ise, "Reactive Power Control for Load Sharing with Virtual Synchronous Generator Control," in Power Electronics and Motion Control Conference (IPEMC), pp. 846–853, 2012.

[12] K. J. P. Macken, M. H. J. Bollen, and R. J. M. Belmans, "Mitigation of voltage dips through distributed generation systems," IEEE Transactions on Industry Applications, vol. 40, no. 6, pp. 1686-1693, 2004.

[13] B. Renders, K. De Gusseme, W. Ryckaert, K. Stockman, L. Vandevelde, M.H.J. Bollen, "Distributed Generation for Mitigating Voltage Dips in Low-Voltage Distribution Grids," IEEE Transactions on Power Delivery, vol. 23, no. 3, pp. 1581-1588, 2008.

[14] B. Renders , W. R. Ryckaert , K. De Gussemé , K. Stockman and L. Vandevelde "Improving the voltage dip immunity of converter-connected distributed generation units", Elsevier Renew. Energy, vol. 33, no. 5, pp.1011 -1018 2008.

[15] M.H.J. Bollen, "Voltage Recovery After Unbalanced and Balanced Voltage Dips in Three-Phase Systems", IEEE Trans. on Power Delivery, Vol. 4, No. 18, pp. 1376–1381, 2003.

[16] M. H. J. Bollen, "Characterization of Voltage Sags Experienced by Three-Phase Adjustable-Speed Drives", IEEE Trans. Power Delivery, Vol. 12, No. 4, pp.1666 -1671, 1997.

[17] J. Alipoor, Y. Miura, T. Ise, " Distributed generation grid integration using virtual synchronous generator with adoptive virtual inertia," in Proc. IEEE Energy Conversion Congress and Exposition (ECCE), pp. 4546-4552, 2013.

Control of Distributed Generation Systems under Unbalanced Voltage Conditions

R. Kabiri   D. G. Holmes   B. P. McGrath

School of Electrical and Computer Engineering
RMIT University, Melbourne, Australia

roozbeh.kabiridehkordi@rmit.edu.au   grahame.holmes@rmit.edu.au   brendan.mcgrath@rmit.edu.au

Abstract — **Managing power delivery from distributed generation systems is challenging when the grid voltages are unbalanced, since the negative sequence voltage causes power oscillations at twice the fundamental grid frequency. Current regulation using sequence components can be used to control these real and reactive power oscillations, and thus help mitigate the unbalanced network voltages. This paper presents a flexible control scheme using double sequence frame current regulators that can readily adjust between eliminating real or reactive power oscillations created by unbalanced grid voltages, or alternatively simply balancing the three phase currents. The mitigation influence of these alternative control strategies on unbalanced grid voltages is then examined for resistive and reactive distribution networks. Simulation results are presented to validate the proposed control strategies and illustrate their influence on unbalanced network voltages.**

Keywords—distributed generation, power oscillation control, unbalanced voltages, double synchronous frame

I. INTRODUCTION

Across the world, the paradigm of an electricity supply grid is changing as increasing levels of renewable generation are connected to the power grid at the distribution network level [1-3]. Since these small-scale distributed generation (DG) sources (PV solar, wind turbines and fuel cells) are generally interfaced through power electronic converters, they can also assist with the operation and control of the electrical grid network by varying both their real and reactive power injection in response to network needs [4-5].

DG systems are commonly interfaced through a power electronic Voltage Source Inverter (VSI), operating in closed loop current regulation [6]. Under balanced network voltage conditions, the real and reactive power injection from such systems is controlled by commanding from the current regulator, a current phasor that has the appropriate magnitude and phase angle with respect to the measured grid connection point voltage. Synchronous frame current regulation is typically used for convenience and ease of implementation [7].

When the network voltages become unbalanced (because of unbalanced loads or short term disturbances), current regulation becomes more challenging, since the negative-sequence component of the unbalanced voltage causes double fundamental frequency oscillations in both the real and reactive power injections. Various strategies have been proposed to address this issue, looking to either balance the three phase AC currents, or to minimise the

real and/or reactive power injection oscillation, depending on the identified target objective [8-13]. However, little work has been reported on the impact on network voltages, of these alternative approaches to unbalanced voltage operation.

This paper presents a novel control strategy that allows a double sequence frame current regulator to continuously adjust between eliminating double fundamental frequency real power oscillations, balancing the inverter three phase currents, and eliminating double fundamental frequency reactive power oscillations. This smooth control capability allows the impact of these control alternatives on unbalanced voltages at the inverter Point-of-Common-Coupling to be readily investigated under different levels of P and Q injection. In particular, this allows the response of grids with more inductive or more resistive line characteristics to be explored. The results provide new understanding as to the influence of real/reactive power ripple mitigation on unbalanced network voltages for grid connected DG systems.

II. THREE-PHASE DG CONVERTER MODEL

The DG converter arrangement used in this paper is a standard three phase VSI, connected to the electrical network at the PCC through an LCL filter. Typically, such systems are controlled by a three level control structure – a high level power control scheme, which feeds into a mid level current regulation system, and which then in turn calculates voltage commands for the lowest level PWM controller. When an exact time domain representation of the converter is not required, an average voltage representation of the lowest level modulation scheme can be used for simplicity - this is the approach used in this paper [14].

For an unbalanced system, there are three possible reference strategies for the higher level controllers. The first approach is to operate in the $\alpha\beta$ stationary frame, using resonant regulators to control the inverter currents since all quantities are sinusoidal [6]. However, this requires four resonant controllers to independantly regulate the positive and negative sequence $\alpha\beta$ frame currents. The second strategy is to operate in the positive synchronous rotating dq reference frame, using simple dc regulators to control the positive sequence currents. However, in this frame the negative sequence currents oscillate at twice the fundamental frequency, and hence two resonant regulators are still required in this rotating frame [12]. The third alternative is to use the positive rotating synchronous frame to control the positive

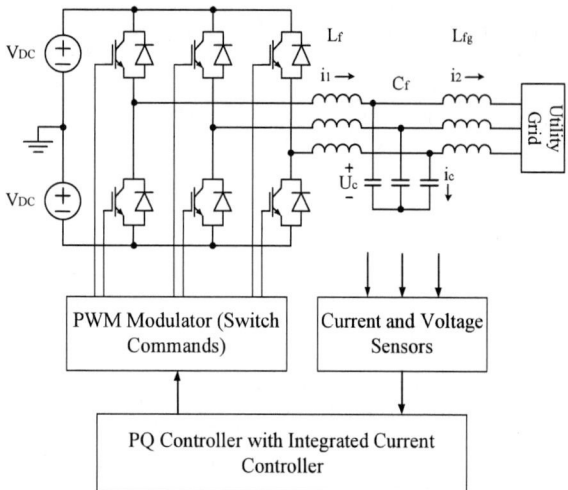

Fig.1. Structure of a three phase VSI used for a DG system.

sequence currents only, and to introduce a negative rotating sequence frame to control the negative sequence currents only. This arrangement requires only two simple PI control structures in each rotating frame of reference, which is straightforward to design and conceptually easy to implement. However, it does require the measured $\alpha\beta$ frame voltages and currents to be separated into positive and negative sequence components before the rotating frame transformation.

For this paper, this third approach has been used, since calculation of the real and reactive power steady state and oscillatory quantities is much easier using "dc" representations of the positive and negative sequence voltages and currents in their respective rotating frames of reference.

Fig. 1 shows the structure of a typical three phase voltage source inverter used to connect a DG system to the electrical network. It consists of a center tapped dc voltage source (the DG energy source) which supplies a switched three phase bridge converter. The converter outputs connect to the utility network via an LCL filter at the point-of-common-coupling. To adequately simulate the dynamic response of such a DG inverter system, it is important to properly represent all three of these converter control systems. This is quite satisfactorily achieved using an average voltage representation, where the switched phase leg voltages are modeled as equivalent continuous time variables [14].

To operate in the synchronous rotating frames, the converter voltages and currents that are measured in the stationary abc frame must be appropriately transformed. This is done by transforming firstly to the stationary $\alpha\beta$ frame using (voltage transformation only shown)

$$[v_{\alpha\beta}] = \begin{bmatrix} 1 & -0.5 & -0.5 \\ 0 & \sqrt{3}/2 & -\sqrt{3}/2 \end{bmatrix} \begin{bmatrix} v_a \\ v_b \\ v_c \end{bmatrix} \quad (1)$$

Next, the positive and negative sequence $\alpha\beta$ quantities are extracted using

$$[v_{\alpha\beta}]^p = \frac{1}{2}\begin{bmatrix} 1 & -i \\ i & 1 \end{bmatrix}[v_{\alpha\beta}]$$
$$[v_{\alpha\beta}]^n = \frac{1}{2}\begin{bmatrix} 1 & i \\ -i & 1 \end{bmatrix}[v_{\alpha\beta}] \quad (2)$$

where "i" denotes a 90° phase shift operator in the time domain. While various methods have been proposed to generate this phase shifted signal, the standard approaches are to use a time delay operator, which has dynamic response limitations, or a Phase Locked Loop (PLL based on a second order generalized integrator [15-17]. This second approach is used in this paper.

Finally, the extracted stationary frame sequence quantities are transformed into their respective positive and negative sequence rotating frames using

$$[v_{dq}]^p = \begin{bmatrix} \cos(\omega t) & \sin(\omega t) \\ -\sin(\omega t) & \cos(\omega t) \end{bmatrix}[v_{\alpha\beta}]^p = T(\omega t)[v_{\alpha\beta}]^p \quad (3)$$

$$[v_{dq}]^n = \begin{bmatrix} \cos(\omega t) & -\sin(\omega t) \\ \sin(\omega t) & \cos(\omega t) \end{bmatrix}[v_{\alpha\beta}]^n = T(-\omega t)[v_{\alpha\beta}]^n \quad (4)$$

Note that the LCL output filter system used with the DG system will introduce a resonance peak into the plant frequency response which can cause current regulation stability problems. While either passive or active damping can be used to manage this instability, active damping using a capacitor current compensation term is the preferred approach [18-19]. For this system, four separate capacitor current feedback signals are required for active damping – one for each of the dq currents in each of the two synchronous rotating frames. Fig. 2 shows the converter and control system structure used in this paper, based on these control concepts.

III. POWER CONTROL FOR THREE-PHASE DG SYSTEM

The primary objective of the top level power control system is to regulate the real and reactive power that is injected into the grid system. The secondary objective under unbalanced voltage conditions is to either 1) eliminate the active power ripple, or 2) eliminate the reactive power ripple, or 3) simply balance the grid injected currents, as required.

The power control aims to generate commanded values for the current regulators that form the next level down control system. For this purpose, these commanded values are calculated based on the target real and reactive power commands P_{ref} and Q_{ref} which are either commanded from a remote controller, or set to constant values for a desired level of PQ control.

The amount of power injected into the grid can be measured using

$$\begin{cases} P = v_d i_d + v_q i_q \\ Q = v_q i_d - v_d i_q \end{cases} \quad (5)$$

where v_{dq} and i_{dq} are the measured grid voltages and currents in the dq frame. For unbalanced PCC voltage conditions, the real and reactive power flows can be calculated from the positive and negative sequence voltage and current components [20] using

$$\begin{bmatrix} P \\ Q \\ P_c \\ P_s \\ Q_c \\ Q_s \end{bmatrix} = \begin{bmatrix} +v_d^p & +v_q^p & +v_d^n & +v_q^n \\ +v_q^p & -v_d^p & +v_q^n & -v_d^n \\ +v_d^n & +v_q^n & +v_d^p & +v_q^p \\ +v_q^n & -v_d^n & -v_q^p & +v_d^p \\ +v_q^n & -v_d^n & +v_q^p & -v_d^p \\ -v_d^n & -v_q^n & +v_d^p & +v_q^p \end{bmatrix} \times \begin{bmatrix} i_d^p \\ i_q^p \\ i_d^n \\ i_q^n \end{bmatrix} \quad (6)$$

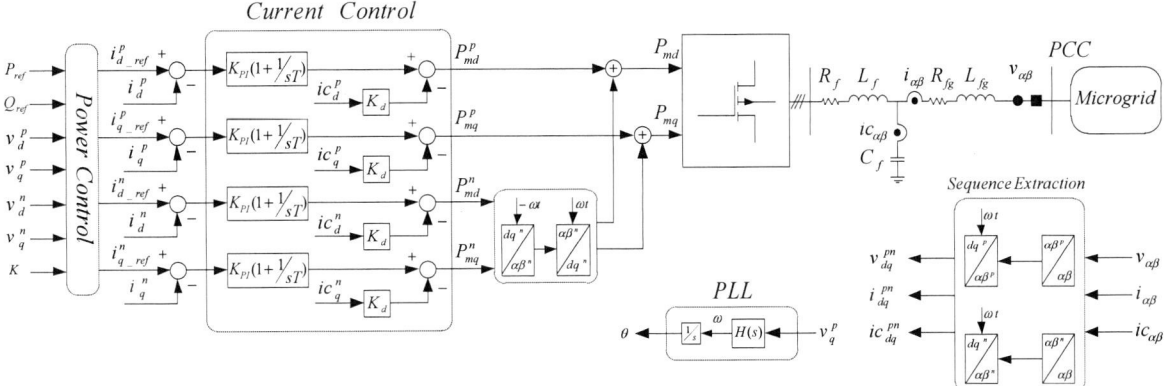

Fig. 2. Proposed DG converter system with double synchronous frame controllers and actively damped LCL filter system

where the P and Q terms in (6) are the average real and reactive powers injected into the grid, and the P_c, P_s, Q_c, Q_s terms define the magnitude of the double fundamental frequency oscillating real and reactive powers as quadrature components referenced to the synchronous rotating frame. Eqn. (6) can then be used to create commanded references for the current regulators from the required grid power injections:

$$
\begin{bmatrix} i_d^p \\ i_q^p \\ i_d^n \\ i_q^n \end{bmatrix}_{ref} =
\begin{bmatrix}
+v_d^p & +v_q^p & +v_d^n & +v_q^n \\
+v_q^p & -v_d^p & +v_q^n & -v_d^n \\
+v_d^n & +v_q^n & +v_d^p & +v_q^p \\
+v_q^n & -v_d^n & -v_q^p & +v_d^p \\
+v_q^n & -v_d^n & +v_q^p & -v_d^p \\
-v_d^n & -v_q^n & +v_d^p & +v_q^p
\end{bmatrix}^{-1}
\times
\begin{bmatrix} P \\ Q \\ P_c \\ P_s \\ Q_c \\ Q_s \end{bmatrix}_{ref}
\tag{7}
$$

By deleting unnecessary rows, inverting the remaining reduced matrix and replacing the calculated powers with known reference values, commanded currents for the three different ripple control strategies can be created as follows:

A. Active Power Ripple Control

For control of active power ripple only, the last two rows of (6) can be deleted, since the reactive power ripple is uncontrolled. The reduced matrix is then inverted, the average P and Q injections are set to known values of P_{ref} and Q_{ref}, and the target quadrature active power ripple magnitudes are set to zero, i.e. $P_c = P_s = 0$, to give:

$$
\begin{bmatrix} i_d^p \\ i_q^p \\ i_d^n \\ i_q^n \end{bmatrix}_{ref} =
\begin{bmatrix}
+v_d^p & +v_q^p & +v_d^n & +v_q^n \\
+v_q^p & -v_d^p & +v_q^n & -v_d^n \\
+v_d^n & +v_q^n & +v_d^p & +v_q^p \\
+v_q^n & -v_d^n & -v_q^p & +v_d^p
\end{bmatrix}^{-1}
\times
\begin{bmatrix} P_{ref} \\ Q_{ref} \\ 0 \\ 0 \end{bmatrix}
\tag{8}
$$

$$
\rightarrow
\begin{bmatrix} i_d^p \\ i_q^p \\ i_d^n \\ i_q^n \end{bmatrix}_{ref} =
\begin{bmatrix}
+v_d^p/D & +v_q^p/E \\
+v_q^p/D & -v_d^p/E \\
-v_d^n/D & +v_q^n/E \\
-v_q^n/D & -v_d^n/E
\end{bmatrix}
\times
\begin{bmatrix} P_{ref} \\ Q_{ref} \end{bmatrix}
\tag{9}
$$

where
$$
\begin{cases}
D = (v_d^p)^2 + (v_q^p)^2 - (v_d^n)^2 - (v_q^n)^2 \\
E = (v_d^p)^2 + (v_q^p)^2 + (v_d^n)^2 + (v_q^n)^2
\end{cases}
\tag{10}
$$

With the reference currents set by commanded average power injections and zero real power oscillations only, reactive power oscillations will now occur, since their

magnitude is uncontrolled. The quadrature component magnitudes of these oscillations can be obtained by combining (9) with the last two lines of (6), to give

$$
\begin{bmatrix} Q_c \\ Q_s \end{bmatrix} = 2
\begin{bmatrix}
\left(\dfrac{v_d^p v_q^n - v_d^n v_q^p}{D}\right) & \left(\dfrac{v_d^n v_d^p + v_q^n v_q^p}{E}\right) \\
-\left(\dfrac{v_d^n v_d^p + v_q^n v_q^p}{D}\right) & \left(\dfrac{v_d^p v_q^n - v_d^n v_q^p}{E}\right)
\end{bmatrix}
\times
\begin{bmatrix} P \\ Q \end{bmatrix}_{ref}
\tag{11}
$$

B. Reactive Power Ripple Control

For control of reactive power ripple only, the middle two rows of (6) can be deleted, since active power ripple is uncontrolled. The reduced matrix is then inverted, the average P and Q injections are set to known values of P_{ref} and Q_{ref}, and the quadrature reactive power ripple magnitudes are set to zero, i.e. $Q_c = Q_s = 0$, to give:

$$
\begin{bmatrix} i_d^p \\ i_q^p \\ i_d^n \\ i_q^n \end{bmatrix}_{ref} =
\begin{bmatrix}
+v_d^p & +v_q^p & +v_d^n & +v_q^n \\
+v_q^p & -v_d^p & +v_q^n & -v_d^n \\
+v_q^n & -v_d^n & +v_q^p & -v_d^p \\
-v_d^n & -v_q^n & +v_d^p & +v_q^p
\end{bmatrix}^{-1}
\times
\begin{bmatrix} P_{ref} \\ Q_{ref} \\ 0 \\ 0 \end{bmatrix}
\tag{12}
$$

$$
\rightarrow
\begin{bmatrix} i_d^p \\ i_q^p \\ i_d^n \\ i_q^n \end{bmatrix}_{ref} =
\begin{bmatrix}
+v_d^p/E & +v_q^p/D \\
+v_q^p/E & -v_d^p/D \\
+v_d^n/E & -v_q^n/D \\
+v_q^n/E & +v_d^n/D
\end{bmatrix}
\times
\begin{bmatrix} P_{ref} \\ Q_{ref} \end{bmatrix}
\tag{13}
$$

Similarly to (11), the quadrature magnitudes of the oscillating real power can be obtained from (13) and (6) as

$$
\begin{bmatrix} P_c \\ P_s \end{bmatrix} = 2
\begin{bmatrix}
\left(\dfrac{v_d^n v_d^p + v_q^n v_q^p}{E}\right) & \left(\dfrac{v_d^n v_q^p - v_d^p v_q^n}{D}\right) \\
\left(\dfrac{v_d^p v_q^n - v_d^n v_q^p}{E}\right) & \left(\dfrac{v_d^n v_d^p + v_q^n v_q^p}{D}\right)
\end{bmatrix}
\times
\begin{bmatrix} P \\ Q \end{bmatrix}_{ref}
\tag{14}
$$

C. Balanced Current Control

For balanced grid currents, only the top left hand four elements of (6) are required, since both the active and reactive power ripple are uncontrolled, and the negative sequence current references are forced to zero, so that:

$$
\begin{bmatrix} i_d^p \\ i_q^p \end{bmatrix}_{ref} =
\begin{bmatrix}
+v_d^p & +v_q^p \\
+v_q^p & -v_d^p
\end{bmatrix}^{-1}
\times
\begin{bmatrix} P_{ref} \\ Q_{ref} \end{bmatrix}, \quad i_{d\,ref}^n = i_{q\,ref}^n = 0
\tag{15}
$$

$$
\rightarrow
\begin{bmatrix} i_d^p \\ i_q^p \end{bmatrix}_{ref} =
\begin{bmatrix}
\dfrac{+v_d^p}{(v_d^p)^2+(v_q^p)^2} & \dfrac{+v_q^p}{(v_d^p)^2+(v_q^p)^2} \\
\dfrac{+v_q^p}{(v_d^p)^2+(v_q^p)^2} & \dfrac{-v_d^p}{(v_d^p)^2+(v_q^p)^2}
\end{bmatrix}
\times
\begin{bmatrix} P_{ref} \\ Q_{ref} \end{bmatrix}
\tag{16}
$$

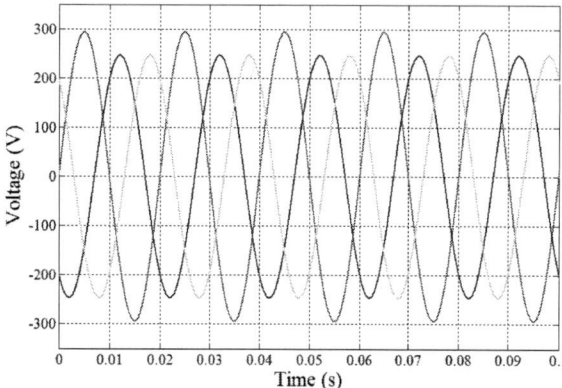

Fig.3. Faulty grid voltages, where two phases sag to 80%.

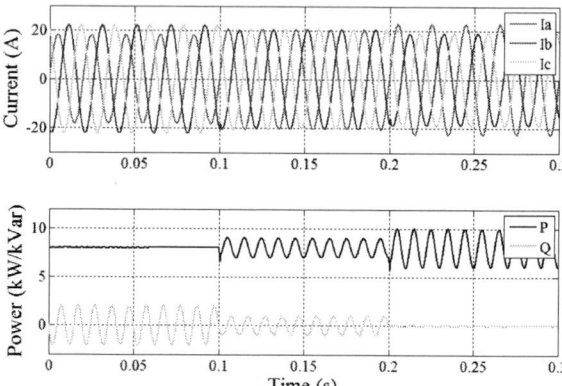

Fig.4. Proposed controller. (upper trace) Three-phase currents. (lower trace) Real and reactive powers.

Balanced currents are obtained at the expense of both active and reactive power oscillations. The oscillating power magnitudes can be calculated by substituting (16) into the last four lines of (6), to give

$$\begin{bmatrix} P_c \\ P_s \\ Q_c \\ Q_s \end{bmatrix} = \begin{bmatrix} \left(\frac{v_d^n v_d^p + v_q^n v_q^p}{(v_d^p)^2+(v_q^p)^2}\right) & \left(\frac{v_d^n v_q^p - v_q^n v_d^p}{(v_d^p)^2+(v_q^p)^2}\right) \\ \left(\frac{v_q^n v_d^p - v_d^n v_q^p}{(v_d^p)^2+(v_q^p)^2}\right) & \left(\frac{v_d^n v_d^p + v_q^n v_q^p}{(v_d^p)^2+(v_q^p)^2}\right) \\ \left(\frac{v_q^n v_d^p - v_d^n v_q^p}{(v_d^p)^2+(v_q^p)^2}\right) & \left(\frac{v_d^n v_d^p + v_q^n v_q^p}{(v_d^p)^2+(v_q^p)^2}\right) \\ -\left(\frac{v_d^n v_d^p + v_q^n v_q^p}{(v_d^p)^2+(v_q^p)^2}\right) & \left(\frac{v_q^n v_d^p - v_d^n v_q^p}{(v_d^p)^2+(v_q^p)^2}\right) \end{bmatrix} \times \begin{bmatrix} P \\ Q \end{bmatrix}_{ref} \quad (17)$$

Comparing the oscillating power magnitudes of (11), (14) and (17), it can be seen that the factor of "2" has disappeared for the case where balanced currents are generated. This is because the power oscillations are shared between both the real and reactive powers for this third case.

D. Combined Power Ripple Control

Eqns (9), (13) and (16) can be combined as follows:

$$\begin{bmatrix} i_d^p \\ i_q^p \\ i_d^n \\ i_q^n \end{bmatrix}_{ref} = \begin{bmatrix} +v_d^p/D' & +v_q^p/E' \\ +v_q^p/D' & -v_d^p/E' \\ -Kv_d^n/D' & +Kv_q^n/E' \\ -Kv_q^n/D' & -Kv_d^n/E' \end{bmatrix} \times \begin{bmatrix} P \\ Q \end{bmatrix}_{ref} \quad (18)$$

where
$$\begin{cases} D' = (v_d^p)^2 + (v_q^p)^2 - K\big((v_d^n)^2 + (v_q^n)^2\big) \\ E' = (v_d^p)^2 + (v_q^p)^2 + K\big((v_d^n)^2 + (v_q^n)^2\big) \end{cases} \quad (19)$$

and K is a real number in the range of $[-1,1]$ that defines the required power ripple control strategy. When $K = +1$, (18) and (19) revert to (9) and (10) to control active power ripple only at the expense of unbalanced grid currents and oscillatory reactive power. Similarly, when $K = -1$, (18) and (19) revert to (12) and (13) to control reactive power ripple only at the expense of unbalanced grid currents and oscillatory active power. Finally, when $K = 0$, (18) and (19) revert to (15) and (16), and the injected grid currents only are balanced.

A voltage unbalance of 20% reduction for two of the three phases (Fig. 3) is now used as the unbalanced grid operating condition to investigate the ripple control strategies presented in this paper. Under this grid condition, the controller performance is evaluated as shown in Fig. 4. For the period $0.0 \leq t < 0.1s$, the

"*Active Power Control*" strategy ($K = 1$) is implemented and no oscillation is present in the real power. As expected, oscillations appear in the reactive power, and the three-phase currents are also not balanced. For the period $0.1 \leq t < 0.2s$, the "*Current Control*" strategy is used with $K = 0$. Balanced currents are then generated to feed into the grid, but with both real and reactive power oscillation as a consequence. Finally, for the period $0.2 \leq t < 0.3s$, K is changed to $K = -1$ to enable the "*Reactive Power Control*" strategy. In contrast to active power control, no reactive power oscillations are now present, but there is now real power oscillation, and once again the grid currents are unbalanced.

IV. POWER FLOW RIPPLE CONTROL AND ITS EFFECT ON GRID VOLTAGE

The combined power ripple control concept has been used to investigate the effect of eliminating the active or the reactive power ripple, or just balancing the grid injected currents, when the DG is feeding into a grid distribution system with a significant voltage unbalance at the PCC.

For this study, as shown in Fig. 5, a DG unit is connected to a grid with different line characteristics (Z_{grid}). Changing the grid characteristic from resistive to inductive, the three different control approaches are then applied to explore their impact on the voltage quality at the point of common coupling.

For this investigation, the voltage unbalanced *n*-factor [21], which is the ratio between negative and positive sequence voltage amplitudes, is used to quantify the system voltage unbalance at the inverter PCC. This factor is given by

$$n = \frac{V^n}{V^p} = \sqrt{\frac{(v_d^n)^2+(v_q^n)^2}{(v_d^p)^2+(v_q^p)^2}} \quad (20)$$

Th n-factor parameter is calculated from measurements at the PCC and is compared against the (constant) n-factor

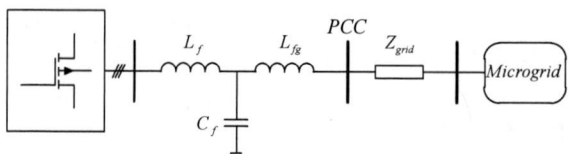

Fig. 5. Three-phase grid-connected DG inverter.

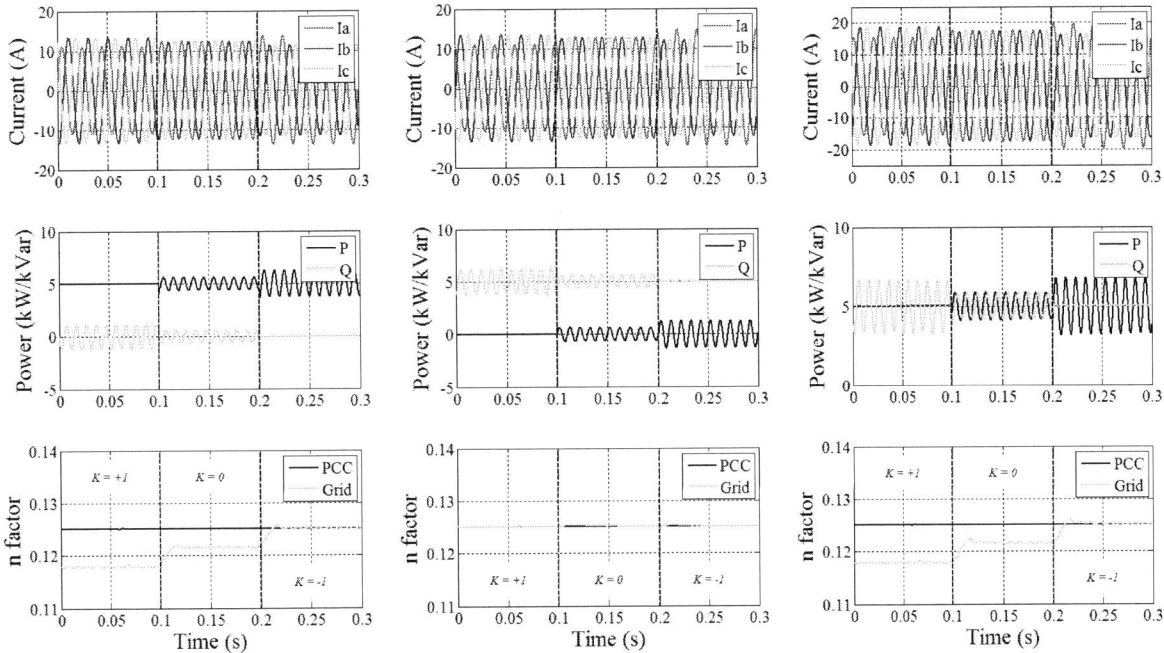

Fig. 6. Simulation results for grid with resistive characteristic.
(left) Real power injection. (middle) Reactive power injection. (right) combined PQ injection.

at the grid source caused by the unbalanced grid supply voltages. Hence, with regard to the different case studies, the impact of the alternative control strategies can be evaluated. Additionally, the voltage magnitudes of the sagged phase voltages are considered for each operating condition to see how well they have been compensated.

For each investigation, the voltage sag at the PCC is determined with no active power ripple, balanced grid current, and no reactive power ripple, respectively, for commanded average power injections of P only, Q only, and a combined PQ command. The power ripple objective is set by varying K over the range $K = +1, 0, -1$ at 0.1sec steps, for each average PQ test condition.

For this paper, results for the power ripple control strategy performance are presented for resistive and inductive grid network impedance, for the three conditions of 5kW average P only injection, 5 kVAr Q only injection, and combined 5kW P and 5 kVAr Q injection. However, the investigation strategy could be readily applied to a wide variety of grid loading and ripple control scenarios.

A. Case I: Grid with resistive characteristic

Fig. 6 shows the results for a grid with a resistive line impedance, where for each PQ condition the ripple power objective is changed at 0.1 second intervals during the simulation, from "no active power ripple", $K = +1$, to "balanced currents", $K = 0$, to "no reactive power ripple", $K = -1$. From this figure, it can be seen that irrespective of the average PQ injection levels, the power ripple control scheme achieves exactly the commanded ripple objective, which confirms the capability of the strategy. Furthermore, it is clear that for this resistive grid network, eliminating real power ripple achieves the lowest n-factor outcome, and hence this ripple objective provides the best grid voltage negative sequence mitigation irrespective of the average power injection conditions.

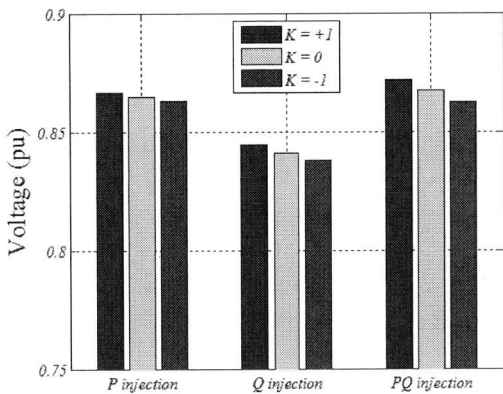

Fig. 7. Voltage magnitude of sagging phases for resistive grid.

Fig. 7 shows how the different ripple objectives influence the absolute voltage magnitude of the sagging phase voltages at the PCC. Once again, for this grid network, real power ripple elimination achieves the best outcome under all average PQ injection conditions.

In addition, in terms of voltage support during faults or unbalanced voltage conditions, active power injection is the most effective solution for voltage rise as well as unbalanced voltage mitigation. Therefore, during abnormal conditions DG units should produce as much active power as possible, and the rest of their capacity should be then be used for reactive power injection.

B. Case II: Grid with inductive characteristic

In this study the grid impedance is considered to be inductive. The simulation results obtained from this case study are shown in Fig. 8, while Fig. 9 presents the voltage amplitude for the sagging phases.

978-1-4799-2706-7/14 $31.00 © 2014 IEEE

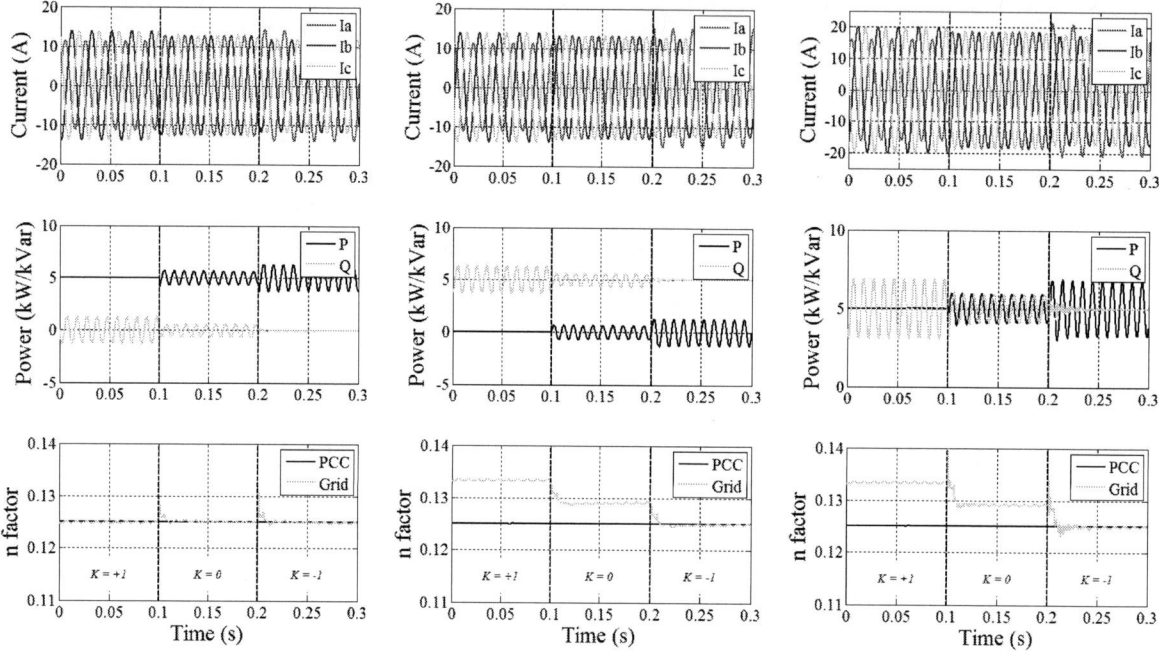

Fig. 8. Simulation results for grid with inductive characteristic.
(left) Real power injection. (middle) Reactive power injection. (right) combined PQ injection.

These results show that injecting active power into the system as shown in Fig. 8 (left) does not have any significant impact on the voltage unbalance factor, while injecting reactive power increases the n-factor unless the reactive power oscillation is regulated ($K = -1$). Therefore, in case of a grid with an inductive characteristic, the best unbalanced voltage mitigation is achieved when reactive power oscillation is regulated. When DG unit is supplying both real and reactive power, the system performance is as shown in Fig. 8 (right), with a voltage unbalance factor that is essentially the same as supplying reactive power only. So, for a grid with an inductive impedance, reactive power support is more dominant compared to active power for grid voltage negative sequence mitigation.

Fig. 9 also shows that for voltage support with an inductive grid impedance (i.e. increasing the sagging voltages), real power injection is still the best alternative. In other words, if real power is being injected into the grid, the best DG performance is seen when $K = +1$. Conversely, with average reactive power injection, the "no reactive power ripple" strategy is best to increase the voltage without increasing the n-factor. However, active power injection is always more effective for voltage support and unbalance voltage mitigation compared to reactive power injection, similar to Case II.

It is also interesting to note that the $K = 0$ strategy (balanced currents) has no significant PCC voltage quality improvement compared to either real or reactive power ripple control. In other words this strategy sits in between the other two, and hence has the least benefit.

C. Case III: Grid with resistive/inductive impedance

The last study case investigates a grid with a mixed resistive/inductive characteristic. For this investigation, the grid impedance has an angle of 45 degree. For active

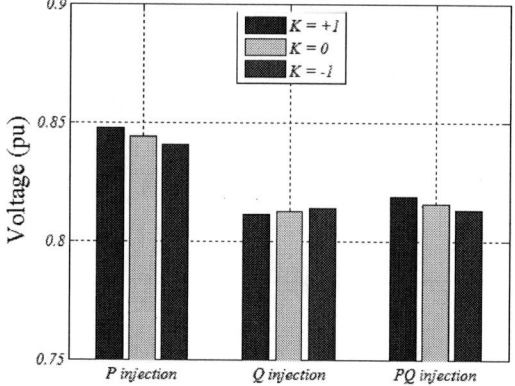

Fig. 9. Voltage magnitude of sagging phases for inductive grid.

average power injection and the three different control strategies, the system performance is similar to the case of a grid with resistive characteristics, as shown in Fig. 10 (left). On the other hand when average reactive power is injected into the grid the results are the same as for reactive power injection in inductive grids, as shown in Fig. 10(middle). Finally, the performance with mixed average power injection is in between the response when the grid impedance is purely resistive or inductive.

These studies also show that regardless of the type of grid impedance, active power injection always achieves the best voltage unbalance mitigation. Furthermore, in order to increase the voltage of the sagged phases, the "no active power ripple" strategy ($K = +1$) is the best approach unless reactive power is injected into an inductive grid.

978-1-4799-2706-7/14 $31.00 © 2014 IEEE   3311

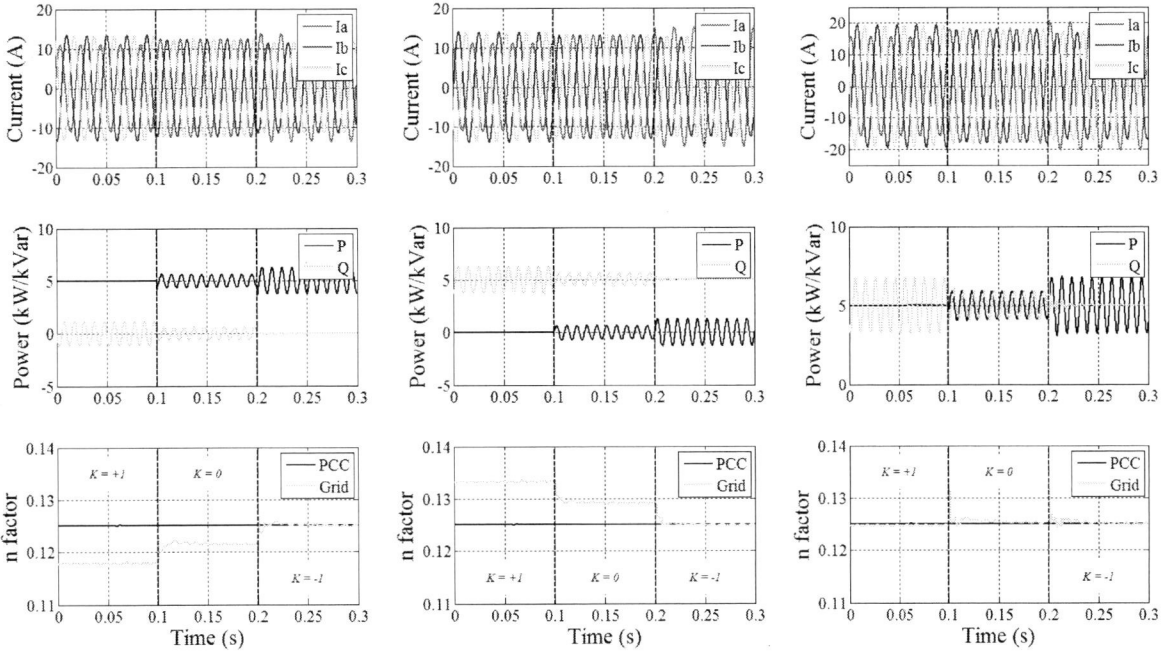

Fig. 10. Simulation results for grid with mixed RL characteristic.
(left) Real power injection. (middle) Reactive power injection. (right) combined PQ injection.

V. CONCLUSION

This paper presents a strategy to adjust between eliminating the real power ripple, the reactive power ripple, or to just balance the grid currents, for a DG inverter operating into a grid network with unbalanced voltages at the PCC. The strategy calculates positive and negative sequence current references from the required average power injections and power ripple objectives, and feeds these references to synchronous frame closed loop current regulators. The system achieves the required power ripple elimination objective under all load conditions with any R/X ratio grid impedance, and allows the mitigation effect of this control strategy on the unbalanced grid voltages at the PCC to be readily investigated. Detailed simulation results are presented to confirm the strategy.

APPENDIX

The controller parameters as well as the parameters of the DG models used in this are given in Table I and II.

ACKNOWLEDGMENT

The authors gratefully acknowledge the financial support of Energy Networks Association Limited, Australia for the research reported in this paper.

REFERENCES

[1] R. H. Lasseter, "Smart Distribution: Coupled Microgrids," Proceedings of the IEEE, vol. 99, pp. 1074-1082, 2011.
[2] "Smart Grid: An Introduction U.S. Department of Energy," 2009.
[3] R. H. Lasseter and P. Paigi, "Microgrid: a conceptual solution," in Power Electronics Specialists Conference, 2004. PESC 04. 2004 IEEE 35th Annual, 2004, pp. 4285-4290 Vol.6.
[4] Carrasco, J.M.; Franquelo, L.G.; Bialasiewicz, J.T.; Galvan, E.; Guisado, R.C.P.; Prats, Ma.A.M.; Leon, J.I.; Moreno-Alfonso, N., "Power-Electronic Systems for the Grid Integration of Renewable Energy Sources: A Survey," Industrial Electronics, IEEE Transactions on , vol.53, no.4, pp.1002,1016, June 2006.

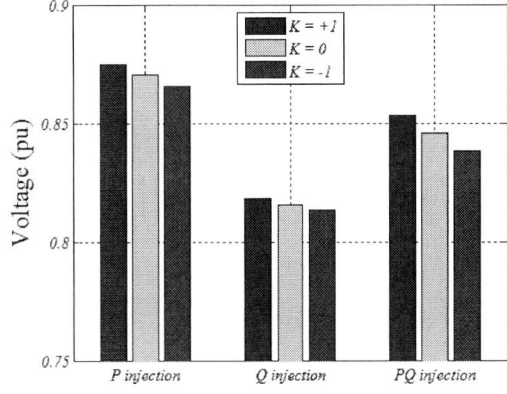

Fig. 11. Voltage magnitude of sagging phases for RL grid.

TABLE II. CONTROLLER PARAMETERS

	Symbol	Value
Proportional gain	K_{PI}	0.08
Time constant	T	0.03
Damping gain	K_d	0.095

TABLE I. SYSTEM PARAMETERS

	Symbol	Nominal Value
Inverter inductor	L_f	6 mH
Grid-side inductor	L_{fg}	2 mH
Filter capacitance	C_f	15 μF
Switching frequency	f_{sw}	5 kHz
Sampling frequency	f_{samp}	10 kHz
Grid voltage	V	400 V (rms)
Grid frequency	f	50 Hz

[5] Barker, P.P.; De Mello, R.W., "Determining the impact of distributed generation on power systems. I. Radial distribution systems," Power Engineering Society Summer Meeting, 2000. IEEE , vol.3, no., pp.1645,1656 vol. 3, 2000.

[6] Zmood, D.N.; Holmes, D.G., "Stationary frame current regulation of PWM inverters with zero steady-state error," Power Electronics, IEEE Transactions on , vol.18, no.3, pp.814,822, May 2003.

[7] Zmood, D.N.; Holmes, D.G., "Improved voltage regulation for current-source inverters," Industry Applications, IEEE Transactions on , vol.37, no.4, pp.1028,1036, Jul/Aug 2001.

[8] P. Rodriguez, A. V. Timbus, R. Teodorescu, M. Liserre, and F. Blaabjerg, "Flexible Active Power Control of Distributed Power Generation Systems During Grid Faults," Industrial Electronics, IEEE Transactions on, vol. 54, pp. 2583-2592, 2007.

[9] Y. A. R. I. Mohamed and E. F. El-Saadany, "A Control Scheme for PWM Voltage-Source Distributed-Generation Inverters for Fast Load-Voltage Regulation and Effective Mitigation of Unbalanced Voltage Disturbances," Industrial Electronics, IEEE Transactions on, vol. 55, pp. 2072-2084, 2008.

[10] W. Fei, J. L. Duarte, and M. A. M. Hendrix, "Pliant Active and Reactive Power Control for Grid-Interactive Converters Under Unbalanced Voltage Dips," Power Electronics, IEEE Transactions on, vol. 26, pp. 1511-1521, 2011.

[11] A. Camacho, M. Castilla, J. Miret, J. C. Vasquez, and E. Alarcon-Gallo, "Flexible Voltage Support Control for Three-Phase Distributed Generation Inverters Under Grid Fault," Industrial Electronics, IEEE Transactions on, vol. 60, pp. 1429-1441, 2013.

[12] I. Etxeberria-Otadui, U. Viscarret, M. Caballero, A. Rufer, and S. Bacha, "New Optimized PWM VSC Control Structures and Strategies under Unbalanced Voltage Transients," Industrial Electronics, IEEE Transactions on, vol. 54, pp. 2902-2914, 2007.

[13] S. Xianwen, W. Yue, H. Weihao, and W. Zhaoan, "Three Reference Frame Control Scheme of 4 wire Grid-connected Inverter for Micro Grid Under Unbalanced Grid Voltage Conditions," in Applied Power Electronics Conference and Exposition, 2009. APEC 2009. Twenty-Fourth Annual IEEE, 2009, pp. 1301-1305.

[14] D. G. Holmes, T. A. Lipo, B. P. McGrath, and W. Y. Kong, "Optimized Design of Stationary Frame Three Phase AC Current Regulators," Power Electronics, IEEE Transactions on, vol. 24, pp. 2417-2426, 2009.

[15] Rodriguez, P.; Teodorescu, R.; Candela, I.; Timbus, A.V.; Liserre, M.; Blaabjerg, F., "New Positive-sequence Voltage Detector for Grid Synchronization of Power Converters under Faulty Grid Conditions," Power Electronics Specialists Conference, 2006. PESC '06. 37th IEEE , vol., no., pp.1,7, 18-22 June 2006.

[16] Ciobotaru, M.; Teodorescu, R.; Blaabjerg, F., "A New Single-Phase PLL Structure Based on Second Order Generalized Integrator," Power Electronics Specialists Conference, 2006. PESC '06. 37th IEEE , vol., no., pp.1,6, 18-22 June 2006.

[17] Rodríguez, P.; Luna, A.; Candela, I.; Mujal, R.; Teodorescu, R.; Blaabjerg, F., "Multiresonant Frequency-Locked Loop for Grid Synchronization of Power Converters Under Distorted Grid Conditions," Industrial Electronics, IEEE Transactions on , vol.58, no.1, pp.127,138, Jan. 2011.

[18] Parker, S.G.; McGrath, B.P.; Holmes, D.G., "Regions of Active Damping Control for LCL Filters," Industry Applications, IEEE Transactions on , vol.50, no.1, pp.424,432, Jan.-Feb. 2014.

[19] J. Dannehl, C. Wessels, and F. W. Fuchs, "Limitations of Voltage-Oriented PI Current Control of Grid-Connected PWM Rectifiers With LCL Filters," *Industrial Electronics, IEEE Transactions on,* vol. 56, pp. 380-388, 2009.

[20] Hong-Seok Song; Kwanghee Nam, "Dual current control scheme for PWM converter under unbalanced input voltage conditions," Industrial Electronics, IEEE Transactions on , vol.46, no.5, pp.953,959, Oct 1999.

[21] von Jouanne, A.; Banerjee, B., "Assessment of voltage unbalance," Power Delivery, IEEE Transactions on , vol.16, no.4, pp.782,790, Oct 2001.

Stability Analysis of Grid-Connected Inverters with LCL-Filter Based on Harmonic Balance and Floquet Theory

Jing Bian, Hong Li, Trillion Q. Zheng
School of Electrical Engineering
Beijing Jiaotong University
Beijing, China
hli@bjtu.edu.cn

Abstract—**The inherent resonance of the grid-connected inverter with LCL-filter can lead to an instability of such a system. Normally, a damping resistor in series with the capacitor in the LCL-filter will be used to suppress the resonance, and the damping resistance is designed roughly based on the Routh-Hurwitz stability criterion. But, in practice, the resistance should be bigger than the designed value to make the inverter stable and reliable, which results in the arbitrariness of the parameter selection. In this paper, a novel method is provided based on the harmonic balance and Floquet theory to analyze the stability of the inverter. In detail, the approximate analytical periodic expressions of the inverter are obtained by using the harmonic balance method; following, the stability of the inverter versus damping resistance is analyzed according to Floquet theory. By comparing with the damping resistance got by Routh-Hurwitz stability criterion, the resistance obtained by the provided stability analysis is more practical, therefore, the provided method can be used to optimize the damping resistance.**

Keywords— Floquet theory, grid-connected inverter with LCL-filter, harmonic balance method, stability analysis

I. INTRODUCTION

The single phase grid-connected inverter with LCL-filter is widely used because of its advantage in harmonic reduction. However, the inherent resonance of LCL-filter can lead to the instability of such a system, which can further cause the increase of harmonics and the failure of the inverter. Passive damping method is a kind of method to suppress the resonance [1-3], which is very practical, but will result in low efficiency of the inverters due to the loss on damping resistor. Hence, finding the appropriate resistance of damping resistor is very important in practical engineering application. Routh-Hurwitz stability criterion is normally applied to analyze the stability of the inverter in literatures by establishing linear control model and deducing characteristic equation of the inverter [4]. A method based on harmonic balance and Floquet theory is introduced to analyze the stability of the single phase grid-connected inverter with LCL-filter, by which a more appropriate resistance of damping resistor is designed, comparing with that got by the method based on Routh-Hurwitz stability criterion. When the damping resistor is the

minimum got by provided method, the inverter can achieve good performance of harmonics. In practice, the provided method avoids the arbitrariness in the damping resistance selection. What's more, the provided method is more suitable to analyze the stability changing trend of the single phase grid-connected inverter with LCL- filter as the change of the parameter in the inverter. Finally, simulation results are given to verify the rationality of harmonic balance and the accuracy of the stability analysis method proposed in this paper.

II. PERIODIC SOLUTIONS OF THE INVERTER BASED ON HARMONIC BALANCE

The schematic diagram of the single phase grid-connected inverter with LCL- filter is illustrated in Fig. 1. The control of grid current is the key to the inverter, because it has impact on the quality of the grid inverter. So the direct current control technology is applied to control the current of the grid side [5].If typical grid current closed-loop control is used directly, the stability of the grid inverter cannot be guaranteed [6].Damping resistor R_C in series with capacitor C is adopted to avoid the phenomenon of instability[7]. So control method is typical direct current control method based on PI. The control approach is not only simple, but also ensures the stability of the single phase grid-connected inverter with LCL-filter in the case of matching parameters.

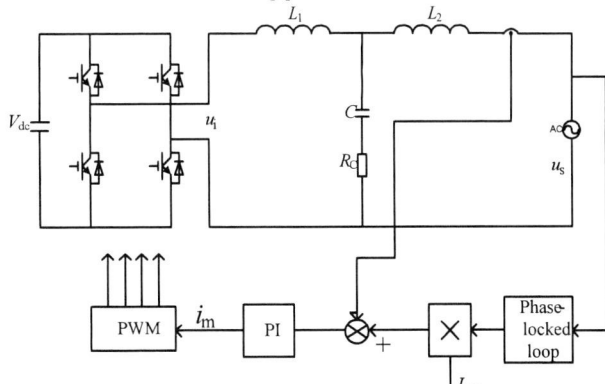

Fig. 1. The schematic diagram of single phase grid –connected inverter with LCL- filter

The single polarity SPWM method is adopted in the paper. The principle of single polarity SPWM technology and the triangle based on regular sampling are illustrated in Fig.2.

$$d = \frac{i_m}{I_{sm}} \tag{1}$$

Where, d is the duty cycle of the single phase grid-connected inverter with LCL- filter. i_m is the signal after PI controller in Fig.1. I_{sm} is the amplitude of the triangle carrier wave.

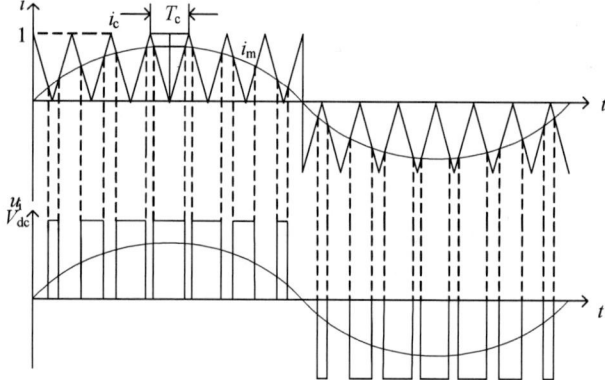

Fig. 2. The principle of single polarity SPWM technology

The voltage after inverter can be equivalent to a time-varying voltage source [8].

$$u_i = dV_{dc} \tag{2}$$

Where, u_i is the voltage after the inverter. V_{dc} is the value of the dc source.

According to the topology of the single phase grid-connected inverter with LCL- filter in this paper, main circuit state equations can be obtained as follows:

$$C\frac{du_C}{dt} = i_1 - i_2$$
$$L_1\frac{di_1}{dt} = -u_C - R_C i_1 + R_C i_2 + u_i \tag{3}$$
$$L_2\frac{di_2}{dt} = u_C + R_C i_1 - R_C i_2 - u_s$$

Where, i_1 and i_2 are separately the current through inductor L_1 and L_2. Likewise, u_C is the voltage across capacitor C. R_C is the damping resistor in series with capacitor in the LCL-filter. u_s is the grid voltage.

According to the control logic of the control circuit in Fig. 1, the transfer function of PI controller is

$$\frac{i_m(s)}{i_{ref}(s) - i_2(s)} = k_p + \frac{k_i}{s} \tag{4}$$

Where, i_{ref} is the given ac current, whose phase is consistent with the grid voltage. Thus, output current of PI controller meets the following differential equation

$$\frac{di_m}{dt} = -k_p\frac{di_2}{dt} + k_p\frac{di_{ref}}{dt} + k_i(i_{ref} - i_2)$$
$$= -\frac{k_p u_C}{L_2} - \frac{k_p R_C}{L_2}i_1 + (\frac{k_p R_C}{L_2} - k_i)i_2 \tag{5}$$
$$+ \frac{k_p}{L_2}u_s + k_i i_{ref} + k_p\frac{di_{ref}}{dt}$$

Summarizing the above equation (1) to (5), the state space average model of the single phase grid-connected inverter with LCL- filter is given in (6),

$$\begin{cases} \dfrac{du_C}{dt} = \dfrac{1}{C}i_1 - \dfrac{1}{C}i_2, \\[2mm] \dfrac{di_1}{dt} = -\dfrac{1}{L_1}u_C - \dfrac{R_C}{L_1}i_1 + \dfrac{R_C}{L_1}i_2 + \dfrac{i_m V_{dc}}{L_1}, \\[2mm] \dfrac{di_2}{dt} = \dfrac{1}{L_2}u_C + \dfrac{R_C}{L_2}i_1 - \dfrac{R_C}{L_2}i_2 - \dfrac{u_s}{L_2}, \\[2mm] \dfrac{di_m}{dt} = -\dfrac{k_p u_C}{L_2} - \dfrac{k_p R_C}{L_2}i_1 + (\dfrac{k_p R_C}{L_2} - k_i)i_2 \\[2mm] \qquad + \dfrac{k_p}{L_2}u_s + k_i i_{ref} + k_p\dfrac{di_{ref}}{dt}. \end{cases} \tag{6}$$

For brevity, let $Y = \begin{bmatrix} u_C & i_1 & i_2 & i_m \end{bmatrix}^T$. Hence, (6) can be rewritten as

$$\dot{Y} = G(t, Y). \tag{7}$$

If the single phase grid-connected inverter with LCL- filter is in stable periodic state, it will meet the principle of energy balance in (8) [9].

$$i_m V_{dc} i_1 = Cu_C\frac{du_C}{dt} + L_1 i_1\frac{di_1}{dt} + u_s i_2$$
$$+ L_2 i_2\frac{di_2}{dt} + R_C(i_1 - i_2)^2 \tag{8}$$

Where, equation (8) contains three state variables in the main circuit and one variable in a control circuit of the single phase grid-connected inverter with LCL- filter. The variables in (8) meet the constraint relationships in (9):

$$\frac{di_m}{dt} = -k_p\frac{di_2}{dt} + k_p\frac{di_{ref}}{dt} + k_i(i_{ref} - i_2),$$
$$L_1\frac{di_1}{dt} = i_m V_{dc} - L_2\frac{di_2}{dt} - u_s, \tag{9}$$
$$\frac{du_C}{dt} = \frac{1}{C}i_1 - \frac{1}{C}i_2$$

Putting the constraint relationships into (8), (8) can be simplified into a single-variable differential equation (10),

$$f(t, i_2) = g(t, i_2). \tag{10}$$

The equation (10) is not only about i_2, but also the derivative of i_2. We let i_2 be expressed in the approximate analytical periodic expressions [9]

$$i_2 = a_0 + \sum_{n=1}^{M} [a_n \cos(n\omega t) + b_n \sin(n\omega t)], \qquad (11)$$

Where, $\omega = 2\pi f$, f is the line frequency. M is the total number of harmonics that a variable contains. Of course, the greater the value of M is, the expressions of i_2 solved by the harmonic balance method is more close to the actual waveform. However, the calculation will increase as the value of M becomes bigger. Thus considering the calculation accuracy and computational complexity, the number $M = 3$ is selected in this paper. Let

$$C_s = [1, \cos(\omega t), \cos(2\omega t), \cos(3\omega t), \\ \sin(\omega t), \sin(2\omega t) \sin(3\omega t)] \qquad (12)$$

$$A = [a_0, a_1, a_2, a_3, b_1, b_2, b_3]^T \qquad (13)$$

Hence,

$$i_2 = C_s A \qquad (14)$$

Since i_2 is a continuous periodic variable of period $1/f$, the equation (14) can be solved in the time interval [0, $1/f$]. Therefore, we can get $2M+1 = 7$ algebraic equations, which contain all the elements of matrix A.

$$\int_0^{1/f} f(a_0, a_1, a_2, a_3, b_1, b_2, b_3)C_s^T dt \\ = \int_0^{1/f} g(a_0, a_1, a_2, a_3, b_1, b_2, b_3)C_s^T dt \qquad (15)$$

According to (15), the coefficients in (13), i.e., a_0, a_n, b_n, can be figured out, then, the steady state solutions $Y_0(t)$ of the variables in (7) can be totally obtained. Now, the stability of the inverter can be analyzed according to Floquet theory using the obtained steady state solution.

III. STABILITY ANALYSIS OF THE INVERTER

In this paper, in order to solve the unstable phenomenon of the single phase grid-connected inverter with LCL- filter due to the resonance, the solution is adding damping resistor in series with capacitor. Thus the damping resistance has important influence on the stability of the single phase grid-connected inverter with LCL- filter. Based on this, the following discussion will be focused on the relationship of the damping resistance and the stability of the single phase grid-connected inverter with LCL- filter.

A. Stability Analysis Based on Floquet Theory

Through the above analysis and calculation, the periodic solution of the single phase grid-connected inverter with LCL-filter will be obtained. Subsequently, the Floquet theory will be used to analyze the stability of the system. The stability of the system can be equal to the stability of the equivalent model (16),

$$\varphi' = J(t, Y_0(t))\varphi \\ \varphi = Y(t) - Y_0(t) \qquad (16) \\ Y_0(t) = [u_{C0} \quad i_{10} \quad i_{20} \quad i_{m0}]^T$$

Where, $J(t, Y_0(t))$ denotes the Jacobian matrix at $Y_0(t)$. According to Floquet theory, the stability of the single phase grid-connected inverter with LCL- filter can be judged by the eigenvalues of the transition matrix deduced from (16) [10-11]. Specific judgment method is as follows: if the absolute values of all eigenvalues are less than 1, then the system is stable; if as long as the absolute value of one eigenvalues is greater than 1, then the system is unstable.

Since the eigenvalues of the transition matrix is related to R_C, the eigenvalues of the transition matrix can be changed with the variation of damping resistor R_C, as shown in Table 1.

TABLE I
CHANGING TREND OF THE EIGENVALUES WITH THE DECREASE OF DAMPING RESISTOR R_C

R_C/Ω	λ 1,2	λ 3,4	state
4.99	-0.1803±0.4648i	-4.8440e-19, 1.2546e-16	stable
4.98	-0.2230±0.5672i	-1.0008e-16, 1.5364e-16	stable
4.97	-0.275±0.6921i	9.6450e-17±1.0717e-16i	stable
4.96	-0.3405±0.8445i	3.2963e-16, -7.8173e-17	stable
4.95	-0.4209±1.0305i	-2.9185e-16, 2.2721e-16	unstable

Based on Table 1, it is obvious that when R_C is larger than 4.95 Ω, the single phase grid-connected inverter with LCL-filter will be stable. Fig. 3 shows the changing trend of eigenvalues of the single phase grid-connected inverter with LCL-filter with the decrease of damping resistance in table 1. The stability state of the single phase grid-connected inverter with LCL-filter can be seen more vividly in Fig.3.

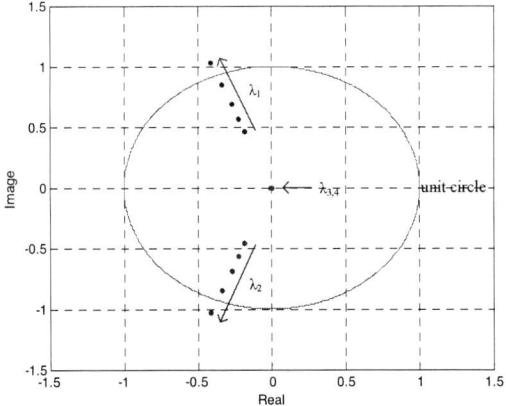

Fig. 3. Movement of the eigenvalues with the decrease of R_C

B. Stability Analysis Based on Routh-Hurwitz stability criterion

As it is shown in Fig.1, the single phase grid-connected inverter with LCL-filter is a nonlinear system, which can be equivalent to a linear system when ignoring some nonlinear factors. Besides, the full bridge of the single phase grid-connected inverter with LCL-filter is generally seen as a proportional gain K_{PWM} [12], namely:

$$K_{PWM} = U_{dc} \qquad (17)$$

Therefore, the linear control model of the single phase grid-connected inverter with LCL-filter is shown in Fig. 4.

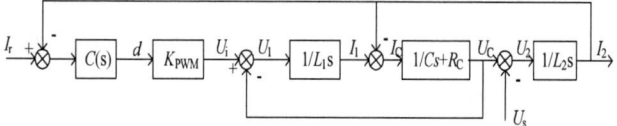

Fig. 4. Linear control model of the single phase grid-connected inverter with LCL-filter

By the linear control model of the single phase grid-connected inverter with LCL-filter, the transfer function of the output current of the single phase grid-connected inverter with LCL-filter is as follows:

$$I_2(s) = \frac{K_{PWM}C(s)}{K_{PWM}C(s)+F(s)} I_r(s) - \frac{L_1Cs^2+1}{K_{PWM}C(s)+F(s)} U_s(s) \qquad (18)$$

Where,

$$C(s) = k_p + \frac{k_i}{s}$$
$$F(s) = L_1L_2Cs^3 + (L_1+L_2)s \qquad (19)$$

Hence, characteristic equation of the single phase grid-connected inverter with LCL-filter is as follows:

$$D(s) = a_0s^4 + a_1s^3 + a_2s^2 + a_3s^1 + a_4 \qquad (20)$$

Where,

$$a_0 = L_1L_2C$$
$$a_1 = RC(L_1+L_2)$$
$$a_2 = L_1 + L_2 + K_{PWM}k_pRC \qquad (21)$$
$$a_3 = K_{PWM}k_p + K_{PWM}k_iRC$$
$$a_4 = K_{PWM}k_i$$

According to the Routh-Hurwitz stability criterion, the stability conditions for the single phase grid-connected inverter with LCL-filter can be obtained as follows [4]:

$$a_i > 0, a_1a_2 - a_0a_3 > 0, a_1a_2 - a_0a_3 - a_1^2a_4 > 0 \quad (22)$$

Namely,

$$R_CC(L_1+L_2)(L_1+L_2+K_{PWM}k_pR_CC)$$
$$-L_1L_2C(K_{PWM}k_p + K_{PWM}k_iR_CC) > 0$$
$$R_CC(L_1+L_2)(L_1+L_2+K_{PWM}k_pR_CC) \qquad (23)$$
$$-L_1L_2C(K_{PWM}k_p + K_{PWM}k_iR_CC)$$
$$-(L_1L_2C)^2K_{PWM}k_i > 0$$

Hence, the lower limit of the damping resistance R_C, which makes the single phase grid-connected inverter with LCL-filter stable is 4.79 Ω obtained from (23).

IV. SIMULATION VERIFICATION

In order to verify the rationality of the periodic solution based on harmonic balance method, the harmonic balance method and Matlab/Simulink simulation method are used to calculate the grid-connected current of the single phase grid-connected inverter with LCL-filter. The simulation result of the grid-connected current is shown in Fig. 5, which is got by Simulink. At the same time, the steady state solution of grid-connected current i_2 can be calculated based on harmonic balance method and also shown in Fig. 5.

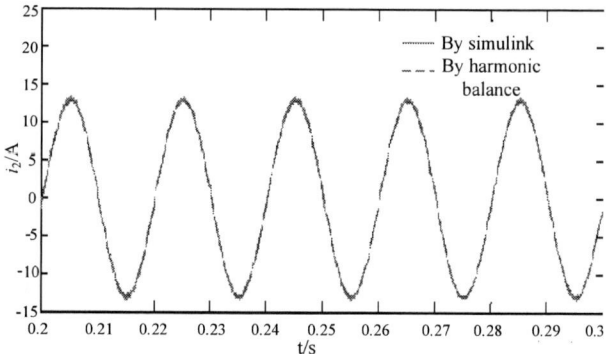

Fig. 5. Waveform of grid-connected current i_2 based on simulink and harmonic balance method

From Fig.5, it can be concluded that the calculation result of grid-connected current using harmonic balance is consistent with the simulation/Simulink result. The only difference is that the simulation solution contains more harmonics. The situation that the waveforms based on the two methods are more similar can be achieved by increasing M. Hence, the rationality of harmonic balance is verified.

As mentioned, the single phase grid-connected inverter with LCL-filter will be in stable state when the R_C is larger than 4.79 Ω based on the Routh-Hurwitz stability criterion, and 4.95Ω based on the harmonic balance and Floquet theory, respectively. Simulation results of the single phase grid inverter with LCL-filter by Matlab/Simulink are shown in Fig. 6 and Fig. 7 when the resistance of R_C is respectively equal to 4.8Ω and 5Ω. The corresponding FFT analysis diagrams under the two situations are shown in Fig. 8 and Fig. 9.

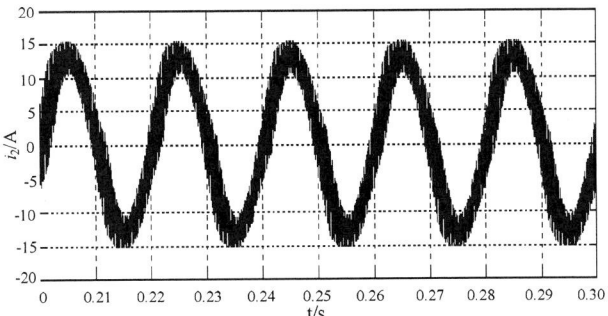

Fig. 6. Waveform of grid-connected current i_2 when R_C is 4.8Ω

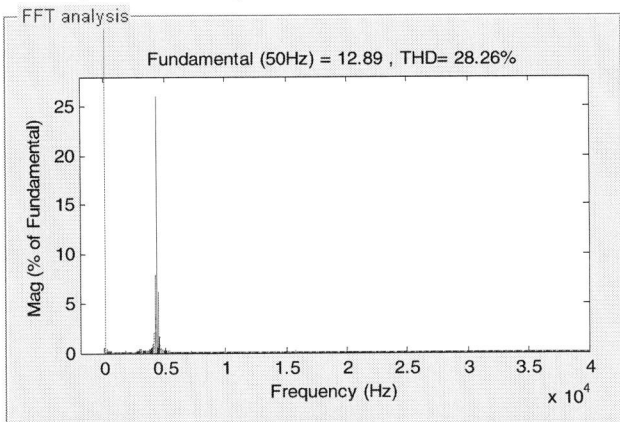

Fig. 7. Waveform of grid-connected current i_2 when R_C is 5Ω

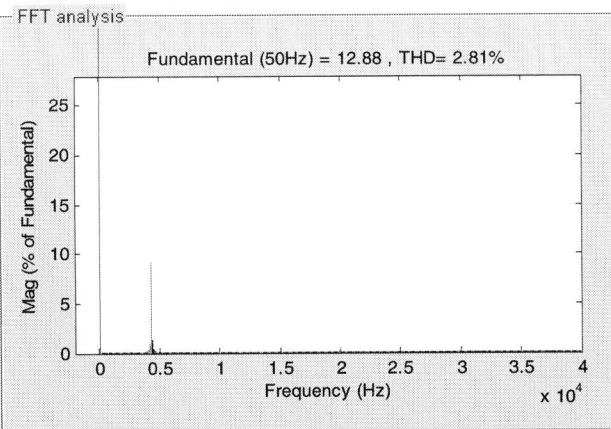

Fig. 8. FFT analysis of grid-connected current i_2 when R_C is 4.8 Ω

Fig. 9. FFT analysis of grid-connected current i_2 when R_C is 5Ω

From Fig.6 to Fig.9, the damping resistor R_C got by the stability analysis method based on harmonic balance and Floquet theory can make the inverter achieve better performance, comparing with that by Routh-Hurwitz stability criterion. In practice, we can choose the R_C easily without keeping the damping resistance in bigger margin. It can be said that a more practical damping resistor R_C can be designed by the stability analysis method based on harmonic balance and Floquet theory.

V. CONCLUSIONS

In this paper, a stability analysis method based on harmonic balance and Floquet theory has been presented, and used to determine the resistance of the damping resistor in the single phase grid-connected inverter with LCL-filter. Different resistances are obtained according to Routh-Hurwitz stability criterion and the stability analysis method proposed in this paper; further, the simulation results show the proposed method is more accurate, and the outputs of the inverter with the damping resistance designed by the proposed method have better harmonic characteristic. Hence, the proposed stability analysis provides a practical parameter design reference.

ACKNOWLEDGMENT

This work was supported by the National Natural Science Foundation of China under Grants 51007004 and 50937001; by the Fundamental Research Funds for the Central Universities under Grant 2012JBM096; by Beijing Natural Science Foundation under Grant 3142015; by Beijing Higher Education Young Elite Teacher Project under Grant YETP0569.

REFERENCES

[1] Zhixiong Huang, Baoyou Xu, Lingfei Shen, and Zhiyong Li, "New current double closed loop control strategy of LCL grid-connected inverter," Power System Protection and Control. vol. 40, pp. 1-5, September 2012.

[2] Rende Zhao, Qiang Zhao, Fang Li, and Ping Wang, "Impact of damping resistance on grid inverter with LCL-filter," Proceedings of the CSU-EPSA. vol. 21, no. 6, pp. 112-116, December 2009.

[3] Timothy CY Wang, Zhihong Ye, Gautam Sinha, Xiaoming Yuan, "Output Filter Design for a Grid-interconnected Three-Phase Inverter," IEEE, Power Electronics Specialist Conference. USA, vol. 2, pp. 779-784, 2003.

[4] Xiaoqiang Guo, Weiyang Wu, Herong Gu, Liqiao Wang, and Qinglin Zhao, "Modeling and stability analysis of direct current control for LCL interfaced grid-connected inverters," Transactions of China Electrotechnical Society. vol. 25, no.3, pp. 102-109, March 2010.

[5] Jianbin Wang, "Research on the technology of low power transformerless grid-connected photovoltaic inverter," Beijing Jiaotong University. 2012.

[6] Zhiying Xu, Aiguo Xu, Shaojun Xie, "Dual-loop current control technique for grid-connected inverter using an LCL filter," Proceedings of the CSEE. vol. 29, pp. 36-41, September 2009.

[7] Chenghui Zhang, Ying Ye, Alian Chen, and Chunshui Du, "Research on grid-connected photovoltaic inverter based on output current control," Transactions of China electrotechnical society. vol. 22, pp. 41-45, August 2007.

[8] Guixin Wang, Yong Kang, and Jian Chen, "Control modeling of a single-phase inverter based on state-space average method," Power Electronics, vol. 38, no.5, pp. 9-12, October 2004.

[9] Yuan Zhang, Hao Zhang, and XikuiMa, "Intermediate-scale instability in one-cycle controlled Cuk power factor correction converter," ACTA PHYSICA SINICA. vol.59 no.12, pp. 8432-8443, December 2010.

[10] R. Seydel, Practical Bifurcation and Stability Analysis: From Equilibrium to Chaos, 3rd ed. New York: Springer, 2010.

[11] Faqiang Wang, Hao Zhang, and Xikui Ma, "Analysis of slow-scale instability in boost PFC converter using the method of harmonic balance and Floquet theory," IEEE Transactions on Circuits and Systems, vol.57, no.2, pp. 405-414, February 2010.

[12] Zhiying Xu, "Research on grid current control technique for grid-connected inverters," Nanjing University of Aeronautics and Astronautics. 2009.

Comparative Evaluation of Passive Damping Topologies for Parallel Grid-Connected Converters with LCL Filters

Remus Beres, Xiongfei Wang, Frede Blaabjerg,
Claus Leth Bak
Department of Energy Technology
Aalborg University
Aalborg, Denmark
rnb(xwa,fbl,clb)@et.aau.dk

Marco Liserre
Institute of Power Electronics and Electrical Drives
Christian-Albrechts University of Kiel
Kiel, Germany
ml@_tf.uni-kiel.de

Abstract – In this paper a comprehensive analysis of three passive damping methods is done under parallel operation of multiple current controlled voltage source converters. One could argue that a well damped LCL filter with no peaking in the output impedance and stable designed controllers will turn into stable status under the parallel operation with other converters that share similar configuration. However, it is shown that this is not always the case, especially because of the coupling between converters and the grid impedance. For the considered ratings, under grid impedance variation, it is found that with grid-side current feedback the stability may be improved in parallel operation while for converter-side feedback, the stability of the current controller is always decreased compared with the single converter case. The proposed stability analysis and experimental tests demonstrates the theoretical analysis.

Keywords – Impedance based stability analysis, LCL filter, passive damping, voltage-source converters.

I. INTRODUCTION

Recently, the trend is to use active damping methods instead of passive damping in order to damp the LCL filter resonance [1-3]. The main reason is to avoid resistive losses and to reduce the component count in the damping circuit, therefore improving the reliability. Furthermore, has been suggested that if the filter resonance frequency is higher than one sixth of the control frequency, a simple current controller with the grid current as feedback is sufficient to control the Voltage Source Converter (VSC) without any additional damping [4]. However, this is essentially true only when the grid impedance is free of resonances. Furthermore, possible resonances or peaks in the grid impedance should be far away from the filter resonance in order to minimize the risk of instability. Therefore, at least for unknown grid impedance, one could say that damping may be used in order to attenuate the filter resonance and ensure the stability of the inverter based system connected to the public grid. The problem becomes even more complicated in the case when multiple VSCs are connected to same point of common coupling (PCC), as both damping and stability of the system changes. This is due the mutual interactions between current controller loops of the parallel VSC and the grid impedance [2]. On

the other hand, it is well known that the use of passive damping methods always improves the stability of the system and the use of more complicated damping topologies can highly decrease the resistive losses [5]. However, no previous work has addressed the passive damping of the LCL resonance under parallel operation of the VSCs. Therefore, in this paper a comprehensive analysis of the passive damping is done under parallel operation of multiple current controlled voltage source converters. Another aspect is to investigate if the stability has direct impact on the size footprint of the filter. This becomes interesting from a filter design point of view. The methodology used to assess the stability of the parallel VSC is based on the impedance-based stability criterion [6] and some considerations given by the authors in [7-9]. The basic diagram of the parallel connected VSCs with LCL filters is illustrated in Fig. 1 where Z_1 is the converter side impedance of the filter, Z_2 is the grid side impedance of the filter, Z_p is the parallel branch of the filter that includes the damping circuit and Z_g denotes the grid impedance.

The content of the paper has the following structure. In Section II, the proposed passive damping methods and corresponding design are presented. An impedance-based stability criterion is described in Section III for both converter-side and grid-side current feedback and it is generalized for the parallel operation of the VSCs. Section IV validates the impedance-based analysis in the frequency domain through time-domain simulations and experimental tests. The last section is reserved for the main findings and contributions.

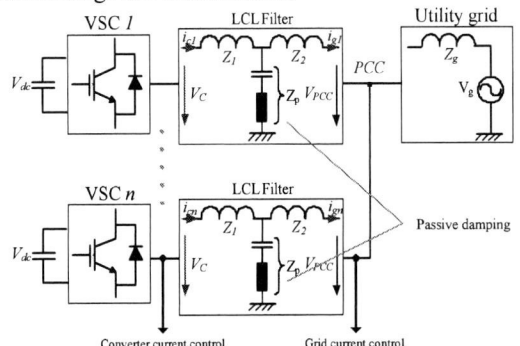

Fig. 1. One line diagram of n-parallel connected VSCs with passively damped LCL filters.

II. PASSIVE DAMPING METHODS

Three passive damping methods are investigated in this work, namely a series resistor with the filter capacitor, a parallel R_dC_d damper and a trap filter combined with an R_dC_d damper as illustrated in Fig. 2. The later topologies provide lower damping losses [10] due to multiple branches in the LCL filter and therefore much lower current in the damping resistor. Another aim for choosing three filter topologies is that they all provide different characteristics, i.e. resonance frequency, attenuation, etc.

A. LCL with damping resistor R_d

The damping resistor in series with the filter capacitor is a very simple passive damping solution. However it provides highest losses and decreases the attenuation of the filter to 40dB/decade after the resonance frequency.

B. LCL with R_dC_d damper

The solution with additional parallel R_dC_d branch increases the attenuation to 60dB/decade at higher frequencies and reduces the resistive losses even if the resulting resistor has much higher value compared with the simple resistor case [11]. Due to high attenuation at higher frequencies this topology is well suited for electromagnetic interference (EMI) filtering.

C. LCL with trap and R_dC_d damper

The introduction of the trap in the LCL filter tuned at the switching frequency has the advantage of reducing the overall size of the filter (by decreasing the grid-side inductor). The reduction in size is given by the selective attenuation functionality of the trap and therefore the contribution of the LCL filter inductors to the switching ripple attenuation is minimized. With a much lower grid side inductor, the resonance frequency of the filter is also increased, resulting in increased bandwidth of the current controllers. The high frequency attenuation in this case is also 40dB/decade as in the series resistor case.

D. Design example

1) PWM inductor: L_1

The current ripple can be used as design parameter used to derive the PWM inductor value. Typical current ripple values adopted in industry are in the range of 10-25% of the rated current, which translates to 20-50% of peak to peak current ripple (I_{pp}). The current ripple expression depends mainly on the modulation strategy used and the given VSC ratings (Table I). For a given PWM strategy, the maximum current ripple can be calculated considering the inductor volt-second balance principle. For the inverter-mode operation and adopting the space vector modulation (SVM) presented in [12], the inductor value can be calculated using (1).

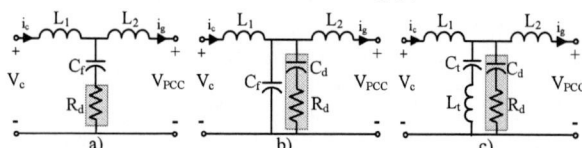

Fig. 2. Proposed passive damping methods: a) series damping resistor with the filter capacitor; b) additional RC damping parallel branch; c) trap and RC damping.

TABLE I
VSC PARAMETERS

Symbol	Meaning	Value
V_g	Output phase RMS voltage	230 V
I_g	Output RMS current	14.5 A
f_1	Grid fundamental frequency	50 Hz
f_{sw}	Switching frequency	10 kHz
V_{dc}	DC voltage	700 V
I_{pp_max}	Converter current ripple	10%

$$L_1 \simeq \frac{V_{dc}}{15 I_{pp\_max} f_{sw}} \quad (1)$$

2) Grid side inductor: L_2

The selection of the inductor on the line side of the LCL filter is a trade-off between the reactive power installed in the filter capacitor C_f and harmonic attenuation. Adopting the same value as the PWM inductor will further limit the filter capacitor within reasonable values (lower than 5%). In the case of a trap configuration and taking into account the VSC ratings given in Table I, the line inductance can be reduced by 5 times compared with two other considered topologies.

3) Parallel impedance components

a) LCL filter: C_f, R_d

The filter capacitor results from condition (2) which ensures proper attenuation around the dominant harmonics. The terms in (2) can be found in Fig. 1 and Fig. 2 while i_d is the harmonic current limits imposed by the standards or by utility grid operators, h is the order of the dominant harmonics around the modulation frequency m_f and ω_1 is the fundamental frequency in rad/s.

$$\left\| \frac{i_g(j\omega)}{i_C(j\omega)} \right\|_{V_{PCC}=0} = \left| \frac{Z_p(j\omega)}{Z_p(j\omega)+Z_2(j\omega)} \right| \geq \left| \frac{i_C(h\omega_1)}{i_d(h\omega_1)} \right|_{\substack{h=m_f\pm2; m_f\pm4; \\ 2m_f\pm1; 2m_f\pm5; 2m_f\pm7; \\ 3m_f\pm2; 3m_f\pm4; 3m_f\pm8}} \quad (2)$$

The resistor is chosen to 30% of the characteristic resistance defined in (3) [13].

$$R_d = 0.3 \sqrt{\frac{L_1 L_2}{C_f (L_1 + L_2)}} \quad (3)$$

b) Additional RC damper: C_f, C_d, R_d

The selection of the damping parameters is done according to [14] while still taking into account the condition in (2) computed with the new damping components. The choice of resistor using [14] will cause the minimum peaking in the output filter admittance.

c) Single-tuned trap and RC damper: C_t, L_t, C_d, R_d

The design starts with individually tuning of the trap at the dominant switching frequency, e.g. the switching frequency in this case using

$$\omega_t = \sqrt{\frac{1}{C_t L_t}} \quad (4)$$

It is worth to mention that in the design process it should be considered the possible detuning due to tolerance of the filter components, aging or variations of the grid fundamental frequency. A good practice is to tune the trap to a lower frequency. The selection of the damping capacitor and resistor can be done based on the required damping considering the guidelines given in [14]. For all three filter topologies about the same damping was achieved in order to minimize the risk of possible resonances with the grid impedance. In addition, because the resonance frequency ω_{res} of the designed filters is higher than the critical frequency ω_{crit} of the controllers which was shown to be one sixth of the sampling frequency (T_s), another set of filter parameters is defined in which the resonance frequency is lower than the critical frequency according to the condition in (5). This can easily be achieved by increasing the filter capacitors values while decreasing the grid side inductance of the LCL filters as can be found in Table II.

$$\omega_{res} = \sqrt{\frac{L_1 + L_2}{C_f L_1 L_2}} \approx \left. \sqrt{\frac{1}{C_f L_1}} \right|_{L_2 \gg L_1} < \omega_{crit} = \frac{T_s}{6} \quad (5)$$

III. IMPEDANCE-BASED STABILITY ANALYSIS

The basis of the stability analysis is the current controller closed loop feedback illustrated in Fig. 3 for converter current feedback and in Fig. 4 for grid current feedback. The LCL filter control admittances (plant) can be written as (6) and (8) for converter current and grid current control, respectively. In addition, the disturbance transfer functions Y_{cp} and Y_{gg} are defined in (7) and (9) where v_P denotes the voltage across the parallel branch of the filter. The other two unknowns transfer functions from Fig. 3 and Fig. 4 are the proportional gain K_p of the current controller tuned with technical optimum criterion and the delay due to digital computation which is about 1.5 of the sampling period T_s, both defined in (10) and (11). The corresponding parallel branch impedances in the filter admittance expressions are depicted in Table III.

$$Y_{cc} = \left. \frac{i_c}{v_c} \right|_{v_P = 0} = \frac{1}{Z_1} \quad (6) \qquad Y_{cp} = \left. \frac{i_c}{v_p} \right|_{v_{PCC} = 0} = \frac{Z_p + Z_2}{Z_2 Z_p} \quad (7)$$

$$Y_{gc} = \left. \frac{i_g}{v_c} \right|_{v_{PCC} = 0} = \frac{Z_p}{Z_1 Z_2 + Z_1 Z_p + Z_2 Z_p} \quad (8)$$

$$Y_{gg} = \left. \frac{-i_g}{v_{PCC}} \right|_{v_c = 0} = \frac{Z_1 + Z_p}{Z_1 Z_2 + Z_1 Z_p + Z_2 Z_p} \quad (9)$$

$$G_d = e^{-1.5 T_s} \quad (10) \qquad G_c = K_p \quad (11)$$

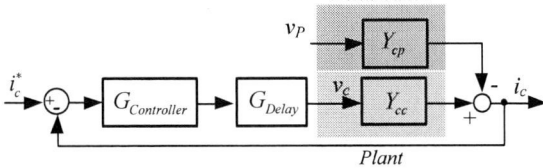

Fig. 3. Closed-loop current control with converter current as feedback.

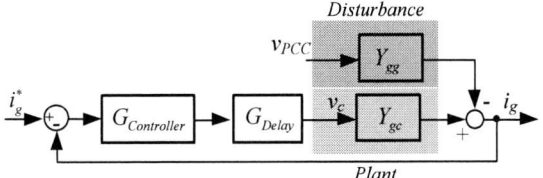

Fig. 4. Closed-loop current control with grid current as feedback.

TABLE II
FILTERS COMPONENTS PARAMETERS IN % (SET I: $F_{LICF} > F_{CRIT}$, SET II: $F_{LICF} < F_{CRIT}$)

	SERIES RESISTOR		RC DAMPER		TRAP AND RC DAMPER	
	SET I	SET II	SET I	SET II	SET I	SET II
L_1	3	3	3	3	3	3
L_2	3	1	3	1	0.6	0.2
C_f	2.35	4.7	1.66	3.3	-	-
R_d	25	12.5	200	100	135	135
C_d	-	-	1.14	2.3	1.15	1.6
C_t	-	-	-	-	2.35	2.9
L_t	-	-	-	-	1	0.05

TABLE III
RELATIONSHIPS FOR THE PARALLEL BRANCH IMPEDANCE Z_P

SERIES RESISTOR	RC DAMPER	TRAP AND RC DAMPER
$\dfrac{1}{sC_f} + R_d$	$\dfrac{1}{sC_f} \left\| \left(\dfrac{1}{sC_d} + R_d \right) \right.$	$\left(sL_t + \dfrac{1}{sC_t} \right) \left\| \left(\dfrac{1}{sC_d} + R_d \right) \right.$

A. Single VSC operation

To assess the stability of the individual VSC the closed loop gains can be analyzed by means of root locus analysis of the filter open loop transfer functions. In addition, the grid impedance should be added to the filter grid side inductance.

B. Parallel VSCs

The main key to assess the stability of n-parallel VSCs is to derive the equivalent admittance seen by the i-th VSC, which is under investigation. In order to simplify the analysis all the converters, controller gains and the corresponding LCL filters are assumed to be identical.

1) Current sensor on the converter side

The open loop and closed loop transfer functions can be written as (12) and (13) where Y_f denotes the filter control admittances, which should be replaced with either Y_{cc} or Y_{gc} as a function of which current is tracked:

$$G_{ol} = K_p G_d Y_f \quad (12) \qquad G_{cl} = \frac{G_{ol}}{1 + G_{ol}} \quad (13)$$

The closed-loop input admittances are resulting as:

$$Y_{dc} = \frac{Y_{cp}}{1 + G_{olc}} \quad (14) \qquad Y_{dg} = \frac{Y_{gg}}{1 + G_{olg}} \quad (15)$$

In Fig. 5 the equivalent closed-loop model of one VSC is derived from Fig. 3 with the specification that the admittance Y_{ce} is the equivalent admittance of the filter combining the closed-loop input admittance Y_{dc} from (14), the parallel impedance Z_P and the grid side impedance of the filter Z_2. In Fig. 6 is illustrated the deriving circuits for the equivalent impedance seen by the i-th VSC provided that all controller parameters, filters are identical. Equalizing all corresponding terms from Fig. 6a it is easy to follow that corresponding equivalent impedance seen by one of the VSC can be derived as n times the grid impedance as shown in Fig. 6c [8]. Therefore, assuming 100 identical VSCs and a grid inductance of *1 mH*, the equivalent inductance seen by one of the VSC is *100 mH*.

$$Z_{ech,c} = n Z_g \quad (16)$$

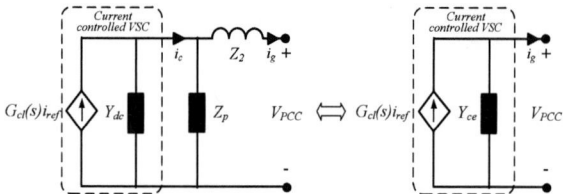

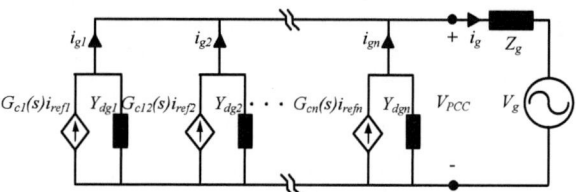

Fig. 5. Equivalent current-source models for the converter side closed loop current control

Fig. 7. Closed-loop model for n-paralleled VSCs with LCL filters (grid current feedback).

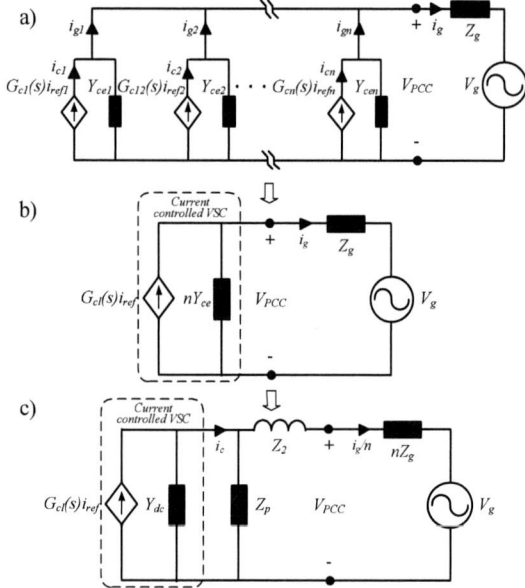

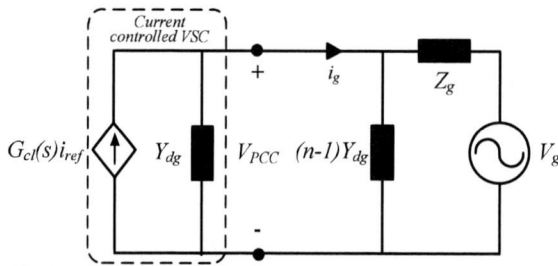

Fig. 8. Equivalent closed-loop model for n-identical paralleled VSCs with LCL filters (grid current feedback).

Fig. 6. Deriving circuits for the equivalent closed-loop model for n-identical paralleled VSCs with LCL filters (converter current feedback).

The formulation in (18) is valid only to find the stability limits of the closed loop poles. The considered approximation in (18) can be validated also by the help of (17). Finding the stability limits for the proportional gain using (18) one could then introduce the obtained values of the proportional gain in (17). For a good match the new obtained gain from (17) should be about the same with one previously obtained by using (18).

To investigate the stability of the system, the equivalent impedance from (16) seen by the i-th VSC should be added to the filter grid side inductance Z_2 of the i-th VSC. Further on, the controllers closed loop poles can be found by root locus analysis.

C. Frequency domain results

The open loop gain transfer functions for the proposed damping methods are illustrated in Fig. 9 for both grid and converter current feedback for set I of values indicated in Table II. For simplicity reasons the plots for the set II of values are not indicated. In Fig. 10 the stability limits for the proportional gain is plotted as a function of the number of parallel VSCs by considering (16) and (18). The values are computed from the root locus analysis. One could also consider to plot the corresponding minor loop gains ($Y_{dg}/Y_{ech,g}$) and to investigate the stability by means of Nyquist plots. However, for the sake of simplicity the root locus analysis method was preferred.

2) Current sensor on the grid side

The closed-loop model for n-paralleled VSCs with LCL filters can be derived from Fig. 4 as illustrated in Fig. 7. The closed loop disturbance transfer function Y_{dg} is depicted in (15). For grid current feedback the equivalent impedance seen the i-th VSC results as:

$$Y_{ech,g} = \frac{1}{Z_g} + (n-1)Y_{dg} \qquad (17)$$

The factor (n-1) results due to the fact that the grid current is controlled. By adding (17) to the LCL filter grid side inductance the stability of the system can be investigated similar with previous cases. It is worth to mention that using (17) the closed loop gain poles specific to the controller gain $K_{p,i}$ of the i-th VSC is only to be investigated while the gains of the of the j-th VSCs ($i \neq j$) are equal and constant. In order to match the proportional gain to be the same for all VSCs as in the converter current feedback case, the closed loop disturbance admittance Y_{dg} can be approximated with the open loop transfer function G_{olg} as indicated in (18):

$$Y_{ech,g} \simeq \frac{1}{Z_g} + (n-1)Y_{olg} \qquad (18)$$

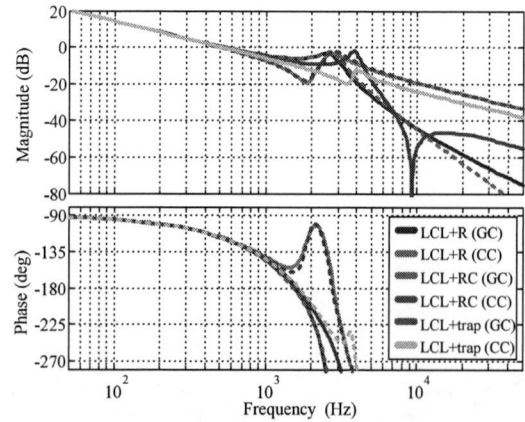

Fig. 9. Open loop transfer functions of the proposed damping methods using Set I parameters from Table II (for grid (GC) and converter current feedback (CC)).

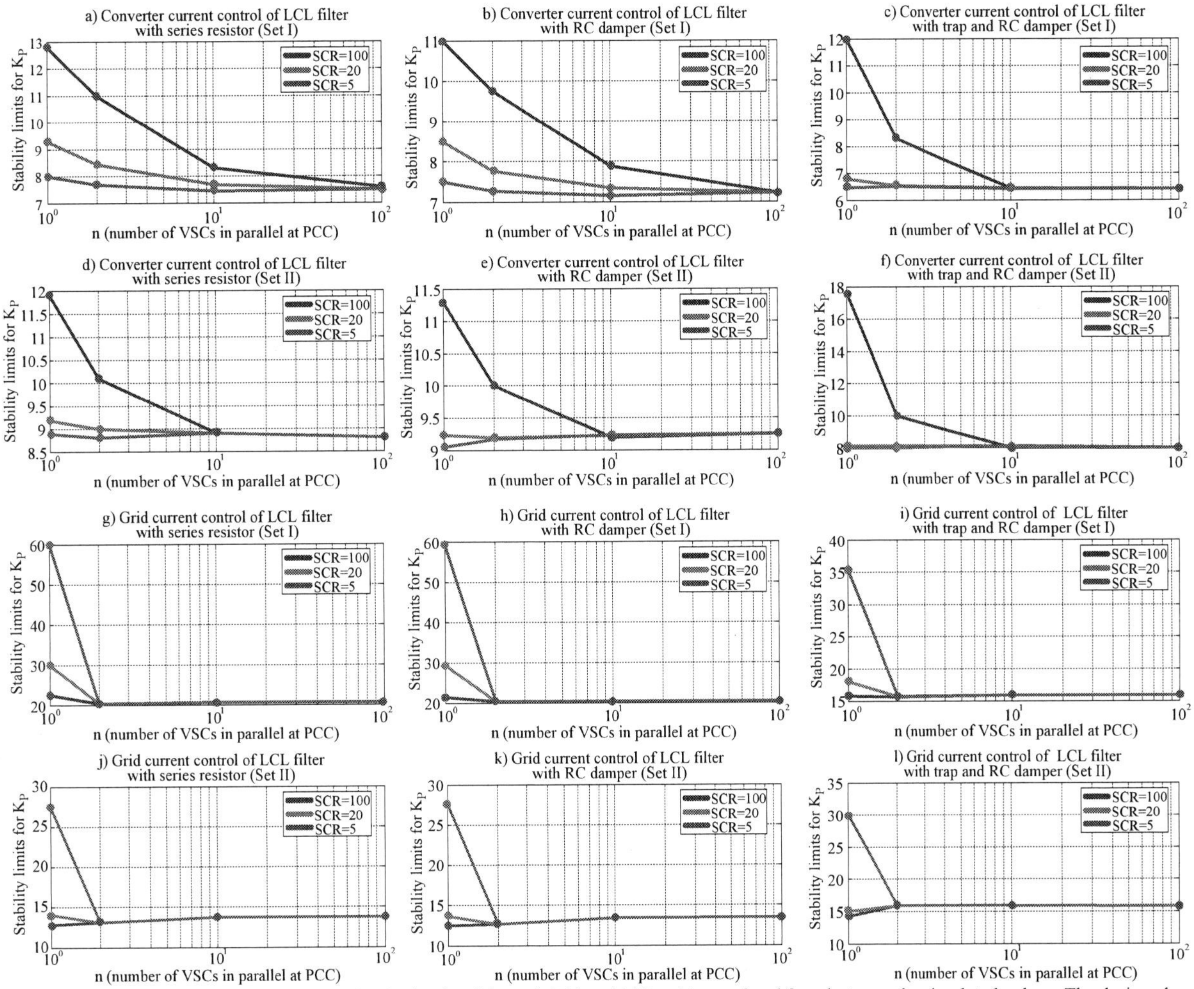

The 2014 International Power Electronics Conference

Fig. 10. Stability limits for the proportional gain simulated for n=1,2,10 and 100 and interpolated for n between the simulated values. The designed proportional gain using technical optimum criterion are: Kp=10 (cases *a, b, g, h*); Kp=6.6 (cases *c, i*); Kp=6 (cases *d, e, j, k*); Kp=5.3 (cases *f, l*).

It is worth to mention that instead of the grid inductance, the Short Circuit Ratio (SCR) was preferred as indicator for the grid impedance. The SCR can be approximated as being inversely proportional with the grid impedance in percent's. For example an SCR of 100 means actually 1% of the base inductance of one VSC.

In Fig. 11 the root loci of the LCL filter with series damping resistor for the case when two identical VSCs are connected to the PCC is illustrated. First, the equivalent impedance seen by one of the VSC is calculated using (18) for a wide range of SCRs, e.g. 20, 100 and 5 respectively. Then the calculated impedance is added to LCL filter grid side inductance of the VSC under study. From the root locus, the closed loop gains can be found to be the same for the three SCRs considered. The stability limits of the proportional gain are then introduced successively in (12), (15) and finally in (17) in order to demonstrate the validity of (18). For the three considered SCRs again the closed loop gains are similar but slightly different than the closed loop gains calculated using (17) especially at higher frequencies. However, their closed loop gain frequency is identical at the frequency of interest which is about 1.5 kHz.

Furthermore, also the stability limits of the proportional gain matches very closely which also is agreement with the gains depicted in Fig. 10, case *g* for *n=2*.

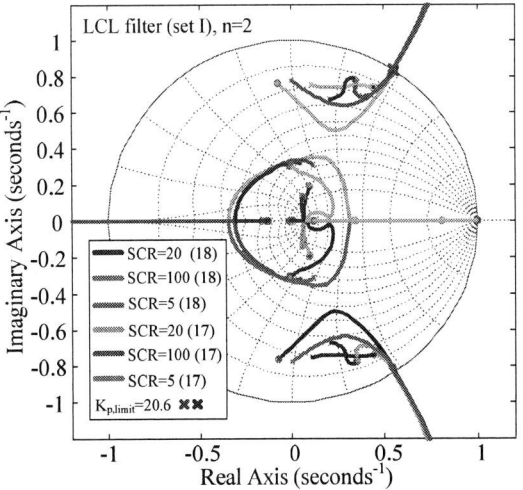

Fig. 11. Root locus of the LCL filter with series resistor using K_p from (18) and introduced in (17) for validation purposes.

978-1-4799-2706-7/14 $31.00 © 2014 IEEE

IV. SIMULATION AND EXPERIMENTAL RESULTS

A. Simulation results

The simulations results are performed in Simulink and PLECS blockset to model the VSCs, filters and the grid. The grid is considered by an equivalent inductor and an ideal voltage source as presented in Fig. 1. For more detailed studies, more realistic models may be used, especially if impedance measurements at PCC are available. In the following the number of the VSCs connected in parallel is limited to two. Furthermore, the VSCs share the same filter and controllers parameters.

In Fig. 12 the time domain simulations are shown for all considered damping methods of the LCL filter for both converter and grid current control feedback. The corresponding gains are also illustrated and were chosen in such a way in order to describe the overall trends for the VSC stability. For instance for case a the stability limits of the proportional gain can be taken from Fig. 10 case g considering n=1 and n=2, for SCR=100. For one VSC the proportional gain limit is 22.5 while for two paralleled VSCs is 20.5. Therefore a value of 21 was chosen in order to show that even the converter is marginally stable in single operation, when in paralleled the stability is decreased and may turn into the unstable operation. Fortunately, the chosen gain exceeds the designed gain tuned with the technical optimum criterion ($K_p=(L_1+L_2)/(3T_s)$) and therefore this will not lead into instability problems in practice. However if we consider the current being sensed on the converter side of the filter for set I of filter parameters the designed gains will not provide enough stability margin for changes in the grid impedance or when the number of the VSCs in parallel increases. It can be concluded that the time domain simulations closely matches the frequency domain analysis illustrated in Fig. 10.

B. Experimental results

The experimental setup used for validation includes two Danfoss FC302 inverters controlled through dSPACE 1006 using P+R current controllers. The DC-link voltage is kept constant from three Delta SM 300-10 DC sources connected in series. Hence, no DC voltage loop is used. The ratings of the setup are identical with the ones illustrated in Table I. Furthermore, between the PCC of the VSCs and the grid there is an isolation transformer. Additional inductances are added in series with the isolation transformer in order to meet SCRs of about 5 and 20 as in the simulations. The SCR of 100 corresponding to 1% grid inductance is omitted for the experiments due to the presence of 3% leakage inductance from the isolation transformer. Furthermore the values of the filter inductors can also vary in a range of ±50% within the 50 Hz – 10 kHz range than those given in Table II due to the magnetic core characteristic and nonlinearities.

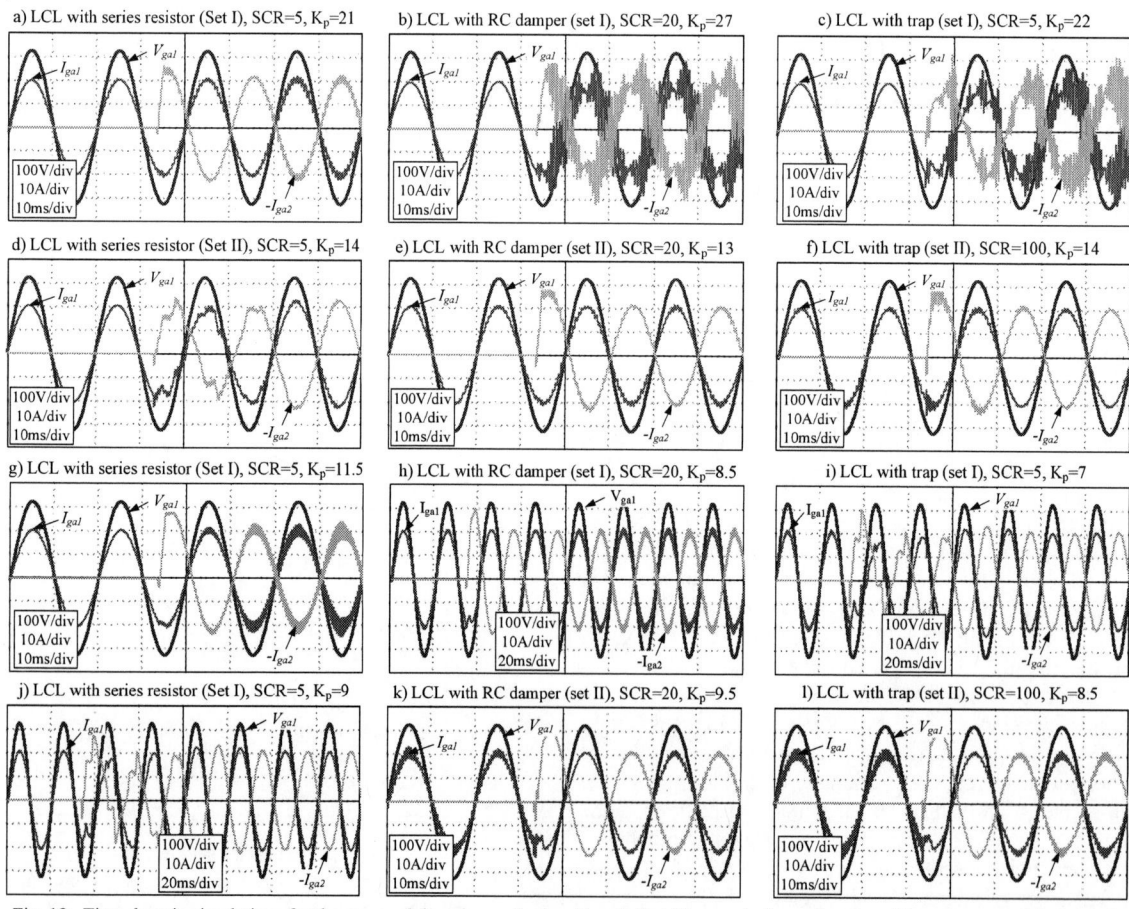

Fig. 12. Time domain simulations for the proposed damping methods with n=1 first 33ms and after n=2 under grid current control (cases *a-f*) and converter current control (cases *g-l*).

1) Influence of the grid inductance

To evaluate the influence of the grid inductance variation on the current controllers gain, the current and voltage waveforms on phase a are shown in Fig. 13 and Fig. 14 for converter and grid side current feedback, respectively. When the converter current is sensed the closed loop gain margin is decreased with increasing the grid inductance for the case when $n=1$ and $n=2$. However, when the grid current is sensed the opposite is true for $n=1$ while for $n=2$ the proportional gain remains about the same. It is obvious that that more stability margin is obtained by adopting the grid side current sensing for the given ratings.

2) Single operation vs. parallel operation of VSCs

A comparison between the single and parallel operation of identical VSCs is illustrated in Fig. 15 including the steady state waveforms for $n=1$.

The steady state waveforms demonstrate the design of the individual VSC for the considered passive damping methods. However there is some data acquisition noise in the voltage sensor as no dominant harmonics are identified using harmonic decomposition. An interesting aspect is that for the set II of filter parameters somehow the opposite conclusions were obtained compared with frequency domain analysis. It can be noticed that the stability margin increases in case of two paralleled VSCs compared with the single case. One reason could be the tolerance of the filter components and slightly differences with the values given in Table II. Another aspect could be the additional damping found in the system and neglected in analysis as well as damping due to the dead time effect. The detailed proportional gains that lead to marginal stability for the experiments are given in Table IV in Appendix.

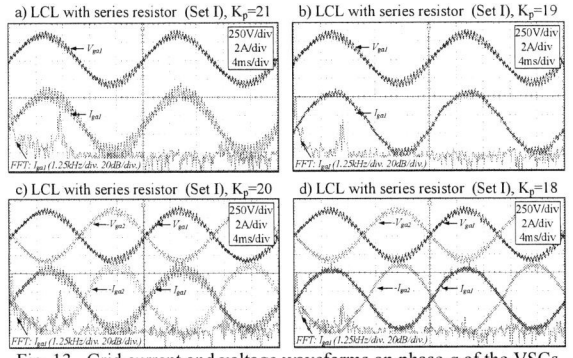

Fig. 13. Grid current and voltage waveforms on phase a of the VSCs for converter current feedback: SCR=20 (case a, c); SCR=5 (case b, d).

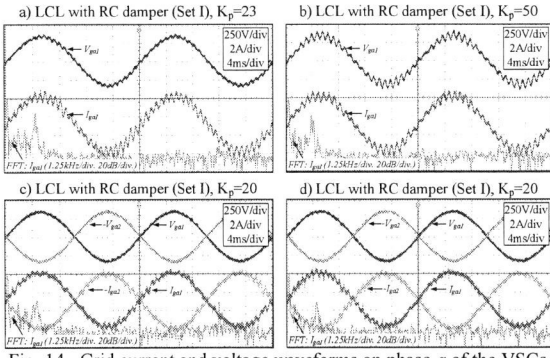

Fig. 14. Grid current and voltage waveforms on phase a of the VSCs for grid current feedback: SCR=20 (cases a, c); SCR=5 (cases b, d).

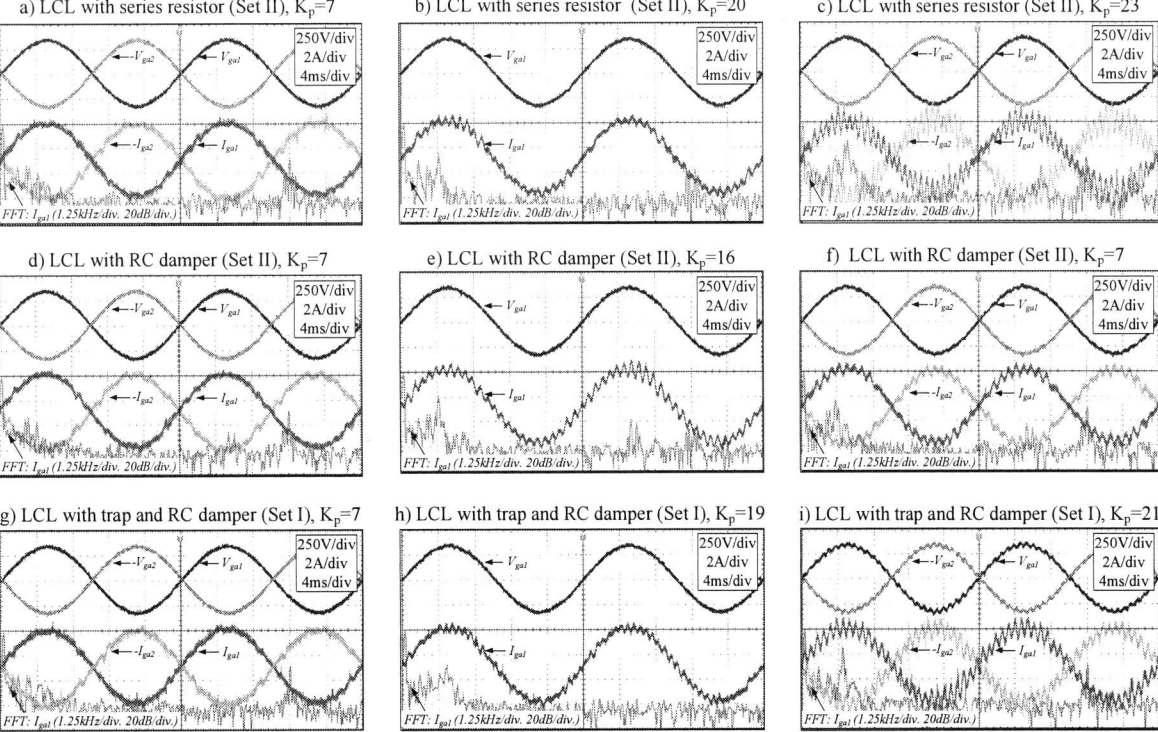

Fig. 15. Grid current and voltage waveforms when the grid current is sensed: steady state waveforms for one VSC and the proportional gain tuned according to the technical optimum criterion (cases a, d, g), proportional gain limits test for one VSC (cases b, e, h) and two VSCS (cases c, f, i).

V. Conclusions

Damping of the LCL resonance becomes even more crucial under parallel operation of the VSCs employed in renewable energy applications. In this paper it was shown that especially for the converter-side current feedback a stable system in single operation could easily be unstable in parallel operation of two VSCs or even under grid impedance variation. For more paralleled VSCs the effect is even worse due to the coupling grid impedance. However, the effect tends to be smoothened with increasing the number of VSCs. For single VSC operation under grid-side current feedback the stability margin for the proportional gain increases with increasing the grid impedance. For the converter-side current feedback the opposite is true, which means that increasing the grid impedance, the stability of the controller decreases. Of course there is not one answer that can resolve all the problems, but as general rule one could decrease the gain of the proportional controller in case of resonances due to impedance interactions between the paralleled VSCs and the grid impedance.

The simulation results match the stability analysis from frequency domain. An experimental setup was built to demonstrate the simulations result. Although there are some slightly differences between the simulation and experimental results the overall trends are the same also in the experiments. The developed stability analysis platform could be easily applied in the filter design process when a mass product filter and the VSC are enclosed in a small-scale wind or solar power inverter with the aim of minimizing the interactions between parallel converters.

Appendix

The stability limits of the proportional gain taken from the experimental measurements are given in Table IV.

TABLE IV
EXPERIMENTAL RESULTS FOR THE PROPORTIONAL GAIN LIMITS (CC – CURENT CONTROL)

| Parallel's VSC | SCR | LCL with series resistor | | | | LCL with RC damper | | | | LCL with trap and RC damper | |
| | | Conv. CC | | Grid CC | | Conv. CC | | Grid CC | | Conv. CC | Grid CC |
		Set I	Set II	Set I	Set II	Set I	Set II	Set I	Set II	Set I	Set I
n=1	20	21	27	32	20	16	21	23	16	19	19
	5	19	23	65	50	16	20	50	33	16	39
n=2	20	20	25	28	23	16	20	20	16	17	21
	5	18	23	28	23	16	20	20	22	16	24

References

[1] J. Dannehl, M. Liserre and F. W. Fuchs, "Filter-Based Active Damping of Voltage Source Converters With LCL-filters," *IEEE Trans. on Industrial Electronics,* vol. 58, pp. 3623-3633, 2011.

[2] X. Wang, F. Blaabjerg, M. Liserre, Z. Chen, J. He and Y. Li, "An active damper for stabilizing power electronics-based AC systems," *IEEE Trans. on Power Electronics,* vol. 29, no. 7, pp.3318-3329, July 2014.

[3] M. Hedayati, A. Acharya and V. John, "Common Mode And Differential Mode Active Damping For PWM Rectifiers," *IEEE Trans. on Power Electronics*, vol.29, no.6, pp.3188-3200, 2014.

[4] S. Parker, B. McGrath and D. Holmes, "Regions of active damping control for LCL filters," *IEEE Proc. on Energy Conversion Congress and Exposition (ECCE), 2012 IEEE,* pp. 53-60, 2012.

[5] R. Pena-Alzola, M. Liserre, F. Blaabjerg, R. Sebastián, J. Dannehl and F. W. Fuchs, "Analysis of the Passive Damping Losses in LCL-filter Based Grid Converters," *IEEE Trans. on Power Electronics,* vol. 28, no. 6, pp.2642-2646, 2013.

[6] J. Sun, "Impedance-based stability criterion for grid-connected inverters," *IEEE Trans. on Power Electronics,* vol. 26, pp. 3075-3078, 2011.

[7] L. Harnefors, M. Bongiorno and S. Lundberg, "Input-admittance calculation and shaping for controlled voltage-source converters," *IEEE Trans. on Industrial Electronics,* vol. 54, pp. 3323-3334, 2007.

[8] J. L. Agorreta, M. Borrega, J. López and L. Marroyo, "Modeling and control of N-paralleled grid-connected inverters with LCL filter coupled due to grid impedance in PV plants," *IEEE Trans. on Power Electronics,* vol. 26, pp. 770-785, 2011.

[9] Xiongfei Wang, F. Blaabjerg, Zhe Chen and Weimin Wu, "Modeling and analysis of harmonic resonance in a power electronics based AC power system," *IEEE Proc. on Energy Conversion Congress and Exposition (ECCE), 2013 IEEE,* pp. 5229-5236, 2013.

[10] R. N. Beres, X. Wang, F. Blaabjerg, M. Liserre and C. L. Bak, " A review of passive filters for grid-connected voltage source converters," *IEEE Proc. on Applied Power Electronics Conference and Exposition, APEC 2014,* 2014.

[11] Weimin Wu, Yuanbin He, Tianhao Tang and F. Blaabjerg, "A New Design Method for the Passive Damped LCL and LLCL Filter-Based Single-Phase Grid-Tied Inverter," *IEEE Trans. on Industrial Electronics,* vol. 60, no. 10, pp. 4339-4350, Oct. 2013.

[12] J. Boys and P. Handley, "Harmonic analysis of space vector modulated PWM waveforms," *IEEE Proc. on Electric Power Applications, ,* vol. 137, pp. 197-204, 1990.

[13] M. Liserre, F. Blaabjerg and S. Hansen, "Design and control of an LCL-filter-based three-phase active rectifier," *IEEE Trans. on Industry Applications,* vol. 41, pp. 1281-1291, 2005.

[14] R. Middlebrook, "Design techniques for preventing input-filter oscillations in switched-mode regulators," *IEEE Proc. on Powercon,* 1978, pp. A3-1.

Study and Implementation of a SEPIC LED Driver with Adjustable Output Voltage

Po-Jung Tseng, Yu-Chen Liu,Yu-Kang Lo and
Huang-Jen Chiu

Department of Electronic Engineering
National Taiwan University of Science and Technology
Taipei City, Taiwan, ROC
Email: D9702201@mail.ntust.edu.tw

Yun-Chu Chiu

Lite-On Technology Corporation
New Taipei City, Taiwan, ROC
Email: Donut.Chiu@liteon.com

Abstract-**This paper presents the study and implementation of an LED driver of which the topology is a SEPIC operated under boundary conduction mode (BCM). The output diode features zero-current switching (ZCS). An Inverse-Buck converter is cascaded at the output for the design of an adjustable output voltage. The stability, efficiency and output ripple of a traditional single-stage converter with wide range of output voltage can be improved. For the dimming circuit, the light sensor and IR sensor are additionally used to detect the brightness and presence of human beings. The studied LED driver can be applied for different numbers of LEDs in a string. And the intelligent dimming system can save the energy and extend the applications. Finally, a 126-W laboratory prototype is implemented and verified with the simulations and theoretical analysis. The input voltage ranges from 90Vac to 130Vac and the output voltage can be changed from 25 V to 45 V. The efficiency can achieve 87% at full load condition.**

I. INTRODUCTION

For the energy saving and the development of green energy industry, Light-Emitting Diode (LED) is widely used in some lighting applications. In traditional LED driver circuit design, they are too many components are used in two stage power supply. These make the control circuit is more complex, the transfer efficiency is lower and the volume is bigger. In addition, the life time of LED is more longer. So the lift time of LED drive is restricted by the bulk electrolytic capacitor. Therefore, single stage converter is used to improve the efficiency, remove the bulk electrolytic and reduce the components and cost. But they are some disadvantages which need to considered. Because the power factor correction and output voltage regulation of single stage converter are considered in the same time. So the output voltage ripple is bigger. And it is not easy to design a adjustable output voltage via single stage converter. In order to solve these problems, the improved SEPIC is proposed. An Inverse-Buck converter is combined to one of the secondary output of SEPIC, and the output of Inverse-Buck converter is cascoded to the other output voltage of SEPIC. Beside, the feedback control is changed and adds adjustable output voltage control to reduce output ripple voltage and achieve adjustable output volatge.

Furthermore, the efficiency performance of LED driver is improved at all output voltage condition. Finally, dimming and current sharing circuit are added.

II. OPERATING PRINCIPLE

Single stage SEPIC

Fig. 1 shows the single stage SEPIC topology [1]-[4], Important waveforms is described in Figs. The operating mode of controller is boundary conduction mode, this makes the main switch has valley switching and the output diode has zero current switching.

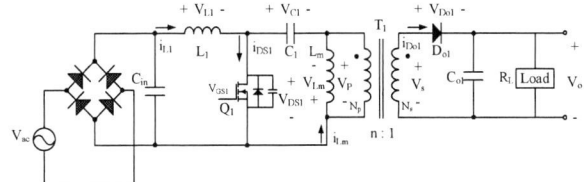

Fig. 1 Single stage SEPIC

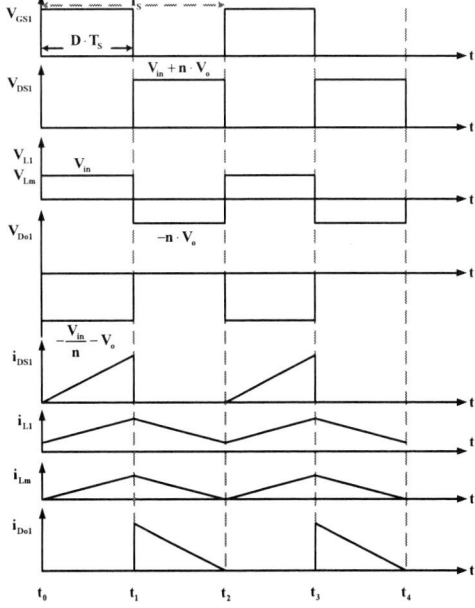

Fig. 2 Key waveforms of single stage SEPIC

Adjustable output voltage circuit

For traditional single stage converter, power factor correction and output voltage regulation are considered. The compensation is designed to filter the twice line frequency which is generated by the rectifier circuit as shown in Fig. 3. In other word, output voltage will have ripple voltage which the frequency is twice as line frequency. It is shown in Fig. 4. It used to change the compensation or increase the output capacitance to decrease the voltage ripple. But this will influence the performance of power factor correction and dynamic response and increase the cost and volume. Therefore, the improved SEPIC is proposed to reduce output voltage ripple and increase the efficiency performance at full input voltage range.

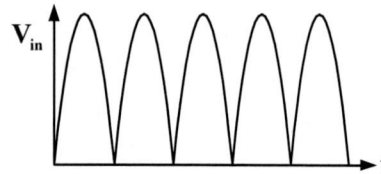

Fig. 3 120Hz rectifier voltage

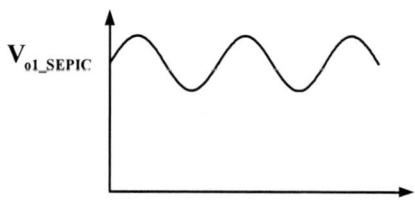

Fig. 4 Output voltage ripple of single stage SEPIC

The proposed improved SEPIC is shown in Fig. 5. V_{o1_SEPIC} and V_{o2_SEPIC} are the output of single stage SEPIC. V_{o1_SEPIC} is the feedback command of single stage SEPIC. V_{o2_SEPIC} is used to be the input voltage of Inverse-Buck converter. Because V_{o1_SEPIC} and V_{o2_SEPIC} provide the power to the load in the same time. This means that the load conditions will increase or decrease in the same time. So V_{o2_SEPIC} doesn't need to fed back to the control circuit of single stage SEPIC. The output voltage V_{o3_Buck} of Inverse-Buck converter is cascoded to the output voltage V_{o1_SEPIC} of single stage SEPIC. The total output voltage V_{o_Total} is the sum of V_{o1_SEPIC} and V_{o3_Buck}. And the feedback signal of Inverse-Buck converter is sensed from the total output voltage V_{o_Total}. When the output voltage ripple of V_{o_Total} is smaller than average output voltage, the output voltage V_{o3_Buck} will increase to keep the total output voltage V_{o_Total} has the same value. When the output voltage ripple of V_{o_Total} is bigger than average output voltage, the output voltage V_{o3_Buck} will decrease to keep the total output voltage V_{o_Total} has the same value. Finally, the output voltage ripple is reduced. The output voltage ripple waveforms are shown in Fig. 6.

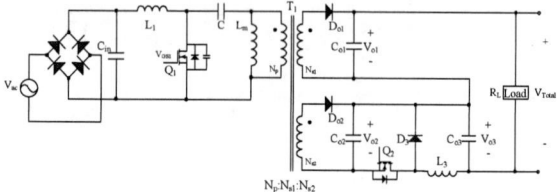

Fig. 5 Proposed improved SEPIC circuit

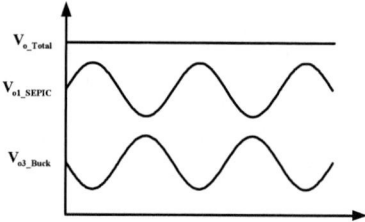

Fig. 6 Output ripple voltage waveforms of improved SEPIC

Fig. 7 shows the function block of proposed improved SEPIC. The single stage SEPIC has two output voltages. One of them is the input voltage of Inverse-Buck converter. The feedback signal is sensed from the total output voltage, and the variable resistor is put in the feedback loop. This makes the output voltage is adjustable via adjust the value of variable resistor.

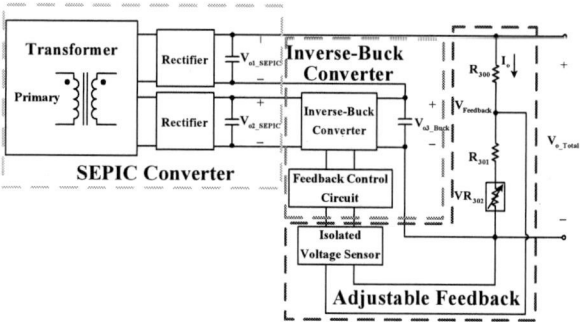

Fig. 7 Function block of proposed improved SEPIC

III. INTELLIGENT DIMMING SYSTEM

Fig. 8 shows the function block of intelligent dimming system. Illumination detector circuit and person detector circuit are added to the dimming circuit. Illumination detector circuit can detect the light in indoor and outdoor to control the dimming circuit. Person detector circuit can turn off the light when no person is detected. In order to make the environment more safety, minimum dimming duty is added. Ultimately, intelligent dimming system can save more energy.

The 2014 International Power Electronics Conference

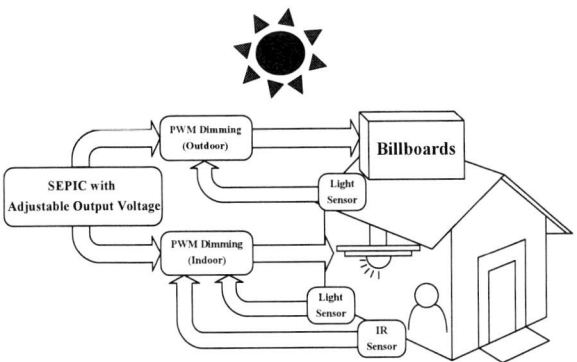

Fig. 8 Function block of intelligent dimming system

PWM dimming and trans-conductance amplifier current sharing circuit [5]-[6] are used in this paper. It adjusts the duty cycle to control the average current of LED. When the LED is turned on, the current of LED is set to the maximum workable current. In order to prevent the visible twinkling light, the operation frequency of PWM is set at 400Hz. Besides, the characteristic of photo resistor is used to detect the illumination. When the light is dark, the resistance is getting higher. When the light is bright, the resistance is getting lower. These can change the feedback voltage of dimming circuit and adjust the illumination. The trans-conductance amplifier is used to do current sharing. When more strings of LED are designed to the lighting system, the more OPAs and transistors are added to achieve current sharing. The controller can be used the same one. The operating current waveform of LED is shown in fig. 9. And the trans-conductance current sharing circuit is shown in fig. 10. Fig. 11 shows the PWM dimming and trans-conductance amplifier current sharing circuit.

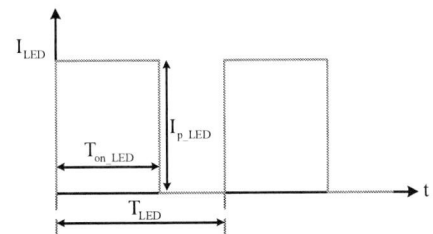

Fig. 9 Operating current waveform of LED

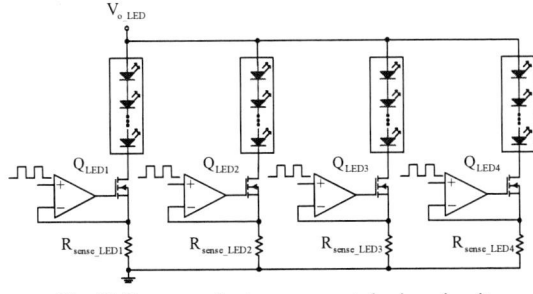

Fig. 10 Trans-conductance current sharing circuit

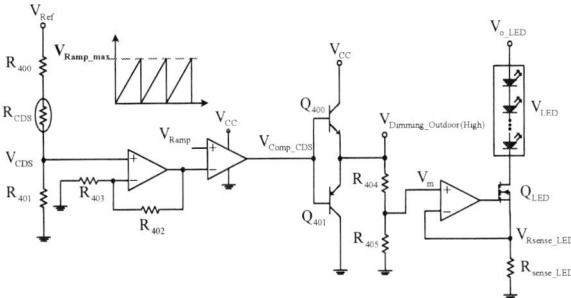

Fig. 11 PWM dimming and current sharing circuit

TL494 is used to design the PWM dimming signal, and the brightness feedback signal V_{CDS} is derived as equation 1.

$$V_{CDS} = V_{Ref} \times \frac{R_{401}}{R_{400} + R_{CDS} + R_{401}} \tag{1}$$

V_{CDS} is compared to Ramp voltage V_{Ramp}, the compared output voltage is V_{Comp_CDS}. And the V_{Comp_CDS} is transferred to $V_{Dimming\_Outdoor}$ via driver circuit. When the $V_{Dimming\_Outdoor}$ is high, the constant current reference voltage V_m is set from R404 and R405, and the R405 is derived as equation 2.

$$R_{405} = R_{404} \times \frac{I_{LED} \times R_{Sense\_LED}}{V_{Dimming\_Outdoor(High)} - I_{LED} \times R_{Sense\_LED}} \tag{2}$$

For normal lighting, the lamp always fully turns on when switch is turned on. And the lamp will not be turn off when someone leaving in a while. The power dissipation is not necessary in this situation. In this paper, the person detector circuit is added to save more energy. And for safety consideration, the minimum dimming signal is added. The minimum on time duty cycle D_{min_D} is derived as equation 3.

$$D_{min\_D} = \frac{V_{Ramp\_max} - V_{min\_D}}{V_{Ramp\_max}} \tag{3}$$

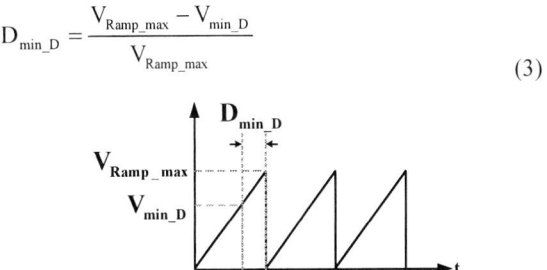

Fig. 12 Triangle waveform of TL494 controller

Finally, minimum indoor dimming signal V_{min_D}, dimming compare signal V_{Comp_CDS}, person detector signal V_{Person} and indoor dimming signal $V_{Dimming\_Indoor}$ are compared, the logic circuit is shown in Fig. 13. And the Key waveforms is shown in Fig. 14.

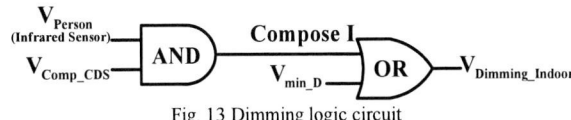

Fig. 13 Dimming logic circuit

978-1-4799-2706-7/14 $31.00 © 2014 IEEE

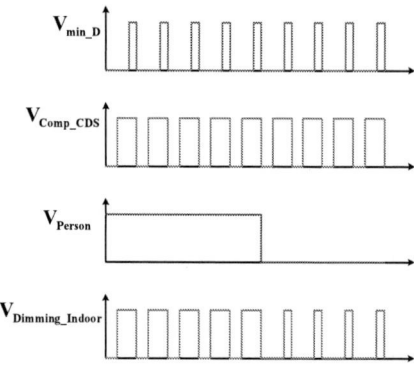

Fig. 14 Operating waveforms of dimming circuit

IV. EXPERIMENTAL RESULTS

In this paper, 126W laboratory prototype for the SEPIC LED driver with Adjustable Output Voltage design is built and tested to verify the studied scheme. The circuit specifications of the prototype converters are listed as below list.

- Input voltage range V_{in} : 90 V_{ac}~130 V_{ac}
- Output voltage V_{o_Total} : 25 V_{dc}~45V_{dc}
- Maximum output current I_{o_max} : 2.8 A
- Maximum output power P_{o_max} : 126 W
- Expected efficiency η : 87 %

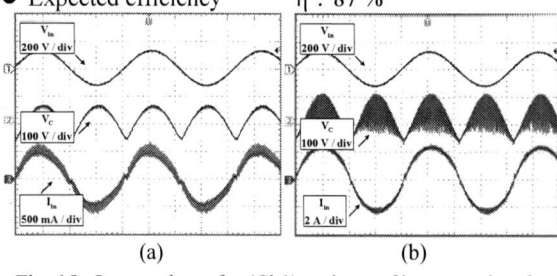

(a)   (b)

Fig. 15 : Input voltage for (Ch1), voltage of bus capacitor for (Ch2) and input current for (Ch3) of SEPIC at (a) 20% load and (b) 100% load (Time : 4 ms/div)

Fig. 15(a) and 15(b) show the input voltage, voltage of bus capacitor and input current of SEPIC at 45Vdc/20% load and 45Vdc/100% load at low-line inputs, respectively. Fig. 16(a) and 16(b) show the output voltage V_{o1_SEPIC} for, output voltage V_{o3_Buck}, total output voltage V_{o_Total} and input voltage at 45V_{dc}-20% load and 45V_{dc}-100% load at low-line inputs, respectively.

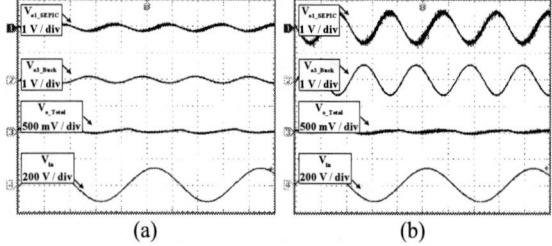

(a)   (b)

Fig. 16 : Output voltage V_{o1_SEPIC} for (Ch1), output voltage V_{o3_Buck} for (Ch2), total output voltage V_{o_Total} for (Ch3) and input voltage for (Ch4) at (a) 20% load and (b) 100% load

(Time : 4 ms/div)

Fig. 17(a) to 17(d) show the minimum indoor dimming signal V_{min_D}, dimming compare signal V_{Comp_CDS}, person detector signal V_{Person} and indoor dimming signal $V_{Dimming\_Indoor}$ at 25% and 100% dimming compare signal, respectively. In Fig. 17, indoor dimming signal is composed of minimum indoor dimming signal, dimming compare signal and person detector signal.

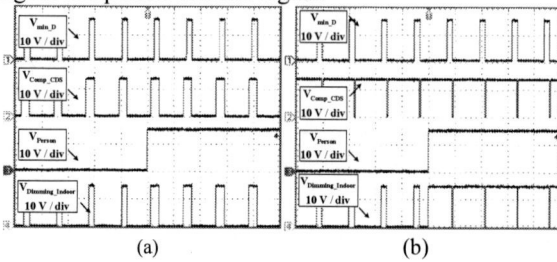

(a)   (b)

Fig. 17 : Minimum indoor dimming signal V_{min_D} for (Ch1), dimming compare signal V_{Comp_CDS} for (Ch2), person detector signal V_{Person} for (Ch3) and indoor dimming signal

$V_{Dimming\_Indoor}$ for (Ch4) at (a) 25% and (b) 100% (Time : 2 ms/div)

Fig. 18 shows the efficiency of proposed circuit at 45V_{dc}-20%/50%/100% load at 90V_{ac}, 110V_{ac} and 130V_{ac} inputs.

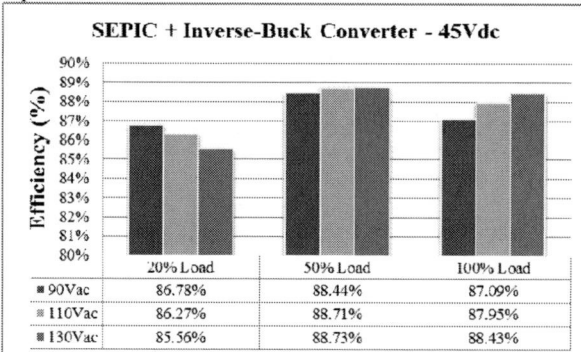

SEPIC + Inverse-Buck Converter - 45Vdc	20% Load	50% Load	100% Load
90Vac	86.78%	88.44%	87.09%
110Vac	86.27%	88.71%	87.95%
130Vac	85.56%	88.73%	88.43%

Fig. 18 : The efficiency curves of proposed circuit at 45 V_{dc} output

Fig. 19 shows the efficiency of proposed circuit at 25V_{dc}-20%/50%/100% load, 35V_{dc}-20%/50%/100% load and 45V_{dc}-20%/50%/100% load at 90V_{ac} input.

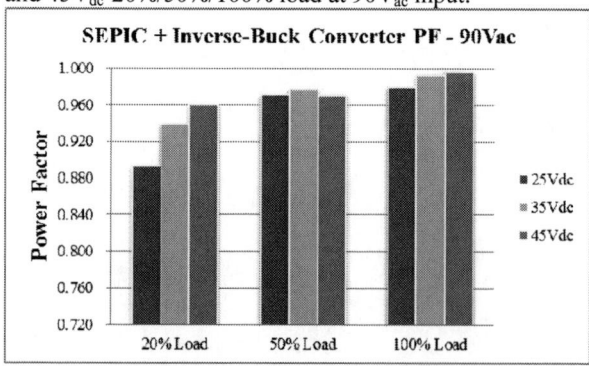

Fig. 19 : The power factor curve of proposed circuit at 90 V_{ac} input

V. CONCLUSION

This paper studies a LED lighting driver that meets the output voltage with adjustable range from 25 V_{dc} to 45 V_{dc}. And the efficiency achieves 87% at full load and high output voltage conditions. The studied circuit improves the efficiency, stability and the output voltage ripple problem of single stage converter via cascade the inverse-buck converter in the secondary side. The DC output voltage V_{o1_SEPIC} of the single stage SEPIC converter is set at 20 V. And the DC output voltage V_{o2_SEPIC} of the single stage SEPIC converter is set at 30 V to supply the inverse-buck converter. The duty utilization of transformer is increasing via cascade output voltage which improves the efficiency at low output voltage condition. And the intelligent dimming circuit is added to save the energy. The proposed topology is verified by the measured results of a laboratory prototype.

REFERENCES

[1] F. Zhang, J. Xu, P. Yang, M. Qin, and H. Yu, "High-Efficiency Capacitive Idling SEPIC PFC Converter with Varying Reference Voltage for Wide Range of Load Variations," Proc. International Conference on Communications Circuits and Systems, pp.536-540, July 2010.

[2] [9] J. M. Kwon, W. Y. Choi, J. J. Lee, E. H. Kim, and B. H. Kwon, "Continuous Conduction Mode SEPIC Converter with Low Reverse-Recovery Loss for Power Factor Correction," IEE Proceedings on Electric Power Applications, vol.153, no.5, pp.673-681, Sept. 2006.

[3] [10] N. Li, X. Lin-Shi, P. Lefranc, E. Godoy, A. Jaafar, and B. Allard, "Universal DC-DC Converter Using SEPIC," Proc. 14th European Conference on Power Electronics and Applications, pp.1-10, Aug. 30 - Sept. 1, 2011.

[4] [11] C. Jingquan and J. Chang, "Analysis and Design of SEPIC Converter in Boundary Conduction Mode for Universal-Line Power Factor Correction Applications," Proc. IEEE 32nd Annual Power Electronics Specialists Conference, vol.2, pp.742-747, 2001.

[5] [12] H. M. Huang, S. H. Twu, S. J. Cheng, and H. J. Chiu, "A Single-Stage SEPIC PFC Converter for Multiple Lighting LED Lamps," Proc. 4th IEEE International Symposium on Electronic Design Test and Applications, pp.15-19, Jan. 2008.

[6] M. Ali, M. Orabi, M. E. Ahmed, and A. El-Aroudi, "Design Consideration of Modified SEPIC Converter for LED Lamp Driver," Proc. 2nd IEEE International Symposium on Power Electronics for Distributed Generation Systems, pp.394-399, June 2010.

An Interleaved Single-Stage LLC Resonant Converter Used for Multi-Channel LED Driving

Chien-Hsuan Chang, Chun-An Cheng, Masahito Jinno, and Hung-Liang Cheng
Department of Electrical Engineering, I-Shou University,
Kaohsiung City 84001, Taiwan
e-mail: chchang@isu.edu.tw

Abstract— In order to simplify circuit complexity, reduce cost and improve system efficiency, this paper proposes an interleaved single-stage LLC resonant converter to drive multi-channel light emitting diodes (LEDs) with high power-factor and current-sharing features. Two derivative buck-boost typed power-factor-correctors (PFCs) with interleaving operation are integrated with an LLC resonant converter to form the presented circuit. The proposed single-stage converter has the features of simple circuit structure, balanced switch currents, low dc-bus voltage, low input current harmonics, and high system reliability. With proper analysis and design, the power switches can turn on with zero-voltage-switching (ZVS), and the output rectifier diodes can turn off with zero-current-switching (ZCS), which results in low switching losses. Additionally, by connecting the primary windings of transformers in series, the current-sharing among all the LED channels can be achieved naturally. Operation principles and circuit analysis are addressed. Finally, a laboratory prototype with 220 V_{rms} utility-line voltage is implemented correspondingly to verify the validity of the theoretical predictions and the feasibility of the proposed circuit.

Keywords— *interleave; single-stage; power-factor correction; LED driving.*

I. INTRODUCTION

Nowadays, light emitting diodes (LEDs) have been applied in the applications of display backlighting, indoor lighting, street lighting, and traffic lighting, because they have the advantages of dc working voltage, high luminescent efficiency, short ignition time, high reliability and pollution free, etc.[1], [2]. In the LED lighting applications, multi-channel structure is very popular to achieve the requirement of high illumination. Figure 1 shows the typical configuration of a multi-channel LED driver, in which there are three stages connected in cascade to achieve high power-factor and current-sharing among all LED channels [3]. The power-factor corrector (PFC) in first stage is used to suppress current harmonics on utility-line, and the dc-dc converter in second stage provides regulated voltage to drive LEDs. Besides, the third stage controls all LED currents to achieve current-sharing. This structure has independent control in each stage so that the functions of power-factor correction, line regulation, load regulation, and current-sharing can be designed optimally. However, the three-stages structure has the disadvantages of complicated

structure, large size, high cost and low efficiency.

In order to improve these shortcomings, several single-stage and high power-factor converters are developed by operating the front-end PFCs in discontinuous conduction mode (DCM) and sharing the active switches with the back-end dc-dc converters [4]-[6]. Furthermore, the current-sharing of LED channels can be naturally achieved by employing multiple transformers and connecting their primary windings in series [7], [8]. Figure 2 shows the block diagram of a single-stage, multi-channel LED driver with high power-factor and current-sharing features.

Since the LLC resonant converter has no light-load regulation issue and low circulating energy [9], [10], it has been widely adopted as the back-end converter to develop single-stage and high power-factor converters with zero-voltage-switching (ZVS) operation [11]-[16]. Due to the merits of simple structure and control, boost converter is the most popular topology of PFC semi-stage among these single-stage converters. However, their dc-bus voltage should be at least twice the peak of utility-line voltage to achieve high power-factor [11], [12]. High voltage stresses of switching components result in high conduction losses and low efficiency. Besides, because of DCM operation, lots of input high-frequency harmonics are induced by the pulse currents. Furthermore, the PFC semi-stage shares only one power switch of the LLC resonant converter so that the current stresses and power

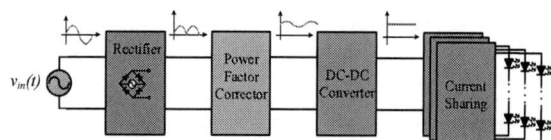

Fig. 1. The block diagram of a three-stage, multi-channel LED driver.

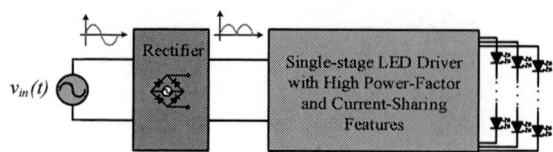

Fig. 2. The block diagram of a single-stage, multi-channel LED driver.

dissipations of the half-bridge switches are extremely different. High current-stress of the shared switch will significantly impact its reliability. Therefore, the literatures [15], [16] have proposed the interleaved single-stage converters via the integration of two interleaved boost typed PFCs and an LLC resonant converter. They have the features of balanced current stresses of power switches and low input current harmonics.

This paper proposes an interleaved single-stage LLC resonant LED driver to further reduce dc-bus voltage. Two derivative buck-boost typed PFCs [14] with interleaved operation are integrated with an LLC resonant converter to form the proposed circuit. The system size and cost can be reduced by saving power switches and control circuits. Since the dc-bus voltage is reduced, the proposed circuit is more suitable for the applications with high utility-line voltage. The current stresses and power dissipations of the half-bridge switches are close and balanced, leading to simple circuit design and high system reliability. Additionally, the input current harmonics can be reduced by the interleaved operation of PFCs so that the size of input low-pass filter can be miniaturized. Both the power switches can turn on with ZVS, and all the output rectifier diodes can turn off with zero-current-switching (ZCS), leading to high system efficiency. Further, considering LED's nonlinear output

characteristics, piecewise linear model and voltage-controlled approach is used in the proposed LED driver. In order to further simplify the circuit structure, this paper proposes to employ multiple transformers and connect their primary windings in series [7], [8]. The current-sharing among all LED channels can be naturally achieved without extra converters.

II. CIRCUIT CONFIGURATION

A high power-factor and multi-channel LED driver, as shown in Fig. 3(a), consists of two derivative buck-boost typed PFCs with interleaved operation and an LLC resonant converter in cascade connection. The two coupled inductors are employed instead of single-winding inductors to accomplish buck-boost conversion. The switches can be integrated by applying the synchronous switch technique [17] to simplify the driving circuit as a single-stage converter. The switches S_{B1} and S_{L1} share a common node of drain-drain so that they can be replaced with an inverted T-type synchronous switch called S_1. Similarly, since the switches S_{B2} and S_{L2} share a common node of source-source, they are replaced with a T-type synchronous switch called S_2. The proposed interleaved, single-stage LLC resonant LED driver can be generated as shown in Fig. 3(b). The diodes D_5 *and* D_6 are used to block the current going from ac utility-line into the

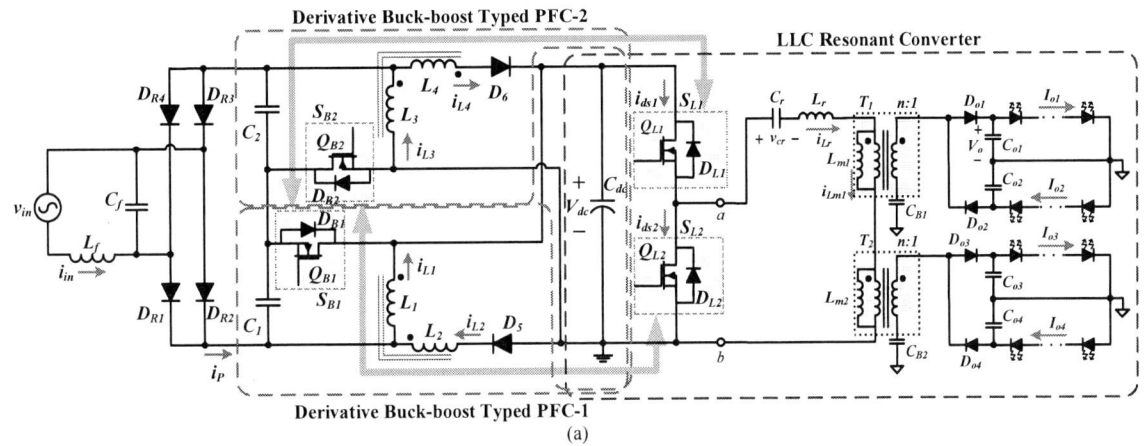

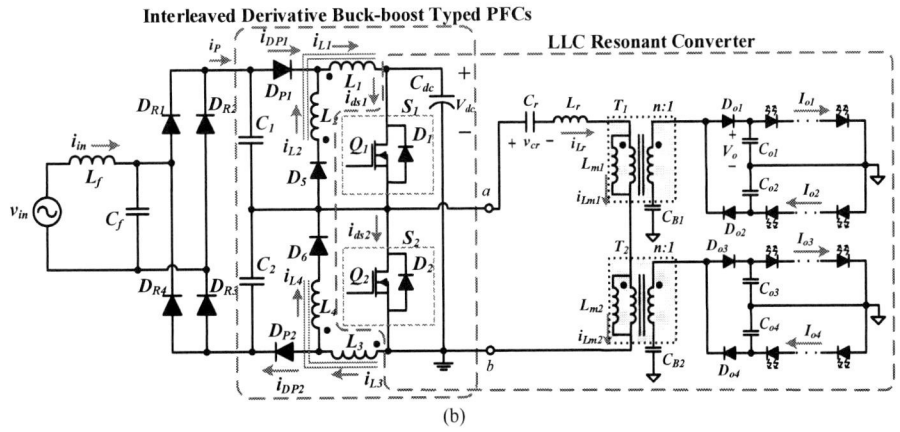

Fig. 3. Circuit derivation of the proposed single-stage LED driver.

inductors L_2 and L_4. Besides, the diodes D_{p1} and D_{p2} can prevent the inductor currents going back the input capacitors C_1 and C_2.

As shown in Fig. 3(b), each switch S_1 (S_2) is composed of an MOSFET Q_1 (Q_2) and its intrinsic anti-parallel diode D_1 (D_2). The resonant tank is formed by the resonant capacitor C_r, the resonant inductor L_r and the magnetizing inductors L_{m1} (L_{m2}) of transformers T_1 (T_2). By conducting S_1 and S_2 alternately, a symmetrical square waveform with the magnitude of $V_{dc}/2$ can be obtained in the input terminal of the resonant tank, where V_{dc} is the dc-bus voltage. The output voltage V_o of one LED channel can be regulated by modulating the switching frequency f_S.

The two derivative buck-boost typed PFCs are embedded in the proposed circuit diagram and share the power switches S_1 and S_2. With accurate designs of the two coupled inductors and dc-bus voltage V_{dc}, the inductor currents can operate in DCM to achieve high power-factor without additional current-shaped controller. The input high-frequency current harmonics can be filtered by the low-pass filter (L_f and C_f).

Since the input voltage of each PFC is half of the utility-line voltage, the peak current of each coupled inductor and the dc-bus voltage will also be half. Therefore, the input current harmonics are reduced, and the power switches with lower voltage-rating can be adopted to decrease their conduction losses. Besides, the currents of two coupled inductors are close, and flow through one of the half-bridge power switches respectively. The current stresses of power switches should be the same, which leads to simple circuit-design and high system-reliability.

III. OPERATION PRINCIPLE

Figure 4 shows the theoretical waveforms of the proposed single-stage LED driver, in which there are twelve operation modes within one switching period. Referring to the equivalent circuits shown in Fig. 5, the operation principles of each mode are introduced in the following. In order to simplify circuits, the input voltage v_{in}, the low pass filter L_f and C_f, the bridge rectifier, and the capacitors C_1, C_2 are replaced with two rectified voltages V_{rec1} and V_{rec2}.

A. Mode 1

This mode starts when the switch S_2 is turned on. The equivalent circuit is shown in Fig. 5(a). Before the gate signal of S_2 is applied, the resonant current i_{Lr} flowed through anti-paralleled diode D_2 and released energy to the resonant capacitor C_r, which provides ZVS operation for S_2 turn-on. The voltage across L_{m1} (L_{m2}) is clamped to $2nV_o$ with reverse polarity so that it current i_{Lm1} (i_{Lm2}) decreases linearly. The energy stored in L_{m1} and L_{m2} will be released through D_{o2} and D_{o4} to LEDs. In addition, because the voltage across the inductor L_3 is V_{rec2}, the current i_{L3} starts to increase linearly from zero. The dc-bus voltage V_{dc} cross on the inductors L_1 and L_2 in series is with reverse polarity so that both L_1 and L_2 discharge to

the dc-bus capacitor C_{dc}. When i_{Lr} reaches zero, this mode ends.

B. Mode 2

As shown in Fig. 5(b), i_{Lr} becomes negative, so the energy transfers from C_r to the resonant inductor L_r. L_{m1} and L_{m2} remain releasing energy to output LEDs until their currents reach zero. Since the voltages across the coupled inductors are the same as previous mode, i_{L1} and i_{L2} remain decreasing, and i_{L3} remains increasing. When i_{Lm1} and i_{Lm2} reach zero, this mode ends.

C. Mode 3

The voltage across L_{m1} (L_{m2}) is the same as previous mode so that i_{Lm1} and i_{Lm2} remain decreasing and become negative. Both C_r and L_r release energy to output LEDs simultaneously. The equivalent circuit is shown in Fig. 5(c).

D. Mode 4

This mode starts when the inductor currents i_{L1} and i_{L2} become zero. The operations of the currents i_{Lm1}, i_{Lm2} and i_{L3} remain the same as previous mode. Since i_{L1} and i_{L2} reach zero before the switch S_1 turns on again, the feature of high power-factor can be achieved. The equivalent circuit is shown in Fig. 5(d).

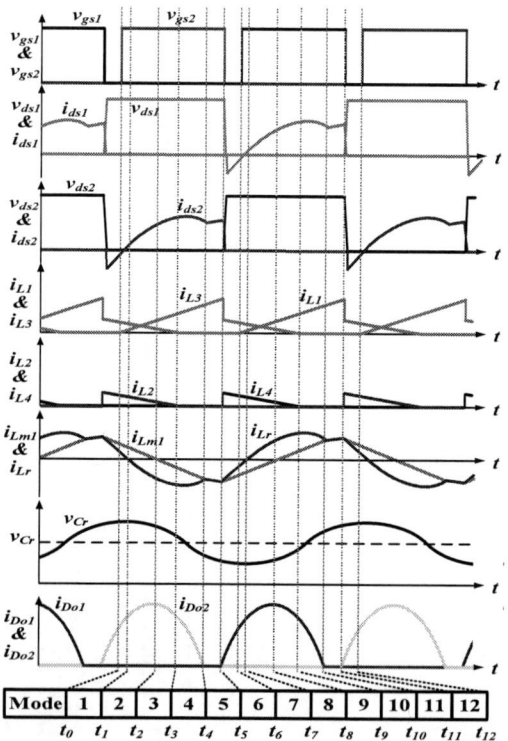

Fig. 4. Theoretical waveforms of the proposed single-stage LED driver.

E. Mode 5

This mode starts when the two inductor currents i_{Lr} and i_{Lm} are equal. Currents circulating through the secondary diodes D_{o2} and D_{o4} naturally decrease to zero so that they can turn off under ZCS. There is no voltage spike occurred by diode reverse-recovery. Because The voltage across L_{m1} (L_{m2}) is no longer clamped to $2nV_o$ with reverse polarity, L_{m1} and L_{m2} are in series with L_r, and participate in the resonance with C_r. Fig. 5(e) shows the equivalent circuit of this mode. The current i_{L3} is increasing continuously, and this mode ends when S_2 is turned off.

F. Mode 6

As shown in Fig. 5(f), while S_2 is turning off, the resonant current i_{Lr} is discharging C_{oss1} and charging C_{oss2}, simultaneously. At the moment of v_{ds1} decreasing to zero, the resonant current i_{Lr} flows through the anti-paralleled diode D_1, which provides ZVS operation for S_1 turn-on. At the same time, the secondary rectifier diodes D_{o1} and D_{o3} turn on. The voltage across L_{m1} (L_{m2}) is clamped to $2nV_o$ so that i_{Lm1} (i_{Lm2}) becomes linear increasing. L_{m1} and L_{m2} are separated from the resonance with L_r and C_r. When the gate signal v_{gs1} is applied to force S_1 turning on under ZVS, this mode ends and enters the other half cycle with symmetrical operation principles.

At the instants of the switch S_2 turning-off ($t = t_5$), the diode D_6 is forward biased and the inductor current i_{L3} flows through the inductor L_4 into the dc-bus capacitor C_{dc}. The magneto motive force in the coupled inductors should keep balance in ampere turns; hence,

$$N_3 i_{L3}(t_5^-) = (N_3 + N_4) i_{L3}(t_5^+) \qquad (1)$$

where N_3 and N_4 are the turn numbers of the coupled inductors L_3 and L_4, and $i_{L3}(t_5^-)$ and $i_{L3}(t_5^+)$ are the current i_{L3} at the instants before and after the switch S_2 turning-off. The inductor current i_{L3} drops dramatically at this instant. The voltage across L_3 and L_4 in series is V_{dc} with reverse polarity so that both i_{L3} and i_{L4} decrease linearly and release their energy to C_{dc}.

G. Modes 7~12

For the LLC resonant semi-stage, the operation principles of next half cycle (Modes 7~12) are symmetrical with the previous six modes mentioned above. When the gate signal v_{gs2} is applied again to force S_2 turning on under ZVS at $t = t_{12}$, the circuit operation goes to Mode 1 of the next switching cycle.

For the PFCs semi-stage, i_{L3} and i_{L4} remain decreasing linearly. In order to achieve high power-factor, they should reach zero before S_2 turns on again. Besides, due to the interleaved operation of two PFCs, their input inductor currents i_{L1} and i_{L3} are symmetrical and flow through the active switches S_1 and S_2, which leads to balanced current-stresses and low conduction-losses.

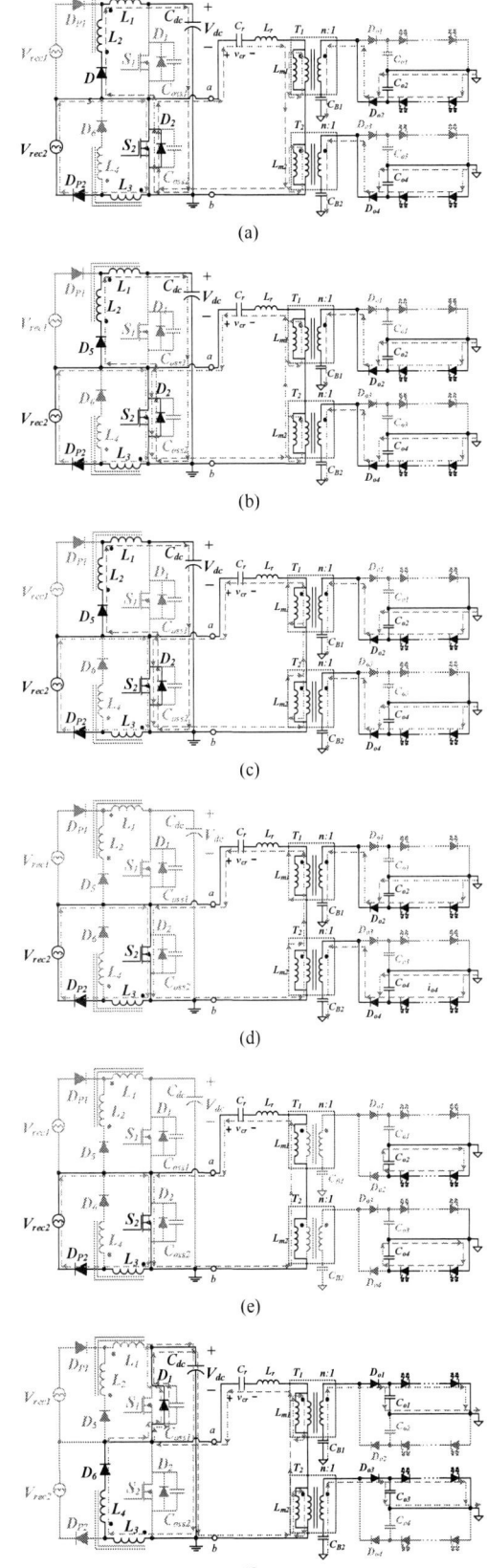

(a)

(b)

(c)

(d)

(e)

(f)

Fig. 5. Equivalent circuits of the proposed single-stage LED driver.

IV. CIRCUIT ANALYSIS

Although the derivative buck-boost PFCs and the LLC resonant converter share the active switches S_1 and S_2, their functions are remained. Therefore, the proposed single-stage LED driver can be divided into the PFCs semi-stage and the LLC resonant semi-stage, and then analyzed separately.

A. Derivative Buck-Boost Typed PFCs

As shown in Fig. 3(b), there are two derivative buck-boost typed PFCs with interleaved operation. The coupled inductors are employed instead of the single buck-boost inductor to accomplish buck-boost conversion. The input utility-line voltage $v_{in}(t)$ is sinusoidal and expressed as

$$v_{in}(t) = V_m \sin(2\pi f_{line} t) \tag{2}$$

where V_m and f_{line} are the peak value and frequency of utility-line voltage, respectively. Theoretically, if the capacitors C_1 and C_2 have the same capacitance, their voltages $V_{rec1}(t)$ and $V_{rec2}(t)$ will be equal and can be expressed as

$$V_{rec1}(t) = V_{rec2}(t) = \frac{1}{2} \times V_m \cdot \left| \sin(2\pi f_{line} t) \right| \tag{3}$$

While the switch S_1 is turning on, the voltage across the inductor L_1 is $V_{rec1}(t)$. Because the switching frequency f_S is much higher than line frequency f_{line}, the voltage $V_{rec1}(t)$ could be regarded as constant within one high-frequency switching period T_S. Therefore, the peak value of diode currents $i_{DP1}(t)$ and $i_{DP2}(t)$ can be expressed as

$$i_{DP1,peak}(t) = i_{DP2,peak}(t) = \frac{1}{2} \times \frac{V_m \cdot \left| \sin(2\pi f_{line} t) \right|}{L_{pf}} \times D T_s \tag{4}$$

where L_{pf} is the inductance of L_1 and L_3, and D is the duty ratio for the switches S_1 and S_2. Thus, the peak value of the rectified input current $i_{p,peak}(t)$ can be expressed as

$$\begin{aligned} i_{p,peak}(t) &= i_{DP1,peak}(t) + i_{DP2,peak}(t) \\ &= \frac{V_m \cdot \left| \sin(2\pi f_{line} t) \right|}{L_{pf}} \times D T_s \end{aligned} \tag{5}$$

By filtering the high-frequency components of $i_p(t)$, the input current $i_{in}(t)$ is equal to the average of $i_p(t)$ over one switching period T_S and can be expressed as

$$i_{in}(t) = \frac{1}{T_S} \int_0^{T_S} i_p(t) dt = \frac{V_m D^2 T_S}{2 L_{pf}} \times \sin(2\pi f_{line} t) \tag{6}$$

In (6), if T_S and D are fixed within one utility-line period T_{line}, $i_{in}(t)$ will be sinusoidal and in phase with utility-line voltage. Thus, a high power-factor can be achieved. Fig. 6 shows the concept waveforms of $i_{DP1,peak}(t)$ and $i_{in}(t)$. By the way, the input power P_{in} can be obtained from the average power over one T_{line}.

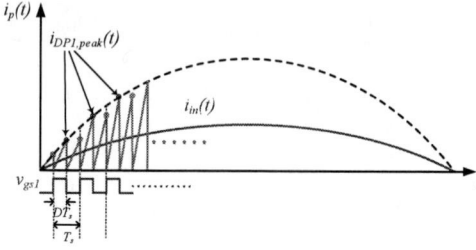

Fig. 6. Concept waveforms of input current and inductor current.

$$P_{in} = \frac{D^2 T_S}{2 L_{pf}} \frac{1}{T_{line}} \int_0^{T_{line}} V_m^2 \cdot \sin^2(2\pi f_{line} t) dt = \frac{V_m^2 D^2 T_S}{4 L_{pf}} \tag{7}$$

While the switch S_1 is turning off, the diode D_{P1} must be reverse bias to accomplish the buck-boost conversion. Therefore, the voltage across L_2 should be designed as higher than half of peak line voltage and can be expressed as

$$\left| v_{L2} \right| = \left(\frac{\alpha}{1+\alpha} \right) \times V_{dc} \geq \frac{V_m}{2} \tag{8}$$

where α is the turn ratio of L_1 and L_2. Besides, according to the volt-second balance principle, V_{dc} should match the following equation to ensure that the buck-boost typed PFC can operate in DCM to achieve high power-factor

$$V_{dc} \geq \frac{V_m}{2} \times \left(\frac{D}{1-D} \right) \times \left(\frac{1 + 2k\alpha + \alpha^2}{1 + k\alpha} \right) \tag{9}$$

where k is the coupling coefficient of the coupled inductor.

B. LLC Resonant Converter

Because the inductances of L_{m1} and L_{m2} are only slightly higher than that of L_r, they may participate in the resonance with L_r and C_r to change the characteristics of resonant tank. The equivalent circuit of the LLC resonant converter can be depicted as shown in Fig. 7, where $2L_m$ is the equivalent inductance of L_{m1} and L_{m2} in series. A linear model, including the cut-in voltage V_r and the forward resistance r_d, is used for the presentation of one LED channel. Therefore, the equivalent ac resistance R_i in primary side can be expressed as

$$R_i = 8 n^2 r_d / \pi^2 \tag{10}$$

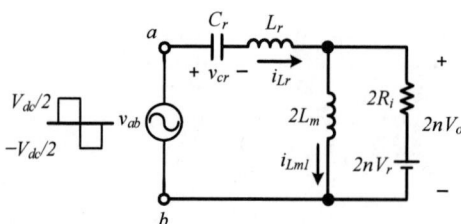

Fig. 7. Equivalent circuit of the LLC resonant semi-stage.

By conducting the switches S_1 and S_2 alternately, a symmetrical square waveform with the magnitude of $V_{dc}/2$ can be obtained in the input terminal of resonant tank, where V_{dc} is the dc-bus voltage. The dc voltage gain of the LLC resonant converter can be determined as follows:

$$M(f_S) = \left| \frac{V_o}{V_{dc}} \right|$$

$$= \frac{1}{4n \cdot \sqrt{(1+A)^2 \left(1 - (\frac{f_L}{f_S})^2\right)^2 + \frac{1}{Q_L^2}(\frac{f_S}{f_L}\frac{A}{1+A} - \frac{f_L}{f_S})^2}} \quad (11)$$

where the inductance ratio A, the second resonant frequency f_L, and the load quality factor Q_L are defined as

$$A = \frac{L_r}{2L_m} \quad (12)$$

$$f_L = \frac{\omega_L}{2\pi} = \frac{1}{2\pi\sqrt{(L_r + 2L_m) \cdot C_r}} \quad (13)$$

$$Q_L = 2R_i \times \sqrt{C_r / (L_r + 2L_m)} = 2R_i \cdot 2\pi f_L \cdot C_r \quad (14)$$

In this paper, the parameters V_r and r_d of the LED model are obtained from real measurements. Fig. 8 shows the measured V-I characteristics of one LED channel. Once these parameters are determined by linear approximation, the relation between output voltage V_o and current I_o can be expressed as

$$V_o = V_r + I_o \times r_d \quad (15)$$

V. EXPERIMENTAL RESULTS

The proposed single-stage LED driver with a 220 V_{rms} utility-line voltage is taken as an illustrative example. According to previous analyses, the key component parameters, such as L_{pf}, C_r, L_r, and L_m, can be calculated and determined. The electrical specifications and component parameters are summarized in Table I.

TABLE I
SPECIFICATIONS AND PARAMETERS OF THE PROTOTYPE

ELECTRICAL SPECIFICATIONS	
Input Voltage, v_{in}	220 V_{rms}, 60 Hz
DC-bus Voltage, V_{dc}	400 V_{DC}
LED Channel Voltage, V_o	23 V_{DC}
LED Channel Current, I_o	1 A
Output Power, P_o	92 W
COMPONENT PARAMETERS	
Input Inductor, L_{pf}	245 μH
Turn Ratio of Coupled Inductor	1.2
Resonant Inductor, L_r	234 μH
Magnetizing Inductors, L_{m1} & L_{m2}	584.5 μH
Resonant Capacitor, C_r	10 nF
Transformer Turn Ratio, n:1	6:1

Figure 9 shows the measured voltage and current waveforms of the switches S_1 and S_2 at full-load and peak line-voltage conditions. As can be seen, both S_1 and S_2 can turn on with ZVS. Besides, because two PFCs operate with interleaving, the switch currents i_{ds1} and i_{ds2} are similar. Therefore, their current stresses and losses are close and balanced, leading to simple design of switching components and high reliability.

Figure 10 shows the measured waveforms of the diode currents i_{Do1} and i_{Do2}. They circulate through the secondary rectifier diodes and naturally decrease to zero before the power switches turning on. The rectifier diodes can turn off under ZCS condition, which can improve efficiency and avoid the voltage spike caused by diode reverse-recovery. Fig. 11 shows the measured efficiency curve, where the highest efficiency is up to 90.3%.

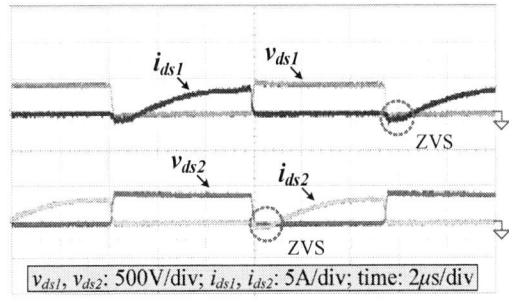

v_{ds1}, v_{ds2}: 500V/div; i_{ds1}, i_{ds2}: 5A/div; time: 2μs/div

Fig. 9. Measured voltage and current waveforms of the switches S_1 and S_2 at full-load operation and peak line-voltage conditions.

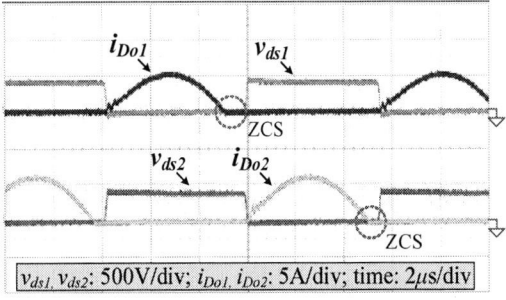

v_{ds1}, v_{ds2}: 500V/div; i_{Do1}, i_{Do2}: 5A/div; time: 2μs/div

Fig. 10. Measured current waveforms of the diodes D_{o1} and D_{o2} at full-load operation.

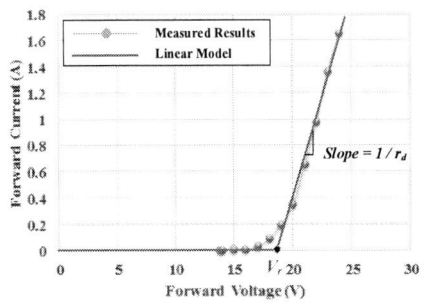

Fig. 8 The V-I characteristics of one LED channel.

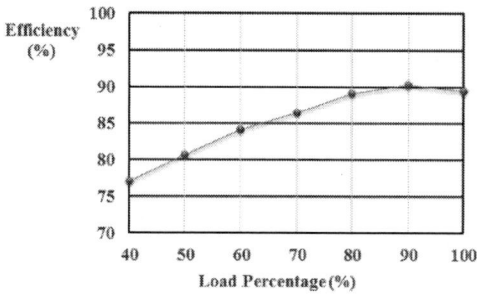

Fig. 11. Measured efficiency curve of the proposed LED driver.

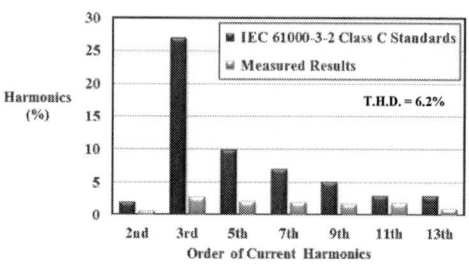

Fig. 14. Measured input current harmonics.

Figure 12 shows the measured waveforms of the coupled-inductor currents i_{L1} and i_{L3} at full-load and peak line-voltage conditions. As can be seen, both i_{L1} and i_{L3} operate in DCM so that high power-factor can be achieved naturally. The interleaved operation of i_{L1} and i_{L3} indicates that the input current harmonics can be suppressed. Besides, at the instants of the switches turning-off, i_{L1} (i_{L3}) drops dramatically and flows through the inductor L_2 (L_4) to release its energy to the dc-bus capacitor C_{dc}. The measured waveforms of utility-line voltage and input current are shown in Fig. 13. The input current is sinusoidal and in phase with the utility-line voltage. The measured power factor is 0.998, and the total harmonic distortion (T.H.D.) of input current is 6.2%. Fig. 14 shows the measured input current harmonics compared with the IEC 61000-3-2 Class C standards, in which all measured results are fully compliant with the requirements.

The measured currents of all LED channels under normal condition are shown in Fig. 15(a), in which all currents are very close to 1 A within 0.3% difference. In order to further verify the current-sharing capability, the LED channel no.3 is set to have ±5% cut-in voltage (V_r) difference with LED channel no.1. The measured LED currents i_{o1} and i_{o3} are shown in Fig. 15(b) and 15(c), respectively. Because the current i_{o1} is with sensing and feedback control, it is exactly 1 A at all conditions. The current i_{o3} is 996 mA and 1009 mA, respectively. The current difference is within 1%, which indicates that good current-sharing can be achieved by the proposed LED driver. Further, Fig. 15(d) shows the measured LED currents i_{o1} and i_{o3}, while one LED short failure occurs on the LED channel no.3. As can be seen, the current i_{o1} is not influenced and still controlled in 1 A. The current i_{o3} recovers to 1020 mA within 30 ms, which verifies that good current-sharing can be maintained even though short failure occurs. All measured results are summarized in Table II for comparison.

VI. CONCLUSION

By sharing power switches and employing multiple transformers, an interleaved single-stage LLC resonant converter has been successfully used for driving multi-channel LEDs with high power-factor and current-sharing features. Two derivative buck-boost typed PFCs with interleaved operation are integrated with an LLC resonant dc-dc converter to form the proposed circuit. The circuit structure are simplified due to saving switches and controllers. ZVS turn-on operation of primary switches and ZCS turn-off operation of rectifier diodes result in improving system efficiency. Experimental results of an illustrative example have really proved that high power-factor and good current-sharing can be achieved by the proposed multi-channel LED driver.

ACKNOWLEDGMENT

This work was supported by the National Science Council of Taiwan, under grant contract NSC 102-2221-E-214-023.

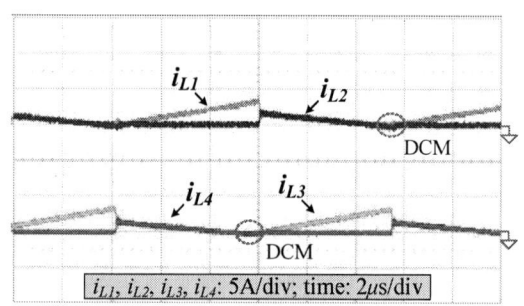

Fig. 12. Measured waveforms of the coupled-inductor currents i_{L1}, i_{L2}, i_{L3} and i_{L4} at full-load and peak line-voltage conditions.

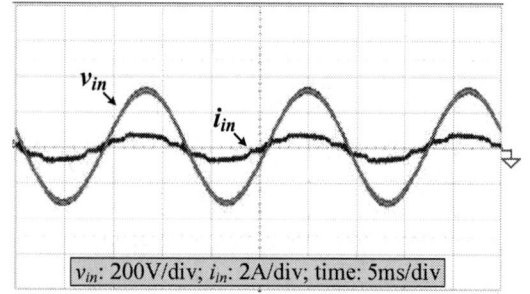

Fig. 13. Measured waveforms of utility-line voltage and input current at full-load operation.

TABLE II
COMPARISONS OF MEASURED LED CURRENTS

Test condition of i_{o3}	i_{o1} (mA)	i_{o3} (mA)	Difference (%)
normal	1003	1002	-0.1%
+5% V_r	1001	996	-0.5%
-5% V_r	1002	1009	+0.7%
short failure	1003	1020	+1.7%

978-1-4799-2706-7/14 $31.00 © 2014 IEEE

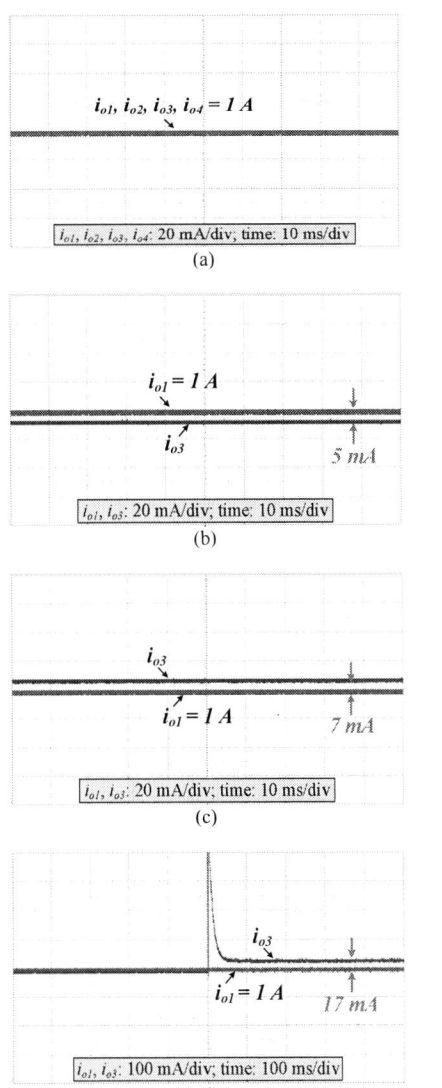

Fig. 15. Measured LED currents under (a) normal condition, (b) +5% V_r, (c) -5% V_r and (d) short failure.

REFERENCES

[1] G. Harbers, S. J. Bierhuizen, and M. R. Krames, "Performance of high power light emitting diodes in display illumination applications," *IEEE/OSA J. Disp. Technol.*, vol. 3, no. 2, pp. 98-109, Jun. 2007.

[2] M. Rico-Secades, A. J. Calleja, J. Ribas, E. L. Corominas, J. M. Alonso, J. Cardesín, and J. García-García, "Evaluation of a Low-Cost Permanent Emergency Lighting System Based on High-Efficiency LEDs," *IEEE Trans. on Ind. Appl.*, vol. 41, no. 5, pp. 1386-1390, Sept./Oct. 2005.

[3] C.-C. Chen, C.-Y. Wu, Y.-M. Chen, and T.-F. Wu, "Sequential color LED backlight driving system for LCD panels," *IEEE Trans. Power Electron.*, vol.22, no. 3, pp. 919-925, May 2007.

[4] H.-J. Chiu, and S.-J. Cheng, "LED Backlight Driving System for Large-Scale LCD Panels," *IEEE Transactions on Ind. Electron.*, vol. 54, no. 5, pp. 2751-2760, 2007.

[5] D. G. Lamar, J. S. Zuniga, A. R. Alonso, M. R. Gonzalez, and M. M. H. Alvarez, "A Very Simple Control Strategy for Power Factor Correctors Driving High-Brightness LEDs," *IEEE Trans. Power Electron.*, vol. 24, no. 8, pp. 2032-2042, 2009.

[6] Y.-C. Chuang, Y.-L. Ke, H.-S. Chuang, and C.-C. Hu, "Single-Stage Power-Factor-Correction Circuit with Flyback Converter to Drive LEDs for Lighting Applications," in *Conf. Rec. 2010 IEEE Ind. Appl. Conf. 39th IAS Annual Meeting*, 2010, pp. 1-9.

[7] S. Ji, H. Wu, X. Ren, and F. C. Lee, "Multi-channel constant current (MC³) LED driver," in *Proc. the IEEE Applied Power Electronics Conference and Exposition*, March 2011, pp. 718-722.

[8] H. Wu, S. Ji, F. C. Lee and X. Wu, "Multi-channel constant current (MC³) LLC resonant LED rriver," in *Proc. the 2011 IEEE Energy Conversion Congress and Exposition*, Sept. 2011, pp. 2568-2575.

[9] B. Yang, F. C. Lee, A. J. Zhang, and G. Huang, "LLC resonant converter for front end DC/DC conversion," in *Proc. IEEE Appl. Power Electron. Conf. Expo.*, Mar. 2002, vol. 2, pp. 1108-1112.

[10] G. Ivensky, S. Bronshtein, and A. Abramovitz, "Approximate analysis of resonant LLC DC-DC converter," *IEEE Trans. Power Electron.*, vol. 26, no. 11, pp. 3274-3284, Nov. 2011.

[11] C.-M. Lai and K.-K. Shyu, "A single-stage AC/DC converter based on zero voltage switching LLC resonant topology," *IET Electr. Power Appl.*, vol. 1, no. 5, pp. 743-752, Sep. 2007.

[12] S.-Y. Chen, Z. R. Li, and C.-L. Chen, "Analysis and design of single-stage AC/DC LLC resonant converter," *IEEE Trans. Ind. Electron.*, vol. 59, no. 3, pp. 1538-1544, Mar. 2012.

[13] C.-H. Chang, H.-Y. Chen, C.-T. Cho, and J.-Y. Chiu, "A novel single-stage LLC resonant AC-DC converter with power factor correction feature," in *Proc. 6th IEEE Conf. on Ind. Electron. and Appl. (ICIEA)*, Jun. 2011, pp. 2191-2196.

[14] C.-H. Chang, C.-A. Cheng, H.-L. Cheng, and H.-Y. Chen, "A single-stage high-power-factor LLC resonant AC-DC converter for light emitting diode driving", *Advanced Science Letters*, vol. 8, pp. 451-456, Apr. 2012.

[15] X. Zhang, W. Wang, and D. Xu, " A novel interleaved single-stage AC/DC converter with high power factor and ZVS characteristic," in *Proc. IEEE 2010 International Conf. on Elec. Mach. and Syst. (ICEMS)*, Oct. 2010, pp. 249-254.

[16] C.-A. Cheng and C.-H. Yen, "A single-stage driver for high power LEDs," in *Proc. 6th IEEE Conf. on Ind. Electron. and Appl. (ICIEA)*, Jun. 2011, pp. 2666-2671.

[17] T.-F. Wu and T.-H. Yu, "Unified approach to developing single-stage power converters," *IEEE Trans. Aerosp. Electron. Syst.*, vol. 34, no. 1, pp. 211-223, Jan. 1998.

978-1-4799-2706-7/14 $31.00 © 2014 IEEE

A Novel Type of Wireless V2H System with Bidirectional Resonant Single-Ended Inverter

Hiroki Fukuoka
Department of Electrical and
Electronic Systems Engineering
Osaka Institute of Technology
Osaka, Japan,
e1310096@st.oit.ac.jp

Yuichi Iga
Department of Electrical and
Electronic Systems Engineering
Osaka Institute of Technology
Osaka, Japan,
iga19890929@gmail.com

Hideki Omori
Department of Electrical and
Electronic Systems Engineering
Osaka Institute of Technology
Osaka, Japan,
omori@ee.oit.ac.jp

Tosimitsu Morizane
Department of Electrical and
Electronic Systems Engineering
Osaka Institute of Technology
Osaka, Japan,
igayu1@cemlab.ee.oit.ac.jp

Noriyuki Kimura
Department of Electrical and
Electronic Systems Engineering
Osaka Institute of Technology
Osaka, Japan,
igayu1@cemlab.ee.oit.ac.jp

Mutuo Nakaoka
Industrial E.E. Application Research Institute
Kyungnam University
Masan, Korea
nakaoka@pe-news1.ee.yamaguchi-u.ac.jp

Abstract— Electric vehicles (EV) offer promise as an effective solution to environmental problems. One of the keys to their successful diffusion is the provision of adequate battery charging infrastructure. In order to create a charging infrastructure by installing equipment in such as locations as carports in private homes, the wireless battery charging system is very suitable. EVs can be used in smart house systems to supplement the energy storage. This vehicle to home (V2H) system essentially requires a bidirectional power transfer feature between the EV and home. This paper presents a new bidirectional inductive power transfer (IPT) system for wireless V2H with simplest components and low cost aiming at wide diffusion for home use. Proposed is a novel type of bidirectional wireless EV charging system with an efficient and compact type single-ended quasi-resonant high-frequency inverter for V2H.

Keywords—single-ended inverter; vehicle to home; wireless EV charger; resonant IPT; bidirectional

I. INTRODUCTION

In recent years, Electric Vehicles (EV) which are highly efficient as well as do not create air pollution, and offer promise as an effective solution to environmental problems with great advances of power electronic technology.

One of the key issues to their successful and wide diffusion is the provision of adequate battery charging infrastructure. In order to create a battery charging infrastructure by installing equipment in such as locations as carports in private homes, the inductive power transfer (IPT)-wireless battery charging system is indispensable for spread. The wireless battery charging system eliminates the use of power cables with plug. Merely by parking the car in a designated spot, the battery can be charged. It is a promising system for wider diffusion because it is easy and safe to use for a broad range of users including the elderly.

Usually as wireless EV charging power supply topologies have been half-bridge, push-pull, full bridge, boost half-bridge

and boost full-bridge circuit configuration with DC-DC converters.[1][2][3][4] The authors have previously put into practice the cost-effective and high-efficiency single-ended quasi-resonant soft-switching inverter.[5]

And although EVs are primarily consider as a method of clean transport, they can also be used in smart house systems to supplement the energy storage. This vehicle to home (V2H) system essentially requires a bidirectional power transfer feature between the EV and home.

This paper presents a new system with the simplest components and low cost aiming at wide diffusion for home use. From a practical point of view, simple high-frequency inverter circuit topologies have to be effectively selected in accordance with specific cost effective applications. Proposed is a novel type of wireless EV charging system based on IPT technology with an efficient and compact type single-ended quasi-resonant high-frequency inverter. The single-ended quasi-resonant high-frequency inverter, which can operate in the frequency range from 20-30 kHz under a self-excited ZVS control and its zero voltage crossing detector of resonant capacitor voltage and a PFM controlled power regulation scheme, is evaluated from an experimental point of view. Furthermore transfer power and efficiency have been successfully improved by pick-up circuit with resonant component. The output power and the efficiency of a proposed system are successfully improved by a resonant IPT circuit.

Therefore, proposed is a new system configuration by a single-ended inverter with a resonant IPT circuit for bidirectional power transfer. And a result of feasibility study by simulation is indicated for V2H.

II. EV CHARGING SYSTEM DESCRIPTION

An IPT-based wireless EV charging system is schematically illustrated in Fig.1. And Fig.2 shows a schematic total system with a single-ended quasi-resonant high-frequency inverter. The system is mainly composed of a single-phase diode D_3

rectifier with a L_3-C_3 filter, a single-ended quasi-resonant high-frequency inverter operating with a ZVS-PFM power regulation scheme in the frequency range from 20-30kHz, resonant capacitor with the primary coil L_1 which is loosely coupled to the pick-up coil L_2 as load side, a single-phase diode D_2 rectifier with a L_4-C_2 filter, L_5-C_4 filter connected with the battery bank of EV, and a specific power regulation control circuit due to the self-excited timing signal processing.

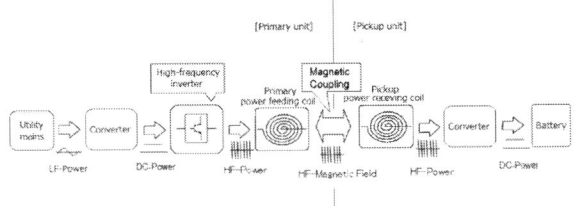

Fig.1. A schematic system configuration of wireless EV battery charger

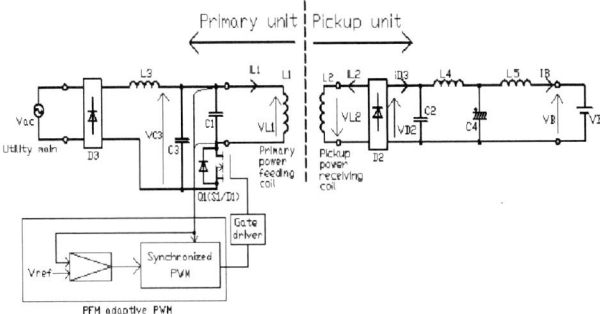

Fig.2. A proposed power conversion conditioning circuit with low-cost IPT-wireless battery charger using single-ended quasi-resonant high-frequency inverter link.

III.   EXPERIMENTAL RESULTS

A.   Wireless Coupling Coil Units

It is noted that the primary-side power feeding coil unit L_1 is loosely coupled to the pick-up coil L_2 of the secondary side power receiving coil unit in the high-frequency resonant link DC-DC converter for battery charging power supply.

In actual, these planar and circular coils in the primary and secondary-side are assembled by the power litz wires in order to reduce the power losses due to the eddy current skin effect.

Two contactless planar and circular coil units with ferrite core sheets; power sending coil and power receiving coil are depicted in Fig.3, including trially-produced planar coils.

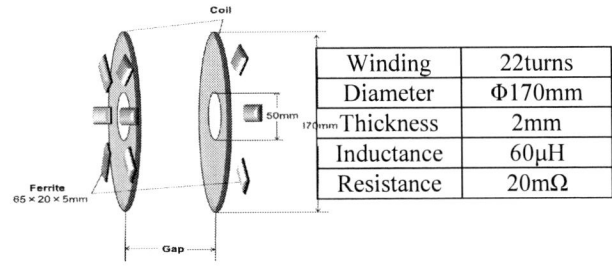

Winding	22turns
Diameter	Φ170mm
Thickness	2mm
Inductance	60µH
Resistance	20mΩ

Fig.3 A wireless electro-magnetic structure with two symmetrical planar and circular type coil units and design specifications.

As a matter of fact, some circuit parameters; mutual inductance M and magnetic coupling coefficient k of two planar spiral coils can depend upon the gap distance variable. Primary inductance L_1 and secondary inductance L_2 is not depending on gap.

B.   Circuit State Equation

In two planar spiral coils shown in Fig.3, circuit equations are given by

$$\left. \begin{array}{l} v_{L1} = r_1 i_1 + L_1 \dfrac{di_{L1}}{dt} + M \dfrac{di_{L2}}{dt} \\[2mm] v_{L2} = r_2 i_2 + M \dfrac{di_{L1}}{dt} + L_2 \dfrac{di_{L2}}{dt} \end{array} \right\} \quad (1)$$

Where,
L_1, L_2 ; self-inductance of the primary-side power feeding coil and secondary side power receiving coil.
r_1, r_2 ; internal resistances of each planar coil.
k ; electromagnetic coupling coefficient of two planar spiral coils.
M ; mutual inductance between L_1 and L_2, which is influenced upon gap distance variable.

The output power in the case of non-resonant type is approximately represented by the following Equation (2):

$$P_{nr} = \frac{|V_{L2}|^2}{R_0} = \frac{k^2}{R_0 + \omega^2 L^2 (1-k^2)^2 \frac{1}{R_0}} V_i^2 \quad (2)$$

And the output power in the case of resonant type is approximately represented by the following Equation (3):

$$P_r = \frac{k^2}{R_0 \{1 - \omega^2 LC (1-k^2)\}^2 + \omega^2 L^2 \frac{(1-k^2)^2}{R_0}} v_i^2 \quad (3)$$

P_r is higher than P_{nr} by the effect of resonant as shown in following Equation (4).

$$\frac{P_r}{P_{nr}} = \frac{R_0 + \omega^2 L^2 \frac{(1-k^2)^2}{R_0}}{\omega^2 L^2 \frac{(1-k^2)^2}{R_0}} = \left\{\frac{R_0}{\omega L (1-k^2)}\right\}^2 + 1 > 1 \quad (4)$$

C.   Design Specifications and System Parameters

Fig.4 shows measured waveforms in a proposed EV charging system, and the self-excited ZVS control scheme well functions.

In particular, in single-ended quasi-resonant inverter link DC-DC converter, note that the maximum voltage applied to the active switch Q_1 (S_1/D_1) becomes relatively high because of quasi-resonant operation.

The 2014 International Power Electronics Conference

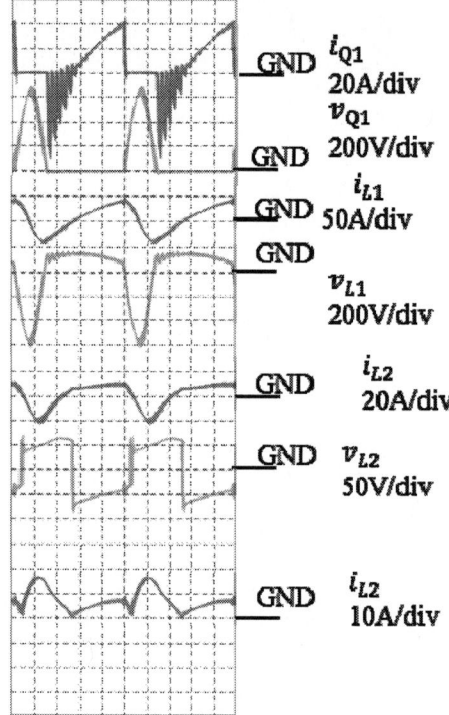

Fig.4 Measured voltage and current waveforms of quasi-resonant high-frequency inverter. (10μs/div)

D. Steady-State Operating Performances

The resonant peak in the primary-side power feeding or power sending coil is approximately kept constant by means of the PWM adaptive PFM control strategy on the basis of instantaneous sensing of resonant tank capacitor.

The variable frequency related pulse width Ton determined through the self-excited pulse processing based on the resonant tank voltage sensor is adjusted to provide the output DC voltage or battery voltage.

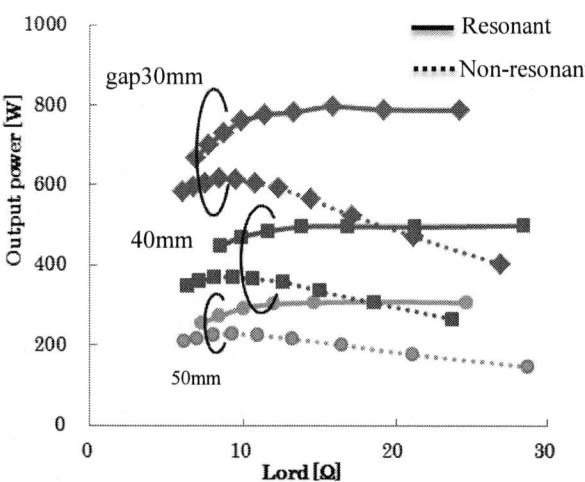

Fig.5 Observed output power vs. lord characteristics.

Fig.5 shows measured value of output power of non-resonant case and resonant case vs. load characteristics for gap distance from 30 to 50mm. The output power with resonant pick-up circuit is 1.5-2 times higher than that with non-resonant one.

Fig.6 shows IPT-efficiency of a proposed system. The efficiency of resonant-IPT is 10% higher than that of non-resonant one. The efficiency is calculated from experimental input power to power feeding unit and output power from power receiving unit. These characteristics of Fig.5 and 6 are explained by Eq. (1) and (2).

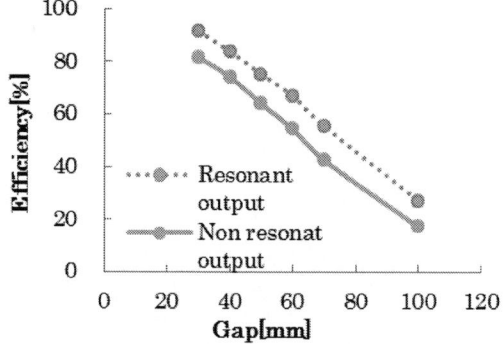

Fig.6 Measured efficiency for gap distances between two symmetrical planar coils.

A prototype of the EV equipped with pick-up coil unit of the resonant IPT-wireless charging system and power feeding coil unit is shown in Fig7. The wireless charging system can charge 1kWh in 90minutes as a feasibility study result as shown in Fig8.

Fig.7. A prototype of the EV equipped with developed charging system.

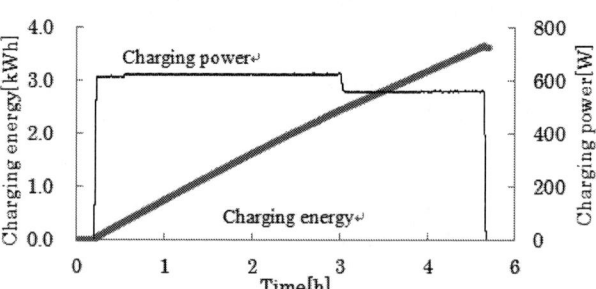

Fig.8. A feasibility study result of developed charging system.

978-1-4799-2706-7/14 $31.00 © 2014 IEEE

IV. A WIRELESS V2H SYSTEM WITH BIDIRECTIONAL RESONANT SINGLE-ENDED INVERTER

A new system configuration by a single-ended inverter with a resonant IPT circuit for bidirectional power transfer is shown in Fig.9. This system is applied to the smart house as V2H system. (see Fig.10)

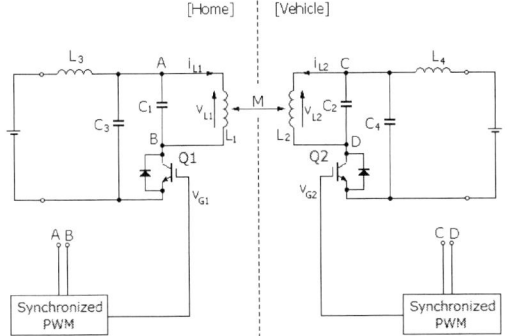

Fig.9. Bidirectional circuit using quasi-resonant high-frequency inverter

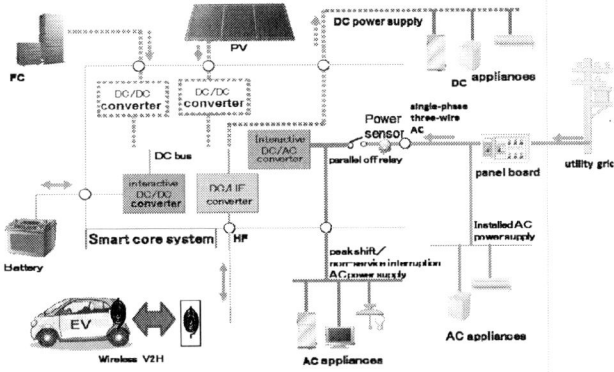

Fig.10. A smart house construction including wireless V2H

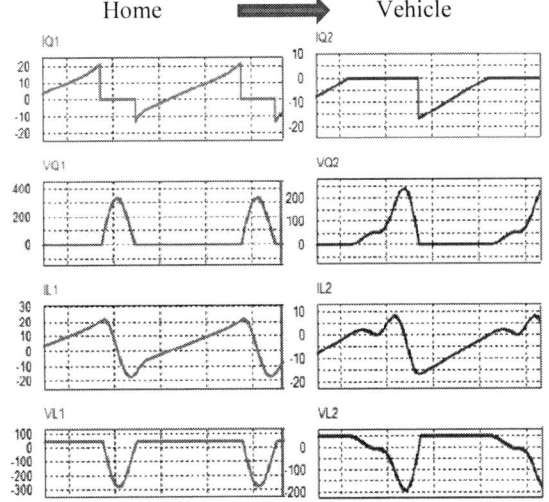

Fig.11. Simulation waveform of Bidirectional circuit using quasi-resonant high-frequency inverter (Home to Vehicle)

In the mode of using EV battery energy, a vehicle-side single-ended inverter produces high frequency power which is transferred to home side circuit by resonant IPT. Capacitance C_2 in vehicle-side circuit operates as resonant capacitance for ZVS transition and capacitance C_1 in home-side circuit operates as resonant capacitance for resonant IPT. Fig.11 shows operating waveforms in this vehicle to home mode. The active switch Q_2 is turned off and turned on with ZVS transition. And the active switch Q_1 is kept off, so a home side circuit operates as rectifier with pick-up resonant circuit. This system operates as forward converter.

On the other hand, in the EV charging mode, a home-side single-ended inverter produces high frequency power which is transferred to vehicle side circuit by resonant IPT. Capacitance C_1 in home-side circuit operates as resonant capacitance for ZVS transition and capacitance C_2 in vehicle-side circuit operates as resonant capacitance for resonant IPT.

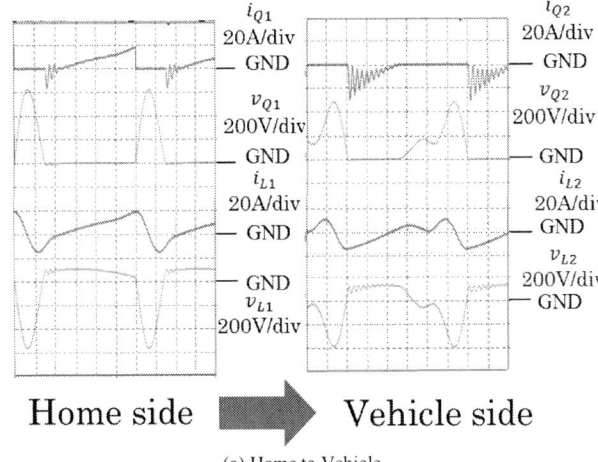

(a) Home to Vehicle

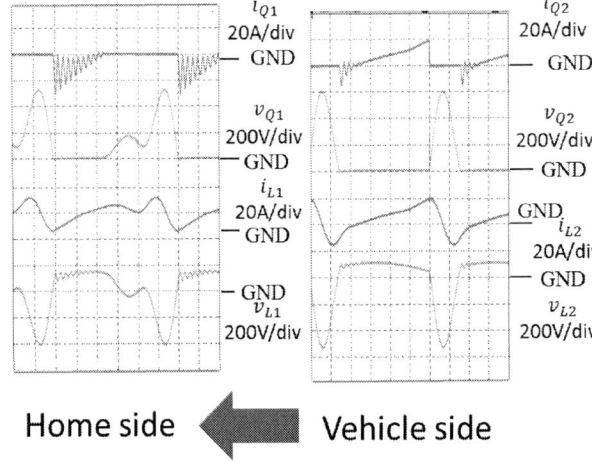

(b) Vehicle to Home

Fig.12. Measured waveforms of bidirectional circuit using quasi-resonant high-frequency inverter (10us/div)

Fig.12 shows measured operating waveforms of proposed wireless EV charging system for V2H system. These waveforms are same as above mentioned simulation waveforms. The ZVS control scheme works well. Measured waveforms of vehicle to home mode also show the reverse operation scheme well function.

Fig.13 shows output power characteristics of home to vehicle mode. The output power decreases depending on gap distance, and the output power can be controlled by conduction time T_{ON}.

Home to Vehicle and Vehicle to Home operation have same characteristics.

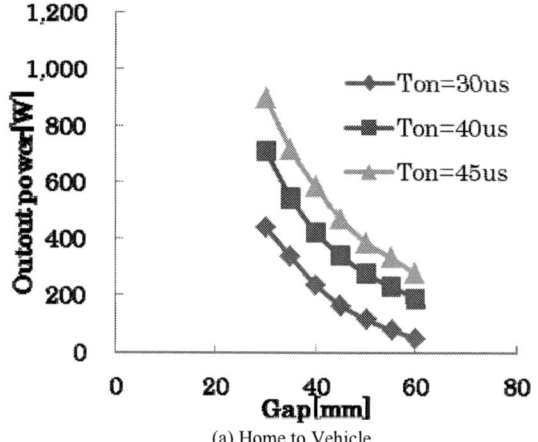

(a) Home to Vehicle

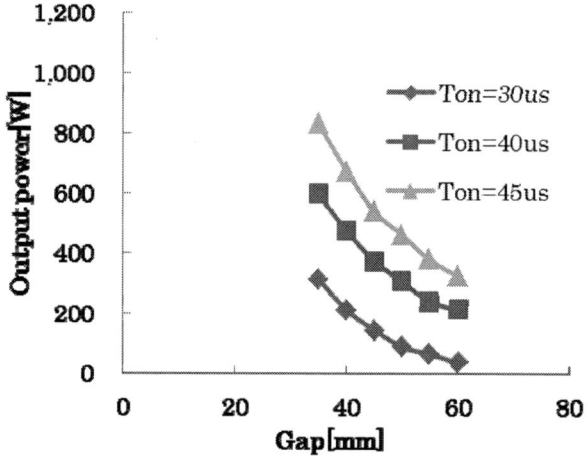

(b) Vehicle to Home

Fig.13. Observed output power vs. gap distances of Bidirectional circuit using quasi-resonant high-frequency inverter

V. CONCLUSION

Presented has been a wireless EV charging system based on IPT technology, with a single-ended quasi-resonant high-frequency inverter. The single-ended high-frequency inverter can operate efficiently under a self-excited ZVS control and a PFM adaptive PWM power regulation scheme.

The output power with resonant pick-up circuit was 1.5-2 times higher than that with non-resonant one. And the efficiency of resonant-IPT was 10% higher than that of non-resonant one.

Proposed has been a new system configuration by a single-ended inverter with a resonant IPT circuit for bidirectional power transfer. And a result of feasibility study for V2H by simulation was indicated.

Acknowledgment

This work was supported by JSPS Grants-in-Aid for Scientific Research Grant Number 25420272.

REFERENCES

[1] Duleepa J. Thrimawithana, Udaya K. Madawala, Yu Shi, "Design of a Bi-Directional Inverter for a Wireless V2G System", *Proc. IEEE International Conf. on Sustainable Energy Technologies, ICSET,* Kandy, Sri Lanka, December 2010

[2] Jin Huh, Wooyoung Lee, Gyu-Hyeong Cho Byunghun Lee, Chun-Taek Rim, "Characterization of Novel Inductive Power Transfer Systems for On-Line Electric Vehicles", *Proc. Annual IEEE Applied Power Electronics Conference and Exposition, APEC,* Texas, USA, March 2011

[3] Shingo Machino, Masahiro Kozako, Katsuhiko Harada, Masayuki Hikita, Kazuyuki Hotta, Yoshinori Kataoka," Construction and Characteristics of Wireless Resonance Type Inductive Power Supply", *Proc. Japan Industry Applications Society Conf.* Okinawa, Japan, August 2011, CD-ROM, Paper No 2-13

[4] Toshihiro Kai, Throngnumchai Kraisorn, Yuusuke Minagawa, "A Study on Receiver Circuit Topology of Non-contact Charger for Electric Vehicle", *Proc. Japan Industry Applications Society Conf.* Okinawa, Japan, August 2011, CD-ROM, Paper No.2-16

[5] Hideki Omori , Mutsuo Nakaoka, "Generic Circuit Topologies and Their Performance Evaluations of Single-Ended Resonant High-Frequency Inverters for Induction-Heated Cooking Appliances", *Trans. on IEE Japan,* pp. 150-159, Vol. 117-D, No. 2, 1997

[6] Omori H., Iga Y. Morizane T., Kimura N., Nakagawa K., Nakaoka M., " A Novel Wireless EV Charger using SiC Single-Ended Quasi-Resonant Inverter for Home Use ", *International Power Electronics and Motion Control Conference and Exposition,* Novisad, Serbia, September, 2012

[7] Y. Iga, H. Omori, T. Morizane, N. Kimura, Y. Nakamura, M. Nakaoka, " New IPT-Wireless EV Charger using Single-Ended Quasi-Resonant Converter with Power Factor Correction ", *International Conference on Renewable Energy Research and Applications,* Nagasaki, Japan, November, 2012

[8] T. Miyoshi, H. Omori, G. Maeda, "Reduction of Magnetic Flux Leakage from an Induction Heating Range", *IEEE Trans on IA, VOL. IA-19, NO.4,* July/August, 1983

[9] Hideki Omori, Yuichi Iga, Hiroki Hukuoka, Tosimitsu Morizane, Noriyuki Kimura, Kunio Nakagawa, Yoshimichi Nakamura, Mutuo Nakaoka, "A New Bidirectional Resonant IPT EV Charging System with Single-Ended Inverter for Wireless V2H", *International Conference on Electrical Drives and Power Electronics,* Dubrovnik, Croatia, October , 2013

Design and Implementation of an Interleaved BCM Boost PFC Control IC

Kuan-Hsien Chou, Tsorng-Juu Liang, *Senior Member, IEEE*, Kai-Hui Chen, and Ji-Shiang Lee

Advanced Optoelectronic Technology Center (AOTC)/ Green Energy Electronics Research Center (GREERC)

Department of Electrical Engineering, National Cheng Kung University, Tainan, Taiwan

Email: tjliang@mail.ncku.edu.tw

Abstract— **In this paper, a novel interleaved BCM boost PFC control IC is designed and implemented. This contoller not only retains the advantage of low swtiching loss of BCM control but also increases the power rating of converter. The voltage mode turn-on instant phase-shift of master-slave control is adopted in the proposed controller. Conventionally, this control scheme may occasionally operate in CCM due to pertubations. To solve this problem, the paper proposes a novel phase-shifter and utilizes the second zero current detector to catch the zero current switching signal cooperated with the phase-shift signal to make the slave converter operate under BCM.**

Index Terms—**interleaved, BCM control, power factor correction**

I. INTRODUCTION

Generally, interleaved control can be classified into two categories: open-loop master-slave control method [7][8] and PLL-based closed-loop method [9][10]. In open-loop master-slave control, the master converter is at stand-alone BCM operation and the slave converter's gate signal is partially control by the master converter to achieve interleaved control. It means the slave converter is synchronized by a time delay equal to half the switching period of the master converter previous switching cycle, which is usually generated from a phase-shifter. However, the slave converter is easy to be influenced by disturbance and enters CCM. In closed-loop control, the phase difference between two channels is measured and by adjusting the turn-off instant or the on-time of the slave converter based on the detected phase error can achieve the interleaved control, but it will slow down the response in encountering disturbance.

In this paper, a novel open-loop master-slave BCM control method for dual boost PFC is proposed. This controller owns a phase-shifter and the second zero current detector cooperating with a proposed function block to confirm the slave converter operating at interleaved BCM control in case of disturbance.

II. ANALYSIS OF TYPICAL INTERLEAVED CONVERTER

Fig. 1 shows the simplified interleaved boost PFC converter.

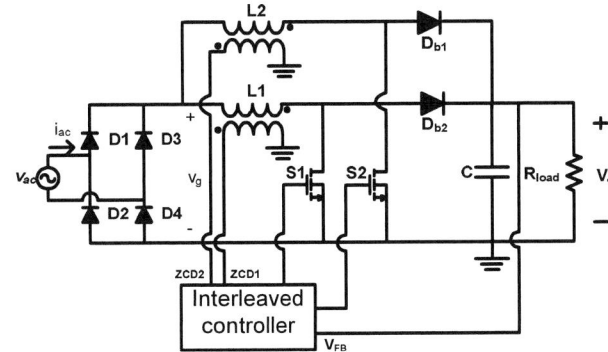

Fig. 1 Interleaved boost PFC converter

Mode I (t = t$_0$ ~ t$_1$): In this mode, S1 is turned on and S2 keeps on at t = t0. L1 and L2 store energy and Cout transfers energy to the output. When S2 is off, this mode ends.

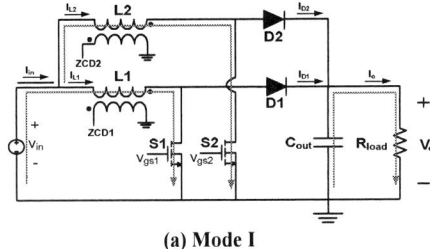

(a) Mode I

Mode II (t = t$_1$ ~ t$_2$): In this mode, S$_1$ keeps on and S$_2$ is off. L$_1$ stores energy but L$_2$ releases energy to the output. This mode ends when current in L$_2$ becomes zero.

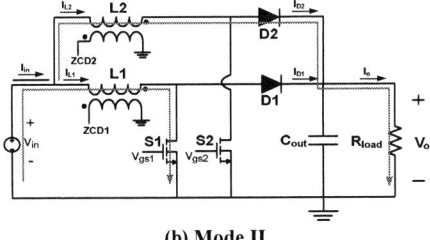

(b) Mode II

Mode III (t = t$_2$ ~ t$_3$): This mode acts as Mode I. The only difference is the energy in L$_2$ is larger than L$_1$ at Mode I, but this situation is reverse at Mode III. When S$_1$ is off, this mode ends.

978-1-4799-2706-7/14 $31.00 © 2014 IEEE

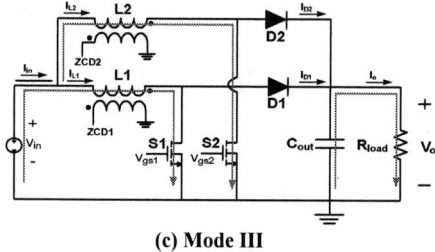

(c) Mode III

Mode IV (t = t₃ ~ t₄): In this mode, L_1 releases energy and L_2 stores energy. This mode ends when L_1 runs out of stored energy and next switching cycle begins.

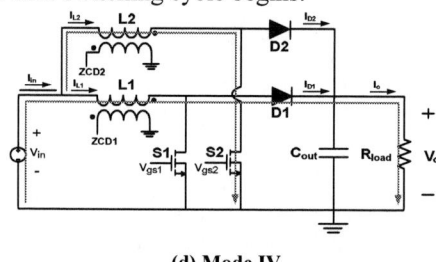

(d) Mode IV

Fig. 2 The operation of interleaved boost PFC converter

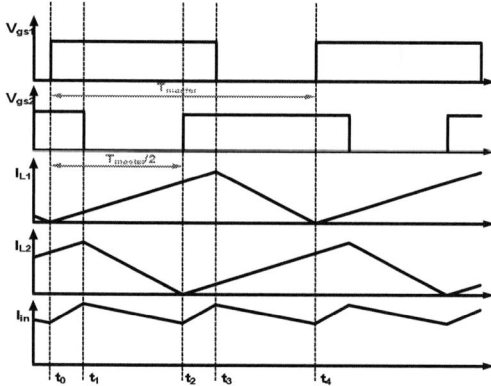

Fig. 3 Typical interleaved key waveforms

III. ANALYSIS OF OPEN-LOOP VOLTAGE-MODE INTERLEAVED CONTROLLER

Open-loop interleaved controller with voltage-mode [11][12] is a popular choice because its control block is easy to design and implement and there is no need to sense any current signal so power loss on series sensing resistors can be removed. Fig. 4 and 5 show the simplified voltage-mode open-loop interleaved control blocks with turn-on instant and turn-off instant phase-shift synchronization and relative waveforms.

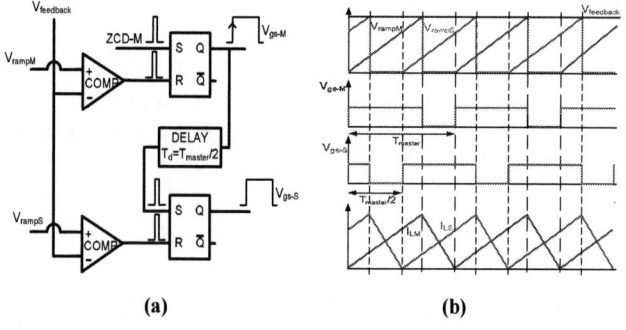

Fig. 4 (a) Simplified voltage-mode turn-on instant phase-shift controller (b) relative waveforms in normal operation and without inductor mismatch

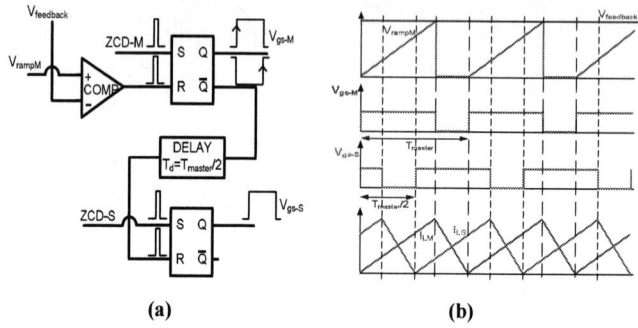

(a)   **(b)**

Fig. 5 (a) Simplified voltage-mode turn-off instant phase-shift controller (b) relative waveforms in normal operation and without inductor mismatch

Fig. 4 shows voltage-mode turn-on instant phase-shift controller. This controller owns only one ZCD to detect the master converter's zero current instant and trigger on V_{gsM}, but the turn-on signal of slave converter is decided by the phase-shift signal. V_{rampM} and V_{rampS} determine the on-time of V_{gsM} and V_{gsS} and these two are always the same in normal operation. Unfortunately, the slave converter may accidentally enter CCM in case of disturbance as shown in Fig. 6. And Fig. 5 is the voltage-mode turn-off instant phase-shift controller. There are two ZCD signals to ensure V_{gsM} and V_{gsS} turned on at BCM but the turn-off signal of slave converter is determined by phase-shift signal after V_{gsM} is off. Due to only one V_{ramp} to keep the on-time of master converter, the on-time of slave converter may be affected and different from the master's once there are some perturbations in slave converter. Fig. 7 shows this control method can retain stable when D > 0.5 but the slave inductor current will be unstable when D < 0.5.

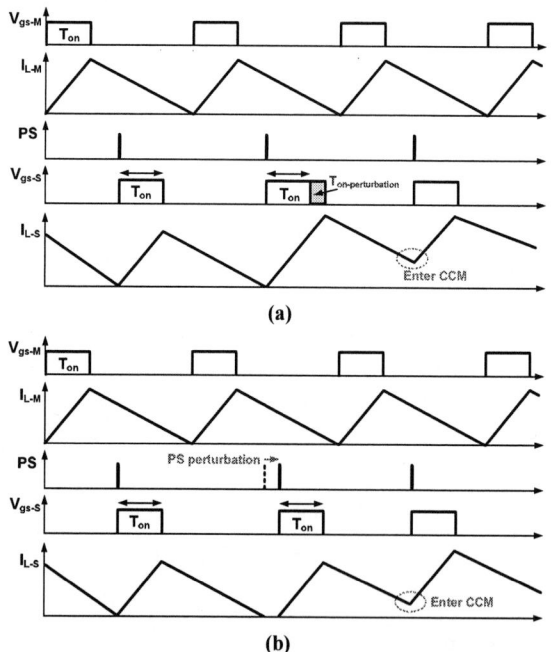

(a)

(b)

Fig. 6 Perturbed slave current waveforms for turn-on instant phase-shift synchronization (a) perturbation on slave turn-on time (b) perturbation on phase-shift signal

978-1-4799-2706-7/14 $31.00 © 2014 IEEE

The 2014 International Power Electronics Conference

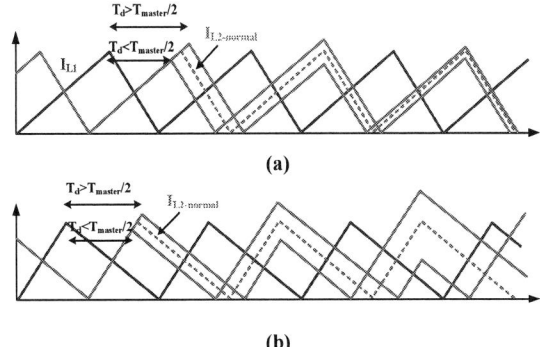

(a)

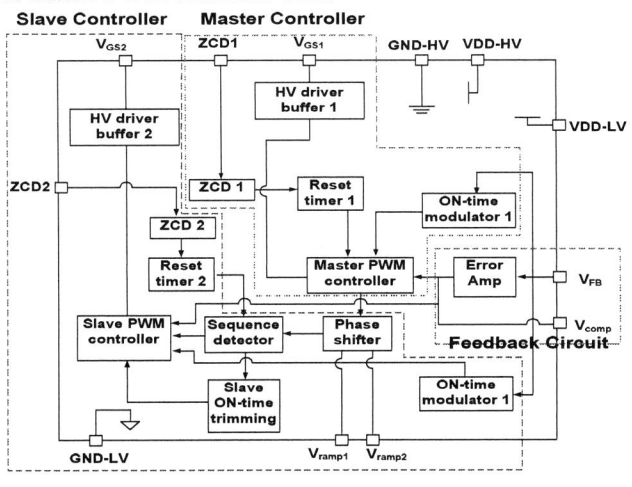

(b)

Fig. 7 Perturbed slave current waveforms for turn-off instant phase-shift synchronization (a) waveform at duty cycle > 0.5 (b) duty cycle < 0.5

IV. CIRCUIT DESIGN AND ANALYSIS OF THE PROPOSED INTERLEAVED CONTROLLER

This paper proposes a novel open-loop voltage-mode interleaved control method with turn-on instant phase-shift synchronization. Because turn-on instant phase-shift synchronization is adopted, the proposed controller rightly breaks the limit that the duty cycle must exceed 50%. Fig. 8 shows the overall control block. The proposed controller owns ZCD not only at the master converter but also at the slave converter. The second ZCD signal can cooperate with the phase-shift signal to assure the slave converter to operate at interleaved BCM as well and immune to any perturbations.

A. Overview of the Proposed Controller

First, the master ZCD signal (ZCDM) is detected and sent into the master PWM controller and the master gate signal (V_{GSM}) is triggered high. And after a voltage ramp generated inside the PWM controller tops the error signal V_{EA}, V_{GSM} is pulled low and awaits the next ZCD signal. On the other hand, V_{GSM} is also sent into the proposed phase-shifter. After half of the master switching cycle from the turn-on instant of V_{GSM}, the phase-shifter will generate a phase-shift signal (PS). Afterwards, PS and slave ZCD signal (ZCDS) are sent into the proposed sequence detector. After PS and ZCDS are both detected by sequence detector, the turn-on signal for slave converter is generated and outputted to the slave PWM controller, which will produce the slave gate signal (V_{GSS}) as the master PWM controller does.

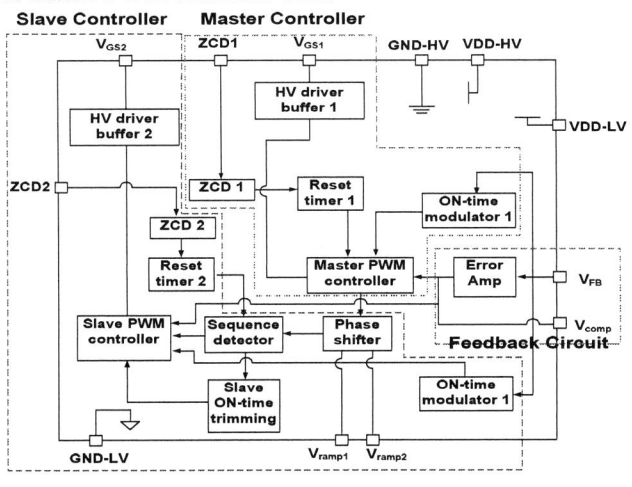

Fig. 8 Simplified block diagram of the proposed controller

B. Detailed Introductions to Utilized Blocks

1. Master/slave PWM controller

This control block is the main controller to generate the switch gate signal V_{GSM} and V_{GSS}. As shown in Fig. 9, when

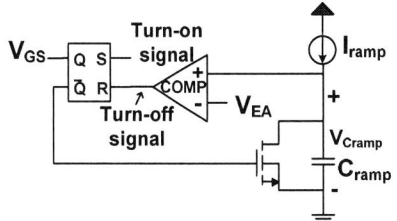

Fig. 9 PWM controller

the SR-latch receives turn-on signal (ZCD or PS signal), V_{GS} is triggered high and C_{ramp} starts charged. At the moment of V_{Cramp} equal to V_{EA}, V_{GS} is triggered low and C_{ramp} is discharged until the next turn-on signal arrives. The relative waveforms are shown as Fig. 10.

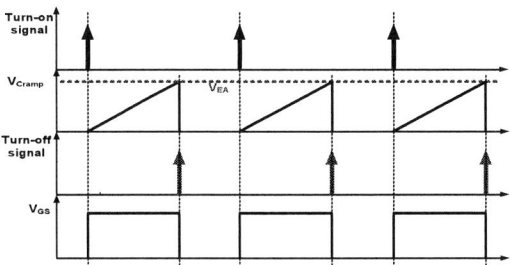

Fig. 10 Relative waveforms of PWM controller

2. Phase-shifter

Phase-shifter is the core of an open-loop interleaved controller as shown in Fig. 11. When the turn-on instant of V_{GSM} arrives, C_{PS1} and C_{PS2} start charged by I_{PS1} and I_{PS2} whose current ratio is 1:2. Additionally, the peak voltage on C_{PS1} is sampled to be the reference for voltage ramp on C_{PS2}. When voltage ramp on C_{PS2} tops the sampled reference voltage represents the delay after half of the master prior switching cycle so that the phase-shifter signal (PS) is derived. The relative waveforms are shown in Fig. 12.

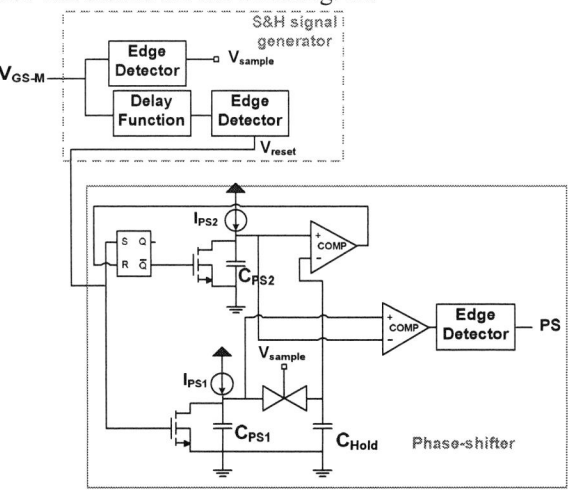

Fig. 11 Proposed phase-shifter

978-1-4799-2706-7/14 $31.00 © 2014 IEEE

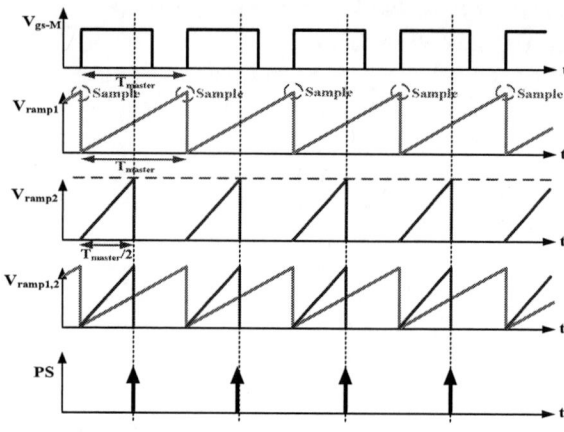

Fig. 12 Relative waveforms of phase-shifter

3. Sequence detector

For conventional turn-on instant phase-shift controller, only PS determines the slave turn-on, and hence it may cause the CCM problem. Thus, this controller proposes the sequence detector cooperating with the slave ZCD (ZCDS) to deal with this problem. As shown in Fig. 13, PS and ZCDS are both the input signal for this block. This block owns two main functions.

(a) Generating slave turn-on signal

The sequence detector will not generate the turn-on signal for slave converter until it catches both the PS and ZCDS signals so that the proposed controller can ensure slave BCM operation. That is to say, the later-coming signal dominates the turn-on signal.

(b) Trimming the slave on-time

The sequence detector also detects the arrival order between PS and ZCDS signals. The relative waveforms are shown as Fig. 14. If ZCDS comes at the same time with PS or earlier than PS, it means the slave converter is operated at perfect 180^0 phase shift because the turn-on instant of slave is dominated by PS. On the contrary, if ZCDS comes later and dominates the turn-on signal, the sequence detector will generate a signal (V_{trim}) to alert the slave PWN controller to slightly shrink the on-time until the domination of turn-on signal is returned to PS.

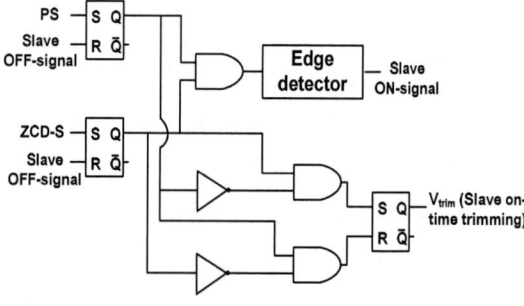

Fig. 13 Sequence detector

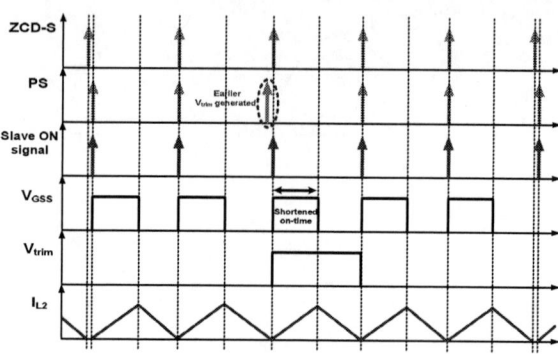

Fig. 14 Relative waveforms of sequence detector

V. SIMULATION AND EXPERIMENTAL RESULTS

To verify the feasibility of the proposed interleaved controller, this controller with a dual-boost PFC converter is simulated completely with the following specifications. And every block waveform is presented individually.

◆ Operating frequency: 25~50 kHz
◆ Input voltage: 90~264 V_{rms}
◆ Output voltage: 400 V_{dc}
◆ Output power: 320 W

3. PWM controller

The simulation and experimental waveform of PWM controller is shown in Fig. 15 and 16, respectively. The $V_{c,ramp}$'s duration represents the on-time of V_{GS}. And the on-time can be modulated with the slope or error signal (V_{EA}).

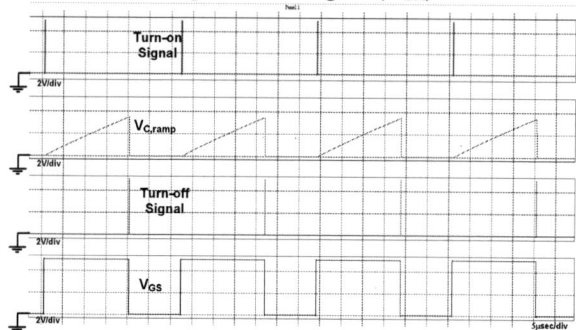

Fig. 15 Simulation waveforms of PWM controller

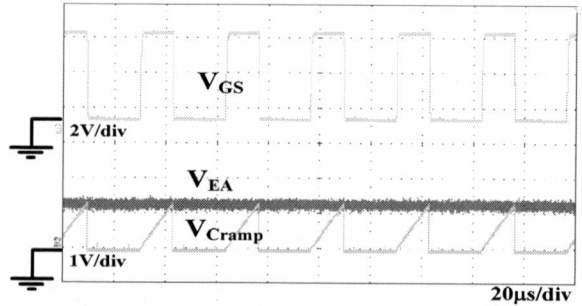

Fig. 16 Experimental waveforms of PWM controller

4. Phase-shifter

Fig. 17 reveals how to generate the phase-shift signal with the proposed phase-shifter and Fig. 18 are the control signals for this phase-shifter. V_{GSM} and V_{GSS} are shown in Fig. 19 to demonstrate the phase-shift accuracy. The experimental waveforms of proposed phase-shifter are shown in Fig. 20.

978-1-4799-2706-7/14 $31.00 © 2014 IEEE

The 2014 International Power Electronics Conference

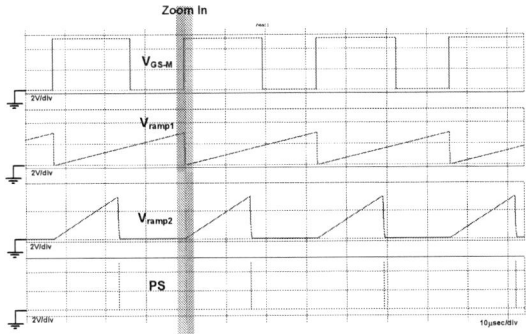

Fig. 17 Waveforms of proposed phase-shifter

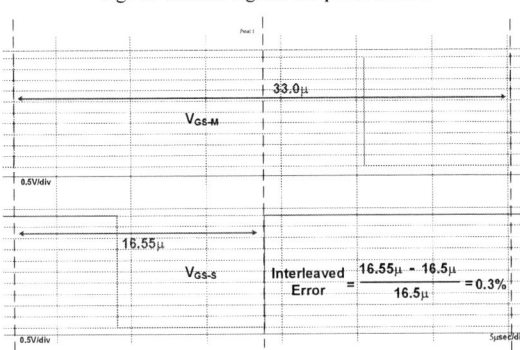

Fig. 18 Control signals for phase-shifter

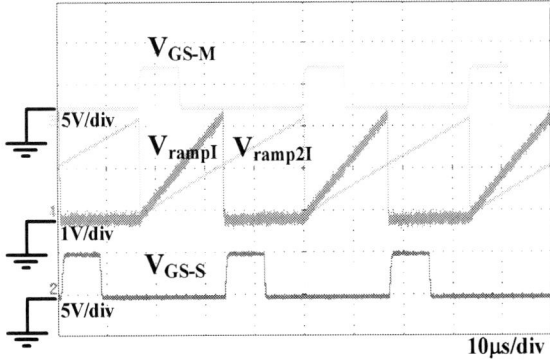

Fig. 19 The interleaved accuracy between V_{GSM} and V_{GSS}

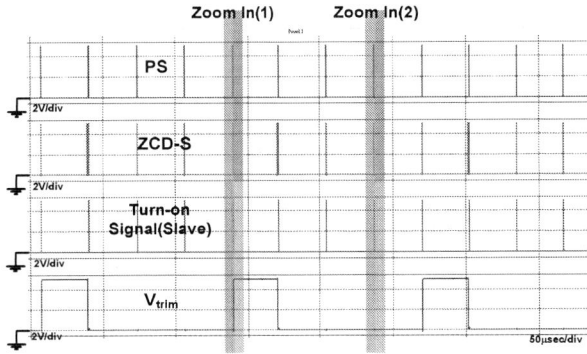

Fig. 21 Waveforms of sequence detector

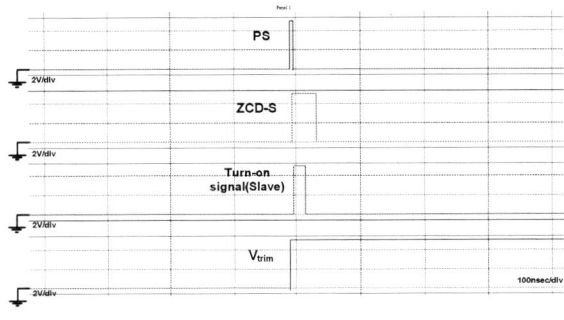

(a)

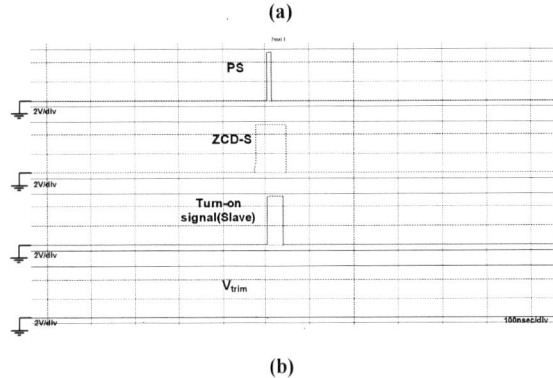

(b)

Fig. 22 Zoom-in waveforms when **(a)** ZCDS is later **(b)** PS is later.

The followings are the system waveforms of the interleaved boost PFC converter. Fig. 23 shows the input voltage (V_{in}=110 V_{rms}) and current (I_{in}, I_{Lm1}, I_{Lm2}) waveforms at rated power 320 W and Fig. 24 are the zoom-in current waveforms derived at the peak of I_{Lm1}, I_{Lm2}. The peak of I_{in} is merely 5.8 A, and the max input current ripple is 2.2 A in this condition.

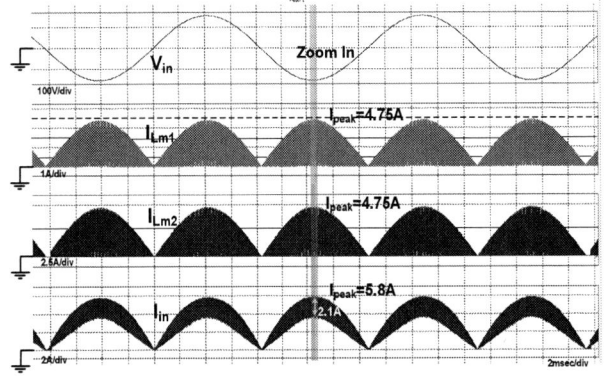

Fig. 23 Input voltage and current waveforms at V_{in}= 110 V_{rms}

Fig. 20 Experimental waveforms of proposed phase-shifter

5. Sequence detector

Fig. 21 shows the waveforms of sequence detector. Fig. 22(a), (b) reveal V_{trim} is high if ZCDS comes later but low if PS is later.

978-1-4799-2706-7/14 $31.00 © 2014 IEEE

The 2014 International Power Electronics Conference

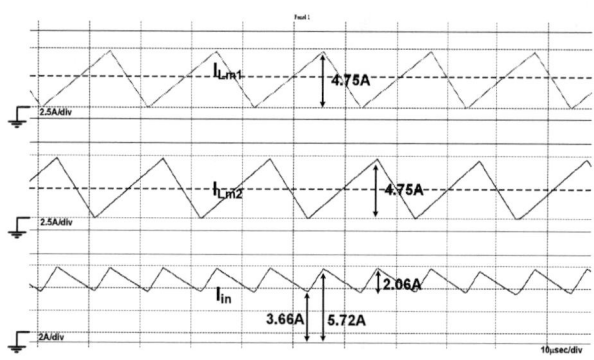

Fig. 24 Zoom-in input current and inductor current waveforms

Fig. 25 shows input current waveforms at 220 Vrms input voltage. It can be found that Iin is totally different from that in Fig. 23. It is because the duty cycle varies widely at 220 Vrms. At zone1, because Vin is low, the duty should be more than 0.5 to afford the high voltage gain. At zone2, Vin is around 200 V so that duty is about 0.5 and meanwhile the ripple cancelling performance is the best. At zone3, the duty is less than 0.5 and gradually decreases because Vin is getting higher, and it can be known that the ripple will become higher again. This situation does not happen at 110 Vrms because the duty is always more than 0.5.

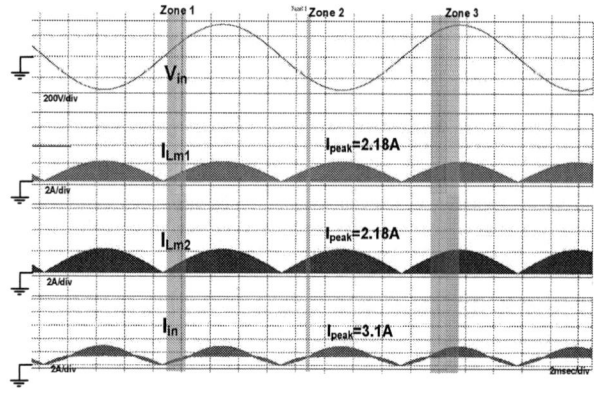

Fig. 25 Input voltage and current waveforms at V_{in}= 220 V_{rms}

VI. CONCLUSIONS

The dual-boost PFC converter with novel open-loop interleaved controller is implemented in this paper. It has the advantages of low switching loss, low current ripple, and increasing power rating. The proposed controller eliminates the turn-on instant phase-shift problem, and unlike the turn-off instant phase-shift controller, it can work normally in utility input voltage. In addition, this controller still retains BCM operation in case of inductor mismatch. The simulation results show the proposed controller can achieve accurate phase-shift and effectively cancel the input current ripple in wide-range input voltage.

ACKNOWLEDGEMENT

The authors gratefully acknowledge financial support from the Ministry of Science and Technology, Taiwan under project No. 102-3113-P-006-020-, and chip implementation support from National Chip Implementation Center, Taiwan.

REFERENCES

[1] M. J. Kocher and R. L. Steigerwald, "An AC-to-DC converter with high quality input waveforms," *IEEE Trans. on Industry Applications*, vol. IA-19, no. 4, July 1983, pp. 586-599.

[2] R. Redl, "Power electronics and electromagnetic compatibility," *Proc. of IEEE Power Electronics Specialists Conference, PESC'96*, pp. 15-21, June 1996.

[3] J. S. Lai and D. Chen, "Design considerations for power factor correction boost converter operating at the boundary of continuous conduction mode and discontinuous conduction mode," *IEEE Applied Power Electronics Conference*, pp. 267-273, Mar. 1993.

[4] "L6561 Power Factor Corrector Controller," STMicroelectronics, 2003.

[5] "FAN9611 / FAN9612 Interleaved Dual BCM PFC Controllers," Fairchild, Apr. 2011.

[6] "UCC28060 Dual Phase Transition-Mode PFC Controller," Texas Instruments, May 2007.

[7] Laszlo Huber, Brian T. Irving, Claudio Adragna, and Milan M. Jovanović, "Implementation of Open-Loop Control for Interleaved DCM/CCM Boundary Boost PFC Converters," *IEEE Applied Power Electronics Conference and Exposition*, pp. 1010-1016, Feb. 2008.

[8] Laszlo Huber, Brian T. Irving, and Milan M. Jovanović, *"Open-Loop Control Methods for Interleaved DCM/CCM Boundary Boost PFC Converters," IEEE Trans. on Power Electronics, vol. 23, no. 4, pp. 1649-1657, July 2008*

[9] Laszlo Huber, Brian T. Irving, and Milan M. Jovanović, "Closed-Loop Control Methods for Interleaved DCM/CCM Boundary Boost PFC Converters," *IEEE Applied Power Electronics Conference and Exposition*, pp. 991-997, Feb. 2009.

[10] X. J. Xu, W. Liu, Alex Q. Huang, "Two-Phase Interleaved Critical Mode PFC Boost Converter with Closed Loop Interleaving Strategy," *IEEE Trans. on Power Electronics, vol. 24, no. 12, pp. 3003-3013, Dec. 2009.*

[11] J. R. Tsai, T. F. Wu, C. Y. Wu, Y. M. Chen, M. C. Lee, "Interleaving Phase Shifters for Critical-Mode Boost PFC," *IEEE Trans. on Power Electronics, vol. 23, no. 3, pp. 1348-1357, May 2008.*

[12] T. F. Wu, J. R. Tsai, Y. M. Chen and Z. H. Tsai, "Integrated Circuits of a PFC Controller for Interleaved Critical-Mode Boost Converters," *IEEE Applied Power Electronics Conference*, pp. 1347-1350, Feb. 2007.

[13] Hangseok Choi, "Novel Adaptive Master-Slave Method for Interleaved Boundary Conduction Mode (BCM) PFC Converters," *IEEE Applied Power Electronics Conference and Exposition*, pp. 36-41, Feb. 2010.

Low Capacitive Inductors for Fast Switching Devices in Active Power Factor Correction Applications

Juan C. Hernandez, Lars P. Petersen, Michael A. E. Andersen

Dept. Electrical Engineering
Technical University of Denmark
Oersteds Plads, 349. Kongens Lyngby, Denmark

jchbo@elektro.dtu.dk   lpet@elektro.dtu.dk   ma@elektro.dtu.dk

*Abstract-*This paper examines different winding strategies for reduced capacitance inductors in active power factor correction circuits (PFC). The effect of the parasitic capacitance is analyzed from an electro magnetic compatibility (EMI) and efficiency point of views. The purpose of this work is to investigate different winding approaches and identify suitable solutions for high switching frequency/high speed transition PFC designs. A low parasitic capacitance PCB based inductor design is proposed to address the challenges imposed by high switching frequency PFC Boost converters

Keywords— Parasitic capacitance, PFC, boundary conduction mode (BCM), high frequency.

I. INTRODUCTION

Increased switching frequency operation in power electronics converters permit achieving high power densities due to the size reduction of the energy storage elements in the circuit. Resonant converters have been a common approach to overcome the reduced switching speeds in the active devices, making possible to mitigate or completely eliminate the switching losses present in hard switching topologies. On the other hand, a new era in power electronics is approaching and starting to be a reality with the introduction of wide bandgap devices based on silicon carbide (SiC) and gallium nitride (GaN) materials. The increased electrical strength and electrical conductivity in these materials allows for a reduction of the switch die size, consequently reducing the device parasitic capacitance, which directly increases the achievable switching speed of these devices. Previous works have proven the advantages of the utilization of wide bandgap devices, [1], [2]. However, increased hard switched converter operating switching frequencies requires more attention to be put into the printed circuit board (PCB) and magnetic components design in order to minimize the introduction of parasitic inductances and capacitances into the circuit.

The work presented in this paper focuses on the reduction of the parasitic capacitances of the PFC input inductor. Some research has already been performed on parasitic capacitance calculations based on analytical models [3], [4], [5] and finite element analysis [6]. The solution presented in [7] and [8] proposes a new winding strategy for reduction of inductors self-capacitance for SiC based power converters. A model for calculating the parasitic capacitance is addressed, and finally the effect of the reduction of this parasitic component is analyzed based on several measurements performed with different SiC devices. Some research has already been performed on the inductor self-capacitance effect on boost PFC's EMI performance. This work is part of a larger research project where the goal is to successfully introduce and take advantage of wide band gap devices for single phase PFC converters. It is the authors' opinion that this implicit means that the switching frequency must be increased in order to fully take advantage of these new devices. The justification of this paper is the work done on comparing different winding strategies in terms of EMI, switching and conduction loss. Furthermore, a novel winding strategy based on PCB manufacturing for reduced cost inductors with low ac resistance and self-capacitance is proposed. The proposed solution addresses the requirements for inductor designs in high frequency boost derived PFC circuits.

II. INDUCTOR PARALLEL CAPACITANCE

The inductor behavior at very high frequencies is clearly dominated by the parasitic capacitance effect and other non-ideal behavior. The component impedance can be approximated based on lumped parasitic models, which can be simplified to the model shown in Fig. 1.

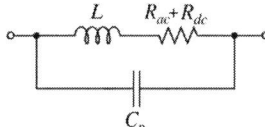

Fig. 1 Inductor equivalent simplified model

This model, presented in [9], includes the windings dc and ac resistances, which will effectively affect impedance curve quality factor at the component resonant frequencies. A parallel resistor, modeling the inductor core loss, is included in [10]. A more complex model based on impedance measurement fitting is presented in [11] and [12], in order to take into account very high frequency parasitic effects.

As presented in [7] and [8], if the capacitance has to be minimized, the layer to layer, the first turn to last turn and the turn to core capacitances represent the mayor contribution to the final capacitance of the inductor. This is due to the fact that even if the turn to turn capacitance is larger than the last ones, they will be interconnected in series minimizing its effect.

978-1-4799-2706-7/14 $31.00 © 2014 IEEE

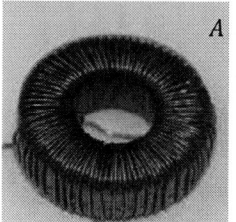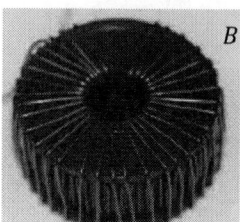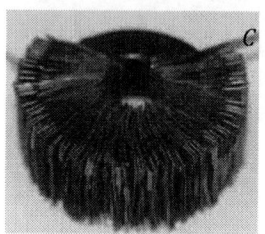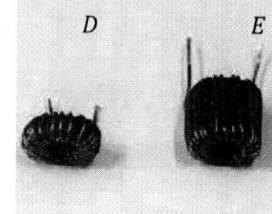

Fig. 2. Implemented inductor prototypes. A-Conventional double layer, B-Double layer with separator, C-Copper foil, Small size low capacitance inductors D-25 μH and E-69 μH.

III. PROTOTYPE IMPLEMENTATION

In order to compare the performance of different winding strategies, a toroidal Kool Mu core from Magnetics is selected. The same copper cross section and number of turns is used in all the implemented prototypes, to perform a fair comparison between the different winding structures. Toroidal cores are selected because they provide a large winding area compared to the core volume and represent a low cost solution in PFC inductor implementation. A first prototype is implemented using a conventional two layer structure that will present very large capacitance due to the layer-to- layer capacitance contribution. The selected core is 0077439A7 Kool Mµ 60 from Magnetics®. The winding is implemented using 94 turns of AWG 18 coated cable obtaining an inductance value of $1.2\ mH$. The well-known progressive winding or sectioned bobbin techniques [13] for reducing self-capacitance are not considered because of the difficulty of implementation in toroidal shaped cores. Instead, based on the work presented in [7] and [8], a two layer toroidal core with a layer to layer separator is implemented for layer-to-layer capacitance reduction. Moreover, a gap is introduced from first to last turn in each of the layers for reduction of the parasitic capacitance. Finally, in order to evaluate the feasibility of introducing PCB windings for this design, a copper foil implementation is selected where the copper cross section is adjusted to match the AWG18 cross section. This structure will present a relatively large turn-to-turn capacitance due to the increased area of the equivalent capacitance plates as shown in (1). Where ε_r is the relative permittivity, A is the plate area and d is the distance between plates.

$$C = \varepsilon_r \cdot \varepsilon_0 \cdot \frac{A}{d} \qquad (1)$$

However this structure will not present any of the critical layer-to-layer or first turn to last turn contribution due to the fact that the increased fill factor will allow implementing the same amount of turns in a single layer structure with a large gap between first and last turn. Moreover, this structure presents a reduced turn to core capacitance contribution respect to the conventional windings because of the reduced A/d ratio. To finalize the comparison, as suggested in [14], the effect of adding a small inductor with very low capacitance in series with a multilayer high capacitance design is analyzed by constructing two single layer toroids using one core and three stacked cores Magnetics® Kool Mµ 125 0077350A7 with an inductance value of 25.9 and

69.9 μH respectively. The implemented prototypes are presented in Fig. 2 and the impedance measurements results are shown in Fig. 3 and Fig. 4.

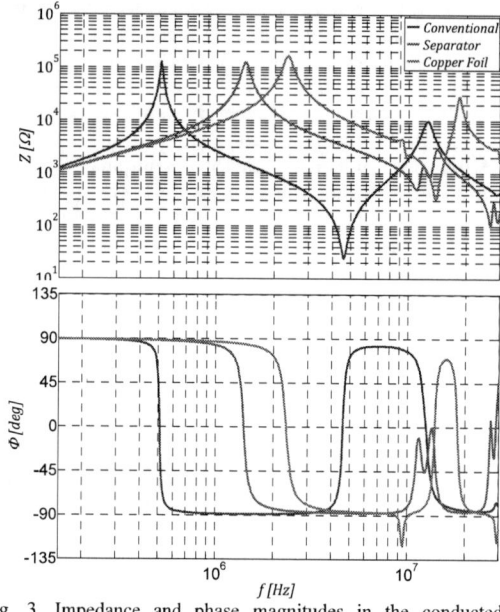

Fig. 3. Impedance and phase magnitudes in the conducted EMI frequency range ($150\ kHz - 30MHz$) for the conventional (blue), separator (green) and copper foil (red) inductors.

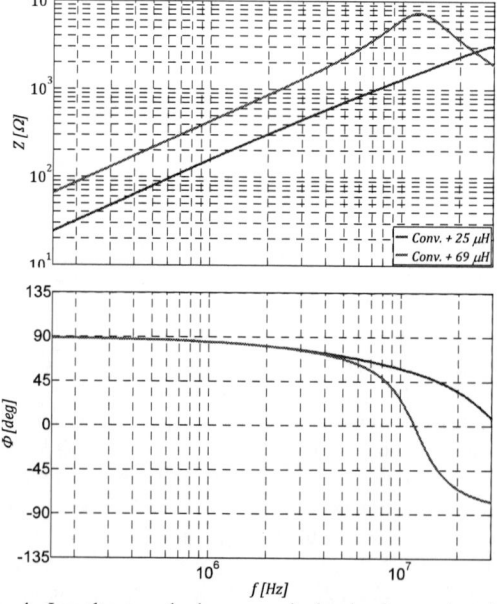

Fig. 4. Impedance and phase magnitudes in the conducted EMI frequency range ($150\ kHz - 30MHz$) for the small size inductors 25 μH (blue), 69 μH (green).

978-1-4799-2706-7/14 $31.00 © 2014 IEEE

The parallel capacitance is calculated from the measured parallel resonant frequency and inductance value. The obtained values are shown in Table I.

TABLE I
PROTOTYPES' INDUCTANCE AND CAPACITANCE VALUES

Prototype	L [mH]	f_0 [MHz]	C_p [pF]
Conventional	1.21	0.51	80.5
Separator	1.16	1.41	11
Copper Foil	1.15	2.34	4
25 μH	0.0259	32.5	1
69 μH	0.0698	11.8	2.7

As it can be observed form Table I, the two layer structure with separator has seven times lower capacitance than the conventional structure due to the reduced layer to layer capacitance and the inserted gap between the first and last turn in each of the layers. The copper foil implementation reaches a capacitance level twenty times lower than the conventional structure. Even with a much larger turn to turn capacitance this structure achieves the smallest parasitic capacitance value. Finally, the low size implemented inductors present a quasi-ideal behavior up to 30 MHz with a parasitic capacitance of 1 and 2.7 pF respectively.

IV. CONDUCTED EMI MEASUREMENTS

The EMI performance of the different implemented prototypes is analyzed using an ac-dc converter evaluation board from Texas Instruments PMP669 (Fig. 5) where the dc-dc conversion power stage has been disabled and the input EMI filter completely removed.

Fig. 5. AC-DC converter with conventional PFC. Evaluation board from Texas Instruments PMP669 MB

The EMI measurement is performed using a Two-line V-Network (LISN) with the converter operating @ $f_{sw} = 98\ kHz$ $V_{ac} = 230\ V_{rms}$ and $P_o = 200\ W$. Fig. 6 shows the measurement result for the three main different prototypes. Fig. 7 shows the effect of adding the small inductor in series with the conventional two layer inductor.

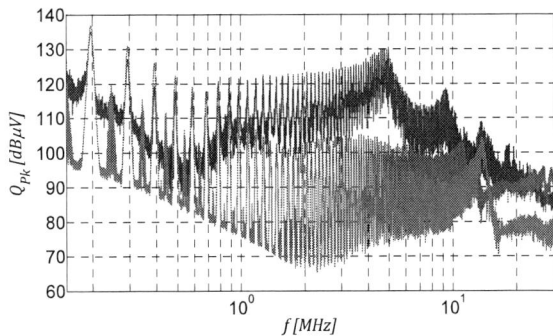

Fig. 6. EMI measurement using LISN network from 150 kHz to 30 MHz @ 230 V_{rms} and 200 W for the conventional (blue), separator (green) and copper foil (red) implemented inductors.

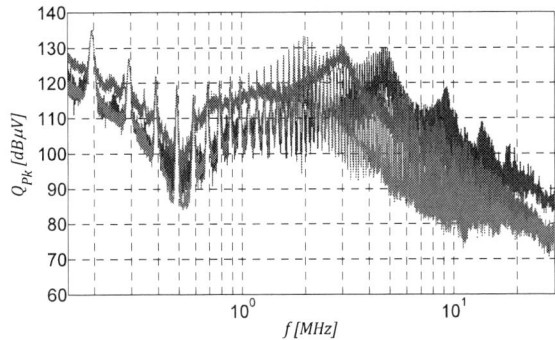

Fig. 7. EMI measurement using LISN network from 150 kHz to 30 MHz @ 230 V_{rms} and 200 W .Conventional (blue), conventional+ 25 μH (green), and conventional+ 68 μH (red).

The capacitance reduction obtained in the two layers with separator and the copper foil implementations provide a significant reduction in conducted EMI ss it can be observed in Fig. 6. The conventional structure shows high amplitude harmonics around 5.5 MHz which corresponds to the location of the minimum impedance measured for this prototype (Fig. 3). Fig. 7 shows the small effect of adding a small series inductance in series with the conventional two layer inductor. In fact, as it can be observed, the high frequency noise will be reduced due to the increased impedance in this area. On the other hand, at low frequencies, the introduction of this inductance will reduce the frequency of the minimum inductor impedance, increasing the propagated noise due to the higher amplitude of the switching frequency harmonics and the increased quality factor at this resonant frequency due to the reduced ac resistance.

V. EVALUATION OF THE IMPACT ON THE SWITCHING AND CONDUCTION LOSS

After comparing the different configurations in terms of conducted EMI, an efficiency related comparison is performed using a low inductive double pulse tester (DPT) shown in Fig. 8. A small die size 600V super-junction device FCD9N60N from Fairchild Semiconductor is used in combination with a 600V SiC diode IDD10SSG60C from Infineon Technologies.

The 2014 International Power Electronics Conference

Fig. 8. Implemented double pulse tester prototype.

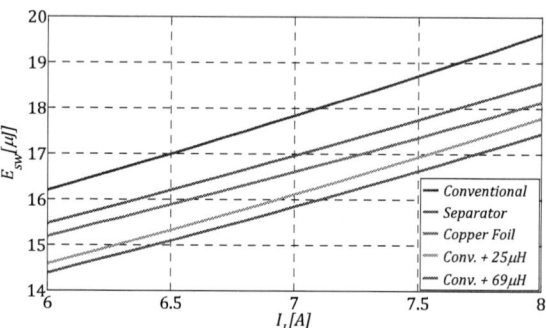

Fig. 11. MOSFET switching energy as a function of the inductor current level for the different inductor prototypes.

Fig. 9 shows the DPT switching waveforms for the conventional inductor (green current) and the copper foil prototypes for an inductor current level of 6A and a bus voltage of 400V. Fig. 10 shows the switching waveforms for the conventional two layer inductor (green current) in series with the $25\,\mu H$ (purple) and the $69\,\mu H$ (blue) prototypes.

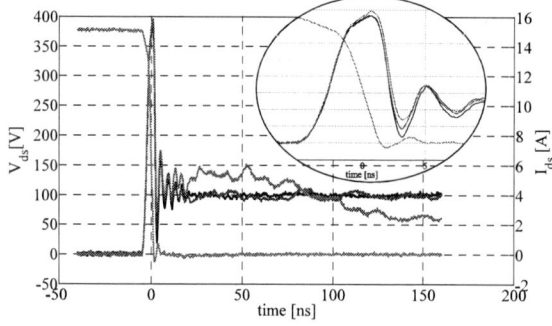

Fig. 9. Switching waveform comparison between the conventional (green current waveform) and the copper foil (blue current waveform) inductor prototypes.

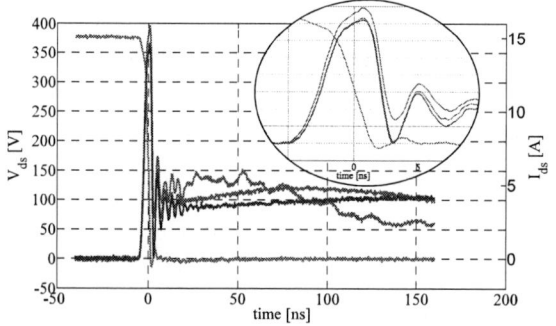

Fig. 10. Switching waveform comparison between the conventional (green current waveform) and conventional with series $25\,\mu H$ (purple) and $69\,\mu H$ (blue) inductor prototypes.

After performing a switching energy loss extraction, the turn on energy dissipated in the MOSFET is plotted as a function of the inductor current level (Fig. 11). As it can be observed, a small difference is obtained in the MOSFET turn on loss due to the inductor parasitic capacitance effect. Acording to the difference in calculated capacitance value and according to (2), the difference in energy loss from the standard double layer and the copper foil inductors should be at least $6\,\mu J$

$$E = \frac{1}{2} \cdot C \cdot V^2 \qquad (2)$$

However, this difference from the measurement to the calculation can be easily explained by looking at Fig. 9 and Fig. 10. As it can be observed the dissipated energy in the MOSFET before the drain to source voltage collapses to zero varies very little between the different measurements. This is due to the fact that the parasitic capacitance will not be charged on this small subinterval. Instead this capacitance will resonate with the parasitic inductance formed by the inductor interconnection and will finalize the charge long time after the switch has completed the switching transition. It can be concluded that the charge of the parasitic capacitance will not create a large increment in the MOSFET switching loss but it will increase the conduction losses in the inductor because of the presence of a high frequency resonant current that will be damped by the component ac resistance. Furthermore, there is also a risk that these resonances can couple through parasitic capacitances to the converter structure and generate common mode noise source further challenging the input EMI filter. Finally, as it can be observed in Fig. 10 the inclusion of the small size inductors in series with the standard double layer inductor will effectively reduce the frequency of this resonance minimizing the joule losses in the circuit because of the reduced ac resistance effect at lower frequencies.

978-1-4799-2706-7/14 $31.00 © 2014 IEEE

The 2014 International Power Electronics Conference

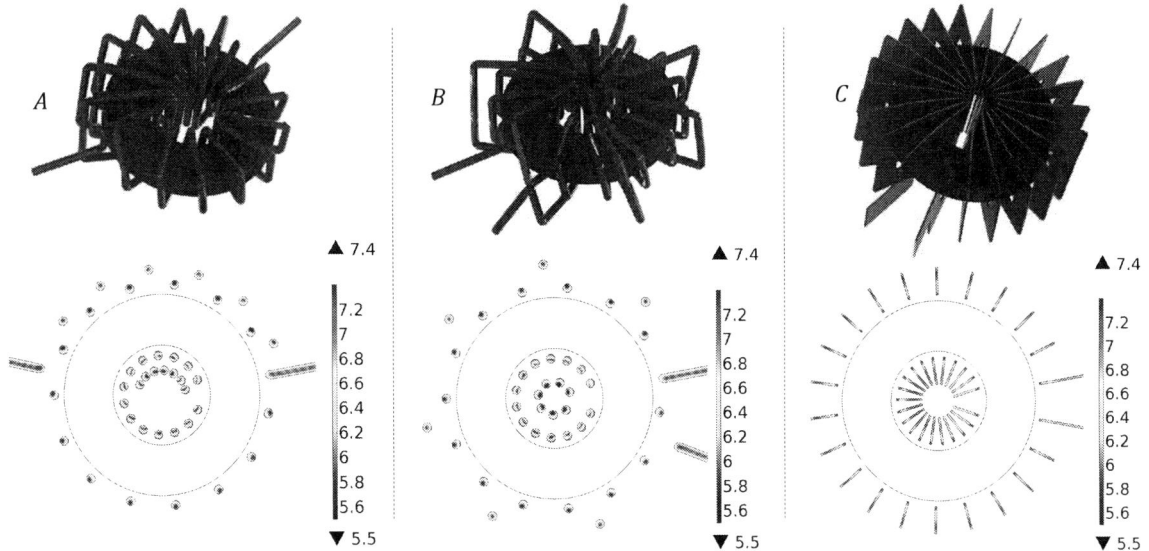

Fig. 12. Simulated inductor structures and 2d plots of the windings current density. A-Conventional double layer, B-Double layer with separator, C-Copper foil.

In order to complete the analysis of the different designs is important to evaluate the ac resistance of the different structures. This is a very important parameter in PFC applications with boundary conduction mode (BCM) operation where the inductor current presents a large high frequency component. Ac resistance measurement of inductors is a difficult task because when the measurement is performed, the magnetic material losses are included in the measurement and they are difficult to separate from each other. Another possibility is to perform an analytical calculation based on Dowell equations (3) where h is the copper thickness amd δ is the skin depth.

$$\frac{R_{ac,m}}{R_{dc,m}} = \frac{\xi}{2}\left[\frac{sinh\ \xi + sin\ \xi}{cosh\ \xi - cos\ \xi} + (2m-1)^2\frac{sinh\ \xi - sin\ \xi}{cosh\ \xi + cos\ \xi}\right]$$

$$\xi = \frac{h}{\delta} \qquad\qquad \delta = \frac{7.5}{\sqrt{f}} \quad [cm] \qquad (3)$$

However even considering that this work takes into account skin and proximity effects, it will not provide a correct solution in this specific problem where a 3d structure is evaluated with a variable distance between turns and layers. Instead, a finite element analysis (FEA) is performed by constructing a 3d model of the different analyzed winding strategies. The size of the simulated inductors is reduced to minimize the complexity of the solution. The same copper cross section $A = 0.56\ mm^2$ is used in the simulated structures. The three implemented models are shown in Fig. 12. A simulation is performed where the ac resistance is calculated for different frequencies up to $400\ kHz$. The windings are meshed to take into account the skin effect. Fig. 12 shows a current density plot of the three different structures at a frequency of $400\ kHz$ represented with a logarithmic colorbar. The skin effect can be easily appreciated in the inductor terminals in the conventional and the separator structures. All the structures present a higher current density near the core due to the high concentration of flux lines in the core proximity. The proximity effect of the second layer can be seen in the inner part of the toroid in the conventional structure where the first layer presents a high current density area in the close proximity to the second layer windings.

The obtained ac resistance for the different structures is shown in Fig. 13. All the structures present a similar ac resistance value at $100\ Hz$ of around $18\ m\Omega$. The conventional structure ac resistance value increases up to $1.54\ \Omega$ compared with $1.07\ \Omega$ and $0.63\ \Omega$ for the separator and copper foil structures respectively.

Finally, the effect of the parasitic capacitance in the converter efficiency is analyzed by testing the copper foil and the conventional implemented inductor prototypes in a PFC stage operating in BCM. The power stage is a modular PFC design operated with a superjunction 600V MOSFET 65R230C7 and a SiC 600V diode IDD04SG60C. The MOSFET is driven by a high current 9A driver FAN3122 and controlled by a BCM controller FAN7930B. Fig. 14 shows the implemented PFC power stage and Fig. 15 shows the measured converter efficiency as a function of the converter output power for a constant dc input voltage $V_{in} = 200V$ and an output voltage $V_{out} = 375V$.

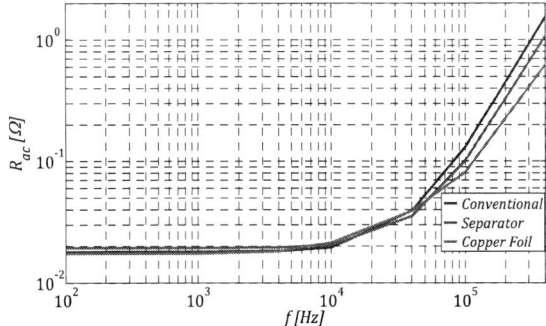

Fig. 13 Simulated ac resistances of the different analyzed structures

978-1-4799-2706-7/14 $31.00 © 2014 IEEE

Fig. 14 Implemented modular PFC power stage

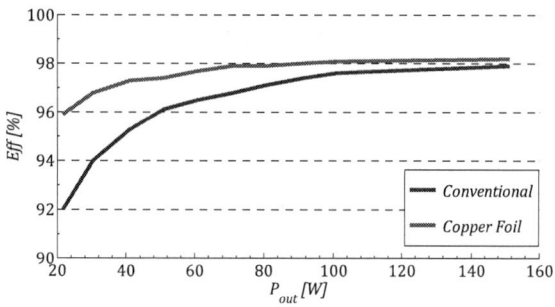

Fig. 15 Measured efficiency as a function of the converter output power for $V_{in} = 200\,V$ and $V_{out} = 375\,V$.

The effect of the capacitance can be appreciated at very low power levels. Under this situation, the converter switching frequency is increased up to $215\,kHz$. The difference in power loss between the two solutions is $0.96W$. With the operating voltage levels, the inductor parasitic capacitor changes its voltage from $200V$ to $-175V$. Taking into account the measured capacitance for the two prototypes, this change in voltage corresponds to a dissipated energy difference of $2.7148\,\mu J$ which corresponds to a power loss difference of $0.58\,W$. Therefore the remaining power loss difference is attributed to ac resistance difference between the two structures. As it can be seen, as the converter output power increases the two efficiency measurements get closer because the converter switching frequency is reduced down to $48\,kHz$ at $150\,W$ output power, minimizing the ac resistance difference between the prototypes.

VI. CONCLUSIONS

This paper analyzes different inductor winding structures focusing on parasitic capacitance reduction of the component. The parasitic capacitance effects are analyzed from conducted EMI and efficiency point of views. Different solutions for reduced capacitance effects are evaluated. The ac resistance of the different structures is evaluated together with the capacitance because it has a large impact on PFC converters efficiency operating in BCM mode. A copper foil winding structure is proposed with very low parasitic capacitance and ac resistance

compared to the conventional structures. This foil winding structure is similar to using PCB windings in a U core or E core structure. Using a single layer configuration can be very effective in reducing the parasitic capacitance mitigating EMI conducted and radiated problems and improving the converter efficiency. Moreover high frequency PFC converters operating in BCM will benefit from a reduced winding ac resistance

REFERENCES

[1] A.M. Abou-Alfotouh, A.V. Radun, Hsueh-Rong Chang, and C. Winterhalter, "A 1-MHz hard-switched silicon carbide DC-DC converter," *IEEE Transactions on Power Electronics*, vol. 21, no. 4, pp. 880-889, 2006.

[2] K.S. Boutros, S. Chandrasekaran, W.B. Luo, and V. Mehrotra, "GaN Switching Devices for High-Frequency, KW Power Conversion," in *IEEE International Symposium on Power Semiconductor Devices and IC's (ISPSD)*, Naples, 2006.

[3] A. Massarini and M.K. Kazimierczuk, "Self-capacitance of inductors," *IEEE Transactions on Power Electronics*, vol. 12, no. 4, pp. 671-676, 1997.

[4] Wenhua Tan, X. Margueron, and N. Idir, "Analytical modeling of parasitic capacitances for a planar common mode inductor in EMI filters," in *International Power Electronics and Motion Control Conference (EPE/PEMC)*, Novi Sad, 2012.

[5] C.K. Lee, Y.P. Su, and S.Y.R. Hui, "Printed Spiral Winding Inductor With Wide Frequency Bandwidth," *IEEE Transactions on Power Electronics*, vol. 26, no. 10, pp. 2936-2945, 2011.

[6] Wang Shishan, Liu Zeyuan, and Xing Yan, "Extraction of parasitic capacitance for toroidal ferrite core inductor," in *IEEE Conference Industrial Electronics and Applications (ICIEA)*, Taichung, 2010.

[7] Mariusz Zdanowski, J. Rabkowski, K. Kostov, and H. Peter-Nee, "The role of the parasitic capacitance of the inductor in boost converters with normally-on SiC JFETs," in *International Power Electronics and Motion Control Conference (IPEMC)*, Harbin, 2012.

[8] M. Zdanowski, K. Kostov, J. Rabkowski, R. Barlik, and H. Nee, "Design and Evaluation of Reduced Self-Capacitance Inductor in DC/DC Converters with Fast-Switching SiC Transistors," *IEEE Transactions on Power Electronics*, vol. PP, no. 99, p. 1, 2013.

[9] M. Bartoli, A. Reatti, and M.K. Kazimierczuk, "Modelling iron-powder inductors at high frequencies," in *Conference Record of the 1994 IEEE Industry Applications Society Annual Meeting*, Denver,CO, 1994.

[10] Shuo Wang, Fred.C. Lee, and J.D. van Wyk, "Inductor winding capacitance cancellation using mutual capacitance concept for noise reduction application," *IEEE Transactions on Electromagnetic Compatibility*, vol. 48, no. 2, pp. 311-318, 2006.

[11] Liyu Yang, "Modeling and Characterization of a PFC Converter in the Medium and High Frequency Ranges for Predicting the Conducted EMI," Virginia, MSc 2003.

[12] J.R.R. Zientarski, R. Piveta, M. Iensen, J.R. Pinheiro, and H.L. Hey, "A design methodology for optimizing the volume in single-layer inductors applied to PFC boost converters," in *Power Electronics Conference, 2009. COBEP '09.Brazilian*, Bonito-Mato Grosso do Sul, 2009.

[13] Marty Brown, *Power Supply Cookbook 2nd ed. pg. 55*. Woburn: Newnes, 2001.

[14] M. Roeber M. Seitz. (2005, August) Squeeze more performance out of toroidal inductors. Power Electronics. [Online]. www.powerelectronics.com

Temperature-Robust LC^3 LED Driver with Low THD, High Efficiency, and Long Life

Eun S. Lee[1], Bo H. Choi[1], Jun P. Cheon[1], Bong C. Kim[2], and Chun T. Rim[1]

[1]Dept. of Nuclear and Quantum Eng., KAIST, Daejeon 305-701, Korea
[2]Optomind Inc., Suwon, Korea

Abstract— A new type of passive LED driver that can reduce total harmonic distortion (THD) significantly by LC parallel resonance is proposed in this paper. Using an inductor and three capacitors, called LC^3, novel characteristics such as high efficiency and power factor (PF) with extremely long life time are achieved. The proposed LED driver has a temperature-robust characteristic because its power is hardly changed by temperature, selecting the number of LED in series n_s appropriately so that the LED power variation due to temperature change in LED can be zero. For analyzing LED power of the proposed LED driver, the phasor transformation technique is applied, which is firstly applied to a non-linear diode rectifier modeling. Nevertheless this non-linear switching, the proposed analyses agreed well with simulation and experiment results. A prototype LED driver showed very high power efficiency of 96.7 % at 60 W, meeting high PF of 0.95 and low THD of 10.1 %, though a reasonably small filter was used.

Keywords— *Passive LED driver, temperature-robust, LC parallel resonance, long lifetime, high efficiency, phasor transformation.*

I. INTRODUCTION

Since LEDs have higher efficacy than fluorescent lamps with much longer lifetimes [1]-[3] now, conventional lamps are being replaced with LEDs. The change of lamps to the LEDs, however, is slowed down because of the high cost of LEDs and their drivers as well as the short lifetime of the drivers [4]. It is quite important for LED drivers to have a longer lifetime than LEDs so that the overall LED lamps can operate for more than 50,000 hours. LED drivers, in general, should be able to regulate or control LED power level, guarantee high input power factor (PF) and low total harmonic distortion (THD). For example, the PF should be greater than 0.9 for LED power ≥ 5 W according to Energy Star Requirement for Solid [5], and the THD should be satisfied by the IEC 61000-3-2 class C standard [6]. Conventionally, switch-mode-power-supply (SMPS)-type LED drivers, which typically include active switches, control ICs, and passive circuit elements, have been widely used to meet the requirements [7]-[9]. They usually provide LEDs with constant currents by PWM switching techniques with a feedback control. The SMPS-type LED drivers, however, typically have low power efficiencies, which result in the large power loss and high junction temperature of the switching devices.

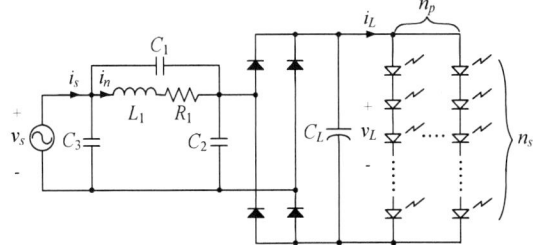

Fig. 1. The proposed LC^3 LED driver, which is quite insensitive to LED temperature, meeting PF and THD requirements.

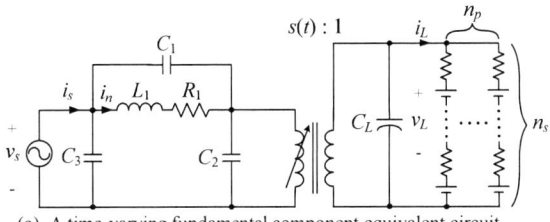

(a) A time-varying fundamental component equivalent circuit.

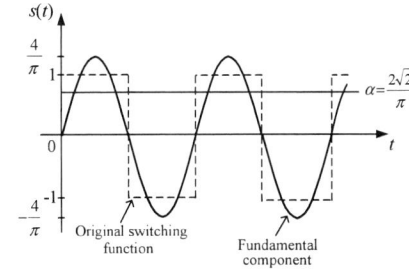

(b) Switching waveforms of a time-varying transformer.

Fig. 2. Large signal AC equivalent circuit of Fig. 1 replacing each LED with a linear approximated model.

Together with DC electrolytic capacitors, this high junction temperature shortens the lifetime of the conventional LED drivers.

In order to overcome these drawbacks, passive-type LED drivers that consist of just passive components such as inductors, capacitors, and diodes have been recently proposed [10]-[15]. These passive LED drivers mentioned so far usually have long lifetimes and high power efficiency because passive circuit elements are quite robust and free from high frequency switching loss. In some cases, however, PF and THD may not be satisfied by the use of passive elements only [10]-[11]. In addition, the LED power may change due to high

978-1-4799-2706-7/14 $31.00 © 2014 IEEE

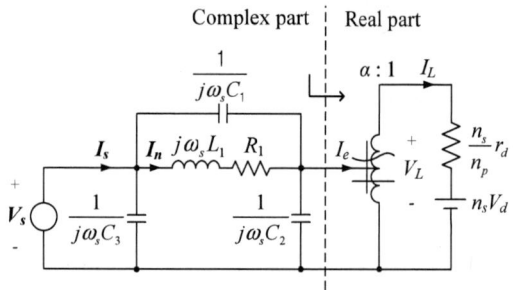

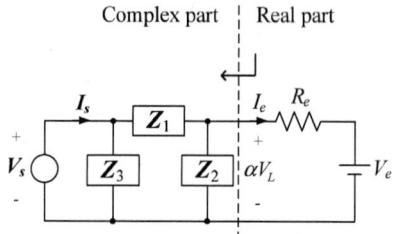

Fig. 3. A static phasor circuit, where rectified current I_L is always real.

(a) Equivalent static circuit of Fig. 3 eliminating the auto-transformer.

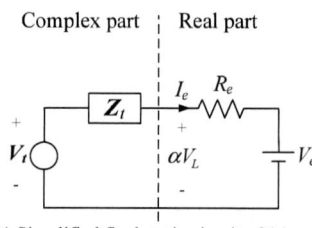

(b) Simplified final static circuit of (a).

Fig. 4. A simplified final static circuit of the proposed LED driver.

operating temperature and variation of line voltage because there is no means of regulating the LED power.

In this paper, compact passive LC^3 LED driver which satisfies low THD and high PF by using LC parallel resonance is newly proposed, as shown in Fig. 1. An optimum number of LED array in series is appropriately selected such that the LED power becomes temperature-robust. The proposed LED driver was analyzed by the powerful phasor transformation techniques [16]-[21], which can be applied to not only general power converters but also special cases of it such as wireless power transfer system [22]-[24]. Especially, the application to non-linear switching case is the first time in this paper and experiments for an LED driver of 60 W showed good agreements with simulations and designs.

II. DESIGN OF THE PROPOSED LED DRIVER

The proposed LED driver is illustrated in Fig. 1, where the source side filters L_1, C_1, and C_2 together with the number of LED array in series n_s determine the line PF, THD in source current i_s, and LED load power P_L. The source side capacitor C_3 is just for PF compensation whereas the load side electrolyte capacitor C_L is for smoothing the LED array voltage v_L. The LED array is composed of n_p number of parallel sub-arrays, as shown in Fig. 1. The steady state behaviors of the proposed LED driver will be analyzed in this section by neglecting high order switching harmonics. The internal resistance of L_1 is considered as R_1 but the ESRs of capacitors are omitted from consideration throughout this paper. The characteristics of all the LEDs are assumed to be identical and temperature distribution over the LED array is even.

A. Circuit Modeling of the Proposed LED Driver

As shown in Fig. 2, each LED in an array can be replaced with a dynamic resistance r_d and a DC voltage source V_d [9]. According to the general equivalence of a converter with a time-varying transformer [16]-[17], the full bridge diode rectifier can be replaced with a time-varying transformer, whose turn-ratio is a switching function $s(t)$ at the line frequency [17], as shown in Fig. 2(a) and (b). Only the fundamental components of voltages and currents of the circuit are considered and all harmonics are neglected from consideration in this paper.

In order to analyze the proposed circuit, the well-known phasor transformation [16]-[21] is applied to the time-varying fundamental component equivalent circuit of Fig. 2, as shown in Fig. 3. Then a time-invariant complex circuit with an auto-transformer of a time-invariant turn-ratio α [16] is obtained. A simplified static circuit, neglecting dynamic behavior, is derived for the static analysis, as shown in Fig. 3. In the phasor transformation, the rectified current at the DC output side should include only real value in general and should not contain any imaginary component [18]. Therefore, the input current of the auto-transformer I_e ($= I_L/\alpha$) should be also real in case the phase reference is tuned to the diode rectifier. In the phasor transformation, the selection of a reference phase may be arbitrary; hence, the phase of the rectifier is set to zero for simplicity of analysis and the phase of source voltage ϕ_s is unknown at the moment.

Throughout this paper, the bridge diode is assumed to be lossless and ideal, where α is almost 0.90 for the continuous conduction mode (CCM) [16].

The remaining work is to find the source voltage V_s of a complex variable so that I_e can be real, where the magnitude V_s is given and the phase ϕ_s should be determined.

$$V_s \equiv V_s e^{j\phi_s} \qquad (1)$$

By reflecting the turn-ratio of auto-transformer to an equivalent circuit seen from left part, Fig. 3 can be converted to Fig. 4(a). Z_1, Z_2, Z_3, R_e, and V_e are defined as follows:

$$Z_1 = (j\omega_s L_1 + R_1) // \frac{1}{j\omega_s C_1}, \quad Z_2 = \frac{1}{j\omega_s C_2}, \quad Z_3 = \frac{1}{j\omega_s C_3} \qquad (2)$$

978-1-4799-2706-7/14 $31.00 © 2014 IEEE   3359

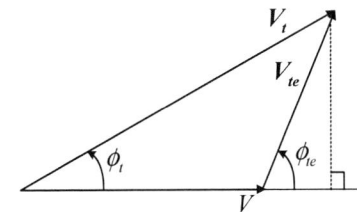

(a) Phasor diagram of V_e, V_t, and V_{te} of Fig. 4(b).

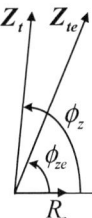

(b) Phasor diagram of R_e, Z_t, and Z_{te} of Fig. 4(b).

Fig. 5. Phasor diagram of the simplified static circuit of Fig. 4(b).

$$R_e = \alpha^2 \frac{n_s}{n_p} r_d , \quad V_e = \alpha n_s V_d . \quad (3)$$

By applying the Thevenin's theorem to the left part of the circuit of Fig. 4(a), a simplified static circuit is finally obtained, as shown in Fig. 4(b), where V_t and Z_t are derived as follows:

$$V_t = V_s \frac{Z_2}{Z_1 + Z_2} \equiv V_s e^{j\phi_s} \cdot \gamma e^{j\phi_\gamma} = V_t e^{j\phi_t} \quad (4)$$

$$V_t = \gamma V_s , \quad \phi_t = \phi_s + \phi_\gamma \quad (5)$$

$$Z_t = Z_1 // Z_2 = \frac{R_1 + j\omega_s L_1}{1 - \omega_s^2 L_1 (C_1 + C_2) + j\omega_s R_1 (C_1 + C_2)} . \quad (6)$$

In (4) and (5), γ and ϕ_γ are defined as follows:

$$\gamma = \left| \frac{Z_1}{Z_1 + Z_2} \right| = \sqrt{\frac{(1 - \omega_s^2 L_1 C_1)^2 + (\omega_s C_1 R_1)^2}{\{1 - \omega_s^2 L_1 (C_1 + C_2)\}^2 + \omega_s^2 R_1^2 (C_1 + C_2)^2}} \quad (7)$$

$$\phi_\gamma = \tan^{-1} \frac{\omega_s [L_1 \{1 - \omega_s^2 L_1 (C_1 + C_2)\} - R_1^2 (C_1 + C_2)]}{R_1} . \quad (8)$$

According to (5), ϕ_s can be obtained if ϕ_t is determined, where ϕ_γ is readily calculated from (8) for the given L_1, R_1, C_1, and C_2. The phasor diagram of Fig. 4(b) for satisfying the condition that I_e has only real part is illustrated in Fig. 5, where ϕ_{te} should be the same as ϕ_{ze} in order to meet the following condition:

$$I_e = \frac{V_t - V_e}{Z_t + R_e} \equiv \frac{V_{te}}{Z_{te}} \equiv \frac{V_{te} e^{j\phi_{te}}}{Z_{te} e^{j\phi_{ze}}} = \frac{V_{te}}{Z_{te}} \quad \text{for} \quad \phi_{te} = \phi_{ze} \quad (9)$$

where ϕ_{te} is the phase of subtracting from V_t to V_e and ϕ_{ze} is the phase of sum of Z_t and R_e, which are defined as follows:

$$\phi_{te} = \phi_{ze} = \tan^{-1} \frac{\omega_s [L_1 \{1 - \omega_s^2 L_1 (C_1 + C_2)\} - R_1^2 (C_1 + C_2)]}{R_1 + [\{1 - \omega_s^2 L_1 (C_1 + C_2)\}^2 + \omega_s^2 R_1^2 (C_1 + C_2)^2] R_e} \quad (10)$$

In (10), ϕ_{ze} is already given; hence, ϕ_{te} can be accordingly determined. A relationship on ϕ_t can be obtained from Fig. 5(a) as follows:

$$\cos \phi_t = \frac{V_t^2 + V_e^2 - V_{te}^2}{2 V_t V_e} \quad (11)$$

where V_{te} is determined from the following equation.

$$V_t^2 = (V_{te} \sin \phi_{te})^2 + (V_e + V_{te} \cos \phi_{te})^2$$
$$\Rightarrow V_{te} = -V_e \cos \phi_{te} + \sqrt{V_t^2 - V_e^2 \sin^2 \phi_{te}} . \quad (12)$$

From (10) - (12), finally ϕ_t is determined as follows:

$$\therefore \phi_t = \cos^{-1} \left(\frac{V_e}{V_t} \sin^2 \phi_{te} + \cos \phi_{te} \sqrt{1 - \left(\frac{V_e}{V_t} \sin \phi_{te} \right)^2} \right) \quad (13)$$

B. Static Analysis of LED Power & Efficiency

I_e can be derived by (9) as follows:

$$I_e = \frac{V_t - V_e}{Z_t + R_e} = \text{Re} \left\{ \frac{V_t e^{j\phi_t} - V_e}{Z_r + jZ_i + R_e} \right\}$$
$$\Rightarrow \therefore I_e = \frac{(V_t \cos \phi_t - V_e)(Z_r + R_e) + V_t Z_i \sin \phi_t}{(Z_r + R_e)^2 + Z_i^2} \quad (14)$$

where Z_r is real part of Z_t and Z_i is imaginary part of Z_t, which are defined as follows:

$$Z_r = \text{Re}\{Z_t\}, \quad Z_i = \text{Im}\{Z_t\} \quad (15)$$

From (14) and (15), LED power P_L can be calculated as follows:

$$P_L = I_e^2 R_e + V_e I_e \quad (16)$$

978-1-4799-2706-7/14 $31.00 © 2014 IEEE

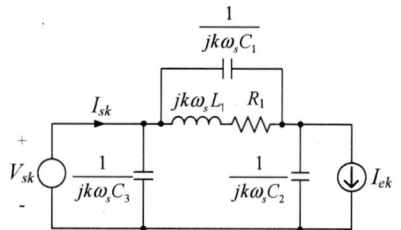

Fig. 6. Equivalent static circuit considering harmonic current I_{ek} by LEDs.

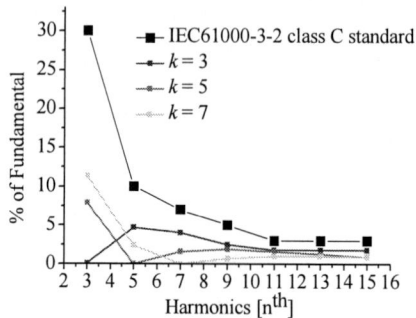

Fig. 7. Simulation result of harmonics for $L_1 = 0.9$ H.

In addition to P_L, the proposed LED driver has the losses of P_n and P_d, which are ohmic loss of L_1 and conduction loss of the bridge diode, respectively. L_1 current I_n, P_n, and P_d can be calculated as follows:

$$I_n = I_e \cdot \left| \frac{Z_2}{Z_1 + Z_2} \right| \cdot \left| \frac{\frac{1}{j\omega_s C_1}}{j\omega_s L_1 + R_1 + \frac{1}{j\omega_s C_1}} \right| \quad (17)$$

$$= \frac{I_e \gamma}{\sqrt{(1 - \omega_s^2 L_1 C_1)^2 + \omega_s^2 R_1^2 C_1^2}}$$

$$P_n = I_n^2 R_1 = \frac{I_e^2 \gamma^2 R_1}{(1 - \omega_s^2 L_1 C_1)^2 + \omega_s^2 R_1^2 C_1^2} \quad (18)$$

$$P_d = I_e^2 \cdot (2R_d) = 2I_e^2 R_d \quad (19)$$

where R_d is diode resistance of the bridge diode.

By (18) and (19), total power efficiency of the proposed LED driver can be derived as follows:

$$\therefore \eta = \frac{P_L}{P_s} = \frac{P_L}{P_L + P_n + P_d} \cdot \quad (20)$$

C. Harmonic Reduction by LC Parallel Resonance

When load is connected to LEDs, load current is operated in discontinuous conduction mode (DCM) due to the DC voltage of LEDs V_d. For this reason, LED current is operated in DCM, which causes harmonic current I_{ek}, as shown in Fig. 6. Especially, it generates the 3rd and 5th harmonics, whose components are dominant more than others. This issue is important for LED driver and I_{ek} should not be flown to I_{sk} to improve THD. LC parallel resonance, composed of L_1 and C_1, is adopted so that corresponding impedance can be infinite. By this characteristic, I_{ek} can be filtered to C_2 and source harmonic current I_{sk} can be drastically reduced by selecting corresponding harmonic degree k. The transfer function G_I, which indicates the ratio of I_{sk} and I_{ek}, is expressed as follows:

$$G_I(jk\omega_s) = \frac{I_{sk}(jk\omega_s)}{I_{ek}(jk\omega_s)} = \left| \frac{Z_{2k}}{Z_{1k} + Z_{2k}} \right| \quad (21)$$

where Z_{1k} and Z_{2k} are defined as follows:

$$Z_{1k} = (jk\omega_s L_1 + R_1) // \frac{1}{jk\omega_s C_1}, \quad Z_{2k} = \frac{1}{jk\omega_s C_2} \cdot \quad (22)$$

To reduce THD significantly, dominant harmonic degree such as 3rd and 5th should be considered and harmonic degree of parallel resonance k_p can be set by appropriate values of L_1 and C_1. The value of C_2 is assumed to be same as that of C_1 and parallel resonance frequency f_{rp}, determined by L_1 and C_1, is expressed as follows:

$$f_{rp} = k_p f_s = \frac{1}{2\pi \sqrt{L_1 C_1}} \quad (23)$$

$$C_1 = C_2 = \frac{1}{k_p^2 \omega_s^2 L_1} \quad (\omega_s = 2\pi f_s). \quad (24)$$

To verify the function of LC resonance, L_1, R_1, C_L, n_s, and n_p are chosen as 0.9 H, 12 Ω, 100 μF, 65, 4, respectively. Because an LED array is equivalent to r_d and V_d, these values are selected as 3.8 Ω and 2.7 V considering typical operating point of LED current. Diode threshold voltage and diode resistance in the bridge diode are 1 V and 1 Ω. Given this condition, simulation result of harmonics performed by PSIM simulation tool is shown in Fig. 7. By setting proper k, corresponding harmonics are greatly reduced. In (23), k_p is selected 4.0 to reduce the dominant harmonic degree of 3rd and 5th in this paper.

III. EXPERIMENTAL VERIFICATIONS

To verify the proposed LED driver by simulation and experiment, 0.9 H of L_1 having $R_1 = 12$ Ω is selected. From (24), both C_1 and C_2 are selected as 0.47 μF for $k_p = 4.0$, and C_L is chosen as 100 μF. n_p is selected as 4 considering 75 mA of normal current rating of LEDs and α is chosen as 0.88 in all of n_s range for simplicity. For experimental measurement, WT1600 and WT310 are used to measure LED power, PF, and THD.

The 2014 International Power Electronics Conference

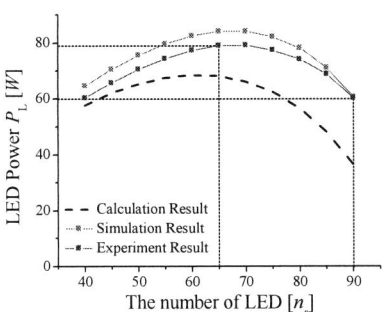

Fig. 8. The LED power versus n_s.

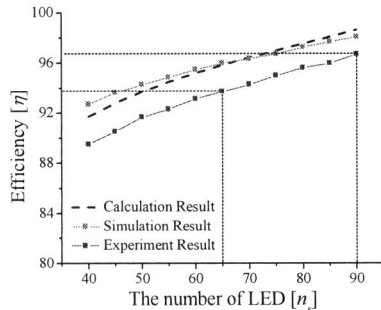

Fig. 9. The efficiency versus n_s.

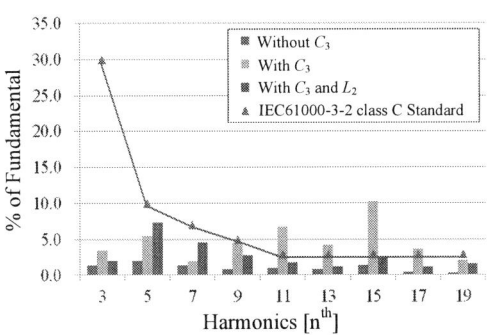

Fig. 10. Experimental result of harmonic characteristics at $n_s = 65$.

LED power and the power efficiency of the proposed LED driver can be calculated by (16) and (20), and calculation, simulation, and experimental results are shown in Fig. 8 and Fig. 9. When an LED array operates, LEDs temperature would be increased, which cannot ensure the constant LED power due to the characteristics of LEDs depending on LEDs temperature. LED power, corresponds to the brightness of LEDs, should not be changed even though the characteristics of LEDs change. In Fig. 8, experiment result has a maximum point at $n_s = 65$ that LED power variation is very insensitive with respect to n_s because the change of r_d and V_d caused by LEDs temperature corresponds to that of n_s. By setting a maximum point to the operating point of LEDs, an LED array can be operated without any change of the brightness of LEDs compared to other point of n_s.

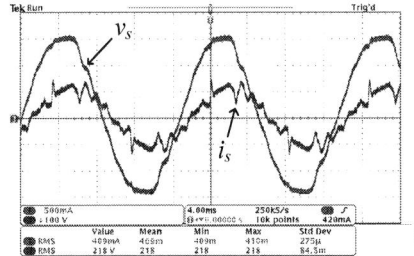

(a) Source voltage and current when C_3 is used for PF.

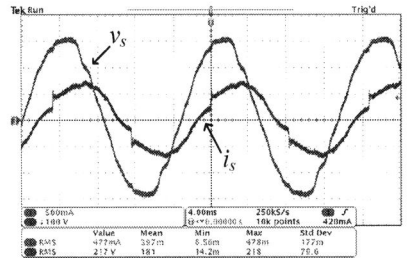

(b) Source voltage and current when C_3 and L_2 are used for PF.

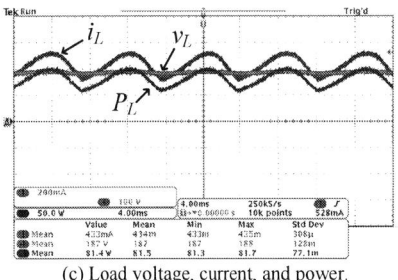

(c) Load voltage, current, and power.

Fig. 11. The experimental waveforms at $n_s = 65$.

On the other hands, desired LED power can be set by selecting appropriate n_s. Especially, to achieve extremely high efficiency, large n_s of operating point is required because less current of an LED array reduce inductor loss caused by R_1. For instance, when the proposed LED driver is operated at $n_s = 90$, LED power is 60 W and a very high efficiency of 96.7 % can be achieved, as shown in Fig. 9. Since R_1 was designed to have quite large value, the efficiency can be more improved when R_1 is designed to have small value.

The discrepancy of simulation and experiment is that r_d and V_d of real LED model is not constant, but highly depends on the operating current. The discrepancy between calculation and experimental results comes from non-linear characteristic of auto-transformer because the proposed circuit for LED driver is non-linear switching case and α is assumed to be constant in all of n_s range. Even though it has some discrepancy, the tendency of LED power can be predicted and observed through calculation results.

When operating point of LEDs is set to $n_s = 65$ for temperature-robust, PF was measured as 0.811, whose value is not satisfied with IEC61000-3-2 class C standard. For PF compensation, compensation capacitor should be used and 1.8 μF of C_3 is selected to satisfy PF = 0.95 at $n_s = 65$. However, over the 9th of harmonics, especially the 11th and 15th of harmonics, are measured as very high

978-1-4799-2706-7/14 $31.00 © 2014 IEEE

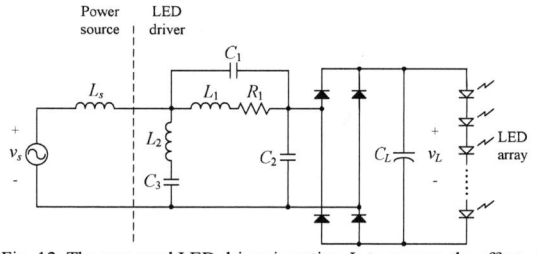

Fig. 12. The proposed LED driver inserting L_2 to remove the effect of source harmonics and L_s.

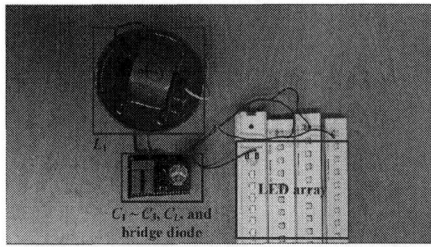

(a) The prototype of the proposed LED driver.

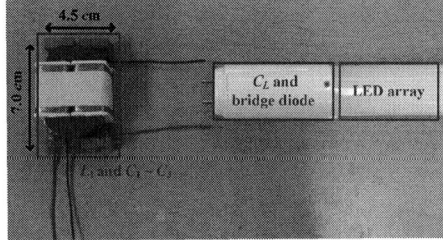

(b) The product of the proposed LED driver.

Fig. 13. Prototype and product of the proposed LED driver without L_2.

value when C_3 is used, as shown in the green bar of Fig. 10, and its waveforms of v_s and i_s are shown in Fig 11(a). It is because over the 9th of harmonics are originally high and it can be confirmed by measuring harmonic characteristics of the condition, where only resistance load is connected to power source, as described in the red bar of Fig. 10. The results showed that the 9th, 11th, 13th, and 15th of fundamental frequency were measured as 0.8 %, 1.0 %, 0.9 %, and 1.4 %, respectively. Moreover, due to source line inductance L_s that basically exists in power source, C_3 is resonated with L_s that is normally 10 mH, which makes corresponding harmonics much higher.

For this remedy, L_2 is inevitably connected to C_3 in series in order to remove the effect of L_s, as shown in Fig. 12. In this case, L_2 is chosen to move the resonant frequency determined by L_2 and C_3 to 4th of fundamental because there is no 4th harmonic; hence, $L_2 = 240$ mH that has 17 Ω of internal resistance. By inserting additional inductor L_2, over 9th of harmonics are greatly reduced, as shown in the purple bar of Fig. 10 and its waveforms of v_s, i_s, v_L, i_L, and P_L are shown in Fig. 11(b) and Fig. 11(c). In addition, all of harmonics are finally satisfied with IEC61000-3-2 class C standard; THD was measured as 10.1 %. L_2 is just used in order to suppress the source harmonic effect and is not mandatory to this circuit, but is necessary in the case that harmonics in the power source and L_s are originally high. The effect of L_2 on PF can be

negligible because C_3 is large enough to compensate leading current and to ignore L_2 inductance. The prototype of the proposed LED driver without L_2 is shown in Fig 13(a), which is quite bulky for practical use. However, the proposed LED driver can be manufactured by decreasing the size of L_1 and integrating C_L and bridge diode into LED tube for the product, which is compact in size and cheap to be implemented in Fig. 13(b).

IV. CONCLUSIONS

A passive-type LED driver that has just one inductor and three capacitors called LC^3 is newly proposed in this paper. By using passive components only, 10.1 % of low THD, 0.95 of high PF, and 96.7 % of a very high efficiency for 60 W with extremely long lifetime are achieved. In addition, LEDs can be operated at temperature-robust point that the power variation is zero, which robustness is beneficial to operate LEDs without any change of the LED power. This characteristic is ultimate solution of the applications, which requires very high efficiency and long life LEDs operation such as parking garage or emergency stairs. Even though the proposed LED driver is non-linear switching case, the characteristics of the proposed LED driver were analyzed by phasor transformation technique, which results are fairly good agreement with simulations and experiments.

ACKNOWLEDGMENT

This work was supported by the Korea Micro Energy Grid (K-MEG) of the Korea institute of Energy Technology Evaluation and Planning (KETEP) grant funded by the Korea government Ministry of Trade, Industry and Energy (MOTIE). (No. 2011T100100024).

REFERENCES

[1] M.O. Holcomb et al., "The LED lightbulb: are we there yet? progress and challenges for solid state illumination," *Conference on Lasers and Electro-Optics, 2003. CLEO '03*, June 2003.

[2] Jeff Y. Tsao, "Solid-state lighting: lamps, chips, and materials for tomorrow", *IEEE Circuits & Devices*, May-Jun. 2004, pp. 28-37.

[3] Daniel A. Steigerwald et al., "Illumination with solid state lighting technology", *IEEE journal of Selected Topics in Quantum electronics*, Apr. 2002, pp. 310-320.

[4] Sung-Soo Hong et al., "A new cost-effective current-balancing multi-channel LED driver for a large screen LCD backlight units," *Journal of Power Electronics (JPE)*, vol. 10, no. 4, pp. 351-356, July 2010.

[5] ENERGY STAR program requirements for solid state lighting luminaires. Eligibility Criteria – Version 1.1 (2008). [Online].

[6] IEC 61000-3-2 class C standard, Limits for harmonic current emissions (equipment input current ≤ 16A per phase), IEC, 2009

[7] Heinz van der Broeckl et al., "Power driver topologies and control schemes for LEDs," *Applied Power Electronics Conference and Exposition (APEC), 2007*, Feb. 2007, pp. 1319-1325.

[8] Aguilar, D. and Henze, C.P., "LED driver circuit with inherent PFC," *Applied Power Electronics Conference and Exposition (APEC), 2010*, Feb. 2010, pp. 605-610.

[9] Liu Yu and Jinming Yang, "The topologies of white LED lamps' power drivers," *Power Electronics Systems and Applications, 2009. PESA*, May 2009, pp. 1-6.

[10] Seoul Semiconductor Co., Ltd, "Luminous device," Korea patent, reg. no. 10-1142939, Apr. 2012.

[11] B. H. Lee, H. J. Kim, B. C. Kim, and Chun T. Rim, "The development of low-cost AC power LED driver using capacitor," *2010 Summer KIPE Conference*, pp. 426-427.

[12] Optomind Inc., "Direct connection-type LED lighting apparatus" Korea patent, apply no. 10-2012-0094901, Aug. 2012.

[13] S. Y. (Ron) Hui, Si Nan Li, Xue Hui Tao, Wu Chen, and W. M. Ng, "A novel passive offline LED driver with long lifetime," *IEEE Trans. on Power Electronics*, vol. 25, no. 10, pp.2665-2672, Oct. 2012.

[14] W. Chen, "A comparative study on the circuit topologies for offline passive light-emitting diode (LED) drivers with long lifetime & high efficiency," *IEEE Energy Conversion Congress and Exposition*, pp. 724-730, Sep. 2010.

[15] B. H. Lee, H. J. Kim, and Chun T. Rim, "Robust passive LED driver compatible with conventional rapid-start ballast," *IEEE Trans. on Power Electronics*, vol. 26, no. 12, pp.3694-3706, Dec. 2011.

[16] Chun T. Rim and G. H. Cho, "Phasor transformation and its application to the DC/AC analyses of frequency/phase controlled series resonant converters (SRC)," *IEEE Trans. on Power Electronics*, vol. 5, no. 2, pp. 201-211, Apr. 1990.

[17] Chun T. Rim, D. Y. Hu, and G. H. Cho, "Transformers as equivalent circuits for switches: General proofs and D-Q transformation-based analysis," *IEEE Trans. on Industry Application*, vol. 26, no. 4, pp. 777-785, July/Aug. 1990.

[18] Chun T. Rim, "Unified general phasor transformation for AC converters," *IEEE Trans. on Power Electronics*, vol. 26, no. 9, pp. 2465-2475, Sep. 2011.

[19] C. B. Park, S. W. Lee, and Chun T. Rim, "Static and dynamic analyses of three-phase rectifier with LC input filter by Laplace phasor transformation," *IEEE Energy Conversion Congress and Exposition*, Sep. 2012, pp. 1570-1577.

[20] Sungwoo Lee, Bohwan Choi, and Chun T. Rim, "Dynamics Characterization of the inductive power transfer system for online electric vehicles by laplace phasor transform," *IEEE Trans. on Power Electronics*, vol. 28, no. 12, pp. 5902-5909, Mar. 2013.

[21] C. T. Rim, and G. H. Cho, "New approach to analysis of quantum rectifier-inverters," *IEEE Electronic Letters*, vol. 25, no. 25, pp. 1744-1745, Dec. 1989.

[22] J. Huh, W. Y. Lee, S. Y. Choi, G, H, Cho and C. T. Rim, "Frequency-domain circuit model and analysis of coupled magnetic resonance systems," *Journal of Power Electronics,* vol. 13, no. 2, Mar. 2013.

[23] Suyong Choi, J. Huh, S. Lee, and C. T. Rim, "New cross-segmented power supply rails for roadway powered electric vehicles," *IEEE Trans. on Power Electronics*, vol. 28, no. 12, pp. 5832-5841, Dec. 2013.

[24] Changbyung Park, Sungwoo Lee, Gyu-Hyeong Cho, Su-Yong Choi, and Chun T. Rim, "Two-dimensional inductive power transfer system for mobile robots using evenly displaced multiple pick-ups," *IEEE Trans. on Industry Applications*, vol. 50, no. 1, pp.558-565, June 2013.

The 2014 International Power Electronics Conference

Optimizing Repulsive Lorentz Forces for a Levitating Induction Cooker

Claudius M. Zingerli, Thomas Nussbaumer, Johann W. Kolar

ETH Zürich, Power Electronic Systems Laboratory

Physikstrasse 3, 8092 Zürich, Switzerland

zingerli@lem.ee.ethz.ch

Abstract—**In this paper we study a novel way of induction cooking. Traditionally, an alternating magnetic field is used to induce currents in a ferromagnetic pan to heat foods. We use non-ferromagnetic materials and optimize the design for high repulsive forces in order to levitate the pan while simultaneously heating it and its contents. Our approach is simulation-based to study the influence of different parameters and to perform multidimensional analyses for high force versus power or loss goals. Finally, an experimental prototype has been realized and successfully operated.**

Index Terms—**Magnetic levitation, modeling, eddy currents, induction motors, induction cooking, home appliances, education**

I. MOTIVATION

The proliferation of induction cookers is constantly growing. Compared to traditional electrical stoves, quick reaction time and – depending on the source – high energy efficiency speak for them [1]. In a typical induction cooker, a flat induction coil creates an alternating magnetic field. The field induces a current flow in the conducting pan on top of the coil [2]. This leads to losses in the pan caused by its electrical resistance, heating up the pan and its contents. The currents in the two coupled coils (excitation coil and pan) also have a force effect. In this publication, we investigate if these forces can be used to levitate the pan. Further, we study design variants targeting stable levitation and concentration of the losses in the pan and keep them away from the excitation – or bearing – coil(s). A design variant is shown in Figure 1.

While newer developments [3] override this restriction, in conventional induction stoves the base of the pan must typically contain ferromagnetic material. Such materials conduct the magnetic field well and concentrate the magnetic flux. This primarily leads to a lower skin depth (higher resistance) as well as to a reduction of the flux path and consequently to an attracting force owing to the reduced reluctance. The current flow in the base of the pan is opposed to the one in

the excitation coil, why a repulsive force is created (Lorentz force). The minimization of this force is often the goal of optimizations in applications like metal production [4]. In our investigation, we follow a completely different approach: that of maximizing Lorentz forces in order to allow the pan to levitate so that an observer may witness an air gap between the excitation/cooktop surface and the pan.

Generally, high excitation currents lead to high repulsive forces. But there are limits on the excitation current because the removal of ohmic losses in the excitation coil becomes more and more challenging. Thus one criterion for optimization is to maximize the quotient of losses in the pan versus losses in the excitation coil. A further criterion is the maximization of the bearing force at a certain loss level.

To simplify the construction, solely repulsive magnetic fields shall be used. Additionally, no parts shall be added to the pan (permanent magnets, coils, back iron). For this reason, active, top or radially acting magnetic bearing designs will not be considered.

Figure 1: Design variant of a levitated induction cooker.

978-1-4799-2706-7/14 $31.00 © 2014 IEEE

II. PREVIOUS WORKS

The principle involved in using eddy currents for heating and levitation is not new. Previous works include applications on contactlessly melt metals [5] and many educational exhibits. We use an eddy current bearing in one of our lectures to demonstrate a very simple magnetic bearing where the levitated object gets very hot quickly. In science and technology museums one can often find an exhibit with a levitating aluminum ring on top of a coil connected to the mains [6, 7].

What is new about our device is that we want to apply said principle for cooking such as a demonstration of induction or as a show effect.

III. ANALYSIS

The shape of the pan has a great influence on the force development. The strongest vertical forces (high axial stiffness) act on surfaces orthogonal to the coil axis, but these surfaces cause no radial stiffness; therefore, a pan consisting solely of such surfaces (e.g. a sheet) would simply slip out and fall down. On the other hand, a spherically shaped pan offers a positive radial stiffness but, despite the tendency to right itself, it has a zero tilting stiffness. A flat bowl/plate shaped pan offers a positive axial, radial and tilting stiffness and is thus researched.

For an efficient excitation, the excitation coil is operated in a (series) resonant circuit. As the pan moves, the inductance changes and hence, the resonant frequency. Experiments have shown that without any extra means of damping, the levitated pan would oscillate and eventually touch the coil surface or even fall out. We adaptively control the excitation frequency by tracking the coil current.

Assuming that the major part of the flux in the excitation coil induces a current in the pan and that the pan is thick enough to shield the field, an increase of the coil radius leads to an increase of the current in the pan and – at a constant air gap field strength – to a proportional increase of the Lorentz force acting on the pan.

At a fixed geometry and air gap, the coupling between excitation and pan is constant. A higher excitation current then leads to a proportionally higher magnetic air gap field and current in the pan. The Lorentz force acting on the pan is proportional to the current in the pan and air gap flux density which is again proportional to the air gap field. Therefore this force should increase with the excitation current squared.

IV. SIMULATION

To verify the feasibility of the concept and to optimize the design, a simulative approach has been chosen. Due to the rotational symmetry, all simulations were 2D FEM-based in a cylindrical coordinate system in the eddy current domain. All dimensions were parameterized in order to allow variation and tuning runs, since with today's computer power, multidimensional searches could be performed within only hours (for example frequency versus coil diameter and pan thickness).

Figure 2 shows the simulation model used for a simple cylindrical pan. The mesh in the levitated object (Pan) was manually adjusted to include multiple elements in the material thickness.

The coil below was excited with a fixed amplitude AC current. This constant current excitation was used because in the experiment the number of turns and maximum amplifier current are fixed while independent of the inductance, the resonant tank can always be tuned to push the voltage amplitude high enough to drive that current.

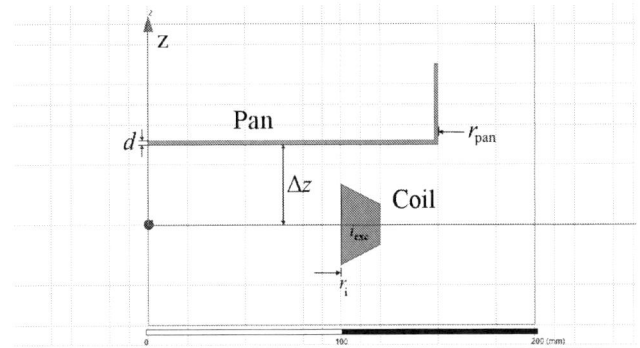

Figure 2: Model of a magnetically levitated induction cooker. The vertical (z) axis shows the rotational symmetry. The excitation coil is stranded and supplied with a fixed amplitude AC current.

As expected, the following proportionalities of the repelling force F were confirmed:

$$
\begin{aligned}
F \quad &\propto \quad r_\mathrm{i} \qquad \text{Inner coil radius } r_\mathrm{i} \\
&\propto \quad i_\mathrm{exc}^2 \qquad \text{Ampere turns } i_\mathrm{exc} \text{ of excitation} \\
&\propto \quad \cos(\alpha) \qquad \text{Pan wall angle } \alpha \\
&\propto \quad 1/\Delta z \qquad \text{Vertical distance } \Delta z \text{ between excitation} \\
&\qquad\qquad\qquad \text{coil center and pan}
\end{aligned}
$$

Figure 3 shows that the repelling force acting on the pan rises proportionally to the excitation coil radius. If the pan is

978-1-4799-2706-7/14 $31.00 © 2014 IEEE

The 2014 International Power Electronics Conference

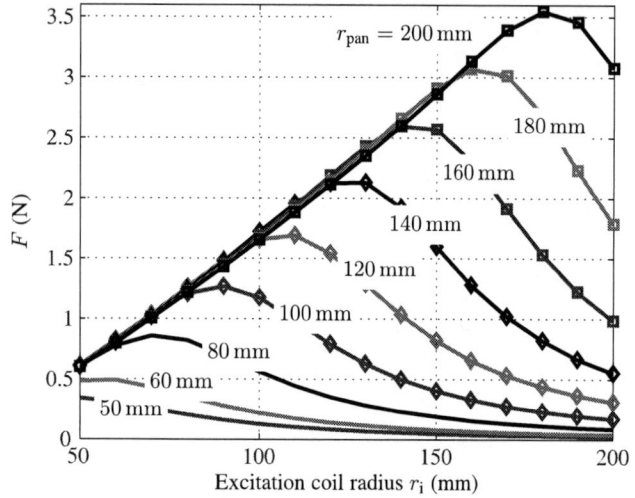

Figure 3: Simulated force F as a function of the excitation coil radius r_i at different pan radii r_{pan} (flat aluminum pan, $i_{exc} = 1000\,\text{A} \cdot \text{turns}$, $h_{pan} = 33\,\text{mm}$, radial excitation coil thickness: 20 mm).

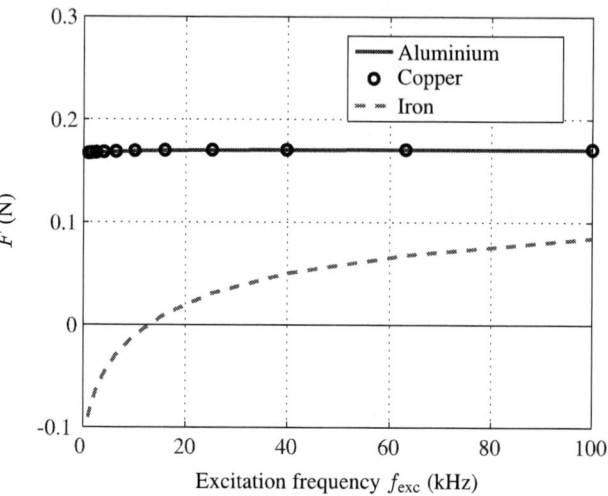

Figure 4: Force F as a function of excitation frequency f_{exc} at different pan materials ($i_{exc} = 500\,\text{A} \cdot \text{turns}$, $h_{pan} = 40\,\text{mm}$, $r_i = 50\,\text{mm}$). Aluminum and copper show no significant difference, while iron generally generates lower repulsive forces, yet (at low frequencies) produces attracting forces.

smaller than the excitation coil, the force starts to decrease again, so matched pan and coil sizes are important.

Good conducting materials (Al, Cu) lead to repulsive forces with little dependency on f_{exc} (Figure 4). The forces are similar in both materials, but aluminum being about three times lighter than copper, the air gap will be wider at the same current. Iron (Fe-1010) reveals a strong dependency on f_{exc} in the studied frequency range, even showing attracting forces when remanence dominates at frequencies below 12 kHz. Due to the high resistivity of iron, the force versus heating power quotient is also low.

In Figure 5, the dependence of the force on the excitation frequency f_{exc} and the pan thickness d is shown. The highest repelling force is generated with a pan thickness of about 4/7 of the corresponding skin depth. This finding has been tested against different materials (Aluminum, Copper and Magnesium) and frequencies ($f_{exc} = 1\,\text{kHz}...100\,\text{kHz}$). If the pan is thinner, the force decreases rapidly. If the pan is thicker, the decrease is less. At a constant excitation current, however, the heating power rises fast, hence, depending on the application – heating power versus levitation – an optimum can be found.

Studies on the influence of the excitation current i_{exc} and levitation height Δz at two different coil radii $r_i = \{50\,\text{mm}, 100\,\text{mm}\}$ were performed (Figure 6 and Figure 7). While in both designs, the force increases with the current squared, the smaller coil only allows for levitated loads in a low range, even at higher currents or small gaps.

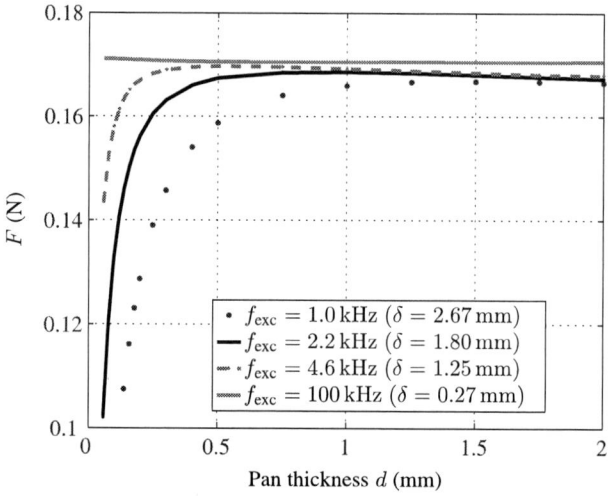

Figure 5: Force F as a function of the pan thickness d at different excitation frequencies f_{exc} ($i_{exc} = 500\,\text{A} \cdot \text{turns}$, $h_{pan} = 40\,\text{mm}$, $r_i = 50\,\text{mm}$, aluminum pan). At low frequencies ($< 30\,\text{kHz}$) a force maximum can be found near $d = 4/7 \cdot \delta$ with the skin depth δ.

The decreasing force with increasing gap size Δz (Figure 8) can be approximated by a hyperbolic curve $1/h_{pan}$. The decrease is due to the smaller coupling which eventually leads to less ohmic losses in the levitated pan than in the excitation coil at high distances (Figure 9). In the range studied, the loss distribution is solely a function of the geometry, independent of the electrical excitation.

978-1-4799-2706-7/14 $31.00 © 2014 IEEE

The 2014 International Power Electronics Conference

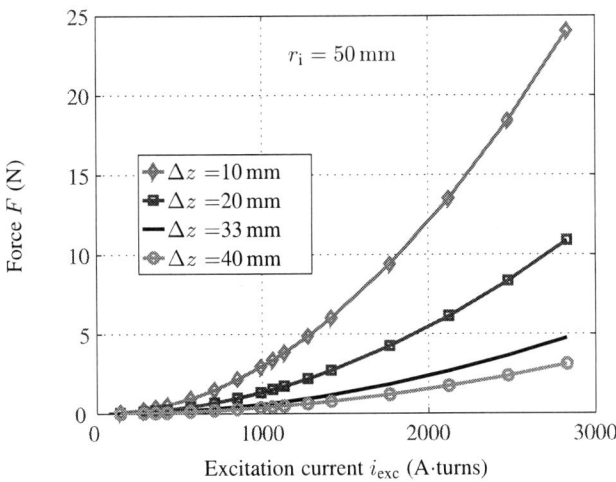

Figure 6: Simulated force F as a function of the excitation current i_{exc} at different gaps Δz ($r_i = 50\,\mathrm{mm}$).

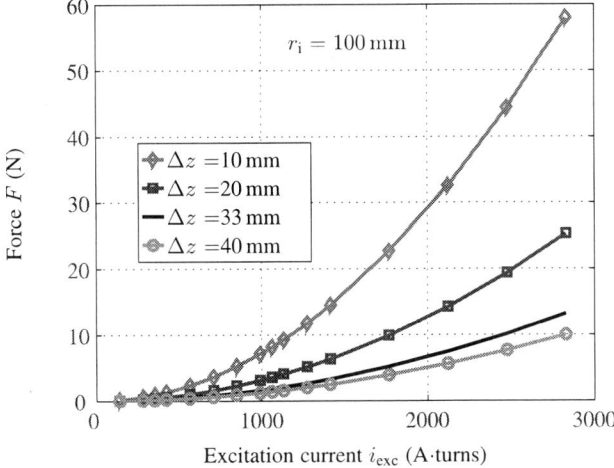

Figure 7: Simulated force F as a function of the excitation current i_{exc} at different gaps Δz ($r_i = 100\,\mathrm{mm}$).

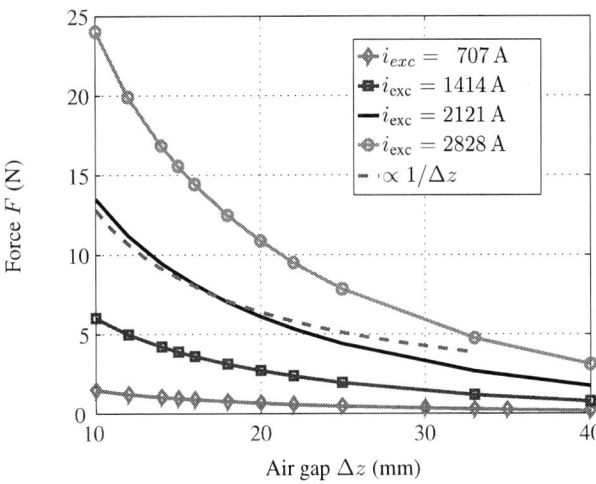

Figure 8: Simulated force F as a function of the air gap Δz at different excitation currents i_{exc} ($r_i = 50\,\mathrm{mm}$, $f_{exc} = 10\,\mathrm{kHz}$).

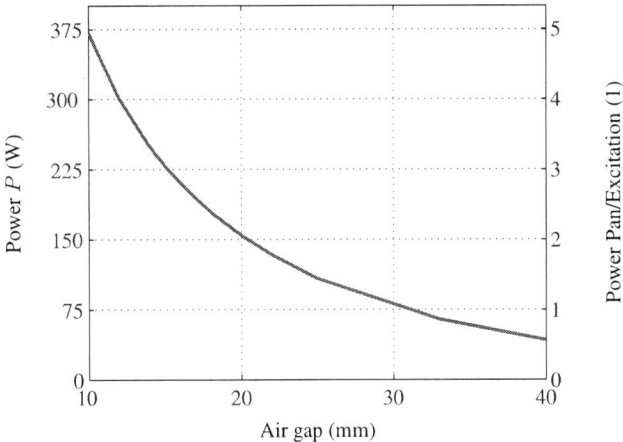

Figure 9: Simulated ohmic losses in the pan at $i_{exc} = 1000\,\mathrm{A} \cdot \mathrm{turns}$. The loss ratio pan/excitation goes below one at wider air gaps h_{pan} ($r_i = 50\,\mathrm{mm}$, $f_{exc} = 10\,\mathrm{kHz}$).

To give an example, levitating a load of $m = 3\,\mathrm{kg}$ (Typical caquelon pan and fondue contents for a party of four), using a coil with an inner radius of $r_i = 100\,\mathrm{mm}$, at a level of $\Delta z = 10\,\mathrm{mm}$, needs a current of $i_{exc} = 2\,\mathrm{kA} \cdot \mathrm{turns}$. This current produces ohmic losses on the pan of $P = 1.5\,\mathrm{kW}$ which is sufficient for cooking applications.

V. EXPERIMENT

A small-scale prototype was built and successfully operated. Levitation and heating have been demonstrated but at a much smaller scale than conventional induction cookers. Table I shows some key data of the first prototype. In Figure 10, the experimental setup is shown: on the left side, the trapezoidal shaped coil with an aluminium test pan on top. On the

right side are series resonance capacitors, voltage and current probes. The setup was powered from a commercial square wave voltage source (48 V, 10 Arms, DC...150 kHz, PWM).

Figure 11 shows the electrical characteristics of the excitation coil modelled as a series connected R-L circuit. The bearing coil being made of liz wire, there is little change in the series resistance R_s up to an excitation frequency $f_{exc} = 50\,\mathrm{kHz}$). Starting at about $f_{exc} = 1\,\mathrm{kHz}$, the magnetic field of the excitation coil is completely shielded by the pan making the inductance and resistance stay constant relative to the unloaded coil. Since the quality factor of the coil

$$Q_L = \frac{2\pi \cdot f_{exc} \cdot L_s}{R_s}$$

is very high (at $f_{exc} = 18\,\mathrm{kHz}$: $Q_L = 350$ without load,

$Q_L = 150$ with a plate and $Q_L = 90$ with a pan at $\Delta z = 33\,\mathrm{mm}$), owing to the low damping of the large air gap, a high voltage gain in the resonant circuit results, shown in Figure 12.

Table I: Key parameters of the experimental setup

Parameter	Value	
Configuration	series resonance	
Coil radius	r_i	$= 50\,\mathrm{mm}$
Number of turns	n	$= 99$
Unloaded coil inductance	L_0	$= 1.245\,\mathrm{mH}$
Operating frequency	f_{exc}	$= 16.7\,\mathrm{kHz}$

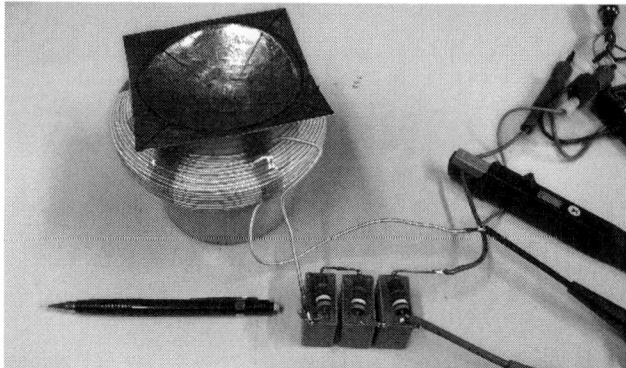

Figure 10: Prototype setup of a magnetically levitated induction cooker.

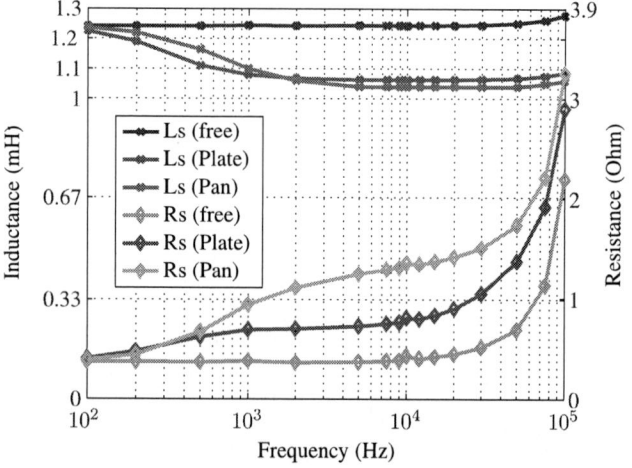

Figure 11: Series inductance L_s and resistance R_s of the coil as a function of the excitation frequency, operating in free space, loaded with a flat pan "Plate" and round pan (21 mm deep, 92 mm radius of the depression).

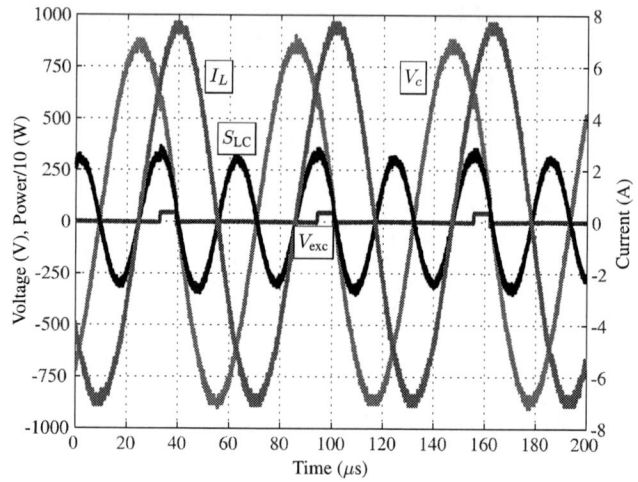

Figure 12: Measured voltage, apparent power and currents in the setup. Due to the big air gap, the quality factor of the resonant circuit is high, resulting in a high voltage V_c.

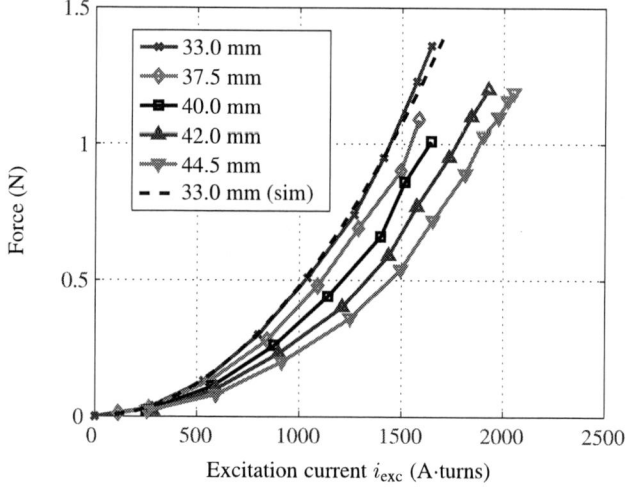

Figure 13: Force F as a function of the excitation current i_{exc} normalized to one turn at different gaps (Δz) comparing measured and simulated values.

The simulated forces in Figure 6 predict the measured values very good (Figure 13, the prototype has a lower limitation on Δz of 33mm).

VI. DISCUSSION

To achieve a big air gap (a high levitation level), high excitation currents are needed ($F \propto i_{exc}^2/\Delta z$ for gaps $\Delta z \ll$ diameter), which inherently lead to an elevated heating power. If this power is not needed the current has to be reduced and the gap will become smaller with the levitation effect being less apparent.

High excitation currents also lead to high losses in the excitation coil that have to be dealt with. While, in a conventional induction cooker, the excess heat from the excitation coil can be dumped into the pan to limit the coil's temperature to the pan's, this cannot easily be done here because of the weak thermal coupling between pan and excitation coil during levitation. We presume that an active (fan based) cooling of the excitation coil can hardly be avoided.

The repulsive force between excitation coil and prototype pan is quite small for cooking applications: as a result, heavier or bigger pans cannot be levitated without increasing excitation current and heating power to very high levels. Simulations of a bigger model were performed (Figure 7, $r_i = 100\,\text{mm}$). At bigger radii and excitation currents or smaller air gaps, significantly higher forces can be reached, but there are at least two downsides: First, the force only increases linearly with the radius, while the weight of the pan generally scales by volume, radius at the power of three. Second, simple air cooling of the excitation coil can only be performed at the coil surface which scales with the radius squared, so cooling of bigger coils or higher currents will become challenging.

The simple construction has the further disadvantage of low stiffness compared to actively controlled, PM-based or attracting magnetic bearings, which could be disadvantageous in relation to external disturbance forces (stirring, adding or removing material from the pan).

VII. CONCLUSIONS AND OUTLOOK

It has been shown that levitating a pan on an induction cooker is not only feasible, but can also be realized, be it as a small prototype or a typically sized pan including contents. The heating power is coupled to the repulsive force and is generally quite high. To increase the force-to-power quotient, the application of permanent magnet or reluctance based passive and active magnetic bearings should be studied.

Currently, the design is rotationally symmetric. Similar to an induction motor, further coils could be added to create a rotating magnetic field in order to spin the levitated pan. The lifting and driving fields could further be superimposed on the control circuit to allow for a simpler construction with fewer coils.

By using position sensors or applying one of the several published approaches for sensor-less control, an active control circuit could be realized to enhance the stiffness of the bearing including axial, tilting and radial movement. While the first do not need any additional coils, control of the later axes asks for independent control by e.g. the coils of the drive.

The current design has the advantage of boosting the heating power the closer the pan gets to the excitation coil. If there is less material in the pan (for example due to evaporation, boiling or removal), the gap is widened: this results in a lower heating power proving a passively safe behavior. Given that power is not reduced to zero with an empty pan, however, additional means of supervision are necessary, which may be implemented using a contactless IR-detector as proposed in [2].

As it is, state of the art levitating induction cookers open up new and exciting possibilities in Fondue event catering or culinary experiences which will not fail to amaze your guests.

REFERENCES

[1] H. Koertzen, J. Van Wyk, and J. Ferreira, "Design of the half-bridge, series resonant converter for induction cooking," in *Power Electronics Specialists Conference, 1995. PESC '95 Record., 26th Annual IEEE*, vol. 2, 1995, pp. 729–735 vol.2.

[2] P. H. Peters, "A portable cool-surface induction cooking appliance," *Industry Applications, IEEE Transactions on*, vol. IA-10, no. 6, pp. 814–822, 1974.

[3] A. Fujita, H. Sadakata, I. Hirota, H. Omori, and M. Nakaoka, "Latest developments of high-frequency series load resonant inverter type built-in cooktops for induction heated all metallic appliances," in *Power Electronics and Motion Control Conference, 2009. IPEMC '09. IEEE 6th International*, 2009, pp. 2537–2544.

[4] S. Hosseini, A. Kashtiban, and G. Alizadeh, "Particle swarm optimization and finite-element based approach for induction heating cooker design," in *SICE-ICASE, 2006. International Joint Conference*, 2006, pp. 4624–4627.

[5] E. C. Okress, D. M. Wroughton, C. Comenetz, P. N. Brace, and J. C. K. Kelly, "Electromagnetic levitation of solid and molten metals," *Journal of Applied Physics*, vol. 23, p. 545, 1952.

[6] University of Illinois, Urbana-Campaign Engineering Open House, "Floating frying pan," March 2007. [Online]. Available: http://www.youtube.com/watch?v=LJzCy8hhnZY

[7] Crealev Magnetic Levitation Technology, TSC Group, "Hovering cooking pan," January 2012. [Online]. Available: http://www.youtube.com/watch?v=CMURy4AgZPA

The 2014 International Power Electronics Conference

Design of a Modular Resonant Converter for 25kV-8A DC Power Supply of RF Cavities

Daniel Siemaszko, Serge Pittet, Davide Aguglia, Louis de Mallac

EPC - Electronic Power Converters Group

CERN - European Organisation for Nuclear Research

Geneva - Switzerland

daniel.siemaszko@a3.epfl.ch

Abstract—**This paper describes the design of an LCC resonant tank that is to be used in a $300kW$ modular power converter for an application in particle accelerators. The aim of the power supply is to provide a so-called RF cavity with $25kV$ DC and a current up to $8A$. Modular approach is chosen for redundancy and availability. Each module is composed of a resonant tank with a step-up Medium Frequency Transformer with its secondary windings put is series. Full gain is ensured in the full operation frequency range considering component value inaccuracies and ageing of capacitors. The paper focusses on the design of the resonant tank and its implementation in a $100kW$ power module prototype.**

Keywords—High Voltage DC Power Supply, LCC Resonant Converter, Modular Converter.

I. INTRODUCTION

CERN, the European Organization for Nuclear Research, has recently built the Large Hadron Collider (LHC), a superconducting circular particle accelerator designed for colliding two beams of protons at 7 TeV energy. The LHC is housed in a 27 km circumference tunnel, about 100m underground. It is actively used for physics since 2010. In order to improve the performance of the LHC, and in particular the rate of collisions observed in the detectors (luminosity), the accelerator complex supplying the LHC with particles must also be improved. Two linear accelerators (Linac3 and Linac4) and four synchrotrons (LEIR, PSB, PS and SPS) are concerned. It is the purpose of the LHC Injectors Upgrade (LIU) project to implement in these accelerators the modifications and upgrades required for the LHC to reach its enhanced luminosity goal.

From the power electronics point of view, some of the installed equipment requires an extensive redesign for fulfilling the new requirements. For the purpose of powering so-called RF cavities in the kV range, the resonant converter topology with a step-up MF-Transformer (Medium Frequency) is being given extensive study for eventually becoming a new family of medium power converter at CERN. The foreseen power converter is being designed for supplying $25kV$ to a tetrode, or a vacuum tube which is seen from the power converter point of view as current sources drawing up to $8A$. The tetrode is generating RF power for a cavity which purpose is to accelerate particles. It requires medium precision, low output noise with embedded medium precision current

and voltage measurement. High efficiency and high reliability is a requirement together with galvanic isolation between mains.

II. DESCRIPTION OF FORESEEN POWER SUPPLY

The converter design is a modular structure with three resonant converters operated in parallel. For reaching high voltages, each of the modules is connected to a multi-winding step-up MF-Transformer which secondary windings are put in series. The converter has two distinctive parts, namely the low-voltage resonant converter at the primary of the step-up multi-winding MF-Transformer, and the high voltage part containing the multi-winding transformer itself and passive rectifiers put in series for reaching $25kV$.

A. Specifications

The characteristics of the power supply must fulfil specifications as depicted in Fig. 1. The output voltage can vary in the range of $[15; 20]kV$ within $200ms$, no matter the state of the output current source. The current can vary in the range of $[0; 8]A$ within $5ms$, no matter the required output voltage.

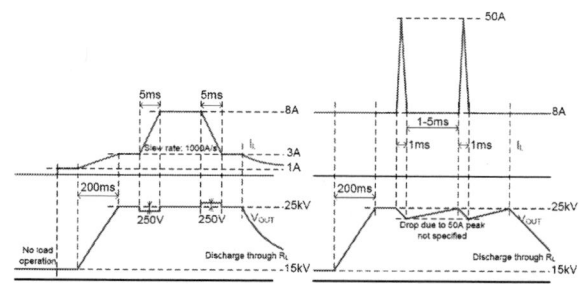

Fig. 1. General specifications for the 25kV-8A DC Power Supply

Stability of the output voltage is the most important feature of this unidirectional power supply. The maximal voltage variation is $250V$. Also, the tetrode may create some short current peaks up to $50A$ with a length up to $1ms$. The voltage drop due to those current peaks should be contained within the specified limits. Since the specified dynamics of the converters are too slow for fulfilling this specification alone, an output capacitor bank filters all perturbation to maintain the voltage stability.

978-1-4799-2706-7/14 $31.00 © 2014 IEEE

B. General structure

The designed power supply is depicted in Fig. 2. It consists of a modular redundant DC/DC power converter. For ensuring reliability and availability of the supply, the converter can go on operate with two modules only. The input DC-link is supplied by a diode bridge rectifier and its 6th harmonic is filtered by a double stage LC filter. The output load is a capacitor with the tetrode supplying RF power to the cavities.

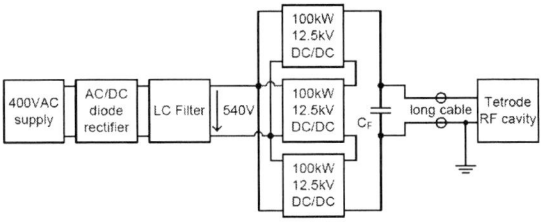

Fig. 2. General structure of the modular resonant power converter for 25kV-8A DC Power Supply

The full $100kW$ DC/DC power module is detailed in Fig. 3. It consists of an IGBT bridge and a resonant tank tuned for working in the range of $[-3;+3]dB$, within frequencies in the range of $[15;20]kHz$. Resonant technology has been chosen for overall efficiency and soft-switching capabilities [1]. A step-up MF-Transformer with multiple secondary windings connect diode rectifiers put in series for building up the $25kV$ voltage. Knowing that the primary transformer voltage in in the range $[270;1080]V$, the overall transformer ratio ν_{TRAFO} is about 12.

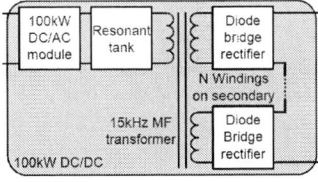

Fig. 3. General structure of the 100kW DC/DC module

C. Low voltage primary side

One the primary side of the MF-Transformer, the circuit is a proper LCC resonant converter as depicted in Fig. 4 operating in the range of $[15;20]kHz$. The design of the resonant tank is totally decoupled from the MF-Transformer design, allowing parallel studies. The passive resonant elements, namely the series capacitor C_S, the parallel capacitor C_P, and the series inductor L_S are all discrete elements. The passive parasitic elements of the MF-Transformer are not used in this design for allowing better flexibility in the choice of resonant tank elements' values. For the study of the resonant tank, the transformer, secondary side rectifier and the tetrode load can be brought on the primary side as one equivalent load R_P.

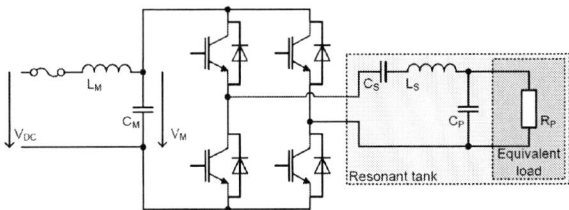

Fig. 4. Structure of the studied resonant converter

D. High Voltage secondary side

On the primary side of the MF-Transformer, the voltage has a pure sinusoidal form due to the resonance. The transformer contains multi windings on the secondary side, all with a one to one ratio, each of which is connected to a diode rectifier. The rectifiers are put in series for reaching the desired voltage, in this case, half of the specified voltage of $25kV$. The transformer elements are designed in such a way that its overall bandwidth is far higher than the resonant tank, at least one decade. The diodes must fulfil strong reliability requirements, at the given operation point. It must withstand twice the winding maximal voltage and switch to full load current at the operating frequency. This part is not detailed in this paper.

III. DESIGN OF RESONANT TANK

Two resonant tank structures were considered as depicted in Fig. 5. The first, LCLC [2], [3], uses the passive elements of the MF-Transformer. This has the advantage of minimizing the number of discrete elements. Beside, the parallel inductor L_P maintains high pass filter with a cut-off frequency that is independent from the load. However, the resonant frequency on which one want to operate is dependant of the parallel capacitor C_P which is very difficult to design in a MF-Transformer. Since one does not want to add capacitors in each secondary winding, the simpler LCC [4], [5] solution with discrete elements on the primary was chosen. The MF-Transformer needs therefore to be designed with a cut-off frequency that is at least one decade higher that the resonant frequency of the resonant tank.

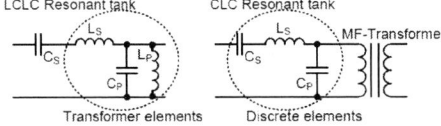

Fig. 5. Two considered resonant tank structures

A. Equivalent circuit and load

The secondary side of the transformer is a diode rectifier with a capacitor and a load R_L. As seen from the resonant tank, the equivalent load is a simple resistor R'_P calculated as in Equ. 3 [6]. As seen on the primary side, the equivalent load resistance R_P is calculated as in Equ. 4.

$$\frac{V_{OUT}}{V_{IN}} = \frac{sR_PC_S}{1 + s(R_SC_S + R_PC_S + R_PC_P) + s^2C_S(L_S + R_SR_PC_P) + s^3L_SR_PC_PC_S} \quad (1)$$

$$\frac{V_{IN}}{I_{IN}} = \frac{1 + s(R_SC_S + R_PC_S + R_PC_P) + s^2C_S(L_S + R_SR_PC_P) + s^3L_SR_PC_PC_S}{sC_S + s^2C_SC_PR_P} \quad (2)$$

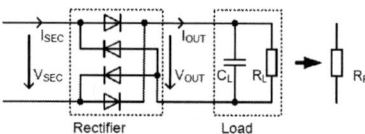

Fig. 6. Equivalent resistive load for output stage

$$R'_P = \frac{8}{\pi^2}R_L \quad (3)$$

$$R_P = \left(\frac{1}{\nu_{TRAFO}}\right)^2 \frac{8}{\pi^2}R_L$$

$$= \left(\frac{1}{12}\right)^2 \frac{8}{\pi^2}\frac{12500}{8} = 9.45\Omega \quad (4)$$

The final system that is being studied and prototyped for validation purpose, is a $100kW$ IGBT bridge with an LCC resonant tank as depicted in Fig. 4. The design of the resonant tank is totally decoupled from the MF-Transformer design, allowing parallel studies. For experimental validation purpose, the equivalent load R_P can be brought as a simple power resistance.

B. Dimensioning of the LCC resonant tank

The studied LCC resonant tank structure is depicted in Fig. 8. The series equivalent impedance Z_S, given by Equ. 5, is given as a function of series discrete capacitor C_S, series discrete inductor L_S with its line resistor R_S. The equivalent parallel impedance Z_P, given by Equ. 6, is a function of parallel discrete capacitor C_P and the equivalent load R_P. The voltage gain is given by Equ. 7 and detailed in Equ. 1. The input impedance as seen from the IGBT bridge is given by Equ. 8 and detailed in Equ. 2.

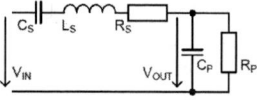

Fig. 8. Structure of the LCC resonant tank

$$Z_S = \frac{1 + sR_SC_S + s^2L_SC_S}{sC_S} \quad (5)$$

$$Z_P = \frac{R_P}{1 + sR_PC_P} \quad (6)$$

$$\frac{V_{OUT}}{V_{IN}} = \frac{Z_P}{Z_S + Z_P} \quad (7)$$

$$\frac{V_{IN}}{I_{IN}} = Z_S + Z_P \quad (8)$$

Some simple design rules are drawn from the behaviour of the overall transfer function. The series inductor L_S limits the resonant current drawn from the IGBT bridge and lead the rule given by Equ. 9. Then, setting the resonance frequency, the parallel capacitor C_P is chosen as a function of L_S as in Equ. 10. Finally, the cut-off frequency of the high-pass filter given by $[C_S; R_P]$ must be lower than the resonant frequency given by $[L_S; C_P]$, leading to the rule given by Equ. 11.

$$L_S > \frac{V_{IN}\nu_{TRAFO}V_{OUT}}{8f_{SW}P_{NOM}} \quad (9)$$

$$C_P = \frac{1}{(2\pi f_0)^2 L_S} \quad (10)$$

$$C_S > \frac{\sqrt{L_SC_P}}{R_P} \quad (11)$$

The full transfer function of the modelled LCC resonant tank, with its designed values is illustrated in Fig. 7a. the characteristics are drawn for full load and no load operation. One can see the effect of the load on the cut-off frequency of the high-pass filter set by $[C_S; R_P]$, the most important is that the resonance frequency is not affected by the filter at the highest load. For harmonic filtering purpose, the operation range is set on the falling characteristic of the transfer function.

A detailed view on the operation range, given by Fig. 7b shows the difference in the transfer function between full load and no load operation. The $+3dB$ gain is reached at $15khz$ at full load, and some $17kHz$ at no load. This shows that the full voltage range is reached at any load condition in the operation frequency range of $[15; 20]kHz$. It is important to keep the impedance curves over the minimal impedance set to limit the semiconductor currents.

C. Effect of parasitic components

When the resonant tank is built, one needs to consider the parasitic elements that may disturb the overall operation. As illustrated in Fig. 10, the main parasitic effect comes from the parallel capacitor C_P, its equivalent series resistor R_{ESR} and equivalent line inductor L_C. The resistor is often given by constructors and should not be higher than the $m\Omega$ range. as for the parasitic inductor, it most often come from the cabling which add a unavoidable $100nH$ in series with the capacitor.

The behaviour of this new modelled system is illustrated in Fig. 9a as a function of the value of the parasitic line inductor. One can see its effect in the higher frequencies, and the limit in inductor value to ensure in order

The 2014 International Power Electronics Conference

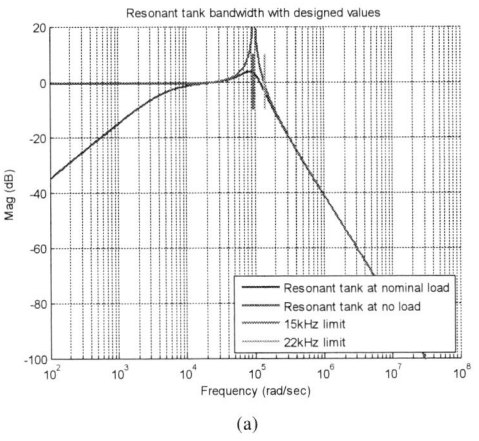
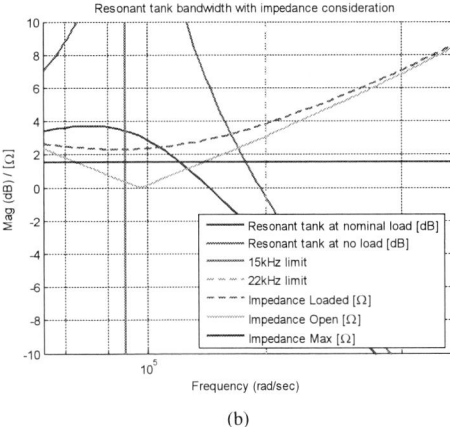

Fig. 7. Transfer function of resonant tank with designed values (a) and zoom on the operating frequencies with tank impedance (b).

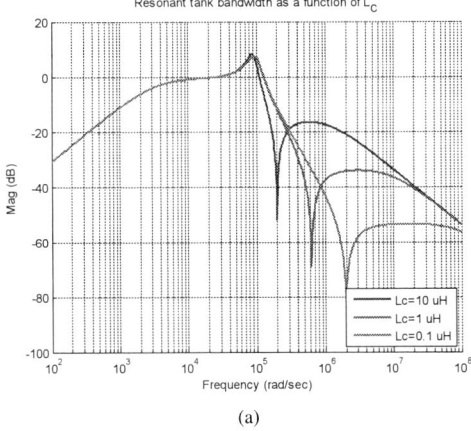
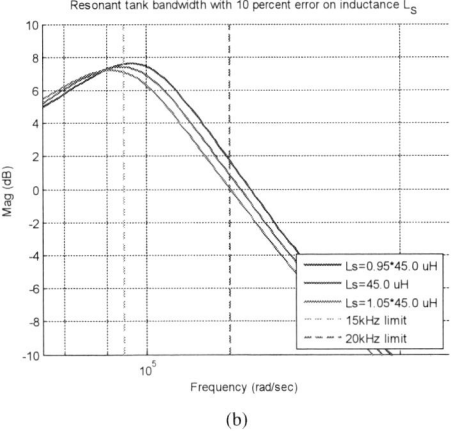

Fig. 9. Transfer function of resonant LCC tank considering parasitic components (a) and 10% error on series inductor value (b).

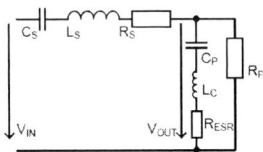

Fig. 10. Structure of the LCC resonant tank with parasitic elements

to assure at least $40dB$ attenuation in the whole high-frequency range. This result however, gives confidence in the model and film capacitor technology. One can see also, that the transfer function in the operation range is not affected by the line inductor, is maintained under $1\mu H$.

D. Inaccuracy in component values and ageing

The control and the resonant tank design should consider inaccuracies in the component values. Namely in the parallel capacitor C_P and the series inductor L_S. First inaccuracy comes in the value itself of the components. It is difficult to build a power inductor with better accuracy than 5% on the whole frequency range as well as on the parallel capacitor C_P. Fig. 9b shows the transfer function

of the tank with $\pm 5\%$ inaccuracy on the inductor, a similar curve is reached by varying the capacitor value. The capacitor lifetime is defined by a range in its value of 2%. This means that the parallel capacitor C_P will lose 2% of its value when reaching its end of lifetime, shifting slowly the transfer function to the higher frequencies.

E. Meeting the Specifications

The designed $100kW$ module must fulfil the specification in output voltage and current, namely $12.5kV$-$8A$ on the secondary side of the step-up transformer. An investigation is made to assess if a given load current can be reached at a given voltage (meaning resonant tank gain) and for which frequency. The final values of the resonant tank must consider the imprecision on the value of the resonant tank elements, the ageing of the capacitors. The worst case appears when the supply network faces 10% dip in all three phases, in which case the resonant tank must reach $3.3dB$ at maximal load and minimal frequency of $15kHz$. The extreme values of the resonant tank elements, namely maximum de-rating against maximum over-rating, are computed in order to

978-1-4799-2706-7/14 $31.00 © 2014 IEEE

The 2014 International Power Electronics Conference

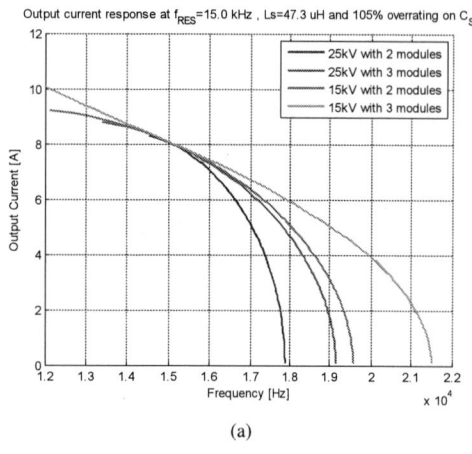
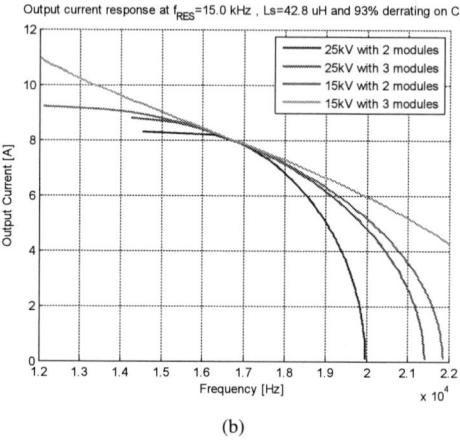

(a)   (b)

Fig. 11. Load current as a function of operating frequency for meeting the output voltage, with maximum overrating of resonant tank elements (a) and maximum derating (b).

set the main resonant frequency in such a way that all operation points can be reached.

The values of the elements of the resonant tank as given in Table I allow reaching the minimum operation point of 8 A with two working modules and a maximum overrating of resonant tank elements of 5%, as illustrated on Fig. 11a. The effect of a maximum dc-rating of 5% and the ageing of 2% of resonant tank elements is illustrated on Fig. 11b. The minimum frequency of 15 kHz is reached at maximum output current.

TABLE I.   DESIGNED VALUES FOR THE LCC RESONANT TANK

Component	Value
Series Inductor L_S	$45\mu F$
Parallel capacitor C_P	$2.5\mu F$
Series capacitor C_S	$30\mu F$
Load condition R_P	7.1Ω

IV.   CONTROL AND MODULATION

The circuit is simulated and prototyped for the validation of the control principles and the overall design. The validation is performed on a one module circuit with a transformer containing only one secondary winding. Two types of controllers are tested for the maintaining of the output voltage inducing naturally two types of modulation schemes which have an obvious effect on the overall switching mechanisms and the harmonics as seen from the load side as well as the grid side.

A. Implemented Circuit

The studied circuit described by Fig. 12 is a single module structure. The input stage is a three phase diode rectifier with two filters in series for reducing the harmonic effect from the grid to the load, as well as reducing the injection of the high frequency harmonic content of the resonant converter to the grid. The line filter set to reduce de $300Hz$ oscillation is a $C - 4C$ structure with optimized dumping for avoiding oscillations while facing negative impedance due to the constant load. The

module filter is set to dump high frequency oscillations as well as for avoiding perturbations from one module to another (when three of them are connected in parallel). The resonant tank is connected to a MF-Transformer of a ratio 1 to 12. For simulation purpose and prototype verification, only one secondary winding is connected to a diode bridge rectifier, with one capacitor for dumping high frequency oscillations. The load is a current source which value changes from 1 to $8A$.

B. Switching mechanisms

The resonant tank is operated above the resonance frequency, which means that the switches are operated with soft turn-ON and hard turn-OFF. When the switching frequency gets closer from the resonant frequency, the current to turn-OFF gets lower, reducing switching losses, but also reducing the (dis-)charging time of capacitor snubbers. Switching mechanisms have two different behaviours, depending on whether the modulation is done on two levels or three levels.

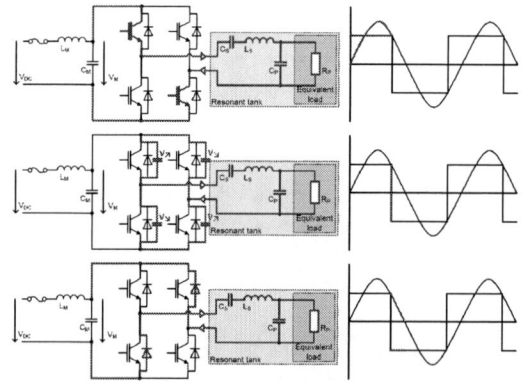

Fig. 14.   Switching mechanisms for positive current and two levels modulation.

The two levels modulation is illustrated on Fig. 14 for positive current. When applying positive voltage, the two opposite switches of the H-bridge are conducting.

978-1-4799-2706-7/14 $31.00 © 2014 IEEE   3375

The 2014 International Power Electronics Conference

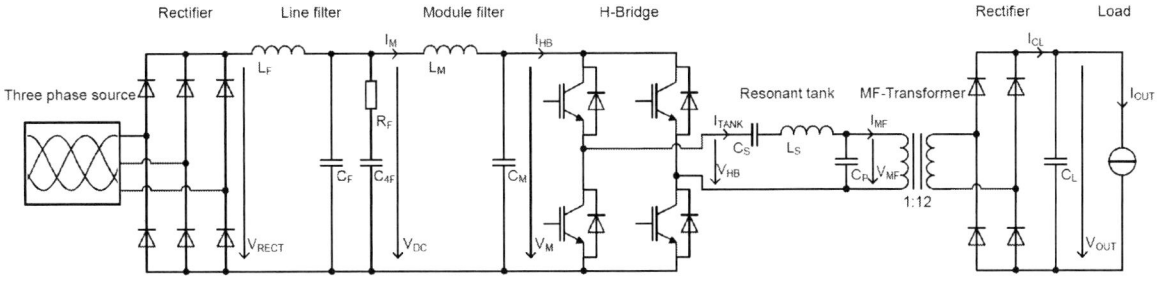

Fig. 12. Full simulated converter model of the 100kW resonant module.

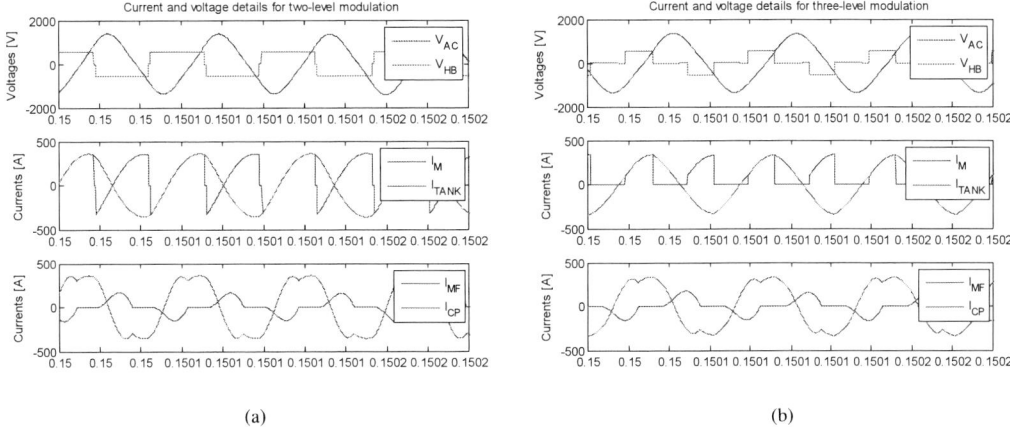

Fig. 13. Details on currents and voltages in resonant tank, with two levels modulation (a), three levels modulation (b).

When switching OFF, the current from the resonant tank naturally charges the snubber capacitors of the previously conducting switches, while discharging the snubber capacitors of the switches to be turned-ON. When the charge is finished, the diodes take the current, and the active switches are turned-ON at zero current.

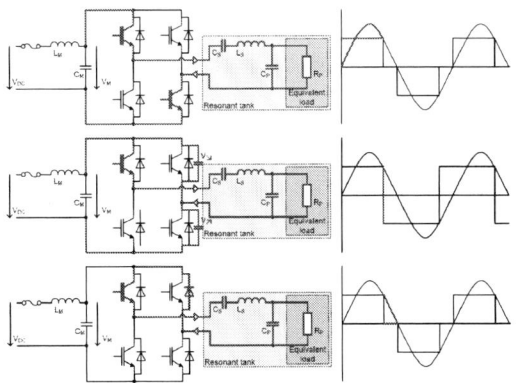

Fig. 15. Switching mechanisms for positive current and three levels modulation.

The three levels modulation is illustrated on Fig. 15 for positive current. When applying positive voltage, the two opposite switches of the H-bridge are conducting. When switching OFF the devices of one leg only, again the current from the resonant tank naturally charges the snubber capacitors of the previously conducting switch,

while discharging the snubber capacitors of the switch to be turned-ON [7]. When the charge is finished, the diodes take the current, and the two active switches are turned-ON at zero current.

The main differences between the two modulation schemes are illustrated on Fig. 13a for the two level case and on Fig. 13b for the three level case. The current I_M taken from the module capacitor C_M and the grid is much lower in the three-phase case because of the bigger free-wheeling time left to the resonant tank. The switching frequency operation is much closer to the resonant frequency, meaning that less current is needed to actually maintain the resonance. With two levels modulation, the operating frequency gets far from the resonant frequency at lower loads, allowing a better switching of snubber capacitors, but much higher current to maintain the resonance.

C. Frequency control

The two levels modulation is associated with the frequency control of the resonant tank. A simple PI controller maintains the output DC voltage V_{OUT} independently of the load. The controller corrects the output voltage by adjusting the switching frequency whose dynamics must respond the the load dynamics. Fig. 16a shows the simulated behaviour of the converter for a current step of I_{OUT} in the output load. The voltage is maintained within the acceptable limits of $\pm 10V$, with

978-1-4799-2706-7/14 $31.00 © 2014 IEEE

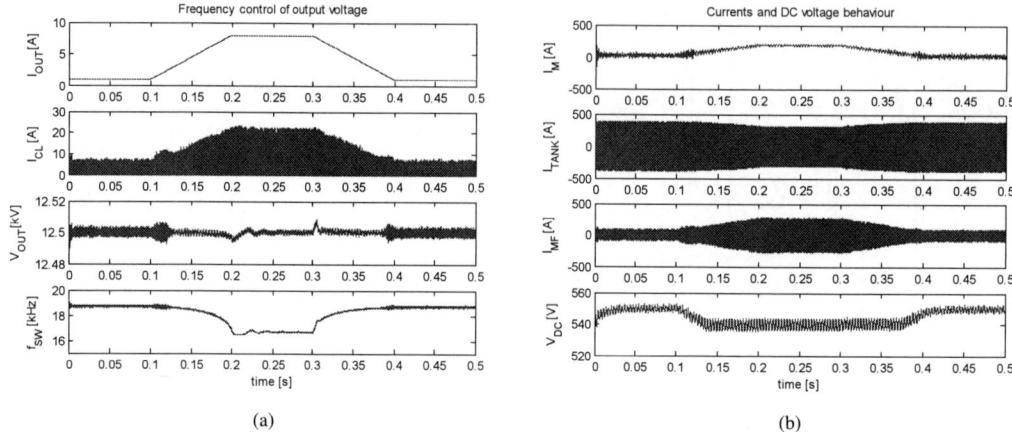

Fig. 16. System behaviour with frequency control, output values and control variable (a) input values (b).

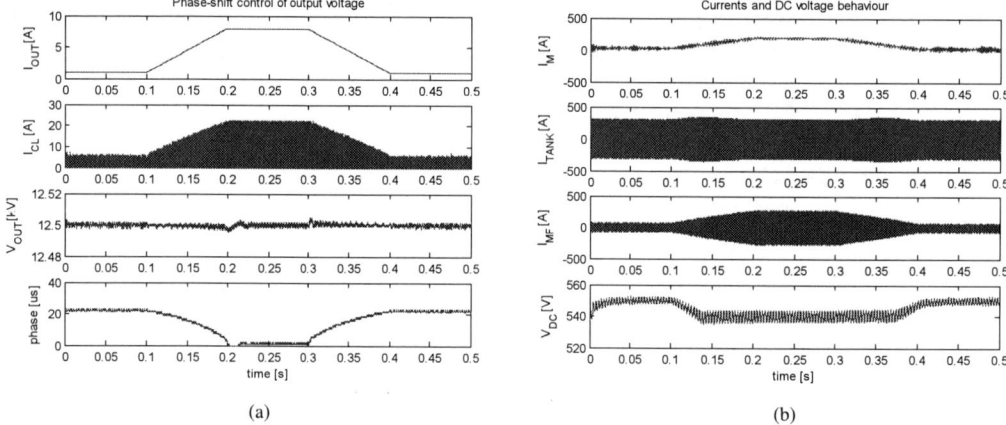

Fig. 17. System behaviour with phase-shift control, output values and control variable (a) input values (b).

transients due to the current variations. The parameters of the PI controller were set with successive iterations, it is therefore difficult to assess whether the optimal values were found. However, the specifications given for this converter are already met. Fig. 16b show the variations in current and voltage of input values, namely the DC link Voltage V_{DC} and the current I_M actually taken from the grid. This shows that most of the current in the resonant tank is actually circulating reactive power and the grid only provides the losses when the load is zero.

D. Phase-shift control

The three levels modulation is associated with the amplitude control of the voltage applied at the resonant tank. The operating switching frequency is constant and the output voltage V_{OUT} is corrected by a simple PI controller that adjusts the phase-shift between the two legs of the H-bridge. Fig. 17a shows the simulated behaviour of the converter for the same I_{OUT} current step as previously applied with the frequency control. The controller parameters are also set with successive iterations and the transient phenomena seem similar if not better than the frequency control. Moreover, the harmonic content of the

output DC link, especially at lower loads, is definitely better. This also reflects on the harmonic content in current and voltage of input values as seen in Fig. 17b.

V. IMPLEMENTATION OF THE 100kW RESONANT POWER MODULE

For validation purpose, a 100kW prototype with one module only is being constructed. The elements are selected with the objective of 20 years lifetime. The most sensitive part being the semiconductors themselves, a careful look is given on thermal cycling. The resonant tank elements are selected keeping in mind their imprecision in value and finally the overall integration of the module is given.

A. Semiconductors

The semiconductor switches are four IGBTs *CM600HA-24A* in individual cases from *Mitsubishi - Powerex*. Considering soft turn-ON and hard turn-OFF at maximum current being the worst operation point, and that for no load there is still half of the current flowing in the resonant tank, the thermal cycling over $1.2s$ cycle can be predicted as in Table II.

TABLE II.   THERMAL CYCLING CALCULATIONS FOR IGBTS

Peak current	$400A$	$200A$
RMS current	$140A$	$70A$
Collector emitter forward voltage	$1.3V$	$1V$
Diode forward votlage	$1.4V$	$1.2V$
E turn ON	$5mJ$	$5mJ$
E turn OFF	$45mJ$	$25mJ$
E reverse recovery	$10mJ$	$10mJ$
Switching losses	$1200W$	$800W$
Conduction losses	$182W$	$70W$
Total loss per IGBT	$1382W$	$870W$
sink temperature	$40^\circ C$	
R_{th} case-to-sink	$0.015K/W$	
R_{th} junction-to-case	$0.034K/W$	
case temperature	$61^\circ C$	$53^\circ C$
Junction temperature	$107^\circ C$	$83^\circ C$
Thermal cycling (1s)	$24^\circ C$	

B. Resonant tank

The values of the resonant tank elements are given in Table I. The series capacitor C_S is the commercial model *FAI66F0306K* from *AVX-TPC* and the parallel capacitor C_P is a custom model also manufactured by *AVX-TPC*. The series inductor L_S is manufactured by *Transrail - Boige et Vignal*, the dimensions for such powers reach the dimensions of $0.3 \times 0.2 \times 0.3$ in meters and some $40kg$. For this prototype, three different values were asked, therefore its dimensions are even bigger, as one can see on Fig. 18. The other components can be easily integrated in a module.

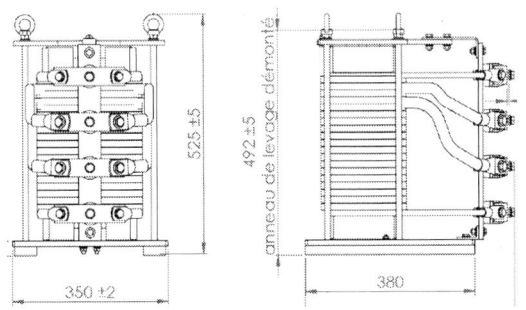

Fig. 18.   Schematic of the tapped inductor ordered for prototyping purpose.

C. Integration

The integration plan is given in Fig. 19. The module capacitor C_M is as close as possible from the semiconductors. A part from the series resonant inductor L_S all

elements are implemented in the $100kW$ module. The cooling of semiconductors is done by demineralized water with a circuit designed for removing up to $6kW$ with a difference in temperature of $10^\circ C$.

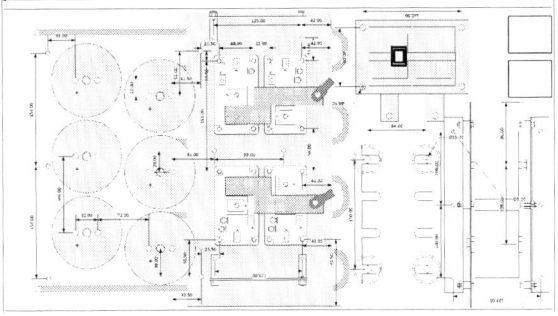

Fig. 19.   Implementation of the 100kW resonant module

VI.   CONCLUSIONS

This paper presents design guideline for building a $100kW$ resonant power module that is to be used in a modular resonant power converter for a $300kW$ application. The first 100kW prototype is about to be finished. It will allow to verify control principles and switching mechanisms in the semiconductors before implementing the full converter.

REFERENCES

[1] Kazimierczuk, M.K.; Czarkowski D., "Resonant Power converters", second edition, Willey, 2011.

[2] Shafiei, N.; Pahlevaninezhad, M.; Farzanehfard, H.; Motahari, S.R., "Analysis and Implementation of a Fixed-Frequency LCLC Resonant Converter With Capacitive Output Filter, "Industrial Electronics, IEEE Transactions on , vol.58, no.10, pp.4773,4782, Oct. 2011.

[3] Ang, Y.A.; Bingham, C.M.; Foster, M.P.; Stone, D.A.; Howe, D., "Design oriented analysis of fourth-order LCLC converters with capacitive output filter," Electric Power Applications, IEE Proceedings - , vol.152, no.2, pp.310,322, 4 March 2005.

[4] Rui Yang; HongFa Ding; Yun Xu; Lei Yao; YingMeng Xiang, "An Analytical Steady-State Model of LCC type Series-Parallel Resonant Converter With Capacitive Output Filter," Power Electronics, IEEE Transactions on , vol.29, no.1, pp.328,338, Jan. 2014.

[5] Martin-Ramos, J.A.; Villegas Saiz, P.J.; Pernia, A.M.; Diaz, J.; Martinez, J.A., "Optimal Control of a High-Voltage Power Supply Based on the PRC-LCC Topology With a Capacitor as Output Filter," Industry Applications, IEEE Transactions on , vol.49, no.5, pp.2323,2329, Sept.-Oct. 2013.

[6] Erickson, R.W.; Maksimovic, D., "Fundamentals of Power Electronics", Chapter 19, Springer, 2001.

[7] Ranstad, P.; Nee, H-P; Linner, J., "A novel control strategy applied to the series loaded resonant converter," Power Electronics and Applications, 2005 European Conference on.

The 2014 International Power Electronics Conference

A Novel Transformer-less Interleaved Four-Phase High Step-down DC Converter with Low Switch Voltage Stress

Ching-Tasi Pan,
Department of Electrical
Engineering, National
TsingHua University,
Hsinchiu 30013, Taiwan,
ctpan@ee.nthu.edu.tw.

Chen-Feng Chuang
Department of Electrical
Engineering, National
TsingHua University,
Hsinchiu 30013, Taiwan,
thomas3628f3@gamil.com

Chia-Chi Chu
Department of Electrical
Engineering, National
TsingHua University,
Hsinchiu 30013, Taiwan,
ccchu@ee.nthu.edu.tw

Hao-Chien Cheng
Hardware R&D Engineer,
ASUSTeK Computer Inc,
Taipei, Taiwan
daryldemi@hotmail.com

Abstract- In this paper, a novel transformer-less interleaved four-phase high step-down conversion ratio dc–dc converter with low switch voltage stress is proposed. In the proposed converter, the new capacitors switching circuits are combined with interleaved four-phase buck converter in order to get a high step-down conversion ratio without adopting an extreme short duty ratio. Based on the capacitive voltage division, the main objectives of the capacitors switching circuits in the converter are both storing energy in the blocking capacitors for increasing the voltage conversion ratio and reducing voltage stresses of active switches. This will allow one to choose lower voltage rating MOSFETs to reduce both switching and conduction losses, and the overall efficiency is consequently improved. In addition, due to the charge balance of the blocking capacitor, the converter features automatic uniform current sharing characteristic of the interleaved phases without adding extra circuitry or complex control methods. The operating principle and steady-state analyses of the voltage gain are discussed. Finally, a 400-V input voltage, 24-V output voltage, and 500W output power prototype circuit is implemented in the laboratory to verify the performance.

Keywords : **Low switch voltage stress, high step-down converter, uniform current sharing characteristic**

I. INTRODUCTION

Recently the high-performance dc–dc converters have been called for the increasing high step-down ratios with high output current rating applications, such as VRMs of CPU boards and battery chargers, and distributed power systems [1]–[3]. For non-isolation applications with low output current ripple requirement, an interleaved buck converter (IBC) has received a lot of attention due to its simple structure and low control complexity.

However, in the conventional interleaved buck converter shown in Fig. 1(a), due to active switches devices suffer from the input voltage, high-voltage devices rated above the input voltage should be applied. High-voltage-rated devices are generally with poor characteristics such as high cost, large on-resistance, large voltage drop, and severe reverse recovery, etc. These limit the switching frequency of the converter and impact the power density improvement. For high-input and low-output voltage regulation applications, pursuing higher power density and better dynamics, it is required operating at higher switching frequencies [4] that will increase both switching and conduction losses. Consequently, the efficiency is further deteriorated. Also, it experiences an extremely short duty

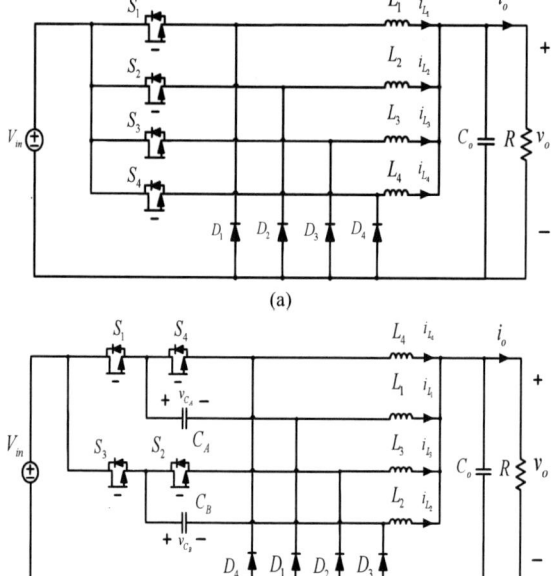

(a)

(b)

Fig. 1. Configuration of (a) conventional interleaved buck converter
(b) four-phase extended duty ratio interleaved buck converter

cycle in the case of high-input and low-output voltage applications.

To overcome the drawbacks of the conventional interleaved buck converter (IBC), a new extended duty ratio multiphase topology has been proposed [5]-[7] is shown in Fig. 1(b). Extended duty ratio (ExtD) mechanisms are very efficient input voltage dividers which reduce the switching voltage and associated losses. However, it cannot achieve automatic uniform current sharing characteristic of the interleaved four-phase and the voltages stress of switches and diodes devices are remains rather high.

In this paper, a novel transformer-less interleaved four-phase high step-down conversion ratio dc–dc converter with low switch voltage stress is proposed. In the proposed converter, two capacitors are series-charged by input voltage and parallel-discharged by a new four-phase interleaved buck converter for providing a much higher step-down conversion ratio without adopting an extreme short duty cycle. Based on the capacitive voltage division, the main objectives of the new voltage-divider circuit in the

978-1-4799-2706-7/14 $31.00 © 2014 IEEE   3379

converter are both storing energy in the blocking capacitors for increasing the step-down conversion ratio and reducing voltage stresses of active switches. As a result, the proposed converter topology possesses the low switch voltage stress characteristic. This will allow one to choose lower voltage rating MOSFETs to reduce both switching and conduction losses, and the overall efficiency is consequently improved. Moreover, due to the charge balance of the blocking capacitor, the converter features automatic uniform current sharing characteristic of the interleaved phases without adding extra circuitry or complex control methods.

The remaining contents of this paper may be outlined as follows. First, the novel circuit topology and operation principle are given in section II. Then the corresponding steady state analysis is made in section III to provide some basic converter characteristics. A prototype is then constructed and experimental results are then presented in section IV for demonstrating the merits and validity of the proposed converter. Finally, some conclusions are offered in the last section.

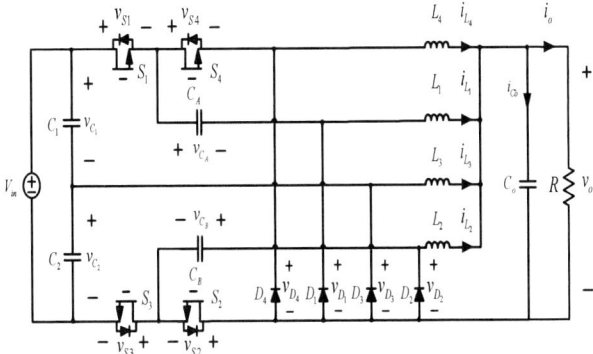

Fig. 2. The circuit configuration of the proposed converter

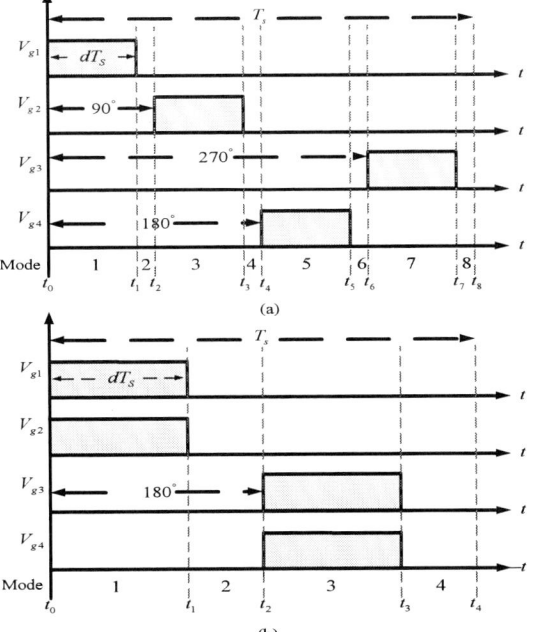

(a)

(b)

Fig. 3. Gate signals of the converter: (a) $0 < D < 0.25$ (b) $0.25 < D < 0.5$

II. OPERATING PRINCIPLE

The proposed novel transformer-less interleaved four-phase high step-down converter is shown in Fig. 2, which is derived from two-phase extended duty ratio interleaved buck converter in [7]. In order to further reduce input current ripple and output voltage ripple, the converter is divided into four-phase small inductors via interleaved operation to minimize those ripples. From Fig. 2 one can see that the proposed converter consists of four inductors, four active power switches, four diodes and four capacitors. The main objectives of the new voltage-divider circuit are twofold. First, they are used to store energy as usual. Second, based on the capacitive voltage division principle, they are used to reduce the voltage stress of active switches as well as increasing the step-down conversion ratio as will be obvious from later explanation. For the theoretical analysis, it will be considered that input and output voltages are ripple free, and all devices are ideal. The system is under steady state and operating in continuous conduction mode (CCM) with duty ratio being less than 0.5 for high step-down conversion ratio purpose. As the main objective is to obtain same step-down conversion ratio and automatic current sharing characteristics can be achieved when the duty cycle is less than 0.5, the converter operating duty ratio $D < 0.25$ [Fig. 3(a)] and duty ratio $D > 0.25$ [Fig. 3(b)] can be controlled by four-phase interleaved and two phase interleaved scheme respectively; hence, the illustrations for the operating modes of the proposed converter and steady-state analysis are made as follows for these two cases, respectively. Referring to the gate signals shown in Fig. 3, the corresponding operating modes of the proposed converter when $0 < D < 0.25$ (Fig. 4) are:

1) **Mode 1** ($t_0 < t \leq t_1$): In this operation mode, switch S_1 is turned on, switch S_2, S_3 and S_4 remain off. Hence, Diode D_1 becomes turned off and diode D_2, D_3 and D_4 remain on. The corresponding equivalent circuit is shown in Fig. 4(a). From Fig. 4(a) it is seen that the stored energy of C_1 is discharged to C_A, L_1, and output load and current i_{L2}, i_{L3} and i_{L4} are freewheeling through D_2, D_3 and D_4 respectively. The V_{L2}, V_{L3} and V_{L4} are equal to $-V_{CO}$, and hence, i_{L2}, i_{L3} and i_{L4} decrease linearly. The voltage across diode D_1 is clamped to V_{C1} minus V_{CA}. The voltage across switch S_3 is clamped to V_{C2} minus V_{CB} and the voltage across the switch S_2 and S_4 are clamped to V_{CB} and V_{C1} respectively.

2) **Mode 2, 4, 6, 8** [$t_{k-1} < t \leq t_k$, $k \in \{2, 4, 6, 8\}$]: For this operation mode, switch S_1, S_2, S_3 and S_4 are off. The corresponding equivalent circuit is shown in Fig. 4(b). From Fig. 4(b), one can see that i_{L1}, i_{L2}, i_{L3} and i_{L4} are freewheeling through D_1, D_2, D_3 and D_4 respectively. All V_{L1}, V_{L2}, V_{L3} and V_{L4} are equal to $-V_{CO}$, and hence, i_{L1}, i_{L2}, i_{L3} and i_{L4} decrease linearly. During this mode, the voltage across S_1, namely V_{S1}, is equal to the difference of V_{C1} and V_{CA}, and V_{S2} is clamped at V_{CB}. Similarly, the voltage across S_3, namely V_{S3}, is equal to the difference of V_{C2} and V_{CB}, and V_{S4} is clamped at V_{CA}.

3) **Mode 3** ($t_2 < t \leq t_3$): During this mode, D_2 becomes turned off while S_2 is turned on. The corresponding equivalent circuit is shown in Fig. 4(c). From Fig. 4(c) one can see that the stored energy of C_B is discharged to L_2 and output load and i_{L1}, i_{L3} and i_{L4} are freewheeling through D_1, D_3, and D_4 respectively. The inductor L_1,

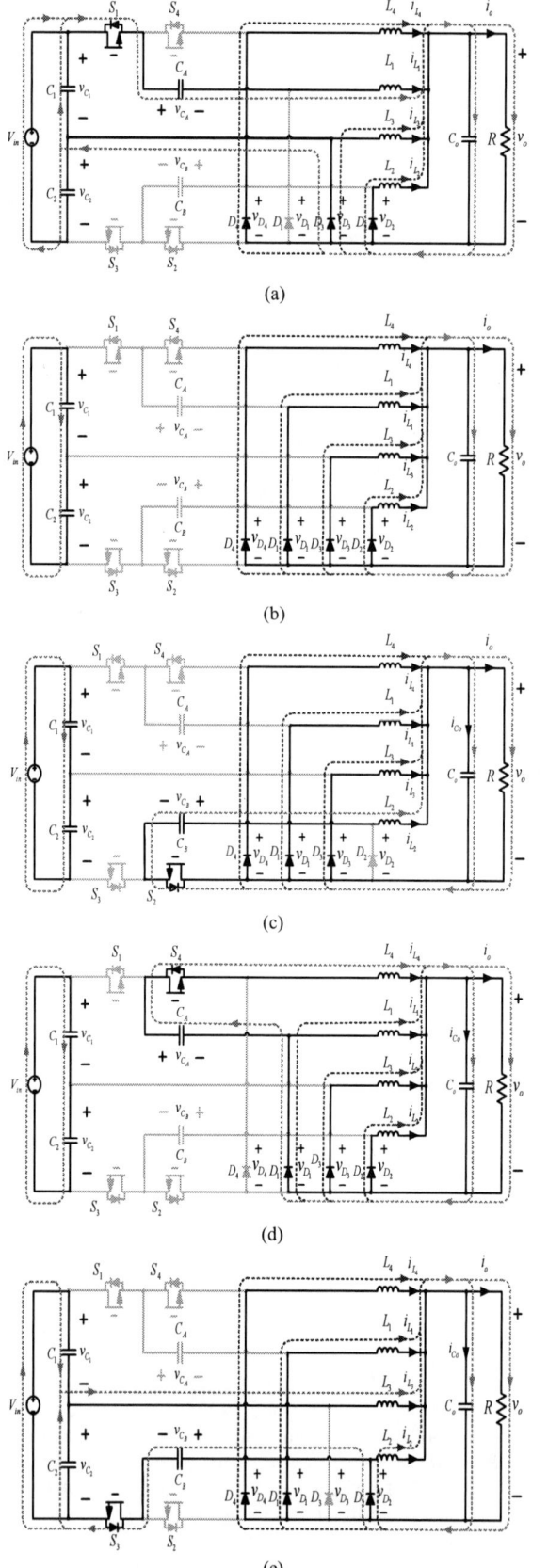

Fig. 4. Equivalent circuits of the proposed novel converter topology for each operating modes when $0 < D < 0.25$: (a) mode 1, (b) mode 2, 4, 6, 8, (c) mode 3,(d) mode 5, and (e) mode 7

L_3, and L_4 are releasing energy to output load. The voltage across diode D_2 is clamped to V_{CB}. The voltage across switch S_1 is clamped to V_{C1} minus V_{CA} and the voltage across the switch S_3 and S_4 are clamped to V_{C2} and V_{CA} respectively.

4) **Mode 5** ($t_4 < t \leq t_5$): During this mode, D_4 becomes turned off while S_4 is turned on. The corresponding equivalent circuit is shown in Fig. 4(d). From Fig. 4(d) one can see that the stored energy of C_A is discharged to L_4 and output load and i_{L1}, i_{L2} and i_{L3} are freewheeling through D_1, D_2, and D_3 respectively. The inductor L_1, L_2, and L_3 are releasing energy to output load. The voltage across diodes D_4 is clamped to V_{CA}. The voltage across switch S_1 is clamped to V_{C1} minus V_{CA} and the voltage across the switch S_2 and S_3 are clamped to V_{CB} and V_{C2} minus V_{CB} respectively.

5) **Mode 7** ($t_6 < t \leq t_7$): During this mode, D_3 becomes turned off while S_3 is turned on. The corresponding equivalent circuit is shown in Fig. 4(e). From Fig. 4(e) it is seen that the stored energy of C_2 is discharged to C_B, L_3, and output load and i_{L1}, i_{L2} and i_{L4} are freewheeling through D_1, D_2 and D_4 respectively. All V_{L1}, V_{L2} and V_{L4} are equal to $-V_{CO}$, and hence, i_{L1}, i_{L2} and i_{L4} decrease linearly. The voltage across diode D_3 is clamped to V_{C2} minus V_{CB}. The voltage across switch S_1 is clamped to V_{in} minus V_{CA} $+V_{CB}$ and the voltage across the switch S_2 and S_4 are clamped to V_{CB} and V_{CA} respectively.

Similarly, referring to the gate signals shown in Fig. 2(b), the corresponding operating modes of the proposed converter when $0.25 < D < 0.5$ (Fig. 5) are:

1) **Mode 1** ($t_0 < t \leq t_1$): For this operation mode, the switch of S_1 and S_2 are turned on. The corresponding equivalent circuit is shown in Fig. 5(a). From Fig. 5(a) one can see that during this mode current i_{L3} and i_{L4} are freewheeling through D_3 and D_4 respectively, and L_3 and L_4 are releasing energy to output load. The stored energy of C_1 is discharged to C_A, L_1, and output load and the stored energy of C_B is discharged to L_2 and output load. Therefore, the voltage across diode D_1 and D_2, are clamped to V_{C1} minus V_{CA} and V_{CB} respectively. The voltage across the switch S_3 and S_4 are clamped to V_{C2} and V_{C1} respectively.

2) **Mode 2** ($t_1 < t \leq t_2$): For this operation mode, the switch of S_1, S_2, S_3 and S_4 are off. The corresponding equivalent circuit is shown in Fig. 5(b). From Fig. 5(b), one can see that i_{L1}, i_{L2}, i_{L3} and i_{L4} are freewheeling through D_1, D_2, D_3 and D_4 respectively. All V_{L1}, V_{L2}, V_{L3} and V_{L4} are equal to $-V_{CO}$, and hence, i_{L1}, i_{L2}, i_{L3} and i_{L4} decrease linearly. During this mode, the voltage across S_1, namely V_{S1}, is equal to the difference of V_{C1} and V_{CA}, and V_{S2} is clamped at V_{CB}. Similarly, the voltage across S_3, namely V_{S3}, is equal to the difference of V_{C2} and V_{CB}, and V_{S4} is clamped at V_{CA}.

3) **Mode 3** ($t_2 < t \leq t_3$): For this operation mode, the switch of S_3 and S_4 are turned on. The corresponding equivalent circuit is shown in Fig. 5(c). From Fig. 5(c) one can see that during this mode current i_{L1} and i_{L2} are freewheeling through D_1 and D_2 respectively, and L_1 and L_2 are releasing energy to output load. The stored energy of C_2 is discharged to C_B, L_3, and output load and the stored energy of C_A is discharged to L_4 and output load.

Therefore, the voltage across diode D_3 and D_4, are clamped to V_{C2} minus V_{CB} and V_{CA} respectively. The voltage across the switch S_1 and S_2 are clamped to V_{in} minus $V_{CA}+V_{CB}$ and V_{CB} respectively.

4) **Mode 4** ($t_3 < t \le t_4$): For this operation mode, as can be observed from Fig. 3(b), all switches are turned off, D_1, D_2, D_3, D_4 are all on. The corresponding equivalent circuit turns out to be the same as Fig. 5(b).

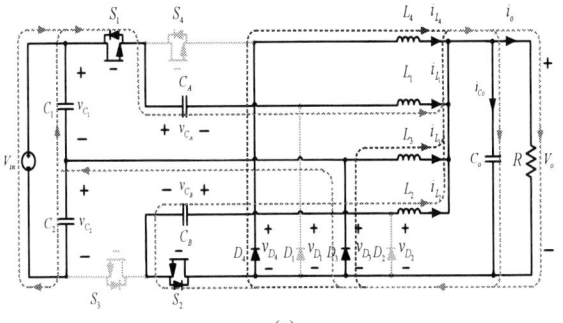

(a)

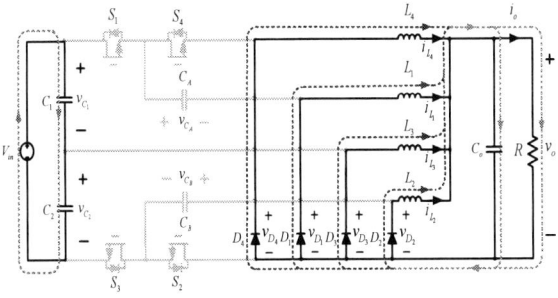

(b)

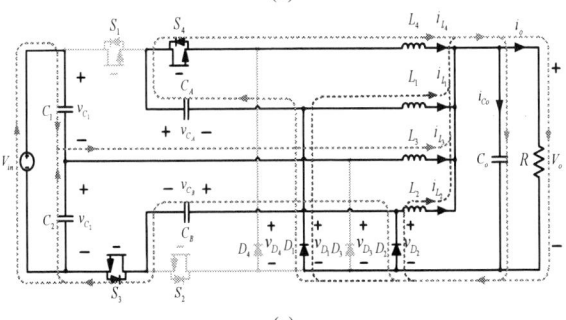

(c)

Fig. 5. Equivalent circuits of the proposed novel converter topology for each operating modes when $0.25 < D < 0.5$: (a) mode 1 (b) mode 2, 4 (c) mode 3.

III. STEADY-STATE ANALYSIS

In order to simplify the circuit analysis of the proposed converter, some assumptions are made as follows.

(1) All components are ideal components.

(2) The capacitors are sufficiently large such that the voltages across them can consider as constant. Also, assume that $C_1=C_2$, $C_A=C_B$.

(3) The system is under steady state and is operating in continuous conduction mode (CCM) and with duty ratio being less than 0.5 for high step-down conversion ratio purpose.

A. Conversion ratio

Referring to Fig. 4(a) -4(e) or Fig. 5(a)-5(c), from the volt–second relationship of inductor L_1-L_4 one can obtain the following relations:

$$(V_{C1} - V_{CA} - V_O)D - V_O(1 - D) = 0 \tag{1}$$

$$(V_{CB} - V_O)D - V_O(1 - D) = 0 \tag{2}$$

$$(V_{C2} - V_{CB} - V_O)D - (1 - D)V_O = 0 \tag{3}$$

$$(V_{CA} - V_O)D - (1 - D)V_O = 0 \tag{4}$$

Also from the equation (1) and (3), voltage V_{C1} and V_{C2} can be derived as follow by substituting the V_{CA} and V_{CB} solutions of (2) and (4).

$$V_{C1} = \frac{2}{D}V_O \tag{5}$$

$$V_{C2} = \frac{2}{D}V_O \tag{6}$$

It follows from (5) and (6) that the output voltage can be obtained as below.

$$V_{in} = V_{C1} + V_{C2} = \frac{4}{D}V_O \tag{7}$$

Thus, the step-down conversion ratio of the proposed converter can be obtained as follows:

$$M_{Step-dwon} = \frac{V_O}{V_{in}} = \frac{D}{4} \tag{8}$$

B. Voltage Stresses on Semiconductor Components

To simplify the voltage stress analyses of the components of the proposed converter, the voltage ripples on the capacitors are ignored. From Fig. 4(a) - 4(e) or Fig. 5(a)-5(c), one can see that the voltage stresses diodes D_1- D_4 can be obtained directly as shown in the following equation.

$$V_{D1,max} = V_{D2,max} = V_{D3,max} = V_{D4,max} = \frac{V_{in}}{4} \tag{9}$$

From (9), one can see that the voltage stress of diodes of the proposed converter is equal to one fourth of the input voltage. Hence, the proposed converter enables one to adopt lower voltage rating devices to further reduce conduction losses.

As can be observed from the equivalent circuits in Fig. 4(a) - 4(e) or Fig. 5(a)-5(c), the open circuit voltage stress of switches S_1, S_2, S_3 and S_4 can be obtained directly as shown in (10).

$$V_{S1,max} = V_{S3,max} = V_{S4,max} = \frac{V_{in}}{2}, \; V_{S2,max} = \frac{V_{in}}{4} \tag{10}$$

In fact, one can see from (10) that the maximum resulting voltage stress of switch S_1, S_3 and S_4 is equal to $V_{in}/2$, and the maximum voltage stress of switch S_2 is equal to $V_{in}/4$. Hence, the proposed converter enables one to adopt lower voltage rating switches to further reduce both switching and conduction losses.

C. Characteristic of uniform input inductor current sharing

By using the state space averaging technique one can get the averaged state equations quite straightforward as follows:

$$0 = \frac{D(-I_{L1} + I_{L3})}{C_1 + C_2} \tag{11}$$

$$0 = \frac{D(I_{L1} - I_{L4})}{C_A} \tag{12}$$

978-1-4799-2706-7/14 $31.00 © 2014 IEEE   3382

$$0 = \frac{D\left(-I_{L2} + I_{L3}\right)}{C_B} \tag{13}$$

$$0 = \frac{\left(I_{L1} + I_{L2} + I_{L3} + I_{L4}\right)R - V_{C_O}}{RC_O} \tag{14}$$

Where $I_o = \dfrac{V_{C_O}}{R}$

It follows from the equation (11) - (14) that one can get the corresponding DC solutions as follows:

$$I_{L1} = I_{L2} = I_{L3} = I_{L4} = \frac{I_O}{4} \tag{15}$$

From (15) one can see that the proposed converter possesses the inherent automatic uniform current sharing capability.

D. Output Current Ripples

Fig. 6 shows the output current and inductor currents of the proposed converter. The output current consists of inductor current i_{L1}, i_{L2}, i_{L3}, and i_{L4}. The waveform of the output current can be obtained by the waveforms of the four inductor currents. Thus, adding the inductor current i_{L1}, i_{L2}, i_{L3} and i_{L4} the output current ripple can be derived as follows:

$$\Delta i_{ripple} = \begin{cases} D(1-4D)\dfrac{V_{in}}{4Lf_S} & 0 < D < 0.25 \\[2mm] D(1-2D)\dfrac{V_{in}}{2Lf_S} & 0.25 < D < 0.5 \end{cases} \tag{16}$$

It follows from (16) that the output current ripple of the proposed converter is affected by the input voltage, inductor value, switching duty ratio, and switching frequency. The normalized output current ripple of the proposed converter is given as (17).

$$\Delta i_{ripple}^N = \begin{cases} M\left(1-16M\right) & 0 < D < 0.25 \\[2mm] 2M\left(1-4M\right) & 0.25 < D < 0.5 \end{cases} \tag{17}$$

E. Performance Comparison

For demonstrating the performance of the proposed converter, the proposed converter is compared with conventional IBC and extended duty ratio IBC as shown in Table I.

Table I summarize the conversion ratio and normalized voltage stress of active as well as passive switches for reference. For comparison, the voltage stress is normalized by the input voltage V_{in}, the conversion ratio M, the normalized switch stresses and the normalized output diode stresses of the conventional IBC and the extended duty ratio IBC are also shown in the same table to provide better view.

It is seen from Table I that the proposed converter can achieve higher step-down conversion ratio than that of the other two IBC converters. Therefore, the proposed converter is rather suitable for use in applications requiring high step-down conversion ratio. From Table I, one can see

that the proposed converter can achieve the lowest voltage stress for the active switches. Also, it is seen that the proposed converter can achieve the lowest voltage stress for the diodes. As a result, one can expect that with proper design the proposed converter can adopt switch components with lower voltage ratings to achieve higher efficiency.

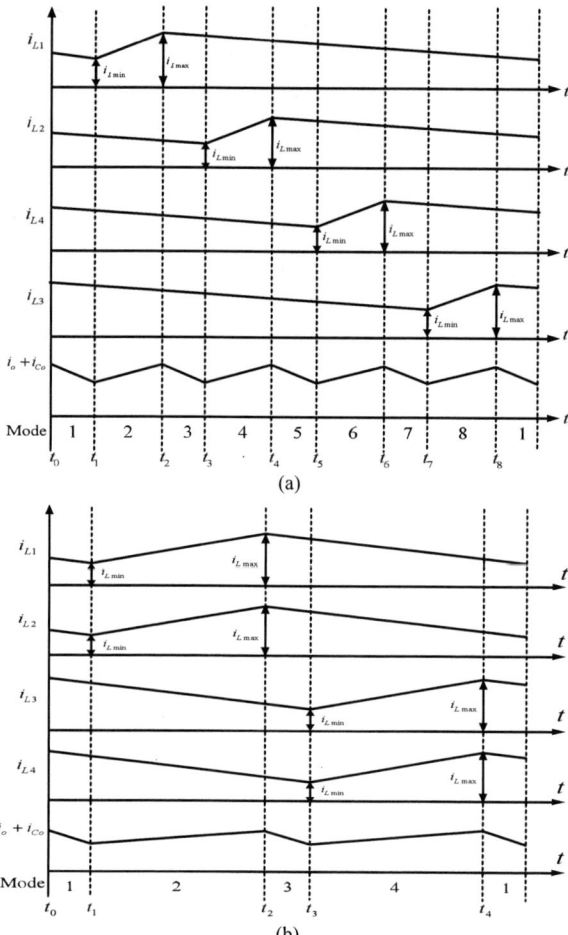

Fig. 6. Inductor currents and output current waveforms of the proposed converter during (a) 0<D<0.25 (b) 0.25<D<0.5 operation conditions

Table I. Comparison of the steady state characteristics

Gain/stress	Conventional IBC	Extended duty ratio IBC	Proposed converter	
Conversion Ratio M	D	D/2	D/4	
Voltage stress of switches $S_{1,3}$	1	1/2	S_1	1/2
			S_3	
Voltage stress of switches $S_{2,4}$	1	1	S_2	1/4
			S_4	1/2
Voltage stress of diodes	1	1/2	1/4	
Automatic uniform current sharing	No	No	Yes	

IV. EXPERIMENTAL RESULTS

To facilitate understanding the merits and serve as a verification of the feasibility of the proposed converter, a prototype with 400V input, 24V output, and 500W rating is constructed as shown in Fig. 2 is chosen. The switching frequency is chosen to be 40 kHz, the duty ratios of S_1, S_2, S_3 and S_4 equal to 0.24 and the corresponding component parameters are listed in Table II for reference. Due to the low switch voltage stress of the proposed converter, three power MOSFETs rating of 250V and conductive resistance of 27 mΩ, namely IXFH100N25P and one power MOSFET rating of 150V and conductive resistance of 13 mΩ, namely IXFH150N15P are adopted. Similarly, four diodes with low forward voltage, namely DSSK60-02A are chosen.

Inductor design: As mentioned above specifications, the voltage conversion ratio is 0.06, as calculated by (8). Next, consider the steady-state inductor currents in (15), the rated output current is calculated to be 20.83A. Moreover, 12% of the full-load inductor current, i.e., 2.5A, can be chosen as peak to peak ripples current. Therefore, the inductor operating in the CCM is

$$L = \frac{v_o}{\Delta I_L}(1-D)T_S = \frac{24(1-0.24)25\mu}{2.5} = 182.4\mu H \quad (18)$$

According to magnetic powder core data sheet provided of CSC, we choose toroid powder core which part number is CM467060. Using magnetic design formulas from data sheet, the design ensures that the inductor operates in the CCM when the load is greater than 120W. In fact, a $250uH$ inductor is chosen in the implementation.

Output capacitor design: As to output capacitances C_o, the peak-to-peak output voltage ripple is considered to be less than 100 mV, then the output capacitor is

$$C_o = \frac{\Delta Q}{\Delta V_o} = \frac{\Delta I_{OL}}{8\Delta V_o f_S} \approx 0.12\mu F \quad (19)$$

To ensure sufficient energy and hold-up time are provided for the post-stage. Therefore, the output capacitor of 220uF is selected in the circuit implementation.

The interleaved structure can effectively increase the switching frequency and reduce the output ripples as well as the size of the energy storage inductors. Fig. 7 shows the four-phase inductor current waveforms of the experimental results. Since output current is sum of four-phase inductor current, it is obviously that with four-phase interleaving control, both output current ripples and switch conduction losses can be reduced.

Table II. COMPONENT PARAMETERS OF THE PROTOTYPE SYSTEM

Components	Specification
Inductors (L_1, L_2, L_3, L_4)	CM467060, 250µH
Active Switches (S_1, S_3, S_4)	IXFH100N25P, 250V, $R_{ds(on)}$=27mΩ
Active Switch S_2,	IXFH150N15P, 150V, $R_{ds(on)}$=13mΩ
Blocking capacitors (C_A, C_B)	10uF/250V(R_c=4.6 mΩ)
Input capacitors (C_1, C_2)	250uF/250V(R_c=44 mΩ)
Output capacitors C_o	220uF/250V(R_c=44 mΩ)
Power diodes (D_1, D_2, D_3, D_4)	DSSK60-02A

The experimental waveforms of the input voltage and output voltage are shown in Fig. 8. The measured input

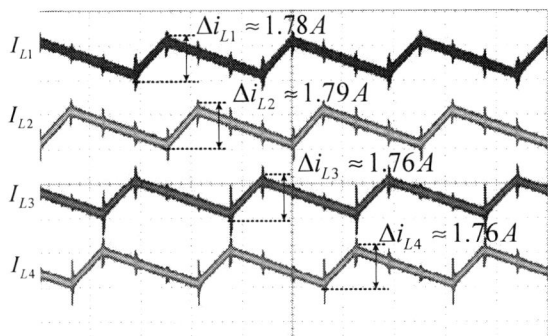

Fig. 7. Waveforms of inductors current I_{L1}, I_{L2}, I_{L3} and I_{L4}

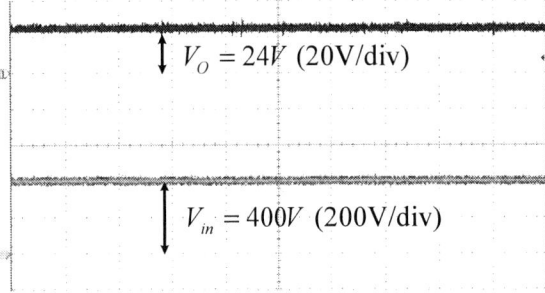

Fig. 8. Waveforms of Input voltage and output voltage

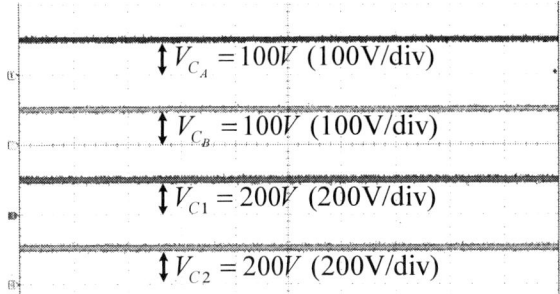

Fig. 9. Waveforms of the blocking capacitors and output capacitors voltages

voltage is 400 V, and the output voltage is 24 V.

To check the validity of the capacitor voltage stress, waveforms of input capacitors and blocking capacitors are recorded as shown in Fig. 9. From Fig. 9 one can see that, with the proposed converter, the voltage stresses of the input capacitors and blocking capacitors are indeed equal to one half and one fourth of the input voltage respectively.

Similarly, to check the correctness of (9), experiments are made and the results are shown in Fig. 10. From Fig. 10 one can observe that the voltage stress of the diodes is equal to one fourth of the input voltage.

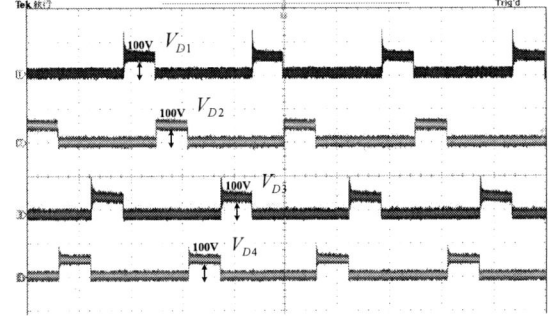

Fig. 10. Waveforms of the voltage stress of V_{D1}, V_{D2}, V_{D3} and V_{D4}

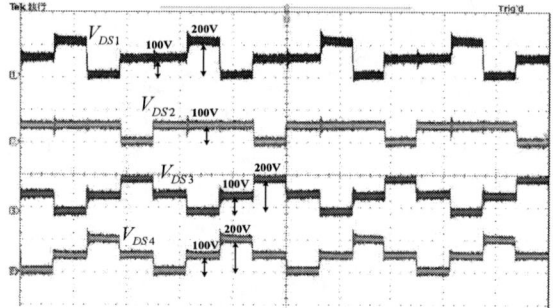

Fig. 11. Waveforms of the voltage stress of V_{Ds1}, V_{Ds2}, V_{Ds3} and V_{Ds4}

The experimental voltage waveforms of active switches are shown in Fig. 11, which indicates that the maximum voltage cross diodes V_{DS1}, V_{DS3} and V_{DS4} equals 200 V which is indeed equal to one-half of the input voltage. The maximum voltage cross switch V_{DS2} is 100 V which is equal to one fourth of the input voltage as expected.

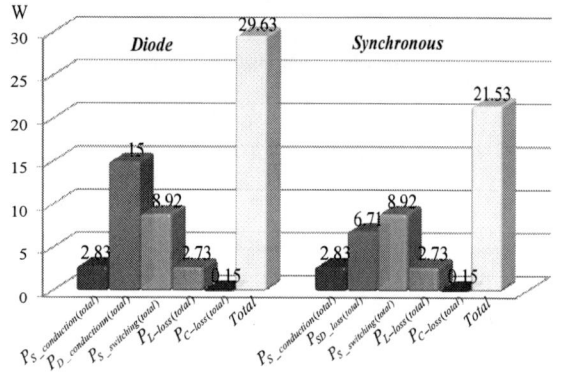

Fig.12. Loss analysis of the proposed converter at full load

Fig.12 shows the corresponding loss analysis of the proposed converter at full load as an illustration. By analyzing the power losses distribution, it can be concluded that the major losses come from the active switches, the diodes, and the output inductors. From Fig. 12, it can also be seen that the conduction loss of diodes is obviously larger than other component losses. In order to further increase the converter efficiency, one can adopt the synchronous rectifier technology, namely replace four diodes with four MOSFETs. The corresponding loss analysis with synchronous rectifier component also is shown in Fig. 12.

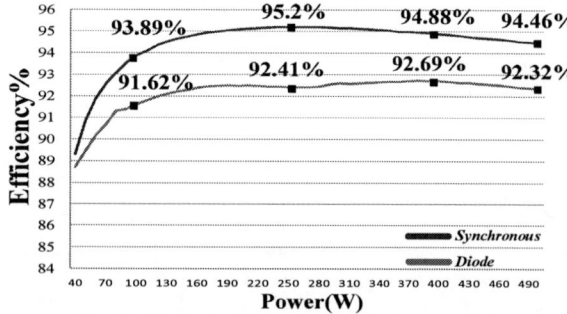

Fig. 13. Measured efficiency of the proposed converter

A precise power meter (YOKOGAWA-WT3000) was used to measure the efficiency of the proposed converter. Fig. 13 shows the measured efficiency curve of the proposed high step-down converter. The measured full-

load efficiency is 92.32% and the maximum efficiency is 92.69%. In order to further increase the converter efficiency, one can replace four diodes with four MOSFETs. The corresponding measured efficiency curve is also shown in Fig. 13. From Fig. 13, it can be seen that the measured full-load efficiency of the proposed converter with synchronous rectifier is 94.6%, and the maximum efficiency is nearly 95.2%.

V. CONCLUSION

In this paper, a novel transformer-less interleaved four-phase high step-down conversion ratio dc–dc converter with low switch voltage stress is proposed. In the proposed converter, the new capacitors switching circuits are combined with interleaved four-phase buck converter in order to get a high step-down conversion ratio without adopting an extreme short duty ratio. Based on the capacitive voltage division, the main objectives of the capacitors switching circuits in the converter are both storing energy in the blocking capacitors for increasing the voltage conversion ratio and reducing voltage stresses of active switches. As a result, the proposed converter topology possesses the low switch voltage stress characteristic. This will allow one to choose lower voltage rating MOSFETs to reduce both switching and conduction losses, and the overall efficiency is consequently improved. In addition, due to the charge balance of the blocking capacitor, the converter features automatic uniform current sharing characteristic of the interleaved phases without adding extra circuitry or complex control methods. The operating principle and steady-state analyses of the voltage gain are discussed. Finally, a 400-V input voltage, 24-V output voltage, and 500W output power prototype circuit is implemented in the laboratory to verify the performance.

ACKNOWLEDGMENT

This work was sponsored by the National Science Council of Taiwan, R.O.C. under Grant NSC-101-2221-E-007-107-MY2.

REFERENCES

[1] Dylan Dah-Chuan Lu, and Vassilios G. Agelidis, "Photovoltaic-Battery-Powered DC Bus System for Common Portable Electronic Devices," IEEE Trans. Power Electronics, vol. 24 no. 3 pp. 849-855, March. 2009.

[2] Kai Sun, Li Zhang, Yan Xing, and Josep M. Guerrero, "A Distributed Control Strategy Based on DC Bus Signaling for Modular Photovoltaic Generation Systems With Battery Energy Storage," IEEE Trans. Industrial Electronics, vol. 26 no. 10 pp. 3032-3045, Oct. 2010.

[3] Majid Pahlevaninezhad, Josef Drobnik, Praveen K. Jain, and Alireza Bakhshai, "A Load Adaptive Control Approach for a Zero-Voltage-Switching DC/DC Converter Used for Electric Vehicles," IEEE Trans. Industrial Electronics, vol. 59 no. 2 pp. 920-933, Feb. 2012.

[4] X. Du and H. M. Tai, "Double-frequency buck converter," IEEE Trans. Ind. Electron., vol. 56, no. 54, pp. 1690–1698, May 2009.

[5] Y. Jang, M. M. JovanoviD , and Y. Panov, "Multiphase Buck Converters with Extended Duty Cycle," in Proc. IEEE Applied Power Electronics Confw (APEC'06), pp.38-44, Mar. 2006.

[6] B. Oraw and R. Ayyanar, "Small signal modeling and control design for new extended duty ratio, interleaved multiphase synchronous buck converter," in Proc. INTELEC, pp. 1–8 Feb. 2006.

[7] Il-Oun Lee, Shin-Young Cho, and Gun-Woo Moon, "Interleaved buck Converter having low switching losses and improved step-down conversion ratio," IEEE Trans. Power Electronics, vol. 27 no. 8 pp. 3664-3675, Aug. 2012.

Efficiency Improvement of Power Supply with Transient Current Circuit Using Digital Control

Haruomi Takashita, Masahito Shoyama
Faculty of Information Science and Electrical Engineering
Kyushu University
Fukuoka, Japan
takashita@ckt.ees.kyushu-u.ac.jp

Yu Yonezawa, Yoshiyasu Nakashima
ICT System Laboratories
Fujitsu Laboratories
Kawasaki, Japan

Abstract. **Optimal techniques for power supply with transient current circuit are introduced in this paper. The circuit reduces switching loss of buck converter's main switch effectively. However, in our previous researches, it could be operated with high efficiency only at heavy loading. The techniques to get rid of this disadvantage are proposed and the validity of it is verified through experiments.**

Keywords— Soft switching, Zero current switching, Transient current Power supply, Digital control

I. INTRODUCTION

Recent years, there is a tendency of miniaturization in electronics, and also it is required for the internal power supply. Making switching frequency high is used for miniaturization usually, but this method increases the switching-loss. On the other hand, the demand of energy saving has been demanded also. Because of this, to decrease power loss and to get high efficiency is important for power supply. Soft switching is one way which reduces switching loss. There is a circuit called "Quasi-resonance converter" as a typical example of a circuit for soft switching. This circuit can reduce switching-loss making resonance of the current or the voltage in switch element by an inductor and a capacitor. This circuit can reduce switching-loss but this technique makes the current or the voltage of the switch more stressful. Because of this, some switch elements which have adequate rating must be used and this increases conduction loss. As a countermeasure to this problem of this quasi-resonant converter, there is one technique of soft switching which called "Partial resonance." This way makes resonant only at a period of time; switched on or switched off. This resonate the current or voltage of switch during a little section so there is lesser stress unlike quasi-resonance converter.

In this paper, buck converter with transient current circuit[1] (TCC) [2] [3] is introduced. This circuit is for partial resonance. The authors got improvement of the efficiency using the circuit at heave loading. However, the low efficiency at light loading was mentioned for the matter. In this research, improvement of the efficiency at light loading is achieved by redesign of TCC and adapting digital control.

II. BUCK CONVERTER WITH TRANSIENT CURRENT CIRCUIT

In this section, buck converter with transient current circuit is introduced. Transient current circuit is introduced in Section-A. The behaviors of the buck converter with TCC are described in Section-B. The drive waveforms of the buck converter with TCC are introduced in Section-C. Then, two specific states of this circuit are explained in Section-D and E.

A. Transient Currnet Circuit(TCC)

The buck converter is shown in Fig.1. In this circuit, large heat occur in main switch. TCC is a circuit which can reduce switching loss when it is applied in the buck converter as shown in Fig.2. TCC can reduce switching loss of high-side switch S_m effectively since it makes zero-current switching operation during turn-off S_m. The power conversion efficiency of the buck converter with TCC is higher than that of the conventional one. TCC is the circuit which is surrounded by blue dotted line in Fig.2. This circuit includes two sections. The first one is operated as a main circuit as buck converter. It is consisted with main switch S_m, low-side switch S_D, inductor L, capacitor C. Then, the second one is operated as an auxiliary circuit. This part is just TCC and, it is consisted with two switches S_{a1} and S_{a2}, two diodes D_{a1} and D_{a2}, inductor L_a and capacitor C_a. The resonance occurred between parasitic inductor L_p and C_a brings zero-current switching to S_m.

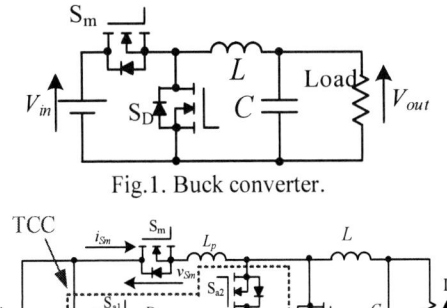

Fig.1. Buck converter.

Fig.2. Buck converter with TCC.

B. Circuit Behavior

TCC is operated at only two periods: charge and discharge capacitor C_a. The other operation is same as the conventional buck converter. During discharging time, resonance occurs between L_p and C_a. In Fig.3, the operation states of this circuit are shown. $V_{GS\text{-}Sm}$, $V_{GS\text{-}SD}$, $V_{GS\text{-}Sa1}$ and $V_{GS\text{-}Sa2}$ are the voltage between gate and source of four switches. The letter i_{La}, v_{Ca}, v_{Sm} and i_{Sm} are electric current and voltage shown in Fig.2.

The operations of the state of each circuit state are shown below. (1) ~ (8) and the schematics Fig4.1~4.8 are one-to-one relationship.

(1): S_m is turned off, and S_D is turned on. During this state, S_{a1} and D_{a1} are turned on. The current flows to C_a and L_a.

(2): In this state, S_{a1} and D_{a1} are turned off. The energy stored in L_a flows to C_a, so C_a get charged until required voltage.

(3): When operation of TCC finishes, and TCC is separated from the buck converter by keeping the switches S_{a1} and S_{a2} OFF.

(4): In this state, S_m is turned on, and S_D is turned off. During this state, this circuit operates the same as conventional buck converter because of separation of the auxiliary circuit.

(5): This state starts from the time S_{a2} is switched on and finishes when S_m is switched off. As S_{a2} is turned on, the resonance occurs. During this state, zero-current switching can be achieved, so that switching loss are reduced.

(6), (7): Electron stored in C_a is discharged to load side. S_{a2} is turned off after voltage of C_a becomes 0V.

(8): Because Sa2 is turned off at state-(7), the auxiliary circuit is separated from the buck converter. This state is same as conventional buck converter.

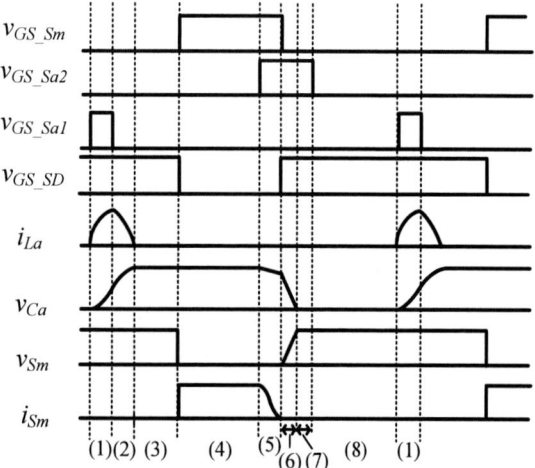

Fig.3. Key waveforms of the buck converter with TCC shown in Fig.2.

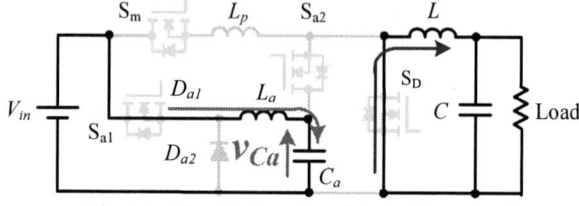

Fig.4.1. Circuit diagram of state (1).

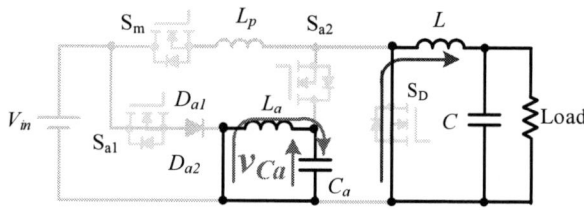

Fig.4.2. Circuit diagram of state (2).

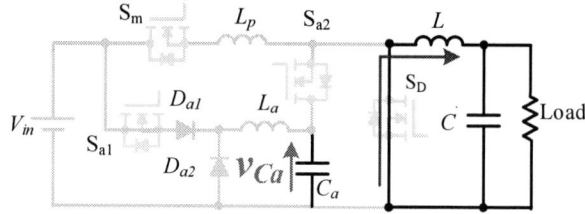

Fig.4.3. Circuit diagram of state (3).

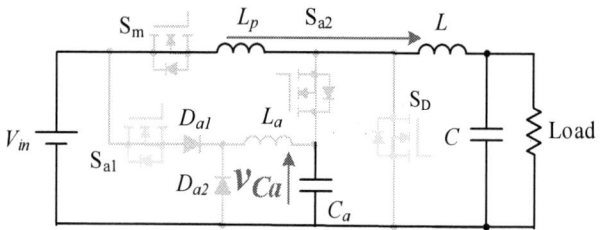

Fig.4.4. Circuit diagram of state (4).

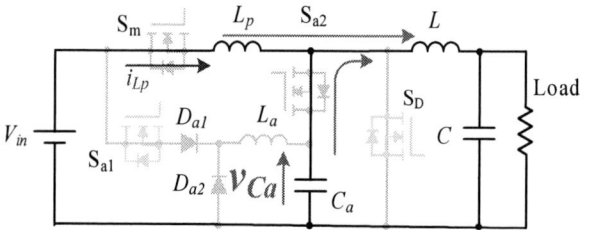

Fig.4.5. Circuit diagram of state (5).

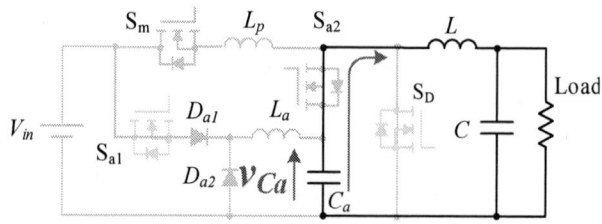

Fig.4.6. Circuit diagram of state (6).

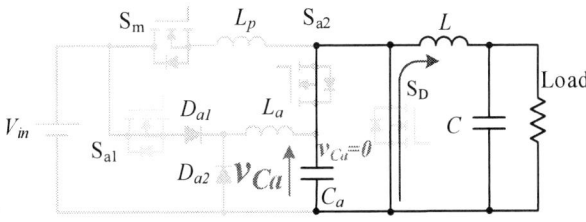

Fig.4.7. Circuit diagram of state (7).

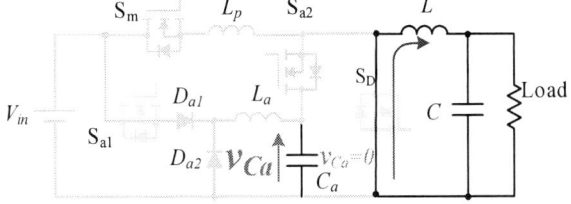

Fig.4.8. Circuit diagram of state (8).

C. Drive Waveforms

Appropriate control each four switches is required for operating the buck converter with TCC as the circuit operation shown in Section II.B. G-S waveforms of each switch are shown in fig.5. This figure says that these four switches have some time setting called T1~T6. T1 is the term during S_{a1} is turned-on. T2 is the time from turning-off S_m to turning-off S_{a2}. T3 is the time from turning-off S_{a2} to turning-on S_{a1}. T5 is the time from turning-on S_{a2} to turning-off S_m. T4 and T6 are the dead time of S_m and S_D.

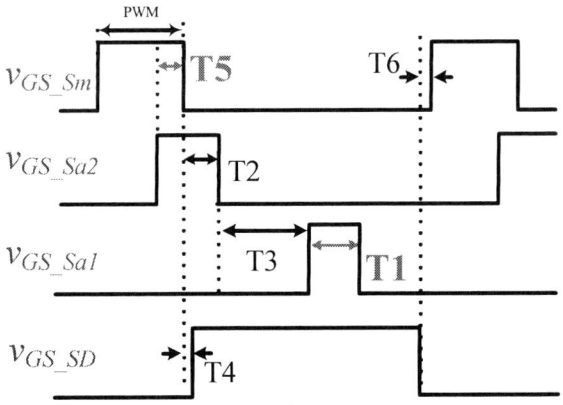

Fig.5. Drive Waveforms.

D. Capacitor Discharge Time

Then we describe about capacitor discharge time. The waveforms from state (4) to state (8) are shown in Fig.6.a. Especially, it is called capacitor discharge time from state (5) to state (7). This discharging time is important for transient current circuit. This reason is that zero-current switching of S_{um} is performed during this period.

At state (7), the current of L_p is resonates, and this resonant current is given by formula (1) and (2).

$$i_{Lp}(t) = I_o - \omega_b C_a \left(v_{ca}(t_a) - V_{in}\right) \sin \omega_b t \tag{1}$$

$$\omega_b = \frac{1}{\sqrt{L_p C_a}} \tag{2}$$

To achieve zero-current switching correctly, it is needed that the current of L_p become $i_{Lp} < 0$ when S_m is turned off. Therefore, it must be satisfied the following formula (3).

$$v_{ca}(t) \geq \frac{I_o}{\omega_b C_a} + V_{in} \tag{3}$$

Then, to reduce the switching loss efficiently, it is required that S_m is turned-off at the best timing which the current go through S_m becomes 0A. T5 is the time to decide the timing. The fixed waveforms of iLp and v_{DS_Sm} are shown in Fig.7. When T5 is not adjusted, the waveforms become as Fig.7.a. This state occur useless loss. When T5 is adjusted, the waveforms become as Fig.7.c. The best value of T5 which decide the good timing is calculated in formula (4).

$$T_4 = \frac{1}{\omega_b} \sin^{-1} \left(\frac{I_o}{\omega_b C_a \left(v_{Ca}(t_a) - V_{in}\right)} \right) \tag{4}$$

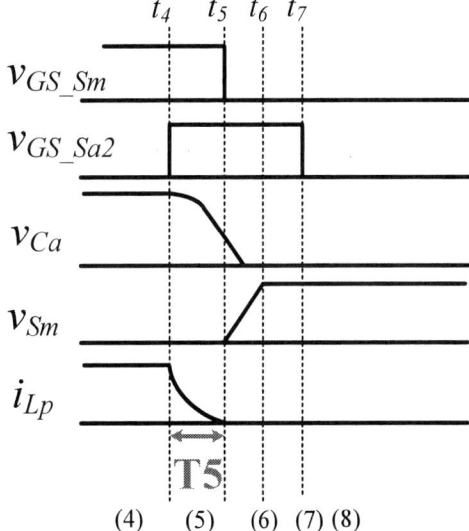

Fig.6.a Waveforms during capacitor discharge time.

E. Capacitor Charge Time

We describe about capacitor charge time. The waveforms from state (1) to state (4) are shown in Fig.6.b. Especially, it is called capacitor discharge time from state (1) to state (2). The role of this discharging time is the preparation period to decide the voltage of C_a. Voltage across C_a is a key parameter of the transient current circuit. Whether ZCS could be achieved or not at the transient current circuit depends on the voltage across C_a. Because of this, the voltage is needed to charge to satisfy the formula (3).

During state1, S_{a1} is on. Therefore, the current of L_a is given by formula (5)

$$i_{La-1} = \omega_a C_a V_{in} \sin \omega_a t \qquad (5)$$

$$\omega_a = \frac{1}{\sqrt{L_a C_a}} \qquad (6)$$

Then, the current i_{Lp} is shown in formula (7)

$$i_{La-2} = \omega_a C_a V_{in} \{\cos \omega_a (t - t_1) - \sin \omega_a (t - t_1)\} \qquad (7)$$

The value of "t1" is decided by the length of T1. Ohe basis of the formula (6) and (8), the voltage of C_a is given by (8)

$$
\begin{aligned}
v_{Ca} &= \frac{1}{C_a} \int i_{La} dt \\
&= \frac{1}{C_a} \int_0^{t_1} i_{La-1} dt + \frac{1}{C_a} \int_{t_1}^{t_2} i_{La-2} dt \\
&= V_{in} \{-\cos \omega_a T_4 + \cos \omega_a (t_2 - t_1) + \sin \omega_a (t_2 - t_1)\}
\end{aligned} \qquad (8)
$$

Then, the operation for effective reduction of the switching loss is required. This way is to occur resonant which goes to minus with charging C_a by minimum energy. The condition is decided by the length of T1. If the length is too long, the waveforms of i_{Lp} and v_{DS_sm} become as Fig.7.a. In this case, the amplitude is increased. The current goes to C_a during S_{a1} in turned-on and it makes needless conduction loss. The waveforms adjusted T1 are shown in Fig.7.b. In this situation, the amplitude is decreased. The best value of T1 is given by formula (9). This formula satisfy the condition that the formula (1) is zero.

$$v_{ca}(t) = \frac{I_o}{\omega_b C_a} + V_{in} \qquad (9)$$

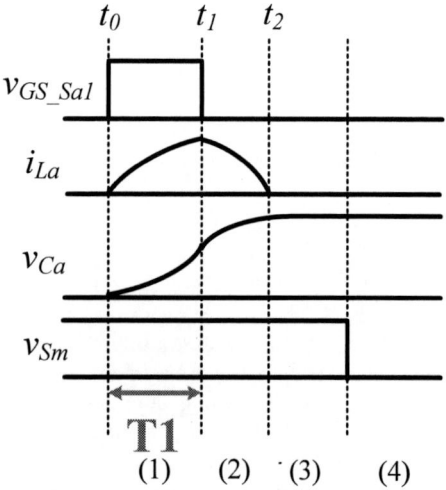

Fig.6.b. Waveforms during capacitor charge time.

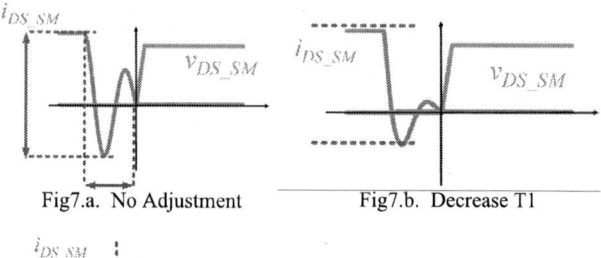

Fig7.a. No Adjustment   Fig7.b. Decrease T1

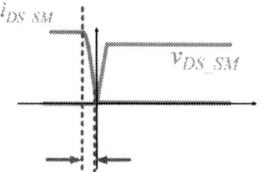

Fig7.c. Increase T5

Fig.7. Relationships among T1, T5 and the buck converter with TCC.

III. EXPERIMENTS

In this section, it is written for the experiment. The conditions of the experiments are said in Section-A. These conditions are decided on the basis of Section-III. The result of the simulation is said in Section-B. It is recognized that TCC can make zero-current switching operation when TCC is used in buck converter. The result of the experiments is said in Section-B. In Section-C, it is recognized that using TCC can increase the efficiency of buck converter if the elements and the control of the circuit is designed properly.

A. Conditions of the experiments

In this paper, two converters, one is the normal buck converter with digital control and the other is the buck converter with TCC using digital control, are used. The conditions are given in TABLE I and TABLE II. On the basis of Section II, the relationships about the voltage and the capacity of C_a and output current of this circuit is given in Fig.8. This result is calculated with the value of L_p is 15nH. When the capacity of C_a is large, the special changes of the voltage are not found, but the case that capacitor is small, the output current affects the value of the voltage of C_a strongly. This means that to adjust T1 according to output current makes the efficiency the best.

The way to choose the capacitor C_a is said here. When the capacity becomes larger, the current flowing through TCC becomes large also. The conduction loss in TCC is given by (10)

$$P_{loss\_TCC} = (R_{Sa1} + R_{La}) \int_0^{t_1} i_{La-1}{}^2 dt + R_{La} \int_{t_1}^{t_2} i_{La-2}{}^2 dt \qquad (10)$$

It is confirmed that it is better the value of L_a is set to large and C_a is set to small. However, C_a must be set with on the basis of the frequency of resonant which is decided by the value of parasitic inductor of S_m.

In this research, L_a is holed in 100uF and 200uH, C_a is chosen from 0.1, 0.47, 1.0, 2.2, 4.7 and 10nF.

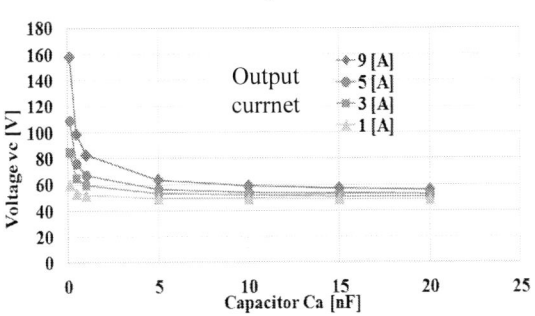

Fig.8. Relationships among Voltage of Capacitor and Capacity of Capacitor.

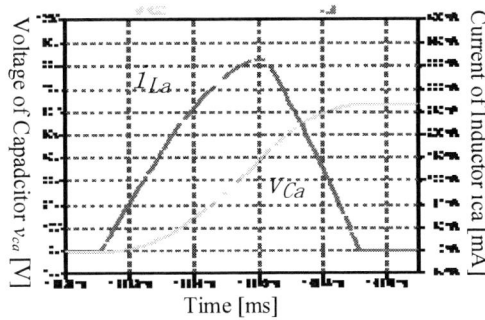

Fig.9. Current La and Voltage Ca in TCC.

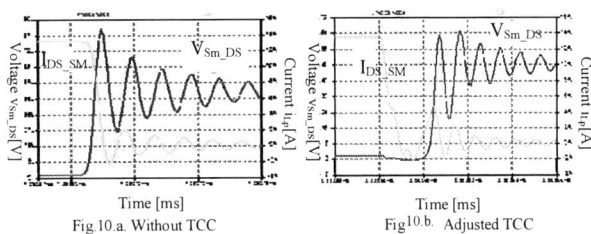

Fig.10.a. Without TCC   Fig10.b. Adjusted TCC

Fig.10. Current and Voltage about S_m.

TABLE I

THE PARAMETERS OF THE POWER SUPPLY WITH TCC.

Symbol	Meaning	Value
V_{in}	Input voltage	48V
V_{out}	Output voltage	12V
I_{out}	Output current	1~9A
f_{SW}	Switching frequency	200kHz
L	Output filter inductor	10μH
C	Output filter capacitor	470μF
L_a	Auxiliary inductor	100・200μH
C_a	Auxiliary capacitor	Variable

TABLE II

THE PARAMETERS OF T1 ~ T6.

Symbol	No adjustment	After adjustment
T1	1500ns	Variable
T2	200ns	200ns
T3	1500ns	1500ns
T4	200ns	200ns
T5	60ns	Variable
T6	200ns	200ns

B. Simulation

Simulation was performed on the same circuit conditions that are presented in Section-A. In this case, L_a is set to 100[μH] and C_a is set to 4.7[μF]. The conditions which are not presented in Section-A are said in TABLE III. The result of the simulation is shown in Fig.9 and Fig.10. In Fig.9, V_{ca} is the voltage of C_a and i_{La} *is the current of* L_a. In Fig.10, V_{Sm_DS} is the voltage between drain and source of S_m and I_{DS_Sm} is the current which flow through S_m. In this simulation, the proper operation of zero current switching and reduction of switching loss is achieved by adjustment of T1 and T4 properly.

TABLE III

THE CONDITIONS OF SIMULATION.

Symbol	Meaning	After adjustment
L_a	Auxiliary inductor	100μH
C_a	Auxiliary capacitor	4.7μF
T1	Time	960nsec
T5	Time	7.5nsec

C. Experiments

From the above results, C_a and L_a is set as TABLE IV. It is needed not only to drive switch but to set the time T1~T6 properly for the optimal operation, when TCC is used in buck converter. In this research, to simplify the control of this circuit, digital control is used. Because this control used here enable T1~T6 to set the value on the program, there is no need to control the circuit operation by each situation without changing circuit elements. The result is shown in Fig.11. It says that the smaller C_a raises the efficiency and the larger inductor L_a also raises the efficiency. Nevertheless, the efficiency is reduced if C_a is too small. This reason is that too small capacitor makes the frequency higher and soft-switching can't be occurred well. The result of the efficiency on the condition fixed L_a to 100uH and C_a to 0.47uF is shown in Fig.12. In this condition, the current flowing through the inductor L_p in TCC goes the peak when T1 is 1080ns. At this time, C_a has been overcharged more than necessary so that extra loss occur in TCC. In the condition, in order to obtain the best efficiency, T1 is required to set 900nsec to charge C_a properly. The validity of the theoretical value is confirmed from this result. To make the value of T1 longer than this increase the loss and the efficiency of the circuit will be reduced.

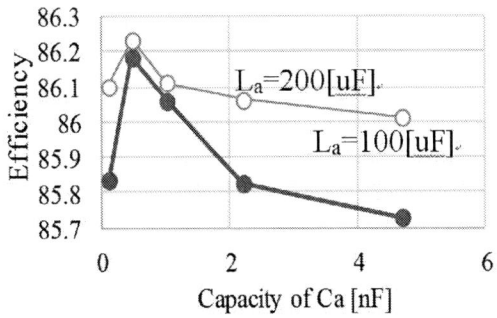

Fig.11. Efficiency characteristic about C_a and L_a.

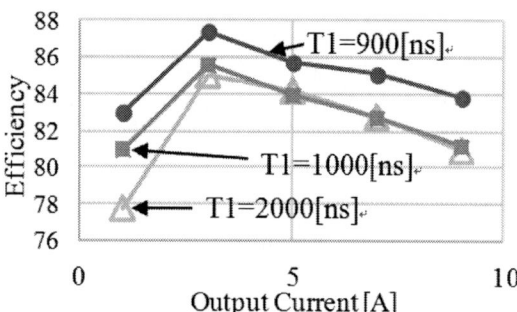

Fig.12. Efficiency characteristic about T1.

Based on the above results, the each power conversion of three converters is compared. The first one is a normal buck converter. The second one is a buck converter with TCC without adjustment. The last one is a buck converter with TCC with adjustment. The last one is designed in order to operate the special converter in the best condition, L_a is set to 200uH and C_a is set to 0.47nF; the value as small as possible. T1 is set by calculation and T5 is fine-tuned by experiment. These conditions are shown in TABLE IV. Then, the results of this experiment are shown in Fig13.

To compare the adjusted converter with unadjusted one, The efficiency in right load is raised 1%.

TABLE IV
THE CONDITIONS OF IMPROVEMENT OF EFFICIENCY.

Circuit Type	La	Ca	T1	T4
Buck	No Data	No Data	No Data	No Data
Unadjusted	100uH	10nF	60ns	1500ns
Adjusted	200uH	0.47nF	2ns	500ns

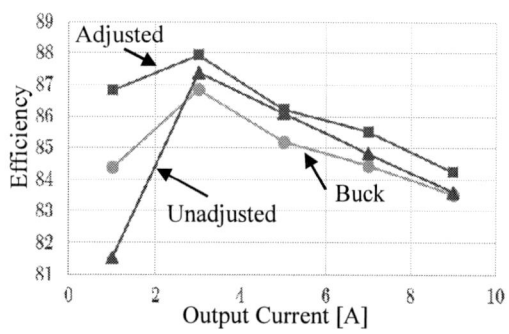

Fig.13. Comparison of Efficiency both before adjustment and after adjustment.

IV. CONCLUSIONS

In this paper, how to use power supply with TCC effectively is introduced. The most important things are design TCC properly and adjustment of T1 and T5. To decrease the capacity of the capacitor C_a in TCC and increase the value of the inductor L_a make the efficiency of the buck converter with TCC higher. T1 and T5 are adjusted by calculation and experiments. The validity of the adjustment is verified through last experiments. The circuit evidently reduces switching-loss of buck converter switch effectively more than the previous circuit which is introduced in the references.

V. INFORMATION

REFERENCES

[1] S. Hiromitsu, M. Ogawa, S. Abe, M. Shoyama, Y. Yonezawa, N. Mishima, "Transient Current Circuit for Switching Loss Reduction in Buck Bonverters," *IEICE technical report*, vol. 109, no. 371, pp. 1-5, Jan. 2010.

[2] R. Liu, M. Shoyama, Y. Yonezawa, N. Mishima, "Loss Analysis of Buck Converter with Transient Current Circuit," *IEICE technical report*, vol. 111, no. 400, pp. 59-63, Jan. 2012.

[3] K. Harada, T. Ninomiya, B. Ko, "The fundamentals of switched-mode converters," *Corona Publishing*, Feb 1992.

Ultra High Step-Down Converter

Y. T. Yau[1], *Member, IEEE*, and K. I. Hwu[2], *Member, IEEE*

Department of Electrical Engineering, National Taipei University of Technology, Taipei, Taiwan

E-mail[1]: tsmc35@yahoo.com.tw, E-mail[2]: eaglehwu@ntut.edu.tw

Abstract– **In this paper, An ultra high step-down converter is presented, which combines one coupling inductor and one energy-transferring capacitor. Therefore, the corresponding voltage conversion ratio can be much lower than that of the traditional buck converter. As compared with the traditional buck converter, there are two key merits in the proposed converter. One demerit is that the voltage conversion ratio of this converter does not have nonlinearity characteristics, and the other is that if one switch fails or is abnormally controlled, and at the same time any other switch is made turned on, then the high voltage does not appear in the output terminal, so the output load can be protected. In this study, brief theoretical deductions and some experimental results are given to verify the feasibility and effectiveness of the proposed converter.**

Keywords—Buck converter, coupling inductor, energy-transferring capacitor, step-down converter.

I. INTRODUCTION

As generally recognized, the step-down converter is widely used in powering the digital circuit. The 48V voltage bus is generally used in the communication system and is generated from the AC-DC converter in the network communication room. Such a 48V voltage bus is to be down to 3.3V or less via a DC-DC converter, that is, the corresponding pulse-width-modulated (PWM) duty cycle would be down to 10% or less, thereby causing the control design to be not easy enough and the accompanying switching loss to be relatively high. Therefore, up to now, the two-stage step-down structure has been widely used. In order to power the CPU, the DRAM and the hard disk, the first-stage transfers 48V to 12V so as to power the point of load (POL), and then the second-stage transfers 12V to 3.3V, 2.5V, 1.8V, 1.5V, 1.2V or 1V via the POL.

The methods described in [1]-[4] are based on the two-stage buck converter so as to realize the high step-down voltage gain. But these methods need more active switches and components, more PWM gate driving circuits and signals, etc. And the overall efficiency is the product of the efficiencies of two stages. Hence, the two-stage buck converter is not widely used today. The method described in [5] is based on a single-stage high-efficiency open-loop bus converter. The voltage 48V is stepped down to 12V, and then the voltage 3.3V or lower is created from the POL, so as to feed the load. Basically, this topology also belongs to a two-stage converter. In addition, such a bus converter needs four active switches and two magnetic devices, and if the second-stage buck converter is taken into account, then there are six active switches, three magnetic devices, one additional control IC to be needed. The methods described in [6]-[11] are based on multiple voltage regulators connected in parallel and coupling inductors, and accordingly by using interleaved PWM signals, the ratio of the input to the output can be improved. As compared to the

traditional buck converter, under the same input and output voltages, this converter can operate with a larger duty cycle so as to reduce the switching loss. But this structure needs two phases operating simultaneously, so it is suitable for applications with high output currents. The methods described in [7]-[19] are based on coupling inductors so as to achieve high step-down output. These circuits are simple but leakage inductances would create high voltage spikes to tend to break down switches, and hence passive snubbers are required to avoid tending to destroy switches, thereby reducing the corresponding efficiency. Although the active snubbers presented by the literatures [19]-[21] can reduce the leakage energy, the accompanying circuits are too complex. In [6][13][22]-[25], the number of switches and magnetic devices are too many, and hence the corresponding circuits are too complex, so the resulting cost is high and hence they are not suitable for middle-power applications. In [7][8][11][13][18][20][22][23][26][27], the switches require floating gate drivers instead of low-cost high-bridge gate drivers. In this case, if the pulse transformer is adopted, the PCB space becomes relatively large, leading to difficulty in applications. The method described in [4] possesses the floating output, resulting in application limitation. The method described in [9][12]-[14][16][20][21][28]-[30] possesses the nonlinearity of voltage conversion ratio, thereby making the corresponding controllers difficult in design. The method described in [31] needs many inductors, capacitors and diode arrays, leading to too many components and relatively low efficiency.

Based on the aforementioned, a novel ultra high step-down converter is presented herein. As compared with the traditional buck converter, there are three advantages in this converter as follows: (i) the proposed converter has one additional coupling inductor, one additional grounded switch and one additional capacitor with small capacitance, and can be driven using existing buck PWM control ICs; (ii) the voltage conversion ratio of this converter does not have nonlinearity characteristics; (iii) if one switch fails or is abnormally controlled, and meanwhile any other switch is made turned on, then the high voltage does not appear in the output terminal, so the output load can be protected.

II. PROPOSED CIRCUIT CONFIGURATION

Fig. 1 shows the proposed converter, which contains three switches Q_1, Q_2 and Q_3, one energy-transferring capacitor C_B and one output capacitor C_O, and one coupling inductor composed of the primary winding N_1 and the secondary winding N_2. Moreover, Q_1 and Q_2 are driven simultaneously with both gates connected together, and hence, although there are three switches in this circuit, only one half-gate driver is needed to drive them.

978-1-4799-2706-7/14 $31.00 © 2014 IEEE

Also, in Fig. 1, there are some symbols to be given as follows: (i) the currents flowing through Q_1, Q_2 and Q_3 are signified by I_{ds1}, I_{ds2} and I_{ds3}, respectively; (ii) the voltage across C_B is denoted by V_{CB}; (iii) the currents flowing through N_1 and N_2 are represented by I_{N1} and I_{N2}, respectively; and (iv) the input and output voltages are indicated by V_{in} and V_O, respectively.

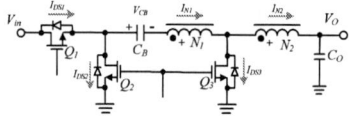

Fig. 1. Proposed step-down converter.

III. BASIC OPERATING PRINCIPLES

In this circuit, there are seven operating states, to be described as follows.

1) State 1: As shown in Fig. 2, the switch Q_1 is turned on but the switches Q_2 and Q_3 are turned off. During this state, the windings N_1 and N_2 of the coupling inductor are magnetized, and hence the currents in the windings N_1 and N_2, i.e., I_{N1} and I_{N2}, are increasing slowly. Therefore,

$$v_{N1} = \left(V_{in} - V_{CB} - V_O\right) \cdot \left(\frac{N_2}{N_1 + N_2}\right) \quad (1)$$

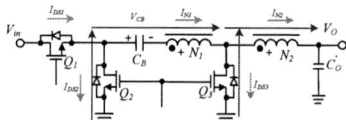

Fig. 2. Current flow in state 1.

2) State 2: As shown in Fig. 3, the switch Q_1 is turned off, and the switches Q_2 and Q_3 are still turned off. Since this state is one dead time over one PWM cycle and the currents in the coupling inductor keep continuous, the body diodes of the switches Q_2 and Q_3 are turned on.

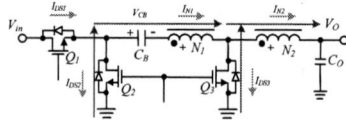

Fig. 3. Current flow in state 2.

3) State 3: As shown in Fig. 4, the switches Q_1, Q_2 and Q_3 are all still turned off, but the body diodes of the switches Q_2 and Q_3 are still turned on. As soon as the energy in the leakage inductance of the winding N_1, released to the output terminal, falls to zero, the proceeds to state 4.

Fig. 4. Current flow in state 3.

4) State 4: As shown in Fig. 5, the switch Q_1 is still turned off but the switches Q_2 and Q_3 are turned on with zero voltage switching (ZVS). During this state, the energy-transferring capacitor C_B magnetizes the winding N_1 in the opposite direction and the energy stored in the coupling inductor is transferred to the output terminal via the winding N_2 based on transformer behavior. Hence, the current in the winding N_1, i.e., I_{N1}, is increasing in the opposite direction, and the current in the winding N_2, i.e., I_{N2}, is increasing, also. Due to the turns ratio, the voltage across C_B can be expressed to be

$$v_{N1} = V_{CB} = V_O \cdot \left(\frac{N_1}{N_2}\right) \quad (2)$$

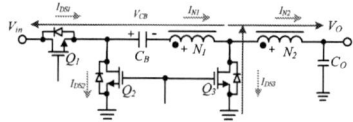

Fig. 5. Current flow in state 4.

5) State 5: As shown in Fig. 6, the switch Q_1 is still turned off, and the switches Q_2 and Q_3 are turned off. Since this state is the other dead time over one PWM cycle, and the currents in the coupling inductor keep continuous, the body diodes of the switches Q_1 and Q_3 are turned on.

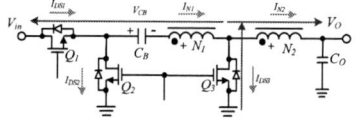

Fig. 6. Current flow in state 5.

6) State 6: As shown in Fig. 7, the switches Q_1, Q_2 and Q_3 are all still turned off, but the body diodes of the switches Q_1 and Q_3 are still turned on. As soon as the energy in the leakage inductance of the winding N_1, demagnetized to the output terminal in the opposite direction, falls to zero, the state goes to state 7.

Fig. 7. Current flow in state 6.

7) State 7: As shown in Fig. 8, the switch Q_1 is turned on with ZVS, but the switches Q_2 and Q_3 are still turned off. During this state, the input voltage V_{in} charges the energy-transferring capacitor C_B. At the same time, the current in the winding N_1, i.e., I_{N1}, is smaller than the current in the winding N_2, i.e., I_{N2}, the body diode of the switch Q_3 is still turned on. As soon as the current I_{N1} is equal to the current I_{N2}, the current in the body of the switch Q_3 stops flowing, the operation state goes back to state 1, and the next cycle is repeated.

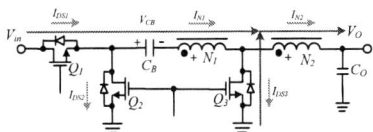

Fig. 8. Current flow in state 7.

Based on (1) and (2) and the volt-second balance of the coupling inductor, the following equation can be obtained as follows:

$$D \cdot \left(V_{in} - V_{CB} - V_O \right) \cdot \left(\frac{N_1}{N_1 + N_2} \right) = \left(1 - D \right) \cdot \left(\frac{N_1}{N_2} \right) \cdot V_O \qquad (3)$$

Finally, the voltage conversion ratio of the proposed converter can be represented by

$$\frac{V_O}{V_{in}} = D \cdot \left(\frac{N_2}{N_1 + N_2} \right) \qquad (4)$$

From (4), it can be seen that the voltage conversion ratio of the proposed converter can be adjusted not only by the duty cycle but also by the primary and secondary turns.

IV. CIRCUIT SPECIFICATIONS AND PARAMETERS

The following describes the circuit specifications and parameters: (i) the input voltage is 48V; (ii) the output voltage is 3.3V; (iii) the rated output current is 15A; (iv) the switching frequency is 100kHz; (iv) the coupling inductor, with the primary winding N_1 of 24Turns, the secondary winding N_2 of 8Turns and the primary inductance of 86µH, adopts a 106-M125 MPP core with the inductance coefficient of 157nH/N^2; (v) two 10µF/16V TDK MLCC capacitors, connected in parallel, are selected for the capacitor C_b; (vi) one 1800µF/16V Rubycon electrolytic capacitor is chosen for the capacitor C_O; (vii) the product name of the switches Q_1 and Q_2, made by NXP Co., is PHB34NQ10T with the voltage stress of 100V and the turn-on resistance of 10mΩ; (viii) the product name of the switch Q_3, made by NXP Co., is PHD96NQ3LT with the voltage stress of 25V and the turn-on resistance of 6mΩ; and (ix) the product name of the gate driver is HIP2101 made by Intersil Co.

V. EXPERIMENTAL RESULTS

Figs. 9 to 11 show the waveforms relevant to the rated load. Fig. 9 shows the gate driving signals V_{gs1} and V_{gs2} for the switches Q_1 and Q_2, respectively, and the currents I_{N1} and I_{N2} for the windings N_1 and N_2, respectively. Fig. 10 shows the positive edge of V_{gs1} and the corresponding V_{ds1}, and the negative edge of V_{gs2} and the corresponding V_{ds2}. Fig. 11 shows the negative edge of V_{gs1} and the corresponding V_{ds1}, and the positive edge of V_{gs2} and the corresponding V_{ds2}. From these figures mentioned above, it can be seen that the proposed converter can operate well to some extent, and the switches Q_1 and Q_2 possess ZVS turn-on and blocking times of about 100ns between them. Fig. 12 shows the load transient response due to step load change from 50% to 100% load. Fig. 13 shows the

load transient response due to step load change from 100% to 50% load. From Figs. 12 and 14, it can be seen that the corresponding undershoot or overshoot is about 320mV with the recovery time of about 500µs. Fig. 14 shows the curve of efficiency versus load current. From Fig. 14, it can be seen the efficiency all over the load range is above 84.86% and can be up to 95.78%. The efficiency at rated load is 88.79%.

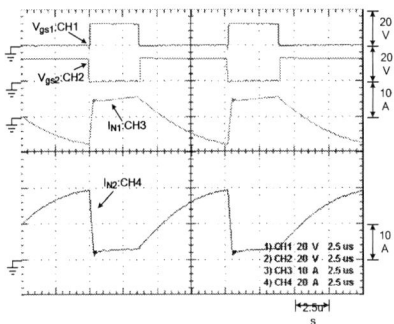

Fig. 9. Measured waveforms at rated load: (1) V_{gs1}; (2) V_{gs2}; (3) I_{N1}; (4) I_{N2}.

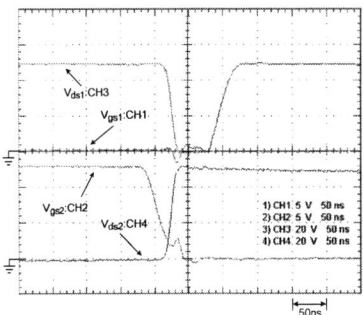

Fig. 10. Measured waveforms at rated load for positive edge of V_{gs1} and negative edge of V_{gs2}: (1) V_{gs1}; (2) V_{gs2}; (3) V_{ds1}; (4) V_{ds2}.

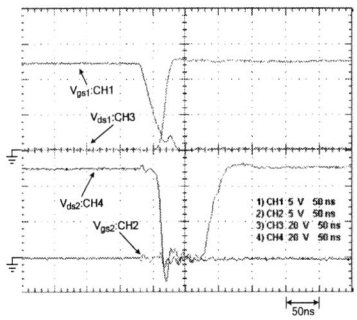

Fig. 11. Measured waveforms at rated load for negative edge of V_{gs1} and positive edge of V_{gs2}: (1) V_{gs1}; (2) V_{gs2}; (3) V_{ds1}; (4) V_{ds2}.

The 2014 International Power Electronics Conference

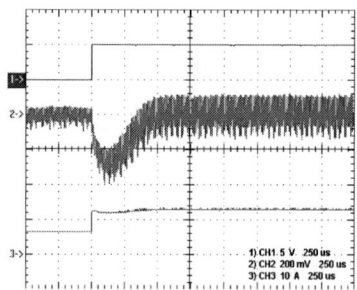

Fig. 12. Load transient responses from 50% to 100%:
(1) LOAD_EN; (2) V_o; (3) I_o.

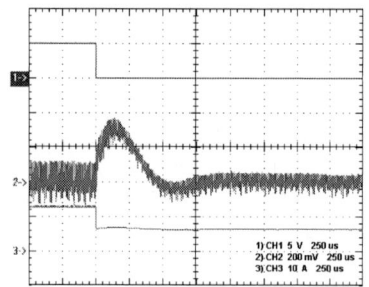

Fig. 13. Load transient responses from 100% to 50%:
(1) LOAD_EN; (2) V_o; (3) I_o.

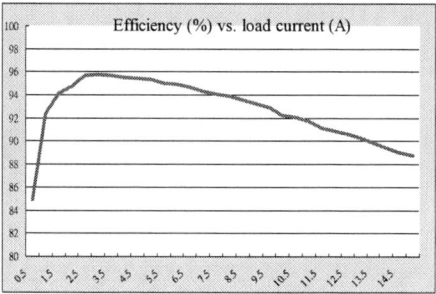

Fig. 14. Efficiency versus load current.

VII. CONCLUSION

An ultra high step-down converter is presented herein. By combining one coupling inductor and one energy-transferring capacitor, the corresponding voltage conversion ratio can be much lower than that of the traditional buck converter. As compared with the traditional buck converter, there are three merits in this converter as following: (i) the proposed converter has one additional coupling inductor, one additional grounded switch and one additional capacitor with small capacitance, and can be driven using existing buck PWM control ICs; (ii) the voltage conversion ratio of this converter does not have nonlinearity characteristics; (iii) if one switch fails or is abnormally controlled, and in the meantime any other switch is made turned on, then the high voltage does not appear in the output terminal, so the output load can be protected. To sum up, the structure of the proposed converter is quite simple and very suitable for industrial applications.

REFERENCES

[1] Yuancheng Ren, Ming Xu, Kaiwei Yao, Yu Meng, F. C. Lee, Jinghong Guo and Y. Ren, "Two-stage approach for 12 V VR," *in Proc. IEEE APEC Conf.*, vol. 2, pp. 1306-1312, 2004.

[2] Yuancheng Ren, Ming Xu, Kaiwei Yao and F. C. Lee, "Two-stage 48V power pod exploration for 64-bit microprocessor," *in Proc. IEEE APEC Conf.*, vol. 1, pp. 426-431, 2003.

[3] Yuancheng Ren, Ming Xu, Yu Meng and F. C. Lee, "12V VR efficiency improvement based on two-stage approach and a novel gate driver," *in Proc. IEEE PESC Conf.*, pp. 2635-2641, 2005.

[4] Hong Mao, J.A. Abu-Qahouq, Shiguo Luo and I. Batarseh, "Zero-voltage-switching (ZVS) two-stage approaches with output current sharing for 48 V input DC-DC converter," *in Proc. IEEE APEC Conf.*, vol. 2, pp. 1078-1082, 2004.

[5] K. I. Hwu and Y.T. Yau ,"Resonant voltage divider with bidirectional operation and startup considered," *IEEE Trans. Power Electron.*, vol. 27, no. 4, pp. 1996-2006, Apr. 2012.

[6] Wuhua Li, Jianguo Xiao, Jiande Wu, Jun Liu and Xiangning He, "Application summarization of coupled Inductors in DC-DC converters, " *in Proc. IEEE APEC Conf.*, pp. 1487-1492, 2009.

[7] Wuhua Li and Xiangning He, "A family of interleaved DC-DC converters deduced from a basic cell with winding-cross-coupled inductors (WCCIs) for high step-up or step-down conversions ," *IEEE Trans. Power Electron.*, vol. 23, no. 4, pp. 1791-1801, Jul. 2008.

[8] Il-Oun Lee, Shin-Young Cho and Gun-Woo Moon, "Interleaved buck converter having low switching losses and improved step-down conversion ratio," *IEEE Trans. Power Electron.*, vol. 27, no. 8, pp. 3664-3675, Aug. 2012.

[9] Cheng-Tao Tsai and Chih-Lung Shen, "Interleaved soft-switching buck converter with coupled inductors," in Proc. IEEE ICSET Conf., pp. 877-882, 2008.

[10] Zhiliang Zhang, E. Meyer, Yan-Fei Liu and P.C. Sen, "A non-isolated ZVS self-driven current tripler topology for low voltage and high current applications," in Proc. IEEE ECCE Conf., pp. 1983-1990, 2009.

[11] Yungtaek Jang, M. M. Jovanovic and Y. Panov, "Multiphase buck converters with extended duty cycle," *in Proc. IEEE APEC Conf.*, pp. 38-44, 2006.

[12] D. A. Grant, Y. Darroman and J. Suter, "Synthesis of tapped-inductor switched-mode converters," *IEEE Trans. Power Electron.*, vol. 22, no. 5, pp. 1964-1969, Sep. 2007.

[13] Sheng Ye, W. Eberle and Yan-Fei Liu, "A novel non-isolated full bridge topology for VRM applications," *IEEE Trans. Power Electron.*, vol. 23, no. 1, pp. 427-437, Jan. 2008.

[14] Hao Cheng, K. M. Smedley and A. Abramovitz, "A wide-input–wide-output (WIWO) DC–DC converter," *IEEE Trans. Power Electron.*, vol. 25, no. 2, pp. 280-289, Feb. 2010.

[15] M. Batarseh, Xiangcheng Wang and I. Batarseh, "Non-isolated half bridge buck based converter for VRM application," in *Proc. IEEE PESC*, 2007, pp. 2393-2398.

[16] K. Nishijima, D. Ishida, K. Harada, T. Nabeshima, T. Sato and T. Nakano, "A novel two-phase buck converter with two cores and four windings," *in Proc. IEEE INTELEC Conf.*, pp. 861-866, 2007.

[17] Kaiwei Yao, Yuancheng Ren, Jia Wei, Ming Xu and F. C. Lee, "A family of buck-type DC-DC converters with autotransformers," *in Proc. IEEE APEC Conf.*, vol. 1, pp. 114-120, 2003.

[18] Zhihua Yang, Sheng Ye and Yan-Fei Liu, "A New Transformer-Based Non-Isolated Topology Optimized for VRM Application," in *Proc. IEEE PESC*, 2005, pp. 447-453.

[19] M. H. Vafaie, E. Adib and H. Farzanehfard , "A self powered gate drive circuit for tapped inductor buck converter," in *Proc. IEEE PEDSTC*, 2012, pp. 379-384.

[20] P. Xu, J. Wei and F. C. Lee, "The active-clamp couple-buck converter-a novel high efficiency voltage regulator modules," *in Proc. IEEE APEC Conf.*, vol. 1, pp. 252-257, 2001.

[21] B.-R. Lin, J.-J. Chen and F.-Y. Hsieh, "Analysis and implementation of a bidirectional converter with high conversion ratio," in *Proc. IEEE ICIT*, 2008, pp. 1-6.

[22] Zhiliang Zhang, W. Eberle, Yan-Fei Liu and P.C. Sen, "A novel non-isolated ZVS asymmetrical buck converter for 12 V voltage regulators," in Proc. IEEE PESC Conf., pp. 974-978, 2008.

[23] Sheng Ye, E. Meyer, Yan-Fei Liu and Liu Xiao Dong, "A novel non-isolated two-phase full bridge topology for VRM applications," in Proc. IEEE APEC Conf., pp. 24-30, 2008.

[24] Zhiliang Zhang, E. Meyer, Yan-Fei Liu and P. C. Sen, "A 1-MHz, 12-V ZVS nonisolated full-bridge VRM with gate energy recovery," *IEEE Trans. Power Electron.*, vol. 25, no. 3, pp. 624-636, Mar. 2010.

[25] Hyosang Jang, Taeyoung Ahn and Byungcho Choi, "New half-bridge dc-to-dc converters for wide input voltage applications," *in Proc. IEEE INTELEC Conf.*, pp. 1-6, 2009.

[26] Ling Gu, Wenjing Cao, Ke Jin and Xinbo Ruan, "A family of switching capacitor regulators," *in Proc. IEEE ECCE Conf.*, pp. 3370-3376, 2011.

[27] Daocheng Huang, Xinke Wu and F.C. Lee , "Novel non-isolated LLC resonant converters," *in Proc. IEEE APEC Conf.*, pp. 1373-1380, 2012.

[28] Zhiliang Zhang, W. Eberle, Yan-Fei Liu, P.C. Sen , "A nonisolated ZVS asymmetrical buck voltage regulator module with direct energy transfer," *IEEE Trans. Ind. Electron.*, vol. 56, no. 8, pp. 3096-3105, Aug. 2009.

[29] K. W. E. Cheng, "Tapped inductor for switched-mode power converters," *in Proc. IEEE ICPESA Conf.*, pp. 14-20, 2006.

[30] Zhihua Yang, Sheng Ye and Yanfei Liu, "A novel nonisolated half bridge DC-DC converter," *in Proc. IEEE APEC Conf.*, vol. 1, pp. 301-307, 2005.

[31] J. Leyva-Ramos, L. H. Diaz-Saldierna and M. G. Ortiz-Lopez, "Control of high-step down voltage converters for voltage regulator modules," *in Proc. IEEE CCE Conf.*, pp. 1-6, 2011.

978-1-4799-2706-7/14 $31.00 © 2014 IEEE

Digital Control of PWM Inverter using Ultra High Speed Network for Feedback Signals with Communication Disturbance Observer based on Rocket I/O Protocol

Ryo Saito, Kazuo Tsuchida, Tomoki Yokoyama
Tokyo Denki University
5, Asahicho, Senju, Adachiku, Tokyo, Japan
E-mail : yoko@fr.dendai.ac.jp

Abstract—An new network communication system for power electronics controller to transmit the feedback signal from the sensors using FPGA based hardware controller was proposed. Based on RocketI/O network controller, customized Rocket I/O protocol was implemented to achieve high speed feedback communication between two FPGA controller. To compensate the network delay of the feedback signal, communication disturbance observer is combined to the digital control method of PWM inverter system.

I. INTRODUCTION

To construct the power electronics equipments, various sensors are necessary for the measurement of the state variables and these signals have to feedback to control the system. These signals are connected by metal wire from the sensors to the control board, and the signals are used for the control system. In the case of rather large capability power electronics system, multi-cell or multi-unit system is generally applied and the number of feedback signals becomes huge, so many wire cables are necessary to build up the system. This result is the prevention of the reduction of the manufacturing cost and the complexity of the system.

On the other hand, progress of network technology makes it possible to realize the real time high speed communication system to apply to the power electronics system. It is expected that the data feedback of power electronics will shift to the digital data communication using network from the conventional transfer technique of analog signal in the future. In order to realize the transform of the feedback signal using network communication in practical use, it is necessary to consider the trade-off of the merit (reduction of the wire) and the demerit (restriction of transmission speed, the quality and safety of communication pathway).

In this research, the network communication method for power electronics is proposed and verified using FPGA based hardware controller. The ultra high speed control communication system for power electronics based on Rocket I/O protocol is designed and implemented to FPGA chip [1]. And the power electronics controller was also implemented to the same FPGA chip. Using FPGA, all the protocol processing is implemented in the hardware, the network transmission speed becomes maximum, the protocol processing is processed by the high-speed control circuit. Through the experiments, the accurate operation of the data communications were verified, and the transfer delay of the data is observed around 0.56μ seconds. Communication disturbance observer was introduced to compensate the network delay of the feedback signal. Network disturbance was defined to simulate the network conditions. The first-order and the second-order network disturbance were used to derive the disturbance observer. The simulations of the inverter control system based on the voltage feedback is carried out based on the proposed feedback system, and the proper operation was confirmed.

II. NETWORK DESIGN OF THE CONTROLLER

A. System Configuration

Fig. 1 shows the system configuration. Board 1 treats the A/D interface and transfer the converted digital data into the packet then pass to the Rocket I/O transceiver. Board 2 treats the packet receiving and take out the feedback data to the inverter control block, then the digital control and communication disturbance observer is applied to the system. In each board, all the calculation were implemented in one FPGA chip, and all the functions for the network communication were processed in FPGA chip.

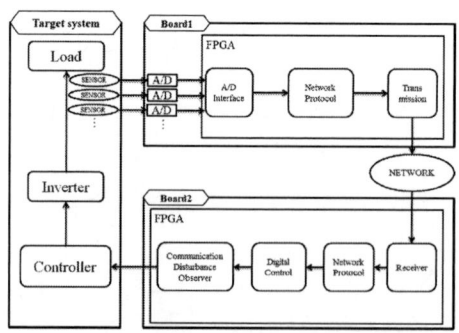

Fig. 1. Configuration of the Network Feedback Controller

B. RocketI/O Transceiver

In this research, all the control system is built on the FPGA board. The FPGA board is equipped with the Rocket I/O transceiver, and the Rocket I/O transceiver is connected to the FPGA directly. The Virtex-5 FPGA XC5V-LX110T-PCIEXP is used as a FPGA based hardware controller. The proposed system communicates all the feedback data through the Rocket I/O transceiver.

In the Virtex5 chip, the Rocket I/O transceiver is implemented in the chip itself. PHY structure is used to communicate with the FPGA through the physical layer, so the control block can access to Rocket I/O network directly in the FPGA chip.

1) Transmission Function and Receiving Function: The transmission function and receiving function was applied to the transmitter of the Rocket I/O transceiver. These function were configured by the PCS layer and the PMA layer. The data coding, decoding and the control of a signal integrity were performed in the PCS layer. Parallel to serial conversion of the data is carried out in the PMA layer. Fig.2 shows a block diagram of a transmitter.

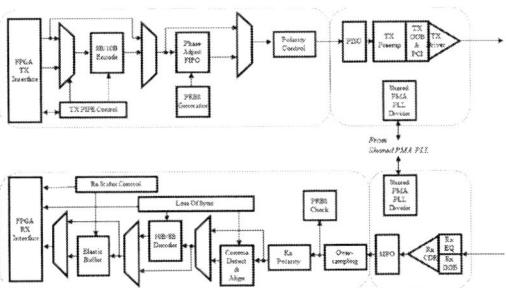

Fig. 2.   Transmission Block and Receiving Block

2) Communication Speed and Construction of the FPGA Module: The theoretical value of the transmitting processing time was calculated from the VHDL code as follows.

- Transmitting processing time : 238[$nsec$]

- Receiving processing time : 319 [$nsec$]

The total time of the theoretical value for the transmitting process and the receiving process becomes 557.6[$nsec$].

C. Deadbeat Control with Disturbance Current Compensation for PWM inverter[7]

1) Modeling: Fig.3 shows the system configuration of the proposed method. The state equation of the continuous time model becomes as (1).

$$\dot{x} = A_C x + B_C u_{db} \tag{1}$$

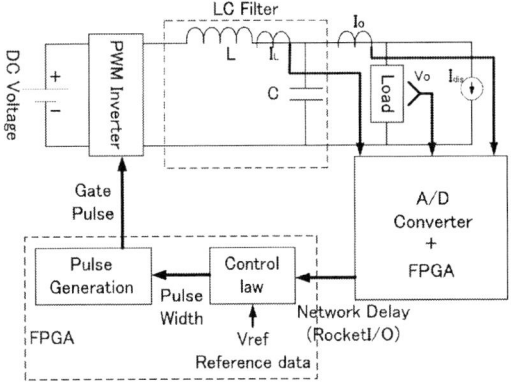

Fig. 3.   Single Phase PWM Inverter

$$, where \quad x = \begin{bmatrix} V_O & i_L & i_{dis} \end{bmatrix}^T,$$

$$A_C = \begin{bmatrix} -1/CR & 1/C & -1/C \\ -1/L & 0 & 0 \\ 0 & 0 & 0 \end{bmatrix},$$

$$B_C = \begin{bmatrix} 0 & 1/L & 0 \end{bmatrix}^T.$$

Discrete-time state equation of (1) become as follows.

$$x[k+1] = A_d x[k] + B_d u_{db}[k] \tag{2}$$

$$, where \quad A_d = e^{A_C T} = \begin{bmatrix} F_{11} & F_{12} & F_{13} \\ F_{21} & F_{22} & F_{23} \\ 0 & 0 & 1 \end{bmatrix},$$

$$B_d = e^{\frac{A_C T}{2}} B_C E = \begin{bmatrix} g_1 & g_2 & 0 \end{bmatrix}^T,$$

$$T = sampling \ time.$$

In this method, the disturbance current i_{dis} was defined to compensate the disturbance from the nominal conditions. i_{dis} can be calculated as the following equation.

$$i_{dis}(k) = i_O(k) - V_O(k)/R \tag{3}$$

In (2), replacing $V_O(k+1)$ as the voltage reference, the pulse width for the next sampling instant can be obtained as follows.

$$u(k) = \frac{V_{o\_ref}(k+1) - F_{11}V_o(k) - F_{12}i_L(k) - F_{13}i_{dis}(k)}{g_1} \tag{4}$$

D. Variable sampling quasi multirate deadbeat control[1]

Variable sampling quasi multirate deadbeat control (VSQMDB) was also applied to verify the experimental system. In the this method, two kinds of modeling methods are used as shown in Fig.4. Fig.5 shows the sampling procedure of the proposed method. T_u is the sampling interval. T_1 is the start point of the carrier, and T_6 is the end point of the carrier. First, the state variables are sampled at the beginning of a carrier period T_1, and a pulse width is calculated using the centered pulse model in Fig.5(a). Next the state variables are sampled in the center of a carrier period T_2, and a pulse width is re-calculated using the left sided pulse model in Fig.5(b). Next, if the re-calculated pulse is longer than the instant T_3, the state variables are sampled at instant T_3, which is the center of T_2 and T_6, and re-calculate the pulse width using the left sided pulse method. In the same way, at the instant T_4 and T_5, re-calculation is carried out if the re-calculated pulse width was longer than that sampling instant.

To improve the robustness for the parameter variations, quasi multirate deadbeat control method is also combined to the proposed method(VSQMDB). In the multirate control method, two pulse width are derived in (10) and (11). To avoid the sampling frequency becomes half of the carrier frequency, two pulse width are combined to one pulse as shown in Fig.6[7]. When the sampling instant is T_1, the centered pulse method in Fig.6(a) is applied, and in the case of T_2,T_3,T_4 and T_5, the left sided pulse method in Fig.6(b) is applied respectively.

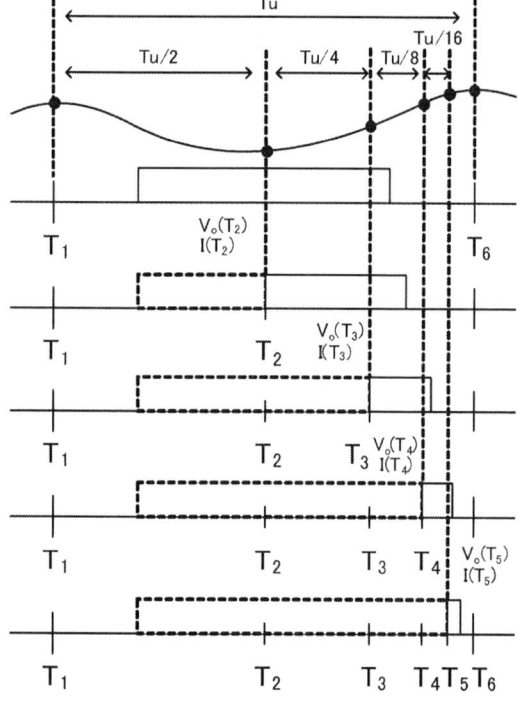

Fig. 4. VSDB

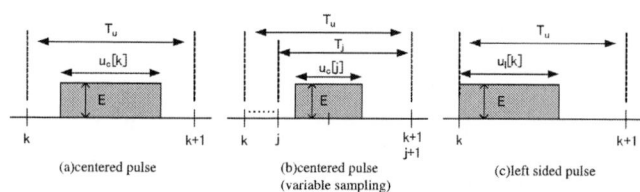

Fig. 5. Discrete time model

Here, the sampling interval is T_u, and the pulse is outputted in the center of the carrier interval as shown in Fig.5(a), A_d and B_d becomes as follows[2],

$$A_d = e^{A_c T_u} \quad B_d = e^{\frac{A_c T_u}{2}} B_c E. \tag{5}$$

When a pulse is outputted from a sampling point as shown in Fig.5(b), A_{dl} and B_{dl} becomes as follows[8],

$$A_{dl} = e^{A_c T_u} \quad B_{dl} = e^{A_c T_u} B_c E. \tag{6}$$

From (2) and (6), the state equation of the next sampling instant becomes (7). Using (2) and (7), the multirate state equation (8) can be derived for the centered pulse model.

$$x[k+2] = A_d x[k+1] + B_d u[k+1] \tag{7}$$

$$x[k+2] = A_d^2 x[k] + [A_d B_d, B_d][u_c[k], u_c[k+1]]^T \tag{8}$$

Using (2) and (6), the multirate state equation (9) for the left sided pulse model can be derived.

$$x[k+2] = A_{dl}^2 x[k] + [A_{dl} B_{dl}, B_{dl}][u_l[k], u_l[k+1]]^T \tag{9}$$

(8) and (9) can be rewritten in the sampling period as (10) and (11).

$$x[i+1] = A_m x[i] + B_m u_c[i] \tag{10}$$

,where $u_c[i] = [u_c[k], u_c[k+1]]^T$,
$A_m := A_d^2, B_m := [A_d B_d, B_d]$.

$$x[i+1] = A_{ml} x[i] + B_{ml} u_l[i] \tag{11}$$

,where $u_l[i] = [u_l[k], u_l[k+1]]^T$,
$A_{ml} := A_{dl}^2, B_{ml} := [A_{dl} B_{dl}, B_{dl}]$.

$$u_c[i] = B_m^{-1} x[i+1] - B_m^{-1} A_m x[i] \tag{12}$$

$$u_l[k+1] = B_{ml}^{-1} x[k+2] - B_{ml}^{-1} A_{ml} x[k+1] \tag{13}$$

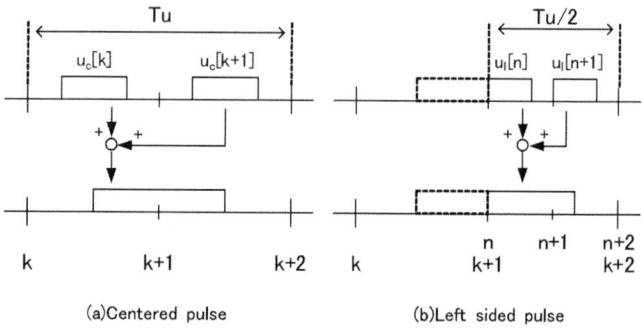

Fig. 6. VSQMDB

This control method re-calculates pulse width immediately after a sampling, and it needs to be reflected in a pulse output. For example when carrier frequency is 2[kHz], sampling frequency is 31.25[μs], therefore high operation speed is required. And so FPGA is applied to a controller, and operation is ended to less than 1[μs] by conversion time 800[ns] of an AD converter and parallel operation of FPGA. Consequently the operation result is reflected immediately in the pulse after sampling[1].

E. Implementation of Rocket I/O system

In the proposed system, Rocket I/O was used to realize ultra high speed feedback data transmission. Data format of Rocket I/O for feedback system is indicated in Fig.7.

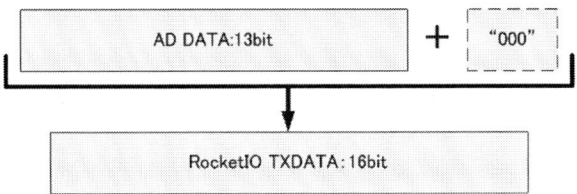

Fig. 7. Data format of TXDATA

13bit A/D converter is applied in this system, 16bit data band is prepared to construct data packet. VHDL module configuration of the board 1 (A/D interface and packet generation) was indicated in Fig. 8. Rocket I/O module treats the packet generation procedure and A/D module treats the interface between FPGA and A/D converter. Fig.9 shows the RTL of the Rocket I/O module and A/D module.

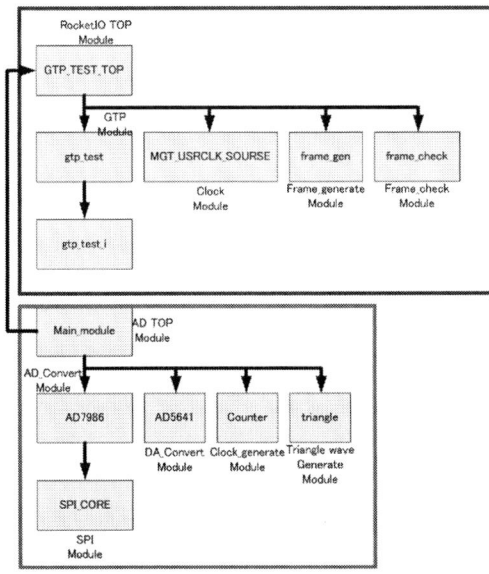

Fig. 8. VHDL Module Configuration of Rocket I/O

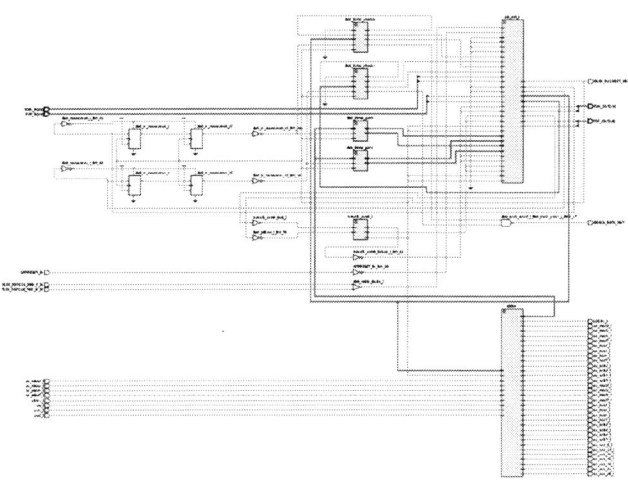

Fig. 9. RTL diagram of Rocket I/O and A/D control module

F. Experimental result of Rocket I/O

Fig. 10 shows the experimental packet data of Rocket I/O using the proposed system. Data transmission between the board 1 and the board 2 was confirmed.

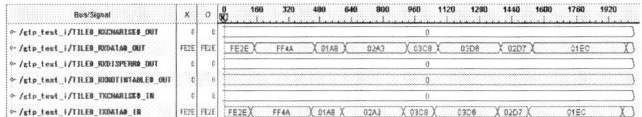

Fig. 10. Transmitting and Receiving Data

G. Communication Disturbance Observer (CDOB)

In the case that the network is used for the feedback signal, the network delay is exist in the feedback line and should be compensated. Communication disturbance observer (CDOB) was proposed to compensate the network delay in which the communication delay was treated as the disturbance[3]. In this paper, CDOB was adopted to Rocket I/O protocol to compensate the network delay.

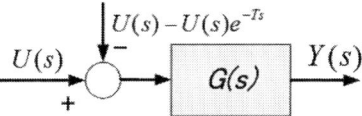

Fig. 11. Network Disturbance

Fig.11 indicates the concept of network disturbance (ND) and Fig.12 shows the control block diagram of the network disturbance. First-order CDOB and second-order CDOB were considered to assume that the ND is the polynomial of time domain t. ND was defined as follows.

$$d_{net}(t) = u(t) - u(t - T) \tag{14}$$

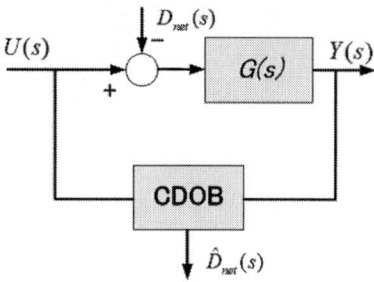

Fig. 12.   Network Disturbance

$$\dot{d}_{net}(t) = 0 \qquad (15)$$

$$\ddot{d}_{net}(t) = 0 \qquad (16)$$

(14) is the first-order CDOB and (15) is the second-order CDOB. ND was treated as the vector in (17).

$$d_{net}(t) = \begin{bmatrix} d_{net}(t) \\ \dot{d}_{net}(t) \end{bmatrix} \qquad (17)$$

The state equation of the proposed PWM inverter system becomes as follows.

$$\dot{x} = A_C x + B_C u \qquad (18)$$

$$, where \quad x = \begin{bmatrix} V_O & i_L & i_{dis} \end{bmatrix}^T,$$

$$A_C = \begin{bmatrix} -1/CR & 1/C & -1/C \\ -1/L & 0 & 0 \\ 0 & 0 & 0 \end{bmatrix},$$

$$B_C = \begin{bmatrix} 0 & 1/L & 0 \end{bmatrix}^T.$$

Here, these equations can be combined, the CDOB of the proposed system can be described as follows. h is the observer gain.

$$\begin{bmatrix} \dot{\hat{x}}(t) \\ \dot{\hat{d}}_{net}(t) \end{bmatrix} \begin{bmatrix} A & B \\ O_{2 \times n} & E \end{bmatrix} \begin{bmatrix} \hat{x}(t) \\ \hat{d}_{net}(t) \end{bmatrix}$$
$$+ \begin{bmatrix} b \\ o_{2 \times 1} \end{bmatrix} u(t) + h(y(t) - \hat{y}(t)) \qquad (19)$$

$$\hat{y}(t) = \begin{bmatrix} c & o_{1 \times 2} \end{bmatrix} \begin{bmatrix} \hat{x}(t) \\ \hat{d}_{net}(t) \end{bmatrix} \qquad (20)$$

Then the ND can be estimated as follows in the first-order CDOB and the second-order CDOB respectively. g_{net} is the cut-off frequency gain of the low pass filter for the first-order CDOB and k_1 and k_2 is the cut-off frequency gain of the low pass filter for the second-order CDOB respectively.

$$\hat{D}_{net}(s) = \frac{g_{net}}{s + g_{net}} D_{net}(s) \qquad (21)$$

$$\hat{D}_{net}(s) = \frac{k_1}{s^2 + k_2 s + k_1} D_{net}(s) \qquad (22)$$

$$V_O(t) = d_{net}(t) - V_O e^{-Ts} \qquad (23)$$

III.   SIMULATION

Simulations were carried out to evaluate the effectiveness of the CDOB,the following conditions were used to compare the output characteristics.

TABLE I.   SYSTEM PARAMETERS

Output voltage frequency [Hz]	50
Output voltage amplitude [V_{rms}]	100
Input dc voltage [V]	180
Sampling Time [ms]	0.5
Simulation Time [ms]	0.2
L	4.930 [mH]
C	34.94 [μF]
Load Conditions	
Resistive load 9.1[Ω]	
Constant Delay : 1[ms],557.6[ms],1[μs]	
Variable Delay : 0~0.1[ms]	

THD and the steady state error of the output voltage are used as the criteria for evaluation. Table I shows the system parameters for the simulations.

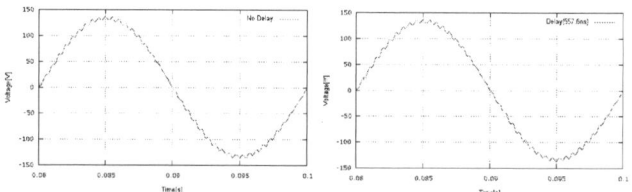

Fig. 13.   Output Voltage Waveforms
(a) no network delay (b) with network delay (557.6[$nsec$])

Fig.13 (a) shows the output voltage for the deadbeat control method without network delay and (b) shows the output voltage with 557.6 ns network delay. In this case, deadbeat control without CDOB was applied.

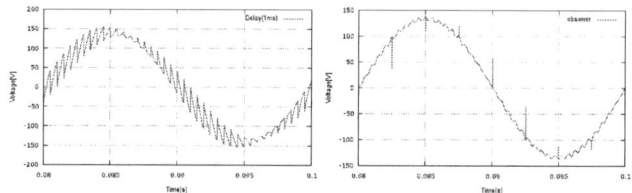

Fig. 14.   Output Voltage Waveforms
(a) network delay (1[$msec$]) without CDOB (b) network delay(1[$msec$]) with CDOB

Fig.14 (a) and (b) show the output voltage for 1ms network delay condition with and without CDOB respectively. In such a severe conditions, CDOB compensates network delay and superior characteristics was obtained. In the Rocket I/O, maximum network communication delay is insured within 1

TABLE II.   SIMULATION RESULTS

	RMS[V]	SSE[%]	THD[%]
No Delay	94.36	-5.64	2.58
Constant Delay without CDOB(1ms)	98.47	-1.53	14.04
Constant Delay with CDOB(1ms)	94.37	-5.63	2.67
Constant Delay without CDOB(1us)	94.37	-5.63	2.58
Constant Delay with CDOB(1us)	94.36	-5.63	2.58
Variable Delay without CDOB(0[ms] → 20[ms])	91.77	-8.23	46.11
Variable Delay with CDOB(0[ms] → 20[ms])	94.36	-5.64	2.58

μs. Fig.15 (a) and (b) shows the output voltage for $1\mu s$ delay conditions with and without CDOB respectively.

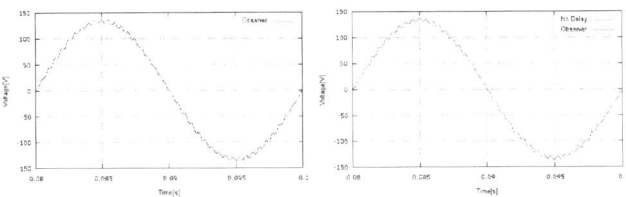

Fig. 15.   Output Voltage Waveforms
(a) constant delay ($1[\mu s]$) without CDOB (b) constant delay ($1[\mu s]$) with CDOB

Fig.16 (a) and (b) shows the output voltage comparison for variable delay conditions which varied from 0ms to 20ms in 200ms.

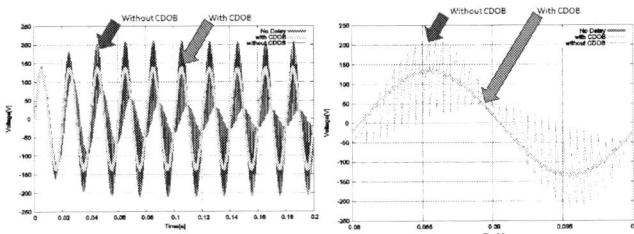

Fig. 16.   Output Voltage Waveforms
(a) Variable delay($0[ms] \to 20[ms]$) (b) Enlargement Waveforms($0[ms] \to 1[ms]$)

In TableII, the simulation results were summarized. In the case that the network delay is 1ms, the superior characteristics of CDOB was verified.

As the network delay of Rocket I/O is very small, the affection of the network delay is not so large, but the effectiveness of CDOB was also confirmed. In the case of variable delay conditions, the effectiveness of CDOB is very clear.

IV. CONCLUSION

An ultra fast network feedback system using Rocket I/O protocol was implemented to the digital control system of PWM inverter using FPGA based hardware controller. Deadbeat control was applied and the communication disturbance observer was combined to compensate the network delay of the feedback signal and superior characteristics was verified.Also the operation of the ultra fast data feedback system using Rocket I/O was experimentally confirmed.

REFERENCES

[1] S.Tahara, T.Fujii, T.Yokoyama : "Variable Sampling Quasi Multirate Deadbeat Control Method for Single Phase PWM Inverter in Low Carrier Frequency", PCC Nagoya 2007

[2] K.Natori,R.Oboe, and K.Ohnishi : "Stability Analysis and Practical Design Procedure of Time Delayed Control Systems with Communication Disturbance Observer",IEEE Transactions on Industrial Informatics, Vol.4,No.3,pp.185-197(2008)

[3] K.Natori, R.Kubo, T.Shimono, K.Ohnishi : "Time-Delay Compensation by Communication Disturbance Observers of Different Orders in Bilateral Teleoperation Systems", IEEJ 2009 pp.353-362

[4] N.Doi, T.Saito, T.Yokoyama : "Verification of the load sharing characteristics in Autonomous Decentralized UPS system using FPGA based Hardware Controller", EPE-PEMC 2008

[5] T.Ishioka, N.Doi, T.Yokoyama "Experimental Verification of Autonomous Decentralized UPS system with Instantaneous Power Detection using FPGA based Hardware Controller", ECCE 2009

[6] S.Kojima, T.Ishioka, T.Yokoyama, "A study of Communication System for Power Electronics Controller using FPGA based Hardware Controller", IPEC 2010

[7] H.Uchida, T.Yokoyama, "1MHz Variable Multi Sampling Digital Control of Single Phase PWM Inverter using FPGA based Hardware Controller", ECCE 2012

[8] K.Tsuchida, T.Yokoyama : "Verification of high-speed network communication system using FPGA", JIASC 2012

[9] "XILINX Corporation" http://japan.xilinx.com/

100 kHz DC Chopper Digitally Gate Controlled with Partial Turn- off Switching Using SiC-MOSFET and FPGA

Yukinori Tsuruta
Electrical and Computer Engineering
Yokohama National University
Yokohama, Japan
tsuruta@ynu.ac.jp

Atsuo Kawamura
Electrical and Computer Engineering
Yokohama National University
Yokohama, Japan
kawamura@ynu.ac.jp

Abstract— **This paper describes high efficiency chopper applying SiC-MOSFETs to soft switching method with partial turn-off. This chopper basically operates by digitally gate control with partial turn-off switching using FPGA under high frequency of 100 kHz and 25 kHz. For confirming effectiveness by partial turn-off and applying SiC-MOSFETs, the 8kW rating test, efficiency measurement and principal operational test have been completed and efficiency of 98.4 % at 25 kHz was obtained.**

Keywords— *SiC-MOSFET, Partial Turn-off, Boost chopper, FPGA.*

I. INTRODUCTION

Improvement of efficiency on DC-DC converter have been studied in many public civil area sectors such as home electric appliances, transportation, electric power and so on in the past. Various type of circuit methods have been proposed until now. In 1994 2 switch partial resonant scheme of ZVT converter was proposed [1]. This type can obtain high efficiency without increasing the current stress of main switch due to the current flow of auxiliary circuit not passing through main switch. Recent development of electric drive in transportation, especially in the automobile field is remarkable. The design target of DC-DC converter already reaches 98-99 % [2]. In 2005 authors proposed SAZZ (snubber assisted zero voltage zero current transition) chopper and improvement of efficiency reaches 99 % under condition of 25 kHz, 8 kW in 2009 [3]. However when power converter is applied to high power application, tail loss increase due to turning off high current. The behavior of tail current under soft switching was studied in [4]. In waveforms during turn off period under ZVS soft switching condition increase of tail current was observed. The authors proposed tail loss cancel circuit (TLC) to reduce power dissipation resulted from tail loss during turn off period [5] [6]. However efficiency was not obtained as expected due to the increase of auxiliary circuit loss by over-damping cancel current. Recent years in industry application power converter using new material of Silicon Carbide (SiC) is studied by hard switching [7][8]. Then SiC application to ZVS soft switching converter is also studied [9] [10].

Partial turn off for high power chopper using SiC-MOSFET by new TLC topology is discussed through 8 kW load test as follows in this paper.

II. NEW SiC TLC-II TYPE CHOPPER

A. Circuit configuration

Fig.1 and Table. I show new proposal TLC-II type soft switching dc chopper using SiC-MOSFETs. E_1 is input DC power source. L_1 is main DC reactor. SL_1 is saturable reactor to avoid occurrence of surge voltage in resonant auxiliary circuit. The output part is composed of output diode D_5, output capacitor C_{out}, and load resistor R_L. Discharge current through the resonant loop $C_2(+)-L_2-C_{in}(+)-C_{in}(-)-S_1(D)-D_1-S_2-D_4-C_2(-)$ cancels main collector current through main switch S_1 (including tail current).

TABLE I.
FUNDAMENTAL SPECIFICATION

Input	200 VDC		
Output	400 VDC		
Output power	8 kW		
Frequency	100 kHz	25 kHz	
Circuit type	TLC-II		
Control	Digital control by FPGA		
Cooling	With ventilating fans		
S_1	50MT060WH, 600V, 50A		IR
S_2	CMF20120D, 1200V, 32A×3P		CREE
D_1	CSD20060, 600V, 20A×3P		CREE
D_4	CSD20060, 600V, 20A×4P		CREE
D_5	CSD20060, 600V, 20A×10P		CREE
D_2, D_6, D_8	USR100PP12A, 1200V, 50×2A		

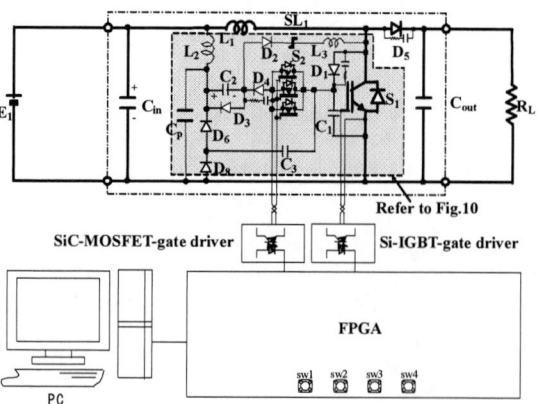

Fig. 1. New tail loss cancel circuit (TLC) using SiC-MOSFETs.

B. Principle of operation

Fig. 2 shows the basic operating waveforms of TLC-II chopper at steady state and equivalent circuits of each 5 operation modes are illustrated in Fig. 3.

1) Mode 1: Pre-turn-on and Turn-on with regenerating Auxiliary switch S_2 turns on at t = -t_1. Snubber capacitor voltage V_{C1} change from plus voltage to zero in sinusoidal waveform. At that instant auxiliary switch S_2 turns on as ZCS shown in Fig.2. With very short time delay, at t = t_0, main switch S_1 turns on. Diode D_5 turns off and regenerating diode D_4 turns on. Storage energy of auxiliary reactor L_2 is under regenerating state as shown in Fig.3 (a).

2) Mode 2: On state with charging C_2 At t = t_1, capacitor C_2 which cancels tail loss of main IGBT S_1 starts to be charged as shown in Fig.3(b).

3) Mode 3: On state Reactor L_1 current builds up as shown in Fig.3 (c).

4) Mode 4:turn-off with canceling tail current of main switch S_1 At t = t_4 very short time before main switch S_1 turns off auxiliary switch S_2 turns on again, charge stored in C_2 which cancels main current discharge in L_2C_2 resonant mode as shown in Fig.3 (d). At t = t_5 when this L_2C_2 resonant discharging current reaches at peak value, main switch S_1 is turned off. At t = t_6 when current through auxiliary switch S_2 becomes zero after decreasing like sinusoidal half wave, auxiliary switch S_2 is turned off. By this operation in mode 4 main current through main switch S_1 is canceled and turned off.

5) Mode 5:off-state Switches are all in off-state as shown in Fig.3 (e).

C. Simulation

New TLC-II circuit was simulated by PSIM 9.13 Service Pack 1. Simulation circuit is shown in Fig.4. Simulation result shown in Fig.5 is voltage $V(S_1)$ across switch S_1, current $I(S_1)$ through switch S_1 and current $I(S_2)$ through auxiliary switch S_2 at 8 kW, a 200V input and a 400V output, 100 kHz.

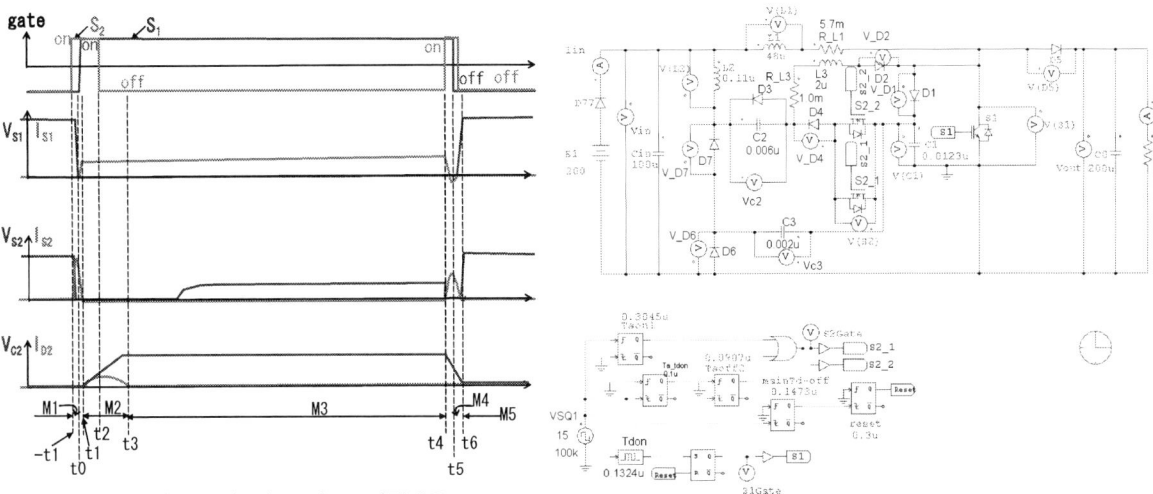

Fig. 2. Basic operational waveforms of TLC-II.

Fig. 4. Simulation circuit by PSIM 9.1.3 Service Pack 1.

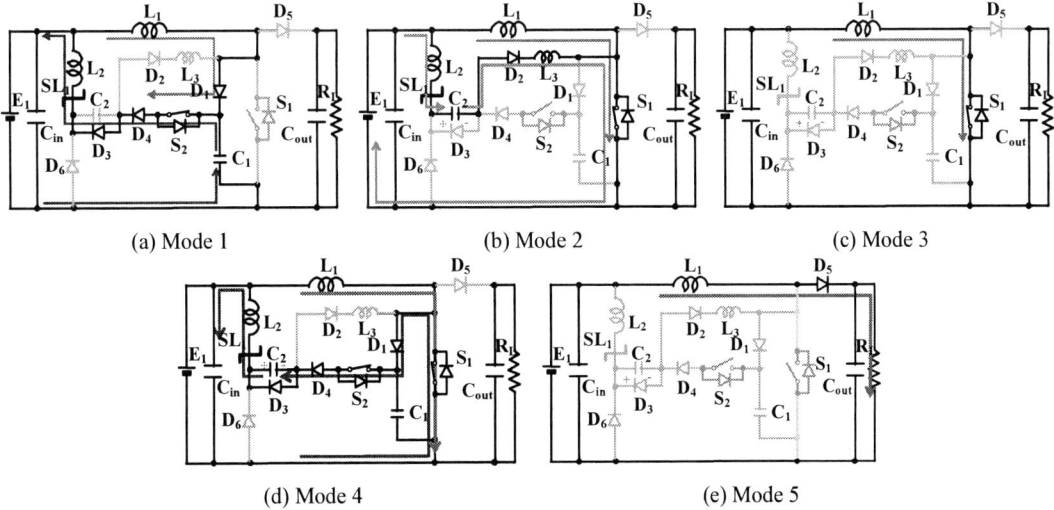

(a) Mode 1   (b) Mode 2   (c) Mode 3

(d) Mode 4   (e) Mode 5

Fig. 3. Equivalent circuits in 5 operational modes.

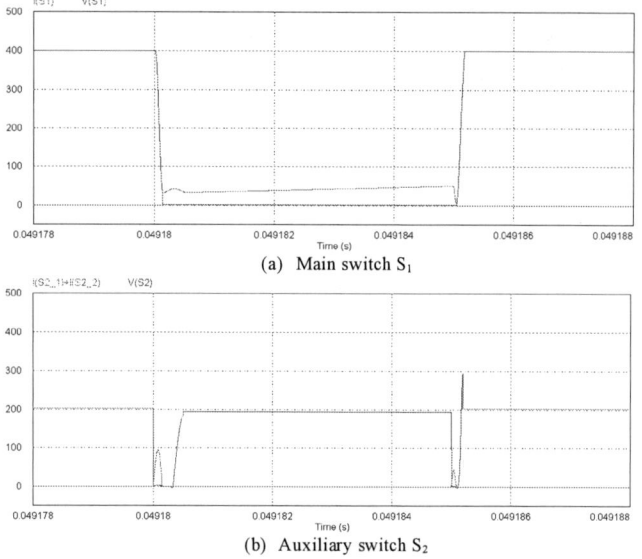

(a) Main switch S_1

(b) Auxiliary switch S_2

Fig. 5 Waveforms of voltage across switches and current through switches at 8 kW, 200 V / 400 V, 100 kHz.

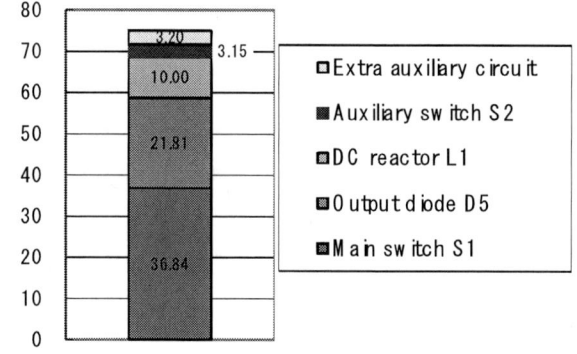

Fig. 6 Theory Limit Loss break down at 8 kW, 200 V / 400 V, 100 kHz by PSIM 9.1.3 Service Pack 1.

Fig.6 shows theory limit loss breakdown at 8 kW, 200 V/400 V, 100 kHz by PSIM 9.1.3 Service Pack 1. Efficiency of 99 % with main circuit loss of 62.65 W and auxiliary circuit loss of 6.35 W is resulted from on state loss only considered by theory limit.

III. BUILD OF DIGITALIZING THE GATE CONTROL FOR TLC-II TOPOLOGY USING FPGA

New digital gate control for TLC-II topology was built by programing with NIOS Development Board Cyclone II. All digital gate timing for new TLC-II topology is set by flow chart as shown in FIG.7. Setting values shown in Table. II are corresponding to gate timing chart shown in Fig. 8. Design of FPGA programing was obtained by gate time chart as shown in Fig.9.

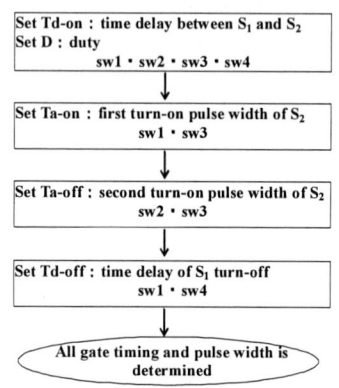

Fig. 7. Flow chart of digitalizing gate pulse setting using FPGA.

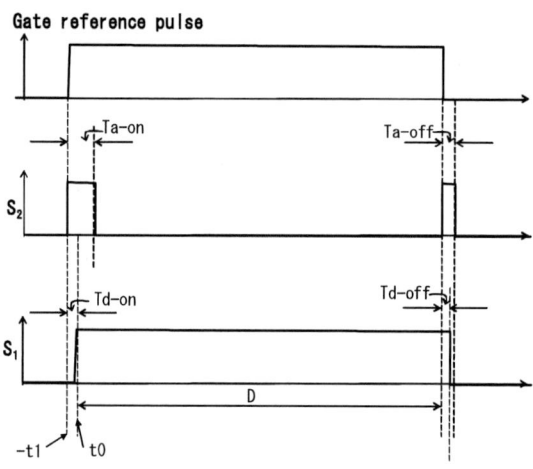

Fig. 8. Digital gate timing chart for new TLC-II.

TABLE II
Digital gate timing set by FPGA

$T_{d\text{-}on}$	Delay time: auxiliary S_2– main S_1	[ns]	150
$T_{a\text{-}on}$	auxiliary S_2 first on interval	[ns]	550
$T_{d\text{-}off}$	Delay time: auxiliary S_2 on– main S_1 off	[ns]	100
$T_{a\text{-}off}$	auxiliary S_2 second on interval	[us]	400
D	Duty	[unit]	0.48

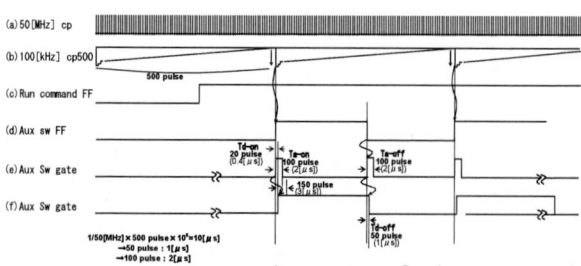

Fig.9 Digital gate time chart for FPGA programing (100[kHz] duty D=0.5=5[μs]/10[μs])

IV. EXPERIMENTAL EVALUATION AND DISCUSSION

A. Prototype model for experiment

8 kW, 100 kHz and 25 kHz prototype model for experiment was made as shown in Fig.10. It is a new proposal TLC-II type soft switching dc chopper using SiC-MOSFETs. Rating parameters are as follows.

- Output ratings: Vout=400 V, Pout=8 kW
- Switching frequency =100 kHz and 25 kHz
- Circuit parameters: C_1=16 nF, L_1=48 μH(at 100 kHz) and 200μH(at 25 kHz), L_3=2 μH, C_2=15 nF, C_3=2.35 μF

B. Evaluation of circuit operational verification

Principal circuit operation was verified by rating test. Fig.11 shows S_1 and S_2 waveforms under rating condition and soft switching by partial turn off is proved to be achieved. Corresponding to Fig.5(a), Fig.12 (a) shows voltage and current waveforms of main switch S_1 at 100 kHz. Corresponding to Fig.5(b), Fig.12(b) shows voltage and current waveforms of auxiliary switch S_2 at 100 kHz. In figures from 11 and 12 experimental result nearly corresponds to simulation results. So normal operation of prototype was confirmed.

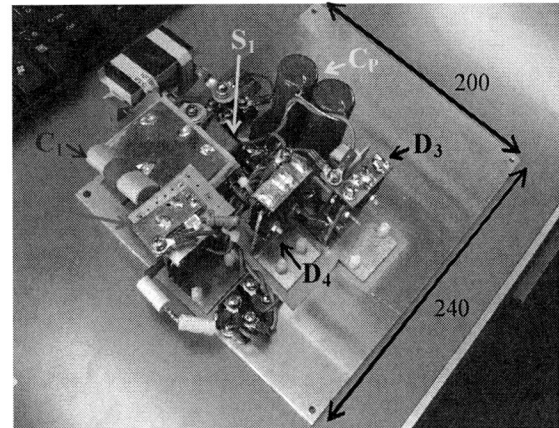

Fig. 10. Exterior of 8 kW prototype model of new SiC-TLC-II

C. Comparison of circuit operation between 25 kHz and 100 kHz

Fig.13 shows the voltage and current waveforms of main switch S_1 at 25 kHz corresponding to Fig.11 at 100 kHz. Reduction of switching frequency from 100 kHz to 25 kHz results in the improvement of efficiency from 97.2 % to 98.4%.

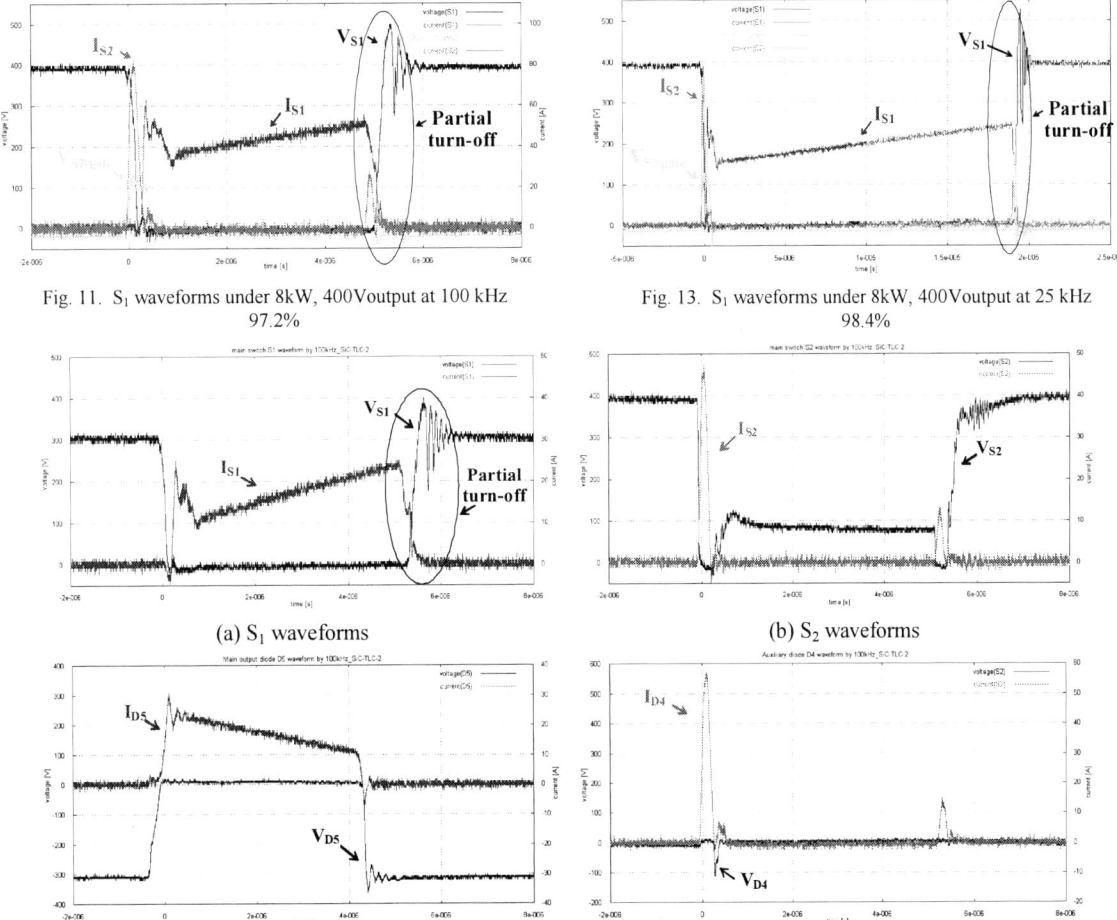

Fig. 11. S_1 waveforms under 8kW, 400Voutput at 100 kHz
97.2%

Fig. 13. S_1 waveforms under 8kW, 400Voutput at 25 kHz
98.4%

(a) S_1 waveforms

(b) S_2 waveforms

(c) D_5 waveforms

(d) D_4 waveforms

Fig. 12. Operating waveforms at 100 kHz

D. Loss breakdown

Figures from 14 to 25 were measured at 100 kHz, 8 kW, 200 V/400V. Fig.14 shows turn off loss of 29.5 W in main switch S_1. Fig.15 shows turn on loss of 20.5 W and turn off loss of 9 W in auxiliary switch S_2.

On state loss 40.3 W and recovery loss of 25.2 W in output diode D_5 are shown in Fig.16. Figures from 17 to 22 were loss breakdown of auxiliary diodes D_1 - D_8.

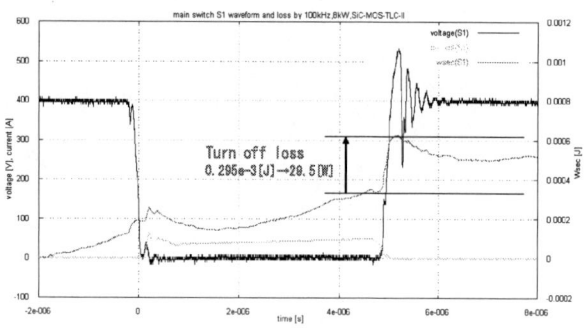

Fig. 14. 8 kW main switch S_1 loss breakdown

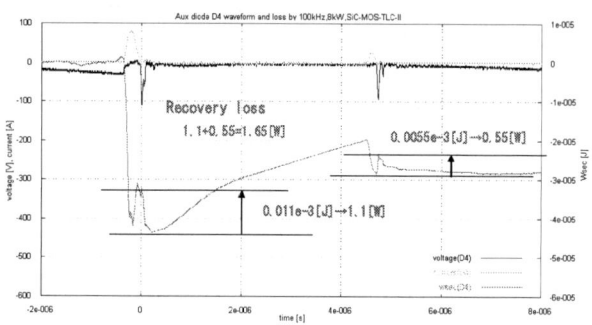

Fig. 18. 8 kW auxiliary diode D_3 loss breakdown

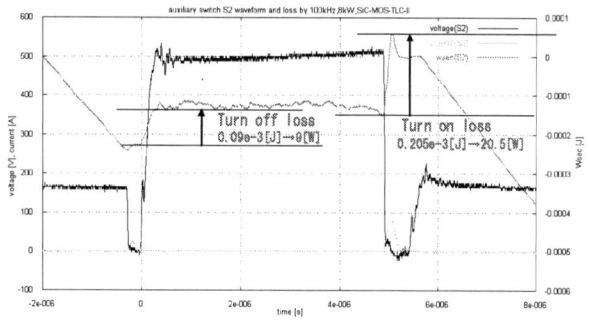

Fig. 15. 8 kW aux switch S_2 loss breakdown

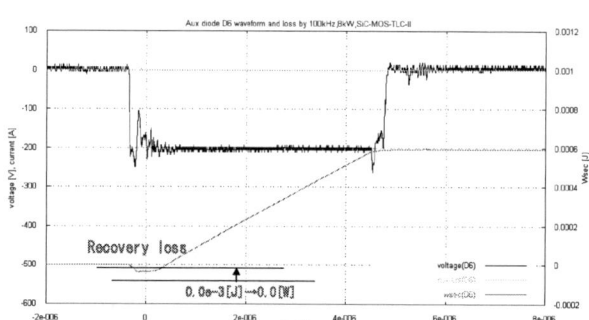

Fig. 19. 8 kW auxiliary diode D_2 loss breakdown

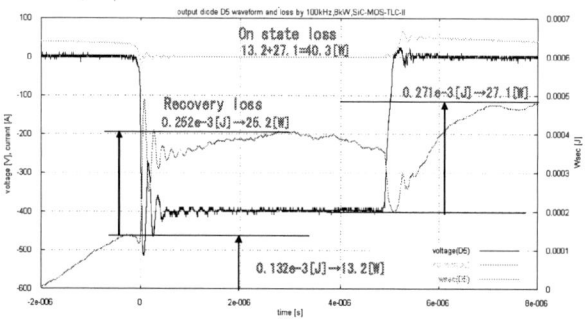

Fig. 16. 8 kW output diode D_5 loss breakdown

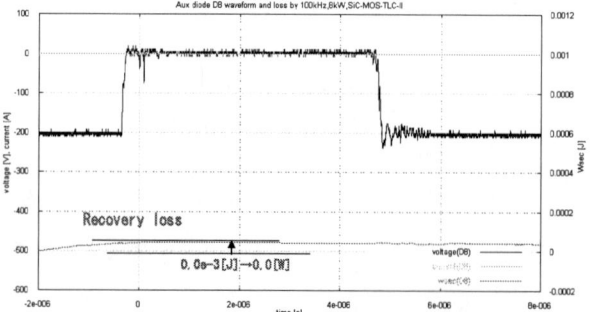

Fig. 20. 8 kW output diode D_6 loss breakdown

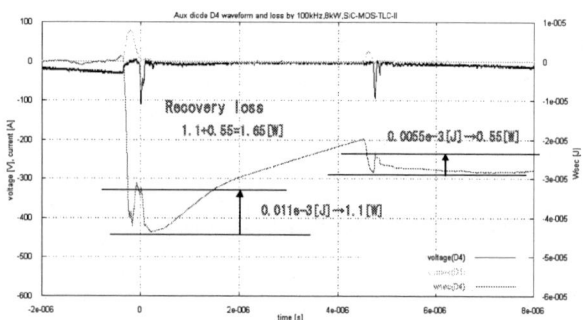

Fig. 17. 8 kW auxiliary diode D_4 loss breakdown

Fig. 21. 8 kW auxiliary diode D_8 loss breakdown

Recovery loss of auxiliary circuit is mainly shown from Fig.17 to Fig.22. Figures from 23 to 25 show CR snubber loss in diode D_1, D_4, and D_5.

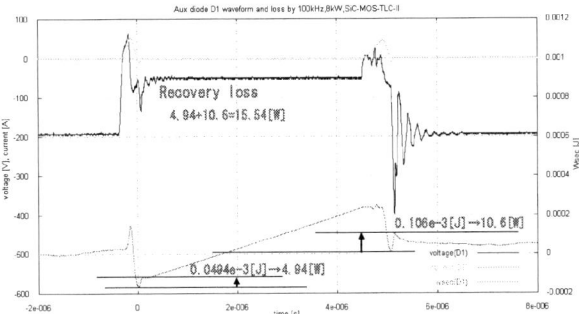

Fig. 22. 8 kW snubber diode D_1 loss breakdown

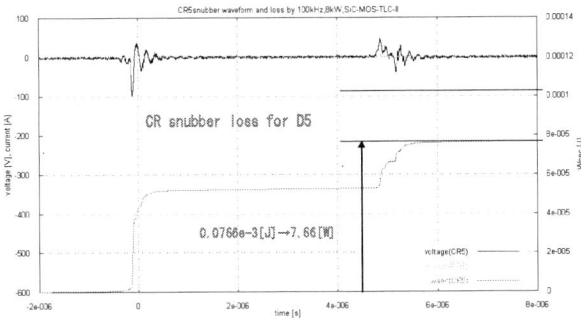

Fig. 23. 8 kW CR snubber for D_5 loss breakdown

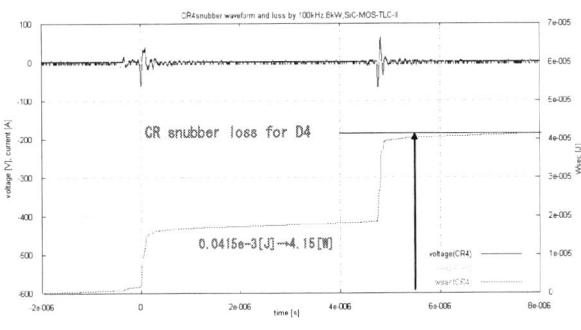

Fig. 24. 8 kW CR snubber for D_4 loss breakdown

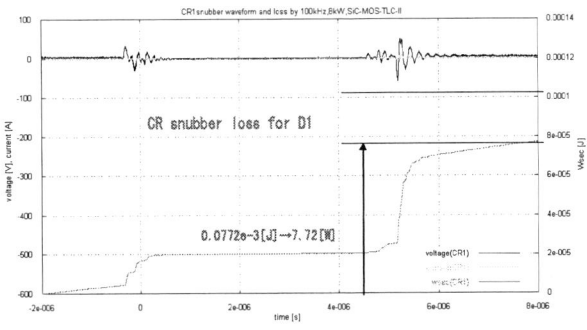

Fig. 25. 8 kW CR snubber for D_1 loss breakdown

E.   Breakdown evaluation of the total loss

Fig.26 and Fig.27 show comparison of total loss breakdown classified by circuit elements at 100 kHz and 25 kHz. Loss reduction rate by reduction of frequency from 100 kHz to 25 kHz proved 41 %(=1-129.01/219.79).

Fig.28 and Fig.29 show comparison of total loss breakdown classified by loss factors at 100 kHz and 25 kHz. Recovery loss reduction of auxiliary circuit proved from 57.35 W to 14.24 W. Table. III shows 100 kHz vs.25 kHz-8 kW SiC-TLC-II loss breakdown.

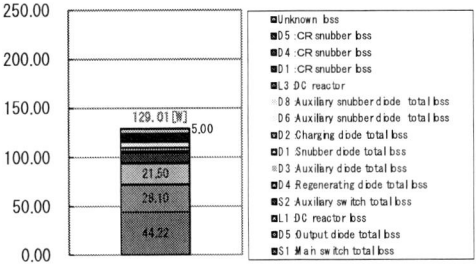

Fig. 26. Total loss breakdown classified by circuit elements at 25 kHz, 8 kW

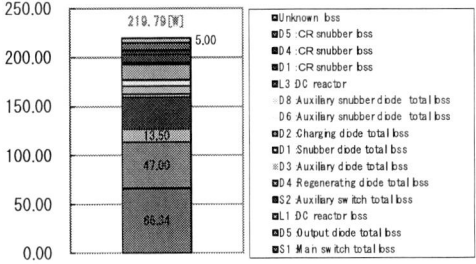

Fig. 27. Total loss breakdown classified by circuit elements at 100 kHz, 8 kW

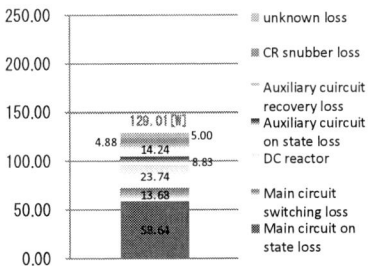

Fig. 28. Total loss breakdown classified by loss factors at 25 kHz, 8 kW

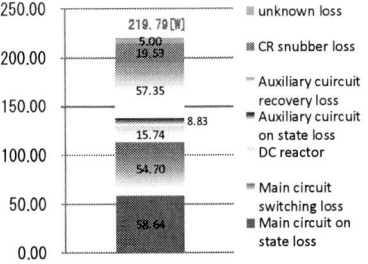

Fig. 29. Total loss breakdown classified by loss factors at 100 kHz, 8 kW

TABLE III

100 kHz vs.25 kHz-8 kW SiC-TLC-II loss breakdown

Circuit element	100[kHz]	25[kHz]	notes
S1 :Main switch turn on bss	0	0	
S1 :Main switch on state bss	36.84	36.84	PSIM
S1 :Main switch turn off bss	29.5	7.375	
1 S1 :Main switch total bss	66.34	44.22	
D5 :Output diode turn on bss	0.00	0.00	
D5 :Output diode on state bss	21.80	21.80	PSIM
D5 :Output diode recovery bss	25.20	6.30	
2 D5 :Output diode total bss	47.00	28.10	
3 L1 :DC reactor bss	13.50	21.50	IPEMC-2009
S2 :Auxiliary switch turn on bss	20.50	5.13	
S2 :Auxiliary switch on state bss	3.15	3.15	PSIM
S2 :Auxiliary switch turn off bss	9.00	2.25	
4 S2 :Auxiliary switch total bss	32.65	10.53	
D4 :Regenerating diode turn on bss	0.00	0.00	
D4 :Regenerating diode on state bss	1.60	1.60	PSIM
D4 :Regenerating diode recovery bss	1.65	0.41	
5 D4 :Regenerating diode total bss	3.25	2.01	
D3 :Auxiliary diode turn on bss	0.00	0.00	
D3 :Auxiliary diode on state bss	1.40	1.40	PSIM
D3 :Auxiliary diode recovery bss	6.77	1.69	
6 D3 :Auxiliary diode total bss	8.17	3.09	
D1 :Snubber diode turn on bss	0.00	0.00	
D1 :Snubber diode on state bss	2.13	2.13	PSIM
D1 :Snubber diode recovery bss	3.89	3.89	
7 D1 :Snubber diode total bss	6.02	6.02	
D2 :Charging diode turn on bss	0.00	0.00	
D2 :Charging diode on state bss	0.29	0.29	PSIM
D2 :Charging diode recovery bss	15.54	0.88	
8 D2 :Charging diode total bss	15.83	1.17	
D6 :Auxiliary snubber diode turn on bss	0.00	0.00	
D6 :Auxiliary snubber diode on state bss	0.13	0.13	PSIM
D6 :Auxiliary snubber diode recovery bss	0.00	0.00	
9 D6 :Auxiliary snubber diode total bss	0.13	0.13	
D8 :Auxiliary snubber diode turn on bss	0.00	0.00	
D8 :Auxiliary snubber diode on state bss	0.13	0.13	PSIM
D8 :Auxiliary snubber diode recovery bss	0.00	0.00	
10 D8 :Auxiliary snubber diode total bss	0.13	0.13	
11 L3 :DC reactor	2.24	2.24	L2 :None
12 D1 :CR snubber bss	7.72	1.93	
13 D4 :CR snubber bss	4.15	1.04	
14 D5 :CR snubber bss	7.66	1.92	
15 Unknown bss	5.00	5.00	unknown bss
16 Total	219.79	129.01	
17 Input power[kW]	8.22	8.13	
18 Output power[kW]	8.00	8.00	
19 Efficiency[%]	97.33	98.41	

F. Efficiency measurement

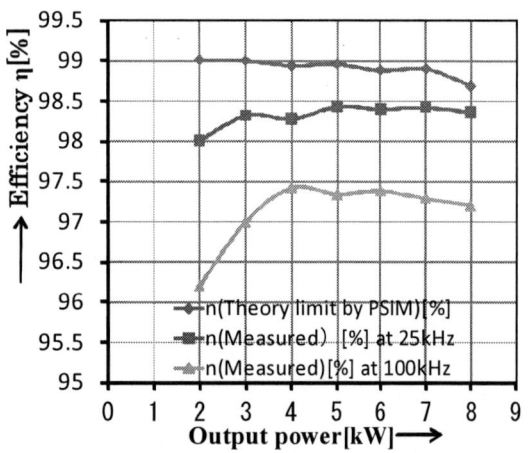

Fig. 30. Efficiency (measured), efficiency (PSIM) vs. output power

Optimization of circuit resonant parameter was verified by means of repetition of changing circuit constant and gate timing and pulses. Fig.30 shows comparisons between measurements and theory limit. Efficiency was measured every ΔP_{out}=1 kW step until rating power of 8 kW. Total efficiency was measured by means of 3390 Hioki digital power meter and CT6863 current sensor with 0.16 % accuracy. Output power of 8 kW with efficiency of 97.2 % was obtained under frequency of 100 kHz and 98.4 % was at 25 kHz.

V. CONCLUSIONS

This time we studied applying SiC-MOSFETs to TLC-II topology as soft switching method with partial TLC(Tail Loss Cancel). To confirm effectiveness DC-DC converter circuit (TLC-II) using SiC-MOSFETs was fabricated and new digital gate control for TLC-II topology was built by programing using FPGA. Normal operation of prototype was confirmed and output power of 8 kW with efficiency of 97.2 % at 100 kHz and 98.4 % at 25 kHz.

REFERENCES

[1] G. Hua, C.S. Leu, and F.C. Lee, "Novel zero-voltage-transition PWM converters", *IEEE Trans. on Power Electronics*, Vol.9, NO.2, pp.213-219, 1994

[2] W.Yu, and J.Lai, "Ultra High Efficiency Bidirectional DC-DC Converter With Multi-Frequency Pulse Width Modulation", APEC08, pp.1079-1084, 2008

[3] Y. Tsuruta, M. Pavlovsky, and A. Kawamura, "Very High Efficiency SAZZ Chopper Using High Speed IGBT," Proc. of 2009 IEEE 6th International Power Electronics and Motion Control Conference (IPEMC 2009), pp. 573-579, 2009

[4] Malay Trivedi, et al: "Internal Dynamics of IGBT Under Zero-Voltage and Zero-Current Switching Conditions", IEEE Transactions on Electron Devices, Vol.46, pp.1274-1282, 1999

[5] Y. Tsuruta and A. Kawamura: "Mid-Power SAZZ Chopper with Switched Tail Loss Cancel Circuit", *Proc. of IEEE International Power Electronics Conference -ECCE ASIA-*, IPEC-Sapporo2010, 23I1-4, pp.1195-1201 (2010)

[6] Y. Tsuruta, A. Kawamura, "Evaluation of Partial Turn off Dynamics for High Power Chopper using IGBT", 15th European Conference on Power Electronics and Applications (EPE2013), pp.1-8, 2013

[7] B. Eckard, A. Hofman, S. Zeltner, and M. Maerz, "Automotive Powertrain DC-DC Converter with 25kW/dm³ by using SiC Diodes," 4th International Conference on Integration of Power Electronics System, 2006, pp.1-6

[8] R. Ohama, I. Yuzurihara, and A. Takayanagi, "DC-DC converter with SiC Devices Considering Distribution System and Control Circuit," IEICE Technical Report, 111(161), 2011, pp.7-11

[9] A. Kadavelugu, V. Baliga, S. Bhattacharya, M. Das, and A. Agarwal, "Zero Voltage Switching Performance of 1200V SiC MOSFET, 1200V Silicon IGBT and 900V CoolMOS MOSFET," Proc. of the IEEE Energy Conversion Congress and Exposition (ECCE), pp. 1819-1826, 2011

[10] Y. Tsuruta, A. Kawamura, "Zero Voltage Switched Chopper with SiC-MOSFETs", IEEE Energy Conversion Congress & EXPO Conference (ECCE2013), pp. 5553-5559, 2013

The 2014 International Power Electronics Conference

Variable carrier deadbeat control with digital hysteresis method using SoC-FPGA for utility interactive inverter

Shunsuke Ohashi, Morito Yoshida, Tomoki Yokoyama
Tokyo Denki University
5, Asahicho, Senju, Adachiku, Tokyo, Japan
E-mail : yoko@fr.dendai.ac.jp

Abstract—**Variable carrier frequency control of PWM inverter based on deadbeat control using SoC-FPGA based hardware controller was proposed. The inductor current was controlled using digital hysteresis method and the output current was controlled using deadbeat control with disturbancecompensator. As the result, the control accuracy and the efficiency of the inverter can be adjustable with superior tracking accuracy to the reference current. Verifications were carried out through simulations and experiments.**

Keywords—SoC-FPGA, utlity interactive inverter, deadbeat control, Variable carrier frequency control

I. INTRODUCTION

The distributed power system like PV generation, wind power generation and fuel cells generation has much attention in the worldwide. In these generation systems, the utility interactive inverter is necessary to connect the DC power source to the utility line. In the near future, many distributed power systems will be connected to the utility line in the local area. Therefore, a demand of the control accuracy to the utility interactive inverter will also become strict more than the present ability. Also the demand of the high efficiency utility interactive inverter is increased.Deadbeat control is the digital control approach to realize precise control accuracy to the inverter control. To adopt to the various grid condition of each connecting point of the inverter, authors proposed a single-rate deadbeat control with disturbance compensation method to achieve the robust control and very superior characteristics was obtained[6]. In the conventional deadbeat control method, the carrier frequency is fixed to design the control method because of the limitation of the ability of the control processors to derive the control gain in each carrier period. So the control gain was calculated before the operation and stored as one of the control parameters in the memory of the controller.

On the other hand, SoC-FPGA chip was announced and released in the commercial use [10],[11]. SoC-FPGA is the advanced FPGA chip, which include CPU hardware core within the FPGA chip. CPU hardware core of SoC-FPGA has full spec features of high performance CPU, and the data transmission rate between CPU block and FPGA block achieves over 100Gbit/s, so the very fast calculation capability can be realized.

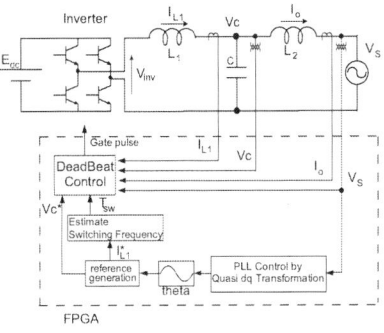

Fig. 1. Configuration of utility interactive inverter system

Using SoC-FPGA based hardware controller, the real time control gain calculation of deadbeat control can be realized, which result in the realization of variable carrier frequency control of PWM inverter based on deadbeat control method. It is well known that the efficiency of the inverter is much depends on the carrier frequency. So if the carrier frequency of the deadbeat control can be appropriately selected in one sinusoidal cycle, the control accuracy and the efficiency of the inverter can be adjustable by the operator. In this paper, a variable carrier frequency deadbeat control with hysteresis band and PLL control based on a 1MHz sampling with quasi dq transformation were combined. The control system was implemented using SoC-FPGA based hardware controller, parallel processing controller can be realized with the cooperation of FPGA block and CPU core block, and all the calculation circuit, PLL control circuit, reference generation circuit, analog-to-digital interface circuit, digital-to-analog interface circuit and pulse pattern generation circuit can be implemented in one SoC-FPGA chip. The characteristics of the control accuracy and the efficiency of the inverter were verified through simulations and experiments. The superior characteristics of the proposed method was verified.

II. UTLITY INTERACTIVE INVERTER SYSTEM

Fig.1 shows the diagram of the proposed system configuration. The PWM inverter bridge with the DC voltage E_{dc} outputs V_i, then connected to the output filter L_1 and C, then connected to the utility voltage V_s via a nominal inductance L_2. The inductor current I_{L1}, the capacitor voltage V_c, the output current I_o and the utility voltage

978-1-4799-2706-7/14 $31.00 © 2014 IEEE   3410

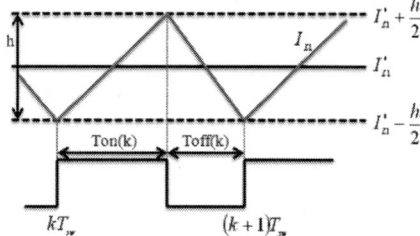

Fig. 2.   Inductor Current moduration with Hysteresis band

V_s are measured as the state variables respectively. The output current I_o is depend on the difference voltage between the inverter output voltage V_c and the utility voltage V_s. As the utility voltage is stable, so if the output voltage V_c is determined, the output current can be controlled, and the current regulation depends on the parameter of the inductor L_2.

III.   CONTROL METHOD

In the proposed method, to decide the carrier frequency $T_c(k)$, the inductor current I_{L1} is controlled within the hysteresis band width h. The derived pulse width to control the output current I_o matches to the reference using $T_c(k)$.

A. REFERENCE GENERATION

The capacitor voltage V_c is controlled to ensure the output current I_o equals to the current reference I_o^* for the next sampling instant. The output current I_o becomes;

$$\frac{d}{dt}I_o = \frac{V_c - V_s}{L_2} \tag{1}$$

As T_s is sampling period, sampling data $I_o(k)$ can be rewritten as follows.

$$
\begin{aligned}
I_o(k+1) &= I_o(k) + T_s \dot{I}_o(k) \tag{2}\\
&= I_o(k) + \frac{T_s}{L_2}(V_c(k) - V_s(k)) \tag{3}
\end{aligned}
$$

Here, $I_o(k+1)$ and $V_c(k)$ represent $I_o^*(k)$, $V_c^*(k)$ respectively. The capacitor voltage reference V_c^* becomes;

$$V_c^*(k) = \frac{L_2}{T_s}(I_o^*(k) - I_o(k)) + V_s(k) \tag{4}$$

From $I_o^*(k)$ and $V_c^*(k)$, $I_{L1}^*(k)$ becomes;

$$I_{L1}^*(k) = I_o^*(k) + C * \dot{V}_c^*(k) \tag{5}$$

B. DECISION of CARRIER FREQUENCY

To decide the carrier frequency, the inductor current I_{L1} is controlled within the hysteresis band width h. The inverter output voltage is assumed $\pm E_{dc}[V]$ in the duration of T_{on}. And the inverter output voltage $0[V]$ is assumed in the duration of T_{off}. The inductor current I_{L1} is assumed as shown in Fig.2. The inductor current I_{L1} becomes;

$$\frac{d}{dt}I_{L1} = \frac{V_{inv} - V_c}{L_1} \tag{6}$$

The inductor current $I_{L1}(k) + \Delta I_{L1}(T_{on}(k))$ becomes;

$$
\begin{aligned}
I_{L1}(k) + \Delta I_{L1}(T_{on}(k)) &= I_{L1}(k) + T_{on}\dot{I}_{L1}(k) \tag{7}\\
&= I_{L1}(k) + T_{on}(k)\frac{V_{inv}(k) - V_c(k)}{L_1} \tag{8}
\end{aligned}
$$

It is assumed that the inductor current I_{L1} equals to the boundary of the hysteresis band in $T_{on}(k)$. (8) can be rewritten as follows.

$$I_{L1}^*(k) + \frac{h}{2} = I_{L1}(k) + T_{on}(k)\frac{V_{inv}(k) - V_c(k)}{L_1} \tag{9}$$

$T_{on}(k)$ can be rewritten as follows.

$$T_{on}(k) = \frac{L_1}{V_{inv}(k) - V_c(k)}(I_{L1}^*(k) + h/2 - I_{L1}(k)) \tag{10}$$

Here, the inverter voltage V_{inv} is $\pm E_{dc}$. The inductor current $I_{L1}(k+1)$ becomes;

$$I_{L1}(k+1) = I_{L1}(k) + \Delta I_{L1}(T_{on}(k)) + T_{off}(k)\frac{-V_{inv}(k) - V_c(k)}{L_1}$$

It is assumed that the inductor current I_{L1} equals to the hysteresis band in $k+1$. $I_{L1}(k+1)$ can be rewritten as follows.

$$I_{L1}^*(k) - \frac{h}{2} = I_{L1}^*(k) + \frac{h}{2} + T_{off}(k)\frac{-V_{inv}(k) - V_c(k)}{L_1} \tag{11}$$

$T_{off}(k)$ becomes;

$$T_{off}(k) = -\frac{hL_1}{V_{inv}(k) + V_c(k)} \tag{12}$$

Here, it is assumed that $V_c \approx V_s$. T_{on} and T_{off} can be rewritten as follows.

$$
\begin{aligned}
T_{on}(k) &= \frac{L_1}{V_{inv}(k) - V_s(k)}(I_{L1}^*(k) + \frac{h}{2} - I_{L1}(k))\\
T_{off}(k) &= -\frac{hL_1}{V_{inv}(k) + V_c(k)}
\end{aligned}
$$

The carrier period $T_c(k)$ and the carrier frequency $f_c(k)$ become;

$$
\begin{aligned}
T_c(k) &= |T_{on}(k)| + |T_{off}(k)| \tag{13}\\
f_c(k) &= \frac{1}{T_c(k)} \tag{14}
\end{aligned}
$$

To prevent the rapid frequency change of the $f_c(k)$, the first order low-pass filter is adopted to derive the $f_c(k)$.

C. VARIABLE CARRIER DEADBEAT CONTROL (VCDB)

The state equation of the utility interactive system in Fig.1 becomes;

$$\dot{x}_{v2} = A_{v2}x_{v2} + B_{v2_1}u_v + B_{v2_2}V_s \tag{15}$$

$$x_{v2} = \begin{bmatrix} I_{L1} \\ V_c \\ I_o \end{bmatrix} \quad A_{v2} = \begin{bmatrix} 0 & -\frac{1}{L_1} & 0 \\ \frac{1}{C} & 0 & -\frac{1}{C} \\ 0 & \frac{1}{L_2} & 0 \end{bmatrix}, B_{v2_1} = \begin{bmatrix} \frac{1}{L_1} \\ 0 \\ 0 \end{bmatrix}$$

$$B_{v2_2} = \begin{bmatrix} 0 \\ 0 \\ -\frac{1}{L_2} \end{bmatrix}, u_{v2} = [V_i].$$

The state equation for discrete time system using carrier period $T_c(k)$ based on PWM hold method[1] becomes;

$$x_{v2}(k+1) = F_{v2}x_{v2}(k) + G_{v2}\Delta T(k) + H_{v2}V_s(k) \tag{16}$$

$$F_{v2} = e^{A_{v2}T_c(k)} = \begin{bmatrix} f_{v2_{11}} & f_{v2_{12}} & f_{v2_{13}} \\ f_{v2_{21}} & f_{v2_{22}} & f_{v2_{23}} \\ f_{v2_{31}} & f_{v2_{32}} & f_{v2_{33}} \end{bmatrix},$$

$$G_{v2} = \begin{bmatrix} g_{v2_{11}} \\ g_{v2_{21}} \\ g_{v2_{31}} \end{bmatrix}, H_{v2} = \begin{bmatrix} h_{v2_{11}} \\ h_{v2_{21}} \\ h_{v2_{31}} \end{bmatrix}.$$

From (16), the derived pulse width to control the output current I_o matches to the reference becomes;

$$
\begin{aligned}
\Delta T(k) &= (V_c^*(k) - f_{v2_{21}}I_{L1}(k) - f_{v2_{22}}V_c(k)\\
&\quad - f_{v2_{23}}I_o(k) - h_{v2_{21}}V_s(k))/g_{v2_{21}} \tag{17}
\end{aligned}
$$

TABLE I. SIMULATION PARAMETER(FOR 100kHz BASED INVERTER)

E_{dc}	200V	L_1	0.1mH(0.314%)
V_s	$100V_{rms}$,50Hz	C	1μF(0.314%)
Sampling	1MHz	L_2	0.5mH(1.57%)

TABLE II. SWITCHING DEVICE PARAMETER(COOLMOS)

	Parameter	Typ	Unit
$R_{DS_{on}}$	Drain-to-Source On-Resistance	0.09	Ω
T_r	Rise Time	5	ns
T_f	Fall Time	5	ns

TABLE III. SWITCHING DEVICE PARAMETER(IRFB3306PBF)

	Parameter	Typ	Unit
$R_{DS_{on}}$	Drain-to-Source On-Resistance	0.33	Ω
T_r	Rise Time	76	ns
T_f	Fall Time	77	ns

D. VARIABLE CARRIER DEADBEAT CONTROL WITH DISTURBANCE COMPERASTION (VCDBDC)

The inverter output voltage V_c is depend on the condition of the PWM inverter bridge voltage V_i and the utility voltage V_s. V_i is the PWM pulse waveforms and the V_s is a continuous sinusoidal waveform in the normal utility line. In the proposed method, the V_i and the V_s are divided in different terms in (18), the state equation becomes as follows.

$$\dot{x_v} = A_v x_v + B_{v1} u_v + B_{v2} V_s,$$

$$x_v = \begin{bmatrix} I_{L1} \\ V_c \\ I_o \\ I_{dis} \\ \dot{I}_{dis} \end{bmatrix}, A_v = \begin{bmatrix} 0 & -1/L_1 & 0 & 0 & 0 \\ 1/C & 0 & -1/C & 0 & 1/C \\ 0 & 1/L_2 & 0 & 0 & 0 \\ 0 & 0 & 0 & 0 & 0 \\ 0 & 0 & 0 & 1 & 0 \end{bmatrix},$$

$$B_{v1} = \begin{bmatrix} 1/L_1 \\ 0 \\ 0 \\ 0 \\ 0 \end{bmatrix}, B_{v2} = \begin{bmatrix} 0 \\ 0 \\ -1/L_2 \\ 0 \\ 0 \end{bmatrix}, u_v = [V_i]. \quad (18)$$

The state equation for discrete time system becomes (19).

$$x_v(k+1) = F_v x_v(k) + G_v u_v(k) + H_v V_s(k) \quad (19)$$

$$F_v = e^{A_v T_c}, H_v = \int_0^{T_c} e^{A_v \tau} B_{v2} d\tau \quad (20)$$

$$G_v = \int_{(T_c - T_p)/2}^{(T_c + T_p)/2} e^{A_v (T_c - \tau)} B_{v1} d\tau u_v(k) = [\Delta T(k)] \quad (21)$$

T_c is the control period, T_p is the inverter output pulse width. Finally, the derived pulse width to contrtol the output current I_o matches to the reference becomes as (22).

$$\Delta T(k) = (V_c^* - f_{v21}I_{L1}(k) - f_{v22}V_c(k) - f_{v23}I_o(k) -$$
$$f_{v24}I_{dis} - f_{v25}\dot{I}_{dis} - h_{v21}V_s(k))/g_{v21} \quad (22)$$

IV. SIMULATION

A. Variations of carrier frequency

Simulations were carried out for the proposed method in the case of a 100kHz carrier based inverter. System parameters for the simulations were indicated in TABLE VI. The simulation results are shown in Fig.8 and Fig.9 respectively. A total harmonic distortion (THD) and inverter loss (P_{loss}) were used as an evaluation index. P_{loss} is the sum of the steady loss and the switching loss due to the switching device. THD and P_{loss} have a trade-off relationship, the trend of THD and P_{loss} were plotted by changing the hysteresis band width h. TABLE II and TABLE III show switching device parameter used in simulations. Fig.5 shows the trend of THD and P_{loss}. From these simulation results, the control accuracy and the efficiency of the inverter can be controlled by the hysteresis band h. The efficiency of the inverter is improved by increasing the hysteresis band, and the control accuracy is improved by decreasing the hysteresis

band. So the operator can select the adequate parameter h to derive the desired control accuracy and the efficiency of the inverter.

B. Variation range of carrier frequency

In the proposed method, the upper limit and the lower limit of carrier frequency can be settled by the operator. First, the upper limit was settled to 100kHz and the lower limit was settled to 20kHz. Next, the simulations were carried out in the case that the upper limit was settled to 5kHz and the lower limit was settled to 1kHz. In Fig.6 and Fig.7, in the case that the hysteresis band was settled to 140% and 50% were indicated respectively. The parameter for the simulation was summarized in TABLE IV. In these cases, it was confirmed that in the lower carrier frequency range, the controllability and the efficiency can be adjusted by the hysteresis band setting.

C. Comparison of VCDB and VCDBDC

Simulations were carried out for the proposed method in the case of a VCDB and VCDBDC. System parameters for the simulations were indicated in TABLE VI. The simulation results are shown in Fig.8, Fig.9, Fig.10 and Fig.11 respectively. A total harmonic distortion (THD) and inverter loss (P_{loss}) were used as an evaluation index. P_{loss} is the sum of the steady loss and the switching loss due to the switching device. THD and P_{loss} have a trade-off relationship, the trend of THD and P_{loss} were plotted by changing the hysteresis band width h. TABLE II shows

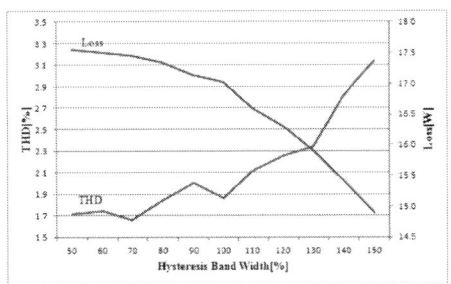

Fig. 5. THD and Inverter Loss vs h(100kHz carrier Inverter)

TABLE IV. SIMULATION PARAMETER((FOR UPPER LIMIT 5KHZ,LOWER LIMIT 1KHZ)

E_{dc}	200V	L_1	3.0mH(9.42%)
V_s	$100V_{rms}$,50Hz	C	13.0μF(48.4%)
Sampling	1MHz	L_2	15.4mH(4.08%)

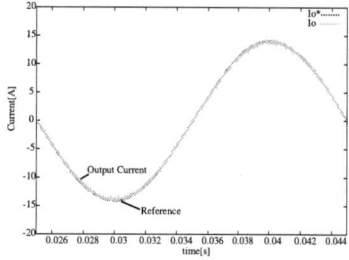 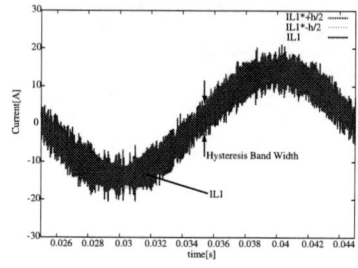

Fig. 3. Simulation result(Upper limit 100kHz,Lower limit 20kHz):Hysteresis Band h=50%(Left: Output Current I_o Right:Inductor Current I_{L1})

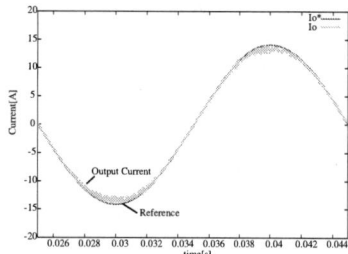 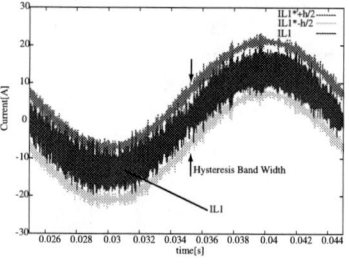

Fig. 4. Simulation result(Upper limit 100kHz,Lower limit 20kHz):Hysteresis Band h=140%(Left: Output Current I_o Right:Inductor Current I_{L1})

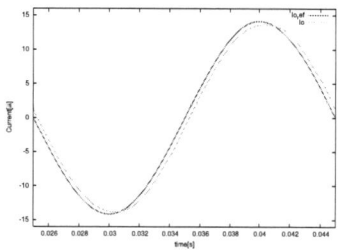 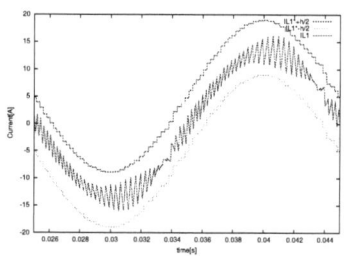

Fig. 6. Simulation result(Upper limit 5kHz,Lower limit 1kHz):Hysteresis Band h=100%(Left: Output Current I_o Right:Inductor Current I_{L1})

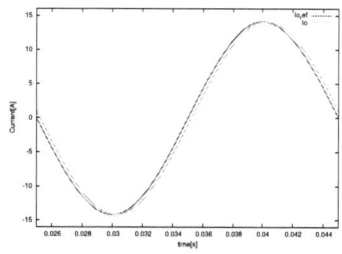 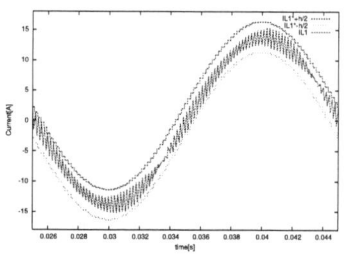

Fig. 7. Simulation result(Upper limit 5kHz,Lower limit 1kHz):Hysteresis Band h=50%(Left: Output Current I_o Right:Inductor Current I_{L1})

TABLE V.   SIMULATION RESULT

Upper limit[Hz]	Lower limit[Hz]	Hysteresis Band[A]	THD[%]	Error[%]	Loss[W]
100k	20k	14	2.8	3.40	15.4
100k	20k	5	1.7	1.92	17.6
5k	1k	14	7.1	16.7	9.59
5k	1k	5	0.31	6.59	11.4

TABLE VI.   SIMULATION PARAMETER

E_{dc}	200V	L_1	0.1mH(0.314%)
V_s	100V_{rms},50Hz	C	1μF(0.314%)
Sampling	1MHz	L_2	0.5mH(1.57%)

switching device parameter used in simulations. Fig.12

and Fig.13 show the trend of THD and P_{loss}. From these simulation results, the control accuracy and the efficiency of the inverter can be controlled by the hysteresis band h. The efficiency of the inverter is improved by increasing the hysteresis band, and the control accuracy is improved by decreasing the hysteresis band. So the operator can select the adequate parameter h to derive the desired control accuracy and the efficiency of the inverter.

D. Response of amplitude with an instantaneous voltage drop

Simulation were carried out for VCDB and VCDBDC to verify instantaneous voltage, the grid voltage was stepped down to 20% of nominal voltage for two cycle duration. The simulation result are shown in Fig. 14

The 2014 International Power Electronics Conference

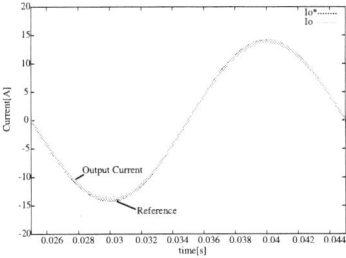

Output Current I_o

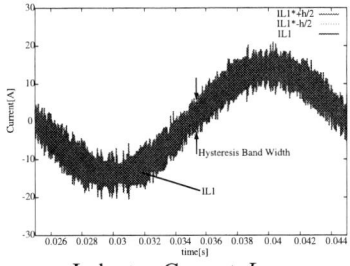

Inductor Current I_{L1}

Fig. 8.   Simulation result(VCDB):Hysteresis Band h=50%

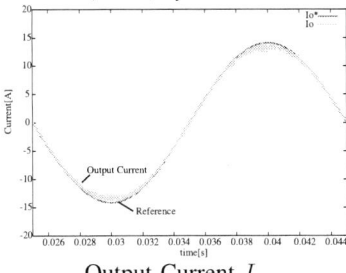

Output Current I_o

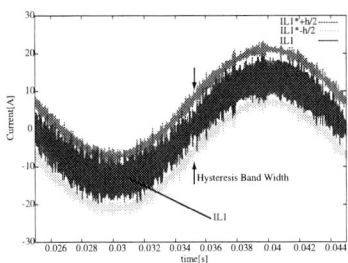

Inductor Current I_{L1}

Fig. 9.   Simulation result(VCDB):Hysteresis Band h=140%

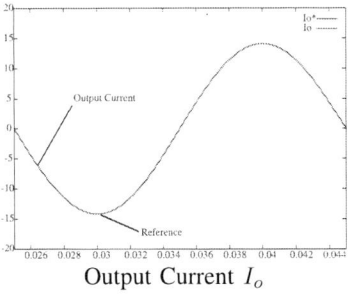

Output Current I_o

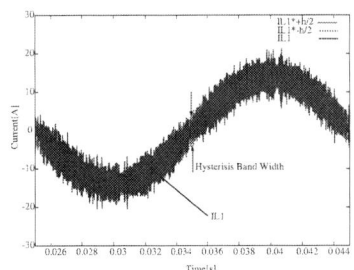

Inductor Current I_{L1}

Fig. 10.   Simulation result(VCDBDC):Hysteresis Band h=50%

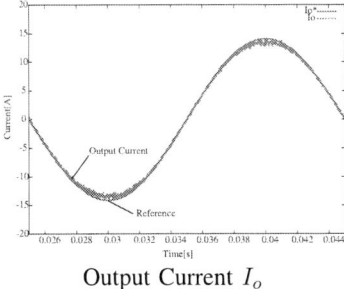

Output Current I_o

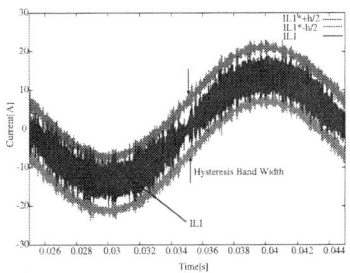

Inductor Current I_{L1}

Fig. 11.   Simulation result(VCDBDC):Hysteresis Band h=140%

and Fig. 15. The output current followed reference value smoothly while the voltage fluctuation for each control method.

V. EXPERIMENTS

The Zynq-7000 integrates a feature-rich dual-core ARM Cortex-A9 based processing system (PS) and 28 nm Xilinx programmable logic (PL) in a single device. The ARM Cortex-A9 CPUs are the heart of the PS and also include on-chip memory, external memory interfaces, and a rich set of peripheral connectivity interfaces[9]. Ultra fast calculation capability ensures the realization of

variable carrier deadbeat control which requires the re-calculation of the control gain in every sampling period. The system block diagram of SoC-FPGA based controller is described in Fig.16. Fig.17 shows the configuration of SoC-FPGA based hardware controller and Fig.18 shows the block diagram of the carrier frequency decision module in SoC-FPGA. Fig.19 shows the experimental results for VCDB in the case that the hysteresis band h is settled to 50% and Fig. 20 shows in the case that the hysteresis band h is settled to 140%. It is confirmed that the carrier is varied due to the variable carrier method and the output current matches to the reference current smoothly in both

978-1-4799-2706-7/14 $31.00 © 2014 IEEE   3414

The 2014 International Power Electronics Conference

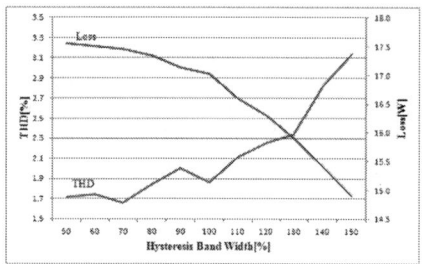

Fig. 12. THD and Inverter Loss vs h (VCDBDC)

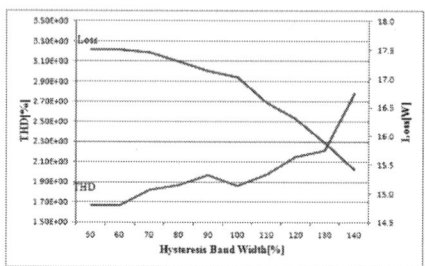

Fig. 13. THD and Inverter Loss vs h (VCDBDC)

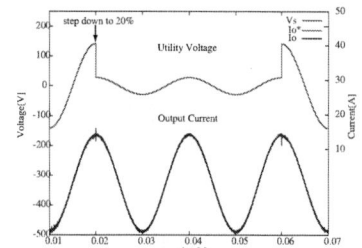

Output Current I_o and Utillty Voltage V_s

Fig. 14. Simulation result for instantaneous voltage drop(VCDB)

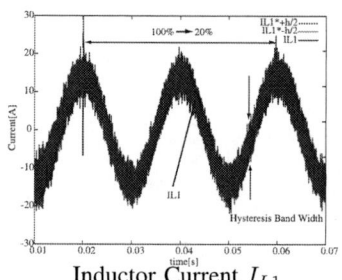

Inductor Current I_{L1}

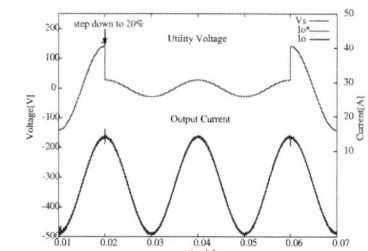

Output Current I_o and Utillty Voltage V_s

Fig. 15. Simulation result for instantaneous voltage drop(VCDBDC)

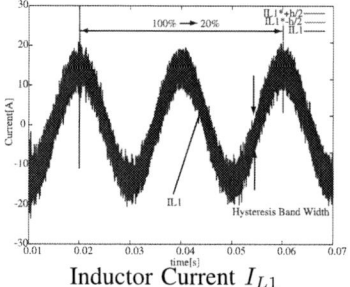

Inductor Current I_{L1}

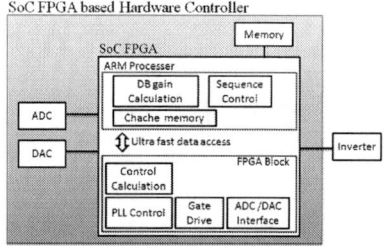

Fig. 16. System Block Diagram of SoC-FPGA based Hardware Controller

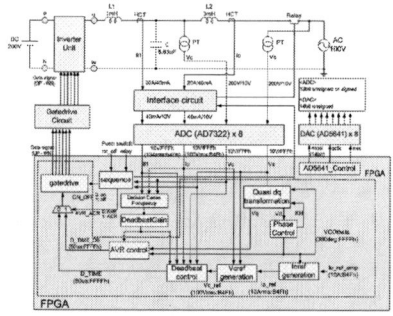

Fig. 17. Configuration of SoC-FPGA based hardware controller

TABLE VII. EXPERIMENTAL RESULT(THD[%])

Hysteresis band width	Experiment
50%	3.2
140%	3.8

cases. Table. VII shows the THD of the output current vs.the hysteresis band h. The waveform of the inductor current was varied by changing the hysteresis band h, and the THD of the output current was varied same as the simulations.

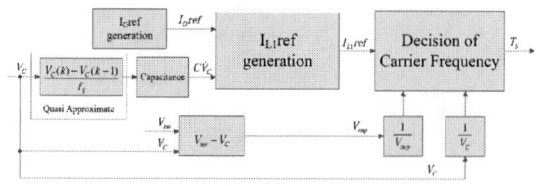

Fig. 18. Block diagram of Carrier Frequency Decision

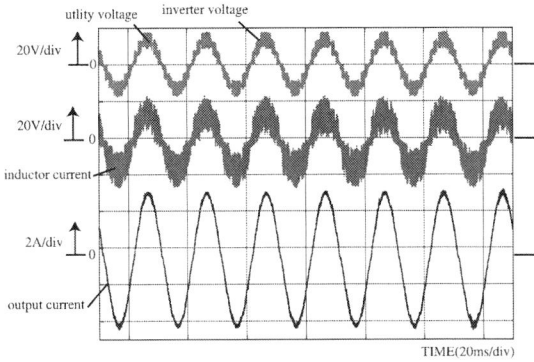

Fig. 19.　Experimental waveforms for VCDB : Hysteresis band h = 50%

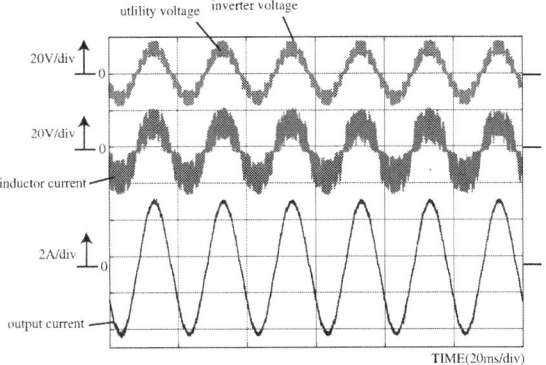

Fig. 20.　Experimental waveforms for VCDB : Hysteresis band h = 140%

VI.　CONCLUSION

A variable carrier frequency deadbeat control with hysteresis band was proposed for the utility interactive inverter. The characteristics of the control accuracy and efficiency of the inverter were verified through simulations and experiments. It is confirmed that the control accuracy and the efficiency of the inverter can be adjustable by setting the hysteresis band width, which is effective in practical use.

REFERENCES

[1]　K.P. Gokhale, A. Kawamura, R.G. Hoft, "Dead Beat Microprocessor Control of PWM Inverter for Sinusoidal Output Waveform Synthesis", IEEE Trans. on Ind. Appl. Vol.IA-23, No.3, Sep/Oct. pp.901-910(1987).

[2]　A.Kawamura, T.Haneyoshi: "Deadbeat Controlled PWM Inverter with Parameter Estimation Using Only Voltage Sensor" IEEEtrans. on PE, Vol.9, No.5, 1988

[3]　T.Haneyoshi, A.Kawamura: "Waveform Compensation of PWM Inverter with Cyclic Fluctuating Loads" IEEEtrans. on INDUSTRY APPLICATIONS, Vol.24, No.4, 1988

[4]　E.Shimada, K.Aoki, T.Komiyama, T.Yokoyama: "Implementation of Single Phase Utility Interactive Inverter using FPGA based Hardware Controller", EPE 2005 T14-2

[5]　T.Saigusa, T.Yokoyama: "Digital Control Method for 100kHz Single Phase Utility Interactive Inverter with FPGA based Hardware Controller", EPE-PEMC 2010

[6]　Yuichi Hanashima, Tomoki Yokoyama :" Fault Ride Through Capability of 100kHz Single Phase Utility Interactive Inverter with FPGA based Hardware Controller ", EPE-PEMC 2012

[7]　Morito Yoshida, Tomoki Yokoyama :" Variable Carrier Frequency Deadbeat Control with Hysteresis Band using SoC-FPGA for Utility Interactive Inverter ", EPE-PEMC 2013

[8]　Walter Stefanutti, Paolo Mattavelli:"Fully Digital Hysteresis Modulation With Swiching-Time Prediction",IEEE TRANSACTIONS ON INDUSTRY APPLICATIONS, VOL.42,NO.3,MAY/JUNE 2006

[9]　Xilinx Inc, "Zynq-7000 All Programmable SoC Overview" http://japan.xilinx.com/support/documentation/zynq-7000.htm

[10]　ALTERA Corporation, http://www.altera.co.jp/

[11]　Xilinx Inc,http://japan.xilinx.com/

A Space Vector Modulation Strategy for Three-Level Operation Based on Dual Two-Level Voltage Source Inverters

Yuttana Kumsuwan
Department of Electrical Engineering
Faculty of Engineering, Chiang Mai University
Chiang Mai, Thailand
yt@eng.cmu.ac.th

Watcharin Srirattanawichaikul
Department of Electrical Engineering
Faculty of Engineering, Chiang Mai University
Chiang Mai, Thailand
watcharin.cmu@gmail.com

Abstract— **This paper presents a space vector modulation (SVM) technique for a three-level operation based on dual two-level voltage source inverter (VSI). This inverter topology uses two standard two-level VSI placed in parallel using a three-phase transformer to connect them together. This paper proposes a general modulation algorithm for dual two-level VSI based on the basic two-level SVM technique. The proposed inverter produces five-level line-to-line voltage waveforms, which is able to generate voltage levels similar to a three-level inverter. The practicality and performance of the modulation strategy for the proposed inverter topology has been verified via simulation and experiment.**

Keywords— *Dual two-level voltage source inverter, three-level voltage source inverter, space vector modulation, multilevel inverter.*

I. INTRODUCTION

Multilevel converter topologies have recently been increasingly applied in medium-voltage and high-power industrial applications such as active power filters, adjustable-speed drive and renewable energy generation. The multilevel converters have advantages of a high power rating, a low device voltage rating, a high quality output waveforms associated with reduced voltage/current harmonic distortions and higher efficiency when compared to the standard two-level voltage source converters [1].

There are generally commercial classified topologies of multilevel converters available with diode-clamped converters, cascaded H-bridge converters, and flying-capacitor converters [2]. Among the various multilevel converter topologies, the most popular topology in high power applications is the three-level neutral point clamped (NPC) VSI, which is able to five-level line-to-line voltage. The advantage of the three-level NPC VSI topology is that the power switches and the dc-link capacitors have to endure only one-half of the dc-link voltage. In addition, the converter can deal with double voltage and power value more than a standard two-level VSI with the same switching frequency. However, this converter has some drawbacks, such as additional clamping diodes and complexity to the modulation method. Furthermore, the performance of the

three-level NPC VSI depends on the ability to maintain stability of the voltage balance of the dc-link neutral point, which also needs partial voltage balancing control.

The three-level operation using the dual two-level VSI topology has been extensively researched. This inverter configuration uses two standard two-level VSI in parallel using transformers or coupled inductors to connect them together. Reference [3] presents the multilevel operation with a dual two-level inverter feeding a three-phase machine with open-end windings based on the use of two insolated dc supplies. This paper has been focusing on the development of a modulation technique for regulating the power sharing between the two dc sources. However, the switching scheme of this inverter is complex. In [4], the three-limb coupled inductor operation for paralleled three-phase voltage sourced multi-level PWM inverters was proposed. This paper presents a modified discontinuous PWM method that eliminates these voltage differences.

In this paper, a SVM strategy for three-level operation based on dual two-level voltage source inverter is proposed. The modulation algorithm is based on the basic two-level SVM, the following advantages are achieved.

1) The proposed topology can generate voltage levels similar to a three-level NPC VSI.

2) The proposed topology is powered by two isolated dc sources of equal/unequal voltage. It doesn't have the voltage balancing problem as well as the three-level NPC VSI.

3) The proposed modulation method can simplify the algorithm and can be easily implemented by using DSP/micro controller board, which supports only two-level modulation control.

4) Ideal for closed loops and also for using during dynamic operations.

Finally, experimental waveforms and simulation results are shown to highlight the modulation technique for three-level operation based on dual two-level VSI.

978-1-4799-2706-7/14 $31.00 © 2014 IEEE

II. THREE-LEVEL OPERATION BASED ON DUAL TWO-LEVEL VOLTAGE SOURCE INVERTER TOPOLOGY

The dual two-level VSI for the three-level operation is schematically shown in Fig. 1. The proposed inverter topology consists of two standard two-level VSI (upper and lower VSI) in parallel using three-phase transformer to connect them together, which is powered by two isolated dc sources of equal/unequal voltage. Each output voltage of the dual two-level VSI is connected to the primary winding. The secondary winding supplies a wye-connected into the three-phase load, with isolated neutral, which generates the five-level line-to-line voltages. This topology was chosen as it provides a relatively simple modification to a two-level VSI, but can be used to verify the operation of the three-level inverter.

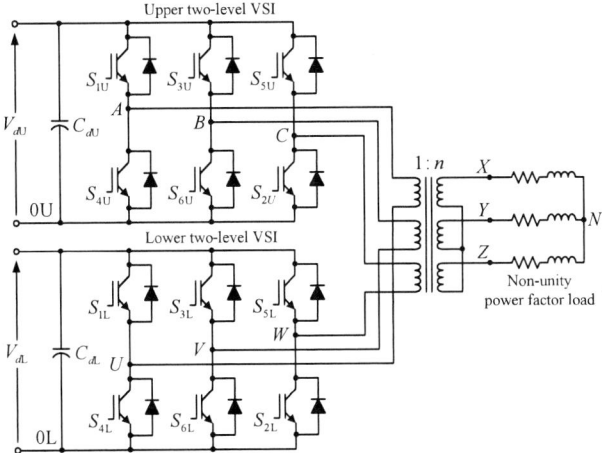

Fig. 1. Simplified schematic of the proposed three-level operation based on dual two-level voltage source inverter.

From the schematic in Fig. 1, the output line-to-line voltages $v_{LL,U}, v_{LL,L}$ of the dual two-level VSI, upper and lower VSI, can be expressed as

$$
v_{LL,U} = \begin{bmatrix} v_{AB} \\ v_{BC} \\ v_{CA} \end{bmatrix} = \begin{bmatrix} v_{A0U} - v_{B0U} \\ v_{B0U} - v_{C0U} \\ v_{C0U} - v_{A0U} \end{bmatrix} \tag{1}
$$

$$
v_{LL,L} = \begin{bmatrix} v_{UV} \\ v_{VW} \\ v_{WU} \end{bmatrix} = \begin{bmatrix} v_{U0L} - v_{V0L} \\ v_{V0L} - v_{W0L} \\ v_{W0L} - v_{U0L} \end{bmatrix} \tag{2}
$$

where $v_{A0U}, v_{B0U}, v_{C0U}$ and $v_{U0L}, v_{V0L}, v_{W0L}$ are the pole voltages of upper and lower VSI, respectively. v_{AB}, v_{BC}, v_{CA} and v_{UV}, v_{VW}, v_{WU} are the line-to-line voltages of upper and lower VSI, respectively.

For a three-phase star-connected balanced load, the line-to-line secondary voltages of the dual two-level VSI for a three-level operation are defined as

$$
\begin{bmatrix} v_{XY} \\ v_{YZ} \\ v_{ZX} \end{bmatrix} = n \begin{bmatrix} v_{AB} - v_{UV} \\ v_{BC} - v_{VW} \\ v_{CA} - v_{WU} \end{bmatrix} \tag{3}
$$

where v_{XY}, v_{YZ}, v_{ZX} are the line-to-line voltages of load, n is turn ratio of the three-phase transformer.

III. SPACE VECTOR MODULATION STRATEGY FOR THREE-LEVEL OPERATION BASED ON DUAL TWO-LEVEL VOLTAGE SOURCE INVERTERS

A. Stationary Space Vector

Space vector modulation strategy is most popularly used for digital control of the standard two-level VSI due to its flexibility and good performance. This section presents the principle of the SVM strategy for the dual two-level VSI for the three-level operation. In this technique, the two voltage references $\vec{V}_{U,ref}, \vec{V}_{L,ref}$ are represented by a unique space vector rotating on a hexagon, which is divided into six sectors (I to VI). The space vector diagram for the dual two-level VSI is shown in Fig. 2. The operation of this inverter for the three-level operation on this plane is made up of twelve active voltage vectors, $\vec{V}_{U1}$ to $\vec{V}_{U6}$ for upper VSI and $\vec{V}_{L1}$ to $\vec{V}_{L6}$ for lower VSI, and two null voltage vectors $\vec{V}_{U0}$ and $\vec{V}_{L0}$ at the center of the hexagon, which depends on the on/off status of its six switches [000]-[111]. There are fourteen possible combinations of switching states in the dual two-level VSI for a three-level operation as listed in Table I.

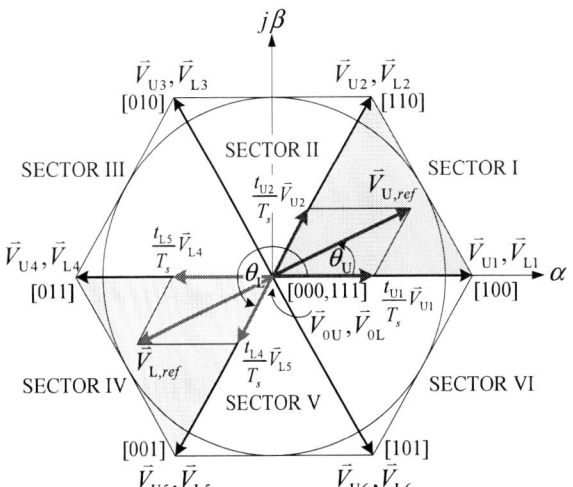

Fig. 2. Space vector diagram of the dual two-level VSI for a three-level operation.

TABLE I
SPACE VECTORS, SWITCHING STATES AND ON-STATE SWITCHES OF THE
DUAL TWO-LEVEL VSI FOR A THREE-LEVEL OPERATION

Space Vector		Switching State	On-state Switch	Vector Definition
Zero Vector	$\vec{V}_{U0}$	[111]	S_{U1}, S_{U3}, S_{U5}	0
		[000]	S_{U4}, S_{U6}, S_{U2}	
	$\vec{V}_{L0}$	[111]	S_{L1}, S_{L3}, S_{L5}	0
		[000]	S_{L4}, S_{L6}, S_{L2}	
Active Vector	$\vec{V}_{U1}$	[100]	S_{U1}, S_{U6}, S_{U2}	$(2/3)V_{dU}e^{j0}$
	$\vec{V}_{U2}$	[110]	S_{U1}, S_{U3}, S_{U2}	$(2/3)V_{dU}e^{j\pi/3}$
	$\vec{V}_{U3}$	[010]	S_{U4}, S_{U3}, S_{U2}	$(2/3)V_{dU}e^{j2\pi/3}$
	$\vec{V}_{U4}$	[011]	S_{U4}, S_{U3}, S_{U5}	$(2/3)V_{dU}e^{j\pi}$
	$\vec{V}_{U5}$	[001]	S_{U4}, S_{U6}, S_{U5}	$(2/3)V_{dU}e^{j4\pi/3}$
	$\vec{V}_{U6}$	[101]	S_{U1}, S_{U6}, S_{U5}	$(2/3)V_{dU}e^{j5\pi/3}$
	$\vec{V}_{L1}$	[100]	S_{L1}, S_{L6}, S_{L2}	$(2/3)V_{dL}e^{j0}$
	$\vec{V}_{L2}$	[110]	S_{L1}, S_{L3}, S_{L2}	$(2/3)V_{dL}e^{j\pi/3}$
	$\vec{V}_{L3}$	[010]	S_{L4}, S_{L3}, S_{L2}	$(2/3)V_{dL}e^{j2\pi/3}$
	$\vec{V}_{L4}$	[011]	S_{L4}, S_{L3}, S_{L5}	$(2/3)V_{dL}e^{j\pi}$
	$\vec{V}_{L5}$	[001]	S_{L4}, S_{L6}, S_{L5}	$(2/3)V_{dL}e^{j4\pi/3}$
	$\vec{V}_{L6}$	[101]	S_{L1}, S_{L6}, S_{L5}	$(2/3)V_{dL}e^{j5\pi/3}$

Using the space vector representation in Fig. 2, the two output voltage vectors $\vec{V}_U, \vec{V}_L$ of the dual two-level VSI are determined regarding to the location of the two voltage references $\vec{V}_{U,ref}, \vec{V}_{L,ref}$ in the space vector diagram. By transforming the three-phase voltages into two-phase voltages in stationary reference frame $\alpha\beta$, the two voltage vectors can be expressed as

$$\vec{V}_U = V_{U,ref}e^{j\theta_U} = V_{U\alpha} + jV_{U\beta} = \frac{2}{3}V_{dU}e^{j(k-1)\frac{\pi}{3}} \quad (4)$$

$$\vec{V}_L = V_{L,ref}e^{j\theta_L} = V_{L\alpha} + jV_{L\beta} = -\frac{2}{3}V_{dL}e^{j(k-1)\frac{\pi}{3}} \quad (5)$$

where $V_{U\alpha}$ and $V_{U\beta}$ are the α and β components of $V_{U,ref}$, $V_{L\alpha}$ and $V_{L\beta}$ are the α and β components of $V_{L,ref}$, θ_U is are the angular displacement between $V_{U,ref}$ and α-axis of $\alpha\beta$ plane, the θ_L is are the angular displacement between $V_{L,ref}$ and α-axis of $\alpha\beta$ plane, V_{dU} and V_{dL} are the dc-link voltage of upper and lower VSI, and k is the number of sector the reference space vector ($k = 1, 2, ..., 6$).

Therefore, the total voltage vectors $\vec{V}_O$ of the proposed dual two-level VSI can be expressed as

$$\vec{V}_O = n(\vec{V}_U - \vec{V}_L) = \frac{2}{3}n(V_{dU} + V_{dL})e^{j(k-1)\frac{\pi}{3}} \quad (6)$$

From (6), the output total voltage of the proposed inverter topology $|V_O|$ depends on the magnitude of two dc-link voltages of the upper and lower VSI V_{dU}, V_{dL} and the turn ratio of the three-phase transformer n.

B.   Dwell Time Calculation and Modulation Index of Space Vector Modulation

The dwell times for the stationary vectors are the times for which each of the state space vectors is utilized. The objective of the proposed SVM technique is to generate a voltage with an average value equal to the two reference voltages $\vec{V}_{U,ref}, \vec{V}_{L,ref}$ using the summation of three vectors nearest to voltage references during a sampling time period T_{sw}.

Assuming that the sampling time period T_{sw} is sufficiently small, the two reference voltage vectors $\vec{V}_{U,ref}, \vec{V}_{L,ref}$ can be considered constant during. By using sector I and IV as an example, the calculation flow of the duel two-level SVM algorithm can be clearly demonstrated. Considering the reference voltage $\vec{V}_{U,ref}$ falls into sector I and the reference voltage $\vec{V}_{L,ref}$ falls into sector IV as shown in Fig. 2, it can be synthesized by $\vec{V}_{U1}, \vec{V}_{U2}, \vec{V}_{U0}$ and $\vec{V}_{L4}, \vec{V}_{L5}, \vec{V}_{L0}$ for upper and lower reference voltage vectors, respectively. Therefore, the equations for on time of the voltage vectors for upper and lower reference voltage vectors can be given as

$$\vec{V}_{U,ref}T_{sw} = \vec{V}_{U1}t_{U1} + \vec{V}_{U2}t_{U2} + \vec{V}_{U0}t_{U0}$$
$$T_{sw} = t_{U1} + t_{U2} + t_{U0} \quad (7)$$

$$\vec{V}_{L,ref}T_{sw} = \vec{V}_{L4}t_{L5} + \vec{V}_{L5}t_{L4} + \vec{V}_{L0}t_{L0}$$
$$T_{sw} = t_{L1} + t_{L2} + t_{L0} \quad (8)$$

where t_{U0}, t_{U1} and t_{U2} are supposed to be normalized by the sampling time period T_{sw} of upper VSI and t_{L0}, t_{L4} and t_{L5} are supposed to be normalized by the sampling time period T_{sw} of lower VSI.

From above expressions, the switching time intervals of voltage vectors can be calculated as follows:

$$\begin{bmatrix} t_{Uk} \\ t_{Uk+1} \end{bmatrix} = \frac{\sqrt{3}V_{U,ref}T_{sw}}{V_{dU}} \begin{bmatrix} \sin\left(\frac{k\pi}{3}\right) & -\cos\left(\frac{k\pi}{3}\right) \\ -\sin\left(\frac{(k-1)\pi}{3}\right) & \cos\left(\frac{(k-1)\pi}{3}\right) \end{bmatrix} \begin{bmatrix} \cos(\theta_U) \\ \sin(\theta_U) \end{bmatrix} \quad (9)$$

$$\begin{bmatrix} t_{Lk} \\ t_{Lk+1} \end{bmatrix} = \frac{\sqrt{3}V_{L,ref}T_{sw}}{V_{dL}} \begin{bmatrix} \sin\left(\frac{k\pi}{3}\right) & -\cos\left(\frac{k\pi}{3}\right) \\ -\sin\left(\frac{(k-1)\pi}{3}\right) & \cos\left(\frac{(k-1)\pi}{3}\right) \end{bmatrix} \begin{bmatrix} \cos(\theta_L) \\ \sin(\theta_L) \end{bmatrix} \quad (10)$$

Equation (9) and (10) can be also expressed in term of modulation index as follows:

$$m_{\mathrm{U}} = \frac{\sqrt{3}V_{\mathrm{U},ref}}{V_{d\mathrm{U}}}$$

$$m_{\mathrm{L}} = \frac{\sqrt{3}V_{\mathrm{L},ref}}{V_{d\mathrm{L}}} \qquad (11)$$

where m_U is the modulation index of upper VSI and m_L is the modulation index of lower VSI, which the modulation index of upper and lower VSI for the SVM strategy are in the range of $0 \le m_{\mathrm{U}}, m_{\mathrm{L}} \le 1.0$.

The line-to-line voltage (rms) of the upper and lower VSI $V_{LL,\mathrm{U}}, V_{LL,\mathrm{L}}$ produced by the SVM strategy can be expressed as

$$V_{LL,\mathrm{U}} = \frac{m_{\mathrm{U}}V_{d\mathrm{U}}}{\sqrt{2}}$$

$$V_{LL,\mathrm{L}} = \frac{m_{\mathrm{L}}V_{d\mathrm{L}}}{\sqrt{2}} \qquad (12)$$

In (12), the maximum fundamental line-to-line voltage (rms) is around 15.5% higher than that for the sinusoidal PWM strategy. Finally, the output line-to-line voltages of the dual two-level VSI for a three-level operation can be concluded

$$V_{O,1} = \frac{n}{\sqrt{2}}\left(m_{\mathrm{U}}V_{d\mathrm{U}} + m_{\mathrm{L}}V_{d\mathrm{L}}\right) \qquad (13)$$

C. Switching Sequence Arrangement

The switching sequence of the dual two-level VSI for a three-level operation also needs attention. Generally the switching sequence that minimizes the switching states, minimum switching losses or minimum total harmonic distortion is favorable. Table II shows the switching sequences for the two reference voltage vectors $\vec{V}_{\mathrm{U},ref}, \vec{V}_{\mathrm{L},ref}$ residing in all six sectors.

TABLE II
SWITCHING SEQUENCE OF THE DUAL TWO-LEVEL VSI
FOR A THREE-LEVEL OPERATION

Sector	Switching Sequence						
I	$\vec{V}_0$	$\vec{V}_1$	$\vec{V}_2$	$\vec{V}_0$	$\vec{V}_2$	$\vec{V}_1$	$\vec{V}_0$
	000	100	110	111	110	100	000
II	$\vec{V}_0$	$\vec{V}_3$	$\vec{V}_2$	$\vec{V}_0$	$\vec{V}_2$	$\vec{V}_3$	$\vec{V}_0$
	000	010	110	111	110	010	000
III	$\vec{V}_0$	$\vec{V}_3$	$\vec{V}_4$	$\vec{V}_0$	$\vec{V}_4$	$\vec{V}_3$	$\vec{V}_0$
	000	010	011	111	011	010	000
IV	$\vec{V}_0$	$\vec{V}_5$	$\vec{V}_4$	$\vec{V}_0$	$\vec{V}_4$	$\vec{V}_5$	$\vec{V}_0$
	000	001	101	111	101	001	000
VI	$\vec{V}_0$	$\vec{V}_5$	$\vec{V}_6$	$\vec{V}_0$	$\vec{V}_6$	$\vec{V}_5$	$\vec{V}_0$
	000	001	101	111	101	001	000
V	$\vec{V}_0$	$\vec{V}_1$	$\vec{V}_6$	$\vec{V}_0$	$\vec{V}_6$	$\vec{V}_1$	$\vec{V}_0$
	000	100	101	111	101	100	000

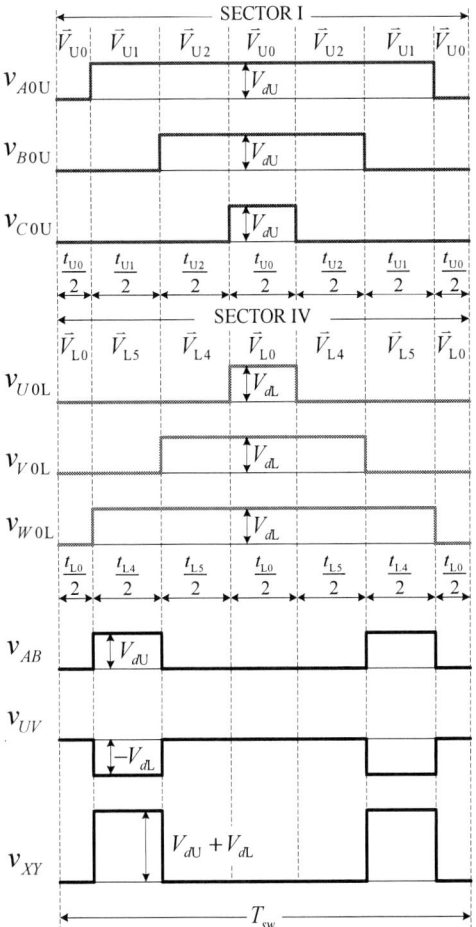

Fig. 3. Switching sequence in proposed SVM strategy for three-level operation based on dual two-level VSI in sector I for upper VSI and sector IV for lower VSI.

For example, a case in Fig. 3, a typical switching sequence and inverter output voltage waveforms for the two voltage references $\vec{V}_{\mathrm{U},ref}, \vec{V}_{\mathrm{L},ref}$ are placed into sector I and sector IV (Fig. 1), where the upper and lower VSI are synthesized by the voltage vectors $\vec{V}_{\mathrm{U}0}, \vec{V}_{\mathrm{U}1}, \vec{V}_{\mathrm{U}2}$ and voltage vectors $\vec{V}_{\mathrm{L}0}, \vec{V}_{\mathrm{L}4}, \vec{V}_{\mathrm{L}5}$, respectively. Based on the switching sequence, the line-to-line voltage waveforms of the upper and lower VSI v_{AB}, v_{UV} in a sampling period T_{sw} are illustrated. It can be note that the waveforms of load line-to-line voltage produced by two different voltage of the dual two-level VSI.

IV. SIMULATION RESULTS

This section presents the simulation results that have been obtained using the SVM technique in the dual two-level VSI for a three-level operation. The performance of the proposed inverter topology is simulated with the Matlab/Simulink software. The detailed system specifications and parameters are the same as those used in the experimental, which are given in Section V.

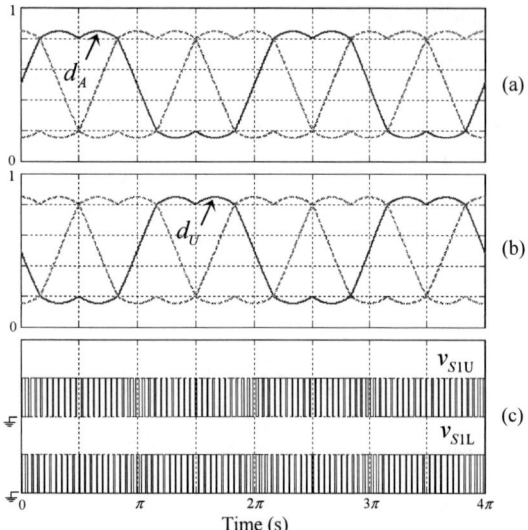

Fig. 4. Simulated waveforms of duty cycles and the gating pulses for power switches (a) duty cycles of upper VSI d_A, d_B, d_C, (b) duty cycles of lower VSI d_U, d_V, d_W, and (c) the gating pulses for power switches v_{S1U}, v_{S1L}.

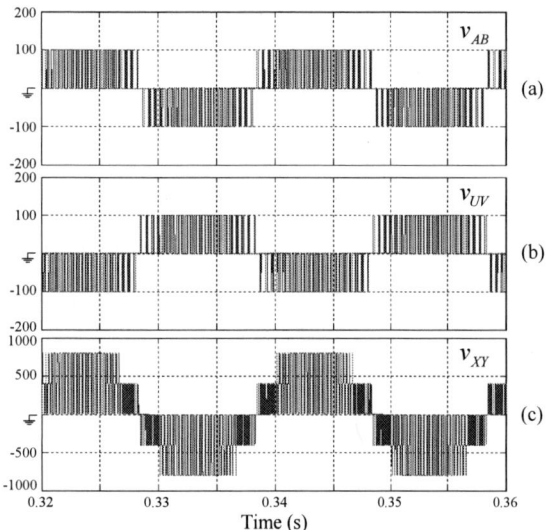

Fig. 5. Simulated waveforms of steady-state operation at modulation index $m_U = m_L = 0.8$ (a) line-to-line voltage of upper VSI v_{AB}, (b) line-to-line voltage of lower VSI v_{UV}, and (c) line-to-line voltage of load v_{XY}.

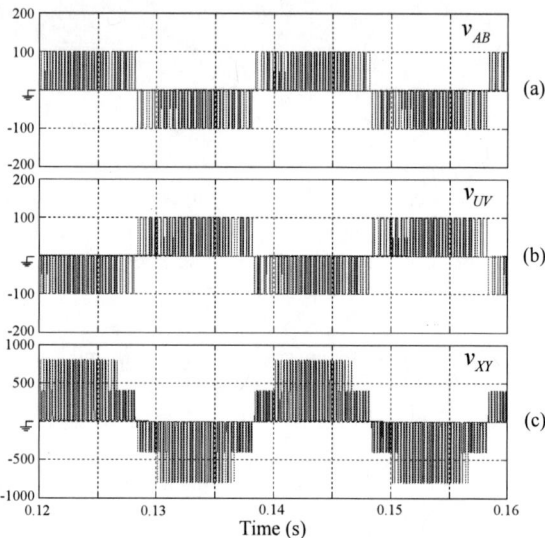

Fig. 6. Simulated waveforms of steady-state operation at modulation index $m_U = m_L = 0.4$ (a) line-to-line voltage of upper VSI v_{AB}, (b) line-to-line voltage of lower VSI v_{UV}, and (c) line-to-line voltage of load v_{XY}.

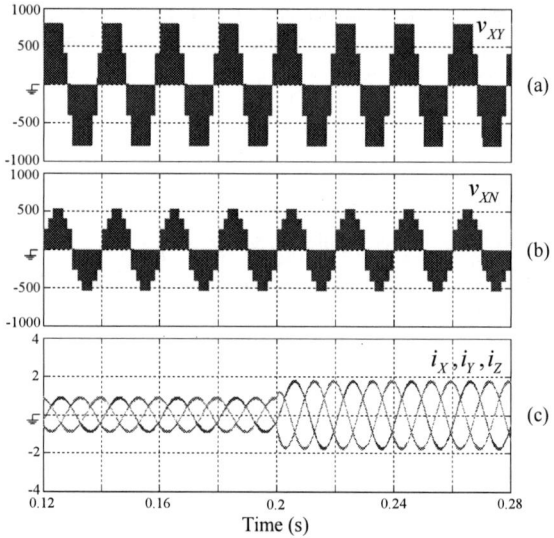

Fig. 7. Simulated waveforms of dynamic response operation for a unit step modulation index reference (a) line-to-line voltage of load v_{XY}, (b) phase voltage of load v_{XN}, and (c) phase current of load i_X, i_Y, i_Z.

By averaging the output of the SVM technique in the dual two-level VSI for a three-level operation, Fig. 4 shows the simulated waveforms of duty cycles and the gating pulses for power switches. Instead of providing the gate signal as a pulse in each switching cycle, the SVM generates the duty cycles in terms of d_A, d_B, d_C for the upper VSI and d_U, d_V, d_W for the lower VSI, as shown in Fig. 4 (a) and (b), respectively. It can be seen that the average duty cycles are continuous. The gating pulses of the power switches v_{S1U}, v_{S1L} for leg A and U in the dual two-level VSI are represented in Fig. 4 (c), which operated the switching frequency of 2 kHz.

The output line-to-line voltage waveforms of the dual two-level VSI for a three-level operation are given in Fig. 5 and Fig. 6 with both high and low modulation indexes in steady-state conditions. Fig. 5 shows the simulated line-to-line voltage waveforms with the high modulation index $m_U = m_L = 0.8$. The line-to-line voltage of upper and lower inverters v_{AB}, v_{UV} and the line-to-line voltage of load v_{XY} are illustrated in Fig. 5 (a)-(c), respectively. The two inverters generate the same three voltage levels but phase shifted by $180°$ and the five voltage levels on the line-to-line voltage of load at high modulation indexes.

The 2014 International Power Electronics Conference

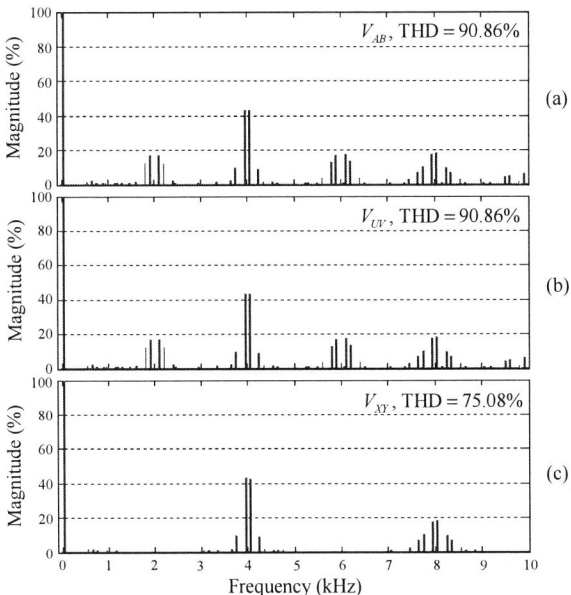

Fig. 8. Normalized harmonic spectrum of the voltage at modulation index $m_U = m_L = 0.8$ (a) line-to-line voltage spectrum of upper VSI, (b) line-to-line voltage spectrum of lower VSI, and (c) line-to-line voltage spectrum of load.

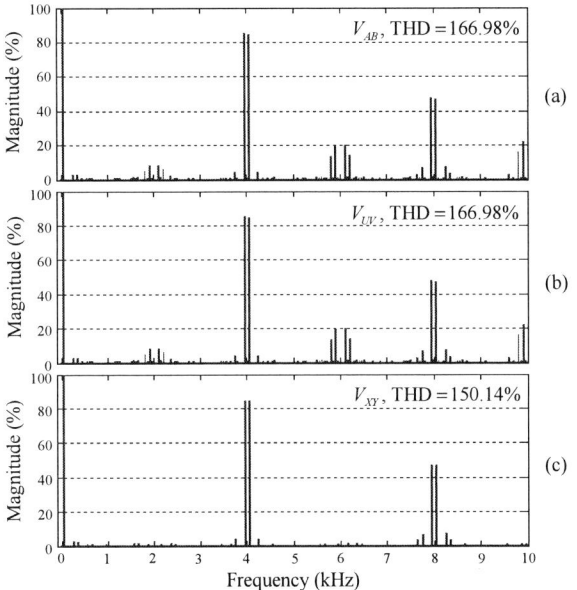

Fig. 9. Normalized harmonic spectrum of the voltage at modulation index $m_U = m_L = 0.4$ (a) line-to-line voltage spectrum of upper VSI, (b) line-to-line voltage spectrum of lower VSI, and (c) line-to-line voltage spectrum of load.

Fig. 6 shows the simulated line-to-line voltage waveforms with the low modulation index $m_U = m_L = 0.4$. The line-to-line voltages of upper and lower inverters v_{AB}, v_{UV} are illustrated in Fig. 6 (a) and (b), respectively, which generate the same three voltage levels. In Fig. 6 (c), the five voltage levels are achieved on the line-to-line voltage of load. Fig. 7

illustrates the simulated waveforms of the dynamic response operation for a unit step modulation index. The modulation index reference increased starting from low to high value (0.4 to 0.8) for the unit step operation in 0.2s. As can be seen from simulated waveform, the output voltage and current waveform proves that smooth for the dynamic response operation. The line-to-line voltage v_{XY} has five voltage levels at high and low modulation index and the waveform of load currents i_X, i_Y, i_Z are close to sinusoidal and low amount of harmonic distortion, which is due to the elimination of low-order harmonics by the modulation technique and the filtering effect of the load inductance.

The normalized harmonic spectrum of the dual two-level VSI for a three-level operation are given in Fig. 8 and Fig. 9. The harmonic spectrums of line-to-line voltage are shown in Fig. 8 (a)-(c). The dual two-level inverters operates under the condition of $m_U = m_L = 0.8$ and $f_{sw} = 2$ kHz. The total harmonic distortion (THD) of the line-to-line voltages with the high modulation index are v_{AB}, v_{UV} = 90.86% and v_{XY} = 75.08%. Fig. 9 shows the harmonic spectrum of voltage under the condition of $m_U = m_L = 0.4$. The THD of v_{AB}, v_{UV} and v_{XY} is 166.98% and 150.14%, respectively.

Fig. 10 shows the THD profile of the line-to-line voltage schemes for the dual two-level VSI for a three-level operation modulated by proposed SVM strategy. The THD is evaluated in the whole linear modulation index ranges (0.1-1.15) with incremental step value of 0.1. The THD of output line-to-line voltage between upper and lower inverters v_{AB}, v_{UV} are the same as that of its counterpart. Further, the THD of line-to-line voltage of load v_{XY} produced by the proposed modulation technique is lower the line-to-line voltage between of upper and lower inverters in all range.

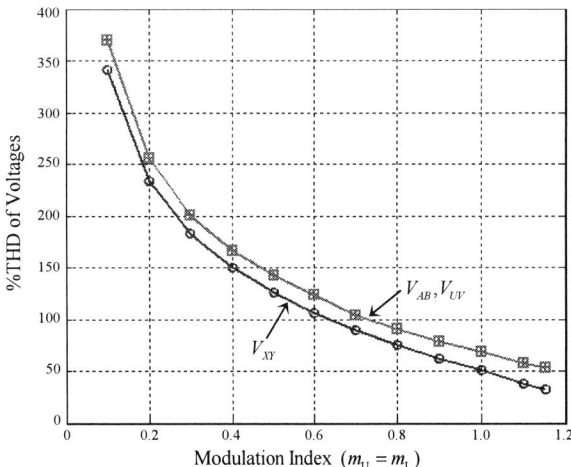

Fig. 10. THD profile of the line-to-line voltage schemes for the dual two-level VSI for a three-level operation.

V. EXPERIMENTAL RESULTS

A prototype system was built and tested in the laboratory. The values of the parameters are the same as those used in the simulation. The parameters that were used for simulation are listed as follows: the nominal dc-link voltage $V_{dU}, V_{dL} = 100\text{V}$, the dc-link capacitor $C_{dU}, C_{dL} = 3300 \ \mu\text{F}$, the fundamental frequency $f_1 = 50\text{Hz}$, the switching frequency $f_{SW} = 2$ kHz, and the turn ratio of three-phase transformer $n = 4$. A three-phase wye-connected the non-unity power factor load with a 1.2kVA, 380V, 50Hz, 0.95 lagging power factor was used in the experimental verification. The photograph of the experimental prototype of the proposed dual two-level VSI is shown in Fig. 11.

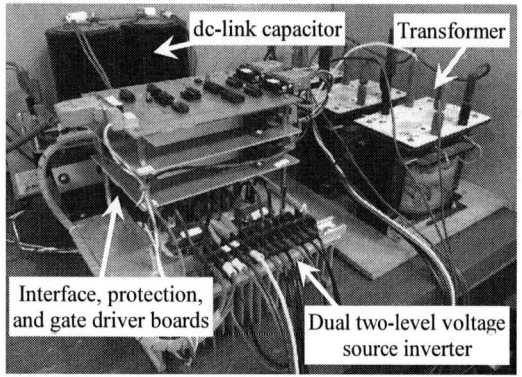

Fig. 11. Photograph of the experimental prototype.

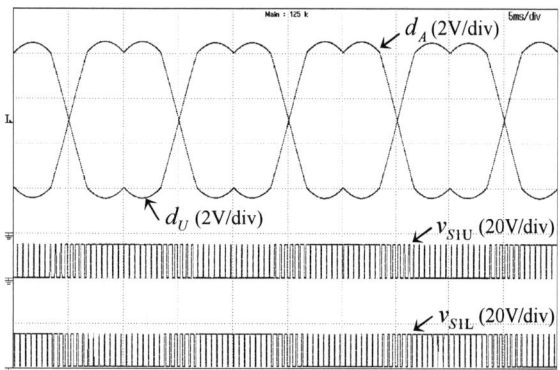

Fig. 12. Experimental waveforms of duty cycles d_A, d_U and the gating pulses for power switches v_{S1U}, v_{S1L}.

Fig. 12, top, shows measured waveforms of duty cycles d_A, d_U of the upper and lower VSI at both modulation indexes $m_U = m_L = 0.8$. From this result, it can be seen that the duty cycle d_A is the same as d_U, which are the same amplitude and frequency, but phase shifted of π radian (out of phase). The gating pulses of the power switches v_{S1U}, v_{S1L} for leg A and U in the dual two-level VSI are represented in Fig. 12 (middle and bottom).

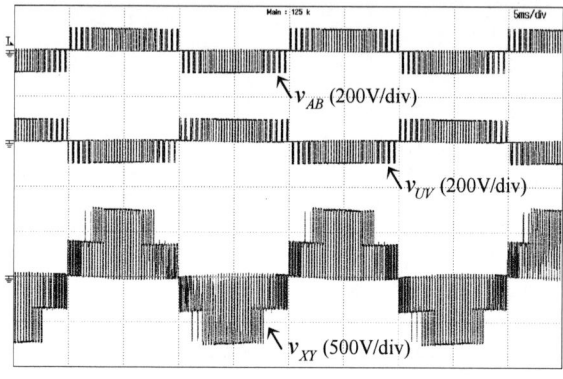

(a)

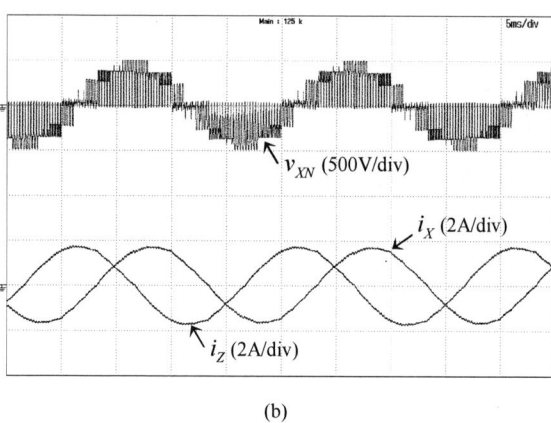

(b)

Fig. 13. Experimental waveforms of steady state operation at modulation index $m_U = m_L = 0.8$: (a) line-to-line voltage of upper and lower VSI v_{AB}, v_{UV} and line-to-line voltage of load v_{XY} (b) phase voltage of load v_{XN} and phase currents of load i_X, i_Z.

The proposed inverter voltage waveforms with both high $(m_U = m_L = 0.8)$ and low $(m_U = m_L = 0.4)$ modulation indexes in steady-state operation are shown in Fig. 13 and Fig. 14.

This waveforms in Fig. 13 (a) show, from top to bottom, the line-to-line voltages of upper and lower VSI v_{AB}, v_{UV} and the line-to-line voltage of load v_{XY}. The modulation indexes for the inverters are both set at 0.8 and the dc-link voltage is maintained at 100 V. It can be seen that the load line-to-line voltage v_{XY} of the inverter under steady-state operation is generated by the five-level voltage. The fundamental of the load line-to-line voltage $v_{XY,1}$ is 420 V (rms). Fig. 13 (b) shows measured waveforms of the phase voltage v_{XN}, and phase currents i_X, i_Z with non-unity power factor load. It can be seen that the phase voltage under steady-state operation generated by the seven-level voltage and low ripple current.

The 2014 International Power Electronics Conference

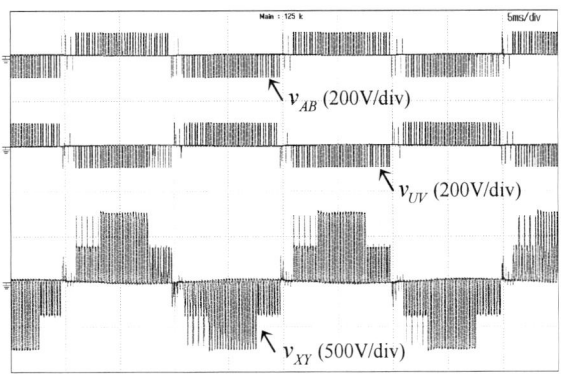

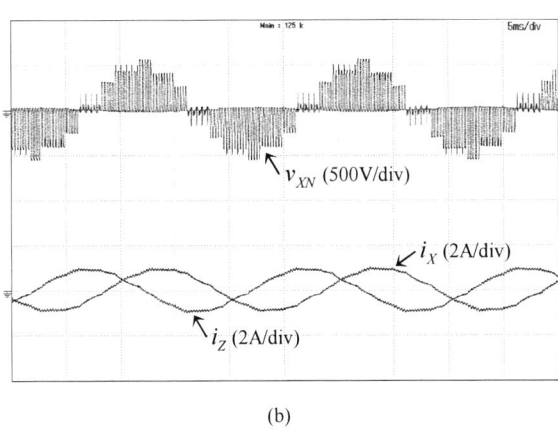

(a)

(b)

Fig. 14. Experimental waveforms of steady state operation at modulation index $m_U = m_L = 0.4$: (a) line-to-line voltage of upper and lower VSI v_{AB}, v_{UV} and line-to-line voltage of load v_{XY} (b) phase voltage of load v_{XN} and phase currents of load i_X, i_Z.

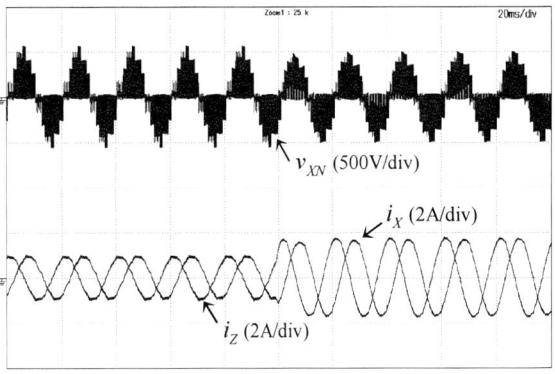

Fig. 15. Experimental waveforms of dynamic response operation for a unit step modulation index reference: phase voltage of load v_{XN}, and phase current of load i_X, i_Z.

VI. CONCLUSIONS

A SVM strategy for three-level operation based on dual two-level VSI topology was proposed in this paper. The topology uses two standard two-level VSI in parallel with a three-phase transformer to connect them together. The proposed inverter produced five-level line-to-line voltage waveform, which is equal to three-level NPC VSI. The modulation algorithm is based on the basic two-level SVM. It can simplify the algorithm and can be easily implemented. The feasibility and reliability of the proposed modulation method for the dual two-level VSI has been verified by simulation and experiment. Experimental results confirmed the good performance with quality to the output voltages and the output phase currents in the dynamic and steady-state operations under high and low modulation indices. As stated above, the proposed topology is suitable for high power medium voltage applications.

ACKNOWLEDGMENT

The work was supported by the Faculty of Engineering, Chiang Mai University.

REFERENCES

[1] J. Rodríguez, S. Bernet, B. Wu, J. O. Pontt, and S. Kouro, "Multi-level voltage-source-converters topologies for industrial medium-voltage drives," *IEEE Trans. Ind. Electron.*, vol. 54, no. 6, pp. 2930-2945, Dec. 2007.

[2] J. Rodríguez, J. Lai, and F. Peng, "Multilevel inverters: A survey of topologies, controls and applications," *IEEE Trans. Ind. Electron.*, vol. 49, no. 4, pp. 724–738, Aug. 2002.

[3] D. Casadei, G. Grandi, A. Lega, and C. Rossi, "Multilevel operation and input power balancing for a dual two-level inverter with insulated dc sources," *IEEE Trans. Ind. Applicat.*, vol. 44, no. 6, pp. 1815-1824, Nov./Dec. 2008.

[4] J. Ewanchuk and J. Salmon, "Three-limb coupled inductor operation for paralleled multi-level three-phase voltage sourced inverters," *IEEE Trans. Ind. Electron.*, vol. 60, no. 5, pp. 1979-1988, May 2013.

Similarly, Fig. 14 also shows the measured waveforms in steady-state operation. During the test, the dc voltage was maintained at 100V by the modulation index for upper and lower inverter is 0.4. From these results in Fig. 14 (a), it can be shown that the line-to-line voltages of upper and lower VSI v_{AB}, v_{UV} and the line-to-line voltage of load v_{XY} that the line-to-line voltage of load is generated by the five-level voltage. The fundamental of the load line-to-line voltage $v_{XY,1}$ is 240 V (rms). Fig. 14 (b) shows measured waveforms of the phase voltage v_{XN}, and phase currents of the load i_X, i_Z. The output voltage and current in this case is reduced to half of the rated voltage.

Fig. 15 shows the dynamic response operation of the measured waveforms of the phase voltage v_{XN}, and phase currents i_X, i_Z of the load when modulation index reference has a unit step increases from 0.4 to 0.8. It demonstrates that the phase voltage waveform proves smooth output voltage, which is generated by the seven-level voltage at high and low modulation index.

Investigation on the Parallel Operation of All-GaN Power Module and Thermal Performance Evaluation

Stone Cheng*, Po-Chien Chou

Dept. of Mechanical Engineering, National Chiao-Tung University, Hsinchu, Taiwan, ROC

*stonecheng@mail.nctu.edu.tw

Abstract- **This paper presents a 270-V, 56-A GaN power module with three AlN substrates are prepared for the module. Each substrate is composed of three parallel connected GaN chips which incorporates six 2-A AlGaN/GaN-on-Si high electron mobility transistors (HEMTs) cells. The devices are wire-bonded in parallel connection to increase the power rating. The packaged GaN HEMTs exhibit the pulsed drain current of 0.435 A/mm. Both DC and pulsed current-voltage (I_D-V_{DS}) characteristics are measured for different connection and sizes of devices, at various power densities, pulse lengths, and duty factors. The static parameters of threshold voltage and leakage currents were extracted to show how these parameters would scale as the devices are paralleled. Performance of multiple chip GaN power module package and thermal evaluation is studied. Experimental results demonstrated the ability to parallel nine GaN HEMTs die together and to verify the current sharing during the dynamic switching to attain high-current capacities.**

Keywords- GaN HEMTs, Thermal Management, Cascode circuit.

I. INTRODUCTION

In this work, the AlGaN/GaN HEMT structure is shown in Fig.1 [1,2]. The chip consists of 6 cells; each cell has 10 gate fingers with 2-μm width and 500-μm length. The entire GaN based module was designed to reduce the thermal stack by using thermally conductivity materials and eliminating the interfacial thermal resistance of the die-attach layer. Figure 2 shows that each module contains an array of 24 GaN HEMT cells arranged in two rows of four chips. The chips are bonded to the DBC with sintering of silver nano-particles and then wire-bonded to the electrical interconnect board. The DBC material provides good coefficients of thermal expansion (CTE) matching and an efficient heat transfer path for removing heat from GaN HEMTs [3,4].

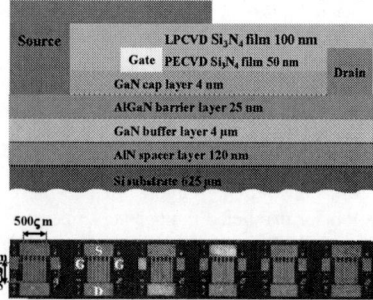

Fig. 1. Cross-sectional structures of fabricated AlGaN/GaN HEMT with source field plate structure. The total width of the gate periphery is 30 mm (6 cells; gate length, 2 μm; gate width, 500 μm).

Fig. 2. Test module contains four GaN HEMT chips, mounted on a common source pad. Parallel drain leads extend to the upper and gate leads to the lower. The circuit layout is designed and optimized for sorting of the parallel-connected HEMTs.

II. CHARACTERISTIC OF PARALLELED GAN-HEMT DEVICES

Individual cells were tested to verify voltage blocking capability, gate leakage current, gate threshold voltage characteristics, and the forward I–V characteristics. Figure 3(a) shows the room temperature on-state characteristic of four 2 A, 400 V GaN chips that have been mounted on an AlN package substrate. The I_D–V_{DS} forward characteristics of each substrate were determined by individually pulsing currents through each GaN chips. The current pulses were limited in duration to minimize any self-heating in the chips that were calculated to be less than 0.4 °C at the maximum current pulse. Ten cells (C1~C10) are selected from these chips for parallel testing. I_D–V_{DS} curve of each GaN cell was tested in the same chopper circuit with the gate-source voltage swinging from 0 to -10 V to obtain its dynamic characteristics. The specific on-resistance is shown to vary between 0.685 Ω-mm² to 0.77 Ω-mm² with V_{GS} = 0 V and V_{DS} = 1 V, which corresponds to the active device area of 0.25 mm². Threshold voltage (V_{th}) is an important device parameter that can be determined by linear extrapolation of the transfer characteristic (I_D-V_{GS}). The individual transfer curves at room temperature were measured to verify the threshold voltage among parallel devices, as shown in Fig. 3(b). The drain current is plotted on a linear scale as a function of the gate voltage. The threshold voltages extracted from the transfer curves are between -6.43 V to -6.62 V. Each devices show the small drain current which flows in the 2DEG channel below threshold, and because of the threshold voltage mismatch (~190 mV) among the HEMT devices, a small difference in the drain current is observed. Parallel operation is an important feature of GaN HEMT cells in a power module. The static electrical characteristics of the prototype module were measured at room temperature.

The 2014 International Power Electronics Conference

Comparison of drain currents between eight parallel GaN cells with the sum of each GaN cell shows no significant difference, which is shown in Fig. 3(c). The calculated individual device currents ranged from 2 to 2.2 A. This demonstrates good static current sharing with an unbalance of less than 5 % among the eight GaN cells. The static current sharing between the cells should be nearly uniform to avoid an overloading of the devices due to an excessive current. To balance the current between all GaN HEMT modules, the on-state characteristics have been extracted to match as closely as possible to each other. The parallel-connected module is found to be stable under forward bias conditions with no significant change in characteristics observed.

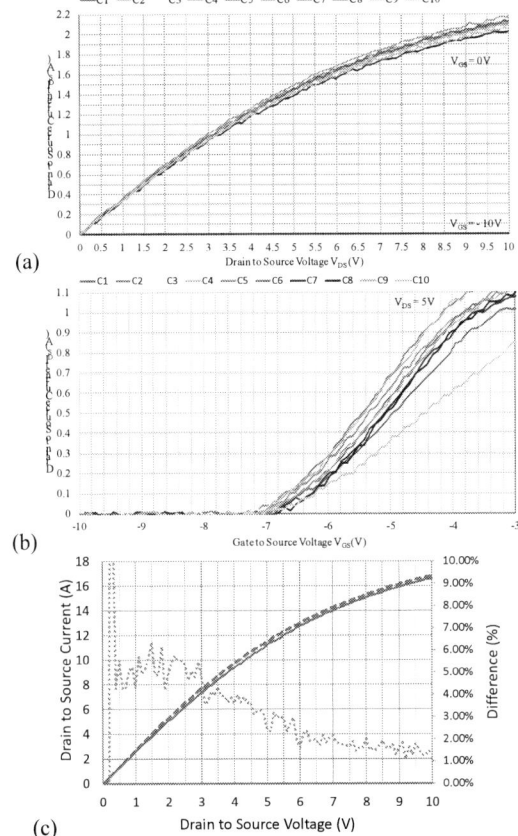

(a)

(b)

(c)

Fig. 3. (a) I_D-V_{DS} curve of each GaN cell. All cells have similar characteristics. (b) I_D-V_{GS} characteristics of each GaN cell. (c) Drain current comparison between eight GaN cells parallel connected with the sum of each GaN cell.

Device operations in the off-state were analyzed by measuring breakdown voltage, which was defined as the drain-source voltage at a constant drain current of 5 mA. The breakdown performance of the device was measured with a Keithley 2657A high power source meter. The drain-source voltage constantly increased when V_{GS} was fixed at -10 V. Figure 4(a) shows the forward blocking characteristics at a blocking voltage of 425 V and a leakage current of 1mA/mm. Individual cells were tested to verify on-state characteristics, voltage blocking, and to measure gate leakage current. Four HEMT cells with device leakage current within 100nA were then selected for analysis in a parallel connection. Figure 4(b) shows

the hard breakdown I-V curve over time and the avalanche breakdown voltage. Since the leakage current density is limited to 1 mA/mm to avoid damaging the devices on reaching breakdown point, the drain current is then dropped to 2µA after breakdown. The parallel GaN cells showed an off-state breakdown voltage of 270V, and the drain leakage current showed an upward trend to 3 µA before breakdown. These static characteristics were matched before transient response testing in order to maximize dynamic current sharing.

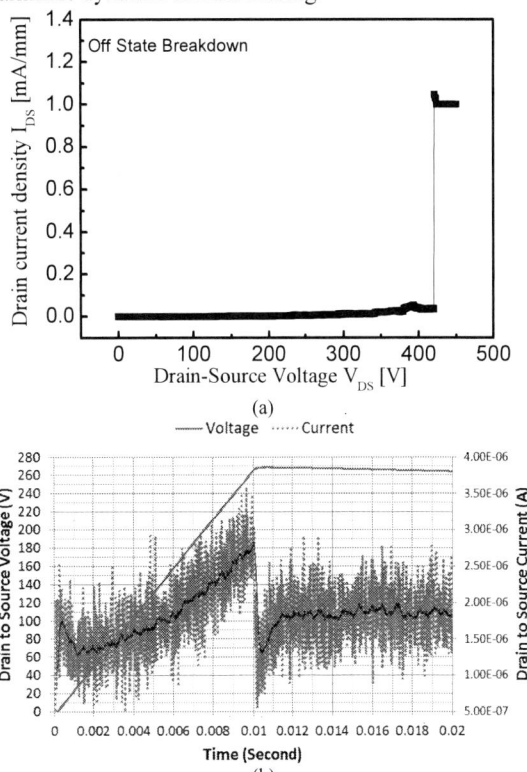

(a)

(b)

Fig. 4. The GaN HEMT reverse voltage blocking characterization. (a) Off-state breakdown voltage was 425 V for single GaN HEMT. The V_{GS} was fixed at -10 V. (b) Off-state breakdown voltage for four parallel GaN HEMTs was 270 V. Drain current leakage started from 1.5 µA and increased to 3µA at 270 V breakdown voltages. The V_{GS} was fixed at -10 V.

III. STATIC AND TRANSIENT BEHAVIOR OF GAN HEMTS ACROSS DIFFERENT TEMPERATURES

To quantize the effect of temperature on the performance of both single and paralleled cells, a set of GaN HEMTs was characterized at different base-plate temperatures. The module is placed inside a thermostatic chamber of which the ambient temperature (T_A) is set to 25 °C to 175 °C. For GaN HEMT, 175 °C is a good rule for maximum channel temperature. At each temperature, the DC I-V curves were measured, and maximum drain current ($I_{D\ max}$), on resistance (R_{ON}), transconductance (g_m), and pinch-off voltage (V_P) were recorded. The experiment architecture uses Keithley 2601A to operate gate terminal and Keithley 2651A to operate drain terminal for electrical measurement. These electrical parameters are measured by drain and gate bias simultaneously in submicrosecond time scale. At this

978-1-4799-2706-7/14 $31.00 © 2014 IEEE

condition, there is dissipated negligible power, and the channel temperature is the same as the ambient temperature (assuming thermal conductivities of both GaN and Si are linear over the 25 °C to 175 °C temperature range). The measured cells were chosen to have the approximate transfer characteristics (I_D-V_{DS} and I_D-V_{GS} curves), to facilitate comparison, and ensure that the cells can be matched within the range of impedances available from the individual die-level characterization. In each case, the least-squares curve fit of experimental data are shown. The static and dynamic characteristics are extracted to show how these parameters would scale as the HEMTs are paralleled.

A. Maximum drain current ($I_{D\,max}$) and on resistance (R_{ON}) measurements

The forward I_D-V_{DS} characteristics of GaN HEMTs were obtained for both single and paralleled HEMT cells at different temperatures from 25 °C to 175 °C. Figure 5 shows the ON resistance for a single HEMT cell and for two paralleled HEMT cells over a temperature range from 25 °C to 175 °C. The $I_{D\,max}$ decreased with temperature, as expected. The change in $I_{D\,max}$ with temperature is -0.42 %/°C for single HEMT cell and -0.84 %/°C for two paralleled HEMT cells. It has positive temperature coefficient, which means with the temperature increasing, $I_{D\,max}$ decreases and R_{ON} increases. This feature makes it easier for parallel operation. The paralleled HEMT cells have almost half the resistance as the single HEMT cell at the same ambient temperature. The R_{ON} with temperature, as expected, is at a slope of 1.64 %/°C for single HEMT cell and 0.87 %/°C for two paralleled HEMT cells.

B. Pinch-off voltage (V_P) and peak transconductance (R_{ON}) measurements

Figure 6 shows the I_D-V_{GS} characteristics of single and paralleled HEMTs. I_D-V_{GS} plots were obtained by sweeping V_{GS} from -10 V to V_{GS} (max = 0 V) with V_{DS} set to 5 V. The HEMTs are normally depletion-mode (normally-on) devices, and the gate must be maintained at a negative voltage to keep the device off. The pinch-off voltage shift increases with increasing ambient temperature for both the single and paralleled devices, as shown in Fig. 7. The V_P, has been evaluated by extrapolating the I_D current, in the I_D versus V_{GS} graph, measured as when I_D = 1 % of room temperature $I_{D\,max}$, V_{DS} = 10 V, and taking the corresponding V_{GS} value as V_P. The reason V_P was chosen at 1 % $I_{D\,max}$ (20 mA), in place of the typical 1 mA/mm leakage current, was that the leakage current has strong temperature dependence due to the presence of the presence of defects. The change in V_P with temperature is 0.13 %/°C and 0.16 %/°C for the two HEMT cells. The results obtained with the definition of V_P show a weak dependence on temperature in parallel operation. It increases with temperature at 0.07 %/°C for the paralleled HEMTs and almost 0.03 %/°C mismatch between the two cells. The measured values of the peak transconductance (g_m) versus temperature are shown in Fig. 8. The figure shows that the temperature dependence of g_m is significant and consistently negative for single and parallel measurements. The change in g_m with temperature is -0.23 %/°C for single HEMT cell and -0.39 %/°C for two paralleled HEMT cells. The paralleled HEMT cells exhibit a higher g_m than the single HEMT cell. This increase in g_m will improve the current handling capability of the HEMT cell as they are paralleled.

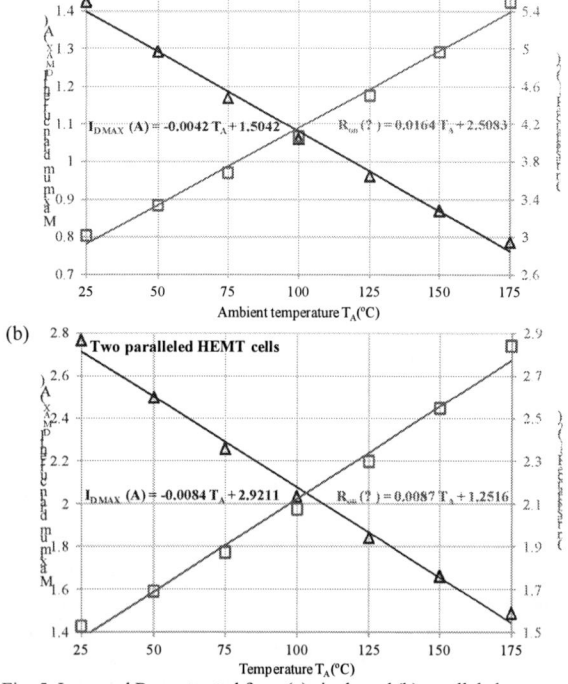

Fig. 5. $I_{D\,max}$ and R_{ON} extracted from (a) single and (b) paralleled HEMTs over a temperature range T_A (25 °C–175 °C). As temperature increases, $I_{D\,max}$ decreases and R_{ON} increases.

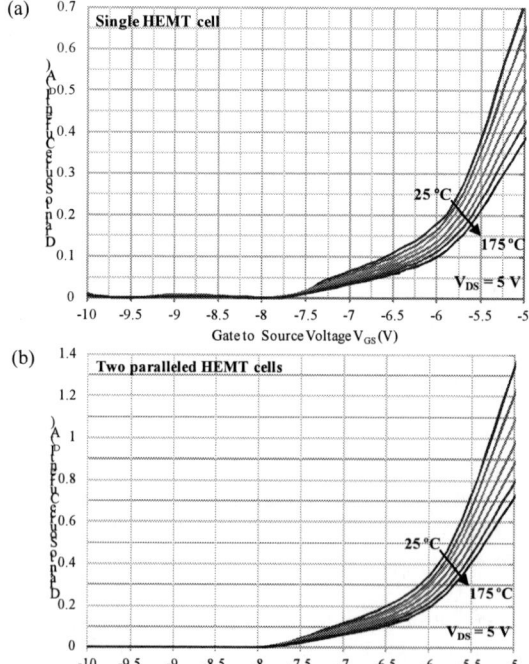

Fig. 6. Transfer characteristics (I_D-V_{GS}) of both (a) single and (b) two paralleled HEMT cells.

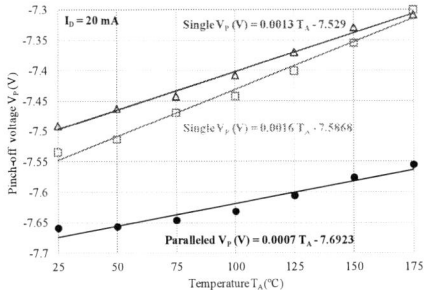

Fig. 7. Variation of pinch-off voltage across temperature range. The pinch-off voltage exhibited a positive shift at increasing temperatures.

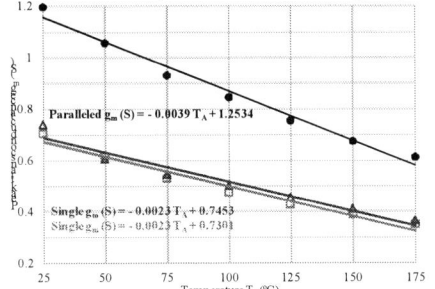

Fig. 8. Dependence of transconductance on temperature. The transconductance decreases with increase in temperature for both the single and paralleled cells.

IV. DYNAMIC CHARACTERIZATION AND ANALYSIS OF PARALLELED HEMT CELLS

The static current characteristics of HEMT devices are not enough for HEMT-based power module design and operation. The switching performance is evaluated with the circuit shown in Fig. 9 and tested with gate pulse measurement. The operating temperature is swept from room temperature 25 °C to 175 °C with placing inside a thermostatic chamber, which is same as used in static I/V characterization. Switching is performed up to 100 kHz and 20% duty cycle. Considering HEMT gate breakdown and pinch-off voltages, and their variation in two paralleled HEMT cells, $V_{GS} = 0$ V is chosen for turn-on and $V_{GS} = -10$ V is used for turn-off when V_{DS} is kept at 5 V. In the test, for two paralleled HEMTs, a RC circuit made with a parallel resistor (R_G) and capacitor (C_G) is used to attenuate the parasitic oscillations. The parasitic oscillation comes from the different package parasites on each HEMT loop. The suitable R_G can aid fast switching while avoiding shoot-through and oscillation problems caused by high dv/dt and di/dt. The external R_G in the range of 10 Ω~33 Ω and the external C_G in the range of 1~100 nF is tried to ensure a high dynamic gate current during transients. A 22 Ω resistor and a 10 nF capacitor are optimally chosen for a parallel RC network, which can increase the synchronous switching capability of the HEMT cells and reduce the risk of over heat.

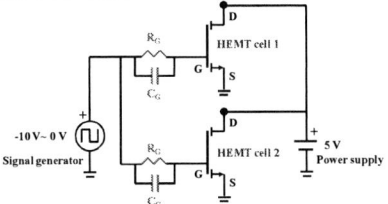

Fig. 9. Schematic of the drive circuit.

Like the SiC MOSFET, GaN HEMT devices have a positive temperature coefficient for resistance and allow for parallel use [5]. If a HEMT cell shares increased drain current (I_D) initially, it heats up faster. The resistance (R_{ON}) increased and causes current to shift to the other cell in parallel. The dynamic switching behaviors during the parallel operation of two HEMT cells are shown in Fig. 9 and Fig. 10, respectively. It may be noticed that the current sharing among the HEMT cells is equitable under initial transient conditions during turn-on and turn-off. There appears to be minimal mismatching and no apparent problem is foreseen. It has been found that an external RC network connected at each HEMT gate eliminates parasitic oscillation while minimizing switching losses. The figure shows a very close match of the waveforms during turn-on. However, there appears to be minimal current mismatching at turn-off, as shown in Fig. 10(b) and Fig. 11(b), until the characteristics of each cell dominate when approaching steady state, possibly owing to some undesirable parasites and device parameter mismatches [6].

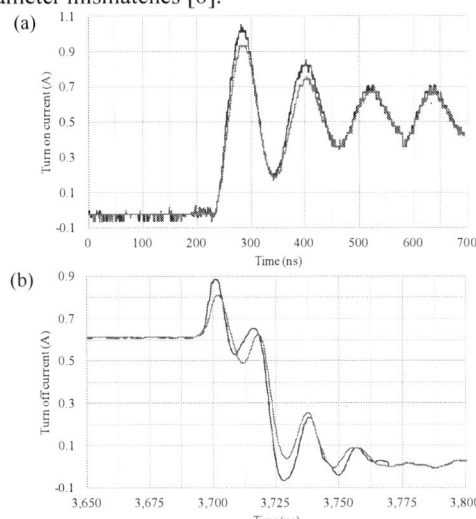

Fig. 10. (a) Turn-on and (b) turn-off waveforms of two paralleled HEMT cells at room temperature 25 °C.

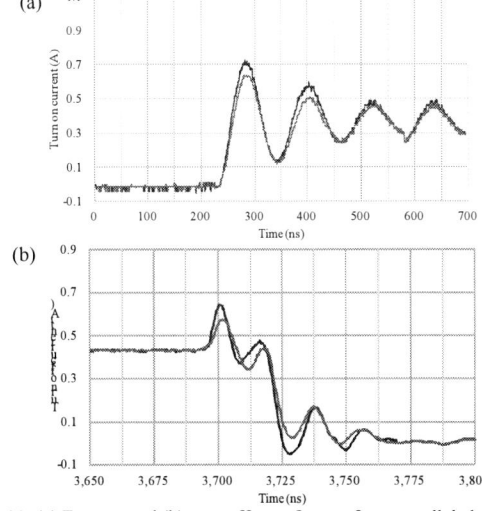

Fig. 11. (a) Turn-on and (b) turn-off waveforms of two paralleled HEMT cells at operating temperature 175 °C.

Figure 12 shows the repetitive switching of the two paralleled HEMT cells at room temperature 25 ℃. The high degree of current sharing achieved by the HEMT pair is clearly illustrated in this figure. The current waveforms are superimposed and they are so identical that there is only a 0.2 % mismatch in current between the HEMTs. This clearly shows that the series resistors are used to equalize the total resistance in each parallel branch to promote current sharing. Even though the current is shared evenly between the paralleled cells, the improvement in current sharing comes with the disadvantage of higher power consumption in the series resistors and lower efficiency power conversion. This also illustrates the influence of package parasites for high-speed switching devices. Parasitic oscillation between parallel HEMTs is unacceptable because of greatly reduced reliability and possibly increased switching losses. To avoid such parasitic oscillations, a symmetrical layout is designed for balancing gate loop parasites to ensure appropriate current sharing among the cells. The experimental results are satisfactory and demonstrate the feasibility of the parallel operation of discrete GaN HEMTs, capable of current sharing of two paralleled HEMT cells.

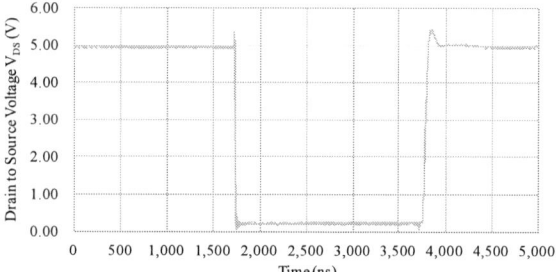

Fig. 12. The switching waveform of two paralleled HEMT cells.

V. CASCODE CIRCUIT OF GAN HEMT

Normally-off operation characteristic is required in industrial power drive circuits to satisfy fail-safe criteria and simple gate drive configuration. The normally-off GaN device, as compared to the conventional HEMT, is preferred in this aspect. A popular solution proposed to address the normally-off requirement is a cascode configuration consisting of a high-voltage, normally-on GaN HEMTs device and a low-voltage silicon MOSFET [7]. By avoiding direct focus on threshold voltage of GaN transistor structures, this approach provides one of the simplest and fastest way to deliver a normally-off GaN product, as shown in Fig. 13(a). The cascode configuration requires the integration of a low on-resistance E-mode MOS device. The LV-MOSFET die including the diode measured 4.05 x 4.05 mm. To maximize the switching performance of the GaN device, every effort was made to allow the current flow to be vertical in the 2 transistors, cascode connected, die stack in the TO-3P package, and every effort was taken to allow for parallel connections wherever possible, thus minimizing resistance and inductance inside the package. The I-V characteristics of a packaged LV-MOSFET and 80 mm GaN HEMT, Fig. 13(b), are shown in Figure 14.

(a)   (b)

Fig. 13. The cascade structure of a single GaN device and LV-MOSFET on TO-3P lead frame. (b) 80 mm GaN device: 500 μm (gate width) x 160 HEMTs on silicon substrate.

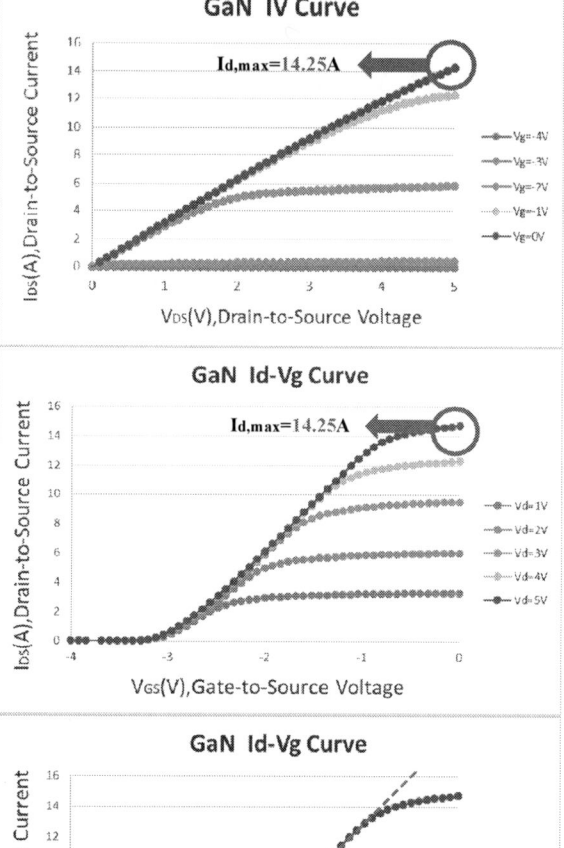

(a) The I-V characteristics of 80mm GaN HEMTs.

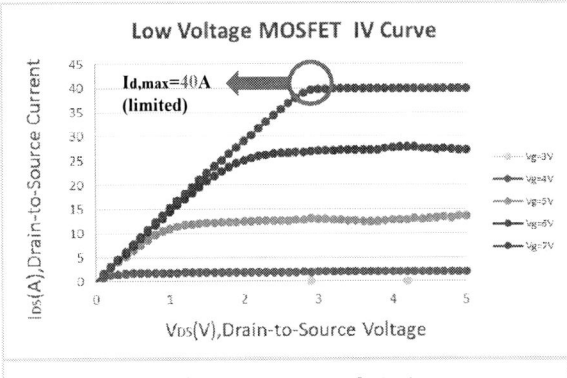

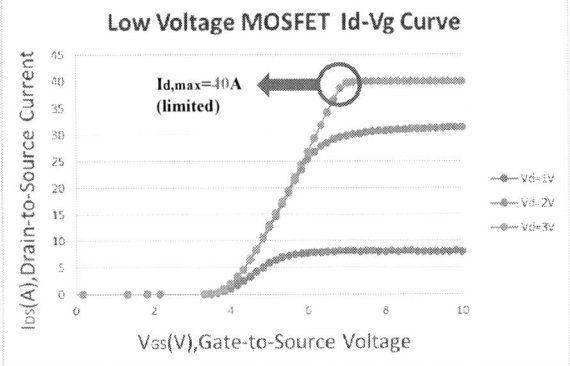

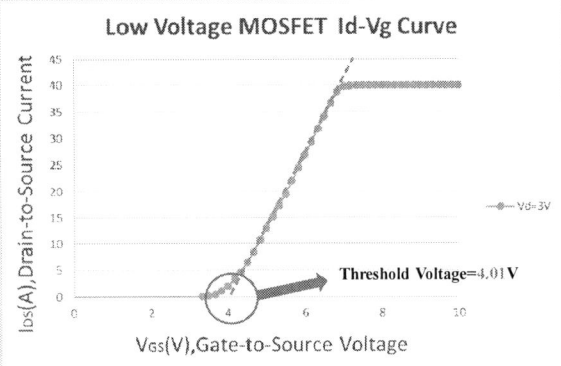

(b) The I-V characteristics of LV-MOSFET.

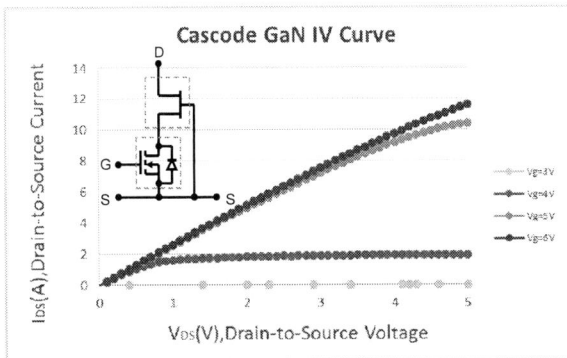

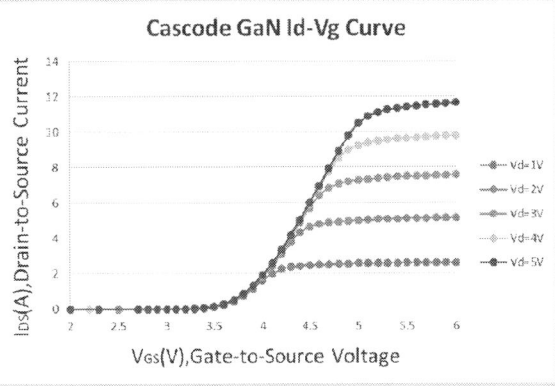

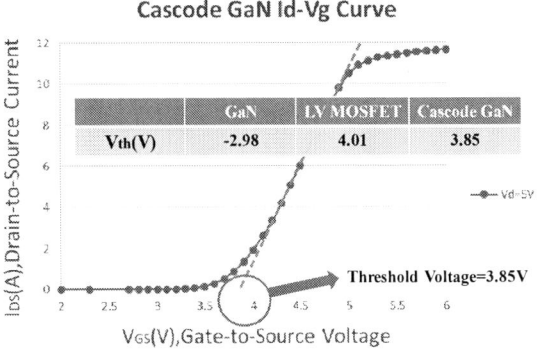

(c) The I-V characteristics of a packaged LV-MOSFET and 80 mm GaN HEMT.

Fig. 14. The I-V characteristics of (a) 80mm GaN HEMTS, (b) LV-MOSFET, and (c)packaged LV-MOSFET and 80 mm GaN HEMT.

VI. CONCLUSIONS

To validate the use of GaN technology in high power applications, there is a need to demonstrate a power module that has multiple paralleled GaN HEMT devices. In this paper, a 270-V, 56-A all-GaN power module is fabricated and characterized. The individual GaN device was presented at a lower current rating of 2 A per cell. The design requirements for this particular module were focused on a parallel operation and current sharing characteristics. Both DC and pulsed current-voltage (I_D-V_{DS}) characteristics are measured and sorted for different connection and sizes of devices. Performance of multichip GaN power module package is studied. The static current sharing between the paralleled devices was obtained to show the effect of device differences and package on the device performance. Experimental results demonstrated the ability to parallel nine GaN HEMTs die together and to verify the high current rating feasibility during the dynamic switching to attain high power capacities.

REFERENCES

[1] Stone Cheng, Po-Chien Chou, "Novel packaging design for high-power GaN-on-Si high electron mobility transistors (HEMTs)," *International Journal of Thermal Sciences*, vol. 66, pp.63-70, 2012.

[2] C.-T. Chang, H.T. Hsu, E. Y. Chang, Y.-C. Chen, H.-D. Trinh and K.J. Chen: "Normally-off operation AlGaN/GaN MOS-HEMT with high threshold voltage," *Electron. Lett.*, vol. 46, no. 18, pp. 1280-1281, 2010.

[3] Robert A. Wood and Thomas E. Salem, "Evaluation of a 1200-V, 800-A All-SiC Dual Module," *IEEE Trans. Power Electron.*, vol. 26, no. 9, pp. 2504-2511, Sep. 2011.

[4] Michael J. Palmer, R. Wayne Johnson, Tracy Autry, Rizal Aguirre, Victor Lee, and James D. Scofield, "Silicon Carbide Power Modules for High-Temperature Applications," *IEEE Trans. Components, Packaging and Manufacturing Technology*, vol. 2, no. 2, pp. 208-216, Feb. 2012.

[5] Madhu Chinthavali1, Puqi Ning, Yutian Cui, Leon M. Tolbert, "Investigation on the Parallel Operation of Discrete SiC BJTs and JFETs," *26th Annual IEEE Applied Power Electronics Conference and Exposition* (APEC), pp.1076-1083, Mar. 2011.

[6] T. Laurent, R. Sharma, J. Torres, P. Nouvel, S. Blin, C. Palermo, L. Varani, Y. Cordier, M. Chmielowska, J.-P. Faurie, and B. Beaumont, "Measurement of Pulsed Current–Voltage Characteristics of AlGaN/GaN HEMTs from Room Temperature to 15 K," in *ACTA PHYSICA POLONICA A*, vol. 119, no. 2, , pp.196-198, Feb. 2011.

[7] T. MacElwee, J. Roberts, H. Lafontaine, I. Scott, G. Klowak, and L. Yushyna, "Characterization and Performance of D-mode GaN HEMT Transistor Used in a Cascode Configuration," Electrochemical Science and Technology, Transactions, 58 (4) 167-177 (2013).

Full Silicon Carbide Boost Chopper Module for High Frequency and High Temperature Operation

Sami Pettersson, Slavo Kicin, Toni Holm, Enea Bianda, and Francisco Canales

ABB Switzerland Ltd.

Corporate Research

Baden-Daettwil, Switzerland

sami.pettersson@ch.abb.com

Abstract—This paper presents a 1200-V 20-A full silicon carbide boost chopper module designed for high temperature and high frequency operation. The developed module is based on one silicon carbide metal oxide semiconductor field effect transistor chip and two parallel connected silicon carbide schottky diode chips manufactured by Cree, Inc. The static and dynamic characteristics of the module have been experimentally determined and its performance tested in a 2-kW boost converter. The test results show that the developed module performs well and is able to provide a good conversion efficiency even at high switching frequency.

Keywords—*silicon carbide, MOSFET, module, boost converter*

I. INTRODUCTION

Considering the state-of-the-art semiconductor and packaging technology, the silicon based insulated gate bipolar transistor (IGBT) is the dominant device and used in most of today's industrial power electronic converters, especially in low-voltage applications. The latest generation IGBT based power modules can handle junction temperatures up to 175 °C, but the higher the junction temperature, the worse the system efficiency due to the increasing conduction and switching losses of the IGBTs [1]. The same applies to the silicon based metal oxide semiconductor field effect transistor (MOSFET) technology. Especially the on-state resistance of the superjunction MOSFETs increases significantly with temperature [2].

Commercial power electronic converters are specified for a certain operating temperature range within which they should be able to operate continuously at nominal power. In addition to the normal operating conditions there may be other application specific requirements, such as a short-time overload capability, which has to be taken into account in the cooling system design. Therefore the maximum junction temperature is typically designed to be much lower than the limit set by the semiconductor modules.

Wide band gap materials, e.g. silicon carbide (SiC) and gallium nitride (GaN), are expected to be the next generation materials for power semiconductor devices. Their superior material properties compared to standard silicon enable implementation of power electronic converters with improved efficiency and power density. In addition, SiC and GaN based devices may be operated at higher temperatures than the standard silicon based devices without excessively compromising their performance, which reduces the thermal management requirements of the converter.

However, faster switching speeds and higher operation temperatures have to be taken into account in the system design so that a good system performance can be achieved. To ensure a proper switching behavior the current commutation loops should be short and low inductive. As for high temperature operation, a suitable packaging concept is required for the semiconductor devices in terms of packaging materials as well as module assembly techniques [3].

Several full SiC power modules have been demonstrated over the years. First, SiC junction field effect transistors (JFET) ([4],[5]) and SiC bipolar junction transistors (BJT) ([6]) were used in the modules. However, the BJT and normally-off type JFET require a large gate current during on-state, which increases the power consumption of the gate driver. In the case of the normally-on type JFET a negative gate voltage is needed to turn off the device. Therefore the device is conducting when no auxiliary power is available, e.g. during the system startup and in case of auxiliary power supply failure, which has to be taken into account in the system design.

The SiC MOSFET has matured rapidly during the last few years. Because it is a normally-off type device and easy to drive, it is often preferred to the JFET and the BJT. Various kinds of SiC MOSFET based power modules have been presented, e.g. in [7]–[12].

This paper presents a 1200-V 20-A full SiC boost chopper module designed for high frequency and high temperature operation. The objective was to design and implement a packaging concept enabling fast switching speeds and a continuous operation at 200 °C junction temperature. The module is based on the SiC MOSFET and SiC schottky diode chips manufactured by Cree, Inc. The on-state resistance and the switching energies of the module were experimentally characterized. In addition, the module was tested in a 2-kW boost type converter

978-1-4799-2706-7/14 $31.00 © 2014 IEEE

demonstrator to evaluate the performance of the module in a real application. Section II briefly explains the design and development processes as well as the structure of the module. In the first part of Section III, the results of the static and dynamic characterization of the module are shown. The second part describes the designed and implemented boost converter demonstrator based on the developed SiC module and presents the experimental results.

II. FULL SiC BOOST CHOPPER MODULE

In order to assess the benefits of SiC in various power electronic applications, the parts of the module (substrate, housing etc.) were designed to be flexibly used in various module configurations, e.g. single switch, half bridge and chopper. The module design is also very compact to minimize the internal inductance so that a good performance at high switching speeds can be achieved. If further improvement in fast switching capability is required, the universal substrate can be replaced by the substrate for the specific module configuration. The module layout was optimized using the ANSYS Q3D software.

Additionally, the materials and assembly technologies were chosen with the aim to operate the modules at junction temperatures up to 200 °C [13]. The module presented in this paper contains one 1200-V 80-mΩ SiC MOSFET and two 1200-V 10-A SiC schottky diode dies. The SiC dies were attached to the thermally robust Si_3N_4 substrate by silver sintering (Fig. 1a). The pins for electrical connection of the module were attached using solder preforms with the melting temperature above 280 °C (Fig. 1b). Figure 1b also shows the attachment of the thermocouple to the MOSFET surface for a direct measurement of the junction temperature. The housing was machined from polyether ether ketone which is suitable for continuous operation up to 250 °C. The finalized module is shown in Fig. 1c. The dimensions of the module are $41 \times 29 \times 9$ mm^3.

The module has no baseplate and the substrate is directly attached to the heatsink during operation. Therefore the thermal resistance is very small and in combination with the high temperature capability of the module allows to reach high current densities.

III. EXPERIMENTAL TESTING

A. Static and Dynamic Characterization

To evaluate the performance of the developed SiC module, the on-state resistances and the switching energies were measured using a high power curve tracer and a double-pulse test setup with inductive load. The measured on-state resistance of the SiC MOSFET is presented in Fig. 2a and the forward voltage of the SiC boost diode in Fig. 2b. For comparison, the measurement results of a commercial dual boost chopper module based on SiC MOSFET and SiC schottky diode chips are also presented. Because each boost converter unit in the commercial module consists of two MOSFET chips and

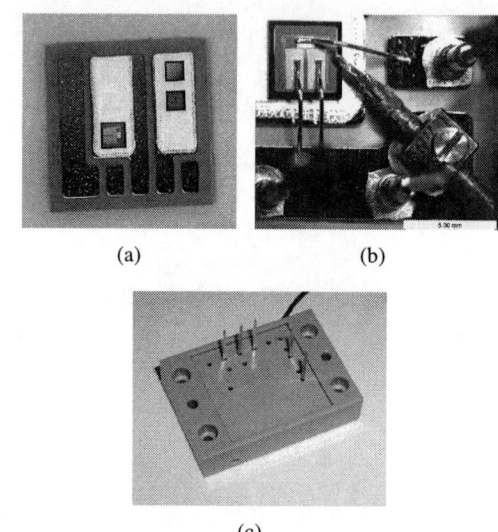

Fig. 1: (a) Si_3N_4 substrate with sintered SiC dies. The area of the substrate is 26×26 mm^2.
(b) Detailed view on the thermometry integrated in the module. The thermocouple is directly attached to the surface of the MOSFET. The MOSFET area is 4×4 mm^2. The soldered interconnections of the pins are also visible.
(c) Finalized 1200-V 20-A SiC boost chopper module.

four boost diode chips, the results are shown for one MOSFET chip and two diode chips for a fair comparison.

Regarding the switching energies of the modules, the measurement results of the developed module are shown in Fig. 2c and those of the commercial module in Fig. 2d. These include the turn-on and turn-off energies of the MOSFET (E_{on} and E_{off}) as well as the boost diode turn-off energy (E_{rr}). The results show that the developed module has a good overall performance. One noticeable difference between the modules is that the on-state resistance of the developed module is considerably smaller at high junction temperatures. However, the main reason for this is that the chips used in the commercial module are from a different manufacturer. The packaging has only a small effect on the on-state resistance.

The impact of junction temperature on the switching energies of the developed module is shown in Fig. 2e. It can be seen that both the MOSFET and the diode turn-off losses are larger but the MOSFET turn-on losses are smaller at 150 °C junction temperature compared to those at 25 °C. Therefore, unlike in the case of IGBTs, the total switching losses of the SiC MOSFETs are not significantly changing at higher junction temperatures. In fact, they are actually slightly smaller at 150 °C junction temperature (725 μJ) than at 25 °C (779 μJ) at 20-A nominal current.

As shown in Fig. 2a, the on-state resistance of the developed module increases ca. 15 % when the junction temperature increases from 25 °C to 150 °C. It has a negative temperature coefficient at lower junction temperatures until ca. 70 °C and a positive one after that, which is an interesting and noteworthy characteristic of this SiC MOSFET chip.

B. SiC Boost Converter Demonstrator

In order to test the performance of the developed module in a real application, a simple boost converter demonstrator was designed and implemented. A voltage pre-regulator for a single-phase photovoltaic (PV) inverter was selected as an application example [14]. The schematic of the boost converter is shown in Fig. 3a and a photograph of the demonstrator in Fig. 3b.

Typical commercial transformerless single-phase and three-phase PV string inverters available on the market operate at switching frequencies between 15 and 20 kHz. This frequency range is at the high end of the hearing range of human ear, which helps to keep the audible noise caused by the pulse-width modulation at a comfortable level. This is beneficial because PV string inverters are commonly used in commercial and residential buildings. Regarding the converter performance, in the case of multilevel converter topologies a fairly good balance between the power density and the conversion efficiency can be achieved in this frequency range. Nowadays many commercial PV string inverters contain SiC diodes to even further improve the conversion efficiency.

In the case of the developed full SiC module, switching frequencies up to 100 kHz were considered. In this range both the conversion efficiency and the power density of the converter were estimated to be good, and the fundamental switching frequency is still below the electromagnetic interference (EMI) frequency range ($\geq$ 150 kHz). However, the multiples of the switching frequency may still have an impact on the EMI filter de-

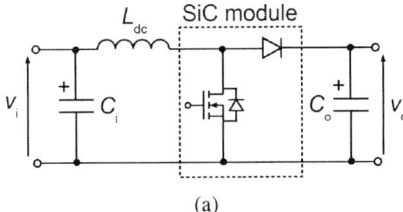

(a)

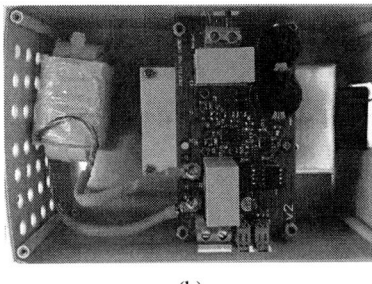

(b)

Fig. 3: Schematic (a) and a photograph (b) of the SiC boost converter demonstrator.

TABLE I: BOOST CONVERTER SPECIFICATIONS

Parameter	Value
Input voltage (V_{in})	$\geq$ 125 V
Maximum input current	16 A
Output voltage (V_{out})	400 V
Rated output power	2 kW
Switching frequency (f_{sw})	25, 50, 75 kHz
Gate voltage of MOSFET (V_{GS})	20 V / -3 V
Gate resistor of MOSFET (R_G)	0 Ω

sign in commercial systems. Three switching frequencies (25, 50 and 75 kHz) were selected for the experimental tests.

The specifications of the demonstrator are listed in Table I. The boost converter is controlled in an open-loop manner by generating the gate control signal with the desired duty ratio using a signal generator. A negative gate voltage is used to limit the SiC MOSFET leakage current in off-state. The cooling system was implemented using a standard extruded aluminum heatsink with the dimensions of $40 \times 65 \times 100$ mm^3. Even though there is a 40-mm cooling fan attached to the heatsink, only a natural convection based cooling was used in the experimental tests.

The boost inductors were implemented using the POWERLITE® AMCC type cores. Four different designs, referred to as L1, L2, L3 and L4, were tested in the boost converter demonstrator to evaluate the effect of the inductor current ripple on the conversion efficiency and the thermal behavior of the inductors. The inductor specifications are presented in Table II. All the inductors were wound using a litz wire consisting of 40 strands with 0.355-mm diameter.

At the beginning of each test run the converter was operated at full power until the heatsink temperature reached 100 °C before the measurements were started. Overall one test run consists of 42 operating points, i.e. two input voltages (125 V and 250 V), three switching frequencies and seven output powers (5 %, 10 %, 20 %, 30 %, 50 %, 75 % and 100 % of the rated output power). The input voltage was generated by a direct voltage power supply and the converter output was connected to a Chroma 63204 programmable dc electronic load. The efficiencies were measured using a Yokogawa WT3000 high precision power analyzer.

Figure 4 presents the measured current and voltage waveforms of the demonstrator operating under the worst-case conditions (2-kW output power and 125-V input voltage) at 50-kHz switching frequency with inductor L4. The test results show that both the developed SiC module and the boost converter demonstrator are performing well. There are no significant oscillations in the current and voltage waveforms in Fig. 4 and the overshoot in the MOSFET drain-source voltage is only ca. 10 %, which indicate that the inductance in the current commutation path is very small. This can be considered to be a good

The 2014 International Power Electronics Conference

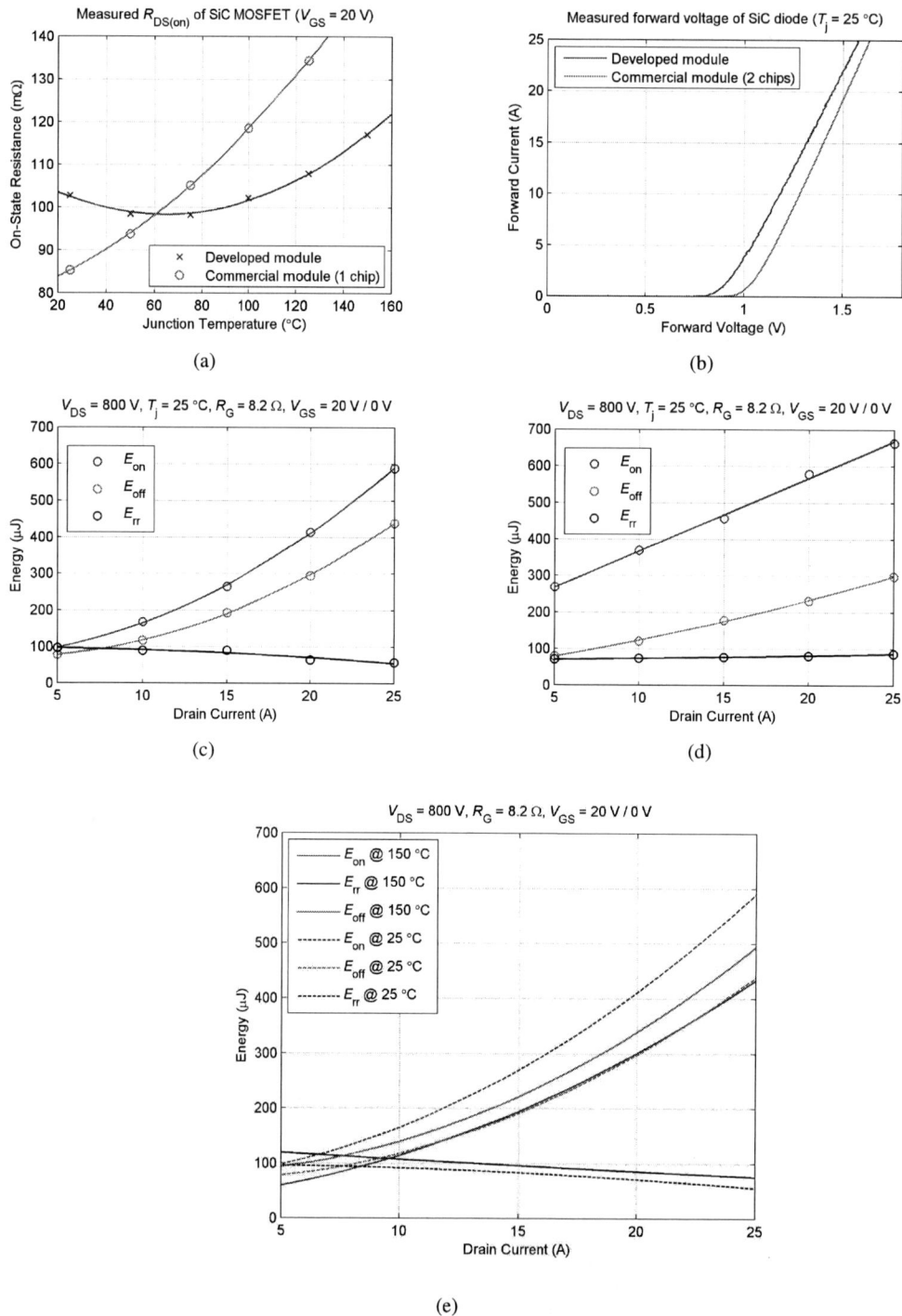

(a)

(b)

(c)

(d)

(e)

Fig. 2: The measured on-state resistance of the SiC MOSFET at 20-V gate voltage (a), the measured forward voltage of the SiC diode at 25 °C (b), the measured switching energies of the developed module at 25 °C (c), the measured switching energies of the commercial module at 25 °C (d), and the impact of junction temperature on the switching energies of the developed module (e).

978-1-4799-2706-7/14 $31.00 © 2014 IEEE

TABLE II: INDUCTOR SPECIFICATIONS

Parameter	L1	L2	L3	L4
Winding current density (A/mm^2)	4	4	4	4
Nominal direct current (A)	16	16	16	16
Maximum peak current (A)	20	20	20	20
Nominal inductance (µH)	1900	740	730	560
Number of turns	56	44	40	38
Winding dc resistance (mΩ)	34	27	35	25
Physical air gap (mm)	0.5	0.5	0.5	0.5
Core type	AMCC-40	AMCC-16B	AMCC-25	AMCC-16B

result since there is no external gate resistor to limit the switching speed.

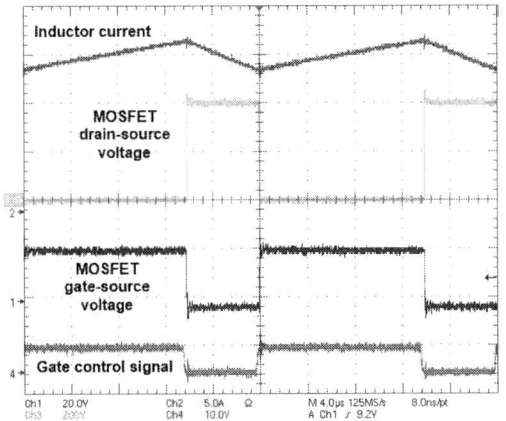

Fig. 4: Measured current and voltage waveforms of the SiC boost converter demonstrator operating at 50-kHz switching frequency with inductor L4 (2-kW output power and 125-V input voltage).

Regarding the conversion efficiency, Fig. 5 presents the measured efficiencies with all four different inductors at three different switching frequencies and two input voltages. The impact of the inductance value on the conversion efficiency is the most significant at 25-kHz switching frequency, as can be seen in Figs. 5a and 5b. The conversion efficiency is clearly better in certain operating points with the largest inductance value (L1), i.e. with the smallest inductor current ripple, than with the other inductors. However, the situation gets more even at 50 and 75-kHz switching frequencies (Figs. 5c–5f) when the current ripple is small with all of the four inductors, which results in smaller inductor core losses. It can be noticed in Figs. 5d and 5f that the efficiency curve of L4 behaves slightly differently at lower output powers than the others. The reason is that the boundary between the discontinuous and continuous conduction modes is at higher power with L4 than in the case of the other inductors, which creates a knee-like shape in the efficiency curve.

Because PV inverters do not constantly operate at full power, a weighted efficiency is usually preferred when denoting and comparing conversion efficiencies of PV inverters. The weighted European efficiency is defined as follows [15]:

$$\eta_{\text{euro}} = 0.03 \cdot \eta_{5\%} + 0.06 \cdot \eta_{10\%} + 0.13 \cdot \eta_{20\%}$$
$$+ 0.1 \cdot \eta_{30\%} + 0.48 \cdot \eta_{50\%} + 0.2 \cdot \eta_{100\%} \quad (1)$$

where $\eta_{n\%}$ is the conversion efficiency at $n\%$ of the inverter rated power. The European efficiencies of the SiC boost converter demonstrator are summarized in Table III. As can be seen, the inductors have a very small impact on the European efficiency in this case. This is because the efficiencies at 50 % and 100 % powers have the largest weight coefficients in the European efficiency and the efficiencies of the SiC boost converter demonstrator with each inductor in these operating points are fairly similar, as shown in Fig. 5.

As in this case the European efficiency is not significantly affected by the inductance, it makes sense to use a small core with a short wire to minimize the inductor volume, the material cost and the dc resistance of the winding. However, a small core has a small surface area and therefore a limited heat transfer capability, which easily results in a large temperature rise, especially in the case of a natural convection based cooling. To evaluate the thermal behavior of the inductors, the temperature of the inductors was measured between the bobbin and the core during the test runs. Inductor L1 with the largest core volume and inductance had the lowest temperature, about 70 °C when the ambient temperature was approximately 30 °C. In the case of inductors L2 and L3 the temperature stayed between 70 °C and 80 °C. The highest temperatures were recorded for inductor L4. The temperature of L4 was generally between 90 °C and 100 °C at maximum input current.

TABLE III: EUROPEAN EFFICIENCIES OF THE SIC BOOST CONVERTER DEMOSTRATOR

V_{in}, f_{sw}	L1	L2	L3	L4
125 V, 25 kHz	97.7 %	97.6 %	97.4 %	97.5 %
125 V, 50 kHz	97.5 %	97.4 %	97.3 %	97.3 %
125 V, 75 kHz	97.2 %	97.1 %	97.1 %	97.0 %
250 V, 25 kHz	98.5 %	98.5 %	98.4 %	98.3 %
250 V, 50 kHz	98.5 %	98.4 %	98.4 %	98.3 %
250 V, 75 kHz	98.3 %	98.2 %	98.2 %	98.2 %
Average	97.95 %	97.87 %	97.80 %	97.77 %

The 2014 International Power Electronics Conference

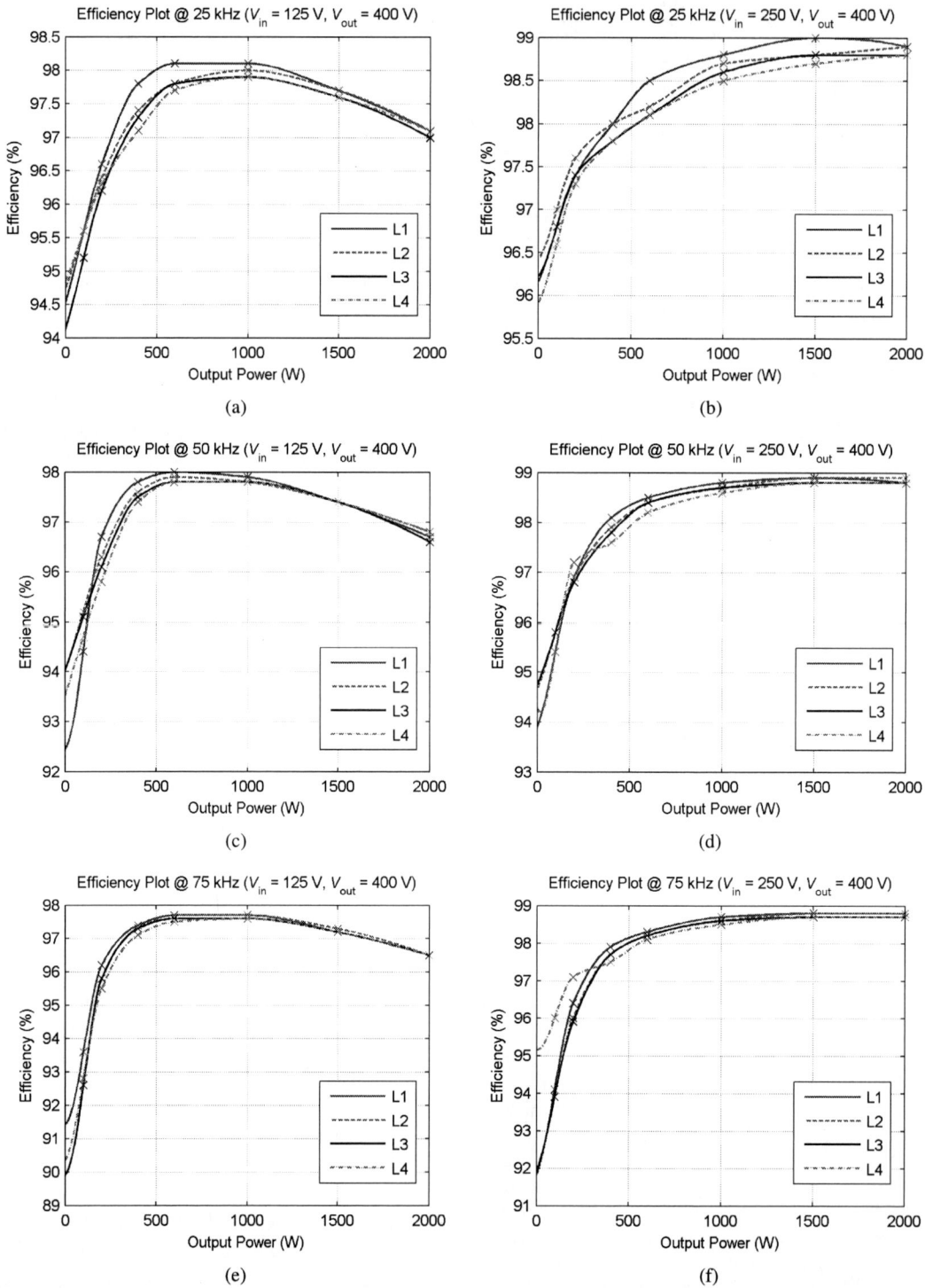

Fig. 5: Measured efficiencies of the SiC boost converter demonstrator with different inductors: $f_{sw} = 25$ kHz and $V_{in} = 125$ V (a), $f_{sw} = 25$ kHz and $V_{in} = 250$ V (b), $f_{sw} = 50$ kHz and $V_{in} = 125$ V (c), $f_{sw} = 50$ kHz and $V_{in} = 250$ V (d), $f_{sw} = 75$ kHz and $V_{in} = 125$ V (e), $f_{sw} = 75$ kHz and $V_{in} = 250$ V (f).

978-1-4799-2706-7/14 $31.00 © 2014 IEEE

The behavior of the conversion efficiency at different switching frequencies with inductor L4 is illustrated as a function of input voltage and output power in Fig. 6. As can be seen, increasing switching frequency has the most noticeable effect on the conversion efficiency in the lower right corner of the efficiency graphs when the input voltage is low and the output power high, i.e. at large input currents. Because the conduction losses are practically independent of the switching frequency (unless the current ripple changes very dramatically), the main cause for the efficiency reduction is the higher switching losses. The larger the current, the larger the switching energies, which means that when the switching frequency is doubled or tripled, the impact is the most significant in this region.

As a final test, the SiC boost converter demonstrator was operated with output powers up to 200 % of the rated power. The input voltage was set at 250 V in order that the maximum input current was not exceeded. The test was run with 400-V and 500-V output voltages at 50-kHz switching frequency using inductor L4. The measured efficiencies presented in Fig. 7 show that the developed SiC module provides a good conversion efficiency also at higher output powers. In the case of transformerless grid connected three-phase PV inverters and single-phase PV inverters based on a half-bridge inverter topology, a large input-output voltage boost ratio may be required to guarantee the proper operation of the dc-ac stage. Higher output voltages result in higher switching losses, but owing to the good switching characteristics of the SiC module the efficiency reduction is not very significant when comparing the efficiencies at 400-V and 500-V output voltages in Fig. 7. This is also affected by the fact that when the output power is constant the boost diode current is smaller at higher output voltage, which reduces the boost diode losses and therefore compensates at least part of the higher switching losses.

IV. CONCLUSION

This paper presented a 1200-V 20-A boost chopper type power module based on silicon carbide power semiconductor devices. The module was built using the first generation SiC MOSFET and SiC schottky diode chips from Cree, Inc. The design objective was a low internal inductance to guarantee a good performance at high switching speeds and a thermal capability for operation up to 200 °C junction temperature.

The static and dynamic performances of the module were experimentally characterized and the results compared with those of a commercial SiC boost chopper module. The results showed that the performance of the developed module is good, also at higher junction temperatures.

To evaluate the performance of the developed module in a real application, a 2-kW boost converter demonstrator was designed and implemented. The boost converter was run using only a natural convection based cooling at switching frequencies between 25 and 75 kHz without an external gate resistor to limit the switching speed of

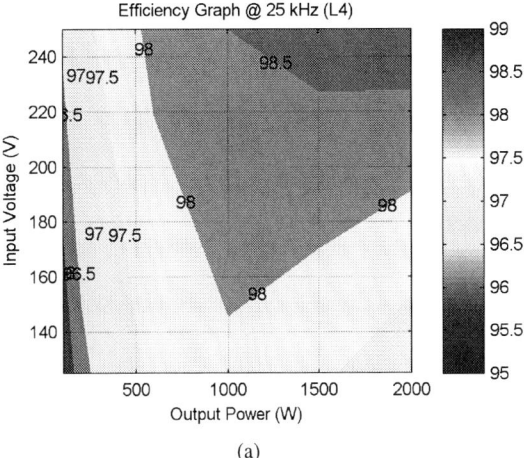

(a)

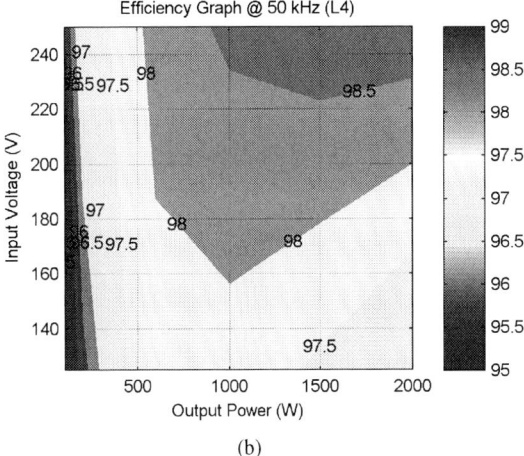

(b)

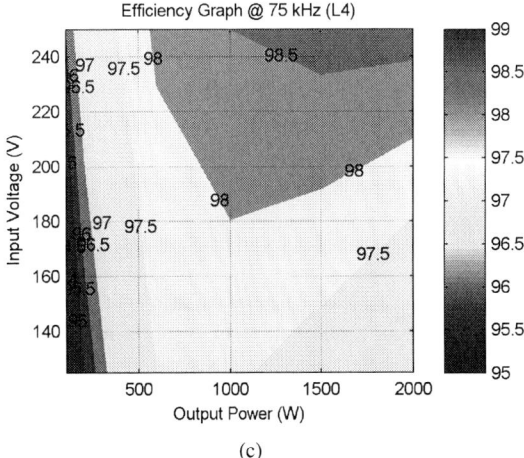

(c)

Fig. 6: Efficiency graphs of the SiC boost converter demonstrator with inductor L4 and 400-V output voltage: 25-kHz switching frequency (a), 50-kHz switching frequency (b), and 75-kHz switching frequency (c).

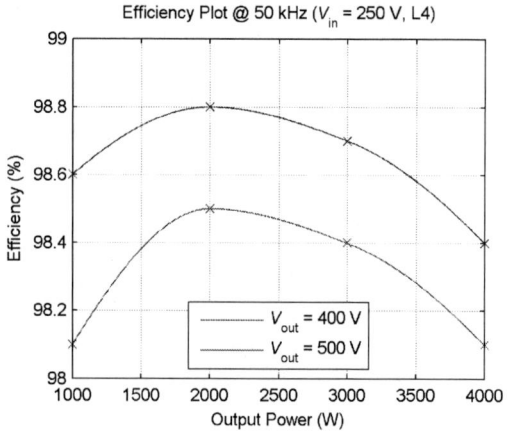

Fig. 7: Measured efficiencies of the SiC boost converter demonstrator with output power up to 200 % of the rated power ($f_{sw} = 50$ kHz, $V_{in} = 250$ V, L4).

the SiC MOSFET. The measured switching waveforms showed no excessive oscillations and the conversion efficiency was found promising.

Regarding the future work, there are still several tasks that have not been accomplished so far:

- Thermal resistance of the module has not been determined. Therefore the cooling system for the SiC boost converter demonstrator could not be optimized.

- The thermocouple inside the module has not been put to use and tested, because of which the chip temperature could not be measured during the test runs.

- EMI measurements have not been carried out to evaluate the impact of fast switching speeds on the EMI filtering requirements.

REFERENCES

[1] Infineon Technologies AG, *Datasheet of FF50R12RT4 dual-pack IGBT module*, Nov. 2013.

[2] Infineon Technologies AG, *Datasheet of IPW60R070C6 Cool-MOS™ power transistor*, Feb. 2010.

[3] P. Ning, F. Wang, and K. D. T. Ngo, "High-temperature SiC power module electrical evaluation procedure," *Power Electronics, IEEE Transactions on*, vol. 26, no. 11, pp. 3079–3083, 2011.

[4] H.-R. Chang, E. Hanna, and A. Radun, "Development and demonstration of silicon carbide (SiC) motor drive inverter modules," in *Power Electronics Specialist Conference, 2003. PESC '03. 2003 IEEE 34th Annual*, vol. 1, pp. 211–216, 2003.

[5] J. M. Hornberger, S. D. Mounce, R. Schupbach, A. Lostetter, and H. Mantooth, "High-temperature silicon carbide (SiC) power switches in multichip power module (MCPM) applications," in *Industry Applications Conference, 2005. Fourtieth IAS Annual Meeting. Conference Record of the 2005*, vol. 1, pp. 393–398, 2005.

[6] D. Katsis, B. Geil, T. Griffin, G. Koebke, S. Kaplan, G. Ovrebo, and S. Bayne, "Silicon carbide power semiconductor module development for a high temperature 10 kW ac drive," in *Industry Applications Conference, 2005. Fourtieth IAS Annual Meeting. Conference Record of the 2005*, vol. 1, pp. 399–403 Vol. 1, 2005.

[7] T. Salem, D. Urciuoli, R. Green, and G. Ovrebo, "High-temperature high-power operation of a 100 A SiC DMOSFET module," in *Applied Power Electronics Conference and Exposition, 2009. APEC 2009. Twenty-Fourth Annual IEEE*, pp. 653–657, 2009.

[8] D. Urciuoli, R. Green, A. Lelis, and D. Ibitayo, "Performance of a dual, 1200 V, 400 A, silicon-carbide power MOSFET module," in *Energy Conversion Congress and Exposition (ECCE), 2010 IEEE*, pp. 3303–3310, 2010.

[9] R. Wood and T. Salem, "Evaluation of a 1200-V, 800-A all-SiC dual module," *Power Electronics, IEEE Transactions on*, vol. 26, no. 9, pp. 2504–2511, 2011.

[10] T. Funaki, M. Sasagawa, and T. Nakamura, "Multi-chip SiC DMOSFET half-bridge power module for high temperature operation," in *Applied Power Electronics Conference and Exposition (APEC), 2012 Twenty-Seventh Annual IEEE*, pp. 2525–2529, 2012.

[11] Z. Chen, Y. Yao, D. Boroyevich, K. Ngo, P. Mattavelli, and K. Rajashekara, "A 1200 V, 60 A SiC MOSFET multi-chip phase-leg module for high-temperature, high-frequency applications," in *2013 CPES Annual Conference*, 2013.

[12] Y. Hinata, M. Horio, Y. Ikeda, R. Yamada, and Y. Takahashi, "Full SiC power module with advanced structure and its solar inverter application," in *Applied Power Electronics Conference and Exposition (APEC), 2013 Twenty-Eighth Annual IEEE*, pp. 604–607, 2013.

[13] S. Kicin and J. Hamers, "Assessment of selected materials and assembly technologies for PE modules with the capability to operate at high temperatures," in *15th European Conference on Power Electronics and Applications (EPE'13)*, 2013.

[14] S. Kjaer, J. Pedersen, and F. Blaabjerg, "A review of single-phase grid-connected inverters for photovoltaic modules," *Industry Applications, IEEE Transactions on*, vol. 41, no. 5, pp. 1292–1306, 2005.

[15] H. Haeberlin, "Evolution of inverters for grid connected PV-systems from 1989 to 2000," in *17th European Photovoltaic Solar Energy Conference*, Oct. 2001.

Development of Ultrahigh Voltage SiC Power Devices

Kenji Fukuda[1], Dai Okamoto[1], Shinsuke Harada[1],Yasunori Tanaka[1], Yoshiyuki Yonezawa[1],
Tadayoshi Deguchi[1], Shuji Katakami[1,2], Hitoshi Ishimori[1], Shinji Takasu[1], Manabu Arai[1,2],
Kensuke Takenaka [1], Hiroyuki Fujisawa[1], ManabuTakei[1,2], Kazushi Matsumoto[1], Naoyuki Ohse[1],
Mina Ryo[1], Chiharu Ota[4], Kazuto Takao[4], Makoto Mizukami[4], Tomohisa Kato[1],
Toru Izumi[5], Toshihiko Hayashi[5], Koji Nakayama[5], Katsunori Asano[5], Hajime Okumura[1],Tsunenobu Kimoto[6]

[1]Advanced Power Electronics Research Center, National Institute of Advanced Industrial Science and Technology, AIST Tsukuba
Central 2, 1-1-1 Umezono, Tsukuba, Ibaraki 305-8568, Japan, k-fukuda@aist.go.jp
[2]New Japan Radio Co. Ltd., 2-1-1 Fukuoka, Fujimino, Saitama 356-8510, Japan
[3]Corporate R&D Headquarters, Fuji Electric Co. Ltd., 1 Fujimachi, Hino, Tokyo, Japan
[4] Toshiba Corporation Corporate Research & Development Center 1,
Komukai Toshiba-cho, Saiwai-ku, Kawasaki 212-8582 Japan
[5] Power Enginnering R&D Center,Kansai Electric Power Co., Inc.,3-11-20 Nakoji, Amagasaki, Hyogo, Japan
[6]Department of Electronic Science and Engineering Kyoto University, A1-301 Katsura, Nishikyo, Kyoto 615-8510, Japan

Abstract— Ultrahigh voltage SiC devices and their package technology were investigated. As a result, we have succeeded in creating a 13kV level PiN diode without forward voltage degradation by using 4° off substrates and a 15kV level p-channel IGBT and 16kV level n-channel IGBT with a low differential specific on-resistance ($R_{diff,on}$). Moreover, the results reveal that the nano-tech resin, improved resin and Si_3N_4 DBC substrate are the best materials for package at high temperature and ultrahigh voltage.

Keywords— SiC, ultrahigh voltage, PiN, n-IGBT, p-IGBT, package technology, resin, ceramics DBC substrate

I. INTRODUCTION

Ultrahigh voltage SiC devices must be necessary to realize power electric components such as SVC (Static Var Compensator), SVR (Step Voltage Regulator), LBC(Loop Balance Controller), STATCOM(Static Synchronous Compensator), which help create smart grid systems and HVDC (High Voltage Direct Current) transmission systems for a low carbon emission society. Because the breakdown electric field of SiC is 10 times higher than that of Si, the thickness of the drift layer of the SiC power devices can be 1/10. This thin drift layer leads the low on resistance. Hence, SiC is suitable for ultrahigh voltage devices with blocking voltages higher than 10 kV and low on-resistance[1]. Moreover, package technology for power modules with ultahigh voltage SiC devices must be developed. In this study, fabrications method and evaluation results of SiC PiN diodes, p-channel IGBTs, n-channel IGBTs and their package technologies are reported.

This research is supported by a grant from the Japan Society for the Promotion of Science (JSPS) through the "Funding Program for World-Leading Innovative R&D on Science and Technology (FIRST Program)," initiated by the Council for Science and Technology Policy (CSTP). Part of this work has been implemented under a joint research project of Tsukuba power Electronics Constellations (TPEC). We would like to acknowledge Tokyo Electron Limited for supporting thick epitaxial wafer.

II. >10KV SiC PiN DIODE [2],[3]

A. Device structure

Figure 1 shows the cross section of fabricated SiC PiN diode with a two-zone junction termination extension (JTE). The JTE structures were formed by Al^+ ion implantation at 500 °C followed by activation annealing in Ar at 1620 °C. Al/Ni and Ni/Ti were employed as ohmic contacts on the anode and cathode. The doping density and thickness of the n-drift layer are 5×10^{14} cm^{-3} and 120 μm, respectively.

B. Electrical characteristics

At first, the effect of the off-angle of a substrate on forward degradation was investigated as shown in figures 2 and 3. There is a large forward degradation for the PiN diode fabricated on the 8° off substrate, which is caused by basal plane dislocations. On the other hand, in the case of 4° off substrate, there are very few the basal plane dislocations and the forward degradation is sufficiently suppressed. Figure 4 shows the temperature dependence of forward characteristics of large capacity PiN diode with $R_{diff,on}$ = 12 mΩ cm^2, and there is forward current higher than 40 A at the voltage of 5 V at room temperature. As temperature rises, the current increases. This is due to the increase of the carrier life time in the drift layer. The blocking voltage of the fabricated PiN diode was 13.7 kV as shown in figure 5. Figure 6 shows the temperature dependence of the reverse recovery characteristics measured using the package technology that will be described in section IV. This result reveals that the reverse recovery

The 2014 International Power Electronics Conference

current increases at high temperature, which is also caused by the increase in the carrier life time.

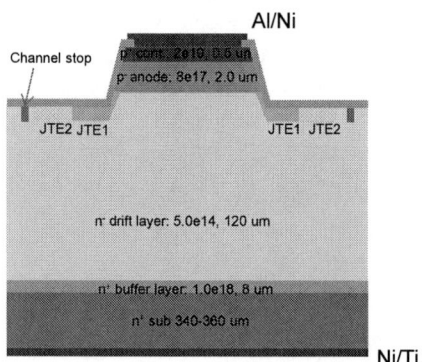

Fig.1. Schematic cross section of fabricated PiN diode.

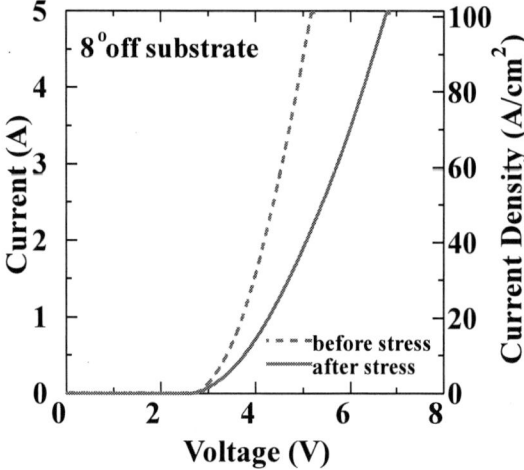

Fig.2. Forward degradation phenomena of PiN diode fabricated on 8° off substrate

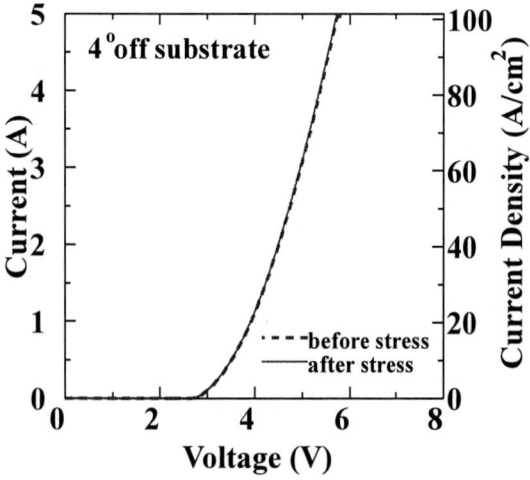

Fig.3. Forward degradation phenomena of PiN diode fabricated on 4° off substrate

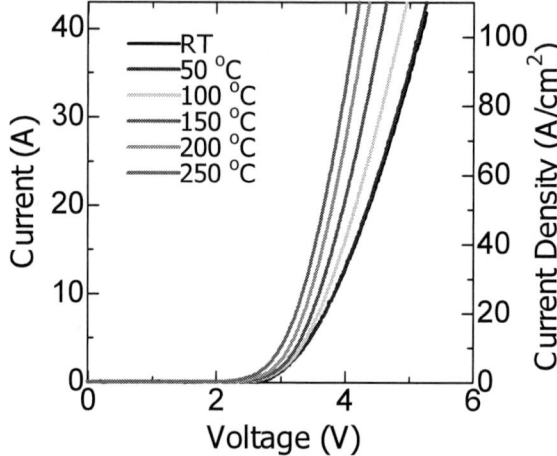

Fig. 4. Forward characteristics of fabricated large capacity PiN diode. Chip size is 8x8mm^2

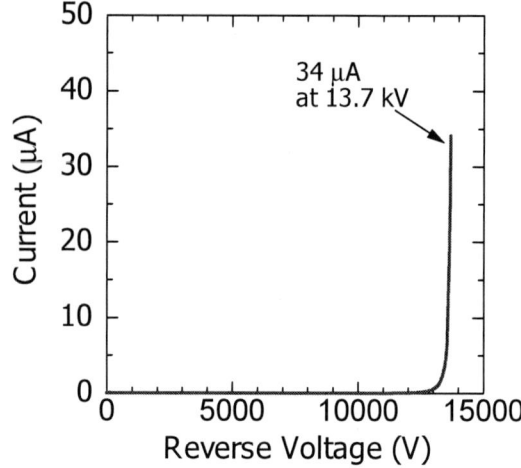

Fig.5. Reverse characteristics of fabricated large capacity PiN diode. Chip size is 8x8mm^2.

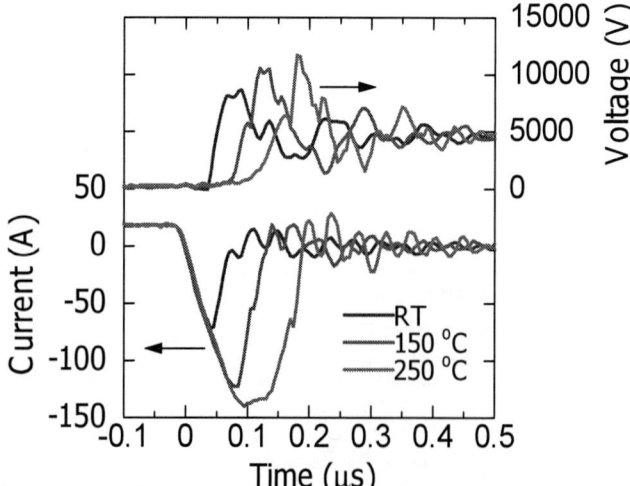

Fig. 6. Reverse recovery characteristics of SiC-PiN diode.

978-1-4799-2706-7/14 $31.00 © 2014 IEEE   3441

III. P-CHANNEL SiC-IGBT[4]

A. Device structure

P-channel SiC-IGBT can be easily manufactured without complicated device processes involving backside-grinding and ion implantation to eliminate the effects of high resistivity p-type substrates for the n-channel SiC-IGBT. Therefore, at first, the development of p-channel IGBT was carried out. Figure 7 shows a schematic cross section of a fabricated 4H-SiC p-channel IGBT. The device utilizes a 152μm-thick p-type drift layer, with a 5.2×10^{14} cm^{-3} doping density, and a 1.0-μm-thick p-type buffer layer with a 1.0×10^{17} cm^{-3} doping density. A p-type charge storage layer (CSL) and an n-well with retrograde profiles were formed by ion implantation of Al and N ions, taking into account the fact that the channel mobility is strongly dependent on the doping density. A p-type JFET region with a box profile (1.5×10^{16} cm^{-3}) was formed by additional Al ion implantation. All the implants were activated at 1650 °C for 10 min. Detailed process flow afterwards are described elsewhere [5,6]. A double-zone JTE was formed around the device periphery to obtain a high blocking voltage. The JTE length and doping concentration were designed using a TCAD simulation. The total JTE length and the inner JTE doping density were 500 μm and $1.1 \sim 1.2 \times 10^{13}$ cm^{-2}, respectively.

B. Electrical characteristics

Figure 8 shows the gate voltage dependence of field-effect mobility (μ_{FE}) for p-channel MOSFETs with various gate oxides [5]. The 50-nm-thick gate oxides were grown in dry O_2 atmosphere at 1200 °C followed by wet re-oxidation annealing at 1100 °C, 950°C and N_2O annealing at 1200 °C. As a result, the highest μ_{FE} peak of 13.5 cm^2/Vs can be obtained by gate oxide grown in dry O_2 atmosphere at 1200 °C followed by wet re-oxidation annealing at 1100 °C. This gate oxide condition was applied to fabrication of the 13kV level p-IGBTs. Figure 9 shows Temperature dependence of the μ_{FE} and threshold voltage (V_{th}) of the p-channel SiC-MOSFET. The μ_{FE} reveals the highest value at around 75 °C. The V_{th} decreases with increasing temperature. However, the V_{th} is -6V at even 200°C, the p-channel MOSFET is normally-off type FET at high temperature.

Figure 10 shows the photograph of the top view of fabricated p-IGBT. The chip size, active area, and L_{JFET} of the device were 2.9 mm × 2.9 mm, 0.022 cm^2, and 3 μm, respectively. Figure 11 shows the blocking characteristics of p-IGBTs (active: 1.0×10^{-3} cm^2). The gate was grounded to the emitter for the measurement of the blocking voltage at room temperature (R.T.). A maximum blocking voltage of over 15 kV (the power supply limit) was achieved for an L_{JFET} of 2 μm for a p-IGBT with a low leakage current density of 0.4 μA/cm^2. Figure 12 shows the pulsed on-state collector current ($-I_C$) versus the collector voltage ($-V_C$) of a SiC p-IGBT at RT and 250 °C. The low forward voltage drop (V_F) and $R_{on,diff}$ were observed to be −8.5 V and 33 mΩ-cm^2 (I_C = 2.25 A, J_C = −100 A/cm^2, V_G = −20 V) by increasing the

temperature to 250 °C, whereas these were high ($R_{on,diff}$ = 148 mΩ cm^2) at R.T. The strong temperature dependence could exist because the CSL was formed by ion implantation. That is, at low temperatures, the on-state characteristics of the p-IGBT can be affected by lower lifetimes due to crystal damage by ion implantation, which decreases the effect of conductivity modulation. In addition, low carrier lifetimes of 0.75 μs at RT could be also one reason for the high V_F and $R_{on,diff}$ at RT.

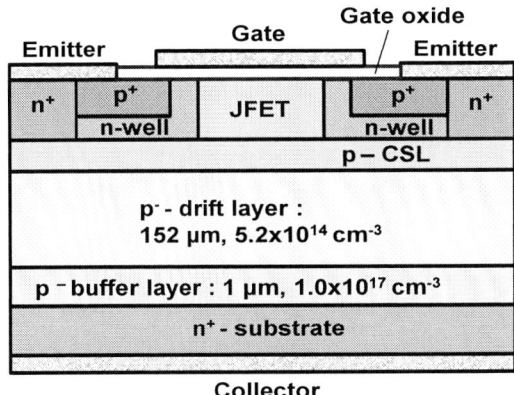

Fig. 7. Schematic cross section of a fabricated 4H-SiC p-channel IGBT.

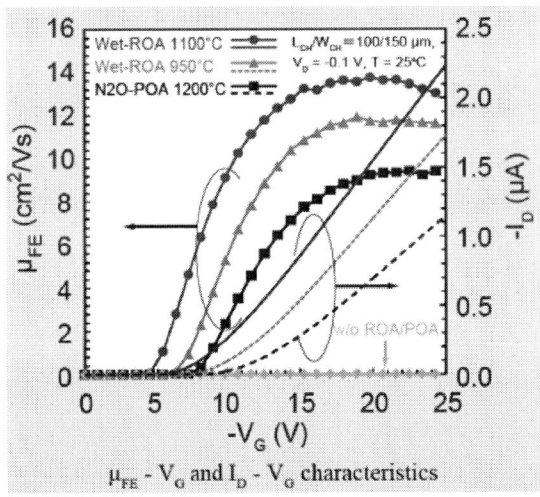

Fig. 8. Field-effect mobility as a function of Vg of the SiC-MOSFET with the gate oxide grown under various conditions.

The 2014 International Power Electronics Conference

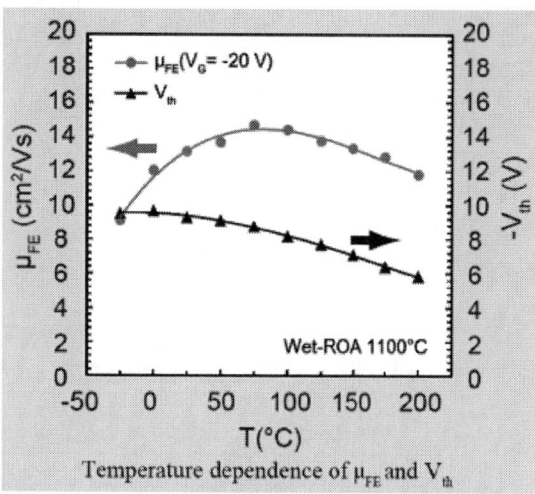

Fig. 9. Temperature dependence of field-effect mobility and threshold voltage of the p-channel SiC-MOSFET with the gate oxide grown by dry oxidation following by re-oxidation in wet atmosphere.

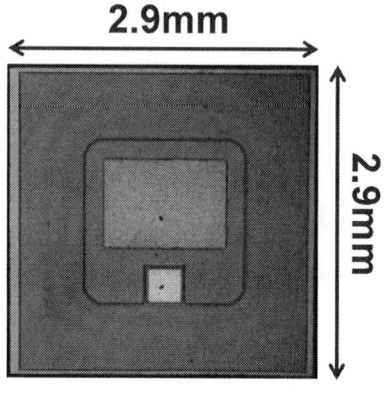

Fig. 10. Photograph of top view of the fabricated p-channel IGBT.

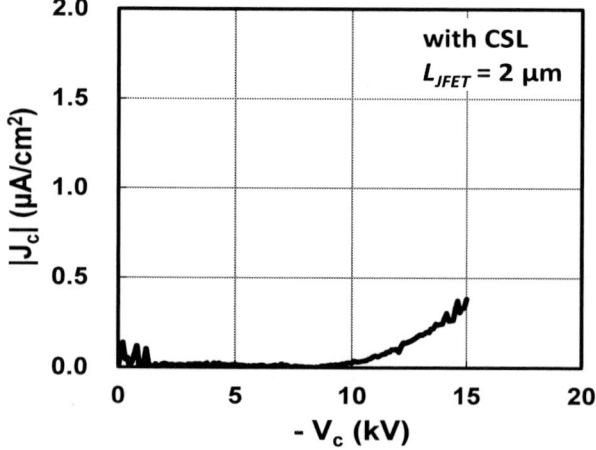

Fig. 11. Blocking characteristics of a fabricated p-channel IGBT.

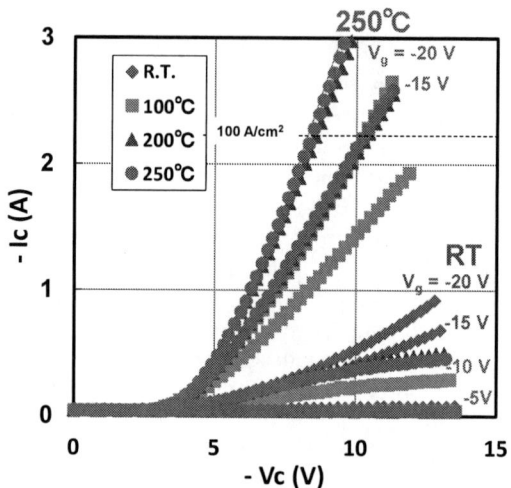

Fig. 12. Forward characteristics of a fabricated 4H-SiC p-channel IGBT.

IV. N-CHANNEL SiC-IGBT[6]

A. Device structure and fabrication flow

Fig. 13 shows a schematic cross section of the fabricated IE-IGBT. In order to improve the low channel mobility in SiC-MOSFETs, we proposed that implantation and epitaxial MOSFET (IE-MOSFET) using the 4H-SiC (000-1) carbon face, which has a channel mobility greater than 100 cm^2/Vs, considerably higher than that achievable with the 4H-SiC (0001) silicon face (20–30 cm^2/Vs). The bottom and top of the p-well of IE-MOSFET are formed by ion implantation and epitaxial growth, respectively. Thus, the surface of the p-well is a smooth and channel mobility is higher as compared to the conventional double implantation type MOSFET. Fig. 14 shows the process flow of a flip-type (000-1) carbon face substrate. Firstly, an n-type drift layer thicker (4×10^{14}cm^{-3}) than 150 μm with nitrogen doping was grown on the (0001) silicon face of an n^{++} substrate. Then, a buffer layer was grown. After the p$^+$ collector with an aluminum-doped layer was formed, the p^{++} substrate layer (2×10^{19} cm^{-3}) was grown to a thickness of over 100 μm. Then, we turned the substrate over (flip), removed the n^{++} substrate, and polished the surface using the CMP process. A charge storage layer (CSL) with a nitrogen concentration of 1.5×10^{16} cm^{-3} was grown for increasing the charge storage effect in the drift layer. The device structure was optimized by TCAD simulation. In particular, the IE-IGBTs were fabricated with the JFET width from 1.2 to 3 μm in order to investigate the JFET width dependence of blocking voltage and V$_f$. The p$^+$-base, aluminum ions (Al$^+$) were implanted for forming the bottom of the p-well. Then, a 0.5μm-thick p$^-$ epitaxial layer was grown as the top p$^-$ layer. The JFET region was formed by nitrogen ion (N$^+$) implantation into the p$^-$ layer. A two-zone junction termination extension (JTE) structure was used for termination. The total termination length and the inner JTE

978-1-4799-2706-7/14 $31.00 © 2014 IEEE 3443

The 2014 International Power Electronics Conference

dose are 750 μm and 1.4 x 10^{13} cm^{-2}, respectively. The gate oxide was grown by dry oxidation following by post oxidation annealing in hydrogen rich atmosphere. Finally, emitters and collectors were formed.

B. Electrical characteristics

Figure 15 shows the JFET width dependence of the blocking voltage. The blocking voltage decreases with increasing the JFET width. The V$_f$ is almost independent of JFET width. Therefore, the JFET width of 1.6 μm was selected in order to get more than 15kV the blocking voltage. Figure 16 shows the blocking characteristics of fabricated the 5.3mm x 5.3 mm IE-IGBT. The blocking voltage is 16.5kV. Figure 17 shows the pulsed on-state I-V characteristics of the 5.3mm x 5.3 mm IE-IGBT. The differential-specific on–resistance was 15 mΩcm^2 and the V$_f$ at 100 A/cm^2 is 5.2 V. It is reported that long oxidation times and C implantation are effective for enhancing carrier lifetimes to more than 10 μs . Moreover, If the process temperature is controlled under low one. The V$_f$ of IE-IGBTs might be low, and it reveals good electrical performance. Figure 18 shows the V$_{th}$ shift at gate voltages of +30 V and -30 V at 200 °C. Those are approximately 0.15 V and nearly 0 V, respectively, after 1000s. The novel wet gate oxidation process carried out in hydrogen rich atmosphere sufficiently suppressed the V$_{th}$ shift. This might mean that hydrogen atoms can effectively terminate interface states in SiO$_2$/SiC.

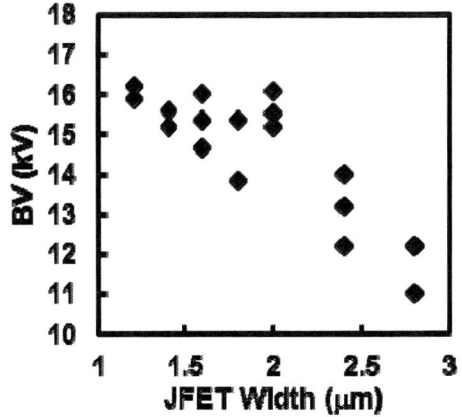

Fig. 15. Blocking voltage of the fabricated TEG of IE-IGBTs as a function of JFET width.

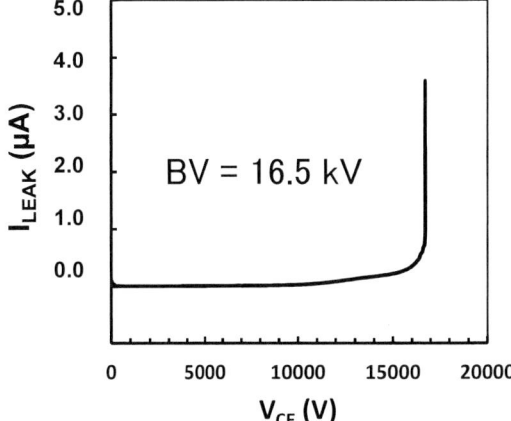

Fig.16. Blocking voltage of the fabricated IE-IGBT.

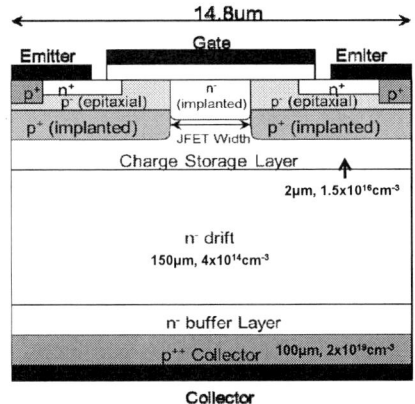

Fig. 13. Schematic cross section of fabricated IE-IGBT.

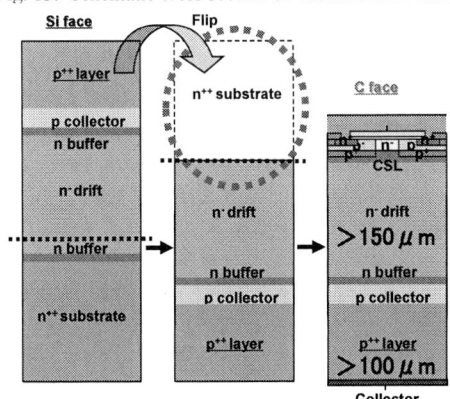

Fig 14. Fabrication process flow of flip-type IE-IGBT.

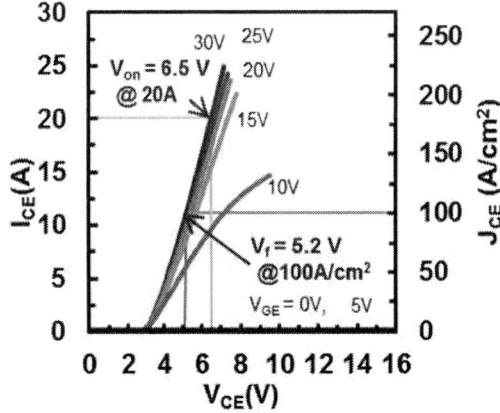

Fig. 17. Forward characteristics of fabricated IE-IGBT.

978-1-4799-2706-7/14 $31.00 © 2014 IEEE

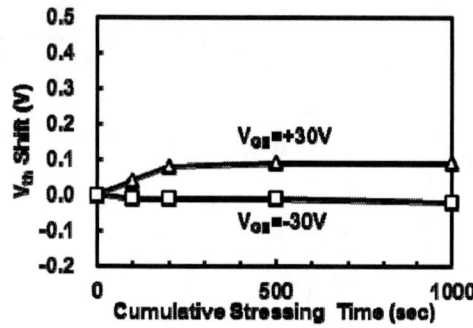

Fig. 18. Vth shift as a function stress time under the gate voltage $=\pm30\text{V}$ at 200 °C.

IV. PACKAGE TECHNOLOGY

A. Resin[7]

Figure 19 shows an Si_3N_4 substrate with two electrodes which used to evaluate resins. Three types of resin were evaluated in order to investigate the long-term stability of packages with high heat-resistant resins for ultrahigh voltage SiC devices at high temperature. Resin A and Resin B were produced by ADEKA Corporation. Resin C is the commercial available one. The electrical insulation characteristics for the resins were measured by applying a high voltage of 20 kV between the electrodes at room temperature. Resin was mounted to the substrate with a thickness of 5 mm. The gas was removed from the resins in a vacuum, and then the resins were thermally cured. Figure 20 shows the ratio of the number of samples with a blocking voltage higher than 20 kV to the total number of the samples as a function of the storage time at 250 °C. There are no samples that can have the blocking voltage higher than 20 kV more than 200 h for samples covered with Resin C. On the other hand, all samples covered with Resin A and Resin B can keep the blocking voltage higher than 20 kV at least until 500 h. The storage time dependence of the penetration length was measured in order to clarify the reason as shown in Figure 21. The penetration length becomes short because of the hardening of the resin, which is generally caused by oxidation. The penetration length of Resin C is the shortest among the three resins. This means that the reduction of the blocking voltage is caused by occurrence of cracks due to the strong hardening of resins.

Fig. 19. Optical photograph of an Si_3N_4 substrate.

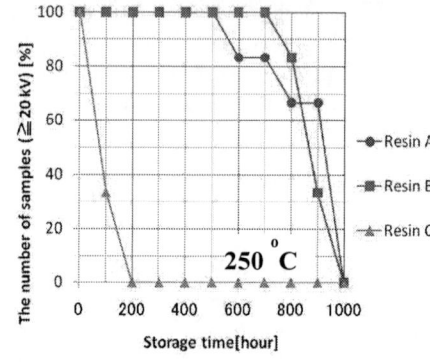

Fig. 20. The ratio of the number of samples with a blocking voltage of more than 20 kV to the total number of the samples.

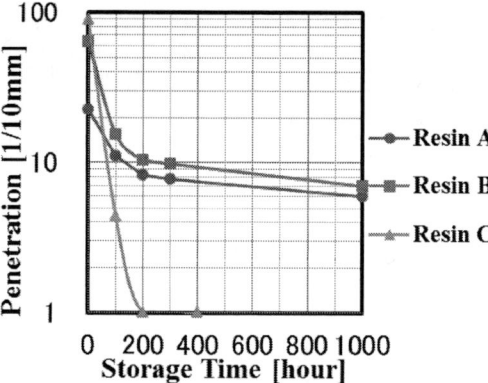

Fig. 21. Storage time dependence of penetration of three resins.

B. Ceramics DBC substrate[8]

Figure 22 and Table I show an optical photograph of the prototype package and the designed parameters of the fabricated packages, respectively. Figure 23 shows the leakage current of sample A with an Si_3N_4 DBC substrate as a function of the applied voltage at room temperature and 250 °C. The leakage current of sample A does not increase up to 20 kV at room temperature and 250 °C, which means it is suitable for high temperature operation. On the contrary, the leakage current of sample B with a AlN DBC substrate at room temperature is very low, however, at 250 °C, the leakage current increases around 0 V. The thermal conductivity of AlN is higher than that of Si_3N_4, which indicates an advantage as an insulator of heat spread for high-power SiC devices operating at high temperature. However, the AlN is not suitable as the DBC substrate of package of ultrahigh voltage SiC devices at high temperature because of large leakage current at 250 °C. Therefore, the package for an ultrahigh voltage SiC PiN diode was fabricated using a high heat-resistant resin as mentioned in session A and a Si_3N_4 DBC substrate. Finally, we succeeded in the measurement of the reverse recovery waveforms of the SiC PiN diode with a reverse bias voltage of 10 kV, as shown in figure 24.

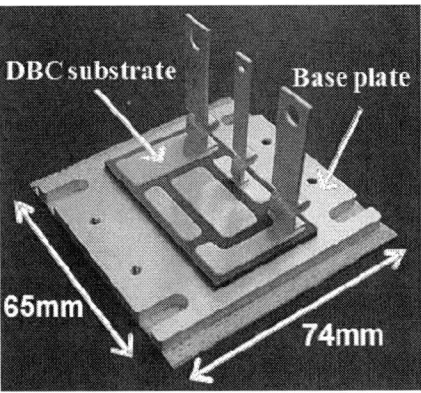

Fig. 22. Prototype package of 13kV level ultrahigh voltage SiC devices.

TABLE I

DESIGNED PARAMETERS OF FABRICATED PACKAGE

	DBC substrate		Base plate	
	Ceramics	thickness (mm)	Metal	thickness (mm)
Sample A	Si_3N_4	1.0	W	4.0
Sample B	AlN	1.5	Cu-Mo	4.0

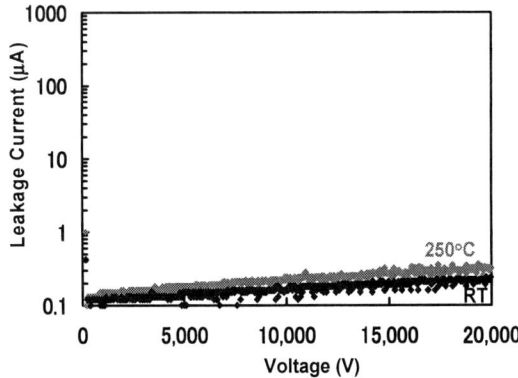

Fig. 23. Leakage current of sample A as a function of applied voltage at room temperature and 250 °C.

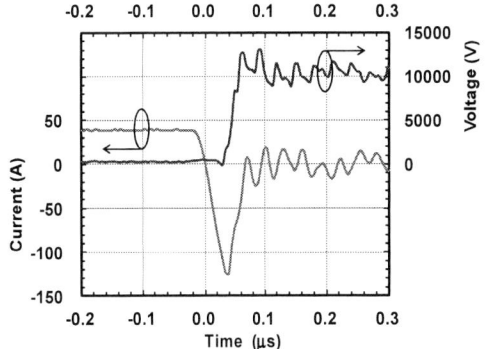

Fig. 24. Reverse recovery waveforms of the ultra-high-voltage SiC diode module with reverse bias voltage of 10 kV.

V. CONCLUSION

The ultrahigh voltage SiC bipolar devices of 13kV class PiN diodes and p-/n-channel IGBTs were fabricated, and their properties were characterized. The PiN diode succeeded in a large capacity of more than 40 A with no forward degradation. The P-channel IGBT revealed that $R_{diff,on}$ improved with increasing temperature, becoming 33 $m\Omega cm^2$ at 250 °C, which was caused by an increase in the carrier lifetime and an enhancement of the charge storage effect by the CSL. The flip type n-channel IE-IGBT with 20A and the 16kV blocking voltage was demonstrated. The $R_{diff,on}$ = 15 $m\Omega cm^2$ and V_f=5.2V, which is good performance. Moreover, a package using high heat-resistant resin and a Si_3N_4 DBC substrate was demonstrated. This technology can be used for a smart grid system and HVDC transmission systems, which will lead to a low carbon emission society.

REFERENCES

[1] S. Ryu, et al., Materials Science Forum Vols. 717-720, p.1135 (2012)

[2] D. Okamoto et al., "Development of High-Voltage 4H-SiC PiN Diodes on 4° and 8° Off-Axis Substrates" Mat. Sci. Forum 740-742, (2013) 907.

[3] D. Okamoto et al., "13kV, 20A 4H-SiC PiN Diodes for Power System Applications" the proceedings of ICSCRM2013.

[4] T. Deguchi et al., "Effect of Current-Spreading Layer Formed by Ion Implantation on the Electrical Properties of High-Voltage 4H-SiC p-channel IGBTs" to be presented in ICSCRM 2013.

[5] S. Katakami et al., "Fabrication of a P-channel SiC-IGBT with High Channel Mobility" Materials Science Forum Vols. 740-742 (2013) 958-961.

[6] Y. Yonezawa et al., "Low Vf and highly reliable 16 kV ultrahigh voltage SiC flip-type n-channel implantation and epitaxial IGBT", in Proc. of International Electron Devices Meeting (IEDM), 2013 IEEE International, 2013, pp. 6.6.1–6.6M2013.

[7] T. Hayashi ,T.Izumi, T.Hemmi and K.Asano "Insulating Properties of Package for Ultrahigh-Voltage, High-Temperature devices" Mat. Sci. Forum 740-742, (2013) 1036.

[8] T.Izumi, T.Hemmi, T.Hayashi and K.Asano, "Stability Investigation of High Heat-resistant resin unser High Temperature for Ultlahigh Blocking voltage SiC devices" Mat. Sci. Forum 740-742, (2013) 669.

The 2014 International Power Electronics Conference

High Switching Performance of 1.7kV, 50A SiC Power MOSFET over Si IGBT for Advanced Power Conversion Applications

Samir Hazra, Ankan De, and Subhashish Bhattacharya
FREEDM Systems Center
North Carolina State University
Raleigh, NC, USA
Email: {shazra, ade, sbhatta4}@ncsu.edu

Lin Cheng, John Palmour, Marcelo Schupbach,
Brett Hull, and Scott Allen
Power R&D, Cree, Inc.
4600 Silicon Dr., Durham, NC 27703, USA

Abstract—**Silicon Carbide (SiC) has wider band gap compared to Silicon (Si) and hence MOSFET made in SiC has considerably lower drift region resistance, which is a significant resistive component in high-voltage power devices. Due to low on-state resistance combined with its inherently low switching loss, SiC MOSFET is an excellent candidate for high power converter design. With its lower power loss and operation capability at higher switching frequency, power converters based on SiC MOSFETs can offer much improved efficiency and compact size compared to those using Si IGBTs. In this paper, we report switching performance of a new 1.7kV, 50A SiC MOSFET; designed and developed by Cree, Inc. Hard-switching losses of the SiC MOSFETs with different circuit parameters and operating conditions are measured and compared with the 1.7kV, 50A Si IGBTs, using the same test setup. Switching performance of the 1.7kV SiC MOSFET and 1.7kV SiC Schottky diode connected in series are also evaluated under a zero current switching (ZCS) condition and important findings are reported.**

Keywords—*SiC MOSFET, Si IGBT, Si BiMOSFET, device characterization, hard-switching, ZCS, switching losses.*

I. INTRODUCTION

With the advent of SiC technology, MOSFET ON-state resistance is reduced drastically and hence MOSFET device manufacturing for higher voltage and current has been encouraging [1], [2]. For high power converter application, even with hard-switched topology that is free of snubbers, SiC MOS-FETs provide various advantages over the well-established Si IGBTs. Due to multi-fold lower switching losses compared to Si IGBTs, SiC MOSFET based power converters can be operated at higher switching frequencies and higher junction temperatures. Therefore, the size of the passive filter elements in the circuit can be greatly reduced. Use of high-power SiC devices in Dual Active Bridge (DAB) [3] based high power density application is also beneficial due to the size reduction and efficiency improvement. Also, reverse recovery energy loss of SiC Schottky diode is almost zero compared to higher loss of Si-PiN diodes. The efficiency improvement of a hybrid device combining Si IGBT and SiC Schottky diode is evaluated in [4] and the products are commercially available for use in the market (Powerex, QID1210005). Detailed characterization

of 1.2 kV, 100 A SiC MOSFET module is reported in [5]. The two-Level Voltage Source Converter (2L-VSC) loss using such a SiC power module is compared with 1.2 kV, 100 A Si IGBT [5]. It is shown that the 1.2 kV, 100 A SiC MOSFET based converter can achieve higher efficiency and considerable reduction in the size of the heat sink.

With these new 1.7 kV, 50 A SiC MOSFETs, dc bus voltage of a 2L-VSC can go up to 1.2 kV which enables an extensive use of these SiC devices in applications such as wind, solar and motor drives at higher voltages with higher efficiencies. In this paper, the new 1.7 kV, 50 A SiC MOSFET is characterized and its switching losses are analyzed using different gate resistances at different temperatures. Turn ON and OFF dv/dt and di/dts are captured at different operating conditions and reported. Conduction and switching losses of the 1.7 kV, 50 A SiC MOSFET are compared with Si IGBTs for the similar I-V ratings. The conduction and switching losses of a 1.7 kV, 50 A Si IGBT (5SMY 12G1721) from ABB [6] and a 1.7 kV, 42 A BiMOSFET (IXBH42N170) from IXYS [7] are measured and compared with those of the SiC MOSFET reported in this work.

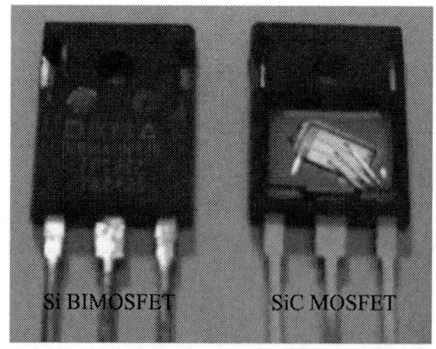

Fig. 1.   1.7 kV, 42 A Si BiMOSFET and 1.7 kV, 50 A SiC MOSFET.

The Si BiMOSFET is a hybrid device which is made by integrating an IGBT and MOSFET to optimize the switching loss and forward voltage drop [8]. As shown in Fig.1, both the 1.7 kV, 42 A BiMOSFET and the SiC MOSFET are packaged in TO247. The Si IGBT die characteristic data are taken

978-1-4799-2706-7/14 $31.00 © 2014 IEEE

The 2014 International Power Electronics Conference

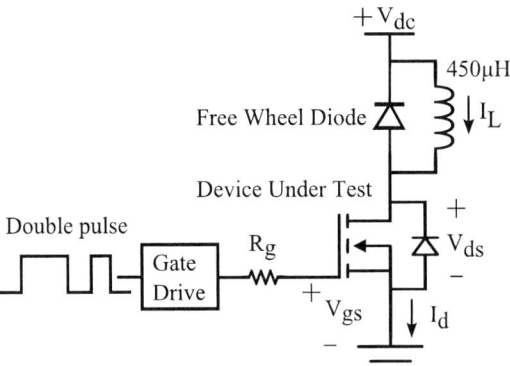

Fig. 2. Double pulse test circuit for device characterization under hard-switched condition.

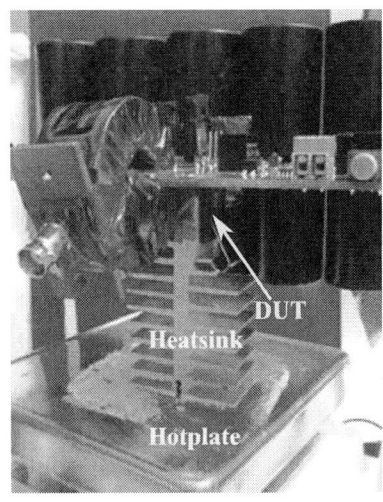

Fig. 4. Hard-switching characterization setup with device heating arrangements.

from the published data sheet [6] for comparison. While the Si BiMOSFET characteristics are evaluated in the same test circuit for a direct comparison of those of the SiC MOSFET.

A reverse voltage blocking switch made out of $1.7\,\text{kV}, 50\,\text{A}$ SiC MOSFET and $1.7\,\text{kV}, 50\,\text{A}$ SiC Schottky diode in series, is characterized for zero current switching operation. The diode is tested for soft turn-OFF. This test ensures almost zero loss of the SiC devices compared to non zero loss of the Si IGBTs which shows a voltage bump during soft turn-ON [9], [10]. Impact of circuit parasitic elements on device voltage stress in soft-switched circuit is also analyzed.

II. HARD-SWITCHING CHARACTERIZATION CIRCUIT

To characterize the switching performance of the $1.7\,\text{kV}, 50\,\text{A}$ SiC MOSFET under hard-switched condition, a double pulse (DP) test-bed is set up [5]. The equivalent test circuit is shown in Fig.2. The inductor used in the DP circuit is designed carefully to minimize terminal capacitance which can cause device current ringing or over-shooting during turn ON switching transient. For simplicity, an air core inductor is designed to handle higher current without saturation. Although, the number of turns increases for same inductance value for the air core, the conductor diameter can be smaller because

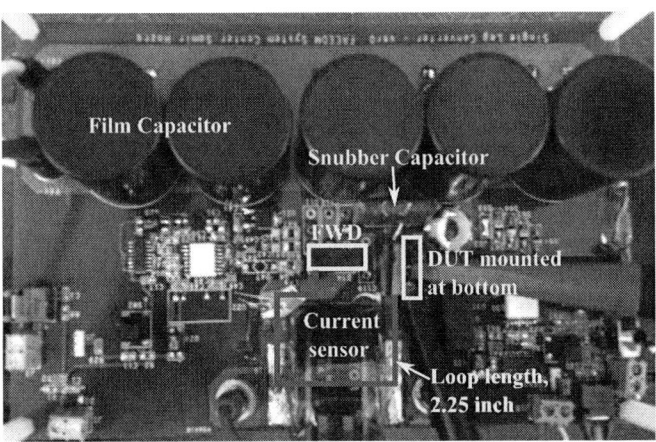

Fig. 3. Hard-switching characterization circuit board.

it is not required to be designed for high rms current but for handling current for very short time ($\sim 5 - 6\,\text{ms}$). However, time constant ($L/R = 4.5\,\text{ms}$) of the inductor should be high enough compared to the OFF period between first and second pulse ($\sim 10\,\mu\text{s}$) so that freewheeling current remains almost constant.

The $1.7\,\text{kV}, 50\,\text{A}$ SiC MOSFET and $1.7\,\text{kV}, 42\,\text{A}$ BiMOS-FET are inserted in the circuit as the device under test (DUT) and $1.7\,\text{kV}$ SiC Schottky diode is used to free wheel the inductor current as shown in Fig.2. The hardware test circuit board is shown in Fig.3. A current probe by Pearson Electronics (Model-3972) is used to measure the switch current. The total length of the current commutation path is $\sim 2.5\,\text{inches}$ resulting in $\sim 15\,\text{nH}$ of stray inductance. The device is mounted from bottom of the board as shown in Fig.4, to facilitate heating of the device for characterization under certain junction temperature. The device is attached to a heat sink which sits on a hot plate. By heating the hot plate, device case temperature can be brought to a desired steady state value. By placing hotplate temperature sensor near the device case the temperature can be controlled. Device case temperature is an indicative of the junction temperature in this DP testing condition as the energy loss for two pulses at the junction of the device is negligible to effect any further temperature rise.

III. HARD-SWITCHING CHARACTERIZATION OF 1.7KV, 50A SIC MOSFET

The $1.7\,\text{kV}, 50\,\text{A}$ SiC MOSFET is tested with the dc bus voltage of $1.2\,\text{kV}$ and maximum switch current of $40\,\text{A}$. At the end of the first pulse the inductor current rises to desired value and the switch is tested for turn-OFF characteristics. Inductor current free wheels through SiC Schottky diode till the start of the second pulse when the device is tested for hard-switching turn-ON. Gate voltage for SiC MOSFET is $+20\,\text{V}$ during turn-ON condition and $-5\,\text{V}$ during turn-OFF condition.

In Fig.5, the turn-ON characteristics of the device with

978-1-4799-2706-7/14 $31.00 © 2014 IEEE   3448

The 2014 International Power Electronics Conference

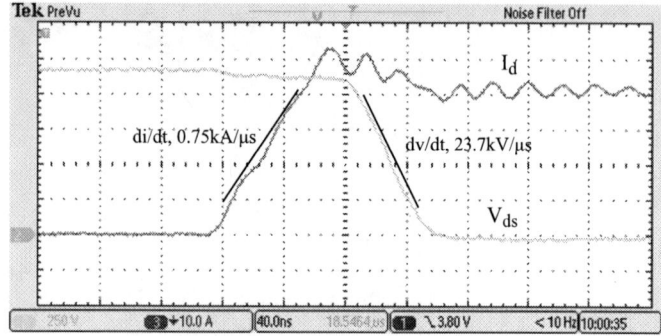

Fig. 5. Turn-ON characteristics with $R_g = 20\,\Omega$ at T_j=30 °C, Measured E_{on} = 4.52 mJ, Scale: V_{ds} =250 V/div, I_d = 10 A/div , Time = 40 ns/div.

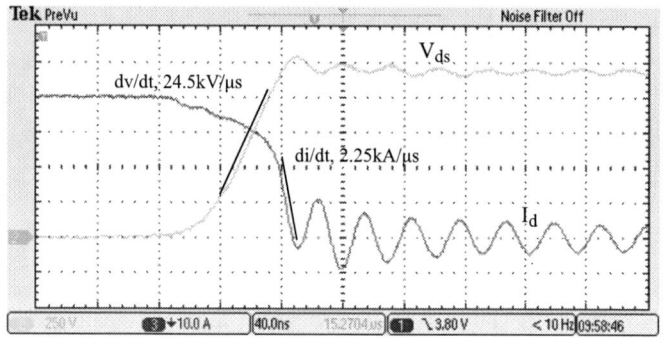

Fig. 6. Turn-OFF characteristics with $R_g = 20\,\Omega$ at T_j=30 °C, Measured E_{off} = 1.11 mJ, Scale: V_{ds} =250 V/div, I_d = 10 A/div , Time = 40 ns/div.

$20\,\Omega$ gate resistance is shown. Switching energy is measured by integrating the product of device voltage (V_{ds}) and device current (I_d). Measured energy loss (E_{on}) during turn-ON transient is 4.52 mJ. turn-OFF behavior of the device is shown in Fig.6 and measured turn-OFF loss (E_{off}) is 1.11 mJ. The turn-ON and OFF dv/dt are found to be $23.7\,\text{kV}/\mu s$ and $24.5\,\text{kV}/\mu s$ respectively. The turn-ON and OFF di/dt are $0.75\,\text{kA}/\mu s$ and $2.25\,\text{kA}/\mu s$ respectively.

Switching energy loss as a function of the switching current at dc bus voltage of $1.2\,\text{kV}$ is shown in Fig.7. This data can be

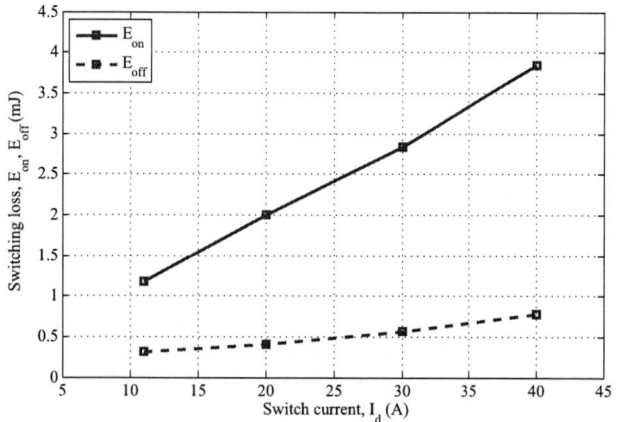

Fig. 7. Switching loss at different switch current with $R_g = 15\,\Omega$ at T_j=30 °C and dc bus voltage of $1.2\,\text{kV}$.

utilized for power loss computation of a hard-switched voltage source converter using $1.7\,\text{kV}, 50\,\text{A}$ SiC MOSFET. Both turn-ON and OFF gate resistance is kept at $15\,\Omega$ and the test is carried out at an ambient temperature of $30\,°\text{C}$.

Fig.8 shows the switching loss as a function of the dc bus voltage at $40\,\text{A}$ of switching current and a gate resistance of $15\,\Omega$ at ambient temperature of $30\,°\text{C}$. This data can be used to compute power losses in a hard-switched current source converter.

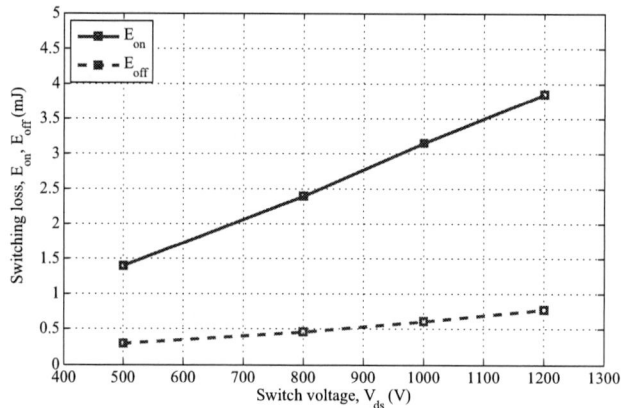

Fig. 8. Switching loss at different switch voltage with $R_g = 15\,\Omega$, T_j=30 °C and switch current I_d of 40A.

Switching energy loss with different gate resistances is shown in Fig.9. Switching loss increases with gate resistance mainly due to slower dv/dt and di/dt.

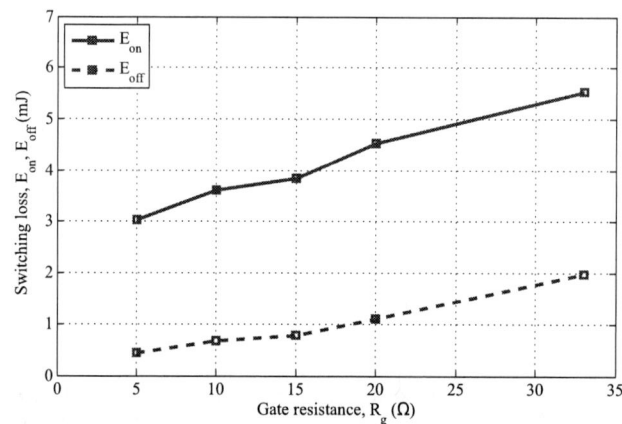

Fig. 9. Switching loss at different gate resistance, R_g with T_j=30 °C, V_{dc}=1.2kV and I_d=40A.

However, with lower gate resistance the dv/dt and di/dt in the converter increase which leads to an increment of common mode current and EMI issues in the system [11]. The turn-ON and OFF dv/dt variation versus gate resistance is plotted in Fig.10. Also, the turn-ON and OFF di/dt variation is plotted with the change of gate resistance as shown in Fig.11. The maximum dv/dt and di/dt found to be $36\,\text{kV}/\mu s$ and $3\,\text{kA}/\mu s$ respectively with the lowest gate resistance of $5\,\Omega$ used. It can be seen that the turn-ON and OFF dv/dt are almost same at different gate resistance. Whereas, turn-OFF

The 2014 International Power Electronics Conference

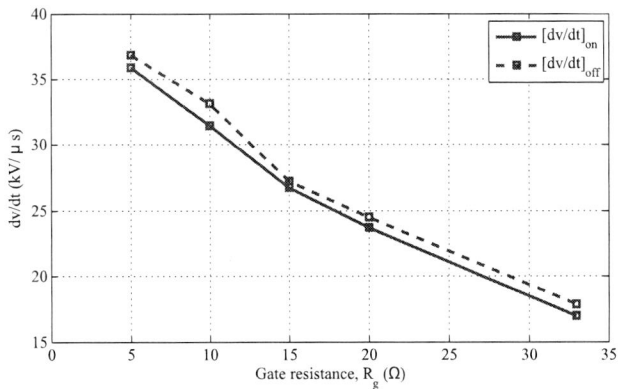

Fig. 10. Turn-ON and OFF dv/dt with different gate resistance, R_g at T_j=30 °C.

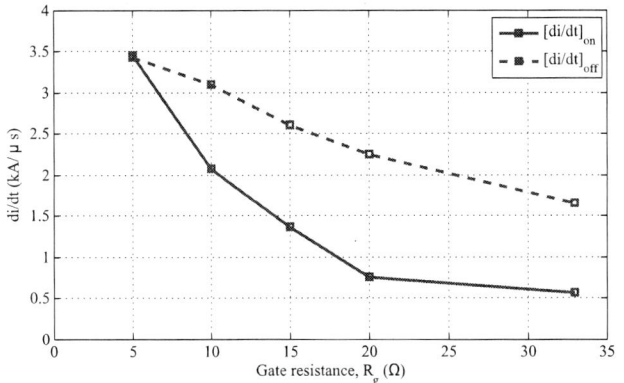

Fig. 11. Turn-ON and OFF di/dt with different gate resistance, R_g at T_j=30 °C.

di/dt is much higher compared to turn-ON di/dt with higher gate resistances. However, both turn-ON and turn-OFF di/dt increases with the decrease of gate resistance.

High dv/dt can increase the common mode current flow in the system. Common mode current flowing into the gate drive circuit through the coupling capacitance of the isolation transformer can cause significant voltage drop across the gate resistance and hence it can turn-ON a switch when it is in OFF state. In a circuit, dv/dt of one switch can cause problem for

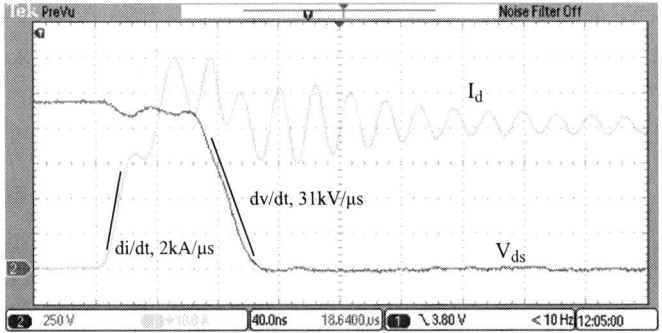

Fig. 12. Turn-ON characteristics with $R_g = 10\,\Omega$ at T_j=30 °C, Measured E_{on} = 3.6 mJ, Scale: V_{ds} =250 V/div, I_d = 10 A/div , Time = 40 ns/div.

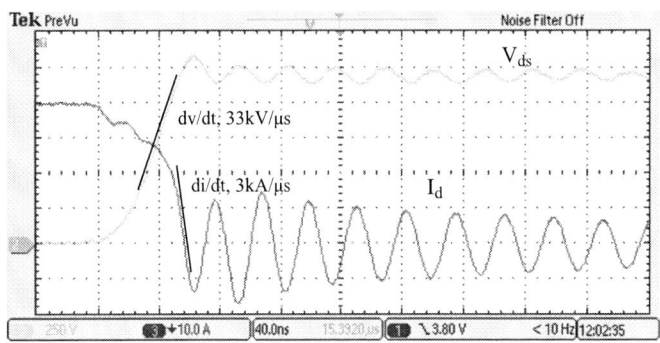

Fig. 13. Turn-OFF characteristics with $R_g = 10\,\Omega$ at T_j=30 °C, Measured E_{off} = 0.67 mJ, Scale: V_{ds} =250 V/div, I_d = 10 A/div , Time = 40 ns/div.

other switches. With the increase in di/dt, the current ringing increases resulting in significant voltage drop across common source inductor of the device. This forces the gate voltage to oscillate and if gate oscillation reaches gate threshold voltage then device turns ON. Therefore, to control turn-OFF di/dt OFF state gate resistance can be increased, which does not increase the turn-OFF loss significantly as the turn-OFF loss much lower compared to turn-ON loss of the SiC MOSFET.

From the switching energy loss and dv/dt and di/dt variations with respect to gate resistance, it is clear that a tradeoff between energy loss and acceptable dv/dt and di/dt values can be achieved by selecting ON and OFF state gate resistances properly.

In Fig.12 and Fig.13 the switch turn-ON and OFF behavior with gate resistance of $10\,\Omega$ are shown. It can be seen that the current ringing and voltage overshoot have increased significantly. However, in an actual converter the ringing can be controlled better by minimizing the current commutation loop. Here; in double pulse circuit, the insertion of a current probe increased the commutation loop length and introduced around 15 nH of inductance. Also, by placing low ESL snubber capacitor close to the devices can minimize the commutation loop length and can reduce the ringing substantially [5].

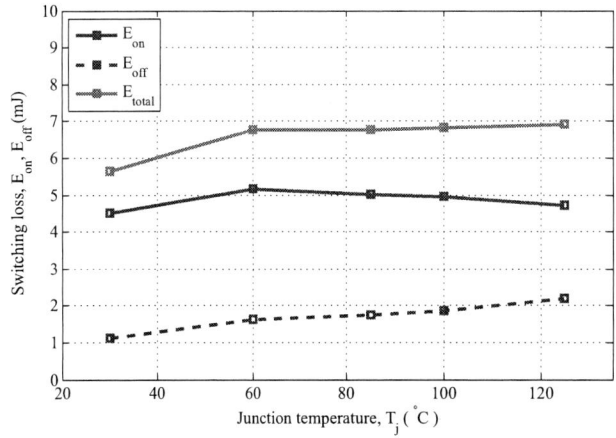

Fig. 14. E_{on} and E_{off} measured with $R_g = 20\,\Omega$ at $V_{ds} = 1.2\,kV$, $I_d = 40\,A$ at different junction temperatures, T_j.

Temperature dependence of switching losses are plotted in Fig.14. Turn-ON loss goes down slightly whereas turn-OFF loss increases with the increase of junction temperature. However, the total switching loss remains almost constant with temperature variation.

IV. HARD-SWITCHING CHARACTERIZATION OF 1.7KV, 42A SI BIMOSFET

BiMOSFET from IXYS is a Si device which combines the strength of IGBT and MOSFET. It is constructed by keeping MOSFET and IGBT in its structure [8]. Hence, it has an intrinsic diode in it. The BiMOSFET has lower switching losses than that of an IGBT but higher than a MOSFET. The forward voltage drop is lower than a MOSFET but higher than that of an IGBT. Therefore, BiMOSFET can be used for slightly higher switching frequency than IGBT [8]. In this work we chose BiMOSFET for comparison to show further multi-fold switching performance improvement by SiC MOSFET.

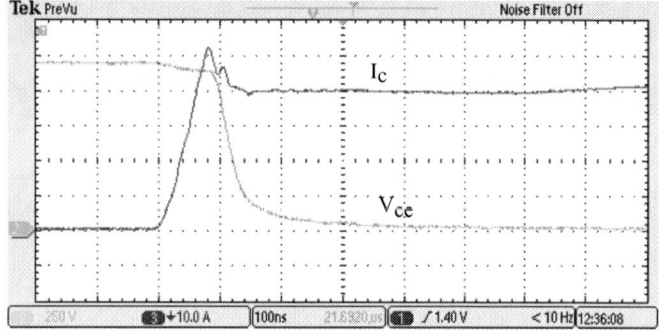

Fig. 15. Turn-ON characteristics with $R_g = 10\,\Omega$ at T_j=30 °C, Measured E_{on} = 4.81 mJ, Scale: V_{ce} =250 V/div, I_d = 10 A/div , Time = 100 ns/div.

Switching losses of a 1.7 kV, 42 A BiMOSFET is evaluated using the same circuit setup for a direct comparison with those of SiC MOSFET. The gate voltage applied to BiMOSFET is from $-15\,$V to $+15\,$V. Figs.15-16 show the BiMOSFET turn-ON and OFF characteristics using a gate resistance of $10\,\Omega$. Turn-ON transient is comparatively faster than the turn-OFF transient due to tail current effect. The turn-ON and turn-OFF transient times are 200 ns and 2 μs respectively as shown in Figs.15-16. Hence, the turn-OFF loss of the BiMOSFET, is

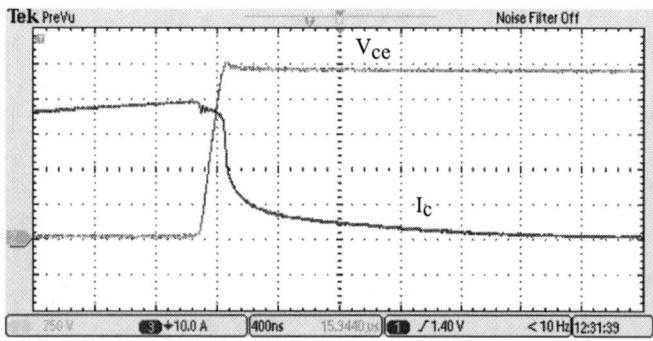

Fig. 16. Turn-OFF characteristics with $R_g = 10\,\Omega$ at T_j=30 °C, Measured E_{on} = 16.62 mJ, Scale: V_{ce} =250 V/div, I_d = 10 A/div , Time = 400 ns/div.

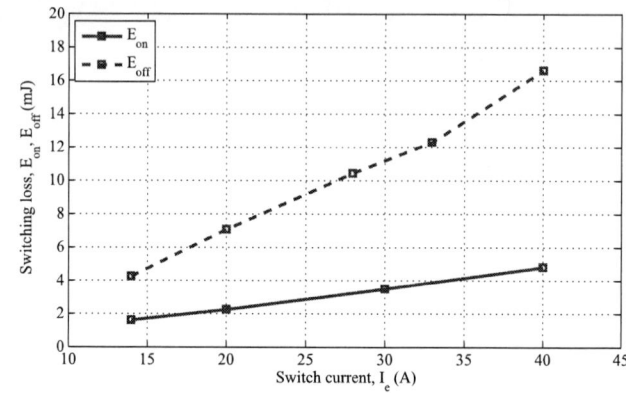

Fig. 17. Switching loss at different switch current with $R_g = 10\,\Omega$, T_j=30 °C and dc bus voltage (V_{dc}) of 1.2kV.

much higher compared to turn-ON loss like an IGBT due to current tailing.

The BiMOSFET switching loss is plotted as a function of switch current in Fig.17. The switching losses increase with the increase of switching current.

Switching losses of BiMOSFET is characterized with different gate resistances and the result is plotted in Fig.18. It is found that the turn-OFF loss is much higher than the turn-ON loss and remains almost unaffected by the change of gate resistance. Turn-ON loss increases with the increase of gate resistance. From Fig.18, it can be seen that the total loss does not decrease considerably with the decrease of gate resistance.

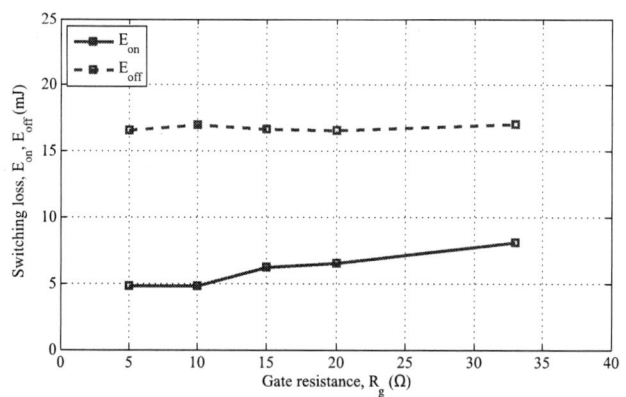

Fig. 18. Switching loss at different gate resistance, R_g with T_j=30 °C, V_{dc}=1.2kV and I_c=40A.

V. PERFORMANCE COMPARISON OF 1.7KV 50A SIC MOSFET, 1.7KV 50A SI IGBT AND 1.7KV 42A BIMOSFET

In this section the 1.7 kV, 50 A SiC MOSFET performance is compared with 1.7 kV, 50 A Si IGBT from ABB, 1.7 kV, 42 A BiMOSFET from IXYS. Si IGBT data is taken from its datasheet (5SMY 12G1721) [6]. The BiMOSFET performance data is evaluated from the double pulse circuit used for characterization of the SiC MOSFET.

The 2014 International Power Electronics Conference

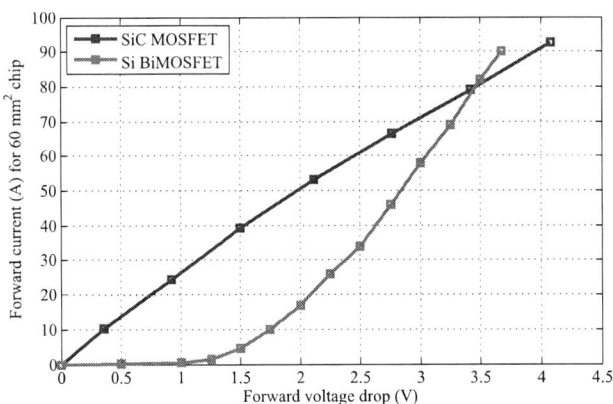

Fig. 19. 1.7 kV, 50 A SiC MOSFET and 1.7 kV, 42 A Si BiMOSFET V-I characteristics for same chip size (60 mm^2) at T_j=125 °C.

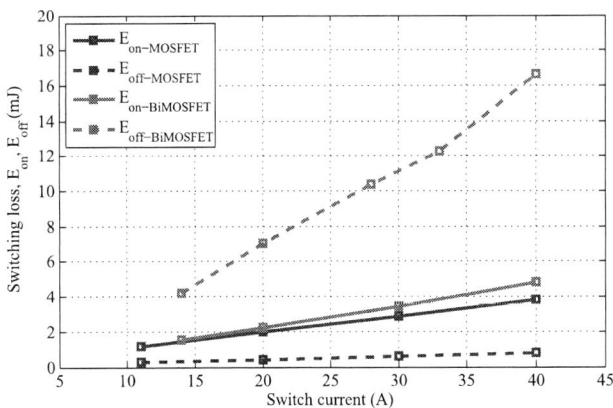

Fig. 20. 1.7kV, 50A SiC MOSFET (R_g= 15 Ω) and 1.7kV, 42A Si BiMOSFET (R_g= 10 Ω) switching loss comparison at 1.2kV, 40A.

Generally, due to conductivity modulation of drift region in bi-polar device, the on-state voltage drop of 1.7 kV, 50 A Si IGBT is quite low [12]. However, with the use of a wide band gap material like SiC, for the same break down voltage the drift region width of MOSFET is smaller. Hence, it offers considerably lower on-state voltage drop than the Si devices with the same chip sizes at the same current rating [8].

In Fig.19, the V-I characteristics of a 1.7 kV, 50 A SiC MOSFET at 125 °C are given. The V-I characteristics of 1.7 kV, 42 A Si BiMOSFET at 125 °C [7] also plotted in the same figure. The 1.7 kV, 50 A SiC MOSFET chip area is 30 mm^2 whereas the chip area of 1.7 kV, 42 A BiMOSFET is 60 mm^2. The forward characteristics are normalized for same chip area to have a better comparison. It can be seen that the SiC MOSFET has lower forward voltage drop for same chip size. The 1.7 kV, 50 A Si IGBT by ABB has a voltage drop of 3 V at 70 A at 125 °C for chip area normalized to 60 mm^2, which is almost same as SiC MOSFET at 70 A.

It also should be noted that ON-state voltage drop of SiC MOSFET at lower switching current is much lower than that of either Si IGBT or Si BiMOSFET, which is an indicative of higher low load efficiency of the power converters based on SiC MOSFET.

The switching losses in the SiC MOSFET and Si BiMOSFET are compared in Fig.20. It can be seen that the turn-OFF loss of BiMOSFET is much higher than SiC MOSFET. Also, the turn-ON and OFF switching losses of Si IGBT from ABB with gate resistance of 22 Ω are 14 mJ and 20 mJ respectively at the switching current and voltage of 50 A and 900 V respectively. Therefore, Si IGBT by ABB has higher switching loss compared to BiMOSFET, which is expected. So the total switching loss of SiC MOSFET from Cree, Inc. are significantly lower than that of both Si IGBT from ABB and BiMOSFET from IXYS.

From conduction loss analysis with normalized chip area, it is evident that SiC MOSFET has marginally lower ON-state voltage drop compared to both Si IGBT and Si BiMOSFET at rated current and it only improves at lower current. However, bigger gain comes in terms of switching loss reduction of

the power converters if SiC MOSFET is used. Hence, the efficiency of the converter increases and it reduces the heatsink size. Also, switching frequency of converter can be increased to higher value to design the system with smaller passive elements.

VI. SOFT-SWITCHING CHARACTERIZATION OF 1.7KV MOSFET AND 1.7KV SIC SCHOTTKY DIODE

Although the switching loss of SiC MOSFET is already very low under hard-switching operations, it can be more efficient to use SiC MOSFET in a soft-switched power converter [13]. A SiC MOSFET in series with a SiC Schottky diode forms a reverse voltage blocking switch, which can be used in current source converter. Here, such combination of MOSFET and diode is tested for zero current turn-ON and OFF of the switch and soft turn-OFF of the diode. The switching energy losses and device voltage and current stress in such circuit are characterized.

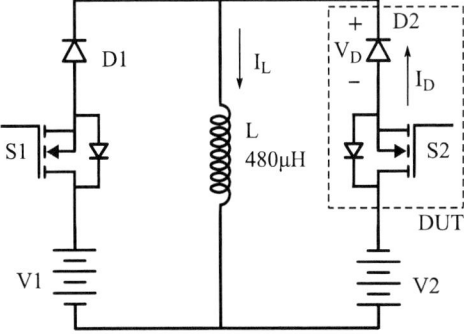

Fig. 21. Circuit for ZCS characterization of reverse voltage blocking switch.

The circuit shown in Fig.21 is used to test the device under zero current switching condition. The circuit operation for ZCS testing is depicted in Fig.22. When both $S1$ and $S2$ are in OFF condition diode $D2$ blocks voltage $V2$. The switch $S1$ is turned ON at t_1 and the current through the inductor (L) rises linearly. Blocking voltage of $D2$ increases to ($V1 + V2$). The switch $S2$ is turned ON at t_2 and this is zero current

978-1-4799-2706-7/14 $31.00 © 2014 IEEE   3452

The 2014 International Power Electronics Conference

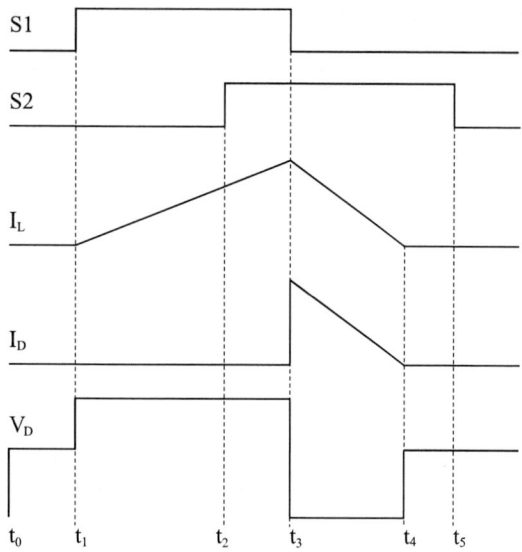

Fig. 22.   ZCS switching sequence and circuit operation.

turn-ON. At t_3 switch $S1$ is turned OFF and the current in the inductor starts flowing through $V2$, $S2$ and $D2$. The diode voltage drops to zero. There is no change in SiC MOSFET voltage as the current starts flowing through it. If the switch is Si IGBT then it shows a voltage bump when current starts flowing through it [9], [10]. Inductor current drops linearly as it feeds the power into voltage source $V2$. After the inductor current reaches zero the diode $D2$ turns off at t_4 and blocks the voltage $V2$. This is soft turn-OFF for the diode. After the diode is turned OFF the MOSFET $S2$ is turned OFF at t_5. This is again zero current turn-OFF for the MOSFET.

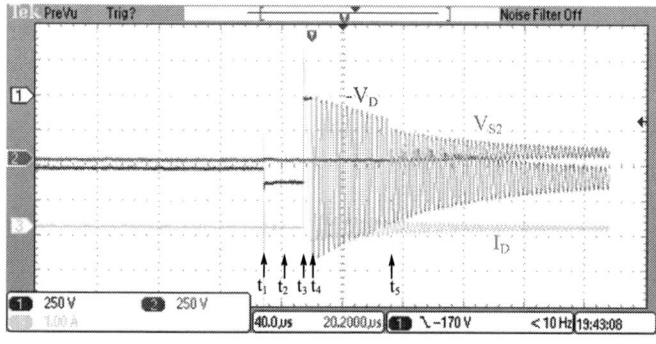

Fig. 23.   Device characteristics under ZCS operation.

Fig.23 shows the device characteristics under ZCS. It can be seen that the MOSFET turns ON at t_2 at zero current and loss is zero a MOSFET voltage V_{s2} remains zero. It also turns OFF at zero current at t_5. Fig.24 shows the zoomed view of diode turn-OFF which is a soft turn-OFF. It can be seen that the diode voltage rises only after current through it goes to zero. Hence, there is no reverse recovery loss.

Diode voltage oscillates after diode turns OFF. After diode turn-OFF terminal parasitic capacitance of the inductor (L), MOSFET and diode capacitance resonate with stray induc- tance in the loop. After MOSFET $S2$ is turned OFF, its

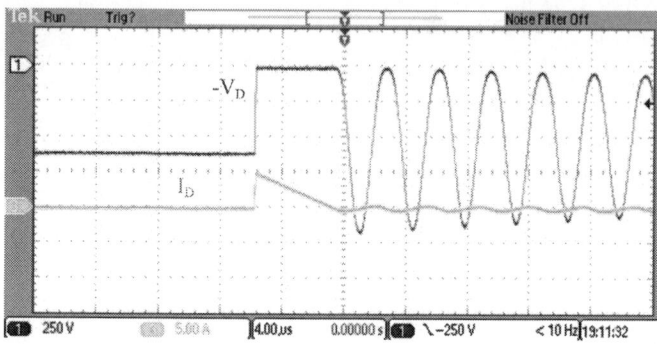

Fig. 24.   Diode soft turn-OFF; turn-OFF loss is zero.

output capacitance starts resonating and the terminal voltage oscillation appears as seen in Fig.23. With the increase of $V2$ the magnitude of oscillation increases. With the increase of $V2$ turn-OFF di/dt of i_D increases which increases the initial energy in the resonating circuit. In Fig.25 and 26 the resonating current and voltage oscillation are compared with the magnitude of $V2$. With higher magnitude of $V2$, turn-OFF di/dt increases and oscillation peak increases. This oscillation can die down due to circuit resistances. However, it disappears at the start of new cycle if it does not decay before that.

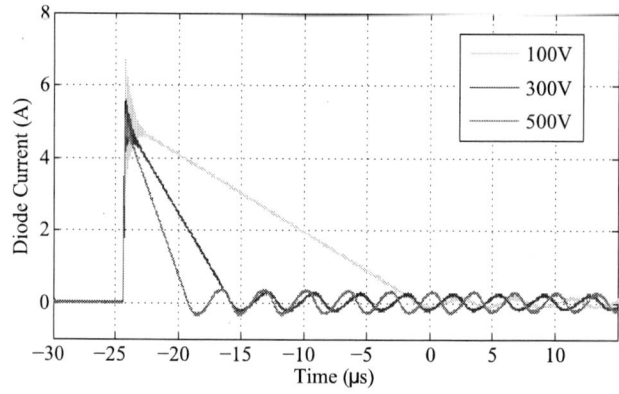

Fig. 25.   Variation of turn-OFF di/dt of the diode current with respect to blocking voltage $V2$.

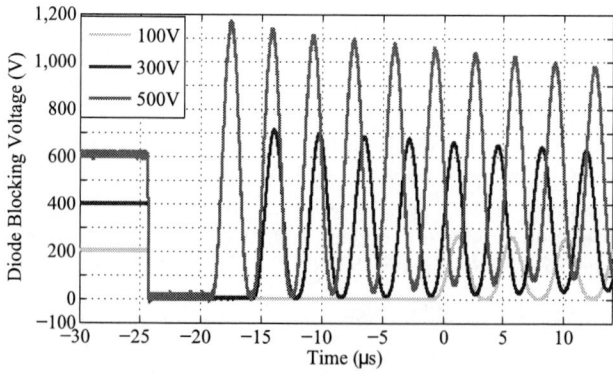

Fig. 26.   Variation of diode voltage oscillation with respect to blocking voltage, $V2$.

978-1-4799-2706-7/14 $31.00 © 2014 IEEE   3453

The 2014 International Power Electronics Conference

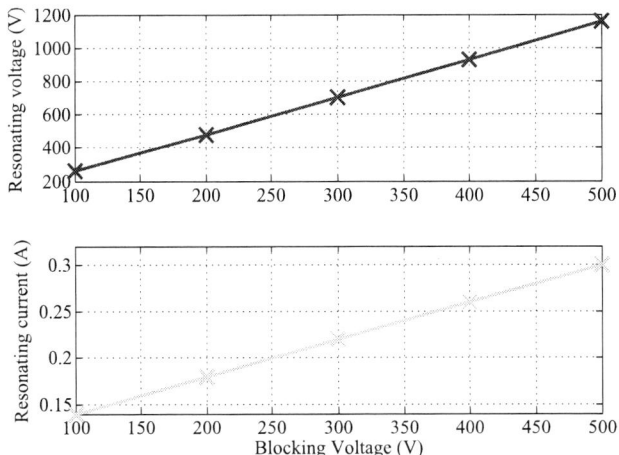

Fig. 27. Variation of peak ringing of voltage and current as a function of blocking voltage, $V2$.

In Fig.27, the variation of peak oscillation voltage and current is plotted against the blocking voltage, $V2$. The voltage oscillation can be stressful for the MOSFET and diode and it may require overrating the diode. However, careful design can minimize the terminal parasitic capacitance of the inductor and stray inductance in the circuit. Voltage oscillation can be minimized with better circuit design.

VII. CONCLUSION

In this paper, a new $1.7\,\text{kV}, 50\,\text{A}$ SiC MOSFET from Cree, Inc., is characterized under hard-switched conditions. The switching behaviors are analyzed and switching losses are evaluated as a function of gate resistances and junction temperature at different dc bus voltage and switch current. Switching losses increase with the increase of gate resistance. The dv/dt and di/dt in the circuit increases with the decrease of gate resistance. Maximum dv/dt and di/dt found to be $36\,\text{kV}/\mu\text{s}$ and $3\,\text{kA}/\mu\text{s}$ respectively with the lowest gate resistance of $5\,\Omega$ used. Total switching losses of the SiC MOSFET remains almost unchanged with the variation of its junction temperature.

Conduction and switching losses of a $1.7\,\text{kV}, 50\,\text{A}$ Si IGBT from ABB and a $1.7\,\text{kV}, 42\,\text{A}$ Si BiMOSFET are compared with those of the SiC MOSFET. The BiMOSFET is characterized in the same test setup to extract the switching loss data for direct comparison with SiC MOSFET. The $1.7\,\text{kV}, 50\,\text{A}$ SiC MOSFET is showing superior performance and expected to achieve higher converter efficiency even at higher operating switching frequency. The reverse voltage blocking switch made from the combination of SiC MOSFET and SiC Schottky diode in series, is tested under zero current switching condition and found to have negligible power losses.

ACKNOWLEDGMENT

This work made use of FREEDM ERC shared facilities supported by National Science Foundation under Award Number EEC-0812121.

REFERENCES

[1] B. Callanan, "Application considerations for Silicon Carbide MOSFETs," *Appl. Notes*, CREE Inc, Morrisville, NC, Jan. 2011.

[2] M. K. Das, "SiC MOSFET module replaces up to 3x higher current Si IGBT modules in Voltage Souce Inverter application," *Bodo's Power Systems®*, pp. 22-24, Feb. 2013.

[3] M. H. Kheraluwala, R. W. Gascoigne, D. M. Divan, and E. D. Baumann, "Performance Characterization of a High-Power Dual Active Bridge dc-to-dc Converter," *IEEE Trans. on Ind. Appl.*, vol. 28, no. 6, pp. 1294-1301, Nov./Dec. 1992.

[4] B. Ozpineci, M. S. Chinthavali, L. M. Tolbert, A. S. Kashyap, and H. A. Mantooth, "A 55-kW Three-Phase Inverter with Si IGBTs and SiC Schottky Diodes," *IEEE Trans. on Ind. Appl.*, vol. 45, no. 1, pp. 278-285, Jan./Feb. 2009.

[5] S. Hazra, S. Madhusoodhanan, G. K. Moghaddam, K. Hatua, and S. Bhattacharya, "Design considerations and performance evaluation of 1200 V, 100 A SiC MOSFET based converter for high power density application," in *Proc. of IEEE Energy Conversion Congress and Exposition (ECCE)*, Denver, USA, 2013, pp. 4278-4285.

[6] http://www05.abb.com/global/scot/scot256.nsf/veritydisplay/dd4a5c760ecd594cc1257ada0031dd53/$file/5SMY%2012G1721%205SYA%201324-02%2011%2002.pdf

[7] http://ixdev.ixys.com/DataSheet/DS98710C(IXBH-BT42N170).pdf

[8] R. E. Locher and A. D. Pathak, "Use of BiMOSFETs in modern radar transmitters," in *Proc. of Power Electronics and Drive Systems*, vol. 2, Oct. 2001.

[9] A. De, S. Roy, S. Bhattacharya, and D. M. Divan, "Performance Analysis and Characterization of Current Switch under Reverse Voltage Commutation, Overlap Voltage Bump and Zero Current Switching," in *Proc. of IEEE Applied Power Electronics Conference and Exposition (APEC)*, Long Beach, California, USA, 2013, pp. 2429-2435.

[10] A. De, S. Roy, S. Bhattacharya, and D. M. Divan, "Characterization and performance comparison of reverse blocking SiC and Si based switch," in *Workshop on Wide Bandgap Power Devices and Applications (WiPDA)*, Columbus, Ohio, 2013, pp. 80-83.

[11] S. Ogasawara, and H. Akagi, "Modeling and damping of high-frequency leakage currents in PWM inverter-fed AC motor drive systems," *IEEE Trans. Ind. Appl.*, vol. 32, no. 5, pp. 11051114, 1996.

[12] B. Jayant Baliga, Fundamentals of Power Semiconductor Devices, Springer: New York, 2008.

[13] A. Kadavelugu, V. Baliga, S. Bhattacharya, M. Das and A. Agarwal, "Zero voltage switching performance of 1200V SiC MOSFET, 1200V silicon IGBT and 900V CoolMOS MOSFET," in IEEE Energy Conversion Congress and Exposition (ECCE), Phoenix, AZ, 2011.

978-1-4799-2706-7/14 $31.00 © 2014 IEEE

The 2014 International Power Electronics Conference

Control Method for Five Level Converter with Common Flying Capacitors to Avoid Voltage Level Skip

Wei Yan, Hui Zhang, Kazuya Ogura
R & D Division
Meiden Singapore
Singapore
hui.zhang@meidensg.com.sg

Shota Urushibata
Research & Development Group
Meidensha Corporation
Numazu, Japan

Abstract— A multilevel converter topology with common flying capacitors (CFC) was proposed recently. It has the advantage that fewer components are required, thus lower cost and higher power density are possible to be realized. However, due to the common circuit shared among different phases, voltage level skips are easy to be caused, which will make the system suffer surge voltage and could be a threat to system insulation. Especially in motor drive application, a costly output filter is always used to protect motor from the impact of voltage level skip. In order to solve this issue and avoid using bulky output filter, this paper proposes a predictive control method for this CFC converter. By using MATLAB/Simulink, it was verified that voltage level skip can be avoided and performance is improved.

Keywords— common flying capacitor, five level converter, voltage level skip.

I. Introduction

Multilevel converters are playing a very important role in power conversion applications nowadays, especially in megawatt and medium voltage application. The attractive features of multilevel converters against conventional two-level converters are lower voltage distortion, lower voltage stress etc [1][2]. In the past two decades multilevel topologies like neutral point clamped converter (NPC) [3], flying capacitors converter [4] and cascaded H-bridge converter [5] have been proposed. Coming into 21st century some advanced and featured multilevel topologies such as active neutral point clamped converter (ANPC) [6] and modular multilevel cascade converter (MMCC) [7] were proposed. Recently proposed common flying capacitor converter (CFC) [2] as shown in Fig. 1 has fewer components, lower cost and higher power density.

In this paper further research has been done to solve voltage level skip [8] issue existing in CFC converter. In motor control system, voltage level skip in line-to-line voltage waveform is a big challenge for motor insulation system, due to surge voltage caused. It may decay insulation level and reduce its life span. Normally, a bulky output filter is needed to protect a motor from surge voltage effectively. However, cost and converter volume will increase accordingly. This paper explains

why voltage level skip happens in CFC and presents how to solve it through a control approach instead of using output filters. For verification, simulation in MATLAB/Simulink was carried out.

II. Five-Level CFC Converter

A CFC topology comprises a common and several selector circuits [2]. The common circuit shared by different phases implies that fewer components are required. As shown in Fig.1, a five-level CFC converter, as the minimum system of this topology, contains a two-stage common circuit shared by all three phase selector circuits, which reduces the number of switching devices and flying capacitors to 22 and 2, compared to 24 and 3 in ANPC five-level converter. However, this common circuit also makes three phases related to each other physically, which consequently results in some vectors unavailable in 5-level space vector diagram [2].

Based on SVM (Space Vector Modulation) [8] applied on five-level topology, the whole region is divided into six sectors, while each sector can be further divided into 16 triangles, as illustrated in Fig.2. The space vectors marked by dots are named according to g-h coordination system [9], such as V_{00} and V_{40}, etc. A 'voltage code'

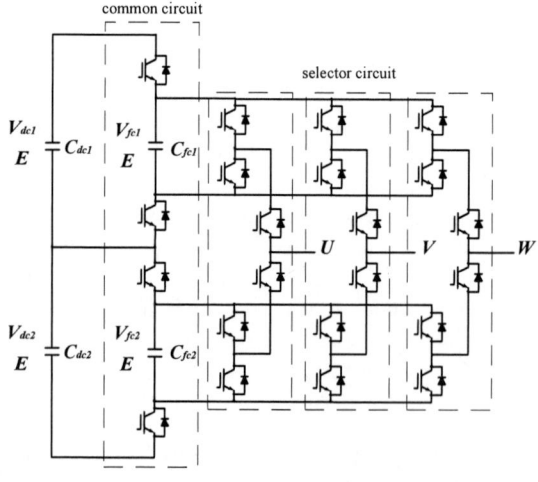

Fig. 1 A five-level CFC converter

The 2014 International Power Electronics Conference

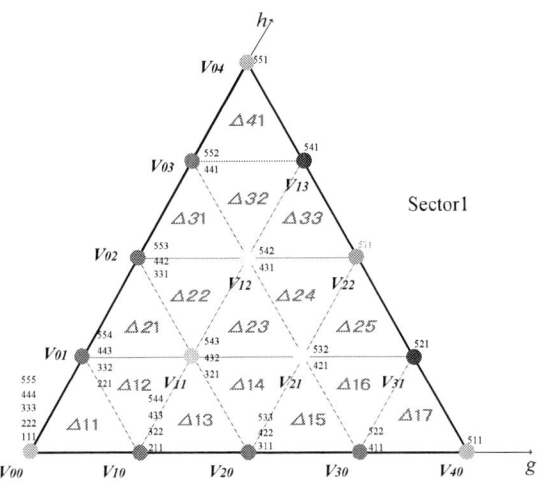

Fig. 2 Space vectors in sector1

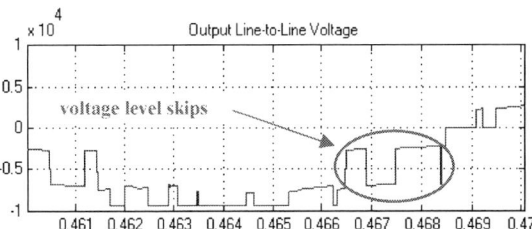

Fig. 3 Zoomed-in output line-to-line voltage with level skips

defined as a three-digit number, from 1 to 5, is introduced to indicate voltage levels in three phases, where '1' represents '-2E', '2' for '-E', and so on (assuming that voltage of DC capacitors in DC link and flying capacitors in common circuit are '2E' and 'E' respectively). It can be seen that most of vectors include more than one 'voltage code', and among these vectors, V_{22} is unavailable to be output by 5-level CFC converter [2].

III. Voltage Level Skip Issue

Voltage level skip is a phenomenon that more than one voltage level is skipped in output voltage, for example, from '-3E' jump to '-E' directly as illustrated in Fig.3. In fact, it is due to more than one pair of IGBTs switching at the same time. For CFC topologies, voltage level skip normally happens:

A. When unavailable vector is involved

As mentioned, one of constrains of CFC is the unavailable vector in each sector. Like V_{22} in sector 1, to replace it, V_{13} and V_{31} can be selected [2]. Process ① in Fig.4 shows how to avoid V_{22} by selecting 421→521→541→532. However, voltage skip appears during transitions from '521' to '541' and from '541' to '532.

B. During transitions between certain adjacent vectors

In this case voltage level skip happens during dead time of certain transaction, for example 522→421, ② in Fig. 4.

Fig.5 illustrates such kind of situation. In Fig. 5 IGBTs in circles keep turned-on; IGBTs in shadows switch from 'ON' to 'OFF'; while IGBTs in rectangles from 'OFF' to 'ON'. During dead time of this transition, after IGBTs in shadows turned off, line-to-line voltage (UV) changes to 4E. Then after dead time finished and IGBTs in rectangles turned on, line-to-line voltage (UV) steps down to 2E directly, thus level skip happens. Obviously, there are three pairs of IGBTs switched at the same time in this case.

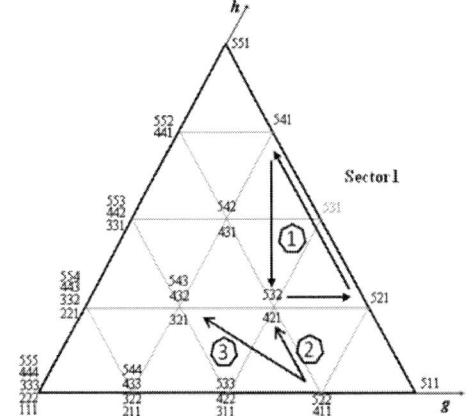

Fig. 4 Causes of voltage level skips

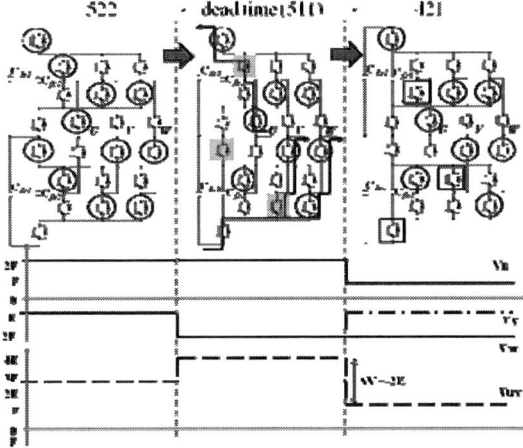

Fig. 5 Level skip caused by dead time

C. During transitions between distant vectors

This could happen between any of two switching patterns in distant vectors even without dead time, for example 522→432, ③ in Fig. 4.

Certainly the issue of voltage level skip can be solved by introducing an effective output filter, but cost and converter volume will be problems. Therefore if we can settle it at control stage, CFC system will be more competitive.

978-1-4799-2706-7/14 $31.00 © 2014 IEEE   3456

IV. Control Scheme

A. Replace unavailable vector

For the case of III.A, a special control stratagem is adopted. When voltage reference $\vec{V}_{ref}$ is in any triangle of 24, 25 or 33, it can be expressed by (1). It means recalculation can be done in the big triangle (Fig. 6) formed by four small triangles 23, 24, 25 and 33. Meanwhile in sector 1 (Fig.2), we can obtain (2). Substitute (2) into (1), equation (3) is deduced, from which it can be seen that voltage level skip issue caused by unavailable vector can be settled by rotation of $\vec{V}_{31} \rightarrow \vec{V}_{21} \rightarrow \vec{V}_{12} \rightarrow \vec{V}_{13}$.

$$\vec{V}_{ref} \cdot T_s = \vec{V}_{31} \cdot T_1 + \vec{V}_{11} \cdot T_2 + \vec{V}_{13} \cdot T_3, \quad (1)$$

$$\vec{V}_{11} = \tfrac{1}{3}\vec{V}_{21} + \tfrac{1}{3}\vec{V}_{12} \quad (2)$$

$$\vec{V}_{ref} \cdot T_s = \vec{V}_{31} \cdot T_1 + \tfrac{1}{3}\vec{V}_{21} \cdot T_2 + \tfrac{1}{3}\vec{V}_{12} \cdot T_2 + \vec{V}_{13} \cdot T_3, \quad (3)$$

where T_s is sampling time, T_1, T_2 and T_3 are executive time of corresponding vectors.

B. Make sequence table

Avoiding voltage level skip is in fact to avoid more than one pair of IGBTs switching at the same time. Before further explanation, we introduce a new term called switching pattern. It is the numbers in Fig.7 and in Fig. 8 besides of voltage code, which represents different current paths under this voltage code. Voltage code only indicates voltage level of each phase, but according to the behavior of charge or discharge flying capacitors, there are more than one switching patterns in some voltage codes. For example, there are two switching patterns, 71 and 72, under voltage code 411 (Fig.8), and they are charge and discharge upper flying capacitor respectively [2].

In Fig.7 a state-flow is presented, in which all vectors and switching patterns in sector 1 are demonstrated. While lines between switching patterns indicate all transitions in which only one pair of IGBTs switching is involved.

While observing transition 522→421 in Fig.8 again, there are only two three-step sequences, 73→44→82 and 73→42→81, which do not involve any voltage level skip. Thus we must make a sequence table in each triangle based on Fig.7 so as to avoid level skip under all operational conditions. However, we also notice that the table size can be infinite if there are no limits on number of steps. For example if pattern 73 and 81 are chosen as start and stop switching patterns in triangle 15, there are sequences like 73→42→81, 73→44→73→42→81, 73→44→82→44→73→42→81, and so on, where no level skip exists. In simulation after trade-off between switching losses and DC and FC ripples, number of steps is capped at 7. In table 1, it lists some of the switching sequences in triangle 15 where sequences are limited at five steps and switching patterns are shown as bold numbers and followed by how they affect DC and FC voltage.

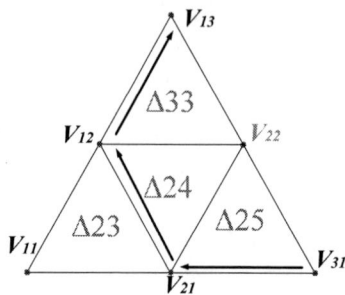

Fig. 6 Solution for unavailable vector

Fig. 7 State-flow chart without voltage level skips

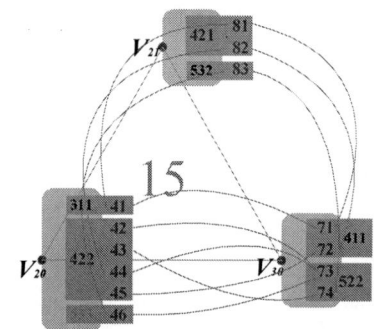

Fig. 8 Zoomed-in state-flow chart in triangle 15

C. Select sequence

Upon sequence selection, predictive control [10][11] is adopted to minimize DC and FC ripples. Predictive control is an approach of using system model to predict future behavior of controlled variables. It is very popular in motor drive control and many algorithms are developed in the past three decades. We choose DC and FC voltage errors as predicting variables. Firstly, we assume load current is constant within one sampling period, since sampling frequency is much higher than system frequency. Therefore equation (4) can be obtained for a fixed switching pattern.

Table 1 Switching sequence table of triangle 15

NO	SP1	FC1	FC2	DC	SP2	FC1	FC2	DC	SP3	FC1	FC2	DC	SP4	FC1	FC2	DC	SP5	FC1	FC2	DC	ΔV_{fc1}	ΔV_{fc2}	ΔV_{dc}	j
1	41	NC	NC	+u	71	-u	NC	+u	81	-u	-v	+u	71	-u	NC	+u	41	NC	NC	+u	-0.743	0.212	1.06	1.724
2	42	-u	+u	+u	73	NC	+u	NC	42	-u	+u	+u	81	-u	-v	+u	71	-u	NC	+u	-0.902	0.69	0.902	2.104
3	43	-u	-u	NC	71	-u	NC	+u	41	NC	NC	+u	71	-u	NC	+u	81	-u	-v	+u	-0.902	0.0531	0.902	1.631
4	44	+u	+u	NC	73	NC	+u	NC	42	-u	+u	+u	81	-u	-v	+u	42	-u	+u	+u	-0.531	0.849	0.637	1.409
5	45	+u	-u	-u	72	+u	NC	NC	82	+u	-v	NC	44	+u	+u	NC	73	NC	+u	NC	0.902	0.372	-0.159	0.978
6	46	NC	NC	-u	73	NC	+u	NC	42	-u	+u	+u	81	-u	-v	+u	-	-	-	-	-0.584	0.69	0.425	0.997
7	71	-u	NC	+u	41	NC	NC	+u	71	-u	NC	+u	81	-u	-v	+u	42	-u	+u	+u	-0.902	0.372	1.06	2.079
8	72	+u	NC	NC	82	+u	-v	NC	72	+u	NC	NC	41	NC	NC	+u	71	-u	NC	+u	0.531	0.212	0.425	0.507
9	73	NC	+u	NC	42	-u	+u	+u	73	NC	+u	NC	42	-u	+u	+u	81	-u	-v	+u	-0.743	0.849	0.743	1.825
10	74	NC	-u	-u	43	-u	-u	NC	74	NC	-u	-u	83	NC	w	-u	74	NC	-u	-u	-0.318	-0.849	-0.743	1.375
11	81	-u	-v	+u	42	-u	+u	+u	81	-u	-v	+u	71	-u	NC	+u	41	NC	NC	+u	-0.902	0.372	1.06	2.079
12	82	+u	-v	NC	44	+u	+u	NC	73	NC	+u	NC	42	-u	+u	+u	73	NC	+u	NC	0.425	0.849	0.159	0.927
13	83	NC	w	-u	46	NC	NC	-u	73	NC	+u	NC	46	NC	NC	-u	83	NC	w	-u	0	0.106	-0.743	0.563

......

1. SP1 (SP2, SP3, SP4, SP5) is switching pattern 1 (switching pattern 2, switching pattern 3, switching pattern 4, switching pattern 5).
2. FC1 (FC2, DC) means how this switching pattern affects upper flying capacitor (low flying capacitor, neutral point).
3. ΔV_{fc1} (ΔV_{fc2}, ΔV_{dc}) means how much upper flying capacitor (low flying capacitor, neutral point) voltage fluctuates with this switching pattern.
4. NC is no charge.
5. +u (+v, +w) means charged by current of phase u (v, w); -u (-v, -w) means discharged by current of phase u (v, w).

$$\Delta V_{SP} = \frac{I}{2\pi f \cdot C} \cdot \Delta t \qquad (4)$$

where ΔV_{SP} is capacitor voltage error at the end of switching pattern; I is load current; and Δt is executive time of switching pattern.

And for a certain sequence, voltage error at the end of this sequence can be calculated by (5).

$$\Delta V_{SQ} = \Delta V_{SP1} + \Delta V_{SP2} + \Delta V_{SP3} + \Delta V_{SP4} + \Delta V_{SP5} \quad (5)$$

Secondly, calculate capacitors voltage errors (DC and FC) of each possible sequence in one triangle based on equation (4) and (5). Finally minimize cost function j (equation (6)) to select proper sequence.

$$j = \Delta V_{SQ,fc1}^2 + \Delta V_{SQ,fc2}^2 + k \cdot \Delta V_{SQ,dc}^2 \qquad (6)$$

where k is balance gain for ΔV_{fc} and ΔV_{dc};

Thus neutral point voltage and FC voltage can be controlled.

In order to have a clear picture of selection, an example of triangle 15 is explained. In this example, assume that three-phase currents are 100A, -50A and -50A respectively. Sampling frequency is 1kHz and executive time of each vector are 0.3ms (V_{20}), 0.3ms (V_{30}) and 0.4ms (V_{21}). For easy calculation, executive time is equally shared by switching patterns within a same vector. For instance, in sequence no.5 in table1, there are two V_{20} switching patterns (44 and 45), two V_{30} switching patterns (72 and 73) and one V_{21} switching pattern (82). By sharing equally, switching time of each switching pattern is 0.15ms (44), 0.15ms (45), 0.15ms (72), 0.15ms (73) and 0.4ms (82). By applying equation (4) and (5) to a 50Hz system with all capacitors selected as 300μF, $\Delta V_{SQ,fc}$ and $\Delta V_{SQ,dc}$ can be calculated. And finally cost function j can be gotten by equation (6). Among results the minimum j is 0.507 (k=1), so sequence no.8 should be selected for minimum ripple.

V. Simulation Results

Simulation for the five-level CFC inverter in MATLAB/Simulink was carried out to verify the proposed control approach. For comparison study, the control approach is implemented in the same 1MVA, 6.6kV motor driving system as [2] and under the same conditions of V/F control, constant sampling rate of 1kHz and 300μF as both DC and FC capacitance. Voltage of FC is initialized as 0.

Fig.9 shows output line-to-line voltage waveform in steady state with modulation index of 0.8, which means voltage reference in most outer band, in order to get a nine-level line-to-line voltage waveform. Fig.10 is a zoomed-in waveform of Fig.9, showing voltage level skip is successful avoided. DC and FC voltage are presented

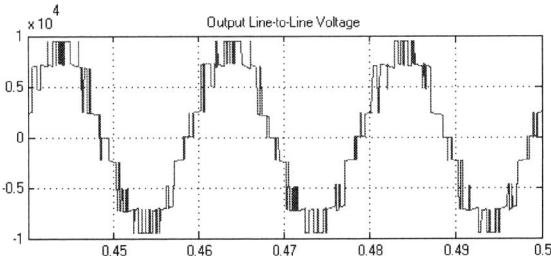

Fig. 9 Output line-to-line voltage (steady state)

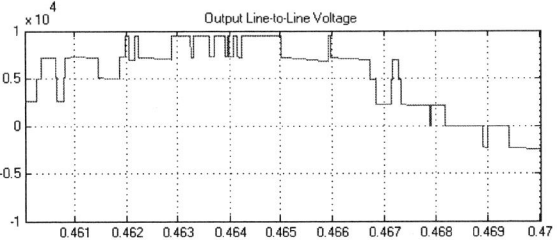

Fig. 10 Zoom-in output line-to-line voltage

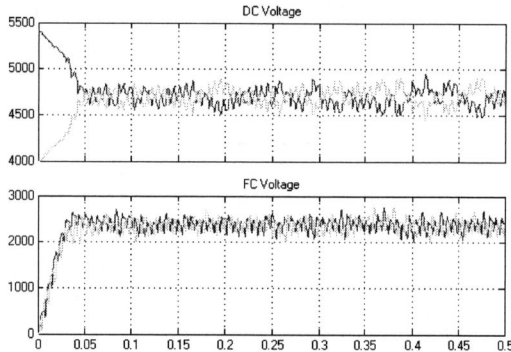

Fig. 11 DC and FC voltage waveforms

in Fig.11. It is also noted that DC and FC voltage balancing control are not affected.

VI. Conclusions

A control method for a five-level CFC converter is proposed and verified in a motor drive system by MATLAB/Simulink simulation. With this method voltage level skips can be effectively avoided. Therefore output filter can be designed smaller which directly affects cost and system volume.

References

[1] Jose Rodriguez, Jih-Sheng Lai, and Fang Zheng Peng, "Multilevel Inverters: A Survey of Topologies, Controls, and Applications" IEEE Trans. Ind. Electron., vol. 49, no. 4, pp. 724-738, Aug. 2002

[2] Hui Zhang, Wei Yan, Kazuya Ogura and Shota Urushibata, "A Multilevel Converter Topology with Common Flying Capacitors" ECCE 2013, Denver, USA

[3] H. Miyazaki, H. Fukumoto, S. Sugiyama, M. Tachikawa and N. Azusawa, "Neutral-Point-Clamped Inverter with Parallel Driving of IGBT's for Industrial Applications" IEEE Trans. on Ind. Applications, vol. 36, no. 1, pp. 146-151, Jan./Feb. 2000

[4] Jih-Sheng Lai and Fang Zheng Peng, "Multilevel Converter – A New Breed of Power Converter" 30th IAS Annual Meeting, 8-12 Oct. 1995, Orlando

[5] Leon M. Tolbert, Fang Z. Peng and Thomas G. Habetler, "Multilevel Inverter for Electric Vehicle Application" Power Electronics in Transportation, 22-23 Oct 1998, Dearborn

[6] P. Barbosa, P. Steimer, J. Sternke, L. Meysenc, M. Winkelnkemper and N. Celanovic, "Active Neutral-Point-Clamped Multilevel Converters," in PECS, 2005, June. 16, Recife

[7] H. Akagi, "Classification, Terminology, and Application of the Modular Multilevel Cascade Converter (MMCC)" IPEC, 21-24 June 2010, Sappro

[8] M. Trabelsi, L. Ben-Brahim, T. Yokayama, A. Kawamura, R. Kurosawa and T. Yoshino, "An Improved SVPWM Method for Multilevel Inverters" 15th International Power Electronics and Motion Control Conference, EPE-PEMC 2012 ECCE Europe, Novi Sad, Serbia

[9] N. Celanovic and D. Boroyevich, "A Fast Space-Vector Modulation Algorithm for Multilevel Three-Phase Converters," IEEE Trans. Ind. Appl., vol. 37, no. 2, pp. 637–641, Mar./Apr. 2001.

[10] Arne Linder, Rahul Kanchan, Ralph Kennel and Peter Stolze, "Model-Based Predictive Control of Electric Drives" Chapter 4 pp. 17-16

[11] P. Cortes, M.P. Kazmierkowski, R.M. Kennel, D.E. Quevedo and J. Rodriguez, "Predictive Control in Power Electronics and Drives" IEEE Transactions on Industrial Electronics , vol.55, no.12, pp. 4312-4324, Dec.2008

978-1-4799-2706-7/14 $31.00 © 2014 IEEE

The 2014 International Power Electronics Conference

Low-Complexity Analytical Approximations of Switching Frequency Harmonics of 3-Phase N-Level Voltage-Source PWM Converters

Ralph M. Burkart and Johann W. Kolar
Power Electronic Systems Laboratory
ETH Zurich, Physikstrasse 3
Zurich, 8092, Switzerland
burkart@lem.ee.ethz.ch

Abstract—**In this paper, analytical calculations of the switching frequency harmonics of PWM-controlled 3-phase voltage-source DC/AC power converters are presented. Detailed knowledge of the modulation-dependent spectra of such systems is highly valuable for the EMI filter design and the estimation of the high-frequency losses in the filter components and the load. The paper presents a formula for the calculation of the switching frequency harmonics which is simple and highly versatile. It can be applied to converters with arbitrary numbers of voltage levels N and in combination with a broad variety of modulation signals. All calculations are verified by means of simulations and show a high accuracy over a wide range of input parameters. Finally, an EMI filter design example demonstrates the practical application and benefits of the derived formula. The MATLAB code used for the calculations is provided at the end of the paper.**

I. INTRODUCTION

Switched-mode pulse-width-modulated (PWM) 3-phase power converters with a voltage-type DC-link are the backbone of a wide range of today's power electronics applications, such as AC motor drive systems, grid-connected converters in renewable energy systems and uninterruptible power supplies (UPS). Besides generating the fundamental voltages which drive the desired fundamental currents, such systems also generate undesired high-frequency (HF) voltage harmonics as an inherent result of the PWM switched-mode operation. These HF harmonics often have a crucial impact on the system design: on the one hand, the harmonic spectrum determines the required EMI filter attenuation in grid-connected applications and for motor drive inverters. On the other hand, the resulting HF currents, i.e. the ripple currents cause additional losses in the filter components and the load (e.g. electrical machines). Consequently, a detailed knowledge of the converter's harmonic spectrum is a basic requirement for the appropriate design of the full converter system. It is furthermore particularly important if accurate efficiency estimations for low-load operation are required where the HF losses usually become increasingly dominant compared to the losses associated with the fundamental current.

A simple approach to compute the switching frequency harmonics of a PWM-controlled converter is by means of applying the fast Fourier transform (FFT) to the respective simulated or measured waveforms. However, the accuracy of the FFT is limited by the number of available sampling points which potentially results in a high computational effort [1]. As a consequence, the use of versatile yet simple analytical

methods can offer advantages if a systematic insight regarding the spectra of different converter topologies, modulation signals and modulation depths must be gained.

A wide variety of methods exist to analytically describe the harmonics of PWM signals. While decomposition techniques [2], implicit descriptions [3] and Kapteyn [4] or Lagrange [5] series have been reported, by far the most widely used method in literature is the double-Fourier-series (DFS) analysis based on 3D geometric wall-model representations of the PWM signal. The method was originally presented in [6] and [7] and later adopted for different 3-phase current- and voltage-source PWM converters employing various PWM schemes. [1], [8]–[12]. An extensive overview of the formulas obtained from this method can be found in [13].

The above mentioned methods have the common benefit of yielding exact analytical closed-form results. The main drawback, however, lies in the requirement of deriving new formulas for each combination of topology and type of modulation signal. These derivations and the results are involved and the mathematical effort increases quickly with increasing number of voltage levels and more demanding modulation signals. Generalized formulas exist either for N-level converters with restriction to a pure sinusoidal modulation signal [10], [13], or for converters with 2-level characteristics only but arbitrary modulation signals [1]. These results are even more intricate and cannot readily be applied. Eventually, all methods generate formulas including infinite double sums of Bessel functions which considerably impedes the numerical evaluation.

An alternative but not widely used approach for the analytical calculation of switching frequency harmonics is presented in [14]. This method is based on a subsequent local and global integration and is thus called local-global-integral (LGI) method for the remainder of this work. The LGI method overcomes most of the above stated drawbacks. In particular, the resulting formula can take into account a broad variety of modulation signals and consists of a simple integral which is numerically easy to evaluate. However, the LGI method assumes an infinitely high number z of switching intervals within a period of the output voltage fundamental and is thus, in contrast to most other methods, not exact in the case of real converter systems with finite switching frequencies. Nevertheless, sufficient accuracy can still be obtained in most practical cases, i.e. for converters with $z \geq 30$.

The LGI method has so far only been applied to a 2-level

978-1-4799-2706-7/14 $31.00 © 2014 IEEE

The 2014 International Power Electronics Conference

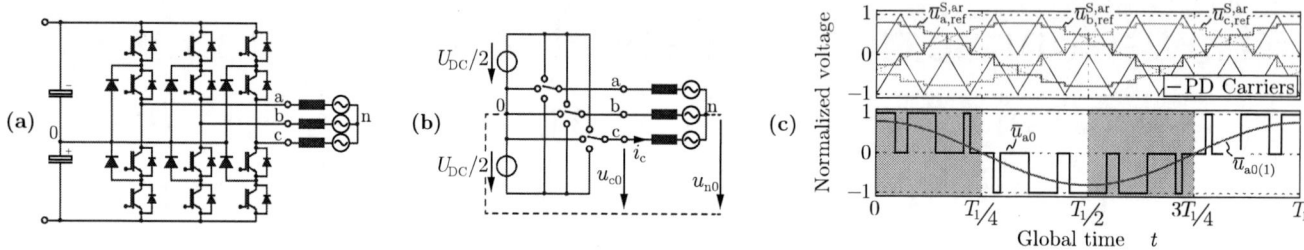

Fig. 1: 3-phase 3-level voltage-source DC/AC power converter with sinusoidal PWM. **(a)** Typical practical implementation. **(b)** Idealized model used for the harmonic analysis in this work. **(c)** Asymmetric regular-sampling PWM with synchronized phase disposition (PD) carrier waves, sampled sinusoidal reference signals $\overline{u}_{p,\text{ref}}^{\text{S,ar}}$, $p = \{a,b,c\}$ and resulting idealized output voltage pattern $\overline{u}_{a0}$ and fundamental harmonic $\overline{u}_{a0(1)}$. Due to the underlying assumptions in this work, the converter output voltages feature a quarter-wave symmetry, i.e. $\overline{u}_{p0}(t) = \overline{u}_{p0}(-t) = -\overline{u}_{p0}(t+T_1/2)$.

converter in [14]. Moreover, a detailed analysis of the accuracy and limitations of the result is missing yet. In this paper, it is shown that the LGI method can also be applied to a 3-level converter with only moderate mathematical effort and yielding a similarly simple formula with high accuracy (**Sec. II**). In a second step, the LGI method is used to derive an advanced but still simple and versatile universal formula which describes the voltage and current harmonics of a general N-level converter with almost arbitrary modulation signal. Finally, an example of application is given which demonstrates how the formula can be used to design an EMI filter (**Sec. IV**). The **Appendix** contains the MATLAB implementation of the general N-level converter formula derived in this work.

II. LGI METHOD FOR 3-LEVEL CONVERTERS

In this section, the LGI method presented in [14] is adopted for a 3-level 3-phase voltage-source converter as shown in **Fig. 1(a)**. **Sec. II-A** gives a short overview of possible PWM schemes for this topology. After discussing the underlying assumptions in **Sec. II-B**, the derivation of the formula to calculate the switching frequency voltage harmonics is presented in **Sec. II-C** and **Sec. II-D**. **Sec. II-E** compares the obtained LGI-based result to the existing DFS-based result and **Sec. II-F** presents a verification of the derived formula and investigates its accuracy and the useful parameter ranges.

For the remainder of this work, $T_1 = 1/f_1$ denotes the fundamental period and $T_{\text{sw}} = 1/f_{\text{sw}}$ the switching period. Where a distinction between the two different periods is necessary, the position within T_1 is described by the (macroscopic) global time t while the (microscopic) local time t_μ is used to refer to the position within a switching interval $[-T_{\text{sw}}/2, T_{\text{sw}}/2]$, placed symmetrically around t. Normalized voltages are denoted by overlines, $\overline{u} = u/(U_{\text{DC}}/2)$. p stands for one of the three phases $\{a,b,c\}$ with corresponding phase shifts $\phi = \{0, -2\pi/3, 2\pi/3\}$. $u_{p,\text{ref}}$ denotes an arbitrary modulation signal while $u_{p,\text{ref}}^{\text{S}}$ is a specific modulation signal, e.g. $u_{p,\text{ref}}^{\text{S}}$ denoting a pure sinusoidal signal. The term "reference signal" is equivalently used for "modulation signal".

A. PWM Schemes for 3-Level Converters

In this section, PWM schemes employing different carrier signals and sampling techniques are discussed, while the discussion of different modulation signals follows in **Sec. II-D**.

The most basic idea of PWM employed in DC/AC power converters is to generate trains of rectangular switched voltage pulses u_{p0} whose fundamental $u_{p0(1)}(t)$ is equal to a given sinusoidal modulation signal, i.e. the reference signal $u_{p,\text{ref}}^{\text{S}}$ (cf. **Fig. 1**),

$$u_{p0(1)}(t) \equiv u_{p,\text{ref}}^{\text{S}}(t) \equiv \hat{U}_{\text{ref}} \cos\left[2\pi f_1 t + \phi\right] . \tag{1}$$

This can be achieved by determining the switching instants through intersection of the reference signal with a HF carrier signal as depicted in **Fig. 2**. Commonly used carriers are periodic sawtooth or triangular signals. However, only triangular carriers are considered here since sawtooth carriers generally lead to a higher harmonic distortion of the generated output u_{p0} [13]. For a 3-level topology, 2 carriers are required which can be implemented with either phase disposition (PD) or phase opposition disposition (POD) as shown in **Fig. 2(a)**. This work focuses on PD carriers rather than POD carriers due to the better harmonic performance that can be achieved in three phase systems [10], [13]. Finally, using a digitally sampled reference curve $\overline{u}_{p,\text{ref}}^{\text{ar}}$ (asymmetric regular sampling PWM) rather than a continuous signal $\overline{u}_{p,\text{ref}}^{\text{n}}$ (natural sampling PWM) is better suited for the digital controllers employed in most modern converters. Regular sampling PWM generates slightly different voltage pulses (and thus different harmonic amplitudes) when compared to natural sampling PWM as depicted in **Fig. 2(b)**. The switching rules, however, are independent from the type of sampling or the reference signal waveform,

$$\overline{u}_{p0}(t) = \begin{cases} 1 & \overline{u}_{p,\text{ref}}(t) \geq \text{carrier}_{\text{high}}(t) \\ 0 & \text{carrier}_{\text{low}}(t) \leq \overline{u}_{p,\text{ref}}(t) < \text{carrier}_{\text{high}}(t) \\ -1 & \overline{u}_{p,\text{ref}}(t) < \text{carrier}_{\text{low}}(t) \end{cases} . \tag{2}$$

Whether natural or asymmetric regular sampling is used becomes irrelevant at very high switching frequency ratios z due to the vanishing differences between the generated voltage pulses and switching instants. As high values of z are assumed for all analytical calculations throughout this work, the sampling type is not indicated anymore for the remainder of this paper. For all simulations used to evaluate the calculated results, a PWM scheme with PD carriers and regular sampling is employed (**Fig. 1(c)**).

A typical spectrum of the output voltages u_{p0}, employing the selected PWM scheme with triangular PD carriers, can be seen in **Fig. 3**. The spectrum contains harmonic groups at integer multiples m of the carrier frequency f_{sw}. Within these groups, the individual harmonics are integer multiples k of the fundamental frequency f_1 distant from each other. Based on this observation, the harmonic order q of a particular harmonic $\hat{U}_{p0(q)}$ can uniquely be expressed by

$$q = mz + k, \quad m \in \mathbb{N}, \ k \in \mathbb{Z} \ \big| \ |k| \leq \lfloor z/2 \rfloor , \tag{3}$$

where $\lfloor * \rfloor$ denotes the floor function.

General remark: although this work is based on the consideration of triangle-carrier-based PWM, the presented analysis is also valid for space-vector-based modulation concepts. This is due to the existence of equivalent reference signals in a triangle-carrier-based PWM scheme which can be used to represent any space-vector modulation scheme [15].

B. Assumptions

In order to successfully apply the LGI method, the following assumptions must be made:

(i) The converter output PWM phase voltages $u_{p0}(t)$ are ideal rectangular, even and periodical with T_1,

$$u_{p0}(t) = u_{p0}(-t) = u_{p0}(t \pm nT_1) , \quad n \in \mathbb{N}_0 . \quad (4)$$

(ii) The ratio z between switching frequency f_{sw} and fundamental frequency f_1 is assumed to be very high, i.e. theoretically tending towards infinity,

$$z = \frac{f_{sw}}{f_1} \to \infty . \quad (5)$$

(iii) The frequency ratio z is odd and a multiple value of 3,

$$z = 6n + 3 , \quad n \in \mathbb{N}_0 . \quad (6)$$

Assumption (i) implies the consideration of a converter with idealized switches and a constant DC-link voltage source U_{DC} as depicted in **Fig. 1(b)**. Assumption (ii) is required for the mathematical derivations. In practice, however, for the considered asymmetric regular sampling $z \geq 30$ is sufficient to achieve accurate results (see **Sec. II-F**, [14]). The presented formula in this work can in theory be applied for arbitrary values of $z \geq 30$. However, an odd frequency ratio is assumed in (iii) as this guarantees odd harmonics only and avoids a DC offset of the output phase voltages (cf. **Fig. 3**). Even harmonics and DC offsets are generally undesired in practice, which is also reflected by the more stringent limits for such harmonics in grid harmonics standards such as [16]–[18]. The restriction to multiples of 3 in (iii) allows for synchronized, symmetric PWM of the three phases {a,b,c} with the same PD carrier signals for all references (cf. **Fig. 1(c)**). Assumption (iii) in conjunction with (i) furthermore implies output signals with quarter-wave symmetry (cf. **Fig. 1(c)**) and associated spectra that contain odd order cosine terms only,

$$u_{p0}(t) = \sum \hat{U}_{p0(q)} \cos\left[q(2\pi f_1 t + \phi)\right], \; q{=}2n{+}1, \; n \in \mathbb{N}_0. \quad (7)$$

Note that due to the symmetric PWM the amplitudes $\hat{U}_{p0(q)}$ are identical in all three phases. Furthermore, there are no phase shifts other than $\pm\pi$ between the harmonics of the individual phases which is taken into account by negative harmonic amplitudes $\hat{U}_{p0(q)}$. Finally, the consideration of only even output signals u_{p0} does not represent a restriction in practice as it can always be achieved by appropriately defining the coordinates.

C. Analytical Derivations

In a first step of the LGI method, a local Fourier analysis within a single switching period is performed. This result can later be used for a global analysis of the full fundamental period. For reasons of symmetry (assumption (iii)), it is sufficient to perform the analysis for phase p = a only.

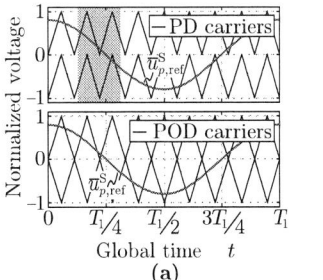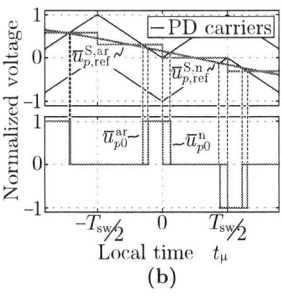

Fig. 2: PWM schemes for 3-level converters. **(a)** HF triangular carriers with phase disposition (PD) and phase opposition disposition (POD), sinusoidal reference signal $\overline{u}_{p,ref}^{S}$. **(b)** Detail of **(a)** (marked gray) showing the generation of the switching time instants by intersection of the reference signal with the HF PD carriers. The digitally sampled reference $\overline{u}_{p,ref}^{S,ar}$ used for asymmetric regular sampling PWM results in slightly different voltage pulses than the continuous reference $\overline{u}_{p,ref}^{S,n}$ used for natural sampling PWM.

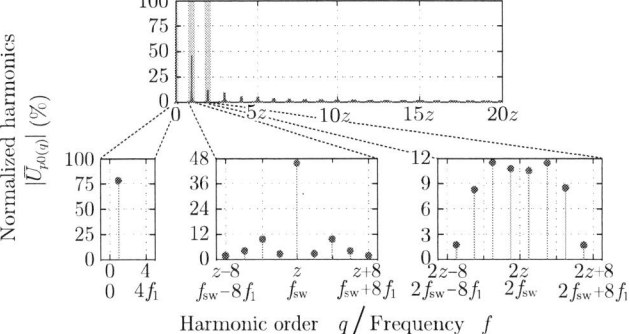

Fig. 3: Typical spectrum of the output voltages $\overline{u}_{p0}$ for high frequency ratios z and a PWM scheme with triangular PD carriers. The spectrum contains baseband low-frequency (LF) harmonics and harmonic groups centered around integer multiples m of the carrier frequency, i.e. the switching frequency f_{sw}. For high values of z, the baseband harmonics represent the harmonic content of the used reference signal (here pure sinusoidal reference $\overline{u}_{p,ref}^{S}$ and thus one harmonic at the fundamental frequency f_1). Within the harmonic groups at integer multiples m of f_{sw}, sideband harmonics occur at integer multiples k of the fundamental frequency f_1 distant from the respective center frequency $mf_{sw} = mzf_1$. Thus, for $z = \frac{f_{sw}}{f_1}$, the harmonic order $q = mz + k$ can be defined with reference to f_1. As a result of the considered triangular PD carriers only sidebands with k even around odd m and with k odd around even m occur. Consequently, if z is odd, only odd order harmonics occur.

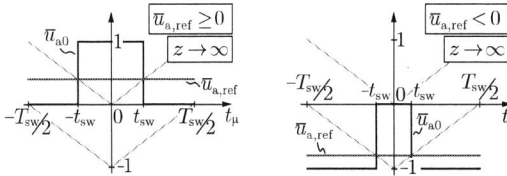

Fig. 4: Normalized output voltage pattern $\overline{u}_{a0}(t_\mu)$ within a single switching period T_{sw} depending on the sign of the reference signal $u_{a,ref}(t)$. Due to the assumed infinite switching frequency ratio z, the considered time interval is very short and thus the reference signal does not change, $\overline{u}_{a,ref}(t) = \text{const.}$ for $t_\mu \in [-T_{sw}/2, T_{sw}/2]$. As a consequence, the switching instants are symmetric around $t_\mu = 0$ and $\overline{u}_{a0}(t_\mu) = \overline{u}_{a0}(-t_\mu)$.

1) Local Analysis:
The voltage pattern within a single switching period can be seen in **Fig. 4**. Due to assumption (ii), $z \to \infty$, the reference value $\overline{u}_{a,ref}$ is constant within the infinitely short switching period. As a result, the output voltage $\overline{u}_{a0}$ is symmetric around $t_\mu = 0$ with the switching times

The 2014 International Power Electronics Conference

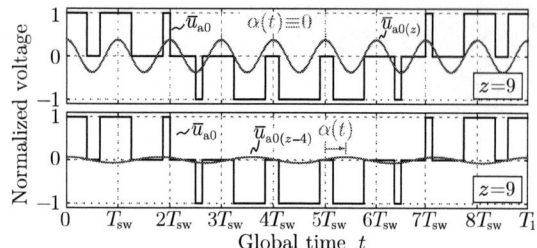

Fig. 5: Normalized output voltage $\overline{u}_{a0}(t)$ and its 9th and 5th harmonic $\overline{u}_{a0(9)}(t)$ and $\overline{u}_{a0(5)}(t)$, respectively, for $z=9$. While the carrier harmonics at integer multiples of z ($q=mz$) stay in phase with respect to the consecutive switching periods, the sideband harmonics with $q=mz+k$, $k \neq 0$, experience a phase shift $\alpha(t)$. The low frequency ratio $z=9<30$ has been chosen for the purpose of a clearer illustration.

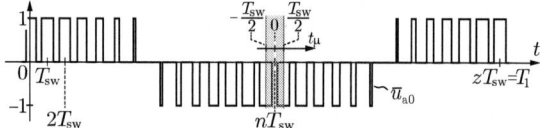

Fig. 6: Summation (13) of the z local Fourier integrals (12) at the positions $t=nT_{sw}$, $n=\{1,..,z\}$, over the fundamental period $T_1=zT_{sw}$. For $z \to \infty$, $T_{sw} \to dt$, and $nT_{sw} \to t$. As consequence, the sum (13) can be expressed as the integral (16).

$$|t_{sw}(t)| = \begin{cases} \dfrac{T_{sw}}{2}\overline{u}_{a,ref}(t) & \overline{u}_{a,ref}(t) \geq 0 \\ \dfrac{T_{sw}}{2}\left(1+\overline{u}_{a,ref}(t)\right) & \overline{u}_{a,ref}(t) < 0 \end{cases} . \quad (8)$$

The local Fourier analysis within a single switching period over the local time t_μ can now be carried out using

$$\hat{U}_{a0(q)}[t] = \frac{U_{DC}}{T_{sw}}\int_{-\frac{T_{sw}}{2}}^{\frac{T_{sw}}{2}}\overline{u}_{a0}(t,t_\mu)\cos\left[q2\pi f_1 t_\mu - \alpha(t)\right]dt_\mu . \quad (9)$$

Although the periods $T_q = T_{mz+k}$ with $k \neq 0$ of the sideband harmonics do not equal integer multiples of the switching period T_{sw}, the difference becomes negligibly small for $z \to \infty$ and finite k,

$$qf_1 = (mz+k)f_1 \overset{z \gg k}{\approx} mzf_1 = \frac{m}{T_{sw}} , \quad (10)$$

and hence (9) can be used without restrictions for both the carrier harmonics ($k=0$) and the sideband harmonics ($k\neq 0$). Note that finite values of k are guaranteed by definition (3) of q, where $|k|$ is minimized with respect to mz which consequently also minimizes the approximation error in (10).

Over the course of the fundamental period, the sideband harmonics experience a phase shift $\alpha(t)$ with respect to the individual switching periods,

$$\alpha(t) = k2\pi f_1 t . \quad (11)$$

This phase shift is illustrated in **Fig. 5** for the case of a carrier harmonic at a multiple of the switching frequency ($k=0$) where the phase shift is always zero and for a sideband harmonic ($k\neq 0$) with non-zero phase shift.

Using (8), (10) and (11), the integral (9) can be solved,

$$\hat{U}_{a0(q)}[t] = \frac{U_{DC}}{m\pi}\cos\left[k2\pi f_1 t\right]\sin\left[m\pi\overline{u}_{a,ref}(t)\right]$$
$$\cdot \begin{cases} 1 & \overline{u}_{a,ref}(t) \geq 0 \\ (-1)^m & \overline{u}_{a,ref}(t) < 0 \end{cases} . \quad (12)$$

2) Global Analysis: In order to obtain the global harmonics $\hat{U}_{a0(q)}$ of the entire fundamental period $T_1 = zT_{sw}$, i.e. the actual output voltage spectrum, the local results (12) of the z individual switching intervals must be summed up and weighted accordingly (cf. **Fig. 6**),

$$\hat{U}_{a0(q)} = \frac{1}{zT_{sw}}\sum_{n=1}^{z}\hat{U}_{a0(q)}[nT_{sw}] \cdot T_{sw} . \quad (13)$$

For an infinite switching frequency ratio $z \to \infty$,

$$\frac{T_1}{z} = T_{sw} \overset{z \to \infty}{\longrightarrow} dt , \quad (14)$$

$$n\frac{T_1}{z} = nT_{sw} \overset{z \to \infty}{\longrightarrow} t , \quad (15)$$

the local time interval T_{sw} is converted into a global time differential dt and the discrete points of time nT_{sw} merge into the continuous global time t. As a consequence, the sum in (13) can be expressed as a global integral,

$$\hat{U}_{a0(q)} \overset{z \to \infty}{=} \frac{1}{T_1}\int_0^{T_1}\hat{U}_{a0(q)}[t]dt . \quad (16)$$

The above integral (16) again illustrates how the contributions of the individual switching periods for a specific harmonic are averaged over the fundamental period. This local-global integral method (LGI) assuming infinite switching frequency ratios z can also be found in [19], [20] for the calculation of the semiconductor currents of PWM converters. Substitution of $\hat{U}_{a0(q)}[t]$ in (16) with (12) yields,

$$\hat{U}_{a0(q)} = \frac{2\left[1-(-1)^{m+k}\right]U_{DC}}{m\pi T_1}$$
$$\cdot \int_0^{\frac{T_1}{4}}\cos\left[k2\pi f_1 t\right]\sin\left[m\pi\overline{u}_{a,ref}(t)\right]dt , \quad (17)$$

where the quarter-wave symmetry of $\overline{u}_{a0}$ was exploited.

3) Final Result: Eq. (17) can be further simplified as it does not depend on T_1. In order to see this, the substitution

$$\beta = 2\pi f_1 t , \quad (18)$$

is used to obtain the final result

$$\hat{U}_{a0(q)} = \frac{\left[1-(-1)^{m+k}\right]U_{DC}}{m\pi^2}$$
$$\cdot \int_0^{\frac{\pi}{2}}\cos\left[k\beta\right]\sin\left[m\pi\overline{u}_{a,ref}(\beta)\right]d\beta ,$$

$$q = mz+k, \ m \in \mathbb{N}, \ k \in \mathbb{Z} \ \big| \ |k| \leq \lfloor z/2 \rfloor , \quad (19)$$

where $\overline{u}_{a,ref}(\beta)$ is the reference signal $\overline{u}_{a,ref}(t)$ with its period normalized to 2π. Inspection of (19) results in the following observations:

- The effective amplitude of a particular harmonic with order q is the result of (infinite) overlapping sidebands. E.g. for $z=33$, both $(m,k)=(1,2)$ and $(m,k)=(2,-31)$ contribute to the harmonic at $q=35$.
- For continuous and bounded reference signals $\overline{u}_{a,ref}(t)$, the integral in (19) yields decreasing values for increasing ratios $|k|/m$. Moreover, the values become less accurate due to approximation (10).
- By means of the unique definition (3) of q which minimizes $|k|$ with respect to mz, only the most significant and

most accurate sideband contribution (m, z) is considered in (19).

- Note again that only the amplitudes of the harmonics are approximated in (19) whereas the resulting phase shifts (0 or $\pm\pi$) are exact.

Further mathematical conversions of (19), i.e. the analytical evaluation of the integral is only possible with a given reference $\overline{u}_{a,ref}$ and would lead to more complex expressions containing Bessel functions as known from [8]–[10]. Instead, direct numerical integration is preferred as shown in the next section.

D. Evaluation for Different Reference Signals

Eq. (19) is the result in its most general form which can easily be evaluated numerically for almost any given reference $\overline{u}_{a,ref}$. Assumptions (i) and (iii) imply the class of permissible reference signals $\overline{u}_{p,ref}$ that generate output voltage waveforms u_{p0} with the assumed properties. Any even reference signal consisting of only odd order cosine terms is possible,

$$\overline{u}_{p,ref}(\beta) = \sum \hat{U}_{ref(q)} \cos\left[q(\beta + \phi)\right], \; q{=}2n{+}1, \; n \in \mathbb{N}_0 \,. \quad (20)$$

All permissible reference signals feature quarter-wave symmetry. Besides the standard pure sinusoidal reference,

$$\overline{u}_{p,ref}^{S}(\beta) = M \cos\left[\beta + \phi\right] \,, \quad (21)$$

with $M = \frac{U_{p0(1)}}{U_{DC/2}}$ being the modulation depth, other common and widely applied references belong to this class. Examples are the sinusoidal PWM with third harmonic injection,

$$\overline{u}_{p,ref}^{S3}(\beta) = M\left(\cos\left[\beta + \phi\right] - \frac{1}{6}\cos\left[3\beta\right]\right) \,, \quad (22)$$

or the symmetric PWM,

$$\overline{u}_{p,ref}^{SY}(\beta) = \begin{cases} \frac{\sqrt{3}}{2}M\cos\left[\beta - \frac{\pi}{6} + \phi\right] & 0 \le \beta < \frac{\pi}{3} \\ \frac{3}{2}M\cos\left[\beta + \phi\right] & \frac{\pi}{3} \le \beta < \frac{\pi}{2} \end{cases} \,, \quad (23)$$

which allow for an increased linear range of the modulation depth ($M_{max} = 2/\sqrt{3}$ instead of $M_{max} = 1$ for pure sinusoidal PWM). The 60° flat-top PWM,

$$\overline{u}_{p,ref}^{FT}(\beta) = \begin{cases} 1 & 0 \le \beta < \frac{\pi}{6} \\ -1 + \sqrt{3}M\cos\left[\beta - \frac{\pi}{6} + \phi\right] & \frac{\pi}{6} \le \beta < \frac{\pi}{2} \end{cases} \,. \quad (24)$$

can be employed to additionally decrease the switching losses at the expense of a richer harmonic spectrum. Further information on the above types of reference signals can be found in [15].

Eq. (19) can now be evaluated when substituting $\overline{u}_{a,ref}$ by a concrete reference signal, e.g. the pure sinusoidal reference. E.g. for $q = 1 \cdot z + 4$, $M = 0.8$ we obtain (cf. **Fig. 3**),

$$\hat{U}_{a0(1\cdot z+2)} = \frac{2U_{DC}}{\pi^2}\int_0^{\frac{\pi}{2}} \cos\left[2\beta\right]\sin\left[1\cdot\pi\cdot0.8\cdot\cos\left[\beta\right]\right]\mathrm{d}\beta$$

$$\approx -0.093\frac{U_{DC}}{2} \,. \quad (25)$$

E. Comparison to the Double-Fourier-Series Method

Applying the widely used double-Fourier-series (DFS) method instead of the LGI method used in this work, the following formula for the calculation of the switching frequency harmonics $\hat{U}_{a0(q)}^{DFS}$ of a 3-level converter employing PD carriers and natural sampling can be obtained [13],

$$\hat{U}_{a0(q)}^{DFS} = \sum_{\substack{mz+k=q \\ m\in\mathbb{N}, k\in\mathbb{Z}}} \frac{U_{DC}}{m\pi}\left(1 - \cos\left[(m+k)\pi\right]\right)$$

$$\cdot \left[\begin{array}{c} J_k\left[m\pi M\right]\sin\left[k\frac{\pi}{2}\right] \\ + \frac{4}{\pi}\sum_{\substack{n=1 \\ 2n-1\neq|k|}}^{\infty} J_{2n-1}\left[m\pi M\right]\frac{(2n-1)\cos\left[k\frac{\pi}{2}\right]}{(2n-1+k)(2n-1-k)} \end{array}\right] \,. \quad (26)$$

Note that an appropriate formula for asymmetric regular sampling could not be found in the literature. By inspection of (26) several observations can be made:

- Formula (26) is clearly more complex than the LGI-method-based formula (19). Furthermore, in contrast to (19), (26) is only valid for a pure sinusoidal reference signal and different formulas would thus be required in case of other references.
- The result (26) is exact in theory. However, the numerical evaluation is involved due to the present Bessel functions and infinite sums. Especially the latter can only be approximated in practice. Therefore, on the one hand, the DFS method does not yield exact results in practice and on the other hand, a careful error analysis is required as for the LGI method (cf. **Sec. II-F**).
- A fundamental difference between the DFS-based and the LGI-based result lies in the first summation in (26). This summation takes into account the overlapping sideband contributions of all possible combinations (m, z), $m \in \mathbb{N}, k \in \mathbb{Z}$ with $mz + k = q$. In contrast, the LGI-based formula (19) neglects this summation as this results in only small errors if large z are assumed.

In [13], a simplified and more flexible result of the DFS method is presented for 2-level converters. It contains, similar to the LGI-based result presented in this paper, an unevaluated integral instead of Bessel functions and can be used for several reference signals. However, no such result is presented for 3-level and N-level converters and the complexity of the corresponding derivation and possible result remains unclear.

F. Verification and Error Analysis

Fig. 7 compares the significant, i.e. dominant harmonics occurring at the switching frequency multiples $m = \{1, 5, 15\}$, calculated by means of (19) and by means of a simulation-based FFT for the flat-top PWM (24) and $z = 243$. The figure proofs the validity of the formula and that a high accuracy can be obtained.

A systematic and detailed analysis of the achievable accuracy of the presented formula was performed. However, due to reasons of brevity, only the results of this analysis can be presented here.

In a first step, a lower limit $z = z_{min}$ was searched for which assumption (ii), $z\to\infty$, is sufficiently met yielding high accuracy of the calculated harmonics. As for 2-level systems [14], it was found that

$$z \geq z_{\min} = 30 \, , \tag{27}$$

is sufficient yielding amplitude errors below 10 %.

By inspection of **Fig. 7**, it can be observed that the errors tend to increase with increasing harmonic order. A detailed analysis reveals that the total energy of the calculated harmonics decreases when compared to the FFT-based harmonics as the occurring overlapping at high frequencies is not taken into account (cf. **Sec. II-E**). Therefore, in a second step, it was investigated up to what harmonic orders the result still provides sufficient accuracy. It was found that the maximum permissible value of m increases for higher values of z. If the energy deviation between each calculated harmonic group to the corresponding FFT-based results is limited to $< 20\,\%$, the upper limit for m,

$$m_{\max}(z) = \begin{cases} z/5 & \overline{u}_{p,\mathrm{ref}} \in \{\overline{u}_{p,\mathrm{ref}}^{\mathrm{S}}, \overline{u}_{p,\mathrm{ref}}^{\mathrm{S3}}\} \\ z/6 & \overline{u}_{p,\mathrm{ref}} \in \{\overline{u}_{p,\mathrm{ref}}^{\mathrm{SY}}, \overline{u}_{p,\mathrm{ref}}^{\mathrm{FT}}\} \end{cases} , \tag{28}$$

can be found.

Based on the above analysis it can thus be concluded that the obtained formula can be applied to most modern 50/60 Hz grid-connected converters which normally readily fulfill $z \geq z_{\min}$ ($f_{\mathrm{sw}} \geq 1.5/1.8\,\mathrm{kHz}$). Moreover, the defined upper bound in this section mostly represents no serious practical constraint for the application of the result of this paper, as in many applications only the dominant low-order harmonics are of interest.

III. GENERALIZATION FOR N-LEVEL CONVERTERS

The derivations for the 3-level converter shown in **Sec. II** can be generalized with little effort to obtain a universal formula for N-level topologies as depicted in **Fig. 8(a)**. Since the approach and ideas of the derivation are similar to those already shown, for reasons of brevity only the result is presented here,

$$\hat{U}_{\mathrm{a0}(q)} = \frac{2\left[1 - (-1)^{m+k}\right] U_{\mathrm{DC}}}{(N-1)m\pi^2} \int_0^{\frac{\pi}{2}} \cos\left[k\beta\right]$$
$$\cdot \sin\left[\frac{m\pi}{2}\left(1 + N - 2N^*\left[N, \overline{u}_{\mathrm{a,ref}}(\beta)\right] + (N-1)\overline{u}_{\mathrm{a,ref}}(\beta)\right)\right] \mathrm{d}\beta \, . \tag{29}$$

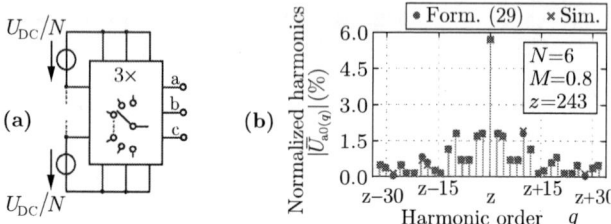

Fig. 8: Calculation of the output voltage spectrum of N-level topologies. **(a)** General three-phase N-level voltage source converter. **(b)** Comparison of the calculated spectra of a 6-level converter using the proposed universal formula (29) and FFT-based values from a simulation.

The function

$$N^*\left[N, \overline{u}\right] = 1 + \left\lfloor \frac{(1 + \overline{u})(N-1)}{2} \right\rfloor , \tag{30}$$

identifies the relevant carrier signal N^* amongst the $(N-1)$ available PD carriers (cf. **Fig. 2(a)**) as a function of the local value of the reference voltage $\overline{u}_{\mathrm{a,ref}}(\beta)$. Formula (29) represents a highly versatile result which can be evaluated for both arbitrary number of voltage levels as well as all reference signals defined in (20). The formula is verified by the example shown in **Fig. 8(b)**.

IV. EXAMPLE OF APPLICATION

In this section, it is demonstrated how to use the derived formulas in a practical EMI filter design example. **Fig. 9** shows a typical 10 kW solar inverter used in residential applications connected to the public mains. All specifications are listed in **Tab. I**. The input DC/DC boost converter ensures a stable DC-link voltage of $U_{\mathrm{DC}} = 650$ V which implies a constant modulation depth of $M = 1.0$ if a constant grid line-voltage of $\hat{U}_{\mathrm{g}} = (230 \cdot \sqrt{2})$ V is assumed.

Depending on the geographic location, different standards concerning EMI must be fulfilled. For residential solar applications in European countries, IEC 61000-6-3 [21] is a typical standard which ultimately requires compliance with the Class B quasi-peak (QP) EMI limits as stated in CISPR 14 [22]. As a consequence, the output filter of the inverter must be accordingly dimensioned. For simplicity reasons, only the DM

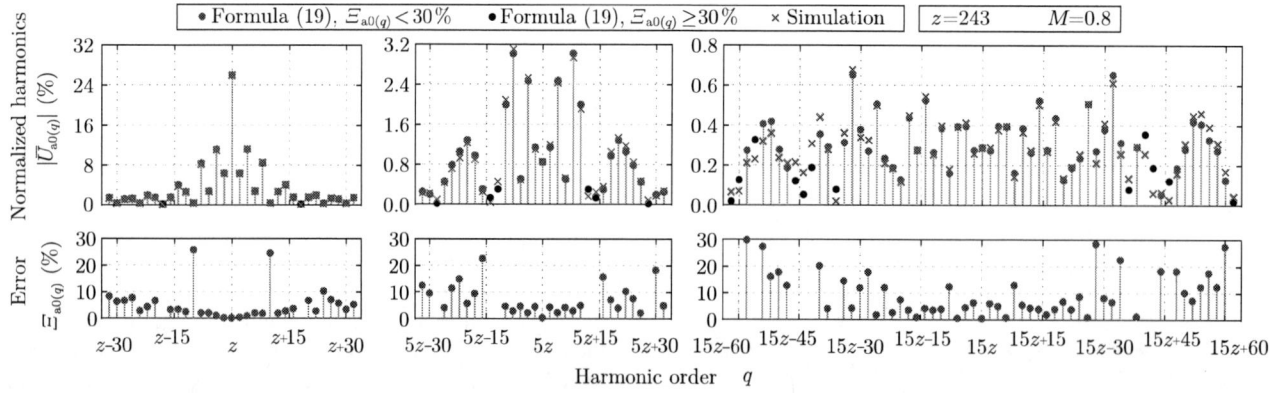

Fig. 7: Verification of (19) for the flat-top PWM (24) by means of a comparison to a simulation-based FFT analysis. It can be observed that at higher multiples m of the switching frequency the growing number of significant sideband harmonics have a lower amplitude but are more widely distributed around the respective center order harmonic at $q = mz$. A high accuracy is achieved in general while large individual errors $\Xi_{\mathrm{a0}(q)} > 30\,\%$ predominantly occur at only non-significant harmonics (marked black). It can be observed, that for increasing harmonic orders the individual errors tend to increase. This is a result of the harmonic groups which start to overlap on the edges at high values of m. As indicated in (28), the process of overlapping occurs already at lower values of m if z is low, as the harmonic groups are located closer to each other than for high values of m.

The 2014 International Power Electronics Conference

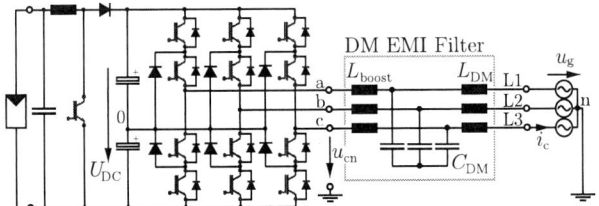

Fig. 9: Typical solar inverter connected to the grounded public mains. The DM EMI filter together with the CM filter (not shown) must be designed so as to fulfill EMI limits such as CISPR 14, Class B. The specifications of this converter are listed in **Tab. I**.

Tab. I: Specifications of the example solar inverter depicted in **Fig. 9**.

General	Rated power	P_r	10 kW
	DC-link voltage	U_{DC}	650 V
	Grid line-voltage	$\hat{U}_g$	$(230 \cdot \sqrt{2})$ V
	Grid frequency	$f_g = f_1$	50 Hz
Modulation	Modulation signal	$u_{p,ref}(\beta)$	$u_{p,ref}^{S3}(\beta)$
	Modulation depth	M	1.0
	Switching frequency	f_{sw}	12.15 kHz
	Frequency ratio	z	243
DM filter	EMI standard	IEC 61000-6-3 (CISPR 14, Class B)	
	Boost inductance	L_{boost}	300 µH
	DM inductance	L_{DM}	30 µH
	DM capacitance	C_{DM}	65 µF

filter is designed here, while in reality additional CM filter components must be employed to attenuate the CM emissions.

According to CISPR 16 [23], the QP EMI noise level of a signal must be determined by means of the following procedure:

1.) Filtering of the signal with a bandpass filter (RBW filter) with 9 kHz ($= 2 \cdot 90 \cdot f_1$) window width and a filter center frequency of 150 kHz.
2.) Application of the obtained filtered signal u_{RBW} to the predefined QP detector circuit as depicted in **Fig. 10(a)** which provides the steady-state QP emission level u_{QP} at the output.
3.) Sweep of the RWB filter center frequency from 150 kHz to 30 MHz and continuous repetition of steps 1.) and 2.) to obtain the frequency-dependent QP emissions $u_{QP}(f)$.

If a calculation of the EMI emissions is preferred over using an actual hardware prototype and appropriate measurement equipment, either a simulation or, more flexibly, the formulas derived in this paper can be used. For the given application

in **Fig. 9**, formula (19) was used to create a table of the harmonics of the output voltages u_{pn} as shown in **Fig. 10(a)**. The described filtering process of step 1.) was then performed by synthesizing the filtered time-domain signal u_{RBW} using the calculated DM harmonics within the sweeping RBW filter window. Step 2.) was performed by means of an equivalent mathematical representation of the QP detector circuit [24]. Note that the (mostly imperative) distinction between DM and CM noise can easily be done in the frequency-domain of this approach and normally requires extra effort in a simulation.

Fig. 10(b) shows the calculated harmonic DM spectrum of the voltages u_{pn} at the ports {a,b,c} before the filter. Furthermore, the determined QP emission level u_{QP} is depicted. The Class B limits do not decrease beyond 1 MHz, which is why the critical frequency for the filter design can be identified in **Fig. 10(b)** at $f_{crit}=158$ kHz. The inspection of the plot reveals a required DM attenuation A_{DM} of 75 dB at this frequency. Since the shown Class B limits apply to the total DM and CM emissions, a common practice is to increase the required DM attenuation by 6 dB to obtain a margin for the subsequent CM filter design [24]. The resulting filtered spectrum is measured at the interface {L1,L2,L3} between converter and mains by means of a line impedance stabilization network (LISN, [23]). This measurement circuit has an approximate inner resistance of $R_{LISN} = 50\,\Omega$ for frequencies above 150 kHz. The total attenuation of the DM filter and LISN can thus be approximated with

$$\frac{1}{A_{DM}(f)} = \left| \frac{i_p(f)}{u_{pn}(f)} \right| \cdot R_{LISN} \approx \frac{R_{LISN}}{L_{boost}C_{DM}L_{DM}(2\pi f)^3} , \quad (31)$$

if linear filter components are assumed. Using (31) and the component values in **Tab. I**, the desired attenuation

$$A_{DM}(f_{crit} = 158\,\text{kHz}) \approx 1.14 \cdot 10^4 \approx 81.14\,\text{dB} , \quad (32)$$

results. **Fig. 10(c)** shows the calculated attenuated DM spectrum as well as the respective emission levels u_{QP}. It can be seen that the Class B limits are clearly met with a margin of approximately 11 dB. The discrepancy to the targeted 6 dB is due to the approximation in the filter transfer function (31). The validity and quality of the approach based on the derived formula in this work is proven by the high accuracy when compared to the simulated emission levels. The error is around 2 dB at the upper limit $m_{max} = 48 \hat{=} 583$ kHz as defined in **Sec. II-F** and reaches only 3 dB at 1 MHz ($\hat{=} m \approx 82$).

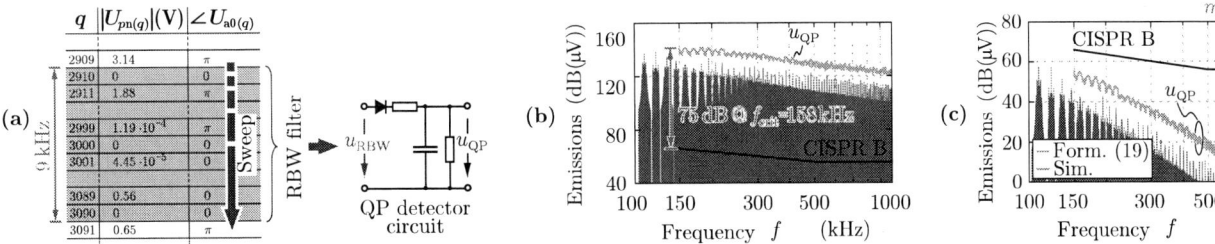

Fig. 10: Analysis and design steps of the DM EMI filter of the application shown in **Fig. 9**. **(a)** Calculation of the quasi-peak (QP) emission levels at the converter output ports {a,b,c}. The calculated DM harmonics ($q \neq 3n$) within the bandpass RBW filter with window width of 9 kHz are used to synthesize the filtered signal u_{RBW} in time domain, which can then be applied to the QP detector circuit with defined charge and discharge constants. The resulting steady-state voltage across the capacitor is the measured QP emission level u_{QP}. This procedure is repeated while sweeping the center frequency of the RBW filter from 150 kHz ($\hat{=} q=3000$) to 30 MHz. **(b)** Calculated spectrum of **(a)**, determined DM emission levels u_{QP} and permissible QP emission levels according to CISPR 14, Class B [22]. A minimum damping of 75 dB must be achieved at the critical frequency $f_{crit}=158$ kHz. **(c)** Calculated emissions including the filter at the ports {L1,L2,L3} and comparison to the emission values obtained with the FFT-based spectrum from a simulation. The error remains lower than 2 dB up to the upper limit $m_{max}=48 \hat{=} 583$ kHz as defined in **Sec. II-F** and reaches 3 dB at 1 MHz ($\hat{=} m \approx 82$).

V. CONCLUSION

In this paper, an analytical formula for the approximate calculation of the switching frequency harmonics of idealized PWM-controlled 3-phase 3-level voltage-source DC/AC power converters was presented. The formula is based on the assumption of an infinitely high ratio z between switching and output fundamental frequency. This assumption is, however, fulfilled to a sufficient degree by most modern converter systems, allowing for accurate calculations of the spectra. An in-depth error analysis has shown that $z \geq 30$ is sufficient, while good accuracy can be obtained up to high harmonic orders. The simple formula which can be evaluated for almost arbitrary modulation signals is thus an attractive alternative to exact methods, whose demanding derivations yield complex formulas which can mostly be evaluated for a single modulation signal only.

The presented method was further used to derive a universal yet still simple formula for general N-level converters and again arbitrary modulation signals. Furthermore, it is shown how the effect of non-linear, current-dependent inductors in the output impedance can be incorporated for systems with decoupled phases or, equivalently, single-phase systems.

A design example of an EMI filter demonstrated the usefulness and accuracy of the derived formulas. The MATLAB code presented in the **Appendix** can be used to calculate most of the presented results.

Future work could include a more comprehensive error analysis. Particularly the impact of varying the number of voltage levels N on the accuracy in general, z_{min} and m_{max} could be investigated. Finally, more advanced formulas could be derived with the LGI method which incorporate other non-ideal effects besides non-linear inductors. Examples are switching delays, finite switching speeds of the semiconductors and a varying DC-link voltage, which all lead to a distortion of the assumed ideal rectangular voltage pulse trains.

APPENDIX
MATLAB CODE FOR THE OUTPUT VOLTAGE SPECTRUM OF AN N-LEVEL CONVERTER

The derived formulas in this work are most suitably implemented in MATLAB using the Symbolic Math Toolbox. This enables a clear and flexible code while benefiting from MATLAB's efficient and accurate numeric methods. Here, the MATLAB code for the generalized N-level formula (29) is presented.

First the desired modulation function must be defined, such as the flat-top reference,

```
uref = @(M,beta)...
   (1.).*        (beta >= 0 & beta <= pi/6)+...
   (-1.+sqrt(3.).*M.*cos(beta-pi./6)).*...
                 (beta > pi/6 & beta <= pi/2)+...
   (1.-sqrt(3.).*M.*cos(-beta+5.*pi./6)).*...
                 (beta > pi/2 & beta <= 5*pi/6)+...
   (-1).*        (beta >= 5*pi/6 & beta <= pi);
```

Other modulation signals can be coded analogously. In a next step, the integrand of the N-level formula (29) and (30) are defined,

```
N_ast = @(N,M,beta)...
   1+floor((uref(M,beta)+1)./(2/(N-1)));

int_NLevel = @(UDC,N,m,k,M,beta)...
   2.*(1-(-1)^(m+k)).*UDC/((N-1).*m.*pi^2).*...
```

```
   cos(k.*beta).*sin(1/2.*m.*pi.*...
   (1+N-2.*N_ast(N,M,beta)+(-1+N).*uref(M,beta)));
```

An arbitrary harmonic can now be calculated using MATLAB's `integral()` method,

```
UDC=1000; N=3; m=1; k=0; M=1.;

U_mz_k = integral(@(beta)...
   int_NLevel(UDC,N,m,k,M,beta) ,0,pi/2)
```

REFERENCES

[1] M. Odavic, M. Sumner, P. Zanchetta, and J. Clare, "A Theoretical Analysis of the Harmonic Content of PWM Waveforms for Multiple-Frequency Modulators," *IEEE Trans. PE*, vol. 25, no. 1, pp. 131–141, 2010.

[2] J. T. Boys and P. G. Handley, "Harmonic Analysis of Space Vector Modulated PWM Waveforms," *Proc. IEE*, vol. 137, no. 4, pp. 197–204, 1990.

[3] R. Guinee and C. Lyden, "A Novel Single Fourier Series Technique for the Simulation and Analysis of Asynchronous Pulse Width Modulation," in *Proc. APEC*, vol. 1, pp. 123–128 vol.1, 1998.

[4] G. Fedele and D. Frascino, "Spectral Analysis of a Class of DC-AC PWM Inverters by Kapteyn Series," *IEEE Trans. PE*, vol. 25, no. 4, pp. 839–849, 2010.

[5] D. Kostic, Z. Avramovic, and N. Ciric, "A New Approach to Theoretical Analysis of Harmonic Content of PWM Waveforms of Single- and Multiple-Frequency Modulators," *IEEE Trans. PE*, vol. 28, no. 10, pp. 4557–4567, 2013.

[6] W. R. Bennett, "New Results in the Calculation of Modulation Products," *Bell System Technical Journal*, vol. 12, no. 2, pp. 228–243, Apr. 1933.

[7] H. Black, *Modulation Theory*, 1st ed. Van Nostrand Reinhold Company, 1953.

[8] S. Bowes and B. Bird, "Novel Approach to the Analysis and Synthesis of Modulation Processes in Power Convertors," *Proc. IEE*, vol. 122, no. 5, pp. 507–513, 1975.

[9] S. Bowes, "New Sinusoidal Pulsewidth-Modulated Invertor," *Proc. IEE*, vol. 122, no. 11, pp. 1279–1285, 1975.

[10] G. Carrara, S. Gardella, M. Marchesoni, R. Salutari, and G. Sciutto, "A New Multilevel PWM Method: A Theoretical Analysis," *IEEE Trans. PE*, vol. 7, no. 3, pp. 497–505, 1992.

[11] J. F. Moynihan, M. Egan, and J. M. D. Murphy, "Theoretical Spectra of Space-Vector-Modulated Waveforms," *Proc. IEE EPA*, vol. 145, no. 1, pp. 17–24, 1998.

[12] M. Bierhoff, F. Fuchs, and S. Pischke, "Theoretical Output Current Spectra of Three Phase Current Source Converters," in *Proc. PEA*, pp. 9 pp.–P.9, 2005.

[13] D. G. Holmes and T. A. Lipo, *Pulse Width Modulation for Power Converters*. New York: Wiley, 2003.

[14] H. van der Broeck, "Analysis of the Voltage Harmonics of PWM Voltage Fed Inverters Using High Switching Frequencies and Different Modulation Functions," *ETEP*, vol. 2, no. 6, pp. 341–350, Nov./Dez. 1992.

[15] K. Zhou and D. Wang, "Relationship Between Space-Vector Modulation and Three-Phase Carrier-Based PWM: A Comprehensive Analysis," *IEEE Trans. IE*, vol. 49, no. 1, pp. 186–196, 2002.

[16] *Electromagnetic Compatibility (EMC) – Part 3-12: Limits – Limits for Harmonic Currents Produced by Equipment Connected to Public Low-voltage Systems with Input Current >16 A and ≤75 A per Phase*, IEC 61000-6-4, 2011.

[17] *Recommended Practices and Requirements for Harmonic Control in Electrical Power Systems*, IEEE 519, 1992.

[18] *Technical Guideline Generating Plants Connected to the Medium-Voltage Network*, BDEW Std., 2008.

[19] J. W. Kolar, H. Ertl., and F. C. Zach, "Calculation of the Passive and Active Component Stress of Three-Phase PWM Converter Systems with High Pulse Rate," in *Proc. EPE*, pp. 1303–1311, 1989.

[20] J. Kolar, H. Ertl, and F. C. Zach, "Influence of the Modulation Method on the Conduction and Switching Losses of a PWM Converter System," *IEEE Trans. IA*, vol. 27, no. 6, pp. 1063–1075, 1991.

[21] *Electromagnetic Compatibility (EMC) – Part 6-3: Generic Standards – Emission Standard for Residential, Commercial and Light-Industrial Environments*, IEC 61000-6-3, 2006.

[22] *Electromagnetic Compatibility – Requirements for Household Appliances, Electric Tools and Similar Apparatus*, CISPR 14, 2008.

[23] *Specification for Radio Disturbance and Immunity Measuring Apparatus and Methods*, CISPR 16, 2010.

[24] M. L. Heldwein, "EMC Filtering of Three-Phase PWM Converters," Ph.D. dissertation, ETH Zurich, 2007.

The 2014 International Power Electronics Conference

Dynamic Voltage Balancing Algorithm for Modular Multilevel Converter with Three-Level Flying Capacitor Submodules

Apparao Dekka$^\diamond$, Bin Wu$^\diamond$ and Navid R. Zargari†

$^\diamond$ Dept. of Electrical and Computer Engineering, Ryerson University, Toronto, ON M5B 2K3, Canada
† Medium-Voltage Drive R&D, Rockwell Automation, Cambridge, ON N1R 5X1, Canada
E-mail: adekka@ee.ryerson.ca, bwu@ee.ryerson.ca and nrzargari@ra.rockwell.com

Abstract—The modular multilevel converter (MMC) has several three-level flying capacitor (3L-FC) submodules in cascade. To balance the submodule capacitors voltage, this paper describes a dynamic voltage balancing algorithm based on the carrier phase shifted pulse width modulation (CPS-PWM) scheme. The submodules are controlled based on the instantaneous value of capacitor voltage and the direction of current. For the selection of the submodules, the maximum and minimum voltage submodule selection logics are designed. These voltage logics are generating an index number for each submodule based on the relative comparison of capacitors voltage. Finally the switching state of the submodules is generated by comparing the submodule index number with the dynamic reference index number. The performance of the proposed voltage balancing algorithm at different operating conditions is evaluated on 6kV/2MVA MMC system with the MATLAB simulation and the corresponding results are presented. In addition, the performance comparison of MMC with 3L-FC and conventional two-level half bridge (2L-HB) submodules is presented.

Index Terms—Capacitor voltage balancing, circulating currents, flying capacitor (FC) converter, modular multilevel converter (MMC), pulse width modulation (PWM)

I. INTRODUCTION

The increase in the production rate and the economy of scale are demands the high power, multilevel converter systems [1]–[3]. In the last few decades, various multilevel converter topologies are investigated and are presented in [1]–[3]. The most popular multilevel converter topologies are (i) neutral point clamped (NPC) converter, (ii) flying capacitor (FC) converter and (iii) cascade H-bridge converter (CHB) [1]–[3]. The application of these converter topologies at high voltage, high power level is restricted by the semiconductor ratings, extra number of devices, assembly of the converter, dc link capacitors and the dc voltage balancing capability etc [1]–[3].

In recent years, the modular multilevel converter has gained more attention for high voltage applications due to its inherent features such as modular structure, common dc bus terminals and the possibility of back-back operation etc [4]–[8]. The modular multilevel converter is first introduced in [8]–[10] for high voltage direct current (HVDC) applications. The research on MMC is continuously growing in the areas of medium voltage motor drives [6], medium voltage static compensator [11], solar photo voltaic application [7] etc.

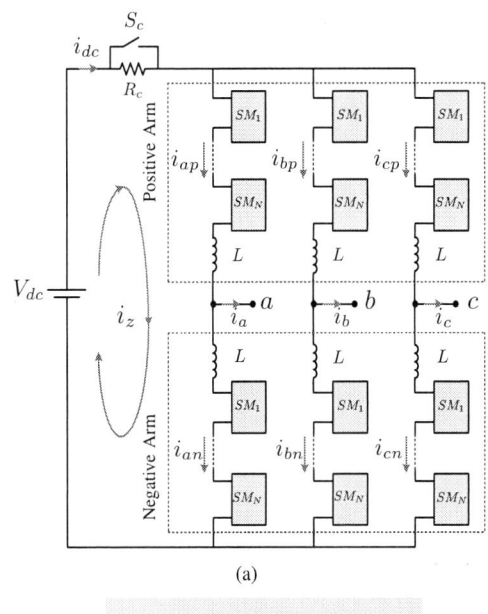

(a)

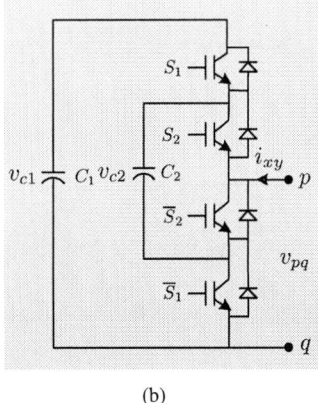

(b)

Figure 1. (a) Power circuit of three phase MMC (b) Three level half bridge flying capacitor converter

Various configurations of MMC and their applications are discussed in [4]–[8]. The generalized structure of the modular multilevel converter is shown in Fig. 1(a), uses two-level half bridge (2L-HB) converter as submodule. To meet the required operating voltage, these submodules are connected in cascade.

978-1-4799-2706-7/14 $31.00 © 2014 IEEE

However it is also possible to use the well established multilevel converter technology in MMC. The three-level flying capacitor (3L-FC) submodule is considered as an alternative for two-level half bridge (2L-HB) submodule, which is shown in Fig. 1(b). With any submodule technology, the voltage balancing of floating capacitors is necessary. In the literature, various voltage balancing algorithms are discussed for MMC with 2L-HB submodules [7], [12]–[16]. These algorithms are facing the following difficulties such as complexity in the implementation for higher number of submodules, accuracy and the requirement of converter parameters, requirement of huge memory to store the mapping process and the stability issues etc.

In this paper a new generalized voltage balancing algorithm is proposed for modular multilevel converter. The proposed algorithm is also applicable to the conventional two-level half bridge (2L-HB) submodules. The balancing algorithm does not require the sorting technique to select the submodules, which will reduce the burden on the controller. Here a simple submodule selection logics are designed based on the relative comparison of normalized submodule capacitors voltage. The final switching state for the submodules is generated by dynamically. With the present approach, the implementation and the extending to higher number of modules is quite easy. In addition, a simple modulation stage based on the carrier phase shifted pulse width modulation (CPS-PWM) scheme is presented for modular multilevel converter. The output of the modulation stage is a normalized reference PWM waveform, which resembles the actual output voltage waveform of the converter. The effectiveness of the proposed algorithm is verified at different operating conditions by means of MATLAB simulation and the results are presented. The main features of the proposed modulation stage and balancing algorithms are:

- It avoids the extra computations such as calculation of required number of on-state cells at every switching instant
- In addition to the voltage balancing, it can generate the predefined voltage waveform at the output of the converter
- Implementation and extending to the higher number of modules per arm is easy
- It does not require the sorting technique for the selection of submodules

In addition to the voltage balancing, the circulating currents is a one of the major issue in modular multilevel converter. The second harmonic current component is a dominant component in the circulating currents and its magnitude is depends on the modulation index (M), fundamental frequency (F) and power factor (PF) [17]–[21]. These circulating currents have the severe impact on the performance of the modular multilevel converter, especially the harmonic distortion, device power losses and the converter efficiency are deteriorates. The circulating currents are minimized by using buffer inductors [17]. However, the size of the buffer inductors is small so its impact

Table I
SWITCHING STATES OF 3L-FC SUBMODULE

S_1	S_2	$\overline{S}_1$	$\overline{S}_2$	v_{pq}	$i_{xy} > 0$	$i_{xy} \leq 0$
1	1	0	0	v_{c1}	$v_{c1}\uparrow, v_{c2}\approx$	$v_{c1}\downarrow, v_{c2}\approx$
0	1	1	0	v_{c2}	$v_{c1}\approx, v_{c2}\uparrow$	$v_{c1}\approx, v_{c2}\downarrow$
1	0	0	1	$v_{c1} - v_{c2}$	$v_{c1}\uparrow, v_{c2}\downarrow$	$v_{c1}\downarrow, v_{c2}\uparrow$
0	0	1	1	0	$v_{c1}\approx, v_{c2}\approx$	$v_{c1}\approx, v_{c2}\approx$

on the circulating currents is limited. The control method using resonant current controller is presented in [18]. In this method, the current controller is designed to eliminate the dominant harmonic component from the circulating currents. The control method based on the synchronous reference frame theory with the simple proportional-integral (PI) controller is discussed in [21]. However, the circulating currents can be minimized by using properly designed multilevel submodules. In this paper, the 3L-FC submodule is proposed to minimize the circulating currents. By selecting the proper capacitance value of the floating capacitors, the voltage ripples and the magnitude of circulating currents are significantly minimized. To study the effectiveness of the 3L-FC submodule, a performance comparison between the MMC with 3L-FC and 2L-HB submodules is presented.

The paper is organized as follows: Section II provides the basic structure and the operation details of the MMC. The section III gives the modulation scheme and the voltage balancing algorithm for MMC. In section IV, the effectiveness of voltage balancing algorithm is presented. In addition, the performance comparison of MMC with 3L-FC and 2L-HB submodules at the different operating power factors is discussed. Finally conclusions are given in section V.

II. MODULAR MULTILEVEL CONVERTER

A. Basic Structure

The generalized three phase structure of the modular multilevel converter is shown in Fig. 1(a), which is realized by using two three-phase CHB structures connected their output terminals through buffer inductors (L_{xy}). Where x represents the phase ($x = a, b, c$) and y represents the arm ($y = p, n$). The output terminal (x) is divided the each phase into two arms, positive arm (p) and negative arm (n), and each arm consists of N-submodules and a buffer inductor (L_{xy}). The structure of the submodule is shown in Fig. 1(b), which is a three-level flying capacitor (3L-FC) topology. Each submodule is further divided into two cells, where S_1, $\overline{S}_1$ and C_1 are considered as cell-1, and S_2, $\overline{S}_2$ and capacitor C_2 are considered as cell-2. Each 3L-FC submodule has two floating capacitors C_1 and C_2 with a voltages of v_{c1} and v_{c2} respectively. The outer capacitor voltage (v_{c1}) is twice that of inner capacitor voltage (v_{c2}). The charging and discharging of these floating capacitors is depends on the direction of arm current and the corresponding switching states. The Table I shows the output voltage levels

978-1-4799-2706-7/14 $31.00 © 2014 IEEE

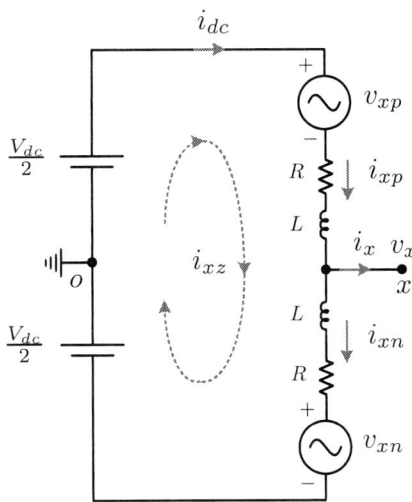

Figure 2. Single phase equivalent circuit of MMC

and the variation of capacitors voltage in 3L-FC submodule for different switching states.

B. Principle of Operation

Depends on the switching state, the 3L-FC submodule can generate a different voltage levels at the output. So each submodule can be modeled as a controlled voltage source and each arm contains N-controlled voltage sources in cascade. The output arm voltage is equal to the sum of N-controlled voltage sources. For the analysis, the MMC system can be further simplified and the single phase equivalent circuit is shown in Fig. 2. Where V_{dc} and i_{dc} are the total dc bus voltage and dc current, L and R are the buffer inductance and their equivalent resistance, v_x and i_x are the converter output voltage and output line current. The output voltage of the cascaded submodules in positive and negative arms are represented as a AC voltage source of v_{xp} and v_{xn} respectively. The required voltage at the output (v_{xo}) is generated by controlling the voltage sources of positive (v_{xp}) and negative arms (v_{xn}). From the equivalent circuit shown in Fig.2, the current flowing through the positive and negative arms are given as

$$i_{xp} = i_{xz} + \frac{i_x}{2} \tag{1}$$

$$i_{xn} = i_{xz} - \frac{i_x}{2} \tag{2}$$

where i_{xp} and i_{xn} are the current flowing through the positive and negative arms respectively, i_{xz} is the circulating current and i_x is the output current flowing through the load. By adding the equation (1) and (2), the circulating current flowing through the phase is given as

$$i_{xz} = \frac{i_{xp} + i_{xn}}{2} \tag{3}$$

By subtracting the equation (2) from (1), then the output current is becomes

$$i_x = i_{xp} - i_{xn} \tag{4}$$

From the equivalent circuit shown in Fig. 2, the output phase voltage with respect to the dc bus midpoint is expressed as follows:

$$v_{xo} = \frac{1}{2}\left[v_{xn} - v_{xp} + L\frac{d(i_{xn} - i_{xp})}{dt} + R(i_{xn} - i_{xp})\right] \tag{5}$$

substitute the equation (4) in (5), then the final output voltage is become as follows

$$v_{xo} = \frac{1}{2}\left[v_{xn} - v_{xp} - L\frac{di_x}{dt} - Ri_x\right] \tag{6}$$

The voltage across the buffer inductor due to the circulating currents is given as

$$L\frac{di_{xz}}{dt} + Ri_{xz} = \frac{1}{2}(V_{dc} - v_{xp} - v_{xn}) \tag{7}$$

From the equations (6) and (7), the output voltage and the circulating currents are realized by controlling the positive and negative arm voltages. Based on the above analysis, the positive and negative arm voltages are deducted as

$$\begin{aligned} v_{xp} &= \frac{V_{dc}}{2} - v_x \\ v_{xn} &= \frac{V_{dc}}{2} + v_x \end{aligned} \tag{8}$$

Where v_x is the output phase voltage, which is in the following form:

$$v_x = \frac{1}{2}V_{dc} * M * sin(\omega_0 t + \theta_x) \tag{9}$$

where M is the modulation index ($0 \leq M \leq 1$), θ_x is the initial output phase angle and ω_0 is the output angular frequency. The equation (8) shows, each arm in MMC can be controlled independently by using individual modulating signals. Even though, the above analysis and the conclusions are deducted based on the single phase equivalent circuit, but this can be applicable to the MMC system with the any number of phases.

III. MODULATION SCHEME AND VOLTAGE BALANCING ALGORITHM

A. Modulation Scheme for MMC

The number of voltage levels at the output of the MMC is depends on the phase shift between the triangular carrier signals and the number of submodules. The 3L-FC submodule is considered as an alternative to the 2L-HB submodule. Each 3L-FC submodule can be controlled by using two phase shifted triangular carrier signals [12], so for $2N$-submodules per phase requires a $4N$-triangular carrier signals. The $4N$-triangular carriers are disposed with a phase shift of $\frac{360°}{4N}$ and the carriers which belongs to the same arm are disposed with a phase shift of $\frac{360°}{2N}$. With the above carrier arrangement, the modular multilevel converter with N-3L-FC submodules per arm can generate a $2N + 1$ voltage levels at the output. In addition to the carrier signals, the PWM modulator is requires a reference modulating signal. In modular multilevel converter, each phase can be divided into two arms and each arm can be controlled independently. So it requires a two reference modulating signals, one for the positive arm (v_{xp}^{ref}) and another

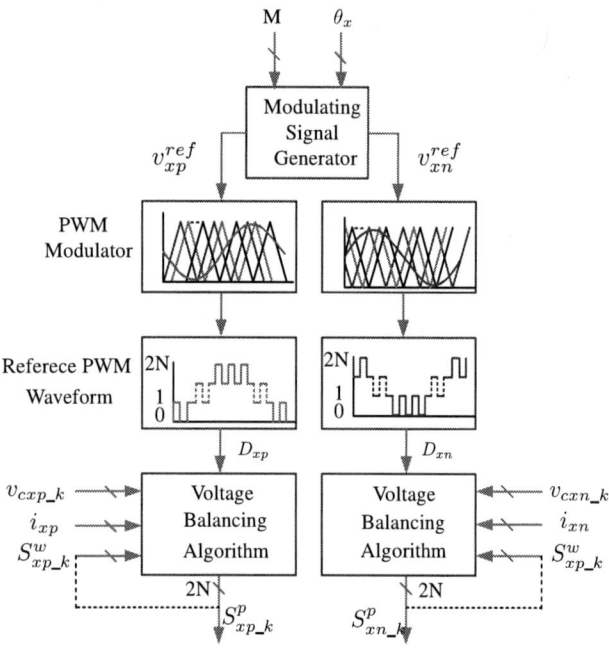

Figure 3. Single phase based pulse width modulator for MMC

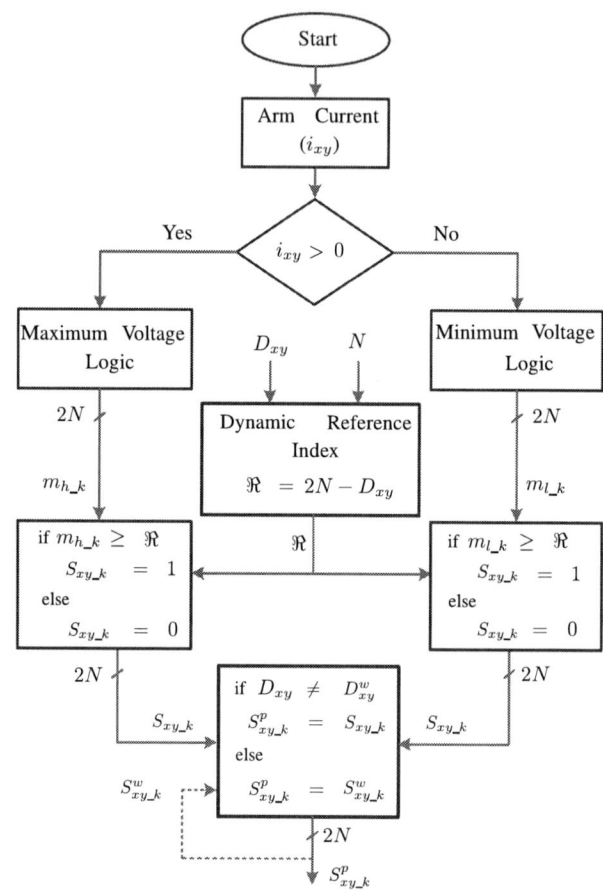

Figure 4. Flow chart of voltage balancing algorithm for MMC

for the negative arm (v_{xn}^{ref}), where x represents the phase ($x = a$, b, c). The individual reference modulating signals are generated by adding the fixed offset to the reference output phase voltage. The reference modulating signals for each arm is given as

$$v_{xp}^{ref} = \frac{V_{dc}}{2} - v_x$$
$$v_{xn}^{ref} = \frac{V_{dc}}{2} + v_x \tag{10}$$

where v_x is the reference output phase voltage, which is given in the equation (9). The structure of the PWM modulator is shown in Fig. 3, where the reference modulating signal for each arm is generated corresponding to the modulation index (M) and the initial phase angle (θ_x). The reference modulation signal of the arm is compared with the each triangular carrier signal of the corresponding arm. The output of each comparator is summed up, which gives the normalized reference PWM waveform (D_{xy}), where y represents the arm ($y = p$, n). The reference PWM waveform (D_{xy}) represents the normalized voltage levels of 0, 1, 2, ..., 2N, which is also equal to the required number of on-state cells. If D_{xy}=2N, then the required number of on-state cells is equal to $2N$. The output of PWM modulator is a reference PWM waveform, which resembles the actual output voltage waveform of the converter. The normalized PWM waveform is given as the input to the voltage balancing algorithm. The voltage balancing algorithm will select and apply the required switching state to the submodule based on the required number of on-state cells, instantaneous value of submodule capacitors voltage and the direction of arm current.

B. Voltage Balancing Algorithm

The balancing of submodule capacitors voltage is achieved either by using external controllers or by using the redundant switching states. The complexity in the designing of the external controllers for voltage balancing is increases with the number of submodules. Also this method is involves the stability problems due to the interaction between the controllers [21]. In contrast, the voltage balancing with the redundancy switching states is the easiest method. However, the number of redundant switching states is drastically increases with number of submodules. The selection of required switching state from such a high redundancy is difficult and requires a sophisticated method. Also this method is involves the huge memory to store the redundancy switching states. To minimize the complexity and the requirement of memory, this paper is describes a generalized voltage balancing algorithm based on the dynamic switching state selection approach. The proposed approach is easily implemented in the digital controllers like DSP/FPGA, computational less complex and easily extendable to the higher number of submodules. The voltage balancing algorithm is mainly involves the following steps:

- step-1: Calculation of required number of submodules
- step-2: Selection of submodules
- step-3: Generation of switching state of submodules

The 2014 International Power Electronics Conference

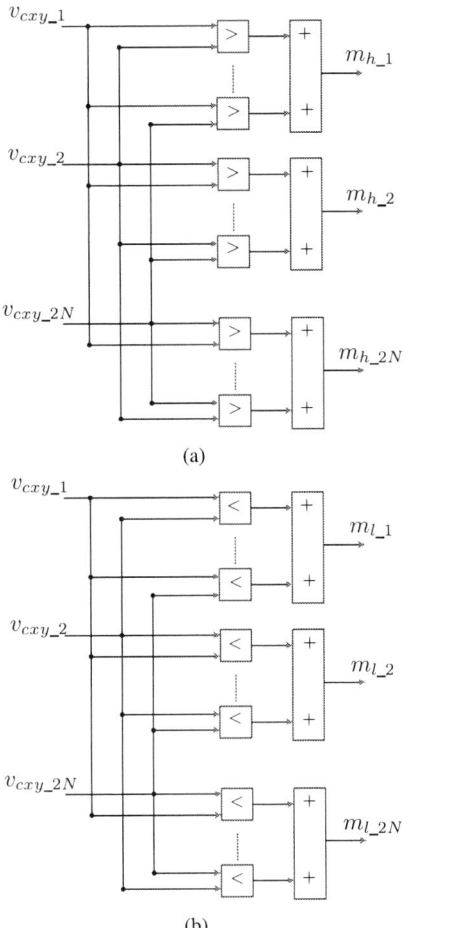

(a)

(b)

Figure 5. Submodule selection logics: (a) Maximum voltage logic (b) Minimum voltage logic

- step-4: Duty cycles of submodules
- step-5: Generation of gating signals for submodules

However, with the proposed structure of the PWM modulator, some of the above steps (step-1 and 4) can be eliminated from the voltage balancing algorithm. The output of the PWM modulator is a normalized PWM waveform, which provides the information of required number of submodules and the duty cycles of the selected submodules. This will further simplify the computational complexity of the voltage balancing algorithm. The flow chart of the proposed voltage balancing algorithm is shown in Fig. 4. To select the submodules, the algorithm requires the direction of arm current and the instantaneous value of capacitors voltage. Based on the direction of arm current, the algorithm will select the maximum or minimum voltage submodules selection logic and are shown in Fig. 5. The selection logics are requires the instantaneous value of capacitors voltage as input. Each submodule has two floating capacitors with the different voltages. Each capacitor voltage is normalized w.r.t their corresponding nominal capacitor voltage and is given as input to the submodule selection logics. With this approach, the modular structure of selection logics is maintained and also it can be extendable to any number and

the any type of submodules without any major changes. In submodule selection logic, each normalized capacitor voltage is compared with the other normalized capacitor voltages. The output of the each comparator is added together and the resultant output is the index number of the corresponding cell (m_{h_k} or m_{l_k}). Similarly, the relative comparison between the other capacitors voltages is executed and the corresponding index numbers are provided by the submodule selection logics. The index number of each cell represents the priority order of that particular cell. The switching state for the each cell is generated by comparing the cell index number with the dynamic reference index number ($\Re$). If cell index number (m_{h_k} or m_{l_k}) is greater than or equal to the reference index number ($\Re$) then the corresponding cell is turned on. The dynamic reference index number is generated based on the equation (11)

$$\Re = 2N - D_{xy} \qquad (11)$$

where N is the number of submodules per arm and D_{xy} is the reference PWM waveform. From the equation (11) the variation of reference index number is opposite and proportional to the D_{xy}. But the variation of cell index number is proportional to the capacitors voltage, so any change in the capacitor voltage will effect the m_{h_k} or m_{l_k}. This variation will further increase the device switching frequency, but the average switching frequency of the converter is remain constant. To minimize the device switching frequency, the present voltage level (D_{xy}) is compared with the previous voltage level (D_{xy}^w). If the present and previous voltage levels are equal then the algorithm will apply the previous switching state ($S_{xy\_k}^w$) to the converter. By using this approach the device switching losses can be minimized. The proposed method is computationally less complex and can be easily extended to any number of submodules per arm. Also the selection of redundant switching states is simple.

IV. SIMULATION RESULTS

A. Dynamic Performance of Voltage Balancing Algorithm

The modular multilevel converter is designed for 6kV, 2MW power level. The system shown in Fig. 1(a) is considered for the simulation study with the 3L-FC submodules. To validate the effectiveness of the proposed balancing algorithm, the simulations are carried out by using MATLAB/Simulink under different operating conditions with the parameters given in Table II. Before applying the gating signals, the floating capacitors are needs to be pre-charged to avoid the inrush currents. Generally the floating capacitors are pre-charged by using current limiting resistor (R_c) as shown in Fig. 1(a). However the present study is mainly focused on the steady state performance of the modular multilevel converter, so the start-up process is not discussed. The effectiveness of the proposed algorithm is presented for worst case scenario, where the initial capacitors voltage of all the submodules is not equal. The results for phase-a is presented with modulation index M=0.99 and fundamental frequency F=60 Hz in Fig. 6. At t = 0 sec, the balancing algorithm is enabled. The

978-1-4799-2706-7/14 $31.00 © 2014 IEEE   3472

The 2014 International Power Electronics Conference

TABLE II
SIMULATION PARAMETERS

Parameter	SI
DC voltage (V_{dc})	10 kV
No. of submodules per arm (N)	5
Submodule capacitance (C_1 & C_2)	2.2 μF
Capacitor voltage (v_{c1} & v_{c2})	2 kV & 1 kV
Buffer inductor (L_{xy})	5 mH
Carrier frequency (F_c)	200 Hz
Line-Line voltage (v_{ab})	6 kV
Power (P)	2 MW
Frequency (F)	60 Hz

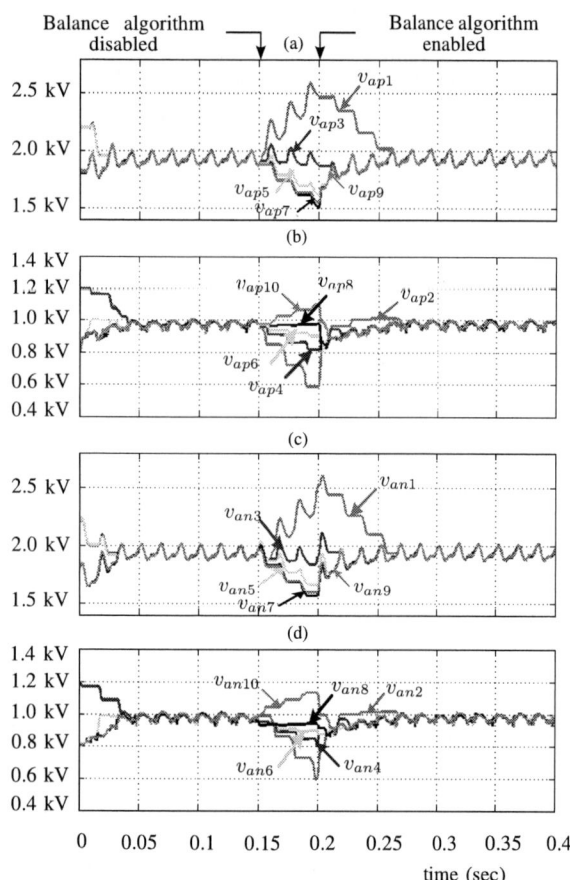

Figure 6. Simulation results of phase-a: (a) Positive arm outer capacitor voltages (b) Positive arm flying capacitor voltages (c) Negative arm outer capacitor voltages (d) Negative arm flying capacitor voltages

capacitor voltages are reached to its nominal value within three fundamental cycles. The balancing algorithm is disabled at t = 0.15 sec and the capacitor voltages are diverging from its nominal value. As the duty cycles of each cell is different, so the variation in capacitors voltage is not equal. Again at t = 0.2 sec, the balancing algorithm is enabled and the capacitor voltages are reached to its nominal value within three cycles.

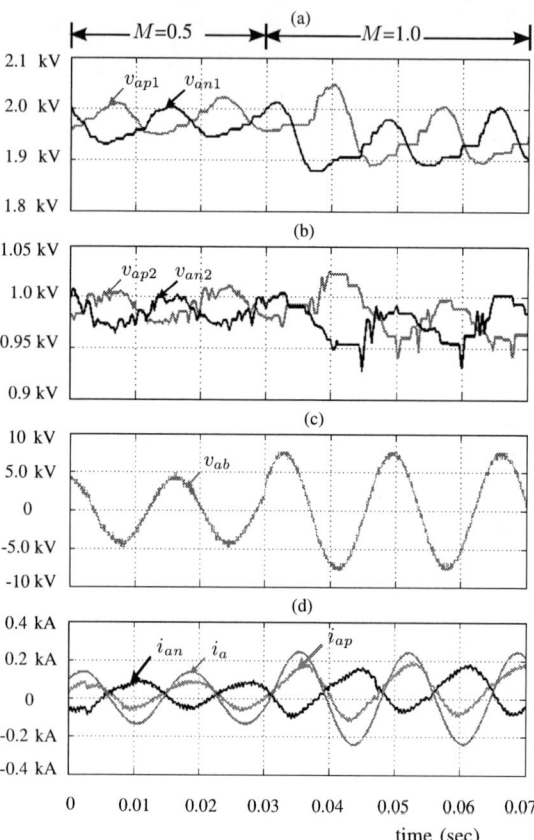

Figure 7. Simulation results of phase-a: (a) Positive and negative arm outer capacitor voltage (b) Positive and negative arm flying capacitor voltage (c) Line-Line voltage waveform (d) Arm currents and output line current waveform

Also the effect of load variation on the performance of the balancing algorithm is studied and the results are shown in Fig. 7. Here the load variation is implemented by changing the modulation index (M). During this study, initially the converter is operating at steady state with the modulation index of M=0.5 and the capacitors voltage is perfectly balanced and maintained at its nominal value. At t=0.03 sec, the modulation index is changed from 0.5 to 1.0. During the transient, the change in the capacitors voltage is insignificant and is less than 1 %. The capacitors voltage is reached to the steady state within one fundamental cycle. But the voltage ripples are high as compared to M=0.5, due to the increase in the load current. From the current and voltage waveforms, it is seen that there is no abrupt change in the magnitudes. From the results, it is clear that the proposed algorithm is effectively balancing the capacitor voltages at its nominal value without effecting the converter and load performance.

B. Performance of MMC with the Novel Submodules

This section presents the performance of the MMC with the novel submodules interms of the magnitude of circulating currents w.r.t the load power factor. This paper proposes the 3L-FC submodule to improve the MMC performance by minimizing the magnitude of circulating currents. To further

978-1-4799-2706-7/14 $31.00 © 2014 IEEE

The 2014 International Power Electronics Conference

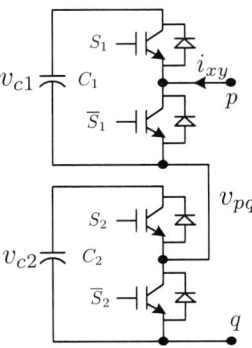

Figure 8. Cascade connection of HB module

comparison, the 3L-FC submodule is reconfigured into the cascade connection of two 2L-HB submodules which is shown in Fig. 8. The cascade connection of 2L-HB is also generates the three output voltage levels as that of single 3L-FC submodule. However, the effect of switching states on the capacitors voltage is quite different from 3L-FC submodule. Various switching states and the capacitors voltage variation of cascade 2L-HB submodule is given in Table III. With this approach, the N-3L-FC submodules per arm is equal to the $2N$-2L-HB submodule. In overall, the MMC will generates the same number of output voltage levels with both the submodule technologies. Also the device reverse voltage blocking capability and the number of semiconductor devices per arm is equal.

The proposed PWM modulator and the voltage balancing algorithm is applied to the MMC with the 2L-HB submodules. In MMC, the magnitude of circulating currents and capacitor voltage ripples are depends on the load power factor. So the performance of modular multilevel converter is evaluated at different load power factors using MATLAB/Simulink. The simulation results are presented for the worst case scenario, where the load power factor is zero (supplying reactive power) with the modulation index M=0.99 and the fundamental frequency F=60 Hz. Under zero power factor operation, the capacitor voltage ripples are very high as compared to the unity power factor operation. So, the zero power factor operation is an ideal condition to evaluate the performance of MMC with the novel submodules. The simulation results of MMC with 3L-FC and 2L-HB submodules for zero power factor operation (supplying pure reactive power to the load)

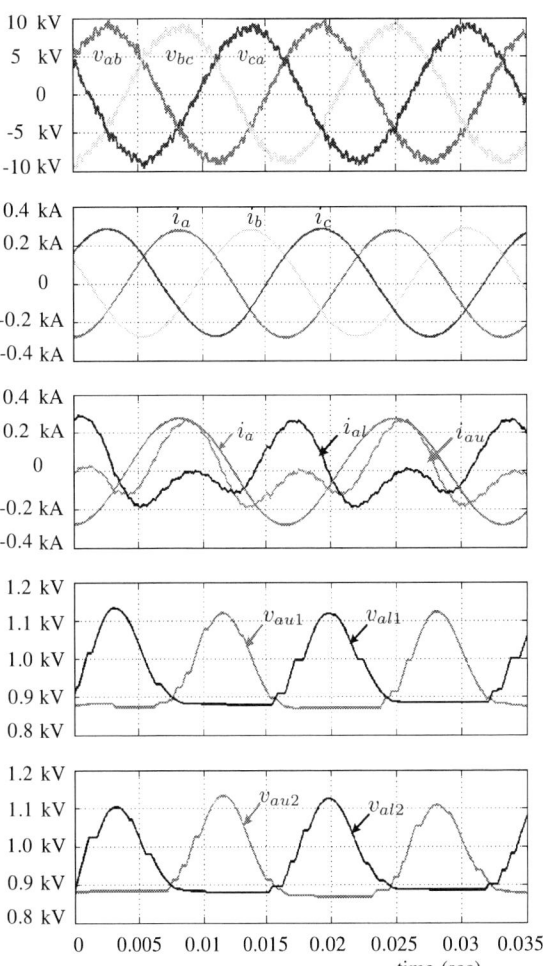

Figure 9. Performance of MMC with HB modules at PF=0

is shown in Fig. 9 and Fig. 10 respectively. The result shows, the MMC with 3L-FC system has the lowest voltage ripples and better shape of arm current as compared to the MMC with 2L-HB submodule. The circulating currents is mainly contains the second harmonic component and its magnitude can be significantly minimized by using 3L-FC submodules. Hence, the 3L-FC submodules can helps in improving the efficiency and reliability of the MMC converter.

V. CONCLUSION

In this paper, a new structure of CPS-PWM modulator along with a generalized capacitor voltage balancing algorithm is presented for MMC. The proposed balancing algorithm is computationally less complex and it is easily extended to the any number of submodules per arm, as it does not require the sorting technique. The submodules selection logic based on the normalized capacitors voltage is presented. This approach is helps in the maintaining the modular structure of the control algorithm and also the implementation of proposed algorithm to the any type of submodules with any voltage rating is easy. The switching state of the submodule is generated by dynamically, so the requirement of memory is less

Table III
SWITCHING STATES OF HB MODULE FOR THREE LEVEL OPERATION

S_1	S_2	$\overline{S}_1$	$\overline{S}_2$	v_{pq}	$i_{xy} > 0$	$i_{xy} \leq 0$
1	1	0	0	$v_{c1}+v_{c2}$	$v_{c1}\uparrow, v_{c2}\uparrow$	$v_{c1}\downarrow, v_{c2}\downarrow$
0	1	1	0	v_{c2}	$v_{c1}\approx, v_{c2}\uparrow$	$v_{c1}\approx, v_{c2}\downarrow$
1	0	0	1	v_{c1}	$v_{c1}\uparrow, v_{c2}\approx$	$v_{c1}\downarrow, v_{c2}\approx$
0	0	1	1	0	$v_{c1}\approx, v_{c2}\approx$	$v_{c1}\approx, v_{c2}\approx$

978-1-4799-2706-7/14 $31.00 © 2014 IEEE

The 2014 International Power Electronics Conference

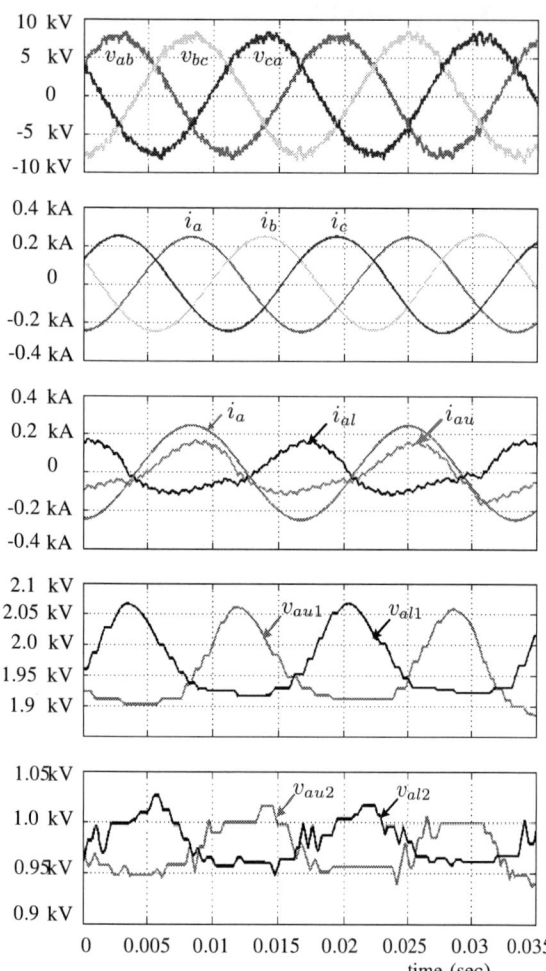

Figure 10. Performance of MMC with FC modules at PF=0

as compared to the other voltage balancing algorithms. The dynamic performance of the balancing algorithm is verified under different operating conditions. The result shows, the capacitors voltage is effectively maintained at its nominal value.

In addition, the performance of MMC with the novel submodules is presented interms of the magnitude of the circulating currents w.r.t the load power factor. The proposed 3L-FC submodules are reconfigured into the cascade connection of 2L-HB submodules. The results shows, the MMC performance can be greatly improved with the use of 3L-FC submodules. Also the circulating currents are significantly minimized. This will further improves the efficiency and reliability of the converter. The above advantages are makes the 3L-FC submodule as a potential candidate for modular multilevel converter.

REFERENCES

[1] H. Abu-Rub, J. Holtz, J. Rodriguez, and G. Baoming, "Medium-voltage multilevel converters-state of the art, challenges, and requirements in industrial applications," *IEEE Trans. Ind. Electron.*, vol. 57, no. 8, pp. 2581–2596, 2010.

[2] S. Kouro, M. Malinowski, K. Gopakumar, J. Pou, L. Franquelo, B. Wu, J. Rodriguez, M. Perez, and J. Leon, "Recent advances and industrial applications of multilevel converters," *IEEE Trans. Ind. Electron.*, vol. 57, no. 8, pp. 2553–2580, 2010.

[3] J. Rodriguez, S. Bernet, B. Wu, J. Pontt, and S. Kouro, "Multilevel voltage-source-converter topologies for industrial medium-voltage drives," *IEEE Trans. Ind. Electron.*, vol. 54, no. 6, pp. 2930–2945, 2007.

[4] H. Akagi, "Classification, terminology, and application of the modular multilevel cascade converter (mmcc)," *IEEE Trans. Power Electron.*, vol. 26, no. 11, pp. 3119–3130, 2011.

[5] M. Guan and Z. Xu, "Modeling and control of a modular multilevel converter-based hvdc system under unbalanced grid conditions," *IEEE Trans. Power Electron.*, vol. 27, no. 12, pp. 4858–4867, 2012.

[6] M. Hagiwara, K. Nishimura, and H. Akagi, "A medium-voltage motor drive with a modular multilevel pwm inverter," *IEEE Trans. Power Electron.*, vol. 25, no. 7, pp. 1786–1799, 2010.

[7] J. Mei, B. Xiao, K. Shen, L. Tolbert, and J. Y. Zheng, "Modular multilevel inverter with new modulation method and its application to photovoltaic grid-connected generator," *IEEE Trans. Power Electron.*, vol. 28, no. 11, pp. 5063–5073, 2013.

[8] A. Lesnicar and R. Marquardt, "An innovative modular multilevel converter topology suitable for a wide power range," in *Power Tech Conference Proceedings, 2003 IEEE Bologna*, vol. 3, 2003, pp. 6 pp. Vol.3–.

[9] M. Glinka and R. Marquardt, "A new ac/ac-multilevel converter family applied to a single-phase converter," in *Power Electronics and Drive Systems, 2003. PEDS 2003. The Fifth International Conference on*, vol. 1, 2003, pp. 16–23 Vol.1.

[10] S. Allebrod, R. Hamerski, and R. Marquardt, "New transformerless, scalable modular multilevel converters for hvdc-transmission," in *Power Electronics Specialists Conference, 2008. PESC 2008. IEEE*, 2008, pp. 174–179.

[11] M. Hagiwara, R. Maeda, and H. Akagi, "Negative-sequence reactivepower control by a pwm statcom based on a modular multilevel cascade converter (mmcc-sdbc)," *IEEE Trans. Ind. Appl.*, vol. 48, no. 2, pp. 720–729, 2012.

[12] M. Hagiwara and H. Akagi, "Control and experiment of pulsewidthmodulated modular multilevel converters," *IEEE Trans. Power Electron.*, vol. 24, no. 7, pp. 1737–1746, 2009.

[13] L. Angquist, A. Antonopoulos, D. Siemaszko, K. Ilves, M. Vasiladiotis, and H.-P. Nee, "Open-loop control of modular multilevel converters using estimation of stored energy," *IEEE Trans. Ind. Appl.*, vol. 47, no. 6, pp. 2516–2524, 2011.

[14] G. Adam, O. Anaya-Lara, G. M. Burt, D. Telford, B. Williams, and J. McDonald, "Modular multilevel inverter: Pulse width modulation and capacitor balancing technique," *IET Power Electr.*, vol. 3, no. 5, pp. 702–715, 2010.

[15] S. Rohner, S. Bernet, M. Hiller, and R. Sommer, "Modulation, losses, and semiconductor requirements of modular multilevel converters," *IEEE Trans. Ind. Electron.*, vol. 57, no. 8, pp. 2633–2642, 2010.

[16] E. Solas, G. Abad, J. Barrena, S. Aurtenetxea, A. Carcar, and L. Zajac, "Modular multilevel converter with different submodule concepts - part i: Capacitor voltage balancing method," *IEEE Trans. Ind. Electron.*, vol. 60, no. 10, pp. 4525–4535, 2013.

[17] Q. Tu, Z. Xu, H. Huang, and J. Zhang, "Parameter design principle of the arm inductor in modular multilevel converter based hvdc," in *Power System Technology (POWERCON), 2010 International Conference on*, 2010, pp. 1–6.

[18] Z. Li, P. Wang, Z. Chu, H. Zhu, Y. Luo, and Y. Li, "An inner current suppressing method for modular multilevel converters," *IEEE Power Electr. Letters*, vol. 28, no. 11, pp. 4873–4879, 2013.

[19] M. Zhang, L. Huang, W. Yao, and Z. Lu, "Circulating harmonic current elimination of a cps-pwm-based modular multilevel converter with a plug-in repetitive controller," *IEEE Trans. Power Electron.*, vol. 29, no. 4, pp. 2083–2097, 2014.

[20] J.-W. Moon, C.-S. Kim, J.-W. Park, D.-W. Kang, and J.-M. Kim, "Circulating current control in mmc under the unbalanced voltage," *IEEE Trans. Power Del.*, vol. 28, no. 3, pp. 1952–1959, 2013.

[21] Q. Tu, Z. Xu, and L. Xu, "Reduced switching-frequency modulation and circulating current suppression for modular multilevel converters," *IEEE Trans. Power Del.*, vol. 26, no. 3, pp. 2009–2017, 2011.

Modular Medium Voltage Drive for Demanding Applications

Drazen Dujic, Jonas Wahlstroem, Juan Alberto Marrero Sosa, Dominik Fritz

ABB Switzerland Ltd, Medium Voltage Drives, Turgi, Switzerland

Abstract-Nowadays, several state-of-the-art multilevel converter topologies are used for numerous industrial medium-voltage (MV) high-power variable-speed drive applications. The 3-level neutral point clamped (3L-NPC) topology is one of those and is often used as a benchmark for various alternatives. In this paper, a 3L-NPC medium voltage special-purpose drive based on an integrated gate commutated thyristor (IGCT) is presented, highlighting the design paradigms followed in order to achieve highest performances at maximum possible flexibility. Modular hardware design supported by highly dynamic direct torque control, fulfills the needs of the demanding applications.

Keywords— DTC, Medium Voltage Drives, NPC.

I. INTRODUCTION

Nowadays, multilevel converter topologies are well established on the market and serve medium voltage high-power applications [1]. A diverse range of industrial applications, equally those with low or high performance requirements, is not particularly (if at all) sensitive to the type or topology of the converter used, as long as the requested requirements can be met and the availability of the system can be guaranteed. Various multilevel converter topologies have so far established themselves as mature technology, e.g. 3L-NPC, cascaded H-Bridges (CHB), 5L-Active NPC, flying capacitor topology, to name a few [2-4]. This is greatly supported and enabled by the development of high-power semiconductor devices for high power applications, such as Integrated Gate Commutated Thyristors (IGCTs) [5,6] and Insulated Gate Bipolar Transistors (IGBTs) and the IGBT clone, the so called Injection Enhanced Gate Transistor (IEGT) [7].

In addition to the state-of the-art converter topologies mentioned above, a fairly quick and successful commercialization of the modular multilevel converter (MMC/M2LC) [8] opened new possibilities, and options regarding voltage scaling during the design, although penetration into the drives market is minor, at this stage. Modularity, inherently built into the MMC concept, has often been cited as a great advantage of this topology. Reality is, that the concept of modularity has been present for a long time in state-of-the-art MV drives, both at the level of the Power Electronic Building Block (PEBB) [9], and the level of basic functional converter units.

This paper presents a modular and scalable 3.3kV medium voltage converter, based on the 3L-NPC topology, successfully deployed in demanding industrial applications where high dynamic performances are required. Thanks to the modular design, the converter can be scaled in powers between 3MVA and 36MVA. At the same time, the converter can serve single-motor and also multi-motor drives, as will be discussed hereinafter.

II. APPLICATION REQUIREMENTS

In order to limit the scope and length of the paper, demanding high power metal industry applications are taken as basis for the discussion. Such applications include: different types of rolling mills (hot strip, plate, steckel, tandem, wire rod, seamless, profile), winders, un-winders, crop shears, etc. In many of these applications it is common to use multiple motors in cyclic operation with frequent accelerations and decelerations (motoring and regeneration modes) accompanied with high torque and wide speed range requirements (up to several times of the rated torque and speed). At the same time, installation space tends to be limited, with converters often located away from the motors requiring lengthy cabling. Powers involved vary in the range from several MWs up to several tens of MWs, depending on the actual type of mill.

The presence of multiple motors that often operates in different modes (motoring or braking) offer great potentials for energy savings on system level: the regeneration energy can be captured, eventually stored or channeled and used immediately in a part of the installation where it is needed. To achieve this in installations which are based on multiple single-motor MV drives, every drive has to be a four-quadrant one. This allows the excess energy to reach the grid, from where it will be drawn by another drive in need. However, the multiple conversion stages in the path increase the overall losses in the system.

To reduce the losses in these types of installations, it becomes advantageous to connect several MV converters (inverters) to the same MV DC link that is fed from the network side either through an active rectifier, or in some cases even through a diode rectifier. This greatly improves the overall system efficiency and also reduces installation footprint, as discussed hereinafter.

Fig. 1. Multi-motor drive example - Profile mill.

III. POWER ELECTRONIC BUILDING BLOCK

The IGCT is a monolithic gate-controlled turn-off semiconductor device which turns off like a transistor, but conducts like a thyristor with the lowest conduction losses [6]. Since its commercial introduction, in the late 90's, the IGCT has established itself in the high-power and high-current applications due to its low on-state and turn-on losses (related to the circuit design), high surge current capabilities, possible switching frequencies up to 1kHz, and simple mechanical integration.

For a 3.3kV inverter unit (INU) based on a 3L-NPC topology, 4.5kV IGCTs are needed. In contrast to IGBT devices, IGCT indeed behaves as a switch and does not modulate transition speed, neither during turn-on nor during turn-off. Thus, to control (limit) the di/dt during diode turn-off and fault current, a clamp inductor L is needed per half DC-link.

A principal schematic of an IGCT based 3L-NPC inverter unit (INU) is shown in Fig.2. Only one phase is shown in detail, including the well-known 3L-NPC leg having four asymmetric IGCTs (thus four separate free-wheeling diodes), two neutral point diodes and a separate RCD clamp circuit per half DC-link. The clamp inductors L, per each half of DC-link, are shown as well.

The practical realization of one phase module of a 12MVA (13MVA peak) INU is shown in Fig.3. The mechanical arrangement is compact and ensures that the required forces are applied to semiconductor devices within a stack. Cooling plates with de-ionized water are in direct contact with semiconductor devices and are also connected to the bus-bar system towards the DC-link, clamp inductor L and AC terminal of a phase. Low-power signal interfaces include only power supply for the IGCT gate unit and fiber-optical interfaces towards the control hardware.

Thanks to the compact design, the phase stack can be easily manipulated by one person, and in the unlikely event of a failure, can be replaced in less than 30 minutes without any special tools. Quick couplings on the cooling system connections ensures a quick and leak-free separation from the rest of a drive system. This ensures minimal downtime for the plant.

Fig. 3. 12MVA (13MVA peak) 3L-NPC single-phase stack.

Compactness of a phase module positively impacts the realization of the 3-phase inverter unit, which is packed inside the cabinet of size 1.5m x 1.1m x 2m with resulting volume of 3.3m³. Considering the power rating of 12MVA, this yields an impressive power density[1] for the INU that is reaching almost 3.5MVA/m³.

Considering that the INU cabinet has no extra cooling system to complement the water-cooled phase stacks, an additional protection for the IGCTs has been realized by implementing an SW thermal model. Properties are demonstrated in Fig.4, for a cyclic operation with preload (25s-35s) followed by overload (35s-40s) and continuous operation afterwards (40s-50s).

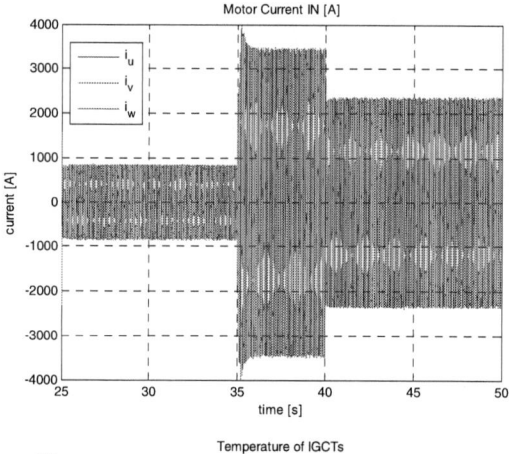

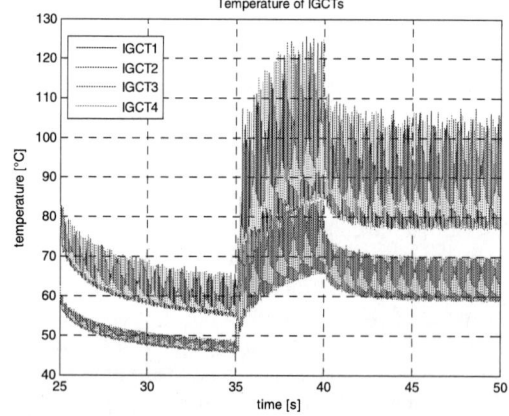

Fig. 4. Thermal simulations: motor current (top; temperature of IGCTs (bottom) .

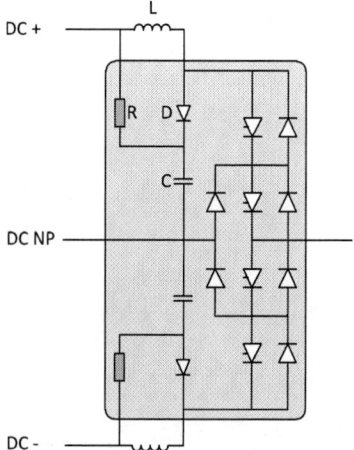

Fig. 2. Principal schematic of the phase leg of an IGCT based 3L-NPC inverter unit.

---

[1] Please note that number refers only to the INU, without considering DC link capacitors or the cooling system.

IV. MODULARITY AND SCALABILITY

To meet the variety of requirements defined by the actual industrial process, a modular approach has been followed during the design of the MV drive (converter). A sufficient number of functional units, that perform major functions has been identified and common electrical and mechanical interfaces are defined, that allow various configurations of a drive line-up. Basic functional units are:

- Conversion Units:
  - LSU – Line Supply Unit
  - ARU – Active Rectifier Unit
  - INU – Inverter Unit
  - EXU – Excitation Unit
- Energy Storage Unit:
  - CBU – Capacitor Bank Unit
- Control and Interface Units:
  - COU – Control Unit
  - TEU – Terminal Unit
  - WCU – Water Cooling Unit
- Protection Units:
  - VLU – Voltage Limiter Unit
  - RBU – Resistor Braking Unit
  - BCU – Braking Chopper Unit
  - DIU – Discharging Unit
- Optional Units:
  - IFU – Input Filter Unit
  - IRU – Input Reactor Unit
  - ISU – Isolator Unit

A modular solution allows the combination of both standardized and engineered units into one converter in order to achieve maximum flexibility and quality.

Conversion units are those units processing the power. The connection to the grid can be established either using passive diode rectification LSU (6, 12, or 24 pulse) or ARU when regeneration is required. The ARU and INU are topologically identical, difference being only in the function they perform. For a drive with synchronous machines (SM), an EXU can be realized either as Direct EXU (DC) or Brushless EXU (AC). Intermediate energy storage is realized by means of CBU, whose ratings are determined by the installed power of a drive. Rated DC-link voltage allows for a 3.3kV INU output voltage. Internal and external control and power interfaces are handled by means of COU, TEU and WCU.

In some special cases, some of the protection units are needed either to deal with a sudden change of operating modes imposed by the application, or to handle excess of energy when it is not possible to feed it back to the grid. All protection units are in essence DC chopper units with different energy and power capability. Another differentiation comes from the location of the braking resistors: installed locally inside the cabinet for VLU and RBU, and installed externally for the BCU. In case that the application does not require a chopper unit, a DIU is installed in the drive line-up to allow a fast and active discharging of the DC link.

Finally, when necessary and depending on the grid strength and conditions at the customer's site, a LCL filter, the IFU, may be used in combination with an ARU, or an input reactor, the IRU is used in combination with an LSU in order to guarantee that network harmonic limits are respected as defined in standards, e.g., IEC 61000-2-4 or recommended by IEEE-519. The ISU is used in case of redundant drives to connect or disconnect DC links of different CBUs.

An example of a four-quadrant single-motor drive is shown in Fig.5. It is one of the simpler configurations (the simplest would be two-quadrant drive with LSU instead of ARU). With rated output voltage of 3.3kV, scaling of basic conversion units is done by current resulting in portfolio of ARU/INU with rated powers of: 3, 5, 7, 9, or 12MVA. When the application requirements exceed these levels, paralleling of ARU/INU units is possible in order to achieve higher powers.

This is illustrated in Fig.6 for a four-quadrant multi-motor drive with power a rating higher than ratings of basic conversion units. The connection to the grid is established by means of three ARU connected in parallel to the same DC link of CBU. Five INUs are connected to the same CBU, whereas two INUs are used to supply two induction motors (IMs) and the remaining three INUs are hard paralleled to supply a high power rated synchronous machine (SM). In addition, there is an EXU providing supply to the field winding of synchronous machine.

Depending on the actual type and/or power of the machine, it can be that a separate INU supplies one set of three-phase winding of a multi-winding machine, instead of hard paralleling for single three-phase winding (as shown in Fig.6). Alternatively, five INUs could be supplying five motors.

Fig. 5. ACS 6000 single-motor drive combining functional units (from left to right): ARU, COU/TEU, INU, CBU, VLU and WCU.

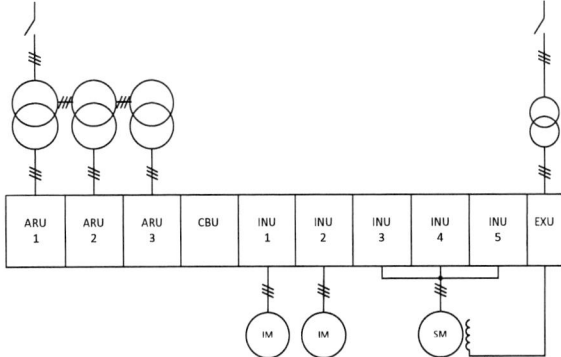

Fig. 6. Multi-motor drive example including three ARUs and five INUs supplying two induction motors and one synchronous motor.

Presently, up to eight conversion units can be connected within a multi-motor drive. This provides enough versatility to address the most demanding applications and, more importantly, offers great advantages on the system level, in terms of reduced footprint of the installation, reduced cabling and improved efficiency.

V. Control and Protection

Using fig.6 as an example the connection to the network is realized using three double-winding transformers, from which two are often with an open-end primary winding. Secondary windings are phase shifted accordingly, usually 0°, +20° and -20°. This 18-pulse transformer connection eliminates certain odd order harmonics, leaving critical harmonics of order $18k\pm1(k=1,2,3,..)$ for control considerations.

Vector control principles applied to the ARU are using modulators based on the Optimized Pulse Pattern (OPP) principles and are working with low switching frequencies. Depending on the number of additional harmonics (N-1) that need to be eliminated, N switching angles are determined through an optimization process over the modulation index range of interest. Typical switching frequencies are 3, 5, 7 or 9 times the fundamental frequency (50Hz), or respectively, 150Hz, 250Hz, 350Hz, 450Hz. For example, with 18-pulse ARU and OPP with N=5, the first critical harmonic in the system is 53rd. The input power factor can be controlled as well to meet installation requirements regarding reactive power.

In the INU, an enhanced version for the 3L-NPC topology of Direct Torque Control (DTC) is used. DTC for controlling induction and synchronous motors has been well explained in the literature [10-12] and its description is out of the scope of this paper. Main advantages of DTC can be summarized in [13, 14]:

- accurate and fast torque response
- robust control
- low motor parameters sensitivity

It is in demanding high power metal industry applications where DTC stands out from other motor control techniques, e.g. Field Oriented Control (FOC).

An example of the torque response at different speeds is shown in Fig.7. The motor used is a low speed synchronous motor (7MW, 30rpm) for hot-rolling mills. Data were acquired from a Hardware-In-the-Loop (HIL) simulator where the complete ACS 6000 control software and hardware is developed and tested (Fig.8). Despite of being a high-power medium-voltage drive, the torque response can be compared to that of low voltage drives in terms of rise time.

The fast and linear torque response of DTC allows aggressive tuning of the speed controller with minimum overshoot. This is a major advantage in hot-rolling mills because the speed drop after the slab-in can be minimized. This is shown in Fig.9 and Fig.10, where a real slab-in and slab-out process was simulated in the HIL for a hot-rolling mill stand with top and bottom motors (2 x 7MW synchronous motors).

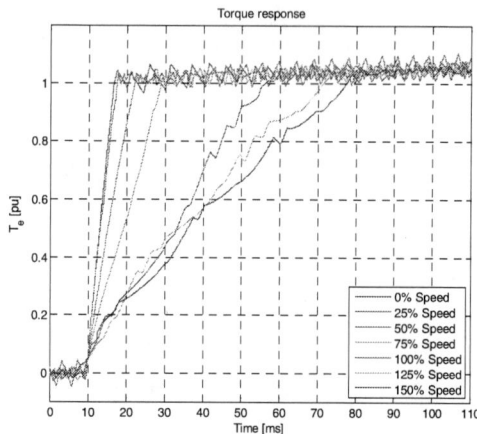

Fig. 7. INU DTC torque response at different speeds. Speed forced, 1 p.u. torque step. Synchronous motor data: 7MW, 30rpm.

Fig. 8. HIL simulator for modular multi-motor MV Drive.

With regard to the protection concept, the ACS 6000 medium voltage drive features a "fuse-less" design, thus increasing the availability and avoiding nuisance trips. Thanks to the high surge current capabilities of the IGCT devices, HW protection relies on the principle of "protection firing", also known as "firing through". Basically, a concept similar to thyristor crowbar across DC link, but realized with existing LSU/ARU/INU functional units.

Depending on the origin and severity of the fault, several control SW protection functions could be activated in order to safely isolate the fault and prevent major damage or disturbance to the industrial process. For example, in the case of multi-motor drive and failure on one of the motors, a pulse removal can be applied to the INU of that motor, to selectively isolate the motor, without interruption of the rest of the system. For a high severity faults and as a last resort, control SW may activate "protection firing" and turn-on all the switches in the drive, thus effectively creating short-circuit (SC) conditions that are sustained for long enough to trip the main-circuit-breaker (MCB). The electrical and mechanical design of a drive has been designed in accordance with expected stresses during these conditions.

978-1-4799-2706-7/14 $31.00 © 2014 IEEE

The 2014 International Power Electronics Conference

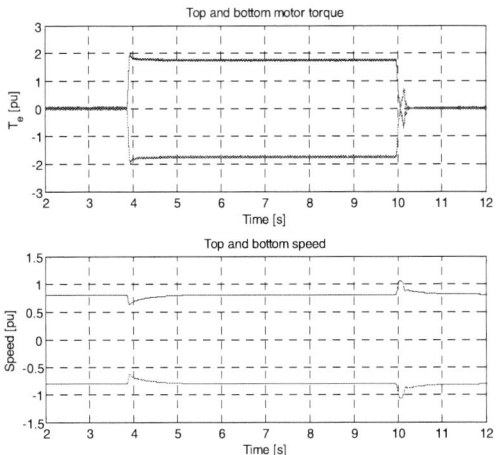

Fig. 9. Hot rolling mill slab-in and slab-out process simulation in HIL. Top and bottom motors torque and speed (2 x 7MW synchronous motors).

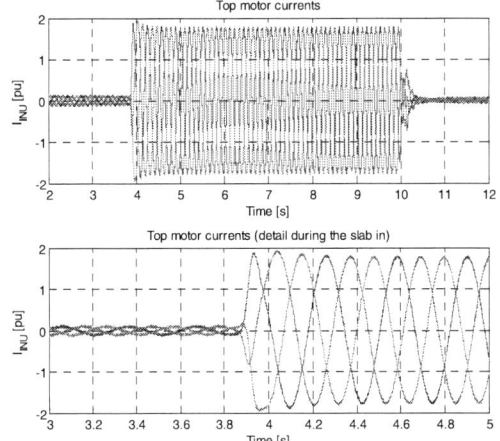

Fig. 10. Hot rolling mill slab-in and slab-out process simulation in HIL. Currents of a top motor.

VI. EXPERIMENTAL RESULTS

To test and verify performance of the power electronic hardware and that of control software beyond the laboratory environment, testing under real operating and loading conditions is occasionally needed. To be able to carry out such experiments, an ACS 6000 18MVA back-to-back test configuration has been realized, as shown in Fig.11. It features two four-quadrant drives, connected to the same supply line and feeding two mechanically coupled MV induction machines. Usually, tests are carried with energy circulating between two ACS 6000 drives, but when needed an additional machine can be connected and two ACS 6000 drives are then used for loading.

With ARU and the use of optimized pulse patterns, network friendly operation with low distortion is achieved, what can be seen from Fig.12, for the 9MVA unit. When demanded by the application or installation conditions, the input factor can be controlled to compensate for the reactive power.

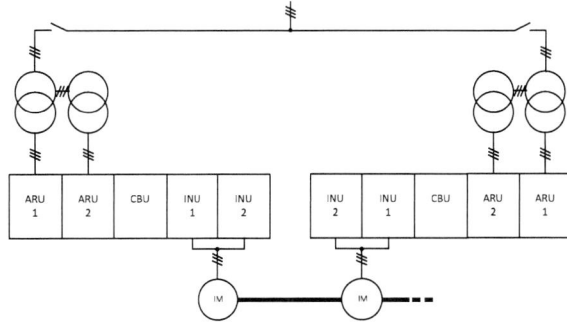

Fig. 11. ACS 6000 test stand.

The output line voltage of the INU, shown in Fig.13 with 5-level voltage waveform, is distinctive for 3L-NPC topology. Switching frequency is maintained around an average value (usually below or up to 250Hz), taking into account the allowed thermal limits of IGCTs. Protection of the semiconductors is based on a thermal model within the control software. It can be seen that the INU switching pattern does not follow a regular PWM pattern, since it is generated by the DTC. This provides extremely high dynamics regarding the torque and speed rise times, outperforming other control methods often used for medium voltage drives.

Finally, electrical losses and efficiency of the INU have been measured and results are presented in Table 1. Under rated operating conditions of the 12MVA unit, the largest portion of losses is associated with semiconductors losses (~60%). The remaining losses are coming from the remaining electrical components (~40%). The Joule losses associated with copper bus-bars are rather negligible. Overall efficiency of the INU, under rated operating conditions is rather high and around 99.2% - 99.3%.

TABLE 1: LOSSES OF THE 9MVA AND 12MVA INUs

Components		9MVA	12MVA
Semiconductors	[kW]	39	53
El. Components	[kW]	32.5	35.7
Total Losses	[kW]	71.5	88.7
Efficiency	**[%]**	**99.2**	**99.3**

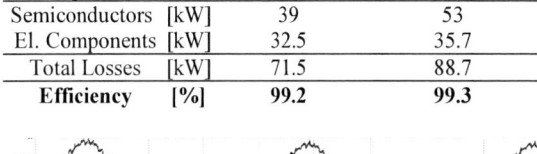
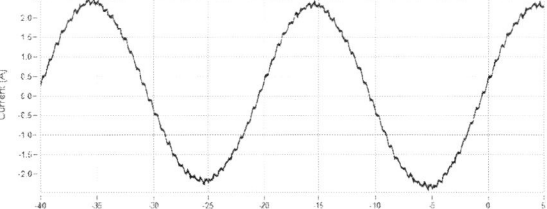

Fig. 12. ARU input current (50Hz grid)

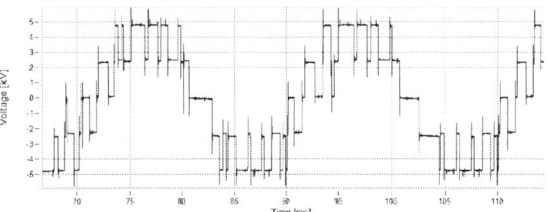

Fig. 13. INU output line voltage.

978-1-4799-2706-7/14 $31.00 © 2014 IEEE

VII. CONCLUSION

Nowadays, the term modular multilevel converter is mainly used in conjunction MMCs. Reality is that the concept of modularity and the ability to configure the converter to match application requirements is not new in industrial applications. This is described in this paper, using the ACS 6000, a 3.3kV IGCT based 3-L NPC medium voltage converter, as an example. Using the pre-defined functional units, a number of drive line-ups can be assembled to address the variety of system architectures. The strengths of this approach become especially apparent when it comes to multi-motor drives, resulting in several advantages, as discussed in the paper.

Combining this approach with high performance control algorithms provides a solution that is able to respond to the most demanding applications, both statically and dynamically. The metals industry is a typical example where the discussed features and advantages are fully translated into benefits for the plant owners and operators.

To conclude, Fig.14 illustrates an actual line-up of ACS 6000 multi-motor drive, rated at 36MVA for five large synchronous motors with a total length of 26 meters. Often, in this kind of installations, total power rating of the ARUs is lower than the actual sum of the power ratings of the INUs. This is application specific since the braking energy coming from any of the motors is immediately consumed by another motor connected to the same DC link through its own INU. Thus, in addition to the overall high performances delivered to the application, great energy savings are achieved simultaneously.

ACKNOWLEDGMENT

The authors would like to express their sincere gratitude to Dr. Adrian Zuckerberger (A2Z Consulting Ltd) for his valuable contributions to the work presented in this paper.

REFERENCES

[1] J.Rodriguez, S.Bernet, Bin Wu, J.O.Pontt, S.Kouro, "Multilevel Voltage-Source-Converter Topologies for Industrial Medium-Voltage Drives," *IEEE Trans. on Industrial Electronics*, vol.54, no.6, pp.2930,2945, Dec. 2007.

[2] A.Nabae, I.Takahashi, H.Akagi, "A New Neutral-Point-Clamped PWM Inverter," *IEEE Trans. on Industry Applications*, vol.IA-17, no.5, pp.518,523, Sep. 1981.

[3] J-S.Lai, F.Z.Peng, "Multilevel converters-a new breed of power converters," *IEEE Thirtieth IAS Annual Meeting, IAS '95.*, vol.3, pp.2348-2356, Oct. 1995.

[4] F.Kieferndorf, M.Basler, L.A.Serpa, J.-H.Fabian, A.Coccia, G.A.Scheuer, "A new medium voltage drive system based on ANPC-5L technology," *IEEE International Conference on Industrial Technology (ICIT)*, pp.643,649, Mar. 2010.

[5] A.Zuckerberger, E.Suter, C.Schaub, A.Klett, P.K.Steimer, "Design, simulation and realization of high power NPC converters equipped with IGCTs," *The 33rd IEEE IAS Annual Meeting Ind. Appl. Conf.*, vol.2, pp.865-872, vol.2, 1998.

[6] P.K.Steimer, O.Apeldoorn, E.Carroll, "IGCT devices-applications and future opportunities," *IEEE Power Engineering Society Summer Meeting*, vol.2, pp.1223-1228, 2000.

[7] M.A.Mamun, D.Yoshizawa, M.Mukunoki, "Performance evaluation of a large capacity 3-level IEGT inverter," *IEEE ECCE Asia Downunder*, pp.201-207, Jun. 2013.

[8] A.Lesnicar, R.Marquardt, "An innovative modular multilevel converter topology suitable for a wide power range," *IEEE Power Tech Conference Proceedings*, vol.3, pp.6 Jun. 2003.

[9] P.K.Steimer, "Power electronics building blocks - a platform-based approach to power electronics," *IEEE Power Engineering Society General Meeting*, vol.3, pp.1360-1365 Jul. 2003.

[10] I.Takahashi, T.Noguchi, "A new quick response and high efficiency control strategy for the induction motor," *IEEE Trans. Ind. Appl.*, vol. IA-22, no.2, pp. 820-827, Sep./Oct. 1986.

[11] I.Takahashi, Y.Ohmori, "High-performance direct torque control of an induction motor," *IEEE Trans. Ind. Appl.*, vol. 25, no.2, pp. 257-264, Mar./Apr. 1989.

[12] P.Tiitinen, P.Pohjalainen, J.Lalu, "Next generation motion control method: Direct torque control (DTC)," *EPE Journal*, vol. 5, no. 1, Mar. 1995.

[13] B.Wu, *High-Power Converters and AC Drives*, Wiley-Interscience, 2006.

[14] P.Vas, *Sensorless Vector and Direct Torque Control*, Oxford University Press, 1998.

Fig. 14. ACS 6000 four-quadrant 36MVA multi-motor drive for five synchronous motor, consisting of 23 functional units (from left to right): WCU1, TEU, ARU1, ARU2, COU0, ARU3, COU1, INU1, CBU1, CBU2, VLU, INU2, COU2, INU3, COU3, INU4, INU5, WCU2, EXU1, EXU2, EXU3, EXU4, EXU5.

The 2014 International Power Electronics Conference

Asymmetrical Fault Ride-through of Three-phase PV Systems using Four-wire Dc-ac Converters

Shivkumar Iyer, Bin Wu
Dept. of Electrical and Computer
Engineering, Ryerson University
Toronto, Ontario, Canada
shivkiyer@ieee.org, bwu@ee.ryerson.ca

Yunwei Li
Dept. of Electrical and Computer
Engineering, University of Alberta
Edmonton, Alberta, Canada
yunwei.li@ualberta.ca

B.N. Singh
Hydro One Inc.
Toronto, Ontario, Canada
bob.singh@hydroone.ca

Abstract—This paper examines the fault ride-through strategy of a three-phase PV system with respect to compliance to the German Transmission Code. Specifically, the paper examines the case of a single line-ground fault which occurs predominantly in the distribution system. The paper proposes the injection of reactive power on a per-phase basis to the faulty phases while maintaining active power injection to the healthy phases thereby avoiding power generation deficit. Simulation results from a study performed in MATLAB/Simulink with the SimPowerSystem toolbox present the unbalanced nature of currents during an unbalanced fault. An experimental setup has been shown and described that has been used to implement the strategy. Experimental results have been provided to validate the simulation results.

Keywords—Grid compliance, four-wire dc-ac converter, phase locked loop, current control.

I. INTRODUCTION

Renewable energy (RE) based distributed generators (DGs) are emerging as a viable solution to meeting the electrical power demand. Many utility companies all over the world have set targets for integrating a certain level of RE based DGs particularly solar photovoltaic (PV) and Wind Turbine Generators (WTGs) within the coming decade. These steps would significantly decrease the carbon emissions and promote greener energy. However, the integration of a significant level of intermittent and fluctuating DGs distributed over the power distribution and transmission system poses significant operational challenges for utility companies. Moreover, this has also led to utility companies defining the roles to be played by DGs during normal and abnormal operation of the grid.

Grid compliance and regulations regarding the operation of DGs applies to normal (steady-state) and abnormal (transient) conditions. During normal operations, the DGs are required to inject balanced currents having a THD lesser than a specified level. The abnormal conditions are short or long duration voltage dips that accompany faults on one or more phases and frequency fluctuations away from the nominal frequency. This paper deals specifically with voltage dips caused due to faults in the distribution system.

This paper refers to the German Transmission Code 2007 [1] that lays down guidelines for renewable based DGs on their performance during abnormal conditions. The code specifies the voltage profile with respect to time

during which the DG is expected to remain connected. Moreover, the code also specifies the nature of support to be provided by the DG during the voltage dip. For example, the code specifies that for the first 150 ms after the fault, the DG must remain connected even if the voltage dips to zero. Subsequent to that, the DG may be required to ride through for voltages higher than 30% and lower than 70%. The DG is required to ride through for dips higher than 70%. Moreover, the DG must provide reactive current during the voltage dip. For a dip of up to 10%, the DG is not required to inject a reactive component of current. For dips larger than 10%, the reactive current injected by the DG rises linearly up to its kVA rating. For voltage dips greater than 50%, the DG does not have to supply active power and only needs to provide reactive power support.

This paper includes several aspects of PV interconnection to a three-phase four-wire grid. Grid interconnection of PV panels requires accurate information of the phase angle of the grid. Therefore, the paper emphasises on a single-phase Phase Locked Loop (PLL) which is a modified version of the PLL presented in [2], [3]. A review of different current controllers for dc-ac converters that forms the second stage of the conversion process can be found in [4]. The paper has used the inverter with inductor-capacitor-filter (LCL) for experimental results. Several manuscripts have been published related to the control of inverters with LCL filters [5]. A review of fault handling algorithms for DGs is presented in [6].

The outline of the paper is as follows. Section II describes the topology and control of the dc-ac converter used to interface the PV panel to the three-phase four-wire grid. Section III presents simulation results of a voltage dip on one phase following a single line-ground fault. Section IV provides experimental results. Section V provides a discussion of the simulation results with respect to the results presented in literature. Section VI concludes the paper.

II. SYSTEM DESCRIPTION

Fig. 1 shows the converter topology and the associated control used to interface the PV panel to a three-phase four-wire grid. The two-stage converter consists of a dc-dc boost converter followed by a three-phase four-wire dc-ac converter. The capacitors C_{pv} and C_{dc} are the electrolytic decoupling and dc link capacitors

978-1-4799-2706-7/14 $31.00 © 2014 IEEE

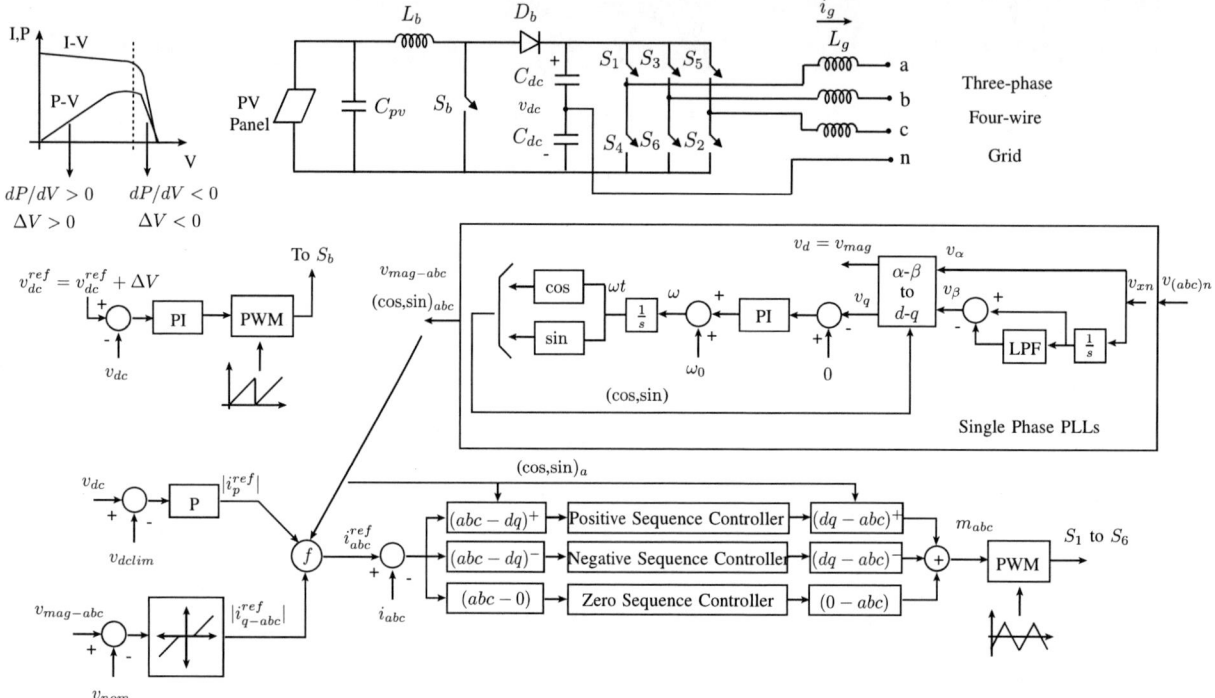

Fig. 1. Converter topology

respectively. The switch S_b of the dc-dc converter is an Insulated Gate Bipolar Transistor (IGBT). The switches S_1 to S_6 are IGBTs with their associated anti-parallel diodes. The switching frequency of both converters is 5 kHz. The simulation consider the dynamics of the PV panel and the associated dc-dc converter as well as the dc-ac converter. The output filter of the dc-ac converter has been considered to be a simple inductor filter. However, the grid has been considered to be a constant three-phase ac source. The experimental results assume the PV panel and the dc-dc converter as a fixed dc bus with a constant voltage. The output filter of the dc-ac converter has been considered to be an LCL filter and the superior performance has been shown in the experimental results. However, the grid has been emulated by another three-phase four-wire dc-ac converter and therefore, the impact of the fault ride-through strategy on the grid has been shown. The control strategy is described as follows:

A. PV panel and maximum power point tracking

The top left corner of Fig. 1 shows the I-V and P-V characteristics of the PV panel being simulated. [7] describes in detail the simulation of the I-V characteristics taking as inputs the incident solar radiation (G) and the temperature (T). This paper has adopted the current source, diode and resistor model proposed in [7] and a description of the same will be omitted here. On the I-V characteristic in Fig. 1 is superimposed the P-V characteristic showing the maximum power point. The maximum power point tracking (MPPT) algorithm can be found in [8]. The MPPT algorithm is based on the fact that $\frac{dP}{dV} = 0$ at the MPP while it is positive in the constant current region on the left and is negative in the constant voltage region on the right as shown in

Fig. 1. Therefore, the perturb-and-observe method has been adopted to add an incremental voltage ΔV to the dc link voltage reference v_{dc}^{ref} based on the computed value of $\frac{dP}{dV}$.

B. Dc-dc converter control

Fig. 1 shows the topology of the boost converter. The dc link voltage v_{dc} is the output voltage of the boost converter whose reference v_{dc}^{ref} is set by the MPPT algorithm. A Proportional-Integral (PI) regulates v_{dc} by varying the duty ratio d of the switch S_b [9].

C. Phase locked loop (PLL)

The PLL is shown below the circuit in Fig. 1. The PLL is applied to a three-phase four-wire system and receives as input the three phase voltages with respect to the neutral v_{an}, v_{bn}, v_{cn}. However, instead of the conventional three-phase PLL where a d-q transformation is performed on the three voltages, in this paper, PLL is applied to each phase for the following two reasons:

- The emphasis of this paper is the ability to deal with unbalanced faults. To accurately detect dips in the voltage of a single phase, the magnitude of each phase is required instead of a positive sequence and negative sequence component of the three-phase voltages.

- Unbalanced faults are often accompanied by the phase angle of the faulted phase experiencing a jump particularly when the R/X ratio of the fault is different from the R/X ratio of the grid. Three single-phase PLLs will ensure phase angle locking during faulted conditions.

The 2014 International Power Electronics Conference

TABLE I. SIMULATION PARAMETERS AND SYSTEM CONDITIONS

	Topology	Control
Grid	$V_{L-L} = 208$ V (rms), frequency $= 60$ Hz	PLL: LPF cut-off frequency $\omega_0 = 20$ rad/s
		$K_p = 0.75$, $K_i = 1$
PV panel	$V_{oc} = 120$ V and $I_{sc} = 16$ A	MPPT: $\Delta V = 0.5$ V
Dc-dc converter	$C_{pv} = 470$ μF, $L_b = 250$ μH, $C_{dc} = 1100$ μF	$K_p = 0.025$, $K_i = 0.005$
Dc-ac converter		Ac current generation:
	$L_g = 2.5$ mH	$v_{dc}^{lim} = 350$ V, $K_p = 0.05$
		$K_p^+ = 0.0075$, $K_i^+ = 0.5$
		$K_p^- = 0.0075$, $K_i^- = 0.5$
		$K_p^0 = 0.02$, $K_i^0 = 0.75$

For each phase voltage (shown as v_{xn} in Fig. 1), an integrator followed by a Low Pass Filter (LPF) creates a quadrature component of the signal with the LPF removing the dc offset that may be generated by the integrator [3]. The original signal along with the quadrature component generated form the v_α-v_β components that are transformed to the synchronously rotating frame of reference. From this point on, the PLL is similar to a conventional PLL [2], where the objective is for the d axis of the synchronous reference frame to be aligned with the original signal by forcing the q component to 0. The three single-phase PLLs generate three sets of $(\cos\omega t, \sin\omega t)$ for phase a, b and c respectively with ω being the angular frequency of the grid as shown in Fig. 1.

D. Grid current reference generation

The grid currents contain two components - the active power component and the reactive power component as shown in Fig. 1. The active power component $|i_p^{ref}|$ is generated with respect to the deviation of the dc link voltage v_{dc} from a lower limiting value v_{dc}^{lim} by a Proportional (P) controller. The difference $v_{dc} - v_{dc}^{lim}$ is proportional to the power fed by the PV panel through the dc-dc converter. The reactive power component is generated on a per-phase basis $|i_{q-abc}^{ref}|$ as a function of the difference between the phase-neutral voltage magnitudes $v_{mag-abc}$ and the nominal voltage magnitude v_{nom} as shown in Fig. 1. Details of the requirement of reactive power support to be provided by DGs during voltage sags can be found in [1]. For a deviation of upto $\pm 10\%$ from the nominal voltage, there is no reactive current injection requirement after which the reactive current component increases linearly.

The signals of $|i_p^{ref}|$ and $|i_{q-abc}^{ref}|$ are fed to a function block f shown in Fig. 1. This block f receives the current component magnitude signals and combines them with the cosine and sine templates of the respective phases. Additionally, the block f also receives the individual phase voltage magnitude information and if a particular phase voltage dips below 50%, the active component of current in that phase is set to zero and the phase current is purely reactive. Another operation which is not shown in Fig. 1 is performed upon detection of a voltage dip. The MPPT algorithm is discontinued by setting v_{dc}^{ref} to a constant to prevent the dc link voltage from rising during the dip.

E. Current controller

The errors e_{abc} between the references i_{abc}^{ref} and the actual measured grid currents i_{abc} are fed to the current controller as shown in Fig. 1. The current references i_{abc}^{ref} will be balanced sinusoids during normal operation but will be unbalanced sinusoids with a zero sequence component during faults. Therefore, the current controller consists of three different controllers - positive, negative and zero sequence controller.

The positive sequence transformation is as follows:

$$\begin{bmatrix} e_d^+ & e_q^+ \end{bmatrix}' = \mathbf{T_1 T_2} \begin{bmatrix} e_a & e_b & e_c \end{bmatrix}' \quad (1)$$

where $'$ indicates the matrix transpose operation and the matrices $\mathbf{T_1}$ and $\mathbf{T_2}$ are:

$$\mathbf{T_1} = \begin{bmatrix} \cos\omega t & \sin\omega t \\ -\sin\omega t & \cos\omega t \end{bmatrix} \ \& \ \mathbf{T_2} = \frac{2}{3}\begin{bmatrix} 1 & -\frac{1}{2} & -\frac{1}{2} \\ 0 & \frac{\sqrt{3}}{2} & -\frac{\sqrt{3}}{2} \end{bmatrix}$$

The negative sequence transformation is as follows:

$$\begin{bmatrix} e_d^- & e_q^- \end{bmatrix}' = \mathbf{T_1 T_3} \begin{bmatrix} e_a & e_b & e_c \end{bmatrix}' \quad (2)$$

where the matrix $\mathbf{T_3}$ is:

$$\mathbf{T_3} = \frac{2}{3}\begin{bmatrix} 1 & -\frac{1}{2} & -\frac{1}{2} \\ 0 & -\frac{\sqrt{3}}{2} & \frac{\sqrt{3}}{2} \end{bmatrix}$$

The zero sequence transformation is:

$$e^0 = \frac{2}{3}\begin{bmatrix} \frac{1}{2} & \frac{1}{2} & \frac{1}{2} \end{bmatrix}\begin{bmatrix} e_a & e_b & e_c \end{bmatrix}' \quad (3)$$

The three separate error vector signals are fed to three different PI controllers to produce three sets of modulation signal vectors - $\begin{bmatrix} m_d^+ & m_q^+ \end{bmatrix}'$, $\begin{bmatrix} m_d^- & m_q^- \end{bmatrix}'$ and m^0. These are subjected to inverse transformation as follows:

$$\mathbf{m_{abc}^+} = \begin{bmatrix} m_a^+ & m_b^+ & m_c^+ \end{bmatrix}' = \frac{3}{2}\mathbf{T_2}'\mathbf{T_1}'\begin{bmatrix} m_d^+ & m_q^+ \end{bmatrix}'$$

$$\mathbf{m_{abc}^-} = \begin{bmatrix} m_a^- & m_b^- & m_c^- \end{bmatrix}' = \frac{3}{2}\mathbf{T_3}'\mathbf{T_1}'\begin{bmatrix} m_d^- & m_q^- \end{bmatrix}'$$

$$\mathbf{m_{abc}^0} = \begin{bmatrix} m_a^0 & m_b^0 & m_c^0 \end{bmatrix}' = \begin{bmatrix} \frac{1}{2} & \frac{1}{2} & \frac{1}{2} \end{bmatrix}m_0$$

$$(4)$$

The final modulation signals are:

$$\mathbf{m_{abc}} = \begin{bmatrix} m_a & m_b & m_c \end{bmatrix}' = \mathbf{m_{abc}^+} + \mathbf{m_{abc}^-} + \mathbf{m_{abc}^0} \quad (5)$$

The modulation signals m_a, m_b, m_c and their negative sum $-(m_a + m_b + m_c)$ are fed to three sine-triangle PWM blocks to generate switching signals for the switches S_1 to S_6.

III. SIMULATION RESULTS

The circuit and associated control in Fig. 1 are simulated using the SimPowerSystems blockset in MATLAB/Simulink. Table. I lists the system conditions and the simulation parameters. The purpose behind the simulation results is to show how the control philosophy described in Section II affects the currents injected into the grid during an unbalanced sag. The unbalanced sag is a 70% sag on phase b that occurs at $t = 2$s.

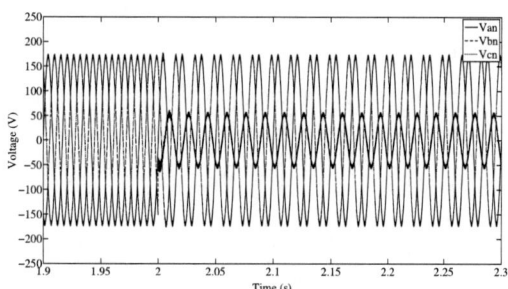

Fig. 2.   Grid voltages during the sag

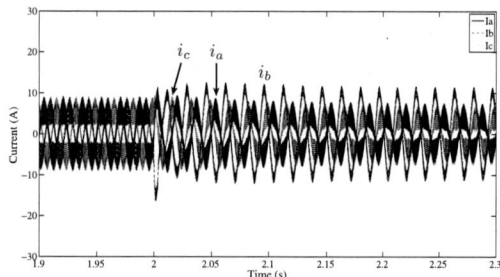

Fig. 3.   Grid currents during the sag

Fig. 2 shows the measured grid voltages at the PCC. Phase b voltage dips at $t = 2$s to 30% of the nominal value. Fig. 3 shows the currents injected into the grid. The grid currents contain the switching ripple generated by the dc-ac converter. These high-frequency harmonics can be removed by a line filter or by choosing an L-C-L filter instead of a L filter as will be shown in the experimental results. Prior to the fault, the currents are balanced and sinusoidal. Immediately following the fault, the currents undergo a cycle of transients due to the sudden change of the grid current references as described in Section II. Steady state is attained within 3 cycles of the transient. At steady state during the faulted condition, the phase a and c currents have the same magnitude and are displaced by 120°. However, phase b currents have a different magnitude as calculated by the reactive power support requirement. Moreover, the current i_b in phase b will lag the voltage v_{bn} by 90°.

Figs. 4, 5 and 6 show the PV panel voltage, panel current and dc link voltage before and during the fault. Prior to the occurrence of the fault, maximum power was being extracted from the fault with PV panel current being close to the short circuit current of 16 A. The perturbations in V_{pv} and I_{pv} are due to the perturb-and-observe nature of the MPPT algorithm. Before the sag,

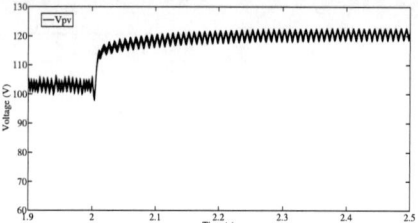

Fig. 4.   PV panel voltage during the sag

v_{dc}^{ref} is set by the MPPT algorithm. However, when a fault is detected, reference v_{dc}^{ref} is clamped to 400 V and the MPPT is discontinued. This is to prevent a sharp rise in the dc link voltage due to excess power from the PV panel being fed into the dc link but which is not injected into the grid due to the change in the current references. After steady state is reached within 3 cycles, V_{pv} and I_{pv} settle to new values. However, significantly large oscillations of angular frequency 2ω can be observed in V_{pv} and I_{pv} due to the unbalanced currents being injected into the grid.

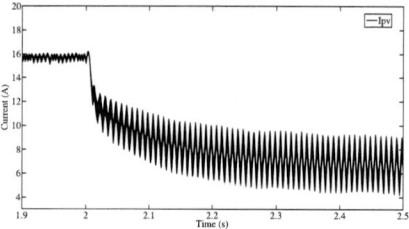

Fig. 5.   PV panel current during the sag

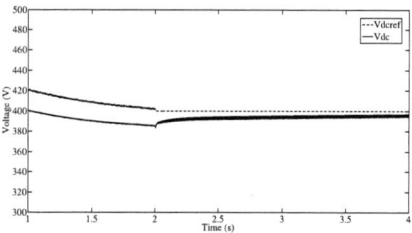

Fig. 6.   Dc link voltage during sag

IV. EXPERIMENTAL RESULTS

Fig. 7 shows the experimental setup used to validate the concept of asymmetrical fault ride-through while Fig. 8 shows the schematic of the experimental setup. From Figs. 7 and 8, the experimental setup consists of two dc-ac converters with LCL filters. A description of the setup from left to right is as follows. The capacitors and inductors form the LCL filter of the first dc-ac converter that is being used to emulate the grid. By using a dc-ac converter controlled in voltage feedback mode, an asymmetrical sag of any magnitude and phase jump can be emulated. Following the LCL filter comes the dc-ac converter. Fig. 9 shows the dc-ac converter that has been fabricated. Continuing with Fig. 7, following the grid emulator is the dc-ac converter which emulates the DG interface. Following the second dc-ac converter is the LCL filter of the DG interface. The LCL filter of

The 2014 International Power Electronics Conference

Fig. 7.   Photograph of the experimental setup

each dc-ac converter consists of six 1 mH inductors and twelve 50 μF capacitors. Therefore, an LCL filter with the inductors $L_f = L_g = 1$ mH and the filter capacitors $C_f = 200\mu$F are being used. From Fig. 8, it can be observed that the two dc-ac converter are connected together to form a Point of Common Coupling (PCC). A three-phase resistive load has been connected to the PCC. However, in general, the PCC is the bus at which consumers are connected and the voltage of the PCC is the voltage affecting the connected loads.

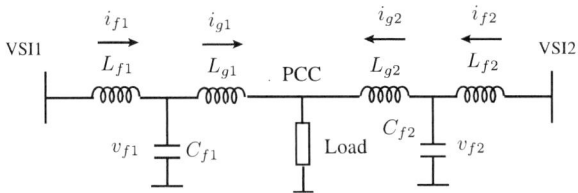

Fig. 8.   Schematic of the experimental setup

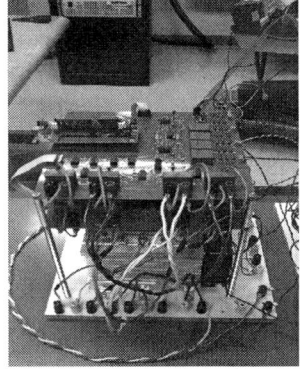

Fig. 9.   Photograph of the dc-ac converter

The power device used is a six-IGBT Intelligent Power Module (IPM) from Infineon - IFS150V12P4 - with devices rated for 1200V, 150A. These IPMs have in-build gate drivers and also provide fault feedback for desaturation of the IGBTs due to overcurrent and a failure of the power supply to the gate drivers. The power device mounted on a heatsink has been closely integrated with the sensor and controller circuit which has been mounted directly above the power device so as to minimize noise. The control is being implemented on a Texas Instruments floating point Digital Signal Processor TMS320F28335 Delfino Experimental Kit. The current and voltage sensors are hall effect transducers from LEM.

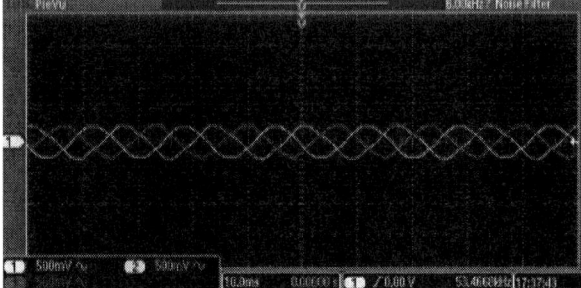

Fig. 10.   Grid currents in steady state

The current transducers are LA 55P with a maximum range of 55A and the voltage sensors are LV 25P with a maximum range of 1000V.

Fig. 10 shows the currents i_g injected into the grid emulator by the DG interface. The currents are seen to be balanced sinusoids with a THD measured to be approximately 3.0%. The currents are seen to have some oscillations due to the LCL filter resosnance. These oscillations have been damped actively in the control strategy rather than using resistors in the LCL filter which will decrease the efficiency of the overall system. However, advanced controllers have been proposed in literature for control of dc-ac converters with LCL filters [5]. The grid emulator has been used to produce a sag of 30% in phase b of its output voltage. It should be noted here that the sag is produced across the filter capacitors C_f of the dc-ac converter LCL filter. However, the grid has been emulated with its associated line impedance in the form of the final grid inductor L_g. The DG interface measures the grid voltage while synchronizing and applying the fault ride-through algorithm. In this experimental setup, measurement of the grid voltages is possible due to the proximity of the converters. However, in a practical system, the grid voltage will have to be estimated from the measured PCC voltages.

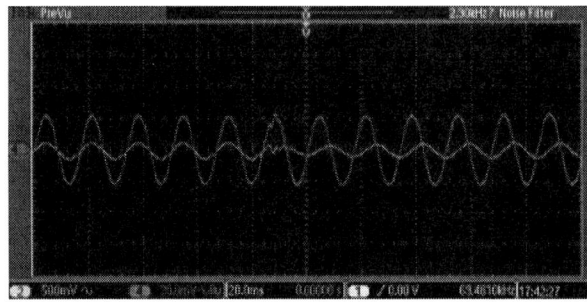

Fig. 11.   Grid currents and voltage during the sag in phase b

Fig. 11 shows the PCC voltage of phase b and the current injected into the PCC by the DG interface in phase b. Before the fault, the current in phase b is observed to be in phase with the phase b PCC voltage as the DG injects only active power into the healthy grid. However, upon detecting the sag in the grid voltage, the phase b currents are seen to lag the PCC voltage by 90° thereby injecting reactive power into the grid. As a result, the PCC voltage

978-1-4799-2706-7/14 $31.00 © 2014 IEEE   3486

The 2014 International Power Electronics Conference

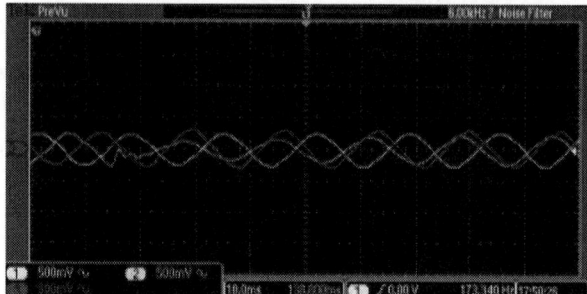

Fig. 12.   Grid currents in transient

Fig. 13.   Grid currents during asymmetrical sag

is seen to remain practically constant during the sag. It should be noted that the reason for this mitigation in the fault is the impedance of the grid. This impedance of the grid will vary with respect to the strength of the grid that is decided by the length of the feeder following the transformer to which the loads and the DG are connected.

Fig. 12 shows the currents injected by the DG into the PCC during the sag. As can be the seen the balanced currents before the fault become severely unbalanced during the fault with phase b injecting purely reactive power into the grid. It is to be noted that the four-wire dc-ac converter is essential for the purpose of fault ride-through in a three-phase four-wire system as zero sequence currents are inevitable when fault ride-through is applied on a per-phase basis. Fig. 13 shows the currents injected by the DG interface into the PCC during the fault. The currents continue to maintain a THD of approximately 3.2% indicating that the controller continues to perform effectively while supplying a zero sequence current.

V.   DISCUSSION

This paper has proposed a strategy by which a DG connected to a distribution system can continue to operate during an asymmetrical fault and also can provide a certain degree of relief in terms of reactive power support to the system. However, at this stage, the operational impacts of such an operational strategy needs to be considered. In general, at low voltages of 208V, 400V, 690V or other standard voltage levels used in different parts of the world most distribution system feeding consumers are radial systems. Therefore, most system operators will expect the DG to disconnect during a fault in order that the protection system should be allowed to de-energize the fault so that the fault clears upon reclosure. The

possibility of DGs remaining connected during a fault could result in the DG feeding the fault thereby causing a temporary fault to turn into a permanent fault.

The present trend of increasing DG penetration in the low voltage distribution system by allowing consumers to fit PV panels on their rooftops and inject power into the grid will result in a significant change in the current distribution of the grid. Moreover, if these DGs were to form a significant proportion of the total power produced, the disconnection of all DGs at the low voltage system could cause significant stability problems upon fault clearance when power generation will be insufficient. Therefore, in a distribution system, the location of the fault is of importance. In the case of a fault occuring on a downstream feeder, DGs connected upstream could continue to remain connected and supply consumers who have not been affected by the operation of the grid protection system. This would improve the power balance of the grid and also in case the DG has the capacity to support the grid, would be able to provide relief to the grid in terms of voltage support.

This paper has examined the ride-through strategy in terms of reference current generation and current control. However, it has not considered the issue of locating the fault in the distribution system with respect to the point of the connection of the DG. Identification of the location of the fault would be similar to anti-islanding techniques that have been proposed. However, the difference would be that the system to be considered would be far more complex that most of the techniques available in literature. This is due to the fact that not only loads with varying characteristics in terms of frequency response are connected to the distribution system but also other DGs implementing different anti-islanding strategies may significantly alter the effective frequency-response of the distribution system as viewed by the DG.

The paper has examined the fault ride-through concept with respect to the compliance regulation imposed on larger DGs such as wind farms connected at the transmission level. These DGs are required to inject reactive power when a voltage sag occurs. However, in the low voltage distribution system where the R/X ratio may vary significantly, injecting reactive power may cause the grid voltage to fluctuate. The experimental results show that the PCC voltage is almost unaffected by a 30% sag in the grid voltage due to reactive power injection. This is due to the fairly large reactance of the grid which is emulated by the grid dc-ac converter. However, this proves that further studies are required in distribution systems for the mode of operation of DGs during faults.

VI.   CONCLUSIONS

This paper has examined a PV system interfaced to a four-wire distribution system with specific emphasis on the ability to ride-through unbalanced faults. The importance of fault ride-through appears when the penetration of DG increases in the distribution system to an significant extent. In such a case, the disconnection of all DGs as is usually required by the utility in the distribution system may result in a power deficit. Furthermore, to enable reactive power support to be provided only to faulty phases and not to healthy phases, a four-wire dc-ac converter has been used. The control philosophy has been modified by using three single-phase PLLs for improved fault detection and also better synchronization to individual phases in the case of phase angle jumps in the faulty phases. The paper presents experimental results for dc-ac converters with LCL filter showing the low THD in the currents injected into the grid. Moreover, the four-wire dc-ac converter has been shown to be capable of injecting a current with a large zero sequence component.

ACKNOWLEDGEMENT

The authors would like to thank Hydro One Inc. and Natural Sciences and Engineering Research Council of Canada (NSERC) for their support in this project.

REFERENCES

[1] E. Troester, "New german grid codes for connecting PV systems to the medium voltage power grid," in *2nd International Workshop on Concentrating Photovoltaic Power Plants: Optical Design, Production, Grid Connection*, March 2009, pp. 1–4.

[2] V. Kaura and V. Blasko, "Operation of a phase locked loop system under distorted utility conditions," *IEEE Transactions on Industry Applications*, vol. 33, no. 1, pp. 58–63, January/February 1997.

[3] R. M. S. Filho, P. F. Seixas, P. C. Cortizo, L. A. B. Torres, and A. F. Souza, "Comparison of three single-phase PLL algorithms for UPS applications," *IEEE Transactions on Industrial Electronics*, vol. 55, no. 8, pp. 2923–2932, August 2008.

[4] A. Timbus, M. Liserre, R. Teodorescu, P. Rodriguez, and F. Blaabjerg, "Evaluation of current controllers for distributed power generation systems," *IEEE Transactions on Power Electronics*, vol. 24, no. 3, pp. 654–664, March 2009.

[5] M. Liserre, F. Blaabjerg, and S. Hansen, "Design and control of an lcl-filter-based three-phase active rectifier," *IEEE Transactions on Industrial Applications*, vol. 41, no. 5, pp. 1281–1291, September/October 2005.

[6] P. Rodriguez, A. V. Timbus, R. Teodorescu, M. Liserre, and F. Blaabjerg, "Flexible active power control of distributed power generation systems during grid faults," *IEEE Transactions on Industrial Electronics*, vol. 54, no. 5, pp. 2583–2592, October 2007.

[7] M. G. Villalva, J. R. Gazoli, and E. R. Filho, "Comprehensive approach to modeling and simulation of photovoltaic arrays," *IEEE Transactions on Power Electronics*, vol. 24, no. 5, pp. 1198–1208, May 2009.

[8] E. Koutroulis, K. Kalaitzakis, and N. C. Voulgaris, "Development of a microcontroller-based, photovoltaic maximum power point tracking control system," *IEEE Transactions on Power Electronics*, vol. 16, no. 1, pp. 46–54, January 2001.

[9] S. J. Chiang, K. T. Chang, and C. Y. Yen, "Residential photovoltaic energy storage system," *IEEE Transactions on Industrial Electronics*, vol. 45, no. 3, pp. 385–394, June 1998.

Operation Mode Analysis for Solving the Partial Shadow in a Novel PV Power Generation System

Qi Zhang, Xiangdong Sun, Yanru Zhong , Lie Guo
Department of Electrical Engineering
Xi'an University of Technology
Xi'an, China
Zhangqi_hero@163.com，sxd1030@163.com

Mikihiko Matsui
Department of Electrical Engineering
Tokyo Polytechnic University
Atsugi, Japan
matsui@ee.t-kougei.ac.jp

Abstract—A photovoltaic (PV) module integrated converter (PVMIC) composed of a PV panel and its parallel-connected bidirectional flyback converter is proposed for solving the partial shadow problems. The PVMIC includes sleeping mode, energy feedback mode, bypass-diode mode and DC module mode. Based on PVMIC, the novel PV generation system adjust the operating modes to accommodate the different shadow level and thus to ensure maximum power output of the PV array. In this approach, the PV system has higher conversion efficiency, higher PV output power and enhanced operation functions in the condition of partial shadow. The feasibility of proposed method is verified by the simulations and experimental results.

Keywords—PVMIC, bidirectional flyback converter, partial shadow problem, DC module.

I. Introduction

The PV power generation system is always influenced by its internal parameters as well as natural environments during the generating operation. Those such as the shadows generated by the dust, tall buildings, clouds, and the parameter mismatch caused by different PV panel orientations and the different PV cell aging degrees are all called partial shading problem. Those shading problems have the potential impact on the following aspects.

1) All series-connected PV modules deviate from the maximum power point (MPP), and the substantial decrease in the output power is inevitably.

2) If there are not any paralleled bypass diodes equipped in PV panels, the probability of hot spot effect will increase significantly, which will cause irreparable damage for PV cells, thereby affecting the operation of PV system.

3) To prevent the hot spot effect from the physical nature, the bypass diodes are paralleled equipped in each PV panels. However, it will be appear multiple MPP for

---

This work was supported by Specialized Research Fund for the Doctoral Program of Higher Education（20116118110006）, Specialized Fund of Shaanxi Education Department（12JK0567、2013JK0999）, Industrialization Projects（2012JC17）, Specialized Fund for Key Discipline Construction of Shaanxi（105-5X1201）, Startup Fund of Doctor research（105-211207）.

the PV power-voltage (*P-V*) output characteristics curve under the unbalanced PV panels' illumination, and the traditional MPP tracking (MPPT) algorithm becomes invalid in this situation.

From the algorithms perspective to solve the shadow problem, the essence is to get the real maximum output power point from multi-point power condition. Many works focus on the algorithms research, such as [1-2], which provides a complex algorithm to find the optimal PV operating point, while the method is not easy to realize by traditional low digital processors. From the circuit structure perspective, each panels equipped with an electrical converter as a PV module unit to deliver the power respectively, and the panels in the partial shadowed PV array can work in their own maximum power point. Circuit topologies and corresponding control algorithms service for the partial shadow problem are provided in [3-7].

II. NEW PHOTOVOLTAIC POWER GENERATION SYSTEM

A. Circuit Topology

Fig.1 (a) shows the schematic diagram of PVMIC, and its simplified schematic is presented in Fig.1 (b). A bidirectional flyback circuit is chosen exactly for the parallel connected power converter of the PVMIC due to the unit power level is typically less than 300W. It can be seen in Fig.2, series-connected PV string together with PVMICs formats the DC bus voltage through a reverse charge diode for the next power converter stage or power loads. Energy of the DC bus capacity feed from input of the flyback circuit back to each panel's interface. When power demand increases, the parallel branch should be considered as shown in Figure 3, which has multiple series PV strings with PVMICs' circuits. These series and parallel connected PVMICs can be viewed as a whole power sources.

B. New PV String Operation Modes

The global MPP for the whole PV source is realized by next power stage. There are several operating modes for the PVMICs. As for single-string system, the

The 2014 International Power Electronics Conference

operation includes sleeping mode, energy feedback mode, bypass diode mode and DC module mode. For multi-string system and supplying the power for three-phase inverter power load, the operation includes sleeping mode, energy feedback mode, and bypass diode mode. Due to the same operating principle, the single-string system is taken as an example for the operation modes' elaboration.

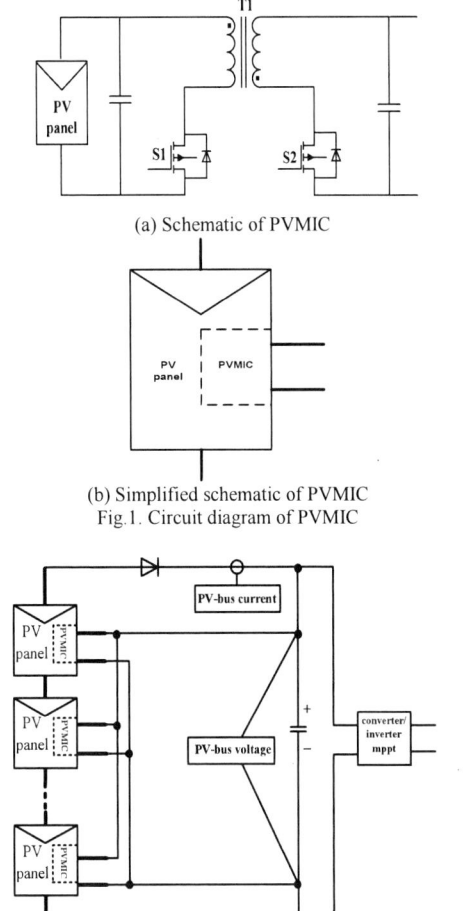

(a) Schematic of PVMIC

(b) Simplified schematic of PVMIC
Fig.1. Circuit diagram of PVMIC

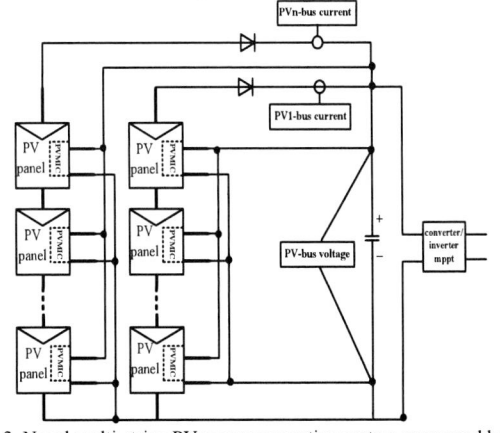

Fig.2. Novel single-string PV power generation system composed by PVMICs

Fig.3. Novel multi-string PV power generation system composed by PVMICs

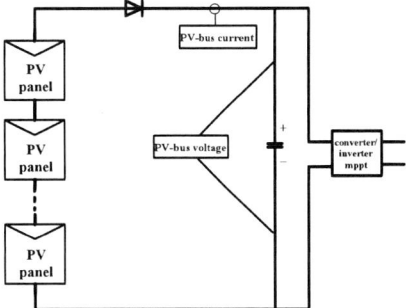

Fig.4(a). Equivalent circuit of sleeping mode

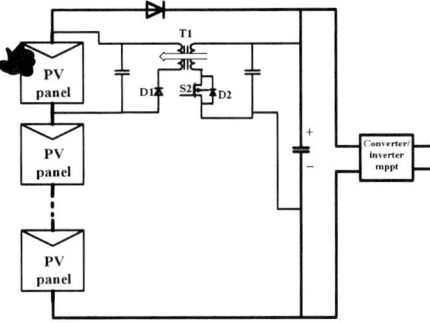

Fig.4(b). Equivalent circuit of energy feedback mode

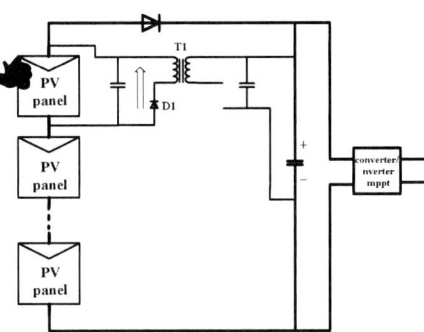

Fig.4(c). Equivalent circuit of bypass diode mode

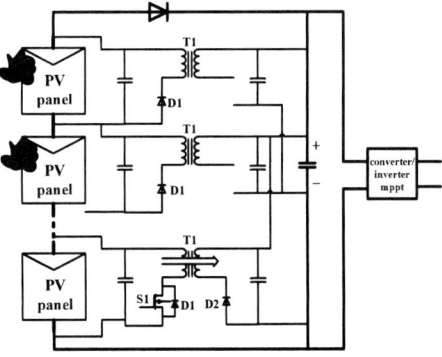

Fig.4(d). Equivalent circuit of DC module mode
Fig.4 Operation mode diagram of PV generation power system

C. New PV String Operation Modes

The global MPP for the whole PV source is realized by next power stage. There are several operating modes for the PVMICs. As for single-string system, the operation includes sleeping mode, energy feedback mode, bypass diode mode and DC module mode. For multi-

978-1-4799-2706-7/14 $31.00 © 2014 IEEE   3490

string system and supplying the power for three-phase inverter power load, the operation includes sleeping mode, energy feedback mode, and bypass diode mode. Due to the same operating principle, the single-string system is taken as an example for the operation modes' elaboration.

（1）Sleeping mode: When illumination balanced and PV parameters consistent, flyback of PVMIC does not work and there is hardly any power loss in the PVMICs. This situation is completely the same as the normal PV power generation system which is shown in Figure.4 (a).

（2）Energy feedback mode: When a certain PV panel is shadowed, its corresponding PVMIC operates to compensate the shadowed panel's terminal voltage back to the maximum power point voltage value. However, the other normal PVMICs are still working in the sleeping mode to ensure the high system efficiency. This mode diagram is shown in Fig.4 (b).

（3）Bypass diode mode: It is shown in Fig.4 (c), when a PV panel is serious shadowed and its output voltage collapse, the corresponding PVMIC is invalid, and the diode connected the transformer primary side provide a current path to clamp the panel's terminal voltage and then prevent hot spot effect.

（4）DC module mode: If numerous panels seriously shadowed and operate in bypass diode mode, the PV source cannot satisfy the power demand of the next power stage and unable to support the DC bus voltage. In this situation, the panels without shadow problem operate in DC module mode, which is the same as a traditional parallel-connected DC module mode. Fig.4 (d) illustrates the DC module mode.

D. Partial shadow analysis

PV panels are composed by numerous series or parallel connected PV cells. Assuming that the characteristics of all the PV cells are consistent, include temperature and illumination intensity, the mathematical model of the PV panel can be described as (1).

$$I_{pv} = I_{sc} - I_{do}\left[\exp\left(\frac{q}{nAkT_c}V_{pv}\right) - 1\right] \quad (1)$$

Where, I_{pv} and V_{pv} are the output current and voltage of PV panel, I_{sc} is the panel's short-circuit current, and I_{do} is the diode reverse saturation current. q is the electron charge, $A=1.0{\sim}3.0$, k is a Boltzmann's constant, n is a series-connected coefficient of PV cells, and I_{sc} is decided by (2), where, h_t is temperature coefficient expressed by (3). T_c is a absolute temperature, T_{co} is the absolute standard temperature, φ is illumination intensity, and φ_o is standard value.

$$I_{sc} = I_{sco} - \left[1 + h_t\left(T_c - T_{co}\right)\right]\frac{\phi}{\phi_o} \quad (2)$$

$$h_t = 6.4 \times 10^{-4}\left(K^{-1}\right) \quad (3)$$

Without considering the temperature factor, (1) can be rewritten as (4).

$$I_{pv} = I_{sc} - \varepsilon\left[\exp\left(\xi V_{pv}\right) - 1\right] \quad (4)$$

where, $\varepsilon = I_{do}$, $\xi = q/nAkT_c$.

PV panel's terminal voltage is obtained by (5), and its output power is described as (6).

$$V_{pv} = \frac{\ln\left(\dfrac{I_{sc} - I_{pv}}{\varepsilon} + 1\right)}{\xi} \quad (5)$$

$$P_{pv} = V_{pv} \cdot I_{pv} \quad (6)$$

If $dP_{pv}/dV_{pv} = 0$, $V_{pv} = V_{mpp}$, the voltage of MPP is expressed by (7), the current I_{mpp} of MPP is obtained from (1) and (7).

$$\left(\xi V_{mpp} + 1\right)\exp(\xi V_{mpp}) = \frac{I_{sc}}{\varepsilon} + 1 \quad (7)$$

Take the series connection of PV1 and PV2 as an example, which is shown in Fig.5. The PVMICs do not work, and two PV panels have the same circuit parameters, their output voltages and currents are V_{pv1}, V_{pv2}, I_{pv1} and I_{pv2}, respectively. Therefore, the following equation (8) is always valid. Where, V_s is the PV-bus voltage and I_s is the PV-bus current.

$$\begin{cases} I_{pv1} = I_{pv2} = I_s \\ V_{pv1} + V_{pv2} = V_s \end{cases} \quad (8)$$

The differential voltage between PV1 and PV2 is calculated by (9).

$$\Delta V = V_{pv1} - V_{pv2} = \frac{\ln\left(\dfrac{I_{sc1} - I_{pv1}}{\varepsilon} + 1\right) - \ln\left(\dfrac{I_{sc2} - I_{pv2}}{\varepsilon} + 1\right)}{\xi} = \frac{\ln\left(\dfrac{I_{sc1} - I_s + \varepsilon}{I_{sc2} - I_s + \varepsilon}\right)}{\xi} \quad (9)$$

If $I_{sc1} = I_{sc2}$, $V_{pv1} = V_{pv2}$ and $\Delta V = 0$, which means that two PV panels have the same output voltages without the shadow. If only PV2 is shadowed, $I_{sc1} > I_{sc2}$ and $V_{pv1} > V_{pv2}$, $\Delta V > 0$. Therefore, we can determine whether the shadow exists by measuring the terminal voltage of each PV panel. When all terminal voltages are almost the same, the PVMICs never work, and the sleeping mode is operated.

Supposing $\Delta I_{sc} = I_{sc1} - I_{sc2} \geq 0$, we can deduce equation (10) from equation (4) and equation (8).

$$V_{pv1} = \frac{1}{\xi}\ln\left[\exp\left(\xi V_{pv2}\right) + \frac{\Delta I_{sc}}{\varepsilon}\right] \quad (10)$$

It is seen from (4) and (10) that the PV array's output voltage, current and power are expressed as follows.

$$V_s = V_{pv2} + \frac{1}{\xi}\ln\left[\exp\left(\xi V_{pv2}\right) + \frac{\Delta I_{sc}}{\varepsilon}\right] = f\left(V_{pv2}\right) \quad (11)$$

$$I_s = I_{pv2} = I_{sc2} - \varepsilon\left[\exp\left(\xi f^{-1}(V_s)\right) - 1\right] \quad (12)$$

$$P_s = V_s I_s = V_s I_{sc2} - \varepsilon V_s\left[\exp\left(\xi f^{-1}(V_s)\right) - 1\right] \quad (13)$$

Since $V_{pv1} > 0$, the following relationships are known from (8), (10) and (11).

$$\begin{cases} V_{pv2} < 0, \quad V_s < \dfrac{1}{\xi}\ln\left(1 + \dfrac{\Delta I_{sc}}{\varepsilon}\right) \\[2mm] V_{pv2} = 0, \quad V_s = \dfrac{1}{\xi}\ln\left(1 + \dfrac{\Delta I_{sc}}{\varepsilon}\right) \\[2mm] V_{pv2} > 0, \quad V_s > \dfrac{1}{\xi}\ln\left(1 + \dfrac{\Delta I_{sc}}{\varepsilon}\right) \end{cases} \quad (14)$$

On the basis of the above analysis, if V_s satisfies the equation of (15), the diode D1 is reversed, and D2 conducted, which has been defined as the bypass diode mode for PV2.

$$0 < V_s < \frac{1}{\xi}\ln\left(1 + \frac{\Delta I_{sc}}{\varepsilon}\right) \tag{15}$$

Then, if V_{mpp1} is the maximum power point voltage of PV1, and it satisfies the relationship shown in Equation (16), output power of PV string monotonically increases. If V_{mpp1} satisfies (17), the PV string has only one MPP which is the MPP of PV1.

$$V_{mpp1} \geq \frac{1}{\xi}\ln\left(1 + \frac{\Delta I_{sc}}{\varepsilon}\right) \tag{16}$$

$$0 < V_{mpp1} < \frac{1}{\xi}\ln\left(1 + \frac{\Delta I_{sc}}{\varepsilon}\right) \tag{17}$$

V_{oc} is the open-circuit voltage of PV string. If V_s meets the equation of (18), both D1 and D2 are all reversed.

$$\frac{1}{\xi}\ln\left(1 + \frac{\Delta I_{sc}}{\varepsilon}\right) < V_s < V_{oc} \tag{18}$$

$$\frac{dP_s}{dV_s} = I_s + V_s\frac{dI_s}{dV_s} \tag{19}$$

$$= I_s - \varepsilon\xi V_s \exp[\xi g^{-1}(V_s)]\cdot[g^{-1}(V_s)]'$$

$$\frac{d^2 P_s}{dV_s^2} = -2\varepsilon\xi V_s \exp[\xi g^{-1}(V_s)]\cdot[g^{-1}(V_s)]' - \varepsilon\xi V_s \exp[\xi g^{-1}(V_s)]\cdot[g^{-1}(V_s)]'' \tag{20}$$

$$- \varepsilon\xi^2 V_s \exp[\xi g^{-1}(V_s)]\cdot[g^{-1}(V_s)']^2 < 0$$

The first-order and second-order partial derivatives are listed by (13), equations (19) and (20) are given. Therefore, equation (21) is deduced.

$$\left.\frac{dP_s}{dV_s}\right|_{V_s = \frac{1}{\xi}\ln\left(1 + \frac{\Delta I_{sc}}{\varepsilon}\right)} = I_{sc2} - \varepsilon\ln\left(1 + \frac{\Delta I_{sc}}{\varepsilon}\right)\frac{\Delta I_{sc}/\varepsilon + 1}{\Delta I_{sc}/\varepsilon + 2} \tag{21}$$

When equation (22) is satisfied, it is known from (19) and (20) that the PV string has a unique MPP. Otherwise, the PV string has two MPPs.

$$I_{sc2} > \varepsilon\ln\left(1 + \frac{\Delta I_{sc}}{\varepsilon}\right)\frac{\Delta I_{sc}/\varepsilon + 1}{\Delta I_{sc}/\varepsilon + 2} \tag{22}$$

E. Work condition of operation modes

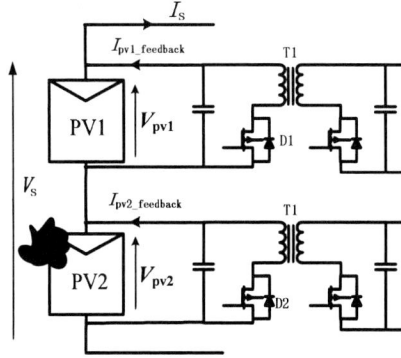

Fig.5 Schematic diagram of two series-connected PV panels

Take the series connection of PV1 and PV2 shown in Fig.5 as an example. When equation (16) is satisfied, the system may operate in energy feedback mode. At this situation, PVMIC according to PV2 operates to compensate the current which reduced by the shadow. Due to the compensation that PV-bus current is always the optimal current and each PV panel operates at the MPP. Ignoring the temperature difference, equation (22) is given, where, I_{mpp}, $I_{mpp\text{-}shading}$ are the optimal currents of PV1 and PV2, respectively, V_{mpp1} and V_{mpp2} are the optimal voltages of PV1 and PV2.

$$\begin{cases} I_s = I_{pv1} = I_{mpp} = I_{sc1} - \varepsilon\left[\exp\left(\xi V_{mpp1}\right) - 1\right] \\ I_{pv2} = I_{mpp-shading} = I_{sc2} - \varepsilon\left[\exp\left(\xi V_{mpp2}\right) - 1\right] \\ I_s - I_{pv2} = I_{mpp} - I_{mpp-shading} = \Delta I_{sc} \end{cases} \tag{22}$$

Thus, the total power generated by PV string is expressed by equation (23).

$$P_s = V_s I_s \tag{23}$$

the total power generated by PV module is expressed by equation (24).

$$P_{mpp} = P_{pv1} + P_{pv2} = V_{mpp1}I_{mpp} + V_{mpp2}I_{mpp-shading} \tag{24}$$

The compensation power to PV2 is obtained by (25).

$$P_{pv2-feedback} = V_{mpp2}I_{pv2-feedback} = V_{mpp2}\left(I_{mpp} - I_{mpp-shading}\right) \tag{25}$$

It is seen from (22)~(25) that the power delivered to load is described by (26) when the PVMIC of PV2 works.

$$P_{load} = P_s - P_{pv2-feedback} = P_{mpp} \tag{26}$$

Based on the above analysis, work conditions of PVMIC are summarized as follows.

1) Sleeping mode: if all the terminal voltages of series-connected PV panels are the same basically, the PVMICs never act, the system works in the sleeping mode.

2) Bypass diode mode: if terminal voltage of a PV panel collapses, the corresponding PVMIC does not work and the parallel diode provides a current path for the shadowed panel.

3) Energy feedback mode: When a certain PV panel is shadowed , the system works in the energy feedback mode.

4) DC module mode: if the PV-bus voltage is lower than the required voltage, the system also operates in the DC module mode.

Similarly, the analysis method is suitable to PV string including multiple series-connected PV panels. the following conclusions are obtained. Shadows degree are observed by detecting the PV panels' terminal voltage. If no shadow occurs, the system works in the sleeping mode. If partial shadow happens, the operation mode is decided according to the shadow degree, if the shadow is mild, energy feedback mode is chosen; while the shadow is serious, bypass diode mode is selected; and if there are many PV panels operate in the bypass diode mode, DC module mode is used to realize the power conversion.

III. SIMULATION RESULTS

Simulation model of a novel single-string PV system in the PSIM software is established. Three PVMICs are connected in series, and the boost chopper is the public converter which delivers the power to the resistor load. Parameters of PV panel are shown in Table 1. The illumination is the ratio of its standard value $1000W/m^2$.

（1）Sleeping mode

Fig.6 shows the waveforms of illumination degree which are defined as L1, L2 and L3. Each panel's output current, voltage, power are also listed in sequence. Three panels basically operate at the MPP.

（2）Energy feedback mode

Illumination of PV1 changes from $1000W/m^2$ to $800W/m^2$ at 0.6s, and illuminations of PV2 and PV3 are still $1000W/m^2$. Fig.7 shows that PV1 is compensated by its PVMIC, and each panel works in their own MPP. It is obvious that PV2 and PV3 are not affected by the shadow.

（3）Bypass diode mode:

It is seen from fig.8, illumination of PV1 changes from $1000W/m^2$ to $100W/m^2$, which means that PV1 is serious shadowed. The corresponding PVMIC is in bypass diode mode, and the flyback converter does not work. Terminal voltage of PV1 is 0V due to diode clamping, while other two PV panels still work at their own MPP.

（4）DC module mode:

In Fig.9, illuminations of PV1 and PV2 changes from $1000W/m^2$ to $100W/m^2$, and illumination of PV3 is still $1000W/m^2$. PV1 and PV2 are all operated in bypass diode mode, and the corresponding flyback converter of PV3 operates in DC module mode to stabilize the PV-bus voltage.

IV. EXPERIMENTAL RESULTS

A single-string PV power generation system is established, experimental circuit and parameters are the same as those of simulation. PV panels are defined as PV1, PV2 and PV3.

TABLE.1
PV MODULE PARAMETERS

V_{oc}	44.4V
I_{sc}	5.15A
V_{mpp}	35.6V
I_{mpp}	4.78A
P_{mpp}	170W

Fig.6 Simulation waveforms of sleeping mode

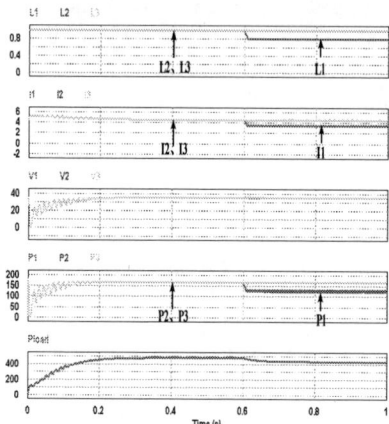

Fig.7 Simulation waveforms of energy feedback mode

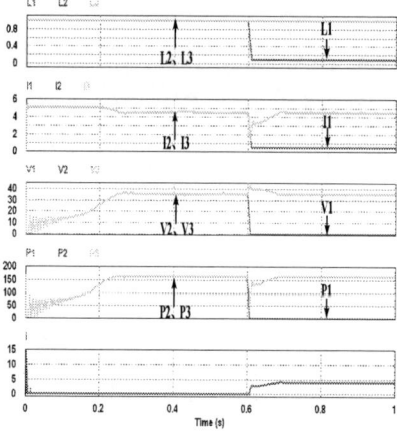

Fig.8 Simulation waveforms of bypass diode mode

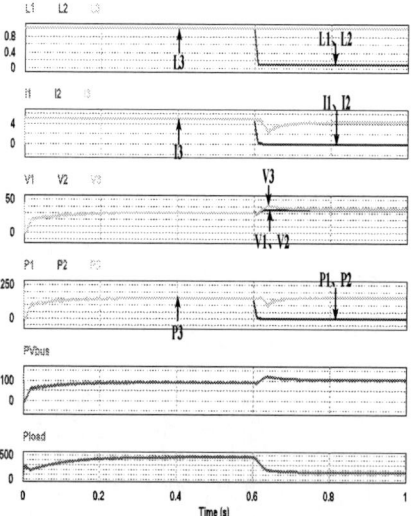

Fig.9 Simulation waveforms of DC module mode

Fig.10 shows the experimental results of sleeping mode, CH1, CH2 and CH4 are output voltages of PV1, PV2 and PV3 respectively, and CH3 is the PV-bus current. It can be seen from Fig.10 that three PV panels almost work at the MPP.

The 2014 International Power Electronics Conference

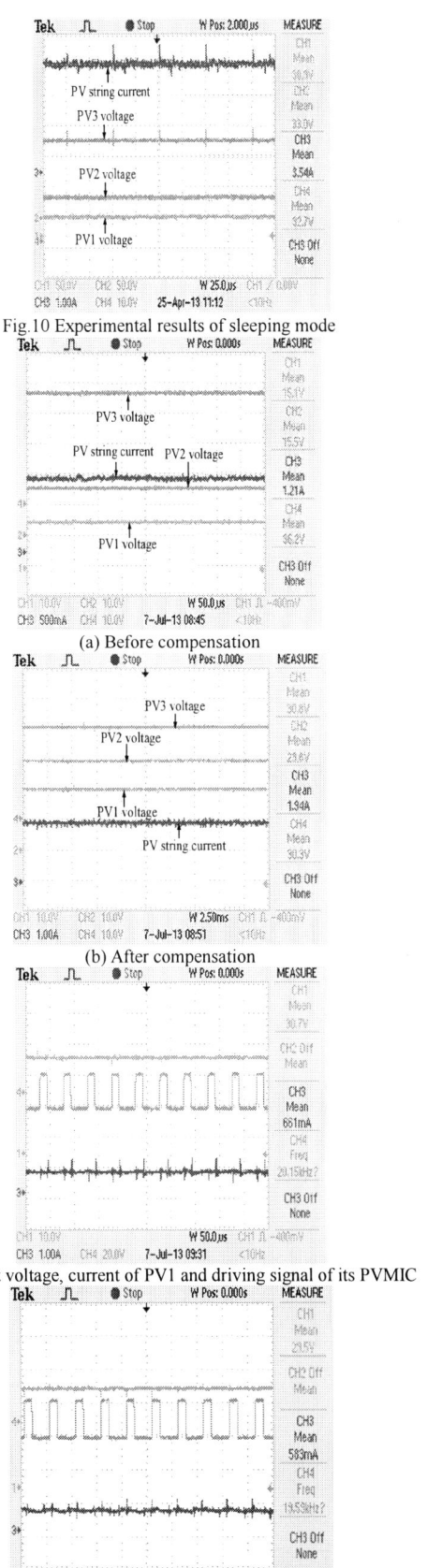

Fig.10 Experimental results of sleeping mode

(a) Before compensation

(b) After compensation

(c) Output voltage, current of PV1 and driving signal of its PVMIC

(d) Output voltage, current of PV2 and driving signal of its PVMIC
Fig.11 Experimental result of energy feedback mode

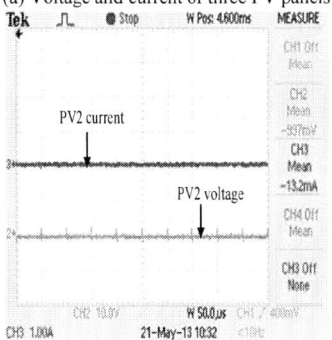

(a) Voltage and current of three PV panels

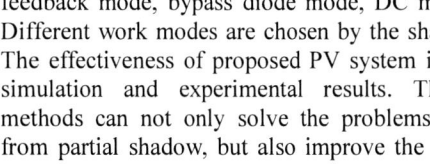

(b) Voltage and current of PV2
Fig.12 Experimental results of bypass diode mode

Fig.11 shows the experimental results in energy feedback mode when PV1 and PV2 are affected by the shadow. Fig.11 (a) and (b) show the experimental results before and after compensation, respectively. CH1, CH2 and CH4 are output voltages of PV1, PV2 and PV3 respectively, and CH3 is the PV-bus current. Fig.11 (c) shows PV1's output voltage, output current and driving signal of the corresponding flyback converter. Fig.11 (d) demonstrates those of PV2. It is known that three PV panels deviate from their MPP and the system has serious power loss before compensation. After the system works in the energy feedback mode, the corresponding flyback converter of PV1 and PV2 compensate the PV-bus current. As a result, three PV panels almost operate at the MPP and system output power increases.

Fig.12 is experimental results of bypass diode mode. PV2 is serious shadowed. It is seen from Fig.12, PV2 works in bypass diode mode and its voltage is clamped by the diode voltage, PV1 and PV3 almost operate at the MPP.

Because the feedback PV-bus voltage is replaced by a DC voltage source in the experiment, DC module mode is not realized

CONCLUSIONS

A PVMIC is proposed for solving partial shading problem. The PVMIC includes sleeping mode, energy feedback mode, bypass diode mode, DC module mode. Different work modes are chosen by the shadow degree. The effectiveness of proposed PV system is verified by simulation and experimental results. The proposed methods can not only solve the problems which arise from partial shadow, but also improve the efficiency of power generation system.

978-1-4799-2706-7/14 $31.00 © 2014 IEEE   3494

References

[1] Toshihiko Noguchi, Shigenori Togashi, Ryo Nakamoto, "Short-current pulse-based maximum power point tracking method for multiple photovoltaic and converter module system,"*IEEE Trans. On Industrial Electronics*, vol.49,no.1,pp:.217-222, 2002.

[2] Solodovnik,E.V., Shengyi Liu, Dougal, R.A, "Power controller design for maximum power tracking in solar installations,"*Trans. On Power Electronics*, vol.19.no.5,pp.1295-1304,2004.

[3] Toshihisa Shimizu, Osamu Hashimoto, Gunji Kimura, "A Novel High-Performance Utility-Interactive Photovoltaic Inverter System," *IEEE Trans. on Power Electronics*, vol.18,no.2,pp.704-711, 2003.

[4] Toshihisa Shimizu, Masaki Hirakata, Tomoya Kamezawa, Hisao Watanabe, "Generation Control Circuit for Photovoltaic Modules,"*IEEE Trans. on Power Electronics*,vol.16,no.3,pp.293-300,2001.

[5] Z. Salam, M. Z. Ramli, "A Simple Circuit to Improve the Power Yield of PV Array During Partial Shading ,"*Energy Conversion Congress and Exposition (ECCE), IEEE*, pp.1622-1626, 2012.

[6] Abdalla, J. Corda, and L. Zhang, 2012:Multilevel DC-Link Inverter and Control Algorithm to Overcome the PV Partial Shading,"*IEEE Trans. on Power Electronics*,vol.28,no.1,pp.14-18,2013.

[7] Riad KADRI, Jean+Paul GAUBERT, Gérard CHAMPENOIS, "New Converter Topology to Improve Performance of Photovoltaic Power Generation System Under Shading Conditions," *Proceedings of the 2011 International Conference on Power Engineering, Energy and Electrical Drives*, Spain,2011.

The 2014 International Power Electronics Conference

Analysis of Partial Power Processing Distributed MPPT for a PV Powered Electric Aircraft

Ahmad Diab Marzouk[1], Sébastien Fournier-Bidoz[2], Jessica Yablecki[2],
Kenneth McLean[2], and Olivier Trescases[1]

[1]University of Toronto, 10 King's College Road, Toronto, ON, M5S 3G4, Canada
[2]Solar Ship Inc., 366 Adelaide St., Toronto, ON, M5A 3X9, Canada
E-mail: ot@ece.utoronto.ca

Abstract—Habitations in remote areas around the world lack basic infrastructure to achieve an efficient supply chain. Over 90% of roads are unpaved and fuel infrastructure is scarce. The Solarship, a hybrid between a bush plane and airship, was conceived to address this problem. It is a buoyant low-altitude aircraft with an electric power train and wing mounted photovoltaic array. Fully electric operation requires efficient lightweight power electronics to achieve a minimum range of 200 km carrying a 200 kg payload. A detailed system model is developed to explore the impact of wingspan, flight speed and drag coefficient on the flight range. A Partial Power Processing (PPP) converter based on the bidirectional Ćuk topology is demonstrated for this application. Due to the PPP concept, the converter is rated for only 26% of 2.7 kW generated PV power. The rating is optimized based on the battery and photovoltaic array voltage ranges. The experimental prototype uses Silicon Carbide MOSFETS and achieves a system efficiency of up to 99.3% with an effective specific power of 5.23 kW/kg.

I. INTRODUCTION

Canada, Africa, Australasia, Brazil, and Northern Eurasia represent the world's largest commodity regions and have combined remote areas of over 50 million square km, with over 90% unpaved roads and very limited fuel infrastructure. The world's largest commodity economies have gaps in their ability to move cargo. To move goods from California to Toronto by truck costs less than 10 cents per ton/km, while moving goods from Bujumbura to the eastern Congo costs more than $10 per ton/km. Solarship is a buoyantly-assisted airplane made possible through a confluence of modern developments, as shown in Fig. 1. Advanced aerodynamics, synthetic textile laminates, smart power electronics, lightweight batteries, and high-efficiency photovoltaics are the enabling technologies for a practical solar aircraft.

The Solarship aims to address the economic and logistical barriers that prevent adequate supply delivery to remote regions around the globe by (1) reducing the cost of transport, (2) enabling movement in-and-out of areas where other transport methods are ineffective due to lack of fuel and runways, and (3) ensuring cold chain storage and distribution. The Solarship is a hybrid between a bush plane and an airship. The added buoyancy from the Helium filled wing increases the payload, while the heavier-than-air design eliminates the need for expensive anchors.

(a)

(b)

Fig. 1. (a) 11 Meter prototype during test flight. (b) Solarship for future design.

One of the greatest advantages compared to standard aircraft is the ability to land in a small area the size of a soccer field. The simplified Solarship electrical architecture is shown in Fig. 2. The architecture, which is similar to ground based Electric Vehicles (EVs) [1], consists of a central battery pack, two electric motors driven by inverters, and a set of dc-dc converters for performing Distributed Maximum Power Point Tracking (DMPPT) on the wing-mounted PV array.

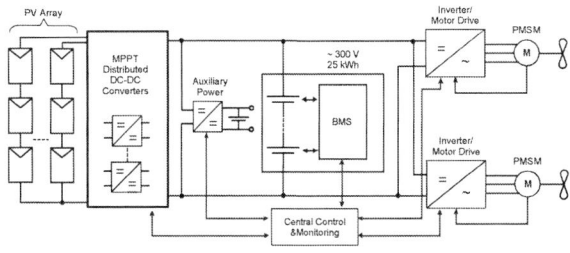

Fig. 2. Solarship electrical system overview.

Optimizing the Solarship design is a multi-disciplinary challenge, due to the strong influence of the wingspan on the lift capability, the available area for PV power harvesting and the required electrical systems. The minimum short-term requirement of the aircraft design is to achieve a 200 km electric range with a 200 kg payload for a 2014 field-trial in Canada.

A DMPPT Partial Power Processing (PPP) converter approach is considered for this weight-sensitive aerospace application. The objective of PPP is to reduce the power rating of the dc-dc converter, and thus reduce the mass of heatsinks and magnetic components. The objectives of this paper are to 1) report the results of system-level optimization of the Solarship for future designs, based on the stated range and payload objectives, 2) investigate the impact of PV voltage drift on the DMPPT PPP converter, and 3) demonstrate a lightweight PPP converter. The power rating of the PPP based dc-dc converter is highly sensitive to the system level design, which is not treated in the literature thus far.

II. PARTIAL POWER PROCESSING

The PPP concept is outlined in Fig. 3 for a single PV string, where V_{PV} and I_{PV} are the PV string voltage and current, respectively, V_{BATT} is the battery bus voltage, η_P is the converter efficiency, I_P is the current at the battery port, and ΔV is the voltage at the secondary port of the PPP converter. Converters based on PPP have been previously proposed in [2] for PV systems in battery backup systems, and in [3] as a PV array regulator for spacecraft applications. An experimental PPP prototype with both buck and boost capabilities is presented in [4] for a fuel-cell application. While the PPP concept has been successfully demonstrated in the literature, the impact of voltage drift on ΔV, and the system-level design considerations for PV applications have not been treated. The processed power of the dc-dc converter, P_P,

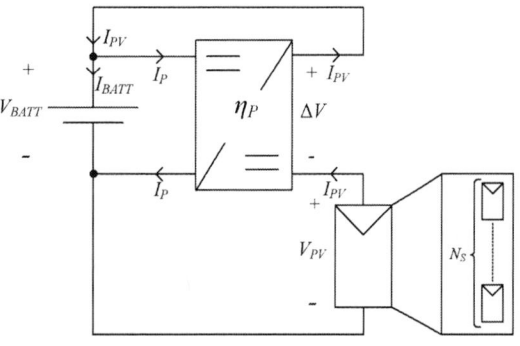

Fig. 3. Partial power processing dc-dc converter connected between a PV string and a battery.

is proportional to the difference between the battery and PV voltages,

$$P_P = \Delta V \cdot I_{PV}, \qquad (1)$$

which implies that for sufficiently low ΔV, the processed power can be minimized compared to the full PV power.

A. Modes of Operation and System Efficiency

In order to minimize the power rating of the dc-dc converter in Fig. 3, the converter needs to operate both in buck and boost modes, since the battery voltage may be greater or less than the PV voltage. Buck mode operation is shown in Fig. 4. The arrow above the converter indicates the direction of power transfer. In this mode, V_{PV} is given by,

$$V_{PV} = V_{BATT} - \Delta V, \qquad (2)$$

where V_{PV} is less than V_{BATT}.

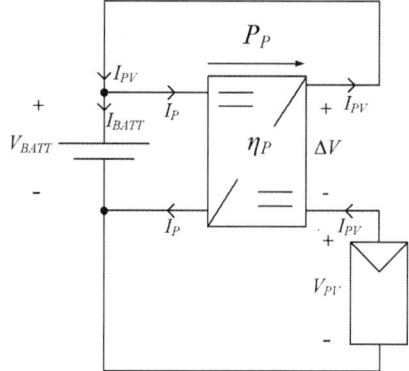

Fig. 4. Buck mode in the PPP converter, $V_{BATT} > V_{PV}$.

Boost mode operation is shown in Fig. 5. In this case,

$$V_{PV} = V_{BATT} + \Delta V, \qquad (3)$$

where V_{PV} is greater than V_{BATT}. In boost mode, the direction of power transfer in the dc-dc converter is reversed. Note that in both modes, power is transferred

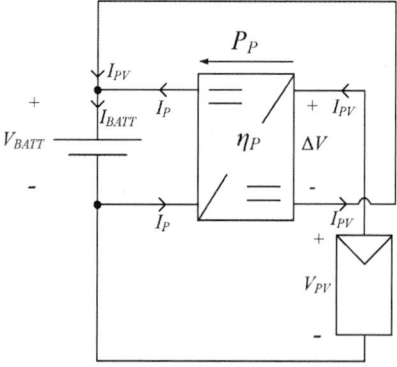

Fig. 5. Boost mode in the PPP converter, $V_{BATT} < V_{PV}$.

from the PV array to the battery, since I_P is less than I_{PV}, hence I_{BATT} is positive. The system efficiency, η_{sys}, can be expressed as a function of PPP converter efficiency, η_P, in both modes [2]. For buck mode,

$$
\begin{aligned}
\eta_{sys} = \frac{P_{BATT}}{P_{PV}} &= \frac{V_{BATT}(I_{PV} - I_P)}{V_{PV}I_{PV}} \\
&= \frac{1 - \frac{\Delta V}{\eta_P V_{BATT}}}{1 - \frac{\Delta V}{V_{BATT}}}.
\end{aligned}
\tag{4}
$$

For boost mode,

$$
\begin{aligned}
\eta_{sys} = \frac{P_{BATT}}{P_{PV}} &= \frac{V_{BATT}(I_{PV} + I_P)}{V_{PV}I_{PV}} \\
&= \frac{1 + \eta_P \frac{\Delta V}{V_{BATT}}}{1 + \frac{\Delta V}{V_{BATT}}}.
\end{aligned}
\tag{5}
$$

From (4) and (5), it is clear that for a small ratio of $\Delta V/V_{BATT}$, when the PV and battery voltages are nearly identical, the *system efficiency* is not sensitive to the converter efficiency, η_P, as shown in Fig. 6.

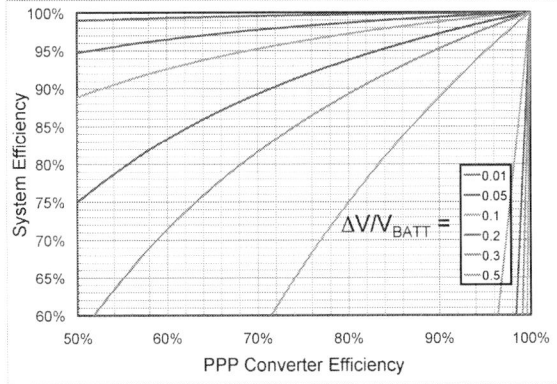

(a)

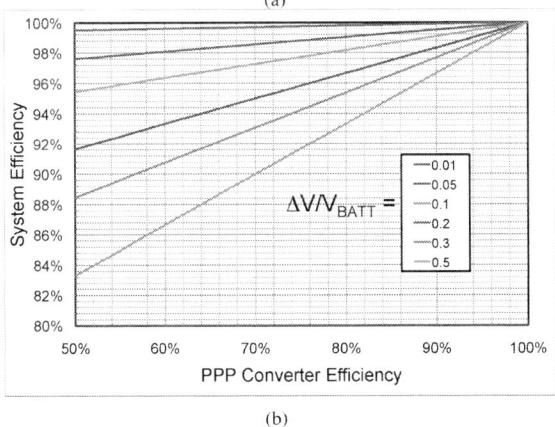

(b)

Fig. 6. System efficiency using a PPP scheme in (a) buck mode and (b) boost mode.

B. Converter Topology

The operation described in Section II-A is realized by a four quadrant isolated converter. Four quadrant operation is necessary since current must flow in both directions, and the converter must be capable of bipolar voltage output.

It is possible to realize the above requirements by starting with an isolated bidirectional converter, modified

to achieve bipolar operation. The Dual Active Bridge (DAB) proposed in [5] is suitable, however it has eight switches, while soft-switching is lost both at light-load and when operating over a wide voltage range [5]. The bidirectional LLC resonant converter [6] is capable of achieving soft-switching under light-load conditions, however it also contains eight switches and achieving buck mode under a wide operating range is a challenge due to variable frequency operation. The isolated Ćuk converter shown in Fig. 7 is capable of bidirectional operation, contains only two low-side switches, and operates at fixed frequency. At the same time it has three magnetic components and may require external snubbers. The Ćuk converter is chosen in this work due to its reduced number of high-frequency switches and simpler magnetics design compared to the full-bridge design in [4]. The magnetic component size is reduced by operating at a high switching frequency, enabled by Silicon Carbide (SiC) MOSFETS. Continuous current in both inductors reduces the size of the input and output capacitors. Duty cycle control is used to achieve MPPT, while the conversion ratio, $M = \Delta V/V_{BATT}$, is ideally independent of the load condition in Continuous Conduction Mode (CCM),

$$
M = \frac{n_2}{n_1} \frac{D}{1-D}.
\tag{6}
$$

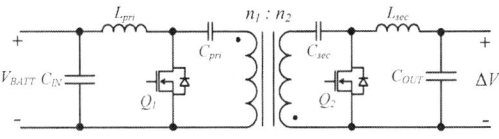

Fig. 7. Isolated bidirectional Ćuk converter.

In order to achieve a bipolar output, an additional bridge is used at the secondary side, similar to the unfolder in single-stage PV microinverters [7], as shown in Fig. 8. The unfolder can actively invert ΔV in boost mode. The unfolder is realized by a bridge of four bidirectional blocking switches as shown in Fig. 9. Only two sets of switches are enabled in each mode: (S_1, S_4) in buck mode, (S_2, S_3) in boost mode. Given that ΔV changes sign very slowly based on irradiance and battery voltage fluctuations, the low-frequency unfolder only contributes to conduction losses. It is therefore recommended to use low R_{on} and low V_f devices. Moreover, the bridge provides an additional safety disconnect feature for the PV array, which is why it is connected on the secondary side. The complete PPP topology is shown in Fig. 10.

The design procedure for the Ćuk converter is well covered in the literature [8] and not repeated here. If ΔV is smaller then V_{BATT}, the input current, voltage stress, and inductor voltage swings decrease in the converter. This allows the reduction or elimination of any required snubbers in the converter, use of smaller inductors, and higher FOM switches.

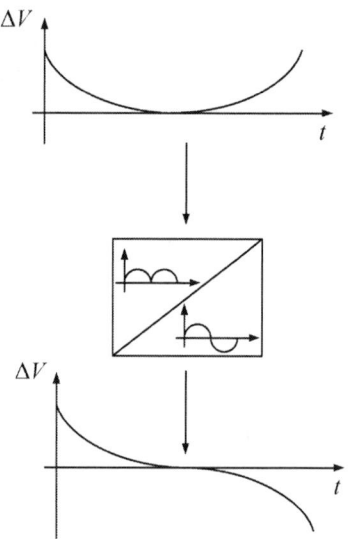

Fig. 8.   Unfolder concept to achieve bipolar output.

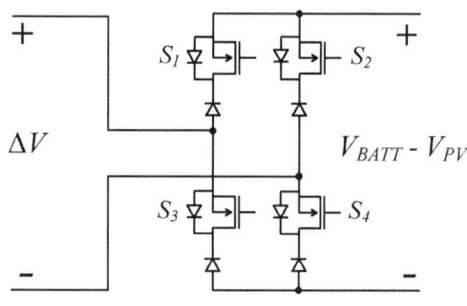

Fig. 9.   Unfolder implementation.

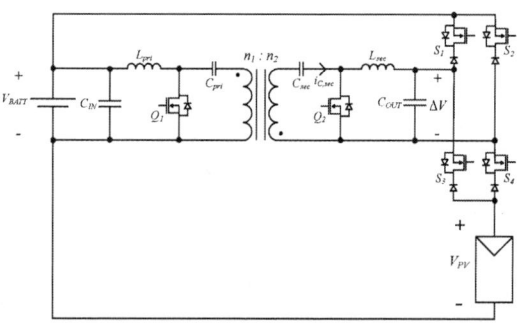

Fig. 10.   Complete PPP converter topology.

C. Rating the PPP Converter

The PPP converter rating, P_r, is closely related to its weight and volume. For the Solarship application, P_P is defined in (1), where V_{BATT} is a strong function of the State of Charge (SOC) [9], and V_{PV} is a strong function

of cell temperature [10]. Two worst-case conditions for P_P are defined assuming MPPT operation,

$$P_{P,1} = (V_{BATT,max} - V_{MPP,min})I_{PV} \qquad (7)$$

$$P_{P,2} = (V_{MPP,max} - V_{BATT,min})I_{PV}, \qquad (8)$$

where $V_{BATT,max}$ and $V_{BATT,min}$ are the maximum and minimum battery voltage, respectively, $V_{MPP,max}$ is the maximum PV MPP voltage at the maximum irradiance and minimum temperature, G_{max} and T_{min}, respectively, $V_{MPP,min}$ is the minimum PV MPP voltage at G_{max} and T_{max}. P_r is then given by

$$P_r = max\left\{P_{P,1}, P_{P,2}\right\}, \qquad (9)$$

and depends heavily on the PV environmental characteristics.

III.   SOLARSHIP SYSTEM MODEL

A detailed parametrized system model was produced to predict the Solarship behavior for varying wingspan and flight speed, under time-varying irradiance and temperature conditions. This model accurately captures the performance of the electrical and mechanical subsystems, including losses.

A. Electrical Model

The electrical model consists of the PV array and battery, with the load power modeled according to the flight conditions. The PV array modeling is based on [11]. The power of the array is dictated by the available surface area on the solar ship wing. A time varying irradiance profile is fed into the model during the transient simulation. For a typical day, the profile is symmetrical around solar noon, assuming an optimal departure time. The profile is obtained using the method in [12], where the hourly irradiance is estimated from historical monthly averaged clear day data for the mean day of the particular month. Historical data was used, since this is an empirical value and captures environmental conditions in the geographical location of interest, which is Thunder Bay, Canada. The cell temperature, T_c, is modeled according to [13]:

$$T_c - T_a = R_{th}G, \qquad (10)$$

where T_a is the ambient temperature, R_{th} is a thermal coefficient and G is the irradiance. The battery is modeled according to [9]. For each design iteration, battery specifications are determined based on the Specific Energy (SE) in kWh/kg, the allowable battery mass from the mechanical design, and the nominal battery voltage.

B. Mechanical Model

The wingspan and mechanical structure of the Solarship dictate the available PV area and battery mass. The solar panels are placed on the wing of the Solarship. The surface area available for solar panels, A_{solar}, changes as a square function of the wingspan, W,

$$A_{solar} \propto W^2. \qquad (11)$$

The lift, L, and battery mass, M_{Batt}, of the Solarship are related by

$$M_{Batt} \propto L \propto W^3 \propto V_f^2. \tag{12}$$

The cubic function is a result of the increase in Helium volume available in the Solarship and the square function is a result of the drag that must be overcome by the electric motor. The load power, P_{Load}, is determined by the drag force, F_D, which is a function of the aircraft speed, V_f, drag coefficient, C_d, air density, ρ, and surface area of the Solarship, A_s:

$$P_{Load} \propto F_D = \frac{1}{2}\rho C_d V_f^2 A_s. \tag{13}$$

IV. SIMULATION RESULTS

The electrical and mechanical equations are incorporated within a MATLAB Simulink model. The short-term objective is to achieve a minimum flight range and payload of 200 km and 200 kg, respectively. The aircraft travels at a constant speed until one of two conditions is met: the target flight range is achieved, or the battery reaches its minimum voltage, in which case the simulation is halted.

A. System Design Analysis

A simulation study was carried out to analyze the relationship between C_d, W, V_f, and flight range achieved. The simulation was iterated for several values of W and V_f at a fixed C_d and a transient simulation was performed for each point. The resulting range is plotted for every coordinate of (W, V_f).

The results of this iteration are shown in Fig. 11 for drag coefficients $C_{d,1}$, $C_{d,2}$, where $C_{d,1} < C_{d,2}$. The mission objective of 200 km is achieved with increases in W and V_f. This is because wing area and lift capability are increased, subsequently allowing more solar power and carrying capacity for batteries. The impact of the drag coefficient can also be seen from Fig. 11, where the minimum target range can be achieved for smaller W at $C_{d,1}$.

B. PPP Converter Rating for Solarship Application

The power rating of the converter, P_r, is dependent on ΔV. The goal is to determine the limits of V_{MPP} at the expected temperature and irradiance conditions. The results of the limits for V_{MPP} are listed in Table I. The battery voltage limits are $V_{BATT,max} = 400$ V and $V_{BATT,min} = 288$ V. Using (7) to (9), the power rating of the PPP converter is therefore determined:

$$P_{P,1} = 687.4\,W, \tag{14}$$

$$P_{P,2} = 583.1\,W, \tag{15}$$

$$P_r = max\{P_{P,1}, P_{P,2}\} = 687.4\,W. \tag{16}$$

Knowing P_r, the percentage of power processed compared to a full power processing converter is:

$$\frac{P_r}{P_{MPP,max}} = 0.26, \tag{17}$$

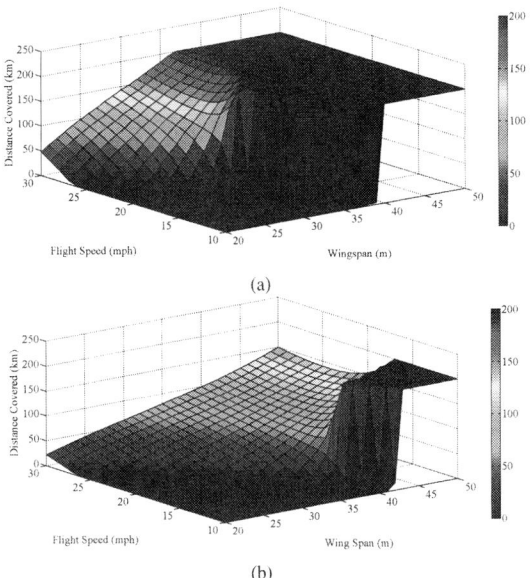

(a)

(b)

Fig. 11. Range analysis at (a) $C_{d,1}$ and (b) $C_{d,2}$.

where (17) indicates that the PPP converter power is rated to 26% of the full input power from the PV string.

TABLE I. ENVIRONMENTAL CONDITIONS AND V_{MPP} RANGE

G	T_a	R_{th}	V_{MPP}	I_{MPP}	P_{MPP}
1200 W/m^2	10 °C	0	370 V	7.16 A	2.64 kW
1200 W/m^2	35 °C	0.03	301 V	6.94 A	2.09 kW

V. EXPERIMENTAL RESULTS

A prototype was built to demonstrate the PPP converter, as shown in Fig. 12. The prototype specifications are listed in Table II. The converter weighs 516 g and contains no electrolytic capacitors for high reliability. The Ćuk converter is controlled using an on-board microcontroller. Silicon Carbide MOSFETS rated for 1.2 kV are used for Q_1 and Q_2, eliminating the need for external snubbers.

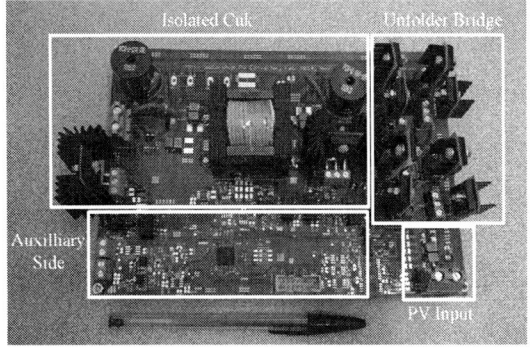

Fig. 12. Converter prototype.

The PPP converter was tested at different levels of ΔV at both extremes of the battery voltage range. The

TABLE II. PPP CONVERTER PROTOTYPE SPECIFICATIONS

Parameter	Value	Unit
Maximum Battery Voltage, $V_{BATT,max}$	400	V
Minimum Battery Voltage, $V_{BATT,min}$	288	V
Maximum PPP output voltage ΔV_{max}	120	V
Minimum PPP output voltage ΔV_{min}	20	V
Switching Frequency, f_s	200	kHz
Turns ratio, n_1/n_2	2	
Primary-side Inductance, L_{pri}	274	μH
Secondary-side Inductance, L_{sec}	100	μH
Primary-referred Magnetizing Inductance, L_m	2	mH
Leakage Inductance, L_{lk}	1.76	μH
Input Capacitance, C_{IN}	4.4	μF
Output Capacitance, C_{OUT}	6.6	μF
Primary-side Capacitance, C_{pri}	4.4	μF
Secondary-side Capacitance, C_{sec}	6.6	μF

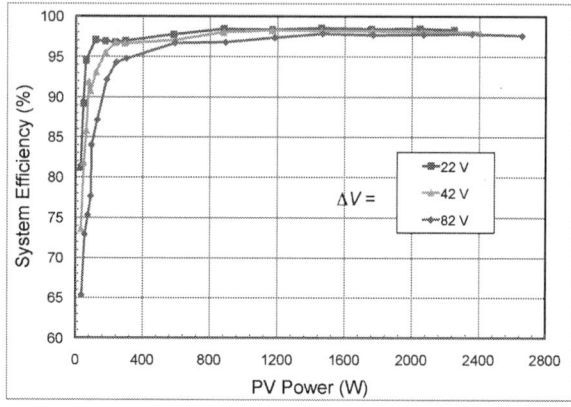

Fig. 14. PPP converter efficiency in boost mode at the minimum battery voltage, $V_{BATT,min}$ = 288 V.

converter operates in the forward and reverse direction for buck and boost modes, respectively. The measured system efficiency for buck and boost modes is shown in Fig. 13 and Fig. 14, respectively. The efficiency drops for increasing ΔV due to the higher processed power. The light-load efficiency could be further improved by operating the Ćuk converter in burst-mode below P_{PV} = 400 W. The system efficiency is 98.8% in buck mode for a maximum power of P_{PV} = 2.7 kW and minimum ΔV = 30 V. This corresponds to a processed power of P_P = 250 W. The efficiency in the reverse direction for ΔV = 82 V is 97.6% at P_{PV} = 2.7 kW. This corresponds to a processed power of P_P = 602 W.

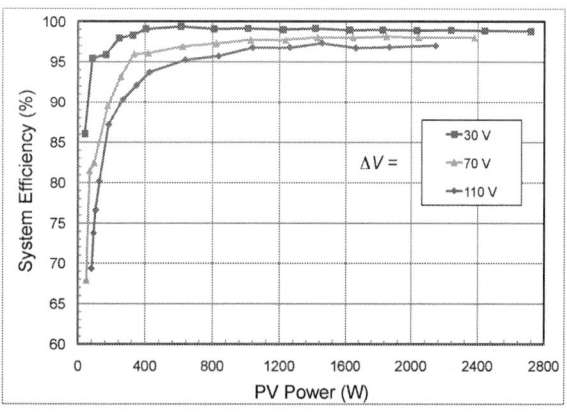

Fig. 13. PPP converter efficiency in buck mode at the maximum battery voltage, $V_{BATT,max}$ = 400 V.

The steady-state waveforms of the converter are shown in Fig. 15, where the output voltage, ΔV, the secondary side capacitor current, $i_{C,sec}$, and the primary switching node, $V_{DS,pri}$, are shown at input power P_{PV} = 2.7 kW in buck mode at a typical value of ΔV = 30 V. The overshoot on the primary side MOSFET drain is within its 1.2 kV rating, and snubbers are not required.

A thermal image of the converter is shown in Fig. 16 for ΔV = 30 V, P_{PV} = 2.7 kW, and output current

$I_{L,sec}$ = 7.33 A. The image shows that the warmest components are the diodes in the unfolder bridge, while the transformer core temperature is below 60 °C.

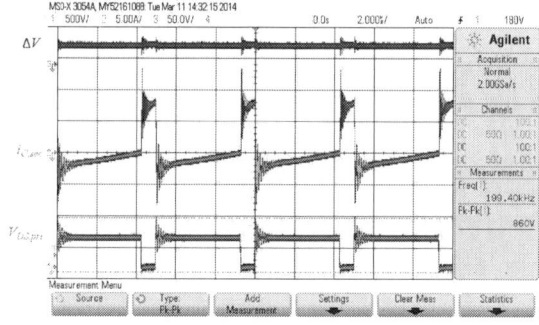

Fig. 15. Steady-state waveforms of ΔV, $i_{C,sec}$, and $V_{DS,pri}$ under ΔV = 30V, $I_{L,sec}$ = 7.33 A.

Fig. 16. Converter temperature distribution for ΔV = 30 V, P_{PV} = 2.7 kW.

VI. CONCLUSION

The partial power processing converter concept can achieve a high power density for a string-level PV converter in this mass sensitive aerospace application. The system-level model for the Solarship is useful to optimize the battery and PV systems, while minimizing the processed power for the DMPPT dc-dc converters. The experimental prototype demonstrates that the isolated bidirectional Ćuk converter with an added output bridge is a promising candidate for this PPP application. The string-level power converter achieves an effective specific power of 5.23 kW/kg and maintains an efficiency above 95% for a broad range of conditions. Further work is required to optimize the dynamic transition between buck and boost modes.

ACKNOWLEDGEMENT

This work was supported by the Ontario Centres of Excellence, the Natural Sciences and Engineering Research Council of Canada, the Canadian Foundation for Innovation and the Ontario Research Fund.

REFERENCES

[1] M. Lukasiewycz, S. Steinhorst, S. Andalam, F. Sagstetter, P. Waszecki, W. Chang, M. Kauer, P. Mundhenk, S. Shanker, S. Fahmy, and S. Chakraborty, "System architecture and software design for electric vehicles," in *Design Automation Conference (DAC), 2013 50th ACM / EDAC / IEEE*, 2013, pp. 1–6.

[2] D. Snyman and J. H. R. Enslin, "Novel technique for improved power conversion efficiency in pv systems with battery back-up," in *Telecommunications Energy Conference, 1991. INTELEC '91., 13th International*, 1991, pp. 86–91.

[3] R. Button, "An advanced photovoltaic array regulator module," in *Energy Conversion Engineering Conference, 1996. IECEC 96., Proceedings of the 31st Intersociety*, vol. 1, 1996, pp. 519–524 vol.1.

[4] A. G. Birchenough, "A High Efficiency DC Bus Regulator / RPC for Spacecraft Applications," in *Space Technology and Applications*, ser. American Institute of Physics Conference Series, M. S. El-Genk, Ed., vol. 699, Feb. 2004, pp. 606–613.

[5] M. Kheraluwala, R. Gascoigne, D. Divan, and E. Baumann, "Performance characterization of a high-power dual active bridge dc-to-dc converter," *Industry Applications, IEEE Transactions on*, vol. 28, no. 6, pp. 1294–1301, 1992.

[6] J.-H. Jung, H.-S. Kim, J.-H. Kim, M.-H. Ryu, and J.-W. Baek, "High efficiency bidirectional llc resonant converter for 380v dc power distribution system using digital control scheme," in *Applied Power Electronics Conference and Exposition (APEC), 2012 Twenty-Seventh Annual IEEE*, Feb 2012, pp. 532–538.

[7] M. Joshi, E. Shoubaki, R. Amarin, B. Modick, and J. Enslin, "A high-efficiency resonant solar micro-inverter," in *Power Electronics and Applications (EPE 2011), Proceedings of the 2011-14th European Conference on*, Aug 2011, pp. 1–10.

[8] R. Erickson and D. Maksimovic, *Fundamentals of Power Electronics*, ser. Power electronics. Springer, 2001.

[9] O. Tremblay, L.-A. Dessaint, and A.-I. Dekkiche, "A generic battery model for the dynamic simulation of hybrid electric vehicles," in *Vehicle Power and Propulsion Conference, 2007. VPPC 2007. IEEE*, 2007, pp. 284–289.

[10] M. Zaman, S. Poshtkouhi, V. Palaniappan, K. Li, H. Bergveld, S. Myskorg, and O. Trescases, "Distributed power-management architecture for a low-profile concentrating-pv system," in *Power Electronics and Motion Control Conference (EPE/PEMC), 2012 15th International*, 2012, pp. LS2d.4–1–LS2d.4–8.

[11] M. Villalva, J. Gazoli, and E. Filho, "Comprehensive approach to modeling and simulation of photovoltaic arrays," *Power Electronics, IEEE Transactions on*, vol. 24, no. 5, pp. 1198–1208, 2009.

[12] J. A. Duffie and W. A. Beckman, *Solar Engineering of Thermal Processes*. Hoboken, New Jersey: John Wiley and Sons, Inc., 2013.

[13] W. Maranda and M. Piotrowicz, "Extraction of thermal model parameters for field-installed photovoltaic module," in *Microelectronics Proceedings (MIEL), 2010 27th International Conference on*, 2010, pp. 153–156.

Impacts of Rectifier Circuit Loads on Islanding Detection of Photovoltaic Systems

Yoshiaki Yoshida
Dept. of Electrical Systems Engineering
Hiroshima Institute of Technology
Hiroshima, Japan
y.yoshida.bc@it-hiroshima.ac.jp

Hirokazu Suzuki
Dept. of Electrical Engineering & Information Systems
The University of Tokyo
Tokyo, Japan
suzuki@p-front.t.u-tokyo.ac.jp

Abstract— Simulation analysis of the islanding tests with the rectifier circuit load is performed. The rectifier circuit load has the impacts to the harmonic voltage after the islanding operation, and reduces the frequency and fundamental voltage greater than if no rectifier circuit loads. It is probable that total harmonic voltage distortion occurs more than 2V in a balanced three-phase circuit, if the ratio of the rectifier circuit load is at least 10%. However, the changes in the harmonic voltage after the islanding operation are dependent on the difference between the harmonic voltage included in the system power supply and the harmonic voltage by the rectifier circuit load. The characteristics on the islanding detection time are clarified in case of using the new active islanding detection method that is called "a frequency feedback method with step reactive power injection".

Keywords— photovoltaic system, islanding detection, harmonic voltage, rectifier circuit load

I. INTRODUCTION

When a large penetration of the distributed generators such as photovoltaic (PV) systems is growing up in distribution systems, it is very important for PV systems to quickly detect islanding caused by power system fault in order to ensure electrical safety.

New Energy and Industrial Technology Development Organization (NEDO) in Japan developed a new active islanding detection method for the multiple-inverter of PV system [1]. This method is called "a frequency feedback method with step reactive power injection (FFB)". This method has been used to Japanese standard as an islanding detection technique for the residential PV systems. When the generation power of PV system is nearly equal to the load power, the grid system frequency and fundamental voltage do not little change after islanding. Thus, when the frequency change is within 0.01Hz, after detecting the change (> 1%) of total harmonic voltage (*THD*), the step injection of reactive power is conducted in the direction of reducing the frequency. However, it is not clear whether the change of *THD* is greater than 1% (2V) after islanding.

So far, it has been said that the nonlinear magnetic characteristics of the pole transformer cause mainly the increase of the harmonic voltage after islanding. However, when the number of connected PV systems is increased, the load impedance for the higher harmonics is decreased because the load power equal to the PV generation power also increases. So, there is a possibility that the magnetic characteristics of the pole transformer can't cause *THD* that is greater than 2V [2].

The parallel connection circuit of the resistance and the regenerative load (motor load or resonance circuit) has been normally used as the load circuit for the islanding test [3]. These load circuits slightly emit the harmonic current. On the other hand, recently, the general household loads have many rectifier circuits that emit a lot of harmonic current. In past research [4], the frequency characteristics during islanding have been studied using a rectifier circuit. But the harmonic voltage characteristics during islanding have not well been clarified.

In this study, the simulation analysis of the islanding tests with the rectifier circuit load is performed. And we clarify that the rectifier circuit load has the influence to the harmonic voltage after islanding operation.

II. SIMULATION METHOD

A. Islanding test circuit

Fig. 1 shows the proposed circuit for the islanding test. The parallel connection circuits of the resistance and the resonance circuit have been normally used as the load circuit for the islanding test. We add the rectifier circuit load to it.

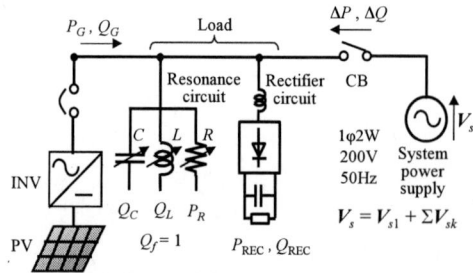

Fig. 1. Proposed circuit for the islanding test.

When PV inverter has been connected to the power system, the following equations are given.

$$P_G + \Delta P = P_R + P_{REC} \equiv P_{Load} \tag{1}$$

$$Q_G + \Delta Q = Q_L + Q_C + Q_{REC} \tag{2}$$

$$Q_L = Q_f \cdot P_G \tag{3}$$

$$\alpha = \frac{P_{REC}}{P_R + P_{REC}} \times 100 \tag{4}$$

Where, ΔP and ΔQ are the active and reactive power flow at the circuit breaker (CB). P_G and Q_G are the active and reactive power generated by PV inverter. P_R is the active power of the resistor R. Q_L is the reactive power of the inductor L. Q_C is the reactive power of the capacitor C. P_{REC} and Q_{REC} are the active and reactive power of the rectifier circuit load. P_{Load} is the total active power of the loads. Q_f is the quality factor (= 1). α is the rate of P_{REC} for P_{Load}. The sign of the lag load is positive. The sign of the lead load is negative.

Particularly, when ΔP, ΔQ and α are 0%, Q_f is given by the following equation.

$$Q_f = R\sqrt{\frac{C}{L}} \tag{5}$$

B. Test cases

Table I indicates the test cases. Case A is the normal tests that change the power flow of ΔP and ΔQ. The test of 25 points is carried out by the "△" marks of Fig. 2.

TABLE I
SIMULATION TEST CASES

Case	α [%]※1	ΔP, ΔQ [%]	V_{sk} [V] ※2 ($k = 5$)
A	2 points (0% , 25%)	25 points (All combination for 0%, ±5%, ±10)	0 V
B	6 points (0%, 5%, 10%, 15%, 20%, 25%)	1 point (At the power flow, frequency and fundamental voltage hardly change after islanding.)	0 V
C	5 points (0%, 5%, 10%, 15%, 20%)		0 - 14 V∠ 0°

※1) $\alpha = P_{REC}/(P_R + P_{REC}) \times 100$

※2) V_{sk}: Harmonic voltage included in system power supply

Case B is the non-detection-zone (NDZ) test. At NDZ, the frequency and the fundamental voltage hardly change after the islanding operation. The tests of 6 points are carried out by the "○" marks of Fig. 2. It means that the power flow of ΔP and ΔQ separated from 0% according to the increase of α, because the rectifier circuit load is non-linear load. In case A and case B as shown in Table II, the islanding detection functions are masked because the main aim of this study is to clarify the islanding phenomenon. In case C, FFB is used to evaluate the islanding detection function. We consider the influence of the harmonic voltage that is included in the system power supply, and will clarify the characteristics on the islanding detection time. The voltage in the system power

supply is defined as follow.

$$V_s = V_{s1} + \sum V_{sk} \tag{6}$$

Where, V_s is the voltage in the system power supply. V_{s1} is the fundamental wave component of V_s. V_{sk} is the k-th order harmonic component of V_s. For the sake of simplicity, we will consider the fifth order harmonic voltage that is included most in the distribution systems.

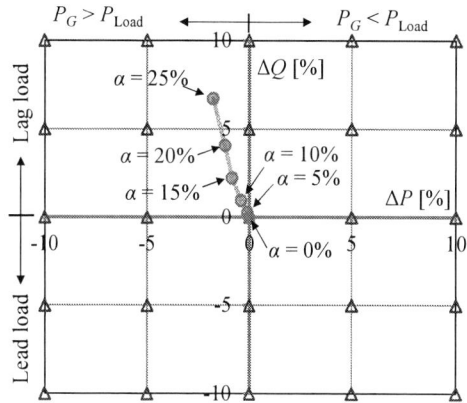

△ : Test points for Case A (25 points)

● : Test points for Case B (6 points)

Fig. 2. Islanding test points on ΔP-ΔQ plane.

C. PV inverter model

Fig. 3 shows the block diagram of the simulation model for the single-phase grid-connected inverter [5]. To simplify the model, we treat the inverter as a current source that has control system, because the inverter for PV system is generally used the fast current controlled type. Table II indicates the specifications of PV inverter. In case A and case B, the reference of the active signal (Q^*) is zero because the active islanding function is masked.

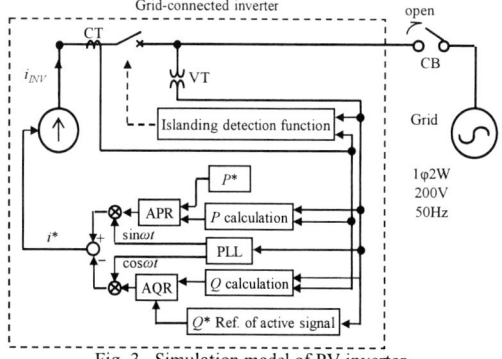

Fig. 3. Simulation model of PV inverter.

Fig. 4 shows the block diagram of FFB [6]. The grid-connected inverter applies the positive feedback to the reactive power Q in the direction of increasing the frequency change. FFB is similar to Sandia Frequency Shift (SFS) [7] in that the positive feedback function is used. The step injection function of the reactive power is different from SFS. When the frequency change $|\Delta f|$ is within 0.01 Hz, after detecting the changes of the total harmonic distortion voltage ΔTHD or the fundamental voltage ΔV_1, the step injection of the reactive power Q is conducted in direction of reducing the frequency. This step injection technique can make the NDZ very smaller. When the shift pattern of Δf is above the threshold value, the inverter detects the islanding operation and is disconnected from the power system.

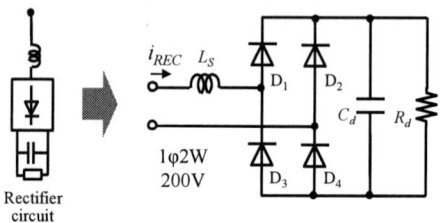

Fig. 5. Simulation model of rectifier circuit load.

TABLE II
SPECIFICATIONS OF PV INVERTER

Iverter type	Voltage type currrent controlled method
Active islanding detection method	mask (Case A/B)
	FFB (Case C)
Passive islanding detection method	mask
Rated output voltage	AC 200V
Eletrical system	1φ2W
Rated frequency	50 Hz
Power factor	1.0
Rated output current	20 Arms (at 200V)
Rated output power	4.0 kW

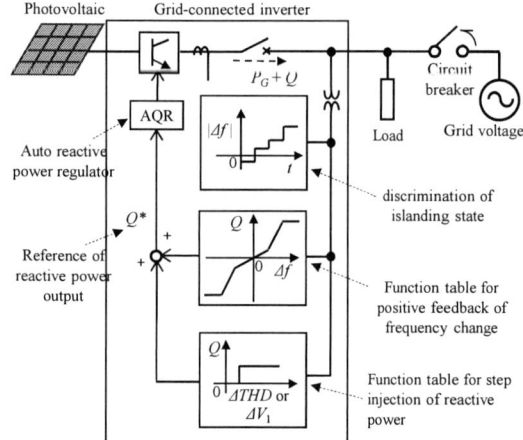

Fig. 4. Block diagram of frequency feedback method with step reactive power injection (FFB).

D. Rectifier circuit load model

Fig. 5 shows the rectifier circuit load. We treat the capacitor input type rectifier circuit that is often used in the general household loads. Table III indicates the circuit specification. The rated active power per one unit is 200 W.

Fig. 6 illustrates the waveforms of AC input voltage and current. Table IV shows the harmonic components of the input current. By adjusting the value of the reactance L_S, the harmonic current is smaller than the maximum of class D determined by JIS C 61000-3-2 [8]. The increase of the rectifier circuit load is carried by the increase of the unit number of parallel connection. For example, P_{REC} of 5 units is equal to 1000W. The harmonic currents are increased almost linearly according to the increase of α.

TABLE III
SPECIFICATIONS OF RECTIFIER CIRCUIT LOAD

Rated active power	P_{REC}	200 [W]
Rated reactive power	Q_{REC}	61.4 [var] (Lag)
Rated voltage	-	200 [V]
Rated frequency	-	50 [Hz]
Displacement power factor	$\cos\theta_1$	0.956 (Lag)
Total power factor	$\cos\theta$	0.712 (Lag)
Reactance for harmonic suppression	L_S	15.0 [mH]
Smoothing capacitor	C_d	0.2623 [uF]
DC resistance	R_d	343.2 [Ω]

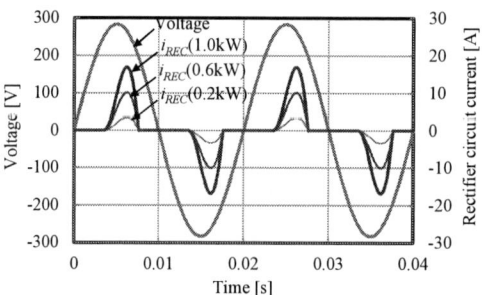

Fig. 6. Waveform of rectifier circuit load.

TABLE IV
HARMONIC CURRENT OF RECTIFIER CIRCUIT LOAD
(IN CASE OF P_{REC} = 200W/UNIT)

Harmonic current	Value [Arms]	Rate [%]	Maximum rate of class D [%]
I_1	1.046	100	100
I_3	0.8018	76.6	78.2
I_5	0.4471	42.7	43.7
I_7	0.1617	15.5	23.0
I_9	0.08396	8.0	11.5
I_{11}	0.07139	6.8	8.05
I_{13}	0.03927	3.8	6.81

III. SIMULATION RESULTS AND DISCUSSION

A. Simulation results of Case A

Table V indicates the frequency analytical result after 0.5 s from the start of islanding. Similarly, Table VI shows the fundamental voltage analytical result. Table V shows that in all the test points, the frequency in the case that α is 25% are lower than the case that α is 0%. The fundamental voltage analytical result is also same.

978-1-4799-2706-7/14 $31.00 © 2014 IEEE

TABLE V
FREQUENCY ANALYTICAL RESULT
AFTER 0.5S FROM THE START OF ISLANDING OPERATION

Reactive power (ΔQ)		Active power (ΔP)				
		-10%	-5%	0%	+5%	+10%
+10%	$\alpha = 0\%$	52.70Hz	52.70Hz	52.70Hz	52.70Hz	52.70Hz
	$\alpha = 25\%$	50.89Hz	50.84Hz	50.80Hz	50.76Hz	50.72Hz
+5%	$\alpha = 0\%$	51.29Hz	51.29Hz	51.29Hz	51.29Hz	51.29Hz
	$\alpha = 25\%$	49.66Hz	49.62Hz	49.58Hz	49.54Hz	49.50Hz
0%	$\alpha = 0\%$	50.00Hz	50.00Hz	50.00Hz	50.00Hz	50.00Hz
	$\alpha = 25\%$	48.52Hz	48.48Hz	48.45Hz	48.41Hz	48.37Hz
-5%	$\alpha = 0\%$	48.79Hz	48.79Hz	48.79Hz	48.79Hz	48.79Hz
	$\alpha = 25\%$	47.45Hz	47.42Hz	47.39Hz	47.36Hz	47.32Hz
-10%	$\alpha = 0\%$	47.67Hz	47.67Hz	47.67Hz	47.67Hz	47.67Hz
	$\alpha = 25\%$	46.46Hz	46.43Hz	46.40Hz	46.37Hz	46.34Hz

TABLE VI
FUNDAMENTAL VOLTAGE ANALYTICAL RESULT
AFTER 0.5S FROM THE START OF ISLANDING OPERATION

Reactive power (ΔQ)		Active power (ΔP)				
		-10%	-5%	0%	+5%	+10%
+10%	$\alpha = 0\%$	211.31V	205.68V	200.64V	195.69V	191.23V
	$\alpha = 25\%$	208.64V	203.04V	198.17V	193.54V	189.32V
+5%	$\alpha = 0\%$	211.04V	205.29V	200.10V	195.18V	190.98V
	$\alpha = 25\%$	208.36V	203.02V	198.02V	193.36V	189.18V
0%	$\alpha = 0\%$	210.8V	204.94V	199.98V	195.38V	190.66V
	$\alpha = 25\%$	208.77V	203.36V	198.34V	193.66V	189.34V
-5%	$\alpha = 0\%$	211.11V	205.22V	200.01V	195.18V	190.75V
	$\alpha = 25\%$	208.9V	203.49V	198.46V	193.82V	189.46V
-10%	$\alpha = 0\%$	211.42V	205.57V	200.37V	195.59V	191.03V
	$\alpha = 25\%$	209.32V	203.91V	198.90V	194.24V	190.03V

TABLE VII
HARMONIC VOLTAGE ANALYTICAL RESULT
AFTER 0.5S FROM THE START OF ISLANDING OPERATION

Reactive power (ΔQ)		Active power (ΔP)				
		-10%	-5%	0%	+5%	+10%
+10%	V_{THD}	17.77V	17.20V	16.70V	16.21V	15.75V
	V_3	16.65V	16.11V	15.64V	15.18V	14.75V
	V_5	5.94V	5.75V	5.59V	5.43V	5.28V
	V_7	1.85V	1.78V	1.72V	1.67V	1.61V
+5%	V_{THD}	17.45V	16.88V	16.39V	15.90V	15.46V
	V_3	16.33V	15.80V	15.34V	14.88V	14.47V
	V_5	5.87V	5.67V	5.52V	5.36V	5.21V
	V_7	1.86V	1.78V	1.73V	1.67V	1.61V
0%	V_{THD}	17.15V	16.63V	16.13V	15.65V	15.21V
	V_3	16.04V	15.55V	15.08V	14.63V	14.22V
	V_5	5.79V	5.62V	5.46V	5.30V	5.15V
	V_7	1.85V	1.80V	1.74V	1.68V	1.62V
-5%	V_{THD}	16.88V	16.35V	15.87V	15.39V	14.97V
	V_3	15.77V	15.28V	14.82V	14.38V	13.98V
	V_5	5.73V	5.55V	5.39V	5.23V	5.09V
	V_7	1.86V	1.79V	1.74V	1.68V	1.63V
-10%	V_{THD}	16.61V	16.10V	15.62V	15.18V	14.77V
	V_3	15.50V	15.03V	14.58V	14.17V	13.78V
	V_5	5.66V	5.49V	5.33V	5.18V	5.05V
	V_7	1.86V	1.80V	1.74V	1.69V	1.64V

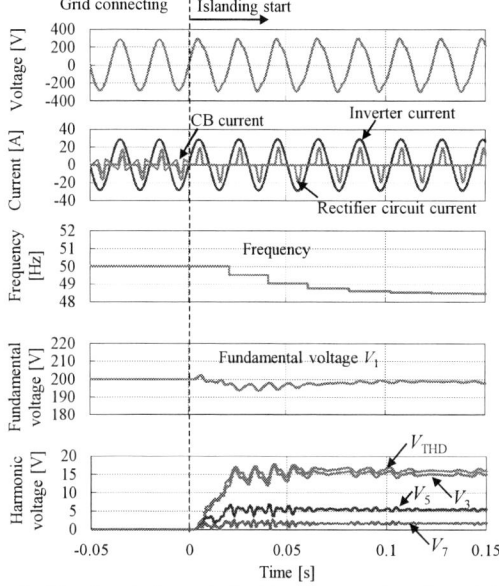

Fig. 7. Waveform before and after islanding operation when α is 25% (ΔP and ΔQ are 0%).

Particularly, we focus on the case that ΔP and ΔQ are 0%. When α is 0%, the frequency and the fundamental voltage hardly change. When α is 25%, the frequency is lowered to 48.45Hz, and the fundamental voltage is lowered to 198.34V. The reason may be considered that the fundamental component of active power of the rectifier circuit load after islanding was increased a little, and considered that the fundamental component of reactive power of the rectifier circuit load after islanding was decreased in the direction of lead load.

Table VII indicates the analytical result in harmonic voltage after 0.5 s from the start of islanding in the case that α is 25%. This shows that in all the test points, the harmonic voltages have occurred significantly by the rectifier circuit load. So, it is possible that by capturing these harmonic changes may be made to detect the islanding without being affected by the power flow at the circuit breaker.

Fig. 7 shows the waveform before and after islanding operation when α is 25% (ΔP and ΔQ are 0%). After islanding operation, the voltage waveform has become the distorted wave, and the harmonic voltage has increased rapidly, and THD has reached to 8.1 %. The peak current of the rectifier circuit has increased a little by the influence of the distorted wave voltage.

B. Simulation results of Case B

Fig. 8 shows the waveform before after islanding operation when α is 25% (ΔP is -1.8% and ΔQ is +6.7%). After islanding operation, the frequency and the fundamental voltage hardly change and the voltage waveform has become the distorted wave, and the harmonic voltage has increased rapidly, and THD has reached to 8.3 %. The peak current of the rectifier circuit has increased a little by the influence of the distorted wave voltage.

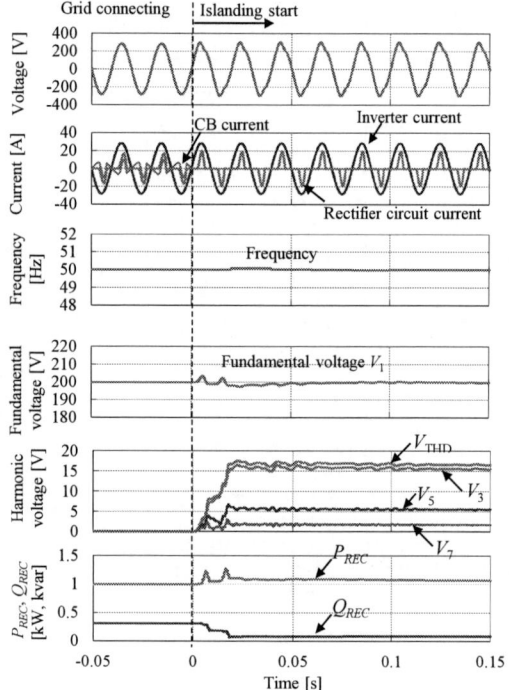

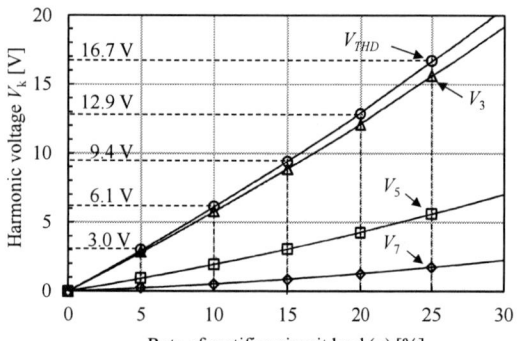

Fig. 8. Waveform before and after islanding operation when α is 25% (ΔP is -1.8% and ΔQ is +6.7%).

Fig. 9. Analytical result of the harmonic voltage after 0.5 s from the start of islanding.

Fig. 9 indicates the analytical result of the harmonic voltage after 0.5 s from the start of islanding. With the increase of α, the harmonic voltage is also increased. The 3rd harmonic voltage shows the maximum value in the harmonic voltages. In a case of the balanced circuit of three phase three wire, the 3rd harmonic voltage does not occur because the 3rd harmonic current circulates in the delta circuit. But When α is greater than 10%, the calculated THD value by using the 5th and the 7th harmonic voltage is greater than 2.0V. So, if the harmonic voltages that are included in the system power supply are small enough to be ignored, the step injection function of the reactive power will be easy to act. However, when the harmonic voltage is contained richly in the system power supply, the change in harmonic voltage before and after the islanding operation will be almost zero or minus. The influence will be analyzed by the next case C.

C. Simulation results of Case C

Fig. 10 shows the characteristics on the islanding detection time against the harmonic voltage that is included in the system power supply. Since the simulation has forcibly stopped after 1 second from the start of the islanding operation, the maximum duration of an islanding operation is 1 second.

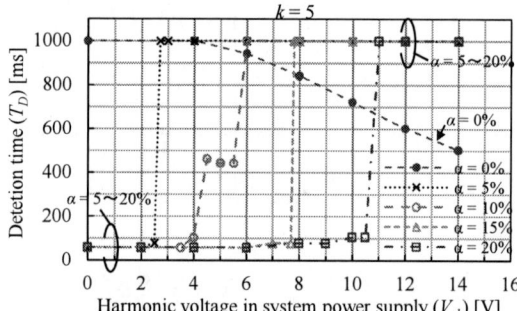

Fig. 10. Characteristics on detection time against harmonic voltage that is included in the system power supply.

When α is zero, it means that the islanding operation continues for more than 1 second under the condition that the V_{sk} is 4V or less. If V_{sk} exceeds 4V, the detection time by frequency feedback function is fast gradually. When V_{sk} increases, it will affect the power flow at the circuit breaker before the islanding operation. It is considered that it led to a slight change in the frequency after the islanding operation and the action of the frequency feedback has been easy.

If α is from 5% to 20% and V_{sk} is below a certain threshold respectively, the PV inverter can detect the islanding operation at high speed. However, the islanding operation is continued for more than 1 second if V_{sk} exceeds a certain threshold. This is close to the value of the V_{THD} shown in Fig. 9 for each of α. In other words, since ΔTHD is an area of less than 2V, the step injection function does not work. It is considered that frequency feedback function also did not work because the change in frequency is less than 0.01Hz.

Fig. 11 shows an example of waveform that has led to the islanding detection by the step injection function. The condition is that α is 20% and V_{sk} is 10V. As shown in Fig. 9, V_{THD} reaches to 12.9V, so ΔTHD is 2.9V at the maximum. Since ΔTHD exceeds 2V, the step injection function has worked at time 40.2ms. Then, frequency feedback function is working at high speed with time 60.7ms. As a consequence, the islanding operation has been detected at time 105ms.

Then, Fig. 12 shows an example of waveform that the step injection function did not work and the islanding operation could not be detected. The condition is that α is 20% and V_{sk} is 12V. The maximum of ΔTHD is 0.9V. So, the step injection function did not work. Since the change in frequency is almost zero, the islanding operation was continued for 1 second or more.

978-1-4799-2706-7/14 $31.00 © 2014 IEEE

The 2014 International Power Electronics Conference

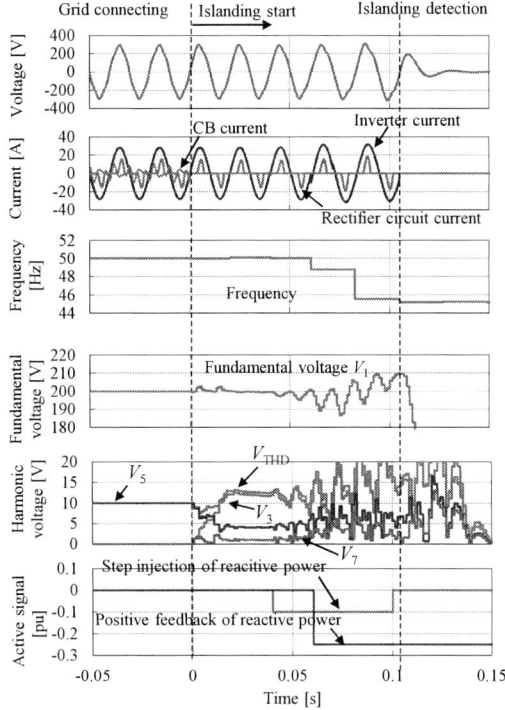

Fig. 11. Waveform before and after islanding operation when α is 20% and V_{sk} is 10V$\angle$0° (ΔP is -1.2% and ΔQ is +4.1%).

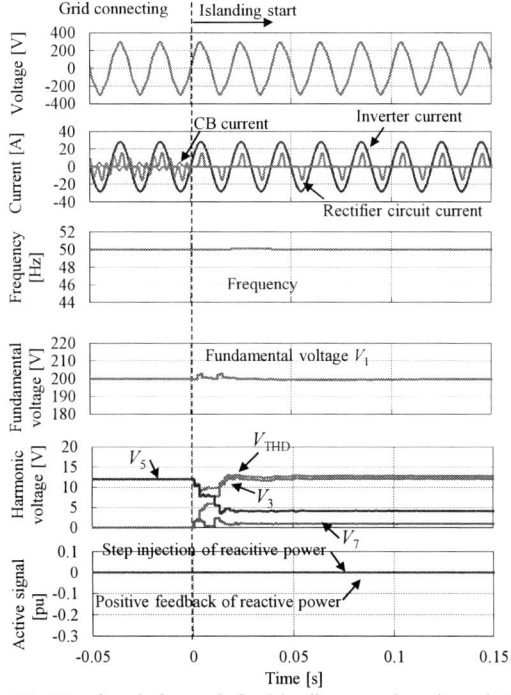

Fig. 12. Waveform before and after islanding operation when α is 20% and V_{sk} is 12V$\angle$0° (ΔP is -1.2% and ΔQ is +4.1%).

IV. CONCLUSIONS

In this study, the simulation analysis of the islanding tests with the rectifier circuit load has been performed. After the islanding operation, the rectifier circuit reduces the frequency and fundamental voltage greater than if no

rectifier circuit loads. As the rectifier circuit load increases, the harmonic voltage after the islanding operation has increased. It is probable that *THD* occurs more than 2V in a balanced three-phase circuit, if α is at least 10%.

However, the changes in the harmonic voltage after the islanding operation are dependent on the difference between the harmonic voltage included in the system power supply and the harmonic voltage caused by the rectifier circuit load. When the difference is equal to 2V or more, the islanding operation can be detected at a high speed by the frequency feedback function or step injection function. On the other hand, if the difference is below 2V, it is probable that the islanding operation is continued for more than 1 second. These results will provide an important guidance for detecting the islanding operation quickly and reliably.

In the future work, it is necessary to examine in consideration of the line impedance. And the experimental verification is needed.

REFERENCES

[1] The Japan Electrical Manufacturers' Association: JEM1498, A frequency feedback method with step reactive power injection (Standard active islanding detection scheme of utility-interactive photovoltaic power conditioners), 2012 (in Japanese).

[2] Yoshiaki Yoshida, Koji Fujiwara, Yoshiyuki Ishihara, Hirokazu Suzuki, "Analysis of Islanding-Prevention by Detecting Harmonic Voltage Considering Nonlinear Magnetizing Characteristics of Pole Transformer," Journal of International Council on Electrical Engineering Vol. 3, No. 1, pp.12-19, 2013.

[3] IEC 62116 Ed.1: "Test procedure of islanding prevention measures for utility interconnected photovoltaic inverters," 2008.

[4] The Japan Electrical Manufacturers' Association: "Establishment of islanding detection technology," 2009 http://www.jema-net.or.jp/Japanese/res/hokoku2009.pdf.

[5] H. Kobayashi, M. Itou, "Development of Transient Analysis Model of Grid Interconnected PCS for Photovoltaic Power Generation," CRIEPI REPORT, R07027, pp. 2-4, 2008 (in Japanese).

[6] Japan Electric Association: "Grid-interconnection Code JEAC 9701-2012", Ohmsha, 2013 (in Japanese).

[7] W. Bower, M. Ropp, SANDIA REPORT, "Evaluation of Islanding Detection Methods for Utility-Interactive Inverters in Photovoltaic Systems," SAND2002-3591, pp. 31-32, 2002.

[8] JICS C 61000-3-2: Electromagnetic compatibility (EMC)-Part 3-2: Limits-Limits for harmonic current emissions (equipment input current $\leq$ 20 A per phase), 2011 (in Japanese).

Induction motor made of SMC

Masayuki Morimoto and Mamiko Inamori
Department of Electrical and Electronic Engineering
Tokai University
4-1-1, Kita-kaname, Hiratsuka, Japan
E-Mail: morimoto@ieee.org

Abstract— In this presentation the development of an induction motor of which stator core is made of molded SMC (Soft Magnetic Composite) will be presented. The molded core shows low iron loss compared to the iron loss of machined core. Iron loss of machined core is large because of the heat at the machining of bulk SMC. Moreover, by using SMC, 3D shaped core is easily constructed. The teeth of the stator core extended in order to enlarge the pole surface area in the same outer volume of the motor.

Keywords— *iron loss, efficiency, distributed windings, magnetic pole surface.*

I. INTRODUCTION

The resource crisis of rare earth materials has broken out in motor industry in the world. Rare earth material, as the name shown, is "rare" in the globe. Nowadays, many motors for broad application use rare earth permanent magnet (PM). The efficiency of PM motors is high even in partial load condition or at low speed operation.

Nd-Fe-B magnet is mainly used for such motors. High efficiency and high power density can be easily realized by using Nd-Fe-B magnet as the rotating field magnet. Almost all hybrid electric vehicles in the market use several kilograms of Nd-Fe-B magnet in their traction motor and generator. The number of HEV and EV are growing rapidly year by year. Therefore, not only motor industries but automobile industries should consider about how to reduce rare-earth magnet and what is the alternative to the PM motors.

After 1990th, the developing effort of the motor technology is mainly on PM motors because almost all newly designed motors are PM synchronous motor. New motor technology of design and manufacturing is applied to PM motors one after another. An induction motor is going to be thought as the past technology. The author believes that the improvement can be realized by the application of new technology to the induction motor.

Recent advances of SMC materials can be realized motors made of SMC[9]-[13]. All motors proposed in the literatures are PM machines. Because the literatures say that the assistance of magnetic flux by permanent magnet is necessary to attain high performance of the motor [13]. That means SMC is not suitable for electro magnet motor such as induction motor.

The authors proposed that one solution for rare-earth-free motor is the induction motor made of Magnetic Composite (SMC). SMC is compressed iron powder covered by insulation coating which is easily to form by press molding[1]-[8].

The author has already reported the increase of iron loss by machining SMC core [14], however, the motor performance did not report at that time. So, in this paper, the performance improvement of induction motor made of molded SMC will be described. The stator core made of SMC has three dimensional shape. By using molded core, efficiency and maximum output of experimental induction motor increases from that of the motor using trial production of machined SMC core.

II. DIVIDED COREE WITH DISTRIBUTED WINDINGS

In this study, SMC material used for the stator core is MBS318 (Diamet Corporation). The major advantage of SMC material is that the core can be made by press-molding process. This expresses that mass production is easy, however, larger press machine is required if the size of SMC component is large. Therefore, the core should be separated into small parts, which is the divided core.

Divided core is becoming popular in PM synchronous motors with direct concentrated windings. However, distributed winding is required for induction motor. Novel distributed winding method for divided core will be presented in this paper.

Temporary assembled divided core is shown in Fig.1. The stator core is divided into 48 parts, they are 24 tooth and 24 yokes.

Fig.1.Temporary assembled divided core

Proposed assembling procedure of the stator is shown in Fig.2. The procedure is as follows;

(1) Wind coils at proper bobbin. Form as the required shape (See Fig.3).

(2) Teeth are attached to the inner circular attachment, which diameter is the size of inside of the stator.

(3) Pre-formed coil is inserted between teeth from outside.

(4) Yoke is glued to teeth from outside after all the coils are inserted.

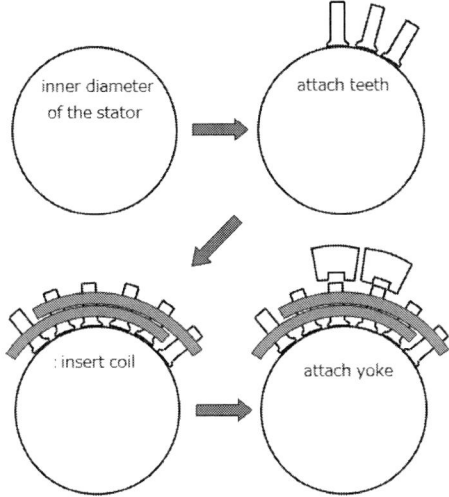

Fig.2 The assembling procedure

By the proposed assembling procedure, all the parts of the stator can be assembled from outside. This new procedure is easier than that of the conventional procedure all from inside of the stator.

Fig.3. Pre-formed coil

III. THREE DIMENSIONAL IRON CORE

The most advantageous characteristics of SMC is its freedom of the shape. In this study, 3D shaped core of the induction motor is realized. Only the pole surface of the teeth is extended at the end of the stator. The extended teeth are shown in Fig.4.

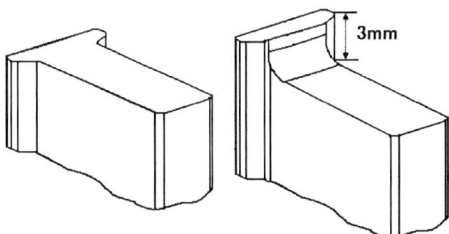

Fig.4. Extended teeth and conventional teeth

Extended teeth realizes increase of the pole surface by using same shape of the coil. The advantage of our 3D shaped scheme is that the outer size of the motor is same. The scheme is shown in Fig.5. The stack height (this naming is strange in our case) is same, though, the magnetic pole surface of the air gap is large. The shape and length of the coil are same. The difference between 2D shape and 3D shape is only the shape of the teeth. But, the end of the pole surface of the teeth is extended 3mm.

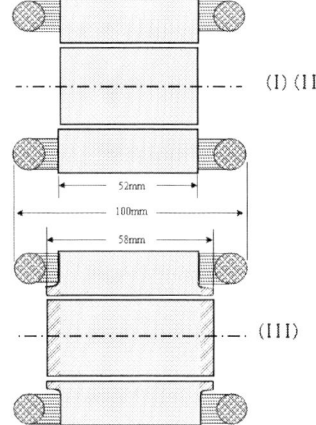

Fig. 5. The advantage of 3D shaped core

able.1 Specifications of experimental machine

By using 3D shaped core, pole surface of the motor increase 10% ($29144\text{mm}^2 \Rightarrow 32507\text{mm}^2$). Although, the external volume of the motor is same. The specification of experimental 3D motor is shown together with that of 2D SMC motor in Table1. The experimental performance improvement of 3D motor from 2D motor is shown in Fig.6 [14]. The performance improvement is less than calculated value because of the increase of iron loss.

The experimental core used in this 3D motor is made by machining. The heat generated at the machining degrades the iron loss characteristics of the core. The insulation layer of the powder at the machined surface area has broken. The surface resistance became almost zero. Therefore, all the surface of the machined core is conductive. All surface act as a path of eddy current.

Therefore, in order to attain high performance, the molding of the core is required.

Table.1 Specifications of 2D and 3Dmachines

	Names	2D (conventional shape)	3D (extended teeth)
Rated value	Number of poles	4	
	Rated voltage [V]	200	
Stator core	Inner diameter [mm]	90	
	Outer diameter[mm]	145	
	Axial length of the core[mm]	52	58
	Number of slots	24	
	Core material	SMC	
	Number of turns	72	
Rotor core	Inner diameter [mm]	22	
	Outer diameter[mm]	89.4	
	Axial length of the core[mm]	52	58
	Number of slots	44	38
	Core material	50A800 (Lamination)	
	Rotor conductor	Copper	

Fig.8 The molded stator core with extended teeth

The iron loss of the molded core is shown in Fig.9, compared to that of machined core. The iron loss of molded core is about half of that of machined core. And iron loss of molded core is proportional to the frequency, on the contrary, iron loss of machined core is proportional to the square of the frequency.

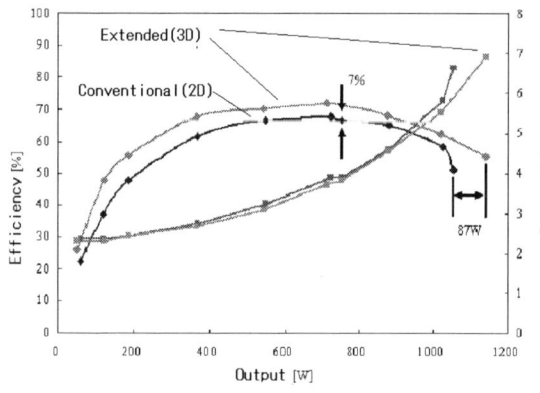

Fig. 6 Load characteristics of experimental 3D motor

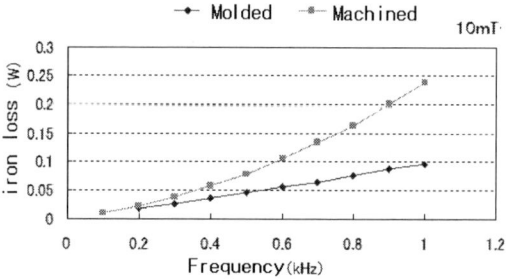

Fig.9 Iron loss of the molded core

Iron loss analysis of the core is shown in Fig. 10. Eddy current loss of molded core has been changed into almost zero. Therefore, most of measured iron loss of molded core is hysteresis loss. This is intrinsic property of SMC iron loss.

IV. PRESS MOLDING

In order to improve the increase of iron loss by machining, experimental core is manufactured by press molding. Fig 7 shows the pictures of molded yoke and teeth. Fig.8 shows the temporary assembled stator core.

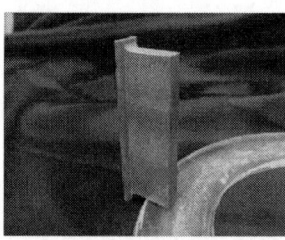 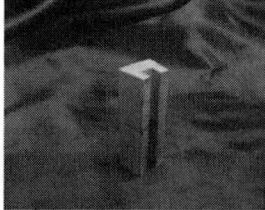

Fig. 7 molded teeth and yoke

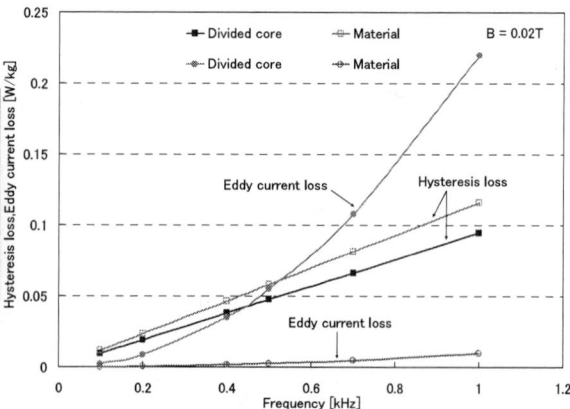

Fig. 10 Iron loss analysis of the cores

V MOLDED MOTOR

Molded core is assembled as the motor. Calculated performance of the molded motor is shown in Fig.9[15]. The maximum output power increase. The efficiency increases.

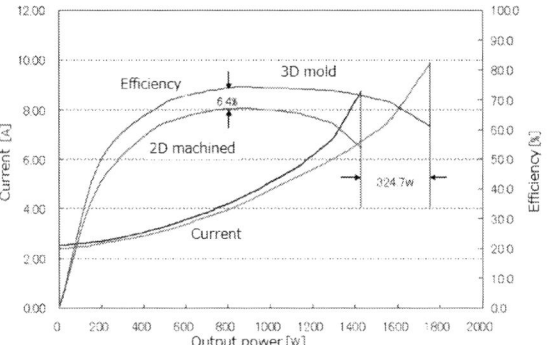

Fig.11 Calculated performance of molded motor

VI CONCLUSION

In this presentation, an induction motor of which stator core is made of SMC is presented. The followings are summary of the results.

➢ The performance of an induction motor will be improved if the design of the motor is adequate to the property of SMC.

➢ SMC application to the motors should be used the divided core from the limitation of press process.

➢ Three dimensional design of the core is easily realized by using SMC. Three dimensional design opens new aspect of the motor design.

➢ Molding of SMC is necessary even if the production is trial. Because the machining of SMC degrades iron loss characteristics of SMC

➢ Because of mechanical problem, experimental performance of the molded motor could not shown in this manuscript.

ACKNOELEDGEMENT

Part of this work is supported by New Energy and Industrial Technology Development Organization (NEDO), Japan.

REFERENCES

[1] T. Fukuda, M. Morimoto, "Load characteristics of Induction Motor Made of SMC", The 11th International Conference on Electric Machines and Systems, CMP-02 (2008.)

[2] M.Morimoto, "Rare Earth less Traction Motor for Electric Vehicle", EVS24, The 24th International Battery, Hybrid and Fuel Cell Electric Vehicle Symposium & Exhibition(2009)

[3] M.Morimoto, "Induction motor made of Iron Powder Core", EPE 2009, European Power Electronics Conference, DS2.11, No.0074(2009).

[4] Y.Sasaki, M.Morimoto, ""The Inverter Drive loss of Induction Motor Made by Soft Magnetic Composite", PEDS 2009, Power Electronics and Drive Systems, No. 453 (2009)

[5] M. Morimoto, "Induction Motor made of Iron powder Core", Proc. of IPEC-Sapporo 2010, p1814(2010).

[6] M. Morimoto, "Iron loss of non rare earth traction motor for EVs", VPPC2010, RT-6-1-5 (2010).

[7] N.Tsuchiya and M. Morimoto, "Iron Loss Characteristics of Induction Motor Made of Soft Magnetic Composite (SMC) ", Proc. of ICEMS 2010, ISM-11(2010).

[8] M.Morimoto, "Performance Improvement of Induction Motor made of Three Dimensional Shaped SMC Core", PEDS 2011, No.192 (2011).

[9] Y. G. Guo, J. G. Zhu, P. A. Watterson and W. Wu, "Comparative Study of 3-D Flux Electrical Machines With Soft Magnetic Composite Cores", *IEEE Transactions on Industry Applications*, Vol. 39, No. 6, November / December, 2003.

[10] A. Jack, B. Mecrow, P. Dickinson, D. Stephenson, J. Burdess, N. Fawcett and J. Evans, "Permanent-Magnet Machines with Powdered Iron Cores and Prepressed Windings", *IEEE Transactions on Industry Applications*, Vol. 36, No. 4, July / August, 2000.

[11] HUANG Yunkai, HU Qiansheng, ZHU Jianguo, GUO Youguang, LIN Zhiwei, "Design and Analysis of a High-Speed Claw Pole Motor With Soft Magnetic Composite Core", *IEEE Transaction on Magnetics,* Vol.43, No.6, Page.2492-2494 (2007).

[12] T. Kosaka, N. Matsui, I. Izumi, "Less Rare-Earth Magnet-High Power Density Hybrid Excitation Motor Designed for Hybrid Electric Vehicle Drives", EPE 2009, European Power Electronics Conference, DS2.10, No.0772(2009)

[13] H. Amano, Y. Enomoto, M. Ito, R. Masaki, M. Matsuzawa and M. Mita, "Characteristics of a Permanent-Magnet Synchronous Motor with a Dual-Molding Permanent-Magnet Rotor", *IEEE Power Engineering Society General Meeting*, 2007.

[14] M.Morimoto, "Efficiency Improvement of Induction Motor by 3-D Core Made of SMC "EPE-PEMC 2012 ECCE Europe, LS1b.3, (2012).

[15] M.Morimoto and M.Inamori, "Performance improvement of Induction Motor by press molded SMC Core", EPE'13 ECCE Europe, 0096 (2013).

Estimation and Comparison of the Windage Loss of a 60 kW Switched Reluctance Motor for Hybrid Electric Vehicles

Kyohei Kiyota, Takeo Kakishima, and Akira Chiba
Department of Electrical and Electronic Engineering
Tokyo Institute of Technology
S3-1, 2-12-1, Ookayama, Meguro, Tokyo, 152-8552, Japan
kiyota.k@belm.ee.titech.ac.jp

Abstract— A Switched Reluctance Motor (SRM) has been designed with outer dimensions, maximum torque, operating area, and maximum efficiency identical to the rare-earth PM motor used in the Toyota Prius. However, an SRM has salient poles in the rotor; thus, additional windage loss is generated. In this paper, estimation of the windage loss of the SRM is carried out. Then, a comparison to the machine test result is carried out. The influence of the windage loss is identified.

Keywords— *Efficiency, Hybrid electrical vehicle, Swiched reluctance motor, Windage loss.*

I.   INTRODUCTION

Hybrid electrical vehicles (HEVs) have become more popular in the past decade because they have reduced the exhaust gas emissions of automobiles. There are considerable demands for developing electronic motors with high torque density, high efficiency and low cost for HEVs, plug-in HEVs, and pure EVs [1]. Interior Permanent Magnet Synchronous Motors (IPMSMs) are the most popular electric motors [2]. However, the IPMSM contains Nd-Fe-B permanent magnets with rare-earth material such as Neodymium and Dysprosium. At their peak, the price of Nd and Dy increased to 36 and 43 times, respectively, above that of 7 years ago. Although the prices have been decreased since then to six and ten time, the cost and the supply of rare-earth permanent magnets have been recognized as a problem for HEV mass production.

Therefore, there are high demands of a rare-earth-free-motor or a less-rare-earth-motor for the next generation HEVs. Several projects have investigated substitute motors [3]. A switched reluctance motor (SRM) is one of the possible solutions. There are several advantages in SRMs such as a simple structure, low cost, rotor robustness and possible operation in high temperatures or high rotational speeds. The SRMs need no permanent magnet, but only low loss silicon steel and stator winding are needed.

SRMs have been developed extensively around the world for automotive propulsion [4]-[13]. In mass production HEVs, a maximum torque density of 35-45

Nm/l is required for 18 seconds and longer with a current density of 20-26 A/mm^2. The highest efficiency of 96% and a constant power operation is required up to the five times of the base speed [14].

A SRM has been designed with the identical outer diameter and axial length with respect to the IPMSM employed in 2009 Toyota Prius [15]. It has been shown that the torque density, the operation area, and the maximum efficiency can be competitive to the IPMSM in FEM analysis. A test machine has been built and the maximum torque test results have been reported [16] [17]. However, a SRM has salient poles in the rotor; thus, the windage loss caused by the salient poles is not negligible at the high speed area.

In this paper, estimation of the windage loss of the SRM is carried out by a simple calculation and machine test. Then, the windage loss influence on the motor efficiency is carried out.

II.   SPECIFICATION OF REFERENCE IPMSM AND DESIGNED SRM

The test motor described in this paper is based on the design previously presented in [15]. Thus, only an outline and important characteristics are described in this section.

Fig. 1 shows cross sectional views of the reference IPMSM and the designed SRM. The outer diameter and the axial length of the IPMSM and the SRM are identical values. The outer diameter of the stator core is 264 mm. The axial length is 108 mm, which is the sum of the stator iron stack length and both coil-end lengths. To improve torque density, there are 18 stator poles and 12 rotor poles in the SRM. Note that the SRM employs 6.5% high silicon steel characterized by low iron loss to improve the efficiency.

Table I summarizes specifications of the reference IPMSM and the tested SRM. The maximum axial length of the test machine is 112 mm, which is 3.7% longer than the designed value because a part of the coil-end length is larger than the designed value. In the IPMSM, the maximum shaft output power is 60 kW or slightly less in the speed range from 2768 r/min to 13 900 r/min. The

SRM can provide 60 kW in the same speed range, moreover, the SRM can provide 100 kW output power from 5400 r/min to 13 900 r/min under the DC voltage and rms current limits of 650 V and 138 A, respectively. When the design process is started, the rms current of the IPMSM at the maximum torque is estimated to be 141 A; thus, the SRM is designed to have the maximum rms current of 138 A. Afterwards, the rms current of the IPMSM is found to be as high as 170 A. Because the PM magnetization is inferior in high temperature, the required rms current is about 170 A. In the SRM, temperature dependence is not very apparent. In the IPMSM, the measured efficiency is reported to be around 96% [17]. The maximum efficiency of the SRM is also 96%, but this does not include the bearing loss and windage loss.

III. ESTIMATE THE WINDAGE LOSS

Table II shows the rotor dimensions of the SRM. These rotor dimensions are designed to maximize the maximum output torque at the base speed of 2768 r/min.

In the SRM, there are the electrical losses such as the copper loss and iron loss and the mechanical losses such as the bearing loss and the windage losses caused by the salient poles and cylindrical rotor surface. In this paper, the windage losses are calculated based on the reference [19].

A. Calculation

The Reynolds number R_e of the SRM is given as

$$R_e = \frac{Rt\omega}{\mu/\rho} \tag{1}$$

where, ω is the motor angular velocity, μ is the air viscosity of 18 μPa·s, and ρ is the air density as 1.20 kg/m³. Note that the air temperature is set at a room temperature. Then, the drag coefficient C_d caused at the rotor surface is calculated by the following equation:

$$\frac{1}{\sqrt{C_d}} = 2.04 + 1.768\ln(R_e\sqrt{C_d}) \tag{2}$$

Fig. 2 shows the variation of the Reynolds number and the drag coefficient versus the rotational speed. The Reynolds number is proportional to the rotational speed, while the drag coefficient is decreased.

The windage loss W_{wc} of the cylindrical shape rotor is given as

$$W_{wc} = \pi C_d \rho R^4 \omega^3 L \tag{3}$$

Then, the additional windage loss W_{wp} caused by the salient poles is given as

$$W_{wp} = (K-1)\pi C_d \rho R^4 \omega^3 L \tag{4}$$

where, K is the salient pole coefficient given by the rotor outer radius and the rotor pole depth as

$$K = 8.5(H/R) + 2.2 \approx 3.97 \tag{5}$$

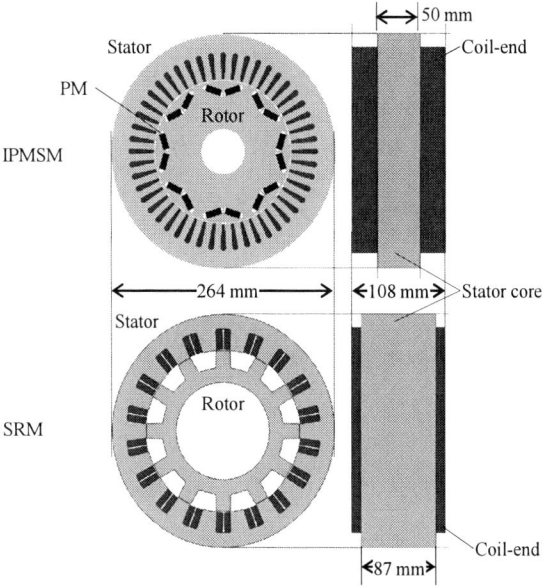

Fig. 1. Cross section of the reference IPMSM and the designed SRM.

TABLE I
COMPARISON OF THE IPMSM AND THE SRM.

	Referenced IPMSM	Tested SRM
Outer diameter	264 mm	264 mm
Axial length	108 mm	112 mm
DC side voltage	650 V	650 V
Maximum peak current*	240 A	240 A
Maximum rms current*	170 A[18]	138A
Maximum torque*	207 Nm	209 Nm
Maximum power*	60 kW	109 kW
60 kW output range*	2768 – 13 900 r/min	2768 – 13 900 r/min
100 kW output range*	N/A	5400 – 13 900 r/min
Maximum efficiency	96%[14]	96%

*Short time

TABLE II.
ROTOR PARAMETER OF THE DESIGNED SRM.

Rotor outer radius, R [m]	91×10^{-3}
Rotor axial length, L [m]	87×10^{-3}
Airgap length, t [m]	0.5×10^{-3}
Rotor pole depth, H [m]	19×10^{-3}

Thus, the salient poles of the SRM generate 2.97 times higher windage loss with respect to that of a cylindrical rotor motor which has the same rotor radius and axial length. Fig. 3 shows the variations of the windage loss of the salient pole rotor and the cylindrical shape rotor. The shaded area of Fig. 3 indicates the additional windage loss caused by the salient poles. The windage loss of the salient pole rotor is 1765 W at the maximum rotational speed of 13 900 r/min, while that of the cylindrical shape rotor is 444 W; thus, the windage loss caused by the salient poles is 1321 W at the maximum rotational speed, which is 1.3% of the maximum output power of 100 kW. It is found that the efficiency at the maximum output power of the SRM is decreased by 1.3% purely due to the

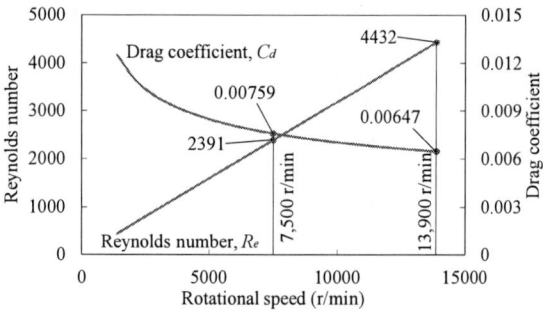

Fig. 2. Reynolds number and drag coefficient of the SRM.

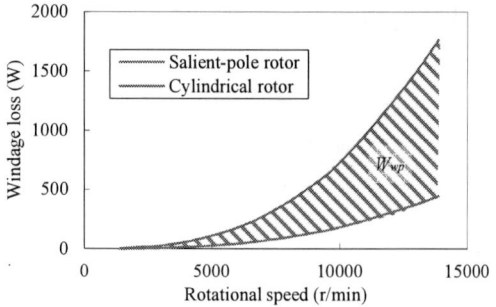

Fig. 3. Calculated windage loss of the cylindrical rotor and the salient-pole rotor.

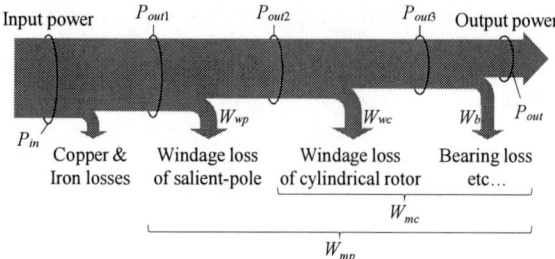

Fig. 4. Power flow from the input to the output.

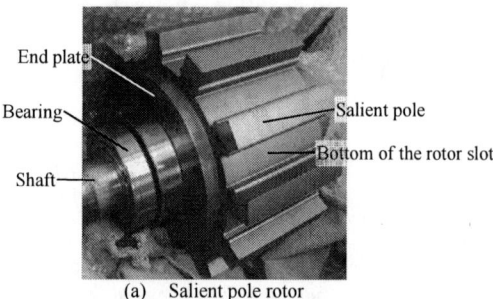

(a)   Salient pole rotor

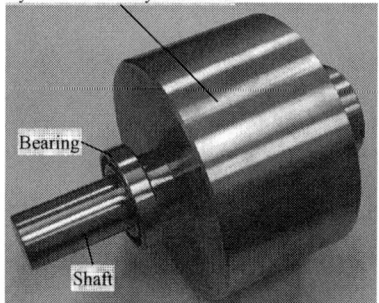

(b)   Cylindrical-shape dummy rotor

Fig. 5. Fabricated SRM and dummy rotor.

windage loss caused by the salient poles of the rotor. The influence of the windage loss will be apparent at the low output region.

B.   Machine Test Condition

Fig. 4 shows the power flow from the electrical input power to the shaft output power. The windage losses W_{wp} and W_{wc} caused by salient poles and cylindrical rotor, respectively, are given as

$$W_{wp} = P_{out1} - P_{out2} \qquad (6)$$

$$W_{wc} = P_{out2} - P_{out3} \qquad (7)$$

The bearing loss W_b is given as

$$W_b = P_{out3} - P_{out0} \qquad (8)$$

In the experiment, the windage loss W_{wp} of the salient-poles is estimated by fabricating a cylindrical shape rotor which has the identical outer diameter and axial length.

Figs. 5 (a) and (b) show the rotors of the SRM. The length of the rotor is 87.15 mm. The outer diameter is 182 mm. The outer diameter of the end plates and bottom of the rotor slot of the salient pole rotor are 130 mm and 144 mm, respectively; thus, no shrouds are assumed, so axial gas flow is possible at both ends of the rotor. The mechanical loss W_{mp} of the salient pole rotor is given as

$$W_{mp} = W_{wp} + W_{wc} + W_b \qquad (9)$$

Fig. 5 also shows the cylindrical-shape dummy rotor with the same outer diameter and bearings as the test

SRM rotor. Carbon steel is used for the rotor core. The mechanical loss W_{mc} of the cylindrical-shape dummy rotor is given as

$$W_{mc} = W_{wc} + W_b \qquad (10)$$

From (9) and (10), W_{wp} is the difference between the mechanical loss of the salient pole rotor and that of the cylindrical rotor. Note that the weight of the dummy rotor shaft and the SRM rotor shaft are 20.9 kg and 16.6 kg, respectively; thus, measured W_{mc} includes additional bearing loss caused by the increased weight. The bearing loss is estimated as proportional to the rotational speed. The correction of the bearing loss increase is described in the later section.

Fig. 6 shows an experimental system condition for the mechanical loss measurements. A test bed has a high speed but rather low torque induction generator with a torque and speed transducer (HBM T10F). The mechanical loss of the test SRM and the cylindrical dummy rotor are the product of the measured torque and

The 2014 International Power Electronics Conference

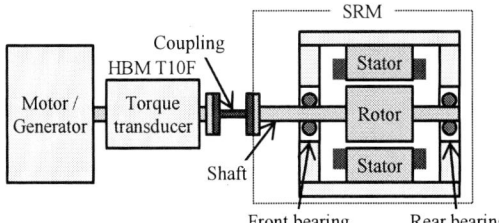

Fig. 6. System configuration for the mechanical loss.

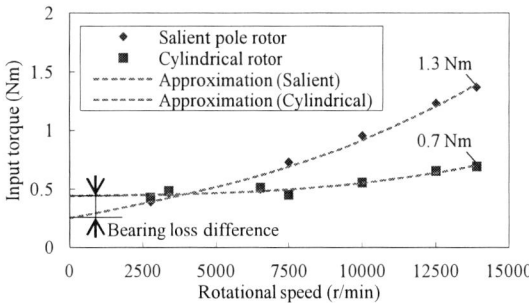

Fig. 7. Measured no-load torque of the salient-pole rotor and the cylindrical rotor.

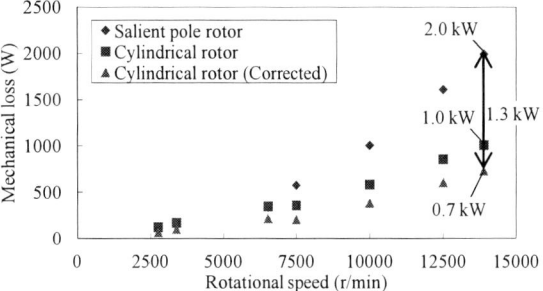

Fig. 8. Mechanical loss of the salient-pole rotor and the cylindrical rotor.

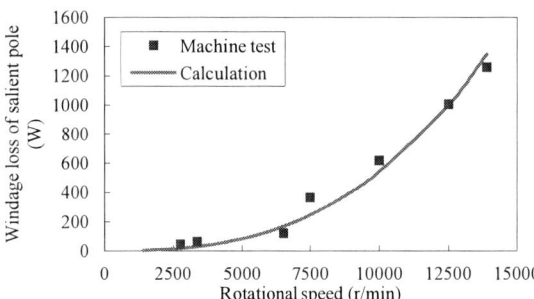

Fig. 9. Calculation results and measurement results of the windage loss of the salient-pole.

the rotational angular speed when the load machine drives the shaft with no excitation in the test SRM. This mechanical loss consists of bearing loss and windage loss. For the measurement of the cylindrical dummy rotor, the bearings are changed with the rotor shaft; thus, the measured mechanical loss includes the error of bearing condition. About a 20% variation in the mechanical loss results depending on the bearing condition; thus, the no-load torque is measured every 0.1 second and averaged over 60 seconds Note that the torque transducers have ±0.2% error with respect to the full-scale torque. The full-scale of the torque detectors are both 200 Nm; thus, the torque measurement error caused by the torque detector ΔT is ±0.4 Nm. It is also noted that both the inner diameter of the front bearing and the rear bearing of the SRM are 45 mm; these are 10 and 5 mm larger with respect to those of the IPMSM because SRM has been developed for only test purpose of electrical performance evaluation. Thus, the bearing loss of the SRM may be rather high with respect to that of the IPMSM.

C. Machine Test Result

Fig. 7 shows the measured torque of the salient pole rotor and the cylindrical dummy rotor. Note that this mechanical loss includes the windage loss caused by the salient poles of the SRM. The cylindrical dummy rotor has the identical outer diameter and bearings; thus, the mechanical loss torque increase is mainly caused by the windage loss of the salient poles. At the low rotational speed of 2768 r/min, on the other hand, the torque of the salient pole rotor is slightly lower than that of the cylindrical rotor because of the 40% weight increase of the cylindrical rotor. Fig. 7 also shows the torque approximation T_a of two rotors; those are given as

$$T_a(v) = T_{a3}v^3 + T_{a1}v + T_{a0} \qquad (11)$$

where, T_{a3}, T_{a1} and T_{a0} are coefficients and v is the rotational speed. From (3) and (4), the windage losses are zero at the zero rotational speed. The bearing loss torque is given as the constant coefficient T_{a0}. The constant coefficient T_{a0} of the salient pole rotor and the cylindrical shape rotor are 0.25 Nm and 0.44 Nm, respectively; thus, the bearing loss torque increase ΔT_b is 0.19 Nm. The bearing loss of the cylindrical rotor is increased by around 75% because the weight increase and bearing condition.

Fig. 8 shows the mechanical loss of the two rotors. The measured mechanical loss of the salient pole rotor and the cylindrical rotor are 2.0 kW and 1.0 kW at the maximum rotational speed of 13 900 r/min, respectively. Fig. 8 also shows the corrected mechanical loss of the cylindrical rotor. In the correction, the increased torque $\Delta T_b = 0.19$ Nm of the bearings is removed. The corrected mechanical loss of the cylindrical rotor is 0.7 kW at 13 900 r/min. The difference of the mechanical loss at the maximum rotational speed is 1.3 kW; thus, the windage loss caused by the salient poles of the SRM is identified as 1.3 kW at 13 900 r/min.

Fig. 9 compares the calculated and measured result of the windage loss of the salient pole rotor. The calculated windage loss of the salient poles is 1320 W at the maximum rotational speed, whereas the measurement result is 1.3 kW. The measurement results are well

978-1-4799-2706-7/14 $31.00 © 2014 IEEE   3516

confirmed at the all rotational speed. Note that the bearing loss is assumed to be linear as the rotational speed, however, it is not linear in general; thus, the measurement result may include the bearing loss at the high speed region. It is also note that the torque transducer has ±0.2% error with respect to the full-scale torque. The full-scale of the torque detectors are both 200 Nm; thus, the torque measurement error caused by the torque detector is ±0.4 Nm; thus, the measurement error is around ±582 W at 13 900 r/min. The calculated value and the estimated windage loss are within the error range; thus, the windage and bearing losses of 1.3 kW and 0.7 kW are considered in the later section.

IV. CORRECTION OF THE EFFICIENCY

In the test machine, forced cooling is not implemented. In general, an oil-cooling or a water jacket is provided, however, this test machine has only natural cooling; thus, measurements are carried out at a coil-end temperature of 75 °C, which is the identical temperature at the prediction stage.

The measured efficiency η_m is calculated as

$$\eta_m = \frac{P_o}{P_i} \qquad (12)$$

where, P_o is the output shaft power and P_i is the input electrical power. This is the lowest efficiency including bearing and windage losses.

To see the efficiency decrease caused by the windage loss W_{wp} of the salient pole, the efficiency η_w excluding the windage loss is calculated as below,

$$\eta_w = \frac{P_o + W_{wp}}{P_i} \qquad (14)$$

To see the efficiency decrease caused by the mechanical loss W_{mp} which is the sum of the windage losses and bearing loss as previously shown in Fig. 4, the efficiency η_e, is calculated as

$$\eta_e = \frac{P_o + W_{mp}}{P_i} \qquad (13)$$

This efficiency is assumed to have a good correspondence to the analysis prediction.

Fig. 10 shows the efficiencies η_m, η_e and η_w at the maximum rotational speed of 13 900 r/min. The maximum efficiency of η_m is 89.5%, whereas those of η_e and η_w are 95.0% and 92.6%, respectively. We can observe an efficiency decrease at 18 kW that is close to the maximum efficiency point at the analytical prediction. At 18 kW output power efficiency decreases caused by the windage loss of the salient poles is 6.5%. A 4.7% efficiency decrease is caused by bearing loss and cylindrical rotor windage loss. In Total, the efficiency is decreased by 11.2%. On the other hand, the efficiency decrease is only 1.1% at the output power of 100-kW. The influence of efficiency decrease is rather apparent at low rotational speed.

Fig. 11 shows the efficiencies η_m, η_e and η_w at the middle rotational speed of 7500 r/min. The maximum efficiency of η_e is 96.4% that is the maximum efficiency of the SRM, whereas those of η_w and η_m are 95.7% and 94.9%, respectively. Let us examine the efficiency decrease in the region about the maximum efficiency, near 36 kW of output power. At 36 kW output power, the efficiency decreases caused by the windage loss of the salient poles is 0.8%. Note that the 0.7% efficiency decrease is caused by bearing loss and cylindrical rotor windage loss. The influence of mechanical loss is limited

Fig. 12 shows the efficiencies η_m, η_e and η_w at the base speed of 2768 r/min. The efficiency decrease from the windage loss of the salient poles is low with respect to that of the other mechanical losses. The efficiency η_e has a maximum value of 95.7% at the output of 8.8 kW, whereas those of η_w and η_m are 94.5% and 94.1%, respectively. The influence of windage loss is limited.

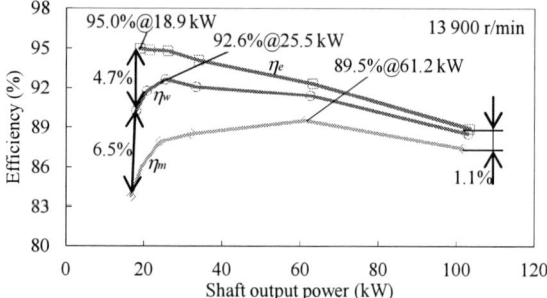

Fig. 10. Mechanical loss influence on the efficiency at 13 900 r/min.

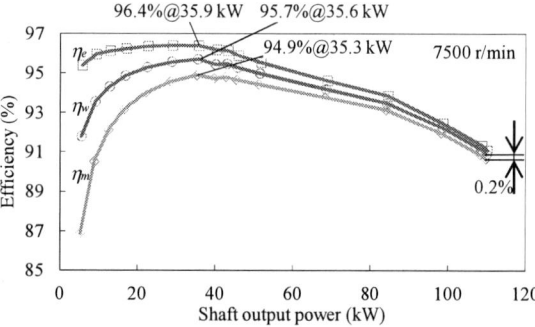

Fig. 11. Mechanical loss influence on the efficiency at 7500 r/min.

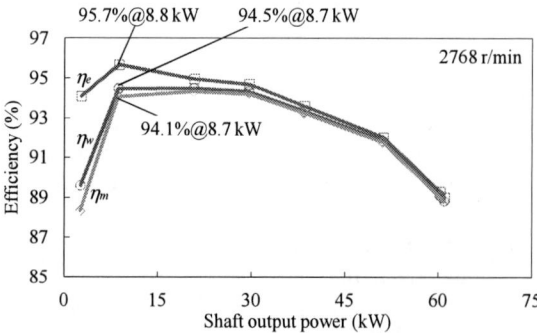

Fig. 12. Mechanical loss influence on the efficiency at 2768 r/min.

Note that the output power of 10 kW or lower is a main operation region of HEVs; thus, the influence of the windage loss may be not significant in city mode driving schedules.

V. CONCLUSIONS

In this paper, estimation of the windage loss of the SRM is carried out by calculation and machine testing. It is found that windage loss caused by the salient pole is 1.3 kW at the speed of 13 900 r/min. In addition, the decrease of the motor efficiency caused by salient pole windage loss is as high as 6.5% at 18 kW output. The efficiency decrease is significant at low output power, although it is only 1.1% at 100 kW. The reduction of the windage loss remains for a future project.

ACKNOWLEDGMENT

The authors would like to acknowledge the test machine fabrication by Mr. Saito in the Motion System Tech. They also are grateful for an FPGA controller and SR inverter fabricated by the Myway Plus Corporation.

REFERENCES

[1] P.B. Reddy, A.M. EL-Refaie, Huh Kum-Kang, J.K. Tangudu, T.M. Jahns, "Comparison of Interior and Surface PM Machines Equipped With Fractional-Slot Concentrated Windings for Hybrid Traction Applications", *IEEE Transactions on Energy Conversion*, vol.27, no.3, pp.593-602, Sept. 2012

[2] K. Yamamoto, "The development trend of a next-generation car and the its propulsion motor", 2010 *International Conference on Electrical Machines and Systems*, Oct. 10-13, 2011, pp.25-31.

[3] Akira Chiba, Nobukazu Hoshi, Masatsugu Takemoto, Satoshi Ogasawara, Shigeo Morimoto, Masayuki Sanada, Takashi Kosaka, "Rare-Earth-Free AC Motors -an Alternative Approach Advances", *Keynote Lecture, International Electric Machines and Drives Conference*, May 15-18, Niagara Falls, Canada, 2011.

[4] John M. Miller, Allan R. Gale, Patrick J. McCleer, Franco Leonardi, Jeffrey H. Lang, "Starter-Alternator for Hybrid Electric Vehicle: Comparison of Induction and Variable Reluctance Machines and Drives", *Conference Record of IEEE Industrial Applications Society Annual Meeting*, VOL. 1, pp.513-523, 1998.

[5] Khwaja M. Rahman, Babak Fahimi, G. Suresh, Anandan Velayutham Rajarathnam, M. Ehsani, "Advantages of Switched Reluctance Motor Applications to EV and HEV: Design and Control Issues", *IEEE Transactions on Industry Applications*, VOL. 36, NO. 1, pp.111-121, 2000.

[6] Shuanghong Wang, Qionghua Zhan, Zhiyuan Ma, Libing Zhou, "Implementation of a 50-kW Four-Phase Switched Reluctance Motor Drive System for Hybrid Electric Vehicle", *IEEE Transactions on Magnetics*, VOL. 41, NO. 1, pp.501-504, 2005.

[7] Nigel Schofield, Stephen A. Long, David Howe, Mike McClelland, "Design of a Switched Reluctance Machine for Extended Speed Operation", *IEEE Transactions on Industry Applications*, VOL. 45, NO. 1, pp.116-122, 2009.

[8] X.D. Xue, K.W.E. Cheng, T.W. Ng, N.C. Cheung, "Multi-Objective Optimization Design of In-Wheel Switched Reluctance Motors in Electric Vehicles," *IEEE Transactions on Industrial Electronics*, vol.57, no.9, pp.2980-2987, Sept. 2010

[9] Martin D. Hennen, Markus Niessen, Christian Heyers, Helge J. Brauer, Rik W. De Doncker, "Development and Control of an Integrated and Distributed Inverter for a Fault Tolerant Five-Phase Switched Reluctance Traction Drive," *IEEE Transactions on Industrial Electronics*, vol.27, no.2, pp.547-554, Feb. 2012.

[10] Kim C.S., Geunhong Lee, Kwanghyun Lee, Jaesung Lee, Yongwan Cho, Jang Hyeok Won, Hongchul Shin, Changhwan Choi, Han Kyung Bae, "Design of Π core and Π2 core PM-aided switched reluctance motors," 2012 IEEE *International Electric Vehicle Conference*, pp.1-6, 4-8 March 2012.

[11] M. Takeno, A. Chiba, N. Hoshi, S. Ogasawara, M. Takemoto, M. A. Rahman, "Test Results and Torque Improvement of the 50kW Switched Reluctance Motor Designed for Hybrid Electric Vehicles", *Transactions on Industry Applications*, vol.48, no.4, pp.1327-1334, July-Aug. 2012.

[12] A. Labak, N. C. Kar, "Designing and Prototyping a Novel Five-Phase Pancake-Shaped Axial-Flux SRM for Electric Vehicle Application Through Dynamic FEA Incorporating Flux-Tube Modeling", *Transactions on Industry Applications*, vol.49, no.3, pp.1276-1288, May-June 2013

[13] B. Bilgin, A. Emadi, M. Krishnamurthy, "Comprehensive Evaluation of the Dynamic Performance of a 6/10 SRM for Traction Application in PHEVs", *Transactions on Industrial Electronics*, vol.60, no.7, pp.2564-2575, July 2013

[14] T. A. Burress, S. L. Campbell, C. L. Coomer, C. W. Ayers, A. A. Wereszczak, J. P. Cunningham, L. D. Marlino, L. E. Seiber, H. T. Lin, "Evaluation of the 2010 Toyota Prius Hybrid Synergy Drive System", ORNL/TM-2010/253, 2010

[15] K. Kiyota, A. Chiba, "Design of Switched Reluctance Motor Competitive to 60-kW IPMSM in Third-Generation Hybrid Electric Vehicle", *IEEE Transactions on Industry Applications*, vol.48, no.6, pp.2303-2309, Nov.-Dec. 2012.

[16] K. Kiyota, T. Kakishima, H. Sugimoto, A. Chiba, "Comparison of the Test Result and 3D-FEM Analysis at the Knee Point of a 60 kW SRM for a HEV", *IEEE Transactions on Magnetics*, vol.49, no.5, pp.2291-2294, May 2013.

[17] K. Kiyota, T. Kakishima, A. Chiba, "Comparison of Test Result and Design Stage Prediction of Switched Reluctance Motor Competitive to 60 kW Rare-earth PM Motor", *IEEE Transactions on Industrial Electronics*, IEEE Early Access Articles.

[18] R. Mizutani, "Technical Feature and Subjects of Traction Motors for HEV/EV", *The Institute of Electrical Engineers of Japan Vehicle Technology*, RM-13-026, pp.67-70, 2013

[19] J.E.Vrancik, "Prediction of windage power loss in alternators", *NASA Technical Note* D-4849, 1968.

Development of High-power PMASynRM Using Ferrite Magnets for Reducing Rare-earth Material Use

Masayuki Sanada, Shigeo Morimoto and Yukinori Inoue

Dept. of Electrical and Information Systems,
Osaka Prefecture University
1-1 Gakuencho, Naka-ku, Sakai, Osaka, Japan
sanada@eis.osakafu-u.ac.jp

Abstract— Recently, the use of permanent magnet (PM) motors has expanded remarkably as a result of improvements in the performance of the rare-earth PM motor. However, it is desirable to reduce the use of rare-earth materials, which are an important aspect of the high-performance PM motor, because of the high cost and the unpredictability involved in procuring such materials. The performance of motors that use rare-earth materials has reached a very high level, one which would not be easy to match without such materials. In this paper, we introduce novel structures for a high-power density PM-assisted synchronous reluctance motor (PMASynRM) that uses a ferrite PM. The performance of a PMASynRM is evaluated based on the 2-D finite element method and through an experiment using a prototype machine. The analysis results reveal that the proposed PMASynRM has the same power density and equivalent efficiency as conventional rare-earth PM synchronous motors for use in vehicle applications. The results of this study indicate that the proposed structure achieves both high-power and high-efficiency.

Keywords— *PMASynRM, ferrite magnet, power density, efficiency*

I. INTRODUCTION

Recently, the use of permanent magnet (PM) motors has expanded remarkably. Hybrid electric vehicles (HEVs) and electric vehicles (EVs) use PM synchronous motors (PMSMs) that contain Nd-Fe-B rare-earth PMs. This is because PMSMs that use rare-earth PMs have high efficiencies and high power densities. However, it is desirable to reduce the use of rare-earth materials, which are an important aspect of the high-performance PM motor, because of their high cost and the unpredictability involved in procuring such materials [1]. The performance of motors that use rare-earth materials has reached a very high level, which would not be easy to match without such materials [2]-[5]. A PM-assisted synchronous reluctance motor (PMASynRM) that use a ferrite PM is considered to be a candidate as a high-performance motor that does not use a rare-earth PM [6]-[9].

In this paper, the motor structure of a high-power

PMASynRM with a ferrite PM is proposed. The performance of the PMASynRMs is evaluated based on the 2-D finite element method and an experiment using a prototype machine. In addition, the demagnetization behavior of the PM is discussed [10]-[12]. The results indicate that a ferrite PMASynRM with a performance equal to that of a rare-earth PM motor can be achieved.

II. ROTOR DESIGN OF PMASYNRM WITH FERRITE PERMANENT MAGNET

The rotor structures of a PMASynRM with a ferrite PM for high power density are shown in this section. Here, when the rotor is designed considering only the torque and output characteristics, the PMs in the rotor may be irreversibly demagnetized because ferrite PMs are less robust than rare-earth PMs [13]. Therefore, the basic concept of the rotor design to reduce the demagnetization of PM [10] is first described.

A. Concept of Rotor design to Reduce Demagnetization

An example of the demagnetization curve of a ferrite PM is shown in Fig. 1. In this paper, irreversible demagnetization was determined as follows. When the flux density in the PM is lower than the flux density of the knee, the magnet is assumed to have suffered irreversible demagnetization. The ratio of irreversible demagnetization is calculated base on the magnet volume of irreversible demagnetization with respect to the total magnet volume [9],[14]. This ratio is referred to as the demagnetization rate. For the ferrite PMs used in this study, a knee point appears on the demagnetization curve at low temperatures. Thus, the criterion for the flux density of irreversible demagnetization is 0.15 T at $-20°C$. In order to assume the most severe conditions, the current phase angle β is set to 90° (β is defined as the leading electrical angle with respect to the q-axis).

A number of rotor designs that are intended to prevent the irreversible demagnetization of the ferrite PM are shown in Fig. 2. These figures show one pole part of the rotor of 6-pole machine as an example. The type-A rotor

The 2014 International Power Electronics Conference

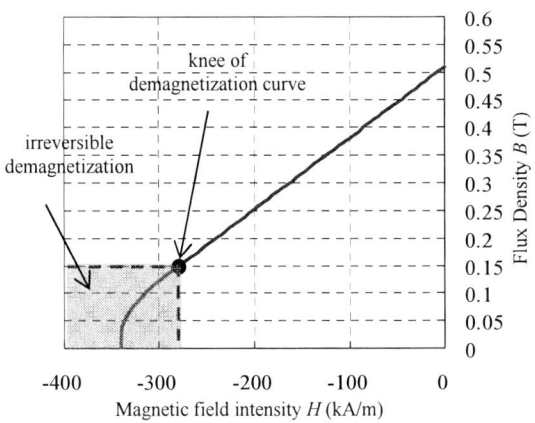

Fig. 1. Demagnetization curve of a ferrite PM.

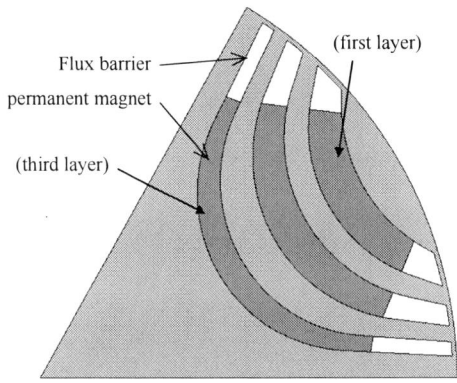

(a) Type-A rotor using a thick PM in a shallow layer

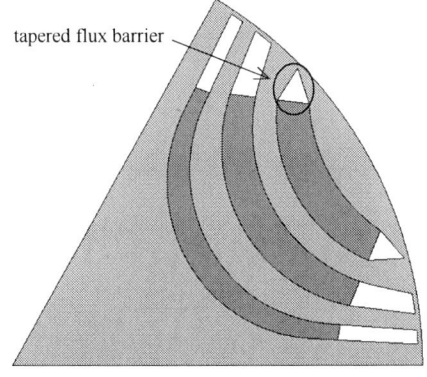

(a) Type-B rotor with a tapered flux barrier shape

Fig. 2. Rotor design to reduce demagnetization of the PM.

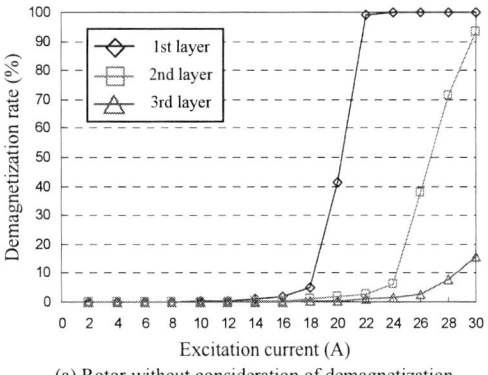

(a) Rotor without consideration of demagnetization

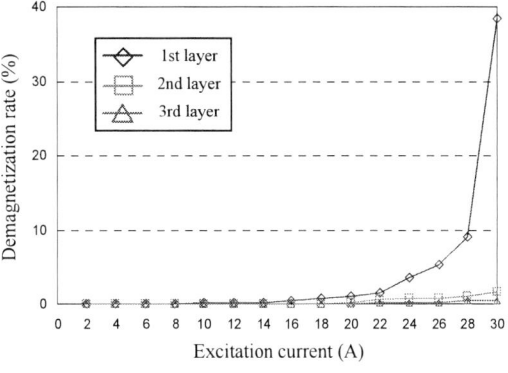

(b) Type-A rotor

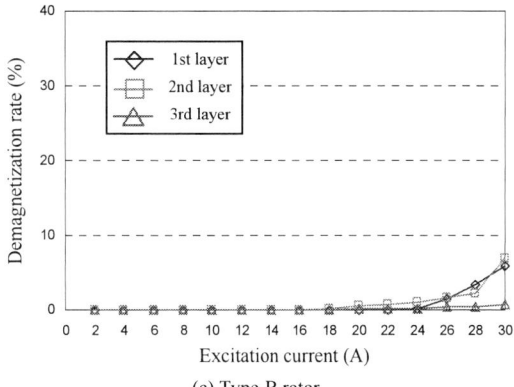

(c) Type-B rotor

Fig. 3. Demagnetization rate characteristics.

design uses a magnet with greater thickness in the shallow layer, as compared to a rotor designed without consideration of demagnetization. Unlike the type-A rotor, the type-B rotor has a tapered flux barrier in the first layer.

The demagnetization rate characteristics at −20°C are shown in Fig. 3. If the rotor structure without consideration of demagnetization is used, the demagnetization rate of the first layer PM reaches 100% for a current of only 22 A, which is 2.2 times the rated current, as shown in Fig. 3(a). On the other hand, the irreversible demagnetization is improved in the type-A

rotor structure, as shown in Fig. 3(b). In the structure of the type-A motor, the demagnetization rate is improved by increasing the volume of the PM and by making the shallow layer of the PM thicker. The demagnetization rate of the first-layer PM of the type-A motor is still approximately 40% at 30 A, which is three times the rated current, and the improvement is therefore insufficient. In the structure of the type-B motor, which has a tapered flux barrier in the first layer, the demagnetization flux easily flows through the rotor surface. Therefore, the demagnetization rate in the first layer is greatly improved, as shown in Fig. 3(c). The demagnetization rate of the first-layer PM of the type-B motor is approximately 7% at 30 A.

As a result, it was revealed that the demagnetization

978-1-4799-2706-7/14 $31.00 © 2014 IEEE

The 2014 International Power Electronics Conference

rate is greatly improved by increasing the PM thickness and by tapering the flux barrier [9],[10].

B. Analysis Models

The structure and specifications of the 20-kW and 50-kW PMASynRMs with ferrite PMs are shown in Fig. 4 and Table I, respectively [11],[12]. The rotor of the 20-kW model shown in Fig. 4(b) has a center rib in order to provide mechanical strength in the high-speed region and to counter the effect of the armature reaction on the PM. In the rotor of the 50-kW model shown in Fig. 4(d), flat magnets are used in consideration of the costs of the magnets and their manufacture. Moreover, the flux barriers of the first and second layers are tapered in order to reduce the effect of irreversible demagnetization, as described in previous section. In both rotor structures, the demagnetization rate is 9% or less for a current of three times of the rated current. The stator features eight poles, 48 slots and a distributed winding.

A ferrite PM with a coercive force of 342 kA/m at 20°C is chosen for use in this study. The rated current density and the maximum current density of the motor are assumed to be 7.5 and 15 A/mm^2, respectively.

III. ANALYSIS RESULTS

The maximum torque characteristics of the 20-kW PMASynRM are shown in Fig. 5. The current was set to 50 A, 100 A, and 133.3 A, which correspond to current densities of 7.5 A/mm^2, 15 A/mm^2, and 20 A/mm^2, respectively. As shown in Fig. 5, the reluctance torque is approximately 81%, 89%, and 89% of the total torque at 7.5 A/mm^2, 15 A/mm^2, and 20 A/mm^2, respectively. The maximum torque characteristics of the 50-kW PMASynRM are shown in Fig. 6. The current was set to 137.5 A and 275 A, which correspond to current densities of 7.5 A/mm^2 and 15 A/mm^2, respectively. As shown in Fig. 6, the reluctance torque is approximately 84% and 92% of the total torque at 7.5 A/mm^2 and 15 A/mm^2, respectively.

Figures 5 and 6 also show that the PMASynRM with ferrite PMs produces reluctance torque, whereas the permanent magnet assists in torque production.

The torque and output power vs. speed relationships are shown in Fig. 7. Here, the voltage limit V_{am} was set to 306 V, and the limit of the current density was set to 7.5 A/mm^2. With the proposed rotor structures, the output power exceed 20 kW and 50 kW at the rated currents of the 20-kW and 50-kW motors, respectively, as shown in Fig. 7. The output power densities (power/core volume) of both PMASynRMs with ferrite magnets exceed 10

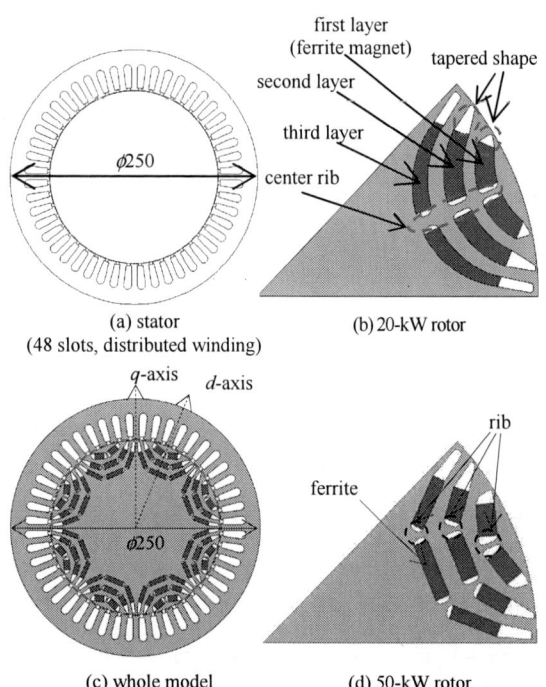

Fig. 4. Analysis models of the PMASynRM.

TABLE I
SPECIFICATIONS OF PROPOSED PMASYNRMS

Item (Unit)	Value	
Output Power	20kW	50kW
Number of poles	8	
Stator diameter (mm)	250	
Rotor diameter (mm)	169.1	
Stack length (mm)	40	100
Air gap length (mm)	0.9	
Current density (A/mm^2)	7.5, 15	
Coercive force of PM (kA/m)	342	
Winding resistance (Ω)	0.0604	0.0118

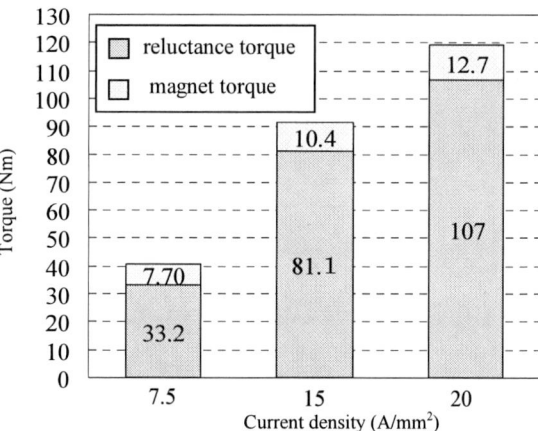

Fig. 5. Maximum torque characteristics.

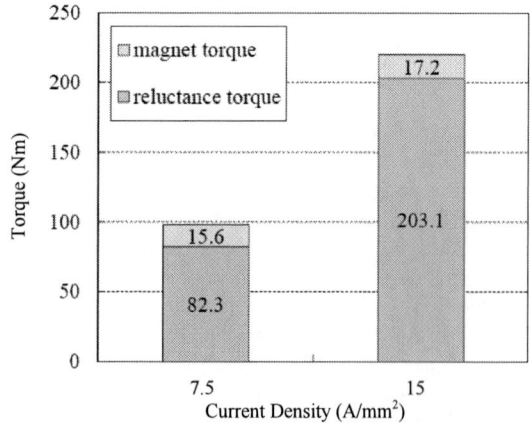

Fig. 6. Maximum torque characteristics.

978-1-4799-2706-7/14 $31.00 © 2014 IEEE

The 2014 International Power Electronics Conference

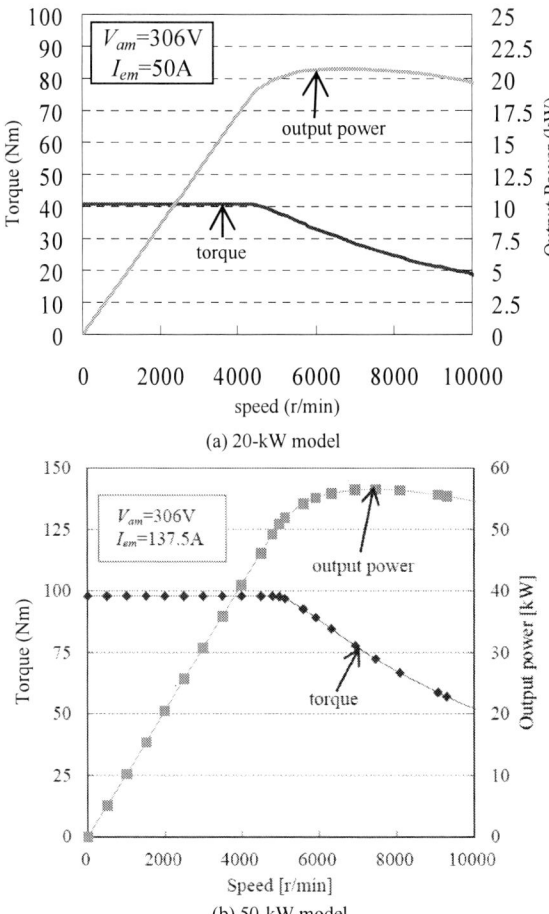

(a) 20-kW model

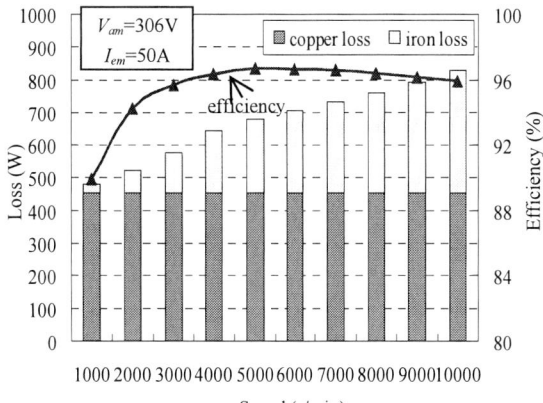

(a) 20-kW model

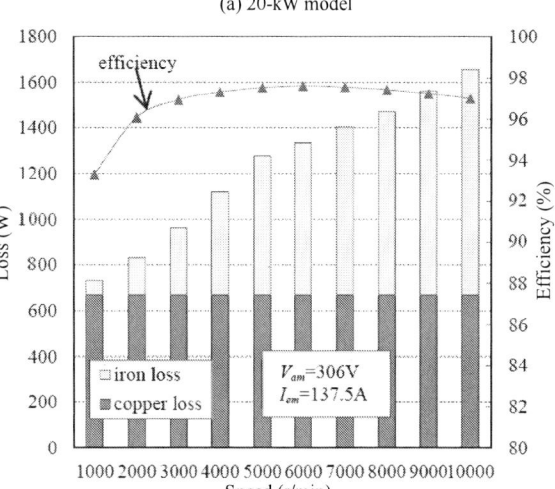

(b) 50-kW model

Fig. 7. Torque and power versus speed characteristics.

(b) 50-kW model

Fig. 8. Loss and efficiency versus speed characteristics.

kW/L, which is approximately equal to that of the rare-earth PMSM used in the commercial HEV in 2003.

The loss and efficiency vs. speed relationships under the operating conditions in Fig. 7 are shown in Fig. 8. The iron loss, copper loss, and efficiency were calculated using the following equations:

$$W_{hi} = \sum_{e=1}^{n_{elem}} \left\{ \sum_{k=1}^{n} a(B_k) \times f_k \right\} \times V_e \qquad (1)$$

$$W_{ei} = \sum_{e=1}^{n_{elem}} \left\{ \sum_{k=1}^{n} b(B_k, \ f_k) \times f_k^2 \right\} \times V_e \qquad (2)$$

$$W_i = W_{hi} + W_{ei} \qquad (3)$$

$$W_c = 3 R_a I_e^2 \qquad (4)$$

$$\eta = \frac{\omega T - W_i}{\omega T + W_c} \times 100 \qquad (5)$$

where W_{hi} is the hysteresis loss (W), W_{ei} is the eddy current loss (W), W_i is the iron loss (W), n_{elem} is the number of elements in the FEM analysis model, e is the element number, f_k is the harmonic order frequency (Hz), k is the harmonic order, V_e is the volume of the e-th element (m^3), B_k is the amplitude of the flux density harmonic k in the radial and tangential directions of the e-th element (T), $a(B_k)$ and $b(B_k, f_k)$ are evaluated based on the loss curve of the material data, W_c is the copper loss

(W), R_a is the armature winding resistance (Ω), I_e is the phase current RMS value (A), η is the efficiency (%), and ω is the mechanical angular speed (rad/s).

As shown in Fig. 8, the copper loss is constant because the phase current is constant and the iron loss gradually increases with speed. The maximum efficiencies of the 20-kW and 50-kW models under the rated current condition (7.5 A/mm^2) are 96.7% and 97.6%, respectively. These characteristics demonstrate that high

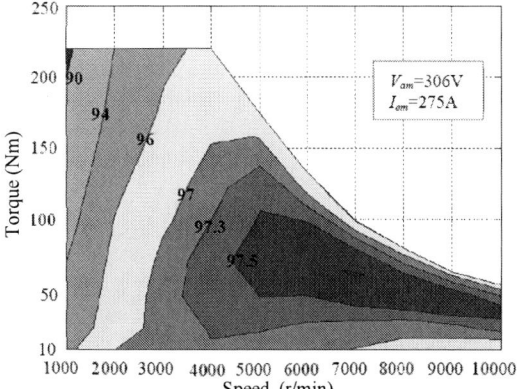

Fig. 9. Efficiency map of the 50-kW PMASynRM.

efficiency can be achieved in PMASynRMs by using ferrite PMs.

An efficiency map of the 50-kW model is shown in Fig. 9, where the maximum efficiency is 97.7% at 62.4 Nm and 7,000 r/min. In addition, an efficiency of over 90% is achieved over a wide range. This efficiency is sufficiently high compared to that of a PMSM with a rare-earth PM.

IV. EXPERIMENTAL RESULTS

A test motor based on the 50-kW model was constructed, and its characteristics were measured. The specifications of the tested motor are listed in Table II, and photographs of its rotor and stator are shown in Fig. 10. From Table II, the tested motor is different from the analytical model with respect to several aspects.

The stack length of the tested motor is 20 mm, which is 20% of that of the analysis model. Thus, in order to maintain the magnet-motive-force and current density, the number of turns per slot and the rated current of the tested motor are set to be 22 and 25 A, respectively. In order to evaluate the analysis model and the prototype motor under the same conditions, the analytical results are converted based on the specifications of the tested motor. Figure 11 shows the experimental setup, which is

a standard method of measuring motor characteristics using a load motor and a torque transducer. The torque transducer was a SS2100 (Ono Sokki Co. Ltd.), and the load motor was a 10-kW induction motor (Fuji Electric Co. Ltd.). The tested motor was driven by a PWM inverter (My Way Plus Co.) with a carrier frequency of 10 kHz. The maximum speed during the measurement was limited to 6,000 r/min for safety.

The torque characteristics are shown in Fig. 12. The measured maximum torque at 25 A is approximately 8% smaller than the analytical results. This difference is thought to be due to incomplete magnetization of the PM, and the fact that the flux leakage occurs at the edge of the rotor. The measured Ψ_a is 0.0270 Wb, which is approximately 7.5% smaller than the analytically obtained Ψ_a (= 0.0292 Wb). The output power vs. speed characteristics under the maximum power control is shown in Fig. 13. As shown in Fig. 13, the measured maximum output power is 7.3 kW, which is smaller than the analytical result (9.6 kW). In particular, the large difference in the high-speed region is due to the lower power factor caused by the inductance increment resulting from the flux leakage and the small Ψ_a.

The efficiency and loss are shown in Fig. 14. The armature resistance of the tested motor (0.280 Ω) is larger than the analytical value (0.261 Ω). Therefore, the measured copper loss is higher than the analytical value.

TABLE II
SPECIFICATIONS OF THE TESTED PMASYNRM

Item (Unit)	Value
Number of poles	8
Stator diameter (mm)	250
Rotor diameter (mm)	169.1
Stack length (mm]	20
Air gap length (mm)	0.9
Current (A)	12.5, 25
Current density (A/mm^2)	7.5, 15
Number of turns per slot	22
Coercive force of PM (kA/m)	342
Winding resistance (Ω)	0.280

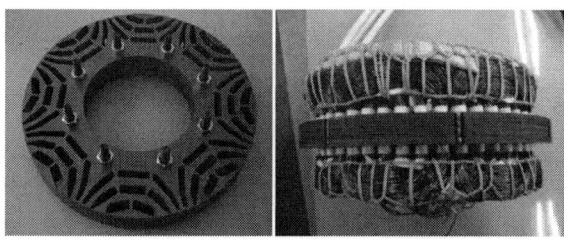

(a) Rotor   (b) Stator

Fig. 10. Photographs of the tested machine.

Fig. 11. Experimental setup.

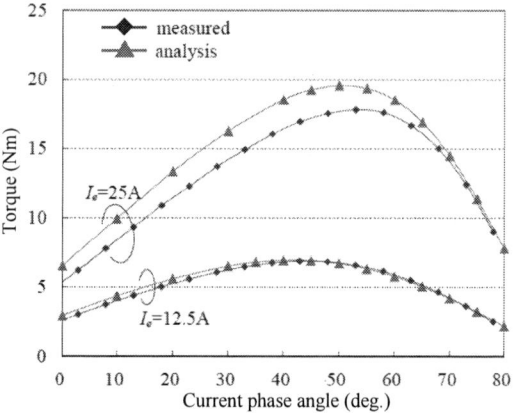

Fig. 12. Torque versus current phase characteristics.

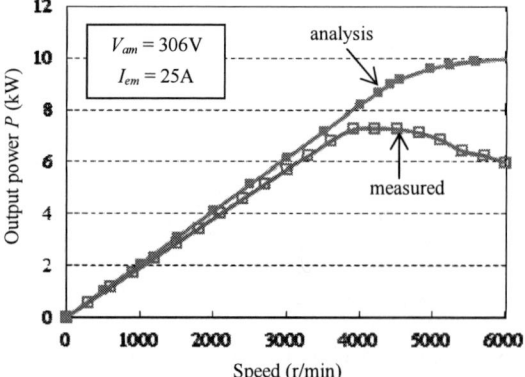

Fig. 13. Power versus speed characteristics.

The 2014 International Power Electronics Conference

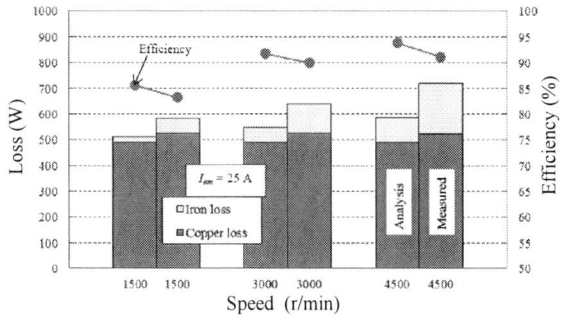

Fig. 14. Efficiency and loss characteristics.

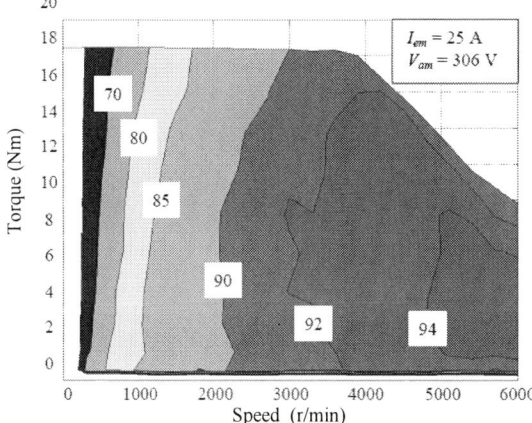

Fig. 15. Experimental efficiency map.

The measured iron loss is also higher than the analytical value. This is because the iron losses due to the carrier frequency components and mechanical loss are included in the measurement results. As shown in Fig. 14, at 4,500 r/min, the analytical efficiency is 93.9 % and the measured efficiency is 91.1 %. The copper loss is very large compared to the iron loss because the ratio of the coil end winding is large, as shown in Fig. 10(b). A significant portion of the windings does not produce torque. If the stack length becomes large, the ratio of the coil end windings becomes small, and the motor has a reasonable balance between copper loss and iron loss. Thus, the efficiency characteristics will be improved.

An efficiency map of the experiment is shown in Fig. 15, which indicates that the maximum efficiency is 94.7 % at 6.17 Nm and 6,000 r/min. In addition, an efficiency of over 90 % is achieved over a wide range, and at high-speeds in particular.

V. CONCLUSIONS

Permanent-magnet-assisted synchronous reluctance motors with ferrite magnets for high-power applications were designed, and their characteristics were examined. By incorporating ribs, a PMASynRM with ferrite PMs was designed in order to satisfy the mechanical strength requirements in the high-speed region and to resist demagnetization. The output power density, 10 kW/L, was as high as that for the rare-earth PMSM used in the commercial HEV in 2003. The proposed PMASynRMs

achieved an efficiency of over 90% over a wide operating range with a maximum efficiency of 97.7%.

The experimental results of this study reveal that the tested PMASynRM has a maximum output power of 7.3 kW at a rated current of 25 A, which corresponds to a power density of 7.4 kW/L at the rated current. The tested PMASynRM achieved an efficiency of over 90 % over a wide operating range with a maximum efficiency of 94.7 %. Furthermore, assuming a stack length of the tested machine of 100 mm, the efficiencies were estimated based on the experimental results. The maximum efficiency was found to be 96.9 % at 30.9 Nm and 6,000 r/min, and an efficiency of over 90 % was achieved over a wide range. These results indicate that if the stack length of the tested machine is 100 mm, the motor has a reasonable balance between copper loss and iron loss, and the efficiency characteristics are improved.

REFERENCES

[1] T. Naruta, Y. Akiyama, Y. Niwa and D. Uneyama: "A Study of BLDC motor design and its cost trend for rare earth materials", *Proc. of JIASC2007*, Vol. 3, pp. 261-264, 2007.

[2] M. Kamiya: "Development of Traction Drive Motors for the Toyota Hybrid System", *Proc. of IPEC-Niigata 2005*, pp. 1474-1481, 2005.

[3] P. Niazi, H. A. Toliyat, D. H. Cheong, J. C. Kim: "A Low-Cost and Efficient Permanent-Magnet-Assisted Synchronous Reluctance Motor Drive", *IEEE Transactions on Industry Applications*, Vol. 43, No. 2, pp. 542-550, 2007.

[4] E. Armando, P. Guglielmi, M. Pastorelli, G. Pellegrino, A. Vagati: "Accurate Magnetic Modelling and Performance Analysis of IPM-PMASR Motors", *Proc. of IEEE-IAS Ann. Meet.*, pp. 133-140, 2007.

[5] D. G. Dorrell, A. M. Knight, M. Popescu: "Performance Improvement in High-Performance Brushless Rare-Earth Magnet Motors for Hybrid Vehicles by use of High Flux-Density Steel", *IEEE Transactions on Magnetics*, Vol. 47, No. 10, pp. 3016-3019, 2011.

[6] N. Matsui, A. Chiba and Y. Takeda: "Electric Machines Employing Reluctance Torque", *T. IEE Japan*, Vol. 114-D, No. 9, pp. 824-832, 1994.

[7] M. Morimoto, N. Matsui and Y. Takeda: "Recent Advances of Reluctance Motors", *T. IEE Japan*, Vol. 119-D, No. 10, pp. 1145-1148, 1994.

[8] S. Morimoto, M. Sanada and Y. Takeda: "Performance of PM Assisted Synchronous Reluctance Motor for High-Efficiency and Wide Constant Power Operation", *Proc. of IEEE IAS Ann. Meet.*, pp. 509-514, 2000.

[9] M. Sanada, Y. Inoue, S. Morimoto: "Structure and Characteristics of High-Performance PMASynRM with Ferrite Magnets", *Electrical Engineering in Japan*, Vol. 187, No. 1, pp. 42-50, 2014.

[10] M. Sanada, Y. Inoue, S. Morimoto: "Rotor Structure for Reducing Demagnetization of Magnet in a Ferrite-PMASynRM and its Characteristics", *Proc. of ECCE2011*, pp. 4189-4194, 2011.

[11] S. Ooi, S. Morimoto, M. Sanada, Y. Inoue: "Performance Evaluation of a High-Power-Density PMASynRM With Ferrite Magnets", *IEEE Transactions on Industry Applications*, Vol. 49, No. 3, pp. 1308-1315, 2013.

[12] M. Obata, S. Morimoto, M. Sanada, Y. Inoue: "High-performance PMASynRM with Ferrite Magnet for EV/HEV Applications", *Proc. of EPE'13*, pp. 1-9 (CD-ROM), 2013.

[13] A. Yamagiwa, K. Aota, Y. Sanga, H. Takabayashi, M. Natsumeda: "Demagnetization analysis of IPMSM using FEM", *Papers of Technical Meeting on Rotating Machinery, IEE Japan*, RM-03-41, pp. 43-48, 2003.

[14] K. Murata, M. Sanada, S. Morimoto, Y. Takeda: "Torque Performance and Magnet Arrangement for IPMSM Considering Demagnetizing Ability", *Proc. of ICEMS2006*, pp. 1-4 (CD-ROM), 2006.

The 2014 International Power Electronics Conference

Consideration of 10kW In-Wheel Type Axial-Gap Motor Using Ferrite Permanent Magnets

Kodai Sone, Masatsugu Takemoto, Satoshi Ogasawara
Graduate School of Information Science and Technology
Hokkaido University
Sapporo, Japan

Kenichi Takezaki, Wataru Hino
Dynax Corporation
Chitose, Japan

Abstract— In general, the in-wheel type permanent magnet synchronous motor (PMSM) for electric city commuters uses powerful rare earth permanent magnets (PMs). However, the employment of the rare earth PMs should be reduced due to high prices. Therefore, it is important to develop an in-wheel PMSM that does not use the rare earth PMs. We proposed 5kW in-wheel type axial gap motor that can generate high torque density and has features such as a coreless rotor structure with ferrite PMs and a reduction gearbox in the inner side of the stator. And we have presented details of experimental results on a prototype. So, in this paper, in order to achieve further higher power output, we considered a 10 kW in-wheel motor with the proposed structure. The motor with 10-kW output was examined by means of using 3D-FEM and experimental results in order to attain the further higher power as a rare earth free motor.

Keywords— *Permanent magnet synchronous motors, electric city commuters, axial gap motors, coreless rotor strucutre, ferrite permanent magnets.*

I. INTRODUCTION

Currently, electric vehicles (EVs) are expected as one solution to global warming and energy problems. However, their inability to travel long distances with a single charge and their high battery price may prevent the spread of EVs. Therefore, electric city commuters have attracted much attention because they only need to travel short distances and have a small battery size.

Because electric city commuters are one- or two-seated small electric vehicles, the interior space of electric city commuters is limited, so use of an in-wheel motor is desirable to make more effective use of the narrow interior. The in-wheel type permanent magnet synchronous motors (PMSMs) generally utilize powerful rare earth permanent magnets (PMs) in order to attain high-performance features such as high output, high efficiency, and small size [1]–[2]. However, the employment of the rare earth PMs should be reduced due to high prices.

Therefore, a low-cost in-wheel motor for electric city commuters, which uses ferrite PMs instead of rare earth ones, has been proposed by our research team. Residual magnetic flux density and coercive force of ferrite PMs are as low as 30% those of rare earth PMs, so decreased torque becomes a serious problem. Owing to solving this

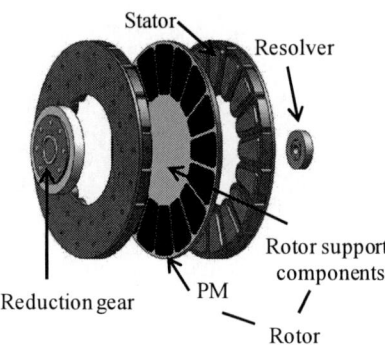

Fig. 1. Outline of the proposed motor.

problem, we adopt an axial gap configuration that can achieve high torque density [1]–[7] and a coreless rotor structure that is effective in maximizing magnetic torque. It is expectable to attain high efficiency by reducing iron loss because of low magnetic flux density reduced by this rotor structure. Moreover, in order to utilize the limited wheel space effectively, a reduction gearbox and a resolver is inserted into the inner side of the stator.

In previous papers [6]–[7], we introduced this motor structure with these features, and reported the operational characteristics of the proposed motor from the results of experimentation on a prototype in 5kW size. It was indicated that the proposed motor can satisfy the demands of an in-wheel motor for use in electric city commuters, despite using ferrite PMs.

In this paper, in order to achieve further higher power output, we considered a 10 kW size motor. As an initial design, a 16-pole/18-slot (16p18s) configuration similar to the 5 kW size was designed. To examine the fundamental characteristics of the 10 kW size motor, we also produced and tested a 10 kW size prototype. Although the 16p18s configuration was satisfactory performance in the 5 kW size, it is confirmed from experimental and analysis results that this configuration generates a large eddy-current losses within a rotor support component in a high output motor of 10 kW. Therefore, we research 16-pole/24-slot (16p24s) configuration which can suppress the eddy-current loss. This paper reports these studies using the experimental results and the analysis results of a three-dimensional (3D) finite element analysis (FEA), and shows that the

978-1-4799-2706-7/14 $31.00 © 2014 IEEE   3525

The 2014 International Power Electronics Conference

TABLE I
TARGET VALUE OF THE PROPOSED MOTOR

Maximum torque	73 Nm
Maximum power output	10 kW
Maximum phase voltage amplitude	Less than 110 V
Maximum current density	11.9 Arms/mm^2
Cogging torque	Less than 3% of average torque
Base speed	1300 rpm
Maximum speed	5000 rpm
Outer diameter	294 mm
Total motor axial length	Less than 66 mm
Slot fill factor of coil	Less than 55 %
Air gap	1.5 mm

(a) Front view

(b) Side view

(c) Rotor

Fig. 2. Photographs of the proposed motor prototype.

16p24s configuration can satisfy the properties demanded by the in-wheel motor of an electric city commuter.

II. INITIAL DESIGN OF 16P18S CONFIGURATION

A. Basic structure

Fig. 1 shows an overall view of the 10 kW proposed motor with 16p18s configuration similar to the 5 kW size. The proposed motor is an internal-rotor-external-stator type, in which one rotor is sandwiched by two stators. The rotor is structured by building only ferrite PMs into a non-magnetic stainless steel rotor support component in order to maximize magnetic torque. We call this a "coreless rotor structure." The coreless rotor structure has the following three main advantages: (I) it is possible to prevent irreversible demagnetization in the

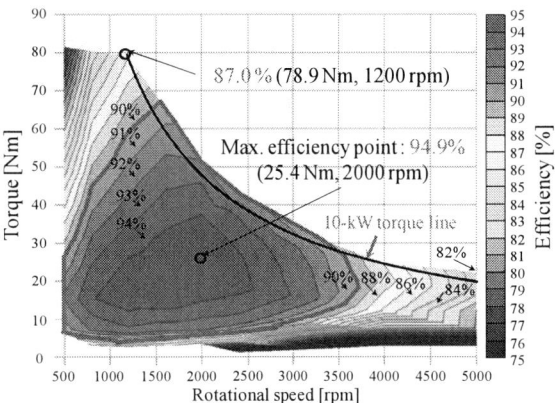

Fig.3. Efficiency map of the proposed motor with 16p18s configuration.

ferrite PMs; (II) when a rotor is eccentric to the axial direction, unbalanced electromagnetic force acting on the rotor can be reduced; and (III) iron loss can be suppressed because this rotor structure can lower magnetic flux density in the stator. Moreover, as compared to the rare earth PM, the iron loss generated in the ferrite PM is small in high-speed region. So, higher efficiency can be expected in the proposed motor. In addition, to meet the required torque characteristics of the electric city commuter, a reduction gearbox with a gear ratio of 5:1 is inserted inside the stator of one side as shown in Fig. 1. Conversely, a resolver is inserted into the stator of other side. The proposed motor thus utilizes the limited wheel space effectively.

Table I shows the target value of the 10 kW proposed motor. The dimensions of each part of the designed motor are optimized by repeatedly performing 3D FEA until the target values in Table I are satisfied. Fig. 2(a) and (b) shows the exterior of a prototype. Although the outer diameter of 10 kW motor is increased as compared to 5 kW motor, the axial length of 10 kW motor is reduced. Fig. 2(c) shows the rotor in a 10 kW size prototype of the proposed motor, demonstrating the coreless rotor structure. The ferrite PMs are fixed into the rotor support component with an adhesive.

B. Experimental result of 16p18s configuration

Fig. 3 shows an efficiency map of the 10 kW size prototype with the 16p18s configuration shown in Fig. 1 under the full operation range. A load test was performed under various current densities, current phases, and rotational speeds. This efficiency map was obtained by correcting bearing loss. The solid line in Fig. 3 denotes the torque curve, which realizes a constant power output of 10 kW at rotational speeds from 1200 to 5000 rpm. When performing the experiments, the DC voltage of the inverter was constant at the battery voltage of 200 V. The experimental results establish that the proposed motor can achieve a constant power output operation of 10 kW by means of field weakening control.

As shown in Fig. 3, the area surrounded by a red line

978-1-4799-2706-7/14 $31.00 © 2014 IEEE   3526

The 2014 International Power Electronics Conference

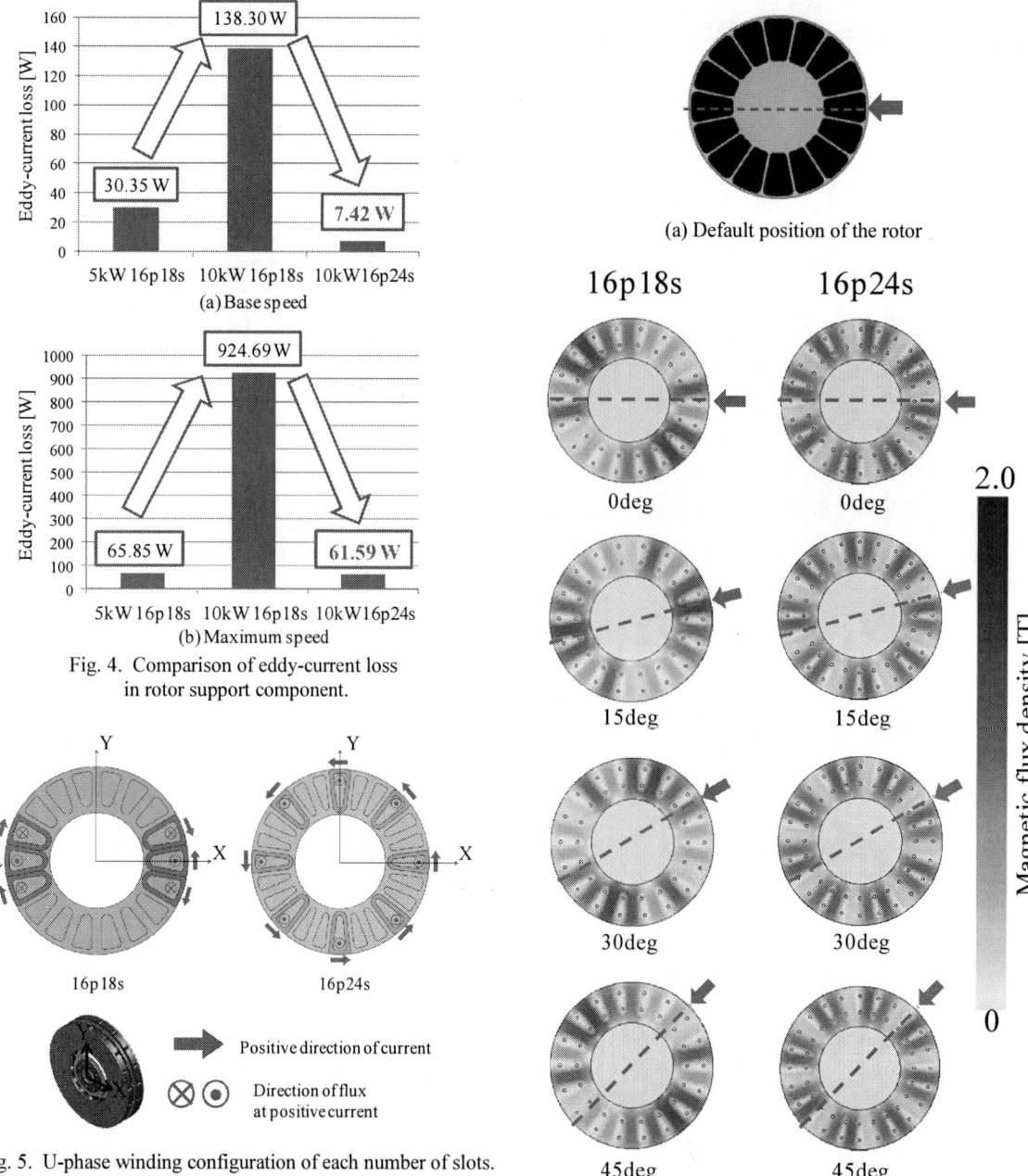

Fig. 4. Comparison of eddy-current loss in rotor support component.

(a) Default position of the rotor

(b) Magnetic flux density

Fig. 6. Comparison of magnetic flux density between 16p18s and 16p24s configuration.

Fig. 5. U-phase winding configuration of each number of slots.

indicates that the efficiency is 90% or more, and the maximum efficiency is 94.9% at 2000 rpm. The 10 kW size prototype has good characteristics in the region of up to about 3500rpm. However, the eddy-current loss in the rotor support component known in Fig. 1 and Fig. 2(c) is largely generated, and the efficiency is decreased in the high-torque region and high-speed region. It is essential to suppress the large eddy-current loss in the rotor support component in order to put the 10 kW motor in practical use. This problem was coped with in the following section.

III. COMPARISON OF THE EDDY-CURRENT LOSS

In this section, the eddy-current loss within the rotor support component is discussed with the analysis results

of 3D FEA. Fig. 4 shows a comparison of the eddy-current losses in the rotor support component at the maximum power output in each motor size at base speed and maximum speed, respectively. The eddy-current loss of 10 kW motor in the rotor support component increases significantly in comparison with 5kW motor. It is found that the eddy-current loss of 10 kW motor at base speed is 138.30 W, 4 times more than that of 5 kW motor and that the loss of 10 kW motor at maximum speed is 924.69 W, 14 times more than that of 5 kW motor. This caused the decline in the efficiency at the high torque output and the high-speed region as shown in Fig. 3. The eddy-

The 2014 International Power Electronics Conference

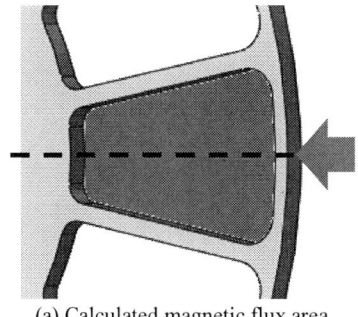

(a) Calculated magnetic flux area

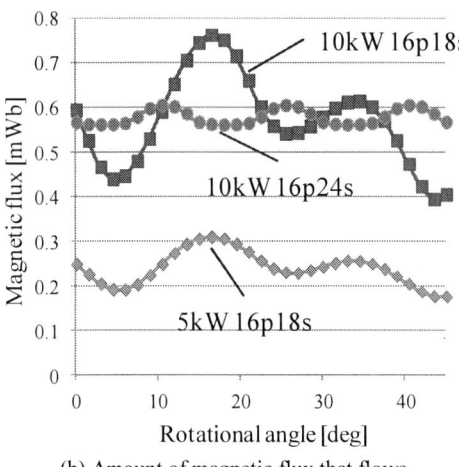

(b) Amount of magnetic flux that flows
into the specified area

Fig. 7. Changes of the amount of magnetic flux that flows into
the specified area versus rotational angle.

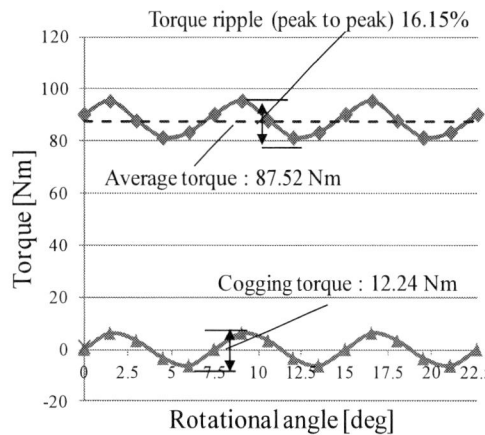

Fig.8. Changes in instantaneous and cogging torques with
rotational angle at a maximum current density of 11.9
Arms/mm^2 in 16p24s configuration.

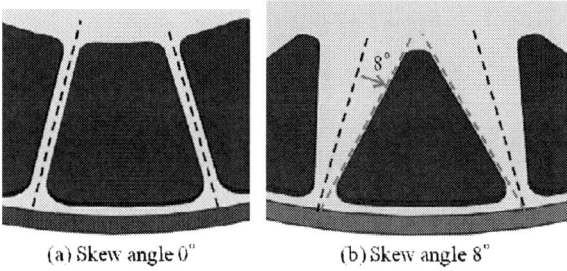

(a) Skew angle 0°   (b) Skew angle 8°

Fig. 9. Comparison of skew magnet.

current loss generated in the rotor support component
also becomes a serious problem in terms of heat
generation of the rotor which is hard to cooling.

Therefore, we have studied 16p24s configuration in
order to suppress the eddy-current loss. Fig.5 shows the
U-phase winding configurations of each number of slots.
As shown in Fig. 4, the eddy-current loss in the rotor
support component is reduced to 7.42 W at base speed
and 61.59 W at maximum speed considerably as
compared with the 16p18s configuration by the use of
16p24s configuration which is common for 16-pole
motor.

We consider about this cause. Fig. 6 shows the flux
density distributions of the stator core of each
configuration under the maximum current density at each
rotation angle. Fig. 6 (a) shows the initial position of the
rotor at the rotation angle of 0°. Fig. 6 (b) shows where
the initial position of a rotor comes by the arrow. As
shown Fig. 6, magnetic flux density distribution in the
16p24s configuration varies in synchronization with the
rotation of the rotor. The magnetic flux density
distribution and the initial position of the rotor marked
are the same positional relationship at any rotation angle.
This indicates that the change of the magnetic flux
density is small. On the other hand, the magnetic flux
density distribution in the 16p18s configuration does not

change in synchronization with the rotation of the rotor.
As one example, in the initial position of the rotor, the
magnetic flux density is high at 15° and low at 45°. It can
be seen that the magnetic flux density changes severely.

Fig. 7 represents clearly this phenomenon. Fig. 7
shows the interlinkage magnetic flux in a specified area
of the rotor support component under the maximum
current density at each rotation angle. Fig. 7 (b) shows
comparison of the amount of interlinkage magnetic flux
which flows into the area shown in Fig. 7 (a) in 5 kW
size of 16p18s configuration, 10 kW size of 16p18s
configuration, and 10 kW size of 16p24s configuration,
respectively. As shown in Fig. 7, the change in the
interlinkage magnetic flux is small in 10 kW size of
16p24s configuration. On the other hand, it is obvious
that the change of that is large in 10 kW size of 16p18s
configuration. A eddy current flows in the direction
which cancels this large interlinkage magnetic flux
change in the rotor support component, and the eddy-
current loss is increased.

IV. DESIGN FOR 16P24S CONFIGURATION

A. Magnet skewing

Fig. 8 shows the instantaneous torque waveform at the
maximum current density of 11.9 Arms/mm2 and the
cogging torque waveform under no torque load. It was
possible to reduce the eddy-current loss by adopting

The 2014 International Power Electronics Conference

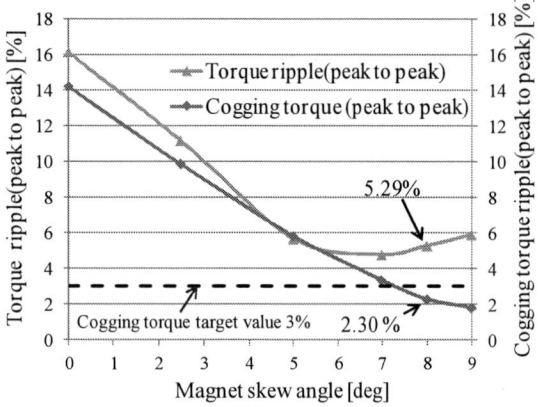

Fig. 10. Changes in torque ripple and cogging torques with magnet skew angle.

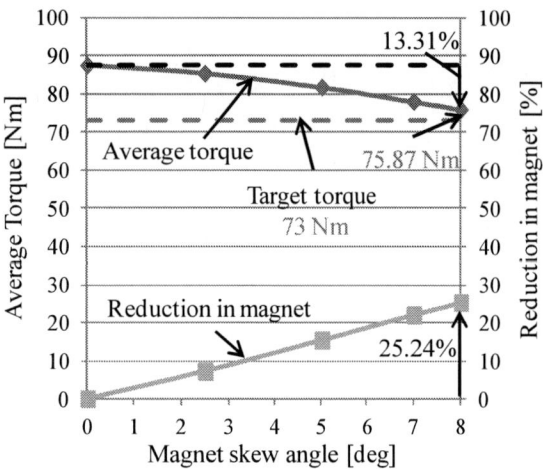

Fig. 11. Changes in average torque and reduction in magnet with magnet skew angle.

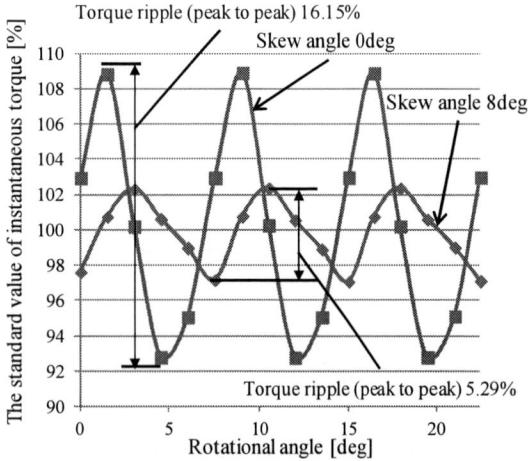

Fig. 12. Comparison of the standard value of instantaneous torque between skew angle of 0° and 8°.

16p24s configuration. However, both the torque ripple and the cogging torque has become a large value. Then, the magnet skewing is applied for torque ripple and the

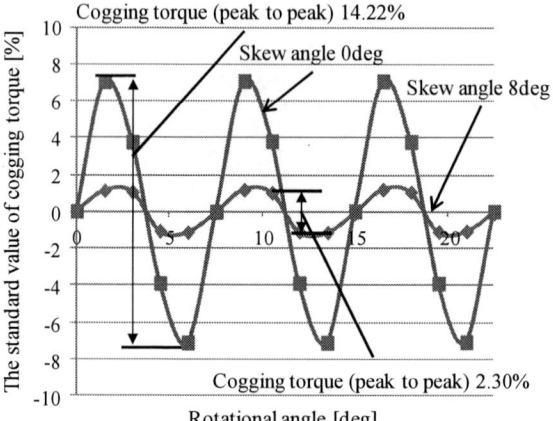

Fig. 13. Comparison of the standard value of cogging torque between skew 0° and 8°.

cogging torque reduction [8]–[9]. Fig. 9 shows rotor structure with magnet skew angle of 0° and 8°, respectively.

Fig. 10 shows the transition of the cogging torque and torque ripple versus magnet skew angle. Both the values are derived from the proportion of peak to peak to the maximum average torque in each skew angle. As the magnet skew angle increases, both the torque ripple and the cogging torque decreases. In the magnet skew angle of 8°, it has checked that the percentage of the cogging torque to the maximum average torque was 2.3% at less than the target value of 3.0%. When the magnet skew angle surpasses more than 7°, the torque ripple starts to slightly increase. However, the magnet skew angle of 8° was set to the optimal value in order to prioritize the target value of the cogging torque.

It should be noted that magnet skew causes the reduction in the magnet volume. The reduction in the amount of magnet linked to a decreased torque directly. Fig. 11 shows the transition of the rate of decrease in the magnet and the average torque versus the magnet skew angle. The magnet is reduced with increasing the skew angle, the average torque decreases. However, in the skew angle 8° to achieve the cogging torque of the target value, the proposed motor achieved 75.87 Nm in excess of the target value of 73 Nm. The rate of decrease in the average torque is 13.31% compared with the rate of decrease in magnet of 25.24%. It is below a magnet decrement. So, the average torque per unit of magnets is rather increasing.

Fig. 12 shows comparison of the standardized value of instantaneous torque between magnet skew angle of 0° and 8°. In Fig. 12, the maximum average torque of each magnet skew angle is standardized as 100% in order to compare simply. The torque ripple ratios of the magnet skew angle of 0° and 8° are 16.15% and 5.29%, respectively. It is known that the torque ripple can be reduced by the magnet skewing. Fig. 13 shows the ratio of the instantaneous cogging torque to the maximum average torque between magnet skew angle of 0° and 8°. The cogging torque also decreases from 14.22% to 2.30%

978-1-4799-2706-7/14 $31.00 © 2014 IEEE

The 2014 International Power Electronics Conference

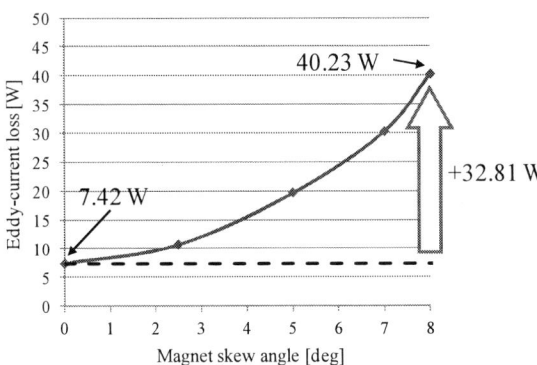

Fig. 14. Change of eddy-current loss versus magnet skew angle.

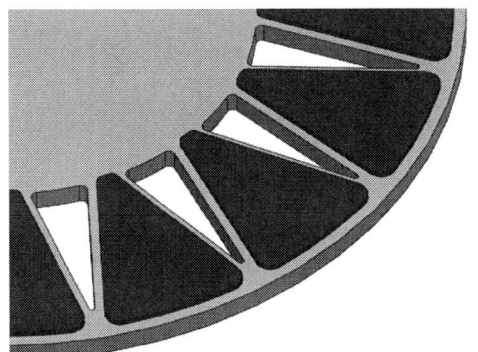

Fig. 15. Rotor after opening the hole.

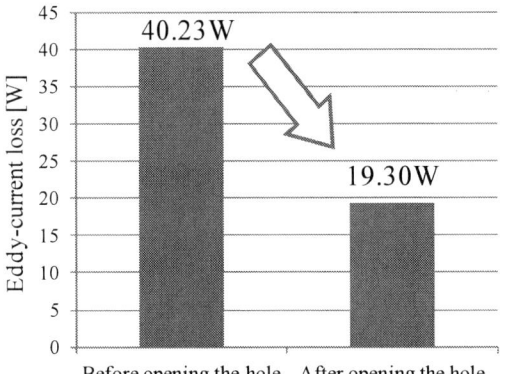

Fig. 16. Comparison of eddy-current loss before and after opening the hole.

meeting the required specification of 3% in the skew angle of 8°.

B. Countermeasure of the eddy-current loss

In the previous section, we have succeeded in suppressing the torque ripple and cogging torque which are a problem in 16p24s configuration by skewing the rotor magnets. However, the area of the rotor support component is increased as shown in Fig. 9, as the magnet skew angle is increased. This means that the portion where eddy-current loss occurs is increased. Fig. 14

shows the change of the eddy-current loss versus the magnet skew angle. The eddy-current loss increases to 40.23 W at the optimum skew angle of 8°. Then, we try to reduce the eddy-current loss by making a hole in the rotor support component as shown in Fig. 15. Fig. 16 shows the eddy-current loss in the rotor support component before and after opening the hole. By opening the hole, we succeeded in reducing the eddy-current loss from 40.23 W to 19.30 W. The decrement as the whole loss may be small, but considering the temperature rise of the rotor itself, which is difficult to cool, opening the hole has an instant effect because the losses in the rotor is reduced by approximately 50%.

It should be noted here that opening the hole in the rotor support component causes the fall of durability to stress. We therefore used SolidWorks 3D CAD software to investigate the durability to stress caused by rotating the rotor. The analysis is conducted by using the rotational speed of 6500 rpm which is 1.3-fold maximum rotational speed. A maximum von Mises stress is 199.50 MPa. The 18Cr-8Ni austenitic steel used for the rotor support component has the lowest yield strength at 206 MPa. The safety factor for the rotational speed is therefore considered sufficiently high at about 1.3.

V. CONCLUSIONS

This paper introduced a 10 kW in-wheel axial gap motor with coreless rotor structure using ferrite permanent magnets for an electric city commuter. Firstly, as an initial design, we investigated the 16p18s configuration. It was confirmed from experimental and analysis results that the 16p18s configuration causes a large eddy-current loss within a rotor in a high output motor of 10 kW. Nextly, we therefore researched the 16p24s configuration which can suppress the eddy-current loss by indicating the change in the flux linkage and the magnetic flux density distribution. The 16p24s configuration is superior in terms of reducing the eddy-current loss, but cogging torque and torque ripple are large. So, in the 16p24s configuration, the magnet skewing is applied for ripple reduction. Although the eddy-current loss increased with the increase in volume of the rotor support component by magnet skewing, it was coped with by opening the hole in the rotor support component.

REFERENCES

[1] Y.-P. Yang and D. S. Chuang, "Optimal design and control of a wheel motor for electric passenger cars", *IEEE Trans. Magn.*, vol. 43, no. 1, pp. 51-61, 2007.

[2] C. Versele, Z. De Greve, F. Vallee, R. Hanuise, O. Deblecker, M. Delhaye, and J. Lobry, "Analytical design of an axial flux permanent magnet in-wheel synchronous motor for electric vehicle", *Proc. of the 13th European Conference on Power Electronics and Applications* (EPE '09), 9 pages (CD-ROM), 2009.

[3] M. Aydin, S. Huang, and T. A. Lipo, "Torque Quality and Comparison of Internal and External Rotor Axial Flux Surface-Magnet Disc Machines", *IEEE Trans. Ind. Electronics*, vol. 53, No. 3, pp. 822-830, 2006.

[4] W. Fei, P. Luk, and K. Jinupun, "A new axial flux permanent magnet Segmented-Armature-Torus machine for in-wheel direct drive applications", *Proc. of the 2008 IEEE Power Electronics Specialists Conf.* (PESC 2008), pp.2197-2202, 2008.

[5] K. Rahman, N. Patel, T. Ward, J. Nagashima, F. Caricchi, and F. Crescimbini, "Application of direct drive wheel motor for fuel cell electric and hybrid electric vehicle propulsion system", *Conf. Record of the 2004 IEEE IAS Annual Meeting*, pp1420-1426, 2004.

[6] K. Sone, M. Takemoto, S. Ogasawara, K. Takezaki, and H. Akiyama, "A Ferrite PM In-Wheel Motor without Rare Earth Materials for Electric City Commuters", *IEEE Trans. Magn.*, vol. 48, no. 11, pp. 2961-2964, 2012.

[7] K. Sone, M. Takemoto, S. Ogasawara, K. Takezaki, and W. Hino, "Operation Characteristics of Ferrite Permanent Magnet In-Wheel Axial-Gap Motor with Coreless Rotor Structure for Electric City Commuters", *Proc. of the 2013 IEEE Energy Conversion Congress and Exposition* (ECCE 2013), pp. 3186-3193, 2013.

[8] F. Caricchi, F. Capponi, F. Crescimbini, and L. Solero, "Experimental study on reducing cogging torque and core power loss in axial-flux permanent magnet machines with slotted winding", *Conf. Record of the 2002 IEEE IAS Annual Meeting*, pp. 1295-1302, 2002.

[9] A. B. Letelier, D. A. Conzalez, J. A. Tapia, R. Wallace, and M. A.Valenzuela, "Cogging torque reduction in an axial flux PM machinevia stator slot displacement and skewing", *IEEE Trans. Ind. Appl.*, vol.43, no. 3, pp. 685-693, 2007.

Power Control Method for Multi-Parallel DC Distribution System Through the Equivalent Circuit Model

Seok-Jin. Hong, Soo-Cheol. Shin, Hee-Jun. Lee,
Chung-Yuen Won*

College of Information & Communication Engineering
SungKyunKwan University
Suwon, Republic of Korea
gujjim@naver.com

Taeck-Kie. Lee

Department of Electronic Engineering
HanKyong National University
Anseong, Republic of Korea

Abstract— **In this paper, power control method is proposed when power converters are configured in parallel for increasing power capacity. Each module and load is transformed by equivalent circuit model. And power is controlled through the equation. Power control method can be controlled power of each module and reduced circulate current. Proposed power control method is applied to power converter modules for DC distribution system and verified through simulation result and experiment.**

Keywords— *DC distribution, Circulating current, Parallel configuration, Parallel System, Power Control*

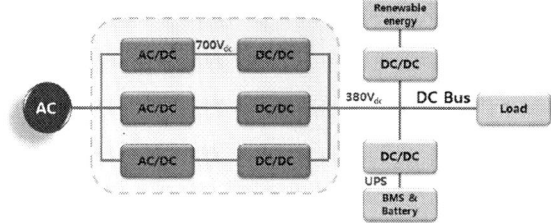

Fig. 1. Schematic of DC distribution system

I.   INTRODUCTION

Demand of DC power has been increased because supply of digital load has rapidly spread. But it has been normal that power generation facilities produce AC power directly by generator. So, to supply power to digital load, power conversion loss is increased by lots of power converting.

Also, as digital load has been increased, interest to renewable energy has been increased. This causes that DC power source which has DC output is increasing, too. But DC power must be conversed to AC power for DC power source making a connection to normal power or supplying power to load. That makes more reduction of low voltage utilization of renewable energy. Thus, DC distribution, which can supply power to digital load, and make an efficient linkage to renewable energy, is attended. [1]

Several countries studied about DC distribution early have already built a test bed applied to commercial buildings. Like this, study about DC distribution is being accelerated.

System for applying to commercial building requires large power capacity. Figure 1 shows configuration of proposed system. If DC distribution is applied to commercial building as shown in Figure 1, it can make easy linkage to ESS, and renewable energy and minimize loss by power conversion through integrating power conversion device. On this wise, the system that requires large power capacity like this, generally used modular power converter and parallel operation. But system

comprised of parallel, has many advantage more than single module such as radiant heat, standardization, , ease of maintenance, reliability, performance, volume reduction, etc. But it requires additional power control method. Power control methods that have been used are UOVD(Uniform Output Voltage Distribution) Control[2], Droop control[3], PI power control, etc. However like these methods require certain constraints(unconditionally equal output, demand of look–up table, defined certain constant) and these are sensitive to environmental changes, slow response. In particular, if the specification of voltage fluctuations is strict, the existing power control methods applying to system is more difficult.

Thus, in this paper, DC distribution system is comprised of the plurality of modules and power control method through the equivalent circuit model is proposed. If proposed power control method is applied, the system is able to have a high reliability and distribute power freely. Also, proposed power control method is reduced complexity of processer because it is using simple mathematical equation. Through simulation, we examined proposed validity of power control method.

II.   PROPOSED DC DISTRIBUTION SYSTEM

Fig.2 shows the schematic of DC distribution system that proposed control method is applied. In the configuration of Figure 1, it is schematic about AC/DC large power converter which is responsible the main power supply. In large power system, configuration of

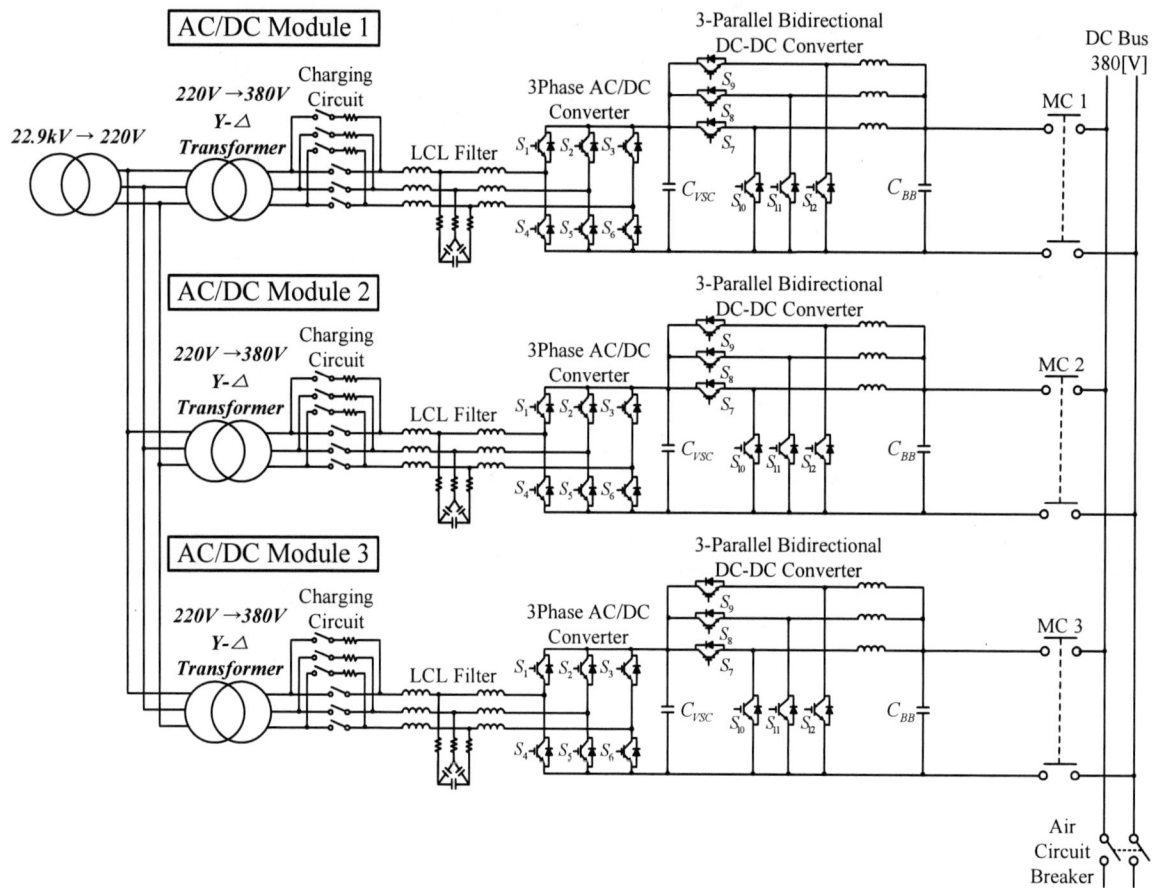

Fig. 2. Schematic of DC distribution system

modular power conversion device has a lot of advantages. So, entire system is configured in modular for easy maintenance, and power converter is configured in a module that considered radiant heat, standardization, maintenance, and volume size. And the system is configured same module of three in parallel to increase capacity of entire system. Maximum capacity of each module is 50[kW] and each module has bi-directional power flow.

A. System configuration

In Fig. 2, system of large capacity, such as operation of the modular power conversion devices in parallel is more efficient. Thus entire system is configured to modular power converter in parallel. So the system can be maintained easy via modularized power converter.

Commercial facilities in Korea generally use 3-phase 380[V_{rms}], 60[Hz] Y-Δ wiring transformer. Therefore, entire system is configured to take advantage of existing power facilities. It can make easy extension via detaching each module. Also DC bus voltage is 380[V_{dc}].[4] Thus, Input AC voltage is 380[V_{rms}], 60[Hz] and output voltage of module(DC bus voltage) is 380[V_{dc}].

When output DC capacitor of 3-phase AC/DC PWM converter is empty, overcurrent flows to the capacitor

when the initial start-up. In order to prevent this, circuit has charging time with 3τ.

And to reduce harmonics by switching of 3-phase AC/DC PWM converter, LCL filter is applied between 3-phase AC/DC PWM converter and system. [5][6]

Output DC-link voltage of 3-phase AC/DC PWM converter is controlled to 700[Vdc] considering inductance of system voltage and LCL filter.

And bi-directional DC/DC converter is connected in series with AC/DC PWM converter for stepping-down 700[Vdc] to 380[Vdc]. Bi-directional DC/DC converter is applied 3-parallel non-isolated Buck-Boost DC/DC converter. The current flowing to inductor can be smaller than generally used DC/DC converter because this topology can reduce the volume of inductor [7].

Configured system as Fig.1, more than 2 power converters connected in parallel control each DC voltage. But system configured as such, difference may be occurred to controlling voltage by measurement error of sensor, error of MCU gain value, line impedance, etc. when each components control voltage. This difference of voltage can occur circulating current between modules that share the AC grid. Especially, between modules that share the AC grid, larger circulating currents can be occurred. Circulating currents increases loss of the total system, drop system utilization, and worsen power quality because circulating currents is similar to reactive power. [7] Thus, we inhibited circulating current to control the power of parallel module by using (1). [8]

B. Control of Each Module

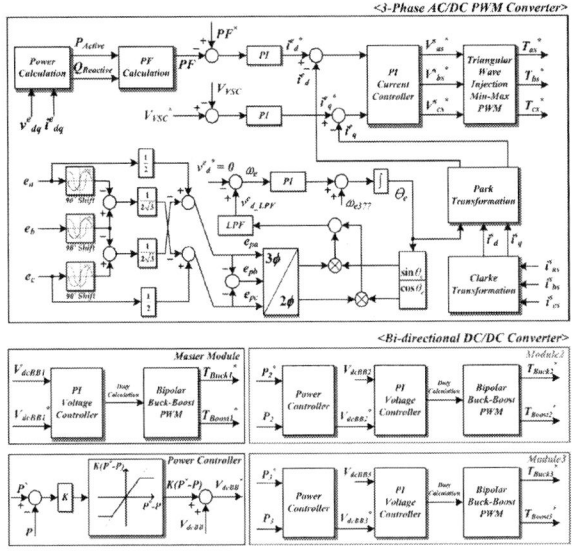

Fig.3. Control Blocks of Entire System

At the proposed DC distribution system, each module is controlled by control blocks of entire system shown as Fig. 3. The AC/DC PWM converter of each module controls voltage of DC link 700[Vdc] through vector control based on the grid voltage angle. Positive phase sequence component extraction algorithm has been applied for analyzing the unbalance phenomenon that can be occurred at grid voltage. Also PLL has been applied for real time tracking of voltage angle. Control block diagram of Fig.3 contains positive phase sequence component extraction algorithm, power factor control, PLL and SVPWM. In addition, basically bi-directional DC/DC converters(Buck-Boost Converter) of each module always control DC bus voltage steadily to 380[Vdc]. Voltage and power control is conducted for voltage control of DC/DC converter.

III. PROPOSED DC DISTRIBUTION SYSTEM

A. System configuration

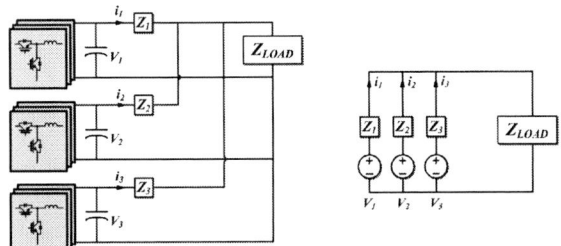

Fig.4. Equivalent Circuit of Multi-Parallel DC Distribution System

When equivalent circuit is converted considering line impedance of each module from entire circuit diagram in Fig.2, it can be expressed as Fig.4. By the equivalent circuit diagram of Fig.4, the voltage output from each module can be expressed as follows.

$$
\begin{aligned}
V_1 &= (Z_1 + Z_L)i_1 + Z_L i_2 + Z_L i_3 \\
V_2 &= (Z_2 + Z_L)i_2 + Z_L i_1 + Z_L i_3 \\
V_3 &= (Z_3 + Z_L)i_3 + Z_L i_1 + Z_L i_2
\end{aligned}
\tag{1}
$$

The output of performed voltage controller at each module is current reference. So, following equation (2) is mutual simultaneous equation of equation (1). And this is deployed output current from each module.

$$
\begin{aligned}
i_1 &= \frac{V_2 Z_L (Z_3 + Z_L) + V_3 Z_L (Z_2 + Z_L) - Z_L^2 (V_2 + V_3)}{Z_1 Z_2 (Z_3 + Z_L) + Z_3 Z_L (Z_1 + Z_2)} \\
i_2 &= \frac{Z_3 (V_2 - Z_L i_1) + Z_L (V_2 - V_3)}{Z_2 Z_3 + Z_2 Z_L + Z_3 Z_L} \\
i_3 &= \frac{Z_2 (V_3 - Z_L i_1) + Z_L (V_3 - V_2)}{Z_2 Z_3 + Z_2 Z_L + Z_3 Z_L}
\end{aligned}
\tag{2}
$$

By equation (2), we can find that output current of each module is decided by line impedance, load, output voltage of each module. Thus, output current can be controlled by controlling output voltage of each module, and controlling current in consistent DC-BUS voltage is same as controlling power. In practice, if voltage source of DC output is connected in parallel, large current flows in small voltage difference. Thus, large current can be controlled under almost consistent DC-BUS voltage in proposed system. Proposing system in this paper, AC/DC power conversion module 1 keeps voltage of DC-BUS consistently, and the rest AC/DC power conversion module 2, and 3 control the output by varying voltage slightly connected in parallel. As making slight voltage difference between each module from equation (2), output current can be controlled, so output power of each module can be controlled. Thus, only simple control method can control output power from each module without using of complex controller.

B. Power Control Using Power Reference Point Tracking Method

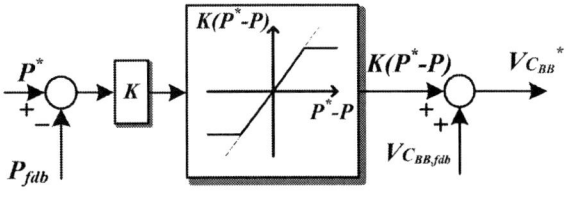

Fig.5. Control Blocks of Entire System

Fig.5. is a simple power controller. Power reference point tracking (PRPT) power controller follows power reference using error between power reference and actual power. Error between power reference and actual power can be limited under certain range via consistent constant. This value becomes output value of power controller added to actual voltage, and inputs to voltage controller as a reference value.

$$V_{dcBB}^* = V_{dcBB} + K(P^* - P) \qquad (3)$$

Output of power controller is decided using equation (3). Reference value of voltage controller is output by multiplying constant K by difference between reference power and actual power. This control method is similar to MPPT control method of photovoltaic system. If reference power is larger than actual outputting power, larger value than present voltage value will be selected as output of controller, if actual outputting power is larger than reference power, smaller value than present voltage value will be selected as output of controller, and the value will be used as input of voltage controller.

IV. SIMULATION

Table 1 shows the fundamental physical constant value that applied to simulation

TABLE I
FUNDAMENTAL PHYSICAL CONSTANTS

Symbol		Meaning	Value
V_{in}		Input Voltage	$380\ V_{rms}\ /\ 60\ Hz$
$V_{DC\ Link}$		DC Link Voltage	$700\ V_{dc}$
f_{sw}		Switching Frequency	$5\ kHz$
$V_{DC\ Bus}$		DC Bus Voltage	$380\ V_{dc}$
LCL Filter	L	Converter Side Inductor	$700\ \mu H$
	C_f	Filter Capacitor	$50\ \mu F$
	L_g	Grid Side Inductor	$120\ \mu H$
Z_{LOAD}		Total Load	$150\ kW$

Input voltage is AC 380[V_{rms}] / 60[Hz] and AC/DC PWM converter control DC link voltage to 700[V_{dc}]. This voltage is controlled 380[V_{dc}] DC bus voltage by 3-parellel bi-directional DC/DC converter. Basis of parameters of Table 1 is applied to the system of simulation using Power SIM. Total load that is applied to simulation is 50[kW] and simulation is conducted in some situation.

Figure 6 shows the result of simulation of entire system through the proposed power control method. Fig.0 shows power is controlled by current controlling under same voltage.

Figure 6-(a) shows the waveform of load that 10[kW] each step increasing to 150[kW]. Each module shares load gradationally to 50[kW]. Through the waveform, we could find each module can output current differently by power command, and supply same power 1/3 equally to load.

Figure 6-(b) shows the waveform of load that 50[kW]

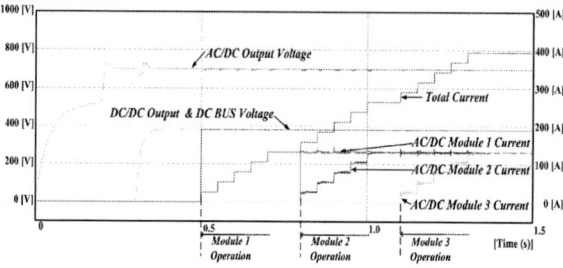

(a)   10[kW] Load is increased to step

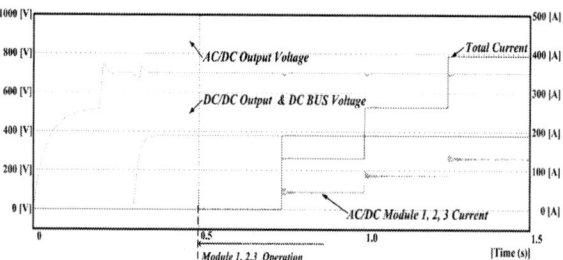

(b)   50[kW] Load is increased to step

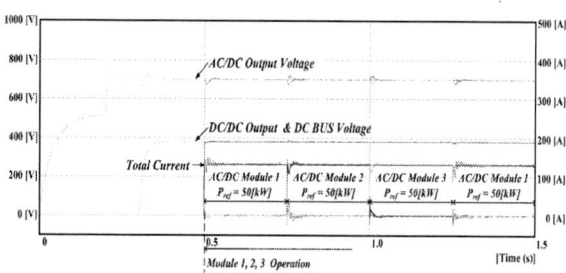

(c)   The waveform of power control shows 50[kW] Load

Fig 6. Simulation result of proposed system

each step increasing to 150[kW]. Each module supplies same power to load 1/3 equally.

Figure 6-(c) shows the waveform of load that 50[kW] alternating sharing. When each module has to be detached due to failure/repair etc., control as Fig.5-(c) is required.

When proposed control method is applied to system, stable power supply and free power control in DC distributed system are possible.

V. EXPERIMENTAL RESULT

Fig. 7 is hardware configuration of proposed DC distribution system. Each power conversion module locates inside of stack, and output of each module is connected to DC-BUS in parallel. DC-BUS and load are separated by ACB(Air Circuit Breaker). Each power converter controls using TMS320F28335, and master controller, controls totally, is additionally configured. Even rated capacity of each module is designed as 50[kW], test was conducted reducing rating of each power conversion module due to limit of load SET. Load SET, comprised 8 loads of 7.2[kW] in parallel, increases gradationally.

(a) Front

(b) Back & Right Side

Fig.7. Hardware Configuration

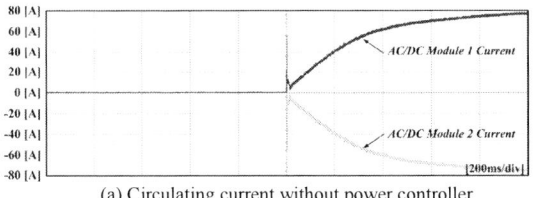

(a) Circulating current without power controller

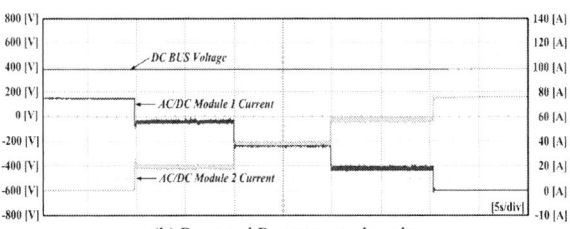

(b) Proposed Power control apply

Fig. 8. Experimental result of proposed power control method

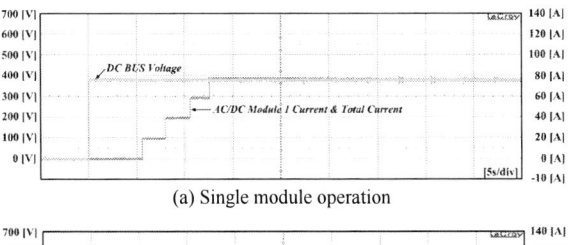

(a) Single module operation

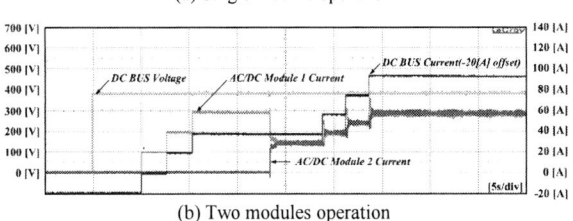

(b) Two modules operation

Fig. 9. Experimental result of single and parallel operation

Fig.8.-(a). is result of not applying proposing power control method. In case of not applying proposing power control method, we can verify generation of circulating current because of slight voltage difference between each module.

Fig.8.-(b) is result of applying proposing power control method. In case of applying proposing power control method, we can verify that outputting power keeps consistently following reference, and circulating current is not generated.

Fig.9. is result of single module operation and 2 modules operation in parallel. Proposed power control method is applied on both tests. We can verify that power control is conducted with minimizing circulating current as test result shown in Fig.6.-(b).

VI. CONCLUSIONS

In this paper, in case of connecting power converters in parallel for extension of system capacity, necessary simple power control method for circulating current and power controlling is proposed. Through the proposed power control method, DC distribution system is able to control power freely and hold high quality of distribution voltage having stable and low voltage fluctuation. The efficiency of power controller is verified via simulation and test, through this, reliability and expandability of large capacity parallel link system is expected.

ACKNOWLEDGMENT

This work was supported by the National Research Foundation of Korea(NRF) grant funded by the Korea government(MEST) (No. 2012T100100064)

This work was supported by the technology Innovation of the Korea Institute of Energy Technology Evaluation and planning(2011T100100025) grant funded by the Korea government ministry of Knowledge Economy(No. 2011T100100025)

REFERENCES

[1] Annabelle Pratt, Pavan Kurnar, Tomm V.Aldridge, "Evaluation of 400V DC Distribution in Telco and Data Centers to Improve Energy Efficiency", IEEE Telecommunications Energy Conference, pp.32-39, 2007.

[2] Siri, K, "Uniform Current/Voltage-Sharing for Intercon -nected DC-DC Converters", Aerospace Conference, 2007 IEEE, pp. 1 - 17, 2007.

[3] Guerrero, J. M, "Hierarchical Control of Droop-Controlled AC and DC Microgrids - A General Approach Toward Standardization", Industrial Electronics, IEEE Transac -tions on, Volume. 58, Issue. 1, pp. 158 - 172, 2011.

[4] Daniel Nilsson, Ambra Sannino, "Efficiency analysis of low- and medium- voltage dc distribution systems", IEEE PES, Vol.2, 2004, pp. 2315-2321

[5] Marco Liserre, Frede Blaabjerg, Steffan Hansen, "Design and Control of LCL-Filter-Based Three-Phase Active Rectifier", IEEE Transactions On Industry Applications, Vol. 41, No. 5, September/October. 2005.

[6] A.A. Rockhill, Marco Liserre, Remus Teodorescu, Pedro Rodriguez, "Grid-Filter Design for a Multimegawatt Medium-Voltage Voltage-Source Inverter", IEEE Transactions On Industrial Electronics, Vol.58, No. 4, April 2011.

[7] Chi-Hwan Choi, Soo-Cheol Shin, Hee-Jun Lee, Chul-Ho Jung, Hak-Seong Kim, Chung-Yuen Won, "Parallel System of

Bidirectional DC/DC Converter for Improvement of Transient Response in DC Distribution System for Building Applications", IEEE Vehicle Power and Propulsion Conference(VPPC), Oct. 2012.

[8] L. Asiminoaei, E. Aeloiza, P. N. Enjeti, and F. Blaabjerg, "Shunt active-power-filter topology based on parallel interleaved inverters," IEEE Trans.Ind.Electron., vol. 55, no. 3, pp. 1175-1189, Mar. 2008.

[9] Chul-Ho Jung, Soo-Cheol Shin, Hee-Jun Lee, Tae-Bok Jung, Chung-Yuen Won, Young-Real Kim, "Control Method for Reduction Circulating Current in Parallel Operation of DC Distribution System for Building Applications", IEEE Vehicle Power and Propulsion Conference(VPPC), Oct. 2012.

[10] J. W. Choi, S. K. Sul, "Fast Current Controller in Three-phase AC/DC Boost Converter Using d-q Axis Crosscoupling," *IEEE Trans. Power Electorn.*, Vol. 13, No. 1, pp.179-185, Jan. 1998.

A Communication-Less Distributed Voltage Control Strategy for a Multi-Bus AC Islanded Microgrid

Yanbo Wang, Yongdong Tan
Electrical Engineering School
Southwest Jiaotong University
Chengdu, China
ywa@et.aau.dk, ydtan@home.swjtu.edu.cn

Zhe Chen, Xiongfei Wang ,Yanjun Tian
Department of Energy Technology
Aalborg University
Aalborg, Denmark
zch@et.aau.dk, xwa@et.aau.dk, yti@et.aau.dk

Abstract— **This paper presents a communication-less distributed voltage control strategy for a multi-bus AC islanded microgrid. First, a Kalman Filter-based network voltage estimator is proposed to obtain voltage responses without communication links in the presence of load disturbances. Then, a voltage optimal controller using MPC (Model Predictive Control) are developed to implement voltage optimal control. The contributions of this paper are demonstrated: (1) The proposed voltage estimator can dynamically obtain network voltage responses just through local voltage and current associated with each DG (Distributed Generator) unit rather than relying on any communication facilities; (2) In contrast to the conventional PI-based voltage control method, the proposed voltage control strategy can implement not only offset-free control for single-bus, but also optimal control for multi-bus. Thus the flexibility and reliability is improved for islanded microgrids due to communication-less operation. The simulations and experimental results are presented to validate the proposed distributed voltage control strategy.**

Keywords— *distributed voltage control, voltage estimator, communication-less, islanded microgrid.*

I. INTRODUCTION

The expansion of distributed generations has brought new emerging power system concepts and technologies like microgrids[1] and virtual power plants. A microgrid is designed to integrate a cluster of distributed generators, loads and energy storing devices, which can be operated either in a grid-connection mode or in an islanded mode according to the power system condition [2]. During the islanded operation, generally speaking, the droop control methods [3-4] are adopted to automatically assign the generated active power and reactive power among DG (Distributed Generation) units in parallel operation without any communication facilities. However, the network voltages consequently drop [5] since droop controller reduces voltage to track the increased reactive power under uncertain load disturbances. The voltage deviations will result in a poor performance in voltage-dependent load regulation [6] at steady state.

Therefore, one of essential concerns in droop control-based islanded microgrids is to restore the voltage deviations in the presence of load disturbances. A large number of voltage control methods such as the

centralized voltage control [7] and decentralized voltage control [8] have been developed to deal with this issue.

A potential function method for centralized voltage control is proposed in [9], where dynamic set points are updated using communication system within the microgrids. A hierarchical control structure is proposed, where a PI-based secondary voltage controller is developed to perform voltage restoration [10]. It is obviously noted that, in these cases, the voltage control strategies for islanded microgrids, to large extent, rely on the communication links to acquire the network voltage information and to send the control commands. To improve flexibility and reliability for islanded operations, the distributed voltage control schemes have also attracted many attentions recently. A distributed secondary voltage control method based on distributed cooperative control of multi-agent systems [8] is developed, where the one-way communication links are required to exchange information among neighboring agents. A second control layer, making up voltage deviation caused by the droop controller, is presented in [11]. A distributed secondary control scheme performing voltage restoration and reactive power sharing is proposed in [12].

To sum up, the conventional voltage control methods are subject to the following problems: (1).The critical communication links are necessary to obtain voltage responses and send control commands, which results in an expensive and low reliable microgrid. In particular, when various DG units and loads are located far away from each other, the complicated communication links are no longer attractive for such voltage control techniques. (2) The conventional PI-based voltage control methods only can perform voltage control for the single bus. In other words, they are difficult to implement the global voltage optimal for a multi-bus islanded microgrid. Hence, it would be significant to avoid communication facilities for improving the flexibility and reliability of voltage control system.

To overcome the aforementioned challenges, this paper presents a communication-less distributed voltage control strategy for an islanded microgrid that is dominated by multiple inverter-interfaced DG units. The term "communication-less distributed" means that the voltage control is implemented in each DG unit without the communication links. First, a Kalman Filter-based network voltage estimator is employed to obtain voltage responses. Furthermore, a dynamic voltage optimal controller using MPC (Model Predictive Control) is developed. The contribution of this paper is demonstrated as follows: (1) The proposed voltage estimator can dynamically obtain network voltages just through local voltage and current associated with each DG unit rather than the conventional communication facilities; (2) In contrast to the conventional PI-based voltage control methods, the proposed voltage control strategy can implement not only offset-free control for the specified single bus, but also optimal control for multi-bus.

II. THE DRAWBACKS OF THE CONVENTIONAL VOLTAGE CONTROL STRATEGY

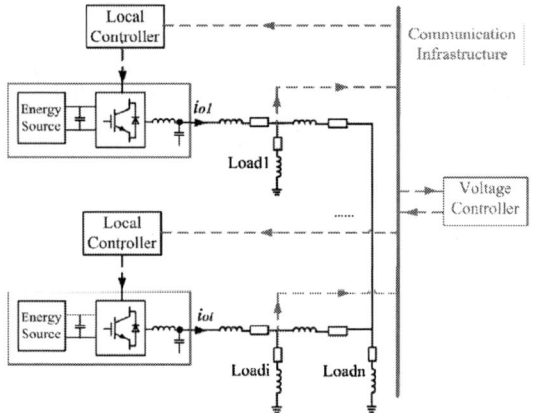

Fig.1.The centralized voltage control method.

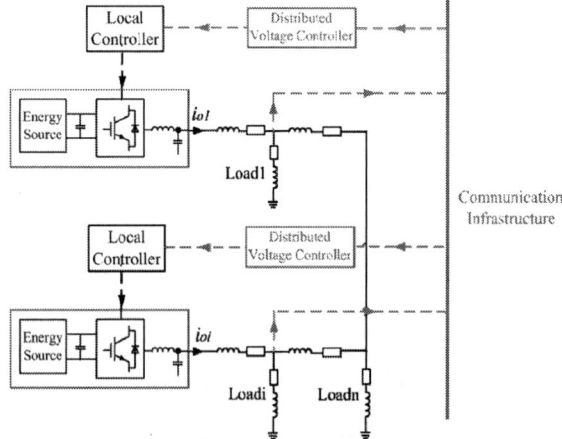

Fig.2.The distributed voltage control method.

To perform power sharing without communication links, the droop controllers [3-4] will reduce reference voltage to track reactive power in the presence of load disturbances. Hence, network voltages will also drop due to inherent droop effect. To compensate for the voltage deviations, a few voltage control approaches are adopted to deal with the issue.

In this section, the drawbacks of conventional voltage control methods for an islanded microgrid are reviewed, including centralized voltage control [7],[13] and distributed voltage control [8],[12],[14].

Fig.1 depicts the centralized voltage control methods. The communication facilities are developed to obtain system voltage responses for supporting the operation of the centralized voltage controller. Also, the voltage control commands are sent to local controller by the communication links. Fig.2 illustrates the distributed voltage control approaches. Compared with the centralized voltage control approaches, distributed voltage controllers are able to perform voltage restoration quickly and locally. However, the communication facilities are indispensable to support system operation for either the centralized voltage controller or the distributed voltage controller. Once communication faults or data loss happen, these control approaches will be out of work.

Therefore, this paper presents a communication-less voltage controller, which is able to perform both offset-free control for single-bus and optimal control for multi-bus without communication links.

III. THE PROPOSED COMMUNICATION-LESS DISTRIBUTED VOLTAGE CONTROL STRATEGY

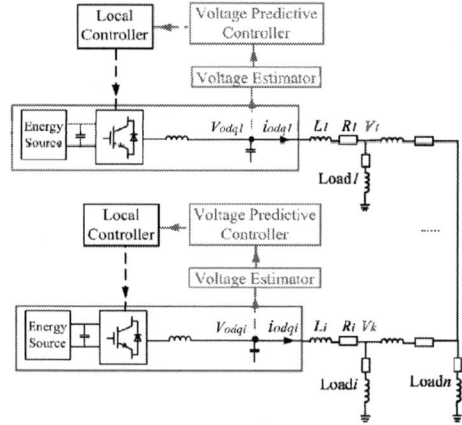

Fig.3.The proposed communication-less distributed voltage control strategy.

As depicted in Fig.3, the proposed communication-less voltage control strategy consists of voltage estimator and voltage predictive controller, respectively. The former is developed to estimate dynamically network voltages just through voltage and current of each DG unit, and the latter is responsible for computing voltage control commands according to the estimated information. Ultimately, the voltage control commands are carried out through local droop controller.

A. The Small Signal Model of An Islanded Microgrid

In fact, the proposed voltage estimator is based on the small signal model to evaluate voltage dynamics, which is able to reconstruct voltage responses just through local voltage and current associated with each DG unit. In this section, the small signal model of an islanded microgrid is first developed to produce voltage responses in the presence of load disturbances. The previous small signal models have been proposed in [15]-[16]. A small signal model with consideration of network voltage responses and loads characteristic is also derived in our previous work [17]. The instantaneous power of ith DG unit [16] can be represented in d-q rotating frame as (1) and (2):

$$p_i = V_{odi}i_{odi} + V_{oqi}i_{oqi} \qquad (1)$$

$$q_i = V_{odi}i_{oqi} - V_{oqi}i_{odi} \qquad (2)$$

where V_{odqi}, i_{odqi} are output voltage and current of ith DG unit on individual d-q frame. p_i, q_i are instantaneous active and reactive power. And the average active power and reactive power are formulated from instantaneous powers passing one-order low-pass filters as (3) and (4):

$$P_i = \frac{\omega_c}{s + \omega_c} p_i \qquad (3)$$

$$Q_i = \frac{\omega_c}{s + \omega_c} q_i \qquad (4)$$

Then, the conventional active power-frequency (P-f) and reactive power-voltage (Q-V) droop control method [16] can be represented as (5) and (6), respectively.

$$\omega_i = \omega^* - m_{pi}P_i \qquad (5)$$

$$V_{odi} = V^* - n_{qi}Q_i \qquad (6)$$

where P_i, Q_i are average active and reactive power; m_{pi}, n_{qi} are droop coefficient of active power and reactive power; ω_i is rotating angle frequency of ith DG unit;

Further, the Q-V droop control method with consideration of secondary control can be rewritten as (7).

$$V_{odi} - u_{ci} = V^* - n_{qi}Q_i \qquad (7)$$

The currents of DG unit [16] can be represented as:

$$\frac{di_{odi}}{dt} = -\frac{R_i}{L_i}i_{odi} - \omega i_{oqi} + \frac{1}{L_i}V_{odi} - \frac{1}{L_i}V_{dk} \qquad (8)$$

$$\frac{di_{oqi}}{dt} = -\frac{R_i}{L_i}i_{oqi} + \omega i_{odi} + \frac{1}{L_i}V_{oqi} - \frac{1}{L_i}V_{qk} \qquad (9)$$

The overall small signal dynamics of each DG unit can be modelled combing and linearizing (1)-(9) as follows:

$$\Delta \dot{x}_{invi} = A_{invi}\Delta x_{invi} + B_{invi}\Delta V_k + B_{invc}\Delta u_{ci} \qquad (10)$$

where $\Delta x_{invi} = [\Delta\delta_i, \Delta P_i, \Delta Q_i, \Delta V_{odi}, \Delta i_{odi}, \Delta i_{oqi}]^T$, $\Delta V_k = [\Delta V_{dk} \quad \Delta V_{qk}]^T$ is bus voltages. Δu_{ci} is voltage control input. A_{invi}, B_{invi} B_{invc} are parameters matrixes of each DG unit.

Similarly, the small signal dynamics of loads and network also can be modelled in a state space form. The detail modelling procedures of DG units, load dynamics and network dynamics can be referred in [15]-[17]. Finally, the overall model including inverters, loads and network dynamics and disturbances can be represented as

$$\Delta \dot{x} = A\Delta x + B\Delta i_{dis} + B_c\Delta u_c \qquad (11)$$

$$\Delta y = C\Delta x + D\Delta i_{dis}$$

where Δx is overall state vector of the whole microgrid, $\Delta x = [\Delta x_{inv1} \cdots \Delta x_{invi}, \Delta i_{line1} \cdots \Delta i_{linei}, \Delta i_{load1} \cdots \Delta i_{loadi}], \Delta i_{dis}$ are unmeasured load disturbances [17]. Δu_c are voltage control inputs of all the DG units. A, B, B_c, C, D are system parameters matrixes.

An essential step for the proposed voltage controller is determining discrete time model. The discrete time model can be obtained from the small signal model aforementioned as (12)

$$\Delta x(k+1) = A_d\Delta x(k) + B_d\Delta i_{dis}(k) + B_{cd}\Delta u_c(k) \qquad (12)$$

$$\Delta y(k) = C_d\Delta x(k) + D_d\Delta i_{dis}(k) \qquad (13)$$

The parameters matrixes A, B, B_c, C, D are obtained though Euler discretization. To permit a simpler description, sign Δ in increment function is omitted in next contents.

B. The Network Voltage Estimator.

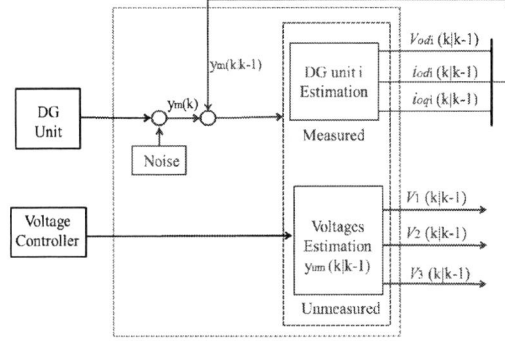

Fig.4. The operation principle of the voltage estimator

To improve flexibility and reliability of islanded operations, a Kalman Filter-based network voltage estimator is first proposed to obtain voltage responses rather than relying on the conventional communication links. The discrete time model is adopted to generate states responses and support voltage estimation. Even though the small signal model is available, the model mismatch resulting from uncertain disturbances is inevitable in real plant. To reject model mismatch and improve system robustness, disturbance models are augmented to original plant model according to the internal model principle [18]. In addition, the measurement noises are also disturbance sources for voltages estimation. To imitate sensor measurement noise, the noise dynamics should be considered in voltages estimation process. Thus, an overall system states combining full states, disturbance states and measurement noises can be represented as (14):

$$x_T(k+1) = A_ox_T(k) + B_ou_c(k) + W_ow_T(k) \qquad (14)$$

Meanwhile, the measured output can be shown as (15):

$$y_m(k) = C_ox_T(k) + D_ow_T(k) \qquad (15)$$

where $x_T = [x \quad x_{dis} \quad x_m]^T$ is augmented states, including system states, disturbance states and noises states. A_o, B_o C_o, D_o, W_o are system parameters matrixes. $w_T(k)$ is disturbance and noise vector.

The principle of network voltage estimator is depicted as Fig.4. The inputs of the proposed estimator are local state vector $y_m(k)$ as well as voltage control commands, where all the states can be updated and revised continuously via these measured information. Also, the outputs of it are estimated states vector $y_m(k|k-1) = [V_{odi}(k|k-1), i_{odi}(k|k-1), i_{oqi}(k|k-1)]^T$ and the estimated voltages vector.

In fact, the proposed estimator is a steady-state Kalman Filter which implements an estimate for the augmented states and unknown network voltages by measured local states. With assumption that (C_o, A_o) is detectable for overall model (14)-(15), full state estimation equation could be given by

$$x_T(k|k) = x_T(k|k-1) + K(y_m(k) - y_m(k|k-1)) \quad (16)$$

K is the Kalman gain [19], which is the solution of Ricatti matrix equation. The measured outputs of the estimator are the voltage and current of each DG unit while unmeasured outputs are network voltages. Furthermore, the estimated state update and output update equation are given from (14) and (15) as:

$$x_T(k+1|k) = A_o x_T(k|k) + B_o u_c(k) \quad (17)$$

$$y_m(k|k-1) = C_o x_T(k|k-1) \quad (18)$$

These estimated states are updated and revised continuously via update of measured voltage and current of each DG unit. Then, state estimation equation can be rewritten combing (16)-(18) as:

$$x_T(k+1|k) = A_T x_T(k|k-1) + B_T y_m(k) + B_o u_c(k) \quad (19)$$

where $A_T = A_o - A_o * K * C_o, B_T = A_o * K$.

Note that the output equation of the proposed estimator consists of two parts as (20) and (21), including the local voltage and current (measured and estimated) and network voltages (unmeasured and estimated)

$$y_m(k|k-1) = C_o x_T(k|k-1) \quad (20)$$

$$y_{um}(k|k-1) = C_{oum} x_T(k|k-1) \quad (21)$$

where C_o, C_{oum} is the measured and unmeasured output matrix of the estimator. The further details about the estimator principle also can be referred in our previous work [17].

C. The Dynamic Voltage Controller.

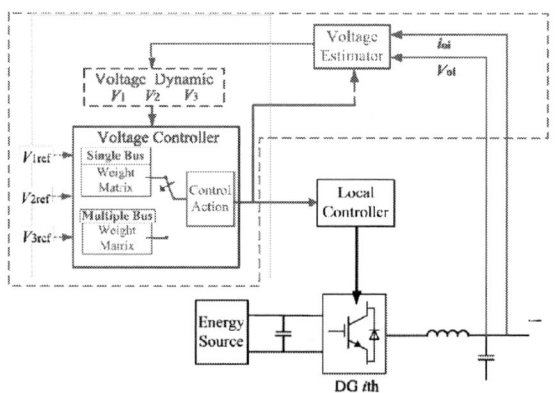

Fig.5 Block diagram of the proposed dynamic voltage controller.

Once network voltage responses are obtained from the proposed estimator, the voltage controller will compute optimal control commands according to the estimated information. In practice, the desired behavior of the voltage controller is obtained by computing an optimal cost function that minimizes voltage error. When computing is finished, voltage controller sends control commands to local controller and operates with the control commands until next sampling update.

Periodically, the voltage controller obtains new voltages estimation because of measurement feedback and revises its otherwise control plan. Then, voltage control commands are provided to compensate for errors between estimated voltages and references. And the control commands can be computed via solving an optimization cost function, which is formulated as (13)

$$\min_{\Delta u(k|k),\cdots,\Delta u(m-1+k|k),\varepsilon} J = \left\{ \sum_{i=0}^{p-1} \left((\sum_{j=0}^{n_b} w_b V_{bj}(k+i+1|k) - V_{bjref})^2 + \sum_{j=1}^{n_c} \left| w_c \Delta u_{cj}(k+i|k) \right|^2 + \rho \varepsilon^2 \right) \right\}$$

(13)

Subject to

$$\Delta u_{cj\min}(i) - \varepsilon V_j^{\Delta u_c}{}_{\min}(i) \le \Delta u_{cj}(k+i|k) \le \Delta u_{cj\max}(i) - \varepsilon V_j^{\Delta u_c}{}_{\max}(i)$$

In the cost function, ε is slack variable for softening constrains, w_b, w_c are weights for network voltage errors and control increments; Δu_{cj} is control increment; $\Delta u_{cj,\min}, \Delta u_{cj,\max}$ are low/upper bounds of control increments; p is prediction horizon; m is control horizon; n_b is number of bus, n_c is number of voltage control inputs; $V_j(k+i+1|k)$ denotes the voltage estimations for time $k+i+1$ based on the measured information available at time k; V_{jref} are reference voltages, which are set to 0 (initial equilibrium state defined in origin). Finally, the voltage control commands are given to local controller at the next instant. The process repeats independently by the proposed voltage controller of each DG unit.

(1) The Voltage Offset-free Control for Single Bus.

As is known, it's impossible to hold all the bus voltages at set points due to the inherent circuit configuration. When there are several voltages to be controlled, it is necessary to set priority so that controller can hold the most important voltage at its set point, allowing others float within an accepted range. To implement offset-free control for a specified single bus, the cost function is defined to minimize the voltage error. Therefore, weight for the specified bus should be far larger than that of others. In the case analyzed later, the control objective is to hold bus3 voltage at its point.

(2) The Voltage Optimal Control for Multi-Bus.

Even though it is impossible to hold all the voltages at set points, optimal control for multi-bus still can be performed. It means that all the voltages to be controlled have same priorities so that controller can force all the voltages go back to original state. Of course, there indeed exist steady-state errors at several buses. For the control objective, voltage weights are set to the same value. As shown in Fig5, the voltage controller can be switched between the two control objectives.

For the proposed voltage controller, another important concern is weight penalty for voltage control increment in cost function, which is used to avoid reactive power fluctuation due to aggressive update of control commands. The relationship between the voltage control commands and reactive power dynamic has been well established as (7). Note that the weight coefficients have a dramatic influence on the closed-loop performance, which will be analyzed in details later. Besides, another important parameter in the cost function is the prediction horizon *p*. In the implementation of simulations and experiments, it has been chosen that p=10. Also, it is the control horizon that is an essential parameter associated with control commands but not occur in cost function. In the implementation, the control horizon has been chosen that m=2 in the proposed controller, which means control inputs are executed during the time span from k to k+2, where k is the sampling instant.

The communication-less distributed voltage control strategy had an attractive advantage that the local voltage controller of each DG unit is completely independent without communication links, thus it provides flexibility and reliability due to communication-less operation.

D. The Closed-loop Dynamic Performance Analysis.

To further validate the relationship between weight coefficients in controller and dynamic performance, the closed-loop dynamic performance is analyzed in details. In the case analyzed, the closed-loop dynamic performance of the voltage controller is investigated by checking the positions of closed-loop poles when adjusting weights in proposed cost function.

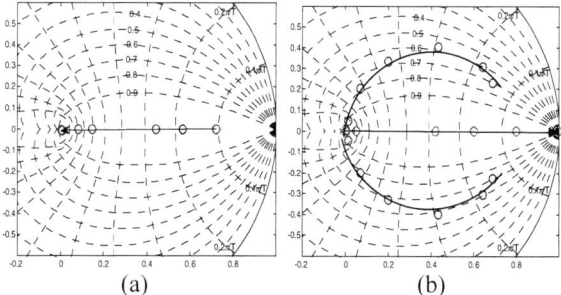

(a)   (b)

Fig.6.The poles map for weight variation of DG1 controller. (a) The bus3 voltage weight w1= 50, 80, 150, 200, 500, 800, 1000. (b) The control increment weight w2= 0.1, 1, 5, 10, 20, 50, 200.

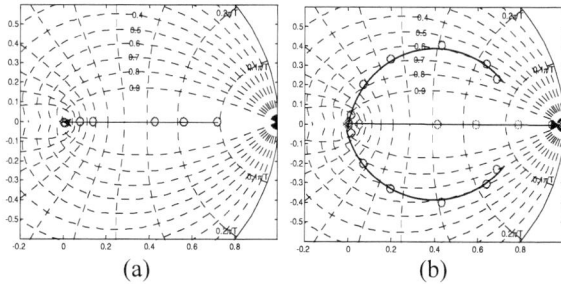

(a)   (b)

Fig.7. The poles map for weight variation of DG2 controller. (a) The bus3 voltage weight w3= 50, 80, 150, 200, 500, 800, 1000. (b) The control increment weight w4= 0.1, 1, 5, 10, 20, 50, 200.

As depicted in Fig.6, the closed-loop poles are plotted when modifying bus3 voltage weight of DG1 controller in the proposed cost function. Note that 22 poles can appear in pole map since the whole system has 22 orders, but just a real pole is sensitive highly to the weight variation, where it moves towards origin as bus3 voltage weight increases. Also, dynamic performance for modifying voltage increment weight of DG1 controller is illustrated in Fig.6.(b), where one real pole and one complex conjugate pole pair is sensitive for the weight

variation. With the increase of voltage control increment weight, the real pole and conjugate pole pair is driven to move from origin towards on inside unit cycle, slowing down the speed response. Similarly, as illustrated in Fig.7, dynamic variation of a real closed-loop pole is shown when bus3 voltage weight of DG2 controller is increased. It can be observed that the real pole moves towards origin, thus system had a much faster dynamic response. Further, a real pole, together with a conjugate pole pair varies from original point towards one inside unit cycle, shown in Fig.7.(b).

The analysis conclusions can be drawn that (1) A real pole is sensitive highly to variation of bus3 voltage weight. With the increase of it, dynamic response can be improved; (2) A real pole, along with a conjugate pole pair, is sensitive to variation of voltage control increment. The less aggressive voltage control increment, the slower dynamic response.

IV. SIMULATION VERIFICATION

The proposed dynamic voltage control strategy has been verified in MATLAB/Simulink simulations and experiments for a three phase 50 Hz prototype islanded microgrid. As shown in Fig.8, the system setup is composed of two inverters and three loads. In order to verify effectiveness of the proposed network voltage estimator and dynamic voltage controller, two cases for different control objectives are conducted repectively.

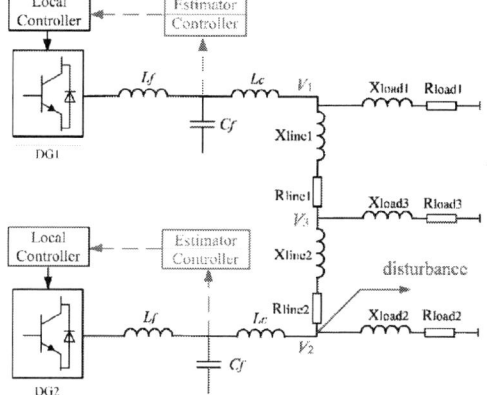

Fig.8.The simulation and experiment setup

TABLE I
PARAMETERS FOR SIMULATION AND EXPERIMENT

Parameters	Value	Parameters	Value
inverter rate	10kVA	Lc	1.8mH
fs	10k	Rline1+jXline1	0.2+j0.628
Lf	1.5mH	Rline2+jXline2	0.2+j0.628
Cf	25μF	w1/w3	1000
mp1/mp2	2.5e-5/1e-4	w2/w4	0.1
nq1/nq2	1e-3/1e-3		

A. The Network Voltage Estimation.

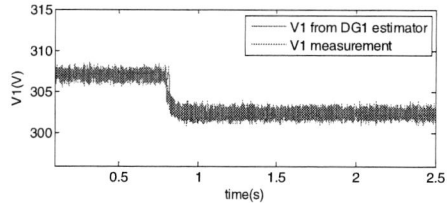

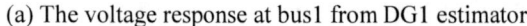
(a) The voltage response at bus1 from DG1 estimator.

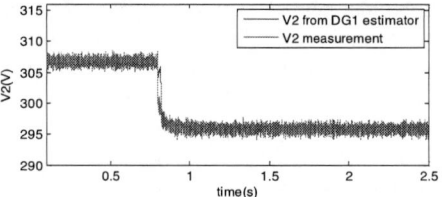
(b) The voltage response at bus2 from DG1 estimator.

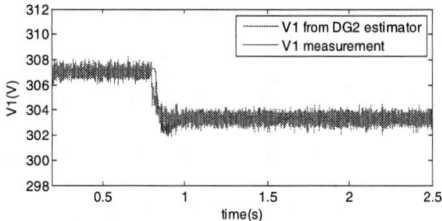
(c) The voltage response at bus1 from DG2 estimator.

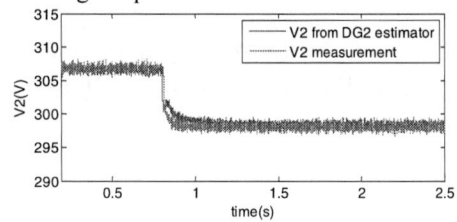
(d) The voltage response at bus2 from DG2 estimator.

Fig.9 The simulation results for voltage responses estimation

To validate the voltage responses from the proposed estimator, a RL disturbance load (L=50mH, R=10 Ω) is exerted at bus2.

Fig.9.(a)-(b) demonstrates voltage estimation results from DG1 voltage estimator, where the voltage responses under load disturbance are obtained from proposed estimator (blue curves) and simulation measurement (green curves) respectively. Similarly, Fig.9.(c)-(d) illustrates the network voltage dynamics from DG2 estimator. It can be observed that voltages decrease in the presence of the load disturbance due to inherent droop effect. Also, the proposed estimators can track network voltage responses effectively at steady state. Thus, the proposed estimators can provide an effective solution to obtain network voltage responses rather than relying on the conventional communication links.

B. The Voltage Control for Single Bus.

In the case analyzed, the unique control objective for the proposed voltage controller is to hold bus3 voltage at steady-state set point. Thus, weight on bus 3 is set to 1 and others are set to 0.01, which means bus 3 has a top priority compared with others. The same disturbance load is exerted at bus2. Fig.10 illustrates the voltage responses at the different buses. It can be observed that network voltages drop (blue curves) due to inherent droop effect in the presence of load disturbance. When the proposed voltage controller is activated, the bus3 voltage is brought to original state as shown in Fig.10.(c). Instead, bus voltages at bus1 and bus2 appear the steady-state deviations due to less weights (green curves).

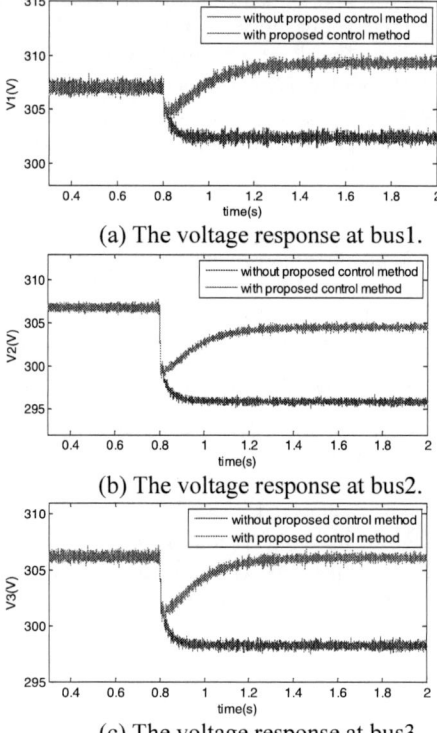
(a) The voltage response at bus1.

(b) The voltage response at bus2.

(c) The voltage response at bus3.

Fig.10. The simulation results of voltage responses with and without the proposed control method for single-bus.

C. The Voltage Control for Multi-Bus.

To validate the proposed voltage control approach, another case about voltage optimal for multi-bus is conducted, where the voltage control objective is to drive all the buses return to their set points in the presence of load disturbance. It's impossible to hold all the bus voltages at set points accurately due to inherent circuit configuration. Thus, a little steady-state errors are permitted for different buses. Therefore, all the voltage weight coefficients in the controller are set to same value 1, which means each bus has the same priority.

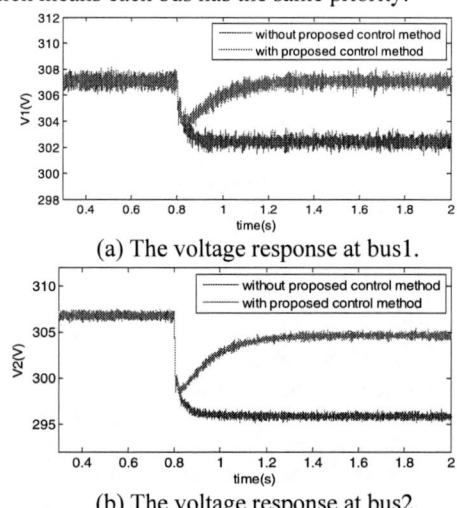
(a) The voltage response at bus1.

(b) The voltage response at bus2.

The 2014 International Power Electronics Conference

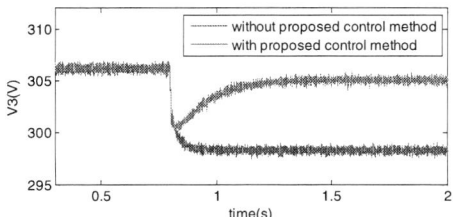

(c) The voltage response at bus3.

Fig.11 The simulation results of voltage responses with and without the proposed control method for multi-bus.

Fig.11 depicts the simulation results of the proposed controller for multi-bus. It can be seen that the voltages at all the buses are forced to go back to their original states when the proposed voltage optimal controller is activated. Of course, it indeed exists little steady-state errors because of the inherent circuit configuration.

V.   EXPERIMENTS VERIFICATION

To further verify effectiveness of the proposed voltage control scheme, the accompanying experiments are conducted. The whole platform of the islanded microgrid is controlled by dSPACE 1006.

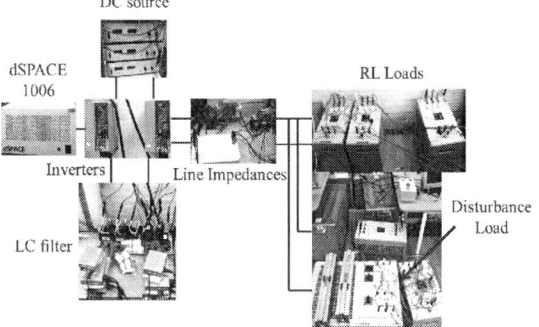

Fig.12. The hardware photo of the experiment setup.

A. Network Voltage Estimation.

In the experiment setup, a disturbance load (R=10 Ω, L=70mH) is exerted at bus2 at 4s.

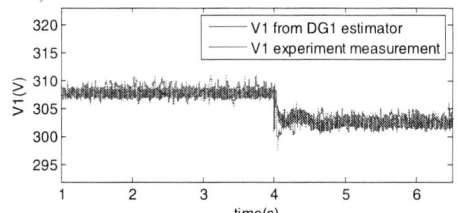

(a) The voltage response at bus1 from DG1 estimator.

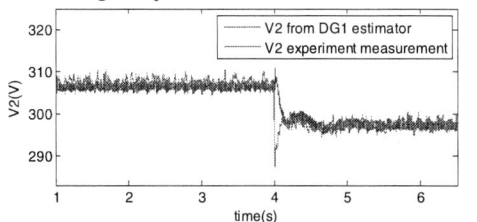

(b) The voltage response at bus2 from DG1 estimator.

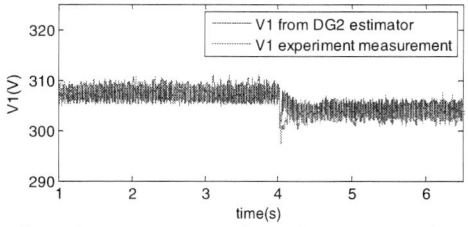

(c) The voltage response at bus3 from DG2 estimator.

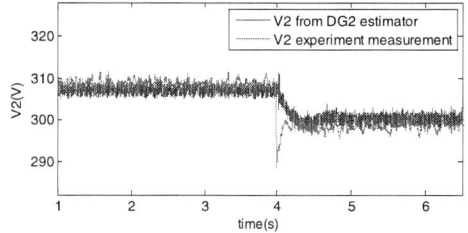

(d) The voltage response at bus3 from DG2 estimator.

Fig.13. voltages estimation results

It can be seen that the proposed voltages estimator of two DG units can track dynamically change of voltages at bus1 and bus2 as illustrated in Fig.13.(a)-(d). Thus, the experimental results, along with simulation results, show that the proposed voltage estimators are able to estimate system voltage responses effectively.

B. The Voltage Control for Single Bus

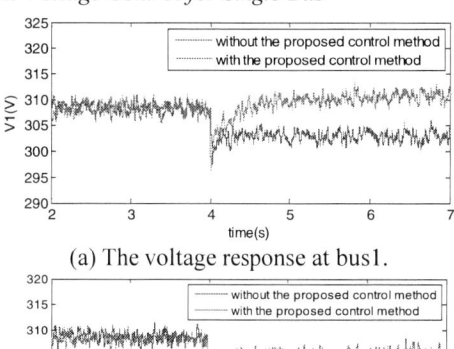

(a) The voltage response at bus1.

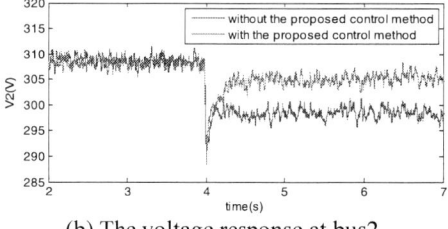

(b) The voltage response at bus2.

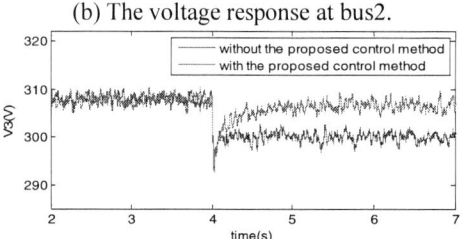

(c) The voltage response at bus3.

Fig.14. The experiment results of voltage responses with and without the proposed control method for single-bus.

Also, in the accompanying experiment, a disturbance load is exerted at bus2. Fig.14 depicts the voltage responses from experimental setup. The experiment results show that the proposed voltage controller bring bus3 voltage to original state accurately in the presence of disturbance load. It can be seen obviously the matched

978-1-4799-2706-7/14 $31.00 © 2014 IEEE   3544

results between simulations and experiments. Hence, the correctness and effectiveness of the proposed voltage control approach is confirmed.

C. The Voltage Control for Multi-Bus.

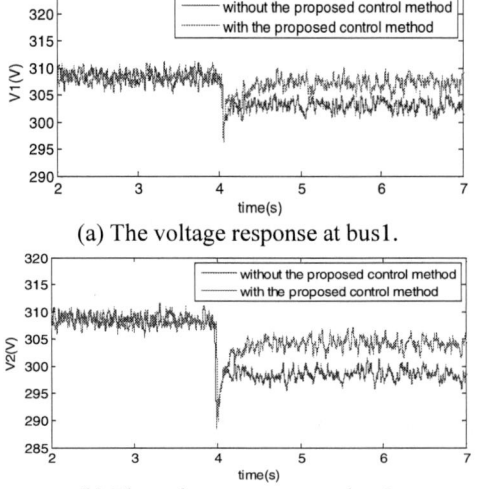

(a) The voltage response at bus1.

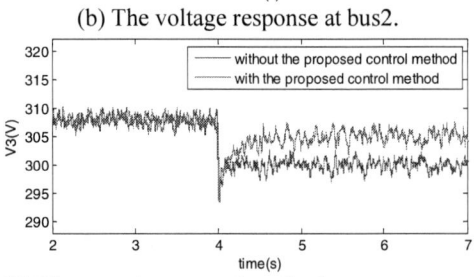

(b) The voltage response at bus2.

Fig.15. The experiment results of voltage responses with and without the proposed control method for multi-bus.

As can be seen from Fig.15, the results obtained from experiment shows that the voltages are brought toward their original states when the proposed optimal controller is enabled at 4s. Hence, the proposed voltage optimal controller for multi-bus is also verified.

I. CONCLUSIONS

In this paper, a communication-less distributed dynamic voltage control strategy was proposed for a droop control-based multi-bus islanded microgrid. First, a local network voltage estimator has been proposed to obtain voltage responses without communication links. The voltage estimator can accurately estimate network voltages just through local voltage and current associated with each DG unit. Second, a voltage optimal controller was developed to perform different voltage control objectives, including voltage offset-free control for single-bus and optimal control for multi-bus. The results obtained from simulations and experimental results verify the correctness and effectiveness of the proposed control strategy. Therefore, the proposed voltage control strategy improves significantly the flexibility and reliability of islanded microgrid due to communication-less operation.

ACKNOWLEDGMENT

The authors would like to thank for funding from China scholarship Council and the Danish Council for

Strategic Research for providing the financial support for the project "Development of a Secure, Economic and Environmentally-friendly Modern Power Systems" (DSF 09-067255).

REFERENCES

[1] N. Hatziargyriou, H. Asano, R. Iravani, and C. Marnay, " Microgrids," IEEE Power Energy Mag., vol.5, no. 4, pp.78-94, Jul./Aug.2007.

[2] Y.W Li, C.N Kao. "An accurate power control strategy for power-electronics-interfaced distributed generation units operating in a low-voltage multibus microgrid," IEEE Trans. Power Electronics, vol.25, pp. 6-15, Jan.2010.

[3] Qing-Chang Zhong. "Robust Droop controller for Accurate Proportional Load Sharing Among Inverters Operated in Parallel," IEEE Trans. Industry Electronics,vol.60,pp. 1281-1290.APR.2013.

[4] X.Wang, J.M. Guerrero, Z. Chen, F. Blaabjerg. "Distributed energy resources in grid interactive AC microgrids," The 2nd IEEE international symposium on power electronics for distributed generation systems. 2010. pp.806-812.

[5] Il-Yop Chung, Wenxin Liu, David A. Cartes, Emmanuel G.Collins. " Control Methods of Inverter-Interfaced Distributed Generators in a Microgrid System," IEEE Trans. Industry Applications, vol.46,pp. 1078-1088. MAY/JUNE.2010.

[6] M. Hua, H. Hu, Y. Xing, J. M. Guerrero. "Multilayer control for inverters in parallel operation without intercommunications," IEEE Trans Power Electronics.vol.27, pp. 3651-3663, Aug. 2012.

[7] J. Rocabert, A. Luna, F. Blaabjerg, P. Rodriguez. "Control of power converters in AC microgrids," IEEE Trans Power Electronics.vol.27, pp. 4734-4749, Nov, 2012.

[8] A. Bidram, A. Davoudi, FL Lewis, JM Guerrero. "Distributed cooperative secondary control of microgrids using feedback linearization," IEEE Trans. Power System, vol.28, pp. 3462-3470, Aug.2013.

[9] A.Mehrizi-Sani and R. Iravani, "Potential-function based control of a microgrid in islanded and grid-connected models," IEEE Trans. Power Syst., vol. 25, pp. 1883-1891, Nov. 2010.

[10] J. M. Guerrero, J. C. Vasquez, J. Matas, L. G. de Vicuna, and M. Castilla, "Hierarchical control of droop-controlled AC and DC microgrids: A general approach toward standardization," IEEE Trans. Ind. Electron., vol. 58, no. 1, pp. 158–172, Jan. 2011.

[11] M. Hua, H. Hu, Y. Xing, J. M. Guerrero. "Multilayer control for inverters in parallel operation without intercommunications," IEEE Trans Power Electronics.vol.27, pp. 3651-3663, Aug. 2012.

[12] Q Shafiee, Josep M. Guerrero , etc. "Distributed secondary control for islanded microgrid - a novel approach," IEEE Trans. Power Electronics, vol.29, pp. 1018-1031, Feb.2014.

[13] Tine L. Vandoorn, Clara M.Ionescu, Jeroen D.M. "Theoretical Analysis and Experimental Validation of Single-Phase Direct Versus Cascade Voltage Control in Islanded Microgrids," IEEE Trans. Ind Electronics.vol.60, pp.789-798, Feb.2013.

[14] A.H. Etemadi, R. Iravani. "A decentralized robust control strategy for multi-Der microgrids- Part I : fundamental concepts," IEEE Trans. power delivery, vol.27, pp. 1843-1853, Oct.2012.

[15] E.A.A.Coelho, P.C.Cortizo, P.F.D.Garcia. "Small-signal stability for parallel-connected inverters in stand-alone AC supply systems," IEEE Trans. Ind Applications, vol.38, pp. 533–542, Mar. 2002.

[16] Nagaraju Pogaku. "Modeling, analysis and testing of autonomous operation of an inverter-based microgrid," IEEE Trans. Power Electronics, vol.22, pp. 613-625, Mar.2007.

[17] Y.B Wang, Y.J. Tian, X.F Wang, Z Chen. "Kalman-Filter-Based state estimation for system information exchange in a multi-bus islanded microgrid," The 7th IET international conference on Power Electronics, Machines and Drives. Apr, 8-10. 2014.

[18] U. Maedar, F.Borrelli, M. Morari. "Linear offset-free model predictive control," Automatica, 45, pp. 2214-2222, 2009.

[19] Greg Welch, Gary Bishop. "An Introduction to the Kalman Filter," (1995)

An Enhanced Load Power Sharing Strategy for Low-voltage Microgrids Based on Inverse-droop Control Method

Yixin Zhu, *Student Member, IEEE*, Fang Zhuo, *Member, IEEE*, Baoquan Liu, *Student Member,* Hao Yi*, *Student Member, IEEE*

School of Electrical Engineering, Xi'an Jiaotong University
Xi'an, China
* Corresponding author Email: zhuyixin.1987@stu.xjtu.edu.cn

Abstract—it is important for an autonomous microgrid to share the load demand properly among multiple distributed generation (DG) units. Normally, the traditional ω-P and E-Q droop control method is used for its "plug and play" feature. However, when employed in a low-voltage microgrid, the conventional droop control is subject to the power coupling and steady-state reactive power sharing errors. Furthermore, the complex microgrid configurations often make the reactive power sharing more challenging. In this paper, an enhanced power sharing strategy is proposed based on inverse-droop control, which can behave well in low-voltage microgrids but has real power sharing errors. So a synchronous regulation process for real power sharing is added to inverse-droop control. After the regulation starts, an integration term is added to keep the well shared reactive power, and the real power errors are eliminated through the frequency regulation, just like the work of traditional droop control. Finally, the Matlab simulation results validate the feasibility of the proposed strategy.

Keywords—*distributed generation (DG); microgrid; power sharing; inverse-droop control; low-bandwidth communication.*

I. INTRODUCTION

With the increased concerns on environment and energy consumption, more renewable energy sources (RES) or microsources such as photovoltaic cells, small wind turbines, and microturbines are integrated into the power grid in the form of distributed generation (DG) units, which are normally interfaced to the grid through power electronics converters [1]. On the other hand, the high penetration of these power electronics-based DG units will introduces a few issues, such as system resonance, protection interface, etc. In order to deal with these problems, the microgrid concept has been proposed [2]-[7]. The microgrid coordinates the operation manners of those multiple DG units, and integrates them as a controllable unit which is connected to the power grid through a point of common coupling (PCC). Microgrid can operate in two modes: during grid-connected mode, it can be controlled as a power node to regulate the power flow of power grid; and during island mode, it offers high reliability power supply to the critical loads.

During island mode, making the load demand be properly shared is an important issue that should be researched. Conventionally, the frequency-real power and voltage magnitude-reactive power droop control method is adopted, which has "plug and play" feature and realizes power sharing autonomously. However, the droop control governed microgrid is prone to have some power control stability problems when used in a low-voltage microgrid, where the DG feeders are mainly resistive as listed in Table Ⅰ. It can also be seen that the real power sharing at the steady state is always accurate while the reactive power sharing is sensitive to the impacts of mismatched feeder impedance. Moreover, the existence of local loads and the networked microgrid configurations often further aggravate reactive power sharing problems.

TABLE I
UNIT FEEDER IMPEDANCE UNDER DIFFERENT VOLTAGE CLASSES

Voltage Classes	R (ohm/1km)	X (ohm/1km)	Impedance ratio
Low voltage	0.642	0.083	7.7
Mid voltage	0.161	0.190	0.85
High voltage	0.060	0.191	0.31

A few improved methods were proposed to solve these power control issues. In [8]-[11], the predominant virtual output inductor is placed at the DG output terminal, which can help preventing the power control instability [12]-[15]. However, this approach cannot deal with the reactive power sharing problem at the same time. In [16], based on virtual inductor control, the feeder voltage drop is compensated by estimating the feeder voltage, but this method was invalid in complex microgrids. According to the low-voltage microgrid characteristics, the ω-Q and E-P inverse droop control method was proposed, which can eliminate the reactive power sharing error but sacrificed the real power sharing accuracy. It is worth noting that these aforementioned methods were developed based on simplified microgrid configurations. Indeed, due to the "plug and play" concept, DG units and loads may change with time and this time-varying feature makes power sharing more difficult. In [17], an enhanced power sharing strategy was developed, which estimated the reactive power control error through injecting small real power disturbances, and eliminated the error through a slow integration term. It should be noted that this one-way communication method does not affect the "plug and play" feature.

In this paper, in order to realize power sharing for a low-voltage microgrid, a synchronous regulation process is added to the inverse droop control. This regulation process is activated by a low-bandwidth synchronization signal from the microgrid control center, and the "plug and play" feature of DG units will not be affected. Before the regulation starts, the reactive power demand is accurately shared among DG units. Then the proposed method uses this reactive power information during the synchronous regulation stage. The reactive power output is fixed through an integration term, and the real power output is regulated by frequency regulation just as droop control works. After this, a new inverse-droop control state is formed, both the real and reactive power are well shared. Note that the proposed accurate power control method is effective for microgrids with all types of configurations and load locations, and it does not need the detailed microgrid structural information. Matlab simulation results are provided to verify the proposed load demand sharing method.

II. TECHNICAL WORK PREPARATION

As mentioned before, in islanded mode, the loads must be properly shared by multiple DG units. Conventionally, the frequency and voltage magnitude droop control is adopted, which aims to achieve microgrid power sharing in a decentralized manner. This method is based on the flow of real power and reactive power between two nodes separated by a feeder line as:

$$P = \frac{U}{R^2 + X^2}(UR - ER\cos\delta + EX\sin\delta) \tag{1}$$

$$Q = \frac{U}{R^2 + X^2}(UX - EX\cos\delta - ER\sin\delta) \tag{2}$$

Where R and X are the resistor and reactance of the line impedance between the two nodes, U and E are the voltage magnitudes of the power output node and the terminal node, δ is the phase angle difference between the two nodes. For mainly inductive line impedance condition, and considering the δ is typically small, the expressions (1) and (2) can be simplified as:

$$P = \frac{UE\sin\delta}{X} \tag{3}$$

$$Q = \frac{U(U - E)}{X} = \frac{U}{X}\Delta E \tag{4}$$

Therefore, the output real power is proportional to δ and the output reactive power is proportional to ΔE. So the real power flow from DG unit can be regulated by changing the DG voltage frequency, and the reactive power flow can be regulated by changing the DG voltage magnitude. That is the basic control of traditional droop control method. However, in a low voltage microgrid, where feeders are mainly resistive, the expressions (1)

and (2) can be simplified as:

$$P = \frac{U}{R}(U - E\cos\delta) = \frac{U}{R}\Delta E \tag{5}$$

$$Q = -\frac{UE\sin\delta}{R} \tag{6}$$

That means for traditional droop control, increasing the real power output will decrease the reactive power output; and increasing reactive power output will increase the real power output at the same time. To restrain this power coupling in low-voltage condition, the ω-Q and E-P inverse-droop control can be adopted, which exchanges the power control relationship compared with traditional droop control.

Mathematically, the ω-Q and E-P characteristics can be expressed as:

$$\omega_i = \omega_0 + D_{\omega i} \cdot Q_i \tag{7}$$

$$U_i = U_0 - D_{ui} \cdot P_i \tag{8}$$

Where ω_0 and U_0 are the nominal values of DG angular frequency and DG voltage magnitude, P_i and Q_i are the measured real and reactive powers after the first-order low-pass filtering (LPF), $D_{\omega i}$ and D_{ui} are the inverse-droop control slopes, which can be set as:

$$D_{\omega i} = \frac{\omega_{max} - \omega_{min}}{Q_{i\_max} - Q_{i\_min}} \tag{9}$$

$$D_{Ui} = \frac{E_{max} - E_{min}}{P_{i\_max} - P_{i\_min}} \tag{10}$$

Where $\{E_{max}, E_{min}\}$ are the maximum and minimum voltage magnitude of microgrid; $\{\omega_{max}, \omega_{min}\}$ are the maximum and minimum operation frequency of microgrid; $\{P_{i\_max}, Q_{i\_max}\}$ and $\{P_{i\_min}, Q_{i\_min}\}$ are the maximum and minimum power of DGi.

According to (7) and (8), it can be observed that DG units will output reactive power proportional to their capacities due to the same frequency. However, the real power outputs will not follow the desired relationship due to the unequal voltage drop on feeders.

Compared with traditional droop control method, the inverse-droop control has a better stability in low voltage microgrids. However, it has real power sharing errors due to the mismatched feeder impedance, just like the reactive power sharing condition in droop control.

From the above analysis, it can be found that the system frequency is a communication link for power sharing. Therefore, if the frequency regulation is well utilized, the power sharing problem can be well solved.

III. PROPOSED POWER SHARING STRATEGY

A. Operation of Microgrid

Fig. 1 illustrates the configuration of a microgrid. As shown, the microgrid is composed of a number of DG units and loads. Each DG unit is interfaced to the microgrid with an inverter, and the inverters are connected to the common ac bus through their respective feeders. For the focus of this paper is the fundamental real and reactive power control, nonlinear loads are not considered in the microgrid. According to the operation requirements, the microgrid can be connected (grid-connected mode) or disconnected (islanding mode) from the main grid by controlling the static transfer switch (STS) at the point of common coupling (PCC). To realize the proposed power sharing strategy, a synchronous compensation process for reactive power sharing error elimination is added to inverse-droop control. The microgrid control center (MGCC) sends low-bandwidth synchronization signals to each DG unit to activate the extra compensation.

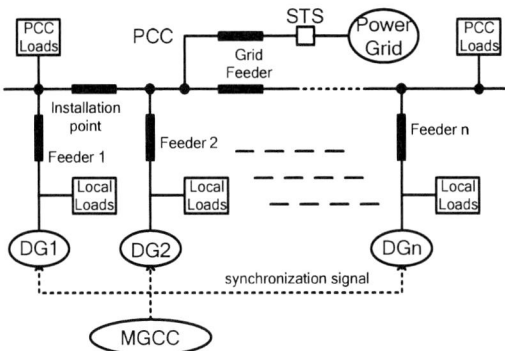

Fig. 1. Simplified configuration of the microgrid.

Note that only one-way communication from the central controller is needed for starting the DG regulation with a synchronized manner. The intercommunication among DG units is not necessary, so that the "plug-and-play" feature will not be affected.

Since developing the circuit model-based reactive power sharing error compensation strategy is difficult. Therefore, the objective of this section is to develop an enhanced compensation method that can eliminate the reactive power sharing errors without knowing the detailed microgrid configuration. This feature is very important to achieve the "plug-and-play" operation of DG units and loads in the microgrid. To simplify the analysis, the DG units discussed here are all identical.

B. Proposed Compensation Control

The enhanced power control strategy is realized through two stages. For most of the time, DG units adopt inverse-droop control method, and this is the first stage; after receiving the synchronization signal, the real power compensation is started, and this is the second stage. Finally, the DG units will automatically return to stage 1, waiting for the next synchronization signal. The whole operation time sequence of the proposed method is shown in Fig. 2.

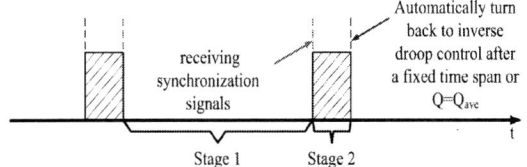

Fig. 2. Time sequence of the proposed method.

1) Stage 1: Inverse Droop Control Method

Before reaching the synchronization signal, the inverse droop controller (7) and (8) are adopted for initial load power sharing. Meanwhile, the DG local controller monitors the status of the compensation flag dispatched from the microgrid central controller. During this stage, the steady-state averaged reactive power (Q_{ave}) shall be measured and saved for use in Stage 2. As the cutoff frequency of LPFs cannot be made too low to get the ripple-free value Q_{ave} due to the consideration of system stability, a moving average filter (MAF) is added after the LPF to further filter out the power ripples as shown in Fig. 3.

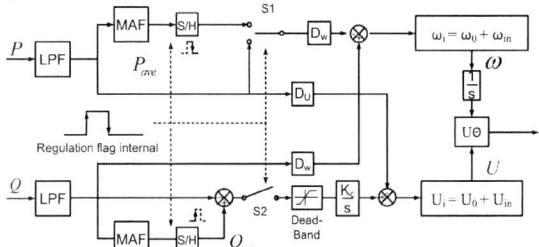

Fig. 3. The control diagram of the proposed control strategy.

2) Stage 2: Synchronized Compensation

During this stage, an integration term is added to reactive power control, and this will make the reactive power output track the Q_{ave} value saved in Stage 1. To compensate the real power sharing error, the real power is regulated through frequency control. The new control relationships can be expressed as:

$$\omega_i = \omega_0 + D_{\omega x} \cdot Q - D_{\omega x} \cdot P \tag{11}$$

$$U_i = U_{0i} - D_{ui} \cdot P + \frac{Kc}{s}(Q_{ave} - Q_i) \tag{12}$$

where Kc is the integral gain, which is selected to be the same for all the DG units. For a single DG unit, once its reactive power output equals to Q_{ave} and its real power output comes into steady state, the compensation will be stopped, and the input channel of the integration term will be cut off. Then a new inverse-droop control state is obtained as:

$$\omega_i = \omega_0 + D_{\omega x} \cdot Q - D_{\omega x} \cdot P_{ave} = \omega_0' + D_{\omega x} \cdot Q \tag{13}$$

$$U_i = U_{0i} - D_{ui} \cdot P + \int Kc \cdot (Q_{ave} - Q) = U_{0i}' - D_{ui} \cdot P \tag{14}$$

Where P_{ave} is the averaged real power saved at the end of Stage 2. Once the reactive power channel of voltage magnitude control disconnected, the integral effect is left, added to the previous P-E droop characteristic. Similarly to the voltage frequency control, the saved P_{ave} value is added to the Q-ω inverse-droop control characteristic. After this compensation, the nominal frequency value ω_0 and voltage magnitude value U_0 are changed. Due to the same P_{ave} value, each DG unit will still have a same ω_0 value. After the new inverse-droop control state being formed, both the real and reactive power will be well shared. It should be noted that the load variation during this stage will affect the validity of the proposed strategy. So the time interval between two synchronization signals should be properly selected to avoid this situation. Once such situation happens, the power sharing will recover after receiving the periodical reset signal.

IV. SMALL-SIGNAL MODELING AND ANALYSIS

During Stage 1, DG units employ traditional inverse droop control method. During Stage 2, the proposed regulation controller (11) and (12) are adopted. In order to investigate the stability and transient performances of DG units while using the proposed method, small-signal analysis methods can be applied.

According to (1) and (2), the real and reactive power variations according to DG voltage disturbances can be obtained in (15) and (16):

$$\Delta P_o = \left(\frac{\partial P_o}{\partial \theta}\right) \cdot \Delta\theta + \left(\frac{\partial P_o}{\partial U}\right) \cdot \Delta U = k_{P\theta} \cdot \Delta\theta + k_{PU} \cdot \Delta U \tag{15}$$

$$\Delta Q_o = \left(\frac{\partial Q_o}{\partial \theta}\right) \cdot \Delta\theta + \left(\frac{\partial Q_o}{\partial U}\right) \cdot \Delta U = k_{Q\theta} \cdot \Delta\theta + k_{QU} \cdot \Delta U \tag{16}$$

Where the $k_{P\theta}$, k_{PU}, $k_{Q\theta}$, k_{QU} represent the power flow sensitivity to voltage angle and magnitude regulation.

When there are some power fluctuations during Stage 1, by expending the regulation method, the small-signal response of the DG voltage can be expressed as:

$$\Delta\omega = D_\omega \cdot \Delta Q = \frac{D_\omega}{\tau s + 1} \cdot \Delta Q_o \tag{17}$$

$$\Delta U = -D_U \cdot \Delta P = \frac{-D_U}{\tau s + 1} \cdot \Delta P_o \tag{18}$$

where τ is the time constant of LPFs used in the power calculation.

Considering that $\Delta\theta$ is the integration of $\Delta\omega$, by a simple manipulation on (15) to (18), the dynamic performance of the DG unit during Stage 1 yields the following matrix equation as:

$$\left(\begin{bmatrix} s(\tau s + 1) & 0 \\ 0 & s(\tau s + 1) \end{bmatrix} + \begin{bmatrix} -D_\omega & 0 \\ 0 & D_U \cdot s \end{bmatrix} \cdot \begin{bmatrix} k_{Q\theta} & k_{QU} \\ k_{P\theta} & k_{PU} \end{bmatrix}\right) \cdot \begin{bmatrix} \Delta\theta \\ \Delta U \end{bmatrix} = 0$$

Then the closed-loop characteristic equation at the matrix equation can be obtained in:

$$s^3 \Delta\theta + A_1 \cdot s^2 \Delta\theta + B_1 \cdot s\Delta\theta + C_1 \Delta\theta = 0$$

where A, B, and C are represent follow expressions:

$$A_1 = \frac{2\tau + K_{PU} D_U \tau}{\tau^2}$$

$$B_1 = \frac{1 + K_{PU} D_U - D_\omega K_{Q\theta} \tau}{\tau^2}$$

$$C_1 = \frac{-D_\omega K_{Q\theta} - K_{PU} D_U D_\omega K_{Q\theta} + K_{QU} D_U D_\omega K_{P\theta}}{\tau^2}$$

By a similar way, the dynamic performance of traditional droop control method can be obtained:

$$\left(\begin{bmatrix} s(\tau s + 1) & 0 \\ 0 & s(\tau s + 1) \end{bmatrix} + \begin{bmatrix} D_\omega & 0 \\ 0 & D_U \cdot s \end{bmatrix} \cdot \begin{bmatrix} k_{P\theta} & k_{PU} \\ k_{Q\theta} & k_{QU} \end{bmatrix}\right) \cdot \begin{bmatrix} \Delta\theta \\ \Delta U \end{bmatrix} = 0$$

Then the corresponding closed-loop characteristic equation at the matrix equation can be obtained:

$$s^3 \Delta\theta + A_2 \cdot s^2 \Delta\theta + B_2 \cdot s\Delta\theta + C_2 \Delta\theta = 0$$

Where A, B, and C are represent follow expressions:

$$A_2 = \frac{2\tau + K_{QU} D_U \tau}{\tau^2}$$

$$B_2 = \frac{1 + K_{QU} D_U + D_\omega K_{P\theta} \tau}{\tau^2}$$

$$C_2 = \frac{D_\omega K_{P\theta} + K_{QU} D_U D_\omega K_{P\theta} - K_{PU} D_U D_\omega K_{Q\theta}}{\tau^2}$$

The DG unit circuit parameters are selected to be the same as in the simulation, and are listed in Table I. Fig.4 shows the performance of the system using the inverse-droop control method, where the reactive power inverse droop gain D_ω is fixed to 0.00001 while the real power gain D_U increasing. As illustrated, it is a three order system and the dynamic performance of the system is mainly determined by the dominate pole λ_1 and λ_2. It can be seen that the positions of the dominate poles are not sensitive to the variations of D_U value. So the calculated value of D_U from (10) 0.000125 is proper for the inverse droop control system. Fig.5 shows the performance of the system using the traditional droop control method, where the real power droop gain D_ω is fixed and the reactive power gain D_U changes.

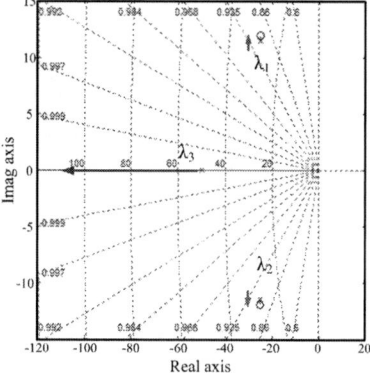

Fig. 4. Family of root locus diagram with the inverse-droop controller: $D_\omega = 0.00001$, and D_U increasing.

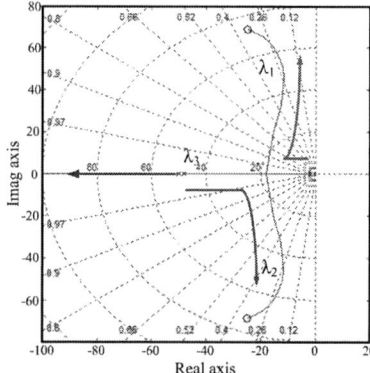

Fig. 5. Family of root locus diagram with the traditional droop controller: $D_\omega = 0.00001$, and D_U increasing.

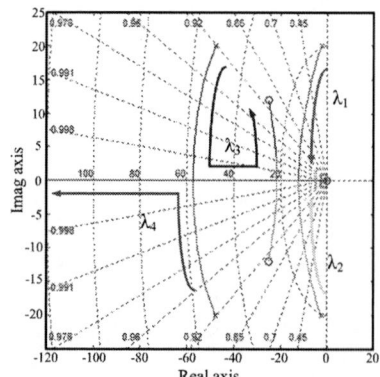

Fig. 6. Family of root locus diagram with the proposed regulation controller: $D_\omega = 0.00001$, $K_c = 0.004$ and D_U increasing.

It can be found that the traditional droop control is much more sensitive to the D_u value compared with inverse-droop control. So it can be conclude that in a low-voltage microgrid, where feeders are mainly resistive, the inverse-droop control system is more stability. Normally, the traditional droop control can adopt virtual impedance to improve its performance in low-voltage microgrid.

When DG unit operates in Stage 2, the response of the system to different D_u is also depicted in Fig. 6. During this stage, the dynamic performance of the DG unit yields the following matrix equation as:

$$\left(\begin{bmatrix} s(\tau s+1) & 0 \\ 0 & s(\tau s+1) \end{bmatrix} + \begin{bmatrix} -D_\omega & D_\omega \\ K_c & D_U \cdot s \end{bmatrix} \cdot \begin{bmatrix} k_{Q\theta} & k_{QU} \\ k_{P\theta} & k_{PU} \end{bmatrix} \right) \cdot \begin{bmatrix} \Delta\theta \\ \Delta U \end{bmatrix} = 0$$

The closed-loop characteristic equation can be obtained in:

$$s^4 \Delta\theta + A \cdot s^3 \Delta\theta + B \cdot s^2 \Delta\theta + C \cdot s\Delta\theta + D\Delta\theta = 0$$

Where A, B, C and D are represent the follow expressions:

$$A_3 = \frac{2\tau + K_{PU}D_U\tau}{\tau^2}$$

$$B_3 = \frac{1 + K_{PU}D_U - D_\omega K_{Q\theta}\tau + D_\omega K_{P\theta}\tau + K_c K_{QU}\tau}{\tau^2}$$

$$C_3 = \frac{K_c K_{QU} - D_\omega K_{Q\theta} + D_\omega K_{P\theta} - K_{Q\theta}D_U D_\omega K_{PU} + K_{QU}D_U D_\omega K_{P\theta}}{\tau^2}$$

$$D_3 = \frac{D_\omega K_{P\theta}K_c K_{QU} - K_{PU}K_c D_\omega K_{Q\theta}}{\tau^2}$$

In contrast to the conclusions from Fig.5, the DG unit becomes a fourth order system in this case. Once again, the performance of system is evaluated by the dominate pole approximation, and the selected value for D_u (0.000125) is still proper.

Similarly, the response of the system to the integration gain K_c variation is examined in Fig. 7. It can also be observed that the stability and damping of the system is sensitive to change of the integration gain. To maintain proper stability and damping features, the selected gain K_c is 0.004.

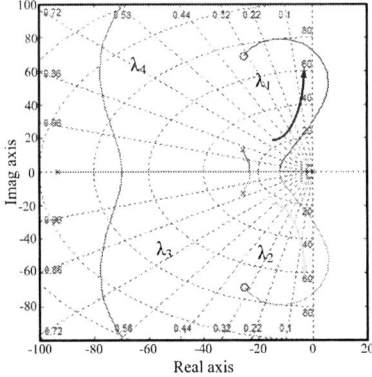

Fig. 7. Family of root locus diagram with the proposed regulation controller: $D_\omega = 0.00001$, $D_U = 0.000125$ and K_C increasing.

From the detailed small-signal analysis, it can be concluded that the stability and damping performance of the microgrid is affected during the regulation process. However, as will be demonstrated in the next section, the

well-designed control parameter can maintain satisfied stability and damping performance during both steady-state operation and regulation transients.

V. SIMULATION VERIFICATION

A networked microgrid model has been established using MATLAB/Simulink. As shown in Fig.9, the simulated microgrid is composed of three identical DG units and five linear loads. With the same power rating, three DG units shall share the load equally. The detailed configuration of the DG unit is presented in Fig.10, where an LC filter is placed between the IGBT bridge outputs and the DG feeders. The DG line current and filter capacitor voltage are measured to calculate the real and reactive powers. In addition, the well-known multi-loop voltage controller is employed to track the reference voltage. The circuit and control parameters of the DG unit are listed in Table Ⅱ.

TABLE Ⅱ
PARAMETERS OF THE SIMULATION SYSTEM

	Parameters	Values
Microgrid Parameter	Rated RMS voltage (L-L)	380V 50Hz
	Total Loads	155kw 110kvar
DG unit Parameter	Frequency droop D_ω	0.00001
	Magnitude droop D_U	0.000125
	Integral gain K_c	0.004
	LPF time constant τ	0.02s
	Saturation value (Q_{ave}-Q)	500var
	Rated output real power	50kw
	Rated output reactive power	50kvar
	Max output real power	100kw
	Max output reactive power	100kvar

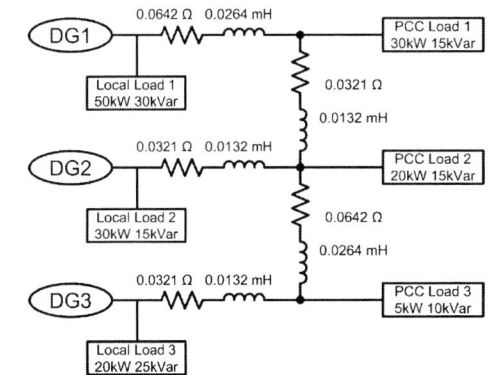

Fig. 8. Networked microgrid in the simulation.

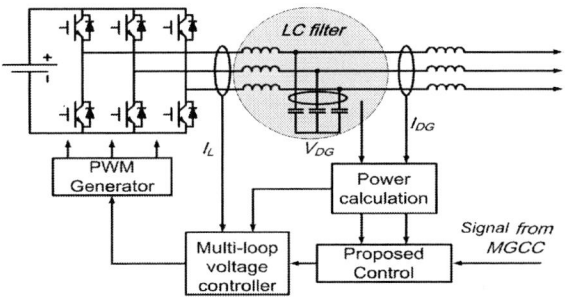

Fig. 9. Configuration of DG unit.

Fig. 10 demonstrates the real power flow of the DG units. Due to the unequal voltage drops on networked microgrid feeders, there are significant sharing errors of real power with the original inverse droop control method. On the other hand, the synchronization regulation method starting at 4.0 s can effectively adjust the real power sharing error to almost zero. Fig. 11 shows the reactive power flow of these DG units. Before the regulation, the reactive power is evenly shared with the inverse-droop control method. When the synchronization regulation is enabled at 4.0 s, due to the transient real and reactive power coupling introduced by (15) and (16), there are some disturbances in the reactive power. However, the output reactive power goes back to the original value after 1.3 s, and it can be observed that the switch process is seamless. Fig. 12 and Fig. 13 show the associated DG line currents. With the original inverse-droop control method, the magnitude and phase of DG currents are not the same as illustrated in Fig. 12. Consistent with the power sharing analysis, the DG line currents in Fig. 13 are almost identical after the proposed compensation.

The voltage magnitudes at different locations are also obtained. Fig.14 shows the changes of DG unit voltage magnitudes during this process. In order to realize equal real power sharing, these voltages have small deviations during Stage 2. This is because the unequal voltage drops on the feeders are compensated by the DG units. The voltage magnitudes at the installation points (see Fig. 8) are also obtained in Fig. 15. Similarly, these voltage magnitudes also have slight deviations during the regulation.

To test the sensitivity of the proposed regulation method to synchronization accuracy, a 0.1 s delay is intentionally added to the regulation process of DG1, and the simulation results are shown in Figs. 16 and 17.

Compared to the case in Fig.10 and Fig.11, it can be observed that the compensation performance is close to the situation with-out any delay. Although the reactive power sharing has some disturbances after the compensation, the transient sharing errors are limited in a tolerable range. The results are acceptable for most of the microgrid applications.

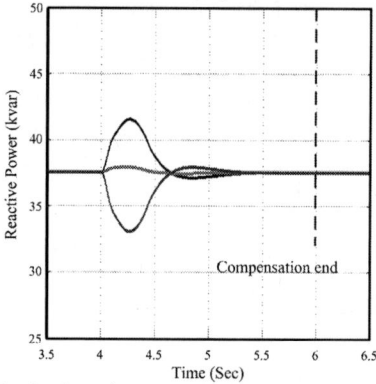

Fig. 10. Simulated real power sharing performance in a network microgrid with the proposed method.

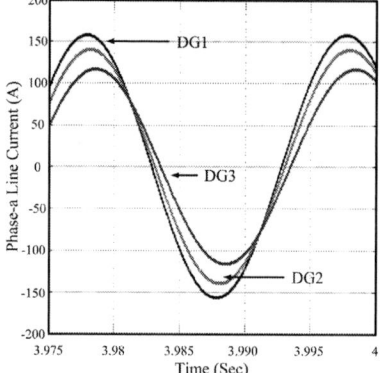

Fig. 11. Simulated reactive power sharing performance in a network microgrid with the proposed method.

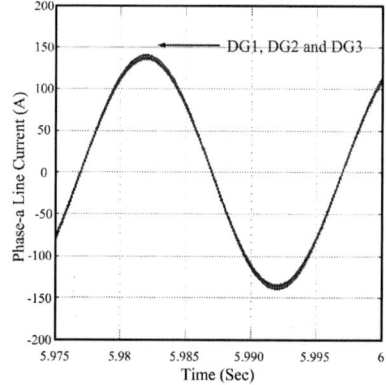

Fig. 12. Simulated DG currents before compensation.

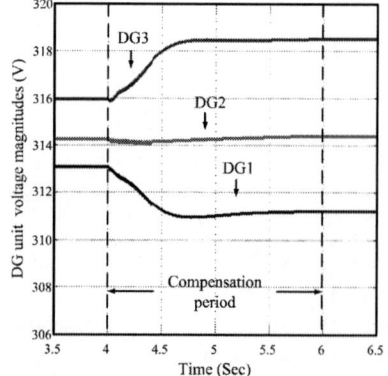

Fig. 13. Simulated DG currents after compensation.

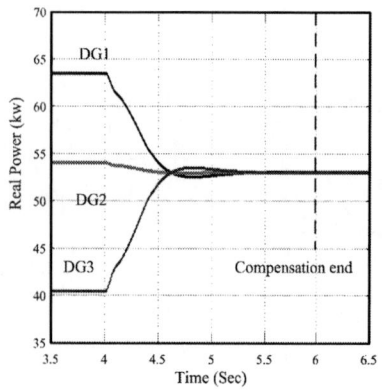

978-1-4799-2706-7/14 $31.00 © 2014 IEEE

Fig. 14. Simulated DG voltage magnitudes.

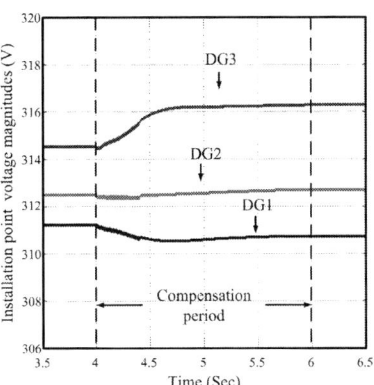

Fig. 15. Simulated installation points voltage magnitudes.

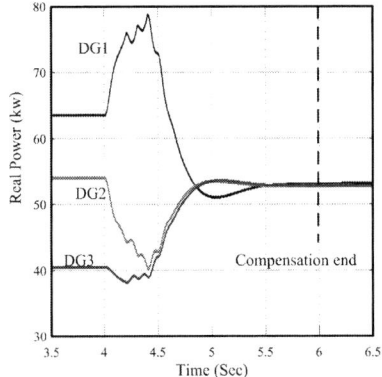

Fig. 16. Simulated real power sharing performance in a network microgrid (0.1 s delay of synchronization regulation in DG 1).

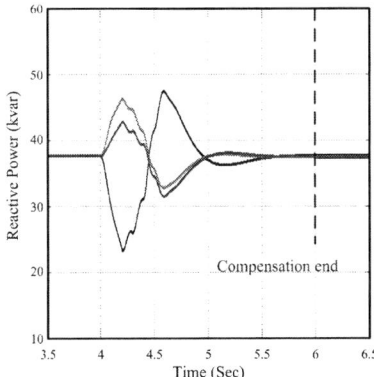

Fig. 17. Simulated reactive power sharing performance in a network microgrid (0.1 s delay of synchronization regulation in DG 1).

The simulation results demonstrate that the proposed method is useful to solve load demand sharing problem in low voltage microgrids without knowing the detailed microgrid structural information.

VI. CONCLUSION

In this paper, an improved microgrid power sharing strategy was proposed. The method injects a real-reactive power transient coupling term during the inverse-droop control period. Then the identified errors of power sharing are compensated by using a slow integral term in

DG local control. Finally, accurate power sharing can be achieved, and the "plug and play" feature is reserved. In addition, the proposed method is not sensitive to the microgrid configurations, which is especially suitable for a complex mesh or networked microgrid.

REFERENCES

[1] J. M. Carrasco, L. G. Franquelo, J. T. Bialasiewicz, E. Galvan, R. C. P. Guisado, M, A, M, Prats, J, I, Leon, and N, Moreno, Alfonso, "Power electronic systems for the grid integration of renewable energy sources: A survey," *IEEE Trans. Power Electron.*, vol. 53, no. 4, pp. 1002-1016, Aug. 2006.

[2] R. Lasseter, "Microgrids," presented at the Power. Eng. Soc, Winter Meet., 2002.

[3] M. Barnes, J. Kondoh, H. Asano, J. Oyarzabal, G. Ventakaramanan, R. Lasseter, N. Hatziargyriou, and T. Green, "Real-world microgrids—An overview," in *Proc. 2007 IEEE Int. Conf. Syst. Syst. Eng.*, pp. 1-8.

[4] Y. W. Li, D. M. Vilathgamuwa, P. C. Loh, "Design, analysis, and real-time testing of a controller for multibus microgrid system," *IEEE Trans. Power Electron.*, vol. 19, no.5, pp. 1195-1204, Sep. 2004.

[5] N. Pogaku, M. Prodanovic, and T. C. Green, "Modeling, analysis and testing of autonomous operation of an inverter-based microgrid," *IEEE Trans. Power Electron.*, vol. 22, no. 2, pp. 613-625, Mar. 2007.

[6] F. Katiraei and M. R. Iravani, "Control and protection of power electronics interfaced distributed generation systems in a customer-driven microgrid," *IEEE Trans. Power Syst.*, vol. 21, no. 4, pp. 1821-1831, Nov. 2006.

[7] F. Z. Peng, Y. W. Li, and L. M. Tolbert, "Control and protection of power electronics interfaced distributed generation systems in a customer-driven microgrid," in *Proc. 2009 IEEE Power Eng. Soc. Gen. Meet.*, pp. 1-8.

[8] C. T. Lee, C. C. Chu and P. T. Cheng, "A new droop control method for the autonomous operation of distributed energy resource interface converters," in *Proc. 2010 IEEE Energy Convers. Congr. Expo.*, pp. 702-709.

[9] J. M. Guerrero, L. G. Vicuna, J. Matas, M. Castilla, and J. Miret, "Output impedance design of parallel-connected UPS inverters with wireless load sharing control," *IEEE Trans. Ind. Eectron.*, vol. 52, no. 4, pp. 1126-1135, Aug. 2005.

[10] J. M. Guerrero, L. G. Vicuna, J. Matas, M. Castilla, and J. Miret, "A wireless controller to enhance dynamic performance of parallel inverters in distributed generation systems," *IEEE Trans. Power Eectron.*, vol. 19, no. 4, pp. 1205-1213, Sep. 2004.

[11] Y. Li and Y. W. Li, "Decoupled power control of DG units in a power electronics interfaced low voltage microgrid," in *Proc. 2009 IEEE IPEMC*, pp. 2490-2496.

[12] E. Barklund, N. Pogaku, M. Prodanovic, C. Hernandez-Aramburo, and T. C. Green, "Energy management in autonomous microgrid using stability-constrained droop control of inverters," *IEEE Trans. Power Electron.*, vol. 23, no. 5, pp. 2346-2352, Sep. 2008.

[13] Y. A.-R. I. Mohamed and E. F. El-Saadany, "Adaptive decentralized droop controller to preserve power sharing stability of paralleled inverters in distributed generation microgrids," *IEEE Trans. Power Electron.*, vol. 23, no. 6, pp. 2806-2816, Nov. 2008.

[14] J. He and Y. W. Li, "Analysis, design and implementation of virtual impedance for power electronics interfaced distributed generation," *IEEE Trans. Ind. Appl.*, vol. 47, no. 6, pp. 2525-2538, Nov./Dec. 2011.

[15] E. A. A. Coelho, P. C. Cortizo, and P. F. D. Garcia, "Small-signal stability for parallel-connected inverters in stand-alone AC supply systems," *IEEE Trans. Ind. Appl.*, vol. 38, no. 2, pp. 533-542, Mar./Apr. 2002.

[16] Y. W. Li, and C. N. Kao, "An accurate power control strategy for power electronics-interfaced distributed generation units operation in a low voltage multibus microgrid," *IEEE Trans. Power Electron.*, vol. 24, no. 12, pp. 2977-2988, Dec. 2009.

[17] J. W. He, and Y. W. Li, "An Enhanced Microgrid Load Demand Sharing Strategy," *IEEE Trans. Power Electron.*, vol. 27, no. 9, pp. 3984-3995, Sep. 2012.

Adding Virtual Resistance in Source Side Converters for Stabilization of Cascaded Connected Two Stage Converter Systems with Constant Power Loads in DC Microgrids

Mingfei Wu
School of Electrical and Information Engineering
The University of Sydney
Sydney, Australia
mingfei.wu@sydney.edu.au

Dylan D. C. Lu
School of Electrical and Information Engineering
The University of Sydney
Sydney, Australia
dylan.lu@sydney.edu.au

Abstract— **In DC microgrids, the operation of the constant power load (CPL) and its associated input filter is very likely poorly damped if no damping method is applied. Active damping methods can significantly improve the power efficiency comparing with the passive damping methods. This paper proposes an active damping method which builds a virtual resistance in the source side converter, and it can effectively stabilize the DC microgrids with CPLs. The advantage of this active damping method is that the stabilization effect is from the source side converters, and therefore, there is no need to sacrifice the transient performance of the CPL. Simulation and experimental results are reported to verify the effectiveness of the proposed idea.**

Keywords— *Active damping methods, Constant power loads, DC microgrids*

I. INTRODUCTION

In DC microgrid, different loads require different voltage levels. A cascaded connected two stage converters may be needed for a load. The architecture of the DC microgrid is shown in Fig. 1 and the configuration of the cascaded connected two stage converter system is shown in Fig. 2(a). Similar to other DC electric power systems, in DC microgrid, point of load (POL) converters are used to regulate their output voltages. From the system level of view, these converters behave as constant power loads (CPLs). Meanwhile, the source stage converter is also tightly regulated and operates like a constant voltage source. In addition, in DC microgrids an LC input filter is necessary for reduction of input current ripple and noise. The simplified model for the cascaded system is shown in Fig. 2(b).

The operation of CPLs and their associated input filters is very likely to be unstable. In order to ensure the stability of the system, in the small signal analysis, impedance inequality criteria must be fulfilled. Passive damping methods are one kind of approach of reshaping the output impedance of the input filter by adding extra physical resistor [1]. The main drawback of passive damping methods is that the extra physical resistor causes extra power loss.

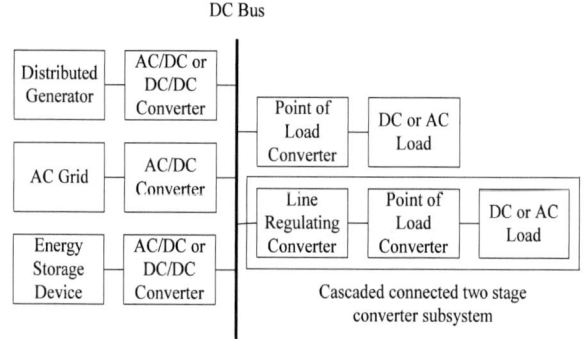

Fig.1 The architecture of DC microgrid with cascaded connected two stage converter subsystem

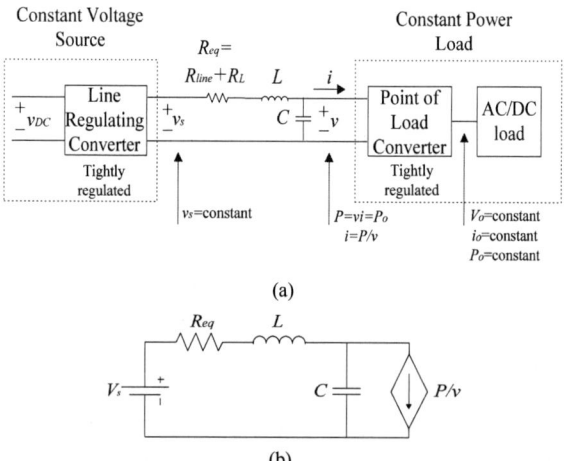

Fig. 2 (a) The configuration of cascaded connected two-stage converter system, (b) Simplified model

Recently, in order to overcome this drawback, several active damping methods which stabilize the system by control technologies were proposed [2-7]. However, many of these methods are based on the cascaded system without LC input filter [2, 5-7]. The instability can also occur between the CPL and the output filter of the source

978-1-4799-2706-7/14 $31.00 © 2014 IEEE

stage converter if the controller is not well designed. This cascaded system without input filter between two stage converters is popular in some DC electric power system in a small space, for example, electric vehicles [2]. However, in a relatively larger space or longer distance between each power stage of the cascaded system, for example, in a DC microgrid for a floor of rooms or a building, this cascaded system without another stage of filter is not applicable and neither the previously proposed active damping methods. There are also a few active damping methods for the DC microgrid with LC input filter. Pierre Magne, *et al* proposed an active damping method by adding compensating current to the load [8]. i.e. during the transient performances such as change of DC bus voltage, the input power of CPL will be added a compensation part instead of keeping constant, which effect the performance of the CPL. X. Liu, A. Forsyth and A. Cross proposed a similar method by adding compensating current, which has the same disadvantage [9]. In DC microgrids, it has a characteristic that the source side converters are also tightly regulated and have similar bandwidth as the POL converters, for example, the switching mode rectifier for power conversion from AC grid or distributed generators. Therefore, we can use this characteristic to stabilize the DC microgrid with CPLs from the source stage converters. This paper will show how we can build virtual resistance from source side converter to stabilize the DC microgrid with CPL. In Section II, stability analysis of cascaded system in DC microgrid consisting of LC input filters and CPLs is studied and then in Section III, the active damping method by adding virtual resistance in the source side converter is proposed for stabilization of the cascaded system with CPL and experiment results are reported in Section IV to verify the analytic results. Finally, the conclusion is presented in Section V.

II. STABILITY ANALYSIS

The differential equations presenting the operation of the system shown in Fig. 2(b) are displayed in (1).

$$\begin{cases} \frac{di}{dt} = -\frac{R_{eq}}{L}i - \frac{1}{L}u + \frac{1}{L}V_s \\ \frac{du}{dt} = \frac{1}{C}i - \frac{P}{Cu} \end{cases} \quad (1)$$

where, i is the inductor current and u is the capacitor voltage and they are all non-negative values. From (1), it can be found that the system is nonlinear. The value of R_{eq} determines stability of the system. A proper value of R_{eq} can well damp the oscillation and stabilize the system, otherwise, the system is unstable. The phase portrait of the system with $V_s = 10V, P = 10W, L = 500\mu H, C = 220\mu F$ and $R_{eq} = 0\Omega$ is shown in Fig. 3 [5].

According to this Fig., the system operates along the limit-circle and has a considerable oscillation with any non-negative initial conditions. It can also be found that the constant power load *v-i* characteristic inside the limit circle is closed to a straight line. i.e. the negative incremental impedance change slightly during the operation in the limit circle. Therefore, we can make a linearization of the nonlinear component $\frac{P}{u}$ in the equation

(1) around its equilibrium point. Hence, a linear time invariant state space representation can be obtained

$$\begin{bmatrix} \frac{di}{dt} \\ \frac{du}{dt} \end{bmatrix} = \begin{bmatrix} -\frac{R_{eq}}{L} & -\frac{1}{L} \\ \frac{1}{C} & \frac{1}{CR} \end{bmatrix} \begin{bmatrix} i \\ u \end{bmatrix} + \begin{bmatrix} \frac{1}{L} \\ 0 \end{bmatrix} V_s \quad (2)$$

where R is the instantaneous impedance of the CPL in the above small signal model. In order to make the two roots of the equation

$$det(sI - A) = 0 \quad (3)$$

where $A = \begin{bmatrix} -\frac{R_{eq}}{L} & -\frac{1}{L} \\ \frac{1}{C} & \frac{1}{CR} \end{bmatrix}$, locate in right half plane, R_{eq} must fulfill the following inequality as shown in (4).

$$\frac{L}{CR} < R_{eq} < R \quad (4)$$

In addition, in order to get a fast settle time, the real part of the roots $\frac{1}{2CR} - \frac{R_{eq}}{2L}$ avoids being small. i.e. the R_{eq} is much larger than $\frac{L}{CR}$. We can use this method to get a suitable R_{eq}.

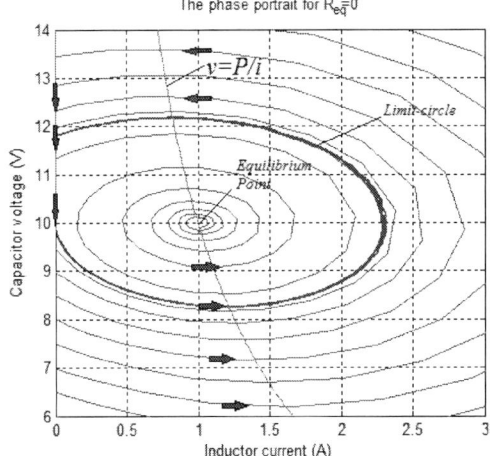

Fig. 3 The phase portrait of the nonlinear system

III. PROPOSED ACTIVE DAMPING METHOD

From the analysis in Section II, a sufficient large resistor R_{eq} can stabilize the system. However, it causes extra power loss if increasing the resistance of the physical resistor R_{eq}. In DC microgrids, the source side converters are usually tightly regulated. Therefore, these source side converters can mimic a virtual resistance to stabilize the CPL with its LC input filter. The schematic diagram of the cascade system and the proposed active damping method is shown in Fig. 5. A buck converter is used as the source side converter as an example. This active damping method can also be applied in other DC/DC converters. In the buck converter, the output current is measured and then multiplied by a virtual resistance and feedback to the voltage control. Therefore, in the bandwidth of the buck converter, it performs as a Thevenin equivalent circuit. i.e. a constant voltage source series connected with a resistor R_{eq}. Therefore, we can make the R_{eq} as shown in the Fig. 5 fulfill the stability criterion as shown in (4) to stabilize the cascaded system. An LTspice simulation is made based on Fig. 5. The parameters of the buck converter are $V_{in} = 15V$, $V_{ref} = 9V$, $L_s = 150\mu H$, $C_s = 330\mu F$ and $R_{eq} = 1\Omega$ and the parameters of the LC input filter are $L_f = 500\mu H$ and $C_f = 220\mu F$. The physical resistance of the power line and the inductor is 0.05Ω. The series equivalent resistance of the input filter capacitor is 0.17Ω.

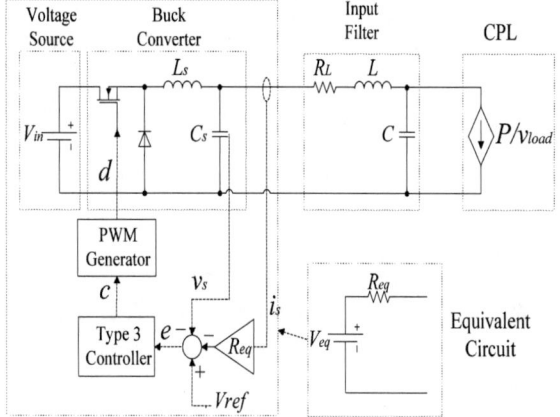

Fig. 5 The schematic of the cascaded system and the proposed active damping method

The simulation results are shown in Fig. 6. The inductor current and the capacitor voltage of the LC input filter will oscillate uniformly as shown in Fig. 6(a) if there is no damping method applied. In comparison, the oscillation of the current and voltage is well damped and system is stabilized with the proposed active damping method as shown in Fig. 6(b). It can also be found in Fig. 6(b) the theory of this control method. That is the inductor current i_s of the LC input filter is measured and multiplied by a virtual resistance R_{eq}. The output voltage v_s is measured and controlled to track the reference voltage $V_{ref} - i_s R_{eq}$ instead of a constant voltage reference. The new reference value will vary according to the value of i_s and hence produce the equivalent function of adding physical resistor R_{eq}.

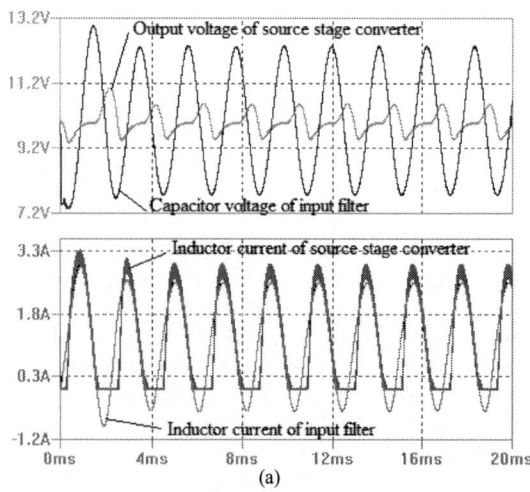

(a)

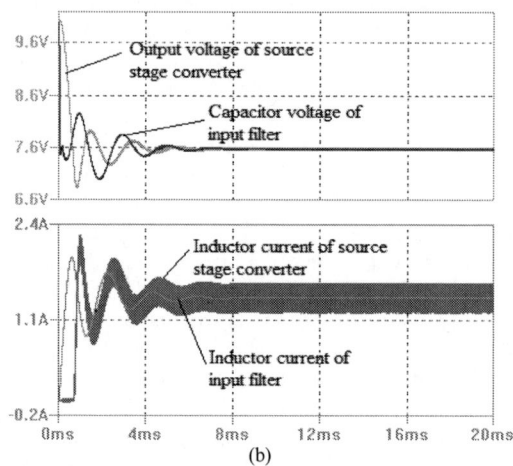

(b)

Fig. 6 The simulation result of output voltage of buck converter, capacitor voltage of input filter and inductor current of input filter in the system (a) without damping (b) with active damping.

IV. EXPERIMENTAL RESULTS

The parameters in the experimental circuit are the same as the simulation circuit. The experimental circuit is shown in Fig. 7. The CPL is implemented by an analogue CPL model which was designed in [2]. The experimental results are shown in Fig. 8 and 9. In Fig. 8(a), the system operates without damping control and there is an oscillation when we start to supply power to the CPL and Fig. 8(b) shows the current and voltage and power waveforms of the CPL, from which we can see that the power of CPL is almost constant. Fig. 9 shows the damped inductor current and capacitor voltage of the input filter with the proposed active damping method.

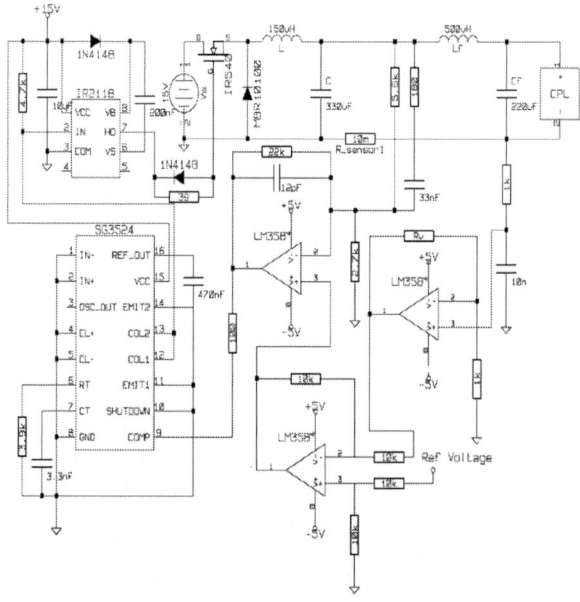

Fig. 7 The experimental circuit for the verification of the proposed active damping method

The 2014 International Power Electronics Conference

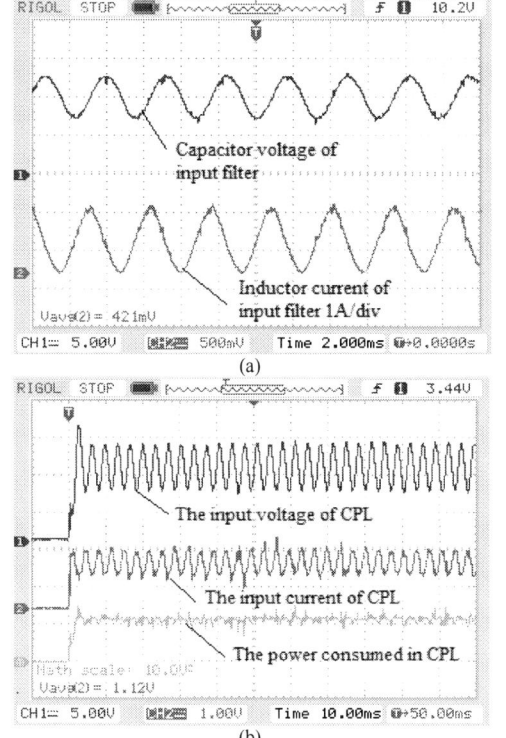

(a)

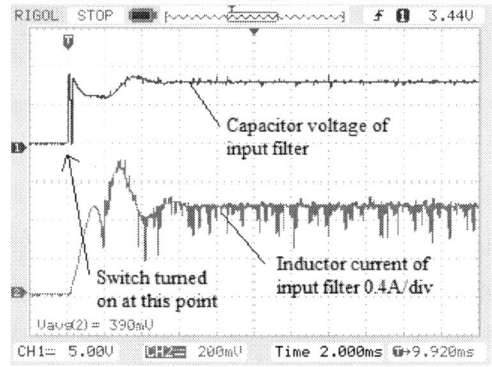

(b)

Fig. 8 (a) The waveform of the capacitor voltage and inductor current of the input filter without damping method (b) the waveform of the input voltage and input current and input power of the CPL without damping method

Fig. 9 The waveform of the capacitor voltage and inductor current of the input filter with the proposed active damping method

V. CONCLUSION

This paper proposed a new active damping method for stabilization of the cascaded connected system in DC microgrids with CPL. The proposed method is designed by building a virtual resistance at the source side converter. The main advantage of this proposed method is that there is no compensating current into the load and hence it does not affect the performance of the CPL during the change of the power condition in DC bus, for example the DC bus voltage. Simulations and experimental results verify the effectiveness of the proposed active damping method. The future work will focus on the application of the proposed active damping method for the stabilization of DC microgrids with multiple parallel connected CPLs.

REFERENCES

[1] M. Cespedes, X. Lei, and S. Jian, "Constant-Power Load System Stabilization by Passive Damping," *Power Electronics, IEEE Transactions on,* vol. 26, pp. 1832-1836, 2011.

[2] A. M. Rahimi and A. Emadi, "Active Damping in DC/DC Power Electronic Converters: A Novel Method to Overcome the Problems of Constant Power Loads," *Industrial Electronics, IEEE Transactions on,* vol. 56, pp. 1428-1439, 2009.

[3] P. Magne, B. Nahid-Mobarakeh, and S. Pierfederici, "General Active Global Stabilization of Multiloads DC-Power Networks," *Power Electronics, IEEE Transactions on,* vol. 27, pp. 1788-1798, 2012.

[4] W. Du, J. Zhang, Y. Zhang, and Z. Qian, "Stability Criterion for Cascaded System With Constant Power Load," *Power Electronics, IEEE Transactions on,* vol. 28, pp. 1843-1851, 2013.

[5] A. Kwasinski and C. N. Onwuchekwa, "Dynamic Behavior and Stabilization of DC Microgrids With Instantaneous Constant-Power Loads," *Power Electronics, IEEE Transactions on,* vol. 26, pp. 822-834, 2011.

[6] C. H. Rivetta, A. Emadi, G. A. Williamson, R. Jayabalan, and B. Fahimi, "Analysis and control of a buck DC-DC converter operating with constant power load in sea and undersea vehicles," *Industry Applications, IEEE Transactions on,* vol. 42, pp. 559-572, 2006.

[7] A. M. Rahimi, G. A. Williamson, and A. Emadi, "Loop-Cancellation Technique: A Novel Nonlinear Feedback to Overcome the Destabilizing Effect of Constant-Power Loads," *Vehicular Technology, IEEE Transactions on,* vol. 59, pp. 650-661, 2010.

[8] P. Magne, D. Marx, B. Nahid-Mobarakeh, and S. Pierfederici, "Large-Signal Stabilization of a DC-Link Supplying a Constant Power Load Using a Virtual Capacitor: Impact on the Domain of Attraction," *Industry Applications, IEEE Transactions on,* vol. 48, pp. 878-887, 2012.

[9] L. Xinyun, A. J. Forsyth, and A. M. Cross, "Negative Input-Resistance Compensator for a Constant Power Load," *Industrial Electronics, IEEE Transactions on,* vol. 54, pp. 3188-3196, 2007.

978-1-4799-2706-7/14 $31.00 © 2014 IEEE

The 2014 International Power Electronics Conference

Expansion of Operating Range and Improvement of Torque Response of PMSM Drive by Using Model Predictive Control

Abstract—This paper describes expansion of operating range and improvement of torque response of Permanent Magnet Synchronous Motor (PMSM) drive by using Model Predictive Control (MPC). Controller based on MPC selects the voltage vector of inverter by predicting and evaluating future states (torcue, current) of PMSM. Proposed torque control system of PMSM based on MPC makes it possible to drive PMSM in overmodulation mode (including rectangular-wave mode) of inverter in addition to linear mode . In this paper, performance of proposed method is shown by simulation that proposed MPC control is compared and evaluated with high gain conventional PI-based vector control about expansion of operating range and improvement of torque response.

I. INTRODUCTION

Permanent Magnet Synchronous Motor (PMSM) has been used in a wide area because of high efficiency and hight power per volume and weight. Recently, PMSM is applied to Electric Vehicles (EV) and Hybrid Electric Vehicles (HEV). In these applications, expansion of operating range and improvement torque response is important. These performance realizes easily by increase of DC-link voltage. However, increase of DC-link voltage causes a new problem such as system volume, weight, and cost.

Therefore overmodulation mode (including rectangular-wave mode) of inverter has been used mainly to realizes these performance. This method makes it possible to increase the upper limit of the inverter output voltage. On the other hand, the inverter output voltage contains low order harmonic components which make current control loop unstable. Conventionally, the controller in overmodulation mode is switched to another one, such as voltage phase controller[1]. However, transient response cased by switching these controllers becomes a new problem.

To overcome this problem, we have proposed a torque control system of PMSM based on Model Predictive Control (MPC)[2] which makes it possible to drive PMSM both in linear amplitude mode of inverter, that modulation ratio is less than 1, (linear mode) and in overmodulation mode (including rectangular-wave mode). Proposed controller directly selects the switch states of inverter instead of the voltage reference[3]. In this sense, it may be similar to Direct Torque Control (DTC)[4], Space Vector Modulation(SVM). However, our proposed method have the advantage which means that is able to decide the output with prediction of the future current and torque behavior by using the inverter switching

state and mathematical model of PMSM. It makes inverter output voltage appropriate even in the transient state as the steady state.

In this paper, simulation result shows the proposed system realizes expansion of operating range and improvement of torque response, which is compared with simulation results of conventional PI-based vector control.

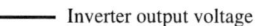
——— Inverter output voltage
– – – – Fundamental frequency component of inverter output voltage

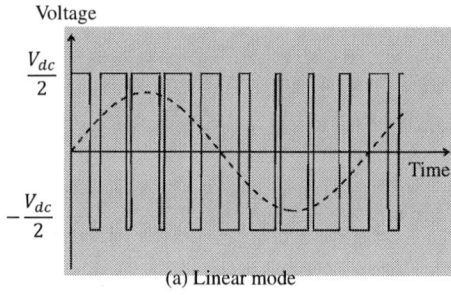
(a) Linear mode

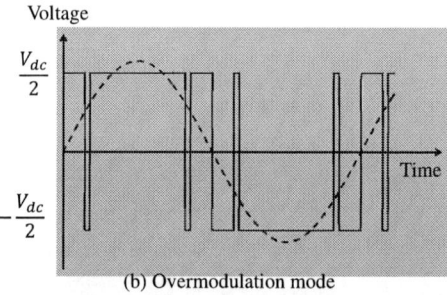
(b) Overmodulation mode

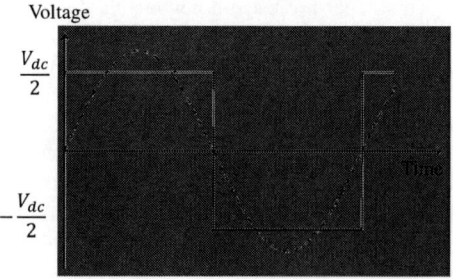
(c) Overmodulation mode (Rectangular wave mode)

Fig. 1.   Modualtion mode of PWM inverter

978-1-4799-2706-7/14 $31.00 © 2014 IEEE   3557

II. TORQUE CONTROL SYSTEM BASED ON MPC

Proposed torque control system of PMSM based on MPC is shown in Fig 2. Current reference generator, for current control just at steady state, consists of Maximum Torque per Ampere control (MTPA), Flux Weakening control (FW), and so on. In this following section, the switching controller based on MPC (hereafter MPC controller) in Fig. 2 is described.

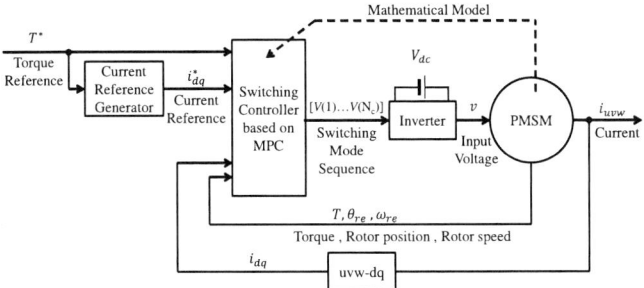

Fig. 2. Torque control system based on MPC

A. MPC controller and output voltage of inverter

This paper assumes that PMSM is driven by two-level voltage source PWM inverter as shown in Fig. 3. In this type of inverter, there are eight different switching states determined by the combination of the switching state at each phase. Each switching state corresponds to each of inverter output voltage vectors (V_0 to V_7) shown in Fig. 4.

Conventional current control system generates the voltage reference without considering the inverter. Switching control of inverter that is based on the reference is done by some PWM methods which are completely independent of the current control. However, in overmodulation mode, the inverter output is not able to follow the reference because of nonlinear amplification of voltage and low order high harmonic components. These problems have a negative influence on current response and control performance.

On the other hand, MPC controller directly selects one of eight voltage vectors under the limitation of DC-link voltage and switching states. Because of this, proposed system is able to consider the inverter which has the property of outputting spatial discrete voltage vector by searching and selecting a suitable voltage vector from among eight voltage vectors.

B. Selection of voltage vector based on MPC

This section shows how MPC controller determine the voltage vector. First, the finite-time period from $n = 0$ to $n = N_p$ (N_p is finite positive integer.) is defined as the prediction period as shown in Fig. 5 Next, the finite sequence of the inverter voltage vectors (V_0 to V_7) in the prediction horizon (the total time for prediction) $V_p^{(k)}$ is shown as follows:

$$V_p^{(k)} = [V^{(k)}(1), V^{(k)}(2), \cdots, V^{(k)}(N_p)] \quad (1)$$
$$V^{(k)}(n) \in V_0 \sim V_7$$

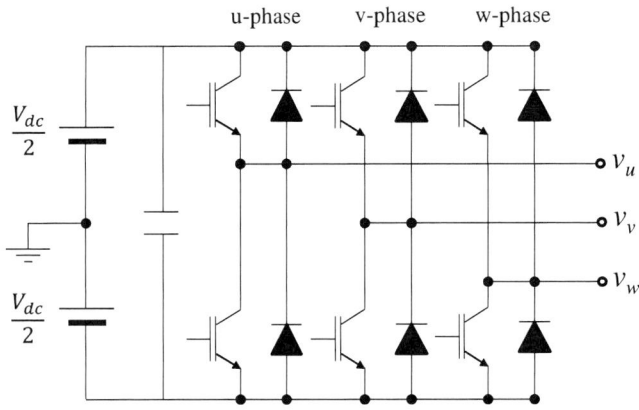

Fig. 3. Two-level voltage source PWM inverter

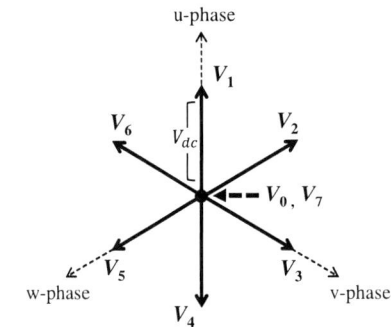

Fig. 4. Inverter output voltage vector

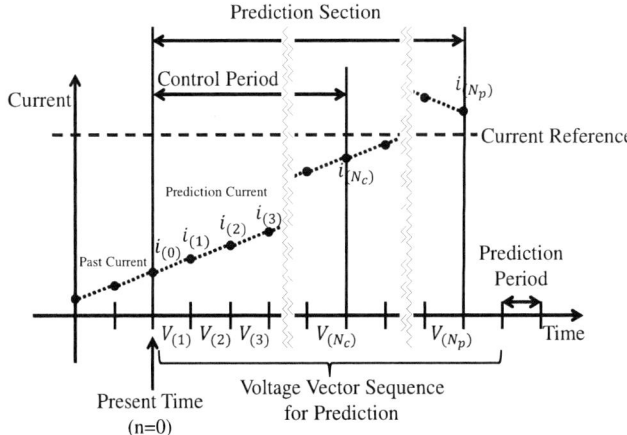

Fig. 5. Prediction of the current behavior in MPC controller

The superscript (k) denotes that $V_p^{(k)}$ is k-th sequence of all possible sequences. The coordinate transformation using the rotor position eventually yields the voltage vector sequence in $d-q$ axis ($[v_{dq}^{(k)}(0), v_{dq}^{(k)}(1), \cdots, v_{dq}^{(k)}(N_p-1)]$). Then, assuming that the rotor speed is constant in the prediction horizon, MPC controller predicts current behavior and torque behavior by using a mathematical model of PMSM. The state equation of PMSM is given as follows:

$$\frac{d}{dt}\boldsymbol{i_{dq}} = \underbrace{\begin{bmatrix} -\frac{R}{L_d} & \frac{\omega_{re}L_q}{L_d} \\ -\frac{\omega_{re}L_d}{L_q} & -\frac{R}{L_q} \end{bmatrix}}_{\boldsymbol{A}} \boldsymbol{i_{dq}}$$

$$+ \underbrace{\begin{bmatrix} \frac{1}{L_d} & 0 \\ 0 & \frac{1}{L_q} \end{bmatrix}}_{\boldsymbol{B}} (\boldsymbol{v_{dq}} - \underbrace{\begin{bmatrix} 0 \\ \frac{\omega_{re}K_E}{L_q} \end{bmatrix}}_{\boldsymbol{e}}) \quad (2)$$

where $\boldsymbol{i_{dq}}(= [i_d\, i_q]^T)$ and $\boldsymbol{v_{dq}}(= [v_d\, v_q]^T)$ are the $d-q$ axis vectors which denote the stator currents and the stator voltages. ω_{re} is the electric rotor speed. R is the resistance and L_d, L_q are the d-axis and q-axis inductances of a stator winding. K_E is the EMF constant.

The discrete-time state equation of PMSM is given as follows:

$$\boldsymbol{i_{dq}}(n+1) = \underbrace{exp(\boldsymbol{A}t_s)}_{A'}\boldsymbol{i_{dq}}(n)$$

$$+ \underbrace{\int_0^{t_s} exp(\boldsymbol{A}\tau)d\tau\boldsymbol{B}}_{B'}(\boldsymbol{v_{dq}}(n) - \boldsymbol{e_{dq}}) \quad (3)$$

where t_s is prediction period. Relationship between t_s and control period T_c is given as follows:

$$t_s = T_c/N_c \quad (4)$$

where N_c is finite positive integer, which defines a ratio of prediction period to control one. In proposed controller, upper limit of average switching frequency of inverter is restricted by T_c and phase resolution of inverter voltage is decided by N_c.

The variable n is the discrete-time instant and $n = 0$ points the present time. The equation (3) and the voltage sequence yield the sequence of the predicted current behavior in the prediction horizon $I_p^{(k)}$ as follows:

$$\boldsymbol{I_p}^{(k)} = [\boldsymbol{i_dq}^{(k)}(0), \boldsymbol{i_dq}^{(k)}(1), \cdots, \boldsymbol{i_dq}^{(k)}(N_p)] \quad (5)$$

If the predicted current behavior can be obtained, the torque behavior in the future can be predicted. Torque equation used in this paper is as follows:

$$T = P_n(K_E + (L_d - L_q)i_d)i_q \quad (6)$$

where Pn is a number of pole pairs.

Finally, the evaluation function (For example, the equation (8) is used in linear mode.) is calculated from the predicted current sequence $\boldsymbol{i_dq}(n)$, torque sequence $T(n)$, the current

reference $\boldsymbol{i_dq}^*$ and torque reference T^*. After prediction and the evaluation mentioned above are done to all possible voltage vector sequences, MPC controller picks up k which minimizes the value of the evaluation function and outputs the N_c step voltage vector sequence to inverter. This process is repeated at every control period.

$$V_{out} = [V^{(k)}(1), V^{(k)}(2), \cdots, V^{(k)}(N_c)] \quad (7)$$

$$J = W_T \sum_{n=1}^{N_p} |T^* - T(n)|$$

$$+ W_{i_d}\sum_{n=1}^{N_p} |i_d^* - i_d(n)| + W_{i_q}\sum_{n=1}^{N_p} |i_q^* - i_q(n)| \quad (8)$$

C. Prediction method according to modulation rate

In overmodulation mode (including rectangular-wave mode), inverter output voltage contains low order harmonic components. As the result, MPC controller is not able to select appropriate voltage vector by using the method which is only considered for linear mode of inverter.

To solve this problem, the prediction horizon must be expanded to the cycle of current ripple and the evaluation function must be changed to conform the average of current to current reference(equation (10)).

On the other hand, the prediction horizon expanded increases calculation cost. Therefore, it is important to reduce the number of pattern for prediction and evaluation.

In this paper, the number of pattern is reduced by changing searching method in rectangular-wave mode. In rectangular-wave mode, voltage vector of inverter changes six times in the electric rotor cycle. In other words, the switching state changes only once in cycle of current ripple. Therefore, it is enough to search only switching timing of voltage vector. Present voltage vector is selected by the voltage reference described in equation (9). For example , in figure 6, selected present voltage vector is V_2. Therefore, it is necessary to determine changing timing from V_2 to V_3.

As the result, it is possible to reduce the calculation cost of prediction in rectangular-wave mode.

$$v_{ref} = \begin{bmatrix} R & -\omega_{re}L_q \\ \omega_{re}L_d & R \end{bmatrix} i_{ref} + \begin{bmatrix} 0 \\ \omega_{re}K_e \end{bmatrix} \quad (9)$$

$$J = W_T |\sum_{n=1}^{N_p}(T^* - T(n))|$$

$$+ W_{i_d}|\sum_{n=1}^{N_p}(i_d^* - i_d(n))| + W_{i_q}|\sum_{n=1}^{N_p}(i_q^* - i_q(n))| \quad (10)$$

The 2014 International Power Electronics Conference

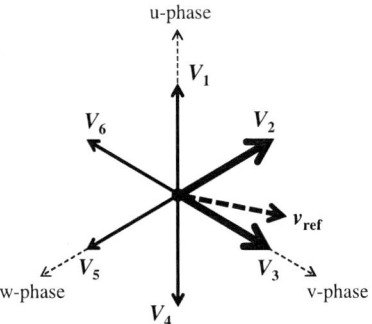

Fig. 6.   Selection of voltage vectors

III.   SIMULATION

This section shows the performance of MPC control system in simulation. The parameters of PMSM and inverter are shown in TABLE I. The operating points on Speed-Torque map is shown in Fig.7. At expect the Point 3, inverter is operated in linear mode and at the Point 3, it must be operated in rectangular-wave mode. The current references are calculated based on MTPA. The rotor speed is controlled as constant by load motor.

Fig. 8 and 11 show the simulation results of torque step response by using conventional PI controller and MPC controller. TABLE II and III shown parameter of PI controller and MPC controller respectively. First, proposed torque control system is able to use rectangular-wave mode in steady state in Point 3. Next, torque response is improved in Point 1 to Point 2, Point 2 to Point 1, Point 1 to Point 4 and Point 4 to Point 1. Among them, the response improvement in Point 1 to Point 2 and Point 4 to Point 1 is caused by using overmodulation mode in transient state. Moreover, response between linear mode and overmodulation mode is stable in Point 2 to Point 3 and Point 3 to Point2.

Fig. 9 and 10 shows simulation results of high gain PI-based vector control in same drive pattern. By high gain of PI controller, transient response is improved as same as proposed MPC controller in Point1 to Point2, Point2 to Point3 and Point4 to Point1 . Operating point, however, become shrunk in overmodulation mode. Moreover, torque response by MPC controller is fast in Point 2 to Point 1 , Point 3 to Point 2 and Point 1 to Point 4. Fig. 12, Fig. 13 and Fig 14 shows torque response and transitions of voltage vector. In comparison to PI controller, Proposed MPC controller selects voltage vector which decrease torque (Fig. 15) positively in transient state.

IV.   CONCLUSION

This paper described expansion operating range and improvement of torque response of PMSM by using Model Predictive Control MPC. Proposed torque control system of PMSM based on MPC is able to use inverter overmodulation mode (including rectangular-wave mode) in addition to linear mode. As the result, Inverter output voltage increases in transient state in addition to steady state. Simulation result showed the proposed torque control system realizes expansion of operating range and improvement of torque response, is compared and evaluated with high gain conventional PI-based vector control.

REFERENCES

[1] H.Nakai, H.Ohtani, E.Satoh, Y.Inaguma, ”Development and Testing of the Torque Control for the Permanent-Magnet Synchronous Motor”, IEEE Trans. Industrial Electronics, Vol.52, No.3, pp. 800-806, 2005.

[2] M.Kadota, S.Lerdudomsak, S.Doki, S.Okuma, ”A Novel Current Control System of IPMSM Operating at High Speed Based on Model Predictive Control”, The 4th Power Conversion Conference, 2007.

[3] Zhixun Ma, Saeid Saeidi, Ralph Kennel , ”Continuous Set Nonlinear Model Predictive Control for PMSM Drives”, EPE’13 ECCE Europe, 2013.

[4] G.S. Buja, M.P. kazmierkowski, ”Direct Torque Control of PWM inverter-Fed AC motors-A survey”, IEEE trans. Industrial Electronics, Vol.51, No.4, pp.744-757, 2004.

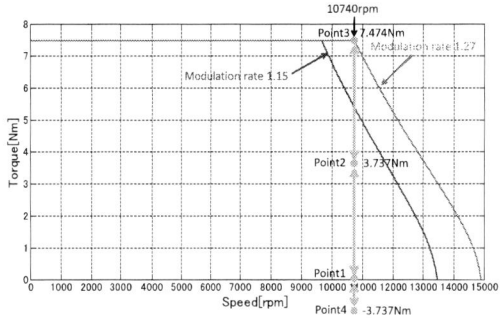

Fig. 7.   Operating points in simulation

TABLE I.   PARAMETERS OF PMSM AND INVERTER

Resistance(R)	0.13 Ω
Inductance(d-axis) (L_d)	0.12 mH
Inductance(q-axis) (L_q)	0.47 mH
EMF constant (K_E)	0.02 V/(rad/s)
Number of pole pairs(P_n)	6
DC-link voltage (V_{DC})	240 V
Switching frequency limit(f_{clim})	25 kHz

TABLE II.   PARAMETERS OF PI CONTROLLER

Control period	$20\mu s$
Cut-off frequency	2000,5000,10000 rad/s
Anti-windup scheme	Stop integration
Voltage limiter	Minimum phase error method
Voltage modulation	3rd harmonic injection PWM

TABLE III.   PARAMETERS OF CONTROLLER BASE ON MPC

Control period (T_c)	$20\mu s$
Prediction horizon(expect in Point 3) (T_p)	$20\mu s$
Prediction horizon(in Point3) (T_p)	$165\mu s$
Prediction period(t_s)	$1\mu s$

The 2014 International Power Electronics Conference

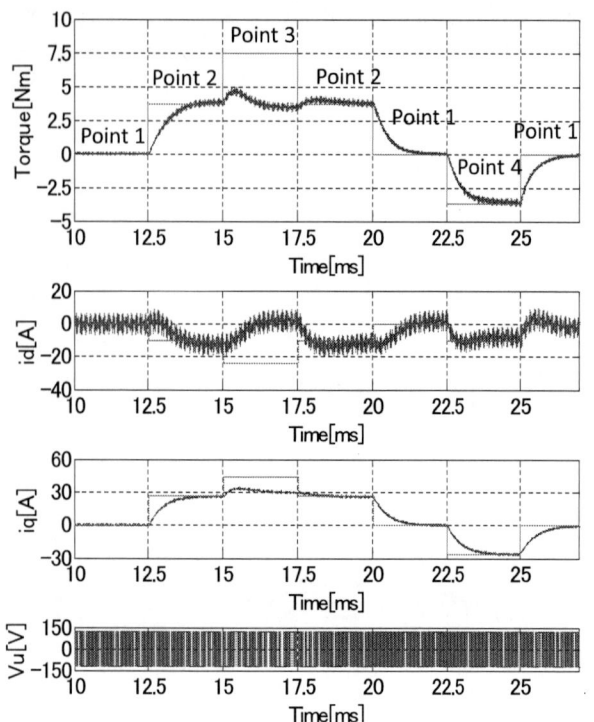

Fig. 8.   Torque and current response of conventional torque control system(2000rad/s)

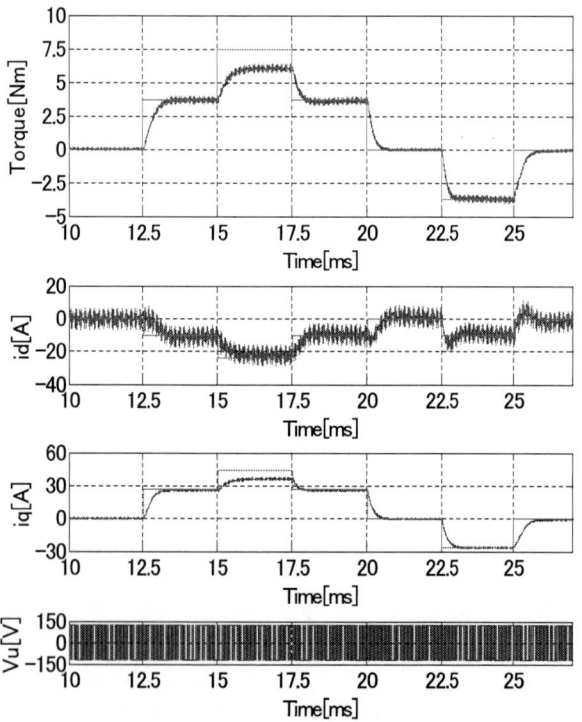

Fig. 9.   Torque and current response of conventional torque control system(5000rad/s)

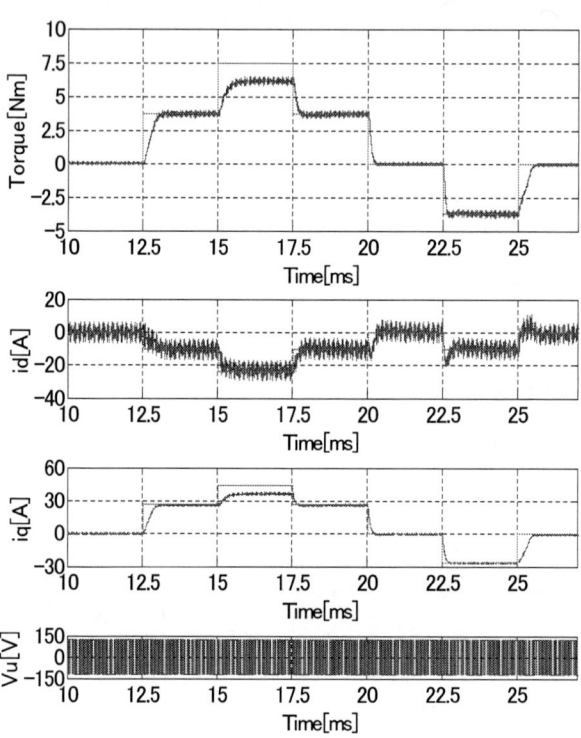

Fig. 10.   Torque and current response of conventional torque control system(10000rad/s)

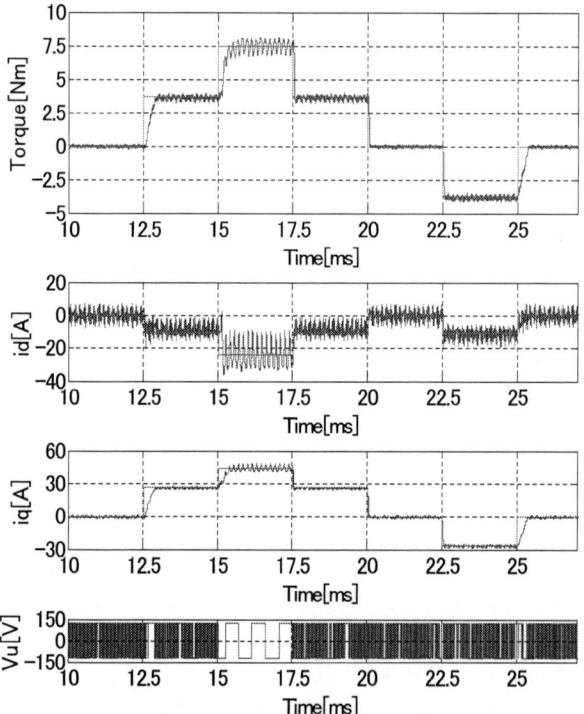

Fig. 11.   Torque and current response of proposed control system based on MPC

978-1-4799-2706-7/14 $31.00 © 2014 IEEE   3561

The 2014 International Power Electronics Conference

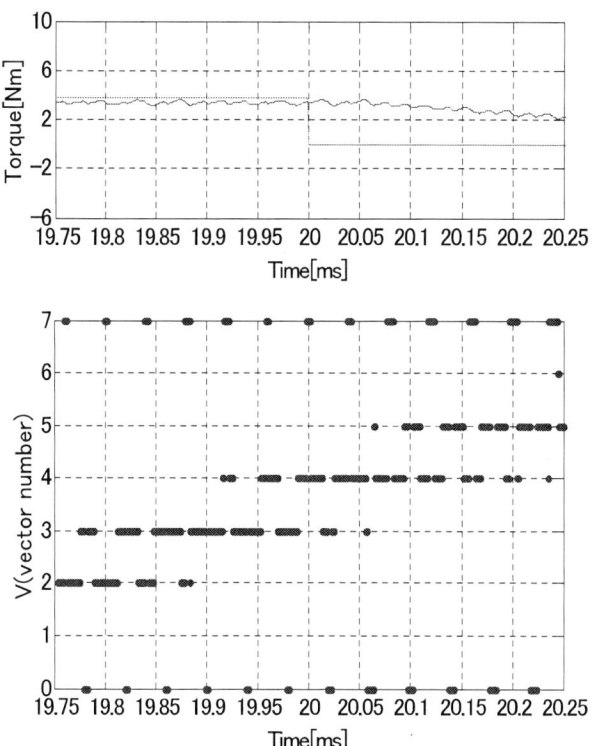

Fig. 12. Torque response and transition of voltage vector of conventional control system(2000rad/s)

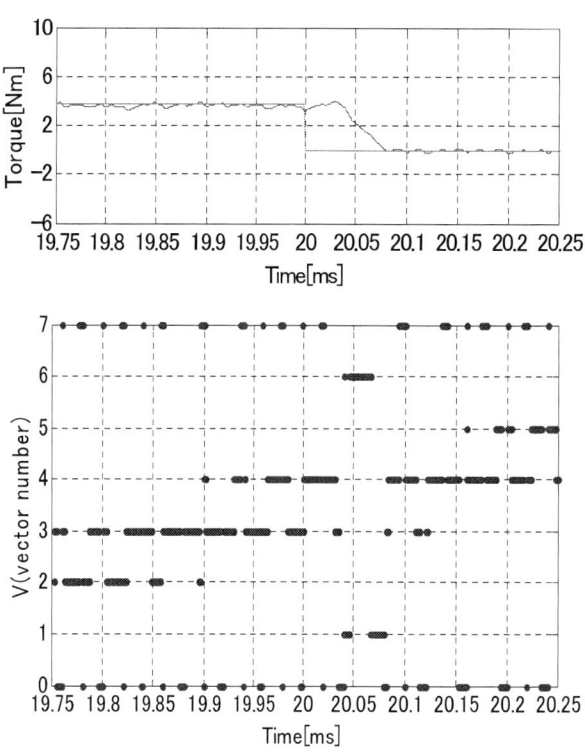

Fig. 14. Torque response and transition of voltage vector of proposed control system based on MPC

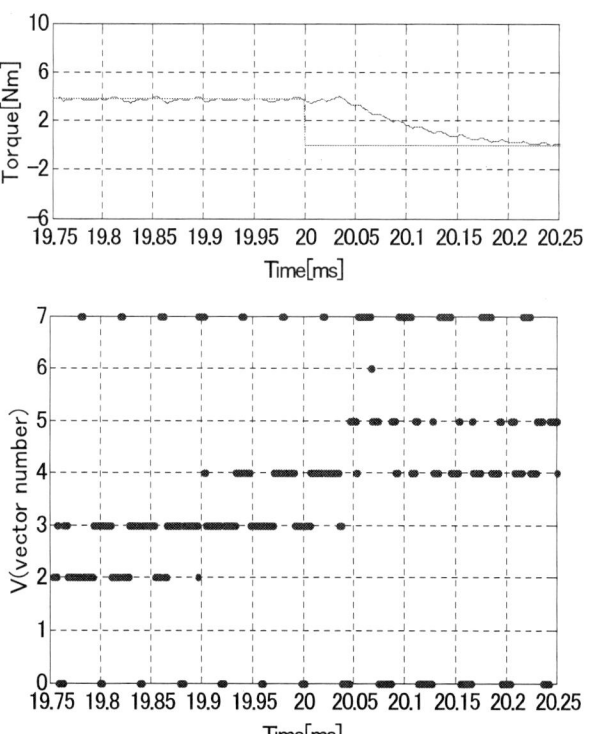

Fig. 13. Torque response and transition of voltage vector of conventional control system(10000rad/s)

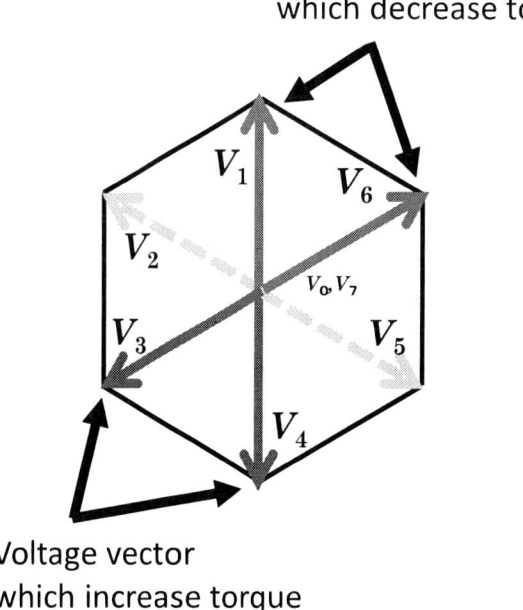

Fig. 15. Relationship between voltage vector and torque

Nonlinear Model Predictive Torque Control of a Load Commutated Inverter and Synchronous Machine

Stefan Almér, Thomas Besselmann and Joachim Ferreau

ABB Corporate Research

Segelhofstrasse 1K, 5405 Baden-Dättwil, Aargau, SWITZERLAND

Email: stefan.almer@ch.abb.com

Abstract—**The paper at hand considers the design of a controller for torque regulation of a variable speed synchronous machine fed by a line commutated rectifier and a load commutated inverter. The control approach is model predictive control where a constrained optimal control problem is solved to minimize the deviation of the torque from its reference. The dynamic model of the converters and drive is nonlinear and considers both the rectifier and inverter firing angles as control input. Controlling both firing angles simultaneously, as opposed to in a cascaded manner, implies improved potential for dynamic performance and disturbance rejection. In particular, the controller handles voltage dips on the line better than a conventional PI controller. The nonlinear MPC solution is implemented through on-line optimization. The optimization algorithm has been implemented on an embedded system and has been shown to execute sufficiently fast for the targeted control frequency.**

I. INTRODUCTION

In recent years there has been considerable interest in model predictive torque and/or speed control of variable speed electric machines, see *e.g.*, [1]–[4]. However, the focus of most, if not all, research has been on voltage source converter topologies. In the present paper we consider a synchronous machine fed by *current source* converters. To the best of our knowledge, model predictive control (MPC) has not been applied to this type of system prior to this paper.

The paper considers a variable speed synchronous machine connected to the grid via a line commutated rectifier and a load commutated inverter (LCI) [5], and considers the design of a torque controller. The work is motivated by gas compression plants which are often situated in remote locations and operate under weak grid conditions. Our goal is to design an improved torque controller which can keep delivering torque during partial loss of grid voltage (so-called brown-outs).

The control approach is set in the framework of MPC. The MPC formulation considers both the rectifier and inverter firing angles as control inputs and stabilizes the DC link current and rotor fluxes while tracking the torque reference. The fact that the rectifier and inverter angles are controlled "simultaneously" rather than in a cascaded fashion implies that the controller has potential for higher dynamic performance compared to conventional

approaches. In particular, there is increased potential to deliver torque during brown-outs.

The literature on MPC of power electronics is to a large extent focused on so-called finite control set MPC (FCS-MPC), see *e.g.*, [6]. In FCS-MPC, the control input is restricted to a finite set of values and the MPC problem is solved by enumerating all possible combinations of the input over the prediction horizon. The FCS-MPC approach has a number of drawbacks, including very short prediction horizon, chattering and unpredictable and time varying switching frequency, see [7] for an extensive discussion. The drawbacks of FCS-MPC are mitigated by considering a continuous control variable, such as a duty-cycle or a firing angle, which is mapped to switching action through a modulator. The control approach outlined in the present paper belongs to the later class of methods which considers a continuous control variable.

Implementing the model predictive controller requires to solve a constrained nonlinear, nonconvex optimization problem in real-time. This is a challenging task as our application asks for sampling times of one millisecond or less and the embedded computing power is limited. Solving nonlinear MPC problems in such a situation requires both a careful problem formulation and highly efficient, state-of-the-art optimization algorithms. In this paper, we follow the promising approach of auto-generating customized nonlinear MPC algorithms that are tailored to the problem at hand based on a symbolic problem formulation as proposed in [8].

The paper is organized as follows. In Section II we briefly describe the synchronous machine and the load-commutated inverter. Afterwards a dynamic model of this system is presented in Section III. The developed model predictive torque controller is outlined in Section IV, whereas the state estimation is described in Section V. Section VI contains simulation results with the proposed controller. Finally conclusions are drawn in Section VII.

II. CURRENT SOURCE CONVERTERS AND SYNCHRONOUS MACHINE

The paper considers a variable speed drive system composed of line commutated rectifiers, inductive DC link, load commutated inverters and a synchronous machine, see Fig. 1. The considered configuration of the

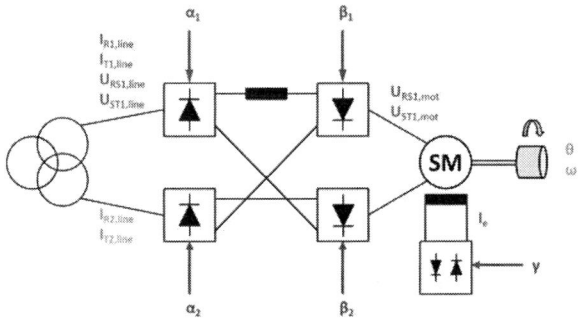

Fig. 1. Variable speed drive system comprised of line commutated rectifier, load commutated inverter and synchronous machine.

converter is a 12/12 pulse setting, whereas the proposed scheme can easily be adapted to other configurations. The shown rectifier and inverter blocks consist of six pulse thyristor bridges. This type of drive systems are suitable for high power applications in the range of several ten's of Megawatts such as high speed compressors or rolling mills.

We note that the synchronous machine has two sets of three phase windings. Each winding pair is connected to its own inverter and both are physically displaced 30° from each other.

The control inputs (signals to be manipulated by the controller) are the firing angles α_1, α_2 of the rectifiers and β_1, β_2 of the inverters. Furthermore, the excitation flux is controlled by an excitation voltage v.

III. DYNAMIC MODEL

The rotor excitation flux varies considerably slower than the other states of the system. We therefore control the excitation flux with a slower outer loop and the design of this control loop is not discussed in this paper. The control variable (excitation voltage) v discussed above is therefore not considered in the sequel and the excitation flux is treated as a parameter.

For reduced computational complexity we impose that both rectifiers apply the same firing angles and that both inverters also apply the same firing angles. Thus, the control input is

$$\alpha \ : \ \text{rectifier firing angle}$$
$$\beta \ : \ \text{inverter firing angle}$$

and we impose $\alpha_1 = \alpha_2 = \alpha$ and $\beta_1 = \beta_2 = \beta$.

The state of the system consists of the DC ink current and machine damper windings fluxes:

$$\psi_D \ : \ \text{damper winding flux, d-component}$$
$$\psi_Q \ : \ \text{damper winding flux, q-component}$$
$$i_{DC} \ : \ \text{DC link current.}$$

Using an adaptation of the model of the dual-three phase synchronous machine described in [9] and using an averaged approximation of how the stator currents depend

on the inverter firing angle, the dynamics of the damper winding fluxes

$$\Psi := \begin{bmatrix} \psi_D & \psi_Q \end{bmatrix}'$$

are modelled as

$$\frac{d}{dt}\Psi = A\Psi + Bi_{DC}\begin{bmatrix} \cos(-\beta + \Delta) \\ \sin(-\beta + \Delta) \end{bmatrix} + F\psi_f, \quad (1)$$

where ψ_f is the excitation flux, Δ is the stator voltage angle in the dq-frame and where A, B, F are constant matrices.

The DC link current dynamics are described by

$$\frac{d}{dt}i_{DC} = \frac{1}{L_{DC}}\Big(- R_{DC}i_{DC} + u_{rec,1} + u_{rec,2}$$
$$- u_{inv,1} - u_{inv,2}\Big) \quad (2)$$

where L_{DC}, R_{DC} are the inductance and parasitic resistance of the inductor and where $u_{rec,i}$, $u_{inv,i}$ are the DC voltages of the rectifier and inverter bridges respectively. We adopt an averaged model to describe the relation between the AC and DC side voltages of the rectifier and inverter. Neglecting the switching and commutation intervals we have

$$u_{rec,i} \approx kU_L\cos(\alpha), \quad u_{inv,i} \approx kU_{M,i}\cos(\beta) \quad (3)$$

where k is a constant, U_L is the amplitude of the line voltages and $U_{M,i}$ are the amplitudes of the stator winding voltages. The line voltage amplitude U_L is a parameter in the MPC problem formulation. The stator voltage amplitudes $U_{M,i}$ of the motor are a (nonlinear) function of the system state. The equations (1)-(3) comprise the dynamic model of the synchronous machine, converters and DC link used in the MPC problem formulation.

A. Torque Expression

The MPC problem formulation penalizes the deviation of the torque from a given reference and we therefore need an expression for the torque. The torque is given by

$$T = (\psi_{d,1}i_{q,1} + \psi_{d,2}i_{q,2} - \psi_{q,1}i_{d,1} - \psi_{q,2}i_{d,2})$$

where $\psi_{d,i}$, $\psi_{q,i}$, $i_{d,i}$, $i_{q,i}$ are the stator fluxes and currents. Using the flux linkage equations and an averaged approximation of how the stator currents depend on the inverter firing angle, the torque can be expressed as a nonlinear function of the system state.

IV. MODEL PREDICTIVE CONTROLLER

At each sampling time, the model predictive controller takes an estimate of the system state as initial condition and minimizes a finite time horizon cost integral subject to the dynamic constraints of the system and constraints on the state and input. The cost criterion is

$$J := \int_{kt_s}^{kt_s+t_p} (T - T_{ref})^2 dt \quad (4)$$

where t_s is the sampling period, t_p is the prediction horizon length, T_{ref} is the torque reference.

A. Controller Implementation and On-Line Solution

In order to minimize criterion (4), the corresponding optimal control problem first needs to be discretized in time. If the dynamic model would be linear, one would only need to perform this problem discretization once before the actual runtime of the controller. In that case, the only computational effort to be performed on-line would be solving a convex quadratic programming (QP) problem. Recent years have seen a rapid development of on-line QP solvers that are able to solve such kind of linearized problems in the milli- or even microsecond range on embedded hardware, see *e.g.*, [10]–[13]. Since our model of the drive comprises nonlinear dynamics, we are forced to discretize the optimal control problem on-line at each sampling instant. Along with this discretization in time, we also compute first-order derivatives of the state trajectory with respect to the initial state value and the control moves along the horizon (also called sensitivities). In doing so, we obtain a discrete-time linearization of the optimal control problem, which corresponds to a convex quadratic programming (QP) problem. Finally, we eliminate all state variables from the QP formulation to arrive at a small-scale, dense QP problem. The QP is solved by an adapted variant of the on-line QP solver qpOASES [10]. This procedure to solve nonlinear MPC problems is known as real-time iteration scheme with Gauss-Newton approximation of the second-order derivatives [14].

In order to obtain a highly efficient implementation of the nonlinear MPC algorithm sketched above, we make use of the code generation functionality of the ACADO Toolkit [8], [15]. This software takes a symbolic formulation of the control problem and allows the user to automatically generate customized nonlinear MPC algorithms that are tailored to the specific problem structure. The resulting C code is self-contained, highly optimized and able to run on embedded computing hardware. In our case, the nonlinear MPC controller is running on ABB's controller AC 800PEC, which is based on a 32-bit Power PC processor with a clock speed of up to 600 MHz and also includes an FPGA and a 64-bit IEEE floating point unit. On this platform, the controller has been shown to execute in less than 1 millisecond.

V. State Estimation

The MPC assumes that measurements or estimates of the entire system state are fed to the controller at each sampling time. In the present paper we do *not* assume that the state is measured. Instead, the state is estimated by an observer. The input to the observer corresponds to quantities which are typically measured in an industrial application. This section outlines how the estimates are obtained.

The available measurements are indicated in Figure 1. While the stator current voltages $u_{\alpha\beta} = [u_\alpha, u_\beta]^T$ are available directly, the DC link current i_{DC} and the stator winding currents $i_{\alpha\beta} = [i_\alpha, i_\beta]^T$ can be deduced from the line side currents and the switching positions of the thyristor bridges.

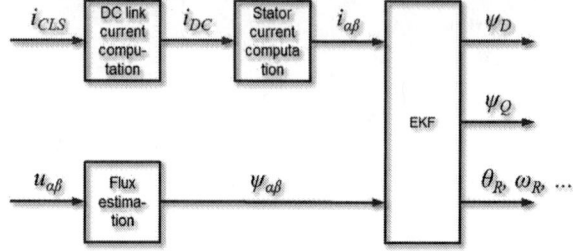

Fig. 2. Structure of the state estimator: Flux estimation and an Extended Kalman Filter.

Figure 2 depicts the structure of the state estimation which consists of two parts; a stator flux estimation and an extended Kalman filter (EKF), [16]. The reason for this separation is that the flux estimator runs at a higher sampling rate in order to increase the estimation accuracy.

The stator flux estimator takes the stator winding currents and voltages in $\alpha\beta$-coordinates and uses the so-called voltage model,

$$ u_{\alpha\beta} = R i_{\alpha\beta} + \frac{d}{dt}\psi_{\alpha\beta}, $$

to compute an estimate of the stator flux $\psi_{\alpha\beta} = [\psi_\alpha, \psi_\beta]^T$. Using this model, the stator flux is computed by integration.

The stator flux and the stator currents are the inputs to the EKF. The EKF uses a dynamic model of the synchronous machine, similar to the one described in Section III, to estimate the remaining states of the system, in particular the damper winding fluxes ψ_D, ψ_Q.

VI. Performance Evaluation

The performance of the controller and observer is evaluated in simulation using a high-fidelity Simulink model. The model implements a complex grid model including transformers and harmonic filters. The rectifier, inverter and DC link are implemented using SimPower components and the synchronous machine is represented by a continuous time model derived from first principles. The Simulink model also includes a model of the load which is a rotating mass. The machine considered has rated power of 12 MW and the controller operates with a sampling period of 1 millisecond.

In the sections below we consider the dynamic performance of the MPC controller and compare it to a state of the art PI-based solution. The prediction horizon of the MPC controller is 20 milliseconds. All signals are represented in the per-unit system.

A. Reference Tracking

We consider the system at nominal steady state and apply steps in the torque reference. The reference, torque and firing angles are shown in Fig. 3 where the left column corresponds to the MPC controller and the right corresponds to the PI controller. It can be seen that both the MPC and PI controller achieve good tracking without

The 2014 International Power Electronics Conference

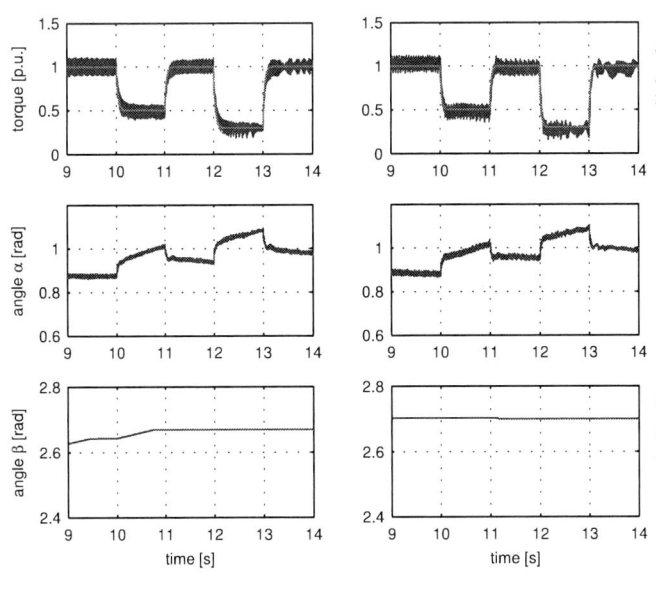

Fig. 3. System response to steps in the torque reference. Left: MPC controller. Right: PI controller.

overshoot or oscillations. The MPC controller exhibits slightly less variation of the rectifier firing angle α.

B. Line Voltage Drop

We consider the system at nominal steady state and apply steps in the line voltage amplitude. The resulting supply voltage amplitude, torque, DC link current and firing angles are shown in Fig. 4 and 5.

The voltage steps considered have a duration of 0.1 seconds and are spaced one second apart. We consider a sequence of increasingly larger drops. The voltage magnitudes are 0.8 p.u., 0.6 p.u., 0.4 p.u. and 0.2 p.u. (see Fig. 4). This kind of disturbances has been reported to occur when the phases of the transmission line briefly touch due to strong wind.

From Fig. 4 it is clear to see that the voltage steps in the source induces high frequency oscillations in the voltage amplitude seen by the rectifier. This is mainly because the harmonic filters which are present at the side are excited. The results show that the MPC controller manages to keep delivering torque for voltage drops down to 0.4 p.u. For lower drops, the torque goes to zero, but the system is not destabilized and the torque returns to the reference when the line voltage returns.

The MPC controller is superior to the PI controller which trips already at the first voltage drop down to 0.8 p.u. (see Fig. 5). The PI controller handles the step down in line voltage, but when the voltage returns, the system trips due to a large over-current in the DC-link.

VII. CONCLUSION

The paper considered nonlinear model predictive control for torque regulation of a synchronous machine supplied by current source converters. The MPC formulation does not impose a cascaded control structure, but uses

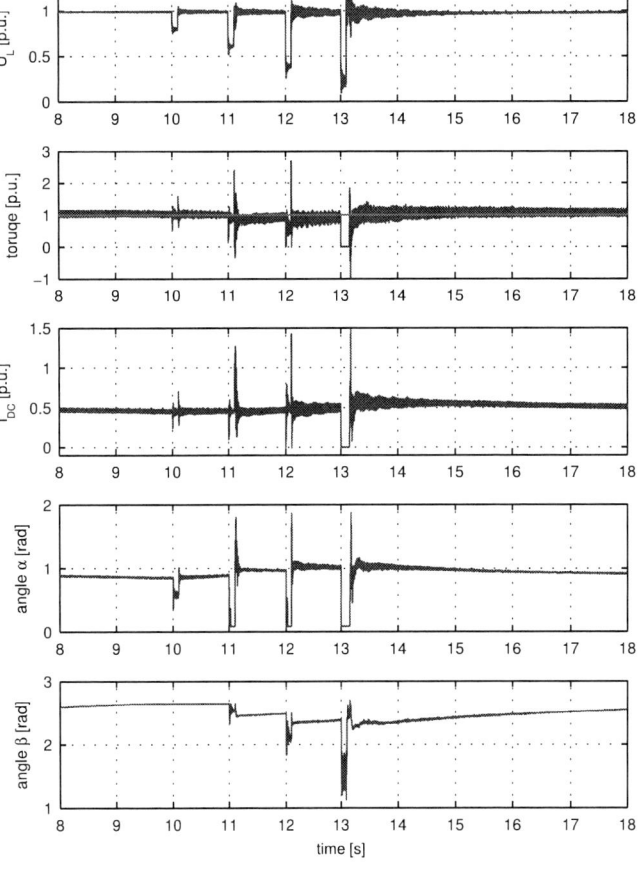

Fig. 4. MPC controller: Response to sudden change in line voltage amplitude.

both the rectifier and inverter angles simultaneously to stabilize the system state and control the torque. This implies increased potential to stabilize the system and reject disturbances. Simulations indeed show that the controller can track the torque reference in the presence of large line voltage drops where a traditional PI controller fails. Thus, the proposed controller increases the system ability of ride-through of brown-outs. Future work could include experimental implementation and development of dynamic models with lower complexity.

REFERENCES

[1] S. Mariéthoz, A. Domahidi, and M. Morari. Sensorless explicit model predictive control of permanent magnet synchronous motors. In *IEEE International Conference on Electric Machines and Drives*, pages 1250–1257, 2009.

[2] F. Morel, J.-M. Retif, Xuefang Lin-Shi, and C. Valentin. Permanent Magnet Synchronous Machine Hybrid Torque Control. *Industrial Electronics, IEEE Transactions on*, 55(2):501–511, 2008.

[3] P. Cortes, M.P. Kazmierkowski, R.M. Kennel, D.E. Quevedo, and J. Rodriguez. Predictive control in power electronics and drives. *Industrial Electronics, IEEE Transactions on*, 55(12):4312–4324, 2008.

[4] T. Geyer, G.A. Beccuti, G. Papafotiou, and M. Morari. Model predictive direct torque control of permanent magnet synchronous motors. In *Energy Conversion Congress and Exposition (ECCE), 2010 IEEE*, pages 199–206, 2010.

The 2014 International Power Electronics Conference

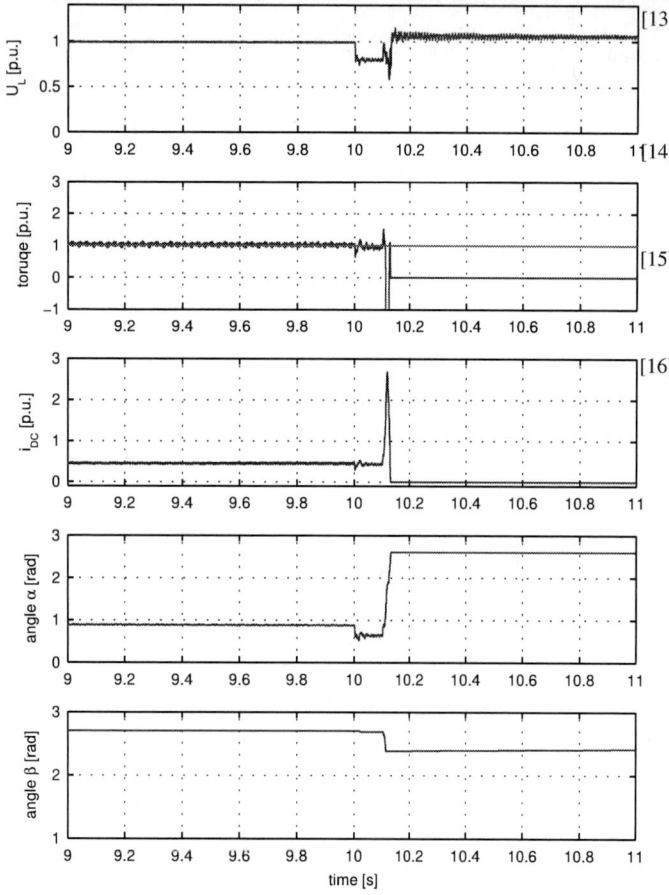

Fig. 5. PI controller: Response to sudden change in line voltage amplitude.

[5] A. B. Plunkett and F. G. Turnbull. Load-Commutated Inverter/Synchronous Motor Drive Without a Shaft Position Sensor. *IEEE Transactions on Industry Applications*, IA-15(1):63–71, 1979.

[6] S. Kouro, P. Cortes, R. Vargas, U. Ammann, and J. Rodriguez. Model Predictive Control - A Simple and Powerful Method to Control Power Converters. *IEEE Transactions on Industrial Electronics*, 56(6):1826–1838, June 2009.

[7] S. Almér, S. Mariéthoz, and M. Morari. Sampled Data Model Predictive Control of a Voltage Source Inverter for Reduced Harmonic Distortion. *IEEE Transactions on Control Systems Technology*, 21(5):1907–1915, Sept 2013.

[8] B. Houska, H.J. Ferreau, and M. Diehl. An Auto-Generated Real-Time Iteration Algorithm for Nonlinear MPC in the Microsecond Range. *Automatica*, 47(10):2279–2285, 2011.

[9] J. Kaukonen. *Salient Pole Synchronous Machine Modelling in an Industrial Direct Torque Controlled Drive Application*. Phd thesis, Lappeenranta University of Technology, March 1999.

[10] H. J. Ferreau, H. G. Bock, and M. Diehl. An online active set strategy to overcome the limitations of explicit MPC. *International Journal of Robust and Nonlinear Control*, 18(8):816–830, 2008.

[11] J. Mattingley and S. Boyd. *Convex Optimization in Signal Processing and Communications*, chapter Automatic Code Generation for Real-Time Convex Optimization. Cambridge University Press, 2009.

[12] S. Richter, S. Mariéthoz, and M. Morari. High-Speed Online MPC Based on a Fast Gradient Method Applied to Power Converter Control. In *Proceedings of the American Control Conference*, pages 4737–4743, Baltimore, MD, USA, 2010.

[13] A. Domahidi, A. Zgraggen, M.N. Zeilinger, M. Morari, and C.N. Jones. Efficient interior point methods for multistage problems arising in receding horizon control. In *IEEE Conference on Decision and Control (CDC)*, pages 668 – 674, Maui, HI, USA, December 2012.

[14] M. Diehl, H.G. Bock, J.P. Schlöder, R. Findeisen, Z. Nagy, and F. Allgöwer. Real-time optimization and Nonlinear Model Predictive Control of Processes governed by differential-algebraic equations. *J. Proc. Contr.*, 12(4):577–585, 2002.

[15] B. Houska, H.J. Ferreau, and M. Diehl. ACADO Toolkit – An Open Source Framework for Automatic Control and Dynamic Optimization. *Optimal Control Applications and Methods*, 32(3):298–312, 2011.

[16] B.D.O. Anderson and J. Moore. *Optimal Filtering*. Prentice-Hall, Englewood Cliffs, N.J, 1979.

Model predictive current control for PMSM considering number of switching operations

Tadanao Zanma, Yuji Yasumura and KangZhi Liu
Graduate School of Engineering
Chiba University, Japan
Email: {zanma, yuji.yasumura}@chiba-u.jp, kzliu@faculty.chiba-u.jp

Abstract—**This paper presents a model predictive-based current control for PMSMs. In our approach, a reference voltage to an inverter can be strictly treated as a discrete voltage vector so that the output saturation of the inverter is avoided. Various constraints can be also taken into consideration in our approach. Both tolerance ranges for control variables and number of switching operations are incorporated into a single objective function in order to consider a trade-off between current ripple and switching operations. The effectiveness of the proposed method is verified through simulations and experiments.**

Keywords—Model predictive control, PMSM, Optimal current control.

I. INTRODUCTION

Several adjustable speed ac motors have been used in various fields with recent advancements in the field of power electronics. In particular, permanent magnet synchronous motors (PMSMs) have received considerable attention, owing to their advantages of low excitation loss, high power factor, high torque per current, and high efficiency. In fact, PMSMs have widely adopted for commercial applications such as electric and hybrid vehicles. Pulse width modulation (PWM) is usually applied in order to control the stator winding current in the PMSM. The switching vector in the inverter is dependent on the PWM carrier frequency. The efficiency of the PMSM drive improves if the carrier frequency is high, but this is not the case owing to the inherent trade-off between control performance and switching loss.

In voltage saturation or overmodulation mode, a voltage reference generated in a controller cannot be realized in the inverter, which may considerably degrade control performance. The positive use of the overmodulation to derive the output voltage of the inverter generates non-sinusoidal output in the inverter. Field weakening control[1], [2] is well known as an extension method of operation area of the PMSM, The approach achieves the extension by modifying the reference in the current constraint even the use of the approach imposes on a limitation of link-voltage in the inverter. Therefore, it is necessary to integrate the current control system taking into the output constraint in the inverter consideration[3]. The use of the overmodulation mode in the inverter augments operation regions of PMSMs Specifically, it is possible to increase the voltage generated in the inverter corresponding to the fundamental element of the three-phase voltage reference by the overmodulation

PWM and one-pulse mode subject to the constant dc-link voltage. Moreover, an amplitude compensation has been proposed for the inverter when it is regarded as a nonlinear amplifier[4]. In a current stabilization[5], harmonic current is estimated using the PMSM model and eliminated from the feedback current. Other reference voltage modification methods have been proposed for voltage saturation appearing in transient response[6], [7], [8]. However, most existing methods need mode switching and complicated preadjustment or computation, which makes its implementation intractable.

In this paper, inverter-driven motor control systems can be regarded as hybrid system in the sense that both continuous variables such as current and discrete voltage vectors. For such a system, model predictive control (MPC)[9] has been receiving much attention as an effective synthesis. In MPC, an optimal control input can be derived subject to constraints on input and/or state, which is also applied to the considered system even if input is discrete. Exploiting the feature above, we have presented a model predictive-based current control[10], [11] or direct torque control[12], [13] for the PMSM. In our earlier works, optimality is ensured in the sense that an evaluation of a given objective function is always minimized while satisfying some imposed constraints. In this paper, we propose a model predictive current control for the PMSM in order to achieve higher performance in the overmodulation region with no additional control part. In contrast to conventional approaches, the discrete output voltage is explicitly treated in our approach, which can avoid the output saturation mentioned above. Moreover, switching of control modes are not required even in the overmodulation, and various constraints can be taken into consideration. Both tolerance ranges for control variables and number of switching operation are considered in an objective function so that the trade-off between current ripple and switching operation is solved. The effectiveness of the proposed method is verified through simulations and experiments.

II. PRELIMINARIES

This section describes the mathematical model of the PMSM, the inverter and its output from the viewpoint of hybrid nature and MPC as a design method.

A. Mathematical model of PMSM

The notations used in this paper are listed in Table I.

978-1-4799-2706-7/14 $31.00 © 2014 IEEE

TABLE I.   NOTATION.

$i_{dq} = \begin{bmatrix} i_d & i_q \end{bmatrix}'$	stator current vector in dq-axis
$v_{dq} = \begin{bmatrix} v_d & v_q \end{bmatrix}'$	stator voltage vector in dq-axis
R	stator winding resistance
L_d, L_q	stator winding inductance of dq-axis
K_e	electromotive force constant
P_n	number of pole pairs
ω_{re}	electric angular velocity of rotor
θ_{re}	electrical angle of rotor

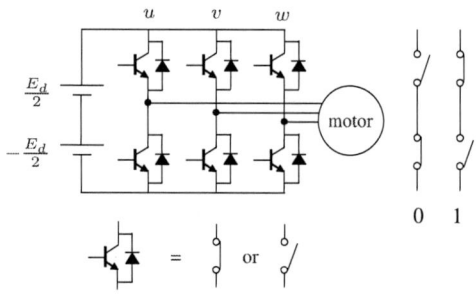

Fig. 1.   Three-phase voltage source inverter.

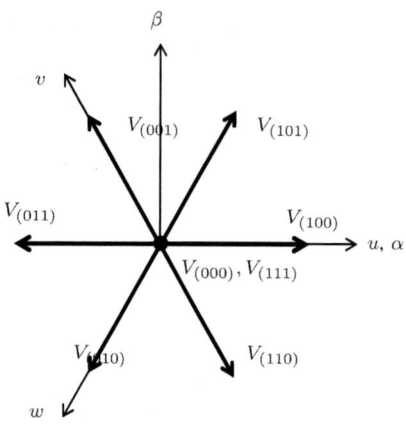

Fig. 2.   Space vectors.

We obtain the following state equation with respect to dq-current of the PMSM.

$$\frac{d}{dt} i_{dq} = \begin{bmatrix} -\frac{R}{L_d} & \omega_{\text{re}} \frac{L_q}{L_d} \\ -\omega_{\text{re}} \frac{L_d}{L_q} & -\frac{R}{L_q} \end{bmatrix} i_{dq} + \begin{bmatrix} \frac{1}{L_d} & 0 \\ 0 & \frac{1}{L_q} \end{bmatrix} v_{dq} + \begin{bmatrix} 0 \\ -\frac{K_e \omega_{\text{re}}}{L_q} \end{bmatrix}. \tag{1}$$

Note that i_d and i_q in Eq. (1) are controllable by v_d and v_q. Each current vector is controlled by the vector control.

B. Three-phase voltage source inverter

In this paper, as shown in Fig. 1, the state when the upper arm is conducted in each phase is denoted by 1 whereas the opposite is denoted by 0. Note that both the arms are not conducted simultaneously to ensure protection of the devices. The output voltage in each phase with the virtual neutral point is limited to $\frac{E_d}{2}$ or $-\frac{E_d}{2}$, for the dc-link voltage E_d. Then, the instantaneous output voltage of the inverter is confined to the $2^3 = 8$ switching patterns. Each instantaneous inverter switching is shown in Fig. 2. The subscript of the vector in Fig. 2 corresponds to u, v and w phases in sequence. For example, $V_{(000)} = \frac{E_d}{2} \begin{bmatrix} -1 & -1 & -1 \end{bmatrix}'$, $V_{(100)} = \frac{E_d}{2} \begin{bmatrix} +1 & -1 & -1 \end{bmatrix}'$, and so forth. Formally,

$$\{V_{(000)}, \ldots, V_{(111)}\} = \left\{ \frac{E_d}{2} s : s \in \{-1, 1\}^3 \right\}. \tag{2}$$

In Eq. (1), v_{dq} is given as $v_{dq} = C_{uvw}^{dq}(\theta_{\text{re}}) v_{uvw}$ for $v_{uvw} \in \{V_{(000)}, \ldots, V_{(111)}\}$. The transformation matrix $C_{uvw}^{dq}(\theta_{\text{re}})$ for transforming uvw to dq is defined as follows.

$$C_{uvw}^{dq}(\theta_{\text{re}}) = \sqrt{\frac{2}{3}} \begin{bmatrix} \cos\theta_{\text{re}} & \cos(\theta_{\text{re}} - \frac{2}{3}\pi) & \cos(\theta_{\text{re}} + \frac{2}{3}\pi) \\ -\sin\theta_{\text{re}} & -\sin(\theta_{\text{re}} - \frac{2}{3}\pi) & -\sin(\theta_{\text{re}} + \frac{2}{3}\pi) \end{bmatrix}. \tag{3}$$

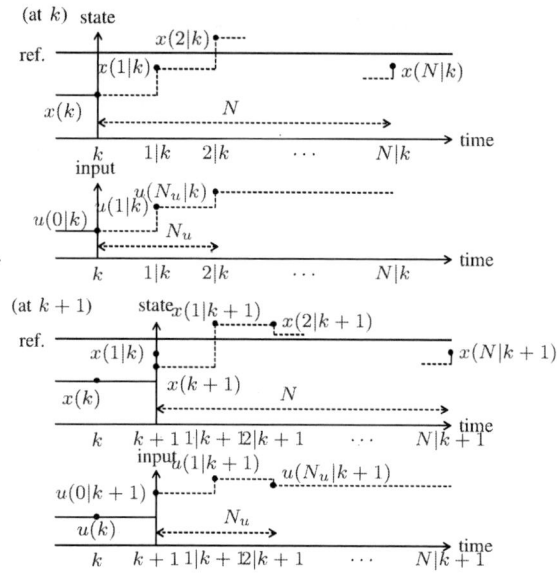

Fig. 3.   Illustrative outline of MPC.

C. Model Predictive Control

In model predictive control (MPC)[14], [15], N-step-ahead behavior of controlled variables is predicted using the system model. An objective function is usually set so that the predictive values approach the reference. In MPC, various constraints can be also taken into account while minimizing an evaluation of the given objective function.

Fig. 3 shows an illustrative outline of the MPC. In Fig. 3, $x(k)$ is the state measured at k, while $x(j|k)$ ($j = 0, \ldots, N-1$) denotes the predicted state j steps ahead from k. The positive integer N is called the predictive horizon. The aim is to find input $u(k)$ at k. The optimal input sequence is calculated $\{u^*(0|k), u^*(1|k), \ldots, u^*(N_u - 1|k)\}$ over $[k, k + N_u]$ (generally, $N_u \leq N$) which minimizes the evaluation of the objective function subject to the constraints. As for the optimal input sequence, only the first entry of $u^*(0|k)$ is applied to the system as $u(k)$ and is held until the

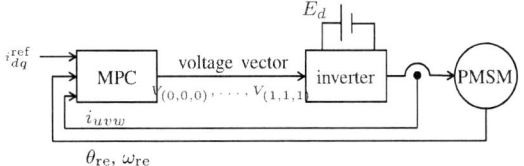

Fig. 4. Configuration of model predictive based current control system.

next instance of sampling, as shown at the bottom of Fig. 3. The process is iterated at each instance of sampling receding the predictive horizon.

Even if the prediction is not perfectly correct owing to modeling and/or measurement errors, the prediction is recalculated according to the latest measurements at each instance sampling. Therefore, MPC functions well as long as the predicted model is admissible.

III. MODEL PREDICTIVE BASED CURRENT CONTROL

This section presents a model predictive-based current control for the PMSM.

A. System Configuration

The proposed control system configuration is shown in Fig. 4. In the proposed method, the command to the inverter is different from that by the conventional PWM method. The switching mode in the proposed system is determined so that an estimation of a given objective function is minimized while satisfying constraints. In the system, a discrete voltage vector is selected from $\{V_{(0,0,0)}, \ldots, V_{(1,1,1)}\}$ at each sampling instance. This is achieved by using the predictive state i_{dq}, θ_{re}, ω_{re} in Fig. 4. A key feature of MPC current controller is summarized as follows. 1) The controller derives an optimal switching vector sequence subject to given constraints, 2) The inverter outputs the optimal switching vector, and 3) No mode-switching in controller is required at all operating points.

Predictive variables are derived on the basis of the mathematical model the PMSM. The discrete-time state-space model of the PMSM is given as

$$i_{dq}(k+1) = A(\omega_{\mathrm{re}})i_{dq}(k) + Bv_{dq}(k) + f(\omega_{\mathrm{re}}), \quad (4)$$

where A, B and f are obtained by a proper discrete transformation of Eq. (1).

B. Optimization

In the proposed approach, the inverter switching command is directly determined. The optimization problem is formally described as follows.

$$\{s^*(0|k), \ldots, s^*(N-1|k)\}$$
$$= \operatorname*{argmin}_{\{s(0|k), \ldots, s(N-1|k)\}} J(i_{dq}(k)) \qquad (5)$$

subject to

$$\begin{cases} s(-1|k) = s(k-1), \quad i_{dq}(0|k) = i_{dq}(k), \\ \theta_{\mathrm{re}}(j|k) = \theta_{\mathrm{re}}(k) + j\omega_{\mathrm{re}}T_s, \\ i_{dq}(j+1|k) = A(\omega_{\mathrm{re}})i_{dq}(j|k) + Bv_{dq}(j|k) + f(\omega_{\mathrm{re}}), \\ v_{dq}(j|k) = C_{uvw}^{dq}(\theta_{\mathrm{re}}(j|k))\frac{E_d}{2}s(j|k), \\ s(j|k) \in \{-1, 1\}^3, \\ (j = 1, \ldots, N). \end{cases}$$
$$(6)$$

Let us discuss two objective functions as $J(i_{dq}(k))$ in Eq. (5).

First, consider the error between the current reference i_{dq}^{ref} and the predictive current. The current reference is assumed to be constant in prediction. Then, the following objective function is considered.

$$J_1(i_{dq}(k)) = \sum_{j=0}^{N-1} ||i_{dq}^{\mathrm{ref}} - i_{dq}(j+1|k)||_Q^2, \qquad (7)$$

where, $Q \succ 0$ is the weighting matrix.

Second, recall that switching operations can be taken into account in our approach. In other words, unnecessary switching operations can be removed by considering the trade-off between switching and current ripple. Then, the following objective function is defined.

$$J_2(i_{dq}(k)) = \sum_{j=0}^{N-1} |s(j|k) - s(j-1|k)|$$
$$+ Q_d p_d(j|k) + Q_q p_q(j|k), \qquad (8)$$
$$p_\star(j|k) = \begin{cases} 0 & \text{if } |i_\star^{\mathrm{ref}} - i_\star(j|k)| \leq \varepsilon_\star, \\ (i_\star^{\mathrm{ref}} - i_\star(j|k))^2 & \text{if } |i_\star^{\mathrm{ref}} - i_\star(j|k)| > \varepsilon_\star, \end{cases}$$
$$(9)$$
$$(\star : d, q),$$

where $\varepsilon_\star$ ($\star : d, q$) is the tolerance bound for the current reference, and $Q_\star \succeq 0$ is the weight, respectively.

The first term on the right side of Eq. (8) accounts for the total number of switching operations of switching devices. It should be noted that the number of switching is a nonnegative integer at most three at each sampling instance.

The last two terms are zero as long as the predictive current value is within the tolerance bound. However, they are active if the predictive value is out of the tolerance.

In Eq. (5), $J_1(\cdot)$ in Eq. (7) or $J_2(\cdot)$ in Eq. (8), is adopted as $J(\cdot)$. The effectiveness dependent on the employed objective function will be compared in the next section.

TABLE II.   PARAMETERS OF IPMSM.

rated power	1.5[kW]
resistance R	0.681[Ω]
inductance (d-axis) L_d	0.0100[H]
inductance (d-axis) L_q	0.0152[H]
e.m.f. constant K_e	0.2832[Vs/rad]
number of pole pairs P_n	3

TABLE III.   PARAMETERS OF CONTROLLER.

average switching frequency	18[kHz]
sampling period T_s	50[μs]
prediction horizon N	1
J_1 in Eq. (7)	$Q = \text{diag}(1,\ 1)$
J_2 in Eq. (8)	$Q_d = \frac{4}{\varepsilon_d},\ Q_q = \frac{4}{\varepsilon_q}$

IV.   SIMULATIONS AND EXPERIMENTS

In this section, the two proposed methods are compared with the conventional vector control through simulations and experiments.

In Eq. (4), $A(\omega_{\mathrm{re}})$, B and $f(\omega_{\mathrm{re}})$ are given as the first order approximation of the Taylor expansion of Eq. (1). Specifically, they are given as $A(\omega_{\mathrm{re}}) = \begin{bmatrix} 1 - \frac{RT_s}{L_d} & \omega_{\mathrm{re}}\frac{L_q T_s}{L_d} \\ -\omega_{\mathrm{re}}\frac{L_d T_s}{L_q} & 1 - \frac{RT_s}{L_q} \end{bmatrix}$, $B = \begin{bmatrix} \frac{T_s}{L_d} & 0 \\ 0 & \frac{T_s}{L_q} \end{bmatrix}$, and $f(\omega_{\mathrm{re}}) = \begin{bmatrix} 0 \\ -\frac{K_e \omega_{\mathrm{re}} T_s}{L_q} \end{bmatrix}$.

A. Condition

Consider that the operating point changes from the linear to the overmodulation regions. The voltage is saturated in the transition. The average switching frequency in the proposed method is adjusted to that in the conventional method for equitable condition.

The parameters of the motor and the proposed current controller are listed in Tables II and III, respectively. In the proposed method, a voltage vector fed during the sampling period is constant, whereas that in the conventional vector control is settled by the reference. Hence, it is difficult to compare the control performance of the proposed and conventional methods under the same condition. In the vector control, which is composed of PI control and decoupling control, the career frequency is 3 [kHz]. Then, the average switching frequency is 18 [kHz]. In contrast, in the proposed method, the average switching frequency is not explicitly determined since the switching is not necessarily required at each sampling instance. Then, the control period is set to be approximately 18 [kHz] for $N_p = 1$ by setting the sampling time is 15 [μs] through preliminary simulations.

Recall that $J_2(\cdot)$ in Eq. (8) has adjustable parameters ε_d and ε_q associated with the tolerance bounds. For comparison, we set $\varepsilon_d = \varepsilon_q = 0.25, 0.15$. In addition, note that the dimension of the first term in Eq. (8) is different from any of those of the rests. Thus, the second and three terms are normalized in order to balance contribution of each term. Recall that the first term at one sampling instance takes at most three as mentioned

TABLE IV.   OPERATING POINTS FOR TRANSITION

	OP 1	OP 2
speed	500[rpm]	500[rpm]
torque reference	1.0[Nm]	3.0[Nm]
current reference (d-axis) i_d^{ref}	-0.0254[A]	-0.2261[A]
current reference (q-axis) i_q^{ref}	1.176[A]	3.516[A]

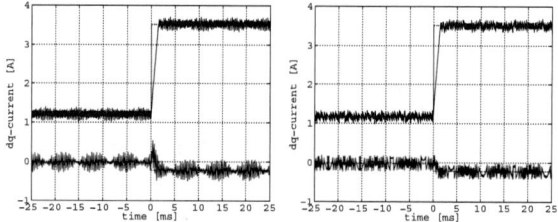

(a) conventional (PWM) (left) and (b) Proposed (J_1) (right) .

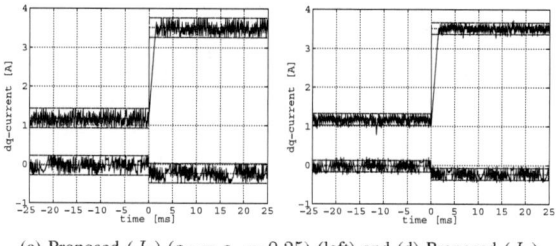

(c) Proposed (J_2) ($\varepsilon_d = \varepsilon_q = 0.25$) (left) and (d) Proposed (J_2) ($\varepsilon_d = \varepsilon_q = 0.15$) (right).

Fig. 5.   Current response (simulation).

earlier. Then, it is desirable that evaluations of the last two terms are about three for fair comparison. For the purpose, $Q_d = \frac{4}{\varepsilon_d}$ and $Q_q = \frac{4}{\varepsilon_q}$ in Eq. (8) in the simulations and experiments.

Operating point changes from the linear to the overmodulation regions, which are listed in Table IV.

B. Simulation results

The proposed method are compared with the conventional method from the perspective of the dq-current response and the average switching frequency. In the simulations, the dc-link voltage is set as 100[V].

Fig. 5 shows simulation results of the step response of the dq-current by the conventional vector control and the proposed method. In Fig. 5, the solid and the dashed lines are the response and the reference, respectively. The operating point is switched from OP 1 to OP 2 at 0[ms] in Fig. 5. In the result of the conventional method shown in Fig. 5(a), the ripple and overshoot can be seen, especially in the d-axis current at the stepwise transition. The result can be explained quantitatively below. In the vector control, the inverter is driven in the one-pulse mode in order to track to the reference voltage at the transient response. This yields pulsating current since the output voltage from the inverter includes a large harmonic component. Hence, the dq-current ripple increases in especially OP 2 as shown in Fig. 5(a),

In the proposed method, the ripple, overshoot and settling time are improved, especially in the d-axis current,

TABLE V. AVERAGE SWITCHING FREQUENCY (SIMULATION).

objective function	average switching frequency [kHz]
J_1	22.50
J_2 ($\varepsilon_d = \varepsilon_q = 0.25$)	14.68 (-37.9%)
J_2 ($\varepsilon_d = \varepsilon_q = 0.15$)	21.35 (-8.9%)

as shown in the result of the proposed method in Fig. 5 (b). In the vector control, it is necessary to generate the reference voltage considering harmonic voltage in the one-pulse mode and an appropriate limitation for the reference voltage in the saturation mode, in order to improve the performance of transient current response. On the contrary, the proposed method can achieve better current response without any particular compensation.

Let us see the effects by the different objective functions in the proposed method. In Fig. 5(c) and (d), which are the results by the objective function $J_2(\cdot)$, the transient current response is not so different from that by the vector control. In addition, the ripple in steady-state is kept within the specified tolerance. Compared with the conventional method, the average switching frequencies with respect are reduced by 37.9% and 8.9% as the tolerance bounds are narrower, respectively, as listed in Table V.

Next, let us see the result by J_2 ($\varepsilon_d = \varepsilon_q = 0.15$) shown in Fig. 5(d). In the plot, it can be observed that the q-axis current is beyond the tolerance bound at about -12, 7 and 14[ms] though the stepwise response is similar to the other plots. When J_2 is adopted as the objective function, if the current is within the tolerance bound, the number of switching operations are preferentially considered, otherwise, the current is controlled into the bound at the expense of switching operations. In the case that the tolerance bound is narrower, it is more difficult to keep the current within the bound with less number of switching operations. As a result, the error between the reference and the measured current is considered, which results in the violation of the tolerance bound.

C. Experimental results

The main control unit in experiment is composed of PE-Expert3 (Myway Plus Corp.). The dead-time is set as $5.0[\mu s]$. It is necessary to set an equal average switching frequency in the proposed and the conventional methods for a fair comparison. By trial and errors, the career frequency is set 3[kHz] in the conventional method. The other parameters are the same as those in the simulation.

Fig. 6 shows the dq-current responses using J_1 as the objective function in the proposed approach under the same condition as the simulation except the dead-time. Compared with the plot of Fig. 5 (b), the rising time of the dq-current response is slower as shown in Fig. 6. As the reason, various causes can be considered such as the dead-time, lack or perturbation of the dc-link voltage. Through some experiments, it is found that increase of the dc-link voltage contributes to the rising time. In the subsequent experimental results, the dc-link voltage is set to 110[V] so that the rising time is similar to that in the plot of Fig. 5 (b)

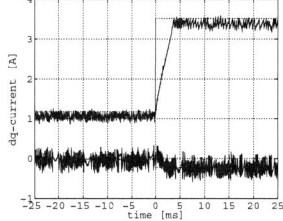

Proposed (J_1), $E_d = 100$[V].

Fig. 6. Current response (experiment).

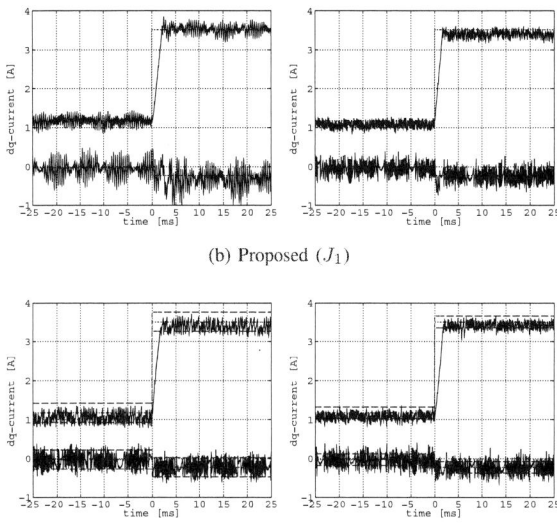

(b) Proposed (J_1)

(e) Proposed (J_2) ($\varepsilon_d = \varepsilon_q = 0.15$).

Fig. 7. Current response (experiment).

Fig. 7 shows the dq-current responses. As shown in Fig. 7(a), the ripple and overshoot can bee seen in each current for the stepwise reference. In addition, the d-axis current response responds in opposite side for the reference. This is because that the decoupling control does not function as desired The change of the q-axis voltage command is larger than that of the q-axis voltage for stepwise references. In other words, the d-axis voltage command becomes relatively small. As a result, the voltage vector command is weighted in the q-axis current. In such a case, the decoupling components of d-axis voltage is not compensated appropriately, which degrades the current response.

We can see the steady-state errors in Fig. 7(a) after the stepwise transition. The ratios of the steady-state errors are about 0.20 and 0.05 in the dq-current, respectively. In contrast, in the proposed method, it can be seen that the dq-current track to their references quickly with no overshoot in the transient response. These results demonstrate the effectiveness of the proposed model predictive-based current control method experimentally.

Let us see the results by the different objective functions $J_1(\cdot)$ and $J_2(\cdot)$ in the proposed method. As shown in Fig. 7(c), the ripple in steady-state and response at the stepwise transition are almost the same as the result

TABLE VI. AVERAGE SWITCHING FREQUENCY.

objective function	average switching frequency [kHz]
$J_1(\cdot)$	19.20
$J_2(\cdot)$, ($\varepsilon_d = \varepsilon_q = 0.25$)	14.90 (-22.4% to $J_1(\cdot)$)
$J_2(\cdot)$, ($\varepsilon_d = \varepsilon_q = 0.15$)	17.90 (-6.8% to $J_1(\cdot)$)

shown in Fig. 7(b). For the other result shown in Fig. 7(d), the transient response is similar to the result shown in Fig. 7(c). The average switching frequencies of each method is listed in Table VI. Table VI indicates that the control performance by $J_2(\cdot)$ is similar to that by $J_1(\cdot)$ while decreasing the number of switching operations in the inverter. However, focusing on the steady-state in Fig. 7 after the stepwise change, we can see the steady-state errors. The modeling error and/or parameter variation such as resistance and inductances can be considered as the possible reasons.

V. CONCLUSION

In this paper, we proposed a model predictive current control for permanent magnet synchronous motors. In the proposed approach, both the number of switching operations and current control were considered together, which were not realized in conventional PWM methods. In the proposed approach, two objective functions were presented; one considered the current behavior only and the other took the number of switching operations into consideration while keeping the current within a tolerance bound.

Compared with the conventional PWM method, the effectiveness of the proposed method was verified through simulations and experiments. Both the simulations and experiments showed that the proposed approach reduced the number of switching operations while achieving the same current response as the conventional PWM method.

Future subjects include implementation of the proposed approach for a longer prediction horizon, which requires the use of an efficient algorithm for optimization calculation. In addition, the offset in the torque response is dependent on the speed of the motor. To reduce the offset, we are presently developing an appropriate setting for the reference values.

REFERENCES

[1] S. Morimoto, M. Sanada, and Y. Takeda, "Wide-speed operation of interior permanent magnet synchronous motors with high-performance current regulator," *IEEE Trans. on Industrial Applications*, vol. 30, no. 4, pp. 920–926, 1994.

[2] Q. Liu, M. A. Jabbar, and A. M. Khambadkone, "Design optimization of interior permanent magnet synchronous motors for wide- speed operation," *Proc. of the 4th IEEE International Conference Power Electronics and Drive Systems*, vol. 2, pp. 475–478, 2001.

[3] S. Morimoto, Y. Takeda, T. Hirasa, and K. Taniguchi, "Expansion of operating limits for permanent magnet motor by currentvector control considering inverter capacity," *IEEE Trans. on Industrial Applications*, vol. 26, no. 5, pp. 866–871, 1990.

[4] E. A. Hava, R. J.Kerkman, and T. A. Lipo, "Carrier-based pwm-vsi overmodulation strategies : Analysis, comparison and design," *IEEE Trans. on Power Electronics*, vol. 13, no. 4, pp. 674–689, 1998.

[5] S. Lerdudomsak, M. Kadota, S. Doki, and S. Okuma, "Harmonic currents estimation and compensation method for current control system of ipmsm in overmodulation range," *Proc. of the 2007 Power Conversion Conference (PCC nagoya 2007)*, pp. 1320–1326, 2007.

[6] J. Jung and K. Nam, "A dynamic decoupling control scheme for high-speed operation of induction motors," *IEEE Trans. on Industrial Electronics*, vol. 46, no. 1, pp. 100–110, 1999.

[7] J. K. Seok, J. S. Kim, and S. K. Sul, "Overmodulation strategy for high-performance torque control," *IEEE Trans. on Power Electron*, vol. 13, no. 4, pp. 786–792, 1998.

[8] S. Lerdudomsak, M. Kadota, S. Doki, and S. Okuma, "Novel techniques for fast torque response of ipmsm based on space-vector control method in voltage saturation region," *Proc. of IEEE the 33rd Annual Conference of Industrial Electronics Society (IECON)*, pp. 1015–1020, 2007.

[9] J. B. Rawlings, "Tutorial overview of model predictive control," *IEEE Control System Magazine*, vol. 20, no. 3, pp. 38–52, 2000.

[10] A. Imura, T. Takahashi, M. Fujitsuna, T. Zanma, and M. Ishida, "Instantaneous-current control of PMSM using MPC: Frequency analysis based on sinusoidal correlation," *Proc. of IEEE the 37th Annual Conference of Industrial Electronics Society (IECON)*, pp. 3551–3556, 2011.

[11] A. Imura, T. Takahashi, M. Fujitsuna, T. Zanma, and S. Doki, "Refinement of inverter model considering dead-time for performance improvement in predictive instantaneous current control," *IEEJ Trans. on Electrical and Electronic Engineering*, vol. 9, no. 1, pp. 83–89, 2014.

[12] M. Hagino, T. Zanma, and M. Ishida, "Optimal direct torque control for pmsm based on model predictive control," *Proc. of the 2011-14th European Conference on Power Electronics and Applications (EPE 2011)*, p. (electronic medium), 2011.

[13] T. Zanma, M. Kawasaki, K. Z. Liu, M. Hagino, and A. Imura, "Optimal direct torque control for pmsm based on model predictive control," *IEEJ Journal of Industry Applications*, vol. 3, no. 2, pp. 121–130, 2014.

[14] J. B. Rawlings, "Tutorial: model predictive control technology," *Proc. of the American Control Conference*, vol. 1, pp. 662–676, 1999.

[15] J. Maciejowski, *Predictive Control with Constraints*. Prentice Hall, 2001.

Predictive Indirect Matrix Converter Fed Torque Ripple Minimization with Weighting Factor Optimization

Muslem Uddin, Saad Mekhilef
Power Electronics and Renewable
Energy Research Laboratory
(PEARL), Department of Electrical
Engineering, University of Malaya
50603 Kuala Lumpur, Malaysia
muslem_eee04@siswa.um.edu.my
saad@um.edu.my

Marco Rivera
Department of Industrial
Technologies
Universidad de Talca
Curico, Chile
marcoesteban@gmail.com

Jose Rodriguez
Departamento de Electrónica
Universidad Técnica Federico Santa
María (USM)
Chile
jrp@usm.cl

Abstract—**Predictive control is a powerful and promising control algorithm in the control of power converter and electrical machine drive's system. The system performance depends on the selection of weighting factor in the cost function. Therefore, this paper proposed a weighting factor optimization method to reduce the torque ripple of induction motor fed by an indirect matrix converter. Also, predictive torque and flux control with conventional weighting factor is being investigated in this paper and is compared with the proposed optimum weighting factor based predictive control algorithm. The introduced weighting factor optimization method in predictive control algorithm is validated through simulation and shows potential tracking of variables and control with their corresponding references and consequently minimizes the torque ripple compare to the conventional weighting factor based predictive control method.**

Keywords—Indirect Matrix Converter, Induction Motor, Predictive Control, Torque Ripple, Weighting Factor.

I. INTRODUCTION

Diverse control targets, variables and constraints can be included in a single cost function at predictive control algorithm and be controlled with the basis of priority control factor is known as weighting factor. For the variables with same nature in the cost function no need to set the weighting factor but when target variables are in different nature (different order of magnitude and different unit) in a single cost function then weighting factor selection become as a great issue for the system stability. Till the date in literature no analytical or numerical methods or control design theories to adjust the weighting factor and currently they are determined with the iterative evaluation method [1]. Though this procedure is extensively used to adjust the weighting factor and potential performances can be attained but this is quietly approximated. Therefore, for the best performance of the system, the optimized weighting factor is needed. In recent past, a weighting factor optimization method is applied for torque ripple reduction fed by three phase voltage source inverter (VSI) in [2] with good performance and predictive two-level inverter fed induction motor control strategy with

weighting factor look up table and divide control interval have been investigated in [3]. In [4], a ranking approach based multi objective optimization has been proposed to replace the single cost function at the predictive horizon which allows the predictive control of torque and flux without weighting factors. Therefore, in this investigation, predictive control algorithm has been used to control torque and flux of the induction motor precisely and is specially focused on indirect matrix converter fed weighting factor optimization method to minimize the torque ripple and control the flux of the induction motor. There are different types of AC-AC converter in the power converter field. The cyclo-converter is one of them which transfer the power without any intermediate energy storage devices with a significant amount of harmonic contents at the output frequency due to the commutations and these harmonics cannot be filtered by the load inductance. Also, direct matrix converter (DMC) is another AC-AC converter without DC-link storage device but its control strategy is so complex. On the other hand, indirect matrix converter is an AC-AC converter which has been proposed to remove all the demerits stated above. The most important improvement of this topology is the simplicity and less complex in the control compared to DMC and allowing secure commutation of the system [5] without particular sensing devices as needed for DMC [6]. Furthermore, the indirect matrix converter is with a longer life span and size becomes more compact. Recently, some investigations with the multilevel inverter have been investigated in [7, 8] with digital control method.

Some recent investigations have been carried out with indirect matrix converter, such as: predictive torque and flux control with unity power factor [9], current control [10], imposed sinusoidal source and load current [11], current control with filter resonance mitigation in [12]. Also, predictive control applications with three phase VSI in [13, 14], active front end rectifier control with unity displacement in [15] and for matrix converter (MC) have been investigated in [16, 17]. Furthermore, a comprehensive review on MC have been elaborately presented in [18]. Also, a three-stage 18-level

978-1-4799-2706-7/14 $31.00 © 2014 IEEE

The 2014 International Power Electronics Conference

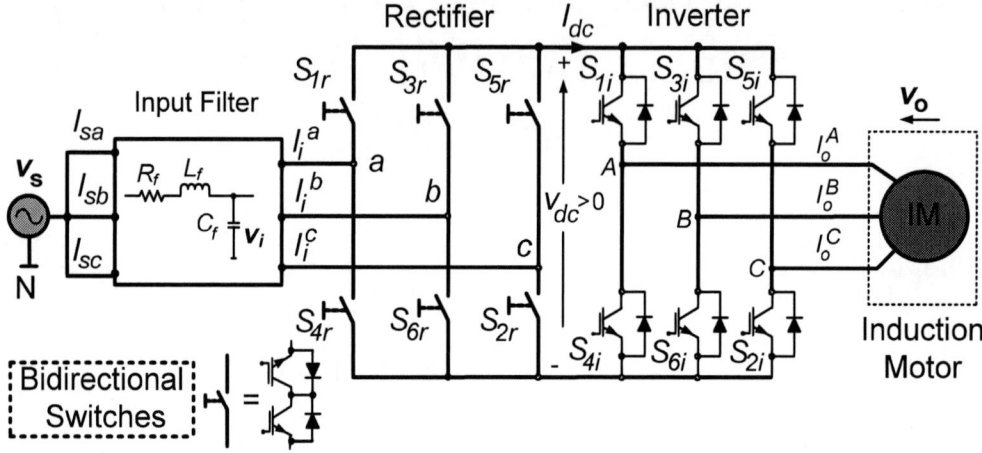

Fig. 1. Induction motor fed indirect matrix converter topology.

hybrid inverter circuit and its innovative control method have been presented in [19, 20].

This paper is organized in the following manner: Section II is related to the mathematical modeling of the indirect matrix converter topology of the system. Section III present the proposed predictive torque ripple reduction and flux control algorithm with weighting factor optimization method. Section IV states, verification results and discussion of the proposed investigation to reduce the torque ripple corresponding to conventional weighting factor based predictive control algorithm and finally, a fruitful conclusion is drawn in section V.

II. INDIRECT MATRIX CONVERTER TOPOLOGY

The Fig. 1 shows the topology of the indirect matrix converter (IMC) which consists of rectifier and inverter part. The IMC has 24 possible switching states that are utilizes in the predictive control algorithm to select the best actuation for the converter. The modeling equations of the indirect matrix converter are given in below:

$$V_{dc} = \begin{bmatrix} S_{1r} - S_{4r} & S_{3r} - S_{6r} & S_{5r} - S_{2r} \end{bmatrix} V_i \quad (1)$$

$$I_{ri} = \begin{bmatrix} S_{1r} - S_{4r} \\ S_{3r} - S_{6r} \\ S_{5r} - S_{2r} \end{bmatrix} I_{dc} \quad (2)$$

$$V_o = \begin{bmatrix} S_{1i} - S_{4i} \\ S_{3i} - S_{6i} \\ S_{5i} - S_{2i} \end{bmatrix} V_{dc} \quad (3)$$

$$I_{dc} = \begin{bmatrix} S_{1i} & S_{3i} & S_{5i} \end{bmatrix} I_o \quad (4)$$

where rectifier switching states $= S_{1r}$ to S_{6r}; input voltage $V_i = \begin{bmatrix} V_i^a & V_i^b & V_i^c \end{bmatrix}^T$; rectifier input current $I_{ri} = \begin{bmatrix} I_i^a & I_i^b & I_i^c \end{bmatrix}^T$; output voltage $V_o = \begin{bmatrix} V_o^A & V_o^B & V_o^C \end{bmatrix}^T$; output current $I_o = \begin{bmatrix} I_o^A & I_o^B & I_o^C \end{bmatrix}^T$.

III. PREDICTIVE CONTROL ALGORITHM

Predictive control algorithm uses the finite number of valid switching states of the power converter. The proposed scheme maintains the predictive values closed to their respective references at the end of the sampling instant and maintain positive DC-link voltage between the rectifier and inverter stages which eliminates the extra usage of the large energy storage devices. Hence reduces the size and increase the life span of the converter. This proposed optimum weighting factor based predictive control algorithm and scheme are presented in Fig. 2 and Fig. 3, respectively. Predictive controller satisfies all the aforementioned constraints by using the following five steps:

Steps 1: Supply voltage V_s^k, input voltage V_i^k, stator current I_o^k and speed ω^k of the induction motor are measured in the kth sampling instant.

Step 2: PI controller is used to set nominal torque T_{nom} from the error signal between the measured and reference speeds of the induction motor where reference speed ω_{ref} is known value.

Step 3: Stator reference flux ψ_{ref} is a given value and a flux estimator has been used to estimate the stator and rotor flux.

Step 4: For each valid switching states of indirect matrix converter, values of torque T_e^{k+1} and stator flux ψ_s^{k+1} are predicted in the next sampling period (k+1).

Step 5: All the predictive values are compared with their respective references and determine the cost functions for all possible switching states based on conventional weighting factor and with imposed optimized weighting factor. The switching state corresponds to the minimum cost function is selected in the next sampling time period to actuate the converter.

978-1-4799-2706-7/14 $31.00 © 2014 IEEE

The 2014 International Power Electronics Conference

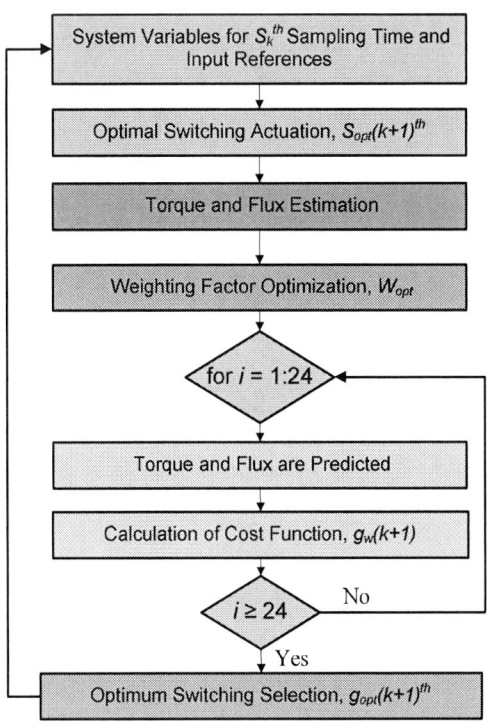

Fig. 2. Predictive torque and flux control algorithm.

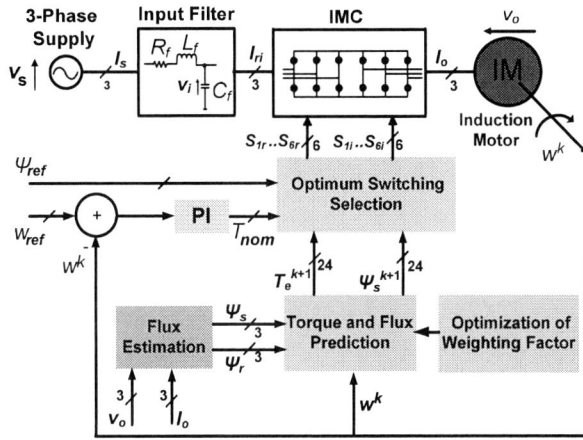

Fig. 3. Control scheme with weighting factor optimization.

A. Torque and Flux Prediction

The predictive flux, current and torque equations are given in equations (5)-(7) respectively.

$$\psi_s^{k+1} = \psi_s^k + V_o^{k+1} T_s - R_s I_o^{k+1} T_s \tag{5}$$

$$I_o^{k+1} = \psi_r^k (V_o^{k+1} + (\tau_r k_r - jk_r \omega^k)) \frac{T_s}{\sigma L_s} + I_o^k (1 - \frac{T_s r_\sigma}{\sigma L_s}) T_s \tag{6}$$

here, $r_\sigma = R_s + R_r k_r^2$, $\tau_r = \dfrac{L_r}{R_r}$, $k_s = \dfrac{L_m}{L_s}$, $k_r = \dfrac{L_m}{L_r}$ and

$\sigma = 1 - k_s k_r$.

$$T_e^{k+1} = \frac{3}{2} p(\psi_s^{k+1} \times I_o^{k+1}) \tag{7}$$

where R_s and R_r are the stator and rotor resistance respectively.

B. Weighting Factor Optimization Method

Torque ripple of the induction motor can be derived in the following

$$T_r^2 = \frac{1}{T_s} \int_0^{T_s} (D_T + m_1 t)^2 dt \tag{8}$$

where, $D_T = T_e - T_{nom}$, T_r = Torque ripple, T_s = Sampling time, T_e = Torque of induction motor, T_{nom} = Nominal torque and m_1 = Ascending torque slope and is given in below:

$$m_1 = -(\frac{R_s}{\sigma L_s} + \frac{R_r}{\sigma L_r}) T_e + K[(V_{oq} \psi_{rd} - V_{od} \psi_{rq}) - \omega^k (\psi_{sd} \psi_{rd} + \psi_{sq} \psi_{rq})] \tag{9}$$

where V_{od} = Stator voltage α-axis component; V_{oq} = Stator voltage β-axis component and $\sigma = 1 - k_r k_s = 1 - \dfrac{L_m^2}{L_r L_s}$,

$$K = \frac{3}{2} p$$

The simplified torque ripples can be represented as follows:

$$T_r^2 = \frac{1}{T_s} \int_0^{T_s} (D_T^2 + m_1^2 t^2 + 2m_1 t D_T) dt \tag{10}$$

The first derivative of torque ripple to the weighting factor has to be set to zero in order to find the weighting factor that minimizes the ripple of torque. The relation is as follows:

$$T_r^2 = T_r^2(m_1), \quad m_1 = m_1(\overline{V}_o), \quad \overline{V}_o = \overline{V}_o(W_{opt})$$

Therefore, $T_r^2 = T_r^2(W_{opt})$ \hfill (11)

To find the optimum weighting factor in the cost function the derivative of the torque ripple must be zero. Therefore,

$$\frac{dT_r^2}{dW_{opt}} = 0 \tag{12}$$

By solving the equation (12), optimum weighting factor becomes,

$$G = \frac{3D_T}{2KT_s} + \omega^k (\psi_{sd} \psi_{rd} + \psi_{sq} \psi_{rq}) + \frac{1}{K}(\frac{R_s}{\sigma L_s} + \frac{R_r}{\sigma L_r}) T_e \tag{13}$$

$$W_{opt} = \frac{\beta_2 \psi_{rd} - \beta_1 \psi_{rq}}{G + \psi_{rq}(\alpha_1 + V_{od,k-1}) - \psi_{rd}(\alpha_2 + V_{oq,k-1})} \tag{14}$$

Equation (14) represents the optimum weighting factor.

C. Cost Function Calculation

The cost functions used in conventional and optimized weighting factor based predictive control are given in (15) and (16) respectively.

$$g^{k+1} = X_1 |(T_e^{k+1} - T_{nom})| + X_2 |(\psi_s^{k+1} - \psi_{ref})| \tag{15}$$

$$g_w^{k+1} = \frac{1}{2}(|T_e^{k+1} - T_{nom}|^2 + W_{opt} ||\psi_s^{k+1}|^2 - |\psi_{ref}|^2|^2) \tag{16}$$

where, X_1, X_2 are the conventional weighting factors and W_{opt}, is the optimized weighting factor.

978-1-4799-2706-7/14 $31.00 © 2014 IEEE   3576

The 2014 International Power Electronics Conference

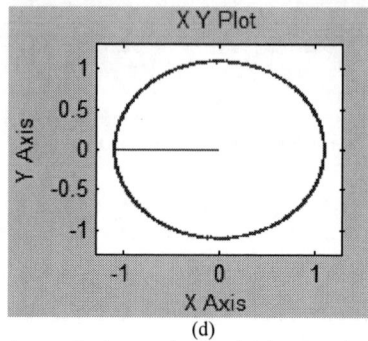

(d)

IV.   RESULTS AND DISCUSSION

The predictive control of induction motor fed by an indirect matrix converter is verified in MATLAB Simulink environment to justify the performance of the proposed optimization of weighting factor. The parameters used in verifications are given in Table I and the simulations has been carried out with sampling time, T_s = 20 μs. In this investigation, two cases are analyzed, First, the validation of the predictive control algorithm is performed with conventional weighting factor, while in the second case, an optimization method is adopted to select the optimized weighting factor for the predictive control algorithm. In both cases, the induction motor starts at 0.01s without any load torque, varying the reference speed from 0 to 35 rad/s and the torque is limited to 7 N-m. A load torque of 3 N-m is applied at time of 0.4s and a reverse torque at 0.5s is applied to change the speed in the reverse direction from 35 rad/s to -35 rad/s. In this investigation, stator reference flux has been assumed as 1.1 Wb in all the verifications. The speed controller generates torque references at transients which is different from zero and be appreciated as a good tracker of speed given in Figs. 4(a) and 5(a), of torque indicated in Figs. 4(b) and 5(b), of stator

Fig. 4. Verification results (conventional weighting factor): (a) Motor speed, (ω^k) [rad/s] and reference speed, (ω_{ref}) [rad/s]; (b) Motor torque, (T_e) [N-m] and Nominal torque, (T_{nom}) [N-m]; (c) Stator flux, (ψ_s) [Wb] and reference flux, (ψ_{ref}) [Wb]; (d) α-β presentation for the stator flux, (ψ_s) [Wb].

flux depicted in Figs. 4(c) and 5(c) in both the conventional and proposed optimum weigthing factor based predictive control schemes respectively. Also the stator flux α-β representations are plotted in Figs. 4(d) and 5(d) for the both cases respectively and follows the reference magnitude of around 1.1 Wb accurately.

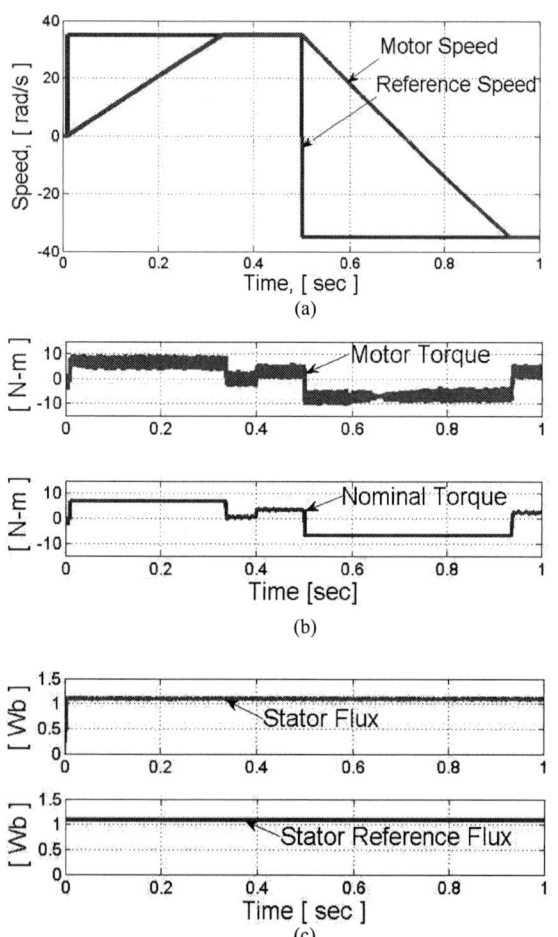

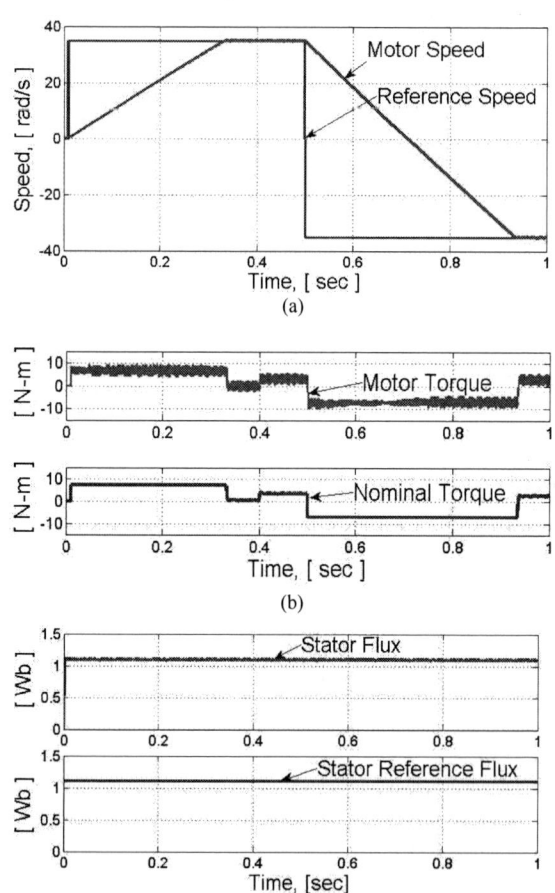

The 2014 International Power Electronics Conference

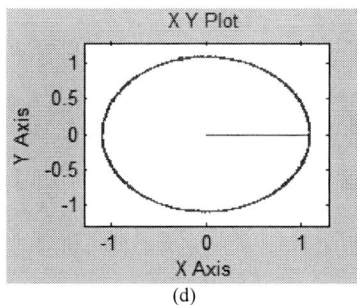

(d)

Fig. 5. Verification results (optimized weighting factor): (a) Motor speed, (ω^k) [rad/s] and reference speed, (ω_{ref}) [rad/s]; (b) Motor torque, (T_e) [N-m] and Nominal torque, (T_{nom}) [N-m]; (c) Stator flux, (ψ_s) [Wb] and reference flux, (ψ_{ref}) [Wb]; (d) α-β presentation for the stator flux, (ψ_s) [Wb].

A. Torque Ripple Reduction in Forward Speed Region

The Fig. 6(a) shows the maximum value of the torque ripples is 5 N-m and the minimum is 1.85 N-m in a certain time range of verification. Therefore the difference between the maximum and minimum peak value of the torque ripples is 3.15 Nm for the predictive method based on conventional weighting factor. On the other side, for the proposed weighting factor optimization method, the maximum and minimum peaks are in between 4 N-m and 2 N-m respectively in the same time interval in Fig. 6(b) and the difference of the values is only 2 N-m. Therefore, the proposed weighting factor optimization method has reduced the torque ripples (3.15 - 2.0) N-m or, 1.15 N-m corresponding to conventional predictive control algorithm in the forward speed region.

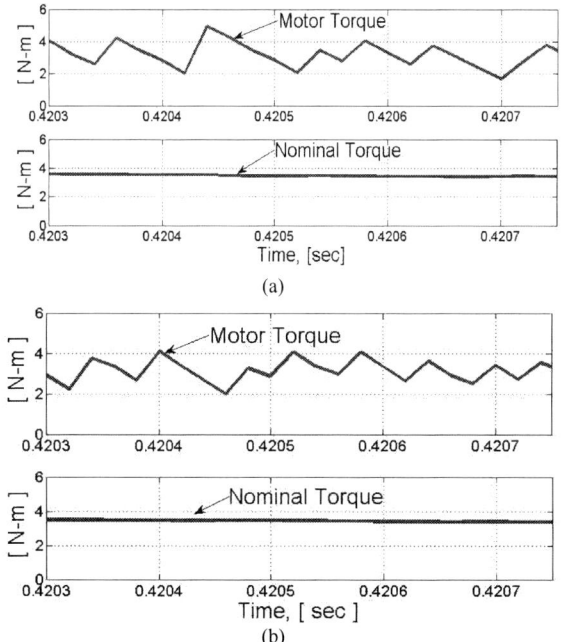

Fig. 6. Torque ripples (N-m) (forward speed): (a) Using conventional weighting factor; (b) With imposed weighting factor optimization.

B. Torque Ripple Reduction in Reverse Speed Region

The Fig. 7(a) implies that, the maximum value of the torque ripples is 5.5 N-m and the minimum is 1.1 N-m in a certain range of verification. Therefore the difference between the maximum and minimum value of the torque ripples is 4.4 N-m for the method of conventional weighting factor. On the other hand, for the proposed weighting factor optimization method, the maximum and the minimum values of the torque ripples are in between 3.6 N-m and 2.2 N-m respectively in the same time interval and the difference of the values is only 1.4 N-m. Hence, the proposed weighting factor optimization method has reduced the torque ripples by (4.4–1.4) N-m, or 3.0 N-m in the reverse speed corresponding to conventional weighting factor based predictive control.

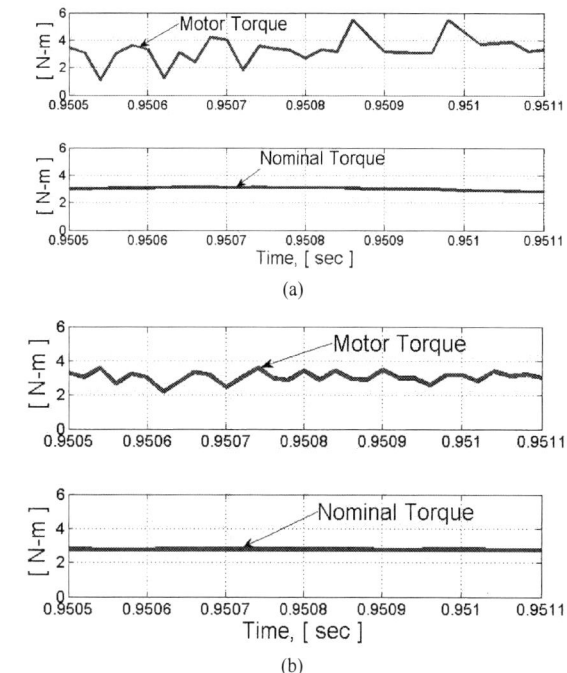

Fig.7. Torque ripples (N-m) (reverse speed): (a) Using conventional weighting factor; (b) With imposed weighting factor optimization.

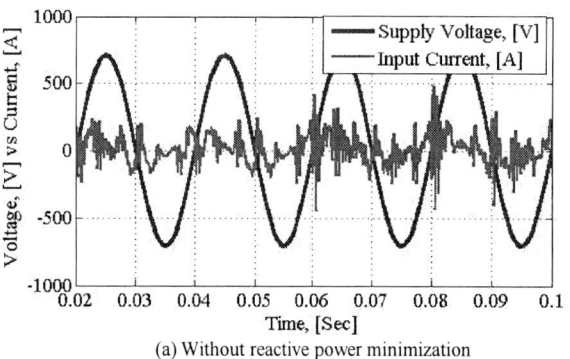

(a) Without reactive power minimization

978-1-4799-2706-7/14 $31.00 © 2014 IEEE

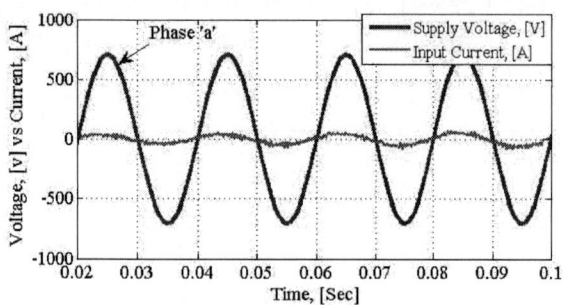

(b) With reactive power minimization.

Fig. 8. Supply voltage, V_s^a [V] vs input current, I_i^a [A]: (a) Without reactive power minimization; (b) With reactive power minimization.

TABLE I
PARAMETERS USED IN SIMULATION

Symbol	Meaning	Value
T_s	Sampling time	20 μs
V_s	Supply phase voltage (RMS)	500 V
f_s	Supply frequency	50 Hz
R_f	Input filter resistance	0.5 Ω
L_f	Input filter inductance	400 μH
C_f	Input filter capacitance	21 μF
ω_{ref}	Reference speed	35 rad/s
T_{nom}	Nominal torque	7 N-m
R_s	Stator resistance	1.35 Ω
L_s	Stator inductance	0.2861 mH
R_r	Rotor resistance	7.2037 Ω
L_r	Rotor inductance	0.2861 mH
L_m	Mutual inductance	0.2822 mH
p	Number of poles	2
X_1	Weighting factor	63
X_2	Weighting factor	13500

In the indirect matrix converter, input side current is sinosoidal which is shown in Fig. 8. The Fig. 8 shows the relation between the supply voltage and input current to the converter.

From the result it is clear that in Fig. 8(a), a chaoitic behavior is observed in input current of indirect matrix converter regardless of reactive power compensation in the cost function and this reactive power can be minimized by adding a reactive power minimization term as the investigation of [9] and the result is shown in Fig. 8(b).

Consequently, at the input side current becomes more sinusoidal as well as unity power factor is maintained properly.

V. CONCLUSIONS

Predictive control algorithm with conventional weighting factor and proposed weighting factor optimization methods are presented in this paper in order to reduce the torque ripple and flux control of the induction motor fed by an indirect matrix converter. The system behavior is highly changeable with the values of the weighting factor in the cost function. So, this paper is highlighted with an imposed optimized weighting factor calculation method to reduce the torque ripple of the induction motor corresponding to the conventional predictive control scheme. Finally, the proposed weighting factor optimization method used in the cost function of the predictive control algorithm significantly improves the torque ripple of induction motor with potential control.

APPENDIX

A. Torque Slope Calculation

The torque for the induction motor can also be determined by the following relation,

$$T_e = T = K(\psi_{sq}\psi_{rd} - \psi_{sd}\psi_{rq}) \tag{17}$$

First derivative of (17) implies the slope of the torque as follows,

$$\frac{dT}{dt} = K(\frac{d}{dt}\psi_{sq}*\psi_{rd} + \frac{d}{dt}\psi_{rd}*\psi_{sq} - \frac{d}{dt}\psi_{sd}*\psi_{rq} - \frac{d}{dt}\psi_{rq}*\psi_{sd}) \tag{18}$$

Considering the induction motor dynamic model from equation (18),

$$\frac{dT}{dt} = K[(V_{sq} + W_2)*\psi_{rd} + W_3*\psi_{sq} - (V_{sd} + W_1)*\psi_{rq} - W_4*\psi_{sd}] \tag{19a}$$

$$\begin{bmatrix} W_1 \\ W_2 \\ W_3 \\ W_4 \end{bmatrix} = \begin{bmatrix} Q_{11} & Q_{12} & Q_{13} & Q_{14} \\ Q_{21} & Q_{22} & Q_{23} & Q_{24} \\ Q_{31} & Q_{32} & Q_{33} & Q_{34} \\ Q_{41} & Q_{42} & Q_{43} & Q_{44} \end{bmatrix} \cdot \begin{bmatrix} \psi_{sd} \\ \psi_{sq} \\ \psi_{rd} \\ \psi_{rq} \end{bmatrix} \tag{19b}$$

where,

$$Q_{11} = Q_{22} = -\frac{R_s}{\sigma L_s}, Q_{13} = Q_{24} = \frac{R_s(1-\sigma)}{\sigma L_m}, Q_{31} = Q_{42} = \frac{R_r(1-\sigma)}{\sigma L_m},$$

$$Q_{21} = Q_{12} = Q_{32} = Q_{41} = Q_{14} = Q_{23} = 0 \quad , \quad Q_{34} = -Q_{43} = \omega^k \quad ,$$

$$Q_{33} = Q_{44} = -\frac{R_r}{\sigma L_r}, \text{ So, the ascending slope of the torque can}$$

be depicted with the following equation,

$$\frac{dT}{dt} = K[(V_{sq}\psi_{rd} - V_{sd}\psi_{rq}) + \omega^k(\psi_{sd}\psi_{rd} + \psi_{sq}\psi_{rq})] - (\frac{R_s}{\sigma L_s} + \frac{R_r}{\sigma L_r})T_e = m_1 \tag{20}$$

where, $K = \frac{3}{2}p$, $\sigma = 1 - k_r k_s = 1 - \frac{L_m^2}{L_r L_s}$

B. Relationship Between Weighting Factor and Stator Voltage

Applying the Taylor expansion around the nominal values in equation (17) to express the model predictive variables in a linear manner as follows:

$$T_e = T_{nom} + K(\psi_{rd}^o\Delta\psi_{sq} + \psi_{sq}^o\Delta\psi_{rd} - \psi_{sd}^o\Delta\psi_{rq} - \psi_{rq}^o\Delta\psi_{sd}) \tag{21}$$

$$|\psi_s|^2 = |\psi_{ref}|^2 + 2\psi_{sd}^o\Delta\psi_{sd} + 2\psi_{sq}^o\Delta\psi_{sq} \tag{22}$$

Hence, (16) becomes,

$$g^{k+1} = \frac{1}{2}[(\Delta T)^2 + W_{opt}(\Delta|\psi_s|^2)^2]$$ (23)

Torque and flux displacements are related to stator and rotor flux as follows:

$$Y(t_k) = G.X(t_n)$$ (24a)

$$Y = \begin{bmatrix} \Delta T_e \\ \Delta|\psi_s|^2 \end{bmatrix}, G = \begin{bmatrix} -K\psi_{rq}^o & K\psi_{rd}^o & K\psi_{sq}^o & -K\psi_{sd}^o \\ 2\psi_{sd}^o & 2\psi_{sq}^o & 0 & 0 \end{bmatrix}, X = \begin{bmatrix} \Delta\psi_{sd} \\ \Delta\psi_{sq} \\ \Delta\psi_{rd} \\ \Delta\psi_{sq} \end{bmatrix}$$ (24b)

On the other case, in stationary reference frame induction motor discrete model can be described as below:

$$X(t_{k+1}) = R.X(t_k) + S.U(t_k)$$ (25a)

$$X = \begin{bmatrix} \Delta\psi_{sd} \\ \Delta\psi_{sq} \\ \Delta\psi_{rd} \\ \Delta\psi_{sq} \end{bmatrix} U = \begin{bmatrix} \Delta V_{od} \\ \Delta V_{oq} \end{bmatrix}, S = \begin{bmatrix} 1 & 0 \\ 0 & 1 \\ 0 & 0 \\ 0 & 0 \end{bmatrix}$$ (25b)

$$R = T_s \begin{bmatrix} \frac{1}{T_s} + Q_{11} & Q_{12} & Q_{13} & Q_{14} \\ Q_{21} & \frac{1}{T_s} + Q_{22} & Q_{23} & Q_{24} \\ Q_{31} & Q_{32} & \frac{1}{T_s} + Q_{33} & Q_{34} \\ Q_{41} & Q_{42} & Q_{43} & \frac{1}{T_s} + Q_{44} \end{bmatrix}$$ (25c)

The parameters regarding the matrix R have been introduced before in (19b). For discrete nature of the system considering in the next sampling periods the before mentioned equations gives:

$$Y(t_{k+1}) = G.X(t_{k+1}) = G.R.X(t_k) + G.S.U(t_k)$$ (26)

The appropriate input vector satisfies the following set of equations

$$\frac{\delta}{\delta\Delta V_{od}} g_w = 0$$ (27)

$$\frac{\delta}{\delta\Delta V_{oq}} g_w = 0$$

From equations (16), (26) and (27) lead to the following voltage displacement

$$\Delta V_{od} = \alpha_1 + \frac{\beta_1}{W_{opt}}$$ (28a)

$$\Delta V_{oq} = \alpha_2 + \frac{\beta_2}{W_{opt}}$$ (28b)

where, $\alpha_1 = \dfrac{-f_1(e_{22}^2 e_{11} - e_{12}e_{21}e_{22})}{(e_{11}e_{22} + e_{12}e_{21})^2}$, $\beta_1 = \dfrac{-f_2(e_{12}^2 e_{21} - e_{11}e_{22}e_{12})}{(e_{11}e_{22} + e_{12}e_{21})^2}$,

$$\alpha_2 = \frac{-f_1(e_{21}^2 e_{12} - e_{11}e_{21}e_{22})}{(e_{11}e_{22} + e_{12}e_{21})^2}, \beta_2 = \frac{-f_2(e_{11}^2 e_{22} - e_{11}e_{12}e_{21})}{(e_{11}e_{22} + e_{12}e_{21})^2}$$ (29a)

$$\begin{bmatrix} f_1 \\ f_2 \end{bmatrix} = G.R \begin{bmatrix} \Delta\psi_{sd} \\ \Delta\psi_{sq} \\ \Delta\psi_{rd} \\ \Delta\psi_{sq} \end{bmatrix}$$ (29b)

$$\begin{bmatrix} e_{11} & e_{12} \\ e_{21} & e_{22} \end{bmatrix} = \begin{bmatrix} -KT_s\psi_{rq}^o & KT_s\psi_{rd}^o \\ 2T_s\psi_{sd}^o & 2T_s\psi_{sq}^o \end{bmatrix}$$ (29c)

These equations are expressed the stator voltage and weighting factor relationship.

C. Optimization of Weighting Factor

Let, derivative of the torque ripples with respect to weighting factor equals to zero.

$$\frac{dT_r^2}{dW_{opt}} = \frac{d}{dW_{opt}}(m_1^2 \cdot \frac{T_s^2}{3} + m_1 D_T T_s + D_T^2) = 0$$ (30)

Therefore, $\dfrac{dm_1}{dW_{opt}} = 0$ (31)

And, $(m_1 \cdot \dfrac{2}{3}T_s^2 + D_T T_s) = 0$ (32a)

$$m_1 = -\frac{3D_T}{2T_s}$$ (32b)

From equation (9) and (31) the relation obtained as below;

$$\frac{dm_1}{dW_{opt}} = \frac{d}{dW_{opt}}[K\{(V_{oq}\psi_{rd} - V_{od}\psi_{rq}) + \omega^k(\psi_{sd}\psi_{rd} + \psi_{sq}\psi_{rq})\} - (\frac{R_s}{\sigma L_s} + \frac{R_r}{\sigma L_r})T_e]$$ (33a)

It gives $\psi_{rd}\dfrac{d}{dW_{opt}}V_{oq} = \psi_{rq}\dfrac{d}{dW_{opt}}V_{od}$ (33b)

The derivatives of the stator voltage to the weighting factor are obtained as:

$$\frac{d}{dW_{opt}}V_{od} = \frac{d}{dW_{opt}}(V_{od}^o + \Delta V_{od}) = \frac{d}{dW_{opt}}(\Delta V_{od}) = -\frac{\beta_1}{W_{opt}^2}$$ (34a)

$$\frac{d}{dW_{opt}}V_{oq} = \frac{d}{dW_{opt}}(V_{oq}^o + \Delta V_{oq}) = \frac{d}{dW_{opt}}(\Delta V_{oq}) = -\frac{\beta_2}{W_{opt}^2}$$ (34b)

As a result, equation (31) is not a suitable equation to calculate the optimized weighting factor parameter because of each cancel from both sides of the equation (33). Therefore, equations (32a) and (32b) are the best criterion for weighting factor optimization. Therefore,

$$-\frac{3D_T}{2T_s} = K[(V_{oq}\psi_{rd} - V_{od}\psi_{rq}) - \omega^k(\psi_{sd}\psi_{rd} + \psi_{sq}\psi_{rq})] - (\frac{R_s}{\sigma L_s} + \frac{R_r}{\sigma L_r})T_e$$ (35)

where, $G = V_{oq}\psi_{rd} - V_{od}\psi_{rq}$ (36)

The equation (36) is the criterion for weighting factor optimization. Combining the equations (28) and (36) optimized weighting factor can be determined.

ACKNOWLEDGMENT

The authors wish to thank the financial support from the University of Malaya through HIR-MOHE project UM.C/HIR/MOHE/ENG/17.

REFERENCES

[1] P. Cortes, S. Kouro, B. La Rocca, R. Vargas, J. Rodriguez, J. I. Leon, S. Vazquez, and L. G. Franquelo, "Guidelines for weighting factors design in model predictive control of power converters and drives," in *Proc. IEEE International Conf. on Ind. Tech.*, Gippsland, VIC, 2009, pp. 1-7.

[2] S. A. Davari, D. A. Khaburi, and R. Kennel, "An Improved FCS–MPC Algorithm for an Induction Motor With an Imposed Optimized Weighting Factor," *IEEE Trans. Power Elec.*, vol. 27, pp. 1540-1551, 2012.

[3] S. A. Davari, D. A. Khaburi, and R. Kennel, "Using a weighting factor table for FCS-MPC of induction motors with extended prediction horizon," in *Proc. IEEE 38th Annual Conference on Industrial Electronics Society*, 2012, pp. 2086-2091.

[4] C. Rojas, J. Rodriguez, F. Villarroel, J. Espinoza, C. Silva, and M. Trincado, "Predictive Torque and Flux Control Without Weighting Factors," *IEEE Trans. Ind. Elec.*, vol. 60, pp. 681-690, 2013.

[5] J. W. Kolar, F. Schafmeister, S. D. Round, and H. Ertl, "Novel three-phase AC–AC sparse matrix converters," *Power Electronics, IEEE Transactions on*, vol. 22, pp. 1649-1661, 2007.

[6] P. W. Wheeler, J. C. Clare, L. Empringham, M. Bland, and M. Apap, "Gate drive level intelligence and current sensing for matrix converter current commutation," *Industrial Electronics, IEEE Transactions on*, vol. 49, pp. 382-389, 2002.

[7] S. Mekhilef, M. AbdulKadir, and Z. Salam, "Digital Control of Three Phase Three-Stage Hybrid Multilevel Inverter," *IEEE Trans. Ind. Inf.*, vol. 9, pp. 719-727, 2013.

[8] S. Mekhilef and A. Kadir, "Voltage control of three-stage hybrid multilevel inverter using vector transformation," *IEEE Trans. Power Elec.*, vol. 25, pp. 2599-2606, 2010.

[9] S. Muslem Uddin, S. Mekhilef, M. Rivera, and J. Rodriguez, "A FCS-MPC of an induction motor fed by indirect matrix converter with unity power factor control," in *Proc. 8th IEEE Conference on Industrial Electronics and Applications (ICIEA)* Melbourne, Australia, 2013, pp. 1769-1774.

[10] P. Correa, J. Rodríguez, M. Rivera, J. R. Espinoza, and J. W. Kolar, "Predictive control of an indirect matrix converter," *IEEE Trans. Ind. Elec.*, vol. 56, pp. 1847-1853, 2009.

[11] M. Rivera, J. Rodriguez, J. R. Espinoza, T. Friedli, J. W. Kolar, A. Wilson, and C. A. Rojas, "Imposed sinusoidal source and load currents for an indirect matrix converter," *IEEE Trans. Ind. Elec.*, vol. 59, pp. 3427-3435, 2012.

[12] M. Rivera, J. Rodriguez, B. Wu, J. R. Espinoza, and C. A. Rojas, "Current control for an indirect matrix converter with filter resonance mitigation," *IEEE Trans. Ind. Elec.*, vol. 59, pp. 71-79, 2012.

[13] R. Kennel and A. Linder, "Predictive control of inverter supplied electrical drives," in *Proc. IEEE 31st Annual Power Electronics Specialists Conf.*, Galway, 2000, pp. 761-766.

[14] J. Rodriguez, J. Pontt, C. A. Silva, P. Correa, P. Lezana, P. Cortés, and U. Ammann, "Predictive current control of a voltage source inverter," *IEEE Trans. Ind. Elec.*, vol. 54, pp. 495-503, 2007.

[15] S. Muslem Uddin, P. Akter, S. Mekhilef, M. Mubin, M. Rivera, and J. Rodriguez, "Model predictive control of an active front end rectifier with unity displacement factor," in *Proc. IEEE International Conference on Circuits and Systems (ICCAS)* Kuala Lumpur, Malaysia, 2013, pp. 81-85.

[16] M. Rivera, J. Rodriguez, P. W. Wheeler, C. A. Rojas, A. Wilson, and J. R. Espinoza, "Control of a matrix converter with imposed sinusoidal source currents," *IEEE Trans. Ind. Elec.*, vol. 59, pp. 1939-1949, 2012.

[17] M. Rivera, R. Vargas, J. Espinoza, J. Rodriguez, P. Wheeler, and C. Silva, "Current control in matrix converters connected to polluted AC voltage supplies," in *Proc. IEEE Power Electronics Specialists Conf.*, 2008, pp. 412-417.

[18] J. Rodriguez, M. Rivera, J. W. Kolar, and P. W. Wheeler, "A review of control and modulation methods for matrix converters," *IEEE Trans. Ind. Elec.*, vol. 59, pp. 58-70, 2012.

[19] S. Mekhilef and M. A. Kadir, "Novel vector control method for three-stage hybrid cascaded multilevel inverter," *IEEE Trans. Ind. Elec.*, vol. 58, pp. 1339-1349, 2011.

[20] A. Kadir, S. Mekhilef, and H. Ping, "Voltage vector control of a hybrid three-stage 18-level inverter by vector decomposition," *Power Electronics, IET*, vol. 3, pp. 601-611, 2010.

High-Power Density Hybrid Converter Topologies for Low-Power Dc-Dc SMPS

Aleksandar Radić, S.M. Ahssanuzzaman, Behzad Mahdavikhah, and Aleksandar Prodić
Laboratory for Power Management and Integrated Switch-Mode Power Supplies
ECE Department, University of Toronto, CANADA
{e-mail: prodic@ece.utoronto.ca}

Abstract-**This paper gives a review of several emerging dc-dc converter topologies that combine capacitor-based and inductive converters in single hybrid converter structures. It is shown that, compared to the conventional topologies, the hybrid buck converters allow for a drastic reduction of the inductive components while minimizing switching losses and improving the overall power processing efficiency. Therefore, the hybrid converters result in a higher power density. As examples, buck with merged capacitive divider, a two-phase interleaved buck, and a differential buck-based multi-output power module for mobile applications are shown. The presented converters have up to four times smaller inductor volume and, at the same time, about 12% lower losses.**

I. INTRODUCTION

Power management systems of modern portable devices, computers, and numerous other applications incorporate a large number of low-power switch-mode power supplies, processing power from a fraction of watt to several hundreds of watts. The reactive components of these SMPS, especially inductors, take a significant portion of the overall device weight and volume. In numerous applications, they occupy much more than 25% of the overall device volume [1], and, as such, are a large obstacle to further system minimization. Also, the power processing efficiency of these SMPS is usually significantly smaller than that of the converters processing more power, increasing cooling requirements and, in battery-powered application, affecting operating time.

A number of different methods for minimizing the volume of the reactive components [2]-[9], have been proposed in the past. Generally, those can be divided in the frequency based and topological solutions. The frequency based methods [2] increase effective switching frequency of the converter to minimize the filter

requirements at the expense of larger switching losses.

On the topological side, arguably, among the most interesting are switched-capacitor (SC) converters [3]-[7], that eliminate the inductive components. The SC converters show advantages in terms of power density and power processing efficiency for certain fixed voltage conversion ratios. However, the absence of the inductor in those structures causes voltage regulation problems and negatively affects power processing efficiency (or the system volume) in applications where the conversion ratio is not fixed [4]. Therefore, the use of SC in typical applications of interest, where the load changes frequently and input voltage is not constant, is fairly limited. To eliminate the previously mentioned problems, cascade connections of a SC and a conventional buck (Fig.1) have been proposed in [7]-[9]. The cascaded topologies eliminate the voltage regulation problem and have much smaller inductor than the conventional buck. However, these solutions often increase the resistance of in the conduction path and, consequently, suffer from increased conduction losses. The conventional cascaded solutions also require a relatively bulky intermediate capacitor (C_{sc} of Fig.1) for energy transfer and balancing of the capacitor cells.

The main goal of this paper is to review several recently emerged hybrid converter topologies [10]-[13] that eliminate the drawbacks of the previous two-stage solution. As, shown in Fig.1, *the hybrid converters merge the capacitive and inductive converter in a single structure such that components are shared between them and/or the need for a bulky intermediate capacitor is eliminated.*

In the following sections, the principle of operations of three types of topologies that perform commonly required functions in power management system of interest are reviewed. Namely, a buck converter with a merged

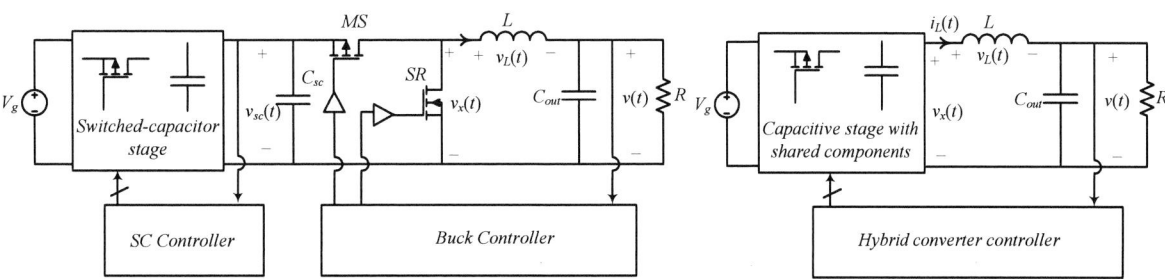

Fig.1. Cascaded switched-capacitor (SC) and buck converter topology (left) and hybrid converter (right).

capacitive divider [10], a two-phase interleaved buck [11], and a multi-output power management module utilizing differential buck connection [12] are shown. In comparison with the conventional buck based solutions the hybrid converters reviewed here require much smaller inductor and have better power processing efficiency. Both of these result in a significant increase in the power density of the hybrid structures. To achieve these advantages, all of the presented hybrid converters utilize the principle of reduced inductor volt-second product, which is reviewed in the following section.

As demonstrated in the other related work [14],[15], the hybrid conversion principles reviewed in this paper apply not only for buck-based converter topologies but also for boost-based and other indirect energy transfer converters, used in applications such as rectifiers [14] and inverters [15].

II. MINIMIZING THE INDUCTOR AND SWITCHING LOSSES THROUGH VOLTAGE SWING REDUCTION

To analyze the fundamental principle of hybrid converters operation, we can start from the expression for the steady-state inductor current ripple [2]. For a general inductive converter operating at a constant switching frequency f_{sw}, with a duty ratio D the inductor ripple can be expressed as

$$\Delta I_{ripple} = \frac{|V_{L\_on}|}{2L}\frac{D}{f_{sw}} \quad , \quad (1)$$

or

$$\Delta I_{ripple} = \frac{|V_{L\_off}|}{2L}\frac{1-D}{f_{sw}} \quad , \quad (2)$$

where, V_{L_low} and V_{L_high} are the voltages applied across the inductor during the on and off state of the main

converter switch, respectively, and L is the inductance value. For a conventional buck the two values are:

$$V_{L\_on} = V_g - V \quad \text{and} \quad (3)$$
$$V_{L\_off} = -V,$$

where V_g and V are the converter input and output voltage values, respectively.

To minimize the ripple amplitude and, therefore, reduce the inductor, most commonly the switching frequency is increased. The main drawback of this solution is that, at the same time, switching losses and the inductor core losses are increased negatively affecting the overall power processing efficiency.

In hybrid and cascaded topologies of Fig.1, the inductor is reduced by minimizing V_{L_on} or/and V_{L_off}, resulting in consequent reduction of the inductor volt-second product. In the following section it will be shown that the reduction of the V_{L_low} or/and V_{L_high} values can be achieved without the use of bulky intermediate capacitors existing in the straightforward cascaded solutions [7]-[9]. It will also be demonstrated that the hybrid architectures not only allows for a significant minimization of the inductors but also reduce the voltage stress across the switches, minimizing the switching losses and, at the same time, allowing for minimization of the conduction losses.

III. BUCK CONVERTER WITH MERGED ACTIVE CAPACITIVE DIVIDER

The buck converter with merged capacitive divider [10],[11] is shown in Fig.2, together with equivalent circuits describing its operation. This converter can also be viewed as a topology obtained through a source-load inversion [2] of a three level boost converter [15]. The converter consists of the active capacitive divider, four

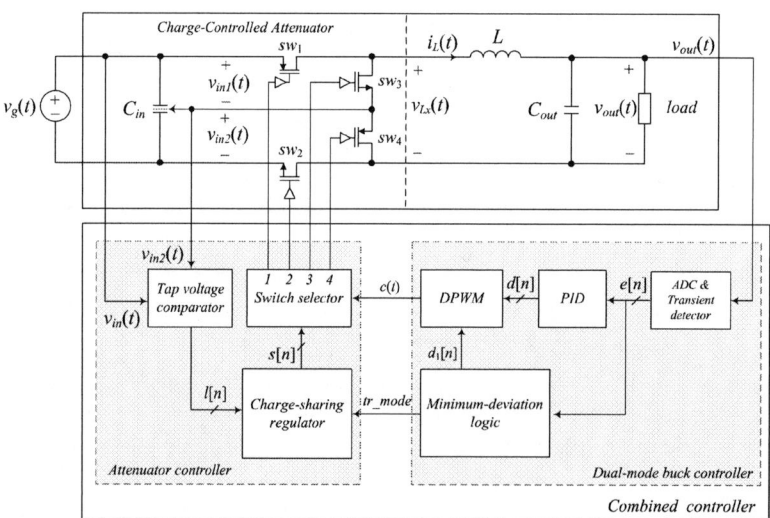

Fig.2. Buck converter with merged capacitive divider [10].

switches, and the downstream buck stage. Here, as shown in Fig. 2, the downstream portion of the converter and the active capacitive divider C_{in} share the same switches. To eliminate the need for a balancing and intermediate capacitors existing in conventional SC converters, in this topology, the buck inductor is used for the balancing of the capacitor cells.

The operation of the converter can be understood by looking at the equivalent circuits of Fig.3. When the divider capacitors are balanced, i.e. voltage $V_g/2$ is across both of them. The switching sequence is performed over two switching periods, where $T_s=1/f_{sw}$ is the switching period. The converter passes through the following modes $a - b/d - c - b/d$, such that, as described in [10], the mode a starts at the beginning of the switching period, mode c starts at T_s, and both of these modes last for DT_s time. Ideally, for perfectly matched components and the duty ratio value, this operation results in the same amount of charge taken from both capacitors and equal voltage sharing. In the case of a mismatch causing unbalanced voltages, the sequence is altered such that more current is taken from the capacitor with higher voltage, until the balance is regained.

It can be seen that in this converter $V_{L\_on} = (V_g/2 - V)$ is lower than that of the conventional buck, resulting in a reduction of volt-second product and a consequent reduction of the inductor value.

By looking at Fig.2 it can also be seen that the blocking voltage of all is $V_g/2$, i.e. half of the value of the conventional buck. This reduction of the voltage stress allows for a significant reduction of switching losses and for the use of a transistors with lower voltage rating that for the same amount of silicon used have about four times smaller on resistance [16]. As a result, a drastic reduction of both switching and conduction losses with this topology is possible.

A. Extension to the two–phase topology

Fig.4 shows extension of the buck with merged capacitive divider to a two phase case [12], where, again,

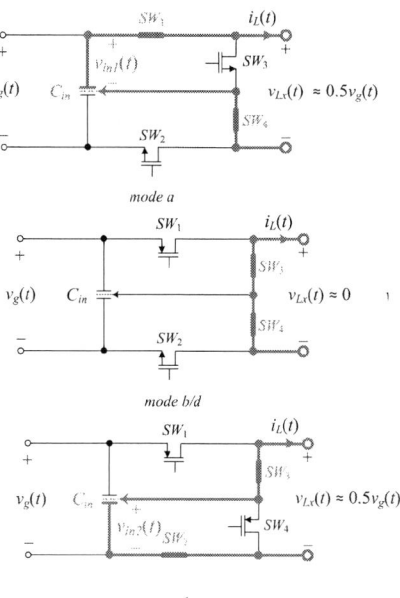

Fig.3. Modes of operation of the buck converter with a merged capacitive divider.

inductors of the downstream portion of the converter are balancing the capacitor divider taps. Operating modes of the converter are shown in Fig.5. In this case, the charging of the capacitive divider is performed through the input filter and Q_1, during the period when the SR switch of the upper phase (transistor Q_9) is turned on. The discharging, and balancing of the capacitors is performed with the inductors of the corresponding buck phases. In this case again, V_{L_on} in both converter phases is reduced to $V_g/2 - V$ and the blocking voltages of all components are reduced to a half of those needed in conventional topologies. This topology also provides inherent equal current sharing between two phases [7] eliminating the need for current balancing circuits.

In the implementation shown in Fig.4 the charging of

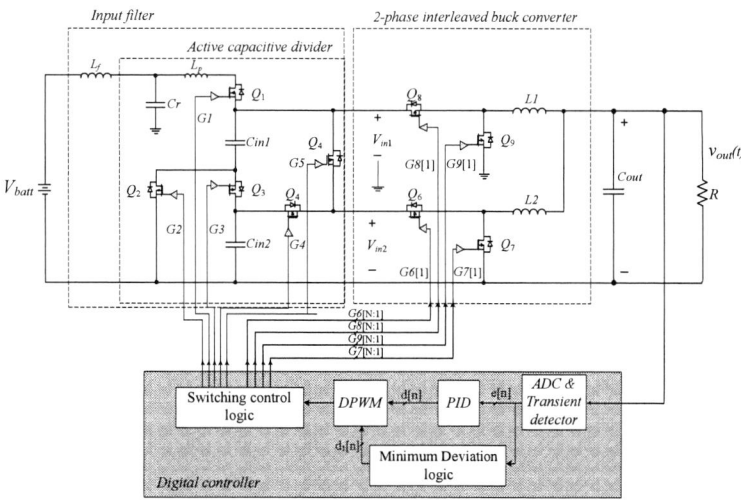

Fig.4. Two phase buck converter with merged capacitive attenuator [12].

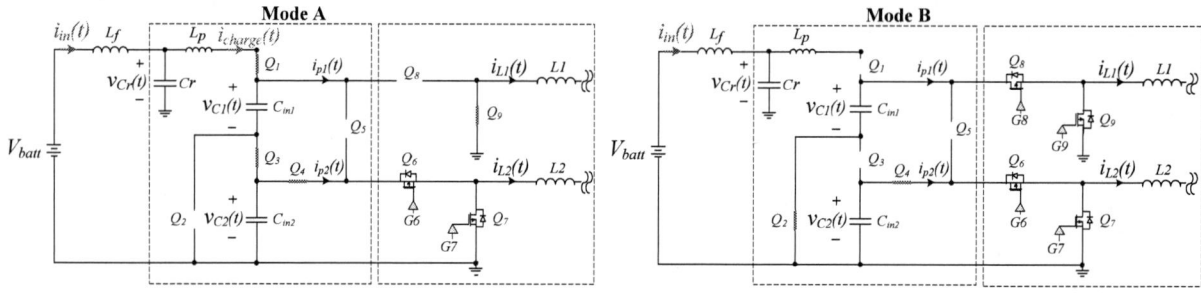

Fig. 5. Operating modes of a two phase buck with merged capacitive divider: adiabatic charging of the capacitors (left) discharging and balancing (right).

both capacitors is performed through Q_1 and the resonant tank L_r-Cr, which provides adiabatic capacitor charging and, therefore, minimizes the losses of Q_1.

It should be noted that even though both of the presented topologies require a larger number of switches than the conventional ones, the total conduction losses are not increased, due lower voltage rating of the switches and, consequently, smaller on resistance values.

IV. MULTI OUTPUT HYBRID POWER MANAGEMENT BLOCK BASED ON DIFFERENTIAL BUCK CONNECTIONS

Hybrid converter topologies can also be utilized at the system level, to increase the power density of power management systems in portable applications.

A typical architecture of a power management system for a battery powered applications is shown in Fig.6. The system of consists of a multiple buck converters, supplied by a bus voltage. The buck converters provide well-regulated voltages for the dedicated functional blocks, such as various data processors and memory. In these applications the reactive components take a significant portion of the overall device volume that depending on the application can vary between 12% and 80% of the overall device volume [1], [11].

Fig.7 shows a hybrid power management structure. In this architecture the sharing of the components is performed such that the capacitive string of the front-end multi-output switched-capacitor converter (MoSC) also acts as the input filter capacitor [2] of the downstream buck converter, i.e. replaces C_{bus} of Fig.6. The downstream buck converters are connected differentially to the output taps of front-end stage. The tap voltages are set such that at the terminals of the buck converters the voltages are slightly higher and slightly lower than the desired output voltage, rather than changing the switching node voltage between the full bus value and the ground. As a result a large reduction in V_{L_on} and $V_{L\text{-}off}$ values (volt-second products) is obtained and, at the same time, the blocking voltages of the switches are reduced.

A practical implementation of this topology for a typical battery-powered application is shown in Fig.8, and the inductor value reductions in comparison with a conventional architecture (Fig.6) operating with the same

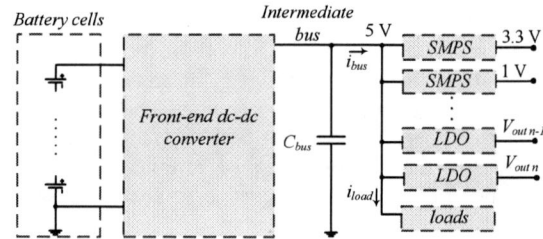

Fig.6. Conventional power management system of a battery-powered device.

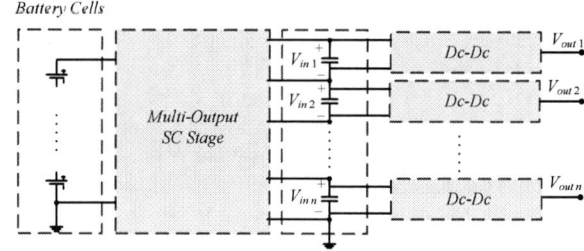

Fig.7. Multi-output switched capacitor (MoSC) based hybrid converter (power management block) based on differential buck converter stages.

input and output voltages are shown in Table 1. The table compares efficiencies and the volumes of the components for two cases, when the buck converters of the hybrid architecture operate at the same frequencies as those of the conventional system and when those converters operate at two times larger frequencies.

It can be seen that for the operation of both system at the same switching frequency the inductance values are reduced up to 50% and the reduction in switching losses of up to 56% is achieved. The table also shows that for the case when a portion of the savings on the switching losses is traded for operation at the higher frequency the inductor can be reduced by up to four times while maintaining significantly smaller switching losses.

Results of efficiency comparison measurements for a 1 V output obtained from a 2-cell Li-ion battery-powered system are shown in Fig.9. The results confirm that with the MoSC hybrid topology significant efficiency improvements are obtained throughout the whole

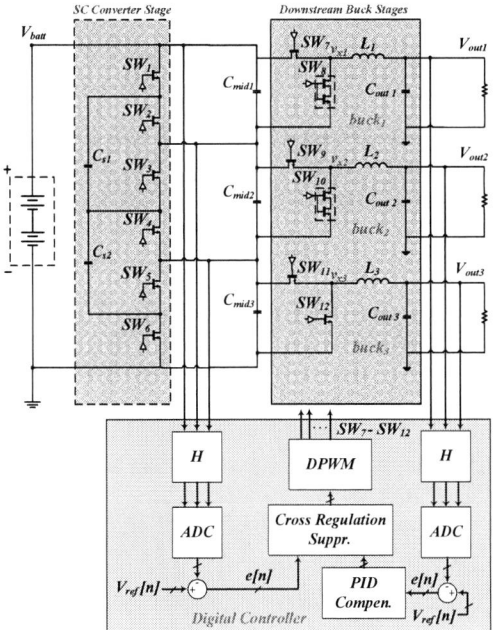

Fig.8. Multi-output hybrid converter based on differential buck converter stages [13].

operating range confirming that both volume reduction and efficiency improvements are obtained with the hybrid architecture.

TABLE I
INDUCTOR VOLUME AND SWITCHING LOSS REDUCTION

	$V_{s\ norm}$	L_{norm}	$P_{sw\ norm}$	$f_{sw\ norm}$
Dual-input buck 3.3 V	0.44	0.49	0.44	1
Dual-input buck 1 V	0.44	0.68	0.44	1
Dual-input buck 3.3 V (incr. f_{sw})	0.44	0.25	0.88	2
Dual-input buck 1 V (incr. f_{sw})	0.44	0.34	0.88	2

V. CONCLUSIONS

A review of three hybrid topologies combining a switched-capacitor (SC) and buck converter in single structures is given. It is shown that by sharing components and utilizing the SC to reduce inductor volt-second product drastic reduction of inductive components can be achieved and, at the same time, significant power processing efficiency improvements obtained.

REFERENCES

[1] "DELPHI DNT04 2.4~5.5Vin/0.75~3.63Vo/3A Non-Isolated Point of Load datasheet," Delta Electronics Inc.
[2] R. W. Erickson and D. Maksimovic, *Fundamentals of Power Electronics*.: Springer Media Inc., 2001.
[3] M. D. Seeman, "A design methodology for switched-capacitor dc-dc converters", Ph.D. thesis, University of California at Berkeley, 2009.

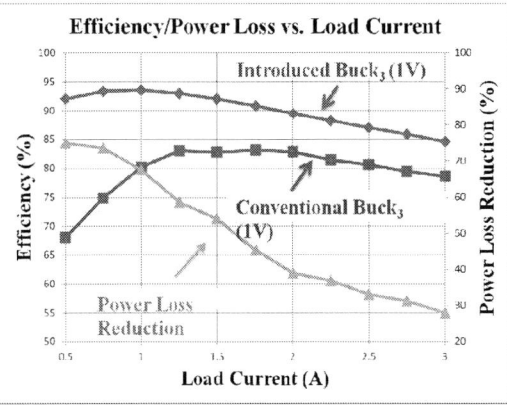

Fig.9. Efficiency comparisons of the dual-input buck of the MoSC based architecture and the conventional downstream converter.

[4] M.D. Seeman, and S. R. Sanders, "Analysis and optimization of switched-capacitor dc-dc converters," *IEEE Trans. on Power Electron.*, vol. 23, pp. 841 – 851, Mar. 2008.
[5] D. Maksimovic, and S. Dhar, "Switched-capacitor dc-dc converters for low-power on-chip applications," in *Proc. IEEE Power Electronics Specialists Conf.*, 1999, pp. 54-59. Aug. 1999.
[6] M. Shoyama, T. Naka, and T. Ninomiya. "Resonant switched capacitor converter with high efficiency," in *Proc. IEEE Power Electronics Specialists Conf.*, 2004, pp. 3780-3786, June 2004.
[7] J. Sun, M. Xu, Y. Ying, and F. C. Lee, "High power density, high efficiency system two-stage power architecture for laptop computers," in *Proc. IEEE Power Electronics Specialists Conf.*, 2006, pp. 1-7, June 2006.
[8] J. Sun, M. Xu, and F. C. Lee, "Transient analysis of the novel voltage divider," in *Proc. IEEE Applied Power Electronics Conf.*, 2007, pp. 550–556. Feb. 2007.
[9] R.C.N. Pilawa-Podgurski, D.M. Giuliano, and D.J. Perreault, "Merged two-stage power converter architecture with soft charging switched-capacitor energy transfer," *in Proc. IEEE Power Electronics Specialist Conf.*, 2008, pp. 4008-4015, June 2008.
[10] A. Radić, A. Prodić, "Buck Converter With MergedActive Capacitive Attenuation," IEEE Trans. on Power Electronics, March 2012,
[11] A. Radić,, "Practical Volume Reduction Strategies for Low-Power Switch-Mode Power Supplies," Ph.D. Thesis, University of Toronto, 2013. .
[12] B. Mahdavikhah, P. Jain, A. Prodić, "Digitally controlled multi-phase buck-converter with merged capacitive attenuator," in *Proc. IEEE (APEC)*, 2012.
[13] S.M. Ahsanuzzaman, J. Blackman, T. Mcrea and A.Prodić, "A multi-output low-volume power management module for portable battery-powered applications," in Proc. *IEEE APEC)*, 2013.
[14] B. Mahdavikhah, R. DiCecco, and A.Prodić, "Low-volume PFC rectifier based on non-symmetric multi-level boost converter," in Proc. IEEE Applied Power Electronics Conference (APEC), 2013
[15] T.A. Meynard, H. Foch, "Multi-level conversion: high voltage choppers and voltage-source inverters," in Proc. IEEE PESC '92, pp.397-403 vol.1.
[16] B.J. Baliga, "Fundamentals of Power Semiconductor Devices," NewYork, NY: Springer, 2008.

The 2014 International Power Electronics Conference

Coupled Inductor Based Current-Fed Switched Inverter for Low Voltage Renewable Interface

Soumya Shubhra Nag and Santanu Kumar Mishra

Department of Electrical Engineering, Indian Institute of Technology, Kanpur, U.P., India

E-mail:- soumyasn@iitk.ac.in and santanum@iitk.ac.in

Abstract: **This paper presents a novel coupled inductor based high boost inverter topology which can be utilized in low voltage renewable systems where high voltage step-up is needed to interface with 110 V/220 V AC systems. The proposed inverter possesses high boost ability with superior EMI immunity compared to a traditional voltage source inverter (VSI). Unlike the traditional VSI, the proposed inverter does not need dead-time circuit for its switching signals as it utilizes shoot-through state of the inverter in its single-stage configuration. Insertion of shoot-through state also helps it to achieve high boost operation essential for renewable energy applications. The proposed inverter is derived from Current-Fed Switched Inverter topology. Apart from topology derivation, this paper describes the steady-state analysis of the inverter and establishes the relation between input, DC-link, and AC output. An experimental prototype is built to validate the proposed inverter circuit. A 220 V (rms) AC is obtained from 52 V DC input to demonstrate its boost mode of operation.**

Index Terms — **ZSI, SBI, Coupled inductor, Shoot through state, EMI immunity.**

I. INTRODUCTION

Voltage source inverters are widely used in UPS, motor drives, grid connected and stand-alone renewable systems, etc. The main limitations of traditional VSI are:

1) The output AC voltage cannot be more than its input DC voltage as VSI is a buck inverter. Due to this reason a DC-DC boost converter stage is needed prior to the VSI to achieve step-up DC-AC inversion when the input DC voltage is limited like in the case of solar PV, fuel cell, etc. Commercially available solar PV panel voltage ranges from 12 V to 48 V typically whereas for fuel cells, it is typically between 24 V to 56 V. For this reason, a high step-up inversion is needed to connect the renewable sources to 110 V/ 240 V AC systems which cannot be obtained from a VSI.

2) The upper and lower switching devices of any leg of the VSI cannot be turned on simultaneously thus requiring for a dead-time circuit which in turn contributes to waveform distortion. Although, adding dead-time in the switching signals cannot alleviate the chances of mis-gating or shoot-through due to spurious signals or EMI noise [1].

To eliminate these drawbacks of VSI, inverters like Z-Source Inverter (ZSI) [1], Quasi-ZSI (q-ZSI) [2], Switched Boost Inverter (SBI) [3]-[5], Boost-Derived Hybrid Converter (BDHC) [6], Trans-ZSI (T-ZSI) [7], etc., were proposed. These new-age inverters present single stage DC-AC inversion with high boost capability and utilize the shoot-through phenomenon in the inverter legs to provide superior EMI immunity. In the lines of these inverters, Current-Fed Switched Inverter (CFSI) was proposed [8]-[9] which

provided high gain (same as ZSI) with the low passive component count. Due to the presence of input inductor, CFSI provided continuous input current property which is necessary for renewable applications. In all of the above mentioned inverters, shoot-through state imposes some restriction on the modulation index which limits them to achieve high overall input DC to output AC gain. Thus, in recent years, there is a constant effort among the researchers to increase the overall DC-to-AC conversion ratio of these shoot-through inverters by

a) Modifying the pulse width modulation scheme so that the constraint on modulation index can be minimized. It has resulted in invention of new modulation techniques like Constant Boost Control, Maximum Boost Control schemes, etc.

b) Improving the boost factor (input DC-to-inverter input gain) of the inverters by using either passive network (switched capacitor, switched inductor etc.) or magnetic (tapped inductor, coupled inductor etc.) network.

Nevertheless, inverters with low component count, continuous input current, low device stress are always an attractive option owing to their high efficiency, ease of integration with renewable sources, low device cost and device footprint.

This paper presents a coupled inductor based high boost inverter topology derived from Current-Fed Switched Inverter (CFSI) which is named as Coupled Inductor based Current-Fed Switched Inverter (Trans-CFSI) as it utilizes energy transfer through the transformer action of the coupled inductor to achieve high boost. Like SBI, the proposed inverter uses an active network between the DC input and inverter bridge with one LC-filter pair. In the next section, CFSI topology is reviewed. Derivation of Trans-CFSI topology from CFSI is discussed in section III along with its steady-state characteristics. In section IV, the PWM control scheme of Trans-CFSI is described. The proposed inverter is verified with experimental results in section V. Section VI presents some conclusions.

II. REVIEW OF CFSI TOPOLOGY

The circuit schematic of Current-Fed Switched Inverter (CFSI) is shown in Fig. 1 (a). CFSI provides high-boost operation similar to ZSI and q-ZSI utilizing the shoot through state of the inverter legs. The operating states of the CFSI can be broadly categorized into i) Shoot through state and ii) Non-Shoot through state, the later can be further be divided into active state (power interval of the inverter) and zero state (zero interval of the inverter). The equivalent circuit of the CFSI is shown in Fig. 1 (b). In the shoot through interval (or duty

978-1-4799-2706-7/14 $31.00 © 2014 IEEE

interval D) switch S is turned on along with both the switches of any inverter leg. In this interval source V_g and capacitor C_o charges inductor L together. In non-shoot through interval ((1-D) interval or D' interval), switches S is turned off which forces diodes D_a and D_b to turn on, and the inductor charges C_o and power is delivered to the AC-load through the inverter. The equivalent circuits of CFSI in D and D' intervals are shown in Fig. 1 (c). From Fig. 1 (c), the voltage across inductor L in one switching period of T_s is given by (1) (assuming small ripple approximation) from which the boost conversion ratio of CFSI can be obtained as shown in (2).

$$v_L = \begin{vmatrix} V_g + V_c & During\ D.T_s \\ V_g - V_c & During\ (1-D).T_s \end{vmatrix} \quad (1)$$

$$B_{CFSI} = \frac{V_c}{V_g} = \frac{1}{1-2D} \quad (2)$$

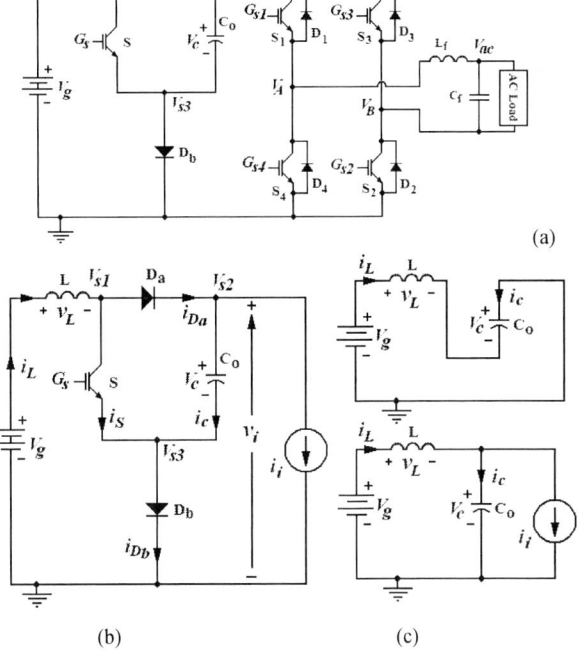

(a)

(b) (c)

Fig. 1. Schematic of (a) Current-Fed Switched Inverter (CFSI), (b) equivalent circuit of CFSI, and (c) equivalent circuit of CFSI in shoot though and non-shoot through state.

Although CFSI provides high boost output, use of shoot through state restricts the modulation index to a value always less than (1-D) in simple boost control method. This also imposes higher stress on the inverter switches. In the next section, a coupled inductor based CFSI topology (Trans-CFSI) will be derived which will mitigate the problems of CFSI as stated above.

III. DEVELOPEMENT OF TRANS-CFSI TOPOLOGY

The coupled inductor based CFSI topology, namely Trans-CFSI, is shown in Fig. 2. It utilizes energy transfer through the transformer action of the coupled inductor to achieve high voltage boost which depends on the turns-ratio $n_1{:}n_2$.

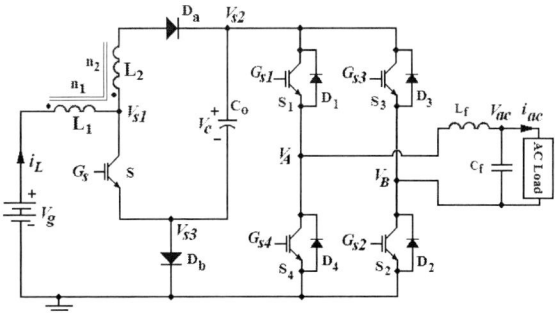

Fig. 2. Schematic of Trans-CFSI topology

The equivalent circuit diagram of Trans-CFSI is shown in Fig. 3 which is obtained by replacing the two windings of the coupled inductor with an ideal transformer and a mutual inductance L_m on the primary side.

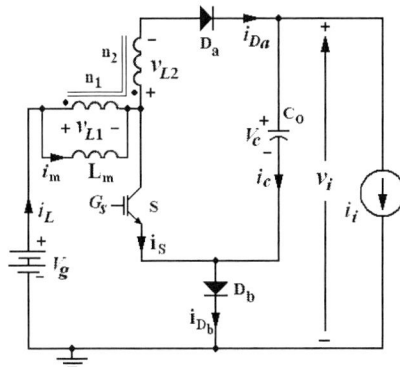

Fig. 3. Equivalent circuit of Trans-CFSI.

In the shoot through duty interval (D interval), switch S is turned on with the inverter leg being shorted while both the diodes remain reverse biased (as shown in Fig. 4 (a)). The inductor voltages in this interval can be written as in (3). In the non-shoot through duty interval ((1-D) interval), switch S is turned off and the inverter operates either in active or zero state. In this interval both the diodes remain in conduction (as shown in Fig. 4 (b)). The inductor voltages in (1-D) interval can be written as in (4).

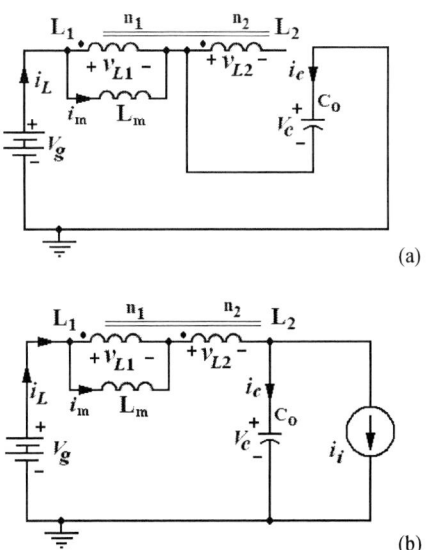

(a)

(b)

Fig. 4. Equivalent circuit of Trans-CFSI in (a) D-interval (shoot-through state), and (b) in (1-D)-interval (non-shoot through state).

$$v_{L1} = v_M = V_g + V_c$$

$$v_{L2} = \frac{n_2}{n_1} v_{L1} = \frac{n_2}{n_1}(V_g + V_c) \quad \bigg| \quad During\ D.T_s \qquad (3)$$

$$v_{L1} + v_{L2} = v_L = V_g - V_c$$

$$v_M = \frac{n_1}{n_1 + n_2}(V_g - V_c) \quad \bigg| \quad During\ (1-D).T_s \qquad (4)$$

Applying volt-second balance [10] using (3) and (4), the boost factor of Trans-CFSI can be obtained as,

$$(V_g + V_c).D + \left(\frac{n_1}{n_1 + n_2}(V_g - V_c)\right).(1-D) = 0$$

$$\Rightarrow B_{Trans-CFSI} \quad \frac{V_c}{V_g} = \frac{1+nD}{1-(2+n)D} \qquad (5)$$

Where coupled inductor turns-ratio $n = n_1:n_2$. The switch-node voltages of Trans-CFSI, viz. V_{s1}, V_{s2}, V_{s3} are shown Fig. 5 (a) in along with shoot through switching signal and input current (current through the primary winding of the coupled inductor) of the inverter. The boost factor (V_c/V_g) of Trans-CFSI for different values of coupled inductor turns-ratio is plotted in Fig. 5 (b).

IV. PWM CONTROL SCHEME OF TRANS-CFSI

To incorporate shoot-through state in the PWM control, the traditional PWM technique for VSI is modified accordingly. The modified PWM scheme for Trans-CFSI is developed based on the traditional sine-triangle PWM with unipolar voltage switching. Sinusoidal modulation signals $V_m(t)$ and $-V_m(t)$ and high frequency carrier signal $V_{tri}(t)$ are shown in Fig. 6 (a).

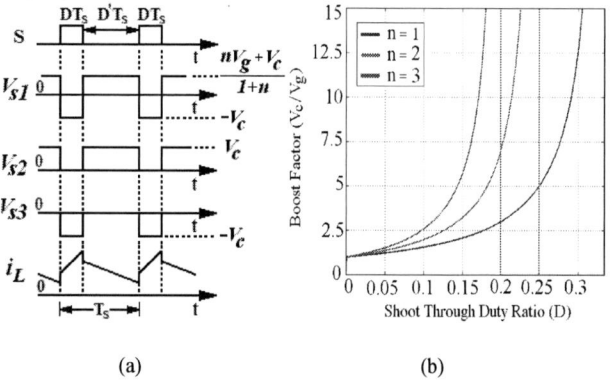

(a)   (b)

Fig. 5. (a) Switch-node voltages along with shoot through duty and inverter input current of Trans-CFSI, and (b) boost factor of Trans-CFSI.

Shoot-through constant voltages V_{ST} and $-V_{ST}$, and Gate signals G_S, G_{S1}, G_{S2}, G_{S3}, G_{S4} of the modified modulation scheme for positive and negative half cycles of the sinusoidal modulation signal $V_m(t)$ is shown in Fig. 6 (a). Shoot-through signals $ST1$ and $ST2$ are generated by comparing V_{ST} and $-V_{ST}$ with carrier signal as shown in Fig. 6 (b) and Fig. 6 (c) for the positive and negative half-cycle of the modulation signal, respectively. In this half cycle, the shoot-through interval is inserted in the gate signals G_{S3} and G_{S4}. Gate signals G_{S3}, G_{S4}, and G_S are generated using the following logic.

$$G_{S3} = \overline{G_{S2}} + ST1 \ ; \ G_{S4} = \overline{G_{S1}} + ST2 \ ; \ G_S = ST1 + ST2$$

Likewise, in the negative half-cycle ($V_m(t) < 0$) of the modulation signal, gate signals G_{S3} and G_{S4} are generated by comparing the sinusoidal modulation signals $-V_m(t)$ and $V_m(t)$ with carrier signal $V_{tri}(t)$. The shoot-through interval is inserted in gate signals G_{S1} and G_{S2}. Gate signals G_{S1}, G_{S2}, and G_S are generated using the following logic equation.

$$G_{S1} = \overline{G_{S4}} + ST1 \ ; \ G_{S2} = \overline{G_{S3}} + ST2 \ ; \ G_S = ST1 + ST2$$

The PWM control signals G_{s1}, G_{s2}, G_{s3}, and G_{s4} are shown in Fig. 6 (d) for the positive half-cycle of $V_m(t)$.

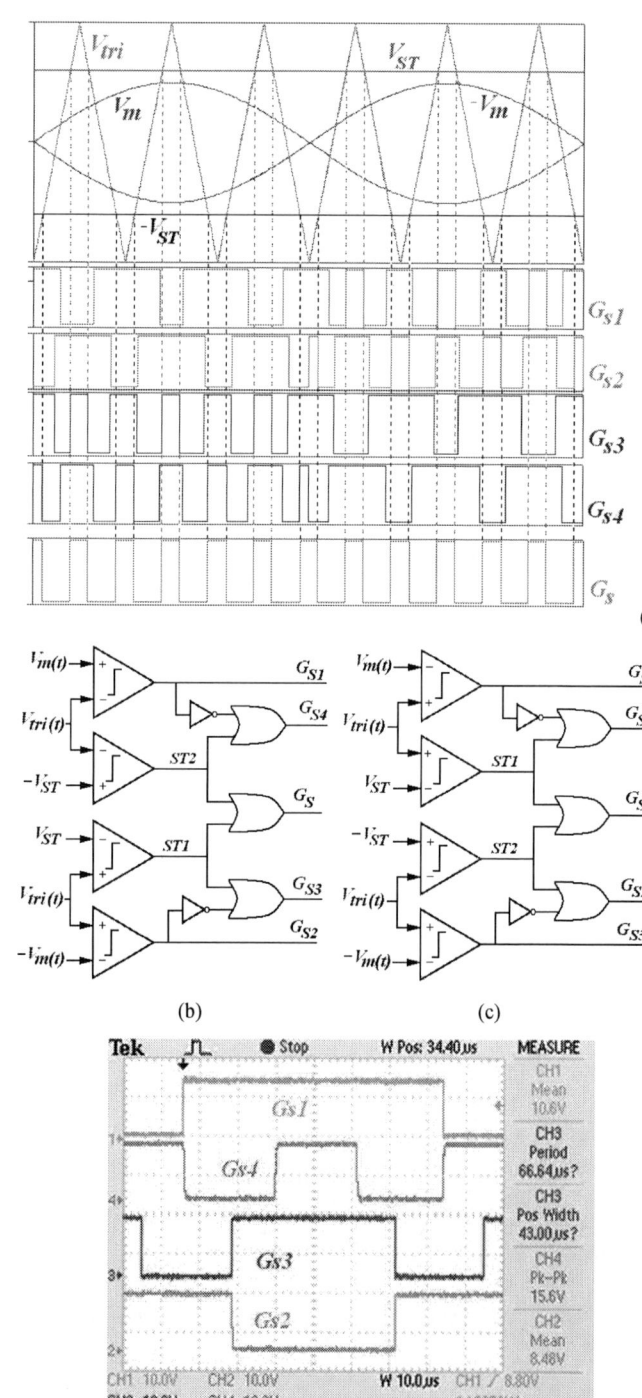

Fig. 6. (a) Generation of PWM control signals, (b) PWM control scheme when $V_m(t) > 0$, (c) PWM control scheme when $V_m(t) < 0$, and (d) PWM control signals for the positive half-cycle of the modulation signal.

The mathematical relation between D and V_{ST} can be written as,

$$D = 1 - (V_{ST} / \hat{V}_{tri}) \qquad (6)$$

In order to ensure that the shoot-through duty D interval doesn't overlap with the power interval of the inverter, D should be chosen such that

$$D \leq 1 - M \qquad (7)$$

Where M is the modulation index of the inverter. The relation between the AC output and DC input (V_g) is

$$V_{ACpeak} = B_{Trans-CFSI} \times M \times V_g = M \times \frac{1+nD}{1-(2+n)D} V_g \qquad (8)$$

V. EXPERIMENTAL VERIFICATION

A prototype is built to test the proposed Trans-CFSI topology. The PWM control for the inverter is developed in Texas Instruments TMS320F28335 DSP. The design specifications are tabulated in Table I.

TABLE I

PROTOTYPE SPECIFICATION

Parameter/ Component	Attributes
Modulation Frequency (f_m)	50 Hz
Carrier Frequency (f_{tri})	25 kHz
Coupled Inductor (L$_1$, L$_2$, n)	1.78 mH, 1.78 mH, 1
Capacitor (C$_o$)	940 uF
Output Filter Inductor (L$_f$)	4.6 mH
Output Filter Capacitor (C$_f$)	10 μF
Output Power	225 W

Steady-State Operation

Fig. 7 (a) shows converter switch node voltages along with the DC-link capacitor voltage. Fig. 7 (b) shows the inverter circuit operation of Trans-CFSI in boost mode where a 110 V AC output (at 50 Hz) is obtained from 230 V DC-link voltage with input voltage of 28 V.

Fig. 8 (a) shows the high boost operation of Trans-CFSI where a 220 V AC output is obtained from an input voltage of 52 V DC at D=0.285, M=0.68, n=1 with DC-link capacitor voltage (V_c) of 450 V. The output power is 225 Watt at unity power factor (resistive load). Steady-state operation of Trans-CFSI in buck mode is shown in Fig. 8 (b) where 36 V AC output is generated from an input voltage of 60 V DC with DC-link capacitor voltage (V_c) of 59 V with D=0, M=0.87, n=1.

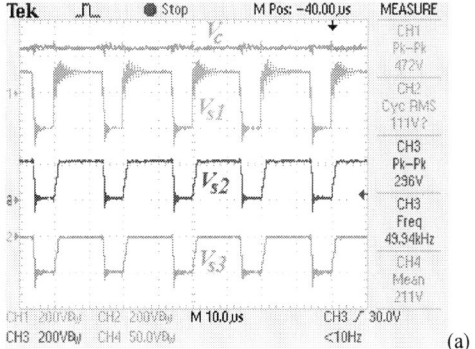

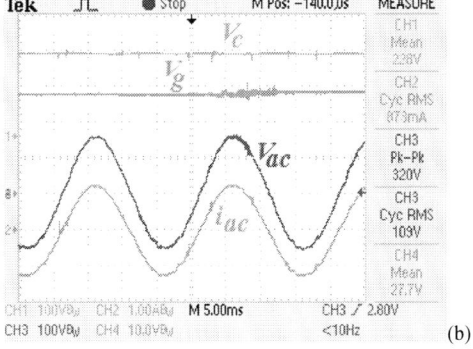

Fig. 7. (a) Steady-state switch-node voltages and DC-link voltage (V_c) of Trans-CFSI, and (b) steady-state output AC voltage (V_{ac}) and output current (I_{ac}) with DC-link voltage (V_c) and DC-input voltage (V_g) for 110 V AC systems.

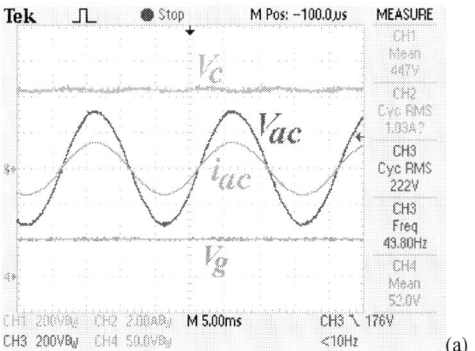

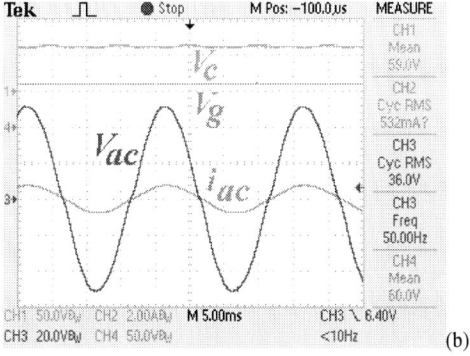

Fig. 8. Steady-state operation of Trans-CFSI in (a) high-boost mode: DC-link capacitor voltage (V_c), input DC voltage (V_g), output AC voltage (V_{ac}), and output AC current (i_{ac}) waveforms for 220 V AC systems with D=0.285, M=0.68, n=1, (b) buck mode: DC-link capacitor voltage (V_c), input DC voltage (V_g), output AC voltage (V_{ac}), and output AC current (i_{ac}) waveforms with D=0, M=0.87, n=1.

Effect of DC-link Fault Due to EMI

Due to the presence of shoot-through state, Trans-CFSI provides better EMI immunity than traditional VSI. To prove the EMI immunity of Trans-CFSI, an EMI noise of 1-ms duration is applied to the inverter bridge which makes all the gate switching signals high. The operation of the circuit under this test is shown in Fig. 9 (a). From the result, it can be seen that the output AC voltage and DC-Link capacitor voltage remains at their previous voltage levels and does not respond to the noise. Although the input current, i_{in}, (i_{in} is the input current drawn by the converter after placing an input filter capacitor) of the inverter rises to about 4 times of the normal value, it comes back to steady-state within a short period of time (3-ms). The Trans-CFSI circuit is also tested for a long duration DC-bus fault test with 80 ms fault duration which is shown in Fig. 9 (b). It can be seen from the test result that there is no damage to the inverter and the circuit regains steady-state after the fault is cleared.

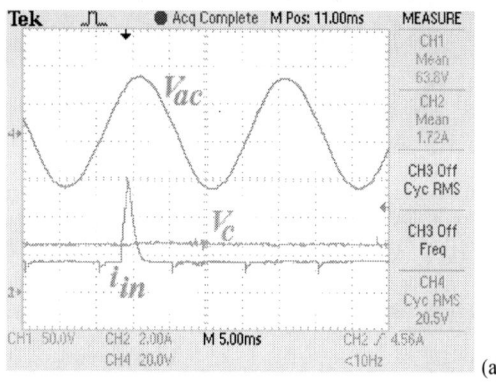

(a)

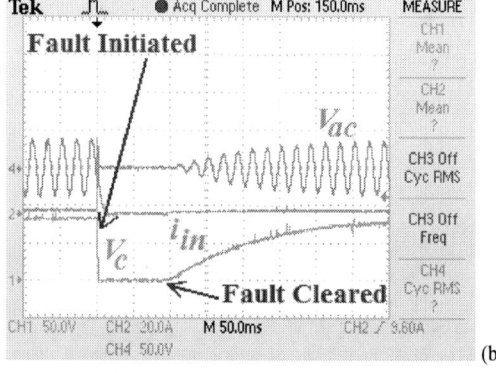

(b)

Fig. 9. (a) EMI testing on Tran-CFSI with 1 ms duration EMI pulse, and (b) DC-bus fault testing on Trans-CFSI with 80 ms fault duration.

VI.   CONCLUSION

This paper proposed a coupled inductor based high boost inverter, named Trans-CFSI, which exhibits improved EMI noise immunity similar to the ZSI, SBI etc. The high gain of the inverter is obtained by the transformer action of the coupled inductor and insertion of shoot-through state. In this paper the development of Trans-CFSI topology is described in details along with its steady-state characteristics and PWM switching scheme. The proposed inverter is tested on a laboratory prototype and verified. The inverter is also tested for EMI and DC-bus fault which shows that the inverter shows EMI immunity and can sustain DC-bus fault.

ACKNOWLEDGEMENT

This work was supported by Science and Engineering Research Board (SERB), Govt. of India under grant no. SR/S3/EECE/0187/2012.

VII.   REFERENCES

[1] F.Z. Peng, "Z-Source Inverter," *IEEE Transaction on Industrial Applications*, Volume. 39, No. 2, 2003, pp. 504-510.

[2] J. Anderson, F.Z. Peng, "Four Quasi-Z-Source Inverters" in *IEEE Power Electronics Specialists Conference, PESC* 2008, pp. 2743 – 2749.

[3] S. Mishra, R. Adda, and A. Joshi, "Inverse Watkins–Johnson Topology-Based Inverter," *IEEE Transactions on Power Electronics*, Volume: 27, Issue: 3, 2012, pp. 1066 – 1070.

[4] R. Adda, O. Ray, S. Mishra, and A. Joshi, "Synchronous Reference Frame Based Control of Switched Boost Inverter for Standalone DC Nanogrid Applications," *IEEE Transactions on Power Electronics*, Volume: 28, Issue: 3, 2013, pp. 1219 - 1233.

[5] A. Ravindranath, S. Mishra, and A. Joshi, "Analysis and PWM Control of Switched Boost Inverter," *IEEE Trans. Ind. Electron.*, vol. 60, no. 12, pp. 5593–5602, Dec. 2013.

[6] Olive Ray and Santanu Mishra, "Boost-Derived Hybrid Converter with Simultaneous DC and AC Outputs," accepted for publication in *IEEE Trans. on Indus. Applications*, 2013. DOI 10.1109/TIA.2013.2271874.

[7] Wei Qian, F. Z. Peng, H. Cha, "Trans-Z-Source Inverters" *IEEE Transactions on Power Electronics*, Volume: 26, Issue: 12, 2011, pp. 3453 – 3463.

[8] Soumya Shubhra Nag and Santanu Mishra, "Current–Fed Switched Inverter," *IEEE Early Access Article, IEEE Transactions on Indus. Electronics*, 2014, DOI 10.1109/TIE.2013.2289907.

[9] Soumya Shubhra Nag, Ravindranath Adda, Olive Ray, Santanu Mishra, "Current-Fed Switched Inverter Based Hybrid Topology for DC Nanogrid Application, " in *39th Annual Conference of IEEE Industrial Electronics Society, IECON, 2013*, pp. 7146-7151.

[10] R. W. Erickson and D. Maksimovic, Fundamentals of Power Electronics, 2nd Edition, Springer science+ business media Inc., NY, 2001.

The 2014 International Power Electronics Conference

A Semi-Isolated Multi-Input Converter for Hybrid PV/Wind Power Charger System

Cheng-Wei Chen, Kun-Hung Chen, and Yaow-Ming Chen
Electric Energy Processing Research Laboratory
Department of Electrical Engineering
National Taiwan University, Taipei, Taiwan
f98921020@ntu.edu.tw

Abstract— **The objective of this paper is to propose a semi-isolated multi-input converter (S-MIC) for hybrid PV/wind power charger system which can simplify the power system, reduce the cost and deliver continuous power with higher reliability. The S-MIC consists of a forward-type isolated pulsating voltage source cell (I-PVSC) and a SEPIC prime converter that can realize the maximum power point tracking (MPPT) function for each PV/wind source. Due to the semi-isolated configuration, the proposed S-MIC can adopt PV/wind power sources with larger operation voltage difference. In this paper, the operational principle of the proposed S-MIC is explained and the small-signal ac model is developed. Computer simulations and prototype circuit experimental results are presented to verify the performance of the S-MIC.**

Keywords — Maximum power point tracking (MPPT), multi-input converter (MIC), renewable energy, small-signal ac model.

I. INTRODUCTION

Nowadays, renewable energy sources such as solar panels or wind turbines have been developed rapidly [1], [2]. However, because of their intermittent feature, different types of renewable energy sources should be integrated together in order to deliver less fluctuated and more reliable energy to the load [3], [4].

Generally, various sources with individual power converters can be connected in parallel at the load port to draw the maximum power as well as to increase the power reliability. However, in order to increase the flexibility of power expansion and the utilization of the power converter, different types of multi-input converters (MICs) have been proposed [5]-[16].

The MIC can deliver power from different energy sources to the load simultaneously or individually. A systematic approach to develop the circuit topologies of the MIC have been proposed [17]. In [17], the concept of pulsating source cell (PSC), including pulsating voltage source cell (PVSC) and pulsating current source cell (PCSC), was first proposed and rules of connecting them with other dc-dc converters were established. For example, a buck-type PVSC inserted into a buck–boost prime converter was reported in [9]. The power drawn from a low-voltage dc source and a high-voltage one is delivered to a common dc load. A two-loop control

strategy for a buck-type PVSC SEPIC prime converter has been proposed in [18], providing the small signal ac model developed for dc battery sources and a dc load.

However, the application of MICs for hybrid PV/wind power charger system is a challenge since the output voltage of the battery and PV module are usually quite low and the rectified output voltage of a wind turbine using permanent magnet synchronous generator (PMSG) is be much higher. The use of transformer with high voltage step up/down feature becomes a promising solution to solve this problem. Therefore, based on [17], a semi-isolated multi-input converter (S-MIC) consists of a forward-type isolated pulsating voltage source cell (I-PVSC) and a SEPIC prime converter is proposed in this paper to realize the maximum power point tracking (MPPT) function for each PV/wind source.

In this paper, the operational principle of the proposed S-MIC for hybrid PV/wind power charger system is explained and the small-signal ac model is developed. Computer simulations and prototype circuit experimental results are presented to verify the performance of the S-MIC.

II. THE PROPOSED S-MIC

The schematic diagram of the proposed S-MIC, which consists of a forward-type PVSC and a SEPIC prime converter, for the hybrid PV/wind battery charger system is shown in Fig. 1. The input voltage sources V_{PV} and V_{Wind} are the output of the PV panel and the rectified output voltage of a wind turbine, respectively.

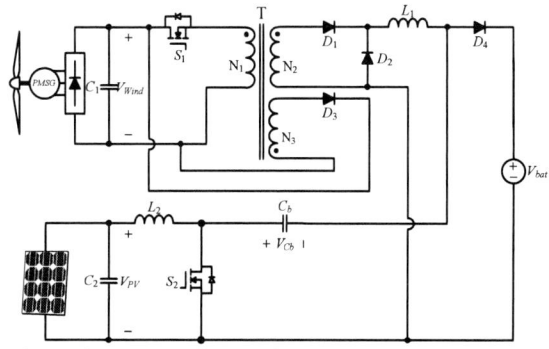

Fig. 1. The schematic diagram of the proposed S-MIC for the hybrid PV/wind battery charger system.

978-1-4799-2706-7/14 $31.00 © 2014 IEEE

By controlling the switches S_1 and S_2, both sources can achieve their maximum power with appropriate MPPT algorithm, individually or simultaneously.

According to the status of the power switches, there are four different operation modes which can be explained as follows.

Mode I (S_1:On / S_2:On)

The equivalent circuit for Mode I is shown in Fig. 2(a) in which S_1, D_1 and S_2 are turned on, and D_2, D_3 and D_4 are turned off with reverse-biased voltages. The input source V_{i1} and the capacitor C_b will charge the inductor L_1 while the input source V_{i2} will charge the inductor L_2. In this operation mode, the transformer will step down the input source V_{i1} to charge the energy storage components, inductor L_1, while the input source V_{i2} will charge the energy storage components, inductor L_2. The energy stored in capacitor C_b will be transfer inductor L_1 too. Both of L_1 and L_2 will provide the electric energy for the battery during other operation mode.

Mode II (S_1:On / S_2:Off)

The equivalent circuit of Mode II is shown in Fig. 2(b). In Mode II, the power switch S_1 is turned on and S_2 is turned off. Diodes D_1 and D_4 are turned on but D_2 and D_3 are reverse biased. Because switch S_2 is turned off so the power diode D_4 will provide a current path for inductor current i_{L2}. The capacitor C_b is charged by inductor current i_{L2}, too. In this mode, both of the input source sources will transfer electric energy into the proposed double-input dc/dc converter, simultaneously.

Mode III (S_1:Off / S_2:On)

Fig. 2(c) shows the equivalent circuit for Mode III where the power switch S_1 is off and S_2 is on. Also, the power diode D_1 and D_4 are reverse biased as an open circuit, and D_2 and D_3 are forward biased as a short circuit. In this operation mode, the transformer will be reset via the reset winding and D_3. The input voltage source V_{i1} stop providing energy and the input source V_{i2} will charge the energy storage components, inductor L_2. The capacitor C_b will transfer its stored energy into inductor L_2.

Mode IV (S_1:Off / S_2:Off)

The equivalent circuit for Mode IV is shown in Fig. 2(d). Both of the power switches S_1 and S_2 are turned off. Power diodes D_2 and D_4 will provide the current path for the inductor current i_{L1} and i_{L2}. In this mode, the input source V_{i1} is disconnected from the proposed double-input converter and the transformer will be reset. The input source V_{i2} will transfer electric energy into the load via the proposed double-input dc/dc converter. The electric energy stored in L_1 will be released into the load, too.

The steady-state input-output voltage relationship of the proposed S-MIC can be derived based the from the volt-second balance theory. The duty ratios d_1 and d_2 for S_1 and S_2, respectively, can be different depending on the status of the input voltage sources.

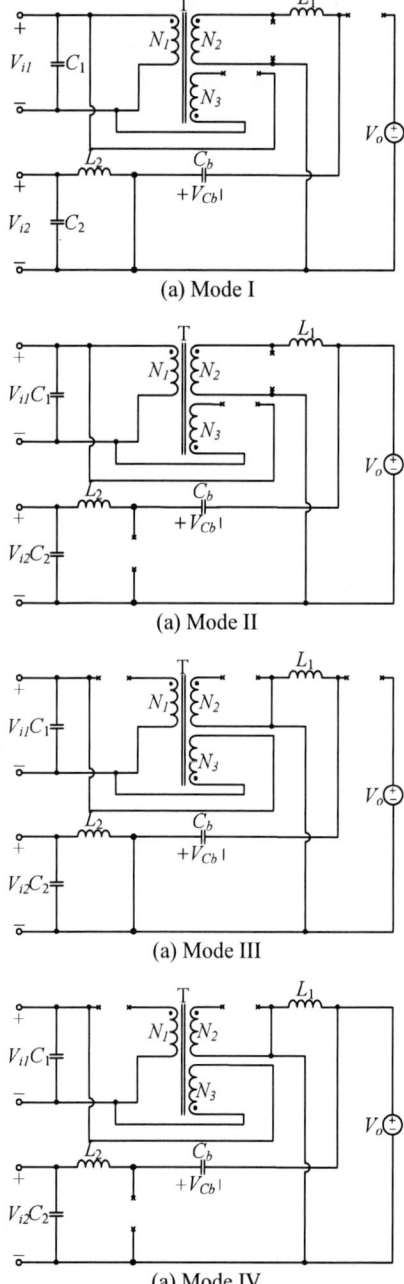

(a) Mode I

(a) Mode II

(a) Mode III

(a) Mode IV

Fig. 2. The equivalent circuits of the proposed S-MIC under different operation modes.

When $d_1 > d_2$, the equivalent operation circuit of the proposed converter during one switching cycle will follow the sequence of Modes I, II, and IV. By applying the volt–second balance theorem on the inductor L_1 and L_2, the following equations can be obtained:

$$d_2 T_s (N V_{i1} + V_{Cb}) + (d_1 - d_2) T_s (N V_{i1} - V_o)$$
$$-(1 - d_1) T_s V_o = 0 \qquad (1)$$

$$d_2 T_s V_{i2} + (d_1 - d_2) T_s (V_{i2} - V_{Cb} - V_o)$$
$$+(1 - d_1) T_s (V_{i2} - V_{Cb} - V_o) = 0 \qquad (2)$$

where T_S is the switching period and N is the turn ratio defined as N_2/N_1.

From (1) and (2), the output voltage expression can be obtained as:

$$V_o = NV_{i1}d_1 + \frac{V_{i2}d_2}{(1-d_2)} \quad . \tag{3}$$

If $d_2 > d_1$, the equivalent operation circuit of the proposed converter in one switching cycle will follow another sequence of Modes I, III, and IV. By applying the volt-second balance theory, the same output voltage expression as shown in (3) can be obtained. As shown in (3), if any one of the voltage sources is not available, the proposed S-MIC can still achieve the desired output voltage. Eq. (3) also reveal the advantage of the semi-isolated configuration under high voltage-transfer-ratio applications, since the power switches can be operated under moderate duty cycle with less current stress and higher efficiency.

III. CIRCUIT MODELING AND CONTROLLER DESIGN

A. The small signal ac model

The output characteristic of the rectified output voltage of a PMSG wind turbine is similar to a PV panel. Therefore, under steady state, both the wind turbine and the PV panel can be represented by a voltage source V_{g1} V_{g2} and a series resistor R_{eq1} and R_{eq2}, respectively, with values determined by the actual operating condition.

By using the state space averaging method, the small-signal ac models of the proposed forward-SEPIC hybrid PV/wind battery charger can be represented in the matrix form as follows:

$$\dot{\tilde{x}} = A\tilde{x} + B\tilde{u}$$
$$\tilde{y} = C\tilde{x} + D\tilde{u} \tag{4}$$

where $\tilde{x}$ is the state-variable, $\tilde{u}$ is the input vector, and $\tilde{y}$ is the output vector. For the proposed S-MIC, these parameters can be defined as:

$$[x] = [v_{Wind} \quad v_{PV} \quad i_{L1} \quad i_{L2} \quad v_{Cb}]^T ;$$

$$[u] = [d_1 \quad d_2 \quad v_{bat}]^T ;$$

$$A = \begin{bmatrix} -\dfrac{1}{C_1 R_{eq1}} & 0 & -\dfrac{ND_1}{C_1} & 0 & 0 \\ 0 & -\dfrac{1}{C_2 R_{eq2}} & 0 & -\dfrac{1}{C_2} & 0 \\ \dfrac{ND_1}{L_1} & 0 & 0 & 0 & -\dfrac{D_2}{L_1} \\ 0 & \dfrac{1}{L_2} & 0 & 0 & \dfrac{1-D_2}{L_2} \\ 0 & 0 & \dfrac{D_2}{C_b} & \dfrac{D_2-1}{C_b} & 0 \end{bmatrix} ;$$

$$B = \begin{bmatrix} \dfrac{NI_{L1}}{C_1} & 0 & 0 \\ 0 & 0 & 0 \\ -\dfrac{NV_{PV1}}{L_1} & \dfrac{V_{Cb}-V_{bat}}{L_1} & \dfrac{D_2-1}{L_1} \\ 0 & \dfrac{V_{Cb}-V_{bat}}{L_2} & \dfrac{D_2-1}{L_2} \\ 0 & -\dfrac{I_{L1}+I_{L2}}{C_b} & 0 \end{bmatrix} ;$$

$$C = \begin{bmatrix} 1 & 0 & 0 & 0 & 0 \\ 0 & 1 & 0 & 0 & 0 \\ 0 & 0 & 0 & 0 & 0 \\ 0 & 0 & 0 & 0 & 0 \\ 0 & 0 & 0 & 0 & 0 \end{bmatrix} ; \quad D = 0 . \tag{5}$$

The transfer function matrix of the converter can be obtained from the small-signal ac model as follows:

$$G = C(sI - A)^{-1}B + D . \tag{6}$$

According to the number of control variables and (6), the transfer function matrixes are obtained as follows:

$$\underbrace{\begin{bmatrix} v_{Wind} \\ v_{PV} \end{bmatrix}}_{y} = \underbrace{\begin{bmatrix} G_{11} & G_{12} & G_{13} \\ G_{21} & G_{22} & G_{23} \end{bmatrix}}_{G} \underbrace{\begin{bmatrix} d_1 \\ d_2 \\ v_{bat} \end{bmatrix}}_{u} \tag{7}$$

where component G_{ij} represents the transfer function between y_i and u_j.

B. Controller design

Fig. 3 shows the control block diagram of the proposed S-MIC, where two loop gains, $T_1(s)$ and $T_2(s)$, are measured at the designated location. Due to the interaction between the two loops, the design of the compensator A_1 and A_2 is no longer as straight-forward as it used to be for a single input converter. However, the complexity can be reduced if $T_X(s)$ and $T_Y(s)$ are both larger than unity, which can be achieved easily with proper design before the crossover frequency. Equation (7) and (8) shows the simplified loop gains and the decoupled control-to-input transfer functions G_1 and G_2.

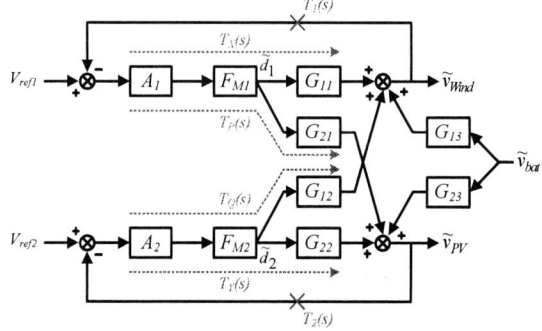

Fig. 3. The control block diagram of the proposed S-MIC.

$$T_1(s) = T_X(s) - \frac{T_P(s) \cdot T_Q(s)}{1 + T_Y(s)}$$

$$\approx \underbrace{\left(G_{11} - \frac{G_{12} \cdot G_{21}}{G_{22}} \right)}_{G_1} \cdot F_{M1} \cdot A_1(s) \tag{8}$$

$$T_2(s) = T_Y(s) - \frac{T_P(s) \cdot T_Q(s)}{1 + T_X(s)}$$

$$\approx \underbrace{\left(G_{22} - \frac{G_{12} \cdot G_{21}}{G_{11}} \right)}_{G_2} \cdot F_{M2} \cdot A_2(s) \tag{9}$$

With the small signal ac model derived in the previous section and the specifications listed in Table I. The bode plots of G_1 and G_2 can be obtained as shown in Fig. 4 and Fig. 5.

Fig. 4 and Fig. 5 show that G_1 can be considered as a transfer function with a double pole and a zero at higher frequency, while G_2 has a single double-pole. In both cases, the desirable loop gain have to meet the stability criterion with wide band width and high gain at low frequency in order to eliminate the steady state error and to increase the transient response. Therefore, a type-III compensator with two zeros and three poles is introduced to the control loop.

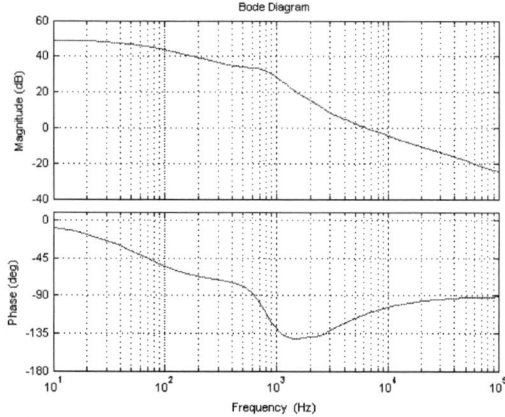

Fig. 4. The bode plot of G_1.

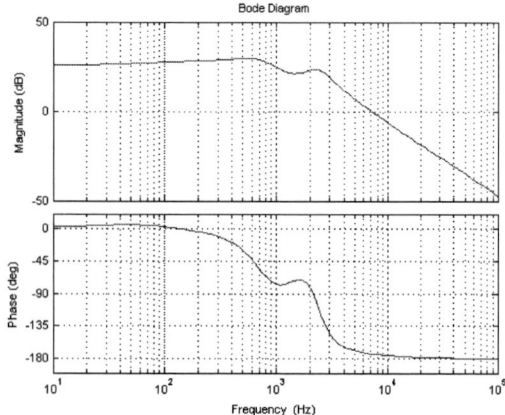

Fig. 5. The bode plot of G_2.

Table I Specifications of the prototype circuit.

Symbol	Meaning	Value
$V_{PV,mpp}$	MPP voltage of PV	35V
$I_{PV,mpp}$	MPP current of PV	3.33A
$V_{wind,mpp}$	Rectified MPP voltage of wind turbine	75V
$I_{wind,mpp}$	Rectified MPP current of wind turbine	2A
V_{bat}	Battery voltage	48V
L_1	Output inductor	400μH
L_2	Input inductor	400μH
C_1	Input capacitor	200μF
C_2	Input capacitor	10μF
C_b	Capacitor	15μF
V_{Cb}	Voltage of C_b	2V
f_{sw}	Switching Frequency	50kHz
N	Turn Ratio of the transformer N_2/N_1	0.667

The two zeros are placed around the double poles to reduce the excessive phase lag. One pole is placed at the origin to increase the magnitude of the loop gain at low frequency, another pole is designed to cancel out the zero in G_1, and the other pole is used to attenuate the switching noise. Since G_2 doesn't have a zero to be cancelled out, the additional pole in the compensator can be used to attenuate the switching noise.

IV. COMPUTER SIMULATIONS ADN EXPERIMENTAL RESULTS

To verify the feasibility and performance of the proposed S-MIC for the hybrid PV/Wind power charger system, both computer simulations and hardware experiments with specifications listed in Table I are carried out. The hardware prototype has been built and tested to verify the accuracy of the small-signal model and the performance of the proposed S-MIC.

Based on the controller design criterion proposed in the previous section, two different compensators A_1 and A_2 with crossover frequency at 1 kHz are designed as:

$$A_1(s) = \frac{1.16e^8 s^2 + 8.04e^{11} s + 4.59e^{14}}{6.12s^3 + 8.21e^5 s^2 + 1.97e^{10} s}$$

$$A_2(s) = \frac{2.89e^7 s^2 + 4.9e^{11} s + 1.6e^{15}}{3.98s^3 + 5.04e^6 s^2 + 5e^{10} s} \quad . \tag{10}$$

The circuit simulation software PSIM is adopted to obtain the simulated frequency-domain responses of the proposed S-MIC. The software Mathcad is adopted to draw the bode plots of those data obtained from the derived mathematical equations, computer simulations and experimental measurements. The PV simulator, Agilent E4360A, is used to simulate both the PV panel and the rectified output voltage of the wind turbine. The MPPT function and the compensator are realized by a micro controller unit (MCU) dsPIC33FJ16GS502.

The 2014 International Power Electronics Conference

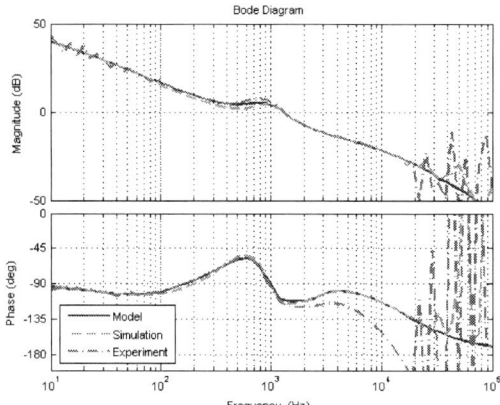

Fig. 6. The bode plot of T_1.

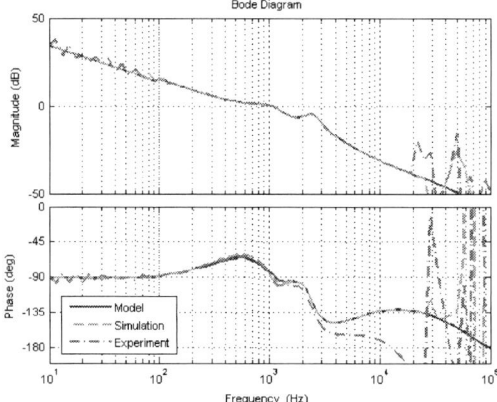

Fig. 7. The bode plot of T_2.

Fig. 6 and Fig. 7 show the bode plots of the loop gain with the designed compensator listed in (10). Different bode plots are obtained from the derived small-signal model, the computer simulation, and the experimental measurement. It should be mentioned that the measured bode plots of the prototype circuit are obtained by using the frequency response analyzer AP-300. The results show that the loop gain of the derived small-signal model is agree with those obtained by computer simulation or hardware experiment. However, due to the digital controller implementation of the prototype circuit, some phase lag caused by the one switching cycle time delay appears in the experiment results.

Fig. 8 shows the hardware measured waveform of the prototype S-MIC for the hybrid PV/wind battery charger. The result shows that both sources can achieve their maximum power point without affecting each other.

It is well-known that renewable energies have intermittent feature. Therefore, to verify the transient response of the proposed S-MIC with the ability to deliver power to the battery either individually or simultaneously, different tests are performed. The input voltage of the PV module or the wind turbine will be forced to turn off for a period of time then resume on afterward.

Fig. 9 shows the measured input voltages and current when wind turbine has sudden power off and on changes.

It can be clearly seen that the operation of the other input terminal, PV module, is not affected by the sudden change of the wind power. It also reveals the proposed S-MIC can transfer power from two different sources either simultaneously or individually, without affecting each other's operation.

Similar experimental waveforms are shown in Fig. 10 where the PV module has sudden power off and on change. Again, the operation at the other input terminal, the wind turbine, is not affected by the sudden change of the PV power. The experimental results verify the performance of the controller for the proposed S-MIC.

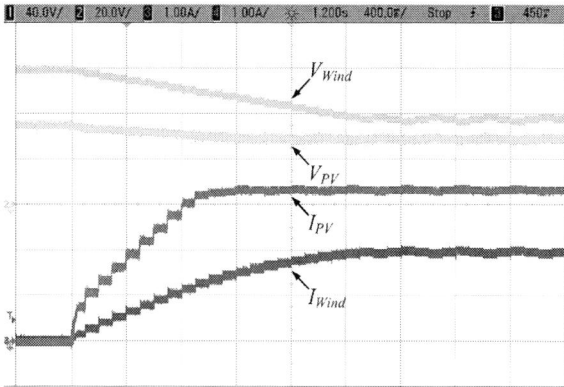

Fig. 8. The MPPT feature for both PV module and Wind turbine.

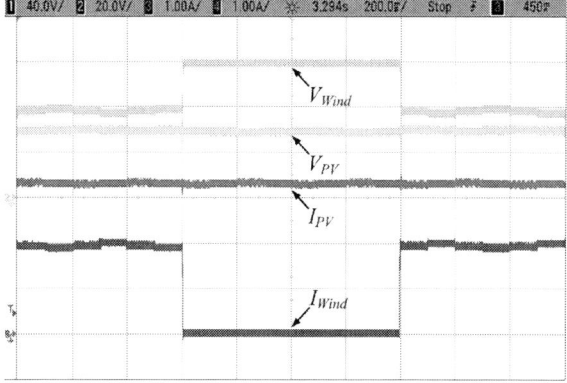

Fig. 9. Sudden power off and on of the wind turbine.

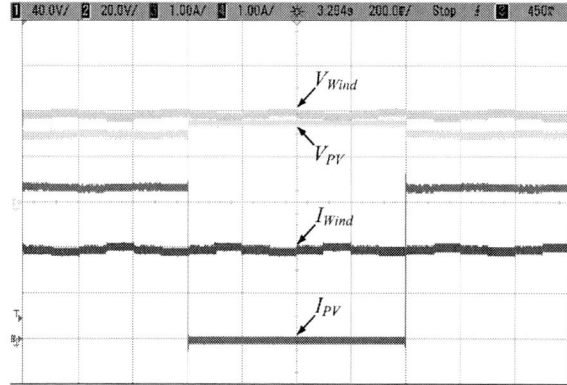

Fig. 10. Sudden power off and on of the PV module.

978-1-4799-2706-7/14 $31.00 © 2014 IEEE

V. CONCLUSIONS

This paper proposed a S-MIC for the hybrid PV/wind power charger system which can simplify the power system, reduce the cost, deliver continuous power with higher reliability and overcome high voltage-transfer-ratio problems. The operation principle of the proposed S-MIC is introduced and the small signal ac model is developed. Computer simulations and prototype hardware circuit experimental results are presented to verify the accuracy of the proposed small signal model and the performance of controller for the proposed S-MIC for hybrid PV/wind battery charger system.

REFERENCES

[1] Q. Li and P. Wolfs, "A review of the single phase photovoltaic module integrated converter topologies with three different DC link configurations," *IEEE Trans. Power Electron.*, vol.23, no.3, pp.1320-1333, May 2008.

[2] S.M. Muyeen, R. Takahashi, T. Murata, and J. Tamura, "A variable speed wind turbine control strategy to meet wind farm grid code requirements," *IEEE Trans. Power Electron.*, vol.25, no.1, pp.331-340, Feb. 2010.

[3] Z. Qian, O. Abdel-Rahman, and I. Batarseh, "An integrated four-port DC/DC converter for renewable energy applications," *IEEE Trans. Power Electron.*, vol.25, no.7, pp.1877-1887, July 2010.

[4] X. Li and A.K.S. Bhat, "Analysis and design of high-frequency isolated dual-bridge series resonant DC/DC Converter," *IEEE Trans. Power Electron.*, vol.25, no.4, pp.850-862, April 2010.

[5] Y.-M. Chen, Y.-C. Liu, and F.-Y. Wu, "Multi-input DC/DC converter based on the multiwinding transformer for renewable energy applications," *IEEE Trans. Ind. Appl.*, vol.38, no.4, pp.1096-1104, Jul/Aug 2002.

[6] Y.-M. Chen, Y.-C. Liu, and F.-Y. Wu, "Multiinput converter with power factor correction, maximum power point tracking, and ripple-free input currents," *IEEE Trans. Power Electron.*, vol.19, no.3, pp. 631- 639, May 2004.

[7] N. Benavides and P. Chapman, "Object oriented modeling of a multiple-input multiple-output flyback converter in dymola," *IEEE Workshop on Computers in Power Electronics*, pp.156-160, 2004.

[8] H. Matsuo, W. Lin, F. Kurokawa, T. Shigemizu, and N. Watanabe, "Characteristics of the multi-input dc-dc converter," *IEEE Trans. Industrial Electronics*, vol.51, no.3, pp.625-631, June 2004.

[9] Y.-M. Chen, Y.-C. Liu, and S.-H. Lin, "Double-input PWM DC/DC converter for high-/low-voltage sources," *IEEE Trans. Industrial Electronics,* vol.53, no.5, pp.1538-1545, Oct. 2006.

[10] H. Tao, A. Kotsopoulos, J. Duarte, and M. Hendrix, "Family of multiport bidirectional dc-dc converters," *IEE Proceedings – Electric Power Applications,* vol. 153, no. 3, May 2006.

[11] R. Zhao and A. Kwasinski, "Multiple-input single ended primary inductor converter (SEPIC) converter for distributed generation applications," in Proc. *IEEE Energy Conversion Congress and Exposition*, 2009, pp.1847-1854.

[12] A. Kwasinski, "Identification of feasible topologies for multiple-input DC–DC converters," *IEEE Trans. Power Electron.*, vol.24, no.3, pp.856-861, March 2009.

[13] A. Khaligh, J. Cao, and Y.-J. Lee, "A multiple-input DC–DC converter topology," *IEEE Trans. Power Electron.*, vol.24, no.3, pp.862-868, March 2009.

[14] Z. Qian, O. Abdel-Rahman, H. Al-Atrash, and I. Batarseh, "Modeling and control of three-port DC/DC converter interface for satellite applications," *IEEE Trans. Power Electron.*, vol.25, no.3, pp.637-649, March 2010.

[15] C. Zhao, S.D. Round, and J.W. Kolar, "An isolated three-port bidirectional DC-DC converter with decoupled power flow management," *IEEE Trans. Power Electron.*, vol.23, no.5, pp.2443-2453, Sep. 2008.

[16] H. Tao, J.L. Duarte, and M.A.M. Hendrix, "Three-port triple-half-bridge bidirectional converter with zero-voltage switching," *IEEE Trans. Power Electron.*, vol.23, no.2, pp.782-792, March 2008.

[17] Y.-C. Liu and Y.-M. Chen, "A Systematic Approach to Synthesizing Multi-Input DC-DC Converters," *IEEE Trans. Power Electron.*, vol. 24, pp. 116-127, Jan. 2009.

[18] M. Veerachary, "Two-Loop Controlled Buck–SEPIC Converter for Input Source Power Management," *IEEE Trans. Industrial Electronics*, vol. 59, no.11, pp. 4075-4087, Nov. 2012

HFL PV Micro-Inverter with Front-End Current-Fed Converter and Half-Wave Cycloconverter

D. R. Nayanasiri, D. M. Vilathgamuwa
School of Electrical and Electronic Engineering
Nanyang Technological University
Singapore
mdul0001@e.ntu.edu.sg, emahinda@ntu.edu.sg

D. L. Maskell
School of Computer Engineering
Nanyang Technological University
Singapore
asdouglas@ntu.edu.sg

Abstract— A high-frequency-link micro inverter is proposed with a front-end dual inductor push-pull converter and a grid-connected half-wave cycloconverter. Pulse width modulation is used to control the front-end converter and phase shift modulation is used at the back-end converter to obtain grid synchronized output current. A series resonant circuit and high-frequency transformer are used to interface the front-end and the back-end converters. The operation of the proposed micro-inverter in grid-connected mode is validated using MATLAB/Simpower simulation. Experimental results are provided to further validate the operation.

Keywords— *high-ferqunecy-link micro-inverter, dual inductor push-pull converter, half-wave cycloconverter, phase shift modulation, pulse width modulation.*

I. INTRODUCTION

Photovoltaic (PV) energy has gained popularity as a sustainable renewable energy source during fast few decades due to its low energy payback time (EPBT), energy return factor (ERF) and carbon footprint [1], [2]. Distributed PV energy conversion architectures are commonly used to overcome drawbacks in centralized architectures due to shading and module mismatch problems [3]. Micro-inverters (AC modules) and micro-converters (DC-optimizers) are used to convert PV generated DC power into suitable form (AC or DC) in distributed architectures. Micro-inverters help to directly integrate PV modules to the local grid (or load) without an additional power converter. High-frequency operation is used to reduce the weight and size of the micro-inverters due to small passive component sizes. High-frequency micro-inverters are categorized as dc-link less, pseudo dc-link and dc-link based on the nature and availability of the DC-link in the micro-inverter[4]. DC-link less (or high-frequency-link) micro-inverters have gained research interest due to their double stage power conversion [5].

High-frequency-link (HFL) micro-inverters consist of two power conversion stages. The front-end converter is used to obtain high-frequency AC current at the HFL and the back-end converter modulates that high-frequency current to obtain a low frequency output current. The front-end converter can be either a voltage-fed converter or current-fed converter. Front-end voltage-fed converter based HFL micro-inverters were proposed in [6], [7] and [8]. However, front-end current-fed converter based HFL

micro-inverters are not common in the research literature. Cycloconverters are used as the back-end converter in HFL micro-inverters, operating in either half-wave or full-wave modes. Resonant circuits are used in these topologies in order to realize soft switching (zero-voltage-switching and zero-current-switching) at the front-end and back-end converters.

Different power modulation schemes were proposed for HFL micro-inverters which are based on resonant circuits. Pulse-width modulation and phase shift modulation were used to control voltage-fed front-end converters. But frequency modulation based power modulation schemes are not commonly used due to an unpredictable noise spectrum, more complex filtering of the output voltage and poor utilization of magnetic components. Moreover, pulse-width modulation is used in the front-end current-fed converters [9]. Phase-shift based modulation schemes were proposed for the back-end cycloconverter of the HFL micro-inverters.

A front-end current-fed converter based HFL micro-inverter is proposed in this paper with a series resonant circuit and grid-connected half-wave cycloconverter. The operation of the proposed micro-inverter is analyzed and explained using MATLAB/Simpower simulations. Experimental results are used to further validate the operation of the proposed micro-inverter.

II. THE PROPOSED MICRO-INVERTER

The proposed HFL micro-inverter with a front-end current-fed converter, a series resonant tank and a back-end half-wave cycloconverter is shown in Fig. 1. The front-end converter consist of a dual inductor push-pull converter [10] which is obtained by applying the duality principle [11] to a voltage-fed half-bridge converter [12]. The front-end converter is interfaced to the series resonant tank (L and C) through a parallel capacitor (C_1) at the output of the front-end converter. The series resonant circuit can be placed in either the primary side or secondary side of the isolation transformer. However, the size of the resonant inductor can be significantly reduced if the resonant circuit is placed in the primary side of the isolation transformer. Therefore, a resonant capacitor can be placed in the secondary side with proper scaling. The high-frequency transformer isolates the micro-inverter from the load and provides passive voltage step-up. The secondary side of the high-frequency transformer is connected to the half-wave cycloconverter

978-1-4799-2706-7/14 $31.00 © 2014 IEEE

through a DC blocking capacitor. This capacitor helps to block the DC magnetizing current of the isolation transformer. A capacitive filter (C_f) is used to remove high frequency current ripple in the output current (i_{ohf}).

Pulse width modulation is used to control the operation at the maximum power point (MPP) of the PV module. The phase shift between resonant current (i_L) and input voltage of the half-wave cycloconverter (v_{cyc}) is modulated to obtain grid synchronized output current (i_o). This phase shift modulation is realized at the grid-connected half-wave cycloconverter.

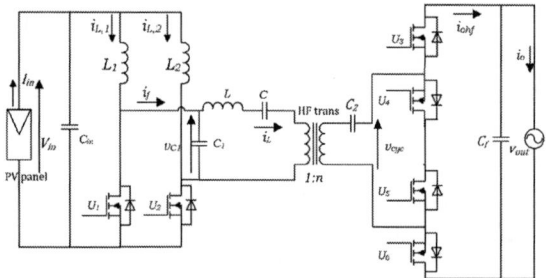

Fig. 1. The proposed HFL micro-inverter.

Therefore the proposed micro-inverter has the following advantages:
- Reduced number of power switches due to two power conversion stages.
- High power density because of the reduced number of semiconductor devices and the small size of the passive components due to high-frequency operation.
- Small transformer size due to a lower turn ratio compared to voltage-fed topologies.
- Lower electromagnetic interference and efficient magnetic utilization due to a fixed frequency operation.
- Reduced input buffer capacitor requirement compared to voltage-fed front-end converter based topologies due to the current-fed operation.
- Reduced voltage surges across the drain to source of the primary side power switches due to the inclusion of the transformer leakage inductance in the series resonant circuit.

The voltage across the primary side power switch is the capacitor voltage (v_{C1}) when the switches are turned-off. Hence the primary side power switch blocking voltage should be large enough to prevent possible failure.

III. OPERATION AND ANALYSIS OF THE HFL MICRO-INVERTER

The proposed micro-inverter is analyzed in grid-connected mode to explain its operation. The following assumptions are used throughout this analysis. 1.) L_1 and L_2 are large enough to assume continuous input current: 2.) The switching frequency (f_s) is high enough to assume constant grid voltage over one switching cycle: 3.) High quality factor in the series resonant circuit to assume sinusoidal current through inductor (L). The switching frequency (f_s) can be selected higher or lower than resonant frequency (f_r) of the series resonant circuit. In our analysis, we assume that the resonant frequency is

greater than the switching frequency, that is ($f_s > f_r$). The operation of the proposed micro-inverter has a completely different behavior if the switching frequency (f_s) is lower than the resonant frequency (f_r). However, we do not consider this mode of operation here. Moreover, there should be a constant switching frequency (f_s) to have proper operation and better utilization of the magnetic components.

The ideal waveforms to explain the operation of the HFL micro-inverter are shown in Fig. 2. The duty ratios of the front-end converter power switches are both selected as 0.75, but are phase shifted by 180°. The control signals of the half-wave cycloconverter power switches that need to have ZVS at turn-on are shown in Fig. 2 with a 0.5 duty ratio and the required dead times.

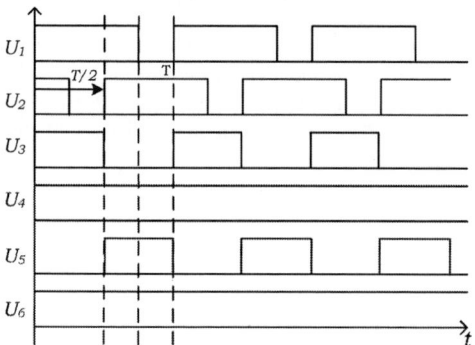

Fig. 2. Power switch control signals of the micro-inverter when $v_{out} > 0$.

A. The operation of the front-end current-fed converter

The operation of the front-end current-fed dual inductor push-pull converter can be analyzed by taking the parallel capacitor (C_1), series resonant circuit and dual inductor push-pull converter as a single unit. The series resonant circuit can be considered as a perfect sinusoidal current source which has an output as shown in Fig. 3. This sinusoidal current results from the second order filter behavior of the series resonant circuit on its input, which is the voltage across the parallel capacitor (C_1).

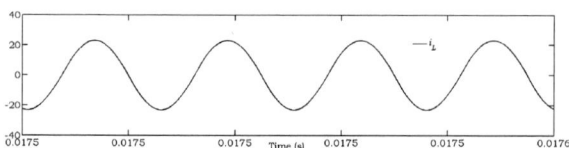

Fig. 3. Resonant current (A) in the high-frequency-link.

The parallel capacitor (C_1) voltage (v_{C1}) and current (i_{C1}) are shown in Fig. 4. The voltage across capacitor (C_1) increases with time when either primary side power switch is turned-off. This is a result of the input inductors releasing their energy stored during the corresponding power switch turn-on period. The parallel capacitor current (i_{C1}) reaches zero when the capacitor is fully charged. Then, the energy stored in the capacitor is released into the series resonant tank within the same power switch turn-off period, as shown in Fig. 4. Hence, the resultant current in the intermediate link (i_f, which is shown in Fig. 1) has a sudden dip, as shown in Fig. 5, which impacts the power switch currents. Thus, it is evident that the capacitor (C_1) is not part of the resonant tank, although the capacitor (C_1) and the resonant tank operate as a single unit. Therefore the operation of the front-end dual inductor push-pull converter with series resonant tank can be explained using Figs. 6 and 7.

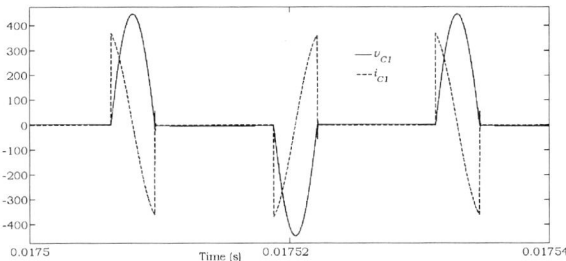

Fig. 4. Parallel capacitor voltage (V_{C1} in V) and scaled (1:20) current (i_{C1} in A).

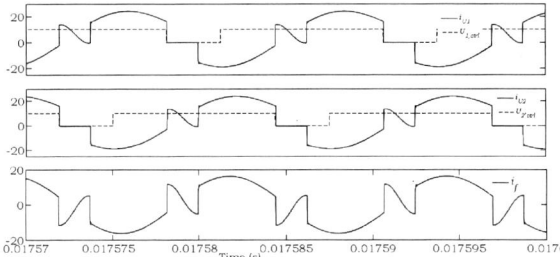

Fig. 5. Current (A) through power switch U_1 and U_2, and current i_f.

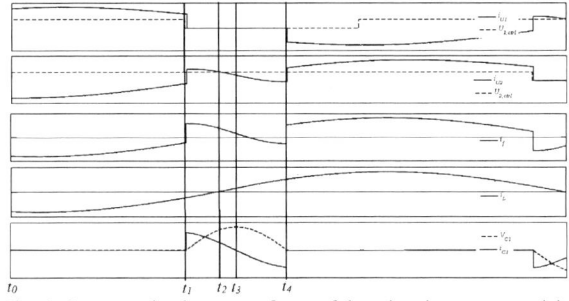

Fig. 6. Current and voltage waveforms of the micro-inverter to explain the operation of the front-end current-fed converter with series resonant tank.

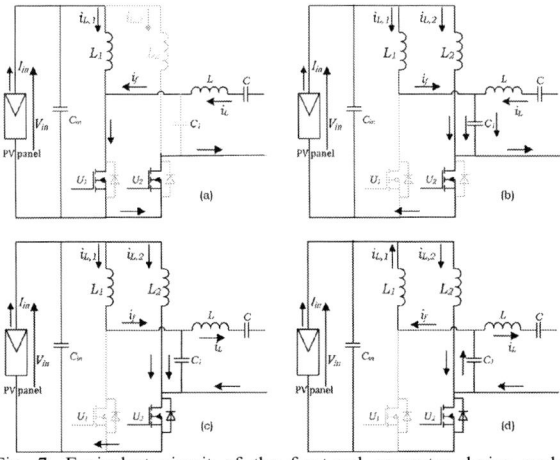

Fig. 7. Equivalent circuit of the front-end converter during each subinterval shown in Fig. 6.

Interval 1 ($t_0<t<t_1$; Fig. 7(a)): During this period both power switches are turned-on. There is a positive current through input inductor L_1 and U_1, hence input power is stored in L_1. The resonant current is negative during this period and flows in the direction shown in Fig. 7(a). Hence there is a negative current though power switch U_2 as shown in Fig. 6.

Interval 2 ($t_1<t<t_2$; Fig. 7(b)): This period begins when U_1 is turned-off. But U_2 is still turned-on. The current direction through U_2 suddenly changes at $t=t_1$, as shown in Fig. 7(b). Inductor L_1 releases its stored energy during this process to capacitor (C_1) and hence the voltage across C_1 (v_{C1}) increases and current (i_{C1}) reduces with time. As a result, the direction of the current i_f changes suddenly and becomes positive when $t=t_1$. The resonant current is still negative and flows in the same direction as in *interval 1*.

Interval 3 ($t_2<t<t_3$; Fig. 7(c)): This interval begins when the resonant current changes its direction from negative to positive. Inductor L_1 still transfers its stored power into C_1 while L_2 stores power from the input. This interval ends when i_{C1} reaches zero and v_{C1} reaches its peak value.

Interval 4 ($t_3<t<t_4$; Fig. 7(d)): During this period i_{C1} changes its direction and v_{C1} begins to decrease. The current i_f changes its direction at the same time. It is evident that capacitor C_1 releases its stored energy into the series resonant tank and the input inductor. Therefore v_{C1} reaches zero and i_{C1} reaches its maximum negative value when $t=t_4$. As a result, current through U_1 and U_2 suddenly attain higher values and allow resonant current to flow through them after $t=t_4$.

B. The operation of the half-wave cycloconverter

The operation of the half-wave cycloconverter can be analyzed by assuming it has a current source at its input port, as shown in Fig. 8. The half-wave cycloconverter modulates the high-frequency sinusoidal input current to obtain grid synchronized 50Hz output current. The switching control signals of the half-wave cycloconverter during the positive half-cycle of the grid voltage (v_{out}) are shown in Fig. 2. Power switches U_4 and U_6 are completely turned-on and U_3 and U_4 are switched during the positive half cycle of the grid voltage ($v_{out}>0$). Subsequently, power switches U_3 and U_5 are completely turned-on and U_4 and U_6 are switched when ($v_{out}<0$). Therefore, the critical waveforms needed to analyze operation of the half-wave cycloconverter are shown in Fig. 9. The switching behavior of the half-wave cycloconverter is elaborated in [13].

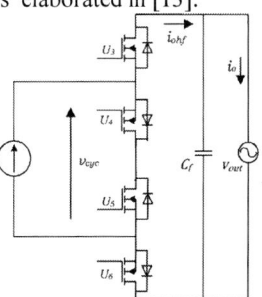

Fig. 8. Equivalent circuit of the half-wave cycloconverter.

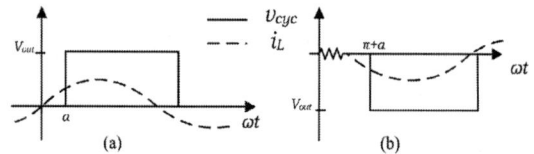

Fig. 9. Half-wave cycloconverter input voltage and secondary side resonant current when (a) $v_{out}>0$ (b) $v_{out}<0$.

The phase shift (α) shown in Fig. 9 needs to modulate with respect to the secondary side resonant current (i_L) to obtain the grid synchronized current. The phase shift (α) can be linearly modulated in order to obtain a low distortion output current by assuming nearly sinusoidal behavior in the resonant current.

C. Analysis of switching behavior

Switching loss is a common problem in high-frequency power converters. Voltage surge at turn-off of the primary side power switches is another common problem in dual and single inductor push-pull converters, due to the leakage inductance of the isolation transformer. The leakage inductance of the isolation transformer is included into the resonant inductor of the proposed topology. Hence switching behavior of the power switches can be analyzed by considering the presence of the series resonant tank and the parallel capacitor (C_1) across the output of the dual inductor push-pull converter. The drain to source voltage (V_{DS}) and power switch control signals of the primary side power switches are shown in Fig. 10.

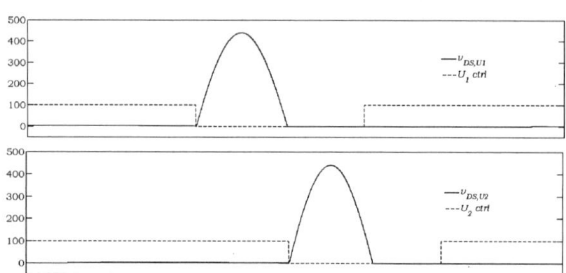

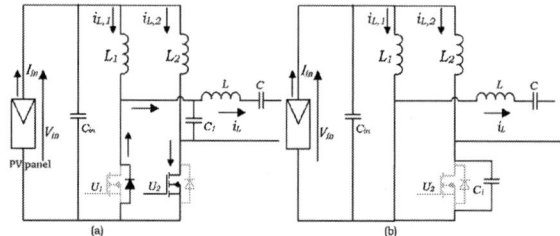

Fig. 10. Drain to source voltage and power switch control signals.

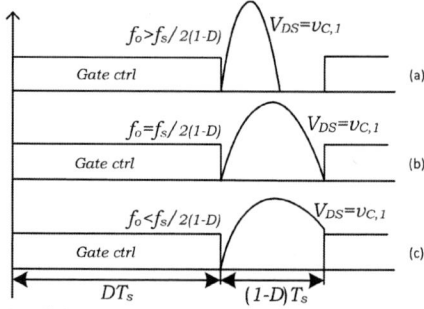

Fig. 11. Equivalent circuit when (a) U_1 turned-on, and (b) U_2 turned-off.

The switching behavior at turn-off of the primary side power switches can be analyzed using the equivalent circuit at the turn-off of U_2, as shown in Fig. 11(b). The drain to source voltage does not suddenly increase when power switch U_2 is turned-off due to the parallel capacitor (C_1). Hence there is a reduced voltage across the power switch. As a result there is insignificant power loss at turn-off of the power switch. Fig. 11(b) shows that the input inductor (L_2) and capacitor (C_1) creates a second

order LC circuit. A similar equivalent circuit can be obtained to analyze the operation of input inductor (L_1) and capacitor (C_1) due to the symmetrical nature of the front-end current-fed dual inductor push-pull converter. Hence the capacitor current (i_{C1}) and voltage (v_{C1}) can be represented as:

$$i_{C,1} = I_0 \cos(\omega_0 t) \qquad (1)$$
$$v_{C,1} = V_0 \sin(\omega_0 t) \qquad (2)$$

where $\omega_0 = 1 \Big/ \sqrt{0.5L_n C_1}$, n= 1, 2.

The behavior of the capacitor current (i_{C1}) and voltage (v_{C1}) needs to be analyzed to better understand not just the switching losses at turn on and turn off, but also to determine device ratings of the primary side power switches. It can be further analyzed using Fig. 12 which shows the behavior of the capacitor voltage (v_{C1}) with respect to the switching frequency (f_s) and resonant frequency (f_o) of the LC circuit.

Fig. 12. Parallel capacitor (C_1) voltage with gate control signal of primary side power switch.

Minimum switching losses can be obtained when the resonant frequency (f_o) of the LC circuit is less than or equal to $f_s/2(1-D)$. Then, there is zero voltage across the power switch when it is to be turned-on. The inductance of the input inductors (L_1 and L_2) are based on the switching frequency and required input current ripple. Hence the resonant frequency of the LC circuit depends on the parallel capacitor (C_1) which needs to be carefully chosen in order to prevent losses at larger duty ratios. Moreover, Figs. 10 and 15 clearly show that there a significant reduction in the peak voltage of the v_{C1} when the resonant frequency (f_o) is less than or equal to $f_s/2(1-D)$, compared to the case when it is greater.

There is a negative current through either primary side power switch when each is about to be turned-on, as shown in Fig. 5. Hence the current should flow through the parallel diode across the power switch, as shown in Fig. 11(a), resulting in an insignificant voltage across the power switch equal to the forward voltage drop of the diode. As a result there is reduced switching loss at turn-on of the power switches. However, conducting current for a long time through the body diode of the power MOSFETs results in higher losses. Therefore, an additional parallel diode can be placed across the primary side power MOSFETs in order to prevent such losses. This condition is only valid for the cases which are shown in Figs. 12(a) and (b).

IV. SIMULATION AND EXPERIMENTAL RESULTS

A. The proposed controller

The overall structure of the proposed micro-inverter is shown in Fig. 13. The maximum power point tracker output is used to obtain the required reference to generate pulse width control signals of the front-end converter. Two loops are used in the front-end controller: the outer current loop and the inner voltage loop. The fast inner loop controls the maximum power point and the slower outer loop tracks irradiance change variations.

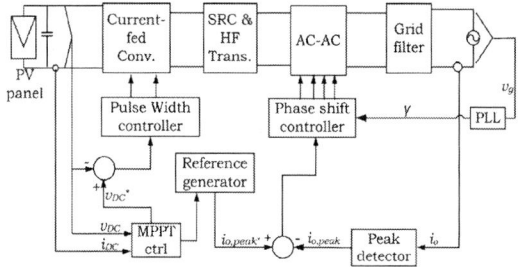

Fig. 13. The overall structure of the proposed HFL micro-inverter.

The controller of the back-end converter is used to generate the required phase shift control signals of the half-wave cycloconverter. A phase locked loop (PLL) tracks the grid voltage and generates the phase angle (γ). This phase angle is fed into the phase shift controller, along with the current reference signal, to generate the gate control signal of the half-wave cycloconverter.

B. Simulation results

The proposed HFL micro-inverter is simulated using the MATLAB/Simpower block set to validate its operation in the grid connected mode with the parameters shown in Table I. Fig. 14 shows the grid synchronized output current of the micro-inverter with in phase grid voltage. This result verifies the effectiveness of the proposed linear phase shift based power modulation method used to obtain low frequency output current from the HFL sinusoidal current.

TABLE I.
COMPONENT PARAMETERS OF THE SIMULATION

Component	Value
Input inductor (L_1 and L_2)	$25\mu H$
Input capacitor (C_{in})	$400\mu F$
Parallel capacitor (C_1)	$20nF$
Resonant inductor (L)	$319\mu H$
Resonant capacitor (C)	$13.05nF$
Transformer turns ratio (n)	1:4
Input voltage (V_{in})	$40V$
Switching frequency (f_s)	$80kHz$
Output voltage (V_{out})	$230V/50Hz$

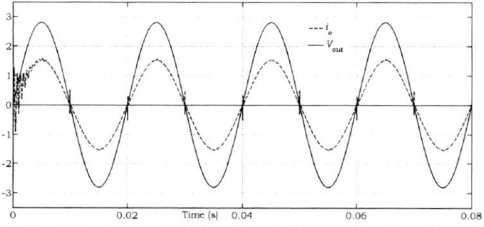

Fig. 14. Output current (A) and scaled (160:1) grid voltage (V) of the micro-inverter.

The parallel capacitor (C_1) value shown in Table I is used to obtain the condition shown in Fig. 12(a). The capacitor voltage (v_{C1}), along with the primary side power switches control signals, are shown in Fig. 10. To achieve the conditions shown in Figs. 12(b) and (c) the capacitor values are changed to $78nF$ and $120nF$, respectively. The simulation results under these conditions are shown in Fig. 15 and clearly validate the analysis of the LC network at the front-end converter in section III(C).

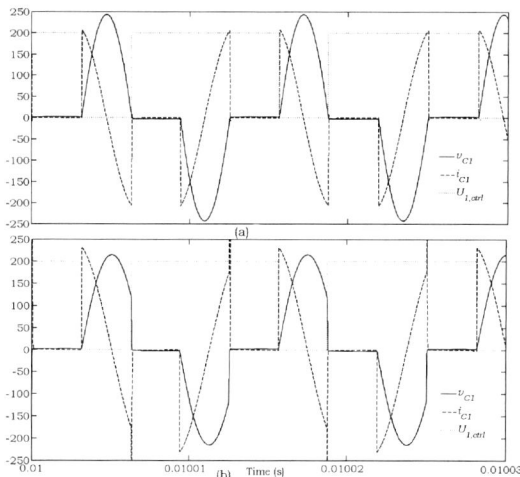

Fig. 15. Parallel capacitor voltage (v_{C1} in V), scaled (1:10) current (i_{C1} in A) and scaled (1:200) U_1 gate control signal when C_1 is (a) 78nF, and (b) 120nF.

C. Experimental results

A prototype of the proposed HFL micro-inverter is built in the laboratory, using parameters shown in Table II, to further validate its operation. Open loop control was used in this experiment and the gating control signals are generated using a $TMS320f28335$ DSP controller. The proposed micro-inverter is tested using a resistive load.

TABLE II.
COMPONENT PARAMETERS OF THE EXPERIMENT SETUP

Component	Value
Input inductor (L_1 and L_2)	$283\mu H$
Parallel capacitor (C_1)	$40nF$
Resonant inductor (L)	$1.10mH$
Resonant capacitor (C)	$44.2nF$
Transformer turns ratio (n)	1:3
Input voltage (V_{in})	$18V$
Switching frequency (f_s)	$25kHz$
Load	220Ω

The output current and load terminal voltage of the micro-inverter, along with the AC source voltage used to simulate grid voltage, is shown in Fig. 16. The proposed power modulation method of the half-wave cycloconverter was able to generate output current with a relatively low total harmonic distortion, as shown in Fig. 16.

Fig. 17 shows the gate control signal ($V_{GS,U1}$) and drain to source voltage ($V_{DS,U1}$) of power switch U_1 with two different parallel capacitor values. The drain to source voltage has a behavior similar to the theoretical analysis. Fig. 17 shows insignificant voltage surges in the drain to source voltage when the power switch is to be turned-off. It is evident that there are insignificant switching losses at the turn-off instant. Moreover there is zero voltage across the power switch when it is to be turned-on. Hence there is ZVS at turn-on of the primary side power switches.

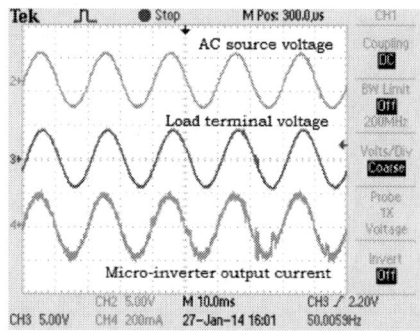

Fig. 16. The micro-inverter output current, load terminal voltage and reference AC source voltage.

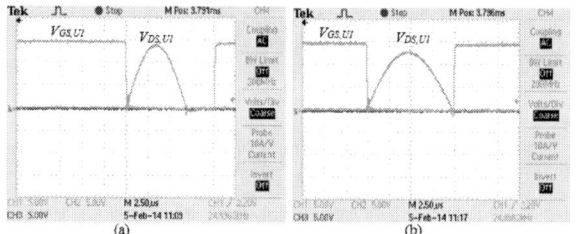

Fig. 17. Gate control signal ($V_{GS,U1}$) and drain to source voltage ($V_{DS,U1}$) of U_1 when the parallel capacitor C_1 is (a) $40nF$, and (b) $104nF$.

V. DERIVED TOPOLOGY

The proposed HFL micro-inverter consists of a half-wave cycloconverter at the back-end which can be replaced by a full wave cycloconverter, as shown in Fig. 18. However, while this micro-inverter has full-wave operation it has an increased number power switches. A similar switching control strategy, to the half-wave cycloconverter, can be used with this micro-inverter. In this case, two back to back switches can be controlled using the same switching control signal. Additionally, the size of the output capacitor and DC blocking capacitor can be significantly reduced due to full-wave operation.

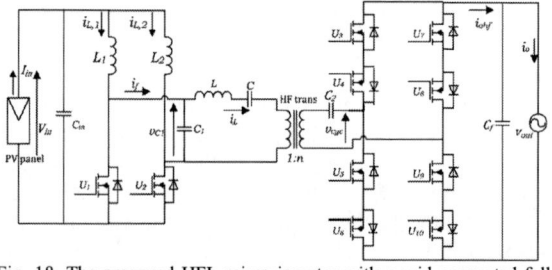

Fig. 18. The proposed HFL micro-inverter with a grid connected full-wave cycloconverter.

VI. CONCLUSIONS

A HFL micro-inverter is proposed with a front-end current-fed dual inductor push-pull converter and a grid-connected half-wave cycloconverter. The front-end converter is interfaced to the half-wave cycloconverter through the parallel capacitor, series resonant circuit and high-frequency transformer. A pulse width modulation and phase shift modulation based power modulation scheme is proposed for the HFL micro-inverter. The operation of the front-end converter with parallel capacitor and series resonant tank is analyzed. Behavior of the parallel capacitor is further investigated to better understand the switching behavior and device parameters of the power switches of the front-end converter. The operation of the half-wave cycloconverter is explained. Simulation and experimental results are provided to validate operation of the proposed micro-inverter. Lastly, a HFL micro-inverter using a grid-connected full-wave cycloconverter is proposed as an extension of the proposed HFL micro-inverter.

REFERENCES

[1] S. A. Halasah, D. Pearlmutter, and D. Feuermann, "Field installation versus local integration of photovoltaic systems and their effect on energy evaluation metrics," *Energy Policy*, vol. 52, pp. 462–471, Jan. 2013.

[2] M. J. (Mariska) de Wild-Scholten, "Energy payback time and carbon footprint of commercial photovoltaic systems," *Sol. Energy Mater. Sol. Cells*, vol. 119, pp. 296–305, Dec. 2013.

[3] L. Zhang, K. Sun, Y. Xing, L. Feng, and H. Ge, "A Modular Grid-Connected Photovoltaic Generation System Based on DC Bus," *IEEE Trans. Power Electron.*, vol. 26, no. 2, pp. 523–531, Feb. 2011.

[4] Q. Li and P. Wolfs, "A review of the single phase photovoltaic module integrated converter topologies with three different DC link configurations," *IEEE Trans. Power Electron.*, vol. 23, no. 3, pp. 1320–1333, 2008.

[5] S. B. Kjaer, J. K. Pedersen, and F. Blaabjerg, "A review of single-phase grid-connected inverters for photovoltaic modules," *IIEEE Trans. Ind. Appl.*, vol. 41, no. 5, pp. 1292–1306, 2005.

[6] A. Trubitsyn, B. J. Pierquet, A. K. Hayman, G. E. Gamache, C. R. Sullivan, and D. J. Perreault, "High-efficiency inverter for photovoltaic applications," *Proc. IEEE ECCE*, no. 1, pp. 2803–2810, 2010.

[7] H. Krishnaswami, "Photovoltaic microinverter using single-stage isolated high-frequency link series resonant topology," in *Proc. IEEE ECCE*, 2011, vol. 1, no. 1, pp. 495–500.

[8] D. R. Nayanasiri, D. M. Vilathgamuwa, and D. L. Maskell, "Half-Wave Cycloconverter-Based Photovoltaic Microinverter TopologyWith Phase-Shift PowerModulation," *IEEE Trans. Power Electron.*, vol. 28, no. 6, pp. 2700–2710, 2013.

[9] M. Kazimierczuk and D. Czarkowski, "Resonant power converters," 2nd ed., John Wiley & Sons, 2011, p. 311.

[10] W. de A. Filho and I. Barbi, "A comparison between two current-fed push-pull DC-DC converters-analysis, design and experimentation," in *in Proc. IEEE INTELEC*, 1996, vol. 00, pp. 313–320.

[11] S. D. Freeland, "Techniques for the practical application of duality to power circuits," *IEEE Trans. Power Electron.*, vol. 7, no. 2, 1997.

[12] P. J. Wolfs, "A current-sourced DC-DC converter derived via the duality principle from the half-bridge converter," *IEEE Trans. Ind. Electron.*, vol. 40, no. 1, pp. 139–144, 1993.

[13] D. R. Nayanasiri, D. M. Vilathgamuwa, and D. L. Maskell, "High-Frequency-Link Micro-Inverter with Front-End Current-Fed Half-Bridge Boost Converter and Half- Wave Cycloconverter," in *proc. IEEE IECON*, 2013, pp. 6987 – 6992.

Comprehensive Study about Stability Issues of Multi-module Distributed System

Fangcheng LIU, Jinjun LIU, Haodong ZHANG, Danhong XUE, Qinyun DOU

Power Electronic and Renewable Energy Research Center
School of Electrical Engineering, Xi'an Jiaotong University
Xi'an, Shaanxi, China

Abstract—when the stability issues of multi-module distributed DC power system are analyzed, the physical understandings of terminal characteristics of each module should be focused. According to the proposed analysis method, all the sub-modules are divided into two groups based on the different terminal property types: impedance (Z) type and admittance (Y) type. Equivalent circuits of total system are established to analyze the stability issues and mathematical equations of equivalent circuit are derived. Based on the mathematical equations, two generalized stability criteria for multi-module distributed system have been proposed. In this paper, comparisons between the two stability criteria are made and comprehensive study about stability issues in multi-module system is made through the comparison.

Keywords—admittance; impedance; multi-module system; stability criteria

I. INTRODUCTION

Compared to traditional power conversion system, multi-module distributed system has the benefits as follows: 1) it can provide the possibility to coordinate several sources and loads; 2) the power and voltage rating can be easily increased by connecting different module in cascade or parallel; 3) the production and maintenance are more easily due to the standardized design. Based on these benefits, multi-module distributed system has been widely adopted in practical applications.

However, the dynamic characteristics of each sub-module in the system are quite different, thus the interaction among different sub-modules will deteriorate the performance of the total system. Although each sub-module is stable in standalone mode, the total system may be unstable due to the interaction.

The stability issues of multi-module system have been discussed in some researches [1]-[4]. Following the impedance-based stability criterion in cascade DC system [5], the multi-module system is divided into two groups: source group and load group. The interaction between source group and load group is considered. The equivalent output impedance of source group and the equivalent input impedance of load group are utilized to judge the system stability.

In recent years, it has been discovered that the stability

This work was supported in part by the National Basic Research Program (973 Program) of China under Project 2009CB219705 and in part by the State Key Laboratory of Electrical Insulation and Power Equipment under the Project EIPE09109.

issues of cascade system are not relying on the distinction of source and load any more [6]-[8]. Instead, the terminal property of each sub-module turns out to be the key point in the stability analysis. According to this recognition, the study about the terminal characteristics of sub-module is carried out [9]. Based on the study about the behavior of sub-module, two generalized stability criteria for multi-module distributed system have been proposed [10][11].

In this paper, comparisons between two proposed stability criteria have been made. Based on the comparison, a comprehensive understanding about the stability issues of multi-module system is illustrated. To deliver the full idea about in this topic, this paper is organized as follows: the terminal characteristics of single sub-module are studied in Section II. The equivalent circuit of total system is drawn and the stability issues are analyzed and two stability criteria are derived in Section III. Comparisons between these two criteria are made in Section IV. Brief experimental evidences are provided to in Section V.

II. TERMINAL CHARACTERISTICS OF SUB-MODULE

When the electrical interaction is analyzed, external behavior of each sub-module is of more interest than its internal loop stability.

In DC system, there are only two variables at common terminal: current and voltage. Hence, the relationship between terminal current and voltage describes the terminal characteristic of each module. The terminal impedance or admittance of the converter is the linearized model of terminal behavior around certain operating point actually. When the terminal property of active module is considered, it is clear that the terminal characteristics do not only depend on power stage but also on the controller as shown in Fig.1.

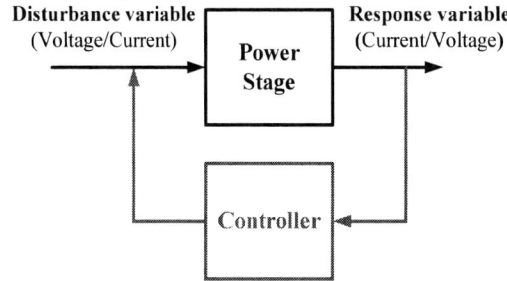

Figure 1. Structure of active module

It is very important to distinguish disturbance variable from response variable in controller design. In normal conditions, only one kind of the ideal sources is appropriate at the input or output terminal of active converter due to the limitation in topology design. For example, if a capacitor is connected in parallel with an ideal voltage source, the effect of capacitor is attenuated by the voltage source and the capacitor may be damaged due to the overcurrent when there is a steep change in the voltage source. Therefore, the capacitor should avoid being connected with ideal voltage source in parallel directly. For a similar reason, the inductor should avoid being connected with ideal current source in series directly. According to the type of available source at the common terminal, the active converter can be classified into two groups: current-fed (CF) converter and voltage-fed (VF) converter.

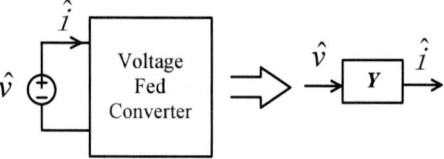

Figure 2. Terminal characteristic of VF converter

In VF converter, the terminal voltage is treated as the disturbance variable, while the terminal current is treated as the response variable [12][13]. Thus, the terminal characteristic is equivalent to admittance (Y) as shown in Fig.2. Since the VF converter should be stable in standalone mode, no RHP is allowed in terminal admittance.

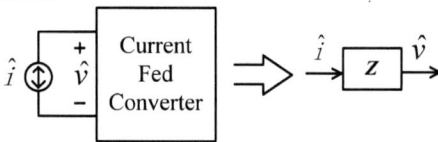

Figure 3. Terminal characteristic of CF converter

There are still CF converters in practical applications as discussed in [6][14]. The terminal current should be treated as the disturbance variable and the terminal voltage should be treated as the response variable in CF converter. The terminal characteristic of CF converter is equivalent to impedance (Z) as shown in Fig.3.

It is easy to draw the equivalent circuit of each module according to the basic circuit theory when the type of terminal property is fixed [15]. The terminal property of common bus terminal is more attractive in the stability analysis, so only the characteristics of common bus terminal of each module are concerned.

The Thevenin equivalent circuit is used to model the terminal characteristics of Z type module while Y type module is represented by the Norton equivalent circuit in general cases. This set makes the following analysis clear. Otherwise, if the Norton equivalent circuit is used for Z module where the terminal voltage is regulated tightly, an infinite current source is needed because the terminal impedance is almost zero at 0 Hz. However, the infinite source is not acceptable in both practical application and theoretical analysis.

III. STABILITY ANALYSIS

A multi-module system is shown in Fig.4. All the Z type modules are represented by the Thevenin equivalent circuit and all the Y type modules are represented by the Norton equivalent circuit.

The simplified equivalent circuit of the total system is shown in Fig. 5. All the Z type modules are replaced by an equivalent Z type module and all the Y type modules are replaced by an equivalent Y type module. This equivalent circuit is similar to the circuit which is shown in the stability analysis of cascade system [16].

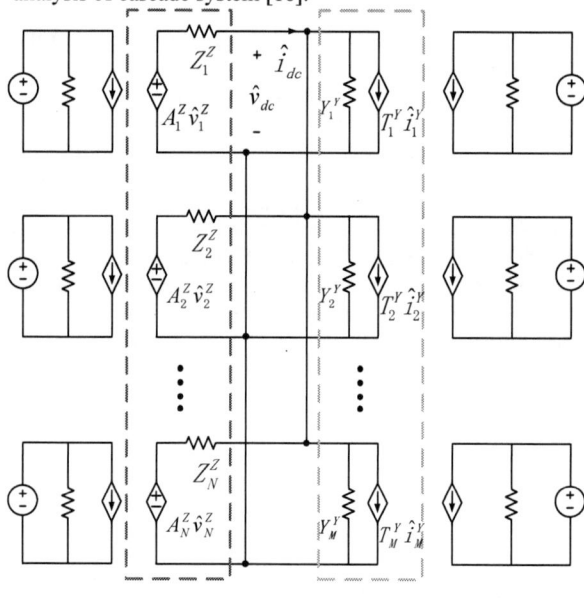

N impedance modules   **M admittance modules**

Figure 4. Multi-module distributed system

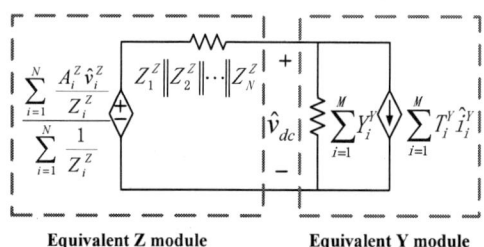

Equivalent Z module   **Equivalent Y module**

Figure 5. Equivalent circuit of multi-module system

Based on the equivalent circuit, two stability criteria can be derived.

A. Stability Criterion 1[10]

The sub-modules are divided into two groups and the mathematical expression of DC bus voltage can be expressed as (1).

$$\hat{v}_{dc} = \frac{\dfrac{\sum_{i=1}^{N} \dfrac{A_i^Z \hat{v}_i^Z}{Z_i^Z}}{\sum_{i=1}^{N} \dfrac{1}{Z_i^Z}} \cdot (Z_1^Z \| Z_2^Z \| \cdots \| Z_N^Z) - (Z_1^Z \| Z_2^Z \| \cdots \| Z_N^Z) \cdot \sum_{i=1}^{M} T_i^Y \hat{i}_i^Y}{1 + (Z_1^Z \| Z_2^Z \| \cdots \| Z_N^Z) \cdot \sum_{i=1}^{M} Y_i^Y} \quad (1)$$

Since each sub-module is stable in standalone mode, it is clear that there should be no RHP in the numerator in (1). Then, the stability of the total system depends on the numbers of RHZ in the denominator in (1).

In practical application, detailed models of some modules are unknown sometimes. In that case, it is impossible to derive the mathematical transfer function. However, the terminal characteristics of these modules can be estimated according to the measured data by network analyzer. When the measured data are analyzed, graphic stability analysis methods such as Nyquist stability criterion are more suitable.

When Nyquist stability criterion is applied to (1), it is clear that the system stability can be assessed from T_m in (2):

$$T_m = (Z_1^Z \| Z_2^Z \| \cdots \| Z_N^Z) \cdot \sum_{i=1}^{M} Y_i^Y \qquad (2)$$

The basic equation of Nyquist stability criterion is shown in (3). When the numbers of times that open-loop frequency response T_m encircle the point (-1, 0) in counter-clockwise direction equals the numbers of RHP in T_m, the system is stable [16].

$$N_{Tm} = RHZ(1 + T_m) - RHP(T_m) \qquad (3)$$

Where, N_{Tm} is the numbers of clockwise encirclements of the critical point (-1, 0), $RHZ(1+T_m)$ is numbers of RHZ of $1+T_m$, and $RHP(T_m)$ is the numbers of RHP of T_m.

It is clear that the numbers of $RHZ(1+T_m)$ is wanted because it influences the stability of total system directly.

Moreover, N_{Tm} can be drawn based on the measured data. Therefore, the numbers of $RHP(T_m)$ should be estimated before the Nyquist criterion is applied to T_m.

As shown in (2), T_m has two parts: equivalent impedance of all the Z type modules ($Z_{eq} = Z_1^Z \| Z_2^Z \| \cdots \| Z_N^Z$) and equivalent admittance of all the Y type modules ($Y_{eq} = \sum_{i=1}^{M} Y_i^Y$). Since all the modules are required to be stable in standalone mode, there is no RHP in the equivalent admittance of all Y type modules. Hence, only the numbers of RHP in the equivalent impedance of all Z type modules should be estimated.

Then, the expression of Z_{eq} is shown in (4):

$$Z_{eq} = Z_1^Z \| Z_2^Z \| \cdots \| Z_N^Z = \frac{\prod_{i=1}^{N} Z_i^Z}{\sum_{i} \left(\frac{1}{Z_i^Z} \cdot \prod_{j=1}^{N} Z_j^Z \right)} \qquad (4)$$

Due to the stability requirement for sub-module in standalone mode, there is no RHP in the numerator. Therefore, only the numbers of RHZ in the denominator need to be estimated.

Similarly, the Nyquist criterion is applied to the denominator of Z_{eq} to assess the numbers of RHZ in $\sum_{i}^{n} \frac{1}{Z_i} \cdot \prod_{j=1}^{n} Z_j$.

$$N_{\sum_i (\frac{1}{Z_i^Z} \cdot \prod_{j=1}^{N} Z_j^Z)} = RHZ(\sum_i (\frac{1}{Z_i^Z} \cdot \prod_{j=1}^{N} Z_j^Z)) - RHP(\sum_i (\frac{1}{Z_i^Z} \cdot \prod_{j=1}^{N} Z_j^Z)) \qquad (5)$$

The definitions of each item in (5) are similar to the definitions in (3). However, the critical point is changed to (0, 0) due to the mathematical application. Meanwhile, the numbers of RHP in $\sum_{i}^{n} \frac{1}{Z_i} \cdot \prod_{j=1}^{n} Z_j$ is also zero. Therefore, equation (5) can be simplified as (6).

$$N_{\sum_i (\frac{1}{Z_i^Z} \cdot \prod_{j=1}^{N} Z_j^Z)} = RHZ(\sum_i^N (\frac{1}{Z_i^Z} \cdot \prod_{j=1}^{N} Z_j^Z)) \qquad (6)$$

Based on the analysis above, a two-step stability criterion is proposed:

1) Draw the trajectory of $\sum_{i}^{n} \frac{1}{Z_i} \cdot \prod_{j=1}^{n} Z_j$ (which is the denominator of $Z_1^Z \| Z_2^Z \| \cdots \| Z_N^Z$) in s-plane. The times that trajectory encircle the point (0,0) in clockwise direction are equal to the numbers of RHP in Z_{eq};

2) Draw the trajectory of $Z_{eq} \cdot Y_{eq}$ in s-plane. If the times that trajectory encircle the point (-1, 0) in counter-clockwise direction are equal to to the numbers of RHP in Z_{eq}, then the system is stable; otherwise the system is unstable.

The first step can be explained as the assessment of the interaction among all the Z type modules. If there is only one Z type module in the system, there is no interaction and the denominator of equation (6) is 1. Then, the first step can be neglected and only the second step is needed. The second step is used to assess the interaction between the equivalent Z module and equivalent Y module.

B. Stability Criterion 2[11]

All the sub-modules can be treated as connected in parallel. Then, the small signal expression of DC bus voltage can also be expressed as (7).

$$\hat{v}_{dc} = \frac{\prod_{i=1}^{N} Z_i^Z \cdot (\sum_{i=1}^{N} \frac{A_i^Z \hat{v}_i^Z}{Z_i^Z} - \sum_{i=1}^{M} T_i^Y \hat{i}_i^Y)}{\prod_{i=1}^{N} Z_i^Z \cdot (\sum_{i=1}^{N} \frac{1}{Z_i^Z} + \sum_{i=1}^{M} Y_i^Y)} \qquad (7)$$

Similar to the analysis procedure in Criterion 1, Cauchy theorem is employed to develop the stability criterion.

The stability issues of total system can be expressed as (8) after some simplification steps.

$$N = RHZ(\sum_{i=1}^{N} \frac{1}{Z_i^Z} + \sum_{i=1}^{M} Y_i^Y) - RHZ(\prod_{i=1}^{N} Z_i^Z) \qquad (8)$$

Where, N is the numbers of clockwise encirclements of the critical point (0, 0) of the total sum of the admittance of each module.

Based on the analysis above, a two-step stability criterion is proposed:

1) Draw the trajectory of Z_i^Z individually in s-plane, the times that trajectory encircle the point (0,0) in

clockwise direction is the numbers of poles of

$$\sum_{i=1}^{N} \frac{1}{Z_i^Z} + \sum_{i=1}^{M} Y_i^Y \text{ in RHP;}$$

2) Draw the trajectory of the total sum of the admittance of each sub-module $\sum_{i=1}^{N} \frac{1}{Z_i^Z} + \sum_{i=1}^{M} Y_i^Y$ in s-plane, if the times that trajectory encircle the point (0, 0) in counter-clockwise direction equals to the numbers got in step 1, then the system is stable, otherwise the system is unstable.

IV. COMPARISIONS BETWEEN STBAILITY CRITERIA

The validities of criterion 1 and 2 in stability assessment have been proved in [10] and [11]. In some cases, the stability criterion is used to design the parameters to realize specific gain margin and phase margin in addition to stability assessment. In classical control theory, the definitions of gain margin (GM) and phase margin (PM) relate to the distance between system trajectory and critical point (-1, 0), as shown in Fig.6.

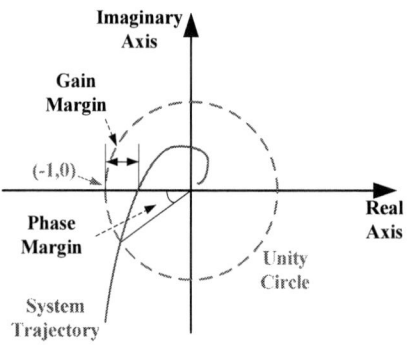

Figure 6. Definitions of Gain Margin and Phase Margin

It is clear that the second step of criterion 1 is based on the Nyquist criterion and the critical point is (-1, 0). However, the second step of criterion 2 is based on Cauchy theorem and the critical point is (0, 0). Therefore, the criterion 2 is not suitable for parameter design unless the definitions of GM and PM are modified with the point (0, 0).

There are also big differences between these two criteria in some applications. For example, an individual sub-module A will be added to an existing system B as shown in Fig.7. Both the individual sub-module A and the system B stable before integration. The stability status of integrated system needs to be assessed before the integration.

If the detailed structure of the system B is known, then both criterion 1 and 2 can be used to assess the stability. However, the effect of criterion 1 and 2 are different if the detailed structure of the system B is unknown.

It should be noted that the criterion 1 is based on the concept that the total system can be divided into equivalent Z type group and equivalent Y type group. If the detailed system is unknown, then it is impossible to

divide the total system into two equivalent groups. Therefore, the criterion 1 will fail in this case.

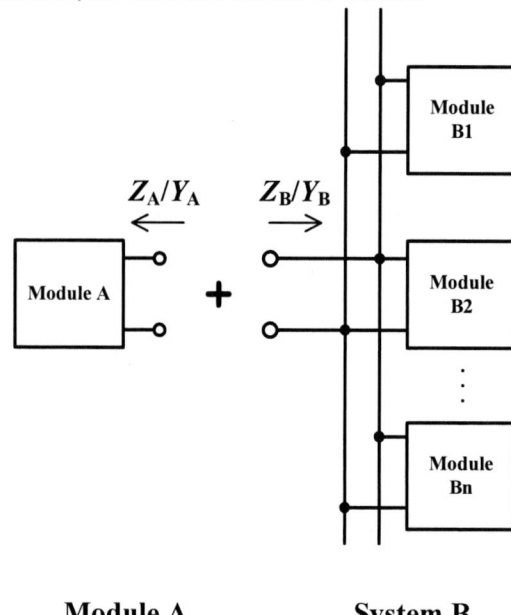

Module A **System B**

Figure 7. Application of System Integration

However, the criterion 2 is based on another concept that all the sub-modules are connected in parallel. Therefore, if the system B is stable, then it can be derived that there is no RHZ in the sum of equivalent admittance of each sub-module. Since the admittance is the reciprocal of impedance, it can be explained as that there is no RHP in the equivalent impedance of system B. Based on the terminal characteristics defining method in Section II, the system B can be regarded as a Z type module. Then, the system integration problem turns into the stability issue of a two module system. If module A is a Y type module, then the system can be regarded as a Z+Y type system. If module A is a Z type module, then the system can be regarded as a Z+Z type system. Therefore, the stability criterion of two module system in [17] can be used to solve stability problem caused by system integration.

Based on above discussion, it is clear that criterion 1 is more advantageous in system parameter design and criterion 2 is more useful in system integration application.

V. EXPERIMENTAL VERIFICATION

To verify the discussion about criterion 2 in system integration application, a three-module system is established in the lab. The photo of test bench and is shown in Fig.8. The system is the same as studied in [10] and [11]. However, the analysis method is different. Sub-module 1 and sub-module 2 are combined together to form system B. The sub-module 3 is regarded as module A.

According to the analysis in Section IV, the system B can be regarded as a Z type module. In addition, the module A is a Y type module. Therefore, the total system is a Z+Y system. Then, the stability of total system can be assessed by counting the clockwise encirclement that the trajectory of $Y_A \cdot Z_B$ encircle point (-1,0).

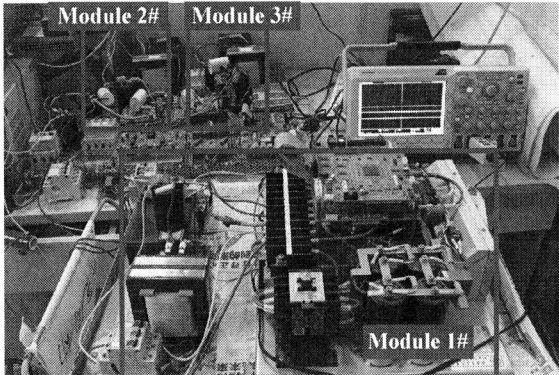

Figure 8. Test bench

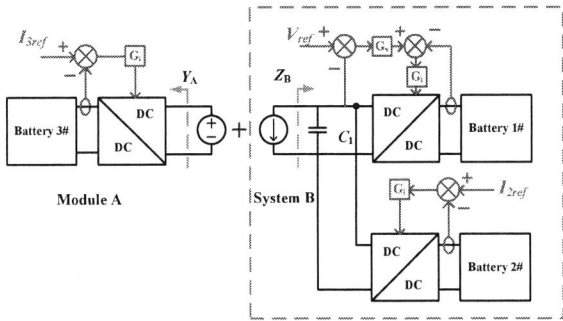

Figure 9. System Structure

The control parameters are modified to form a stable case and an unstable case to verify the stability criterion.

The parameters in the first group experiment are shown in Table I. The input current of each battery is positive.

TABLE I. PARAMETERS IN THE FIRST GROUP

Reference Values	
V_{ref}	30 V
I_{2ref}	3.75 A
I_{3ref}	3.75 A
Output limit of voltage regulator (upper / lower)	15 / -15
Output limit of current regulator (upper / lower)	0.8 / 0.2
Circuit Elements	
C_1	1.5 mF
L_1, L_2, L_3	1 mH, 1 mH, 1 mH
Control Parameters in Stable Case	
Voltage Regulator in Module 1	1+40/s
Current Regulator in Module 1	0.1+40/s
Current Regulator in Module 2	1+40/s
Current Regulator in Module 3	1+40/s
Control Parameters in Unstable Case	
Voltage Regulator in Module 1	0.01+10/s
Current Regulator in Module 1	0.1+40/s
Current Regulator in Module 2	2*(s+200)/(s-200)
Current Regulator in Module 3	2*(s+200)/(s-200)

A. Stable case

The terminal characteristics of each sub-module are measured by using the Agilent network analyzer E5061B. The measured data are drawn in Matlab as shown in Fig.10.

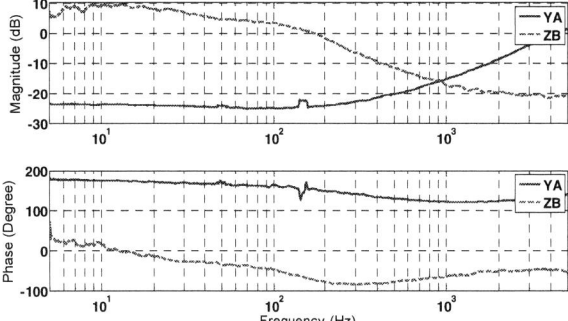

Figure 10. Bode plots of terminal characteristics for sub-modules

Then, the trajectory of $Y_A \cdot Z_B$ is drawn based on the measured data, as shown in Fig.11. It can be seen that the trajectory is far away from point (-1,0). There is no encirclement around point (-1,0). Therefore, the system should be stable.

The time domain waveforms of total system are shown in Fig.12. It is clear that the system is stable and this is consistent with the stability assessment.

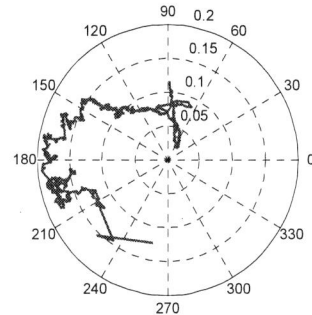

Figure 11. Trajectory of $Y_A \cdot Z_B$ in s-plane

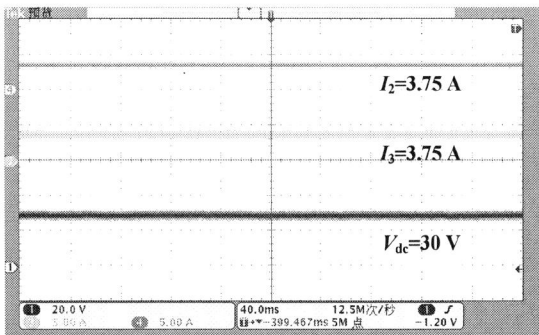

Figure 12. Time domain waveforms of stable case

B. Unstable case

Then, the control parameters are modified to adjust the stability of total system. The terminal characteristics of each sub-module are also measured and the measured data are drawn in Matlab as shown in Fig.13.

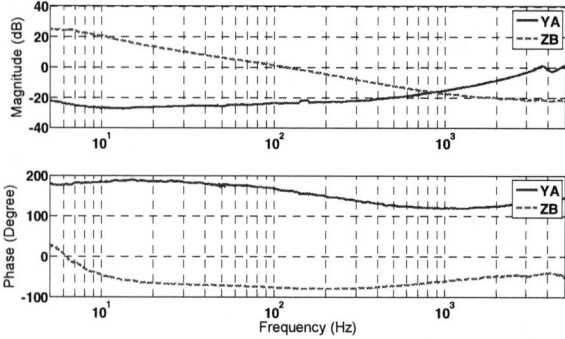

Figure 13. Bode plots of terminal characteristics for sub-modules

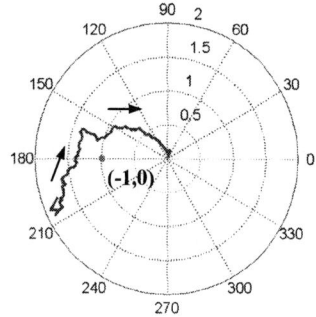

Figure 14. Trajectory of $Y_A \cdot Z_B$ in s-plane

Then, the trajectory of $Y_A \cdot Z_B$ is drawn based on the measured data, as shown in Fig.14. It can be seen that the trajectory encircle point (-1,0) once in clockwise direction. Therefore, the system should be unstable.

The time domain waveforms of total system are shown in Fig.15. It is clear that the system is unstable at designed operating point when the soft start ends, and this result is consistent with the stability assessment.

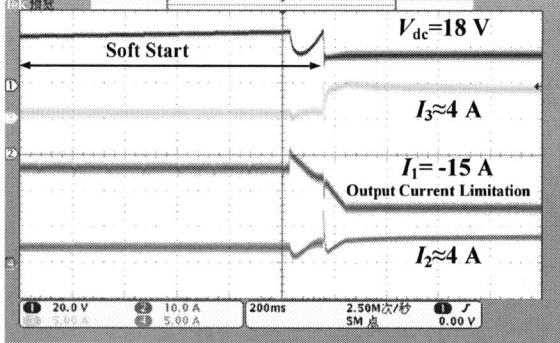

Figure 15. Time domain waveforms of unstable case

VI. CONCLUSIONS

Comparisons between two existing generalized stability criteria of multi-module distributed system have been made. It is clear that criterion 2 is more useful in system integration application. This is because that criterion 1 is developed based on a partial recognition about the system behavior. Meanwhile, the understanding about the system behavior in criterion 2 is clearer. In distributed multi-module system, the bus voltage for each sub-module is the same, thus all the sub-modules should be regarded as in parallel connection. This is the key point

of system behavior study and all the other existing stability criteria are just equivalent mathematical tools.

It should be noted that there are some potential problems due to the non-ideality in practical application. The measuring range and resolution of the network analyzer will influence the accuracy in measurement. The noise in measurement will also affect the accuracy. These practical problems may lead to wrong stability assessment when the trajectory is very close to the critical point so that it is very hard to make sure whether the trajectory encircle the critical point or not. In the study of stability of cascade system, this problem is solved by setting different forbidden regions to keep the trajectory off the critical point. Future work will focus on the design of forbidden region in multi-module application, as well as the design of sub-module control.

REFERENCES

[1] Y.Panov, J.Rajagopalan, F.C.Lee, "Analysis and design of N paralleled DC-DC converters with master-slave current-sharing control," *Applied Power Electronics Conference and Exposition, 1997. APEC '97 Conference Proceedings 1997, Twelfth Annual*, vol.1, pp.436-442, 23-27 Feb 1997.

[2] X.Xie, S.Yuan, J.Zhang, Z.Qian, "Analysis and Design of N paralleled DC/DC Modules with Current-Sharing Control," *Power Electronics Specialists Conference, 2006. PESC '06. 37th IEEE* , pp.1-4, 18-22 June 2006.

[3] V.J.Thottuvelil, G.C.Verghese, "Analysis and control design of paralleled DC/DC converters with current sharing," *IEEE Trans. Power Electron*, vol.13, no.4, pp.635-644, Jul 1998.

[4] D.Hou, J.Liu, H.Wang, W.Huang, "The stability analysis and determination of multi-module distributed power electronic systems," *Power Electronics for Distributed Generation Systems (PEDG), 2010 2nd IEEE International Symposium on*, pp.577-583, 16-18 June 2010.

[5] R.D.Middlebrook, "Design techniques for preventing input-filter oscillations in switched-mode regulators," *in Proc. Powercon 5*, 1978, pp. A3-1–A3-16.

[6] T.Suntio, J.Leppaaho, J.Huusari, L.Nousiainen, "Issues on Solar-Generator Interfacing With Current-Fed MPP-Tracking Converters," *IEEE Trans. Power Electron*, vol.25, no.9, pp.2409-2419, Sept. 2010.

[7] J.Sun, "Impedance-Based Stability Criterion for Grid-Connected Inverters," *IEEE Trans. Power Electron*, vol.26, no.11, pp.3075-3078, Nov. 2011.

[8] F.Liu, J.Liu, B.Zhang, H.Zhang, S.U.Hasan, S.Zhou, "Unified stability criterion of bidirectional power flow cascade system," *Applied Power Electronics Conference and Exposition (APEC), 2013 Twenty-Eighth Annual IEEE*, pp.2618-2623, 17-21 March 2013.

[9] F.Liu, J.Liu, B.Zhang, H.Zhang, S.U.Hasan, "General impedance/admittance stability criterion for cascade system," *ECCE Asia Downunder (ECCE Asia), 2013 IEEE*, pp.422-428, 3-6 June 2013.

[10] F.Liu, J.Liu, H.Zhang and D.Xue, "Generalized Stability Criterion for Multi-module Distributed DC System", *Journal of Power Electronics*, vol. 14, no. 1, January 2014.

[11] F.Liu, J.Liu, H.Zhang and D.Xue, "Terminal Admittance Based Stability Criterion for Multi-module DC Distributed System," *Applied Power Electronics Conference and Exposition (APEC), 2014 Twenty-Ninth Annual IEE*E, March 2014

[12] R.W.Erickson, D.Maksimovic, "Fundamentals of power electronics", 2nd edition. New York: Kluwer, 2001.

[13] R.D.Middlebrook, S.Cuk. "A general unified approach to modelling switching-converter power stages." *In Power Electronics Specialists Conference*, vol. 1, pp. 18-34. 1976.

[14] A.Capel, J.C.Marpinard, J.Jalade, M.Valentin, "Current Fed and Voltage Fed Switching DC/DC Converters - Steady State and Dynamic Models their Applications in Space Technology," *Telecommunications Energy Conference, 1983.* INTELEC '83. Fifth International, pp.421-430, 18-21 Oct. 1983.

978-1-4799-2706-7/14 $31.00 © 2014 IEEE

[15] Alexander C.K., M.N.O. Sadiku, "Fundamentals of Electric Circuits," 2005: McGraw-Hill Higher Education.

[16] M.Driels," Linear Control Systems Engineering", Tsinghua University Press,2000.

[17] F. Liu, J. Liu, B. Zhang, H. Zhang, S.U.Hasan and L. Zhou, "Stability issues of Z+Z or Y+Y type cascade system," *Energy Conversion Congress and Exposition (ECCE), 2013 IEEE* , pp.434,441, 15-19 Sept. 2013

Characteristics Study of Neural Network Aided Digital Control for DC-DC Converter

Hidenori Maruta*, Masashi Motomura†, Fujio Kurokawa*

*Division of Electrical Engineering and Computer Science, Faculty of Engineering, Nagasaki University

†Graduate School of Engineering, Nagasaki University

*† 1-14, Bunkyo–machi, Nagasaki, 852–8521, Japan

*e-mail: {hmaruta, fkurokaw}@nagasaki-u.ac.jp

Abstract—This paper studies the digital controlled dc–dc converter which employs the neural network in coordination with PID control. Especially, we focus on the problem of over-compensation to the transient response of the output voltage in the dc–dc converter by the neural network control. Our previous studies suggested that the neural network control is effective to improve the transient response of the output voltage. Therefore, it can contribute to the low power consumption of power supplies for information and communication systems. However, the neural network aided control has a problem that it tends to be overcompensation, i.e., the compensation effect of the neural network becomes too much and badly affects to control the plant. The overcompensation causes from a delay time in sensing and calculation to determine the timing and duration of the neural network control itself. To address the problem of overcompensation, our previous studies investigated methods to devise the timing and duration of neural network control. However, since they need some pre–determined parameters which are optimized to the specific target converter to avoid the effect of delay, the implementation of it tends to be complicated one in practical use. In this study, we adopt the neural network which predicts parameters to improve the transient response and control timing and duration of itself. Therefore, it does not need any pre-determined parameters for timing and duration control. From the simulation result for evaluation, it is confirmed that the presented method has a superior transient characteristics compared with conventional method.

Keywords—neural network, digital control, dc-dc converter

I. INTRODUCTION

The global request to reduce the energy consumption is common for the information and communication systems. The power supply for them has a potential to reduce the consumption and its control method becomes more important to contribute such purpose. To realize the low energy consumption in power supply for such systems, dc–dc converters used in them have a very important role. For such purposes, the digital control method for dc–dc converters are widely focused since they have advantages in energy management and providing effective control which can not be realized by analog one[1][2][3].

One of the methods which cannot be realized by the analog control is a neural network based one[4][5][6][7] since it needs to be realized in the digital way using some integral circuit systems such as DSP, FPGA and so forth. We have previously

presented a neural network based control method for dc–dc converters which can improve the transient response effectively and contribute the low energy consumption of them[8]. In the previous method, the transient response is compensated by the effect of neural network control, however, its effect becomes too much when the timing and duration of neural network control is not optimal. To address the problem, we studied to devise the timing and duration to avoid the overcompensation to design them by employing pre–determined control parameters[9][10]. However, since it needs to determine those parameters for the specific target converter, we require more functional method for the timing and duration control of the neural network control.

In this study, we present a neural network aided digital control method for dc–dc converters which can address the problem of the overcompensation by the neural network control itself. The presented method works in coordination with a conventional PID control and improves the transient response by modification of the reference in the PID control in the same manner as our previous study[8]. The neural network is adopted to modify the reference value in the PID control and predict the optimal timing and duration of itself simultaneously. Therefore, the presented method does not need any pre–determined parameters for timing and duration control used in the neural network control. This can contribute to the effective implementation of the presented method in practical use.

For evaluation, we examine the transient characteristics of presented method compared with conventional PID control and the e_o sensed neural network control. From the results, it is revealed that the presented method have superior transient characteristics compared with other methods.

II. OPERATION PRINCIPLE

In this section, we present a neural network control method which works in coordination with conventional PID control.

Figure 1 shows the buck-type dc–dc converter used in this paper as a case study. Symbols represent circuit parameters as follows; E_i is an input voltage; E_o^* is a desired output voltage; C is a output capacitor; L is an inductor; T_r represent a switch; D represents a diode; R represents a load, respectively.

The 2014 International Power Electronics Conference

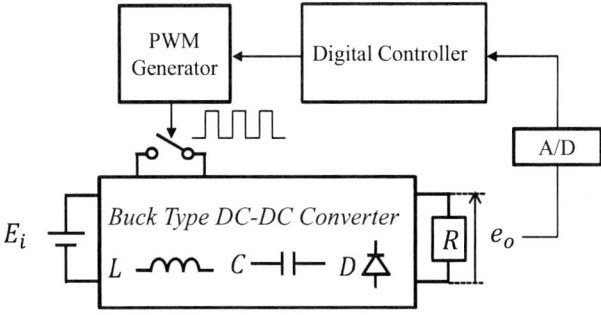

Fig. 1.   Digitally controlled buck converter.

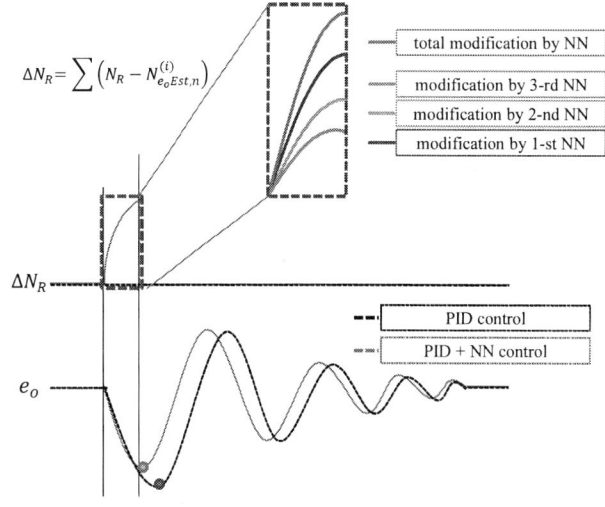

Fig. 2.   Reference modification by neural network control[8].

As same manner as our previous study[8], the neural network works in coordination with conventional PID control. The conventional PID control to calculates the on–time duration for digital PWM signal $N_{T_{on}}$ is expressed as Eq. (1):

$$
\begin{aligned}
N_{T_{on},n} =\; & N_B - K_P \left(N_{e_o,n-1} - N_R \right) \\
& - K_I \sum \left(N_{e_o,n-1} - N_R \right) \\
& - K_D \left(N_{e_o,n-1} - N_{e_o,n-2} \right) \quad (1)
\end{aligned}
$$

where K_P, K_I and K_D are the proportional, integral and differential coefficients; N_R is the fixed reference value and N_B is the bias term. We also mention that the suffix n denotes the n-th switching period in this study.

It is well known that, in the PID control, the proportional (P) term is dominant for the improvement of overshoot compensation in the transient response. Therefore, the neural network modifies the reference value in the P term in Eq. (1) to improve the transient response of the output voltage. It works to reduce the difference between e_o and the original reference value (desired output voltage) N_R.

Here we describe how the neural network improves the transient response in detail.

At first, there is no reference modification by the neural network, i.e., only the PID control works. The control parameters K_P, K_I and K_D in the PID control are tuned and remains as same values after this tuning.

To perform as a function to modify the reference value in the P term, the neural network needs to be trained. Therefore, we need to obtain the output voltage $\{e_{o,n}|n\}$ in the transient state as training data and train the neural network which can predict the forthcoming e_o using the former data. The neural network in this study is the three-layered neural network which has three input nodes, six hidden nodes and two output nodes. This neural network is trained to predict $\widehat{e_{o,n}}$ and $\widehat{e_{o,n+1}}$ from its three former data $e_{o,n-1}, e_{o,n-2}, e_{o,n-3}$. After the training, this (first) neural network modifies N_R in the P term using $\widehat{e_{o,n}}$ as follows;

$$
\begin{aligned}
N_{T_{on},n} =\; & N_B - K_P \{ N_{e_o,n-1} - (N_R + \Delta N_R) \} \\
& - K_I \sum \left(N_{e_o,n-1} - N_R \right) \\
& - K_D \left(N_{e_o,n-1} - N_{e_o,n-2} \right) \quad (2)
\end{aligned}
$$

$$
\Delta N_R = N_R - N^{(1)}_{e_o Est,n} \quad (3)
$$

where $N^{(1)}_{e_o Est,n}$ is the digital value of $\widehat{e_{o,n}}$. The remained $\widehat{e_{o,n+1}}$ is used for timing and duration control, which is discussed in later. This modification proceeds to suppress the first undershoot as shown in Fig. 2 since the suppression of the first undershoot is effective for the improvement of whole transient response[8].

Using the PID control and the first trained neural network control, we construct the second neural network as same manner as the first one.

This training and construction process repeatedly proceeds until the enough improvement of the transient response is obtained. After M times procedures of the neural network training, the reference modification ΔN_R becomes Eq. (4);

$$
\Delta N_R = \sum_{i=1}^{M} \left(N_R - N^{(i)}_{e_o Est,n} \right) \quad (4)
$$

It is known that popular feedback control methods, including the PID control, need the optimal control parameter setting to satisfy the superior transient response and not to cause the overcompensation phenomenon. The neural network control has also the problem of overcompensation.

Here we discuss this remained problem about timing and duration of the neural network control described above.

The previously devised timing and duration of the neural network control shown in Fig. 2 [8] is effective, however, it has difficulty about the peak–point detection of undershoot. Figure 3 shows the peak–point difference which causes the bad effect for the neural network control. As shown in Fig. 3, when the neural network control is performed, the transient response varies and the peak point of undershoot also varies. From this reason, the ideal control timing for neural network is affected and varied, and the difference of peak time Δ_t

978-1-4799-2706-7/14 $31.00 © 2014 IEEE   3612

The 2014 International Power Electronics Conference

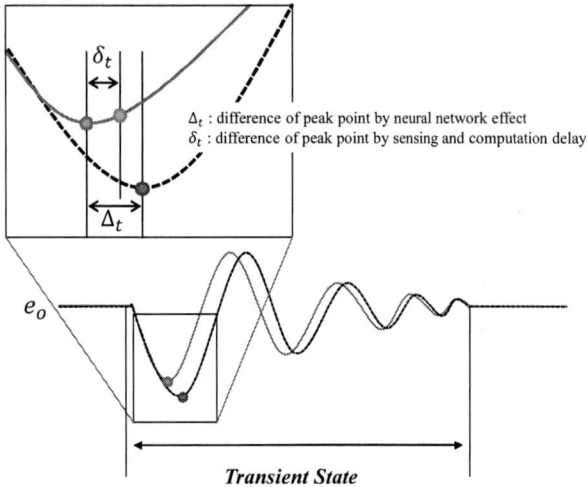

Fig. 3. Two types of peak difference which badly affects to neural network control.

TABLE I.   CIRCUIT PARAMETERS OF DC–DC CONVERTER USED IN SIMULATION STUDY.

Circuit Parameter	Value
Input Voltage	E_i =20 V
Desired Output Voltage	E_o^* = 5 V
Output Capacitor	C=940 μF
Inductor	L =189 μH
Switching frequency	f_S=100 kHz

appears. Moreover, the control timing is also affected from the sensing delay. Since the peak point is detected by the sensed output voltage e_o, there is a delay time δ_t caused from the sensing and computation process. These differences causes the *overcompensation* by neural network control and the bad effect to improve the transient response[8].

To address the problem of optimal peak point detection, we already presented the method which uses pre–determined timing and control duration for the specific target of dc-dc converter[9] [10] . This method can improve the overcompensation effect by the neural network control, however, it has a problem that it tends to be complicated in a practical use by the consideration of the pre–determined parameters in the duration control. Therefore, we present a method to predict the peak point to obtain optimal timing and duration of the neural network control.

The neural network is trained to predict $\widehat{e_{o,n}}$, $\widehat{e_{o,n+1}}$ as mentioned before. Therefore, it is possible to predict the optimal peak point by these values as follows. When the $\Delta_{e_o,n+1} = \widehat{e_{o,n}} - \widehat{e_{o,n+1}}$ changes its sign, we can consider that the output voltage e_o reaches its peak of the undershoot at between the n-th switching period and the $n+1$-th switching period. Then, it is the optimal timing for the neural network control to stop performing.

After M times training, we obtain M neural networks which are combined to modify the reference value as shown in Eq. (4). The timing and duration control is done by M-th neural network. This neural network control procedure is shown in Fig. 4.

III.   SIMULATION STUDY

To evaluate the presented method, we compare it with the conventional PID control and the neural network control which

uses e_o sensed timing and duration control[8] . The simulation software PSIM is used for this simulation study.

The circuit parameters for this evaluation are shown in Table I.

In the simulation experiments, our presented method is compared to the conventional PID control method and the neural network control method which uses the sensed output voltage for timing and duration control to perform the neural network control term. For evaluation, the transient characteristics of each method are compared when the step change in load R is 25Ω (0.2A) to 5Ω (1A), which is the change from the Continuous Current Mode (CCM) to CCM. The optimal gain parameters for PID control are set to be $K_P = 4$, $K_I = 0.015$ and $K_D = 4$ respectively. These control parameters are determined as the transient response of the output voltage are optimized by the PID controller. As mentioned in the previous section, these parameters are remained same values in the training and control of the neural network, which works to modify the reference value in the P term of the PID controller.

To obtain the neural network controller, the training of the neural network proceeds repeatedly three times ($M = 3$), which is the best case in the e_o sensed neural network control[8]. In the presented method, the number of (trained) neural network is also three ($M = 3$) for evaluation.

Figures 5 through 7 show waveforms of the transient response obtained by the conventional PID control, e_o sensed neural network control and presented neural network control respectively.

Table II summarizes results of each control method. In this table, NN_3 means that the neural network training is repeated three times, i.e., it is the case of $M = 3$ in Eq. (4).

From these results, it is revealed that the presented method improves the transient characteristics of the output voltage effectively compared with other conventional methods. Therefore, we can confirm that the presented method can improve the transient response and contribute to the implementation in practical use simultaneously.

IV.   CONCLUSION

In this study, we show characteristics of the neural network aided control method for digitally controlled dc–dc converters. The neural network control works in coordination with the

978-1-4799-2706-7/14 $31.00 © 2014 IEEE

The 2014 International Power Electronics Conference

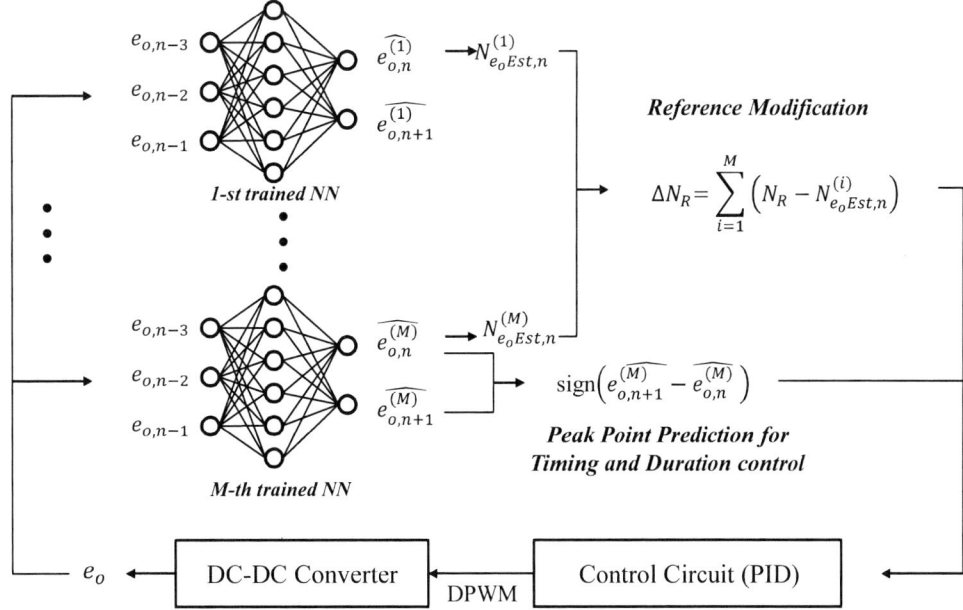

Fig. 4.   Reference modification and peak point prediction by neural network control.

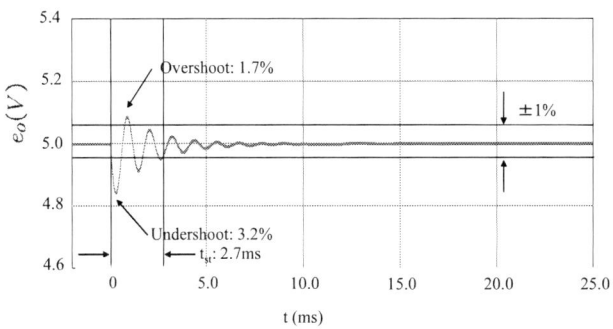

Fig. 5.   Waveform of transient response with conventional PID control.

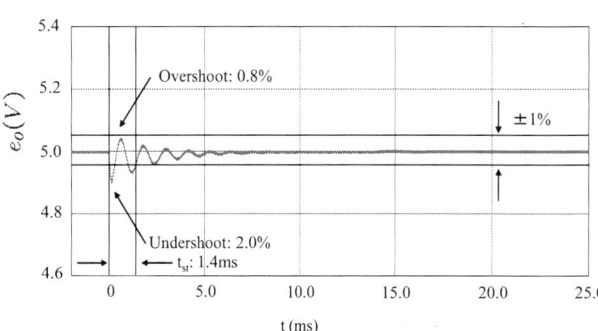

Fig. 6.   Waveform of transient response with e_o sensed neural network control.

TABLE II.   SUMMARY OF SIMULATION RESULTS.

Control method	Undershoot (%)	Overshoot (%)	Settling time within $\pm 1\%$(ms.)
PID	3.2	1.7	2.7
PID+NN_3 (e_o sensed[8])	2.0	0.8	1.4
PID+NN_3 (Presented)	1.8	0.44	1.4

conventional PID control and modifies the reference value in the PID control to improve the transient response of output voltage. The presented method have advantages in which the neural network can control the reference value for compensation and timing and duration of itself simultaneously. Therefore, it does not need any pre–determined parameters for neural network control and it is more suitable for practical use and design.

From the simulation results, we can confirm that the presented method has the superior transient characteristics compared with conventional method.

The neural network aided control method can be adopted in any type of converter in the same manner as presented in this paper. It can provide easy tuning which does not need any complicated determination of the control parameters to improve the transient response. Therefore, we expect that the presented method can realize high performance dc–dc converters and contribute to reduce the energy consumption of information and communication systems effectively.

ACKNOWLEDGMENT

This work was partially supported by JSPS KAKENHI Grant Number 24360112.

978-1-4799-2706-7/14 $31.00 © 2014 IEEE

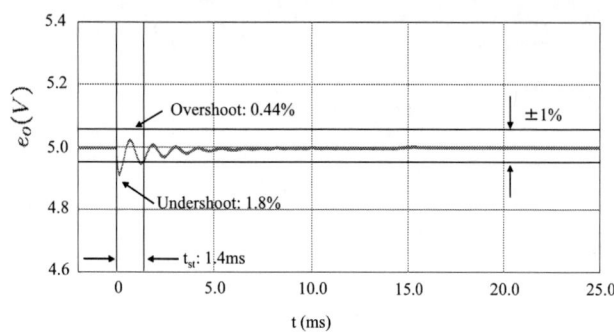

Fig. 7. Waveform of transient response with presented neural network control.

REFERENCES

[1] S. Buso, P. Mattavelli: Digital Control in Power Electronics, Morgan & Claypool, 2006.

[2] D. Maksimovic, R. Zane, R. Erickson: "Impact of digital control in power electronics, " Proc. of 2004 International Symposium on Power Semiconductor Devices & ICs, pp. 13-22, 2004.

[3] P. T. Krein: "Digital control generations digital controls for power electronics through the third generation," Proc. of the IEEE PEDS, pp. 1-5, 2007.

[4] B. R. Lin, R. G. Hoft: "A Neural Network Controller for Switching Power Converters," IEEE PESC '93 Record, pp. 887–892, 1993.

[5] H. C. Chan, K. T. Chau, C. C. Chan: "Power electronics converter control based on neural network and fuzzy logic methods," IEEE PESC '93 Record, pp. 900–906, 1993.

[6] K. T. Chau, C. C. Chan: "Real–Time Implementation of an On–Line Trained Neural Network Controller for Power Electronics Converters," IEEE PESC '94 Record, pp. 321–327, 1994.

[7] R. J. Wai, L. C. Shih: "Adaptive Fuzzy-Neural-Network Design for Voltage Tracking Control of a DC–DC Boost Converter," IEEE Transactions on Power Electronics, Vol. 27, pp. 2104–2115, 2012.

[8] H. Maruta, M. Motomura, K. Ueno, F. Kurokawa: "Reference modification control DC-DC converter with neural network predictor, " Proc. 2012 IEEE 13th Workshop on Control and Modeling for Power Electronics (COMPEL), pp. 1–4, 2012.

[9] H. Maruta, M. Motomura, F. Kurokawa: "Effect of Time-Duration of Neural Network Based Control on Transient Response of DC-DC Converter," Proc. 10th International Conference on Power Electronics and Drive Systems (PEDS), pp. 250–255, 2013.

[10] H. Maruta, M. Motomura, F. Kurokawa: "A Novel Timing Control Method for Neural Network Based Digitally Controlled DC-DC Converter," Proc. 15th European Conference on Power Electronics and Applications (EPE), pp. 1–8, 2013.

The 2014 International Power Electronics Conference

Zero Current Switching Current-fed Parallel Resonant Push-pull (CFPRPP) Converter

Radha Sree Krishna Moorthy
Electrical and Computer Engineering
National University of Singapore
Singapore 117583
a0107273@nus.edu.sg

Akshay Kumar Rathore, *Senior Member, IEEE*
Electrical and Computer Engineering
National University of Singapore
Singapore 117583
eleakr@nus.edu.sg

Abstract— Current-fed parallel resonant push-pull (CFPRPP) converter has been analyzed and designed. The topology maintains zero current switching (ZCS) of the semiconductor devices. Variable frequency modulation has been adopted for load voltage regulation and to achieve ZCS over wide input voltage variation. Soft switching permits to raise the device switching frequency resulting in compact and light converter. At fixed input voltage, ZCS and load voltage are load independent. Natural clamping of devices by ZCS is achieved using parallel resonance. Push-pull converters have only two devices with common ground with source requiring simple gate drive circuitry. A comprehensive study of the proposed converter including steady-state analysis and design has been performed. Also, a 500 W prototype was developed in the laboratory to evaluate the performance of the converter. Experimental results have been demonstrated to verify the claims.

Keywords— Zero current switching, Current-fed converter, Push-pull topology, Parallel resonance.

I. INTRODUCTION

The liaison between human population's growth and energy demand has been the driving force behind the idea of micro-grid [1]. Given the unpredictable nature of the renewable energy sources, the use of energy storage systems (ESS) becomes indispensable. In order to interface the low voltage ESS to high voltage dc bus it is necessary to rely on step up dc/dc converters. Thus dc/dc converter with good performance characteristics along with compactness, light weight and high efficiency is a general requirement for battery, solar, fuel cells, and uninterruptible power supply (UPS) applications.

Among various dc/dc converter topologies reported in literature, isolated topologies are preferred when high voltage conversion ratio is required [2-3]. While current-fed non-isolated converters can allow voltage conversion ratio up to 10, isolated current-fed topologies permit voltage conversion ratio beyond 10. Further, current-fed converters are meritorious over voltage-fed converters as discussed in [4-6] for low voltage high current applications requiring high voltage gain. These converters have low input current ripple, reduced peak and circulating currents, negligible rectifier diode ringing, no duty cycle loss etc. [7]. One of the major problems associated with current-fed converters is the large device voltage overshoot at turn-off. This is

alleviated by employing dissipative or regenerative snubber. However, these snubbers affect the converter efficiency significantly [8]. Active-clamping permits energy recovery and aids in zero voltage switching (ZVS), thereby showing high efficiency [9]. But these merits come at the cost of large film capacitor and floating high side devices [10].

Resonant converters may be a plausible option because they utilize circuit parasitics or resonant tank to assist in soft-switching of semiconductor devices. This aspect makes them suitable for isolated applications as the leakage inductance and parasitic capacitance of high frequency (HF) transformer if not utilized effectively may give rise to large voltage and current spikes and ringing [11-12]. Load independent CLL resonant converter topology operating at constant frequency has been proposed in [13-14]. But to account for the non-idealities of the converter components and for input voltage variation, variable frequency modulation may be necessary. On the other hand, higher voltage gain is obtained by resonance. Also, these converters can be operated at higher switching frequency with higher power density [15-17]. Advantages of parallel resonant converter for voltage regulator type applications have been discussed in [18].

A current-fed full-bridge resonant converter has been proposed in [19]. But, push-pull topology is attractive owing to only two devices with common ground with source simplifying gate driving requirement. [20-23] provide a glimpse on how ZVS and ZCS can be achieved in push-pull converter using resonant tank.

In this paper, a current-fed parallel resonant push-pull (CFPRPP) dc/dc converter, shown in Fig. 1, has been analyzed and designed. The resonance owing to transformer leakage inductance and parallel capacitance to achieve ZCS turn-off of the device has been implemented. The proposed converter operates with ZCS of switches and rectifier diodes reducing switching losses. Natural voltage clamping (NVC) and ZCS eliminate the need for additional active-clamp or snubber enhancing the converter efficiency. Section II elaborates the steady-state operation and analysis of the proposed converter. The converter design has been described in detail in Section III. Section IV demonstrates the experimental results obtained from a laboratory prototype of 500 W.

978-1-4799-2706-7/14 $31.00 © 2014 IEEE

The 2014 International Power Electronics Conference

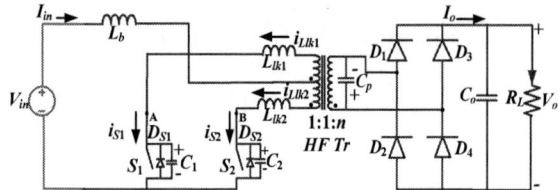

Fig. 1. Proposed CFPRPP converter.

II. STEADY-STATE OPERATION AND ANALYSIS

The steady-state operation and analysis of the CFPRPP converter are explained in this Section. Transformer leakage inductance L_{lk} and external parallel capacitor C_p form the resonant tank for ZCS operation of the primary switches.

The following assumptions are made to study and understand the analysis of the converter: a) Input boost inductor L_b is large to maintain a constant current through it. b) All the semiconductor devices are ideal and lossless. c) Magnetizing inductance is infinitely large. d) Leakage inductances of the transformer are equal i.e., $L_{lk1}=L_{lk2}=L_{lk}$.

The control of power transferred is achieved by variable frequency modulation with a constant duty cycle for input voltage variation. The gating signals for the switches are phase shifted by 180° with a duty cycle D greater than 50%. The steady-state operating waveforms are shown in Fig. 2.

The analysis has been carried out for the first half cycle and for the next half cycle these intervals repeat in the same sequence with other symmetrical devices conducting. The equivalent circuits depicting the different modes of operation of the converter are laid out in Fig. 3. Analytical equations derived facilitate in determining the values and ratings of components and theoretical converter performance.

A. Interval 1 (Fig. 3(a): $t_0 < t < t_1$)

During this interval, switch S_2 and diodes D_1 and D_4 are conducting and the source energy is transferred to the load. Constant current I_{in} flows through L_{lk2}. The parallel capacitor's voltage is clamped at $-V_o$ which is an attribute of parallel resonant circuits. The transformer primary voltage v_{AB} is $2V_o/n$. The equations for this interval are given by

$$v_{Cp}(t) = -V_o \tag{1}$$

$$i_o(t) = \frac{I_{in}}{n} \tag{2}$$

B. Interval 2 (Fig. 3(b): $t_1 < t < t_2$)

At $t = t_1$, the switch S_1 is turned on. The device capacitor C_1 discharges through S_1 in a short interval of time.

C. Interval 3 (Fig. 3(c): $t_2 < t < t_3$)

The switches S_1, S_2 and the diodes D_1 and D_4 are conducting while the diodes D_2 and D_3 are blocking the output voltage V_o. Positive voltage appears across L_{lk1} and hence i_{Llk1} starts increasing in the positive direction with the slope of V_o/nL_{lk}, while the current i_{Llk2} starts decreasing with the same slope. The current flowing through the diodes D_1 and D_4 are also decreasing. Current

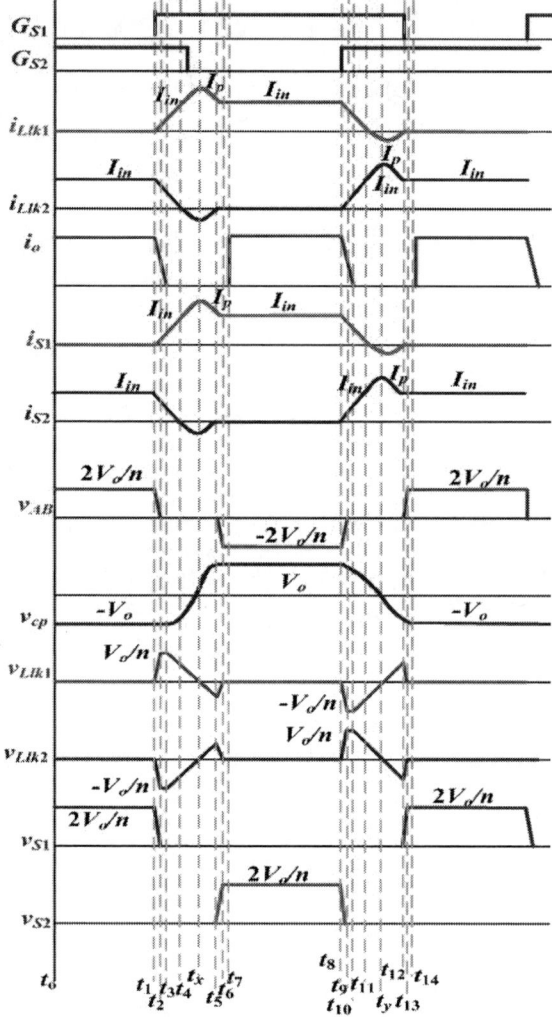

Fig. 2. Steady-state operating waveforms of the proposed converter during different intervals of operation.

through the leakage inductances or switches is given as

$$i_{Llk1}(t) = i_{S1}(t) = \frac{V_o/n}{L_{lk1}}(t - t_2) \tag{3}$$

$$i_{Llk2}(t) = i_{S2}(t) = -\frac{V_o/n}{L_{lk2}}(t - t_2) + I_{in} \tag{4}$$

Where, I_{in} is the current flowing through the boost inductor L_b.

$$i_o(t) = -\frac{1}{n}(i_{Llk2}(t)) = \left[\frac{V_o}{n^2 L_{lk2}}(t - t_2) - \frac{I_{in}}{n}\right] \tag{5}$$

The voltage across the primary of the transformer $v_{AB} = 0$. At the end of this interval, the current flowing through the leakage inductors are equal can be given as,

$$I_{Llk1} = I_{Llk2} = \frac{I_{in}}{2} \tag{6}$$

Therefore, the net current flowing through the secondary circuit becomes zero and the diodes D_1 and D_4 get commutated naturally. The duration of this interval is given by

$$T_{32} = (t_3 - t_2) = \frac{nL_{lk1}I_{in}}{2V_o} \tag{7}$$

978-1-4799-2706-7/14 $31.00 © 2014 IEEE   3617

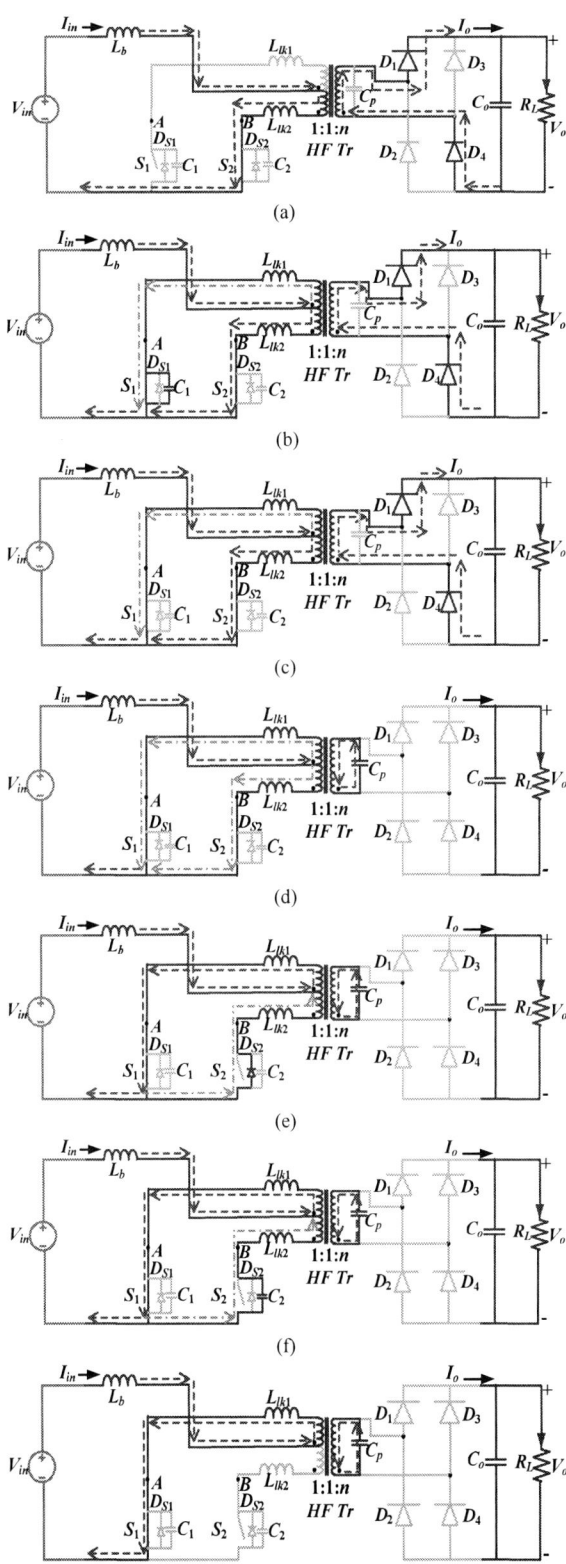

Fig. 3. Equivalent circuits representing the different intervals of operation of the proposed converter.

D. *Interval 4 (Fig. 3(d): $t_3 < t < t_4$)*

During this interval, the current flowing through S_2 and

S_1 decreases and increases respectively in a resonant fashion. The equations governing this interval are as follows:

$$i_{S1}(t) = i_{Llk1}(t) = \frac{I_{in}}{2} + \frac{V_0}{nZ_r}\sin(2\pi f_r(t-t_3)) \quad (8)$$

$$i_{S2}(t) = i_{Llk2}(t) = \frac{I_{in}}{2} - \frac{V_0}{nZ_r}\sin(2\pi f_r(t-t_3)) \quad (9)$$

The resonant frequency and the characteristic impedance are given by

$$f_r = \frac{1}{2\pi\sqrt{L_{eq}C'_p}} \quad (10)$$

$$Z_r = \sqrt{L_{eq}C'_p} \quad (11)$$

Where, the equivalent inductance, $L_{eq} = L_{lk1} + L_{lk2}$ and C'_p is the parallel capacitance reflected on the primary side. The voltage across capacitor C_p is given by

$$V_{Cp} = -V_0\cos(2\pi f_r(t-t_3)) \quad (12)$$

At the end of the interval, current i_{S2} reduces to zero naturally. $i_{S2} = 0$. The duration of this interval is given by

$$T_{43} = (t_4 - t_3) = \frac{1}{2\pi f_r}\sin^{-1}\left(\frac{I_{in}nZ_r}{2V_0}\right) \quad (13)$$

E. *Interval 5 (Fig. 3(e): $t_4 < t < t_5$)*

During this interval, S_1 and the body diode D_{S2} go into conduction. The switch S_2 can now be turned off with ZCS. At $t = t_x$, the current i_{Llk1} reaches the maximum I_p and voltage across the parallel capacitor reaches zero and starts increasing. The pivotal condition for ZCS to be achieved if $I_p > I_{in}$ is

$$I_p = |i_{Llk1}(t)|_{max} = \frac{I_{in}}{2} + \frac{V_0}{nZ_r} \geq I_{in} \quad (14)$$

The duration T_{53} is given by

$$T_{53} = (t_5 - t_3) = \frac{\pi - 2\pi f_r(t_4 - t_3)}{2\pi f_r} \quad (15)$$

The resonant period ends with this interval and the current i_{Llk1} becomes equal to I_{in}. Switch S_2 is turned-off with ZCS and i_{DS2} commutates naturally. $i_{DS2} = 0$ and $i_{Llk1} = I_{in}$.

F. *Interval 6 (Fig. 3(f): $t_5 < t < t_6$)*

In this mode, switch S_2 has been turned off with ZCS and the device capacitor C_2 gets charged and switch voltage builds up from 0 to $2V_0/n$.

G. *Interval 7 (Fig. 3(g): $t_6 < t < t_7$)*

In this mode switch S_1 is conducting and the current i_{Llk1} becomes equal to I_{in} and the parallel capacitor C_p gets charged by the boost inductor current I_{in} during this interval. The voltage across the primary of the transformer, v_{AB} is $-2V_0/n$. $i_{S1}(t) = I_{in}$ and $i_{S2}(t) = 0$. The capacitor voltage is given by

$$v_{Cp} = -V_0\cos(2\pi f_r T_{42}) + \frac{I_{in}}{nC_p}(t-t_6) \quad (16)$$

At the end of this interval, the parallel capacitor is fully charged to V_0 and $v_{AB} = -2V_0/n$. The diodes D_2 and D_3 get forward biased and go into conduction. The duration of this interval is given by

$$T_{76} = (t_7 - t_6) = \frac{C_p' V_0 \left(1 + \cos\left(2\pi f_r T_{53}\right)\right)}{n I_{in}} \quad (17)$$

The duration of the 7 intervals t_0 to t_7 is equal to half the switching cycle. Therefore,

$$T_{70} = (t_7 - t_0) = \frac{T_s}{2} \quad (18)$$

III. CONVERTER DESIGN

In this Section, the design of the proposed converter has been illustrated for the following specifications: Input Voltage V_{in} = 38 to 50 V; Nominal voltage = 48 V; Output Voltage V_o = 380 V; Output Power P_o = 500 W; Maximum value of switching frequency f_s = 200 kHz. Defining the parameters normalized frequency f_n, characteristic impedance Z_r and voltage gain M simplifies the design procedure and also provides a better insight on the converter performance.

A. Maximum device voltage

The maximum voltage across the primary switches, rectifier diodes and parallel capacitor C_p are respectively given by

$$V_{sw,max} = \frac{2V_0}{n} \quad (19)$$

$$V_{D,max} = V_o \quad (20)$$

$$V_{Cp,max} = V_o \quad (21)$$

From (19), it's clearly seen that the maximum voltage appearing across the switch is dependent on turns ratio n.

B. Duty Ratio

The minimum and the maximum value of duty cycle required for variable frequency modulation can be obtained by equating T_{41} and T_{51} with the overlap period $(D-0.5)T_s$ respectively and is given by

$$D_{min} = 0.5 + \frac{n I_{in} L_{lk} f_s}{2V_0} + \frac{f_n}{2\pi} * \sin^{-1}\left(\frac{n I_{in} Z_r}{2V_0}\right) \quad (22)$$

$$D_{max} = 0.5 + \frac{n I_{in} L_{lk} f_s}{2V_0} + \frac{f_n}{2\pi} * \sin^{-1}\left(\frac{n I_{in} Z_r}{2V_0}\right) + \frac{f_n}{2} \quad (23)$$

Where, f_n is the normalized frequency and is defined as the ratio of switching frequency f_s to the resonant frequency f_r. To ensure ZCS, the duty ratio selected should be such that

$$D_{min} \leq D \leq D_{max} \quad (24)$$

The converter can be operated with ZCS for the designed input conditions and load variations if the selected duty cycle lies within this range.

C. RMS current through the switch

The rms current through the primary switch and the leakage inductance is given by

$$I_{sw,rms} = I_{Llk,rms} =$$

$$\left(I_{in}^2 * \left(\frac{1}{2} + f_n\left[\frac{11}{48} * \frac{1}{\pi(x - 0.5)} + \frac{43}{24} * \frac{1}{\pi} * \sin^{-1}\left(\frac{1}{2x - 1}\right)\right]\right)\right)^{1/2} \quad (25)$$

Where, x is the ratio of peak current to input current ($x = I_p/I_{in}$). The rms current through the switch is a function of the normalized frequency and the ratio of peak to input current.

D. Relationship between output and input voltage

Voltage gain of the converter is given by

$$M = \frac{V_0}{V_{in}} = \frac{n\eta}{\left(1 - f_n\left[1 + \frac{x}{4\pi} - \frac{1}{\pi}\sin^{-1}\left(\frac{x}{2}\right) + \frac{1 - \sqrt{1 - x^2/4}}{x\pi}\right]\right)} \quad (26)$$

For $x > 1.3$, (26) can be reduced as

$$M = \frac{V_o}{V_{in}} = \frac{n\eta}{1 - f_n} \quad (27)$$

Assuming $\eta = 1$, the voltage gain of the converter is a function of n and f_n as depicted in Fig. 4.

E. Selection of turns ratio of the transformer

From Fig. 4, it should be noticed that the voltage gain M depends on the turns ratio of the HF transformer and the normalized frequency f_n. These two parameters have a pronounced effect on the voltage and the current stress. It is therefore necessary to optimize these two parameters to minimize the conduction losses. Lower conduction losses in the primary circuit should be the quintessential criteria for selecting the turns ratio as the converter efficiency primarily depends on the losses in the switches on the primary side owing to high current.

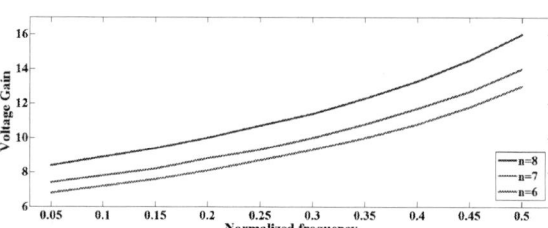

Fig. 4. Variation of voltage gain M with normalized frequency f_n.

Another parameter of paramount importance is duty ratio D. Current-fed converters allow higher duty ratio to reduce the transformer turns ratio for the same voltage gain. But reduction in n increases the switch voltage forcing to employ devices with high on-state resistance increasing the conduction losses. The duty ratio of CFPRPP converter has been kept low to confine the switch voltage to lower value. Even though there is a slight increase in the rms current, the switch on-state resistance of the latter gets halved when compared to the former reducing the conduction losses significantly.

Taking into account all the issues discussed above the turns ratio has been chosen as 1:1:6.5. Having chosen n, f_n can be computed from (27) for variations in the input voltage. For this design example, the switching frequency band is 82 kHz up to 200 kHz for V_{in} = 50 V down to 38 V under full load.

F. Condition for ZCS

The vital condition to be met for ZCS is $I_p > I_{in}$. In order to achieve ZCS for wide input voltage variations, the parameters of the resonant tank should be diligently designed. This imposes a constraint on the characteristic impedance at resonance Z_r and the constraint is given by

$$Z_r < \frac{2V_{in,\min} R_{FL}}{nV_o} \tag{28}$$

Where, $V_{in,min}$ is the minimum input voltage and R_{FL} is the resistance at full load condition. Z_r is a critical factor which determines the range of ZCS operation with input voltage. Smaller the Z_r wider is the range of ZCS operation and higher is the current stress as the ratio of I_p/I_{in} increases. Hence a trade-off between the range of operation and the current stress is compulsory for the selection of Z_r.

G. L and C values

The values of L and C constituting the tank circuit can be obtained by solving (10) and (11). For this particular design example Z_r was taken as 5.363 to achieve ZCS over the entire range of input voltage variation while keeping the current stress as low as possible. The values of L and C obtained are 1.5 µH and 2.6nF respectively.

H. Body diode conduction time

The duration for which the body diode of the primary switches conduct leading to ZCS is given by,

$$T_{DS} = \frac{\pi - 4\pi f_r T_{43}}{2\pi f_r} \tag{29}$$

IV. SIMULATION AND EXPERIMENTAL RESULTS

The proposed converter was simulated using PSIM 9.1 for the specifications and design values of components. The simulation results are presented in Fig. 5 for various operating conditions of input voltage and load. It should be observed from Fig. 5(a) and (b) that the simulation waveforms match closely with operating waveforms shown in Fig. 2. Effect of parallel resonance should be noticed. It has resulted into ZCS turn-off the devices and natural device voltage clamping for the given design range of input voltage and load variation.

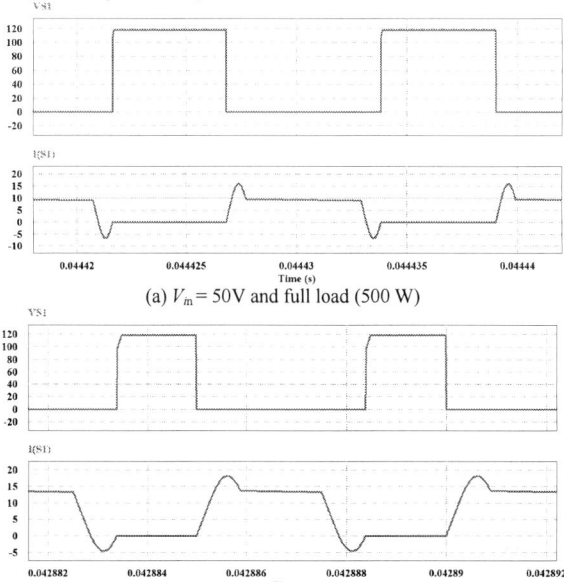

(a) V_{in} = 50V and full load (500 W)

(b) V_{in} = 38V and full load (500 W)

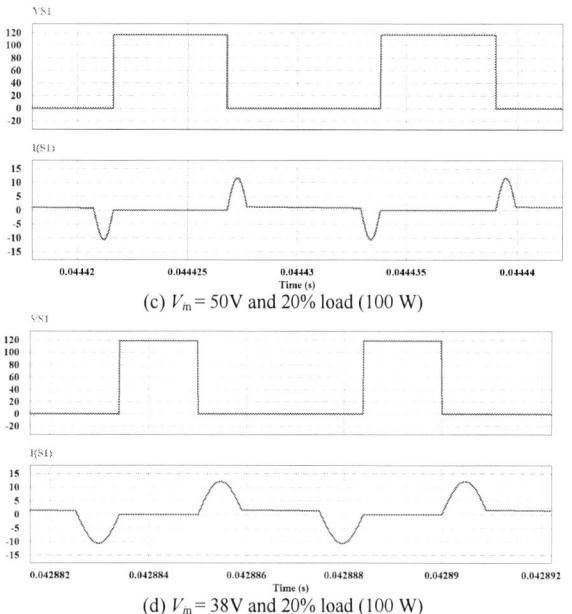

(c) V_{in} = 50V and 20% load (100 W)

(d) V_{in} = 38V and 20% load (100 W)

Fig. 5. Simulation results of the CFPRPP converter.

The developed laboratory prototype rated at 500W is shown in Fig. 6. The details of the prototype are listed in Table I. The gating signals were generated by a Xilinx Spartan-6 FPGA. Driver IR2181 was used for gate driver circuit. The leakage inductance of each primary winding is 796.4nH and 941nH. Therefore, to meet the design value, external inductors of 0.734µH and 0.540µH, respectively, were connected in series with the transformer primary windings as mentioned in Table I.

Experimental results for V_{in} = 50V at rated load (500W) are shown in Fig. 7. It should be observed that experimental results match closely with analytically predicted steady-state operating waveforms shown in Fig. 2. It verifies the steady-state operation and analysis.

Fig. 7(a) shows the input inductors' current waveforms. To meet the design requirement (L_b = 50 µH that could carry input current of 13.16 A), two PCV-2-104-10L (100 µH, 10 A) were used and connected in parallel. Two inductors share equal current ($I_{in}/2$) and are in phase. The inductors' current has low ripple and ripple frequency is 2x switching frequency.

Fig. 6. Experimental setup of the proposed CFPRPP converter.

TABLE I PARAMETERS OF THE LABORATORY PROTOTYPE

Components	Parameters
Boost Inductor L_b	PCV-2-104-10L; 100µH; 10.1A 2 inductors in parallel.
Primary switches $S_1 \sim S_2$	IRFB4127PbF; 200V; 76A; $R_{ds,on}$ = 17mΩ
External series inductances	TDK5901PC40Zcore, 0.734µH and 0.540µH
External parallel Capacitance C_p	2.4nF, 1000V ceramic capacitor
HF Transformer	TX55/32/18-3E27 ferrite core, Primary turns, N_1=4; secondary turns N_2=26; Leakage inductances referred to the primary =796.4nH and 941nH respectively; Parasitic capacitance in the secondary= 0.18nF
Input Capacitors	4.7 mF, 50V electrolytic & 0.68µF, 400V HF film capacitor
Output Capacitors	100 µF, 400V electrolytic capacitor & 2.2µF, 400V HF film capacitor

Fig. 7(b) shows the gate signal v_{gs1}, switch voltage v_{ds1}, and switch current i_{S1} waveforms of switch S_1. It should be noticed that switch current i_{S1} naturally decreases to zero due to resonance and later its body diode takes over leading to ZCS. The gate signal v_{gs1} is removed after switch current i_{S1} is zero resulting in its ZCS turn-off. Device voltage v_{ds1} rises after i_{S1} = 0 and v_{gs1} = 0 causing NVC without overshoot (voltage spike or ringing) and clamped at $2V_o/n$.

Fig. 7(c) presents the same for switch S_2. Similar ZCS, and NVC are obtained for S_2 180° after commutation of switch S_1. These waveforms (Fig. 7(b) and 7(c)) clearly demonstrate the expected performance and confirm the claims. Soft-switching of devices and natural device voltage clamping is obtained.

Fig. 7(d) presents the transformer primary voltage v_{AB} and the two primary currents (leakage inductance currents) i_{Llk1} and i_{Llk2}. Transformer voltage is bipolar showing two switch voltages and is balanced with no dc offset. The two primary currents are the same as corresponding switch currents shown in Fig. 7(b) and 7(c), which is clear from the circuit (Fig. 1) and operating waveforms (Fig. 2).

Fig. 7(e) shows the secondary side waveforms, i.e., current through and voltage across rectifier diodes. Zero-current turn-off of the diodes should be observed with diode voltage rising to output voltage V_o after diode commutation. Use of SiC Schottky diodes enabled smooth turn off with no reverse recovery and ringing.

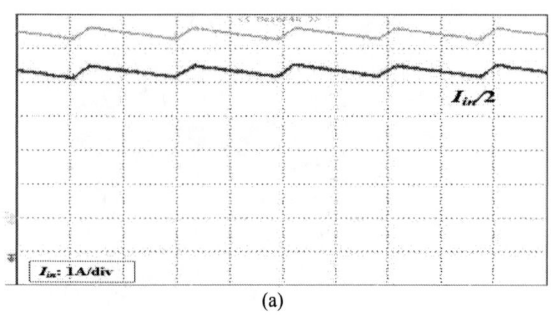

(a)

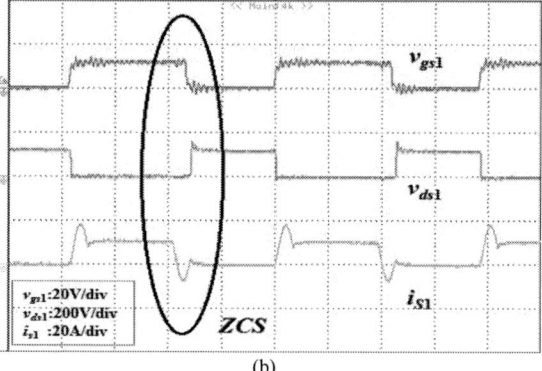

(b)

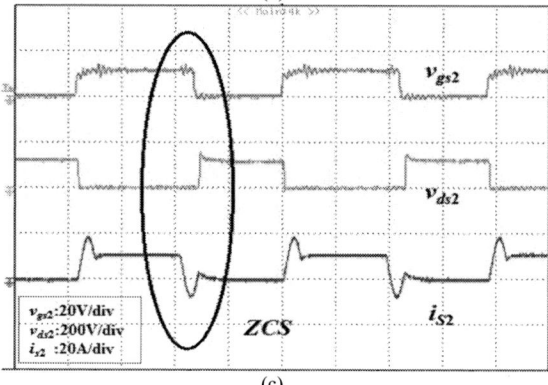

(c)

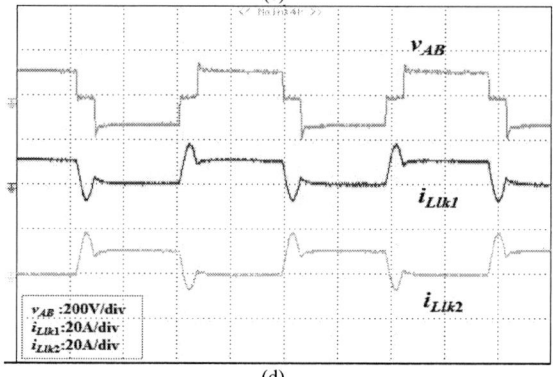

(d)

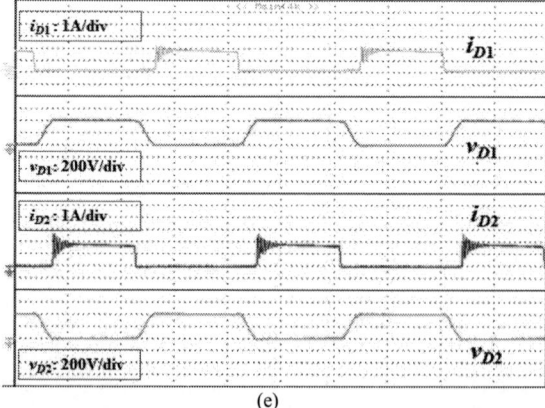

(e)

Fig. 7. Experimental results for V_{in} = 50 V at full load (x axis: 2µs/div). (a) Boost inductor current I_{in}, (b) Gate to source voltage v_{gs1} and drain to source voltage v_{ds1} of S_1 and current through switch S_1, (c) Gate to source voltage v_{gs2} and drain to source voltage v_{ds2} of S_2 and current through switch S_2, (d) Voltage v_{AB} across transformer primary and current through the leakage inductors L_{lk1} and L_{lk2}, (e) Voltage across diodes D_1 and D_2 and current through them.

The 2014 International Power Electronics Conference

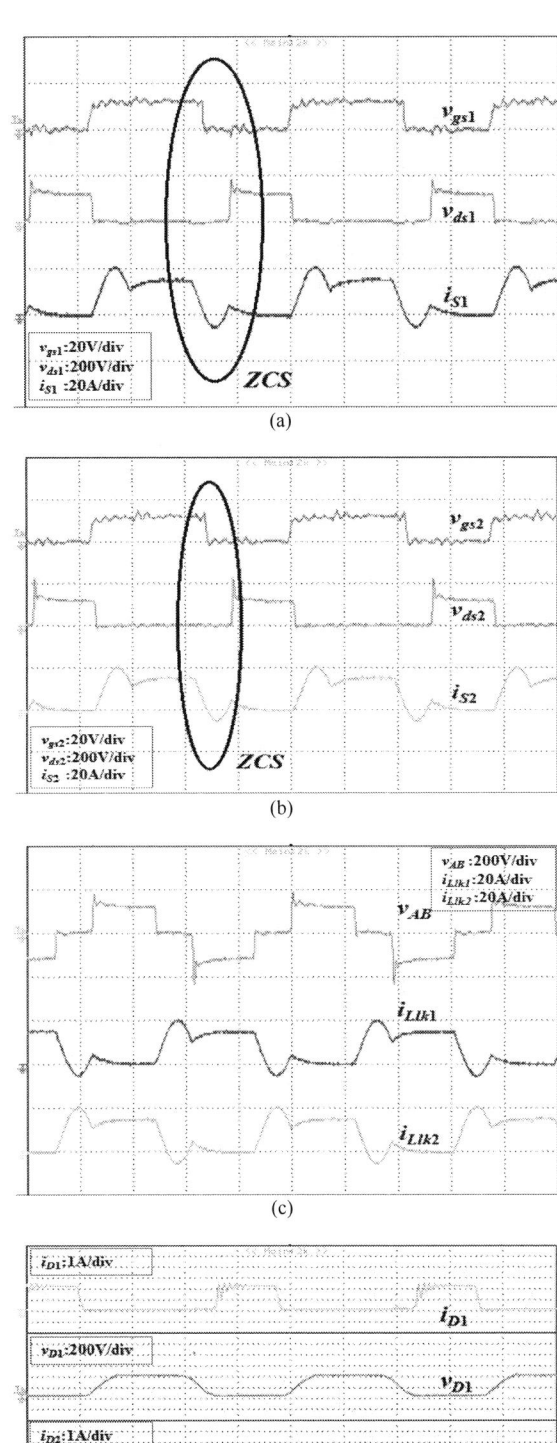

(a)

(b)

(c)

(d)

Fig. 8. Experimental results for V_{in} = 38 V at full load (x axis: 1μs/div). (a) Gate to source voltage v_{gs1} and drain to source voltage v_{ds1} of S_1 and current through switch S_1, (b) Gate to source voltage v_{gs2} and drain to source voltage v_{ds2} of S_2 and current through switch S_2, (c) Voltage v_{AB} across transformer primary and current through leakage inductors L_{lk1} and L_{lk2}, (d) Voltage across diodes D_1 and D_2 and current through them.

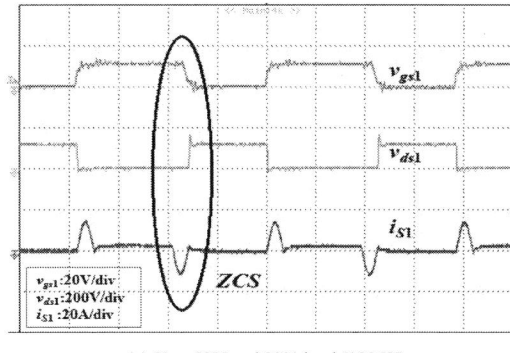

(a) V_{in} = 50V and 20% load (100 W)

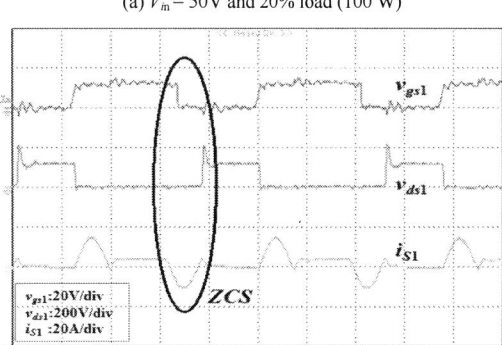

(b) V_{in} = 38V and 20% load (100 W)

Fig. 9. Experimental waveforms for light load conditions.

Experimental results for (1) V_{in} = 38 V, rated load (500 W) at f_s = 200 kHz, (2) V_{in} = 50 V, 20% load (100 W) at f_s = 77 kHz, (3) V_{in} = 38 V, 20% load (100 W) at f_s = 195 kHz are shown in Figs. 8, 9(a), and 9(b) respectively. Duty cycle is kept constant at D = 0.57.

It is clear that the attributes ZCS and NVC are preserved even under variable input voltage. It should be noted that under light load conditions, the ratio of I_p/I_{in} is higher than rated load. The same has also been noticed through simulation waveforms in Fig. 5. In addition, the switching frequency variation is not much sensitive to the load at fixed source voltage. The variation in device switching frequency is observed at input voltage variation mainly. With input voltage variation, ZCS and NVC are maintained by varying the device switching frequency. The claimed soft-switching and NVC of devices have been demonstrated and validated experimentally.

The variation in converter efficiency with load is plotted in Fig. 10 for two extreme input voltages. The measurements are taken and recorded on proof-of-concept circuit and therefore, the values can be improved with selection of suitable components, surface mount devices with low on-state resistance, paralleling of devices optimized PCB design, etc. The converter operates with the highest efficiency of about 94% for V_{in} = 50 V full load. Minimum efficiency of 87% is obtained at V_{in} = 38V 10% load (100W).

Loss distribution in the converter for 38 V and 50 V input voltages are shown in Fig. 11. At low input voltage, both the current through the primary circuit and the device switching frequency is high. At higher voltage, constant and circulating losses dominate.

978-1-4799-2706-7/14 $31.00 © 2014 IEEE

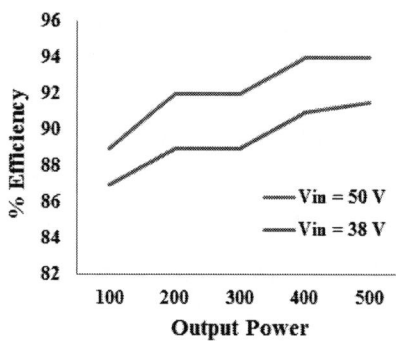

Fig. 10. Plot showing the variation in efficiency with output power for V_{in} = 50 V and V_{in} = 38 V.

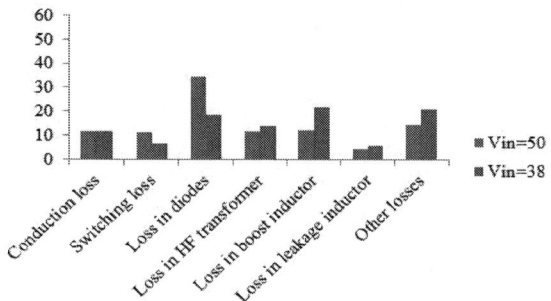

Fig. 11. Loss distribution in the converter for 50V and 38V input.

V. CONCLUSIONS

A novel ZCS current-fed parallel resonant push-pull (CFPRPP) converter has been proposed. The proposed topology can maintain ZCS for designed variation of input voltage and load variations. In addition, the device voltage is naturally clamped by ZCS using parallel resonance. Even though current-fed converter requires a large input inductor, it is reliable compared to large electrolytic capacitor in voltage-fed converter. In addition, it brings merits of reduction in the peak and the circulating currents for low voltage high current applications along with reduced transformer requirement and built-in short circuit protection. Low power experimental prototype has been tested to validate the claims, proposed operation and design. With high switching frequency and ZCS of the semiconductor devices, the proposed converter achieves higher power density. The proposed converter is suitable for applications that require high voltage gain with reduced number of switching devices.

REFERENCES

[1] P. Biczel, "Power electronic converters in DC microgrid," in *Proc. Compatibility in Power Electronics*, 2007, pp. 1-6.

[2] T.-F. Wu, Y.-C. Chen, J.-G. Yang, and C.-L. Kuo, "Isolated bidirectional full-bridge DC–DC converter with a flyback snubber," *IEEE Trans. Power Electron.*, vol. 25, no. 7, pp. 1915-1922, Jul. 2010.

[3] P. Xuewei and A. Rathore, "Novel Bidirectional Snubberless Naturally Commutated Soft-switching Current-fed Full-bridge Isolated DC/DC Converter for Fuel Cell Vehicles," *IEEE Trans. Industrial Electronics*, doi: 10.1109/TIE.2013.2271599.

[4] A. K. Rathore, A. K. S. Bhat and R. Oruganti, "A comparison of soft switched dc-dc converters for fuel-cell utility-interface

applications," in *Proc. Power Convers. Conf.*, Nagoya, Japan, Apr. 2007, pp. 588–594.

[5] M. Mohr, and F.-W. Fuchs, "Voltage fed and current fed full bridge converter for the use in three phase grid connected fuel cell systems," in *Proc. IEEE Int. Power Electron. Motion Control Conf.*, 2006, pp. 1-7.

[6] F. Krismer, J. Biela, and J. Kolar, "A comparative evaluation of isolated bi-directional DC/DC converters with wide input and output voltage range," *Fourtieth IAS Annual Meeting in Industrial Applications Conference*, 2005, vol. 1, pp. 599-606.

[7] H. Ma, L. Chen, and Z. Bai, "An active-clamping current-fed push-pull converter for vehicle inverter application and resonance analysis," *IEEE ISIE*, 2012, pp. 160-165.

[8] F. J. Nome, and I. Barbi, "A ZVS Clamping mode-current-fed push-pull dc-dc converter," in *Proc. ISIE*, 1998, vol. 2, pp. 617-621.

[9] A. K. Rathore, K. S. Bhat and R. Oruganti, "Analysis, design and experimental results of wide range ZVS active-clamped L-L type current-fed DC/DC converter for fuel cells to utility interface," *IEEE Trans. Ind. Electron.*, vol. 59, no. 1, pp. 473-483, Jan. 2012.

[10] A. K. Rathore and U. R. Prasanna, "Analysis, design, and experimental results of novel snubberless bidirectional naturally clamped ZCS/ZVS current-fed half-bridge dc/dc converter for fuel cell vehicles," *IEEE Trans. on Industrial Electronics*, vol. 60, no. 10, pp. 4482-4491, Oct. 2013.

[11] S. Jalbrzykowski, and T. Citko, "Current-fed resonant full-bridge boost DC/AC/DC converter," *IEEE Trans. Industrial Electronics*, vol. 55, no. 3, pp. 1198-1205, Mar. 2008.

[12] R. Chen, R. Lin, T. Liang, J. Chen, and K. Tseng, "Current-fed full-bridge boost converter with zero current switching for high voltage applications," *Fourtieth IAS Annual Meeting in Industrial Applications Conference*, 2005, vol. 3, pp. 2000-2006.

[13] C. Chakraborty, M.Ishida and T.Hori, "Performance, Design and Pulse Width Control of a CLL Resonant DC/DC Converter Operating at Constant Frequency in the Lagging Power Factor Mode," *IEEE- PEDS'99*, Hongkong, pp.767-772.

[14] D. J. Tschirhart and P. K. Jain, "A CLL resonant asymmetrical pulsewidth-modulated converter with improved efficiency," " *IEEE Trans. Industrial Electronics*, vol. 55, no. 1, pp. 114–122, 2008.

[15] J. Ying, Q. Zhu, H. Lin, and Z. Wu, "A zero-voltage-switching (ZVS) push-pull DC/DC converter for UPS," in *Proc. 5th Int. Power Electronics and Drive systems Conf.*, 2003, vol.2, pp. 1495-1499.

[16] C.-L. Chu, and C.-H. Li, "Analysis and design of a current-fed zero-voltage-switching and zero-current-switching CL-resonant push-pull dc-dc converter," *Power Electronics, IET*, vol. 2, no. 4, pp. 456-465, 2009.

[17] J. S. Glaser, and J. M. Rivas, "A 500 W push-pull dc-dc power converter with a 30 MHz switching frequency," in *Proc. IEEE APEC*, 2010, pp. 654-661.

[18] C. Chakraborty, M. Ishida, and T. Hori, "Performance and design of an L-C-L converter for voltage regulator type applications," *Trans. IEE Jpn.*, vol. 119-D, no. 6, pp. 848–856, 1999.

[19] R.-Y. Chen, T.-J. Liang, J.-F. Chen, R.-L. Lin, and K.-C. Tseng, "Study and implementation of a current-fed full-bridge boost DC–DC converter with zero-current switching for high-voltage applications," *IEEE Trans. Industry Applications*, vol. 44, no. 4, pp. 1218-1226, Jul./ Aug. 2008.

[20] I. Boonyaroonate, and S. Mori, "A new ZVCS resonant push-pull DC/DC converter topology," in *Proc. IEEE-APEC*, 2002, pp. 1097-1100.

[21] M. J. Ryan, W. E. Brumsickle, D. M. Divan, and R. D. Lorenz, "A new ZVS LCL-resonant push-pull DC-DC converter topology," *IEEE Trans. Industry Applications*, vol. 34, no. 5, pp. 1164-1174, Sept./Oct. 1998.

[22] W. Chen, Z. Lu, X. Zhang, and S. Ye, "A novel ZVS step-up push-pull type isolated LLC series resonant dc-dc converter for UPS systems and its topology variations," in *Proc. 23rd IEEE-APEC*, 2008, pp. 1073-1078.

[23] D. Edry, and S. Ben-Yaakov, "Capacitive-loaded push-pull parallel-resonant converter," *IEEE Trans. Aerospace and Electronic Systems*, vol. 29, no. 4, pp. 1287-1296, Oct. 1993.

Characteristics of Transmission Carrier in a New Wire Communication System by the Use of High-Ripple DC-DC Converter

Akihiko Katsuki
Next Generation Switching Power Circuit Course
Nagasaki University
1-14, Bunkyo-machi, Nagasaki, 852-8521 Japan
e-mail: katsuki@nagasaki-u.ac.jp

Kosuke Morita, Kazufumi Masutomo
Dept. of Computer Science and Electronics
Kyushu Institute of Technology
680-4, Kawazu, Iizuka-shi, Fukuoka, 820-8502 Japan
e-mail: katsuki-lab@cse.kyutech.ac.jp

Tatsuya Mizuki, Kohei Shibahara
Dept. of Computer Science and Electronics
Kyushu Institute of Technology
680-4, Kawazu, Iizuka-shi, Fukuoka, 820-8502 Japan
e-mail: katsuki-lab@cse.kyutech.ac.jp

Shigetaka Maeyama
Technical Center, TDK Corporation
2-15-7, Higashi-Ohwada, Ichikawa-shi, Chiba, 272-8558
Japan
e-mail: maeyama2@jp.tdk.com

Abstract— **In the newly proposed wire communication system, converters are utilized as transmitter as well as power supply. Signal transmission is performed by the use of output voltage ripple. At transmission of a message, the ripple is intentionally increased and modulated by the signal. In this paper, the output filter capacitor of the dc-dc converter is changed according to reception or transmission. At reception, output voltage ripple is made small enough. At transmission, output voltage ripple is enlarged and used as carrier. Output voltage ripple of a buck-type dc-dc converter having small capacitance of the output capacitor becomes almost sinusoidal, when the duty ratio is 0.5. From view point of impedance matching, the output impedance is investigated. Magnitude of ripple voltage is varied by several reasons. Therefore, digital FM (Frequency Modulation), for example, FSK (Frequency-Shift Keying) can be recommended.**

Keywords— *DC-DC converter, Ripple modulation, Transmission, Wire communication system*

I. INTRODUCTION

Most wire communication systems including telephone system and data communication system utilize communication lines as power lines. In the conventional systems, a main power unit supplies all the power consumed in input/output terminal devices [1]-[3]. These systems have common problems in reliability and adaptability to alterations in the system. If the main power supply stops operating, the whole system breaks down. Power rating of the main power-supply should be determined after consideration to the number of terminals. Excess of terminals over the estimated number may leads to an exchange of the main power supply to avoid system down.

To remove these disadvantages, we have proposed a novel wire communication system [4]. Main part of this system is constructed by the use of switched-mode dc-dc converters. A converter is utilized as power supply having the function of transmission. Signal transmission is performed by modulation for the output voltage ripple. These converters construct a parallel-connected power supply system.

With respect to the power capacity of total system, we assume that the total power rating of power supply system is selected to be more than the total consumed power. In this case, this combination makes a parallel redundant system. The system reliability and availability are much increased in comparison with conventional communication systems. Failure in small number of terminals leads to no system down. Repair or maintenance can be implemented with the system active. When we increase the number of powered terminals, the number of powering terminals should be increased. In this case, there is no need to worry about system down because of shortage in power.

Though the speed of signal transmission is not so high in comparison with optic fiber systems, metallic wire systems can deliver not only communication signals but also dc power. This system having simple configuration is suitable for controlling electronic equipment and so on.

In [4]-[6], signal transmission by the way of analog or digital AM (Amplitude Modulation) was discussed. In [7], characteristics of modulation and demodulation by digital FM (Frequency Modulation) were presented. In this paper, the output impedance and the others are investigated on a transmitter by the use of buck-type dc-dc converter.

II. NEW WIRE COMMUNICATION SYSTEM

Fig. 1 and Fig. 2 represent a configuration of the conventional system and that of a new system, respectively. In many applications, commercial ac is used as power source of the system. In Fig. 2, we introduce two types of terminals, that is, the powering terminal and the powered terminal.

The former is powered by commercial ac and supplies dc power to the communication lines. The powering terminal has a low-power switched-mode ac-dc converter. These converters construct a parallel-connected power supply system. For example, powering terminal can be used as built-in controller of equipment powered by commercial ac.

On the other hand, the latter is powered from the communication lines. The powered terminals are parallel-connected load. Powered terminal can be utilized as remote controller of above-mentioned equipment, built-in controller of device having sensor and the others.

III. TERMINALS FOR COMMUNICATION

With respect to the new wire communication system, Fig. 3 shows an example of operation in powering terminal. In powering terminals, the dc power is supplied to the communication lines. Each powering terminal has a power supply of low-power rating. Uninterruptible power supply (UPS) including battery should be placed against ac power failure. About switches, symbols "T" and "R" represent "Transmission" and "Reception," respectively. In the case of Fig. 3, communication signal is transmitted by FM.

Next, operation of powered terminal is introduced in Fig. 4. The dc power is received from the communication lines. Arrows with "DC" and "AC" show flow of the dc power-supply current and that of the ac communication signal, respectively.

At the state of standby, all terminals are at reception and are waiting a signal. Usually, one terminal becomes at transmission and the others are kept at reception, because controllers of the terminals give directions.

About communication signal, all terminals connected to communication lines are parallel-connected ac load. To prevent this impedance from decreasing, high input impedance against communication signal.

In the new wire communication system, bi-directional electronic choke [8] is desirable because the same circuit can be applicable for both the powering terminal and the powered terminal. Terminals 1-1' are connected to the communication lines. In the electronic choke, the ac communication signal is input to terminals 1-1' and output from terminals 3-3'. The dc power supply current flows from the terminal 2 to the terminal 1 in the powering terminal. On the other hand, dc current flows from the terminal 1 to the terminal 2 in the powered terminal.

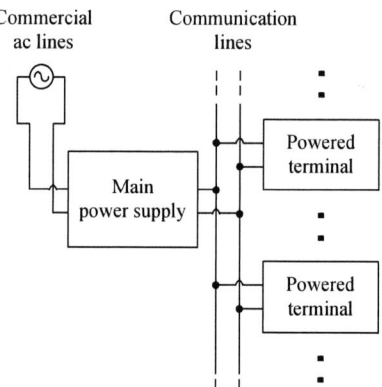

Fig. 1. Conventional wire communication system.

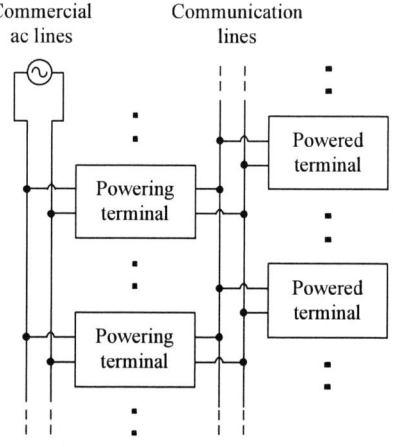

Fig. 2. Proposed wire communication system.

The 2014 International Power Electronics Conference

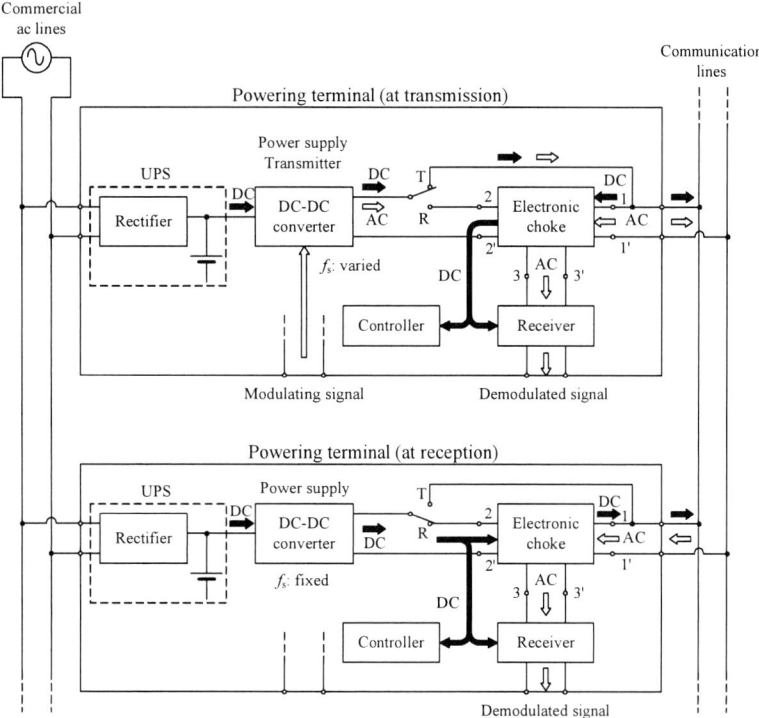

Fig. 3. Operation of powering terminal in the proposed wire communication system.

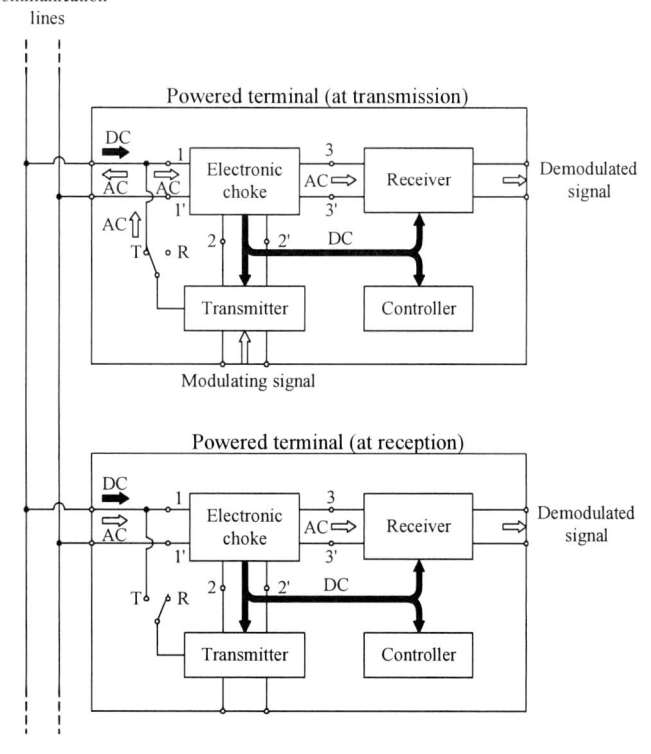

Fig. 4. Operation of powered terminal in the proposed wire communication system.

978-1-4799-2706-7/14 $31.00 © 2014 IEEE

The 2014 International Power Electronics Conference

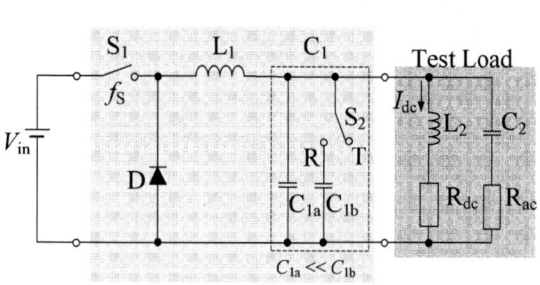

Fig. 5. Experimental circuit for transmitter.

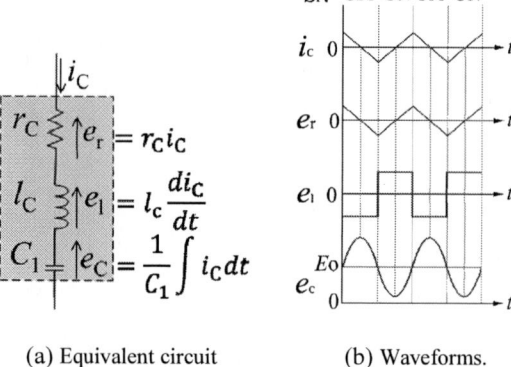

(a) Equivalent circuit
Of the capacitor C_1.

(b) Waveforms.

Fig. 6. Equivalent circuit and waveforms
on the capacitor C_1.

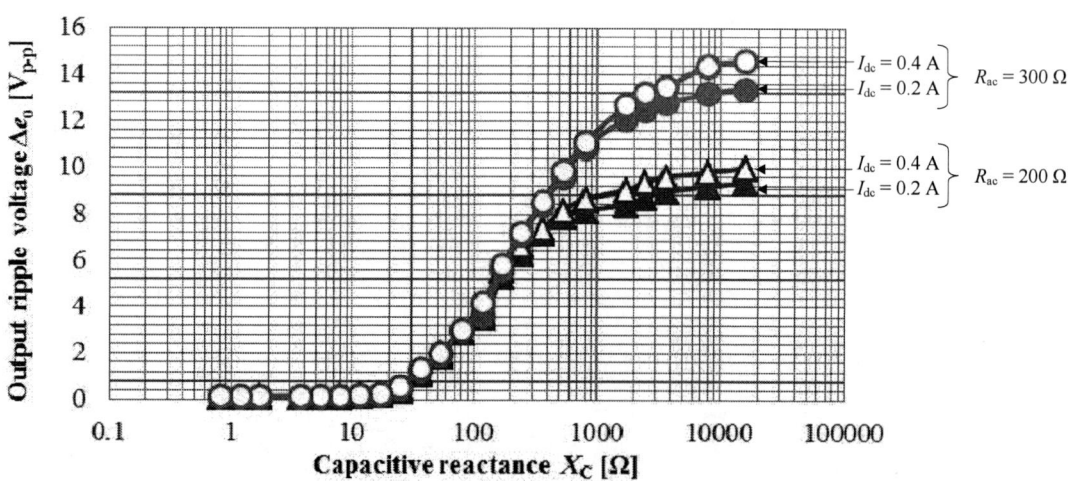

Fig. 7. Relations between the amplitude of output ripple voltage Δe_o
and the reactance X_C of capacitor C_1, where Δe_o is the voltage across
the resistor R_{ac}.

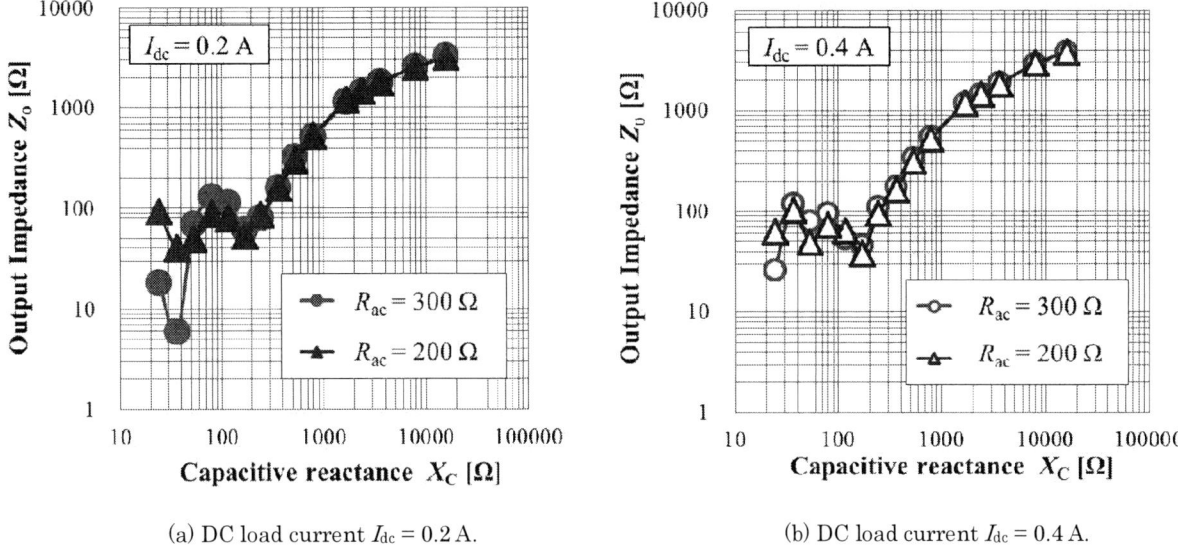

(a) DC load current $I_{dc} = 0.2$ A.

(b) DC load current $I_{dc} = 0.4$ A.

Fig. 8. Relations between the absolute value of output impedance Z_o and the reactance X_C of capacitor C_1.

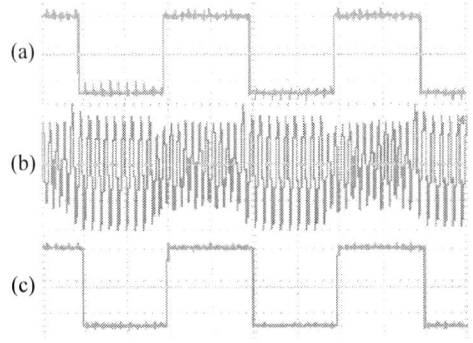

Fig. 9. Example of observed FSK waveforms; (a) Modulating signal, (b) FM wave, and (c) Demodulated signal. (Vertical: 5 V/div., Horizontal: 25 μs/div.)

IV. TRANSMITTER

Fig. 5 shows experimental circuit of transmitter. As transmitter, a buck-type dc-dc converter was used. At reception, the switch S_2 is connected to the capacitor C_{1b} of high capacitance. The output voltage ripple is small, since the total capacitance C_1 (= $C_{1a} + C_{1b}$) of filter capacitor is high, where $C_{1a} \ll C_{1b}$.

On the other hand, at transmission, the switch S_2 is open and the capacitance C_1 equals C_{1a}. The output ripple voltage is much increased. When C_1 is small, it should be noted that waveform of the output ripple voltage becomes near sinusoidal as shown in Fig. 6(a) and (b).

In order to transmit a digital modulating signal, we change the switching frequency f_S of the main switch S_1 between two values (i.e., Frequency-Shift Keying (FSK)). Actually, analog FM can be also possible.

V. EXPERIMENTAL RESULTS

A. Magnitude of output ripple voltage

Fig. 7 presents relations between the amplitude of output ripple voltage Δe_o and the reactance X_C of capacitor C_1. Ideally, $X_C = 1/(2\pi f_S C_1)$, where C_1 is the capacitance of C_1.

In this wire communication system, the ac load R_{ac} equals to the characteristic impedance of cable and then the parallel input impedance of terminals. The characteristic impedance is $50 \sim 75\ \Omega$ (in the coaxial cables), or $200 \sim 300\ \Omega$ (in parallel feeders). In our applications, parallel feeder is mainly used. The reason is that an alteration of wiring can be easily performed and the cost is low.

So as to obtain suitable magnitude of output ripple voltage at the range of $R_{ac} = 200 \sim 300\ \Omega$, the capacitance C_1 at transmission should be selected. In this experiment, $C_1 = 0.0015\ \mu\text{F}$ and $f_S = 200$ kHz, therefore X_C equals about $530\ \Omega$. At these conditions, waveform of the output ripple voltage is almost sinusoidal. According to the dc load current I_{dc}, the magnitude of output ripple voltage Δe_o varies a little. However, there exists no problem by employment of FM transmission.

With an extremely small capacitance C_1, at the right side of Fig. 7, the waveform of output ripple voltage Δe_o is widely different from sinusoidal wave.

On the other hand, the left side of Fig. 7 corresponds to the conditions at reception. For example, when $C_1 = 100\ \mu\text{F}$ and $f_S = 200$ kHz, X_C equals about $8\ \text{m}\Omega$. The output ripple voltage is very small. However, waveform of the ripple becomes non-sinusoidal.

B. Output impedance

Fig. 8 shows frequency characteristics on the absolute value of output impedance Z_o. Fig. 8(a) and (b) are the data when the dc load current $I_{dc} = 0.2$ A and 0.4 A, respectively. When the reactance X_C is about $50 \sim 500\ \Omega$, waveform of the output voltage ripple is almost sinusoidal. Out of this range in X_C, measurement errors increase.

We can see that the absolute value of output impedance Z_o is almost independent of the dc load current I_{dc} as well as the ac load R_{ac}. Other parameters are as follows: $L_1 = 2.2$ mH, $L_2 = 2.5$ H, $C_2 = 1.5\ \mu\text{F}$, $V_{in} = 48$ V.

C. Example of modulation by FSK

FSK is a method of digital modulation. Fig. 9 presents an example of observed waveforms on digital FM transmission and reception by FSK. In the circuit of Fig. 5, the switching frequency f_S was changed according to the value of digital modulating wave.

Fig. 9(a) is the modulating signal (a symmetrical square wave, the rate frequency: 10 kHz). Fig. 9(b) shows the FM wave which is the output ripple voltage (the voltage across the ac load resistor R_{ac}) of dc-dc converter. FSK modulation was performed under the condition of the mark frequency: 240 kHz and the space frequency: 200 kHz.

In Fig. 9(b), voltage variation with the rate frequency is observed. After filtering by high-pass filter and demodulating by PLL (Phase-Locked Loop) IC, we can obtain the demodulated signal as Fig. 9(c). The demodulated signal is similar to the modulating signal. The time lag is negligibly small.

VI. CONCLUSION

It was made clear that impedance matching between the output impedance of converter and the characteristic impedance of communication cable is possible mainly by suitable selection of the capacitance C_1 at transmission. The magnitude of output ripple voltage Δe_o can be adjusted also by the inductance L_1. By the use of FSK modulation, this circuit can be utilized for a digital signal transmitter.

REFERENCES

[1] J. Pest, "Circuit for Supplying Direct Current to a Telephone Line," *United States Patent*, 3649769, March 1972.

[2] P. H. Sutterlin, W. R. Bemiss, and G. M. Hey, "Data Communication Network Providing Power and Massage Information," *United States Patent*, 5148144, September 1992.

[3] H. Ott and H. Zierhut, "Transmission System," *European Patent*, 365696B1, February 1995. (in German)

[4] A. Katsuki, M. Matsushima, and N. Takimoto, "Wire Communication System Utilizing Ripple-Modulated DC-DC Converter as Signal Transmitter and Power Supply," *Proceedings of INTELEC'99 (The 21st International Telecommunications Energy Conference)*, No. 13-1, June 1999.

[5] A. Katsuki, S. Ogura, and M. Matsushima, "Ripple Characteristics against Inductive Load of a High-ESR-Filter-Capacitor DC-DC Converter Utilized in New Wire Communication System," *Proceedings of IPEC'00 (The 4th International Power Electronics Conference)*, pp. 108-117, April 2000.

[6] A. Katsuki and M. Matsushima, "Signal Transmission Characteristics of Inductively Loaded High-Ripple DC-DC Converter Used in a New Wire Communication System," *Proceedings of INTELEC'00 (The 22nd International Telecommunications Energy Conference)*, pp. 741-748, September 2000.

[7] A. Katsuki, T. Matsumoto, T. Eto, and Y. Hashimoto, "Digital-FM Transmission and Reception in a New Wire-Communication System That Utilizes DC-DC Converter as Transmitter," *Proceedings of INTELEC'03 (The 25th International Telecommunications Energy Conference)*, pp. 615-622, October 2003.

[8] A. Katsuki, T. Matsumoto, S. Watanabe, and M. Fukunaga, "Ripple Reduction Characteristics in the Parallel-Connected DC Power Distribution System Constructed by New Terminals for Wire Communication," *Proceedings of IPEC'05 (The 5th International Power Electronics Conference)*, pp. 2134-2141, April 2005.

5MHz PWM-Controlled Current-Mode Resonant DC-DC Converter Using GaN-FETs

Akinori Hariya/Nagasaki University
Graduate School of Engineering
Nagasaki Univ.
Nagasaki, Japan
bb52312202@cc.nagasaki-u.ac.jp

Ken Matsuura/TDK-Lambda Corporation
Advanced Development Dept.
TDK-Lambda
Nigata, Japan
matsuken@jp.tdk-lambda.com

Hiroshige Yanagi / TDK-Lambda Corporation
Advanced Development Dept.
TDK-Lambda
Nigata, Japan
yanagi@jp.tdk-lambda.com

Satoshi Tomioka / TDK-Lambda Corporation
Advanced Development Dept.
TDK-Lambda
Nigata, Japan
s.tomioka@jp.tdk-lambda.com

Yoichi Ishizuka/Nagasaki University
Graduate School of Engineering
Nagasaki Univ.
Nagasaki, Japan
isy2@nagasaki-u.ac.jp

Tamotsu Ninomiya /
The International Centre for the Study of
East Asian Development
Electronics Research Group for Sustainability
ICSEAD
Kitakyushu, Japan
t_ninomiya@icsead.or.jp

Abstract— In this paper, the method of the realization of a MHz level switching frequency DC-DC converter for high power-density is presented. For high power-density, Gallium Nitride field effect transistor (GaN-FET) and current-mode resonant DC-DC converter are adopted. In addition, the proposed pulse width modulation (PWM) control method which is suitable for the isolated current-mode resonant DC-DC converter operated at MHz level switching frequency, and the novel primary-side zero voltage switching (ZVS) turn on method for the proposed PWM control are presented.

Some experiments have been done with 5MHz isolated DC-DC converter which has GaN-FET, and the total volume of the circuit is 16.14cm^3. With the proposed PWM control method, input voltage range is 36-44V, and maximum load current range is 8A at V_i = 44V. The primary-side ZVS turn on is confirmed, and the maximum power-efficiency is 89.4%.

Keywords— High Switching Frequency, PWM Control, Current-Mode Resonant DC-DC Converter, GaN-FET

I. INTRODUCTION

Recently, high power-efficiency and high power-density DC-DC converters have been required in information and communication technology (ICT) equipment. Corresponding to the requirement of high power-density, the increase in the switching frequency of these converters has been considered to be one of key technologies. In particular, MHz level switching frequency contributes significantly to downsizing.

However, by adopting high switching frequency, power loss such as switching loss and gate driving loss increases. To solve this problem, the current-mode resonant DC-DC converter is effective because this converter can reduce switching loss. Also, GaN-FETs are suitable for high switching frequency operation as semiconductor switches because of low gate driving loss. Therefore, we have developed a 5MHz current-mode resonant DC-DC converter with GaN-FETs [1].

Current-mode resonant DC-DC converters are usually controlled by pulse frequency modulation (PFM). However, PFM control is hard to design to control the output voltage at MHz level switching frequency operation. The details are going to be described in the section III.

To solve the issue, this paper presents a novel PWM control method for the current-mode resonant DC-DC converter for MHz level switching frequency. This converter topology is same as the conventional current-mode resonant DC-DC converter with synchronous rectification. The feature of the converter is controlling the output voltage without any additional components. By using transformer's leakage inductance and secondary-side synchronously rectifying switches, the novel control method for boost conversion is realized.

In the previous researches, some PWM-controlled current-mode resonant DC-DC converters have been presented [2-4]. For example, the method of additional auxiliary circuits for regulating output voltage [2-3], and the method of controlling the duty ratio of primary-side switches [4] have been proposed. However, these methods need some additional components for regulating output voltage.

978-1-4799-2706-7/14 $31.00 © 2014 IEEE

On the other hand, the advantages of the proposed method are controlling secondary-side switches and no additional components.

To accomplish the primary-side ZVS turn on, small magnetizing inductance L_m have been used in current-mode resonant DC-DC converter generally. However, small L_m leads to increase of primary-side current which is cause of reducing power-efficiency. Therefore, to accomplish the ZVS operation in the proposed method, phase-shift between primary and secondary-side switches which control resonant current is adopted. As a result, this proposed converter maintains primary-side ZVS turn on.

The targets of the study are to obtain the high performance which is the small volume, 36-75V or 42-53V of input voltage range, 10A of maximum load current range, the realization of primary-side ZVS turn on, and high power-efficiency.

In the section II, the approach of the realization of DC-DC converter operated at MHz level switching frequency is described. In the section III, the issue of the conventional PFM-controlled current-mode resonant DC-DC converter in MHz level operation is revealed. In the section IV, the proposed PWM-controlled current-mode resonant DC-DC converter is explained. In the section V, the experimental results are revealed.

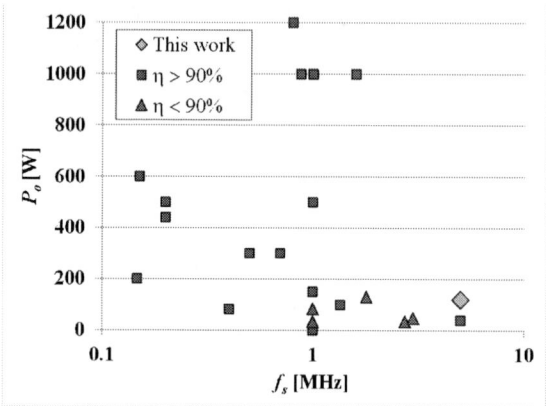

Fig. 1. The performance of isolated DC-DC converters in the previous papers.

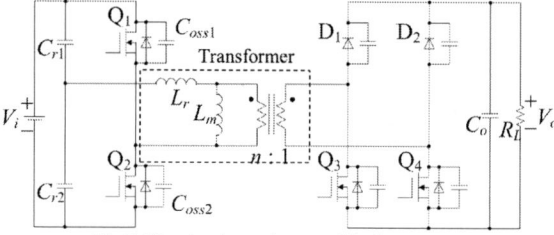

Fig 2. The circuit topology used in the paper.

II. THE APPROACH OF THE REALIZATION OF DC-DC CONVERTER OPERATED AT MHZ LEVEL SWITCHING FREQUENCY

For miniaturization of the DC-DC converter, the increase in the switching frequency of these converters is considered to be one of key technologies. Therefore, this study challenges 5MHz of switching frequency at 120W of output power. Studies that satisfy both the frequency and the output power have not been challenged so far, as shown in Fig. 1. To suppress increasing core power loss with high switching frequency, NiZn ferrite core is used in the study because this core material is suitable for high frequency operation [5]. In addition, to decrease parasitic inductance, multilayer printed circuit board (PCB) and planar transformer like [6], [7] is adopted.

However, by adopting high switching frequency, power loss such as switching loss and gate driving loss increases dramatically. The switching loss P_{sw} is expressed by

$$P_{sw} = C_{oss}V_{DS}^2 f_s \qquad (1)$$

where C_{oss} is output capacitance of FET, V_{DS} is drain-to-source voltage of FET, and f_s is switching frequency. The gate driving loss P_d is expressed by

$$P_d = Q_G V_{GS} f_s \qquad (2)$$

where Q_G is gate charge, and V_{GS} is gate-to-source voltage of FET. From eqs. (1) and (2), these power losses are proportional to f_s.

For reducing gate driving loss, GaN-FET is adopted in the study because GaN-FET has low Q_G and be driven at low gate-source voltage. From these reasons, GaN-FET can realize low P_d. Also, another researches have been proved utilizing GaN-FET for high switching frequency DC-DC converter is very practical [8-10]. Some literatures [11-13] prove that GaN devices are more effective than silicon devices.

However, because GaN-FETs have low gate-to-source threshold voltage V_{th}, low maximum gate-to-source voltage V_{GSS}, and high source-to-drain voltage V_{SD}, it is difficult to drive these FETs. Therefore, suitable drive circuit for GaN-FET is needed. Some literatures show that driver LM5113 is suitable for driving GaN-FETs [14], [15].

Furthermore, current-mode resonant DC-DC converter topology featuring primary-side ZVS turn on operation is adopted because switching loss can be suppressed. The circuit topology is shown in Fig. 2.

III. THE ISSUE OF THE CONVENTIONAL PFM-CONTROLLED CURRENT-MODE RESONANT DC-DC CONVERTER IN MHZ LEVEL OPERATION

To reduce the switching loss, the current-mode resonant DC-DC converter is widely used. Also, other researches have been proved using current-mode resonant DC-DC converter for high power-efficiency is very practical [16-19]. Generally, the current-mode resonant DC-DC converter is controlled by PFM control which varies switching frequency. PFM control has been valid for the control of the resonant DC-DC converters in kHz level switching operation so far.

To show the static characteristics, the definitions of the converter are

$$M = (2nV_o)/V_i , \qquad (3)$$
$$\kappa = L_m/L_r , \qquad (4)$$
$$F = f_s/f_0 , \qquad (5)$$

$$f_0 = 1 \Big/ \left(2\pi \sqrt{L_r C_r} \right), \qquad (6)$$

$$\omega_0 = 2\pi f_0, \qquad (7)$$

$$Q = Z_0 / R_{ac}, \qquad (8)$$

$$R_{ac} = \left(8n^2 R_L \right) \Big/ \pi^2 \qquad (9)$$

and

$$Z_0 = \sqrt{L_r / C_r} \qquad (10)$$

where turn ratio n, magnetizing inductance L_m, resonant inductance L_r, switching frequency f_s, resonant capacitance C_r and load resistance R_L.

From literatures, voltage conversion ratio M of PFM-controlled current-mode resonant DC-DC converter is written by

$$M = \cfrac{1}{\sqrt{\left(1 + \cfrac{1}{\kappa}\left(1 - \cfrac{1}{F^2} \right) \right)^2 + Q^2 \left(F - \cfrac{1}{F} \right)^2}} \qquad (11)$$

From eq. (11), the static characteristics of PFM controlled current-mode resonant DC-DC converter can be leaded. Two examples of the static characteristics are shown in Fig. 3. One is $V_i = 48\text{V}$, $n = 2.2$, $L_m = 200\text{nH}$, $L_r = 100\text{nH}$, $C_r = 10\text{nF}$, $Z_0 = 3.16$, $f_0 = 5.04\text{MHz}$ and $\kappa = 2$. The other is $V_i = 48\text{V}$, $n = 2.2$, $L_m = 200\text{nH}$, $L_r = 10\text{nH}$, $C_r = 80\text{nF}$, $Z_0 = 0.354$, $f_0 = 5.63\text{MHz}$ and $\kappa = 20$.

It is assumed that input voltage range is from 42 to 53V at $R_L = 1.2\Omega$ and $V_o = 12\text{V}$. In case of $L_r = 100\text{nH}$, for realization of the range, F is changed from 0.81 to 0.99 (f_s is changed from 4.06 to 4.99MHz). In contrast, in case of $L_r = 10\text{nH}$, for realization of the range, F is changed from 0.38 to 0.92 (f_s is changed from 2.16 to 5.17MHz).

The difference between two parameters can be confirmed. If switching frequency changes widely for controlling output voltage, the large noise filter will be needed. Therefore, the DC-DC converter is prevented from downsizing by large noise filter. If switching frequency changes narrow for controlling output voltage, the large resonant inductance is needed for controlling.

As a result, in MHz level operation for the miniaturization of the DC-DC converter, it is shown that PFM control is hard to be designed.

IV. THE PROPOSED PWM-CONTROLLED CURRENT-MODE RESONANT DC-DC CONVERTER

The proposed current-mode resonant DC-DC converter can be controlled at fixed switching frequency. In addition, for controlling output voltage, this method need not any additional components.

A. The Circuit Topology

The circuit topology is based on a half-bridge type current-mode resonant DC-DC converter as shown in Fig. 2. The primary-side is the half-bridge topology. Q_1 and Q_2 are driven in 50% duty ratio, alternatively. C_{oss1} and C_{oss2} are parasitic capacitance of primary-side switches

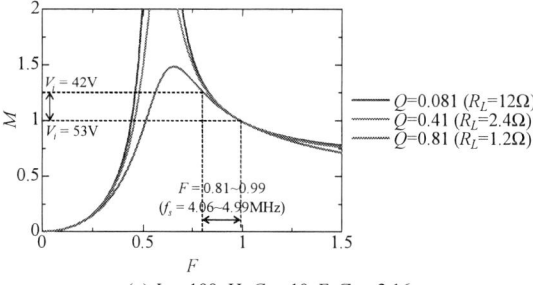

(a) $L_r = 100\text{nH}$, $C_r = 10\text{nF}$, $Z_0 = 3.16$.

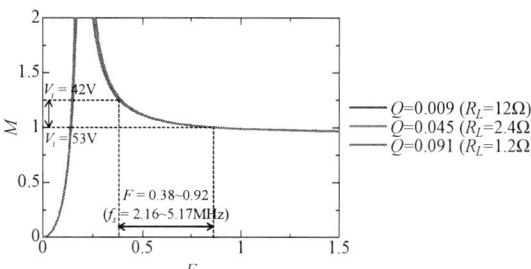

(b) $L_r = 10\text{nH}$, $C_r = 80\text{nF}$, $Z_0 = 0.354$.

Fig 3. The static characteristics of LLC resonant DC-DC converter with the conventional PFM control method.

Q_1 and Q_2. C_{r1} and C_{r2} are the resonant capacitors which have same capacitances and also make averaged voltage of v_c to a half of the input. The inside of the broken line is the magnetic transformer which equivalently indicated that L_r is leakage inductance, and L_m is the transformer's magnetizing inductance. The turn ratio is $n : 1$. L_r is used as the resonant inductance. The secondary-side is the full-bridge topology composed with diodes D_1 and D_2 for high-side arm switches, and transistors Q_3 and Q_4 for low-side arm switches.

B. The Principle of the Proposed PWM Control Method

To simplify analysis of the circuit operation, the following assumptions are made:

- FETs are treated an ideal switch;
- The body diodes of the primary-side FET are neglected;
- The output capacitances of the primary-side FETs are constant during operation, satisfying $C_{oss1} = C_{oss2}$, $C_{oss} = C_{oss1} + C_{oss2}$;
- Resonant capacitances are satisfied $C_{r1} = C_{r2}$, $C_r = C_{r1} + C_{r2}$;
- The forward voltage drop and the parasitic capacitance of the secondary-side diodes are neglected;
- The output capacitance and the body diodes of the secondary-side FETs are neglected;
- The output voltage is constant;

The output voltage can be controlled with changing the duty ratio of Q_3 and Q_4, simultaneously. When the duty ratio is less than 0.5, the circuit is operated as well as conventional current resonance circuit. When the duty ratio is more than 0.5, the circuit is operated in the proposed operation.

The circuit can be separated into 5 states in the proposed operation with the switch combination as shown in TABLE I. The operational waveforms are shown in Fig. 4. The equivalent circuits for each state of a half switching term are shown in Fig. 5. In this figure, the switches drawn with weak colors represent OFF, and red line represents current flow. The definitions of the duty ratio D is followed as

$$D = T_{on}/T_S \; , \tag{12}$$

$$D = 1/2 + D_1 + D_4 + D_5 \; , \tag{13}$$

$$D_1 + D_2 + D_3 + D_4 + D_5 = 1/2 \tag{14}$$

and

$$\begin{cases} D \le 0.5 \cdots\cdots\text{(conventional operation)} \\ D > 0.5 \cdots\cdots\text{(proposed operation)} \end{cases} \tag{15}$$

where T_s is the switching period, and T_{on} is the on-term of Q_3 and Q_4. $D_1 \sim D_5$ are the duty ratio of the each state. The

definitions of the resonant are followed;

$$\omega_1 = 1/\sqrt{(L_r + L_m)C_r} \; , \tag{16}$$

$$Z_1 = \sqrt{(L_r + L_m)/C_r} \tag{17}$$

and

$$\omega_{oss} = 1/\sqrt{L_r C_{oss}} \tag{18}$$

The definitions of the initial value of the variable are followed;

$$v_{c1}(0) = V_i/2 - V_c \tag{19}$$

and

$$i_{r1}(0) = I_r \tag{20}$$

The description for each state is described below.

State 1 $(0 < t < D_1 T_s)$:

In this state, t_1 is defined as $t_1 = t$. The primary-side switch Q_1 is turned ON. Also, the secondary-side switch-

TABLE I. CIRCUIT OPERATION STATES

State	FET				Diode	
	Q_1	Q_2	Q_3	Q_4	D_1	D_2
State 1	ON	OFF	ON	ON	OFF	OFF
State 2	ON	OFF	ON	OFF	OFF	ON
State 3	ON	OFF	ON	OFF	OFF	OFF
State 4	ON	OFF	ON	ON	OFF	OFF
State 5	OFF	OFF	ON	ON	OFF	OFF

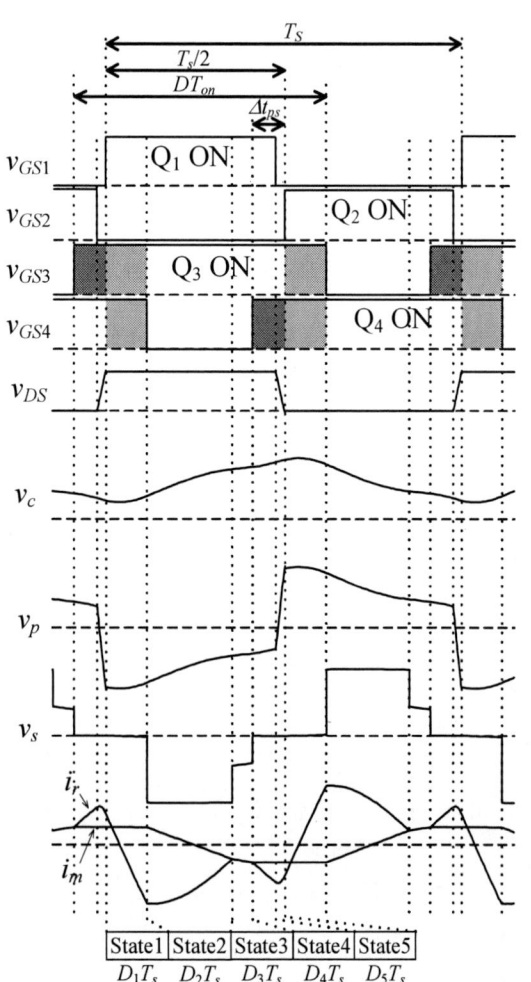

Fig. 4. The operational waveforms of the proposed PWM control.

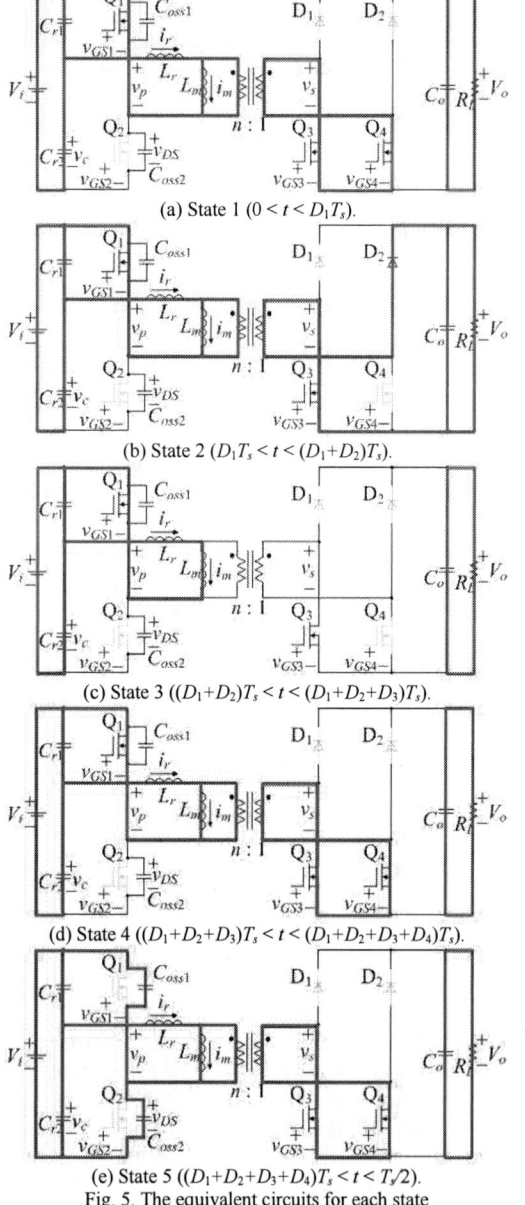

(a) State 1 $(0 < t < D_1 T_s)$.

(b) State 2 $(D_1 T_s < t < (D_1 + D_2)T_s)$.

(c) State 3 $((D_1 + D_2)T_s < t < (D_1 + D_2 + D_3)T_s)$.

(d) State 4 $((D_1 + D_2 + D_3)T_s < t < (D_1 + D_2 + D_3 + D_4)T_s)$.

(e) State 5 $((D_1 + D_2 + D_3 + D_4)T_s < t < T_s/2)$.

Fig. 5. The equivalent circuits for each state of the proposed PWM method.

978-1-4799-2706-7/14 $31.00 © 2014 IEEE

es both Q_3 and Q_4 are turned ON. Q_3 and Q_4 are overlapped as indicated with the light gray area in Fig. 4. The resonant inductance L_r is magnetized by i_r for boosting output voltage. From the figure, $v_{c1}(t_1)$ and $i_{r1}(t_1)$ are become

$$v_{c1}(t_1) = V_i - \{(V_i/2 + V_c)\cos(\omega_0 t_1) + Z_0 I_r \sin(\omega_0 t_1)\} \quad (21)$$

and

$$i_{r1}(t_1) = -1/Z_0 (V_i/2 + V_c)\sin(\omega_0 t_1) + I_r \cos(\omega_0 t_1). \quad (22)$$

The final values of the state 1 are

$$V_{c2} = v_{c1}(D_1 T_S) \quad (23)$$

and

$$I_{r2} = i_{r1}(D_1 T_S). \quad (24)$$

State 2 ($D_1 T_s < t < (D_1 + D_2) T_s$):

In this state, t_2 is defined as $t_2 = t - D_1 T_s$. After Q_4 is turned OFF, the direction of the voltage applied to D_2 is inverted, and diode of D_2 becomes ON. The inductance current which is magnetized in state 1 flow through diode D_2 and switch Q_3, to the load. From the figure, $v_{c2}(t_2)$ and $i_{r2}(t_2)$ are become

$$v_{c2}(t_2) = V_i - nV_o + (V_{c2} - V_i + nV_o)\cos(\omega_0 t_2) - Z_0 I_{r2} \sin(\omega_0 t_2) \quad (25)$$

and

$$i_{r2}(t_2) = 1/Z_0 (V_{c2} - V_i + nV_o)\sin(\omega_0 t_2) + I_{r2}\cos(\omega_0 t_2). \quad (26)$$

The final values of the state 2 are

$$V_{c3} = v_{c2}(D_2 T_S) \quad (27)$$

and

$$I_{r3} = i_{r2}(D_2 T_S). \quad (28)$$

State 3 ($(D_1+D_2)T_s < t < (D_1+D_2+D_3)T_s$):

In this state, t_3 is defined as $t_3 = t - D_1 T_s - D_2 T_s$. The direction of the diode D_2 current is inverted, and diode of D_2 becomes OFF. In the state, resonant current i_r equal to magnetizing current i_m. From the figure, $v_{c3}(t_3)$ and $i_{r3}(t_3)$ are become

$$v_{c3}(t_3) = V_i + (V_{c3} - V_i)\cos(\omega_1 t_3) - Z_1 I_{r3} \sin(\omega_1 t_3) \quad (29)$$

and

$$i_{r3}(t_3) = 1/Z_1 (V_{c3} - V_i)\sin(\omega_1 t_3) + I_{r3}\cos(\omega_1 t_3). \quad (30)$$

The final values of the state 3 are

$$V_{c4} = v_{c3}(D_3 T_S) \quad (31)$$

and

$$I_{r4} = i_{r3}(D_3 T_S). \quad (32)$$

State 4 ($(D_1+D_2+D_3)T_s < t < (D_1+D_2+D_3+D_4)T_s$):

In this state, t_4 is defined as $t_4 = t - D_1 T_s - D_2 T_s - D_3 T_s$. This state is similar to state 1. In the state, the resonant inductance L_r is magnetized by i_r for primary-side ZVS turn on. Q_3 and Q_4 are overlapped as indicated with the dark gray

area in Fig. 4. From the figure, $v_{c4}(t_4)$ and $i_{r4}(t_4)$ are become

$$v_{c4}(t_4) = V_i + (V_{c4} - V_i)\cos(\omega_0 t_4) - Z_0 I_{r4} \sin(\omega_0 t_4) \quad (33)$$

and

$$i_{r4}(t_4) = 1/Z_0 (V_{c4} - V_i)\sin(\omega_0 t_4) + I_{r4}\cos(\omega_0 t_4). \quad (34)$$

The final values of the state 4 are

$$V_{c5} = v_{c4}(D_4 T_S) \quad (35)$$

and

$$I_{r5} = i_{r4}(D_4 T_S). \quad (36)$$

State 5 ($(D_1+D_2+D_3+D_4)T_s < t < T_s/2$):

In this state, t_5 is defined as $t_5 = t - D_1 T_s - D_2 T_s - D_3 T_s - D_4 T_s$. The primary-side switch Q_1 is turned OFF. All primary-side switches are turned OFF, called dead-time. The parasitic capacitor C_{oss1} of Q_1 is discharged by a half of resonant inductance current i_r. Q_3 and Q_4 are overlapped as indicated with the dark gray area in Fig. 4. From the figure, $v_{c5}(t_5)$, $v_{DS5}(t_5)$ and $i_{r5}(t_5)$ are become

$$v_{c5}(t_5) = V_{c5} - \frac{\omega_0^2 (V_{c5} - V_i)}{\omega_0^2 + \omega_{oss}^2}$$
$$+ \frac{\omega_0^2 (V_{c5} - V_i)}{\omega_0^2 + \omega_{oss}^2}\cos\left(\sqrt{\omega_0^2 + \omega_{oss}^2} \cdot t_5\right) \quad ,$$
$$- \frac{I_{r5}}{C_r \sqrt{\omega_0^2 + \omega_{oss}^2}}\sin\left(\sqrt{\omega_0^2 + \omega_{oss}^2} \cdot t_5\right) \quad (37)$$

$$v_{DS5}(t_5) = \frac{\omega_{oss}^2 (V_{c5} - V_i)}{\omega_0^2 + \omega_{oss}^2} + V_i$$
$$- \frac{\omega_{oss}^2 (V_{c5} - V_i)}{\omega_0^2 + \omega_{oss}^2}\cos\left(\sqrt{\omega_0^2 + \omega_{oss}^2} \cdot t_5\right)$$
$$+ \frac{I_{r5}}{C_{oss}\sqrt{\omega_0^2 + \omega_{oss}^2}}\sin\left(\sqrt{\omega_0^2 + \omega_{oss}^2} \cdot t_5\right) \quad (38)$$

and

$$i_{r5}(t_5) = \frac{V_{c5} - V_i}{L_r \sqrt{\omega_0^2 + \omega_{oss}^2}} \cdot \sin\left(\sqrt{\omega_0^2 + \omega_{oss}^2} \cdot t_5\right)$$
$$+ I_{r5}\cos\left(\sqrt{\omega_0^2 + \omega_{oss}^2} \cdot t_5\right). \quad (39)$$

C. The Method for Achieving Primary-side ZVS Turn On in the Proposed PWM Control

To accomplish the primary-side ZVS turn on, small magnetizing inductance L_m have been used generally. However, small L_m leads to increase of primary-side current which is cause of reducing power-efficiency. Therefore, to accomplish the ZVS operation in the proposed converter, phase-shift between primary and secondary-side switches which control resonant current is adopted. As shown in Fig. 4, the proposed operation has phase-shift between primary and secondary-side switches. Δt_{ps} is the time length of the phase-shift. Without the phase-shift, Δt_{ps} = 0ns, the previous state of the dead-time becomes discontinuous current state. With this situation, initial current cannot be charged enough for ZVS because of secondary parasitic capacitance. Therefore, even with the long dead-time term, ZVS cannot be achieved. With the phase-shift, the problem of the initial current can be solved. From eqs. (34) and (36), the initial current of the dead-time, I_{r5} is

$$I_{r5} = 1/Z_0 (V_{c4} - V_i)\sin(\omega_0 D_4 T_S) + I_{r4} \cos(\omega_0 D_4 T_S). \quad (40)$$

The term of the phase-shift becomes

$$\Delta t_{ps} = D_4 T_S + D_5 T_S. \quad (41)$$

In eq. (41), as defined as the $D_5 T_s$ is fixed, the term of state 4 becomes larger with the increase of Δt_{ps}. I_{r5} is related to Δt_{ps} almost linearly. With the enough amount of I_{r5}, the amount of electrical charge q_{r5} which flows resonant inductance in dead-time becomes larger. For an example, in the conditions of C_{oss} = 900pF, V_i = 36V, R_L = 1.5ohm, $D_5 T_s$ = 5ns, the relation of Δt_{ps} and q_{r5} are shown in Fig. 6. The electrical charge q_{r5} can be calculated with

$$q_{r5} = \int_0^{D_5 T_S} i_{r5}(t)dt \quad (42)$$

For achieving ZVS, q_{r5} has to be larger than the amount of the electrical charge of the parasitic capacitance of the switch as

$$|q_{r5}| > q_{oss} \quad (43)$$

$$\text{where } q_{oss} = C_{oss} V_i \quad (44)$$

From the figure, it can be seen that the phase-shift Δt_{ps} between primary and secondary-side switches are valid for ZVS operation for the proposed PWM control.

D. The Advantage of the Proposed Method at MHz Level Operation

Similar to section III, two examples of the static characteristics of the proposed PWM controlled current-mode resonant DC-DC converter are shown in Fig. 7.

It is assumed that input voltage range is from 42 to 53V at R_L = 1.2Ω and V_o = 12V. In case of L_r= 100nH, for realization of the range, D is changed from 0.5 to 0.61. In contrast, in case of L_r= 10nH, for realization of the range, D is changed from 0.5 to 0.55. By comparing between two parameters, the case of L_r= 10nH can control in narrow duty ratio than the other one. Therefore, the proposed PWM control method is seem to be suitable for the miniaturization of the DC-DC converter because the output voltage can be controlled with small resonant inductance.

V. EXPERIMENTAL RESULTS

In this section, it can be confirmed the difference between the experimental results of the proposed method and the targets of the study.

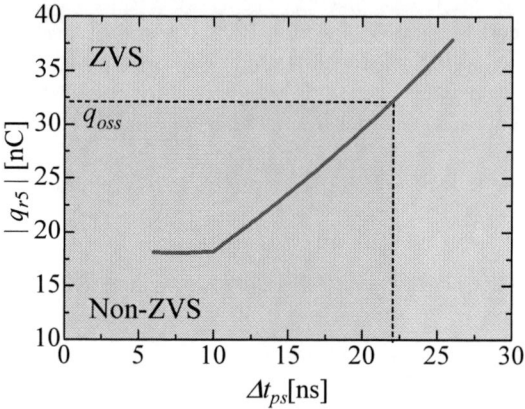

Fig 6. The effect of the phase-shift Δt_{ps} vs $|q_{r5}|$.

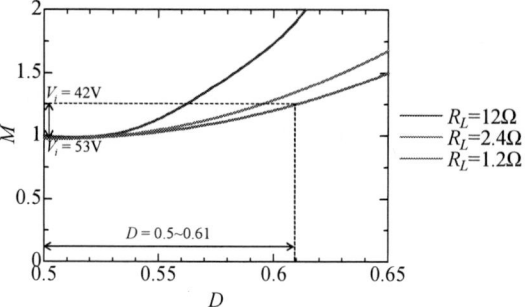

(a) L_r = 100nH, C_r = 10nF, Z_0 = 3.16.

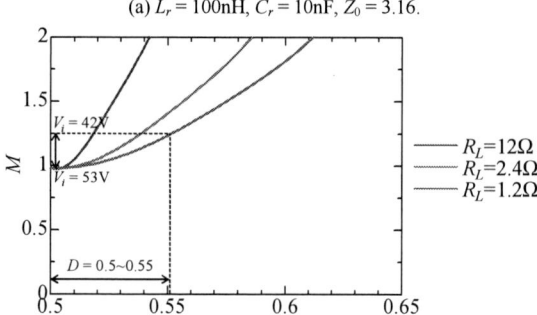

(b) L_r = 10nH, C_r = 80nF, Z_0 = 0.354.

Fig 7. The static characteristics of resonant DC-DC converter with the proposed PWM control method.

TABLE II. EXPERIMENTAL PARAMETERS

Specifications	Value
Output reference voltage: V_o	12V
Transformer ratio n : 1	2 : 1
Switching frequency: f_s	5MHz
Resonant frequency: f_r	4.98MHz
Transformer leakage inductance: L_r	33nH
Transformer magnetizing inductance: L_m	200nH
Resonant capacitor: C_{r1}, C_{r2}	15.5nF
Output capacitor: C_o	18.8μF

A. The Experimental Conditions

As a prototype digital controller, field programmable gate array (FPGA) Cyclone IV is used, which generates individual gate signal for each switch. The on-term of the gate signals are manually changed with software. The resolution of the gate signals is 1nano second.

Some experiments have been carried out with parameters as shown in TABLE II, and 12V of constant output voltage with open loop control. Components used in the experiment are shown in TABLE III.

The main circuit of the proposed PWM-controlled 5MHz DC-DC converter with GaN-FETs is shown in Fig. 8. The total volume of the main circuit of the proposed 5MHz DC-DC converter is 16.14cm^3.

B. The Performance of the Proposed Method

The open loop static characteristics of the 5MHz PWM-controlled DC-DC converter are exhibited as shown in Fig. 9. From the figure, it can be seen that voltage transfer ratio M is controlled by duty ratio D, and input voltage range can be achieved 36-44V. The difference between the performance of the proposed method and the targets can be confirmed. The reason is that the ideal transformer turn ratio n is 2.2, however, the actual n is 2. Planar transformer cannot be changed n easily because the winding is incorporate into the substrate.

The waveforms of $I_o = 0.89A$ and $I_o = 6.3A$ with the phase-shift are shown in Fig. 10. The blue bar of dead-time shows achieving primary-side ZVS turn on and Δt_{ps} is effective for primary-side ZVS operation.

The power-efficiency of the DC-DC converter at $V_o = 12V$ is shown in Fig. 11. The maximum power-efficiency is 89.4%, and maximum load current is 8A at $V_i = 44V$. Also, at low input voltage 44V or 36V, it can be seen that the power-efficiency is relatively low, and maximum load current is low. The reason is that large duty ratio is needed at low input voltage and large output current. The large duty ratio leads to large peak current which is cause of power dissipation. ZVS operation has been confirmed in the range of the experimental conditions.

The temperature distribution of the DC-DC converter as shown in Fig. 12, has been taken of the breadboard at $V_i = 44V$ and $I_o = 8A$. From the results, the temperature of the secondary-side is still in high level. The reason is that large duty ratio leads to the hard-switching of secondary-side switches, and large conduction loss of diodes is happened. On the other hand, the primary-side temperature is relatively low because of achieving ZVS turn on. Therefore, in the future, the hard-switching of secondary-side switches will be improved.

TABLE III. EXPERIMENTAL COMPONENTS

Name	Manufacture	Part Name/ Material
Primary side GaN-FET	EPC	EPC2001
Secondary side GaN-FET	EPC	EPC2015
FET Driver	TEXAS INSTRUMENTS	LM5113
Diode	DIODES	PDS1040L
Transformer Core Material	TDK	NiZn Ferrite Core
Resonant Capacitor	TDK	C1608C0G1H392J
Input Capacitor	TDK	C3216X7R1H105K
Output Capacitor	TDK	C2012X7R1E475M
FPGA	Terasic (ALTERA)	DE0-nano (Cyclone IV)
Isolator	TEXAS INSTRUMENTS	ISO722

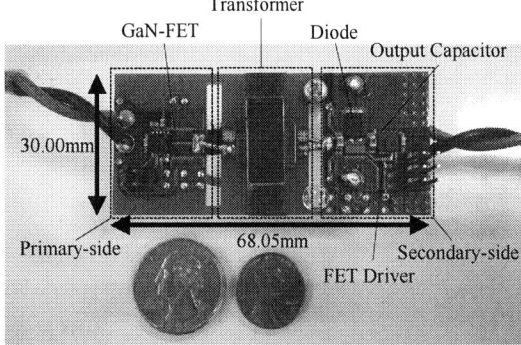

Fig. 8. The main circuit of the DC-DC converter.

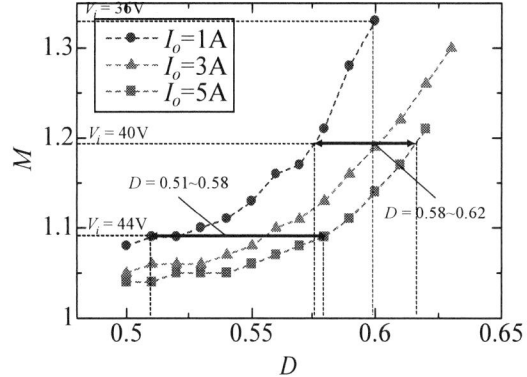

Fig. 9. The open loop static characteristics of resonant DC-DC converter with the proposed PWM control method.

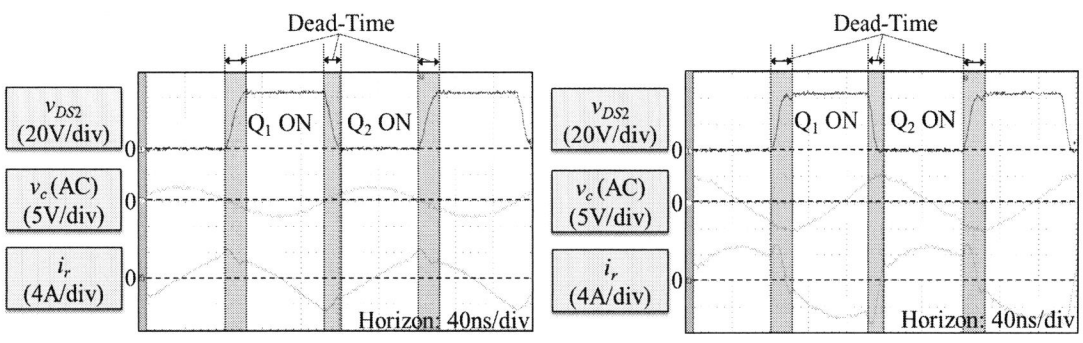

(a) $V_i = 44V$, $V_o = 12V$, $I_o = 0.89A$, $D = 0.53$, $\Delta t_{ps} = 3ns$.   (a) $V_i = 44V$, $V_o = 12V$, $I_o = 6.3A$, $D = 0.6$, $\Delta t_{ps} = 13ns$.

Fig. 10. Experimental waveforms of resonant DC-DC converter with the proposed PWM control method.

VI. CONCLUSIONS

In this paper, the method of the realization of a MHz level switching frequency DC-DC converter for high power-density is described. Furthermore, the novel PWM control method and achieving ZVS operation method for the current-mode resonant DC-DC converter in MHz level operation has been proposed.

The targets of the study is to obtain the high performance which is the small volume, 36-75V or 42-53V of input voltage range, 10A of maximum load current range, the realization of primary-side ZVS turn on, and high power-efficiency.

Some experiments have been done with 5MHz isolated DC-DC converter which has GaN-FET, and the total volume of the circuit is 16.14cm^3. With the proposed PWM control method, input voltage range is 36-44V, and maximum load current range is 8A at 44V. The primary-side ZVS turn on is confirmed, and the maximum power-efficiency is 89.4%.

As the future work, to accomplish the targets of the study, detailed analysis of the proposed method with phase-shift, realization of the wide control range and feedback control by digital controller are under considerations.

ACKNOWLEDGMENT

The authors would like to thank Mr. Shohei Iwasaki of Nagasaki University Technical staff, for him technical support.

REFERENCES

[1] K. Matsuura, H. Yanagi, S. Tomioka and T. Ninomiya, "Power-Density Development of a 5MHz-Switching DC-DC Converter," IEEE APEC 2012, pp.2326-2332.

[2] T. Zaitsu, T. Ninomiya, M. Shoyama, and H. Tanaka, "PWM-Controlled Current-Mode Resonant Converter Using an Active-Clamp Technique," IEEE PESC1996, pp.89-93.

[3] W. J. Lee, S. W. Choi, C. E. Kim, and G. W. Moon, "A New PWM-Controlled Quasi-Resonant Converter for a High Efficiency PDP Sustaining Power Module," IEEE Trans. Power Electron., vol. 23, no.4, 2008, pp.1782-1790.

[4] X. Xu, A. M. Khambadkone, T. M. Leong, and R. Oruganti, "A 1-MHz Zero-Voltage-Switching Asymmetrical Half-Bridge DC/DC Converter: Analysis and Design," IEEE Trans. Power Electron., vol. 21, no. 1, 2006, pp. 105-113.

[5] Q. Li, M. Lim, J. Sun, A. Ball, Y. Ying, F. C. Lee, and K. D. T. Ngo, "Technology Road Map for High Frequency Integrated DC-DC Converter," IEEE APEC 2010, pp. 533-539.

[6] N. Huapeng, P. Yunqing, Y. Xu, W. Laili, and W. Zhaoan, "Design of High Power Density DC Bus Converter Based on LLC Resonant Converter with Synchronous Rectifier," IEEE IPEMC 2009, pp. 540-543.

[7] Z. Gong, Q. Chen, X. Yang, B. Yuan, W. Feng, and Z. Wang, "Design of High Power Density DC-DC Converter Based on Embedded Passive Substrate," IEEE PESC 2008, pp. 273-277.

[8] S. Ji, D. Reusch and F. C. Lee, "High-Frequency High Power Density 3-D Integrated Gallium-Nitride-Based Point of Load Module Design," IEEE Trans. Power Electron., vol. 28, no. 9, 2013, pp.4216-4226.

[9] J. Delaine, P. O. Jeannin, D. Frey and K. Guepratte, "Improvement of GaN Transistors Working Conditions to Increase Efficiency of A 100W DC-DC Converter," IEEE APEC 2013, pp.656-663.

[10] Y. Wang, W. Kim, Z. Zhang, J. Calata, and K. D. T Ngo, "Experience with 1 to 3 Megahertz Power Conversion Using eGaN FETs," IEEE APEC 2013, pp.532-539.

[11] W. Zhang, Y. Long, Y. Cui, F. Wang, L. M. Tolbert, B. J. Blalock, S. Henning, J. Moses, and R. Dean, "Impact of Planner Transformer Winding Capacitance on Si-based and GaN-based LLC Resonant Converter," IEEE APEC 2013, pp.1668-1674.

[12] W. Zhang, Y. Long, Z. Zhang, F. Wang, L. M. Tolbert, B. J. Blalock, S. Henning, C. Wilson, and R. Dean, "Evaluation and Comparison of Silicon and Gallium Nitride Power Transistors in LLC Resonant Converter," IEEE ECCE 2012, pp. 1362-1366.

[13] D. Reusch, and F. C. Lee, "High Frequency Isolated Bus Converter with Gallium Nitride Transistors and Integrated Transformer," IEEE ECCE 2012, pp. 3895-3902.

[14] Y. Xi, M. Chen, K. Nielson, and R. Bell, "Optimization of the Drive Circuit for Enhancement Mode Power GaN FETs in DC-DC Converters," IEEE APEC 2012, pp.2467-2471.

[15] "5A, 100V Half-Bridge Gate Driver for Enhancement Mode GaN FETs," Texas Instruments Datasheet.

[16] C. Oeder, and T. Duerbaum, "ZVS Investigation of LLC Converters Based on FHA Assumptions," IEEE APEC 2013, pp. 2643-2648.

[17] R. L. Steigerwald, "A Comparison of Half-Bridge Resonant Converter Topologies," IEEE Trans. Power Electron., vol. 3, no. 2. 1988, pp. 174-182.

[18] S. Yang, M. Shoyama, and S. Abe, "Design of Low-Profile LLC Resonant Converter for Low Transformer Loss," IEEE TENCON 2010, pp. 1301-1306.

[19] B. Lu, W. Liu, Y. Liang, F. C. Lee, and J. D. Van Wyk, "Optimal Design Methodology for LLC Resonant Converter," IEEE APEC 2006, pp. 533-538.

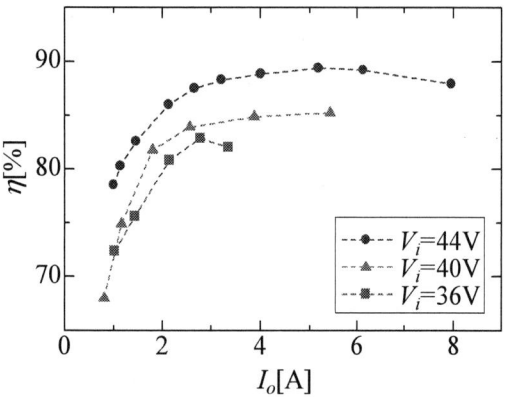

Fig. 11. The power-efficiency of resonant DC-DC converter with the proposed PWM control method at V_o = 12V.

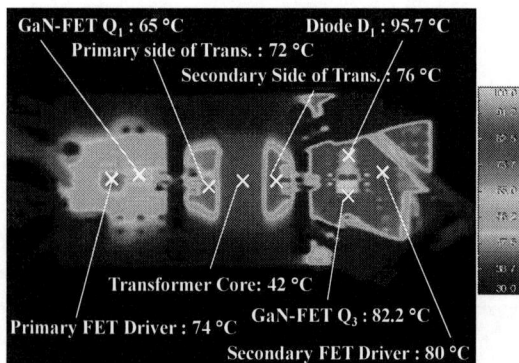

Fig. 12. The temperature distribution of resonant DC-DC converter with the proposed PWM control method at V_i = 44V, V_o = 12V, I_o = 8A.

Design and Performance Evaluation of Digital Control for LLC Series Resonant Dc-to-Dc Converters

Syam Kumar Pidaparthy and Byungcho Choi
School of Electronics Engineering, Kyungpook National University
Daegu, South Korea
syamkumar.537@gmail.com

Jinhaeng Jang
LG Electronics
Pyeongtaek, South Korea
jinhaeng.jang@lge.com

Abstract—**This paper presents the design and performance evaluation of the digital control adapted to an LLC series resonant dc-to-dc converter operating with wide input and load variations. The main theme of this paper is to highlight the advantage of using the push-pull mode of digital pulse width modulation (DPWM) operation over the complementary mode operation for the implementation of digital control to the resonant converters using a low cost digital signal controller (DSC). Conventionally, the complementary mode of DPWM operation is used for resonant converter. However, the push-pull mode of DPWM operation is chosen for the resonant converter circuit to improve closed-loop performance. Design procedures and performance evaluation are illustrated for both DPWM operational modes using an experimental 150 W LLC converter.**

Keywords—*Complementary mode operation, control design, digital control, digital pulse width modulation (DPWM), push-pull mode operation, LLC series resonant dc-to-dc converter, performance evaluation.*

I. INTRODUCTION

Digital control has emerged as a viable alternative to the conventional analog control of PWM and resonant dc-to-dc converters. The digital control of PWM converters has been extensively studied in past researches, yielding many useful results [1]-[5]. In comparison, the application potential of the digital control to resonant converters [6] is not fully explored yet.

In [6], the selection of required frequency resolution of the hardware is given only for the series-resonant parallel-loaded converter to eliminate limit-cycles. But, it is not applicable for the LLC series resonant converter due to lack of proper modeling technique, which adds to more difficulties in the implementation of digital control for the system. And also, [6] demonstrates only the resolution effects on limit cycles, mainly due to the analog-to-digital converter (ADC) and DPWM, yet it needs to address the quantization effects due to the compensator and time delay problem for the resonant converter. The more time delay in the system can degrade the closed-loop performance. In practice, the system requires higher ADC sampling rates which reduces the time delay, however, we need to consider the cost of hardware implementation and its computational time delay. For a low cost DSCs like dsPIC33FJ16GS [12], the ADC sampling frequency is limited to the switching frequency due to the higher computational delay.

In practice, the complementary mode operation [6], [12] is used for the resonant converter, where the ADC sampling frequency is equal to the switching frequency. However, to improve the ADC sampling frequency, the push-pull mode of DPWM operation is selected to this application circuit. Accordingly, the push-pull mode has the ADC sampling rate as two times the switching frequency. The increment in the ADC sampling rate notably decreases the time delay and improves the transient response characteristics. But, it decreases the frequency resolution to half of the resolution of complementary mode operation that causes limit cycle oscillations. However, it will be shown that with sufficient frequency resolutions, the push-pull mode of operation offers good closed-loop performance than that of the complementary mode operation.

Challenges to the digital control of resonant converters include variations in both the small-signal gain of a counter-based digitally controlled oscillator (DCO) and small-signal dynamics of power stage. The small-signal DCO gain varies as a quadratic function of the switching frequency. The power stage dynamics are sensitively affected by operational conditions [8]-[11].

This paper addresses the design and performance evaluation of the digital control for LLC series resonant dc-to-dc converters. This paper adopts the standard emulation method for digital compensation design. The effect of ADC sampling frequency, the quantization effect of compensator, and time delay problem are analyzed in a qualitative manner by comparing the closed-loop performances of both complementary and push-pull DPWM operational modes. The design procedures and performance evaluation are illustrated using an experimental 150 W LLC series resonant converter. This paper will show that the digital control could offer the performance that favorably compares with the best case of the conventional analog control when the variations in DCO gain and power stage dynamics are all correctly incorporated into

---

This research was supported by the MSIP (Ministry of Science, ICT & Future Planning), Korea, under the C-ITRC (Convergence Information Technology Research Center) support program (NIPA-2013-H0401-13-1005) supervised by the NIPA(National IT Industry Promotion Agency).

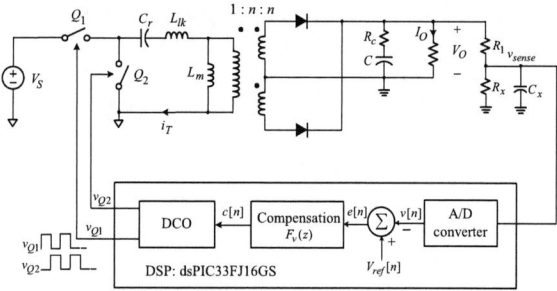

Fig. 1. Experimental 150 W LLC series resonant dc-to-dc converter with digital control: $V_S = 340 - 390$ V, $V_O = 24$ V, $I_O = 1 - 6$ A, $C_r = 47$nF, $L_{lk} = 160\mu$H, $L_m = 1.24$mH, $n = 0.14$, $C = 2$ mF, $R_c = 6$ mΩ, $R_1 = 1.75$ kΩ, $R_2 = 250$ Ω, $C_x = 2.2$ nF. The ADC resolution is 3.22 mV and the DCO counter period is 1.04 ns.

the design.

II. DIGITALLY-CONTROLLED LLC CONVERTER AND COMPENSATION DESIGN

A. Converter Configuration

Fig. 1 is a simplified circuit diagram of the experimental LLC converter employing a digital control. As a typical case of off-line power supplies, the converter receives the input voltage of $V_S = 340 - 390$ V and produces $V_O = 24$ V output at the load current of $I_O = 1 - 6$ A. The output voltage is sensed by an anti-aliasing low-pass filter and fed to a 16-bit DSP: dsPIC33FJ16GS from Microchip [12]. The 10-bit ADC inside the DSP accepts the sensed output voltage in the range of $v_{sense} = 0 - 3.3$ V. The digital compensation $F_v(z)$ receives the error signal $e[n]$ and provides the input signal $c[n]$ for the DCO. The DCO operates with a counter-based DPWM scheme [4]. The DCO outputs derive two MOSFET switches, Q_1 and Q_2, in a push-pull mode of operation [12].

B. Power Stage Dynamics

The earlier studies [8]-[11] proved that small-signal dynamics of LLC converters vary substantially as operational conditions are altered. The major result of these analyses is shown in Fig. 2 in conjunction with the small-signal dynamics of the experimental LLC converter. Fig. 2(a) depicts the pole/zero trajectory of the frequency-to-output transfer function with respect to the input voltage variation. The trajectory portraits the locations of the poles and zero as the input voltage decreases from $V_S = 390$ V, represented by **Point B** in Fig. 2, to $V_S = 340$ V, denoted as **Point A**, while delivering the load current of $I_O = 6$ A. The detailed analysis for the locations of poles and zero for the points A and B are given in [8]-[11].

Fig. 2(b) displays the measured Bode plot of the control-to-output transfer function, G_{vc}, at these operating points. The transfer function exhibits notable changes in the exact same manner as predicted from the pole/zero trajectory in Fig. 2(a). It can be observed that point B

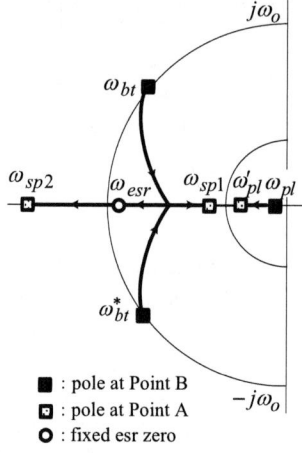

■ : pole at Point B
□ : pole at Point A
O : fixed esr zero

(a)

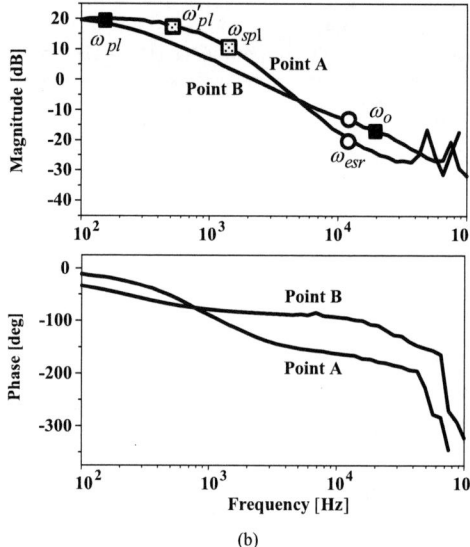

(b)

Fig. 2. Power stage dynamics of experimental LLC converter. (a) Pole/zero trajectory of frequency-to-output transfer function. (b) Bode plot of control-to-output transfer function: $\omega_o = |\omega_{bt}| = |\omega_{bt}^*|$.

exhibits first-order system behavior. Whereas point A illustrates the second-order behavior with rapid-declining phase characteristics, with the presence of the two neighboring low-frequency poles, ω_{pl}' and ω_{sp1}. Accordingly, point A exhibits the worst small-signal dynamics, thus being a logical target for the compensation design. From Fig. 2(b), the control-to-output transfer function at point A can be approximated up to mid-frequencies as

$$G_{vc}(s)|_A = K_{vco}K_{vf_A}\frac{1 + \dfrac{s}{\omega_{esr}}}{\left(1 + \dfrac{s}{\omega_{pl}'}\right)\left(1 + \dfrac{s}{\omega_{sp1}}\right)} \quad (1)$$

where K_{vco} is the gain of voltage controlled oscillator (VCO) at point A, $K_{vco} = 2\pi \cdot 6.10 \times 10^4$ (rad/s)/V, K_{vf_A} is the low-frequency gain of the frequency-to-output transfer function at point A. It is determined from

The 2014 International Power Electronics Conference

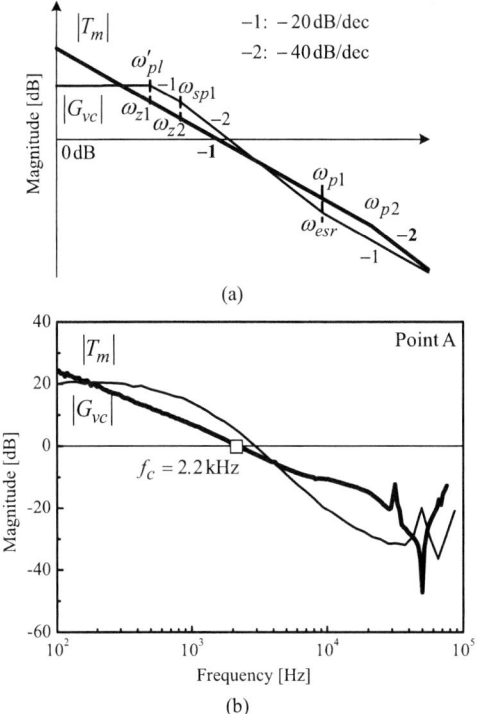

(a)

(b)

Fig. 3. Digital voltage feedback compensation design and loop gain at point A. (a) Design strategy. (b) Design verification.

the Fig. 2(b) as

$$20\,log(K_{vco}K_{vf_A}) = 20 \text{ dB}$$
$$\Rightarrow K_{vf_A} = 2.62 \times 10^{-5} \text{ V/(rad/s)} \quad (2)$$

the esr zero is calculated as, $\omega_{esr} = 1/(CR_c) = 0.75 \times 10^5$ rad/s. The location of the low-frequency poles, ω'_{pl} and ω_{sp1}, are estimated as 5×10^3 rad/s and 7.0×10^3 rad/s respectively from Fig. 2(b).

C. Analog Compensation Formulation

An analog compensation is initially formulated and later converted to the digital compensation using the emulation method. One practical design strategy at the presence of variations in small-signal dynamics is to calibrate the compensation design for the worst-case operational condition. As illustrated in the above, the compensation is formulated on basis of G_{vc} at point A. An earlier publication [8] showed that a three-pole two-zero compensation provides good dynamic performance for the entire operational region

$$F_v(s) = \frac{K_v(1 + s/\omega_{z1})(1 + s/\omega_{z2})}{s(1 + s/\omega_{p1})(1 + s/\omega_{p2})} \quad (3)$$

Fig. 3(a) shows asymptotic plots for the control-to-output transfer function $|G_{vc}|$ and the loop gain, $|T_m|$ at point A. Fig. 3(a) is constructed with the following design procedure

- The compensation zeros are located at the low-frequency poles, $\omega_{z1} = \omega'_{pl} = 5 \times 10^3$ rad/s and $\omega_{z2} = \omega_{sp1} = 7 \times 10^3$ rad/s.

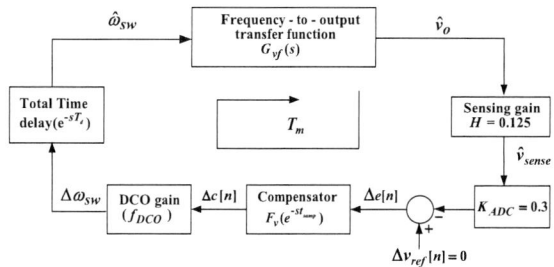

Fig. 4. Small-signal block diagram of digitally-controlled LLC converter with voltage mode control.

- The first compensation pole is placed at the esr zero, $\omega_{p1} = \omega_{esr} = 0.75 \times 10^5$ rad/s, whereas the second pole is located at $\omega_{p2} = 0.95 \times 10^5$ rad/s.

- Integrator gain K_v is adjusted to attain the loop gain crossover frequency at high frequencies while securing a sufficient phase margin.

For the practical design purposes, the loop gain crossover frequency, ω_{cr}, which gives sufficient phase margin throughout the operational region is first selected and the integrator gain K_v is determined later. Accordingly, $\omega_{cr} = 2\pi \cdot 2.3 \times 10^3$ rad/s is chosen. Referring to the asymptotic plot of the loop gain, $|T_m|$ as illustrated in Fig. 3(a), the following relation is derived

$$20\,log\frac{K_{m_A}}{\omega_{cr}} = 0 \text{ dB} \quad (4)$$

where K_{m_A} is the DC gain of loop gain $|T_m|$ at point A. From the small-signal block diagram of digitally-controlled LLC converter shown in Fig. 4, it is expressed as

$$K_{m_A} = HK_{ADC}K_v f_{DCO_A} K_{vf_A} \quad (5)$$

Substituting (5) in (4) gives

$$K_v = \frac{\omega_{cr}}{HK_{ADC}f_{DCO_A}K_{vf_A}} \quad (6)$$

where, $H = 0.125$ is the low-frequency gain of the anti-aliasing filter. $K_{ADC} = 0.3$ V^{-1} is the ADC gain based on [5]. f_{DCO_A} is the small-signal DCO gain at point A. The evaluation of f_{DCO} is given in detail in the following section. From Fig. 6(c), the value f_{DCO} that corresponds to point A for the push-pull mode operation is found to be $2\pi \cdot 1.50 \times 10^5$ and $K_{vf_A} = 2.62 \times 10^{-5}$ was determined in (2). By incorporating the values of these above parameters in (6), the integrator gain for the digitally-controlled LLC converter is calculated as

$$K_v = \frac{2\pi \cdot 2.3 \times 10^3}{0.125 \cdot 0.3 \cdot 2\pi \cdot 1.50 \times 10^5\, 2.62 \times 10^{-5}}$$
$$= 1.56 \times 10^4$$

The above design guidelines formulates the following s-domain function

$$F_v(s) = \frac{1.56 \times 10^4 (1 + s/5 \times 10^3)(1 + s/7 \times 10^3)}{s(1 + s/0.75 \times 10^5)(1 + s/0.95 \times 10^5)} \quad (7)$$

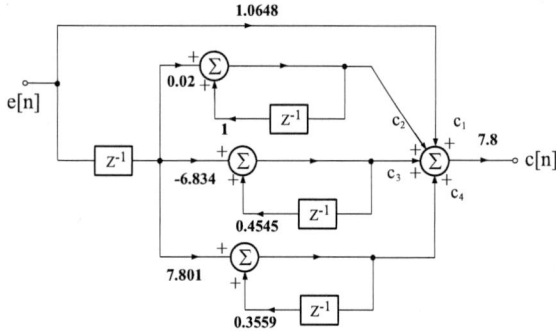

Fig. 5. Parallel form realization of digital voltage feedback compensator.

It can be selected as a reference for the compensation design through the emulation method. Although designed for Point A, the resulting feedback compensation offers a stable operation with satisfactory closed-loop performance for the entire operational region.

D. Conversion to Digital Compensation

The analog compensation (7) is converted into the digital compensation through the bilinear transform using MATLAB software. For the push-pull mode operation, the ADC sampling period is determined as $t_{samp} = 1/(2f_{sw})$ where f_{sw} is the switching frequency at point A. The bilinear-transformed z-domain expression of (7) is determined as

$$
\begin{aligned}
F_v(z) &= \left. F_v(s) \right|_{s = \frac{z-1}{z+1} \frac{2}{t_{samp}}} \\
&= \frac{8.3058(z+1)(z-0.9512)(z-0.9324)}{(z-1)(z-0.4545)(z-0.3559)} \quad (8)
\end{aligned}
$$

and later converted into the parallel structure

$$
\begin{aligned}
&F_v(z) \\
&= 7.8 \left(1.0648 + \frac{0.02}{z-1} - \frac{6.834}{z-0.4545} + \frac{7.801}{z-0.3559} \right) \\
&\hspace{10cm} (9)
\end{aligned}
$$

Fig. 5 is the simulation block diagram of (9), which can readily be converted into the C-code programming. Fig. 3(b) shows the experimental Bode plots for $|G_{vc}|$ and $|T_m|$ of the prototype LLC converter operating at point A. This figure validates the theoretical design procedure shown in Fig. 3(a).

III. PRACTICAL CONSIDERATIONS IN THE IMPLEMENTATION OF DIGITAL CONTROL FOR LLC RESONANT CONVERTER

In the previous section, the digital voltage feedback compensator is designed and verified at point A. However, the implementation of the digital control for the resonant converter can bring certain issues when operating point shifts. Mainly, those are limit-cycle oscillations and time delay. The limit-cycle oscillation is caused by the quantization effects of ADC and DCO and integrator gain of the compensator. The time delay originates from the ADC

process, computation, DPWM generation and etc. These problems can degrade the closed-loop performance. The closed-loop performance can be improved by reducing the time delay and eliminating the limit-cycles. This is done with the proper incorporation of variations in the DCO gain with respect to the variations in the power stage dynamics into the design.

A. Digitally Controlled Oscillator and Small-Signal Gain

Fig. 6 shows the two different DPWM operational modes of DCO namely complementary mode and push-pull mode. In commercial dSPICs, the DCO is used from the DPWM unit [6] and [12], where the complementary DPWM operational mode is generally used for the resonant converters. Conventionally, the ADC samples the error signal once in one switching period due to the higher computational time and the limitation exist in the micro controllers. As a result, it has the time delay of one switching period that gives degraded closed-loop performance. In order to reduce the time delay, the push-pull mode of operation is chosen to increase the ADC sampling rate. The detailed explanation about these DPWM modes are given in [12]. It can be seen from Fig. 6, the ADC samples the sensing signal once in one switching period for the complementary mode operation and twice in one switching period for the push-pull mode operation. Therefore, the sampling rate for the push-pull mode operation becomes two times the switching frequency.

Fig. 6(a) and (b) show the timing diagram of a counter-based DCO for the complementary and push-pull mode operations. The DCO receives the command $c[n]$ from the digital compensation and multiplies it with a constant G_{DCO}, to yield an intermediate variable $N_C = c[n]G_{DCO}$, for complementary mode operation and $N_P = 0.5N_C = 0.5c[n]G_{DCO}$, for push-pull mode operation. The variable N_C or N_P is then multiplied with a prefixed DCO clock, t_{clk}, to drive the MOSFETs at the frequency of

$$
f_{sw} = \begin{cases}
\dfrac{1}{N_C t_{clk}} = \dfrac{1}{c[n]G_{DCO}t_{clk}} \\
\quad : \text{complementary mode operation} \\[4pt]
\dfrac{1}{N_P t_{clk}} = \dfrac{1}{0.5c[n]G_{DCO}t_{clk}} \\
\quad : \text{push-pull mode operation}
\end{cases} \quad (10)
$$

The small-signal DCO gain can be evaluated by taking the derivative of (10)

$$
\begin{aligned}
f_{DCO} &\equiv \frac{\Delta f_{sw}}{\Delta c[n]} \\
&= \begin{cases}
-\dfrac{1}{t_{clk}G_{DCO}}\dfrac{1}{c[n]^2} = -f_{sw}^2 G_{DCO}t_{clk} \\
\quad : \text{complementary mode operation} \\[4pt]
-\dfrac{2}{t_{clk}G_{DCO}}\dfrac{1}{c[n]^2} = -2f_{sw}^2 G_{DCO}t_{clk} \\
\quad : \text{push-pull mode operation}
\end{cases}
\end{aligned} \quad (11)
$$

The 2014 International Power Electronics Conference

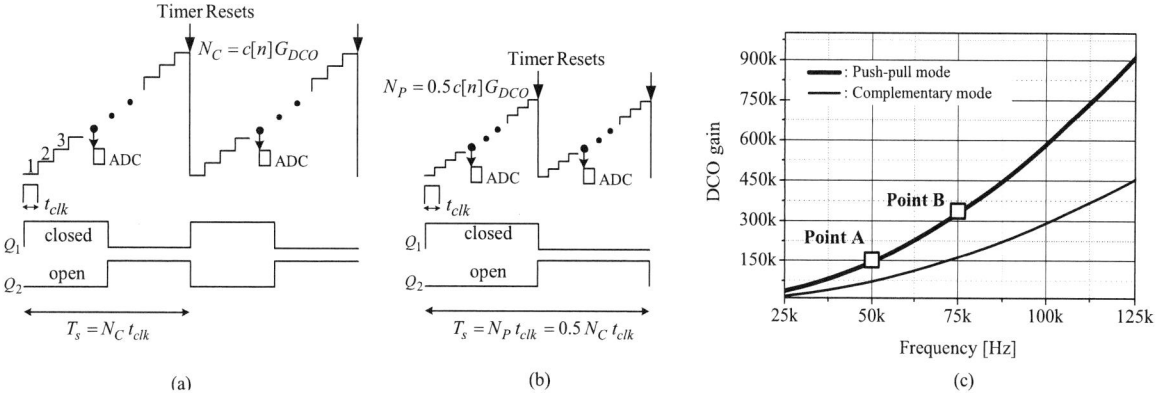

Fig. 6. DPWM operational modes of DCO. (a) Complementary mode operation. (b) Push-pull mode operation. (c) Small-signal gain of DPWM operational modes of DCO.

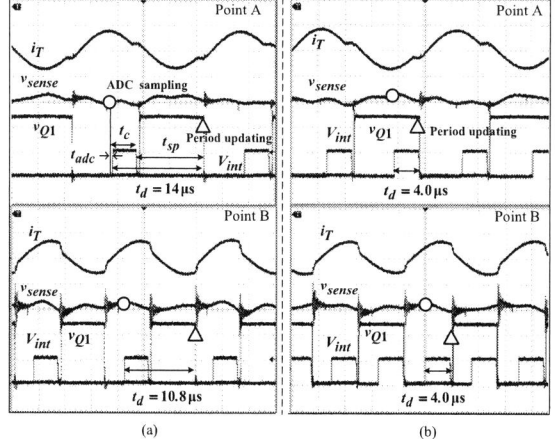

Fig. 7. Power stage and control waveforms at points A and B for the selected DPWM operational modes. (a) Complementary mode operation . (b) Push-pull mode operation.

From (11), it can be observed that the DCO gain increases as the square of the switching frequency. And also, the DCO gain of the push-pull mode operation is twice to that of the complementary mode operation. On the contrary, the DPWM frequency resolution of push-pull mode operation is decreased to the half to that of the complementary mode operation. Fig. 6(c) is the magnitude plot of the DCO gain of the converter for the given DPWM operational modes, where the DCO gains at points A and B are shown for the push-pull mode of operation. It can be observed that the DCO gain has a non-uniform characteristics. However, it has a potential merit when compared to the constant small-signal VCO gain of analog control case and it will be presented in the next section.

B. Comparison of closed-loop performances of the complementary and push-pull DPWM operational modes

Fig. 7 depicts the power stage and control waveforms for the both operational DCO modes. The tank current i_T, output of the anti-aliasing filter v_{sense}, upper MOSFET drive signal v_{Q1}, and ADC interrupt signal, v_{int}, are

shown in Fig. 7. As it can be seen, the ADC samples v_{sense} twice in one switching period and once in one switching mode for the push-pull and complementary mode operations respectively. The total time delay, T_d, in the digital control loop is approximated to

$$T_d \approx t_d + t_{ZOH} \qquad (12)$$

t_d is the time period between the ADC sampling instant to the next period update is shown in Fig. 7. It mainly includes ADC conversion time, $t_{ADC} = 2t_{clk}$, $t_c = 4\ \mu s$ is the computational delay, and t_{sp} is the spare time. t_{ZOH} is the zero-order-hold (ZOH) equivalent delay in the DPWM. The time delay caused by the ZOH is given by [7]

$$
\begin{aligned}
t_{ZOH} &\equiv DT_s - \frac{floor(ND)}{N}T_s \\
&\approx \begin{cases} 0.5T_s : \text{complementary mode operation} \\ 0 \qquad : \text{push-pull mode operation} \end{cases}
\end{aligned}
$$

$$(13)$$

From (12), (13), and Fig. 7, it can be concluded that the complementary mode operation has the larger time delay than the push-pull mode operation at both operating points A and B. Hence, the push-pull mode operation improves the closed-loop performance for the entire operational region. However, it is reported earlier that the frequency resolution is decreased for this mode of operation. Nonetheless, if the complementary mode operation is chosen to get fine frequency resolution but it can also produce either limit-cycle oscillations or degraded closed-loop performance. And, it is illustrated in both the frequency-domain characteristics and time-domain characteristics at points A and B in Figs. 8 and 9.

Fig. 8 shows the loop gain characteristics of the selected DPWM operational modes for the both points A and B. Fig. 9 compares the step-load transient response characteristics at points A and B for the given DPWM operational modes. Initially, the compensator is designed for the push-pull mode operation at point A, which has the integrator gain as $K_v = 1.56 \times 10^4$. If the compensator

978-1-4799-2706-7/14 $31.00 © 2014 IEEE

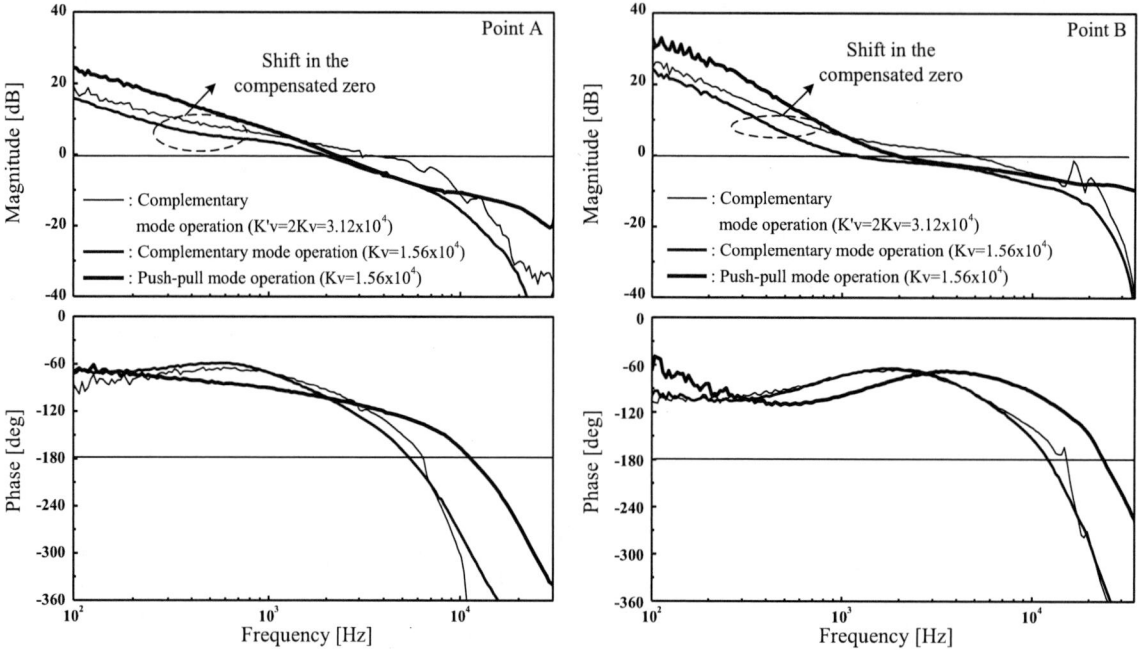

Fig. 8. Comparison of loop gain characteristics for the selected DPWM operational modes at points A and B.

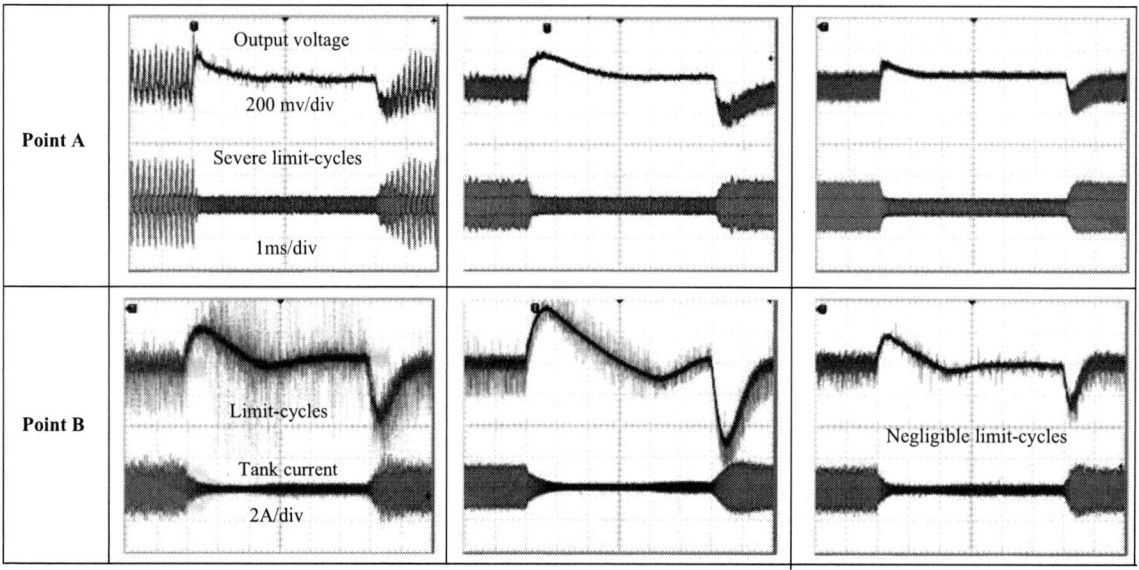

Fig. 9. Comparison of step-load transient responses at points A and B for the selected DPWM operational modes. (a) Complementary mode operation with $K_v' = 2K_v = 3.12 \times 10^4$. (b) Complementary mode operation with $K_v = 1.56 \times 10^4$. (c) Push-pull mode operation with $K_v = 1.56 \times 10^4$.

is designed for the complementary mode operation to get the fine frequency resolutions at the same designed conditions, the integrator gain becomes double to the push-pull mode operation, $K_v' = 2K_v = 3.12 \times 10^4$. Even though, the converter has increased DCO resolution for the complementary mode operation, it can generate the limit-cycle oscillations due to increment in the integrator gain [3]-[4] and [6].

When the converter is operated with this complementary mode of operation, the loop gain characteristics violates the designed loop gain shaping strategy. And it can be seen in Fig. 8, the shift in the designed compensated zero in the loop gain characteristics due to the quantization of compensator coefficients. As a result, it has low DC gain of the loop gain along with the high loop gain crossover frequencies and low phase margins at both the operating points A and B, when compared to the push-pull mode operation. Therefore, the converter

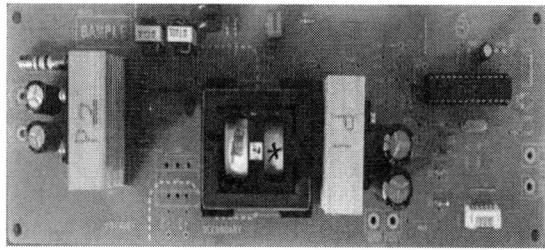

Fig. 10. Photograph of the prototype converter.

shows the severe limit-cycle oscillations at both points A and B as shown in Fig. 9(a).

In order to eliminate the limit-cycles oscillations for the complementary mode operation in the system, the integrator gain is decreased to $K_v = 1.56 \times 10^4$. It is obvious that the decrement in the integrator gain gives lower loop gain crossover frequencies for the system. Nevertheless, this case also violates the loop gain shaping strategy as depicted in Fig. 8. In respect to this, the transient response characteristics for the both operating points A and B show the degraded performance when compared to the push-pull mode operation as illustrated in Fig. 9 (b) and (c).

It can be seen from Fig. 8, both the cases of complementary mode operation has a difference in the DC gain of the loop gain characteristics with similar loop gain structure and violates the designed loop gain strategy. Moreover, they exhibit larger phase lag characteristics at high frequencies when compared to the push-pull mode operation. It is expected that it is mainly because of ADC samples the sensing signal one time in one switching period in the complementary mode. On the other hand, the push-pull mode operation, which has the ADC sampling frequency twice the switching frequency, displays the exact designed loop gain shaping strategy as already shown in Section II. As illustrated in above, the time delay is considerably reduced for the push-pull mode operation. Hence, it shows designed loop gain crossover frequencies and better phase characteristics than the both cases of complementary mode operations as shown in Fig. 8. So, it ultimately improves the step load transient response with negligible limit-cycle oscillations as shown in 9(c).

From the above qualitative analysis, it can be concluded that the complementary mode operation which is designed to get the better closed-loop performance shows the limit-cycle oscillations. However, to eliminate the limit-cycle oscillations, the closed-loop performance need to be sacrificed. On the other hand, with the sufficient DCO frequency resolutions at the given operating frequencies, the push-pull mode operation can be useful to improve the transient performance with negligible limit-cycles oscillation problem.

IV. PERFORMANCE OF DIGITALLY-CONTROLLED LLC CONVERTER

The experimental results are obtained from the 150 W digitally-controlled LLC resonant converter. The proto-type hardware is shown in Fig. 10. The performance of the digitally-controlled LLC converter is measured at points A and B, and compared with the case of the conventional analog control.

Loop Gain: In the top row of Fig. 11, the loop gain characteristics of the digital control are compared with those of the analog control. The analog control is optimally designed following the design procedure proposed in [8]. At point A, the digital control produces larger phase lag at high frequencies. This is a well-known consequential effect of the unavoidable time delay in the digital control. Except for the additional phase lag, the digital control nearly duplicates the loop gain characteristics of the best case of the analog control.

At point B, the loop gain of the digital control exhibits a 7 dB magnitude boost, compared with that of the analog control. The magnitude boost is an outcome of the non-uniform DCO gain. As shown in Fig. 6(c), the DCO gain at point B is larger than the DCO gain at point A by a factor of 2.25. The larger DCO gain raises the loop gain magnitude of the digital control by $20 \log 2.25 = 7$ dB , compared with the case of the analog control where the VCO gain is a constant. The above analysis hints the potential merit of the digital control with a counter-based DCO. The non-uniformity of the DCO gain can be exploited to compensate for undesired changes in power stage dynamics.

Step Load Response: The bottom waveforms in Fig. 11 show the transient response of the output voltage and tank current in response to $I_O = 6\,\text{A} \Rightarrow 1\,\text{A} \Rightarrow 6\,\text{A}$ changes in the load current. Two control schemes produce largely the same responses at point A. However, at point B, the digital control shows much superior transient responses due to the improved loop gain characteristics.

V. CONCLUSIONS

This paper has presented the design and performance evaluation of the digital control adapted to an LLC series resonant dc-to-dc converter operating with wide input and load variations. It is demonstrated that, the push-pull mode of operation significantly improved the closed-loop performance of the converter than that of the complementary mode. This paper showed that, when all the design considerations are correctly addressed, the digital control could achieve the performance that exceeds the best case of the conventional analog control.

REFERENCES

[1] Z. Lukic, K. Wang, and A. Prodic, "High-frequency digital controller for dc-dc converters based on multi-bit $\Sigma - \Delta$ pulse-width modulation," IEEE APEC 2005, pp.35-40.

[2] D. Maksimovic, and R. Zane,"Small-signal discrete-time modeling of digitally controlled DC-DC converter," IEEE COMPEL Workshop, Rensselaer Polytechnic Institute, Troy, NY, July 16-19, 2006.

[3] A. V. Peterchev, S. R. Sanders, "Quantization resolution and limit cycling in digitally controlled PWM converters," IEEE Trans. Power Electron., vol.18, no.1, pp. 301-308, Jan. 2003.

[4] H. Peng, A. Prodic, E. Alaecon, and D. Maksimovic,"Modeling of quantization effects in digitally controlled DC-DC converters," IEEE Trans. Power Electron., vol. 22, no. 1, Jan. 2007.

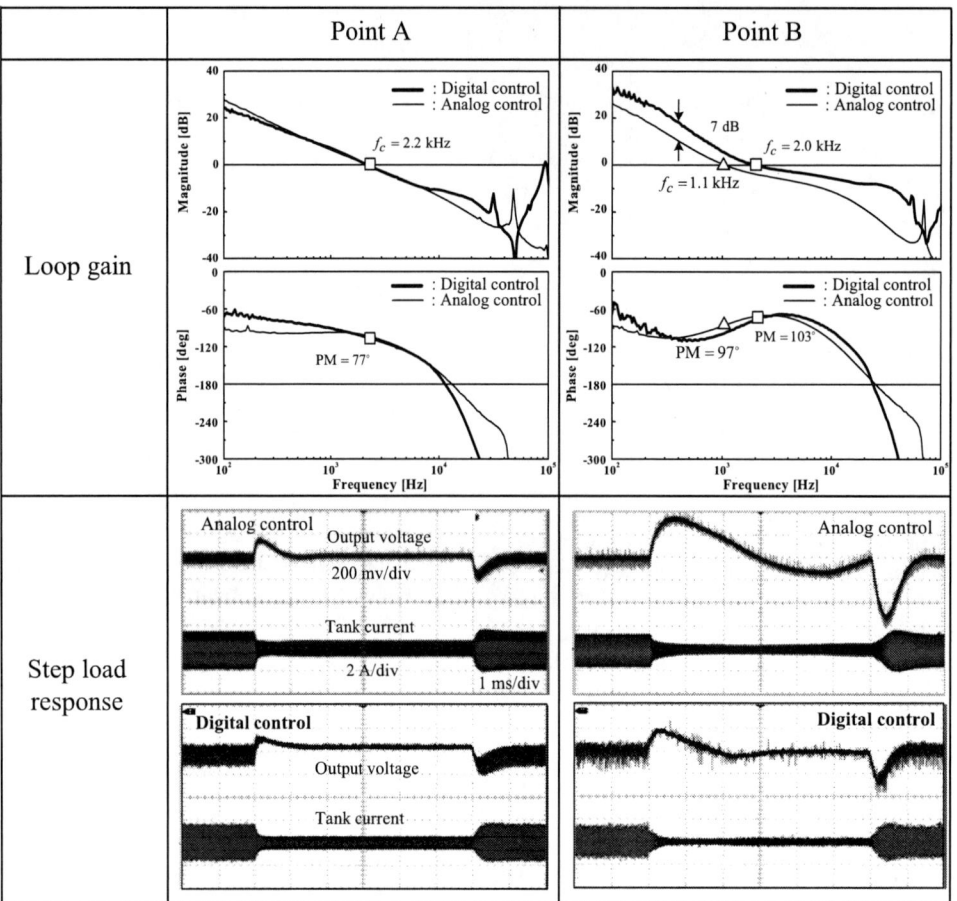

Fig. 11. Performance of digitally-controlled experimental LLC converter.

[5] Al-Atrash, H.; Batarseh, I.,"Digital Controller Design for a Practicing Power Electronics Engineer, "IEEE Applied Power Electronics Conference, APEC 2007, Feb. 25 2007-March 1 2007, pp.34-41.

[6] M. M. Peretz, S. B. Yaakov, "Digital control of resonant converters : resolution effects on limit cycles," IEEE Trans. Power Electron., vol.25, no.6, pp. 1652-1661, June 2010.

[7] L. Corradini and P. Mattavelli "Analysis of multiple sampling technique for digitally controlled dc-dc converters", Proc. IEEE Power Electronics Specialists Conf. (PESC), pp.2410 -2415, 2006.

[8] J. Jang, M. Joung, B. Choi, S. Hong, and S. Lee, "Dynamic analysis and control design of optocoupler-isolated LLC series resonant converters with wide input and load variations," IET Power Electron., vol.5, pp.755-764, June 2012.

[9] C.H. Park, S.H. Cho, J. Jang, S. K. Pidaparthy, T. Y. Ahn, and B. Choi "Average Current Mode Control for LLC Series Resonant DC-to-DC Converters," JPE, vol.14, no.1, pp.40-47, 2014.

[10] Bo Yang,"Topology investigation for front end DC/DC Power Conversion for Distributed Power Systems," Ph. D. dissertation, Virginia Polytechnic Institute and State University, Blacksburg, VA., 2003.

[11] Bo Yang and F.C. Lee,"Small-signal analysis for LLC resonant converters," CPES Seminar 2003, S7.3, pp.144-149.

[12] "dsPIC33F family-high-performance 16-bit digital signal controllers," DS70165A, Microchip Technology Inc.

Experimental Verification of Noiseless Sampling for Buck Chopper Circuit with Current Control

Shun Takeuchi and Keiji Wada
Department of Electrical and Electronics Engineering
Tokyo Metropolitan University, 1-1 Minami Oosawa, Hachioji, Tokyo, JAPAN
kj-wada@tmu.ac.jp

Abstract—**In digitally controlled circuits for power converter circuits, sampling data is important because a digital control circuit is operated on the basis of these data. If the sampled values have been affected by switching noise from the power circuit, the control stability of the circuit would be disturbed. This paper proposes a noiseless sampling method for both synchronous sampling and multisampling, which can sample a value without being affected by noise. The synchronous sampling method may be affected by noise depending on the duty ratio of the circuit. The noiseless sampling method for synchronous sampling changes the timing of the sampling to a position that is less susceptible to noise. The noiseless sampling method for multisampling does not obtain the data immediately after turn-on and -off switching. The control circuit can avoid switching noise by using noiseless sampling, which leads to a disturbance in the control circuit and enhances the robustness of the circuit when applying the multisampling method. Experimental results are presented to verify that the current control of the proposed sampling methods is not disturbed by noise.**

I. INTRODUCTION

Power converter circuits are composed of a power circuit and control circuit. A digital controller such as a digital signal processor (DSP) and field-programmable gate array (FPGA) as the control circuit is widely used to control the power circuit[1]. In addition, the feedback controls of an inductor current and a capacitor voltage are applied to the power converter circuit. In this case, the currents and voltages are converted via these sensors and AD converters. Therefore, it is important to set the sampling-points and -frequency properly because these values may influence the circuit operation on the basis of the digital control. Generally, a synchronous sampling method is used to digitally control the power converter circuit[2]-[4]. The method samples one or two times per switching cycle and synchronizes with the carrier waveform. The synchronous sampling method is commonly used because it can sample the mean value without switching ripple components. Moreover, this sampling method is unaffected by the noise component under turn-off and -on operations; however, it may be affected by noise when the digitally controlled circuit operates at an extremely high or low duty ratio[5].

Recently, researches on multisampling methods has reported on the improvement in the tracking performance with respect to the reference value and the enhancement of the

robustness of the converter circuits[6]-[9]. The control time delay of multisampling is lower than that of synchronous sampling; however, the computational speed of the digital controller of the multisampling method should be higher. In addition, the effect of the switching noise on the circuit operation has never been sufficiently examined when applying a multisampling method.

This paper discusses the relationship between the sampling points and the control performance of a buck chopper circuit with current control. In this case, this experiment does not include an anti-aliasing filter before the AD converter[10][11]. When the conventional sampling method is applied, unstable operations will be demonstrated in case of high- or low-duty operation of the current control by the experimental results. This paper proposes a noiseless sampling method for the synchronous sampling and multisampling methods. The validity of the proposed method is verified, and a buck

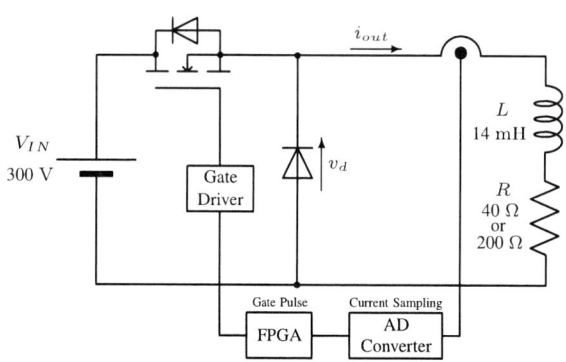

(a) Buck chopper circuit.

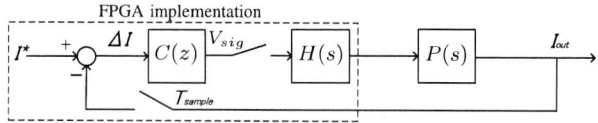

(b) Control block diagram.

Fig. 1. Experimental circuit configuration.

The 2014 International Power Electronics Conference

TABLE I.   SPECIFICATIONS OF THE DIGITAL CIRCUIT AND CURRENT SENSOR

FPGA	CycloneII EP2C8Q208C8N
FPGA reference clock	100 MHz
A/D Converter	LTC1412, 12 bit, Parallel 3 MS/s
Number of bits	12 bit
Maximum sampling rate	3 Msps
AD conversion time	less than 300 ns
Current transducer	LEM, LA 55-P, 0 to 200kHz

chopper circuit rated at 300 V, an FPGA, and a high-speed AD converter are applied[12].

II.   CIRCUIT CONDITIONS

Fig. 1(a) shows the buck chopper circuit used in the experiment, and Fig. 1(b) is the control block diagram based on the z-transformation. A detailed explanation of the sampling method will be explained in section III. Table I shows the specifications of the FPGA and current sensor.

Here, the controller $C(z)$ is the proportional control and is denoted with

$$C(z) = K_p. \tag{1}$$

The proportional gain K_p is set to $K_p = 256$ V/A in the following experiment. By setting this value, the output current reaches a steady state in one step delay in the case of synchronous sampling once per cycle. The zero-order hold $H(s)$ associated with digital control is expressed by

$$H(s) = \frac{1 - e^{-sT_s}}{s}, \tag{2}$$

where T_s is the sampling period of the AD converter. The RL load circuit is expressed as follows:

$$P(s) = \frac{1}{Ls + R}. \tag{3}$$

The closed-loop pulse transfer function $G(z)$ is obtained by using the z-transformation of the digital control system in Fig. 1(b):

$$G(z) = \frac{I_{out}(z)}{I_{out}^*(z)} = \frac{\frac{K_p}{R}(1 - e^{-\frac{RT_s}{L}})}{z - e^{-\frac{RT_s}{L}} + \frac{K_p}{R}(1 - e^{-\frac{RT_s}{L}})}. \tag{4}$$

Moreover, the steady-state current $I_{out}(\infty)$ for the step response is given by

$$I_{out}(\infty) = \lim_{z \to 1}(z - 1)I_{out}^*(z) = \frac{K_p}{R + K_p}I_{out}^*. \tag{5}$$

III.   CONVENTIONAL SAMPLING METHOD

A. Synchronous Sampling

Fig. 2 shows the sampling points and triangular carrier waveforms of the two types of synchronous sampling methods at low duty ratio. Fig. 2(a) shows the positive-peak sampling method that samples the current at the positive peaks of the carrier waveform. This sampling method can sample the current that does not include noise and ripple components in the

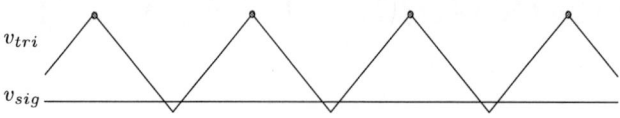

(a) Positive-peak synchronous sampling.

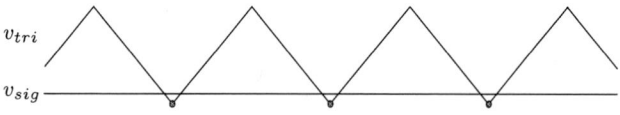

(b) Negative-peak synchronous sampling.

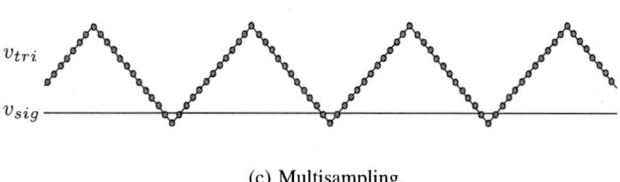

(c) Multisampling.

Fig. 2.   Conventional sampling methods verified in this paper.

vicinity of the switching operation at a low duty ratio. Fig. 2(b) shows the negative-peak sampling method that is almost the same sampling method as the positive-peak sampling method. This sampling method can obtain the discrete inductor current at each sampling point that does not include noise and ripple components in the vicinity of the switching operation at a high duty ratio. When the duty ratio is comparatively low, as shown in Fig. 2, the sampling value may be affected by switching noise, causing an inaccurate output current. Here, the control stability and steady-state value are the same for both sampling methods because the time delay of the controller is also the same value.

Generally, synchronous sampling that samples both peaks of a triangular waveform is widely applied to a digitally controlled circuit, but this paper discusses the experimental

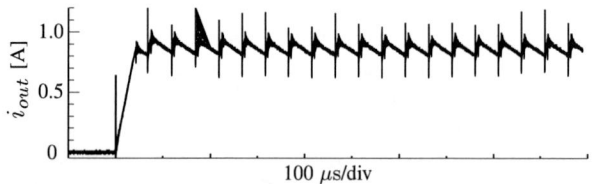

Fig. 3.   Step response waveform for the positive-peak synchronous sampling method.

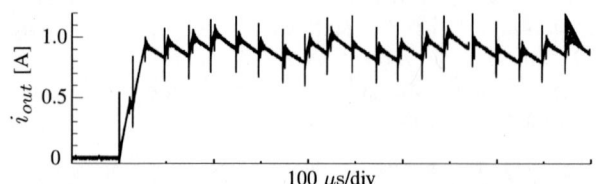

Fig. 4.   Step response waveform for the negative-peak synchronous sampling method ($I^* = 1$).

978-1-4799-2706-7/14 $31.00 © 2014 IEEE

results for both positive- and negative-peak sampling in order to clarify the relationship between the duty ratio and the sampling point.

B. Multisampling

A multisampling method that samples a continuous inductor current multiple times per switching cycle is shown in Fig. 2(c). In this experiment, the sampling frequency of multisampling is set to 800 kHz; thus, it samples the current 40 times per switching cycle, and the time delay of the control is therefore smaller than that of the conventional control.

C. Experimental Results for the Conventional Sampling Method

The DC voltage of the chopper circuit is set to 300 V, and the switching frequency of the buck chopper is set to 20 kHz in the following experiment. The load resistance R in the experiment for the step response is set to 40 Ω. In this case, the duty ratio at steady state of the chopper circuit is 11.5%. On the other hand, the current reference I^* is set to 2.6 A under the ramp signal, and the load resistance R is set to 200 Ω. The duty ratio of the circuit continuously varies from 0% to 97%.

Fig. 3 shows the step response of the digitally controlled current-mode buck chopper with the positive-peak synchronous sampling method. The current waveform changes to steady state by the one-sample delay. The buck chopper circuit achieved stable current control in spite of the switching noise from the converter circuit.

For positive-peak sampling, the sampling value of the digital controller does not contain switching noise. As a result, the buck chopper circuit can control the stable inductor current. Fig. 4 shows the step response of a digitally controlled current-mode buck chopper with the negative-peak synchronous sampling method. However, Fig. 4 indicates that the inductor current exhibits unstable operation; thus, the inductor current fluctuates. The difference between the waveforms in Figs. 3 and 4 is only the sampling positions of the digital controller; however, the inductor current waveforms are quite different. In this circuit condition with a low duty ratio, the sampling values are affected by the noise component that causes unstable current output for the negative sampling point.

Figs. 5 and 6 show the ramp response of a digitally controlled current-mode buck chopper with the positive- and negative-peak synchronous sampling methods, respectively. The duty ratio of the chopper circuit changes from 0% to 97%. For a high duty ratio, the current waveform of the positive-peak sampling method is disturbed by the switching noise, and the current could not be controlled at a constant value. In contrast, the waveform of the negative-peak sampling method is disturbed by the noise when the duty ratio of the chopper circuit is low. As a result, the current control that applies conventional synchronous sampling methods may be influenced by the switching noise depending on the duty ratio.

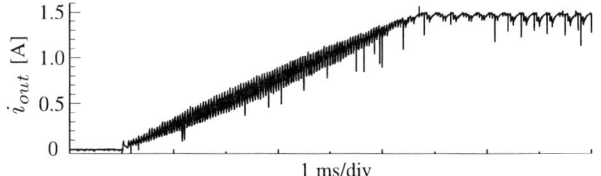

Fig. 5. Ramp response waveform for the positive-peak synchronous sampling method.

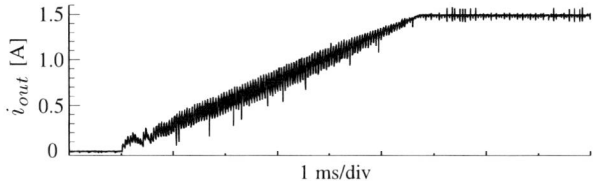

Fig. 6. Ramp response waveform for the negative-peak synchronous sampling method.

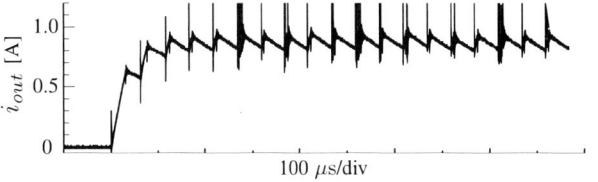

Fig. 7. Experimental waveform for the multisampling method ($I^* = 1$).

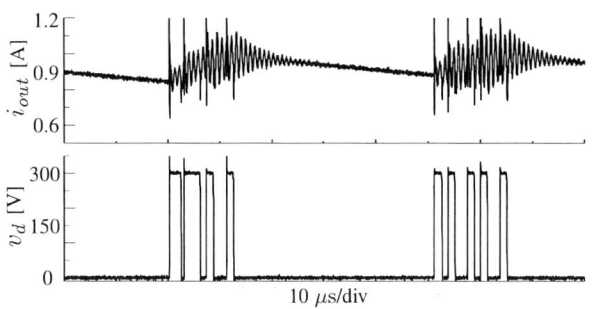

Fig. 8. Enlarged experimental waveforms for the multisampling method ($I^* = 1$).

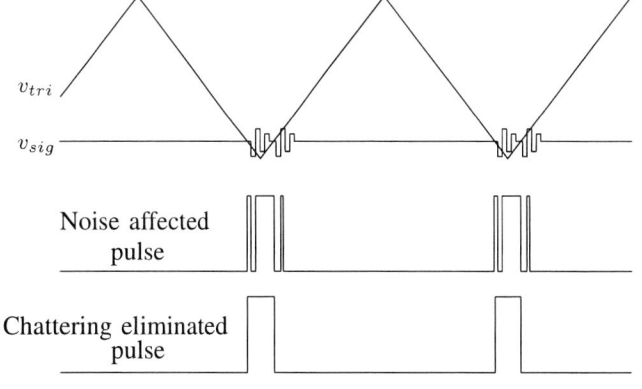

Fig. 9. PWM pulse implemented in a multi-switching prevention program.

Fig. 7 shows the experimental results when the multisampling method is applied. From these experimental results, the inductor current contains some noise because the digital controller is influenced by the switching noise, as the digital controller samples the inductor current every 1.25 μs. Fig. 8

shows the enlarged experimental waveforms for multisampling operation that show the output current i_{out} and the diode voltage v_d. From the experimental waveforms, the MOSFET is turned-on and turned-off even more than expected. The reference signal, which is affected by noise, intersects with the triangular waveform multiple times. These multiple switching operations will increase the switching losses and noise.

1) Multi-switching prevention method: In general, a multi-switching prevention method is applied to converter circuit. Fig. 9 shows the carrier v_{tri}, signal v_s, and PWM pulse waveforms. The signal wave is oscillating, but this is a representation of the influence of noise immediately after switching. If v_s oscillates according to the influence of noise after the switching operation, v_s intersects v_{tri} multiple times and consequently produces a noise-affected pulse. Thus, it is possible to produce a chattering-eliminated pulse by adding the condition that v_s and v_{tri} are only compared when the signal intersects the rising and falling slopes of the triangular waveform the first time in each switching period.

Fig. 10 shows the experimental results when multisampling with the multi-switching prevention method are applied. By adopting the multi-switching prevention method, it is possible to realize stable current control without the effects of noise for the reference current $I^*=1$ A and the duty ratio of 11.5%.

D. Low Duty Ratio Operation

Fig. 11 shows the experimental results when the reference current $I^*=0.25$ A, and the duty ratio is 2.88%. The output current waveform is distorted and unstable. From the results in Figs. 10 and 11, the effect of noise on the sampling values is remarkable when the duty ratio of the circuit is low. Fig. 12 shows the enlarged inductor current waveforms for the positive-peak synchronous sampling method. The output current ripple is given by

$$\Delta I_{out} = \frac{V_{IN}}{L}(1-d)dT_{sw}. \qquad (6)$$

ΔI_{out} is maximized when the duty ratio is 50%. From Figs. 12(a), 12(b), and 12(c), resonance noise occurs independent of the output current. Therefore, the resonance noise becomes larger versus ΔI_{out}, and the effect of noise increases if the duty ratio of the circuit is closer to 0% or 100%. Thus, the multi-switching prevention method is not sufficient when the duty ratio of the circuit is high or low.

IV. NOISELESS SAMPLING METHOD

A. Noiseless Sampling Method for Synchronous Sampling

Fig. 13(a) shows the noiseless sampling method for synchronous sampling. This sampling method changes the sampling point depending on the duty ratio of the circuit.

When the duty ratio is less than 50%, the digital controller obtains the inductor current at the positive peak. On the other hand, the digital controller obtains the inductor current at the negative peak of the triangular waveform when the duty ratio is greater than 50%. As a result, the noiseless sampling method

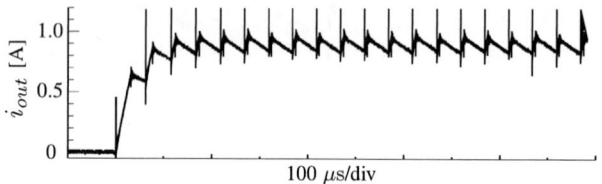

Fig. 10.   Experimental waveform of I^*=1 A implemented with a multi-switching prevention program.

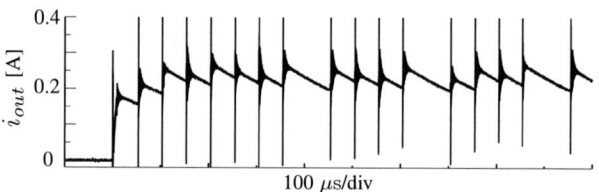

Fig. 11.   Experimental waveform of I^*=0.25 A implemented with a multi-switching prevention program.

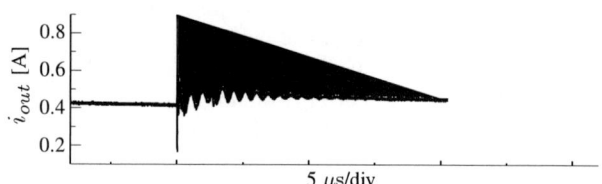

(a) Enlarged current waveform of $I^* = 0.5$ A with a duty ratio of 5.77%.

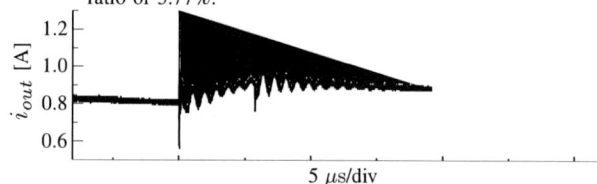

(b) Enlarged current waveform of $I^* = 1$ A with a duty ratio of 11.5%.

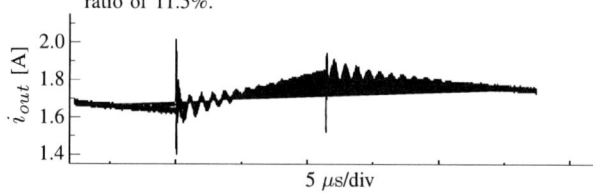

(c) Enlarged current waveform of $I^* = 2$ A with a duty ratio of 23.1%.

Fig. 12.   Enlarged current waveforms.

can possibly avoid switching noise for more than 1/4 of a cycle of the switching period. Therefore, this sampling method can obtain the inductor current that does not include the noise and ripple components in the vicinity of switching operations for all duty ratios.

B. Noiseless Sampling Method for Multisampling

Fig. 13(b) shows a diagram of the noiseless sampling method for multisampling. This method does not take the inductor current quickly after switching during the period of T_{away}. This T_{away} period is set after the turn-on and -off of the MOSFET in order to avoid the switching noise.

1) Upper limit of T_{away}: From the previous section, the influence of noise on the sampling value increases when the duty ratio of the circuit becomes closer to a duty ratio of 0%. In this section, the determination of T_{away} is described for a given reference current I^* of 1 A. From Fig. 12(b), the resonance noise after switching operation remains for approximately 6 μs; in order to avoid this noise, T_{away} is set to 6.5 μs.

If T_{away} is longer than the switching period T_{sw}, the digital circuit cannot sample the current in each switching period, and the digital circuit can not calculate appropriate v_{sig}. The upper limit of T_{away} can be obtained as follows. In the buck chopper circuit, both T_{on} and T_{off} are at minimum values when the duty ratio is 50%. Therefore, the digital circuit cannot sample the current in each switching period when T_{away} is too large close to the duty ratio of 50%. In a transient state, the duty ratio of the circuit may exceeds 50%, even when the step response of $I^* = 1$ A, and the duty ratio is 11.5%. Therefore, considering the sampling cycle T_s, it is necessary to set T_{away} within the limit of

$$T_{away} < 0.5T_{sw} - T_s. \tag{7}$$

2) Conditions for average current sampling with noiseless multisampling: If the digital circuit samples the current during both the on and off periods at least once in the switching cycle, it can sample the average output current in one switching cycle. This is because the digital circuit samples a value $\Delta I_{out}/2$ greater than the average output current during the on period, and it samples a value $\Delta I_{out}/2$ less than the average output current during the off period. The conditions for T_{away} are given by

$$T_{away} < dT_{sw} - T_s \quad (d < 0.5), \tag{8}$$

$$T_{away} < (1-d)T_{sw} - T_s \quad (d > 0.5). \tag{9}$$

For example, the digital circuit can sample the current at least once during both the on and off periods for $I^* = 1$ A and a duty ratio of 11.5% ($d = 0.115$) according to (8) if T_{away} is set to less than 4.5 μs.

3) Corrections to the theoretical average current: In these experiments, the digital circuit cannot sample the average output current and does not match (5) if the T_{away} is set within the range in (8). The average output current can be approximated as follows in (10) when the T_{away} is set within the range in (8). The digital circuit cannot sample the average current and therefore controls the circuit using a value including the current ripple component when T_{away} is set within the range in (8). In the following, the average output current in (10) is presented when the digital circuit does not sample the current during the on period by applying the noiseless multisampling method. The sampling data just before the on period considering the current ripple is given by

$$I'_{out} \approx I_{out} - \frac{\Delta I_{out}}{2}. \tag{10}$$

As depicted in Fig. 13(b), the sampling value I'_{out} is held during the on period. Therefore, the average output current is

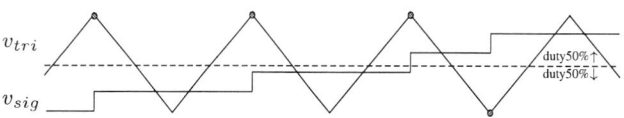

(a) Noiseless sampling method for synchronous sampling.

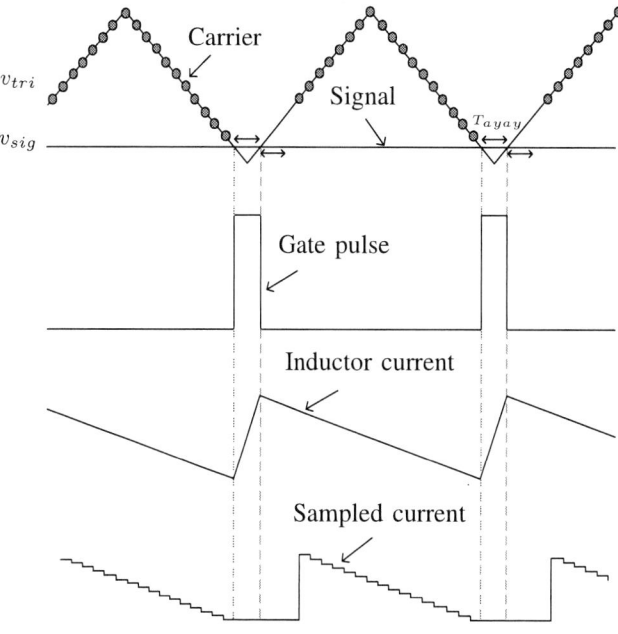

(b) Noiseless sampling method for multisampling.

Fig. 13.   Proposed sampling methods verified in this study.

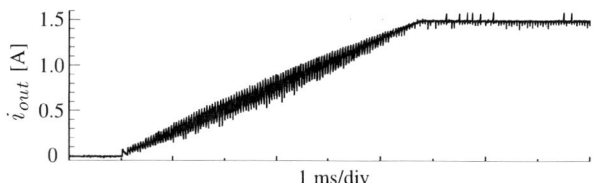

Fig. 14.   Ramp response waveform for the noiseless sampling method for synchronous sampling.

determined by I'_{out}. By changing I_{out} in Fig. 1 to I'_{out}, the output average current is given by

$$I_{out}(\infty) = \frac{K_p}{R + K_p}I^*_{out} + \frac{\Delta I_{out}}{2} \quad (d < 0.5). \tag{11}$$

C. Experimental Results of the Proposed Sampling Method

Fig. 14 shows the ramp response of a digitally controlled current-mode buck chopper with the noiseless sampling method for synchronous sampling. In this case, the inductor current can be controlled for both lower and higher duty ratios without any switching noise component. From the experimental results, it is confirmed that the digital controller can take the inductor current that is not superimposed with the noise component.

The 2014 International Power Electronics Conference

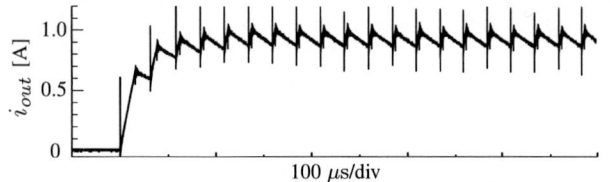

Fig. 15. Experimental waveform for the noiseless sampling method for multisampling ($I^* = 1$).

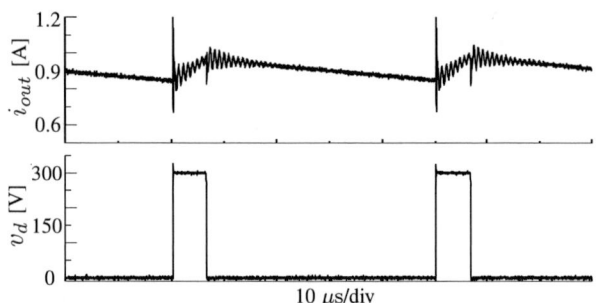

Fig. 16. Enlarged view of the experimental waveforms for the noiseless sampling method for multisampling ($I^* = 1$).

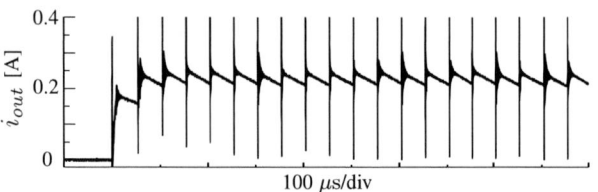

Fig. 17. Experimental waveform for I^*=0.25 A for the noiseless sampling method.

Fig. 15 shows the step response for $I^* = 1$ A and the duty ratio of 11.5% with the noiseless sampling method for multisampling.

By implementing the proposed method, the buck chopper circuit achieved stable current control because it samples the current during the period that is not affected by noise. Fig. 16 shows an enlarged diagram of the experimental waveform for noiseless multisampling. By implementing noiseless multisampling, the output current can be controlled without being affected by the switching noise. The average output current is 924 mA, and this is different from the theoretical value of 865 mA from (5). This is because T_{away} does not satisfy (8); consequently, the digital circuit does not sample the average output current. In this case, the corrected average output current of 920 mA of (11) is applied.

D. Low Duty Ratio Operation

Fig. 17 shows the step response for $I^* = 0.25$ A and a duty ratio of 2.88% with the noiseless sampling method for multisampling. The average output current is 241 mA, and the corrected average current from (11) is 244 mA; therefore, this experimental result is equal to the theoretical value. From the comparison of Figs. 11 and 17, stable current control is achieved without being affected by noise at a lower duty ratio by adopting the noiseless multisampling method.

V. CONCLUSION

This paper has proposed noiseless sampling methods for both synchronous sampling and multisampling. It is confirmed that the behavior of a digitally controlled circuit is influenced by noise when it samples the inductor current relatively quickly after switching. The noiseless sampling method can avoid a malfunction of the circuit for both synchronous sampling and multisampling. By adopting noiseless multisampling method, power converter circuit achieved stable current control at the duty ratio of 2.88% The experimental results rated at 300 V and 1 A confirmed the validity of the proposed sampling method.

REFERENCES

[1] A. Kawamura, H. Fujimoto and T. Yokoyama, "Survey on the real time digital feedback control of PWM inverter and the extension to multi-rate sampling and FPGA based inverter control," IEEE 33rd Annual Industrial Electronics Society, IECON, pp. 2044-2051 (2007)

[2] P. Jintakosonwit, H. Fujita, and H. Akagi, "Control and performance of a fully-digital-controlled shunt active filter for installation on a power distribution system," *IEEE Transactions on Power Electronics,* vol. 17, no. 1, pp. 132-140 (2002)

[3] T. Ômae, K.Kubo, and M.Watanabe, "Microprocessor-based direct digital control for inverter drives, " (in Japanese), Hitachi Rev. vol. 65, no. 4, pp. 267-272 (1983)

[4] D. Sype, K. Gusseme, A. Bossche, J. Melkebeek, "A sampling algorithm for digitally controlled boost PFC converters," *IEEE Transactions on Power Electronics,* vol. 19, no. 3, pp. 649-657 (2004)

[5] S. Maekawa, Y. Hasegawa, N. Suzuki, H. Kubota, "Three-Phase Current Reproduction Method using the DC Current Sensor Capable of Suppressing a Harmonic Noise," IEEJ Transactions on Industry Applications, vol. 134, no. 1, pp. 96-105 (2014)

[6] X. Zhang, and S. W. Joseph, "Study of multisampled multilevel inverters to improve control performance." *IEEE Transactions on Power Electronics,* vol. 27, no. 11, pp. 4409-4416 (2012)

[7] S. Tahara, F. Tabuchi, and T. Yokoyama, "1MHz variable sampling quasi multi-rate deadbeat control for single phase PWM inverter in low carrier frequency," *IEEE Power Electronics Specialists Conference,* pp. 2512-2516 (2008)

[8] L. Corradini, M. Paolo, and S. Stefano, "Elimination of Sampling-Induced Dead Bands in Multiple-Sampled Pulsewidth Modulators for DC-DC Converters," *IEEE Transactions on Power Electronics,* vol. 24 no. 11, pp. 2661-2665 (2009)

[9] R. Sakai, H. Fujita. "A multi-sampling and multi-update Control Method for Active Power Filters ," Technical Meeting on Semiconductor Power Converter, IEE Japan, no. SPC-11-038, pp. 77-82 (2011)

[10] Z. Feng and X. Wei, "Using PWM Output as a Digital-to-Analog Converter on DSP," 2010 International Conference on System Science, Engineering Design and Manufacturing Informatization (ICSEM), pp.278-281 (2010)

[11] H. Funato, T. Mori, T. Igarashi, S. Ogasawara, F. Okazaki and Y. Hirota, "Optimization of Switching Transient Waveform to Reduce Harmonics in Selective Frequency Bands," IEEJ Industry Applications, vol. 2, no. 3, pp. 161-169 (2013)

[12] H. Akiyama, K. Wada and T. Shimizu, "Development of an FPGA Board with AD Converters for Power Electronics Circuit," Annual Conference of IEEJ, Industry Applications Society, (2011)

Control Characteristics Improvement of Full-Bridge DC-DC Converter with Snubber Capacitor

Kazuhide Domoto, Yoichi Ishizuka
Graduate School of Engineering, Nagasaki University
1-14 Bunkyo-machi, Nagasaki, Japan 852-8521

Seiya Abe , Tamotsu Ninomiya
International Centre for the Study of East Asian
Development
1-8 Hibikino, Wakamatsu-ku, Kitakyushu, Japan 808-0135

Abstract— **A high-voltage full bridge dc-dc converter has a big problem of high-voltage surge occurrence across secondary-side rectifying diodes. In order to solve this problem, a simple solution with a snubber capacitor has been proposed. This technique provides a prominent effect of surge-voltage suppression. On the other hand, it deteriorates the control characteristics of this converter. In this paper, the control characteristics are analyzed by the state-space averaging method. Based on this analytical result, the improvement of the control characteristics is proposed. Some experiments confirm the effectiveness of the proposed technique.**

Keywords— Full-Bridge DC-DC Converter, Surge Voltage, Snubber Capacitor, Control Characeristics

I. INTRODUCTION

Recently, the rapid growth of internet traffic has increased the number of ICT equipments in data center. Therefore, the electric power consumption in data center has been increased. The energy saving techniques are required in date center.

The conventional data center utilizes the power distribution system based on the combined power lines of AC and 48VDC. In case of AC system, the number of conversion stages is larger, and then its efficiency is lower. In case of 48V DC system, the efficiency is higher due to smaller number of conversion stage. However, heavy current requires thicker power cables, and then flexibility of installation is limited.

In order to solve these difficulties, the power distribution system using High-Voltage Direct-Current (HVDC) e.g., 400V has been recently researched[1~2]. This HVDC system has several advantages such as higher efficiency due to smaller number of conversion stages, and easier installation due to finer power cables. In this system, a semiconductor circuit breaker and a high power-density isolated DC-DC converter are strongly demanded.

A full-bridge converter is considered as the isolated DC-DC converter. This full-bridge converter has some problems such that a large surge voltage occurs across the secondary side rectifying diodes[3].

In the conventional lower-output-voltage converters, RC snubber circuits were usually used[4~7]. However, in higher-output-voltage converters, power loss at snubber resister becomes much larger, and then this method is impractical. Thus, another surge snubber is needed. Recently, a simple surge snubber using a capacitor only has been proposed in the previous papers[8~9], and its prominent effect on the surge suppression has been confirmed experimentally.

However, this snubber capacitor severally deteriorates the control characteristics of the converter.

In order to solve the above problem, this paper proposes a new technique of utilizing the transformer leakage inductance. The control characteristics of the Full-Bridge Converter with Snubber Capacitor (FB-SC) are analyzed, and the effect of the leakage inductance on the control characteristics improvement is experimentally confirmed.

II. FULL-BRIDGE DC-DC CONVERTER WITH SNUBBER CAPACITOR

A block diagram of the HVDC power distribution system is shown in Fig. 1. The first stage consists of a power-factor-correction (PFC) converter for the harmonic reduction of input current waveform. The subsequent DC-DC converter operates for the dc-isolation and the voltage regulation. A circuit topology of the FB-SC is shown in Fig. 2, where the output-voltage regulation is performed by the phase-shift driving.

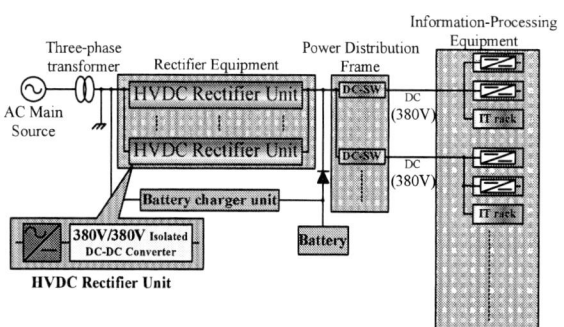

Fig. 1 HVDC distribution system

The surge suppression effect of this topology has been confirmed in reference. However, this snubber capacitor severally deteriorates the control characteristics of the converter. In ideal case, the output voltage can be derived as following equation;

$$V_{CS} = \frac{V_{in}}{n} = V_O \qquad (1)$$

From Eq. (1), the output voltage is decided by snubber capacitor voltage V_{CS}, and the output voltage cannot be controlled by the phase-shift driving.

In order to control the output voltage, the output voltage variation has been experimentally confirmed by the influence of leakage inductance so far[10].

Here, the effect of the leakage inductance on the control characteristics improvement is investigated. The relationship between the leakage inductance and the control characteristics are analyzed in next chapter.

III. ANALYTICAL RESULTS

Figure 3 shows the key waveforms of the FB-SC. From these waveforms, the operating states can be divided into 7 states in a half switching cycle. Figure 4 shows the each operating state of this circuit.

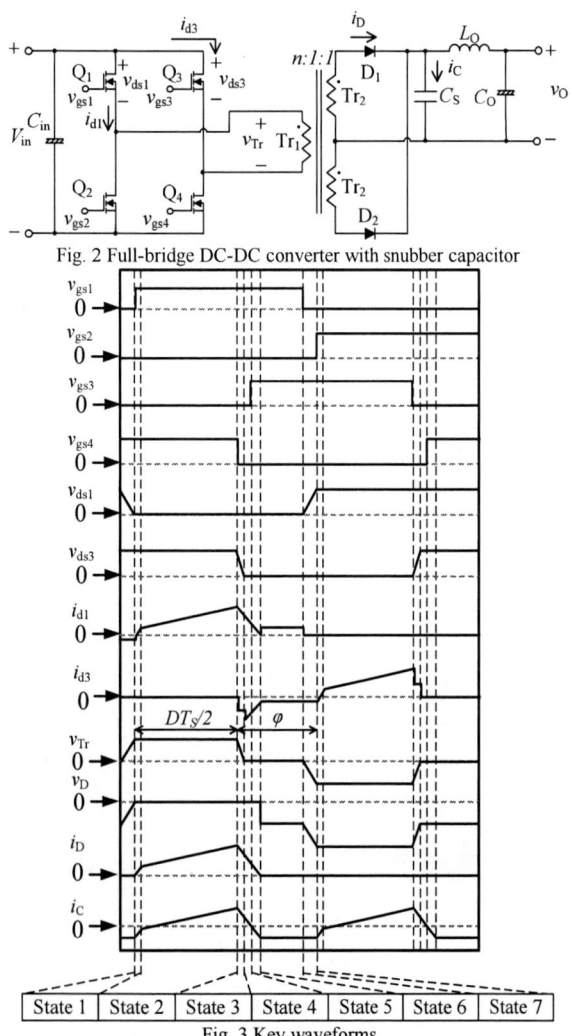

Fig. 2 Full-bridge DC-DC converter with snubber capacitor

Fig. 3 Key waveforms

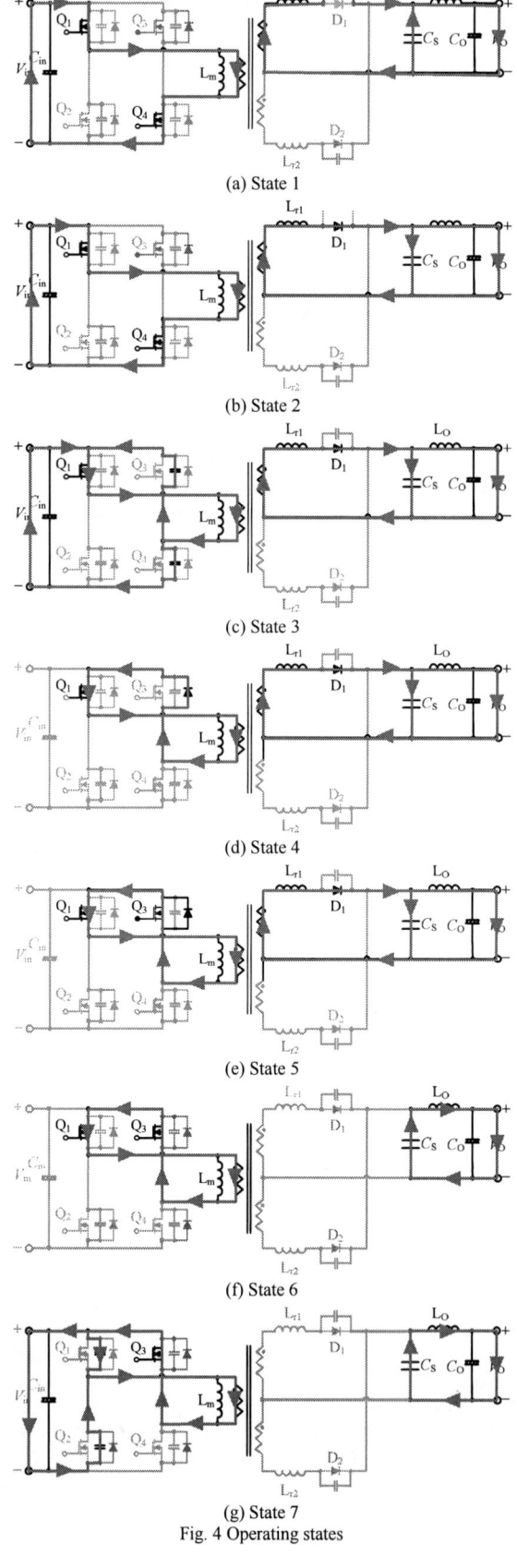

(a) State 1

(b) State 2

(c) State 3

(d) State 4

(e) State 5

(f) State 6

(g) State 7

Fig. 4 Operating states

The property of the circuit operations are shown as follows;

State 1: This state begins when Switch Q_1 turned on. During this period, the electric charge of secondary side diode's junction capacitance is discharged to output side. This state will be continued until the electric charge reaches zero.

State 2: Diode D_1 is turned on, and the power is transferred from primary side to the load. During this period, the applied voltage across the transformer is equal to input voltage V_{in}, and the energy of leakage inductance is stored by the secondary side current. This state will be end when the switch Q_4 turns off.

State 3: The current flowing through the switch Q_3 discharges parasitic capacitor. At that same time, the current flowing through the switch Q_4 charges parasitic capacitor. The energy stored in the leakage inductance is transferred to the load. This state will be end when the voltage V_{ds1} reaches zero and the voltage V_{ds4} reaches input voltage.

State 4: The body diode of switch Q_3 is turned on by the stored energy of the leakage inductance.

State 5: This state begins the switch Q_3 turns on under the ZVS condition. During this period, the current flowing through the switch Q_3 is negative because of the stored energy of the leakage inductance. This state will be end when the secondary side diode current reaches zero (stored energy of leakage inductance reaches zero).

State 6: During this period, the primary side current is circulating through switch Q_1 and Q_3. The secondary side rectifying diode D_1 is turned off. The output power is provided from snubber capacitor. This state will be end when the switch Q_1 turns off.

State 7: This state begins the switch Q_1 turns off. The current flowing through the switch Q_2 discharges parasitic capacitor. At that same time, the current flowing through the switch Q_1 charges parasitic capacitor. This state will be end when the voltage V_{ds2} reaches zero and the voltage V_{ds1} reaches input voltage.

The voltage conversion ratio can be derived by means of the extended state-space averaging method[11]. In order to simplify the state analysis, the primary side is modeled to the just voltage source, and then the parameters are converted to the secondary side. Based on this, the analytical models can be divided into 4 states by the considering of secondary side states as shown in Fig. 5. Where, the state 1 can be neglected because of very short interval. However, the state 1 is needed to calculate initial value. From these equivalent circuits, the state equations are derived as follows;

$$
\begin{cases}
\dfrac{d\hat{v}_S}{dt} = -\dfrac{D^2 T_S V_{in}/n}{4 C_S L_{r1}} + \dfrac{D^2 T_S V_{in}^2/n^2}{4 C_S L_{r1} \hat{v}_S} \\
\qquad\quad + \dfrac{L_{r1}}{C_S T_S \hat{v}_S} I_{D0}^2 + \dfrac{D V_{in}/n}{C_S \hat{v}_S} I_{D0} - \dfrac{\hat{i}_L}{C_S} \\
\dfrac{d\hat{i}_L}{dt} = \dfrac{-r_L \hat{i}_L + \hat{v}_S - \hat{v}_O}{L_O} \\
\dfrac{d\hat{v}_O}{dt} = \dfrac{\hat{i}_L - I_O}{C_O}
\end{cases}
\tag{2}
$$

Where, the duty ratio D has been defined by means of φ which is the phase difference between switch Q_1 and Q_4 as following equation;

$$
D = 1 - \frac{\varphi}{180}
\tag{3}
$$

Moreover, I_{D0} is the initial value of the diode current. In the steady state, the left side of Eq. (2) is equal to zero. Thus, the output voltage is derived as follows;

$$
V_O = \frac{V_{in}}{n} + \frac{\dfrac{I_{D0}^2 L_{r1}}{T_S I_O} - \dfrac{V_{in}}{n}\left(1 - D\dfrac{I_{D0}}{I_O}\right)}{1 + \dfrac{D^2 T_S V_{in}}{4 n L_{r1} I_O}} - r_L I_O
\tag{4}
$$

In order to evaluate Eq. (4), the estimation of initial value is required.

A. Initial value I_{D0}

The initial value is equal to the diode current at the end of state 1. Estimating the initial value I_{D0}, the transient analysis of state 1 is necessary. The equivalent circuit of state 1 is shown in Fig. 5(a). From this equivalent circuit, the diode current i_D can be derived from following equations.

$$
i_D(t) = \frac{V_{in}}{n}\sqrt{\frac{C_d}{L_{r1}}} \sin\frac{t}{\sqrt{C_d L_{r1}}}
\tag{5}
$$

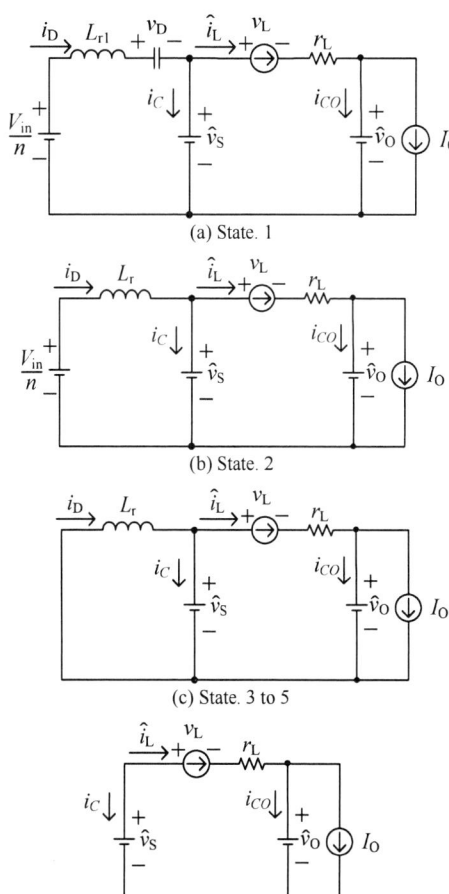

(a) State. 1

(b) State. 2

(c) State. 3 to 5

(d) State. 6 to 7

Fig. 5 Analyzed model

Further, the diode voltage v_D is derived as follows:

$$v_D(t) = \frac{V_{in}}{n}\left(1 - \cos\frac{t}{\sqrt{C_d L_{r1}}}\right) - V_O \qquad (6)$$

When the diode voltage reaches zero, the diode current is equal to initial value i_{D0}.

From Eq. (5) and (6), the initial value I_{D0} is derived as follows;

$$I_{D0} = \frac{V_{in}}{n}\sqrt{\frac{C_d}{L_{r1}}\left\{1 - \left(1 - n\frac{V_O}{V_{in}}\right)^2\right\}} \qquad (7)$$

The initial value depends on the output voltage as shown in Eq. (7). Therefore, the initial value is increasing with increasing the output voltage. When the output voltage is equal to the input voltage, the initial value becomes maximum value as shown in Eq. (8).

$$I_{D0} = \frac{V_{in}}{n}\sqrt{\frac{C_d}{L_{r1}}} \qquad (8)$$

From Eq. (4) and (7), the output voltage is obtained. Further, the output voltage approximately expressed by neglecting the infinitesimal terms as follows;

$$V_O = \frac{V_{in}}{n}\left(1 - \frac{1 - \frac{DV_{in}}{nI_O}\sqrt{\frac{C_d}{L_{r1}}\left\{1 - \left(1 - n\frac{V_O}{V_{in}}\right)^2\right\}}}{1 + \frac{D^2 T_S V_{in}}{4 L_{r1} I_O}}\right) \qquad (9)$$

From Eq. (9), the output voltage depends on the leakage inductance. The control characteristics may be improved by means of the leakage inductance.

B. The control characteristics of FB-SC

The circuit parameters and specifications are shown in Table. I. The voltage difference between input and output voltage becomes small in this case as shown in Table. I. When voltage difference is smaller, the initial value reaches maximum value of the I_{D0} as shown Eq. (8). Therefore, the maximum value of I_{D0} utilizes as the initial value. In this case, the output voltage becomes as follows in approximately;

$$V_O = \frac{V_{in}}{n}\left(1 - \frac{1 - \frac{DV_{in}}{nI_O}\sqrt{\frac{C_d}{L_{r1}}}}{1 + \frac{D^2 T_S V_{in}}{4 L_{r1} I_O}}\right) \qquad (10)$$

Figure. 6 shows the output voltage characteristics against the duty ratio. The output voltage variation becomes larger with large leakage inductance and its variation becomes smaller with large duty ratio. Moreover, the voltage variation becomes small when the leakage inductance is smaller. When the leakage inductance is 1µH, the output voltage is not able to be controlled by the duty ratio independently on load condition as shown in Fig. 6.

Furthermore, the output voltage variation depends on the load current. When the load current is lower, the output voltage variation becomes small. Therefore, the large leakage inductance is needed to obtain the wide control range in light load case.

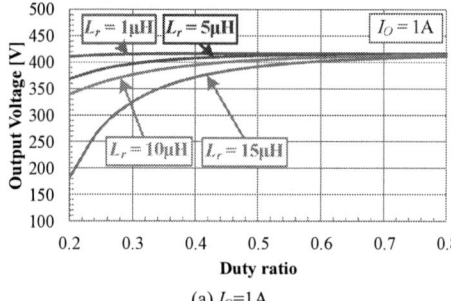

(a) I_O=1A

(b) I_O=2A

Fig. 6 Control characteristics

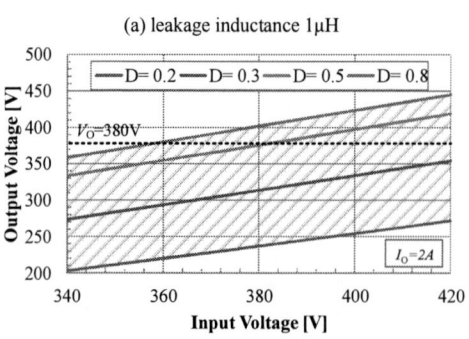

(a) leakage inductance 1µH

(b) leakage inductance 10µH

Fig. 7 Line characteristics

TABLE I   The circuit parameters and specification

Description		Value
Input Voltage		340V to 420V
Output Voltage		380V
Load Current		0.2A to 2.5A
MOSFET		SPP20N60S5 650V/20A
Diode		C2D10120 1200V/10A
Transformer	L_m	450µH
	L_τ	2µH
	Turn ratio	28:30:30
Snubber Capacitor		330nF
Output indutor		300µH
Smoothing capacitor		47µF
Switching frequency		200kHz

Figure 7 shows the relationship between the input and output voltage. The output voltage is almost linearly dependent on the input voltage. In the case of the small leakage inductance, the output voltage variation width by the duty ratio becomes narrow as shown in Fig. 7(a). On the other hand, when the leakage inductance is larger, the output voltage variation width by the duty ratio becomes wider as shown in Fig. 7(b).

Figure 8 shows the load characteristics with various duty ratio. In the case of the small leakage inductance, the output voltage independent on the duty ratio as shown in Fig. 8 (a). When the leakage inductance is lager, the control range of the output voltage becomes wider.

From the these analytical results, it is clarified that the control characteristics is improved by larger leakage inductance.

IV. EXPERIMENTAL RESULTS

In order to examine the validity of the analytical results, the experimental measurements are necessary. The prototype evaluation board has been implemented, and the circuit parameters and specifications are utilized the same as the analytical works as shown in Table I. Here, the external inductances are installed in secondary side instead of leakage inductance.

Figure 9 shows experimental key waveforms. A little ringing is confirmed. However, the experimental key waveforms agree with theoretical key waveforms.

Figure 10 shows the output voltage characteristics against the duty ratio. When the leakage inductance is $2\mu H$, the output voltage cannot be controlled by the duty ratio as shown Fig. 10(a). On the other hand, the output voltage can be controlled by the duty ratio when the leakage inductance is $12\mu H$ as shown Fig. 10(b).

The output voltage can be controlled by the duty ratio between 0.2 and 0.4 when the load current is 1A as shown in Fig. 10 (a). On the other hand, when the load current is 2A, the output voltage can be controlled by the duty ratio between 0.2 and 0.6 as shown in Fig. 10 (b). From these results, it is clarified that the control range depends on not only leakage inductance but also load current. Therefore, the design of this converter is necessary to consider the above dependence.

Figure 12 shows the relationship between the input and output voltage. The output voltage is almost linearly dependent on the input voltage. When the leakage inductance is $2\mu H$, the output voltage is not able to control as shown in Fig. 11(a). On the other hand, when the leakage inductance is $10\mu H$, the output voltage can be controlled as shown in Fig. 11(b).

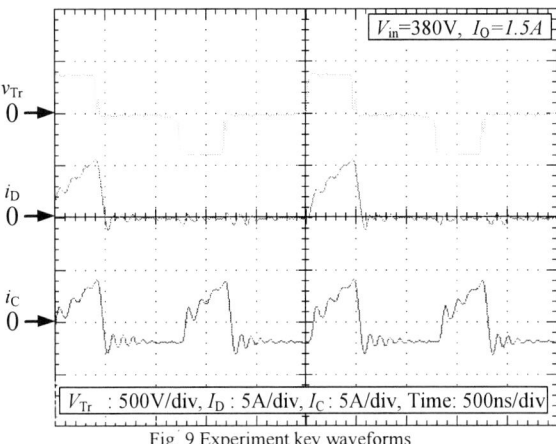

Fig. 9 Experiment key waveforms

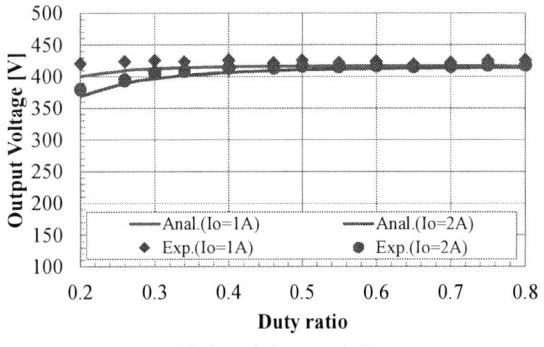

(a) leakage inductance $2\mu H$

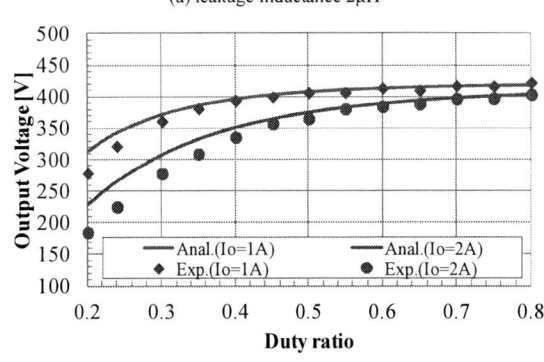

(b) leakage inductance $12\mu H$
Fig. 10 Control characteristics

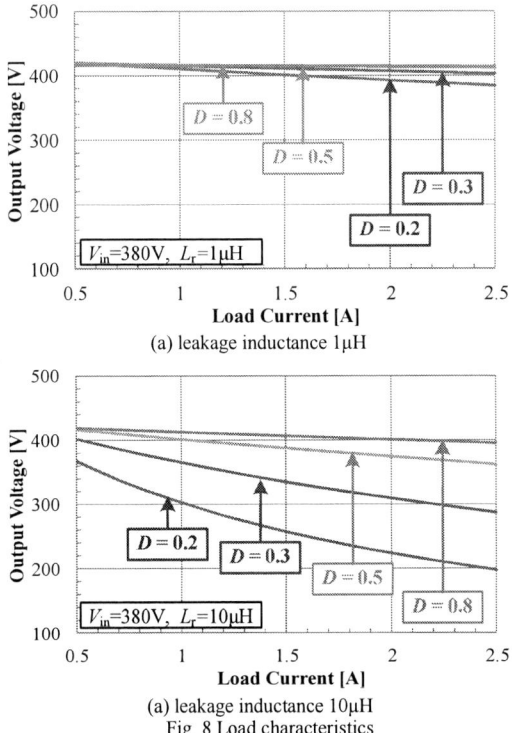

(a) leakage inductance $1\mu H$

(a) leakage inductance $10\mu H$
Fig. 8 Load characteristics

978-1-4799-2706-7/14 $31.00 © 2014 IEEE

Figure 12 shows the load characteristics with various duty ratio. The output voltage decreases with decreasing the load current. The output voltage variation is changed depending on the leakage inductance. When the leakage inductance is 2μH, the output voltage independent on the duty ratio as shown in Fig. 12 (a). When the leakage inductance is 12μH, the control range of the output voltage becomes wider as shown in Fig. 12 (b).

These experimental results are agreed well analytical results. Moreover, the control characteristics improvement by mean of leakage inductance is confirmed by analytical and experimental.

V. CONCLUSIONS

This paper proposed a new technique of utilizing the transformer leakage inductance in order to improve the control characteristics of the Full-Bridge DC-DC converter with snubber capacitor. The effect of the leakage inductance on the control characteristics was clarified by analytical and experimental. As a result, it was confirmed that control range depends on not only leakage inductance but also load current. The design of leakage inductance for wide control range is necessary to consider the load condition.

The design technique of the leakage inductance will be discussed in the future work.

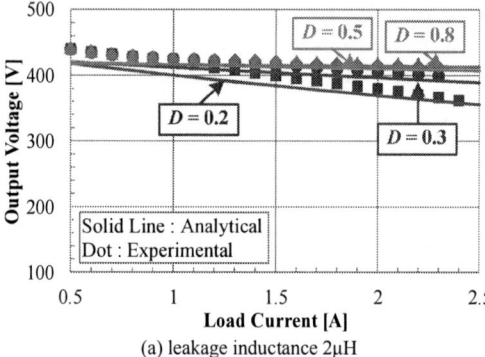

(a) leakage inductance 2μH

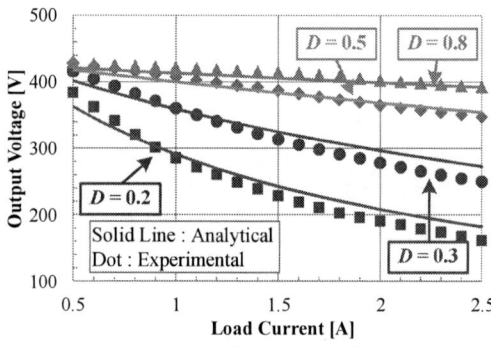

(b) leakage inductance 12μH
Fig. 12 Load characteristics

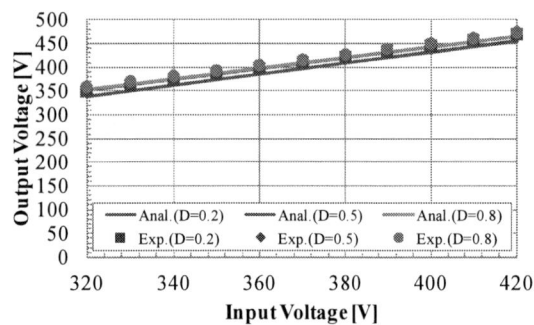

(a) leakage inductance 2μH

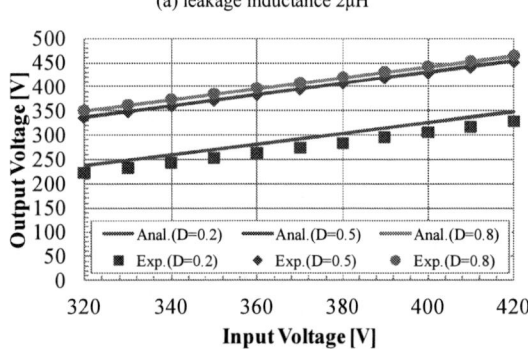

(b) leakage inductance 12μH
Fig. 11 Line characteristics

REFERENCES

[1] A.Matsumoto, A.Fukui, T.Takeda, M.Yamasaki: "Development of 400Vdc Power Distribution System and 400Vdc Output Rectifier," Proc. IEEE INTELEC'09, Sess.PA2-1, pp.68–73, Oct. 2009.

[2] U.Badstuebner, J.Biela, J.W.Kolar:" An Optimized, 99% Efficient, 5kW, Phase-Shift PWM DC-DC Converter for Data Center and Telecom Applications," IPEC'10 Record, pp.626-634, June. 2010.

[3] K.Domoto, et al:" Surge Analysis and Snubber Design for a Full-Bridge Isolated DC-DC Converter in HVDC Power Distribution Systems," IEEJ Trans. IA, Vol.133, No.12, pp.1171-1178, 2013 (in Japanese)

[4] A.Jangwanitlert, J.C.Balda.:"Phase-shifted PWM full-bridge DC-DC converters for automotive applications: reduction of ringing voltages," Proc. IEEE Power Electronics in Transportation, pp.111-115, Oct.2004.

[5] K.Yoshida: "ZVS active clamp full-bridge conveter with CRD snubber circuit," IEICE Technical Report, Vol.101, No.38, pp.43-48,EE2001-5, May 2001 (in Japanese)

[6] K.Orikawa J.Itoh :"Principle of Surge Voltage of a Rectifier in Isolated DC-DC Converters and Snubber Circuit Dsign Method," IEEJ Trans. IA, Vol.133, No.3, pp.350-359, 2013 (in Japanese)

[7] M.Hirokawa, T.Ninomiya : "Non-Dissipative Snubber for Rectifying Diodes in a High-Power DC-DC Converter," IEEJ Trans. IA, Vol.125, No.4, pp.366-371, 2005 (in Japanese).

[8] R.Simanjorang, H.Yamaguchi, H.Ohashi, K.Nakao, T.Ninomiya, S.Abe, M.Kaga, A.Fukui:" High-Efficiency High-Power DC-DC Converter for Energy and Space Saving of Power-Supply System in a Data Center," APEC'11 Record, pp.600-605, Mar. 2011.

[9] H.Kawano (Toyota industries corporation) :"Switching Power Supply circuit," JPA 2006-230075 (2005.02.16)

[10] K.Domoto, T.Ninomiya, Y.Ishizuka, S.Abe, M.Kaga, R.Simanjyorang, H.Yamaguchi:"Surge snubber design for high power-density DC-DC converters in HVDC power distribution systems," Proc. IEEE ICRERA'12, pp1-5,Nov.2012

[11] K. Harada, T. Ninomiya:"On the Mechanism of Switching-Noise Generarion and its Suppression Techniques in a DC-to-DC Converter," IECE Trans. Vol.62, No.12, pp.795-802,1979.

DCM Control Method of Boost Converter based on Conventional CCM Control

Le Hoai Nam, Koji Orikawa, Jun-ichi Itoh
Department of Electrical, Electronics and Information Engineering
Nagaoka University of Technology
Nagaoka, Japan
lehoainam@stn.nagaokaut.ac.jp

Abstract— This paper proposes current control method for DCM to achieve same control performance as CCM. The proposed control loop is designed based on the conventional current control loop in CCM. By using only one PI controller and introduction of two correction factors, the proposed control method can control both CCM and DCM current exactly to design values. In this paper, the operation of the proposed control method is confirmed by simulation and experiment. The simulation results of the current response in both CCM and DCM agree with design values. In the experimental results, due to the delay in feedback, error 11.3% in rise time and error 2.5 point in overshoot occur. Furthermore, the voltage regulation experiment with the proposed current controller as a inner control loop is conducted. As load varies, the recovery time for the output voltage regulation is about 20 ms and the overshoot – undershoot voltage is below 3%.

I. INTRODUCTION

The research in electric power conversion has grown rapidly in recent years due to energy crisis. DC-DC converters present in most power conversion systems or devices such as: fuel cell systems in mobile devices [1], photovoltaic(PV) systems [2], lithium-ion battery in hybrid vehicles (HEVs) [3][4], smart power system for servers [5], e.t.c. Because the minimization of those DC-DC converters not only increases the mobility of mobile devices but also simplifies the layout for PV systems or battery systems in HEVs, many studies have focused on shrinking the DC-DC converter circuit as much as possible. Besides, passive components such as inductors and capacitors account for the majority of the volume of DC-DC converters, so the miniaturization of these passive components leads to compact DC-DC converter generation. One of methods to minimize these passive components is to switch the circuit at high frequency [6]. However, switching at high frequency leads to the increase of frequency-dependent losses such as switching loss in MOSFET or recovery loss in diode [7]. The increase of conversion loss accounts for larger cooling sink, and this makes the minimization of DC-DC converters further difficult. In order to eliminate the increase of loss when switching at high frequency, the concept of resonant converters has been proposed [8]. Some researchers have introduced DC-DC converters operating at frequencies of hundreds of MHz, while maintaining the power conversion efficiency high [9][10]. However, in those resonant converter, because switch's

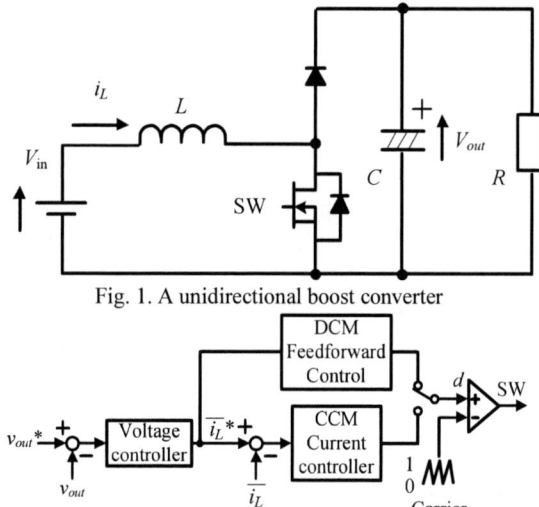

Fig. 1. A unidirectional boost converter

Fig. 2. Conventional CCM/DCM control method.

duty is limited in narrow range, the problems such as the uncontrollability of output voltage or the overheat when resonant converter operates in boot mode [11], are difficult to be solved. This limits the practical applications for resonant converters. On the other hand, the minimization of the inductor can be accomplished by the utility of Discontinuous Current Mode (DCM).

Fig. 1 shows a unidirectional boost converter which is generally applied in PV system to regulate the output voltage while preventing the power from flowing back into the input side. When the inductor is minimized, the ripple of the input current becomes higher, and this leads to the discontinuity of input current. This can be understood that only a small amount of energy can be stored in a minimized inductor, so all stored energy in inductor will be released before a new switching period begins. In order to design controller for DCM similar to the conventional Continuous Current Mode (CCM), many researches have been conducted [12][13][14], but a specific method to accomplish this has not been discussed.

Furthermore, the inductor current alternates between CCM and DCM depending on the variation of the load, and this likely leads to the unstable operation of the converter. Many CCM/DCM control methods have been studied in past years to deal with the nonlinearity in DCM, such as Adaptive Tuning method [15], Bi-frequency Pulse-Train Control technique [16], Peak Current Mode Bifrequency Control Technique [17], Current-Mode

Synthetic Control Technique [18], Switching-Frequency Control [19], e.t.c [20][21][22]. However, those control methods either lead to the complication of the control system, or the increase in cost for the auxiliary circuit such as the variable frequency pulse generator.

Fig. 2 shows a conventional control system which composes of an Automatic Voltage Regulator (AVR) and two Automatic Current Regulators (ACR) [23]. Many designers prefer to utilize PI controller for both AVR and ACR because of its simple design. In conventional inductor current PI control method, the PI parameters for CCM can be designed simply based on the linear CCM transfer function. On the other hand, due to the nonlinear characteristic of the DCM transfer function, the current control loop in DCM is designed based on the reduced-order model [12][23]. This leads to the feed forward current control in DCM. As a result, the DCM current response cannot be controlled as same as the CCM current response. In this case, DCM current control is effective in only critical conditions.

This paper proposes a control method in which only one PI controller is necessary to regulate both CCM and DCM. By introduction two correction factors into the conventional inner current control loop in CCM, the current response of DCM is controlled exactly to that of CCM. Besides, it is not necessary to detect the current mode. In this paper, first, the principle of the proposed CCM-DCM control method is described. Second, the introduction of two correction factors α_{DCM} and K_{DCM} are described. Then, the transition between CCM and DCM is explained. Finally, the operation of the boost converter which the proposed control method is applied, is confirmed by simulation and experiment.

II. PROPOSED CCM-DCM CONTROL SCHEME

A. Conventional inner current loop design in CCM

Fig. 3 shows typical boost converter configuration. The average small signal modeling technique is used to model the boost converter for the inner current control loop design [6]. In this chapter, the output voltage is assumed to be constant, because the response of the inductor current is much higher than that of the output voltage in ACR design step.

Fig. 4 shows inductor current waveform in DCM, where d and d' denote the duty ratio of the first and the second interval respectively. Equation based on the average model of the boost converter in CCM and DCM is given by (1),

$$L\frac{d\bar{i}_L}{dt} = dV_{in} + d'\left(V_{in} - V_{out}\right) \dots\dots\dots\dots\dots\dots\dots(1)$$

where V_{in} is the input voltage, V_{out} is the output voltage, $\bar{i}_L$ is the average inductor current during a switching period T_{sw}, and L is the inductance. In case of CCM, the relationship between d and d' is given by (2).

$$d + d' = 1 \dots\dots\dots\dots\dots\dots\dots\dots\dots\dots\dots\dots\dots\dots(2)$$

Substituting (2) into (1), (3) is obtained.

$$L\frac{d\bar{i}_L}{dt} = V_{in} - V_{out} + dV_{out} \dots\dots\dots\dots\dots\dots\dots\dots\dots(3)$$

Here, the averaged control to inductor current transfer

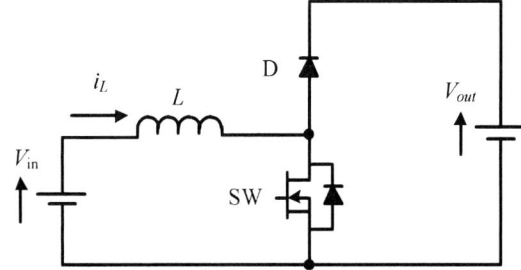

Fig. 3. Typical boost converter configuration in ACR design

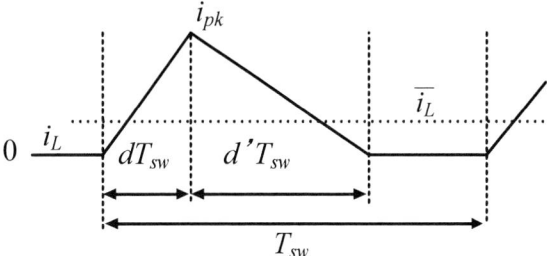

Fig. 4. Inductor current waveform in DCM

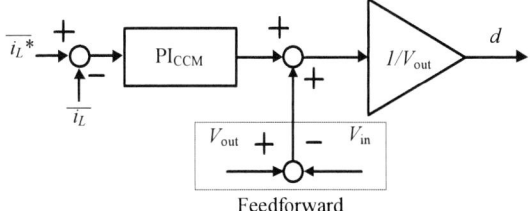

Fig. 5. Conventional inner current loop in CCM

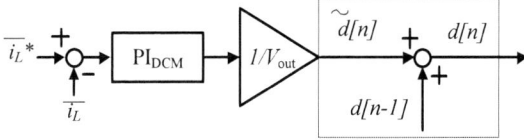

Fig. 6. Proposed inner current loop in DCM

function in CCM is given by (4),

$$G_{CCM}(s) = \frac{\bar{i}_L(s)}{d(s)} = \frac{V_{out}}{sL} \dots\dots\dots\dots\dots\dots\dots\dots\dots\dots(4)$$

Fig. 5 shows ACR-CCM designed based on (3) and (4). The integral period T_i and proportional gain K_p of PI_{CCM} controller are designed based on the second-order standard form (5),

$$G_{2nd-order-delay}(s) = \frac{\omega_n^2}{s^2 + 2\zeta\omega_n s + \omega_n^2} \dots\dots\dots\dots\dots\dots(5)$$

where ω_n is the natural angular frequency, ζ is the damping factor, and both of which are designed to achieve the desired inductor current response. The closed loop transfer function of ACR-CCM is derived by (6).

$$H_{CCM-ACR}(s) = \frac{\bar{i}_L(s)}{\bar{i}_L{}^*(s)} = \frac{\frac{K_p}{LT_i}\left(1 + sT_i\right)}{s^2 + \frac{K_p}{L}s + \frac{K_p}{LT_i}} \dots\dots\dots\dots(6)$$

In order to make the design of PI parameters simple, a low pass filter whose role is to filter the command current $\bar{i}_L{}^*$ is necessary for matching (5) and (6). However, a low pass filter is not essentially needed.

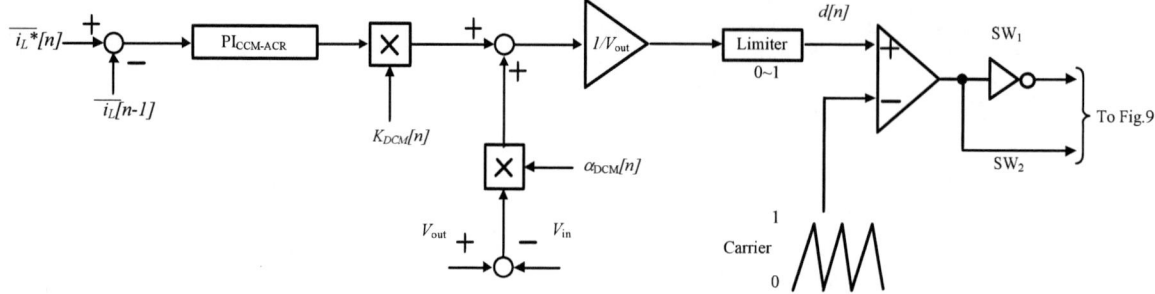

Fig. 7. Configuration of the proposed CCM/DCM current controller.

Using a low pass filter depends on designers. The PI parameters are then obtained by matching (5) and (6).

$$K_p = 2\xi\omega_n L \quad\text{..(7)}$$

$$T_i = \frac{2\xi}{\omega_n} \quad\text{..(8)}$$

B. Small signal modeling in DCM

In case of DCM, the relationship between d and d' is given by (9).

$$d + d' < 1 \quad\text{..(9)}$$

In order to derive the averaged control to inductor current transfer function in DCM, it is necessary to represent the second interval d' in Fig. 4 as a function of the first interval d. Based on [12] and the waveform shown in Fig. 4, this can be done as follow

$$\bar{i}_L = \frac{i_{pk}}{2}(d + d') \quad\text{..(10)}$$

where i_{pk} is the inductor peak current during a switching period T_{sw}. Neglect the voltage drop by the switch, the voltage across the inductor during the first interval d can be assumed as V_{in}. The inductor peak current can be written as

$$i_{pk} = \frac{V_{in}}{L}dT_{sw} \quad\text{..(11)}$$

Substituting (11) into (10) and solving the resulting equation for d' yields

$$d' = \frac{2L\bar{i}_L}{dT_{sw}V_{in}} - d \quad\text{..(12)}$$

Substituting (12) into (1), (13) is obtained.

$$L\frac{d\bar{i}_L}{dt} = \frac{2L\bar{i}_L}{dT_{sw}}\left(\frac{V_{in} - V_{out}}{V_{in}}\right) + dV_{out} \quad\text{...................(13)}$$

In case of DCM, the averaged control to inductor current transfer function in DCM cannot be derived simply from (13) because of its nonlinearity compared to (3). Therefore, (13) is linearized at steady point, then the averaged control to inductor current transfer function in DCM can be derived by linearized equation (14)

$$G_{DCM}(s) = \frac{\tilde{i}_L(s)}{\tilde{d}(s)} = \frac{V_{out}}{L}\frac{2}{s + 2\sqrt{\dfrac{(V_{out}/V_{in}-1)V_{out}}{2LI_L T_{sw}}}} \quad\text{........(14)}$$

where $\tilde{i}_L$ is the small signal of inductor current, $\tilde{d}$ is the small signal of duty, I_L is the value of the inductor current at steady point. Eq. (14) indicates that not only the gain of the DCM transfer function differs from that of the

CCM transfer function, but also the control target changes from the full values $\bar{i}_L$ and d in CCM into the small values $\tilde{i}_L$ and $\tilde{d}$ in DCM. Therefore, in order to achieve the DCM current control by the conventional CCM-ACR, both the control gain and the control target of the PI_{CCM} controller needs to be modified. The modification of control target and control gain is achieved by introducing two corrector factors α_{DCM} and K_{DCM} into conventional current control loop in CCM.

C. Proposed CCM/DCM current control loop

1) Derivation of the first correction factor α_{DCM}

The modification of the control target from CCM to DCM is achieved by compensating the output of the CCM feedforward part using the correction factor α_{DCM} that given in (16) [22]. The procedure to calculate this correction factor α_{DCM} is as follows.

When considering the control system in z-domain, the control target of DCM $\tilde{d}[n]$ is given by

$$\tilde{d}[n] = d[n] - d[n-1] \quad\text{...............................(15)}$$

where $d[n]$ is the output duty to control inductor current, $d[n-1]$ is the value of the duty in one sampling period before. Eq. (15) indicates that, the output duty is the sum between the value of the duty in one sampling period before and the difference duty controlled by the PI controller in DCM.

Fig. 6 shows the proposed DCM current control loop. The objective to change Fig. 5 into Fig. 6, is achieved by multiplying the output of the CCM feedforward part with the correction factor α_{DCM}.

$$\alpha_{DCM}[n] = \frac{V_{out}}{V_{out} - V_{in}}d[n-1] \quad\text{...............................(16)}$$

2) Derivation of the second correction factor K_{DCM}

The modification of the control gain from CCM to DCM is achieved by compensating the output of the PI_{CCM} using the correction factor K_{DCM} that given by (21). The procedure to calculate this correction factor K_{DCM} is as follows.

The discretized control gains (17), (18) are obtained by discretizing (4) and (14) in sampling period T_{sw}. Note that the sampling period equals to the switching period.

$$|G_{CCM}(z)| = \frac{V_{out}T_{sw}}{L} \quad\text{..(17)}$$

$$|G_{DCM}(z)| = \frac{V_{out}}{L}\sqrt{\frac{2LI_L T_{sw}}{(V_{out}/V_{in}-1)V_{out}}} \quad\text{........................(18)}$$

In order to change the control gain from $|G_{CCM}(z)|$ to

$|G_{DCM}(z)|$, a correction factor K_{DCM} is multiplied with the output of the PI$_{CCM}$ controller. K_{DCM} is given by (19).

$$K_{DCM} = \frac{|G_{CCM}(z)|}{|G_{DCM}(z)|} = \sqrt{\frac{(V_{out}/V_{in}-1)V_{out}T_{sw}}{2LI_L}} \quad \ldots\ldots\ldots\ldots \quad (19)$$

The K_{DCM} in (19) is simplified by using the equation of I_L at steady state as below [12]. Note that (20) is obtained by simply let the inductor current differential part $\dfrac{d\bar{i}_L}{dt}$ in (13) be zero.

$$I_L = \bar{i}_L[n-1] = \frac{d^2[n-1]T_{sw}V_{out}V_{in}}{2(V_{out}-V_{in})L} \quad \ldots\ldots\ldots\ldots\ldots\ldots..\ldots \quad (20)$$

Substituting (20) into (19), K_{DCM} is also given by (21).

$$K_{DCM}[n] = \frac{V_{out}-V_{in}}{V_{in}}\frac{1}{d[n-1]} \quad \ldots\ldots\ldots\ldots\ldots..\ldots\ldots\ldots \quad (21)$$

Fig. 7 shows the proposed CCM-DCM current control loop, which controls both CCM and DCM by using only one PI$_{CCM}$ and two correction factors α_{DCM} and K_{DCM}.

Table I shows the values of two correction factors in case of CCM and DCM.

D. Transition between CCM and DCM

As mentioned above, the proposed control method can achieve both CCM and DCM control without detect the inductor current. The approach for this can be explained as follow. From Table I , both correction factors α_{DCM} and K_{DCM} have to be set to 1 when the circuit operates in CCM and be set to (16) , (21) respectively in DCM.

Fig. 8 shows two correction factors α_{DCM} and K_{DCM} as functions of duty d. Let consider the transition of α_{DCM} first. As shown in Fig. 8, the transition point falls right into the boundary between CCM and DCM, which means α_{DCM} can trade between CCM value and DCM value without detecting the circuit operating mode. This can be explained as when the circuit operates in CCM , the duty d follows as

$$d = \frac{V_{out}-V_{in}}{V_{out}} \quad \ldots\ldots\ldots\ldots\ldots..\ldots\ldots\ldots\ldots\ldots\ldots\ldots..(22)$$

Neglect the parasitic characteristic of switch and diode, (22) is always established whenever the circuit operates in CCM. Substituting (22) into (16), it is understood that α_{DCM} is set to 1 because of the steady operation of CCM.

On the other hand, as shown in Fig. 8, K_{DCM} do not change into 1 automatically when the circuit has already switched from DCM to CCM. However, because α_{DCM} can change states automatically, the transition of K_{DCM} is accomplished by observing the value of α_{DCM}. To be specific, whenever α_{DCM} becomes 1, K_{DCM} is also set to 1, otherwise K_{DCM} follows as (21). Because α_{DCM} can trade values between CCM and DCM without detecting current mode, the transition of K_{DCM} based on α_{DCM} state is also accomplished without sensing which mode the circuit is operating.

III. SIMULATION RESULTS

Simulation are carried out to verify the proposed CCM/DCM control method.

Fig. 9 shows the simulation system. The switching device in the upper arm SW$_1$ is used when the boost

TABLE I
VALUES OF TWO CORRECTION FACTOR IN CCM AND DCM

Current Mode	CCM	DCM
α_{DCM}	1	$V_{out}d[n-1]/(V_{out}-V_{in})$
K_{DCM}	1	$(V_{out}-V_{in})/(V_{in}d[n-1])$

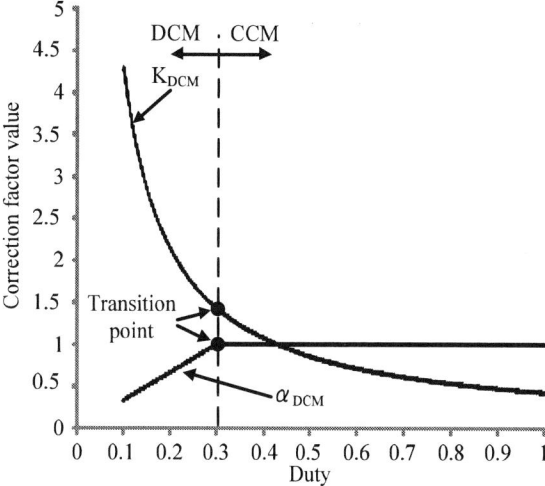

Fig. 8. Relationship between duty and correction factors at steady point

Fig. 9. Boost converter for ACR Simulation and Experiment.

TABLE II
SIMULATION AND EXPERIMENTAL PARAMETERS

Input voltage V_{in}	70 V
Output voltage V_{out}	100 V
Rated output power P_{out}	56 W
Boost chopper inductor L	360 µH
Switching frequency f_{sw}	20 kHz
Sampling frequency f_{samp}	20 kHz
Damping factor ζ ACR	0.7
Natural angular frequency ω_n ACR	2500/3000/3500 rad/s

converter operates in CCM. When the DCM operation is tested, the switch SW$_1$ is always kept at off-state.

Table II lists the system parameters and ratings. In order to avoid the appearance of the delay time in feedback, the sampling of the average inductor current is achieved by simply taking advantage of the periodic average block in simulator.

Fig. 10 shows the simulation results of the step response. Fig. 10(a) and Fig. 10(b) show the current waveform of CCM and DCM respectively. Fig. 10(c) shows the step response of the average inductor current in both CCM and DCM. It can be concluded from Fig. 10(c) that the DCM current response is the same as the CCM current response. In order to verify the operation of the proposed CCM-DCM control method in many cases, the measurement of the overshoot and the rise time of the current response are conducted while varying the natural

The 2014 International Power Electronics Conference

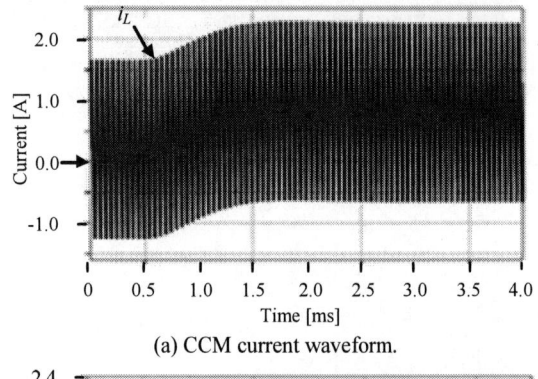

(a) CCM current waveform.

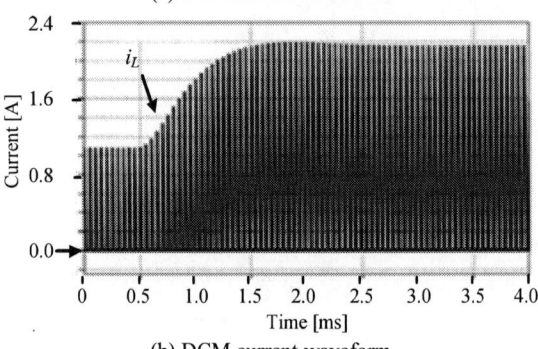

(b) DCM current waveform.

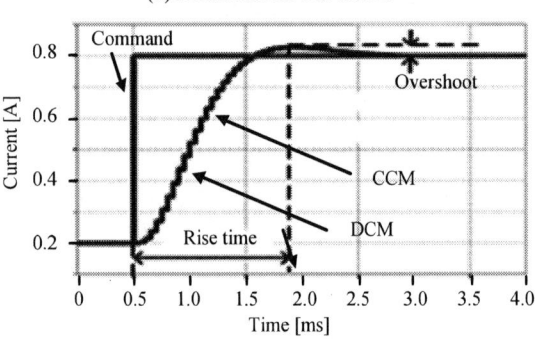

(c) CCM and DCM average current response
Fig. 10. Simulation results of step response.

frequency of the PI_{CCM} controller.

Fig. 11 shows the overshoot and the rise time of the current response in simulation while varying the natural frequency of the PI_{CCM} controller. It is confirmed that the rise time and the overshoot of both CCM and DCM current are agreed with the design values.

IV. EXPERIMENTAL RESULTS

The experimental conditions are the same as those of the simulation as shown in Table II. In experiment, the average inductor current is sampled by using a Low Pass Filter (LPF) with cutoff frequency of 2 kHz. Note that the switch SW_1 is always kept at off-state in case of DCM.

Fig. 12 and Fig.13 show the experimental step response of the inductor current response at $\omega_n = 3000$ rad/s and $\omega_n = 3500$ rad/s, respectively. Both the inductor current and the average values of the inductor current in CCM and DCM are demonstrated in same waveform. There is a slight difference in the shape of the current response between CCM and DCM. This difference can be explained as follows. The state of SW_1 is different between CCM and DCM. As a result, the on-resistance in

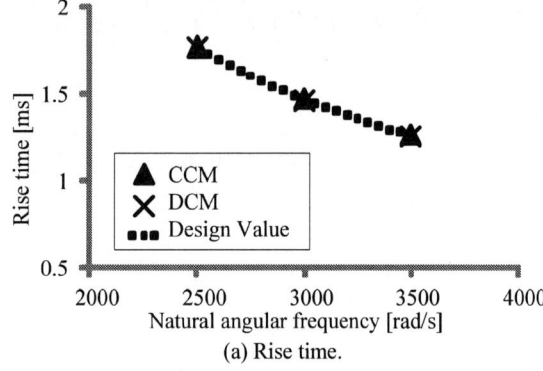

(a) Rise time.

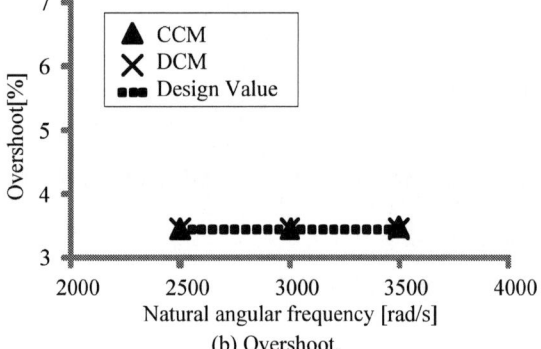

(b) Overshoot.
Fig. 11. Rise time and overshoot of the current response
in simulation.

MOSFET and the voltage drop by diode change the shape of the current response.

Fig. 14 shows the overshoot and the rise time of the current response in experiment while varying the natural frequency of the PI_{CCM} controller. There are errors between CCM/DCM current response and the design values. In the rise time results, the maximum error is 11.3%, while it is 2.5 point in the overshoot results. It can be seen from Fig. 13 that in both CCM response and DCM response, the experimental results of the rise time tend to become smaller than the design values, while the experimental results of the overshoot tend to become higher than design values. This error tendency can be explained as follows. When a LPF is used to obtain the average values of any signal, the time constant in LPF becomes the delay time between the input and output of LPF. This delay time increases the total delay time in the feedback of the controller. As a result, the rise time of the system response becomes shorter, while the overshoot becomes higher than the design values. Next step is to confirm the validity of the proposed CCM/DCM current controller in the output voltage regulation system.

Fig. 15 shows the boost converter whose the output voltage needs to be regulated for proper operation . Note that the upper switch is changed into a diode in order to let the circuit operate in DCM at light load. This also prevents the current flowing back into the source, which damages the source in such PV systems. Besides, the PI parameters in AVR are as follows

$$K_p = 2\xi\omega_n C \dots\dots\dots\dots\dots\dots\dots\dots\dots\dots(23)$$

$$T_i = \frac{2\xi}{\omega_n} \dots\dots\dots\dots\dots\dots\dots\dots\dots\dots\dots(24)$$

Table III lists the system parameters and ratings in

978-1-4799-2706-7/14 $31.00 © 2014 IEEE

The 2014 International Power Electronics Conference

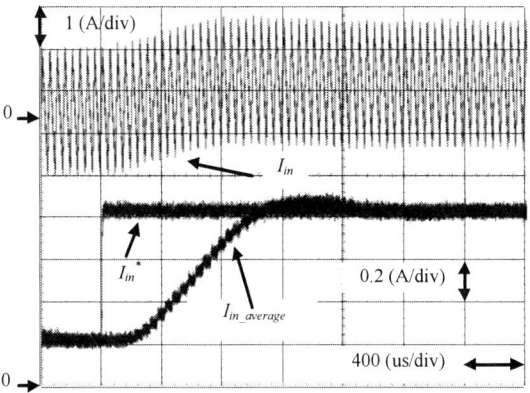

(a) CCM current waveform and average current response.

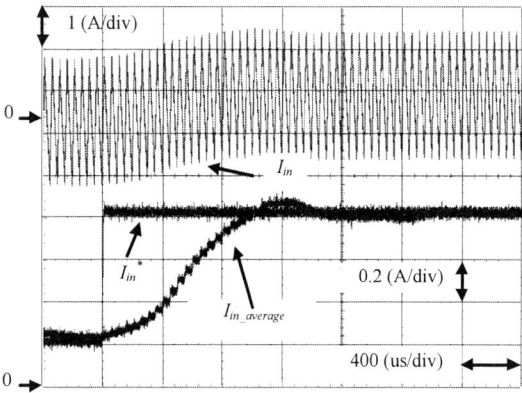

(a) CCM current waveform and average current response.

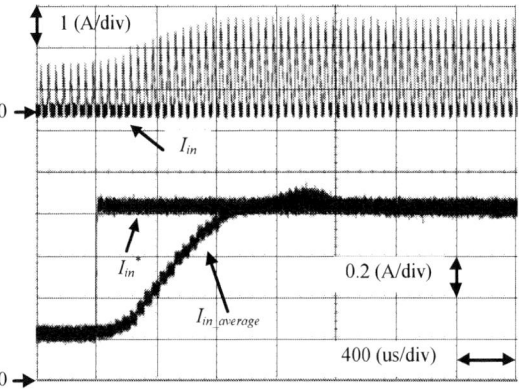

(b) DCM current waveform and average current response.
Fig. 12. Experimental results of step response
($\omega_n = 3000$ rad/s)

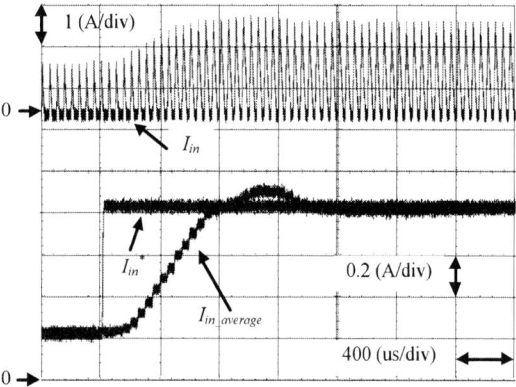

(b) DCM current waveform and average current response.
Fig. 13. Experimental results of step response
($\omega_n = 3500$ rad/s)

AVR experiment. Note that the natural angular frequency in AVR should be designed smaller at least 10 times than that of ACR in order to achieve the stable operation of the control system.

Fig. 16 shows the experiment results of load transient response between CCM (heavy load) and DCM(light load). Fig. 16(a) shows the output voltage response and the input current response when load varies from 40% in DCM to 100% in CCM. The recovery time for the output voltage regulation is about 20 ms and the undershoot voltage is 2 V, which is below 3% of the output voltage. Fig. 16(b) shows the output voltage response and the input current response when load varies from 100% in CCM to 40% in DCM. The recovery time for the output voltage regulation is about 20 ms and the overshoot voltage is 2 V, which is below 3% of the output voltage. There are two things needs to be noted in Fig. 16. First, when load varies the input voltage also slightly changes. In the experiment, a three-phase diode bridge rectifier and an input side smoothing capacitor are applied as the voltage source, so when the load varies, the voltage drop by the diode bridge and the smoothing capacitor also changes. This results in the variation of the input voltage when the load varies. Second, the input current quite oscillates not only in DCM but also in CCM. This is explained that, the accuracy of the average input current sampling mainly depends on three factors: the ripple of the input current, the speed of the D/A converter, and the time constant of LPF. According with the minimization

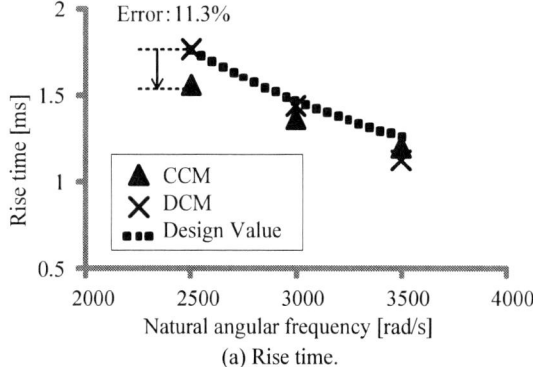

(a) Rise time.

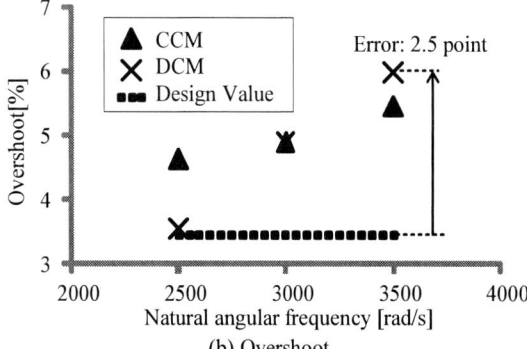

(b) Overshoot.
Fig. 14. Rise time and overshoot of the current response in experiment

of the boost chopper inductor, the ripple current becomes higher. The average value of the input current needs to be sampled in order to achieve the proper operation of ACR. Therefore, a LPF is applied in order to obtain average value of the input current. However, this LPF introduces a considerable delay time into the feedback, while the mismatch in each sampling point in the D/A converter lead to the feedback into ACR is just a delayed value of the actual current. This results in the oscillation of the input current. However, in both cases, the output voltage is regulated at the command. On the other hand, in order to avoid the oscillation of duty in CCM which might cause the current mode determination inaccuracy, instead of comparing with 1, α_{DCM} is compared with 0.9 to determine the state of K_{DCM}. This achieves the CCM/DCM current control without detecting the input current or the parameters of the boost converter circuit.

Fig. 17 shows the experiment results of load transient at light load. Fig. 17(a)(b) shows the output voltage response and the input current response when load varies from 20% in DCM to 40% in CCM and vice versa. The recovery time for the output voltage regulation is about 15 ms and the overshoot voltage is least than 1 V, which is below 2% of the output voltage. Both Fig. 16 and Fig. 17 confirm the validity of the proposed CCM/DCM current control method.

V. CONCLUSION

In this paper, a control method for both CCM and DCM was proposed. In the proposed method, the control of CCM and DCM current is achieved by using only one PI controller and introduction of two correction factors. Besides, the CCM/DCM current control can be achieved without sensing the input current or any circuit parameters. Therefore, the simplification of the CCM/DCM control system is possible. The validity of the proposed current control is confirmed by both simulation and experiment.

Future works will be focused on designing the minimized inductor and evaluating the conversion efficiency.

REFERENCES

[1] Koji Orikawa, Jun-ichi Itoh: "A comparison of the series-parallel compensation type DC-DC converters using both a fuel cell and a battery", IEEE ECCE 2010, pp.1428-1435, 2010

[2] Sachin Jain, Vivek Agarwal: "A Single-Stage Grid Connected Inverter Topology for Solar PV Systems With Maximum Power Point Tracking", IEEE Vol.22, No.5, pp.1928-1940, 2007

[3] Liqin Ni, Dean J. Patterson, Jerry L. Hudgins: "High Power Current Sensorless Bidirectional 16-Phase Interleaved DC-DC Converter for Hybrid Vehicle Application", IEEE Vol.27, No.3, pp.1141 – 1151, 2012

[4] Koichi Matsu-ura, Koji Orikawa, Jun-ichi Itoh: "Reduction of a boost inductance using a switched capacitor DC-DC converter", IEEE ECCE 2011, pp.1315 – 1322, 2011

[5] Satoshi Miyawaki, Jun-ichi Itoh, Kazuki Iwaya: "Experimental investigation and loss calculation for a bi-directional isolated DC/DC converter using series voltage compensation", APEC 2013, pp.1931-1938, 2013

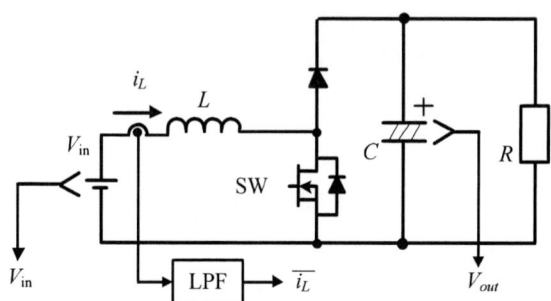

Fig. 15. Boost converter for AVR Experiment.

TABLE III
EXPERIMENTAL PARAMETERS

Input voltage V_{in}	40 V
Output voltage V_{out}	70 V
Rated output power P_{out}	49 W
Boost chopper inductor L	180 µH
Smoothing Capacitor C	680 µH
Switching frequency f_{sw}	50 kHz
Sampling frequency f_{samp}	50 kHz
LPF cutoff frequency f_{cutoff}	5 kHz
Damping factor ζ ACR	0.7
Natural angular frequency ω_n ACR	3000 rad/s
Damping factor ζ AVR	0.7
Natural angular frequency ω_n AVR	300 rad/s

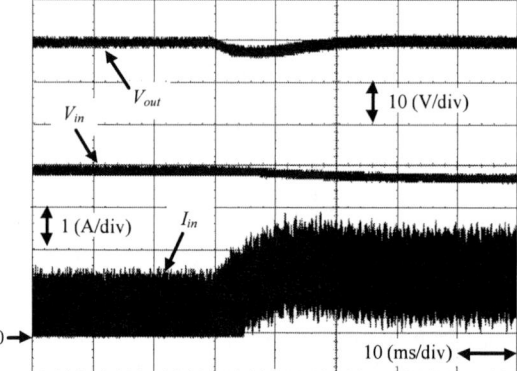

(a) From 40% load in DCM to 100% load in CCM.

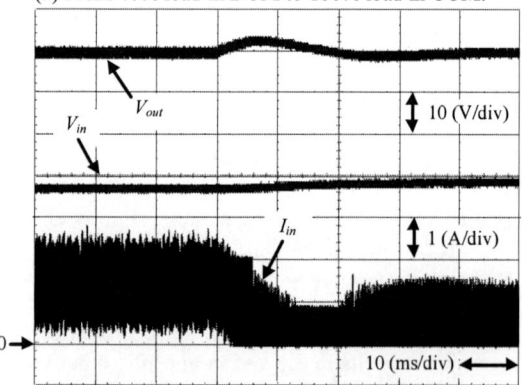

(b) From 100% load in CCM to 40% load in DCM.

Fig. 16. Experimental results of load transient response from 40% load in DCM to 100% load in CCM and vice versa. With the proposed current control method, the input current smoothly alternates between CCM and DCM, so this ensures the stability of the output voltage regulation when load varies.

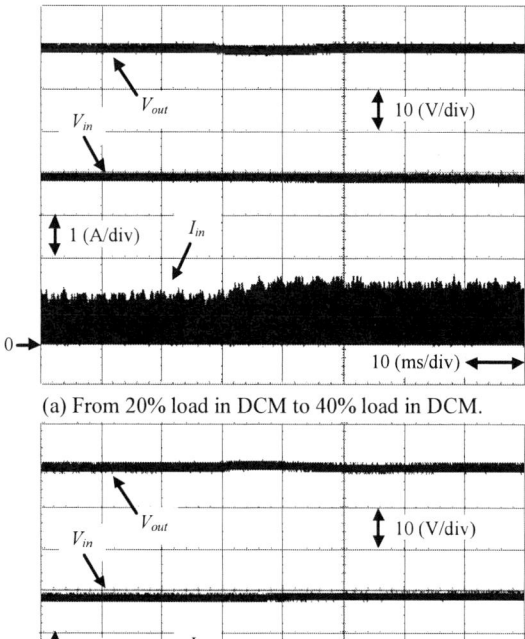

(a) From 20% load in DCM to 40% load in DCM.

(b) From 40% load in DCM to 20% load in DCM.

Fig. 17. Experimental results of load transient response from 20% load in DCM to 40% load in DCM and vice versa. With the proposed current control method, the input current in DCM is controlled exactly to the conventional CCM, so this ensures the stability of the output voltage when load varies.

[6] Robert W. Erickson and Dragan Maksimovic, Fundamentals of Power Electronics, 2nd ed. New York: Springer-Verlag, 2001

[7] Peter Haaf, Jon Harper: "Understanding Diode Reverse Recovery and its Effect on Switching Losses", Fairchild Power Seminar 2007, ppA.23-A.33, 2007

[8] Juan M. Rivas, Olivia Leitermann, Yehui Han, David J. Perreault: "A Very High Frequency dc-dc Converter Based on a Class Φ2 Resonant Inverter", IEEE PESC 2008, pp.1657-1666, 2008

[9] Juan M. Rivas, David Jackson, Olivia Leitermann, Anthony D. Sagneri, Yehui Han, David J. Perreault: "Design Considerations for Very High Frequency dc-dc Converters", IEEE PESC 2006, pp.1-11, 2006

[10] Robert C. N. Pilawa-Podgurski, Anthony D. Sagneri, Juan M. Rivas, David I. Anderson, David J. Perreault: "Very-High-Frequency Resonant Boost Converters", IEEE Vol.24, No.6,pp.1654-1665, 2009

[11] Anthony D. Sagneri, David I. Anderson, David J. Perreault: "Optimization of Integrated Transistors for Very High Frequency DC-DC Converters", IEEE Vol.28, No.7, pp.3614-3626, 2013

[12] Jian Sun, Daniel M. Mitchell, Matthew F. Greuel, Philip T. Krein, Richard M. Bass: "Averaged Modeling of PWM Converters Operating in Discontinuous Conduction Mode", IEEE Vol.16, No.4, pp.482-492, 2001

[13] Miao Zhu, Fang Lin Luo: "Remaining Inductor Current Phenomena of Complex DC–DC Converters in Discontinuous Conduction Mode General Concepts and Case Study", IEEE Vol.23, No.2, pp.1014-1019, 2008

[14] Xiaotian Zhang, Joseph W. Spencer: "Analysis of Boost PFC Converters Operating in the Discontinuous Conduction Mode", IEEE Vol.26, No.12, pp.3621-3628, 2011

[15] Jeffey Morroni, Luca Corradini, Regan Zane, Dragan Maksimovic: "Adaptive Tuning of Switched-Mode Power Supplies Operating in Discontinuous and Continuous Conduction Modes", IEEE Vol.24, No.11, pp.2603-2611, 2009

[16] Jianping Xu, Jinping Wang: "Bifrequency Pulse-Train Control Technique for Switching DC–DC Converters Operating in DCM", IEEE Vol.58, No.8, pp.3658-3667, 2011

[17] Jinping Wang, Jianping Xu: "Peak Current Mode Bifrequency Control Technique for Switching DC–DC Converters in DCM With Fast Transient Response and Low EMI", IEEE Vol.27, No.4, pp.1876-1884, 2012

[18] Yi-Ping Su, Yean-Kuo Luo, Yi-Chun Chen, Ke-Horng Chen: "Current-Mode Synthetic Control Technique for High-Efficiency DC–DC Boost Converters Over a Wide Load Range", IEEE Vol.PP, pp.1-12, 2013

[19] Yu-Kang Lo, Jing-Yuan Lin, Sheng-Yuan Ou: "Switching-Frequency Control for Regulated Discontinuous-Conduction-Mode Boost Rectifiers", IEEE Vol.54, No.2, pp760-768, 2007

[20] Jong-Won Shin, Bo-Hyung Cho: "Digitally Implemented Average Current-Mode Control in Discontinuous Conduction Mode PFC Rectifier", IEEE Vol.27, No.7, pp.3363-3373, 2012

[21] Koen De Gussemé, David M. Van de Sype, Alex P. M. Van den Bossche, Jan A. Melkebeek: "Digitally Controlled Boost Power-Factor-Correction Converters Operating in Both Continuous and Discontinuous Conduction Mode", IEEE Vol.52, No.1, pp.88-97, 2005

[22] Shu Fan Lim, Ashwin M. Khambadkone: "A Simple Digital DCM Control Scheme for Boost PFC Operating in Both CCM and DCM", IEEE Vol.47, No.4, pp.1802-1812, 2011

[23] Tai-Sik Hwang, Sun-Yeul Park: "Seamless Boost Converter Control Under the Critical Boundary Condition for a Fuel Cell Power Conditioning System", IEEE Vol.27, No.8, pp.3616-3626, 2012

The 2014 International Power Electronics Conference

Technical Assessment of Load Commutation Switch in Hybrid HVDC Breaker

Arman Hassanpoor
School of Electrical Engineering
KTH Royal Institute of Technology
Stockholm, Sweden
arman.hassanpoor@ee.kth.se

Jürgen Häfner
Grid Systems
ABB AB
Ludvika, Sweden
jurgen.hafner@se.abb.com

Björn Jacobson
Grid Systems
ABB AB
Beijing, China
bjorn.jacobson@cn.abb.com

Abstract—The development of a large scale high voltage direct current (HVDC) power grid requires a reliable, fast and low-loss circuit breaker. The load commutation switch (LCS) is an essential part of ABB's 1200 MW hybrid HVDC breaker concept which builds up a low-loss conducting path for the load current. The technical requirements for the LCS are expressed in this paper by studying the operation principle of the hybrid HVDC breaker. The voltage stress over the LCS is calculated and simulated based on a dc grid with 320kV and 2kA rated voltage and current. A system model of the hybrid HVDC breaker is developed in PSCAD/EMTDC software to study the design criteria for snubber circuit and arrester blocks. It is observed that conventional snubber circuits are not suitable for a bidirectional hybrid HVDC breaker as the current of snubber capacitors prevent the fast interruption action. A modified snubber circuit is proposed in this paper along with two more alternatives for the load commutation switch to overcome this problem. Moreover, the power loss model for a semiconductor device is discussed in this paper based on the 4.5 kV StakPak IGBT. The model is used to calculate the conduction power losses for different LCS topologies. Ultimately, a matrix of 3x3 IGBT modules is selected to provide a reliable LCS design which can handle several internal fault cases with no interruption of operation. A full-scale prototype has been constructed and tested in ABB HVDC Center, Ludvika, Sweden. The experimental test results are also included in the paper in order to verify the calculation and simulation study.

Keywords—HVDC Grid, Solid state circuit breaker, Semiconductor breaker

I. INTRODUCTION

Transporting power electricity from distant renewable resources to load centers with high efficiency and low visibility environmental impact, are important features which can be achieved with a high voltage direct current (HVDC) grid. The evolution of the power electricity network leads to a demand for advanced power transmission methods and more dc grids. The key equipment for developing the dc grid and dc switchyards is the dc circuit breaker. The lack of a fast and reliable dc breaker has always been an issue preventing the development of dc grids but this obstacle has now been removed by introducing the hybrid HVDC breaker in [1]. The hybrid HVDC breaker is a stand-alone system which can independently interrupt the dc current.

The hybrid HVDC breaker consists of three essential

components; load commutation switch (LCS), ultra fast disconnector (UFD) and Main breaker. The hybrid HVDC breaker is a proactive component which is able to interrupt the load current either during fault cases or normal steady state conditions. Therefore, it should be able to handle both fault current and steady state current in a multi-terminal dc grid. In the hybrid HVDC breaker there is a bypass current path over the Main breaker in which the UFD and LCS are placed in series. The load current flows through the bypass path during normal operation, while turning off the LCS commutates the current to the Main breaker path; thereafter, LCS is exposed to the Main breaker voltage drop, during the UFD opening time. Since the LCS is conducting the load current, it is necessary to consider the possible failure modes and design a highly reliable bi-directional topology for this switch.

Following the introduction Section, the operation principle of the hybrid HVDC breaker is discussed in Section II. Section III presents system modeling, design specification and different LCS topologies. This Section conclude the most reliable, fast and low-loss LCS through a detailed system study. Section IV is devoted to the experimental results from tests on a full-scale prototype of LCS.

II. HYBRID HVDC BREAKER OPERATION PRINCIPLE

The concept of hybrid HVDC breaker was first proposed by Jürgen Häfner in [1] . Fig. 1 demonstrates the operation principle of the hybrid HVDC breaker. The load current flows through the UFD and LCS during normal operation. The green line in Fig. 1(a) shows the load current path during this time. In the case of any fault, the conducting current will increase proportionally to the fault current. The fault case mode can be triggered according to grid requirements and protection scenarios. In this paper, an increase of 20% in the breaker current above the nominal current will trigger the fault case mode and consequently, the LCS turn-off signal will be initiated after a time delay. The fault type can be determined during this time delay. So, as Fig. 1(b) shows, the fault current goes through the LCS for a certain period of time. As the LCS is switched off, the current will be commutated to the Main breaker path and the UFD will isolate one side of the LCS. At this time the Main breaker can interrupt the fault current and complete the current interruption. Fig. 1(c) illustrates the commutated current in the Main breaker path. In the end, the

978-1-4799-2706-7/14 $31.00 © 2014 IEEE

fault energy will be absorbed by the arrester across the Main breaker, Fig. 1(d). Note that a current limiting reactor is placed in series to the LCS and Main breaker. The purpose of this reactor is to limit the current derivative during faults. The corresponding current interruption sequence is demonstrated in Fig. 2. Referring to Fig. 2, the fault occurred at t=6ms when the line carried a current I_c of 2kA. At the time when the fault is established (20% above the nominal current in this example, I_c=2.4kA), the LCS is turned off and the current is commutated to the Main breaker branch. Thereafter, the UFD can isolate the LCS in the system and enable Main breaker operation. The Main breaker can either immediately interrupt the current or limit the current based on the protection logics. However, Fig. 2 shows the case that a fault current of 9kA is interrupted by the Main breaker.

III. SYSTEM STUDY

The LCS design criteria can be addressed by a detailed study of stresses to this switch in a system model. A simulation model is described in this section which includes a simplified network model along with the hybrid HVDC breaker model. The network is modeled as a dc voltage source and a resistive load to derive the maximum current of 2 kA in normal operation mode. Since the most severe fault would be a fault at the breaker terminals, then zero fault impedance has been considered for the worst case scenario in the simulation study. However, different topologies for LCS are introduced later in this Section and evaluated for different power loss generation.

A. System Modeling

Fig. 3 demonstrates the studied model of hybrid HVDC breaker including a simplified dc grid, Main breaker, UFD, LCS and arrester banks. The hybrid HVDC breaker is evaluated in a 320kV dc grid for a rated current of 2kA. This grid is represented by a stiff dc source of 320kV along with a resistive load. The current limiting reactor is selected in a way to limit the current rise rate to 3.5kA/ms for a low inductance fault at dc switchyard terminals. As is shown in [1], a dc reactor of 100mH can fulfill this requirement assuming a breaking time of less than 5ms. The Main breaker consists of four 80kV modules which are able to interrupt the current in either direction, employing 40 IGBT modules in each direction. Each module also utilizes an arrester bank to limit the maximum voltage over the corresponding module. Since the internal behavior of the Main breaker is not the subject of this study, it is sufficient to model the Main breaker using 160 IGBTs in one direction and 160 Diodes in the other direction while an arrester bank is positioned over all semiconductor devices. Furthermore, the equivalent turn-off snubber is available for the Main breaker.

The UFD is modeled as an ideal switch which can be operated only at zero current. A delay time of 2ms is introduced in the operation of this switch in order to model the mechanical time delay [2]. The UFD isolates the LCS from the voltage over the Main breaker. Hence, the voltage rating requirement of LCS is relatively low. In this study, the LCS is

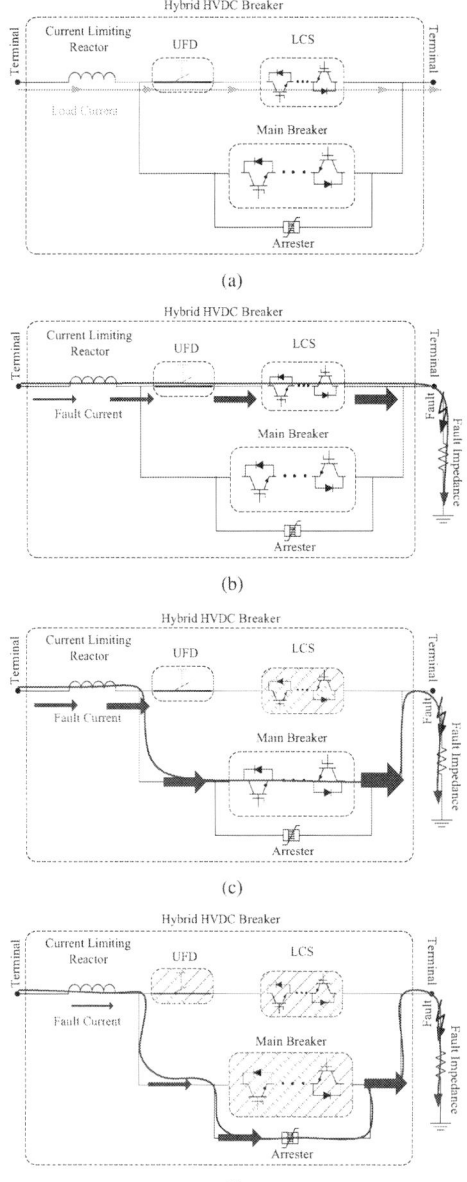

Fig. 1. Hybrid HVDC breaker operation principle (a) normal load current path, (b) fault initiates operation, (c) LCS interrupts and commutates current to Main breaker, (d) Main breaker interrupts and commutates current to the arrester

initially modeled as two IGBT modules which are connected in opposite directions. The inductance L_s links the bypass path and the Main breaker path; this is the stray inductance of the mechanical structure. Since the LCS is the main object under study, the values of the dominating parasitic electrical elements of the mechanical structures and components are roughly estimated and implemented in the simulated model.

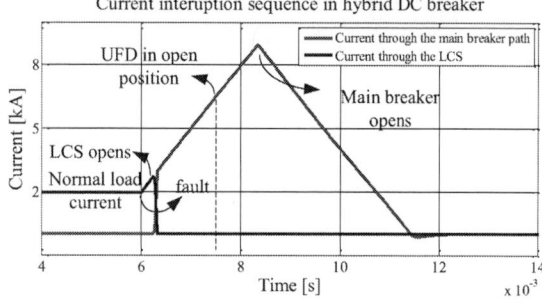

Fig. 2. Current interruption sequence in hybrid HVDC breaker

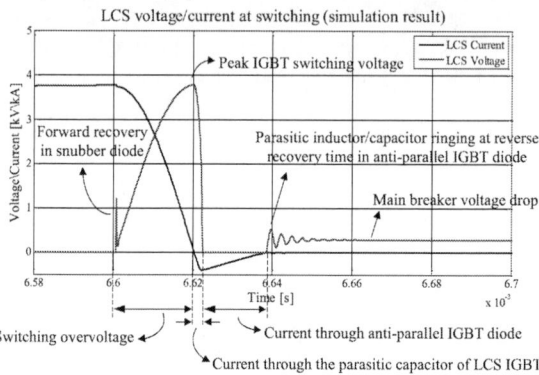

Fig. 4. LCS voltage stress during the fault current interruption interval

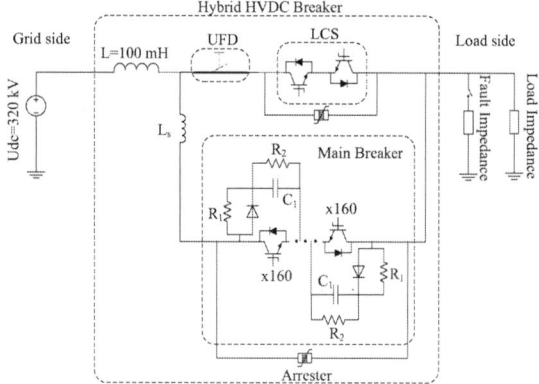

Fig. 3. Simulation model for hybrid HVDC breaker

B. Design Specification

Fig. 4 shows the voltage stress over the LCS during the interruption interval. This is the time interval when the load current is sequentially commutating from the bypass path to the Main breaker path and then arrester path, in order to complete the interruption. Referring to Fig. 4, the turn-off signal triggers the LCS at $t=6.6$ms and the IGBT switching voltage is exposed over the LCS at this moment. The switching voltage depends on the number of series IGBTs, snubber circuit and stray inductance between the LCS and the Main breaker. The larger the stray inductance, the longer the commutation time and, consequently, the larger the voltage overshoot over the LCS. This shows how mechanical design affects the voltage stress over the LCS. The assumed stray inductance and snubber circuit introduce a peak voltage of 3.9kV in the case of one IGBT module in the LCS. The parasitic capacitor over the LCS IGBTs completes a loop through both the Main breaker and the UFD at the snubber diode reverse recovery time. At this moment, the negative IGBT voltage drop makes the anti-parallel diode forward biased. This diode will continue conducting until the positive voltage drop over the Main breaker appears over the LCS IGBT and makes the anti-parallel diode reverse biased. The parasitic stray inductances and capacitors in the bus-bars and cables create an oscillation at the reverse recovery time of the anti-parallel diode. While the fault current totally commutates to the Main breaker path,

the forward voltage drop in the Main breaker would appear over the LCS until the UFD isolates the LCS. Additionally, to protect the LCS against any undesirable over-stress, an arrester bank has been placed over the LCS.

In order to lower the dv/dt at switching time, a resistor-capacitor-diode (RCD) turn-off snubber is placed at each IGBT module in LCS. Standard RCD snubber is shown in Fig. 5. R_s is the turn-on resistor which controls the discharging behaviour of the snubber capacitor, C_s, at turn on time. The diode D_s bypasses R_s during the turn-off period in order to reduce the IGBT stress and charge the snubber capacitor as fast as possible. However, using the conventional RCD snubber for LCS prevents fast UFD action, because the snubber capacitor C_s has a low resistive path to be discharged through the R_s, UFD and Main breaker during the breaking action. The discharging current goes through the UFD and precludes fast opening action. The current through the UFD and the Main breaker is plotted in Fig. 6 and the discharging path is highlighted in Fig. 5. Three different alternatives are proposed in order to overcome the discharging problem by the conventional RCD snubber. Any design should maintain the following requirements for LCS:

- Bi-directional current interruption

- High system reliability

- Minimize the component's stress

- Several switching actions of LCS in a short period of time

- Simple mechanical structure

1) Alternative one (reversal connection): Using a diode in series with the LCS will prevent the snubber capacitor to be discharged through the Main breaker. In this case the standard RCD snubber can be used while the same configuration should be used as an anti-parallel connected LCS to enable bi-directional operation of the hybrid HVDC breaker. However, a resistor is used in parallel to the snubber capacitor to discharge the capacitor with a large time constant. This facilitates a more secure maintenance period and lowers risk to personnel.

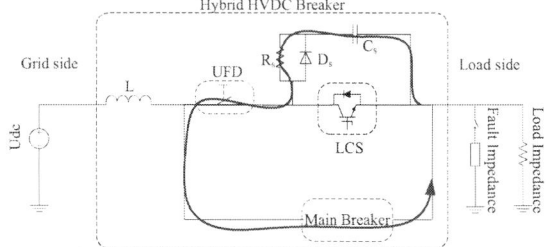

Fig. 5. Standard RCD snubber (the red line highlights the discharging problem with the standard snubber circuit)

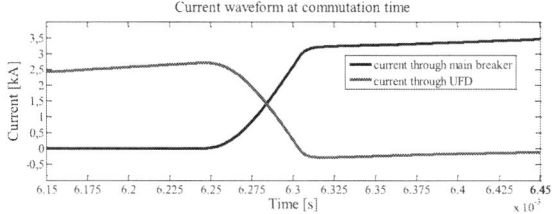

Fig. 6. Current through the Main breaker path and bypass path

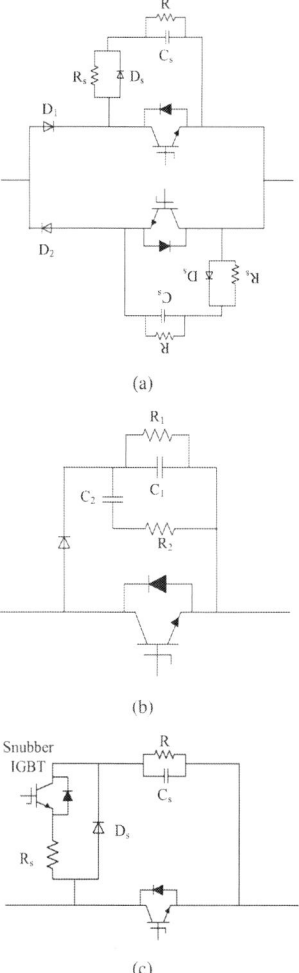

Fig. 7. (a) Alternative one (series diode connection), (b) Alternative two (modified snubber circuit), (c) Alternative three (snubber IGBT)

Fig. 7(a) illustrates the topology for this alternative. Another advantage of this topology is that only one IGBT stack will conduct the current in each direction so the conduction losses only include IGBT losses and no diode loss in the anti-parallel IGBT diodes. The diode power loss is only generated through the series diodes D_1 and D_2 in this alternative.

2) Alternative two (modified snubber circuit): A modified snubber circuit is presented in Fig. 7(b). During the turn-off period, C_1 is charged and shapes the switching voltage over the corresponding IGBT. While the current commutation is accomplished, the stored charge in C_1 will be shared by parallel capacitor C_2 through R_2. Thereafter, C_1 can handle another switching action in the corresponding IGBT. The capacitance of C_2 needs to be at least three times higher than C_1 and R_2 resistance also needs to be much smaller than R_1. Moreover, R_1 ensures capacitor discharge with a relatively large time constant. A typical value for this snubber circuit is as follows:

- C_1=7.5μF, C_2=20μF
- R_1=10kΩ, R_2=10Ω

3) Alternative three (snubber IGBT): In order to unify the mechanical design along with the full functionality of the turn-off snubber, an IGBT module can be placed in the snubber circuit according to Fig. 7(c). All LCS IGBTs, including snubber IGBTs, are switched simultaneously; therefore, no additional firing pulse control logic needs to be implemented for the snubber IGBT. The snubber IGBT will block the discharging current through the snubber resistance at turn-off while it forms a discharging path through the snubber resistance at the time that LCS IGBT is switched on again. This feature allows fast turn-on turn-off actions for LCS with no excessive over voltage formation over the IGBT modules [3].

C. LCS Topology

Regardless which of the above-mentioned snubber configurations is used, the load commutation switch continuously conducts the load current in steady-state operation. In this subsection, a reliable and robust design of the LCS is addressed by providing an additional conducting path and also decreasing the stresses on each internal component.

Additional conducting paths can be provided by paralleling the semiconductor devices in the LCS, while series connection results in a lower voltage stress on each device. Therefore, the LCS reliability can be improved through paralleling and series connection of semiconductor devices. On the other hand, each semiconductor device contributes to power losses and also increase the total system complexity, so an optimized design for the LCS is a trade-off between system reliability, power loss and system complexity. However, ten different LCS topologies have been studied and evaluated regarding the component stresses and losses. These cases are depicted in

The 2014 International Power Electronics Conference

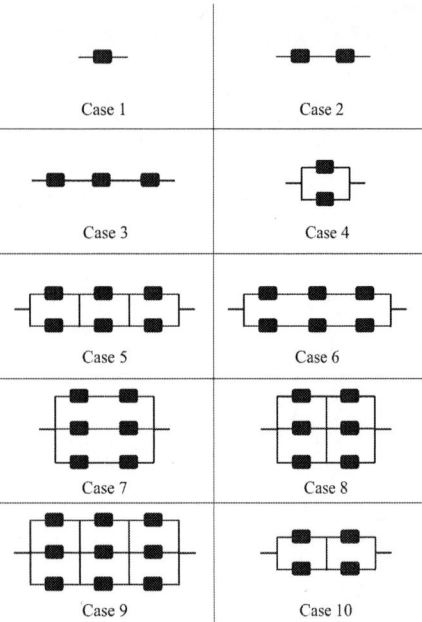

Fig. 8. Topologies for the load commutation switch

TABLE I. COMPONENT LOSSES FOR DIFFERENT TOPOLOGIES

Case no.	Nominal component current	Max. IGBT conduction loss	Max. diode conduction loss
1	2.7 kA	6.9 kW	5.8 kW
2	2.7 kA	6.9 kW	5.8 kW
3	2.7 kA	6.9 kW	5.8 kW
4	1.4 kA	2.5 kW	2.2 kW
5	1.4 kA	2.5 kW	2.2 kW
6	1.4 kA	2.5 kW	2.2 kW
7	0.9 kA	1.5 kW	1.3 kW
8	0.9 kA	1.5 kW	1.3 kW
9	0.9 kA	1.5 kW	1.3 kW
10	1.4 kA	2.5 kW	2.2 kW

TABLE II. REQUIREMENT SPECIFICATIONS FOR LOAD COMMUTATION SWITCH

Requirements	Dimensions	Values
Nominal current	A_{DC}	2000
Maximum continuous current	A_{DC}	2600
Maximum current level during commutation	kA_{peak}	8
Maximum voltage stress after commutation	kV_{peak}	4.8
Maximum transient voltage stress during commutation	kV_{peak}	9
Repetitive switching capability within one operating cycle (turn off / turn on)		10

Fig. 8. The semiconductor devices and electrical connections are represented by black blocks and plain lines in this Figure. Power losses are calculated utilizing the 4.5kV StakPak IGBT module [4]. IGBT conduction loss can be approximated by modeling the IGBT on-state zero current collector-emitter voltage (V_{CE0}) as a dc source voltage and a collector-emitter on-state resistance (R_{CE0}) in series:

$$v_{CE}(t) = V_{CE0} + R_{CE0}i_{CE}(t) \tag{1}$$

where, $v_{CE}(t)$ and $i_{CE}(t)$ are the instantaneous collector-emitter voltage and current, respectively. Similarly, the instantaneous voltage drop over the anti-parallel diode ($v_D(t)$) can be approximated by employing the forward voltage drop (FVD) and diode on-resistance (R_{on}):

$$v_D(t) = FVD + R_{on}i_D(t) \tag{2}$$

in which, $i_D(t)$ is the instantaneous diode current. Furthermore, the instantaneous IGBT and diode conduction loss ($p_{CE}(t)$ and $p_D(t)$) can be, respectively, formulated as:

$$p_{CE}(t) = v_{CE}(t)i_{CE}(t)$$
$$p_D(t) = v_D(t)i_D(t) \tag{3}$$

while substituting (1) and (2) in (3) expresses the instantaneous power losses with respect to instantaneous device current:

$$p_{CE}(t) = V_{CE0}i_{CE}(t) + R_{CE0}i_{CE}^2(t)$$
$$p_D(t) = FVDi_D(t) + R_{on}i_D^2(t). \tag{4}$$

The maximum conduction losses in each IGBT module and anti-parallel diode were calculated by implementing (4) in a PSCAD simulation model utilizing the 4.5kV StakPak IGBT parameters for V_{CE0}, R_{CE0}, FVD and R_{on}. Results are available in Table I for various LCS topologies. The LCS

requirement specifications including the design margins are summarized in Table II which determines the final design of the LCS according to topology number 9. Two parallel semiconductor devices are required for 2.6kA continuous current while a third one ensures redundancy in this set. A second set of three paralleled devices is required to allow a voltage drop across the LCS which is large enough to ensure short circuit failure mode (SCFM) of a failed device in the Main breaker. A third set of three paralleled devices is needed for redundancy if one of the other series devices fails. However, in the case of device failure in any parallel set, there are two more alternative current paths which can handle the load current. In order to have full operation availability, the cooling system is designed based on the maximum power loss in Case 10. In this case all back-up devices in the LCS are fully used. In addition to all above-mentioned redundancies in the LCS, an arrester block will protect the LCS against any kind of unexpected stresses which might occur in internal or external failure cases. This arrester block limits the maximum over-voltage stress to 4.5kV in the case of 4.5kV StakPak IGBT module utilization. The final stack design of the LCS is illustrated in Fig. 9. The stack design brings full modularity and easy maintenance. Parallel connections are accomplished through flexible busbars between each set of three series devices while insulator plates separate the series sets.

IV. EXPERIMENTAL TEST RESULTS

An LCS full-scale prototype has been built according to the design specification in Section III-C. The designed LCS has been studied in a test circuit including the Main breaker, UFD

978-1-4799-2706-7/14 $31.00 © 2014 IEEE

The 2014 International Power Electronics Conference

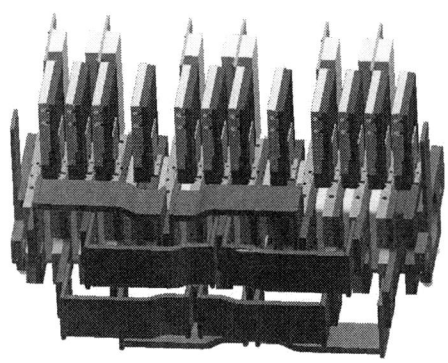

Fig. 9. Stack design for load commutation switch

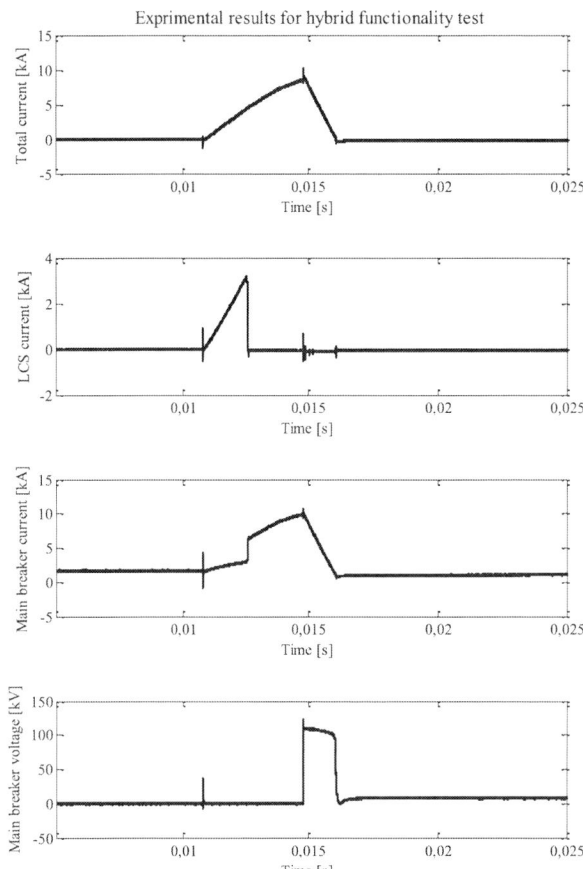

Fig. 11. Experimental results for hybrid functionality test

and arresters, to illustrate the stresses of the LCS components. The test sequence is similar to the introduced functional logic of the hybrid HVDC breaker in Section II. However, the fault current in this test setup is generated by discharging the capacitor banks and its rate of change (di/dt) is regulated by reactors in the loop.

A. Test Setup

Fig. 10 illustrates the hybrid HVDC breaker test circuit. The capacitor bank $C1$ builds up the desired dc voltage level of 40kV, supplied by a ±150kV outdoor dc switchyard. Initially, the breaker $Q11$ charges the capacitor bank $C1$. Later, the dc switchyard will be isolated by opening the breaker $Q11$. Thereafter, the capacitor bank $C1$ discharges through the reactor $L1$ and test object, when the spark gap $Q5$ triggers the test. The variable reactor $L1$ can be manipulated to achieve different rates of rise of the current di/dt. The object under test including Main breaker, UFD, LCS and arrester bank is represented by $Q41$, $Q42$, $Q43$ and $F1$, respectively in Fig. 10. The optical current measurement devices are also integrated in the test circuit to measure different quantities during the test procedure. $T1$ measures the total discharging current, $T2$ measures the current through LCS and $T3$ measures the arrester current. The breakers $Q21$, $Q22$, $Q25$ and $Q26$ are protective breakers ensuring circuit grounding for safety purposes.

B. Test Results

The results from tests are illustrated in Fig. 11. The spark gap $Q5$ initiates the test at $t_0 = 0.0112$s. In this test, the discharging current is shared between LCS and the Main breaker path according to resistance. Note that only one Main breaker stack (80kV module) is used in this test, otherwise there would not be any current flow through the Main breaker path in normal operation of the hybrid HVDC breaker. However, the LCS is turned off when the conducting current reaches 3.2kA. At this moment, the current commutates to the Main breaker path and, as is shown in Fig. 11, the LCS current becomes zero. A delay time of $250\mu s$ is assumed for commutation time, therefore, the UFD can be opened after this time delay. Finally, the total current of 9 kA is interrupted

by turning off the Main breaker. Fig. 11 shows currents in different paths of the hybrid HVDC breaker and also voltage stress over the Main breaker during the test interval.

An expanded view of the LCS IGBT module voltage stress is available in Fig. 12a. Comparing the test result in this Figure with the simulation result in Fig. 4 validates the simulation model with an acceptable error margin. Similar to Fig. 4, the first overshoot in Fig. 12a is due to the forward recovery in the snubber diode. The maximum voltage of 2.7 kV is governed by snubber circuit, LCS topology and stray inductances in the commutation loop. Hereafter, the zero voltage over the IGBT modules in the LCS explain the current flowing through the anti-parallel diode of the IGBT module. And finally, the voltage oscillation appears at the reverse recovery time for the anti-parallel diode which is due to the parasitic components between the LCS and the Main breaker. Fig. 12b shows the expanded view for the LCS current at the switching time which matches the simulation results illustrated in Fig. 4.

V. CONCLUSION

The load commutation switch (LCS) is one of the key parts of the hybrid HVDC breaker concept. The LCS continuously conducts the load current in the normal steady-state operation

978-1-4799-2706-7/14 $31.00 © 2014 IEEE

The 2014 International Power Electronics Conference

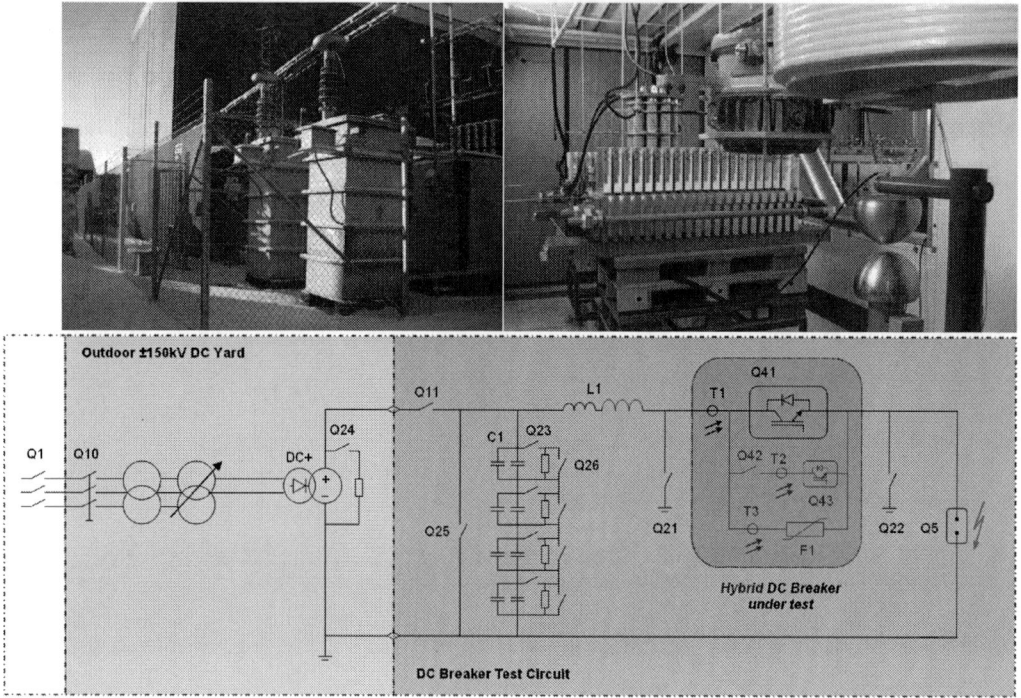

Fig. 10. Hybrid HVDC breaker test circuit

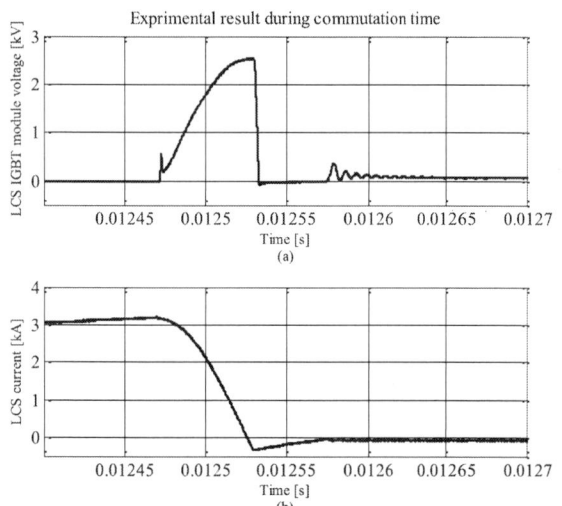

Fig. 12. Load commutation switch voltage and current during commutation time

of the hybrid HVDC breaker. In this paper, the operating principle of the hybrid HVDC breaker is analysed primarily towards developing a simulation model of the breaker. Specific requirements for the LCS has been introduced in order to make it fast, reliable and robust commutation switch of LCS. It is shown that conventional snubber circuits prevent fast operation of the LCS. Therefore, three alternatives are introduced in order to overcome this problem. On the other

hand, ten different internal topologies for LCS, including different number of conducting paths, are evaluated regarding the generated power loss, component stress and reliability factors. Finally, an LCS full-scale prototype has been built and tested in a high-voltage test setup along with the other parts of the hybrid HVDC breaker. The test was conducted to illustrate the maximum component stresses over the LCS during current interrupting event in the hybrid HVDC breaker. The functionality test of the hybrid HVDC breaker does not only confirm the reliable mechanical design but also verifies the detailed simulation model of the system. The presented LCS design brings modularity and easy maintenance as well as fast, low-loss and reliable operation of the hybrid HVDC breaker.

ACKNOWLEDGMENT

The authors gratefully acknowledge the contributions of ABB HVDC test group in Ludvika, Sweden.

REFERENCES

[1] J. Häfner and B. Jacobson, "Proactive hybrid hvdc breakers - a key innovation for reliable hvdc grid," in *Cigrésymposium*, Bologna, Italy, 13-15 September 2011.

[2] P. Skarby and U. Steiger, "An ultra-fast disconnecting switch for a hybrid hvdc breaker - a technical breakthrough," in *Cigrésymposium*, Alberta, Canada, 9-11 September 2013.

[3] A. Hassanpoor and J. Häfner, "Hvdc hybrid circuit breaker with snubber circuit," Patent WO/2013/071 980 A1, 05 23, 2013.

[4] "5SNA2000K450300 data sheet," ABB Semiconductor, Lenzburg, Switzerland, March 2013. [Online]. Available: http://www.abb.com

978-1-4799-2706-7/14 $31.00 © 2014 IEEE

The 2014 International Power Electronics Conference

Control of Hexagonal Modular Multilevel Converter for 3-phase BTB System

Shin-ichi Hamasaki, Kazuki Okamura, Takashi Tsubakidani, Mineo Tsuji
Division of Electrical Engineering and Computer Science,
Nagasaki University,
Nagasaki, Japan
hama-s@nagasaki-u.ac.jp

Abstract— The Modular Multilevel Converter (MMC) with cascaded full bridge cells is available for AC/AC transmission system without transformers in high voltage line. This research applied a Hexagonal MMC (H-MMC) to 3-phase AC-AC back-to-back (BTB) system. The H-MMC has six arm modules with floating capacitors and is required to regulate the input-output power and capacitor voltages by controlling currents in arm modules. Theory of the power flow of H-MMC is analyzed and the control scheme of the power flow and capacitor voltage regulation is proposed. Currents in the arm modules are controlled by the internal model principle and H-MMC can operate appropriately. Effectiveness of the proposed control method is presented in simulation and experiment.

Keywords— *modular multilevel converter, AC/AC converter, back to back system, power flow control*

I. INTRODUCTION

In recent years, energy conservation problems have been important due to CO_2 emission, abandoning nuclear generation and so on. In such situation, renewable energy systems have been focused and a lot of photovoltaic and wind power generation systems have introduced all over the world. However renewable energy generations give influence of voltage fluctuation to power grid because of their unstable output and reverse power flow. Therefore it is important to maintain the voltage of power line. This importance will increase by promotion of the electricity liberalization. In general, AC/AC transmission system is introduced for power flow control. Many kinds of AC/AC system such as back-to-back (BTB) system, cycloconverter and matrix converter are developed and investigated.

On the other hands, several kinds of multilevel converter for high voltage and less harmonics are investigated. Conventional multilevel converter has transformers on DC bus, which is one of the reasons of heavy and large system. A modular multilevel converter (MMC)[1]-[9] is one of the transformer-less converter and applies to high voltage and high power conversion. The MMC consists of cascaded switching device modules with floating capacitors.

The multilevel converter such as the MMC can improve waveform and reduce harmonics by multilevel voltage output. The triple star type MMC[6] and the

hexagonal type MMC[7][8] are constructed as a 3-phase AC/AC converter. The hexagonal MMC (H-MMC) can be applied to power flow control system instead of BTB system. In addition, this is less flexibility than the triple star type MMC and conventional multilevel BTB system, but is able to realize to reduce a number of cells compared with those systems. This H-MMC topology is expected for a wide range of fields such as a wind power converter, a grid connection converter and high voltage motor drive.

In this paper, H-MMC is investigated and applied to BTB system for a power flow controller. Theory of the power flow of H-MMC is analyzed and the control scheme of the power flow and capacitor voltage regulation for H-MMC is proposed. Effectiveness of the proposed control is shown by simulation and experiment in typical cases.

II. CONFIGURATION OF CIRCUIT

Fig.1 shows the configuration of H-MMC. The H-MMC has 6 arm-modules connected in hexagonal ring shape. These modules consist of multiple cascade full bridge cells which have floating capacitors. Each arm module has a buffer reactor for preventing short circuit and current regulation. The number of cells in an arm-module is 2 as shown in Fig.1(a). Number of these can be increased depending on usage situation. In Fig.1(a), terminals of R,S,T are primary side of power line and terminals of U,V,W are secondary side. Primary power via R,S,T can be flowed to floating capacitors and secondary side via UVW. This configuration means that RST-delta connection and UVW-delta connection overlap each other. Thus each arm module has primary and secondary line-to-line current.

The power on primary side, the power on secondary side and all the voltages of floating capacitors in cells are required to regulate by controlling output of the modules. When the arm current is controlled in case of different frequencies between primary and secondary side, it is necessary that two frequency components are controlled at the same time. In this research, the internal model principle is applied to the current control to obtain a good current regulation.

978-1-4799-2706-7/14 $31.00 © 2014 IEEE

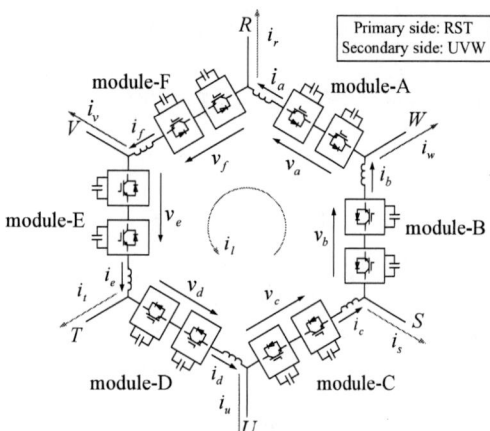

(a) Summary of H-MMC circuit

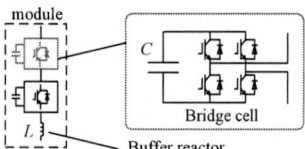

(b) Configuration of module and cell

Fig. 1. Circuit configuration of H-MMC

III. CONTROL OF HEXAGONAL TYPE MMC

When H-MMC is applied to a power converter, it is able to convert frequency and voltage arbitrarily without transformer. At this time, currents of each arm module must be controlled to be non-sinusoidal waveform, which is due to overlap line-to-line currents of primary and secondary. It is important that every capacitor voltage is managed to be constant and arm currents are exactly controlled without steady state error. In this study, capacitor voltages are kept by current control based on analysis the power flow in H-MMC. When the line-to-line current i_{rs} is current for primary node R-S and the line-to-line current i_{vw} is current for secondary node X-Y, current of arm-module A is calculated in (1).

$$i_a = i_{rs} + i_{vw} \qquad (1)$$

Other currents of modules are obtained the same as (1).

A. PQ-coordinate transformation

The control of primary and secondary power is realized by the pq-coordinate transform. In particular, the primary power control is handled in the pq-current in order to regulate all the floating capacitor voltages. Thereafter the pq-quantity is transformed to 3-phase quantity and the primary 3-phase currents are controlled together with the secondary 3-phase currents by the arm modules.

The pq-transform is defined as following equation.

$$\begin{bmatrix} F_p \\ F_q \end{bmatrix} = \sqrt{\frac{2}{3}} \begin{bmatrix} \sin \omega_0 t & \sin\left(\omega_0 t - \frac{2\pi}{3}\right) & \sin\left(\omega_0 t - \frac{4\pi}{3}\right) \\ \cos \omega_0 t & \cos\left(\omega_0 t - \frac{2\pi}{3}\right) & \cos\left(\omega_0 t - \frac{4\pi}{3}\right) \end{bmatrix} \begin{bmatrix} f_u \\ f_v \\ f_w \end{bmatrix} \qquad (2)$$

,where F_p and F_q are active and reactive component respectively. f_u, f_v and f_w are 3-phase components. ω_0 is an angular frequency the same as u, v, w-phase one.

B. Power flow control in secondary side

Power in secondary side can be controlled to arbitrarily value by using the pq-theory. The pq currents in secondary side are converted to 3-phase line-to-line currents with voluntary frequency, which are controlled by the arm modules.

C. Power flow control in primary side

Power in primary side can be controlled in accordance with the secondary power by the pq-theory. In this control, however, the primary power is also controlled to maintain each capacitor voltage. The pq-current references in primary side are calculated from capacitor voltages in modules because there is relevance between the primary power and the capacitor voltage. Then power flows involved in arm-modules are analyzed. The average power of each arm-module is constructed by overlap of primary and secondary power, which is defined as P_A, P_B, P_C, P_D, P_E and P_F, respectively. These average powers are calculated by the following equations.

$$P_{ACE} = P_A = P_C = P_E$$
$$= \frac{V_1 I_1}{\sqrt{3}} \cos(\phi_1 - \frac{\pi}{6}) - \frac{V_2 I_2}{\sqrt{3}} \cos(\phi_2 - \frac{5\pi}{6}) \qquad (3)$$

$$P_{BDF} = P_B = P_D = P_F$$
$$= -\frac{V_1 I_1}{\sqrt{3}} \cos(\phi_1 - \frac{5\pi}{6}) + \frac{V_2 I_2}{\sqrt{3}} \cos(\phi_2 - \frac{\pi}{6}) \qquad (4)$$

,where V_1 and I_1 are RMS volumes in primary side, V_2 and I_2 are RMS volumes in secondary side. ϕ_1 is the power factor angle in primary side, and ϕ_2 is the power factor angle in secondary side.

And the average power is calculated by (5).

$$P_X = \frac{1}{T} \int_{t-T}^{t} p_x \, dt = \frac{1}{T} \int_{t-T}^{t} v_x i_x \, dt \qquad (5)$$

$$(X = A\text{-}F, x = a\text{-}f)$$

(3) and (4) show that modules A,C,E and modules B,D,F have the same average power P_{ACE} and P_{BDF} respectively. Following equations show sum of P_{ACE} and P_{BDF}.

$$P_{ACE} + P_{BDF} = V_1 I_1 \cos\phi_1 + V_2 I_2 \cos\phi_2 \qquad (6)$$

(6) shows the fluctuation of energy in H-MMC. If (6) is positive, all the capacitor voltages increase because the energy is stored in H-MMC. Otherwise, all the capacitor voltages decrease because the energy is emitted from H-MMC. Therefore, it is desirable that (6) is maintained to 0 in order to operate H-MMC exactly at steady state. In this case, (7) is obtained.

$$V_1 I_1 \cos\phi_1 = -V_2 I_2 \cos\phi_2 \qquad (7)$$

(7) means that input active power in primary side is equal to the output active power in secondary side. When (7) is satisfied, the average voltage of all the capacitors is maintained to be constant.

From the above, the average value of all capacitor voltage is controlled by the primary active power, and the average power difference is controlled by the primary reactive power respectively. The following control strategies are given by considering relevance between capacitor voltage and average power.

$$I_{p1}^* = (K_{P1} + \frac{K_{I1}}{s})(\bar{v}_C - v_C^*) - \frac{V_2}{V_1}I_{p2}^* \qquad (8)$$

,where $\bar{v}_C$ is average value of all the capacitor voltages. In (8), first term of right-hand side means PI control of capacitor voltage and second term means feed-forward term depending on the active power in secondary side. On the other hand, I_{q1}^* is given voluntarily. In usual, I_{q1}^* is set to 0 in order to keep the power factor 1. The pq-currents are converted to three-phase currents and determine the line-to-line currents in primary side.

D. Loop current control

The power control on primary side is able to balance the capacitor voltages between modules A, C, E and B, D, F. Voltage balance in all the arm-modules is realized by a loop current control. The loop current has two frequencies of primary side and secondary side. The loop current can be calculated in (9).

$$i_l = (i_a + i_b + i_c + i_d + i_e + i_f)/6 \qquad (9)$$

The loop current can be separated in three components. First one is current depending on primary current, second one is current depending on secondary current and the last is pure loop current in the ring of H-MMC.

The first loop current reference is given by the following equation.

$$
\begin{aligned}
i_{l1}^* = &-K_L(\bar{v}_C - \bar{v}_{C-RS})\sin\omega_1 t \\
&-K_L(\bar{v}_C - \bar{v}_{C-ST})\sin(\omega_1 t - \frac{2}{3}\pi) \qquad (10) \\
&-K_L(\bar{v}_C - \bar{v}_{C-TR})\sin(\omega_1 t + \frac{2}{3}\pi)
\end{aligned}
$$

,where ω_1 is primary side angular frequency and $\bar{v}_{C-XY}$ (X, $Y = R$, S, T) is average of capacitor voltages between node X and Y. For example, $\bar{v}_{C-RS} = (\bar{v}_{CA} + \bar{v}_{CB})/2$ is derived from Fig.1.

The second loop current reference is given by the following equation.

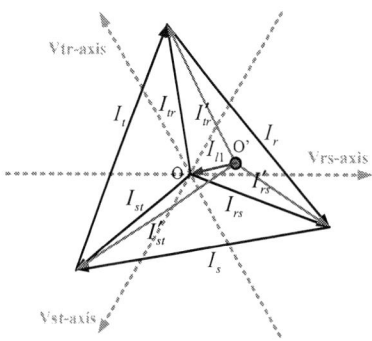

Fig. 2. current vector plot (in primary side)

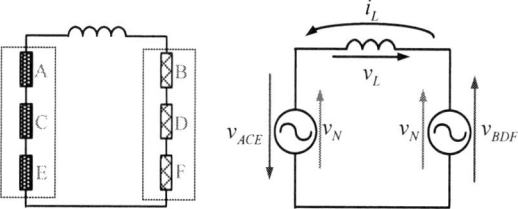

Fig. 3. Equivalent circuit for pure loop current control.

$$
\begin{aligned}
i_{l2}^* = &-K_L(\bar{v}_C - \bar{v}_{C-UV})\sin\omega_2 t \\
&-K_L(\bar{v}_C - \bar{v}_{C-VW})\sin(\omega_2 t - \frac{2}{3}\pi) \qquad (11) \\
&-K_L(\bar{v}_C - \bar{v}_{C-WU})\sin(\omega_2 t + \frac{2}{3}\pi)
\end{aligned}
$$

,where ω_2 is secondary side angular frequency and $\bar{v}_{C-XY}$ (X, $Y = U$, V, W) is average of capacitor voltages between node X and Y. For example, $\bar{v}_{C-UV} = (\bar{v}_{CD} + \bar{v}_{CE})/2$ is derived from Fig.1.

Fig.2 shows a current vector plot at the case. This figure shows the primary side and the same principle can be applied to the secondary side. By the loop current control, the neutral point is intentionally displaced by (10) and (11). The displaced difference of the neutral point is the same as the loop current. In this way, the loop current control can adjust voltage balance among modules by unbalancing three-phase line-to-line current. Note that this unbalance affects only line-to-line current in H-MMC. It means that phase currents of before and after the conversion are not affected by this unbalance.

In addition to the above controls, pure loop current control between modules is introduced. This scheme is able to accommodate active power between modules A, C, E and B, D, F.

Fig.3 shows an equivalent circuit between modules A, C, E and modules B, D, F. Sum of voltages of A, C, E and voltages of B, D, F are regarded as voltage source v_{ACE} and v_{BDF} respectively. This is assumed for pure loop current control. To obtain circulate current in ring of H-MMC, additional current and voltage references are given in (12) and (13) respectively.

$$i_L^* = K_{DI}(v_{C\_BDF} - v_{C\_ACE}) \qquad (12)$$

$$v_L^* = \left| K_{DV} \left(v_{C\_BDF} - v_{C\_ACE} \right) \right| \qquad (13)$$

The total loop current reference is calculated by (14).

$$i_l^* = i_{l1}^* + i_{l2}^* + i_L^* \qquad (14)$$

(14) is added to current reference of each arm module and (13) is added to voltage reference of arm module B,D,F and subtracted to voltage reference of arm module A,C,E.

E. Internal model principle

Power conversion and module voltage balance are controlled by regulating currents in H-MMC. Currents of modules are expressed by the following equations.

$$i_a^* = i_{rs}^* + i_{vw}^* + i_l^* \qquad (15)$$

Other current references of modules are obtained the same as (15). Each module current has both frequencies in primary and secondary side. Therefore it is difficult to regulate the current of arm-modules accurately.

In this research, the module current is directly controlled by applying the internal model principle. The internal model principle has a merit that the applied controller can realize current control without steady error. It is defined that the transfer function of control system has mathematical model of reference signal. $G_C(s)$ is the transfer function of controller which includes two sinusoidal function models.

$$G_C(s) = \frac{K_{Sc1}s}{s^2 + \omega_1^2} + \frac{K_{Sc2}s}{s^2 + \omega_2^2} \qquad (16)$$

Based on (16), Control strategy of current controller of each module is shown in (17).

$$
\begin{aligned}
v_x^* = &\, K_{Pc}(i_x^* - i_x) \\
&+ \left(\frac{K_{Sc1}\omega_1^2 s}{s^2 + 2\zeta\omega_1 s + \omega_1^2} + \frac{K_{Sc2}\omega_2^2 s}{s^2 + 2\zeta\omega_2 s + \omega_2^2} \right)(i_x^* - i_x)
\end{aligned}
\qquad (17)
$$

$$(x = a\text{-}f)$$

First term in the right side is a proportional control, and second term is a control of the internal model principle with damping. The damping coefficient is valid at transient, but is 0 in steady state. The controller has sinusoidal internal model of primary and secondary frequencies.

F. Voltage balance in each module

Voltage balance in each module is introduced[1]. Because unbalance in each arm module occurs due to variations of individual differences of capacitance, switching delay and so on. The balancing control is given by (18).

$$v_{B-xn}^* = -K_B \left(\overline{v}_{Cx} - v_{C-xn} \right) i_x \qquad (18)$$

$$\overline{v}_{Cx} = \frac{\sum_{n=1}^{N} v_{C-xn}}{N} \qquad (19)$$

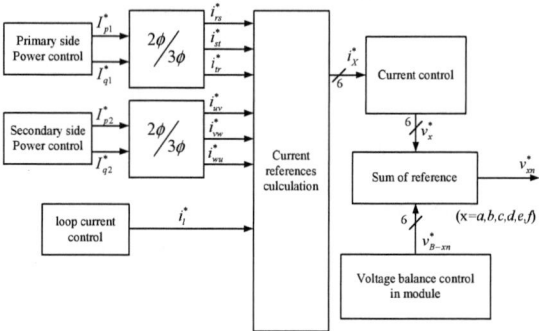

Fig. 4. Configuration of total control system

where (19) is average value of cell capacitor voltages in an arm module and N is the number of cells. $v_{C\text{-}xn}$ is capacitor voltage, $v_{B\text{-}xn}^*$ is voltage reference value for balancing capacitor voltage in each module. Index '-xn' means n-th cell in x-module. This control can adjust voltage balance among cells in an arm module because of exchange of power among cells. Voltage reference values in (13), (17) and (18) are summed up and output from each cell.

The configuration of overall control system is shown in Fig.4.

IV. SIMULATION

Simulation is performed to verify the proposed control methods by using interconnection circuit of power line in Fig.5. Table I shows the parameters of the circuit, and table II shows the control gains. In this research, the H-MMC has 2 full-bridge cells per an arm module, and each cell is operated the uni-pole PWM switching with shifted triangular wave of which phase is shifted according to the number of modules. The line voltage becomes 9-levels in case of 4 cells in line-to-line modules. The H-MMC can realize multilevel voltage output by this PWM switching pattern.

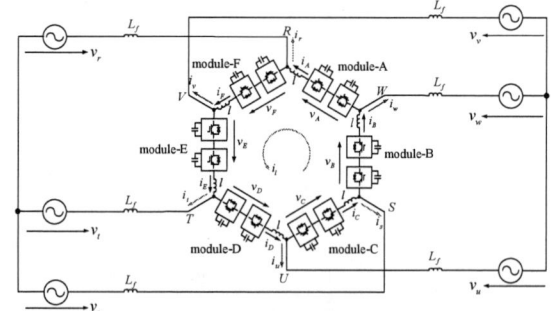

Fig. 5. Simulation circuit

TABLE I
PARAMETERS OF CIRCUIT

V_1	200Vrms	V_2	200Vrms
l	5mH	C	5000µF
v_c^*	200V	L_f	3mH

The 2014 International Power Electronics Conference

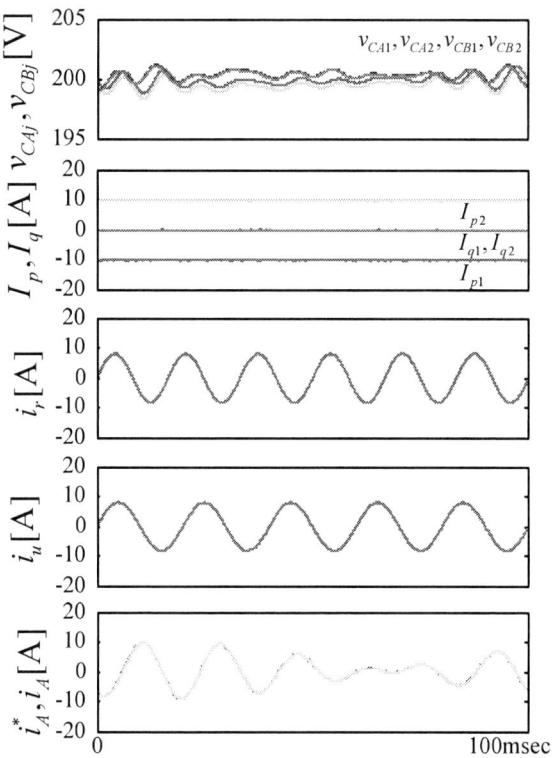

Fig. 6. Simulation result (f_1: 60Hz, f_2: 50Hz)

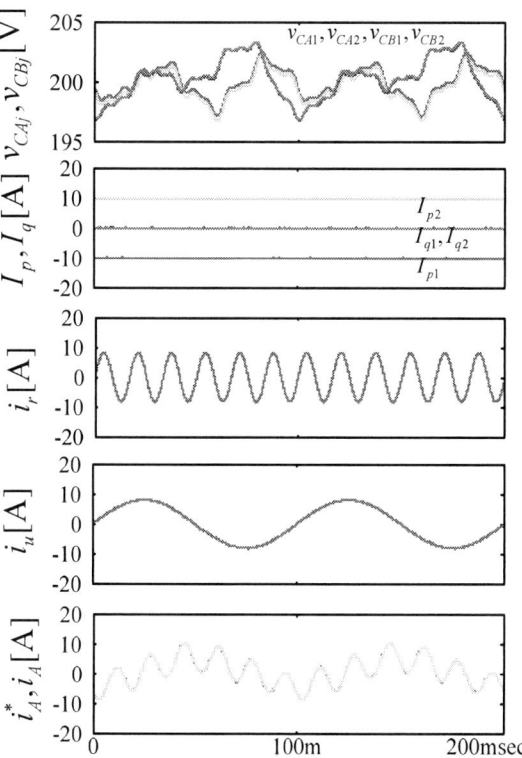

Fig. 7. Simulation result (f_1: 60Hz, f_2: 10Hz)

TABLE II
GAINS OF CONTROL

K_{P1}	1.5	K_{I1}	5.0
K_{Sc1}	0.1	K_{Sc2}	0.1
K_{Pc}	40	K_L	0.5
K_{DI}	1.5	K_{DV}	15
K_B	0.1		

Fig.6 shows a simulation result that input frequency is 60Hz and output frequency is 50Hz. It is confirmed that I_{p1} and I_{p2} are the same absolute value 10A but opposite sign at steady state. This means that received active power of H-MMC from primary side is the same value as the supplied active power to secondary side. Average voltage of all floating capacitors is maintained to be constant value around 200V. On the other hand, current of primary side, secondary side and arm module can be outputted accurately. Especially, current of arm module is precisely regulated by the internal model principle in spite of mixed frequency.

Fig. 7 shows a simulation result that input frequency is 60Hz and output frequency is 10Hz. The good result is obtained like Fig.6. These simulation results show that H-MMC is able to convert exactly regardless of frequency and voltage. The H-MMC can be applied to grid connection converter as a power flow controller and a power conditioner for wind generators.

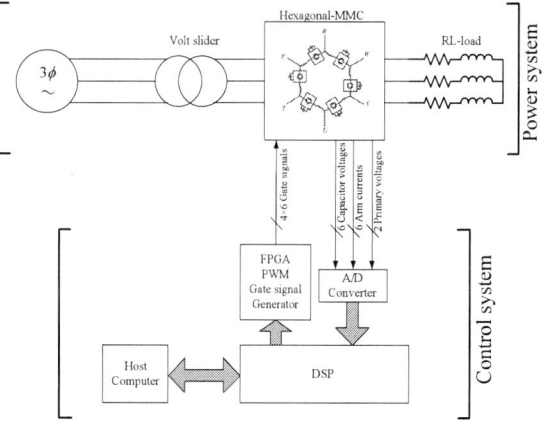

Fig. 8. Experimental system

TABLE III
PARAMETERS OF EXPERIMENTAL CIRCUIT

V_1	12Vrms	V_2	12Vrms
l	1.5mH	C	3300μF
R	22Ω	L	10mH
v_C^*	30V		

V. EXPERIMENT

Experimental system is constructed for a basic test of H-MMC as shown in Fig.8. This system has single cell per arm module and power scale of circuit is small in Table III. All the controls are executed by the DSP (TI : TMS320C6713-225) and PWM of the shifted triangular

978-1-4799-2706-7/14 $31.00 © 2014 IEEE

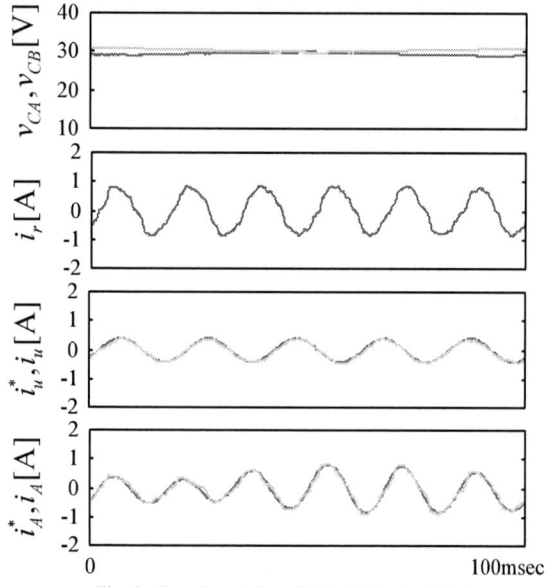

Fig. 9. Experimental result (f_1: 60Hz, f_2: 50Hz)

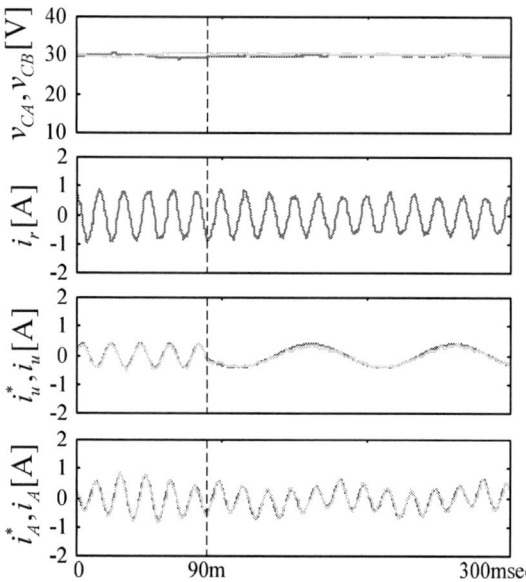

Fig. 10. Experimental result (f_1: 60Hz, f_2: 50 to 10Hz)

capacitor keeps constant around 30V without influence of change.

Experimental results clarify that the proposed control method for H-MMC is able to work correctly and has good performance for power flow and voltage regulation of floating capacitors.

VI. CONCLUSIONS

This paper investigated a new control of Hexagonal MMC for 3-phase BTB system. The H-MMC is applied to grid connection system and proposed how to control the power flow in primary and secondary side and the floating capacitor voltage of the H-MMC. In the H-MMC, sinusoidal currents with mixed frequency are controlled directly by using the internal model principle controller. The simulation and the experiment demonstrated that the proposed control method is able to control exactly and have a good performance. We intend to increase the number of cell of the experimental system and test in the near future.

ACKNOWLEDGMENT

This work was supported by JSPS KAKENHI (Grant-in-Aid for Young Scientists(B)) Grant Number 24760237.

REFERENCES

[1] M. Hagiwara, H. Akagi, "Control and Experiment of Pulsewidth-Modulated Modular Multilevel Converters", *IEEE Trans. on PE*, Vol.24, No.7, pp.178-1746, 2009.
[2] M. Hagiwara, R. Maeda, and H. Akagi, "Theoretical Analysis and Control of the Modular Multilevel Cascade Converter Based on Double-Star Chopper-Cells (MMCC-DSCC)"(in Japanese), *IEEJ Trans. on IA*, Vol.16, No,.1, pp.8142, 2011.
[3] M. Glinka, R. Marquardt, "A New AC/AC Multilevel Converter Family", *IEEE Trans. on IE*, Vol.52, No.3, pp. 66169, 2005.
[4] R.Marquardt,"Modular Multilevel Converter:An universal concept for HVDC-Networks and extended DC-Bus-applications",*The Internationa Power Electronics Conference(IPEC)*,pp.502-507,2010.
[5] H. Fujita, M. Hagiwara, H. Akagi, "Power Flow Analysis and DC-Capacitor Voltage Regulation for the MMC-DSCC"(in Japanese), *IEEJ Trans. on IA*, Vol.17, No.6, pp.659-665, 2012.
[6] W. Kawamura, H. Akagi, "Control Strategy of a Modular Multi-level Cascade Converter Based on Triple-Star Bridge-Cells (MMCC-TSBC)"(in Japanese), *The Institute of Electrical Engineers of Japan*, 1-14, pp.241-246, 2012
[7] L. Baruschka, A. Mertens, "A new 3-phase AC/AC modular multilevel converter with six branches in hexagonal configuration", *Energy Conversion Congress and Exposition (ECCE)*, pp.4005-4012, 2011
[8] T. Hosaka, K Akiba and H Fujita, "A Unified Power-flow controller Consisting of Six Cascaded H-Bridge Converters" (in Japanese), *The Institute of Electrical Engineers of Japan*, 1-35, Vol.1, pp.233-236, 2011
[9] S. Hamasaki, K. Okamura, M. Tsuji, "Power Flow Control of Modular Multilevel Converter based on Double-Star Bridge Cells Applying to Grid Connection", *Journal of ICEMS*,Vol.2, No.2, pp.248-255, 2013
[10] S. Fukuda, T. Yoda "A Current Control Method for Active Filters Using Sinusoidal Internal Model" (in Japanese), *T.IEEJ*, Vol.120-D, No.12, pp.111-146, 2000
[11] S. Hamasaki, M. Tsuji, E. Yamada "A Study on Power Flow Control for Distributed Generator with EDLC", *SYMPOSIUM ON POWER ELECTRONICS, ELECTRICAL DRIVES, AUTOMATION AND MOTION (SPEEDAM)*, Vol.1, pp.1502-1507, 2010.

wave is executed by the FPGA (Xilinx : XC3S1500-4FG456C). Switching frequency of each cell is 10kHz and sampling period of DSP is 100µs in the experiment.

Fig.9 shows an experimental result that input frequency is 60Hz and output frequency is 50Hz. Voltage of each floating capacitor is maintained to be constant value around 30V. Current of primary side, secondary side and arm module can be outputted accurately. Current of arm module is precisely regulated by the internal model principle in spite of mixed frequency. This behavior is very similar to the simulation.

Fig.10 shows an experimental result that input frequency is 60Hz and output frequency changes from 50Hz to 10Hz by step. Output current in secondary side changes smoothly and regulated correctly. Voltage of

A Synthesized Capacitors Voltage Control for Modular Multilevel Converter in HVDC Application

Rongfeng Yang, Shunke Sui, Binbin Li, Wei Wang, Dianguo Xu

School of electrical engineering and automation
Harbin Institute of Technology, Harbin, China
yrf@hit.edu.cn, suishunke@126.com, libinbinhit@126.com, wangwei602@hit.edu.cn, Xudiang@hit.edu.cn

Abstract—**The capacitors voltage balance is one of the major problems in modular multilevel converter control which has not been resolved perfectly. Based on energy transmission analysis through steady state equations, this paper proposed a synthesized balance control scheme that involves leg voltage, arm voltage, individual voltage balance control and circular current control. Comparing with other methods, the proposed scheme incorporates the leg balance control into circular current control, and employs the differential mode voltage to balance the upper and lower arm energy. The proposed method can maintain capacitors voltage balance under different power factor conditions with few harmonics in circular current and dc line voltage. The proposed strategy has been verified by detailed simulation results.**

Keywords— *capacitors voltage balance, circular current, Modular multilevel converter (MMC).*

I. INTRODUCTION

The modular multilevel converter (MMC) is a prominent topology for high voltage and high power application[1,3,4,16]. Comparing with other multilevel converters, this type of converter not only has less switch stress, few clamped diodes and capacitors, but also has modular structure that makes the design convenient, controllable and reliable. The double-start connected modular multilevel converter with chopper cells (DSCC) has the capability of 4-quadrant operation with positive and negative sequence control ability[1], thus it has been widely implemented in various industry application, such as high power motor drive, high voltage static var compensation (STATCOM) and high voltage direct current transmission (HVDC).

The configuration of this converter includes different number of legs (phases), meanwhile each leg is constructed by upper and lower arms, and each arm is series-connected by several chopper cells, as shown in Fig.1. The modular multilevel converter is first presented in [2], and later has been successfully realized in industry on HVDC by Siemens, which urged the researchers to

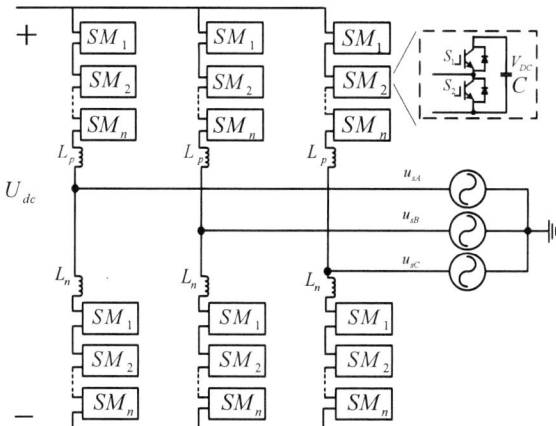

Fig.1 Modular multilevel converter configuration.

concentrate on this type of converter. Many literatures have been proposed in recent years focusing on MMC modulation, capacitors voltage balance and circular current control.

The modulation methods have firstly been studied. The SVPWM, PDPWM[11,12,8], NLC[4-6] and PSPWM[9,10,13,14] are commonly adopted in multilevel converters. The SVPWM exhibits excellent harmonics performance and high DC voltage utilization ratio, which has widely employed in low voltage and two level converters. However, its calculation should get tremendously complex as the level number increases, which constrains its application in multilevel converters. Thus more researchers employ carrier comparing based modulation methods, such as PDPWM and PSPWM. Implemented in modular multilevel converters, the PSPWM method has similar on-off switching decision procedure for each power cell. Moreover, each power switch has equal switching frequency and power loss, so it specially suits modular multilevel converters. But the PSPWM cannot maintain the number of the active cells constant, which generates harmonics in circular current and calls for special harmonics suppressing control[5].

The floating capacitors voltage balance is a major issue for modular multilevel converters. The nonideal device parameters and working condition deviation should bring with power cells capacitors voltage imbalance, and influence the system normal operation. The imbalance could be categorized as individual voltage,

---

[1] 1. Project 51237002 supported by National Natural Science Foundation of China.
2. Project DREM2012001 Sponsored by grants from the Power Electronics Science and Education Development Program of Delta Environmental & Educational Foundation.

arm voltage and leg voltage imbalance[18, 21].

The abundant redundant states of multilevel converters provide flexible measures to balance the individual voltage. So the first type of individual balance control methods is to sort the capacitors voltage, and decide each cell appropriate switching state according to the charge or discharge status[3-10]. [11,12] changes carrier wave bias to realize balance effect based on the similar consideration. This type method can obtain satisfied balance effect despite the varied modulation, but has the disadvantage that the devices switching frequency cannot reserve equal. Another individual balance control is to plus the regulation voltage to each cell's reference voltage[13,14]. This method would not change the arm output voltage thus has no influence on other system control but could adjust each cell's instantaneous energy storage effectively.

The leg voltage imbalance and arm voltage imbalance comparatively has less been studied in proposed literatures. The arm voltage could reserve balance under steady state[15,16,22] thus its imbalance has less been researched, however, the deviation caused by transient change may deteriorate the system performance that must be suppressed. [17] proposed an arm balance scheme but had not enough demonstration about its effect on circular current. [19] proposed a novel leg and arm balance control method based on active and reactive positive sequence components control of circular current. This method is complex and must be done in two rotating reference frame which limits its application field. [21]proposed a comprehensive balance control for leg, arm and individual imbalance. However, the arm balance control parameters is complex to calculate, and the consequence influence on dc voltage ripple is not discussed. [16,18,20]analyzed the energy transmission relationship and employed one similar measure like[19], but did not used in HVDC system .

The circular current is another special phenomenon for modular multilevel converter. As the DC terminals are connected together, energy could flow from DC line to each leg. Meanwhile there exists energy flow between legs when the leg voltage is imbalance, which brings with circular current harmonics[23,24]. The circular current harmonics should reduce system maximum power ratio and deteriorate system performance. The circular suppressing measure will affect the balance control, however, [9,25,26] only proposed circular current suppression methods but did not analyze the cooperation with voltage balance control.

This paper analyzed the converter energy flow in HVDC application and proposed a novel comprehensive voltage balance control strategy. Section II deduced the circuit equivalent model and the general control rules. Section III provided the power flowing equations. Section IV illustrated the voltage balance control strategy, including leg, arm and individual voltage balance. One circular current real-time control method was proposed to balance the leg voltage. Being distinct from previous

methods, the differential mode voltage was employed to balance the arm voltage with negligible influence on AC side to avoid conflicts with dc ripple control. Section V presented the system current control incorporating with balance control. Section VI verified the proposed methods through simulation results.

II. MODULAR MULTILEVEL CONVERTER MODEL AND ITS CONTROL

The multilevel converter is simplified as Fig.2. Each arm is thought as one voltage source converter (VSC), and the VSC output voltage is assumed to trace the reference voltage after modulation control. Analyzing one submodule (SM) of the circuit, as depicted in Fig.1, if switch S1 is turned on and switch S2 is turned off, the SM generates $+V_{dc}$ voltage, else if switch S1 is turned off and switch S2 is turned on, the SM generates 0 voltage.

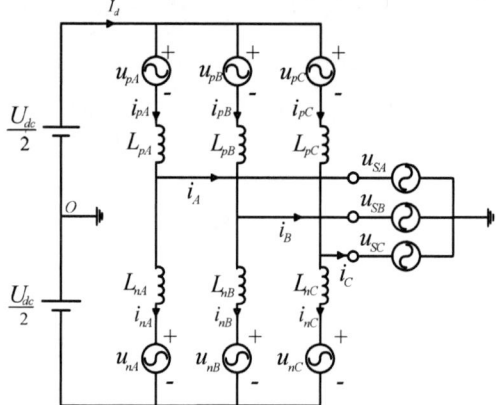

Fig.2 Simplified equivalent model.

With averaged model, the upper arm voltage is

$$u_{pj} = M * \sum_{i=1}^{N} V_i \qquad (1)$$

Where $M = \dfrac{u_{pref}}{\sum\limits_{i=1}^{N} V_i}$, N is the number of SMs in the

upper arm, u_{pref} is the setting reference voltage of upper arm. V_i is the capacitor voltage in this arm. The lower arm has the similar express as (1).

Applying Kirchhoff's voltage law (KVL), the VSC voltage and current relationship is expressed as,

$$\begin{cases} u_{pj} = \dfrac{U_{dc}}{2} - L\dfrac{di_{pj}}{dt} - u_{Sj} \\[3mm] u_{nj} = \dfrac{U_{dc}}{2} - L\dfrac{di_{nj}}{dt} + u_{Sj} \end{cases} \quad j=\text{A, B or C} \qquad (2)$$

Where i_{pj} and i_{nj} are j-phase upper and lower arm current, u_{sj} is the supply voltage.

The circular current i_{zj} and AC side current i_j could be expressed as (3), and they are also common mode and differential mode current respectively.

$$\begin{cases} i_{zj} = \dfrac{i_{pj} + i_{nj}}{2} = i_{commj} \\[3mm] i_j = i_{nj} - i_{pj} = i_{diffj} \end{cases} \qquad (3)$$

From (2) and (3), it could be deduced that

$$\begin{cases} u_{pj} + u_{nj} = U_{dc} - L\dfrac{d(i_{pj} + i_{nj})}{dt} \\ u_{nj} - u_{pj} = 2u_{sj} - L\dfrac{d(i_{nj} - i_{pj})}{dt} = u_o \end{cases} \quad (4)$$

Uo is the phase output voltage on AC side. Define the common mode and differential mode voltage as,

$$\begin{cases} u_{commj} = \dfrac{u_{pj} + u_{nj}}{2} \\ u_{diffj} = u_{nj} - u_{pj} \end{cases} \quad (5)$$

Substituting (5) into (4) yields

$$\begin{cases} u_{commj} = \dfrac{U_{dc}}{2} - L\dfrac{d(i_{commj})}{dt} \\ u_{diffj} = 2u_{sj} - L\dfrac{d(i_{diffj})}{dt} \end{cases} \quad (6)$$

It is evident that the circular current and the AC side current can be control respectively by the differential and common mode voltage.

III. CONVERTER ENERGY STORAGE ANALYSIS OF STEADY STATE

The SMs capacitors voltage has significant relationship with the power flowing throughout the converter. There exists power exchange between the DC and AC terminals, and the integrated power yields the energy storage in SMs.

From eq.(2) and eq.(3), neglecting the electromotive voltage on inductor, the upper and lower arms instantaneous power is

$$\begin{cases} P_{pj} = u_{pj}i_{pj} = \dfrac{U_{dc}}{2}i_{zj} - u_{Sj}i_{zj} - \dfrac{U_{dc}}{2}\dfrac{i_j}{2} + u_{Sj}\dfrac{i_j}{2} \\ P_{nj} = u_{nj}i_{nj} = \dfrac{U_{dc}}{2}i_{zj} + u_{Sj}i_{zj} + \dfrac{U_{dc}}{2}\dfrac{i_j}{2} + u_{Sj}\dfrac{i_j}{2} \end{cases} \quad (7)$$

In (7), the first component is the energy from the DC terminal, and the last component is the energy from the AC terminal. Under steady state, the circular current i_{zj} and the DC voltage U_{dc} are constant, meanwhile the supply voltage u_{sj} and supply current i_j are sinusoidal, so the average energy exchange in one period is

$$\int_0^{2\pi} P_{nj} = \int_0^{2\pi} P_{pj} = \int_0^{2\pi}\dfrac{U_{dc}}{2}i_{zj} + \int_0^{2\pi} u_{Sj}\dfrac{i_j}{2} \quad (8)$$

From (8), the energy is only exchanged between the DC and AC terminals in steady state. However, the energy should fluctuate that the upper arm and the lower arm energy has opposite directional variability during the period as expressed in (7).

IV. SMS CAPACITORS VOLTAGE BALANCE CONTROL

As the SMs' capacitors are float, different SMs' voltage should deviate because of the nonideal circuit parameters. Furthermore, the working condition will not always maintain equivalent like steady state analyzed in section III. The slight energy exchanging difference will result in irreversible imbalance without balance control. So it is necessary to take effective measures to prevent the imbalance at the beginning occurrence. The imbalance phenomenon includes legs, arms and individual SM capacitor imbalance[18,21].

A. Leg voltage balance control.

The leg voltage is the sum of the capacitors voltage in one leg,

$$\begin{aligned} V_{legj} &= V_{pj} + V_{nj} \\ V_{pj} &= \sum_{i=1}^{N} V_{ij} \\ V_{nj} &= \sum_{i=N+1}^{2N} V_{ij} \end{aligned} \quad (9)$$

V_{pj} is the upper arm capacitors voltage, V_{nj} is the lower arm capacitors voltage. The leg voltage reflects the power stored in one phase, namely the total energy of the upper and lower arms. With (7), the instantaneous leg energy is expressed as

$$\begin{aligned} \int P_j &= \int (P_{pj} + P_{nj}) = \int (U_{dc}i_{zj} + u_{Sj}i_j) \\ &= \int (U_{dc}i_{zj} + U_{Sj}I_j\cos\varphi) \end{aligned} \quad (10)$$

Here U_{sj} and I_j denote the amplitude of the supply voltage and current respectively, φ is the power factor angle. Eq.(10) means there are two measures to change the energy storage in one leg: changing the AC side or DC side power flow. To change the AC side power flow, it has to change the converter AC side voltage, which will bring with nonequivalent voltage on the converter output, thus influences the AC side decoupled control. So the better way is to adjust the DC side energy flow. However, because of the capacitors voltage oscillation and its consequent circular current, many system control strategies have circular current suppress tackles, which should conflict with the leg voltage balance control. To resolve this problem, a novel idea comprising circular current control and leg voltage balance control is proposed as in fig.3. The difference of leg voltage generates additional reference circular current, and the wanted circular current is obtained by the inner current control. The compensated voltage u_{LBj} will be added to converter common mode voltage later.

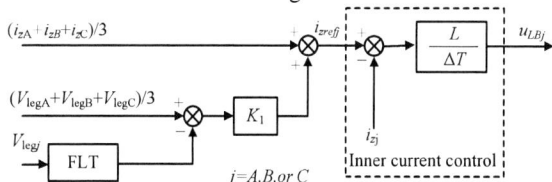

Fig.3 Leg voltage balance control.

B. Arm voltage balance control.

The arm capacitors voltage derives from the upper and lower VSC operation, and the energy transformation of steady state is expressed in (7). It is obvious that the average power of P_{pj} and P_{nj} is constant, so the arm voltage could retain balance[22] under ideal working condition. However, as the actual system is nonideal, the small parameter deviation should bring with further accumulative imbalance, so the arm voltage balance control is a requisite to ensure system reliable and stable.

Substituting P_{pj} and P_{nj} from (7) yields the arm power difference in one period,

$$\int_0^{2\pi}(P_{nj}-P_{pj})=\int_0^{2\pi}(2u_{sj}i_{zj}+\frac{U_{dc}}{2}i_j) \qquad (11)$$

If (11) has none-zero value, it could be employed to balance the arm energy. To avoid changing the converter output AC voltage, the conventional arm balance control is to adjust the common mode voltage[18-21]. This method adds an alternate component in phase of supply voltage u_{sj} to the circular current i_{zj}, so the item $2u_{sj}i_{zj}$ could generate active power and balance the arm voltage. This method works well in one-phase system, but in 3-phase system, the arms imbalance degree is different, so the regulation of 3-phase is asymmetric. Assuming each arm circular current has same DC component I_{ldc} and alternate component with different amplitude I_1, I_2, and I_3, the total circular current is,

$$I_{leg}=I_{legA}+I_{legB}+I_{legC}$$
$$=3I_{ldc}+I_1\sin\omega t+I_2\sin(\omega t-\frac{2}{3}\pi)+I_3\sin(\omega t+\frac{2}{3}\pi) \qquad (12)$$

It means that the total circular current will not maintain constant, which leads to system DC voltage disturbance in one period, thus deteriorates the system performance, even causes arm imbalance failure.

So this paper proposed another compensate method without inducing energy disturbance on the DC side. Considering the $u_{sj}i_{zj}$ in (11), adding a dc component to u_{sj} also results in the none-zero value of the average arm power difference. A novel arm balance scheme is depicted in fig.4, where if $i_{zj}>0$, $K_4=1$, else $K_4=-1$.

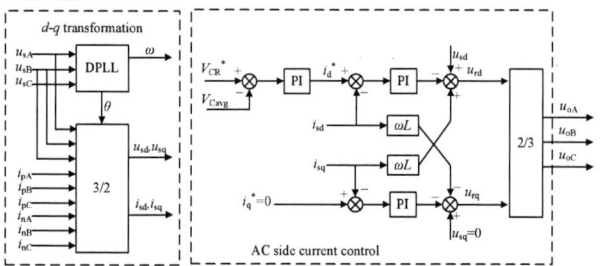

Fig.4 Arm voltage balance control.

After leg and arm balance control, the upper and lower reference voltages are

$$\begin{cases}u_{pj}=\frac{1}{2}U_{dc}-u_o-u_{aBj}+u_{LBj}\\[2mm]u_{nj}=\frac{1}{2}U_{dc}+u_o+u_{aBj}+u_{LBj}\end{cases} \qquad (13)$$

Here u_o is the AC side current control results, u_{Lbj} and u_{aBj} are the leg and arm balance control regulation respectively. The dc component in u_{sj} and its balance effect also could be explained as follows. If the upper capacitors voltage is larger than the lower one, and this leg is releasing energy through circular current, increasing the upper arm output voltage or decreasing the lower arm output voltage could reduce the gap between the energy stored in the upper and lower arms, thus the arm capacitors voltage is balanced.

C. Individual balance control.

The individual voltage control is applied in each arm. If the arm is charging, increasing the chopper cell's output voltage which has smaller capacitor voltage should increase the charging energy, and vice verse. For example, Assuming the average arm capacitor voltage is

V_{arm}^* (V_{pj} or V_{nj}), and each chopper cell's capacitor voltage is V_1, V_2, ..., V_N, the chopper cell's output voltage is regulated as follows,

$$u_{idvBi}=K_5(V_i-V_{arm}^*)$$
$$K_5=\begin{cases}+|K_5| & i_{arm}>0\\-|K_5| & i_{arm}<0\end{cases} \qquad (14)$$

Here, $i=1,\ldots,N$, and u_{idvBi} is the regulation for individual balance control. i_{arm} is the arm current, and it could be i_{pj} or i_{nj}. It is notable that the total arm output voltage will not change because

$$\sum_{i=1}^N u_{idvBi}=K_5\sum_{i=1}^N(V_i-V_{arm}^*)$$
$$=K_5(\sum_{i=1}^N V_i-NV_{arm}^*)=0 \qquad (15)$$

So this individual balance control will not change the converter output voltage, and it will not conflict with other control schemes.

V. SYSTEM CONTROL STRATEGY

In HVDC application, the converter absorbs energy from the AC supply source and transmits it to DC load. The control block diagram that absorbs active power is shown in fig.5, which generates converter AC side output voltage through decoupled scheme. The V_{CR}^* is reference capacitors voltage, and V_{Cavg} is the actual average value of all cells' capacitors voltage. The difference of V_{CR}^* and V_{Cavg} is used to generate active current reference. The reactive current reference is set to zero functioning as rectifier.

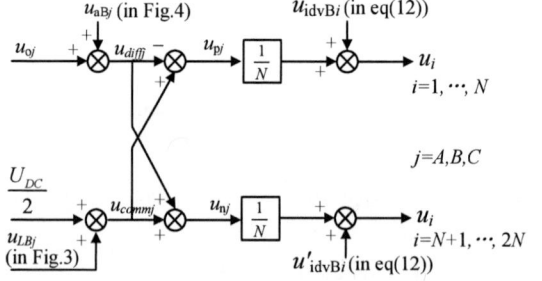

Fig. 5 AC side current decoupled control.

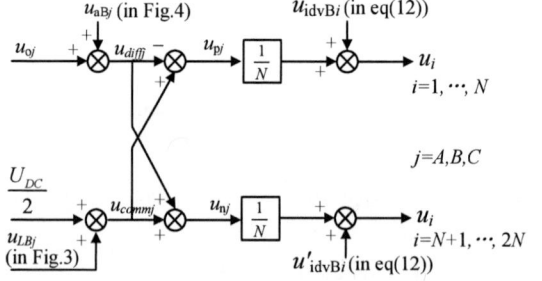

Fig.6 Balance control scheme.

To control the circular current and balance the capacitors voltage, the discussed regulation has been added as shown in fig.6, which includes circular current control, leg voltage balance control (u_{LBj}), arm balance control (u_{aBj}) and individual balance control (u_{idvBi}).

After calculating voltage command of each chopper cell u_i, it is normalized by each cell's capacitor voltage V_i, then after comparing with the phase shifted triangle waveform, the switches states are determined.

VI. SIMULATION

The simulation is carried out with Matlab/Simulink tools. The circuit configuration is displayed in Fig.1, and the parameters of the circuit are listed in Talbe I. The system control strategy is exhibited in Fig.5 and Fig.6. The PSPWM is employed so the converter equivalent switching frequency is 12kHz (each cell works at 1kHz and there are 12 cells each leg), which is helpful to suppress switching harmonics.

TABLE I. SYSTEM PARAMETERST

Parameter	Symbol	Value
Supply voltage	V_s	2.5kV
Cells number of each arm	N	6
DC-voltage reference	U_{DCref}	5000V
DC capacitor capacitance	C	5600μF
Unit switching frequency	f_s	1kHz
Reactor	L	5mH
Rated Current	Ir	150A
DC load	R_L	50Ω

A. Steady state simulation results

Fig.7 shows the simulation waveforms under steady state condition. From Fig.7(b), the supply current is symmetric and sinusoidal, which means the AC side current control works well. The capacitors voltage always maintains balance. From Fig.7(c), the 3-leg average capacitors voltage V_{legj} is balance, and the voltage oscillates following the supply current. However, there are inevitable AC components in each capacitor's voltage V_{ci} including fundamental frequency (50Hz) and 2nd-order (100Hz) harmonic components, as shown in Fig.7(d), which displays all 12 cells' capacitors voltage waveform of leg A. The circular current i_z in Fig.7(e) has been excellently controlled as the three-phase circular current is equal with only dc components and high frequency switching harmonics, meanwhile the DC voltage U_{DC} in Fig.7(f) is always keeping with the set value. The good performance of AC and DC side control illustrates that these two side controls have been decoupled correctly through differential and common mode control, as expressed in (6).

B. Transient state simulation results

Fig.8 shows the simulation waveforms under transient state condition. The capacitors initial voltage is disordered, and there are leg, arm and individual imbalance simultaneously which is used to verify different balance control strategy. The average capacitors voltages of three legs are 900V, 775V, 975V respectively, and the capacitors voltages in A-phase upper and lower arms are 800V, 900V, or 1000V. From Fig.8(c)(d), it displays that the imbalance is gradually reduced, and all the capacitors voltage gets balance after 0.3s.

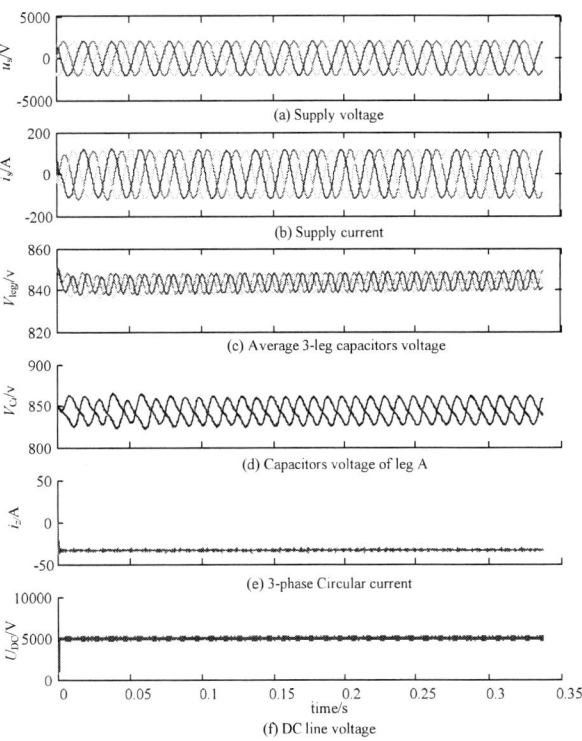

Fig.7 Simulation waveforms in steady state.

Fig.8 Simulation waveforms in transient state.

During this process, the three-phase circular currents are not equal thus the difference is used to compensate the leg imbalance. Meanwhile the supply current is no longer symmetric because of the differential mode voltage caused by arm balance control. However, the slight symmetric has not disturbed the system regularly running and disappears after the arm balanced. As the arm imbalance is generally small under normal working conditions, the arm balance control influence on other control schemes can be neglected.

Fig.9 shows the performance under different power factors, the i_q^* in fig.5 is set as 75A. The balance effect is similar with the results of i_q^* being 0 in Fig.8, which means this balance scheme will not be effected by the reactive power.

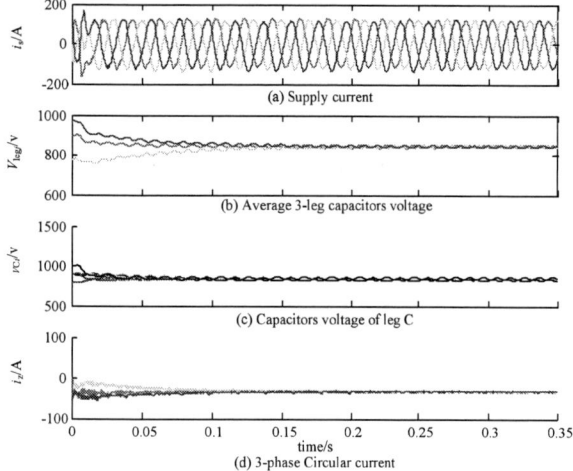

Fig.9 Simulation waveforms with non-zero power factor.

VII. CONCLUSION

This paper proposed one novel capacitor voltage balance scheme being suitable for HVDC application. The novelty of this balance scheme involves one leg balance control incorporating with circular current control, and the arm balance control based on converter differential mode voltage compensation. The arm balance control should slightly change the ac side voltage. However, as the arm capacitors voltage deviation is small under normal condition, the balance control influence on converter ac side voltage is negligible.

The simulation results verified the proposed methods. The balance process is effective and stable. The circular current is always under control with few harmonics, and the dc line voltage is keeping constant during balance procedure. The excellent performance demonstrates that the proposed strategy is suitable for HVDC application.

REFERENCES

[1] H. Akagi, "Classification, Terminology, and Application of the Modular Multilevel Cascade Converter (MMCC)," *IEEE Transactions on Power Electronics*, vol. 26, no. 11, pp.3119-3130, 2011.

[2] R. Marquardt, A. Lesnicar, "A new modular voltage source invertertopology," *Conf. Rec. Eur. Conf. Power Electron. Appl.*, pp. 1–10, 2003.

[3] E. Solas, G. Abad, J.A. Barrena, S. Aurtenetxea, et al, "Modular Multilevel Converter With Different Submodule Concepts—Part I: Capacitor Voltage Balancing Method," *IEEE Transactions on Industrial Electronics*, vol. 60, no. 10, pp. 4525-4535, 2013.

[4] J. Peralta, H. Saad, S. Dennetiere, J. Mahseredjian, et al, "Detailed and Averaged Models for a 401-Level MMC–HVDC System", *IEEE Transactions on Power Delivery*, vol. 27, no. 3, pp. 1501-1508, 2012.

[5] Zixin Li, Ping Wang, Haibin Zhu, Zunfang Chu, et al. "An Improved Pulse Width Modulation Method for Chopper-Cell-

Based Modular Multilevel Converters," *IEEE Transactions on Power Electronics*, vol. 27, no. 8, pp. 3472-3481, 2012.

[6] Gum Tae Son, Hee-Jin Lee, Tae Sik Nam, Yong-Ho Chung, et al. "Design and Control of a Modular Multilevel HVDC Converter With Redundant Power Modules for Noninterruptible Energy Transfer," *IEEE Transactions on Power Delivery*, vol. 27, no. 3, pp. 1611-1619, 2012.

[7] Y. Zhang, G.P. Adam, T.C. Lim, S.J. Finney, et al, "Analysis of modular multilevel converter capacitor voltage balancing based on phase voltage redundant states," *IET Power Electronics*, vol.5, no. 6, pp. 726-738, 2012.

[8] M. Saeedifard, R. Iravani, "Dynamic Performance of a Modular Multilevel Back-to-Back HVDC System," *IEEE Transactions on Power Delivery*, vol. 25, no. 4, pp. 2903-2912, 2010.

[9] Qingrui Tu, Zheng Xu, Lie Xu, "Reduced Switching-Frequency Modulation and Circulating Current Suppression for Modular Multilevel Converters," *IEEE Transactions on Power Delivery*, vol. 26, no. 3, pp. 2009-2017, 2011.

[10] Fujin Deng, Zhe Chen, "A Control Method for Voltage Balancing in Modular Multilevel Converters", *IEEE Transactions on Power Electronics*, vol. 29, no. 1, pp. 66 – 76, 2014.

[11] Jun Mei, Ke Shen, Bailu Xiao, L.M. Tolbert, et al, "A New Selective Loop Bias Mapping Phase Disposition PWM With Dynamic Voltage Balance Capability for Modular Multilevel Converter," *IEEE Transactions on Industrial Electronics*, vol. 61, no. 2, pp. 798-807, 2014.

[12] Jun Mei, Bailu Xiao, Ke Shen, L.M. Tolbert, et al, "Modular Multilevel Inverter with New Modulation Method and Its Application to Photovoltaic Grid-Connected Generator," *IEEE Transactions on Power Electronics*, vol. 28, no. 11, pp. 5063-5073, 2013.

[13] D. Montesinos-Miracle, M. Massot-Campos, J. Bergas-Jane, S. Galceran-Arellano, et al, "Design and Control of a Modular Multilevel DC/DC Converter for Regenerative Applications," *IEEE Transactions on Power Electronics*, vol. 28, no. 8, pp. 3970-3979, 2013.

[14] M. Hagiwara, H. Akagi, "Control and Experiment of Pulsewidth-Modulated Modular Multilevel Converters," *IEEE Transactions on Power Electronics*, vol. 24, no. 7, pp. 1737-1746, 2009.

[15] A. Antonopoulos, L. Angquist, L. Harnefors, K. Ilves, et al. "Global Asymptotic Stability of Modular Multilevel Converters," *IEEE Transactions on Industrial Electronics*, vol. 61, no. 2, pp. 603-612, 2014.

[16] M. Hagiwara, H. Akagi, "A battery energy storage system with a modular push-pull PWM converter," *IEEE ECCE'2012*, pp. 747-754, 2012.

[17] R.F. Lizana, M.A. Perez, J. Rodriguez, "DC voltage balance control in a modular multilevel cascaded converter," *2012 IEEE International Symposium on Industrial Electronics (ISIE)*, pp. 1973-1978, 2012.

[18] J. Pou, S. Ceballos, G. Konstantinou, G.J. Capella, et al, "Control strategy to balance operation of parallel connected legs of modular multilevel converters," *2013 IEEE International Symposium on Industrial Electronics (ISIE)*, pp. 1-7, 2013.

[19] G. Bergna, E. Berne, P. Egrot, P. Lefranc, et al, "An Energy-Based Controller for HVDC Modular Multilevel Converter in Decoupled Double Synchronous Reference Frame for Voltage Oscillation Reduction," *IEEE Transactions on Industrial Electronics*, vol. 60, no. 6, pp. 2360-2371, 2013.

[20] G. Casadei, R. Teodorescu, C. Vlad, L. Zarri, "Analysis of dynamic behavior of Modular Multilevel Converters: Modeling and control," *2012 16th International Conference on System Theory, Control and Computing (ICSTCC)*, pp. 1-6, 2012.

[21] M. Hagiwara, R. Maeda, H. Akagi, "Control and Analysis of the Modular Multilevel Cascade Converter Based on Double-Star Chopper-Cells (MMCC-DSCC)," *IEEE Transactions on Power Electronics*, vol. 26, no. 6, pp. 1649-1658, 2011.

[22] Li Xiaoqian, Song Qiang, Li Jianguo, Liu Wenhua, "Capacitor voltage balancing control based on CPS-PWM of Modular Multilevel Converter," *2011 IEEE Energy Conversion Congress and Exposition (ECCE)*, 4029-4034, 2011.

[23] S. Ceballos, J. Pou, Choi Sanghun, M. Saeedifard, et al, "Analysis of voltage balancing limits in modular multilevel converters," *37th Annual Conference on IEEE Industrial Electronics Society (IECON)*, pp. 4397-4402, 2011.

[24] K. Ilves, A. Antonopoulos, Staffan Norrga, H.-P. Nee, "Steady-State Analysis of Interaction Between Harmonic Components of Arm and Line Quantities of Modular Multilevel Converters," *IEEE Transactions on Power Electronics*, vol. 27, no. 1, pp. 57-68, 2012.

[25] Zixin Li, Ping Wang, Zunfang Chu, Haibin Zhu, "An Inner Current Suppressing Method for Modular Multilevel Converters," *IEEE Transactions on Power Electronics*, vol. 28, no. 11, pp. 4873-4879, 2013.

[26] Ji-Woo Moon, Chun-Sung Kim, Jung-Woo Park, Dea-Wook Kang, et al, "Circulating Current Control in MMC Under the Unbalanced Voltage," *IEEE Transactions on Power Delivery*, vol. 28, no. 3, 1952-1959, 2013.

The 2014 International Power Electronics Conference

Operating Phase and Frequency Selection of Low Frequency AC Transmission System Using Cycloconverters

Pichetjamroen Achara*and Toshifumi Ise*

* Division of Electric, Electronic and Information Engineering,
Osaka University,
2-1 Yamada-oka, Suita, Osaka, Japan
achara@pe.eei.eng.osaka-u.ac.jp

Abstract-This paper describes the characteristics of low frequency ac transmission system using cycloconverters with three configurations. The purpose is to evaluate and choose the most suitable operating frequency and number of phases for this system. The new control scheme is also provided for all operating transmission system. Cable and filter selection are also included in this paper.

Keywords—LFAC, Cycloconverter, Power transmission, AC transmission

I.   INTRODUCTION

To support large capacity of electric power user, existing technology of power transmission on the same or longer distance still cannot use full performance in electrical engineering. Not only stepped up voltage level but also reduced transmitting frequency can improve the performance of transmission system. The conventional ac transmission system which is operating at 50 or 60 Hz has some weak points. At this range of frequency, it can cause the problem due to large charging current of cable capacitance, which causes large losses. HVDC, high voltage dc transmission system, is one of the alternatives because it is no need to consider about frequency. However, it still has difficulties of control for multi-terminal system in case of line commutated converters because of complicated design[1]. In terms of application on superconducting cable, it can achieve large power capacity and low-loss power transmission. However, there still have problem with ac losses, which proportional to frequency square. With frequency reduction in low frequency ac transmission system, it can decrease ac losses that can be lead to heat problem to superconducting cable[2].

To achieve these purposes without much adjustment to the conventional transmission system, low frequency ac transmission system is proposed in this paper.

In 2000s, LFAC, low frequency ac transmission system, has been recently proposed [3]-[6] by using cycloconverter to lower the grid frequency to small value. Even the transmitting frequency is lower, but it still be an ac system. So that the exiting technologies and available power system components such as transmission line

design and protection systems that are used in 60 Hz conventional transmission system is can be applied to LFAC system. In terms of control scheme for this LFAC transmission system by using cycloconverter, the load capacitive in cable transmission line caused current ripple around zero-cross point. It is difficult to decide positive or negative operation of cycloconverter. To solve this problem three control methods have been proposed to control power flow between each terminal [6]. Fundamental component of output current, positive-phase sequence current control and the balancing control, these three control methods can stable operation of cycloconverters in capacitive load with the third method has the fastest system responding because it is feedforward control.

With LFAC transmission system concept, the experimental on fractional frequency transmission system has been performed by using cycloconverter as frequency converter only on receiving end side to step up frequency from 50/3 Hz to utility grid [7]. This can confirm one of the applications of LFAC system at a reference low frequency transmission.

However, on existing research, with the three-phase 6-pulse cycloconverter applied to LFAC, large harmonics component existing and large filters are necessary to suppress them. it still lack of the application on various types of system configuration and also the information to find the suitable operating frequency and cable length in this low frequency ac transmission system.

In this paper, the low frequency ac transmission system is proposed in more details and various types of system information that do not include in any others reference papers. At first, the configuration of three types of low frequency ac transmission system by using 12-pulse cycloconverter and different number of operating phase will be shown and explained. Next, the new control method that can control power flow and limit over current to protect commutation failure will be discussed here. On the system analysis part, the information between frequency with cable length and voltage level, the effect from frequency to system response are also shown here on low frequency side. On the last part,

978-1-4799-2706-7/14 $31.00 © 2014 IEEE

simulation results of power flow control and power reversal are shown. With the proposed control method for LFAC transmission system, circuit configurations and control schemes performance are investigated by PSCAD/EMTDC simulations.

II. SYSTEM CONFIGURATION

A. Single-phase system

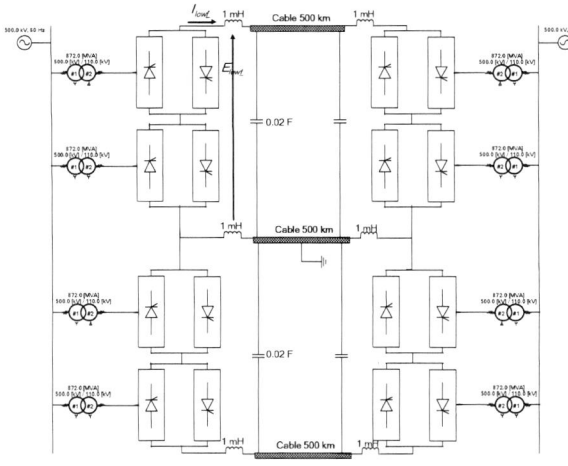

Fig.1 Configuration of single-phase low frequency ac transmission system

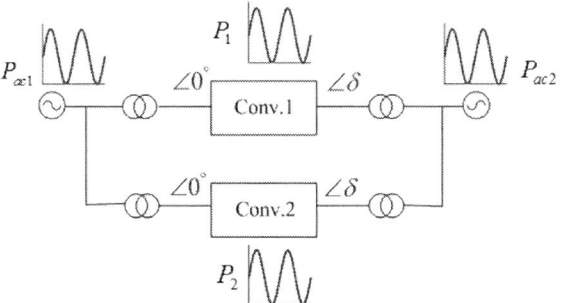

Fig.2 Power fluctuation of single-phase low frequency ac transmission system

Fig. 1 shows configuration of single-phase for low frequency ac transmission system. Each side consists of three phase (Y-Δ) and (Y- Y) transformers and two sets of 12-pulse ac-ac converter connected to power cable. At sending end, the cycloconverter works as frequency converter changing from utility ac input voltage from 60 Hz to low frequency during transmitting power via the cables. At receiving end, another set of cycloconverter works as frequency converter by changing low frequency to 60 Hz of ac voltage. From this single-phase circuit configuration, it can be operated with same phase to transfer power between terminals as shown in Fig.2.

B. Two-phase system

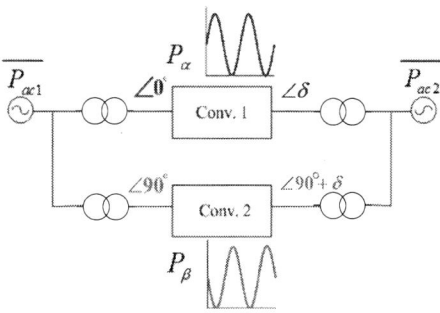

Fig.3 Power fluctuation of two-phase of low frequency ac transmission system

As proposed circuit configuration from single-phase system, the same configuration in Fig.1 can be also used for two-phase system with 90° phase differences between each phase. There are two sets of 12-pulse cycloconverter, the upper part called α-phase and the lower part called β-phase, with same rating, then total amount of transmitting power can be equally shared in each phase. Fig.3 shows the phase operation for this type of transmission system. P_{ac1} and P_{ac2} represent total active power on line frequency side. Power fluctuation on each operating phase of low frequency ac transmission system can be eliminated with this operation. The transmitting power on line frequency side can keep constant. That is one of the advantages from this two-phase configuration.

C. Three-phase system

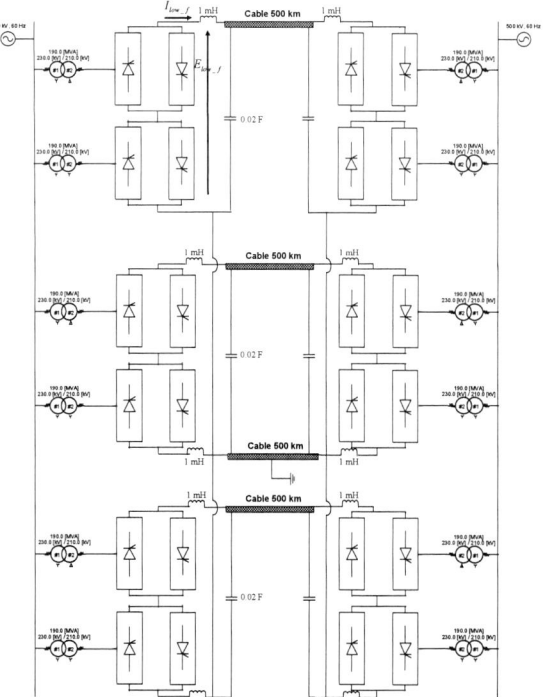

Fig.4 Configuration of three-phase low frequency ac transmission

Fig.4 shows three-phase low frequency ac transmission system configuration. This system has the configuration similar to the above two-phase transmission system. Its configuration is added one more phase from two-phase configuration. The three-phase system configuration can be described into phase-a, phase-b and phase-c. Each phase has 120° phase difference among them. It is also connected with the same cable parameter as single-phase and two-phase system.

From all three configurations, they will be used to investigate on system characteristics for evaluating merit and drawback for operating in low frequency ac transmission system.

III. SYSTEM CONTROL

The discontinuous current condition because of the capacitive characteristics in the cable transmission line around zero cross point can cause problem to converter operation. The controller could be difficult to decide the right control unit to converters. Fig. 5 shows the output voltage and current waveform on low frequency side affected by large capacitive. To avoid this problem, the non-circulate cycloconverter type is chosen with the new proposed control scheme. This section describes an adequate control scheme for this low frequency ac transmission system.

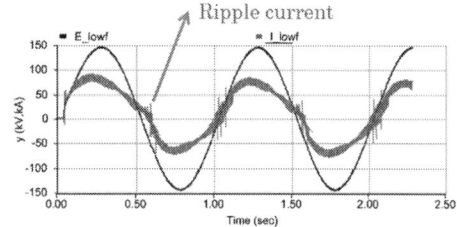

Fig.5 Large capacitive effects

A. Concept of control scheme

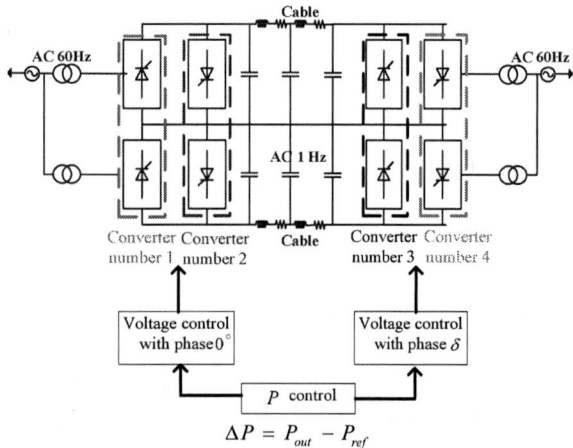

Fig.6 Layout of control system for low frequency ac transmission

Fig. 6 shows the layout of control scheme that is applied to this LFAC transmission system. The control scheme is designed by using voltage control with current limiter, gamma control and power control scheme.

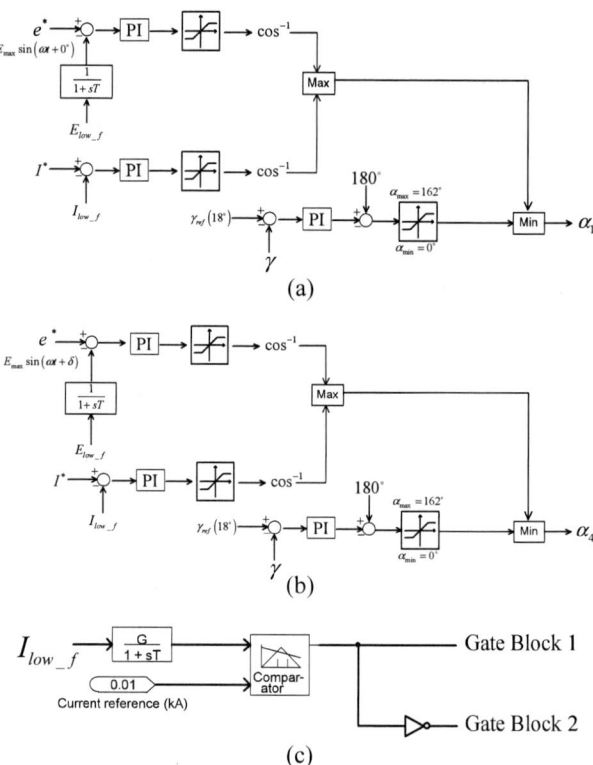

Fig.7 Control block of low frequency transmission system

$$P = \frac{V_s V_R}{X} \cdot \sin \delta \qquad (1)$$

where, V_S : the sending end voltage

V_R: the sending end voltage

X : line reactance

δ : transmitting angl

Firstly, control strategies for cycloconverters in both terminals are considered. As indicated by equation(1), transmitting power can be controlled by using phase difference or transmitting angle (δ) between V_S and V_R. The voltage control scheme with current limiter is used, in case of over current, the current limiter control will operate to protect damage to devices in the system by limited firing angle to thyristors. Extinction angle (γ) control is applied to avoid commutation failure for rectifier as shown in Fig. 7(a). In the case of inverter, the control scheme is similar to rectifier as shown in Fig. 7(b). In addition, during the current zero cross point, gate block control scheme as shown in Fig. 7(c) is applied to every converter.

Power control is used to generate phase difference δ between two terminals following reference power, and then voltage control works with the phase difference angle to generate firing angle to all converters.

Fig.8 shows the control block of transmitting power control. This power control scheme is applied to control the amount of transmitting power by comparing to power references to calculate the value of phase difference δ between two terminals following each converter operation mode.

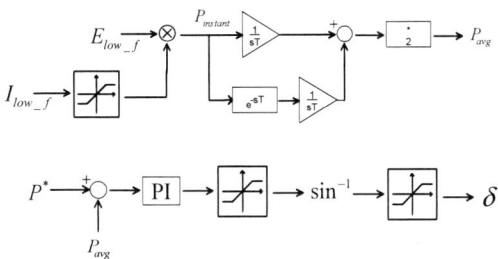

Fig.8 Control block of transmitting power control

B. Current control

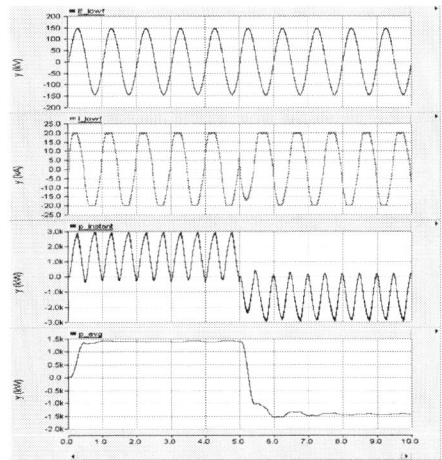

Fig.9 Simulation result for current limiter control

As, mentioned in section IIIA, the proposed control scheme includes current limiter and gamma control. First, current limiter will be explained. During the period of transition mode of cycloconverter can cause any spike current. This can malfunction to devices in the system so that current limiter is used for protection. From the control scheme as shown in Fig.7, the current limiter is working by comparing the rating current with measure current from low frequency side to generate firing angle to cycloconverter. The current limiter can limit over current by limit firing angle to cycloconverter. Fig. 9 shows the result of current limiter control that has current limit at 20 kA on low frequency side. E_lowf waveform is transmitting low frequency ac voltage, following reference value from control scheme. I_lowf waveform is transmitting current on low frequency side, which is limited at 20 kA due to current limiter. p_instant

waveform is calculation from control block in Fig.8 and p_avg waveform is also calculated from moving average control in Fig. 8 by comparing with power reference. From the result, it can consider that the current limiter control can work properly.

C. Constant extinction angle control (γ)

For line commutated converters, in practice, it has some delay angle due to commutation reactance. Constant extinction angle control(γ) is used to avoid commutation failure. This puts a limit on the maximum angle of firing angle(α) following equation (2):

$$\alpha = \pi - (u + \gamma_0) \qquad (2)$$

As shown in Fig.7, this gamma control works by comparing gamma reference with gamma measure, then minimum firing angle is chosen to be gating pulse for cycloconverters. Fig. 10 shows the simulation result of gamma control that cannot exceed limit, which means the control scheme works well. This control scheme can be applied to all single-phase, two-phase and three-phase system.

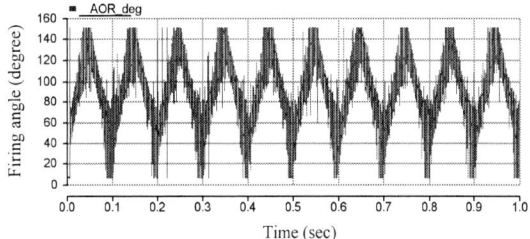

Fig. 10 Simulation result for constant extinction angle control

IV. CABLE AND FILTER DESIGN

To simulate all proposed system configurations, the simulation model at operating maximum voltage level at 500kV, and the capacity is set maximum to 1400 MW(1 pu). The length of cable is varying from 50 km to 500 km and transmission frequency is set at 1 Hz and 10 Hz for the simulation and 1 Hz, 10 Hz and 60 Hz for calculation. Cable model as the nominal model is shown in Fig.11

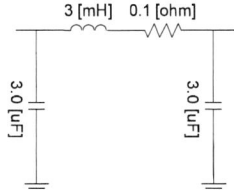

Fig.11 Cable model for the simulation

This cable model is including in every proposed low frequency transmission system configurations. The cable model can be calculated from following equations with D = 139.2 and d = 112.8:

$$C = \frac{0.02413\varepsilon_s}{\log_{10} D/d}$$
$$= 0.6 \quad [\mu F/km]$$
$$L = 0.05 + 0.4605 \log_{10} \frac{D}{r} \quad [mH/km]$$
$$= 0.3 \quad [mH/km]$$
$$R_{ac} = 0.0164 \quad [\Omega/km]$$

; D = Outside diameter of insulation
d = Inside diameter of insulation

Fig.11 shows a set of 10 km cable length, in simulation, we can connect this model to represent any cable length. Three configurations of low frequency ac transmission system, which are single-phase, two-phase and three-phase, are simulated by using these cable parameters to study system behavior for each configuration on LFAC transmission system.

As for LC filter design, the design procedure is aim to filter harmonics on low frequency side. It can be computed base on [9], which maximum transmitting power (P) is 1400 MW, reactive power (Q) is 1000 MVar. There is a linear relation between L and C, on low frequency side of system configuration as shown in Fig.1, according to C = aL+b and tan(ϕ) = Q/P. Based on this equation, we can determine the pair of (L,C) as shown in Fig. 12.

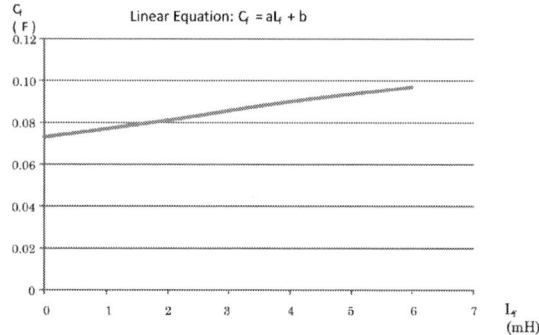

Fig.12 Pairs of LC filter on low frequency side

As for the filter design, these results will be used for numerical simulation to obtain the most suitable number of phase and operating frequency in low frequency ac transmission system.

V. PHASE AND FREQUENCY SELECTION

In this section, the analysis is divided into two sections. First, the analysis in different voltage level and transmitting frequency on low frequency side by transmission capacity is explained. Another is about characteristics on frequency and time response.

A. Analysis on transmission frequency with voltage level and cable length

In this section, maximum transmitting active power is shown to find the optimum point of operating frequency and voltage level with cable length. The amount of transmitting active power to cable length can be expressed as (3).

$$P = \sqrt{S_{th}^2 - \left(2\pi f C l V_l^2\right)^2} \qquad (3)$$

$$S_{th} = \sqrt{3} I_i V_l$$

where; P = maximum transmission power
f = transmitting frequency
C = cable capacitance
l = cable length
V_l = voltage level
I_i = rated current

It is simulated by using cable length from 50 km to 500 km with several voltage levels at 66 kV, 150 kV, 275 kV and 500 kV and transmission frequency is used as a parameter at 1, 10 Hz for simulation and 1, 10 and 60 Hz for calculation by (3). The transmission power cable is modeled by using single-core XLPE underground power cable from ABB[8].

Table-I : Parameters of XLPE underground power cable

	66kV	150kV	275kV	500kV
Maximum voltage (kV)	72.5	170	300	550
Rated current (A)	1730	1705	1705	1705
Cross section (mm²)	2000	2000	2000	2000
Inductance (mH/km)	0.45	0.47	0.49	0.51
Capacitance (µF/km)	0.52	0.31	0.23	0.19

The relationship between sending end active power P and maximum transmission distance is calculated using parameters from Table-I. These parameters are using in cable model as shown in Fig.11 calculated by using eq. (3) to obtain the transmission power in each voltage level.

In order to confirm with real application, with same parameters is simulated by PSCAD/EMTDC. The comparison results can be shown in Fig. 13. It shows that at 60 Hz, it can transfer power with the shortest length of cable. In the contrary with 10 Hz and 1 Hz, they can transfer power far more than 60 Hz in terms of cable length at every voltage level. Calculation results are shown in dash line and simulation results are shown in normal line on every voltage level. (µF/km)

Consider the result in terms of voltage level, the transmission capacity in low voltage level becomes not much different among several frequencies. However, the differences become noticeable in higher voltage level. When the transmission system operates in higher voltage level, lower frequency has much more transmission capacity comparing with conventional frequency.

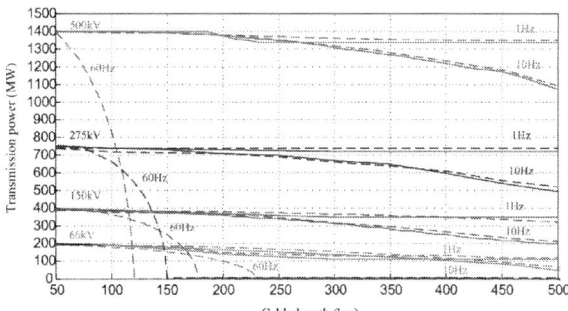

Fig.13 Transmission power to cable length

B. Analysis on transmission frequency with time response

Another study in this section is considering on time response with varying transmission frequency. Time response analysis of the system is performed in this section to choose a suitable range of operating frequency. With the proposed system and control method, simulation is carried out by using only single-phase system for easily understanding. Parameter of this simulated system is frequency by choosing 1 Hz as very low frequency, 10 Hz and 20 Hz as medium frequency for low frequency ac transmission system.

Fig. 14 shows waveform on low frequency side, the reference frequency is 1 Hz, 10 Hz and 20 Hz and maximum of voltage reference is 500 kV. In this simulation, we need to transmitting power at 300 MW and has power reversal at 2.5 s.

In this simulation results, some ripple appears on transmission frequency at 10 Hz and more effect at 20 Hz because frequency is constant value in control scheme, which is designed at 1 Hz. However, in this part only frequency and system time response are considered without changing any parameters in control scheme.

As the result, when power flow direction suddenly changes, we can see that 1 Hz system has the longest time delay but 10 Hz system has only 0.1 second before power can stabilize again. On 20 Hz system, it shows small delayed time before power can stabilize to reference point. From the simulation results, in terms of time response and frequency, for very low transmitting frequency such as 1 Hz has very slow time response that is a drawback to power transmission system. However, if we consider higher frequency such as 10 Hz, the time response is much better than 1 Hz, and very close to the time response in the conventional ac transmission system.

As the result, when power flow direction suddenly changes, we can see that 1 Hz system has the longest time delay but 10 Hz system has only 0.1 second before power can stabilize again. On 20 Hz system, it shows small delayed time before power can stabilize to reference point.

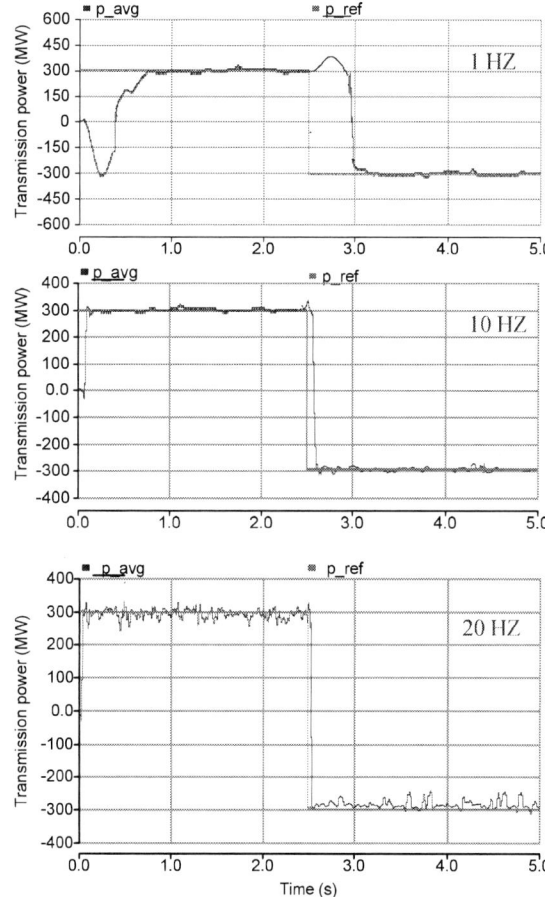

Fig.14 Comparison between transmission frequency and time response results on three operating frequencies.

From the simulation results, in terms of time response and frequency, for very low transmitting frequency such as 1 Hz has very slow time response that is a drawback to power transmission system. However, if we consider higher frequency such as 10 Hz, the time response is much better than 1 Hz, and very close to the time response in the conventional ac transmission system.

The ripple that occurs on Fig.14 at transmitting frequency 20 Hz causes by LC filter on low frequency side using 10 Hz as base frequency. To eliminate this ripple, pair of LC filter should be recalculate as mentioned in section IV.

VI. SIMULATION RESULTS OF 2-PHASE, 10-Hz OPERATION

From three configurations of low frequency ac transmission system, two-phase system is chosen because it meets the requirement of transmitting power distance more than 500 km, and it also has quick system responding compared to single-phase system.

To eliminate fluctuation on line frequency side. With 90° difference of phase angle in voltage reference on low frequency side, it can lead to 180° phase difference of instantaneous power in each phase. This way of the operation can solve power fluctuation on line frequency side without using three-phase system which contains many devices.

Since, this LFAC system requires power transmission distance more than 500 km with minimum losses in cable, the lowest transmission frequency is required. With the additional information from system time response, 10 Hz is chosen from the small delay compared to transmission frequency at 1 Hz.

Table-II : Parameters of simulated system

LFAC system	500kV, 60 Hz,
Transmitting frequency	10 Hz
Transformer	500kV/110kV
Max. power transfer	1400 MW

To confirm the application on this two-phase system, the simulation is performed with parameters in Table-II, the simulation results on low frequency and line frequency side are considered.

A. Two-phase system on low frequency side

Reference frequency is using at 10 Hz, transmitting power for α-phase is 700 MW and total transmitting power is 1400 MW. Fig.15 shows waveforms of voltage E_lowf_alpha, current I_lowf_alpha, instantaneous power p_instant_alpha and average power of α-phase, p_avg_alpha on low frequency side, respectively. The green line is the reference value of transmitting power with power reversal at t = 1. With this simulation results, it can be confirmed that the new control scheme works well with two-phase system configuration.

B. Two-phase system on line frequency side

The merit for choosing two-phase system is no fluctuation on line frequency side. This simulation results can confirm as shown in Fig.16 and Fig.17. Fig.17 shows the results on line frequency side of single-phase system with the fluctuation problem on active (P_{ac1}) and reactive power (Q_{ac1}). However, with the two-phase system operation, the results on line frequency side become no fluctuation on active (P_{ac2}) and reactive power (Q_{ac2}) as shown in Fig.16.

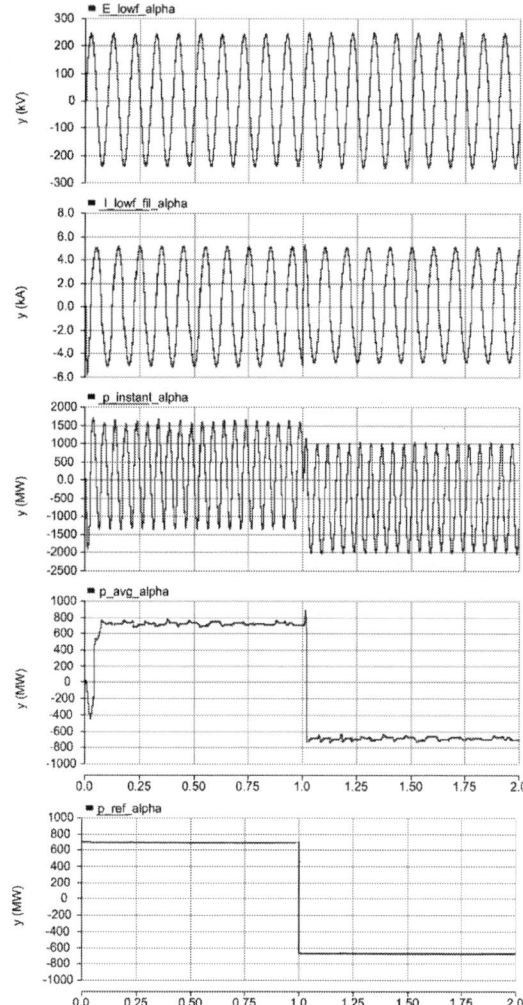

Fig. 15 Simulation result of two-phase low frequency ac transmission system

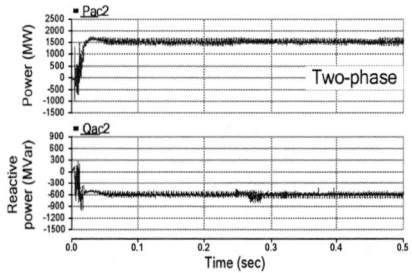

Fig. 16 Simulation results on line frequency side of two-phase system

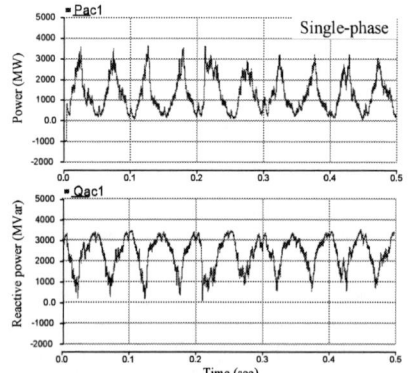

Fig. 17 Simulation results on line frequency side of single-phase system

VII. CONCLUSION

The number of phase and operating frequency on low frequency ac transmission system by using cycloconverter with the application of new control scheme on three types of transmission configurations; single-phase, two-phase and three-phase systems has been already presented. There are three main important points to be concluded here as follows:

1. The new control scheme can control power flow and limit over current, which can cause the damage to low frequency ac transmission system with cycloconverter

2. Two-phase of low frequency ac transmission system was chosen because of no fluctuation on line frequency side. It can be operated same as three-phase system but less number of devices.

3. As the simulation results, all three configurations were tested by power flow control and power reversal. All configuration system can operate properly with the proposed control scheme. The transmitting frequency at 10 Hz was chosen because it can transmit power more than 500 km, which is suitable for the system target.

From the view point of response time, this frequency has quicker responding than 1 Hz but slightly slower than 20 Hz only 0.05 sec delay, which can be assumed no differences.

With this study, it can lead us to consider in more details of choosing the proper operating number of phase and frequency for LFAC transmission system. The applications on this system such as multi-terminal system application should be investigated to improve and obtain the best performance of the transmission system.

References

[1] E. Prieto-Araujo, F. D. Bianchi, A. Junyent-Ferré, and O. Gomis-Bellmunt, "Methodology for droop control dynamic analysis of multiterminal VSC-HVDC grids for offshore wind farms," IEEE Trans. Power Del., vol. 26, no. 4, pp. 2476–2485, Oct. 2011.

[2] Cau, F. ; Fusion Technol., EPFL-CRPP, Villigen, Switzerland; Bruzzone, P.,"AC Loss Measurements in CICC With Different Aspect Ratio," IEEE Trans. Applied Superconductivity, vol. 19, no. 3, pp. 2383-2386, June 2009.

[3] Y. Cho, G. J. Cokkinides, and A. P. Meliopoulos, "Time domain simulation of a three-phase cycloconverter for LFAC transmission systems,"presented at the IEEE Power Energy Soc. Transm. Distrib. Conf. Expo., Orlando, FL, May 2012.

[4] Tsuyoshi Funaki, Kenji Matsuura, "Proposition of a Low Frequency AC Power Cable Transmission System", IEEJ Vol 120-B, No.8/9, 2000, PP.1077-1083.

[5] Tsuyoshi Funaki, Kenji Matsuura, "Feasibility of the Low Frequency AC Transmission", Power Engineering Society Winter Meeting,2000,PP.2693-2698

[6] Ryosuke Nakagawa, Tsuyoshi Funaki and Kenji Matsuura, "Installation and Control of Cycloconverter to Low Frequency AC Power Cable Transmission", Power Conversion Conference, 2002, PP.1417-1422

[7] X. Wang, C. Cao, and Z. Zhou, "Experiment on fractional frequency transmission system," IEEE Trans. Power Syst., vol. 21, no. 1, pp.372–377, Feb. 2006.

[8] ABB, XLPE Underground Cable System User's Guide [Online]. Available: http://www.abb.com

[9] B. R. Pelly, Thyristor Phase-Controlled Converters and Cycloconverters. New York: Wiley, 1971.

Fast Acting DC Circuit Breaker for HVDC Transmission Line Based on DC/DC Chopper

Liangyi Tang[1], Student Member, IEEE, Bin Wu[2], Fellow Member, IEEE, Venkata Yaramasu[2], Member, IEEE,
Weirong Chen[1], Member, IEEE, and Hussain S. Athab[2], Senior Member, IEEE
[1] Southwest Jiaotong University, Chengdu, Sichuan, China
[2] Ryerson University, Toronto, Ontario, Canada
Email address: tangliangyu1985@gmail.com

Abstract—**In this paper, two fast acting dc circuit breakers are investigated to interrupt the short-circuit faults in HVDC transmission lines. These circuit breakers are realized using dc/dc choppers which can also perform step-up and/or step-down functions. Compared to the existing topologies, the number of active switches in the proposed configurations are minimized to achieve lower initial cost and higher conversion efficiency. A double closed-loop controller based on PI regulators is designed to regulate the inductor current and output voltage of the converters. An auxiliary controller is also proposed to ensure proper shut-down of the chopper in the event of short-circuit faults. The simulation and experimental results are presented to validate the proposed converter topologies and control scheme.**

Keywords—dc/dc power conversion, HVDC circuit breakers, HVDC transmission lines, Power electronics, Power system faults.

I. INTRODUCTION

As prices of fossil fuels rise and as concern about the dangers of global warming grows, electricity production from renewable energy sources such as wind and photovoltaic (PV) energy conversion systems is a subject of great interest nowadays. Many megawatt-level PV plants and offshore wind farms have been installed around the globe [1]-[2]. The high voltage direct current (HVDC) transmission is a promising solution to integrate the power generated from these renewable energy sources into the existing power grid. The technological advancements in power electronics have enabled the lower cost and higher performance operation for HVDC transmission lines [3]-[5].

Earlier HVDC transmission lines were mainly configured as two terminal systems with limited applications, and recently there has been significant interest for multi-terminal HVDC and HVDC grids [6]. They allow integration of many energy sources and the tapping of industrial and consumer loads [7]-[9]. A typical multi-terminal HVDC system with such a flexible power integration and allocation is shown in Fig. 1. The important technical and operational requirements for multi-terminal HVDC system are summarized below:
- Depending on the different power requirements, the tapped loads demand step-up or step-down of the transmission-level dc voltages. When the tapped loads need more power, the output voltage should be forced to step-up, and when some of the tapped loads shut down, the output voltage should be forced to decrease. For example, a 200 kV dc voltage can be stepped-up to 300 kV or stepped-down to 100 kV.
- In the event of short-circuit fault at the low voltage terminal of the HVDC transmission line, the main HVDC transmission lines should be isolated from the fault instantly to ensure the whole grid safety [10]. To fulfill the above two important requirements, dc/dc choppers as depicted in Fig. 1 are commonly used. They perform step-up and step-down functions in addition to the dc circuit breaker (DCCB) functionality.

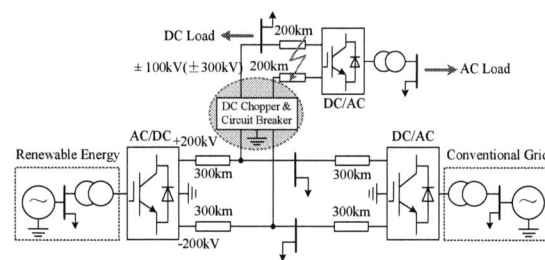

Fig. 1. Multi-terminal HVDC system with a dc circuit breaker (dc/dc chopper).

Various dc circuit breakers are researched for HVDC systems such as: mechanical DCCB [11], solid-state DCCB [12], hybrid DCCB [13] and power electronic converter based DCCB [14]-[21]. Compared between these DCCB's in comprehensive performance including isolation time, control complexity, costs and losses, the power electronic converter based DCCB is found to be the best. Recently many scholarly works have been carried-out to integrate the circuit breaker function into the dc/dc choppers [14]-[21].

The circuit breaker illustrated in Fig. 1 is specially designed for the fault clearance, and it can also perform the step-up and/or step-down functions. A multi-functional four-switch dc/dc chopper with short-circuit protection functionality has been analyzed for high power application in [19] and this topology is shown in Fig. 2. Comparing with the mechanical DCCB and the hybrid DCCB, this configuration has better performance. However, this configuration is costly due to the greater number of active switches which are realized by MV-IGBT's or IGCT's. Many active switches are connected in series and in parallel to handle the high voltage level and high peak fault current. For example, the test system showed in Fig. 2 needs a string with 134 IGBT's （4.5 kV, 650 A） in series, and 3 strings in parallel. Moreover, compared with diodes, the price of IGBT's or IGCT's is much more expensive. The control complexity is also greater as this converter requires more gating signals. In addition, many equalizing resistors and buffer devices are needed to ensure equal voltage among IGBT's. A modified converter configuration is described in [20] with one active switch group replaced by a diode group and this topology is shown in Fig. 3. These two choppers can offer bidirectional power flow. But, most of the ac or dc tapped loads do not need to provide power back to the main HVDC lines [8]-[9]. In addition, the cost of the DCCB is very high because of the huge number of the active switches and control devices. In order to reduce cost and complexity of DCCB, two new converter configurations are proposed in this paper. The proposed configurations deal with the unidirectional power flow and make them more suitable for the HVDC systems. The proposed configurations use less number of active switches compared to the existing ones, and thus they decrease the overall cost, control complexity and power losses. These choppers can step-up and/or step-down the transmission-level dc voltages in addition to the circuit breaker functionality.

The control scheme development is as important as of the converter configuration. In [20], double-loop control algorithm has been proposed for the operation of DCCB. However, this control scheme cannot interrupt and isolate the short-circuit faults completely. To solve this issue, a novel auxiliary controller is also proposed for proper shut-down of the chopper in the event of short-circuit faults. The proposed converter configurations and the control scheme are verified through MATLAB simulations and dSPACE DS1103 real-time controller based experiments.

II. PROPOSED CONVERTER TOPOLOGIES

A bipolar dc/dc chopper shown in Fig. 2a was described in [18], where a four-switch chopper shown in Fig. 2b is connected in cascade to achieve bipolar configuration.

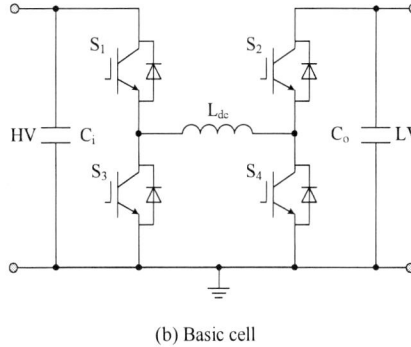

(b) Basic cell

Fig. 2. The four-switch dc/dc choppers employed in HVDC systems.

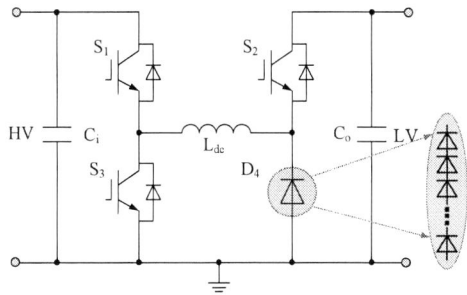

Fig. 3. Three-switch topology [20].

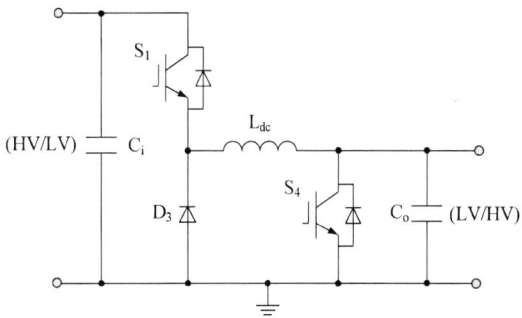

(a) Two-switch topology

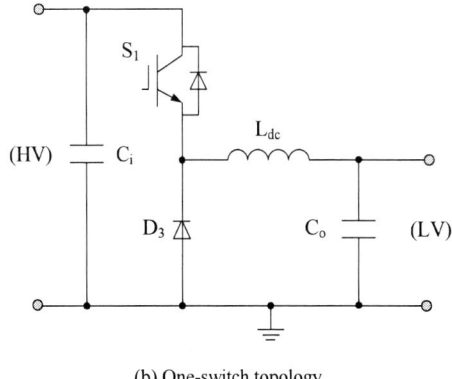

(b) One-switch topology

Fig. 4. Proposed dc/dc choppers with minimal number of active switches.

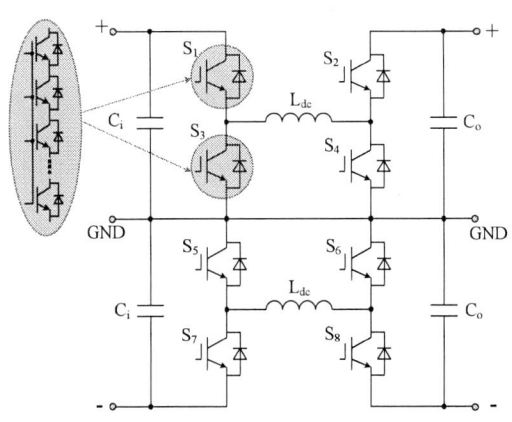

(a) Bipolar dc/dc chopper

In this paper, the proposed converter topologies for circuit breaker function are shown in Fig. 4. Two-switch topology is shown in Fig. 4a and One-switch topology is shown in Fig. 4b. The two new converter topologies can perform step-up and/or step-down functions in addition to the dc circuit breaker function. Compared to the work presented in [20], one more IGBT group is replaced by a diode group as shown in Fig. 4a. It leads to further reduction in the cost and complexity of the system. This two-switch topology can perform step-up or step-down functions. Another configuration with only one active switch is proposed as shown in Fig. 4b for step-down operation. Compared to the topologies presented in [19] and [20], the proposed topologies possesses less active switches, and thus the cost and switching losses are further reduced while the system reliability is enhanced due to the use of only one group of active switches.

The operating principle of proposed two-active-switch topology is discussed in Fig. 5. During turn-on period, the active switches S_1 and S_4 are conducting and the energy is stored in dc inductor L_{dc}. The energy stored in L_{dc} is discharged to the load through D_2 and D_3 during turn-off period. The mode of operation considering continuous inductor current is shown in Fig. 5c. During one switching period Ts, the relation between the input and output voltage can be obtained as

$$V_o = \frac{D}{1-D}V_i \tag{1}$$

From (1) it can be noticed that the operation of the proposed topology is similar to the buck-boost converter, where the output voltage (V_o) is higher than the input voltage (V_i) when $D>0.5$, and V_o is lower than V_i when $D<0.5$. The output voltage equals input voltage when $D=0.5$.

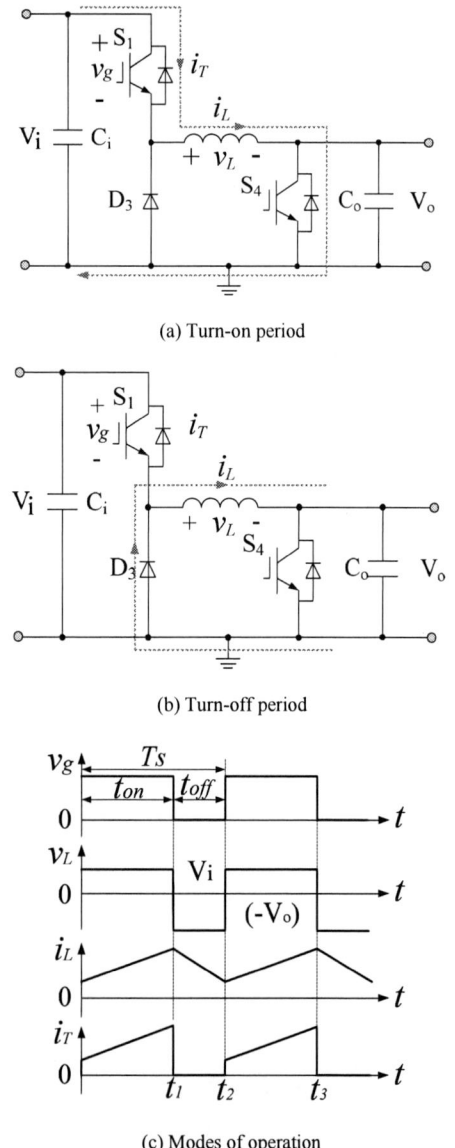

(a) Turn-on period

(b) Turn-off period

(c) Modes of operation

Fig. 5. Operating principle of proposed two-switch topology (circuit breaker).

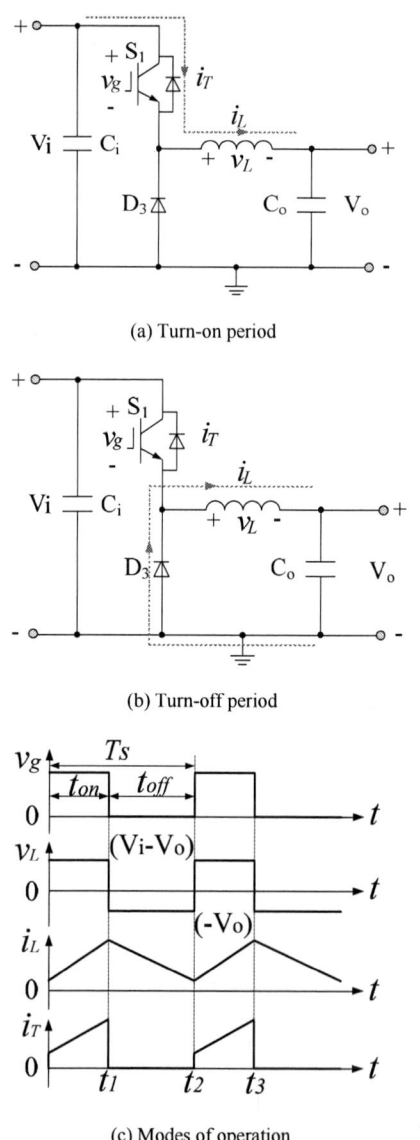

(a) Turn-on period

(b) Turn-off period

(c) Modes of operation

Fig. 6. Operating principle of proposed one-switch topology (circuit breaker).

The operating principle of the one-active-switch topology is shown in Fig. 6. During the turn-on period, the energy is stored in L_{dc} through S_1 and D_2. During turn-off period, the L_{dc} energy is discharged through D_2 and D_3, which is similar to the two-switch topology. The input and output voltages can be related as

$$V_o = DV_i \qquad (0 \le D \le 1) \qquad (2)$$

It can be noticed from (2) that similar to the buck converter, the output voltage of the converter is a fraction of the input voltage.

III. PROPOSED CONTROL SCHEME

The conventional double closed-loop controller is shown in Fig. 7, where the outer-loop controls the output voltage (V_o) of the converter, while the inner-loop controls the inductor current (i_L). This control scheme cannot distinguish the normal operation and recovery period from the faulty operation of the transmission lines and thus produces gating signals even during the short-circuit faults. This is an undesirable characteristic of the classical double closed-loop controller.

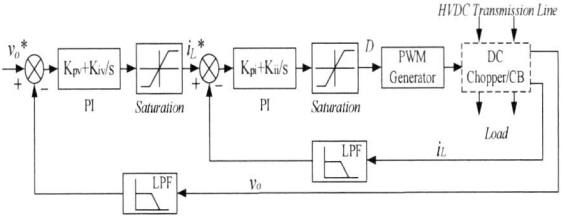

Fig. 7. Classical double closed-loop control scheme.

To overcome this problem, a novel auxiliary controller is proposed as shown in Fig. 8 to be used in conjunction with the classical controller. The auxiliary controller performs zero crossing detection on the duty cycle, and when the duty cycle value falls below zero, it sends a trigger signal to a two-port switch. With the trigger signal being applied, the two-port switch produces the zero duty cycle value throughout the operation of converter. When the fault is completely cleared, the trigger signal can be disabled to restore normal operation. This algorithm is applicable for the proposed converters and also for the conventional converters in [19]-[20].

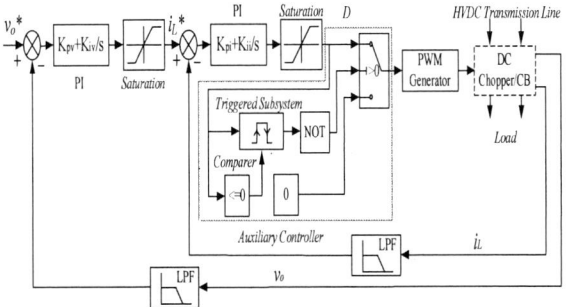

Fig. 8. Proposed double closed-loop control scheme with additional auxiliary controller.

IV. SIMULATION RESULTS

In order to verify the proposed control scheme and converters, simulations were carried out on a high-voltage system using MATLAB/Simulink software. The parameters of the system are shown in Appendix-Table A.1. The performance of the proposed converter is also compared with four-switch converter introduced in Fig. 2 [19]. The value of the dc inductor L_{dc} is selected such that the converter operates in continuous conduction mode with 5% ripple in the inductor current. The value of the output filter capacitor C_o is chosen such that the output voltage contains 10% ripple under the rated operating conditions of the converter. The transmission line resistance R_t and inductance L_t are also considered in the simulation tests. A 200 km transmission line is chosen with an inductance value of 0.11155 mH/km and resistance value of 0.014 Ω /km [22]. The values of R_t and L_t used in the simulation tests are 3 Ω and 20 mH. In all the tests, the converter is assumed to be operating in steady-state before the occurrence of short-circuit fault.

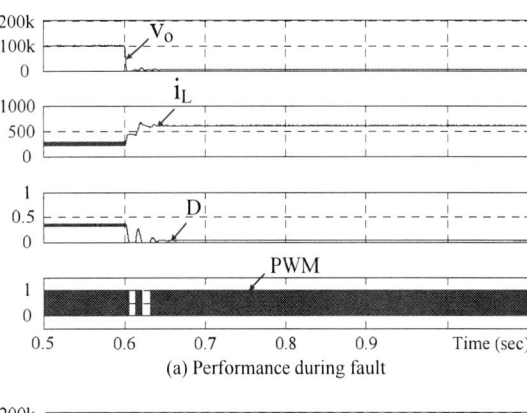

(a) Performance during fault

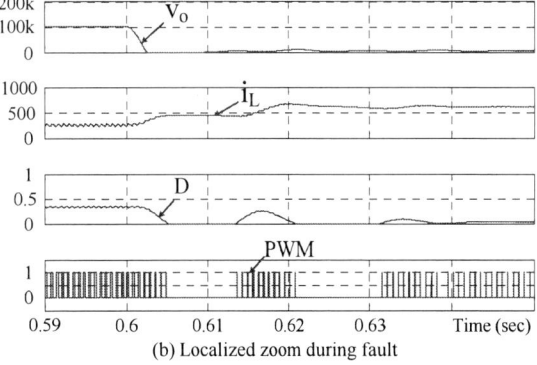

(b) Localized zoom during fault

Fig. 9. Simulation results for conventional four-switch topology without the auxiliary controller.

To validate the auxiliary controller and to demonstrate its significance, a test with conventional double closed-loop controller (without the auxiliary controller) has been carried out with the conventional four-switch converter, the results of which are presented in Fig. 9. The step-down operation from 200 kV to 100 kV is considered with a duty cycle value of less than 0.5. A short-circuit fault is applied at the output of the converter at 0.6 s. When a short-circuit fault occurs, the output voltage sharply decreases to zero. The double closed-loop controller forces the converter to turn-off rapidly by bringing the duty cycle D to zero. In the Fig. 9, the duty cycle increases again after about 10 ms because of the double closed-loop adjustment, which makes the switches turn-on and the current increases again. So, there is an oscillation in the circuit. Because the short-circuit fault is still in circuit, the output voltage v_o cannot be recovered back to 100 kV. The inductor current is maintained at a value higher than its rated value. This is an undesirable operation in HVDC systems as it might cause subsequent faults and destroy the whole system.

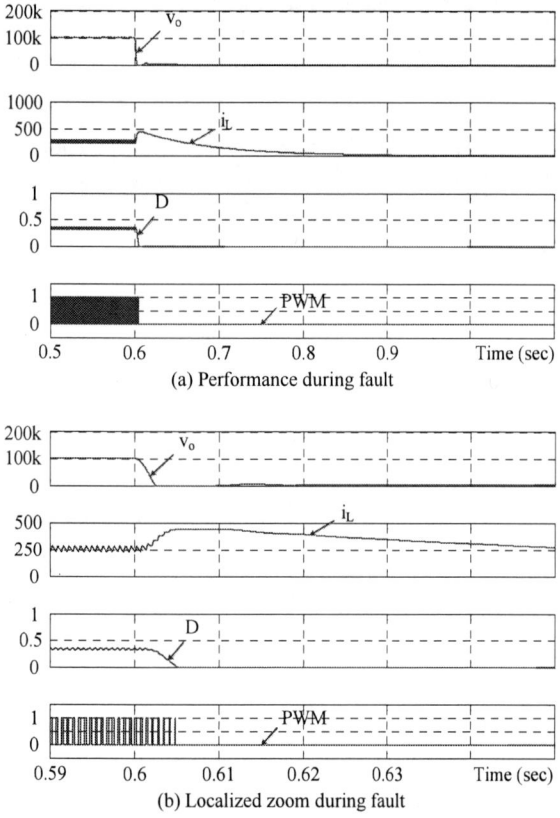

(a) Performance during fault

(b) Localized zoom during fault

Fig. 10. Simulation results for conventional four-switch topology with the auxiliary controller.

The results with the auxiliary controller are shown in Fig. 10. The peak value of the fault current is also reduced from 600 A to 400 A approximately compared to the previous case (Fig. 9) because of the instant cut off the short-circuit fault. In addition, the auxiliary controller ensures that the converter is completely shut-down after the duty cycle becomes zero. The response time of the converter is measured as the time between the occurrence of fault and the instant the duty cycle becomes zero. In this test, a fast response time of 6.5 ms has been achieved. This time is very low compared to the mechanical/hybrid breakers and thus the fault current is limited to a safe value without employing any other current limiters. After the fault is cleared by manual intervention, the system can be recovered back to normal operation by resetting

the auxiliary controller. The proposed auxiliary controller can be applied to any dc circuit breakers based on power electronic converters which are already described in literatures or for the proposed topologies. The analysis presented here also applies for step-up operation of the circuit breakers.

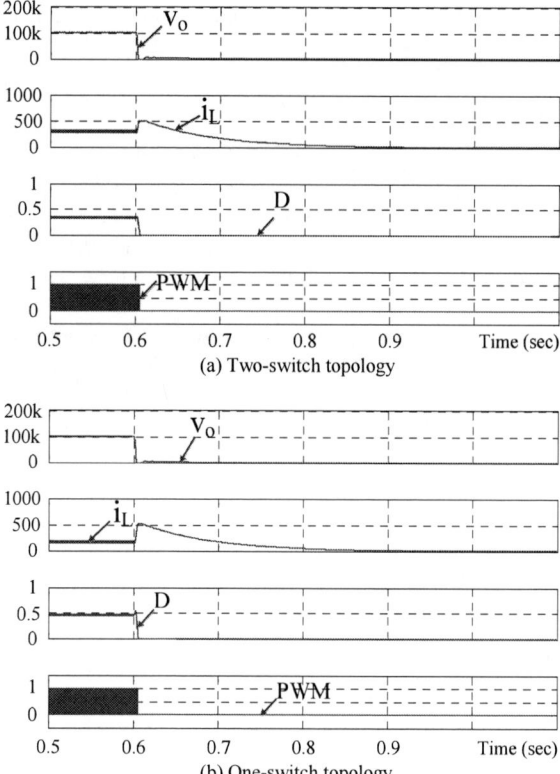

(a) Two-switch topology

(b) One-switch topology

Fig. 11. Simulation results for proposed topologies with auxiliary controller during step-down.

The performance of the proposed topologies during step-down operation is shown in Fig. 11. Similar to the four-switch topology presented in Fig. 10, these converters operate at a duty cycle value less than 0.5. The response time of the converters during the short-circuit faults is almost same as that of four-switch topology. The results validate the proposed topologies with reduced number of active switches for circuit breakers in step-down operation.

The comparison between the four-switch converter topology and proposed two-switch topology is presented in Fig. 12 for the step-up operation. The output voltage reference is set at 300 kV and thus both the converters operate at a duty cycle higher than 0.5. When the short-circuit fault occurs at 0.6 s, both the converters completely turned-off with almost identical response times. These results prove that with the auxiliary controller, the proposed converter configurations can perform better dc circuit breaker performance. Compared with the four-switch topology, the proposed converter configurations employ less number of active switches.

The 2014 International Power Electronics Conference

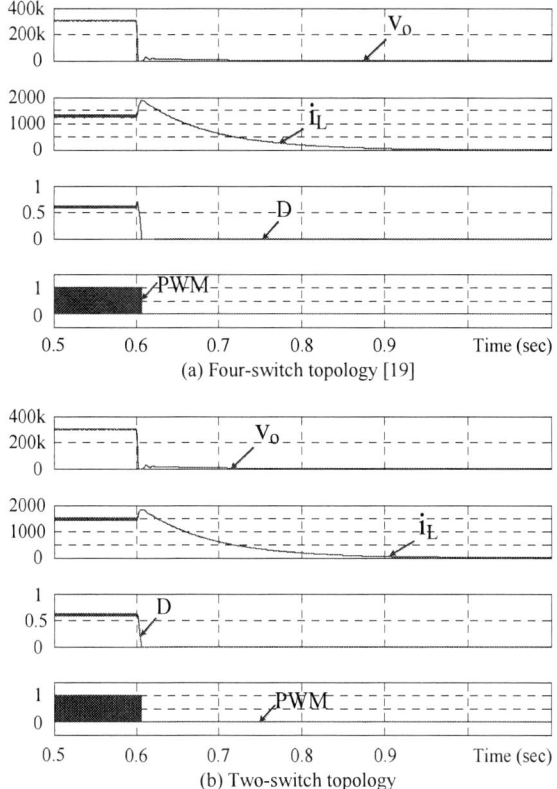

(a) Four-switch topology [19]

(b) Two-switch topology

Fig. 12. Simulation results for the conventional and proposed topology during step-up with auxiliary controller.

V. EXPERIMENTAL RESULTS

The experimental results with the proposed topologies during the step-down and step-up operations present similar performances to the four-switch topology. The experimental results for a four-switch converter during step-down operation with the conventional double closed-loop controller (without the auxiliary controller) are shown in Fig. 13a, where the response of the system is different from the high-power simulations presented in the previous section. But it carries same meaning that the classical controller initiates the gating signals when the fault is still in the circuit. The low-power simulation results are presented in Fig. 13b to better compare to the experimental results.

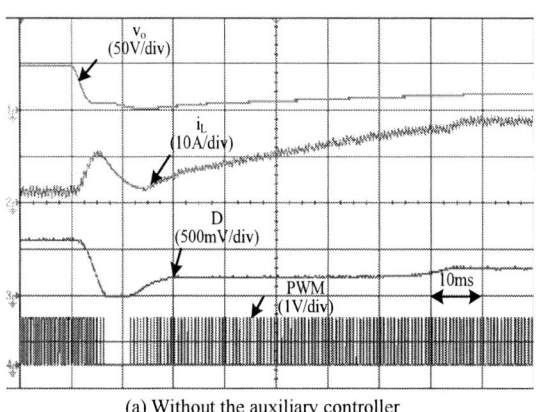

(a) Without the auxiliary controller

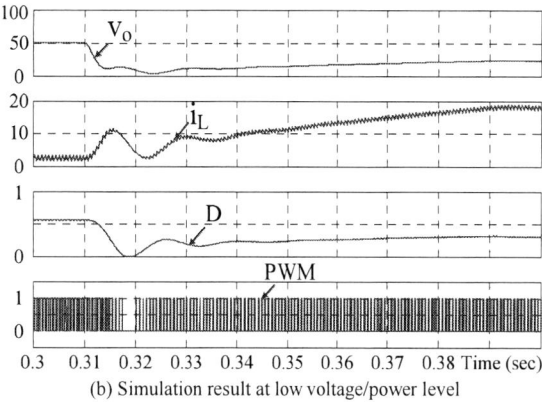

(b) Simulation result at low voltage/power level

Fig. 13. Performance of conventional four-switch topology without the auxiliary controller.

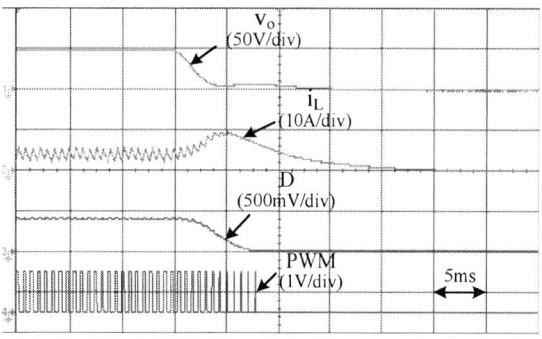

(a) Two-switch topology

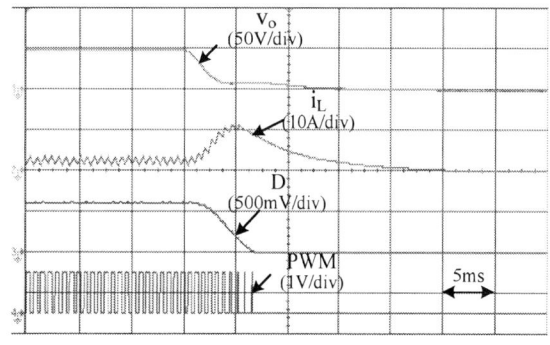

(b) One-switch topology

Fig. 14. Experimental results for proposed topologies with auxiliary controller during step-down.

The experimental results with the proposed topologies during the step-down operation are presented in Fig. 14. The results with two- and one-switch topologies are presented in Figs. 14a and 14b, respectively. These converters present similar performance in step-down operation and in disabling the gating signals during short-circuit faults. The experimental results presented here validate the proposed converter configurations and control scheme. The experimental results are very similar to those of the low power simulation results.

978-1-4799-2706-7/14 $31.00 © 2014 IEEE

VI. CONCLUSION

This paper proposed two new power converter topologies for dc circuit breakers based on dc/dc choppers for unidirectional power flow applications. The proposed converters perform multiple functions such as step-up and/or step-down operation and dc circuit breaker. The proposed converters exhibit similar performance compared to the conventional topologies during normal and abnormal conditions, but with less number of active switches. The proposed topologies offer great reduction in the cost, complexity and switching losses of the HVDC systems. The proposed auxiliary controller ensures proper shout-down of the converter during short-circuit faults in the power system. The response time of the proposed controller is very fast compared to the mechanical/solid-state breakers, and the solutions provided in this work are very promising for implementation in the practical HVDC systems.

APPENDIX

Parameters of the simulation and experimental tests are summarized in Table A.1.

Table A.1 Parameters Used in the Simulation and Experimental Tests

Variable	Description	Simulation	Experimental
V_i	Input dc Voltage (V)	200k	100
V_o	Output dc Voltage (V)	100k/300k	50/150
L_{dc}	dc Inductor (mH)	800	10
C_i	Input dc-link Capacitance (μF)	22	1000
C_o	Output dc-link Capacitance (μF)	47	1500
R_t	Transmission Resistance (Ω)	3	0.5
L_t	Transmission Inductance (mH)	20	1
R_o	Load Resistance (Ω) (Step-down/Step-up)	600	20
P_o	Output Power (W) (Step-down/Step-up)	16.7M/150M	125/1125
f_{sw}	Switch Frequency (Hz)	1500	1500
K_{pv}, K_{iv}	V_o PI Parameters	0.05, 50	0.01, 30
K_{pi}, K_{ii}	i_L PI Parameters	0.2, 600	0.3, 300

ACKNOWLEDGMENT

The authors are thankful for the support from the Laboratory for Electric Drive Applications and Research (LEDAR) and Centre for Urban Energy (CUE) of Ryerson University in Canada.

REFERENCES

[1] Renewable Energy Policy Network for the 21st Century (REN21), "Renewables global status report," 2013, available at: http://www.ren21.net, accessed on Aug. 2013.

[2] Global Wind Energy Council (GWEC), "Wind report," 2012, available at: http://www.gwec.net, accessed on Aug. 2013.

[3] De Doncker, R.W.; Meyer, C.; Lenke, R.U.; Mura, F., "Power Electronics for Future Utility Applications," Power Electronics and Drive Systems, 2007. PEDS '07. 7th International Conference on, pp. K-1-K-8, 27-30 Nov., 2007.

[4] Franck, C.M., "HVDC Circuit Breakers: A Review Identifying Future Research Needs," Power Delivery, IEEE Transactions on, vol. 26, no. 2, pp. 998-1007, Apr., 2011.

[5] Carstensen, C.; Biela, J., "10kV/30kA unipolar arbitrary voltage source for Hardware-in-the-Loop Simulation Systems for HVDC circuit breakers," Power Electronics and Applications (EPE 2011), Proceedings of the 2011-14th European Conference on, pp. 1-10, Sep., 2011.

[6] Salloum, A.; Heydt, G.T., "Innovative HVDC connections in power transmission systems," Transmission and Distribution Conference and Exposition (T&D), 2012 IEEE PES, pp. 1-8, 7-10 May 2012.

[7] Ekstrom, A.; Lamell, P., "HVDC tapping station: power tapping from a DC transmission line to a local AC network," AC and DC Power Transmission, 1991, International Conference on, pp. 126-131, 17-20 Sep., 1991.

[8] Pietersen, D., "Tapping small amounts of power from HVDC transmission lines using parallel cascaded converters," AFRICON, 2004. 7th AFRICON Conference in Africa, vol. 2, pp. 691-696, 15-17 Sep., 2004.

[9] Chetty L, Ljumba N. "Rural electrification using overhead HVDC transmission lines." Energize, no. 2, pp. 56-64, 2012.

[10] Bachmann, B.; Mauthe, G.; Ruoss, E.; Lips, H. P.; Porter, J.; Vithayathil, J., "Development of a 500kV Air blast HVDC Circuit Breaker," Power Apparatus and Systems, IEEE Transactions on, vol. PAS-104, no. 9, pp. 2460-2466, Sep., 1985.

[11] Steurer, M.; Frohlich, K.; Holaus, W.; Kaltenegger, K., "A novel hybrid current-limiting circuit breaker for medium voltage: principle and test results," Power Delivery, IEEE Transactions on, vol. 18, no. 2, pp. 460-467, Apr., 2003.

[12] Novello, L.; Gaio, E.; Piovan, R., "Feasibility Study of a Hybrid Mechanical-Static DC Circuit Breaker for Superconducting Magnet Protection," Applied Superconductivity, IEEE Transactions on, vol. 19, no. 2, pp. 76-83, Apr., 2009.

[13] Meyer, C.; Schroder, S.; De Doncker, R.W., "Solid-state circuit breakers and current limiters for medium-voltage systems having distributed power systems," Power Electronics, IEEE Transactions on, vol. 19, no. 5, pp. 1333-1340, Sep., 2004.

[14] Jovcic, D.; Bin Wu, "Fast fault current interruption on high-power DC networks," Power and Energy Society General Meeting, 2010 IEEE, pp. 1-6, 25-29 Jul., 2010.

[15] Jovcic, D.; Ooi, B. -T, "Developing DC Transmission Networks Using DC Transformers," Power Delivery, IEEE Transactions on, vol. 25, no. 4, pp. 2535-2543, Oct., 2010.

[16] Jovcic, D.; Ooi, B.T., "Theoretical aspects of fault isolation on high-power direct current lines using resonant direct current/direct current converters," Generation, Transmission & Distribution, IET, vol. 5, no. 2, pp. 153-160, Feb., 2011.

[17] Soong, T.; Lehn, P., "A transformerless high boost DC-DC converter for use in medium / high voltage applications," IECON 2012 - 38th Annual Conference on IEEE Industrial Electronics Society, pp.174-179, 25-28 Oct., 2012.

[18] Cairoli, P.; Kondratiev, I.; Dougal, R. A., "Coordinated Control of the Bus Tie Switches and Power Supply Converters for Fault Protection in DC Microgrids," Power Electronics, IEEE Transactions on, vol. 28, no. 4, pp. 2037-2047, Apr., 2013.

[19] Zhan, C.; Smith, C.; Crane, A.; Bullock, A.; Grieve, D., "DC transmission and distribution system for a large offshore wind farm," AC and DC Power Transmission, 2010. ACDC. 9th IET International Conference on, pp. 1-5, 19-21 Oct., 2010.

[20] Hajian, M.; Jovcic, D.; Bin Wu, "Evaluation of Semiconductor Based Methods for Fault Isolation on High Voltage DC Grids," Smart Grid, IEEE Transactions on, vol. 4, no. 2, pp. 1171-1179, Jun., 2013.

[21] Corzine, K.A.; Ashton, R.W., "A New Z-Source DC Circuit Breaker," Power Electronics, IEEE Transactions on, vol. 27, no. 6, pp. 2796-2804, Jun., 2012.

[22] Shukr, M.; Thomas, D. W P; Zanchetta, P., "VSC-HVDC transmission line faults location using active line impedance estimation," Energy Conference and Exhibition (ENERGYCON), 2012 IEEE International, pp. 244-248, 9-12 Sep., 2012.

1700V Si-IGBT and SiC-SBD Hybrid Module for AC690V Inverter system

Haining Wang, O.Ikawa, S.Miyashita, T.Nishimura and S.Igarashi

Application technology department, Fuji Electric Co., Ltd
4-18-1 Tsukama, Matsumoto, Nagano 390-0821, Japan
Emai:wang-haining@fujielectric.co.jp

Abstract— **The 1700V/400A hybrid module is consisted of Si-IGBTs and SiC-SBDs mounted in a general 2in1 package. Because the reverse recovery current of SiC-SBD is very small to be an unipolar device, reverse recovery and turn-on losses at 400A on the hybrid module are 83% and 38% lower than that of the conventional all Si module, respectively. Therefore there is a further advantage of hybrid module at high frequency operation. Radiation noise on hybrid module becomes higher with increasing collector current, but the peak value of the noise from hybrid module is almost same as the all Si module if the collector current is less than 300A. In AC690V PWM inverter, the total power dissipation of hybrid module is 8% lower at 1 kHz and 29% lower at 10 kHz compare to the all Si module. Therefore the 1700V hybrid module is useful as a power module for an AC690V high efficiency inverter system such as wind power generation system and high voltage solar power generation system. This paper reports about the static and dynamic characteristics and the radiation noise measurement results on the 400A/1700V hybrid module.**

Keywords — *SiC-SBD, 1700V hybrid module, radiation noise, FFT.*

I. INTRODUCTION

"Smart grids" are promoted as next generation power infrastructures. Several dispersed power generators such as solar and wind power generation systems are included in parts of smart grids, and these systems need high efficiency inverters for connecting to general transmission lines. Also, the high efficiency inverters are needed from general-purpose facilities such as power supplies and air-conditioning equipments. Recently, performance progress of silicon (Si) power devices is almost saturated because of the material property. And then, silicon carbide (SiC) is focused as one of materials for next generation power devices. SiC have more superior potentials than Si in band-gap, breakdown field strength and saturated drift velocity. These advantages provide power device to high avalanche voltage, low on-state voltage and high temperature operation. Therefore several companies provide commercially discrete SiC schottky barrier diodes (SBDs). Then, as a first step of power module included in SiC device, we developed 50-100A/600V and 35-50A/1200V hybrid module exchanged Si pn diodes (PNDs) to SiC-SBDs[1][2][3][4][5]. Next we developed 400A/1700V 2in1 Module for higher voltage application such as

AC690V inverter systems for wind turbine and high voltage solar generator. This paper reports about static, dynamic characteristics and radiation noise measurement results on the 400A/1700V hybrid module.

II. ASSEMBLED

Figure 1 shows the outline picture of the 1700V/400A 2in1 hybrid module. The SiC-SBD was made on the collaboration between Fuji Electric and National Institute of Advanced Industrial Science and Technology (AIST)[6]. "V-series" IGBT die, which is the latest generation Si-IGBT in Fuji Electric [7] [8], is used for the IGBT. One arm on the hybrid module is consisted of one Si-IGBT die and several SiC-SBD dies are connected in parallel as free wheeling diode. The package size is W110mm x D80mm x H30mm. And also, as a reference for comparative evaluations, a 1700V/400A module with Si-IGBT and Si-FWD was assembled exactly the same package.

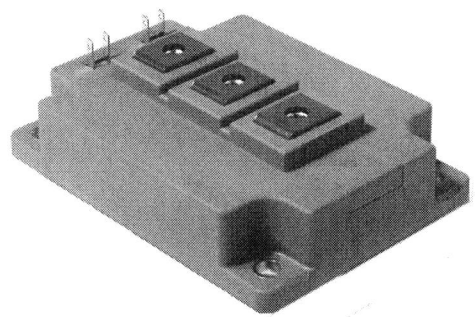

80mm×110mm×30mm

Fig.1. 1700V, 400A 2in1 Hybrid module

III. STATIC CHARACTERISTICS

Figure 2 shows the forward I-V characteristics of free wheeling diodes of the 1700V/400A hybrid module and the 1700V/400A all Si module at 25°C, 125°C and 150°C. The hybrid module has linear slope because a drift layer of SiC-SBD works as a resistance. On the other hand, the all Si module has parabolic slope to occur conductivity modulation at drift layer of Si-PND. Figure 3 shows temperature characteristics comparison of the forward voltage drop at current 400A.The forward voltage drop of the hybrid module is lower than the Si module in <110°C region. Also, the positive temperature coefficient of the hybrid module is stronger than that of the all Si module. This feature is the one of the advantage of the hybrid module when the modules are connected in parallel for making a high power system.

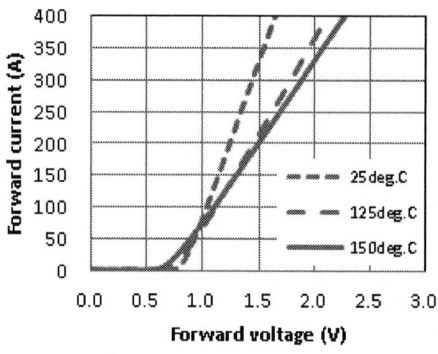

(a) Hybrid module

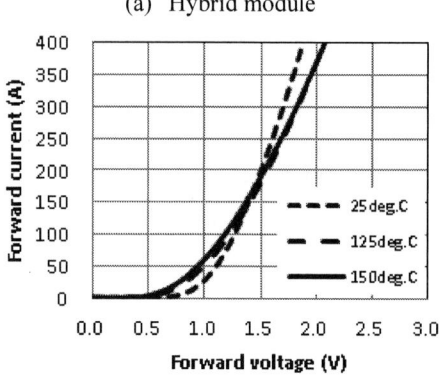

(b) All Si module

Fig.2. Forward voltage characteristics

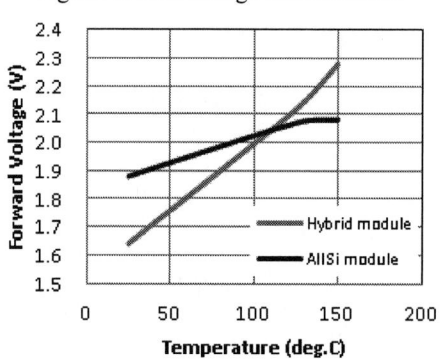

Fig.3. Temperature characteristics of the forward voltage

IV. Switching behavior

We evaluated comparisons of switching behavior between the hybrid module and the all Si module under the following conditions; DC bus voltage VDC = 900 V, gate voltage VGE = ±15 V, external turn on gate

resistance Rg(on) = 3.4 Ω, external turn off gate resistance Rg(off) = 1.8Ω, junction temperature Tj = 125°C.

1. Turn-on Behavior

Turn-on behavior comparison is shown in Figure 4. Fig.4 (a) shows the turn-on loss of the hybrid module and the Si module. Turn-on loss of the hybrid module is lower than that of the all Si module and the difference becomes wider with increasing collector current. At current Ic = 400 A, the turn-on loss decreases by 38% compared to the all Si module. The both modules have the same Si IGBT dies but the FWD is changed from Si-PND to SiC-SBDs. It is the reason that the peak collector current becomes smaller on the hybrid module and the turn-on loss is reduced as shown in Fig.4 (b) (c).

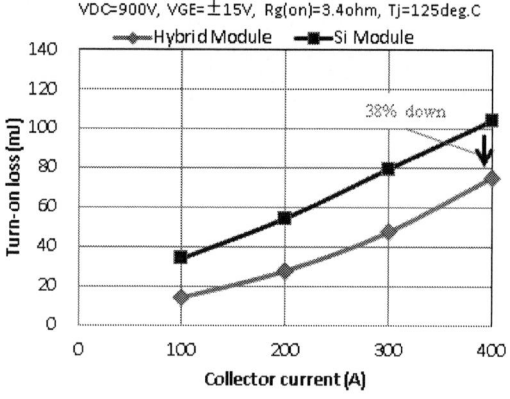

(a) Eon-Ic

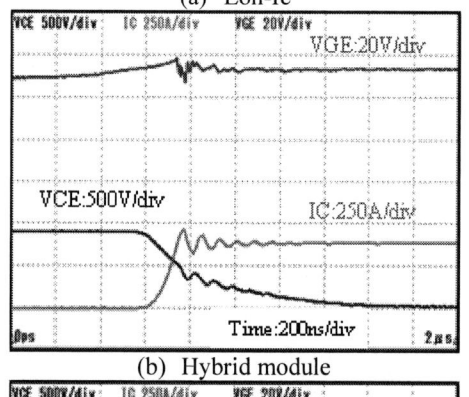

(b) Hybrid module

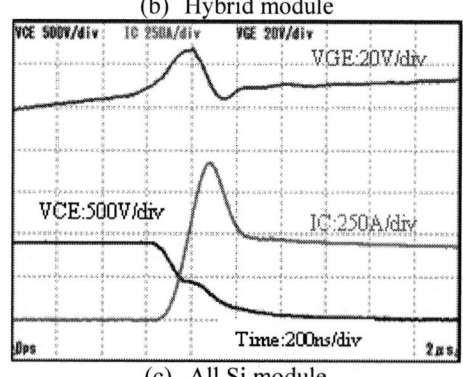

(c) All Si module

Fig.4. Comparison of turn on behavior

2. Reverse Recovery Behavior

Reverse recovery behavior comparison is shown in Fig.5. Reverse recovery loss of the hybrid module at 400A is 83% lower than that of the all Si module. The peak of the reverse recovery current of the hybrid module is very small because the SiC-SBD, which is unipolar device, has little storage of minority carrier. The actual switching waveforms are shown in Fig.5 (b) and (c). There is an oscillation during the reverse recovery switching on the hybrid module. As compared the hybrid module to the all Si module, the reverse recovery di/dt at low current is higher and the junction capacitance is larger because the SiC-SBD has low reverse recovery charge and very thin drift layer. These characteristics generate the oscillation to switch with the high di/dt at RLC resonance circuit configured with stray resistance, stray inductance and the junction capacitance.

3. Turn-off Behavior

Figure 6 shows a comparison of turn-off behavior. Fig.6 (a) shows turn-off comparison and Fig.6 (b) and (c) show actual turn-off waveforms. Turn-off loss at 400A of the hybrid module is 8% lower than that of the all Si module. The difference of the turn-off loss is due to a smaller spike voltage on the hybrid module is lower than on the all Si module, which is 35V lower at 400A. It is the reason that the voltage drop at forward recovery of the hybrid module is lower than that of the all Si module.

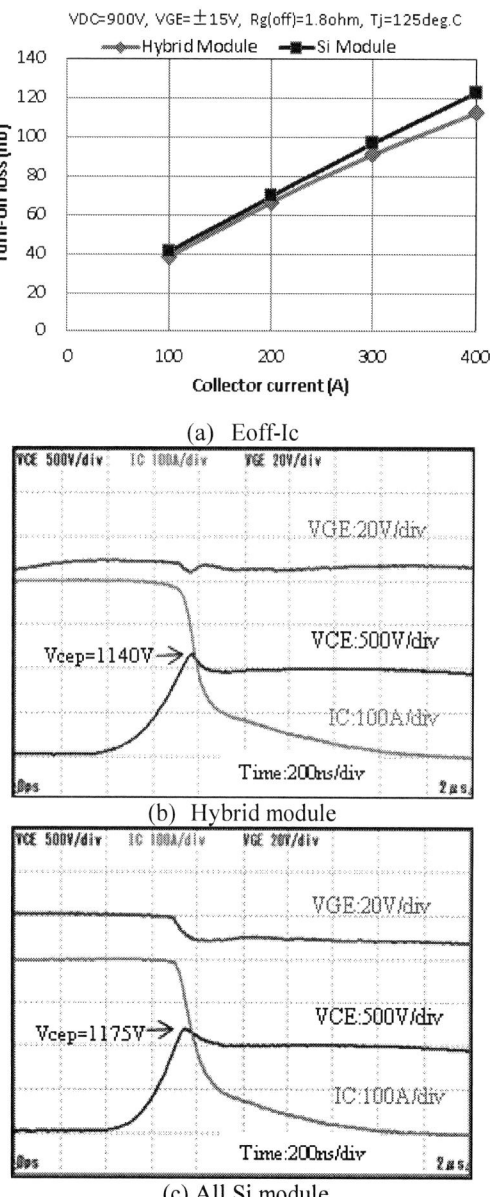

(a) Eoff-Ic

(b) Hybrid module

(c) All Si module

Fig.6. Comparison of turn-off behavior

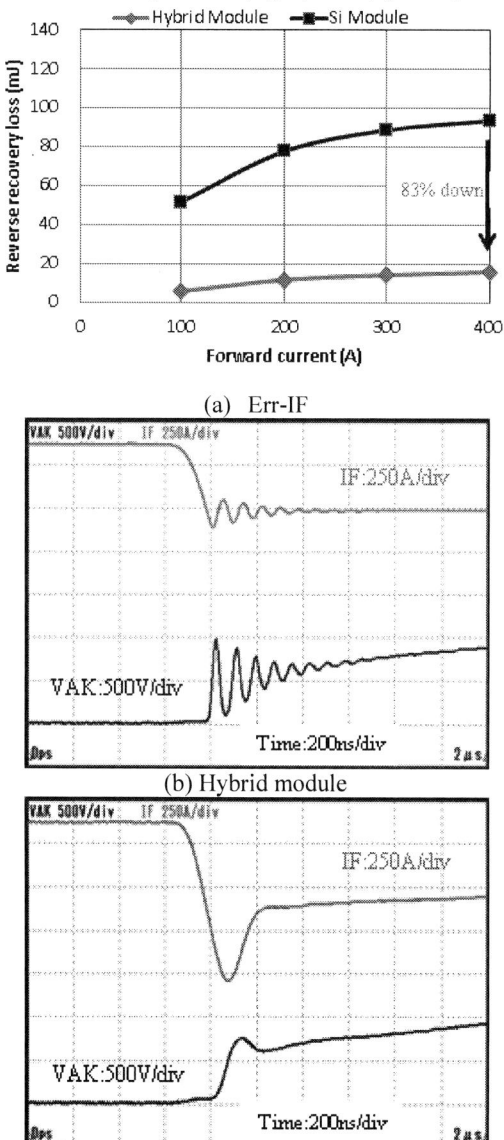

(a) Err-IF

(b) Hybrid module

(c) All Si module

Fig.5. Comparison of reverse recovery behavior

4. Radiation noise

We investigated radiation noise from the hybrid module by a spectrum which is got by doing Fourier transformation (FFT) for search coil voltage. Fig. 7 shows the measurement method, the voltage Vs is the induced voltage across the search coil by radiation noise from the module. And the switching conditions are the followings:

DC bus voltage VDC = 900 V, gate bias voltage VGE = ±15 V, external turn-on resistance Rg(on) = 3.4 Ω, external turn-off resistance Rg(off) = 1.8 Ω, junction temperature Tj = 25°C.

Fig.8. shows FFT spectrums at 30A and 300A switching on the hybrid module. In the region of over 30MHz, the peak amplitude of the FFT spectrum at Ic=300A switching is higher than at 30A.

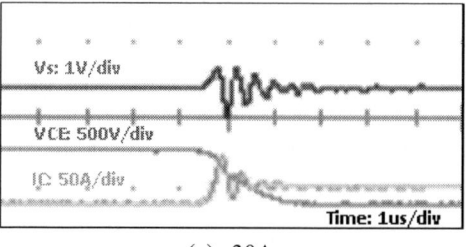

(a) 30A

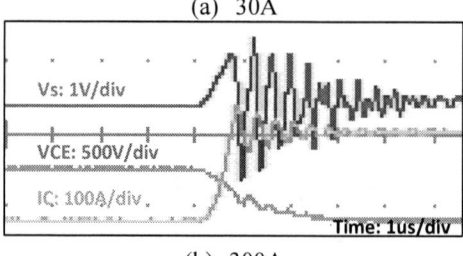

(b) 300A

Fig.9. Turn on waveform of Hybrid module

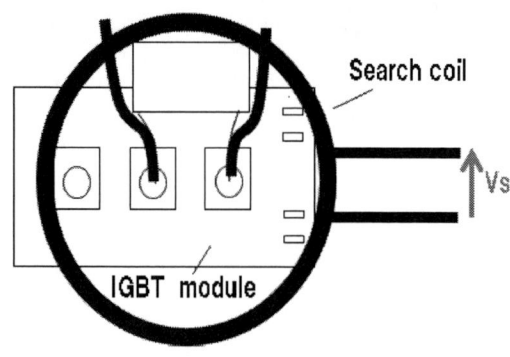

Fig.7. Radiation noise measurement method

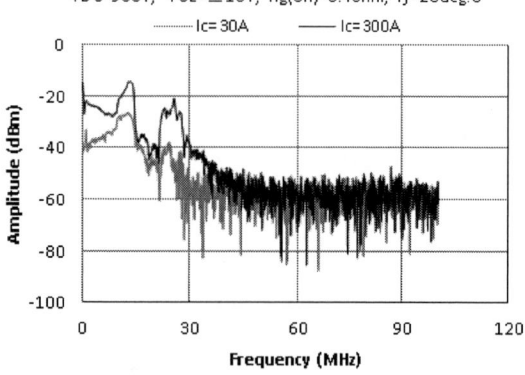

Fig.8. Radiation noise on Hybrid module

Fig.9. shows the turn on waveform of the hybrid module at Ic=30A and Ic=300A. Vs is a voltage of the search coil. The oscillation amplitude of Vs at Ic=300A is larger than Ic=30A.Therefore it is considered that the large amplitude oscillation caused high radiation noise level in >30MHz region.

Fig.10. shows a comparison of current dependence of radiation noise between the hybrid module and conventional all Si module. The hybrid module shows different trends from all Si module, the noise peak level becomes higher with increasing collector current. In the current range less than Ic=300A (ex. AC690V 45kW inverter system), the maximum value of the noise peak level of the hybrid module is almost same than that of all Si module at Ic=30A.

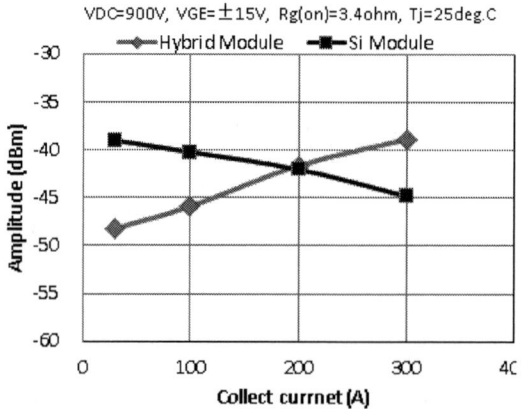

Fig.10. Current dependence of radiation noise

V. CALCULATION OF POWER DISSIPATION

We calculated dissipation power losses of power module for AC 690V PWM Inverter such as wind power generation and high voltage solar generator with "Vdcmax=1500V". Fig.11. shows the calculation results of power dissipation with several carrier frequencies under the following conditions; three arm PWM modulation, DC bus voltage VDC = 900 V, output phase current Io = 230 Arms, output frequency Fo = 50 Hz, power factor cos φ = 0.9, modulation ratio λ = 1.0. Total power dissipation of the hybrid module is 29% lower than that of the all Si module at 10 kHz. At 1 kHz switching frequency, the hybrid module is 8% lower loss compared to the all Si module. The ratio of switching losses in the total dissipation increases as higher carrier frequencies. Therefore hybrid module has the further advantage at high frequency operation.

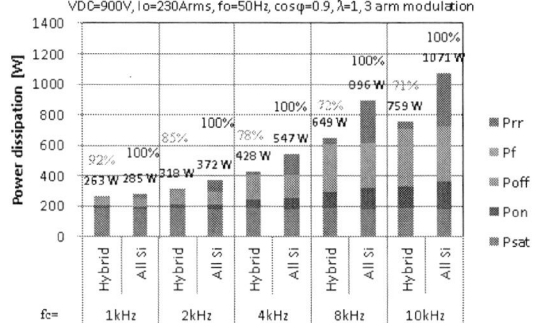

Fig.11. Power dissipation

VI. Conclusions

It is confirmed that the reverse recovery loss of 1700V/400A hybrid module is 83% lower than that of the conventional all Si module. The turn-on loss is also 38% lower compared to the all Si module. The turn-off spike voltage of the hybrid module is lower than the all Si module because of decreasing the forward recovery voltage drop.

The current dependence of radiation noise was measured. The peak level of FFT spectrum in the region over 30 MHz at Ic=300A switching is higher than at Ic=30A switching, but in the current range less than 300A (ex. AC690V 45kW inverter system), the maximum value of the noise peak value of the hybrid module is almost same as the all Si module.

A power dissipation of the hybrid module was calculated for AC 690V PWM inverter system such as wind power generation system and high voltage solar power generation system. The total power dissipation of hybrid module is 8% and 29% lower than the all Si module at 1 kHz and 10 kHz, respectively.

As a conclusion, the newly developed 1700V/400A hybrid module is useful as a power module for high efficiency AC690V inverter system because it has a high efficiency and the peak level of the radiation noise is almost same as the conventional all Si power module..

REFERENCES

[1] Toshiyuki Miyanagi, et al,," Newly Development Hybrid Module by Si-IGBT and SiC-SBD", Proceeding of PCIM Asia, 2012

[2] T. Tsuji, A. Kinoshita, N. Iwamuro, K. Fukuda, T. Tsuyuki, and H.Kimura, "Experimental demonstration of 1200V SiC-SBDs with lower forward voltage drop at high temperature," Proceedings of ICSCRM 2011, Tu-P-27, September 2011

[3] S. Harada, Y. Hoshi, Y. Harada, T. Tsuji, A. Kinoshita, M. Okamoto, Y. Makifuchi, Y. Kawada, K. Imamura, M. Gotoh, T. Tawara, S. Nakamata, T. Sakai, F. Imai, N. Ohse, M. Ryo, A. Tanaka, K. Tezuka,T. Tsuyuki, S. Shimizu, N. Iwamuro, Y. Sakai, H. Kimura, K. Fukuda, and H. Okumura, " High performance SiC-IEMOSFET/SBD module," Proceedings of ICSCRM 2011, MO-3A-1, September, 2011, pp. 52-57.H.Shigekane, et al. " Macro-Trend and A Future Expectation of Innovations in Power Electronics and Power Devices" . Proceedings of IPEMC'09. 2009.

[4] N. Ikeda, Y. Niiyama, H. Kambayashi, Y. Sato, T. Nomura, S. Kato, and S. Yoshida, "GaN Power Transistors on Si Substrates for Switching Applications," Proceedings of the IEEE, vol.98, July 2010, pp.1151-1161.

[5] H. Mine, Y. Matsumoto, R. Yamada, K. Mino, H. Kimura "Characteristics of the Power Electronics Equipments Applying the SiC Power Devices," IEEE Conference on Power Engineering and Renewable Energy 2012. 3-5. July 2012, Bali, Indonesia.

[6] A.Kinoshita et al, "Experimental Demonstration of 1200V SiC-SBDs with Lower Forward Voltage at High Temperature" in Proceedings of 19th SiC and Related Wide Band gap Semiconductors, the Japan Society of Applied Physics, 2010 (in Japanese)

[7] Y.Kobayashi, "The new concept IGBT-PIM with the 6th generation V-IGBT chip technology", in PCIM 2007

[8] S.Igarashi, "SiC-Device for High-Efficiency Power Supplies", Proceeding of SPEED, 2013

The 2014 International Power Electronics Conference

Switching Simulation of SiC High-Power Module with Low Parasitic Inductance

Takashi Yamamoto, Kohei Hasegawa, and Masaaki Ishida
Corporate Manufacturing Engineering Center
Toshiba Corporation
Yokohama, Japan

Kazuto Takao
Corporate Research & Development Center
Toshiba Corporation
Kawasaki, Japan

Abstract— In this paper, we present a simulation technique for a high-power module using silicon carbide (SiC) phase leg units. Each phase leg unit has parasitic inductances that are caused by bonding wires and patterns on a substrate in the phase leg unit. The parasitic inductances are extracted using a quasi-electromagnetic simulation tool. We also estimate mutual inductances between the parasitic inductances in the high-power module. We fabricate a high-power module that consists of six phase leg units with low parasitic inductance. Simulation results show a spike voltage of 156 V and a total parasitic inductance of 7.1 nH. Measurement results indicate a spike voltage of 154 V and a total parasitic inductance of 8.1 nH. These results show that the proposed technique is appropriate for predicting the switching performances of the high-power modules.

Keywords— Silicon carbide (SiC) devices, power modules, parastic inductance, simulation.

I. Introduction

High-power modules have been used for high efficiency power conversion systems in many areas such as trains and electric vehicles. Silicon carbide (SiC) devices are the most effective candidates for high-power modules because they have a high breakdown voltage, a high current density and a low resistance. Moreover, high-power modules based on SiC devices have the potential for high-frequency operation because of the high electron saturation velocity of SiC devices.

However, when the parasitic inductances in a high-power module are large, the short time variation of the current under high-frequency operation causes a high spike voltage [1]-[4]. Thus, it is necessary to reduce the parasitic inductances in high-power modules. The conventional technique for extracting parasitic inductances has been to measure a fabricated module. However, the parasitic inductances should be determined and evaluated before the fabrication.

In this study, we fabricate a high-power module with six phase leg units. The layout of the parts and patterns on a substrate is determined to reduce the parasitic inductances in each phase leg unit. The parasitic inductances of the bonding wires and patterns on the

substrate including the mutual inductances are calculated using a quasi-electromagnetic (EM) simulation tool. We also simulate the switching operation of the high-power module using a simulation parameter integrated circuit with emphasis (HSPICE). Device models of a field-effect transistor (FET) and a diode are made by optimizing the HSPICE parameter.

II. Concept of High-Power Module

Figure 1 shows a basic schematic diagram of the high-power module, which consists of FETs (Q1, Q2), diodes (D1, D2) and a capacitor (C). In Fig. 1, the P-terminal, AC-terminal and N-terminal exhibit a DC positive node, an AC node and a DC negative node, respectively. The gates (G1, G2) of the two FETs control the switching operation of the high-power module. If an inductance for a load is connected between the P-terminal and AC-terminal, the current flows through the load and Q2 when Q1 is in the off-state and Q2 is in the on-state. When Q1 is in the on-state and Q2 is in the off-state, the current flows through the load and D1. In the switching operation without parasitic inductance, a spike voltage does not occur between the AC-terminal and N-terminal.

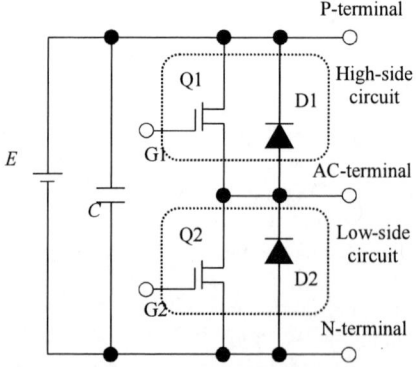

Fig. 1. Basic schematic diagram of high-power module.

978-1-4799-2706-7/14 $31.00 © 2014 IEEE   3707

However, the high-power module has a parasitic inductance generated by the bonding wires and patterns on the substrate. For example, a bonding wire is connected between each FET and each pattern on the substrate. These wires and patterns cause the parasitic inductance. Figure 2 shows a schematic diagram of a real high-power module. When the high-power module is carrying out the switching operation, the spike voltages (V_{sp}) between the AC-terminal and N-terminal are given by as below

$$V_{sp} = L_p \frac{di}{dt}, \qquad (1)$$

where L_p is the parasitic inductance and i is the current. Therefore, a low parasitic inductance is advantageous for the high-power module.

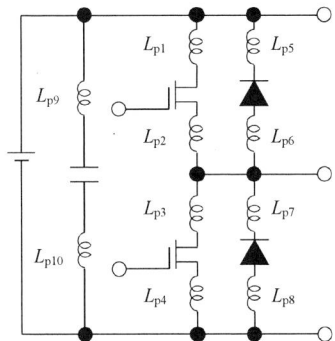

Fig. 2. Schematic diagram of high-power module.

The resonant frequency (F_r) of the high-power module is given by

$$F_r = \frac{1}{2\pi\sqrt{L_p C_{total}}}, \qquad (2)$$

where C_{total} is the total capacitance of the high-power module. Thus, the total parasitic inductance of the high-power module is calculated using eq. (2).

III. Phase Leg Unit

Figure 3 shows a three-dimensional (3-D) model of the high-power module that uses phase leg units with the FETs and diodes fabricated on the SiC wafer. The 3-D model in Fig. 3 shows half the model, which is divided at the referential plane. The layout of parts and patterns on the substrate results in low parasitic inductances in each phase leg unit. A current of 360 A and a breakdown voltage of 1.2 kV are assumed for the high-power module. In this module, there are six phase leg units, with four FETs and four diodes used in each phase leg unit. Each phase leg unit has a current of 60 A and a breakdown voltage of 1.2 kV. We assume that each FET and diode

are CPMF1200S080B and CPW41200S020B model by Cree, respectively. The continuous drain current of the FET is 30 A at a gate-source voltage of 20 V and a junction temperature of 150 °C. Thus, a current of 60 A is attained by connecting the FETs in parallel.

We assume that the FETs and diodes are mounted on a substrate. The substrate thickness is 0.32 mm and the copper pattern thickness is 0.3 mm. The FETs and diodes are connected to the substrate using bonding wires with diameters of 150 μm and 350 μm. Three bonding wires are used to connect the FETs and diodes to the substrate in order to maintain the current capacitance.

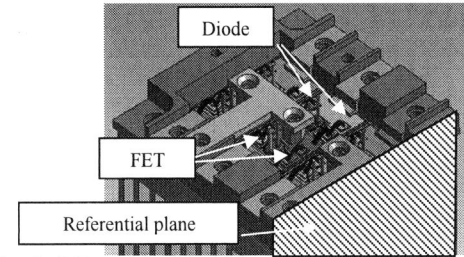

Fig. 3. 3-D model of the high-power module.

IV. Extraction of Parasitic Inductances

We extract the parasitic inductances of the bonding wires and the patterns on the substrate using a Q3D extractor. Such extractors are widely used in studies on EMC.

The 3-D model of the Q3D extractor of the phase leg unit is shown in Fig. 4. This figure shows half the model divided at the referential plane. The patterns on the substrate and bonding wires are assumed to be made of copper in the analysis.

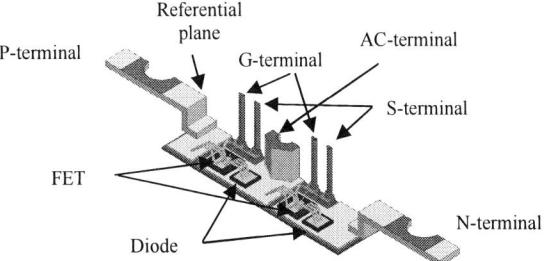

Fig. 4. 3-D model of the Q3D extractor of the phase leg unit.

The equivalent circuit of the half phase leg unit considering the parasitic inductances is shown in Fig. 5. The parasitic inductances of the bonding wires and the patterns on the substrate are from 0.3 to 4.4 nH. The mutual inductances of the three wires are 1.0 to 2.3 nH.

The mutual effects between the phase leg units are also calculated.

The 2014 International Power Electronics Conference

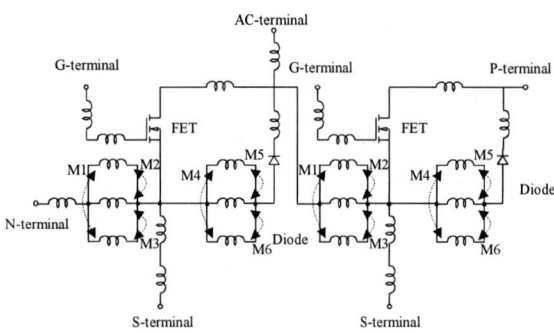

Fig. 5. Equivalent circuit of half phase leg unit.

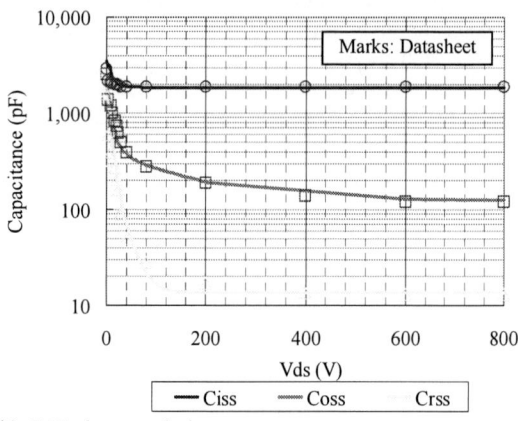

(b) C-V characteristics
Fig. 6. Simulation results for CPMF1200S080B FETs.

V. Circuit Simulation

Simulation models of the SiC devices such as the FETs and diodes are made using the HSPICE. To make simulation models of the FETs and diodes, we use a BSIM3 (Berkeley short-channel IGFET model), which is an n-channel MOS (metal-oxide-semiconductor) FET model in the HSPICE. For the diodes, we use general diode parameters in the HSPICE. We make the FET model by fitting an I-V characteristic and a C-V characteristic written on a datasheet of the CPMF1200S080B FET. The diode model is made from a forward voltage and a repetitive peak reverse voltage written on a datasheet of the CPW41200S020B diode.

The simulation results for the I-V and C-V characteristic of the FETs are shown in Fig. 6, which are found to match those from the datasheet.

The simulation results for the CPW41200S020B diodes are shown in Fig. 7, which are in good agreement with the experimental results.

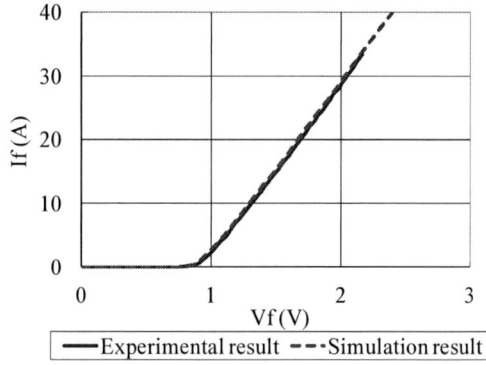

(a) I-V characteristics

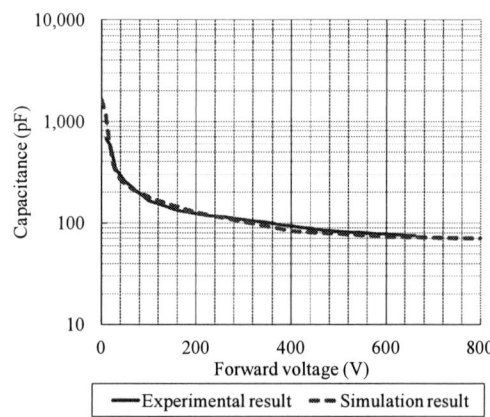

(b) C-V characteristics
Fig. 7. Simulation results for CPW41200S020B diodes.

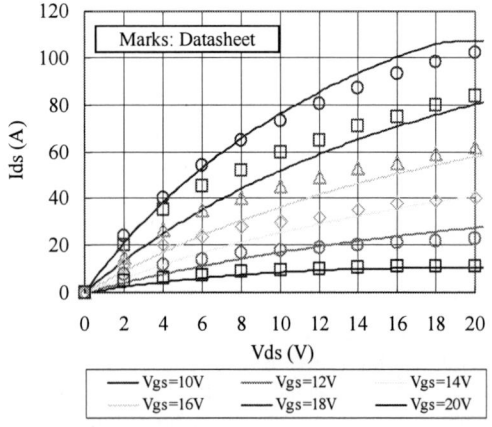

(a) I-V characteristics
Fig. 6. Simulation results for CPMF1200S080B FETs.

The switching operation of the high-power module is simulated using the SiC device models. The simulation model of the high-power module is shown in Fig. 8. The simulation model consists of six phase leg units, which are connected in parallel. To increase the impedance of the voltage source, which is set at 600 V, a high inductance is connected in series to the simulation model of the voltage source. This inductance is set at 1 H. Also, a current source of 360 A is used instead of the load

978-1-4799-2706-7/14 $31.00 © 2014 IEEE

impedance. The parasitic inductance of a film capacitor is measured to be 27.8 nH by an Agilent Technologies impedance analyzer (Model 4294A). To observe the operation in the off-state, the gate and source terminals of the high-side circuit are short-circuited. Moreover, the gate voltage of the low-side circuit is set at 20 V and 0 V in the on-state and off-state, respectively.

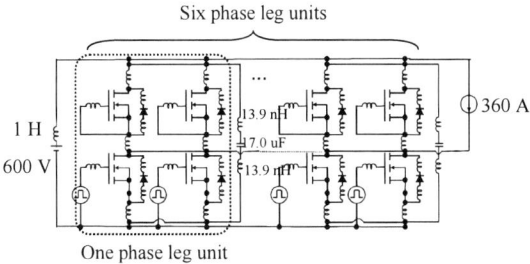

Fig. 8. Simulation model of high-power module.

The simulation result of the switching operation is shown in Fig. 9. In this figure, the spike voltage is 156 V, and one cycle of the resonance of the spike voltage is 25.9 ns.

The C-V characteristic of the FET obtained by simulation gives a drain capacitance of 131 pF at Vds=600 V. The diode capacitance at Vd=600 V is 69.5 pF. Therefore, the total capacitance of the high-power module is 2,406 pF. The total parasitic inductance of the high-power module is then calculated to be 7.1 nH using eq. (2).

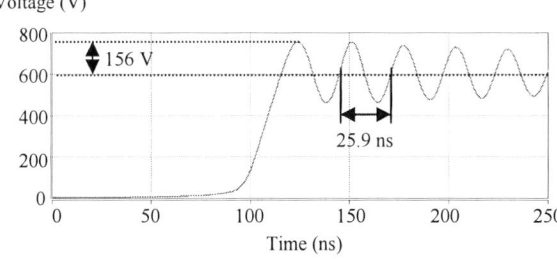

Fig. 9. Simulation result of switching operation.

VI. Experimental Results

The fabricated high-power module consisting of six phase leg units is shown in Fig. 10. The substrate is silicon nitride (SiN). The high-power module is connected to the voltage source and the load inductance, which are set at 600 V and 800 μH, respectively. The gate voltages of the on-state and off-state are set at 20 V and 0 V, respectively.

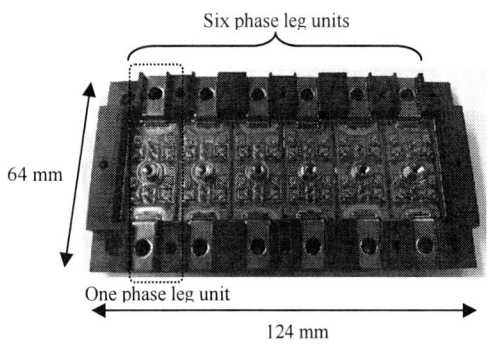

Fig. 10. Fabricated high-power module.

The drain voltage of the low-side FET is shown in Fig. 11, which is measured by a general oscilloscope. In Fig. 11, the spike voltage and one cycle of the resonance of the spike voltage are 154 V and 26.3 ns, respectively. The total capacitance of the fabricated high-power module is calculated to be 2,160 pF from the datasheets of the CPMF1200S080B FET and CPW41200S020B diode. The parasitic inductance of the fabricated high-power module is the obtained to be 8.1 nH using eq. (2).

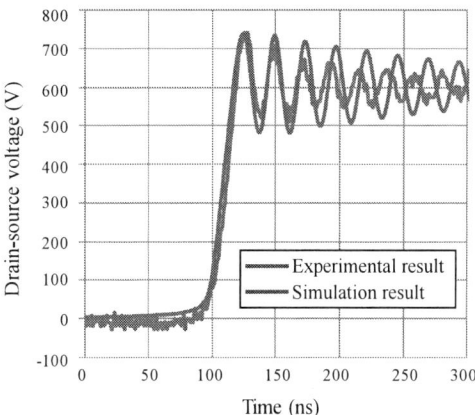

Fig. 11. Experimental and simulation results for switching operation.

VII. Conclusions

The total parasitic inductance of a high-power module was calculated using a 3-D simulation tool. The parasitic inductances included the mutual effects between the bonding wires, the patterns of the substrate and the phase leg units. The simulation results for the spike voltage and the total parasitic inductance corresponded closely with the experimental results. This shows that the extracted inductances are correct values and that the proposed simulation technique is useful for designing high-power modules.

References

[1] K. Haehre, R. Kling, and W. Heering, "Design of a soft-switching inverter operating in the MHz-range based on SiC MOSFETs," *PCIM Europe 2013*, pp. 707-713, May 2013

[2] L. Abbatelli, G. Catalisano, B. Rubino, and S. Buonomo, "1200V SiC MOSFET and N-off SiC JFET performances and driving in high power-high frequency power converter," *PCIM Europe 2013*, pp. 982-989, May 2013

[3] J. Fabre, P. Ladoux, and M. Piton, "Characterization of SiC MOSFET dual modules for future use in railway traction chains," *PCIM Europe 2012*, pp. 539-546, May 2012

[4] S. Liebig, M. Nuber, J. Engstler, and A. Engler, "Characterization and evaluation of 1700V SiC-MOSFET modules for use in an active power filter in aviation," *PCIM Europe 2012*, pp. 952-958, May 2012

The 2014 International Power Electronics Conference

Switching Performance of Parallel-Connected Power Modules with SiC MOSFETs

Juan Colmenares, Dimosthenis Peftitsis, Hans-Peter Nee
Laboratory of Electrical Energy Conversion
KTH Royal Institute of Technology
Stockholm, Sweden

Jacek Rabkowski
Institute of Control and Industrial Electronics
Warsaw University of Technology
Warsaw, Poland

Abstract- **Parallel connection of silicon carbide power modules is a possible solution in order to reach higher current ratings. Nevertheless, it must be done appropriately to ensure a feasible operation of the parallel-connected power modules. High switching speeds are desired in order to achieve high efficiencies in medium and high-power applications but parasitic elements may give rise to a non-uniform current sharing during turn-on and turn-off, leading to non-uniformly distributed switching losses. This paper presents the switching performance of parallel-connected power modules populated with several silicon carbide metal-oxide-semiconductor field-effect-transistors chips. It is experimentally shown that turn-on and turn-off switching times of approximately 50 ns and 100 ns, respectively, can be reached, while an acceptably uniform transient current sharing is obtained. Moreover, based on the obtained results, an efficiency of approximately 99.35% for a three-phase converter rated at 312 kVA with a switching frequency of 20 kHz can be estimated.**

Keywords—Silicone Carbide, Gate Driver, Metal-Oxide-Semiconductor Field-Effect-Transistors, Power Module.

I. INTRODUCTION

Silicon carbide (SiC) power devices have several advantages compare to the existing silicon (Si) counterparts. This could lead to a potential replacement of the semiconductor components in several power electronics applications [1]-[8]. The three main advantages of SiC compared to Si are: higher maximum operating temperature, higher efficiency, and higher possible switching frequency [5]. High efficiency is one of the main targets when medium or high-power converters are considered. By keeping both conduction and switching losses under certain levels it is feasible to reach efficiencies exceeding 99 % [4], [6]. Due to the low on-state resistance of SiC devices, the conduction losses condition could easily be fulfilled. On the other hand, the switching losses condition mainly depends on the switching speed. Fast switching speeds should be achieved with the least oscillative performance.

Nevertheless, for power ratings beyond a few tens of kVA the existing current ratings of the available devices may be insufficient. A possible solution in order to reach higher power levels is to parallel-connect either several discrete single-chip devices or to build power modules populated with several chips. In [9]-[12] deep investigations of parallel-connection of several discrete SiC power switches are presented. In order to ensure the

feasibility of the parallel connection there are two main factors to consider. These factors are, first the placement of the devices in the circuit layout and second, to keep track of the spreads in the device parameters [10]-[12]. In particular, the switching performance is affected by the parasitic inductances between the leg connections of the discrete device packages. Moreover, for modules with high power ratings special driver designs may be necessary in order to ensure an acceptable switching performance. This is mainly due to the parasitic components such as stray inductances and high values of Miller capacitance caused by the parallel connection of several chips [13].

Another possibility to reach even higher power levels, which has not been studied before is the parallel connection of power modules. At the present time, SiC power modules populated with parallel-connected metal-oxide-semiconductor field-effect-transistors (MOSFETs) are mass produced and they are available on the market. Several studies have been performed using SiC MOSFET power modules evaluating the switching performance, high-temperature and high-frequency operation [13]-[18].

In contrast to mentioned investigations, this paper presents the switching performance of parallel-connected SiC MOSFET power modules employed in a phase-leg configuration. A fast switching performance and a good transient and steady-state current sharing have been achieved. A description of the module and gate driver used in this application, are presented in Section II. Section III describes the experimental setup, while recorded experimental results and power-loss and efficiency estimations for a 312 kVA three-phase inverter are shown in Section IV. Finally, Section V gives the conclusions from this work.

II. DESCRIPTION OF THE MODULE AND DRIVER

The half-bridge SiC power module used in this paper (Cree, Inc. CAS100H12AM1) has the ratings of 1200 V and 168 A, at room temperature (RT). It is built with five parallel-connected single-chip SiC power MOSFETs, and five antiparallel SiC Schottky diodes per switch position. The on-state resistance at RT is approximately 16 mΩ. Due to the placement of the chips inside the module a proper way to represent a power module switch position is shown in Fig. 1. L_D, L_S, and L_G represent the total parasitic inductance of the drain, the source and, the gate leads, respectively. These parasitic inductances affect the

978-1-4799-2706-7/14 $31.00 © 2014 IEEE

switching performance when a parallel-connection is targeted [9]-[12]. Specifically, the distribution of parasitic inductances in the devices could contribute as a cause for a non-uniform current sharing during the switching transient, which, in turn causes a non-uniform distribution of the switching losses.

In order to drive this power module, a simple, totem-pole gate driver is used (Fig.1). The switching speeds, and therefore the switching losses, are controlled by means of adjusting the parameters of the gate-drive unit. The design of the printed circuit board (PCB) was done in such a way to reduce the stray inductances caused by the gate-source loop. The turn-on and turn-off current paths are created in a symmetrical way. Another important property for the driver is the ability to supply high current peaks for a short time during the transients. This would help to ensure the very short switching times. Fig. 1 shows the schematic of the gate-drive unit used for the SiC power module. Through D_1 and R_1, the turn-on current, represented by the solid red arrow in Fig. 1, is controlled. On the other hand, the turn-off current shown as the dashed red line in Fig. 1 is controlled through D_2 and R_2.

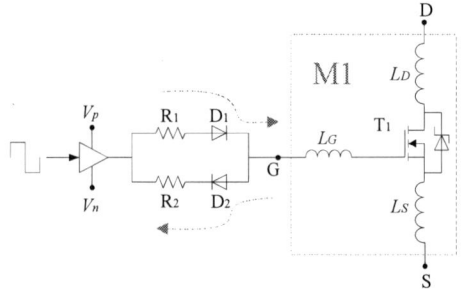

Fig. 1. Schematic diagram of the gate driver and combined schematic diagram of a single-switch position in the power module.

III. DESCRIPTION OF THE EXPERIMENTAL SETUP

In order to reduce the on-state resistance, i.e. the conduction losses, it was decided to connect several power modules in parallel, without any sorting criteria. Fig. 2(a) illustrates the experimental setup which consists of a capacitor C that is in parallel with a voltage supply V_{DC}, and a load, which is an air coil inductor, L. As mentioned above, the SiC power modules are configured as half-bridges. For expected operating conditions the antiparallel diodes only conduct during the blanking time, which is approximately 2 µs, while after this time the reverse current flows through the channels of the SiC MOSFETs in the upper switch position. Table I summarizes the parameters of the experimental setup.

TABLE I
PARAMETERS OF THE EXPERIMENTAL SET-UP

Parameter	Value
DC supply voltage, V_{DC}	700 V
Inductance, L	70 µH
Capacitance, C	560 µF

In order to ensure equally-distributed stray inductances between the various connections in the experimental setup, a symmetrical placement of the power modules

was used. This is possible due to U-shape bus bars as shown in Fig. 2(b). Positive and negative bus bars are placed on top of each other, and therefore the stray inductances introduced by them are partially cancelled with respect to each other. Besides that, distributed capacitors are connected evenly among the bus bars, reducing inductance of the switching loops. Moreover, the gate drivers are connected directly to the gate pins of the power modules. A photograph of the Double Pulse Test (DPT) setup (single phase leg) is shown in Fig. 2(c). Several experiments were performed and the results showing the switching performance are presented in the next section.

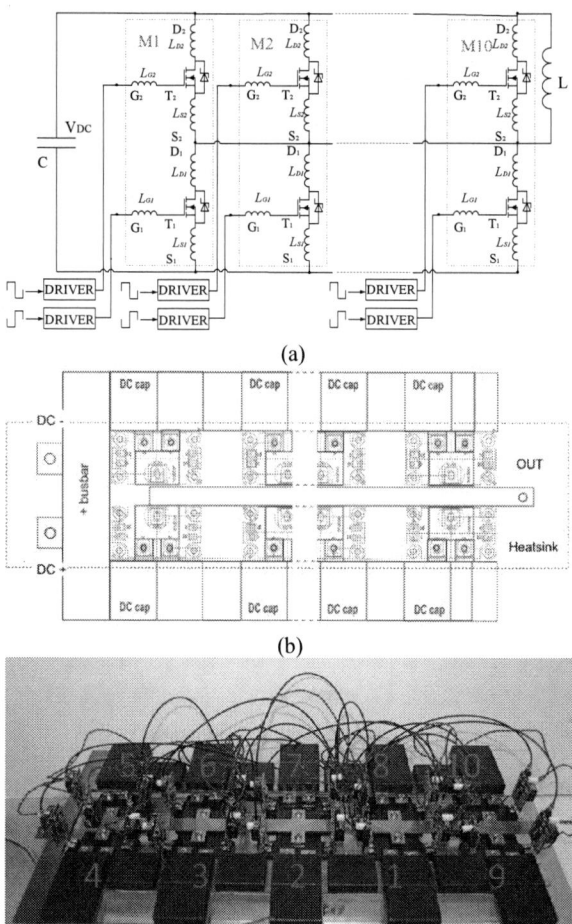

(a)

(b)

(c)

Fig. 2. (a) Schematic diagram of the double pulse test setup, (b) Layout of the bus bars, (c) Photograph of the double-pulse test setup prototype.

IV. EXPERIMENTAL RESULTS

Switching performance of the SiC MOSFET power modules was investigated by number of DPT measurements. It be noted that the definition used for the rise and fall times is the time required for the voltage and currents to rise or fall from 10% to 90%, or vice versa. The current is measured using Rogowski coils. The voltage is measured using a Tektronix P5100 high-voltage probe. The delays of the voltage and current probes were adjusted in order to compute the switching loss correctly. A data acquisition system is used for these tests, National Instrument® PXI-5105, with 8 simultaneously sampled channels, 12-bit vertical resolution, 60 MS/s real-time sampling rate and, 60 MHz analog bandwidth.

A. Transient Performance

The switching performance of the phase-leg, built with ten parallel-connected SiC power modules, has been investigated using a standard double-pulse test (DPT). The devices under test (DUT) are the lower switch positions, while the upper switches only operate as freewheeling diodes. In this paper, V_{DC} was set to 700 V. Fig. 3 shows the steady-state current sharing of the power modules. It is observed that modules No. 9 and No. 10, are taking significantly higher currents than the other modules. The reason for this is the lower stray inductance between the mid-point of the modules and the connection to the inductor. The rest of the modules share the static current with a maximum difference of 25%.

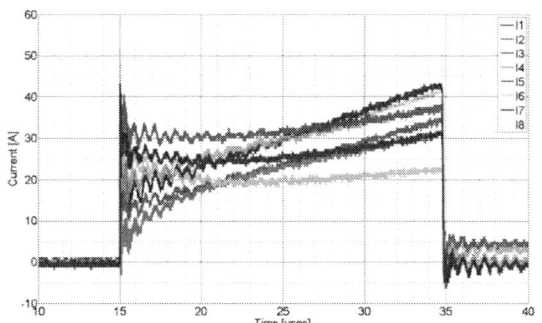

Fig. 3. Current sharing of the SiC MOSFETs. Measured drain-source current of each SiC power module.

Fig. 4 shows the turn-on transient. The current waveforms of the 10 parallel-connected power modules are shown as well as the voltage waveform is shown in Fig. 4. The turn-ON process takes approximately 75 ns. During this period the current sharing is acceptably uniform.

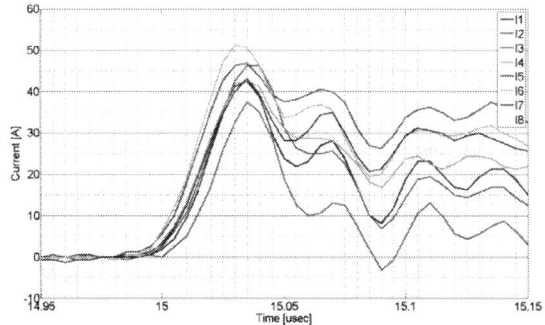

Fig. 4. Turn-ON transient of the SiC MOSFETs. Measured drain-source current of each SiC power module, Measured drain-source voltage.

Fig. 5 shows the turn-OFF transient, where the current waveforms of the parallel-connected power modules and voltage waveform are shown. The turn-off process takes

approximately 50 ns. The turn-off current sharing is uniform for all the modules except for No. 2 and No. 8, which conduct a higher current for a certain period of time, as shown in Fig. 5. It is believed that this overshoot of the current is due to the difference in the switching speeds of the drivers. This unequal transient current sharing will cause a difference between the turn-off losses in the modules.

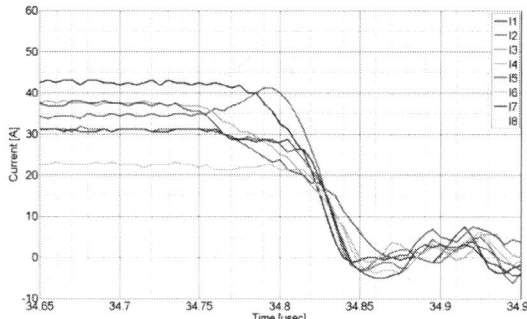

Fig. 5. Turn-OFF transient of the SiC MOSFETs. (a) Measured drain-source current of each SiC power module, (b) Measured drain-source voltage.

B. Steady-State Operation

For the experimental investigation regarding the steady-state operation, the single phase-leg setup prototype was connected in a step-down dc-dc converter configuration. The load consists of an inductor of approximately 100 µH was connected in series with an output resistor of 1.5 Ω. During the test, V_{DC} was increased until the output current equals 70 A. If a duty cycle of 0.18 is considered, the input voltage, V_{DC}, was set to approximately 450 V. A total output power of 5.7 kW was reached, which is 5.7 % of the nominal power. Figs. 6 and 7 shows the results in steady-state operation, where it can be noted that the converter is operating in continuous conduction mode as expected. The steady-state current is approximately 7 A per module. It is the hypothesis of the authors that the unequal distribution of the steady-state currents is due to spread in the device parameters, such as on-state resistance of the MOSFETs. Moreover, the high-frequency resistance of the bus-bars should also be considered as possible reason. It is also noted that the ripple current of the inductor is approximately 40 A.

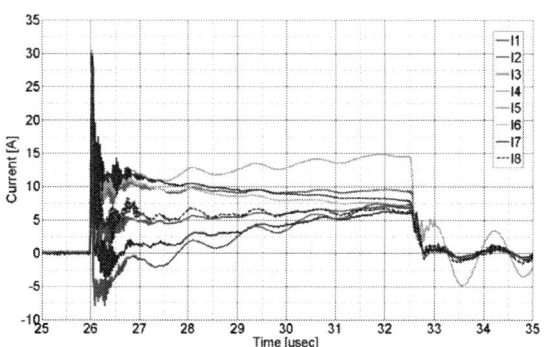

Fig. 6. Current sharing in steady-state of the phase-leg. Measured drain-source current of each SiC power module.

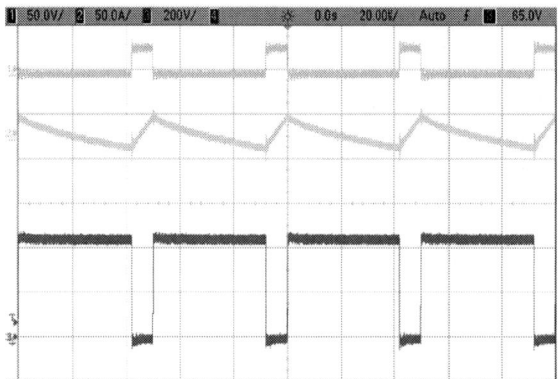

Fig 7. Steady-state operation of the phase-leg test set up prototype. Measured drain-source voltage of the SiC MOSFETs, (purple line, 200 V/div), measured gate-source voltage of the SiC MOSFETs, (yellow line, 50 V/div), ripple current of the inductor, (green color, 50 A/div), (time-base 20 us/div).

A second test was also performed. In this case, the data acquisition system was not used due to its saturation limit. During the test, V_{DC} was increased until the output current reached 90 A. If a duty cycle of 0.18 is considered, the input voltage, V_{DC}, was set for approximately 600 V. A total output power of 9.9 kW was reached, which is approximately 10 % of the nominal power. Fig. 8 shows the results on steady-state operation, where it can be noted that the converter is operating in continuous conduction mode as expected. The current sharing of modules No. 2 and No. 6 is observed. The steady state current is approximately 9 A. It is also noted that the ripple current of the inductor is approximately 50 A.

Fig. 9(a) shows the turn-ON transient. The current waveforms of the parallel-connected power modules and the drain-source voltage waveform are shown. The turn-on process takes approximately 50 ns. During this period the current sharing is acceptably uniform. Moreover, the high-frequency resistance of the bus-bars should also be considered as possible reason. Fig. 9(b) shows the turn-OFF transient. The turn-OFF process takes approximately 100 ns. The turn-OFF current sharing is sufficiently uniform for the modules.

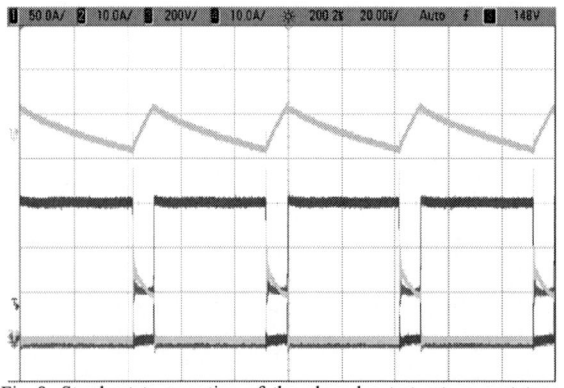

Fig 8. Steady-state operation of the phase-leg test set up prototype. Measured drain-source voltage of the SiC MOSFETs, (purple line, 200 V/div), drain current of the SiC MOSFET, M_2 (pink color, 10 A/div), drain current of the SiC MOSFET, M_6 (green color, 10 A/div), ripple current of the inductor, (yellow color, 50 A/div), (time-base 50 ns/div).

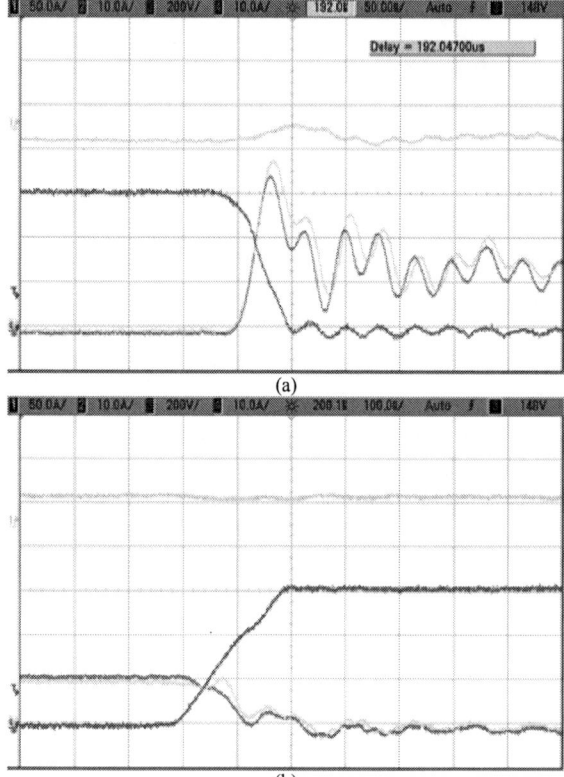

(a)

(b)

Fig. 9. Turn-ON (a) and turn-OFF (b) transient of the phase-leg test set up prototype. Measured drain-source voltage of the SiC MOSFETs, (purple line, 200 V/div), drain current of the SiC MOSFET, M_2 (pink color, 10 A/div), drain current of the SiC MOSFET, M_6 (green color, 10 A/div), ripple current of the inductor, (yellow color, 50 A/div), (time-base 50 ns/div).

During the third test the load resistor was removed, in order to reach higher currents. Duty cycle was set to 0.1 and using the air coil inductor as load itself, V_{DC} was increased until the output current reached 450 A, i.e. nominal output current. Fig. 9 shows the results on steady-state operation, where it can be noted that the converter is operating in continuous conduction mode as expected. The current sharing of modules No. 2, No. 4, No. 6 and No. 8 is observed. The steady state current is approximately 45 A per module. With this higher current, it is possible to notice the difference in the on-state resistances between the modules. Specifically, steady state current in module No. 8 is 40 A.

Fig. 10(a) shows the turn-ON transient. The current waveforms of the parallel-connected power modules are shown. The turn-on process takes approximately 50 ns. During this period the current sharing is even more uniform than the previous results. Fig. 9(b) shows the turn-OFF transient. The turn-OFF process takes approximately 50 ns. The turn-OFF current sharing is also more uniform than the previous cases. It is the hypothesis of the authors that the delay of the current in the module No. 2 could be due to several reasons, not investigated in this paper. The difference in the switching speeds of the drivers or the spread in the device parameters, such as stray inductance in the modules should be considered as possible reason. Moreover, a difference of response time among the Rogowski coils used in the measurements could also be a reason for this effect.

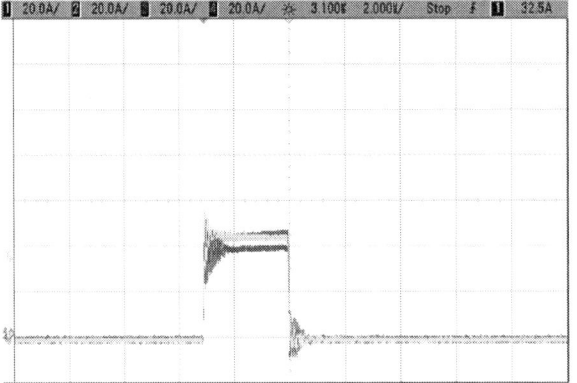

Fig 10. Steady-state operation of the phase-leg test set up prototype. Measured drain current of the SiC MOSFETs, M_8 (purple line, 20 A/div), drain current of the SiC MOSFET, M_2 (pink color, 20 A/div), drain current of the SiC MOSFET, M_6 (green color, 20 A/div), drain current of the SiC MOSFET, M_4 (yellow color, 20 A/div), (time-base 50 ns/div).

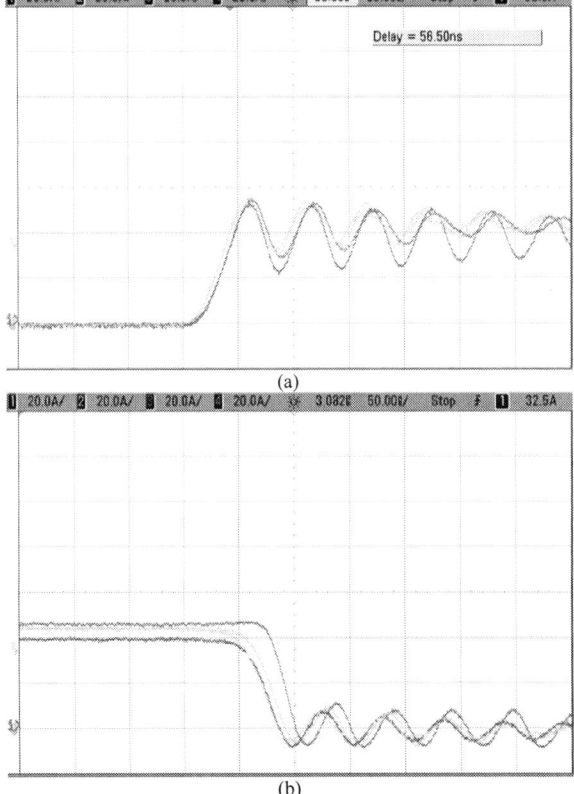

Fig. 11. Turn-ON (a) and turn-OFF (b) transient of the phase-leg test set up prototype. Measured drain current of the SiC MOSFETs, M_8 (purple line, 20 A/div), drain current of the SiC MOSFET, M_2 (pink color, 20 A/div), drain current of the SiC MOSFET, M_6 (green color, 20 A/div), drain current of the SiC MOSFET, M_4 (yellow color, 20 A/div), (time-base 50 ns/div).

C. Thermal Analysis

Considering the last experimental set-up with the

higher current and thermal analysis was performed. Due to physical constraints in the prototype, two thermal camera images were recorded from each side of the prototype (Fig. 12) in order to ensure that steady-state conditions were reached. The steady-state conditions were reached and Fig. 12 shows infrared images with the temperatures of the modules during the operation. Table II shows the average temperature in each module. The images were analyzed using FLIR Quick Report software, from the infrared camera. The average temperature of the modules is 38.73˚C, and the maximun difference of temperature between two modules is 1.5˚C, which is between modules No. 4 and No. 6. From these results, it is concluded that the loss distribution among the modules is sufficiently uniform.

TABLE II. TEMPERATURE ANALYSIS DURING DC-DC CONVERTER OPERATION

Average Temperature of the Modules			
Module 1	38.8 ˚C	Module 6	37.8 ˚C
Module 2	38.9 ˚C	Module 7	38.6 ˚C
Module 3	39.0 ˚C	Module 8	38.7 ˚C
Module 4	39.3 ˚C	Module 9	39.2 ˚C
Module 5	38.5 ˚C	Module 10	38.5 ˚C

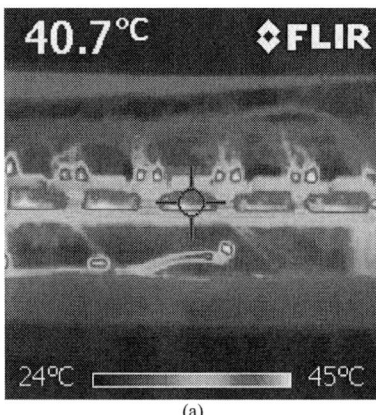

(a)

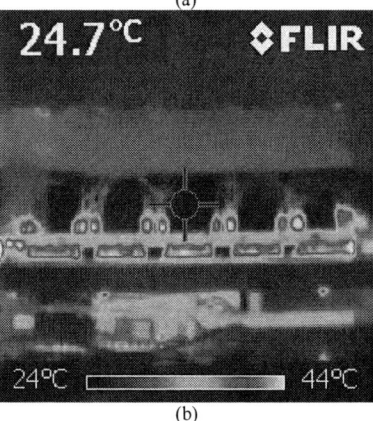

(b)

Fig. 12. Infrared camera picture of the dc-dc converter during operation at 10% of the rated power (a) right side and (b) left side.

D. Loss Estimation

From the measurements presented above, it is found that the switching losses are approximately 0.65 mJ per module (600 V DC and 9 A per module). Assuming a three-phase, two-level voltage source converter (VSC) rated at 312 kVA (700 V DC and 450 A RMS) an estimation of the total power losses, as well as, the efficiency has been performed. If an on-state resistance of 1.8 mΩ at 100° C and a switching frequency of 20 kHz are considered, the corresponding conduction and switching losses were found to be 972 W and 1.21 kW using Eqn. (1) and (2), respectively. The switching loss estimation was performed using MATLAB®.

$$P_{cd} = \left(I_{rms}\right)^2 R_{DSon} \times 3 \qquad (1)$$

$$P_{sw} = \frac{I_{RMS}}{I_{TEST}} \frac{V_{DC}}{V_{TEST}} \frac{f_{sw}(E_{on} + E_{off})}{\pi} \times 6 \qquad (2)$$

Moreover, the converter efficiency is found to be approximately 99.35% which has been calculated using Eqn. (3), at the rated power.

$$\eta = 100 \times \frac{P_n - (P_{sw} + P_{cd})}{P_n} \qquad (3)$$

V. CONCLUSIONS

The switching performance of ten parallel-connected power modules populated with SiC MOSFETs has been investigated. This investigation considered not only the transient current sharing, but also the differences in the steady-state current sharing of the power modules. A standard double-pulse setup was used and low-impedance gate drivers were employed for the experimental procedure. Electrical and thermal measurements have been performed operating the phase-leg as a dc-dc converter, revealing a sufficiently uniform distribution of the losses. Turn-ON and turn-OFF times of approximately 50 ns and 50 ns, respectively, were recorded, which results in an average switching energy of 0.65 mJ per module. This is a result of a combination of parallel connection, and high switching speed. Moreover, a closer investigation of the loss distribution reveals that the distribution of the switching losses is acceptably uniform. More uniform distribution of the currents among the modules was achieved at higher currents, during transients and steady-state operation. It is believed that the positive temperature coefficient of the on-state resistances of the MOSFETs will contribute towards a more uniform loss distribution under operation of the modules in a converter. Based on the recorded data, an efficiency of approximately 99.35% is expected for a three-phase VSC rated at 312 kVA.

REFERENCES

[1] B. Wrzecionko, J. Biela, and J.W. Kolar, "SiC power semiconductors in HEVs: Influence of junction temperature on power density, chip utilization and efficiency ", 35th Annual Conference of IEEE Industrial Electronics IECON '09, 2009, pp. 3834 – 3841.

[2] D. Bortis, B. Wrzecionko, and J.W. Kolar, "A 120°C ambient temperature forced air-cooled normally-off SiC JFET automotive inverter system," 2011 Twenty-Sixth Annual IEEE Applied Power

Electronics Conference and Exposition (APEC), 2011, pp.1282-1289.

[3] H. Zhang, L.M. Tolbert, and B. Ozpineci, "Impact of SiC Devices on Hybrid Electric and Plug-in Hybrid Electric Vehicles," IEEE Industry Applications Society Annual Meeting, IAS, 2008, pp.1-5.

[4] D. Peftitsis, G. Tolstoy, A. Antonopoulos, J. Rabkowski, J.-K. Lim, M. Bakowski, L. Ängquist, and H.-P. Nee, "High-Power Modular Multilevel Converters With SiC JFETs," IEEE Transactions on Power Electronics, vol. 27, no. 1, pp. 28-36, January, 2012.

[5] J. Rabkowski, D. Peftitsis, and H.-P. Nee, "Silicon carbide power transistors − A new era in power electronics is initiated," IEEE Industrial Electronics Magazine, vol. 6, no. 2, June 2012.

[6] J. Rabkowski, D. Peftitsis, and H.-P. Nee, "Design Steps Toward a 40-kVA SiC JFET Inverter With Natural-Convection Cooling and an Efficiency Exceeding 99.5%," IEEE Transactions on Industry Applications, vol.49, no.4, pp.1589,1598, July-Aug. 2013.

[7] T. Evans, T. Hanada, Y. Nakano, T. Nakamura, "Development of SiC power devices and modules for automotive motor drive use," Future of Electron Devices, Kansai (IMFEDK), 2013 IEEE International Meeting for , pp.116,117, 5-6 June 2013.

[8] Y. Hinata, M. Horio, Y. Ikeda, R. Yamada, Y. Takahashi, "Full SiC power module with advanced structure and its solar inverter application," Applied Power Electronics Conference and Exposition (APEC), 2013 Twenty-Eighth Annual IEEE , pp.604,607, 17-21 March 2013.

[9] M. Chinthavali, P. Ning, Y. Cui, and L. Tolbert; "Investigation on the Parallel Operation of Discrete SiC BJTs and JFETs", Twenty-Sixth Annual IEEE Applied Power Electronics Conference and Exposition (APEC), 2011.

[10] D. Peftitsis, R. Baburske, J. Rabkowski, J. Lutz, G. Tolstoy, H.-P. Nee, "Challenges Regarding Parallel Connection of SiC JFETs," IEEE Transactions on Power Electronics, vol.28, no.3, pp.1449,1463, March 2013.

[11] J. Rabkowski, D. Peftitsis, H.-P. Nee, "Parallel-operation of Discrete SiC BJTs in a 6kW/250kHz dc/dc Boost Converter," accepted for publication on IEEE Transactions on Power Electronics.

[12] J.-K. Lim, D. Peftitsis, J. Rabkowski, M. Bakowski and H.-P. Nee, "Modeling of the impact of parameter spread on the switching performance of parallel-connected SiC VJFETs", 9th European Conference on Silicon Carbide and Related Materials 2012, ECSCRM 2012, St. Petersburg, 2-6 September 2012.

[13] J. Colmenares, D. Peftitsis, J. Rabkowski, D. Sadik, H.-P. Nee, "Dual-Function Gate Driver for a Power Module with SiC Junction Field-Effect Transistors," accepted for publication on IEEE Transactions on Power Electronics.

[14] T. Funaki, M. Sasagawa, T. Nakamura, "Multi-chip SiC DMOSFET half-bridge power module for high temperature operation," Applied Power Electronics Conference and Exposition (APEC), 2012 Twenty-Seventh Annual IEEE , pp.2525,2529, 5-9 Feb. 2012.

[15] D. Urciuoli, R. Green, A. Lelis, D. Ibitayo, "Performance of a dual, 1200 V, 400 A, silicon-carbide power MOSFET module," Energy Conversion Congress and Exposition (ECCE), 2010 IEEE , pp.3303,3310, 12-16 Sept. 2010.

[16] M. Horio, Y. Iizuka, Y. Ikeda, E. Mochizuki, Y. Takahashi, "Ultra compact and high reliable SiC MOSFET power module with 200°C operating capability," Power Semiconductor Devices and ICs (ISPSD), 2012 24th International Symposium on , pp.81,84, 3-7 June 2012.

[17] R.A Wood, D.P. Urciuoli, T.E. Salem, R. Green, "Reverse conduction of a 100 A SiC DMOSFET module in high-power applications," Applied Power Electronics Conference and Exposition (APEC), 2010 Twenty-Fifth Annual IEEE , pp.1568,1571, 21-25 Feb. 2010.

[18] M.K. Das, C. Capell, D.E. Grider, R. Raju, M. Schutten, J. Nasadoski, S. Leslie, J. Ostop, A. Hefner, "10 kV, 120 A SiC half H-bridge power MOSFET modules suitable for high frequency, medium voltage applications," Energy Conversion Congress and Exposition (ECCE), 2011 IEEE , pp.2689,2692, 17-22 Sept. 2011.

The 2014 International Power Electronics Conference

Built-In Reliability Design of a High-Frequency SiC MOSFET Power Module

Jianfeng Li, Emre Gurpinar, Saul Lopez-Arevalo,
Alberto Castellazzi

Power Electronics, Machines and Control Group
The University of Nottingham
Nottingham, UK
alberto.castellazzi@nottingham.ac.uk

Liam Mills

Semelab
TT-Electronics
Lutterworth, UK
liam.mills@semelab-tt.com

Abstract— **A high frequency SiC MOSFET-based three-phase, 2-level power module has been designed, simulated, assembled and tested. The design followed a built-in reliability approach, involving extensive finite-element simulation based analysis of the electro-thermo-mechanical strain and stress affecting the switch during both manufacturing and operation: structural simulations were carried out to identify the materials, geometry and sizes of constituent parts which would maximize reliability. Following hardware development, functional tests were carried out, showing that the module is suitable for high switching frequency operation without impairing efficiency, thus enabling a considerable reduction of system-level size and weight.**

Keywords— *SiC MOSFET, multi-chip power modules, reliability.*

I. POWER MODULE DESIGN

A 1.2 kV SiC-MOSFET based three-phase half-bridge power module was designed, specifically for avionic applications, where improved gravimetric and volumetric power density and efficiency are highly appreciated.

The structure of one phase switch is shown in Fig. 1, with indication of assembly layers and location of interconnection pins. Each switch allocates two SiC MOSFETs, which are soldered on the substrate top metallization (Sub_Metal_Top in Fig. 1 b)) together with the power and drive terminals (S-Pins); the substrate bottom metallization is soldered onto the module base-plate. All the solder joints are made of eutectic tin-silver (Sn-3.5Ag) solder of constant 100 µm thickness. Because this module does not target specifically high-temperature applications, aluminum (Al) bond-wire were used for interconnections to the device top-side, with diameter of 125 µm.

Two types of substrates commonly used in the construction of power modules have been considered, namely Aluminum-Nitride (AlN) and Silicon-Nitride (Si_3N_4). Table I summarizes the main characteristics of interest for the two materials: Si_3N_4 has higher flexural strength than AlN and is thus typically processed in thinner layers than AlN for use in building power modules; this also implies that, although Si_3N_4 has a

lower thermal conductivity than AlN, the equivalent thermal resistance of the substrates is comparable if the difference in thickness is taken into account. Here, both experiment and simulations are used to inform the choice on most reliable solution.

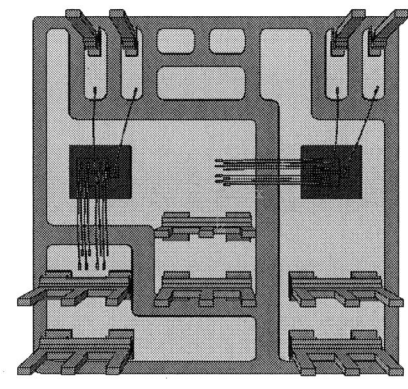

a)

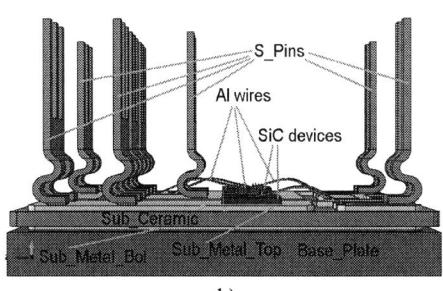

b)

Fig. 1. One-phase SiC switch: a) top view, b) side view.

TABLE I
THERMO-MECHANICAL CHARACTERISTICS OF SUBSTRATE CERAMICS

	Thermal Conductivity (W/m·k)	Flexural Strength (MPa)	Fracture Toughness (MPam1/2)	CTE (ppm/degree-C)
AlN	170	400	2.7	4.5
Si_3N_4	60	850	5.0	2.7

Preliminary experimental testing was carried out on batches of the two types of assemblies, with different thicknesses of copper tracks deposited onto them. The tiles were subjected to passive thermal cycling and Cu tracks detachment/peel-off was used as an indication of failure. The results are summarized in Table II and give a clear indication that Si_3N_4 tiles can withstand a much higher number of cycles before degradation and failure intervenes. Also, it is interesting to note that the thicker the Cu tracks, the shorter the number of cycles to failure.

TABLE II
SIMULATION CASES WITH CONSIDERATION OF SUBSTRATES AND BUMP SHAPES

		Cu Plate Thickness			
		0.2t	0.3t	0.4t	0.5t
Ceramic Substrate Material	AlN 0.635t	300	200	Failure	Failure
	Si3N4 0.32t	>3000	>3000	>3000	>3000

Condition: -40deg.C (30 min.) to 125deg.C (30 min.)
Failure: Cu plate detached after brazing

Finite element (FE) modeling and simulation are employed to investigate the effects of the type and the thickness of substrate and base plate used on the entity of the maximum thermo-mechanical stress and its distribution. Simulations target the analysis of the thermal and thermo-mechanical performance of the assembled power module. The considered simulation cases are listed in Table III.

TABLE III
SIMULATION CASES WITH CONSIDERATION OF SUBSTRATES AND BUMP SHAPES

No	Sub Ceramic Thickness	Sub Metal Bot Thickness	Sub Metal Top Thickness	Base Plate Thickness
A1	1 mm AlN	0.3 mm Cu	0.4 mm Cu	3 mm AlSiC-9
A2	1 mm AlN	0.3 mm Cu	0.4 mm Cu	3 mm AlSiC-12
A3	1 mm AlN	0.3 mm Cu	0.4 mm Cu	2 mm AlSiC-9
A4	0.635 mm AlN	0.3 mm Cu	0.4 mm Cu	2 mm AlSiC-9
A5	0.635 mm AlN	0.2 mm Cu	0.3 mm Cu	2 mm AlSiC-9
A6	0.635 mm AlN	0.2 mm Cu	0.3 mm Cu	2 mm AlSiC-9
B1	0.3 mm Si_3N_4	0.3 mm Cu	0.4 mm Cu	3 mm AlSiC-9
B2	0.3 mm Si_3N_4	0.3 mm Cu	0.4 mm Cu	3 mm AlSiC-12
B3	0.3 mm Si_3N_4	0.3 mm Cu	0.4 mm Cu	2 mm AlSiC-9
B4	0.3 mm Si_3N_4	0.2 mm Cu	0.3 mm Cu	3 mm AlSiC-9
B5	0.635 mm Si_3N_4	0.3 mm Cu	0.4 mm Cu	2 mm AlSiC-9
B6	0.635 mm Si_3N_4	0.2 mm Cu	0.3 mm Cu	2 mm AlSiC-9
B7	0.635 mm Si_3N_4	0.3 mm Cu	0.3 mm Cu	2 mm AlSiC-9

The first one is 0.4mm and 0.3mm thick active brazed Cu on both sides of 1 mm or 0.635mm thick AlN tile. The second one is 0.4 mm and 0.3mm thick or 0.3mm and 0.2mm active brazed Cu on both sides of 0.3mm or 0.635mm thick Si_3N_4 tile, respectively. Also, two types

of base plate materials have been considered: AlSiC-9 and AlSiC-12 composites. For each of the two composites, two thicknesses of 3 mm and 2 mm for the Base_Plate were selected and compared in order to reduce the stress/strain accumulation within the solder joints.

The characteristics of the material used for the solder layer, Lead (Pb)-Tin (Sn)-Silver (Ag) solder of the composition 62Pb36Sn2Ag, are summarized in Table IV.

TABLE IV
SOLDER PROPERTIES

Liquidus Temperature (°C)	179
Density (gm/cm3)	8.41
Electrical Conductivity (1.72μΩ-cm (% of IACS))	11.9
Thermal Conductivity @85°C (W/cm-°C)	0.5
Thermal Coefficient of Expansion @20°C (PPM/°C)	27
Tensile Strength (PSI)	7000
Shear Strength (PSI)	7540
Specific Heat Capacity (J/g-°C)	0.167

Modeling and simulation were carried out using well-established commercially available tools (e.g., Abaqus, Ansys). Fig. 2 a) shows the numerical structural model (meshed) of the switch: the largest brick element is 0.5×0.5×1.5mm, and the smallest is 0.5×0.25×0.025mm. Here, the largest brick element is 0.5×0.5×1.5mm, and the smallest is 0.5×0.25×0.025mm. S4R shell elements of 0.5×0.5mm or 0.5×0.25mm in size were also used to discretize the nickel-phosphorous (NiP) finishing on the surfaces of the substrates, and in both the Al and Ni metallization on two sides of the MOSFETs respectively.

The assembly was first subjected to a predefined temperature profile of 221°C down to 25°C for 3 min to simulate the stress and strain developed during the reflow process. In this stage, all the Al wire bonds were deactivated and therefore no strain/stress was developed on them. Power losses of one MOSFET are shown in Fig. 2 b). These losses were considered as heating sources to simulate the thermal performance of the assembly during a realistic mission profile. They were also taken in account in the heat exchange boundary condition. For this boundary condition, a heat exchange coefficient of 5000W m^{-2} K^{-1} was applied to the bottom surface of the base plate, corresponding to typical heat-sink cooling conditions in our lab experiments. The temperature field obtained from the thermal simulation was used as input to simulate the development of stress/strain in the assembly during two cycles of the mission profile. Thermal and mechanical properties of all the materials used in the thermo-mechanical simulation were taken from either references [1-3] or the datasheets of the materials commercially available. Chaboche's plastic model was

The 2014 International Power Electronics Conference

used to describe the mechanical properties of both the Cu and Al. The properties of Al were applied to both the Al wire bonds and the Al metallization on the front side of the MOSFETs. Anand's creep model was used to describe the mechanical properties of the Sn-3.5Ag solder alloy.

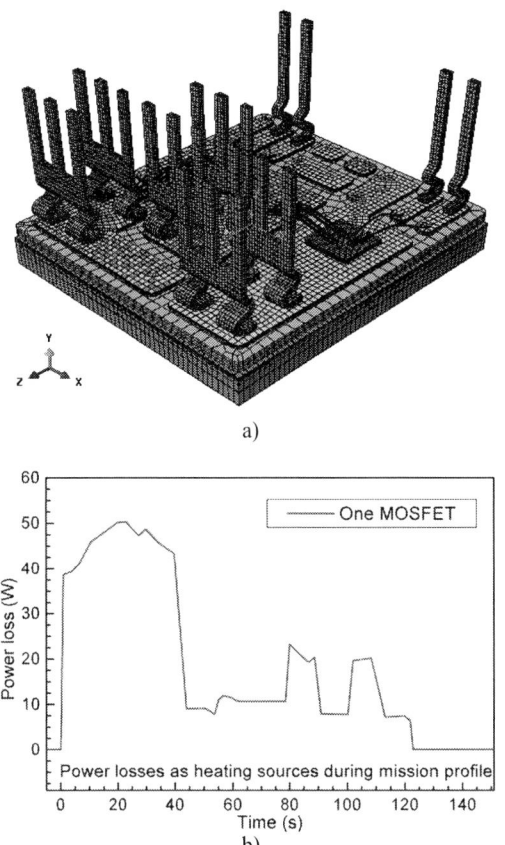

Fig.2. a) Representative meshing system to discretize the designed switch module. b) Power loss of one MOSFET during a realistic mission profile.

II. THERMO-MECHANICAL SIMULATION

For all simulation cases during the mission profile, the highest junction temperature was observed on one MOSFET at 23.01s. The simulation shown is mainly concerned with the stress and strain developed in all solder joints. Simulation results reveal that the maximum von Mises stress in the solder joint between the substrate and the base plate are larger than those in the two solder joints used for attaching the two SiC MOSFETs on the substrate. The maximum creep strain accumulation in the former solder joint is significantly larger than that in the latter two solder joints after the first cycle of the mission profile. However, and depending on the simulation cases, the maximum creep strain accumulation in the former solder joint may be higher or lower than that in the latter two solder joints after the second cycle of the mission profile.

Fig. 3 shows the von Mises stresses and creep strains obtained from simulations in all the solder joints for case A1 after two cycles of the mission profiles. Table V compares the simulation cases for the highest junction temperatures of the MOSFETs and the maximum von Mises stress and creep strain developed in the solder joint between the substrate and the base plate. In Table V, $T_{J,max}$ is for the highest junction temperature of the MOSFETs during the mission profile; S_{max0}, S_{max1} and S_{max2} represent the maximum von Mises stress and ε_{max0}, ε_{max1} and ε_{max2} represent the maximum creep strain in the solder joint of the as-reflowed assembly and the same one after one and two cycles of the mission profile; and $\Delta\varepsilon_{max1}=\varepsilon_{max1}-\varepsilon_{max0}$, $\Delta\varepsilon_{max2}=\varepsilon_{max2}-\varepsilon_{max1}$.

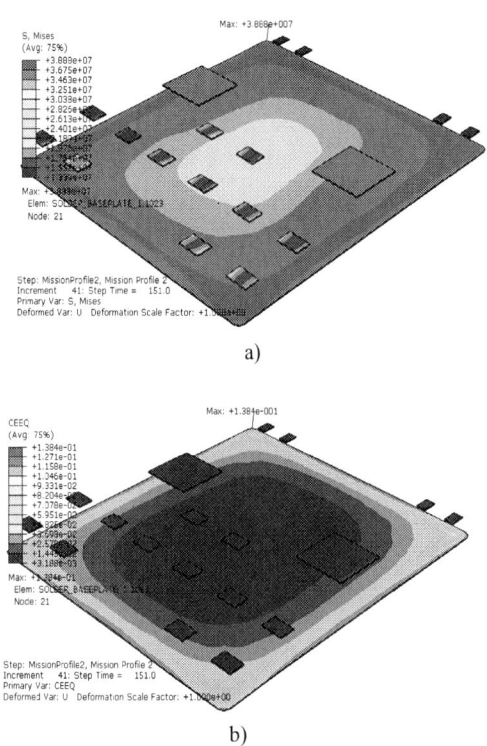

Fig. 3. Simulation results of (a) von Mises stresses and (b) creep strains within all solder joints for simulation case A1 after two cycles of the mission profile.

According to these results, the best design is achieved when the base-plate is 2mm thick AlSiC-9 and the substrate is AlN-based with 0.635mm thick AlN tile (case A4). In this case, the highest junction temperature of the MOSFET is only 0.6°C higher than that in simulation case A1; however, the maximum creep strains and creep strain accumulations dominating the reliability of the critical solder joint in the module would be significantly lower than those in the other simulation cases.

In view of both experimental and simulation results, however, and more particularly due to the relatively small difference in the stress and strain values for the

978-1-4799-2706-7/14 $31.00 © 2014 IEEE   3720

TABLE V

COMPARISON OF THE HIGHEST JUNCTION TEMPERATURES OF THE MOSFETs AND THE MAXIMUM VON MISES STRESS, CREEP STRAIN AND CREEP STRAIN ACCUMULATION IN THE SOLDER JOINT BETWEEN THE SUBSTRATE AND BASE PLATE FOR THE 13 SIMULATION CASES

Case	$T_{J,MAX}$ (K)	S_{max0} (MPa)	S_{max1} (MPa)	S_{MAX2} (MPa)	ε_{MAX0} (%)	ε_{MAX1} (%)	ε_{MAX2} (%)	$\Delta\varepsilon_{MAX1}$ (%)	$\Delta\varepsilon_{MAX2}$ (%)
A1	363.2	42.6	39.5	38.9	12.65	13.40	13.84	0.75	0.44
A2	363.9	44.2	41.6	40.7	18.84	19.67	20.17	0.83	0.50
A3	363.8	41.0	38.2	37.6	6.87	7.25	7.52	0.37	0.27
A4	363.8	41.3	37.9	37.4	6.83	7.10	7.23	0.27	0.13
A5	365.6	41.3	38.6	38.1	7.65	8.13	8.41	0.48	0.28
A6	364.8	40.8	37.9	37.4	7.07	7.35	7.49	0.28	0.14
B1	366.4	42.6	39.1	38.4	10.96	11.80	12.26	0.84	0.46
B2	367.4	44.2	40.5	39.9	16.79	17.66	18.12	0.87	0.46
B3	367.6	41.3	38.1	37.6	7.32	7.70	8.01	0.38	0.31
B4	368.7	43.0	39.6	41.2	12.10	12.88	13.32	0.78	0.44
B5	370.6	42.5	39.0	38.4	8.98	9.61	9.98	0.63	0.37
B6	373.4	42.4	39.5	39.0	10.25	10.88	11.27	0.63	0.39
B7	372.6	42.5	39.0	38.4	9.17	9.81	10.19	0.64	0.38

simulation results using AlN and Si_3N_4 (e.g., case A1 in comparison with case B3), Si3N4 substrates were retained as the best solution for built-in reliability module development.

III. HARDWARE DEVELOPMENT AND TEST

A. Development

Fig. 4 a) shows details of the assembled switches and module. The module of Fig. 4 b) was later filled with a silicone-based dielectric material to provide the required isolation and insulation. Finally it was fully encapsulated by using a plastic lid leading to the final product of Fig. 4 c). The final module as assembled for both functional and reliability testing weighs 126.5g ±0.5g. In view of the specified goal of lightweight development, this is a very satisfactory result, competitive with presently upcoming commercial solutions. A noteworthy feature of the module is the complete separation of the power and driving terminals and current loops. All power terminals were finally implemented with single connectors, doubled to reduce parasitic inductance and enable high switching frequency.

The module has been subjected to both functional testing in inverter operation and to temperature cycles for technology validation. Here, the results of functional testing are reported to demonstrate the high frequency capability and good thermal performance. Reliability tests are still ongoing at the time of submission.

B. Test

A flexible test setup was designed to test a number of modules that have been manufactured. The setup comprises of four electrolytic capacitors (350V, 3.3mF each) with discharge and balancing resistors (47kΩ each), and the SiC module power pins are mounted directly onto the power plane. The schematic is shown in Fig. 5. Fig. 6 shows a picture of the test setup.

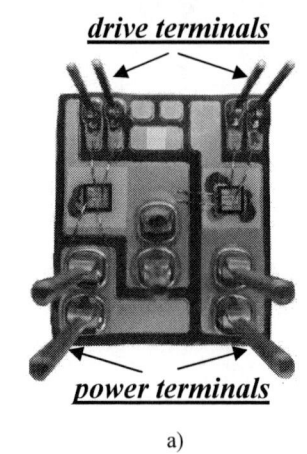

a)

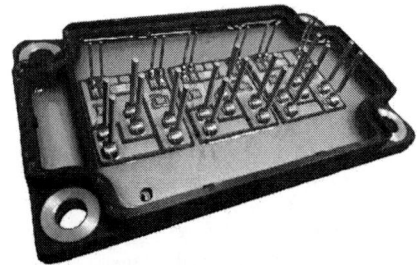

b)

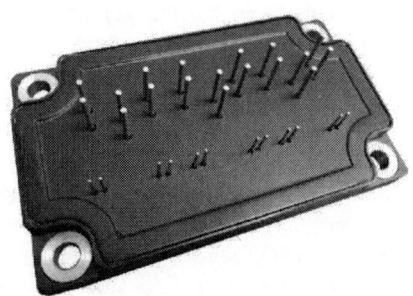

c)

Fig. 4. Assembled switches a), full power module b) and encapsulated device c).

The 2014 International Power Electronics Conference

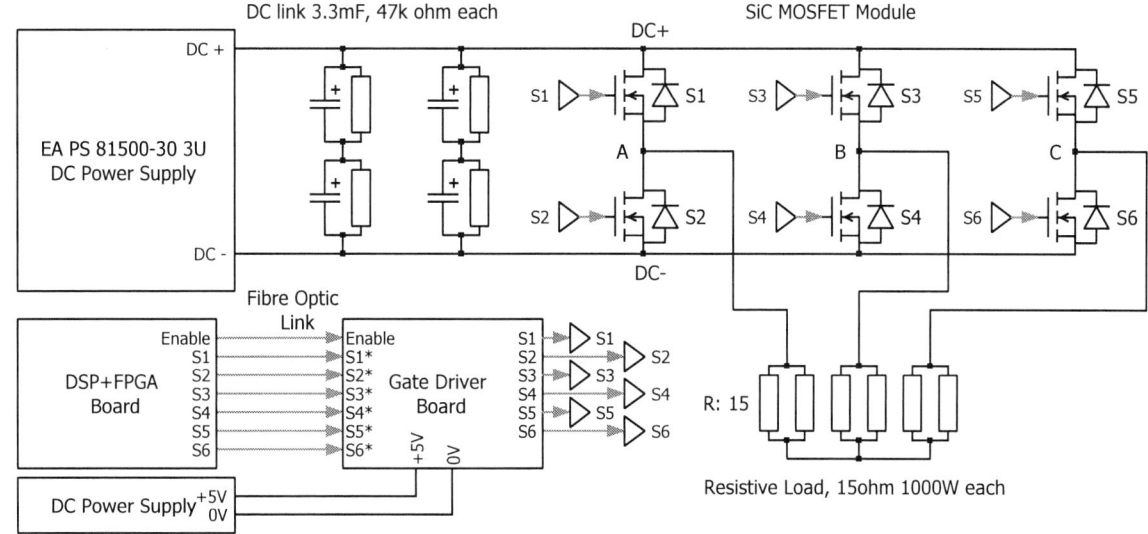

Fig. 5. Schematic of the test circuit.

Fig. 6. Photograph of the test set-up.

The inverter was tested in open loop; the necessary PWM drive signals are generated by a DSP and FPGA control platform with programmable dead-time values. The PWM signals are transmitted to a gate driver board, which is mounted directly onto the SiC module gate pins, through fiber optic cables in order to minimize the noise interference at high switching

frequencies. An isolated gate drive IC supplies +22V to turn-on and -3.1V to turn-off each individual SiC MOSFET. The input DC link voltage is 540V to represent the DC bus in avionic systems. The inverter was tested at full load (6kW) and half load (3kW) conditions. The output inductive filter was 4mH. Output phase-to-neutral RMS voltage is set to 115V with 50Hz fundamental frequency. Tests were run at increasing switching frequencies: 5, 10, 20, 40, 60, 80 and 100 kHz. Fig. 7 shows the three phase current components for a switching frequency of 5 kHz.

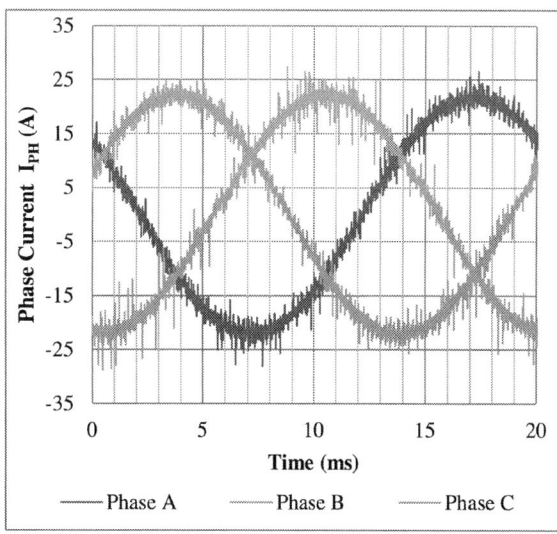

Fig. 7. Phase current waveforms at a switching frequency of 5 kHz.

The output phase voltage and current, and the drain-source (V_{ds}) and gate-source voltage (V_{gs}) waveforms of each SiC MOSFET were also captured. The turn-on and turn-off waveforms of the SiC MOSFET at full load and 100 kHz switching frequency are shown in Fig. 8 a) and b), respectively. These waveforms show that the turn-on and turn-off times of the power switches are less than 60ns at full load.

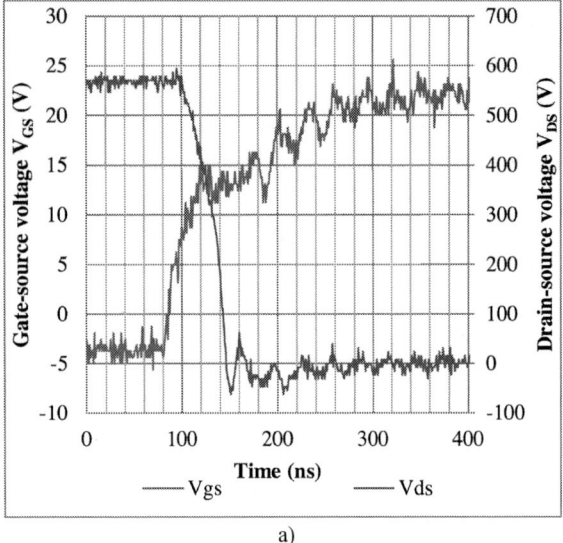

a)

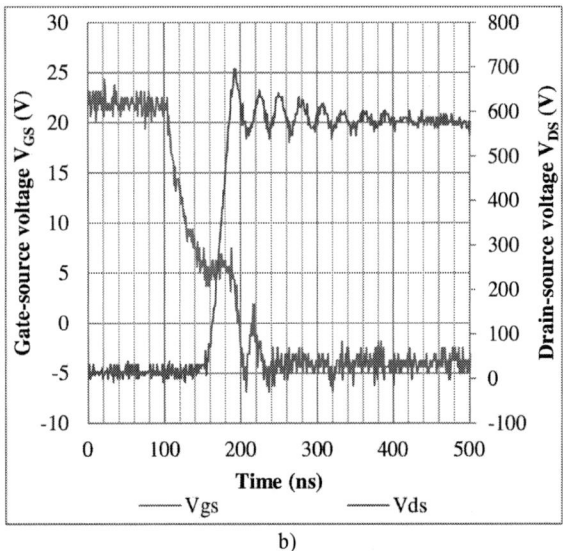

b)

Fig. 8. Turn-on, a), and turn-off waveforms, b), at f_{sw}=100 kHz, full-load.

In a final test, we removed altogether the inductive filter from the load. Fig. 9 shows the output current waveforms of all three phases at full load and 100 kHz switching frequency in this condition. Although the load at the output of the converter is purely resistive, due to leakage inductance at the output cables and resistors, and high switching frequency, the output phase currents become continuous. Clearly, the ripple is not contained, but these measurements give a very clear indication of the dramatic reduction in inverter volume and size which can be enabled by the high-frequency SiC power module.

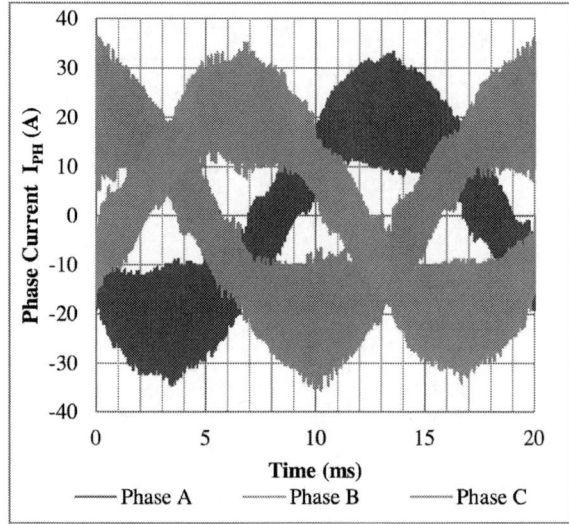

Fig. 9. Output current waveforms at f_s=100 kHz, full-load.

Finally, Fig. 10 reports the measured efficiency at full load as a function of the switching frequency. A decrease of 2.5% (i.e., 150 W) is measured for a twentyfold increase in the switching frequency (from 5 to 100 kHz), which is directly attributed to the SiC device characteristics [4].

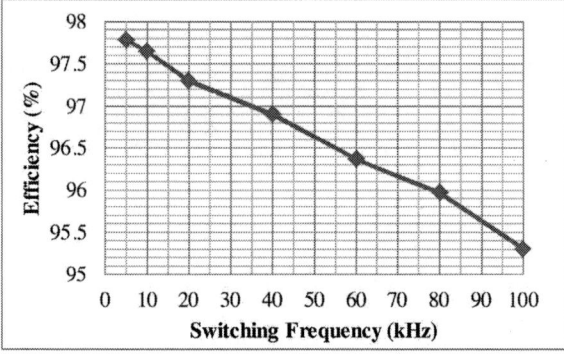

Fig. 10. Efficiency curve of power module at 6 kW output power.

To complete experimental characterization, the SiC power module was dismounted from the heat-sink and the temperature at the baseplate was measured under different operational conditions. To avoid failure, the maximum allowed baseplate temperature was around 90 °C. This corresponded to the following test conditions: P_{OUT}=2.5 kW at f_s=5kHz and at f_s=40 and P_{OUT}=1.5kW 80 kHz. The baseplate temperature distribution corresponding to these test conditions is given in Fig. 11 a), b) and c). All results highlight a very even temperature distribution, with maximum at the center in line with the expectations, and the capability of the module to work properly and efficiently with noticeable base-plate temperatures, that is, minimum cooling requirements, at high switching frequencies and power levels.

IV. CONCLUSION

In this paper, the design, the thermo-mechanical modeling using finite element analysis and, simulation and experimental results of a lightweight SiC MOSFET module are presented. Simulation results using finite element suggests that the best design is obtained when the Base_Plate is based on AlSiC-9 with a thickness of 2mm and the substrate is AlN-based with 0.635mm thick AlN tile. The module has been tested successfully at full and half load operating conditions using a switching frequency of up to 100 kHz. Experimental results show that the designed power module successfully operates in a wide switching frequency range with high-efficiency at full-load and half-load operating conditions, making it ideal for all applications which prize reduction of system size and weight (e.g., avionic, automotive). In view of the dramatic reduction in output filter size that Such frequency increase enable, the developed SiC power module must be regarded as a solution enabling optimum trade-off choices in inverter development.

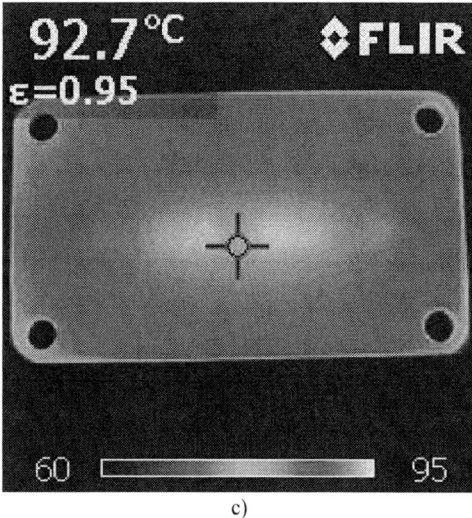

c)

Fig. 11. Infrared thermal mapping of the module baseplate under different operational conditions: P_{OUT}=2.5 kW and f_s=5kHz, a); P_{OUT}=1.5kW at f_s=40 kHz, b); P_{OUT}=1.5kW at f_s=80 kHz, c).

REFERENCES

[1] A. Zeanh et al., "Thermo mechanical modelling and reliability study of an IGBT module for an aeronautical application." *In Proc. 9th Int. Conf. on Thermal, Mechanical and Multiphysics Simulation and Experiments in Micro-Electronics and Micro-Systems*, EuroSimE 2008.

[2] A. Zeanh et al., "Lifetime and reliability assessment of AlN substrates used in harsh aeronautic environments power switch modules", *Advanced Materials Research* vol. 112, pp 113-127, 2010.

[3] W.W. Sheng, R.P. Colino, "Power Electronic Modules: Design and Manufacture", CRC Press LLC, Florida, USA, 2005.

[4] Hui Zhang; Tolbert, L.M.; Jung Hee Han; Chinthavali, M.S.; Barlow, F., "18 kW three phase inverter system using hermetically sealed SiC phase-leg power modules," *Applied Power Electronics Conference and Exposition (APEC), 2010 Twenty-Fifth Annual IEEE*, pp.1108,1112, 21-25 Feb. 2010.

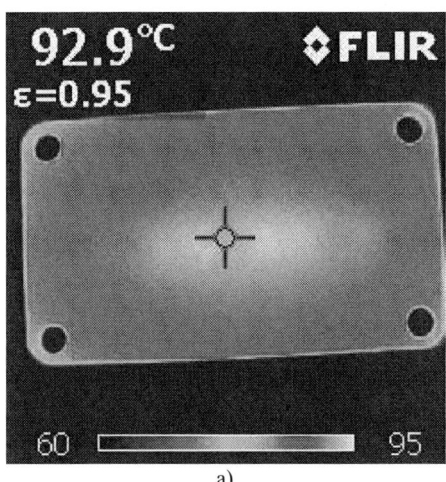

a)

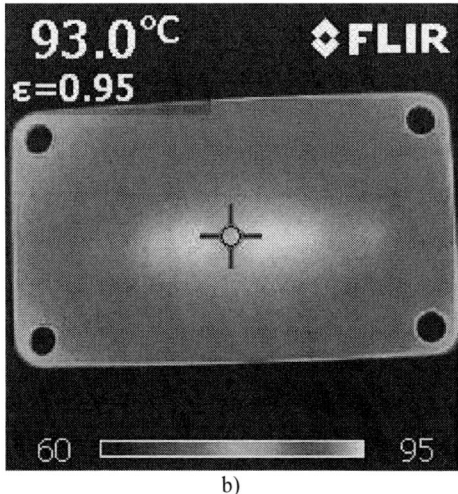

b)

The 2014 International Power Electronics Conference

978-1-4799-2706-7/14 $31.00 © 2014 IEEE

The 2014 International Power Electronics Conference

Experimental Switching Frequency Limits of 15 kV SiC N-IGBT Module

Arun Kadavelugu and Subhashish Bhattacharya
NSF FREEDM Systems Center
Department of Electrical and Computer Engineering
North Carolina State University
Raleigh, USA

Sei-Hyung Ryu, Edward Van Brunt and Dave Grider
Cree, Inc.
Durham, USA

Scott Leslie
Powerex, Inc.
Youngwood, USA

Abstract— **This paper presents extensive experimental switching characteristics of a state-of-the-art 15 kV SiC N-IGBT (0.32 cm^2 active area) up to 10 kV, 10 A and 175°C. The influence of the thermal resistance of the module package, cooling mechanism, and the increased energy loss with temperature are investigated for determining the switching frequency limits of the IGBT. Detailed FEM analysis is conducted for extracting the thermal resistance of each layer in the 15 kV module from the IGBT junction to the base plate, and then down to the ambient. Using this thermal information and the experimental switching data, the inductive switching frequency limits are analytically evaluated for liquid and air cooling cases with 660 W/cm^2 and 550 W/cm^2 power dissipation densities respectively, considering 150°C as maximum junction temperature. The air cooling power dissipation density of the 15 kV IGBT is experimentally validated using a dc-dc boost converter at 10 kV, 6.4 kW output and 550 W/cm^2 under steady state operating conditions. The gate resistances used for the entire experiments are $R_{G(ON)} = 20\ \Omega$ and $R_{G(OFF)} = 10\ \Omega$.**

Keywords— 15 kV SiC IGBT, hard-switching capability, thermal resisistance, turn-off energy loss with temperature

I. INTRODUCTION

The high power conversion efficiency coupled with possibility of using simple two-level voltage source converter topologies at medium voltage level make the 15 kV SiC IGBTs promising for smart-grid power conversion systems and medium voltage drives [1], [2]. It is shown in [3] - [5], that the field-stop buffer layer thickness and doping concentration of the SiC IGBT can be designed for either better conduction or switching characteristics. In the present study, a 5 μm buffer layer thick 15 kV N-IGBT module, shown in Fig. 1, has been experimentally evaluated for its hard-switching frequency capability. The present work also supplements the simulation studies on 15 kV IGBT frequency capability reported in [6].

The 15 kV SiC N-IGBT has active area of 0.32 cm^2. The high power handling requirement of the 15 kV IGBTs with such small chip size result in high power dissipation densities even at moderate switching frequencies. The rise in switching energy loss with temperature due to increased injection [3], [7] adds

significant thermal stress on the IGBT. This increased energy loss coupled with high junction temperature enforced by the package thermal resistance increases power dissipation at the IGBT junction and could drive the IGBT into thermal runaway. Therefore, thermal resistances offered by different layers of the IGBT module in the thermal path between the device junction and heat sink are evaluated. The following sections present switching results, package thermal parameters and the hard switching frequency capability (thermal limits) of the 15 kV SiC N-IGBT.

II. EXPERIMENTAL ENERGY LOSS EVALUATION

The forward conduction characteristics of the 15 kV, 5 μm field-stop buffer layer, SiC N-IGBT (average data of sixteen IGBT modules) are shown in Fig. 2. These results are presented in [7], but are provided again in this paper for complete reference. The IGBT conduction drop is about 4.6 V, 5.6 and 7.1 V at 5 A, 10 A and 20 A respectively. The leakage current at 150°C is found to be less than 150 μA at 12 kV. The switching losses evaluated under purely inductive, partly inductive and resistive load conditions using double pulse test circuit at 10 kV and temperature up to 175°C are presented in detail in this section. The energy loss has been evaluated for different load conditions to determine the IGBT behavior under all the practical applications.

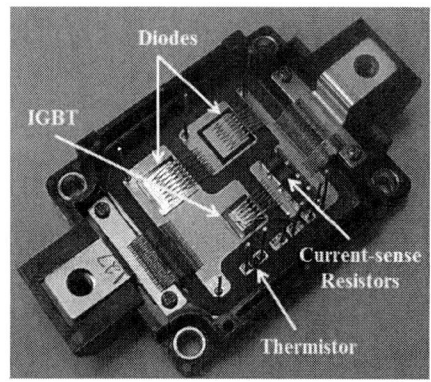

Fig. 1: The 15 kV, SiC N-IGBT Co-pack Module.
(IGBT designed and built by Cree; Packaged by Powerex)

The 2014 International Power Electronics Conference

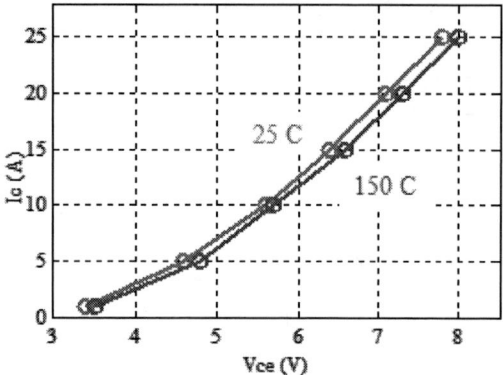

Fig. 2: Forward characteristics of 15 kV, 5 μm buffer layer IGBT at 25°C and 150°C (V_{GE} = 20 V).

A. Clamped-Inductive Load Switching Conditions

The clamped inductive (or fully hard-switched) characteristics are most widely provided data for power semiconductor devices to cover majority of applications. The freewheeling diode used in the switching tests is 20 kV JBS diode module (2x10 kV, 10 A diodes in series). The load is a single-layer, 8 mH inductor (two 4 mH inductors in series) with ultralow (< 10 pF) inter-winding capacitance and saturation current of 30 A. Test results are shown at 10 kV with varying current and temperature.

Fig. 3 shows turn-on switching loss (E_{ON}) and turn-off switching loss (E_{OFF}) at 25°C, at 10 kV, 2 A to 10 A. Unlike the turn-off loss, the turn-on loss is strongly dependent on the current. Fig. 4 shows the turn-off transitions at 10 kV, 5 A, and 10 A at 25°C. The duration of the 10 A transition is about 60 % of that of the 5 A transition, resulting in weaker dependency of energy loss with current. The temperature dependency of the turn-off loss is shown in Fig. 5. The energy loss is increased to a factor of three from 25°C to 175°C, due to increased injection at higher temperature resulting in significantly larger amounts of charge in the drift region. As shown in Fig. 6, the larger amount of charge to be removed at higher temperature is slowing down the voltage rise (before punch through) which in consequence is resulting in higher energy loss.

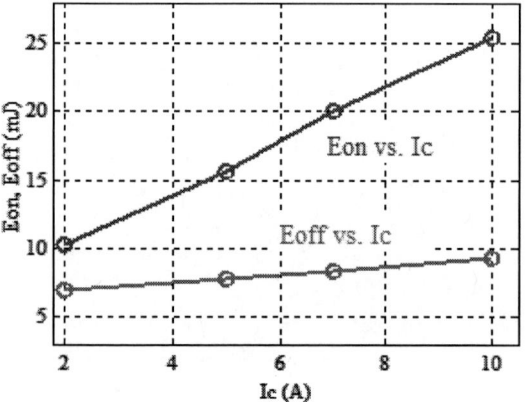

Fig. 3: Turn-on and Turn-off energy loss values with current variation at 10 kV and 25°C, under clamped inductive load.

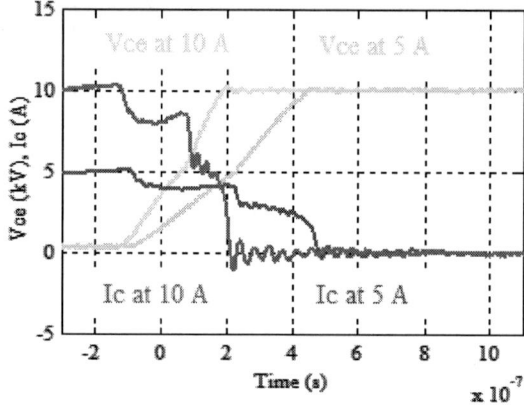

Fig. 4: Turn-off switching transitions at 10 kV, 25°C, 5 A and 10 A, under clamped inductive load conditions.

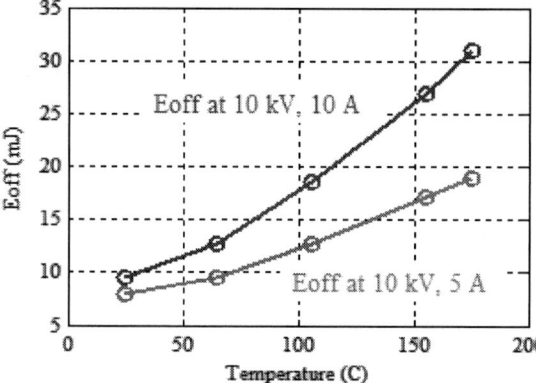

Fig. 5: Turn-off switching loss variation with temperature at 5 A and 10 A at 10 kV, under clamped inductive load conditions.

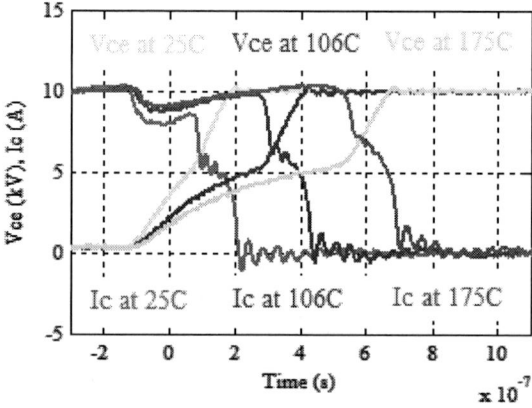

Fig. 6: Turn-off switching transitions with temperature at 10 kV and 10 A, under clamped inductive load conditions.

The turn-off loss at different temperature values is summarized in Table I. The hot plate was allowed to reach steady state for an hour before characterizing the switching transient at a particular temperature. The difference in reading between the hot plate display and the thermistor mounted in the IGBT module (shown in Fig. 1) are provided in the table. The thermistor, rated for

150°C, temperature values are taken as reference for the entire study (lower values) in this paper. As shown in [6], the turn-on energy loss slightly decreases with increase in temperature. This is due to increase in the IGBT gain resulting from increase in MOS channel mobility at high temperature. During the experiments, the high di/dt during turn-on had detrimental effect on the hot plate operation which prohibited turn-on characterization at high temperature. Therefore, the decrease in turn-on loss with temperature is neglected in this paper, providing the most pessimistic (lowest) frequency values for the 15 kV SiC IGBT. This, however, is not affecting the accurate determination of safe power dissipation density (W/cm^2) limits, as explained in the next sections.

B. Partly Inductive Load Switching Conditions

A 1 kΩ resistance in series with the 8 mH inductor is used as the load. Based on the turn-on and turn-off transitions shown in Fig. 7 and Fig. 8 respectively, at 10 kV and 5 A, the RL load is behaving almost as the purely inductive load. The reason for this behavior is the current stiff nature of the RL load (like the L load) determining the sequence of voltage and current transitions. However, the turn-off transition is slightly slower due to decrease in load current because of the inductor energy dissipation in the 1 kΩ resistance during the switching transition.

TABLE I: TURN-OFF ENERGY LOSS VALUES AT 10 KV, 5 A AND 10 A AT DIFFERENT TEMPERATURE UNDER PURELY INDUCTIVE LOAD

Hot-plate Setting	Thermistor Reading	E_{OFF} at 10 kV, 5 A	E_{OFF} at 10 kV, 10 A
25°C	26°C (Measured)	8.0 mJ	9.6 mJ
75°C	65°C (Measured)	9.6 mJ	12.7 mJ
125°C	106°C (Measured)	12.7 mJ	18.6 mJ
175°C	155°C (Measured)	17.2 mJ	27.0 mJ
195°C	175°C (Estimated)	19.0 mJ	31.0 mJ

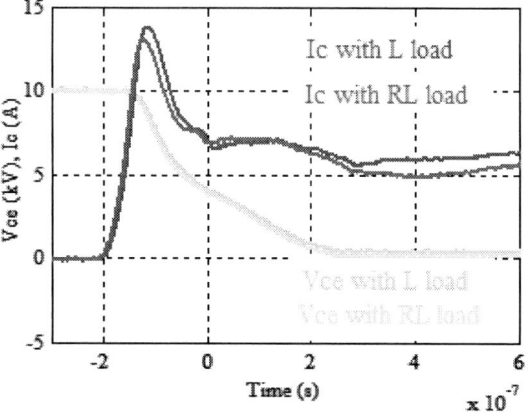

Fig. 7: Comparison of turn-on transitions with L and RL loads at 10 kV and 5 A, at 25°C.

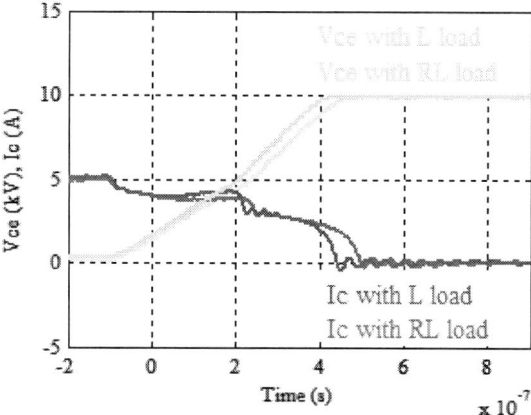

Fig. 8: Comparison of turn-off transitions with L and RL loads at 10 kV and 5 A, at 25°C.

C. Resistive Load Switching Conditions

This switching characterization is done with 2 kΩ wire-wound resistor (with stray inductance of 0.3 mH) for 5 A and 10 kV. The tests at other load currents are conducted by appropriately scaling the load resistance values for the desired current value. Unlike the case for RL load, the R load presents significantly different turn-on and turn-off characteristics in comparison to the purely L load characteristics. The current spike is significantly reduced due to zero current turn-on, and better damping of the LC resonance. The resistive turn-on and turn-off transitions at 10 kV, 5 A and 25°C are compared with corresponding inductive characteristics in Fig. 9 and Fig. 10 respectively. The turn-on and turn-off energy loss comparison with the different loads is provided in Fig. 11 and Fig. 12 respectively. The L and RL loads have almost same amount of energy loss during turn-on and turn-off transitions. The zero current turn-on, effect of LC resonant current spike, and the actual load current rise only after the IGBT voltage dropped to zero resulting in same energy loss at different currents with the R load. The energy dissipation (during turn-off transition) in the R load reduces IGBT current and thus turn-off loss, in comparison to that with L or RL load.

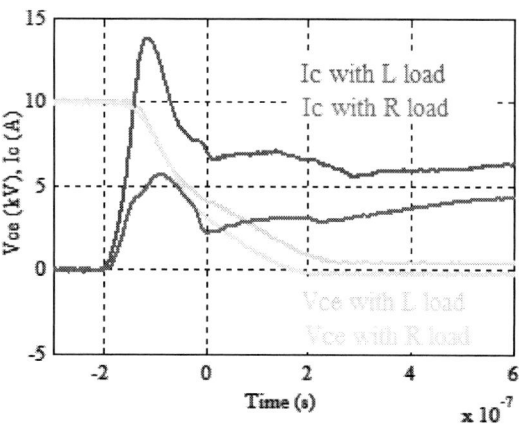

Fig. 9: Comparison of turn-on transitions with L and R loads at 10 kV and 5 A, at 25°C.

The resistive load does not find wide real-time application. However, the characteristics are provided to show behavior of the IGBT under all possible loads. The heating applications where pure R load conditions are seen, are realized by using a series capacitor (for resonance) to cancel out the impedance offered by the stray inductance of the coils. It should be noted that the characteristics shown above, for all the load conditions, differ from the ideal transitions [8], due to capacitance of the IGBT (C_{CE}) and free-wheeling diode.

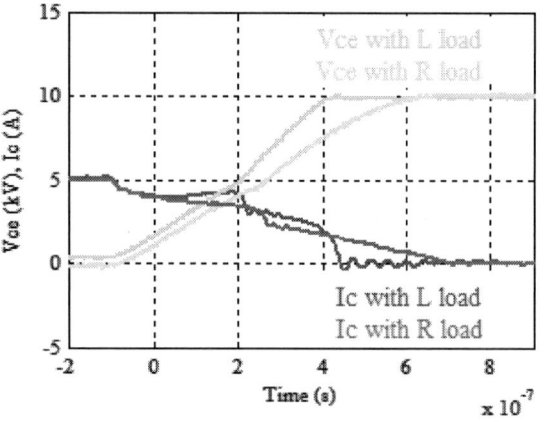

Fig. 10: Comparison of turn-off transitions with L and R loads at 10 kV and 5 A, at 25°C.

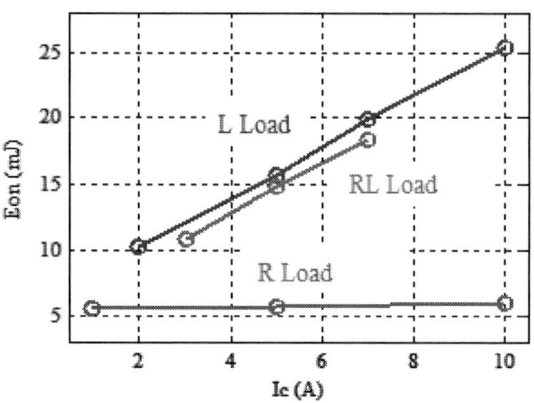

Fig. 11: Comparison of turn-on energy loss with L, RL and R loads at 10 kV and 25°C.

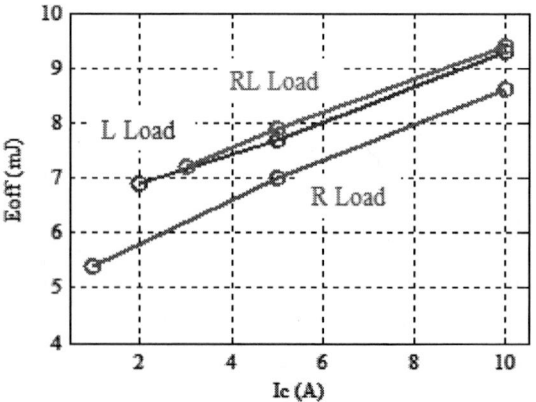

Fig. 12: Comparison of turn-off energy loss with L, RL and R loads at 10 kV and 25°C.

The double pulse test setup used for characterizing the 15 kV IGBT is shown in Fig. 13. The hot plate used for heating the IGBT is seen in the picture. The IGBT collector voltage is measured using a 75 MHz, 20 kV Tektronix (P6015A) probe. The current measurement is done using a 20 MHz pearson coil. The 500 MHz oscilloscope is set at 150 MHz for the voltage measurement and 20 MHz for the current measurement.

III. THERMAL RESISTANCE OF THE 15 KV SiC N-IGBT MODULE

In practical power converter applications, the IGBT power dissipation (determined by the operating current and switching frequency) is limited by the ability to transfer the generated heat from the device junction to the ambient. A low thermal resistance would not increase device junction temperature considerably because of low thermal drop from the power dissipation. The thermal resistance from the junction to the ambient is contributed by the module package, the thermal interface compound and the external heat sinking (cooling) mechanism. As the effective thermal resistance is pivotal in determining the switching frequency (thermal) limits, a detailed investigation is conducted on understanding the different components contributing to the effective thermal resistance of the state-of-the-art 15 kV SiC IGBT Module.

The base plate of the 15 kV IGBT has AlN isolation layer rated for 20 kV dc. The AlN isolation and several other packaging layers contributing to the thermal resistance from the IGBT junction to the ambient are shown in Fig. 14. Thermal simulations have been performed by incorporating thermal conductivity values of the different layers to determine the temperature drop across the entire thermal path for a power dissipation of 125 W by the IGBT. The IGBT junction temperature is found to be 67.2°C for 125 W power dissipation (the heat sink is constrained to 0°C), indicating thermal resistance of 0.54°C/W from the IGBT junction to the bottom of the heat-sink. The corresponding temperature distribution profile of the IGBT module is shown in Fig. 15, along with the temperature drop across each layer in Table II.

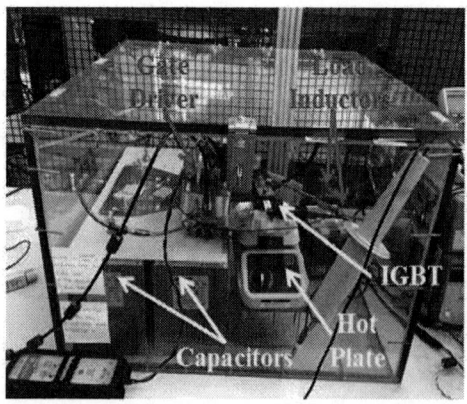

Fig. 13: The 12 kV double-pulse test setup used for characterizing the 15 kV SiC N-IGBT Module

978-1-4799-2706-7/14 $31.00 © 2014 IEEE

The thermal conductivity of the AlN layer is taken as 135 W/mK, corresponding to 175°C. It should be noted that the thermal resistance of the heat sink taken in this analysis corresponds to the liquid cooling mechanism. These thermal resistance values are the basis for determining the IGBT safe operating frequency limits presented in the next section.

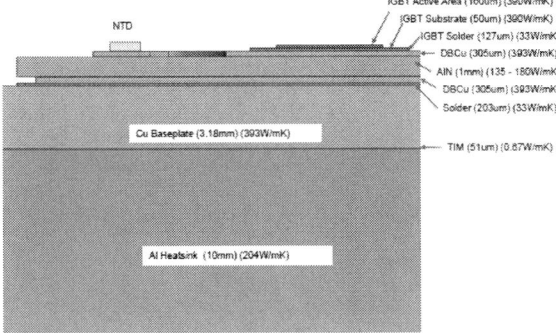

Fig. 14: The different layers in the IGBT module package contributing to the thermal resistance.

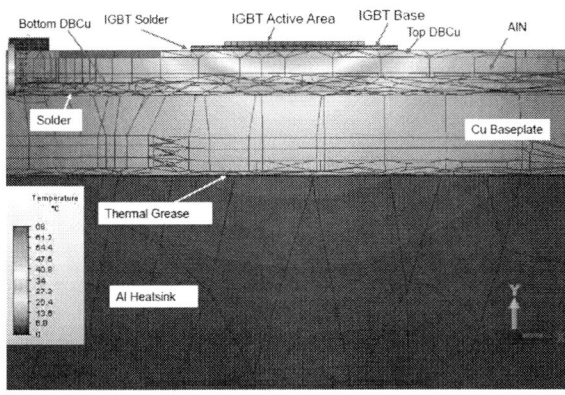

Fig. 15: Simulation result showing the temperature profile across the cross-section of the IGBT module for 125 W power dissipation.

TABLE II: THE TEMPERATURE DROPS ACROSS EACH LAYER IN THE THERMAL PATH OF THE 15 kV SiC IGBT MODULE PACKAGE

Layer	Temperature at Top of Layer (°C)	Temperature drop across Layer (°C)	% Temp drop
IGBT Active	67.2	0.8	1.2 %
IGBT Base	66.4	0.2	0.3 %
IGBT Solder	66.2	13.1	19.5 %
Top DBCu	53.1	1.8	2.7 %
AlN	51.3	19.8	29.5 %
Bottom DBCu	31.5	1.0	1.5 %
Substrate Solder	30.5	5.3	7.9 %
Baseplate	25.2	7.0	10.4 %
Thermal Grease	18.2	12.0	17.8 %
Heat sink	6.25	6.3	9.3 %
Bottom of Heat sink	0	-	-

IV. ANALYTICAL EVALUATION OF IGBT FREQEUNCY LIMITS BASED ON THE EXPERIMENTAL DATA

The 6.5 kV Si IGBTs have maximum operating junction temperature of 125°C [9], [10]. Also, the 1200 V SiC MOSFETs can be safely operated up to 150°C [11]. The 10 kV SiC MOSFET is shown to be operating at 175°C [12]. In this study, 150°C is assumed as the maximum operating temperature for the 15 kV SiC N-IGBT during its operation in a power converter system. The junction power dissipation (or W/cm^2) is limited by this maximum operating temperature and the thermal resistance from the junction to the ambient. Table III shows the IGBT parameters for which the thermal limits are evaluated. The nominal operating voltage in a power converter system is chosen as 66 % of the maximum blocking voltage of 15 kV. Also, the frequency limits are being evaluated for two different current densities as shown in the table. The thermal resistance information for the liquid cooling case is taken from the analysis presented in the earlier section. For the case of air cooling, the heat sink thermal resistance is scaled up appropriately based on the heat sink datasheet used in the boost converter demonstration presented in section V.

Fig. 16 shows 6.2 kHz as the frequency limit for the case of 10 kV, 5 A with liquid cooling mechanism. It is can be seen that at 6.2 kHz, the junction temperature reaches stable operating point of 150°C, where the power loss curve intersects the thermal resistance line. The power loss curves are obtained by curve fitting the turn-off energy loss versus temperature (T) data at 10 kV and 5 A, given in (1). The conduction duty cycle is taken as 50 % and the clamped inductive load switching loss data at 10 kV are used. The reduction of turn-on loss with temperature is ignored, as mentioned earlier, and the conduction loss data used corresponds to 150°C. The thermal resistance line, showing the steady state junction temperature with respect to power dissipation, given in equation (2) is shown in Fig. 15 (red line). At a particular switching frequency, the IGBT power dissipation increases with the junction temperature. Then, the junction temperature rise increases the power dissipation further, which in turn, increases the junction temperature. This process is repeated till the IGBT junction finally settles down to the steady temperature determined by the intersection of temperature dependent power dissipation curve with the thermal resistance line. Based on equation (2), the power dissipation (density) is not influenced due to neglecting the turn-on loss variation with temperature.

$$E_{OFF}(T) = -2 \times 10^{-9} T^3 + 8.2 \times 10^{-7} T^2 - 2.1 \times 10^{-5} T + 0.008. \quad (1)$$

$$T_J = (P_D \times R_{TH}) + R_A. \quad (2)$$

Where, T_J is the IGBT junction temperature, P_D is the power dissipation, R_{TH} is the total thermal resistance and R_A is the ambient temperature.

TABLE III: PARAMETERS OF THE 15 KV SIC N-IGBT

Parameter	Value
IGBT rated blocking voltage	15 kV
Nominal operating voltage	10 kV
Nominal operating current	5 A (15 A/cm^2) 10 A (30 A/cm^2)
Active area	0.32 cm^2
Maximum operating junction temperature	150°C
Thermal resistance from junction to the ambient (liquid cooling)	0.54 °C/W
Thermal resistance from junction to the ambient (air cooling)	0.65 °C/W
Ambient Temperature, R_A	35 °C

The hard frequency limit is 6.2 kHz for 10 kV, 5 A operation with liquid cooling, as shown in Fig. 16. The intersection of the power dissipation curve at 6.2 kHz and the thermal line at 150°C (taken as the maximum operating junction temperature) is seen in the figure. The frequency limit for 10 kV, 10 A is 3.9 kHz, with liquid cooling, as shown in Fig. 17. The 10 kV, 10 A turn-off energy loss versus temperature (T) used from the curve fitting is given in (3). The frequency limits with air cooling at 10 kV, 5 A and 10 A are shown in Fig. 18 and Fig. 19 respectively. The thermal resistance line is derived by using the R_{TH} value corresponding to the air cooling, shown in Table III. The summary of frequency limits for all these cases is provided in Table IV. In moving from liquid cooling to air cooling, the power dissipation density is nominally reduced from 660 W/cm^2 to 550 W/cm^2. This emphasizes the role of the package thermal resistances (rather than the cooling mechanism) in exploiting the SiC material for efficient and high power density power conversion.

$$E_{OFF}\,(T) = -2.2 \times 10^{-9}T^3 + 1.1 \times 10^{-6}T^2 - 8.2 \times 10^{-6}T + 0.0091. \qquad (3)$$

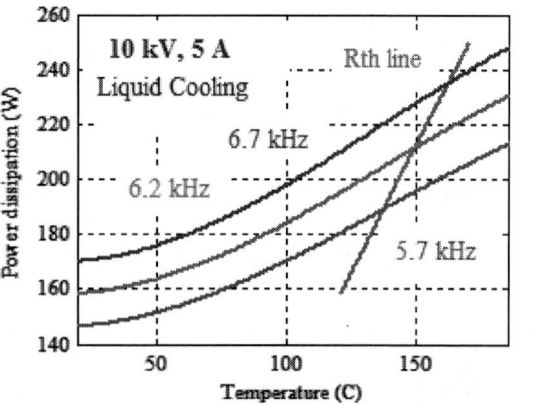

Fig. 16: Frequency limit curves at 10 kV, 5 A (50 % conduction duty cycle) with liquid cooling.

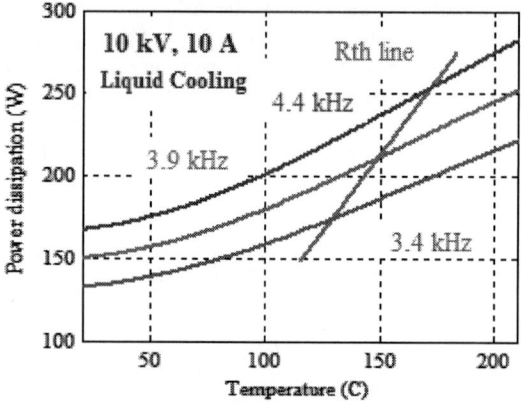

Fig. 17: Frequency limit curves at 10 kV, 10 A (50 % conduction duty cycle) with liquid cooling.

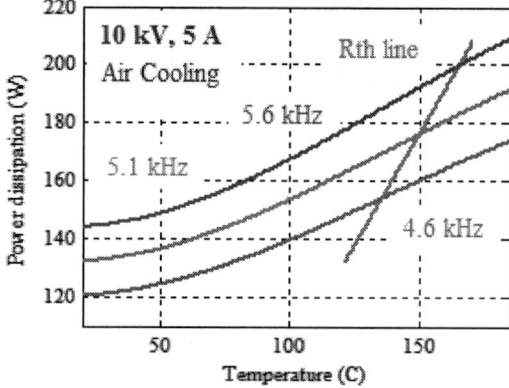

Fig. 18: Frequency limit curves at 10 kV, 5 A (50 % conduction duty cycle) with air cooling.

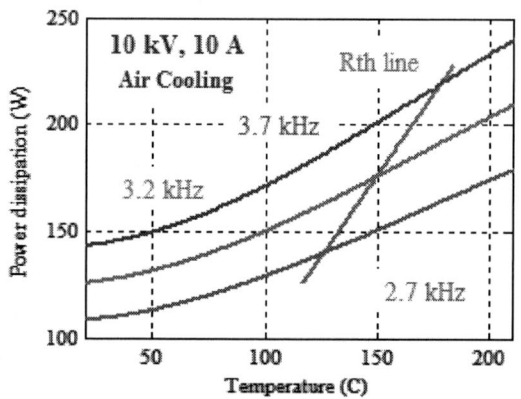

Fig. 19: Frequency limit curves at 10 kV, 10 A (50 % conduction duty cycle) with air cooling.

V. EXPERIMENTAL DEMONSTRATION OF THE IGBT FREQUENCY CAPABILITY ON A 10 KV BOOST CONVERTER

This section presents experimental demonstration of the analytically derived frequency limits provided in the previous section. A dc-dc boost converter topology [13] has been chosen for demonstration, to emulate the hard-switching frequency limits. The converter is operated at

TABLE IV: SUMMARY OF HARD SWITCHING FREQUENCY LIMITS OF THE 15 KV SiC N-IGBT

Parameter	Value
Frequency limits at 10 kV, 5 A	6.2 kHz (liquid cooling) 5.1 kHz (air cooling)
Frequency limits at 10 kV, 10 A	3.9 kHz (liquid cooling) 3.2 kHz (air cooling)
Power dissipation density	660 W/cm2 (liquid cooling) 550 W/cm2 (air cooling)
Maximum operating junction Temperature	150°C
Ambient Temperature (Assumed)	35°C

10 kV output, satisfying the IGBT operating voltage outlined in the previous section. The input voltage is chosen as 2 kV. Therefore, the conduction losses are scaled to 80 % to exactly determine the power dissipation under these conditions. Also, only air cooled condition is validated, due to ease of building and implementing the converter system.

The boost converter is operated at 6.7 kHz at 10 kV output as shown in Fig. 20. The IGBT turn-on current is 1.6 A and the turn-off current is 5 A, with inductor of 78 mH used in the converter. The turn-off and turn-on transitions measured at the end of one hour duration of the boost converter operation are shown in Fig. 21 and Fig. 22 respectively. The conduction loss is calculated for 80 % duty with the ramp rise. The junction temperature is determined from the instant at which IGBT enters punch-through (the point at which the slope of voltage changes as shown in Fig. 6) during turn-off transition. In Fig. 6, it is shown that the duration it takes to reach punch-through is increasing with junction temperature. This fact is used to estimate the junction temperature accurately, by matching the voltage transitions with the hot-plate based double-pulse switching tests. This method of determining temperature is used because of its accuracy and measurement safety at high voltage. Also, the double-pulse tests have been re-conducted with 78 mH (in place of the originally used 8 mH) to match the waveforms for better accuracy, for the temperature estimation.

Based on this method, the junction temperature is found to be 146°C at the end of one hour of the boost converter operation. The corresponding voltage rise curves, from the converter startup instant to one hour operation, are shown in Fig. 23. It is seen that the junction temperature increased negligibly from 30 minute to one hour instants. The details of the heat sink used are given in [14]. It should be noted that the IGBT and diode modules are mounted on the same heat sink. The IGBT power dissipation is 176 W at 6.7 kHz, as summarized in Table V. It is 550 W/cm^2, same as analytically shown in the previous section. It should be noted that that the ambient and the corresponding junction temperature are slightly lower in the demonstration.

TABLE V: SUMMARY OF BOOST CONVERTER DEMONSTRATION

Parameter	Value
Input Voltage, Output Voltage	2 kV, 10 kV
Switching Frequency	6.7 kHz
Filter Inductor	78 mH (Q = 308 at 5 kHz)
Input Power	6.4 kW
Cooling Mechanism	Forced air convection
Conduction Loss (80 % duty)	12 W
Turn-on Loss (10 kV, 1.6 A)	31 W (at 6.7 kHz)
Turn-off Loss (10 kV, 5 A)	133 W (at 6.7 kHz)
Total Power Loss	176 W
IGBT Power Dissipation Density	550 W/cm^2
Thermal Resistance (Junction to Ambient)	0.65 °C/W
IGBT Junction Temperature (After operating for one hour)	146°C
Ambient Temperature (Measured)	32°C

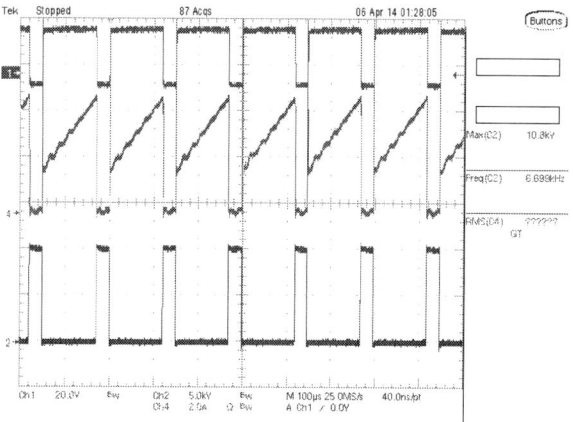

Fig. 20: The boost converter operation at 6.7 kHz, with 2 kV input and 10 kV output, at 6.4 kW. (Ch1: V_{GE}, Ch2: V_{CE}, Ch4: Inductor current)

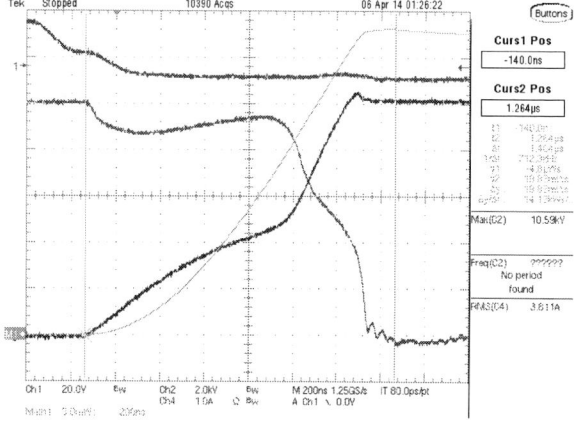

Fig. 21: The turn-off transient during the 6.7 kHz operation of the boost converter at 10 kV, 6.4 kW output (taken after 60 minutes).

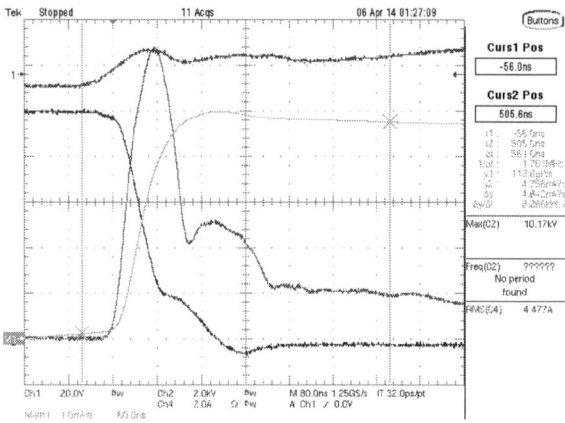

Fig. 22: The turn-on transient during the 6.7 kHz operation of the boost converter at 10 kV, 6.4 kW output (taken after 60 minutes).

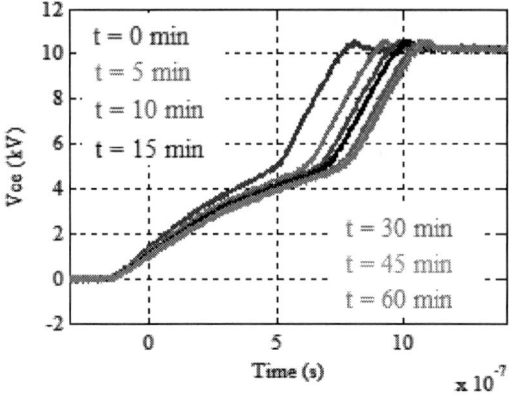

Fig. 23: The turn-off voltage curves (indicating temperature settling) during the 10 kV, 6.7 kHz boost converter demonstration.

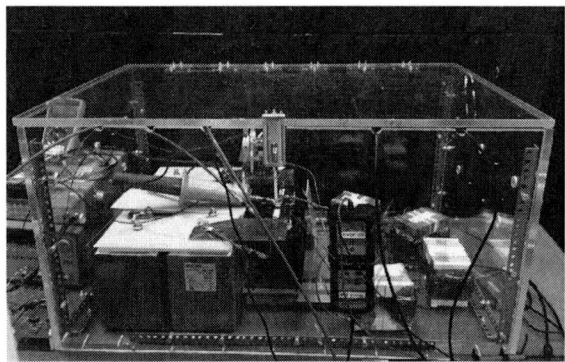

Fig. 24: The test setup of the 10 kV output dc-dc boost converter, demonstrating 15 kV SiC IGBT at 550 W/cm^2 with air cooling.

VI. CONCLUSIONS

The objective of this work is to demonstrate the switching capability of the state-of-the-art 15 kV SiC N-IGBT. As the fundamental means of realizing that, the 15 kV IGBT is thoroughly characterized for 10 kV switching, at different current densities and temperature. Extensive FEM simulations are carried out to determine thermal resistance of each layer in the IGBT module

package. Based on the thermal resistance information extracted and the experimental switching data presented, the hard switching frequency limits of the IGBT are analytically evaluated. At 10 kV, 5 A, the IGBT can be operated up to 6.2 kHz, while limiting the junction temperature to 150°C, with liquid cooling. The corresponding limit for 10 A operation is 3.9 kHz. With air cooling, the IGBT can be switched at 5.1 kHz at 10 kV and 5 A. The corresponding limit for 10 A is 3.2 kHz. The corresponding power dissipation densities for liquid and air cooling are 660 W/cm^2 and 550 W/cm^2. The analytical frequency limits may increase further if the turn-on loss reduction with temperature is considered. However, the power dissipation densities are going to remain the same. Finally, the air cooling limits of the 15 kV SiC N-IGBT module are demonstrated on a dc-dc boost converter operating at 10 kV and 550 W/cm^2.

ACKNOWLEDGMENT

This work is sponsored by ARPA-E Contract #DE-AR0000110, monitored by Dr. Timothy Heidel.

This work made use of the FREEDM ERC shared facilities supported by the National Science Foundation under Award Number EEC-0812121.

REFERENCES

[1] A. K. Agarwal, "An overview of SiC power devices," *International Conf. on Power, Control and Embedded Systems (ICPCES)*, 2010, pp. 1-4.

[2] Q. Zhang, C. Jonas, S. Ryu, A. Agarwal and J. Palmour, "Design and fabrications of high voltage IGBTs on 4H-SiC," *IEEE Int'l Symposium on Power Semiconductor Devices and IC's (ISPSD)*, 2006, pp. 1-4.

[3] S. Ryu, et al., "High performance, ultra high voltage 4H-SiC IGBTs," *IEEE Energy Conversion Congress and Exposition (ECCE)*, 2012, pp. 3603-3608.

[4] T. Tamaki, G. G. Walden, Y. Sui, and J. A. Cooper, "Optimization of on-state and switching performances for 15–20-kV 4H-SiC IGBTs," *IEEE Trans. Electron Devices*, vol. 55, no. 8, pp. 1920–1927, Aug. 2008.

[5] Meng-Chia Lee, Xing Huang, A. Q. Huang and Edward Van Brunt, "An analytical investigation of the effect of varied buffer layer designs on the turn-off speed for 4H-SiC IGBTs," *IEEE Workshop on Wide Bandgap Power Devices and Application (WiPDA)*, 2013, pp. 44-47.

[6] W. Sung, J. Wang, A. Q. Huang, and B. J. Baliga, "Design and investigation of frequency capability of 15kV 4H-SiC IGBT," *IEEE Int'l Symposium on Power Semiconductor Devices and IC's (ISPSD)*, 2009, pp 271-274.

[7] A. Kadavelugu, et al, "Characterization of 15 kV SiC n-IGBT and its Application Considerations for High Power Converters," *IEEE Energy Conversion Congress and Exposition (ECCE)* 2013, pp. 2528-2535.

[8] B. J. Baliga, "Fundamentals of power semiconductor devices," *New York: Springer-Verlag*, 2008.

[9] Datasheet of 6.5 kV Hi Pak ABB IGBT, part no. 5SNA 0400J650100.

[10] Datasheet of 6.5 kV Infineon IGBT, part no. FZ250R65KE3.

[11] Datasheet of 1200 V, Cree, SiC MOSFET, part no. C2M0080120D.

[12] Jun Wang, et al., "10 kV SiC MOSFET based boost converter," *IEEE Transactions on Industrial Applications*, Vol 45, Issue 6, 2009, pp. 2056-2063.

[13] R. W. Erickson and D. Maksimovic, "Fundamentals of Power Electronics," *Springer*, 2001.

[14] Datasheet of Extruded Heat Sinks, Wakefield Engineering, part no. 392-300AB.

The 2014 International Power Electronics Conference

Selection of Suitable Carrier-Based PWM Method for Modular Multilevel Converter

Barış ÇİFTÇİ[1,4], Feyzullah ERTÜRK[2] and Ahmet M. HAVA[3]

[1,2,3]Electrical and Electronics Engineering Department, METU, 06531, Çankaya, Ankara, Turkey
[4]Defense Systems Technologies, ASELSAN A.Ş., 06370, Yenimahalle, Ankara, Turkey
[1]baris.ciftci@metu.edu.tr   [2]feyzullah.erturk@metu.edu.tr   [3]hava@metu.edu.tr

Abstract- **The Modular Multilevel Converter (M2C) has become the preferred topology for use in high power applications in recent years. This paper gives guiding information for M2C design, in terms of carrier-based PWM methods. Phase and level shift methods are considered. The enhancement due to zero sequence signal injection is also quantified. It is found that output voltage characteristics are highly dependent on the PWM method used. Therefore, selection of the switching method is a critical design issue in M2C studies. Performance evaluation and comparisons are based on Matlab generated switching pulse patterns and the results are validated by computer simulations.**

Keywords— carrier-based PWM, harmonics, modular multilevel converter, WTHD

I. INTRODUCTION

The Modular Multilevel Converter (MMLC, M2LC, M2C), topology emerged about a decade ago [1] and has gained a high popularity in research and development studies due to its outstanding properties. M2C operates in a wide range of voltage and power rating, and it is especially evolved through multi MW applications. Fig. 1 shows a double-star configured M2C. In the figure, "SM" stands for chopper cell "sub-module" which is the main constituent of M2C, and it is detailed in Fig. 2.

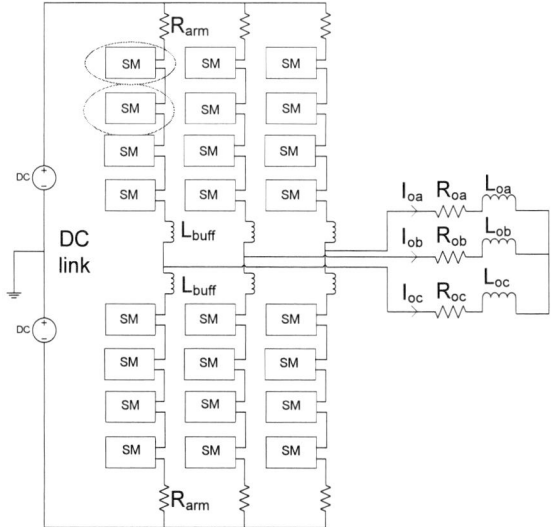

Fig. 1. Double-star configured M2C.

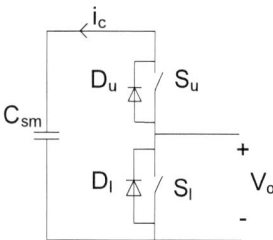

Fig. 2. Chopper cell sub-module.

Design of power electronic converters involves power converter hardware design, controller design, and PWM method design stages. Individual design stages as well as the total design involve numerous stringent criteria such as cost, efficiency, harmonics performance, control complexity, and so on. Of these, the PWM method involves variety of switching strategies; and for a given hardware design and specified performance task, matching a suitable PWM method is a challenging mission. For M2C this issue is even more pronounced as large number of modules brings a wider range of switching strategies, and as the capacitor voltage balancing, hence circulating current control are two fundamental issues related to the PWM methods.

Due to satisfactory performance and ease of implementation, carrier based PWM methods are widely considered for M2C. For M2C switching, *level-shift (LS, sub-harmonic)* and *phase-shift (PS, carrier disposition)* carrier-based PWM methods are used. LS method branches into *phase disposition (PD), phase opposition disposition (POD)* and *alternative phase opposition disposition (APOD)* methods. For PWM switching, if there are N sub-modules per arm in the converter, N carriers are used with a (generally) sinusoidal reference. Implementation of LS methods is based on distributing the carriers contiguously on the whole V_{dc} band in vertically shifted manner; whereas that of PS method is based on distributing them in a phase (horizontally) shifted manner. Carriers of the PD method are all in phase. For POD method, positive and negative carriers are 180° phase shifted. For APOD method, consecutive carriers are 180° phase shifted according to each other. For the PS method, on the other hand, carriers are phase shifted by 360°/N. More detailed information about

978-1-4799-2706-7/14 $31.00 © 2014 IEEE   3734

carrier-based PWM methods and their implementation to the converter can be found in [2]-[5].

In the literature, there are several studies that consider carrier-based PWM methods for M2C switching. One of the earliest studies, [2], introduces LS PWM methods and compares their performances. Carrier-based PWM methods comparisons for specific m_a (=A_o/A_c modulation index) and m_f (=f_c/f_o carrier to fundamental frequency ratio) values are given in [3] and [4]. Another work, [5], also rates sinusoidal PWM methods, for specific m_a and m_f values in terms of harmonics performances.

This paper, different from previous studies, analyzes the PWM methods for M2C in terms of output voltage and current characteristics, sub-module capacitor voltage balancing characteristics and input current characteristics, throughout broad m_a and m_f ranges under the provision of *"equal switching count"* principle. It is a well-known fact for a power converter, higher switching count implies higher output voltage waveform quality, but at the cost of increasing switching losses, thus reduced converter efficiency. In particular, for M2C converters with high power rating, energy efficiency is a very important characteristic. Therefore, a fair output waveform quality comparison of carrier-based PWM methods for M2C should be based on "equal total count of semiconductor switching per leg in a period of output voltage" principle.

The study is conducted in four main steps. First, converter output waveform characteristics are analyzed for different PWM methods. Switching pulse patterns for different m_a and m_f values are generated. Using these pulse patterns, normalized phase and line-to-line voltages are obtained. Harmonic spectrums of line and phase voltages and V_n/n components are analyzed and *"weighted total harmonic distortion (WTHD)"* values of line voltage waveforms are calculated in (1). Different from classical THD calculation, output line voltage WTHD calculation involves the effect of orders of harmonics. It is defined in percent as:

$$\text{WTHD} = \frac{\sqrt{\sum_{i=2}^{n}\left(\frac{V_{LLi}}{i}\right)^2}}{V_{LL1}} \times 100 \qquad (1)$$

where i is the harmonic order and V_{LLi} is the amplitude of the ith line-to-line harmonic voltage. It is important that, V_n/n harmonics and WTHD values are irrespective from the parameters of the topology used, giving a general result. Second, methods are compared according to their sub-module capacitor balancing capabilities. Third, converter input current characteristics are analyzed. Finally, these studies are validated by computer simulations with an M2C for an 8 MW system.

II. PERFORMANCE COMPARISONS FOR OUTPUT VOLTAGE AND CURRENTS

Exploring the carrier and reference waveforms of different carrier-based PWM methods, it is inferred that for the same m_a and m_f values, the PS method has much more total count of carrier and reference crossings (thus switching) than LS methods. Total count of switching in

a phase leg for all the LS and PS PWM methods are computed for different systems with different number of carriers. It is observed that, irrespective of how many carriers are used, the PS method has a total count of switching which is approximately *number of carriers* times that of LS methods. Hence, in order to obtain equal switching count, LS methods should be implemented with an m_f value which is *number of carriers* times that of PS method. This work is conducted in the light of mentioned information.

Double-star chopper cell topology with four sub-modules in each arm, as shown in Fig. 1, is used for PWM pulse pattern generation using Matlab. Performances of different methods are investigated using various m_a and m_f values, keeping total count of switching in a phase leg equal to each other for each set for the same m_a value. In Figs. 3, 6, 9 and 11 switching pulses generated by different modulation methods (PS, PD, POD, and APOD, respectively) in *one period of reference signal (1/50Hz=20ms)* are shown. For simplicity and readability, two of the generated pulse patterns (corresponding to red and blue circled sub-modules in Fig. 1) out of four for the whole phase arm, are shown. One of the pulse patterns (red one) is scaled by two in order to distinguish among each other. In the light of Figs. 3, 6, 9, and 11 it can be concluded, PS method is capable of producing balanced and homogeneous switching functions in terms of duration, switching instants and count of pulses, whereas LS methods produce highly heterogeneous switching functions which could lead to unbalanced usage of semiconductors, high stress in terms of thermal conditions, unequal charge/discharge of sub-module capacitors and therefore high circulation currents [6]. (In fact, because of this unbalance, purest form of LS methods cannot provide a stable circuit without implementing a switching pulse balancing mechanism such as carrier rotation methods. In part III, more detailed information will be given.) Using these switching pulse patterns, normalized (*1 unit of voltage* for each level) output line voltages of a 5-level (9-level line-to-line) M2C system, operated with m_a=0.9, are generated for PS, PD, POD and APOD methods, respectively. Due to lack of space, only the waveforms of PS and PD methods are shown in Figs. 4 and 7. The output line voltages are all well balanced and quarter wave symmetrical; but there exist differences in terms of transitions in voltage steps and voltage harmonics. Investigating Figs. 4 and 7, for PS method, transitions from a voltage level to another may occur as four units (e.g. from level 8 to level 4) as well as two units (e.g. from level 6 to level 4). Actually, this phenomenon is the same for POD and APOD methods, also. On the other hand, for PD method, the transitions always occur as two units. These arguments are validated also by voltage waveforms of methods with different m_a values. Therefore, it is inferred that PD method gives a modulation function like unipolar switched PWM; whereas PS, POD and APOD methods give functions like bipolar switched PWM. This difference revealed its effects on the line voltage harmonics.

As is known, current harmonics are the source of various power quality problems such as heating in the equipment, torque pulsations in motor drives, and heating and waveform distortion problems in grid connected inverters. Current harmonics are directly dependent on voltage harmonics divided by the load impedance. Therefore, for inductive loaded systems, which are the case for most AC motor drive and utility interface applications, low order harmonics are much more problematic, since high order ones are filtered greatly by the load inductance. In the light of this fact, line voltage harmonics are weighted based on their orders and V_n/n terms are calculated as in the WTHD definition. Weighted output line voltage harmonic spectrums can be seen in Figs. 5, 8, 10 and 12. Investigating these figures, it is inferred that dominant harmonics of LS methods are centered about their carrier frequency, while that of PS method is centered about *number of carriers times its own carrier frequency,* catching up the frequency of the carriers of LS methods. Thus, the converter switching frequencies of different methods are approximately equal. Nevertheless, PD method, as a result of its unipolar characterized switching function, has much smaller harmonics magnitude compared to other methods. This advantage makes the PD method favorable, in terms of output harmonics performance. Fig. 13 shows weighted line voltage harmonic spectrum of PD method also for $m_a=0.2$ and $m_a=0.6$. For $m_a=0.2$, again the harmonics centered about carrier frequency are the dominant ones; while for $m_a=0.6$, harmonics about two times the carrier frequency as well as carrier frequency, are dominant.

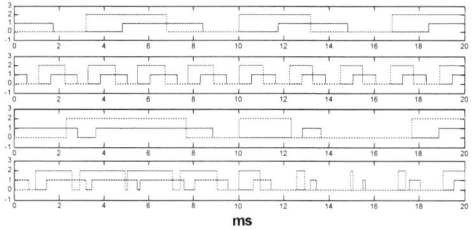

Fig. 3. Pulse patterns of PS for ($m_a=0.1$, $m_f=3$), ($m_a=0.1$, $m_f=9$), ($m_a=0.9$, $m_f=3$) and ($m_a=0.9$, $m_f=9$) from the top to the bottom.

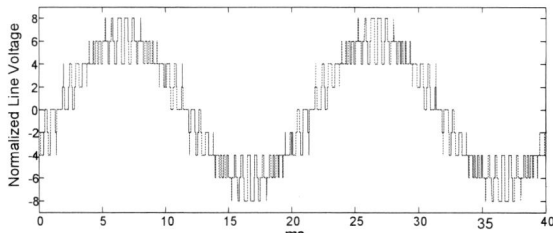

Fig. 4. Normalized line voltage of PS for $m_a=0.9$, $m_f=9$.

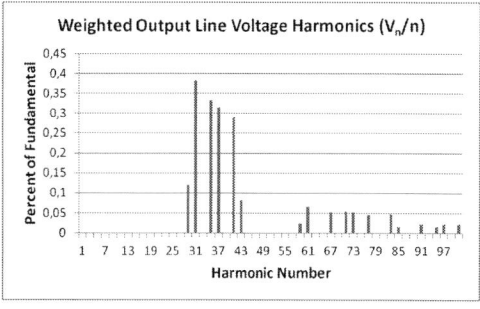

Fig. 5. Weighted line voltage harmonics of PS for $m_a=0.9$, $m_f=9$.

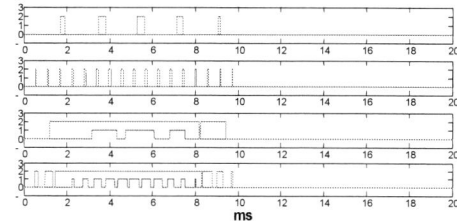

Fig. 6. Pulse patterns of PD for ($m_a=0.1$, $m_f=12$), ($m_a=0.1$, $m_f=35$), ($m_a=0.9$, $m_f=12$) and ($m_a=0.9$, $m_f=35$) from the top to the bottom.

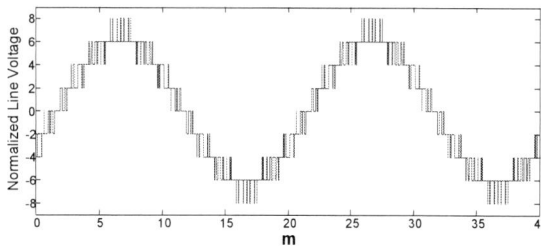

Fig. 7. Normalized line voltage of PD for $m_a=0.9$, $m_f=35$.

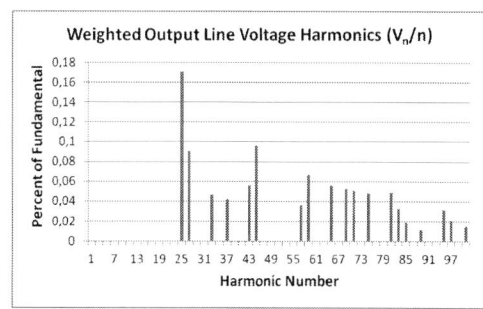

Fig. 8. Weighted line voltage harmonics of PD for $m_a=0.9$, $m_f=35$.

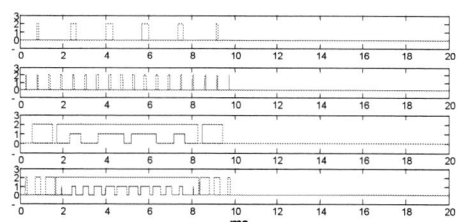

Fig. 9. Pulse patterns of POD for ($m_a=0.1$, $m_f=12$), ($m_a=0.1$, $m_f=36$), ($m_a=0.9$, $m_f=12$) and ($m_a=0.9$, $m_f=36$) from the top to the bottom.

The 2014 International Power Electronics Conference

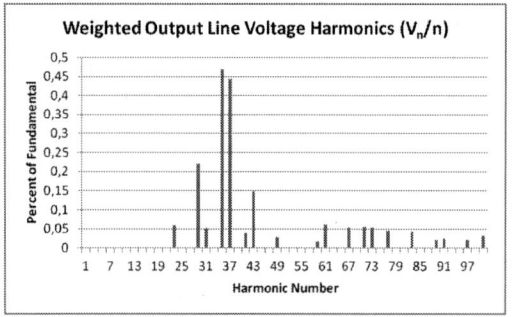

Fig. 10. Weighted line voltage harmonics of POD for m_a=0.9, m_f=36

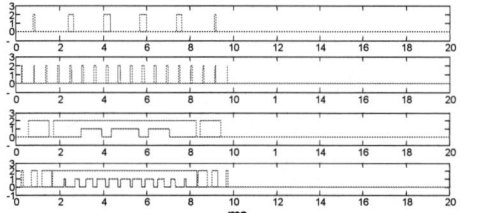

Fig. 11. Pulse patterns of APOD for (m_a=0.1, m_f=12), (m_a=0.1, m_f=36), (m_a=0.9, m_f=12) and (m_a=0.9, m_f=36) from the top to the bottom.

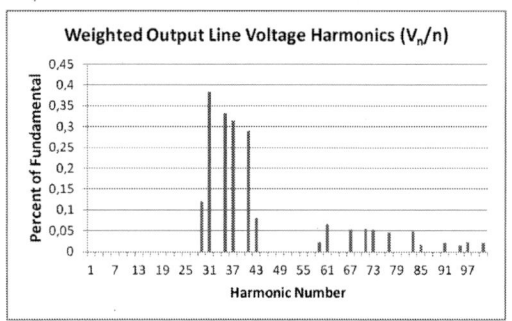

Fig. 12. Weighted line voltage harmonics of APOD for m_a=0.9, m_f=36.

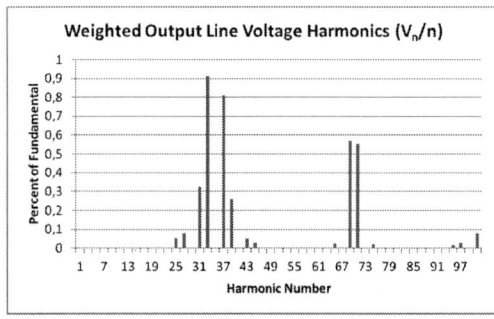

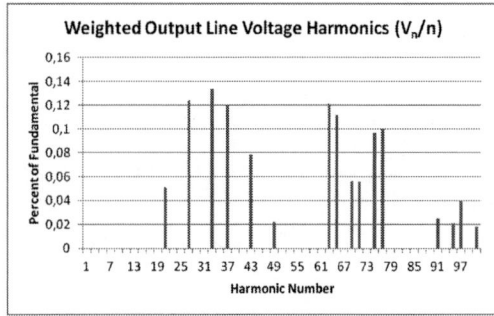

Fig. 13. Weighted line voltage harmonics of PD for m_a=0.2 and m_a=0.6, m_f=35.

Output harmonics performance of different methods can be evaluated by an analytical method, WTHD. As mentioned in part I, WTHD values of the four carrier-based PWM methods are calculated for varying m_a and varying m_f separately. Fig. 14 shows WTHD values for fixed m_f and varying m_a. Fig. 15, on the other hand, shows WTHD values for fixed m_a and varying m_f. Total count of switching throughout changing m_a or m_f ranges in Figs. 14 and 15 are the same for different PWM methods. It is important to note that, for Figs. 14 and 15, m_f values on the chart is valid for PS method, whereas that of LS methods are number of carrier (namely four) times shown m_f values.

Analyzing Fig. 14, it is clear that as m_f value (which also implies total count of switching) increases, WTHD value decreases for the same m_a value, since the dominant voltage harmonics go to higher frequency range. Performances of POD, APOD and PS methods are similar especially at low modulation indices. PD method, on the other hand, has a significant superiority on WTHD performance at low modulation indices. At higher modulation indices, from m_a=0.5 and up, methods have more diverse WTHD performances, but PD method has generally the best performance again. Nonetheless, operating at higher m_f values, performances of different methods become close to each other, also verified by Fig. 15, which shows that the methods' performances differ especially at low m_f values, say up to five. Therefore, it is important for a M2C being operated with the right PWM method when keeping m_f value, and so switching loss, is lower. PD method again has generally, except for a few operation points, better characteristics throughout the m_f range. Consequently, both of the WTHD graphs highlight PD that it would be the right method for the converter considering output voltage harmonics, with equal switching count provision.

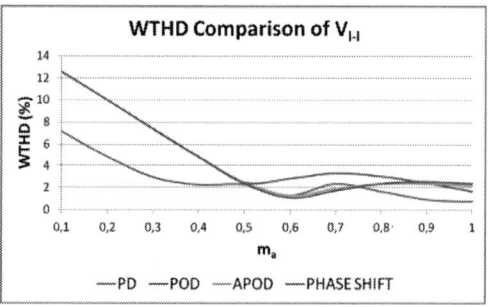

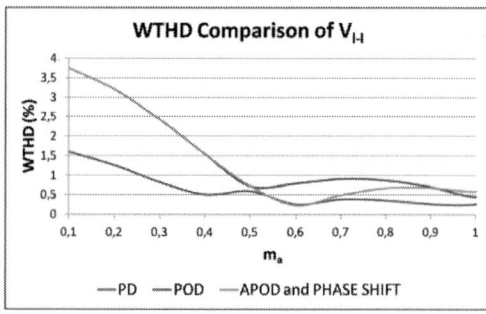

Fig. 14. WTHD values for changing m_a and fixed m_f=3 and m_f=9 at the top and at the bottom.

The 2014 International Power Electronics Conference

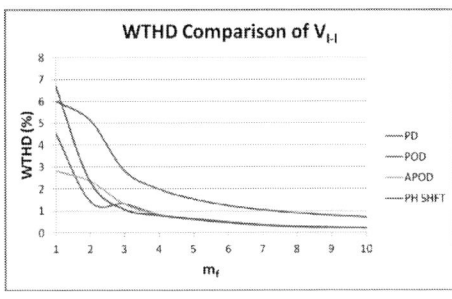

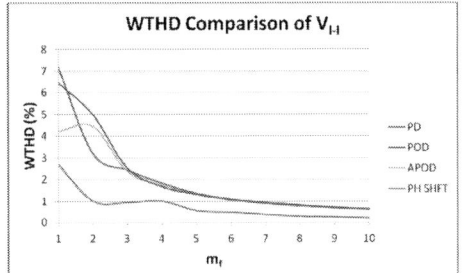

Fig. 15. WTHD values for changing m_f and fixed m_a=0.6 and m_a=0.9 at the top and at the bottom.

Two well-known zero sequence signal injection methods, space-vector PWM (SVPWM) and discontinuous PWM1 (DPWM1), are implemented on PD method. Performance of PD method with SPWM, SVPWM and DPWM1 modulations are investigated. Fig. 16 compares three methods' line voltage WTHD characteristics. It is again clear that after a certain m_f value, say eight, it is not distinguishable which PWM method is used, but for low m_f values, WTHD performances differ. SVPWM method gives WTHD under 1% even at fundamental frequency. Moreover, DPWM1 method has much lower, approximately 0.75 times, total switching count throughout the m_f range than that of two other methods while keeping its WTHD performance close others.

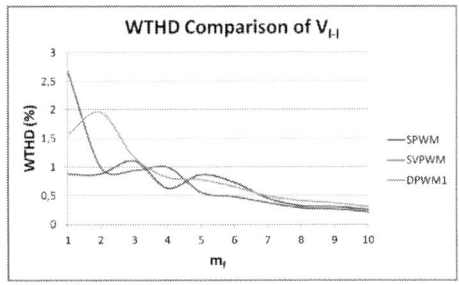

Fig. 16. WTHD values for different PWM methods for changing m_f and fixed m_a=0.9.

III. PERFORMANCE COMPARISONS FOR SUB-MODULE CAPACITOR VOLTAGE BALANCE

One of the fundamental objects of control mechanism for M2C is sub-module capacitor voltage balancing. Various capacitor voltage balancing control methods are introduced in the literature [7]. In this study, performances of carrier based PWM methods with their purest form will be compared. Sub-module capacitor voltages in the steady state are directly dependant on the

charged/discharged energy to/from the capacitor. In order to observe the fundamental energy balance for different methods, simulations are done with dc voltage sources as sub-module capacitors and power integrals are calculated for these dc sources. Power integrals show the total energy sourced, by which energy (so does capacitor voltage) charge/discharge balance of sub-modules can be judged. Fig. 17 shows upper arm sub-module dc sources power integrals for PD and PS method, vertical axis being the joules and horizontal being time as 20ms/div. Power integrals for POD and APOD methods are very similar to that of PD method. Power integrals for PD method are greatly differing from each other. This difference is because of unbalanced and heterogeneous switching pulse patterns as mentioned in part I. Evidently, with these greatly differing power integrals, if sub-module capacitors were used, their voltages would become unstable, which also causes great circulating currents and the circuit would be unstable. In order to equate these power integrals, switching pulse pattern balancing methods, such as carrier rotation method, should be implemented for LS methods [8]. By this way, at the end of each carrier rotation period, power integrals of different sub-modules in either arm could be equated. Fig.18 shows upper arm sub-module dc sources power integrals for PD method implemented with carrier rotation method of 80ms period. PS method, on the other hand, has naturally well balanced sub-module power integrals due to its homogeneous and similar pulse patterns for different sub-modules as explained in part I. Comparing the power integrals for PS and PD with carrier rotation, PS is clearly superior to PD. Although PD has balance at the end of the carrier rotation periods, it has significant unbalances during the period which could lead to higher capacitor voltage ripple and hence higher harmonics in circulating current [6]. As a result, in terms of capacitor voltage balancing and circulation current minimization, PS method has a natural superiority to LS methods all due to their switching pulse patterns.

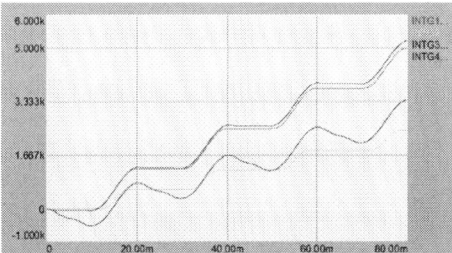

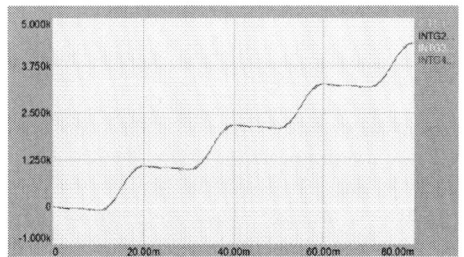

Fig. 17. Upper arm sub-modules power integrals for PD at the top and PS at the bottom.

978-1-4799-2706-7/14 $31.00 © 2014 IEEE   3738

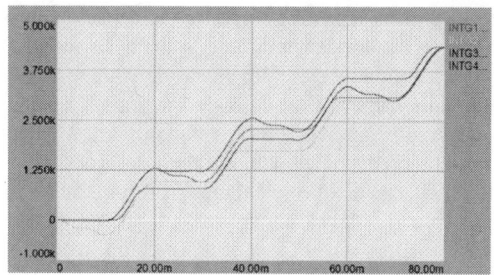

Fig. 18. Upper arm sub-modules power integrals for PD implemented with carrier rotation method of 80ms period.

IV. INPUT CURRENT PERFORMANCE COMPARISONS

The dc bus current and its harmonics spectrum are two determining factors in voltage source inverters design, because of dc bus voltage ripple and capacitor losses caused by capacitor ESR and dielectric loss. Thus, it is important to obtain the dc bus ripple current characteristics for an effective inverter design. From M2C view point, as shown in Fig. 1, there is no capacitor which faces the whole dc link voltage, V_{dc}; but the sub-module capacitors are the dc bus capacitors which are charged to the voltage level of V_{dc}/N. Current flowing through the sub-module capacitors of M2C is defined in [6] for the upper arm and lower arm capacitors, respectively in the following.

$$i_{cu} = \frac{1}{2}\left(i_{dc} + i_{a1} + \sum_{n=1}^{\infty} i_n - m_a cos(wt)i_{dc} - m_a cos(wt)i_{a1} - m_a cos(wt)\sum_{n=1}^{\infty} i_n \right) \quad (2)$$

$$i_{cl} = \frac{1}{2}\left(i_{dc} - i_{a1} + \sum_{n=1}^{\infty} i_n + m_a cos(wt)i_{dc} - m_a cos(wt)i_{a1} + m_a cos(wt)\sum_{n=1}^{\infty} i_n \right) \quad (3)$$

where i_{dc} is related with the average power transfer to the load, i_{a1} is half of the output current and i_n is the nth order harmonic in the circulating current. For a fixed load current and load angle, i_{dc} and i_{a1} are constant and independent from the switching method. Harmonic currents, i_n, on the other hand are switching method dependent.

In this study, in order to observe the characteristic current harmonics of different modulation methods, computer simulations are conducted using a M2C as in Fig. 1 with dc voltage sources in place of sub-module capacitors and 700A_{rms} ac current sources as load. As in the previous studies, equal switching count principle is respected (m_a=0.9, m_f=35 for PD and m_f=9 for PS). For LS methods, two cases are investigated: with and without carrier rotation method. Sub-module dc link current harmonics of PD and PS methods are shown in Fig. 19. Current harmonics spectra are different for PD and PS methods and they are directly dependent on the carrier frequency. Since sub-module capacitors are replaced by dc voltage sources, fundamental frequency component of 250A_{peak} is seen on the figures. Table I shows the rms values of input current and their harmonics. Among all, second harmonic is dominant for all the methods. Actually, this is the dominant component of circulating current of a phase leg. Harmonics below 500Hz have the same value for PD and PS methods. If carrier rotation is applied to PD, sub-harmonics at the carrier rotation frequency (12.5Hz) and its multiples are observed. This

phenomenon deteriorates the frequency spectrum and increases rms values. The low frequency harmonics are the most problematic ones for sub-module capacitors since the impedance (and also ESR) of the capacitors are inversely proportional with frequency. For capacitors, $V=I/(\omega C)$ which means high currents with low frequency causes high voltage ripple.

TABLE I: INPUT CURRENT CHARACTERISTICS

	<500Hz	<1kHz	>1kHz	Total	Dominant Harm.
PD with CR (A)	209	209	71	221	79 (2nd)
PD (A)	199	199	28	200	84 (2nd)
PS (A)	199	206	80	221	79 (2nd)

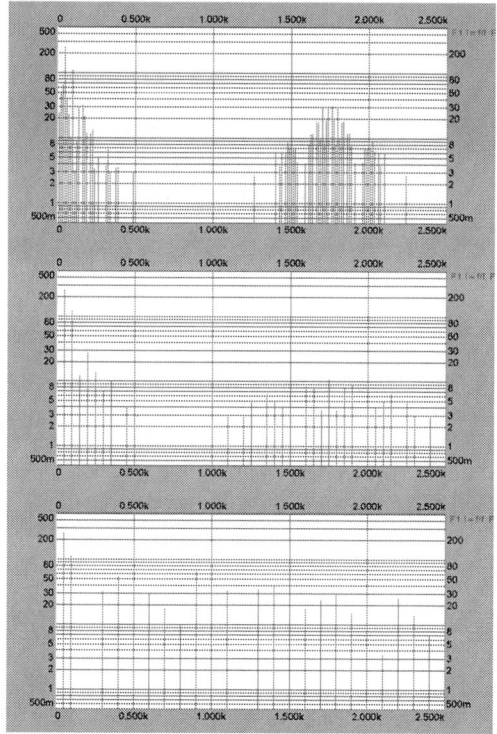

Fig. 19. Sub-module dc source current harmonics of PD (with and without carrier rotation) and PS from the top to the bottom.

The rms value of above mentioned harmonics, I_{inhrms}, is an important data for capacitor design and loss calculations. Using I_{inhrms}, dc link current ripple factor K_{Iin} is defined in [9] as:

$$K_{Iin} = \frac{I_{inhrms}^2}{I_{1rms}^2} \quad (4)$$

where I_{1rms} is the load current fundamental component rms value. K_{Iin} for different m_a and load angle values are calculated using the results of above mentioned simulations. It is seen that, although different modulation methods have different current harmonics spectrum, their rms values are equal to each other; in other words, they are independent of modulation method, for the same m_a and load angle value. In Fig. 20, K_{Iin} graph valid for all LS (with carrier rotation) and PS methods can be seen. K_{Iin} is inversely proportional with m_a and proportional with load angle.

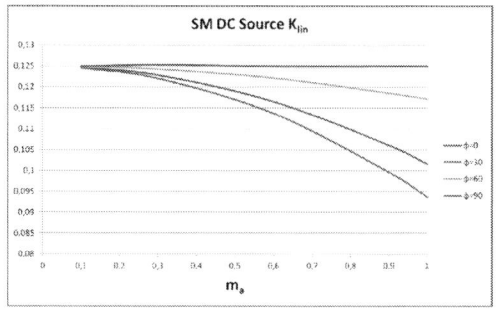

Fig. 20. Sub-module dc source normalized current harmonics valid for all PD, POD, APOD and PS methods.

V. SIMULATION RESULTS

A three-phase, double-star, chopper-cell circuit as in Fig. 1 is simulated with LS and PS modulation methods. SPWM method is implemented and the total count of switching in a phase leg is kept approximately equal to each other for each simulation. Simulation circuit parameters are listed on Table II. First, open loop control is applied without any sub-module capacitor voltage balancing control other than carrier rotation techniques for LS methods as explained in Section III. However, higher circulating currents and instabilities are observed due to poor capacitor voltage balance. Then, direct modulation is implemented with sorting algorithm as sub-module capacitor voltage balancing control [7]. Table III summarizes the simulation results. PD method, as expected from Section III, has the lowest line voltage and current THD. Capacitor voltage sorting mechanism provides stable and parallel capacitor voltage ripple performance for different PWM methods. Peak-to-peak ripple of voltages is about 10% of nominal and nearly same for all PWM methods. Fig. 21 shows the capacitors voltages for PD and PS methods. K_{lin} values for sub-module capacitors are close to each other for POD, APOD and PS; whereas, PD method has lower K_{lin} value. Output line voltages (blue) and phase currents (red) are illustrated in Fig. 22 for all the methods (since load angle is about 30°, line voltage and phase current are almost in phase). The voltage waveforms are as expected from the waveforms in Section II. As a result, the simulation results are consistent with the findings of previous parts.

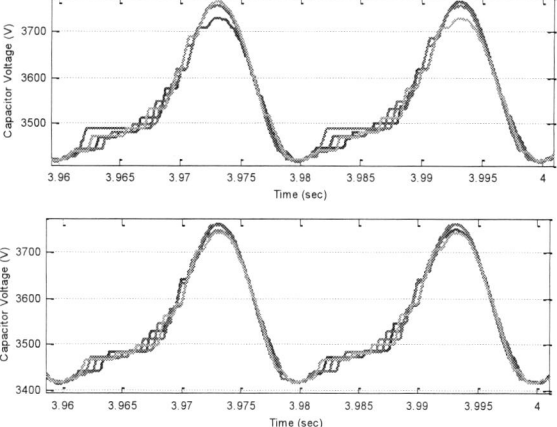

Fig. 21. Sub-module capacitor voltages in an arm for the PD (top) and PS (bottom) methods.

TABLE II: SIMULATION SYSTEM PARAMETERS

Symbol	Meaning	Value
P_{rated}	Rated power	8MW
N	Sub-modules per arm	4
V_g	Line-line voltage	$7.8kV_{rms}$
I_g	Phase current	$700A_{rms}$
f_g	Fundamental frequency	50Hz
V_{DC}	DC bus voltage	14.4kV
f_{sw}	Switching frequency	450Hz for PS 1750Hz for LS
R_g	Load resistance	5.5Ω
L_g	Load inductance	9.9mH (cos φ=0.87)
L_{buf}	Buffer inductance	3mH (0.15 pu)
R_{arm}	Arm resistance	$50m\Omega$
C_{sm}	Sub-module capacitance	4.7mF
m_a	Modulation index	0.9

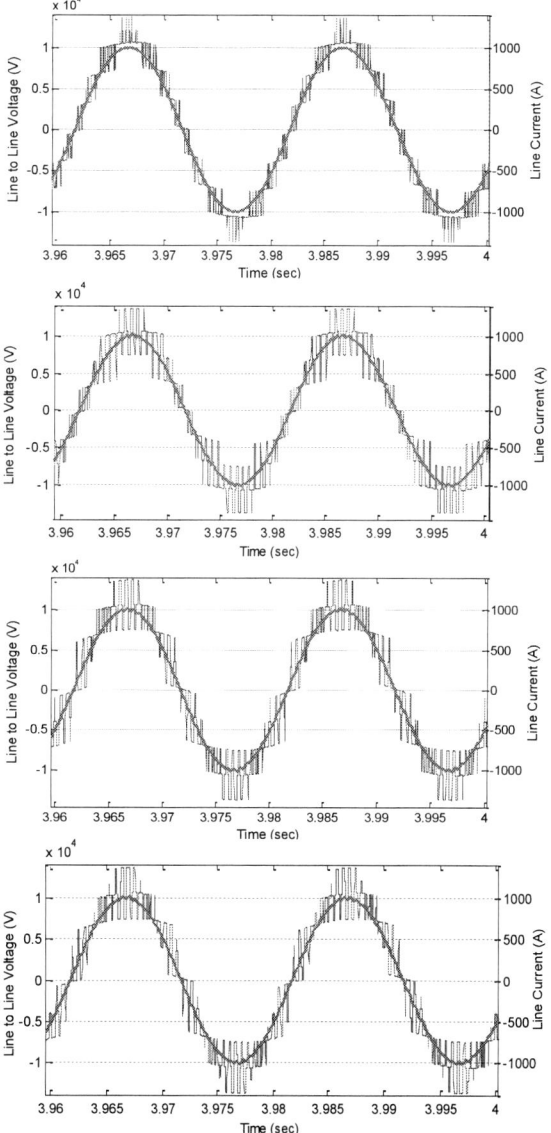

Fig. 22. Line-to-line voltage (V_{AB}) and line current (I_A) with the PD, POD, APOD and PS methods from top to bottom.

In order to test the validity of findings in previous sections under circulating current control, the circulation current suppression method in [10] is implemented. It is shown in Fig. 23 that the method is successful in eliminating the 2nd harmonic circulating current, and as a consequence the conduction losses of circuit elements can significantly be reduced. The results in Table IV validate the arguments of previous sections under circulating current control. They are parallel to the results of Table III. Nonetheless, the method lowers voltage THD by 1% and capacitor voltage ripple by 2% of all PWM methods. Capacitor current ripple factor K_{Iin} is decreased significantly, and the difference between K_{Iin} values of PD and other methods is diminished.

TABLE III: RESULTS OF PWM METHODS WITH CAPACITOR VOLTAGE SORTING ALGORITHM

	PD	POD	APOD	PS
Line Voltage THD (%)	15.14	26.20	25.16	24.92
Current THD (%)	0.56	1.32	1.28	1.24
Capacitor voltage ripple (% of V_{DC}/N)	9.72	9,58	9,72	9,58
Circ. current (DC) (A)	193.3	194.7	193.6	194.2
Circulating current (A) (2nd harmonic - peak)	136.0	136.3	135.6	136.0
K_{Iin} for SM capacitors	0.118	0.130	0.126	0.129

TABLE IV: RESULTS OF PWM METHODS WITH CAPACITOR VOLTAGE SORTING AND CIRCULATING CURRENT CONTROL

	PD	POD	APOD	PS
Line Volt. THD (%)	13.44	25.27	24.18	23.98
Current THD (%)	0.53	1.33	1.27	1.23
Capacitor voltage ripple (% of V_{DC}/N)	7.64	7.64	7.5	7.5
Circ. current (DC) (A)	192.8	192.8	192.8	193
Circulating current (A) (2nd harmonic - peak)	5	2.5	4.7	3.6
K_{Iin} for SM capacitors	0.087	0.090	0.089	0.090

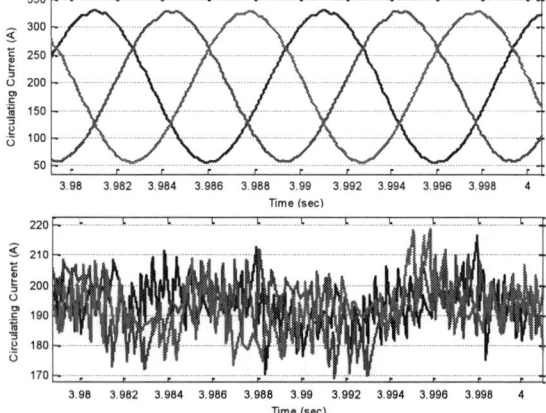

Fig. 23. Circulating current: uncontrolled (top) and controlled (bottom) for PD method.

VI. CONCLUSION

This paper stands as a guide for M2C design with carrier-based PWM methods, based on input and output harmonics and sub-module capacitor voltage balancing performance. Comparing different carrier-based PWM

methods with equal total switching count per leg principle in terms of output harmonic performances, the best method may vary depending on m_a and m_f values. However, PD method generally has superiority over other methods. Also, zero sequence signal injection methods can refine harmonic distortion for low m_f values (SVPWM method) and decrease total count of switching (and the associated switching losses) in a phase leg (DPWM1 method). Input harmonics performances of carrier based PWM methods are similar in terms of dc link ripple current factor, K_{Iin}. However, for LS methods, low frequency harmonics of sub-module capacitor current are higher than that of PS method. Simulation results show that, it is not feasible to operate M2C without implementing a capacitor voltage balancing feedback control algorithm. Simulations with capacitor voltage sorting algorithm validate the superiority of PD over other methods for output harmonics performance. From the point of sub-module capacitor voltage ripples and circulating currents, no method has significant superiority. The sorting algorithm provides similar performances for different methods. After implementing circulation current control method, the results are consistent with the results of previous parts. In conclusion, PD method stands out as a better alternative among carrier based PWM methods for M2C switching because of its superior output harmonics performance.

REFERENCES

[1] A. Lesnicar and R. Marquardt, "An innovative modular multilevel converter topology suitable for a wide power range," in *Proc. IEEE Bologna Power Tech.*, vol.3, 2003.

[2] G. Carrara, S. Gardella, M. Marchesoni, R. Salutari, and G. Sciutto, "A new multilevel PWM method: a theoretical analysis," *IEEE Trans. Power Electron.*, vol. 7, no. 3, pp. 497–505, 1992.

[3] B. P. McGrath and D. G. Holmes, "A comparison of multicarrier PWM strategies for cascaded and neutral point clamped multilevel inverters," in *Proc. IEEE PESC, 2000*, pp. 674–679.

[4] B. P. McGrath, and D. G. Holmes, "Multicarrier PWM strategies for multilevel inverters," *IEEE Trans. Ind. Electron.*, vol. 49, no. 4, pp. 858–867, 2002.

[5] G. S. Konstantinou and V. G. Agelidis, "Performance evaluation of half-bridge cascaded multilevel converters operated with multicarrier sinusoidal PWM techniques," in *2009 4th IEEE Conf. Ind. Electron. Appl.*, pp. 3399–3404, May 2009.

[6] K. Ilves, A. Antonopoulos, S. Norrga, and H.-P. Nee, "Steady-state analysis of interaction between harmonic components of arm and line quantities of modular multilevel converters," *IEEE Trans. Power Electron.*, vol. 27, no. 1, pp. 57–68, Jan. 2012.

[7] D. Siemaszko, A. Antonopoulos, K. Ilves, M. Vasiladiotis, L. Ängquist, and H.-P. Nee, "Evaluation of control and modulation methods for modular multilevel converters," in *Proc. IPEC, 2010*, pp. 746–753.

[8] S. Sedghi, A. Dastfan, and A. Ahmadyard, "A new multilevel carrier based pulse width modulation method for modular multilevel inverter," in *Proc. ICPE/ECCE, Jeju, Korea, 2011*, pp. 1432-1439.

[9] A. M. Hava, R. J. Kerkman, and T. A. Lipo, "Simple analytical and graphical methods for carrier based PWM-VSI drives", *IEEE Trans. Power Electron.*, vol. 14, pp. 49–61, Jan. 1999.

[10] Q. Tu, Z. Xu, and L. Xu, "Reduced switching-frequency modulation and circulating current suppression for modular multilevel converters," *IEEE Trans. Power Del.*, vol. 26, no. 3, pp.2009-2017, Jul. 2011.

Control and Experiment of a 380-V, 15-kW Motor Drive Using Modular Multilevel Cascade Converter Based on Triple-Star Bridge Cells (MMCC-TSBC)

Wataru Kawamura, *Student Member, IEEE*, Makoto Hagiwara, *Member, IEEE*,
and Hirofumi Akagi, *Fellow, IEEE*
Department of Electrical and Electronic Engineering
Tokyo Institute of Technology
Tokyo, JAPAN

Abstract—This paper focuses on an application of the modular multilevel cascade converter based on triple-star bridge-cells (MMCC-TSBC) to adjustable-speed motor drives. The TSBC is a direct ac-to-ac power converter capable of bidirectional power flow with three-phase sinusoidal input and output currents. Therefore, it is suitable to medium-voltage high-power motor drives requiring regenerative braking. However, the TSBC suffers from such a serious problem that capacitor-voltage fluctuation becomes larger as the motor (output) frequency is closer to the supply (input) frequency. Recent research has solved this problem by using a new control method. However, no experimental results applied this control method have been presented. The aim of this paper is to show experimental verification applying the new control method when the output frequency is close to the input frequency. A specially-designed downscaled system rated at 400 V and 15 kW is used to confirm the validity and effectiveness of the new control method.

Keywords—Direct ac-to-ac power conversion, medium-voltage motor drives, modular multilevel cascade converters.

I. INTRODUCTION

The modular multilevel cascade converter (MMCC) family has been attracting attention to power electronics and power system engineers because it has been showing considerable promise as one of the next-generation high-power high-voltage/medium-voltage power converters. Generally, it is applicable to grid connections and/or motor drives. However, the selection of the best one out of the family members depends on its application and voltage/current ratings because each family member has its own character in terms of performance and viability.

This paper focuses on an application of the modular multilevel cascade converter based on triple-star bridge-cells (MMCC-TSBC) to adjustable-speed motor drives. For the sake of simplicity, the MMCC-TSBC is referred to as the TSBC. The circuit configuration of the TSBC is characterized by three sets of star-connected subconverters, each of which consists of three clusters. Each cluster consists of a cascade connection of multiple single-phase full-bridge (H-bridge) converter cells. The TSBC is also called as a "modular matrix converter," the naming of which stems from an expansion of the so-called "matrix converter." The TSBC is more suitable to motor drives requiring regenerative braking rather than grid connections. The reason is that it has the capability of drawing and feeding three-phase sinusoidal currents with any power factor at both supply (input) and motor (output) sides. Since no limitation

exists theoretically in the count of cascaded bridge cells per cluster, it is easy to design and construct high-power medium-voltage converters based on modular structure.

Basic research on the TSBC or the modular matrix converter is going on [1]–[10]. However, the TSBC has a serious problem that capacitor-voltage fluctuation becomes more serious as the motor (output) frequency is close to the supply (input) frequency. For this reason, it is said that the TSBC cannot achieve stable operation in a broad range of frequency control.

In order to solve this problem, the authors has already proposed a new control method in [11]. This control method is characterized by controlling the circulating currents flowing only inside the TSBC so as to cancel out the low-frequency components included in each dc-capacitor voltage. However, no experimental results applied this control method have been presented. Moreover, the proposed mean-voltage balancing control of dc capacitors would make itself unstable because the circulating currents for the balancing control compose a low-frequency power component as a kind of side effect.

The aim of this paper is to show experimental verification of the new control method when the output frequency is close to the input frequency as well as to improve the mean-voltage balancing control which can deal with a broad motor frequency range. A specially-designed downscaled system rated at 400 V and 15 kW is used to confirm the validity and effectiveness of the new control method.

II. CIRCUIT CONFIGURATION

Fig. 1 shows the detailed circuit configuration for a TSBC with four bridge cells per cluster to drive a three-phase motor. The TSBC consists of three star-connected subconverters, each of which has three clusters. Each cluster consists of a stack of four cascaded bridge cells and an ac inductor L. The three neutral points of the three subconverters are connected to the three motor terminals. Each bridge cell consists of four IGBTs and a dc capacitor equipped with a dc voltage sensor, forming a single-phase full-bridge or "H-bridge." Note that neither equipment nor component is connected to the dc side of each bridge cell except for the dc capacitor and the dc voltage sensor. The TSBC shown in Fig. 1 has 36 bridge cells and 144 IGBTs.

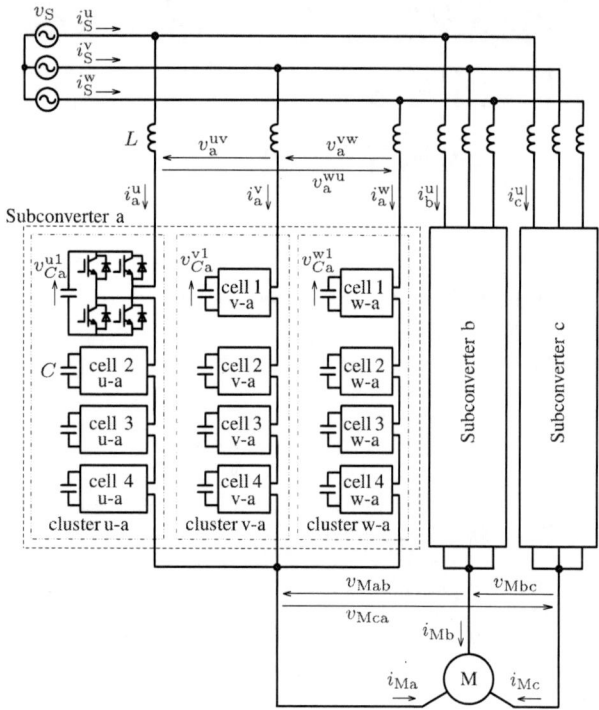

Fig. 1. Circuit configuration of the MMCC-TSBC with four bridge cells per cluster.

III. THE BASIC THEORY OF THE TSBC

A. Cluster Current

The most basic control makes both one third of the supply current and one third of the motor current flow in each cluster. However, a circulating current should be controlled to regulate the mean dc voltage of each dc capacitor and to mitigate ac voltage fluctuation. For this reason, when attention is paid to the cluster current i_a^u, the following equation is obtained

$$i_a^u = \frac{1}{3}i_S^u + \frac{1}{3}i_{Ma} + i_{Za}^u, \quad (1)$$

where i_S^u is the u-phase supply current, i_{Ma} is the a-phase motor current, and i_{Za}^u is the circulating current flowing into the cluster u-a. Note that the circulating current flows only inside the TSBC so that it is independent of both supply and motor currents [12]. Therefore, i_{Za}^u should satisfy the following equations.

$$i_{Za}^u + i_{Za}^v + i_{Za}^w = 0 \quad (2)$$

$$i_{Za}^u + i_{Zb}^u + i_{Zc}^u = 0. \quad (3)$$

Note that the control method for cluster currents has already been proposed in [12], which is applied to the experiments in sections VI and VII.

B. Capacitor-Voltage Fluctuation

Referring to [13] enables to analyze ac voltage fluctuation of each dc capacitor with the following three assumptions:

- All the bridge cells in one cluster produce the same ac voltage.

- The voltage at neutral point of the motor with respect to neutral point of the supply, i.e., the zero-sequence voltage v_N, is always controlled to zero.

- The effect of the ac inductor L on the ac voltage is ignored.

When attention is paid to the bridge cell numbered 1 in the cluster u-a of Fig. 1, the following equation exists

$$p_{a1}^u = \frac{1}{n}(v_S^u - v_{Ma})i_a^u, \quad (4)$$

where p_{a1}^u is an instantaneous active power at the ac side of the bridge cell numbered 1 in the cluster u-a, v_S^u is the u-phase supply voltage, v_{Ma} is the a-phase motor voltage.

The following relationship exists between p_{a1}^u and the capacitor voltage of the bridge cell numbered 1 in the cluster u-a v_{Ca}^{u1},

$$v_{Ca}^{u1} \simeq \frac{1}{CV_C} \int p_{a1}^u dt + V_C, \quad (5)$$

where V_C is the dc component of v_{Ca}^{u1}. Equation (5) suggests that a lower frequency component contained in p_{a1}^u causes a larger capacitor-voltage fluctuation.

IV. AC-VOLTAGE FLUCTUATIONS IN DC CAPACITORS AND MITIGATION

A. Fluctuation in No-Mitigation Control

The supply voltage and current, and the motor voltage and current are assumed as

$$v_S^u = \sqrt{\frac{2}{3}}V_S \sin\theta_S \quad (6)$$

$$i_S^u = \sqrt{2}I_S \sin(\theta_S + \phi_S) \quad (7)$$

$$v_{Ma} = \sqrt{\frac{2}{3}}V_M \sin\theta_M \quad (8)$$

$$i_{Ma} = \sqrt{2}I_M \sin(\theta_M + \phi_M), \quad (9)$$

where

$$\theta_S = 2\pi f_S t \quad (10)$$

$$\theta_M = 2\pi f_M t + \delta_M. \quad (11)$$

Note that V_S and V_M are the supply and motor line-to-line rms voltages, respectively, I_S and I_M are the supply and motor rms currents, respectively, ϕ_S and ϕ_M are the supply and motor phase angle, respectively, and δ_M is the motor initial phase with respect to the supply initial phase at $t = 0$.

When i_{Za}^u is equal to zero, (4) results in

$$p_a^{u1} = \frac{1}{3\sqrt{3}n}(V_S I_S \cos\phi_S - V_M I_M \cos\phi_M$$
$$+ V_M I_M \cos(2\theta_M + \phi_M)$$
$$+ V_S I_M \cos(\theta_S - \theta_M - \phi_M)$$
$$- V_M I_S \cos(\theta_S - \theta_M + \phi_S)) + p_h, \quad (12)$$

where p_h is the high-frequency power consisting of $2f_S$ and $f_S + f_M$ components. The first and second terms on the right-hand side in (12) cancel out each other because

$$P = \sqrt{3}V_S I_S \cos\phi_S = \sqrt{3}V_M I_M \cos\phi_M. \quad (13)$$

On the other hand, the third, fourth, and fifth terms may cause a larger fluctuation because they can contain low-frequency

978-1-4799-2706-7/14 $31.00 © 2014 IEEE

power components. The third term, which has $2f_M$ frequency component, will cause the large fluctuation of the dc-capacitors in the motor low-speed, while the fourth and fifth terms, which have $f_S - f_M$ frequency component, will make the problem when the motor frequency gets close to the supply frequency.

B. Mitigation for Low Output Frequency (presented in [12])

In order to reduce the fluctuation caused by a low motor frequency, the circulating current given by the following equation is used,

$$i_{Za}^{u} = -\frac{\sqrt{2}I_S\cos\phi_S}{3\cos\phi_M}\sin(\theta_S + 2\theta_M + \phi_M). \quad (14)$$

This circulating current forms an additional power p_{aZ}^{u1} as follows:

$$\begin{aligned}
p_{aZ}^{u1} &= v_a^{u1}i_{Za}^{u} \\
&= -\frac{V_S I_S \cos\phi_S}{3\sqrt{3}\cos\phi_M}\cos(2\theta_M + \phi_M) + p_{hZ}, \quad (15)
\end{aligned}$$

where p_{hZ} is the high-frequency power consisting of $f_S + f_M$, $2(f_S + f_M)$, and $f_S + 3f_M$ components, which can be neglected. The total power p_a^{u1} is calculated by summing up (12) and (15). Therefore, the first term on the right-hand side in (15) cancel out the third term on the right-hand side in (12) when the relation (13) exists. Then, p_a^{u1} has no low-frequency component so that the capacitor-voltage fluctuation can be mitigated.

C. Mitigation When the Output Frequency is Close to the Input Frequency (presented in [11])

In order to reduce the fluctuation when the motor frequency is close to the supply frequency, the circulating current given by the following equation is used,

$$\begin{aligned}
i_{Za}^{u} &= -\frac{\sqrt{2}}{3}I_S\sin(2\theta_M - \theta_S - \phi_M) \\
&\quad - \frac{\sqrt{2}}{3}I_M\sin(2\theta_S - \theta_M - \phi_S). \quad (16)
\end{aligned}$$

This circulating current forms an additional power p_{aZ}^{u1} as follows:

$$\begin{aligned}
p_{aZ}^{u1} &= v_a^{u1}i_{Za}^{u} \\
&= \frac{1}{3\sqrt{3}n}\left(-V_S I_M\cos(\theta_S - \theta_M - \phi_M)\right. \\
&\quad + V_M I_S\cos(\theta_S - \theta_M + \phi_S) \\
&\quad - V_S I_S\cos(2\theta_S - 2\theta_M + \phi_S) \\
&\quad \left. + V_M I_M\cos(2\theta_S - 2\theta_M - \phi_M)\right) + p_{hZ}, \quad (17)
\end{aligned}$$

where p_{hZ} is the high-frequency power consisting of $2f_M, 2f_S, 3f_S - f_M$, and $3f_M - f_S$ components, which can be neglected. The total power p_a^{u1} is calculated by summing up (12) and (17). Therefore, the first and second terms on the right-hand side in (17) cancel out the third and fourth terms on the right-hand side in (12). The fourth and fifth terms on the right-hand side in (17) also cancel out each other in the following conditions:

1) The relationship of (13) holds.
2) $Q_S = -Q_M$.

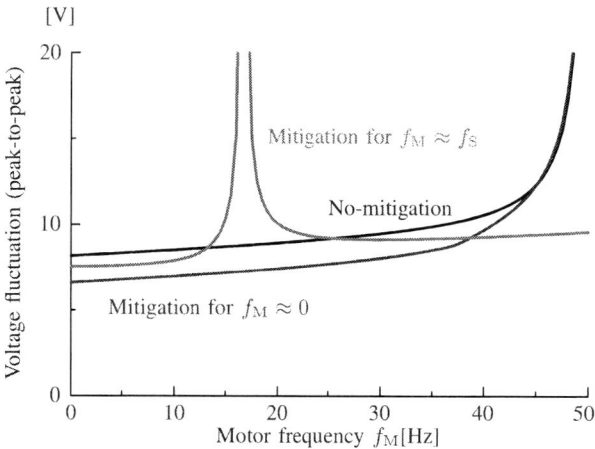

Fig. 2. Capacitor voltage fluctuation in peak-to-peak under 40% torque operation, a unit capacitance constant [14] of $H = 81$ ms, and a dc mean voltage of $V_C = 200$ V.

Here,

$$Q_S = \sqrt{3}V_S I_S\sin\phi_S \quad (18)$$
$$Q_M = \sqrt{3}V_M I_M\sin\phi_M. \quad (19)$$

Note that Q_S and Q_M are the reactive powers of the supply side and the motor side, respectively. The first condition is satisfied in most cases. On the other hand, the second condition is not always satisfied. Hence, the TSBC has to adjust either Q_S or Q_M appropriately (Q_S in most cases). If these two conditions are satisfied, p_a^{u1} has no low-frequency component so that the capacitor-voltage fluctuation can be mitigated. Note that this mitigation is also effective when f_M is exactly equal to f_S.

D. Theoretical Analysis of DC-Capacitor Voltage Fluctuation

Fig. 2 shows theoretical values in peak-to-peak of dc-capacitor voltage fluctuation operated with each control. An assumption of unity-power-factor operation is introduced at both supply and motor sides, to simplify the analysis. Therefore, an induction motor drive would make the actual voltage fluctuation several times larger than Fig. 2. The circuit parameters used in the analysis are shown in Table II , which are same to the experimental downscaled model shown in sections VI and VII. Fig. 2 also assumes 40% torque operation in the whole motor frequency range. Fig. 2 implies that it is required to switch over the three control methods according to motor frequency in order to mitigate the voltage fluctuation. When $f_M < 40$ Hz, the mitigation method for $f_M \approx 0$ is applied, while it is switched over to the mitigation method for $f_M \approx f_S$ after $f_M > 40$ Hz.

V. MEAN-VOLTAGE BALANCING OF DC CAPACITORS

The TSBC is required to regulate and balance the dc mean voltages of all the floating dc capacitors. A few papers have already proposed feedback-based balancing controls [7], [12]. However, some controls could make the dc-capacitor voltage unstable or diverge in a certain motor frequency range. Therefore, the TSBC should switch the balancing control properly as the motor frequency changes.

The dc-mean-voltage balancing control consists of the following three hierarchical subcontrols:

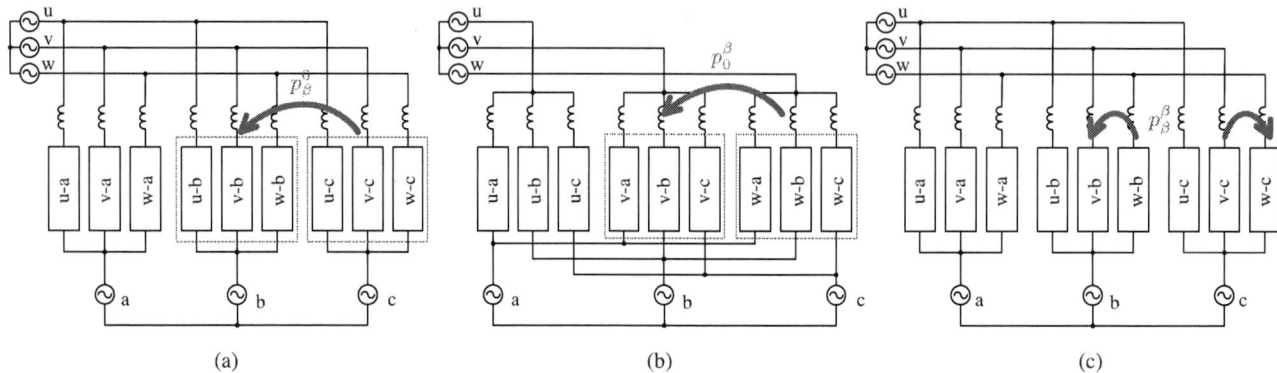

Fig. 3. Power-flow origin and destination of each of p_β^0, p_0^β, and p_β^β. (a) p_β^0. (b) p_0^β. (c) p_β^β.

1) Overall voltage control; it regulates the algebraic-average value of all dc-capacitor voltages $\overline{v}_C$ to the voltage command V_C^*.

2) Cluster-balancing control; it achieves balancing of the algebraic-average values of dc-capacitor voltages among the individual clusters. This control can be divided into an inter-subconverter balancing control and an inner-subconverter balancing control.

3) Individual-balancing control; it achieves voltage balancing of dc capacitors in each cluster.

Overall voltage control and individual-balancing control can be generally applied to all types of MMCC family, and they are effective in a full range of frequency. On the other hand, cluster-balancing control should be adapted to the frequency change.

A. Basic of the Cluster-Balancing Control

The algebraic-average value of the individual dc-capacitor voltages in cluster u-a is defined as $\overline{v}_{Ca}^u$, and those in the other clusters can be defined in the same way. At first, the following voltage matrix $[\overline{v}_{Cabc}^{uvw}]$ is defined.

$$[\overline{v}_{Cabc}^{uvw}] = \begin{bmatrix} \overline{v}_{Ca}^u & \overline{v}_{Ca}^v & \overline{v}_{Ca}^w \\ \overline{v}_{Cb}^u & \overline{v}_{Cb}^v & \overline{v}_{Cb}^w \\ \overline{v}_{Cc}^u & \overline{v}_{Cc}^v & \overline{v}_{Cc}^w \end{bmatrix}. \tag{20}$$

Applying the double $\alpha\beta0$ transformation [12] to (20) gives

$$[\overline{v}_{C\alpha\beta0}^{\alpha\beta0}] = \begin{bmatrix} \overline{v}_{C\alpha}^\alpha & \overline{v}_{C\alpha}^\beta & \overline{v}_{C\alpha}^0 \\ \overline{v}_{C\beta}^\alpha & \overline{v}_{C\beta}^\beta & \overline{v}_{C\beta}^0 \\ \overline{v}_{C0}^\alpha & \overline{v}_{C0}^\beta & \overline{v}_{C0}^0 \end{bmatrix}. \tag{21}$$

If all elements of voltage command matrix $[\overline{v}_{Cabc}^{uvw*}]$ are equal to V_C^*, (21) results in

$$[\overline{v}_{C\alpha\beta0}^{\alpha\beta0*}] = \begin{bmatrix} 0 & 0 & 0 \\ 0 & 0 & 0 \\ 0 & 0 & 3V_C^* \end{bmatrix}. \tag{22}$$

Since $\overline{v}_{C0}^0$ is equal to $3\overline{v}_C$, the regulation of $\overline{v}_{C0}^0$ can be achieved by overall balancing control. The cluster-balancing control has the function of making the other eight voltage elements zero. It is further divided into *inter*-subconverter balancing control and *inner*-subconverter balancing control. The former regulates the two elements of $[\overline{v}_{C\alpha\beta}^0]$ and the two elements of $[\overline{v}_{C0}^{\alpha\beta}]$ to zero, while the latter regulates the four elements of $[\overline{v}_{C\alpha\beta}^{\alpha\beta}]$ to zero.

In order to regulate these eight elements, the power matrix $[p_{abc}^{uvw}]$ is defined as follows, which is almost same as (4):

$$p_u^a = (v_S^u - v_{Ma})i_a^u, \tag{23}$$

and those in the other clusters can be defined in the same way. The relations between these power elements and dc-capacitor voltages are given with the help of (5) as

$$\begin{bmatrix} \overline{v}_{Ca}^u & \overline{v}_{Ca}^v & \overline{v}_{Ca}^w \\ \overline{v}_{Cb}^u & \overline{v}_{Cb}^v & \overline{v}_{Cb}^w \\ \overline{v}_{Cc}^u & \overline{v}_{Cc}^v & \overline{v}_{Cc}^w \end{bmatrix} \simeq \frac{1}{nCV_C} \int \begin{bmatrix} p_a^u & p_a^v & p_a^w \\ p_b^u & p_b^v & p_b^w \\ p_c^u & p_c^v & p_c^w \end{bmatrix} dt$$
$$+ V_C \begin{bmatrix} 1 & 1 & 1 \\ 1 & 1 & 1 \\ 1 & 1 & 1 \end{bmatrix}. \tag{24}$$

The nine power elements of $[p_{\alpha\beta0}^{\alpha\beta0}]$ are related to the nine dc-capacitor voltages of $[\overline{v}_{C\alpha\beta0}^{\alpha\beta0}]$, which are obtained by applying the double $\alpha\beta0$ transformation to (24) as

$$\begin{bmatrix} \overline{v}_{C\alpha}^\alpha & \overline{v}_{C\alpha}^\beta & \overline{v}_{C\alpha}^0 \\ \overline{v}_{C\beta}^\alpha & \overline{v}_{C\beta}^\beta & \overline{v}_{C\beta}^0 \\ \overline{v}_{C0}^\alpha & \overline{v}_{C0}^\beta & \overline{v}_{C0}^0 \end{bmatrix} \simeq \frac{1}{nCV_C} \int \begin{bmatrix} p_\alpha^\alpha & p_\alpha^\beta & p_\alpha^0 \\ p_\beta^\alpha & p_\beta^\beta & p_\beta^0 \\ p_0^\alpha & p_0^\beta & p_0^0 \end{bmatrix} dt$$
$$+ \begin{bmatrix} 0 & 0 & 0 \\ 0 & 0 & 0 \\ 0 & 0 & 3V_C \end{bmatrix}. \tag{25}$$

Equation (25) indicates that $\overline{v}_C (= \overline{v}_{C0}^0/3)$ can be controlled by p_0^0, where p_0^0 corresponds to the power flowing from both supply and motor to the nine clusters. On the other hand, each of the other eight power elements corresponds to the power circulating between one subconverter and another, or between one cluster and another.

Fig. 3 shows the power-flow origin and destination of each of p_β^0, p_0^β, and p_β^β. The power flow of p_β^0 and p_0^β occurs between one subconverter and another, while that of p_β^β happens between one cluster and another. Note that each power flow is bi-directional. Equation (25) suggests that the nine power elements of $[p_{\alpha\beta0}^{\alpha\beta0}]$ can be used to the cluster-balancing control.

1) Inter-Subconverter Balancing Control: There are two options for regulation of each voltage element of $[\overline{v}_{C\alpha\beta}^0]$, $[\overline{v}_{C0}^{\alpha\beta}]$. The following equations intend to balance the $\overline{v}_{C\beta}^0$ element as an example. The voltage element of p_α^0 can be balanced in the same way. The first option for the inter-subconverter balancing control relies on positive-sequence circulating currents as follows:

$$
\begin{cases}
\begin{bmatrix} i_{Za}^{u*} \\ i_{Za}^{v*} \\ i_{Za}^{w*} \end{bmatrix} = \begin{bmatrix} 0 \\ 0 \\ 0 \end{bmatrix} \\[12pt]
\begin{bmatrix} i_{Zb}^{u*} \\ i_{Zb}^{v*} \\ i_{Zb}^{w*} \end{bmatrix} = -K_{0i}\overline{v}_{C\beta}^0 \begin{bmatrix} \sin\theta_S \\ \sin\left(\theta_S - \frac{2}{3}\pi\right) \\ \sin\left(\theta_S + \frac{2}{3}\pi\right) \end{bmatrix} \\[18pt]
\begin{bmatrix} i_{Zc}^{u*} \\ i_{Zc}^{v*} \\ i_{Zc}^{w*} \end{bmatrix} = K_{0i}\overline{v}_{C\beta}^0 \begin{bmatrix} \sin\theta_S \\ \sin\left(\theta_S - \frac{2}{3}\pi\right) \\ \sin\left(\theta_S + \frac{2}{3}\pi\right) \end{bmatrix},
\end{cases}
\tag{26}
$$

where K_{0i} is the proportional control gain. These circulating current commands effect to the power element p_β^0 as follows:

$$
p_\beta^0 = -K_{0i}\overline{v}_{C\beta}^0 V_S.
\tag{27}
$$

This dc power charges/discharges the voltage element $\overline{v}_{C\beta}^0$ with a relation of (25), and achieves the voltage balancing.

The other option relies on the zero-sequence voltage v_N as follows:

$$
v_N^* = -K_{0v}\overline{v}_{C\beta}^0 \cos(\theta_M + \phi_M),
\tag{28}
$$

where K_{0v} is the proportional control gain. This zero-sequence voltage effects to the power element p_β^0 as follows:

$$
p_\beta^0 = -\frac{1}{2}K_{0v}\overline{v}_{C\beta}^0 I_M(1 + \cos 2(\theta_M + \phi_M)).
\tag{29}
$$

This power element contains dc component and ac component with a frequency of $2f_M$. The dc power component can achieve the voltage balancing, while the ac power component causes the voltage fluctuation, which will become larger in case of low motor frequency operation. Therefore, the former option relying on the circulating current is preferable for a broad motor frequency range.

On the other hand, regulation of $[\overline{v}_{C0}^{\alpha\beta}]$ can be achieved by either the circulating currents with a frequency of f_M or the zero-sequence voltage with a frequency of f_S.

2) Inner-Subconverter Balancing Control: Inner-subconverter balancing control has three options. The first option is characterized by using negative-sequence circulating currents as follows:

$$
\begin{cases}
\begin{bmatrix} i_{Za}^{u} \\ i_{Za}^{v} \\ i_{Za}^{w} \end{bmatrix} = \begin{bmatrix} 0 \\ 0 \\ 0 \end{bmatrix} \\[12pt]
\begin{bmatrix} i_{Zb}^{u} \\ i_{Zb}^{v} \\ i_{Zb}^{w} \end{bmatrix} = -K_1\overline{v}_{C\beta}^\beta \begin{bmatrix} \cos\theta_S \\ \cos\left(\theta_S + \frac{2}{3}\pi\right) \\ \cos\left(\theta_S - \frac{2}{3}\pi\right) \end{bmatrix} \\[18pt]
\begin{bmatrix} i_{Zc}^{u} \\ i_{Zc}^{v} \\ i_{Zc}^{w} \end{bmatrix} = K_1\overline{v}_{C\beta}^\beta \begin{bmatrix} \cos\theta_S \\ \cos\left(\theta_S + \frac{2}{3}\pi\right) \\ \cos\left(\theta_S - \frac{2}{3}\pi\right) \end{bmatrix},
\end{cases}
\tag{30}
$$

where K_1 is the proportional control gain. These circulating current commands effect to the power element p_β^β as follows:

$$
p_\beta^\beta = -\frac{1}{\sqrt{2}}K_1\overline{v}_{C\beta}^\beta(V_S - V_M\sin\theta_M\sin\theta_S).
\tag{31}
$$

This power element contains dc component used for the regulation of $\overline{v}_{C\beta}^\beta$, and ac component with a frequency of

TABLE I. OUTLINE OF THE CLUSTER-BALANCING CONTROL.

	v_C	$[i_{\alpha\beta}^{\alpha\beta*}]$	v_N^*	Equiv. gain	Used
Inter-subconv.	$[\overline{v}_{C\alpha\beta}^0]$	f_S	–	$K_{0i}V_S$	✓
		–	f_M	$K_{0v}I_M$	
	$[\overline{v}_{C0}^{\alpha\beta}]$	f_M	–	$K_{0i}V_M$	
		–	f_S	$K_{0v}I_S$	✓
Inner-subconv.	$[\overline{v}_{C\alpha\beta}^\beta]$	f_S	–	K_1V_S	✓ (low f_M)
		f_M	–	K_1V_M	
		dc	dc	$K_{1v}V_N$	✓ (high f_M)

$f_M + f_S$ and $f_M - f_S$. Therefore, this balancing control would get unstable when the motor frequency gets close to the supply frequency.

Another option is almost same as the first option except for using a frequency of f_M. This option also composes the power component with a frequency of $f_M - f_S$, which makes it difficult to use this option when $f_M \approx f_S$.

The other option gives v_N^* as a dc value of V_N, and the dc circulating current as

$$
\begin{cases}
\begin{bmatrix} i_{Za}^{u} \\ i_{Za}^{v} \\ i_{Za}^{w} \end{bmatrix} = \begin{bmatrix} 0 \\ 0 \\ 0 \end{bmatrix} \\[12pt]
\begin{bmatrix} i_{Zb}^{u} \\ i_{Zb}^{v} \\ i_{Zb}^{w} \end{bmatrix} = K_{1v}\overline{v}_{C\beta}^\beta \begin{bmatrix} 0 \\ 1 \\ -1 \end{bmatrix} \\[18pt]
\begin{bmatrix} i_{Zc}^{u} \\ i_{Zc}^{v} \\ i_{Zc}^{w} \end{bmatrix} = K_{1v}\overline{v}_{C\beta}^\beta \begin{bmatrix} 0 \\ -1 \\ 1 \end{bmatrix},
\end{cases}
\tag{32}
$$

where K_{1v} is the proportional control gain. These zero-sequence voltage and circulating currents effect to the power element p_β^β as follows:

$$
p_\beta^\beta = -K_{1v}\overline{v}_{C\beta}^\beta\left(V_N + \frac{1}{\sqrt{2}}V_S\sin\theta_S - \frac{1}{\sqrt{2}}V_M\sin\theta_M\right).
\tag{33}
$$

This power element contains dc component used for the regulation of $\overline{v}_{C\beta}^\beta$, and ac component with a frequency of f_S and f_M. Therefore, it is better to use this option than the others when the motor frequency is close to the supply frequency.

3) Controller Design: Table I summarizes the cluster-balancing control, paying attention to frequencies and equivalent gains used for each subconverter-balancing. Each subconverter-balancing control has two or three options that should be selected from the equivalent gains as well as the ac component caused by the side-effect of the balancing control. The reason is that the equivalent gains indicate which parameters affect the power necessary for balancing more strongly. For example, the two elements of $[\overline{v}_{C\alpha\beta}^0]$ can be controlled by either the circulating currents or the zero-sequence voltage. When the motor currents are small, the former option based on the circulating currents is a better choice in most cases because the latter option based on the zero-sequence voltage cannot generate the power enough to balance the dc-capacitor voltages. Table I also shows the selected options with ticks, which are intended for a motor drive application with a wide-speed range. These selected options will be used for experiments in sections VI and VII.

The 2014 International Power Electronics Conference

Fig. 4. Photo of the experimental system rated at 400 V and 15 kW.

TABLE II. CIRCUIT PARAMETERS OF FIGS. 1, 4 AND 5.

Rated active power	P	15 kW
Supply line-to-line rms voltage	V_S	400 V
Supply line frequency	f_S	50 Hz
AC inductor	L	3.0 mH (8.8%)
DC capacitor of bridge cells	C	1.7 mF
DC-capacitor voltage	V_C	200 V
Unit capacitance constant [14]	H	81 ms
Triangular-carrier frequency	f_C	1 kHz
Switching-ripple frequency	$8f_C$	8 kHz
Dead time		4 μs

Value in () is on a 400-V, 15-kW, and 50-Hz base.

TABLE III. MOTOR PARAMETERS USED IN THIS EXPERIMENT.

Rated output power		15 kW
Rated frequency	f_M	50 Hz
Rated line-to-line rms voltage	V_M	380 V
Rated rotating speed	N_{rm}	1460 min^{-1}
Rated stator rms current	I_1	32 A
Pole-pair number	p	2
Moment of motor inertia	J_M	0.1 kg·m^2
Moment of load inertia	J_L	0.1 kg·m^2

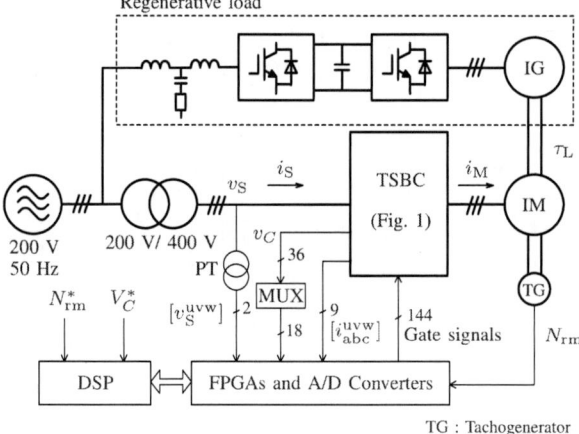

Fig. 5. Overview of the experimental downscaled system rated at 400 V and 15 kW.

VI. EXPERIMENTAL SYSTEM AND CONTROL SYSTEM

A. Experimental System

Fig. 4 shows a photo of the the experimental system rated at 400 V and 15 kW. The main circuit has four units. Three of them are subconverter a, b, and c, and the other is an interface unit playing an important role as the central controller.

Fig. 5 shows an overview of the experimental downscaled system rated at 400 V and 15 kW. The main circuit is the same as that in Fig. 1. The input (supply-side) terminals of the TSBC are connected to the three-phase 200-V and 50-Hz ac mains via a step-up transformer with the secondary voltage of 400 V. Table II summarizes the circuit parameters used in the experiment. The triangular-carrier frequency of each bridge cell was set as $f_C = 1$ kHz, and the dead time of each IGBT was set as 4 μs. The dc-voltage command of each bridge cell was set as 200 V. The unit capacitance constant [14] of the dc capacitors can be calculated as $H = 81$ ms, which is in a practical range.

Table III also summarizes the specifications of a 380-V, 15-kW induction motor tested. The regenerative load in Fig. 5 consists of an induction generator rated at 190 V and 15 kW, and two PWM converters connected back-to-back. The so-called "field-oriented control" is applied to the induction generator, which enables an arbitrary instantaneous torque τ_L to be loaded on the induction motor.

B. Control Circuit

The control circuit consists of one DSP board and two FPGA boards. The A/D converters on the FPGA boards convert detected analog voltage, current, and motor rotating speed signals to digital ones. Since the TSBC has a total of 36 dc capacitors, the use of multiplexers allows the control circuit to reduce the count of the A/D converters by half. Note that the u-phase supply current i_S^u and the a-phase motor current i_{Ma} can be calculated from the detected cluster currents as follows:

$$i_S^u = i_a^u + i_b^u + i_c^u \qquad (34)$$
$$i_{Ma} = i_a^u + i_a^v + i_a^w. \qquad (35)$$

The currents of the other phases can be calculated similarly as (34) and (35).

The DSP board receives the motor rotating speed command N_{rm}^* and the dc-capacitor voltage command V_C^*. Then, the DSP board calculates the voltage commands for the bridge cells, using the detected values (voltages and currents) coming from the FPGA boards. These voltage commands are sent to the FPGA boards, and the FPGA boards produces 144 gate signals finally. This paper adopts the phase-shifted PWM so that the four triangular-carrier signals in each cluster are phase-shifted each other by $45°(= 180°/4)$.

VII. EXPERIMENT

A. Start-up Performance

Fig. 6 shows the experimental start-up performance with 40% torque. Here, N_{rm}^* was increasing from zero to 1500 min^{-1} with a ramp rate of 90 min^{-1}/s. The mitigation for low output frequency was applied when $f_M < 35$ Hz ($N_{rm} \approx 1050$ min^{-1}) and it was switched over to the mitigation for

978-1-4799-2706-7/14 $31.00 © 2014 IEEE

$f_S \approx f_M$ after $f_M > 35$ Hz. The waveforms show that N_{rm} follows its command N_{rm}^* within an acceptable range. The motor current i_{Ma} is sinusoidal even during the start-up period. The peak value of the cluster current i_a^u rapidly increased by almost three times after the mitigation for $f_S \approx f_M$ was applied. The rms value also increased by two times after the mitigation. These differences come from the additional circulating current for the mitigation. The waveforms of v_{Ca}^{u1}, v_{Ca}^{v1}, and v_{Ca}^{w1} show that their dc components are well regulated to the command value of 200 V, and they contains the ac components whose amplitudes are less than 25% of 200 V.

B. Steady-state waveforms

Figs. 7 and 8 show the steady-state experimental waveforms with 40% torque. Fig. 7 shows the waveforms when $N_{rm}^* = 1410$ min^{-1}. The waveforms of i_S^u, i_S^v, and i_S^w are sinusoidal with a supply (line) frequency of 50 Hz, and drawing an amount of reactive power for mitigation of dc-capacitor voltage fluctuation as well as active power. The waveforms of i_{Ma}, i_{Mb}, and i_{Mc} are sinusoidal with a motor frequency of 47.5 Hz. The waveforms of v_{Ca}^{u1}, v_{Ca}^{v1}, and v_{Ca}^{w1} show that their dc components are well regulated to the command value of 200 V. The peak-to-peak voltage of the ac component present in each dc capacitor voltage is 31 V, which is within an acceptable level. Moreover, the ac-voltage component of $f_S - f_M (\approx 2.5$ Hz) is almost mitigated as a result of the mitigation control.

Fig. 8 shows the waveforms when $N_{rm}^* = 1500$ min^{-1}. The waveforms of i_S^u, i_S^v, and i_S^w are sinusoidal with a supply (line) frequency of 50 Hz and i_{Ma}, i_{Mb}, and i_{Mc} are sinusoidal with a motor frequency of 50.6 Hz. The waveforms of v_{Ca}^{u1}, v_{Ca}^{v1}, and v_{Ca}^{w1} show that their dc components are well regulated to the command value of 200 V, and the peak-to-peak voltage of ac components is 28 V. The ac-voltage component of $f_S - f_M (\approx 0.4$ Hz) is almost mitigated as a result of the mitigation control.

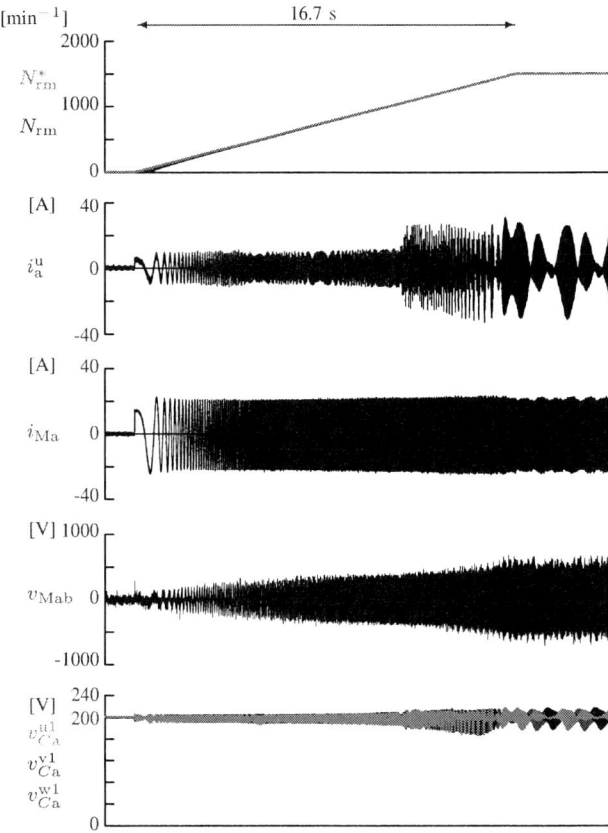

Fig. 6. Experimental start-up performance with 40% torque.

VIII. CONCLUSION

This paper has presented the modular multilevel cascade converter based on triple-star bridge-cells (MMCC-TSBC) with a focus on experimental verification when the motor (output) frequency is close to the supply (input) frequency. It has been shown that new control methods discussed on this paper makes it possible to operate stably even when the motor frequency is close to the supply frequency. A specially-designed downscaled system rated at 400 V and 15 kW has been used to confirm the validity and effectiveness of the new control method.

REFERENCES

[1] R. W. Erickson and O. A. Al-Naseem, "A new family of matrix converters," in *Conf. Rec. IEEE-IECON* 2001, vol. 2, pp. 1515–1520.

[2] S. Angkititrakul and R. W. Erickson, "Control and implementation of a new modular matrix converter," in *Conf. Rec. IEEE-APEC* 2004, vol. 2, pp. 813–819.

[3] S. Angkititrakul and R. W. Erickson, "Capacitor voltage balancing control for a modular matrix converter," in *Conf. Rec. IEEE-APEC* 2006.

[4] C. Oates and G. Mondal, "DC circulating current for capacitor voltage balancing in modular multilevel matrix converter," in *Conf. Rec. IEEE-EPE* 2011.

[5] A. J. Korn, M. Winkelnkemper, P. Steimer, and J. W. Kolar, "Direct modular multilevel converter for gearless low-speed drives," in *Conf. Rec. IEEE-EPE* 2011.

[6] F. Kammerer, J. Kolb, and M. Braun, "A novel cascaded vector control scheme for the modular multilevel matrix converter," in *Conf. Rec. IEEE-IECON* 2011, pp. 1097–1102.

[7] F. Kammerer, J. Kolb, and M. Braun, "Fully decoupled current control and energy balancing of the modular multilevel matrix converter," in *Conf. Rec. IEEE-EPE-PEMC* 2012, LS2a.3.

[8] Y. Miura, T. Mizutani, M. Ito, and T. Ise, "Modular multilevel matrix converter for low frequency ac transmission," in *Conf. Rec. IEEE-PEDS* 2013, pp. 1079–1084.

[9] Y. Miura, T. Mizutani, M. Ito, and T. Ise, "A novel space vector control with capacitor voltage balancing for a multilevel modular matrix converter," in *Conf. Rec. IEEE-ECCE Asia* 2013, pp. 442–448.

[10] Y. Hayashi, T. Takeshita, M. Muneshima, and Y. Tadano, "Independent control of input current and output voltage for modular matrix converter," in *Conf. Rec. IEEE-IECON* 2013, pp. 888–893.

[11] W. Kawamura, M. Hagiwara, and H. Akagi, "A broad range of frequency control for the modular multilevel cascade converter based on triple-star bridge-cells (MMCC-TSBC)," in *Conf. Rec. IEEE-ECCE* 2013, pp. 4014–4021.

[12] W. Kawamura, M. Hagiwara, and H. Akagi, "Control and experiment of a modular multilevel cascade converter based on triple-star bridge cells (MMCC-TSBC)," *IEEE Trans. Ind. Appl.*, to be published.

[13] M. Hagiwara, K. Nishimura, and H. Akagi, "A medium-voltage motor drive with a modular multilevel PWM inverter," *IEEE Trans. Power Electron.*, vol. 25, no. 7, pp. 1786–1799, Jul. 2010.

[14] H. Fujita, S. Tominaga, and H. Akagi, "Analysis and design of a dc voltage-controlled static var compensator using quad-series voltage-source inverters," *IEEE Trans. Ind. Appl.*, vol. 32, no. 4, pp. 970–977, Jul./Aug. 1996.

The 2014 International Power Electronics Conference

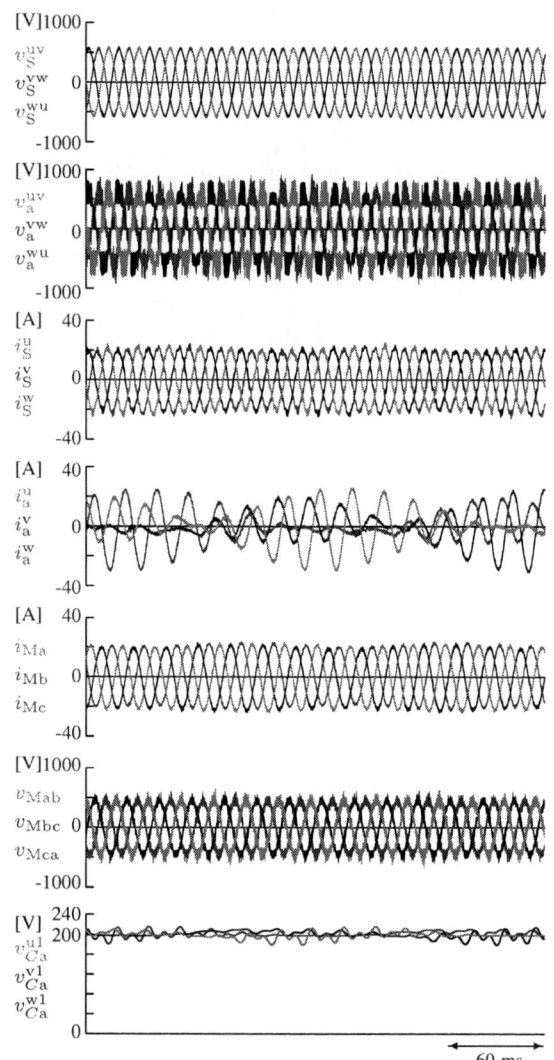

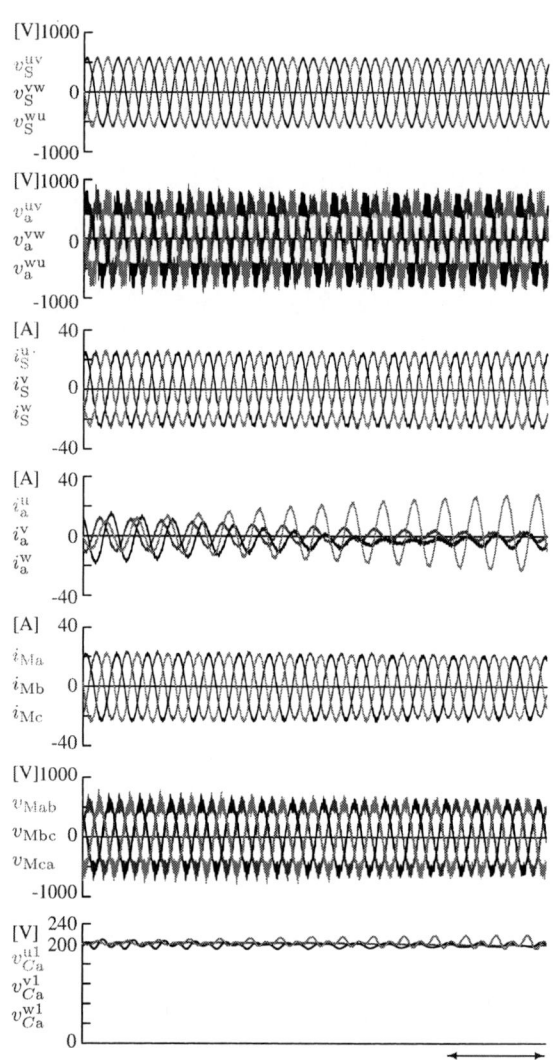

Fig. 7. Experimental waveforms with 40% torque and $N_{\mathrm{rm}}^* = 1410\ \mathrm{min}^{-1}$ ($f_{\mathrm{M}} \approx 47.5\ \mathrm{Hz}$).

Fig. 8. Experimental waveforms with 40% torque and $N_{\mathrm{rm}}^* = 1500\ \mathrm{min}^{-1}$ ($f_{\mathrm{M}} \approx 50.6\ \mathrm{Hz}$).

The 2014 International Power Electronics Conference

A Power Electronic Transformer with Sinusoidal Voltages and Currents using Modular Multilevel Converter

Ashish Kumar Sahoo* and Ned Mohan
Department of Electrical and Computer Engineering
University of Minnesota, Minneapolis, USA
*email−saho0007@umn.edu

Abstract—Power electronic transformer consisting of power converters and a high frequency transformer (HFT) can be used to interface a low voltage machine to a high voltage grid. A modular multilevel converter (MMC) is proposed as the power converter on the high voltage side to generate high frequency, adjustable magnitude sinusoidal voltages. A matrix converter (MC) is used on the low voltage side to synthesize three-phase adjustable frequency PWM AC at the machine terminals. With the leakage inductance of the transformer, a capacitor bank forms an *LC* filter to result in sinusoidal currents through the HFT. With sinusoidal voltages and currents through the transformer, there is significant reduction in transformer losses and also natural commutation of leakage energy is obtained. The magnitude of the output voltage requirement by the machine is met by controlling the output voltage of the MMC on the primary of the HFT, to result in reduced voltage stress and losses in the transformer, secondary side converter and the machine. The operating principle, modulation and control of the proposed PET is validated by simulations in MATLAB/Simulink.

Keywords—High frequency transformer (HFT), modular multilevel converter (MMC), matrix converter (MC), power electronic transformer (PET), pulse-width modulation (PWM)

I. INTRODUCTION

In the modern distribution system, power transformers play an important role in voltage boosting and providing galvanic isolation for protection. Replacement of line frequency transformers with power electronic transformers (PET) results in significant reduction in volume and weight along with added advantages like high power density, voltage regulation, power factor control, on demand reactive power support, etc [1]. This kind of transformer topologies find wide applications in interfacing renewables (solar, wind), traction, medium voltage ASDs, energy storage systems, UPS, mobile substations, etc.

Various PET topologies exist in literature [2]. Most PET configurations involve three stage power conversion [3] [4]. Two stage PET has the secondary side to be a direct AC/AC converter like a cycloconverter [5]. For high voltage and high power applications, PET topologies have to be connected in series/parallel due to limitation of power semiconductor device ratings. In [6], basic building blocks of isolated DC/DC converter made from DC/AC inverter and current source AC/DC converter with a medium frequency transformer are connected

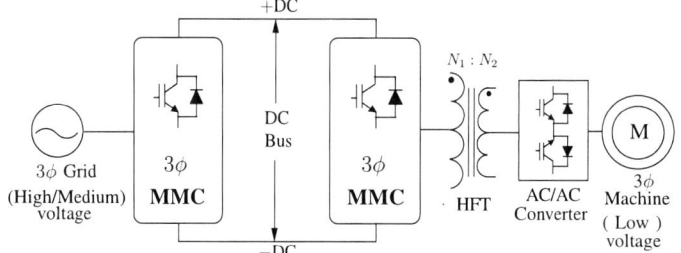

Fig. 1: Proposed MMC based PET

in input series-output parallel configuration. A HVDC transmission line is formed in [7] by series connection of PET modules involving a diode bridge rectifier on the secondary side of the high frequency transformer (HFT) in each module. Use of multilevel topologies like neutral point clamped converter [8] is difficult to scale. Cascaded H-bridge multilevel converter based PET have been widely discussed [9] [10]. Recently the modular multilevel converter (MMC) has emerged as a viable solution for high voltage applications because of its easy scalability to reach high number of voltage levels by simple series connection of submodules, resulting in higher reliability and easy maintenance [11]. Modular converters involving 3-stage power conversion with a dual active bridge and HFT in each module is given in [12]. Use of large number of conversion stages involves intermediate bulky energy storage elements and results in reduced power density and efficiency. Matrix converter based single stage PETs are discussed in [13] [14]. However these topologies cannot be used for high voltage applications and require additional techniques for commutation of leakage energy.

All of the previous modular PETs discussed above for high voltage applications involve a HFT in each module and multiple conversion stages. Also the HFT is subjected to switched voltages/currents which is more lossy. The proposed PET topology involving MMC is shown in Fig. 1. It includes a two stage bidirectional AC/DC and DC/AC stage where the primary is at high or medium voltage and the machine is operated at low voltage. This PET topology can find applications in integrating wind energy generators to a high voltage grid. It can also be used in isolated medium voltage adjustable speed

978-1-4799-2706-7/14 $31.00 © 2014 IEEE 3750

The 2014 International Power Electronics Conference

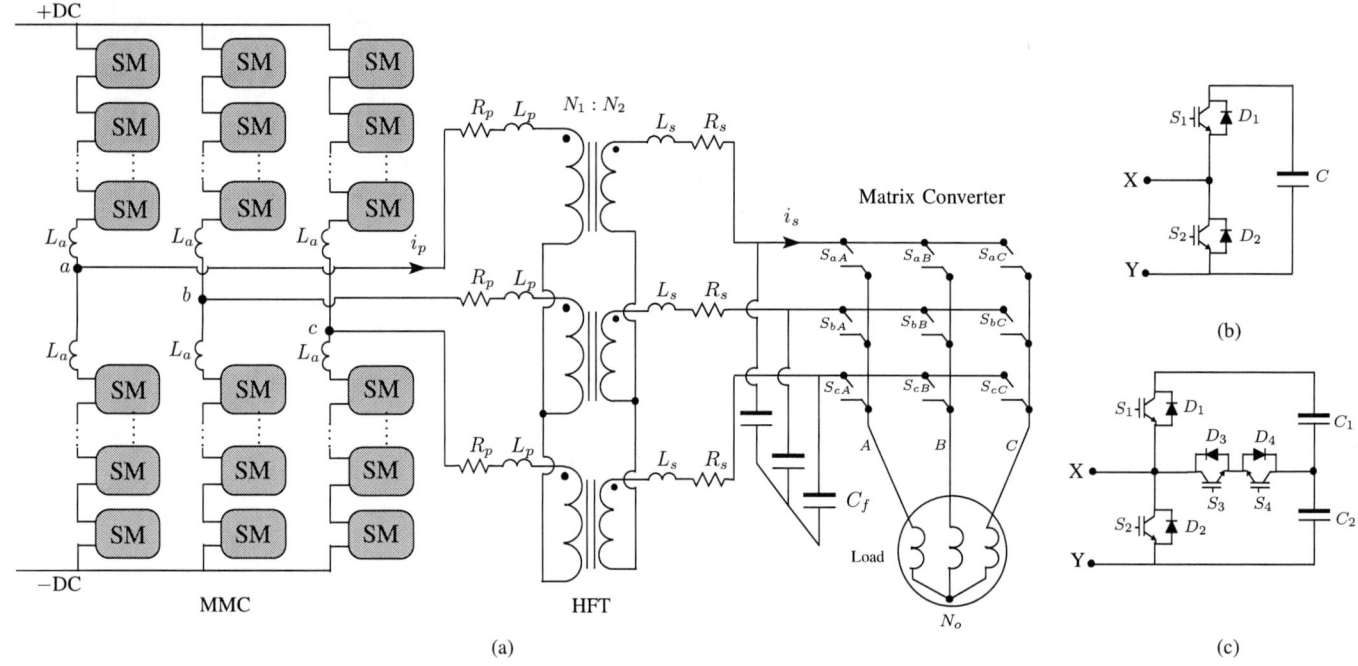

Fig. 2: (a) Proposed power electronic transformer topology, (b) half-bridge submodule (SM), (c) 3-level submodule (SM)

drives. The DC bus could be very short as in the case of onshore wind farms or medium voltage drives. It could also be 100 miles HVDC line for offshore wind farms. The proposed topology results in sinusoidal voltages and currents through the HFT resulting in significant reduction of transformer losses and size. Also due to sinusoidal currents, natural commutation of leakage energy is obtained.

The proposed topology with its working principle is presented in Section II. Section III presents the individual modulation of the primary and secondary side converters, followed by modulation of the combined PET topology. The design of high frequency link filter capacitor is presented in Section IV. The proposed control of primary side MMC to meet the output voltage variation is given in Section V. Key results are presented in Section VI and the paper concludes in Section VII.

II. TOPOLOGY

A partial circuit of Fig. 1 is a single stage DC/AC power electronic transformer topology to interface the high voltage DC bus with the low voltage machine as shown in Fig. 2. To meet the device stress on the high voltage side, a multilevel structure using a modular multilevel converter is proposed. The MMC is modulated to synthesize high frequency near sinusoidal voltages at the primary of the HFT. Another MMC is used at the end of the DC distribution line to interface with the high voltage AC grid as shown in Fig. 1. Use of MMC results in near sinusoidal voltage waveforms at reduced switching frequencies. Each leg of the MMC is made up of two arms consisting of n series connected submodules (SM). Each arm also has a protection choke L_a to limit over-currents

during faults. The internal structure of the submodule can be the conventionally used 2-level half bridge submodule (Fig. 2(b)) or the recently proposed 3-level submodule (Fig. 2(c)). The 3-level submodule has added advantages over the half bridge submodule as it results in nearly half the submodule requirements resulting in a more compact structure and significantly reduced semiconductor losses [15]. The focus of this paper would be on the 3-level submodule structure.

The 3-phase medium/high frequency transformer is made up of three 2-winding transformers with $N_1 : N_2$ turns ratio . Every transformer has leakage inductance's in its winding. L_p, R_p are the primary winding leakage inductance and parasitic resistance. Similarly L_s, R_s are the secondary side leakage inductance and parasitic resistance. Instead of using three 2-winding transformers, a 3-phase high frequency transformer could also be used.

The secondary low voltage side of the HFT is connected to a 3×3 matrix converter (MC). From the sinusoidal voltages produced by the MMC, the matrix converter generates 3-phase PWM AC of desired frequency. It uses an array of controlled bidirectional switches to couple a 3-phase high frequency link with the 3-phase low frequency machine without the need of any intermediate energy storage elements. Because of pulse-width modulation (PWM), the MC injects high frequency switching components into its input current. Hence an LC filter is required. Here the leakage inductance of the transformer is used along with an externally added very small capacitance C_f for filtering action. This results in sinusoidal currents through the HFT. Thus the proposed topology results in sinusoidal voltages and sinusoidal currents through the HFT, thus significantly reducing the transformer losses.

978-1-4799-2706-7/14 $31.00 © 2014 IEEE

The 2014 International Power Electronics Conference

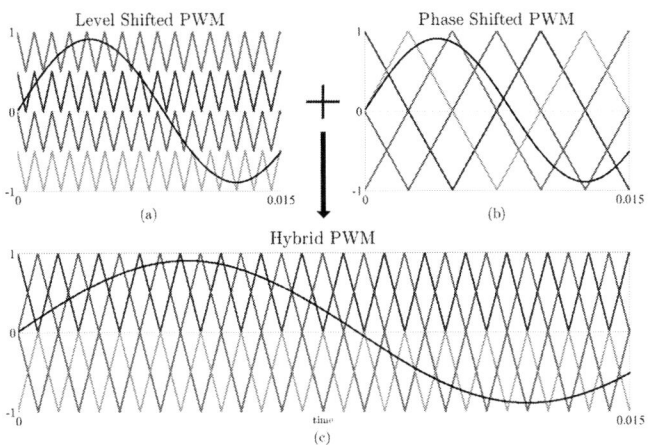

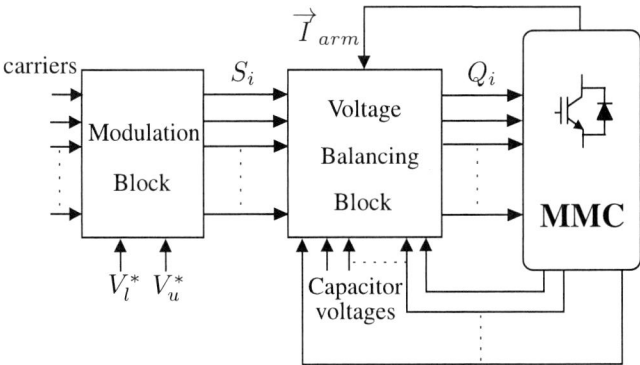

Fig. 3: Hybrid modulation scheme for MMC combining phase shifted and level shifted carriers

Fig. 4: MMC control blocks

III. MODULATION

The modulation of the proposed PET consists of the modulation of the MMC on the primary side and the matrix converter on the secondary side. This section presents a carrier based pulse width modulation approach for the MMC and an indirect space vector modulation approach for the matrix converter.

A. Modulation of primary side MMC

A carrier based PWM scheme is used to modulate the MMC using a combination of phase shifted and level shifted carriers called the hybrid modulation scheme, Fig. 3. The carriers are phase shifted for the submodules in a leg and level shifted for the IGBT's inside each submodule. Every other phase-shifted carrier is applied to a submodule in the upper arm and lower arm respectively. Two level shifted carriers are used for the IGBTs inside each submodule. Two reference signals V_u^* and V_l^* are used in each leg for comparison with the carriers to generate gate signals for the upper arm and lower arm respectively. The output voltage V_{conv}^* of the leg is the difference of the generated voltages V_u^* and V_l^* in the respective arms. A compensation term V_{La}^* is included in the reference signals to account for the drop across the arm inductor L_a.

$$V_u^* = -V_{conv}^* - V_{La}^* + \frac{V_{DC}}{2}$$
$$V_l^* = +V_{conv}^* - V_{La}^* + \frac{V_{DC}}{2} \qquad (1)$$

This modulation results in 3 states of the MMC submodule: FULL-ON state (S_1=1), HALF-ON state (S_3, S_4=1) and BYPASS state (S_2=1). Based on this modulation, unequal capacitance division of $C_1 = C/3$ and $C_2 = 2C/3$ is used which can balance the capacitor voltages naturally in ideal conditions. This is chosen such that the voltage across the two capacitors is maintained constant ($V_C = Q/C = it/C$).

Here V_C is the voltage across a capacitor with charge Q and capacitance value C. The charging/discharging time is t with a current i flowing through it. Since C_2 would be charged or discharged in both the FULL-ON and HALF-ON states for double the time t, its capacitance value is chosen to be double of C_1.

In real life situations, when capacitors degrade or their voltages fluctuate due to disturbances, a voltage balancing technique is used that maintains them constant. The idea is to use the capacitors with the highest voltage when the MMC is on a discharging state, and use the capacitors with the lowest voltage when it is on a charging state. It is based on a sorting algorithm which monitors and sorts the capacitor voltages and then allocates the appropriate gate signal to the corresponding submodule IGBT. This is explained in detail in [16]. The modulation block generates ideal gate pulses S_i which is allocated in the voltage balancing block to the appropriate IGBT. The voltage balancing block then generates the actual gate signals Q_i for each IGBT inside every submodule. The direction of arm currents is taken as a feedback to know if it is a charging or discharging current. The control blocks for the entire MMC modulation is shown in Fig. 4.

B. Modulation of machine connected Matrix Converter

Space vector modulation (SVM) based indirect modulation technique is used to achieve highest possible voltage transfer ratio (0.866) in the matrix converter [17]. In indirect modulation the MC is modulated using two fictitious converters, a current source inverter (CSI) and a voltage source inverter (VSI) connected through a virtual DC-link. The switching states are shown in Fig. 5(a) and Fig. 5(b) respectively. Both of these converters together produce 18 active switching states. For example, state [a b b] in the MC in which a is connected to output phase A and, b is connected to output phase B and C respectively can be implemented by simultaneously applying [a b] and [1 0 0] by the indirect modulation. In this analysis, space vector corresponding to a set of 3-phase quantities x_a, x_b, x_c is a complex vector X as given in (2). The MC converts the high frequency link voltages to low frequency voltages at the machine terminals. Hence the switching frequency of the

978-1-4799-2706-7/14 $31.00 © 2014 IEEE   3752

The 2014 International Power Electronics Conference

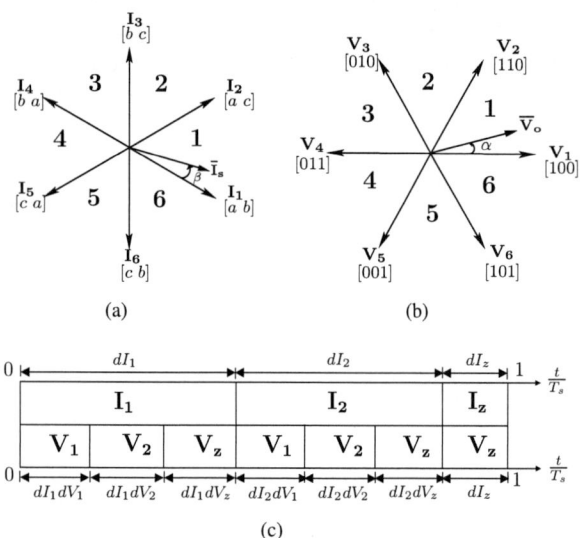

(a)   (b)

(c)

Fig. 5: (a) Current space vectors produced by CSI (b) Voltage space vectors produced by VSI (c) Switching sequence of Matrix converter

matrix converter f_s is set higher than the transformers link frequency f_{link} i.e $f_s > f_{link}$.

$$X = x_a + x_b e^{j2\pi/3} + x_c e^{-j2\pi/3} \tag{2}$$

In one sampling cycle $T_s = 1/f_s$, the reference current vector $\overline{\mathbf{I}}_s$ in Fig. 5(a) and output reference voltage space vector $\overline{\mathbf{V}}_o$ in Fig. 5(b) are generated from two adjacent active vectors and one zero vector, whose duty ratios are given by (3) and (4) respectively. Here m_I is the ratio of the peak of the fundamental component of the input current to the average virtual DC-link current and m_V is the ratio of the peak of the fundamental component of the output voltage to the average virtual DC-link voltage. The switching sequence applied over a sampling cycle is given in Fig. 5(c). The peak of the average output voltage V_o can be written in terms of the peak of the secondary side link voltage V_s as in (5). Similarly the peak of the link current I_s on the secondary of the transformer can be written in terms of the peak of the load current I_o as given in (6) where ϕ_o is the load power factor angle.

$$
\begin{aligned}
dI_1 &= m_I \sin\left(\frac{\pi}{3} - \beta\right) \\
dI_2 &= m_I \sin\beta \\
dI_z &= 1 - dI_1 - dI_2 \\
dV_1 &= \sqrt{3}m_V \sin\left(\frac{\pi}{3} - \alpha\right) \\
dV_2 &= \sqrt{3}m_V \sin\alpha \\
dV_z &= 1 - dV_1 - dV_2
\end{aligned}
\tag{3}
\tag{4}
$$

$$V_o = \frac{3}{2} m_I m_V V_s \tag{5}$$

$$I_s = \frac{3}{2} m_I m_V I_o \cos\phi_o \tag{6}$$

C. Combined modulation of PET

For the proper operation of the PET, both the primary and secondary converters should operate synchronously. The net output voltage generated depends upon the modulation indices of both the converters along with the transformation ratio of the HFT. The peak output voltage synthesized can be represented in terms of the input high voltage DC bus as shown in (7). Here m_p is the modulation index of the primary side MMC and $m_s = m_I m_V$ is the modulation index of the secondary side matrix converter connected to the machine. The high voltage DC-bus current can be represented in terms of the peak output load current as shown in (8).

$$V_o = \frac{3}{2} \frac{N_2}{N_1} m_p m_s V_{dc} \tag{7}$$

$$I_{dc} = \frac{3}{2} \frac{N_2}{N_1} m_p m_s I_o \cos\phi_o \tag{8}$$

IV. HIGH FREQUENCY LINK FILTER DESIGN

Due to pulse-width modulation, the matrix converter injects high frequency switching harmonics over the fundamental link frequency in the secondary current of the HFT. As mentioned in the previous section, the modulation of the matrix converter is done at a switching frequency which is higher than the link frequency. To result in sinusoidal currents flowing through the HFT, a low pass LC filter is designed. The leakage inductance of the transformer forms this filter with an externally added small capacitance. The leakage inductance L_{lkg} is the combination of the secondary side leakage inductance L_s and the primary leakage inductance L_p referred to the secondary side $\left(L_{lkg} = L_p \left(\frac{N_2}{N_1}\right)^2 + L_s\right)$. The leakage resistance is obtained similarly $\left(R_{lkg} = R_p \left(\frac{N_2}{N_1}\right)^2 + R_s\right)$.

The RMS of the input current $\langle i_s \rangle$ of the matrix converter from space vector modulation is derived in [18] and given by (9). The RMS of the switching components $\langle \tilde{i}_s \rangle$ occurring at multiples of the MC switching frequency f_s are given by (10). The per phase equivalent circuit of the MC at its switching frequency is shown in Fig. 6. Assuming all of the ripple components to be at the switching frequency of the matrix converter, a low pass filter can be designed with a corner frequency between the link frequency and switching frequency. The filter capacitor is to be designed such that it takes in most of this higher harmonics ripple current. For an allowable ripple in the secondary side link current $\langle \tilde{i}'_p \rangle = \lambda_1 \left(\frac{I_s}{\sqrt{2}}\right)$ and allowable distortion in the input voltage of the matrix converter

978-1-4799-2706-7/14 $31.00 © 2014 IEEE

$\langle \widetilde{v}_s \rangle = \lambda_2 \left(\dfrac{V_s}{\sqrt{2}} \right)$, the filter capacitor value can be calculated using (11). From the calculated value of C_f and given value of L_{lkg}, the corner frequency of the LC filter is calculated using (12). The leakage resistance R_{lkg} provides the necessary damping at the resonant or corner frequency f_c. The corner frequency should be more than the transformer link frequency but less than the switching frequency of the matrix converter. For a good design, $3f_{link} < f_c < f_s/3$. If the calculated value of f_c is not in the above range, the specifications of λ_1 and λ_2 is varied and the value of C_f is recalculated.

$$\langle i_s \rangle^2_{RMS} = \frac{3\sqrt{3}m_I m_V I_o^2}{\pi^2} \left(\frac{\pi\sqrt{3}}{12} + \frac{3}{8} \right)(1 + \cos 2\phi_o)$$
$$+ \frac{3\sqrt{3}m_I m_V I_o^2}{\pi^2} \left(\frac{\pi}{12} - \frac{\sqrt{3}}{16} \right) \sin 2\phi_o \qquad (9)$$

$$\widetilde{\langle i_s \rangle}^2_{RMS} = \langle i_s \rangle^2_{RMS} - \left(\frac{3}{2\sqrt{2}}m_I m_V I_o \cos\phi_o \right)^2 \qquad (10)$$

$$C_f = \frac{\langle \widetilde{i}_s \rangle_{RMS} - \langle \widetilde{i}'_p \rangle_{RMS}}{\omega_s \langle \widetilde{v}_s \rangle_{RMS}} \qquad (11)$$

$$f_c = \frac{1}{2\pi\sqrt{L_{lkg}C_f}} \qquad (12)$$

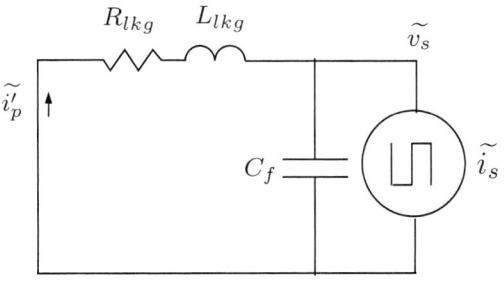

Fig. 6: Per-phase equivalent circuit of the matrix converter with input filter at the switching frequency

V. CONTROL

A rotating machine is a generator in wind applications or an adjustable speed motor drive. Due to variation in rotor speed, the magnitude of voltage produced/required by the machine varies. Conventionally, modulation index of the the AC/AC converter is controlled to adjust this variable magnitude AC generation. The controller G_C generates the reference voltage for the matrix converter to meet the variation in load voltage demand as shown in Fig. 7 (dotted lines). However another solution is to control the modulation index of the MMC at the primary of the transformer as shown in Fig. 7 (solid lines).

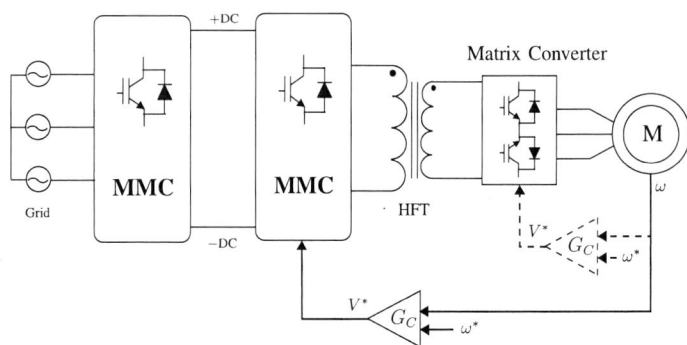

Fig. 7: Control of PET to meet variation in motor speed

The matrix converter is always operated at its full modulation index of 0.866. It only acts as an interface to generate adjustable frequency PWM voltages at the machine terminals. For a reduction in speed to one-half, the voltage requirement by the machine is also reduced to nearly one-half ($V/f \approx$ constant). Hence the MMC is modulated to generate one-half the voltage magnitude at the primary of the HFT. This in turn makes use of just one-half the number of submodules. The purpose of pushing the control to the primary side MMC is to reduce the voltage stress related losses in the transformer, matrix converter and the machine. If the voltage passing through the transformer is one-half, the losses are significantly reduced.

VI. RESULTS

The AC/AC power electronic transformer topology as shown in Fig. 1 is simulated with ideal switches in MAT-LAB/Simulink with parameters as shown in Table 1. It makes use of two 3-phase MMC to interface the high voltage AC grid with the high frequency transformer. Each leg of the MMC consists of eight 3-level submodules. The grid is modeled as an ideal AC source at 11 kV line to line RMS and 60 Hz, with a grid inductance of 5 mH. The DC bus is at 34.5 kV. The HFT has turns ratio 10:1 and an effective leakage inductance of 300 μH, referred to the secondary side. The leakage resistance is 0.1 Ω. The PET feeds a three-phase $R - L$ load with a load power factor of 0.8 at 30 Hz. Following the design in Section IV, for an allowable ripple of 5% in the link current, the designed link filter capacitor comes out to be 8.5 μF.

TABLE I: Table : Simulation Parameters

Parameters	Value
Grid	$V_{g(RMS)}$=11 kV, 60 Hz
Submodule capacitor (MMC)	C_1=1.66 mF, C_2=3.33 mF
Arm inductor L_a	1 mH
Link filter capacitor C_f	8.5 μF
Transformer	10:1, 4 kHz, L_{lkg}=300 μH
Load	20 kW, 0.8 pf, 30 Hz

The 2014 International Power Electronics Conference

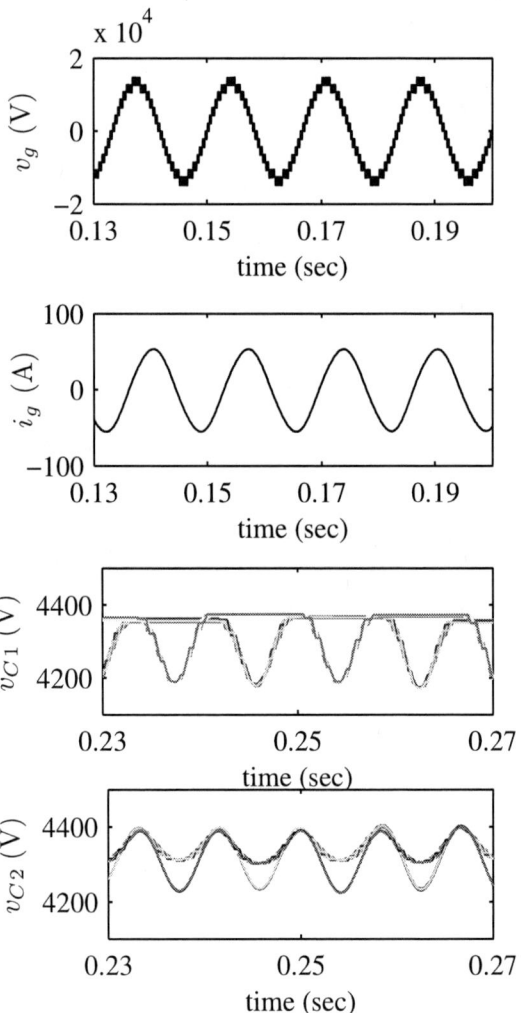

Fig. 8: (top to bottom) Grid phase voltage and line current, balanced capacitor voltages of all capacitors in upper and lower arm for C_1 and C_2

A reduction in output voltage requirement to one-half is met by giving a step change in the modultaion index of the MMC at the primary of the HFT. The high frequency link waveforms are shown in Fig. 9 along with a zoomed view on the right. The waveforms on the left in Fig. 9 shows a clear step change in voltages and currents in the transformers high frequency link when the load changes. The clear zoomed waveforms are shown on the right. As can be seen, the primary voltage of the high frequency transformer as shown in Fig. 9 (top) reaches 15 distinct levels by use of 8 modules of the new 3-level submodule topology, using modulation index $m_p = 0.8$. The switching frequency is set at 8 kHz to result in a single voltage level in every switching cycle of the MMC. The matrix converter injects switched currents at its input which is filtered and only sinusoidal currents pass through the high frequency transformer. This is shown by the sinusoidal currents on the primary side of the HFT, Fig. 9 (second from top). The leakage inductance of the transformer acts as a smoothing reactor to minimize the harmonics in the primary voltage of the HFT. Hence near sinusoidal voltages can be seen on the secondary side of the high frequency transformer. Use of a 10:1 transformer reduces the voltage magnitude by 10 times as can be seen in Fig. 9 (third from top). The high frequency switched currents injected by the matrix converter due to pulse-width modulation are shown in Fig. 9 (bottom).

Because of the proposed control in Section V, the matrix converter is modulated at a full modulation index of 0.866. The output is synthesized at a frequency of 30 Hz. The load parameters are shown in Table I. The corresponding load line to neutral voltage of one phase and line currents are shown in Fig. 10. The change in modulation index of the MMC to half is reflected by a step change in output voltage level of the matrix converter. The control of output voltage magnitude using matrix converter would have resulted in the same voltage levels but use of more number of zero vectors. Instead pushing the control to the primary side of the HFT has resulted in reduced voltage stress related losses in the transformer, matrix converter and the machine.

VII. CONCLUSION

A power electronic transformer based on modular multilevel converter is proposed in this paper which results in sinusoidal voltages and currents through the high frequency transformer. This results in significant reduction in the transformer losses and also natural commutation of leakage energy is obtained. The matrix converter supplying the machine is always operated at its full modulation index. The synthesis of adjustable magnitude voltages at the machine terminals is controlled in the primary side MMC to result in lower voltage stress related losses in the transformer, matrix converter and the machine. The operating principle, modulation and control of the proposed PET along with simulation results have been presented.

Fig. 8(top) and Fig. 8(second from top) shows the grid side voltage and current of the MMC simulated with a modulation index $m = 0.8$ and unity input power factor $\cos \phi_g = 1$. The MMC can reach 17 distinct levels by use of 8 submodules. This is an advantage of using the 3-level submodule topology which can result in double the number of voltage levels as compared to conventionally used half bridge submodules. This can thus result in a more compact converter structure. The voltage balancing is done using the sorting algorithm described in Section III. The balanced capacitor voltages show two distinct balanced waveforms which are the voltages in the upper arm and the lower arm for all the capacitors C_1 and C_2 respectively, Fig. 8(third from top, bottom).

978-1-4799-2706-7/14 $31.00 © 2014 IEEE

The 2014 International Power Electronics Conference

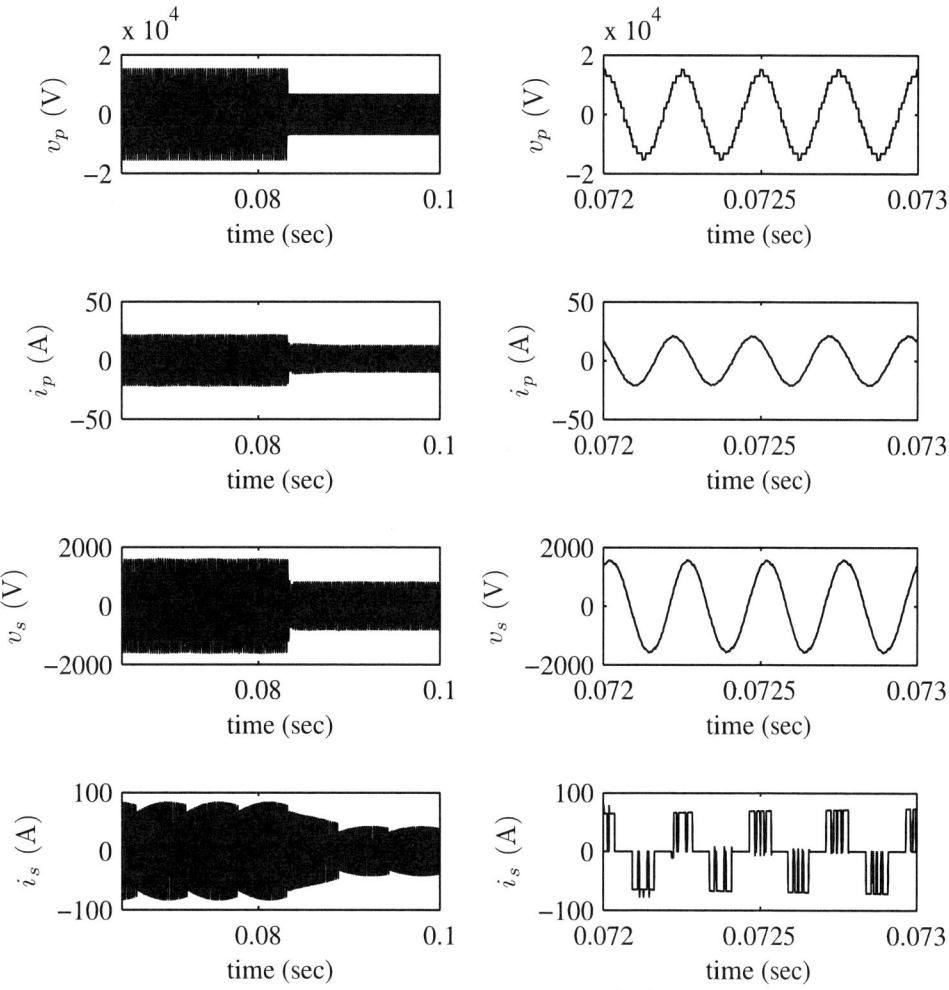

Fig. 9: HFT waveforms (top to bottom) : Primary voltage, primary current, secondary voltage, secondary current along with zoomed version on the right

REFERENCES

[1] X. She, A. Huang, and R. Burgos, "Review of solid-state transformer technologies and their application in power distribution systems," *Emerging and Selected Topics in Power Electronics, IEEE Journal of*, vol. 1, pp. 186–198, 2013.

[2] S. Falcones, X. Mao, and R. Ayyanar, "Topology comparison for solid state transformer implementation," in *Power and Energy Society General Meeting, 2010 IEEE*, 2010, pp. 1–8.

[3] C. G. C. Branco, R. Torrico-Bascope, C. M. T. Cruz, and F. de A Lima, "Proposal of three-phase high-frequency transformer isolation ups topologies for distributed generation applications," *Industrial Electronics, IEEE Transactions on*, vol. 60, no. 4, pp. 1520–1531, 2013.

[4] X. Wang, J. Liu, T. Xu, and X. Wang, "Comparisons of different three-stage three-phase cascaded modular topologies for power electronic transformer," in *Energy Conversion Congress and Exposition (ECCE), 2012 IEEE*, 2012, pp. 1420–1425.

[5] T. Kawabata, K. Honjo, N. Sashida, K. Sanada, and M. Koyama, "High frequency link dc/ac converter with pwm cycloconverter," in *Industry*

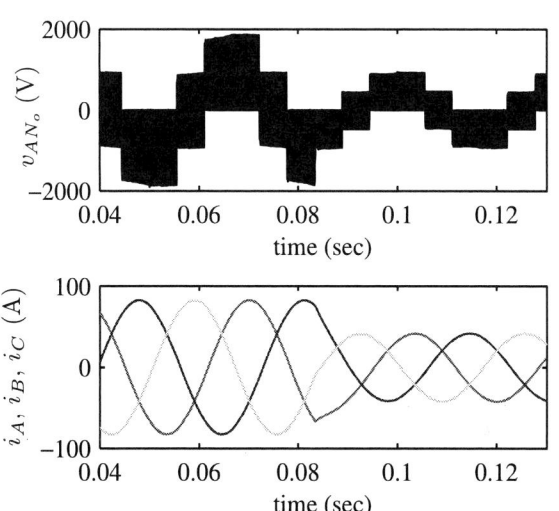

Fig. 10: Load phase voltage (top) and three line currents (bottom)

978-1-4799-2706-7/14 $31.00 © 2014 IEEE

Applications Society Annual Meeting, 1990., Conference Record of the 1990 IEEE, 1990, pp. 1119–1124 vol.2.

[6] M. Carpita, M. Marchesoni, M. Pellerin, and D. Moser, "Multilevel converter for traction applications: Small-scale prototype tests results," *Industrial Electronics, IEEE Transactions on*, vol. 55, no. 5, pp. 2203–2212, 2008.

[7] N. Holtsmark, H. Bahirat, M. Molinas, B. Mork, and H. Hoidalen, "An all-dc offshore wind farm with series-connected turbines: An alternative to the classical parallel ac model?" *Industrial Electronics, IEEE Transactions on*, vol. 60, no. 6, pp. 2420–2428, 2013.

[8] J.-S. Lai, A. Maitra, A. Mansoor, and F. Goodman, "Multilevel intelligent universal transformer for medium voltage applications," in *Industry Applications Conference, 2005. Fourtieth IAS Annual Meeting. Conference Record of the 2005*, vol. 3, 2005, pp. 1893–1899 Vol. 3.

[9] M. Rashed, C. Klumpner, and G. Asher, "High performance multilevel converter topology for interfacing energy storage systems with medium voltage grids," in *IECON 2010 - 36th Annual Conference on IEEE Industrial Electronics Society*, 2010, pp. 1825–1831.

[10] H. Iman-Eini, S. Farhangi, J.-L. Schanen, and J. Aime, "Design of power electronic transformer based on cascaded h-bridge multilevel converter," in *Industrial Electronics, 2007. ISIE 2007. IEEE International Symposium on*, 2007, pp. 877–882.

[11] M. Glinka and R. Marquardt, "A new ac/ac multilevel converter family," *Industrial Electronics, IEEE Transactions on*, vol. 52, no. 3, pp. 662–669, 2005.

[12] J. Shi, W. Gou, H. Yuan, T. Zhao, and A. Huang, "Research on voltage and power balance control for cascaded modular solid-state transformer," *Power Electronics, IEEE Transactions on*, vol. 26, pp. 1154–1166, 2011.

[13] K. Basu, R. Gupta, S. Nath, G. Castelino, K. Mohapatra, and N. Mohan, "Research in matrix-converter based three-phase power-electronic transformers," in *Power Electronics Conference (IPEC), 2010 International*, 2010, pp. 2799–2803.

[14] R. Gupta, K. Mohapatra, and N. Mohan, "A novel three-phase switched multi-winding power electronic transformer," in *Energy Conversion Congress and Exposition. ECCE 2009. IEEE*, 2009, pp. 2696–2703.

[15] A. Sahoo, R. Otero-De-Leon, V. Chandrasekaran, and N. Mohan, "New 3-level submodules for a modular multilevel converter based hvdc system with advanced features," in *Industrial Electronics Society, IECON 2013 - 39th Annual Conference of the IEEE*, Nov 2013, pp. 6269–6274.

[16] A. K. Sahoo and N. Mohan, "Capacitor voltage balancing and an intelligent commutation technique in a new modular multilevel converter based hvdc system," in *Power Electronics, Machines and Drives (PEMD 2014), 7th IET International Conference on*, April 2014, pp. 1–6.

[17] L. Huber and D. Borojevic, "Space vector modulated three-phase to three-phase matrix converter with input power factor correction," *Industry Applications, IEEE Transactions on*, vol. 31, pp. 1234–1246, 1995.

[18] A. Sahoo, K. Basu, and N. Mohan, "Analytical estimation of input rms current ripple and input filter design of matrix converter," in *Power Electronics, Drives and Energy Systems (PEDES), 2012 IEEE International Conference on*, Dec 2012, pp. 1–6.

The 2014 International Power Electronics Conference

Varying and Unequal Carrier Frequency PWM Techniques for Modular Multilevel Converters

Georgios Konstantinou[1] Rosheila Darus[1] Josep Pou[1],[2] Salvador Ceballos[3] and Vassilios G. Agelidis[1]

[1] UNSW Australia, Sydney, NSW, 2052, Australia

[2] Technical University of Catalonia, Catalonia, Spain, [3] Tecnalia Energy, Spain

email: g.konstantinou@unsw.edu.au, r.darus@student.unsw.edu.au, j.pou@unsw.edu.au,
salvador.ceballos@tecnalia.com vassilios.agelidis@unsw.edu.au

Abstract—Carrier-based pulse-width modulation (PWM) strategies are commonly applied to modular multilevel converters (MMC). This paper proposes a modified PWM technique with unequal and varying carrier frequencies between consecutive levels. The variation between two consecutive carriers is calculated in order to maintain a constant average switching frequency regardless of modulation index and number of levels in the output. The proposed method effectively shifts the major harmonic content of output waveforms to higher frequencies at the cost of a slight increase in low order harmonics. Details on the derivation of the carrier frequencies as well as simulation and experimental results of the proposed method demonstrate the viability of the method as an alternative modulation technique for the MMC.

Index Terms—Pulse-width modulation, multilevel converters, modular multilevel converter, carrier frequency

I. INTRODUCTION

Attractive features of the modular multilevel converter (MMC) include its modularity, expandability to higher voltages and a relatively simple voltage balancing task [1]. The topology (Fig. 1) offers significant operational redundancies [2] and multiple sub-module (SM) configurations [3] allowing additional functionalities such as energy storage [4] and dc-fault blocking [5]. The high number of voltage levels significantly reduces the waveform harmonic distortion and makes the reduction of switching losses a high priority task for the operation of the MMC.

The most direct method of reducing the switching losses of the MMC is by modifying the pulse-width modulation (PWM) technique. Various PWM techniques have been proposed for the converter with carrier-based (CB) PWM techniques being the most common ones [6]–[9]. Others include selective harmonic elimination (SHE) [10] and staircase modulation [11], [12]. A modulation technique based on voltage tolerance bands is proposed in [14]. Interleaving between the carriers of the two arms leads to an increase in the number of levels [15] resulting in lower total harmonic distortion (%THD) of the line current and lower capacitors voltage ripples at the cost of higher arm current %THD and harmonics.

SHE-PWM can further reduce the switching frequency of the SMs while providing good harmonic performance [10] and can be applied to fundamental switching frequency schemes [11]. However, the implementation complexity increases sig-

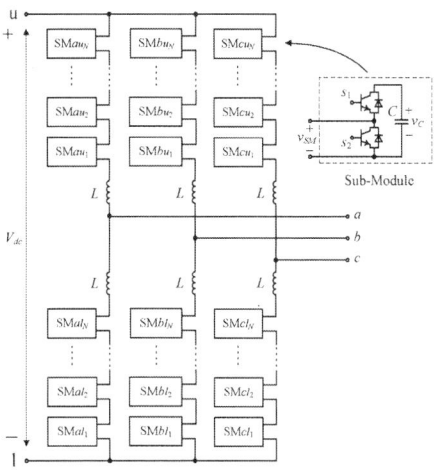

Fig. 1. Circuit configuration of the MMC

nificantly as the number of voltage levels increases. It is, however, a viable alternative for converters with low number of SMs [16], with multiple transitions per level of the converter significantly improving the overall performance [10].

In all previous work focusing on CB-PWM for the MMC, the carrier frequency was assumed constant and equal among the different levels of the waveform. However, this requirement can be relaxed as the switching frequency of the SMs will not be directly affected. One direct effect of the proposed technique is that there is no longer an exact cancellation of switching frequency and side-band harmonics [17] in the output voltage waveform. However, due to the high number of levels the effect on the output voltage spectrum is shown to be minimal and well within the acceptable limits.

The purpose of this paper is to investigate the effect of varying the carrier frequencies among the different levels of the MMC. Different variations in the carrier frequencies are investigated and their effect on the MMC currents and SM capacitor voltages are demonstrated. Unlike other multilevel converter topologies, such an approach is possible with the MMC as the carriers are not directly linked to SMs due to the voltage balancing algorithm that regulates the capacitor

978-1-4799-2706-7/14 $31.00 © 2014 IEEE

The 2014 International Power Electronics Conference

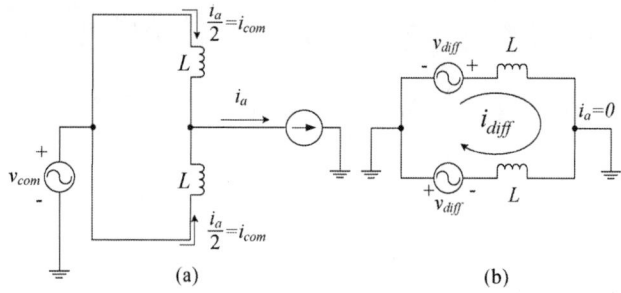

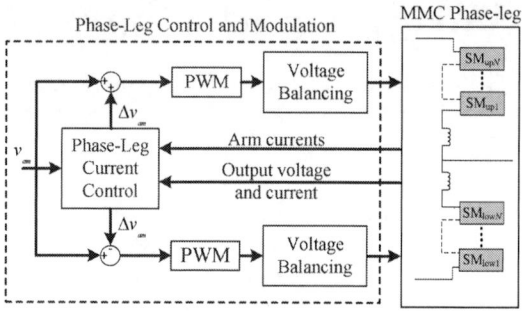

Fig. 2. Equivalent circuits of the MMC: (a) common mode circuit and (b) differential mode circuit.

Fig. 3. Block diagram of the phase-leg current controller.

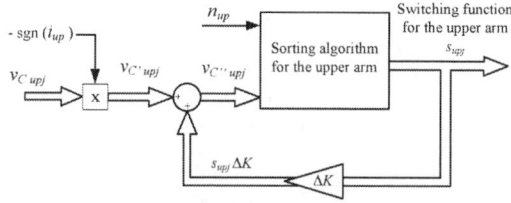

Fig. 4. Implementation of the SM capacitor voltage balancing algorithm with restriction in the number of switching transitions.

voltages. Compared to carriers with constant frequency (typically equal to the average switching frequency of the SMs), the proposed method is shown to reduce the phase voltage %THD and capacitor voltage ripple offering comparable performance and a viable substitute of constant frequency carriers PWM.

The paper is organised as follows. Section II provides a brief description of the topology, fundamentals of circulating current control and the SM capacitor voltage balancing algorithm. An detailed analysis of the proposed variable carrier frequency PWM is provided in Section III. Implementations on simulation models and an experimental prototype, together with a comprehensive comparison of the different proposed techniques, is provided in Section IV. The conclusions of the paper are summarised in Section V.

II. MODULAR MULTILEVEL CONVERTER

A. Topology

An MMC consists of two arms per phase-leg with series connected SMs and also includes one inductor per arm, necessary to limit the dc-fault current and to control the circulating arm currents. Typically, the SMs employ half-bridge structures (Fig. 1) although full-bridge structures [5] provide additional blocking of the dc fault or provide for connection of energy storage. Using the superposition theorem, each arm of the converter can be analysed into two separate and independent circuits describing the common mode (i_{com}) and differential or circulating (i_{diff}) currents [18] as shown in Fig. 2b.

B. Circulating Current Control

Control of the circulating current within the arms of the converter can be achieved without affecting the output current [18]. A differential voltage (Δv_m) is injected to the upper and lower arm modulating signal of the converter with opposite sign, as shown in Fig. 3. The circulating arm current contains a dc component, necessary to maintain the SM capacitors energised, and additional higher order harmonics. These higher order harmonics can be eliminated, leaving only the dc component and reducing the rms value of the circulating current [19], or regulated to fulfil additional control objectives [20] such as capacitor voltage ripple minimisation.

C. Restricted Algorithm for Capacitor Voltage Balancing

A voltage balancing algorithm [12]–[14] is implemented in order to maintain the SM capacitor voltages to the reference value. The algorithm uses the measurements of the capacitor voltages and arm currents in order to determine the SMs that will be connected or disconnected from the arm based on the requirements of the PWM technique. In order to prevent unwanted transitions which increase the switching frequency, a restriction in the number of SMs that are allowed to change state in each switching period is also applied. The implementation of the SM capacitor voltage balancing algorithm for the upper arm of the MMC is given in Fig. 4 and additional details regarding its operation and the effect of the restriction in the switching frequencies of the SMs can be found in [12].

III. PWM FOR THE MMC

A. Carrier-Based PWM

An MMC with N SMs per arm is capable of producing $N+1$ or $2N+1$ levels in the output voltage, depending on whether the switching of the arms is interleaved or not. Although it is possible to directly link the switching of a SM with one particular carrier waveform in phase-shifted (PS) PWM, typically the modulation stage is used to determine transitions between the levels of the waveform and calculate the number of SMs required in each of the arms.

Assuming $N+1$ modulation, N carriers occupy continuous bands within the complete modulation index range between -1 and 1. (Fig. 5a). The carriers are of equal frequency and their phase depends on the choice of PWM technique (PD, POD or APOD [21]). With $N+1$ modulation, out of the $2N$ SMs

978-1-4799-2706-7/14 $31.00 © 2014 IEEE

The 2014 International Power Electronics Conference

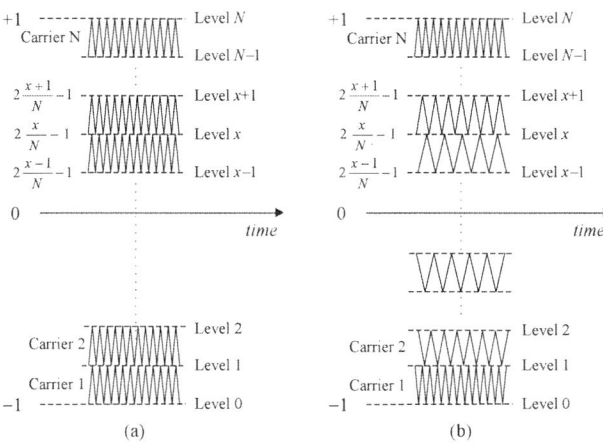

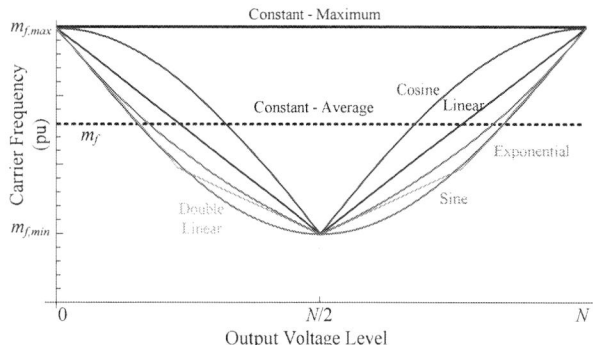

Fig. 6. Possible variations in the normalised carrier frequencies of the MMC.

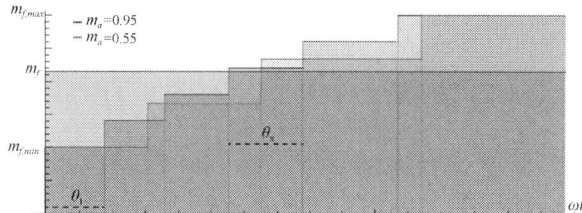

Fig. 5. Level-shifted carriers of a level-shifted PWM strategy (a) PD with constant carrier frequencies (b) POD with unequal carrier frequencies.

Fig. 7. Calculation of carrier switching frequencies for different modulation indices

within each phase-leg, N SMs are connected at any instant defining the output voltage level $x = \{0, 1, ..., N\}$.

In three-phase systems, a third harmonic can be injected in the modulating signal in order to extend the dc-link voltage utilisation. The use of discontinuous PWM through clamping one phase-leg of the MMC to the dc-link for 1/3 of the fundamental period is shown [22] to reduce the switching frequency of the SMs for most of the MMC operating points, while the SM capacitor voltage ripple is also reduced at low modulation indices.

B. Varying and Unequal Carrier Frequency PWM

As one carrier is not directly associated with one SM of the MMC it possible to individually vary the frequency of each carrier without affecting the switching frequency of specific SMs. The frequency of the carriers is varied so that the average switching frequency will be constant at any modulation index, while decreasing the switching frequency of the carriers in the lower levels and increasing the carriers in the higher levels (Fig. 5b). The variation of the carrier frequencies will affect the cancellations of harmonics compared to the case of equal carriers changing the spectrum of the output [17].

A similar variation in the carrier frequencies was discussed in [23] for NPC converters. There, carriers closer to zero were at higher frequency in order to equalise the number of transitions between the levels and provide a further balancing between the losses of the devices in the converter. In the proposed technique, the variation in the carrier frequencies is opposite to that of [23] (from lower to higher as the waveform level increases). This is due to the rate of change of the sinusoidal reference waveform towards the lower levels as well as the lower number of SMs available for selection from the voltage balancing algorithm at higher voltage levels. As a result of the higher switching frequency produced at higher voltage levels, a decrease in the SM capacitor voltage ripple is expected.

The variation of the carrier frequency between the different bands can be selected arbitrarily. Some possible patterns of carrier frequency variation are shown in Fig. 6 (the output voltage level refers to the definitions of Fig. 5), including linear and double linear variations as well as sine, cosine and exponential ones. Carriers 1 and N are always at the highest switching frequency ($m_{f,max}$) while the lowest carrier frequency ($m_{f,min}$) is at the zero crossing of the waveform, between levels ($N/2 - 1, N/2$ and $N/2 + 1$).

The carrier frequencies can be selected constant for all modulation indices (m_a). However, this will result in the target average switching frequency being achieved for one particular m_a and deviating at lower and higher modulation indices. For this reason, the carrier frequencies are not maintained constant but further varied as a function of the modulation index. Assuming an ideal sinusoidal reference $m_a \sin(\omega t)$ with constant switching frequency, the integral of m_f for a quarter of the period will be equal to $\pi m_f/2$. The variation of the carrier frequencies will be selected to produce the same value over the quarter period. The radians for which the reference signal is between levels $2\frac{x}{N} - 1$ and $2\frac{x+1}{N} - 1$ are equal to:

$$\theta_x = \sin^{-1}\left(\frac{x+1}{m_a l(m_a)}\right) - \sin^{-1}\left(\frac{x}{m_a l(m_a)}\right) \qquad (1)$$

where $l(m_a)$ is the number of levels in the output waveform for the particular modulation index m_a. The carriers frequencies are then selected so that the integral of $m_{f,x}$ at any modulation index is equal to $\pi m_f/2$ as demonstrated for three different modulation indices in Fig. 7.

Considering a linear variation of the carriers, the change between two consecutive carrier frequencies with a target average m_f can be written as an equation with two unknowns $m_{f,min}$ and m_a so that

$$\Delta m_f = \frac{\pi(m_f - m_{f,min})}{2\sum_{x=1}^{L(m_a)}(x-1)\theta_x} \tag{2}$$

Δm_f is then calculated by selecting the value of the minimum carrier frequency $m_{f,min}$. Similar calculations can be performed for all possible carrier frequency variations, but these are not considered in this paper, which focuses on the linear carrier frequency variation.

C. Harmonic Analysis

Any PWM modulated waveform, including those of the proposed method, can be written as [17]

$$f(t) = \frac{A_{00}}{2} + \sum_{n=1}^{\infty}[A_{0n}\cos n(\omega_0 t + \theta_0) + B_{0n}\sin n(\omega_0 t + \theta_0)]$$
$$+ \sum_{m=1}^{\infty}[A_{m0}\cos m(\omega_c t + \theta_c) + B_{m0}\sin m(\omega_c t + \theta_c)]$$
$$+ \sum_{m=1}^{\infty}\sum_{\substack{n=-\infty \\ n\neq 0}}^{\infty} \begin{aligned}&[A_{mn}\cos(m(\omega_c t+\theta_c)+n(\omega_0 t+\theta_0)) \\ &+B_{mn}\sin(m(\omega_c t+\theta_c)+n(\omega_0 t+\theta_0))]\end{aligned}$$

$$\tag{3}$$

where

$$C_{mn} = \frac{1}{2\pi^2}\int_{-\pi}^{\pi}\int_{-\pi}^{\pi} f(x,y)e^{j(mx+ny)}dxdy \tag{4}$$

It is, therefore, important to properly define the function $f(x,y)$ within the unity cell that identifies the contours where the function $f(x,y)$ of the output is constant for periodic variations of the fundamental and the carrier frequency [17]. The unity cell is scaled to radians with the x-axis representing the fundamental frequency $x(t) = \omega_0 t + \theta_0$ and the y-axis the carrier waveform frequency $y(t) = \omega_c t + \theta_c$. As there are multiple carrier frequencies in the proposed method, the lowest frequency carrier $m_{f,min}$ is chosed to define the slope $x/y = \omega_c/\omega_0 = m_{f,min}$ of the line which determines the transitions between the levels.

This can be illustrated with a numerical example. Considering a 13-level converter ($N = 12$), we define $m_{f,min} = \frac{m_f}{2}$. The contours of the function $f(x,y)$ are then plotted for the case of constant carrier frequency (Fig. 8a), varying and unequal PD-PWM (Fig. 8b) and varying and unequal POD-PWM (Fig. 8c).

Due to the complexity of the $f(x,y)$ a detailed estimation of the harmonic content through (3) becomes tedious and very long to be of any practical use. Analysis of the harmonic content is preferably performed through modelling and simulation studies of the converter. The variation of $f(x,y)$ as a function of m_a is shown in Figs. 9a-c. At high modulation indices the carriers of the lower band are at lower frequencies, however, as the modulation index decreases, their frequency increases maintaining the average equal to the benchmark

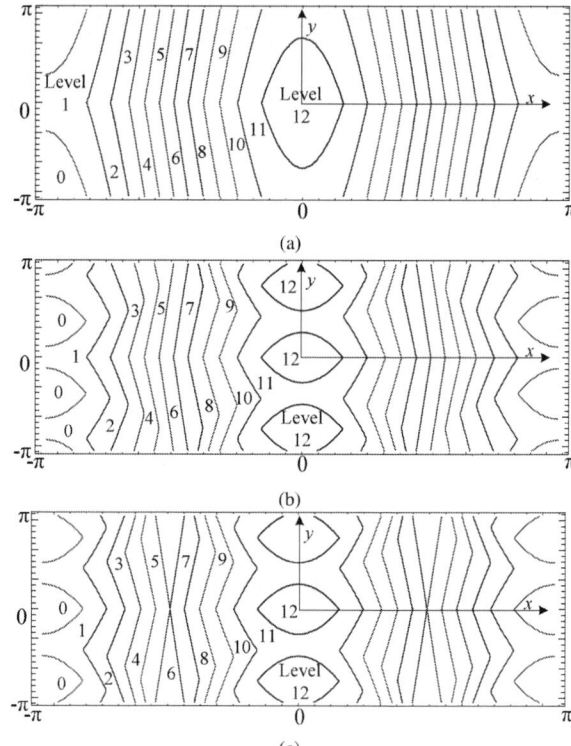

Fig. 8. PWM contour plots of function $f(x,y)$ (a) constant carrier frequencies, (b) PD-PWM with linear variation, (c) POD-PWM with linear variation.

case of constant m_f. The change of carrier frequencies with changing modulation indices occurs for any of the considered variations of Fig. 6.

IV. SIMULATION AND EXPERIMENTAL RESULTS

A. Simulation Results

The proposed PWM is evaluated and compared with the constant carrier frequency through simulation and experimental results. For simplicity and continuity of the analysis, a 13-level converter ($N = 12$) is again used in the simulations. The circulating current controller and restricted voltage balancing algorithm of Sections II.B and II.C are also used to eliminate the higher order harmonics from the circulating arm current.

The target average switching frequency ratio m_f is set equal to 30 (1.5 kHz) while two minimum carrier frequencies $m_{f,min} = 20$ and 10 are selected. Changes in the converter operating point are also considered. The frequency variations, calculated through (2), are summarised in Table I.

In all cases, the circulating current controller forces the circulating current to a dc value while the voltage balancing algorithm maintains the capacitor voltages to the reference value. A quantitative comparison of the results is provided in Table II while Figs. 10a-c provide the harmonic content of the line-to-line voltages for the simulation cases at the same modulation index ($m_a = 0.95$).

Due to the effect of the voltage balancing algorithm and the current controller on the reference signal of each arm, the

The 2014 International Power Electronics Conference

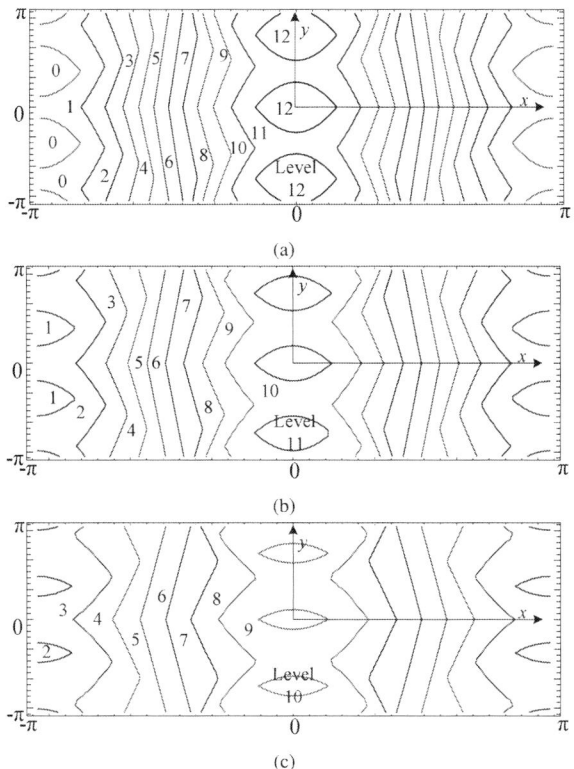

Fig. 9. PD-PWM contour plots of function $f(x,y)$ (a) $m_a = 0.95$, (b) $m_a = 0.75$, (c) $m_a = 0.55$

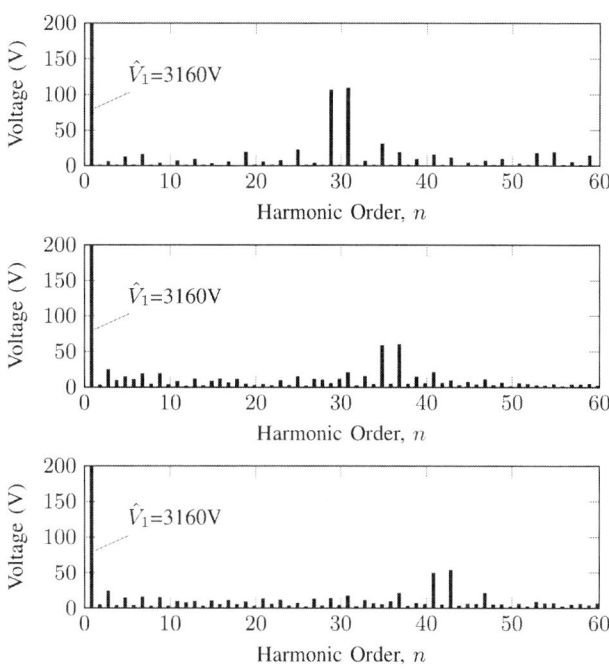

Fig. 10. Harmonic analysis of the line-to-line voltages for variable and constant frequency carriers with the same target switching frequency per device at $m_a = 0.95$, (a) CNT, (b) PD1 and (c) PD2.

TABLE I
CARRIER FREQUENCY RATIOS FOR CONSTANT AND LINEAR VARIATION USED IN SIMULATIONS

Technique	m_{f1}	m_{f2}	m_{f3}	m_{f4}	m_{f5}	m_{f6}
CNT	30	30	30	30	30	30
PD1, $m_a = 0.95$	20	23.19	26.39	29.58	32.77	35.97
PD1, $m_a = 0.75$	20	24.16	28.32	32.48	36.65	–
PD1, $m_a = 0.55$	20	26	32	38	–	–
PD2, $m_a = 0.95$	10	16.38	22.77	29.16	35.55	41.95

TABLE II
COMPARISON OF RESULTS FOR THE DIFFERENT PWM TECHNIQUES

Technique	THD_{ll}	THD_I	$WTHD_{ll}$	f_{sw}	$V_{cap,pp}$
CNT	5.8%	2.7%	0.7%	155Hz	19.2V
PD1, $m_a = 0.95$	6.1%	3.3%	1.13%	185Hz	13.4V
PD1, $m_a = 0.75$	6.7%	4.1%	1.24%	175Hz	12.2V
PD1, $m_a = 0.55$	10.1%	5.3%	1.62%	160Hz	11.1V
PD2, $m_a = 0.95$	5.5%	3.4%	1.01%	155Hz	18.5V

estimation of (2) does not exactly provide an equal average switching frequency between all the methods. As expected, higher switching frequency results in lower SM capacitor voltage ripple while the variation in the carrier frequencies does not affect the rms value of the circulating arm currents. Harmonic cancellations are affected by the use of variable carrier frequencies resulting in slightly increased WTHD values.

However, the main harmonic component of the output waveforms is now located around the highest carrier frequency with only a small increase of lower order harmonics. This observation shows that as the number of SMs in the MMC increases, a variable carrier frequency technique might be more beneficial compared to an approach that considers constant carrier frequencies. A small decrease in the SM capacitor voltage ripple is also observed.

B. Experimental Results

The linear variation of the carrier frequencies is also verified experimentally using a 5-kVA scaled-down single-phase MMC [2], [10] with $N = 5$ SMs per arm. Considering the $N+1$ modulation, the converter will produce six levels in the output. The dc voltage is constant and an RL load is connected between the phase terminal and the dc-side midpoint.

The six-level waveform at the output of the converter limits the carrier variation to only three carrier frequencies defining the PWM pattern. The relatively low number of SMs is somewhat restrictive of the range of available variations that can be considered. The output phase voltage and corresponding harmonic spectrum is shown for the case of equal (Fig. 11a) and unequal carrier frequencies (Fig. 11b). In both implementations the average switching frequency per SM is equal to 300 Hz. The spread in the harmonic spectrum, reduction in the peak values of all harmonics as well as the shift of the main harmonics in the output voltage spectra around the highest carrier frequency can be observed. The current controller forces the differential component to a constant value

978-1-4799-2706-7/14 $31.00 © 2014 IEEE

The 2014 International Power Electronics Conference

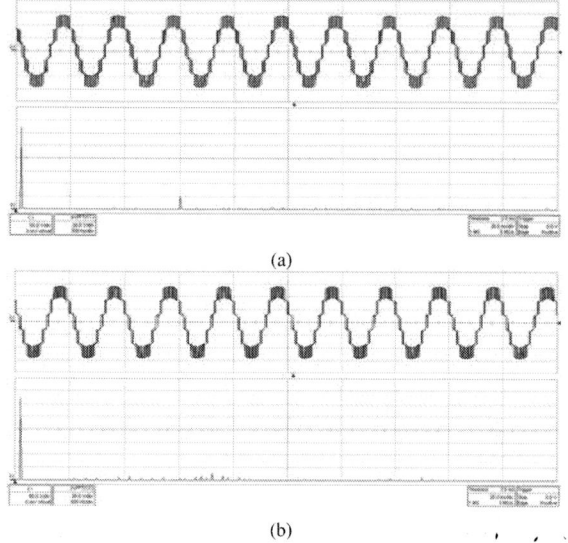

(a)

(b)

Fig. 11. Experimental phase output voltage and associated harmonic spectrum for $m_a = 0.9$ (a) constant carrier frequency $f_c = 1.5kHz$, (b) varying carrier frequency $f_{c1} = 1.05kHz$, $f_{c2} = 1.35kHz$, $f_{c3} = 1.8kHz$

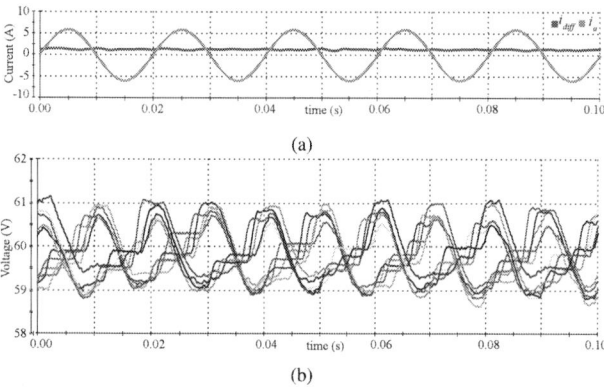

(a)

(b)

Fig. 12. Experimental waveforms: (a) Output and circulating currents within the MMC arm, (b) SM capacitor voltages with varying carrier frequency

(Fig. 12a) while the voltage balancing algorithm maintains SM voltages to the reference value as shown in Fig. 12b.

V. CONCLUSION

The disassociation of carrier waveforms and SMs in MMCs provides additional degrees of freedom and creates multiple opportunities for variations in the PWM patterns and carrier waveforms. This allows for variable carrier frequencies across the different levels while maintaining the average switching frequency constant for all modulation indices. The proposed method achieves this by calculating the carrier frequencies as a function of the modulation index. The variation effectively increases the frequency of the main harmonic content, reduces phase-voltage %THD and capacitor voltage ripples with a small increase in the lower order harmonic due to reduced harmonic cancellations. Simulations and experimental results confirm the viability of the proposed method.

REFERENCES

[1] A. Lesnicar and R. Marquardt, "An innovative modular multilevel converter topology suitable for a wide power range," in *Proc. IEEE Bologna PowerTech Conference*, 23-26 June 2003, Bologna, Italy.

[2] G. Konstantinou, J. Pou, S. Ceballos, and V.G. Agelidis, "Active redundant sub-module configuration in modular multilevel converters," *IEEE Trans. on Power Delivery*, Vol. 28, No. 4, pp. 2333-2341, Oct. 2013.

[3] E. Solas, G. Abad, J. Barrena, S. Aurtenetxea, A. Carcar, and L. Zajac, "Modular multilevel converter with different submodule concepts - Part II: experimental validation and comparison for HVDC applications," *IEEE Trans. Ind. Electron.*, Vol. 60, No. 10, pp. 4536-4545, Oct. 2013.

[4] M. Vasiladiotis and A. Rufer, "Analysis and control of modular multilevel converters with integrated battery energy storage," *IEEE Trans. Power Electron.*, early access, 2014, doi: 10.1109/TPEL.2014.2303297.

[5] M.M.C. Merlin, T.C. Green, P.D. Mitcheson, D.R. Trainer, R. Critchley, W. Crookes, and G. Hassan, "The alternate arm converter: a new hybrid multilevel converter with DC-fault blocking capability," *IEEE Trans. Power. Del.*, Vol. 99, early access, pp. 1-8.

[6] M. Saeedifard and R. Iravani, "Dynamic performance of a modular multilevel back-to-back HVDC system," *IEEE Trans. Power Delivery*, vol. 25, pp. 2903-2912, 2010.

[7] S. Rohner, S. Bernet, M. Hiller, and R. Sommer "Modulation, losses and semiconductor requirements of modular multilevel converters," *IEEE Trans. Ind. Electr.*, 2010, vol. 57, no. 8, pp.2633-2642.

[8] M. Zhang, L. Huang, W. Yao, and Z. Lu, "Circulating harmonic current elimination of a CPS-PWM based modular multilevel converter with plug-in repetitive controller," *IEEE Trans. Power Electr.*, early access.

[9] A. Shojaei, G. Joos, "An improved modulation scheme for harmonic distortion reduction in MMC," in *IEEE PES GM 2012*, pp. 1-7.

[10] G. Konstantinou, M. Ciobotaru, and V.G. Agelidis, "Selective harmonic elimination pulse width modulation of the modular multilevel converter," *IET Power Electronics*, Vol. 6, No. 1, pp. , Jan. 2013.

[11] K. Ilves, A. Antonopoulos, S. Norrga, and H.P. Nee, "A new modulation method for the modular multilevel converter allowing fundamental switching frequency," *IEEE Trans. Power Electron.*, Vol. 27, No. 8, pp. 991-998, Aug. 2012.

[12] R. Darus, J. Pou, G. Konstantinou, S. Ceballos, and V.G. Agelidis, "A modified voltage balancing sorting algorithm for the MMC: evaluation for staircase and phase-disposition PWM", *APEC*, pp. 1-6, Mar. 2014.

[13] F. Deng, Z. Chen, "A control method for voltage balancing in modular multilevel converters,", *IEEE Trans. Power Electr.*, Vol. 29, No. 1, pp. 66-76, Jan 2014.

[14] A. Hassanpoor, K. Ilves, S. Norrga, H.P. Nee, and L. Angquist, "Tolerance band modulation methods for modular multilevel converters," in *EPE 2013*, Lille, pp. 1-10, Sept. 2013.

[15] G. Konstantinou, M. Ciobotaru, and V.G. Agelidis, "Analysis of multi-carrier PWM methods for back-to-back HVDC systems based on modular multilevel converters," *Proc. IEEE IECON*, pp. 4238-4243. Nov. 2011.

[16] A. Hassanpoor, S. Norrga, H.P. Nee, and L. Angquist, "Evaluation of different carrier-based PWM methods for modular multilevel converters in HVDC applications," in *Proc. IECON*, pp. 288-393, Oct. 2012.

[17] T. Lipo and G.D. Holmes, *Pulse Width Modulation for Power Coverters, Principles and Practice*, 1st ed., Hoboken, NJ, John Wiley & Sons, 2003.

[18] R. Darus, J. Pou, G. Konstantinou, S. Ceballos and V.G. Agelidis, "Circulating current control and evaluation of carrier disposition in modular multilevel converters", *IEEE ECCE Asia*, pp. 332-338, Jul. 2013.

[19] Q. Tu, Z. Xu, L. Xu, "Reduced switching-frequency modulation and circulating current suppression for modular multilevel converters," *IEEE Trans. Power Deliv.*, Vol. 26, No. 3, pp. 2009-2017, Mar. 2011.

[20] R. Picas, S. Ceballos, J. Zaragosa, J. Pou, G. Konstantinou, and V.G. Agelidis, "Optimal circulating current harmonics for minimization of the capacitor voltage ripple amplitudes of a modular multilevel converter," in *Proc. IEEE ECCE Asia 2013*, pp. 318 - 324, Jun. 2013.

[21] G. Konstantinou and V. G. Agelidis, "Performance evaluation of half-bridge cascaded multilevel converters operated with multicarrier sinusoidal PWM techniques," in *Proc. IEEE ICIEA 2009*, pp. 3399-3404.

[22] R. Picas, S. Ceballos, J. Zaragoza, J. Pou, G. Konstantinou, and V.G. Agelidis, "Improvements for modular multilevel converter capacitor voltage ripples and power losses with discontinuous modulation," in *Proc. IEEE IECON 2013*, Nov. 2013.

[23] L. M. Tolbert and T. G. Habetler, "Novel multilevel inverter carrier-based PWM method," *IEEE Trans. Ind. Applic.*, Vol.35, No. 5, pp. 1098-1107, Sept.-Oct. 1999.

978-1-4799-2706-7/14 $31.00 © 2014 IEEE

The 2014 International Power Electronics Conference

Comparison of Phase-Shifted and Level-Shifted PWM in the Modular Multilevel Converter

Rosheila Darus[1,2], Georgios Konstantinou[1], Josep Pou[1,3], Salvador Ceballos[4] and Vassilios G. Agelidis[1]

[1] Australian Energy Research Institute (AERI) & School of Electrical Engineering and Telecommunications
The University of New South Wales, Sydney, NSW, 2052, Australia.
[2] Faculty of Electrical Engineering, Universiti Teknologi Mara (UiTM), 40450 Shah Alam, Selangor, Malaysia.
[3]Terrassa Industrial Electronics Group (TIEG), Technical University of Catalonia, Catalonia, Spain.
[4]TECNALIA, Energy Unit, Basque Country, Spain.
email: r.darus@student.unsw.edu.au   g.konstantinou@unsw.edu.au   j.pou@unsw.edu.au   salvador.ceballos@tecnalia.com
vassilios.agelidis@unsw.edu.au

Abstract—**This paper reports a comparison study of different carrier-based PWM techniques applied to the modular multilevel converter. Phase-disposition PWM (PD-PWM) and phase-shifted pulse-width modulation (PS-PWM) with non-interleaving and interleaving are considered in this study. In PS-PWM, two cases are evaluated. In the first case, the particular SMs that have to be activated/deactivated are defined by a voltage balancing algorithm, which is the same one implemented in PD-PWM. In addition, an algorithm to restrict the number of switching SMs is also implemented to reduce the switching frequency of the power devices. In the other case of PS-PWM, each sub-module (SM) has a carrier signal associated to it and capacitor voltage balance is achieved by individual control of its capacitor voltage. The circulating current is controlled to be a dc component in all the cases. Simulation and experimental results are presented to evaluate the quality of the line-to-line output voltages and SM capacitor voltage ripples for the different cases under study.**

Index Terms—**Modular multilevel converter, Modulation technique, Circulating current control, Capacitor voltage balancing**

I. INTRODUCTION

The modular multilevel converter (MMC) is the preferred topology for high voltage (HV) applications because of its modularity and higher voltage ratings. Other advantages of the MMC are the quality of the output voltages, that capacitor voltage balance can be easily implemented, the fault tolerant operation capability, and the elimination of the dc-link capacitor [1]–[4].

The MMC (Fig. 1) consists of a series of N sub-modules (SMs), usually based on the half-bridge circuit, each of them including a capacitor as a voltage source. Configurations with energy storage are also possible [5]. A circulating current exists within each phase-leg of the MMC and has a significant impact on the ratings of the power devices, capacitors voltage ripples and power losses [6]. A circulating current control is necessary to reduce such an impact.

A wide range of modulation techniques can be applied to the MMC, mainly depending on the number of SMs in the phase-legs of the converter. The most common ones are carrier-based PWM (CB-PWM) [7] either phase-shifted PWM (PS-PWM) [8]–[11] or level-shifted PWM (LS-PWM) [11]–[13].

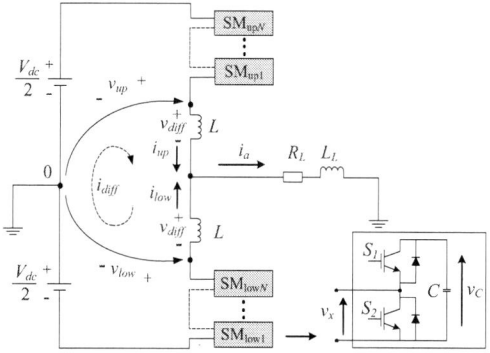

Fig. 1. Phase-leg MMC.

Other techniques include SHE-PWM [14] and band-tolerance modulation based on hysteresis [15]. As a result of its better harmonic performance, phase-disposition PWM (PD-PWM) is the preferred choice of LS-PWM technique in MMCs.

An interesting property of the MMC, owing to its two-arm configuration, is that it can generate two different number of voltage levels for the same number of SMs per arm [11]. The most straightforward way of deriving the additional number of levels is through interleaving of the carriers [10] between the upper and lower arms while maintaining the effective switching frequency of each SM constant.

Capacitor voltage balancing is also required to keep the capacitors voltages at the reference value. A common way to perform capacitor voltage balance is based on sorting the capacitor voltage values from the highest to the lowest or vice versa, and making a decision on the SMs to be activated/deactivated considering the direction of the arm current [16]. A separate voltage balancing algorithm might not be necessary if the modulation technique can provide approximately equal conduction periods [13] per SM (i.e. PS-PWM). Voltage balancing can also be achieved through capacitor voltage estimation and energy averaging.

978-1-4799-2706-7/14 $31.00 © 2014 IEEE   3764

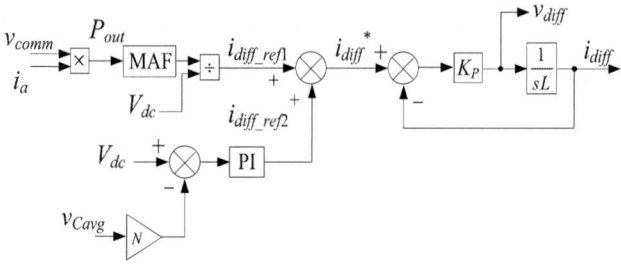

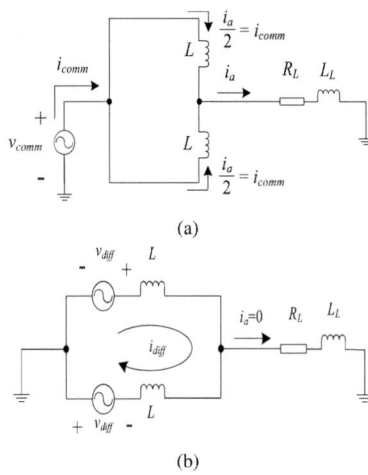

(a)

(b)

Fig. 2. (a) Common mode and (b) Differential mode circuits.

Fig. 3. Circulating current control.

Although CB-PWM techniques have been extensively analysed in the literature, a complete comparison and evaluation when considering carrier interleaving and the voltage balancing algorithms has not been presented. The purpose of this paper is to compare CB-PWM techniques, including the application of carrier interleaving and demonstrate the benefits of the sorting stage in the voltage balancing of the MMC.

The paper is structured as follows. The MMC circuit analysis, circulating current control and modulation techniques used in this study are introduced in Section II. The two capacitor voltage balancing algorithms are presented in Section III, followed by simulation and experimental results and comparison in Section IV. The conclusions of the work are summarised in Section V.

II. MMC, CIRCULATING CURRENT CONTROL AND MODULATION TECHNIQUES

Two independent circuits can be obtained from the analysis of a phase-leg of the MMC (Fig. 1), the common mode and differential mode, as shown in Figs. 2(a) and (b) respectively. The two circuits can be analyzed separately and do not affect each other [17]. The common and differential voltages (v_{comm} and v_{diff}) and currents (i_{comm} and i_{diff}) are:

$$v_{comm} = \frac{v_{up} + v_{low}}{2}, \qquad v_{diff} = \frac{v_{up} - v_{low}}{2} \qquad (1)$$

$$i_{comm} = \frac{i_{up} + i_{low}}{2}, \qquad i_{diff} = \frac{i_{up} - i_{low}}{2}. \qquad (2)$$

The differential (or circulating) current i_{diff} circulates within the arms of the MMC and does not appear in the output. It consists of a dc and ac components [17]. The dc component is essential to keep the phase-leg energised but the ac components can be eliminated, reducing the losses and increasing the efficiency of the MMC.

The circulating current can be controlled by imposing a differential voltage (v_{diff}). Fig. 3 shows a control loop where

the power provided by the phase-leg is calculated and used to estimate the circulating current reference. A moving average filter (MAF) extracts the dc component of the instantaneous power. The average voltage in the SM capacitors of the phase-leg provides additional information to the circulating current reference and is used to regulate the capacitor voltages at the reference value. The final circulating current reference is provided to a control loop based on proportional controller (k_p).

The next step in the control of a phase-leg includes the PWM technique. Its main task is to define the total number of SMs that are activated in the upper and lower arms. Three different carrier dispositions for LS-PWM are usually considered in the literature: (i) PD-PWM, (ii) phase-opposition-disposition PWM (POD-PWM), and (iii) alternating phase-opposition-disposition PWM (APOD) [11]. N additional output voltage levels can be achieved through interleaving the carriers of the upper and lower arm within the same phase-leg. In most cases, the gating signals are not directly derived from the modulation technique as a capacitor voltage balancing stage is usually required.

III. VOLTAGE BALANCING ALGORITHM

Selection of the SMs depends on the SM capacitor voltages and the direction of the arm currents so that the voltages are regulated towards the reference. When the number of activated SMs within an arm increases, the SMs with lowest/highest voltages will be activated if the current direction is such that it charges/discharges the capacitors. However, such an approach may lead to unnecessary switching operations [16] and therefore increased switching losses in the power devices if no additional restriction is taken. This can be avoided if the sorting stage uses virtual voltage values that also depend on whether SM is activated or not [20] effectively reducing the unnecessary transitions and limiting the switching frequency of each SM (Fig. 4). The modified voltages seen by the voltage sorting are expressed with:

$$v'_{Cupj} = -sgn(i_{up}) \times v_{Cupj}, \qquad (3)$$

$$v''_{Cupj} = v'_{Cupj} + s_{upj}\Delta K, \qquad (4)$$

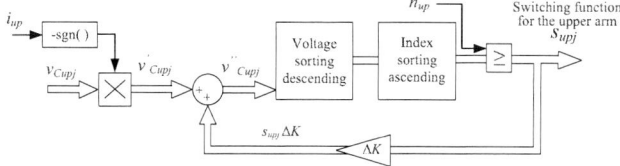

Fig. 4. Restricted voltage balancing algorithm for the upper arm.

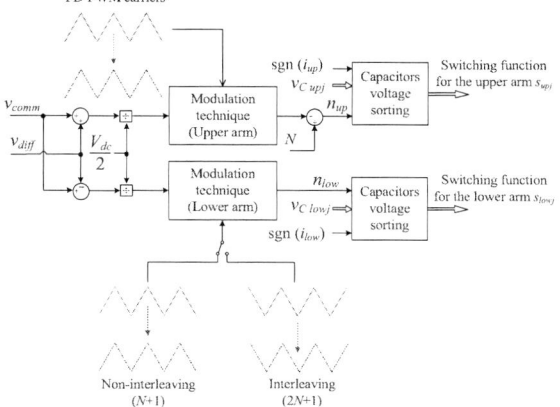

Fig. 5. PD-PWM with non-interleaving and interleaving between the arms with voltage sorting algorithm with N carriers.

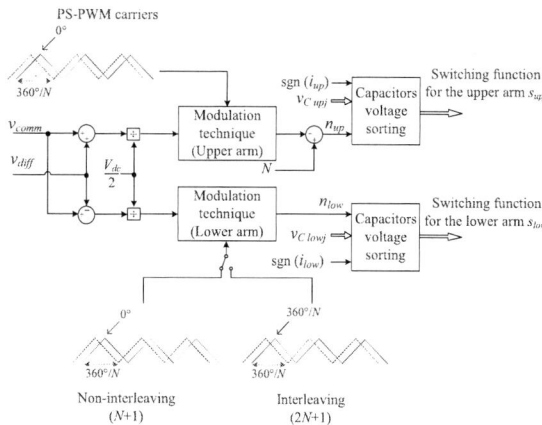

Fig. 6. PS-PWM for non-interleaving and interleaving between the arms with voltage sorting algorithm with N carriers.

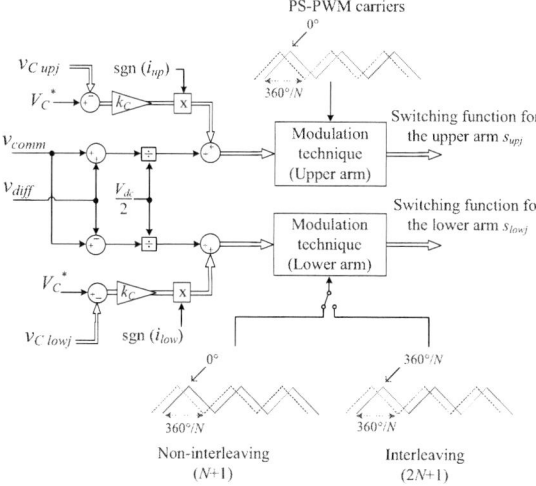

Fig. 7. PS-PWM for non-interleaving and interleaving between the arms with individual capacitor voltage control with N carriers.

A. PD-PWM with Restricted Voltage Balancing Algorithm

The main principles of PD-PWM have been explained in detailed in [7]. In the MMC, the number of carriers is equal to the number of SMs per arm (N) and all the carrier waveforms are in phase [11]. Interleaving of the upper and lower arm can be achieved by phase-shifting the carriers of the upper and lower arm by 180 degrees. The implementation of the current control, PD-PWM stage and capacitor voltage balancing algorithm is shown in Fig. 5.

B. PS-PWM with Restricted Voltage Balancing Algorithm

PS-PWM also requires N carrier waveforms per arm. Unlike LS-PWM, the carriers cover the whole range of modulation indices with consecutive carriers phase-shifted by $2\pi/N$, as shown in Fig. 6. The number of SMs to be activated in the arm is determined by the number of carrier signals that are below the reference signal at any given time. Interleaving is achieved by phase-shifting $2N$ carriers by π/N and distributing them alternatively between the upper and the lower arm, again with N carriers defining the transitions of each arm. The carrier frequency for PS-PWM is selected based on:

$$f_{c(PS-PWM)} = \frac{f_{fc(PD-PWM)}}{N} \qquad (5)$$

so that the average switching frequency of each SM is equal to that of the PD-PWM technique of Subsection III-A.

C. PS-PWM with Individual Capacitor Voltage Control

Alternatively to the sorting stage, voltage balancing can be achieved with PS-PWM where each carrier waveform is directly associated with one particular SM [13]. Additional compensation and feedback loops are also required to modify the individual PWM reference signal driving the capacitor voltages towards the reference value. The implemention of PS-PWM without the sorting algorithm is shown in Fig. 7. Interleaving between the upper and lower arms is similar to that in Subsection III-B. This approach is only valid with PS-PWM as LS-PWM patterns do not provide the required equal conduction times and capacitor voltage balancing cannot be achieved [11].

TABLE I
MMC SIMULATION PARAMETERS

Parameter	Value
Number of SMs per arm, N	10
Dc-link voltage, V_{dc}	6000 V
SM reference voltage, V_C	600 V
SM capacitor, C	1.5 mH
Arm inductors, L	18 mH
Carrier frequency PD-PWM, f_c	3 kHz
Carrier frequency PS-PWM, f_c	300 Hz

IV. SIMULATION AND EXPERIMENTAL RESULTS

A. Simulation Results

The analysis and comparison of the PWM techniques is based on an MMC with ten SMs per arm (N=10) using the parameters of Table I. An average switching frequency of 300 Hz per SM is selected as a balance between producing low switching power losses and the upper SM switching frequency ratio limit of approximately 10, above which no observable gain in harmonic and capacitor voltage ripple is observed [9]. A constant RL load is considered at the output of the MMC.

The capacitor voltages of the upper arm (v_{Cupj}) for all three PWM techniques, both without and with carrier interleaving, are shown in Fig. 8(a)-(f) for an operating point of m_a=0.95. The use of carrier interleaving has a negligible effect in the capacitor voltage ripple which is mainly driven by the average switching frequency of the SMs and the operating point of the converter [21], [22]. As the average SM switching frequency is approximately equal for all techniques, the peak-to-peak SM capacitor voltage ripple (v_{CupjPP}) is also similar at any operating point (Fig. 9).

Despite the fact that all techniques yield the same v_{CupjPP}, the lack of a sorting algorithm results in higher deviations among the SM capacitor voltages of the same arm as seen in Figs. 8 (e) and (f). This is also demonstrated in Fig. 10 where the absolute capacitor voltage deviation from the reference for the PS-PWM with individual capacitor voltage control lies well above the two PWM techniques based on the sorting algorithm.

Interleaving of carriers results in a continuous variation of the number of SMs that is connected to the phase-leg between $N-1$, N and $N+1$ [12]. A direct effect of this variation is observed in the arm currents of PWM techniques with interleaving, resulting in higher harmonic content and, hence, higher RMS value the arm current, slightly increasing the overall losses. The currents of the upper and lower arm under non-interleaving and interleaving PWM techniques are given in Fig. 11

The line-to-line voltages for all six cases for an operating point of m_a=0.95 are presented in Fig. 12. The higher number of output voltage levels for MMCs with carrier interleaving can be observed. Other characteristics of the PWM techniques at this operating point, including the average SM switching

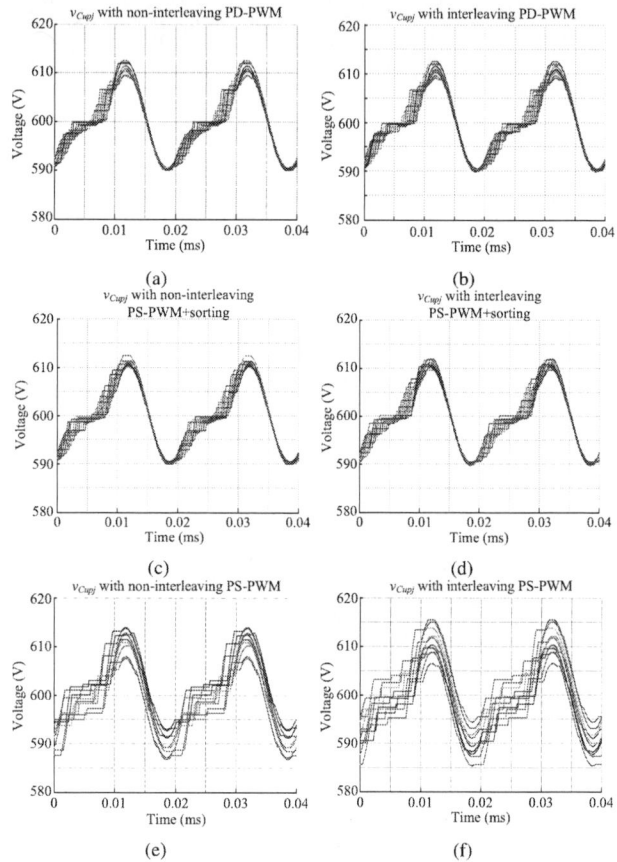

Fig. 8. Capacitors voltages (v_{C_upj}): (a) with non-interleaving PD-PWM, (b) with interleaving PD-PWM, (c) with non-interleaving PS-PWM+sorting, (d) with interleaving PS-PWM+sorting, (e) with non-interleaving PS-PWM and (f) with interleaving PS-PWM.

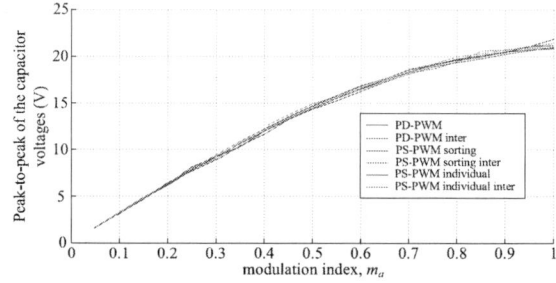

Fig. 9. Peak-to-peak capacitor voltage ripple (v_{CupjPP}) vs modulation index with a constant RL load.

frequency and peak-to-peak capacitor voltage ripple are summarised in Table II. The harmonic spectra of the line-to-line voltages are given in Fig. 13.

The higher number of levels, as a result of carrier interleaving, provides an improved overall performance compared to the non-interleaved PWM techniques albeit with increased arm current harmonics and RMS value. The interleaved techniques exhibit similar %THD for the common mode voltage (Fig. 14) where there is a gap of THD values between non-interleaving

The 2014 International Power Electronics Conference

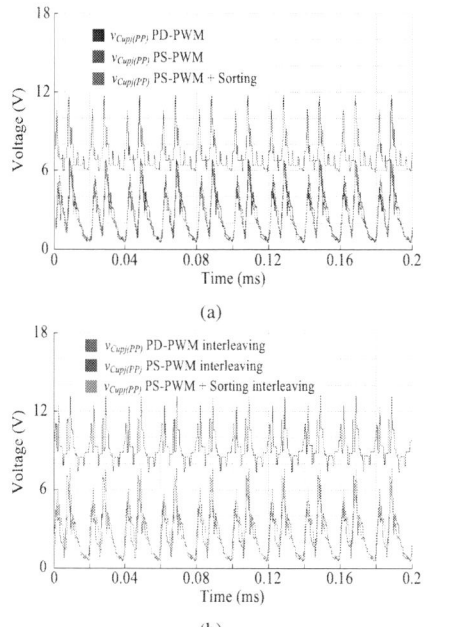

(a)

(b)

Fig. 10. Deviation size of the capacitor voltages in the upper arm ($v_{Cupj_{PP}}$): (a) with carrier non-interleaving and (b) with carrier interleaving.

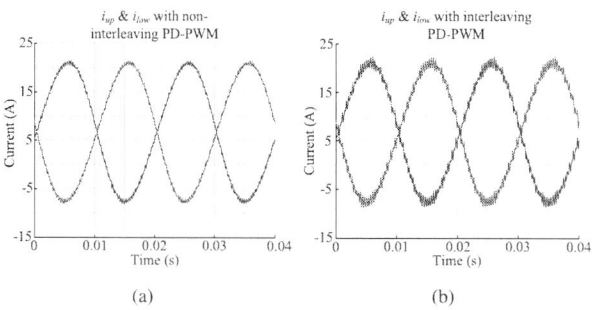

(a) (b)

Fig. 11. Upper and lower arm currents (i_{up} & i_{low}): (a) with non-interleaving PD-PWM and (b) with interleaving PD-PWM.

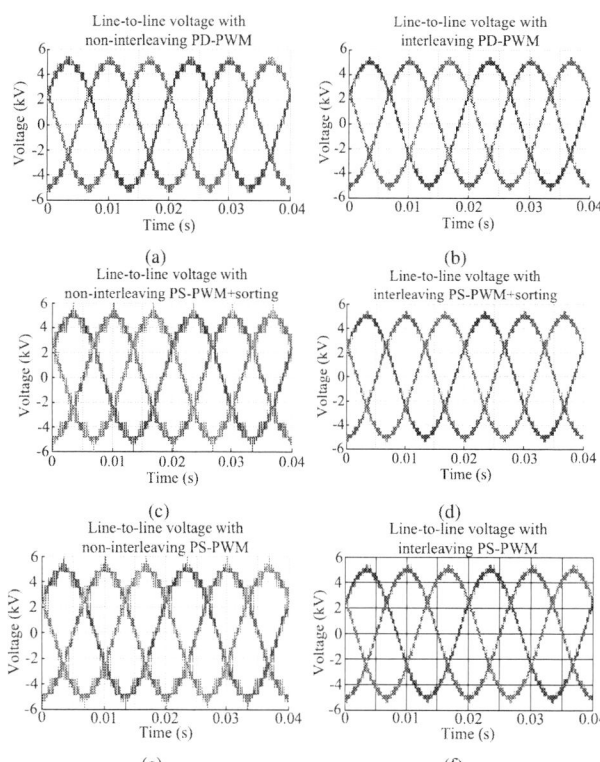

(a) (b)

(c) (d)

(e) (f)

Fig. 12. Line-to-line voltage (v_{LL}): (a) with non-interleaving PD-PWM, (b) with interleaving PD-PWM, (c) with non-interleaving PS-PWM+sorting, (d) with interleaving PS-PWM+sorting, (e) with non-interleaving PS-PWM and (f) with interleaving PS-PWM.

and interleaving, with interleaved producing lower THD. On the other hand, when interleaving between the arms of the converter is not applied, PD-PWM demonstrates a better line-to-line common voltage compared to PS-PWM techniques in terms of %THD and %WTHD as a result of the harmonic cancellation in the line-to line voltages (Fig. 14 and Fig. 15).

B. Experimental Verification

Experimental results are obtained from a laboratory prototype of an MMC phase-leg with five SMs per-arm. Interleaving produces eleven levels ($2N$-1) at the output voltages, while non-interleaving produces six levels only (N-1), as shown in Fig. 16. Interleaving also affects the output current by reducing the current ripple. On the other hand, interleaving increases the ripple in the arm currents and the circulating current (Fig. 17).

PD-PWM with restricted voltage balancing has better control to keep the SM capacitor voltages at the reference value. The peak-to-peak capacitor voltage value is similar either

TABLE II
SIMULATION RESULTS WITH NON-INTERLEAVING AND INTERLEAVING
TECHNIQUE

Modulation technique	$f_c(Hz)$	%THD $v_{LL_{comm}}$	$f_{sw}(Hz)$	Δv_{Cup} (V)
PD-PWM (Non-Inter.)	3000	5.98	345	20.70
PS-PWM (Non-Inter.)	300	8.77	343	21.37
PS-PWM+S (Non-Inter.)	300	8.76	340	20.68
PD-PWM (Interleaving)	3000	4.33	345	20.60
PS-PWM (Interleaving)	300	4.38	350	21.31
PS-PWM+S (Interleaving)	300	4.37	350	20.69

using PD-PWM with restricted voltage balancing or PS-PWM with individual capacitor voltage as shown in Fig. 18. Notice though that the SM capacitor voltages operating with PS-PWM in Fig. 18(b) are more disperse than those in Fig. 18(a) with PD-PWM.

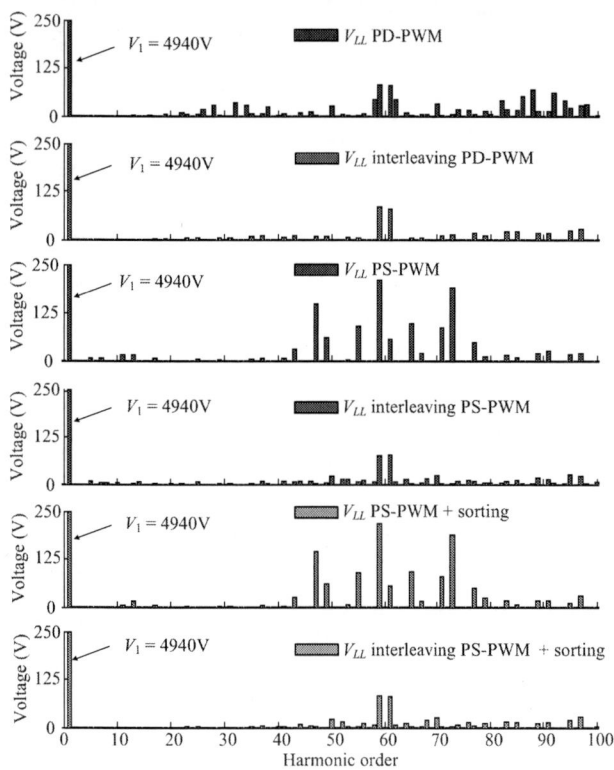

Fig. 13. Harmonic spectra of line-to-line common voltage ($V_{LL_{comm}}$) for all six cases.

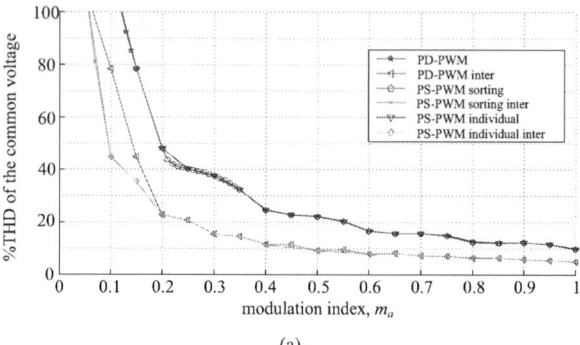

(a)

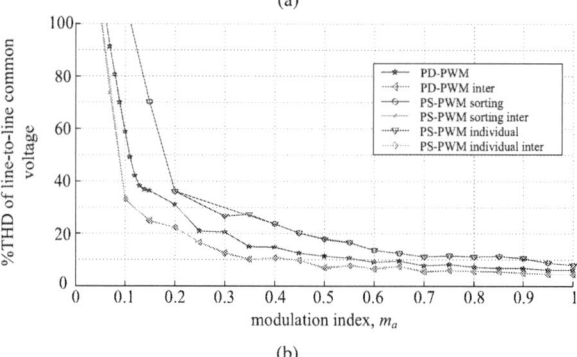

(b)

Fig. 14. Total harmonic distortion: (a) common voltage and (b) line-to-line common voltage.

V. CONCLUSION

n this paper, PD-PWM and PS-PWM have been applied to the MMC and evaluated considering non-interleaving and interleaving. Capacitor voltage balance under PS-PWM has been implemented using individual voltage control and a sorting algorithm like in PD-PWM. It has been shown that the sorting algorithm performs better in regulating the capacitor voltages in both cases PD-PWM and PS-PWM. Although the carriers' frequencies have been selected to produce similar switching frequencies to the power devices, PD-PWM produces better line-to-line output voltage spectra. The application of interleaving between the upper and the lower arms increases the quality of the output voltages and currents but produces more ripples in the arm currents and circulating current. This may force to increase the value of the arm inductors.

VI. ACKNOWLEDGEMENT

This work has been supported by the Ministerio de Economía y Competitividad of Spain under project ENE2012-36871-C02-00.

REFERENCES

[1] A. Lesnicar and R. Marquardt, "An innovative modular multilevel converter topology suitable for a wide power range," in *Proc. IEEE Bolognia Power Tech. Conf., Jun. 2003*.

[2] R. Marquardt, "Modular multilevel converter: An universal concept for HVDC-networks and extended dc-bus-applications," in *Proc. IEEE IPEC*, 2010, pp. 502-507.

[3] R. Marquardt, "Modular multilevel converter topologies with DC-short circuit current limitation," in *Proc. IEEE ICPE* 2011, pp. 1425-1431.

[4] W. Jun, R. Burgos, and D. Boroyevich, "A survey on the modular multilevel converters - modeling, modulation and controls," in *Proc. IEEE ECCE*,2013, pp. 3984-3991.

[5] M. Vasiladiotis and A. Rufer, "Analysis and control of modular multilevel converters with integrated battery energy storage" *IEEE Trans. Power Electron.*, early access, 2014, doi: 10.1109/TPEL.2014.2303297.

[6] J. W. Moon, C. S. Kim, J. W. Park, D. W. Kang, and J. M. Kim, "Circulating currrent control in MMC under the unbalanced voltage," *IEEE Trans. Power Del.*, vol. 28, pp. 1952-1959, Oct. 2013.

[7] D. G. Holmes and T. A. Lipo, *Pulse width modulation for power converters; principles and practice*, IEEE press series on power engineering, 2003.

[8] S. Fan, K. Zhang, J. Xiong, and Y. Xue, "An improved control systemfor modular multilevel converters with new modulation strategy and voltage balancing control," *IEEE Trans. Power Electron.*, vol. doi: 10.1109/TPEL.2014.2304969, 2014.

[9] A. Hassanpoor, S. Norrga, H. Nee, and L. Angquist, "Evaluation of different carrier-based PWM methods for modular multilevel converters for HVDC application," in *Proc. IEEE IECON*, 2012, pp. 388-393.

[10] R. Darus, J. Pou, G. Konstantinou, S. Ceballos, and V. G. Agelidis, "Circulating current control and evaluation of carrier dispositions in modular multilevel converters," in *Proc. IEEE ECCE Asia*, Jun. 2013, pp. 332-338.

[11] G. Konstantinou and V. G. Agelidis, "Performance evaluation of half-bridge cascaded multilevel converters operated with multicarrier sinusoidal PWM techniques," in *Proc. IEEE ICIEA*, 2009, pp. 3399-3404.

[12] G. Konstantinou, M. Ciobotaru, and V. G. Agelidis, "Analysis of multicarrier PWM methods for back-to-back HVDC systems based on modular multilevel converters," in *Proc. IEEE IECON*, 2011, pp. 4391-4396.

[13] M. Hagiwara, and H. Akagi, "Control and experiment of pulsewidth-modulated modular multilevel converters," *IEEE Trans. Power Electron.*, vol. 24, pp. 1737-1746, Jul. 2009.

[14] G. Konstantinou, M. Ciobotaru, V. G. Agelidis, "Selective harmonic elimination pulse width modulation of the modular multilevel converter," *IET Power Electronics*, Vol. 6, No. 1, pp. , Jan. 2013.

The 2014 International Power Electronics Conference

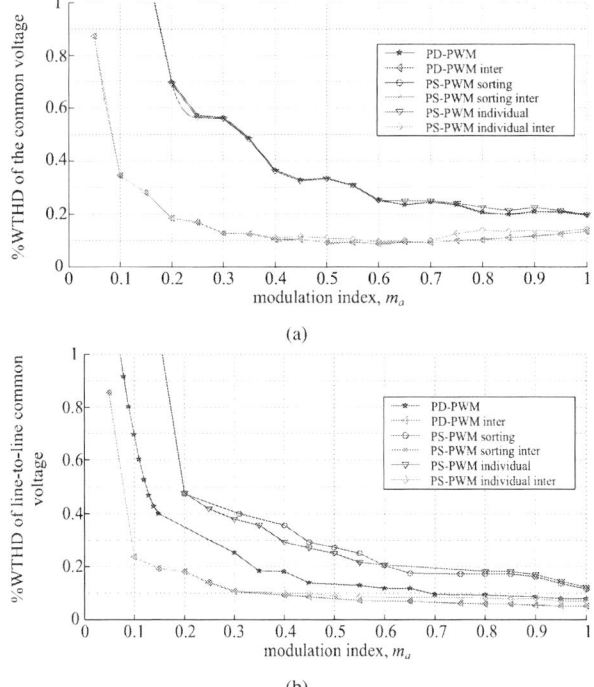

(a)

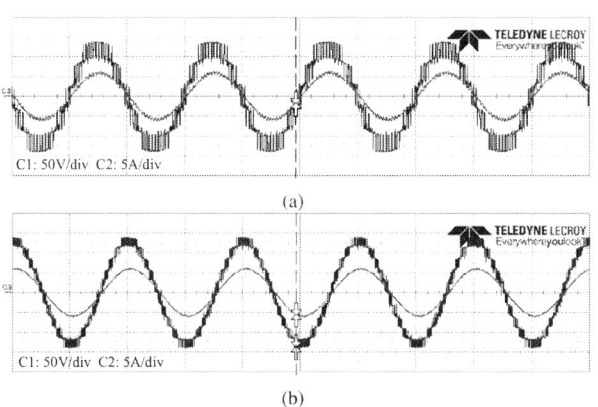

(b)

Fig. 15. Percentage of weighted total harmonic distortion (%WTHD): (a) common voltage and (b) line-to-line common voltage.

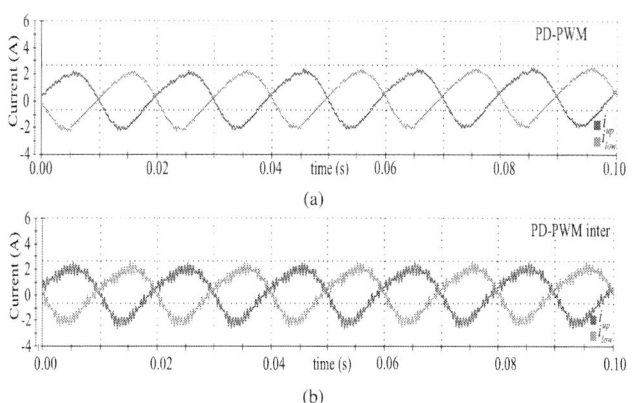

Fig. 17. Arm currents: (a) with non-interleaving PD-PWM and (b) with interleaving PD-PWM.

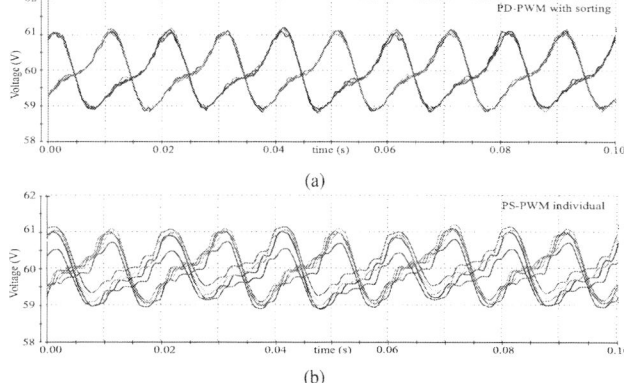

(a)

(b)

Fig. 18. Capacitors voltages: (a) using PD-PWM with restricted voltage balancing and (b) using PS-PWM with individual capacitor voltage control.

(a)

(b)

Fig. 16. Ouput voltage and current with constant RL load: (a) six levels output voltage using non-interleaving PD-PWM and (b) eleven levels output voltage using interleaving PD-PWM.

[15] A. Hassanpoor, L. Angquist, S. Norrga, K. Ilves, H.P. Nee, "Tolerance Band Modulation Methods for Modular Multilevel Converters "*IEEE Trans. Power Electron*, Early Access, 2014,

[16] M. Saeedifard and R. Iravani, "Dynamic performance of a modular multilevel back-to-back HVDC system," *IEEE Trans. Power Del.*, vol. 25, pp. 2903-2912, Oct. 2010.

[17] J. Pou, S. Ceballos, J. Zaragoza, G. Konstantinou, and V. G. Agelidis, "Optimal injection of harmonics in circulating currents of modular multilevel converters," in *Proc. IEEE ISIE*, May 2013, pp. 1-7.

[18] S. Xu and A. Huang, "Circulating current control of double-star chopper-cell modular multilevel converter for HVDC system," in *Proc. IEEE IECON* 2012, pp. 1234-1239.

[19] A. Antonopoulos, L. Angquist, L. Harnefors, K. Ilves, and H. P. Nee, "Stability analysis of modular multilevel converters with open-loop control," *Proc. IEEE IECON*, Nov. 2013, pp. 6316-6321.

[20] R. Darus, J. Pou, G. Konstantinou, S. Ceballos, and V. G. Agelidis, "A modified voltage balancing sorting algorithm for the modular multilevel converter; evaluation for staircase and phase-dispositon PD-PWM," in *Proc. IEEE APEC*, Mar. 2014.

[21] M. Vasiladiotis, N. Cherix and A. Rufer, "Accurate capacitor voltage ripple estimation and current control considerations for grid-connected modular multilevel converters", *IEEE Trans. Power Electron.*, early access, doi: 10.1109/TPEL.2013.2286293, 2013.

[22] S. Ceballos, J. Pou, S. Choi, M. Saeedifard, and V. Agelidis, "Analysis of voltage balancing limits in modular multilevel converters," in *Proc. IEEE Industrial Electronics Conference (IECON'11)*, 7-10 Nov. 2011, Melbourne, Australia, pp. 4397-4402.

A Single-phase Power Conditioner with a Buck–Boost-type Power Decoupling Circuit

Shota Yamaguchi and Toshihisa Shimizu
Dept. of Electrical Engineering
Tokyo Metropolitan University
Tokyo, Japan
shimizut@tmu.ac.jp

Abstract–This paper presents a novel single-phase inverter with a buck–boost-type power decoupling function suitable for PV generation. The proposed buck–boost-type power decoupling circuit enables the reduction of the peak voltage at the DC bus of the PWM inverter. This enables the use of switching devices with lower voltage ratings, and the resultant switching loss and conduction loss of the PWM inverter can be reduced. Another feature of the inverter is that a power decoupling control method based on single-phase p-q theory is applied. The proposed control method provides a low ripple current at the DC input portion, and it enables the use of small long-life capacitors such as film capacitors instead of electric capacitors. The effectiveness of the proposed system is verified by experiment with a 1 kW setup.

Keywords–power conditioner, power decoupling, buck–boost type

I. INTRODUCTION

In recent years, interest in photovoltaic (PV) power generation systems has grown among power electronics engineers because of the current environmental issues. Numerous inverter circuits and control methods have been proposed [1]–[4]. In particular, in a single-phase inverter system, power pulsation with twice the utility frequency appears at the DC input terminal, as shown in Fig. 1 and Eqs. (1) and (2).

$$P_{in} = V_{DC} I_{DC} \qquad (1)$$

$$P_{out} = \frac{1}{2} V_{AC} I_{AC} - \frac{1}{2} V_{AC} I_{AC} \cos 2\omega t \qquad (2)$$

This results in both improper control of the maximum power point tracking (MPPT) of the PV modules (Fig. 2) and distortion of the output current of the single-phase inverter. The second term in Eq. (2) is the pulsated instantaneous power p_{rip}, as shown in Eq. (3).

$$p_{rip} = -\frac{1}{2} V_{AC} I_{AC} \cos 2\omega t \qquad (3)$$

In order to reduce the voltage fluctuation at the input DC bus caused by the power pulsation, electrolytic capacitors with large capacitance have been connected to the DC bus. This method is called passive power decoupling (PPD), and most of the power conditioners on the market have applied the PPD method. However, the lifetime of electrolytic capacitors is relatively short, and it is shortened further when used under high

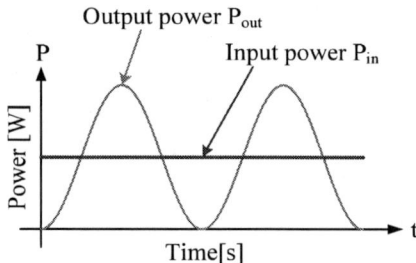

Fig. 1. Input power and output power.

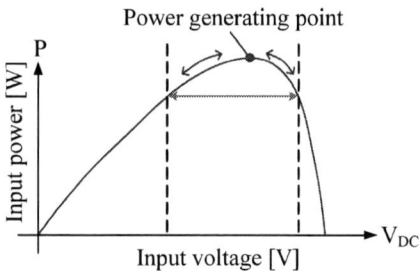

Fig. 2. Characteristics of solar module.

atmospheric temperature conditions. Hence, a serious breakdown may occur due to the short electrolytic capacitor lifetime. The lifetime of the PV panel has been extended to more than 20 years owing to technical innovation. However, most of the power conditioners may not work for more than 20 years without maintenance. So, it is necessary to extend the lifetime of the power conditioner in order to recover the initial cost of the PV power generation systems. Of course, we may be able to use film capacitors instead of the large electrolytic capacitors with the PPD method; however, the volume of the power conditioner must increase because the volume of the film capacitor is more than 20 times larger than that of the electrolytic capacitor.

In order to solve the abovementioned problems, an active power decoupling (APD) concept for a single-phase utility interactive inverter/converter, which can reduce the power pulsation in the input DC capacitors, has been proposed [5]–[8]. A distinctive feature of the APD concept is that the pulsated energy is stored in an additional decoupling capacitor with large voltage fluctuation. The APD circuit is composed of switching devices, a small film capacitor C_X, inductor, and diodes. The instantaneous power being stored in C_X is expressed as follows:

$$P_C = C_X v_X \frac{dv_X}{dt} \qquad (4)$$

where v_X is the voltage of the decoupling capacitor C_X.

A boost chopper circuit has been used to realize the APD function in the single-phase inverter system. However, the peak DC bus voltage at the PWM inverter increases, and switching devices with high voltage ratings have to be used. Hence, the switching loss and conduction loss of the switching devices are increased, and the conversion efficiency of the power conditioner cannot be improved.

In order to overcome this defect, this paper proposes a novel APD circuit and its control method, which provides high-efficiency power conversion. In the proposed APD circuit, both the buck–boost-type APD circuit and the direct power transfer function are applied. The buck–boost-type APD circuit enables storage of the pulsated energy in the APD capacitor with a minimum increase in the DC peak voltage at the inverter portion. Hence, the volume of the APD capacitor can be reduced, and the loss of the PWM inverter circuit can also be minimized. The direct transfer function of the APD reduces the power loss of the APD circuit. Owing to the above contributions, high conversion efficiency and longevity of the proposed power conditioner can be realized.

This paper first describes the concept of the APD method and the main circuit configuration of the proposed power conditioner. Next, a novel control method based on a single-phase p-q theory that the authors have proposed is explained. Finally, the effectiveness of the proposed method is verified through simulation and experiment on a 1 kW experimental setup.

II. MAIN CIRCUIT CONFIGURATION

Figure 3 shows an example of a power conditioner circuit with a boost-type APD circuit, and Fig. 4 shows the proposed power conditioner circuit with a buck–boost-type APD circuit. Table 1 shows the parameters of the proposed circuit. The proposed inverter circuit is composed of a single-phase PWM inverter and an APD circuit, which is indicated by the dotted line. The power decoupling circuit stores the energy caused by the difference in instantaneous power between the input and output terminals of the power conditioner.

Figure 5 shows the power flow of each operation mode. When the instantaneous output power is smaller than the input power, the power decoupling circuit operates in Mode I. The surplus input energy is transferred to the decoupling inductor L_X by turning on switch S_{X1}, and the stored energy is transferred to the decoupling capacitor C_X by turning off switch S_{X1}. Therefore, the voltage of the decoupling capacitor v_X increases. When the output power is larger than the input

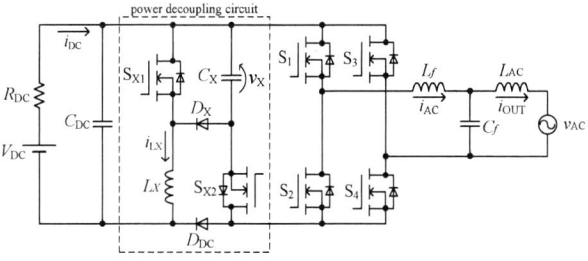

Fig. 3. Power conditioner circuit with boost-type power decoupling circuit.

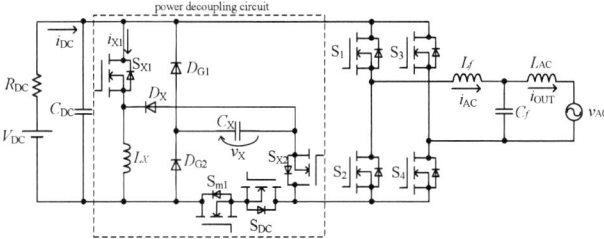

Fig. 4. Power conditioner circuit with buck–boost-type power decoupling circuit.

Table 1. Circuit parameters

Carrier frequency, f_{sw}	20 kHz
Output frequency, f_{AC}	50 Hz
Output voltage, V_{AC}	50 V
Output power, P_{AC}	50 W
DC bus input capacitance, C_{DC}	22 μF
Decoupling capacitance, C_X	50 μF
Inductance of the buck–boost chopper, L_X	500 uH
Inductance of the LC filter, L_f	1 mH
Capacitance of the LC filter, C_f	2.2 μF

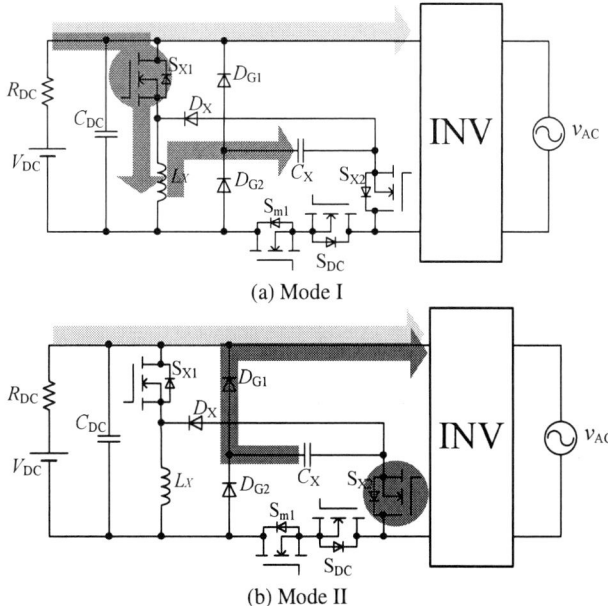

(a) Mode I

(b) Mode II

Fig. 5. Power flow for each operation mode.

power, the power decoupling circuit operates in Mode II. The required AC output energy is supplied both from the input power source V_{DC} and the decoupling capacitor. A part of the required energy is supplied from the input power source while switch S_{X2} is turned off, and the additional energy required for the output power is supplied directory from the decoupling capacitor through switch S_{X2}. Since the stored energy in the decoupling capacitor is supplied directly to the PWM inverter, the loss caused by power transmission can be minimized, and hence, the high conversion efficiency of the power conditioner can be realized.

An additional switch S_{DC} turns off when switch S_{X2} operates to prevent backflow of the current from the decoupling capacitor to the input DC power source when the voltage of the decoupling capacitor is higher than that of the input DC power source. An additional switch S_{m1} is controlled so that current does not flow to the decoupling capacitor when the AC output current flows back.

III. REQUIRED CAPACITANCE AND OPERATING VOLTAGE OF THE DECOUPLING CAPACITOR

In order to store the pulsating power p_{rip} of the single-phase inverter in the power decoupling capacitor, the power p_C stored in the decoupling capacitor needs to have the same value as the pulsated instantaneous power p_{rip}, as shown in Eq. (5),

$$C_X v_X \frac{dv_X}{dt} = -\frac{1}{2} V_{AC} I_{AC} \cos 2\theta \qquad (5)$$

where V_{AC} is the peak value of the utility voltage, I_{AC} is the peak value of the inverter output current, and v_X is the instantaneous voltage of the decoupling capacitor.

The pulsating voltage appearing in the power decoupling capacitor is expressed as follows:

$$v_X(t) = \sqrt{-\frac{V_{AC} I_{AC}}{2\omega C_X} \sin 2\omega t + V_X^{*2}} \qquad (6)$$

where V_X^* is the average voltage of the decoupling capacitor.

Figure 6(a) and (b) shows the relationship between the voltage and the capacitance of the power decoupling capacitor for both the boost-type and buck–boost-type circuits, respectively. Here, the DC bus voltage is V_{DC} =250 V, the output power is P_{OUT} =1 kW, and V_{X_max}, V_{X_min}, and V_X^* represent the maximum voltage, the minimum voltage, and the average voltage of the power decoupling capacitor, respectively.

Figure 7(a) shows the voltage fluctuation in the boost-type power decoupling circuit. We can see that the minimum voltage of the decoupling capacitor v_{X_min1} must be higher than the input voltage V_{DC}, as shown in Eq. (7).

$$v_{X\_min1} = \sqrt{-\frac{V_{AC} I_{AC}}{2\omega C_X} + V_X^{*2}} \geq V_{DC} \qquad (7)$$

Hence, the control command for the average voltage of the decoupling capacitor is expressed as

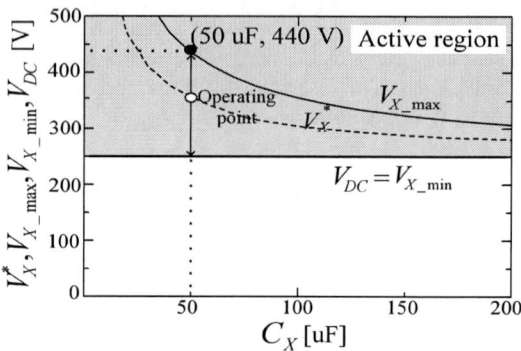

(a)   Boost-type power decoupling circuit

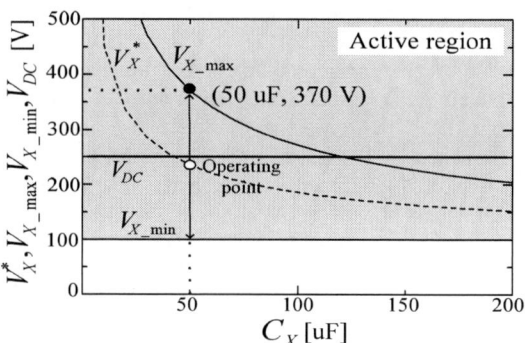

(b) Buck–boost-type power decoupling circuit

Fig. 6. Relationship between the decoupling capacitor and the voltage (V_{DC}=250 V, V_{AC}=100 Vrms, P_{OUT}=1 kW).

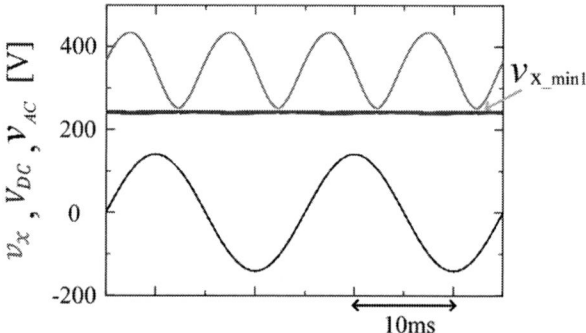

(a) Boost-type power decoupling circuit

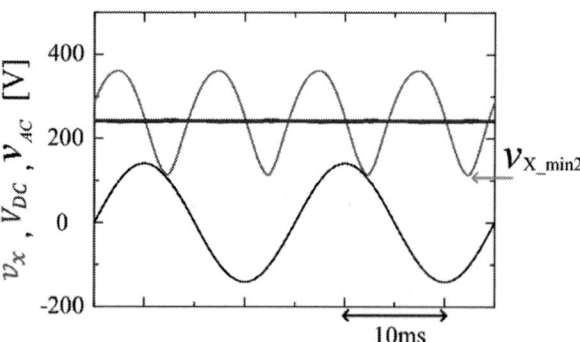

(b) Buck–boost-type power decoupling circuit

Fig. 7. Voltage fluctuation for each type of power decoupling circuit.

$$V_X^* \geq \sqrt{V_{DC}^2 + \frac{V_{AC}I_{AC}}{2\omega C_X}}. \qquad (8)$$

For example, when 50 uF is used for the decoupling capacitance, the maximum voltage appearing at the DC bus of the PWM inverter reaches 440 V. This operating condition is very close to the limitation for a power device with a 600 V voltage rating, and sometimes, a power device with higher voltage rating is required.

Figure 7(b) shows the voltage fluctuation of the buck–boost-type power decoupling circuit. We can see that the minimum voltage of the power decoupling capacitor v_{X_min2} can be reduced below the DC bus voltage, and the limitation of the minimum voltage is determined by the instantaneous AC voltage v_{AC}, as shown in Eq. (9).

$$v_X(t) = \sqrt{-\frac{V_{AC}I_{AC}}{2\omega C_X}\sin 2\omega t + V_X^{*2}} \geq v_{AC} \quad (9)$$

Hence, the control command for the average voltage of the decoupling capacitor is expressed as

$$V_X^* \geq \sqrt{v_{AC}^2 + \frac{V_{AC}I_{AC}}{2\omega C_X}\sin 2\omega t}. \qquad (10)$$

For example, when 50 uF is used for the decoupling capacitance, the maximum voltage appearing at the DC bus of the PWM inverter is reduced to 370 V because the minimum voltage is reduced to 100 V. In this operating condition, a power device with a 600 V voltage rating (or less) can be used very safely, and also, the switching loss of the power device can be reduced. This is a big advantage of the high conversion efficiency of the power decoupling type power converter.

IV. CONTROL METHOD

A. Control of power decoupling circuit

In order to ensure accurate control of the power decoupling circuit, calculation of the pulsated instantaneous power p_{rip}, which is given by Eq. (3), is important. In order to calculate the pulsated power p_{rip}, the single-phase p-q theory that the authors have already proposed [9] is used. A feature of the single-phase p-q theory is that the instantaneous reactive and effective power are calculated in the rotating d-q coordinate system that is synchronized with the utility voltage.

In order to perform the d-q transformation shown in Eq. (11), it is necessary to know two values for the stationary frame: v_α and v_β.

$$\begin{bmatrix} v_d \\ v_q \end{bmatrix} = \begin{bmatrix} \cos \omega t & \sin \omega t \\ -\sin \omega t & \cos \omega t \end{bmatrix} \begin{bmatrix} v_\alpha \\ v_\beta \end{bmatrix} \qquad (11)$$

However, two orthogonal values for single-phase power cannot be defined from a single-phase voltage or current. In this theory, the real voltage of the utility voltage v_{AC} is defined as the orthogonal value of v_α on the α-axis. Also, the other orthogonal value of v_β on the β-axis is given as a virtual value. Since the rotating coordinate system needs to synchronize with the utility voltage, a PLL circuit as shown in Fig. 8 is used. Figure. 9 shows

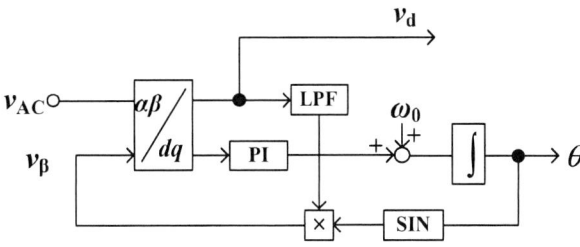

Fig. 8. Block diagram of PLL.

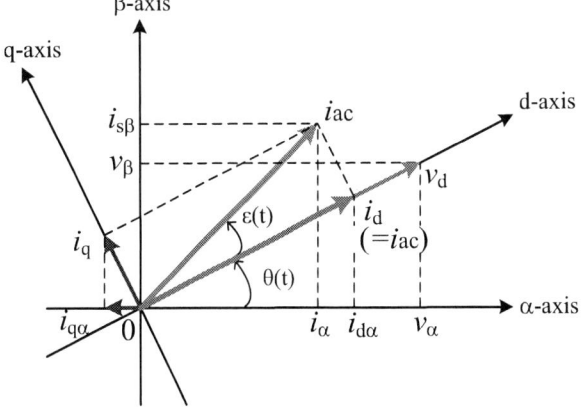

Fig. 9. Single-phase rotating coordinate system.

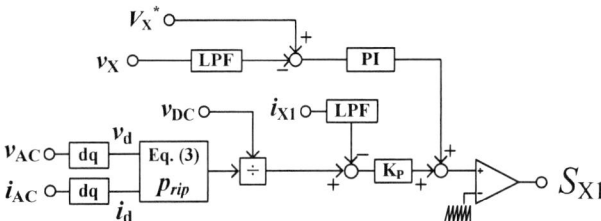

Fig. 10. Block diagram of power decoupling.

the voltage and current values of the single-phase inverter and the virtual voltage and current vectors on the rotating coordinates. In the PLL circuit, an estimated virtual value for v_β is generated using the following equation:

$$v_\beta = v_d \sin \omega t = V_{AC} \sin \omega t. \qquad (12)$$

Hence, we could detect the peak amplitude of the utility voltage V_{AC}, which can be detected from the d-axis value. Since the inverter current i_{AC} is controlled to synchronize with the utility voltage v_{AC}, the current value I_{AC} appears at the d-axis value i_d, in the rotating coordinate system. Once we obtain the voltage and current values on the d-q frame, the instantaneous pulsating power p_{rip}, as shown in Eq. (3), is given. Figure 10 shows a block diagram of the power decoupling control. The power decoupling control circuit is composed of current control for the decoupling inductor i_{X1} and average voltage control for the

978-1-4799-2706-7/14 $31.00 © 2014 IEEE

decoupling capacitor V_X. Hence, the inductor current command $i_{X1}{}^*$ is given by

$$i_{X1}{}^* = \frac{Prip}{V_{DC}} = \frac{v_d i_d}{2V_{DC}}\cos(2\omega t). \qquad (13)$$

The average voltage control command for the decoupling capacitor is added to the power decoupling control command in order to keep the average voltage at the desired value. The gate signal of switch S_{X1} is generated by the PWM device.

B. Output current control

In order to generate the sinusoidal output current, the PWM signal is modified from the conventional one. Figure 11 shows the switching pattern when the output current is positive. While the left arm switches of the inverter (S1, S2) are controlled using PWM with a carrier frequency of 20 kHz, the right arm switches (S3, S4) are controlled to convert the polarity of the output current with a utility frequency of 50 Hz. Switch S4 turns on and S3 turns off while the output current is positive, so their switching losses are reduced. The gating signals of these switches are given based on the operation mode (Modes I and II).

For Mode I, the required output power is supplied from only the input power source. Hence, the output current control and the modulation of the PWM inverter are performed in the same manner as with a conventional PWM inverter. The modulation signal for the switch S_1 is given by

$$\lambda(t) = \frac{V_{AC}\cos\omega t}{V_{DC}} \qquad (14)$$

and the gate signal of switch S_1 and the pulse waveform appearing at the output terminal of the inverter v_{INV} are shown in Fig. 12(a). Please note that the surplus of the input current supplied from the DC power source is stored in the power decoupling capacitor by the proper operation of the power decoupling circuit, and hence, the input current is maintained at a constant value.

In Mode II, the required output power is supplied both from the input power source and the decoupling capacitor. Hence, a modulation that provides both a constant input power and the power decoupling operation has to be applied. In order to realize the above requirement, the time-sharing modulation method is applied.

In the first stage, switches S1 and S4 are turned on, whereas switch S_{X2} is kept off, and part of the required output power is supplied from the input power source. in the conventional PWM inverter, the current flowing from the DC link (i_{DC_link}) fluctuates and is expressed as

$$i_{DC\_link}(t) = \frac{V_{AC}*I_{AC}}{2V_{DC}}(1+\cos2\omega t). \qquad (15)$$

In order to realize a constant DC input current, the modulation signal $\lambda_{DC}(t)$ for switch S1 is modified to

$$\lambda_{DC}(t) = \frac{\lambda(t)}{1+\cos2\omega t} \qquad (16)$$

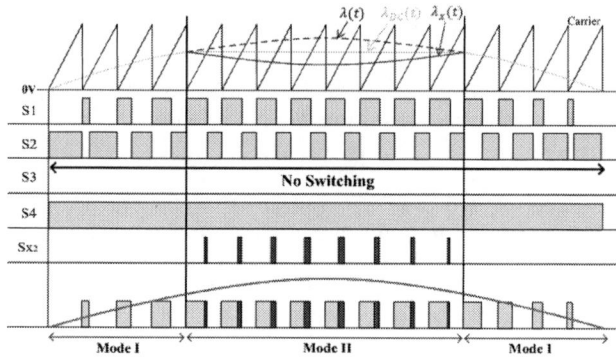

Fig. 11. Switching pattern.

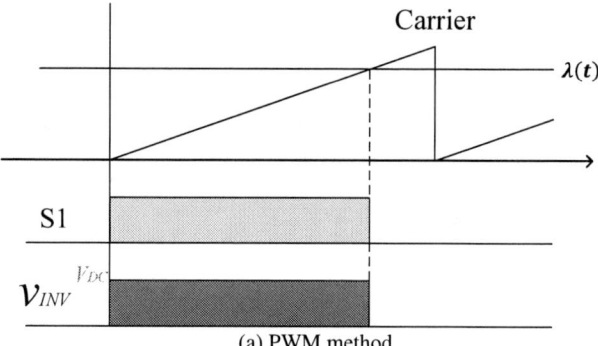

(a) PWM method

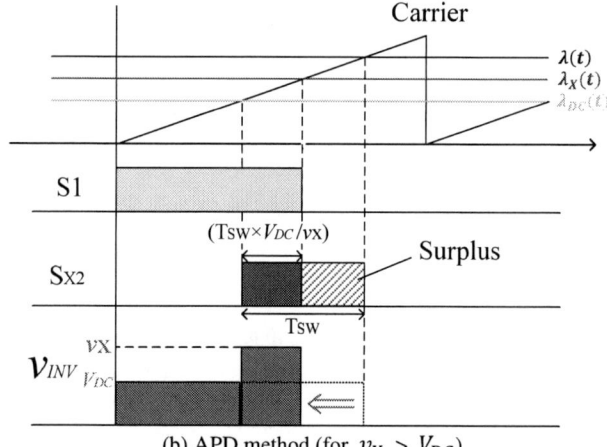

(b) APD method (for $v_X > V_{DC}$)

Fig. 12. Modulation signal and gate pulse.

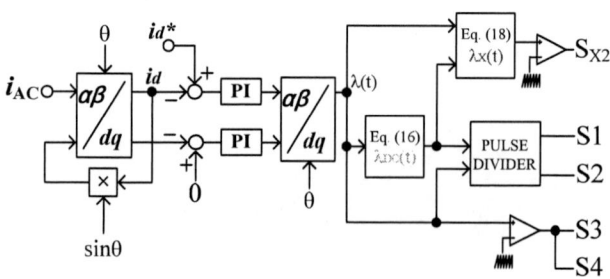

Fig. 13. Block diagram of output current control.

and the resultant DC input current is given as

$$i'_{DC} = \frac{V_{AC} * I_{AC}}{2 V_{DC}}. \qquad (17)$$

Another required output power is supplied from the decoupling capacitor by turning on switch S_{X2}. Hence, the gate signal of switch S_{X2} and the resultant pulse waveform that appeared in the output terminal of the inverter v_{INV} are shown in Fig. 12(b). Since the decoupling voltage v_x is not the same as the DC bus voltage V_{DC}, the original modulation signal $\lambda(t)$ is adjusted based on the variation of the DC bus voltage V_{DC} by the following equation:

$$\lambda_X(t) = \left(\lambda(t) - \lambda_{DC}(t)\right) \times \frac{V_{DC}}{v_x} + \lambda_{DC}(t). \quad (18)$$

Since switch S_{X2} turns off just after switch S_1 is turned off, the switching losses and surge voltage appearing at switch S_{X2} are suppressed.

Based on the above design, a control block diagram of the output current is shown in Fig. 13.

V. EXPERIMENTAL RESULT

Figure 14(a) and (b) shows the simulated results for both the inactive and active condition of the proposed power decoupling method, respectively, where the input voltage is V_{DC}=120 V, the output voltage is v_{AC}=50 V, and the output power is $P_{out} = 100$ W. When the power decoupling circuit is inactive, a large pulsating current (2 A) appears in the DC input current i_{DC}. However, when the APD is activated, the previous pulsating current is reduced to 0.5 A, which is 25% less than the pulsating input current. This contributes to the suppression of the conversion efficiency reduction. A large voltage pulsation with twice the utility frequency appears in the power decoupling capacitor voltage. This means that the pulsating power caused by the AC output power is compensated by the operation of the power decoupling circuit. In addition, the decoupling capacitor voltage reduces to below the DC bus voltage. This result shows that we can use lower voltage ratings for the power devices.

Figure 15(a) and (b) shows the experimental results for both the inactive and active conditions of the power decoupling method, respectively, which are the same conditions as in the simulation results in Fig. 14. The decreasing rate of pulsating input current is 45% before and after power decoupling, which is larger than in the simulation result. This could be reduced more by improving the control method. The total harmonic distortion (THD) of the output current i_{AC} is 4.5% and is sufficient to connect to the grid.

The experimental result for the conversion efficiency is 89% for the conditions in Fig. 15. This could be improved by increasing the output power.

VI. CONCLUSION

This paper presented a novel single-phase inverter with a buck–boost-type power decoupling function. The proposed control method provides a low ripple current at the DC input portion, and it enables the use of small

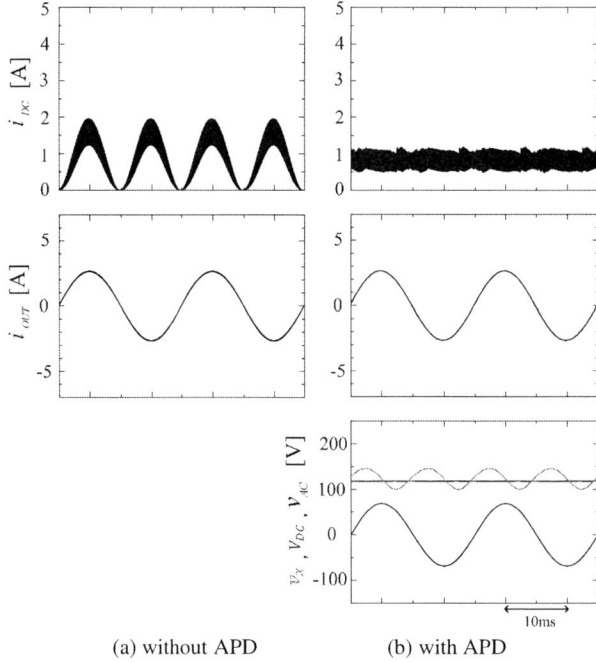

(a) without APD   (b) with APD

Fig. 14. Simulation results.

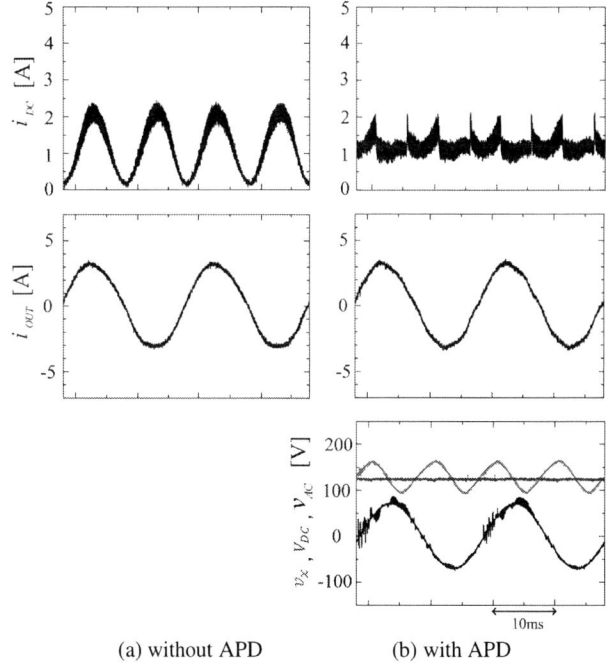

(a) without APD   (b) with APD

Fig. 15. Experimental results.

long-life capacitors such as film capacitors instead of electric capacitors. The effectiveness of the proposed system was verified by experiment with a 1 kW setup.

REFERENCES

[1] D. C. Martins and R. Demonti, "Photovoltaic energy processing for utility connected system," *in Proc. IECON'01*, Vol. 2, pp. 1292-1296, 2001.

[2] T. Shimizu, M. Hirakata, T. Kamezawa, and H. Watanabe, "Generation control circuit for photovoltaic modules," *IEEE Trans. on Power Electronics*, Vol. 16, No. 3, pp. 293-300, 2001.

[3] T. Shimizu, O. Hashimoto, and G. Kimura, "A novel high-performance utility interactive photovoltaic inverter system," *IEEE Trans. on Power Electronics*, Vol. 18, No. 2, pp. 704-711, 2003.

[4] Y. Xue, L. Chang, S. B. Kjaer, J. Bordonau, and T. Shimizu, "Topologies of single-phase inverters for small distributed power generators: an overview," *IEEE Trans. on Power Electronics*, Vol. 19, No. 5, pp. 1305-1314, 2004.

[5] S. B. Kjaer, K. Pederson, and F. Blaabjerg, "A review of single-phase grid-connected inverters for photovoltaic modules," *IEEE Trans. on Power Electronics*, Vol. 41, No. 5, pp. 1292-1306, 2005.

[6] A. Testa, S. De Caro, A. Consoli, and M. Cacciato, "An active current ripple compensation technique in grid connected fuel cell applications," *in ECCE'09*, pp. 2642-2649, 2009.

[7] C. Liu and J. S. Lai, "Low frequency current ripple reduction technique with active control in a fuel cell system with inverter load," *IEEE Trans. on Power Electronics*, Vol. 22, No. 4, pp. 1429-1436, 2007.

[8] S. B. Kjaer and F. Blaabjerg, "A novel single-stage inverter for the ac-module with reduced low frequency ripple penetration," *in Proc. EPE'03*, CDROM, 2003.

[9] Yaow-Ming and Chein-YYao Liao "Three-Port Flyback-Type Single-Phase Micro Inverter With Active Power Decoupling Circuit", *IEEE ECCE*, pp. 501-506 , 2011.

[10] Keisuke Toyama, Keiji Wada, Toshihisa Shimizu, "A Study on The Control Method of The Single-Phase Utility Interactive Inverter with Power Decoupling Function", *IEE-JIASC*, Y-21, 2012.

[11] Keisuke Toyama, Toshihisa Shimizu, "Study on a Control Method of a Single-phase Utility Interactive Inverter with a Power Decoupling Function," *IEEE ECCE USA*, (2013)

The 2014 International Power Electronics Conference

A novel asymmetrical FLC-based MPPT technique for photovoltaic generation system

Yi-Hsun Chiu
Department of Electrical
Engineering, NTUST
Taipei, Taiwan, R.O.C.
D10107203@mail.ntust.edu.tw

Yu-Shan Cheng
Department of Electrical
Engineering, NTUST
Taipei, Taiwan, R.O.C.
D10107206@mail.ntust.edu.tw

Yi-Hua Liu
Department of Electrical
Engineering, NTUST
Taipei, Taiwan, R.O.C.
yhliu@mail.ntust.edu.tw

Shun-Chung Wang
Department of Electrical
Engineering, LUST
Taoyuan, Taiwan, R.O.C.
scwang.hinet@msa.hinet.net

Zong-Zhen Yang
Electric Energy Technology
Division Power Electronics
Department, Green Energy and
Environment Research
Laboratories ICL, ITRI
Hsinchu, Taiwan, R.O.C.
ZZYang@itri.org.tw

Abstract— In this paper, a novel asymmetrical fuzzy-logic-control (FLC) based maximum power point tracking (MPPT) algorithm for photovoltaic (PV) generation system (PGS) is proposed. The proposed method uses the power variation and voltage variation as the inputs, which significantly simplified the computation. Comparing with the conventional perturb and observe (P&O) method, the proposed asymmetrical FLC-based MPPT algorithm can improve the dynamic and steady state performance of the PGS simultaneously. Comparing with the proposed symmetrical FLC-based MPPT method, the transient time and the MPPT tracking accuracy are further improved by 38.1 % and 0.03 %, respectively.

Keywords—fuzzy-logic-control, maximum power point tracking, photovoltaic, photovoltaic generation system(PGS), perturb and observe

I. Introduction

In recent years, with the exhaustion of fossil fuel leading to the ever-increasing cost of energy and the environmental issues invoking public concerns, more and more researches focus on renewable energy systems. Among various renewable energy sources, photovoltaic generation system (PGS) has become one of the most popular green energy systems since it is maintenance-free and environmental friendly. However, to make PGS more competitive in the commercial market, the reduction of cost and size and the improvement of energy efficiency have become significant. Thus, the development of maximum power point tracking (MPPT) technique plays an important role of tracking the available peak power generated by PV panel under any weather conditions. In the previous studies, numerous MPPT methods have been developed and implemented. These techniques perform well under steady weather condition but still exhibit some trade-off between tracking speed and tracking accuracy when insolation changes [1-3]. According to the literatures, fuzzy-logic-control (FLC)-based MPPT can provide good

tracking performance [4-9]. Consequently, the FLC-based MPPT algorithm attracts many research interests.

In this paper, a FLC-based MPPT technique is first developed. Instead of taking an error and an error deviation as the inputs, the proposed method uses the power variation (ΔP) and voltage variation (ΔV) as the inputs, which significantly simplified the computation. Then, to improve dynamic performance, asymmetrical membership function is employed. In this paper, the power stage of the PV system is a boost converter and the proposed algorithm is realized in the dsPIC digital signal controller (DSC) dsPIC33FJ16GS502. Both Experiments and simulations will be conducted to verify the effectiveness of the proposed system. The experimental results shows that proposed FLC based MPPT technique can make the tracking time shorter and enhance the steady state performance in comparison with conventional fixed-step perturb and observe (P&O) method. The proposed algorithm can track maximum power point under any weather conditions and make tracking accuracy higher than 99.8%. The proposed asymmetrical FLC-based method can also reduce the tracking time by 38% comparing to the conventional FLC-based method.

II. System configuration

Fig. 1 shows the block diagram of the proposed system. The developed MPPT algorithm is implemented in a low cost DSC dsPIC33FJ16GS502 from Microchip Corp which provides the required gating signal for the power switch in the converter and gathers data from the signal conditioning circuits. Observing Fig. 1, the whole system can be divided into three parts: power conversion unit, main control unit and data logging system. Detailed descriptions for each unit will be given in the following subsections:

978-1-4799-2706-7/14 $31.00 © 2014 IEEE

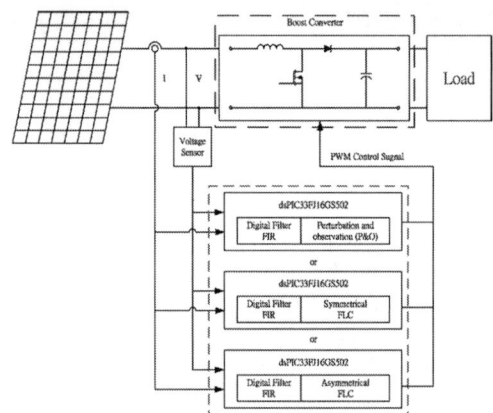

Fig. 1 The block diagram of the proposed system

a. Power conversion unit: To provide the demanded power to the load, a simple boost-type DC-DC converter is employed in this paper. The topology of the power conversion unit is illustrated in Fig. 2. The maximum available PV energy can be transferred to the load through adequate control of the gating signal of the power switch Q. Further discussion will be simplified since both the design and the implementation of the circuit are conventional and well-developed.

The relationship between output voltage and input voltage in boost converter can be expressed as

$$\frac{V_o}{V_{in}} = \frac{1}{1 - Duty} \tag{1}$$

where *Duty* represents the duty cycle. Assuming the conversion efficiency of the boost converter is 100%, relationship between the output current and the input current can be written as

$$\frac{I_o}{I_{in}} = 1 - Duty \tag{2}$$

b. Main control unit: as shown in Fig. 1, main control unit contributes to provide the PWM signal for the boost converter so that maximum power will be obtained from PV panel. The essential PV voltage and current data acquired by DSC from the A/D module will be analyzed and then smoothed out by a 24-order finite impulse response (FIR) filter. The equation of the adopted FIR filter can be expressed as in Eq. (3).

$$Y[n] = \sum_{i=0}^{23} a_i X[n-i] \tag{3}$$

where X is the filter input, Y is the filter output and a_i is the corresponding coefficient of the designed FIR filter.

Once the filtered PV voltage and current are acquired, gating signals can be determined using the developed MPPT controller. In this paper, three different MPPT schemes will be implemented for performance comparison, further description of these MPPT methods will be discussed in section III.

c. Data logging system: It is necessary to have long-term recording of the operating condition to verify the effectiveness of the proposed method. In the study, solar array system TerraSAS DCS80-15 from AMETEK Corp. featuring long-term recording function with the recording

time interval as short as 0.05s is used as the input power source.

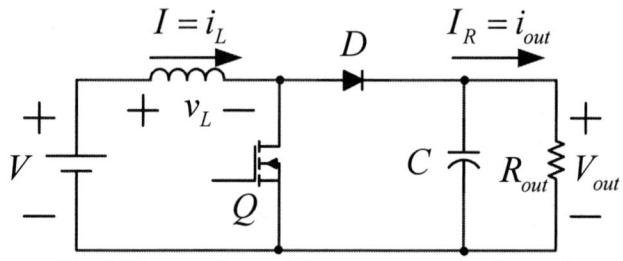

Fig. 2 The topology of the energy conversion unit

III. Derivation of the proposed FLC-based MPPT controller

a. Derivation of the FLC-based MPPT controller

The equivalent circuit of the PV cell can be illustrated in Fig. 3. Also, the I–V characteristic of the PV cell can be mathematically described by Eq. (4)

$$I_{PV} = I_{SC} - I_O \left\{ \exp\left[\frac{q(V + R_S I)}{nkT_K} \right] - 1 \right\} - \frac{V + R_S I}{R_{SH}} \tag{4}$$

where n is the ideality factor, k is the Boltzmann's constant, q is the electron charge, T_K is the temperature in Kelvin, R_S is the equivalent series resistance, R_{SH} is the equivalent shunt resistance and I_{SC}, I_{PV} and I_O are the photogenerated current, panel current and saturation currents, respectively.

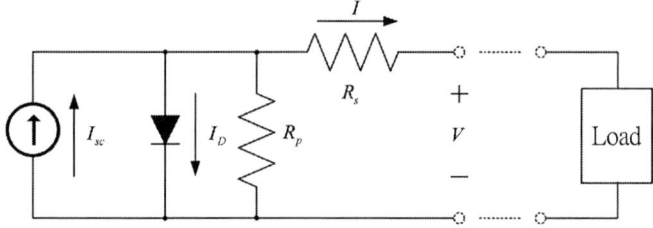

Fig. 3 Equivalent circuit of the PV cell

With the simulation using MATLAB, the I–V curves of the utilized Sanyo VBHN220AA01 solar panel under different irradiation levels can be acquired and shown in Fig. 4(a). In the graph, the calculated maximum power points are represented in circular markers. Fig. 4(b) shows the corresponding P–V curve for 1000 W/m² irradiation level. According to Fig. 4(b), the absolute value of dP/dV of a PV panel varies smoothly and is recommended in [10] as a suitable parameter for determining the variable step size of the incremental conductance algorithm. Therefore, power variation (ΔP_{pv}) and voltage variation (ΔV_{pv}) are adopted as the inputs of the proposed FLC-based MPPT controller.

To realize MPPT, a fuzzy controller is conducted every 20 ms to determine the required perturbation step size. Fig. 5 shows the block diagram of the implemented FLC-based MPPT controller. In this paper, the voltage variation (ΔV_{pv}) and the power variation (ΔP_{pv}) are taken as inputs and duty cycle step size ΔD is chosen as the output. The membership functions (MFs) of the utilized input and output variables for the proposed controller are shown in Fig. 6. Both input and output MFs are in triangular form as illustrated in Fig. 6(a) and Fig. 6(b). For linguistic variables, P represents positive, N represents negative, B, S, and ZE are defined as big, small and zero, respectively.

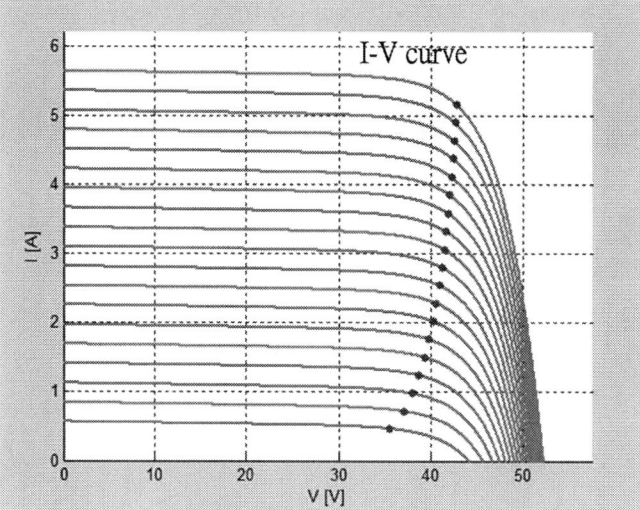

(a) I–V curve

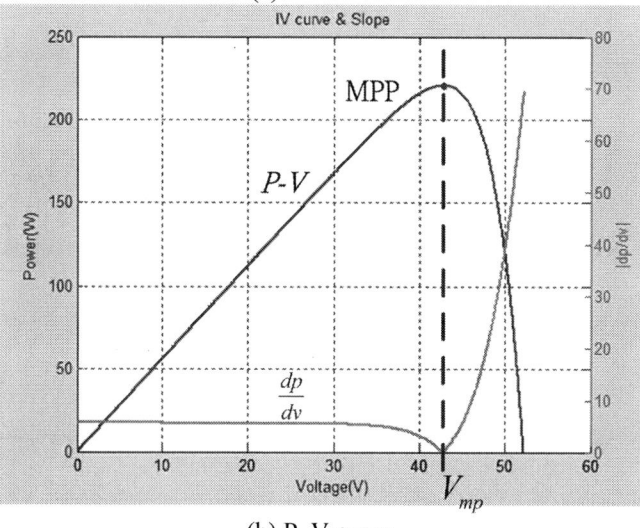

(b) P–V curve

Fig. 4 I–V and P–V curve for the utilized PV panel

Fig. 7 shows the power variation of P-V curve under the fixed voltage perturbation. It can be learned from Fig. 7 that when perturbing a fixed step of voltage on the P-V curve, the operating point on the left side of MPP results in larger power variation while the operating points on the right side of MPP results in smaller power variation. Therefore, according to the characteristic of the P-V curve, the symmetrical MF is replaced by the asymmetrical MF for input variable ΔP_{pv} to improve dynamic performance. In this paper, the utilized asymmetrical ΔP_{pv} MF is shown in Fig. 8. Since each of the input variables ΔV_{pv} and ΔP_{pv} is mapped to 5 different linguistic values, there are 25 different combinations forming the control rules for the proposed system. The further explanations of total sets of rules will be given in section III.b. The defuzzification method used in this paper is the commonly used center of gravity method as shown in Eq. (5).

$$Y_{COG} = \frac{\sum_{i=1}^{n} Y(X_i) X_i}{\sum_{i=1}^{n} Y(X_i)} \tag{5}$$

where Y_i is the inference result of rule i, X_i is the corresponding output of rule i, and Y_{COG} is the output.

b. Derivation of the fuzzy control rule

The design of control rule depends on the relationships between the input and output variables. In this paper, the input variables ΔV_{pv} and ΔP_{pv} are mapped into 5 different linguistic values. Therefore, the rule base of the proposed FLC will consist of 25 different rules. The basic principle of designing these rules is discussed as follows:

As mentioned in section II, a boost converter is employed in the paper. For boost converter, the operating point will move from right to left along the P–V curve when duty cycle increases and vice versa. For the proposed FLC based MPPT system, a positive $\Delta P/\Delta V$ value implies the operating point locates on the left-hand side of MPP. Hence, a negative value of ΔD is required to make the operating point moving toward the MPP. On the other hand, a negative $\Delta P/\Delta V$ value indicates the operating point lies on the right-hand side of MPP. For that case, a positive value of ΔD is required. As a result, the sign of the output variable ΔD can be determined by the sign of the $\Delta P/\Delta V$. Next, the following design procedure will determine the magnitude of the output variable ΔD. A typical P–V curve of a PV panel is illustrated in Fig. 9. From Fig. 9, a larger ΔD is needed to rapidly reach the MPP when the operating point is far away from the MPP. On the contrary, a smaller ΔD can be utilized to reduce the oscillation on steady state when the operating point is close to the MPP. Consequently, the magnitude of the output variable ΔD is determined by the value of $|\Delta P / \Delta V|$. According to these design procedure, a complete set of fuzzy rules for the proposed FLC are summarized in Table 1. In Table 1, darker color stands for larger quantity while lighter color represents smaller quantity. Red implies positive value, blue indicates negative value and white represents zero.

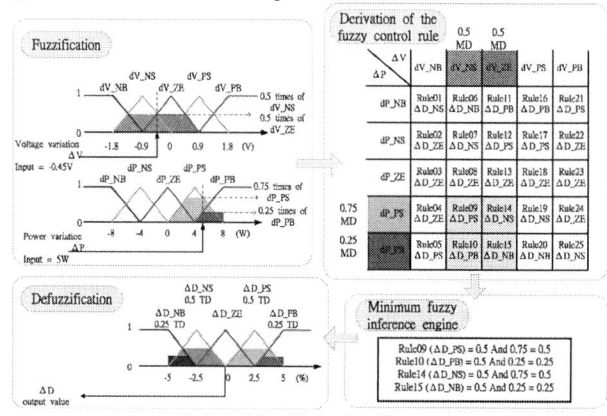

Fig. 5 Block diagram of the implemented FLC-based MPPT controller

The 2014 International Power Electronics Conference

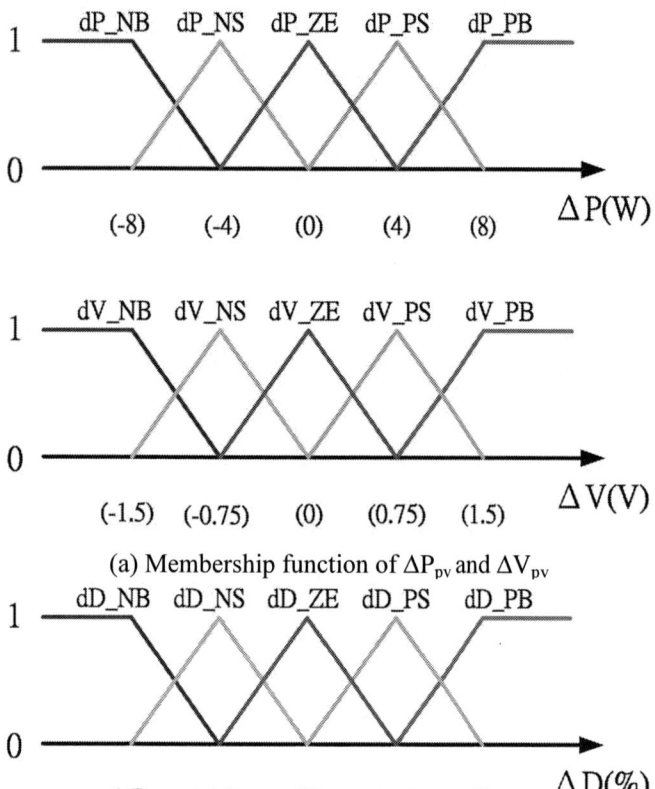

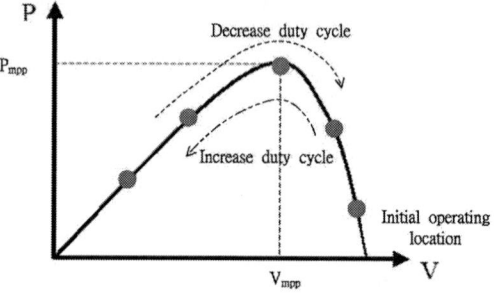

Fig. 9 Typical P–V curve of a solar panel

Table 1 The complete sets of fuzzy rules

Power variation ΔP \ Voltage variation ΔV	dV_NB	dV_NS	dV_ZE	dV_PS	dV_PB
dP_NB	Rule01 ΔD_NS	Rule06 ΔD_NB	Rule11 ΔD_PB	Rule16 ΔD_PB	Rule21 ΔD_PS
dP_NS	Rule02 ΔD_ZE	Rule07 ΔD_NS	Rule12 ΔD_PS	Rule17 ΔD_PS	Rule22 ΔD_ZE
dP_ZE	Rule03 ΔD_ZE	Rule08 ΔD_ZE	Rule13 ΔD_ZE	Rule18 ΔD_ZE	Rule23 ΔD_ZE
dP_PS	Rule04 ΔD_ZE	Rule09 ΔD_PS	Rule14 ΔD_NS	Rule19 ΔD_NS	Rule24 ΔD_ZE
dP_PB	Rule05 ΔD_PS	Rule10 ΔD_PB	Rule15 ΔD_NB	Rule20 ΔD_NB	Rule25 ΔD_NS

(a) Membership function of ΔP_{pv} and ΔV_{pv}

(b) Membership function of ΔD

Fig. 6 Membership functions of the input and output variables
ΔV under the same voltage variation condition

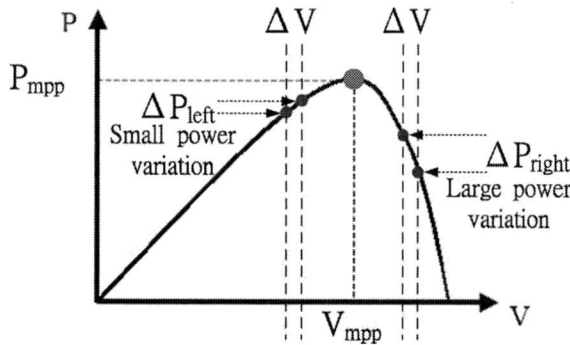

Fig. 7 power variation of P-V curve under the fixed voltage perturbation

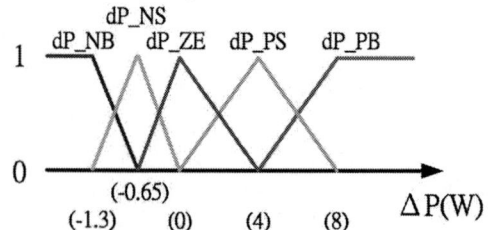

Fig. 8 Membership function of ΔP_{pv} for asymmetrical FLC

IV. Experimental results

In order to validate the proposed FLC-base MPPT controller, a 300W boost converter is implemented. The experiments are carried out with an AMETEK Solar Array Simulator TerraSAS DCS80-15 in SAS mode. The configuration of the utilized PV panel is given in Table 2. According to the system diagram in Fig. 1, the proposed symmetrical FLC-based MPPT method and asymmetrical FLC-based MPPT method are implemented to reveal the effectiveness and correctness in comparison with two P&O MPPT methods with different perturbation step. Here, all algorithms are realized by the low cost DSC dsPIC33FJ16GS502 from Microchip Corp. and each experiment is conducted on the same power circuit to ensure the fairness. Details for the parameters of the proposed MPPT methods are given in Table 3. Due to that the DSC uses a 40MHz oscillator, the required execution time of the P&O MPPT, the symmetrical FLC-based MPPT and the asymmetrical FLC-based MPPT are 1.5 μs, 120 μs and 120 μs, respectively. With the settings of 1000 W/m² solar irradiance and 25°C PV panel temperature, the starting waveform of these four methods can be acquired in Fig. 10. As shown in Fig. 10, the transient time for the two different

978-1-4799-2706-7/14 $31.00 © 2014 IEEE

fixed perturbation step P&O methods are 250 ms (for 5 % fixed step) and 3.13 s (for 0.5 % fixed step), respectively. Also, the transient time for the proposed symmetrical FLC-based MPPT controller and the proposed asymmetrical FLC-based MPPT controller are 1.59 s and 984 ms, respectively. It can be noted that even though the transient time of the proposed method is a little longer than the P&O method using large step size, the oscillation around the MPP for the proposed method is much smaller than that obtained from the P&O method using large step size.

The performances of these four developed methods are summarized in Table 4. The FLC-based MPPT controller not only provides better tracking accuracies than both of the P&O method but also improve the dynamic and steady state performance in the PGS. Moreover, observing Table 4, asymmetrical FLC-based MPPT method has the merit of faster tracking speed comparing to symmetrical FLC-based MPPT method. Additionally, it is a remarkable fact that the only difference between asymmetrical FLC-based MPPT method and symmetrical FLC-based MPPT method is the design of the universe of discourse of ΔP_{pv} MF. Therefore, the implementation complexity of these two methods is the same. To conclude, asymmetrical FLC-based MPPT method can speed up tracking time and increase tracking accuracy over symmetrical FLC-based MPPT without additional calculation.

Table 2 Parameters of the utilized PV panel

PV Model HIT Power 220A			
Maximum Power (P_{max})	220W	Short Circuit Current (I_{sc})	5.65A
Open circuit Voltage (V_{oc})	52.3V	Maximum Power Current (I_{pm})	5.17A
Maximum Power Voltage (V_{pm})	42.7V	Temperature Coefficient (α_v)	-0.336% / °C

Table 3 Parameters of the implemented algorithms

No.	Description	Parameters	Note
1	P&O (0.5%)	Fixed duty cycle 5%	denoted as method 1
2	P&O (5%)	Fixed duty cycle 0.5 %	denoted as method 2
3	Symmetrical FLC	dP_PB=8 W dP_NB=-8 W dV_PB=1.5 V dV_NB=-1.5V	denoted as method 3
4	Asymmetrical FLC	dP_PB=8 W dP_NB=-1.3 W dV_PB=1.5 V dV_NB=-1.5V	denoted as method 4

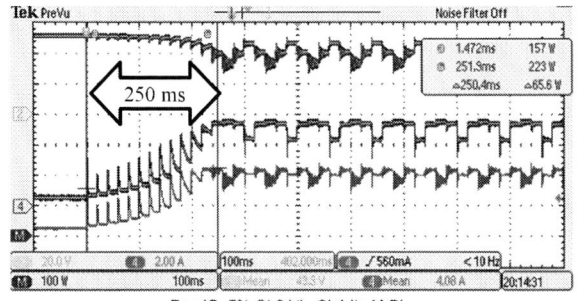

PandO_5%,Ch2:Vin,Ch4:Iin,M:Pin

(CH1：V_{in} 25V/div、CH2：I_{in} 2A/div、CH3：PWM 10V/div、Math：Power (CH1*CH2)100W/div、Time：100 ms)

(a) Measured starting waveform for method 1

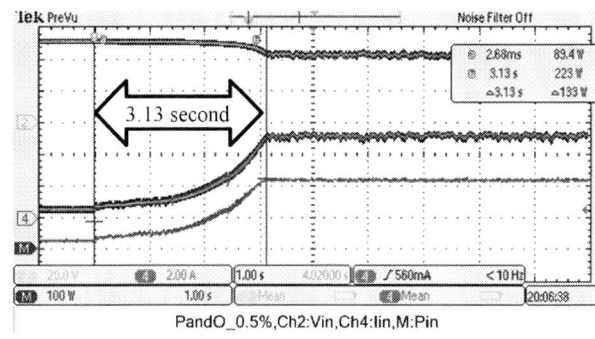

PandO_0.5%,Ch2:Vin,Ch4:Iin,M:Pin

(CH1：V_{in} 25V/div、CH2：I_{in} 2A/div、CH3：PWM 10V/div、Math：Power (CH1*CH2)100W/div、Time：1s)

(b) Measured starting waveform for method 2

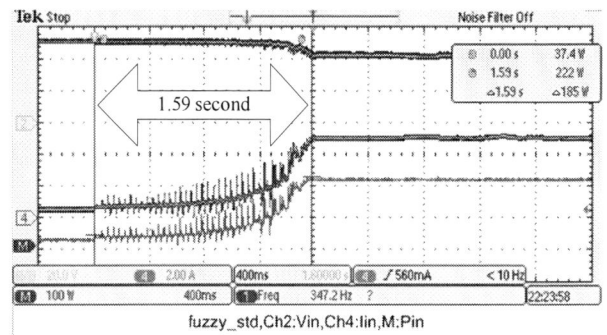

fuzzy_std,Ch2:Vin,Ch4:Iin,M:Pin

(CH1：V_{in} 25V/div、CH2：I_{in} 2A/div、CH3：PWM 10V/div、Math：Power (CH1*CH2)100W/div、Time：400 ms)

(c) Measured starting waveform for method 3

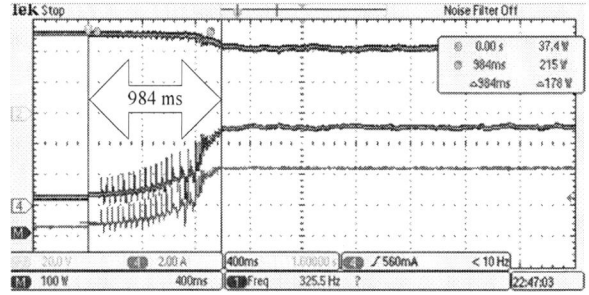

fuzzy_dP_10%,Ch2:Vin,Ch4:Iin,M:Pin

(CH1：V_{in} 25V/div、CH2：I_{in} 2A/div、CH3：PWM 10V/div、Math：Power (CH1*CH2)100W/div、Time：400 ms)

(d) Measured starting waveform for method 4

Fig. 10 Measured starting waveform of four different algorithms

Table 4 Summarized performance of methods

	Average steady state output power	MPPT tracking accuracy	Transient time
Method 1	206.89 W	92.84 %	0.25 s
Method 2	222.45 W	99.82 %	3.13 s
Method 3	222.58 W	99.87 %	1.28 s
Method 4	222.65 W	99.90 %	0.98 s

V. Conclusion

In this paper, a FLC-based MPPT method is proposed. Through the design and implementation, the feasibility and effectiveness of the proposed method is validated. The computation is simplified by adopting power variation and voltage variation as the inputs. Comparing with the conventional P&O method, the proposed MPPT method can satisfactorily deal with the tradeoff between the tracking speed and steady state oscillations. Additionally, an asymmetrical membership function concept is proposed to improve the performance. Comparing with the symmetrical FLC-based MPPT method, the transient time and the MPPT tracking accuracy are further improved by about 38.1 % and 0.03 %, respectively. Experiments are carried out to validate the effectiveness of the proposed method.

ACKNOWLEDGMENT

The authors are indebted to utility-scaled energy storage system and interconnection technology development project from the Bureau of Energy, Ministry of Economic Affairs for supporting this study.

References

[1] T. Esram, and P. L. Chapman, "Comparison of Photovoltaic Array Maximum Power Point Tracking Techniques," IEEE Trans. on Energy Conversion, vol. 22, no. 2, pp.439-449, June. 2007.

[2] A. R. Reisi, M. H. Moradi, and S. Jamasb, "Classification and comparison of maximum power point tracking techniques for photovoltaic system: A review," Renewable and Sustainable Energy Reviews, vol. 19, pp.433-443, Mar. 2013.

[3] K. Ishaque, and Z. Salam, "A review of maximum power point tracking techniques of PV system for uniform insolation and partial shading condition," Renewable and Sustainable Energy Reviews, vol. 19, pp.475-488, Mar. 2013.

[4] A. Messai, A. Mellit, A. Massi Pavan, A. Guessoum, and H. Mekki, "FPGA-based implementation of a fuzzy controller (MPPT) for photovoltaic module," Energy Conversion and Management, vol. 52, no. 7, pp.2695-2704, July. 2011.

[5] M. M. Algazar, H. AL-monier, H. A. EL-halim, and M. E. E. K. Salem, "Maximum power point tracking using fuzzy logic control," International Journal of Electrical Power & Energy Systems, vol. 39, no. 1, pp.21-28, July. 2012.

[6] S. Lalouni, D. Rekioua, T. Rekioua, and E. Matagne, "Fuzzy logic control of stand-alone photovoltaic system with battery storage," Journal of Power Sources, vol. 193, no. 2, pp.899-907, Sep. 2009.

[7] B. N. Alajmi, K. H. Ahmed, G. P. Adam, and B. W. Williams, "Single-Phase Single-Stage Transformer less Grid-Connected PV System," IEEE Trans. on Power Electronics, vol. 28, no. 6, pp.2664-2676, June. 2013.

[8] I.H. Altas, and A.M. Sharaf, "A novel maximum power fuzzy logic controller for photovoltaic solar energy systems," Renewable Energy, vol. 33, no. 3, pp.388-399, Mar. 2008.

[9] C. B. Salah, and M. Ouali, "Comparison of fuzzy logic and neural network in maximum power point tracker for PV systems," Electric Power Systems Research, vol. 81, no. 1, pp.43-50, Jan. 2011.

[10] F. Liu, S. Duan, F. Liu, B. Liu, and Y. Kang, "A Variable Step Size INC MPPT Method for PV Systems," IEEE Trans. on Industrial Electronics, vol. 55, no. 7, pp.2622-2628, July. 2008.

The 2014 International Power Electronics Conference

A novel current link distributed MPPT PV system
– Overall system prototyping and evaluation –

Mikihiko Matsui[*], Toru Sai[*], Akira Kitamura[*], Xiang-Dong Sun[†] and Byung-Gyu Yu[‡]

[*]Faculty of Engineering, Tokyo Polytechnic University, Kanagawa, Japan 243–0297, Email: t.sai@em.t-kougei.ac.jp

[†]Faculty of Automation and Information Engineering, Xi'an University of Technology, Shaanxi, China 710048

[‡]Division of Electrical Electronics and Control Engineering, Kongju National University, Chungnam, Korea 331-717

Abstract—**A novel structure for power maximization of the PV (photovoltaic) system under partially condition is presented. By using CSI (Current Source Inverter) instead of VSI (Voltage Source Inverter), the structure of PV system becomes extremely both more simple and reliable than conventional one. It means that MICs (Module Integrated Converters) can be consisted of buck-only converters and electrolytic capacitors can be removed from the inverter. The experimental results confirmed that the total output power captured the sum of each maximum value of PV under shading conditions.**

I. INTRODUCTION

The distributed MPPT (Maximum Power Point Tracking) can reduce the mismatch and shading losses in PV system. Under partially shaded conditions, the use of distributed MPPT system can recover 10%-30% of annual performance [1]. There are two categories in the distributed MPPT system as shown Fig. 1. Fig. 1(a) shows the AC module type [2]. The AC module type MICs are connected directly to the grid in parallel, thus it is easy to expand the large energy. However, the MICs carry the DC/AC converter in each MIC, thus the costs of MIC increase and the efficiency of DC/AC conversion ratio decrease compare with DC module type. Fig. 1(b) shows the DC module type MICs. The MICs are connected in series and form the string then the top voltage of string is connected to the PWM VSI. Unlike AC module type, as it carries one DC/AC converter, the efficiency performance is higher than AC module type. However, the common MICs of DC module type have three mode circuits that are buck converter, boost converter, and path through mode [3]. Moreover the electrolytic capacitors, that is the most shortest lifetime component of the power electronics circuits, have to connect to DC link node of PWM (Pulse Width Modulation) VSI. To overcome these obstacles, we started thinking about using new semiconductor devices, such as SiC and GaN FET. By using these attractive low resistance and high reverse voltage performance, CSIs might be revived in the power conditioners' industry. By using the CSI instead of VSI, the MICs are greatly simplified and the reliability of the power conditioner is improved.

In section 2 we propose the new distributed MPPT system that is linked by current, and describe the simple two rules for operate the our proposal system. In section 3, the experimental results are shown. The two cascaded MICs are connected to the current load. In section 4, we clarified the current ratio among the parallel connected strings of the proposal system. In section 5, overall system operation is confirmed using PSIM

simulator. Moreover, the experimental results of overall circuits without grid connection are shown. Section 6 concludes this study.

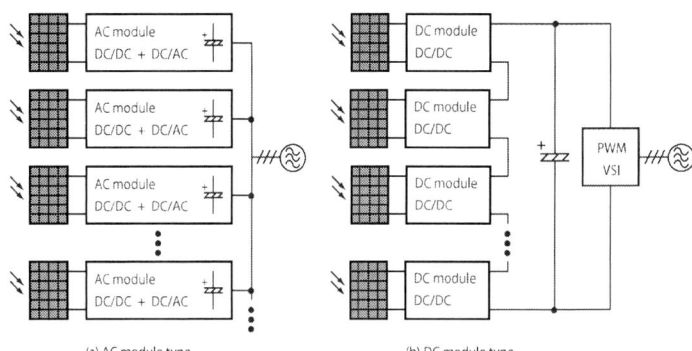

Fig. 1. Typical two different approaches to avoid partial shading problem based on module integrated converter techniques.

II. PROPOSAL SYSTEM

Fig.2 shows the block diagram of proposal system [4]-[6]. The power conditioning system is consist of CSI. Since the input source is current in the CSI, it is possible remove electrolytic capacitors from DC link node. And also the MIC can be consisted of buck-only converter because the input voltage of CSI does not need fixed voltage. MICs' output terminals are connected in series with the controlled current source link of the PCS (Power Conditioning System) for AC grid interfacing. Control rules of this system are quite simple, i.e., (i) each MIC is controlled to maximize its output voltage irrespective of the string current amplitude, (ii) PCS takes a roll of optimizing the amplitude of string current. By proceeding these two rules at the same time, the whole system is led to maximum power operating condition. The PWM CSI has a merit of high efficiency power conversion if the SiC or GaN FET power devices will be applied in the near future.

To assist the understanding of the rule of the buck converter in the propose system, one stage of buck converter is picked up as shown in Fig. 3. Fig. 3(a) shows the block diagram of a single buck converter. The output node V_{out1} of the buck converter is connected to the load resistance R_L and the duty ratio d is controlled by the pulse generator. Fig. 3(b) shows the relationship of V_{out1} and I_{string} when d changed from 0.4 to 0.9. When the duty ratio d=0.9, the buck converter's output

978-1-4799-2706-7/14 $31.00 © 2014 IEEE   3784

The 2014 International Power Electronics Conference

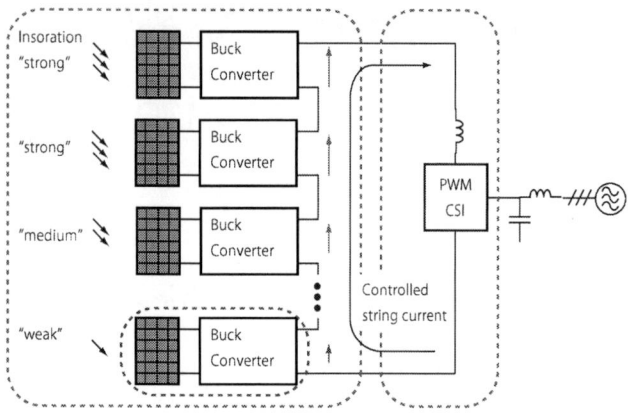

Fig. 2. A new distributed MPPT system proposed in this paper.

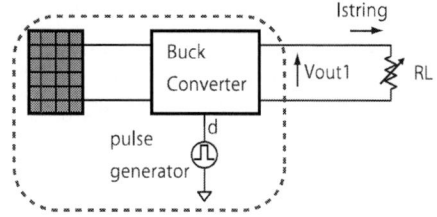

(a) The block diagram of a single buck converter.

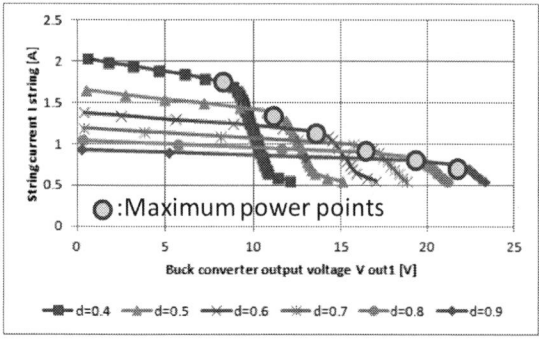

(b) Relationship Vout1 and Isrting when the duty ratio change.

Fig. 3. Duty dependence of a single buck converter's output voltage and string current.

voltage V_{out1} is relatively high and the string current, that is, the output current of buck converter is relatively low. On the other hand, when the duty ratio d=0.4, the buck converter's output voltage V_{out1} is reduced and the string current is increased, because the input power and the output power of the buck converter is ideally constant. By applying MPPT to MICs, MICs can track the maximum power even if duty changed. If the insolation of PV panel becomes week, the buck converter increases its output current to match the string current. As the results, output voltages are reduced. On the contrary if the insolation of PV panels becomes strong, the buck converters reduce their output currents to match the string current. As a result, output voltage increased.

III. EXPERIMENTAL RESULTS

To assess the usefulness of the proposed MIC system, a prototype measurement system has been implemented. Fig. 4 shows the block diagram of the measurement system. The SAS (Solar Array Simulator) supply "soldio" SS-301(Fukushima Electronic) was used instead of solar panel. The buck converters were located between solar array simulators SS-301 and a programmable DC Electric Load 3700 (Array Electronic). The Electric Load is assumed current type inverter and its operation mode is chosen constant current mode. The control circuits are realized by micro computer board SH7144F (Akizuki Denshi) which includes 10-bit AD converters with 12.5MHz CPU clock. The MPPT algorithm is realized by this micro computer program. The voltage (V_{panel}) and current (I_{panel}) of each SAS supply are measured and its nodes are tied to microcomputer. The output powers of PV1, PV2, and the total output node V_{sting} are measured by power analyzer WT500 (Yokogawa Electric).

Fig. 5 shows schematics of the buck converter. The inductor L1 was selected 220uH. The output capacitor was removed from the output node (V_{out1}). The operational frequency of the buck converter was designed about 25kHz for constant current mode operation. To reduce the switching noise, bypass capacitor Cin (47uF) was connected close to drain node of M1. Although the electrolytic capacitor Cin was still remained in this circuit, it can change to film capacitor. The voltage sensor LEM LV25-P and the current sensor LEM LA55-P were connected to measure the output voltage and current of simulated power supply PV1 for P&O (Perturb&Observe) operation. The isolator TLP 250 isolates the micro computer from the buck converter. The supply voltages of the voltage sensor LV-25P and TLP250 were supplied by The DC-DC converter cc1R5-1212. To prevent the reverse current when the illegal start up, schottky-barrier diode D1 was connected between PV1 and the buck converter.

The experimental test-bed is shown in the Fig. 6.

In order to check the total maximum power tracking, three cases of V-P characteristic curves are prepared using SAS supply SS-301. Table 1 shows the three conditions of variable V_{oc} (Open Circuit Voltage) and I_{sc} (Short Circuit Current). The measurements results are shown Fig. 7, Fig. 8, and Fig. 9. The Fig. 7(a), Fig. 8(a), and Fig. 9(a) show the V-P characteristics of SAS supply when the V_{oc} and I_{sc} changed. The Fig. 7(b), Fig. 8(b), and Fig. 9(b) show the confirmation results of our proposed system which definitely captures the maximum power points of total power. Here, the dot lines represent the output power depending on the string current Istring. The solid lines show the total efficiency including the two buck converters. In Fig. 7(b), we can see the maximum power Pout is 40.8W(max) at 1.5A, it corresponds to the sum of 31.1W(max) and 14.1W(max) by considering the efficiency. Likewise, in Fig. 8(b) we can see the maximum power Pout is 37.6W(max) at 1.5 A, it corresponds to the sum of 31.1W(max) and 9.7W(max) by considering the efficiency. Again, in Fig. 9(b) we can see the maximum power Pout is 28.7W(max) at 0.7 A, it corresponds to the sum of 29.7W(max) and 9.7W(max) by considering the efficiency. In addition, the

978-1-4799-2706-7/14 $31.00 © 2014 IEEE

V_{string} of the Case1, Case2, and Case3 are 28.0V, 25.4V, and 42.5V, respectively when the maximum power was obtained. Fig. 10 shows the transient waveform of the Vstring when the system is powered on. We can see that its rise time (10-90%) was about 5 second with no over shoot and a slight oscillation due to P&O.

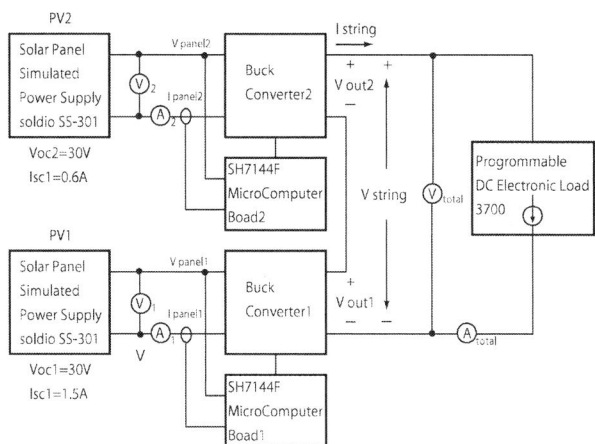

Fig. 4.   Block diagram of the measurement system.

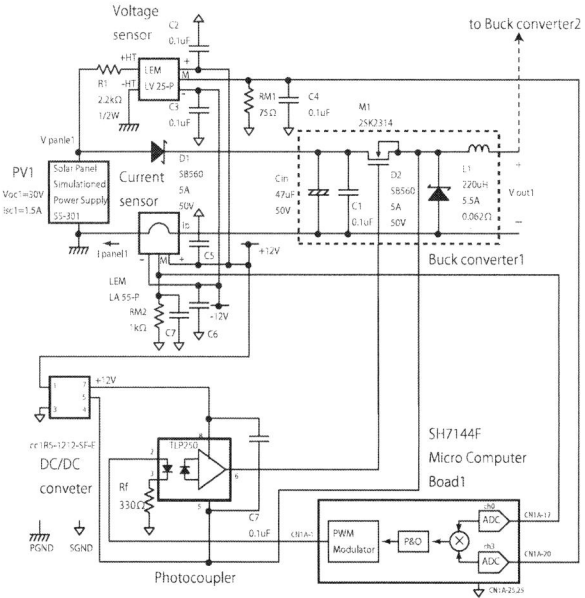

Fig. 5.   Circuit schematics of the buck converter.

TABLE I.   THREE CASES OF VARIATION OF V_{oc} AND I_{oc}

SAS	PV1			PV2		
parameter	V_{oc}	I_{sc}	$P_{w(max)}$	V_{oc}	I_{sc}	$P_{w(max)}$
Case1	30 V	1.5 A	31.1 W	30 V	1.5 A	14.1 W
Case2	30 V	1.5 A	31.1 W	15 V	1.5 A	9.7W
Case3	30 V	0.6 A	17.1 W	30 V	0.6 A	14.1 W

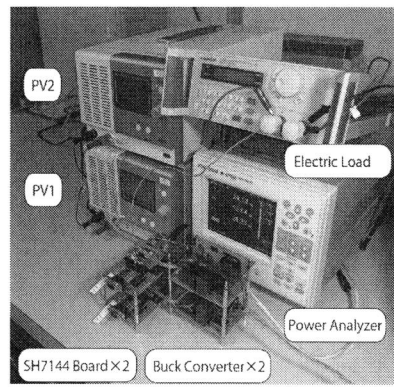

Fig. 6.   Test-bed of the measurement system.

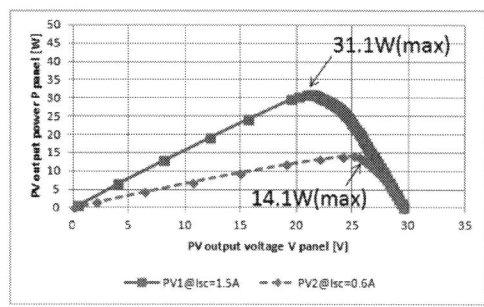

(a) Voc1=Voc2=30V, Isc1=1.5A, Isc2=0.6A

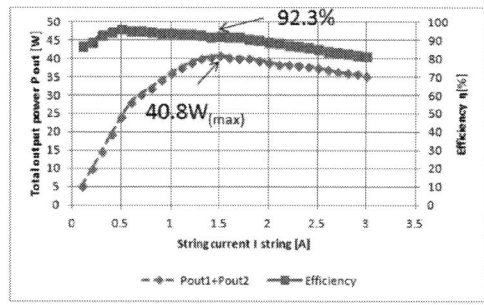

(b) Total output power and efficiency

Fig. 7.   Case1:V-P characteristics(a) and the total output power(b).

IV.   CURRENT SHARING AMONG PARALLELED STRINGS

Fig.11 shows the block diagram of the serial and parallel connection of the proposal system. Here, we considered two cascaded MICs and two parallel strings for easy understanding. Two strings are connected directly at node V_{str_total} without blocking diodes. Because the buck converter carries a reflux diode which can protects reverse current in string. The regulations of current sharing are determined as follows, i.e. (i) obviously, the total current I_{str_total} is the sum of I_{str1} and I_{str2} as given the equation (1), (ii) since the node V_{str_total} is connected directly, the current of I_{str1} and I_{str2} are determined the power ratio, as given equation (2). (iii) The maximum power is obtained when the string current flow over than the $I_{sc(max)}$ of MIC in the each string.

The confirmation of simulation results shows in Fig. 12. Fig. 12 (a) shows the results when all I_{sc} set to 2A. Fig. 12 (b)

The 2014 International Power Electronics Conference

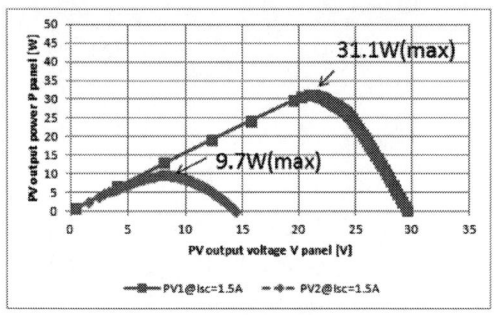

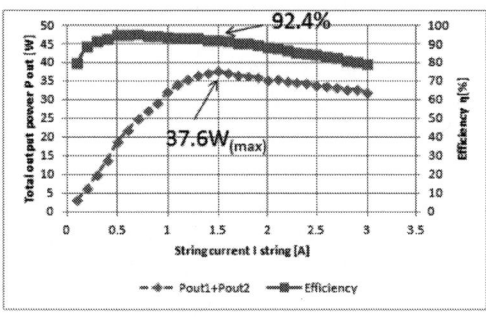

(a) Voc1=30V, Voc2=15V, Isc1=Isc2=1.5A.

(b) Total output power and efficiency.

Fig. 8. Case2:V-P characteristics(a) and the total output power(b).

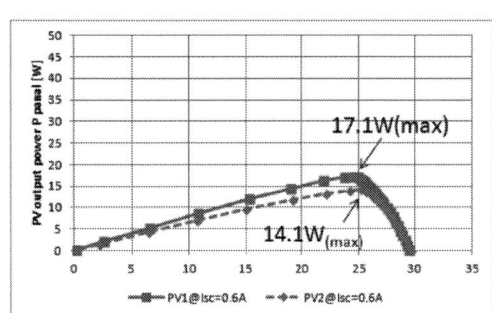

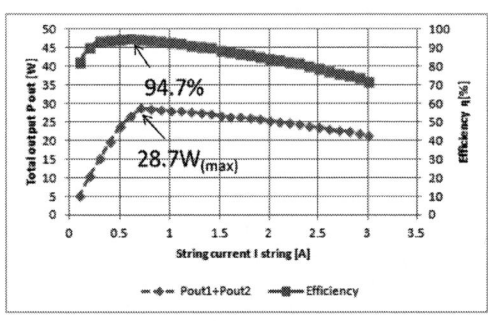

(a) Voc1=Voc2=30V, Isc1=Isc2=0.6A.

(b) Total output power and efficiency.

Fig. 9. Case3:V-P characteristics(a) and the total output power(b).

shows the results when only I_{sc21} set to 0A. Here the all PVs set to the same power, 45.7W. Fig. 12(a) shows the maximum power is obtained when I_{str_total} is 4A and the ratio of I_{str1} and I_{str2} is one. And Fig. 12(b) shows the maximum power is obtained when I_{str_total} is 6A and the ratio of I_{str1} and I_{str2}

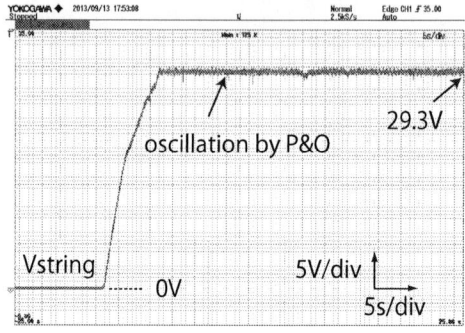

Fig. 10. Output voltage (V_{string}) response when the system power on (Case1: Voc1=Voc2=30V, Isc1=1.5A, Isc2=0.6A).

is two.

$$I_{str\_total} = I_{str1} + I_{str2} \qquad (1)$$

$$\frac{I_{str1}}{I_{str2}} = \frac{P_{11} + P_{12}}{P_{21} + P_{22}} \qquad (2)$$

$$I_{str_n} \geq I_{sc_{nm(max)}} \quad \in \; each \; string$$
$$n = 1, 2 \;\; m = 1, 2 \qquad (3)$$

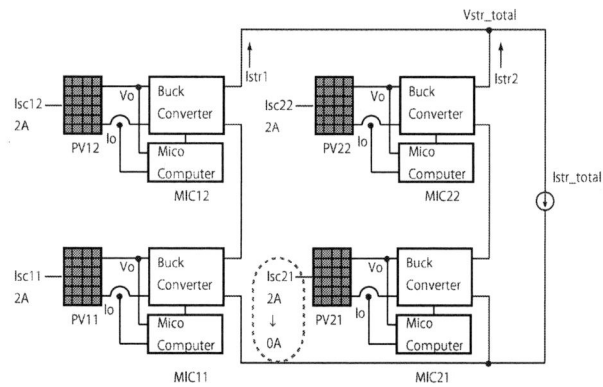

Fig. 11. Block diagram of the serial and parallel connection.

V. OVERALL OPERATION

Fig. 13 shows the overall circuits. Three MICs are connected in series and the MICs and the CSI are tied by the reactor. The PI control block is added to the CSI. In Fig. 13, PVs were consist of behavioral model and bucks were consist of transistor model. We can see that the output current I_s follows the reference current I_{s_ref} in Fig. 14.

Fig. 15 shows the overall experimental circuits without the grid connection. The CSI is consits of SiC FET and SiC diode. The MICs are connected in series under 10V of the open voltage and 0.6A of the short current. Here, the solar panel simulated power supplies are used instead of PV panels. The PI control part was replaced to analog circuits and the reference

978-1-4799-2706-7/14 $31.00 © 2014 IEEE   3787

The 2014 International Power Electronics Conference

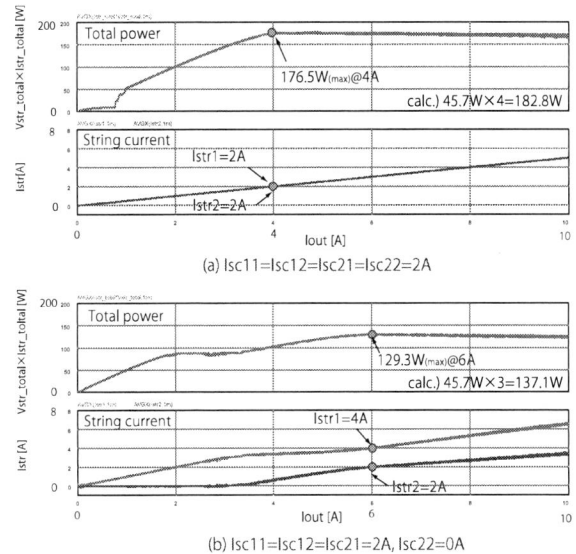

(a) Isc11=Isc12=Isc21=Isc22=2A

(b) Isc11=Isc12=Isc21=2A, Isc22=0A

Fig. 12. Simulation results of total power and string current when the strings are connected in parallel.

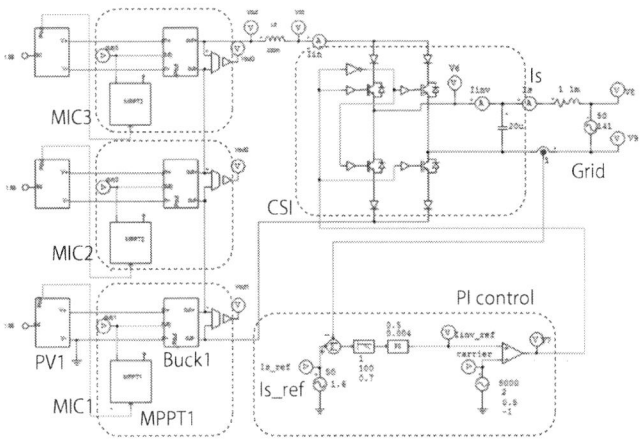

Fig. 13. Overall connection of the MICs and the CSI.

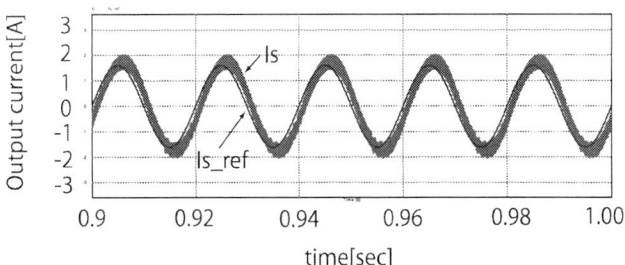

Fig. 14. The simulation results of the CSI output current I_s.

sin wave was applied to positive input node of the OPamp as the I_{s_ref}. The output of the current sensor was applied to the negative input node of the OPamp as the I_s. Fig. 16 shows the experimental results of the Fig. 15. We can see that the input current of CSI is constant and the feedback current node I_s follows the reference current node I_{s_ref}.

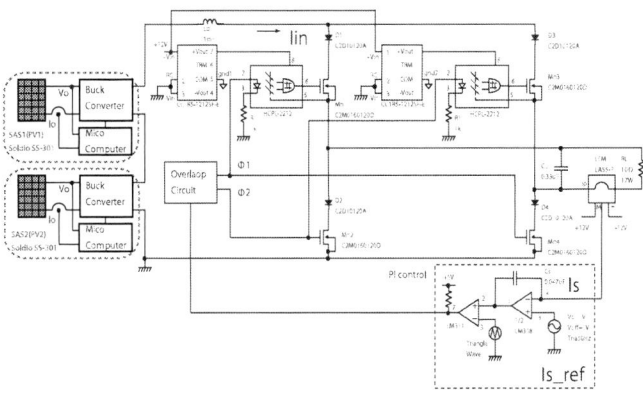

Fig. 15. Overall circuit schematics of PV, MIC and CSI.

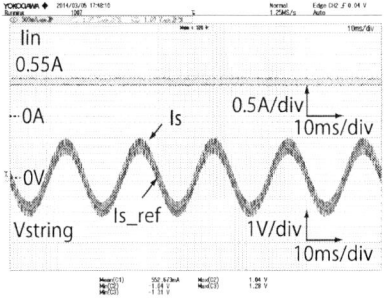

Fig. 16. The experimetal results of the CSI input current I_{in} and output current I_s.

VI. CONCLUSION

The novel idea of the maximization of PV system under partially shaded condition are presented. The principle of our idea is confirmed by experimental results of the two cascaded MICs. And the rules of the current sharing among the paralleled string are revealed. Also the overall operation which includes PV, MIC, and CSI are confirmed.

REFERENCES

[1] Chris Deline, Bill Marion, Jennifer Granata, Sigifredo Gonzalez, "A performance and economic analysis of distributed power electronics in photovoltaic Systems",*Nation renewable energy lab report, NREL Report*, No. TP-5200-50003, Jan. 2011.

[2] Web site of enphase energy: http://enphase.com/.

[3] L. Linares, R. Erickson, S. MacAlpine, and M. Brandemuehl, "Improved energy capture in series string photovoltaics via smart distributed power electronics," *Applied Power Electronics Conference and Exposition (APEC)*, 2009. Twenty-Fourth Annual IEEE, pp. 904 -910, Feb. 2009.

[4] Mikihiko Matsui, Toru Sai, Byung-Gyu Yu and Xiang-Dong Sun, "A new DMPPT technique using buck-type MICs linked with controlled string current", *IEEJ Korea-Japan Joint Technical Workshop on Semiconductor Power Converter*, pp. 106-109, Nov. 2012.

[5] Mikihiko Matsui, Toru Sai, Byung-Gyu Yu, Xiang-Dong Sun, " A new distributed MPPT technique using buck-only MICs linked with controlled string current", *IEEJ Industry Applications Society Conference JIASC2013*, pp. 261-266, August, 2013.

[6] Toru Sai, Mikihiko Matsui, Akira Kitamura, "Next generation controlled current linked modular power conditioning system – Current sharing among paralleled strings –", *IEEJ Japan-Korea Joint Technical Workshop on Semiconductor Power Converter*, pp. 89-90, Oct. 2013.

978-1-4799-2706-7/14 $31.00 © 2014 IEEE

The 2014 International Power Electronics Conference

Power Flow Control and MPPT Parameter Selection for Residential Grid-Connected PV Systems with Battery Storage

Chuenwattanapraniti Chokchai
Department of Electrical Engineering,
Faculty of Engineering, Burapha University
Chonburi, Thailand
chokchai@eng.buu.ac.th

Abstract— **This paper presents a technique to control power flow and a selection of MPPT parameter for grid-connected PV systems with battery storage in order to optimize the use of PV energy in the residential level. The power processor used in such PV systems have to allow the power to flow in all possible directions among 3 elements; PV array, battery and utility grid. Furthermore, to maximize the use of PV energy, the maximum power point (MPP) should be maintained for all conditions. A small PV system has been set up to validate the 3 proposed energy management strategies. The experimental results show the success of power flow control for various control conditions and also illustrate the charge/discharge profile that the battery will be subjected to.**

Keywords— battery storage, grid-connected PV systems, power flow control.

I. INTRODUCTION

Among all of renewable energy technologies, the grid-connected PV market grew most rapidly with a 60 percent annual average growth rates during the period 2002-2006 [1].

Since the photovoltaic energy is concerned in the sustainable development, the market of photovoltaic has been supported by the different incentive policies such as: investment subsidies and feed-in tariffs (FIT).

Once the renewable energy reaches a significant market penetration, this governmental support will be programmed to decrease. The decreasing of FIT progresses directly to the number of PV system installations every year. In the future, with the fast growing of grid connected PV system market like now, the value of PV energy will be decreased certainly as in Germany and Spain nowadays.

Since there is no storage element in almost grid-connected PV systems, all PV energy is injected into utility grid.

In fact, the value of PV energy strongly depends on the feed-in period and other available service. Including battery in grid-connected PV systems provides the possibility to control and optimize the use of PV energy that will give some additional benefits to both the system owner and the distribution systems when sufficient numbers of such PV systems have been installed

especially in the same distribution network [2].

For example, the "Autonomy PV systems" less depend on utility grid [3] and the PV systems which use the weather forecasting data to control the power flow[4].

The power processor used in such PV systems should allow the power to flow in all possible directions among 3 elements; PV array, battery and utility grid and should provide the possibility to operate in 3 modes depending on the control strategy: grid-connected, power factor correction (PFC) or standalone mode.

This paper proposes the structure of power processor ,its power flow control and introduce the proper selection of MPPT parameter in order to meet the desired power management strategies such as "Maximize self-consumption", "Peak shaving" and "Grid power smoothing" and to ensure the maximum power point (MPP) operation for all possible conditions.

II. PV System CONFIGURATION

The proposed grid-connected PV system configuration is depicted in Fig.1. It consists of two bi-directional power converters: a DC-DC and a DC-AC converter. Both are bi-directional power converters.

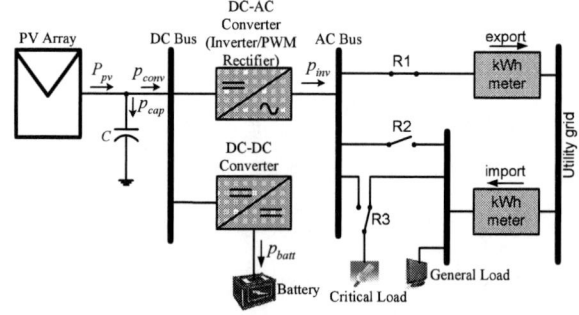

Fig. 1.The proposed grid-connected PV system

A. DC-AC Power Circuit

The DC-AC converter connected between DC and AC bus can function either as the inverter to inject the power from DC side to the AC side (utility grid) or as the PWM

978-1-4799-2706-7/14 $31.00 © 2014 IEEE   3789

rectifier (Power Factor Correction Converter, PFC) to recharge the battery from grid power in some occasions.

The DC-AC power circuit is constructed from the full-bridge topology as shown in Fig.2(a). The converter output is connected to the grid via a filter inductor and a line-frequency transformer. A capacitor is connected with an inductor to form a LC low-pass filter for output voltage filtering in standalone mode. However, this capacitance should be chosen to be as small as possible in order to minimize the phase-shift between the injected current and grid voltage in grid-connected mode. Furthermore, with the small filter capacitance, the current control loop can be designed as a simplified L-filter inverter. This can avoid the instability problem in the controller design of LCL filter inverter that must be special taken care [5].

B. DC-DC Power Circuit

For the DC-DC converter, power can flow from the DC bus to recharge the batteries or to feed the stored energy in battery to DC bus.

Therefore, the power topology of DC-DC converter is a bidirectional 2-quadrant DC chopper (see Fig.2 (b)). It can operate either in buck or boost mode depending on the desired battery current (I^*_{batt}) direction. In buck mode, the power flow from the DC bus to charge the battery, while in boost mode, battery releases its stored energy to DC bus.

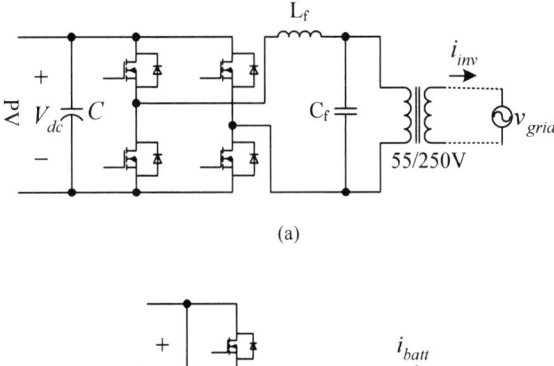

(a)

(b)

Fig. 2. (a) DC-AC and (b) DC-DC power circuit

III. POWER CONVERTERS CONTROL SYSTEM

Fig.3 shows DC-AC converter control system which consists of a current/voltage controller, phase-locked loop (PLL), signal conditioners and gate signal generator. A digital signal processor (DSP) TMS320F2812 was employed for realization of closed loop control while the field-programmable gate array (FPGA) Cyclone EP1C3 manages all logical operations. FPGA running in parallel

with the DSP allows the control system to be closely real-time control.

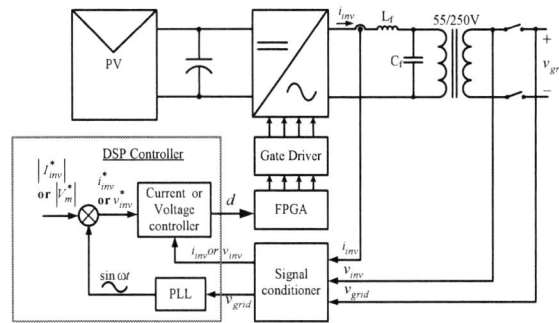

Fig. 3. DC-AC converter's control system

In normal grid condition, the DC-AC converter is controlled in current-controlled mode. The inverter output current (i_{inv}) is regulated to follow a sinusoidal current reference generated from PLL. This reference will be in phase for inverter mode or $180°$ out of phase for PFC mode.

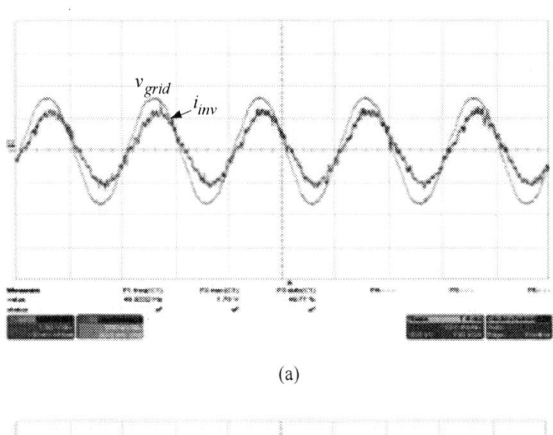

(a)

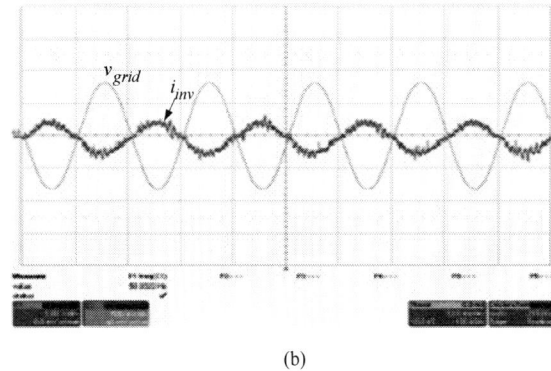

(b)

Fig. 4. Grid voltage and inverter output current in (a) inverter mode (500W) and (b) PFC mode (150W)(Time: 10ms/div, CH1-grid voltage: 200V/div and CH2-injected current: 2.5A/div)

Fig.4 shows the inverter output current for both modes. The inverter was tested with a 105 V constant dc source, 230V 50Hz grid voltage, 33 kHz of switching frequency, 5kHz of sampling frequency and 440 µH output filter inductance.

In grid failure, an inverter will be isolated from utility grid and act as an uninterruptible power supply (voltage-controlled mode) to supply the 230V 50Hz (sinusoidal waveform) to some critical loads.

The DC-DC connected between DC bus and battery can operate in either current or voltage-controlled mode depending on the battery state of charge (SoC). In current controlled mode, the battery current is regulated to follow the battery current command (I_{batt}^*) by the simple PI controller. This command can be a positive sign for charging or a negative for discharging. The voltage controlled mode is used to maintain constant-voltage charging in constant current/constant voltage charging profile (CC/CV charging). By adding outer voltage control loop (dash line in Fig.5), the voltage controlled mode can be achieved.

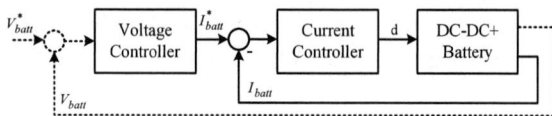

Fig. 5. DC-DC converter's control system

IV. POWER FLOW CONTROL AND MPPT PARAMETER SELECTION

As stated, the technique to control the power flow for 3 energy management strategies: "Maximize self-consumption", "Peak shaving" and "Grid power smoothing" are proposed in this paper.

The amplitude and direction of power flow between each element in the power processor differently depend on the selected strategy, instantaneous load consumption and battery SoC. However, the MPP operation should be maintained for all possible conditions by properly selected control parameter.

For this research, the MPP voltage V_{mpp} which strongly depend on the irradiance and cell temperature is defined based on the well known "Hill-Climbing" algorithm. The PV terminal voltage (v_{dc}) or capacitor voltage (v_c) will track V_{mpp} by unbalancing the energy into the DC bus capacitor. From Fig. 6, the energy flow into capacitor in one cycle of power waveform can be written as:

$$\int_{t_0}^{t_0+T} p_{cap}(t)\, d\tau = \int_{t_0}^{t_0+T} p_{pv}(t)\, d\tau - \int_{t_0}^{t_0+T} p_{conv}(t)\, d\tau \tag{1}$$

where T is the period of p_{conv} waveform, p_{pv} is PV power and p_{cap} is power into a DC bus capacitor.

Assuming 100% of power converter efficiency, the converter input power (p_{conv}) is given by a sum of inverter output power (p_{inv}) and power into battery (p_{batt}) as follows:

$$p_{conv} = p_{inv} + p_{batt} \tag{2}$$

The left side of (1) is the change of stored energy in DC bus capacitor related to the change of capacitor voltage in one cycle as follows:

$$\int_{t_0}^{t_0+T} p_{cap}(t)\, d\tau = \frac{1}{2}C\left\{\left[v_c(t_0+T)\right]^2 - \left[v_c(t_0)\right]^2\right\} \tag{3}$$

From (1) and (3), it can summarize that the unbalance between two energy terms in the right side of (1) will change the capacitor voltage to the other desired level. Of course, only the converter input energy can be controlled.

Equation (2) states that the converter input power (p_{conv}) can be changed via the inverter current reference in RMS (I_{inv}^*) or I_{batt}^* respectively.

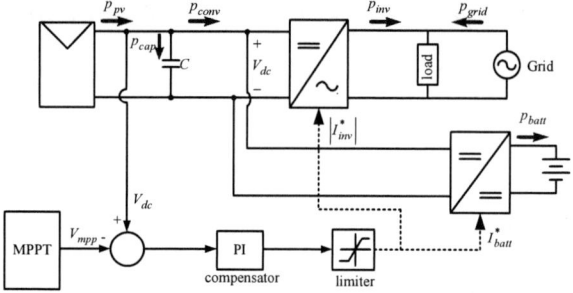

Fig.6. MPPT voltage control loop

A. Maximize Self-Consumption

The principle of this strategy is to consume self-PV generated power on-site if it is possible. Therefore, if there is more than enough of PV generated power, it will be firstly consumed on-site while the excess will be stored in the battery. However, in some situations, it will be fed into utility grid in order to obtain the maximum profit of PV generated power.

Table I presents the control condition and the selection of MPPT control parameters for 3 different battery SoC conditions.

In normal SoC condition (second condition in Table I), battery will be allowed to be either charged or discharged. Therefore, the inverter current reference is fixed to meet the load power requirement (P_{load}) as follows:

$$I_{inv}^* = \frac{P_{load}}{V_{grid}} \tag{4}$$

where P_{load} is the average load power consumption and V_{grid} is the grid voltage in RMS.

For this condition, the DC-DC current reference (I_{batt}^*) has been chosen as a control parameter to track the MPP voltage (V_{mpp}).

First condition in Table I explains the operating condition for undercharge state (when the battery voltage has reached lower SoC limit). In this case, the battery will not be allowed to further discharge. If PV power is less than the load requirement ($P_{pv} < P_{load}$), the best way is to feed all PV power to loads while some parts of load demand is served from utility grid. In this situation, the DC-DC converter will be forced to stop and the PV terminal voltage will reach V_{mpp} by varying the inverter current reference (I_{inv}^*).

On the other hand, if $P_{pv} > P_{load}$, the PV power will be used to supply load and any excess will be stored into the battery. If the recharging process continues until the battery voltage reaches normal SoC (plus small hysteresis

band), it will return to normal condition.

The control action for overcharging condition shown in third condition of Table I is defined similarly to undercharge condition but in the opposite way.

B. Grid Power Smoothing

For this strategy, the inverter current command (I_{inv}^*) will be defined correspondingly to the difference between load power and the predefined smoothing power (P_{smooth}), therefore the inverter current reference is given by

$$I_{inv}^* = \frac{\left(P_{load} - P_{smooth}\right)}{V_{grid}} \tag{5}$$

TABLE I
MPPT CONTROL PARAMETER SELECTION

Battery SoC Condition	Selected Strategy	Condition	MPPT by varying	Fixed Parameter
1.Undercharge	A. Maximize self-consumption	$P_{pv} > P_{load}$	I_{batt}^*	$I_{inv}^* = P_{load}/V_{grid}$
		$P_{pv} \leq P_{load}$	I_{inv}^*	STOP DC-DC
	B. Grid power smoothing	$P_{load} - P_{pv} > P_{smooth}$	I_{inv}^*	STOP DC-DC
		$P_{load} - P_{pv} \leq P_{smooth}$	I_{batt}^* (CHARGE)	$I_{inv}^* = \left(P_{load} - P_{smooth}\right)/V_{grid}$
	C. Peak shaving	$P_{load} > P_{peak}$		
		$P_{load} - P_{pv} > P_{peak}$	I_{inv}^*	STOP DC-DC
		$P_{load} - P_{pv} \leq P_{peak}$	I_{batt}^* (CHARGE)	$I_{inv}^* = \left(P_{load} - P_{peak}\right)/V_{grid}$
		$P_{load} \leq P_{peak}$	I_{batt}^* (CHARGE)	STOP DC-AC
2. Normal	A. Maximize self-consumption	-	I_{batt}^*	$I_{inv}^* = P_{load}/V_{grid}$
	B. Grid power smoothing	-	I_{batt}^*	$I_{inv}^* = \left(P_{load} - P_{smooth}\right)/V_{grid}$
	C. Peak shaving	$P_{load} > P_{peak}$	I_{batt}^*	$I_{inv}^* = \left(P_{load} - P_{peak}\right)/V_{grid}$
		$P_{load} \leq P_{peak}$	I_{batt}^* (CHARGE)	STOP DC-AC
3. Overcharge	A. Maximize self-consumption	$P_{pv} > P_{load}$	I_{inv}^*	STOP DC-DC
		$P_{pv} \leq P_{load}$	I_{batt}^*	$I_{inv}^* = P_{load}/V_{grid}$
	B. Grid power smoothing	$P_{load} - P_{pv} > P_{smooth}$	I_{batt}^* (DISCHARGE)	$I_{inv}^* = \left(P_{load} - P_{smooth}\right)/V_{grid}$
		$P_{load} - P_{pv} \leq P_{smooth}$	I_{inv}^*	STOP DC-DC
	C. Peak shaving	$P_{load} - P_{pv} > P_{peak}$	I_{batt}^* (DISCHARGE)	$I_{inv}^* = \left(P_{load} - P_{peak}\right)/V_{grid}$
		$P_{load} - P_{pv} \leq P_{peak}$	I_{inv}^*	STOP DC-DC

978-1-4799-2706-7/14 $31.00 © 2014 IEEE

The P_{smooth} may be adapted day by day based on several data such as solar irradiance, load forecasting and/or weather forecasting.

The negative sign of I_{inv}^{*} means that an inverter functions as a PFC rectifier drawing the grid power to recharge the battery.

Since I_{inv}^{*} has been fixed, the DC-DC current reference (I_{batt}^{*}) will be chosen as a control parameter to track the MPP voltage (V_{mpp}). In this condition the battery may be charged or discharged depending on the currently available PV power at that time.

Once the battery reaches the overcharge state and if $P_{load} - P_{pv} > P_{smooth}$, it will provide some extra power to fulfill the deficit. In contrast, if $P_{load} - P_{pv} < P_{smooth}$, since the battery is fully charged, the entirely generated power should be fed into grid side to avoid battery overcharge (see third condition in Table I).

When the battery reaches undercharge state and $P_{load} - P_{pv} < P_{smooth}$, the excess PV power will be stored in the battery. However, the grid power smoothing will not succeed if $P_{load} - P_{pv} > P_{smooth}$.

C. Peak Shaving

The concept of this strategy is to use the stored PV energy in battery to suppress the peak demand.

In normal battery state of charge condition, as long as the load demand does not exceed the predefined peak power (P_{peak}), load will be directly supplied from utility grid. The inverter will be forced to stop, all PV generated power will be stored in the battery while the MPP voltage is maintained via the battery current reference (I_{batt}^{*}).

When the peak load occurs, the inverter will be commanded to inject some additional power in order to suppress that peak demand; the inverter current reference for this case is given by

$$I_{inv(fixed)}^{*} = \frac{\left(P_{load} - P_{peak}\right)}{V_{grid}} \qquad (6)$$

For this case, the battery may be charged or discharged depending on the PV power at that time (see condition 2 in Table I).

When the battery voltage has been reached the lower SoC limit (condition 1 in Table I) and if the high peak demand occurs, there is only PV generated power that can be used to suppress that peak demand. Therefore, in some occasions, when the PV power is not enough, the peak elimination will not succeed.

In overcharge state, when $P_{load} - P_{pv} < P_{peak}$, the DC-DC will be forced to stop while the entirely generated PV power will be injected to the grid side.

On the other hand, if the PV power is more than enough ($P_{load} - P_{pv} > P_{peak}$), the excess will be diverted to

recharge the battery.

V. EXPERIMENTAL RESULTS

A small PV system has been set up in order to validate the 3 proposed energy management strategy. It consists of:

- 3 PV modules in series (BP 3165, 3×165Wp at 3× 35.2V_{mmp})
- 5×12V 40Ah (C_5) lead-acid batteries in series (60V and 40Ah or 2.4 kWh battery system)
- power analyzer (Norma D6000 wide-band power analyzer)

In order to prevent the battery from the irreversible damage, the battery state of charge in all experiments will be ranged between 20%-80%. The battery SoC is monitored via the battery terminal voltage. However, this voltage strongly depends on charge and discharge rates. In the experiments, the discharge current rates depends on the chosen energy management strategy. Since it varies between C_{20}-C_{10} rates, hence the 11.6 V which correspond to 20% SoC is defined as the undercharge state.

The charge rates strongly depends on the instantaneous PV power and the selected control strategy. For this research, the average charging rates at C_{10} has been chosen so the battery will reach 80% SoC when the terminal voltages is 13.8 V.

As the defined voltage above, the desired voltage window for the battery system is 58 V for lower limit and 69V for upper limit. The hysteresis band must be added to both sides of these voltage limits in order to eliminate the oscillation. Based on the terminal voltage difference between charge and discharge state, it will be defined as 1 V/battery or 6V for 60V (nominal) battery system.

A. Maximize Self-Consumption

The experimental results for "Maximize self-consumption" are shown in Fig.7. The simulated load profile consists of base consumption of 100W and 300 W for peak load which is applied two times a day: an hour at midday and 2 hours between 19:00-21:00. The initial battery SoC is 40% at the beginning of the experiment.

Between 8:00-10:30, since the PV power is less than the load demand ($P_{pv} < P_{load}$), this causes the battery to be discharged in order to fulfill the load requirement.

The operating condition changes after 10:30 where $P_{pv} > P_{load}$. During this time the excess PV power will be stored into the battery.

During midday where the peak demand occurs, the self consumption will be achieved by the aid of previously stored energy.

In the afternoon, all low power consumption are supplied by PV power while the excess of PV energy not-consumed on-site will be stored in the battery.

For night consumption, since the PV generated power

is null, therefore the load will be supplied by the energy stored during daytime. This stored energy are released until 21:00 where the undercharge state has reached. Thereafter the battery will be forced to stop discharging while the load will be directly supplied from grid power.

It is obvious that after 21:00 there are no energy transfer between each element of system. The battery SoC still approximately 30% at the end of day.

The PV array generates 1783 Wh/day while the total load consumption is 3000 Wh/day. The plentiful solar resource and the 40 % initial SOC at the beginning of experiment provided the long 13 hours of self consumption period in this day.

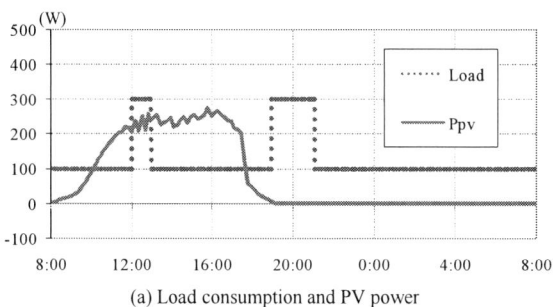

(a) Load consumption and PV power

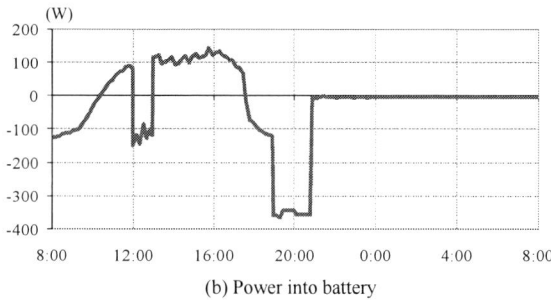

(b) Power into battery

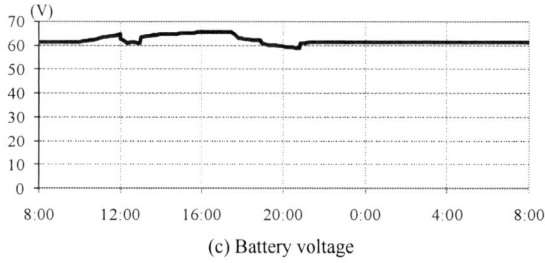

(c) Battery voltage

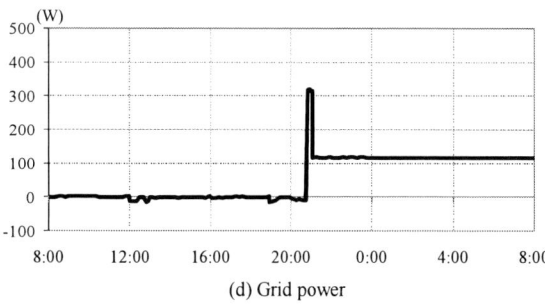

(d) Grid power

Fig. 7. Experimental results for "Maximize self-consumption" strategy

B. Grid Power Smoothing

Fig.8 shows the experimental results for "Grid power smoothing" strategy operating with the 150 W of P_{smooth}.

The base load is defined as 100 W while the 400W of peak load is applied two times a day: three hour around midday and three hours between 18:00-21:00

As can be seen, from 8:00 to 11:00 where the load demand is lower than 150 W, the battery will be charged from 50 W of grid power together with the power generated from PV array. During this period, the battery voltage gradually increases.

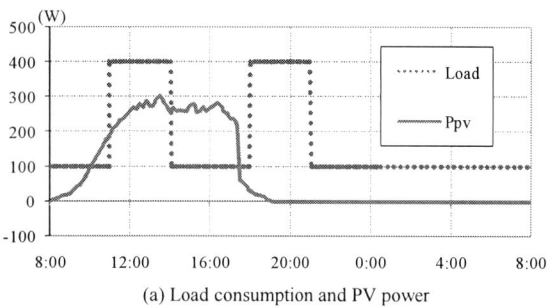

(a) Load consumption and PV power

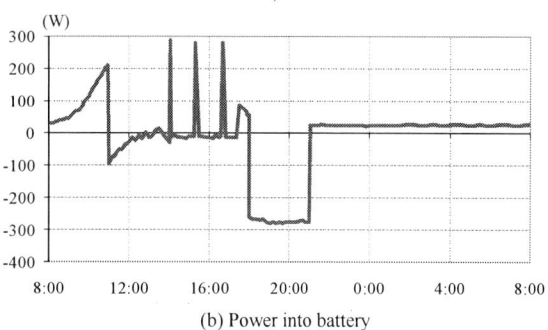

(b) Power into battery

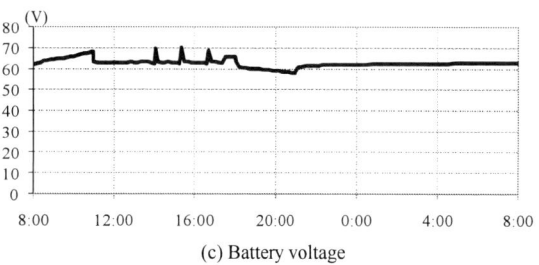

(c) Battery voltage

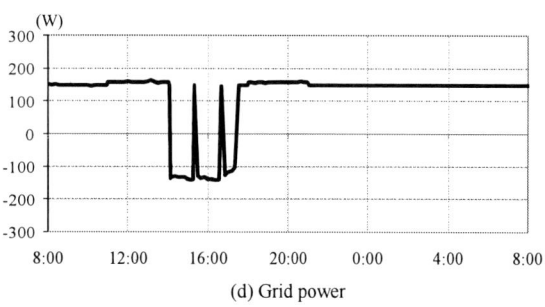

(d) Grid power

Fig. 8. Experimental results for "Grid power smoothing" strategy (150 W of smoothing power)

When the 400 W load is applied during 11:00-14:00, the stored energy in battery together with PV power is used to suppress the grid demand to 150 W.

Between 14:00-18:00, although the consumption is lower than P_{smooth}, the battery is rarely charged since it has reached the overcharge state (69V), therefore all PV power is fed into utility grid. The oscillation between normal and overcharge state occurs during this period which can be eliminated by increasing the hysteresis band.

After sunset, when the high load demand occurs (between 18:00-21:00), it is suppressed by the energy stored in battery during the low consumption period.

C. Peak Shaving

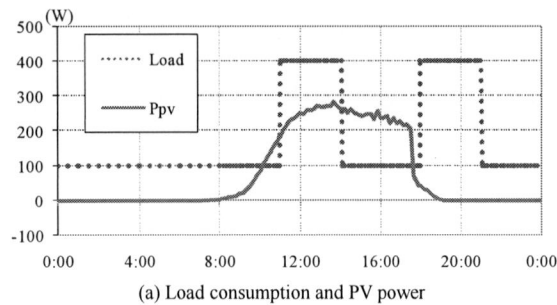

(a) Load consumption and PV power

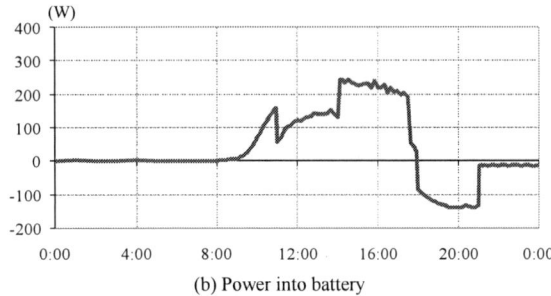

(b) Power into battery

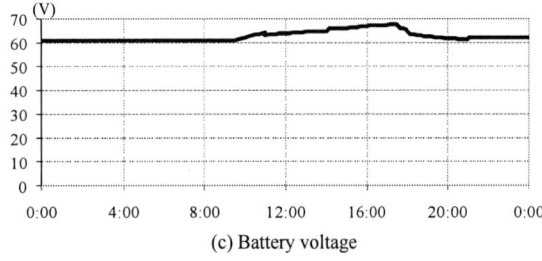

(c) Battery voltage

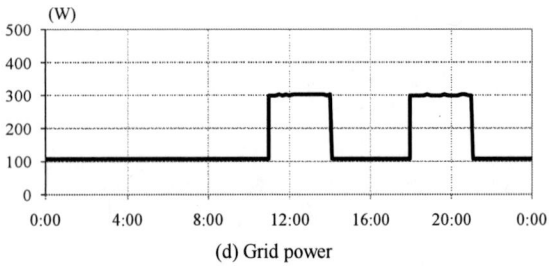

(d) Grid power

Fig. 9. Experimental results for "Peak shaving" strategy
(300 W of predefined peak power)

Like pervious experiment, the base load is defined as 100 W and the 400W of peak load is applied two times a day: three hour around midday and three hours between 18:00-21:00 while the peak power limit is set at 300W (see Fig.9).

As stated, the mainly controlled condition for this strategy is to use all PV power to charge the battery and used it to efface the peak demand.

In the morning, since the load consumption is lower than the defined peak power, all PV generated power is stored into the battery. However, between 11:00-14:00 where the peak load occurs, part of PV power is injected from DC side to grid side of inverter in order to suppress the grid peak demand into the predefined level (300 W).

After 14:00, as $P_{load} < P_{peak}$, all PV generated power is again stored into the battery until 19:00 there is no generation by the PV array, therefore the peak shaving is achieved by this stored energy.

As can be seen in Fig. 9(d), due to plentiful solar resource in this day, the grid power demand is suppressed within the defined peak power limit. The battery voltage at the end of day is higher than that at beginning of day. This means that the battery gains the energy after passing a day.

VI. CONCLUSION

The guideline to control the power flow and the properly selected control parameter to achieve MPP operation for grid-connected PV system with battery storage are presented. This will lead initiated idea for the future research concerned with such PV system. The three control strategies are examined via the small PV system which can also extend to other strategies. Although the experimental results show the success of the power flow control, however, in the view of economic, energy storage is an additional cost. For this reason, in the future work the techno-economical optimization should be applied in order to achieve the optimum system configuration and a suitable control strategy in the future work.

REFERENCES

[1] "Renewable Energy Sources Act (EEG) 2009", Federal Ministry for the Environment, Nature Conservation and Nuclear Safety, 2008.

[2] "Analysis of PV System's Values beyond Energy", Report IEA-PVPS T10-02 , 2008.

[3] K.Kurokawa, S.Wakao, Y.Hayashi, I.Ishii, K.Otani, M.Yamaguchi, T. Ishii, Y. Ono, "Conceptual study on autonomy-enhanced PV Clusters for urban community to meet the Japanese PV2030 Requirements", 20th European Photovoltaic Solar Energy Conference and Exhibition, Barcelona Spain, 2005.

[4] T.Shimada, K.Kurokawa, "Grid-connected photovoltaic systems with battery storages control based on insolation forecasting using weather forecast", Renewable Energy (2006): 228-230.

[5] R. Teodorescu, F. Blaabjerg, "A stable three-phase LCL-filter based active rectifier without damping", Industry Applications Conference 2003, 38th IAS Annual Meeting, 2003, vol.3, pp. 1552-1557.

A Maximum Power Point Tracking Method with Ripple Current Orientation

Chin-Sien Moo, Gwo-Bin Wu

Department of Electrical Engineering
National Sun Yat-sen University
Kaohsiung 80424, Taiwan
mooxx@mail.ee.nsysu.edu.tw

Abstract-This paper proposes a novel maximum power point tracking (MPPT) technique on basis of dynamic characteristics of the solar-cell panel with a power electronic converter. The power converter draws a rippled current from the solar-cell panel with designated circuit parameters. Experimental results indicate that the current is constantly out of phase with the voltage when the peak of the rippled current is less than the averaged current at the maximum power point. On the other hand, a phase deviation implies that the peak has reached the maximum power point current. Accordingly, an MPPT method is developed with a detection circuit designed for observing the occurrence of a phase deviation. The MPPT method is implemented on a laboratory solar power system to demonstrate its effectiveness.

Keywords—Maximum power point tracking (MPPT), Power converter, Rippled current, Solar-cell panel.

I. INTRODUCTION

In last decades, power electronic techniques have been becoming important topics for renewable energy development especially for solar power [1]. For most applications, a power conversion circuit with maximum power point tracking (MPPT) control is required to extract the most available power from the solar-cell panel [2]. Several MPPT algorithms commonly known as hill climbing, perturbation and observation, incremental conductance, and others have been applied to the solar power system [3-6]. These methods are mostly based on data averaging and sampling by using averaged current, voltage and power to develop tracking techniques. The control schemes highlight on different features, such as tracking speed, simplicity in circuits, digital or analog control, and utilization of sensors [7].

When a power electronic converter is attached, a current with high-frequency ripple will be drawn from the solar-cell panel. The ripple can be regarded as a correlated factor for MPPT. A creative method named as ripple correlation control was developed based on the phase commutation between ripples of the current and the power around the maximum power point [8-10]. The relationship between derivatives of the power and the current was employed. This method applies phase observation instead of data averaging and sampling to track the maximum power point rapidly and effectively.

An alternative MPPT method inspired by ripple correlation control is proposed in this research. The method is figured out by scrutinizing the behaviors of the ripple current and voltage from the solar-cell panel. According to the dynamic characteristics of the solar power, a deviation in phases between the output current and voltage shows a transitional shifting as the peak of the output current exceeds the maximum power point current [11]. The phase-shift implies that the phase between the output current and power is not just in commutation but with gradual deflection. Instead of power calculation, the measured voltage and current can be directly used for detecting the phase deviation. Accordingly, the operating point can be oriented to the maximum power point.

II. DYNAMIC CHARACTERISTICS

Experimental results indicate that both dynamic voltage-current (V-I) and power-current (P-I) characteristics are identical with the static ones as the current peak of the solar-cell panel is less than the average current at the maximum power point. Once the peak exceeds the maximum power point current, the output voltage starts to lag with respect to the output current, resulting in a hysteresis in the dynamic curve.

Fig. 1 shows two exemplar cases for a solar-cell panel with a boost converter which outputs essentially a triangular current. In these two cases, the output currents of the solar-cell panel are with a same ripple. In Fig. 1(a), the peak of the rippled current is less than the maximum power point current. Both dynamic V-I and P-I characteristic curves follow the static ones approximately as illustrated in Fig. 2. The voltage decreases as the current increases and vice versa, meaning that the voltage and the current are out of phase.

On the other hand, when the peak of the rippled current is higher than the current at the maximum power point, the measured output waveforms of the solar-cell panel can be different as shown in Fig. 1(b). The valley of the output voltage is no longer coincident with the peak of the rippled current but lags by some degree. As a result, both V-I and P-I characteristic curves do not follow the static ones but turn out to be hysteresis contours as depicted in Fig. 3.

The 2014 International Power Electronics Conference

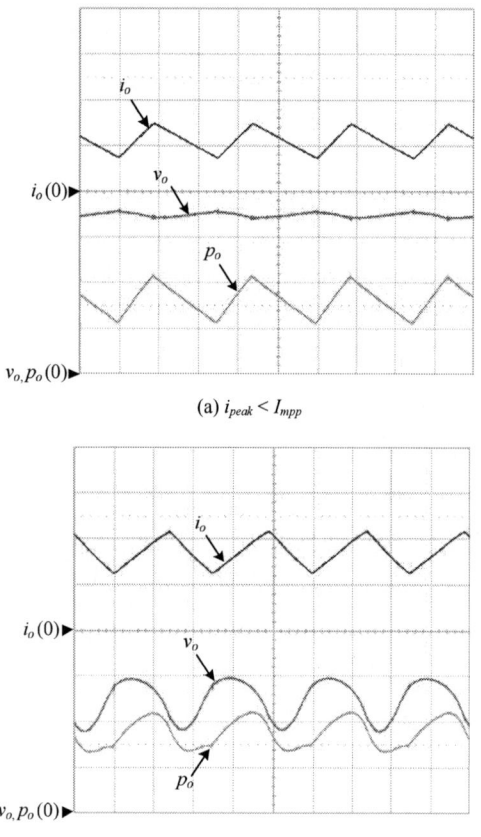

(a) $i_{peak} < I_{mpp}$

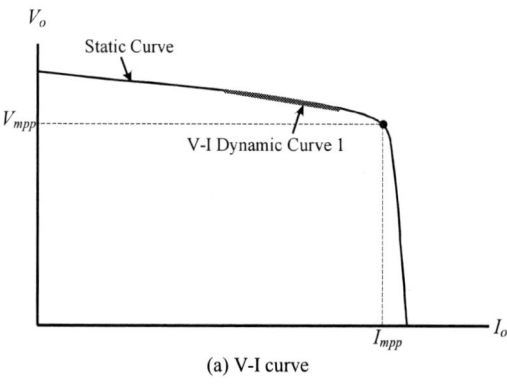

(b) $i_{peak} > I_{mpp}$

(i_o: 0.5 A/div, v_o: 5 V/div, p_o: 6 W/div, time: 20 μs/div)
Fig. 1. Measured waveforms drawn by boost converter.

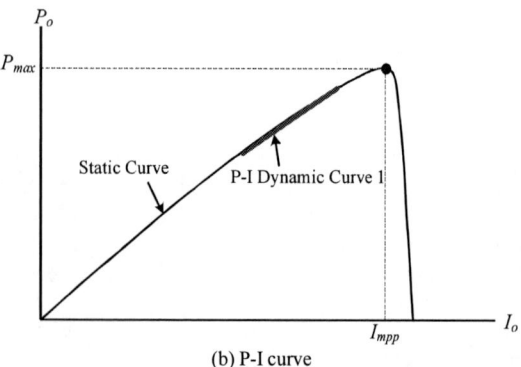

(a) V-I curve

(b) P-I curve

Fig. 2. Static and dynamic characteristic curves ($i_{peak} < I_{mpp}$).

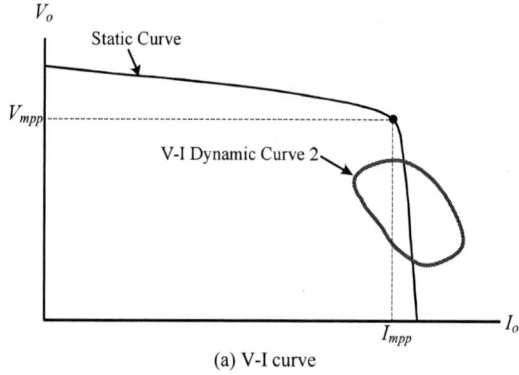

(a) V-I curve

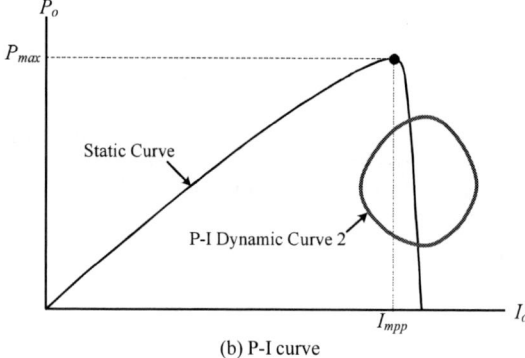

(b) P-I curve

Fig. 3. Static and dynamic characteristic curves ($i_{peak} > I_{mpp}$).

III. PROPOSED MPPT METHOD

As stated above, the dynamic characteristic of the solar-cell panel varies from the static characteristic when the peak of the output current is approximately around the maximum power point current. This phenomenon implies that the maximum power point current can be identified once the valley of the voltage does not coincide with the peak of the current. In other words, the maximum power point current can be simply obtained by means of detecting the instantaneous phase difference between the output voltage and current.

This concept can be evolved into an MPPT method as illustrated in Fig. 4. The method can be divided into two parts. Firstly, the variation of phase difference is detected. Secondly, the averaged current is upraised to match the peak of i_d. Fig. 5 shows the control scenario of the MPPT method.

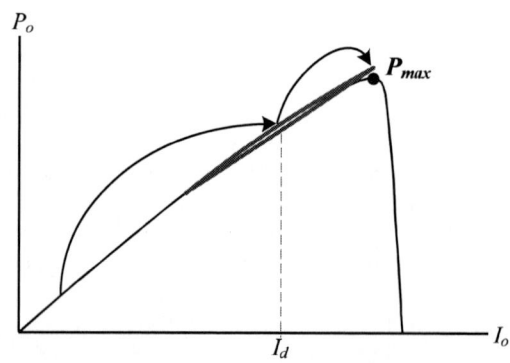

Fig. 4. Illustration of the proposed MPPT method.

978-1-4799-2706-7/14 $31.00 © 2014 IEEE

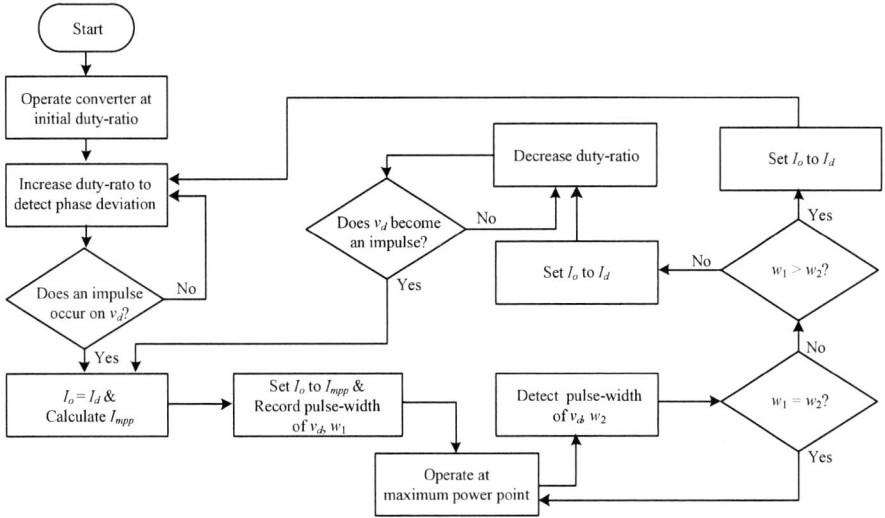

Fig. 5. Control scenario of MPPT.

In the method, a detection circuit is used to scrutinize the phase difference between the output voltage and current. An impulse signal, v_d, is generated once as a phase deviation occurs between them. At this point, an average current I_d with a ripple Δi, is drawn from the solar-cell panel. Then, the average current is upraised to the maximum power point current, I_{mpp}, which can be obtained as

$$I_{mpp} = I_d + \frac{1}{2}\Delta i = i_{peak} \qquad (1)$$

To ensure the solar-cell panel is continuously operated at the maximum power point, the pulse-width of v_d is cyclically examined to observe the variation in phase deviation. The same pulse-width of v_d will be detected when the operation condition remains unchanged. Any change in the pulse-width of v_d implies a change of the operating condition, leading to a deviation of the maximum power point.

IV. PHASE DIFFERENCE DETECTION

The proposed phase detection circuit is shown in Fig. 6. The sensed voltage and current from the solar-cell panel are first reduced in signal levels. Then, the current signal is inversed. The differentiators turn the voltage and current with triangular ripples into quasi-rectangular waveforms. The waveforms are rectified to be rectangular to facilitating the detection of a phase difference. A high state of v_d is recognized by an XOR gate to identify a phase-shift, indicating that the current peak has become higher than the maximum power point current.

Figs. 7 and 8 show two typical detected signals of phase-difference. In Fig. 7, the peak of the output current is less than the maximum power point current. Two signals, v_v and v_i, remain in phase and therefore v_d keeps at 0. In the case of Fig. 8, the peak current just reaches the maximum power point current. The negative edge of v_v falls behind that of v_i, leading to a high state of v_d for a short period, representing that the current peak reaches the maximum power point current at this moment.

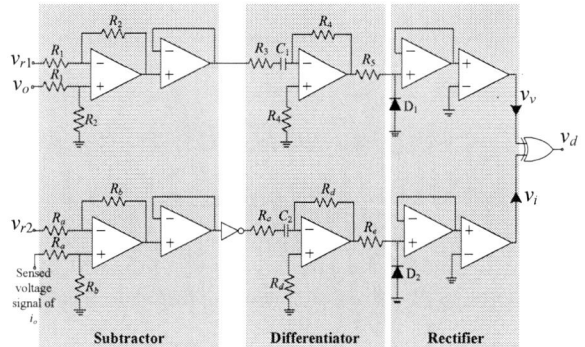

Fig. 6. Phase-difference detection circuit.

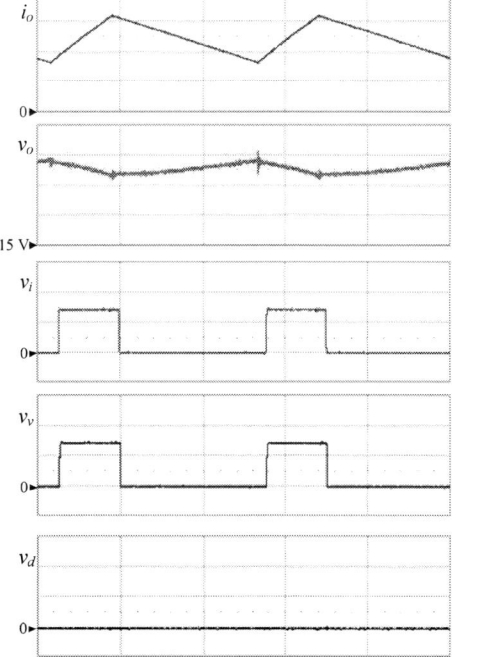

(i_o: 0.2 A/div, v_o: 1 V/div, v_v: 2 V/div, v_i: 2 V/div, v_d: 2 V/div, time: 20 μs/div)

Fig. 7. Detected phase-shift at $i_{peak} < I_{mpp}$.

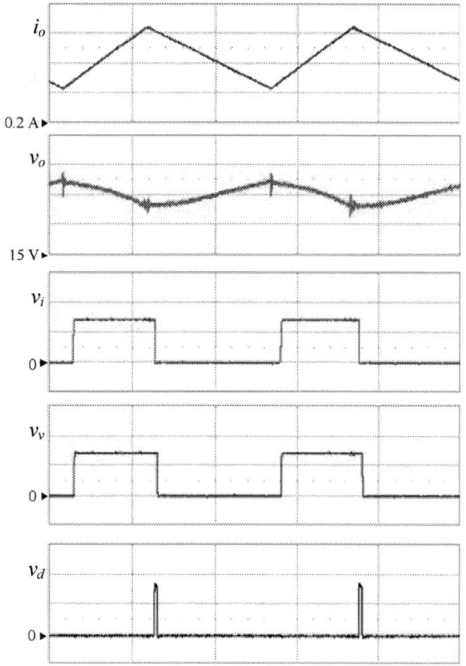

(i_o: 0.2 A/div, v_o: 1 V/div, v_v: 2 V/div, v_i: 2 V/div, v_d: 2 V/div,
time: 20 μs/div)
Fig. 8. Detected phase-shift at $i_{peak} = I_{mpp}$.

V. EXPERIMENTAL TESTS

A solar-cell panel operated with a boost converter is set up for the implementation of the proposed MPPT method. Fig. 9 shows the start-up transient from the detection stage to the steady-state operation at the maximum power point for the experimental result of Case I. At the beginning, the output current and power of the solar-cell panel become higher and higher by increasing the duty-ratio of the boost converter. At the same time, the voltage decreases gradually. The phase difference between the output voltage and current is detected continually during the operating period.

Once a phase difference between the voltage and the inverse of the current has been detected, the maximum power point current is calculated. Then, the duty-ratio of the boost converter is further increased to raise the current and hence the power. Eventually, the solar-cell panel is operated at the MPP.

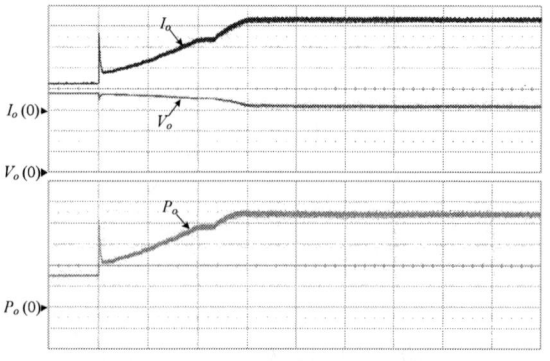

(i_o: 0.2 A/div, v_o: 5 V/div, p_o: 3 W/div, time: 100 ms/div)
Fig. 9. Current, voltage, and power during starting transient (Case I).

Fig. 10 shows the measured waveforms of the output current, voltage and power when the solar-cell panel operating at the maximum power point. The tracked maximum power is 13.43 W with a maximum power point current of 0.84 A.

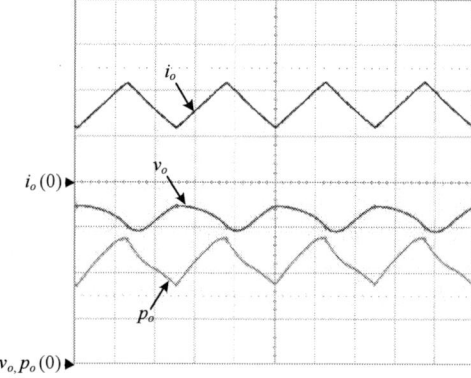

(i_o: 0.5 A/div, v_o: 5 V/div, p_o: 6 W/div, time: 20 μs/div)
Fig. 10. Measured waveforms at the maximum power point.

Fig. 11 shows the experimental result of Case II with a lesser radiation on the solar-cell panel. From the starting to the point where a phase deviation is found, v_d stays always at the low-state. The duty-ratio of the boost converter can be stepped up immediately to increase the output current and power without any intermittent caused by data averaging. When an impulse signal of v_d is recognized by the detection circuit, the current at the maximum power point is calculated as 0.72 A. Consequently, the boost converter is instantly adjusted to obtain an average current at the maximum power point. Then, the solar-cell panel is operated at the maximum power of 11.36 W.

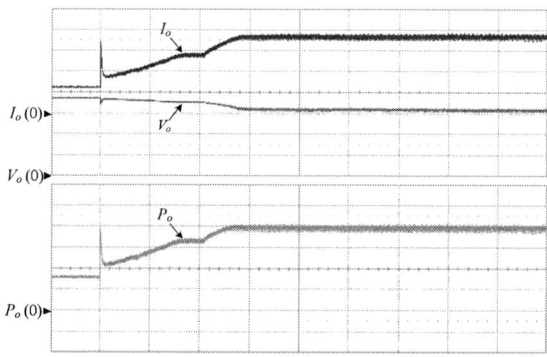

(i_o: 0.2 A/div, v_o: 5 V/div, p_o: 3 W/div, time: 100 ms/div)
Fig. 11. Current, voltage, and power during starting transient (Case II).

Table I summarizes the data of these two experimental tests. In the case of Fig. 9, the phase deviation onsets at the output current of 0.65 A with a ripple of 0.38 A. Accordingly, the current at the maximum power point is calculated as 0.84 A. As compared with the actual maximum power, the tracking accuracy is 99.48 %. For Case II in Fig. 11, the phase deviation is found as the output current is raised up to 0.55 A with a ripple of 0.34 A. As a result, the maximum power point current is obtained as 0.72 A. The tracking accuracy of this case is 99.47 %.

TABLE I
SUMMARIZED DATA OF EXPERIMENTAL TESTS

Experiment	Current at the phase deviation onset (A)	Tracked maximum power point current (A)	Tracked maximum power (W)	Actual maximum power (W)	Tracking accuracy (%)
Case I	0.65 A	0.84 A	13.43 W	13.5 W	99.48 %
Case II	0.55 A	0.72 A	11.36 W	11.42 W	99.47 %

A phase deviation one is found as the solar-cell panel is operated at the maximum power point current since the peak current is much higher than the maximum power point current, leading to a larger hysteresis contour in the dynamic P-I characteristic curve, as shown in Fig. 12. The centers of the dynamic contours represent the average powers of the solar-cell panel with two different operation conditions. Any variation in the phase-shift at these maximum power points reveals a deformation of the dynamic contour and hence a change in the output power. Further adjustment of the duty-ratio of the boost converter has to be made to operate the solar-cell panel at a new maximum power point with a higher power.

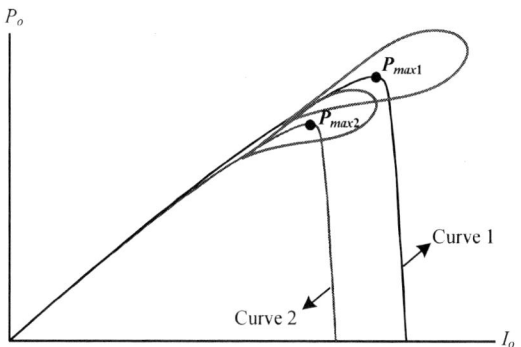

Fig. 12. Maximum power points with dynamic P-I contours.

VI. CONCLUSIONS

A maximum power point tracking technique using the dynamic phase deviation of the rippled current and voltage is introduced and verified for applications of the solar-cell panel with a boost-converter. To realize the observation of the phase deviation, a detection circuit has been designed. An impulse is generated once the peak of the rippled current has reached the current at the maximum power point. The average current at the maximum power point can be obtained in advance. Then, the duty-ratio of the power converter can be adjusted to operate the solar-cell panel with the maximum power point current. This method can track the maximum power point every high-frequency cycle. The effectiveness of the proposed MPPT method has been demonstrated by experiments on a laboratory solar power system.

REFERENCES

[1] S. Rahman, "Green Power: What is it and where can we find it?" *IEEE Power and Energy Magazine*, vol. 1, pp. 30-37, Jan./Feb. 2003.

[2] A. Freitas, F. Antunes, E. Mineiro, A. Lima, S. Daher, and S. Ximenes, "High gain DC-DC converter applied to photovoltaic system with new proposed to MPPT search," in *IEEE EUROCON*, pp. 1081-1085, July 2013.

[3] W. Xiao and W. G. Dunford, "A modified adaptive hill climbing MPPT method for photovoltaic power systems," in *Proc. IEEE PESC*, vol. 3, pp. 1957-1963, June 2004.

[4] F. Liu, Y. Kang, Y. Zhang, and S. Duan, "Comparison of P&O and hill climbing MPPT methods for grid-connected PV converter," in *Proc. IEEE ICIEA*, pp. 804-807, June 2008.

[5] X. Liu and L. A. C. Lopes, "An improved perturbation and observation maximum power point tracking algorithm for PV arrays," in *Proc. IEEE PESC*, vol. 3, pp. 2005-2010, June 2004.

[6] A. Safari and S. Mekhilef, "Simulation and hardware implementation of incremental conductance MPPT with direct control method using Ćuk converter," *IEEE Trans. on Industrial Electronics*, vol. 58, pp. 1154-1161, Apr. 2011.

[7] T. Esram and P. L. Chapman, "Comparison of photovoltaic array maximum power point tracking techniques," *IEEE Trans. on Energy Conversion*, vol. 22, pp. 439-449, June 2007.

[8] P. Midya, P. T. Krein, R. J. Turnbull, R. Reppa, and J. Kimball, "Dynamic maximum power point tracker for photovoltaic applications," in *Proc. IEEE PESC*, vol. 2, pp. 1710-1716, June 1996.

[9] T. Esram, J. W. Kimball, P. T. Krein, P. L. Chapman, and P. Midya, "Dynamic maximum power point tracking of photovoltaic arrays using ripple correlation control," *IEEE Trans. on Power Electronics*, vol. 21, pp. 1282-1291, Sep. 2006.

[10] J. W. Kimball and P. T. Krein, "Digital ripple correlation control for photovoltaic applications," in *Proc. IEEE PESC*, pp. 1690-1694, June 2007.

[11] S. L. Lin, G. B. Wu, W. C. Liu, and C. S. Moo, "Ripple current effect on output power of solar-cell panel," in *Proc. IEEE ICRERA*, pp. 1-5, Nov. 2012.

Output Characteristics of a Surface Permanent Magnet-type Vernier Motor
- Comparison of Test Results and Calculation -

Yasuhiro Kataoka, Masakazu Takayama, Yoshihisa Anazawa

Dept. of Electronics and Information Systems
Akita Prefectural University
84-4 Ebinokuchi, Yurihonjo City, 015-0055, Japan
yasuhiro_kataoka@akita-pu.ac.jp

Yoshitarou Matsushima

Dept. of Electrical and Electronic Engineering
Shizuoka University
3-5-1 Johoku, Naka-ku, Hamamatsu, 432-8561, Japan

Abstract— **This paper describes a surface permanent magnet (SPM)-type vernier motor that can generate high torque at low speeds, and a simple, high-speed method for analyzing its output characteristics is proposed. In this method, if the induced electromotive force, the synchronous reactance, and the iron loss are provided by finite element analysis (FEA), the output characteristics can be calculated immediately using the voltage and torque equations on the γ and δ axes. The proposed method is simpler and faster than the general method that uses FEA only. In this paper, the voltage and torque equations of the SPM-type vernier motor are derived, and the proposed analysis method using these equations is presented. Moreover, the agreement of the results calculated using this method with the measured values confirms a high degree of accuracy.**

Keywords— Vernier motor, Surface permanent magnet type, Calculation method, Output characteristics

I. INTRODUCTION

In industrial machines, such as machine tools, traction machines for elevators, and industrial robots, reduction gears are generally used because the resulting high torque allows for high-precision positioning. However, since reduction gears suffer from disadvantages such as backlash, friction loss, abrasion, noise, and the need for frequent maintenance, direct-drive systems have been developed.

In these systems, since the motor is directly connected to a heavy load, a very high torque must be generated. Although ordinary motors such as stepping motors and brushless DC motors have been used in direct-drive systems, it is difficult for them to precisely control large loads at low speeds because they produce low torque and are subject to torque pulsations. To overcome these issues, vernier motors, which can generate high torque at low speeds and theoretically are not affected by torque pulsations, have been developed [1]–[4]. Such motors are suitable for use in direct-drive systems.

Recently, the increasing energy product of permanent magnets has led to improvements in the performance of ordinary motors. Surface permanent magnet (SPM)-type

vernier motors, which have advantages such as smaller size, have also been developed. However, except for finite element analysis (FEA), a method for analyzing the output characteristics of SPM-type vernier motors has not yet been established. Moreover, FEA requires a huge amount of time due to the modeling and number of calculations involved, so it is difficult to quickly determine the optimal design for a motor using this method.

In this paper, the voltage and torque equations of the SPM-type vernier motor are derived, and an analysis method that combines these equations with simple FEA is proposed. The results of the proposed method agree very well with the measured values for a trial motor. Therefore, it is shown that the output characteristics can be calculated immediately using the proposed method if the induced electromotive force, the synchronous reactance, and the iron loss are provided by FEA. The proposed method is very useful, and its accuracy is excellent.

II. OPERATIONAL PRINCIPLE

The basic configuration and operational principle of the SPM-type vernier motor are shown in this chapter.

A. Basic configuration

In the SPM-type vernier motor, the number of poles P on the primary side and the number of field poles R on the secondary side are different, and based on the vernier principle, the number of slots S can be expressed in terms of P and R as shown in equation (1).

$$S = \frac{R}{2} \pm \frac{P}{2} \tag{1}$$

The two-pole basic composition of the SPM-type vernier motor is shown in Fig. 1. Normal three-phase armature winding is used in the stator, producing two magnetic poles at the gap surface. Around the perimeter of the rotor, ten magnet segments with alternating magneti-

The 2014 International Power Electronics Conference

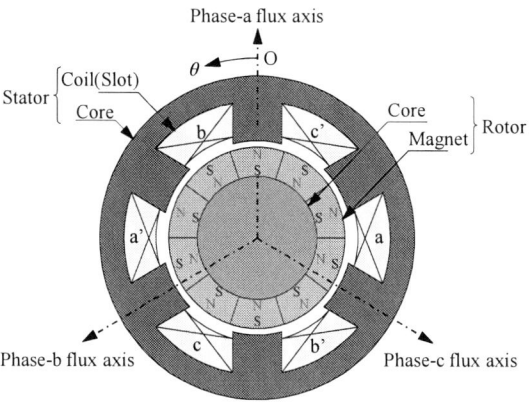

Fig. 1. Two-pole basic composition of SPM-type vernier motor.

zation directions are arranged, producing 10 field poles at the gap surface.

B. Operational principle

The operational principle of the SPM-type vernier motor is shown in Fig. 2, which is a linear representation of Fig. 1. This motor is driven by a three-phase alternating voltage, and a rotating magnetic field is generated in the gap of the motor. In this figure, the magnetomotive force and the magnetic flux density distribution for $i_a = -2i_b = -2i_c$ are shown. The currents of the a phase, b phase, and c phase are denoted by i_a, i_b, and i_c, respectively. Gap permeance pulsates periodically because there are slots and teeth in the stator. Therefore, the magnetomotive force is modulated by the distribution of the gap permeance, and the magnetic flux density of the gap contains the fundamental, 5th, and 7th harmonics.

In this composition, the rotor rotates slowly while synchronizing with the 5th harmonic of the rotating magnetic field because the rotor has 10 magnetic poles. Therefore, the rotor speed is stepped down to 1/5 of the speed of the rotating magnetic field, and the torque is stepped up. Thus, this motor functions as a motor with magnetic gearing. The rotor speed is given by

$$N = \frac{120f}{P} \cdot \frac{P}{R} = \frac{120f}{R} \qquad (2)$$

where N is the synchronous speed (min^{-1}), and f is the power supply frequency (Hz).

III. VOLTAGE AND TORQUE EQUATIONS

The voltage and torque equations of the SPM-type vernier motor are derived in this chapter. An inductance matrix is obtained using evaluation coils for permanent magnets.

A. Derivation of inductance matrix

An equivalent coil conversion for the R poles of permanent magnets is shown in Fig. 3. The permanent magnets in the SPM-type vernier motor can be converted into the equivalent coils shown in Fig. 3(a). The equivalent coils can be separated as shown in Fig. 3(b), and an inductance matrix between each of the coils can be obtained.

The frame of reference of the SPM-type vernier motor is defined in Fig. 4. Here, θ' is defined as the angle formed by the phase-a flux axis and the 1st equivalent coil flux axis, and θ is defined as the angle formed by the phase-a flux axis and a line on the optional coordinate axis. Also, φ is defined as the angle formed by a line on the optional coordinate axis and the 1st equivalent coil flux axis. These relations are given by

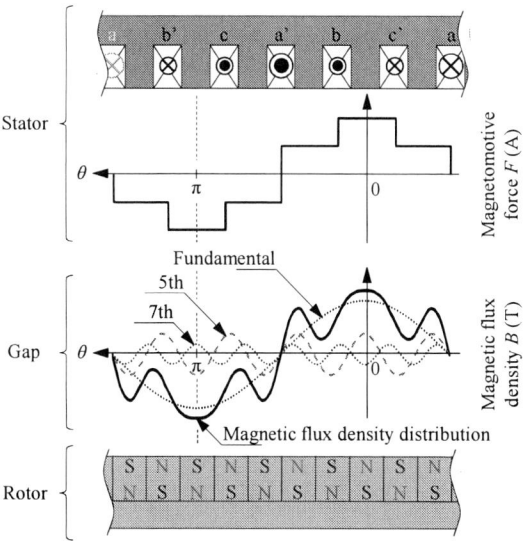

Fig. 2. Operational principle of SPM-type vernier motor.

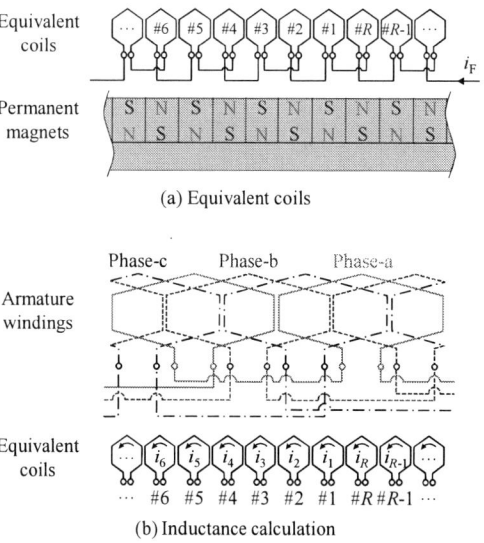

(a) Equivalent coils

(b) Inductance calculation

Fig. 3. Equivalent coil conversion for permanent magnets.

978-1-4799-2706-7/14 $31.00 © 2014 IEEE

$$B(\theta,\theta') = F_{1k}(\theta) \cdot P(\theta,\theta') \qquad (8)$$

According to the above results, the flux linkage of each coil is obtained as follows. First, the flux linkage of the phase-g-th winding on the armature is given by equation (9) if the phase-k-th winding on the armature is excited,

$$\Phi_{1gk} = k_{w1}(v)\, n_1 \int_{\frac{\pi}{2}+(g-1)\alpha}^{\frac{\pi}{2}+(g-1)\alpha} F_{1k}(\theta) \cdot P(\theta,\theta')\, ur\, d\theta \qquad (9)$$

where u is the core length (m), and r is the inner radius of the stator (m). Next, the flux linkage of the phase-g-th winding on the equivalent coil is given by equation (10) if the phase-k-th winding on the armature is excited.

$$\Phi_{Mgk} = k_{w2}(v)\, n_2 \int_{\frac{\pi}{R}+(g-1)\frac{2\pi}{R}}^{\frac{\pi}{R}+(g-1)\frac{2\pi}{R}} F_{1k}(\theta'+\varphi) \cdot P(\theta',\varphi)\, ur\, d\varphi$$
$$(10)$$

Finally, the flux linkage of the phase-g-th winding on the equivalent coil is given by equation (11) if the phase-k-th winding on the equivalent coil is excited.

$$\Phi_{2gk} = k_{w2}(v)\, n_2 \int_{\frac{\pi}{R}+(g-1)\frac{2\pi}{R}}^{\frac{\pi}{R}+(g-1)\frac{2\pi}{R}} F_{2k}(\varphi) \cdot P(\theta',\varphi)\, ur\, d\varphi$$
$$(11)$$

And the inductance matrix for each coils is obtained, and is given by equation (12).

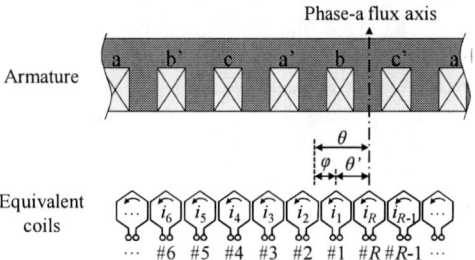

Fig. 4. Frame of reference.

$$\theta = \theta' + \varphi \qquad (3)$$

The magnetomotive force distribution of the phase-k-th winding on the armature is given by equation (4) if a current of I_1 (A) flows to the phase-k-th winding on the armature,

$$F_{1k}(\theta) = \sum_{v=1}^{\infty} \frac{2k_{w1}(v)\, n_1 I_1}{v\pi} \cos v\{\theta - (k-1)\alpha\} \qquad (4)$$

where $k_{w1}(v)$ is the winding factor for v-th order harmonics of the armature winding, n_1 is the number of turns per phase of the armature winding, v is the harmonic order, and α is $2\pi/3$. The winding factor for v-th order harmonics is given by equation (5),

$$k_{w1}(v) = \frac{\sin\left(\frac{v\pi}{2m}\right)}{q\sin\left(\frac{v\pi}{2mq}\right)} \sin v\theta_1 \qquad (5)$$

where q is the number of slots per pole per phase, m is the number of phase, and $2\theta_1$ is the coil pitch expressed as a mechanical angle.

The magnetomotive force distribution of the phase-k-th winding on the equivalent coil is given by equation (6) if a current of I_2 (A) flows to the phase-k-th winding on the equivalent coil,

$$F_{2k}(\varphi) = \sum_{v=1}^{\infty} \frac{2k_{w2}(v)\, n_2 I_2}{v\pi} \cos u\left\{\varphi - (k-1)\frac{2}{R}\pi\right\} \qquad (6)$$

where $k_{w2}(v)$ is the winding factor for v-th order harmonics of the equivalent coil, and n_2 is the number of turns per coil in the equivalent coil.

On the other hand, the gap permeance is given by equation (7),

$$P(\theta,\varphi) = P_{00} + P_{S0}\cos(S\theta) \qquad (7)$$

where P_{00} is the average value in the set of the gap permeance distribution (H), and P_{S0} is the gap permeance pulsating component from the armature slots (H). Moreover, a magnetic flux density distribution $B(\theta,\theta')$ is obtained by multiplying the magnetomotive force distribution $F_{1k}(\theta)$ by the gap permeance distribution $P(\theta,\theta')$ and is given by equation (8).

$$
L = \begin{array}{c}
\\
a \\
b \\
c \\
1 \\
2 \\
\vdots \\
R
\end{array}
\begin{array}{ccc}
a & b & c \\
\left[\begin{array}{ccc}
l_1 + L_0 & -\dfrac{L_0}{2} & -\dfrac{L_0}{2} \\
-\dfrac{L_0}{2} & l_1 + L_0 & -\dfrac{L_0}{2} \\
-\dfrac{L_0}{2} & -\dfrac{L_0}{2} & l_1 + L_0 \\
M_{1a}(\theta') & M_{1b}(\theta') & M_{1c}(\theta') \\
M_{2a}(\theta') & M_{2b}(\theta') & M_{2c}(\theta') \\
\vdots & \vdots & \vdots \\
M_{10a}(\theta') & M_{10b}(\theta') & M_{10c}(\theta')
\end{array}\right. & *
\end{array}
$$

$$
* \begin{array}{cccc}
1 & 2 & \cdots & R \\
\left.\begin{array}{cccc}
M_{a1}(\theta') & M_{a2}(\theta') & \cdots & M_{aR}(\theta') \\
M_{b2}(\theta') & & \cdots & M_{bR}(\theta') \\
M_{c2}(\theta') & M_{c2}(\theta') & \cdots & M_{cR}(\theta') \\
L_{11} & L_{12} & \cdots & L_{1R} \\
L_{21} & L_{22} & \cdots & L_{2R} \\
\vdots & \vdots & \ddots & \vdots \\
L_{R1} & L_{R2} & \cdots & L_{RR}
\end{array}\right]
\end{array} \quad (12)
$$

Next, the relations between the current i_F and the currents from i_1 to i_R are given by equation (13), and a part of connection matrix for the equivalent coils is obtained, and the transposed matrix is shown in equation (14). According to the above results, the connection matrix is given by equation (15).

$$i_F = i_1 = i_3 = \cdots = i_{R-3} = i_{R-1}$$
$$i_F = -i_2 = -i_4 = \cdots = -i_{R-2} = -i_R \tag{13}$$

$$[C]_t = \begin{bmatrix} 1 & -1 & 1 & -1 & \cdots \end{bmatrix} \tag{14}$$

$$\begin{bmatrix} i_a \\ i_b \\ i_c \\ i_{F1} \\ i_{F2} \\ i_{F3} \\ \vdots \\ i_{FR} \end{bmatrix} = \begin{bmatrix} 1 & 0 & 0 & \vdots & 0 \\ 0 & 1 & 0 & \vdots & 0 \\ 0 & 0 & 1 & \vdots & 0 \\ \hdashline 0 & 0 & 0 & \vdots & 1 \\ 0 & 0 & 0 & \vdots & -1 \\ 0 & 0 & 0 & \vdots & 1 \\ \vdots & \vdots & \vdots & \vdots & \vdots \\ 0 & 0 & 0 & \vdots & -1 \end{bmatrix} \begin{bmatrix} i_a \\ i_b \\ i_c \\ i_F \end{bmatrix} \tag{15}$$

The above inductance matrix is converted to an inductance matrix of the SPM-type vernier motor by the connection matrix. The converted inductance matrix of the SPM-type vernier motor is shown in equation (16),

$$[L] = \begin{bmatrix} l_1 + L_0 & & & -\dfrac{L_0}{2} \\ -\dfrac{L_0}{2} & l_1 + L_0 & & \\ -\dfrac{L_0}{2} & & & -\dfrac{L_0}{2} \\ M\cos\dfrac{R\theta'}{2} & & M\cos\dfrac{R(\theta'-\alpha)}{2} & \end{bmatrix} \quad *$$

$$* \quad \begin{bmatrix} -\dfrac{L_0}{2} & & M\cos\dfrac{R\theta'}{2} \\ -\dfrac{L_0}{2} & & M\cos\dfrac{R(\theta'-\alpha)}{2} \\ l_1 + L_0 & & M\cos\dfrac{R(\theta'-2\alpha)}{2} \\ M\cos\dfrac{R(\theta'-2\alpha)}{2} & & l_F + L_F \end{bmatrix} \tag{16}$$

where l_1 is the leakage inductance (H), L_0 is the main inductance (H), M is the mutual inductance (H), and L_F is the self inductance of the equivalent coil.

The main inductance L_0 is shown in equation (17), the mutual inductance M is shown in equation (18), and the self inductance of the equivalent coil L_F is shown in equation (19),

$$L_0 = \sum_{v=1}^{\infty} \left[\frac{4\{k_{w1}(v)n_1\}^2}{v^2\pi} P_{00}ur + \frac{2\{k_{w1}(S+v)n_1\}^2}{v(S+v)\pi} P_{S0}ur \right.$$
$$\left. + \frac{2\{k_{w1}(S-v)n_1\}^2}{v(S-v)\pi} P_{S0}ur \right] \tag{17}$$

$$M = \sum_{v=1}^{\infty} \left[\frac{4k_{w1}(v)k_{w2}(v)n_1n_2}{v^2\pi} P_{00}C(v)ur \right.$$
$$+ \frac{2k_{w1}(v)k_{w2}(S+v)n_1n_2}{v(S+v)\pi} P_{S0}C(S+v)ur$$
$$\left. + \frac{2k_{w1}(v)k_{w2}(S-v)n_1n_2}{v(S-v)\pi} P_{S0}C(S-v)ur \right] R \tag{18}$$

$$L_F = \sum_{v=1}^{\infty} \left[\frac{4\{k_{w2}(v)n_2\}^2}{v^2\pi} P_{00}C(v)^2 ur \right] R^2 \tag{19}$$

where $C(v)$ is the connection factor. Since the field pole R is set to 10, the connection factor $C(v)$ is given by equation (20), and only the 5th harmonics can be realized.

$$C(v) = \frac{2}{5}\sin v\frac{\pi}{2}\left(\cos v\frac{2}{5}\pi - \cos v\frac{\pi}{5} + \frac{1}{2}\right) \tag{20}$$

B. Voltage and torque equations

The voltage equation of the SPM-type vernier motor is given by equation (21),

$$[v] = \{[r] + p[L]\}[i] \tag{21}$$

where $[v]$ is the voltage matrix, $[r]$ is the resistance matrix, and $[i]$ is the current matrix.

The obtained inductance matrix is converted into a new inductance matrix constructed on the γ and δ axes by the three-phase to two-phase conversion, the γ and δ axes conversion, and the commutating conversion. The three-phase to two-phase conversion matrix $[C_1]$ is shown in equation (22), and the γ and δ axes conversion matrix $[C_2]$ is shown in equation (24).

$$[C_1] = \begin{bmatrix} C_1' & 0 \\ 0 & 1 \end{bmatrix} \tag{22}$$

$$[C_1'] = \sqrt{\frac{2}{3}} \begin{bmatrix} \dfrac{1}{\sqrt{2}} & 1 & 0 \\ \dfrac{1}{\sqrt{2}} & \cos\alpha & \sin\alpha \\ \dfrac{1}{\sqrt{2}} & \cos 2\alpha & \sin 2\alpha \end{bmatrix} \tag{23}$$

$$[C_2] = \begin{bmatrix} C_2' & 0 \\ 0 & 1 \end{bmatrix} \tag{24}$$

$$[C_2'] = \begin{bmatrix} \cos\dfrac{R}{2}\theta' & -\sin\dfrac{R}{2}\theta' \\ \sin\dfrac{R}{2}\theta' & \cos\dfrac{R}{2}\theta' \end{bmatrix} \tag{25}$$

According to the above conversions, the voltage equations on the γ and δ axes are given by equation (26).

$$\begin{bmatrix} v_{\gamma} \\ v_{\delta} \\ v_{F} \end{bmatrix} = \begin{bmatrix} r_{a} + p\left(l_{1} + \dfrac{3}{2}L_{0}\right) & -\dfrac{R\omega_{m}}{2}\left(l_{1} + \dfrac{3}{2}L_{0}\right) \\ \dfrac{R\omega_{m}}{2}\left(l_{1} + \dfrac{3}{2}L_{0}\right) & r_{a} + p\left(l_{1} + \dfrac{3}{2}L_{0}\right) \quad * \\ p\sqrt{\dfrac{3}{2}}M & 0 \end{bmatrix}$$

$$* \quad \begin{bmatrix} p\sqrt{\dfrac{3}{2}}M \\ \dfrac{R\omega_{m}}{2}p\sqrt{\dfrac{3}{2}}M \\ r_{F} + pL_{F} \end{bmatrix} \begin{bmatrix} i_{\gamma} \\ i_{\delta} \\ i_{F} \end{bmatrix} \quad (26)$$

In this condition, the torque is given by equation (27),

$$T = \left(\frac{R}{2}\right)\sqrt{\frac{3}{2}}M i_{F} i_{\delta} \qquad (27)$$

where i_{δ} is the current on the δ axis.

Next, the balanced three-phase voltage and the balanced three-phase current at steady state are expressed in equations (28) and (29) if the SPM-type vernier motor is in synchronous operation.

$$\begin{aligned} v_{a} &= \sqrt{2}V\sin(\omega t) \\ v_{b} &= \sqrt{2}V\sin(\omega t - \alpha) \\ v_{c} &= \sqrt{2}V\sin(\omega t - 2\alpha) \end{aligned} \qquad (28)$$

$$\begin{aligned} i_{a} &= \sqrt{2}I\sin(\omega t) \\ i_{b} &= \sqrt{2}I\sin(\omega t - \alpha - \varphi_{1}) \\ i_{c} &= \sqrt{2}I\sin(\omega t - 2\alpha - \varphi_{1}) \end{aligned} \qquad (29)$$

These equations are converted into voltage and current equations on the γ and δ axes and are given by the following equations if the relation of $\omega t = R\theta'/2 + R\delta_{L}/2$ is substituted for the obtained equations:

$$\begin{aligned} v_{\gamma} &= \sqrt{3}V\sin\left(\frac{R}{2}\delta_{L}\right) \\ v_{\delta} &= -\sqrt{3}V\cos\left(\frac{R}{2}\delta_{L}\right) \end{aligned} \qquad (30)$$

$$\begin{aligned} i_{\gamma} &= \sqrt{3}I\sin\left(\frac{R}{2}\delta_{L} - \varphi_{1}\right) \\ v_{\delta} &= -\sqrt{3}I\cos\left(\frac{R}{2}\delta_{L} - \varphi_{1}\right) \end{aligned} \qquad (31)$$

In steady state, these voltages and currents are converted into DC values. Equations (30) and (31) are substituted into (26). Because the differential operator p is given by 0 in the steady state, the voltage equations on the γ and δ axes are given by equations (32) and (33), respectively.

$$V\sin\left(\frac{R}{2}\delta_{L}\right) = -r_{a}I\sin\left(\varphi_{1} - \frac{R}{2}\delta_{L}\right)$$
$$+ X_{S}I\cos\left(\varphi_{1} - \frac{R}{2}\delta_{L}\right) \qquad (32)$$

$$V\cos\left(\frac{R}{2}\delta_{L}\right) = X_{S}I\sin\left(\varphi_{1} - \frac{R}{2}\delta_{L}\right)$$
$$+ r_{a}I\cos\left(\varphi_{1} - \frac{R}{2}\delta_{L}\right) + E_{0} \qquad (33)$$

Solving equations (32) and (33) for the current yields equations (34) and (35), respectively.

$$I\sin\left(\varphi_{1} - \frac{R}{2}\delta_{L}\right) = \frac{V}{Z}\sin\left(\varphi_{S} - \frac{R}{2}\delta_{L}\right)$$
$$- \frac{E_{0}}{Z}\sin\varphi_{S} = -\frac{i_{\gamma}}{\sqrt{3}} \qquad (34)$$

$$I\cos\left(\varphi_{1} - \frac{R}{2}\delta_{L}\right) = \frac{V}{Z}\cos\left(\varphi_{S} - \frac{R}{2}\delta_{L}\right)$$
$$- \frac{E_{0}}{Z}\cos\varphi_{S} = -\frac{i_{\delta}}{\sqrt{3}} \qquad (35)$$

Here, V is the operating voltage (V), r_{a} is the resistance of the armature windings (Ω), $Z = \sqrt{r_{a}^{2} + X_{S}^{2}}$, X_{S} is the synchronous reactance (Ω), $\varphi_{S} = \tan^{-1}(X_{S}/r_{a})$, E_{0} is the induced electromotive force (V), φ_{1} is the power factor angle (rad), $R\delta_{L}/2$ is the power angle (rad), I_{γ} is the γ axis current (A), and I_{δ} is the δ axis current (A).

C. Torque equation

The electromotive force is given by equation (36).

$$E_{0} = -\frac{\omega M i_{F}}{\sqrt{2}} \qquad (36)$$

Here, $\omega = 2\pi f$, and $M_{F}I_{F}$ is the flux linkage (Wb). The torque equation for the SPM-type vernier motor is derived from equations (27), (35), and (36) and is shown in equation (37). In addition, the output equation (38) is derived from the torque equation.

$$T = \frac{3}{\omega}\frac{R}{2}\left\{\frac{VE_{0}}{Z}\cos\left(\varphi_{S} - \frac{R}{2}\delta_{L}\right) - \frac{E_{0}^{2}}{Z}\cos\varphi_{S}\right\} \qquad (37)$$

$$P_{O} = T\frac{2\omega}{R} = 3\left\{\frac{VE_{0}}{Z}\cos\left(\varphi_{S} - \frac{R}{2}\delta_{L}\right) - \frac{E_{0}^{2}}{Z}\cos\varphi_{S}\right\} \qquad (38)$$

The output equation (38) for the SPM-type vernier motor has the same form as that for an ordinary synchronous motor, so the output power can be calculated using the power angle $R\delta_{L}/2$ if the induced electromotive force E_{0} and the synchronous reactance X_{S} are obtained.

978-1-4799-2706-7/14 $31.00 © 2014 IEEE

D. Output characteristics calculation method

An output characteristics calculation method that combines the voltage, the torque equations of the SPM-type vernier motor, and simple FEA is proposed. A flow chart of the proposed calculation method is shown in Fig. 5.

In the first step, the induced electromotive force E_0, the synchronous reactance X_S, and the iron loss P_i are calculated using FEA (JMAG-Designer Ver. 12.1). These results are provided in a short amount of time because the 2D FEA calculation volumes are very small. Moreover, these results are obtained easily without considering a complicated harmonic permeance coefficient.

Next, if several power angles are provided, the output characteristics can be calculated quickly using equations (34), (35), and (38)–(40) in the second step. The power factor pf (%) and the efficiency η (%) are calculated using equations (39) and (40), respectively.

$$pf = \cos\varphi = \frac{1}{VI}\left(\frac{P_O}{3} + r_a I^2\right) \qquad (39)$$

$$\eta = \frac{P_O - P_i}{P_O + 3r_a I^2} \qquad (40)$$

Here, I is the current (A), and P_i is the iron loss (W).

IV. EVALUATION OF OUTPUT CHARACTERISTICS CALCULATION METHOD

The evaluation of the proposed output characteristics calculation method is shown in this chapter. A trial SPM-type vernier motor was prepared, and the calculation and measured results of the trial SPM-type vernier motor are compared in order to validate the proposed calculation method.

A. Specifications of trial motor

The specifications and operating conditions of the trial motor are shown in Table 1, and the one-pole section of the trial motor is shown in Fig. 6. This motor is an inner rotor type. An ordinary concentrated-type three-phase armature winding is arranged in the stator, and the permanent magnets that are magnetized in the direction of the radius are arranged at the surface of the rotor. In these operational conditions, a rotating magnetic field is rotated at a speed of 500 min⁻¹, and the rotor speed is stepped down to 100 min⁻¹.

B. Measurement methods for output characteristics

The induced electromotive force and the synchronous impedance of the trial motor are measured in an open-circuit test and a short-circuit test. A measurement block diagram for the output characteristics of the trial motor is shown in Fig. 7. The electrical input power is measured by a power meter, and the mechanical output power is measured by a speed detector and a torque detector when the vernier motor drives a road of DC generator.

TABLE I
SPECIFICATIONS AND OPERATING CONDITIONS OF TRIAL MOTOR

Item	Unit	Value
Number of slots	-	36
Number of field poles	-	60
Number of poles	-	12
Turns per phase	Turn	336
Packing factor in slot	%	49.2
Coil resistance (Cal., 75 deg)	Ω	0.85
Core length	mm	60
Core material	-	50A290
Residual induction of magnet	T	1.44
Coercive force of magnet	kA/m	927
Operating voltage (line)	V	125
Frequency	Hz	50
Synchronous speed	min⁻¹	100

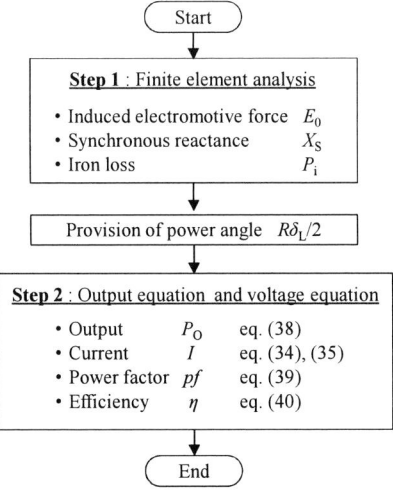

Fig. 5. Flow chart of the proposed analysis method.

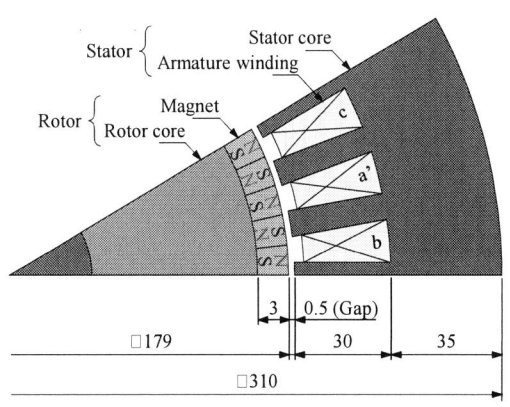

Fig. 6. One-pole section of the trial motor (Units: mm).

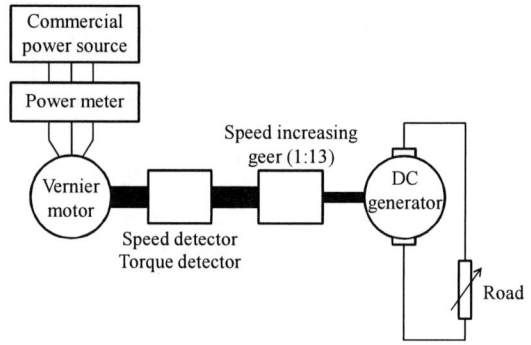

Fig. 7. Output measurement block diagram.

C. Calculation of induced electromotive force and synchronous reactance

At step 1 in the calculation flow, the induced electromotive force, the synchronous reactance, and the iron loss are calculated using FEA. The induced electromotive force E_0 is calculated using 2D FEA, and it is the fundamental wave element in the harmonic analysis for the phase-a voltage waveform. The demagnetized areas of the permanent magnet and the glue layer are also considered in the 2D FEA model. The magnetization model for the permanent magnet is shown in Fig. 8. The arrows indicate the magnetomotive force produced by the permanent magnets. The demagnetized areas have a width of 1/4 of the pole pitch on both sides of the boundaries between the magnets, and the magnetomotive force changes linearly in these areas.

The synchronous reactance X_S is a combination of the reactance x and the coil-end leakage reactance x_{le} and is calculated using equations (41) through (44). The flux linkage ψ without the coil-end leakage flux is calculated using 2D FEA. Since only the coil-end leakage reactance x_{le} cannot be calculated using 2D FEA, it is calculated using a part of Kilgore's equation [8] and is then added to the reactance x. The calculation results for the coil-end leakage reactance are shown in Table II.

$$X_{\mathrm{S}} = x + x_{\mathrm{le}} \tag{41}$$

$$x = \omega L = \omega \frac{\psi}{i} \tag{42}$$

$$x_{\mathrm{le}} = X\lambda_{\mathrm{e}} = X\frac{4}{u}\left(2l_{\mathrm{e2}}+l_{\mathrm{e1}}\right) \tag{43}$$

$$X = 40fum(2p)\left(q\frac{n_{\mathrm{s}}}{c}k_{\mathrm{p}}k_{\mathrm{d}}\right)^{2}\times10^{-8} \tag{44}$$

Here, x is the reactance without the coil-end leakage reactance (Ω), x_{le} is the coil-end leakage reactance (Ω), L is the inductance (H), ψ is the flux linkage without the

coil-end leakage flux (Wb), i is the line current (A), X is the leakage coefficient, λ_e is the leakage permeance of the coil-end (H), $2l_{e2}+l_{e1}$ is the coil-end length (m), u is the core length (m), m is the number of phases, p is the number of pole pairs, q is the number of slots per pole per phase, n_s is the number of coil sides in a slot, c is the number of parallel circuits per phase (=1), k_p is the short-pitch factor (=1), and k_d is the distribution factor (=1).

The calculation results of the 2D FEA are shown in Table III. The calculation results agree well with the measurement results for the induced electromotive force and the synchronous reactance.

D. Calculation of output characteristics

At step 2 in the calculation flow, the output characteristics are calculated by substituting the calculated induced electromotive force and the calculated synchronous reactance for the output calculation equations. A comparison of the calculated and measured output characteristics of the trial motor are shown in Fig. 9. The broken lines indicate the results calculated using the proposed analysis method, and the markers show the test results for the trial motor. The calculated results agree very well with the measured values.

V. CONCLUSIONS

In this paper, the voltage and torque equations for the SPM-type vernier motor are derived, and an analysis method is proposed that combines these equations with

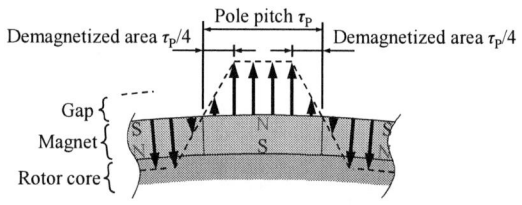

Fig. 8. Magnetization model for permanent magnet.

TABLE II
CALCULATION RESULTS FOR COIL-END LEAKAGE REACTANCE

Item		Unit	Value
Size of coil-end	l_{e1}	mm	55.8
	l_{e2}	mm	18.7
Coil-end leakage reactance x_{le}		Ω	0.84

TABLE III
CALCULATION RESULTS OF 2D FINITE ELEMENT ANALYSIS

Item	Unit	Measurement	Calculation
Induced electromotive force E_0	V	45.4	45.2
Synchronous reactance X_s	Ω	5.84	5.31*
Iron loss P_i	W	–	8.9

* Coil-end leakage reactance is added

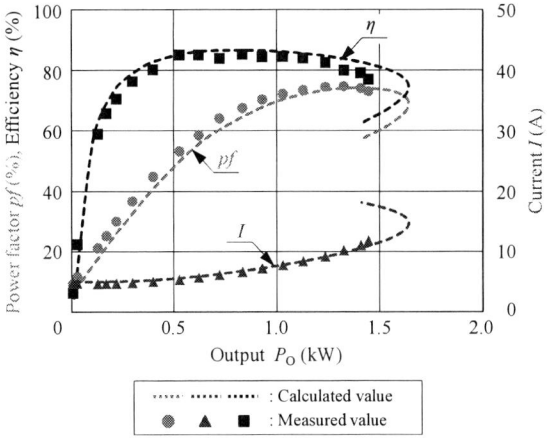

Fig. 9. Comparison of calculated and measured output characteristics.

simple FEA. The results calculated using the proposed analysis method agree very well with the test results for the trial motor. Thus, it has been shown that the proposed method is very useful, and its accuracy is excellent.

Moreover, the effectiveness of the motor design [6] has been confirmed.

In future work, the output power and possible methods for improving the output characteristics will be investigated using this analysis method.

REFERENCES

[1] A. Ishizaki, Y. Shibata, K. Watanabe, and K. Saitoh, "Low-speed High-torque Drive System Applying Vernier Motor Torque," T. IEE Japan, Vol. 111-D, No. 9, pp. 785-793, September, 1991.

[2] A. Ishizaki, T. Tanaka, K. Takasaki, S. Nishikata, and A. Katagiri, "Theory and Torque Characteristics of PM Vernier Motor," T. IEE Japan, Vol. 113-D, No. 10, pp. 1192-1199, October, 1993.

[3] A. Toba, T. Watanabe, Y. Koganei, and H. Ohsawa, "Design and Experimental Evaluation of 5 kW-Surface Permanent Magnet Vernier Machines," T. IEE Japan, Vol. 122-D, No. 2, pp. 162-168, February, 2002.

[4] Y. Anazawa, K. Tajima, A. Kaga, Y. Ito, and H. Isozaki, "Analysis of Vernier Motors Based on the Unified Theory," T. IEE Japan, Vol. 113-D, No. 11, pp. 1317-1323, November, 1993.

[5] IEE Japan, "Design of electric machine," 2nd edition, pp. 331-333 (1982) (in Japanese).

[6] H. Kakihata, Y. Kataoka, M. Takayama, Y. Matshshima, Y. Anazawa, "Design of Surface Permanent Magnet-type Vernier Motor," Journal of International Conference on Electrical Machines and Systems, Vol. 2, No. 2, pp. 127-133, 2013

Topology Optimization for Skew of SPMSM by using Multi-Step Parallel GA

Wataru Kitagawa and Takaharu Takeshita
Nagoya Institute of Technology
Graduate School of Engineering
Gokiso Showa Nagoya , Japan
Kitagawa.wataru@nitech.ac.jp

Abstract— In this paper, it was proposed to the optimization of skew of SPMSM with 3-dimensional finite element method (3D-FEM). The optimization technique was used multi-step parallel Genetic Algorithm (GA) and the allocation of material assignment method. As a result, in useful computation time, It was able to obtain an optimal skew special shape. And, it was able to reduce the cogging torque of more than 97%.

Keywords— IPMSM, 3D-FEM, Optimization, Skew

I. INTRODUCTION

Skew is one of the effective technique for reducing cogging torque. However, it is not able to correctly analyze by the 2-dimensional finite element method, because the skew has 3-dimensional structure. Furthermore, optimization of 3-dimensional model was hard to practical due to limit of computational time and coding. In this paper, it is proposed to the optimization of skew of SPMSM. The optimization technique is used multi-step parallel GA and the allocation of material assignment method.

II. PROPOSED METHOD

The proposed method process is shown in Fig.1. First, material is assigned to the prepared mesh [1]. Analysis region is divided in the grid pattern. In this method is divided by 10th for permanent magnet region of rotor surface to axial direction, and divided by 16th to radial direction. Thus, it have 2 dimensional matrix of 10 - 16. Because the position of the grid points are not changed, the design variable is not necessary remesh the model. And, for a combination optimization problem 0, 1 bit string 160, it is possible to use GA [2-4]. It is possible to change the shape without remeshing, the proposed method is suitable for shape optimization of three-dimensional model. However, it must be finely divided search area in order to obtain a more detailed solution. To solve this problem, using the multi-step method. It is shown multi-step method in Fig.2. First step is to divide the rough searching region. Then, according to the results obtained, Second step is subdivided.

Moreover, to reduce the calculation time, it is used the parallel computing for GA. It is shown parallel GA flowchart in Fig.3. GA is a probabilistic search algorithm. Therefore, so as not to evaluate the same individual, it is checked whether the evaluated individual. In addition, considering the parallel efficiency of the slave, the master distributes the evaluation individuals to slave [5].

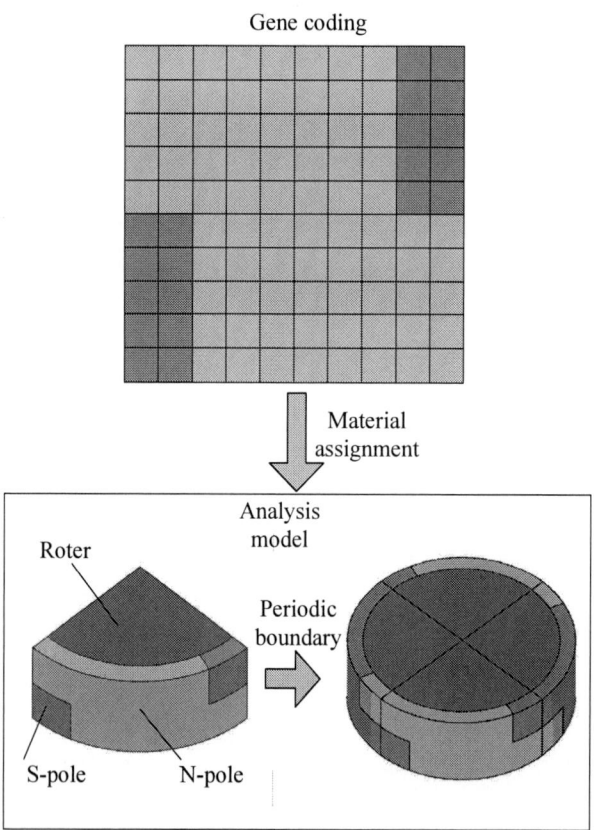

Fig. 1. Material allocation method

The 2014 International Power Electronics Conference

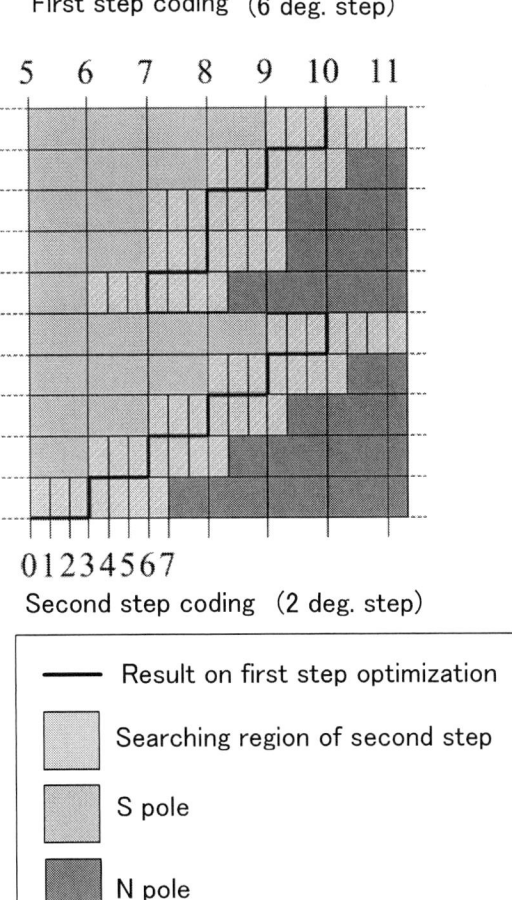

First step coding (6 deg. step)

5 6 7 8 9 10 11

0 1 2 3 4 5 6 7

Second step coding (2 deg. step)

— Result on first step optimization

Searching region of second step

S pole

N pole

Fig. 2. Multi-step method

The "Evaluation check" block in Fig.3 is to prevent re-evaluation. In the population, there are some individuals that are already evaluated in the previous generations because of elitist strategy or result of genetic operations. In order to shortening computing time for main-process, these individuals are skipped the evaluation block.

The "Self mutation" block is one of the genetic operations. In this GA, chromosome coded decimal number can take a value 0-9. Therefore, the initial convergence and falling into local solution is easier to occur than SGA (Simple GA). To maintain the diversity of the population, "Self mutation" is added as new genetic operation. As shown Fig.4, in the self mutation, 1-chromosome is selected at random and calculated the value +1 or -1. This operation uses the characteristic of integer such as decimal number. Its behavior is similar to the local search rather than mutation.

GA has to evaluate and compare two or more mutually independent individuals. Therefore, the "Evaluation" block can be parallelized by distribute individuals to two or more servers, and the computing time is expected to be shortened.

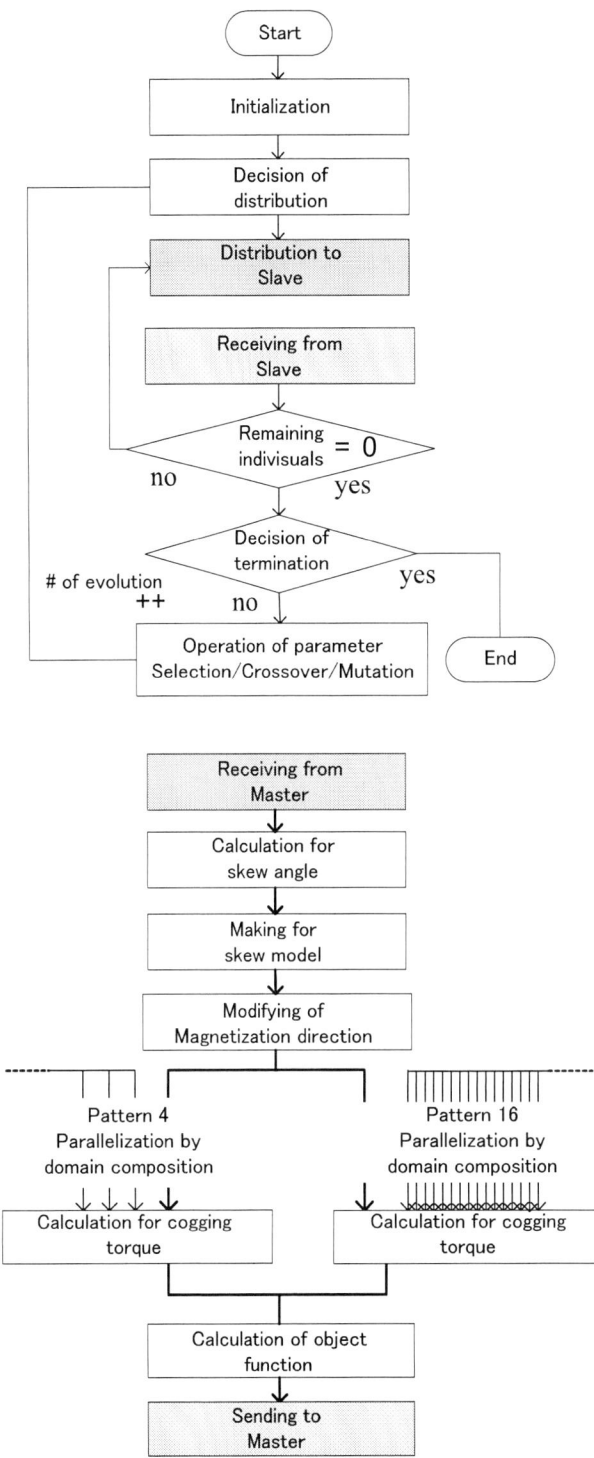

Fig. 3. Flowchart of parallel GA

We use the master/slave type GA for parallelization. As shown Fig.4, there are two scripts: master and slave.

The master performs genetic operations and controls the slave as task manager. The slave performs the evaluation of individuals by the 3-D FEM in several remote servers. The master sends the genetic information

978-1-4799-2706-7/14 $31.00 © 2014 IEEE   3810

of the individual to each slave, and the slave replies the fitness of the individual to the master. Master/slave type GA is simple algorithm, but it is effective method when the computing time becomes long in order to adopt the 3-D FEM for the evaluation.

However, there is a problem in the master/slave type GA when the number of individuals evaluated in a generation is less than the number of slaves. Some slaves don't work in the parallel processing because the evaluation of 1-individual can be performed by only 1-slave. Therefore, it is necessary for effective parallel processing that to be able to share the evaluation of 1-individual with some slaves. If the evaluate target is not transient such as cogging torque or steady torque, the evaluation target can divide in the time domain. In Fig.5 for example, slave#1 need to evaluate only the range of T_1. Thus, the computing time for evaluation of 1-individual can be expected to be shortened to 1/n by evaluating with n-slaves at the same time.

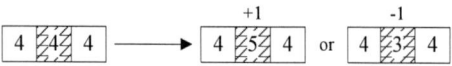

Fig. 3. An example of self mutation

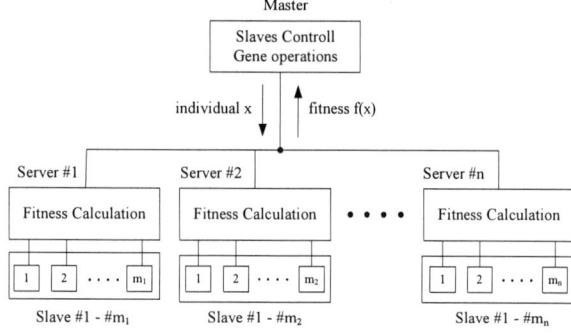

Fig. 4. Parallel processing with master/slave type GA

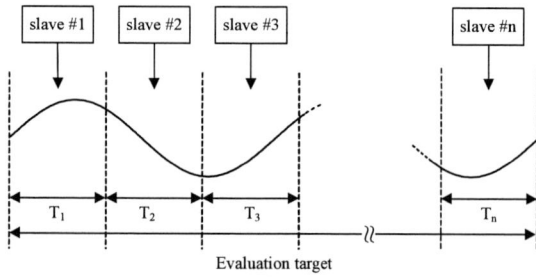

Fig. 5. Parallel evaluation by using several slaves

III. PARAMETER DESIGN

The objective function is used cogging torque which is obtained by 3 dimensional analysis and magnetic flux density of the gap [6-7]. Cogging torque reduction effect can be expected when it is implemented large skew angle in the skew. However, there is a problem flux linkage that the number decreases. Therefore, it is set as a constraint condition flux density in the gap.

$$maximize\ F_i = \frac{100}{W_i}$$

$$W_i = W_C \cdot RATE_C + W_B \cdot (100 - RATE_B) \quad (1)$$

Where, W_C and W_B are weighting function. $RATE_C$ and $RATE_B$ are following equation.

$$RATE_C = \frac{T_{ci}}{T_{c0}}, \qquad RATE_B = \frac{B_i}{B_0} \quad (2)$$

Where, T_{ci} is peak to peak value of cogging torque of ith individual, T_{c0} is peak to peak value of cogging torque of initial model, B_i is average value of magnetic flux density of gap of ith individual and B_0 is average value of magnetic flux density of gap of initial model. Equation (1), to explore the individual while maintaining the flux density in the gap in the initial shape, the cogging torque is reduced. Relationship of this trade-off can be adjusted by a weighting function, it is set as $W_c = 1$, $W_B = 3$ in this case.

IV. ANALYSIS MODEL

The analysis model was using a 3-dimensional model of SPM 3-phase 4-pole 24 slot (Surface Permanent Magnetic) motor. SPM motor has a structure in which pasted a permanent magnet in the rotor surface by Fig.6 (a). The optimization of the 3-dimensional model, was not much practical that from coding limit and the length of the computation time. In this paper, is subjected to a skew in the permanent magnet, to determine such an arrangement to minimize the cogging torque. The motor has a periodic structure. Therefore, it can be analyze as 1/8 model (radial direction 1/4 and axial direction 1/2). Mesh model is shown in Fig.6 (b). Analysis model parameter is shown in Table I.

The 2014 International Power Electronics Conference

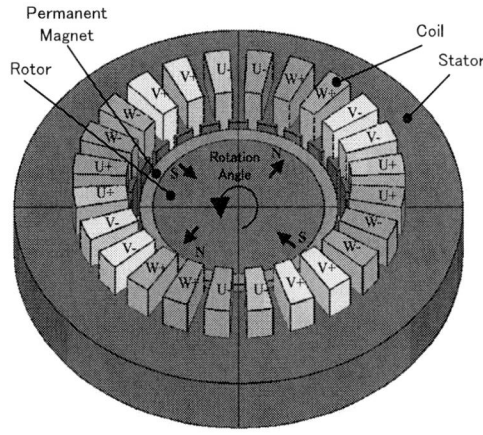

(a) Analysis model (SPMSM)

(b) Mesh of analysis model
Fig. 6. Analysis model

TABLE I
PARAMETER OF ANALYSIS MODEL

Number of element	133680	Radias of stator [mm]	50
Number of node	25058	Radias of rotor [mm]	22
Number of edge	164173	Gap [mm]	2
Unknown value	149556	Thickness [mm]	30
*Computing time [sec/step]	30.1	Switchband wound	Distributed winding
Magnetization [T]	1.0	Turn of coil [turn/phase]	140
Magnetization direction	Parallel	Frequency of power sorce [Hz]	50
Core material (Rotor and Stator)	M-19 (50H250)	Current (rms) [A]	3.0

*CPU: Intel Core-i7 (2.93GHz), Memory: 8GB, Compiler: Intel Compiler ver11.1

V. ANALYSIS RESULTS

The results of optimization shape and cogging torque of 1st step GA is shown in Fig.6 (a) and (b). As the results, skew shape has been created the 2-stage rotor in the vicinity of 40-60 degrees. Cogging torque was reduced by 94%. And, Steady torque waveform and magnetic flux density of gap of 1st step are shown in Fig.7 (a) and (b). The results of optimization shape and cogging torque of 2nd step GA is shown in Fig.8 (a) and (b). There is a difference in the vertical skew angle of the 2nd step than the 1st step model. By skew angle spread, the cogging torque is further reduced. Cogging torque is reduced by about 97% compare to initial model. And, Steady torque waveform and magnetic flux density of gap of 1st step are shown in Fig.9 (a) and (b). Computation time is about 11 hours in each step.

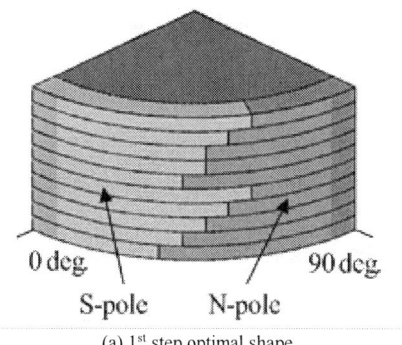

(a) 1st step optimal shape

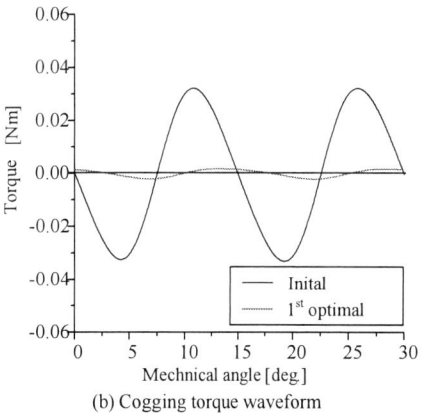

(b) Cogging torque waveform
Fig. 6. 1st step optimization

The 2014 International Power Electronics Conference

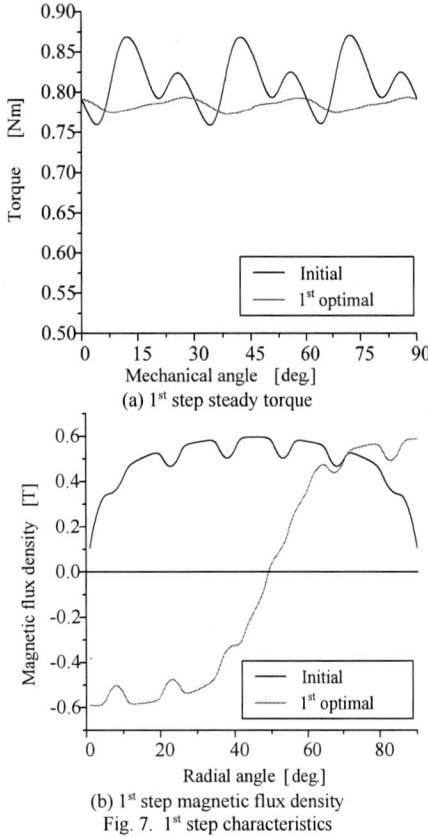

(a) 1st step steady torque

(b) 1st step magnetic flux density
Fig. 7. 1st step characteristics

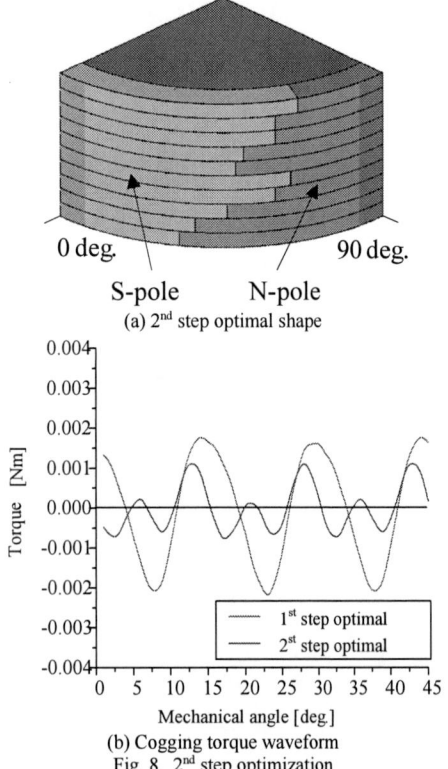

(a) 2nd step optimal shape

(b) Cogging torque waveform
Fig. 8. 2nd step optimization

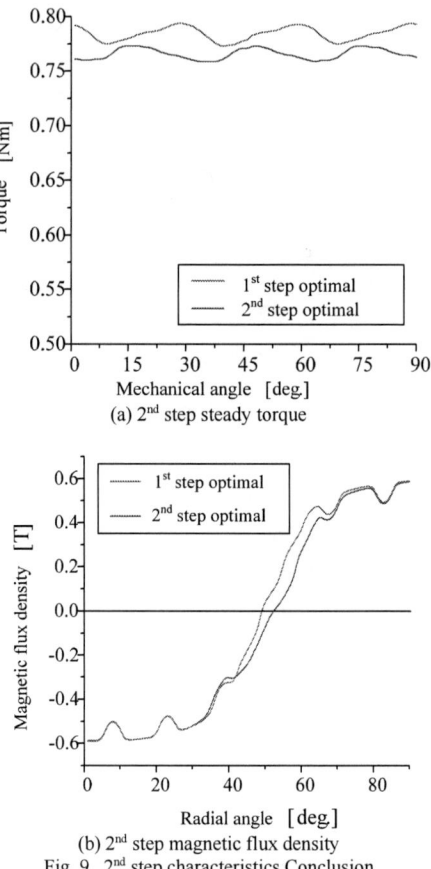

(a) 2nd step steady torque

(b) 2nd step magnetic flux density
Fig. 9. 2nd step characteristics Conclusion

VI. CONCLUSION

In this paper, it was proposed to the optimization of skew of SPMSM. The optimization technique is used multi-step parallel GA and the allocation of material assignment method. As the results, skew shape had been created the 2-stage rotor in the vicinity of 40-60 degrees. Cogging torque was reduced by 94%. And, there was a difference in the vertical skew angle of the 2st step than the 1st step model. By skew angle spread, the cogging torque was further reduced. Cogging torque was reduced by about 97% compare to initial model. Computation time was about 11 hours in each step.

REFERENCES

[1] N. Takahashi, S. Nakazaki and D. Miyagi : Examination of Optimal Design Method of Electromagnetic Shield Using ON/OFF Method, *IEEE Trans.on Magnetics*, Vol.45, No.3, pp.1546-1549, 2009

[2] W. Kitagawa, Y. Kimura and T. Takeshita : A Study on Embedding Genetic Algorithm to Three-Dimensional Finite Element Method by Using Shell Script, *IEEJ Trans. on IA*, Vol.132, No.2, pp.227-232, 2012

[3] D.J. Sim, D.H. Cho, J.S. Chun, H.K. Jung, and T.K. Chung : Efficiency Optimization of Interior Permanent Magnet Synchronous Motor Using Genetic Algorithm, *IEEE Trans. on Magnetics*, Vol.33, No.2, pp.1880- 1883, 2003

[4] Gyu Won Cho, Seok Hyeon Woo, Seung Hun Ji, Ki Bong Jang, Gyu Tak Kim : Optimization of Rotor Shape for Constant Torque Characteristic of IPM Motor, *proceedings of ICEMS 2011*, pp.1-4, 2011

[5] T. Nakano, K. Kawase, T. Yamaguchi, M. Nakamura, N. Nishikawa and H. Uehara : Parallel Computing of Magnetic Field for Rotating Machines Excited on the Earth Simulator, *IEEE Trans. on Magnetics*, vol.46, No.8, pp.3273-3276, 2010

[6] Y. Kawase and S. Ito : New Practical Analysis of Electrical and Electronic Apparatus by 3-D finite element method, *Morikita Publishing Co.*, 1997, (in Japanese)

[7] M. Mori, W. Kitagawa and T. Takeshita : Design of Two-Degree-of-Freedom Electromagnetic Actuator using PMSM and LSM, *proceedings of APSEM*, pp.126-131, 2012

Loss Minimization Design
Using Magnetic Equivalent Circuit
for a Permanent Magnet Synchronous Motor

Daisuke Sato
Department of Electrical Engineering
Nagaoka University of Technology
Nagaoka, Niigata, Japan
dsato@stn.nagaokaut.ac.jp

Jun-ichi Itoh
Department of Electrical Engineering
Nagaoka University of Technology
Nagaoka, Niigata, Japan
itoh@vos.nagaokaut.ac.jp

Abstract— This paper proposes a simple magnetic equivalent circuit using permeance method in order to design rough configurations of the lowest loss motor. First, the iron loss is calculated by the proposed method and compared with the analysis result by finite element method (FEM). As a result, the error of the calculation results between the proposed method and the FEM is 2.9%. Next, the motor losses are calculated by the proposed method when the several specifications are changed. As a result, these results agree in principle with that of the FEM. Therefore, the validity of the proposed method is confirmed. In addition, the method of the motor design is considered in terms of the loss.

Keywords— *Permanent magnet synchronous motor, Loss minimization design, Permeance method, Finite element method*

I. INTRODUCTION

Recently, the permanent magnet synchronous motor (PMSM) is actively researched in order to achieve the high efficiency in the motor drive system. In addition, the PMSM is not only high efficiency but also smaller than the induction motor. Therefore, the PMSM is applied to the electric vehicle and the home electric appliance [1-4].

The analysis of the loss, the torque and the back electromotive force are important in terms of further improvement of the motor. Generally, the characteristics of the motor are analyzed by the finite element method (FEM) [5-7]. However, the FEM spends much analysis time. Moreover, the development of the motor model is required as often as modifying the configuration of the motor. Hence, the FEM is not suitable for the calculation of the optimized solution such as the maximum torque density or the lowest loss point because much time of trial and error is needed to find the optimization point.

On the other hand, the permeance method is used in order to design the motor simply in the past. In the permeance method, a magnetic circuit is replaced with an electric circuit. In this method, the magnetic flux and the magnetomotive force can be expressed as the current and the voltage. Therefore, this method is suitable for the optimized solution because of shorter calculation time, although the calculation accuracy of the permeance

method is less than that of the FEM. This is because the magnetic circuit of the motor is primarily determined as the distributed constant circuit.

The permeance network method has been proposed as one of the motor characteristic analysis using the permeance method [8-11]. This method expresses the stator, the rotor and the air gap on a magnetic circuit. The torque and the back electromotive force can be calculated by calculating the magnetic flux in the air gap from the magnetic circuit. However, the magnetic resistance of the air gap depends on a position of the magnetic pole. Therefore, the construction of the magnetic circuit is complex. On the other hand, the method, which calculates eddy current loss in the permanent magnet, has been proposed [12]. In this method, the eddy current loss can be accurately calculated in comparison with the FEM. However, the calculation process is complicated. In addition, the equation is changed when the shape of the motor is changed.

Generally, first, rough configurations such as size, pole number, teeth width are decided in the design of IPMSM. After that, the details such as flux barrier are decided. Therefore, it is considered that the simple loss calculation is enough in the phase of rough design.

In this paper, the simple magnetic equivalent circuit using permeance method is proposed in order to design the rough configurations of the lowest loss motor. The characteristic of the equivalent circuit is separation between rotor and stator. Therefore, the complex calculations are not needed because the magnetic resistance of air gap can become constant. In addition, if the shape of motor is changed, the magnetic resistance and the magnetomotive force are changed only. Thus, the circuit is not changed.

First, this paper presents the calculation method of flux density in stator core. Next, the loss of IPMSM is calculated and the result is compared with the analysis result of FEM and the measurement result. Finally, the lowest loss motor is designed by the loss calculation in the changing parameter of IPMSM

978-1-4799-2706-7/14 $31.00 © 2014 IEEE

II. MAGNETIC EQUIVALENT CIRCUIT OF PMSM

Generally, in the permeance method, the magnetic circuit is replaced with an equivalent electric direct current circuit. However, it takes the time variation of the magnetic flux density to calculate the iron loss. Therefore, the conventional permeance method is not applied directly. In addition, it is difficult to express as one circuit because the rotor is rolling and the stator is not rolling. Therefore, the magnetic flux in the air gap is calculated by the equivalent circuit of the rotor. Next, the time variation of the magnetic flux density in the teeth and yoke are calculated by the equivalent circuit of the stator. It is noted that the input of the stator circuit is the magnetic flux in the air gap.

Fig. 1 shows the model of the IPMSM and Table I shows the parameters of the IPMSM. The motor is the concentrated winding IPMSM. In addition, the permanent magnets are magnetized in parallel and mounted in the motor. Furthermore, the motor has periodicity every 60 degrees.

Fig. 2 shows the equivalent magnetic circuit of the rotor. It is noted that the equivalent circuit comprises of the half of one pole region. This is because the region of 30 degrees is able to be regarded as symmetry. The magnetic resistance R is given by (1).

$$R = \frac{l}{\mu S} \qquad (1)$$

It is noted that μ is the magnetic permeability, S is the cross section of the material, and l is the length of the material. The magnetic resistance of the rotor core is not taken into consideration because it is sufficiently smaller than the magnetic resistance of the magnet. In addition, the magnetomotive force of the magnet F_m is expressed as

$$F_m = H_e l_m \qquad (2)$$

where H_e is the coercive force of the magnet and l_m is the length of the magnet. Further, the magnetic saturation is expressed by the magnetic resistance R_{sat} which varies by the magnetomotive force because the magnetic saturation occurs in the area between the magnet and the outer of the rotor. Thus, the magnetic flux in the air gap is calculated.

Fig. 3 shows the equivalent magnetic circuit of the stator. This circuit comprises the magnetic resistance of the air gap, the teeth, the yoke and some magnetomotive forces. The magnetomotive force by the flowing current to the armature winding F_i is expressed as

$$F_i = NI_m \sin(\omega t + \theta) \qquad (3)$$

where N is the number of winding turns per the teeth, $I_m \sin(\omega t + \theta)$ is the instantaneous current. In addition, the

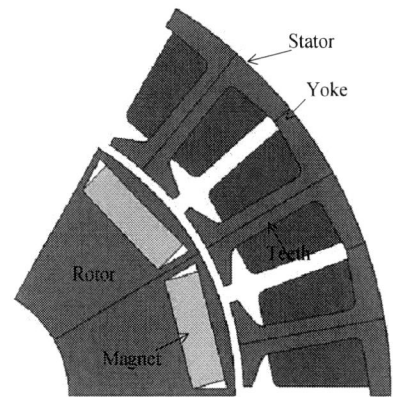

Fig. 1 Concentrated winding IPMSM. The magnets are magnetized in parallel. The motor has periodicity every 60 degrees.

TABLE I	PARAMETERS OF IPMSM.
Core	35H300 / Nippon Steel
Magnet	NMX-41SH / Hitachi Steel
Number of poles	12
Number of slots	18
Coil turns per teeth	11
Outer diameter of stator	100 mm
Inner diameter of stator	66 mm
Outer diameter of rotor	64 mm
Inner diameter of rotor	25 mm
Air gap length	1 mm
Iron stack length	50 mm
Magnet dimensions	$12.5\,\text{mm} \times 25\,\text{mm} \times 4\,\text{mm}$

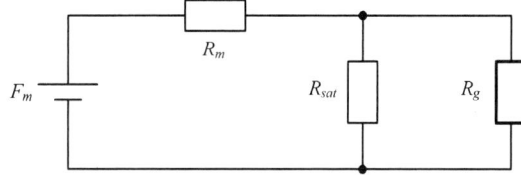

Fig. 2 Magnetic equivalent circuit of rotor. The circuit consists of the half of one pole region. In this circuit, the flux in air gap is calculated.

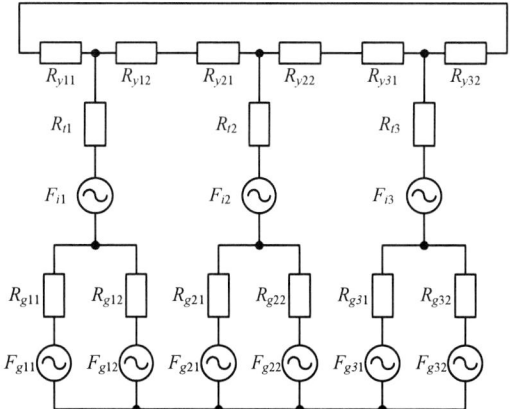

Fig. 3 Magnetic equivalent circuit of stator. The AC magnetomotive force F_g and F_i are set. The magnetic resistance is considered the magnetic saturation. In this circuit, the flux densities in teeth and yoke are calculated.

AC magnetomotive force F_g is set as the element

generated the magnetic flux in the air gap. As a result, the effect of the rolling rotor can be considered.

Fig. 4 shows the waveform of magnetomotive force F_g. it is assumed that F_g is trapezoidal waveform owing to the layout of the magnet. The waveform of F_g is determined from the proportion of the magnet width and the teeth width. The phase of the magnetomotive force ϕ_g is determined from the position of the teeth and expressed as

$$\phi_g = \frac{2\pi}{\theta_p}\theta_t \qquad (4)$$

where θ_t is the position of the teeth and θ_p is the angle of one period.

Further, it is necessary for the magnetic resistance of the teeth and the yoke to consider the magnetic saturation. Therefore, the magnetic resistance depends on the magnetomotive force in the magnetic saturation region. Accordingly, the time variation of the magnetic flux density in the teeth and the yoke are calculated. Moreover, the iron loss in the stator core can be calculated by the magnetic flux density of the teeth and the yoke.

III. CALCULATION OF MAGNETIC FLUX DENSITY AND IRON LOSS

The magnetic flux density and the iron loss are calculated by the permeance method. Furthermore, the calculation results by applying the permeance method are compared with that of the FEM. Table II lists the calculation conditions of the permeance method. The armature current is sinusoidal waveform.

Fig. 5 shows the FEM model of the concentrated winding IPMSM. The magnetic flux density and the iron loss are analyzed by two-dimensional FEM using JMAG Designer (JSOL). It is noted that the number of elements is 6,525 and the time interval per 1 step is 2.71 μs (1024 steps).

Fig. 6 shows the waveform of the magnetic flux density calculated by the permeance method and the FEM. Accordingly, the waveform of the teeth by the permeance method is saturated at 1.75 T. Moreover, it is similar to the waveform of the teeth by the FEM. However, the peak of the magnetic flux density in the yoke by the permeance method is drastically changed. Therefore, the waveform of the yoke by the permeance method is different from the waveform by the FEM.

Fig. 7 shows the harmonic analysis result of the magnetic flux density. From Fig. 7, the fundamental component of the permeance method almost agrees with that of the FEM. However, the error occurs at the odd order harmonic component by the magnetomotive force. Especially, the 3rd order and 9th order components by the permeance method do not occur because the equivalent circuit of the stator is the balanced three-phase circuit. Next, the iron loss in the stator is calculated from the magnetic flux density and the iron loss curve of the

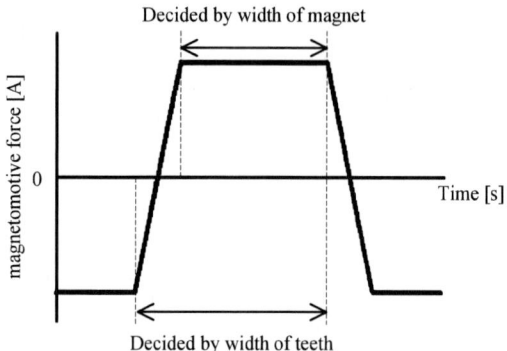

Fig. 4 Waveform of magnetomotive force. The waveform is determined from the proportion of the magnet width and the teeth width.

TABLE II CALCULATION CONDITIONS.

Armature current		18 A$_{rms}$
Motor speed		3600 r/min
Electrical frequency		360 Hz
Magnetic resistance	Teeth	7.0×10^4 A/Wb
	Yoke	5.9×10^4 A/Wb
	Gap	3.4×10^6 A/Wb

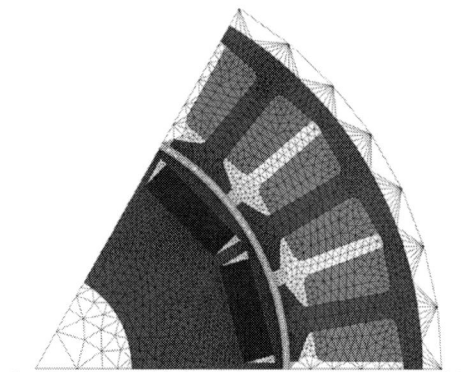

Fig. 5 FEM model of the concentrated winding IPMSM. The number of elements is 6,525. Time interval per 1 step is 2.71 μs (1024 steps).

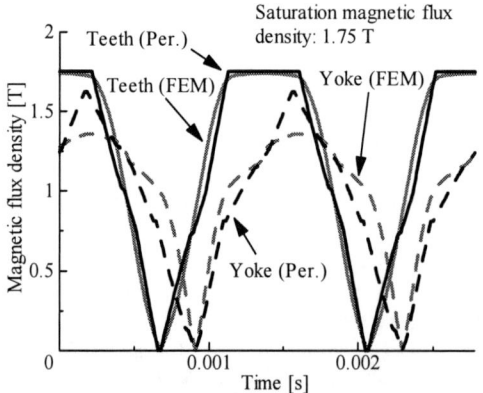

Fig. 6 Waveform of magnetic flux density by permeance method and FEM. The waveform of the teeth by the permeance method is similar to that by the FEM

material.

Fig. 8 shows the iron loss curve of the material

(35H300 / Nippon Steel). As shown in Fig. 8, the iron losses at the required frequency are not shown. For this reason, the iron losses at the required frequency are calculated by the linear interpolation. As the result, it is confirmed that the iron loss in the stator is 30.4 W and the calculated iron loss by the FEM is 31.3 W. The error between the permeance method and the FEM is 2.9%.

IV. COMPARISON WITH IRON LOSS MEASUREMENT RESULT

The motor loss is measured and compared with calculation result in order to confirm the validity of the loss calculation by the proposed method. The IPMSM is typically driven by the inverter. If the carrier frequency is low, the motor loss by harmonic component of the output voltage should be considered. However, the loss can be reduced by the high carrier frequency. In addition, the 3-level inverter can reduce not only the harmonic component of the output voltage but also the switching loss compared with the 2-level inverter. Therefore, in this paper, the motor loss by the carrier harmonics is ignored on the assumption that the IPMSM is driven by the 3-level inverter.

Fig. 9 shows the construction of IPMSM loss measurement system. The parameters of the test motor are shown Fig.1 and Table 1. The 3-level T-type neutral point clamped inverter is applied to IPMSM loss measurement system. In addition, the input power of the motor is measured by the power meter (WT1800/YOKOGAWA) and the shaft power is calculated based on the measurement result of the output torque by the torque meter (TH-2105/Ono sokki).

Fig. 10 shows the measurement result of the motor loss at 3600 r/min. The motor loss is reduced due to the increasing carrier frequency. The reason is that the harmonic component in the output voltage of the inverter is reduced. However, the motor loss is almost unchanged over 7520 Hz. Therefore, the output voltage is equivalent to sinusoidal voltage over 7520 Hz. In this section, the motor loss at 7520 Hz is compared with the calculation results of the proposed method and the FEM because the carrier frequency is ignored.

Next, the iron loss P_{Fe} is separated from the motor loss P_{loss}. The copper loss P_{Cu} and the mechanical loss P_m are expressed by (5) and (6), respectively. The mechanical loss is calculated by the empiric formula of windage loss because it is assumed that the friction loss of bearing is negligible small. Therefore, the iron loss is expressed by (7).

$$P_{Cu} = 3R_a I^2 \qquad (5)$$

$$P_m = 8D(l + 0.15)v_a^2 \qquad (6)$$

$$P_{Fe} = P_{loss} - P_{Cu} - P_m \qquad (7)$$

where, R_a is wire resistance per one-phase, I is root mean

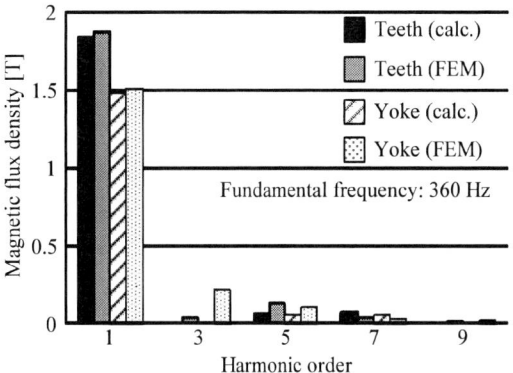

Fig. 7 Harmonic component of magnetic flux density. The fundamental component by applying the permeance method almost agrees with that of the FEM.

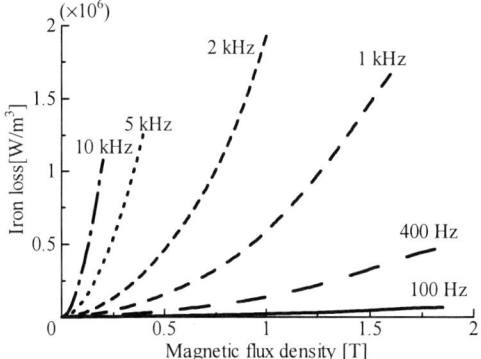

Fig. 8 Iron loss curve of 35H300 (Nippon Steel). The iron losses at the required frequency are calculated by the linear interpolation.

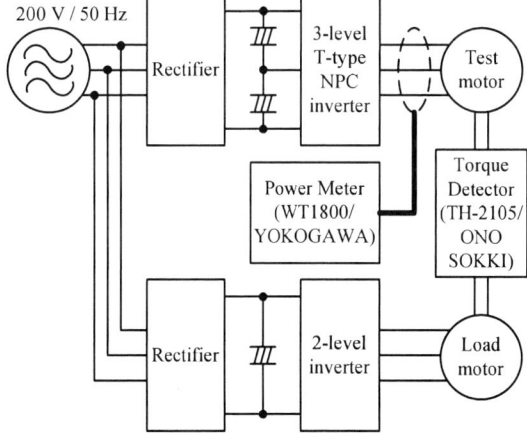

Fig. 9 IPMSM loss measurement system. The test motor is driven by the 3-level T-type neutral point clamped inverter. The input power of the motor is measured by the power meter. The shaft power is calculated based on the measurement result of the output torque.

square value of line current, D is outer diameter of rotor, l is length of rotor and v_a is peripheral speed of rotor.

Fig. 11 shows the measurement and calculation results of the iron loss at 3600 r/min and 2.3 Nm. The error between the calculation result by proposed method and the analysis result by FEM is caused by the loss in the rotor and the magnets. In addition, the error between the analysis result and the measurement result is less than 1%. Therefore, it is assumed that the error is the stray load loss. Thus, the validity of the proposed method is confirmed.

V. MOTOR DESIGN OF LOWEST LOSS

The motor loss is calculated by the proposed method when the mechanical parameter is changed. Moreover, the lowest loss motor is designed based on the calculation results.

Fig. 12 shows the flowchart of the motor design of the lowest loss. 1) Topology of magnetic equivalent circuit is decided based on the configuration of IPMSM. 2) Specifications of the motor are decided. 3) Magnetic resistance and Magnetomotive force in the rotor circuit are calculated. 4) Flux in the air gap is calculated by the equivalent circuit of the rotor. 5) Magnetomotive force is set to the equivalent circuit of the stator from the flux in the air gap. 6) The flux density in stator core is calculated. 7) Iron loss of the stator core is calculated by the flux density and iron loss curve of the core. In addition, copper loss is calculated by the wire resistance and the line current. 8) After, the losses are calculated in several specifications, the lowest loss motor is selected. In this section, the cases of changing pole/slot number, lengths of axial/radial direction and length of teeth are considered.

A. Changing Pole and Slot Number

If the proposed magnetic equivalent circuit is applied to the model changing pole number, the frequency of the magnetomotive force has to be changed only. Therefore, the reconstitution of the circuit is not required. In this section, it is assumed that the ratio between pole number and slot number is constant.

Fig. 13 shows the analysis models of 8 poles 12 slots motor and 16 poles and 24 slots motor. The angles of one period are 90 and 45 degrees, respectively. In addition, the 12 poles 16 slots motor is also calculated and shown in Fig.1. The outer diameter of the rotor and the stator, the lengths of the air gap and the axial direction, the turn number of the coil and the volume of the magnets are constant. The area of slots is changed by the changing slot number. Therefore, the space factor of the coil is constant (50%) and the area of wire (i.e. the resistance of wire) is changed. Moreover, the speed is 3600 r/min. In terms of the output torque, the amplitude of current is adjusted so as to become 3.9 Nm in the FEM. In addition, the amplitude of current in the proposed method is decided by the flux in the air gap.

Fig. 14 shows the motor loss by the changing pole number and slot number based on the constant volume. The motor loss is standardized by the volume. Therefore, if it is assumed that the current density and the magnetic flux density are constant, the calculation result can also be applied to the motor of the different volume. The errors between the proposed method and the FEM are less than 10% in the all models. In addition, the trend of changing loss in the proposed method agrees with that in

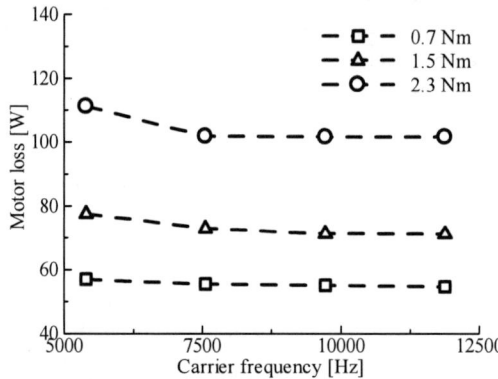

Fig. 10 Measurement result of IPMSM loss at 3600 r/min. The loss is almost unchanged over 7520 Hz because the harmonic component in the output voltage of the inverter is low.

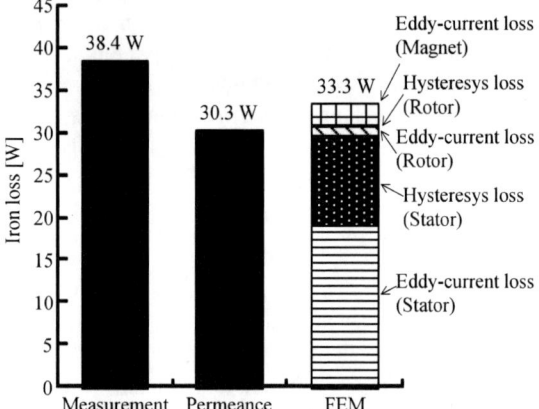

Fig. 11 Measurement result and calculation result by permeance method and FEM of iron loss (3600 r/min, 2.3 Nm). The error between the calculation result by proposed method and the analysis result by FEM is caused by the loss in the rotor and the magnets.

the FEM. Furthermore, in this case, the loss of 12 poles 16 slots motor is the lowest.

The loss of the 12 poles motor is smaller than that of the 8 poles motor. The reason is that the current can be become small under the same output torque by increasing pole number. However, the resistance of wire increases because the area of slot becomes small. Therefore, the loss of the 16 poles motor is larger than that of the 12 poles motor. In addition, the width of the teeth can become large by decreasing pole number and slot number. Therefore, the iron loss decreases by the decreasing flux density. On the other hand, if the pole number and slot number are increased, the iron loss increases because the stator core is saturated magnetically.

B. Changing Lengths of Axial and Radial Direction

Fig. 15 shows the schematic of the changing radial direction length and axial direction length. The calculation conditions are as follows; the motor shown in Fig. 2 is used. The volume of the motor is constant and the axial direction length D and the radius r are changed. The air gap length is set to 1 mm by adjusting the size of the rotor. The amplitude of the armature current is

The 2014 International Power Electronics Conference

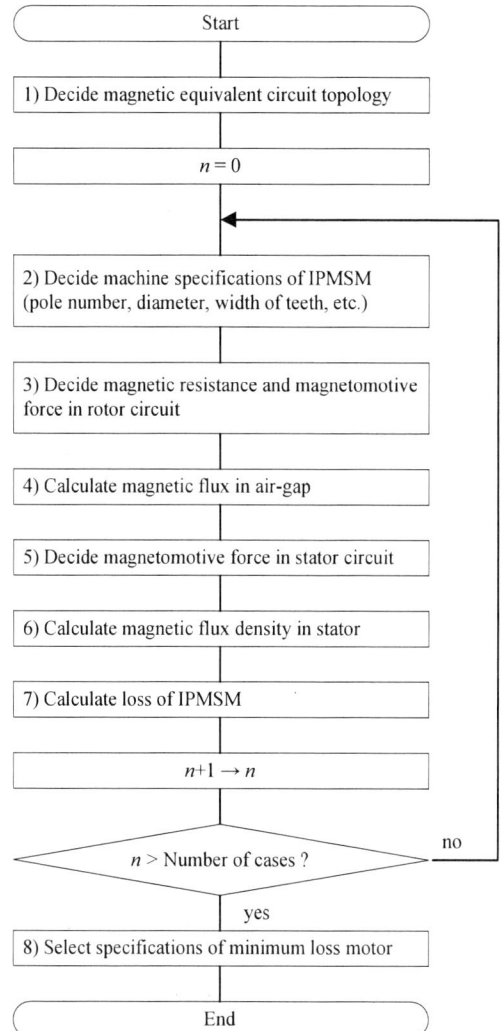

Fig. 12 Flowchart of the motor design of the lowest loss. The lowest loss motor can be designed by the calculating the losses in several specifications.

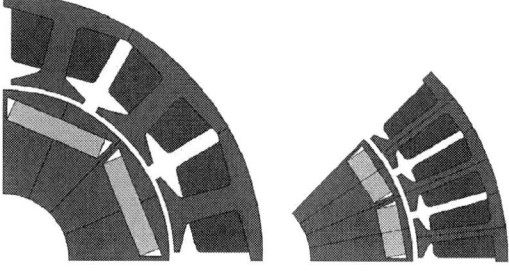

(a) 8 poles 12 slots   (b) 16 poles 24 slots

Fig. 13 IPMSM models of changing pole number. The angles of one period are 90 and 45 degrees, respectively. The outer diameter of the rotor and the stator, the lengths of the air gap and the axial direction, the turn number of the coil and the volume of the magnets are constant.

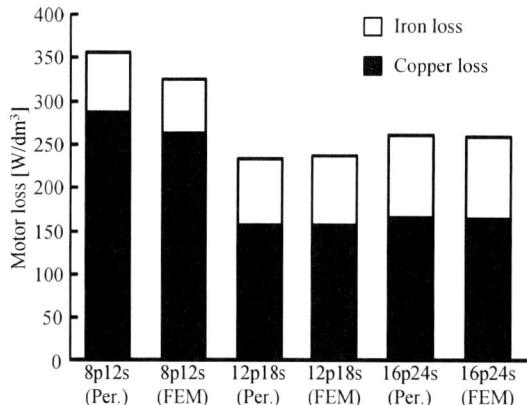

Fig. 14 Motor loss at 3600 r/min and 3.9 Nm by the changing pole number and slot number based on the constant volume. The loss is standardized by the volume. The errors between the proposed method and the FEM are less than 10% in the all models. The loss of 12 poles 16 slots motor is the lowest in this case.

calibrated because the output torque is set to constant. The motor speed is 3600 r/min. The copper loss and the iron loss in the stator are calculated.

Fig. 16 shows the motor loss when the axial direction length and the radius are changed. The radius r is standardized by the axial direction length D and the loss is standardized by the volume of the motor. From the calculation result by the permeance method in Fig. 15, the motor loss decreases when the size is enlarged to the radial direction. In addition, the analysis results by the FEM agree in principle with that of the permeance method. Therefore, the validity of the loss calculation by using the permeance method is confirmed.

If the size is enlarged to the radial direction, the magnetic flux in the air gap decreases because the magnetic resistance of the air gap and the magnet increases. Therefore, the large current is required in order to keep the magnetic flux. However, the wiring resistance can be decreased because the cross section of the coil winding can be enlarged and the length of the coil

winding can become short. Thus, the copper loss can be decreased. On the other hand, if the size is enlarged to the axial direction, the magnetic flux increases in order to decrease the magnetic resistance. However, the magnetic flux density decreases because the cross section of the stator core is enlarged at a great rate. Therefore, the iron loss can be decreased. For these reason, the copper loss and the iron loss are in the relationship of trade-off. As a result, the motor loss is decreased by enlarging to the radial direction because the variation of the copper loss is larger than that of the iron loss. If the current density becomes low, the variation of the iron loss has a significant impact on the motor loss and the lowest loss point exists.

C. Changing Length of Teeth Width

Fig. 17 shows the schematic of the changing teeth width. The losses are calculated when the width of the teeth is changed based on the constant width of the yoke. In addition, the area of the coil is changed by the changing teeth width. Therefore, the space factor of the coil is constant (50%) and the resistance of wire is changed. Moreover, the speed is 3600 r/min and the output torque is 3.9 Nm.

978-1-4799-2706-7/14 $31.00 © 2014 IEEE

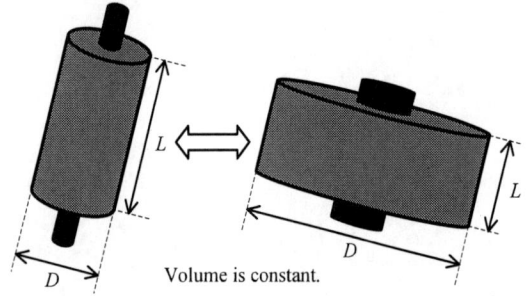

Volume is constant.

Fig. 15 Schematic of the changing radial direction length D and axial direction length L The air gap length is set to 1 mm by adjusting the size of the rotor. The amplitude of the armature current is calibrated because the output torque is set to constant. The motor speed is 3600 r/min. The copper loss and the iron loss in the stator are calculated..

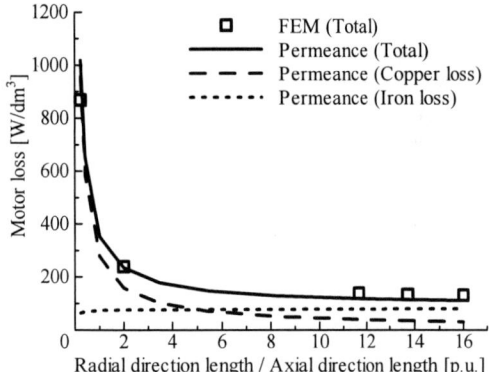

Fig. 16 Motor loss at 3600 r/min and 3.9 Nm when the radial direction length and the axial direction length are changed based on the constant volume. The radius r is standardized by the axial direction length D and the loss is standardized by the volume of the motor. The motor loss decreases when the size is enlarged to the radial direction.

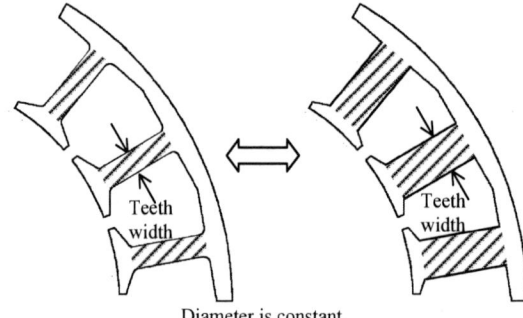

Diameter is constant.

Fig. 17 Schematic of the changing teeth width. The yoke width is constant. The area of the coil is changed by the changing teeth width. The space factor of the coil is constant (50%) and the resistance of wire is changed

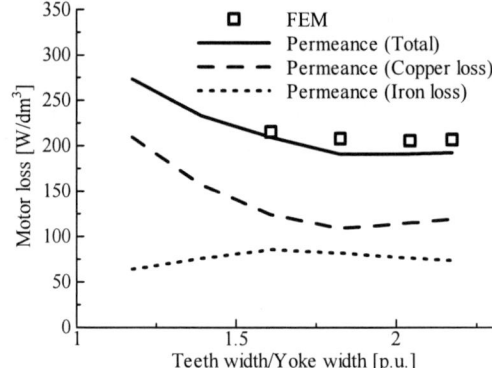

Fig. 18 Motor loss at 3600 r/min and 3.9 Nm when the teeth width is changed based on the constant volume. The loss at 1.8 p.u. is the lowest in the calculation results by the proposed method.

Fig. 18 shows the motor loss when the teeth width is changed based on the constant volume. The teeth width is standardized by the yoke width and the loss is standardized by the volume of the motor. The trend of changing loss in the proposed method agrees with that in the FEM. In addition, when the teeth width is 1.8 p.u., the loss is the lowest in the calculation results by the proposed method. On the other hand, in the results by the FEM, when the width is 2.0 p.u., the loss is the lowest. The difference between the loss at 1.8 p.u. and 2.0 p.u. is very few. Thus, it is considered that the teeth width of minimum loss can be designed by the proposed method.

If the teeth width is become large, the magnetic resistance decreases and the flux in the core increases. Therefore, in less than 1.6 p.u., the iron loss increases due to the increasing flux density of yoke. However, in over 1.6 p.u., the iron loss decreases because the flux density of the teeth decreases. In addition, the flux in the air gap increases by the increasing flux in the core. If the output torque is same, the current can be decreased. Thus, the copper loss can be decreased. However, in over 1.8 p.u., the copper loss increases because the area of coil become small and the resistance of wire become large.

From these results, when the motor parameters are changed, it is confirmed, the validity of loss calculation by the proposed method. Therefore, the minimum loss motor can be designed. For example, if the mechanical strength is considered and the diameter is less than 2 p.u. as a constrained condition, the parameter of the minimum loss motor is as follows: the pole number is 12, the diameter is 2 p.u. and the teeth width is 1.8 p.u..

VI. CONCLUSIONS

This paper proposed the simple magnetic equivalent circuit in order to design the rough configurations of the lowest loss motor. As a result, the error of the iron loss between the proposed method and the FEM is 2.9%. In addition, the motor losses were calculated by the permeance method when several parameters were changed. As a result, these results are almost agreed in principle with the results of the FEM. Therefore the validity of the proposed method was confirmed.

The future works are as follow; first, the eddy current loss in the permanent magnet will be calculated by the permeance method. Second, the motor loss considered the carrier harmonics will be calculated.

REFERENCES

[1] G. Pellegrino, et al. "Performance Comparison Between Surface-Mounted and Interior/min Motor Drives for Electric Vehicle Application", IEEE Trans. Ind. Electron., Vol.59, No.2, 2012.

[2] D. G. Dorrell, A. M. Knight, L. Evans, and M. Popescu, "Analysis and Design Techniques Applied to Hybrid Vehicle Drive Machines—Assessment of Alternative IPM and Induction Motor Topologies", IEEE Transactions on Industrial Electronics, Vol.59, No.10, pp.3690-3699 (2012)

[3] X. Yuan and J. Wang, "Torque Distribution Strategy for a Front- and Rear-Wheel-Driven Electric Vehicle", IEEE Transactions on Vehicular Technology, Vol.61, No.8, pp.3365-3374 (2012).

[4] R. Cao, C. Mi, and M. Cheng, "Quantitative Comparison of Flux-Switching Permanent-Magnet Motors With Interior Permanent Magnet Motor for EV, HEV, and PHEV Applications", IEEE Transactions on Magnetics, Vol.48, No.8, pp.2374-2384 (2012)

[5] T. Okitsu, D. Matsuhashi and K. Muramatsu, "Method for Evaluating the Eddy Current Loss of a Permanent Magnet in a PM Motor Driven by an Inverter Power Supply Using Coupled 2-D and 3-D Finite Element Analyses", IEEE Transactions on Magnetics, Vol.45, No.10, pp.4574-4577 (2009)

[6] K. Yamazaki and N. Fukushima, "Iron-Loss Modeling for Rotating Machines: Comparison Between Bertotti's Three-Term Expression and 3-D Eddy-Current Analysis", IEEE Transactions on Magnetics, Vol.46, No.8, pp.3121-3124 (2010)

[7] Y. Kawase, T. Yamaguchi, T. Umemura, Y. Shibayama, K. Hanaoka, S. Makishima, and K. Kishida, "Effects of carrier frequency of multilevel PWM inverter on electrical loss of interior permanent magnet motor", ICEMS 2009, LS5A-2 (2009)

[8] J. Farooq, S. Srair, A. Djerdir and A. Miraoui, "Use of Permeance Network Method in the Demagnetization Phenomenon Modeling in a Permanent Magnet Motor", IEEE Transactions on Magnetics, Vol.42, No.4, pp.1295-1298 (2006)

[9] B. Sheikh-Ghalavand, S. Vaez-Zadeh and A. H. Isfahani, "An Improved Magnetic Equivalent Circuit Model for Iron-Core Linear Permanent-Magnet Synchronous Motors", IEEE Transactions on Magnetics, Vol.46, No.1, pp.112-120 (2010)

[10] A. R. Tariq, C. E. Nino-Baron, and E. G. Strangas, "Iron and Magnet Losses and Torque Calculation of Interior Permanent Magnet Synchronous Machines Using Magnetic Equivalent Circuit", IEEE Transactions on Magnetics, Vol.46, No.12, pp.4073-4080 (2010)

[11] G. Gotovac, G. Lampic and D. Miljavec "Analytical Model of Permeance Variation Losses in Permanent Magnets of the Multipole Synchronous Machine", IEEE Transactions on Magnetics, Vol.49, No.2, pp.921-928 (2013)

[12] N. Leboeuf, T. Boileau, B. Nahid-Mobarakeh, N. Takorabet, F. Meibody-Tabar, and G. Clerc, "Inductance Calculations in Permanent-Magnet Motors Under Fault Conditions", IEEE Transactions on Magnetics, Vol.48, No.10, pp.2605-2616 (2012)

The Proposal of a New Motor Which Has a High Winding Factor and a High Slot Fill Factor

Shinji Makita
Department of Engineering R&D Center
DENSO CORPORATION
Kariya, Aichi, Japan
shinji_makita@denso.co.jp

Yasuhide Ito, Tomohiro Aoyama
Department of R&D
ASMO CO.,LTD
Kosai, Shizuoka, Japan
yasuhide-itou@asmo.co.jp,
tomohiro-aoyama@asmo.co.jp

Shinji Doki
Department of Electrical Engineering and Computer Science
Nagoya University
Nagoya, Aichi, Japan
doki@nagoya-u.jp

Abstract— In this paper, we propose a new motor which has a high winding factor and a high slot fill factor. A motor with concentrated windings has a high slot fill factor, but a winding factor is low. On the other hand, a motor with distributed windings has a high winding factor, but a slot fill factor is low. To get both good features, the proposal motor has a new distributed winding stator that can be flat band-shaped without dividing windings and can be easy to wind.

Keywords— winding factor, slot fill factor, distributed winding, flat band-shaped core

I. INTRODUCTION

Motors are one of the basic components for household appliances, automobiles and industrial machines and many engineers are pursuing research and development work for downsizing, cost reduction, efficiency increasing, optimization by purpose and productivity improvement.

Now most motors have either concentrated winding stator or distributed winding stator. The former can be a high slot fill factor, because the windings are wound on one tooth, but its winding pitch is 120 degrees electric angle and winding factor is 0.866. A fractional slot combination has a high winding factor than 0.866 [1], but it has problems of large iron loss and flux leakage between adjacent rotor magnets because it has more pole number than a conventional slot combination stator. The latter has a 0.966 to 1.000 winding factor because its winding pitch is more than 120 degrees electric angle. But different phase windings overlap each other in a radial direction, so it is difficult to increase a slot fill factor and to reduce end coil height in comparison to a concentrated winding. To solve this problem, some motor manufacturer divides windings, insert them in stator slots, and connect divided windings by welding [2]. Other insert molding windings in collectively [3]. But these methods limit the number of windings per slot and limits the application use.

In this paper, we propose a new motor which has a high winding factor and a high slot fill factor, and design and estimate a prototype motor. Specifically, the proposal motor has a new distributed winding stator. The stator is manufactured by mounting distributed windings in a flat band-shaped stator core and joining opposite ends of the bent stator together. The slots and teeth layout is different from a conventional distributed winding stator to mount all distributed windings in a flat band-shaped stator core without dividing windings. This technique realizes a higher winding factor at same pole number, and it is not limited by a number of conductors or slot shape.

II. FEATURE OF THE PROPOSED MOTOR

A. Basic Structure and Manufacturing Method

Fig.1 shows a basic structure and a manufacturing method of the proposal motor. Fig1 (a) shows a flat band-shaped stator core, Fig1 (b) shows a stator after jointing opposite ends of the bent stator together. It`s stator teeth width and pitch in a rotational direction are not equal. And there are some spots adjacent windings do not overlap each other in a radial direction. If one of these spots are arranged at the end of the flat-band shaped stator core, it is possible to wind distributed windings without dividing windings.

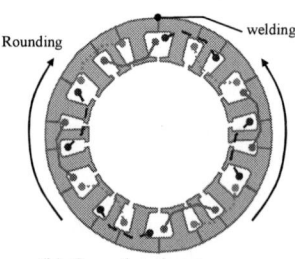

(a) Flat band-shaped stator core with distributed windings

(b) Completed stator core

Fig. 1. Basic structure and a manufacturing method of a proposal motor

This structure makes it possible to enlarge winding pitch up to 120 degrees electrical angle and winding factor. And this manufacturing method makes it possible to raise a slot fill factor and to reduce an axial height of end coils and to raise a yield rate.

978-1-4799-2706-7/14 $31.00 © 2014 IEEE

B. Principle

Next we explain the way of deciding the width and arrangement of the stator teeth in comparison with a normal full pitch distributed winding stator. Fig.2 (a) shows a part of normal full pitch one. A pitch of one tooth is 60 degrees electrical angle and each winding is wound to three teeth, so its winding pitch is 180 degrees electrical angle. Each phase winding is shifted 120 degrees electrical angle each other, and with an unshown one more different phase winding, they configure a three phase motor. In this case, all windings overlap each other in a radial direction, so it is impossible to wind to a flat band-shaped stator core without dividing windings.

Fig2. (b) shows a part of shortening pitch winding stator with same stator teeth in Fig2. (a). In this case, each winding pitch is 120 degrees electrical angle and each winding phase is shifted 30 degrees (one is advanced, the other is delayed). So their short pitch winding factor is 0.866 (=sin (120/2)), distributed winding factor is 0.866 (=cos (30)) and winding factor is the product of them, 0.750.

Fig2 (c) shows a proposed arrangement. The center tooth of Fig2 (b) that is not wound is divided into two and they are embodied next tooth. In this case, each winding pitch is 150 degrees electrical angle, each winding phase is shifted 15 degrees (one is advancing, the other is delayed). So their short pitch winding factor is 0.966 (=sin (150/2)), distributed winding factor is 0.966 (=cos (15)) and winding factor is the product of them, 0.933. And there is a spot that different phase windings are not overlapping each other in the radial direction.

Next we explain the arrangement of this short pitch spot. A deformation of Fig.2 is applied to two phase windings,

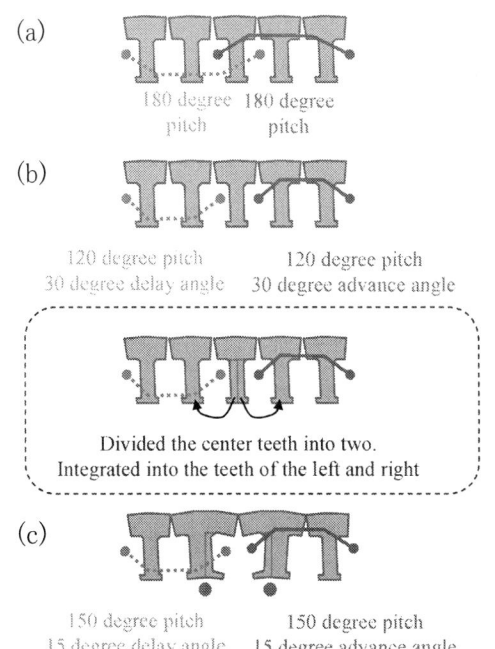

Fig. 2. Proposal arrangement of stator teeth and windings

one winding phase is shifted advance, the other one is shifted delay, so if the number of the short pitch spots is one, the amplitude of three phase windings is different and the phase difference of them is not 120 degrees electrical angle. To ensure the balance of three phase windings, each phase should have at least a pair of advanced winding and delayed winding. In order to do that, there should be three short pitch spots and its phase combination should be different each other.

Fig.3 shows an example of a motor with 8poles. The deformation referred to Fig.2 (c) applies in the positions of 0 degree mechanical angle, 120 degrees mechanical angle, 240 degrees mechanical angle of the 24slots/8poles stator. The deformation in the position of 0 degree mechanical angle makes U-phase winding shifted advance and W-phase winding shifted delayed. The deformation in the position of 120 degrees mechanical angle make V-phase winding shifted advance and U-phase winding shifted delayed. The deformation in the position of 240 degrees mechanical angle make W-phase winding shifted advance and V-phase winding shifted delayed. Each phase has a pair of an advanced winding and a delayed winding, so the all winding phases are the same as 24slots/8poles stator. And the winding amplitude is 0.966 because there are two 180 degrees electrical angle pitch windings, one 150 degrees electrical angle pitch and 15 degrees electrical angle advanced winding and one 150 degrees electrical angle pitch and 15 degrees electrical angle delayed winding. This is a winding factor of the motor in Fig.3.

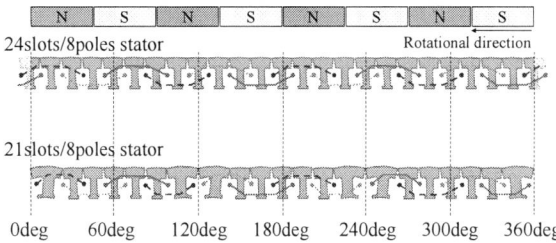

Fig. 3. Deformation from 24slots/8poles stator to 21slots/8poles stator

This motor has the balance of three phase windings, but a cyclic periodicity of the stator teeth and winding arrangement is 360 degrees mechanical angle. So the radial force between a rotor and a stator is unbalanced, and it causes large noise and vibration. To avoid this unbalance, we increase the number of the short pitch spots to six and the same combination spots arrange at the period of 180 degrees mechanical angle each other.

Fig.4 shows an example of a motor with 8poles. The deformation referred to Fig.2 (c) applys in the positions of 60 degrees mechanical angle, 180 degrees mechanical angle, 300 degrees mechanical angle of the 21slots/8poles stator in Fig.3. The phase combination in the position of 60 degrees mechanical angle and the one in the position of 240 degrees mechanical angle is the same. They are in the opposite position each other, so the radial force between a rotor and a stator is balanced. In the same way, the phase combination in the position of 180 degrees mechanical angle and the one in the position of 300 degrees mechanical angle is the same, and the phase combination in the position of 300 degrees mechanical angle and the one in the position of 120 degrees mechanical angle is the same. So three phase balance and radial force balance is ensured. And the winding amplitude is 0.933 because there are two 150 degrees electrical angle pitch and 15 degrees electrical angle advanced winding and two 150 degrees electrical angle pitch and 15 degrees electrical angle delayed winding. This is a winding factor of the motor in Fig.4.

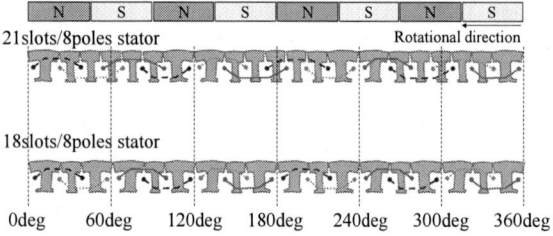

Fig. 4. Deformation from 21slots/8poles stator to 24slots/8poles stator

C. Inspection by FEM Analysis

Fig.5 and Table I show the example of torque and radial force wave forms in comparison with three type of eight poles SPM motor. The average torque ratio of 24slots/8poles machine, 21slots/8poles machine, 18slots/8poles machine is almost 1.000, 0.966, 0.933 and its value is same as each winding factor referred to above. And the radial force is occurring only at 21slots/8poles machines as expected.

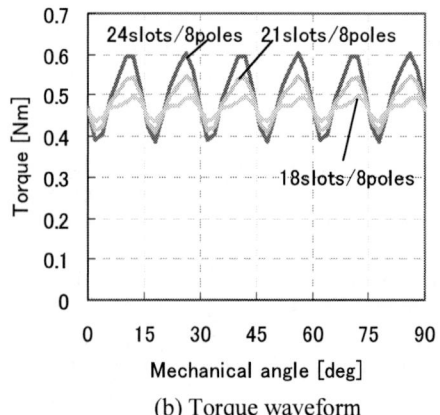

(a) Magnetic density distribution of 24slots/8poles machine, 21slots/8poles machine, 18slots/8poles machine

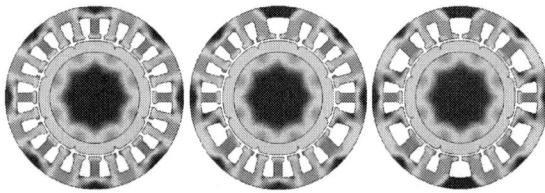

(b) Torque waveform

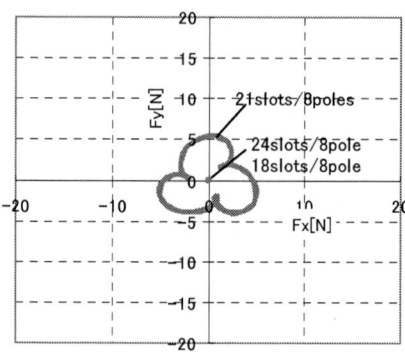

(c) Radial force trajectory

Fig. 5. Torque and radial force waveform of three types of machines

TABLE I
COMPARISON WITH TORQUE AND WINDING FACTOR

Stator	Average Torque [Nm]	Torque Ratio	Winding Factor
24slots/8poles	0.500	1.000	1.000
21slots/8poles	0.484	0.968	0.966
18slots/8poles	0.469	0.936	0.933

We explain the example of 8poles machines, but the number of poles is not limited eight. It is limited over eight and the multiples of four to keep both a three phase balance and a radial force balance. Table II shows a relation among the number of poles, the number of slots and the winding factor. The number of short pitch spots is needed six in spite of the number of poles, So the more the number of poles, the more winding factor.

TABLE II
RELATION OF POLE, SLOT COMBINATION AND WINDING FACTOR

The number of poles	The number of slots	The winding factor
8	18	0.933
12	30	0.955
16	42	0.967
20	54	0.973
24	66	0.978
28	78	0.981

III. THE EVALUATION OF THE PROTOTYPE MOTOR

A. Overview of the Prototype Motor

In consideration of the feature mentioned above, we designed and evaluated a prototype motor with a specification of Table III. Fig.6 shows the process of manufacturing it. Fig.6 (a) is a picture of laminated flat band-shaped stator core, Fig.6 (b) is a picture of the stator core after inserting formed distributed windings, and Fig.6 (c) is a picture of a part of the rounded stator core with windings. Fig.7 shows a picture of the completed stator overview. Its slot fill factor is 54% and end coil height of one side is 6mm in spite of distributed windings

TABLE III
SPECIFICATION OF THE PROTOTYPE MOTOR

Item	Value
The number of poles	8
The number of slots	18
Rotor type	IPM
Stator outer diameter (mm)	40
Rotor diameter (mm)	24
Stack length (mm)	30.5
Air gap length (mm)	0.8
Voltage (V)	13
Rated output (W)	100

(a) Laminated stator core

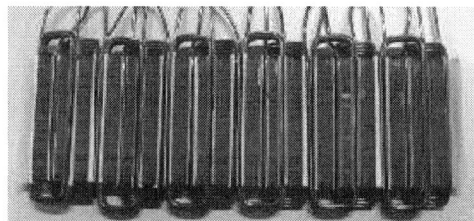

(b) Stator core after inserting formed distributed windings

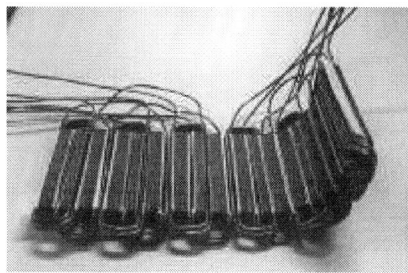

(c) Bent stator core with windings

Fig. 6. Process of manufacturing a prototype motor

Fig. 7. Stator overview of a prototype motor

B. Result of evaluation

Fig.8 shows a measured back EMF of each phase at 1000rpm. It shows three phase amplitude balance and 120 degrees shifted phase each other. And Fig.9 shows a

torque-speed characteristic of the prototype motor. The rated efficiency is 76%.

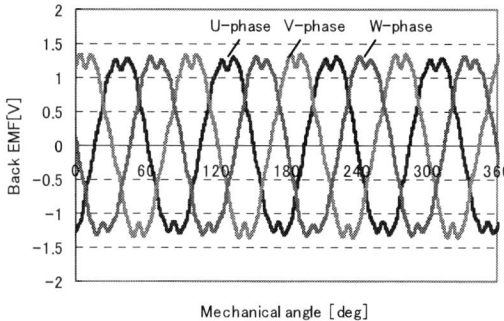

Fig. 8. Measured back EMF waveform (1000rpm)

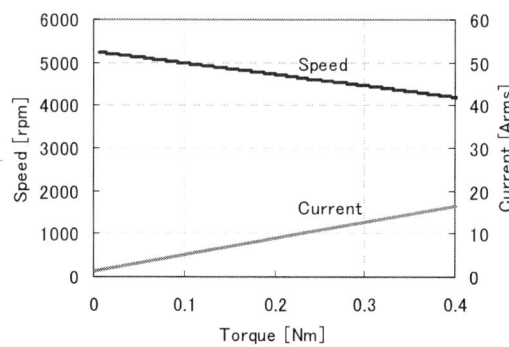

Fig. 9. Measured torque- speed characteristic (DC 13V)

C. Comparison with a conventional motor

In order to check the performance potential of the proposal motor, we compared the physique with the conventional concentrated winding motor of the same performance. In condition that torque, current density of windings, rotor diameter, stator diameter are fixed, we compared the volume of these motors with end coil. Table IV shows the result of the comparison. Although it was an example, the proposal motor has the 6.3% downsizing potential than the conventional motor.

TABLE IV
MOTOR SIZE COMPARISON

Item (Unit)	Proposal	Conventional
Volume [mm^3]	65566	69996
Stator diameter [mm]	47.5	47.5
Rotor diameter [mm]	24	24
Core Axial length [mm]	25	32.5
End coil length [mm]	6	3.5
Number of poles	8	14
Number of slots	18	12
Slot fill factor	0.7	0.6

IV. CONCLUSION

We propose a new motor which has a high winding factor and a high slot fill factor. The stator is manufactured by mounting distributed windings in a flat band-shaped stator core and joining opposite ends of the bent stator together. The slots and teeth layout is different from a conventional distributed winding stator to mount all distributed windings in a flat band-shaped stator core without dividing windings. The feature of the proposal motor is a high winding factor, high slot fill factor and low end coil height.

Furthermore, we design and evaluate a prototype motor. As a result, we checked the 76% rated efficiency, 54% slot fill factor and 6mm end coil height at one side in spite of distributed windings. From now on, we continue to study further characteristic improvement of a proposal motor.

REFERENCES

[1] P. Ponomarev, P. Lindh, J. Pyrhonen, "Effect of Slot-and-Pole Combination on the Leakage Inductance and the Performance of Tooth-Coil Permanent-Magnet Synchronous Machine," *IEEE Trans. On Industrial Electronics*, vol. 60, no. 10, pp. 4310-4317, 2013.

[2] H. Ishikawa, A. Umeda, M. Kohmura, "Development of a More Efficient & Higher Power Generation Technology for Future Electrical Systems," *SAE International Convergence 2000 International Congress on Transportation Electronics, 1999.*

[3] T. Ishigami, Y. Tanaka, H. Homma, "Development of Motor Stator with Rectangular Wire Lap Winding and an Automatic Process for its Production," *IEEJ Journal of Industry Applications*, vol. 132, no. 10, pp. 976-982, 2012.

The 2014 International Power Electronics Conference

Variable Leakage Flux Interior Permanent Magnet Synchronous Machine for Improving Efficiency on Duty Cycle

Masanao Minowa
Student Member, IEEJ
Shibaura Institute of
Technology
Koto-Ku, Tokyo, Japan
e10111@shibaura-it.ac.jp

Hiroki Hijikata
Student Member, IEEJ
Shibaura Institute of
Technology
Koto-Ku, Tokyo, Japan
nb13509@shibaura-it.ac.jp

Kan Akatsu
Member, IEEJ
Shibaura Institute of
Technology
Koto-Ku, Tokyo, Japan
akatsu@sic.shibaura-it.ac.jp

Takashi Kato
Member, IEEJ
Nissan Motor Co., Ltd
EV System Lab.
Atsugi-shi, Kanagawa, Japan
katou-t@mail.nissan.co.jp

Abstract— **Interior permanent magnet synchronous motors (IPMSMs) are used for an electric vehicle (EV) and a hybrid electric vehicle (HEV) due to its high efficiency and high output power density. In particular, the traction motors for the EV and HEV require high efficiency driving in various conditions such as the low load condition and high speed condition. However, it is difficult to change the high efficiency range in the motor, because the magnetomotive force of rotor magnet is constant in conventional IPMSMs. This paper focuses on the leakage flux of the rotor magnet and proposes a variable leakage flux IPMSM (VLF-IPM). The VLF-IPM can change the motor characteristics by simple motor structure drive system because the leakage flux can be controlled by only the armature reaction. This paper shows the principle and the characteristic of VLF-IPM and verifies its effectiveness.**

Keywords—Leakage flux, Traction motor for EV or HEV, Variable characteristics, VLF-IPM

I. INTRODUCTION

A lot of studies have been progressed with the interior permanent magnet synchronous motors (IPMSMs) which are used for an electric vehicle (EV) and hybrid electric vehicle (HEV) due to its high efficiency and high output power density. In particular, the traction motors for the EV and HEV require high efficiency driving in frequently used conditions such as the low torque area and high speed area.

A variable characteristic motor is suitable to meet these requirements because the variable characteristic makes appropriate modification in the ratio of the copper loss and iron loss. Some study of the variable characteristic motor are proposed [1]-[6]. In [1], a type of variable characteristic motor which can mechanically change an air gap length is introduced. The flux density in a stator core is varied by using an actuator. However, the motor drive system becomes larger than the conventional drive system. In [2]-[4], a type of variable characteristic motor which can change the magnetomotive force by the excitation current. The concept of a memory motor has been presented and it uses a special magnet which can optionally select to magnetize or to demagnetize [2]. The pole change

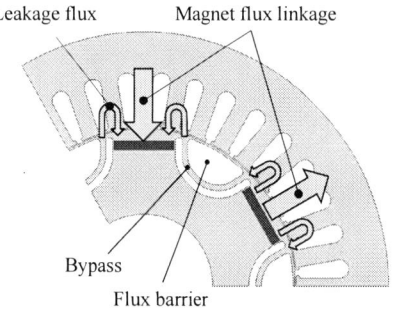

Leakage flux   Magnet flux linkage

Bypass

Flux barrier

(a) Low load condition

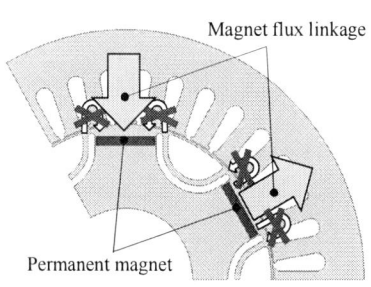

Magnet flux linkage

Permanent magnet

(b) High load condition

Fig. 1 1/3 FEA model of the VLF-IPM.

method is also proposed by the excitation current [3]. However, the current capacity of an inverter is increased because these motors need the excessive armature current for the magnetization or the demagnetization. Moreover, the pole change method is proposed by using additional winding [4]. However, the method requires the large motor capacity and additional inverter circuit. As another idea, there is a concept of using auxiliary winding method [5]. By the field excitation of the auxiliary winding, the motor accomplishes the flux weakening. On the other hand, the disadvantage of these structural approaches is increasing the weight and size. Furthermore, the motor structure becomes complex.

To overcome above disadvantages, this paper proposes a variable leakage flux IPMSM (VLF-IPM). The

978-1-4799-2706-7/14 $31.00 © 2014 IEEE   3828

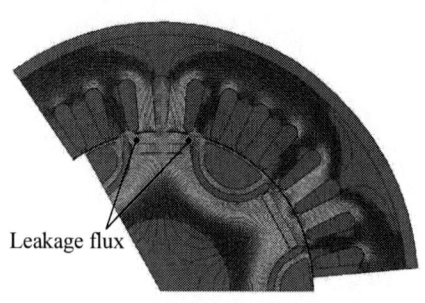

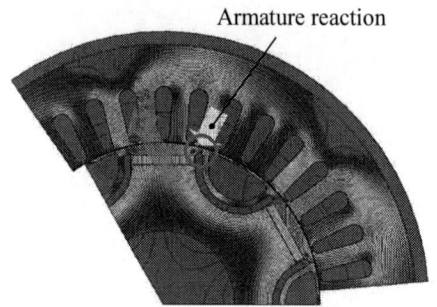

(a) Low load condition

(b) High load condition

Fig. 2 FEA results of the flux density distribution

proposed motor is possible to easily achieve the variable characteristics because the rotor leakage flux is controlled by the only stator armature reaction. As a result, the proposed motor can reduce the magnet flux linkage by increasing the rotor leakage flux in order to reduce the iron loss under the high speed condition. On the other hand, the method can also increase the magnet flux linkage in order to output the large torque by reducing the rotor leakage flux when the motor rotates in low speed condition. Thus, the motor achieves variable characteristics by the simple structure and a conventional 3-phase half-bridge inverter is used.

Although the similar variable characteristics motor has been reported with a type of flux intensified IPM (FIIPM) [7] called the variable flux intensifying IPMSM (VFI-IPM) [8]-[11], the VFI-IPM uses the special magnet which can magnetize or demagnetize optionally by additional d-axis current. In contrast, the proposed VLF-IPM in this paper can achieve the variable characteristic without any additional components by using the conventional Nd-Fe-B magnet. It is note that the proposed machine in this paper is different with the machine in [8] even the machine overview is same.

In this paper, the principle of the motor is firstly described, the variable magnet flux characteristic of the machine is secondary shown and the machine parameters to achieve the maximum variable characteristics are optimized. Adding that a new model which improves the output power is proposed, and finally the paper verifies the efficiency improvement in the high speed region by the finite element analysis (FEA).

II. PRINCIPLE OF VLF-IPM

A. Design Concept for VLF-IPM

The specifications of the VLF-IPM motor are shown in TABLE I. The model includes the distributed winding, large flux barrier, and flux bypass in the rotor. Generally, the IPMSM is designed preferably to reduce the leakage

flux in order to obtain the large torque. On the other hand, the VLF-IPM structure purposely leaks a lot of magnet flux through the flux bypass. The leakage flux is controlled by the armature reaction. Fig. 1 illustrates the relationship between the load conditions and the leakage flux. Fig. 1(a) shows the flux flow in the low load condition. In this condition, the proposed topology can reduce the magnet flux linkage by increasing the rotor leakage flux. As a result, this condition can much reduce the iron loss when the VLF-IPM rotates the high speed. On the other hand, Fig. 1(b) shows the flux flow in the high torque condition. In this condition, the topology can increase the magnet flux linkage by reducing the rotor leakage flux because the flux bypass is saturated by the armature reaction. As a result, this condition can output the large torque at the low speed condition. Therefore, it is found that the VLF-IPM achieves the variable characteristics with simple structure and the only conventional 3-phase half-bridge inverter can be used. Fig. 2 shows the FEA results of the flux density distribution to verify the principle of VLF-IPM. As shown in Fig. 2(a), it is found that the low load condition obtains much leakage flux in the rotor. On the other hand, in the high load condition, lots of flux is distributed in the stator core by reducing the rotor leakage flux.

The difference of the VFI-IPM and VLF-IPM is shown by using the magnetic equivalent circuit in Fig. 3. Where, ψ, F, I, and R_m indicate the magnet flux linkage, magnetomotive force of the permanent magnet, armature current, and magnetic resistance in the flux bypass, respectively. Subscription d and q expresses d–axis and q–axis components. It is said that the VFI-IPM can change the magnetomotive force and the magnet flux linkage by the excitation current. On the other hand, the VLF-IPM can change the leakage flux by the armature reaction.

TABLE I
MOTOR PARAMETERS

Parameters	Value
Maximum current (A_{peak})	90
Gap length (mm)	0.4
Number of poles	6
Number of slots	36
Number of turns per slot	30
Outer rotor diameter (mm)	109.2
Outer stator diameter (mm)	176.0
Magnet material	Nd-Fe-B

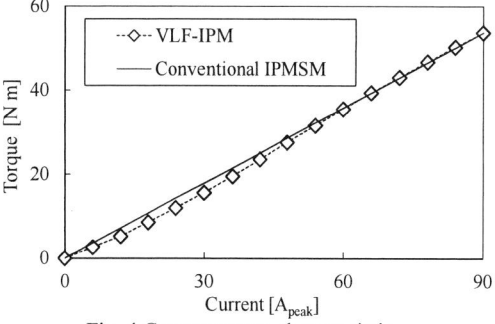

Fig. 4 Current-torque characteristic.

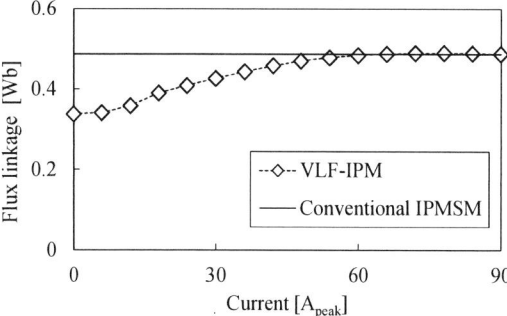

Fig. 5 Current-magnet flux linkage characteristic

B. Magnet Flux Linkage Calculation

This part shows the calculation procedure of the magnet flux linkage by the rotor magnet because it is difficult to divide the magnet flux linkage into the rotor magnet flux and stator armature winding flux. Although the torque is easily calculated if another parameter value such as the inductance is known, it is actually difficult to calculate the inductance without using the magnet flux. Therefore, the magnet flux linkage by the rotor magnet is calculated by following equation without the inductance when the q-axis current is only applied.

$$\psi_d = \frac{T}{P i_q} \quad (1)$$

Fig. 4 shows the current-torque characteristics. The applying current is changed from 0 to $90A_{peak}$ and the current angle is 0 deg. As shown in this graph, it is said that the torque nonlinearly increases with the q-axis current. Fig. 5 shows the current-magnet flux linkage characteristic by using (1) and results of Fig. 4. As shown in this graph, the magnet flux linkage increases about 45 % with the q-axis current. This characteristic is very ideal characteristic for the variable speed machine since the high torque condition with the large q-axis current requests high magnet flux linkage, on the other hand the high speed low torque condition requests low magnet flux linkage to reduce the back-EMF and also the less q-axis current is requested..

III. SIMULATION RESULTS

A. Shape variation for variable range of magnet flux linkage

This section discusses the influence on the variation range of the magnet flux linkage by the width of the flux bypass and depth of the flux barrier. Fig. 6 shows the definition of these parametric studies. The simulation results are summarized in Fig. 7. As shown in this graph, it is found that the variation range of the magnet flux linkage increases with an increasing width of the flux bypass and the depth of the flux barrier. On the other

hand, the maximum magnet flux linkage decreases. It is said that the variation range and the maximum magnet flux linkage is a trade-off.

B. Comparison between the initial model and maximum variable range model

This part compares the current-magnetic flux characteristics of the initial model and maximum variable range model. The maximum variable range model is selected as the largest variation range in Fig. 8. The initial model is similar to the model in Fig. 1. Fig. 9 shows the current-magnet flux linkage characteristics of two models. As shown in this graph, it is found that approximately 100% of the variation range which means the magnet flux linkage increases twice is obtained by the optimization. On the other hand, the maximum magnet flux linkage decreases about 9.3 %. As a result, the maximum variable range model reduces the output torque.

Fig. 10 shows the speed-output power characteristic. It is found that output power of the maximum variable range model is remarkably reduced than the initial model because the magnetic path of q-axis is expanded by having enlarged the depth of the flux barrier as shown in Fig. 11. Since the q-axis magnet flux linkage increases by expanding the q-axis magnetic path, the voltage increases and exceeds the bus voltage of an inverter. It is found that the rotor structure has to be designed with a reduction of the torque and output power in mind. Therefore, the VLF-IPM must be redesigned in order to produce the large torque and output power.

The 2014 International Power Electronics Conference

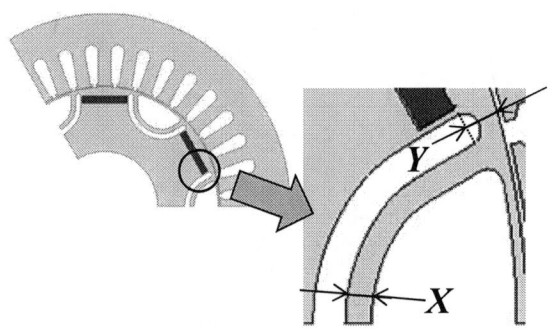

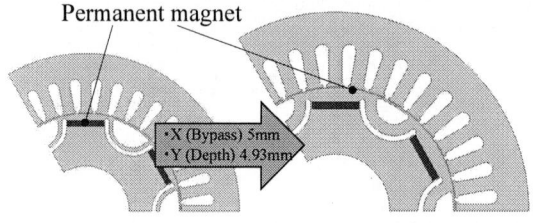

Fig. 6 Design methodologies of the width of the flux bypass and the depth of the flux barrier.

Fig. 8 Maximum variable range model

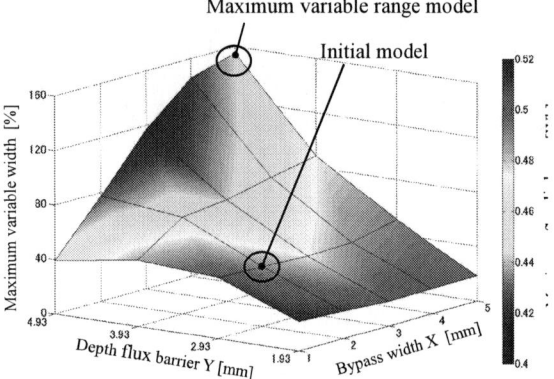

Fig. 7 Summarized simulation results of the relationships among the depth of flux barrier, bypass width, maximum variable width and maximum flux linkage.

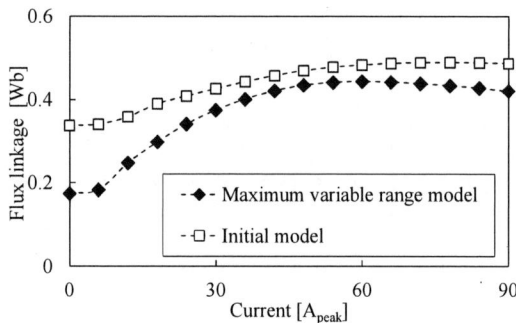

Fig. 9 Current-magnet flux linkage characteristic.

IV. A REALISTIC OPTIMAL DESIGN

A. Proposed a New VLF-IPM Model

This section proposes a new model which increases magnet volume of VLF-IPM (final optimized model) in consideration of the maximum torque and output power. The final optimized model shows in Fig. 12. The specifications of the motor are given in TABLE. II. The magnetic pitch expanded from 72 deg to 129 deg in order to increase the maximum torque. Although the large quantity of the permanent magnet decreases the degree of freedom of the bypass design, the enough variation of the magnet flux linkage is obtained. Moreover, the number of turns is changed from 30 turn to 21 turn with taking into account the rotor structure.

B. Speed-Torque Characteristic

The comparative motor which removes the bypass is shown in Fig. 13. Speed-torque characteristics of with and without bypass model are shown in Fig. 14. It is found that both motors have almost same maximum torque. However, the torque of the comparative model is decreased to zero around 6000 rpm due to the large back-EMF effect even though the comparative model has less

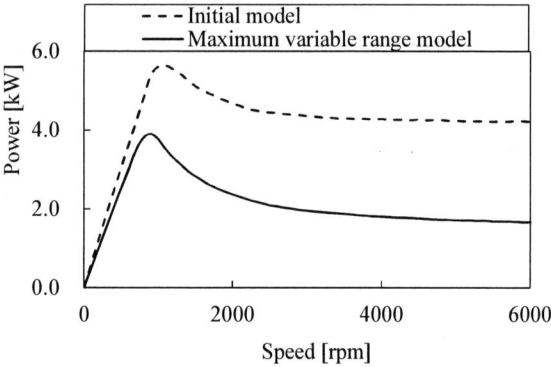

Fig. 10 Speed-output power characteristic.

turn number than the final optimized model. On the other hand, the torque characteristic of the final optimized model can expand the operating area over 6000 rpm due to the variable magnet flux linkage. Fig. 15 shows the speed-current angle characteristics in each model. As shown in this graph, the current angle of the final optimized model is smaller than the comparative model because the final optimized model can easily reduce the d-axis flux linkage through the flux bypass. Therefore, it is found that the d-axis current of the VLF-IPM for flux weakening control can be reduced compared with the comparative model. If the amount of flux weakening current is reduced, the reaction force against magnet can also be reduced. As a result, low cost magnet can be used or thermal durability will be improved.

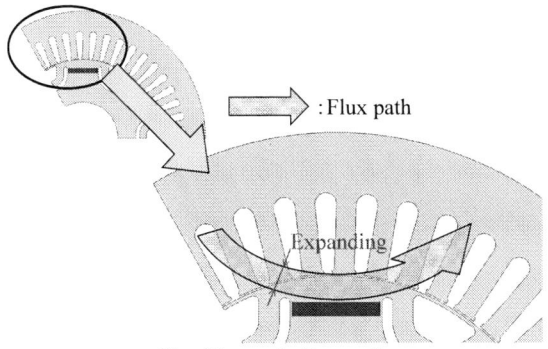

Fig. 11 *q*-axis flux path

TABLE II
PROPOSED MODEL'S PARAMETER

Parameter	Final optimized model	Without bypass model
Maximum current (A_rms)	30	
Magnetic pitch (deg)	129	
Gap length (mm)	0.4	
Number of poles	6	
Number of slots	45	
Number of turns per slot	21	18
Phase resistance (mΩ)	181.8	155.8
Outer rotor diameter (mm)	109.2	
Outer stator diameter (mm)	176.0	
Magnet material	Nd-Fe-B	

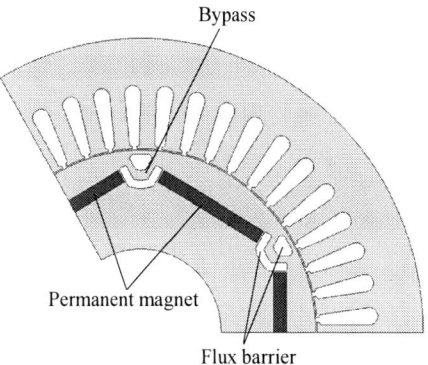

Fig. 12 Final optimized model

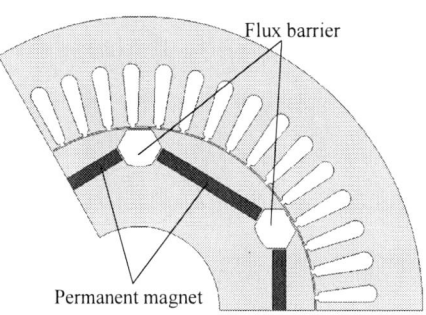

Fig. 13 Comparative model without bypass

The loss analysis at some operating points in the low torque condition is shown in Fig. 16. The final optimized model and comparative model have almost similar output power in each operating point. Since each motor drives without the flux-weakening control in the low torque and low speed condition, the ratio of the iron loss is similar in each machine. On the other hand, in the high speed region, the current and current angle have to be increased in order to use the flux-weakening control because the inductive voltage increases in the condition. Thus the copper loss increases by injecting *d*-axis current. However, the final optimized model requires little *d*-axis current in high speed condition by much leakage flux in the motor. Therefore, the copper loss of the VLF-IPM is reduced in spite of the high speed rotation. In addition, the iron loss of the final optimized model is also reduced in the high speed condition because the flux density in stator teeth is decreased due to the leakage magnet flux. Therefore, the final optimized model achieves high efficiency especially in the high speed area.

The efficiency maps of the final optimized model and comparative model are shown in Fig. 17 and 18, respectively. The maximum efficiency points of the final optimized model and comparative model are 97.24 % at 7 Nm and 2200 rpm, and 97.23 % at 13 Nm and 1900 rpm, respectively. It is found that the high efficiency area of the final optimized model is shifted to the low torque

and high speed condition. Therefore, the proposed motor realizes a demand of a traction motor.

V. CONCLUSION

This paper described the principle of VLF-IPM which can expand the high efficiency area with the flux bypass and large flux barrier. The results of this study reached the following conclusion.

· Some simulation results revealed that the proposed motor has some trade-off among the variation range of the magnet flux linkage, maximum flux linkage, width of the flux bypass and depth of the flux barrier.

· The final optimized model expanded the operating area than the comparative model.

· High efficiency area of the final optimized model shifts to the low torque and the high speed area than the comparative model.

From the above results, it is said that the proposed VLF-IPM motor is suitable for traction motor for an EV and HEV. The experimental results will be shown in the conference.

REFERENCES

[1] T. A. Lipo and M. Aydin; "Field Weakening of Permanent Magnet Machines - Design Approaches," *IEEE Power Electronics and Motion Control Conf., EPE-PEMC'04*, Riga, Latvia, Sep. 2004.

[2] V. Ostovic; "Memory Motors" *IEEE Industry Applications Magazine*, Vol.9, No.1 pp.52-61, Jan./Feb. 2003

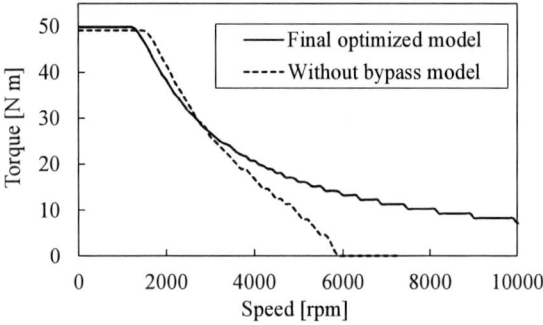

Fig. 14 Speed-torque characteristic

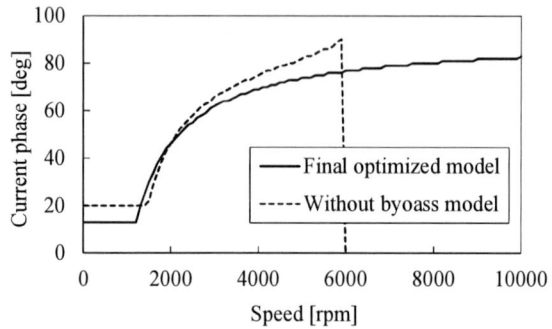

Fig. 15 Speed current angle characteristic

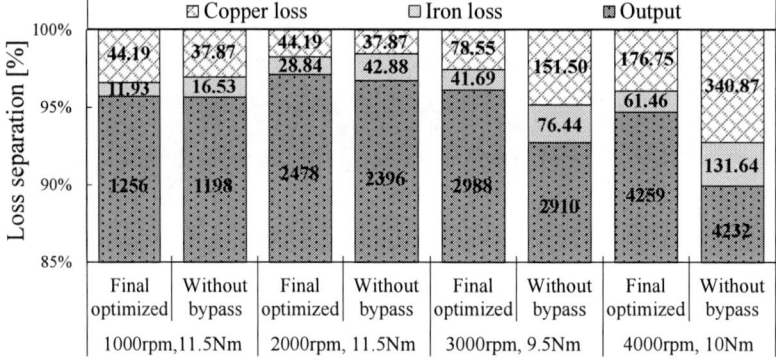

Fig. 16 Classified losses and output power.

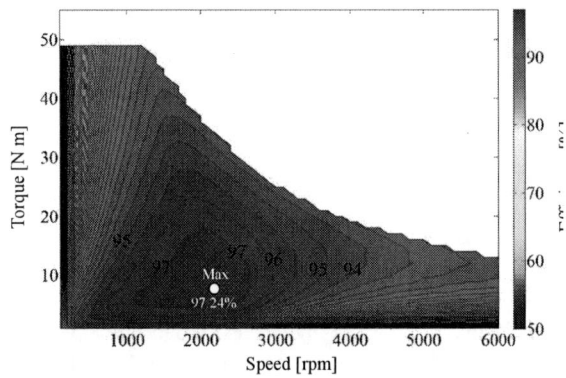

Fig.17 efficiency map of the final optimized model

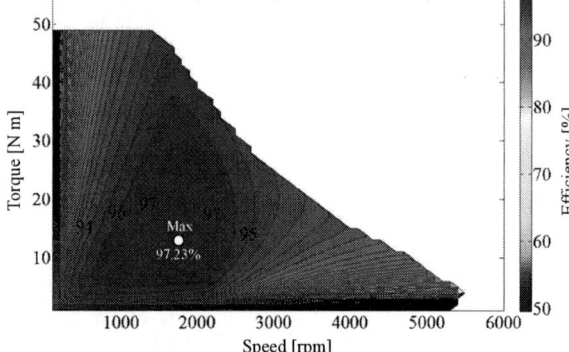

Fig.18 efficiency map of the comparative model

[3] V. Ostovic; "Pole-Changing Permanent-Magnet Machines, " *IEEE Trans. on Industry Applications*, Vol. 38, No. 6, pp. 1493-1499, Nov./Dec. 2002.

[4] K. Sakai and N. Yuzawa; "Permanent Magnet Motor Capable of Pole Changing for High Efficiency," *IEEE Energy Conversion Congress and Exposition (ECCE) 2013*, pp.5064-5071 2013

[5] M. Swamy, T. J. Kume, A. Maemura, and S. Morimoto: "Extended High Speed Operation via Electronic Winding Change Method for AC Motors," *IEEE Trans. on Industry Applications*, Vol.42, No.3 pp.742-752 May-June 2006.

[6] M. Sridharbabu, T. Kosaka, and N. Matsui: "Design Reconsiderations of High Speed Permanent Magnet Hybrid Excitation Motor for Main Spindle Drive in Machine Tools Based on Experimental Results of Prototype Machine," *IEEE Trans. on Magnetics*, Vol. 47, No. 10, pp.4469-4472. Oct. 2011.

[7] K. Akatsu, M. Arimitsu, and S. Wakui; "Design and Control of a Field Intensified Interior Permanent Magnet Synchronous Machine," *IEEJ Trans. on Industry Applications*, Vol. 126-D, No. 7, pp. 827-834, Oct. 2006

[8] T. Kato, N. Limsuwan, C. Yu, K. Akatsu, and R.D. Lorenz, " Rare earth reduction using a novel variable magnetomotive force, flux intensified IPM machine", *IEEE Energy Conversion Congress and Exposition (ECCE) 2012*, pp. 4356-4353, Sep. 2012

[9] T. Fukushige, N. Limsuwan, T. Kato, K. Akatsu, and R.D. Lorenz; "Efficiency Contours and Loss Minimization over a Driving Cycle of a Variable-Flux Flux-Intensifying Interior Permanent Magnet Machine," *IEEE Energy Conversion Congress and Exposition (ECCE) 2013*, pp.591-597 2013

[10] N. Limsuwan, T. Fukushige, K. Akatsu, and R.D. Lorenz; "Design Methodology for Variable-Flux, Flux-Intensifying Interior Permanent Magnet Machines for an Electric-Vehicle-Class Inverter Rating," *IEEE Energy Conversion Congress and Exposition (ECCE) 2013*, pp.1547-1554 2013

[11] C.Y. Yu, T. Fukushige, N. Limsuwan, T. Kato, D. Reigosa, and R.D. Lorenz; "Variable Flux Machine Torque Estimation and Pulsating Torque Mitigation during Magnetization State Manipulation," *IEEE Energy Conversion Congress and Exposition (ECCE) 2013*, pp.852-859 2013

The 2014 International Power Electronics Conference

History and Trends of Converter Technology for DC and AC Transmission in Japan

Teruo Yoshino

Power Electronics Systems Division
Toshiba Mitsubishi-Electric Industrial Systems Corporation
Tokyo, Japan

Abstract-**The paper introduces development history and trends of converter technology for DC and AC transmission in Japan. The high voltage converter development started in 1960s in Japan. The paper describes technology trends in various aspects including semiconductor device, converter insulation, cooling, circuit topology and control. The paper covers the converter applications for HVDC, SVC and STATCOM associated with typical circuit topologies. For the control trends, the sophisticated performance requirements are introduced in addition to the hardware development. Finally, the paper tries to discuss future trends of converter technology in this field.**

Keywords—Frequecy converter station, High voltage DC transmission system, Static Var Compensator, Thyristor valve

I. INTRODUCTION

This paper summarizes trends and development history of high voltage and high power converters for DC and AC transmission systems from viewpoint of various technologies. This paper also tries to discuss the technology trends in future.

Chapter II describes overview on the applications of the high voltage and high power converters for transmission in Japan. Thyristor valves are located at Frequency Converter (FC) stations and High Voltage DC Current transmission (HVDC) stations. At some substations, Static Var Compensators (SVCs) or STATic COMpensators (STATCOMs) are installed. These converters are also briefly introduced.

Chapter III introduces the power semiconductor device development in Japan for high voltage application including conventional thyristor, Light-Triggered Thyristor (LTT), Gate Turn Off thyristor (GTO) followed by Insulated Gate Bipolar Transistor (IGBT), Injection Enhanced insulated Gate Transistor (IEGT) and Gate Commutated Turn-off thyristor (GCT). Impacts to technology trend from the device development are also described.

Chapter IV summarizes insulation and cooling technology trends for high voltage and high capacity converters.

Chapter V introduces the converter circuit topology for HVDC with its mechanical feature. The NPC (neutral point clamped) multi-level self-commutated converter for

STATCOM is also introduced.

Chapter VI introduces the control hardware and performance development. The hardware changed from analog circuit to digital circuit. By taking the HVDC as a typical example, the converter control is introduced for basic functions and for further sophisticated functions in addition to hierarchical structure for the large control system.

Chapter VII briefly introduces technology trends in the power electronics converters applied to other fields including railway substations.

Chapter VIII tries to discuss future trends of the converter technology for transmission application from viewpoints of the converter and the control.

II. HIGH-VOLTAGE AND HIGH-POWER CONVERTERS IN JAPAN

Japan has a unique history in AC power system development. It resulted in two frequency systems with 60Hz and 50Hz. The border line of the frequencies passes near Mr. Fuji. Around the border line, Japan has the FC stations. The sum of the capacity reaches 1200MW as shown in Fig. 1. [1] [2]

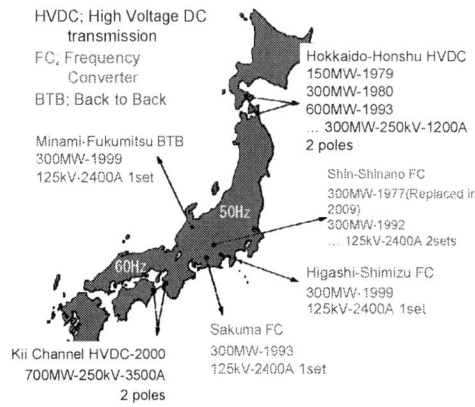

Fig. 1. Frequency converters, HVDC systems and a BTB in Japan.

Japan also has uniqueness in geometry. It is made of many islands. For interconnection between islands, submarine cables are used. For long submarine cable, HVDC technology is applied. The sum of the capacity

978-1-4799-2706-7/14 $31.00 © 2014 IEEE

3834

reaches 2000MW also shown in Fig. 1.

For asynchronous intertie, a back to back converter rated at 300MW is located at Minami-Fukumitsu.

In addition to the DC transmission, reactive power compensators are applied widely to regulate the AC voltage to stabilize the power system. At earlier stage of application, thyristor technology based SVC were applied in Japan. These days, along with capacity increase and performance improvement of the self-commutated converters, STATCOMs plays more popular roles. The high-power self-commutated converter technology is also applied to the power feeder to the electric trains.

III. POWER SEMICONDUCTOR DEVELOPMENT FOR HIGH VOLTAGE AND HIGH POWER CONVERTER

For transmission application, the converters should be rated at several hundreds MW and kV. For such high voltage and high power converters, the power semiconductor devices should also have large ratings of several kA and kV. Taking the capacity of the power semiconductor as the index, the trend of the power semiconductor development is plotted in Fig. 2. The capacity of the semiconductor is simply calculated from the product of its rated voltage and current. Fig. 2 also shows future trends, wide band-gap semiconductors of SiC (Silicon Carbide), under development aiming better performance than the conventional devices based Si (Silicon) for low voltage application.

This chapter focuses on the developments in the LTT and the self-turn-off power semiconductors applied to the large power converters for the power transmission system.

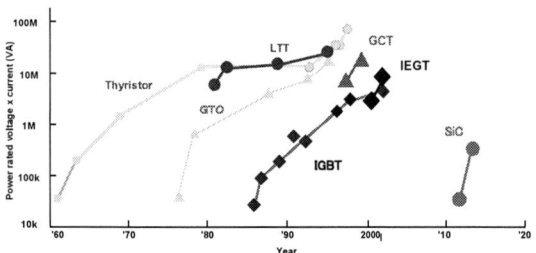

Fig.2. Trends of high power semiconductor development.

A. Light-triggered thyristor

The first semiconductor device with enough capacity applicable to the transmission systems was the thyristor. The thyristor was major semiconductor device in 1960s and 1970s and is still playing major roles in HVDC and in SVC. In 1980s, the development of LTT started.[3] The LTT has various advantages for high voltage application. Almost all DC transmission and thyristor-based SVC projects in Japan use the LTTs.

One of the advantages of LTTs is found in simple circuits at the high voltage level. The electrically-triggered thyristors (ETTs) require gate electronics and its auxiliary power supply circuit. Then, the number of parts used for each thyristor is much larger compared with the LTTs. The LTTs are triggered by the optical gate power

generated in a valve base electronics located at the ground level and no gate electronics is required as shown in Fig. 3. As described in Chapter V, the HVDC converter is made of many series-connected thyristors, the reduction in number of components is one of key factors to improve the converter reliability.[3] The LTTs made an impact worldwide in the high voltage power electronics applications.

The development in thyristor contributed to increase the power density of the thyristor valve for the HVDC converter as shown in Fig. 4. The thyristor valve volume is calculated from the product of width, depth and height of the valve. The spaces for insulation or maintenance are not included. Fig. 4 shows that the power density trend is closely related to the thyristor capacity trend. These simple indices clearly show the development trends of the HVDC converter.

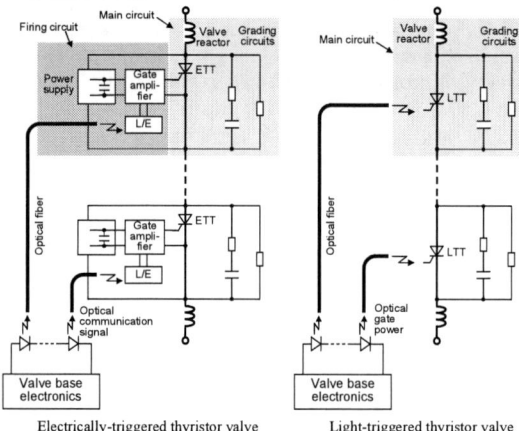

(a)Circuit configuration at high voltage

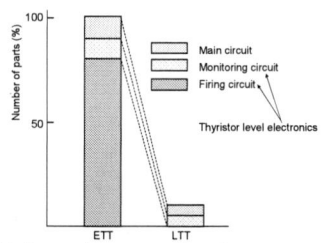

(b) Comparison on number of components

Fig.3. Impacts of LTT to thyristor valves.

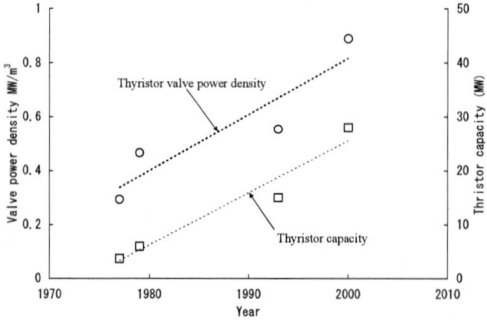

Fig.4. Power density and thyristor capacity of HVDC converter.

B. Self-turn-off power semiconductor

Almost in parallel to the LTT, in late 1970s the GTO was developed as high-capacity self-turn-off device.[5] In 1990s, the GCT has been further developed and shows better characteristics than the GTO.[6] The GTO and GCT are based on the thyristor technology.

Meanwhile, the other branch of power semiconductor device yielded IGBT in late 1980s. The IGBT is originated in the power transistor technology. Then, for higher voltage with lower loss, IEGT has been developed based on IGBT technology.[7]

The first generation of large capacity self-commutated converter was developed based on GTOs. The GTO requires individual snubber circuits to suppress voltage at turn-off and the anode reactor to suppress di/dt at turn-on as shown in Fig. 5(a), which shows a single phase bridge of the two-level converter. After development of the IEGT or the GCT, the self-commutated converter of the second generation has been developed. Fig. 5(b) shows an example for a single phase bridge made of the IEGT. The circuit is much simpler compared with the first generation.

In addition, the efficiency of the converter is improved. The circuit in Fig. 5(a) dissipates larger loss than the circuit in Fig. 5(b). In Fig. 5(a), the snubber capacitor is required to discharge completely when the GTO turns on. Then, the snubber circuit dissipates large loss. On the contrary, in Fig. 5(b), the snubber circuit only absorbs the voltage overshoot when the IEGT turns off. Then, the loss is smaller. Fig. 6 shows the loss comparison between the first and second generation.[8] The loss reduction trend of the self-commutated converters is shown in Fig. 7. Fig. 7 shows points accumulated by surveying the efficiency or loss data published in Japan. [9]

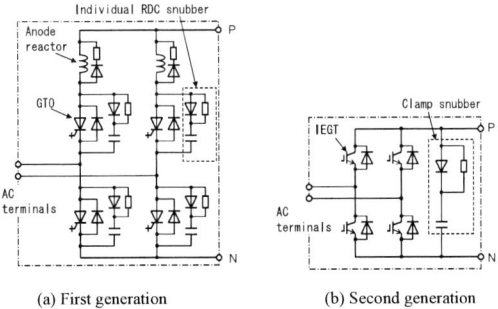

(a) First generation   (b) Second generation

Fig.5. Circuit diagram of first and second generations.

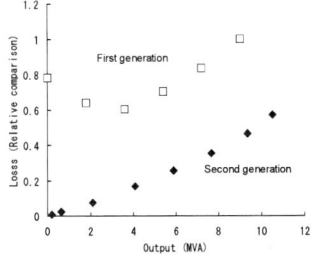

Fig.6. Loss compariton between first and second generations.

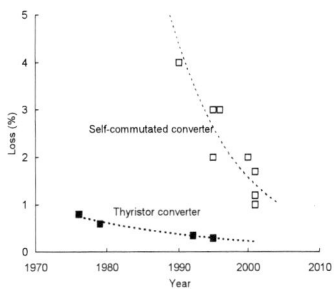

Fig.7. Efficiency trends for high-voltage high-power converters.

IV. INSULATION AND COOLING OF HIGH VOLTAGE AND HIGH POWER CONVERTERS

In Japan, the first field test of HVDC thyristor converter was performed at Sakuma field test station in 1970s.[10][12] It was the one of earliest thyristor valves in the world. There, two types of the thyristor valves were developed. One is insulated by air and cooled by air cooling. The other is oil-immersed. The latter type technology was applied to the first commercial FC installed at Shin-Shinano substation, which started operation in 1977. The structure was similar to the transformer as shown in Fig. 8(a). Hokkaido-Honshu HVDC applied the first type of air-insulation and forced-air-cooling as shown in Fig. 8(b) and started operation in 1978. From 1981 to 1983, in Sakuma again, the air-insulation and water-cooling thyristor valves were tested in the actual FC station.[12] The water cooling system is shown in Fig. 8 (c). After this test, all HVDC converters are made with the air-insulation water-cooling technology in Japan as well as the SVC and the STATCOM.

V. TRENDS OF CONVERTER CIRCUIT TOPOLOGY

The trends of converter circuits are described in two categories. The first one is for the converters for the thyristor based HVDC. The second one is for STATCOM based on the self-turn-off devices.

A. Thyristor valve for HVDC transmission

A typical configuration of the HVDC transmission station is shown in Fig. 9. These days, the 12-pulse topology is popularly used in all of HVDC stations in Japan because of the flexibility in mechanical structure based on air-insulation technology. In contrast to the recent technology, before the replacement, the first Shin-Shinano FC used 6-pulse converters operated in parallel. The FC was based on oil-immersed technology, of which mechanical structure has restrictions to build the valve in the form of double or quadruple structure.

As shown in Fig. 10, a single mechanical structure contains four thyristor valves functioning as arms in the bridges. The mechanical structure is called as the quadruple valve. The valve is further divided to smaller substructures for considering the manufacturing and assembling works. The substructure is called as the thyristor module. In the thyristor module, thyristors are connected in series with its associated circuits including snubber circuits and reactors as shown in Fig. 3(a).

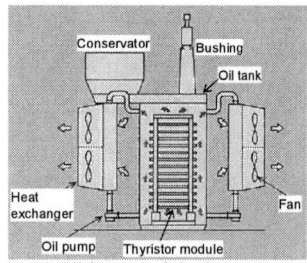

(a) Oil-immerced thyristor valve

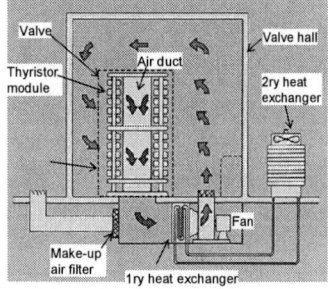

(b) Forced air cooling and air insulation thyristor valve

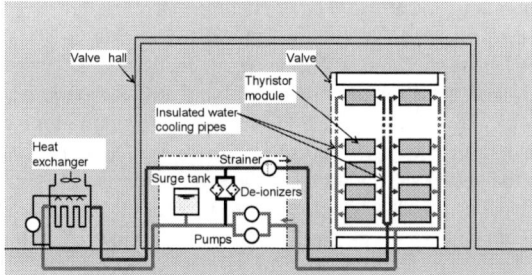

(c) Water cooling and air insulation thyristor valve

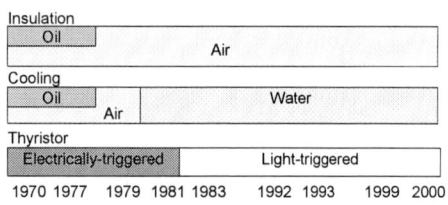

(d) Chronogical trend of HVDC valve technology

Fig. 8. Trends in cooling and insulation technology for high voltage power electronics.

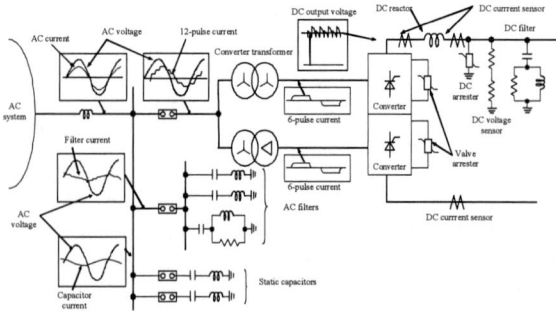

Fig. 9. Typical circuit topology of 12-pulse converter and waveforms for HVDC.

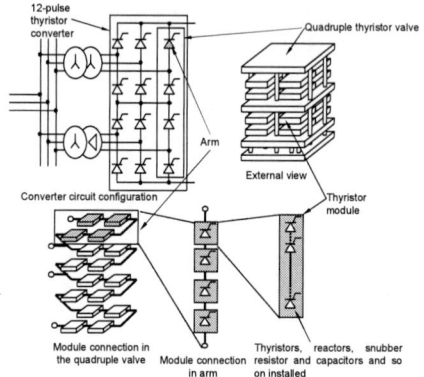

Fig. 10. Thyristor valve structure for 12-pulse converter.

B. Self-commutated converter

Self-commutated converters were developed in early 80s while the GTO ratings started to cover kV and kA range. For AC transmission lines, some of early applications are found for the STATCOM based on the voltage source type converter. One of the pioneers was found at Inuyama Switching Station where a STATCOM rated at 80MVA was installed in 1991.[13] The first STATCOM has been successfully replaced with a new STATCOM of the latest technology.

The first STATCOM was based on the 2-level GTO converter units as shown in Fig. 11(a). The AC output voltages were synthesized through the multistage transformer of eight (8) stages. The converter unit consisted of a voltage source type single-phase bridge, where GTOs are connected in series in an arm as shown in Fig. 11(b). The converter units were controlled with the one-pulse PWM.

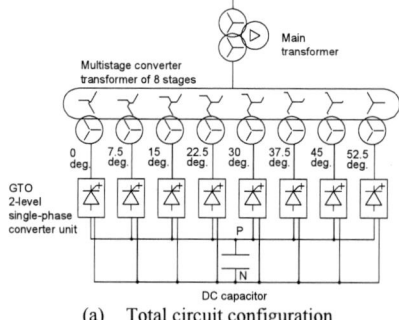

(a)　Total circuit configuration

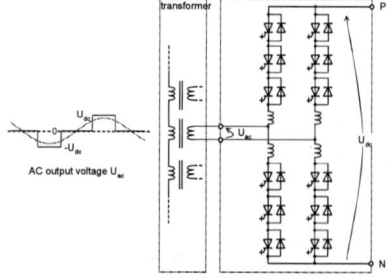

(b)　Configuration of 2-level single-phase converter unit

Fig. 11. Circuit configuration of an early STATCOM.

One of the latest STATCOM is found at Toshin Substation.[14] Fig. 12(a) shows the circuit configuration of the 150MVA unit of STATCOM system rated at 450MVA in total, where the converter units consist of 3-level GCT single-phase bridge as shown in Fig. 12(b). The STATCOM also has a DC overvoltage suppressor.

The 3-level converter outputs higher voltage and better waveforms closer to sinusoidal compared with the 2-level converter output. The 3-level converter offers simpler circuit by reducing the number of stages in the multistage transformer with help of the large-capacity GCT development. The transformer winding structure is also simpler since no phase-shift connections are applied. In the latest STATCOM, only seven (7) stages are required to output 150MVA although the early one had eight (8) stages for 80MVA.

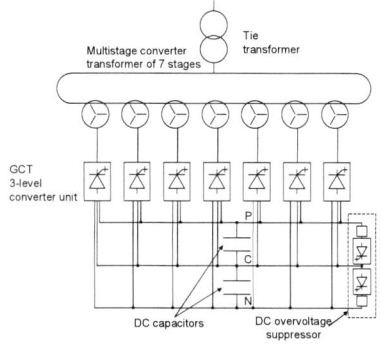

(a) STATCOM with phase-shift multiple winding transformer

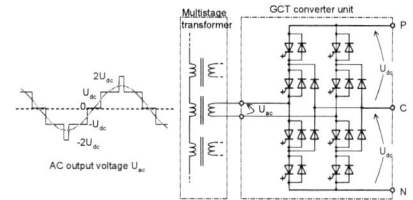

(c)　Configuration of 3-level single-phase converter unit

Fig. 12. Circuit configuration of the latest STATCOM.

VI. CONTROL TECHNOLOGY DEVELOPMENT FOR CONVERTERS

Before describing the converter control, hierarchy of the HVDC control system is introduced to show the position of the converter control. The control system for the SVC or the STATCOM is similar but rather simpler since it does not have DC transmission lines.

A. HVDC control system structure

As shown in Fig. 13, the HVDC station consists of many pieces of equipment and its performance interacts closely with the AC system. Then, the control system is large and takes hierarchical structure as shown in Fig. 13. The converter control has an interface for gate signal multiplication and monitoring of the thyristor valve. The functions are installed in a unit called as the valve base electronics (VBE). They are necessary to handle the hundreds of thyristors are used in the valve. The cooling

system is also large with pumps and many fans. It requires another control. In addition to these, in the AC and DC yards, circuit breakers and switches are installed and controlled individually. The station control and monitoring system takes care of all pieces of equipment in the HVDC station. The station control and monitoring system is further linked to the AC power system control to operate the HVDC as required by the AC power system.

The HVDC control system requires the DC current sensor, DC-CT. Early HVDCs used Kraemer type DC-CT made of two windings. These days, the zero-flux type DC-CTs are widely used in HVDC stations. Furthermore, an optical DC-CT has been developed.[15]

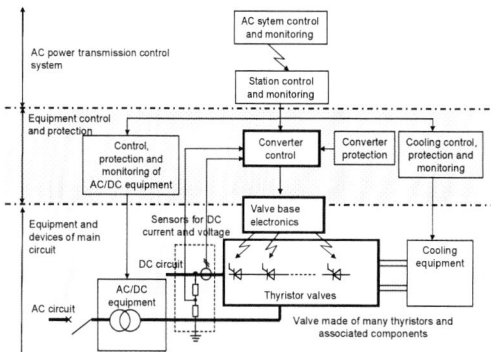

Fig. 13. Simplified control system structure of HVDC station seeing from the converter control.

B. Converter basic control blocks

The basic control blocks in the converter control are shown in Fig. 14. In a two (2) terminal HVDC system, two converter controls operate in coordination. Usually, the sending terminal takes a roll to control the DC current to be Id_{ref} by "ACR", Automatic Current Regulator. The receiving terminal takes the other role to control the DC voltage to be Vd_{ref} by "AVR", Automatic Voltage Regulator. The DC power is fed back to "APR", Automatic Power Regulator" to be kept equal to the power reference Pd_{ref} by adjusting Id_{ref} to the ACR. In case when the AC voltage decreases at the receiving terminal, "AGR", Automatic Gamma Regulator, is selected to keep the gamma enough for successful commutation of the inverter operation. The control signals are exchanged through the communication between the two terminals.

These control blocks are realized by control hardware, of which technical trend is described in the following section.

C. Control hardware technology trend

At the early stage of the converter control, the hardware was based on the analog technology. In 1980s, the development in the personal computer introduced CPU (Central Processing Unit) in the market which had enough performance and could be used for the converter control. The DSPs (Digital Signal Processors) and large capacity memories followed it. They pushed up flexibility and calculation speed in the digital control. Furthermore, large-capacity programmable gate arrays including FPGAs (Field Programmable Gate Arrays) have also developed and the calculation speed is now comparable to the analog control. The digital controls of these days has been much advanced compared with those in 1980s or in early 1990s when the first and second generations of CPU with 8-bit or 16-bit were in the main stream.

In Japan, all of converter control is now based on the digital technology for DC transmission. In the first pole of Hokkaido-Honshu HVDC, the converter control was developed based on analog technology and started operation in 1977. It has been replaced with the latest digital control in 2008. [16]

The digital control trends described above are briefly illustrated in Fig. 15(a).

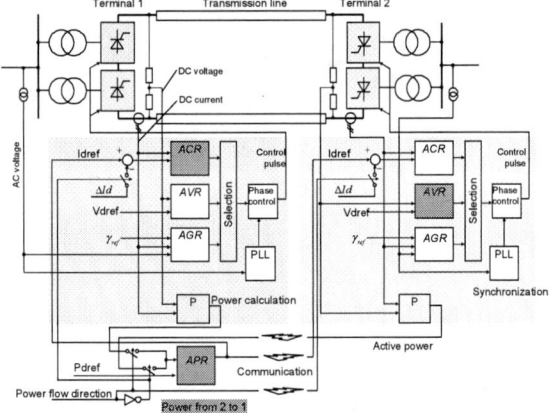

Fig. 14. Basic control block diagram for HVDC transmission.

D. Redundant control system

The digital control has different features in reliability aspect. Considering the feature, redundant control schemes are applied in most of HVDCs in Japan. The duplicated converter control schemes were applied at the same time when the digital control was applied as shown in Fig. 15(b).

The digital control has an advantage to be able to monitor itself on-line with self-diagnostic functions. When a contingency is found in the control unit, the unit is halted to output signals. The halt sequence is performed without interruption since the control pulses are continuously issued to the thyristor valve by the other control unit. The HVDC can continue the power transmission.

E. Advanced control functions

At the early stage, the converters were rather supplemental facilities in the AC power system. However, along with performance improvement of the converter, they begin to be required to play major roles to enhance and to stabilize the AC power system. Then, the converter control should satisfy advanced requirements these days.

Table I lists advanced control functions equipped in HVDC system.[17] The control functions 4), 7) are installed in the converter control.

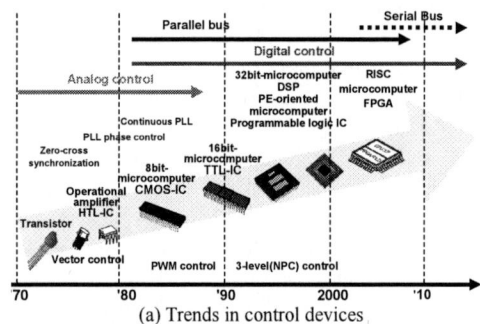

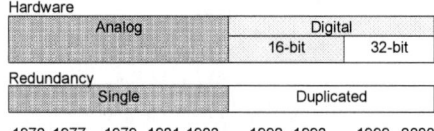

Fig.15. Trends of control for high power converters.

TABLE I. HVDC ADVANCED CONTROL FUNCTIONS

Purpose	Phenomena	Control function
Transient stability improvement	Power unbalance between generation and consumption	1) Emergency power exchange control
	Frequency variation	2) Frequency control
	Power flow swing	3) Power modulation control
	Fast power recovery	4) Converter operation fast recovery control
Overvoltage suppression	Transient overvoltage during and after the AC system fault	5) Reactive device control
Isolated system operation	Coordination with generators	6) Coordinated control with generators
	Subsynchronous torsional interaction	7) SSTI suppression control

VII. HIGH CAPACITY POWER ELECTRONICS APPLICATION TO OTHER FIELDS

The preceding chapters describe the technology trends of the power electronics applied to the transmission systems. The similar trends are found in the other fields applying high-capacity power electronics. This chapter briefly describes the technology trends in the power generation stations and in the railway substations.

A. Power electronics in power generation

In the synchronous generator in the power station, the traditional rectifier is found in the excitation system to feed the DC field current.

As another application for the generation, the static starters for the pumped storage motor-generator was developed based on the thyristor converter in late 1970s.[2] It was almost coincident with the first HVDC and the first FC. The static starters are now used also in the gas turbine power plants.

Then, another sophisticated technology has been developed in late 1980s. It is the AC excitation technology with low-frequency current. The AC excitation enables adjustable speed operation of the motor-generators and offers advantages including higher efficiency. At the first stage, the development was based on the cyclo-converter technology. In 1990s, self-commutated converters have been developed for this application. The development period is synchronized with the STATCOM development. [2]

B. Power electronics in railway substations

Most of the railway feeder systems are fed with DC power converted by conventional rectifiers in Japan. However, the Shinkansen, the bullet train, is fed by AC power. The first Shinkansen line was built between Tokyo and Osaka. As the passenger traffic increases, the electric power consumption has increased in the line. Then, compensators for the voltage variation and the unbalance have been applied to the feeding substations. At the first stage, the thyristor based SVCs were applied in late 1980s. The STATCOMs have been introduced in 1990s.

In 1990s, the RPC, Railway static Power Conditioner, has been developed to regulate the AC voltage balance by exchanging power between the main and the teaser of the Scott-connection transformer, which feeds power to the single phase AC feeder systems.[9]

In 2000s, considering further increase in power demand, FCs are introduced in 50Hz area of the Shinkansen line to increase the active power feeding capacity. These FCs have been developed based on the self-commutated converter technology.[18][19]

VIII. CONVERTER TECHNOLOGY IN FUTURE

This chapter tries to discuss the converter technology trends in future based on the latest development in the circuit topology, control and power device.

A. Circuit topology and control

The modular multi-level converter (MMC) technology is expected to be one of future topologies for HVDC.[20] Some projects have already applied this technology in the world. In order to discuss the features of the MMC, Table II and Fig. 16 are prepared. Table II lists brief comparison between the MMC and the conventional voltage source converter (VSC). The number of the submodule is in the range of hundreds for HVDC rated at hundreds of kV. The hundreds of voltage steps mean that the AC voltage output waveform is very close to sinusoidal. Then, the MMC is expected to show better performance from viewpoint of AC current harmonics.

However, its control is considered to confront many challenges. One of them is the individual control of the capacitor voltage in each submodule. The number of the capacitors is very large and located at very high voltage potential.

One of hopeful technology may be found in the communication field. The mobile phone communication speed is astonishingly increasing on the exponential curve as shown in Fig. 17.[21] In Fig. 17, the vertical axis is scaled with logarithm. The fast serial communication technology can be a practical solution to process many signals going to or coming from the submodules. Without this technology, it might not be possible to realize a practical controller for the MMC.

B. Power semiconductor device

The devices based on SiC (Silicon Carbide) and GaN (Gallium Nitride) are now developed rapidly. As already shown in Fig. 2, the SiC device is now rated at low voltage. It is considered that it takes much time to be applied to the high-voltage converters. One of advantages of the SiC device is faster switching performance than the Si (Silicon) devices. For the low voltage application, the faster switching may offer many benefits.

However, careful consideration is necessary for the application to the high voltage converter. If the SiC device is applied to the conventional converter topology, it will be connected in series to switch hundreds of kV. In this case, the electro-magnetic emission will require careful considerations. The SiC device application to the MMC topology may raise a contradiction where the device has advantage for fast switching although the MMC is based on slow switching operations.

It is a difficult task to discuss the future technology trend based on the past experiences. Years later, revolutionary development might solve the challenges at present.

TABLE II. COMPARISON BETWEEN MMC AND CONVENTIONAL CIRCUIT

Topology	MMC	Conventional VSC
Function	Submodule functioning as a voltage source	Series-connected devices functioning as high-voltage switch
Number of voltage steps	In the range of hundreds	Two or three
Switching frequency	A few per cycle	In the range of thousands per second
Switching loss	Smaller	Larger
DC capacitor	Distributed in submodules	Lumped at DC bus
Difficulty in control	More difficult	Simpler

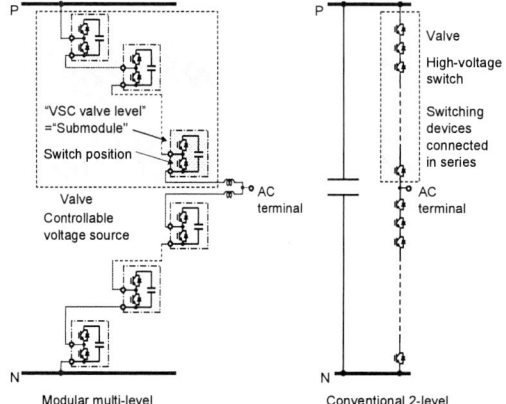

Fig.16. Leg structures for VSC HVDC application.

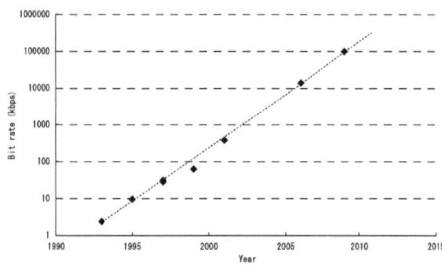

Fig.17. Communication speed trend in mobile phone technology.

IX. CONCLUSION

This paper tries to summaries briefly the technical trends of the power electronics applied to transmission systems in Japan. HVDC and STATCOM technology trends are mainly introduced with the background technologies of the power semiconductors, the insulation, the cooling and the converter circuit structures. The paper also tries to discuss the future technology trends.

It is already passed around 40 years from the first HVDC or the first FC in Japan. However, compared with the history other conventional electrical equipment, the power electronics is considered to be still on the way of development. The technology, which was not possible in the past, is now can be realized along with fast developments in power semiconductor devices and digital devices followed by fast communication technology.

The power electronics is hoped to advance continuously and to contribute further to realize better social infrastructure in future.

REFERENCES

[1] T. Horiuchi, Y. Noro, S. Tanabe, "High-Voltage, Large-Capacity Power Electronics Equipment Applied Widely to Trunk Lines", (in Japanese) Toshiba Review, pp. 8-15, Vol. 55, No. 8, 2000.

[2] Committee of Power Electronics for Electric Power Systems, "Status and design/maintenance guide for Power Electronics for Electric Power Systems", (in Japanese) Technical brochure of Electric Technology Research Association, Vol. 57, No. 2 (2001)

[3] H. Ohashi, T. Tsukakoshi, T. Ogura, & Y. Yamaguchi, "Novel gate structure for high voltage light-triggered thyristor", Japanese J. Appl. Phys.,Vol. 21-1, pp. 91-96, (1982)

[4] S. Kobayashi, T. Takahashi, S. Tanabe, T. Yoshino, T. Horiuchi, T. Senda, "Performance of High Voltage Light-Triggered Thyristor Valve," IEEE Transactions on PAS, vol.PAS-102, no.8, pp.2784,2792, Aug. 1983

[5] M. Kitagawa, Y. Hashiya, "High - Voltage GTO", Toshiba Review, Vol. 43, No. 4, pp. 6-9 (1988)

[6] M. Yamamoto, K. Satoh, T. Nakagawa, A. Kawakami, " GCT (gate commutated turn-off) thyristor and gate drive circuit ", Power Electronics Specialists Conference, 1998. PESC 98 Record, Vol.2, pp. 1711 – 1715, 17-22 May 1998

[7] M. Kitagawa, I. Omura, S. Hasegawa, T. Inoue,A. Nakagawa, "A 4500 V injection enhanced insulated gate bipolar transistor (IEGT) operating in a mode similar to a thyristor", Electron Devices Meeting, 1993. IEDM '93.

[8] M. Tobita, T. Kanai, T. Yoshino, "Development of the 2nd generation SVCS using IEGT," Power Conversion

Conference, 2002. Proceedings of the PCC-Osaka 2002, vol.3, no., pp.1112-1117 2002

[9] Technical Survey Committee of Energy Saving in Static Var Compensator, "Survey on Energy Saving in Static Var Compensator", (in Japanese) IEEJ Technical Report, Vol. 973, (2004)

[10] T. Hayashi et al., "Electric and Electronics Development for 10 Years -4 Development in Power and Energy Field", (in Japanese) Journal of IEE of Japan, pp. 341-347, Vol. 118, No. 6, 1998.

[11] M. Sampei, "Innovation on Energy Power Technology (8) First HVDC Transmission in Japan", (in Japanese) IEE of Japan, Trans on PE, pp. 908-911, Vol.128, No. 7, 2008.

[12] S. Kobayashi, T. Takahashi, "Light-Triggered Thyristor Application to HVDC", (in Japanese) Toshiba Review Vol.38, No. 5. 1983

[13] S. Mori et al., "Development of a large static VAr generator using self-commutated inverters for improving power system stability," IEEE Trans. on Power Systems, Vol.8, No.1, pp.371-377, Feb 1993

[14] T. Fujii, K. Temma, N. Morishima, T. Akedani, T. Shimonosono, H. Harada, "450MVA GCT- STATCOM for stability improvement and over-voltage suppression," Proc. of IPEC2010, pp.1766-1772, 21-24 June 2010

[15] K. Sasaki et al., "Development of DC Optical Current Transformer for Cable Protective Relay in Hokkaido-Honshu HVDC Link", paper presented at ICEE 2009.

[16] T. Murao, K. Shimada, H. Aizawa, "Latest Technologies for Replacement of HVDC systems", (in Japanese) Toshiba Review, pp. 18-21, Vol. 63, No. 12, 2008.

[17] T. Yoshino, Y. Noro, S. Irokawa, "Trends and Development of HVDC Converter Control and Protection", Proc. of IEEJ Annual Meeting 1997, S.23-3 (1997)

[18] K. Kunomura, K. Yoshida, K. Ito, N. Nagayama, M. Otsuki, T. Ishizuka, F. Aoyama, T. Yoshino, "Electronic Frequency Converter", Proc. of IPEC2005 pp.2187-2191, 2-2 April 2005

[19] K. Kunomura, M. Onishi, M. Kai, N. Iio, M. Otsuki, Y. Tsuruma, N. Nakajima, "Electronic frequency converter feeding single-phase circuit for Shinkansen," Proc. of IPEC2010 pp.3136-3143, 21-24 June 2010

[20] A. Lesnicar, R. Marquardt, "An innovative modular multilevel converter topology suitable for a wide power range," Power Tech Conference Proceedings, 2003 IEEE Bologna , Vol.3, no., pp.6, 23-26 June 2003

[21] https://www.nttdocomo.co.jp/info/news_release/report/index.html, "Evolustion of mobile phone communication speed", (in Japanese) NTT Docomo Report No. 40, 2006

Accurate Output Power Control of Converters for Microgrids Based on Local Measurement and Unified Control

Meiqin Mao, Zheng Dong, Yong Ding
Research Center for Photovoltaic System Engineering
Hefei University of Technology
Hefei 230009, China
pvcenter@hfut.edu.cn

Liuchen Chang
Department of Electrical and Computer Engineering
University of New Brunswick
Fredericton, New Brunswick, Canada E3B 5A3
lchang@unb.ca

Abstract—The precise output power control and seamless transit of micro-source converters (MSCs) between grid-tied and islanded operation modes are of great significance for the stability of voltage and frequency, and power flow control in a microgrid. The wireless droop control method (WDCM) is one of the effective methods for MSCs to share loads dynamically and automatically and realize the "plug and play" function. But this method strongly rely on the local voltage and current sampling signals of MSC to dynamically control its output power, therefore the voltage drop induced by the line impedance will cause errors in output power control of MSCs and induce the system unstable. To realize the accurate power tracking and sharing, as well as the automatic seamless transition between grid-tied and islanded modes of microgrid, an accurate output power control method based on local measurement and unified control algorithm is proposed. In this method, two virtual impedances and an on-line estimated power compensation conductance are designed to form a unified controller for MSC, which uses the refined droop control method. The controller is valid for all types of line impedance and loads, besides, the sampling points of voltage and current required are located at the local MSC output terminals, thus increasing the modularity, reliability and flexibility of the controller. Finally, the validity of the improved control scheme is verified by simulation and experiment.

Keywords— Microgrid; droop control; accurate power control; unified controller.

I. INTRODUCTION

A typical microgrid is a single controllable micro electrical network system tightly integrated by local loads, micro-sources and energy storage units through micro-source converters (MSCs). The control algorithm of the MSCs is vital for the realization of microgrid

switching smoothly between grid-tied and islanded modes, accurate power tracking and sharing, and stable voltage and frequency of the system[1].

The traditional controller of a micro-source converter (MSC) needs two different control algorithms for microgrid double-mode operation [2][3]. To guarantee the normal operation of the microgrid, the controller will activate the corresponding algorithm by identifying changes of the operation modes in advance, which usually needs a fast islanding detection method. Therefore, the microgrid performances heavily rely on the speed and accuracy of the island detection method. To address the drawback of above method, the unified controller that could be used at double-modes operation was put forward in references [4][5], where the controller adopts PQ strategy at grid-tied mode, but droop strategy at islanded mode. However, only a single MSC and the inductive line impedance were taken into consideration. In addition, seven close loops at least were used, which makes the design of controller rather complicated. Reference[6] proposed another method for unified control of MSCs by activating or inactivating voltage closed-loop for different operation mode of microgrid, but it needs to keep monitoring the state of microgrid to regulate the voltage closed-loop under different operation mode.

The wireless droop control method (WDCM) [7][8] is one of the effective methods for multi-inverters to share loads dynamically and achieve the automatic transfer between grid-tied and islanded mode, besides, this method contributes to easy system expansion. However, typical WCDM methods don't consider the transmission distance between the output terminal of MSCs and the PCC (point of common coupling) of grid. Therefore, the voltage drop induced by line impedance will cause errors in output power control of MSCs and induce the system unstable. Meanwhile, this conventional frequency and voltage droop control method usually results in real and reactive power coupling, especially in low voltage distribution network. To realize accurate power control, some scholars proposed that the voltage and current signals are sampled at the PCC to eliminate errors in power control caused by the voltage drop of line impedance. Whereas, long communication lines are required by this method, which increases investment and causes sampling signal attenuation and delay. The newly proposed virtual impedance [9] could effectively prevent the coupling between real and reactive power and enhance system stability, however, without proper control algorithm, this method still couldn't realize accurate power control.

To solve the problems above, a unified controller adopting refined droop control method is proposed. In this method, two virtual impedances and an on-line estimated power compensation conductance are designed at the output port of MSC to prevent the coupling between active power and reactive power, and eliminate the errors in power control. Thus, the active and reactive output power of parallel MSCs can be controlled independently and accurately. Meanwhile, the controller could realize the unified control at grid-tied and islanded operation modes of microgrid. Simulation and experiment results verify the correctness of theoretical analysis and validity of the improved control algorithm.

II. CONFIGURATION OF A DG UNIT

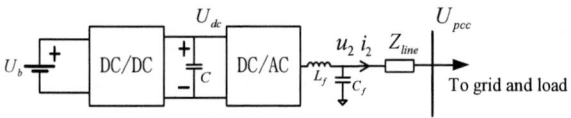

Fig.1 Block diagram of interfacing MSC for battery unit

As shown in Fig.1, the DG unit considered in this paper is for a battery unit in a microgrid. Its power stage includes DC/DC and DC/AC converters with bidirectional power flow, where Z_{line} is equivalent line impedance, inductor L_f and capacitor C_f form the LC filter. The design of DC/AC controller is the focus of the proposed method.

III. BIDIRECTIONAL DC/DC CONTROL ALGORITHM

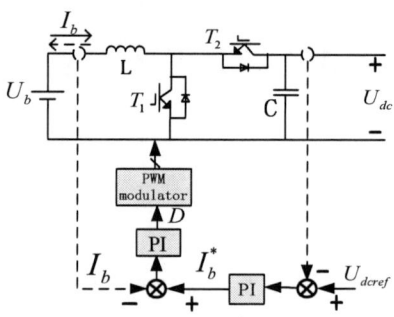

Fig.2 The control method of DC/DC converter

The bidirectional DC/DC converter is shown in Fig.2. This paper adopts outer voltage loop and inner current loop for DC/DC control. The voltage loop is to stabilize the output dc voltage U_{dc}, while the current loop is to improve the stability of system and reduce the regulation time. The battery could be charged and discharged according to the commands from upper energy management unit, and D is duty cycle for power electronic switch T_1 or T_2 in DC/DC converter.

IV. PROPOSED POWER CONTROL ALGORITHM WITH IMPROVED ACCURACY

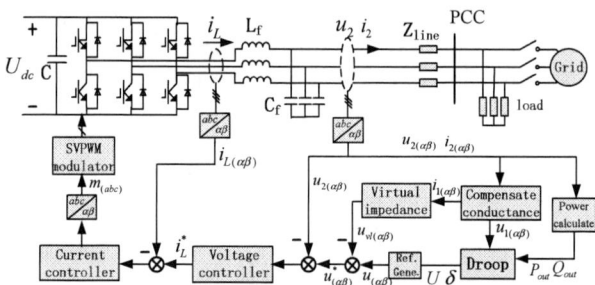

Fig.3 The control scheme for DC/AC converter

In this section, a refined droop control algorithm is proposed, which is based on proper design of the double resistive virtual impedance and a power compensation conductance.

A. Virtual impedance and voltage control loop

Fig. 3 shows the DC/AC converter control scheme. The reference voltage for the converter comes from the real and reactive power droop control loops, which determine the DG output voltage magnitude and frequency. For the voltage control of MSCs, a multi-loop control scheme is implemented, where an inner filter inductor current (i_L) feedback loop is embedded in an outer filter capacitor voltage (u_2) feedback loop. Both the voltage and current controllers are implemented on the stationary frame to avoid the complex frame transformations. In the α-β frame, the reference signals are sinusoidal, and therefore, a PR (proportional-resonant) controller is adopted to ensure the excellent reference tracking, noted for its infinite gain at fundamental frequency. However, for practical use, PR may fail to work, since once the real grid frequency deviates from the designed resonant frequency, PR only have a little gain [2]. Therefore a quasi-PR is adopted with advantages in its much wider bandwidth. The quasi-PR controllers in the form of

$$ G(s) = k_{pu} + \frac{2k_{Iu}\omega_c s}{s^2 + 2\omega_c s + \omega_o^2} \qquad (1) $$

where k_{Pu} is the proportional gain, k_{Iu} is the resonant gain and ω_c is the cutoff frequency for resonant bandwidth control.

Another advantage for the voltage loop adopting quasi-PR regulator is that the inherent equivalent output impedance of MSC is zero. Therefore the equivalent output impedance of MSC only includes the virtual impedance, which will not vary with line parameters and loads.

The resistive output impedance of the MSC is implemented easily by subtracting a proportional term of the output current from the reference voltage (generated from the power loops) to produce the final MSC voltage reference, as illustrated in Fig.4. In contrast with the inductive output impedance, the resistive output impedance could make the overall system more damped and provide automatic harmonic current sharing and decrease THD of output voltage[7][10].

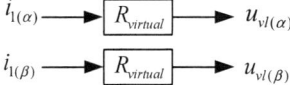

Fig.4 Resistive virtual impedance realization

B. *Accurate power control algorithm*

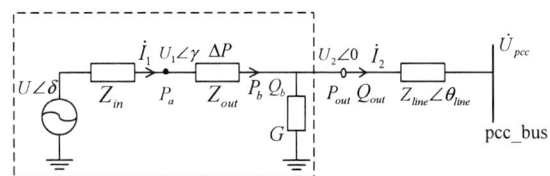

Fig. 5 Improved MSC equivalent circuit

In the proposed method, the sampling points of voltage u_2 and current i_2 are located at MSC output

terminals, and the two virtual impedances and an on-line estimated power compensation conductance are designed. The equivalent circuit is shown in Fig.5.

In Fig.5, Z_{in} and Z_{out} are called inner virtual impedance and outer virtual impedance respectively. Here setting $Z_{in}=R_{in}$, $Z_{out}=R_{out}$ by adding resistive virtual impedance as illustrated in Part A. The sampled voltage u_2 and current i_2 are measured to calculate feedback voltage u_1 and current i_1, and the U_1, U_2, I_1, I_2 are the amplitude of voltage u_1, u_2 and current i_1, i_2 respectively.

As R_{in} and R_{out} are resistive, droop property of amplitude and frequency of voltage u_o is corresponding to P_{out} and Q_{out} respectively. Further when outside virtual impedance R_{out} is large enough, the effects of the line impedance Z_{line} can be neglected, but R_{out} will "consume" part of active output power simultaneously. Therefore, to realize the accurate loads sharing among several MSCs, a power compensation conductance G is introduced to compensate part of the active power consumed by R_{out}.

The controller adopting the same adaptive droop method for both islanded and grid-tied mode is analyzed below in detail.

1) Islanded mode

The control objective of MSCs in parallel to the bus is to share the loads in proportion to their capacity at islanded mode, and the active power control loop and reactive power control loop are shown in Fig. 6.

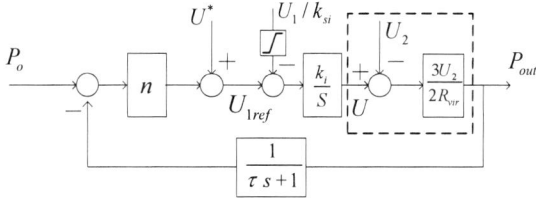

(a) Active power control loop

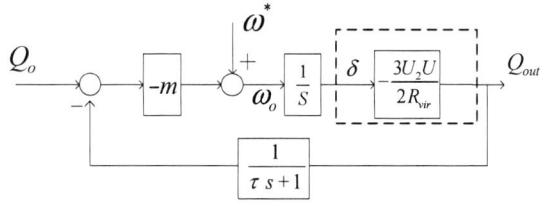

(b) Reactive power control loop

Fig. 6 Control block of droop control method

In Fig.6, n and m are the droop coefficients of active and reactive power droop control, P_o and Q_o are command values of active and reactive power, ω^* and U^* are the basis reference of angle frequency and amplitude of output voltage for droop control. τ is the time constant of first order filter for power feedback. k_{si} is the output voltage compensation coefficient. The parameters of droop control always meet $\omega^* = \omega_g$, $U^* = U_g$, $R_{vir} = R_{1n} + R_{out}$, where ω_g and U_g are the angle frequency and amplitude of grid voltage. To realize accurate power sharing, the MSCs should adopt the same voltage compensation coefficients k_{si}. And the power command values P_{oi} and Q_{oi} of MSCi meet,

$$
\begin{cases}
n_i P_{oi} = n_j P_{oj} \\
m_i Q_{oi} = m_j Q_{oj}
\end{cases}
\tag{2}
$$

In order to eliminate the effect of virtual impedance on active power control, an integral term is added to the active power control loop, as shown in Fig.6(a). In addition, the droop control relationship is derived as,

$$
\begin{cases}
\omega_o = \omega^* + m(Q_{out} - Q_o) \\
U_{1ref} = U^* - n(P_{out} - P_o)
\end{cases}
\tag{3}
$$

It could be proved that, if the outer impedance R_{out} is large enough than line impedance, and R_{outi} meets $R_{outi} / n_i = C$, where C is a constant. Besides, the value of compensation conductance G_i meets [3]:

$$
\mathrm{Re}\left\{\frac{3}{2}\dot{U}_{2i}\left(G_i\dot{U}_{2i}\right)^*\right\} = -\frac{1}{2}\Delta P_i
\tag{4}
$$

Then, the amplitude of feedback voltage u_1 of each MSC has the same value,

$$
U_{11} = U_{12} = \cdots = U_{1k}
\tag{5}
$$

The feedback voltage u_{1i} and current i_{1i} can be calculated from the formulas (6) and (7).

$$
\dot{I}_{1i} = \dot{I}_o + \frac{\dot{U}_{2i}}{G}
\tag{6}
$$

$$
\dot{U}_{1i} = \dot{U}_{oi} + \dot{I}_{1i} R_{outi}
\tag{7}
$$

As the voltage drop of virtual impedance will result in output voltage less than the rated voltage, the voltage compensation coefficient k_{si} is designed to compensate output voltage. A compensation method to regulate the output voltage u_2 to u_g adaptively is shown in Fig.7.

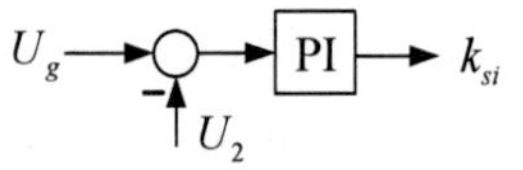

Fig. 7 Output voltage compensation method

When the system is steady, according to the active power loop in Fig.6(a), the voltage U_{1ref} meets,

$$
U_{1ref} = U_1 / k_{si}
\tag{8}
$$

According to derivation above, we know that the value of U_{1refi} of MSCi ($i=1,...,k$) is equal, thus actual output active power of MSCi conforms to

$$
n_1 P_{out1} = n_2 P_{out2} = \cdots = n_k P_{outk}
\tag{9}
$$

where n_i is the droop coefficient of active power of MSCi. When the system is steady, frequency of each MSCi output voltage is at the same value, thus it's derived that

$$
m_1 Q_{out1} = m_2 Q_{out2} = \cdots = m_k Q_{outk}
\tag{10}
$$

Therefore, the MSCs in the microgrid could share active and reactive loads in proportion.

For practical use, the MSC droop coefficient needs to be regulated in inverse proportion to the capacity of DGs. So the DGs could share the active and reactive load in proportion to their capacities.

2) Grid-tied mode

At grid-tied mode, P_o and Q_o are the command value of active and reactive output power for MSC respectively, and their values are determined by the microgrid energy management unit. To track the active and reactive power command at grid-tied mode and avoid algorithm switching, an amplitude limiting is added for the feedback voltage U_1/k_{si}, and its upper limit is the amplitude of grid voltage, as shown in Fig.6 (a).

When system operates at islanded mode, $U_1/k_{si} = U_{1ref} < U_g$, the limit is useless because the feedback quantity is less than the upper limit. While at grid-tied mode, as the MSC discharge, usually $U_2 > U_g$. Therefore, according to Fig.7, k_{si} is smaller than 1($k_{si}<1$), thus $U_1/k_{si} > U_g$, the feedback voltage is limited to the grid voltage. When the battery is charged, $U_2 < U_g$, thus the logical relationship in Fig.7 should be reversed to get a small value of k_{si}. This procedure is realized automatically.

According to Fig.6, as the feedback voltage is limited to grid voltage, the active output power of MSC at grid-tied mode is P_o. The system operates at the same frequency when steady, namely the grid frequency, and therefore reactive power tracks command value Q_o accurately. In this way, both the output active power and reactive power of MSCs can track their command values precisely.

No matter the line impedance is resistive, inductive, capacitive or complex, the derivation is the same. Thus, this method is valid for all kinds of line impedances.

C. *Adaptive modulation of outer virtual impedance*

As the line impedance may change with loads,

network elements and system conditions, the value of outer virtual impedance Z_{out} is adjusted adaptively to the line impedance for better performance of system.

When microgrid is at grid-tied mode, here a real-time equivalent line impedance estimation method is adopted based on online measurement to obtain the equivalent line impedance without injecting disturbance signals into the power supply network:

$$Z_{lineh} = \frac{\dot{U}_{2h} - \dot{U}_{pcch}}{\dot{I}_{2h}} = \frac{\dot{U}_{2h}}{\dot{I}_{2h}} \tag{11}$$

where $\dot{U}_{pcch}$, $\dot{U}_{2h}$ and $\dot{i}_{2h}$ are the grid voltage, output voltage and current of MSC corresponding to h order harmonic, where h is associated with the switching feature of converters. As the grid voltage doesn't contain h order harmonic voltage, thus $\dot{U}_{pcch} = 0$.

We set that the outer impedance Z_{out} meets the following condition [3],

$$\frac{|Z_{out}|}{|Z_{line}|} \geq 10 \tag{12}$$

Then, the equation (5) will be always true. To reduce the voltage drop caused by outer virtual impedance, here set $|Z_{out}| = 10 * |Z_{line}|$.

V. SMALL-SIGNAL MODELING AND ANALYSIS

To analyze the stability and the transient response of the system with the proposed controller, a small-signal model of the MSC system is constructed. According to the Fig.6, as the output impedance is resistive, the active and reactive output powers can be expressed as [8]

$$P_b = \frac{3U_2}{2(R_{in} + R_{out})}(U\cos\delta - U_2)$$
$$Q_b = \frac{-3U_2}{2(R_{in} + R_{out})}U\sin\delta \tag{13}$$

By perturbing (13), and considering the low pass filter which average the instantaneous active and reactive power, it yields to

$$\hat{p}_b = \frac{3}{2(\tau s+1)}\frac{U_2}{R_{in}+R_{out}}(-U\Delta\hat{\delta}+\hat{U})$$
$$\hat{q}_b = -\frac{3}{2(\tau s+1)}\frac{U_2}{R_{in}+R_{out}}(U\hat{\delta}+\hat{U}\Delta) \tag{14}$$

where $^\wedge$ denotes perturbed values，capital letters mean equilibrium point values. In a similar way, we get the perturbed values of active power P_a shown in Fig.6,

$$\hat{p}_a = \frac{3}{2(\tau s+1)}\frac{U_1}{R_{in}}(-U\Delta\hat{\delta}+\hat{U}) \tag{15}$$

As the compensation conductance G only outputs active power $\Delta P / 2$, So, the output power of MSC is

$$P_{out} = P_b + \frac{\Delta P}{2} = \frac{P_a + P_b}{2}$$
$$Q_{out} = Q_b \tag{16}$$

Then the perturbed values of output power are,

$$\hat{p}_{out} = \frac{\hat{p}_a + \hat{p}_b}{2}$$
$$\hat{q}_{out} = \hat{q}_b \tag{17}$$

In addition, formula (3) can be linearized as

$$\hat{\omega} = m\hat{Q}_c$$
$$\hat{U}_{1ref} = -n\hat{P}_c \tag{18}$$

In Fig.6, output of the power controller is the reference voltage U_{1ref} for an integrator with coefficient k_i, the voltage error is fed to a the integrator whose output is U, namely $(U_{1ref} - \frac{U_1}{k_{si}})\frac{k_i}{s} = U$, by perturbing this equation, we obtain

$$\hat{U}_{1ref} = \frac{s}{k_i}\hat{U} \tag{19}$$

By substituting (17) and (19) into (18), and taking $\hat{\omega}$ as $s\hat{\delta}$, the small signal dynamics of parallel system can be obtained as

$$s^4 + As^3 + Bs^2 + Cs + D = 0 \tag{20}$$

Where the coefficients is

$$A = 2/\tau$$
$$B = \frac{1}{\tau}(\frac{1}{\tau} + \frac{3k_i}{4}nK_2 + \frac{3}{2}mK_1U)$$
$$C = \frac{3}{4\tau^2}(K_2nk_i + 2mK_1U)$$
$$D = \frac{9}{8\tau^2}k_imnK_1K_2U(1+\Delta^2) \tag{21}$$
$$K_1 = \frac{U_2}{R_{in} + R_{out}}$$
$$K_2 = \frac{U_2}{R_{in} + R_{out}} + \frac{U_1}{R_{in}}$$

TABLE I

Parameters of MSC

Symbol	Meaning	Value
R_{in}	Inner virtual impedance	2.5 Ω
R_{out}	Outer virtual impedance	6 Ω
R_{line}	Equivalent line resistor	0.2 Ω
L_{line}	Equivalent line inductor	0.2 mH
m	Frequency droop coefficient	0.0004
n	Amplitude droop coefficient	0.0004
τ	Time constant	0.01
R_{load}	Common load resistor	10 Ω
L_{load}	Common load inductor	15 mH
k_i	Integral coefficient of real power loop	100

By using (20), with the parameters listed in Table I, the dynamics and stability of the closed-loop system can be evaluated through the root locus plots shown in Fig.8.

In Fig.8, s_1, s_2, s_3 and s_4 are the four poles of the system, and arrows indicate the evolution of the corresponding poles when the coefficients m and n vary. As droop coefficient n is a small value, poles s_3 and s_4 near the imaginary, they are the dominant poles and determine the dynamic performance of the system. As increasing the coefficient n from 0 to 5×10^{-4}, the conjugated poles s_3 and s_4 go far away from real and imaginary axis, and meet with poles s_1 and s_2, ultimately the system have two real poles s_1, s_2 and two conjugated poles s_3, s_4. When decreasing m, the locus move away from the imaginary axis, making the system response faster. Note that both the virtual impedance R_{in} and R_{out} and the integral coefficient k_i have little effect on the location of the roots in comparison with m and n coefficients. The poles all are in the left half plane of pole-zero map, so the system is always stable.

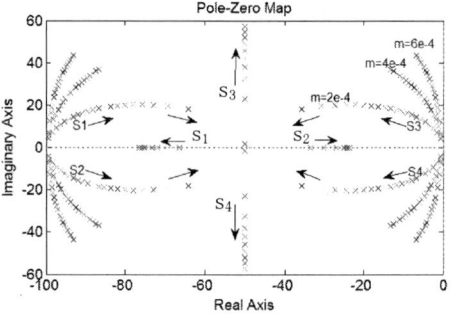

Fig.8 Root locus diagram with $m=2 \times 10^{-4}$, 4×10^{-4} and 6×10^{-4}, for

$$0 < n < 5 \times 10^{-4}.$$

As the system has four eigenvalues, so it could eliminate the steady-state error. For the stability of system, the expression (22) should be met in practical use [1] [9]:

$$Re(s_i) \leq -5 \quad (i = 1, 2, 3, 4) \tag{22}$$

Then, the dynamic performance of the system is mostly determined by conjugated poles s_3 and s_4. To obtain better dynamic performance, we set damping ratio of conjugated poles within the best one [1] [9],

$$1 < \frac{Im(s_j)}{Re(s_j)} < 1.5 \quad (j = 3, 4) \tag{23}$$

Fig.9 The appropriate value range for parameters m and n

According to equation (22) (23), the appropriate value range of m and n is shown in Fig.9. The unit step response of active and reactive output power control loop for corresponding m and n is shown in Fig.10. It shows that the system have quite good dynamic performance with the refined parameters .

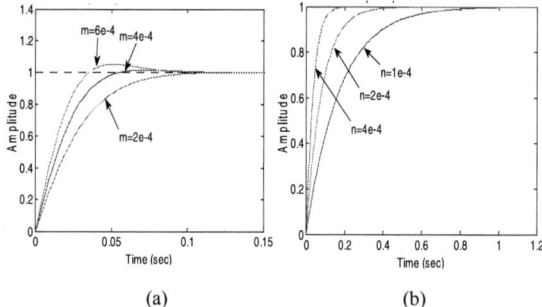

Fig.10 Unit step response of (a)active power, and (b)reactive power

VI. SIMULATIONS AND EXPERIMENTAL VERIFICATION

In order to verify the correctness and effectiveness of the proposed control algorithm, simulations with load variation and modes switching of microgrid with two parallel MSCs are carried out in MATLAB/Simulink. In addition, the power control and seamless transit between grid-tied mode and islanded mode verified in experiment.

The system parameters for the two MSCs are as follows: $R_{out1}=6\Omega$, $R_{out2}=3\Omega$, $n_1=0.0005$, $n_2=0.00025$, $m_1=0.0005$, $m_2=0.001$. (subscript 1 and 2 represent the MSC1 and MSC2)

A. Simulation results with abrupt load change

The system starts at islanded mode, the objective is to control the active output power of MSC1 (P_1) to be half of that of MSC2 (P_2), and the reactive output power of MSC1 (Q_1) is twice of that of MSC2 (Q_2). The microgrid is connected to the grid at 0.4s and the power command values are $P_{o1}=2000W$, $P_{o2}=4000W$, $Q_{o1}=2000var$, $Q_{o2}=1000var$. The load varies at 0.2s and 0.6s. As shown in Fig.11, when the system is steady at islanded mode, $P_1=2000W$, $P_2=4000W$, $Q_1=2000var$, $Q_2=1000var$. At 0.2s, the load varies, the output power becomes $P_1=4000W$, $P_2=8000W$, $Q_1=3800var$, $Q_2=1900var$. At grid-tied mode, the steady output power is $P_1=2000W$, $P_2=4000W$, $Q_1=2000var$, $Q_2=1000var$ and the output power don't vary with the load shift. The simulation results show that output power of inverters is consistent with the proportion set beforehand at islanded mode and track the power command at grid-tied mode despite the loads fluctuation.

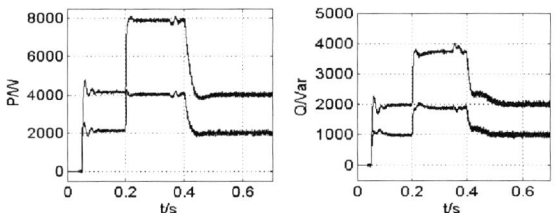

Fig.11 Active and reactive output power of MSCs with loads shifting

B. Multiple transitions process between islanded and grid-tied mode with charge and discharge shift

At first stage, the system operates at islanded mode and the power command value P_{o1}=2000W, P_{o2}=4000W, Q_{o1}=2000var, Q_{o2}=1000var. At 0.25s, the microgrid is connected to the grid. At 0.45s, the command values are changed to P_{o1}=-2000W, P_{o2}=-4000W, Q_1=-2000var, Q_2=-1000var. The system is disconnected from the grid at 0.7s. As shown in Fig.12, at islanded mode, the steady value of output power is P_1=4000W, P_2=8000W, Q_1=3800var, Q_2=1900var. At grid-tied mode, the steady output power is P_1=2000W, P_2=4000W, Q_1=2000var, Q_2=1000var at first, then change to P_1=-2000W, P_2=-4000W, Q_1=-2000var, Q_2=-1000var. When the system returns to islanded mode, the steady value of output power is the same with first stage. This shows that the two MSCs share the load in proportion at islanded mode and follow the charging and discharging command accurately at grid-tied mode. In addition, the proposed method could realize the seamless transit between grid-tied and islanded mode.

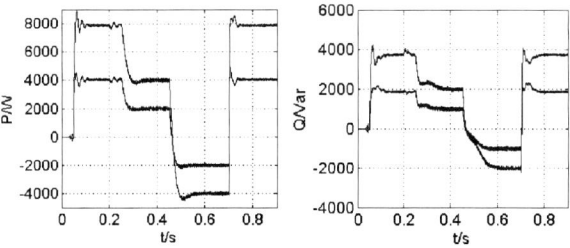

Fig.12 Active and reactive output power of MSCs with mode switching

C. Experiment results for the automatic transit of MSC between grid-tied mode and islanded mode

The rated output phase-to-phase voltage of MSC is set as 80V, and the load resistor is 28 Ω, the active and reactive power command value is P_o=100W, Q_{o1}=20Var respectively. The experiment waveforms of output phase-to-phase voltage and current are shown in Fig.13 when operation modes of microgrid are shifted.

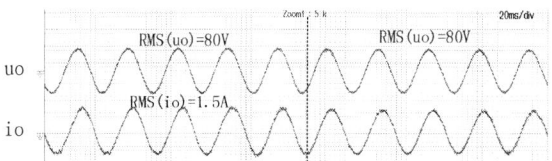

(a) Operation mode transfers from islanded mode to grid-tied mode

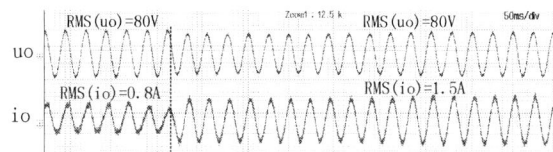

(b) Operation mode transfers from grid-tied mode to islanded mode

Fig.13 Output voltage and current waveforms of MSC

In Fig13, the dashed line indicates the time when operation mode is shifted. It is demonstrated that the proposed control algorithm could realize the automatic seamless transition between different operation mode and stabilize the voltage at the islanded mode.

Fig.14 shows the active output power P and reactive output power Q waveforms of the tested MSC when the operation mode switches and power command values change. At islanded mode, the active output power P=228W, reactive output power Q=0. When the MSC is connected to the grid, its active and reactive output power follow the command value exactly, respectively, as shown in Fig.14(a). As the active power command is increased to P_o=400W, and reactive power to Q_{o1}=100Var, then the corresponding results are shown in Fig.14(b). In addition, with decreasing active power command to P_o=100W and reactive power command remain unchanged, the results is provided in Fig.14(c). The experimental results verify that the output power of MSC could track the power command value precisely. At last, the case disconnecting MSC from the grid is performed as shown in Fig.14(d).

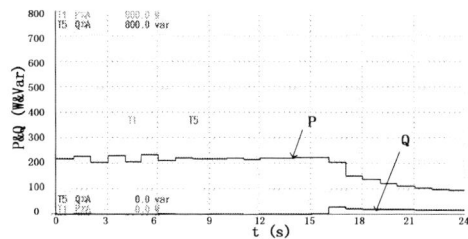

(a) Operation mode transfers from islanded mode to grid-tied mode

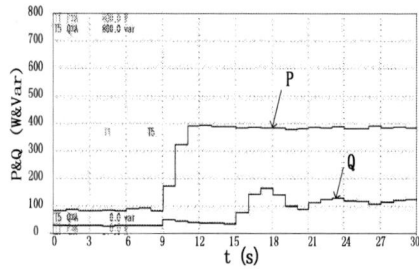

(b) Output power command increases to P_o=400W, Q_o=100Var at grid-tied mode

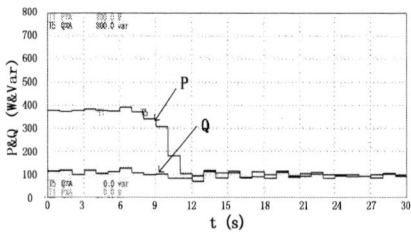

(c) Active output power command decreases to P_o=100W with reactive power command unchanged at grid-tied mode

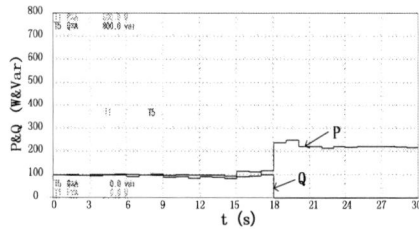

(d) Operation mode transfers from grid-tied mode to islanded mode.

Fig.14 Waveforms of real and reactive output power of MSC

VII. CONCLUSION

This paper proposes a unified controller adopting refined droop method. The control algorithm is useful for multi-MSCs connected in parallel. The simulation and experiment results show the proposed algorithm not only realize accurate power tracking at grid-tied mode and sharing load in proportion to their capacity at islanded mode, but also realize the automatic seamless transition between grid-tied mode and islanded modes of microgrid. The further research will be focused on the situation with abrupt grid voltage.

ACKNOWLEDGEMENT

The authors would like to acknowledge the support partially from 973 Grant Project of MOST (2009CB219708), Grant Project of NSFC (51077033) and Guanggong Provice Introduced Innovation Talent Project (2011N015).

REFERENCES

[1] Yan Du; Guerrero, J.M.; Liuchen Chang; Jianhui Su; Meiqin Mao, "Modeling, analysis, and design of a frequency-droop-based virtual synchronous generator for microgrid applications," ECCE Asia Downunder (ECCE Asia), vol., no., pp.643,649, 3-6 June 2013

[2] Yun Wei Li; ChingNan Kao, "An accurate power control strategy for inverter based distributed generation units operating in a low voltage microgrid," *Energy Conversion Congress and Exposition, 2009. ECCE 2009. IEEE* , vol., no., pp.3363,3370, 20-24 Sept. 2009

[3] Mao Meiqin，Shen Kai，Chang Liuchen. "Accurate Output Power Control of Inverters for Microgrids Based on Local Measurement," IEEE 4th International Symposium on Power Electronics for Distributed Generation Systems,2013.

[4] Yunwei Li; Vilathgamuwa, D.M.; Poh Chiang Loh, "Design, analysis, and real-time testing of a controller for multibus microgrid system," Power Electronics, IEEE Transactions on , vol.19, no.5, pp.1195,1204, Sept. 2004

[5] Delghavi, M.B.; Yazdani, A., "A Unified Control Strategy for Electronically Interfaced Distributed Energy Resources," *Power Delivery, IEEE Transactions on* , vol.27, no.2, pp.803,812, April 2012

[6] Zeng Liu; Jinjun Liu; Yalin Zhao, "A Unified Control Strategy for Three-Phase Inverter in Distributed Generation," *Power Electronics, IEEE Transactions on*, vol.29, no.3, pp.1176,1191, March 2014

[7] Guerrero J M，Vicuna L G，Matas J，et al. A wireless controller to enhance dynamic performance of parallel inverters in distributed generation systems[J]. IEEE Transactions on Power Electronics，2004，19(5)：1408-1413.

[8] Wei Yao; Min Chen; Matas, J.; Guerrero, J.M.; Zhao-ming Qian, "Design and Analysis of the Droop Control Method for Parallel Inverters Considering the Impact of the Complex Impedance on the Power Sharing," Industrial Electronics, IEEE Transactions on , vol.58, no.2, pp.576,588, Feb. 2011

[9] Jinwei He; Yun Wei Li, "Analysis, Design, and Implementation of Virtual Impedance for Power Electronics Interfaced Distributed Generation," *Industry Applications, IEEE Transactions on* , vol.47, no.6, pp.2525,2538, Nov.-Dec. 2011

[10] Guerrero, J.M.; Matas, J.; Luis Garcia de Vicuna; Castilla, M.; Miret, J., "Decentralized Control for Parallel Operation of Distributed Generation Inverters Using Resistive Output Impedance," Industrial Electronics, IEEE Transactions on , vol.54, no.2, pp.994,1004, April 2007

Impedance-Based Analysis of Active Frequency Drift Islanding Detection Method for Grid-Tied Inverter System

Bo Wen, Dushan Boroyevich, Rolando Burgos, and
Zhiyu Shen
Center for Power Electronics Systems (CPES)
Virginia Tech, Blacksburg, VA 24061 USA
wenbo@vt.edu

Paolo Mattavelli
University of Padova, Italy

Abstract— **This paper presents an impedance-based analysis of active frequency drift (AFD) islanding detection method for grid-tied inverter systems. To study the AFD islanding detection method, output impedance of grid-tied inverter is modeled. Islanding detection is analyzed using the interaction between inverter output impedance and load impedance. The analysis further shows that the AFD islanding detection method has the potential to destabilize the grid-connected inverter system when the grid is weak or the size of inverter is large. Experimental results verified the analysis.**

Keywords— Impedance, inverter, islanding detection, stability.

I. INTRODUCTION

Islanding detection is an important function of distributed generation units because of the requirement of grid code compliance [1]. On the other hand, islanding detection function gives distributed generation unit the capability of adjusting its operation mode according to grid condition. For example, the three-phase inverter, as shown in Fig. 1, can work as a current source when it is connected to the grid. It can work as a voltage source to feed the local load (islanding mode) when it is disconnected with grid because of grid faults.

Many islanding detection methods have been developed. Methods that rely on monitoring the changes of grid parameters are defined as passive methods. For example, over/under voltage or frequency detection. Methods generate small perturbations to force the change of grid parameters are defined as active methods. For example, active frequency drift (AFD) islanding detection method forces frequency drifts enough to activate over/under frequency protection when islanding condition happens. Other active methods include grid impedance estimation and negative sequence detection [2]. Among them, AFD methods are popular because of their simplicity, small non-detection zone, and low impact on system power quality. Most AFD methods generate a frequency positive feedback, such as the Sandia frequency shift (SFS) method, which drives the converter system frequency away from the steady-state and detects the islanding event [3-7]. Other methods are

based on the analysis of PLL small-signal stability [8].

Different from traditional analysis of AFD method, this paper proposes an impedance-based analysis which unveils that frequency drift is the consequence of interaction between inverter output impedance and the impedance of local loads. Moreover, impedance-based analysis also shows that, AFD method has the potential to destabilize the grid-connected inverter system when the grid is week or the size of inverter is large.

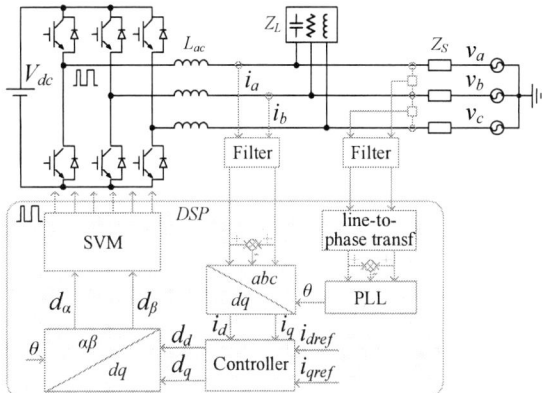

Fig. 1. Three-phase grid-tied PWM inverter system.

The paper is organized as follows: section II discusses the PLL-based and SFS AFD islanding detection methods. Section III discusses the proposed impedance-based analysis. IV shows the experimental results. Section V is the summary.

II. AFD ISLANDING DETECTION METHODS

In this section, the principles of two AFD islanding detection methods are discussed.

A. PLL-Based Method

The PLL-based AFD islanding detection method [8] is shown in Fig. 2. This method uses a typical rotating reference frame PLL scheme with a small-signal feed-forward loop.

This work was supported by the Center for Power Electronics Systems (CPES) Renewable Energy and Nanogrids (REN) mini-consortium.

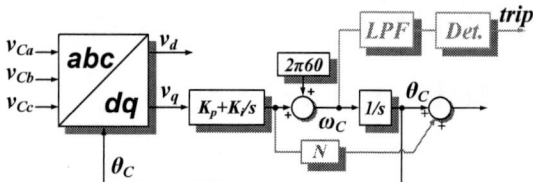

Fig. 2. PLL-based AFD islanding detection method.

Under islanding condition, inverter is controlled as a current source feeding to the local load, as shown in Fig. 3. The terminal voltage v_c of inverter is determined by the current injected to the load and the impedance of the load. This voltage is collected by inverter for PLL to generate frequency and angle to do current control, as shown in Fig. 4. In another word, under islanding condition, inverter is synchronized with the voltage generated by itself. By adjusting the value of N, the pole of PLL can be moved from left half plane to right half plane. When PLL has right half plane pole, it becomes unstable, the output frequency drift away from steady-state value, the bigger the N value is the faster the drift speed is. Islanding event can be detected by over or under frequency detection function of the inverter. Detailed analysis can be found in [8]. This method features a small impact to the inverter system operation as well as an easy implementation.

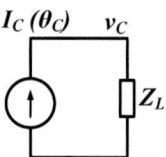

Fig. 3. One line circuit of inverter system under islanding condition.

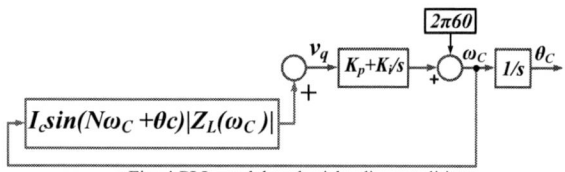

Fig. 4.PLL model under islanding condition.

B. SFS Method

SFS method is one of the positive feedback anti-islanding method. The details of the SFS method can be found from [9, 10] where the inverter reference currents i_{dref} and i_{qref} are processed by a phase angle transformation block to get the new current reference settings i_{dref}^* and i_{qref}^*. The definition of the transformation and the phase angle are:

$$\begin{bmatrix} i_{dref}^* \\ i_{qref}^* \end{bmatrix} = \begin{bmatrix} \cos\theta_f & -\sin\theta_f \\ \sin\theta_f & \cos\theta_f \end{bmatrix} \cdot \begin{bmatrix} i_{dref} \\ i_{qref} \end{bmatrix} \qquad (1)$$

$$\theta_f = \frac{\pi}{2}\left(cf_0 + K(\omega - \omega_0)\right) \qquad (2)$$

The angle θ_f used in the transformation block is calculated from the SFS control, in which ω is the inverter terminal voltage frequency, ω_0 is the grid nominal frequency, K is the positive feedback gain, and

cf_0 is the initial chopping factor. [9]

Positive feedback always tries to perturb the grid-tied inverter system. When the grid is connected, the system is operated stably. However, when islanding event happens, the system is destabilized so that the frequency drift detection unit is tripped.

III.   IMPEDANCE-BASED ANALYSIS OF INVERTER SYSTEM AFD METHODS

Unlike most analyses for AFD islanding detection methods, this paper presents an impedance-based analysis which uses the terminal characteristics of inverter and loads. The advantages of impedance-based analysis is that impedance can be measured, no internal parameters of the inverter are need, and it is easier to address compatibility issue between inverter and the grid by using impedances and Nyquist stability criteria.

Grid-tied inverter can be modeled as a voltage source in series with an output impedance as shown in Fig. 5. Impedance-based analysis studies the islanding event by the interaction between the output impedance of inverter and the input impedance of local load. When the system is connected to the grid, the grid impedance should be considered too; for a stiff grid, the impedance is small, for a weak grid, the impedance is big.

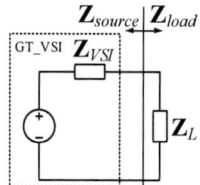

Fig. 5. Impedance-based analysis of inverter system under islanding condition.

A.   Impedance of local load

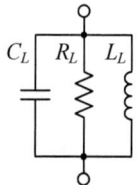

Fig. 6. Load configuration for unintentional islanding test.

Fig. 6 shows the local load configuration (for one line) for unintentional islanding test defined by [1], R_L is equal to 1 pu value of the inverter, which means the real power needed by the resistor is provided only by the inverter. The Q factor of the circuit is equal to 1.0±0.05, when it is equal to 1, the reactive power needed by L_L and C_L are equal to real power provide by the inverter.

B.   Output impedance of inverter with AFD method

Fig. 7 shows the average model of grid-tied inverter with PLL-based AFD method discussed in section II. Notice that the inverter system is divided into two parts, one is in the system d-q domain, and one is in the converter or control system *d-q* domain. The system domain is defined by the grid voltage, while the converter domain is defined by the PLL, which tracks the frequency and angle of the grid voltage to find the position of system domain. In steady-state, the converter domain is aligned with the system domain. When the grid voltage contains small-signal perturbations, which means the system domain is changing, the converter domain is no longer aligned with the system domain because of the PLL dynamics (PI of PLL).

In effect, small-signal perturbations of the system voltage propagate to the PLL output angle, and further to the converter domain current. Through the current controller, the perturbation propagates to the duty ratio and the voltage generated by the inverter power stage and finally to the output current of the inverter. This analysis indicates that the PLL has an impact on the output impedance of the grid-tied inverter.

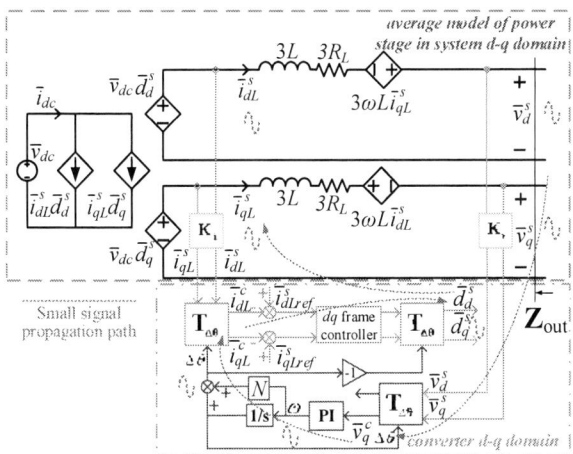

Fig. 7. Average model of grid-tied inverter with AFD method.

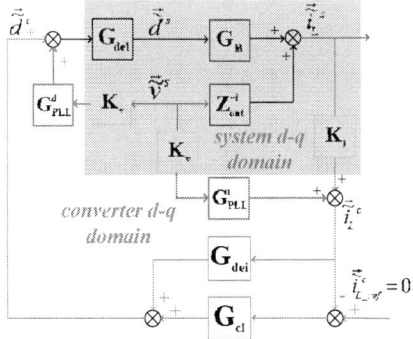

Fig. 8. Transfer function matrix flow chart representation of grid-tied inverter small-signal model with PLL-based AFD method and current feedback control.

Fig. 8 shows the transfer function matrix flow chart representation of grid-tied inverter small-signal model with PLL-based AFD method and current feedback control. $\mathbf{G}_{id}$ is the transfer function matrix from duty ratio $\tilde{d}^s$ to inductor current $\tilde{i}_L^s$, $\mathbf{Z}_{out}$ is the open loop output impedance which can be derived using the small-signal circuit shown in Fig. 9, and the expression is:

$$\mathbf{Z}_{out} = \begin{bmatrix} Z_{dd} & Z_{dq} \\ Z_{qd} & Z_{qq} \end{bmatrix} = \begin{bmatrix} 3Ls + 3R_L & -3\omega L \\ 3\omega L & 3Ls + 3R_L \end{bmatrix} \quad (3)$$

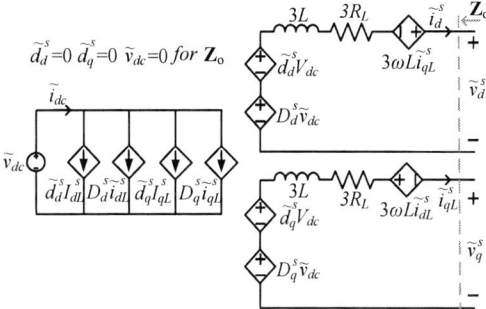

Fig. 9. Average model of grid-tied inverter with AFD method.

In order to model the small-signal propagation path through PLL, transfer function matrices $\mathbf{G}_{PLL}^i$ and $\mathbf{G}_{PLL}^d$ are defined. $\mathbf{G}_{PLL}^i$ models the path from system voltage to current in converter domain. $\mathbf{G}_{PLL}^d$ models the path from system voltage to duty cycle in converter domain. $\mathbf{K}_v$ is the filter for voltage signals and $\mathbf{K}_i$ is for current signals.

Fig. 2 shows the synchronous reference frame (SRF) PLL. the SRF transformation from *abc* to *dq* is:

$$\mathbf{T}_{dq} = \frac{2}{3} \begin{bmatrix} \cos(\theta) & \cos\left(\theta - \frac{2}{3}\pi\right) & \cos\left(\theta + \frac{2}{3}\pi\right) \\ -\sin(\theta) & -\sin\left(\theta - \frac{2}{3}\pi\right) & \sin\left(\theta + \frac{2}{3}\pi\right) \end{bmatrix} \quad (4)$$

Once the power stage is modeled to SRF, the average model of PLL becomes Fig. 10, which shows that PLL is an estimator of system domain voltage vector v_d^s and v_q^s. $T_{\Delta\theta}$ is the rotation matrix shown in (5). $T_{\Delta\theta*}$ is the matrix used for duty cycle and current vectors rotation shown in (6).

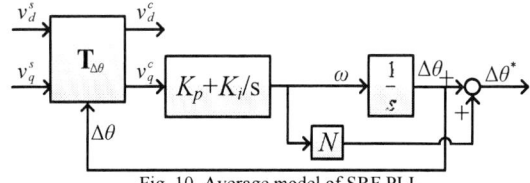

Fig. 10. Average model of SRF PLL.

$$\mathbf{T}_{\Delta\theta} = \begin{bmatrix} \cos(\Delta\theta) & \sin(\Delta\theta) \\ -\sin(\Delta\theta) & \cos(\Delta\theta) \end{bmatrix} \quad (5)$$

$$\mathbf{T}_{\Delta\theta*} = \begin{bmatrix} \cos(\Delta\theta^*) & \sin(\Delta\theta^*) \\ -\sin(\Delta\theta^*) & \cos(\Delta\theta^*) \end{bmatrix} \quad (6)$$

Let

$$tf_{PLL} = K_p + K_i / s \quad (7)$$

Actually, in the grid-tied inverter system, voltage and current vectors are firstly rotated from system domain to converter domain for feedback control, and then duty cycle vector is rotated from converter domain to system domain to give converter power stage command. Their relationship are:

$$\vec{D}^c = \mathbf{T}_{\Delta\theta*}\vec{D}^s, \quad \vec{V}^c = \mathbf{T}_{\Delta\theta}\vec{V}^s, \quad \vec{I}^c = \mathbf{T}_{\Delta\theta}\vec{I}^s \quad (8)$$

Under steady state the angle between vectors in two

domains is zero:

$$\vec{D}^c = \begin{bmatrix} \cos(0) & \sin(0) \\ -\sin(0) & \cos(0) \end{bmatrix} \vec{D}^s \qquad (9)$$

Small signal perturbation can be added to (9) to get:

$$\begin{bmatrix} D_d^c + \tilde{d}_d^c \\ D_q^c + \tilde{d}_q^c \end{bmatrix} = \begin{bmatrix} \cos(0 + \tilde{\theta}^*) & \sin(0 + \tilde{\theta}^*) \\ -\sin(0 + \tilde{\theta}^*) & \cos(0 + \tilde{\theta}^*) \end{bmatrix} \cdot$$
$$\begin{bmatrix} D_d^s + \tilde{d}_d^s \\ D_q^s + \tilde{d}_q^s \end{bmatrix} \qquad (10)$$

By doing approximation of trigonometric functions, and cancel the steady state values. Relationship between converter domain duty cycle and system domain duty cycle can be derived as:

$$\begin{bmatrix} D_d^c + \tilde{d}_d^c \\ D_q^c + \tilde{d}_q^c \end{bmatrix} = \begin{bmatrix} 1 & \tilde{\theta}^* \\ \tilde{\theta}^* & 1 \end{bmatrix} \cdot \begin{bmatrix} D_d^s + \tilde{d}_d^s \\ D_q^s + \tilde{d}_q^s \end{bmatrix} \qquad (11)$$

$$\begin{bmatrix} \tilde{d}_d^c \\ \tilde{d}_q^c \end{bmatrix} \approx \begin{bmatrix} \tilde{d}_d^s + D_q^s \tilde{\theta}^* \\ -D_d^s \tilde{\theta}^* + \tilde{d}_q^s \end{bmatrix} \qquad (12)$$

Since,

$$\tilde{\theta}^* = \tilde{\theta} + N \cdot tf_{PLL} \cdot \tilde{v}_q^c \qquad (13)$$

Then,

$$\begin{bmatrix} \tilde{d}_d^c \\ \tilde{d}_q^c \end{bmatrix} \approx \begin{bmatrix} \tilde{d}_d^s + D_q^s \left(\tilde{\theta} + N \cdot tf_{PLL} \cdot \tilde{v}_q^c \right) \\ -D_d^s \left(\tilde{\theta} + N \cdot tf_{PLL} \cdot \tilde{v}_q^c \right) + \tilde{d}_q^s \end{bmatrix} \qquad (14)$$

By doing similar small signal analysis as above for voltage vectors, relation between PLL output angle and q axis voltage can be derive as [11]:

$$\tilde{v}_q^c = -V_d^s \tilde{\theta} + \tilde{v}_q^s \qquad (15)$$

$$\tilde{\theta} = \frac{tf_{PLL}}{s + V_d^s tf_{PLL}} \tilde{v}_q^s = G_{PLL} \tilde{v}_q^s \qquad (16)$$

Substitute (15) and (16) to (14):

$$\begin{bmatrix} \tilde{d}_d^c \\ \tilde{d}_q^c \end{bmatrix} \approx \begin{bmatrix} \tilde{d}_d^s \\ \tilde{d}_q^s \end{bmatrix} +$$
$$\begin{bmatrix} 0 & D_q^s \left[\left(1 - N tf_{PLL} V_d^s \right) G_{PLL} + N tf_{PLL} \right] \\ 0 & -D_d^s \left[\left(1 - N tf_{PLL} V_d^s \right) G_{PLL} + N tf_{PLL} \right] \end{bmatrix} \begin{bmatrix} \tilde{v}_d^s \\ \tilde{v}_q^s \end{bmatrix} \qquad (17)$$

Then:

$$\mathbf{G}_{PLL}^d = \begin{bmatrix} 0 & -D_q^s \left[\left(1 - N tf_{PLL} V_d^s \right) G_{PLL} + N tf_{PLL} \right] \\ 0 & D_d^s \left[\left(1 - N tf_{PLL} V_d^s \right) G_{PLL} + N tf_{PLL} \right] \end{bmatrix} \qquad (18)$$

Using similar approach, $\mathbf{G}_{PLL}^i$ can be derived as:

$$\mathbf{G}_{PLL}^i = \begin{bmatrix} 0 & I_q^s \left[\left(1 - N tf_{PLL} V_d^s \right) G_{PLL} + N tf_{PLL} \right] \\ 0 & -I_d^s \left[\left(1 - N tf_{PLL} V_d^s \right) G_{PLL} + N tf_{PLL} \right] \end{bmatrix} \qquad (19)$$

Solving the equations represented by Fig. 8, the output impedance of grid-tied inverter system with PLL-based AFD is:

$$\mathbf{Z}_{out\_il\_PLL} = \left(\mathbf{Z}_{out}^{-1} - \mathbf{G}_{id} \mathbf{G}_{dd} \left(\left(\mathbf{G}_{cl} - \mathbf{G}_{del} \right) \mathbf{G}_{PLL}^i - \mathbf{G}_{PLL}^d \right) \mathbf{K}_v \right)^{-1} \cdot$$
$$\left(\mathbf{I} + \mathbf{G}_{id} \mathbf{G}_{del} \left(\mathbf{G}_{cl} - \mathbf{G}_{del} \right) \mathbf{K}_i \right) \qquad (20)$$

TABLE I. PARAMETERS FOR SIMULATION AND CALCULATION

Symbol	Description	Value
V_{dc} (V)	Dc voltage	270
V_d^s (V)	D channel source voltage	99.6
V_q^s (V)	Q channel source voltage	0
f (Hz)	Line frequency	60
L_{ac} (mH)	Ac inductor	1
R_{Lac} (mΩ)	Ac inductor ESR	110
I_{dref} (A)	D channel current reference	-10
I_{qref} (A)	Q channel current reference	0
Kpi	Current controller proportional gain	0.0233
Kii	Current controller integral gain	2.5598
Kp	PLL proportional gain	1
Ki	PLL integral gain	2
f_{sw} (kHz)	Switching frequency	20
R_L (Ω)	Load resistor	10
L_L (mH)	Load inductor	17.4
C_L (μF)	Load capacitor	400

Using the parameters in Table I, Fig. 11 shows the Bode plot of grid-tied inverter output impedance. The results shows that the q-q channel dc impedance is a negative incremental resistance. The magnitude of Z_{qq} equals to the per unit value of inverter system. A higher power rating of the inverter will result in a lower magnitude negative resistance as shown in the Z_{qq}.

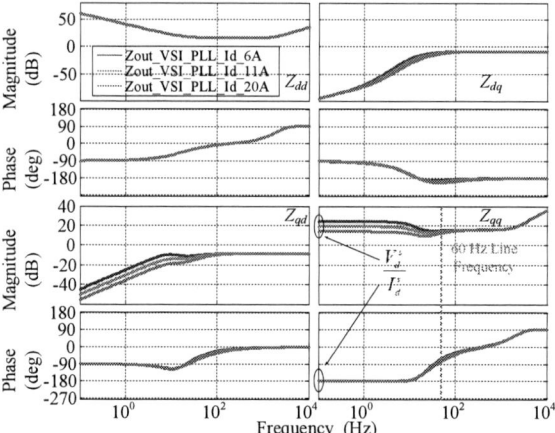

Fig. 11. Output impedance of grid-tied inverter without PLL-based AFD method.

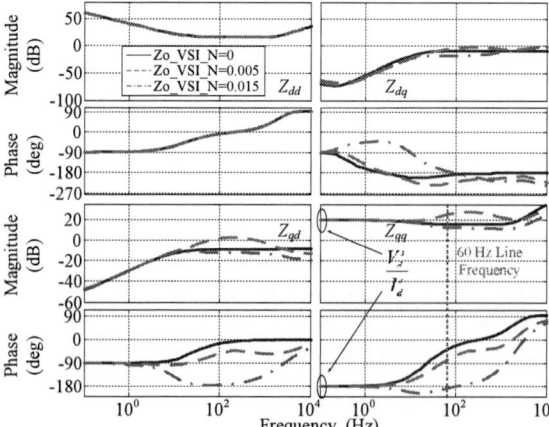

Fig. 12. Output impedance of grid-tied inverter with PLL-based AFD method.

Fig. 12 shows the output impedance of the grid-tied inverter with different feed forward gain N in the PLL-based AFD method. Because of the feed forward loop, the phase of Z_{qq} drops below -180°. The bigger feed forward gain is, the lower the phase of Z_{qq} is.

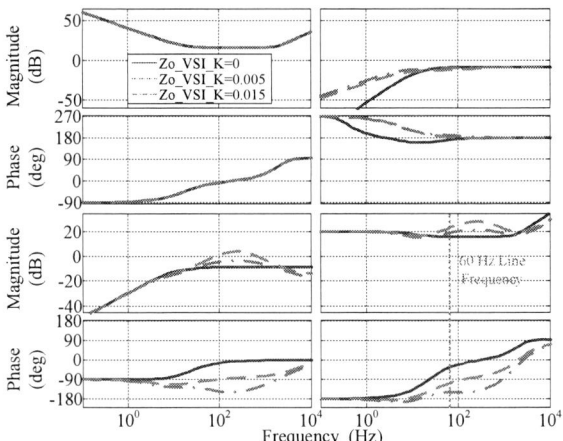

Fig. 13. Output impedance of grid-tied inverter with SFS AFD method.

Fig. 13 shows the output impedance of the grid-tied inverter with different positive feedback gain K in the SFS AFD method. Similar phenomenon can be found that the phase of Z_{qq} drops below -180°. The bigger positive feedback gain K is, the lower the phase of Z_{qq} is.

C. Impedance-based analysis

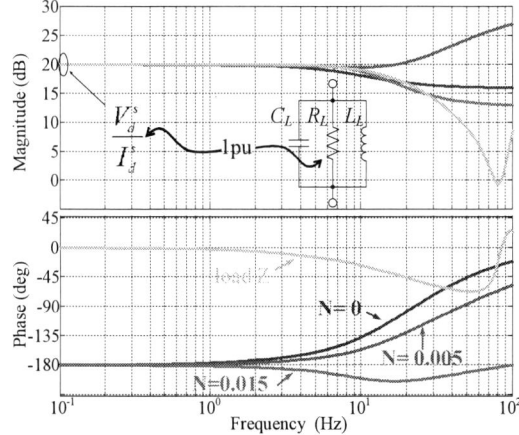

Fig. 14. Impedance interaction in *qq* channel between inverter and local load under islanding condition.

When the output impedance of the grid-tied inverter and the impedance of local load are put together, an impedance interaction is found. As shown in Fig. 14, the dc impedance of the inverter is equal to the dc impedance of the local load. With feed forward gain $N = 0$, the phase difference is 180°; with N bigger than 0.015, the system becomes unstable due to lack of phase margin.

As been reported in [12], grid-tied inverter output impedance has RHP pole. If it is treated as source converter, it is better to use generalized inverse Nyquist (GINC) criteria to study the system stability. Fig. 15

shows the characteristic loci of system return ratio reciprocal which indicates that the system is destabilized with bigger feed forward gain N.

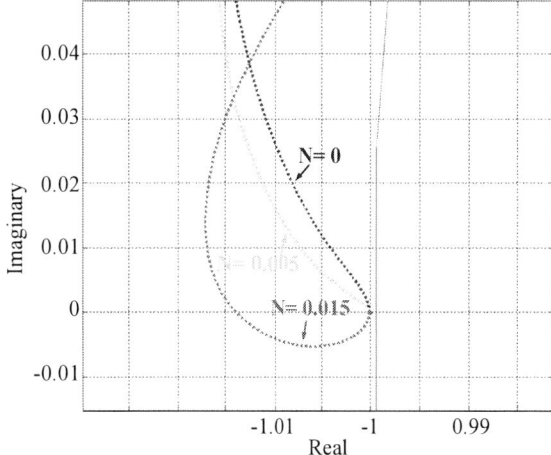

Fig. 15. Characteristic loci of inverter system under islanding condition with different feed forward gain.

IV. EXPERIMENTAL RESULTS

In order to verify the proposed model and analysis, output impedances of a three-phase grid-tied inverter with PLL-based and SFS AFD islanding detection methods are measured with different parameters which is listed in Table II. According to the analysis above, the critical dynamics locates at low frequency range (0.1~10Hz) which the impedance analyzer [13] in the lab cannot measure. In order to show the interesting dynamics, the line frequency is changed to 400 Hz, the bandwidth of PLL is also increased as shown in Table II.

TABLE II. PARAMETERS FOR EXPERIMENTAL VERIFICATION

Symbol	Description	Value
f (Hz)	Line frequency	400
K_p	PLL proportional gain	8.921
K_i	PLL integral gain	3964

Fig. 16 shows the comparison between measurement results and model calculation results for Z_{qq} of grid-tied inverter with and without PLL-based AFD method. Fig. 17 shows the corresponding comparison for SFS AFD method.

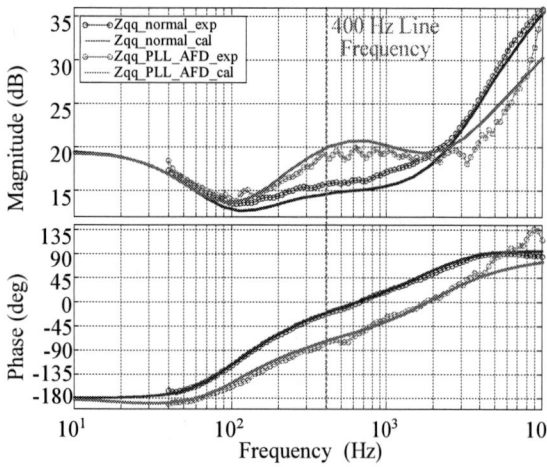

Fig. 16 Measurement results of Z_{qq} for PLL-based AFD method with feed forward gain $N = 0.0005$.

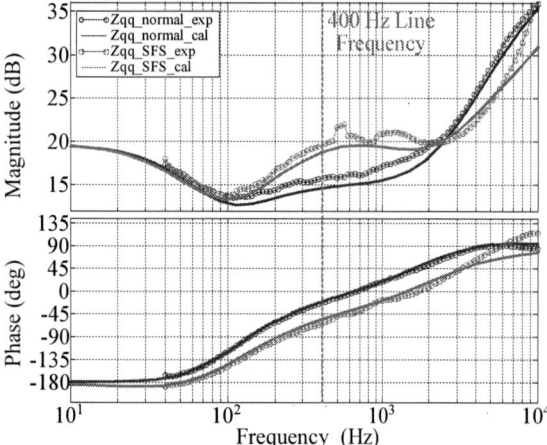

Fig. 17. Measurement results of Z_{qq} for SFS AFD method with positive feedback gain $K = 0.0008$.

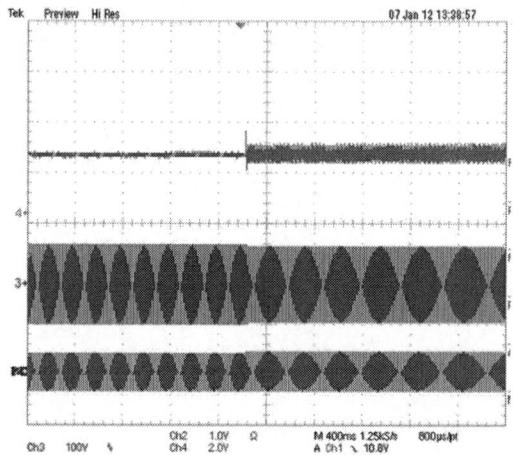

Fig. 18. PLL with feed forward gain $N = 0$ when islanding event happens: terminal voltage (purple) [100 V/DIV], inverter current (blue) [20A/DIV], ω_c (green) [5 Hz/DIV].

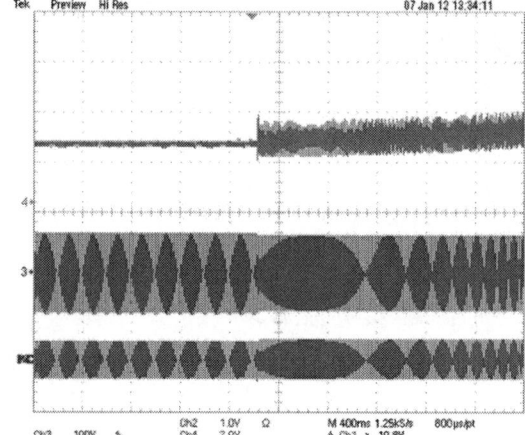

Fig. 19. PLL with feed forward gain $N = 0.015$ when islanding event happens: terminal voltage (purple) [100 V/DIV], inverter current (blue) [20A/DIV], ω_c (green) [5 Hz/DIV].

The time domain experiments are carried out using parameters in Table I. Fig. 18 shows the islanding event with feed forward gain $N = 0$. It shows that the PLL stays at it steady-state value, the system is stable. Both voltage and current does not change due to the matched load condition. Fig. 19 shows the islanding event with N = 0.015. It can be seen that PLL output frequency starts to drift when islanding event happens, the system is unstable.

V. CONCLUSION

This paper presented an impedance-based analysis for active frequency drift islanding detection methods. To do the analysis, the output impedance of a grid-tied inverter was modeled with PLL-based AFD method. The Z_{qq} of the grid-tied inverter output impedance features a negative incremental impedance. Since a grid-tied inverter q-q channel impedance magnitude is equal to the dc impedance magnitude of the local load, impedance interaction happens when islanding event happens. With big feed forward gain N for islanding detection, the phase of Z_{qq} is shown to drop below −180°. Under islanding condition, the inverter system became unstable with a frequency drift away from steady-state. Based on the impedance of the inverter, it could be concluded that the grid-tied inverter with AFD islanding method has the potential to destabilize the grid-connected inverter system when the grid is weak. Experimental results verified the analysis.

REFERENCES

[1] "IEEE Standard for Interconnecting Distributed Resources With Electric Power Systems," *IEEE Std 1547-2003,* pp. 0_1-16, 2003.

[2] R. L. Teodorescu, M.; Rodr?guez, P. , *Grid Converters for Photovoltaic and Wind Power Systems*: Wiley-IEEE Press 2011.

[3] M. E. Ropp, M. Begovic, and A. Rohatgi, "Analysis and performance assessment of the active frequency drift method of islanding prevention," *Energy Conversion, IEEE Transactions on,* vol. 14, pp. 810-816, 1999.

[4] L. A. C. Lopes and S. Huili, "Performance assessment of active frequency drifting islanding detection methods," *Energy Conversion, IEEE Transactions on,* vol. 21, pp. 171-180, 2006.

[5] D. Velasco, C. Trujillo, G. Garcera, and E. Figueres, "An Active Anti-Islanding Method Based on Phase-PLL Perturbation," *Power Electronics, IEEE Transactions on,* vol. 26, pp. 1056-1066, 2011.

[6] A. Yafaoui, W. Bin, and S. Kouro, "Improved Active Frequency Drift Anti-islanding Detection Method for Grid Connected Photovoltaic Systems," *Power Electronics, IEEE Transactions on,* vol. 27, pp. 2367-2375, 2012.

[7] D. Pengwei, Y. Zhihong, E. E. Aponte, J. K. Nelson, and F. Lingling, "Positive-Feedback-Based Active Anti-Islanding Schemes for Inverter-Based Distributed Generators: Basic Principle, Design Guideline and Performance Analysis," *Power Electronics, IEEE Transactions on,* vol. 25, pp. 2941-2948, 2010.

[8] D. Dong, D. Boroyevich, P. Mattavelli, W. Bo, and X. Yaosuo, "Anti-islanding protection in three-phase converters using grid synchronization small-signal stability," in *Energy Conversion Congress and Exposition (ECCE), 2012 IEEE,* 2012, pp. 2712-2718.

[9] W. Xiaoyu, W. Freitas, and X. Wilsun, "Dynamic Non-Detection Zones of Positive Feedback Anti-Islanding Methods for Inverter-Based Distributed Generators," *Power Delivery, IEEE Transactions on,* vol. 26, pp. 1145-1155, 2011.

[10] H. H. Zeineldin and M. M. A. Salama, "Impact of Load Frequency Dependence on the NDZ and Performance of the SFS Islanding Detection Method," *Industrial Electronics, IEEE Transactions on,* vol. 58, pp. 139-146, 2011.

[11] B. Wen, D. Boroyevich, P. Mattavelli, Z. Shen, and R. Burgos, "Influence of phase-locked loop on input admittance of three-phase voltage-source converters," in *Applied Power Electronics Conference and Exposition (APEC), 2013 Twenty-Eighth Annual IEEE,* 2013, pp. 897-904.

[12] B. Wen, D. Boroyevich, P. Mattavelli, Z. Shen, and R. Burgos, "Modeling the Output Impedance Negative Incremental Resistance Behavior of Grid-Tied Inverters," in *Applied Power Electronics Conference and Exposition (APEC), 2014 Twenty-Nineth Annual IEEE,* 2014.

[13] R. B. G. Francis, D. Boroyevich, F. Wang, and K. Karimi, "An algorithm and implementation system for measuring impedance in the D-Q domain," presented at the Energy Conversion Congress and Exposition (ECCE), 2011.

Development of 200-Mvar Class Thyristor Switched Capacitor Supporting Fault Ride-Through

Asuka Ohtake, Fei Zhang, Takafumi Fujimoto, Naoyuki Nakayama
Toshiba Mitsubishi-Electric Industrial Systems Corporation
Tokyo, Japan

Abstract- A static var compensator was developed using a light-triggered thyristor (LTT) with a self-protection function. This paper describes the development of 200-Mvar class thyristor switched capacitors (TSCs) type with LTTs, which are useful for the recovering system voltage after an AC system fault. However, protection of the TSC is achieved not by the self-protection function of the LTT but by an arrester. Therefore, protection coordinated with the arrester is also described. Moreover, the effectiveness of the TSC with the coordinated protection was demonstrated by a simulator test. Finally, the manufacture and operating record of large-capacity SVCs are shown.

Keywords— Fault ride-through, Light-triggered thyristor, Static var compensator, Thyristor switched capacitor

I. INTRODUCTION

Various flexible AC transmission systems (FACTS) with power electronics are applied to electric power systems for stabilizing the transmission system and for reactive power control. In FACTS, static var compensators (SVCs) with thyristors have been used for more than 30 years [1] [2].

Recently, remarkable advances have been made in the development of wind and photovoltaic power generation technologies. These renewable energies were introduced to the electric power grid on a large scale during the last several years. As a result, low voltage ride-through (LVRT) and fault ride-through (FRT) are important aspects in electric power grid for maintaining reliability during a fault.

Especially in cases where a low voltage condition lasts for a long period of time during a system fault in an AC transmission system, it might result in the wind power system trip. Therefore, it is important that the system voltage recovers to a normal level as quickly as possible.

Large-capacity thyristor switched capacitors (TSCs), one type of SVC, are very effective for recovering the system voltage after a system fault. Fig. 1 shows an example of the application of an SVC for connecting to an AC power network.

In the past, we reported the development of large-capacity and high-voltage SVCs for the TCR, the thyristor controlled reactor, with light-triggered thyristors (LTTs) [3]. In this paper, we report the development on 200-Mvar class TSC type with LTTs in terms of the

following aspects:
(1) Function of TSC thyristor valve
(2) Overvoltage protection
(3) Voltage coordination of TSC valve using arresters
(4) Control system of SVC
(5) Simulation results of SVC (TSC)
(6) Operating records of large-capacity SVC

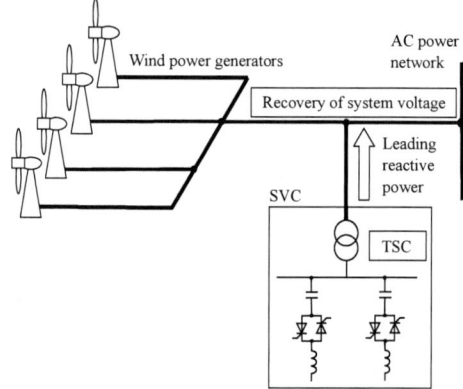

Fig. 1. Application of SVC to AC power network.

II. CONFIGURATION OF SVC WITH THYRISTORS

SVCs with thyristors are categorized into two types: the TSC type, which performs capacitor on/off control and the TCR type, which controls the reactor current by the firing phase. Circuit configurations of these types of SVCs with thyristors are shown in Fig. 2. Pairs of thyristor strings are connected in anti-parallel as components called thyristor valves. This structure is applied to conduct the AC currents for both directions. In addition, a capacitor C_{tsc} and a reactor L_{tsc} are connected to the thyristor valve in the TSC type, whereas reactors L_{tcr} are connected to the thyristor valve in the TCR type.

The TSC type using a capacitor and a reactor can control the leading reactive power step-wise by the capacitor switching. On the other hand, with the TCR type using reactors, it is possible to control the lagging reactive power continuously. However, a harmonic filter should be connected in parallel with the TCR in order to absorb the harmonic current generated by phase control of the thyristors. This harmonic filter is called a fixed capacitor (FC) bank.

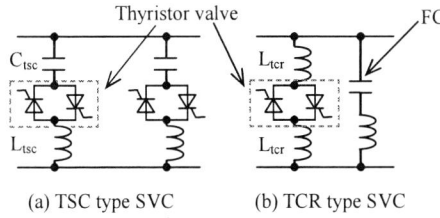

(a) TSC type SVC (b) TCR type SVC

Fig. 2. Circuit configurations of SVCs with thyristors.

III. THYRISTOR VALVE WITH SELF-PROTECTION LIGHT-TRIGGERED THYRISTOR

A LTT with an overvoltage self-protection function protects itself from an overvoltage by self-firing against a forward overvoltage [4]. This function is called voltage break over (VBO) free. The direct light triggering system using LTTs has the following advantages compared with the indirectly light triggering system using an electrically triggered thyristor (ETT).

(a) Higher voltage reliability and simpler maintenance
(b) Better noise immunity
(c) Lower operating limitations by eliminating thyristor level electronics
(d) More compact

By employing the LTT with the self-protection function, the overvoltage-detection circuit can be omitted. Fig. 3 shows circuit diagrams of a thyristor valve with an LTT and a conventional thyristor valve with an ETT. With the LTT, it is possible to develop simpler equipment compared with the ETT type [4]. This is also effective for TCRs.

However, the VBO protection offered by an LTT is not used in the TSC to avoid overvoltage build-up phenomena described later. The overvoltage protection of the TSC is actually carried out by an arrester. Details of arrester protection are shown in Sections V and VI.

Fig. 4 shows a photograph of the developed thyristor valve with LTTs. Separate valve sections for three phases are stacked one upon the other. The maximum thyristor series number is 16. Pairs of thyristor strings connected in anti-parallel are installed in each valve section. The maximum capacity of the developed thyristor valve for

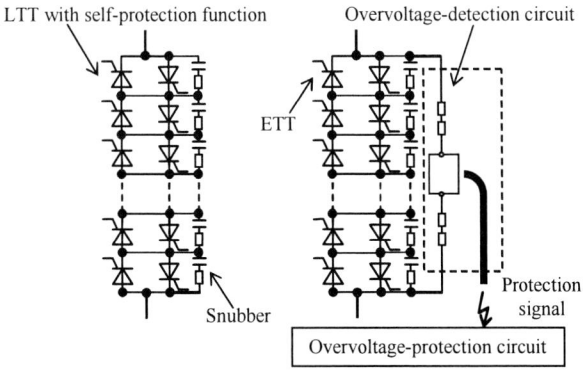

(a) Thyristor valve with LTT (b) Thyristor valve with ETT

Fig. 3. Thyristor valve configurations.

TSC is 345 Mvar. The suitable capacity can be designed for each system by changing the thyristor series number. The specifications of the thyristor valve for TSC are shown in Table I.

TABLE I
SPECIFICATIONS OF THYRISTOR VALVE FOR TSC

Specification Item	TSC
Maximum capacity	345 Mvar
Rated voltage	36 kVrms
Rated current	3200 Arms
Rated frequency	50 / 60 Hz
Number of phases	3 phase
Valve structure	Multiple Valve Unit (MVU)
Thyristor combination	16 series, 2 anti-parallel
Thyristor	Light triggered thyristor
Cooling	Water-cooled
Insulation	Air
Overvoltage Protection	Performed by Arrester

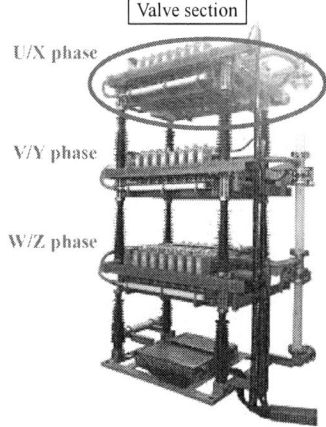

Fig. 4. Photograph of developed thyristor valve with LTTs.

IV. FUNCTION OF TSC THYRISTOR VALVE

A. Gate Logic of TSC

The gate logic of the TSC thyristor valve is shown in Fig. 5. By using a logical AND of the thyristor forward voltage (FV) signal, when the thyristor valve control (TVC) signal (TSC on = 1 / TSC off = 0) becomes true, the gate pulse is generated.

In the case of the ETT, the gate pulse is an electrical signal. Therefore, a long continuous gate pulse can be generated. However, in case of the LTT, an optical gate pulse is generated by a laser diode. A long continuous gate pulse would shorten the life time of the laser diodes. As a solution, the TVC signal is held by a flip-flop circuit during the TSC on operation. With this logic, the gate pulse is generated as soon as the pulse generator receives the forward voltage signal from the thyristor valve.

On the other hand, a high forward voltage (HFV) signal is used for the misfire protection. The details of the HFV will be discussed in Section V.

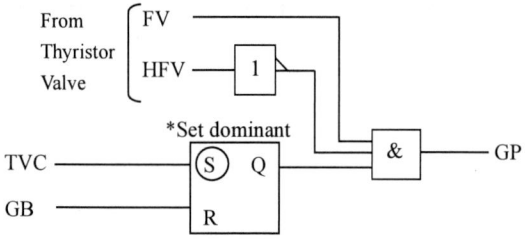

TVC : Thyristor valve control signal (TSC ON/OFF signal)
GB : Gate block signal of SVC
GP : Gate pulse
FV : Forward voltage signal
HFV : High forward voltage signal

Fig. 5. Gate logic configuration of TSC.

B. TSC Normal Operation

In order to compare with a misfire of the TSC, first the normal operation of the TSC with the above gate logic is described.

Fig. 6 shows voltage and current waveforms of a single-phase TSC in the on state. The TSC thyristor valve is operated as a switch. When switching the TSC on, the U phase and X phase of the thyristor valves turn on alternately at each half cycle. The forward voltage detection level is very low. The gate pulses are generated immediately after the thyristor current of the reverse phase becomes zero, because the reactor voltage is applied to the thyristors. Therefore, LTTs operates as they are given continuous gate pulses for conduction period. Then the thyristor valve acts equivalent to the mechanical switch, which continuously conducts the AC current through the capacitor.

Fig. 7 shows voltage and current waveforms of a single-phase TSC in the off state. When switching the TSC off, gate blocking is performed. Even though gate blocking is performed, the TSC cannot switch off until the current conducted through the thyristor becomes zero. Thus, the thyristor valve switches off at the zero-current point after receiving the gate block signal. Furthermore,

the capacitor is charged at the peak voltage when the TSC is off. As a result, the thyristor voltage becomes a superposition of the capacitor DC voltage and the system AC voltage. Because the discharge resistance is usually large, the capacitor voltage is discharged slowly, typically over several minutes.

V.   PROTECTION WITH ARRESTER FOR TSC

Normally, the TSC thyristor valve turns on at nearly the zero voltage point; however, it might turn on under conditions where a high voltage is applied by a false gate pulse. This phenomenon is called a misfire. When the misfire occurs, the voltage charged in the capacitor is discharged, and an overcurrent of several tens of kArms, depending on the capacitor value C_{tsc} and the reactor value L_{tsc}, is conducted through the thyristor valve. The resonance frequency of the TSC capacitor and reactor is usually higher than the system frequency.

Fig. 8 shows voltage and current waveforms when the misfire occurs under the worst-case conditions. In the worst-case conditions, the misfire occurs at the peak voltage just after the TSC switches off. The peak voltage of the system AC voltage defines V_{peak}. When the misfire occurs, a voltage of approximately $2V_{peak}$ is applied to the thyristor valve. After the overcurrent, the TSC switches off, and the capacitor is charged to approximately $3V_{peak}$ in the reverse direction. Furthermore, the superposition of the capacitor DC voltage and the system AC voltage is applied to the thyristor valve. As a result, the peak voltage of the thyristor valve becomes approximately $4V_{peak}$.

In the case where the misfire occurs again at the thyristor voltage of $4V_{peak}$, the subsequent thyristor voltage after the overcurrent will become $6V_{peak}$. As mentioned above, if the misfire occurs at the peak voltage of the thyristor valve after TSC switches off, the capacitor and thyristor voltages become higher and higher due to the repetition of misfires.

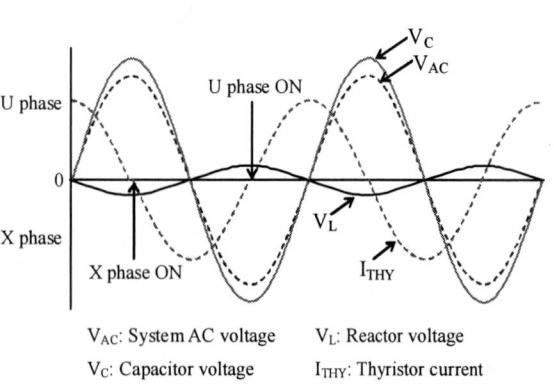

V_{AC}: System AC voltage V_L: Reactor voltage
V_C: Capacitor voltage I_{THY}: Thyristor current

Fig. 6. Voltage and current waveforms of TSC when on.

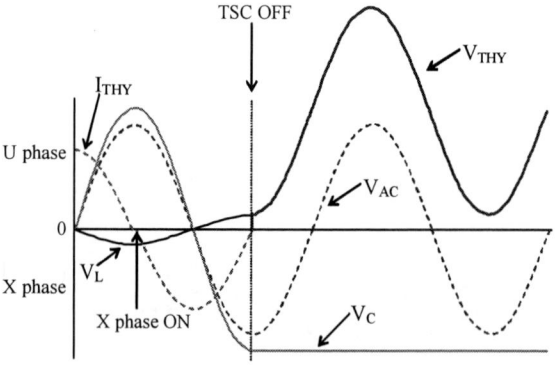

V_{AC}: System AC voltage V_L: Reactor voltage V_{THY}: Thyristor voltage
V_C: Capacitor voltage I_{THY}: Thyristor current

Fig. 7. Voltage and current waveforms of TSC when off.

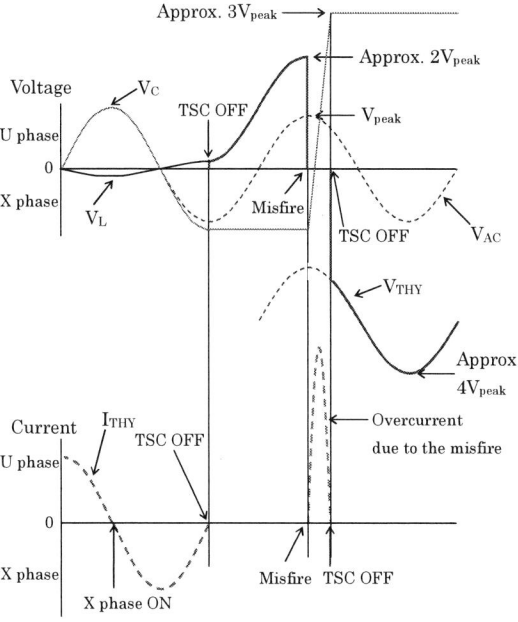

V_AC: System AC voltage   V_L: Reactor voltage   V_THY: Thyristor voltage
V_C: Capacitor voltage   I_THY: Thyristor current   V_peak: Peak of AC voltage

Fig. 8. The voltage and current waveforms when a misfire occurs.

If the overvoltage self-protection function VBO of the LTT is adopted in the TSC, the protection firing will be repeated and the overvoltage is built up in the TSC as described above. Therefore, the overvoltage protection in the TSC is carried out by arresters. The overvoltage above the arrester-protection level is suppressed by the arrester. The protection coordination with the arrester will be described in Section VI.

Additionally, a high forward voltage (HFV) detection circuit is developed for the thyristor valve in order to prevent misfires during high voltage is applied to the valve. The HFV signal is transferred from the thyristor valve to the gate circuit by an optical signal when the voltage across the TSC thyristor valve is higher than the HFV detection level V_hfv. As shown in Fig. 5, generation of the gate pulse is inhibited by the gate circuit while the HFV signal "exists". Therefore, the TSC thyristor valve never turns on at a voltage above the HFV detection level, and misfires are prevented by this protection strategy. Fig. 9 shows voltage and current waveforms during misfire protection using the HFV.

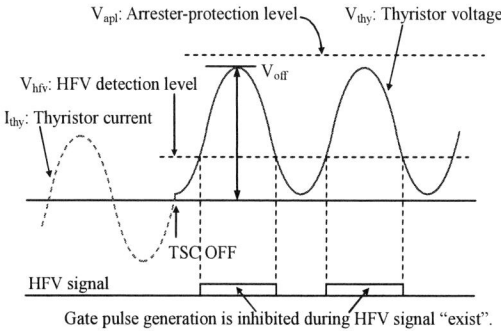

Fig. 9. Misfire protection with a high forward voltage (HFV).

The HFV detection level is set to a non-conducting region of the arrester. Therefore, if a misfire occurs, the current that flows to the arrester is not transferred to the TSC thyristor valve because there is no arrester current.

VI. VOLTAGE COORDINATION OF TSC

When the system voltage increases, voltage control is carried out by switching the TSC off to reduce the capacitive impedance of the system. Therefore, the voltage coordination of the thyristor valve is determined at the peak voltage V_off when the TSC switched off during the overvoltage condition. Voltage coordination based on IEC 61954 (the standard for testing of thyristor valves for static var compensators) [5] is described below. Fig. 10 shows the voltage coordination of the TSC thyristor valve.

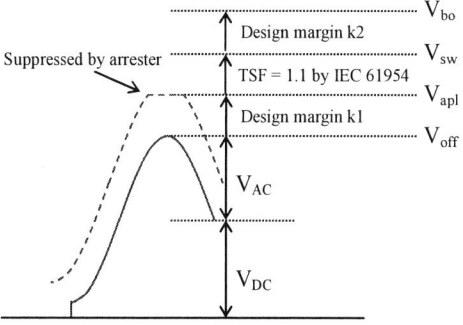

V_DC: DC voltage charged in capacitor at TSC OFF
V_AC: System AC voltage at TSC OFF
V_off: Peak voltage at TSC OFF (maximum overvoltage)
V_apl: Arrester-protection level
V_sw: Switching impulse test level
V_bo: VBO protection level of thyristor valve

Fig. 10. Voltage coordination of TSC thyristor valve.

The arrester-protection level V_{apl} is calculated from V_{off} and the design margin k1 by:

$$V_{apl} = k1 \times V_{off} \qquad (1)$$

The design margin is determined by the arrester capability. According to IEC 61954, the switching impulse test voltage V_{sw} is calculated from the arrester protective level and the test safety factor (TSF) ks = 1.1 by:

$$V_{sw} = ks \times V_{apl} \qquad (2)$$

Therefore, the necessary number of series thyristors is determined so that VBO protection will not occur in the switching impulse test. The required VBO protection level V_{bo} is calculated from the switching impulse test level and the design margin k2 by:

$$V_{bo} = k2 \times V_{sw} \qquad (3)$$

The design margin is determined by the voltage grading of the series thyristors.

VII. CONTROL SYSTEM FOR STATIC VAR COMPENSATOR

The thyristor valves are controlled by a control system. The control system can adjust the output of the SVC to compensate for variations of the system voltage and has a function for protecting the SVC from system faults. Fig.11 shows the typical SVC configuration.

Because the TSC type of SVC, has just ON and OFF states through TSC ON/OFF control, the output of the TSC can be controlled by steps only.

A. Construction of Control System

A typical control system for an SVC includes a human-machine interface (HMI), a local control panel (LCP), a thyristor valve control panel (TVC), and a valve base electronics panel (VBE). A supervisory control and data acquisition (SCADA) is a control and monitor system of the substation. The configuration of the SVC control system is shown in Fig. 12. The functions of the equipments are shown in Table II.

The TVC performs the main control of the SVC and sends control command signals to the VBE, in which the TSC gate logic described in Section IV generates gate pulses for the thyristor valves.

TABLE II
FUNCTIONS OF SVC CONTROL EQUIPMENTS

Equipment	Function
HMI	(1) Interface for operator to operate SVC and monitor the SVC status. (2) Communication with other equipments in substation
LCP	(1) Operation and stop sequence of SVC (2) Connection operation between SVC and main circuit bus
TVC	(1) Automatic voltage regulator (2) TSC ON/OFF control (3) System fault protection
VBE	(1) Gate pulse generation in accordance with the control command signal from TVC (2) Thyristor failure detection of thyristor

B. Control functions

Fundamental SVC control functions include an automatic voltage regulator (AVR), susceptance output limits (Limiter), a slope reactance (X slope), a system voltage detector (V Detector), an SVC current detector (I Detector), a TSC regulator, and so on. An example of the control block diagram for the TSC is shown in Fig. 13.

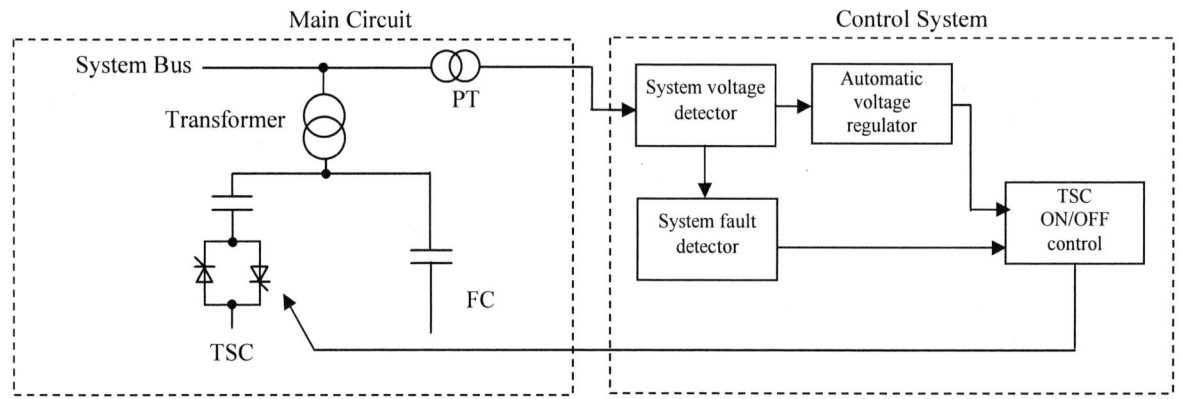

Fig.11. Typical SVC configuration.

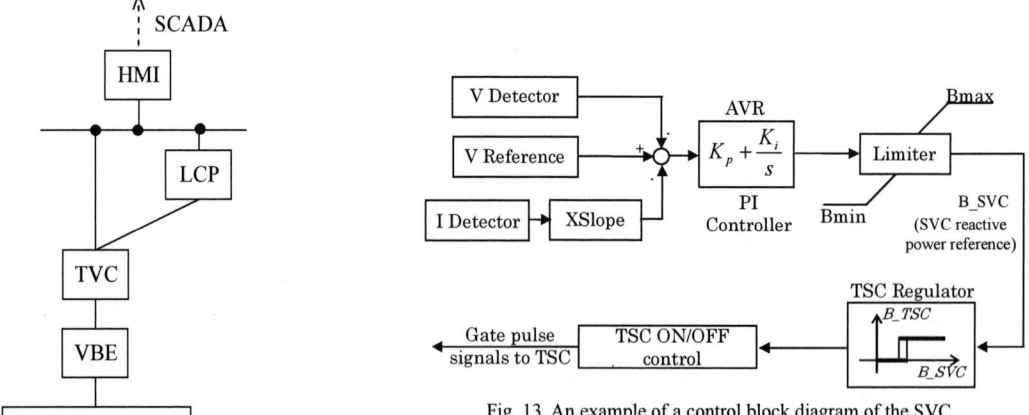

Fig. 12. Configuration of SVC Control system.

Fig. 13. An example of a control block diagram of the SVC control system for the TSC.

978-1-4799-2706-7/14 $31.00 © 2014 IEEE

VIII.   SIMULATOR STUDY

The effectiveness of the TSC at the fault was demonstrated by a simulator test. The simulator test model was constituted of three TSCs (TSC1, TSC2, and TSC3). The ratio of capacities of TSC1, TSC2 and TSC3 were 0.2pu : 0.4pu : 0.4pu respectively. The total capacity of SVC (1.0pu) is more than 200 Mvar. The SVC output could be controlled by 5 steps. The TSC system configuration is shown in Fig. 14. The SVC outputs and TSC on off status are shown in Fig. 15 when the controller switches in the TSCs. The controller has a hysteresis to avoid frequent switching in and out. For the switching out, the different characteristics are used other than shown in Fig. 15.

A.   Simulator Configuration

Simulator tests were performed with the actual SVC controller (HMI, LCP, TVC and VBE), a real time digital simulator (RTDS), and an interface panel. The connection between the actual controller and the simulator is shown in Fig. 16. The main circuits of the SVC and the AC system were simulated in the RTDS, which outputs analogue signals, including system voltage and so on. The actual controller generated gate pulse signals and sent them to the RTDS through the interface panel. Thus, the closed loop control simulation with the actual controller was performed.

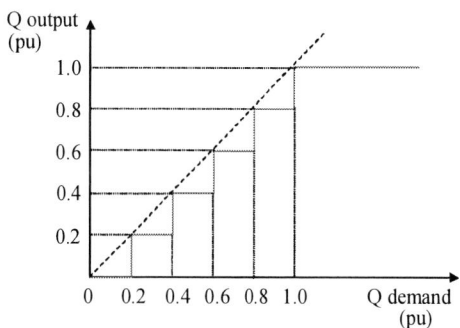

Q demand (pu)	TSC1 (0.2pu)	TSC2 (0.4pu)	TSC3 (0.4pu)
0.2	X	-	-
0.4	-	X	-
0.6	X	X	-
0.8	-	X	X
1.0	X	X	X

Note) "X" means this TSC is operated.

Fig. 15. SVC output and TSC on off status.

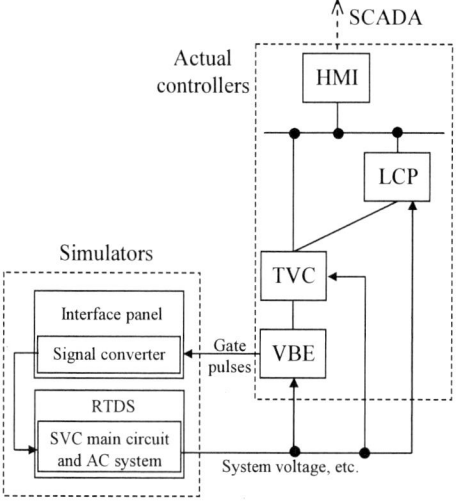

Fig. 16. The connections between the actual controller and the simulator.

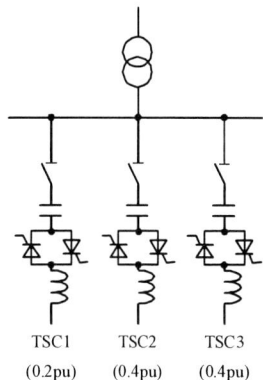

TSC1 (0.2pu)   TSC2 (0.4pu)   TSC3 (0.4pu)

Fig. 14. 200-Mvar class TSC system configuration.

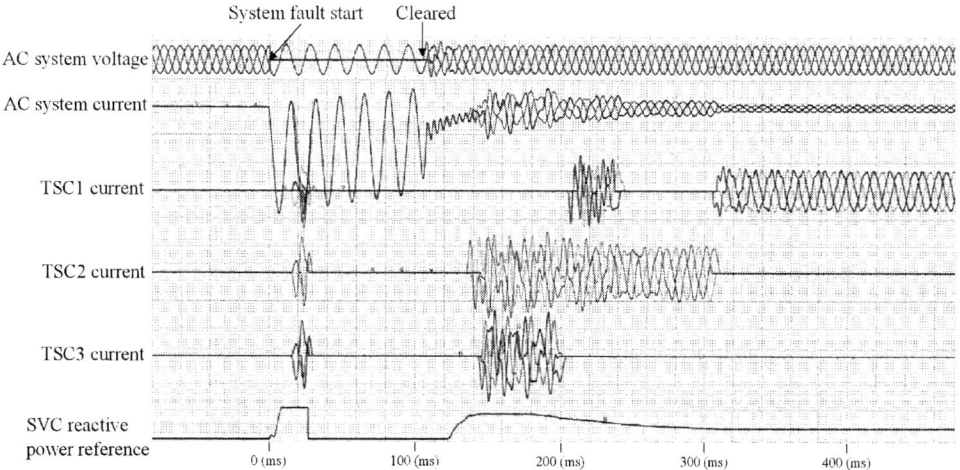

Fig. 17. Simulator test results (phase to phase fault).

B. Results of Simulation

In the simulation, the RTDS simulated various AC system faults. The simulator test results for a phase to phase fault are shown in Fig. 17. During the system fault, first the three TSCs were turned on for a short time to maintain the system voltage and then all TSCs were turned off during a system fault based on the control strategy. Furthermore, after the fault was cleared but the system voltage was still at a low level, TSC2 and TSC3 were turned on to maintain the system voltage according to the SVC reactive power reference. With the recovery of the system voltage, the output of SVC decreased from the full capacity to the minimum capacity step-by-step.

The TSCs can be turned on and off at high speed. As shown in Fig. 17, TSC1 was turned on at 204 ms and off at 240 ms. Furthermore, it was turned on again at 304 ms. This is the advantage for a TSC in comparison with a mechanical switch.

The results of the simulator test demonstrated that the AC system voltage was recovered to a normal level quickly by the TSCs.

IX. OPERATING RECORDS OF SVC

Table III shows the operating records of the large-capacity SVCs manufactured after 2005. 200-Mvar class TSC has been already operated in site. Some systems are constituted with only the TSC type, showing the huge demand for TSCs.

TABLE III
OPERATING RECORDS OF SVC

No.	Total capacity	System
1	330 Mvar	TSC x 3
		TCR x 3
2	260 Mvar	TSC x 2
3	475 Mvar	TSC x 2
		TCR x 1
4	428 Mvar	TSC x 1
		TCR x 1
5	400 Mvar	TSC x 1
		TCR x 1

X. CONCLUSION

In this paper, we described the development of a TSC thyristor valve with LTTs for a large-capacity SVC. The overvoltage protection of the TSC with LTT was performed by an arrester. We showed the behavior during misfires and described the prevention logic using an HFV detection circuit. Furthermore, the design of the voltage coordination considering the arrester-protection level was shown in accordance with IEC 61954. In addition, we also described the control system of the SVC.

The results of simulator tests with the RTDS demonstrated the effectiveness of TSC during a system fault.

TSCs are commonly used for satisfying the criteria of fault ride-through (FRT) and large-capacity TSCs will continue to be applied to electric power grid applications in the future.

REFERENCES

[1] T. Nishikawa et al., "Overvoltage protection of static var compensation system," IEEE Trans. On PAS., Vol. PAS-102, No.7, July (1983).

[2] T. Fujimoto, T. Ikeda, Y. Moriura, F. Aoyama "Recent trends of power electronics apparatus for SVC," The papers of Joint Technical Meeting on Static Apparatus, IEE Japan, Vol.SA-99, No.30-36, pp.13-18, October 1999.

[3] K. Murabayashi, T. Yoshino, "HIGH VOLTAGE, LARGE CAPACITY LIGHT-TRIGGERED THYRISTOR VALVE FOR STATIC VAR COMPENSATORS", Proceedings of The International Conference on POWER CONVERSION and INDUSTRIAL CONTROL, (1986)

[4] T. Fujimoto, M. Imura, and T. Sugiyama, "Development on Static Var Compensator Applying Self-protective Light Triggered Thyristor," 2007 Annual Conference of I.E.E. of Japan, Industry Applications Society, Osaka (Japan), pp. 389-390, August 2007.

[5] Power electronics for electrical transmission and distribution systems - Testing of thyristor valves for static var compensators, IEC 61954 (Edition 2.0 2011-04).

The 2014 International Power Electronics Conference

Detailed Analysis and Design of a Three-Phase Phase-Modular Isolated Matrix-Type PFC Rectifier

Patricio Cortes, Lukas Fässler, Dominik Bortis
and Johann W. Kolar
Power Electronic Systems Laboratory, ETH Zurich
Switzerland
cortes@lem.ee.ethz.ch

Marcelo Silva
Universidad Politécnica de Madrid
Spain

Abstract—Phase-modular isolated PFC rectifiers are an interesting alternative to phase-integrated three-phase rectifiers. The use of a matrix-type converter allows to achieve isolation with a single-stage energy conversion. This paper presents a phase-modular isolated matrix-type rectifier, which can be connected to the mains either in star (Y) or delta (Δ), enabling a wide input voltage range. A detailed analysis of the operating principles and switching behavior of the converter is presented. Then, the design of the active and passive components is discussed considering a 7.5 kW, 400 V output voltage system. Different implementation alternatives are evaluated regarding efficiency and realization effort.

I. INTRODUCTION

In today's power electronic systems, which are powered from the three phase mains, rectifier systems with an active PFC are indispensable in order to comply with directives which specify a maximum level of the noise injected back into the mains [1], [2]. Depending on the underlying application, a galvanic isolation or adaption of the output voltage is needed, which is typically realized with an additional DC/DC-converter. The major disadvantages of this converter cascading are the reduction of the overall efficiency, the large component count and the increasing control complexity. The use of three-phase matrix-type isolated PFC rectifiers has been already proposed as an alternative for single-stage conversion [3]–[6]. In comparison to the conventional two-stage PFC converter systems, the proposed phase-modular indirect matrix-type PFC rectifier system, which can be connected to the mains either in star (Y) or delta (Δ) – enabling a wide input voltage range (IMY/Δ-rectifier) – can perform the PFC functionality and the galvanic isolation in a single-stage (cf. **Fig. 1**).

Due to the phase-modularity of the IMY/Δ-rectifier, according to single-phase rectifier systems, to each phase module only the phase voltage and not the phase-to-phase voltage is applied [7], [8]. Thus, semiconductor devices with a blocking voltage of 600 V can be used, which feature a lower on-state resistance and better switching performance compared to 1200 V-devices, hence leading to even higher efficiency. The phase module, consisting of an input diode rectifier and a full bridge, is basically an indirect single-phase matrix converter and converts the low frequency mains input voltage to a high frequency AC voltage [9]. As already presented in [10], the IMY/Δ-rectifier's full bridges can be controlled with a simple phase-shift modulation where each bridge leg is switched with 50% duty cycle, thus reducing the control complexity compared to conventional two-stage PFC converter system. Furthermore, since the IMY/Δ-rectifier is a buck-type PFC rectifier and due to the missing DC-link capacitors (only a small DC-link capacitor C_{dc} is needed for each phase module during the commutation interval), on the one hand, the inrush currents at startup are strongly reduced since only the input filter capacitors have to be charged. On the other hand, with a buck-type topology the voltage level can be slowly ramped up to the desired output voltage with the

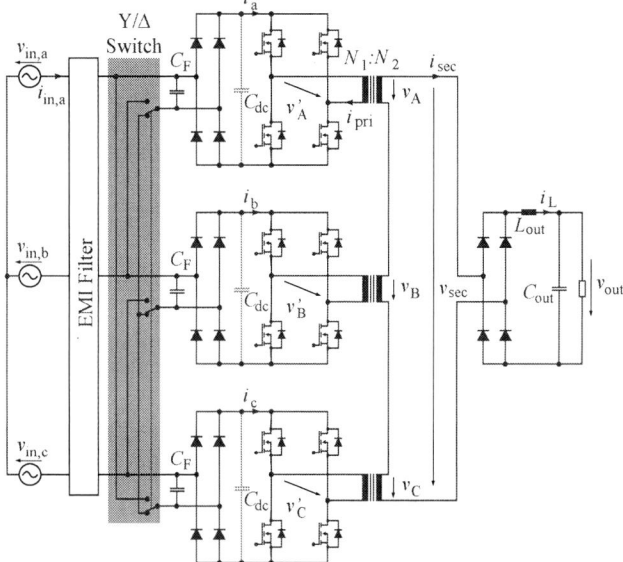

Fig. 1 Circuit diagram of the three-phase phase-modular isolated matrix-type PFC Y/Δ-Rectifier (IMY/Δ-Rectifier).

ability to operate the converter with output current limitation, thus also providing an inherent short-circuit protection.

In this paper a comprehensive analysis of the basic operation, the design and limits of the proposed phase-modular indirect matrix-type PFC rectifier system (IMY/Δ-rectifier) is given.

In **Section II** the basic operation of the IMY/Δ-rectifier is explained. Based on this analysis and the given specifications of **Table I**, in **Section III** the design of a 7.5 kW converter prototype is performed. The selection of the input and output diodes, the switches of the full bridge converters, the design of the input EMI filter and of the output filter are included in this section. The built prototype is finally described in **Section IV**.

TABLE I Electrical specifications of the IMY/Δ-rectifier.

Parameter	Value
Mains voltage, nominal (rms,phase)	230 V
Mains voltage, min-max (rms,phase)	185...306 V
Mains frequency	50 Hz
Output DC voltage	400 V
Nominal power	7.5 kW
Switching frequency	72 kHz

978-1-4799-2706-7/14 $31.00 © 2014 IEEE   3864

The 2014 International Power Electronics Conference

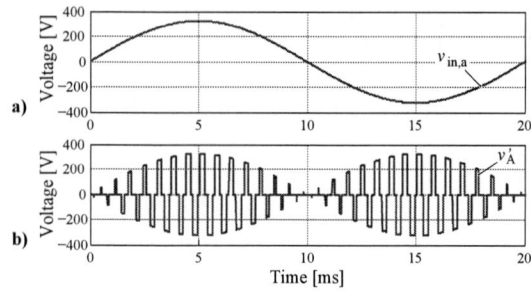

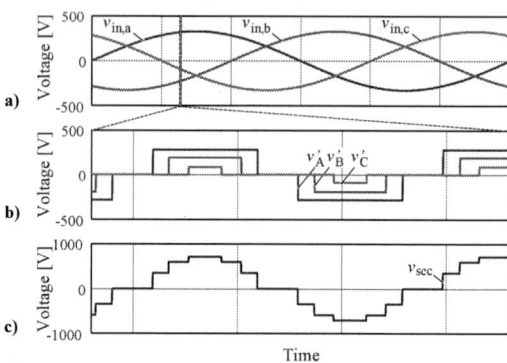

Fig. 2 Schematic waveform of **a)** the phase input voltage $v_{\mathrm{in,a}}$ and **b)** the output voltage v_{A}' of the full bridge of phase A applied to the corresponding single-phase transformer.

II. PRINCIPLE OF OPERATION

The circuit diagram of the proposed phase-modular indirect matrix-type PFC rectifier system (IMY/Δ-rectifier) is shown in **Fig. 1**. As can be noticed, the IMY/Δ-rectifier features a phase-modular structure, whereat each phase module consists of a input diode rectifier and a full bridge. According to single-phase rectifiers, where only the phase-to-neutral instead of the phase-to-phase voltage is processed, 600 V-devices instead of 1200 V-devices can be used in each phase module, which typically have a lower on-state resistance and show an improved switching behavior compared to 1200 V-devices. Thus, lower overall semiconductor losses can be achieved. In addition, depending on the input voltage range, the voltage applied to the phase modules can be adjusted by connecting the IMY/Δ-rectifier either in star or in delta connection to the mains. With an additional relay the IMY/Δ-rectifier can automatically change the type of connection to the mains (cf. **Fig. 1**). In case of a low input voltage, the delta connection is selected in order to apply a higher voltage to each phase module and thus decreasing the input current and the overall conduction losses. Basically, each phase module is an indirect single-phase matrix converter that converts the low frequency mains input voltage to a high frequency AC voltage (**Fig. 2**). As described in [10], each phase module is controlled with a simple phase shift modulation where each bridge leg is switched with a 50% duty cycle, resulting in the high-frequency waveform shown in **Fig. 2 b)**. In order to enable a galvanic isolation, each phase module applies this high frequency AC voltage directly to an individual single-phase transformer.

On the secondary side, the windings of the three single-phase transformers are directly connected in series, thus the output voltages of the individual transformers is summed up. Based on the proposed modulation scheme in [10], where the duty cycle or phase shift respectively is proportional to the corresponding rectified input voltage, a staircase shaped output voltage is generated as shown in **Fig. 3**. Due to the 120° phase displacement between the input voltages $v_{\mathrm{in,a}}$, $v_{\mathrm{in,b}}$ and $v_{\mathrm{in,c}}$ the fundamental component of the secondary voltage v_{sec} shows a constant magnitude over the whole mains period as described in [10]. This high frequency voltage is then rectified by the output diode bridge and filtered by the output inductor L_{out} and capacitor C_{out} in order to provide a constant output voltage v_{out}. Actually, the resulting output voltage v_{out} is equal to the average value of v_{sec},

$$v_{\mathrm{out}} = \overline{|v_{\mathrm{sec}}|} = \frac{3}{2} M \frac{N_2}{N_1} \hat{v}_{\mathrm{in}}. \qquad (1)$$

By only considering the secondary part of the IMY/Δ-rectifier where v_{sec} is the input voltage of this circuit (cf. **Fig. 4**), it can be noticed that the IMY/Δ-rectifier has the same structures as a conventional DC/DC buck converter. Consequently, the IMY/Δ-

Fig. 3 Schematic waveform of **a)** the phase input voltages $v_{\mathrm{in,a}}$, $v_{\mathrm{in,b}}$ and $v_{\mathrm{in,c}}$, **b)** the resulting output voltage of the full bridge v_{A}', v_{B}' and v_{C}' if the symmetric modulation scheme is applied and **c)** the resulting output voltage v_{sec} which is the sum of all three secondary voltages v_{A}, v_{B} and v_{C}.

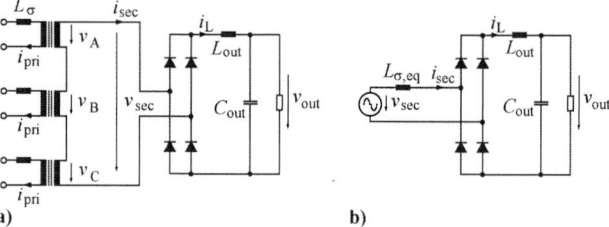

Fig. 4 a) Circuit diagram of the IMY/Δ-rectifier's secondary and **b)** its equivalent circuit which basically is a buck-type converter with input voltage v_{sec}.

rectifier can be controlled with the same duty cycle M, where

$$M = \frac{2}{3} \frac{N_1}{N_2} \frac{v_{\mathrm{out}}}{\hat{v}_{\mathrm{in}}}, \qquad (2)$$

and the same simple control scheme (cascaded control of output current i_{L} and output voltage v_{out}) as can be used for the conventional buck converter.

Assuming a large output inductor L_{out}, the output current i_{L} is almost constant. The same current is also flowing through all secondary windings of the transformers, with the only difference that the current through the secondary windings has a rectangular shape which is then rectified by the output diode bridge. Based on Ampere's law and assuming a large magnetizing inductance, the current in the primary as well as in the full bridge has also a rectangular shape with constant magnitude $\hat{i}_{\mathrm{pri}}$, which is given by the transformer's winding ratio $\frac{N_1}{N_2}$, $\hat{i}_{\mathrm{pri}} = \frac{N_2}{N_1} \hat{i}_{\mathrm{sec}} = \frac{N_2}{N_1} i_{\mathrm{L}}$.

Hence, in order to achieve sinusoidal input currents $i_{\mathrm{in,a}}$, $i_{\mathrm{in,b}}$ and $i_{\mathrm{in,c}}$, which are in phase with the corresponding phase voltages $v_{\mathrm{in,a}}$, $v_{\mathrm{in,b}}$ and $v_{\mathrm{in,c}}$ - thus achieving the required PFC operation - the constant current has to be pulse-width modulated sinusoidally, which means that the duty cycle is proportional to the actual input current or voltage value respectively. The resulting voltages v_{A}', v_{B}' and v_{C}' which are applied to the primary of each single-phase transformer are shown in **Fig. 3 b)**. As can be noticed, based on the modulation scheme proposed in [10], the duty cycles of the three individual phase modules are synchronized in such a way that the resulting output voltages are symmetrically distributed.

For the sake of completeness, it has to be mentioned that it is also possible to slightly shift the input current out of phase with the input voltage in order to enable reactive power compensation.

978-1-4799-2706-7/14 $31.00 © 2014 IEEE

The 2014 International Power Electronics Conference

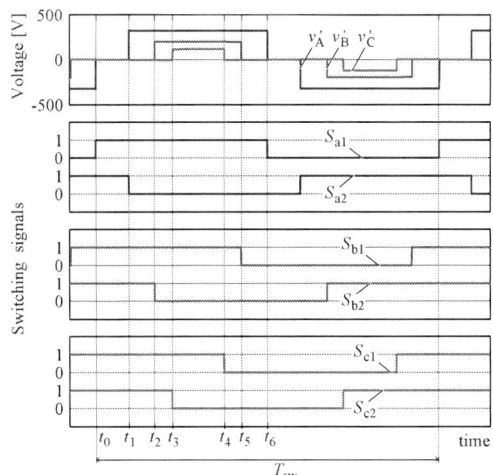

Fig. 5 Resulting voltages v'_A, v'_B and v'_C, which are applied to the individual single-phase transformers and corresponding control signals of the three full bridges for the case $v_{in,a} > v_{in,b} > v_{in,c}$.

III. CONVERTER DESIGN

In order to be able to properly design the IMY/Δ-rectifier with respect to the specifications given in **Table I** and to perform an accurate loss analysis, first the switching behavior of the full bridge of a phase during each switching transition has to be analyzed.

A. Full bridge converter switches

As already mentioned, the bridge legs of each full bridge are switched with a 50 % duty cycle whereat the phase current and the resulting output voltage of the full bridge v'_A, v'_B and v'_C are controlled with the phase shift between the two bridge legs. The control signals of the switches for the three full bridges and the resulting output voltages are illustrated in **Fig. 5** for the case $v_{in,a} > v_{in,b} > v_{in,c}$.

Remark: during the intervals where at least from one of the three full bridges a voltage is applied to the corresponding single-phase transformer, i.e. the full bridge is turned on, the output current impressed by the large output inductor L_{out} is flowing through the secondary windings of all three transformers which are connected in series; accordingly, current is flowing also through the other full bridges, independently of their switching states.

At the beginning of the switching period t_0 the full bridge with the largest input voltage is turned off and all phase modules are now in a freewheeling state, thus no voltage is applied to any transformer's primary winding (cf. **Fig. 6 a)**). Since before that time instant ($t < t_0$) the full bridge with the largest input voltage was conducting the negative load current ($i_{sec} = -i_L$) and the fact that the secondary windings are connected in series, the same negative load current was also flowing through the other full bridges even if they were already in the freewheeling state. If it is assumed that each transformer has a certain leakage inductance L_σ, during the time interval $t_0 - t_1$, the leakage inductances L_σ still force the negative load current to freewheel through the high-side switches S_{1p} and S_{2p} of all full bridges. The corresponding current path during the time interval $t_0 - t_1$ is shown in **Fig. 6 a)** for one full bridge.

At the time instant t_1, the switch S_{2p} of the phase module with the largest input voltage is turned off. Due to the leakage inductance L_σ the current in the primary continues to flow in the same direction, thus the parasitic output capacitance of S_{2p} is charged and the output capacitance of S_{2n} discharges until the current can flow through

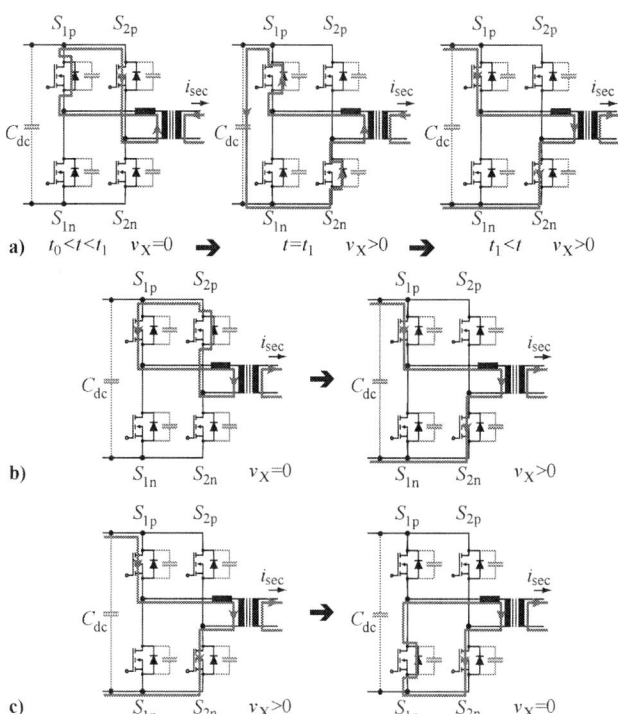

Fig. 6 Current path at different switching transitions: **a)** switching transition at t_1, when the first full bridge is switched from freewheeling to a positive output voltage, **b)** during turn on transitions at t_2 and t_3, when the second and third full bridge are turned on, and **c)** during turn off transitions at t_4, t_5 and t_6, when the full bridges return to the freewheeling state.

the anti-parallel diode of S_{2n} and the small DC-link capacitor (cf. **Fig. 6 a)**). It has to be noticed that the energy stored in the leakage inductance has to be large enough to charge/discharge the output capacitances of the bridge leg. After this commutation, S_{2n} is turned on at zero voltage (ZVS). In addition, the positive input voltage is now fully applied to the leakage inductance L_σ of the transformer, resulting in an almost linear increase of the primary current where the current changes its sign until the positive load current ($i_{sec} = i_L$) is reached.

This can be easily explained with the simplified circuit diagram shown in **Fig. 7**. Before t_1, when $v_{sec} = 0$ V, the current is flowing in negative direction through the leakage inductance $L_{\sigma,eq}$ and the output diodes D_{1n} and D_{2p} (cf. **Fig. 7 a)**). At $t = t_1$, v_{sec} changes to a positive voltage which forces D_{1p} into conduction and then occurs across the leakage inductance $L_{\sigma,eq}$, since the output diode D_{2p} is still conducting, and thus clamps the negative terminal B of v_{sec} to the positive terminal A' of the output rectifier (cf. **Fig. 7 b)**).

Due to the relatively small leakage inductance L_σ, the current rapidly decreases and changes its sign until the load current i_L is flowing in positive direction through the primary and secondary winding of the transformer (cf. **Fig. 7 c)**). Consequently, with the described transition at t_1 and the series connection of the secondary windings, the current direction is also changing in the other phase modules which are still in the freewheeling state (cf. **Fig. 6 b)**, left).

At the time instant t_2, the phase module with the second highest input voltage applies a positive voltage to the corresponding transformer by switching on S_{2n}. Since the current has changed its sign already at t_1, during the switching transition at t_2, the current is forced to

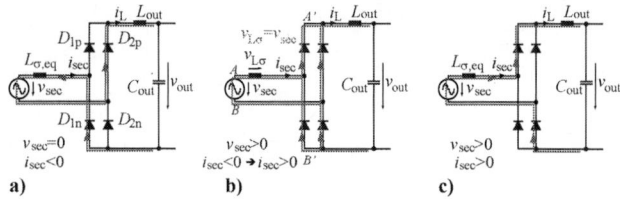

Fig. 7 Equivalent buck-type circuit of the secondary side of the IMY/$\triangle$-Rectifier with the current path **a)** before t_1, when the full bridge is in freewheeling state, **b)** during the switching transition at t_1, when voltage is applied to the leakage inductance and **c)** after t_1, when the current has changed its direction.

commutate from the anti-parallel diode of S_{2p} to the MOSFET S_{2n}, resulting in hard switching with typically high reverse recovery losses (cf. **Fig. 6 b)**). The same hard switching transition occurs also at the time instant t_3, when the phase module with the lowest input voltage is turned on to a positive voltage.

At t_4, t_5 and t_6 the phase modules are again switched off back to the freewheeling state in reverse order, starting with the phase module having the lowest input voltage applied. During these transitions S_{1p} is turned off, thus the current impressed by the leakage inductance L_σ charges/discharges the parasitic output capacitance of the bridge leg until the anti-parallel diode of S_{1n} is conducting; accordingly, the MOSFETs S_{1n} of all three full bridges can be turned on at zero voltage (cf. **Fig. 6 c)**). After t_6 all phase modules are back again in the freewheeling state and the same kind of transitions as described above will take place in the second half of the switching period, when negative voltages are applied to the transformers. In summary, hard switching occurs always during the transition from freewheeling to an active state, but only in the two phase modules which do not block the maximum input voltage within the considered switching interval. In all other switching transitions ZVS is achieved.

Due to the hard switching transitions in the full bridge, MOSFETs with a low reverse recovery charge Q_{rr}, e.g. the CFD2-CoolMOS series from Infineon, have to be used in order to keep the switching losses reasonably low. Unfortunately, such MOSFETs typically feature a higher on-state resistances $R_{ds,on}$ – e.g. compared to MOSFETs of the C7-CoolMOS series (also from Infineon); for example, the on-state resistance $R_{ds,on}$ of the CFD2-MOSFET is approximately two times larger - thus leading to higher conduction losses which also represent a substantial share of the total converter losses as will be explained later.

In order to keep the conduction losses also with the CFD2-CoolMOS low, several MOSFET devices could be connected in parallel. At the same time, however, the reverse recovery charge is proportionally increased with the number of parallel MOSFETs and thus, even if the current per device is lowered, a higher total reverse recovery charge results in higher switching losses. In addition, also the output capacitance is increased linearly, which means that more energy has to be stored in the leakage inductance in order to charge/discharge the output capacitance during the switching transitions and to still achieve soft switching.

Another option to almost avoid any switching losses is to use Si-MOSFET devices in an OR-ing configuration as shown in **Fig. 8**. The series connected Schottky diode D_S prevents the body diode of the Si-MOSFET to conduct current (reverse recovery losses in the body diode are thus avoided), while the anti-parallel silicon carbide (SiC) diode D_P is conducting the current in negative direction through the OR-ing configuration. Consequently, the reverse recovery behavior

Fig. 8 MOSFETs with high reverse recovery charge can be used in an OR-ing configuration with series connected Schottky diode D_S and anti-parallel SiC-diode D_P to prevent high reverse recovery losses.

of the MOSFET's body diode is no more crucial, and instead of the CFD2-CoolMOS now the C7-CoolMOS with lower on-state resistance $R_{ds,on}$ can be used resulting in lower conduction losses. However, the additional conduction losses of the series connected low voltage Schottky diode D_S during turn-on, which are typically small due to the low forward voltage drop of Schottky diodes, and the higher conduction losses of the SiC-diode compared to the MOSFET internal parasitic Si-diode during the freewheeling interval have to be taken into account.

As an alternative to Si-MOSFETs, the utilization of semiconductor devices based on new technologies like SiC-MOSFETs or GaN-transistors would be possible. However, on the one hand SiC-MOSFETs are only available with voltage ratings of 1200 V (CMF20120D, 105 mΩ at $T_j = 150\,^\circ$C, CREE), where the best state of the art products show around three times higher on-state resistances compared to the C7-CoolMOS (IPW65R019C7, 40 mΩ at $T_j = 150\,^\circ$C, Infineon) and on the other hand, 600 V GaN-transistors are not commercially available yet.

As proposed in [10], the control signals of each full bridge are synchronized to each other in such a way that the voltages applied to transformers (v_A, v_B and v_C) are symmetrically distributed, resulting in a symmetric secondary voltage v_{sec} (symmetric modulation). As described above, however, in the first turn on switching transition of the full bridge with the largest input voltage at t_1 - switching from freewheeling state to either a positive or negative voltage - the current direction is changed in all full bridges due to the series connection of the secondary windings, thus resulting in hard switching during the turn on transition of the other two full bridges (cf. **Fig. 6 b)**). In order to avoid these hard switching transitions, the modulation scheme can be modified in such a way that the control signals of the individual full bridges are synchronized to each other with respect to simultaneous turn on transitions, which results in an asymmetric secondary voltage v_{sec} (asymmetric modulation) as shown in **Fig. 9**. Consequently, the freewheeling current impressed by the leakage inductances L_σ will simultaneously charge/discharge the output capacitances of the bridge legs in all three full bridges, thus in each phase module ZVS is achieved during turn on. The turn off transitions are not affected by this modification. It has to be mentioned, that even if the shape of secondary voltage v_{sec} is changed, the output current ripple in the output inductor L_{out} is hardly changed, since the voltage time integral applied to L_{out} stays the same.

With this modified modulation scheme the above stated OR-ing configuration wouldn't be needed anymore and the C7-CoolMOS with low on-state resistance $R_{ds,on}$ could be used, since now in all transitions ZVS is achieved, provided that the energy stored in the leakage inductance ($\hat{E}_{L_\sigma} = 1/2 L_\sigma \cdot \hat{i}_{pri}^2$) is high enough to fully charge/discharge all output capacitances during the simultaneous turn on transition. At low load conditions when the current through the transformer is small, however, the energy stored in the leakage inductance $\hat{E}_{L,\sigma}$ could be too low resulting in hard switching. Even if in that case the currents are low, the total converter losses could be considerably increased and under certain circumstances the MOSFETs

The 2014 International Power Electronics Conference

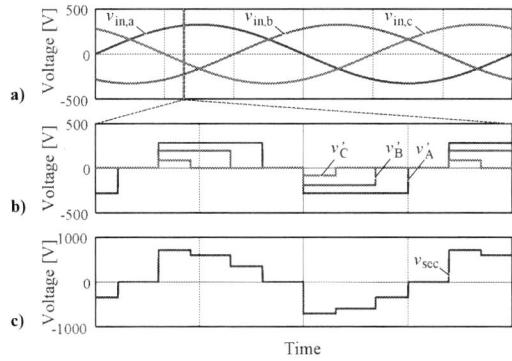

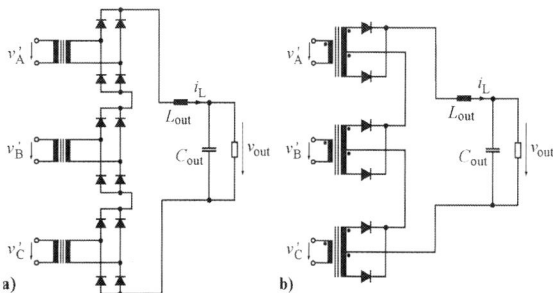

Fig. 10 Possible realization of the IMY/△-rectifier's secondary with **a)** either three separate full bridge output rectifiers or **b)** half bridge output rectifiers and transformers with center-tapped secondary windings.

Fig. 9 Schematic waveform of **a)** the phase input voltages $v_{\mathrm{in,a}}$, $v_{\mathrm{in,b}}$ and $v_{\mathrm{in,c}}$, **b)** the resulting output voltage of the full bridge v'_{A}, v'_{B} and v'_{C} if the asymmetric modulation scheme is applied and **c)** the resulting output voltage v_{sec} which is the sum of all three secondary voltages v_{A}, v_{B} and v_{C}.

could be destroyed.

Therefore, the full bridges of the laboratory prototype are designed with MOSFETs of the C7-CoolMOS series in OR-ing configuration, with the possibility to short circuit the series connected Schottky diode and to experimentally compare the operation as well as the converter efficiency for the built OR-ing configuration with a conventional full bridge under soft switching condition.

B. Input and output diode rectifier

As already mentioned, due to the phase modularity of the IMY/△-rectifier, the voltage stress of all semiconductor devices of the phase modules is defined by the maximum magnitude of the mains phase-to-neutral voltage $\hat{v}_{\mathrm{in}}$ (cf. **Table I**). Thus, for the input rectifier bridges of the IMY/△-rectifier, which are just needed to rectify the low frequency input voltage, Si-diodes with a low forward voltage drop and with a blocking voltage of at least 600 V have to be selected.

The needed blocking voltage capability of the output diode bridge, which rectifies the high-frequency secondary voltage v_{sec}, is basically also given by $\hat{v}_{\mathrm{in}}$, however, the turns ratio of the transformer N_2/N_1 and the superposition of the voltages v_{A}, v_{B} and v_{C} has to be considered. The secondary voltage v_{sec} reaches its maximum when e.g. the input voltage of phase A reaches its maximum $v_{\mathrm{A}} = \hat{v}_{\mathrm{in}}$ and the other two phases B and C are at half the voltage $v_{\mathrm{B}} = v_{\mathrm{C}} = -\frac{1}{2}\hat{v}_{\mathrm{in}}$. Then, the maximum voltage can easily be calculated as $\hat{v}_{\mathrm{sec}} = 2 \cdot N_2/N_1 \cdot \hat{v}_{\mathrm{in,max}}$.

According to the switching behavior of the conventional DC/DC buck converter, due to the hard commutation of the diodes, the output diodes of the IMY/△-rectifier have to be realized either with SiC-diodes for secondary voltages higher than approximately 200 V or with Schottky diodes for voltages below this value.

As already described in detail, if one of the three full bridges is turned on, due to the series connection of the secondary windings the load current also has to flow through all primary windings and all full bridges independent of their switching state. Consequently, this results in high conduction losses in the IMY/△-rectifier, especially if the OR-ing configuration is used, since in the freewheeling state the current can't flow through the MOSFET but has to flow through the anti-parallel SiC-diode with its large forward voltage drop.

In order to be able to reduce the conduction losses in the primary side of the IMY/△-rectifier, the individual single-phase transformers could be decoupled, which could be achieved by extending the modular structure of the IMY/△-rectifier also to the secondary side by using

three series connected individual output rectifiers (cf. **Fig. 10**). With this extension, during the freewheeling interval, the current in each full bridge can be forced to zero independently. Thus, the conduction losses in the primary side of the IMY/△-rectifier can be significantly reduced. Unfortunately, due to the missing freewheeling current during the turn on switching transition (from freewheeling state to either a positive or negative voltage) ZVS is lost. Although the turn on transition can be performed at zero current (ZCS), the stored energy in the MOSFET output capacitance is dissipated.

In addition, if the individual output rectifiers are realized as full wave rectifiers, the conduction losses are increasing, since now the load current is freewheeling through six instead of two diodes (cf. **Fig. 10 a)**). However, the output rectifiers can also be realized as half bridge rectifiers in combination with transformers having a center tapped secondary winding as shown in (cf. **Fig. 10 b)**). Then the load current is only freewheeling through three output diodes, but also has to flow through both secondary windings of the transformer.

C. Single-phase transformers

The single-phase transformers can basically be designed as the transformer of a DC/DC full bridge forward converter, with the only difference that the input voltage is varying in time. Since the duty cycle of the full bridge and the voltage applied to the transformer are proportional to the input voltage v_{in}, the envelope $B_{\mathrm{en}}(t)$ of the flux density in the transformer over one mains period is

$$B_{\mathrm{en}}(t) = \hat{B} \cdot \sin^2(\omega_{\mathrm{in}}t), \tag{3}$$

where the maximum flux density $\hat{B}$ is reached when the maximum input voltage $\hat{v}_{\mathrm{in}}$ is applied to the transformer (with a core cross section A_{core}) during one half period $T_s/2$

$$\hat{B} = \frac{\hat{v}_{\mathrm{in}} \cdot T_s/2}{N_1 \cdot A_{\mathrm{core}}}. \tag{4}$$

Finally, based on (1), the turns ratio of the transformer

$$\frac{N_1}{N_2} = \frac{3}{2} M_{\mathrm{max}} \frac{\hat{v}_{\mathrm{in,min}}}{v_{\mathrm{out}}} \tag{5}$$

can be easily found with the minimum input voltage amplitude $\hat{v}_{\mathrm{in,min}}$ and the maximum modulation index M_{max}, which typically is set around 0.9 in order to leave a certain margin for the control. As already mentioned, if the output rectification is realized with three independent rectifier bridges, the secondary winding of each single-phase transformer has to be realized as a center-tapped winding, however, the turns ratio N_2/N_1 stays the same. Concerning the loss calculation it has to be considered that the voltage applied to the transformer is independent whether one common or three separate

978-1-4799-2706-7/14 $31.00 © 2014 IEEE

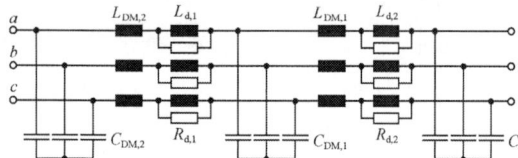

Fig. 11 Circuit diagram of the two-stage EMI input filter.

output rectifier bridges are used. Consequently, for both options the flux excitation as well as the core losses stay the same. The copper losses, however, have to be calculated differently. With one common output rectifier, the same load current with the same rectangular waveform - the pulse width is defined by the phase with the maximum duty cycle - is flowing through all three transformers. With three separate output rectifiers, however, the load current's pulse width is given by the duty cycle of the corresponding full bridge.

D. Input and output filter

The input filter has to be designed in such a way that the EMC directive (CISPR, class B) concerning conducted noise emission in the range of 150 kHz-30 MHz is fulfilled. In order to keep the input filter effort as low as possible, typically the switching frequency is limited below 150 kHz, because then the spectral component at the switching frequency doesn't have to be considered for the input filter design. Unfortunately, buck-type PFC rectifiers feature discontinuous input currents, thus compared to boost-type systems typically a higher filter attenuation is needed resulting in an increased EMI filter effort, e.g. a two-stage filter instead of a single-stage filter has to be built (cf. **Fig. 11**). Due to the discontinuous input currents, usually large filter capacitors C_F are selected which have to be closely placed to the phase module in order to achieve a low commutation inductance. In addition to the filter capacitors C_F also a small dc-link capacitor C_{DC} is provided between the input diode rectifier and the full bridge, which further improves the switching behavior, especially during the turn on transition from freewheeling to either positive or negative voltage (cf. **Fig. 6 a**). However, since the power factor λ is reduced with increasing filter capacitance, the differential mode capacitors shouldn't be selected too large; $\lambda > 0.9$ for the whole input voltage range and a wide output power range is a reasonable value. It should be also noted again that the IMY/$\triangle$-rectifier is capable of slightly shifting the input current out of phase with the input voltage, thus enabling reactive power compensation. Besides the differential mode capacitors, also the common mode capacitors have to be provided, where the capacitance is limited by the maximum ground currents drawn from the mains, which have to be below 3.5 mA.

Based on the specified output voltage and current ripple, the output filter components L_{out} and C_{out} can be designed. Due to the continuous current impressed by the output inductor L_{out}, the current ripple capability of the output capacitor is not as crucial as e.g. with boost-type PFC rectifiers. Assuming a constant output voltage v_{out}, based on the rectified voltage v_{sec} which is applied to the filter inductor, the maximum current ripple is obtained as

$$\Delta i_{L,pp,max} = \frac{\hat{v}_{in}}{L_{out}} \frac{3M}{4f_{sw}} \frac{N_2}{N_1} \left(1 - \frac{\sqrt{3}}{2}M\right). \quad (6)$$

With the given specifications in **Table I**, now the achievable efficiency, and the loss and volume distribution of the IMY/$\triangle$-rectifier are calculated and the following realization options are compared:

- **R1,CFD2 (sym. mod.)**: where the full bridge is realized with MOSFETs of the CFD2-CoolMOS series, which are controlled

based on the symmetric modulation scheme and for the output rectifier only one common diode bridge is used,

- **R1,CFD2 (asym. mod.)**: where compared to **R1,CFD2 (sym. mod.)** only the modulation scheme is changed to the asymmetric modulation,
- **R3,CFD2 (sym. mod.)**: where compared to **R1,CFD2 (sym. mod.)** only the output rectifier is realized with three separate diode bridges and finally,
- **R1,C7+OR-ing**: where the full bridge is realized with MOSFETs of the C7-CoolMOS series, operated in an OR-ing configuration with the symmetric modulation scheme and common diode rectifier.

The comparative evaluation of the mentioned realization options has been done for different design, where per switch also a parallel connection of multiple discrete MOSFET devices (up to four devices per switch) is considered. For the additional diodes in the OR-ing configuration, however, there are no discrete devices placed in parallel, due to the negative temperature coefficient of Schottky diodes; a parallel connection of Si or Schottky diodes should be only used if the diode chips are already available in the same package, thus the chips are thermally well coupled. SiC-diodes, however, feature from a certain current level on a positive temperature coefficient. Therefore, a parallel connection of these diodes would be feasible. Nevertheless, due to the given forward voltage drop, with a parallel connection of diodes the reduction of the conduction losses is limited.

In **Fig. 12** the achievable efficiencies with respect to the resulting converter costs are shown. As can be noticed, for the same costs the realization option **R1,CFD2 (asym. mod.)** shows the highest efficiencies. With four parallel connected MOSFETs of the CFD2-CoolMOS series (IPW65R041CFD, $V_{ds} = 650$ V, $R_{ds,on} = 41$ mΩ) a maximum efficiency of 96.7 % is achieved at nominal input voltage and nominal load. Since for the laboratory prototype an efficiency of at least 96 % is targeted, the design point DP_1 **R1,CFD2 (asym. mod.)**, where two MOSFETs of the CFD2-CoolMOS series (IPW65R041CFD, $V_{ds} = 650$ V, $R_{ds,on} = 41$ mΩ) are connected in parallel per switch, is selected as the reference design (96.0 %) and the costs are normalized with respect to this design.

If the reference design is operated with the symmetric modulation scheme (**R1,CFD2 (sym. mod.)**) at design point DP_2, where two of the six switching transitions result in hard switching, the efficiency is drastically reduced to 95.0 %, which is a difference of 1 % (= 75 W). If per switch even more MOSFETs are connected in parallel, the discrepancy between the two modulations schemes is even more pronounced; while the efficiency of the asymmetric modulation scheme is monotonically increasing, the losses with symmetric modulation are increasing due to the higher recovery losses. This is clearly shown in **Fig. 13**, where the loss distribution of the different designs is given. The only difference in the loss distribution and thus in the efficiency of design point DP_1 and DP_2 is found in the additional switching losses of DP_2. Consequently, the realization option **R1,CFD2 (sym. mod.)** is no more considered.

Another possibility is to built the reference design with three separate half bridge output rectifiers and transformers with center-tapped secondary winding, which should result in reduced conduction losses in the full bridge and in the primary windings of the transformers (**R3,CFD2 (asym. mod.)**). As illustrated in the loss distribution of **Fig. 13**, the conduction losses in the switches can be reduced, however, the conduction losses in the output rectifier are increasing disproportionately, resulting in a lower total converter efficiency and in even higher costs. Surprisingly, the total transformer losses are hardly influenced. This can be explained by the fact that with three

The 2014 International Power Electronics Conference

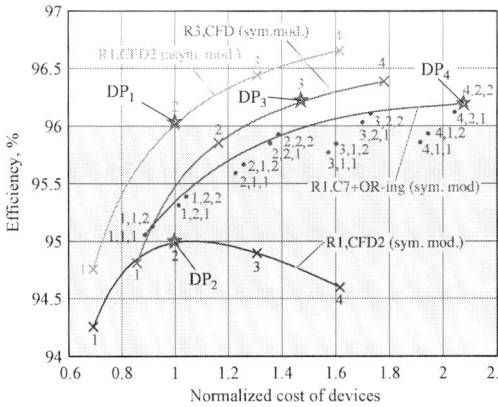

Fig. 12 Comparison of the achievable efficiencies for the different realization options with respect to the resulting converter costs. The numbers indicate how many parallel MOSFETs are used per switch. In the case of the realization option (R1,C7+OR-ing), the number of MOSFETs S, parallel diodes D_P and series diodes D_S is indicated (S,D_P,D_S). The costs are normalized with respect to the reference design (design point DP_1) that achieves an efficiency higher than 96 %.

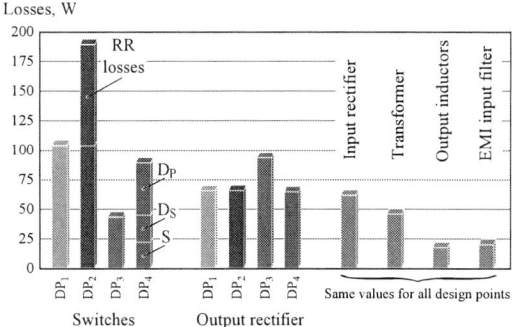

Fig. 13 Loss distribution for the different realization options at the design points DP_1, DP_2, DP_3 and DP_4. In addition to the conduction losses in the MOSFET for DP_2 the reverse recovery losses and for DP_4 the conduction losses in the Schottky diodes D_S and in the SiC-diodes D_P have to be considered.

output rectifiers the losses in the primary winding are effectively reduced, however the losses in the secondary winding are increased by approximately the same amount since now two instead of one secondary winding has to be arranged in the same winding window. Thus, a smaller wire diameter has to be selected which results in higher conduction losses in the secondary windings.

In order to compensate the additional losses in the three output rectifiers and to achieve approximately the same efficiency as with the reference design, now for the design point DP_3 (96.2 %) three instead of two parallel MOSFET are needed per switch (cf. **Fig. 12**). As one might expect, if the output rectifier would be realized as a full bridge rectifier, the conduction losses in the output diodes would be even higher. Such a solution would only be reasonable if Schottky diodes with low forward voltage drop, e.g. at lower output voltages, could be used. Therefore, since for the realization option with three rectifiers **R3,CFD2 (sym. mod.)** the hardware effort and thus the costs ($\sim$ 1.8 times more expensive) are strongly increased, this design is also dropped.

As a fourth option, the full bridges of the IMY/Δ-rectifier can be realized with MOSFETs of the C7-CoolMOS series in OR-ing configuration, thus the reverse recovery losses can be discarded and both modulation schemes can be used (**R1,C7+OR-ing**). Due to the superior low on-state resistance of the C7-MOSFET (IPW65R019C7), the conduction losses in the switch can be drastically reduced. However, this efficiency improvement is again overcompensated by the additional conduction losses in the series Schottky diode and especially in the anti-parallel SiC-diode, which shows a higher forward voltage drop (cf. **Fig. 13**). In design point DP_4 with four parallel connected MOSFETs and two Schottky/SiC-diode per switch a similar efficiency (96.2 %) is achieved as with the reference design.

Although the costs with this realization option are more than doubled compared to the reference design - due to the high number of semiconductor devices - it offers the flexibility to verify the dimensioning and the achievable efficiencies of the different realization options explained above in just one laboratory prototype; e.g. both modulation schemes can be implemented or the series connected Schottky diodes can be shorted in order to verify the reference design point DP_1.

IV. LABORATORY PROTOTYPE

Based on the specifications given in **Table I**, a 7.5 kW laboratory prototype is built with full bridges in OR-ing configuration as described in the previous section.

Each phase module, containing the input diode bridge and the full bridge where discrete devices in D2PAK package are used, is realized on a single-layer insulated metal substrate (IMS) board that improves the heat transfer from the semiconductors to the heatsink. The conduction losses in the input rectifier diodes can be easily calculated based on the average and rms current ratings, which for nominal operating conditions are 4.9 A and 10.24 A, respectively. For the selected rectifier diodes (DSP45-12A), conduction losses of 5 W per diode are expected. Accordingly, the conduction losses in the MOSFETs (IPB65R045C7), Schottky diodes (VBT6045C) and SiC-diode (IDK12G65C5) are found to be 1.8 W, 1.9 W and 3.7 W per switch.

Due to the hard commutation of the output rectifier diodes, SiC Schottky barrier diodes (C4D40120D) are selected, where 16.3 W of conduction losses are expected per diode.

The design of the single-phase transformers is performed with an optimization algorithm, where different cores and turn numbers are evaluated. There, a compromise between volume and losses has to be made. For a selected switching frequency of 72 kHz transformers based on two stacked E55-cores (EPCOS) were realized. The windings with $N_1 = 8$ and $N_2 = 9$ were made of 3 mm litz wire with 840 insulated strands.

Due to the mechanical design, the filter inductor $L_{out} = 2 \cdot 195\,\mu H$ is split in two separate output inductors one of each placed in the positive and negative output rail. Each inductor is also built with two stacked E55-cores, and because of the almost constant output current, the winding is realized with 14 turns of a 2 mm thick solid copper wire. A list of the main components is given in **Table II** and the corresponding volume distribution is shown in **Fig. 14**. More than 50 % of the volume is already captured by the heatsinks, which together with the magnetic components consumes 70.8 %.

The calculated efficiencies of the designed IMY/Δ-rectifier for different input voltages and load conditions is shown in **Fig. 15**. As can be noticed, the efficiency should be maintained over 96 % for all operating conditions. The IMY/Δ-rectifier prototype, on which now the experimental measurements will be performed, is shown in **Fig. 16**.

978-1-4799-2706-7/14 $31.00 © 2014 IEEE   3870

The 2014 International Power Electronics Conference

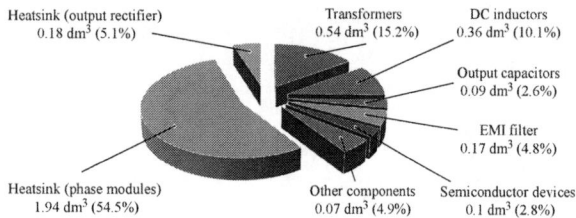

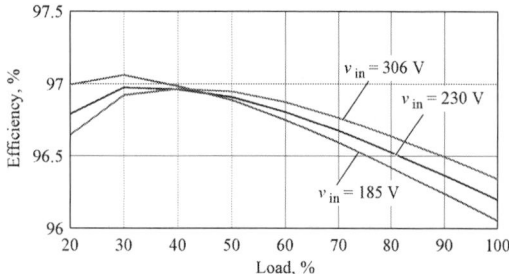

Fig. 14 Volume distribution of the main components.

Fig. 15 Calculated efficiency of the IMY/△-rectifier at different load and input voltage conditions.

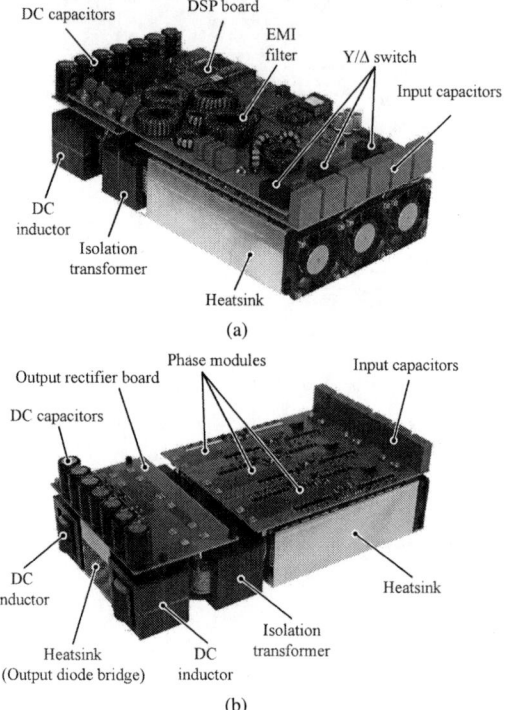

Fig. 16 Realized 7.5 kW IMY/△-rectifier prototype. **a)** fully assembled and **b)** without the main board, in order to show the phase modules and the output rectifier board.

V. CONCLUSIONS

IMY/△-rectifier presents an interesting alternative to three-phase integrated PFC rectifiers. It provides the PFC functionality and the galvanic isolation in a single-stage energy conversion. Due to its phase-modularity, only the phase voltage determining the required blocking capability of the semiconductors, thus 600 V instead of 1200 V semiconductor devices can be used, which typically feature lower on-state resistances and improved switching behavior.

The IMY/△-rectifier can be controlled with a simple phase shift modulation, and due to its buck-type topology the same control structure and duty cycle calculation concept as for a conventional DC/DC buck converter can be implemented. In addition, with the presented modified modulation scheme in each switching transition soft switching (ZVS) can be achieved without any additional circuitry.

In this paper a comprehensive analysis of the basic operation and the design of the IMY/△-rectifier, including the selection of semiconductor devices and magnetic components, is presented. The built laboratory prototype is designed to achieve an efficiency of more than 96 % under different load and input voltage conditions. In the further research, experimental measurements will be performed in order to verify the presented theoretical investigations.

REFERENCES

[1] J. W. Kolar and T. Friedli, "The essence of three-phase PFC rectifier systems," in *IEEE 33rd International Telecommunications Energy Conference (INTELEC)*, 2011, pp. 1–27.

[2] B. Singh, B. Singh, A. Chandra, K. Al-Haddad, A. Pandey, and D. Kothari, "A review of three-phase improved power quality ac-dc converters," *IEEE Transactions on Industrial Electronics*, vol. 51, no. 3, pp. 641–660, 2004.

[3] S. Manias and P. D. Ziogas, "A novel sinewave in ac-to-dc converter with high-frequency transformer isolation," *IEEE Transactions on Industrial Electronics*, vol. IE-32, no. 4, pp. 430–438, 1985.

[4] V. Vlatkovic, D. Borojevic, and F. C. Lee, "A zero-voltage switched, three-phase isolated PWM buck rectifier," *IEEE Transactions on Power Electronics*, vol. 10, no. 2, pp. 148–157, 1995.

[5] J. W. Kolar, U. Drofenik, and F. C. Zach, "VIENNA rectifier II - A novel single-stage high-frequency isolated three-phase PWM rectifier system," in *Conference Proceedings of the 13th Annual Applied Power Electronics Conference and Exposition (APEC)*, vol. 1, 1998, pp. 23–33.

[6] J. W. Kolar, U. Drofenik, H. Ertl, and F. C. Zach, "VIENNA rectifier III - A novel three-phase single-stage buck-derived unity power factor AC-to-DC converter system," in *Conference Proceedings of the Nordic Workshop on Power and Industrial Electronics*, 1998, pp. 9–18.

[7] M. A. de Rooij, J. A. Ferreira, and J. D. Van Wyk, "A three phase, soft switching, transformer isolated, unity power factor front end converter," in *29th Annual IEEE Power Electronics Specialists Conference (PESC)*, vol. 1, 1998, pp. 798–804.

[8] Y. K. E. Ho, S. Hui, and Y. S. Lee, "Characterization of single-stage three-phase power-factor-correction circuit using modular single-phase PWM dc-to-dc converters," *IEEE Transactions on Power Electronics*, vol. 15, no. 1, pp. 62–71, 2000.

[9] J. W. Kolar, T. Friedli, J. Rodriguez, and P. Wheeler, "Review of three-phase PWM AC-AC converter topologies," *IEEE Transactions on Industrial Electronics*, vol. 58, no. 11, pp. 4988–5006, 2011.

[10] P. Cortes, J. Huber, M. Silva, and J. W. Kolar, "New modulation and control scheme for phase-modular isolated matrix-type three-phase ac/dc converter," in *39th Annual Conference of the IEEE Industrial Electronics Society (IECON)*, 2013, pp. 4899–4906.

TABLE II Main components of the IMY/△-rectifier prototype.

Component	Value/details
MOSFETs S	650 V CoolMOS C7 series (IPB65R045C7)
Series diodes D_S	45 V/30 A Schottky barrier (VBT6045C)
Parallel diodes D_P	650 V/12 A SiC Schottky barrier (IDK12G65C5)
Input diodes	1200 V/45 A rectifier diodes (DSP45-12A)
Output diodes	1200 V/54 A SiC diodes (C4D40120D)
Isolation transformers	Stack of two E55 N87 cores, $N_1/N_2 = 8/9$, 3 mm/840 strands Litz wire
Filter inductor L_{out}	2x195 µH, 2 stacked E55 N87 cores, 0.6 mm airgap, 14 turns of 2 mm solid copper wire
Filter capacitor C_{out}	7x82 µF

An Energy Saving Drive Method of an Induction Motor with the Suppression of Sudden Acceleration and Deceleration

Yuji Asano, Kaoru Inoue, Keito Kotera and Toshiji Kato
Department of Electrical Engineering, Doshisha University
Kyotanabe, Kyoto 610-0321, JAPAN
E-mail: dun0307@mail4.doshisha.ac.jp

Abstract—In order to drive the electric machines using the motors efficiently, the energy loss should be minimized during its operation. It has been reported that the design methodology of the optimal torque and rotating speed trajectories to minimize the energy loss of the induction motor (IM) drive systems when the operation time, rotating speed, and rotational angle are given as drive conditions. However, the optimal trajectory may cause sudden acceleration and deceleration. This paper proposes a design methodology of the optimal trajectories for IM drive system by means of the variational method and the S-curve trajectory. The effectiveness of the proposed method will be illustrated by means of some experiments and simulation.

Keywords—Induction Motor, Drive Conditions, Variational Method, S-curve trajectory

I. INTRODUCTION

In order to drive the electric machines using the motors efficiently, the energy loss should be minimized during its operation. The induction motors (IM) are widely used in many applications such as industrial instruments, transportation systems, and so on. The improvement of the efficiency of the existing IM is an important issue to save energy. In the applications of IM, such as transportation systems, conveyers, and elevators, the rotating speed is changed gradually due to the safety of its passengers and packages. The energy loss during the motor operation depends on the waveforms of the motor torque and rotating speed [1], [2]. Then, the optimal torque and rotating speed trajectories of a squirrel-cage induction motor have been proposed by using the variational method [3] under the two drive conditions with respect to the operation time and the rotating speed [1], [4]. The reported optimal trajectory assure that the rotating speed reaches its desired objective speed at the objective time. In the applications, such as transportation systems, elevators, forklift, and conveyers, travel distances should be considered to convey passengers and packages to appropriate objective places. The travel distance is in direct relation to the rotational angle of the motor. The reference [5] have proposed a design methodology of the optimal torque and rotating speed trajectories to minimize the energy loss of the IM drive system when the operation time, rotating speed, and rotational angle are given as the drive conditions. However, the obtained optimal torque trajectory becomes large around the operation start and end time, then it

causes sudden acceleration and deceleration.

This paper proposes a design methodology of the optimal trajectories for IM drive system by means of the variational method and the S-curve trajectories in order to suppress the sudden acceleration and deceleration. The obtained optimal trajectories assure that the rotational angle and rotating speed reach desired objective angle and objective rotating speed at the operation end time, respectively. In the method, trajectories are designed to be smooth in order to suppress the jerk. Jerk is the second derivative of rotating speed, and it is an important factor in both suppressing vibration and achieving high accuracy [6]. In the conveyers and elevators, strong jerk appears at the starting and stopping. The effectiveness of the proposed methods is ensured by comparing the jerk of the variational method and constant torque based on the numerical simulation and experiments.

II. REVIEW OF OPTIMAL TRAJECTORIES DERIVED BY VARIATIONAL METHOD

A. Objective system

Fig. 1 illustrates the basic structure of a three phase induction motor drive system. A mathematical equation of a rotor in IM is shown by

$$ J\dot{\omega} + \xi\omega + T_L = T_e \qquad (1) $$

where J, ξ, ω, T_L, and T_e are the inertia of the rotor, rotational damping coefficient, rotating speed, constant load torque, and motor torque induced by the electrical part of IM, respectively. When an inertial and/or a damping loads are connected to IM by a rigid rotor, the coefficients J and/or ξ change according to the loads. A vector control method is employed in order to control the torque instantaneously.

B. Optimal torque and rotating speed trajectories

The mechanical loss P_{lm} of IM appears due to the rotational damping effect. Then P_{lm} is given by $P_{lm} = \xi\omega^2$. The electric interior loss P_{le} of IM under a vector control becomes the sum of copper losses in the stator and rotor [7] when iron losses are negligible. Then P_{le} is given by $P_{le} = a + bT_e^2$ where constants $a = \frac{R_s}{M^2}\phi_{\gamma r}^2$, $b = \left(\frac{R_s L_r^2}{M^2} + R_r\right)\frac{1}{p^2\phi_{\gamma r}^2}$. Nomenclatures

The 2014 International Power Electronics Conference

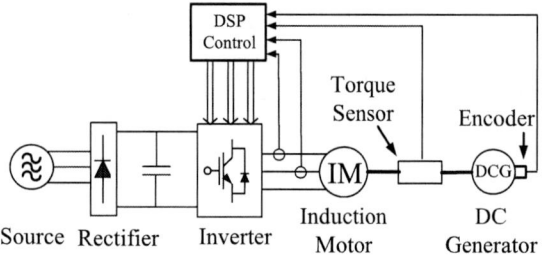

Fig. 1. A basic structure of a three-phase induction motor drive system.

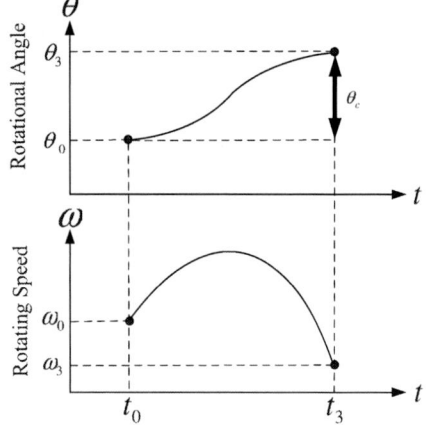

Fig. 2. Schematic diagram of the optimal trajectories to minimize the loss energy during the operation of IM under given 3 drive conditions, the operation time period $[t_0, t_3]$, rotating speed range $[\omega_0, \omega_3]$, rotational angle $[\theta_0, \theta_3]$.

R_s, R_r, M, L_r, p, $\phi_{\gamma r}$, $\phi_{\delta r}$, $i_{\gamma s}$, and $i_{\delta s}$ represent the stator resistance, armature resistance, the mutual inductance, self-inductance of rotor, stator resistance, γ element of rotor flux, δ element of rotor flux, γ element of stator current, and δ element of stator current, respectively. Hence the total loss of IM is given by

$$
\begin{aligned}
P_{la} &= P_{lm} + P_{le} \\
&= bJ^2\dot{\omega}^2 + 2bJ\xi\dot{\omega}\omega + (1+b\xi)\xi\omega^2 + a.
\end{aligned}
\tag{2}
$$

When the motor speed is reduced or accelerated from ω_0 to ω_3 and the rotational angle is changed from θ_0 to θ_3 between the operation time period $[t_0, t_3]$ as shown in Fig. 2, an integrated total loss, i.e. a total loss energy I_{la} is defined as

$$
I_{la} = \int_{t_0}^{t_3} P_{la} dt.
\tag{3}
$$

The aim is to minimize the I_{la} under the subsidiary condition of the rotational angle

$$
\int_{t_0}^{t_3} \omega dt = \theta_3 - \theta_0 = \theta_c
\tag{4}
$$

where θ_c is a given constant value as one of the drive condition. The total loss P_{la} depends on $\dot{\omega}$ and ω, hence the optimal

rotating speed trajectory should be derived to obtain the optimal torque trajectory. To solve the optimization problem with subsidiary condition (4) is equivalent to solving the Euler's equation with Lagrange multiplier λ.

$$
\left\{ \frac{d}{dt}\left(\frac{\partial P_{la}}{\partial \dot{\omega}} \right) - \frac{\partial P_{la}}{\partial \omega} \right\} + \lambda \left\{ \frac{d}{dt}\left(\frac{\partial \omega}{\partial \dot{\omega}} \right) - \frac{\partial \omega}{\partial \omega} \right\} = 0
\tag{5}
$$

By using the variational method, the optimal trajectory of ω minimizes I_{la} is obtained as

$$
\omega = F_1 e^{-\alpha t} + F_2 e^{\alpha t} - \frac{\lambda}{\beta}
\tag{6}
$$

where $\alpha = \sqrt{\frac{2(1+b\xi)\xi}{2bJ^2}}$, $\beta = 2(1+b\xi)\xi$. Consequently, an optimal torque trajectory T_e is obtained from (1) as follows [5].

$$
T_e = (-J\alpha + \xi)F_1 e^{-\alpha t} + (J\alpha + \xi)F_2 e^{\alpha t} - \frac{\xi}{\beta}\lambda
\tag{7}
$$

C. Constants F_1, F_2, and λ

Coefficients F_1, F_2, and λ are determined as follows. Substituting initial rotating speed ω_0 at initial time t_0 and desired objective speed ω_3 at end time t_3 into (6), then the following equations are obtained.

$$
\omega_0 = F_1 e^{-\alpha t_0} + F_2 e^{\alpha t_0} - \frac{\lambda}{\beta}
\tag{8}
$$

$$
\omega_3 = F_1 e^{-\alpha t_3} + F_2 e^{\alpha t_3} - \frac{\lambda}{\beta}
\tag{9}
$$

From the obtained optimal trajectory of rotating speed (6), the subsidiary condition of rotating angle (4) is rewritten as

$$
\begin{aligned}
\theta_c &= -\frac{F_1}{\alpha}\left(e^{-\alpha t_3} - e^{-\alpha t_0} \right) + \frac{F_2}{\alpha}\left(e^{\alpha t_3} - e^{\alpha t_0} \right) \\
&\quad - \frac{\lambda}{\beta}(t_3 - t_0)
\end{aligned}
\tag{10}
$$

From three equations (8), (9), and (10), the coefficients F_1, F_2 and λ are determined. Therefore, the optimal rotating speed, rotating angle, and torque trajectories are derived when the operation time period $[t_0, t_3]$, the rotating speed range $[\omega_0, \omega_3]$ and the rotating angle range $[\theta_0, \theta_3]$ are given as drive conditions.

D. Simulational and experimental results of optimal trajectories by variational method

Simulations and experiments of the obtained optimal trajectory are carried out under the conditions; $\omega_0 = 0$ (rad/s) at $t_0 = 0.05$ (s), $\omega_3 = 0$ (rad/s) at $t_3 = 0.55$ (s), $\theta_c = 30$ (rad). Figure 3 shows the simulational and experimental results in the case of the derived optimal trajectories. The dashed blue lines illustrate the simulation and the solid red lines do experimental results. It is observed that at $t_3 = 0.55$ (s) the rotating speed becomes 0 and the rotational angle reaches 30 (rad). The results show that all the conditions are satisfied.

978-1-4799-2706-7/14 $31.00 © 2014 IEEE

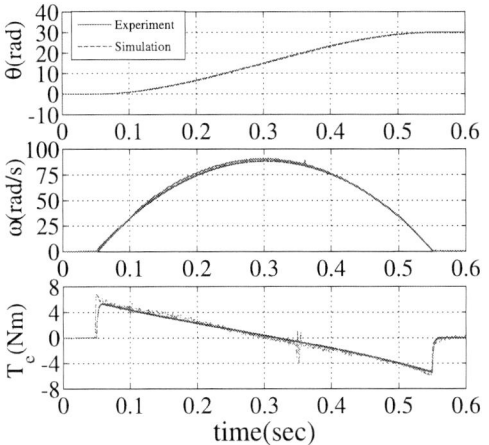

Fig. 3. Simulation and experimental results in the case of the optimal trajectory derived by variational method.

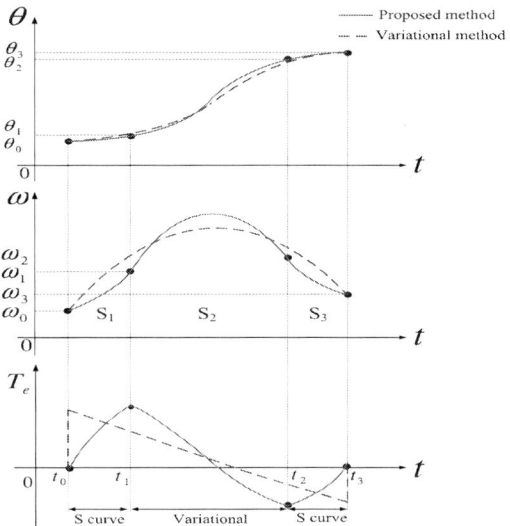

Fig. 4. Schematic diagram of proposed optimal trajectories satisfying the given drive conditions.

III. THE OPTIMAL TRAJECTORY SUPPRESSING SUDDEN ACCELERATION AND DECELERATION

A. S-curve trajectory

The jerk is defined as second derivation of rotating speed. If the acceleration have step form, jerk appears and gives a negative effect to passengers on the transportation systems such as conveyers and elevator [8]. The variational torque trajectory (7) shown in Fig. 3 causes sudden acceleration and deceleration around the operation start and end time. As a result, big jerk appears. To relax the sudden acceleration, S-curve trajectory [9] is adopted . The S-curve trajectory is given by the following equations (11) and (12). Equation (11) represents the acceleration, and equation (12) does the deceleration. In this paper, $\omega' = \frac{\pi}{2t_a}$, $\omega'' = \frac{\pi}{2t_b}$, t_a, t_b, ω_a and ω_b are the parameters of the shape of S-curve trajectory. In this paper, ω_a and ω_b are set as $\omega_a = 90$ (rad/s), $\omega_b = 90$ (rad/s). By using the S-curve trajectories of (11) and (12), the motor accelerates and decelerates gradually around the operation start and the end time.

$$\omega = \omega_a - (\omega_a - \omega_0) \cos(\omega' t) \quad (11)$$

$$\omega = \omega_b - (\omega_b - \omega_3) \sin(\omega'' t) \quad (12)$$

B. Design methodology of optimal trajectories by using S-curve trajectory

Fig. 4 shows the schematic diagram of the optimal motion trajectories by using S-curve trajectory. Because large torque is necessary around the start and end time in case of variational torque trajectory, S-curve trajectory is adopted to suppress the sudden acceleration and deceleration. The point (ω_1, t_1) is the switching point from the S-curve trajectory of acceleration to the variational trajectory. The point (ω_2, t_2) is another

switching point from the variational trajectory to the S-curve trajectory of deceleration. The variational trajectory between the switching points is given by (6). The switching points must be chosen that S-curve trajectory equates to variational trajectory at the switching points. In order to obtain the optimal trajectories by using S-curve trajectory, following equations and the Newton-Raphson iteration method are adopted to derive $t_1, t_2, F_1, F_2,$ and λ. The initial time t_0 and load torque T_L are set as 0 for simplicity.

$$\omega_a - (\omega_a - \omega_0) \cos(\omega' t_1) = F_1 + F_2 - \frac{\lambda}{\beta} \quad (13)$$

$$\omega_b - (\omega_b - \omega_3) \cos(\omega''(t_3 - t_2)) = F_1 e^{-\alpha(t_2 - t_1)} \\ + F_2 e^{\alpha(t_2 - t_1)} - \frac{\lambda}{\beta} \quad (14)$$

$$(\omega_a - \omega_0)\omega' \sin(\omega' t_1) = -\alpha F_1 + \alpha F_2 \quad (15)$$

$$-(\omega_b - \omega_3)\omega'' \sin(\omega''(t_3 - t_2)) = -\alpha F_1 e^{-\alpha(t_2 - t_1)} \\ + \alpha F_2 e^{\alpha(t_2 - t_1)} \quad (16)$$

$$\theta_c = S_1 + S_2 + S_3 \quad (17)$$

where square measure $S_1 = \int_{t_0}^{t_1} \omega_a - (\omega_a - \omega_0) \cos(\omega' t) \, dt$, $S_2 = \int_{t_1}^{t_2} F_1 e^{-\alpha t} + F_2 e^{\alpha t} - \frac{\lambda}{\beta} dt$, $S_3 = \int_{t_2}^{t_3} \omega_b - (\omega_b - \omega_3) \cos(\omega'' t) \, dt$.

Equation (13) shows the rotating speed of S-curve trajectory at t_1 equates to the rotating speed of variational trajectory at t_1. Equation (14) shows the rotating speed of S-curve trajectory at t_2 equates to the rotating speed of variational trajectory at t_2. Equation (15) shows the acceleration of S-curve trajectory at t_1 equates to the acceleration of variational trajectory at t_1. Equation (16) shows the acceleration of S-curve trajectory at t_2 equates to the acceleration of variational trajectory at t_2. Finally, equation (17) is equivalent to (4) in this case.

IV. SIMULATION AND EXPERIMENTAL RESULTS

A. Set up

The motor parameters $J = 0.0073$ (kg·m^2), $\xi = 0.0036$ (Ns), $\alpha = 3.64$, $\beta = 0.00734$ are adopted for analytical discussion and simulation. In the simulations, the Matlab Simulink and SimPowerSystems blocksets are used. The experiments are carried out by using a squirrel-cage induction motor (Kusatsu denki, 750W, 4poles) and a DSP control system (MyWay Geken, PE-ExpertII). The IM is driven from $\omega_0 = 0$ (rad/s) at $t_0 = 0.05$ (s) to $\omega_3 = 0$ (rad/s) at $t_3 = 0.55$ (s) and the rotational displacement angle is $\theta_c = \theta_3 - \theta_0 = 30$ (rad). Fig. 5 shows the control block diagram to let the rotating speed ω follow the obtained optimal rotating speed trajectories. In the figure, ω^* represents the optimal trajectory of the rotating speed (6). The feed-forward input of torque element current $i_{\delta s}{}^*$ is obtained from the optimal trajectory of the motor torque $T_e{}^*$. The reference of the torque element current $i_{\delta sref}$ is compared to the actual motor current $i_{\delta s}$ obtained by a coordinate transformation from three phase AC current, and the error between them is compensated by PI compensator. Then, the voltage $v'_{\delta s}$ is obtained and provided as three phase AC voltages to the induction motor by using the coordinate transformation.

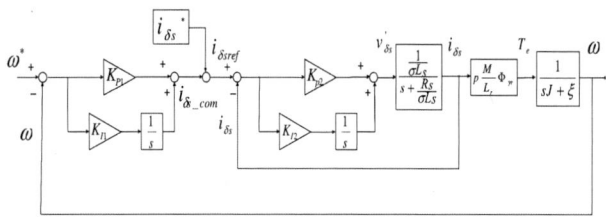

Fig. 5. Rotating speed control diagram in order to follow the obtained optimal trajectories.

B. Effectiveness of the proposed method

S-curve trajectories are set as $t_a = 0.12$ (s) and $t_b = 0.12$ (s). Equations (13) - (17) are solved under the drive conditions, then we obtain $\omega_1 = 25.2$ (rad/s), $\omega_2 = 25.2$ (rad/s), $t_1 = 0.0586$ (s), $t_2 = 0.442$ (s), $F_1 = -300, F_2 = -749$, and $\lambda = -2.94$. Fig. 6 and 7 show the simulation and experimental results by using S-curve trajectory. In the figures, from the top to the bottom, the waveforms of the rotational displacement angle, rotating speed, motor torque, jerk (the time derivative of acceleration) related to comfort and input energy to IM are illustrated. The solid red lines illustrate the proposed optimal case under S-curve trajectory, and the dashed green lines do the constant torque case, the dashed blue lines do the variational trajectories. In the proposed case, the motor torque and the rotating speed are controlled under S-curve trajectory. It is observed that the rotating speed becomes 0

(rad/s) and the rotational displacement angle θ reaches 30 (rad) at $t_3 = 0.55$ (s). The torque and the rotating speed are switched appropriately from/to the variational trajectory to/from the S-curve trajectory at the switching points. The drive conditions are satisfied. Moreover, large torque around the operation start and end time is controlled by means of S-curve trajectory. Consequently, comparing to the constant torque and variational method, jerk is suppressed. Figure 8 shows the input energy comparison of these 3 cases. Comparing to the constant torque case, the input energy of the proposed method to IM is reduced. When t_a and t_b are large, however, the energy becomes larger than the constant torque. These results show that the sudden acceleration and jerk is suppressed and the energy is saved when the S-curve constants t_a and t_b are chosen appropriately. Therefore, it is confirmed that the proposed method is effective by both the simulations and experiments. Figure 9 shows the input energy comparison when t_a is different from t_b of proposed optimal cases. Input energy varies according to t_a and t_b, because the damping effect of rotation works to reduce the speed.

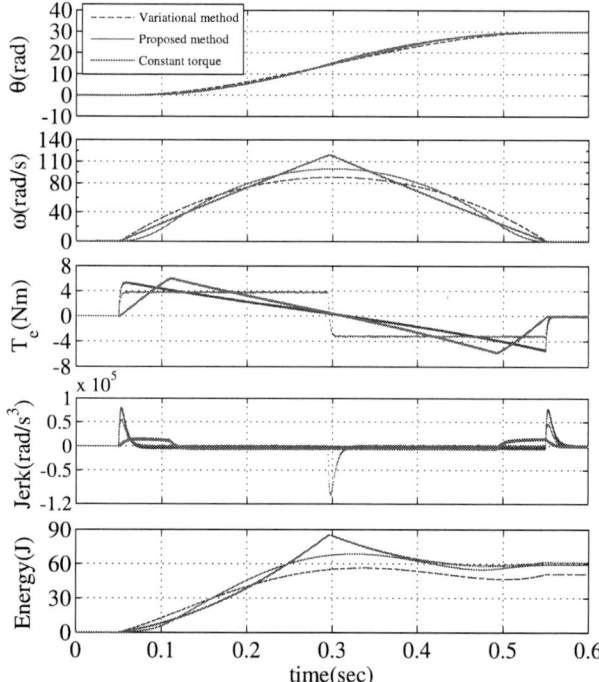

Fig. 6. Waveforms of the proposed method, variational method and constant torque (Simulation).

V. CONCLUDING REMARKS

This paper has proposed a design methodology of the optimal trajectories by using S-curve trajectory. The effectiveness of the proposed methods are illustrated by the numerical simulations and experiments. The proposed method is applicable to

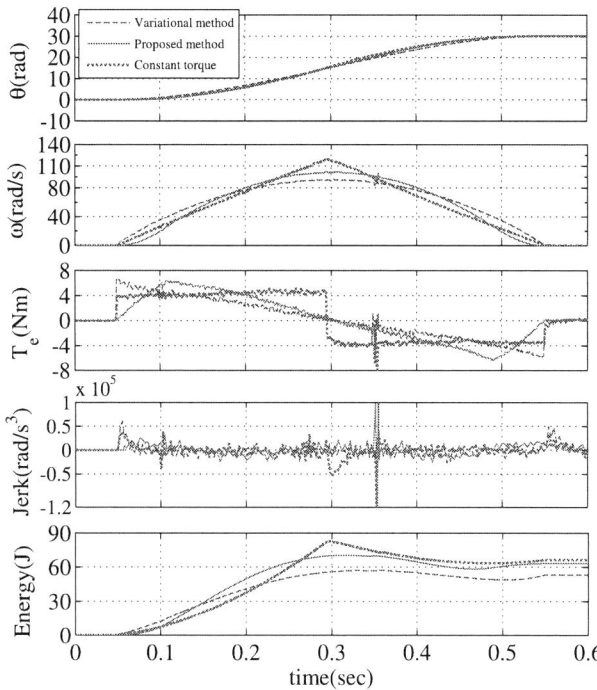

Fig. 7. Waveforms of the proposed method, variational method and constant torque (Experiment).

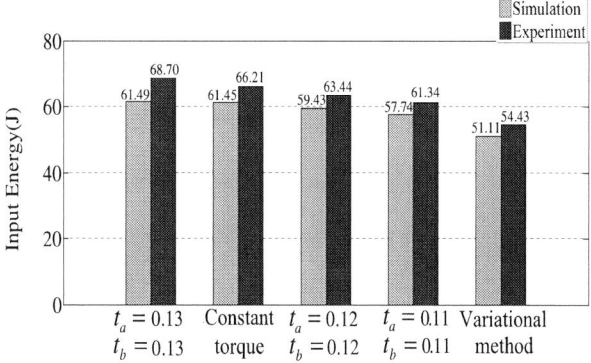

Fig. 8. Input energy of the proposed method ($t_a = t_b$), variational method and constant torque cases.

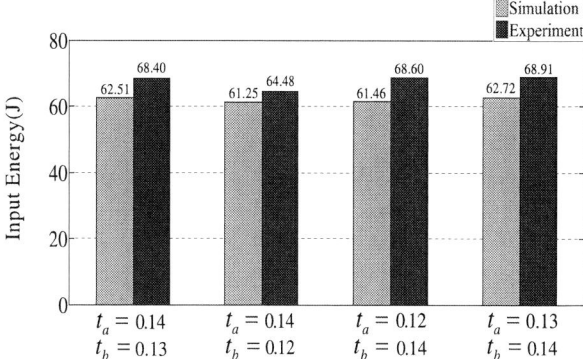

Fig. 9. Input energy of the proposed method for various t_a and t_b

real systems that have large inertia and require the non quick response of rotating speed such that electric trains, elevators, forklift and so on. The analysis of the proposed method is one of the future work when the amplitude limits of the torque and the rotating speed are given.

REFERENCES

[1] K. Inoue, K. Ogata, and T. Kato, "An Efficient Induction Motor Drive Method with a Regenerative Power Storage SystemDriven by an Optimal Torque," in *Proceedings of the 39th IEEE Power Electronics Specialists Conference*, pp.359-364 (2008).

[2] K. Matsuse, T. Yoshizumi, S. Katsuta, and S. Taniguchi, "High-Response Flux Control of Direct-Field-Oriented Induction Motor with High Efficiency Taking Core Loss into Account," *IEEE Transactions on Industry Applications*, Vol.35, No.1, pp.62-69 (1999).

[3] G.B.Arfken and H.J.Weber, *Essential Mathematical Methods for Physicists*, Elsevier (2004).

[4] K. Inoue, K. Ogata, and T. Kato, "An Efficient Power Regeneration and Drive Method of an Induction Motor by means of an Optimal Torque Derived by Variational Method," in *IEEJ Transactions of Industrial Applications*, Vol. 128, No. 9 (2008).

[5] K. Inoue, K. Kotera, and T. Kato, "Optimal Torque Trajectories Minimizing Loss of Induction Motor under Given Condition of Rotational Angle," in *The 3rd Annual IEEE Energy Conversion Congress Exposition*, pp.1734-1738 (2011).

[6] Shinichi Tanaka, Young Min Baek, Naohiro Sugita, Takashi Ueta, Yasuhiro Tamaki and Mamoru Mitsuishi, "Minimum-jerk Trajectory Generation for Master-Slave Robotic System," *The Fourth IEEE RAS/EMBS International Conference on Biomedical Robotics and Biomechatronics*, pp.811-816 (2012).

[7] K. Matsuse, T. Yoshizumi, S. Katsuta, and S. Taniguchi, "High-Response Flux Control of Direct-Field-Oriented Induction Motor with High Efficiency Taking Core Loss into Account," *IEEE Transactions on Industry Applications*, Vol.35, No.1, pp.62-69 (1999).

[8] H. Z. Li, Z. M. Gong, W. Lin and T. Lippa, "Motion profile planning for reduced jerk and vibration residuals," *SIMTech technical reports*, Vol.8, No.1, Jan. - Mar. 2007, pp 32-37.

[9] Phat Minh Ho and Naoki Uchiyama, Shigenori Sano, Kohei Sawada, Atsushi Kato and Takahiro Yonezawa "Simple motion trajectory generation for energy saving of industrial machines," *IEEE International Symposium on System Integration*, pp.476-480 (2012).

The 2014 International Power Electronics Conference

Field Oriented Control of Sensorless Linear Induction Motor Using Matrix Converter

Mahmoud A. Sayed*, Essam Ebaid Mohamed*, Tarek Hassan Mohamed** and Takaharu Takeshita***
*Electrical Engineering Dept., South Valley University, Qena, Egypt
**Electrical Engineering Dept., Aswan University, Aswan, Egypt
***Nagoya Institute of Technology, Nagoya, JAPAN
mahmoud_sayed@ieee.org

Abstract—This paper presents a new controller for the sensorless linear induction motor (LIM) based on field oriented control driven by matrix converter. Conventional PI controlled has been employed for the speed control of mover. Speed estimation of the mover has been employed using Model Reference Adaptive System (MRAS) technique. The PWM switching technique of the matrix converter is designed based on minimum voltage drop switching (MVDS) in order to minimize the converter power loss and harmonic. MVDS switching technique prevents direct switching between minimum and maximum voltage, which reduces the voltage stress on power semiconductor devices. The proposed system has the advantages of high reliability, high efficiency, and low cost due to the elimination of the mechanical speed sensor. The effectiveness of the proposed control technique has been verified using Matlab simulink.

Keywords—Linear Induction Motor (LIM), Field Oriented Control (FOC), sensorless speed control, Model Reference Adaptive System (MRAS), matrix converter, Pulse Width Modulation (PWM).

I. INTRODUCTION

Linear motors are special electrical machines, in which the electrical energy is converted directly into mechanical energy to produce motion in straight line. There are many types of linear motors such as D.C. motors, induction motors, synchronous motors and stepping motors, etc [1], [2]. Among these types, Linear Induction Motor (LIM) is considered one of the most important types of linear motors. Linear induction motor has many desirable performance features, including high-starting thrust force, high-speed operation, simple mechanical construction, no need for a gear between motor and the motion devices, the reduction of mechanical losses and the size of motion devices, silence, and easy maintenance, no backlash and small friction, and suitability for both low speed and high speed applications [3]–[5]. Therefore, LIMs are now widely used in many industrial applications including transportation, conveyor systems, actuators, material handling, pumping of liquid metal, and sliding door closers, robot base movers, office automation, drop towers, elevators etc., with satisfactory performance [6]–[8].

The driving principles of LIM are similar to those of a traditional rotary induction motor (RIM); however, its control characteristics are more complicated since its parameters are time-dependent variables. Conventional

converters are widely used to drive LIM speed using different control techniques such as conventional PI, sliding mode control (SIM), model predictive control (mpc), fuzzy and neural control. Those converters have two stages; AC/DC and DC/AC. Therefore, bulky capacitors must be used as a link between the two stages. This will severely affect the size and durability of the converter. Moreover, conventional converters have high power loss and harmonic contents. Therefore, the performance of single stage drivers for LIM speed control will be better.

Matrix converter is a direct AC/AC bi-directional power flow converter that takes power from AC source and converts it to another AC system with waveforms of different amplitude, frequency, or phase. The matrix converter has recently attracted significant attention among researchers due to its advantages over conventional one. Since it has only one power stage, there is no need for the bulky and lifetime-limited energy-storing electrolytic capacitor that is considered an essential part in the conventional converters based two stages. It also has the ability to control the output voltage magnitude and frequency in addition to operation at unity power factor for any load. Moreover, It provides sinusoidal input and output waveforms, with minimal higher order harmonics and no sub-harmonics. A comprehensive overview of the developments in matrix converter is presented in [9], [10].

This paper presents a new direct AC-AC matrix converter for controlling LIM based on sensorless field oriented control technique. The field orientation principles are used to decouple the mover speed from the secondary flux amplitude. The Model Reference Adaptive System (MRAS) technique is employed to estimate the mover speed [11]. The LIM speed is controlled by using PI controller. The end-effect of LIM is modeled as an external load force dependent on the mover speed [12], [13]. The PWM switching technique for the matrix converter is designed based on minimum voltage drop switching (MVDS) [14], [15] in order to minimize the switching loss and harmonic. Simulation results proved that the proposed controller can be applied successfully to control the LIM speed efficiently.

II. DYNAMIC MODEL OF LINEAR INDUCTION MOTOR

Fig. 1 shows a three-phase LIM that consists of primary and secondary. The primary (mover) is simply a cut

978-1-4799-2706-7/14 $31.00 © 2014 IEEE

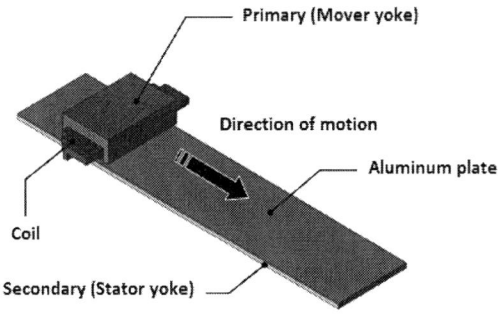

Fig. 1. Schematic diagram of LIM.

open and rolled flat rotary-motor primary. The secondary usually consists of a sheet conductor using aluminum with an iron back for the return path of magnetic flux. The primary and secondary are considered a single-sided LIM. The electrical dynamic model of the LIM is modified from the traditional model of a three phase, Y-connected induction motor in α-β stationary frame and can be described by the following differential equations [16]:

$$Pi_{\alpha s} = -\left(\frac{R_s}{\sigma Ls} + \frac{1-\sigma}{\sigma Tr}\right)i_{\alpha s} + \frac{L_m}{\sigma LsLrTr}\lambda_{\alpha r}$$
$$+ \frac{n_p L_m \pi}{\sigma L_s L_r h}\nu\lambda_{\beta r} + \frac{1}{\sigma L_s}V_{\alpha s} \quad (1)$$

$$Pi_{\beta s} = -\left(\frac{R_s}{\sigma Ls} + \frac{1-\sigma}{\sigma Tr}\right)i_{\beta s} + \frac{L_m}{\sigma LsLrTr}\lambda_{\beta r}$$
$$- \frac{n_p L_m \pi}{\sigma L_s L_r h}\nu\lambda_{\alpha r} + \frac{1}{\sigma L_s}V_{\beta s} \quad (2)$$

$$P\lambda_{\alpha r} = \frac{L_m}{Tr}i_{\alpha s} - \frac{1}{Tr}\lambda_{\alpha r} - \frac{n_p \pi}{h}\nu\lambda_{\beta r} \quad (3)$$

$$P\lambda_{\beta r} = \frac{L_m}{Tr}i_{\beta s} - \frac{1}{Tr}\lambda_{\beta r} - \frac{n_p \pi}{h}\nu\lambda_{\alpha r} \quad (4)$$

$$P\nu = \frac{1}{M}F_e - \frac{D}{M}\nu - \frac{1}{M}F_L \quad (5)$$

where, $T_r = \frac{L_r}{R_r}$ and $\sigma = 1 - \frac{L_m^2}{L_s L_r}$. The longitudinal end-effect is approximated by Taylors series and can be taken as an external load force as shown in the following [12], [13]:

$$F_1 = \theta_1 + \theta_2 V + \theta_3 V^2 \quad (6)$$

where, $\theta_1, \theta_2, and \; \theta_3$ are constants. This end-effect increases with the speed of the mover [17], [18]. Putting F_1 into consideration, (5) can be written as follows:

$$P\nu = \frac{1}{M}F_e - \frac{D}{M}\nu - \frac{1}{M}F_L - \frac{1}{M}F_1 \quad (7)$$

III. INDIRECT FIELD ORIENTED CONTROL OF LIM

In the field oriented control method, the dynamics of the highly coupled nonlinear structure of the induction machine becomes linearized and decoupled. The decoupled relationship is obtained by proper selection of state coordinates, under the hypothesis that the rotor

flux is kept constant [4]. Therefore, the rotor speed is only asymptotically decoupled from the rotor flux, and is linearly related to the torque current only after the rotor flux becomes in the steady state. The flux model of the LIM can be described in the d-q synchronous frame as:

$$P\lambda_{dr} = \frac{L_m}{T_r}i_{ds} - \frac{1}{T_r}\lambda_{dr} + \left(\frac{\pi}{h}\nu_e - \frac{n_p\pi}{h}\nu\right)\lambda_{qr} \quad (8)$$

$$P\lambda_{qr} = \frac{L_m}{T_r}i_{qs} - \frac{1}{T_r}\lambda_{qr} - \left(\frac{\pi}{h}\nu_e - \frac{n_p\pi}{h}\nu\right)\lambda_{dr} \quad (9)$$

In an ideally decoupled induction motor, the secondary flux linkage axis is forced to be aligned with the d-axis, and the field orientation conditions can be applied. It follows that:

$$\lambda_{qr} = 0 \qquad P\lambda_{dr} = P\lambda_{qr} = 0 \quad (10)$$

Using the previous conditions, the desired secondary flux linkage in terms of i_{ds} can be found from 8 as follows:

$$\lambda_{dr} = L_m i_{ds} \quad (11)$$

Moreover, 8 can be combined with 9 and 11 to give the feed-forward slip velocity signal as follows:

$$\nu_{sl} = \frac{\pi}{h}\nu_e - \frac{n_p\pi}{h}\nu = \frac{i_{qs}}{T_r i_{ds}} \quad (12)$$

The electromagnetic force can be described in the d-q synchronous frame as:

$$F_e = k_f\left(\lambda_{dr}i_{qs} - \lambda_{qr}i_{ds}\right) \quad (13)$$

where, k_f is the force constant that can be formulated as follows:

$$k_f = \frac{3n_p L_m \pi}{2L_r h} \quad (14)$$

With the implementation of the field oriented control, 13 can be rewritten using 10 and 11 as:

$$F_e = K_F i_{qs} \qquad (K_F = k_f L_m i_d s) \quad (15)$$

If the d-axis primary current (flux current component) is kept constant at the rated value, the electromagnetic force is directly proportional to the q-axis current, which can be realized via closed loop control. In this case, if the q-axis current (load current component) is rapidly changed in response to the load variation, this will be followed by a rapid change in the motor developed force and the LIM will exhibit a high dynamic performance.

IV. ESTIMATION OF LIM SPEED AND PRIMARY RESISTANCE

The full order adaptive observer for the primary current and secondary flux can be deduced as follows:

$$P\begin{bmatrix}\widehat{i}_s\\\widehat{\lambda}_r\end{bmatrix} = \begin{bmatrix}\widehat{A}_{11} & \widehat{A}_{12}\\A_{21} & \widehat{A}_{22}\end{bmatrix}\begin{bmatrix}\widehat{i}_s\\\widehat{\lambda}_r\end{bmatrix} + \begin{bmatrix}B\\0\end{bmatrix}v_s \quad (16)$$

where the sign $\widehat{}$ signifies the estimated value,

$$i_s = \begin{bmatrix}i_{\alpha s}\\i_{\beta s}\end{bmatrix}, \quad \lambda_r = \begin{bmatrix}\lambda_{\alpha r}\\\lambda_{\beta r}\end{bmatrix}, \quad v_s = \begin{bmatrix}v_{\alpha s}\\v_{\beta s}\end{bmatrix} \quad (17)$$

$$A11 = (\frac{R_s}{\sigma L_s} + \frac{1-\sigma}{\sigma T_r})I \qquad (18)$$

$$A12 = \frac{L_m}{\sigma L_s L_r T_r}I - \frac{n_p L_m \pi v}{\sigma L_s L_r h}J \qquad (19)$$

$$A21 = \frac{L_m}{T_r}I \qquad (20)$$

$$A22 = -\frac{1}{T_r}I + \frac{n_p \pi v}{h}J \qquad (21)$$

$$B = -\frac{1}{\sigma L_s}I, \quad C = \begin{bmatrix} I & 0 \end{bmatrix} \qquad (22)$$

$$I = \begin{bmatrix} 1 & 0 \\ 0 & 1 \end{bmatrix}, \quad j = \begin{bmatrix} 0 & -1 \\ 1 & 0 \end{bmatrix} \qquad (23)$$

Since the primary current can be measured easily, it is selected as the error feedback signal. Subtracting (17) from the actual values given in (1 - 4), assuming $\lambda_r = \widehat{\lambda}_r$, results in:

$$Pe = (A_{11} + G) + (A_{12} - \widehat{A}_{12})\widehat{\lambda}_r \qquad (24)$$

$$e = i_s - \widehat{i}_s \qquad (25)$$

where G is the observer gain matrix. The error between the states i_s and $\widehat{i}_s$ can be used to drive a speed adaptive control mechanism which adjusts the estimated speed. Using the Lyapunov's stability theory [19], a mechanism to adapt the mechanical speed v in addition to the primary resistance R_s from the asymptotic convergence's condition of the current estimation errors can be obtained as follows:

$$\widehat{v} = k_{pv} + \frac{kiv}{S}\left\{\widehat{\lambda}_{\alpha r}(\widehat{i}_{\beta s} - i_{\beta s}) - \widehat{\lambda}_{\beta r}(\widehat{i}_{\alpha s} - i_{\alpha s})\right\} \qquad (26)$$

$$\widehat{R}_s = k_{pR} + \frac{kiR}{S}\left\{\widehat{i}_{\alpha s}(\widehat{i}_{\alpha s} - i_{\alpha s}) - \widehat{i}_{\beta s}(\widehat{i}_{\beta s} - i_{\beta s})\right\} \qquad (27)$$

where (k_{pv}, k_{Iv}, k_{pR}, and k_{IR}) are PI parameters of speed and stator resistance adaptive estimators, respectively.

V. MATRIX CONVERTER MODEL

Fig. 2 shows an analytical model of matrix converter driving LIM. The main circuit of the matrix converter consists of small input filters, which consist of reactors and capacitors, and 9 bi-directional switches that allow any output phase to be connected to any input phase. The bi-directional switches consist of the combination of IGBTs shown in Fig. 2. In this model, the input LC filters, used to eliminate the switching ripples, can be neglected. The loads are expressed by the current sources i_{uv}, i_{vw} and i_{wu}. The source phase voltages e_r, e_s and e_t are given using the line voltage E and the argument θ by :

$$\begin{bmatrix} e_r \\ e_s \\ e_t \end{bmatrix} = \sqrt{\frac{2}{3}}E \begin{bmatrix} \cos\theta \\ \cos(\theta - 2\pi/3) \\ \cos(\theta + 2\pi/3) \end{bmatrix} \qquad (28)$$

$$\theta = \omega t + \varphi_s \qquad (29)$$

where ω and φ_s are the angular frequency and an arbitrary angle of the source voltage, respectively. For the control of the input power-factor, the unit input current references

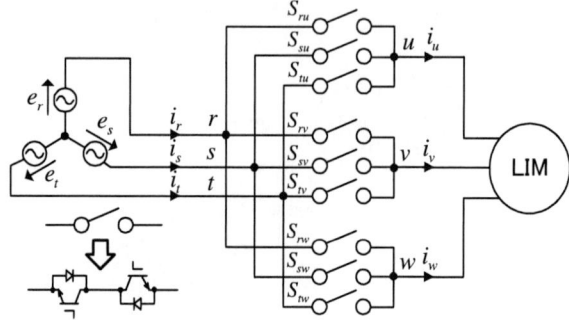

Fig. 2. Matrix converter driving LIM.

i_{ir}^*, i_{is}^* and i_{it}^* are defined using the input power-factor angle reference φ^* as follows;

$$\begin{bmatrix} i_{ir}^* \\ i_{is}^* \\ i_{it}^* \end{bmatrix} = \begin{bmatrix} \cos(\theta + \varphi^*) \\ \cos(\theta + \varphi^* - 2\pi/3) \\ \cos(\theta + \varphi^* + 2\pi/3) \end{bmatrix} \qquad (30)$$

The output phase voltage references v_u^*, v_v^* and v_w^* are given using the line voltage rms value V_L^* and the argument θ_L^* by

$$\begin{bmatrix} v_u^* \\ v_v^* \\ v_w^* \end{bmatrix} = \sqrt{\frac{2}{3}}V_L^* \begin{bmatrix} \cos(\theta_L^*) \\ \cos(\theta_L^* - 2\pi/3) \\ \cos(\theta_L^* + 2\pi/3) \end{bmatrix} \qquad (31)$$

$$\theta_L^* = \omega_L^* t + \varphi_L^* \qquad (32)$$

where ω_L^* and φ_L^* are the angular frequency and an arbitrary angle of the output voltage reference. From (31), the output line voltage references v_{uv}^*, v_{vw}^* and v_{wu}^* are obtained by

$$\begin{bmatrix} v_{uv}^* \\ v_{vw}^* \\ v_{wu}^* \end{bmatrix} = \begin{bmatrix} v_u^* - v_v^* \\ v_v^* - v_w^* \\ v_w^* - v_u^* \end{bmatrix} = \sqrt{2}V_L^* \begin{bmatrix} \cos(\theta_L^* + \pi/6) \\ \cos(\theta_L^* - \pi/2) \\ \cos(\theta_L^* - 7\pi/6) \end{bmatrix} \qquad (33)$$

For realizing the output line voltage v_{uv}, v_{vw}, v_{wu} and the input power-factor $\cos\varphi$ according to the references, the duty cycles $D_{ru} \sim D_{tw}$ of the nine switches $S_{ru} \sim S_{tw}$ and its switching times $T_{ru} \sim T_{tw}$ within the control period T have to be decided.

$$D_{lm} = \frac{T_{lm}}{T} \qquad l = \{r, s, t\} \quad m = \{u, v, w\} \qquad (34)$$

The constraints can be expressed as

$$D_{rm} + D_{sm} + D_{tm} = 1 \qquad m = \{u, v, w\} \qquad (35)$$

VI. MATRIX CONVERTER SWITCHING SCHEME

The modulation scheme of the switching patterns are explained under the input voltages $e_r > e_s > e_t (0 \le \theta \le \pi/3)$ and the output voltage references $v_u^* > v_v^* > v_w^* (0 \le \theta_L \le \pi/3)$. The switching pattern of the matrix converter is shown in Fig. 3. In this scheme, the number of commutations in the control period T is reduced to four. The phase u with the maximum voltage reference v_u^* is connected to the maximum and middle

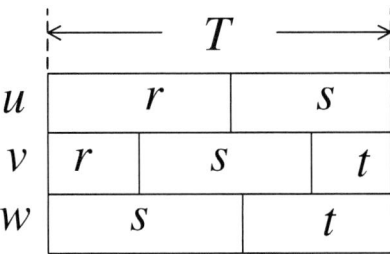

Fig. 3. Proposed switching pattern of the matrix converter.

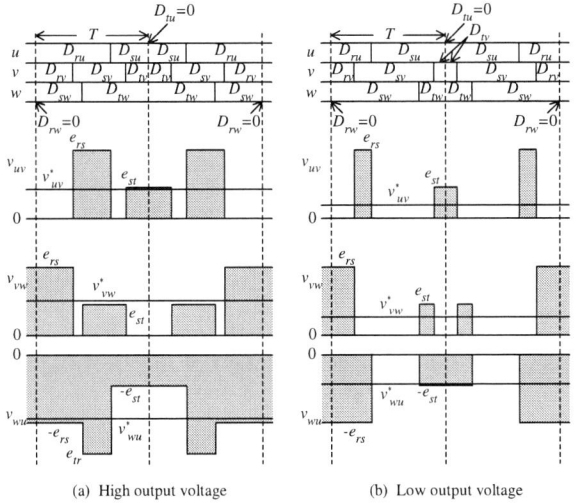

(a) High output voltage   (b) Low output voltage

Fig. 4. Estimated load voltages based on MVDS switching pattern during switching period.

input voltages e_r and e_s only, and not connected to the minimum input voltage e_t. Also, the phase w with the minimum voltage reference v_w^* is connected to the middle and minimum input voltages e_r and e_s, and not connected to the maximum input voltage e_r. Therefore, there is no switching between the maximum and minimum voltages. The middle reference output voltage (v_v^*) is connected to all input voltages, and its value can be realized by switching between the maximum, middle and minimum input voltages. Fig. 4 shows the estimated output voltage of each phase. From the switching pattern and the estimated output voltages, shown in Fig. 3 and Fig. 4, the proposed scheme has no direct commutation between the maximum and minimum input voltages, which reduces the switching loss. Switching duties are determined based on the voltage level of each phase (maximum, middle, and minimum) that can be decided using Table I.

A. Derivation of Duty cycles

Duty cycle calculations is done based on that given in [14], [15]. The switching frequency of the converter is much higher than the fundamental frequency of the input and output voltages and currents. Therefore, during the control period T input and output voltages and currents are assumed to be constant.

1) Duty cycles of maximum and minimum output phase voltages: Duty cycles of all switches connecting phase (u) and phase (w) are calculated based on the equivalent circuit shown in Fig. 5, considering constant voltages and currents in the input and output of the converter. Only two switches are controlling the output phase (u) (S_{ru} and S_{su}), whereas the output phase (w) is controlled by switches (S_{sw} and S_{tw}). The input current can be formulated as a function of the unit reference current, given in (30), as follows:

$$\begin{bmatrix} i_{rwu}^- \\ i_{swu}^- \\ i_{twu}^- \end{bmatrix} = K_{wu} \begin{bmatrix} i_{ir}^* \\ i_{is}^* \\ i_{it}^* \end{bmatrix} \tag{36}$$

The average input power $\bar{P}_{wu}$ during the control period T can be formulated as follows:

$$\bar{P}_{wu} = K_{wu}(e_r i_{ir}^* + e_s i_{is}^* + e_t i_{it}^*) = K_{wu} * P_i \tag{37}$$
$$P_i = e_r i_{ir}^* + e_s i_{is}^* + e_t i_{it}^* \tag{38}$$

Where, the P_i is the unit input power. Assuming ideal converter with equal input and output power, the output current i_{wu} can be formulated as follows:

$$i_{wu} = \frac{K_{wu} P_i}{v_{wu}^*} \tag{39}$$

The duty cycle of the switch S_{ru} can be formulated as follows:

$$D_{ru} = \frac{-i_{rwu}}{i_{wu}} = \frac{-i_{ir}^* v_{wu}^*}{P_i} \tag{40}$$

Since there is no connection between phase (u) and phase (t), the duty cycle of switch S_{tu} is zero ($D_{tu} = 0$). Therefore, based on (35), the duty cycle of the switch S_{su} can be formulated as follows:

$$D_{su} = 1 + \frac{i_{ir}^* v_{wu}^*}{P_i} \tag{41}$$

The phase w is not connected to the maximum input voltage phase r. The input current $\bar{i}_{twu}(= K_{wu} i_{it}^*)$ is equal to the current source i_{wu} through the switch S_{tw}. The duty cycles D_{rw}, D_{sw} and D_{tw} can be derived as follows;

$$D_{rw} = 0 \tag{42}$$
$$D_{tw} = \frac{T_{tw}}{T} = \frac{K_{wu} i_{it}^*}{i_{wu}} = \frac{v_{wu}^* i_{it}^*}{P_i} \tag{43}$$
$$D_{sw} = 1 - D_{rw} - D_{tw} = 1 - \frac{v_{wu}^* i_{it}^*}{P_i} \tag{44}$$

2) Duty cycles of middle voltage phase: Duty cycles of all switches connecting phase (v) with the input phase voltages are calculated based on the equivalent circuit shown in Fig. 6 and Fig. 7. In this case, the input currents $\bar{i}_{ruv}$, $\bar{i}_{suv}$ and $\bar{i}_{tuv}$ are expressed by using the proportional constant value K_{uv} as:

$$\begin{bmatrix} i_{ruv}^- \\ i_{suv}^- \\ i_{tuv}^- \end{bmatrix} = K_{uv} \begin{bmatrix} i_{ir}^* \\ i_{is}^* \\ i_{it}^* \end{bmatrix} \tag{45}$$

The 2014 International Power Electronics Conference

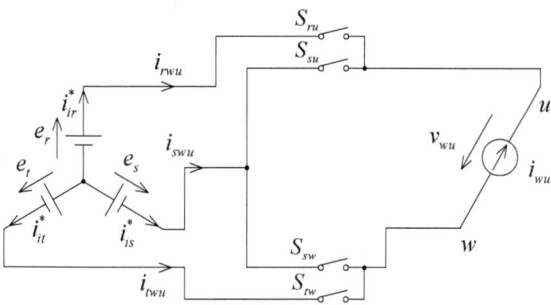

Fig. 5. Equivalent circuit of connecting max-min output phase voltages.

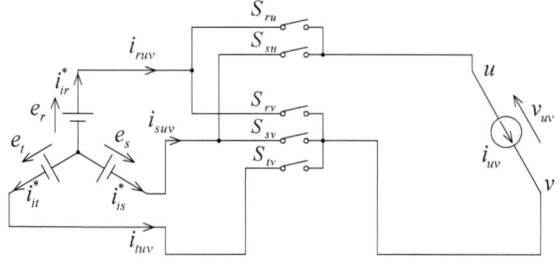

Fig. 6. Equivalent circuit of connecting max-mid output phase voltages.

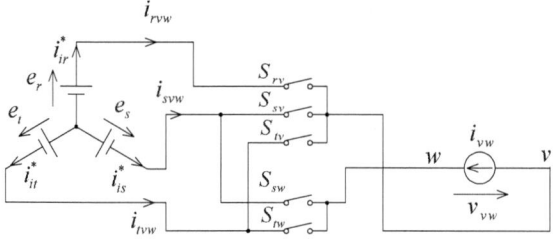

Fig. 7. Equivalent circuit of connecting mid-min output phase voltages.

Since phase (t) is connected directly with phase (v), the duty cycle of switch S_{tv} can be formulated as follows:

$$D_{tv} = \frac{i_{tuv}}{i_{uv}} = -\frac{i_{it}^* v_{uv}^*}{P_i} \quad (46)$$

Fig. 7 shows the connection diagram for the middle output voltage reference phase v and the minimum output voltage reference phase w. The input currents $\bar{i}_{rvw}$, $\bar{i}_{svw}$ and $\bar{i}_{tvw}$ are given by using the unit input current references i_{ir}^*, i_{is}^*, i_{it}^* and the proportional constant value K_{vw} by

$$\begin{bmatrix} \bar{i}_{rvw} \\ i_{svw} \\ i_{tvw} \end{bmatrix} = K_{vw} \begin{bmatrix} i_{ir}^* \\ i_{is}^* \\ i_{it}^* \end{bmatrix} \quad (47)$$

Since phase (r) is connected directly with phase (v), as shown in Fig. 7, the duty cycle of switch S_{rv} can be formulated as follows:

$$D_{rv} = \frac{i_{tvw}}{i_{vw}} = \frac{i_{ir}^* v_{vw}^*}{P_i} \quad (48)$$

TABLE I.   RELATIONSHIP AMONG SOURCE VOLTAGE VALUES

phase angle	source voltage			signs		
θ	e_r	e_s	e_t	e_{rs}	e_{st}	e_{tr}
$0 \sim \pi/3$	e_1	e_2	e_3	+	+	−
$\pi/3 \sim 2\pi/3$	e_2	e_1	e_3	−	+	−
$2\pi/3 \sim \pi$	e_3	e_1	e_2	−	+	+
$\pi \sim 4\pi/3$	e_3	e_2	e_1	−	−	+
$4\pi/3 \sim 5\pi/3$	e_2	e_3	e_1	+	−	+
$5\pi/3 \sim 2\pi$	e_1	e_3	e_2	+	−	−

Therefore, the duty cycle of switch S_{sv} can be formulated based on (35) as follows:

$$D_{sv} = 1 - \left(\frac{i_{ir}^* v_{vw}^*}{P_i} - \frac{i_{it}^* v_{vw}^*}{P_i} \right) \quad (49)$$

B. Matrix converter controllable range

The reference output voltage V_L^* and the input power-factor angle φ^* controllable ranges are given by

$$0 \le V_L^* \le \frac{\sqrt{3}}{2} E |\cos \varphi^*| \quad (50)$$
$$-\pi/6 \le \varphi^* \le \pi/6, \quad 5\pi/6 \le \varphi^* \le 7\pi/6 \quad (51)$$

The maximum value of reference output voltage V_L^* is $\sqrt{3}E/2$ with reference input power factor $\cos \varphi^* = \pm 1$.

VII. SIMULATION RESULT

Fig. 8 shows the system configuration of the LIM driven by matrix converter. The input of matrix converter is connected to symmetrical three-phase voltage source through LC filter. The parameters of the whole system are listed in Table II. The detailed parameters of the used LIM can be found in [20]. Fig. 9 shows the proposed control block diagram of the LIM. Conventional PI controller is used to regulate the LIM speed. The output of the PI speed controller is considered as the reference q-axis LIM current (I_q^*), where as the reference d-axis LIM current is constant value obtained from the motor parameters, ($I_d^* = 0.23A$). Two conventional PI controllers are used to regulate the d-q axis current of the LIM resulting in obtaining the reference d-q axis voltage (V_d^* and V_q^*). Conventional Park/Clark transformation is used to obtain the reference three-phase voltage of the LIM based on the mover position angle, which is obtained from the estimated mover speed. Estimation of LIM speed and its stator resistance has been achieved by using MRAS technique, as shown in Fig. 10. The system is carried out in Matlab/Simulink with the proposed PWM control technique to realize the reference speed of the LIM. The reference speed is 0.8 m/sec. The reference input voltage phase angle φ^* is set at 0° in order to realize unity input power factor. The applied load force on the LIM is changed from 300 N to be 650 N at 0.2 sec.

Fig. 11 shows the simulation waveforms of three-phase input voltage v_s, three-phase input current i_s, three-phase motor current i_L, reference and actual d-q axis currents, reference and actual LIM speed, and load force and electromagnetic developed force in the system shown in Fig. 8. Simulation results prove that the

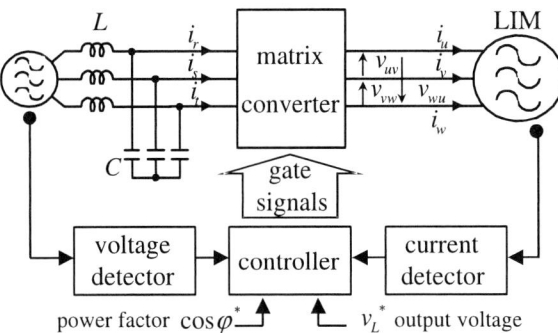

Fig. 8.   Case study of LIM driven by matrix converter.

TABLE II.   LIM SYSTEM PARAMETERS.

Supply and input filter parameters		
Source voltage E, ω	245 V, $2\pi \times 60$ rad/s	
Input filter L, C	3.0 mH, 13.2 μF	
Input power-factor φ^*	0 rad	
LIM		
R_s	5.368 Ω	pole pitch (h)	0.027 m
R_r	3.75 Ω	Mover total mass (m)	2.78 kg
L_s	28.46 mH	Viscous friction and iron loss coefficient (D)	36.045 kg/s
L_r	28.46 mH	Force constant (kf)	593.35 N/Wb.A
L_m	24.19 mH	Rated secondary flux	0.056 Wb
PI controllers		
	Speed Regulator	d-q axis current regulators
K_p	20	10
K_i	350	0.01

reference and actual LIM speed coincide well. At the time of applied load step change, the actual speed reduced by 15 % and coincide with the reference one in 50 *msec*. Moreover, supply and motor currents are sinusoidal and stable.

Fig. 12 shows simulation results of the estimated and actual motor speed using MRAS technique in addition to the estimated and actual stator resistance. It is cleared that estimated and actual LIM speed coincide well. Also, the estimated and actual stator resistance coincide well. However, there are some transient oscillations at the starting of the simulation.

Fig. 13 shows the simulation results of the actual motor phase and line-to-line voltages (v_u and v_{uv}). It is cleared that from the phase voltage there is no direct switching between maximum and minimum input three-phase voltages. Therefore, the line-to-line voltage has periods of no switching between zero and maximum voltage. This results in reducing the switching power losses and the switching noise resulting in high efficiency LIM drive.

Fig. 14 shows a comparison between the simulation results obtained by controlling the LIM using the proposed matrix converter and a conventional three-phase converter. The results prove that the matrix converter provide a faster and smother control response than the conventional converter at starting and at sudden load change. After sudden load change from 300 N to 650 N, the speed control settling time is 50 ms, where as the settling time is 110 ms in case of controlling the LIM using conventional three-phase inverter [21].

According to the simulation results of the whole LIM system driven by matrix converter based MVDS PWM technique, it is cleared that the system is stable and the proposed PWM technique has the ability to control the LIM speed to coincide with the reference value in addition to realization of unity input power factor.

VIII.   CONCLUSION

This paper presented a sensorless speed control of LIM driven by matrix converter using FOC. MRAS technique has been used to estimate the mover speed and the stator resistance of the motor. Conventional

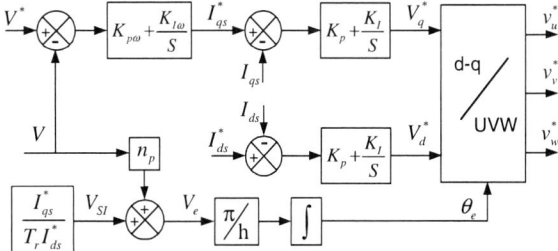

Fig. 9.   Proposed control block diagram.

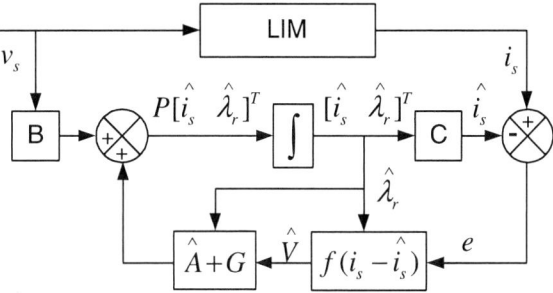

Fig. 10.   Estimation of speed and primary resistance using MRAS.

PI control technique is used to control the matrix converter to regulate the LIM speed. Switching PWM technique has been achieved based on MVDS. According to the reference load voltage, MVDS prevents direct switching between maximum and minimum input voltages. Therefore, the proposed technique has reduced the switching commutation and hence switching loss and harmonics. Reference and actual speed of the LIM agree well. Moreover, estimated and actual speed in addition to estimated and actual stator resistance agree well. The proposed control technique has the ability to control the motor speed even during sudden load change. The Matlab/Simulink results prove that the proposed PWM technique provides an excellent dynamic performance accompanied with an excellent current regulation.

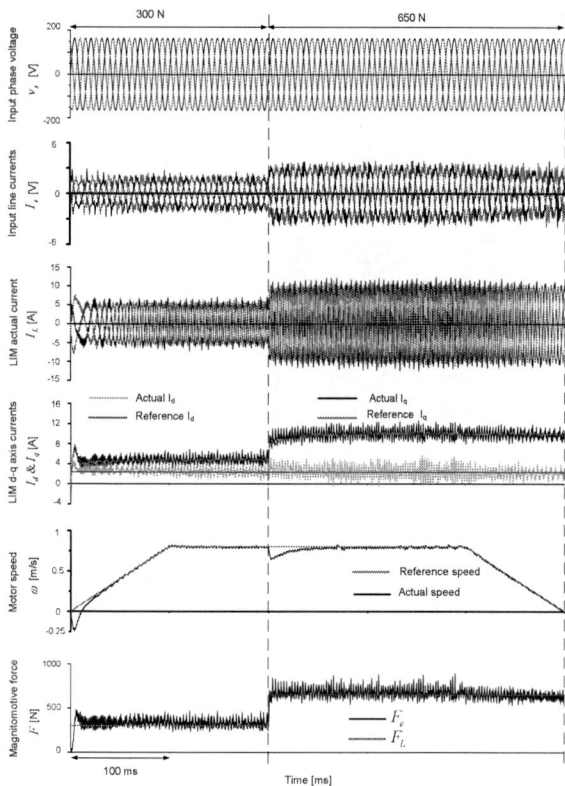

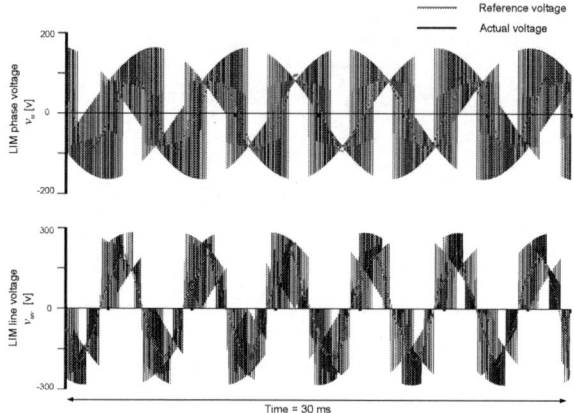

Fig. 13. LIM phase and line-to-line voltages.

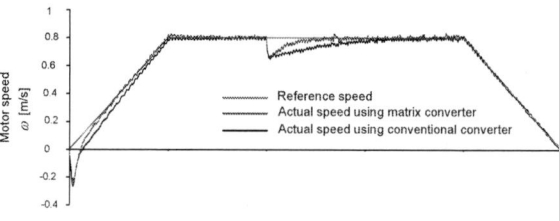

Fig. 11. Simulation results of the LIM driven by the matrix converter.

Fig. 14. Reference and actual LIM speed driven by the proposed matrix converter and conventional converter.

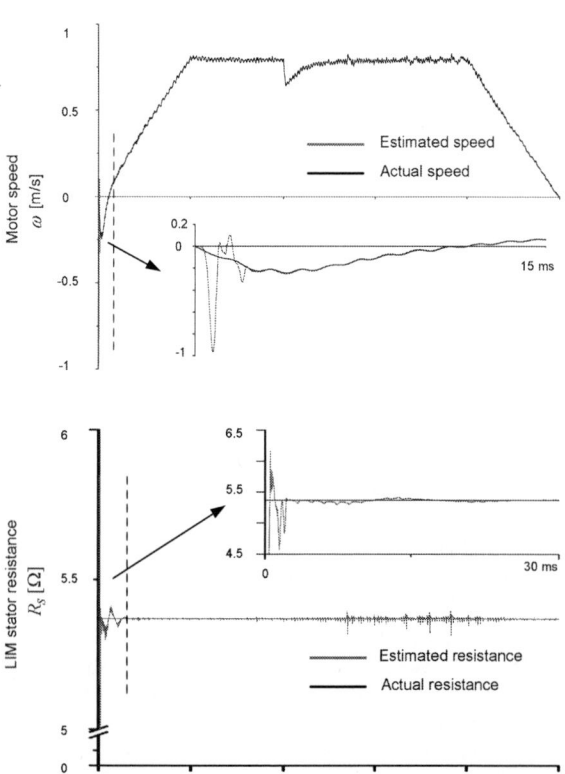

Fig. 12. Estimated and actual values of LIM speed and resistance.

REFERENCES

[1] A. Loukianov, J. Rivera, A. Alanis, and J. Raygoza, "Super-twisting sensorless control of linear induction motors," in *CCE-2012*, pp. 1–5.

[2] B. Bessaih, A. Boucheta, I. Bousserhane, A. Hazzab, and P. Sicard, "Speed control of linear induction motor considering end-effects compensation using rotor time constant estimation," in *SSD-2012*, pp. 1–7.

[3] J. Thomas and A. Hansson, "Speed tracking of a linear induction motor-enumerative nonlinear model predictive control," *IEEE Transactions on Control Systems Technology*, vol. PP, no. 99, pp. 1–1, 2012.

[4] I. Takahashi and Y. Ide, "Decoupling control of thrust and attractive force a lim using a space vector control inverter," in *IEEE IAS Annual Meeting, 1990*, pp. 565–570 vol.1.

[5] A. Y. Alanis, E. N. Sanchez, M. Hernandez-Gonzalez, and L. J. Ricalde, "Discrete-time reduced order neural observer for linear induction motors," in *CIASG-2011*, pp. 1–7.

[6] Z. Zhang, T. Eastham, and G. E. Dawson, "Peak thrust operation of linear induction machines from parameter identification," in *IAS Annual Meeting, IAS '95*, pp. 375–379 vol.1.

[7] F.-J. Lin, C.-K. Chang, and P.-K. Huang, "Fpga-based adaptive backstepping sliding-mode control for linear induction motor drive," *IEEE Transactions on Power Electronics*, vol. 22, no. 4, pp. 1222–1231, 2007.

[8] T. Morizane, K. Tsujikawa, and N. Kimura, "Control of traction and levitation of linear induction motor driven by power source with frequency component synchronous with the motor speed," *IEEE Transactions on Magnetics*, vol. 47, no. 10, pp. 4302–4305, 2011.

[9] P. Wheeler, J. Rodriguez, J. Clare, L. Empringham, and A. Weinstein, "Matrix converters: a technology review," *IEEE Transactions on Industrial Electronics*, vol. 49, pp. 276 –288, apr 2002.

[10] S. Ahmed, A. Iqbal, and H. Abu-Rub, "Generalized duty-ratio-based pulsewidth modulation technique for a three-to-k phase

matrix converter," *IEEE Transactions on Industrial Electronics*, vol. 58, pp. 3925 –3937, sept. 2011.

[11] H. Abu-Rub, M. Khan, A. Iqbal, and S. Ahmed, "Mras-based sensorless control of a five-phase induction motor drive with a predictive adaptive model," in *ISIE-2010*, pp. 3089–3094.

[12] K.-Y. Lian, C.-Y. Hung, C.-S. Chiu, and L.-C. Fu, "Robust adaptive control of linear induction motors with unknown end-effect and secondary resistance," *IEEE Transactions on Energy Conversion*, vol. 23, no. 2, pp. 412–422, 2008.

[13] C.-I. Huang, K.-L. Chen, H.-T. Lee, and L.-C. Fu, "Nonlinear adaptive backstepping motion control of linear induction motor," in *American Control Conference*, pp. 3099–3104.

[14] M. A. Sayed and T. Takeshita, "Novel pwm technique for three-to-five phase matrix converter," in *ICRERA-2013*, pp. 644–649.

[15] H. Shimada and T. Takeshita, "Matrix converter control using direct ac/ac conversion method for reducing output voltage harmonics," in *APEC-2006*, p. 7 pp.

[16] F.-J. Lin and R.-J. Wai, "Robust control using neural network uncertainty observer for linear induction motor servo drive," *IEEE Transactions on Power Electronics*, vol. 17, no. 2, pp. 241–254, 2002.

[17] E. da Silva, C. dos Santos, and J. W. L. Nerys, "Field oriented control of linear induction motor taking into account end-effects," in *AMC-2004*, pp. 689–694.

[18] J.-H. Sung and K. Nam, "A new approach to vector control for a linear induction motor considering end effects," in *IAS-1999*, pp. 2284–2289.

[19] Z. Li, S. Cheng, and K. Cai, "The simulation study of sensorless control for induction motor drives based on mras," in *ICSC 2008*, pp. 235–239.

[20] A. A. Hassan, Y. Sayed, T. Hiyama, and T. H. Mohamed, "Model predictive control of a speed sensorless linear induction motor drive," in *MEPCON-10*, pp. 318–325.

[21] E. E. M. Mohamed, M. A. Sayed, and T. H. Mohamed, "Sliding mode control of linear induction motors using space vector controlled inverter," in *ICRERA-2013*, pp. 650–655.

The 2014 International Power Electronics Conference

A Stator-Equation-Based Reduced-Order Observer for Position-Sensorless Vector Control System of Doubly-Fed Induction Machines

Somrat Smiththisomboon and Surapong Suwankawin

Dept. of Electrical Engineering, Faculty of Engineering, Chulalongkorn University
254 Phayathai Road, Pathumwan, Bangkok, 10330 Thailand. e-mail: surapong.su@eng.chula.ac.th

Abstract— **In this paper, a novel adaptive reduced-order observer is proposed for the position-sensorless drive of doubly-fed induction machines. The reduced-order observer is conducted by the reduced-order model expressed on the holonomic reference frame; the stator equation on stator reference frame. The proposed rotor-position estimation scheme is straightforward and simplicity for implementation without suffering from the DC drift problem arising from pure integration of stator flux. The proposed rotor-position estimation is integrated with the stator-flux vector control. Experimental results of various operating conditions are provided in order to verify the proposed theoretical concepts. The proposed sensorless system can successfully operate in both motor and generator regions, and the performances of sub-synchronous, synchronous and super-synchronous modes are also evaluated.**

Keywords— *Position sensorless, doubly-fed induction machine, adaptive reduced-order observer, stator equation.*

I.   Introduction

Doubly-fed induction machines (DFIMs) are very promising for wind energy conversion systems so far. The rotor-side fed converter attains both engineering and economic aspects. Usually, the vector control with variable speed operation is employed to harvest the maximum power from wind energy conversion. In order to obtain better fault tolerance from encoder's failure and/or malfunction, position-sensorless control scheme is available as a fail-safe function.

Among rotor-position estimation methods made appearance in the literature, the model-based approach is extensively adopted [1]-[8]. In [1]-[3], model reference adaptive system (MRAS) observers are diversified with various error variables, e.g. stator and rotor currents, and the comparative performance were given. Nevertheless, these observers require the stator flux information, the common disadvantage is the DC-offset drift problem caused by the pure integral action of stator-flux calculation. In [4], the rotor current is selected as the error variable, and the estimated rotor flux is calculated through stator flux and torque calculations. This approach not only has the problem from drift problem caused by the stator-flux calculation, but it also needs much more computation than necessary.

The adaptive reduced-order observer is proposed in [5]; the stator flux is considered as the measurable state

variable and the rotor current is the estimated state. There are two drawbacks in this approach; the former is the drift problem from pure integration in stator flux calculation and the latter is that the designed observer's feedback gain required the information of real rotor speed which is unrealistic for the sensorless system; from a theoretical standpoint, using the estimated rotor speed as an alternative is ambiguous. The research works in [6]-[8] avoid the aforementioned DC-offset drift problem. Instead of using pure integration method, the stator-flux observer is used in [6] and the rotor-current error is the error variable for the rotor-position estimation. The observer's gains are employed for the feedback of stator-flux error, and the real stator flux is approximated with doubt. In [7]-[8], rather than the stator flux, the induced voltage is used to calculate the air-gap power. The rotor current and torque; both derived from air-gap power, are chosen as the error variables in [7] and [8] respectively. To compute the air-gap power, these schemes require the additional parameter of core loss which is a barrier to the practice standpoint.

The major disadvantages of the model-based rotor-position estimation methods in the literature can be concluded as follows:
1) DC-offset drift problem arising from pure integration of stator flux [1]-[5].
2) Real rotor speed requirement for the observer's feedback gains caused by inappropriate selection of reference frame for the model [5].
3) Complexity of calculation [4] and more parameters requisition [7]-[8].

In this paper, a novel rotor-position estimation method is proposed. The structure of adaptive reduced-order observer is adopted and the reduced-order model is expressed appropriately on the holonomic reference frame. The proposed scheme offers the following distinguished features:

1) Estimation of rotor position needs no stator flux calculation; the DC-offset drift problem can be avoided.
2) Reduced-order model expressed on holonomic reference frame can exclude the rotor-speed parameter; this helps design the observer's feedback gains without the requirement of real rotor speed.
3) Arrange properly the state variables in the models with stator and rotor current, the computation is naturally simple and requires less parameters.

This work was supported by the Thailand Research Fund (TRF) under Grant RSA 5580039.

978-1-4799-2706-7/14 $31.00 © 2014 IEEE

The contents of the paper are arranged by firstly introducing the model of DFIM on holonomic reference frames. Secondly, the adaptive reduced-order observer based on stator equation is introduced. Next, the proposed rotor-position estimation subsystem and the stator-flux vector control are combined together to assemble the sensorless drive system. Finally, the performances of the drive system are evaluated by the experimental results.

II. Model of Doubly-Fed Induction Machine on Holonomic Reference Frame

Fig. 1 shows the physical structure of DFIM by 2-phase equivalent circuit. Considering the stator windings are fixed on stator reference frame (α - β axis) while the rotor windings are rigidly connected to rotating rotor reference frame (rd – rq axis), the dynamic equations of both windings can be written in (1)-(2). On this so-called holonomic reference frame [9], the dynamic equations are very simple.

A. Dynamic equations of stator/rotor windings

The dynamic equation of stator windings on stator reference frame $\alpha - \beta$ can be expressed as

$$\vec{v}_s = R_s \vec{i}_s + \frac{d\vec{\lambda}_s}{dt} \tag{1}$$

and the dynamic equation of rotor windings on rotating rotor reference frame $dr - qr$ is given in (2)

$$\vec{v}_r' = R_r \vec{i}_r' + \frac{d\vec{\lambda}_r'}{dt} \tag{2}$$

where the stator flux and rotor flux equations can be written as

$$\vec{\lambda}_s = L_s \vec{i}_s + M\left(e^{Jp\theta_r} \vec{i}_r'\right) \tag{3}$$

$$\vec{\lambda}_r' = L_r \vec{i}_r' + M\left(e^{-Jp\theta_r} \vec{i}_s\right) \tag{4}$$

B. Dynamic model of DFIM on holonomic reference frame

Substituting the stator/rotor flux in (1)-(2) with stator/rotor current in (3)-(4), the dynamic model of DFIM on holonomic reference frame is given in (5)-(8)

- Stator equation on stator reference frame

$$\frac{d\vec{i}_s}{dt} = -\frac{R_s}{L_s}\vec{i}_s - \frac{M}{L_s}\frac{d}{dt}\left(e^{Jp\theta_r} \cdot \vec{i}_r'\right) + \frac{\vec{v}_s}{L_s} \tag{5}$$

- Rotor equation on rotating rotor reference frame

$$\frac{d\vec{i}_r'}{dt} = -\frac{R_r}{L_r}\vec{i}_r' - \frac{M}{L_r}\frac{d}{dt}\left(e^{-Jp\theta_r} \cdot \vec{i}_s\right) + \frac{\vec{v}_r'}{L_r} \tag{6}$$

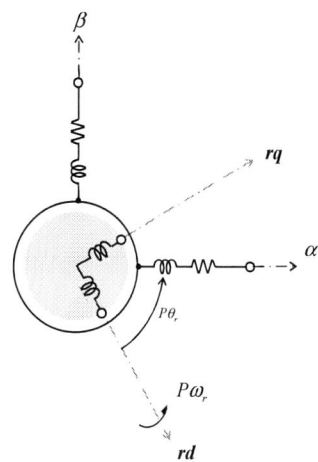

Fig. 1 Model of doubly-fed induction machine.

- Rotor speed and position:

$$\omega_r = \frac{d\theta_r}{dt} \tag{7}$$

- Torque equation:

$$T_e = -P\left(\vec{i}_s \times \vec{\lambda}_s\right) \tag{8}$$

For developing the estimation system; which will be described in the next section, the dynamic models in (5) and (6) have two major advantages; 1) Expressing the model with currents as state variables can avoid the DC-offset drift problem from pure integration of stator flux calculation, and 2) regarding the holonomic reference frame, there exists no rotor speed term in the dynamic equation (5)-(6), this can help design the observer's feedback gains to be independent on the real rotor speed.

III. Novel Rotor-Position Estimations using Adaptive Reduced-Order Observer

The benefit of DFIM compared to other machines is the available of plenty information; voltages and currents on both stator and rotor sides. Considering (5) and (6), the rotor position is the only unknown parameter for position sensorless system. To identify the rotor position, using only stator equation in (5) or rotor equation in (6) is sufficient. In this paper, the estimation scheme is based on the stator equation (5).

A. Stator-Equation-Based Adaptive Reduced-Order Observer

By using the stator equation in (5), the novel adaptive reduced-order observer can be written in (9)-(11) and shown in the block diagram in Fig. 2. The stator current is selected as the error variable for the estimation system. In contrast to [1]-[5], the proposed scheme is straightforward; the estimated stator current can be calculated without stator flux computation beforehand. By comparison with [4], [7]-[8], the calculation is simply carried out with less parameters employment e.g. no need of core loss parameter.

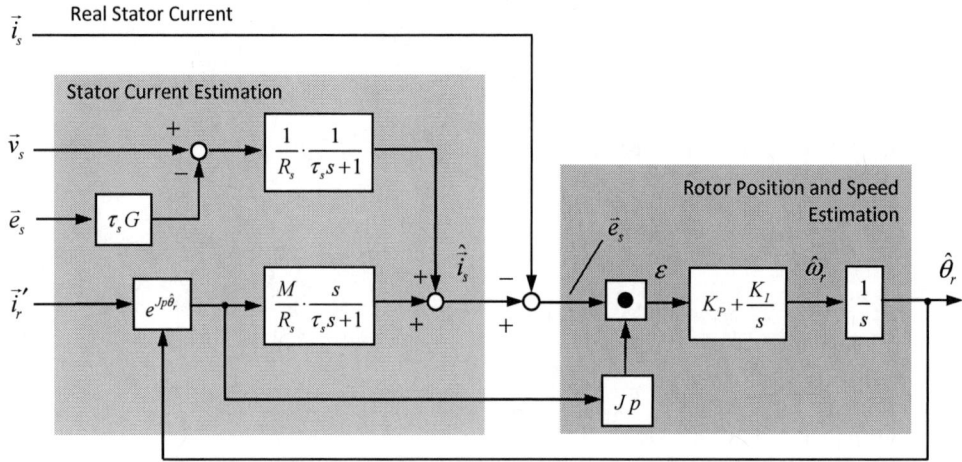

Fig. 2 Rotor-position estimation based on stator equation.

$$\frac{d\hat{\vec{i}}_s}{dt} = -\frac{R_s}{L_s}\hat{\vec{i}}_s - \frac{M}{L_s}\frac{d}{dt}\left(e^{Jp\hat{\theta}_r}\cdot\vec{i}_r'\right) + \frac{\vec{v}_s}{L_s} - G\left(\hat{\vec{i}}_s - \vec{i}_s\right) \quad (9)$$

$$\hat{\omega}_r = \left(K_P + K_I\int dt\right)\cdot\left(Jpe^{Jp\hat{\theta}_r}\cdot\vec{i}_r'\right)^T\left(\hat{\vec{i}}_s - \vec{i}_s\right) \quad (10)$$

$$\hat{\theta}_r = \int\hat{\omega}_r\,dt \quad (11)$$

B. Identifiability Property of Estimation System

The identifiability property of the rotor-speed estimations in (10) is dependent on the regressor vectors $Jpe^{Jp\hat{\theta}_r}\cdot\vec{i}_r'$ which is governed by the magnetizing current and the loading condition. For the conventional grid-connected DFIM drive, the machine usually draws the magnetizing current from the stator side. This means that the regressor vector $Jpe^{Jp\hat{\theta}_r}\cdot\vec{i}_r'$ in (10) will not be available at no-load condition $\vec{i}_r' = 0$ and this causes the estimation system in Fig. 2 loss of identifiability.

Nevertheless, to fulfill the requirements of the recent grid codes, the grid-connected distributed generations should provide the capability of power factor control. This prerequisite leads the DFIM drive to feed the reactive power through the rotor side and the machine will draw the magnetizing current from the rotor side instead. In this regard, the rotor current always exists $\left(\vec{i}_r' \neq 0\right)$ even no-load condition and the estimation in (10) can overcome the loss of identifiability as shown in Fig 3. According to the reactive-power control through the rotor side, the estimation system also has the capability for the stand-alone application.

IV. Stator-Flux Based Vector Control of DFIM

To manipulate the stator flux and torque, the models in (5)-(6) and (8) are expressed on the stator-flux reference frame (d-q axis) as given in (12)-(15) and the torque equation in (16).

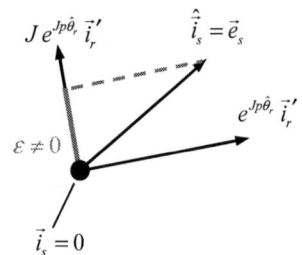

Fig. 3 Vector diagram showing the tuning mechanism of estimation system at no-load condition.

Stator equation on the stator-flux reference frame:

$$\frac{di_o}{dt} = \frac{R_s}{L_s}i_{rd} - \frac{R_s}{L_s}i_o + \frac{v_{sd}}{M} \quad (12)$$

$$\frac{d\theta_o}{dt} = \omega_o = \frac{R_s}{L_s}\frac{i_{rq}}{i_o} + \frac{v_{sq}}{Mi_o} \quad (13)$$

Rotor equation on the stator-flux reference frame:

$$\frac{di_{rd}}{dt} = \frac{1}{\sigma L_r}\left\{-R_r i_{rd} + \omega_s\sigma L_r i_{rq} + \frac{M}{L_s}(R_s i_{sd} - v_{sd}) + v_{rd}\right\} \quad (14)$$

$$\frac{di_{rq}}{dt} = \frac{1}{\sigma L_r}\left\{-R_r i_{rq} - \omega_s\sigma L_r i_{rd} + \frac{M}{L_s}(R_s i_{sq} - v_{sq}) + \frac{M^2}{L_s}p\omega_m i_o + v_{rq}\right\} \quad (15)$$

Torque equation:

$$T_e = -p\frac{M^2}{L_s}i_o i_{rq} \quad (16)$$

Flux and torque of DFIM can be controlled by the rotor current vector; the d-axis component i_{rd} plays the role of controlling the stator-flux magnetizing current i_o (12), while the q-axis component i_{rq} manipulates the induced torque (T_e) (16). It can be seen from the rotor equations (14)-(15) that both d-q axis current is coupled to each other. To control independently each component of the rotor current (i_{rd} , i_{rq}), the decoupling control is introduced in (17)-(18).

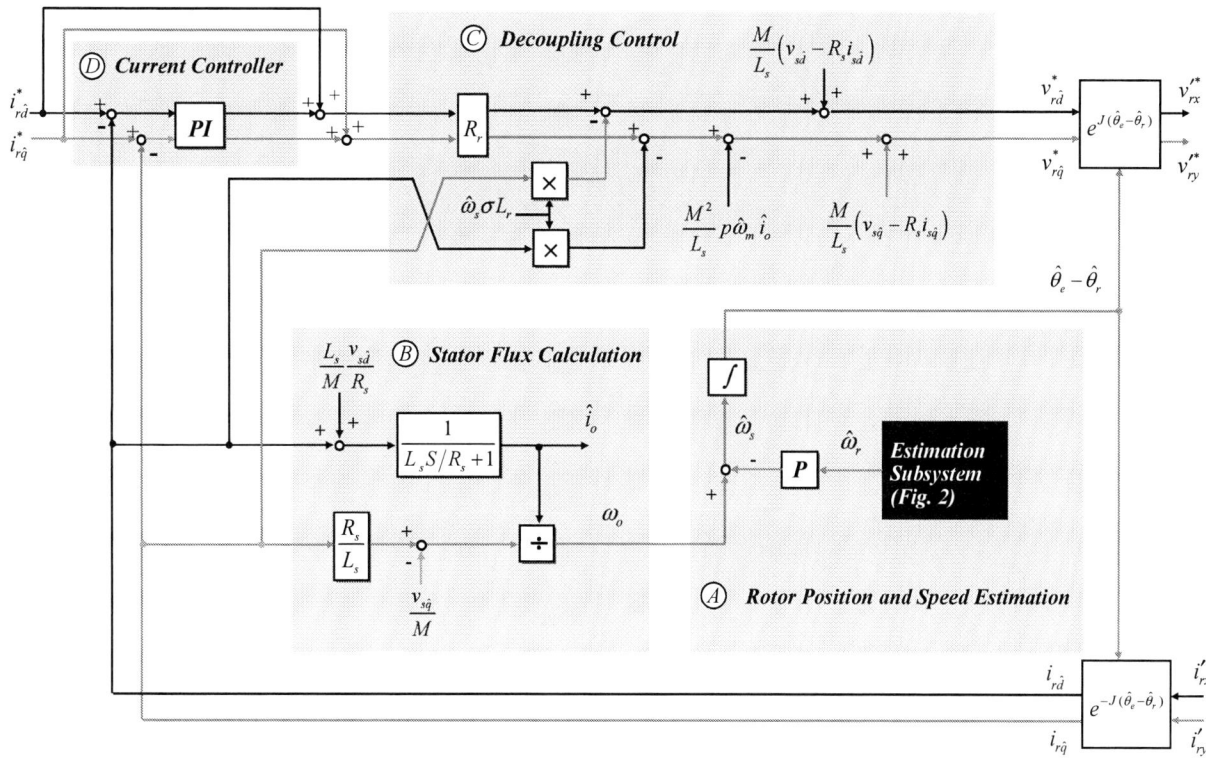

Fig. 4 Position-sensorless vector control of DFIM.

$$v_{rd}^* = \underbrace{R_r i_{rd}^*}_{\text{Feed Forward Term}} \underbrace{-\omega_s \sigma L_r i_{rq} + \frac{M}{L_s}(v_{sd} - R_s i_{sd})}_{\text{Decoupling Voltage Terms}} \quad (17)$$

$$v_{rq}^* = \underbrace{R_r i_{rq}^*}_{\text{Feed Forward Term}} \underbrace{+\omega_s \sigma L_r i_{rd} + \frac{M}{L_s}(v_{sq} - R_s i_{sq}) - \frac{M^2}{L_s}p\omega_r i_o}_{\text{Decoupling Voltage Terms}}$$

$$(18)$$

The commanded rotor voltages (v_{rd}^*, v_{rq}^*) in (17)-(18) are comprised of two terms; the feedforward terms which are assigned by the commanded rotor current (i_{rd}^*, i_{rq}^*) and the decoupling voltage terms. The stator equation (12) helps compute the magnetizing current i_o used in the decoupling control. In addition, the other stator equation (13) calculates the stator-flux position θ_o using in the axis transformation. Fig. 4 shows the position-sensorless vector control system of DFIM. The proposed rotor-position estimation (Fig. 2) is embedded in the stator-flux vector control with current control.

V. Experimental Results

Experiment setup is shown in Fig. 5 and the under test machine's rating and parameters are given in Appendix A. A speed control is created with the senorless controller from Fig. 4, the commanded flux or commanded reactive power can be varied through the commanded d-axis rotor current. The evaluations are conducted with the operating conditions as follows: 1)

motor and generator operations, 2) speed control in synchronous, super-synchronous and sub-synchronous modes, and 3) step change of rotor side's reactive current.

Fig. 6 shows the steady-state response; the rotor position and speed can be well estimated. The position error is inconsiderable and is dependent on the operating conditions. In Fig. 6(a), the utmost position error is roughly 4 degree in the subsynchronous speed region while it is less than 1 degree in synchronous speed and supersynchronous speed regions as shown in Figs. 6(c)-6(e). Figs. 6(a), 6(c) and 6(e) also help confirm the identifiability property at no load. In sequel, using the estimated rotor position and speed, the stator flux can be identified properly and the speed is well regulated. The torque-speed plane in Fig. 6(f) depicts the overall performance of stead-state response; the sensorless system can work properly in sub/super-synchronous modes and motor and generator operations.

The tracking performance is evaluated by the acceleration and deceleration as shown in Fig. 7. In Fig. 7(a), the rotor speed is varied in a wider range which covers ± 30% of slip frequency, the estimated rotor speed and position can track the real ones nicely; the rotor-position error is less than 8 degree during the ramp response of acceleration and deceleration. In Figs. 7(b)-7(d), the rotor speed is varied from synchronous mode to super-synchronous mode and sub-synchronous mode respectively. The satisfactory responses are obtained even though the load is simultaneously taken as shown in Figs. 7(c)-7(d).

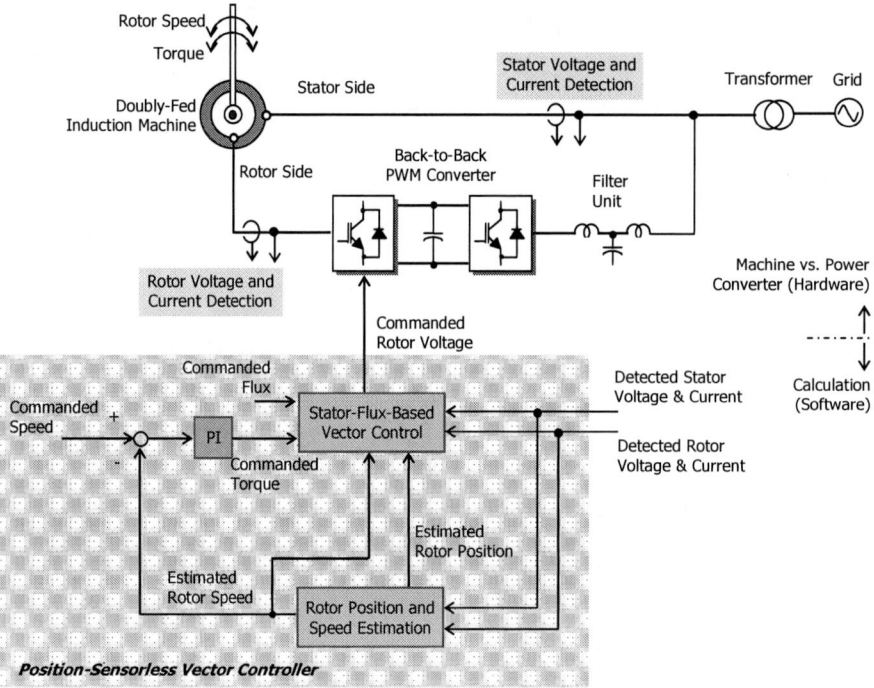

Fig. 5 Experiment setup for evaluation of position-sensorless drive of DFIM.

Fig. 8 demonstrates the capability of reactive current variation; the rotor-position estimation and drive system work properly with the sudden change of the d-axis rotor current. Fig. 9 illustrates the performance against the step change of the load torque for both motoring and generating modes where the commanded speeds are set at synchronous speed, supersynchronous speed and subsynchronous speed. The decoupling control can control the rotor currents $i_{r\hat{d}}$ and $i_{r\hat{q}}$ independently; flux and torque controllability is obtained and the system successfully responds to the step change of load torque.

VI. Conclusion

A novel adaptive reduced-order observer for DFIM sensorless drive is presented in this paper. The salient features of the proposed rotor-position estimation scheme are 1) the estimation systems are straightforward due to the concept of reduced-order model on holonomic reference frame; the stator equation on stator reference frame and, 2) the identification of rotor position is achieved without pure integration of stator flux; the DC drift problem can therefore be avoided. The outcomes of the theoretical concept are verified by the experimental results from various operating conditions.

Appendix

Motor rating and parameters: 4 kW, 50 Hz, 1393 rpm, stator/rotor voltage 400/196 V, stator/rotor current 8.1/13.5 A, $R_s = 1.6\,\Omega$, $R_r = 7.0\,\Omega$, $L_s = 172.6\,mH$, $L_r = 172.6\,mH$, $M = 165.8\,mH$.

Acknowledgment

The authors would like to thank Dr. Somboon Sangwongwanich; department of electrical engineering, Chulalongkorn University, for his valuable academic advice.

References

[1] R.Cardenas, R. Pena, J. Proboste, G. Asher, and J. Clare, "MRAS observer for sensorless control of standalone doubly fed induction generators," *IEEE Trans. Energy Conv.*, vol. 20, no. 4, pp. 710-717, Dec. 2005.

[2] R.Cardenas, R. Pena, J. Proboste, G. Asher, and J. Clare, "Sensorless control of a doubly-fed induction generator using a rotor-current based MRAS observer," *IEEE Trans. Ind. Electron.*, vol. 55, no. 1, pp. 330-339, Jan. 2008.

[3] R.Cardenas, R. Pena, J. Clare, G. Asher, and J. Proboste, "MRAS observers for sensorless control of doubly-fed induction generators," *IEEE Trans. Power Electron.*, vol. 23, no. 3, pp. 1075-1084, May 2008.

[4] F. C. Dezza, G. Foglia, M. F. Iacchetti and R. Perini, "An MRAS observer for sensorless DFIM drives with direct estimation of the torque and flux rotor current components," *IEEE Tran. Power Electron.*, vol. 27, no. 5, pp. 2576-2584, May 2012.

[5] S. Yang and V. Ajjarapu, "A speed-adaptive reduced-order observer for sensorless vector control of doubly fed induction generator-based variable-speed wind turbines," *IEEE Trans. Energy Conv.*, vol. 25, no. 3, pp. 891-900, Sep. 2010.

[6] D. G. Forchetti, G. O. Garcia and M. I. Valla, "Adaptive observer for sensorless control of stand-alone doubly fed induction generator," *IEEE Trans. Ind. Electron.*, vol. 56, no. 10, pp. 4174-4180, Oct. 2009.

[7] G. D. Marques and D. M. Sousa, "Air-gap-power-vector-based sensorless method for DFIG control without flux estimator," *IEEE Trans. Ind. Electron.*, vol. 58, no. 10, pp. 4717-4726, Oct. 2011.

[8] G. D. Marques and D. M. Sousa, "New sensorless rotor position estimator of a DFIG based on torque calculations—stability study," IEEE Trans. Energy Conv., vol. 27, no. 1, pp. 196-203, Mar. 2012.

[9] G. Kron, *Equivalent Circuits of Electric Machinerey*, New York, Dover Publications, 1967.

The 2014 International Power Electronics Conference

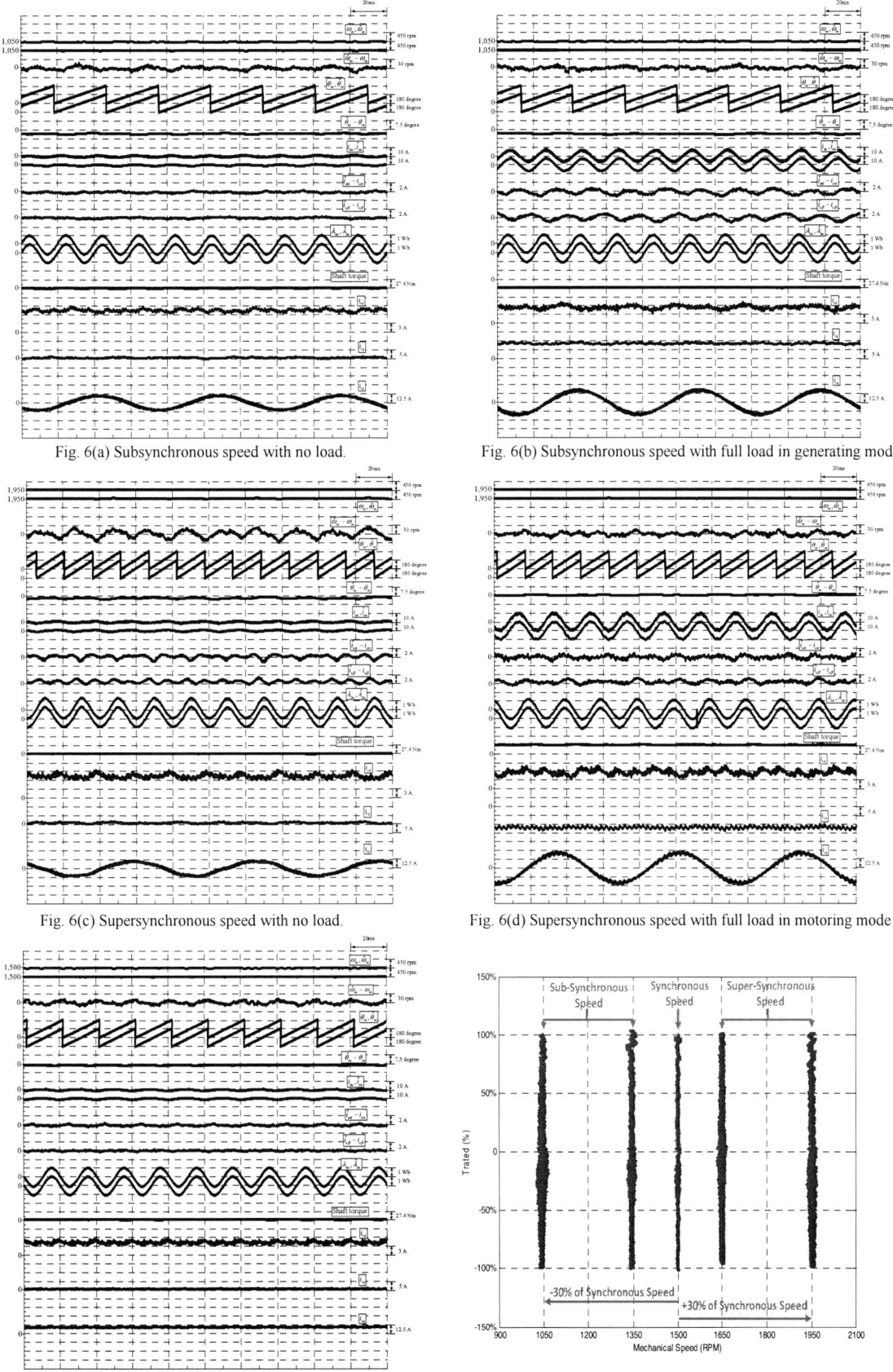

Fig. 6(a) Subsynchronous speed with no load.

Fig. 6(b) Subsynchronous speed with full load in generating mode.

Fig. 6(c) Supersynchronous speed with no load.

Fig. 6(d) Supersynchronous speed with full load in motoring mode.

Fig. 6(e) Synchronous speed with no load.

Fig. 6(f) Torque-speed characteristic.

Fig. 6 Experimental results showing steady-state response.

978-1-4799-2706-7/14 $31.00 © 2014 IEEE

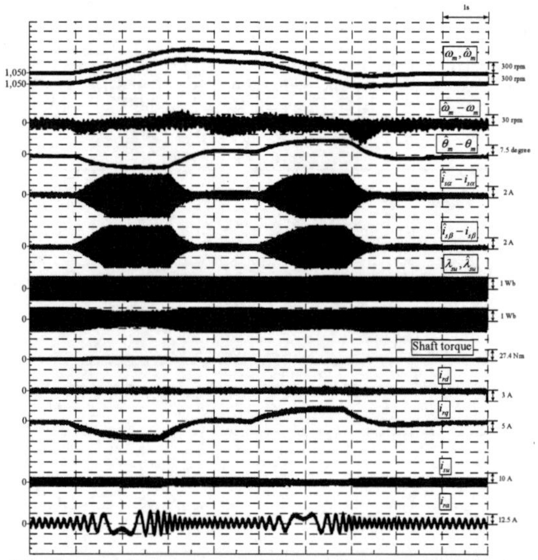

The 2014 International Power Electronics Conference

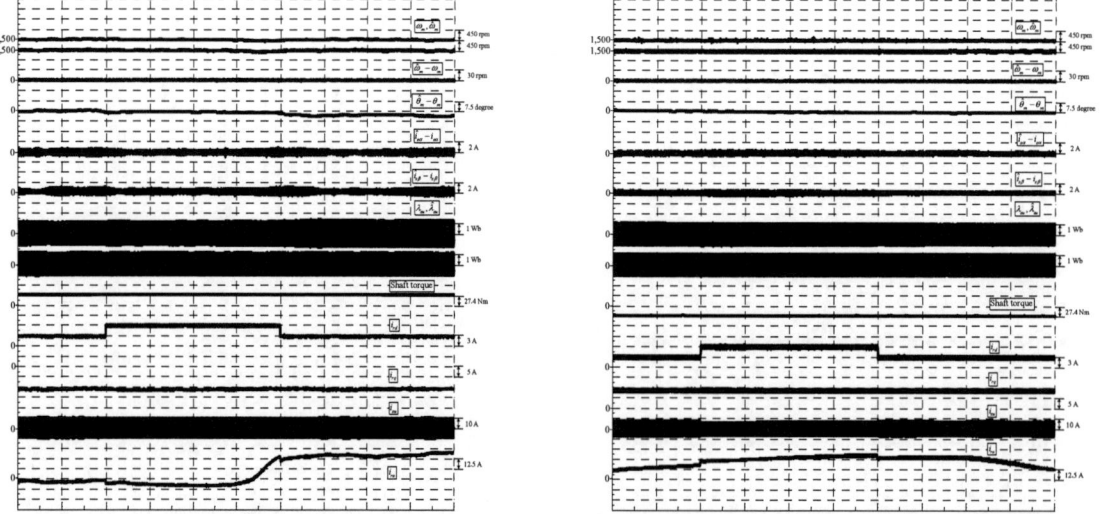

Fig. 7(a) Variable speed range 1,050 – 1,950 rpm with no load.

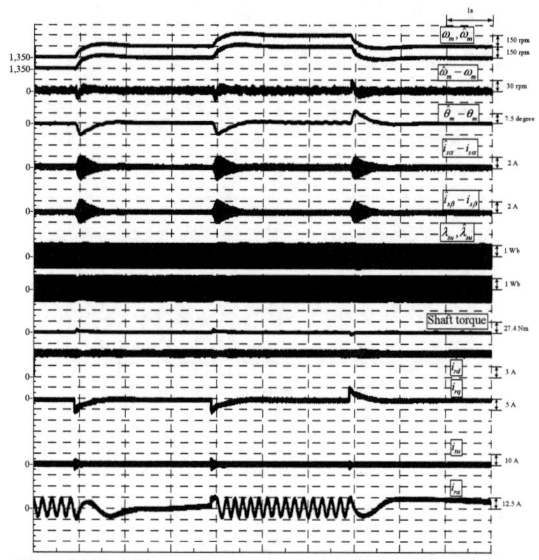

Fig. 7(b) Variable speed range 1,350-1,650 rpm with no load.

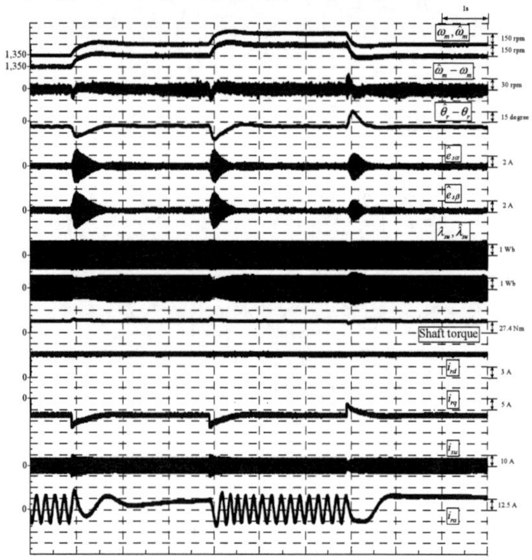

Fig. 7(c) Variable speed range 1,350-1,650 rpm with full load in motoring mode.

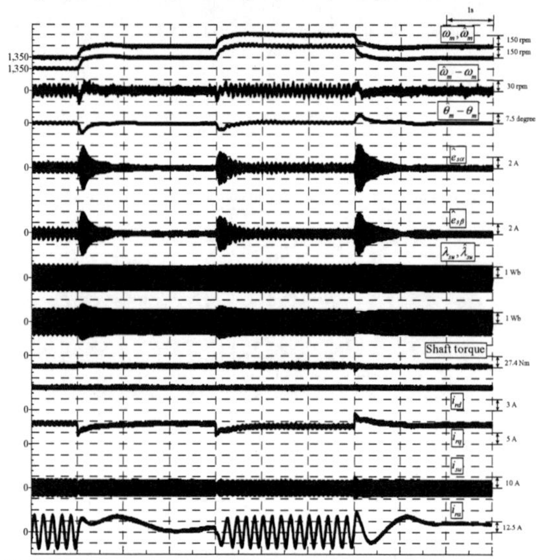

Fig. 7(d) Variable speed range 1,350-1,650 rpm with full load in generating mode.

Fig. 7 Experimental results showing variable speed response.

Fig. 8(a) Synchronous speed with full load in motoring mode.

Fig. 8(b) Synchronous speed with full load in regenerating mode.

Fig. 8 Experimental results showing the step response of changing reactive current on rotor side.

978-1-4799-2706-7/14 $31.00 © 2014 IEEE

The 2014 International Power Electronics Conference

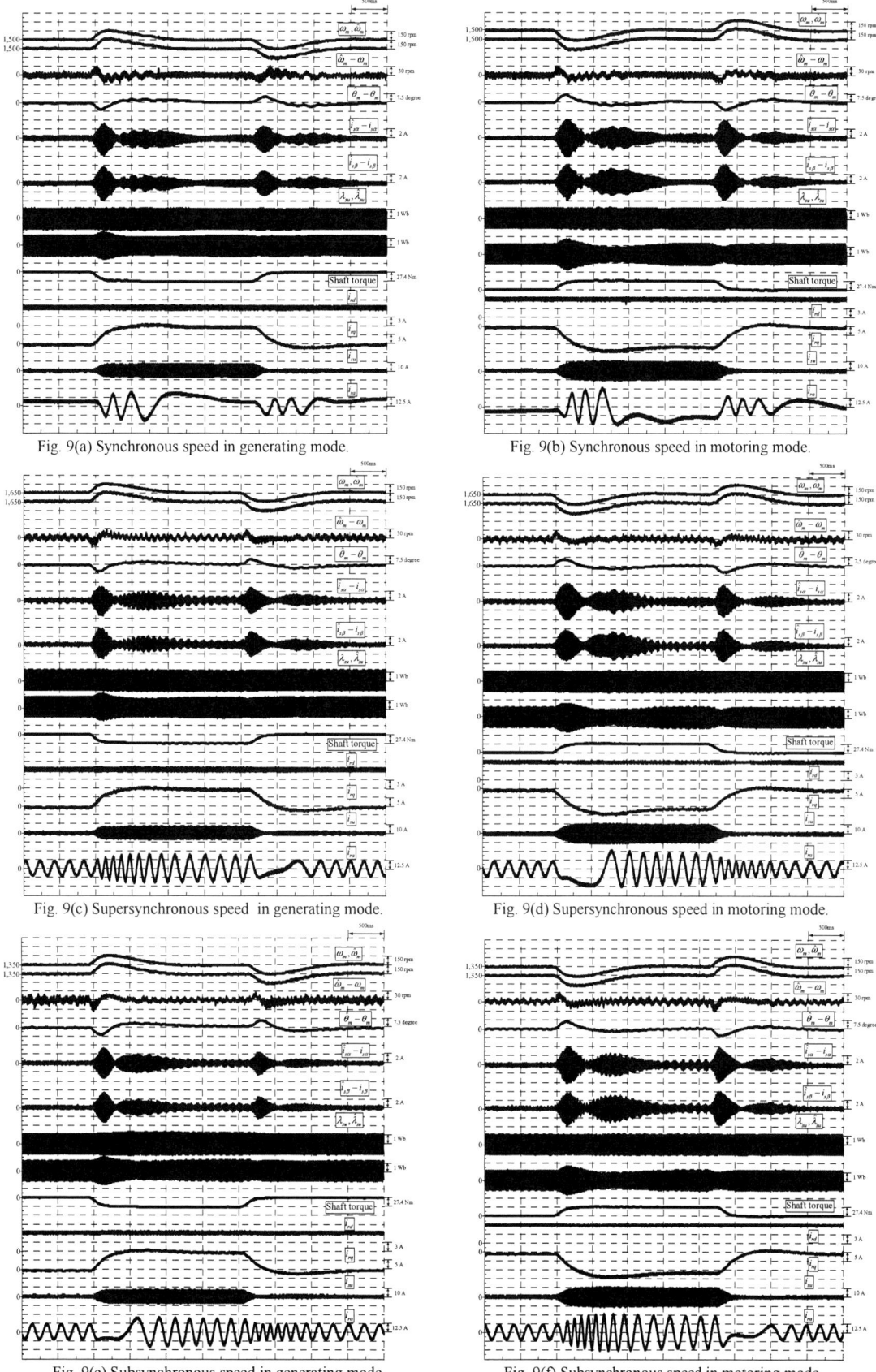

Fig. 9(a) Synchronous speed in generating mode.

Fig. 9(b) Synchronous speed in motoring mode.

Fig. 9(c) Supersynchronous speed in generating mode.

Fig. 9(d) Supersynchronous speed in motoring mode.

Fig. 9(e) Subsynchronous speed in generating mode.

Fig. 9(f) Subsynchronous speed in motoring mode.

Fig. 9 Experimental results showing step-load response.

978-1-4799-2706-7/14 $31.00 © 2014 IEEE

The 2014 International Power Electronics Conference

Input Current Ripple Analysis of Inverter Fed Dual Three-Phase AC Motors

Pekik Argo Dahono and Andri Satria

School of Electrical Engineering and Informatics, Institute of Technology Bandung,
Jl. Ganesa No. 10. Bandung, INDONESIA
padahono@ieee.org

Abstract— **This paper presents an input current ripple analysis of inverter fed dual three-phase induction motors. Four types of modulation strategies are compared. The first is single-carrier with zero phase displaced modulation signals, second is single carrier with 30° phase displaced modulation signals, third is double carrier with zero phase displaced modulation signals, and fourth is double carrier with 30° phase displaced modulation signals. Analysis results show that the third results in the lowest input current ripple. Experimental results are included to show the validity of the proposed analysis method.**

Keywords— Current ripple, inverter, multiphase, PWM.

I.   INTRODUCTION

Multiphase ac motor drives have been proposed for various applications because of some advantages such as lower torque pulsations, higher power density, higher reliability, and lower dc link current ripple[1]-[8]. DC link or inverter input current ripple reduction is very important because it will reduce the size of electrolytic capacitor that is connected on the dc link. Different to three-phase PWM inverters, just a few works on input current ripple analysis of multiphase PWM inverters are available [9]-[11]. In the case of three-phase PWM inverters, it has been shown that the inverter input current ripple cannot be reduced by changing the modulation waveforms. It cannot be reduced also by increasing the switching frequency.

Multiphasing as a method to reduce the inverter input current ripple has been reported in the literature[9]-[11]. The most common phase number is six-phase or specifically it is asymmetrical six-phase. In this case, an AC motor with double three-phase stator winding is used. The two stator windings are usually displaced by 30 electrical degrees in position but can also be located in phase. Each three-phase stator winding is supplied by a three-phase inverter. As there are two stator windings, the system is also commonly called as inverter fed dual AC motor. When the inverter is operated under square-wave mode of operation, it was shown that 30 degree displacement of stator windings results in smaller inverter input current ripple compared to under zero

displacement. On the other hand, no work has shown which phase displacement is better when the inverter is operated under PWM operation.

This paper presents an analysis of input current ripple of inverter fed dual AC motor drives. The inverter is operated under carrier based PWM techniques. Two carrier based PWM techniques are used, the first is single carrier technique and the second is using two identical carrier signals that opposite in phase. Two kinds of asymmetrical six-phase AC motor are used, the first motor has two stator windings with zero phase displacement and the second with 30 degrees phase displacement. Analysis results show that a combination of a six-phase AC motor with zero phase displacement and a PWM technique with two carrier signals results in smallest input current ripple. Simulated and experimental results are included to show the validity of the proposed analysis method.

II.   INVERTER FED DUAL AC MOTORS

The scheme of inverter fed dual AC motors is shown in Fig. 1(a). A common DC voltage source is used to supply both inverters. Each inverter is supplying each three-phase stator winding of the AC motor. The two three-phase stator windings can be displaced by 30 electrical degrees or in phase as shown in Fig. 1(b). The two neutrals are isolated. In the analysis, it is assumed that the DC source voltage is constant and ripple free. The inverter switching devices are assumed as ideal switches.

Various PWM techniques for inverter fed dual three-phase AC motor were proposed [12]-[16]. Carrier based PWM technique is used to control the inverter. In the case of inverter fed dual AC drive system, two identical three-phase reference signals are used. The two references signals are also displaced in phase similarly as the two stator windings of the AC motor. Single and double carrier signals can be used in this applications. If single carrier is used, the two reference signals are compared to a high-frequency carrier signal to determine the ON and OFF signals for the inverter switching devices. If two identical carrier signals with 180 degrees phase displacement are used, each carrier signal is compared to each reference signal to determine the ON

978-1-4799-2706-7/14 $31.00 © 2014 IEEE

The 2014 International Power Electronics Conference

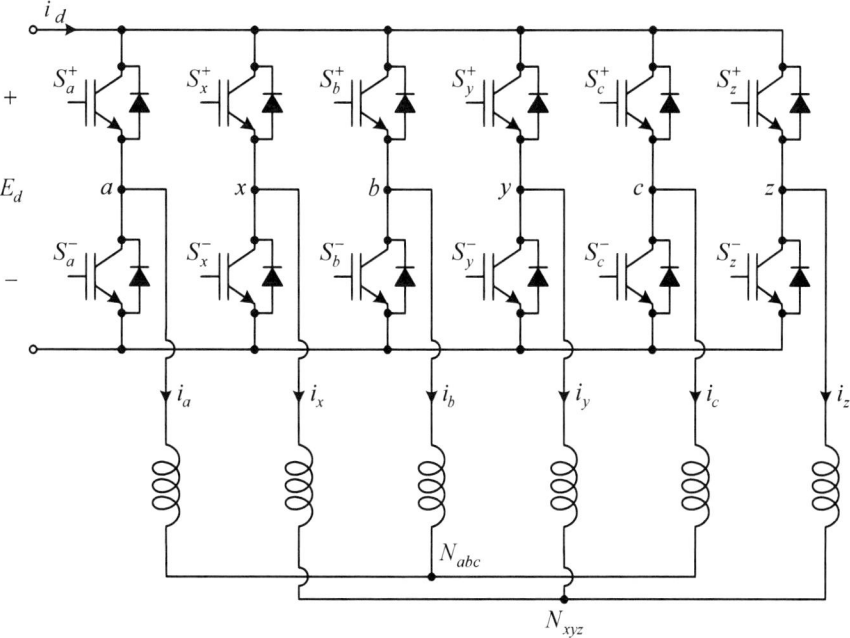

(a) System configuration.

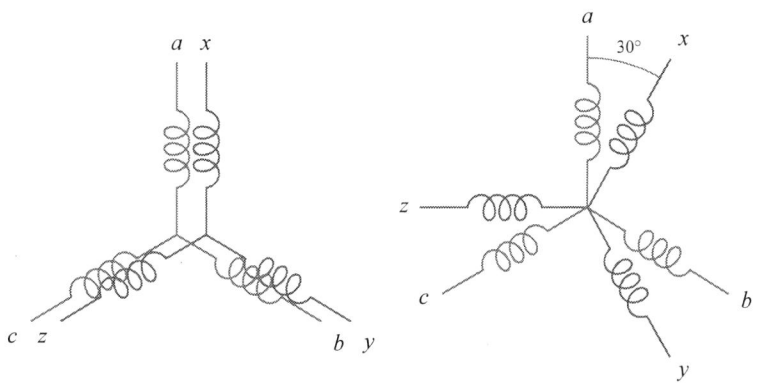

(b) Stator winding connection.

Fig. 1. Scheme of inverter fed dual AC drive system.

and OFF signals of the corresponding inverter switching devices. Thus, for cases are considered here:

i)   Single carrier signal with zero displacement stator windings (SC0).
ii) Single carrier signal with 30 degrees displacement stator windings (SC30).
iii) Double carriers with zero displacement stator windings (DC0).
iv) Double carriers with 30 degrees displacement stator windings (DC30).

Though various reference signals were proposed in the literature, the most commonly used is sinusoidal reference signal. Fig. 2 shows the sinusoidal reference

signals. The two three-phase sinusoidal reference signals can be written as

$$v_a^r = k\sin(\theta)$$
$$v_b^r = k\sin\left(\theta - \tfrac{2\pi}{3}\right)$$
$$v_c^r = k\sin\left(\theta + \tfrac{2\pi}{3}\right)$$
$$v_x^r = k\sin(\theta - \gamma)$$
$$v_y^r = k\sin\left(\theta - \tfrac{2\pi}{3} - \gamma\right)$$
$$v_z^r = k\sin\left(\theta + \tfrac{2\pi}{3} - \gamma\right)$$

(1)

where k is the modulation index, γ is phase angle displacement between the two stator windings, and

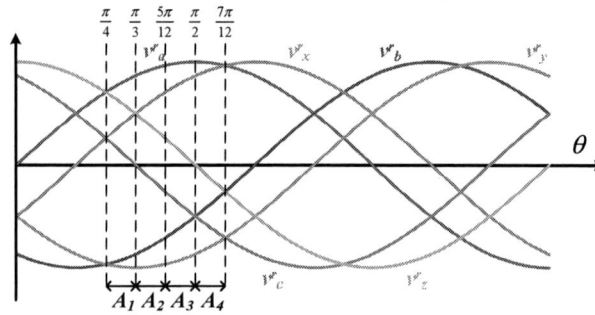

Fig. 2. Reference signal.

$\theta=2\pi ft$ with f is the fundamental output frequency of the inverter.

If the carrier frequency is high, the inverter output currents can be assumed as sinusoidal. The inverter output currents can be written as

$$i_a = \sqrt{2}I_l \sin(\theta - \phi)$$
$$i_b = \sqrt{2}I_l \sin\left(\theta - \phi - \frac{2\pi}{3}\right)$$
$$i_c = \sqrt{2}I_l \sin\left(\theta - \phi + \frac{2\pi}{3}\right)$$
$$i_x = \sqrt{2}I_l \sin(\theta - \phi - \gamma) \qquad (2)$$
$$i_y = \sqrt{2}I_l \sin\left(\theta - \phi - \frac{2\pi}{3} - \gamma\right)$$
$$i_z = \sqrt{2}I_l \sin\left(\theta - \phi + \frac{2\pi}{3} - \gamma\right)$$

where I_l is the rms value of inverter output current and □ is the load power factor angle.

The inverter input current as the function of inverter output currents can be written as

$$i_d = s_a i_a + s_b i_b + s_c i_c + s_x i_x + s_y i_y + s_z i_z \qquad (3)$$

where s_j is the switching state of phase $j(=a, b, c, x, y,$ or $z)$. The value of switching state is unity (zero) when the upper (lower) switching device of the associated phase is receiving an ON signal. The expression (3) will be used as the basis for the analysis that will be presented in the next section.

III. INPUT CURRENT RIPPLE

The reference signals in Fig. 2 are compared to two carrier signals. The resulted switching pattern varies with the fundamental angle θ. If the carrier frequency is much higher than the reference frequency, the reference signals in one carrier period can be assumed as constants. Fig. 3 shows the detailed waveforms over one carrier period in interval A1.

$$\frac{T_0}{T_s} = \frac{1 + v_y^r}{4} \qquad (4)$$

$$\frac{T_1}{T_s} = \frac{-v_a^r - v_y^r}{4} \qquad (5)$$

$$\frac{T_2}{T_s} = \frac{v_a^r - v_c^r}{4} \qquad (6)$$

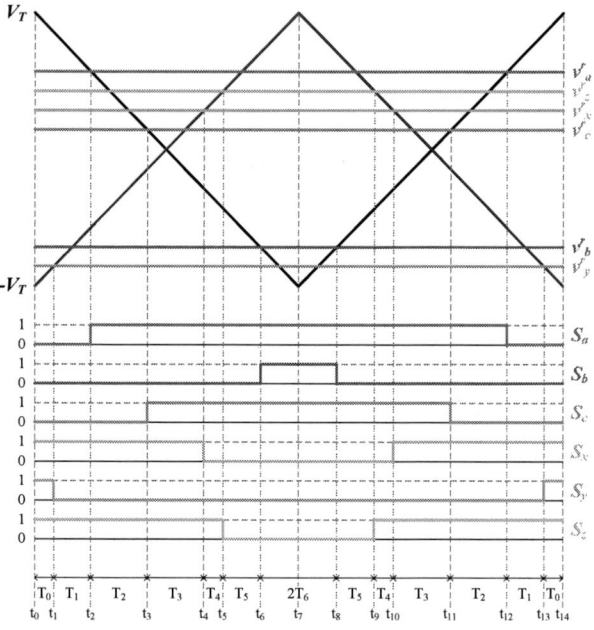

Fig. 3. Detailed waveforms over one carrier period.

The time intervals in Fig. 3 can be calculated as follows

$$\frac{T_3}{T_s} = \frac{v_x^r + v_c^r}{4} \qquad (7)$$

$$\frac{T_4}{T_s} = \frac{v_z^r - v_x^r}{4} \qquad (8)$$

$$\frac{T_5}{T_s} = \frac{-v_b^r - v_z^r}{4} \qquad (9)$$

$$\frac{T_6}{T_s} = \frac{1 + v_b^r}{4} \qquad (10)$$

Based on (3), the inverter input current over one carrier period in Fig. 3 can be written as follow

$$i_d = \begin{cases} 0 & \text{for } t_0 \leq t \leq t_1 \\ -i_y & \text{for } t_1 \leq t \leq t_2 \\ i_a - i_y & \text{for } t_2 \leq t \leq t_3 \\ -i_b - i_y & \text{for } t_3 \leq t \leq t_4 \\ -i_b + i_z & \text{for } t_4 \leq t \leq t_5 \\ -i_b & \text{for } t_5 \leq t \leq t_6 \\ 0 & \text{for } t_6 \leq t \leq t_8 \\ -i_b & \text{for } t_8 \leq t \leq t_9 \\ -i_b + i_z & \text{for } t_9 \leq t \leq t_{10} \\ -i_b + i_z & \text{for } t_{10} \leq t \leq t_{11} \\ -i_b - i_y & \text{for } t_{11} \leq t \leq t_{12} \\ i_a - i_y & \text{for } t_{12} \leq t \leq t_{13} \\ -i_y & \text{for } t_{13} \leq t \leq t_{14} \\ 0 & \text{for } t_{14} \leq t \leq t_{15} \end{cases} \qquad (11)$$

The mean square value during one carrier period of the inverter is

$$I_d^2 = \frac{1}{T_s} \int_{t_o}^{t_o+T_s} i_d^2 dt \tag{12}$$

Substituting (11) into (12), the following can be obtained after performing the integration

$$
\begin{aligned}
I_d^2 = \frac{2T_1}{T_s} i_y^2 + \frac{2T_2}{T_s} \left(i_a - i_y \right)^2 + \frac{2T_3}{T_s} \left(i_b + i_y \right)^2 \\
+ \frac{2T_4}{T_s} \left(i_z - i_b \right)^2 + \frac{2T_5}{T_s} i_b^2
\end{aligned}
\tag{13}
$$

The above method is repeated for intervals A2, A3, and A4. The average of mean square value of inverter input current can be obtained as follow

$$I_{d,avg}^2 = \frac{3}{\pi} \int_{\pi/4}^{7\pi/12} I_d^2 d\theta \tag{14}$$

The ripple component of inverter input current is

$$\tilde{I}_d = \sqrt{I_{d,avg}^2 - I_{dc}^2} \tag{15}$$

where I_{dc} is the dc component of inverter input current. If the inverter losses can be neglected, the dc component of inverter input current can be calculated using the equality between input and output powers. The result is

$$I_{dc} = \frac{3}{\sqrt{2}} k I_l \cos\phi \tag{16}$$

By using (15) and (16) then the ripple component of inverter input current can be obtained as

$$\tilde{I}_d = \frac{I_l}{2} \sqrt{ \frac{ \left[4\left(\sqrt{6}+\sqrt{2}+2\sqrt{3}\right) - 9k\pi \right]\cos^2\phi }{ + 2\left(\sqrt{3}-\sqrt{2}\right)+\sqrt{6} } } \tag{17}$$

The above steps are repeated for other three cases. The results are shown in Table 1.

Fig. 4 shows comparison of inverter input current ripple as a function of modulation index It can be seen that double carrier with zero displacement stator windings results in smaller input current ripple.

IV. EXPERIMENTAL RESULTS

In order to verify the proposed analysis method, a small experimental system was constructed. Locked six-phase induction motors were used in the experiment. The DC voltage was kept constant at 50 Vdc during all experiments. The carrier frequency is 1000 Hz. Inverter switching devices were implemented by using Power MOSFETs.

The load impedance for motor with zero displacement angle is 4.54 Ohm with power factor of 0.5. The load impedance for motor with 30 degrees phase displacement is 5.17 ohm with power factor of 0.47. The inverter input current is measured by using a digital oscilloscope and the results are processed by using a computer to determine the ripple component.

Fig. 5 shows experimental and calculated results of input current ripple as a function of modulation index. The validity of the proposed analysis method can be appreciated from this figure.

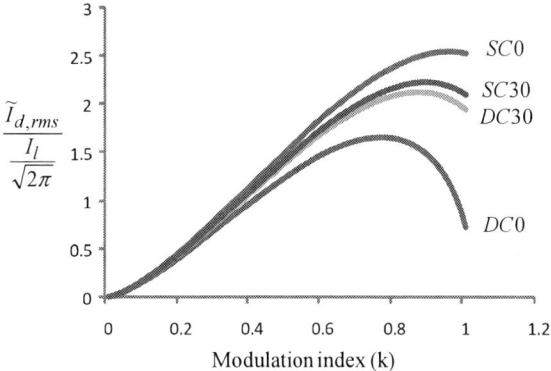

Fig. 4. Input current ripple as a function of modulation index.

V. CONCLUSION

An analysis of input current ripple of inverter fed dual AC drive system has been presented in this paper. It has been shown that a combination of double carrier and zero displacement angle results in smallest input current ripple. Experimental results show the validity of the proposed analysis method. The results are very important for dc filter capacitor specification. Extension of the proposed analysis method for other applications is left for future investigation.

Table 1. Input current ripple.

Type	Input current ripple
SC0	$\dfrac{I_l}{2}\sqrt{\dfrac{2k}{\pi}\left[\left(\sqrt{3}-\dfrac{9}{16}\pi k\right)\cos^2\phi+\dfrac{\sqrt{3}}{4}\right]}$
SC30	$\dfrac{I_l}{2}\sqrt{\dfrac{2k}{\pi}\left[\left(30\sqrt{3}-9\pi k\right)\cos^2\phi+2\left(\sqrt{3}-\sqrt{2}\right)+\sqrt{6}\right]}$
DC0	$\dfrac{I_l}{2}\sqrt{\dfrac{2k}{\pi}\left[\left(16+8\sqrt{3}-9\pi k\right)\cos^2\phi+2\sqrt{3}-2\right]}$
DC30	$\dfrac{I_l}{2}\sqrt{\dfrac{2k}{\pi}\left[\left(4\left(\sqrt{6}+\sqrt{2}+2\sqrt{3}\right)-9\pi k\right)\cos^2\phi+2\left(\sqrt{3}-\sqrt{2}\right)+\sqrt{6}\right]}$

REFERENCES

[1] Jahns, "Improved reliability in solid-state AC drives by means of multiple independent phase drive units," *IEEE Trans. Ind. Appl.*, vol. 16, no. 3, pp. 321-331, 1980.

[2] Klingshirn, "High phase order induction motors," IEEE Trans. Power App. Sys., vol. 102, no. 1, pp. 47-53, 1983.

[3] Singh, "Multiphase induction machine drive research – a survey," *Electr. Power Sys. Res.*, vol. 61, pp. 139-147, 2002.

[4] Williamson and Smith., "Pulsating torque and losses in multiphase induction machines," *IEEE Trans. Ind. Appl.*, vol. 39, no. 4, pp. 986-993, 2003.

[5] Bojoj, et.al., "Dual three-phase induction machine drives control - - as survey," Proc. Int. Power Electr. Conf., pp. 90-99, 2005.

[6] Apsley, et.al., "Induction motor performance as a function of phase number," *IEE Proc. Elec. Power Appl.*, vol. 153, no.6, pp. 898-904, 2006.

[7] Levi, e.al., "Multiphase induction motor drives – a technology status review," *IET Elec. Power App.*, vol. 1, no. 4, pp. 489-516, 2007.

[8] Levi, "Multiphase electric machines for variable-speed applications," *IEEE Trans. Ind. Electr.*,, vol. 55, no. 5, pp. 1893-1908, 2008.

[9] Nelso and Krause, "Induction machine analysis for arbitrary displacement between multiple winding sets," *IEEE Trans. Power App. Sys.*, " pp. 841-848, 1974.

[10] Bojoj, et.al., "Computation and measurements of the DC link current in six-phase voltage source PWM inverters for AC motor drives," Int. Power Conv. Conf., Osaka 2002, pp. 953-958.

[11] Dahono, et.al., "Input ripple analysis of five-phase pulse width modulated inverters," IET Power Electr., vol. 3, no. 5, pp. 716-723, 2010.

[12] Zhao and Lipo, "Space vector PWM control of dual three-phase induction machine using vector space decomposition," IEEE Trans. Ind. Appl., vol. 31, no. 5, pp. 1100-1109, 1995.

[13] Kelly, et.al., "Multiphase space vector pulse width modulation," *IEEE Trans. Eng. Conv.*, vol. 18, no. 2, pp. 259-264, 2003.

[14] Hadiouche, et.al., "Space vector PWM techniques for dual three-phase AC machine : Analysis, performance, evaluation, and DSP implementation, " IEEE Trans. Ind. Appl., vol. 42, no. 4, pp. 1112-1122, 2006.

[15] Zarri, et.al., "Minimization of the power losses in IGBT multiphase inverters with carrier based pulsewidth modulation," IEEE Trans. Ind. Elec., vol. 57, no. 11, pp. 3695-3706, 2010.

[16] Marouani, et.al., A new PWM strategy based on a 24-sector vector space decomposition for a six-phase VSI-Fed dual stator induction motor," IEEE Trans. Ind. Electr., vol. 55, no. 5, pp. 1910-1920, 2008.

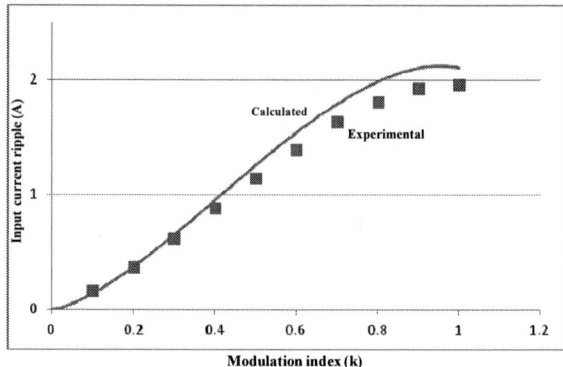

(a) DC0.

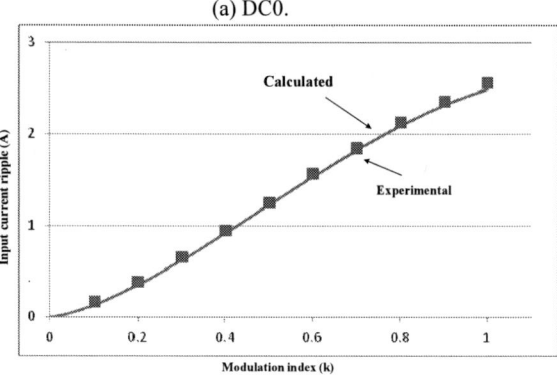

(b) SC30.

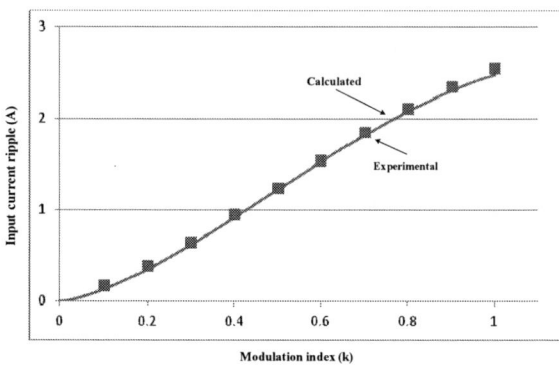

(c) DC30.

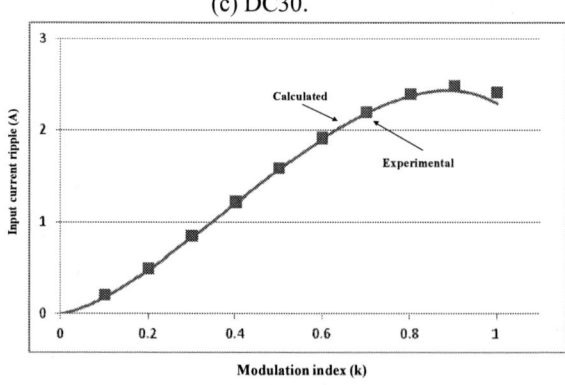

(d) SC0.

Fig. 5. Experimental results.

Offline Extraction of Induction Machine Parameters for Control Strategy Synthesis

Stefan Koschik, Florian Bauer, R. W. De Doncker

Institute for Power Electronics and Electrical Drives (ISEA), RWTH Aachen University

Abstract—**An offline algorithm for induction machine parameter determination from power and torque measurements over several torque-slip curves is presented. Based on a parameter sensitivity function, the algorithm fits the measured data of torque, stator voltage amplitude and power factor to the parameters of an equivalent circuit of the induction machine which takes main inductance saturation, iron losses and skin effect of the rotor into account. The approach is experimentally verified on test bench measurement data.**

I. INTRODUCTION

Equivalent circuits of induction machines with precise parameterization are beneficial for implementation of efficiency optimized control strategies. If the machine parameters are not available from FEM data, it is necessary to determine them from a test bench setup. However, the conventional short circuit and no-load tests do not provide the parameters needed to implement controls for varying main flux path saturation, iron losses or skin effect. Procedures to measure machine parameters from torque versus slip frequency characteristics have been described in [7], though the authors focused on grid-connected motors which operate at high slip frequencies. A similar approach for inverter driven traction drives with focus on the stator current versus flux characteristic was introduced in [5]. In [13], an algorithm based on parameter sensitivity was proposed to fit parameters from machine nameplate data and standard machine tests.

This paper presents an easy-to-implement fitting algorithm, that provides accurate parameter identification from torque versus slip frequency measurements for inverter-driven induction machines. The algorithm evaluates the measured torque, stator voltage amplitude and power factor. The proposed method uses parameter sensitivities of the equivalent circuit to assess model errors during the fitting procedure.

II. MACHINE MODELLING

In this paper, the model shown in Fig. 1 is used, which is an extension of the standard T model for the induction machine [1], [2]. The equivalent circuit is non-linear since main flux path saturation and iron losses are included.

Skin effect is modeled by introducing a parallel *RL* circuit in the rotor, which results in an increased rotor resistance and decreased stray inductance at high slip frequencies. The skin effect parameters $l_{\sigma r}^0$ and $l_{\sigma r}^1$ are chosen such that the rotor stray inductance at no load is equal to the rotor stray inductance

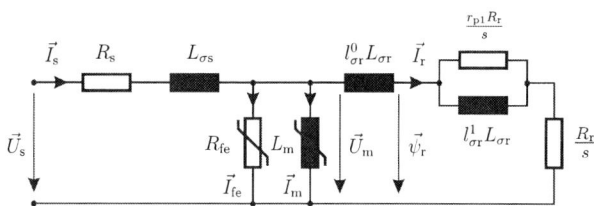

Fig. 1. Steady state equivalent circuit for the induction machine with iron losses, main flux saturation and skin effect

in the standard T model

$$l_{\sigma r}^0 + l_{\sigma r}^1 \overset{!}{=} 1. \tag{1}$$

The effect of saturation and skin effect on the torque versus slip frequency characteristic has been described in [3]. Torque is given by the product of rotor flux linkage $\vec{\psi}_r$ and rotor current $\vec{I}_r$

$$T_m = \frac{3}{2} \cdot p \cdot \Im\left(\vec{\psi}_r^* \vec{I}_r\right). \tag{2}$$

where p is the pole pair number.
Several approaches exist to model the saturation characteristic of the main inductance L_m such as piecewise exponential functions, hyperbolic functions, power functions and polynomials [6], [8]. A polynomial of 3rd order with dependence on the amplitude of the magnetization current $\hat{I}_m$ is used in this paper:

$$L_m(\hat{I}_m) = c_3 \cdot \left(\hat{I}_m\right)^3 + c_2 \cdot \left(\hat{I}_m\right)^2 + c_1 \cdot \hat{I}_m + c_0. \tag{3}$$

In the equivalent circuit, iron losses P_{fe} are represented by the resistance R_{fe} parallel to the main inductance. To describe iron losses, several approaches with varying degrees of accuracy such as Steinmetz equation, generalized Steinmetz equation (GSE), modified Steinmetz equation (MSE) or improved generalized Steinmetz equation (iGSE) have been proposed [4], [9], [10], [11]. For simplicity of measurement and implementation, the Steinmetz equation

$$P_{fe}(f_s, \hat{I}_m) = \frac{3}{2} \cdot k \cdot f_s^a \cdot \hat{I}_m^b \tag{4}$$

is used where the induction $\hat{B}$ is replaced by the magnetizing current $\hat{I}_m$. The coefficient k and exponents a, b depend on material parameters. The iron resistance is determined by the iron losses and therefore varies with stator frequency f_s and magnetization current I_m:

$$R_{fe} = \frac{\left(\omega_s \cdot L_m(\hat{I}_m) \cdot I_m\right)^2}{P_{fe}(\hat{I}_m, f_s)} = \frac{U_m^2}{P_{fe}(\hat{I}_m, f_s)}. \tag{5}$$

A. Operating Point Calculation

The steady state operating points are calculated by solving the non-linear steady state conditions for saturation and iron losses. The stator current amplitude $\hat{I}_s$, the rotor frequency f_m and the slip frequency f_{sl} are input parameters. To obtain the operating point for a given input set of $\{\hat{I}_s, f_m, f_{sl}\}$, the magnetizing current $\hat{I}_m$ is varied from $0\,A$ to $\hat{I}_s$ in i steps. Main inductance, iron losses and iron resistance are calculated according to (3), (4) and (5). Using the values of L_m and R_{fe} for all i calculations, the current in the magnetizing path, designated by $\hat{I}_m^*$, can again be calculated using the current divider equation. The operating point is selected from the calculated operation points by the criterion

$$\min\left(\hat{I}_m^* - \hat{I}_m\right) \tag{6}$$

to account for numeric inaccuracies.

III. MEASUREMENT PROCEDURE

The measurement procedure is divided into three parts. In the first step, stator resistance can be acquired by DC-current measurements or an *LCR* meter. From a second measurement iron losses are determined. From the same measurement data, the saturation characteristic of the main inductance is acquired. For the identification of the stray inductances, rotor resistance and the skin effect parameters, a slip curve measurement is performed. The induction machine is driven to a constant speed by a load machine. A simple PI regulator with low dynamics can be used to control the amplitude of the stator current for a given slip frequency. For all measurements, recording of torque, phase current, voltage and power of the induction machine is required.

A. Iron loss measurement

The iron losses are measured by operating the machine at no-load condition while varying motor speed and stator current amplitude. The active power is measured and the stator winding losses are subtracted

$$P_{fe} = P_{total} - 3 \cdot R_s \cdot I_m^2. \tag{7}$$

The exponents of the loss function a and b are proportional to the logarithm of the Steinmetz equation for a fixed current or a fixed frequency respectively, cf. (4). To determine the proportionality factor k, the calculated parameters a and b are used to calculate the equivalent iron losses with $k = 1$. By dividing the measured losses by the equivalent losses

$$k = \frac{P_{fe,meas}}{\frac{3}{2} \cdot f_s^a \cdot \hat{I}_m^b} \tag{8}$$

the factor k can be determined. This calculation is performed for all measurement points to find a mean value of k.

B. Main inductance measurement

For main inductance characteristic, the same measurement data as for the iron losses can be used since a no-load condition is necessary for the determination of the inductance function. The stator flux can be calculated from the stator current and voltage

$$\Psi_s = \left| \frac{1}{\omega_s} \left(\vec{U}_s - R_s \cdot \vec{I}_s \right) \right|. \tag{9}$$

For increased accuracy, an equivalent iron resistance can be included to accommodate the iron losses. However, the effect on the calculated main inductance is small and will be neglected in the following. Thus, the main inductance is given by

$$L_m = \frac{\Psi_s}{I_s} - L_{\sigma s}. \tag{10}$$

To begin with, it is assumed that $L_{\sigma s} = 0$. The estimation of the stray inductance is done in the fitting algorithm. The values calculated from (10) are fitted to (3) with a least square algorithm.

C. Torque-slip curve measurement

For the extraction of the remaining parameters, a torque versus slip frequency measurement is performed. The induction machine is driven at constant speed. The stator current I_s is kept constant while the slip frequency f_{sl} is varied. This procedure is repeated for different stator current amplitudes up to nominal stator current. The extraction of the desired parameters from the torque versus slip frequency characteristic is described in detail in the next section.

IV. PARAMETER FITTING ALGORITHM

The algorithm is divided into three simple steps. Firstly, for each of the parameters, a start value, a search interval and a step size are defined. The algorithm sequentially varies the parameter within this interval with the given step size and calculates torque, voltage and power factor based on the equivalent circuit. By means of the deviations of calculated and measured values, an optimization index is calculated. If the optimization index improves, the selected parameter value is set as the temporary optimal value. The algorithm then proceeds to the next parameter. After one iteration of all parameters, the selected optimal parameter set is chosen as next start parameters and the algorithm again proceeds to vary all parameters values within the search interval. The algorithm stops if the optimization index is not improved over one cycle of parameter variation.

A. Sensitivity Analysis of Equivalent Circuit Parameters

The optimization index is based on a parameter sensitivity analysis of torque, voltage and power to changes in equivalent circuit parameters. The influence of differential changes of parameters on the characteristic curves is shown in Fig. 2. Several conclusions can be drawn from an analysis of the parameter sensitivity.

In general, the values of rotor resistance, rotor stray inductance

The 2014 International Power Electronics Conference

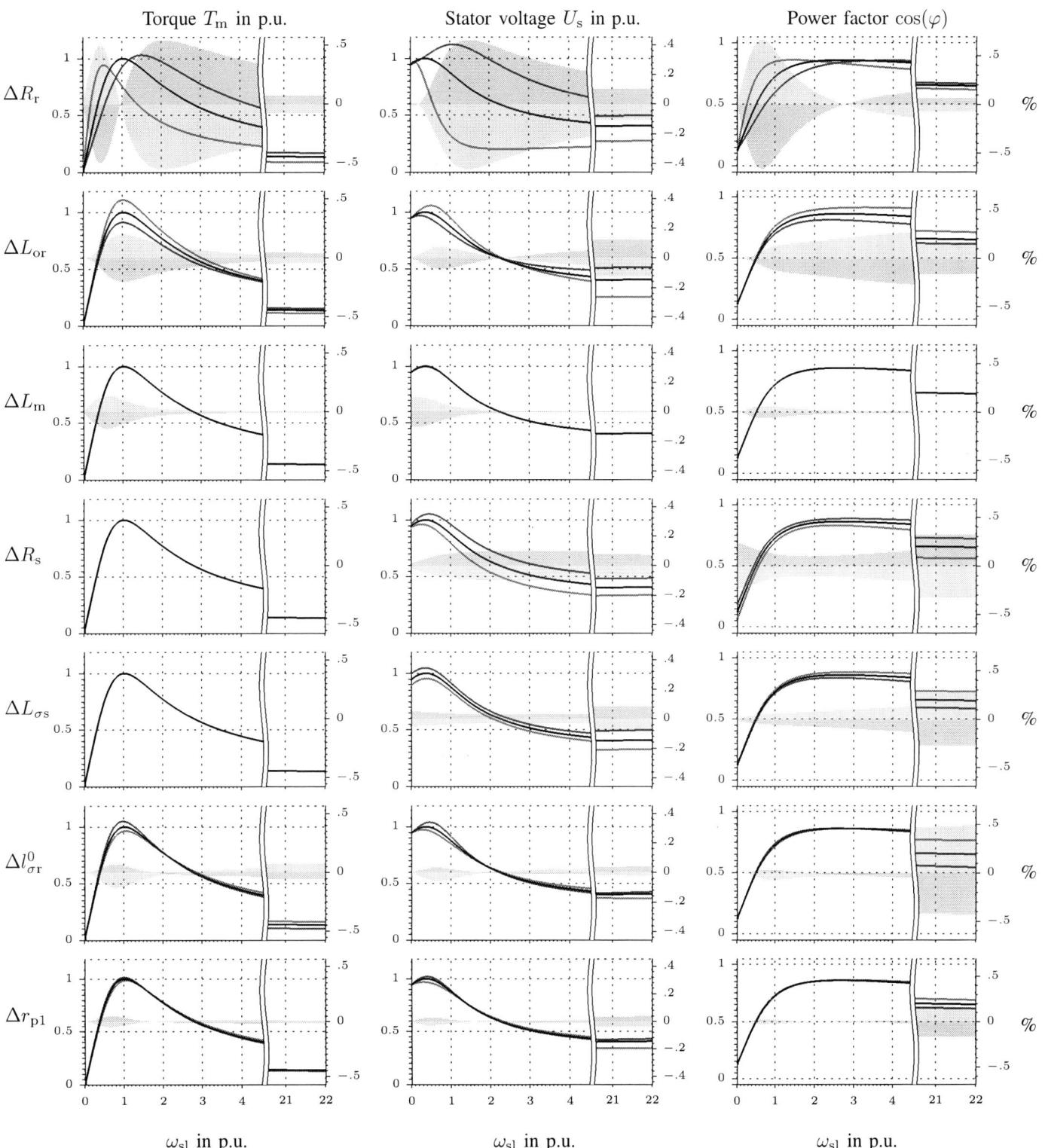

Fig. 2. Parameter sensitivity analysis of the equivalent circuit with regards to torque, phase voltage and power factor. The values for torque and voltage are given per unit. The solid lines show the impact of large deviations in parameters. Each parameter was varied by $\pm 50\,\%$ and the deviations plotted for the increased (red curve) and decreased (green curve) value in comparison with the original value (black curve). For analysis of the impact of small deviations in parameters, each parameter was varied by $\pm 1\,\%$. The shaded areas show the influence of the variations in the values with the axis scaling on the right hand side, where the deviation is given in percent of the nominal value. Parameters from top to bottom: rotor resistance R_r, rotor stray inductance L_or, main inductance L_m, stator resistance R_s, stator stray inductance $L_{\sigma \mathrm{s}}$, skin effect parameters $L_{\sigma \mathrm{r}}^0$, $R_{\mathrm{p}1}$.

978-1-4799-2706-7/14 $31.00 © 2014 IEEE

and main inductance have the highest impact on the machine characteristic. The influence of the main inductance on the curves is only noticeable around the area of maximum torque. It was already noted in [7] that for line start analysis the main inductance saturation has little effect on the machine performance. However, for inverter controlled machines, its effect cannot be neglected.

The deviations of torque and phase due to changes in the rotor resistance are zero at the maximum torque per ampere point. The inversion of sensitivity of the rotor resistance to torque has been discussed in [12]. It is also supported by the discussion of the rotor time constant sensitivity with respect to torque and flux linkage in [15].

In [16], it was presumed that the ratio of stator and rotor stray inductance is negligible as long as the sum of the two remains the same. Considering voltage and power factor this can be validated for high slip frequencies. However, the effect of rotor leakage inductance on torque cannot be compensated by altering the stator leakage inductance.

The parameters of the *RL* skin effect network have the highest impact in the high slip region as might be expected.

Finally, the stator resistance is the only parameter whose sensitivity is speed dependent. The sensitivity to stator resistance decreases towards higher speeds and increases towards lower speeds.

B. Optimization Index

The optimization index for the fitting algorithm is derived from the parameter sensitivity analysis over the sum of errors caused by variations in all parameters x at a given slip frequency

$$\varepsilon(\omega_{\mathrm{sl}}) = \sum_{i=1}^{n} \frac{|f(\omega_{\mathrm{sl}}, x_i) - f(\omega_{\mathrm{sl}}, x_i + \Delta x_i)|}{f^{\mathrm{nom}}}, \quad (11)$$

where n is equal to the number of parameters to be evaluated. In this case the equivalent circuit contains seven parameters. The function $f(\omega_{\mathrm{sl}}, x_i)$ represents torque, voltage or power factor respectively. The functions are scaled to their nominal values. The sensitivity functions scaled to maximum sensitivity are shown in Fig. 3.

For comparison of the overall parameter sensitivity, the

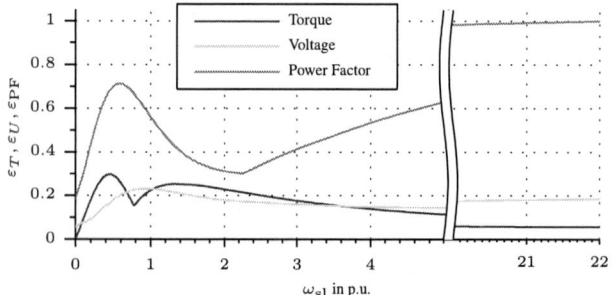

Fig. 3. Sensitivity function for torque, voltage and power factor.

integral of the sensitivity function over slip

$$A_\varepsilon = \int \varepsilon \cdot \mathrm{d}\omega_{\mathrm{sl}} \quad (12)$$

is calculated. The parameter sensitivity is highest with respect to power factor. This tendency increases towards high slip values. For slips up to 4 p.u., the ratio of the areas is $22 : 25 : 53$. At high slips between 20 and 22 p.u., this ratio changes to $5 : 15 : 80$.

For each measured value y_j the algorithm calculates the difference to the theoretical equivalent circuit value y'_j, weights the difference by the nominal value y^{nom} and rounds the value considering the measurement device resolution g_y

$$\Delta y_j = \left\lfloor \frac{|y'_j - y_j|}{g_y \cdot y^{\mathrm{nom}}} \right\rceil \cdot g_y. \quad (13)$$

The errors are scaled with the sensitivity function and added up over all m measurement values

$$\zeta = \sum_{j=1}^{m} \frac{\Delta T_j}{\varepsilon_T} + \sum_{j=1}^{m} \frac{\Delta U_j}{\varepsilon_U} + \sum_{j=1}^{m} \frac{\Delta \mathrm{PF}_j}{\varepsilon_{\mathrm{PF}}} \quad (14)$$

to obtain the optimization index ζ. The algorithm continues to vary parameters as long as ζ improves.

C. Fitting Procedure

To decrease calculation time, the fitting is separated into three steps. At first, the fitting is applied to measurement data at high slips over all current amplitudes to fit values for the skin effect parameters.

In the second step, the fitting is performed at slips below the maximum torque per ampere point to obtain values for ΔL_{m} over different magnetizing currents. These values can be used to correct the saturation curve from the no-load test.

Finally, the fitting is performed for the torque versus slip frequency areas which are relevant for control purposes. Operating points are excluded where the main flux linkage is above 10% of the rated flux linkage. Furthermore, points at high slips which are only reached in field weakening above maximum rated speed are excluded. This encompasses points at high slip and low current amplitude.

V. MEASUREMENT RESULTS

The tests were performed on a 750 W induction motor with a load machine and an analog output torquemeter Lorenz DR-2477 which is connected to a power meter LMG 500 along with the phase currents and phase voltages. Torque, voltage and current accuracy are 0.025 Nm, 25 mV and 0.8 mA, respectively. The test setup along with the nameplate data is shown in Fig. 4.

A. Influence of Temperature

Temperature increase in the machine can have an adverse effect on the fitting procedure because stator and rotor resistance are temperature dependent and change during measurements. To evaluate the effect of temperature increase on the

The 2014 International Power Electronics Conference

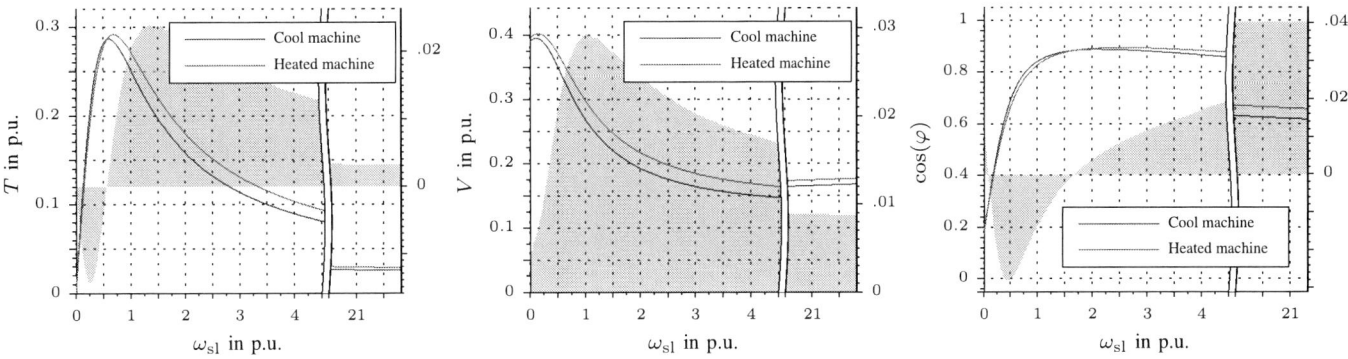

Fig. 5.   Influence of temperature on the machine characteristic. The shaded areas shows the difference between two measurements with the axis scaling given on the right hand side of the plot.

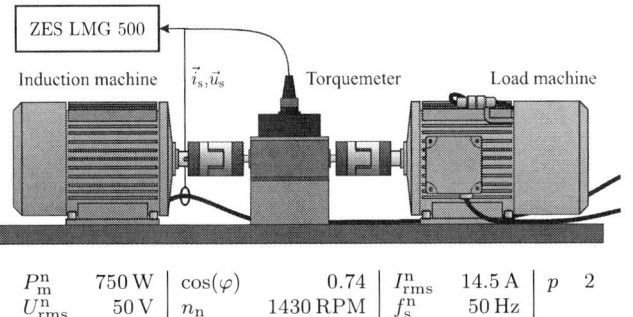

| $P_{\mathrm{m}}^{\mathrm{n}}$ | 750 W | $\cos(\varphi)$ | 0.74 | $I_{\mathrm{rms}}^{\mathrm{n}}$ | 14.5 A | p | 2 |
| $U_{\mathrm{rms}}^{\mathrm{n}}$ | 50 V | n_{n} | 1430 RPM | $f_{\mathrm{s}}^{\mathrm{n}}$ | 50 Hz | | |

Fig. 4.   Test bench setup and nameplate data

machine parameters, two measurement sets were conducted, one on the cool machine, the second on a heated machine. To heat up the stator and rotor of the induction machine, the setup was operated at nominal current and 5x rated slip frequency for half an hour. The results are shown in Fig. 5.

The shaded area shows the deviation between the two measurements with the axis scaling on the right hand side. The similarities of increase in rotor and stator resistance to the sensitivity analysis in Fig. 2 are evident. The increased rotor resistance leads to lower torque at low slips and higher torque at high slips. Simultaneously, the total machine efficiency is increased for the heated machine at slips above the maximum torque per ampere point by about 1.5 %. During the heat up process torque increased by 15 % in total within half an hour. Using the torque equation

$$T_{\mathrm{m}} = \frac{3}{2} \cdot p \cdot \frac{L_{\mathrm{m}}^2}{L_{\mathrm{r}}} \cdot I_{\mathrm{s}}^2 \cdot \frac{\tau \omega_{\mathrm{sl}}}{1 + (\tau \omega_{\mathrm{sl}})^2} \quad (15)$$

with the rotor time constant $\tau = L_{\mathrm{r}}/R_{\mathrm{r}}$ and assuming that $1 << (\tau \omega_{\mathrm{sl}})^2$, the increase in rotor resistance is proportional to the torque increase at high slips

$$R_{\mathrm{r}} \sim \frac{2}{3} \cdot \frac{T_{\mathrm{m}}}{p} \cdot \frac{\omega_{\mathrm{sl}}}{I_{\mathrm{s}}}. \quad (16)$$

Using (16), this signifies an increase of the rotor resistance by 15 % or an equivalent increase of 35 K for an aluminum rotor. The thermal time constant of the rotor $\tau_{\mathrm{r,th}}$ was estimated to 1700 s.

For each current amplitude, 100 measurements were taken which took about 4 seconds per measurement or about 7 minutes for one current amplitude reference. In the fitting process, the rotor resistance is assumed to be constant during measurement cycle of one current amplitude value.

B. Initial Parameter Approximation

For a first approximation of stator resistance and stray inductances, measurements of an *LCR* meter were used. An approximation of the rotor resistance can be calculated from torque measurement at high slip frequency by applying (16).

C. Iron Loss Measurement

Measurement results for the logarithmic plot of the iron losses with fixed current amplitude and varying rotor speed on the one hand and fixed frequency and varying current amplitude at no-load on the other hand can be seen in Fig. 6. On a logarithmic scale, the losses show a linear relation with varying current amplitude and frequency respectively, from which the proportionality factors are extracted. Iron losses account for about 27 % of total losses, which corresponds to 6 % of the mechanical output power at nominal operation.

D. Main Inductance Measurement

Figure 7 shows the fitted curve of the main inductance with the least square fit to the no-load measurement with a polynomial of 3rd order.

The main inductance flux path can saturate due to cross coupling as described in [17]. This effect occurs especially in small power machines with skewed rotor, as in the given test setup. As already mentioned in section IV-A, a deviation in the main inductance value is mainly noticeable in the slip frequency region below the maximum torque per ampere point. To adjust the saturation curve for load dependency and stator stray inductance, the fitting algorithm is applied to measurement points below the maximum torque per ampere point. The fitting algorithm returns an optimal offset value ΔL_{m} for the main inductance value. This value is used to adjust the magnetizing curve at the given magnetizing current level. The procedure is

978-1-4799-2706-7/14 $31.00 © 2014 IEEE

The 2014 International Power Electronics Conference

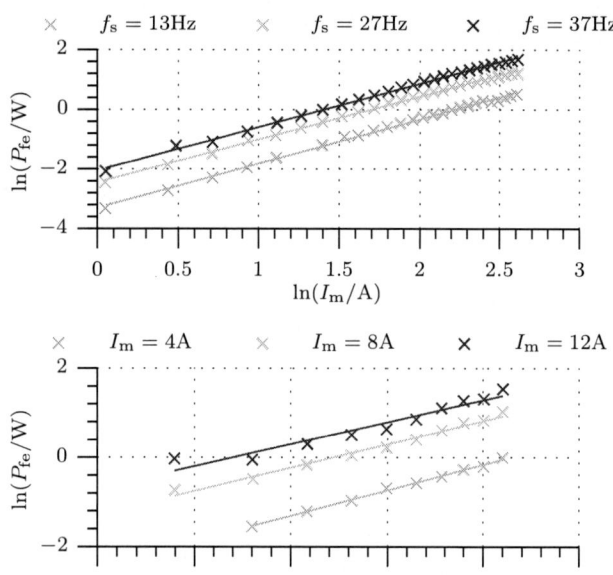

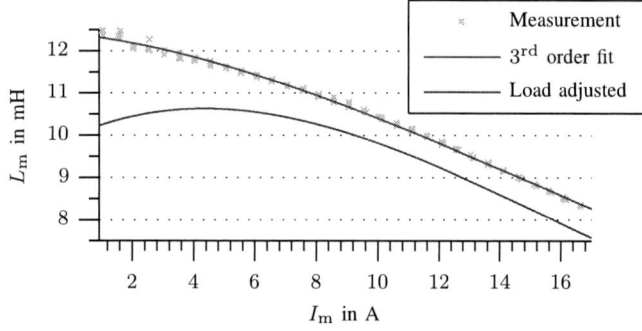

Fig. 6. Iron loss measurement

I_s in A	R_r in mΩ	L_{or} in mH	$L_{\sigma s}$ in mH	$L_{\sigma r}^0$	R_{p1}
4	90.01	1.21	0.42	0.545	1.2
6	91.56	1.11	0.42	0.545	1.2
8	93.12	1.02	0.42	0.545	1.2
10	94.28	1.00	0.42	0.545	1.2
12	95.64	0.97	0.42	0.545	1.2
14	97.39	0.91	0.42	0.545	1.2
16	98.95	0.89	0.42	0.545	1.2
18	101.67	0.84	0.42	0.545	1.2
20	102.64	0.80	0.42	0.545	1.2

Fig. 8. Fitted parameter values.

discussed previously. The estimated value for the rotor stray inductance is comparatively high. Since a change in the rotor stray inductance has a similar effect on torque and voltage characteristic as a change in main inductance, the algorithm will adjust the value of the rotor stray inductance to compensate for parameters deviations in the value of the main inductance. Here, better results might be achieved if a look-up table for the load dependent main saturation characteristic for each load condition is implemented.

Overall, the fitted curves for varying current amplitude match the measured curves well and the error is within 1.5 % of the rated values for all quantities.

VI.   CONCLUSION

A measurement procedure and parameter extraction algorithm for induction machine parameters with non-linearity was presented. The matched parameters agree well with measurements with maximum deviations of 1.5 % of nominal values. Implementation of a sensitivity weight function for the fitting algorithm improved the results of the fitting process. Furthermore, the sensitivity analysis allows insight into the effect of parameter variations on the equivalent circuit which can help as a guideline to implement improved fitting algorithms.

repeated for each torque versus slip frequency measurement to adjust the main inductance curve at different magnetization levels. Another fit is applied to the adjusted curve to receive the final main inductance characteristic.

Fig. 7. Main inductance as a function of magnetizing current

E. Torque-slip curve fitting

The final result of the fitted torque-slip curves at a machine speed of 1000 RPM is shown in Fig. 9. The stator current amplitude was varied from 0 A up to nominal current in 10 % steps of nominal current. In comparison with the standard T model, the point of maximum torque $w_{sl} = R_r/L_m$ is shifted to higher slip frequencies for increased currents due to the effect of saturation. The skin effect causes a slight increase in torque, phase angle and voltage at higher slip frequencies.

The absolute values of the fitted parameters are given in Fig. 8. An increase of rotor resistance and thus rotor temperature of 14 % within 90 minutes of total measurement time is noticeable which is in agreement with the temperature measurement

978-1-4799-2706-7/14 $31.00 © 2014 IEEE

The 2014 International Power Electronics Conference

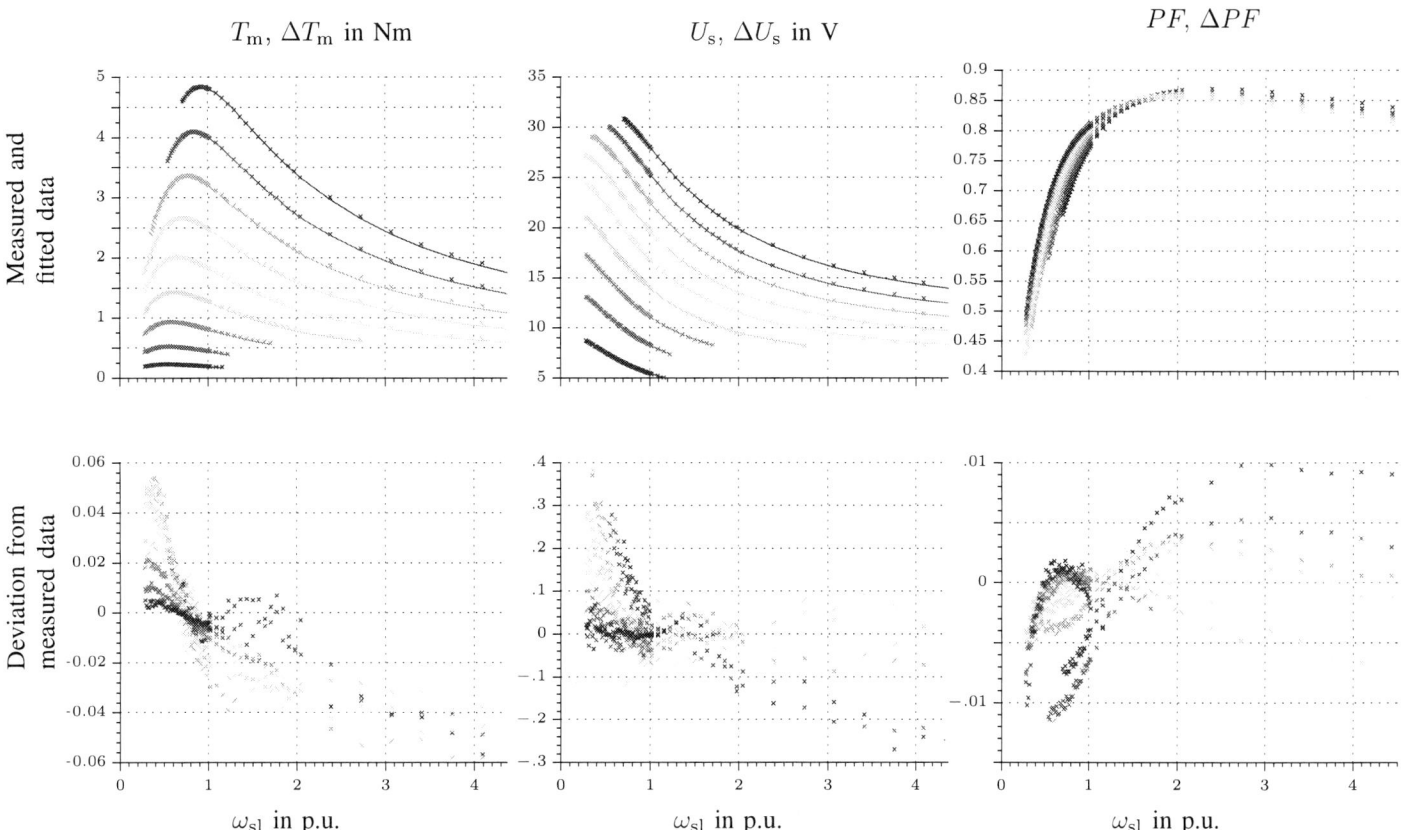

Fig. 9. Curves of torque, stator voltage and power factor versus slip frequency and result of fitting. Solid lines represent values from calculated equivalent circuit.

REFERENCES

[1] De Doncker, R. W.; Pulle, D. W. J.; Veltman, A., "Advanced Electrical Drives: Analysis, Modeling, Control." Springer, 2011.

[2] De Doncker, R. W.; Pulle, D. W. J.; Veltman, A., "Advanced Electrical Drives: Supplementary". Springer, 2011.

[3] R. D. Lorenz and D. W. Novotny, "Saturation effects in field-oriented induction machines," *IEEE Transactions on Industry Applications*, vol. 26, no. 2, pp. 283 289, 1990.

[4] C. P. Steinmetz, "On the law of hysteresis", *Proc. IEEE, vol. 72,* pp. 196221 , Feb. 1984.

[5] N. R. Klaes, "Parameter identification of an induction machine with regard to dependencies on saturation", *IEEE Transactions on Industry Application*, vol. 29, pp. 1135-1140 , Feb. 1993.

[6] Tuovinen, T.; Hinkkanen, M.; Luomi, J., "Modeling of Saturation Due to Main and Leakage Flux Interaction in Induction Machines", *Industry Applications, IEEE Transactions on*, vol.46, no.3, pp.937,945, May-june 2010

[7] Monjo, L.; Corcoles, F.; Pedra, J., "Saturation Effects on Torque- and CurrentSlip Curves of Squirrel-Cage Induction Motors", *Energy Conversion, IEEE Transactions on* , vol. 28, no.1, pp. 243,254, March 2013

[8] Barili, A.; Brambilla, A.; Cottafava, G.; Dallago, E., "A simulation model for the saturable reactor", *Industrial Electronics, IEEE Transactions on*, vol.35, no.2, pp.301,306, May 1988

[9] Muhlethaler, J.; Biela, J.; Kolar, J.W.; Ecklebe, A., "Improved Core-Loss Calculation for Magnetic Components Employed in Power Electronic Systems", *Power Electronics, IEEE Transactions on*, vol.27, no.2, pp.964,973, Feb. 2012

[10] Reinert, J.; Brockmeyer, A.; De Doncker, R. W., "Calculation of losses in ferro- and ferrimagnetic materials based on the modified Steinmetz

equation", *Industry Applications, IEEE Transactions on*, vol.37, no.4, pp.1055,1061, Jul/Aug 2001

[11] Jieli Li; Abdallah, T.; Sullivan, C.R., "Improved calculation of core loss with nonsinusoidal waveforms", Industry Applications Conference, 2001. Thirty-Sixth IAS Annual Meeting. Conference Record of the 2001 IEEE, vol.4, no., pp.2203,2210 vol.4, Sept. 30 2001-Oct. 4 2001

[12] Bastiaensen, C.; Deprez, W.; Symens, W.; Driesen, J., "Parameter Sensitivity and Measurement Uncertainty Propagation in Torque-Estimation Algorithms for Induction Machines", *Instrumentation and Measurement, IEEE Transactions on*, vol.57, no.12, pp.2727,2732, Dec. 2008

[13] Ansuj, S.; Shokooh, F.; Schinzinger, Roland, "Parameter estimation for induction machines based on sensitivity analysis", *Industry Applications, IEEE Transactions on*, vol.25, no.6, pp.1035,1040, Nov/Dec 1989

[14] Sudhoff, S.D.; Aliprantis, D.C.; Kuhn, B.T.; Chapman, P.L., "Experimental characterization procedure for use with an advanced induction machine model", *Energy Conversion, IEEE Transactions on*, vol.18, no.1, pp.48,56, Mar 2003

[15] De Doncker, R. W., "Parameter sensitivity of indirect universal field-oriented controllers", *Power Electronics, IEEE Transactions on*, vol.9, no.4, pp.367,376, Jul 1994

[16] Akaba, M.; Taleb, M.; Rumeli, A., "Improved estimation of induction machine parameters", *Electric Power Systems Research*, vol.34, pp.65-73, 1995

[17] Ranta, M.; Hinkkanen, M., "Online Identification of Parameters Defining the Saturation Characteristics of Induction Machines," *Industry Applications, IEEE Transactions on* , vol.49, no.5, pp.2136,2145, Sept.-Oct. 2013

High Current Planar Transformer for Very High Efficiency Isolated Boost DC-DC Converters

Riccardo Pittini
Technical University of Denmark
Dept. of Electrical Engineering
Oersteds Plads 349
Kgs. Lyngby, Denmark
ripit@elektro.dtu.dk

Zhe Zhang
Technical University of Denmark
Dept. of Electrical Engineering
Oersteds Plads 349
Kgs. Lyngby, Denmark
zz@elektro.dtu.dk

Michael A.E. Andersen
Technical University of Denmark
Dept. of Electrical Engineering
Oersteds Plads 349
Kgs. Lyngby, Denmark
ma@elektro.dtu.dk

Abstract— This paper presents a design and optimization of a high current planar transformer for very high efficiency dc-dc isolated boost converters. The analysis considers different winding arrangements, including very high copper thickness windings. The analysis is focused on the winding ac-resistance and transformer leakage inductance. Design and optimization procedures are validated based on an experimental prototype of a 6 kW dc-dc isolated full bridge boost converter developed on fully planar magnetics. The prototype is rated at 30-80 V 0-80 A on the low voltage side and 700-800 V on the high voltage side with a peak efficiency of 97.8% at 80 V 3.5 kW. Results highlights that thick copper windings can provide good performance at low switching frequencies due to the high transformer filling factor. PCB windings can also provide very high efficiency if stacked in parallel utilizing the transformer winding window in an optimal way.

Keywords— Planar Magnetics, High Current DC-DC Converter, High Efficiency, PCB windings.

I. INTRODUCTION

The major trends in power electronics have been focused on increasing converters efficiency, increasing switching frequencies and power densities, decreasing converter weight and cost. Power semiconductors technology has seen large improvements due to the introduction of new wide bandgap materials and due to the improvements in the conventional silicon technology. However, this is not the case for passive components such as inductors and capacitors. The developments in the magnetic and dielectric materials have been more limited compared to the ones in the power semiconductors. Moreover, passive components represent a significant part in the volume, cost and weight of a power electronic converter.

Planar magnetics [1] have large potential for reducing the magnetic components manufacturing cost, increased component integration and increasing converter power density due to their good thermal performance. Planar transformers are often used for small low power dc-dc converters since the transformer windings are easily integrated on the same PCB with the power semiconductors and the control circuitry [2]. One of the major drawbacks of this type of magnetics comes from its

large stray capacitance which is often unwanted in many commercial applications (e.g. due to increased EMI filter size) [3]. Moreover, for medium power or high current applications it is challenging to achieve high efficiency due to the low transformer window filling factor usually achieved with PCB windings [4].

This paper presents the design approach utilized for a high current planar transformer in a high current high efficiency dc-dc converter. Several winding arrangements are presented and the analysis is focused on winding ac-resistance and on winding leakage inductance. The analysis is validated based on an experimental prototype of a 6 kW dc-dc isolated full bridge boost converter (IFBBC, 30-80 V 0-80 A on the low voltage side and 700-800 V on the high voltage side) achieving a peak efficiency of 97.8 % with power flow from low voltage to high voltage side.

II. PLANAR TRANSFORMER AND PCB TECHNOLOGY

Planar transformers are very suitable for low current applications moreover, creating an interleaved winding structure becomes natural with multilayer PCBs reducing ac-winding resistance and leakage inductance. One of the major challenges in the design of high current PCB windings [5] is the limitation in the PCB manufacturing technology in terms of copper thickness. In order to integrate high currents into PCB windings it is necessary to utilize multilayer PCB and even in this case with conventional copper thicknesses (from 1 oz. to 3 oz. or 35 μm to 105 μm) it is difficult to minimize the winding losses due to the rather low filling factor. After a study of the current production capabilities of several PCB manufacturers it was observed that copper thicknesses up to 6 oz. (210 μm) are easily achievable. Above this value the PCB production cost increases exponentially since it most of the machinery is not capable of handling such high copper thickness.

The following analysis and design are performed on a high current transformer for a dc-dc converter. The analyzed reference cases considers a multilayer PCB (6 layers) with 210 μm of copper thickness and 200 μm of insulation thickness. The multilayer PCB is compared

978-1-4799-2706-7/14 $31.00 © 2014 IEEE

with very high copper thickness custom made PCBs (400 μm and 1500 μm copper bar).

III. DC-DC CONVERTER FOR FUEL CELL APPLICATIONS

The reference design is a 6 kW bidirectional IFBBC for bidirectional solid oxide electrolyzer/fuel cells (SOEC/SOFC) [6]. The converter specifications are defined according to the SOEC/SOFC stack characteristics as specified on Table I. For this topology a complete loss analysis was performed in [7] where it was highlighted how the power loss distribution varies depending on the converter operating point. In the selected topology, the transformer is beneficial since it allows achieving high efficiency with a high converter step-up ratio. However, the transformer design is critical since in the IFBBC the transformer is required to have low leakage inductance in order to minimize the current commutation loop which strongly affects the switching losses on the power semiconductors in the converter low-voltage side [8]. For this reason, in order to achieve high dc-dc conversion efficiency special care has been taken in order to minimize the transformer leakage inductance and ac-resistance.

IV. TRANSFORMER DESIGN AND OPTIMIZATION

Transformer optimization procedures are focused on minimizing the transformer overall losses by minimizing the sum of the winding losses with the core losses [7].

TABLE I
SOFC AND SOEC SPECIFICATIONS FOR DC-DC CONVERTER DESIGN

	SOFC	SOEC	
Low Voltage (LV) side	30-50	50-80	[V]
Current (LV) side	40-0	0-80	[A]
High Voltage (HV) side	700-800	700-800	[V]
Power Rating	~1500	~6000	[W]

However, this paper does not analyze the complete transformer losses but highlights the differences for the analyzed winding topologies. Up-front calculations determined that a good balance between core losses and winding losses is achieved with three primary turns when the converter is operating with a switching frequency of 40 kHz. The analyzed transformer has a turn ratio of 8 defined by the topology, by the converter specifications and by the components ratings. Since the transformer leakage inductance is very critical for the candidate topology the goal was to analyze the trade-offs in the winding arrangements, leakage inductance and winding ac-resistance.

The structures that have been considered for the analysis are presented on Fig. 1: a simple primary-secondary structure (PS) Fig. 1a, a fully interleaved structure (PSPSPS) Fig. 1b, an interleaved structure with very thick windings for primary and secondary (PSPSP) Fig. 1c and a similar structure (PSPSP) only with thinner copper (2*400 μm in parallel, Fig. 1d). This last structure

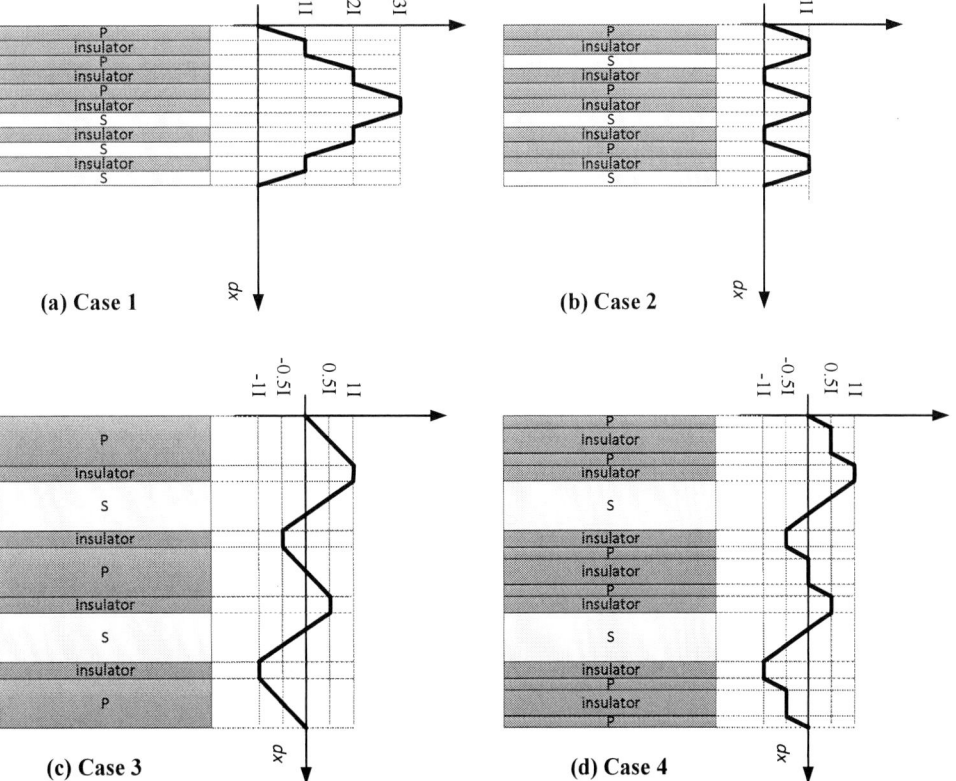

(a) Case 1   (b) Case 2

(c) Case 3   (d) Case 4

Fig. 1 Analysed windings arrangements, Case 1 and Case 2 with PCB copper thickness of 210 μm, Case 3 with very thick copper (1500 μm) and Case 4 with intermediate copper thickness (400 μm two layers in parallel for the primary winding).

aims to highlight the trade-offs between the windings copper thickness and their ac-resistance.

A. Theoretical Background and Analytical Analysis

The four cases are analyzed focusing on the windings ac-resistance and on the leakage inductance for the different configurations. The winding ac-resistance is calculated according to Dowels equations [9]; this takes into account the effect of the skin depth and the proximity effects (1) on the ac-resistance. The proximity effects (2) are included through the magneto-motive force (MMF) of each transformer layer. The totals ac-resistance is computed for each layer of the transformer windings based on (3). The transformer leakage inductance is calculated from the energy stored in the windings and leakage layers (4). This approach considers the MMF varying linearly in the conductor layers and constant in the insulation layers (as presented in [10]), as shown on Fig. 1. In the analyzed cases based on PCB windings the main assumptions are that the primary and secondary windings have the same copper and insulation thicknesses and that the winding window is completely filled.

$$\left(\frac{R_{ac}}{R_{dc}}\right)_{layer} = \frac{\xi}{2}\left[\frac{\sinh(\xi) + \sin(\xi)}{\cosh(\xi) - \cos(\xi)}\right. \\ \left. + (2m-1)^2 \frac{\sinh(\xi) - \sin(\xi)}{\cosh(\xi) + \cos(\xi)}\right] \quad (1)$$

$$m = \frac{MMF(h)}{MMF(h) - MMF(0)} \quad (2)$$

$$R_{ac} = \sum_{layers}\left(\frac{R_{ac}}{R_{dc}}\right)_{layer} \cdot R_{dc,layer} \quad (3)$$

Where $\xi = h/\delta$, h is the conductor height or copper thickness, δ is the skin depth in the conductor and m is >1 for non-interleaved multilayer windings, MMF(h) and MMF(0) are the MMFs values for the analyzed layer. In a similar way the transformer leakage inductance is computed starting from the overall energy stored in the primary and secondary windings and in the

TABLE II
TRANSFORMER LEAKAGE INDUCTANCE

	Leakage	
Case 1	136.8	[nH]
Case 2	19.5	[nH]
Case 3	88.3	[nH]
Case 4	69.1	[nH]

insulation layers, calculated from (4) and (5).

$$E_{leakage} = \frac{1}{2}\mu_0 \int_0^{h_w} H^2 l_w b_w dx \quad (4)$$

$$L_{leak} = \mu_0 \frac{l_w}{b_w} \sum_{i=1}^{n} MMF_{layer}^2 h_{layer} \quad (5)$$

Where H the magnetic field of the layer, h_W is the total winding height, l_W is the transformer mean turn length, b_W is the winding width, MMF and h_{layer} are respectively the magneto motive-force and the height of the considered layer.

Cases three and four require more specific considerations in order to perform a proper analysis. These two cases consider the secondary winding realized with litz wire (two secondary winding sections with 12 turns each) and the primary winding with thick copper foil. The selection of litz wire for the transformer secondary was necessary due to the complexity of realizing a large number of turns with thick copper strands. The litz wire has 30 strands of 0.2 mm in diameter. In this case, different values of ξ has to be used for primary and secondary winding (1). Assuming ideal litz wires (woven and twist), with the ξ coefficeint for the litz wire is given by the litz wire stand diameter over the skin depth (6). This allows to properly calculate the ac-resistance for an ideal litz wire winding.

$$\xi_{litz\ wire} = \frac{d_{litz\ strand}}{\delta} \quad (6)$$

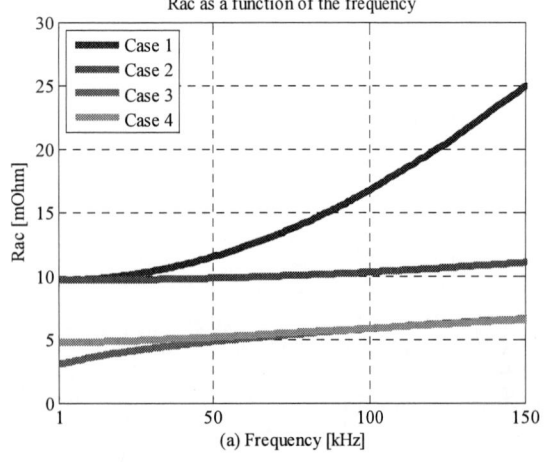
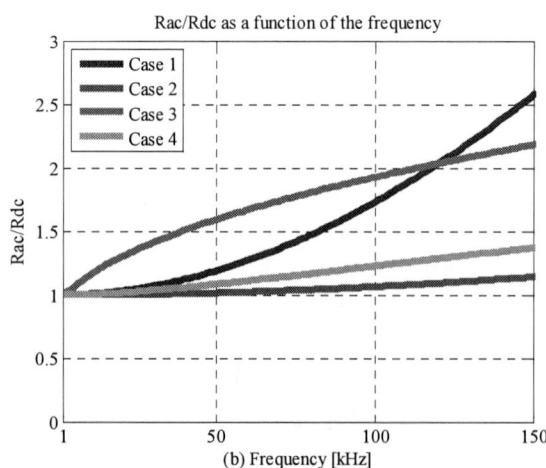

Fig. 2 Results from the analytical calculation of the ac-resistance for the four different cases in the 1-150 kHz range (a) and calculated ac-resistance over dc-resistance for all the analysed cases (b).

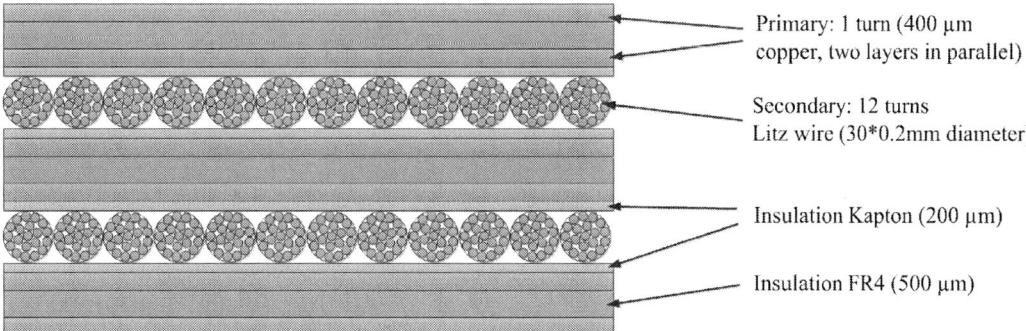

Primary: 1 turn (400 μm copper, two layers in parallel)

Secondary: 12 turns Litz wire (30*0.2mm diameter)

Insulation Kapton (200 μm)

Insulation FR4 (500 μm)

Fig. 3 Winding cross section of the realized transformer prototype according to Case 4, primary winding with thick copper and secondary winding with litz wires.

B. Analysis of the Designs

The cases analyze the stray parameters of different configurations in the 1-150 kHz frequency range; this is the expected operating range of magnetic components in hard-switched multi-kW converters. Results from the analytical analyses of the winding ac-resistance are presented on Fig. 2a and 2b; for the leakage inductance on Table II.

Case 1 and Case 2 (6 layers PCB windings with copper thickness of 200 μm) provide the highest ac-resistance due to the low winding window filling factor in for the planar E64 core pairs. The non-interleaved Case 1 (PS winding structure) suffers of high ac-resistance due to proximity effects, in facts the fully interleaved version Case 2 (PSPSPS winding structure) has minimal variation in ac-resistance over the analyzed spectrum range (Fig. 2a).

Case 3 and Case 4 have similar ac-resistance values even though Case 3 has a significant higher filling factor for the primary winding (a single layer of 1500 μm of copper in Case 3 vs. two layers of 400 μm in Case 4). The difference increases at lower frequencies: ac-resistance of Case 3 is lower than Case 4 due to the increase of the skin depth layer which allows to better utilize the thick copper layers. The increase of ac-resistance over dc-resistance is presented on Fig. 2b. Winding ac-resistance in the extreme thick copper (Case 3) rapidly increases mostly due to the large influence of the skin depth effect. However, even though the frequency increases (e.g. frequencies in the 100-150 kHz range) the thick copper interleaved structures of Case 3 and Case 4 can provide lower ac-resistance than the non-interleaved structure of Case 1.

The structure of Case 2 thanks to the fully interleaving provides the lowest increase in ac-resistance over the analyzed frequency range. However, also Case 4 has a similar gives an increase of ac-resistance close to the fully interleaved Case 2. Therefore, Case 4 represents a good trade-off for achieving a low ac-resistance and minimizing the amount of un-used copper due to due to skin and proximity effects. In fact, the low increase of ac-resistance over the dc-resistance observed in Case 4 is due to the

Transformer leakage inductance is another factor that significantly affects the converter performance. The leakage inductance for the analyzed cases is summarized on Table II. As expected Case 1 has the highest leakage inductance of the analyzed cases due to the large MMF values which give a large contribute in the leakage inductance calculations (5). Case 2 provides the lowers inductance thanks to a full interleaving structure and to the minimal thickness of each layer (both copper and insulation layers). The last two cases have similar leakage inductance however, separated parallel primary winding layers in Case 4 can provide a lower leakage inductance the 1500 μm copper thickness of Case 4.

The most interesting observation can be performed by analyzing the transformer equivalent winding ac-resistance (referred to the primary, Fig. 2a). Even though the windings are realized with thick copper, their ac-resistance at the converter switching frequency (40 kHz) is well below the winding resistance of the Case 1 and Case 2. This is due to the fact that by using a single PCB of 6 layers and with a copper thickness of 6 oz. (210 μm) per layer the transformer filling factor is low and its winding resistance is relatively high compared to the thick copper cases. On the other hand, the PCB windings (6 layers, Case 1 and Case 2) have a finished thickness of ~2.5 mm where the height of the transformer winding window is 10.2 mm (planar E64 pair). This would allow stacking three identical PCB windings (same winding geometry) in order to increase the transformer winding window filling factor and decrease both ac-winding resistance and leakage inductance. With the assumption of equal current distribution in the paralleled layers, it

TABLE III
TRANSFORMER CHARACTERISTICS

Turns ratio	8
Primary	3 turns
Secondary	24 turns
Core	2*E64 pairs
Core material	Magnetics R
Structure	P-S-P-S-P
Primary winding (P)	PCB 2*400 μm
Secondary winding (S)	30*0.2 mm litz wire
Insulation P-S	2*0.1 mm Kapton

(a)

(b)

Fig. 4 Realised transformer prototype according to Case 4 design and specification according to Table III. Primary winding terminations shown on (a) and secondary winding terminations shown on (b).

would result in a reduction by a factor 3 and for both interleaved and non-interleaved case (Case 1 and Case 2). In this theoretical case, the ac-resistance would be reduced below the one achieved in Case 3 and Case 4, as well the leakage inductance.

C. Transformer Prototype Constuction Details and Analysis

In order to validate the performed design transformer was manufactured based on the winding arrangement of Case 4. This case combines the advantages of thick copper foil for the primary windings and minimizes the complexity of the construction by using litz wires for achieving a high number of secondary turns. The design is performed for a converter switching frequency of 40 kHz, as presented in [7].

A detailed cross section of the transformer winding is presented on Fig. 3; the transformer windings have three primary sections interleaved with two secondary sections (PSPSP structure). Each section of the primary winding is realized based on a custom two layer thick copper PCB where each copper layer has a thickness of 400 μm. Each

section of the primary winding has one transformer turn composed by two copper layers in parallel. An insulation layer of FR4 (500 μm) separates the two copper layers (Fig. 3). Each secondary section accommodates 12 secondary turns of 30x0.2 mm litz wire for a total of 24 secondary turns. High voltage insulation between primary and secondary is guaranteed by a 200 μm layer of Kapton. The transformer core is based on two planar E64 core pairs of material R-type from Magnetics. The transformer characteristics and specifications are summarized on Table III.

The realized prototype is presented on Fig. 4a and 4b; where the details of the transformer primary terminations (Fig. 4a) and secondary terminations (Fig. 4b) are highlighted. Case 4 was used as a reference design since it could provide good performance within the analyzed cases and it was possible to realize a custom made laboratory prototype without encountering the manufacturing costs of a single PCB unit. Moreover, calculations in Case 4 were explicitly tuned for the real design case.

A precision impedance analyzer (Agilent 4294A) was

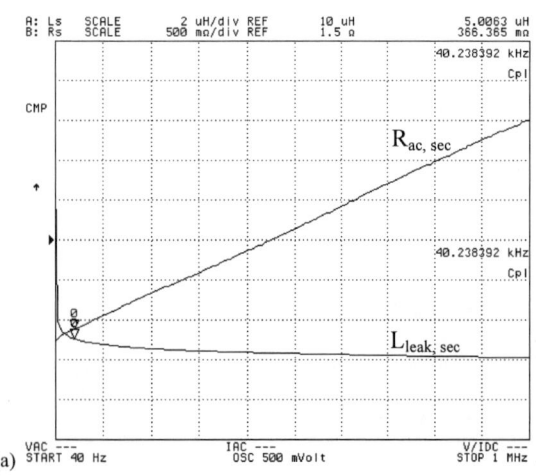
(a)

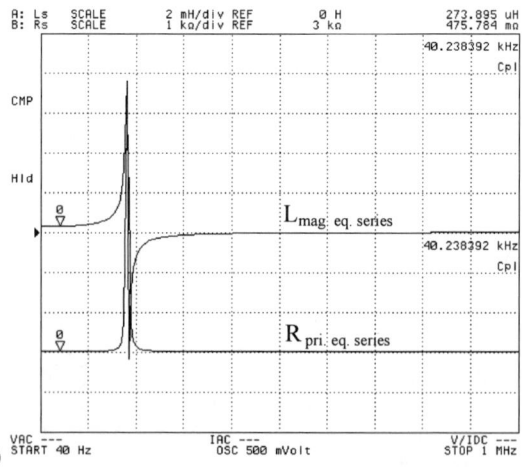
(b)

Fig. 5 Short circuit measurement of the transformer prototype characterized from the secondary side, ac-resistance and leakage inductance at 40 kHz. Open circuit measurement performed from the transformer primary side highlighting the magnetizing inductance at 40 kHz. Transformer characterization in the 40 Hz-1 MHz range. First resonance frequency at 180 kHz.

used to measure the transformer windings resistance and leakage inductance. Transformer parameters have been measured referred to the secondary. This was necessary since the parameters measured from the transformer primary side are very small (in the range of mΩ for the transformer resistance and of nH for the transformer leakage inductance). Measuring the transformer parameters on the secondary side significantly reduced the error due to terminations and allowed the impedance analyzer to operate in a measurement range with better accuracy. In fact, it is possible to observe that the high current screw terminations would introduce several nH of stray inductance.

D. Analysis of the Measurements on the Transformer Prototype

Measurements are performed in short and open circuit conditions in the 40 Hz to 1 MHz band to determine the electrical characteristics of the realized prototype, shown on Fig. 5a and 5b. With the transformer primary shorted by a low stray inductance connection, the total transformer leakage inductance was 5 μH with a total winding resistance of 366 mΩ (measured from the secondary, Fig. 5a). Transformer magnetizing inductance is relatively large compare to the transformer stray parameters, for this reason, the open circuit measurement is performed from the transformer primary side with the transformer secondary side open (Fig. 5b). The transformer magnetizing inductance is 273.9 μH at the converter switching frequency (40 kHz, Fig. 5b). The transformer has a first resonance at a frequency of 180 kHz however, this is not critical case since at this frequency the transformer has high impedance (impedance peak).

The transformer leakage values measured from the secondary side are reflected to the transformer primary side based on the transformer turns ratio Table III obtaining 5.72 mΩ and 78.2 nH. For the realized prototype, the leakage inductance is only 0.0286% of the magnetizing inductance. This value is low compared to regular transformers for dc-dc converters, in fact, leakage inductance is usually 0.1% or more compared to the

magnetizing inductance. This also indicates a very high coupling between primary and secondary of the transformer.

A comparison of the ac-resistance and leakage inductance for the analyzed cases and for the transformer prototype is presented on Fig. 6a and 6b. The calculated ac-resistance at the converter switching frequency is 5.06 mΩ, this leads to an error of about 13%, Fig. 6a. The same error (in this case 13.2%) is observed between the calculated leakage (69.1 nH) inductance and the measured one (78.2 nH), shown on Fig. 6b.

The error observed in the ac-resistance can be explained by analyzing the litz wire structure and the current distribution in the litz wire strands. The calculations for the winding ac-resistance were performed considering a fully transposed litz wire, this is an ideal condition and ensures an optimal current distribution in the litz wire strands (minimum ac-resistance). The transposition of the litz wire strands in the transformer prototype is not ideal; in fact, inner strands are less transposed than the outer strands. This leads to an inhomogeneous current distribution in the litz wire strands and therefore, to a higher ac-resistance. Moreover, ac-resistance calculations for the ideal litz wire neglected the proximity effects of the litz wire inner strands on the outer strands which also increase the ac-resistance.

The error observed in the leakage inductance can be explained by analyzing the construction of the transformer prototype on Fig 4a and 4b. The transformer has not a coil former or a support to contain the windings in a predefined shape. This leads that the windings "relaxed" and filled completely the winding area of the EE64 core pairs. The relaxation led to an increase of the effective secondary winding height (the primary windings were assembled in a glued PCB structure) that gave a slightly larger leakage inductance than the expected calculations. Moreover, also transformer terminations have a large effect when measuring stray inductances in the range of nH.

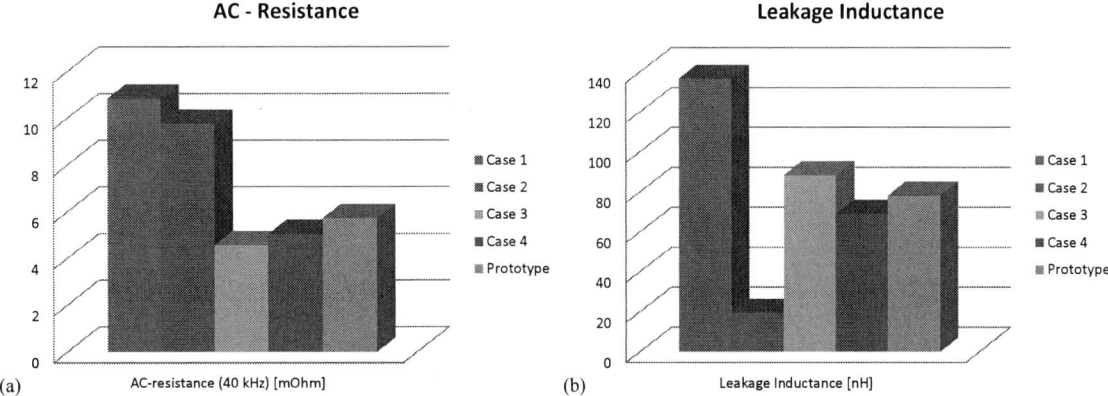

Fig. 6 Picture of the realized converter prototype with highlighted in blue the transformer prototype (a) and complete efficiency characterization with power flow from the low voltage to the high voltage side (b). Darkened area highlights the current limitation area of the converter.

The 2014 International Power Electronics Conference

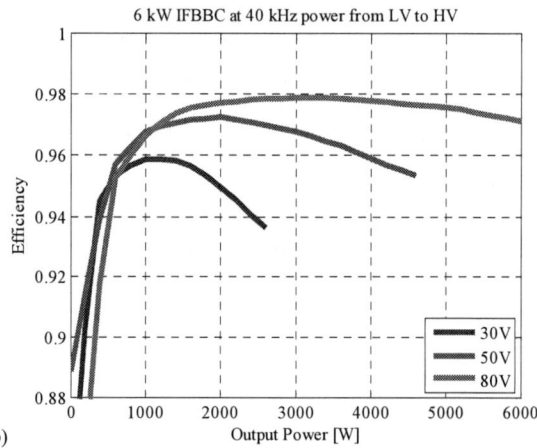

(a)   (b)

Fig. 7 Picture of the realized converter prototype with highlighted in blue the transformer prototype (a) and complete efficiency characterization with power flow from the low voltage to the high voltage side (b). Peak efficiency is 97.8% at the maximum converter input voltage (80 V 40 A). Efficiencies above 97% are already measured at 50 V. Maximum current: 80 A.

V. EXPERIMENTAL RESULTS BASED ON A 6 KW CONVERTER PROTOTYPE

The transformer design and optimization was performed as part of the design process of a high efficiency bidirectional dc-dc converter for fuel cell applications. The realized converter prototype is show Fig. 7a, where it is highlighted the main power transformer. The converter characteristics are defined by the fuel cell characteristics on Table I. The converter low voltage side is characterized by voltage ranged between 30 V up to 80 V with a maximum current of 80 A. The converter was designed for a switching frequency of 40 kHz.

The converter uses 4.1 mΩ MOSFETs on the low voltage side (8 devices arranged two in parallel for each switch in a full bridge) and 1200 V Si IGBTs with anti-parallel 1200 V SiC Schottky diodes on the high voltage side. The boost inductor is also realized with planar KoolMu material (three E6030 core pairs stacked in series) and thick copper PCB windings (custom designed). The transformer prototype was developed with custom made PCBs milled out from a PCB stack for the primary windings (as specified in Case 4). The initial design goal was to achieve a high current planar transformer fully based on PCB windings. However, due to manufacturing complexity and issues encountered during the milling process, the converter secondary windings were realized with litz wires, Table III. Therefore, the analytical analysis has to be updated based on these new constrains.

The transformer leakage inductance has been a major concern since in the selected topology it directly limits the primary MOSFETs current commutation time and therefore, their switching losses [7]. A complete loss analysis of the converter was presented in [7].

On Fig. 7b is presented a complete efficiency characterization of the converter. The design resulted in a dc-dc converter prototype with a peak efficiency of 80 V at ~3.5 kW of 97.8%. At the lowest input voltage (30 V) the peak efficiency was limited to 95.5%. Peak efficiencies are always measured when the current on the converter low voltage side is in the range of 40 A,

The converter is designed for bidirectional power flow however, efficiencies with power flow from the high voltage side to the low voltage side were lower than the ones measured with opposite power flow. The lower efficiency observed in this operating mode was due to the large switching losses of the high voltage power semiconductors (IGBTs). In this other operating mode the maximum measured efficiency was 96.5% at 80 V and 93.4% at 30 V. In this operating mode, a significant improvement in efficiency can be achieved by introducing SiC power semiconductors as replacement for Si IGBTs.

It is interesting to observe the influence of the transformer leakage inductance on the overall converter losses. Based on the four analyzed cases, the converter losses determined by the leakage inductance are calculated at different current levels (40 A, 60 A and 80 A); these are presented on Fig. 8a. It is observed that large leakage inductance values as in Case 1 would introduce large converter losses especially at high current levels (quadratic dependency, as observed in Fig. 8a and as presented in [7]). Case 2, thanks to full interleaving and to the low thickness of each layer, introduces very minimum additional losses in the converter. The realized transformer prototype based on Case 4 introduces an acceptable amount of losses for a converter with target efficiency around 98 %. The leakage inductance introduces a significant amount of losses only at high current levels, as shown on Fig. 8a. The leakage inductance in the transformer causes short avalanche transients of the MOSFETs on the converter low voltage side, as shown on Fig. 8b. The duration of the short avalanche condition depends on the input current level that determines the time interval required to commutate the input current through the transformer primary side windings (low voltage side). In this case, the avalanche interval with an input current 60 A is around 50-55 ns, as shown on Fig. 8b.

978-1-4799-2706-7/14 $31.00 © 2014 IEEE   3911

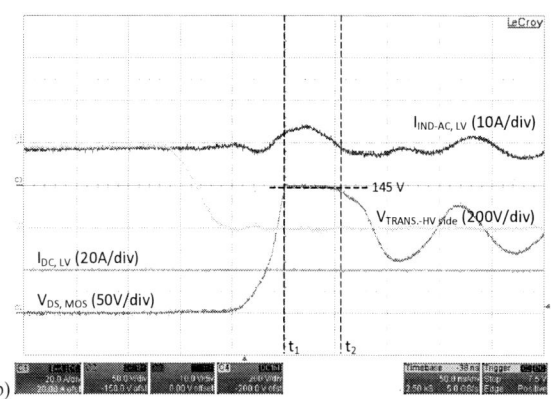

Fig. 8 Converter power loss due to transformer leakage inductance for the different analysed cases and at different current levels, shown on (a). Avalanche interval at 60 A current for the low voltage side MOSFETs, shown on (b). Low voltage side DC-current (Ch.1: 20 A/div), drain to source voltage of the low voltage side MOSFET (Ch.2: 50 V/div), boost inductor AC-current (Ch.3: 10 A/div) and high voltage side transformer voltage (Ch.4: 200 V/div).

VI. CONCLUSIONS

A design and analysis of a high current transformer based on planar magnetics has been presented. The analysis considered different windings structures and different copper thicknesses from 210 µm up to 1500 µm. Results highlighted that the larger copper thicknesses were preferable for achieving low ac-resistances at frequencies ranges around 40 kHz since this type of winding could achieve high filling factor. Cooper foil or thin copper bars become beneficial in transformers for low voltage high current converters especially at frequencies below 100 kHz.

Simple PCB windings struggle to provide the desired transformer performance in high current high efficiency dc-dc converters. However, state of the art PCB technology with copper thicknesses up to 6 oz. (210 µm) and allowing paralleling of multiple PCBs (e.g. three PCBs stacked on top of each other) would also allow to achieve high filling factor and high flexibility for interleaving. This would reduce both ac-resistance and leakage inductance down to levels that are only achievable with copper foil.

A transformer prototype was developed based on thick copper (400 µm) windings. The prototype was utilized in a dc-dc isolated boost converter achieving a peak dc-dc conversion efficiency of 97.8% at the highest input voltage and a peak efficiency of 95.5% at the lowest input voltage.

REFERENCES

[1] Quirke, M.T.; Barrett, J.J.; Hayes, M., "Planar magnetic component technology-a review," IEEE Transactions on Components, Hybrids, and Manufacturing Technology, vol.15, no.5, pp.884,892, Oct 1992.

[2] Ziwei Ouyang; Sen, G.; Thomsen, O.C.; Andersen, M. A E, "Analysis and Design of Fully Integrated Planar Magnetics for Primary–Parallel Isolated Boost Converter," Industrial Electronics, IEEE Transactions on , vol.60, no.2, pp.494,508, Feb. 2013.

[3] Pahlevaninezhad, M.; Das, P.; Djilali, H.; Jain, P.; Bakhshai, A.; Moschopoulos, G., "A novel planar transformer with integrated EMI suppression Y-caps applicable to high frequency high power

DC-DC converters for automotive applications," Twenty-Seventh Annual IEEE Applied Power Electronics Conference and Exposition (APEC 2012), pp.346,351, 5-9 Feb. 2012.

[4] Gunewardena, T.R., "Manufacturing design considerations for planar magnetics," Electrical Insulation and Electrical Manufacturing & Coil Winding Conference Proceedings, 1997,pp.309,311, 22-25 Sep 1997.

[5] Pittini, Riccardo; Zhe Zhang; Ouyang, Ziwei; Andersen, Michael A. E.; Thomsen, Ole C.; "Analysis of planar E+I and ER+I transformers for low-voltage high-current DC/DC converters with focus on winding losses and leakage inductance," 7th International Power Electronics and Motion Control Conference (IPEMC 2012), vol.1, pp.488-493, 2-5 June 2012

[6] Zhang, Zhe; Pittini, Riccardo; Andersen, Michael A.E.; Thomsen, Ole C.; "A Review and Design of Power Electronics Converters for Fuel Cell Hybrid System Applications", Energy Procedia 2012, Volume 20, pp. 301-310.

[7] Pittini, Riccardo; Zhe Zhang; Andersen, Michael A. E.; " Isolated Full Bridge Boost DC-DC Converter Designed for Bidirectional Operation of Fuel Cells/Electrolyzer Cells in Grid-Tie Applications" 15th European Conference on Power Electronics and Applications (EPE-ECCE Europe 2013), 3-5 September 2013.

[8] Nymand, M.; Andersen, M.A.E.; "High-Efficiency Isolated Boost DC–DC Converter for High-Power Low-Voltage Fuel-Cell Applications," IEEE Transactions on Industrial Electronics, vol.57, no.2, pp.505-514, Feb. 2010.

[9] P. L. Dowell, "Effects of eddy currents in transformer windings," Proceedings of IEEE, vol.113, no.8, August 1966, pp. 1387-1394.

[10] Ziwei Ouyang; Thomsen, O.C.; Andersen, M.A.E.; "Optimal Design and Tradeoff Analysis of Planar Transformer in High-Power DC–DC Converters," IEEE Transactions on Industrial Electronics, vol.59, no.7, pp.2800-2810, July 2012.

The 2014 International Power Electronics Conference

High Voltage-Gain Interleaved Boost DC-DC Converter Discarded Electrolytic Capacitor

Quang Trong Nha, Huang-Jen Chiu, Yu-Kang Lo, Pham Phu Hieu
Department of Electronic and Computer Engineering
National Taiwan University of Science and Technology
Taipei, Taiwan, (R.O.C)

Abstract— **This study presents a new topology for alternative power sources, namely interleaved boost converter with current ripple reduction (IBC-CRR), which discards electrolytic capacitors to enlarge the life-span of the converter by using current ripple reduction (CRR) technique. The voltage spikes on switches are eliminated resulting of using of low on-resistance ($R_{DS(ON)}$) MOSFETs to improve efficiency and reducing electromagnetic interference (EMI). Moreover, the proposed topology also provides a high voltage-gain and a continuous conduction mode (CCM) input current operation. These features are suitable for alternative power sources such as PV and fuel-cell. Addition to providing circuit analysis and detailed consideration, two laboratory prototypes of the proposed topology and conventional topology operated under 36-72 V input, 400 V/400 W output were built and inspected to compare their performances. The experimental results shown that the proposed circuit had following superiorities compared with the conventional circuits. A small output voltage ripple was attained with 4.7 µF non-electrolytic capacitors. Because the voltage spikes on MOSFETs were completely eliminated, low $R_{DS(on)}$ MOSFETs were used to improved efficiency. The input current was CCM operation requiring smaller input filtering.**

Keywords— *Current ripple reduction, interleaved boost converter, high voltage-gain converter, non-electrolytic capacitor.*

I. INTRODUCTION

In recent years, the alternative power sources have been commonly used to avoid environmental pollution. Because most of alternative power sources features low DC output voltage. This system requires a high voltage-gain step-up converter to convert to a high DC output voltage. Recently, fuel cell (FC) and photovoltaic (PV) are considered to be two most potential candidates of the alternative energy sources. However, they have several inherent limitations such as: wide output voltage variation, low output voltage, slow load dynamic transience response, and low efficiency under high current-ripple operation [1-3]. Due to having wide-range and low output voltage, the alternative energy sources require high voltage-gain step-up converters. Voltage-fed DC-DC converters with high turn ratio of transformer were presented [4-6] for the alternative energy sources. Those converters suffer from high voltage spike on switches due to presentence of leakage inductance and parasitic capacitance on the power transformer [7-9]. Moreover, the voltage-fed DC-DC

converter characterizes a pulsating input current which harms the alternative power sources and induces electromagnetic interference (EMI) [10-12]. Large electrolytic capacitors are used for pulsating input current filtering. The electrolytic capacitor has several shortcomings such as: large size, poor reliability, large equivalent series resistance (ESR), and limited temperature rating [13-15]. Consequently, converters featuring high voltage-gain, CCM input current operation, and non-electrolytic capacitor are highly recommended for alternative power sources [16, 17].

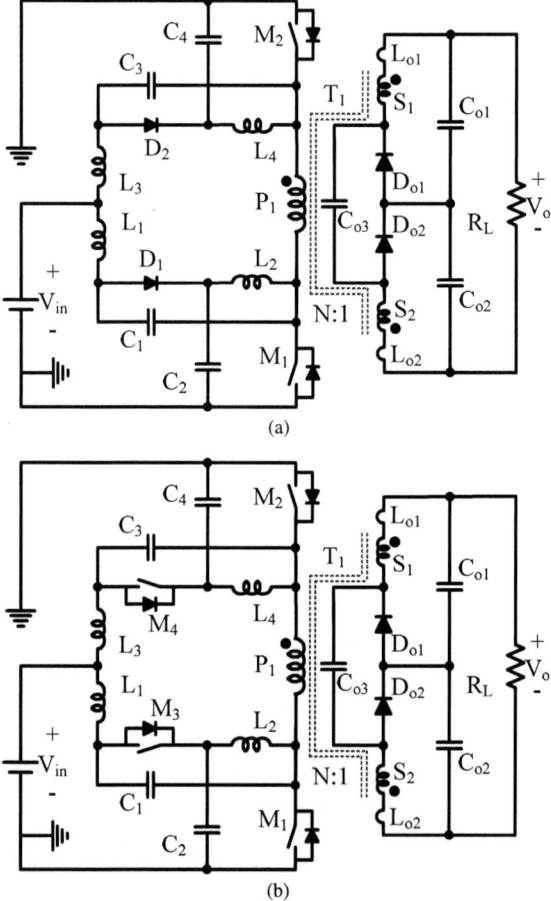

Fig. 1. Proposed converter (a) using diodes and (b) diodes replaced by MOSFETs

978-1-4799-2706-7/14 $31.00 © 2014 IEEE

The 2014 International Power Electronics Conference

Several current-fed converters with voltage-doubled rectifier (VDR), full-wave rectifier (FWR), or center-tapped rectifier (CTR) have been proposed to provide a high voltage-gain and eliminate the usage of electrolytic capacitors for input filtering due to having a CCM input current [18-20]. However, rectification circuit and filter stages are also needed on output side. Employing center-tapped rectifier is an example. High voltage spikes on rectified diodes and high current ripple on output capacitors are occurred. High-cost, high-voltage rating rectified diodes and large electrolytic capacitors still need to use for the output rectification. To discard the usage of electrolytic capacitor, some novel current-fed converters with output current ripple reduction (CRR) have been presented in studies [2, 16, 17]. These configurations feature low parasitic components, CCM input current, voltage-spike-free on the rectified diodes, and non-electrolytic capacitor. However, topologies presented in [2, 17] suffer from high voltage spikes on MOSFETs. High-voltage rating MOSFETs and snubber circuits require. As a result, the overall efficiency will be suffered. In [16], the voltage spikes on MOSFETs have been eliminated by using active clamping method. However, two additional MOSFETs and isolated-driving circuits and signals are required leading to high cost and complexity in circuit design.

Applying CRR approach for output rectification, a high voltage-gain interleaved boost converter with current ripple reduction (IBC-CRR) is presented for alternative power systems. Fig. 1(a) shows circuit diagram of the proposed topology by using diodes D_1 and D_2 on primary side. To improve the efficiency, two diodes (D_1 and D_2) are replaced by two MOSFETs (M_3 and M_4) as shown in Fig. 1(b). Several features, such as high voltage-gain, non-electrolytic capacitor, voltage-spike-free on MOSFETs, and low input current ripple are obtained. In this study, in addition to principle of operation, theoretical analysis, and design considerations, two laboratory prototypes of proposed topology and dual-inductor boost converter with voltage-doubled rectifier (DIBC-VDR) with 36-72 V input, 400 V/400 W output, and 50 kHz switching frequency were built and realized to compare their output-voltage ripple and efficiency performance.

II. OPERATION AND CIRCUIT ANALYSIS

A. Priciple of operation

To simplify circuit analysis, boost inductors (L_1 to L_4) are designed to operate in CCM with small current ripple. The inductor currents (i_{L1} to i_{L4}) are treated to be constant values (I_{L1} to I_{L4}). Due to having similar in circuit operation, only Fig. 1(a) is analyzed in this study. The clamping capacitors (C_1 to C_4) and magnetizing inductance of transformer are assumed to be sufficiently large. Therefore, capacitor voltages (V_{C1} to V_{C4}) are treated to be constant within one switching period. The equivalent circuits corresponding to [t_0-t_1] and [t_1-t_2] intervals are illustrated in Fig. 2(a-b). Some key waveforms are shown in Fig. 3. The time intervals are related to duty cycle D by t_1-t_0=t_3-t_2=DT and t_2-t_1=t_4-t_3=(1/2-D)T. Due to having similar operation, only two time intervals [t_0-t_1] and [t_1-t_2] are analyzed as below:

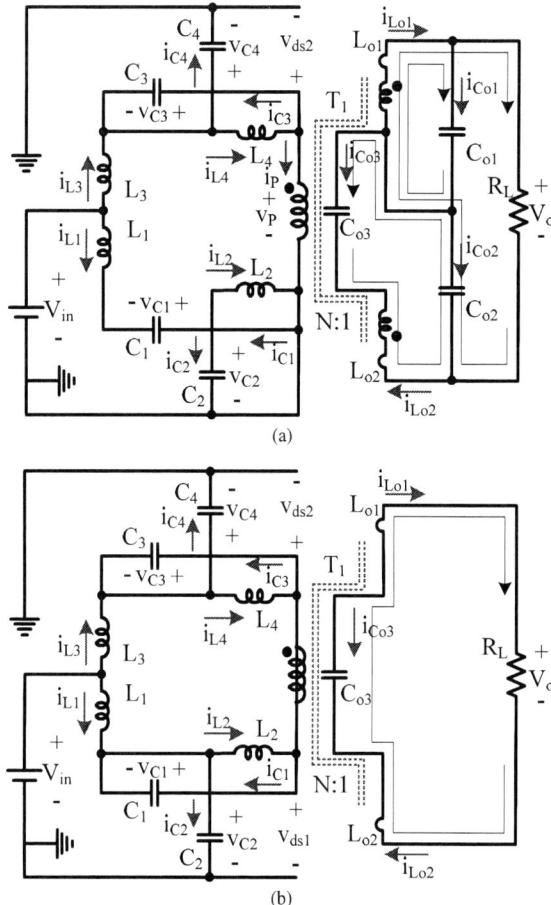

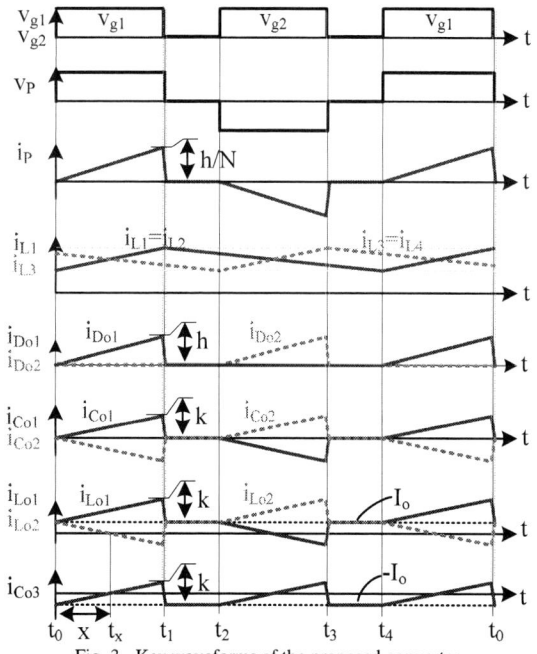

Fig. 2. Equivalent circuits of the proposed converter within (a) [t_0-t_1] and (b) [t_1-t_2] intervals

Fig. 3. Key waveforms of the proposed converter

(a) Stage 1 [$t_0 - t_1$]:

During [t_0-t_1] interval, as indicated in Fig. 2(a), MOSFET M_1 is turned on. MOSFET M_2 is turned off. Diode D_1 turns off. Diode D_2 turns on. Output rectified diode D_{o1} turns on. Diode D_{o2} turns off. Its voltage is clamped by clamping capacitor C_{o3}. The energy is transferred to output via transformer. The output voltage is reflected to the primary winding by amount of $NV_o/2$. This amount is also applied on MOSFET M_2. Moreover, the voltage on this MOSFET is also clamped by capacitors C_3 and C_4 and is calculated by following equation:

$$v_{ds2} = V_{C3} + V_{C4} = NV_o/2 \tag{1}$$

The currents of inductors (L_3 and L_4) linearly decrease by V_{in}, V_{C3}, and V_{C4}. Inductor voltages (v_{L3} and v_{L4}) are calculated as follows:

$$v_{L3} = V_{in} - V_{C4} \tag{2}$$

$$v_{L4} = -V_{C3} \tag{3}$$

Inductor currents (i_{L1} and i_{L2}) linearly increase by V_{in}, V_{C1}, and V_{C2}. Inductor voltages relate to below equations

$$v_{L1} = V_{in} + V_{C1} \tag{4}$$

$$v_{L2} = V_{C2} \tag{5}$$

Clamping current (i_{Co3}) and primary current (I_{Lo2}) relate below

$$i_{Co3} = -i_{Lo2} \tag{6}$$

During [t_0-t_1] interval, the output rectified diode D_{o1} turns on to conduct secondary-winding current (i_{Lo1}) and clamping capacitor current (i_{Co3}). The output rectified diode current (i_{Do1}) linearly increases and reaches peak current (h) at t_1. Therefore, following relationship is found:

$$i_{Do1} = i_{Lo1} + i_{Co3} = i_{Lo1} - i_{Lo2} = \frac{h}{DT}t \tag{7}$$

For transformer T_1, relationship between primary current (i_P) and secondary currents (i_{Lo1} and i_{Lo2}) is described below

$$i_P = \frac{1}{N}(i_{Lo1} - i_{Lo2}) = \frac{1}{N}\frac{h}{DT}t \tag{8}$$

During [t_0-t_1] interval, the output capacitor currents (i_{Co1} and i_{Co2}) are calculated as follows:

$$i_{Co1} = -i_{Co2} = \frac{k}{DT}t \tag{9}$$

The output capacitor current i_{Co1} crosses output current I_o at t_x. [t_x-t_0] time interval is calculated by following equation:

$$t_x - t_0 = x = \frac{I_o}{k}DT \tag{10}$$

Apply Kirchhoff's current law (KCL) for positive output voltage and negative output voltage intersections, the primary-winding currents (i_{Lo1} and i_{Lo2}), output capacitor currents (i_{Co1} and i_{Co2}), and output current (I_o) are related as follows:

$$i_{Lo1} = i_{Co1} + I_o \tag{11}$$

$$i_{Lo2} = i_{Co2} + I_o \tag{12}$$

(b) Stage 2 [$t_1 - t_2$]:

During [t_1-t_2] interval, as shown in Fig. 2(b), MOSFET M_1 is turned off at t_1. MOSFET M_2 keeps turning off. Primary-side diodes D_1 and D_2 turn on. Diodes D_{o1} and D_{o2} turn off. During this time interval, the energy is not transferred to output. Therefore, the primary winding current i_P is zero. Clamping capacitors (C_1-C_4) are discharged via inductors (L_1-L_4). Therefore, clamping capacitor current is equal to inductor current on their conduction path (i_{C1}=I_{L2}, i_{C2}=I_{L1}, i_{C3}=I_{L4}, i_{C4}=I_{L3}). The voltage of L_3 and L_4 are calculated as (2) and (3) respectively. The voltage on MOSFET M_2 is also clamped by capacitors C_3 and C_4 and calculated by (1). Moreover, the voltage on MOSFET M_1 is clamped by capacitors C_1 and C_2 and calculated by following equation:

$$v_{ds1} = V_{C1} + V_{C2} = NV_o/2 \tag{13}$$

The currents of L_1 and L_2 are linearly decreased by V_{in}, V_{C1}, and V_{C2}. Their voltages are calculated by:

$$v_{L1} = V_{in} - V_{C2} \tag{14}$$

$$v_{L2} = -V_{C1} \tag{15}$$

During this time interval, output capacitor currents (i_{Co1} and i_{Co2}) are equal to zero. The clamping capacitor (i_{Co3}) is connected to output capacitors (C_{o1} and C_{o2}) via two leakage inductances (L_{o1} and L_{o2}). The clamping capacitor voltage (V_{Co3}) is calculated as follows:

$$V_{Co3} = V_{Co1} + V_{Co2} = V_o \tag{16}$$

The clamping capacitor (C_{o3}) transfers its energy to the load via leakage inductances (L_{o1} and L_{o2}). Therefore, clamping capacitor current (i_{Co3}), leakage inductance currents (i_{Lo1} and i_{Lo2}), and output current (I_o) are related as:

$$i_{Lo1} = i_{Lo2} = -i_{Co3} = I_o \tag{17}$$

Similarly, during [t_2-t_3] interval, MOSFET M_2 is turned on. Secondary-side diode D_{o2} turns on. The energy is transferred from the input source to output by transformer. The output voltage is reflected to primary winding by -$NV_o/2$ amount. During [t_3-t_4] interval, the energy is not transferred to output. The output current (I_o) is maintained by clamping capacitor C_{o3}. Consequently, the energy stored in the leakage inductances is absorbed by capacitors (C_1-C_4). Moreover, on secondary side, the voltage spikes on secondary-side diodes are clamped by clamping capacitor C_{o3}. Consequently, low voltage MOSFETs and low voltage diodes are able to use to improve conversion efficiency.

The output current I_o is equal to average value of diode currents (i_{Do1} and i_{Do2}). Therefore, peak value of secondary-side diode current is calculated as follows:

$$h = \frac{2I_o}{D} \tag{18}$$

Appling amp-second balance (or charge balance) to clamping capacitor C_{o3}, peak value of output capacitor current (k) is calculated as follows:

$$k = \frac{I_o}{D} \tag{19}$$

Appling volt-second balance to inductors (L_1 and L_2), we have

$$\left(V_{in} + V_{C1}\right)DT + \left(V_{in} - V_{C2}\right)\left(1 - D\right)T = 0 \tag{20}$$

$$V_{C2}DT + \left(-V_{C1}\right)\left(1 - D\right)T = 0 \tag{21}$$

From (20) and (21), the capacitor voltages (V_{C1} to V_{C4}) are calculated as follows:

$$V_{C1} = V_{C3} = V_{in}\frac{D}{1 - 2D} \tag{22}$$

$$V_{C2} = V_{C4} = V_{in}\frac{1 - D}{1 - 2D} \tag{23}$$

From (13), the static voltage-gain is calculated by below equation:

$$\frac{V_o}{V_{in}} = \frac{2}{N}\frac{1}{1 - 2D} \tag{24}$$

From (24), the proposed converter provides a high voltage-gain being suitable for low input voltage, high output voltage applications. The duty cycle is limited at 0.5.

B. Output voltage ripple derivation

In addition to having high voltage-gain, the CRR technique is applied on secondary side to reduce the output voltage ripple. Therefore, electrolytic capacitors are discarded. The circuit principle of operation has been analyzed in section A. In this section, the output voltage ripple is analyzed. Some key waveforms of the proposed topology are shown in Fig. 4 under $L_{o1} > L_{o2}$ assuming. During Δt interval, the secondary winding currents (i_{L1} and i_{L2}) are linearly reduced from their peak value to I_o. The secondary winding current ripples (ΔI_{L1} and ΔI_{L2}) are calculated as follows:

$$\begin{cases} \Delta I_{Lo1} = \dfrac{V_o}{2L_{o1}}\Delta t \\[2mm] \Delta I_{Lo2} = \dfrac{V_o}{2L_{o2}}\Delta t \end{cases} \tag{25}$$

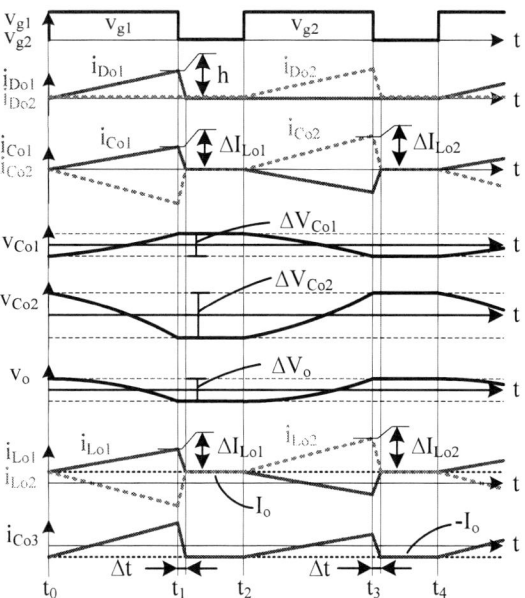

Fig. 4. Key waveforms of the proposed converter with $L_{o1} > L_{o2}$

As mentioned earlier, the output rectified diode D_{o1} turns on during [t_0-t_1] interval to conduct secondary winding currents (i_{Lo1} and i_{Lo2}). The output rectified diode current (i_{Do1}) linearly increases and reaches peak current h at t_1. From (7), the output rectified diode current (i_{Do1}) at t_1 is calculated below:

$$\left(I_o + \Delta I_{Lo1}\right) - \left(I_o - \Delta I_{Lo2}\right) = h \tag{26}$$

Replacing ΔI_{Lo1} and ΔI_{Lo2} from (25) into (26), the result will be found below:

$$\frac{V_o}{2}\left(\frac{1}{L_{o1}} + \frac{1}{L_{o2}}\right)\cdot\Delta t = h \tag{27}$$

By eliminating Δt from (25) and (27), the primary winding current ripples can be obtained as follows:

$$\begin{cases} \Delta I_{Lo1} = \dfrac{L_{o2}}{L_{o1} + L_{o2}}\cdot h \\[2mm] \Delta I_{Lo2} = \dfrac{L_{o1}}{L_{o1} + L_{o2}}\cdot h \end{cases} \tag{28}$$

The output capacitor voltages (v_{Co1} and v_{Co2}) are determined by integrating their current with respect to time. Considering [t_0-t_1] interval, the output capacitor voltages (v_{Co1} and v_{Co2}) are expressed below:

$$v_{Co1}(t) = v_{Co1}(t_0) + \frac{\Delta I_{Lo1}}{2C_{o1}}\frac{D}{T}t^2 \tag{29}$$

$$v_{Co2}(t) = v_{Co2}(t_0) - \frac{\Delta I_{Lo2}}{2C_{o2}}\frac{D}{T}t^2 \tag{30}$$

The voltage-ripple on output capacitor is calculated below

$$\begin{cases} \Delta V_{Co1} = \dfrac{\Delta I_{Lo1}}{2C_{o1}} DT \\[3mm] \Delta V_{Co2} = \dfrac{\Delta I_{Lo2}}{2C_{o2}} DT \end{cases} \qquad (31)$$

As indicated in Fig. 4, the output voltage (V_o) is cancelled by output capacitor voltages (V_{Co1} and V_{Co1}). The output voltage ripple is calculated by below equation:

$$\begin{aligned} \Delta V_o &= \left| \Delta V_{Co1} - \Delta V_{Co2} \right| \\[2mm] &= I_o \left(\frac{1}{L_{o1}+L_{o2}} \right) \left| \frac{L_{o2}}{C_{o1}} - \frac{L_{o1}}{C_{o2}} \right| T \end{aligned} \qquad (32)$$

C. Circuit consideration

The turn ratio of power transformer is chosen by (24) under 36 V input, 400 V output, and 0.41 duty cycle.

The output capacitors (C_{o1} and C_{o2}) are selected by (32). As seen, two small output capacitors can be used to achieve a desire output voltage ripple. However, to avoid resonance caused by output capacitances and leakage inductances, two 4.7 µF film capacitors are chosen in this study.

The input inductances (L_1-L_4) are chosen so that the input current is operated in CCM under high-line, light-load condition. Because the proposed converter characterizes an interleaved boost converter, each inductor currents (I_{L1} to I_{L4}) is therefore ideally equal to one-half average value of the input current. During [t_0-t_1] interval, the boost inductors (L_1-L_4) are chosen by following condition:

$$L_1 = \ldots = L_4 \geq \frac{2V_{in}}{V_o I_o} V_{C2} DT \qquad (33)$$

where V_{C2} is found in (23). The inductances have to be chosen under high-line light-load condition to ensure CCM operation. As diodes (D_1 and D_2) are used for the proposed converter as shown in Fig. 1(a), the inductor currents are zero under DCM operation. However, as these diodes are replaced by MOSFETs (M_3 and M_4), the inductor currents (i_{L1} to i_{L4}) will be continuous and will be negative under DCM operation.

The clamping capacitors (C_1-C_4) should be properly chosen to eliminate the voltage spikes on MOSFETs. By 5% voltage ripple is chosen under low-line, full-load condition. During on-time interval of MOSFET M_1, capacitor voltages (V_{C1} and V_{C2}) are found by (22) and (23). The clamping capacitors are chosen by below conditions:

$$C_1 = C_3 \geq \frac{DT}{0.05V_{C1}} \frac{V_o I_o}{2V_{in}} \qquad (34)$$

$$C_2 = C_4 \geq \frac{DT}{0.05V_{C2}} \frac{V_o I_o}{2V_{in}} \qquad (35)$$

To simplify the circuit consideration, the secondary-winding leakage inductances (L_{o1} and L_{o2}) are assumed to be

identical. Voltage ripple of the clamping capacitor (C_{o3}) is calculated by below equation:

$$\Delta V_{Co3} = \frac{I_o}{C_{o3}} \left(\frac{1}{2} - D \right) T + \frac{k}{C_{o3}} \frac{1}{DT} \frac{x^2}{2} \qquad (36)$$

where x and k values are calculated in (10) and (19), respectively. The voltage ripple of the clamping capacitor C_{o3} is chosen to be equal to 5% of output voltage under low-line, full-load condition. Therefore, the clamping capacitor is calculated by following condition:

$$C_{o3} \geq \frac{I_o}{0.05V_o} \left(\frac{1}{2} - D - \frac{D^2}{2} \right) T \qquad (37)$$

III. EXPERIMENTAL RESULTS

In order to verify the circuit operation and evaluate the performance of the proposed topology, two laboratory prototypes of proposed circuit and dual-inductor boost converter with voltage-doubled rectifier (DIBC-VDR) have been built and inspected to meet following design specifications: 36-72 V input, 400 V/400 W output, and 50 kHz switching frequency. The specifications and main parameters of the designs are shown in TABLE I. To compare the output voltage ripple, 4.7 µF non-electrolytic capacitors were used for the output filtering.

To inspect voltage stress on MOSFETs of dual-inductor boost converter with voltage-doubled rectifier (DIBC-VDR), its experimental waveforms were measured under 36 V input, 400 V/0.6 A output, and 50 kHz switching frequency. As shown in Fig. 5, even only 60% load and 600 V MOSFETs were used for circuit implementation, the voltage spike on MOSFET M_1 (CH3 - v_{ds1}) was 540 V. It is recommended to use snubber circuits to absorb the energy stored in the leakage inductances. As shown in Fig. 6, the experimental results were measured under 36 V input, 400 V/1 A output when resistor-capacitor-diode (R-C-D) snubber circuits were used to eliminate voltage spikes on MOSFETs. As seen, the output voltage ripple (CH4 - v_o) was 4.5 V. Even R-C-D snubber circuits were used; the voltage spikes on MOSFETs were still high (500 V under full-load condition). It is recommended that galvanic isolation between winding layers of transformer and inductors are inserted to avoid arcing.

TABLE I
SPECIFICATIONS AND MAIN PARAMETERS OF THE PROPOSED CIRCUIT

Specifications and Parameters	Proposed converter	DIBC-VDR
Input voltage (V_{in})	36-72 V	
Output (V_o/P_o)	400 V/400 W	
Switching frequency (f_{sw})	50 kHz	
MOSFETs (M_1-M_4)	IXTQ88N28T	45N60C3
Diodes (D_1-D_2)	STTH6004	Not applicable
Transformers (T_1)	Core: EER40 Turn ratio: 32:32	Core: EER40 Turn ratio: 32:32
Capacitors (C_1-C_4)	8.2 µF	Not applicable
Inductors (L_1-L_4)	300 µH	300 µH
Diodes (D_{o1}-D_{o2})	F20L60	
Capacitors (C_{o1}, C_{o2})	4.7 µF	
Capacitor (C_{o3})	1.5 µF	Not applicable

The 2014 International Power Electronics Conference

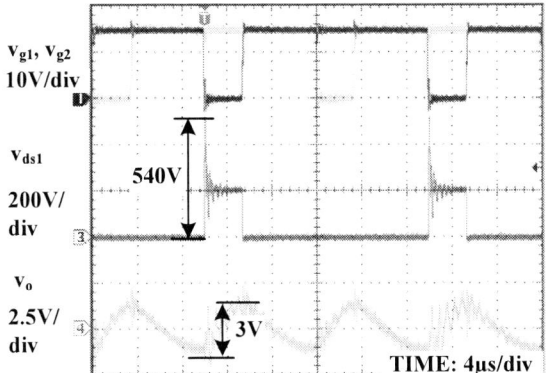

Fig. 5. Experimental waveforms of DIBC-VDR without snubber circuit under 36 V input, 400 V/0.6 A output.

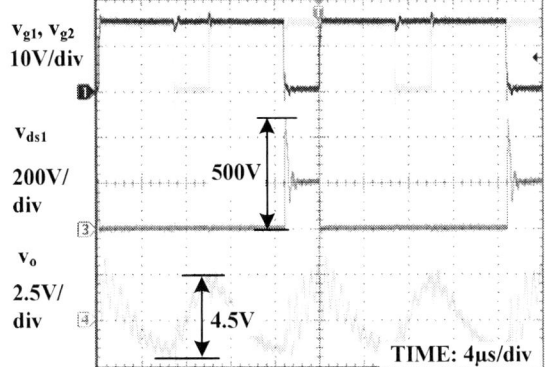

Fig. 6. Experimental waveforms of DIBC-VDR with R-C-D snubber circuit under 36 V input, 400 V/1 A output.

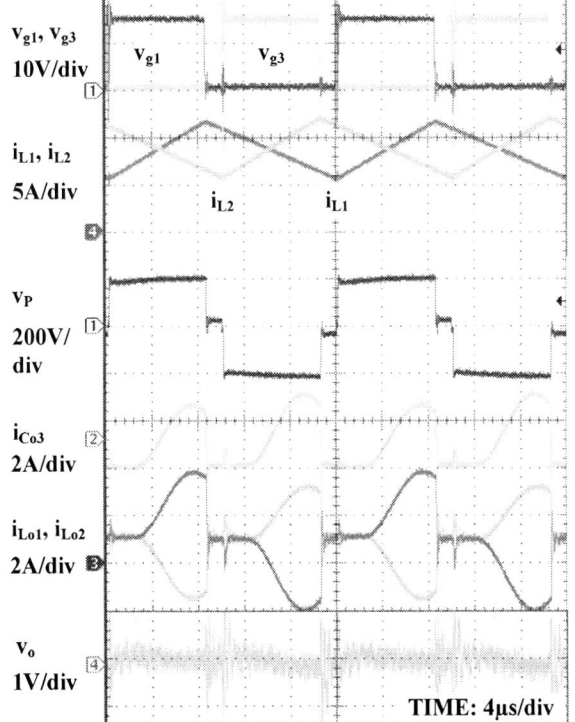

Fig. 7. Waveforms of proposed circuit under 36 V input, 400 V/400 W output

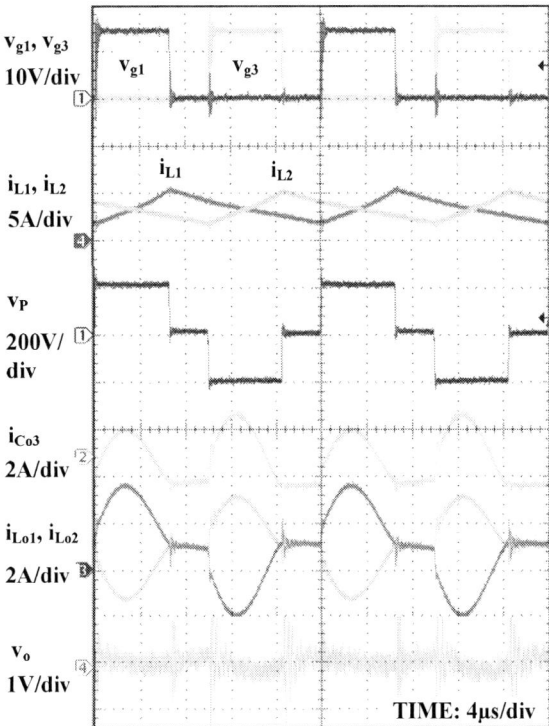

Fig. 8. Waveforms of proposed circuit under 72 V input, 400 V/400 W output

Fig. 7 and Fig. 8 show some key waveforms of the proposed circuit under 36 V input, 400 V/400 W output condition and 72 V input, 400 V/400 W output condition. The experimental results shown that maximun value of the output voltage ripple was 0.4 V when 4.7 μF film-output capacitors were used for output. The voltage spikes on MOSFETs (indicated as v_P waveform) were clamped. The maximun value of the voltage stress on MOSFETs were 210 V. As a result, 250 V MOSFETs with low $R_{DS(ON)}$ are able to use to improve efficience. The inductor currents (i_{L1} and i_{L2} waveforms) were CCM operation. During off-time of output diodes (D_{o1} and D_{o2}), the clamping capacitor current (i_{Co3}) was non-zero value and provided its energy to the load.

Fig. 9 shows measured efficiency of the DIBC-VDR when R-C-D snubber were used to eliminate the voltage spikes on MOFETs. As shown, the maximun efficiency was 93.58%.

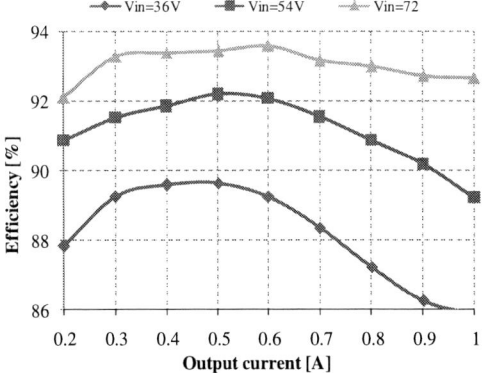

Fig. 9. Measured efficiency of the DIBC-VDR under different conditions

The 2014 International Power Electronics Conference

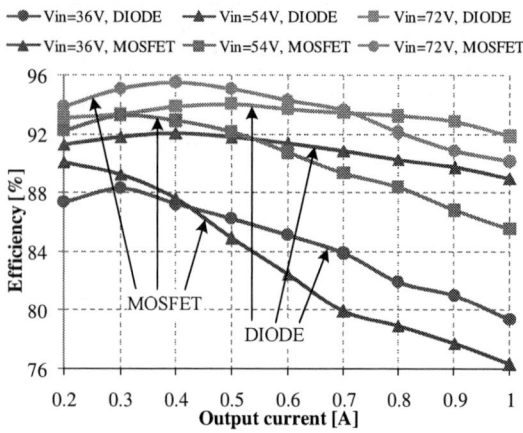

Fig. 10. Measured efficiency of the proposed circuit under different conditions

Fig. 10 shows measured efficiency of proposed circuit when diodes (D_1 and D_2) were used and when two diodes (D_1 and D_2) were replaced by MOSFETs (M_3 and M_4). As shown, the maximun efficiency was 95.54%. Under light-load condition, the efficiency was higher when the MOSFETs were used. However, under heavier load condition, the efficiency tends to be higher when the diodes were used.

IV. CONCLUSIONS

In this paper, a high voltage-gain interleaved boost DC-DC converter featuring non-electrolytic capacitor, voltage-spike-free on MOSFETs, and CCM operation of the input current was presented for alternative energy sources. Two experimental prototypes of the proposed topology and dual-inductor push-pull converter with voltage-doubled rectifier (DIBC-VDR) have been built and tested under 36-72 V input, 400 V/400 W output, 50 kHz switching frequency to compare their performances in order to show the feasibility and superiority of the proposed circuit. The experimental results shown that, under the same operating condition, the proposed converter has obtained very low output voltage ripple compared with the conventional circuit. Additionally, due to obtaining of the voltage spike elimination, low $R_{DS(ON)}$ MOSFETs were used for the proposed topology to improve the efficiency. It was seen that, some additional components were added in new study; however, the proposal could profit from low-voltage, low-cost MOSFETs and lightweight magnetizing components used.

ACKNOWLEDGMENT

The authors would like to acknowledge the financial support of the National Science Council of Taiwan through grant number NSC 100-2628-E-011-009-MY3.

REFERENCES

[1] C. S. Leu and L. Ming-Hui, "A Novel Current-Fed Boost Converter With Ripple Reduction for High-Voltage Conversion Applications," *IEEE Trans. on Ind. Electron.*, vol. 57, pp. 2018-2023, 2010.

[2] C. S. Leu and H. Pin-Yu, "A novel voltage doubler rectifier for high output voltage applications," in *2010 International Power Electronics Conference (IPEC)*, 2010, pp. 2082-2085.

[3] M. Harfman Todorovic, L. Palma, and P. N. Enjeti, "Design of a Wide Input Range DC-DC Converter With a Robust Power Control Scheme Suitable for Fuel Cell Power Conversion," *IEEE Transactions on Industrial Electronics*, vol. 55, pp. 1247-1255, 2008.

[4] J. I. Itoh and F. Hayashi, "Ripple Current Reduction of a Fuel Cell for a Single-Phase Isolated Converter Using a DC Active Filter With a Center Tap," *IEEE Transactions on Power Electronics*, vol. 25, pp. 550-556, 2010.

[5] D. Polenov, H. Mehlich, and J. Lutz, "Requirements for MOSFETs in Fuel Cell Power Conditioning Applications," in *12th International Power Electronics and Motion Control Conference, 2006.*, 2006, pp. 1974-1979.

[6] T. Lixin and S. Gui-Jia, "An Interleaved Reduced-Component-Count Multivoltage Bus DC/DC Converter for Fuel Cell Powered Electric Vehicle Applications," *IEEE Transactions on Industry Applications*, vol. 44, pp. 1638-1644, 2008.

[7] S. C. Kim, S. H. Nam, S. H. Kim, and D. T. Kim, "High power density, high frequency, and high voltage pulse transformer," in *IEEE Conference Record Pulsed Power Plasma Science, 2001.*, 2001, p. 214.

[8] M. A. Perez, C. Blanco, M. Rico, and F. F. Linera, "A new topology for high voltage, high frequency transformers," in *Tenth Annual Applied Power Electronics Conference and Exposition, 1995.*, 1995, pp. 554-559 vol.2.

[9] Y. Bo, Y. Xu, L. Donghao, P. Yunqing, J. Duan, and J. Zhai, "A Current-Fed Multiresonant Converter with Low Circulating Energy and Zero-Current Switching for High Step-Up Power Conversion," *IEEE Transactions on Power Electronics*, vol. 26, pp. 1613-1619, 2011.

[10] G. Fontes, C. Turpin, S. Astier, and T. A. Meynard, "Interactions Between Fuel Cells and Power Converters: Influence of Current Harmonics on a Fuel Cell Stack," *IEEE Transactions on Power Electronics*, vol. 22, pp. 670-678, 2007.

[11] L. Changrong and L. Jih-Sheng, "Low Frequency Current Ripple Reduction Technique With Active Control in a Fuel Cell Power System With Inverter Load," *IEEE Transactions on Power Electronics*, vol. 22, pp. 1429-1436, 2007.

[12] C. S. Leu and Q. T. Nha, "A Half-Bridge Converter With Input Current Ripple Reduction for DC Distribution Systems," *IEEE Transactions on Power Electronics*, vol. 28, pp. 1756-1763.

[13] B. K. Bose and D. K. Kastha, "Electrolytic capacitor elimination in power electronic system by high frequency active filter," in *Conference Record of the 1991 IEEE Industry Applications Society Annual Meeting*, 1991, pp. 869-878 vol.1.

[14] G. Linlin, R. Xinbo, X. Ming, and Y. Kai, "Means of Eliminating Electrolytic Capacitor in AC/DC Power Supplies for LED Lightings," *IEEE Transactions on Power Electronics*, vol. 24, pp. 1399-1408, 2009.

[15] C. Wu and S. Y. R. Hui, "Elimination of an Electrolytic Capacitor in AC/DC Light-Emitting Diode (LED) Driver With High Input Power Factor and Constant Output Current," *IEEE Transactions on Power Electronics*, vol. 27, pp. 1598-1607, 2012.

[16] C. S. Leu, H. Pin-Yu, and L. Ming-Hui, "A Novel Dual-Inductor Boost Converter With Ripple Cancellation for High-Voltage-Gain Applications," *IEEE Trans. on Ind. Electron.*, vol. 58, pp. 1268-1273, 2011.

[17] Y. Bo, Y. Xu, Z. Xiangjun, J. Duan, J. Zhai, and L. Donghao, "Analysis and Design of a High Step-up Current-Fed Multiresonant DC-DC Converter With Low Circulating Energy and Zero-Current Switching for All Active Switches," *IEEE Transactions on Industrial Electronics*, vol. 59, pp. 964-978, 2012.

[18] P. Ki-Bum, M. Gun-Woo, and Y. Myung-Joong, "Two-Transformer Current-Fed Converter With a Simple Auxiliary Circuit for a Wide Duty Range," *IEEE Trans. on Power Electron.*, vol. 26, pp. 1901-1912, 2011.

[19] P. R. Prasanna and A. K. Rathore, "Analysis, Design, and Experimental Results of a Novel Soft-Switching Snubberless Current-Fed Half-Bridge Front-End Converter-Based PV Inverter," *IEEE Trans. on Power Electron.*, vol. 28, pp. 3219-3230, 2013.

[20] W. Tsai-Fu, C. Yung-Chu, Y. Jeng-Gung, and K. Chia-Ling, "Isolated Bidirectional Full-Bridge DC-DC Converter With a Flyback Snubber," *IEEE Trans. on Power Electron.*, vol. 25, pp. 1915-1922, 2010.

Parallel Bi-directional DC-DC Converter for Energy Storage System

Takayuki Ouchi, Akihiko Kanoda
Dept. of Power Electronics Systems Research
Hitachi Research Lab. Hitachi, Ltd.
Hitachi, Ibaraki, Japan
takayuki.ouchi.xp@hitachi.com,
akihiko.kanoda.zz@hitachi.com

Naoya Takahashi
Hitachi Advanced Digital Co., Ltd.
Yokohama, Kanagawa, Japan
naoy-takahashi@hitachi-ad.co.jp

Abstract— A scalable parallel bidirectional DC-DC converter system has been developed and evaluated. Fast and efficient seamless control is achieved by using an H-bridge topology. Although the control structure is simple, four drive modes can be handled. It was extended for use in simple parallel converter management. A parallel function auto-rotation scheme is used to control which units are active on the basis of tracked power. This method achieves high efficiency over a wide load range, especially for light loads. The automatic rotation of which converter acts as the output master distributes the load more evenly, resulting in less load damage. The only hardware required are a power line and communication cable connection. All other functions are provided by software and operate autonomously. A 5-kW prototype converter was constructed and evaluated in a three-unit parallel converter system. The estimated efficiency for 15-kW rated discharge was 97.8% for 333-V battery pack power and 380-V DC-link power.

Keywords—bidirectional DC-DC converter, energy storage system (ESS), parallel converter

I. INTRODUCTION

There is a growing demand for renewable energy systems, such as photovoltaic (PV) and wind turbine systems. One problem with such systems is that the amount of energy generated is uncontrollable. Therefore, battery storage systems are needed to handle the growing gap between supply and demand, because the large gap disturb the stability of grid voltage higher to destroy equipments or lower to be an outage, and a bidirectional converter is a key component in shifting the load.

Various types of bidirectional converters have been reported [1][2]. In particular, parallel-type ones have been developed that reduce the ripple current and improve reliability [3].

In addition, several types of energy storage systems (ESSs), such as electric vehicle (EV) charging infrastructures [4]-[8], have been developed to secure power quality. The output power of an ESS for home or individual use is around 10 kW [9][10].

We have developed a scalable bidirectional converter. It features simple seamless control and achieves fast buck-boost response and smooth charge/discharge mode transition. We also developed an autonomous parallel unit management system. The only hardware required are a power line and communication cable connection. All other functions are provided by software and are completely the same between the master and slave units.

II. FAST AND EFFICIENT BIDIRECTIONAL CONVERTER CIRCUIT CONTROL

An overview of a typical PV system for home or a small building is shown in Fig. 1. A high-voltage (<450 V) DC-link bus line and several energy storage devices, such as batteries, are connected by bidirectional DC-DC converters. The generated PV power is routed through a unidirectional DC-DC converter and the DC-link bus line to the power conditioning system (PCS) or a battery. The routing depends on demand at that time. Since power sent to a battery undergoes charging and discharging, the efficiency of the bidirectional DC-DC converters is very important.

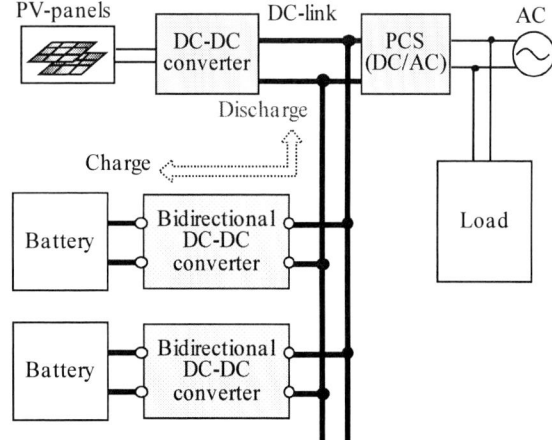

Fig. 1. Typical PV system for home or small building

The rating of the generated power varies because it is determined by the number of PV panels. Likewise, the rated capacity of the batteries depends on the number of cells in each battery. This means that it is important for bidirectional DC-DC converters to be scalable in terms of their rated capacity. In typical home or individual use, the rated power of the PCS is less than 10 kW while in small offices and buildings, it is up to 100 kW. Industrial requirements dictate that a narrow product line-up be able to cover this wide range.

The graph in Fig. 2(a) shows the daily output power transition for a typical bidirectional converter combined with a PV ESS. The output peaks normally appear in the daytime for charging and at nighttime for discharging. The output strength varies because it depends on the weather conditions.

The ratio of converter operation time would be small around the peak power. The general converter efficiency and operation time ratio are plotted against output power in Fig. 2(b). When the load was lighter, the operational time ratio was larger and the efficiency was lower. This means that the ability of bidirectional converters in an ESS to operate efficiently when the load is light is very important.

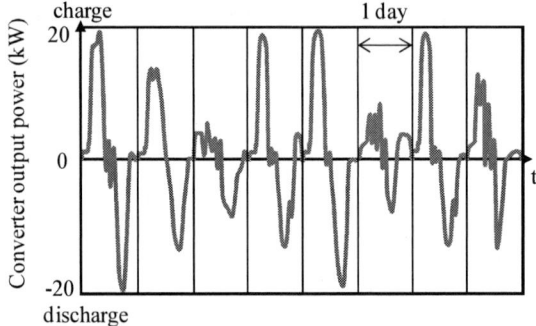

(a) Charge/Discharge characteristics of bidirectional converter

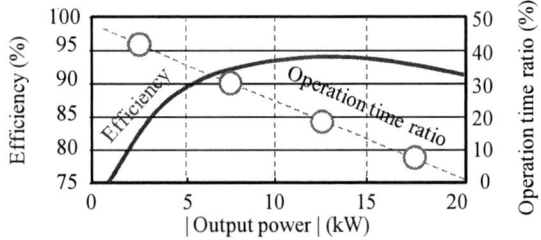

(b) Efficiency and operation time ratio against output power
Fig. 2. Load operation and efficiency of bidirectional converter in PV ESS

Converter efficiency can be increased by increasing the number of battery cells in series. Doing this increases the battery voltage range to around that of the DC-link bus line. The result is a lower peak current and reduced transmission loss in the wire. Although the voltage conversion ratio comes around 1, non-isolated transformer-less topology can be selectable. On the other hand, the relation between battery and DC-link voltage condition which is higher varies, expectable all four-drive mode must be used because the operational voltage range partially overlaps.

Our proposed bidirectional DC-DC converter system has a non-isolated topology, as shown in Fig. 3. This H-bridge bidirectional converter can handle four drive modes, i.e., (1) boost discharge, (2) buck discharge, (3)

boost charge, and (4) buck charge. Two voltages (high-voltage DC link V_{DC} and battery voltage Vbat) and one current (inductor current I_L) are detected in both directions by a hole current transducer. Four main switches (Q1 to Q4) are used to control two sets of insulated gate bipolar transistor (IGBT) modules.

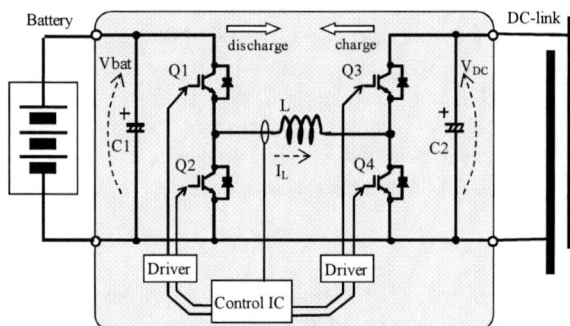

Fig. 3. Proposed bidirectional DC–DC converter system

The duty cycles are calculated using a cascaded control structure (Fig. 4). Its four parallel control blocks enable it to cope with four-mode driving. It selects the one to use on the basis of voltage and current. If a proportional-integral regulator was used, the required convergence time would prevent it from responding quickly during mode transition. Moreover, the overall mode-select and duty-cycle control process would be very complicated.

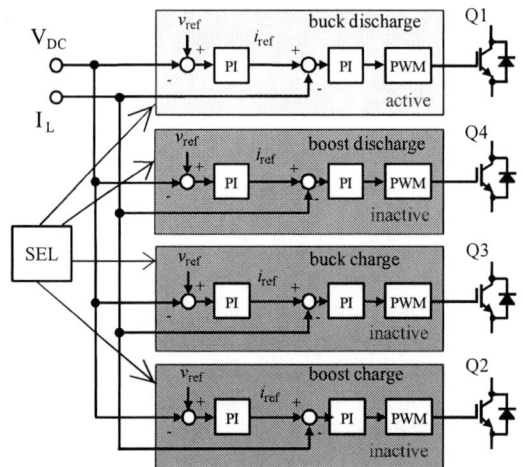

Fig. 4. Basic four-mode parallel-control-block structure of H-bridge converter with mode selector

To avoid these problems, we developed a new control structure. First, to reduce the number of control blocks, we analyzed the output range of the second proportional-integral (PI) regulator (ACR: auto current regulator) and doubled it. In conventional control structures, ACR output value d ranges from 0 to 1 for single-IGBT on-time duty, corresponding to one drive mode. In our structure, d ranges from 0 to 2 for two IGBTs (Q2 and Q3), corresponding to two drive modes (boost and buck charge). The same process was applied to Q1 and Q4, corresponding to buck and boost discharge.

In this scheme, the output value is divided into two ranges, 0 to 1 and 1 to 2. Each division refers to an independent triangular comparative wave to determine the IGBT on-duty. In charge mode, if d is between 0 and 1, Q3 drives by using pulse width modulation (PWM) at on-duty d, and Q2 stops because d is below its duty range (1 to 2). If d is between 1 and 2, Q3 remains in the ON state, and Q2 drives by PWM at on-duty d-1. As a result, the number of control blocks is reduced from four to two.

The symmetric topology of the H-bridge circuit was designed so that switches Q1 and Q2 or Q3 and Q4 cannot both be in the ON state at the same time, which prevents short circuits. That is, the states of Q1 and Q2 (or Q3 and Q4) are either opposite or both OFF. This relation has been applied to a complementary drive for the upper and bottom arm drives in normal operation.

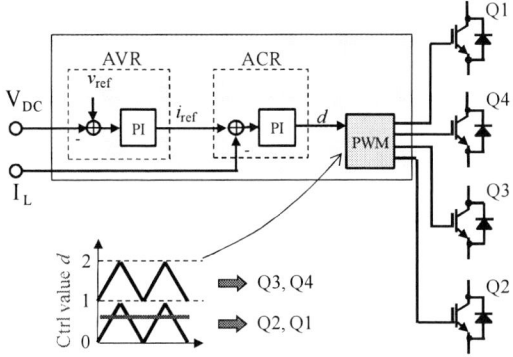

Fig. 5. Advanced unified four-mode control block of H-bridge converter.

The developed advanced control block is very simple, as shown in Fig. 5. It has two cascaded PI regulators, an auto voltage regulator (AVR) and an ACR. Expanded control value d covers 0 to 2. The PWM block receives d and creates four IGBT control pulses and outputs them. The newly introduced two-stage triangular-wave and complementary drive plays no role in the judgment process in the normal duty cycle for boost/buck or charge/discharge mode selection. The unified seamless control system can quickly respond to output power fluctuations.

III. PARALLEL CONVERTER CONTROL AND UNIT OPERATION

The parallel bidirectional converter system takes advantage of scalability. There is a need to combine an ESS with a PCS for a wide rated-power range (5 kW to 100 kW), but the market growth of ESSs has been weak for several years because of feed-in tariffs and other matters. This means that a bidirectional converter is needed that can cover a wide power range with a small line-up. We targeted a line-up of three converters, 5 kW,

10 kW, and 20 kW. In an 8-unit parallel connection system, they can cover 13 points between 5 kW and 100 kW.

A. Parallel Output Current Control

The control system for a conventional parallel converter system is diagrammed in Fig. 6(a). The large shaded rectangles correspond to the controller of each parallel converter (a, b, c,...). The voltage value flows in through a detection block (Det.) and then flows through an AVR and ACR. The output currents are unbalanced because detection errors are inevitable, and each one creates a deviation in the AVR.

To balance the output currents, AVR function is unified to the master unit and others (b, c,...) are masked. The unit acting as the master AVR calculates total N_t units regulation current i_{ref_total}, where N_t is the total number of parallel converters, and transmits i_{ref_nt} to each slave ACR through an RS-485 communication interface.

$$i_{ref\_nt} = i_{ref\_total} / N_t \qquad (1)$$

Each ACR receives i_{ref_nt} and continues the ACR process, as shown in Fig. 6(b). The control current of each converter is defined by (1). A smaller i_{ref_total} means a smaller i_{ref_nt}, which causes the converter to work less efficiently.

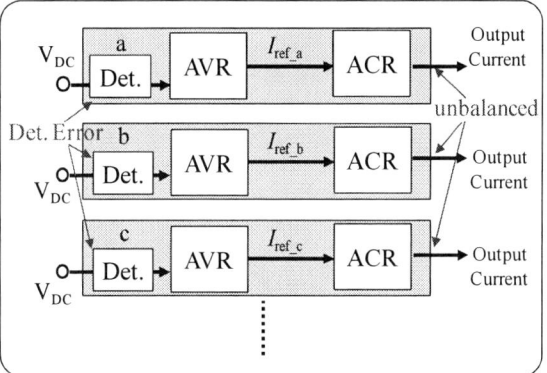

(a) Conventional parallel output control has unbalanced output.

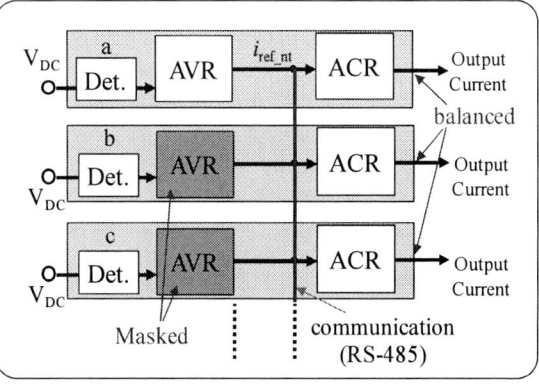

(b) Proposed parallel output control has balanced output.
Fig. 6. Conventional and proposed control systems for parallel converter system

The control structure was therefore extended to improve light-load efficiency. A power calculation process was added to the AVR and is used along with an internal table to set the number of operational units N, which is transmitted to the slave converters along with i_{ref_total}. The slave converters use these data to judge their operational state (ON or OFF). More specifically, they compare N with their self ID (S-ID). (ID management is explained in Sec. III. C.) If N is 2 and S-ID is 2 (i.e., the master ID is 0), the state is set to OFF because only converters with IDs 0 and 1 are active while the others are inactive. If N is 3 and S-ID is 2, the state of the converter is set to ON.

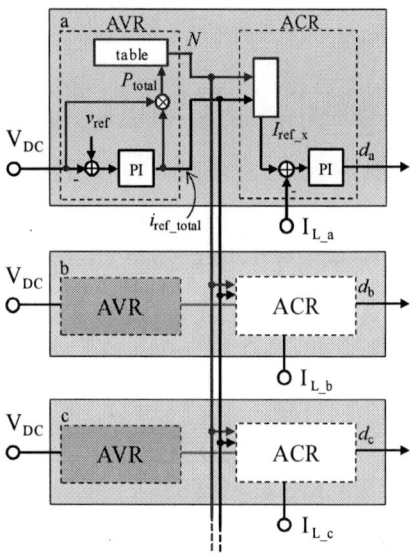

Fig. 7. AVR extension and ACR self-output control.

The extended control structure is depicted in Fig. 7. Each converter receives two pieces of information, i_{ref_total} and N. First, it judges its drive state by comparing its S-ID with N. If the state is active, it calculates its regulation current using

$$i_{ref\_x} = i_{ref\_total} / N \qquad (2)$$

where x is correspondent to each converter a, b, c,...

Each ACR outputs its control value d_x, which is passed to a PWM pulse generation block (not shown in Fig. 7).

This extended control structure features power-tracking active-unit operation. Combined with ID management, the system is self-driven. As shown in Fig. 8, efficiency for a 3-unit parallel 15-kW converter (CNV) system is higher than otherwise in the light-load range.

The threshold power at which each unit is turned on is slightly less than the unit's rated power. To avoid chattering, the threshold values are set slightly different between increase and decrease turn-on unit. The hysteresis of the threshold values was ~0.5 - 1.0 kW in practical use.

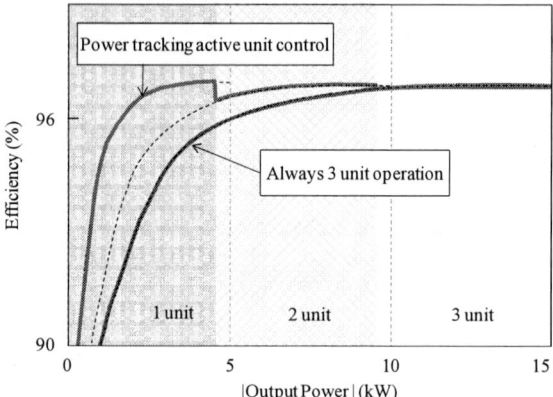

Fig. 8. Use of power-tracking active-unit control improved light-load efficiency.

When the number of active units changes, the control current is continuously changed to prevent current being out of control. The controlled current transition from two active units (a and b) to three units (a, b, and c) and then back to two is illustrated in Fig. 9. The transition time was designed to be less than 2 ms.

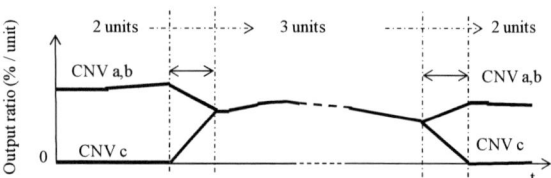

Fig. 9. Control current transition when number of active units changes

B. Load Distribution

As described above, the use of power-tracking active-unit control in our parallel converter system increases light-load efficiency.

The daily output power transition and load allotment with the conventional scheme is depicted in Fig. 10(a). The order in which the converters are turned on is shown on the right; the background colors correspond to the four converters, a, b, c, and d. Peak PV power generation occurs for only a few hours around noon, under sunny conditions. Therefore, converter d is active for a short time while converter a is active for a long time. This unbalanced load degrades the reliability of the converter acting as the master (converter a). The first turn-on master converter in particular is subject to excessive degradation because it has the longest driving time in conventional parallel operation; its electrical chemical capacitors would likely be damaged.

Therefore a scheme has been developed to distribute the load so as to improve total system reliability. The daily output power transition and load allotment with this proposed scheme is depicted in Fig. 10(b). In this scheme, the master and slave assignments are rotated using parallel function auto-rotation (PFR) control.

The 2014 International Power Electronics Conference

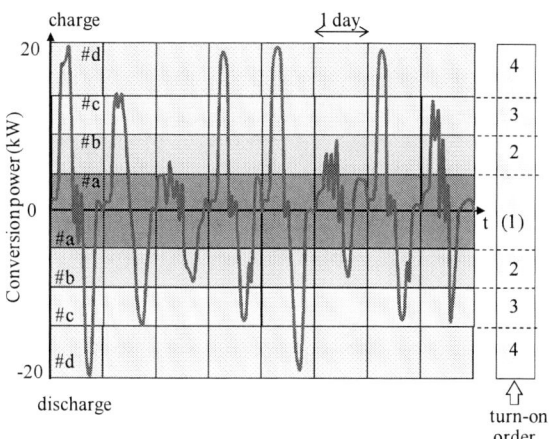

(a) Conventional scheme

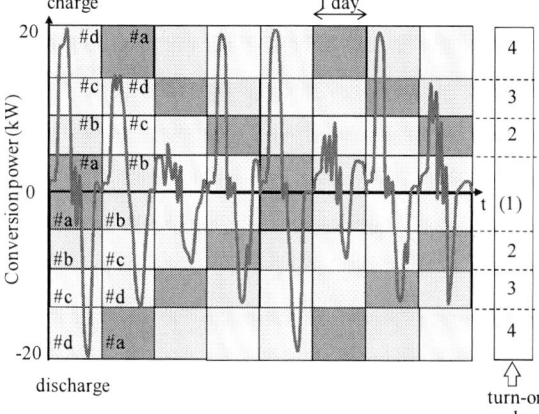

(b) Proposed parallel function auto-rotation scheme
Fig. 10. Daily output power transition and load allotted to each converter (Unit acting as master is shown by dark green.)

C. Parallel Function Auto-rotation

PFR control requires identifying each converter. We therefore define a hardware ID (H-ID) and a software ID (S-ID) for each one. In the initial configuration period, each converter sets its H-ID following connection, with the first one taking an H-ID of 0 and the subsequent ones taking that of the previous one incremented by 1. The unit with an H-ID of 0 functions as the initial hardware master, receiving control signals from the battery pack controller (BPC).

During power conversion, the hardware master handles AVR control. It calculates total regulation current i_{ref_total} and power P_{total}. By referring to the internal table, it determines the number of operation units N. The S-IDs are set in the initial configuration period and are rotated at a triggered timing. The relationships between H-ID and S-ID are listed in Table I, and a diagram of the process is shown in Fig 11.

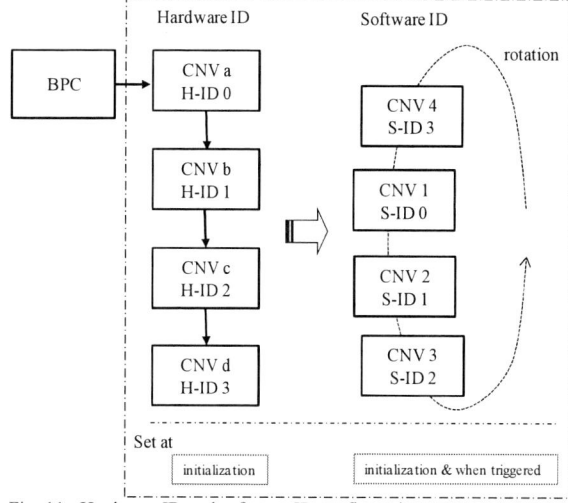

Fig. 11. Hardware-ID and software-ID configuration

TABLE I
RELATIONSHIPS BETWEEN H-IDS AND S-IDS FOR FOUR UNITS

Converter	H-ID	S-ID						
a	0	...	3	2	1	0	3	...
b	1		0	3	2	1	0	
c	2		1	0	3	2	1	
d	3		2	1	0	3	2	

The S-IDs are rotated, for example, when the output power is close to zero, such as in the early morning for most PV ESSs. Triggering S-ID rotation at this time on a daily basis enables the load to be distributed without precise operation time measurement. Since the rotation takes no more than a second, the triggering can easily be set for another preferred time.

D. Summary of Control System and Specifications

A diagram of the parallel control block structure is shown in Fig. 12.

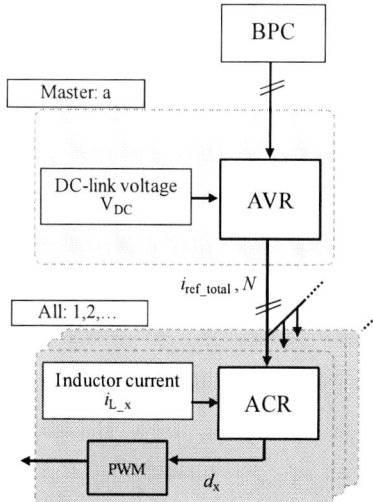

Fig. 12. Diagram of parallel control block structure

The BPC sends ON/OFF signals to the AVR acting as the master converter (converter a), which induces and sends $i_{\text{ref\_total}}$ and N to all ACRs. Each ACR judges its ON/OFF state and controls the output current.

The specifications of the proposed bidirectional DC-DC converter are listed in Table II. It has four operation modes: power compensation for seamless charging and discharging (both with buck and boost power conversion). It has three charge modes: constant current (CC), constant voltage (CV), and pre-charge. There are three types of protection: over-voltage protection (OVP), over-current protection (OCP), and overheat protection. Several other internal protection controls are also applied. The switching frequency of the IGBT drive is 20 kHz.

TABLE II
SPECIFICATIONS OF BIDIRECTIONAL DC–DC CONVERTER

Battery pack voltage	333 V (243–378 V)
DC-link voltage	380 V (350–400 V)
Rated power	5 kW / unit (1–3 units)
Switching frequency	20 kHz
Inductance	600 μH
Operation modes	Power compensation (charge/discharge) Charge only (CC, CV, pre-charge)
Protection	OVP, OCP, Overheat, etc.

The bidirectional rated power is 5 kW for a single unit. An up-to-3-unit 15-kW system with up to 8 parallel connections was tested in this work. The number of parallel connections can easily be expanded.

IV. EXPERIMENTAL RESULTS

The proposed parallel converter system was evaluated using a prototype three-unit system.

A. Prototype 3-unit Parallel Converter System

A single unit of the prototype bidirectional DC-DC converter system is shown in Fig. 13. The rated output power is 5 kW.

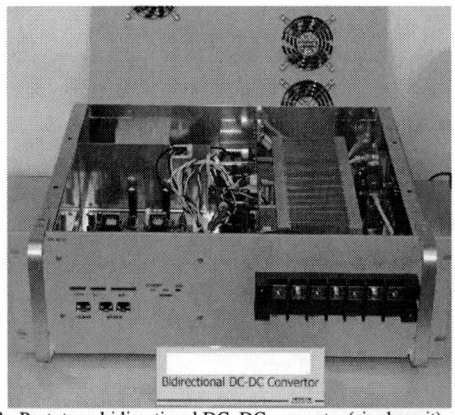

Fig. 13. Prototype bidirectional DC–DC converter (single unit)

The three connection ports on the left are for an upper controller such as a BPC and two parallel loop connections using an RS-485 communication interface. Conventional LAN cables are plugged in for parallel connection. The LED indicators on the right exhibit the operational state, such as wait, drive, and error. Larger terminals on the right side of the unit are for power input.

A block diagram of the prototype parallel evaluation system with power line and communication line connections is shown in Fig. 14. Three bidirectional converters are connected in a loop chain using conventional LAN cable for parallel control. Their hardware and software are completely the same, and any other configuration is needed. The control signals for turn-on and turn-off for each operational mode are received through another line from the BPC. A DC power supply unit works as a battery between 243 and 378 V with up to 9-kW output. Parallel connected electrical loads work as a DC-link bus line between 350 and 400V with up to 7-kW power. Each parallel bidirectional DC-DC converter can handle up to 15-kW power conversion. Under these conditions, it works as discharge condition although most of the measurements were done in compensation mode operation. Operation under other charge modes with suitable settings has also been tested, and the system was found to work well.

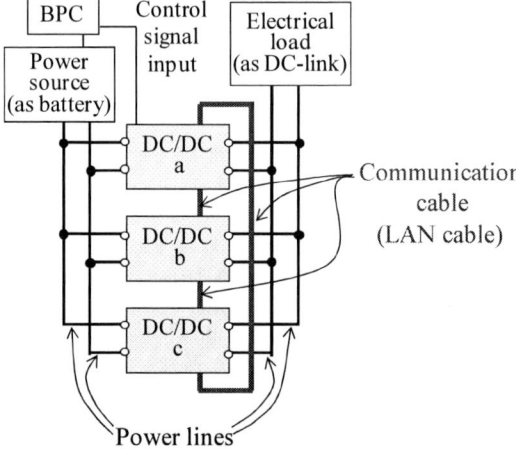

Fig. 14. Block diagram of parallel bidirectional DC–DC converter prototype system with connection lines.

B. Parallel Converter Control by PFR

The four graphs in Fig. 15 show the changes in total output power P_{total} (kW) and inductor current i_L (A) for the three parallel converters with the PFR control scheme. After a standby period from 22 s to 42 s, master unit control is switched from converter a to converter b. The second turn-on unit (1st slave) switches from b to c, and the third turn-on unit (2nd slave) switches from c to a.

The plotted values show that the power-tracking operation worked well. At ~10 s (and ~50 s), the load changed rapidly from 1 kW to 9 kW, and the number of operational converters increased from 1 to 3. The threshold power for increasing the number of active units had been set to 3 kW and 5 kW respectively.

At 22 s, the converters stopped due to receiving a standby signal. The output power and inductor currents of converters a and b immediately dropped, indicating that the "start and stop" process was controlled well. This means that the PFR control scheme worked well: the master and slave assignments were rotated in accordance with the turn-on order. It also means that the power-tracking unit-number control also worked well.

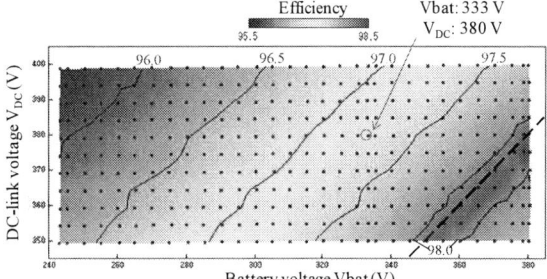

Fig. 16. Efficiency distribution over operational voltage range

We compared the efficiencies for one, two, and three active units in a 3-unit parallel converter system to evaluate the effect of power-tracking active-unit operation. The threshold voltage was modified for this measurement. The voltage condition was rated Vbat 333 V and V_{DC} 380 V. The assumed single 15-kW converter characteristics are shown by the gray solid line in Fig. 17. They are based on the characteristics of a single 5-kW converter. The measurement points were at less than 7 kW due to equipment restrictions.

The efficiency for three active units completely overlapped the assumed 15-kW characteristics for less than 7 kW. This means that efficiency for three units would match the line. At the same output power, efficiency was the highest for one active unit while it was the next highest for two active units.

As shown in Fig. 17, the power-tracking active-unit control using the PFR control scheme improved efficiency over a wide output range. The estimated efficiency at 15-kW rated output power was about 97.8%. This indicates that the efficiency over a 13.3% load range is 96% and the efficiency over a 3.3% load range is 90%.

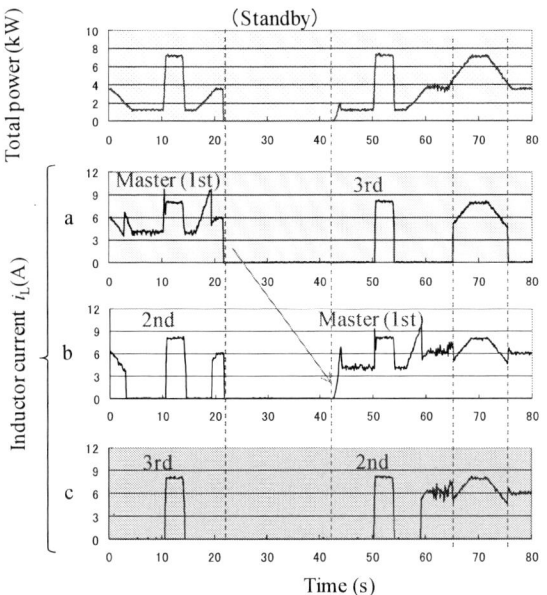

Fig. 15. Changes in total output power and inductor current for three parallel converters with PFR control scheme

C. Efficiency Distribution and Power-tracking Control

The distribution of power conversion efficiency over the operational voltage range of a single converter unit with a 50% load (2.5 kW) in discharge mode is illustrated in Fig. 16. The measured voltage condition is shown in ~5-V steps for DC-link voltage on the vertical axis and for battery voltage on the horizontal axis. Efficiency is indicated by the color correspondence to the color bar at the top of the graph, from 95.5% to 98.5%. The best efficiency conditions were located along the line where the input and output voltages were equivalent. Efficiencies of 96.0 to 98.0% along a 0.5% step line are also indicated in the graph; they were located almost parallel to the equivalent voltage line.

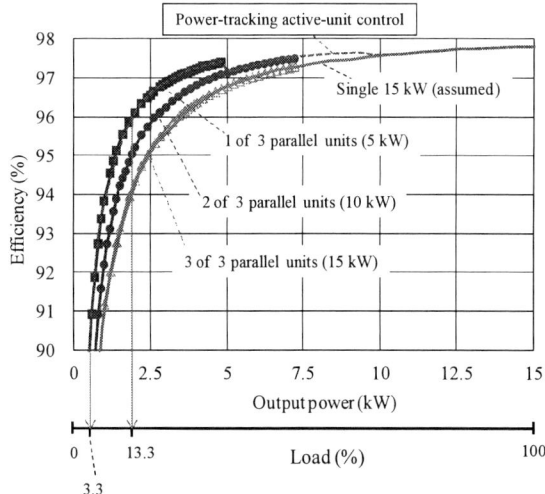

Fig. 17. Dependence of efficiency on number of units and on use of power-tracking control scheme in 3-unit parallel converter system

V. CONCLUSION

A scalable parallel bidirectional DC-DC converter for energy storage systems has been developed and evaluated. Fast and efficient seamless control is achieved by using an H-bridge topology. Although it has a simple control structure, it can handle four drive modes: buck and boost for both charge and discharge. It has been extended to simple and autonomous parallel converter management. The proposed parallel function auto-rotation (PFR) scheme manages power-tracking active-unit control. High efficiency is attained over a wide load range, and efficiency is especially improved in the light-load range. The PFR control scheme rotates the output master function among the converters, thereby distributing the load and reducing load damage. The only hardware needed is a power line and communication cable connection. All other functions are provided by software and operate autonomously. The estimated efficiency of a 5-kW prototype converter in a 3-unit parallel converter system at 15-kW rated discharge was 97.8% for 333-V battery pack power and 380-V DC-link power. This indicates that the efficiency in a 3-unit parallel system over a 13.3% load range is 96% and the efficiency over a 3.3% load range is 90%.

REFERENCES

[1]   C. Citro, A. Luna, J. Rocabert, R. S. Muñoz-Aguilar, I. Candela, and P. Rodriguez, "Overview of Power Processing Structures for Embedding Energy Storage in PV Power Converters," in *IECON 2011 - 37th Annual Conference on IEEE Industrial Electronics Society*, Nov. 2011, pp. 2492-2498.

[2]   M. Young, "The PWM strategy on DC-DC converter," *IEEJ Journal of Industry Applications*, vol. 28, no. 15, pp. 123-129, 1989.

[3]   W. Li, G. Joós, and C. Abbey, "A Parallel Bidirectional DC/DC Converter Topology for Energy Storage System in Wind Application," *Industry Applications Conference, 42nd IAS Annual Meeting*, pp. 179-185, 2007.

[4]   M. Ortuzar, J. Dixon, J. Moreno, "Ultracapacitor-Based Auxiliary Energy System for an Electric Vehicle: Implementation and Evaluation", *IEEE Trans. On Industrial Electronics*, vol. 54, issue 4, Aug.2007, pp. 2147-2156.

[5]   R. M. Schupbachj, C. Bald, "Comparing DC-DC-Converters for Power Management in Hybrid Electric Vehicles," *IEEE 2003 International Electric Machines and Drives Conference*, 2003, vol. 3, pp. 1369-1374.

[6]   J. Czogalla, J. Li, C.R. Sullivan, "Automotive application of Multi-Phase Coupled-Inductor DC-DC Converter," *IEEE 2003 Industry Applications Conference*, 2003, vol. 3, pp. 1524-1529.

[7]   M. Gerber, J. A. Ferreira, N. Seliger, I. W. Hofsajer, "Design and Evaluation of an Automotive Integrated System Module," *IEEE 2005 Industry Applications Conference*, 2005, vol. 2, pp. 1144-1151.

[8]   Y. Du, X. Zhou, S. Bai, S. Lukic, A. Huang, "Review of Non-isolated Bi-directional DC-DC Converters for Plug-in Hybrid Electric Vehicle Charge Station Application at Municipal Parking Decks," *IEEE 2010 Applied Power Electronics Conference*, 2010, pp. 1145-1151, 2010.

[9]   N. M. L. Tan, T. Abe, and H. Akagi, "A 6-kW 2kWh Lithium-Ion Battery Energy Storage System Using a bidirectional Isolated DC-DC Converter," *The 2010 International Power Electronics Conference (IPEC)*, pp. 46-52.

[10] M. Bragard, N. Soltau, R.W. De Donker, and A. Schmiegel, "Design and Implementation of a 5 kW Photovoltaic System with Li-Ion Battery and Additional DC-DC Converter," *IEEE Energy Conversion Congress and Exposition*, pp. 2944-2949, 2010.

Charging Scenario of Serial Battery Power Modules with Buck-Boost Converters

Jhen-Yu Jian, Chu-Shen Chang, Chin-Sien Moo
Department of Electrical Engineering
National Sun Yat-sen University
Kaohsiung, Taiwan
mooxx@mail.ee.nsysu.edu.tw

Hau-Chen Yen
Department of Optoelectronic Engineering
Far East University
Tainan, Taiwan
yenc66@cc.feu.edu.tw

Abstract- **This paper studies the charging scenario of the battery power bank with buck-boost battery power modules (BPMs) connected in series. For the BPMs with serial connection, the charging currents to batteries can be individually scheduled by adjusting the duty-ratios of the associated power converters. During the charging process, those batteries having been completely charged can be isolated from the battery power bank without interrupting the charging operation. To fully utilize the available charger's capacity, a simple charging scenario is proposed according to the state-of-charges (SOCs) of the batteries under the limitations of the allowable charger's power and the maximum battery charging current. Experiments carried out on four buck-boost type BPMs have confirmed the efficient performance of the charging scenario.**

Keywords: Battery, Battery power module (BPM), Charging scenario, State-of-charge (SOC).

I. INTRODUCTION

Recently, battery power has been more and more widely used in our daily life. A dc source with a single-cell battery or a battery packed by a small number of cells was adopted for those designs requiring lower voltages, such as portable electronic devices [1-3]. On the other hand, a power bank with numerous cells in series and/or parallel has to be used to meet the high-voltage and high-power requirements for applications such as electric vehicles and back-up power of uninterruptible power supplies, energy storage of micro-grids in power systems.

When connected in series, batteries are charged and discharged by a same current. Due to intrinsic discrepancies or with different initial states, some batteries in a queue may be over-charged or over-discharged causing a malfunction to the dc power source [4-6]. A conventional solution for this problem is to introduce a battery management system with a charge equalization circuit, leading to an extra cost and additional power losses.

Instead of connecting batteries directly in series, a power bank with the battery power modules (BPMs) was proposed, in which each single-packed battery is equipped with an associated power converter [7-12]. The output terminals can be connected in series to aggregate a high output voltage and in parallel to output a large current. With such a configuration, all battery currents can be scheduled individually through control of the power converters in BPMs. The associated power converter can be bidirectional for both charging and

discharging. This paper focuses on the charging operation of serial BPMs with buck-boost converters. A charging scenario is proposed to charge the batteries in a more efficient manner.

II. BIDIRECTIONAL BUCK-BOOST BPM

Fig. 1 shows the power conversion circuit of a bidirectional buck-boost type BPM. In the bidirectional buck-boost converter, two power MOSFETs are adopted as the active power switches for high frequency operation. The two active power switches, S_B and S_C, which are activated in a mutually complementary manner, play reciprocally the roles of the main switch for power regulation and the auxiliary switch for synchronous rectification. The subscripts "$_B$" and "$_C$" denote that the battery side and the charger side, respectively, where the power switches and the filter capacitors attach with.

The bidirectional buck-boost BPM can be operated at the continuous conduction mode (CCM) or the discontinuous conduction mode (DCM), depending on the designated circuit parameters and the operating power. Fig. 2 shows the operating states for charging. In State 1, the switch S_C is turned on. The capacitor voltage on the charger side is applied on the inductor. The energy from the charger is stored in the inductor current by increasing the current linearly. The stored energy is then delivered to the battery and the output capacitor C_B in State 2. State 3 presents only when the inductor current falls to zero in a

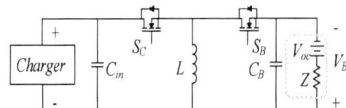

Fig. 1. Bidirectional buck-boost BPM

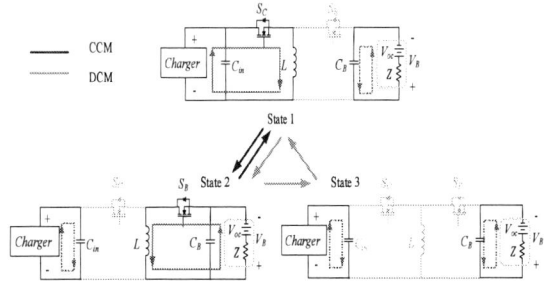

Fig. 2. Operation states of buck-boost BPM

high-frequency switching cycle. In general, the DCM happens when the battery is charged with a small current at the constant-voltage stage. During this state, both two power switches are turned off. Only small currents, which are denoted by dashed lines, present for balancing the capacitor voltages.

III. SERIAL CONFIGURATION OF BPMs

Fig. 3 shows a battery power bank with series configuration of n BPMs. Each BPM is composed of a single-packed battery equipped with a bidirectional buck-boost converter. In the figure, the subscript "i" denotes the i-th BPM. The batteries in BPMs are represented by their internal voltages V_{oci} and equivalent impedances Z_i.

For charging operation, the battery current can be controlled by adjusting the duty-ratio of the buck-boost converter. All BPMs are operated at a same switching frequency with an equal phase-shift to reduce the ripple of the charger's current, i_{chg}. Fig. 4 shows the theoretical waveforms of four BPMs with an equal phase-shift. The ripple of the current from the charger is reduced obviously.

IV. ISOLATION OF BATTERIES

For buck-boost BPMs with series configuration, the batteries can be isolated from the power bank simply by removing the associated gating signal. In other words, those batteries having been fully charged or damaged can be isolated from the BPMs without interrupting the charging operation. In this case, the remaining modules share the available charger's power.

Fig. 5 shows an exemplar case for the isolation of the batteries in BPM$_2$. The isolation can be accomplished by removing the gate signals from the active power switch S_B of the corresponding buck-boost converter. The power supplied to the isolated BPM can transfer to the others for utilizing the allowable charger's power.

V. CIRCUIT OPERATION

The BPMs can be operated individually even though they are connected in series. For charging operation, the charger's voltage, V_{chg}, is equal to the sum of the voltages on the input capacitors of BPMs.

$$V_{in1} + V_{in2} + \cdots + V_{inn} = V_{chg} \tag{1}$$

On the other hand, the average input currents of all BPMs connected in series, which are identical to each other, are essentially equal to the average current I_{chg} drawn from the charger.

$$I_{in1} = I_{in2} = \cdots = I_{inn} = I_{chg} \tag{2}$$

The buck-boost converters in BPMs may be operated at the CCM or the DCM. For the CCM operation, the average battery current is inversely proportional to the duty-ratio of the corresponding buck-boost converter.

$$I_{B1} : I_{B2} : \cdots : I_{Bn} = \frac{1-d_1}{d_1} : \frac{1-d_2}{d_2} : \cdots : \frac{1-d_n}{d_n} \tag{3}$$

where I_{Bi} is the average current charged into the battery and d_i is the duty-ratio of the buck-boost converter in the i-th BPM.

For the DCM operation, the relationship among the average charging currents of BPMs can be expressed as

$$I_{B1} : I_{B2} : \cdots : I_{Bn} = \frac{d_1'}{d_1} : \frac{d_2'}{d_2} : \cdots : \frac{d_n'}{d_n} \tag{4}$$

where d_i' is the time ratio for the inductor current decreasing from its peak to zero in a cycle.

The battery currents can be scheduled by adjusting the duty-ratios of the converters for both CCM and DCM operations.

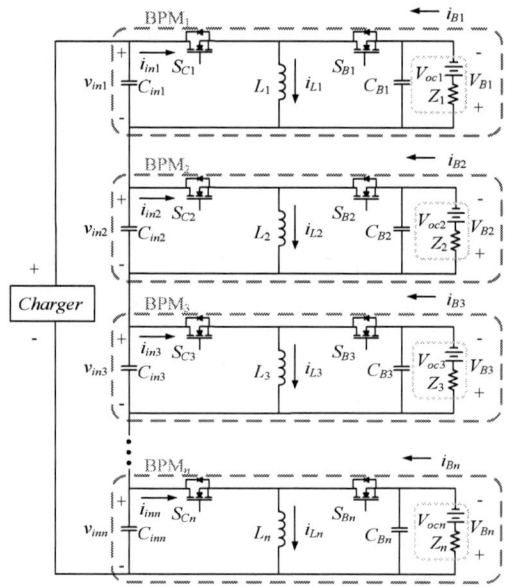

Fig. 3. BPMs with series configuration.

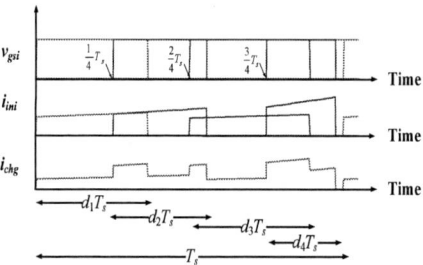

Fig. 4. Theoretical waveforms of BPMs with equal phase-shift.

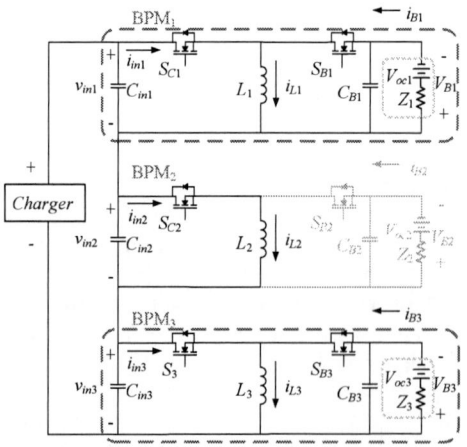

Fig. 5. Isolation of the battery in BPM$_2$.

VI. CHARGING SCENARIO

All batteries in the BPMs of the power bank can be charged at a time through the associated power converters by a dc power source. With the most commonly used constant-current constant-voltage (CC-CV) method charging regime, batteries with lower SOCs are first charged by the CC mode and then changed to the CV mode when the loaded voltages reach a specified threshold. The charging control for fast charging and charge equalization is realized under limitations of the allowable power, P_m, from the charger and the maximum battery charging current, I_{Bm}, specified by the manufacturer.

Fig. 6 illustrates the distribution of the charger's power and the control of the charging scenario. At the beginning, the all batteries are verified to be charged by the CC mode or the CV mode in accordance with the pre-known capacities. The batteries at the CC stage are charged by I_{Bm} from low to high one by one in an order of their capacities until the residual power is not enough to charge a battery with I_{Bm}. When charged by I_{Bm}, these batteries are with different voltages and hence take in different powers. For x batteries charged by I_{Bm}, the residual power, P_r, can be calculated by assuming lossless power conversion in BPMs.

$$P_r = P_m - \sum_{i=1}^{x} \left(I_{Bm} \cdot V_{Bi} \right) \tag{5}$$

where V_{Bi} is the battery voltage of the i-th BPM.

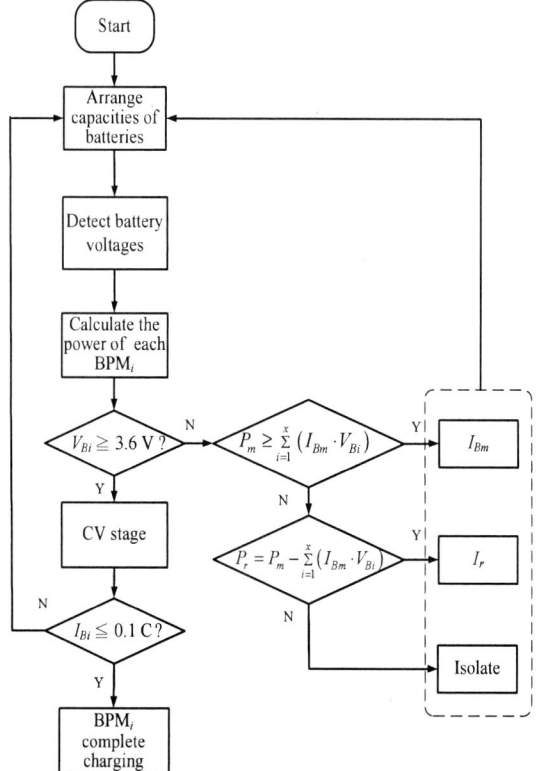

Fig. 6. Control flowchart of charging operation.

Then, the charging currents assigned to the batteries with the next lower capacity can be obtained as

$$I_r = \frac{P_r}{\sum_{i=x+1}^{x+y} V_{Bi}} \tag{6}$$

where y is the number of batteries with the next higher capacity.

Thus far, the available charger's power has been completely used. Therefore, the remaining batteries with higher SOCs are isolated by removing the corresponding gate signals from the active power switches. The batteries with a same capacity will be charged by the same charging current.

For the BPMs in series, the battery currents can be regulated by adjusting the duty-ratios. The battery voltages and currents are detected continuously every second to calculate the charging power and estimate the SOCs. As the BPM operation has been turned into the CV mode, the battery current and thus the power decreases gradually. The charging process is completed when all charging currents have become less than 0.1 C (0.23 A).

VII. EXPERIMENTAL TESTS

The experiments are carried out on a small-scaled laboratory battery power bank with four BPMs in series. Each BPM is formed by a single-cell lithium-ion battery with a buck-boost converter. The nominal voltage and the rated capacity of the batteries are 3.6 V and 2.3 Ah, respectively. Table I lists the parameters of the BPMs. In the experiments, the charger's power is limited at 60 W. The maximum allowable charging current, I_{Bm}, is set at 1 C (2.3 A). The charging operation enters into the CV stage when the battery voltage has reached 3.6 V. The charging cycle is completed as the battery current decreases to be less than 0.1 C (0.23 A).

Figs. 7 and 8 show the experimental results of two test cases. Before testing, all tested batteries are first discharged to a terminal voltage of 2 V, at which the battery is regarded as empty with zero capacity. Then, the batteries are charged to designated capacities.

In Case I, the batteries in three BPMs are initially charged up to capacities of 10.87 %, 21.74 % and 43.47 %, respectively, while the other one remains empty. At the beginning, the empty battery in BPM_1 and the two batteries with lower SOCs in BPM_2 and BPM_3 are

TABLE I
CIRCUIT PARAMETERS

Battery nominal capacity	2.3 Ah
Voltage of CV stage	3.6 V
Battery termination current	0.23 A(0.1 C)
Maximum battery charging current, I_{Bm}	2.3 A
Switching frequency of converter	20 kHz
Inductance of converter, L	200 μH

The 2014 International Power Electronics Conference

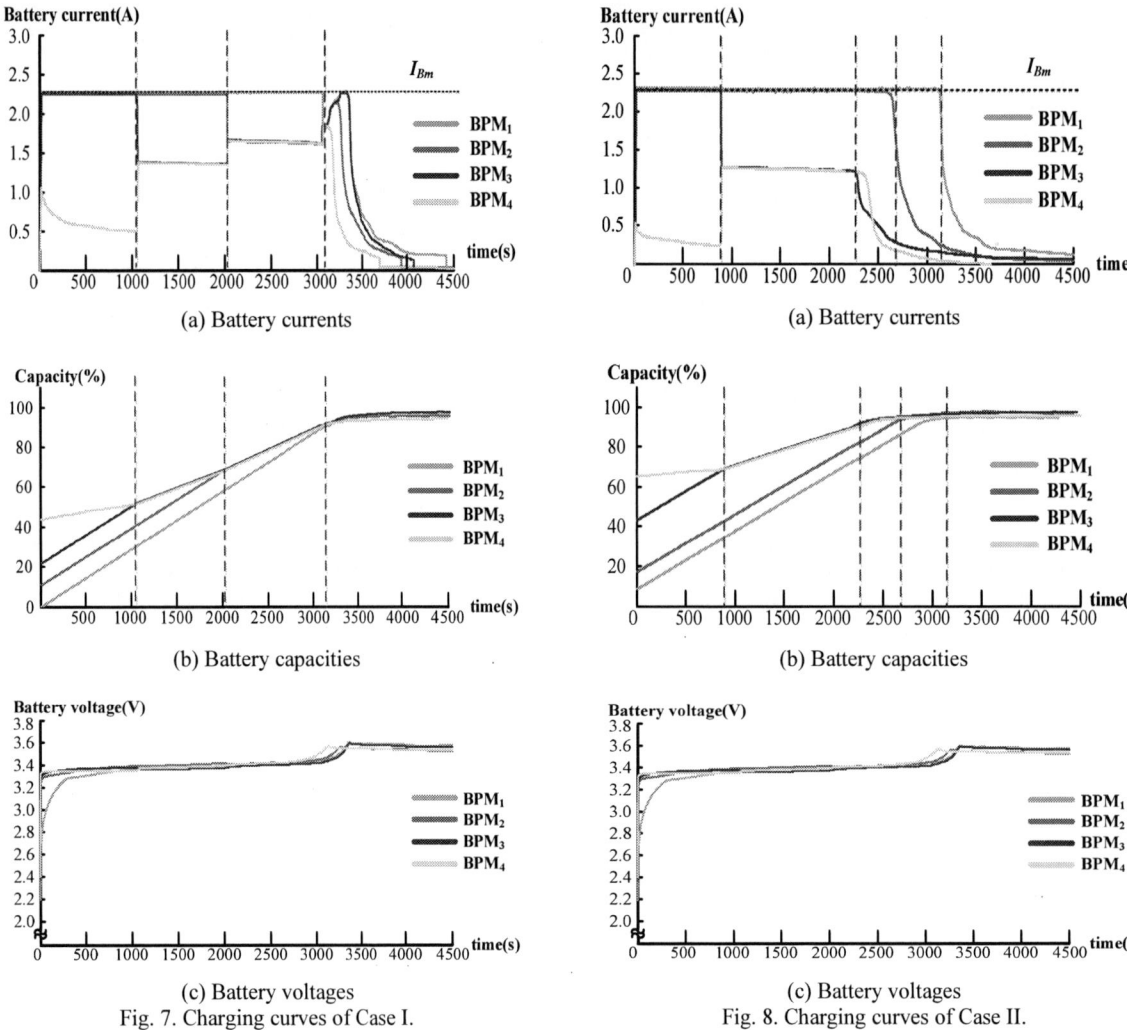

(a) Battery currents

(b) Battery capacities

(c) Battery voltages
Fig. 7. Charging curves of Case I.

(a) Battery currents

(b) Battery capacities

(c) Battery voltages
Fig. 8. Charging curves of Case II.

charged by the I_{Bm}. Limited by the charger's maximum power, the residual power is not enough to charge the battery in BPM$_4$ with I_{Bm}. Therefore, it can only be charged with a current of 1.05 A. However, this current has to be reduced gradually, as the voltages of other batteries increases, drawing more power from the charger. Therefore, the battery current in BPM$_4$ decreases first remarkably and then becomes gradually. At the 1061st second, the battery capacity in BPM$_3$ catches up that in BPM$_4$. Then, the batteries in BPM$_3$ and BPM$_4$ are charged by a same current of 1.45 A, which is much smaller than I_{Bm} but higher than the previous charging current for BPM$_4$. The battery SOC in BPM$_2$ catches up those in BPM$_3$ and BPM$_4$ after 2028 seconds. Then, the charging currents are re-allocated again. At the 3057th second, all batteries have the same capacity and thus with a same charging current smaller than I_{Bm}, which is limited by the charger's maximum power. The BPMs go into the CV stage at different time because of the difference among batteries' intrinsic characteristics and inaccuracy in estimation of batteries' capacities. At the 3116th second, BPM$_4$ enters the CV stage first. The charging power to BPM$_4$ decreases gradually, allowing for a larger charging current to the batteries of other BPMs.

Subsequently, BPM$_2$ goes into the CV stage at the 3220th second. Finally, BPM$_1$ and BPM$_3$ enter the CV stage simul-taneously. The charging cycle is completed when all battery currents decrease to be less than 0.23 A after 4412 seconds. Consequently, the battery capacities of four BPMs are close to each other with 93.48 %, 93.04 %, 93.91 %, and 92.60 %, respectively.

In Case II of Fig. 8, the batteries' capacities in four BPMs are deliberately preset at 8.69 %, 17.39 %, 43.47 %, and 65.21 %, respectively, before being charged. At the beginning, three batteries in BPM$_1$, BPM$_2$ and BPM$_3$ are charged by the maximum current I_{Bm}, while the battery in BPM$_4$ is charged with a current of 0.5 A. At the 810th second, the battery capacity of BPM$_3$ catches up that in BPM$_4$. Then, BPM$_3$ and BPM$_4$ are operated with a same charging current of 1.3 A, and then enter the CV stage after 2250 seconds. The charging current decreases gradually with capacities increases. BPM$_2$ and BPM$_1$ enter the CV stage at the 2650th second and the 3000th second, respectively. The charging cycle is completed when all battery currents decrease to be less than 0.23 A after a charging cycle of 4500 seconds. Eventually, the battery capacities of four BPMs are 92.38 %, 93.28 %, 93.85 %, and 93.38 %, respectively.

978-1-4799-2706-7/14 $31.00 © 2014 IEEE

VIII. Conclusions

The configuration of a battery power bank with buck-boost type battery power modules (BPMs) connected in series has been illustrated. All modules can be operated individually, retaining the advantages of flexible control, easy protection, simple maintenance, and favorable battery power management. A charging scenario is proposed to charge the batteries of BPMs in a more efficient manner. An exemplar battery power bank with four BPMs connected in series has been built and tested. The charging experiments are carried out under the limitations of the allowable charger's power. By adopting the charging scenario, the differences among four battery capacities are effectively reduced during the charging process.

References

[1] F. Feng, R. Lu, G. Wu, and C. Zhu, "A measuring method of available capacity of li-ion series battery pack," in *Proc. IEEE VPPC*, pp. 389-394, Oct. 2012.

[2] J. Kim, J. Shin, C. Chun, and B. H. Cho, "Stable configuration of a li-ion series battery pack based on a screening process for improved voltage/soc balancing," *IEEE Trans. on Power Electronics*, vol. 27, no. 1, pp. 411-424, June 2012.

[3] P. Li, Y. Pan, Y. Ma, and Q. Qin, "Study on an active voltage equalization charge system of a series battery pack," in *Proc. IEEE EMEIT*, vol. 1, pp. 141-144, Aug. 2011.

[4] C. H. Kim, H. S. Park, and G. W. Moon, "A modularized two-stage charge equalization converter for series connected lithium-ion battery strings in an HEV," in *Proc. IEEE PESC*, pp. 992-997, June 2008.

[5] H. Shen, W. Zhu, and W. Chen, "Charge equalization for series connected lithium-ion batteries," in *Proc. IEEE ICEMI*, vol. 4, pp. 1032-1037, Aug. 2009.

[6] F. Wen, J. C. Jiang, W. G. Zhang, and H. P. Guo, "Research on the charge mode of series-connected batteries," in *Proc. IEEE VPPC*, pp. 1-5, Sep. 2008.

[7] Y. C. Hsieh, K. S. Ng, S. P. Chou, and C. S. Moo, "Charge equalization circuit for discharging series-connected batteries with regulated output, " *Journal of the Chinese Institute of Engineers*, vol. 31, no. 6, pp. 1083-1087, Sep. 2008.

[8] C. S. Moo, K. S. Ng, and Y. C. Hsieh, "Parallel operation of battery power modules," *IEEE Trans. on Energy Conversion*, vol. 23, no. 2, pp. 701-707, June 2008.

[9] C. S. Moo, K. S. Ng, and J. S. Hu, "Operation of battery power modules with series output," in *Proc. IEEE ICIT*, pp. 1-6, Feb. 2009.

[10] W. Hong, K. S. Ng, J. H. Hu and C. S. Moo, "Charge equalization of battery power modules in series," in *Proc. IEEE IPEC*, pp. 1568-1572, Nov. 2010.

[11] C. H. Hou, C. T. Yen, T. H. Wu, and C. S. Moo, "A battery power bank of serial battery power modules with buck-boost converters," in *Proc. IEEE PEDS*, pp. 211-216, Apr. 2013.

[12] K. H. Lin, L. R. Yu, C. S. Moo, and C. Y. Juan, "Analysis on parallel operation of boost-type battery power modules," in *Proc. IEEE PEDS*, pp. 809-813, Apr. 2013.

Comparative Thermal Performance Evaluation of SiC MOSFETs and Si MOSFET for 1.2 kW 300 kHz DC-DC Boost Converter as a Solar PV Pre-regulator

Taekyun Kim, Minsoo Jang and Vassilios G. Agelidis
School of Electrical Engineering and Telecommunications
The University of New South Wales (UNSW)
Kensington, Sydney NSW 2052, Australia
taekyun.kim@student.unsw.edu.au

Abstract- **In this paper a comparative thermal performance evaluation of SiC MOSFETs (SiC MOSFET and SiC Z-FET) and Si MOSFET in a high power density of 1.2 kW DC-DC boost converter as a photovoltaic pre-regulator is presented. The good thermal stability of the converter expected from both SiC MOSFETs at high switching frequency up to 300 kHz and high case temperature of 150 °C has been demonstrated by comparing with Si MOSFET. Faster turn-on switching time for SiC MOSFETs, compared with Si MOSFET leads to reduced turn-on switching losses, and consequently the heat sink temperature at 300 kHz switching frequency reaches only at around 50 °C in natural convection. The experimental results given in this paper show that SiC technologies aid in improving reliability of solar PV inverters by enhancing thermal stability of the system.**

I. INTRODUCTION

Increasing demand for more efficient, higher power and operating temperature of power converter looks to the development and availability of wide-band-gap (WBG) semiconductor devices for future direction. As a new generation of WBG power semiconductor devices, silicon carbide (SiC) and gallium nitride (GaN) semiconductor offer the potential to overcome the temperature, frequency and power management limitations of Si semiconductors [1].

GaN has been demonstrated to be superior to Si across a wide range of voltages [2]. However, in terms of technology maturity it seems that the technology achievements with GaN devices have been delayed when compared with the SiC. At the moment, the + 600 V region where SiC and GaN are supposed to compete, it is considered safe for SiC as no GaN device can compete yet [3].

Among many applications, SiC semiconductors have been introduced in solar photovoltaic (PV) applications at research level [4]. SiC technologies can contribute to improved converter performance [5] and reliability of PV energy conversion systems with less thermal management

requirements [6]. The first SiC diodes were introduced in PV inverters in 2001 [7] and in a series of products in 2005 [8], [9]. The first application of SiC Metal Oxide Semiconductor Field Effect Transistor (MOSFET) in PV inverter was reported in 2007 [10]. Now, SiC MOSFET, normally-on/off Junction Field Effect Transistor (JFET) and Bipolar Junction Transistor (BJT) are available in the market. When compared with the high on-resistance R_{DS} of Si MOSFET, SiC MOSFETs become attractive for applications, which are nowadays still dominated by IGBTs. R_{DS} of SiC MOSFET is as low as 60 mΩ [11], and thus is considered well suited for PV inverter development.

The field experience shows that one of the principal reliability issues in solar PV inverter is thermal management and heat extraction mechanisms [12]. In this regard, SiC technologies will aid in improving reliability of solar PV inverters through better heat dissipation and also reducing overall converter design complexity.

The object of this paper is to provide insight into the thermal management in solar PV inverter with SiC technologies by selecting SiC MOSFETs, and to compare its thermal performance with Si MOSFET. No technical paper has provided experimental thermal performance evaluation in terms of switching frequency and temperature, focusing both on switching device itself and the system as well. Fig. 1 shows a non-isolated DC-DC boost converter as a pre-regulator in PV system. Thermal stability of the DC-DC boost converter with SiC MOSFETs is confirmed with high switching frequency test up to 300 kHz. However, it is also witnessed with the high temperature test that the turn-off current waveform of SiC MOSFET (Z-FET) is distorted when the switch temperature reaches 145 °C and above. It is believed to be the SiC MOSFET inherent oxide interface issues at high temperature [13], [14].

978-1-4799-2706-7/14 $31.00 © 2014 IEEE

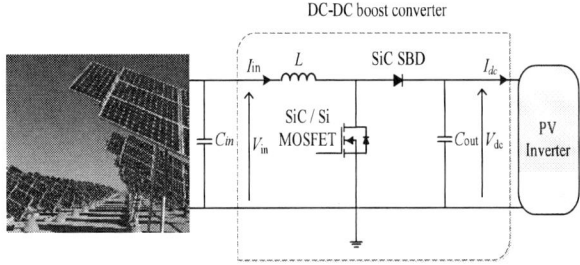

Fig. 1. Pre-regulator DC-DC boost converter in PV energy system.

II. EXPERIMENTAL

Fig. 2 shows the picture of 1.2 kW DC-DC boost converter built for comparative thermal performance test between SiC MOSFETs and Si MOSFET. The specifications of the converter are shown in Table I. The same heat sink with a thermal resistance of 0.5 °C/W in natural convection was used. In this test, similar ratings of SiC MOSFETs (SiC Z-FET and SiC N-channel MOSFET) are selected as the active switches to compare with Si MOSFET. The key parameters of the switches are shown in Table II. The diode used here is SiC Schottky Barrier Diode (SBD) from Semisouth (SDP30S120-T).

TABLE I
SPECIFICATIONS OF A BOOST CONVERTER AS A PHOTOVOLTAIC PRE-REGULATOR

Parameter	Value	Parameter	Value
Input Voltage (V_{in})	240 V	Duty cycle	0.4
Output Voltage (V_{dc})	400 V	External Cooling	None
Output Power (P_{out})	1.2 KW	Switching Frequency	25 ~ 300 kHz
Input Inductor Ripple	20 % OF $I_{IN, AVG}$	Case Temperature	25 ~ 150 °C
Output Voltage Ripple	2 % OF $V_{OUT, AVG}$	Heat sink	0.5 °C/W

Teledyne LeCroy-HDO4054 is used for the accurate measurements and analysis of switching waveforms. Heat sink temperature and switching waveforms without external cooling for both SiC MOSFETs and Si MOSFET were measured by increasing the switching frequency up to 300 kHz in steps of 25 kHz. The case temperature of the switches was slowly increased from 25 °C to 150 °C by a high-temperature test setup (shown on Fig. 3). The same gate driver was used for this test. Fig. 3 shows a high temperature setup using the controlled hot plate connecting to the switches via aluminum plate at the bottom side of the switches. A fan was used to cool the driver board at high frequencies. The designed converter was operated in continuous conduction mode (CCM), open-looped processing 1.2 kW of input power.

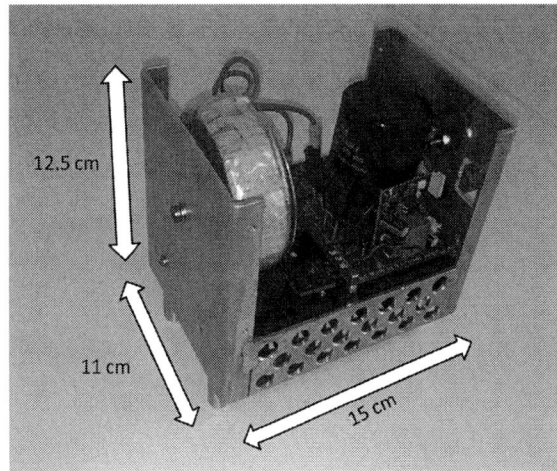

Fig. 2. Non-isolated DC-DC boost converter.

TABLE II
PARAMETERS OF THE EVALUATED DEVICES

Parameter	Devices		
	SiC MOFET (Z-FET)	SiC MOSFET	Si MOSFET
Breakdown Voltage	1200 V	1200 V	1200 V
Rated Current	20 A @100 °C	22 A @100 °C	20 A @100 °C
R_{DS}(on)	0.08 Ω	0.08 Ω	0.53 Ω
Max. Junction temp.	150 °C	150 °C	150 °C
Part No.	C2M0080120D	SCT2080KE	APT28M120B2
Package	TO–247	TO-247	TO-247

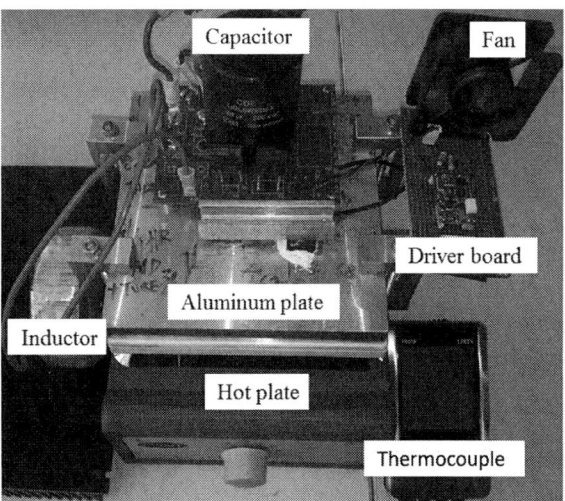

Fig. 3. High-temperature test setup.

A. Switching performance evaluation

Switching waveforms of the devices are compared by increasing switching frequency from 25 to 300 kHz. Note that for Si MOSFET switching frequency only increased up to 150 kHz as the heat sink temperature reached over

100 °C at 150 kHz without a fan to cool it. It is known that switching waveforms during the turn-on and turn-off hardly change with increasing switching frequencies until 300 kHz for SiC MOSFETs and until 150 kHz for Si MOSFET.

However, switching waveforms between the tested switches in the 1.2 kW DC-DC boost converter at 150 kHz switching frequency were different from each other, especially during the turn-on. From Fig. 4, it can be seen that the turn-on time for SiC MOSFETs is over 13 times faster than that of Si MOSFET. Here, the turn-off time is defined as the time current falls to 90 % until the voltage rises to 90 % of its dc value. On the contrary, the turn-on time is defined as the time from the current rises to 10 % of its peak value to the time voltage rises to 90 % of its dc value.

The conduction loss of MOSFET can be determined by

$$P_{S,CON} = I_{Drms}^2 \cdot R_{DSon} (I_{Drms}) \qquad (1)$$

where on-state resistance R_{DSon} can be found in datasheet. From [15], the switching energy losses in MOSFET is estimated by

$$P_{SW} = \tfrac{1}{2} \cdot I_{Drms} \cdot V_{DS} \cdot (t_{off} + t_{on}) \cdot f_{sw} + \tfrac{1}{2} C_{OSS} \cdot V_{DS}^2 \cdot f_{sw} \qquad (2)$$

where C_{OSS} is the output capacitance of the MOSFET.

Similarly, conduction loss of the SiC diode is determined by

$$P_{D,CON} = I_{Frms} \cdot V_F \qquad (3)$$

where forward voltage V_F also can be obtained from the datasheet. Note that reverse recovery loss of SiC SBD is not significant [16], so it is not included in total loss calculation. It is important to mention that the actual results of the measured losses show a great contrast to the calculated results from the equations above as the general equations do not represent the actual losses.

The resultant turn-on switching loss for Si MOSFET is 624.1 µJ, whereas its switching loss for SiC Z-FET and SiC MOSFET only reaches at 22.6 µJ and 94.3 µJ, respectively. Table III summarises measured turn-on and turn-off transient time and its losses for the tested switches.

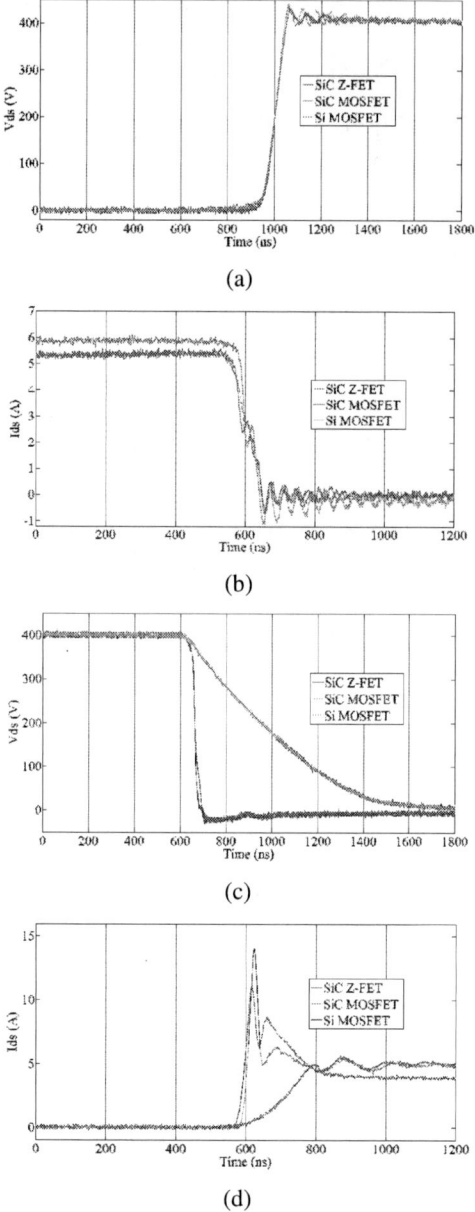

(a)

(b)

(c)

(d)

Fig. 4. Switching waveforms during the turn-on and turn-off for SiC Z-FET, SiC MOSFET and Si MOSFET at 150 kHz switching frequency: (a) and (b) Turn-off waveforms. (c) and (d) Turn-on waveforms.

TABLE III

MEASURED TURN-ON AND TURN-OFF TRANSIENT TIME AND ITS LOSSES OF THE TESTED SWITCHES IN A 1.2 kW DC-DC BOOST CONVERTER AT 150 kHz SWITCHING FREQUENCY

Switch Type	t_{on} (ns)	t_{off} (ns)	Switching loss at t_{on} (W)	Switching loss at t_{off} (W)	Conduction loss (W)
SiC MOSFET (Z-FET)	22.2	31.2	3.5	1.5	1.2
SiC MOSFET	44.9	42.3	15.0	2.6	1.2
Si MOSFET	585.3	50.9	94.0	2.8	7.4

Even if the turn-off current switching transient has approximately the same switching times, it is found from Fig. 4 (c) and (d) that turn-on current switching transient show different transient according to the switches. For example, the turn-on current switching waveform for Si MOSFET is somewhat slower with less oscillatory than SiC MOSFETs. However, as can be seen from Fig.4 (b), high transient speed of SiC MOSFETs generates high di/dt compared to Si MOSFET [17].

The switches are heated up by a high-temperature test set-up via aluminum plate and thermal pad underneath the switches. Switching waveforms for each switch are recorded by increasing case temperature from 25 °C to 150 °C. It is known from this test that switching waveforms are temperature independent for SiC MOSFET, except turn-off current switching waveform of SiC Z-FET. Fig. 5 shows turn-off current transient switching waveforms for SiC Z-FET at the case temperatures of 25 °C, 90 °C and 150 °C. From the technical literature [18], some of the SiC MOSFETs show capability of operating above 200 °C. However, the actual data concerning their reliability are not provided. In this test, it is found that when the case temperature reaches at 145 °C, only the turn-off current waveform of SiC Z-FET is distorted. It is believed that the oxide layer gets affected by this high temperature [13]. [14]. On the other hand, SiC MOSFET (SCT2080KE) does not show any changes of switching waveforms until the case temperature reaches at 150 °C.

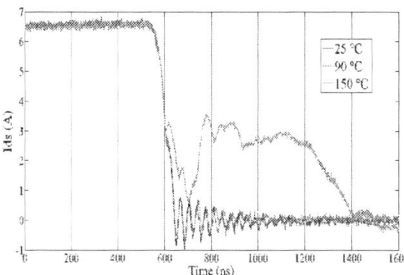

Fig. 5. Turn-off current switching waveforms of SiC Z-FET at different case temperatures.

B. Thermal Performance Evaluation

The heat sink temperatures against the converter switching frequencies are measured by increasing the switching frequency of the tested switches from 25 kHz to 300 kHz in steps of 25 kHz. From Fig. 6, it can be seen that the heat sink temperature for Si MOSFET reaches over 100 °C at 150 kHz. In the case of SiC MOSFETs including SiC MOSFET and SiC Z-FET, however, the heat sink temperature reaches only about 80 °C and 50 °C, respectively.

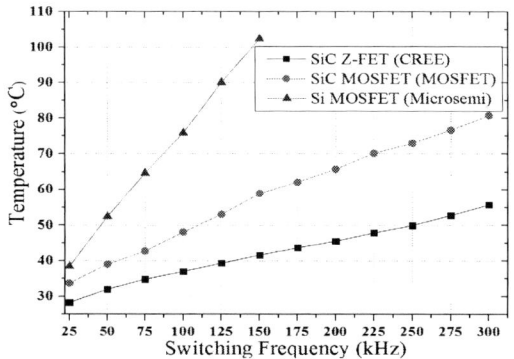

Fig. 6. Heat sink temperature against switching frequency for SiC Z-FET, SiC MOSFET and Si MOSFET.

It is known from this test that heat generated from the SiC Z-FET is relatively much smaller than SiC MOSFET and especially than Si MOSFET. These results provide important insight into the thermal management of the converter, especially into better heat dissipation and reducing overall converter design complexity. For example, when the SiC Z-FET is used in a 1.2 kW DC-DC converter, no external cooling is necessarily required for the switching frequency up to 300 kHz with 0.5 °C/W heat sink. However, if Si MOSFET is used in the same converter system, an external cooling method such as a fan will be required to maintain the system within the allowable operating temperature range even at the lower switching frequencies.

III. CONCLUSION

In this paper, experimental comparison results regarding thermal management of a 1.2 kW DC-DC boost converter as a photovoltaic pre-regulator using SiC Z-FET, SiC MOSFET and Si MOSFET are provided. The good thermal stability of the converter expected from both SiC MOSFETs compared with Si MOSFET at high switching frequency up to 300 kHz has been demonstrated.

SiC Z-FET generate surprisingly less heat at even high switching frequency of 300 kHz, enabling reduced converter design complexity with no external cooling requirements. However it is also found that the turn-off current switching waveform gets distorted when the case temperature of the switch reaches 145 °C, whereas its waveforms for SiC MOSFET and Si MOSFET are not affected by this temperature.

REFERENCES

[1] H. Ohashi, I. Omura, S. Matsumoto, Y. Sato, H. Tadano, and I. Ishii, "Power electronics innovation with next generation advanced power devices," *IEICE Transactions on Communications*, vol. E87-B, pp. 3422-3429, 2004.

[2] N. Ikeda, Y. Niiyama, H. Kambayashi, Y. Sato, and T. Nomura, "GaN power transistors on Si substrates for switching applications," in *Proc. IEEE,* vol. 7, pp. 1151-1161, 2010.

[3] A. Avron and P. Roussel, "SiC Market 2010-2020: 10 year market projection," Technology & Market Report, 2011.

[4] S. K. Singh, F. Guedon, P. J. Garsed, and R. A. McMahon, "Half-bridge SiC inverter for hybrid electric vehicles: Design, development and testing at higher operating temperature," *6th IET Int. Conf. Power Electronics, 2012,* pp. 1-6.

[5] S. Araujo and P. Zacharias, "Perspectives of high-voltage SiC semiconductors in high power conversion systems for wind and photovoltaic sources," in *Proc. 14th Euro. Conf. Power Electron. and Appl., 2011,* pp. 1-10.

[6] L. M. Tolbert, H. Zhang, B. Ozpineci, M. S. Chinthavali, "SiC-based power converters," MRS Spring Meeting, 2008.

[7] P. Frank and B. Bruno "A new high voltage Schottky diode based on silicon-carbide (SiC)," *9th Euro. Conf. Power Electron. and Appl. (EPE), 2001,* pp. 259.

[8] H. Tanaka, T. Hayashi, Y. Shimoida, S. Yamagami, S. Tanimoto, and M. Hoshi, "Ultra-low Von and high voltage 4H-SiC heterojunction diode," in *Proc. 17th Int. Sympo..Power Semiconductor Devices and ICs (ISPSD), 2005,* pp. 287-290.

[9] M. K. Das, B. A. Hull, J. T. Richmond, B. Heath, J. J. Sumakeris, and A. R. Powell, "Ultra high power 10 kV, 50 A SiC PiN diodes," in *Proc. 17th Int. Sympo..Power Semiconductor Devices and ICs (ISPSD)*, 2005, pp. 299-302.

The 2014 International Power Electronics Conference

[10] S. Olivier, B. Bruno, and L. Sascha, "Silicon carbide (SiC) D-MOS for grid-feeding solar- inverters," *12th Euro. Conf. Power Electron. and Appl. (EPE)*, 2007, pp. 1-10.

[11] B. Burger, D. Kranzer, and O. Stalter, "Efficiency improvement of PV-inverters with SiC-DMOSFETs", *Materials Science Forum,* vol. 600-603, pp. 1231-1234, 2009.

[12] N. G. Dhere, "Reliability of PV modules and balance of system components," in *Proc. 31st IEEE Photovolt. Spec. Conf., 2005,* pp. 1570-1576.

[13] K. Sheng, Y. Zhang, M. Su, J. H. Zhao, X. Li, P. Alexandrov, and L. Fursin, "Demonstration of the first SiC power integrated circuit," *Solid State Electron .,* vol. 52, pp. 1636-1646, 2008.

[14] M. Gurfinkel, H. D. Xiong, K. P. Cheung, J. S. Suehle, J. B. Bernstein, Y. Shapira, A. J. Lelis, D. Habersat, and N. Goldsman, "Characterization of transient gate oxide trapping in SiC MOSFETs using fast I-V techniques," *IEEE Trans. Electron Devices,* vol. 55, pp. 2004-2012, 2008.

[15] Z. J. Shen, Y. Xiong, X. Cheng, Y. Fu, and P. Kumar, "Power MOSFET switching loss analysis: a new insight," in *Proc. 41st IAS Annual Meeting*, 2006, pp. 1438-1442.

[16] B. Wrzecionko, D. Bortis, J. Biela, and J. W. Kolar, "Novel AC-coupled gate driver for ultrafast switching of normally off SiC JFETs," *IEEE Trans. Power Electronics,* vol. 27, pp. 3452-3463, 2012.

[17] E. Rondon, F. Morel, C. Vollaire, and J. L. Schanen, "Impact of SiC components on the EMC behaviour of a power electronics converter," *IEEE Energy Conv. Cong. and Expo. (ECCE), 2012,* pp. 4411-4417.

[18] K. Takao, T, Shinohe, S. Harada, K. Fukuda, and H. Ohashi, "Evaluation of a SiC power module using low-on-resistance IEMOSFET and JBS for high power density power converters," in *Proc. IEEE Appl. Power Electron. Conf. Expo, 2010,* pp. 2030-2035.

978-1-4799-2706-7/14 $31.00 © 2014 IEEE

Tolerance Analysis of a Constant-On Time Current-Mode Voltage Regulator with Adaptive Voltage Position Feature

Chih Wei Chen
EE department.
National Taiwan University
Taipei, Taiwan
R00921071@ntu.edu.tw

Dan Chen
EE department.
National Taiwan University
Taipei, Taiwan
chend@cc.ee.ntu.edu.tw

Shin Shiung Wang
Richtek Technology Corporation
Hsinchu, Taiwan
edwin_wang@richtek.com

Abstract — **Voltage regulators (VRs) have been widely used in many computer applications for central processing units (CPUs). In recent years, varieties of constant on-time control schemes have attracted much attention due to its high efficiency under both the light-load and the heavy load condition.**

In this paper, tolerance analysis was conducted for a constant on-time current-mode buck converter. Three methods are used: root-sum squared analysis (RSS), extreme value analysis (EVA) and Monte Carlo analysis. The effects of practical component tolerance and temperature variation are included in the analysis to estimate the converter output load line which is a critical performance for many applications requiring energy-saving adaptive-voltage-position (AVP) feature. A sensitivity analysis is also conducted to assess the necessity of parameter trimming for the analog power control integrated circuits.

Keywords — *tolerance analysis, AVP feature, constant on time, current mode, multi-phase.*

I. INTRODUCTION

The subject of constant on-time controllers for DC-DC converters has been a research topic in recent years. It features relatively high light-load efficiency compared with converters using conventional control scheme while retaining high efficiency under heavy-load condition. The light-load efficiency has become a clear mandate for many applications because most of the devices are operating under light-load condition most of the time.

In this paper, converter tolerance analysis will be conducted for a current-mode constant on time (CMCOT) multi-phase buck converter with adaptive voltage position (AVP) energy-saving feature for computer power applications. The AVP feature is depicted by Fig. 1, in which the converter output load line is required to stay within the tolerance band indicated by the dotted line [1]. The tolerance of the circuit components and the variations of the temperature effects affects the overall output load line tolerance. In this paper, three commonly-used analyses are used: root-sum squared analysis (RSS), extreme value

analysis (EVA), and Monte-Carlo analysis. Sensitivity analysis is also conducted to pinpoint the critical component parameters affecting the performances.

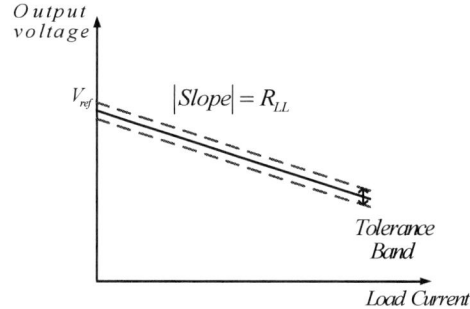

Fig.1 Output load line to depict AVP feature of a DC converter.

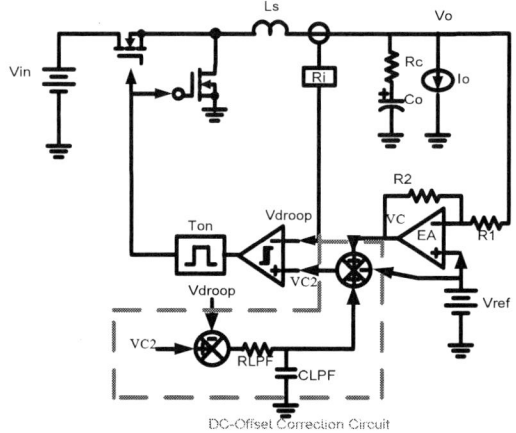

Fig.2 Circuit diagram of OCCMCOT with AVP function

II. DESCRIPTION OF CURRENT-MODE CONSTANT ON-TIME REGULATORS WITH AVP FEATURE

Fig.1 shows the circuit diagram of a CMCOT circuit with AVP feature in which the inductor current is sensed to provide load current information and the output voltage is sensed to achieve necessary output voltage regulation. The switching frequency ripple of both the inductor current and the load voltage are used for duty cycle modulation. When the load current is increased, the output voltage is purposely designed to droop. To achieve AVP function, a low-pass filter $R_{LPF}C_{LPF}$ is used to retrieve the dc information for output offset voltage cancellation [2], that is necessary for ripple-based constant on time converters [3], [4]

In the figure, V_{ref} is the output reference voltage, Rc is the equivalent series resistance (ESR) of the output capacitance Co, A_{vo} is the negative gain provided by the error amplifier (EA), Ri is the current sensing gain to generate a feedback signal V_{droop}. From the arrangement at Node $Vc2$, (1) can be derived. Substituting V_c by $[V_{ref}(1 + A_{vo}) - A_{vo}V_{out}]$, (2) can be derived. Rearranging (2) leads to (3) in which AVP function is built in, and the magnitude of the load line is equal to (Ri/A_{vo}).

$$V_{c2} - V_{droop} + V_c - V_{ref} = V_{c2} \qquad (1)$$

$$V_{ref}(1 + A_{vo}) - V_o A_{vo} = V_{ref} + I_o R_i \qquad (2)$$

$$V_o = V_{ref} - I_L \times \frac{R_i}{A_{vo}} \text{ , where the magnitude of the } \frac{R_i}{A_{vo}} \qquad (3)$$

Is the load line slope R_{LL}

III. TOLERANCE ANALYSIS AND RESULTS

Fig.3 shows the circuit diagram for an OCCMCOT controller in an integrated circuit (IC) implementation for a 4-phase buck converter. From the upper-left corner, the Block

DAC represents a digital-to-analog converter providing the reference voltage (V_{ref}) in Fig.2, EA is an error amplifier, BUFF is a buffer for reducing loading effect of DAC signal. The network inside the dotted line is a negative coefficient compensation (NTC) network used to compensate the temperature effect of the dc resistor (DCR) of the output inductor. Resistor RFB1, RFB2, RCM1 and RCM2 are used for voltage division, Risen is related to the current sensing gain of each phase, EAGM, CS and CSS are trans-conductance amplifiers. There are used to convert voltage signal into current signals which are easier for feedback signals summation. LPF is the low-pass filter $R_{LPF}C_{LPF}$ network. A PWMCP is a comparator for generating a "set" signal to SR latch to achieve pulse-width modulation (PWM). Below the diagram, there are a number of definitions and values of the symbols to be used in the later mathematical analysis.

The relationship between output voltage V_o and the contributing parameters can be derived. It is shown in (4), where the first term is affected by the no-load output voltage setting; the second term is affected by the low pass filter of the offset circuit network, and the third term is affected by the current sensing gain and adaptive voltage positioning specification.

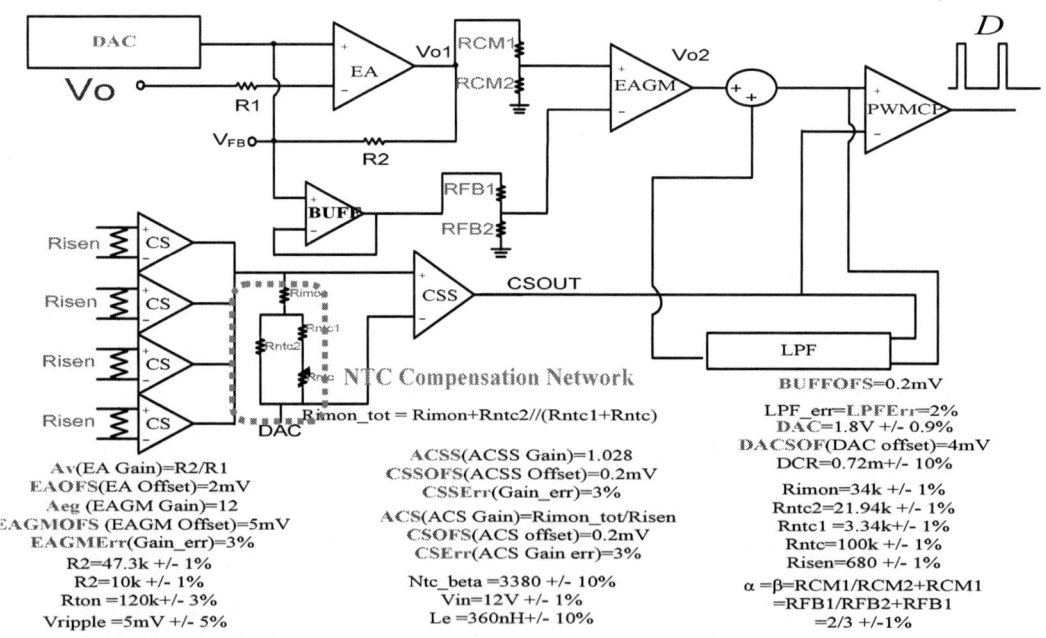

Rimon_tot = Rimon+Rntc2//(Rntc1+Rntc)

Av(EA Gain)=R2/R1
EAOFS(EA Offset)=2mV
Aeg (EAGM Gain)=12
EAGMOFS (EAGM Offset)=5mV
EAGMErr(Gain_err)=3%
R2=47.3k +/- 1%
R2=10k +/- 1%
Rton=120k+/- 3%
Vripple =5mV +/- 5%

ACSS(ACSS Gain)=1.028
CSSOFS(ACSS Offset)=0.2mV
CSSErr(Gain_err)=3%
ACS(ACS Gain)=Rimon_tot/Risen
CSOFS(ACS offset)=0.2mV
CSErr(ACS Gain err)=3%

Ntc_beta =3380 +/- 10%
Vin=12V +/- 1%
Le =360nH+/- 10%

BUFFOFS=0.2mV
LPF_err=LPFErr=2%
DAC=1.8V +/- 0.9%
DACSOF(DAC offset)=4mV
DCR=0.72m+/- 10%

Rimon=34k +/- 1%
Rntc2=21.94k +/- 1%
Rntc1 =3.34k+/- 1%
Rntc=100k +/- 1%
Risen=680 +/- 1%

$\alpha = \beta$=RCM1/RCM2+RCM1
=RFB1/RFB2+RFB1
=2/3 +/-1%

Fig.3 Four-phase IC Implemented circuit of OCCMCOT buck converter

$$V_O = (DAC + DACOFS + EAOFS)(1 + \frac{1}{A_V} - \frac{\beta}{\alpha \times A_V})$$

$$+ \frac{Aeg \times EAGMOFS - Aeg \times \beta \times Buffoffs - \left(\frac{V_{in} - DAC}{2 \times L} \times T_{on} \times DCR \times \left(R_{imon} + \frac{(R_{ntc2})\left(R_{ntc1} + R_{ntc,T} e^{ntc\_beta\left(\frac{1}{T+173} - \frac{1}{303}\right)} \right)}{R_{ntc2} + R_{ntc1} + R_{ntc,T} e^{ntc\_beta\left(\frac{1}{T+173} - \frac{1}{303}\right)}} \right) \times \frac{1}{R_{isen}} \times ACSS + \frac{V_{ripple}}{2} \times A_{vo} \times \alpha \times Aeg \right) \times LPFerr}{Aeg \times \alpha \times A_V}$$

(4)

$$- \frac{ACSS\left[(I_{out1} \times DCR1 + I_{out2} \times DCR2 + I_{out3} \times DCR3 + I_{out4} \times DCR4)[1 + 0.00393 \times (T - 30)] + CSOFS4 + CSOFS3 + CSOFS2 + CSOFS1 \right] \times \left(R_{imon} + \frac{(R_{ntc2})\left(R_{ntc1} + R_{ntc,T} e^{ntc\_beta\left(\frac{1}{T-173} - \frac{1}{303}\right)} \right)}{R_{ntc2} + R_{ntc1} + R_{ntc,T} e^{ntc\_beta\left(\frac{1}{T-173} - \frac{1}{303}\right)}} \right) \times \frac{1}{R_{isen}} + CSSOFS}{Aeg \times \alpha \times A_V}$$

The relationship between output voltage V_o and the contributing parameters can be derived. It is shown in (4), where the first term is affected by the no-load output voltage setting; the second term is affected by the low pass filter of the offset circuit network, and the third term is affected by the current sensing gain and adaptive voltage positioning specification.

From Equation (4), the total output voltage error ΔV_o due to parameter tolerance can be obtained, as shown by (5) and (6), where x's are the contributing parameters and Δx 's are the corresponding tolerances. (5) shows the result obtained by extreme value analysis (EVA), and (6) the root-sum squared (RSS) analysis.

$$\Delta V_{o,error,EVA} = \sum_i \frac{\partial V_o}{\partial x_i} \times \Delta x_i \qquad (5)$$

$$\Delta V_{o,error,RSS} = \sum_i \sqrt{\left(\frac{\partial V_o}{\partial x_i} \times \Delta x_i \right)^2} \qquad (6)$$

Table I shows the result, where the first term in each entry is the nominal value of output voltage and the second term is the output voltage error tolerance of all the parameters indicated in Fig.3. Monte Carlo statistical simulation is used to compare with RSS and EVA. The result shown in Table I can be used to plot the load lines of the converter AVP feature. Fig.1 shows the results. The comparison shows that the RSS underestimates the output voltage error while the EVA overestimates the output voltage error.

Sensitivity analysis was also done and the results are shown in Fig.5, for the top five parameters causing output voltage errors in different load current levels can be observed. From the results, several observations are remarked below. (1) DAC parameter tolerance dominates the output voltage error. (2) DAC's dominance decreases as the load current is increased. (3) DCR effects become increasingly important at high load current level. (4) To meet the specification, the worst condition occurs at full load, and the easiest condition at no load.

TABLE I. Output voltage error ($V_{o,error}$)

	$I_{out} = 0(A)$	$I_{out} = 30(A)$	$I_{out} = 60(A)$	$I_{out} = 100(A)$
Root-sum squared analysis	$1.8 \, V \pm 17.182 \, (mV)$	$1.755 \, V \pm 17.648 \, (mV)$	$1.71 \, V \pm 19.746 \, (mV)$	$1.65 \, V \pm 24.319 \, (mV)$
Extreme value analysis	$1.8 \, V \pm 25.879 \, (mV)$	$1.755 \, V \pm 40.363 \, (mV)$	$1.71 \, V \pm 54.837 \, (mV)$	$1.65 \, V \pm 74.136 \, (mV)$
Monte Carlo analysis	$1.799 \, V \pm 20.389 \, (mV)$	$1.755 \, V \pm 21.279 \, (mV)$	$1.708 \, V \pm 19.746 \, (mV)$	$1.649 \, V \pm 28.395 \, (mV)$

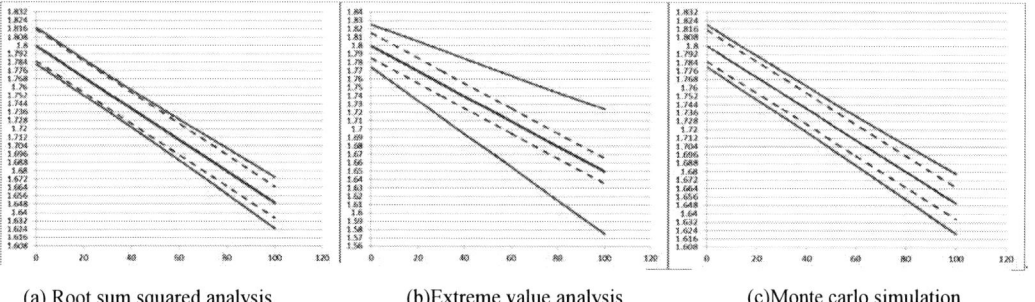

(a) Root sum squared analysis   (b)Extreme value analysis   (c)Monte carlo simulation

Fig.4 The results of tolerance analyses of the converter output load line
Solid line: nominal value
Dotted line: specification (30mV tolerance band)

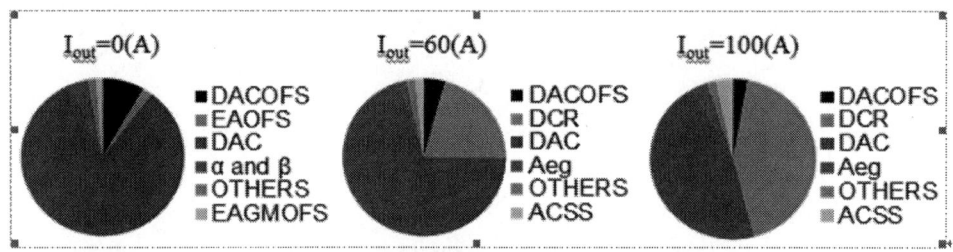

Fig.5 Output voltage error sensitivity to parameter changes
for different load current levels

IV. CONCLUSIONS

Tolerance analyses were conducted for a constant on-time current-mode DC-DC converter with AVP feature for computer voltage regulator supply applications. Practical component tolerances were used to obtain the output load line variation and check against INTEL's specification. It's found that the tolerance of the DAC dominates the output voltage error. Without trimming the DAC offset voltage, it's not possible to meet the specification. The DCR tolerance value of the inductor also plays an important role, especially under heavy load condition. The analytical results presented in the paper provide a useful tool for meeting the specification.

ACKNOWLEDGMENT

This work was supported by a research grant from Richtek Inc. to Taiwan University under Grant No. 102-S-C46. The authors would also like to thank SIMPLIS Technologies, Inc., for supplying SIMPLIS simulation tool.

REFERENCES

[1] Intel, Voltage Regulator-Down (VRD) 11.1

[2] Y.J. Chen, Dan Chen, Y.C. Lin, C.J. Chen, and C.H. Wang,"A Novel Constant On-Time Current-Mode Control Scheme to Achieve Adaptive Voltage Positioning for DC Power Converters,"IEEE Industrial Electronics (IECON), Montreal Canada, Oct. 2012

[3] I-Chieh Wei, Dan Chen, Yu-Cheng Lin, and Ching-Jan Chen,"The Stability Modeling of Ripple-Based Constant On-Time Control Schemes Used in the Converters Operating in DCM,"International Conference on Renewable Energy Research and Applications (ICRERA), Nagasaki, Japan, Nov. 2012

[4] Yu-Cheng Lin, Ching-Jan Chen, Dan Chen, and B. Wang,"A Ripple-Based Constant On-Time Control with Virtual Inductor Current and Offset Cancellation for DC Power Converters" IEEE Trans. on Power Electronics, vol. 27, Issue

[5] Huang, W., Clavette, D.; Schuellein, G.; Crowther, M.; Wallace, J "System Accuracy Analysis of the Multiphase Voltage Regulator Module," APEC, 2003 , Page(s): 731 - 737 vol.2

The 2014 International Power Electronics Conference

FPGA-based Digital-controlled Power Converter Designed with Universal Input Meeting 80 Plus Platinum Efficiency Code and Standby Power Code for Sever Power Applications

Yen-Shin Lai, *Fellow, IEEE*
Center for Power Electronics Technology,
National Taipei University of Technology
Taipei, Taiwan
E-mail: yslai@ntut.edu.tw

Kung-Min Ho
AcBel Polytech Inc.
New Taipei City, Tamsui, Taiwan
E-mail: gmho821@gmail.com

Abstract—**This paper contributes to the design and control of digital-controlled power converter meeting platinum efficiency code and standby power code for sever power applications. Hardware design, control methods and software implementation based upon FPGA will be presented. First, design of hardware and controller is presented, and followed by the loss analysis of power converter to provide a systematic approach to meeting the platinum efficiency code. A feed-forward control for PFC stage with wide range input voltage is presented to provide a *constant* bandwidth despite of input voltage variation. Two control modes for PSFB converter are presented to meet standby power code.**

Software implementation based upon FPGA is then included. Experimental results of FPGA-based digital-controlled power converter are presented. It will be shown the standby power is less than 0.5W and light load efficiency is improved to go up to more than 90% for both low line and high line input voltage. Moreover, it will be confirmed that the platinum efficiency code can be met under 20%, 50% and 100% load conditions. These results fully support the claims and contributions of this paper.

Keywords— Power Factor Corrector, Phase-shift Full-bridge Converter, Server Power, Standby Control

I. INTRODUCTION

Switching power converters have been widely used in development of semiconductor technology and cloud technique, data center and server power applications. In these applications, hundreds to thousands watts power converter with low output voltage and high output current are needed. Several researches for server power have been proposed in [1]-[7]. Two main topics, high efficiency and low standby power losses, are required to follow. Efficiency code has been published in Energy Star - 80 PLUS Certification [8] which requires the efficiency to achieve certain figures under 20%, 50% and 100% rated power conditions. The International Energy Agency (IEA) has proposed a "1-W plan," the participating countries seek to lower standby power less than 1 W in all products by 2010 [9].

Switching power converter consists of AC/DC and DC/DC, in general. Comparisons between single-stage and two-stage topologies of server power have been presented in [10]-[12]. The results show that single-stage topology has

simpler control circuits and power stage. However, its power rating of some key components increases and may result in lower efficiency as compared with those for two-stage topology. For high power and high efficiency applications, two-stage approach is selected in general.

Power factor corrector (PFC) has been widely used for AC/DC conversion while meeting the related harmonics requirement. The standby power control methods of PFC include turned-off of power stage, decrease of switching frequency and burst mode control under no-load condition [12]. Phase-shifted full-bridge (PSFB) converter with current-doubler and synchronous rectifier technique is one of the most popular candidates for DC/DC conversion. Since PSFB converter has the special features of zero-voltage-switching (ZVS) and reduction of current stress of power device. Moreover, PSFB converter with current-doubler and synchronous rectifier can further reduce current ripple and conduction losses. Therefore, PSFB converter with current-doubler and synchronous rectifier is suitable for server power application. The standby power control methods for PSFB include pulse skipping, burst mode and off-time modulation under no-load condition [13]-[15]. In [16], full-bridge switching mode control between PWM and phase-shifted is proposed to improve efficiency under light-load condition. The proposed technique is realized using digital controller rather than discrete integrated circuit. The advantages of digital control include less discrete-component count, no aging issue for the compensator components, higher flexibility and fast time-to-market.

The objective of this paper is therefore to present the design and control of digital-controlled server power which is with universal input voltage and can meet the codes of platinum efficiency and standby power. It is expected that this paper will provide a design reference for the reader interested in the servo power design and its digital implementation.

First, the design of hardware and controller is presented, and followed by the loss analysis of power converter to offer a systematic approach to meeting the platinum efficiency code. A feed-forward control for PFC stage with wide range input voltage is presented in order to provide a constant bandwidth despite of input voltage variation. Two control modes for

978-1-4799-2706-7/14 $31.00 © 2014 IEEE

PSFB converter are presented in order to achieve standby power code. Software implementation based upon FPGA will be included and followed by the experimental results. It will be shown that the standby power is less than 0.5W and light load efficiency is improved to go up to more than 90% for both low line and high line input voltage. Moreover, it will be confirmed that the platinum efficiency code can be met. These results will fully support the design and contributions of this paper.

II. HARDWARE AND CONTROLLER DESIGN OF PFC CONVERTER FOR UNIVERSAL INPUT

A. Hardware design

Figure 1 shows the block diagram for boost PFC converter with average current mode control and feed-forward control. As shown in Fig. 1 the input AC source is converted to DC voltage via front-end diode rectifier and capacitor. Then, a boost inductor is used for output voltage boost by charging energy in it as the MOSFET, "S", is turned on. When the MOSFET, "S", is turned off, DC voltage and charged inductor provides energy to the load and thereby "boosting" the output voltage level. In order to control the output voltage to the reference level, the output voltage is sensed and compared with its reference to give the magnitude of current reference. Moreover, to achieve power factor correction which means to force the input current to be in phase with the input voltage, the input voltage is sensed to be the waveform reference of the current reference.

The input current, I_{in} reaches its maximum value for low line input voltage under rated power condition. Therefore, peak input current of PFC converter can be expressed by:

$$I_{in,peak} = \frac{\sqrt{2} \cdot P_{out}}{\eta_{PFC} \cdot V_{in(\min)}} \qquad (1)$$

Considering ripple current, the inductance of PFC converter can be derived by:

$$L \geq \frac{\sqrt{2} \times V_{in(\min)} \times D_{PFC}}{f_{sw} \times I_{in,ripple}} \qquad (2)$$

Moreover, output capacitance, C_o is designed by hold-up time, T_{HU}. Hold-up time means the time period in which the front-end converter can still provide energy for load to operate while meeting the required specifications after AC line input is lost. In general, the hold-up time requires more than 17 ms [18]. Therefore, output capacitor, C_o can be calculated by:

$$C_o \geq \frac{2 \times P_{out} \times T_{HU}}{V_{bus}^2 - V_{bus,\min}^2} \qquad (3)$$

Where $V_{bus,min}$ is designed by considering turn ratio, duty, output voltage and voltage loss of DC/DC stage as shown in (4).

$$V_{bus,\min} = \frac{N_S}{N_P} \frac{V_{out} + V_{loss}}{D_{DC/DC,stage}} \qquad (4)$$

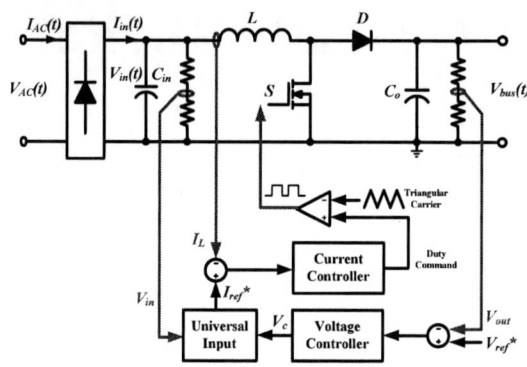

Fig. 1 PFC control, average current mode control with input voltage feed-forward

TABLE I shows designed specifications of PFC converter. According to (1)-(4) and TABLE I, the specifications of boost inductance (L), output capacitance (C_o), power device (S), output diode (D) and bridge rectifiers will be designed and selected. More details of the hardware design are described as follows.

According to (1), the efficiency of PFC converter, η_{PFC} is assumed to be 0.9, $I_{in,peak}$ is 24.44 A under V_{AC} = 90 V_{rms} and P_{out} = 1400 W. Considering 10% of ripple, the peak input current, $I_{in,ripple}$ = 2.444 A. Moreover, duty of boost PFC, D_{PFC} is 0.682 for low line input voltage. Therefore, the inductance of PFC inductor, L, can be determined by (2) to give 748 μH and the related DCR of inductor, R_L = 0.03 Ω (by measurement). The hold-up time of PFC, T_{HU} is assume to be 20 ms and $V_{bus,min}$ is 300 V. Therefore, the capacitance of C_o can be determined by (3) to yield 872 μF and the corresponding equivalent series resistance, R_{ESR} is 37.4 $m\Omega$.

Moreover, the urn-on resistance of a MOSFET, reverse recovery time of output diode and forward voltage drop of the bridge rectifiers should be as small as possible in order to reduce the conduction loss and switching loss. Table II shows the hardware design results of PFC. As shown in Table II, FCH76N60NF power device is selected which is an N-channel low turn-on resistance MOSFET and its gate charge is 230 nC typically. A silicon carbide Schottky diode, C3D20060D, which is with zero reverse recovery current and positive temperature coefficient on forward voltage drop, is picked up for the output diode. Moreover, a low forward voltage drop, 0.76V, single-phase bridge rectifiers, forward voltage drop selected to reduce the forward voltage drop and losses.

Table I Specifications of PFC Converter

Input Voltage, V_{AC}	90~264 V_{rms}/60 Hz
Output Voltage, V_{bus}	400 Vdc
Output Power, P_{out}	1400 W
Switching Frequency, f_{sw}	50 kHz
Inductor current ripple	±10 %
Hold-up time, T_{HU}	20 ms
Efficiency of PFC, η_{PFC}	90 %

The 2014 International Power Electronics Conference

TABLE II HARDWARE DESIGN RESULTS OF PFC

Inductance, L	748 μH
DCR of Inductor, R_L	0.03 Ω
Capacitance, C_o	872 μF
ESR of C_o, R_{ESR}	37.4 $m\Omega$
Power device, S	FCH76N60NF ($R_{ds,on}$=28.7 $m\Omega$, Q_g=230 nC)
Output diode, D	C3D20060D (V_D=1.5 V)
Bridge rectifier	LVB2560 (V_D=0.76 V)

B. Controller design for university input

In order to design controller, the model of PFC represented in s-domain is derived as shown in Fig. 2. The related transfer functions for duty cycle to inductor current and inductor current to output voltage are shown in (5) and (6), respectively.

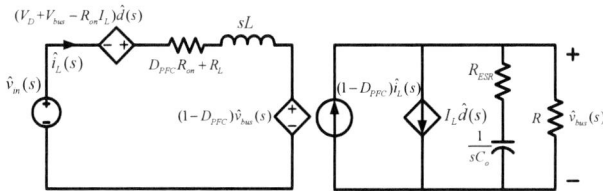

Fig. 2 Model of boost PFC, Laplace domain

$$G_{id}(s) = \frac{sC_o\left[(R_{ESR}+R)(V_D+V_{bus}-R_{on}I_L)+R_{ESR}R(1-D_{PFC})I_L\right]+V_D+V_{bus}-R_{on}I_L+R(1-D_{PFC})I_L}{s^2 LC_o(R_{ESR}+R)+s\left[L+C_o(R_{ESR}+R)(R_L+R_{on}D_{PFC})+C_oR_{ESR}R(1-D_{PFC})^2\right]+R_L+R_{on}D_{PFC}+R(1-D_{PFC})^2} \quad (5)$$

$$G_{oi}(s) = \frac{-s^2 LC_oRR_{esr}I_L+s\left\{C_oR_{esr}R\left[(V_D+V_{bus}-R_{on}I_L)(1-D_{PFC})-(R_L+D_{PFC}R_{on})I_L\right]-LRI_L\right\}+R\left[(V_D+V_{bus}-I_LR_{on})(1-D_{PFC})-(R_L+D_{PFC}R_{on})I_L\right]}{sC_o\left[(R_{esr}+R)(V_D+V_{bus}-R_{on}I_L)+R_{esr}R(1-D_{PFC})I_L\right]+V_D+V_{bus}-R_{on}I_L+R(1-D_{PFC})I_L} \quad (6)$$

For the controller design, the bandwidth of voltage loop as the input voltage changes for boost PFC with and without input voltage feed-forward has been analyzed in [17]. As shown in [17], without feed-forward control, the bandwidth of voltage loop varies as the input voltage changes. And the bandwidth of voltage loop for high line input is increased as it is designed based upon the low line input model. In contrast, the bandwidth of voltage loop for low line input is significantly *reduced* as it is designed based upon the high line input model. It is also shown in our previous results, feed-forward control can achieve constant bandwidth despite the input voltage variation. Fig. 3 shows the block diagram of boost PFC with input voltage feed-forward control. Since both current and voltage controllers are realized by digital controllers, current and voltage are sensed and fed back via AD converters. Based upon the designed hardware shown in TABLE II and PFC models, Fig. 4 shows the Bode plot of plant for current loop with feed-forward control. As shown in Fig. 4, the bandwidth of plant for low line input and high line are the same. Input voltage variation has *no* effect on the bandwidth. Fig. 5 shows the design results for the voltage controller based upon inner loop shown in Fig. 4. As shown in Fig. 5, the bandwidth of voltage loop is constant as the input voltage changes. And the designed results are summarized in TABLE III.

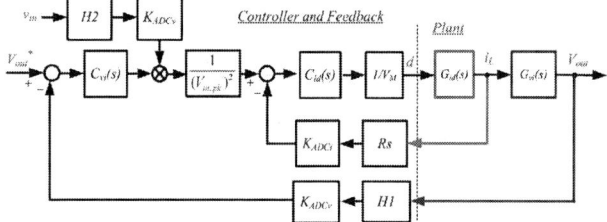

Fig. 3 Block diagram of PFC, with feed-forward control

TABLE III.
DESIGNED RESULTS OF CONTROLLER

	Current Loop	Voltage Loop
Kp	4.4545	43.5859
Ki	63075.72	2296.54
Bandwidth	2.29 kHz	5.1 Hz
Phase Margin	45.7 °	75.5 °

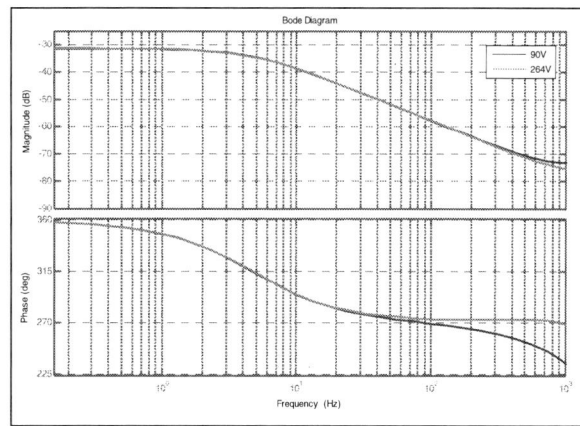

Fig. 4 Bode plot, plant and current loop, no voltage control, input voltage = 90V and 264V

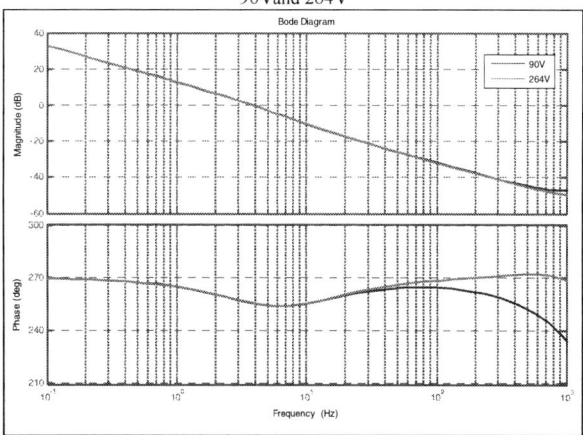

Fig. 5 Bode plot, plant, current loop and voltage control, with feed-forward control, input voltage = 90V and 264V

III. HARDWARE AND CONTROLLER DESIGN OF PHASE-SHIFT FULL-BRIDGE CONVERTER

A. *Hardware design*

Figure 6 shows the block diagram of full-bridge converter with current-doubler and synchronous rectifier. The converter is controlled by phase-shift switching method; therefore, the primary-side power devices can achieve ZVS under certain load condition. The current-doubler topology has been widely used for high power applications due to low current ripple and less conduction loss. Due to lower $R_{ds,on}$ of power device, the conduction losses are smaller than that for diode rectifier, especially for low voltage and high power applications.

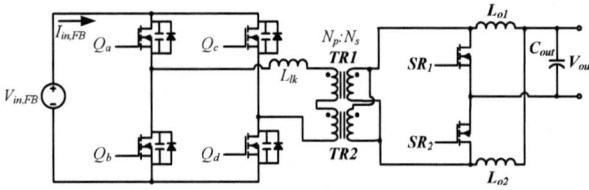

Fig. 6 Block diagram of full-bridge converter with current-doubler and synchronous rectifier

Due to high power application, primary windings of transformer are connected <u>in series</u> and secondary windings of transformer are <u>in parallel</u> in order to <u>reduce loss</u> and thereby increasing the efficiency. The turn number of primary and secondary windings can be expressed by:

$$N_p = \frac{V_{in,FB} \times D_{eff} \times 10^8}{4 \times A_e \times f_s \times B_{max}} \qquad (7)$$

$$N_s \geq \frac{N_p \times \left(V_{out} + V_{out,loss}\right)}{D_{eff,max} \times V_{in,FB}} \qquad (8)$$

Considering ripple current, the inductance of output inductor of PSFB converter, Lo1 and Lo2 can be expressed by:

$$L_{o1} = L_{o2} \geq \frac{V_{out} \times T_s \times \left(1 - D_{FB}\right)}{\Delta I_{out}} \qquad (9)$$

Moreover, for the given voltage ripple and switching frequency, the capacitance of output capacitor of PSFB converter can be derived by:

$$C_{out} \geq \frac{\Delta I_{out}}{\Delta V_{out}} \left(\frac{1}{16 \times f_s} + 65 \times 10^{-6}\right) \qquad (10)$$

The power devices, (Q_a, Q_b, Q_c, Q_d) are turned on in the way of zero voltage switching. To achieve ZVS, resonant inductor, L_{lk} should be large enough in order to draw the energy stored in C_{oss} of power devices. Therefore, resonant inductor, L_{lk}, can be determined by

$$L_{lk} > \frac{2}{I_{in,peak,FB}^2} \frac{4}{3} C_{oss} V_{in,FB}^2 \qquad (11)$$

TABLE IV shows specifications of the designed PSFB converter. According to (7)-(11) and TABLE IV, two

transformers (*TR1*, *TR2*), output inductor (L_{o1}, L_{o2}), output capacitor (C_{out}), power devices f(Q_a, Q_b, Q_c, Q_d), synchronous rectifiers (SR_1, SR_2), resonant inductor (L_{lk}) will be designed as follows.

As shown in Fig. 6, the voltage of primary side of each transformer winding is 200 V. Therefore, EE55 transformer bobbin and PC40 core are selected. Moreover, the maximum flux density, B_{max} is 2730 Gauss, effective area, A_e is 3.54 cm^2 and effective duty of PSFB converter, D_{eff} is 0.45. According to (7), primary-side turn number of transformer, N_p is 10. Similarly, with considering secondary-side loss of transformer and voltage drop, the secondary-side turn number of transformer, N_s is 2 turns which is determined by (8).

Since the rated current of PSFB converter is 100 A and 10% of ripple current is specified, the output inductance of PSFB converter, L_{o1} and L_{o2} can be determined by (9) to give 22 μH. Using Arnold Magnetics Limited-Powder Cores (MS-301125-2) for inductor, the related DCR of inductor, R_L is 5 $m\Omega$ which is derived by measurement. By specifying 5% of output voltage ripple, output capacitance of PSFB converter can be determined by (10) which is 2000 μF and its R_{ESR} is 45 $m\Omega$.

Moreover, turn-on resistance of a MOSFET for being the switch of power device should be small in order to reduce the conduction loss. IPB60R190C6 type MOSFETs with 0.19Ω of turn-on resistance, 63 nC of gate charge capacitance and C_{oss} = 85 pF, are picked up for Q_a, Q_b, Q_c, and Q_d. The power devices of synchronous rectifier (SR_1, SR_2) are IRFB4110PBF. It is a HEXFET Power MOSFET, turn-on resistance is 3.7 $m\Omega$ and gate charge is 150 nC under typical condition. The resonant inductance, L_{lk} can be determined by (11) to give 26.3μH. These hardware designed results for PSFB converter are summary in TABLE V.

TABLE IV DESIGNED SPECIFICATIONS OF PSFB

Input Voltage, $V_{in,FB}$	400 Vdc
Output Voltage, V_{out}	12 Vdc
Output Current, I_{out}	100 A
Output Power, P_{out}	1200 W
Switching Frequency, f_s	50 kHz
Inductor current ripple	±10% of output current
Output voltage ripple	±5%

TABLE V DESIGNED COMPONENTS OF PSFB

Turn ratio of transformer, n	0.1
Inductance, L_{o1}=L_{o2}/ DCR	22 μH/5 mΩ
Capacitance, C_{out}/ ESR, R_{esr}	2000 μF/45 mΩ
Leakage inductor, L_{lk}	26.3μH
Switch of power device, $Qa{\sim}Qd$	IPB60R190C6 ($R_{ds,on}$=0.19 Ω, Q_g=63 nC)
Switch of S.R., $SR1$ and $SR2$	IRFB4110PBF ($R_{ds,on}$=3.7 mΩ, Q_g=150 nC)

B. *Controller design of PSFB converter*

The model of PSFB converter represented in s-domain can be derived as shown in Fig. 7. The related transfer function for duty cycle to output voltage is shown in (12).

$$G_{vd}(s) = \frac{2nV_{in,FB}\left(sC_{out}RR_{esr}+R\right)}{s^2LC_{out}\left(R_{esr}+R\right)+s\left[L+C_{out}\left(R_{esr}+R\right)\left(R_d+R_L\right)+2C_{out}RR_{esr}\right]+R_d+R_L+2R} \quad (12)$$

Fig. 8 shows the block diagram of PSFB. The voltage is sensed and fed back via AD converter to the digital voltage controller. Fig. 9 shows the designed results for the voltage controller based upon plant model and the designed specifications as summarized in TABLE VI.

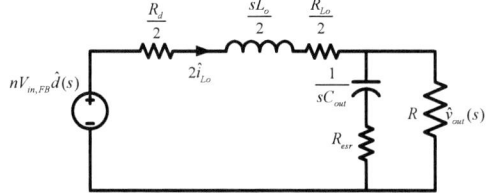

Fig. 7 Model of phase-shift full-bridge converter, Laplace domain

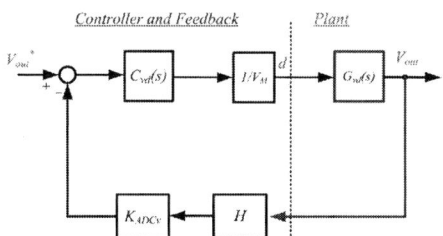

Fig. 8 Block diagram of PSFB

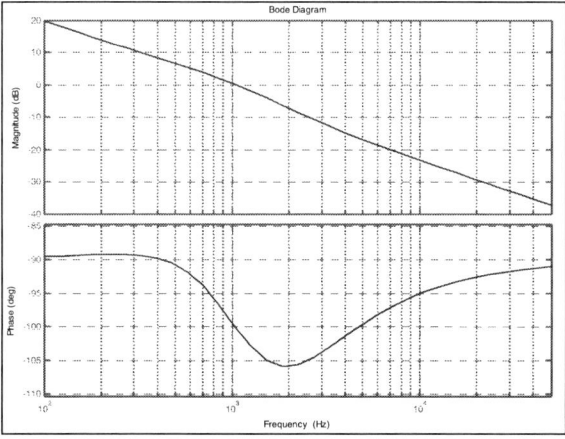

Fig. 9 Bode plot, loop gain with voltage controller

TABLE VI DESIGNED SPECIFICATINOS AND RESULTS OF CONTROLLER

	Voltage Controller
Kp	5.4922
Ki	39274.7
Bandwidth	1.05 kHz
Phase Margin	79.8 °

IV. SWITCHING CONTROL TECHNIQUES UNDER STANDBY CONDITIONS AND POWER LOSS ANALYSIS

A. *Switching Control Technique Under Standby Conditions*

The PSFB converter can achieve ZVS operation to obtain high efficiency. However, under light-load condition, ZVS is difficult to achieve and the duty cycle is small. Therefore, significant circulating loss dominates the losses as shown in our previous research results [16]. To eliminate circulating loss using PWM control method instead of phase-shift control under standby and light-load conditions has been proposed in [16]. To increase the efficiency especially under light load condition, there are three switching control modes for the proposed switching method: phase-shift switching control mode under heavy load condition, PWM switching control mode under light-load condition, and PWM switching with burst-mode control under no-load condition.

Moreover, PFC converter is controlled by burst mode under standby condition in order to reduce standby power losses. And PFC converter is with PWM control based upon average current mode control as load increases greater than 0.8% of rated load.

B. *Power Loss Analysis*

Power loss analysis of components for PFC converter has been presented in [3], [19]. The power loss, $P_{total\_loss,PFC}$, is calculated in half of fundamental period and it can be expressed as

$$P_{total\_loss,PFC} = P_{con\_mos} + P_{con\_diode} + \\ + P_{con\_L} + P_{con\_br} + P_{sw\_mos} + P_{core\_L} \quad (13)$$

The power losses include conduction losses of MOSFET, output diode, inductor, bridge rectifiers, switching loss of MOSFET and core loss of inductor. Fig. 10 and Fig. 11 show power loss analysis of components for PFC under full-load condition for V_{AC}=115 V_{rms} and V_{AC}=230 V_{rms}. As shown in Fig. 10 and Fig. 11, the major power losses are switching loss of MOSFET and core loss of inductor.

Power loss analysis of components for PSFB converter has been presented in [3], [16]. The power loss, $P_{total\_loss,PSFB}$, can be expressed as

$$P_{total\_loss,PSFB} = P_{con\_mos} + P_{con\_tr} + P_{con\_SR} + P_{con\_L} \\ + P_{sw\_mos} + P_{sw\_SR} + P_{cl\_tr} + P_{cl\_L} \quad (14)$$

The power losses of PSFB converter include conduction losses of MOSFETs, transformers, synchronous rectifier and inductor, switching losses of MOSFETs and SRs, and core losses of transformers and inductor. Fig. 12 shows power loss analysis of components for PSFB converter under full-load condition. As shown in Fig. 12, major power losses include conduction loss of transformers, and core loss of transformer and inductor.

According to the power loss analysis results of PFC and PSFB converter, the total power loss of PFC under full-load condition is 65.63 W and 31.2 W for V_{AC} (rms) =115 V and 230 V, respectively. And the total power loss of PSFB converter under full-load condition is 85.79 W. Therefore, the overall efficiency under full-load condition is 88.49% and 90.96% V_{AC} (rms) =115 V and 230 V, respectively.

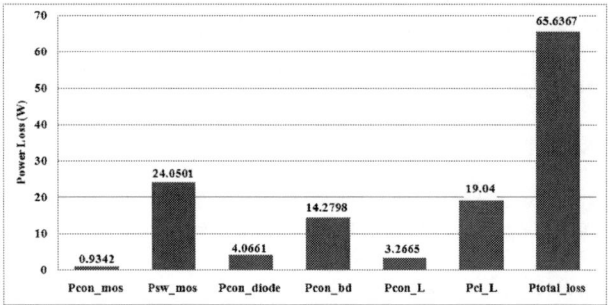

Fig.10 Power loss analysis, PFC converter, full-load condition, V_{AC} = 115 V

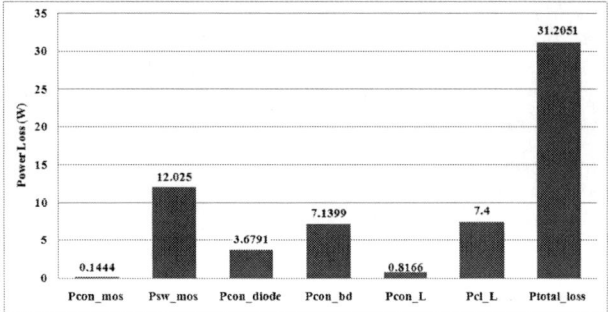

Fig.11 Power loss analysis, PFC converter, full-load condition, V_{AC} = 230 V

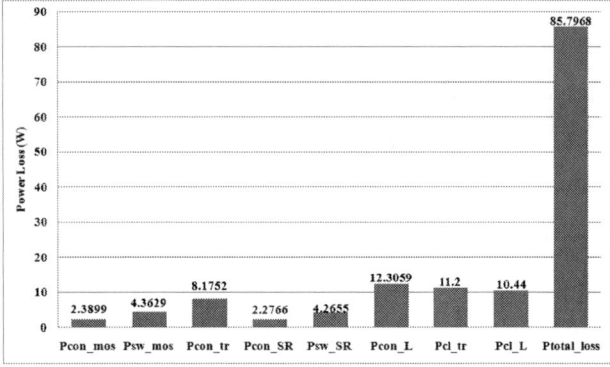

Fig.12 Power loss analysis, PSFB converter, full-load condition

V. FPGA-BASED IMPLEMENTATION

The specifications of the digital-controlled power converter for server applications: the input voltage is 90~264 V AC, the AC/DC output voltage is 400 V DC, the DC/DC output voltage is 12 V DC, the output power is 1200 W, and the switching frequency is 50 kHz. The AD converter is AD7825BN, which is with 8-bit resolution and 0~2V maximum input range. The FPGA is Cyclone II-EP202CF484C8 and its clock rate is 50 M Hz.

Fig. 13 shows the block diagram of the experimental system. As shown in Fig. 13, the FPGA controller is with common ground on the primary side of PSFB converter,

therefore, isolation for the feedback signals of PSFB output voltage and currents are required. Fig. 14 shows the photo of experimental system. In Fig. 14, for PFC converter, two MOSFETs and two output diodes are connected in parallel to reduce the conduction loss and switching loss.

Fig. 15 shows the block diagram of PFC and PSFB for FPGA-based implementation. The system block is composed of frequency multiplier, digital pulse width modulator, AD converters control, standby power control, PFC control loop and PSFB control loop.

The flowchart of PFC and PSFB control realized by Verilog software is shown in Fig. 16. As shown in Fig. 16, the system clock goes up to 150 MHz using frequency multiplier. The digital pulse-width modulator is realized using triangular carrier for PFC and up-count carrier for PSFB. The sampling instants for the AD converters of PFC and PSFB are synchronized.

As the conversion of AD converters is finished, the control loops of PFC and PSFB begin. Both PI controllers of voltage loop of PFC and PSFB are implemented by synchronous state machine [20]-[21]. After that, feed-forward control and current loop controller of PFC and phase-shift control for full-bridge converter are realized. Moreover, the implementation requires about 3050 logic elements in FPGA which is 16% of total logic elements of the chip, Cyclone II-EP202CF484C8.

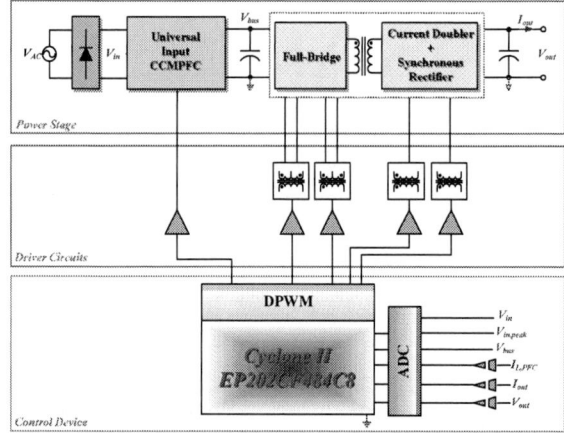

Fig. 13 Block diagram of the experimental system

Fig. 14 Photo of the experimental system

The 2014 International Power Electronics Conference

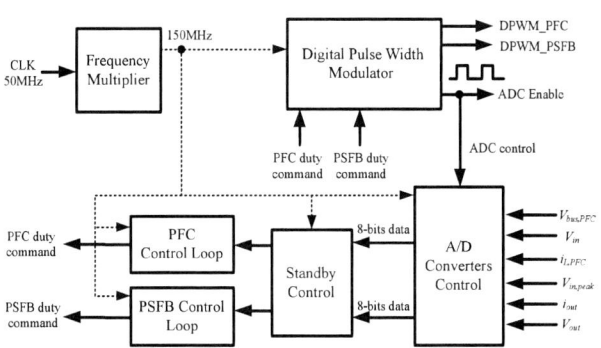

Fig. 15 Block diagram of PFC and PSFB for FPGA-based implementation

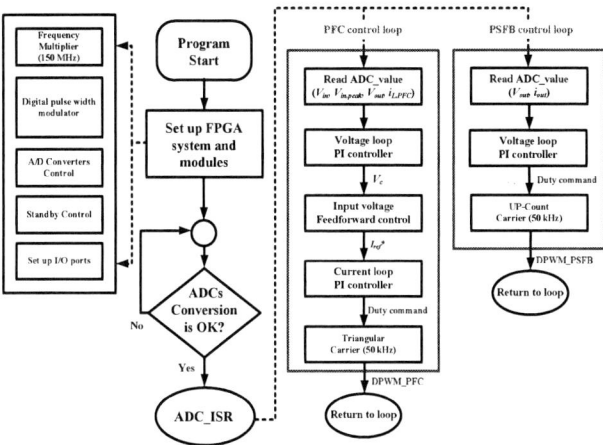

Fig. 16 Flowchart of PFC and PSFB control

VI. EXPERIMENTAL RESULTS

Experimental waveforms for standby power under V_{AC}=115 V_{rms} and V_{AC}=230 V_{rms} conditions are shown in Fig. 17 and Fig. 18. PFC is operated in burst mode control and PSFB converter is operated in burst mode with full-bridge PWM control under no-load condition. The summary of losses for standby control under no-load condition is shown in TABLE VII. All of the losses for standby power are smaller than 0.5 W.

The measured results include input voltage, input current, PFC output voltage and PSFB output voltage under full-load condition. For low line V_{AC}=115 V_{rms}, as shown in Fig. 19, the power factor is greater than 0.998 and efficiency is higher than 89% under 100% load condition. For high line input V_{AC}=230 V_{rms}, as shown in Fig. 20, the power factor is greater than 0.990 and efficiency is higher than 91% under 100% load condition. The inductor current and input voltage is in phase demonstrating good power factor results and good output voltage regulation. The summary of measure results for power factor and measured efficiency are shown in TABLE VIII. According to results of loss analysis, the predicted figures of efficiency are very close to those from measured results. These results fully support the claim and contributions of this paper.

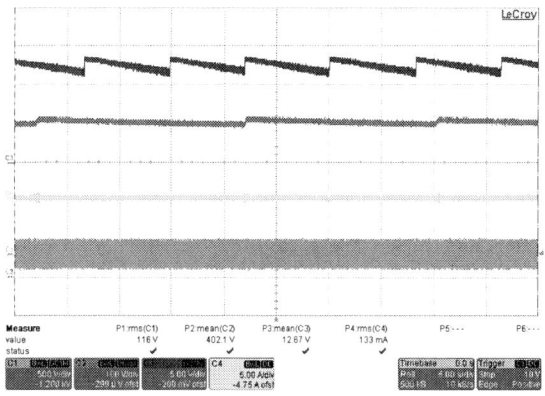

Fig. 17 Experimental waveforms for standby power, V_{AC}=115 V_{rms}, Ch1: Input voltage, Ch2: PFC output voltage, Ch3: PSFB output voltage,Ch4: PFC input current

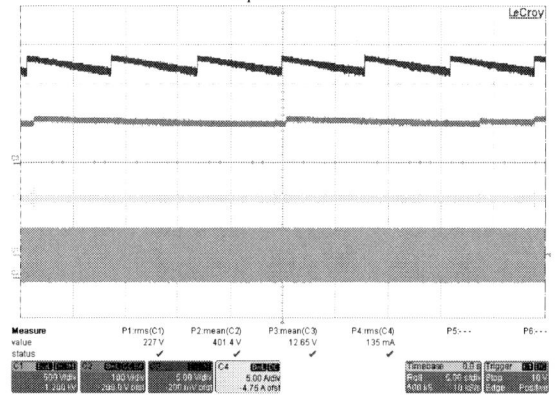

Fig. 18 Experimental waveforms for standby power, V_{AC}=230 V_{rms}, Ch1: Input voltage, Ch2: PFC output voltage, Ch3: PSFB output voltage, Ch4: PFC input current

TABLE VII SUMMARY OF MEASURED LOSSES, WITH STANDBY CONTROL

Input voltage	90 V_{rms}	115 V_{rms}	230 V_{rms}	264 V_{rms}
Standby power	0.042 W	0.069 W	0.237 W	0.34 W
Spec.	< 0.5W			

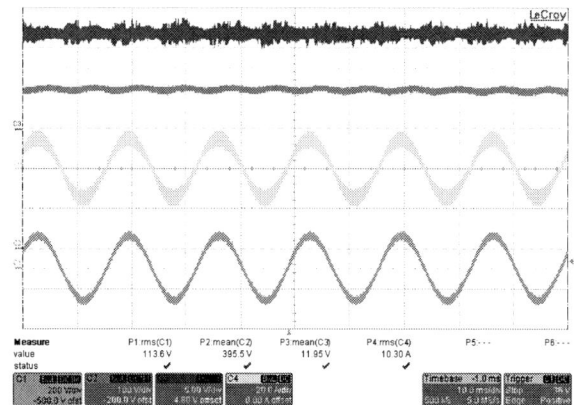

Fig. 19 Experimental results, V_{AC}=115V_{rms}, 100% load, Ch1: Input voltage, Ch2: PFC output voltage, Ch3: PSFB output voltage, Ch4: PFC input current

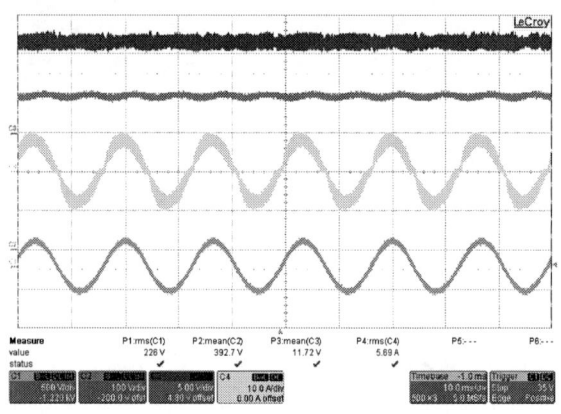

Fig. 20 Experimental results, V_{AC}=230V_{rms}, 100% load, Ch1: Input voltage, Ch2: PFC output voltage, Ch3: PSFB output voltage, Ch4: PFC input current

TABLE VIII. SUMMARY OF MEASURED RESULTS

Input Voltage	Load (%)	Power Factor	Measured Efficiency (%)	80 PLUS Platinum (%)
115V_{rms}	20	0.990	91.60	90
	50	0.997	93.17	92
	100	0.998	89.66	89
230V_{rms}	20	0.937	93.09	90
	50	0.970	94.51	94
	100	0.990	91.64	91

VII. CONCLUSIONS

This paper contributes to the presentation of hardware design, controller design and FPGA implementation for digital-controlled server power which meets the platinum efficiency code and standby power code. Two control modes are presented in this paper in order to achieve standby power code. A control method to achieve constant bandwidth for power factor stage despite of input voltage variation, changing from 90V/AC to 264V/AC. Moreover, hardware and controller design and loss analysis of converter are addressed to achieve platinum efficiency code.

Experimental results show that the standby power is less than 0.5W and light load efficiency is improved to go up to more than 90% for both low line and high line input voltage. Moreover, it will be confirmed that the platinum efficiency code can be met under 20%, 50% and 100% load conditions. These results fully support the claim and contributions of this paper.

ACKNOWLEDGMENT

This work is sponsored by National Science Council, Taiwan. Project number is 99IA01.

REFERENCES

[1] D. Li and X. Ruan, "Comparison of three front-end DC-DC converters for 1200W server power supply," *Proc. of IEEE PESC*, pp. 394-398, June. 2005.

[2] J. Zhang, Y. P. Zou,Y. Zhang and J. Tang, "DSP implementation of digitally controlled SMPS," *Proc. of IEEE IECON*, pp. 1484-1488, Nov. 2007.

[3] D. Y. Kim, Y. D. Kim, K. M. Cho and G. W. Moon, "Adaptive link capacitor voltage control for server power system," *Proc. of IEEE IPEMC*, pp. 505-510, May. 2009.

[4] J. H. Cho, H. W. Seong, S. M. Jung, J. S. Park, G. W. Moon and M. J. Youn, " Implementation of digitally controlled phase shift full bridge converter for server power supply," *Proc. of IEEE ECCE*, pp. 802-809, Spet. 2010.

[5] A. G. J and G. W. Moon, " A high-efficiency three-phase ZVS PWM converter utilizing a positive double-star active rectifier stage for server power supply," *IEEE Trans. on Industrial Electronics*, Vol. 58, No. 8, pp. 3317 – 3329, Aug. 2011.

[6] K. M. Ho, C. A. Yeh and Y. S. Lai, "Novel digital-controlled transition current mode control and duty compensation techniques for interleaved power factor corrector," *IEEE Trans. on Power Electronics*, vol. 25, no. 12, pp. 3085-3094, Dec. 2010.

[7] Y. S. Lai and K. M. Ho," FPGA-based digital-controlled power converter with universal input meeting 80 plus platinum efficiency code and standby power code for sever power applications," *Proc. Int. Conf. ICRERA*, Nov. 2012.

[8] [Online]. Available: http://www.plugloadsolutions.com/

[9] "Standby power use and the IEA "1-watt plan," International Energy Agency, Paris, France, Fact Sheet, 2007.

[10] Z. Jindong, M. M. Jovanovic and F. C. Lee, "Comparison between CCM single-stage and two-stage boost PFC converters," *Proc. of IEEE APEC*, pp. 335-341, Mar. 1999.

[11] C. M. Wang, C. H. Lin, C. H. Liu and T. C. Yang, "High performance single-stage transformer-isolated AC/DC converter," *Proc. of IEEE IPEC*, pp. 131-136, June. 2010.

[12] K. Y. Lee and Y. S. Lai, "Novel circuit design for two-stage AC/DC converter to meet standby power regulations," *IET Power Electronics*, Vol. 2, No. 6, pp. 625-634, Nov. 2009.

[13] B. J. Culpepper and H. Suzuki, "Switching DC-to-DC converter with discontinuous pulse skipping and continuous operating mode without external sense resistor," U.S. Patent 6 396 252, May. 2002.

[14] Y. K. Lo, S. C. Yen, and C. Y. Lin, "A high-efficiency AC-to-DC adaptor with a low standby power consumption," *IEEE Trans. Ind. Electron.*, Vol. 55, No. 2, pp. 963-965, Feb. 2008.

[15] H. S. Choi and D. Y. Huh, "Techniques to minimize power consumption of SMPS in standby mode," *Proc. of IEEE PESC*, pp. 2817-2822, 2005.

[16] B. Y. Chen and Y. S. Lai, "Switching control technique of phase-shift controlled full bridge converter to improve efficiency under light load and standby conditions without additional auxiliary components," *IEEE Trans. on Power Electronics*, Vol. 25, No. 4, pp. 1001-1012, April, 2010.

[17] Y. S. Lai, K. M. Ho and B. Y. Chen, "Bandwidth-based analysis and controller design of power factor corrector for universal applications," *Proc. of IEEE ICIT*, pp.626-631, March, 2012.

[18] Power Supply Design Guideline for 2008 Dual-Socket Servers and Workstations, [Online] Available: https://ssiforum.org/

[19] F. Musavi, W. Eberle and W. G. Dunford, "Efficiency evaluation of single-phase solutions for AC-DC PFC boost converters for plug-in-hybrid electric vehicle battery chargers," *Proc. of IEEE VPPC*, pp.1-6, Sept., 2010.

[20] W. Zhang, Y. F. Liu and B. Wu, "A new duty cycle control strategy for power factor correction and FPGA implementation," *IEEE Trans. on Power Electronics*, vol. 21, no. 6, pp. 1745-1753, Nov. 2006.

[21] E. F. C. Grabovski, T. K. Jappe and S. A. Mussa," Discrete hardware controllers design of a single phase PFC boost converter with FPGA," *Proc. of IET PEMD*, pp.1-5, March, 2012.

Static and Dynamic Analyses of Digital Peak Current Mode DC-DC Converter

Kazuhiro Kajiwara, Fujio Kurokawa and Yuichiro Shibata
Nagasaki University
Nagasaki, Japan
bb52311101@cc.nagasaki-u.ac.jp, fkurokaw@nagasaki-u.ac.jp

Abstract— **The aim of this paper is to present static and dynamic analyses of digital peak current mode dc-dc converter. The current detection circuit of the proposed method consists of RC integrator and comparator as the A-D converter. The proposed method can detect the peak current in real time. Analyses and the transient response of proposed method are discussed in this paper. As a result, the convergence time and undershoot are suppressed by 67% and 34% compared with the conventional digital voltage mode control method, respectively.**

Keywords— dc-dc converter, digital control, peak current mode

I. INTRODUCTION

Recently, the digital control dc-dc converter has come to draw a lot of attention in many applications. Features of digital control method are programmability, miniaturization of control circuit and energy management capability [1]-[6]. The digital control method needs the A-D converter and digital signal processer. They have the delay time from the A-D conversion and processing, and the delay time causes the negative influence on dynamic characteristics [3]-[6].

The peak current mode dc-dc converter is useful to improve the transient response [7]-[9]. It has been generally implemented by the analog control method. It is difficult to implement the digital peak current mode dc-dc converter because the peak current cannot be accurately detected in real time. Even if a fast A-D converter is used, the cost increases. Thus, we have already reported a novel digital peak current mode dc-dc converter capable of peak current detection in real time [9]. Its control circuit uses the RC integrator and comparator as the A-D converter in the current detection. The sampling point of detected current is not fixed and is changed by the digital feedback value of output voltage. Although the transient response of proposed method has been studied, static and dynamic analyses have not been studied yet.

This paper presents static and dynamic analyses of digital peak current mode dc-dc converter using the RC integrator and comparator as the A-D converter for the detected current. It is confirmed that the proposed method shows superior stability and transient response.

II. OPERATION PRINCIPLE

Figure 1 shows the circuit configuration of digital peak current mode dc-dc converter. E_i is the input voltage, e_o is the output voltage, i_S is the switch current, T_r is the main switch, D is the fly wheel diode, L is the energy storage reactor, C_O is the smoothing capacitor, R_O is the load and I_O is the load current. e_S is the voltage detected by the sensing resistor R_S and represented as follows.

$$e_S = A_c R_S i_S \quad (1)$$

where A_c is the gain of ape-amplifier of switch current.

Figure 2 shows the proposed digital control circuit. N_{PID} is calculated value in the PID controller, S_{samp} is the delayed signal generated by the delay circuit, S_{es} represents the signal of peak current detection and S_w is the signal to turn off T_r.

e_o is input to the A-D converter and converted to the digital value N_{eo} as follows.

$$N_{eo} = A_{eo} G_{AD} e_o \quad (2)$$

where A_{eo} is the gain of pre-amplifier and G_{AD} is the gain of the A-D converter. N_{eo} is sent to the PID controller and it calculates the digital feedback value N_{PID} from (3).

$$N_{PID} = N_B - K_P(N_{eo} - N_r) - K_I N_I - K_D N_D \quad (3)$$

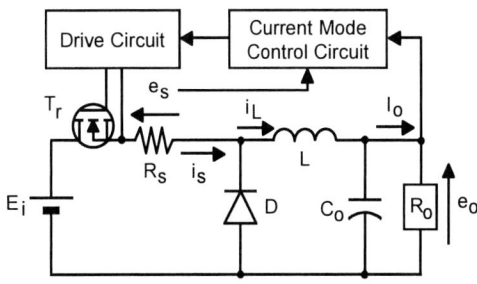

Figure 1. Digital peak current mode dc-dc converter.

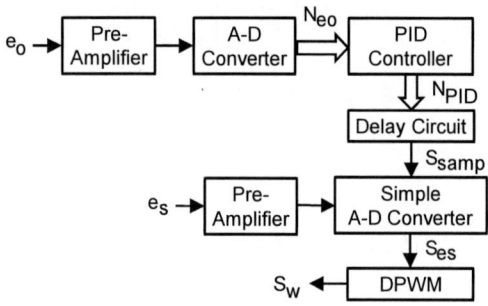

Figure 2. Proposed digital control circuit.

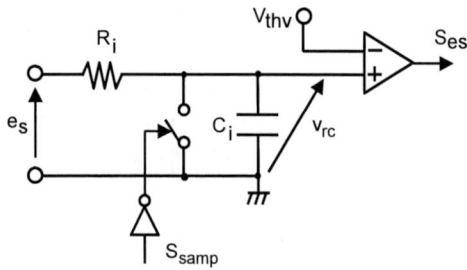

Figure 3. Simple A-D converter using RC integrator.

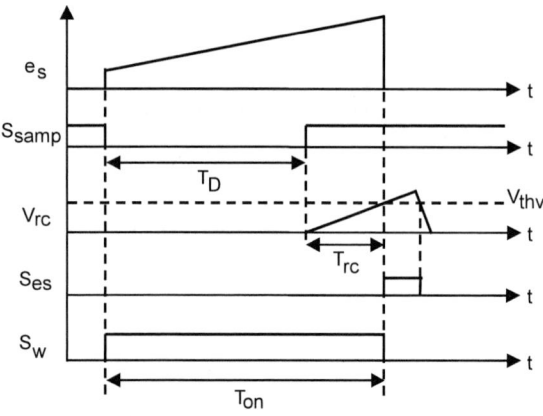

Figure 4. Operation waveforms of peak current detection.

$$N_I = \sum (N_{eo} - N_r) \tag{4}$$

$$N_D = N_{eo,n} - N_{eo,n-1} \tag{5}$$

where N_B is the reference bias, N_r is the desired value of e_o, K_P, K_I and K_D are the proportional, integral, and differential coefficients, respectively. n denotes the n-th switching period.

N_{PID} is fed into the delay circuit. The delay circuit calculates the delay time T_D from (6) and determines the sampling point for the detected current.

$$\frac{T_D}{T_s} = \frac{N_{PID}}{N_{Ts}} \tag{6}$$

where N_{Ts} is the maximum value of A-D converter.

$\Delta T_D(s)/T_s$ is obtained by the Laplace transform as follows;

$$\frac{\Delta T_D(s)}{T_s} = -H(s)\Delta e_o(s)$$
$$= -(H_{Pv} + \frac{H_I}{s} + sH_D)e^{-s\tau_1} \cdot \Delta e_o(s) \tag{7}$$

where τ_1 is the delay time of A-D conversion and processing, and H_{pv}, H_I, H_D are as follows.

$$H_{Pv} = \frac{K_P A_{eo} G_{AD}}{N_{Ts}}$$
$$H_I = \frac{K_I A_{eo} G_{AD}}{N_{Ts} \cdot T_s} \tag{8}$$
$$H_D = \frac{K_D A_{eo} G_{AD} T_s}{N_{Ts}}$$

Figure 3 shows the simple A-D converter using the RC integrator and comparator. The sampling point of i_s is determined by S_{samp}. When v_{rc} reaches the threshold voltage V_{thv}, S_{es} is turned on in the comparator.

Figure 4 illustrates operation waveforms of peak current detection. The peak point of i_s can be detected only by controlling T_D. The on time T_{on} is determined by comparing v_{rc} to V_{thv}. T_{rc} is the time during the current sampling and obtained as follows,

$$T_{rc} = \frac{\tau \cdot V_{thv}}{e_s} \tag{9}$$

where τ is the time constant of RC integrator, that is, $\tau = R_i C_i$.

Furthermore, $\Delta T_{rc}(s)/T_s$ is approximately represented by

$$\frac{\Delta T_{rc}(s)}{T_s} = -H_{Pi}\Delta i_s(s) \tag{10}$$

where

$$H_{Pi} = \frac{\tau \cdot V_{thv}}{A_c R_s T_s} \tag{11}$$

From (6) and (9), T_{on} is represented by (12).

$$T_{on} = T_D + T_{rc} \tag{12}$$

The relationship between the tiny variation ΔT_{on}, Δe_o and Δi_s are represented by (13).

$$\frac{\Delta T_{on}(s)}{T_s} = \frac{\Delta T_D(s)}{T_s} + \frac{\Delta T_{rc}(s)}{T_s}$$
$$= -H(s)\Delta e_o(s) - H_{Pi}\Delta i_s(s) \tag{13}$$

According to [10], the transfer function of proposed digital control dc-dc converter is obtained in Fig. 5.

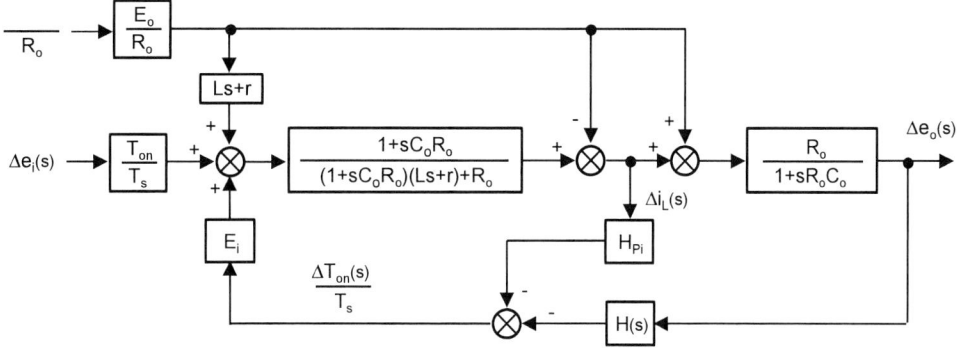

Figure 5. Transfer function of proposed digital control peak current mode dc-dc converter.

III. REGURATION CHARACTERISTICS

In this section, it is discussed that the parameter of RC integrator has an effect on regulation characteristics. E_i is 20V, the desired voltage is 5V, L is 200μH, C_o is 200μF, R_s is 0.05Ω, A_c is 100, A_{eo} is 0.25, G_{AD} is 400, N_{Ts} is 2000 and the A-D converter is 11bits. The switching frequency is 100kHz, that is, the switching period T_s is 10μs. Regulation characteristics are changed by $\tau = R_i C_i$ in Fig. 3. Here, C_i is 110pF, V_{thv} is 0.4V and R_i is changed.

Figure 6 shows regulation characteristics of proposed method. The critical current is 0.1A. The operation range of proposed method is from 0.2A to 1.0A because the current mode dc-dc converter usually switches to the voltage mode control when the load is light. The regulation characteristics become worse with the increase in τ/T_s because T_{rc} becomes large. In this case, τ/T_s should be chosen to less than 0.44 to obtain good regulation characteristics.

IV. TRANSIENT RESPONSE

Figures 7 and 8 show the transient response of conventional digital voltage mode control method and the proposed one. The conventional method only uses the PID control. The step change of I_o is from 0.5A to 1A. K_P is unity, K_I is 0.005 and K_D is unity. τ/T_s is equal to 0.22. The convergence time of e_o to converge within 1% of desired voltage is 2.10ms, the undershoot of e_o is 6.4% and the overshoot of i_L is 29% in Fig. 7. As shown in Fig. 8, the proposed method can improve the boundary stability compared with the conventional one.

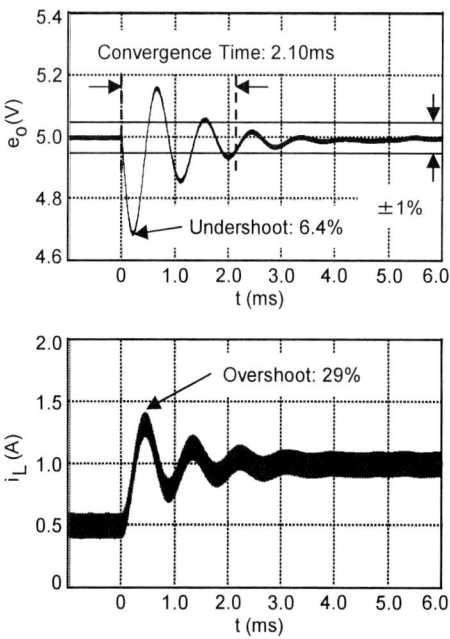

Figure 7. Transient response of conventional method when K_P is unity, K_I is 0.005 and K_D is unity.

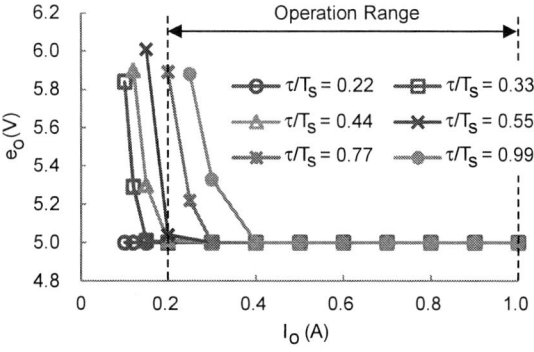

Figure 6. Regulation characteristics of proposed method.

The 2014 International Power Electronics Conference

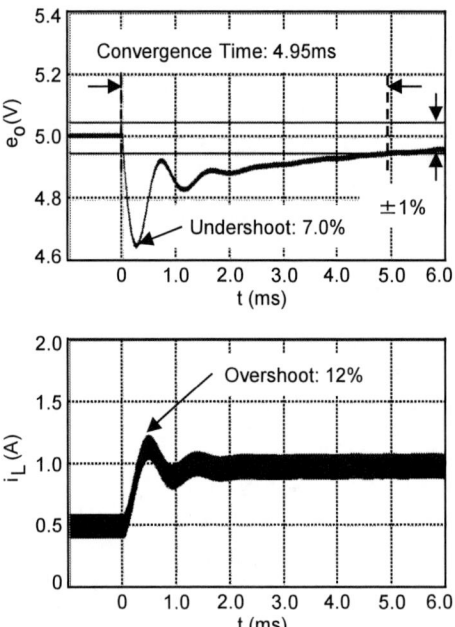

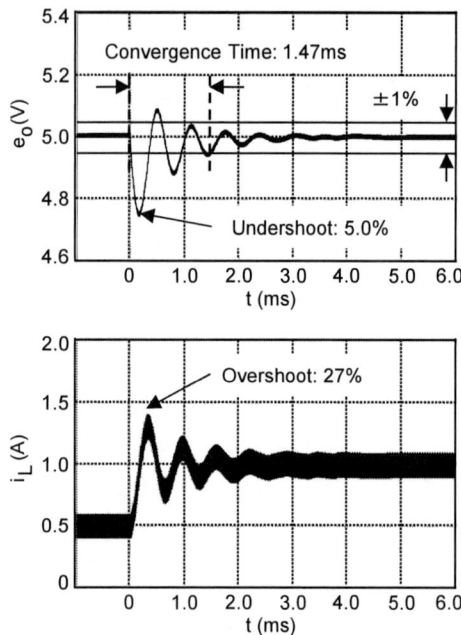

Figure 8. Transient response of proposed method when K_P is unity, K_I is 0.005 and K_D is unity.

Figure 10. Transient response of proposed method when K_P is 3, K_I is 0.04 and K_D is unity.

proposed method can increase K_P and K_I to 3 and 0.04 in Fig. 10 because the more stable performance can be obtained. Moreover, the convergence time and undershoot of e_0 are suppressed to 30% and 22% compared with Fig. 7, respectively. Therefore, the proposed method has a superior transient response.

V. CONCLUSIONS

At first, the operation principle for the peak current detection in real time and the transfer function of proposed digital control circuit were shown in this paper. Second, regulation characteristics of proposed method were discussed when the parameter of RC integrator in the peak current detection circuit was changed. Finally, it was revealed that the proposed method can improve the transient response and stability compared with the conventional method.

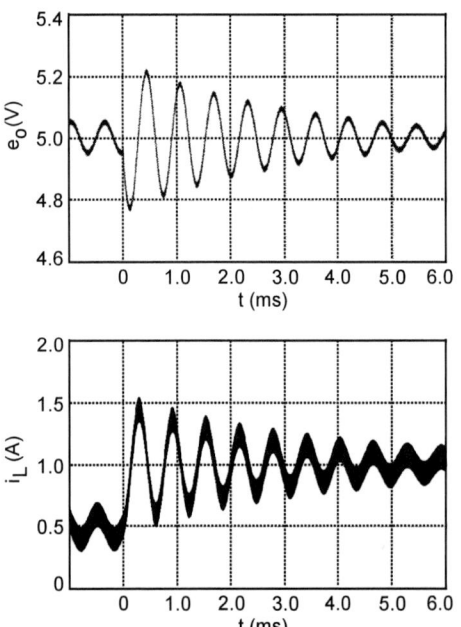

Figure 9. Transient response of conventional method when K_P is 3, K_I is 0.04 and K_D is unity.

ACKNOWLEDGMENT

This work is supported in part by the Grant-in-Aid for Scientific Research (No.24360112) of JSPS (Japan Society for the Promotion of Science) and the Ministry of Education, Science, Sports and Culture.

Figures 9 and 10 show the transient response of conventional and proposed methods when K_P is 3, K_I is 0.04 and K_D is unity. In the conventional method, e_0 and i_L are oscillated because the boundary stability of voltage mode control is not large. On the other hand, the

REFERENCES

[1] T. Sun and D.D. Lu, "Digital time-multiplexing control of single-switch dual-output dc/dc converter," *IEEE Proc. of PEDS*, pp. 828-833, Dec. 2011.

[2] Z. Lukić, N. Rahman and A. Prodić, "Multibit Σ-Δ PWM digital controller IC for dc-dc converters operating at switching frequencies beyond 10 MHz," *IEEE Trans. on IEEE Power Electronics*, vol. 22, no. 5, pp. 1693-1707, Sep. 2007.

[3] K. Ho, C.Yeh and Y. Lai, "Novel digital-controlled transition current-mode control and duty compensation techniques for interleaved power factor corrector," *IEEE Trans. on Power Electronics*, vol. 25, no. 12, pp. 3085-3094, Dec. 2010.

[4] A. Babazadeh and D. Maksimovic, "Hybrid digital adaptive control for fast transient response in synchronous buck dc–dc converters." *IEEE Trans. on Power Electronics*, vol. 24, no. 11, pp. 2625-2638, Nov. 2009.

[5] H. Bae, J. Lee, J. Yang and B. H. Cho, "Digital resistive current (DRC) control for the parallel interleaved dc–dc converters," *IEEE Trans. on Power Electronics*, vol. 23, no. 5, pp. 2465-2476, Sep. 2008.

[6] F. Kurokawa, R. Yoshida, Y. Maeda, T. Takahashi, K. Bansho, T. Tanaka and K. Hirose, "A novel A-D conversion for digital control switching power supply," *IEEE Proc. of ECCE*, pp. 1302-1306, Sept. 2011.

[7] M. Hallworth and S. A. Shirsavar, "Microcontroller-based peak current mode control using digital slope compensation," Trans. on Power Electronics, vol. 27, no. 7, pp. 3340-3351, July 2012.

[8] G. Zhou, J. Xu, C. Mi and Y. Jin, "Transient performance comparison on digital peak current controlled switching dc-dc converters in DCM with different digital pulse-width modulations," Proc. of IEEE IPEMC, pp. 315-319, May. 2009.

[9] F. Kurokawa, K. Kajiwara, Y. Shibata and Y. Yamabe, T. Tanaka and K. Hirose, "A new digital peak current mode dc-dc converter using FPGA delay circuit and simple A-D converter," Proc. of IEEE ECCE, pp. 1698-1702, Sep. 2012.

[10] H. Matsuo, F. Kurokawa and K. Higashi, "Dynamic characteristics of the digitally controlled dc-dc converter," Trans. on Power Electronics, vol. 4, no. 4, pp. 419-426, Oct. 1989.

Extended Discrete Control of Class E Amplifier in order to Achieve Nominal Operation

Tadashi Suetsugu   Xiuqin Wei   Shotaro Kuga

Department of Electronics Engineering and Computer Science
Fukuoka University
Fukuoka, Japan

Abstract—**This paper introduces concept of discrete control method of class E amplifier which keeps nominal operation for varying load resistance. In the proposed method, shunt capacitance and output resonant capacitance are varied electronically by setting status of switches which are connected to capacitances. Proposed method can keep nominal operation, i.e., ZVS and ZDS, both when load resistance is higher than the designed value and lower than the designed value. Operation of proposed circuit is verified with PSIM simulation.**

Keywords—Discrete control, class E amplifier, shunt capacitance, nominal operation, PSIM simulation.

I. INTRODUCTION

The class E amplifier has a feature of high efficiency at high operating frequency higher than 10 MHz. However, this circuit is vulnerable to variations of load impedance. This is because the nominal operation can't be achieved anymore when load impedance is different from the designed value [1]-[6]. In the class E amplifier, relationship among circuit elements and operating waveform is complicated and tightly coupled. Traditionally, frequency modulation (FM) was the most potential candidate of control method for class E amplifier. Because of frequency characteristics of narrow band output filter. However, this control method sacrifices nominal operation. Nominal condition of class E amplifier is often called as 'one point condition' which means all element parameters must coincide with their designed values simultaneously. Otherwise the amplifier can't keep nominal operation, namely high efficiency. The circuit waveform of class E amplifier is intricately related to every element parameters. Therefore, simple control methods such as PWM, PFM and FM, which maneuvers single parameter can't keep nominal operation when load impedance is varied from its designed value [3]-[7]. In order to keep nominal operation when load impedance is varied from the designed value, it is required to maneuver several circuit parameters simultaneously.

In [8]-[10], discrete control methods of class E amplifier were introduced. Methods shown in [8] and [9] are so called DPWM methods, and they control output power by sacrificing nominal condition. In the method shown in [10], plural shunt capacitors are connected in parallel to the transistor. These shunt capacitors can be connected and disconnected by switches. With combination of plural capacitance, required shunt capacitance can be approximately achieved for individual load resistance. This method has ability to maneuver shunt capacitance by electronic signal. But, this method is not enough to cover wide range variation of load resistance. Firstly, it can not keep nominal operation when load resistance is changed. This method only keeps sub-nominal (it is equivalent to ZVS) operation. Therefore, only switching loss can be eliminated but power loss due to transition time of switching device can not be eliminated. This cause a deterioration of efficiency in the high operating frequency. Second, it can not even achieve sub-nominal (ZVS) operation when load resistance is higher than the designed value. In order to achieve sub-nominal operation, the circuit must be operated at less load resistance than the resistance at nominal operation. This means that the highest efficiency can be achieved only at the highest load resistance occurs, although such a situation is not always happen. It is also difficult to predict highest possible load resistance very exactly at the designing stage. The problem of the control method shown in [10] was that only one circuit parameter was maneuvered in order to deal with the load variations.

In this paper, we reviewed operating equation of class E amplifier outside of designed condition. Based on the analytical result, we propose a novel discrete control method to achieve nominal operation in wide range of variation of load resistance. In this method, shunt capacitance and resonant capacitance are varied by connecting and disconnecting plural capacitors in parallel using switches. Note that this method does not maneuver value of resonant inductance which is difficult to do electronically. Proposed method can achieve nominal operation, i.e., ZVS and ZDS, approximately even if the load resistance is higher than the designed value. Of course, the nominal operation can be achieved also if the load resistance is lower than the designed value. Proposed method is designed based on the analytical equation of the voltage waveforms of class E amplifier. Hence, same voltage waveforms are kept even when load resistance is changed from the designed value. Namely, amplitude of output voltage and peak switch voltage are kept to the designed value. This means this method will avoid destruction of switching device from imposing excessive high voltage to the transistor when load resistance is changed significantly. The concept of control operation was verified by the PSIM simulation. The nominal operation, i.e., ZVS and ZDS, can be approximately obtained for several values of load resistance. In every case, switch voltage waveform and output voltage waveform were approximately equal. In this paper, only concept of control method was shown. But, with

978-1-4799-2706-7/14 $31.00 © 2014 IEEE

implementing digital control stage which determines switching combination of shunt capacitance and resonant capacitance, a circuit system which keeps nominal operation for wide range variation of load resistance will be achieved. It will be a next topic of this research.

II. ANALYTICAL RESULT OF CLASS E AMPLIFIER OUTSIDE DESIGNED CONDITION

Figures 1 and 2 show the circuit topology and switch-voltage and output-voltage waveforms of class E amplifier. Steady state operation of class E amplifier was analyzed in [7]. In that analysis, amplitude of output current I_m and input current I_{DD} are described as

$$I_m = -2\frac{U_1}{U_2}I_{DD}, \tag{1}$$

$$I_{DD} = \frac{\pi\omega C_1 R U_1}{\pi^2(1-D)^2 U_3 - U_1 U_2}V_{DD}. \tag{2}$$

where

$$
\begin{aligned}
U_1 &= [2\pi(1-D) + \sin 2\pi D] \\
&\quad \times \cos\phi + (\cos 2\pi D - 1)\sin\phi, \tag{3}
\end{aligned}
$$

$$
\begin{aligned}
U_2 &= [-2\pi(2-D)\cos 2\pi D - \sin 2\pi D]\cos\phi \\
&\quad + [2\pi(2-D)\sin 2\pi D - \cos 2\pi D + 1]\sin\phi, \tag{4}
\end{aligned}
$$

$$U_3 = 2\pi\omega C_1 R + (\cos 2\pi D - 1)[\cos 2\pi D + 2\phi - 1], \tag{5}$$

$$\phi = \pi - \tan^{-1}\frac{A}{B}, \tag{6}$$

$$
\begin{aligned}
A &= [2\pi(1-D) + \sin 2\pi D] \\
&\quad \times [2\pi(D-1) - 0.5\sin 4\pi D + 2\pi\omega C_1 X] \\
&\quad + (\cos 2\pi D - 1)[-2\cos 2\pi D \\
&\quad -1.5 - 0.5\cos 4\pi D - 2\pi\omega C_1 X], \tag{7}
\end{aligned}
$$

$$
\begin{aligned}
B &= [2\pi(1-D) + \sin 2\pi D] \\
&\quad \times [0.5 - 0.5\cos 4\pi D + 2\pi\omega C_1 R] \tag{8} \\
&\quad + (\cos 2\pi D - 1)[-2\sin 2\pi D + 2\pi(D-1) \\
&\quad + 0.5\sin 4\pi D + 2\pi\omega C_1 X], \tag{9}
\end{aligned}
$$

$$X = \omega L - \frac{1}{\omega C}. \tag{10}$$

Switch voltage waveform and output voltage waveform are described by using I_m and I_{DD},

$$v_S = \frac{1}{\omega C_1}[I_D D\omega t + I_m \cos \omega t + \phi - I_m \cos\phi]$$

$$\text{for } \frac{2\pi D}{\omega} < t < \frac{2\pi}{\omega}, \tag{11}$$

$$v_o = RI_m \sin(\omega t + \phi). \tag{12}$$

From the above equations, we can know that the switch voltage v_S is governed only by four parameters: $\omega C_1 R$, $\omega C_1 X$, D, and V_{DD}. Therefore, with constant V_{DD} and D, only $\omega C_1 R$ and $\omega C_1 X$ governs waveform of switch voltage. In order to keep nominal operation, $\omega C_1 R$ and $\omega C_1 X$ should be kept same value as the designed value even if some of their parameters such as C_1, R, and X might change to different values.

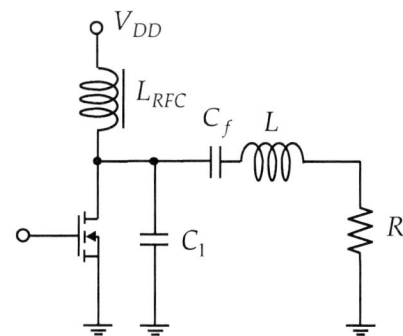

Fig. 1. Basic circuit of the class E amplifier.

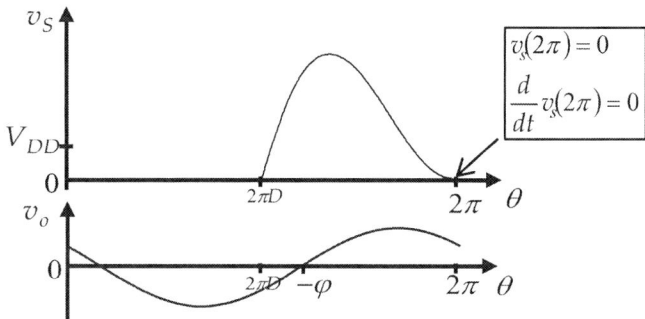

Fig. 2. Waveforms of switch voltage and the output voltage.

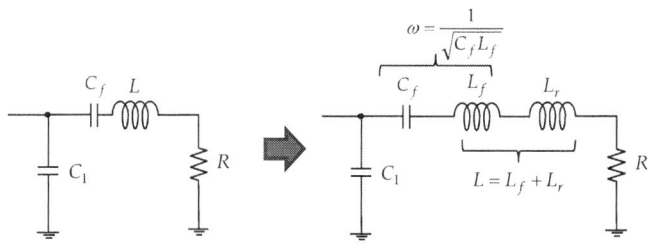

Fig. 3. Output filter circuit at designed condition.

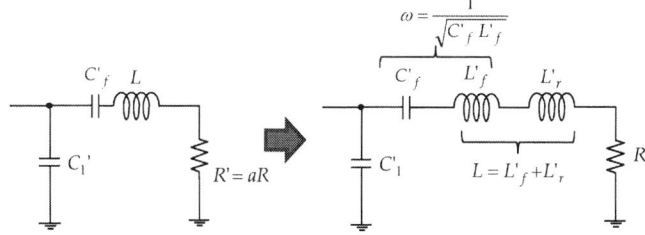

Fig. 4. Output filter circuit outside designed condition.

III. VARIATION OF SHUNT CAPACITANCE AND RESONANT CAPACITANCE TO KEEP NOMINAL OPERATION

Suppose that the class E amplifier satisfies nominal condition at R, C_1, C_f, L_f, and ω (Fig. 3). With some disturbance, the load resistance is supposed to change from its nominal value R to a different value R' (Fig. 4) where

$$R' = aR \qquad \text{for } 0 < a < \infty. \tag{13}$$

In order to keep nominal operation C_1 is changed to C_1', X is changed to X'. In order to keep $\omega C_1 R$ and $\omega C_1 X$ constant,

$$\omega C_1' R' = \omega C_1 R. \tag{14}$$

Because $R' = aR$,

$$C_1' = \frac{C_1}{a}. \tag{15}$$

Nextly,

$$\omega C_1' X' = \omega C_1 X. \tag{16}$$

Substituting (15) into (16),

$$X' = aX. \tag{17}$$

In most cases, it is more difficult to change inductance during operation of circuit than changing capacitance. Hence, in this method, resonant capacitance C_f is changed instead of resonant inductance. The resonant inductance can be equivalently divided into series connected two inductors L_f and L_r

$$L = L_f + L_r, \tag{18}$$

where L_f is at operating frequency

$$\omega = \frac{1}{\sqrt{C_f L_f}}. \tag{19}$$

Hence, L_r takes role of required phase shift reactance for achieving nominal operation

$$X = \omega L_r. \tag{20}$$

Therefore, when the load resistance is R', X should be changed to $X' = aX$. So,

$$L_r' = aL_r. \tag{21}$$

If the total inductance L is constant,

$$L_f' = L - aL_r = L_f + (1 - a)L_r. \tag{22}$$

As L_f has been changed to L_f', C_f must be changed to a new value C_f' in order to be resonant at the operating frequency

$$\omega = \frac{1}{\sqrt{C_f' L_f'}} = \frac{1}{\sqrt{C_f'[L_f + (1 - a)L_r]}}. \tag{23}$$

Using (19) and (23), we have

$$C_f' = \frac{C_f}{1 + (1 - a)\dfrac{L_r}{L_f}}. \tag{24}$$

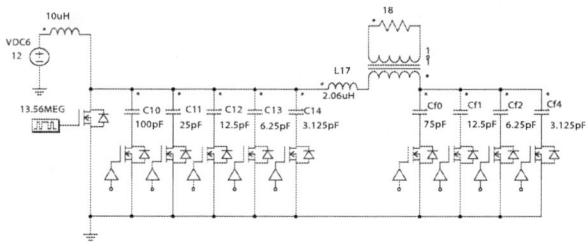

Fig. 5. Proposed class E amplifier.

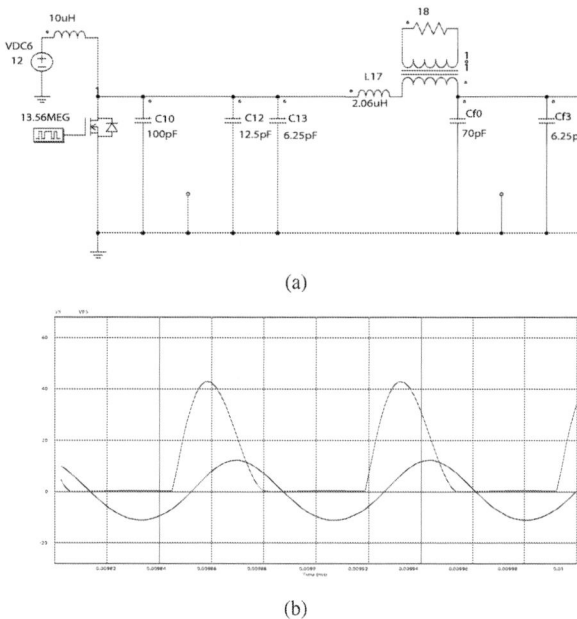

(a)

(b)

Fig. 6. Circuit configuration at $R = 18\ \Omega$ and simulated waveforms.

IV. CIRCUIT CONFIGURATION AND CONCEPT OF CONTROL METHOD

In section above, inductance was kept constant and two capacitances are changed instead. These capacitances are changed by electronic method shown in [10]. Several capacitances are connected in parallel. Each capacitances are connected in series to a switch. With turning on the switch, capacitance is connected to the main circuit. Because these capacitances are connected in parallel, total capacitances are sum of on state capacitances. In order to change two capacitances C_1 and C_f, circuit configuration shown in Fig. 5 is proposed. In this circuit, total capacitances of C_1 and C_f are controlled by combinations of individual control circuit for C_1 and C_f. These control circuits determine combination of switch state based on resistance of load circuit. At this time, we only made concept of control scheme and detail of control circuit is not made yet. Combination of switch state is determined by hand calculation and input to the simulation soft. Hence, realization of control circuit will be necessarily future task.

V. SIMULATION RESULTS

Proposed concept was verified with PSIM simulation. In the simulation, part of capacitors in circuit shown in Fig. 5

The 2014 International Power Electronics Conference

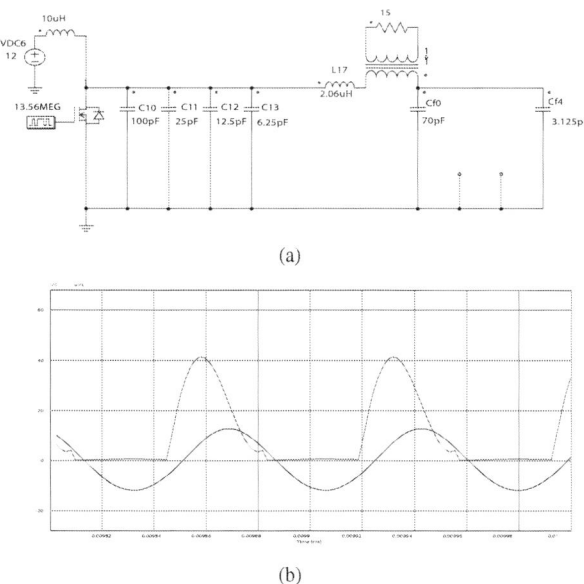

(a)

(b)

Fig. 7. Circuit configuration at $R = 15\ \Omega$ and simulated waveforms.

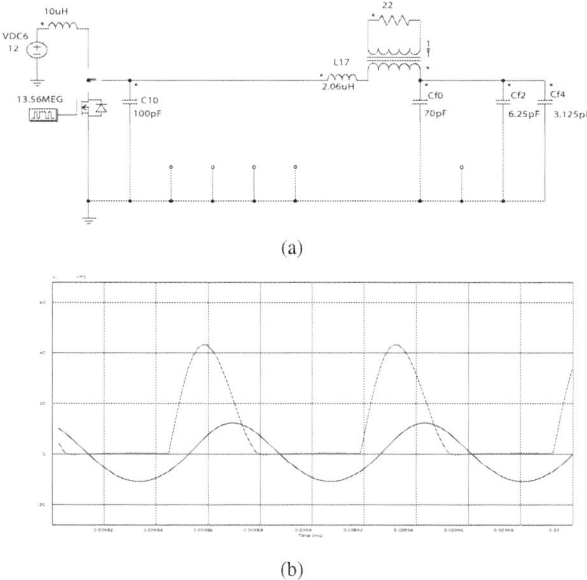

(a)

(b)

Fig. 8. Circuit configuration at $R = 22\ \Omega$ and simulated waveform.

were disconnected manually and simulated. Fig. 6 shows circuit configuration when load resistance is at the degined value $R = 18\ \Omega$. Theoretically required shunt capacitance was $C_1 = 122$ pF and resonant capacitance was $C_f = 75$ pF. As shown in Fig. 6(a), simulated shunt capacitance was $C_1 = 100\ \text{pF} + 12.5\ \text{pF} + 6.25\ \text{pF} = 118.75$ pF, and resonant capacitance was $C_f = 70\ \text{pF} + 6.25\ \text{pF} = 76.25$ pF. Fig. 6(b) shows switch voltage and output voltage waveform of this circuit. It obviously achieved ZVS and ZDS operation. Fig. 7 shows circuit configuration and simulation result of $R = 15\ \Omega$. Simulated shunt capacitance was $C_1 = 143.75$ pF (144 pF in theory) and resonant capacitance was $C_f = 73.125$ pF (73 pF in theory). As shown in Fig. 7(b), it nearly achieved ZVS and ZDS operation. Fig. 8 shows circuit configuration and simulation result of $R = 22\ \Omega$. Simulated shunt capacitance

was $C_1 = 100$ pF (99 pF in theory) and resonant capacitance was $C_f = 79.375$ pF (77 pF in theory). As shown in Fig. 8(b), it nearly achieved ZVS and ZDS operation. It can be easily expected that this method needs smart selection of combination of capacitances in order to achieve precious nominal operation. Further, the simulation shown here did not consider parasitic capacitance of transistor switch. However, this simulation results demonstrates possibility of electronic nominal operation control of class E amplifier. Note that amplitude of output voltage is constant when nominal operation is achieved. Namely, this method also has a feature of fixed output voltage control of class E amplifier over a range of load resistances.

VI. CONCLUSIONS

In this paper, concept of discrete control of class E amplifier which keeps nominal operation for varying load resistance was shown. With PSIM simulation, nominal operation was approximately achieved for a range of load resistance. In the simulation, parasitic capacitance was not taken into account. In the practical circuit, effect of parasitic capacitance will need to be included. Further, this method only considered variation of resistance. Control method for considering variation of reactance will be required for practical system. And, prototyping of digital processing circuit which determines combination of switch states will be future topics.

REFERENCES

[1] N. O. Sokal and A. D. Sokal, "Class E — A new class of high-efficiency tuned single-ended switching power amplifiers," IEEE J. Solid-State Circuits, vol. SC-10, pp. 168-176, June 1975.

[2] F. H. Raab, "Idealized operation of the class E tuned power amplifier," IEEE Trans. Circuits Sys., vol. 24, no. 12, pp. 725-735, Dec. 1977.

[3] F. H. Raab, "Effects of circuit variations on the class E tuned power amplifier," IEEE J. Solid-State Circuits, vol. SC-13, no. 2, pp. 239-247, Apr. 1978.

[4] M. K. Kazimierczuk and D. Czarkowski, Resonant Power Converters, 2nd Ed., IEEE Press/Wiley-Interscience, New York, NY, 2011.

[5] M. Albulet, RF Power Amplifiers, Noble Publishing, 2001.

[6] M. K. Kazimierczuk, RF Power Amplifiers, Wiley and Sons, 2008.

[7] T. Suetsugu and M. K. Kazimierczuk, "Steady-state behavior of class E amplifier outside designed conditions," in Proc. IEEE ISCSA2005, Kobe, Japan, May 2005, pp. 708-711.

[8] F. T. Abu-Nimeh and F. M. Salem, "Digital class E power amplifier for the RF band," in Proc. the 2006 IEEE International Conference on Mechatronics and Automation, Luoyang, China, June 2006, pp. 1548-1552.

[9] J. Ye, Z. Noaker, K. Wakabayashi, T. Yagi, O. Yamamoto, N. Takai, K. Niitsu, K.Kato, T. Ootsuki, I. Akiyama, H. Kobayashi, "Architecture of high-efficiency digitally-controlled class-E power amplifier (written in English)," IEICE Tech Report, ECT-11-41, 2011.

[10] V. Chironi, B. Debaillie, S. D' Amico, A. Baschirotto, J. Craninckx, and M. Ingels, "A digitally modulated class-E polar amplifier in 90 nm CMOS," IEEE Trans. Circuits Syst. I: Regular Papers, vol. 60, no. 4, pp. 918-925, Apr. 2013.

Adaptive Power Efficiency Control by Computer Power Consumption Prediction Using Performance Counters

Shinichi KAWAGUCHI, Toshiaki YACHI
Department of Electrical Engineering
Graduate School of Engineering, Tokyo Univ. of Science
Tokyo, Japan

Abstract— As the number of data centers has grown in pace with the burgeoning expansion in information technology (IT), reducing the power consumption of data centers has become an important social challenge. IT equipment is a primary contributor to data center power consumption and, as such, is an area where power conservation is indispensable. Nevertheless, while the efficiency of the power supply modules integrated into computers has recently seen significant improvements, their overall efficiency generally depends on load rates. This is especially true under low power load conditions, where it is known that efficiency decreases drastically. Recently, power-saving techniques that work by controlling the power module configuration under low power load conditions have been considered, and based on such techniques, further efficiency improvements can be expected by interlocking the real time computer load status with the power supply configuration control. In this study, the performance counters built into the processor of a computer are used to predict power load variations and an equation that predicts server power consumption levels is defined. In a server application experiment utilizing prototype computer hardware and multiple regression analysis, we validated that the equation could precisely predict computer power requirements. Furthermore, the adaptive efficiency control using the power prediction equation is considered. In order to validate the effect, the circuit simulation of the efficiency control was conducted. The simulation results show that power supply efficiency improvements up to 7.2% could be achieved using the model equations to control the efficiency of power supply modules.

Keywords— *adaptive control, efficiency, performance counter, power supply*

I. INTRODUCTION

The establishment of large-scale data centers is globally accelerating in pace with the expansion of information technology. The data center power consumption is rapidly increasing globally, and that is becoming an important social challenge [1]. To date, efforts aimed at reducing data center power consumption have focused on two approaches. The first is reduction of power losses from non-IT-related equipment, such as direct current (DC) power supply systems, etc [2]. the second involves power conservation of IT equipment itself such as servers, routers, etc. For server power consumption reductions, power efficiency improvements

to individual components, such as processors or memory modules, is one approach that is typically taken [3]. The efficiency of server power supply modules has also been improved, with some of the latest versions boasting maximum efficiency levels of 95% or more [4]. However, the efficiency of a power supply module generally decreases under light load conditions. While a high power supply is required if a server is fully engaged in CPU bound data processing, servers generally tend to operate under light power loads for I/O bound server applications because of the large data amounts that are handled and the frequent input/output (I/O) access waits. When reviewing both job types from the viewpoint of power supply efficiency, it can be seen that CPU bound jobs are executed in the high efficiency domain, while I/O bound jobs are executed in the low efficiency domain. Since numerous server applications, such as databases, Web servers, etc. are classified as I/O bound jobs, the efficiency of light load operations is very important. In order to improve power supply efficiency under light load conditions, various approaches have focused on controlling the number of power supply modules or using the phase shedding technology of a DC-DC converter, etc [5] [6]. These techniques aim at changing the maximum efficiency load point by adjusting power supply module configurations. An example of the efficiency improvement is shown in Fig. 1.

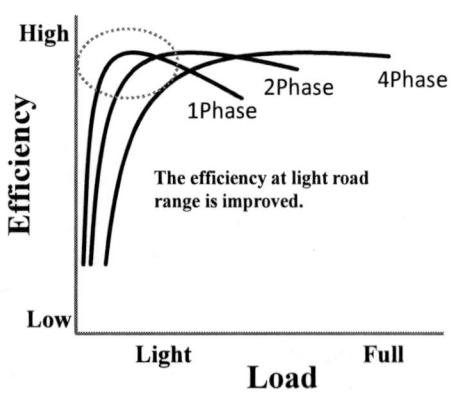

Fig. 1. Efficiency improvements obtained via power supply module configuration control.

The power supply control is shifting from analogue to digital control using field-programmable gate array (FPGA) or digital signal processing (DSP) techniques, which are making more advanced power management possible [7]. These techniques, in turn, can facilitate power efficiency improvements by dynamically coordinating power efficiency controls and server data processing workloads using the digital control functionality. As a result, highly efficient power supply configurations can be created for each load condition and overall power supply efficiency can be improved for all load conditions, including the light load range.

II. PROPOSED POWER SAVING METHOD

Fig. 2 shows the concept of the proposed technique for adaptively interlocking power controls with data processing workloads in order to boost power supply efficiency, which is the subject of this paper.

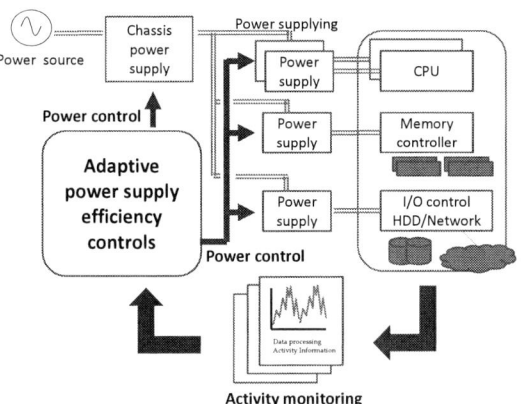

Fig. 2. Concept of adaptive power efficiency controls.

As shown in the figure above, the system aims at achieving the most efficient power supply state by dynamically interlocking the configuration with the constantly changing data processing load. To achieve this, it is first necessary to clarify the optimal server workload monitoring method and confirm its suitability for power efficiency control. Additionally, it is necessary to determine how such power efficiency control processes respond to real-time workload information.

In the computers such as servers, processors are responsible for the largest portion of power consumed. Furthermore, since the power consumption of other data processing components, such as the memory or I/O, is strongly influenced by processor activity, it is clear that the processor activities dominate computer power consumption. The processor power consumption is described in Eq. (1) .

$$P_{cpu} = AfCV^2 + I_{leak}V \tag{1}$$

A : activity rate f : clock frequency

C : gate capacitance V : core voltage

I_{leak} : leak current

Eq. (1) shows that processor power depends on the square of the core voltage, internal frequency, and activity rate. Among those factors, the activity rate and internal frequency are the indicators that show the type of data processing in the CPU and its execution performance. Typical processors usually have integrated performance counters to monitor various events in the processing unit. While those performance counters are normally used to analyze and optimize software execution, they can also be used to gather key information on processor power consumption. To date, several techniques have been proposed for the power consumption monitor using CPU performance counters [8].

In this research, it was investigated the adoption of processor integrated performance counters for the purpose of power supply efficiency controls in computers. Using the event number obtained by the performance counter function, power prediction equation was defined to predict processor power consumption. Then, the application of the power prediction procedure was considered in ordered to improve power supply efficiencies.

The power consumption prediction equation is defined in Eq. (2) below:

$$P_{prediction}(t_k + \delta) = \sum_{i}^{n} a_i \times CNT_i(t_k) + b \tag{2}$$

$P_{prediction}(t_k + \delta)$ Power prediction at δ time after time slot t_k
$CNT_i(t_k)$ Value counted at performance counter i during time slot t_k
a_i Coefficient of value of performance counter i
b Constant value

$CNT_i(t_k)$ is the number of event (i), which is counted in performance counter at the time (t_k). "a_i" is coefficient to the value of performance counter (i) and "b" is intercept. Thus, the power consumption can be predicted at time $t_k+\delta$ using the performance counter number at the time t_k. The information obtained from this equation can be utilized as inputs for power supply efficiency controls.

III. EXPERIMENTAL PROCEDURE

A. Drawing power consumption prediction equation

Fig. 3 shows the computer hardware environment on which performance counter and processor power consumption was measured in the experiments.

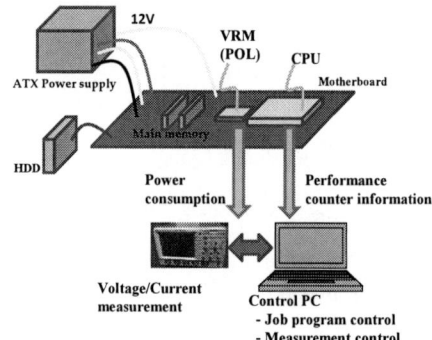

Fig. 3. Experimental computer setup.

The specifications of measured computer hardware are shown in the TABLE I.

TABLE I
COMPUTER CONFIGURATION USED FOR THE
MEASUREMENT

Items	System Features
System configuration	Single Processor System
CPU	Intel Pentium M Processor
CPU frequency	1.86GHz / FSB 533MHz
LSI Process	90nm Process
Core Voltage	1.25V
Thermal Design Power	27W
Main Memory	DDR/1GB
System Software (OS)	Ubuntu 10.04 Linux (Kernel 3.2.0)

A Pentium M 1.86 GHz processor was used in the experiment. The operation system (OS) was LINUX (Kernel 3.2.0). We also developed a device driver that permits the measurement application to control performance counter function. The performance counter can be observed via the CPU's external pin which toggles at internal counter increment. In the experiment, the pin was sampled every 500ns and average of the assertion was calculated every 1ms. Only one performance event of "total number of micro-operation" was used for the performance counter. The event item is the closest one which shows the strongest relation to the power consumption model defined for Pentium IV [9]. The event covers overall processor activities.

TABLE II BENCHMARK PROGRAMS

test name	type	description
pgbench	System (I/O bound)	Simple TPC-B like benchmark of PostgreSQL (ORDBS : object relational database system)
compilebench	Disk (I/O bound)	Aging a filesystem by simulating some of the disk IO common in creating
apache	System (Memory bound)	Test of ab which is the Apache web server benchmark program
ramspeed	System (Memory bound)	Cache and memory performance benchmark program of computer systems
postmark	Disk (I/O bound)	NetApp's PostMark benchmark designed to simulate small-file testing similar to the tasks endured by web and mail servers
sqlite	Disk (I/O bound)	Simple benchmark of SQLite database management system
stresscpu2	Processor (CPU bound)	Series of GROMACS inner loops hand coded in assembly for speed and efficiency on SSE units
idle	Idle	No runnable process is being scheduled on a CPU

Power consumption measurements were conducted at the point of load (POL) where 1.25 V power was supplied to the CPU. Phoronix Test Suite (Phoronix Media) job programs were used as the test server applications when measuring the performance counters and the power consumption data. Seven programs for each of the job execution types, such as CPU bound jobs and I/O bound jobs were selected for the experiment as described in the TABLE II.

Fig. 4 shows the determination process for the Eq. (2) coefficients. Both the performance counter value and power consumption level of the defined time period are obtained when the job program is running. Multiple regression analysis is performed to obtain the coefficients. In this analysis, the performance counter value is the explanatory variable and the CPU power consumption is the response valuable. Fig. 5 shows the process of data collection.

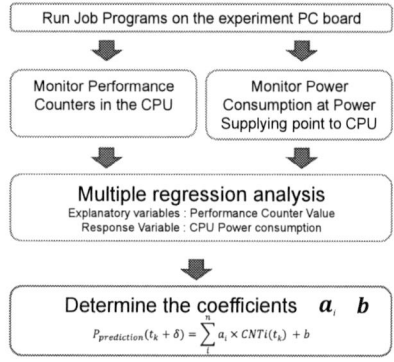

Fig. 4. Process for coefficients determination.

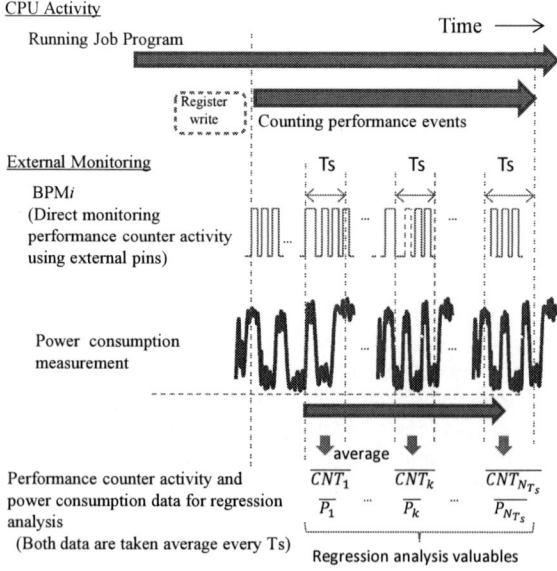

Fig. 5. Data collection process.

B. Simulation of adaptive efficiency controls using power consumption prediction

Fig. 6 shows the diagram of adaptive power efficiency controls which was evaluated in the circuit simulation. The simulation was conducted using OrCAD Pspice simulator. The processor was modeled using the measured data of power consumption and performance counter obtained in the experiments.

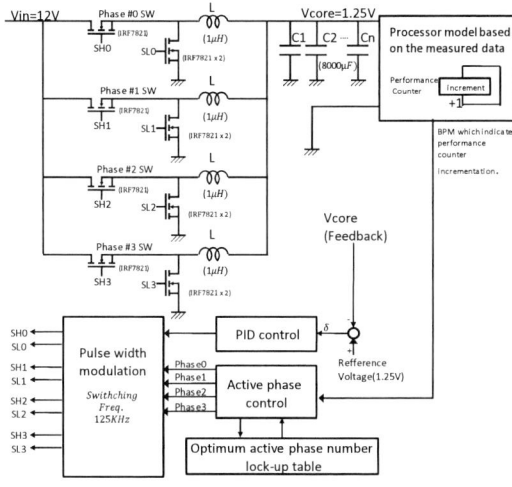

Fig. 6. Diagram of adaptive power efficiency controls which was evaluated in the SPICE simulation.

Obtain CPU performance counter information   (CNT)

Predict CPU power load by the Eq.(2) using the coefficient value (a, b) obtained from the regression analysis in the section A.

$$P_{prediction} = \sum a \times CNT + b$$

Look up the table of optimum active power phase number with the predicted power load rate ($P_{prediction} / P_{max}$).

Output the active phase number (**Phase0 -3**) for the period of T_s.

Fig. 7. Active phase control process.

In the simulation, power source active phase number was controled by the operation flow of Fig. 7. Using the prediction equation which is obtained in the section III (A), the power consumption of the upcoming cycle is predicted and the power supply configuration is optimized to the highest efficiency configuration for the predicted value by phase shedding technique. The active phase control block in the Fig. 6 sends the signals to pulse width modulation block to configure the most efficient power source phase number based on the optimum active phase number look-up table (TABLE III). The look-up table shows the phase numbers which show highest efficiency for power load rate in the efficiency curve which was obtained from circuit simulations(Fig.8.). The phase number selection and power configuration change was done every $1ms$ which was optimum sampling average period in the computer platform[10]. This period is descrived as T_s in the Fig. 5 and Fig. 7.

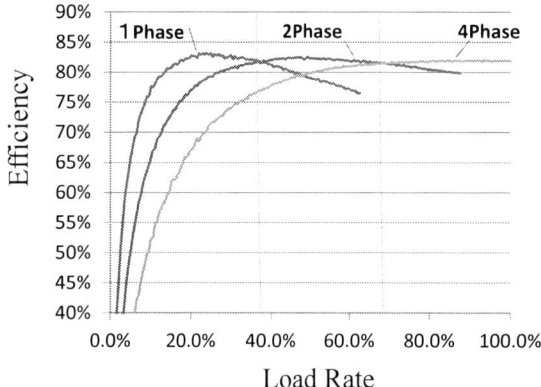

Fig. 8. Power efficiency curve of the simulated circuit.

TABLE III OPTIMUM ACTIVE PHASE LOOK-UP TABLE

predicted power	Activated phase			
load range	PH0	PH1	PH2	PH3
0% – 36%	ON	OFF	OFF	OFF
37% – 68%	ON	OFF	ON	OFF
69% – 100%	ON	ON	ON	ON

IV. RESULTS AND DISCUSSION

A. Comparison between predicted and observed power levels

TABLE IV shows the regression analysis results. As shown in the table, the correlation coefficient of the regression analysis is close to 1. This coefficient shows a strong correlation between the power consumption at t_{k+1} and performance counter value at t_k. TABLE V shows the difference between measured and predicted power consumption for each benchmark job. As the table shows, the prediction equation is properly modeling the power consumption.

TABLE IV
REGRESSION ANALYSIS RESULTS

Cefficient of equation	slope (uops) a		21.35
	intercept b		13.75
Regression statistics	Coefficient of determination R^2		0.855
	Standard error		2.35 (8.4%)
Observations			1222

TABLE V
POWER COMSUMPTION PREDICTION ERROR

	Benchmark job	Power consumption (W)		Prediction error (W)
		Measured	Predicted	
1	pgbench	13.88	13.81	0.07
2	compilebench	17.77	16.54	1.24
3	apache	25.73	22.92	2.81
4	ramspeed	21.30	17.20	4.10
5	postmark	15.92	15.96	−0.04
6	sqlite	12.92	14.09	−1.16
7	stresscpu2	27.07	28.01	−0.94
8	idle	12.95	13.65	−0.70

B. Simulation of the adaptive efficiency control

The circuit simulations of the proposed adaptive efficiency control were conducted. Fig. 9-Fig. 11 show the simulation results of the adaptive controls. The power load predictions are shown in the (a). Optimum active phase numbers (b) are selected in the look-up table (TABLE III) based on the predicted power consumption shown in (a). As shown in (c), the almost constant high efficiency was obtained regardless of the power load change when applying the adaptive efficiency controls. The averages every 1ms period of efficiency improvement are shown in (d). In general, the power supply operates at the maximum configuration if power efficiency control is not applied. The efficiency of the maximum power supply configuration is compared with the result of the adaptive power efficiency control approach. As shown in Fig. 9(d) and Fig. 10(d), efficiency improvements were obtained for both benchmarks. However, little efficiency improvement was obtained for stresscpu2 benchmark as shown in Fig. 11 (d). In this case, the phase control brought little effects since program type of the job is CPU bound and its power load is in the high efficiency domain of maximum power source configuration efficiency curve during the whole job execution.

The average of efficiency improvement ratio for the various job programs by the adaptive efficiency controls are shown in the TABLE VI. As can be seen in the table, there are efficiency improvements for all jobs excepting the stresscpu2. Especially, significant efficiency improvements are obtained for the I/O bound jobs which power consumption are light. As described earlier, the power supply is used at the range of light load when executing those types of jobs. These results show that the adaptive efficiency control approach provides significant improvements during light load power supply conditions. The simulation results also show that power supply efficiency improves when the CPU is in an idle state.

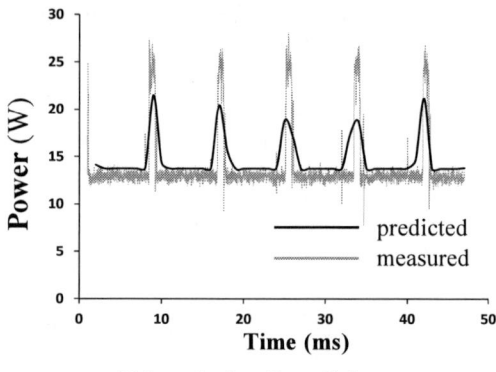

(a) Power load profile prediction

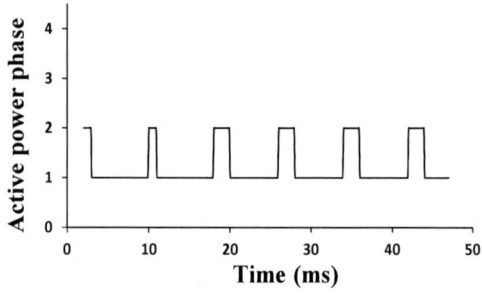

(b) Active phase controls

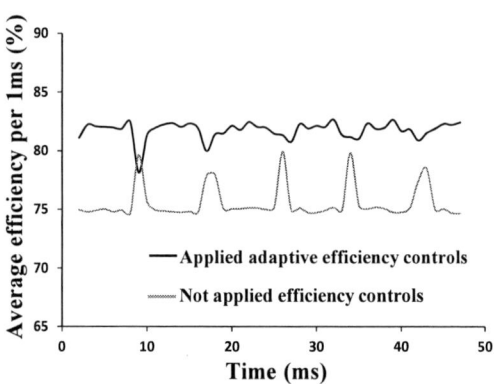

(c) Average efficiency of the power source (every 1ms)

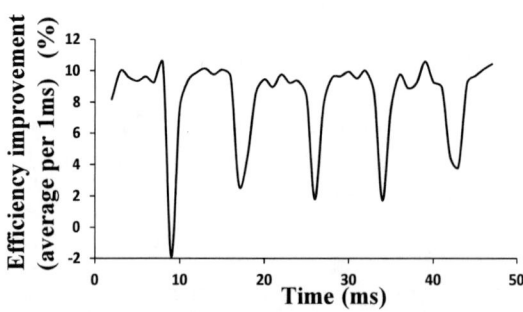

(d) Average of efficiency improvement ratio when applying the adaptive controls (every 1ms)

Fig. 9. Simulation result of adaptive efficiency controls (pgbench).

The 2014 International Power Electronics Conference

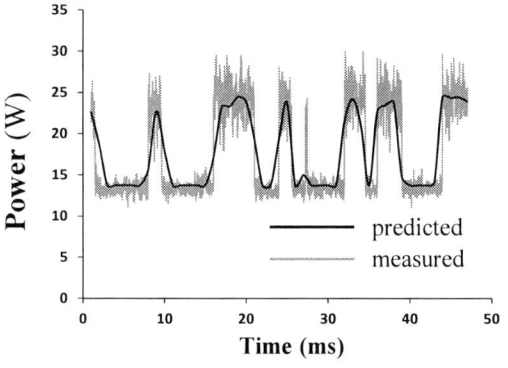

(a) Power load profile prediction

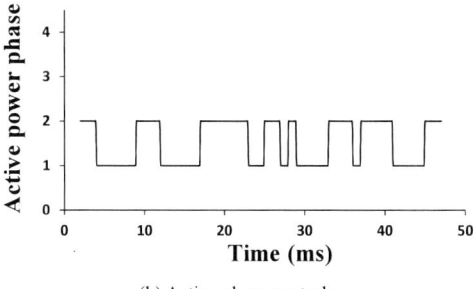

(b) Active phase controls

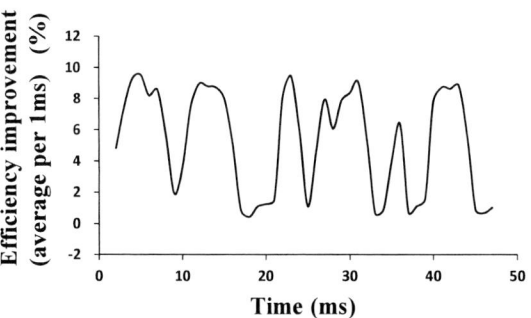

(c) Average efficiency of the power source (every 1ms)

Efficiency improvement (average per 1ms) (%)

(d) Average of efficiency improvement ratio when applying the adaptive controls (every 1ms)

Fig. 10. Simulation result of adaptive efficiency controls (compilebench).

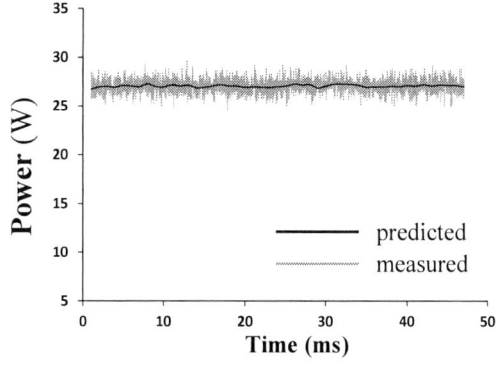

(a) Power load profile prediction

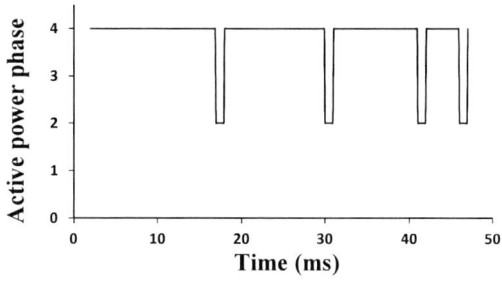

(b) Active phase controls

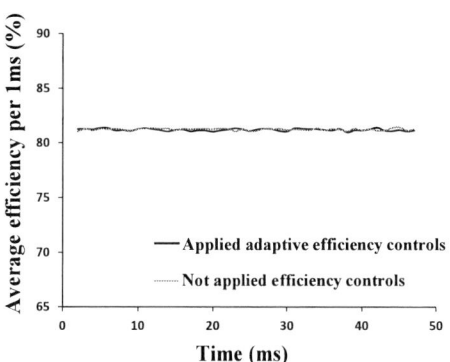

(c) Average efficiency of the power source (every 1ms)

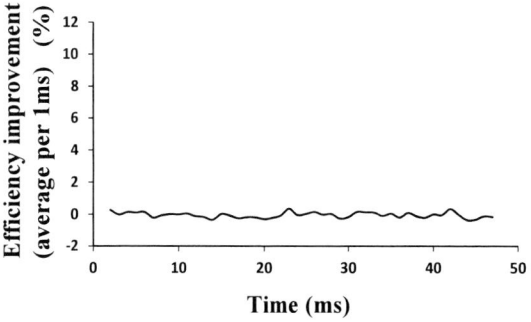

(d) Average of efficiency improvement ratio when applying the adaptive controls (every 1ms)

Fig. 11. Simulation result of adaptive efficiency controls (stresscpu2).

978-1-4799-2706-7/14 $31.00 © 2014 IEEE   3964

TABLE VI
AVERAGE EFFICIENCY IMPROVEMENT RATE
BY THE ADAPTIVE EFFICIENCY CONTROLS

Benchmark job		Power effciency		
		Not applied	Applied	Improvement rate
1	pgbench	75.5%	81.8%	6.2%
2	compilebench	77.6%	81.6%	4.0%
3	apache	80.1%	81.9%	1.8%
4	ramspeed	80.1%	81.9%	1.8%
5	postmark	76.7%	81.8%	5.1%
6	sqlite	74.9%	82.0%	7.1%
7	stresscpu2	81.2%	81.2%	0.0%
8	idle	75.0%	82.2%	7.2%

V. CONCLUSION

This paper proposes a new power conservation approach to dynamical control of CPU power supplies that works by interlocking with real time data processing workloads. The appropriate power consumption prediction method for the power efficiency controls was proposed and a prediction equation that anticipates power consumption levels using the performance counter information was defined. In the adopted approach, the performance counters built into the CPU is used to predict power load changes. Furthermore, applying the equation to the power supply efficiency controls was considered. The circuit simulation of the efficiency controls using the prediction equation was conducted. We validated the approach improves power supply efficiency in some cases of practical server workloads.

ACKNOWLEDGEMENTS

The part of the funding for this work was provided by the Japan MEXT Grants-in-Aid for scientific research (B), No24360112.

REFERENCES

[1] Da Teng, Yingjie Yang, "Power Requirements for Large Data Center", in Proceedings of Intelec'12, Scottsdale, 14.1, 2012.

[2] Tadatoshi Babasaki, Toshimitsu Tanaka, Kaoru Asakura, Yousuke Nozaki and Fujio Kurokawa, "Examination progress and development of HVDC power feeding system", in Proceedings of IEEE Int. Power Electron. Conf., Sapporo, pp. 870–873, Jun. 2010.

[3] Vinay Hanumaiah, Sarma Vrudhula, "Energy-Efficient Operation of Multicore Processors by DVFS, Task Migration, and Active Cooling," IEEE Transactions on Computers, vol. 63, no. 2, pp. 349-360, Feb. 2014.

[4] "80 PLUS Certified Power Supplies and Manufacturers" http://www.plugloadsolutions.com/80PlusPowerSupplies.aspx, Ecova Plug Load Solutions Website, accessed Feb. 2014.

[5] Myungbok Kim, "Two-Phase Interleaved LLC Resonant Converter with Phase Shedding Control", in Proceedings of IEEE Int. Power Electron. Conf., Sapporo, pp. 1642–1645, Jun. 2010.

[6] Jen-Ta Su, Chung-Wen Hung, Tsun-His Chang, Chih-Wen Liu, "A Novel Phase Shedding Scheme for Improved Light Load Efficiency of Interleaved DC/DC Converters", in Proceedings of Applied Power Electronics Conference and Exposition (APEC), Fort Worth, pp.1482-1487, 2011

[7] Fujio Kurokawa, Yuki Maeda, Yuichiro Shibata, Hidenori Maruta, Tsukasa Takahashi, Kouta Bansho, Toru Tanaka and Keiichi Hirose, "A New Digitally Controlled Switching Power Supply for Green IT", in Proceedings of IEEE Int. Power Electron. Conf., Sapporo, pp. 888–891, Jun. 2010.

[8] W. Lloyd Bircher and Lizy K. John, Complete System Power Estimation Using Processor Performance Events, IEEE Transactions on Computers vol. 61, no. 4, pp. 563-577, APRIL 2012.

[9] Shinichi Kawaguchi, Toshiaki Yachi, "New power consumption prediction procedure using performance counters to increase computer power supply efficiency", in Proceedings of Intelec'13, Hamburg, pp.710-715, Oct. 2013.

[10] Shinichi Kawaguchi, Toshiaki Yachi, "Computer Power Supply Efficiency Improvement by Power Consumption Prediction Procedure Using Performance Counters", IEICE Transactions on Communications Vol.E97-B, No.2, pp. 408-415, Feb. 2014.

AUTHOR INDEX

Abe, Kodai 3153
Abe, Seiya 177, 1179, 2216, 2222, 3652
Abe, Shigeru 1109, 1115
Abe, T. 3007
Abe, Takashi 2183, 2189, 3024
Abe, Tomohiko 1575
Abiko, Hiroshi 634
Achara, Pichetjamroen 3687
Adachi, Mitsuo 92
Adhikari, Jeevan 1775
Agelidis, Vassilios G. 1458, 3758, 3764, 3933
Agelidis, Vassilios Georgios 640
Aguglia, Davide 3371
Ahmed, Furqan 480, 790
Ahssanuzzaman, S. M. 3582
Aiso, Kohei 1141
Ajima, Toshiyuki 383, 682
Akagi, Hirofumi 750, 1586, 1761, 2290, 2323, 3742
Akagi, Masataka 629
Akahane, Masashi 2302
Akatsu, Kan 1128, 1141, 1234, 2673, 3828
Aketa, M. 2074
Akira 3784
Akiyama, Satoru 2285
Alemi, Payam 1201
Alipoor, Jaber 3298
Aljankawey, A. S. 2156
Allen, Scott 3447
Almer, Stefan 3563
Ama, Naji Rajai Nasri 2413, 2988, 3278
Amanci, Adrian Z. 1303
Amano, Yuki 1824
Amma, Ryosuke 2027
Anazawa, Yoshihisa 3801
Andersen, Michael A. E. 78, 506, 2842, 3352, 3905
Ando, Itaru 1516
Ando, Masato 1317
Anthon, Alexander 78
An-Yeol Ko 796
Aoki, Mutsumi 2400
Aoyagi, Shigehisa 2451
Aoyama, Fumio 2644
Aoyama, Kohei 2266
Aoyama, Masahiro 1405
Aoyama, Tomohiro 3823
Ara, Takahiro 3044
Arai, Haruki 403
Arai, Manabu 3440
Araki, Jun 1728
Araki, Takahiro 1613
Arata, Masanori 1874
Arikawa, S. 3007
Arimatsu, Kenji 415
Arita, Hideaki 2673
Asai, Inami 123

Asakimori, Koki 1567
Asama, Junichi 988
Asano, Katsunori 3440
Asano, Yoshinari 1997
Asano, Yuji 3872
Ashikaga, Tadashi 1886
Athab, Hussain S. 3695
Atsushi, Manabe 2745
Awaji, Sosuke 3194
Ayano, Hideki 2385
Azuma, M. 1892
Baba, Jumpei 1849
Babasaki, Tadatoshi 1567
Bac Xuan Nguyen 2722
Bafleur, M. 707
Bahman, A. S. 2862
Bahrani, Behrooz 1386
Bak, Claus Leth 3320
Bakran, Mark-M. 2113, 3255
Bang, Deok-Je 2427
Bani Shamseh, Mohammad 2794
Baoquan Liu 1155, 3546
Barater, Davide 433
Barrade, Philippe 1081
Barth, Henry 2881
Basari, Amat A. 3194
Basu, Kaushik 3061
Bauer, Florian 3898
Bauer, Pavol 1193, 3200
Baumgartner, Thomas 1707
Beczckowski, Szymon 2547
Beczkowski, S. M. 2862
Beczkowski, Szymon 2850
Belanger, Jean 2644
Ben Guo 3129
Ben, Hongqi 2318
Beres, Remus 3320
Berhouet, S. 707
Berkouk, El Madjid 560
Bessegato, Luca 1087
Besselmann, Thomas 3563
Bhat, Ashoka K. S. 1721
Bhattacharya, Subhashish 651, 656,
  758, 1626, 2562, 3225, 3286, 3447, 3726
Bianda, Enea 3432
Biela, J. 868
Biela, Jurgen 1788
Bilal, Akin 230
Bin Wu 3482, 3695
Binbin Li 3680
Bizen, Yosio 2983
Blaabjerg, F. 548, 1912, 2862
Blaabjerg, Frede 216, 857, 1529,
  1634, 1801, 2610, 3320
Blank, Frederic 264

AUTHOR INDEX

Bo Wen 944
Bocker, J. 2887
Bocker, Joachim 346, 1501, 1508
Boehm, Andreas 283
Boillat, David O. 1073
Boitier, V. 707
Boroyevich, Dushan 944, 2626, 3850
Bortis, D. 1291
Bortis, Dominik 1309, 2079, 3864
Bosshard, R. 2167
Bosshard, Roman 1904
Boyu Wang 2893
Braz Cardoso, F. 3225
Burger, Niklaus 1386
Burgos, Rolando 944, 2626, 3850
Burkart, Ralph M. 891, 3460
Buticchi, Giampaolo 433
Byoungchang Jung 1185
Byung Moon Han 937
Byung-Geuk Cho 2802
Byung-Gyu Yu 3784
Cai, Zheng-Xiu 429
Canales, Francisco 1043, 3432
Cao, Guoen 2587
Cao, Wei 567
Cao, Yuan 647
Cardoso, Braz J. 3270
Caris, M. L. A. 2954
Carvalho, Eden Luiz 1276
Casolari, Ronaldo Pedro 1276
Castellazzi, A. 2503
Castellazzi, Alberto 433, 2920, 3718
Ceballos, Salvador 3758, 3764
Cha, Honnyong 110, 480, 790
Chai Feng 3129
Chang, C.-H. 2050
Chang, Chien-Hsuan 2523, 3333
Chang, Hsiu-Feng 330
Chang, Kai-Chi 105
Chang, L. 2156
Chang, Yuan-Chih 330, 1832
Changsheng Hu 782
Changwoo Kim 1646
Chao Wang 2950
Chao-Fu Wang 2758
Chattopadhyay, Ritwik 3225
Chen, Ching-Guo 1734
Chen, H. 1471
Chen, Hsin-Chih 1639
Chen, Hung-Chi 2580
Chen, Jiann-Jong 2910
Chen, Jung-Chieh 677
Chen, Min 485
Chen, Qianhong 1425
Chen, Shen-Li 236
Chen, Wei 66, 72
Chen, Wenjie 2950
Chen, Yaow-Ming 3592
Chen, Ying-Zuo 351
Chen, Zhe 3538
Chen-Feng Chuang 3379
Cheng Deng 782
Cheng, Chun-An 2523, 3333
Cheng, Hung-Liang 2523, 3333
Cheng, Po-Tai 1261, 1639
Cheng, Shih-Jen 199, 2593
Cheng, Stone 3425
Cheng-Chieh Yu 2910
Cheng-Wei Chen 3592
Cheol-O Yeon 1738
Cheon, Jun P. 3358
Chia-Chi Chu 3379
Chiang, Hsin-Wei 2100
Chiba, Akira 982, 988, 3513
Chiba, Yoshinori 634
Chien-Yu Lin 2758
Chih Wei Chen 3938
Ching-Hsiang Yang 1639
Ching-Tasi Pan 3379
Ching-Wei Wang 1639
Chiu, Chian-Song 440
Chiu, Huang-Jen 172, 199, 2593, 3328
Chiu, Tse-Wei 440
Cho, Bo-Hyung 2272, 2575
Choi, Bo H. 2232, 3358
Choi, Byungcho 3638
Choi, Hangseok 2575
Choi, Seong-Chon 409
Choi, Sewan 1394, 2247
Choi, Su Y. 1103
Chokchai, Chuenwattanapraniti 3789
Chou, Tzu-Han 208, 421
Chow, T. Paul 2208
Chu, Xi 1322
Chun, Chang Yoon 2272
Chung, Tsung-Yuan 2523
Chung-Chuan Hou 2821
Chung-Yi Lin 3185
Chung-Yuen Won 796, 3532
Chunkag, Viboon 694
Chu-Shen Chang 3928
Ciftci, Baris 3734
Coldevin, Grete H. 1861
Colmenares, Juan 3712
Concari, Carlo 433
Cortes, Patricio 3864
Cortizio, Porfirio C. 3225
Cosovic, Mirsad 1148
Daesu Han 1185
Dahidah, Mohamed S. A. 1283

AUTHOR INDEX

Dahono, Pekik Argo..............................3893
Dai, Wei-Fu 330, 1832
Daikoku, A...1892
Daikoku, Akihiro 2011, 2673
Dan Chen 3938
Darba, Araz..718
D'Arco, Salvatore..................................1544
Darus, Rosheila 3758, 3764
Daskalos, Mike 2330
Davoodnezhad, R. 1482
Dawson, Francis P..................................1303
De Belie, Frederik 718
De Carvalho, Kelly Caroline
Mingorancia 2413, 2618, 2988, 3278
De Doncker, R. W. 3898
De Doncker, Rik W. 736, 2729, 3145
De Haan, Sjoerd 2787
De Mallac, Louis 3371
De Miranda, Rubens Domingos..............1276
De Paula, Helder 3225
De S. Brito, Jose A.................................3225
De Vega, Angel Ruiz 2547, 2850
De, Ankan 651, 2562, 3286, 3447
De, D..2503
De, Dipankar 433
Deguchi, Tadayoshi................................3440
Dehong Xu 782
Dekka, Apparao 3468
Demetriades, Georgios D.......................1220
Deng, Lirong..465
Dianguo Xu 3174, 3680
Dianguo, Xu 341
Diduch, C. P..2156
Dilhac, J-M..707
Diniz, Rogerio Azevedo.........................3270
Doki, Shinji 907, 2445, 3079, 3823
Domoto, Kazuhide 3652
Dong Le...1837
Dong-Hee Lee 994, 2693
Dong-Jing Lee 1452
Dongkook Son 2914
Dongouk Kim..925
Dongwook Kim 2914
Dou, Qinyun 3604
Dowaki, Kiyoshi....................................1207
Do-Yun Kim 796
Drofenik, Uwe.......................................1043
Du Yan 2668
Du, Yimian..1721
Duarte, J. L. 2954
Dujic, Drazen..3476
Durand Estebe, P. 707
Dutta, R...2679
Endo, Takahisa.............................2541, 2977
Eni, E..1912

Enomoto, Toshio....................................2421
Enomoto, Yuji 1997
Erturk, Feyzullah 3734
Eui-Cheol Nho 2763
Fang Zheng 1342
Fang Zhuo......................................1155, 3546
Fang, Xiaocun 335
Fassler, Lukas 3864
Fei Lin 807
Fei Meng...2815
Fei Zhang 3857
Fernandes, B. G. 2433
Ferrari, Bruno Augusto 3278
Ferreau, Joachim...................................3563
Ferreira, J. A. 1935
Ferreira, Jan A. 2787
Figueredo, Ricardo Souza...............2413, 2618
Fletcher, J. 2679
Fletcher, John 2926, 2932
Foo, Gilbert 2722
Fosso, Olav B. 1861
Foureaux, Nicole C................................3225
Fournier-Bidoz, Sebastien 3496
Franca, Gleisson J. 3270
Franceschini, Giovanni 433
Franke, Toke 78
Fritz, Dominik 3476
Frohleke, N. 2887
Fujii, Junji..1654
Fujii, Kansuke 1748
Fujii, Toshiyuki 2663
Fujimoto, Hiroshi 1671, 2421
Fujimoto, Takafumi 3857
Fujimoto, Yasutaka 1685, 1968
Fujisaki, Keisuke 289, 2856, 2874
Fujisawa, Hiroyuki................................3440
Fujita, Hideaki.............1006, 1160, 1350, 2027, 2042
Fujitsuna, Masami..................................3079
Fuketa, Hiroshi 2228
Fukuda, Kenji 3440
Fukuhara, Shuhei 289
Fukumoto, Hisao 724, 730, 3067, 3249
Fukuoka, Hiroki 3341
Fukushima, Kentaro 2189
Fulin Zhou 1050
Funabiki, Shigeyuki 2470
Funaki, Tsuyoshi 1621
Funato, Hirohito............................1728, 2517
Furukawa, Kimihisa...............................383
Furukawa, Tatsuya 724, 730, 3067, 3249
Furukawa, Yutaka 2252
Furuta, R. 2120
Gaing, Zwe-Lee 278
Gao Qiang 3050
Gao, Qiang 614

AUTHOR INDEX

Geng, Hua 543
Gerling, Dieter 774
Ghimire, Pramod 2547, 2850
Giaretta, Antonio Ricardo.......................1276
Goehler, Lutz...2554
Goh Teck Chiang 1028
Gohara, Hiromichi 671
Goto, Akira 130
Goto, Yasuyuki..1490
Goto, Yuichi..1671
Graus, Johannes.......................................270
Grider, Dave 3726
Gruber, Wolfgang....................1691, 1701
Gu, Beom W. 1103
Gueldner, Henry 2554
Guidi, Giuseppe.......................................1544
Gunasekaran, Deepak...............................1342
Guo, Wei...160, 475
Gurpinar, Emre 433, 3718
Ha, Jung-Ik...3140
Hafner, Jurgen 3667
Haga, Hitoshi..................................415, 3153
Haghbin, Saeid 1373
Hagiwara, Makoto 1586, 1761, 2323, 3742
Hahn, Ingo.......................................270, 283
Haining Wang...3702
Haitao Yang...782
Hak-Soo Kim...2763
Hakutou, Takuma.....................................2297
Hama, Ryota 2470
Hamasaki, Shin-Ichi 2775, 3674
Hamasaki, Sin-Ichi 3093
Hamazaki, Yasuhiro 2126
Hanada, T. 2074
Hanamoto, Tsuyoshi 538, 1811
Han-Shin Youn...1743
Hao Huang...2967
Hao Yi 1155, 2960, 3546
Hao, Xiang...2950
Hao-Chien Cheng.....................................3379
Hara, Hidenori 1654, 1898
Harada, Shingo...1671
Harada, Shinsuke......................................3440
Harakawa, Masaya 2638
Hariya, Akinori..3630
Hasegawa, Isamu 1365
Hasegawa, Kohei......................................3707
Hasegawa, Masaru...................183, 907, 2445
Hasegawa, Masataka.................................2212
Hasegawa, Shinya 294, 299, 2972, 3055, 3159, 3162
Hasegawa, Tomonori...............................2126
Hashimoto, Kento 1974
Hashimoto, S 1471
Hashimoto, Seiji......................................3194
Hashimoto, Shizuka 3018

Hassanpoor, Arman 3667
Hatanaka, Ayumu 2285
Hattori, Fumiya 811
Hatua, Kamalesh 758, 1626
Hau-Chen Yen 3928
Hava, Ahmet M. 498, 2034, 3734
Hayase, Masanori 1207
Hayashi, Makoto 1950
Hayashi, Toshihiko 3440
Hayashi, Yusuke 1560
Hayashiya, Hitoshi 1062
Hazeyama, M..1892
Hazra, Samir....................758, 1626, 3447
He, Guofeng 485
Hee-Jun Lee 3532
Hei, Xinhong 647
Hella, Mona 2208
Hermansson, Willy 1220
Hernandez, Juan C 3352
Heung-Geun Kim 790, 2763
Hibino, Shinya 2638
Hidaka, Akira 2216
Hidayat, Nabil M 573, 2529
Higuchi, Shinichi 1522
Higuchi, T. 3007
Higuchi, Tsuyoshi 3024
Hijikata, Hiroki 2673, 3828
Hikita, Masayuki 689
Hinata, Toshifumi 919
Hinkkanen, Marko...................................2489
Hino, Wataru...3525
Hintz, Andrew 2343
Hira, Yuki 730
Hirahara, Hideaki 3044
Hirai, Junji..1974
Hirakawa, Yuki 3024
Hiraki, Eiji..3292
Hirano, Yosei 1956
Hirao, Kuniaki 191, 1365
Hirase, Yuko 1552
Hirokado, K. 146
Hirose, Toshiro 2252
Hirota, Yukitsugu 1728
Hisada, Yoshihiro 3292
Hisato, Hosoyama 2745
Ho, Kung-Min...3942
Hoene, Eckart 2366
Hoffmann, Stefan 2366
Hofmann, Wilfried 2881
Hojo, Masahide..2152
Hojo, Toshiaki 1276
Hokazono, Hiroaki 2870
Holm, Toni 3432
Holmes, D. G............1482, 2019, 2372, 3306
Homma, Hiroshi 1880

AUTHOR INDEX

Hong Li 3314
Hong, Ki-Nam 2598
Hong-Hee Lee 1013
Hongqi Ben 3213
Hori, Yoichi 2421
Horiguchi, Takeshi 2290
Horita, Yasuhisa 1317
Hosaka, Tatsuya 1350
Hoshi, Nobukazu 1242
Hosoyamada, Yu 801
Hou, Chih-Hao 1796
Hou, Jiaxin 526
Hou, Lixiang 577
Hredzak, Branislav 1458
Hsieh, Guan-Cyun 526
Hsieh, Hung-I 526, 2380
Hsieh, Min-Fu 278
Hsieh, Yao-Ching 429, 1796
Hsin-Chih Chen 1261
Hsin-Ping Su 2821
Hu, Jia-Sheng 278
Hu, Shang-Hung 2606
Hu, Sheng 555
Hu, Taiyuan 335
Huang, Hsin-Wei 421
Huang, Jia-Wei 3233
Huang, Lang 2950
Huang, Min 2610
Huang, T. D. 2195
Huang, Wen-Nan 1734
Huang, Zhenhui 647
Huang-Jen Chiu 2758, 2810, 3185, 3913
Huber, Jonas E. 766
Huber, Tobias 1508
Hui Liu 1634
Hui Zhang 1365, 3455
Huisman, H. 2954
Hull, Brett 3447
Huu-Nhan Nguyen 1013
Hwang, Seon-Hwan 2427
Hwang, Yuh-Shyan 2910
Hwu, K. I. 204, 2754, 3190, 3392
Hyoyol Yoo 1646
Ichihara, Junichi 2189
Ichiya, Takahiro 370
Ichiyanagi, Katsuhiro 1490
Ide, Kozo 933
Ieda, Jun 2663
Iga, Yuichi 3341
Igarashi, Kazunori 2983
Igarashi, S. 3702
IIda, Mikiya 2977
IIjima, Ryuji 117
IIjima, Yukihia 2095
Ikawa, O. 2569, 3702

Ikeda, Hidehiro...2476
Ikeda, Masahiro 2183
Ikeda, Tomohiko 1575
Ikeda, Y. 2569
Ikeda, Yoshinari 2870
Il-Kuen Won 796
Ilves, Kalle 1087
Imai, Jun 2470
Imakiire, Akihiro 689
Imamura, Yasutaka 863
Imanishi, Takao 2663
Imaoka, Jun 811, 883, 2497
Inamori, Mamiko 3509
In-Dong Kim 2763
Inomata, Kentaro 1654
Inoue, Kaoru 3872
Inoue, Keita 130
Inoue, M. 1892
Inoue, Tatsuki 363
Inoue, Y. 246, 258, 312, 390
Inoue, Yukinori 324, 356, 363, 370, 2183, 3018, 3519
Irokawa, Shoichi...1357
Ise, Tomofumi 1430
Ise, Toshifumi 1536, 1560, 2632, 3298, 3687
Ishibashi, Makoto 724
Ishida, Koichi 2228
Ishida, M. 146
Ishida, Masaaki 3707
Ishida, Masaki 3162
Ishida, Takahito 634
Ishigami, Takashi 1880
Ishigma, Satoru 403
Ishihara, Chio 1984
Ishihara, Yuji 1135
Ishii, Hirotaka 294
Ishikawa, Hiroki.............................1135, 2183, 2189
Ishikawa, Katsumi............................2140, 2285
Ishikawa, Takeo................................ 252, 1697
Ishimaru, Yusuke 92
Ishimori, Hitoshi...3440
Ishitobi, Manabu 811
Ishizuka, Tomotsugu 2644
Ishizuka, Yoichi 2222, 2252, 2737, 3630, 3652
Isida, Takashi 1950
Itako, Kazutaka 3244
Ito, Yasuhide 3823
Ito, Yoichi 403
Itoh, Hideaki 724, 730, 3067, 3249
Itoh, Jum-Ichi 1943
Itoh, Jun-Chi 1253
Itoh, Junichi 130
Itoh, Jun-Ichi 84, 138, 152, 191, 682,
    1021, 1028, 1095, 1613, 2277, 3659, 3815
Itoh, Tomomichi 850
Itoh, Youichi 415

AUTHOR INDEX

Itoh, Yuki.............................883, 2497
Iwaji, Yoshitaka 2451
Iwakami, Tetsuro 817
Iwasaki, Makoto 1665
Iwasaki, Shinya 2663
Iwata, Tetsuki 403
Iyer, Kartik V 3037, 3061
Iyer, Shivkumar 3482
Izumi, Toru 3440
Jacobson, Bjorn 3667
Jae-Bum Lee 1738
Jaeho Choi 2656
Jae-Hun Jung 2763
Jae-Hyun Kim 1738
Jaesig Kim 2656
Jang, Jinhaeng 3638
Jang, Young-Jin 664
Jang-Hwan Kim 925
Jardini, Jose Antonio 1276
Jauch, Felix 1788
Javed, Riffat 624
Jayoon Kang 1185
Jen-Hao Teng 1452
Jenn-Jong Shieh 3190
Jeon, Jin-Yong 166
Jeong, Seog Y 1103
Jeong, Seon-Yeong 2406
Jeongjoong Kim 1185
Jhen-Yu Jian 3928
Jia Liu 1536
Jia, Y. 1594
Jianfeng Li 3718
Jiang, Dawang 647
Jiang, Maoh-Chin 105
Jiang, W. 1471
Jiang, W. Z. 204, 3190
Jiang, Yongjie 458
Jianhui Su 2668
Jiann-Fuh Chen 2714
Jianwen Zhang 3124
Jih-Hua Yu 2910
Jin Miaoxin 3050
Jin, Miaoxin 614
Jin, Xu 341
Jin, Yasuhiro 3207
Jing Bian 3314
Jing-Hsiao Chen 3233
Jing-Yuan Lin 2758
Jinjun Liu 835
Jinno, Masahito 1781, 3333
Jin-Woo Ahn 994, 2693
Jinyong Zhang 3213
Ji-Shiang Lee 3346
Joebges, Philipp 2729
Jokipii, J. 514, 2240

Jokipii, Juha 1466
Jong Kyou Jeong 937
Jonghyung Park 1185
Jonishi, Akihiro 2302
Jou, Sung-Tak 224
Jung, Hochang 1990
Jung, Jae-Jung 1268
Jung, Sang-Yong 1990
Jung, Yong-Chae 166, 409
Junghum Lee 2656
Junjie Feng 835
Juntao Fei 3168
Juyoung Jang 2656
Kabasawa, Yuichiro 2175
Kabiri, R. 3306
Kadavelugu, Arun 758, 1626, 3726
Kai, Masahiko 1054
Kai-Hui Chen 2750, 3346
Kaipia, T. 587
Kajiwara, Kazuhiro 3950
Kakishima, Takeo 3513
Kalogera, Maria 1193, 3200
Kameshiro, Norifumi 2140
Kamikura, Mamoru 2064
Kamnarn, Uthen 694
Kanagawa, Kinji 2983
Kanai, Yasuyuki 1567
Kanamori, Masaki 2541, 2977
Kaneko, Junji 2745
Kaneko, Yasuyoshi 1109, 1115
Kanematsu, Masato 2421
Kanemoto, Daisuke 2737
Kang, Feel-Soon 2260
Kang, Yong 555
Kanno, Hiroshi 2302
Kano, Yoshiaki 2004, 2457
Kanoda, Akihiko 3920
Kanouda, Akihiko 2058
Kantar, Emre 2034
Kanthaphayao, Yutthana 694
Kari, Mat Nasir 573
Karki, Ujjwal 1342
Karvonen, Andreas 1373
Kasai, Makoto 3194, 3194
Kashihara, Yugo 1943
Kasper, Matthias 2079
Katade, Motohumi 130
Katakami, Shuji 3440
Kataoka, Yasuhiro 3801
Katayama, Noboru 1207, 1227
Kato, Hideaki 2972, 3162
Kato, Koji 403, 415
Kato, Shinji 2175
Kato, Takashi 3828
Kato, Taro 2972

AUTHOR INDEX

Kato, Tomohisa 3440
Kato, Toshiji 2183, 2189, 3872
Kato, Yutaka 2644
Katoh, Kaoru..2285
Katoh, Shuji..850
Katsuki, Akihiko 1575, 3624
Katsura, Seiichiro 1679
Kawachi, Konosuke 863
Kawaguchi, Shinichi 3959
Kawahara, Keiji 1062
Kawakami, Noriko 2095
Kawamura, Atsuo 801, 2266, 2794, 3403
Kawamura, Mitsuhiro 3012
Kawamura, Wataru 3742
Kawano, Daisuke 1671
Kawano, Kenji..883
Kawazoe, Yosuke 2011
Kazuya, Ogura 452
Kempen, S...2887
Kenji, Matsumoto 3218
Kern, Ansgar...712
Khan, Ashraf Ali 110
Khan, Faisal H..2161
Khant, Hlaing Kyi Pyar......................................183
Khomfoi, Surin 2392
Kicin, Slavo...3432
Kihyun Lee 1646
Kikuchi, Takuya 1328
Kim, Bong C..3358
Kim, Chong-Eun 1738, 1743
Kim, Dong-Hun 790
Kim, Dong-Rak 409
Kim, Hee-Jun 2587
Kim, Heung-Geun 110, 480
Kim, Hyejin 2575
Kim, Jae-Hyun 1743
Kim, Jang-Mok 2406
Kim, Ji H..2232
Kim, Ji-Won 2427
Kim, Jonghoon 619
Kim, Minjae..2247
Kim, Seonghye 2260
Kim, Su-Han 480
Kim, Sungmin 1268
Kim, Yong-Jae...1990
Kimoto, Tsunenobu..3440
Kimura, Hiroshi 1920
Kimura, Noriyuki...................1299, 1806, 2183, 3341
Kimura, Shota 883, 2497
Kinouchi, Shin-Ichi....................................750, 2290
Kish, Gregory J..951
Kitabayashi, Tatsuaki 2517
Kitagawa, Wataru 2310, 3809
Kitajima, Jun...1247
Kitazawa, Satochi..1438

Kiyota, Kyohei 3513
Kleinecke, John 2330
Kluge, Andreas..2554
Knott, Arnold.. 506
Kobayashi, H. 2569
Kobayashi, Hiroya 2517
Kobayashi, Ryota 1115
Kobayashi, Takenori 1868
Kobayashi, Y. 2569
Kodama, Takashi 1365
Kogi, Ryosuke 2874
Kogoshi, Sumio 1207, 1227
Kohama, Teruhiko 522, 2781
Kohno, Yusuke 2183
Koiwa, Kazuhiro 84, 130, 1028
Kolar, J. W. 1291, 2167
Kolar, Johann W. 766, 821, 891, 899, 975,
        1073, 1309, 1707, 1904, 2079, 2834, 3365,
        3460, 3864
Komada, Satoshi.. 1974
Komatsu, Wilson 1276, 2413, 2988
Komeda, Shohei 1160
Komiya, Hiroshi 2421
Kon, Saytaro 3263
Kondo, Keiichiro 1438, 2126
Kondo, Seiji 415
Kondo, Takeshi 1365
Kondou, Masahiko 2421
Kono, Y. 2120
Kono, Yasuhiko 2140
Konoto, Masaaki 2189
Konstantinou, Georgios 1458, 3758, 3764
Korner, Olaf..2113
Kosaka, T. 2438
Kosaka, Takashi 1984, 1997
Koschik, Stefan 3145, 3898
Koseki, Takafumi 1334, 2126
Kotegawa, Ryo 1317
Kotera, Keito 3872
Kouki, Matsuse 3134
Kouno, Yusuke 2189
Kounoto, Masaaki 2175
Koyama, Masato... 750
Krafft, Eberhard 2113
Krismer, Florian 2834
Kuan-Hsien Chou 3346
Kubo, H. 395, 1594
Kubo, Hajime..1601
Kubo, Yuji..3134
Kubota, Hisao 919, 1929, 3119, 3134
Kubota, Yutaka 2183
Kudo, Takahiro 1109
Kuga, Shotaro 3955
Kukita, Akio 1444, 2351
Kumagai, Shunji..3194

AUTHOR INDEX

Kumakura, Yoshito....................1715
Kume, Tsuneo....................1898
Kumsuwan, Yuttana....................3417
Kun-Hung Chen....................3592
Kunomura, Ken....................1054
Kuo, Kuan-Yi....................278
Kuperman, A.....................2240
Kurabayashi, Toshiyuki....................1962
Kuribayashi, H.....................2569
Kurihara, Takeshi....................299
Kurihara, Yoshihiro....................1874
Kurita, Nobuyuki....................252, 1697
Kuroda, Y.....................1892
Kurokawa, Fujio....................2108, 3611, 3950
Kusaka, Keisuke....................191
Kusukawa, Jumpei....................2904
Kusunoki, Hironobu....................2330
Kutsuki, Tomohiro....................2064
Kuwahara, Akinobu....................3179
Kuzumaki, Atsuhiko....................1929
Kwasinski, Alexis....................2649
Kwasinski, Andres....................2649
Kwon, Soon-Kurl....................2359
Kyungbae Lim....................2656
Kyungmin Sung....................744
Kyungsub Jung....................1646
Lai, Yen-Shin....................3942
Lamantia, A.....................2503
Lana, A.....................587
Lang, Klaus-Dieter....................2366
Larsson, Tomas....................1220
Laska, Bernd....................2113
Law, Kah Haw....................1283
Le Hoai Nam....................3659
Lee, Chia-Tse....................1639
Lee, Dong-Choon....................1201, 2406
Lee, Eun S.....................1103, 2232, 3358
Lee, Hong-Hee....................2826
Lee, Jae-Bum....................1743
Lee, June-Hee....................493, 532
Lee, June-Seok....................493, 532
Lee, Kyo-Beum....................224, 493, 532
Lee, Min-Hua....................236
Lee, Seong Ryong....................3292
Lee, Shiu-Hui....................1734
Lee, Sung W.....................1103
Lee, Taeck-Kie....................595
Lee, Tzung-Lin....................2606
Lee, Woo-Cheol....................595
Lee, Ya-Ting....................440
Lee, Yuang-Shung....................208, 421
Lehmann, Oliver....................3085
Lehn, Peter W.....................951
Lei, Wanjun....................160, 475
Leibl, Michael....................899

Lelie, Markus....................2729
Leslie, Scott....................3726
Leuenberger, D.....................868
Leuer, Michael....................346
Li Yan....................2899
Li, Ding....................341
Li, Haiqing....................2095
Li, Hong....................2893
Li, Ning....................160, 475
Li, Qian....................2161
Li, Yanxiang....................3002
Lian, K. L.....................2195
Liang Hao....................3174
Liangyi Tang....................3695
Liao, Jhen-Yu....................2580
Lie Guo....................3489
Lin Cheng....................3447
Lin, Chiao-Chien....................3072
Lin, Chia-Yu....................1832
Lin, Chien-Yu....................172
Lin, Chung-Yi....................199, 2593
Lin, Fei....................335, 1322, 2133
Lin, Jing-Yuan....................172
Lin, L.-C.....................2050
Lin, Z. Y.....................1471
Lindberg-Poulsen, Kristian....................2842
Liping Zheng....................1837
Liserre, Marco....................857, 3320
Liu, Baoquan....................577
Liu, Fang....................567
Liu, Fangcheng....................3604
Liu, Fuxin....................458, 2768
Liu, Hanchao....................967
Liu, Jianyu....................614
Liu, Jilong....................66, 72
Liu, Jinjun....................624, 2815, 3604
Liu, Kangzhi....................3568
Liu, Ning....................2156
Liu, Rongqiang....................3099
Liu, Tai-Chun....................105
Liu, Xiankai....................647
Liu, Xiaosheng....................3002
Liu, Yi-Hua....................3233
Liu, Yu-Chen....................199, 2593
Liuchen Chang....................1476, 2668, 3842
Lo, Yu-Kang....................172, 199, 2593
Lobsiger, Yanick....................1309
Loh, Poh Chiang....................216, 1529, 1634, 1801, 2610
Longlong Zhang....................782
Lopez-Arevalo, Saul....................3718
Lovatt, Howard....................2679
Low, K. S.....................446
Lu, Dylan D. C.....................3553
Lu, Kao-Yi....................105
Luo, Guomin....................2145

AUTHOR INDEX

Luthardt, Sven 3029
Ma, K. 548, 2862
Ma, Weigang 647
Madawala, Udaya K. 2722
Madhusoodhanan, Sachin 656, 1626
Maekawa, Sari......................919, 1929
Maemura, Akihiko........................1898
Maeyama, Shigetaka...............1575, 3624
Maezono, Paulo Koiti....................1276
Mahdavikhah, Behzad.....................3582
Mainali, Krishna...................758, 1626
Makaino, Yuki..............................914
Makita, Shinji..........................3823
Mamun, Mostafa 97
Manias, Stefanos N......................1606
Mannen, Tomoyuki........................2042
Manolas, Iakovos 1606
Mao, Meiqin.............................2156
Maret, C................................3239
Marrero Sosa, Juan Alberto 3476
Martinz, Fernando Ortiz......2413, 2988, 3278
Maru, Naoki.............................2285
Marukawa, Yasuhiro......................1984
Marumori, Hiroki........................3055
Maruta, Hidenori........................3611
Marz, Andreas...........................2113
Marzouk, Ahmad Diab.....................3496
Masaki, Kenji...........................2663
Mashino, Masahiro.......................3162
Masic, Semsudin.........................1148
Maskell, D. L...........................3598
Masuda, Hiroyuki...........................92
Masui, Takeshi..........................1317
Masutomo, Kazufumi......................3624
Masuzawa, Hiroshi.......................1054
Masuzawa, Takashi.......................2366
Matakas, Lourenco 2413, 2618, 2988, 3278
Matsubara, Masakatsu....................1874
Matsuda, Katsuhiro........................415
Matsuhashi, Daiki.......................1886
Matsuhashi, Masataka....................1516
Matsui, Hitoshi.........................1586
Matsui, Keiju..............................183
Matsui, Mikihiko........................3489
Matsui, N...............................2438
Matsui, Ryota...........................1128
Matsui, Yoshihiro.......................2385
Matsui, Yoshinobu.......................2745
Matsumoto, Akira........................1560
Matsumoto, Atsushi......................2445
Matsumoto, Kazushi......................3440
Matsumoto, Satoshi......................2216
Matsumoto, Shuhei.......................1929
Matsumoto, Yasushi......................1920
Matsuo, Hirofumi........................1781

Matsuo, Keisuke 1886
Matsuo, Yusuke 1671
Matsuoka, Kazumasa 3207
Matsuoka, Yuji 744
Matsushima, Yoshitarou 3801
Matsushita, Makoto 3012
Matsuura, Kei 1516
Matsuura, Ken 3630
Matsuzaki, Ryohei 1978
Mattavelli, Paolo 3850
Mattsson, A. 587
Mauerer, M. 1291
McGrath, B. P....................1482, 2019, 2372, 3306
McLean, Kenneth 3496
Meiqin Mao....................1476, 2668, 3842
Mekhilef, Saad 560, 3574
Melkebeek, Jan 718
Meng, Fei 624
Meng, Tao 2318
Merahi, Farid 560
Messo, T. 514, 2240
Messo, Tuomas 1466
Mihara, Teruyoshi 1728
Mii, Kenji 2737
Miiura, Yushi 1430
Mikihiko 3784
Mills, Liam 3718
Ming Yang 3174
Mingfei Wu 3553
Mingyan Wang 3129
Mino, Kazuaki 1920
Minoshima, N. 2438
Minowa, Masanao 3828
Minsoo Jang 3933
Mira, Maria C. 506
Mishima, Tomoakzu 2533
Mishra, Santanu 2707
Mishra, Santanu Kumar 3587
Misu, Daisuke 1874, 3012
Mitterhofer, Hubert 1701
Miura, Yushi 1536, 3298
Miyajima, Hiroki 1054
Miyajima, Takayuki 2421
Miyakawa, Takayuki 2421
Miyama, Yoshihiro 2673
Miyashita, S. 3702
Miyawaki, Satoshi 84
Miyazaki, Hideki 383
Miyazaki, Kensuke 601
Miyazaki, Toshimasa 1956
Miyazaki, Yuji 750
Mizoguchi, Takahiro 1660
Mizukami, Makoto 3440
Mizuki, Tatsuya 1575, 3624
Mizuma, Takeshi 2126

AUTHOR INDEX

Mizuno, Takayuki 1886
Mizusaki, Hiroshi 3093
Moballegh, Shiva....................................656
Mochikawa, Hiroshi................................1929
Mochizuki, Eiji............................671, 2870
Mochizuki, K. 2569
Mohamed, Essam Ebaid 3877
Mohamed, Tarek Hassan 3877
Mohan, Ned 1036, 1412, 3037, 3061, 3750
Mohd Arif, Mohd Johari 573
Molinas, Marta 1861
Momose, Fumihiko 671
Moo, Chin-Sien 1796, 3796, 3928
Moon, Dongok 1394
Moon, Gun-Woo............................1738, 1743
Moon, Sang-Ho.......................................224
Moorthy, Radha Sree Krishna.........2087, 3616
Moraes, Lenin 3225
Mori, Tomohiro......................................2983
Morikawa, R. 258
Morimoto, Masayuki 3509
Morimoto, S. 246, 258, 312, 390
Morimoto, Shigeo..............324, 356, 363, 370, 1997, 3018, 3519
Morimoto, Shinya 1654
Morishita, Shin 130
Morita, Hiroshi.......................................1490
Morita, Kazunori 191, 582
Morita, Kosuke 3624
Morita, M 1892
Morizane, Toshimitsu 1299, 1806
Morizane, Tosimitsu 3341
Moroi, Takayuki 3134
Morozumi, Akira 671
Mory, David...2554
Motizuki, Shun 2745
Motoi, Naoki 801, 2266
Motomura, Masashi................................3611
Mouri, Masayuki 1728
Mrak, Branimir 1701
Mukai, Ryosuke 2775
Mukunoki, Makoto 97, 1950
Munk-Nielsen, Stig........................2547, 2850
Murai, Kensuke 1567
Murai, Toshiaki 1122
Murakami, Daichi 1728
Murakami, Kouhei 2385
Murakami, Toshiyuki 1962
Murata, Koji 2108
Murata, Munehiro 1173
Murata, Yuichiro 2064
Musing, Andreas 821
Mustapa, Rijalul Fahmi...........................2529
Muta, Shoichiro 3067
Nag, Soumya Shubhra 3587

Nagai, Shinichiroh..................................811
Nagano, Tetsuaki 2638
Nagano, Tsuyoshi 1253
Nagano, Y. 146
Nagashima, Tomohiro 2175
Nagata, Shun...2252
Nagatomo, Yoshinobu 2663
Nagel, Andreas..2113
Nagura, Hirokazu 2451
Nagy, Istvan 2700
Naitoh, Haruo 1135
Nakagawa, Hidehiko 1552
Nakagawa, Yuki 2533
Nakahara, Mizuki 744, 2511
Nakajima, Yoichiro 403
Nakamura, M. 376
Nakamura, Ritaka 92
Nakamura, Sota 2400
Nakamura, T. 2074
Nakamura, Tatsuya 1575
Nakamurame, Fuminori 2632
Nakanishi, Toshiki 1095
Nakano, Y. 2074
Nakao, Hiroshi..2745
Nakao, Noriya 1128, 1141
Nakaoka, Mutsuo 2359, 2533
Nakaoka, Mutuo 3341
Nakashima, Yoshiyasu 2745, 3386
Nakata, Yuki 138
Nakatsu, Kinya 2904
Nakatsugawa, Junnosuke 2451
Nakayama, Koji 3440
Nakayama, Naoyuki 3857
Nakayama, Yasushi 2290
Nakazawa, Yosuke 1357
Nam, Kwang-Hee 664
Narita, Takayoshi..............294, 299, 3055, 3159
Nashida, N. 2569
Nayanasiri, D. R. 3598
Nee, Hans-Peter 3712
Neubert, Markus 3145
Nguyen, D. 2679
Nguyen, Quoc Khanh 318
Nguyen, Thanh Hai 2406
Nha, Quang Trong...................................3913
Nho Van Nguyen 2826
Nian Heng 843
Nicolae, Ileana-Diana 2996
Nicolae, Marian-Stefan 2996
Nicolae, Petre-Marian 2996
Niijima, Koji 1299, 1806
Nilssen, Robert.......................................1412
Nimura, Tomohiro 3079
Ning Liu 2668

AUTHOR INDEX

Ninomiya, Tamotsu 177, 1179, 2216, 2222, 3630, 3652
Nishida, Katsumi 2359
Nishida, Yasuyuki 2189
Nishikata, Shoji 959
Nishimura, T. 3702
Nishimura, Tomohiro 2870
Nishimura, Yoshitaka 671
Nishio, Haruhiko 2302
Nishioka, Tomoya 2152
Nishisu, Koji 2285
Nishiyama, Noriyoshi.......................................2011
Nishizawa, Shinichi 117, 744
Niu, Ruigen..160, 475
Noda, Koji..2541
Noda, Taku 2175
Noguchi, Kenji..2277
Noguchi, S. 2569
Noguchi, Toshihiko 1173, 1405
Noh, Yong-Su 166
Nomura, Naofumi 1522
Nomura, Shinichi 3218
Nonaka, Hirotaka 2737
Norigoe, Isami 117
Noro, Osamu 1552
Norrga, Staffan 1087
Noto, Yasuo 682
Nozaki, Takahiro 1660
Nozawa, Ryosuke 1115
Nussbaumer, Thomas 975, 3365
Nuutinen, P...587
Oboe, Roberto 1679
O'Byrne, Sean 2926, 2932
Oda, Yoshinori 829
Odawara, Shunya.........................289, 2856, 2874
Ogasawara, Satoshi.................1728, 2977, 3525
Ogashi, Yoshihiro 92
Ogawa, Kazutoshi 2140, 2285
Ogawa, Takashi 2285
Ogura, Kazuya 582, 3455
Ogura, Tsuneo 2068
Oh, Min-Seok 166
Ohara, Shinya 850
Ohashi, Hiromichi.....................................117, 744
Ohashi, Shunsuke 3410
Ohchi, Masashi.....................724, 730, 3067, 3249
Ohishi, Kiyoshi 1247, 1516, 1956, 3153
Ohnishi, Kouhei 1660, 2483
Ohnuma, Takumi 914
Ohnuma, Yoshiya 84
Ohse, Naoyuki 3440
Ohtake, Asuka 3857
Oi, Kazunobu 452
Oi, Takeshi 2290
Oishi, K. 376

Oishi, Koji 3012
Oiwa, Takaaki 988
Ojika, Satoshi 1430
Oka, T. 376
Oka, Toshiaki 2330
Okamoto, Dai 3440
Okamoto, Masayuki 3292
Okamoto, Shoji 2290
Okamura, Kazuki 3674
Okazaki, Fumihiro 1728
Okazaki, Yuhei 1586
Okitsu, Takashi 1886
Okubo, Toshikazu 2058
Okuma, Jun 1978
Okuma, Yasuhiro 2834
Okumura, Hajime 3440
Okuyama, Yoshihiro 1811
Omata, Shinpei 2944
Omi, Masataro 1317
Omori, Hideki.......................1299, 1806, 3341
Omote, Kenichiro 1950
Omura, Mototsugu 1685
Ong, Andrew 2722
Onishi, Mitsuru 1054
Ooishi, Eiji 183
Ooshima, Masahide 1715
Orikawa, Koji 191, 1613, 2277, 3659
Ortiz, G. 1291
Ortiz, Gabriel 1309
Oshima, Ryo 1021
Oshinoya, Yasuo.........294, 299, 2972, 3055, 3159, 3162
Oso, Hiroshi 629
Ota, Chiharu 3440
Ota, Satoru 2095
Otsuki, Midori 1054
Ouchi, Takayuki 3920
Ouyang, Shaodi 624
Ouyang, Ziwei 2842
Ozaki, Takayuki 2638
Ozkan, Ziya 498
Pala, Vipindas 2208
Palmour, John 3447
Pan, Miao 582
Panda, S K 1775
Panda, Sanjib Kumar 1580
Pansier, F. 1935
Papafotiou, Georgios 1606
Papastergiou, Konstantinos 1220
Park, Gyeong-Jae 1990
Park, Junsung 1394
Park, Yongsoon 2598
Parker, S. G. 2019, 2372
Partanen, J. 587
Patel, Dhaval 758, 1626
Pedersen, Kristian Bonderup 2547

AUTHOR INDEX

Peftitsis, Dimosthenis 3712
Peltoniemi, P. 587
Peng Gao 2926, 2932
Peng Wang 3124
Peng Wen 782
Peng, Han 2208
Peretti, L. 3111
Peters, A. 2887
Peters, Wilhelm 1508
Petersen, Lars P. 3352
Petersen, Lars Press 2842
Petrich, Matthias 318
Pettersson, Sami 3432
Pham Phu Hieu 3913
Pidaparthy, Syam Kumar 3638
Piepenbreier, Bernhard 1816, 3029
Ping-Heng Wu 1261
Pires, Igor A. 3225, 3270
Pittet, Serge 3371
Pittini, Riccardo 3905
Po-Chien Chou 3425
Poh Chiang Loh 857
Po-Jung Tseng 2810, 3328
Popa, Lucian-Dinut 2996
Popova, L. 548
Popovic, J. 1935
Poshtkouhi, Shahab 2336
Pou, Josep 3758, 3764
Prasanna, I. V. 1580
Prasanna, U. R. 230, 395, 1594
Prasanna, Udupi. R. 2343
Prodic, Aleksandar 3582
Pyrhonen, J. 548
Qi Zhang 3489
Qu, Lizhi 609
Qunzhan Li 1050
Rabkowski, Jacek 3712
Radic, Aleksandar 3582
Radman, Karlo 1691
Rae-Sung Yu 925
Rahman, F. 2679
Rahman, M. A. 982
Rahman, M. F. 2686
Rajashekara, K. 395, 1594
Rajashekara, Kaushik 230, 2343, 3134
Raju, Siddharth 1036
Ramadan, Husam A. 863
Rambetius, Alexander 270, 3029
Rannestad, Bjorn 2547
Rannested, Bjorn 2850
Rathore, Akshay K. 1775
Rathore, Akshay Kumar 2087, 3616
Ray, Olive 2707
Razik, H. 3239
Reiter, Tomas 774

Ren, Kangle 465
Riffat, Javid 2815
Rikitake, Jungo 2216
Rim, Chun T. 1103, 2232, 3358
Rivera, Marco 3574
Robbins, William P 3037, 3061
Rodriguez, Jose 3574
Rongfeng Yang 3680
Rosekeit, Martin 2729
Roth-Stielow, Jorg 264, 318, 3085
Roy, Sudhin 651, 2562, 3286
Ruan, Xinbo 458, 2768
Ruda, Harry E. 1303
Ruderman, Michael 1665
Rufer, Alfred 1081, 1386
Ryo, Mina 3440
Ryu, Sei-Hyung 3726
Saarakkala, Seppo E. 2489
Saga, Yasunao 1748
Sahoo, Ashish Kumar 3750
Saikusa, H. 3007
Saito, Eiichi 1679
Saito, Katsuhiko 2064
Saito, Ryo 3397
Saito, Ryoji 2189
Saito, Takashi 671
Saitoh, Ryoh 914
Sakaba, Kouichi 3159
Sakai, Kazuto 240
Sakai, Tomoyasu 1490
Sakai, Toshifumi 2451
Sakaino, Sho 1978
Sakimoto, Kenichi 1552
Sakurai, Naoki 2297
Sakurai, Takayasu 2228
Sampath, Prasad K. 2722
Sanada, M. 246, 258, 312, 390
Sanada, Masayuki 324, 356, 363, 370, 3018, 3519
Sand, Kjell 1861
Sariyildiz, Emre 2483
Sasaki, Tomotake 2745
Sasongko, Firman 1761
Sato, Daisuke 3815
Sato, Koji 1671
Satria, Andri 3893
Sauer, Dirk Uwe 2729
Sawada, Tadashi 1122
Sayed, Khairy 2359
Sayed, Mahmoud A. 3877
Schob, Reto. T. 1691
Schon, Andre 3255
Schrittwieser, L. 1291
Schupbach, Marcelo 3447
Schuster, Johannes 3085
Segaran, D. S. 2372

AUTHOR INDEX

Segsa, Karl-Heinz......................................2554
Seilmeier, Markus 1816
Sekiba, Yoichi..2175
Sekisue, Takayuki 2175, 2189
Seo, Gab-Su 2272
Seok-Jin Hong..3532
Seunghoo Song 1646
Seung-Ki Sul 925, 2802
Severson, Eric 1412
Shah, Shahil...843, 967
Shao Zhang 1342
Shaodi Ouyang 2815
Shaofeng Xie 1050
Shaohua Sun 3213
Shaohui Zhong 1155
Shen, Na 582
Shen, Zhiyu 3850
Shenghui Cui 1268
Sheng-Kai Kao 2714
Shi, Hongtao 577
Shi, Rongliang 567
Shibahara, Kohei 1575, 3624
Shibahara, Ryota 2222
Shibanuma, Kenichi 634
Shibata, Yuichiro 3950
Shieh, Hsin-Jang 351
Shieh, Jenn-Jong 204
Shigematsu, Koichi 2183, 2189
Shih, Bing-Jyun 105
Shih, Sheng-Fang 2380
Shih-Jen Cheng 3185
Shimada, Takae 2058
Shimamori, Hiroshi 2745
Shimao, Toshihiro 415
Shimatou, T. 2939
Shimizu, Kyohei 1968
Shimizu, Toshihisa ... 876, 1166, 2944, 2983, 3044, 3771
Shimizu, Toshimasa 1054
Shimode, Daisuke 1122
Shimomura, Junichi 2183
Shimono, Tomoyuki 1685
Shin Shiung Wang 3938
Shin, Hyunhak..110
Shin, Yesl 493
Shinagawa, Syuhei 252
Shinbo, Mitsuo 634
Shindo, Yuji 1552
Shinnaka, Shinji 1824
Shinohara, Atsushi 324
Shinozaki, Ikki 1728
Shinozuka, Yasuhiro...................................2228
Shioda, Masashi 130
Shirakawa, Kazuhiro.........................304, 1379
Shiraki, N. 2120
Shirasawa, Koki 3106

Shishida, Yasuhiro 2644
Shiting Weng 1476
Shixi Hou 3168
Shoeiby, B. 1482
Shoji, Hiroyuki 2058
Shoyama, Masahito 863, 3386
Shu-Hung Liao 1452
Shuitao Yang..1342
Shun-Chung Wang 3233, 3778
Shunke Sui 3680
Shuren Wang 3194
Shu-Wei Kuo 3185
Siemaszko, Daniel 3371
Silva, Marcelo 3864
Silva, Sidelmo M. 3225
Silventoinen, P. 587
Sin, Min-Ho 409
Singh, B. N. 3482
Sintamarean, C. 1912
Sitbon, M. 2240
Sivakumar K 1400
Siwakoti, Yam P. 1801
Skuriat, Robert 2920
Smadi, Issam 1968
Smaka, Senad...1148
Smiththisomboon, Somrat 3885
Sogawa, Yuki 522, 2781
Solomon, Adane Kassa 2920
Sone, Kodai 3525
Song Kejian 640
Song, Z. Q. 2686
Sonoda, Hideki...634
Soo-Cheol Shin 3532
Specht, Andreas 1501
Srirattanawichaikul, Watcharin 3417
Steinert, Daniel 975
Steinke, Gina..1081
Stumpf, Peter 2700
Su, Bonan 614
Su, Hong-Wei...1781
Su, Jianhui 2156
Suetsugu, Tadashi.......................................3955
Sugao, Kazumi 403
Sugimoto, Hiroya 982
Sugiura, Makoto 1135
Suh, Yongsug 1185, 1646
Su-Han Kim 790
Sul, Seung-Ki 1268, 2598
Sumida, Hitoshi 2302
Sun, Jian 843, 967, 2202
Sun, Shaohua 2318
Sun, Wei 609
Sunaga, Keita 3162
Sung, Kyungmin 117, 829
Sunsoon Park 1646

AUTHOR INDEX

Suntio, T. 514, 2240
Suntio, Teuvo 1466
Suryadevara, Rohit 2433
Suto, Kenji 3194, 3194
Suul, Jon Are 1544
Suwankawin, Surapong 3885
Suzuki, Genri 1697
Suzuki, Hirokazu 3503
Suzuki, Katsumi 959
Suzuki, Michiaki 883
Suzuki, Nobuyuki 919, 2541
Suzuki, Ryosuke 2972
Suzuki, Shun 1166
Suzuki, Takashi 1062
Suzuki, Toshiki 907
Svensson, Jan R 1220
Tabira, K. 2939
Tadano, Yugo 1242
Tadokoro, D. 390
Taeck-Kie Lee 3532
Taekyun Kim 3933
Tae-Won Chun 2763
Taga, Hironori 1328
Tajima, G. 2438
Takada, Hiromu 1697
Takagi, Ryo 1328
Takahashi, Akiko 2470
Takahashi, Hiroki 152, 1021
Takahashi, Hirotaka 1068
Takahashi, Hisashi 3106
Takahashi, K. 2569
Takahashi, Naoya 3920
Takahashi, Nobuhiro 3207
Takahashi, Osamu 2285
Takahashi, Takehiro 3207
Takahashi, Yoshikazu 671, 2870
Takamiya, Makoto 2228
Takao, Kazuto 744, 3440, 3707
Takasaki, Mika 2252
Takashita, Haruomi 3386
Takasu, Shinji 3440
Takatsuka, Yushi 1898
Takayama, Masakazu 3801
Takayanagi, Atsushi 2794
Takeda, Kotaro 1654
Takeda, Masashi 801
Takeda, Takashi 1490
Takei, Manabu 3440
Takemoto, Masatsugu 1000, 3525
Takenaka, Kensuke 3440
Takenami, Fumiaki 2737
Takenoiri, S. 2939
Takeshita, Takaharu 123, 601, 2310, 3809, 3877
Takeuchi, Katsutoku 3012
Takeuchi, Shun 3646

Takezaki, Kenichi 3525
Taki, Hiroshi 876, 1379
Takino, Toshiaki 817
Tam Khanh Tu Nguyen 2826
Tamada, Shunsuke 1357
Tamura, Hiroshi 682
Tan, Nadia M. L. 750
Tanabe, Ryo 1234
Tanai, Masanobu 829
Tanaka, Daiki 3179
Tanaka, Junya 240
Tanaka, Kiminori 2222
Tanaka, Koutaro 1006
Tanaka, Seiyu 982
Tanaka, Takahide 2302
Tanaka, Toshihiko 3292
Tanaka, Yasunori 3440
Tanaka, Yuichiro 1880
Tanifuji, Hikaru 1115
Taniguchi, Shun 2465
Tao Meng 3213
Tatsuta, Fujio 959
Tauchi, Yuki 3119
Teng, Jen-Hao 677
Teodorescu, R. 1912
Tera, Takahiro 304, 876
Terabe, Ryosuke 2638
Terao, Yutaka 2644
Teshima, Masato 1068
Thiringer, Torbjorn 1373
Thogersen, Paul 2547
Thogersen, Paul Bach 2850
Tian, Yanjun 3538
Ting, Pangan 677
Ting, Yeh 2787
Tint Soe Win 3292
Toba, Akio 2011
Toda, Hiroaki 1984
Togashi, Ryo 356
Toi, Takahiro 1109
Tokiwa, Tsuyoshi 2977
Tokuda, Hirokazu 2175
Tokumasu, Akira 1379
Tokuyama, Takeshi 2904
Tominaga, Shinji 2290
Tomioka, Satoshi 3630
Tomita, Mutuwo 907
Tonogi, K. 2438
Toru 3784
Tosaka, Shuhei 1207, 1227
Town, Graham E. 1801
Toyoda, Hajime 1560
Tran, Q. V. 446
Trescases, Olivier 2336, 3496
Trillion Zheng 2893

AUTHOR INDEX

Trintis, Ionut 2547
Tripathi, Awneesh 758, 1626
Trompa, Thomas 2554
Tsai, Jiung-Lin 1781
Tsai, Ming-Hsiao 278
Tsan Chen 2810
Tse, Chi K. 1425
Tseng, K. J. 2145
Tsorng-Juu Liang 2750, 3346
Tsubakidani, Takashi 3674
Tsuboi, Yoshiki....................................1160
Tsuboi, Yuichi 92
Tsuchida, Kazuo.................................3397
Tsuda, Junichi 1929
Tsuji, Mineo.....................2775, 3093, 3674
Tsuji, Satoshi 522, 2781
Tsuji, Toshiaki 1978
Tsukakoshi, Masahiko 92, 2330
Tsuruma, Yoshinori.....................1054, 2644
Tsuruta, Hironori.................................629
Tsuruta, Ryoji 1350
Tsuruta, Yukinori 2266, 3403
Tsuyoshi, Hanamoto 2476
Tu, Yunwu 465
Tukiman, Rahayu 2529
Turpin, Santiago 1974
Tuysuz, Arda 1904
Tzou, Ying-Yu 3072
Uddin, Muslem 3574
Ueda, K. 312
Ueda, Tetsuya 403
Ueda, Tetsuzo 2075
Ueda, Yoshinobu 1855
Uemura, Hirofumi 891, 2834
Ukai, Hiroyuki 2400
Umeda, Nobuhiro 2183
Umeno, Masayoshi 183
Umesh B S 1400
Umetani, Kazuhiro 304
Undeland, Tore....................................1412
Uno, Masatoshi...........................1444, 2351
Urushibata, Hiroaki 2290
Urushibata, Shota 1365, 3455
Ushiro, Nobumasa..............................3106
Vaisanen, V...587
Vajk, Istvan 2700
Van Brunt, Edward 3726
Van Wyk, J. D. 1935
Van-Long Tran 1214
Vasiladiotis, Michail 1386
Vasquez-Arnez, Ricardo Leon 1276
Veerasamy, Balaji 2310
Vieto, Ignacio 843
Viinamaki, J. 2240
Vilathgamuwa, D. M. 2722, 3598

Vogt, T. 2887
Wada, Keiji 744, 1379, 2511, 3646
Wahlstroem, Jonas 3476
Wajima, Kiyoshi 1984
Wallscheid, Oliver 1501
Wang Hui 640
Wang, Bin 2133
Wang, Chao-Fu 172
Wang, Fei 470
Wang, Fusheng 465
Wang, H. 1912
Wang, Hengli 72
Wang, Jun 944
Wang, Lingxiang 465
Wang, Lipeng 458
Wang, Xiaojian 2815
Wang, Xinyu 624, 2815
Wang, Xiongfei 216, 1529, 3320, 3538
Wang, Yanbo 3538
Wang, Yong 470
Wang, Yue 160, 475
Wang, Zhao'An 160, 475
Watanabe, Daisuke 988
Watanabe, Hiroki 84
Watanabe, S. 2939
Watanabe, Shoichiro 1334, 2126
Watashima, T. 2939
Wei Jiang 3194
Wei Liu 1050
Wei Wang..3680
Wei Yan 3455
Wei, Guo 2318
Wei, Sun 341
Weili Dai...3168
Weirong Chen 3695
Wei-Ting Hsu 2910
Wen, Bo 3850
Wen, Chao-Kai 677
Wen, Huiqing 702
Wen-Chien Hsu 2714
Wenjie Chen 2967
Wen-Tai Li 677
Wheeler, Pat 2920
Won, Chung-Yuen 166, 409
Wong, Siu-Chung 1425
Wonsuk Choi 2914
Woojin Choi 1214
Wu Mingli 640
Wu, Bin 3468
Wu, Chun-Wei 330, 1832
Wu, G. F. 1471
Wu, Gwo-Bin 3796
Wu, T.-F. 2050
Wu, Tsung-Hsi 1796
Wu, Weimin 2610

AUTHOR INDEX

Wu, Weiyang 582
Wu, Wenlong 470
Wu, Wen-Zhe 429
Wunsch, B. 2167
Xia, Huan 2133
Xiangdong Sun 3489
Xiang-Dong Sun 3784
Xiao, D. 2686
Xiao, Fei 66, 72
Xiao, Shuai 543
Xiaojie You 2893
Xiaojie Zhuang 2638
Xiaolong Ma 835
Xiaomei Song 2967
Xie, Ruiliang 2950
Xiong, Li 230
Xiuqin Wei 3955
Xu Cai 1842, 3124
Xu Dianguo 3050
Xu Yang 2967
Xu, David 3099
Xu, Dehong 485
Xu, Dianguo 609, 614, 3002
Xu, Haizhen 567
Xu, Rong 609
Xue, Danhong 3604
Xuling Chen 2768
Yablecki, Jessica 3496
Yachi, Toshiaki 3959
Yakabe, Seichiro 730
Yamada, Hiroaki 538, 1811
Yamada, Kenji 1898
Yamada, Ryuji 1920
Yamada, Takatoshi 2212
Yamada, Tatsuji 3263
Yamagata, Shinichi 829
Yamagishi, Tatsuya 750
Yamaguchi, Shota 3771
Yamaguchi, Takashi 1242
Yamaichi, Katsuya 2517
Yamaji, Masaharu 2302
Yamamoto, Eiji 1654
Yamamoto, Junichi 177, 1179
Yamamoto, Kenji 3106
Yamamoto, Kichiro 689
Yamamoto, Kohei 1438
Yamamoto, Masayoshi 811, 883, 2497
Yamamoto, Shu 3044
Yamamoto, Takashi 3707
Yamamoto, Yasuhiro 1601
Yamamura, N. 146
Yamanaka, Kenji 2152
Yamanaka, Tatsuya 1207, 1227
Yamanoi, Takashi 1062
Yamashita, Nobuyuki 2983

Yamashita, Shigeharu 2745
Yamazaki, Akira 933
Yan Li 3124
Yan Zhang 835
Yanagi, Hiroshige 3630
Yang Chuan 1962
Yang, Cs 199, 2593, 3185
Yang, Daeki 2247
Yang, Geng 543, 582
Yang, Guorun 66
Yang, Hong-Tzer 2100
Yang, Rongfeng 609
Yang, Shih-Sian 2606
Yang, Sihun 863
Yang, Xu 2950
Yang, Zhongping 335, 1322, 2133
Yanhong, Zhang 452
Yano, Yoshihiro 2775
Yanru Zhong 3489
Yaramasu, Venkata 3695
Yashiro, Daisuke 1974
Yashun Li 782
Yasubayashi, Mikio 183
Yasui, Kazuya 2465
Yasumura, Yuji 3568
Yasuno, Takashi 3179
Yau, Y. T. 2754, 3392
Yazdkhasti, Pegah 2156
Yi, Hao 577
Yi-Chun Lin 2714
Yi-Hsun Chiu 3778
Yi-Hua Liu 3778
Yixin Zhu 1155, 3546
Yizhanyi Tang 2977
Yoda, Kazuyuki 1748
Yokoi, Y. 3007
Yokoi, Yuichi 3024
Yokokura, Yuki 1956
Yokoyama, H. 2120
Yokoyama, Natsuki 2285
Yokoyama, Tomoki 3397, 3410
Yonemori, Ryo 689
Yonezawa, Hikaru 3055
Yonezawa, Yoshiyuki 3440
Yonezawa, Yu 2745, 3386
Yong Ding 1476, 3842
Yong, Yu 341
Yong-Cheol Kwon 925
Yongdong Tan 3538
Yongjae Lee 3140
Yoon, Sung Hyun 2272
Yoshida, Morito 3410
Yoshida, S. 2569
Yoshida, T. 2438
Yoshida, Yoshiaki 3503

AUTHOR INDEX

Yoshikawa, Yuichi 2011
Yoshimoto, Kantaro....................................2421
Yoshimura, Eiji 1552
Yoshino, Teruo 2644, 3834
Yoshino, Yukio 2745
Yoshioka, S. 246
Yoshioka, Takashi....................................1956
Yoshizawa, Daisuke 97, 1950
Young-Do Kim 1738, 1743
Youngjoon Choi 1185
Young-Ryul Kim 796
Yu, Changzhou 567
Yu, F. Y. 1471
Yu, Ling-Chia 208
Yu, Shuai 2318
Yu, Weikai 647
Yu, Yifan 1458
Yu, Yong 567, 609
Yu-Chen Liu 2810, 3328
Yue Chen 2768
Yue, Xiaolong 2960
Yu-Jen Wang 1420
Yu-Kang Lo 2758, 2810, 3185, 3328, 3913
Yuki, Kazuaki 2465
Yukita, Kazuto 1490
Yukutake, Seigo 2140
Yunchang Kwak 2693
Yun-Chu Chiu 3328
Yung-Ching Huang 1452
Yunmei Fang 3168
Yunwei Li 3482
Yura, Masashi 2297
Yu-Shan Cheng 3233, 3778
Yuzurihara, Itsuo 2794
Zaitsu, Toshiyuki 177, 1179
Zanma, Tadanao 3568
Zargari, Navid R. 3468
Zehelein, Matthias....................................3085
Zeliang Shu 1050
Zeljkovic, Sandra 774
Zhang Wei 3050
Zhang Yajing....................................2899
Zhang, Haodong 3604
Zhang, Huiguo 2202
Zhang, Tao 485
Zhang, Wei 614, 1425
Zhang, Xing 465, 567
Zhang, Xuning 2626
Zhang, Yuzhuo 647
Zhang, Zhe....................................78
Zhao, Wei....................................567
Zhao-Qin Guo 1580
Zhe Wang 3129
Zhe Zhang....................................3905
Zheng Dong 3842

Zheng Li 1842
Zheng, T. Q. 807
Zheng, Trillion Q. 2899, 3314
Zhengzhi Han 3124
Zhenyao Xu 994
Zhongping Yang 807
Zhu, B. 395, 1594
Zhu, Honglin 3002
Zhuo, Fang 577, 2960
Zian Qin 857
Zingerli, Claudius M. 3365
Zitouni, Y. 3239
Zong-Zhen Yang....................................3778
Zou, Xudong 555
Zwyssig, Christof....................................1707